10 0497403 7
WITHDRAWN
ROM THE LIBRARY

KU-279-662

Fifth Edition

PLANT PATHOLOGY

Fifth Edition

PLANT PATHOLOGY

GEORGE N. AGRIOS
Department of Plant Pathology
University of Florida

ELSEVIER
ACADEMIC
PRESS

Amsterdam • Boston • Heidelberg • London • New York • Oxford
Paris • San Diego • San Francisco • Singapore • Sydney • Tokyo

Publisher	Dana Dreibelbis
Associate Acquisitions Editor	Kelly D. Sonnack
Project Manager	Troy Lilly
Marketing Manager	Linda Beattie
Cover Design	Eric DeCicco
Composition	SNP Best-set Typesetter Ltd., Hong Kong
Cover Printer	RR Donnelley & Sons Company
Interior Printer	RR Donnelley & Sons Company

Elsevier Academic Press
30 Corporate Drive, Suite 400, Burlington, MA 01803, USA
525 B Street, Suite 1900, San Diego, California 92101-4495, USA
84 Theobald's Road, London WC1X 8RR, UK

This book is printed on acid-free paper. ∞

Library of Congress Cataloging-in-Publication Data

Agrios, George N., 1936–
Plant pathology / George Agrios. — 5th ed.
p. cm.
Includes bibliographical references and index.
ISBN 0-12-044565-4 (hardcover: alk. paper)
1. Plant diseases. I. Title.
SB731.A35 2004
571.9′2 — dc22

2004011924

British Library Cataloguing in Publication Data
A catalogue record for this book is available from the British Library

For all information on all Elsevier Academic Press Publications
visit our Web site at www.books.elsevier.com

Printed in the United States of America
04 05 06 07 08 09 9 8 7 6 5 4 3 2 1

This, the 5th and probably the last edition of *Plant Pathology* by me, is dedicated:

To the memory of my parents, Nikolas and Olga, who, in spite of their limited education, sacrificed everything to give me the most and best education possible.

To the memory of Dr. Walter F. Buchholtz, my major professor at Iowa State University, who challenged me before I had even taught my first lecture to "write my own textbook on Plant Pathology".

To my sisters, Dimitra and Evangelia, who have been there for me forever and who also sacrificed some of their interests for my benefit.

To my wife, Annette, whose love and support have been the most precious things to me throughout our life together, and who helped me in many facets of preparation of this and of previous editions of *Plant Pathology*.

To my daughters-in-law, Betsy and Vivynne, who, by joining our family, added beauty, love, enjoyment, and four wonderful grandchildren.

Finally, to Mark and Maximos, our youngest grandchildren, who, someday, when they read their names in the book, may be reassured of "Granpa's" love for them, and may feel proud of their grandfather.

Contents

Preface xix
Photo credits xxi
About the Author xxiii

part one

GENERAL ASPECTS

chapter one

INTRODUCTION

Prologue: The Issues 4
Plants and Disease 4
The Concept of Disease in Plants 5
Types of Plant Diseases 7
History of Plant Pathology and Early Significant Plant Diseases 8
Introduction 8
Plant Diseases as the Wrath of Gods — Theophrastus 9
Mistletoe Recognized as the First Plant Pathogen 14
Plant Diseases as the Result of Spontaneous Generation 16
Biology and Plant Pathology in Early Renaissance 16
Potato Blight — Deadly Mix of Ignorance and Politics 19
The Expanding Role of Fungi as Causes of Plant Disease 21
The Discovery of the Other Causes of Infectious Diseases 23
Nematodes — Protozoan Myxomycetes — Bacteria — Viruses. Protozoa — Mollicutes — Viroids — Serious Plant Diseases of Unknown Etiology 23
Koch's Postulates 26
Viruses, Viroids, and Prions 27
Losses Caused by Plant Diseases 29
Plant diseases reduce the quantity and quality of plant produce. 29
White, Downey, and Dry Vineyards — Bring on the Bordeaux! 30
Plant diseases may limit the kinds of plants and industries in an area. 32
Chestnuts, Elms, and Coconut Palm Trees — Where have they gone? 32
Plant diseases may make plants poisonous to humans and animals 37
Ergot, Ergotism, and LSD: a Bad Combination 37
Mycotoxins and Mycotoxicoses 39
Plant diseases may cause financial losses. 41
The Insect — Pathogen Connection: Multifaceted and Important 42
Plant Pathology in the 20th Century 45
Early Developments 45

The Descriptive Phase 45
The Experimental Phase 46
The Etiological Phase 46
The Search for Control of Plant Diseases 46
The Main Areas of Progress 47
Chemical Control of Plant Diseases 47
Appearance of Pathogen Races Resistant to Bactericides and Fungicides 48
Public Concern about Chemical Pesticides 48
Alternative Controls for Plant Diseases 49
Interest in the Mechanisms by Which Pathogens Cause Disease 50
The Concept of Genetic Inheritance of Resistance and Pathogenicity 52
Epidemiology of Plant Disease Comes of Age 53
Plant Pathology Today and Future Directions 54
Molecular Plant Pathology 54
Aspects of Applied Plant Pathology 56
Plant Biotechnology — The Promise and the Objections 56
Food Safety 58
Bioterrorism, Agroterrorism, Biological Warfare, etc. Who, What, Why 59
Worldwide Development of Plant Pathology as a Profession 60
International Centers for Agricultural Research 60
Trends in Teaching and Training 61
Plant Disease Clinics 62
The Practice and Practitioners of Plant Pathology 63
Certification of Professional Plant Pathologists 63
Plant Pathology as a Part of Plant Medicine; the Doctor of Plant Medicine Program 64
Plant Pathologists' Contributin to Crops and Society 65
Some Historical and Present Examples of Losses Caused by Plant Diseases 65
Plant Diseases and World Crop Production 65
Crop Losses to Diseases, Insects and Weeds 66
Pesticides and Plant Diseases 69
Basic Procedures in the Diagnosis of Plant Diseases 71
Pathogen or Environment 71
Infectious Diseases 72
Parasitic Higher Plants — Nematodes — Fungi and Bacteria: Fungi — Bacteria and Mollicutes 72
Viruses and Viroids — More than One Pathogen 73
Noninfectious Diseases 73
Identification of a Preciously Unknown Disease: Koch's Postulates 74

chapter two

PARASITISM AND DISEASE DEVELOPMENT

Parasitism and Pathogenicity 77
Host Range of Pathogens 78
Development of Disease in Plants 79
Stages in the Development of Disease: The Disease Cycle 80
Inoculation 80
Inoculation, Types of Inoculum, Sources of Inoculum, Landing or Arrival of Inoculum 80
Prepenetration Phenomena 82
Attachment of Pathogen to Host, Spore Germination and Perception of the Host Surface, Appressorium Formation and Maturation, Recognition between Host and Pathogen, Germination of Spores and Seeds, Hatching of Nematode Eggs 82
Penetration 87
Direct Penetration through Intact Plant Surfaces, Penetration through Wounds, Penetration through Natural Openings 87
Infection 89
Infection, Invasion, Growth and Reproduction of the Pathogen (Colonization) 89
Dissemination of the Pathogen 96
Dissemination by Air, Dissemination by Water, Dissemination by Insects, Mites, Nematodes, and Other Vectors, Dissemination by Pollen, Seed, Transplants, Budwood, and Nursery Stock, Dissemination by Humans 96
Overwintering and/or Oversummering of Pathogens 100
Relationships between Disease Cycles and Epidemics 102

chapter three

EFFECTS OF PATHOGENS ON PLANT PHYSIOLOGICAL FUNCTIONS

Effects of Pathogens on Photosynthesis 106
Effect of Pathogens on Translocation of Water and Nutrients in the Host Plant 106

Interference with Upward Translocation of Water and Inorganic Nutrients 106
Effect on Absorption of Water by Roots 108
Effect on Translocation of Water through the Xylem 108
Effect on Transpiration 108
Interference with the Translocation of Organic Nutrients through the Phloem 113
Effect of Pathogens on Host Plant Respiration 115
Respiration of Diseased Plants 117
Effect of Pathogens on Permeability of Cell Membranes 118
Effects of Pathogens on Transcription and Translation 118
Effect on Transcription 119
Effect on Translation 119
Effect of Pathogens on Plant Growth 119
Effect of Pathogens on Plant Reproduction 121

chapter four
GENETICS OF PLANT DISEASE

Introduction 125
Genes and Disease — Variability in Organisms — Mechanisms of Variability 126
General Mechanisms: Mutation — Recombination — Gene and Genotype Flow among Plant Pathogens — Population Genetics, Genetic Drift, and Selection — Life Cycles — Reproduction — Mating Systems — Out-crossing — Pathogen Fitness 129
Specialized Mechanisms of Variability in Pathogens: Sexual-like Processes in Fungi Heterokaryosis — Parasexualism — Vegetative Incompatibility — Heteroploidy 131
Sexual-like Processes in Bacteria and Horizontal Gene Transfer 132
Genetic Recombination in Viruses 133
Loss of Pathogen Virulence in Culture 133
Stages of Variation in Pathogens 134
Types of Plant Resistance to Pathogens 134
True Resistance: Partial, Quantitative, Polygenic, or Horizontal Resistance — R-Gene Resistance, Monogenic, or Vertical Resistance 136
Disease Escape — Tolerance to Disease 137
Genetics of Virulence in Pathogens and of Resistance in Host Plants 139
The Nature of Resistance to Disease — Pathogenicity Genes in Plant Pathogens 142
Genes Involved in Pathogenesis and Virulence by Pathogens 142
Pathogenicity Genes of Fungi controlling: Production of Infection Structures — Degradation of Cuticle and Cell Wall — Secondary Metabolites — Fungal Toxins — Pathogenicity Signaling Systems 144
Pathogenicity Genes in Plant Pathogenic Bacteria controlling: Adhesion to Plant Surfaces — Secretion Systems — Enzymes that Degrade Cell Walls — Bacterial Toxins as Pathogenicity Factors — Extracellular Polysaccharides as Pathogenicity Factors — Bacterial Regulatory Systems and Networks — Sensing Plant Signaling Components — Other Bacterial Pathogenicity Factors 146
Pathogenicity Genes in Plant Viruses: — Functions Associated with the Coat Protein — Viral Pathogenicity Genes 149
Nematode Pathogenicity Genes 150
Genetics of Resistance through the Hypersensitive Response 151
Pathogen-Derived Elicitors of Defense Responses in Plants 151
Avirulence (avr) Genes: One of the Elicitors of Plant Defense Responses Characteristics of avr Gene-Coded Proteins: — Their Structure and Function Role of *avr* Genes in Pathogenicity and Virulence 154
hrp Genes and the Type III Secretion System 155
Resistance (R) Genes of Plants: Examples of R Genes — How Do R Genes Confer Resistance? – Evolution of R Genes — Other Plant Genes for Resistance to Disease 155
Signal Transduction between Pathogenicity Genes and Resistance Genes: — Signaling and Regulation of Programmed Cell Death — Genes and Signaling in Systemic Acquired Resistance 160
Examples of Molecular Genetics of Selected Plant Diseases: — The Powdery Mildew Disease — Magnaporthe grisea*, the Cause of Rice Blast —*

Fusarium, *the Soilborne Plant Pathogen* — Ustilago maydis *and Corn Smut* 161

Breeding of Resistant Varieties 165

Natural Variability in Plants — Breeding and Variability in Plants — Breeding for Disease Resistance Sources of Genes for Resistance — Techniques Used in Classical Breeding for Resistance — Seed, Pedigree, and Recurrent Selection — Tissue Culture and Genetic Engineering Techniques 165

Genetic Transformation of Plant Cells for Disease Resistance 169

Advantages and Problems in Breeding for Vertical or Horizontal Resistance 169

Vulnerability of Genetically Uniform Crops to Plant Disease Epidemics 170

chapter five

HOW PATHOGENS ATTACK PLANTS

Mechanical Forces Exerted By Pathogens on Host Tissues 177

Chemical Weapons of Pathogens 179

Enzymes in Plant Disease 180

Enzymatic Degradation of Cell Wall Substances 180

Cuticular Wax — Cutin — Pectic Substances — Cellulose — Cross-Linking Glycans (Hemicelluloses) — Suberin — Lignin — Cell Wall Flavonoids — Cell Wall Structural Proteins 180

Enzymatic Degradation of Substances Contained in Plant Cells 189

Proteins — Starch — Lipids 189

Microbial Toxins in Plant Disease 190

Toxins That Affect a Wide Range of Host Plants 190

Tabtoxin — Phaseolotoxin — Tentoxin — Cercosporin — Other Non-Host-Specific Toxins 191

Host-Specific or Host-Selective Toxins 193

Victorin, HV Toxin — T-Toxin [*Cochliobolus (Helminthosporium) heterostrophus* Race T-Toxin] — C-Toxin — *Alternaria alternata* Toxins — Other Host-Specific Toxins 194

Growth Regulators in Plant Disease 196

Detoxification of Low-Molecular Weight Antimicrobial Molecules 201

Promotion of Bacterial Virulence By avr Genes 202

Role of Type III Secretion in Bacterial Pathogenesis 202

Pathogenicity and Virulence Factors in Viruses and Viroids 202

chapter six

HOW PLANTS DEFEND THEMSELVES AGAINST PATHOGENS

Whatever the Plant Defense or Resistance, It Is Controlled by Its Genes 208

Non-host Resistance — Partial, Polygenic, Quantitative, or Horizontal Resistance — Monogenic, R Gene, or Vertical Resistance 208

Preexisting Structural and Chemical Defenses 210

Preexisting Defense Structures 210

Preexisting Chemical Defenses 211

Inhibitors Released by the Plant in Its Environment — Inhibitors Present in Plant Cells before Infection 211

Defense through Lack of Essential Factors 212

Lack of Recognition between Host and Pathogen: Lack of Host Receptors and Sensitive Sites for Toxins — Lack of Essential Substances for the Pathogen 212

Induced Structural and Biochemical Defenses 213

Recognition of the Pathogen by the Host Plant 213

Transmission of the Alarm Signal to Host Defense Providers: Signal Transduction 214

Induced Structural Defenses: Cytoplasmic Defense Reaction — Cell Wall Defense Structures 214

Histological Defense Structures: Formation of Cork Layers — Abscission Layers — Tyloses — Deposition of Gums *215*

Necrotic Structural Defense Reaction: Defense through the Hypersensitive Response 217

Induced Biochemical Defenses in: Non-Host Resistance — In Partial, Quantitative

(Polygenic, General, or Horizontal) Resistance: *Function of Gene Products in Quantitative Resistance — The Mechanisms of Quantitative Resistance — Effect of Temperature on Quantitative Resistance* 217

Induced Biochemical Defenses in the Hypersensitive Response (R Gene) Resistance 221

The Hypersensitive Response: Genes Induced During Early Infection — Functional Analysis of Plant Defense Genes — Classes of R Gene Proteins — Recognition of Avr Proteins of Pathogens by the Host Plant — How Do R and Avr *Gene Products Activate Plant Responses? — Some Examples of Plant Defense through R Genes and Their Matching* Avr *Genes: — The Rice Pi-ta Gene. The Tomato Cf Genes. The Tomato Bs2 Gene. The Arabidopsis RPM1 Gene — The Co-function of Two or More Genes.* 221

Defense Involving Bacterial Type III Effector Proteins 229

Production of Active Oxygen Species, Lipoxygenases, and Disruption of Cell Membranes — Reinforcement of Host Cell Walls with Strengthening Molecules 231

Production of Antimicrobial Substances in Attacked Host Cells — Pathogenesis-Related (PR) Proteins 232

Defense through Production of Secondary Metabolites — Phenolics: — Simple Phenolic Compounds — Toxic Phenolics from Nontoxic Phenolic Glycosides — Role of Phenol-Oxidizing Enzymes in Disease Resistance — Phytoalexins 233

Detoxification of Pathogen Toxins by Plants — Immunization of Plants against Pathogens: Defense through Plantibodies 236

Resistance through prior Exposure to Mutants of Reduced Pathogenicity 237

Systemic Acquired Resistance: — Induction by Artificial Inoculation with Microbes or by Treatment with Chemicals 237

Defense through Genetically Engineering Disease-Resistant Plants 242

With Plant-Derived Genes — With Pathogen-Derived Genes 242

Defense through RNA Silencing by Pathogen-Derived Genes 242

chapter seven

ENVIRONMENTAL EFFECTS ON THE DEVELOPMENT OF INFECTIOUS PLANT DISEASE

Introduction 251

Effect of Temperature 253

Effect of Moisture 257

Effect of Wind 257

Effect of Light 257

Effect of Soil pH and Soil Structure 257

Effect of Host-Plant Nutrition 257

Effect of Herbicides 262

Effect of Air Pollutants 262

chapter eight

PLANT DISEASE EPIDEMIOLOGY

The Elements of an Epidemic 266

Host Factors That Affect the Development of Epidemics 267

Levels of Genetic Resistance or Susceptibility of the Host — Degree of Genetic Uniformity of Host Plants — Type of Crop — Age of Host Plants 267

Pathogen Factors That Affect Development of Epidemics 269

Levels of Virulence — Quantity of Inoculum Near Hosts — Type of Reproduction of the Pathogen — Ecology of the Pathogen — Mode of Spread of the Pathogen 269

Environmental Factors That Affect Development of Epidemics 271

Moisture — Temperature 271

Effect of Human Cultural Practices and Control Measures 272

Site Selection and Preparation — Selection of Propagative Material — Cultural Practices — Disease Control Measures — Introduction of New Pathogens 272

Measurement of Plant Disease and of Yield Loss 273

Patterns of Epidemics — Comparison of Epidemics — Development of Epidemics — Modeling of Plant Disease Epidemics — Computer Simulation of Epidemics 274

Forecasting Plant Disease Epidemics 281

Evaluation of Epidemic Thresholds — Evaluation of Economic Damage Threshold — Assessment of Initial Inoculum and of Disease — Monitoring Weather Factors That Affect Disease Development 281

New Tools in Epidemiology 283

Molecular Tools. GIS. Remote Sensing. Image Analysis. Information Technology 283

Examples of Plant Disease Forecast Systems 285

Forecasts Based on Amount of Initial Inoculum — On Weather Conditions Favoring Development of Secondary Inoculum — On Amounts of Initial and Secondary Inoculum 285

Disease-Warning Systems — Development and Use of Expert Systems in Plant Pathology — Decision Support Systems 289

chapter nine

CONTROL OF PLANT DISEASES

Control Methods that Exclude the Pathogen from the Host 295

Quarantines and Inspections — Crop Certification — Evasion or Avoidance of Pathogen — Use of Pathogen-Free Propagating Material — Pathogen-Free Seed — Pathogen-Free Vegetative Propagating Materials — Exclusion of Pathogens from Plant Surfaces by Epidermal Coatings 295

Control Methods that Eradicate or Reduce Pathogen Inoculum 298

Cultural Methods that Eradicate or Reduce the Inoculum: — Host Eradication — Crop Rotation — Sanitation — Creating Conditions Unfavorable to the Pathogen — Polyethylene Traps and Mulches 300

Biological Methods that Eradicate or Reduce the Inoculum: — Suppressive Soils 303

Reducing Amount of Pathogen Inoculum through Antagonistic Microorganisms Soilborne Pathogens — Aerial Pathogens — Mechanisms of Action — Control through Trap Plants — Control through Antagonistic Plants 305

Physical Methods that Eradicate or Reduce the Inoculum — Control by Heat Treatment — Soil Sterilization by Heat — Soil Solarization Hot-Water Treatment of Propagative Organs — Hot-Air Treatment of Storage Organs Control by Eliminating Certain Light Wavelengths — Drying Stored Grains and Fruit — Disease Control by Radiation — Trench Barriers against Root-transmitted Tree Diseases 310

Chemical Methods that Eradicate or Reduce the Inoculum — Soil Treatment with Chemicals — Fumigation — Disinfestation of Warehouses — Control of Insect Vectors 312

Cross Protection — Induced Resistance: Systemic Acquired Resistance — Plant Defense Activators — Improving the Growing Conditions of Plants — Use of Resistant Varieties 314

Control through Use of Transgenic Plants Transformed for Disease Resistance Transgenic Plants that Tolerate Abiotic Stresses — Transgenic Plants Transformed with: Specific Plant Genes for Resistance — with Genes Coding for Anti-pathogen Compounds — with Nucleic Acids that Lead to Resistance and to Pathogen Gene Silencing — with Combinations of Resistance Genes — Producing Antibodies against the Pathogen — Transgenic Biocontrol Microorganisms 319

Direct Protection of Plants from Pathogens 322
By Biological Controls: — Fungal Antagonists: Heterobasidion (Fomes) annosum *by* Phleviopsis (Peniophora) gigantea *— Chestnut Blight with Hypovirulent Strains of the Pathogen — Soilborne Diseases — Diseases of Aerial Plant Parts with Fungi. — Postharvest Diseases Bacterial Antagonists: Soilborne Diseases — Diseases of Aerial Plant Parts with Bacteria — Postharvest Diseases — with Bacteria of Bacteria-Mediated Frost Injury 328*
Viral Parasites of Plant Pathogens 328
Biological Control of Weeds 328
Direct Protection by Chemicals 329
Methods of Application of Chemicals for Plant Disease Control — Foliage Sprays and Dusts — Seed Treatment — Soil Treatment — Treatment of Tree Wounds — Control of Postharvest Diseases 332
Types of Chemicals Used for Plant Disease Control 338
Inorganic — Inorganic Sulfur Compounds — Carbonate Compounds — Phosphate and Phosphonate Compounds — Film-Forming Compounds 338
Organic Chemicals: Contact Protective Fungicides — Organic Sulfur Compounds: Ditihiocarbamates 339
Systemic Fungicides: — Heterocyclic Compounds — Acylalanines — Benzimidazoles — Oxanthiins — Organophosphate Fungicides — Pyrimidines — Trizoles — Strobilurins or QoI Fungicides — Miscellaneous Systemics 340
Miscellaneous Organic Fungicides — Antibiotics — Petroleum Oils and Plant Oils — Electrolyed Oxidizing Water — Growth Regulators — Nematicides: — Hologenated Hydrocarbons — Organophosphate Nematicides — Isothiocoyanates — Carbamates — Miscellaneous Nematicides 343
Mechanisms of Action of Chemicals Used to Control Plant Diseases — Resistance of Pathogens to Chemicals — Restrictions on Chemical Control of Plant Diseases 345
Integrated Control of Plant Diseases: — In a Perennial Crop — In an Annual Crop 348

part two

SPECIFIC PLANT DISEASES

chapter ten

ENVIRONMENTAL FACTORS THAT CAUSE PLANT DISEASES

Introduction: General Characteristics — Diagnosis — Control 358
Temperature Effects: High-Temperature Effects — Low-Temperature Effects — Low- Temperature Effects on Indoor Plants 358
Moisture Effects: Low Soil Moisture Effects — Low Relative Humidity — High Soil Moisture Effects 365
Inadequate Oxygen 367
Light 368
Air Pollution 368
Air Pollutants and Kinds of Injury to Plants — Main Sources of Air Pollutants — How Air Pollutants Affect Plants — Acid Rain 000
Nutritional Deficiencies in Plants 372
Soil Minerals Toxic to Plants 372
Herbicide Injury 378
Other Improper Agricultural Practices 381
The Often Confused Etiology of Stress Diseases 383

chapter eleven

PLANT DISEASES CAUSED BY FUNGI

Introduction 383
Some Interesting Facts about Fungi (Box) 387
Characteristics of plant pathogenic fungi 388
Morphology — Reproduction — Ecology — Dissemination 388
Classification of Plant Pathogenic Fungi 390

Fungallike Organisms — The True Fungi 391
Identification: Symptoms Caused by Fungi on Plants 397
Isolation of fungi (and Bacteria) 398
Preparing for Isolation — Isolating the Pathogen 398
Life Cycles of Fungi 402
Control of Fungal Diseases of Plants 403
Diseases Caused by Fungallike Organisms 404
Diseases Caused by Myxomycota (Myxomycetes) 404
Diseases Caused by Plasmodiophoromycetes 405
Clubroot of Crucifers 407
Diseases Caused by Oomycetes 409
Pythium Seed Rot, Damping-off, Root Rot, and Soft Rot 410
Phytophthora Diseases 414
Phytophthora Root and Stem Rots — Phytophthoras Declare War on Cultivated Plants and on Native Tree Species (Box) 414
Late Blight of Potatoes 421
The Downy Mildews 427
Introduction — Downy Mildew of Grape 428
Diseases Caused by True Fungi 433
Diseases Caused by Chytridiomycetes 433
Diseases Caused by Zygomycetes 434
Diseases Caused by Ascomycetes and Mitosporic Fungi 439
Sooty molds — Taphrina leaf Curl Diseases — Powdery Mildews 440
Foliar Diseases Caused by Ascomycetes and Deuteromycetes (Mitosporic Fungi) 452
Alternaria Diseases — Cladosporium Diseases — Needle Casts and Blights of Conifers 452
Mycosphaerella Diseases: Banana Leaf Spot or Sigatoka Disease 458
Septoria Diseases — Cercospora Diseases — Rice Blast Disease 460
Cochliobolus, Pyrenophora and Setosphaeria Diseases of Cereals and Grasses 466
Diseases of Corn: — Southern Corn Leaf Blight — Northern Corn Leaf Blight — Northern Corn Leaf Spot 466
Diseases of Rice — Brown Spot Disease of Rice 468
Cochliobolus Diseases of Wheat, Barley, and Other Grasses 469
Crown Rot and Common Root Rot — Spot Blotch of Barley and Wheat 469
Pyrenophora Diseases of Wheat, Barley and Oats 469
Net Blotch of Barley — Barley Stripe — Tan Spot of Wheat 469
Stem and Twig Cankers Caused by Ascomycetes and Deuteromycetes (Mitosporic Fungi) 473
Black Knot of Plum and Cherry — Chestnut Blight — Nectria Canker — Leucostoma Canker 476
Cankers of Forest Trees: — Hypoxylon Canker — Pitch Canker — Butternut Canker — Phomopsis Blight — Seiridium Canker 481
Anthracnose Diseases Caused by Ascomycetes and Deureromycetes (Mitosporic Fungi) 483
Black Spot of Rose 485
Elsinoe Anthracnose and Scab Diseases: — Grape Anthracnose or Bird's-eye Rot — Raspberry Anthracnose — Citrus Scab Diseases — Avocado Scab 486
Colletotrichum Diseases: Colletotrichum Anthracnose Diseases of Annual Plants 487
Anthracnose of Beans — Anthracnose of Cucurbits — Anthracnose or Ripe Rot of Tomato — Onion Anthracnose or Smudge — Strawberry Anthracnose — Anthracnose of Cereals and Grasses 490
Colletotrichum Anthracnoses: A Menace To Tropical Crops (Box) Colletotrichum 491
Bitter Rot of Apple — Ripe Rot of Grape 494
Gnomonia Anthracnose and Leaf Spot Diseases 498
Dogwood Anthracnose 501
Fruit and General Diseases Caused by Ascomycetes and Deuteromycetes (Mitosporic Fungi) 501
Ergot of Cereals and Grasses — Apple Scab — Brown Rot of Stone Fruits — Monoliophthora Pod Rot of Cacao — Botrytis Diseases — Black Rot of Grape — Cucurbit Gummy Stem Blight and Black Rot — Diaporthe, Phomopsis, and Phoma Diseases — Stem Canker of Soybeans — Melanose Disease of Citrus — Phomopsis Diseases — Black Rot of Apple 501

Vascular Wilts Caused by Ascomycetes and Deuteromycetes (Mitosporic Fungi) 522

Fusarium Wilts: Of Tomato — Fusarium or Panama Wilt of Banana 523

Verticillium Wilts 526

Ophiostoma Wilt of Elm Trees: Dutch Elm Disease 528

Ceratocystis Wilts — Oak wilt — Ceratocystis Wilt of Eucalyptus 532

Root and Stem Rots Caused by Ascomycetes and Deuteromycetes (Mitosporic Fungi) 534

Gibberella Diseases — Gibberella Stalk and Ear Rot, and Seedling Blight of Corn 535

Fusarium (Gibberella) Head Blight (FHB) or Scab of Small Grains 535

Fusarium Root and Stem Rots of Non-Grain Crops 538

Take-All of Wheat — Thielavopsis Black Root Rot — Monosporascus Root Rot and Vine Decline of Melons 540

Sclerotinia Diseases: Sclerotinia Diseases of Vegetables and Flowers — Phymatotrichum Root Rot 546

Postharvest Diseases of Plant Products Caused by Ascomycetes and Deuteromycetes 553

Postharvest Decays of Fruits and Vegetables 556

Aspergillus, Penicillium, Rhizopus, and Mucor — Alternaria — Botrytis — Fusarium — Geotrichum — Penicillium — Sclerotinia 556

Control of Postharvest Decays of Fresh Fruits and Vegetables 557

Postharvest Decays of Grain and Legume Seeds 558

Mycotoxins and Mycotoxicoses 559

Aspergillus Toxins — Aflatoxins 559

Fusarium Toxins — Other Aspergillus Toxins and Penicillium Toxins 559

Control of Postharvest Grain Decays 560

Diseases Caused by Basidiomycetes 562

The Rusts — The Smuts — Root and Stem Rots — Wood Rots and Decays — Witches' Broom 562

The Rusts 562

Cereal Rusts — Stem Rust of Wheat and Other Cereals 565

Rusts of Legumes — Bean Rust — Soybean Rust — A Major Threat to a Major Crop (Box) 571

Cedar-Apple Rust — Coffee Rust 574

Rusts of Forest Trees: — White Pine Blister Rust — Fusiform Rust 577

The Smuts 582

Corn Smut — Loose Smut of Cereals — Covered Smut, or Bunt, of Wheat 588

Karnal Bunt of Small Grains–Legitimate Concerns and Political Predicaments (Box) 592

Root and Stem Rots Caused by Basidiomycetes 593

Root and Stem Rot Diseases Caused by the "Sterile Fungi" Rhizoctonia and Sclerotium 593

Rhizoctonia Diseases — Sclerotium Diseases 594

Root Rots of Trees 602

Armillaria Root Rot of Fruit and Forest Trees 602

Wood Rots and Decays Caused by Basidiomycetes 604

Witches' Broom of Cacao 611

chapter twelve

PLANT DISEASES CAUSED BY PROKARYOTES: BACTERIA AND MOLLICUTES

Introduction 616

Plant Diseases Caused by Bacteria 618

Characteristics of Plant Pathogenic Bacteria 618

Morphology — Reproduction — Ecology and Spread — Identification of Bacteria — Symptoms Caused by Bacteria — Control of Bacterial Diseases of Plants 618

Bacterial Spots and Blights 627

Introduction — Wildfire of Tobacco — Bacterial Blights of Bean — Angular Leaf Spot of Cucumber — Angular Leaf Spot or Bacterial Blight of Cotton — Bacterial Leaf Spots and Blights of Cereals and Grasses — Bacterial Spot of Tomato and Pepper — Bacterial Speck of

Tomato — Bacterial Fruit Blotch of Watermelon — Cassava Bacterial Blight — Bacterial Spot of Stone Fruits 627

Bacterial Vascular Wilts 638

Bacterial Wilt of Cucurbits — Fire Blight of Pear and Apple — Southern Bacterial Wilt of Solanaceous Plants — Bacterial Wilt or Moko Disease of Banana — Ring Rot of Potato 639

Bacterial Canker and Wilt of Tomato — Bacterial Wilt (Black Rot) of Crucifers — Stewart's Wilt of Corn 651

Bacterial Soft Rots 656

Bacterial Soft Rots of Vegetables 656

The Incalculable Postharvest Losses from Bacterial (and Fungal) Soft Rots (Box) 660

Bacterial Galls 662

Crown Gall 662

The Crown Gall Bacterium — The Natural Genetic Engineer (Box) 664

Bacterial Cankers 667

Bacterial Canker and Gummosis of Stone Fruit Trees — Citrus Canker 667

Bacterial Scabs 674

Common Scab of Potato 667

Root Nodules of Legumes 675

Xylem-Inhabiting Fastidious Bacteria 678

Pierce's Disease of Grape — Citrus Variegated Chlorosis — Ratoon Stunting of Sugarcane 679

***Phloem-Inhabiting Fastidious Bacteria** 683*

Yellow Vine Disease of Cucurbits — Citrus Greening Disease — Papaya Bunchy Top Disease 684

Plant Diseases Caused By Mollicutes: Phytoplasmas and Spiroplasmas 687

Properties of True Mycoplasmas — Phytoplasmas — Spiroplasmas 688

Examples of Plant Diseases Caused by Mollicutes 691

Aster Yellows — Lethal Yellowing of Coconut Palms — Apple Proliferation — European Stone Fruit Yellows — Ash Yellows — Elm Yellows (Phloem Necrosis) — Peach X-Disease — Pear Decline 691

Spiroplasma Diseases 699

Citrus Stubborn Disease — Corn Stunt Disease 691

chapter thirteen

PLANT DISEASES CAUSED BY PARASITIC HIGHER PLANTS, INVASIVE CLIMBING PLANTS, AND PARASITIC GREEN ALGAE

Introduction — Parasitic Higher Plants 705

Dodder –Witchweed — Broomrapes — Dwarf Mistletoes of Conifers — True or Leafy Mistletoes 706

Invasive Climbing Plants 716

Old World Climbing Fern — Kudzu Vine 717

Parasitic Green Algae: Cephaleuros 719

Plant Diseases Caused by Algae 719

chapter fourteen

PLANT DISEASES CAUSED BY VIRUSES

Introduction 724

Characteristics of Plant Viruses 724

Detection — Morphology — Composition and Structure: Of Viral Protein — Of Viral Nucleic Acid 725

Satellite Viruses and Satellite RNAs 731

The Biological Function of Viral Components: Coding 731

Virus Infection and Virus Synthesis 731

Translocation and Distribution of Viruses in Plants 733

Symptoms Caused by Plant Viruses 734

Physiology of Virus-Infected Plants 737

Transmission of Plant Viruses By: Vegetative Propagation — Sap — Seed — Pollen — Insects — Mites — Nematodes — Fungi — Dodder 737

Epidemiology of Plant Viruses and Viroids 743

Purification of Plant Viruses — Serology of Plant Viruses 743
Nomenclature and Classification of Plant Viruses 747
Detection and Identification of Plant Viruses 751
Economic Importance of Plant Viruses 752
Control of Plant Viruses 753
Diseases Caused by Rigid Rod-Shaped Viruses 757
Diseases Caused by Tobamoviruses: — Tobacco Mosaic 757
The Contribution of Tobacco Mosaic Virus to Biology and Medicine (Box) 757
Diseases Caused by Tobraviruses: — Tobacco Rattle by Furoviruses — by Hordeiviruses — by Pecluviruses — by Pomoviruses — by Benyviruses 758
Diseases Caused by Filamentous Viruses 762
Diseases Caused by Potexviruses — by Carlaviruses — by Capilloviruses and Trichoviruses — by Allexiviruses, Foveaviruses, and Vitiviruses 762
Diseases Caused by Potyviridae 764
Diseases Caused by Potyviruses 764
Bean Common Mosaic and Bean Yellow Mosaic — Lettuce Mosaic — Plum Pox — Papaya Ringspot — Potato Virus Y — Sugarcane Mosaic — Tobacco Etch — Turnip Mosaic — Watermelon Mosaic — Zucchini Yellow Mosaic 767
Diseases Caused by Ipomoviruses, Macluraviruses, Rymoviruses, and Tritimoviruses — by Bymoviruses 773
Diseases Caused by Closteroviridae 774
Diseases Caused by Closteroviruses: — Citrus Tristeza — Beet YellowsDiseases Caused by Criniviruses: — Lettuce Infectious Yellows 774
Diseases Caused by Isometric Single-Stranded RNA Viruses 779
Diseases Caused by Sequiviridae, Genus Waikavirus 779
Rice TungroDiseases Caused by Tombusviridae 779
Diseases Caused by Luteoviridae 781
Barley Yellow Dwarf — Potato Leafroll — Beet Western Yellows 781
Diseases Caused by Monopartite Isometric (+)ssRNA Viruses of Genera Not Yet Assigned to Families 783
Diseases Caused by Comoviridae 784
Diseases Caused by Comoviruses 784
Diseases Caused by Nepoviruses 784
Tomato Ring Spot — Grapevine Fanleaf — Raspberry Ring Spot 785
Diseases Caused by Bromoviridae 787
Diseases Caused by Cucumoviruses 787
Cucumber Mosaic 788
Diseases Caused by Ilarviruses: Prunus Necrotic Ring Spot 790
Diseases Caused by Isometric Double-Stranded RNA Viruses 792
***Diseases Caused by Reoviridae** 792*
Diseases Caused by Negative RNA [(–)ssRNA] Viruses 794
Plant Diseases Caused by Rhabdoviruses 794
Plant Diseases Caused by Tospoviruses 795
Plant Diseases Caused by Tenuiviruses 799
Diseases Caused by Double-Stranded DNA Viruses 801
Diseases Caused by Caulimoviruses and Other Isometric Caulimoviridae 801
Diseases Caused by Badnaviruses 803
Diseases Caused by Single-Stranded DNA Viruses 805
Plant Diseases Caused by Geminiviridae 805
Beet Curly Top — Maize Streak — African Cassava Mosaic — Bean Golden Mosaic — Squash Leaf Curl — Tomato Mottle — Tomato Yellow Leaf Curl 809
Plant Diseases Caused by Isometric Single-Stranded 813
DNA Viruses: The Circoviridae 813
Banana Bunchy Top 814

Coconut Foliar Decay 815

Viroids 816

Plant Diseases Caused by Viroids 816

Taxonomy (Grouping) of Viroids 816

Potato Spindle Tuber — Citrus Exocortis — Coconut Cadang-Cadang 820

chapter fifteen

PLANT DISEASES CAUSED BY NEMATODES

Introduction 826

Characteristics of Plant Pathogenic Nematodes 827

Morphology — Anatomy — Life Cycles — Ecology and Spread — Classification 827

Isolation of Nematodes 831

Isolation of Nematodes from Soil 831

Isolation of Nematodes from Plant Material 832

Symptoms Caused by Nematodes 832

How Nematodes Affect Plants 833

Interrelationships between Nematodes and Other Plant Pathogens 835

Control of Nematodes 836

Important Nematodes and Diseases 838

Root-Knot Nematodes: *Meloidogyne* 838

Cyst Nematodes: *Heterodera* and *Globodera* 842

Soybean Cyst Nematode: Heterodera glycines 843

Sugar Beet Nematode: Heterodera schachtii 846

Potato Cyst Nematode: Globodera rostochiensis and Globodera pallida 847

The Citrus Nematode: *Tylenchulus Semipenetrans* 848

Lesion Nematodes: *Pratylenchus* 849

The Burrowing Nematode: *Radopholus* 853

The Added Significance of Plant Nematodes in the Tropics and Subtropics (Box) 858

Stem and Bulb Nematode: *Ditylenchus* 858

Sting Nematode: *Belonolaimus* 860

Stubby-Root Nematodes: *Paratrichodorus* and *Trichodorus* 863

Seed-Gall Nematodes: *Anguina* 865

Foliar Nematodes: *Aphelenchoides* 867

Pine Wilt and Palm Red Ring Diseases: *Bursaphelenchus* 870

Pine Wilt Nematode: Bursaphelenchus xylophilus 870

Red Ring Nematode: Bursaphelenchus cocophilus 872

chapter sixteen

PLANT DISEASES CAUSED BY FLAGELLATE PROTOZOA

Introduction 875

Nomenclature of Plant Trypanosomatids — Taxonomy — Pathogenicity — Epidemiology and Control of Plant Trypanosomatids 877

Plant Diseases Caused by: 878

Phloem-Restricted Trypanosomatids 878

Phloem Necrosis of Coffee — Hartrot of Coconut Palms — Sudden Wilt (Marchitez Sopresiva) of Oil Palm — Wilt and Decay of Red Ginger 878

Laticifer-Restricted trypanosomatids 882

Empty Root of Cassava 882

Fruit-and Seed-Infecting Trypanosomatids 882

Fruit Trypanosomatids 882

Glossary 887

Index 903

Preface

Since the appearance of the 1st edition of *Plant Pathology* in June 1969, tremendous advances have been made both in the science of plant pathology and in the publishing business. New information published in the monthly plant pathological and related biological journals, as well as in specialized books and annual reviews, was digested and pertinent portions of it were included in each new edition of the book. The worldwide use of the book, in English or in its several translations, also created a need to describe additional diseases affecting crops important to different parts of the world. There has been, therefore, a continuous need to add at least some additional text and more illustrations to the book with as little increase in the size of the book as possible. Fortunately, through the use of computers, tremendous advances have been made in the publishing business, including paper quality and labor costs and, particularly, in the reproducibility and affordability of color photographs and diagrams. Plant diseases and plant pathology come alive when illustrated in full color and it has been the author's dream to have all the figures in color. Add to these advances the interest of the author and of the publishers to spare no effort or expense in the production of this book and you have what we believe is the best book possible for the effective teaching of plant pathology at today's college level worldwide.

To begin with, "Plant Pathology, 5th edition" provides each instructor with all the significant new developments in each area and gives the instructor choices in the type and amount of general concepts material (Chapters 1–9) and of specific diseases (Chapters 10–16) he/she will cover. Each chapter begins with a fairly detailed, well-organized table of contents that can be used by students and instructors as an outline for the chapter. The instructor can also use it to cover parts of it in detail in class while some of the topics are covered briefly and others are assigned to the students as further reading. Each student, however, has all the latest material, well organized and beautifully illustrated, available in a way that is self-explanatory and, with the complete glossary provided, can be understood with minimal effort.

Instructors will have an even greater choice in the kinds of specific diseases one would use in a specific area of the country or of the world where one teaches. While one may want to include the teaching of potato late blight, apple scab, wheat rust, bacterial soft rot, root knot, and some other diseases of general interest, one often also wants to cover diseases of particular interest in the region, both because of their regional importance and because of their availability locally for further study in the classroom and the laboratory. This edition makes this possible by covering and illustrating in full color a wide variety of diseases, some of which are important to the grain plains of the Midwest and the northwestern United States, others to the fruit- and vegetable-producing Pacific and northeastern states, others to the cotton-, peanut-, tobacco-, rice-, and citrus-vegetable producing southern states, and so on. A special effort has also been made to describe and to fully illustrate in full color several diseases of tropical crops important in different parts of the world, such as rice in the Far East, beans in Central and South America, cassava, cacao, and sorghum in Africa, and tropical fruits such as citrus, papaya, coconut, and coffee in the Americas, and so on. Instructors can pick and choose to study, in the classroom and, if possible, in the laboratory, whatever diseases of whichever crops they deem most significant for the particular area and for the ever-shrinking world we all live in.

The overall arrangement of this edition is similar to that of previous editions. However, all aspects of the book have been thoroughly updated and illustrated. Newly discovered diseases and pathogens are described, and changes in pathogen taxonomy and nomenclature are incorporated in the text. Changes or refinements in plant disease epidemiology and new approaches and new materials used for plant disease control are discussed. The chapters on diseases caused by prokaryotes (bacteria and mollicutes), especially the one on diseases caused by plant viruses and viroids, have been revamped due to the large amount of new information published in recent years about such pathogens and diseases. And in all cases, partial tables of contents have been added to each chapter and to its main subdivisions for better clarity and understanding of the arrangement and inclusion of the topics in the appropriate subdivisions. A new feature that has been added to the book is the presentation of a number of topics of special interest in separate boxes. In these, the various topics are approached from a different angle and highlight the importance of the topic whether it has historical, political, or scientific significance. Special attention has also been given to highlighting the historical developments in plant pathology and the scientists or others who contributed significantly to these developments.

As in other recent editions, much of the progress in plant pathology has been in the areas of molecular genetics and its use in developing defenses in plants, against pathogens. Discoveries in basic molecular genetics, particularly discoveries in how plants defend themselves against pathogens and in the development of mechanisms to produce disease resistant plants, receive extensive coverage. It is recognized that some of the included material in Chapters 4 (Genetics of Disease), 5 (How Pathogens Attack Plants), and 6 (How Plants Defend Themselves against Pathogens) may be both too much for students taking plant pathology for the first time and somewhat difficult to follow and comprehend. However, the importance of that material to the future development of plant pathology as a science and its potential future impact on control of plant diseases is so great that its inclusion is considered justified if only to expose and initiate the students to these developments.

There are numerous colleagues to whom I am indebted for suggestions and for providing me with numerous slides or electronic images of plant disease symptoms or plant pathology concepts that are used in the book. Their names are listed in the legend(s) of the figures they gave me and in the list of "Photo Credits." I would particularly like to express my sincere appreciation and thanks to Dr. Ieuan R. Evans of the Agronomy Unit of the Alberta Agriculture, Edmonton, Alberta, Canada, who, as editor of the slide collection of the Western Committee on Plant Disease Control, provided me with hundreds of excellent slides and permission to use them in the book. I also thank Dr. Wen Yuan Song for reviewing the chapter on "How Plants Defend Themselves against Pathogens." Finally, I again thank publicly my wife Annette for the many hours she spent helping me organize, copy, scan, and reorganize the many slides, prints, and diagrams used in this book. Not only did she do it better, she also did it faster than I could have done it.

George N. Agrios
July 2004

Photo Credits

The need for high-quality photographs to include in this book necessitated the request of appropriate photographs from colleagues around the world. All of them responded positively and I am very thankful to all of them. I am particularly indebted to the following individuals and organizations who, although I was asking from them one or a few photographs, sent me those plus all the related or other pertinent photographs that I might want to use in the new edition of the book. Moreover, several of them offered to give me any other photographs they had and which I might want to use.

I am particularly indebted to Dr. Ieuan R. Evans of the Agronomy Unit, Agriculture, Food, and Rural Development of Alberta, Canada, for providing me with several hundreds of slides put together by the Western Committee on Plant Diseases (WCPD) for general use for educational purposes. Those contributing slides through the WCPD include P. K. Basu, Agriculture Canada, Ottawa, Ontario; J. G. N. Davidson, Agric. Canada, Beaverlodge, Alberta; P. Ellis, Agric. Canada, Vancouver, British Columbia; I. R. Evans, Agric. Canada, Edmonton, Alberta; G. Flores, Agric. Canada, Ottawa, Ontario; E. J. Hawn, Agric. Canada, Lethbridge, Alberta; R. J. Howard, Alberta Agriculture, Brooks, Alberta; H. C. Huang, Agriculture Canada, Lethbridge, Alberta; J. E. Hunter, NYAES, Geneva, New York; G. A. Nelson, Agriculture Canada, Lethbridge, Alberta; R. G. Platford, Manitoba Department of Agriculture, Winnipeg, Manitoba; and C. Richard, Agriculture Canada, Sainte-Foy, Quebec.

I am equally indebted to Dr. Gail Wisler, Chair, Plant Pathology Department, University of Florida, for allowing me to use whatever slides of the departmental Plant Disease Clinic would be useful in illustrating the book. Since all of the slides were stamped with the name of Dr. G. W. Simone, and some of them were undoubtedly taken by him while he was an Extension Plant Pathologist in charge of the Plant Disease Clinic in the Department, now retired, I would like to express my thanks to Dr. Simone also.

I am also thankful to several other organizations that gave me permission to use many of their photographs and offered to give me any others I might need. They include the Extension Service of the University of Florida Institute of Food and Agricultural Sciences (UF/IFAS), the American Phytopathological Society, and several United States Department of Agriculture (USDA) Laboratories. I am particularly thankful to the USDA Forest Service along with the University of Georgia who, through "Forestry Images" and "Bugwood Network," provided me with several images of forest tree diseases.

I am particularly indebted to the following colleagues, listed alphabetically, each of whom gave me numerous slides or electronic images and offered to give me as many more of their photographs as I needed: Dr. Eduardo Alves, Federal University of Lavras, Brazil; Dr. Mohammad Babadoost, University of Illinois; Dr.

Edward L. Barnard, Florida Division of Forestry, Forest Health Section; Dr. Benny D. Bruton, USDA, ARS, Lane, Oklahoma; Dr. David J. Chitwood, USDA, Nematology Lab, Beltsville, Maryland; Dr. Daniel R. Cooley, University of Massachusetts; Dr. Danny Coyne, CGIAR, intern. Institute Tropical Agriculture, Ibadan, Nigeria; Richard Cullen, University of Florida; Dr. L. E. Datnoff, University of Florida; Dr. Donald W. Dickson, University of Florida; Dr. Michel Dollet, CIRAD, Montpellier, France; Dr. Michael Ellis, Ohio State University; Mark Gouch, University of Florida; Dr. Edward Hellman, Texas A&M University; Dr. Ernest Hiebert, University of Florida; Dr. Donald L. Hopkins, University of Florida; Jackie Hughes, Intern. Institute of Tropical Agriculture, Ibadan, Nigeria; Dr. Bruce Jaffee, University of California; Dr. Alan L. Jones, Michigan State University; Dr. Daniel E. Legard, University of Florida; Dr. Patrick E. Lipps, Ohio State University; Dr. Don E. Mathre, Montana State University; Dr. Robert J. McGovern, University of Florida; Dr. Robert T. McMillan, Jr., University of Florida; Dr. Charles W. Mims, University of Georgia; Dr. Krishna S. Mohan, University of Idaho; Dr. Lytton John Musselman, American University of Beirut, Lebanon; Dr. Steve Nameth, Ohio State University; Dr. Joe W. Noling, University of Florida; Dr. Kenneth I. Pernezny, University of Florida; Dr. Jay W. Pscheidt, Oregon State University; Dr. H. David Thurston, Cornell University; Dr. James W. Travis and Jo Rytter, Pennsylvania State University; Dr. Tom Van Der Zwet, USDA, retired; Dr. David P. (Pete) Weingartner, University of Florida; and Dr. Tom Zitter, Cornell University.

I am equally thankful to the following colleagues, also listed alphabetically, who provided me with the photographs I requested of them: Dr. Luis Felipe Arauz, Universitad de Costa Rica, San Jose; Dr. Gavin Ash, Charles Sturt University, Australia; Dr. Donald E. Aylor, Connecticut Agric. Experimental Station, New Haven; Dr. Ranajit Bandyopathyay, CGIAR, Nigeria; Dr. George Barron, University of Guelph; Dr. Gwen A. Beattie, Iowa State University; Dr. Dale Bergdahl, University of Vermont; Dr. Ian Breithaupt, AGPP, FAO; Dr. Scott Cameron, International Paper Co.; Dr. Mark Carlton, Iowa State University; Dr. Asita Chatterjee, University of Missouri; Dr. C. M. Christensen (via Dr. Frank Pfleger), University of Minnesota; Dr. William T. Crow, University of Florida; Dr. Howard Davis, Scottish Agricultural Research Institute, UK; Dr. Michael J. Davis, University of Florida; Dr. O. Dooling, USDA Forest Service; Dr. Sharon Douglas, Connecticut Agric. Experimental Station, New Haven; Dr. Robert A. Dunn, University of Florida; Dr. D. Dwinell, USDA Forest Service; Dr. D. M. Elgersma, The Netherlands; Shep Eubanks, University of Florida; Dr. Stephen Ferreira, University of Hawaii; Dr. Catherine Feuillet, University of Zurich; Dr. Robert L. Forster, University of Idaho; Dr. L. Giunchedi, University of Bologna, Italy; Dr. Tim Gottwald, USDA, Ft. Pierce, Florida; Dr. James H. Graham, University of Florida; Dr. Sarah Gurr, Oxford University, UK; Dr. Everett Hansen, Oregon State University; Dr. Mary Ann Hansen, Virginia Tech University; Dr. Thomas C. Harrington, Iowa State University; Dr. Robert Hartzler, Iowa State University; Dr. Robert Harveson, University of Nebraska; Dr. Kenneth D. Hickey, Pennsylvania State University; Dr. Richard B. Hine, University of Arizona; Dr. Molly E. Hoffer, Oregon State University; Dr. Harry Hoitink, Ohio State University; Dr. Tom Isakeit, Texas A&M University; Dr. Ramon Jaime, USDA, New Orleans; Dr. Wojciech Janisiewicz, USDA, Appalachian Fruit Res., West Virginia; Dr. P. Maria Johansson, Plant Pathology and Biocontrol Unit, Sweden; Dr. R. Johnston, USDA; Dr. Robert Johnston, Montana State University; Dr. Linda Kinkel, University of Minnesota; Dr. Jurgen Kranz, University of Giessen, Germany; Dr. Richard F. Lee, University of Florida; Dr. Mark Longstroth, Michigan State University; Dr. Rosemary Loria, Cornell University; Dr. Otis Maloy, Washington State University; Dr. Douglas H. Marin, Banana Development Corp., San Jose, Costa Rica; Dr. Don Maynard, University of Florida; Dr. Patricia McManus, University of Wisconsin; Dr. Glenn Michael, Appalachian Fruit Res., West Virginia; Dr. Themis Michailides, University of California; Dr. Gary Munkvold, Pioneer Hybrid Int., Johnston, Fowa; Dr. Cynthia M. Ocamb, Oregon State University; Dr. Laud A. Ollennou, Cocoa Research Institute, Ghana; Dr. Tapio Palva, University of Helsinki, Finland; Dr. Frank Phleger (for C. M. Christensen), University of Minnesota; Dr. Mary Powelson, Oregon State University; Dr. David F. Ritchie, North Carolina State University; Dr. Chester Roistacher, University of California; Dr. John P. Ross, North Carolina State University; Dr. Randall Rowe, Ohio State University; Dr. Robert Stack, North Dakota State University; Dr. James R. Steadman, University of Nebraska; Dr. Brian J. Steffenson, University of Minnesota; Dr. R. J. Stipes, Virginia Tech University; Dr. Virginia Stockwell, Oregon State University; Dr. Krishna V. Subbarao, University of California; Dr. Pavel Svihra, University of California; Dr. Beth Teviotdale, University of California; Dr. L. W. Timmer, University of Florida; Dr. Greg Tylka, Iowa State University; Dr. S. V. van Vuuren, ARC-ITSC, Nelspruit, South Africa; Dr. John A. Walsh, Horticultural Research Institute, UK; Dr. Robert K. Webster, University of California, Davis; Dr. Wickes Westcott, Clemson University; Dr. Carol Windels, University of Minnesota; Dr. X. B. Yang, Iowa State University; and Dr. Ulrich Zunke, Hamburg, Germany.

About the Author

Professor George N. Agrios was born in Galarinos, Halkidiki, Greece. He received his B.S. degree in horticulture from the Aristotelian University of Thessaloniki, Greece, in 1957, and his Ph.D. degree in plant pathology from Iowa State University in 1960. Following graduation he served 2 years as an officer in the Engineering Corps of the Greek army. In January 1963 he was hired as an assistant professor of plant pathology at the University of Massachusetts at Amherst. His assignment was 50% teaching and 50% research on viral diseases of fruits and vegetables. His teaching included courses in introductory plant pathology, general plant pathology, plant virology, and diseases of florist's crops. His research included studies on epidemiology, genetics, and physiology of viral diseases of apple, cucurbits, pepper, and corn, in which he directed the studies of 25 graduate students and published numerous journal publications. Dr. Agrios was promoted to associate professor in 1969 and to professor in 1976.

In 1969, he published the first edition of the textbook "Plant Pathology" through Academic Press. The book was adopted for plant pathology classes at almost all universities of the United States and Canada and of most other English-speaking countries. The first edition was later followed by the 2nd edition (1978), 3rd edition (1987), and 4th edition (1997). The book was translated into several major languages, including Spanish, Arabic, Chinese, Korean, and Indochinese, and became the standard plant pathology text throughout the world.

In the meantime, Dr. Agrios served on several departmental, college and university committees as well as committees of the northeastern division of the American Phytopathological Society (APS) and of the national APS. He was elected president of the northeastern division (1980) of APS. He was instrumental in founding the APS Press, of which he served as the first editor-in-chief (1984–1987). He was elected vice-president of APS in 1988, serving as vice-president , president-elect, and president (1990 and 1991). In 1988, professor Agrios accepted a position as chairman of the Plant Pathology Department of the University of Florida, overseeing approximately 50 Ph.D. plant pathologist faculty. Half of the faculty were located at the university campus in Gainesville, Florida, while the others worked at 1 of 13 agricultural research centers throughout the state of Florida where they studied all types of diseases of various crops. In 1999, the Florida Board of Regents approved the establishment of the new and unique Doctor of Plant Medicine Program and professor Agrios was appointed its first director. In 2002, Dr. Agrios relinquished his position as chairman of the Plant Pathology Department to concentrate on his duties as director of the Doctor of Plant Medicine Program. In June 2002, however, health reasons forced Dr. Agrios to retire from the University of Florida.

part one

GENERAL ASPECTS

chapter one

INTRODUCTION

PROLOGUE: THE ISSUES
4

PLANTS AND DISEASE
4

HISTORY OF PLANT PATHOLOGY AND EARLY SIGNIFICANT PLANT DISEASES
8

LOSSES CAUSED BY PLANT DISEASES
29

PLANT PATHOLOGY IN THE 20TH CENTURY
46

PLANT PATHOLOGY TODAY AND FUTURE DIRECTIONS
54

WORLDWIDE DEVELOPMENT OF PLANT PATHOLOGY AS A PROFESSION
60

PLANT PATHOLOGY'S CONTRIBUTION TO CROPS AND SOCIETY
65

BASIC PROCEDURES IN THE DIAGNOSIS OF PLANT DISEASES
72

PROLOGUE: THE ISSUES

Plant pathology is a science that studies plant diseases and attempts to improve the chances for survival of plants when they are faced with unfavorable environmental conditions and parasitic microorganisms that cause disease. As such, plant pathology is challenging, interesting, important, and worth studying in its own right. It is also, however, a science that has a practical and noble goal of protecting the food available for humans and animals. Plant diseases, by their presence, prevent the cultivation and growth of food plants in some areas; or food plants may be cultivated and grown but plant diseases may attack them, destroy parts or all of the plants, and reduce much of their produce, i.e., food, before they can be harvested or consumed. In the pursuit of its goal, plant pathology is joined by the sciences of entomology and weed science.

It is conservatively estimated that diseases, insects, and weeds together annually interfere with the production of, or destroy, between 31 and 42% of all crops produced worldwide (Table 1-1). The losses are usually lower in the more developed countries and higher in the developing countries, i.e., countries that need food the most. It has been estimated that of the 36.5% average of total losses, 14.1% are caused by diseases, 10.2% by insects, and 12.2% by weeds.

Considering that 14.1% of the crops are lost to plant diseases alone, the total annual worldwide crop loss from plant diseases is about $220 billion. To these should be added 6–12% losses of crops after harvest, which are particularly high in developing tropical countries where training and resources such as refrigeration are generally lacking. Also, these losses do not include losses caused by environmental factors such as freezes, droughts, air pollutants, nutrient deficiencies, and toxicities.

Although impressive, the aforementioned numbers do not tell the innumerable stories of large populations in many poor countries suffering from malnutrition, hunger, and starvation caused by plant diseases; or of lost income and lost jobs resulting from crops destroyed by plant diseases, forcing people to leave their farms and villages to go to overcrowded cities in search of jobs that would help them survive.

TABLE 1-1
Estimated Annual Crop Losses Worldwide

Attainable crop production (2002 prices)	$1.5 trillion
Actual crop production (–36.5%)	$950 billion
Production without crop protection	$455 billion
Losses prevented by crop protection	$415 billion
Actual annual losses to world crop production	$550 billion
Losses caused by diseases only (14.1%)	$220 billion

Moreover, the need for measures to control plant diseases limits the amount of land available for cultivation each year, limits the kinds of crops that can be grown in fields already contaminated with certain microorganisms, and annually necessitates the use of millions of kilograms of pesticides for treating seeds, fumigating soils, spraying plants, or the postharvest treatment of fruits. Such control measures not only add to the cost of food production, some of them, e.g., crop rotation, necessarily limit the amount of food that can be produced, whereas others add toxic chemicals to the environment. It is therefore the duty and goal of plant pathology to balance all the factors involved so that the maximum amount of food can be produced with the fewest adverse side effects on the people and the environment.

PLANTS AND DISEASE

Plants make up the majority of the earth's living environment as trees, grass, flowers, and so on. Directly or indirectly, plants also make up all the food on which humans and all animals depend. Even the meat, milk, and eggs that we and other carnivores eat come from animals that themselves depend on plants for their food. Plants are the only higher organisms that can convert the energy of sunlight into stored, usable chemical energy in carbohydrates, proteins, and fats. All animals, including humans, depend on these plant substances for survival.

Plants, whether cultivated or wild, grow and produce well as long as the soil provides them with sufficient nutrients and moisture, sufficient light reaches their leaves, and the temperature remains within a certain "normal" range. Plants, however, also get sick. Sick plants grow and produce poorly, they exhibit various types of symptoms, and, often, parts of plants or whole plants die. It is not known whether diseased plants feel pain or discomfort.

The agents that cause disease in plants are the same or very similar to those causing disease in humans and animals. They include pathogenic microorganisms, such as viruses, bacteria, fungi, protozoa, and nematodes, and unfavorable environmental conditions, such as lack or excess of nutrients, moisture, and light, and the presence of toxic chemicals in air or soil. Plants also suffer from competition with other, unwanted plants (weeds), and, of course, they are often damaged by attacks of insects. Plant damage caused by insects, humans, or other animals is not usually included in the study of plant pathology.

Plant pathology is the study of the organisms and of the environmental factors that cause disease in plants; of the mechanisms by which these factors induce disease in plants; and of the methods of preventing or controlling disease and reducing the damage it causes. Plant pathology is for plants largely what medicine is for humans and veterinary medicine is for animals. Each discipline studies the causes, mechanisms, and control of diseases affecting the organisms with which it deals, i.e., plants, humans, and animals, respectively.

Plant pathology is an integrative science and profession that uses and combines the basic knowledge of botany, mycology, bacteriology, virology, nematology, plant anatomy, plant physiology, genetics, molecular biology and genetic engineering, biochemistry, horticulture, agronomy, tissue culture, soil science, forestry, chemistry, physics, meteorology, and many other branches of science. Plant pathology profits from advances in any one of these sciences, and many advances in other sciences have been made in attempts to solve plant pathological problems.

As a science, plant pathology tries to increase our knowledge about plant diseases. At the same time, plant pathology tries to develop methods, equipment, and materials through which plant diseases can be avoided or controlled. Uncontrolled plant diseases may result in less food and higher food prices or in food of poor quality. Diseased plant produce may sometimes be poisonous and unfit for consumption. Some plant diseases may wipe out entire plant species and many affect the beauty and landscape of our environment. Controlling plant disease results in more food of better quality and a more aesthetically pleasing environment, but consumers must pay for costs of materials, equipment, and labor used to control plant diseases and, sometimes, for other less evident costs such as contamination of the environment.

In the last 100 years, the control of plant diseases and other plant pests has depended increasingly on the extensive use of toxic chemicals (pesticides). Controlling plant diseases often necessitates the application of such toxic chemicals not only on plants and plant products that we consume, but also into the soil, where many pathogenic microorganisms live and attack the plant roots. Many of these chemicals have been shown to be toxic to nontarget microorganisms and animals and may be toxic to humans. The short- and long-term costs of environmental contamination on human health and welfare caused by our efforts to control plant diseases (and other pests) are difficult to estimate. Much of modern research in plant pathology aims at finding other environmentally friendly means of controlling plant diseases. The most promising approaches in-clude conventional breeding and genetic engineering of disease-resistant plants, application of disease-suppressing cultural practices, RNA- and gene-silencing techniques, of plant defense-promoting, nontoxic substances, and, to some extent, use of biological agents antagonistic to the microorganisms that cause plant disease.

The challenges for plant pathology are to reduce food losses while improving food quality and, at the same time, safeguarding our environment. As the world population continues to increase while arable land and most other natural resources continue to decrease, and as our environment becomes further congested and stressed, the need for controlling plant diseases effectively and safely will become one of the most basic necessities for feeding the hungry billions of our increasingly overpopulated world.

The Concept of Disease in Plants

Because it is not known whether plants feel pain or discomfort and because, in any case, plants do not speak or otherwise communicate with us, it is difficult to pinpoint exactly when a plant is diseased. It is accepted that a plant is healthy, or normal, when it can carry out its physiological functions to the best of its genetic potential. The meristematic (cambium) cells of a healthy plant divide and differentiate as needed, and different types of specialized cells absorb water and nutrients from the soil; translocate these to all plant parts; carry on photosynthesis, translocate, metabolize, or store the photosynthetic products; and produce seed or other reproductive organs for survival and multiplication. When the ability of the cells of a plant or plant part to carry out one or more of these essential functions is interfered with by either a pathogenic organism or an adverse environmental factor, the activities of the cells are disrupted, altered, or inhibited, the cells malfunction or die, and the plant becomes diseased. At first, the affliction is localized to one or a few cells and is invisible. Soon, however, the reaction becomes more widespread and affected plant parts develop changes visible to the naked eye. These visible changes are the symptoms of the disease. The visible or otherwise measurable adverse changes in a plant, produced in reaction to infection by an organism or to an unfavorable environmental factor, are a measure of the amount of disease in the plant. Disease in plants, then, can be defined as  the series of invisible and visible responses of plant cells and tissues to a pathogenic organism or environmental factor that result in adverse changes in the form, function, or integrity of the plant and may lead to partial impairment or death of plant parts or of the entire plant.

The kinds of cells and tissues that become affected determine the type of physiological function that will be disrupted first (Fig. 1-1). For example, infection of roots may cause roots to rot and make them unable to absorb water and nutrients from the soil; infection of xylem vessels, as happens in vascular wilts and in some cankers, interferes with the translocation of water and minerals to the crown of the plant; infection of the foliage, as happens in leaf spots, blights, rusts, mildews, mosaics, and so on, interferes with photosynthesis; infection of phloem cells in the veins of leaves and in the bark of stems and shoots, as happens in cankers and in diseases caused by viruses, mollicutes, and protozoa, interferes with the downward translocation of photosynthetic products; and infection of flowers and fruits interferes with reproduction. Although infected cells in most diseases are weakened or die, in some diseases, e.g., in crown gall, infected cells are induced to divide much faster (hyperplasia) or to enlarge a great deal more (hypertrophy) than normal cells and to produce

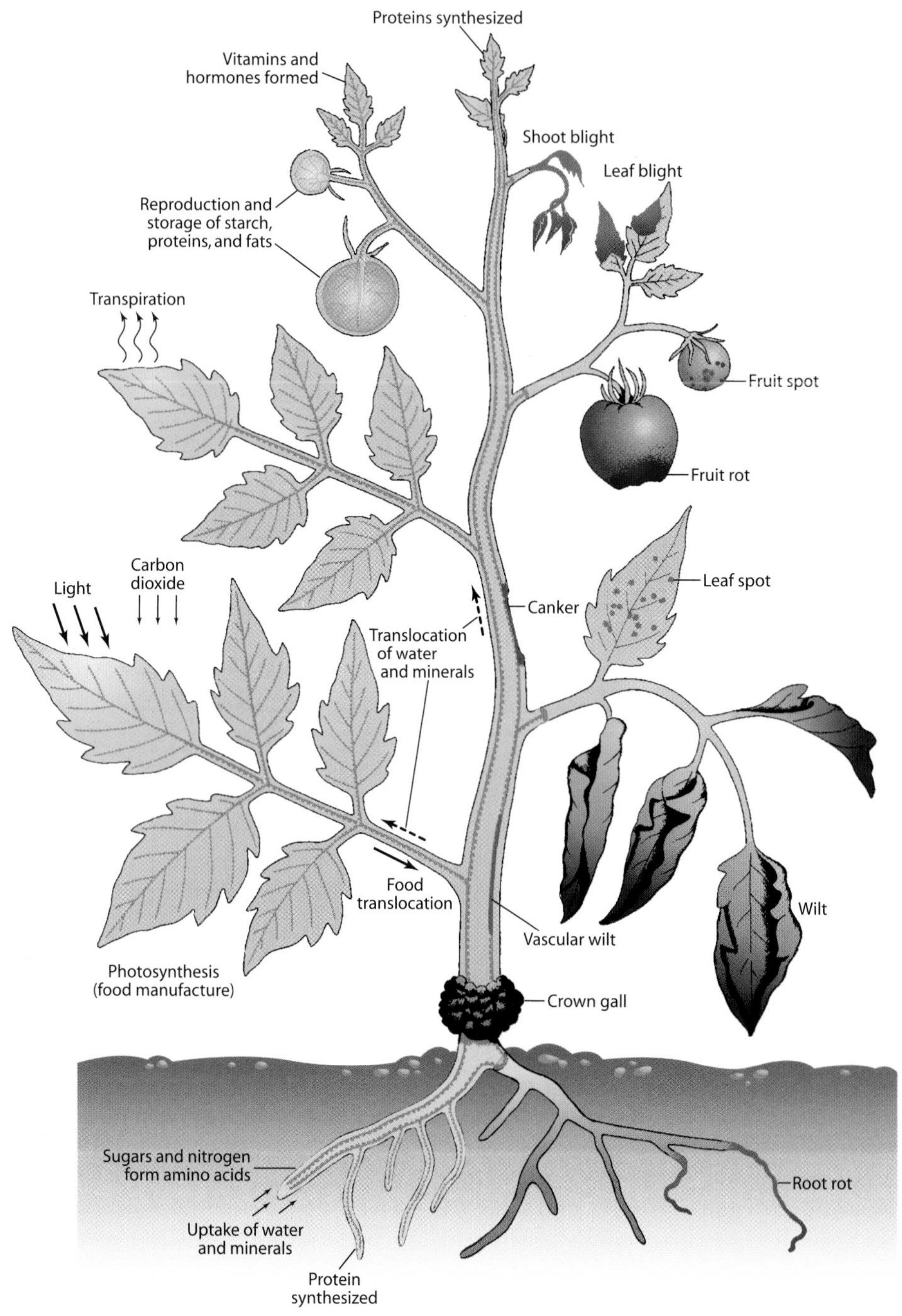

FIGURE 1-1 Schematic representation of the basic functions in a plant (left) and of the kinds of interference with these functions (right) caused by some common types of plant diseases.

abnormal amorphous overgrowths (tumors) or abnormal organs.

Pathogenic microorganisms, i.e., the transmissible biotic (= living) agents that can cause disease and are generally referred to as pathogens, usually cause disease in plants by disturbing the metabolism of plant cells through enzymes, toxins, growth regulators, and other substances they secrete and by absorbing foodstuffs from the host cells for their own use. Some pathogens may also cause disease by growing and multiplying in the xylem or phloem vessels of plants, thereby blocking the upward transportation of water or the downward movement of sugars, respectively, through these tissues. Environmental factors cause disease in plants when abiotic factors, such as temperature, moisture, mineral nutrients, and pollutants, occur at levels above or below a certain range tolerated by the plants.

Types of Plant Diseases

Tens of thousands of diseases affect cultivated and wild plants. On average, each kind of crop plant can be affected by a hundred or more plant diseases. Some pathogens affect only one variety of a plant. Other pathogens affect several dozen or even hundreds of species of plants. Plant diseases are sometimes grouped according to the symptoms they cause (root rots, wilts, leaf spots, blights, rusts, smuts), to the plant organ they affect (root diseases, stem diseases, foliage diseases), or to the types of plants affected (field crop diseases, vegetable diseases, turf diseases, etc.). One useful criterion for grouping diseases is the type of pathogen that causes the disease (see Figs. 1-2 and 1-3). The advantage of such a grouping is that it indicates the cause of the disease, which immediately suggests the probable

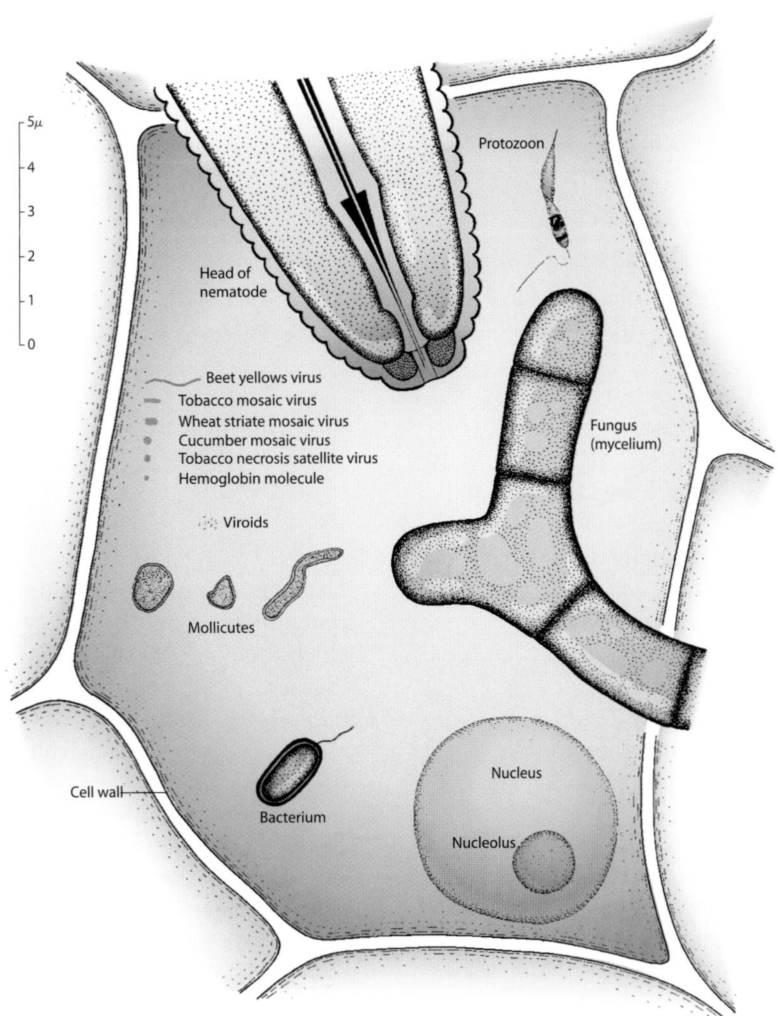

FIGURE 1-2 Schematic diagram of the shapes and sizes of certain plant pathogens in relation to a plant cell. Bacteria, mollicutes, and protozoa are not found in nucleated living plant cells.

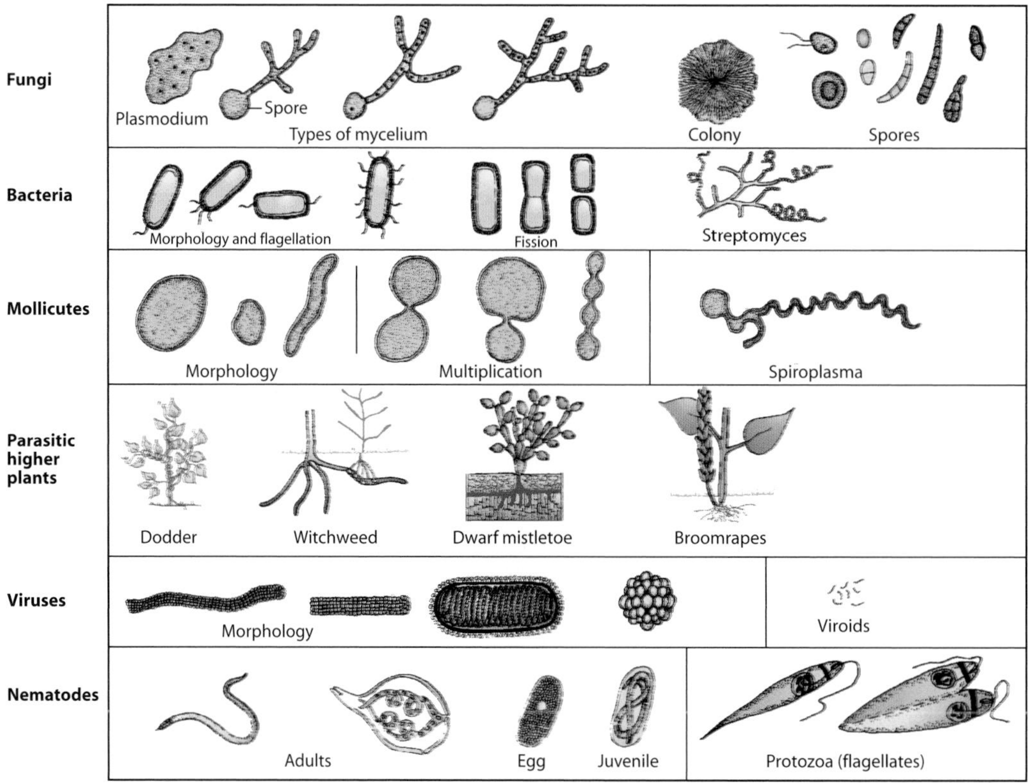

FIGURE 1-3 Morphology and ways of multiplication of some of the groups of plant pathogens.

development and spread of the disease and also possible control measures. On this basis, plant diseases in this text are classified as follows:

I. Infectious, or biotic, plant diseases
   1. Diseases caused by fungi (Figs. 1-4A and 1-4B)
   2. Diseases caused by prokaryotes (bacteria and mollicutes) (Figs. 1-4C and 1-4D)
   3. Diseases caused by parasitic higher plants (Fig. 1-5A) and green algae
   4. Diseases caused by viruses and viroids (Fig. 1-5B)
   5. Diseases caused by nematodes (Fig. 1-5C)
   6. Diseases caused by protozoa (Fig. 1-5D)

II. Noninfectious, or abiotic, plant diseases (Fig. 10-1)
   1. Diseases caused by too low or too high a temperature
   2. Diseases caused by lack or excess of soil moisture
   3. Diseases caused by lack or excess of light
   4. Diseases caused by lack of oxygen
   5. Diseases caused by air pollution
   6. Diseases caused by nutrient deficiencies
   7. Diseases caused by mineral toxicities
   8. Diseases caused by soil acidity or alkalinity (pH)
   9. Diseases caused by toxicity of pesticides
   10. Diseases caused by improper cultural practices

HISTORY OF PLANT PATHOLOGY AND EARLY SIGNIFICANT PLANT DISEASES

Introduction

Even when humans lived as hunters or nomads and their food consisted only of meat or leaves, fruit, and seeds, which they picked wherever they could find them, plant diseases took their toll on hunted animals and on humans. Plant diseases caused leaves and shoots to mildew and blight, and fruit and seeds to rot, thereby forcing humans to keep looking until they could find enough healthy fruit or food plants of some kind to satisfy their hunger. As humans settled down and became farmers, they began growing one or a few kinds of food plants in small plots of land and depended on these plants for their survival throughout the year. It is probable that every year, and in some years more than in others, part of the crop was lost to diseases. In such years food supplies were insufficient and hunger was common. In years when wet weather favored the development of plant diseases, most or all of the crop was

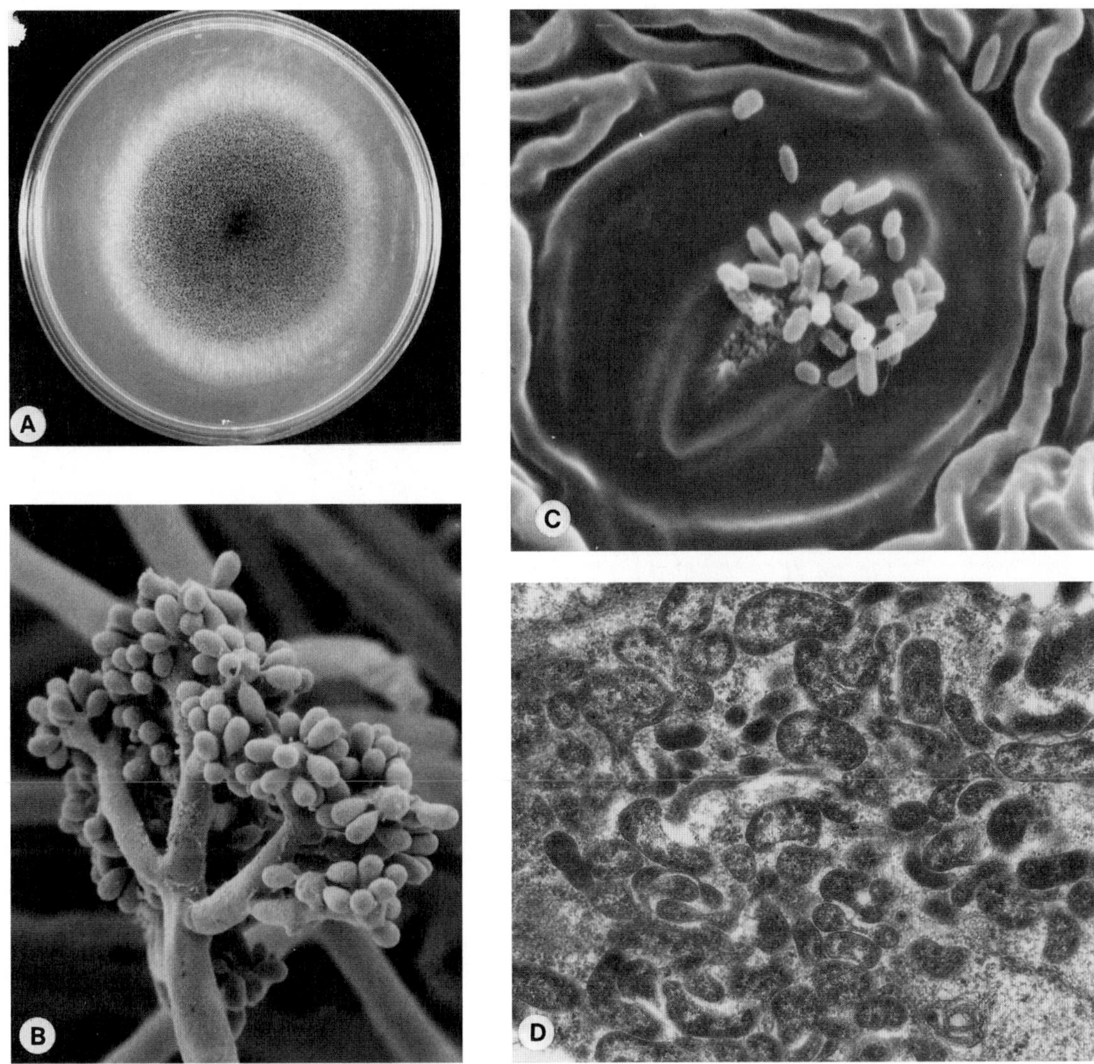

FIGURE 1-4 Three types of pathogenic microorganisms that cause plant diseases. (A) Fungus growing out of a piece of infected plant tissue placed in the center of a culture plate containing nutrient medium. (B) Mycelium and spores of a plant pathogenic fungus (*Botrytis* sp.) (600×). (C) Bacteria at a stoma of a plant leaf (2500×). (D) Phytoplasmas in a phloem cell of a plant (5000×). [Photographs courtesy of (B) M. F. Brown and H. G. Brotzman, (C) L. Mansvelt, I. M. M. Roos, and M. J. Hattingh, and (D) J. W. Worley.]

destroyed and famines resulted, causing immense suffering and probably the death of many humans and animals from starvation. It is not surprising, therefore, that plant diseases are mentioned in some of the oldest books available (Homer, c. 1000 B.C., Old Testament, c. 750 B.C.) and were feared as much as human diseases and war.

BOX 1 Plant diseases as the wrath of gods — theophrastus

The climate and soil of countries around the eastern Mediterranean Sea, from where many of the first records of antiquity came to us, allow the growth and cultivation of many plants. The most important crop plants for the survival of people and of domesticated animals were seed-producing cereals, especially wheat, barley, rye, and oats; and legumes, especially beans, fava beans, chickpeas, and lentils. Fruit trees such as apple, citrus, olives, peaches, and figs, as well as grapes, melons, and squash, were also cultivated. All of these crop plants suffered losses annually due to drought, insects, diseases, and weeds. Because most families grew their own crops and depended on their produce for survival

continued

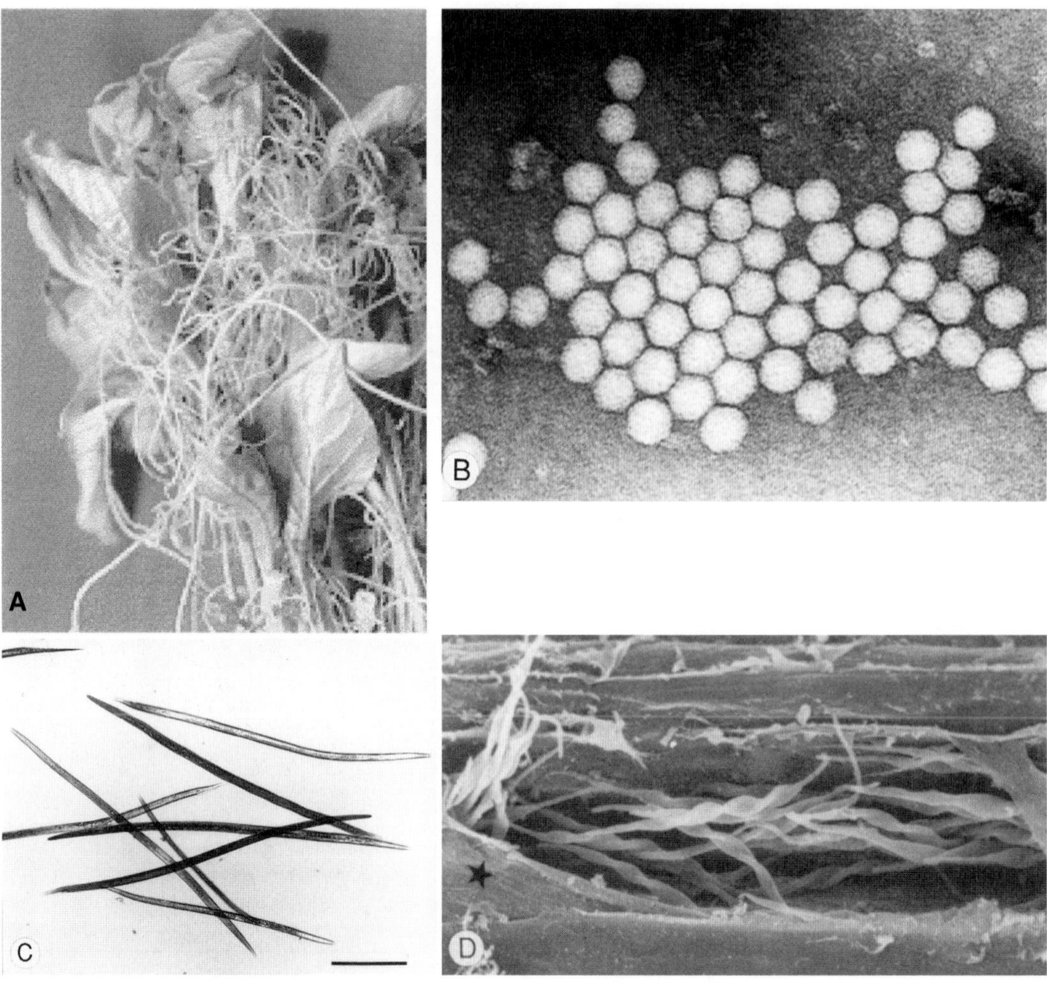

FIGURE 1-5 The other four types of pathogens that cause plant disease. (A) Thread-like parasitic higher plant dodder (*Cuscuta* sp.) parasitizing pepper seedlings. (B) *Tobacco ringspot virus* isolated from infected tobacco plants (200,000×). (C) Plant parasitic nematodes (*Ditylenchus* sp.) isolated from infected onion bulbs (80×). (D) Protozoa (*Phytomonas* spp.) in a phloem cell of an oil palm root (4000×) [Photographs courtesy of (A) G. W. Simone, (C) N. Greco, supplied courtesy R. Inserra, and (D) W. de Sousa].

until the next crop was produced the following year, losses of any amount of crops, regardless of cause, created serious hunger and survival problems for them. Occurrences of mildews (Fig. 1-6, see also pages 448–452), blasts (Figs. 1-7 A, 1-7B, 1-8A and 1-8B, see also pages 582–591), and blights on cereals (Figs. 1-9 A and 1-9B, see also pages 562–571) and legumes (Figs. 1-10A and 1-10B) are mentioned in numerous passages of books of the Old Testament (about 750 B.C.) of the Bible. Blasts, probably the smut diseases, destroyed some or all kernels in a head by replacing them with fungal spores. Blights, probably rusts, weakened the plants and used up the nutrients and water that would fill the kernels, leaving the kernels shriveled and empty (Fig. 1-9B).

Mention of plant diseases is found again in the writings of the Greek philosopher Democritus, who, around 470 B.C., noted plant blights and described a way to control them. It was not, however, until another Greek philosopher, Theophrastus (Fig. 1-11, c. 300 B.C.) made plants and, to a much smaller extent, plant diseases the object of a systematic study. Theophrastus was a pupil of Aristotle and later became his successor in the school. Among others, Theophrastus wrote two books on plants. One, called "*The Nature of Plants*," included chapters on the morphology and anatomy of plants and

FIGURE 1-6 Powdery mildew symptoms on (A) leaves of young wheat plant, (B) cluster of grape berries, (C) lilac leaf, and (D) azalea plant. [Photographs courtesy of (A) G. Munkvold, Iowa State University, (B) E. Hellman, Texas A&M University, and (C and D) S. Nameth, Ohio State University.]

descriptions of wild and cultivated woody plants, perennial herbaceous plants, wild and cultivated vegetable plants, the cereals, which also included legumes, and medicinal plants and their saps. The other book, called "*Reasons of Vegetable Growth*," included chapters on plant propagation from seeds and by grafting, the environmental changes and their effect on plants, cultural practices and their effect on plants, the origin and propagation of cereals, unnatural influences, including diseases and death of plants, and about the odor and the taste of plants. For these works, Theophrastus has been considered the "*father of botany*."

The contributions of Theophrastus to the knowledge about plant diseases are quite limited and influenced by the beliefs of his times. He observed that plant diseases were much more common and severe in lowlands than on hillsides and that some diseases, e.g., rusts, were much more common and severe on cereals than on legumes. In many of the early references, plant diseases were considered to be a curse and a punishment of the people by God for wrongs and sins they had committed. This implied that plant diseases could be avoided if the people would abstain from sin. Nobody, of course, thought that farmers in the lowlands sinned more than those on the hillsides, yet Theophrastus and his contemporaries, being unable to explain plant diseases, believed that God controlled the weather that "brought about" the disease. They believed that

continued

FIGURE 1-7 Loose smut (blast) of (A) barley and (B) wheat caused by the fungus *Ustilago* sp. [Photographs courtesy of (A) P. Thomas and (B) I. Evans, WCPD.]

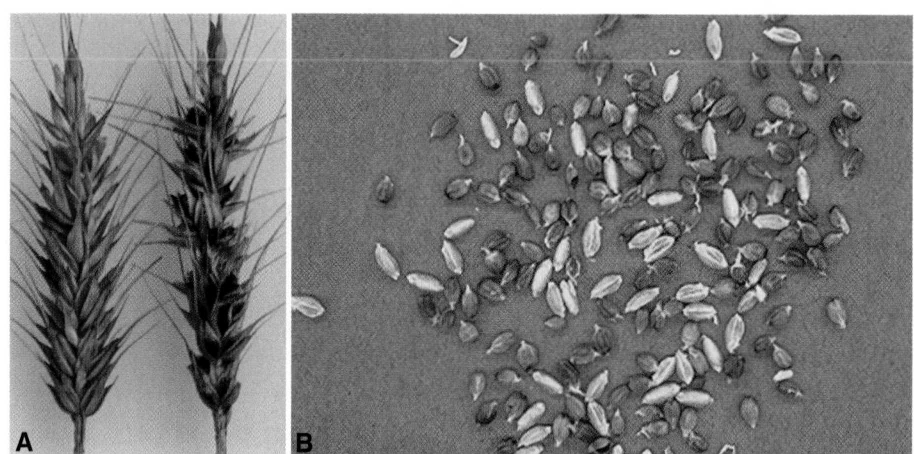

FIGURE 1-8 Cover smut or bunt (blast) of wheat caused by the fungus *Tilletia*. (A) Plant on the left is healthy; plant on the right shows infected, smaller, rounded, black wheat kernels in glumes spread out. (B) Healthy (light colored) and covered smut-infected (dark colored) kernels of wheat. [Photographs courtesy of (A) WCCPD and (B) P. Lipps, Ohio State University.]

FIGURE 1-9 (A) Wheat stems and leaves infected heavily with stem rust of wheat caused by the fungus *Puccinia tritici.* (B) Wheat kernels from rust-infected plants on the left are thin and almost empty of nutrients compared to kernels on the right from a healthy wheat plant, which are plump, full of starch and other nutrients. [Photographs courtesy of (A) CIMMYT and (B) USDA, Cereal Dis. Lab., St. Paul, MN.]

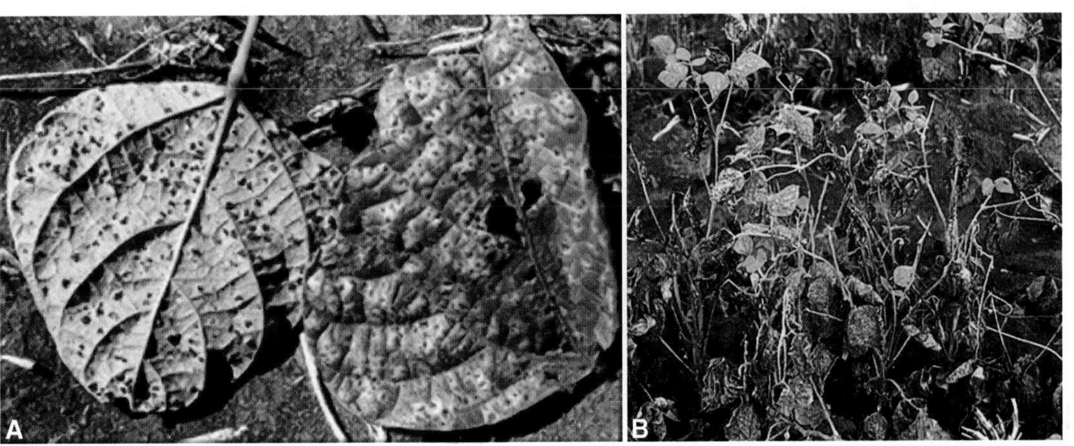

FIGURE 1-10 Close-up of bean rust caused by the fungus *Uromyces appendiculatus.* (A) Rust spots on the upper and lower sides of bean leaves. (B) Rust-infected bean plants in the field showing many leaves killed by the rust and fallen off. [Photographs courtesy of (A) R. G. Platford, WCPD, and (B) J. R. Steadman, University of Nebraska.]

continued

FIGURE 1-11 Theophrastus, the "father of botany."

plant diseases were a manifestation of the wrath of God and, therefore, that avoidance or control of the disease depended on people doing things that would please that same superpower. In the fourth century B.C.; the Romans suffered so much from hunger caused by the repeated destruction of cereal crops by rusts and other diseases that they created a separate god, whom they named Robigus. To please Robigus, the Romans offered prayers and sacrifices in the belief that he would protect them from the dreaded rusts. The Romans even established a special holiday for Robigus, the Robigalia, during which they sacrificed red dogs, foxes, and cows in an attempt to please and pacify Robigus so he would not send the rusts to destroy their crops.

Efforts to control plant diseases were similarly hampered by the lack of information on the causes of disease and by the belief that diseases were manifestations of the wrath of God. Nevertheless, some ancient writers, e.g., Homer (c. 1000 B.C.), mention the therapeutic properties of sulfur on plant diseases, and Democritus (c. 470 B.C.) recommended controlling plant blights by sprinkling plants with the olive grounds left after extraction of the olive oil. Most ancient reports, however, dealt with festivals and sacrifices to thank, please, or appease a god and to keep the god from sending the dreaded rusts, mildews, blasts, or other crop scourges. Very little information on controlling plant diseases was written anywhere for almost 2000 years.

During the two millennia of fatalism, a few important observations were made on the causes and control of plant diseases, but they were not believed by their contemporaries and were completely ignored by the generations that followed. It was not until about A.D. 1200 that a higher plant, the mistletoe, was proposed as a parasite that obtains its food from the host plant, which it makes sick. It was also noted that the host plant can be cured by pruning out the part carrying the mistletoe. Nobody, however, followed up on this important observation.

BOX 2 Mistletoe recognized as the first plant pathogen

Mistletoes are plants that live as parasites on branches of trees (see pages 715) but, for various reasons, they have caught the fancy of people in various cultures and have made a name for themselves way beyond their real properties.

Although mistletoe is the first plant pathogen to be recognized as such and the first pathogen for which a cultural control (by pruning affected branches) was recommended, both by Albertus Magnus (Fig. 1-12A) around 1200 A.D., a great deal more has been fantasized, said, written, and practiced about it than its importance as a pathogen would indicate. Mistletoe, to be sure, both the common or leafy mistletoe (*Viscum* in Europe and elsewhere, *Phoradendron* in North America), which infects many deciduous trees (Figs. 1-12B and 1-12C) and especially the dwarf mistletoes (*Arceuthobium*), which infects conifers, cause considerable damage to trees they infect. In many cases, the evergreen mistletoe plants can be seen clearly after normal leaf fall in the autumn and make up as much as half of the top of the deciduous tree they infect. They generally damage trees by making their trunks and branches swell where they are infected and then break there during windstorms, thereby reducing the surface of the tree and reducing the quality of timber.

Mistletoes, of course, are evergreen parasitic plants that sink their "roots," usually called sinkers or haustoria, into branches of trees. Through the sinkers they absorb all the water and mineral nutrients and most of the organic substances they need from the plant. True

FIGURE 1-12 (A) Albertus Magnus, who recognized the mistletoe as a plant parasite. (B) Tufts of individual mistletoe plants growing on branches of an oak tree in winter. (C) Close-up of a mistletoe plant whose main stems are growing out of the trunk of an oak tree.

mistletoes, however, have well-developed leaves and chlorophyll and carry on photosynthesis and manufacture at least some of the sugars and other organic substances they need. Mistletoe plants produce separate male and female flowers and berry-like fruits containing a single seed. The seeds are coated with a sticky substance and are either forcibly expelled and stick to branches of nearby trees or are eaten by birds but go through their digestive tract and stick to branches on which birds drop them.

The striking visibility of true mistletoes on deciduous trees, and their ability to remain green while their host leaves fall for the winter, excited the imagination of people since the times of the ancient Greeks and inspired many myths and traditions involving the mistletoe plant through the centuries. The plant itself was thought to possess mystical powers and became associated with many folklore customs in many countries. It was thought to bestow life and protect against poison, to act as an aphrodisiac, and to bestow fertility. Mistletoe sprigs placed over house and stable doors or hung from ceilings were believed to ward off witches and evil spirits. The Romans decorated their temples and houses in midwinter with mistletoe to please the gods to whom it was sacred. In Nordic mythology, the mistletoe was sacred to Frigga, the goddess of love, but was used by Loki, the goddess of evil, as an arrow and killed Frigga's son, the god of the summer sun. Frigga managed to revive her son under the mistletoe tree and, in her joy, she kissed everyone who was under the mistletoe tree. But, for its misdeed to her son, she condemned the mistletoe to, be in the future, a parasite and to have no power to cause misfortune, sorrow, or death. She decreed instead that anyone standing under a mistletoe tree was due not only protection from any harm, but also a kiss, a token of peace and love. So, in Scandinavia, mistletoe was thought of as a plant of peace: under the mistletoe, enemies could agree on a truce or feuding spouses could kiss and make up. In England, a ball of mistletoe was decorated with ribbons and ornaments and was hung up at Christmas. If a young lady was standing under the ball, she could not refuse to be kissed or she could not expect to get married the following

continued

year. A couple in love that kiss under the mistletoe is equivalent to promising to marry and a prediction of long life and happiness together. Nowadays, in many parts of Europe and America, a person standing under a ball or even a sprig of mistletoe at Christmastime is inviting to be kissed by members of the opposite gender as a sign of friendship and goodwill. There are, actually, more myths and customs associated with mistletoe. Who would think that a minor parasitic higher plant would excite the imagination of so many others and have so many stories about it.

BOX 3 Plant diseases as the result of spontaneous generation

Following Theophrastus, other than the proposal by Magnus that the mistletoe was a parasite, there was little useful knowledge that was added about plants or about plant diseases for about 2000 years, although there are reports of famines in several parts of the world. Especially bad were outbreaks in north-central Europe of ergotism, a disease of humans and animals caused from eating grains contaminated with parts of the fungus that causes the ergot disease of cereals (see pages 501–504). People continued to associate plant diseases with sin and the wrath of God and therefore were fatalistic about the occurrence of plant diseases, the repeated losses of food, and the hunger and famines that followed. References to the ravages of plant diseases appeared in the writings of several contemporary historians, but little was added to the knowledge about the causes and control of plant diseases. People everywhere believed that plant diseases, as well as human and animal diseases, just happened spontaneously. Whatever was observed on diseased plants or on diseased plant produce was considered to be the product or the result of the disease rather than the cause of it. After the invention of the compound microscope in the mid-1600s, which enabled scientists to see many of the previously invisible microorganisms, scientists, as well as laypeople, became even stronger believers in the spontaneous generation of diseases and of the microorganisms associated with diseased or decaying plant, human, or animal tissues. That is, they came to believe that the mildews, rusts, decay, or other symptoms observed on diseased plants, and any microorganisms found on or in diseased plant parts, were the natural products of diseases that just happened rather than being the cause and effect of the diseases.

Biology and Plant Pathology in Early Renaissance

People continued to suffer from hunger and malnutrition due partially at least to diseases destroying their crops and their fruit. They, however, continued to consider plant diseases as the work and wish of their God and, therefore, an event that could neither be understood nor avoided. In the mid-1600s, however, a group of French farmers noted that wheat rust was always more severe on wheat near barberry bushes than away from them (Fig. 1-13). The farmers thought that the rust was produced by the barberry plants from which it moved to wheat. They, therefore, asked the French government to pass the first plant disease regulatory legislation that would force towns to cut and destroy the barberry bushes to protect the wheat crop.

FIGURE 1-13 A bush of barberry (*Berberis vulgaris*) growing at the edge of a wheat field and helping close the dioecious disease cycle of wheat stem rust disease. The fungus, *Puccinia graminis*, overwinters on barberry on which it produces spores that infect wheat plants near the barberry (see photo) from which then spores of the fungus spread to more wheat plants. (Photograph courtesy of USDA Cereal Dis. Lab., St. Paul, MN.)

In 1670, the French physician Thoullier observed that ergotism or Holy Fire, a serious and often deadly disease of humans in northcentral Europe (see pages 39 and 559), did not spread from one person to another but seemed to be associated with the consumption of ergot-contaminated grains. At about the same time, Robert Hooke, in England, invented the double-lensed (compound) microscope with which he examined thin slices of cork and called its units "cells." Soon after, the Dutchman Antonius van Leeuwenhoek (Fig. 1-14A) improved significantly the lenses and the structure of the microscope and began to examine not only the anatomy of plants, but also the body of filamentous fungi and algae, protozoa, sperm cells, blood cells, and even bacteria. All of these microorganisms, of course, were considered to be produced by whatever organism (animal

FIGURE 1-14 (A) Antonius van Leeuwenhoek. (B) Carl von Linne'. (C) Charles Darwin.

or plant) or medium they happened to be found in and were not thought of as independent, autonomous organisms. In 1735, the Swedish philosopher–botanist Carl von Linne' (Fig. 1-14B) published his main work "*Systema Naturae*," by which he established the diagnosis of plant species and the binomial nomenclature of plants. Linne's species, however, were rigid and were supposed to have remained unchanged since creation. It was not until more than a century later, in 1859, that the Englishman Charles Darwin (Fig. 1-14C) published his book "*The Origin of Species by Means of Natural Selection*" and showed that species of all organisms, plants and animals, evolve over time and adapt to changes in their environment for survival.

The discovery and availability of the microscope, however, sparked significant interest in microscopic fungi and, subsequently, their possible association with plant diseases. In 1729, the Italian botanist Pier Antonio Micheli described many new genera of fungi and illustrated their reproductive structures. He also noted that when placed on freshly cut slices of melon, these structures grew and produced the same kind of fungus that had produced them. He proposed, therefore, that fungi arise from their own spores rather than spontaneously, but because the "spontaneous generation" theory was so imbedded in people's minds, nobody believed Micheli's evidence. Similarly, in 1743, the English scientist Needham observed nematodes inside small, abnormally rounded wheat kernels but he, too, failed to show or suggest that they were the cause of the problem.

In 1755, the Frenchman Tillet, working with smutted wheat, showed that he could increase the number of wheat plants developing covered smut (Figs. 1-8A and 1-8B) by dusting wheat kernels before planting with smut dust, i.e., with smut spores (Fig. 1-15). He also noted that he could reduce the number of smutted wheat plants produced by treating the smut-treated kernels with copper sulfate. Tillet, too, however, did not interpret his experiments properly and, instead of concluding that wheat smut is an infectious plant disease, he believed that it was a poisonous substance contained in the smut dust, rather than the living spores and fungus coming from them, that caused the disease. More than 50 years later, in 1807, Prevost, another Frenchman, repeated both the inoculation experiments and those in which the seeds were treated with copper sulfate, as done by Tillet, and he obtained the same results. In addition, Prevost observed smut spores from untreated and treated wheat seed under the microscope and noticed that those from untreated seed germinated and grew whereas those from treated seed failed to germinate. He, therefore, concluded correctly that it was the smut spores that caused the smut disease in wheat and that the reduced number of smutted wheat plants derived from copper sulfate-treated seed was due to the inhibition of germination of smut spores by the copper sulfate. Prevost's conclusions, however, were not accepted by the French Academy of Sciences because its scientists and other scientists throughout Europe still believed that microorganisms and their spores formed through spontaneous generation and were the result rather than the cause of disease. In 1855, a nematode was observed in galls of cucumber roots, but again they were thought to have appeared there spontaneously. These beliefs continued to be held and expounded by scientists until the early 1860s, when, in 1861–1863, Anton deBary (Fig. 1-16A) proved that potato late blight was caused by a fungus and Louis Pasteur (Fig. 1-16B) proved that microorganisms were produced from preexisting microorganisms and that most infectious diseases were caused by germs. The latter established the "*germ theory of disease*," which changed the way of thinking of scientists and led to tremendous progress. Significant

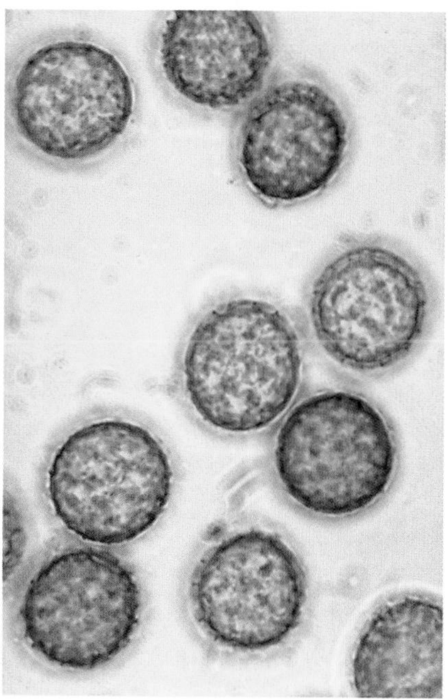

FIGURE 1-15 Teliospores of the fungus *Tilletia*, the cause of the covered smut or bunt of wheat. (Photograph courtesy of M. Babadoost, University of Illinois.)

FIGURE 1-16 (A) Anton deBary. (B) Louis Pasteur. (C) Robert Koch.

impetus to this progress was added by Robert Petri, who developed artificial nutrient media for culturing the microorganisms (Petri dishes), and by Robert Koch (Fig. 1-16C), who established that for proving that a certain microorganism was the cause of a particular infectious disease, certain necessary steps (Koch's postulates) must be carried out and certain conditions must be satisfied.

BOX 4 Potato blight and the irish famine: a deadly mix of ignorance and politics

In about 1800, the potato, which was introduced in Europe from South and Central America around 1570 A.D., was a well-established crop in Ireland. After strong objections against adopting it because (1) it was new and not mentioned in the Bible, (2) it was produced in the ground and, therefore, was unclean, and (3) because parts of it were poisonous, the potato was nevertheless adopted and its cultivation spread rapidly. Adoption of potato cultivation came as a result of it producing much more edible food per unit of land than grain crops, mostly wheat and rye, grown until then. It was adopted also because the ground protected it from the pests and diseases that destroyed aboveground crops and from destruction by the soldiers sent by absentee English landlords to collect overdue land rents.

At that time, most Irish farmers were extremely poor, owned no land, and lived in small windowless, one-room huts. The farmers rented land from absentee English landlords who lived in England, and planted grain and other crops. The yields were poor and, in any case, large portions of them had to be used for paying the exorbitant rent so as to avoid eviction. The Irish farmers also kept small plots of land, usually as small as a quarter of an acre and basically survived the winter with the food they produced on that land. Potato production was greatly favored by the cool, wet climate of Ireland, and the farmers began growing and eating potatoes to the exclusion of other crops and foodstuffs. Irish farmers, therefore, became dependent on potatoes for their sustenance and survival. Lacking proper warehouses, the farmers stored their potato tubers for the winter in shallow ditches in the ground. Periodically, they would open up part of the ditch and remove as many potatoes as they thought they would need for the next few weeks.

The potatoes grew well for many years, free of any serious problems. In the early 1840s, potato crops began to fail to varying extents in several areas of Europe and Ireland. Most of the growing season of 1845 in Ireland was quite favorable for the growth of potato plants and for the formation of tubers. Everything looked as though there would be an excellent yield of potatoes everywhere that year. Then, the weather over northern Europe and Ireland became cloudy, wetter, and cooler and stayed that way for several weeks (Fig. 1-17A). The potato crop, which until then looked so promising, began to show blighted leaves and shoots (Fig. 1-17B), and whole potato plants became blighted and died. In just a few weeks, the potato fields in northern Europe and in Ireland became masses of blighted and rotting vegetation (Fig. 1-17C). The farmers were surprised and worried, especially when they noticed that many of the potatoes still in the ground were rotten and others had rotting areas on their surface (Fig. 1-17D). They did what they could to dig up the healthy-looking potatoes from the affected fields and put them in the ditches to hold them through the winter.

The farmer's worry became horror when later in the fall and winter they began opening the ditches and looking for the potatoes they had put in them at harvest. Alas, instead of potatoes they found only masses of rotting tubers (Figs. 1-17D and 1-17E), totally unfit for consumption by humans or animals. The dependence of Irish farmers on potatoes alone meant that they had nothing else to eat — and neither did any of their neighbors. Hunger (Fig. 1-17F) was quickly followed by starvation, which resulted in the death of many Irish. The famine was exacerbated by the political situation between England and Ireland. The British refused to intervene and help the starving Irish with food for several months after the blight destroyed the potatoes. Eventually, by February of the next year (1846), food, in the form of corn from the United States, began to be imported and made available to the starving poor who paid for it by working on various government construction projects. Unfortunately, the weather in 1846 was again cool and wet, favoring the potato blight, which again spread into and destroyed the potato plants and tubers. Hunger, dysentery, and typhus spread among the farmers again, and more of the survivors emigrated to North America. It is estimated that one and a half million Irish died from hunger, and about as many left Ireland, emigrating mostly to the United States of America.

The cause of the destruction of the potato plants and of the rotting of the potato tubers was, of course, unknown and a mystery to all. The farmers and other simple folk believed it to have been brought about by "the little people," by the devil himself whom they tried to exorcise and chase away by sprinkling holy water in the fields, by locomotives traveling the countryside at devilish speeds of up to 20 miles per hour and discharging electricity harmful to crops they went by, or to have been sent by God as punishment for some unspecified sin they had committed. The more educated doctors and clergy were so convinced of the truth of the theory of spontaneous generation that even when they saw the mildewy fungus growth on affected leaves and on some stems and tubers, they thought that this growth was produced by the dying plant as a result of the rotting rather than the cause of the death and rotting of the plant.

Some of the educated people, however, began to have second thoughts about the situation. Dr. J. Lindley, a professor of botany in London, proposed incorrectly that the plants, during the rains, overabsorbed water through their roots and because they could not get rid

continued

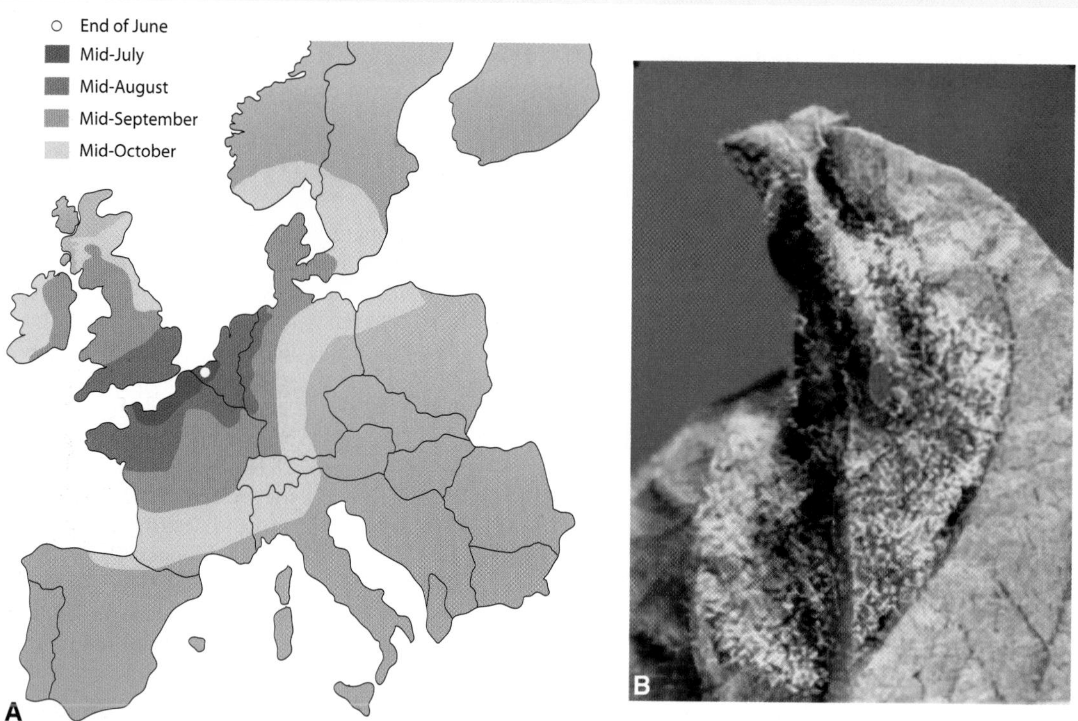

FIGURE 1-17 The late blight of potato and the Irish famine. (A) Itinerary of the advance of the potato blight between June, when the blight was first detected in Belgium, and the end of October 1845, by which time it spread from Italy to Ireland and from Spain to the Scandinavian countries. (B) A young lesion on a potato leaf covered with sporangiophores and sporangiospores of the fungus (oomycete). (C) A potato plant killed completely by the blight (right) next to a healthy-looking resistant plant (left). (D) External and internal appearance of potato tubers infected with the late blight disease. The oomycete is still found near the surface. (E) Advanced invasion and rotting of potato tuber infected with late blight. (F) A period drawing of a family digging for potatoes to avoid starvation during the Irish famine. [Photographs courtesy of (A) W. E. Fry, Cornell University, (B) D. P. Weingartner, University of Florida, (C and D) Cornell University, (E) USDA, and (F) Illustrated London News, 1849.]

of the excess water, their tissues became swollen and rotted. The Reverend Dr. Miles Berkeley, however, noticed that the mold covering potato plants about to rot was a fungus (oomycete) similar but not identical to a fungus he observed on a sick onion. The fungus on potato, however, was identical to a fungus recovered from sick potato plants in northern Europe. Berkeley concluded that this fungus was the cause of the potato blight, but when he proposed it in a letter to a newspaper, it was considered as an incredible and bizarre theory unsupported by facts. The puzzle of what caused blight of potato continued unanswered for 16 years after the 1845 destruction of potatoes by the blight. Finally, in 1861, Anton deBary (Fig. 1-16A) did a simple experiment that proved that the potato blight was caused by a fungus. DeBary simply planted two sets of healthy potatoes, one of which he dusted with spores of the fungus collected from blighted potato plants. When the tubers germinated and began to produce potato plants, the healthy tubers produced healthy plants, whereas the healthy tubers dusted with the spores of the fungus produced plants that became blighted and died. No matter how many times deBary repeated the experiment, only tubers treated with the fungus became infected and produced plants that became infected. Therefore, the fungus, which, we know now, is an oomycete was named *Phytophthora infestans* (*"infectious plant destroyer"* from phyto = plant, phthora = destruction, infestans = infectious), was the cause of the potato blight. DeBary also showed that the fungus did not just reappear from nowhere the following growing season but instead survived the winter in partially infected potato tubers in the field or storage. In the spring, the fungus infected young plants coming from these partially rotten tubers, produced new spores on these plants, and the spores then spread to other cultivated potato plants that were infected and killed. With this experiment deBary actually disproved the **theory of spontaneous generation**, which stated that microorganisms are produced spontaneously by dying and dead plants and animals, and ushered in the **germ theory of disease**. The honor for this proof, however, is reserved for Louis Pasteur, who proved the theories while working with bacteria at about the same time, 1861–1863, that deBary published his work with the potato blight fungus.

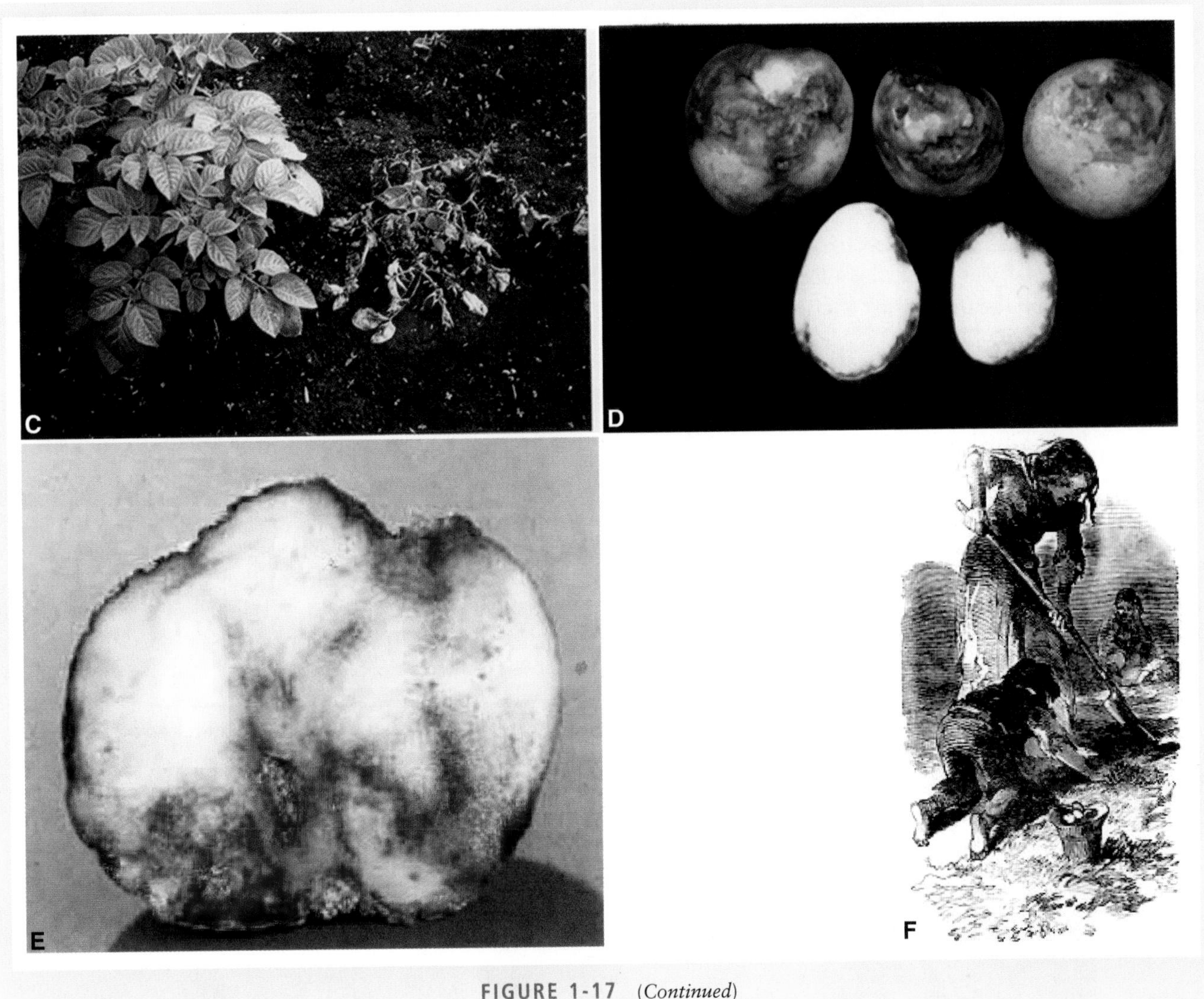

FIGURE 1-17 *(Continued)*

The Expanding Role of Fungi as Causes of Plant Disease

Following the observation by French farmers around the mid-1600s and, independently, by Connecticut farmers in the early 1700s that wheat rust was worse near barberry bushes, the farmers came to believe that barberry fathered the rust, which then moved to wheat. The request by farmers for legislation to force towns to eradicate barberries and in that way to protect the wheat plants from rust followed. At about the same time, spores of the rust fungus were observed with the compound microscope for the first time in England (Hooke, 1667). In Italy, Micheli 60 years later (1729) described many new genera of fungi, illustrated their reproductive structures, and noted that when he placed them on freshly cut slices of melon, these fungal structures generally reproduced the same kind of fungus that produced them. He proposed that fungi arose from their own spores rather than spontaneously, but nobody believed him. New information about plant pathogenic fungi continued to be developed, but most of it was not accepted by the scientists of the time for a long time.

As mentioned previously, in 1755, Tillet in France showed that wheat smut is a contagious plant disease, but even he believed that it was a poisonous substance contained in the smut dust, rather than a living microorganism, that caused the disease. In 1807, Prevost, also in France, repeated and expanded Tillet's experiments and appeared to have demonstrated conclusively that wheat smut was caused by a fungus. His conclusions, however, were not accepted because the scientists were blinded by the belief that microorganisms and their spores were the result rather than the cause of disease. These beliefs continued to be shared and expounded by scientists for at least another 50 years.

The devastating epidemics of late blight of potato in northern Europe, particularly Ireland, in the 1840s not only dramatized the effect of plant diseases on human suffering and survival, but also greatly stimulated interest in their causes and control. In 1861, deBary finally established experimentally beyond criticism that a fungus (*Ph. infestans*) was the cause of the plant disease known as late blight of potato, a disease that closely resembles the downy mildews.

It is, perhaps, worth noting here that it was during those years (1860–1863) that Louis Pasteur proposed, and finally provided irrefutable evidence, that microorganisms arise only from preexisting microorganisms and that fermentation is a biological phenomenon, not just a chemical one. Pasteur's conclusions, however, were not generally accepted for many years afterward. Nevertheless, the proof for involvement of microorganisms (germs) in fermentation and disease signaled the beginning of the end of the theory of spontaneous generation and provided the basis for the germ theory of disease.

Although fungi had already been the object of study by many scientists, proof that they were causing disease in plants greatly increased interest in them. DeBary himself also carried out studies of the smut and rust fungi, of the fungi causing downy mildews, and of the fungus *Sclerotinia*, which induces rotting of vegetables. The German Kühn in the 1870s and later contributed significantly to the studies of infection and development of smut in wheat plants and promoted the development and application of control measures, particularly seed treatment for cereals. Kühn also wrote the first book on plant pathology, "*Diseases of Cultivated Crops, Their Causes and Their Control*," in which he recognized that plant diseases are caused by an unfavorable environment but can also be caused by parasitic organisms such as insects, fungi, and parasitic plants.

During the years of Pasteur and Koch, several scientists also made significant contributions to plant pathology and to biology and medicine. After establishing beyond criticism in 1861 that the potato blight was caused by a fungus, DeBary went on to show conclusively that smut and rust fungi were also the causes and not the results of their respective plant diseases. Moreover, he showed that some rust diseases require two alternate host plants (see Fig. 1-13) to complete their life cycle, e.g., the fungus causing the stem rust of wheat requires wheat and barberry. DeBary also showed (1886) that some fungi induce rotting of vegetables (Fig. 1-18) by secreting substances (enzymes) that diffuse into plant tissues in advance of the pathogen.

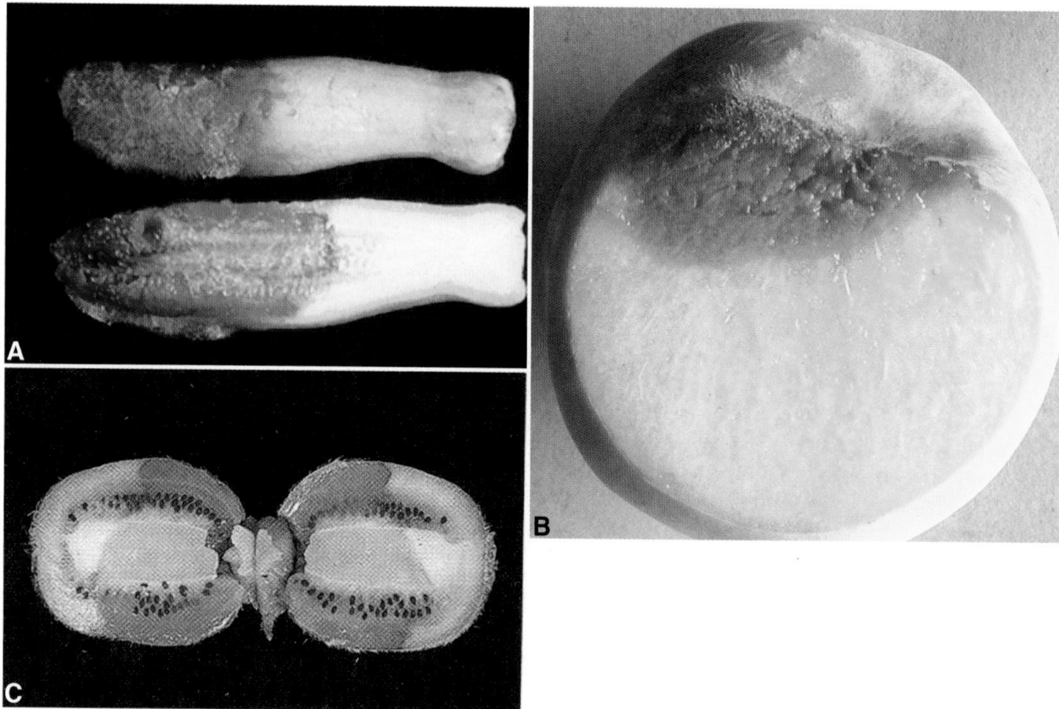

FIGURE 1-18 Infection and advanced internal rotting of summer squash (A) by the fungus *Choanephora*, of peach fruit (B) by the fungus *Rhizopus* sp., and (C) of kiwi fruit by the fungus *Botrytis cinerea*. In all cases, fruit rot is a result of, primarily, pectinolytic enzymes secreted by the fungi and advancing ahead of the mycelium. A small amount of the fungi can be seen on the surface of the fruits. (C) Courtesy of T. Michailides, University of California.

The Discovery of Other Causes of Infectious Diseases

Although Leeuwenhoek first saw microbes with the microscope he invented in 1674, little progress was made toward the concept of microbes as the cause of disease for almost another 200 years. In 1776, Jenner introduced vaccination against the virus-induced smallpox, an extremely infectious and severe disease that used to kill 10 to 20% of those infected, but could only speculate as to its cause and how it worked. In 1861, however, deBary showed that the potato blight was caused by a fungus while Pasteur formulated the germ theory of fermentation. In 1864, Pasteur invented pasteurization and, in 1880, made the first vaccine against the chicken cholera. In the meantime, in 1876, Koch identified the anthrax bacillus, *Bacillus anthracis*, as the first bacterium to cause disease in animals and humans. In addition, in 1887, Koch formulated his rules of disease diagnosis that became known as "Koch's postulates." These rules became the standard procedure for proving that a disease is caused by a bacterium or any other kind of pathogen.

Nematodes

The first report of nematodes associated with a plant disease was made in England by Needham in 1743. He observed nematodes (Fig. 1-19A) within small, abnormally rounded wheat kernels (wheat galls; Fig. 1-19B); however, he did not show or suggest that they were the cause of the disease. It was not until 1855 that a second

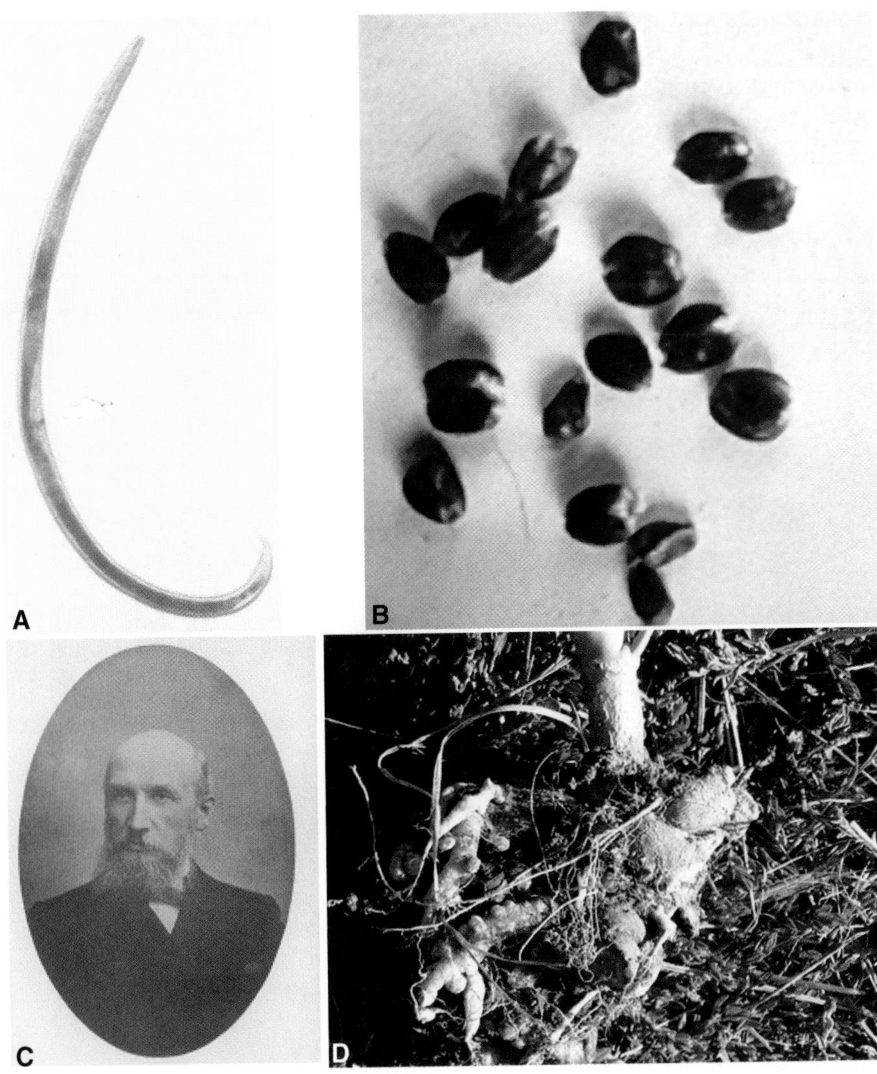

FIGURE 1-19 (A) A typical nematode. (B) Wheat seed galls, each filled with as many as 30,000 nematodes. (C) M. Woronin. (D) Clubroot of cabbage caused by the protozoon *Plasmodiophora brassicae*. [Photographs courtesy of (A and B) USDA Nematology Laboratory, Beltsville, Maryland, and (D) C. M. Ocamp, Oregon State University.]

nematode, the root knot nematode, was observed in cucumber root galls. In the next 4 years two other plant parasitic nematodes, the bulb and stem nematode and the sugarbeet cyst nematode, were reported from infected plant parts. Several more nematodes parasitizing plants were described in the early part of the 20th century by Cobb, who made numerous significant contributions to plant nematology.

Protozoan Myxomycetes

In 1878, Woronin (Fig. 1-19C), in Russia, was the first to show that a plant disease, the clubroot disease of cabbage (Fig. 1-19D), was caused by a fungus that has been shown to be a protozoan plasmodiophoromycete. These are fungus-like, single-celled microorganisms that lack a cell wall and, as a result, produce an amoeba-like body called a plasmodium and zoospores. These microorganisms used to be thought of as lower fungi but are now considered members of a different kingdom, the kingdom protozoa.

Bacteria

Soon after Koch showed that bacteria cause disease in animals and humans, Burrill in Illinois showed, in 1878, that bacteria (Fig. 1-20A) caused the fire blight disease (Fig. 1-20B) of pear and apple. Following Burrill's discovery, several other plant diseases were shown, particularly by Erwin Smith (Fig. 1-20C) of the U.S. Department of Agriculture (USDA), to be caused by bacteria. In the early 1890s, Smith was the first to show that crown gall disease (Fig. 1-20D), which he considered similar to cancerous tumors of humans and animals, was caused by bacteria. Studies of how this bacterium, known as *Agrobacterium tumefaciens*, caused tumors in plants led to the discovery, almost a century later, that whenever the bacterium infects plants

FIGURE 1-20 (A) The fire blight bacterium *Erwinia amylovora*. (B) Fire blight on apple trees. (C) Erwin F. Smith. (D) Crown gall, caused by the bacterium *Agrobacterium tumefaciens*. [Photographs courtesy of (A) Oregon State University, and (B) K. Mohan and (D) R. L. Forster, University of Idaho.]

it transfers part of its DNA to the plant and that the DNA is expressed by the plant as if it were plant DNA (see also pages 624–625). The discovery that the bacterium acts as a natural genetic engineer of plants led to the development of this bacterium so that it could be loaded with, and then transfer to plants, DNA segments coding for desirable characteristics, which formed the basis of biotechnology, especially of plants. As with fungal plant pathogens, however, acceptance of bacteria as causes of disease in plants was slow. For example, as late as 1899, Alfred Fischer, a prominent German botanist, rejected the results of Smith and others who claimed to have seen bacteria in plant cells.

Viruses

At about the same time that more diseases of plants were shown to be caused by bacteria, the Dutchman Adolph Mayer (Fig. 1-21A), in 1886, injected juice obtained from tobacco plant leaves showing various patterns of greenish yellow mosaic (Fig. 1-21B) into healthy tobacco plants and the latter then developed similar mosaic patterns. Because no fungus was present on the plant or in filtered juice, Mayer concluded that the disease was probably caused by bacteria. In 1892, however, Ivanowski showed that whatever caused the tobacco mosaic disease could pass through a filter that retains bacteria, so he concluded that the disease was caused by a toxin secreted by bacteria or, perhaps, by unusually small bacteria that passed through the pores of the filter. In 1898, Beijerinck, by repeating some of these experiments, finally concluded that the tobacco mosaic disease was caused not by a microorganism, but by a "contagious living fluid' " that he called a virus.

No one had any idea, however, what a virus was and what it looked like for another 40 years. The true nature, size, and shape of the virus (Fig. 1-21C) remained unknown for several more decades. In 1935, Stanley added ammonium sulfate to tobacco juice extracted from infected tobacco leaves and obtained as a sediment in the flask a crystalline protein that, when rubbed on tobacco, caused the tobacco mosaic disease. This led him to conclude that the virus was an autocatalytic protein that could multiply within living cells. Although his results and conclusions were later proved incorrect, for his discovery Stanley received a Nobel Prize in Chemistry. In 1936, Bawden and colleagues demonstrated that the crystalline preparations of the virus actually consisted of not only protein, but also a small amount of ribonucleic acid (RNA). The first virus (tobacco mosaic virus) particles were seen with the electron microscope in 1939 by Kausche and colleagues. Finally, in 1956, Gierrer and Schramm showed that the protein could be removed from the virus and that the ribonucleic acid carried all the genetic information that enabled it to cause infection and to reproduce the complete virus. It was shown subsequently that although the nucleic acid of most viruses infecting plants is single-stranded RNA, some viruses have double-stranded RNA, some double-stranded DNA, and some single-stranded DNA.

The search for the cause of the many thousands of plant diseases led to the discovery of at least three more kinds of pathogens and it is likely that others remain to be discovered.

Protozoa

Flagellate trypanosomatid protozoa were observed in the latex-bearing cells of laticiferous plants of the family

FIGURE 1-21 (A) Adolph Mayer. (B) Tobacco leaf showing symptoms of tobacco mosaic. (C) Particles of *tobacco mosaic virus*.

Euphorbiaceae by Lafont in 1909. Such protozoa, however, were thought to parasitize the plant latex without causing disease on the host plant. In 1931, Stahel found flagellates infecting the phloem of coffee trees, causing abnormal phloem formation and wilting of the trees. In 1963, Vermeulen presented convincing evidence of the pathogenicity of flagellates to coffee trees, and in 1976 flagellates were reported to be associated with several diseases of coconut and oil palm trees in South America and in Africa. In recent years, of course, the Myxomycota and the Plasmodiophoromycota, which were previously thought to be fungi, have been transferred to the kingdom protozoa.

Mollicutes (Phytoplasmas)

For nearly 70 years after viruses were discovered, many plant diseases were described that showed symptoms of general yellowing or reddening of the plant or of shoots proliferating and forming structures that resembled witches' brooms. These diseases were thought to be caused by viruses, but no viruses could be found in such plants. In 1967, Doi and colleagues in Japan observed mollicutes, i.e., wall-less mycoplasma-like bodies in the phloem of plants exhibiting yellows and witches' broom symptoms. That same year the same group showed that the mycoplasma-like bodies and symptoms disappeared temporarily when the plants were treated with tetracycline antibiotics. Since then, mycoplasma-like organisms (MLOs) that infect plants have been reclassified as phytoplasmas, and some of them that have helical bodies and can be found in other environments besides plants are known as spiroplasmas.

Viroids

In 1971, studies of the potato spindle tuber disease showed that it was caused by a small, naked, single-stranded, circular molecule of infectious RNA, which was called a viroid (see later). Viroids have been found to be the cause of several dozen plant diseases. Viroids seem to be the smallest infectious nucleic acid molecules. Although more than 40 viroids have been found to infect plants, no viroids have been found that infect animals or humans.

Apparently, however, an even smaller type of infectious agent, called a prion, exists (see later). Prions apparently consist only of a small (~55,000 Da) protein, which is encoded by a chromosomal gene of the host. Prions have been shown to cause the scrapie disease of sheep, "mad cow" disease, and at least three slow-developing degenerative diseases of humans. So far, no prions have been found to infect plants, but there is no obvious reason why they should not.

Serious Plant Diseases of Unknown Etiology

Although pathogens as large and complex as fungi and nematodes or as tiny and simple as viroids and prions have been discovered, there are many severe diseases of plants, particularly of trees, for which we still do not know their real cause, despite years of searching and research. Some of them, such as peach short life in the southeastern United States, waldsterben, or forest decline in central Europe and various forest tree declines in the northeastern and northwestern United States, may be caused by more than one pathogen or by combinations of pathogens and adverse environment. Others, such as citrus blight in Florida and South America, spear rot in oil palm in Suriname and Brazil, and mango malformation in India and other mango-growing countries, seem to have a biotic agent as the primary cause, but the activity of the agent seems to be strongly affected by environmental factors such as soil or temperature. Despite more than 100 years of research on some plant diseases, the causes of these diseases remain unknown.

BOX 5 Koch's postulates

Robert Koch (1843–1910) (Fig. 1-16C) was a medical doctor and a bacteriologist. He was the first to show, in 1876, that anthrax, a disease of sheep and other animals, including humans, was caused by a bacterium that he called *Bacillus anthracis*. He subsequently discovered, in 1882, that tuberculosis and, in 1883, that cholera are each caused by a different bacterium, which led to the general conclusion that each disease is caused by a specific microbe. These experiments confirmed for the first time the germ theory of disease proposed earlier by Louis Pasteur.

Before Koch's experiments, and while Koch himself was carrying out the work on the diseases mentioned earlier, there was confusion and uncertainty about the occurrence and the cause of each disease. Much of the time when bacteria or fungi were isolated from diseased or dead human, animal, or plant tissues, the isolated bacteria or fungi were subsequently shown to be saprophytes, i.e., they coexisted with the microorganism that caused the disease but could not by themselves cause the disease for which they were being considered. Based on his experiences, in 1887, Koch set out the four steps or criteria that must be satisfied before a microorganism isolated from a diseased human, animal, or plant can be considered as the cause of the disease. These four steps, rules, or criteria are known as "Koch's postulates."

1. The suspected causal agent (bacterium or other microorganism) must be present in every diseased organism (e.g., a plant) examined.
2. The suspected causal agent (bacterium, etc.) must be isolated from the diseased host organism (plant) and grown in pure culture.
3. When a pure culture of the suspected causal agent is inoculated into a healthy susceptible host (plant), the host must reproduce the specific disease.
4. The same causal agent must be recovered again from the experimentally inoculated and infected host, i.e., the recovered agent must have the same characteristics as the organism in step 2.

Koch's rules are possible to implement, although not always easy to carry out, with such pathogens as fungi, bacteria, parasitic higher plants, nematodes, most viruses and viroids, and the spiroplasmas. These organisms can be isolated and cultured, or can be purified, and they can then be introduced into the plant to see if they cause the disease. With the other pathogens, however, such as some viruses, phytoplasmas, fastidious phloem-inhabiting bacteria, protozoa, and even some plant pathogenic fungi that are obligate parasites of plants (such as the powdery mildew, downy mildew, and rust fungi), culture or purification of the pathogen is not yet possible and the pathogen often cannot be reintroduced into the plant to reproduce the disease. Thus, with these pathogens, Koch's rules cannot be carried out, and their acceptance as the actual pathogens of the diseases with which they are associated is more or less tentative. In most cases, however, the circumstantial evidence is overwhelming, and it is assumed that further improvement of techniques of isolation, culture, and inoculation of pathogens will someday prove that today's assumptions are justified. However, in the absence of the proof demanded by Koch's rules and as a result of insufficient information, all plant diseases caused by phytoplasmas (e.g., aster yellows) and fastidious vascular bacteria (e.g., Pierce's disease of grape) were for years thought to be caused by viruses.

Despite the difficulties of carrying out Koch's postulates with some causal agents, they have been and continue to be applied, sometimes with certain modifications, in all cases of disease. They have had and continue to have a tremendous effect in deciding and in convincing others that a particular microorganism is the cause of a specific disease. By attempting to carry out Koch's postulates in all newly discovered diseases, a great deal of work with potential saprophytes has been avoided, while, at the same time, doubt and criticism are reduced to a minimum while confidence in and use of the identification increase greatly and quickly.

BOX 6 Viruses, Viroids, and Prions

Although they have been with us forever, we know relatively little about how these pathogen operate. There are many common characteristics among viruses and viroids. The relationship of prions to others is only in their small size but they are contrasted to the other two in that they do not depend on any kind of nucleic acid (RNA or DNA). Viruses cause numerous severe diseases in all types of organisms, have been studied the longest, and we know the most about them. Viroids cause more than 40 diseases in plants, some of them lethal. Prions seem to affect only humans and animals in which they cause degenerative diseases of the brain, such as the recently much publicized "mad cow disease."

Viruses are submicroscopic spherical, rod-shaped, or filamentous entities (organisms) (Figs. 1-22A–1-22C) that consist of only one type of nucleic acid (DNA or RNA). The nucleic acid is surrounded by a coat consisting of one or more kinds of protein molecules. Viruses infect and multiply inside the cells of humans, animals, plants, or other organisms and usually cause disease.

Viroids were discovered by Diener (Fig. 1-22D) and colleagues in 1971 while they were studying the potato spindle tuber disease (Fig. 1-22E). Viroids are the smallest infectious agents that multiply autonomously in plant cells; they consist only of small, circular RNA molecules (Fig. 1-22F) that are too small to code for even one small protein and therefore lack a protein coat. Viroids infect plant cells and are replicated in their nucleus, using the substances and enzymes of plant cells. Viroids infect only plants and in many of them they usually cause disease. Viroids have not yet been detected in any other kind of organism besides plants.

Prions were proposed for the first time in 1972 by Prusiner (Fig. 1-22G) who, for that and subsequent work, received the Nobel Prize in Physiology or Medicine in 1997. Prions are at first normal small protein molecules produced in nerve and other cells of the brain. Prions become pathogenic, i.e., they cannot carry out their normal functions and, instead, have adverse effects on the brain and cause disease. This occurs when prions are forced by conditions in the brain to change shape (Fig. 1-22H). The change in shape signals the onset of infection. Prions are not associated with any nucleic acid. Abnormal prions appear to increase in number and to cause the appearance of amyloid fibrils and plaques, as well as the appearance of small cavities (Fig. 1-22I) in the brain of diseased animals and humans. Prions have not been observed in plants or other organisms.

continued

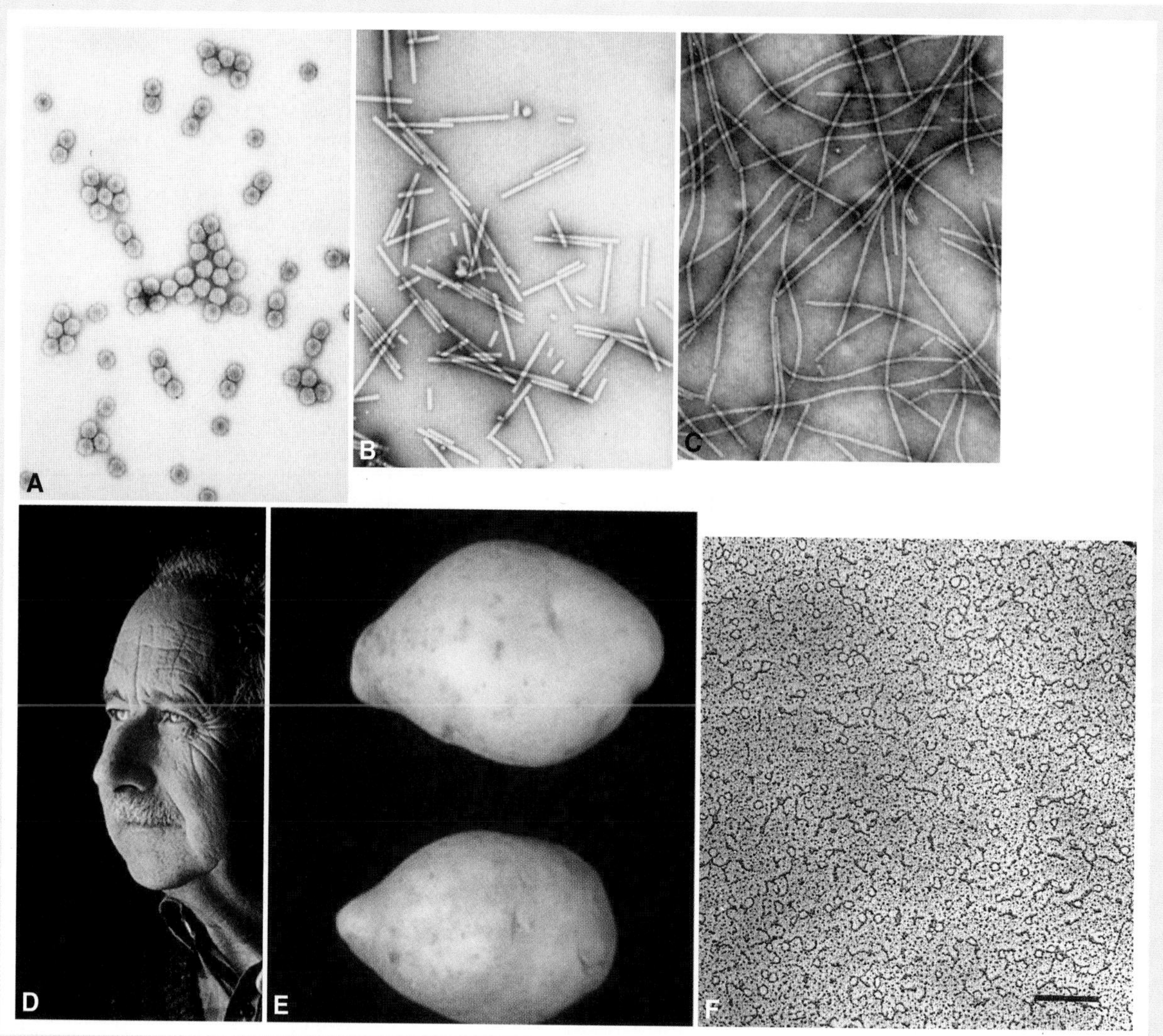

FIGURE 1-22 (A–C) Relative shapes and sizes of plant viruses: spherical, rod shaped, and flexuous. (D) T. O. Diener. (E) Potatoes infected with potato spindle tuber viroid. (F) Circular and linear particles of the coconut cadang-cadang viroid. (G) Stanley Prusiner. (H) Schematic presentation of a normal protein and of a deformed inactive one, i.e., a prion. (I) Plaques in the brain of an animal affected by a prion. [Photographs courtesy of (E) H. D. Thurston, Cornell University, (F) J. W. Randles, University of Adelaide, Australia, and (H and I) S. Prusiner, University of California.]

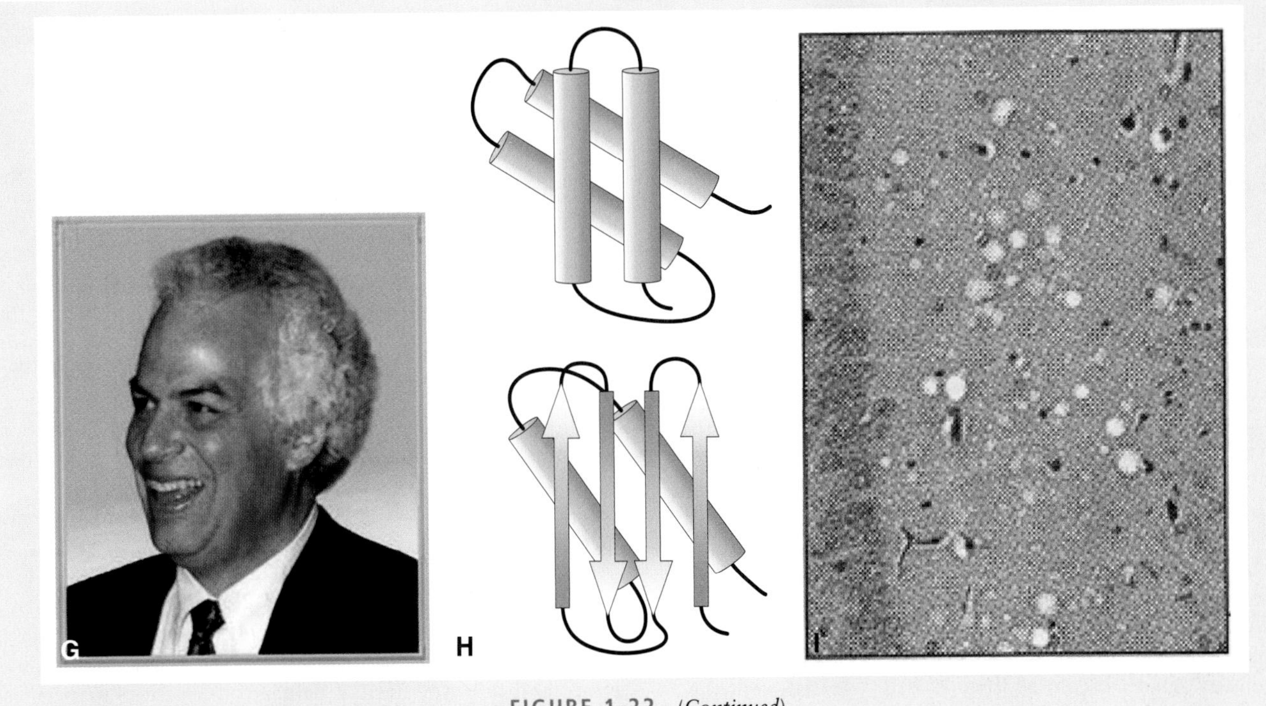

FIGURE 1-22 *(Continued)*

LOSSES CAUSED BY PLANT DISEASES

Plant diseases are of paramount importance to humans because they damage plants and plant products on which humans depend for food, clothing, furniture, the environment, and, in many cases, housing. For millions of people all over the world who still depend on their own plant produce for survival, plant diseases can make the difference between a comfortable life and a life haunted by hunger or even death from starvation. Death from starvation of one and a quarter million Irish people in 1845 and much of the hunger of the underfed millions living in the developing countries today are examples of the consequences of plant diseases. For countries where food is plentiful, plant diseases are significant primarily because they cause economic losses to growers. Plant diseases, however, also result in increased prices of products to consumers; they sometimes cause direct and severe pathological effects on humans and animals that eat diseased plant products; they destroy the beauty of the environment by damaging plants around homes, along streets, in parks, and in forests; and, in trying to control the diseases, people release billions of pounds of toxic pesticides that pollute the water and the environment.

Plant Diseases Reduce the Quantity and Quality of Plant Produce

The kinds and amounts of losses caused by plant diseases vary with the plant or plant product, the pathogen, the locality, the environment, the control measures practiced, and combinations of these factors. The quantity of loss may range from slight to 100%. Plants or plant products may be reduced in quantity by disease in the field, as indeed is the case with most plant diseases, or by disease during storage, as is the case of the rots of stored fruits, vegetables, grains, and fibers. Sometimes, destruction by the disease of some plants or fruits is compensated by greater growth and yield of the remaining plants or fruits as a result of reduced competition. Frequently, severe losses may be incurred by reduction in the quality of plant products. For instance, whereas spots, scabs, blemishes, and blotches on fruit, vegetables, or ornamental plants may have little effect on the quantity produced, the inferior quality of the product may reduce the market value so much that production is unprofitable or a total loss. For example, with apples infected with apple scab, even as little as 5% disease may cut the price in half; with others, e.g., potatoes infected with potato scab, there may be no effect on price in a market with slight scarcity, but there may be a considerable price reduction in years of even minor gluts of produce.

BOX 7 White, dry, and downy vineyards — bordeaux to the rescue!

During the second half of the 1800s, the saying that bad things come in threes found perfect application in the European and particularly the French grape and wine industry. In the 1840s, a condition known to exist on grapes in America but never before observed in Europe appeared first in England and soon after in France: young grape leaves would be covered with spots of white powder (Fig. 1-23A). Later, as the leaf grew in size and age, the white spots would spread and cover most of the leaf. The white mildewy stuff would also get on the berries, which would become dirty gray, wither, and sometimes crack. The condition was called powdery mildew and was later shown to be caused by the fungus *Uncinula necator*. Often, parts of the leaf would turn brown to black and die, while the berries would remain small, discolored (Fig. 1-23B), and unfit for wine production or to be eaten fresh. By 1854, French wine production was reduced by 80% due to the new disease. New grapevines were frantically imported from many countries in the hope that some of them would survive the powdery mildew. Fortunately, at the same time, it was noticed in England that when a mixture of powdered lime and sulfur was dusted on the vines, it significantly protected the leaves and the berries from powdery mildew. This practice became somewhat accepted in France and losses from powdery mildew were reduced significantly.

The early scramble for and importation of foreign vines, however, brought with it a second calamity to the French and European grape and wine industry that was much more disastrous than powdery mildew. In the early 1860s, young leaves on French vines would develop several small galls on the underside (Fig. 1-24A), but then, a few weeks later, all the leaves would turn yellowish to red in early spring and summer and subsequently would wither and fall off (Fig. 1-24B) in July or August. Affected vines produced little or no fruit and the following year they died. The dead, dry leaves gave to the condition the name "phylloxera" (="dryleaf" from the Greek phyllo = leaf, and xera = dry). It was later noted that phylloxera was associated with aphids, some of which fed on the young leaves and induced galls, while many more were found feeding on the roots of grapevines. The aphids not only induced galls on the small roots, they also multiplied quickly and sucked the nutrients out of the roots, killing the roots and, by denying the plant water, caused the leaves to discolor, wither, and fall off. The phylloxera condition was spreading slowly but, in vineyards into which it spread, it had devastating results.

It was determined that phylloxera aphids had probably been brought in from the United States with vines imported for resistance to the powdery mildew problem. The phylloxera aphids, however, did not seem to attack or cause serious damage on American grapevines. So, a new wave of importation of American vines began. These vines were used as rootstocks on which the European varieties were grafted. The degree of resistance of some of the rootstocks to

FIGURE 1-23 Powdery mildew of grape on (A) leaves and (B) grape cluster. White mycelium may cover all green parts, which become dry and brown. (Photographs courtesy of M. A. Ellis, Ohio State University.)

FIGURE 1-24 Phylloxera on grape caused by the grape root aphid. (A) Patch of grapevines showing dry foliage or defoliation due to infection of their roots by the phylloxera aphid. (B) Phylloxera aphids (*Dactylosphaira vitifolia*) feeding on and eventually killing the rootlets of grapevines, thereby causing drying and death of the plants. (Photographs: Queensland Dept. Natural Resources.)

the phylloxera aphids was excellent (Fig. 1-24B) and so the French and other European vineyards could be restored significantly over time.

Unfortunately, however, a third calamity hit the European vineyards while they were just beginning to feel that they had figured out how to escape the destructiveness of phylloxera. In 1878, grape leaves in some French vineyards began to show whitish downy spots on their undersides (Fig. 1-25A), while the upper sides of such leaves corresponding to the underside downy spots became yellow at first and then turned brownish black and died. This condition became known as downy mildew and was shown to be caused by the fungus *Plasmopara viticola*. As the number and size of the spots increased, most or all of the leaf was affected, died, and fell off the vine. Young shoots were also affected, as were young grape clusters, becoming covered with the white downy growth (Fig. 1-25B). Later, they turned brown and eventually shriveled. Berries infected later in the season remained hard compared to healthy ones, exhibited a light green to reddish mottle, and eventually dropped.

The downy mildew spread rapidly within vineyards and from one vineyard to another. It reduced grape yields and quality greatly and killed the young vines in many vineyards. Downy mildew was especially severe and spread the most in cool, rainy weather. Within 5 years of its appearance in France it spread to all the vineyards of that country and into those of adjacent countries. The grape producers in these countries became panicky again. Many scientists showed concern for the problem and interest in finding a solution for it. Some of them used different substances, which they added to the soil or dusted on the vines, trying to protect them from downy mildew. For several years nothing worked. Then one day, the French botany professor Pierre Alexis Millardet (Fig. 1-25C), while walking among the vineyards, noticed that in some of them, the vines of a few rows along the dirt road had a bluish film on their leaves. What was most noteworthy was that these vines seemed to still have all their leaves healthy, whereas vines in rows that did not have the bluish film, the leaves, young twigs, and berry clusters were affected severely by downy mildew (Fig. 1-25D). The owner of the vineyard told him that the bluish film was actually bluestone (copper sulfate), mixed with some hydrated lime to better stick on the leaves. The mixture was sprayed on the vines to create the impression that it was poisonous and in that way to keep passersby from going into his vineyard and taking his grapes. With that information in hand, Millardet went back to his laboratory where he mixed copper sulfate and hydrated lime in various proportions and tried them on downy mildew-affected vines. Finally, in 1885, he found the best combination for the control of downy mildew. This solution (8-8-100) became known as Bordeaux mixture and ushered in the era of control of plant diseases with fungicides. Bordeaux mixture proved to be an excellent fungicide and bactericide and for more than a century was the fungicide used the most throughout the world.

continued

FIGURE 1-25 Downy mildew of grape. Early symptoms on (A) grape leaf and (B) grape cluster. (C) P. Millardet. (D) At left, grapevines exposed to downy mildew but treated with Bordeaux mixture still retain most of their foliage, whereas, at right, unprotected grapes lost almost all of their foliage as a result of downy mildew. [Photographs courtesy of (A) J. Travis and J. Rytter, Pennsylvania State University, (B) University of Georgia, Extension, and (D) G. Ash, Charles Sturt University, Australia.]

Plant Diseases May Limit the Kinds of Plants and Industries in an Area

Plant diseases may limit the kinds of plants that can grow in a large geographic area. For example, the American chestnut was annihilated in North America as a timber tree by the chestnut blight disease, and the American elm is being eliminated as a shade tree by Dutch elm disease.

BOX 8 Familiar trees in the landscape: going, going, gone

In the 19th century, two plant diseases, powdery and downy mildews of grape, and an insect pest of grapes, the phylloxera aphid, each of which alone could have destroyed the European vineyards, spread from North America into Europe. The rediscovery of the use of sulfur against powdery mildew, the discovery of Bordeaux mixture against downy mildew, and the discovery of rootstocks resistant to phylloxera saved the European grape industry in each case. In the 20th century, Europe returned the favor to North America by giving North America two plant diseases, chestnut blight and Dutch elm disease, each of which killed billions of trees, bringing their respective host species to the brink of extinction. Unfortunately, no good control of these diseases exists even to date, and more of the remaining, at least elm trees, continue to be killed. Another disease, lethal yellowing of coconut palms, has spread through several of the Caribbean islands and adjacent countries, the states of Florida and Texas, west Africa, and elsewhere. Lethal yellowing has destroyed the majority of coconut palms in these areas and, like chestnut blight and the Dutch elm disease, it is still impossible or very difficult to control and continues to kill and threaten the remaining trees with extinction.

Chestnut Blight

There was a time not too long ago that in a broad band of land of the United States, several hundred miles in width and extending from the bottom of the states of Georgia and Mississippi to the top of Maine and Michigan and into Ontario, Canada (Fig. 1-26A), that the most common trees in the forests were the majestic American chestnuts (Fig. 1-26B). They provided timber and chestnuts, the latter serving as a source of food for humans and for wildlife, while the trees served as a habitat for wildlife. Both timber and chestnuts provided a source of income for the local people. The trees had been there apparently forever and looked like they would also last forever.

Then something seemingly minor happened. In 1904, the leaves of a few branches of large chestnut trees and a few young trees in the New York zoo began to turn brown and die. Before anyone could figure out what was happening, many more young trees and branches of older ones died, giving the trees a blighted appearance. From there, chestnut blight spread rapidly through eastern North America so that by the 1920s the blight could be found in the entire natural range of the American chestnut tree. By now, scientists in

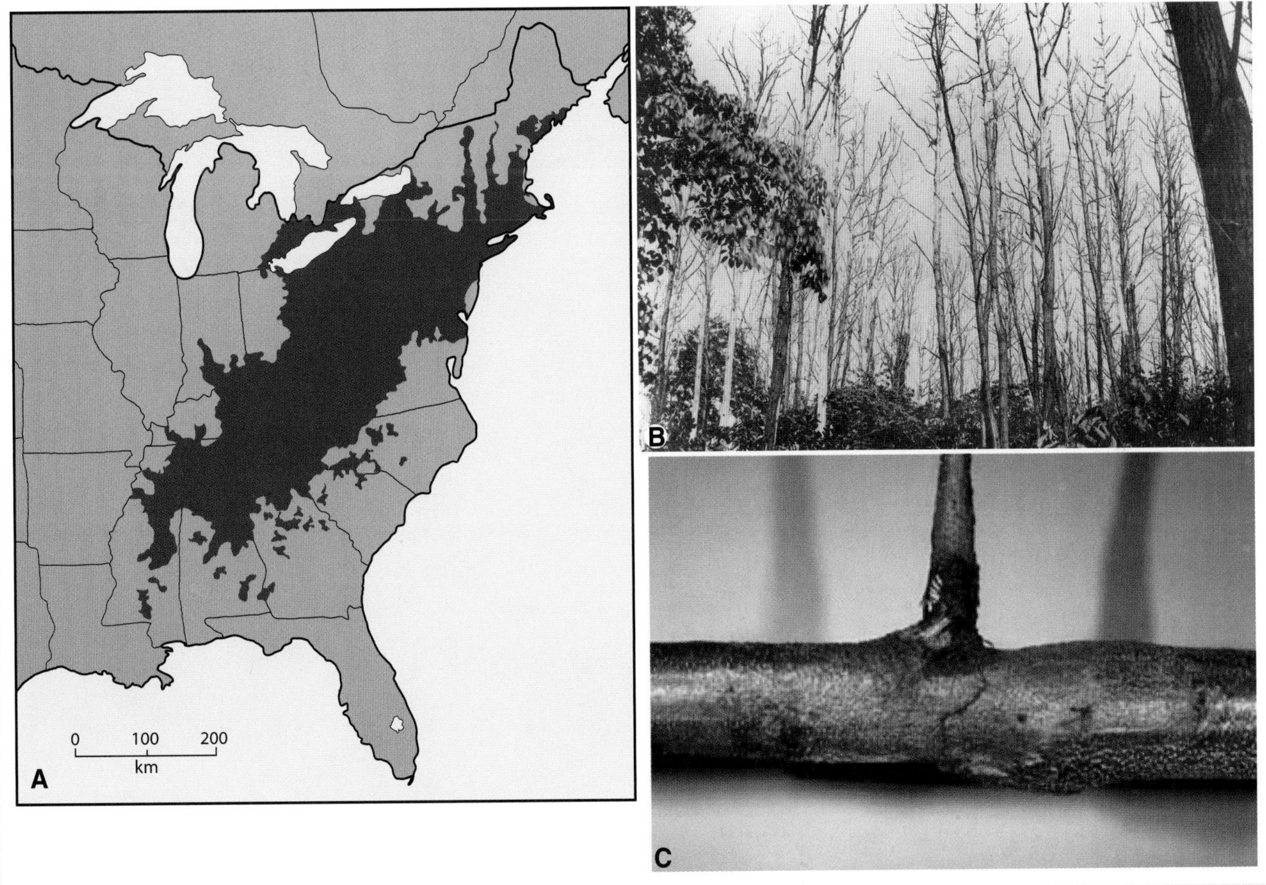

FIGURE 1-26 Chestnut blight. (A) Natural range of American chestnut before the chestnut blight fungus epidemic of 1904–1944. (B) Stand of young, pole-sized chestnut trees devastated by chestnut blight. (C) Chestnut blight canker on trunk of young chestnut tree causing the death of the tree. [Photographs courtesy of (B) W. L. MacDonald, West Virginia University, and (C) R. L. Anderson, U.S. Forest Service.]

continued

general, and plant pathologists in particular, were quite adept at identifying most causes of plant disease, and chestnut blight was quite easy to diagnose. It was soon shown that chestnut blight is caused by a fungus, *Cryphonectria parasitica*. The fungus attacks and kills the bark of branches and of young trees, causing a canker (Fig. 1-26C) that expands along and around the stem, girdling stems at that point and causing the leaves above the canker to wilt and die. Unfortunately, the fungus produces spores that are carried to other branches and trees by wind-blown rain, by insects, and by birds. By the late 1920s, about three and a half billion American chestnut trees had become infected. Infected trees and branches would produce sprouts from areas below the canker and the sprouts would grow without becoming infected until they were 2 to 4 inches in diameter. At some point, and before they produced any fruit, the fungus would attack and kill them too. That way, although the huge original chestnut trees kept producing new sprouts year after year for many years, their killing by the ever-present fungus finally exhausted the trees and they finally died to their roots. Hardly any trees escaped, making chestnuts the first tree to approach extinction in modern times because of a plant disease caused by a fungus.

Dutch Elm Disease

The American elm grows to be a big, gracefully shaped and beautiful vase-like tree that exists naturally mixed with other hardwoods throughout eastern North American forests and extending into the Great Plains. The elm was soon adopted by early homeowners and town settlers in North America and beautified many a street by being planted in rows on both sides of the street. Then, in 1930, a few elm trees in Cleveland, Ohio, began to show wilting, yellowing, and then browning of the leaves of some branches (Fig. 1-27A). The wilted, brown leaves later fell off and the branch appeared defoliated and dead. More branches showed similar symptoms later that year or the following year, and the entire elm tree usually died (Fig. 1-27B) within 1 or a few years. Trees with similar symptoms were soon observed in some east coast states. The disease became known as Dutch elm disease because, although it had been reported from France in 1917, it was the first report from Holland in 1921 that

FIGURE 1-27 Dutch elm disease. (A) Early symptoms of elm tree showing wilting, curling, and browning of leaves of branch infected with the Dutch elm disease fungus. (B) Advanced symptoms of wilt, defoliation, and death of large branches of tree affected with the disease. (C) Dead elm trees along a road, all killed by Dutch elm disease. [Photographs courtesy of (A) R. J. Stipes, Virginia Tech University, (B) R. L. Anderson, U.S. Forest Service, and (C), E. L. Barnard, Florida Forest Service.]

received all the publicity. The Dutch elm disease spread rapidly in North America, crossing the Mississippi River by 1956 and reaching the Pacific coast states by 1973. In its path, the disease has killed the vast majority of yard, park, and street trees (Fig. 1-27C), although quite a few trees in their natural forest habitat are still free of the disease.

Dutch elm disease is caused by the fungus *Ophiostoma ulmi*. The fungus is carried to healthy elm trees by two elm bark beetles that lay their eggs in weakened or dead elm trees or logs, often those killed by the Dutch elm disease. The eggs hatch and produce larvae that form tunnels, and if the tree or logs are infected with the disease, the fungus grows into and produces spores in the tunnels. The adult beetles then emerge covered with spores of the fungus and look for vigorous young elm branches to feed on. While they are feeding and causing hardly any damage to the elm trees, they deposit spores of the fungus in the feeding wound. The spores germinate and produce mycelium and more spores, both of which spread and multiply in the xylem vessels of the tree and cause the vessels to become clogged. Water and minerals cannot move from the root to the shoots and leaves beyond the point of clogging. The shoots and leaves subsequently wilt and die and, eventually, the entire tree dies.

Lethal Yellowing of Coconut Palms

Lethal yellowing-like symptoms on dying palm trees had been included in brief reports from the Cayman Islands, Cuba, and Jamaica even during the 19th century. In 1955, coconut palm trees in the Key West islands of Florida were noticed to drop their coconuts prematurely. Then, the next inflorescence had blackened tips and set no fruit. Soon, first the lower, older leaves and then the next younger leaves turned yellow and then brown and died. Finally, all the leaves and the vegetative bud died (Fig. 1-28A) and the entire top of the tree fell off, leaving the tall palm trunk looking like a telephone pole (Fig. 1-28B). The lethal yellowing disease was first found in mainland Florida in 1971 and killed 15,000 coconut palm trees by1973, 40,000 by 1974, and, by 1975, 75% of the coconut palm trees in Dade County were dead or dying from the disease. Tremendous losses of palm trees occurred in many other countries. For example, in Jamaica, of six million trees counted in 1961, 90% had been killed by lethal yellowing by 1981. Thousands of hectares of palm trees were killed in Mexico and also in Tanzania, more than a million coconut palm trees were killed in Ghana within 30 years, and more than 60,000, about 50% of the palm trees in Togo, were killed by lethal yellowing by 1964.

The lethal yellowing disease is caused by a phytoplasma, which is a kind of bacterium that lacks a cell wall. The phytoplasma lives and multiplies in the phloem sieve elements of palm trees and causes the lethal yellowing symptoms by clogging some of the sieve tubes and interfering with the transportation of organic foodstuffs out of the leaves and also by producing biologically active substances that are toxic and cause the yellowing and death of the leaves, inflorescence, and vegetative bud of coconut trees. The phytoplasma is spread from diseased to healthy trees by a small plant hopper. The plant hopper sucks up juice

FIGURE 1-28 Lethal yellowing of coconut palm trees. (A) Coconut palms at different stages of the disease, with the disease advancing from the lower fronds upward until the apical bud is killed. (B) Telephone pole-like trunks of coconut palms left after trees were killed by the lethal yellows phytoplasma. (Photographs courtesy of University of Florida.)

continued

from the phloem of palm trees and, if the tree is infected with the mycoplasma, the plant hopper sucks up some phytoplasmas also. When the plant hopper lands and feeds on a healthy palm tree, it transmits some of the phytoplasmas it carries into the phloem sieve elements. Once in the phloem cells, the phytoplasmas multiply and move throughout much of the phloem of the tree and cause the tree to develop the symptoms of lethal yellowing and to die.

Oak Wilts and Sudden Death

Oaks have been killed for decades by oak wilt (see page 532) caused by the fungus *Ceratocystis fagacearum*, but its spread and development are slower than the Dutch elm disease of elm. At the same time, the oak population is larger and distributed more widely compared to elm. Recently, different species of *Phytophthora* have been attacking and killing oak trees in California, Oregon, Europe, and elsewhere (see pages 418). The progression of these epidemics is hard to predict, but the loss of thousands of oak trees is certain.

Butternut Canker

Butternut trees are native to eastern North American forests and their wood has been used for furniture and for carving. In 1967, butternut trees in Iowa were observed to have multiple cankers on branches and stems and to subsequently die from the disease. Soon afterward, the disease was found to occur widely in the forests of the southeastern coastal region and was shown to be caused by the imperfect fungus *Sirococcus clavigignenti-juglandacearum*. Contrary to chestnut trees killed by chestnut blight, in butternut trees, the canker fungus infects both young and old trees through wounds. Because butternut trees do not sprout after their stem is killed, they are lost entirely. The disease has spread so rapidly that the US Forest Service estimated that about 80% of the butternut trees in the southeast had been killed by the mid-1990s. The remaining survivors were mostly along the banks of streams and rivers, but most of them were also heavily infected and were not reproducing.

Cypress Canker

Cypress trees (*Cupressus sempervirens*) and other species grow in Mediterranean climates, including California, the Mediterranean, and Persia. For more than three millennia they have been valued as ornamentals for their tall, statuesque, columnar shape, as well as for their wood, which is resistant to woodworms, rots, and decays. Cypress trees are extremely long lived, some of them possibly living for more than 2000 years. Many of the world's centers of civilization, such as the Acropolis of Athens, Olympia, Delphi, Florence, and others, and many of the paintings over the centuries derive much of their classic beauty from the real or painted cypress trees in them.

The first cypress canker outbreak was described in California in the mid-1920s, but the disease apparently existed there for more than 10 years before that. The disease then spread inland across the United States and into South America and, apparently, was transported from there across the oceans into the Mediterranean countries, New Zealand, and South Africa so that by now it is believed to be present in most parts of the world where cypress trees grow. Cypress canker or cypress blight is caused by three species of the fungus *Seiridium*, particularly *S. cardinale*. The fungus produces spores (conidia) that infect twigs and small branches through wounds and causes cankers that kill the twigs and branches. Resin flows out of the cracks of cankers while the foliage of infected twigs and branches turns yellowish to red at first, becoming reddish brown as the twigs die. A noticeable dieback of twigs, branches, and tree tops becomes visible at a distance. Heavily infected trees die. Large numbers and large percentages of cypress trees have been killed by the cypress canker fungus in the last few decades. Spread of the disease among the remaining trees continues, possibly at an accelerated rate. As many as one million cypress trees have been killed in central Italy, which includes Florence, with some groves showing more than 45% tree mortality from cypress canker infections. In some of the Greek islands and in parts of the mainland, 70 to 98% of the cypress trees have been killed by this disease.

The Xylella *Outbreak*

The European grape, *Vitis vinifera*, which provides all high-quality table and wine grapes throughout the world, cannot be grown in the southeastern United States because it is devastated by the indigenous xylem-inhabiting bacterium *Xylella fastidiosa*, the cause of Pierce's disease of grape. The disease had been reported in California in the 1880s, but lack of appropriate vectors, appropriate alternate hosts, and timing of unfavorable weather conditions kept the disease under control. As a result, grapes in California and Texas were free of that enemy but, in 1990, the disease was found in Texas where it has spread widely among the vineyards and has caused heavy losses. In 1998, one of its planthopper vectors and the bacterium causing Pierce's disease were found in vineyards of southern California, threatening not only the grape industry, but also many of the ornamental crops of California. *Xylella* bacteria were expected to do well in the California climate, but the absence of an effective vector of the bacteria provided protection and comfort to its agricultural industry. Now that the bacteria and one of their vectors have been brought together in that state, the California grape industry, and possibly its ornamentals, will probably never be the same again.

Plant diseases may also determine the kinds of agricultural industries and the level of employment in an area by affecting the amount and kind of produce available for local canning or processing. However, plant diseases are also responsible for the creation of new industries that develop chemicals, machinery, and methods to control plant diseases; the annual expenditures to this end amount to billions of dollars in the United States alone.

Plant Diseases May Make Plants Poisonous to Humans and Animals

Some diseases, such as ergot of rye and wheat, make plant products unfit for human or animal consumption by contaminating them with poisonous fruiting structures (Fig. 1-29).

BOX 9 Ergot, ergotism, and LSD: a bad combination

For centuries, if not for millennia, people and domestic animals from northern Spain to Russia, and probably elsewhere, suffered periodically from a variety of symptoms ranging from reddening and blistering of the skin to a burning sensation, to excruciating pain in the lower abdomen, muscle spasms, trembling, shaking, and convulsions, hallucinations and permanent insanity, gangrene and loss of fingers and limbs, and, occasionally, death. As a result of the initial burning sensation afflicted persons felt, the disease became known as "devil's curse," "fire," or "holy fire." In 1093, following a series of years of severe outbreaks of the disease, a religious order was formed in southern France to help those suffering from the disease. Because the patron saint of the order was Saint Anthony, the disease became known as "St. Anthony's fire." The disease varied in severity and occurrence from year to year and appeared to affect poor people more often than the well-to-do.

The disease seems to have existed since ancient times. It was described in China as early as 1100 B.C., in Assyria in 600 B.C., and was reported to severely affect the troops of Julius Caesar in one of his campaigns in France. Actually, France has experienced several serious epidemics of "holy fire," including the well-documented ones of 857, of 994 (which is said to have killed between 20,000 and 50,000 people), and of 1093. It is speculated that the Salem witchcraft trials in Salem, Massachusetts, in 1692, may indeed be the result of the "holy fire" disease caused by the consumption of ergot-contaminated flour. In 1722, 20,000 soldiers of the army of Peter the Great of Russia died from consuming bread made from severely infected wheat. Outbreaks of "holy fire" occurred even during the 20th century. For example, in 1926–1927 in Russia, as many as 10,000 people were affected by the disease, more than 200 cases were reported in 1927 in England, and more than 200 people were affected in 1951 in Provence, France, 32 of them becoming insane and 4 dying, all from eating bread made from ergot-contaminated wheat flour.

St. Anthony's fire is known today as ergotism and is the result of people and animals consuming grain coming from cultivated cereals and wild grasses infected with one of several ergot-producing fungi. Ergot (from the French "argot," which means a spur) is the fruiting structure produced by *Claviceps purpurea* and related fungi in place of the seed of the plant (Figs. 1-29A–1-29D) and contaminates the grain after harvest. Ergot is also the name of the disease of cereals and grasses caused by this and related fungi. Ergot, the plant disease, can reduce grain yields significantly, as each ergot replaces completely the kernel that it infects. Most of the damage to the crop, however, is because it makes the rest of the crop unfit for human or animal consumption unless the ergots are removed.

Ergots contain a number of potent alkaloids and other biologically active compounds that affect primarily the brain and the circulatory system. The best known of the ergot alkaloids is lysergic acid diethylamide, the infamous LSD (Fig. 1-29E) that was widely used as a hallucinogen by the hippie culture of the 1960s. Depending on the weather, the host plant (wheat, rye, barley, etc.) and the species of the ergot-forming fungus, the amount of ergot in the field and in the harvested grain may vary, as does the frequency and severity of the symptoms of ergotism (Figs. 1-29F and 1-29G). Rye, which is often consumed by animals and poor people, is the most frequent host of ergot, whereas wheat, preferred by the rich, is the least frequent host of ergot. The property of ergot alkaloids to constrict blood vessels and cause gangrene in humans and animals that consumed food contaminated with ergot sclerotia was put to good use by doctors and midwifes who used ground ergots at the wound to stop excessive bleeding occurring at childbirth and at severe accidents.

continued

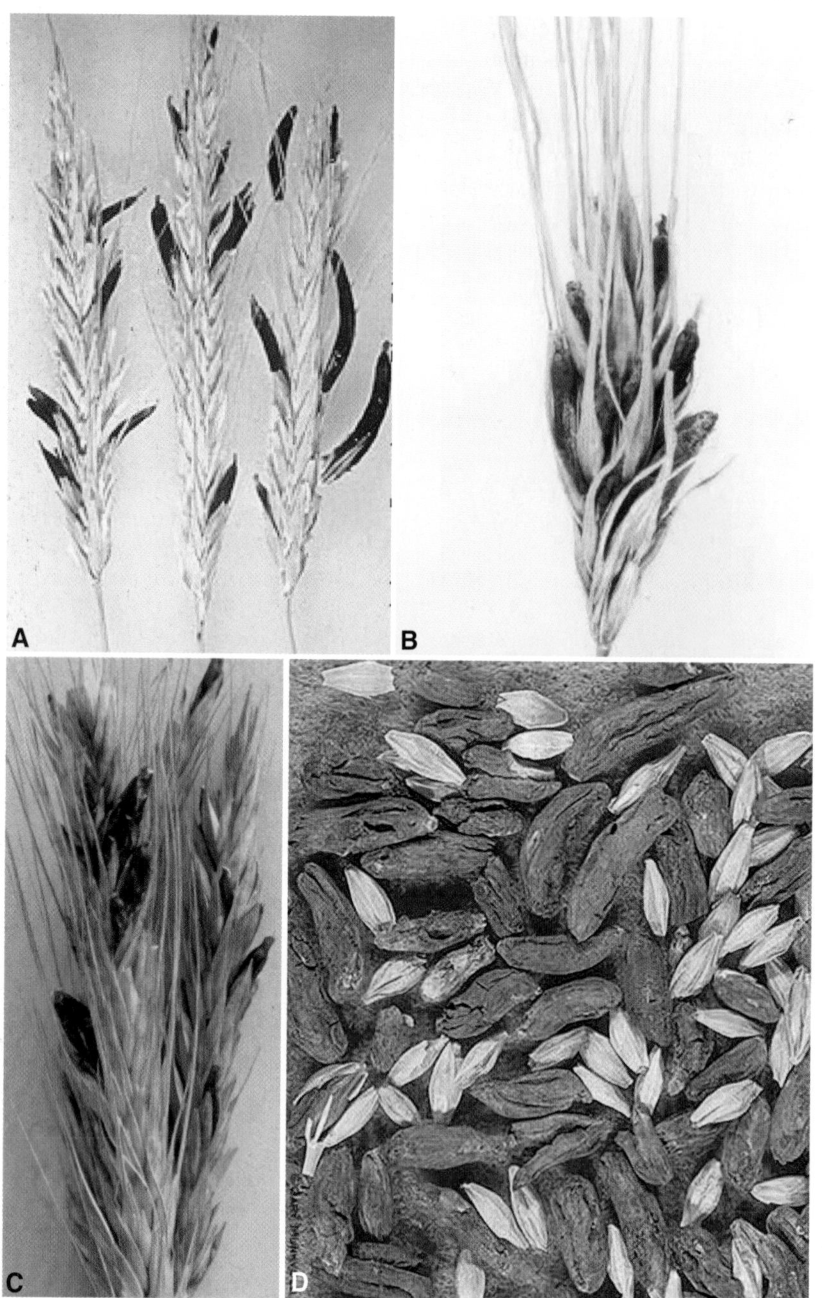

FIGURE 1-29 Ergot of cereals. Ergot sclerotia replacing the kernels in the heads of (A) rye, (B) barley, and (C) wheat. (D) Ergot sclerotia from barley mixed with healthy barley kernels. (E) The chemical formula of LSD found in ergot sclerotia. (F) Calf legs showing hemorrhage caused by consumption of feed containing ergot sclerotia. (G) A sketch of several people, some of whom had become maimed as a result of eating bread containing ground ergot sclerotia. [Photographs courtesy of (A–C) I. R. Evans, WCCPD, (D) G. Munkvold, Iowa State University, (F) Department of Veterinary Science, NDSU, and (G), Breugel, 16th Century, Art History Museum, Vienna.]

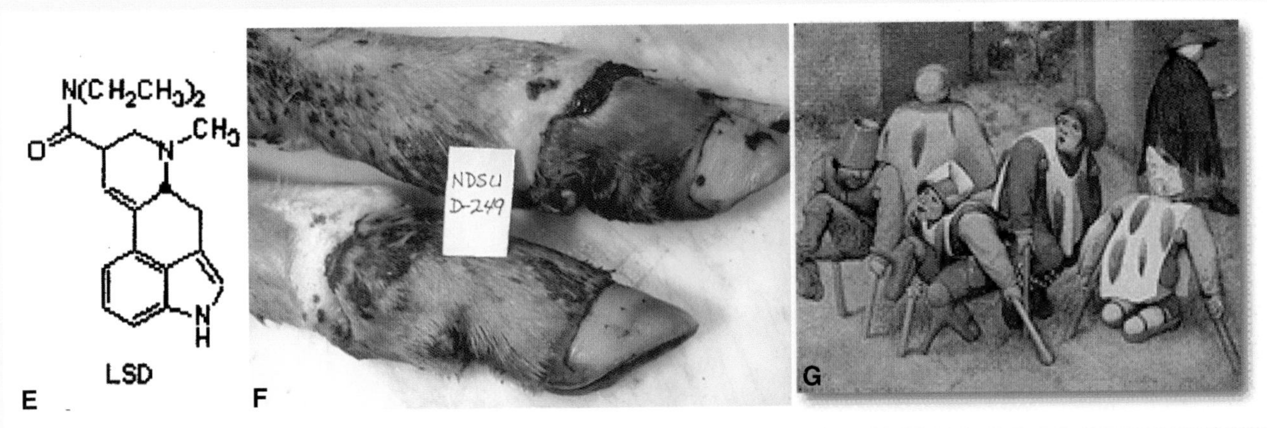

FIGURE 1-29 (*Continued*)

BOX 10 Mycotoxins and mycotoxicoses

Many grains (Figs. 1-30A–1-30D) and sometimes other seeds and also plant products such as bread (Fig. 1-30E), hay, purees, and rotting fruit (Fig. 1-30F) are often infected or contaminated with one or more fungi that produce toxic compounds known as mycotoxins. Animals or humans consuming such products may develop severe diseases of internal organs, the nervous system, and the circulatory system and may die. Also, many pasture grasses are infected with certain endophytic fungi that grow internally in the plant (Fig. 1-30G) and, although they do not seem to seriously damage the grass plants, they produce toxic compounds that cause severe diseases in the wild and domestic animals that eat the plants. Similarly, toxic and sometimes lethal to animals are some grasses whose seeds are infected with bacteria carried there by a nematode; these bacteria are often themselves infected with a virus (bacteriophage) that induces the bacteria to produce compounds very toxic to animals.

Ergotism is an example of a mycotoxicosis caused by food and feed made extremely unhealthy by mycotoxins produced by the fungus *Claviceps purpurea*. Ergotism causes very direct and dramatic symptoms and has been known for many centuries, if not millennia. There have been, however, innumerable other cases in which people or animals became chronically or acutely ill by eating food or feed that contained unsuspected toxic substances. The existence and identity of the toxic substances had remained unknown, the sources of such unsafe food and feed had been little noticed, and the ailments affecting humans and animals remained unexplained. It was not until the 1960s that a severe disease of young turkey birds was shown to be caused by moldy feed and called attention to the importance of mycotoxins in the health of people and animals.

Mycotoxins are toxic fungal metabolites that are released by relatively few but universally present fungi growing on grains, legumes, and nuts. Such produce, especially when harvested while still containing a high percentage of moisture or if it is damaged and stored at relatively high humidity, becomes moldy, i.e., it supports the growth of mycotoxin-producing fungi. Such moldy produce is likely to carry high concentrations of mycotoxins. Several of the mycotoxins are proven carcinogens, may disrupt the immune system, and may retard the growth of animals or humans that consume them. Even very small amounts of mycotoxins bring about the detrimental effect of mycotoxins on the immune system and metabolism of humans and animals, thereby posing a continuous health hazard. At higher concentration, which occur often on moldy produce, mycotoxins cause visible clinical symptoms (mycotoxicoses) in both humans and animals in the form of nervous agitation, dermal and subcutaneous lesions, impaired growth, damage to kidneys and liver, cancer, and others symptoms. Mycotoxins and mycotoxicoses are described in greater detail on page 559–560.

Although the last recorded outbreak of gangrenous ergotism occurred in Ethiopia in 1978, it was not until 1960 that the first general interest in mycotoxicoses was shown when the so-called "turkey X disease" appeared in farm animals in England. It was eventually shown that the disease was caused by feed contaminated with aflatoxins, and when these were shown to cause cancer in the liver of humans and animals, interest in mycotoxins skyrocketed. Aflatoxins are extremely toxic, appear in the milk of animals consuming contaminated feed, attack primarily the liver, and are mutagenic, teratogenic, and carcinogenic. In the last several decades, several outbreaks of aflatoxicosis have occurred in tropical countries where many adults in rural populations often consume moldy corn. Blood examinations in adults and children living in some tropi-

continued

FIGURE 1-30 Mycotoxin-containing plant products infected with mycotoxin-producing fungi. (A) Portion of ear of corn infected with *Aspergillus*. (B) Damaged corn kernels infected heavily with mycotoxin-producing *Gibberella fungi*. Wheat (C) and rye (D) kernels from fields infected heavily with the wheat scab-causing *Fusarium* spp. (E) Bread infected with *Aspergillus*, *Penicillium*, and other fungi. (F) Orange fruit infected with *Penicillium*. (G) Fluorescent mycelium of an endophytic fungus in a grass plant in which it produces mycotoxins. [Photographs courtesy of (A) P. Lipps, Ohio State University, (B) R. W. Stack, North Dakota State University, (C and D) WCCPD, and (G) A. DeLucca, USDA.]

cal areas and showing various symptoms of varying intensity have revealed the presence of aflatoxins in them, with significant seasonal variations.

In addition to aflatoxins produced by the two aforementioned species of *Aspergillus*, several other equally toxic mycotoxins, e.g., ochratoxins, are produced by these and by other species of *Aspergillus*, by *Penicillium*, and by other fungi. Ochratoxins occur in cereals, coffee, bread, and in many preserved foods of animal origin. About 20,000 people in the northern Balkans seem to be suffering from diseases caused by chronic exposure to ochratoxin. Poisoning from moldy sugar cane is caused by a mycotoxin produced by species of *Arthrinium*, and in one rural area in China it affected more than 800 persons who had ingested moldy sugar cane. *Aspergillus* and *Penicillium* are extremely common in nature and are almost always present to some extent in any feed and in most foods. Aflatoxins are the most common mycotoxins, but even more potent mycotoxins, e.g., patulin, roquefortin C, and others, are also produced by species and strains of *Penicillium*.

A number of potent mycotoxins, the trichothecins, are produced by several species of *Fusarium* and, to a lesser extent, by species of *Trichoderma*, *Trichothecium*, *Myrothecium*, and *Stachybotrys*. The most common trichothecin is deoxynivalenol, also known as vomitoxin. Another type of mycotoxin, zearalenone, is produced by somewhat different species of *Fusarium* (*F. graminearum*). Vomitoxin and zearalenone often occur together, especially in scabby wheat and in corn infected with *Gibberella* ear rot, but they have also been found in moldy rice, cottonseed, flour, barley, malt, beer, and other foods. In addition to humans, vomitoxin and zearalenone affect cattle, swine, chickens and other birds, cats, dogs, and fish. Individuals fed contaminated food or feed over a period respond by vomiting, refusal to eat, suppression of their immune system, diarrhea, loss of weight, and low milk production in the case of cows. A still different group of mycotoxins, called fumonisins, are produced by *Fusarium verticillioides* (*F. moniliforme, F. proliferatum*) and related species, primarily in corn and corn-based products. Fumonisins affect all or most of the animals affected by the other *Fusarium* toxins but they also affect and are particularly toxic to horses. In horses, low concentrations of fumonisins cause liquefaction of the brain, resulting in the "blind staggers" and "crazy horse disease" in which horses display blindness, head butting and pressing, constant circling and being agitated, and finally die. In swine, fumonisin attacks the heart and the respiratory system, in which it causes swellings, and it also causes lesions in the liver and pancreas. In humans, fumonisins have been linked to cancer. In the last 10 years, outbreaks of fumonisins in feed or food have been reported in several states from Arizona to Virginia and from South Carolina to the upper Midwest and in some Canadian provinces.

In most of the cases just mentioned, most of the damage is caused by the mycotoxins in food or feed consumed by humans and animals. However, for people and animals spending considerable time surrounded by moldy food or feed, there is the added danger of directly breathing spores of these fungi. It is not clear how detrimental to their health this is, but humans and animals, especially horses, exposed to spores of *Stachybotrys chartarum* develop irritation of the mouth, throat and nose, shock, skin necrosis, decrease in leukocytes, hemorrhage, nervous disorder, and death. *Stachybotrys* grows on straw and feed and on moist surfaces on walls and in air-conditioning ducts and is considered one of the most important causes of the "sick building syndrome."

Plant Diseases May Cause Financial Losses

In addition to direct losses in yield and quality, financial losses from plant diseases can arise in many ways. Farmers may have to plant varieties or species of plants that are resistant to disease but are less productive, more costly, or commercially less profitable than other varieties. They may have to spray or otherwise control a disease, thus incurring expenses for chemicals, machinery, storage space, and labor. Shippers may have to provide refrigerated warehouses and transportation vehicles, thereby increasing expenses. Plant diseases may limit the time during which products can be kept fresh and healthy, thus forcing growers to sell during a short period of time when products are abundant and prices are low. Healthy and diseased plant products may need to be separated from one another to avoid spreading of the disease, thus increasing handling costs.

The cost of controlling plant diseases, as well as lost productivity, is a loss attributable to diseases. Some plant diseases can be controlled almost entirely by one or another method, thus resulting in financial losses only to the amount of the cost of the control. Sometimes, however, this cost may be almost as high as, or even higher than, the return expected from the crop, as in the case of certain diseases of small grains. For other diseases, no effective control measures are yet known, and only a combination of cultural practices and the use of somewhat resistant varieties makes it possible to raise a crop. For most plant diseases, however, as long as we still have chemical pesticides, practical controls are available, although some losses may be incurred, despite the control measures taken. In these cases, the benefits from the control applied are generally much greater than the combined direct losses from the disease and the indirect losses due to expenses for control.

Despite the variety of types and sizes of financial losses that may be caused by plant diseases, well-informed farmers who use the best combinations of available resistant varieties and proper cultural, biological, and chemical control practices not only manage to produce a good crop in years of severe disease outbreaks, but may also obtain much greater economic benefits from increased prices after other farmers suffer severe crop losses.

BOX 11 The insect-pathogen connection: multifafaceted and important

Insects and similar organisms, such as mites and nematodes, are involved intimately and commonly in the facilitation, initiation, and development of many biotic and abiotic plant diseases. Some insects, e.g., gall-forming aphids and some mites, cause disease-like conditions in plants on which they feed. The importance of insect involvement in the development of pathogen-induced plant disease is so great that it can hardly be exaggerated. For some reason, however, it does not receive sufficient coverage in textbooks and in courses of plant pathology. Insects become involved in disease development in plants primarily through the following four types of action. (1) Insects visit infected plant organs oozing bacteria or fungal spores or plants covered with fungal spores, become smeared with bacteria or spores, and, quite passively, transfer them to other plants where they might cause disease. (2) They cause wounds on plant organs (leaves, fruit, shoots, branches, stems, roots) on which they feed or deposit their eggs and these allow pathogens, primarily fungi and bacteria, to enter the plant. (3) By feeding on plants, especially perennial ones, insects weaken them and make them more vulnerable to attack by some pathogenic fungi. (4) Insects act as vectors of certain pathogens, including a few fungi and bacteria, many viruses, and all phytoplasmas and protozoa. Insects carry these pathogens from diseased to healthy plants where they initiate new disease. These pathogens depend totally on insects for transmission, i.e., in the absence of the insect vectors there is no spread of the pathogen and no new diseased plants.

The first type of incidental transfer of bacteria or fungal spores to other plants or organs where they might cause disease probably involves many types of crawling, walking, or flying insects, such as flies (Figs. 1-31A and 1-31B). Some insects walk through or feed on flower nectar, as, for example, do bees (Fig. 1-31C)) in pear blossoms infected with the fireblight (Fig. 1-31D) bacterium, or on sugars released in infected areas, such as cankers, on stems, or spots or powdery and downy mildews on leaves, or on spots on fruit still on the tree or after harvest. Such insects may include different types of fruit flies, aphids, leafhoppers, beetles, ants, and many others.

Numerous insects feed and cause feeding wounds on various plant organs, e.g., fruits and roots, and several insects cause wounds when they deposit their eggs into such organs. Fungal and, sometimes, bacterial pathogens, such as the soft rot bacterium of potatoes and many other fleshy organs, are facilitated greatly in entering these organs through the wounds made by the insects. For example, the plum curculio beetle (Fig. 1-31E) creates wounds on fruit (Fig.1-31F) during ovipositing. The increased number of entry points for the fungus made on the fruit by insects makes it possible for fungi such as those causing brown rot of pome and stone fruits to be much more damaging in orchards where insect control is poor.

When insects feed on roots, leaves, or shoots of plants, especially perennial ones, the plants not only are wounded in numerous places and allow plant pathogenic fungi and bacteria to enter through the wounds and cause disease, they are also weakened greatly, especially in their ability to mobilize their defenses against pathogens and to protect themselves from becoming diseased. This situation is commonly observed on trees whose roots have been damaged by insects or have been defoliated by insects. In such trees, cankers or root rots, caused by fungi that are normally weak pathogens, develop much more rapidly and cause severe damage or may even kill the entire tree, something that would not have happened in the absence of the damage.

The fourth way in which insects influence the development of disease in plants is by forming close associations with certain pathogens. In such specific insect/pathogen associations, transmission and spread of certain pathogens from diseased to healthy plants depend almost entirely on the availability and involvement of one or a few specific insect vectors. For example, the corn flea beetle (Fig. 1-32A) is the main vector of the bacteria causing bacterial wilt of corn (Fig. 1-32B), whereas the striped and spotted cucumber beetles (Fig. 1-32C) are the main vectors of the cucurbit wilt bacteria (Fig. 1-32D). Similarly, without the vectoring ability of two species of elm bark beetles (Fig. 1-32E) , Dutch elm disease (Fig. 1-32F), which is caused by a fungus, would not possibly occur. Certain insects have also formed symbiotic associations with phloem-inhabiting bacteria such as the citrus greening disease bacteria; with specific xylem-inhabiting bacteria, e.g., the planthoppers that transmit the bacterium that causes Pierce's disease of grapevines; with the xylem-inhabiting nematode causing pine wilt; and with phloem-inhabiting plant pathogenic protozoa causing wilt diseases in coffee and palm trees.

The association of certain insects with specific pathogens, however, has reached its greatest frequency with the plant pathogenic phloem-inhabiting phytoplasmas that cause the yellows, proliferation, and decline diseases of numerous plants (e.g., aster yellows, apple proliferation, coconut palm lethal yellowing), and also with many of the phloem-inhabiting plant viruses. Phytoplasmas are transmitted by the closely related leafhoppers, plant hoppers, and psyllid insects.

Plant viruses, however, are transmitted by one or a few species belonging to the following groups of insects: aphids (Fig. 1-33A) transmit a large number of viruses, such as potato virus Y (Fig. 1-

FIGURE 1-31 Examples of insects helping spread plant diseases. Common flies (A) help spread fruit diseases such as brown rot of cherries (B). Bees (C) help spread diseases, such as fire blight of apple and pear (D). Curculio weevil (E) makes holes when ovipositing on fruit (F) that allow fruit-rotting fungi to enter the fruit. [Photographs courtesy of (A and C) University of Florida, (B) J. W. Pscheidt, Oregon State University, (D) T. Van Der Zwet, and (E and F) Clemson University.]

33B); leafhoppers and planthoppers (Fig. 1-33C) vector numerous viruses, such as the rice grassy stunt virus (Fig. 1-33D) (as well as phytoplasmas, spiroplasmas, and xylem and phloem-inhabiting bacteria); and whiteflies (Fig. 1-33E) vector geminiviruses, such as tomato yellow leaf curl virus (Fig. 1-33F). Other specific virus vectors include certain thrips, beetles, and mealybugs. The mechanisms of transmission of viruses by their insect vectors vary considerably. Although all phytoplasmas and most viruses transmitted by leafhoppers are taken up by the insect vector, circulated internally in its body, and multiply in some of its organs before they are injected into the phloem of new hosts, in many of the viruses, especially those transmitted by aphids, the virus is carried on or in the stylet of the vector and through it is deposited in phloem or parenchyma cells of the new host plant.

continued

FIGURE 1-32 Examples of insects serving as specific vectors of many important bacterial and fungal diseases. The corn flea beetle (A) is the vector of Stewart's wilt of corn (B). The striped cucumber beetle (C) is one of two vectors of bacterial wilt off cucurbits (D). The elm bark beetle (E) is one of two vectors of Dutch elm disease (F). [Photographs courtesy of (A) G. Munkvold and (B) M. Carlton, both Iowa State University, (C and D) Clemson University, (E) U.S. Forest Service, and (F) Minnesota Department of Natural Resource Archives.]

FIGURE 1-33 Examples of insects serving as specific vectors of viruses. Aphids (A) are the most important specific vector of numerous plant viruses such as *potato virus* Y (B). Leafhoppers and related planthoppers (C) are specific vectors for many viruses, such as *grassy stunt virus* (D) and also for phytoplasmas and xylem- and phloem-limited fastidious bacteria. Whiteflies (E) are the specific vectors of many devastating viruses, such as the *tomato yellow leaf curl geminivirus* (F). [Photographs courtesy of (A, B, E, and F) University of Florida and (C and D) H. Hibino.]

PLANT PATHOLOGY IN THE 20TH CENTURY

Early Developments

The Descriptive Phase

As agriculturists, botanists, naturalists, and other scientists, such as physicians, became aware of and familiar with the existence of plant disease and with some of the causes of plant disease, reports began to be published in scientific, popular, and semipopular journals describing numerous plant diseases on a variety of agricultural and ornamental plants. The availability of improved magnifying lenses and of microscopes made possible the detection and description of many fungi, nematodes, and, later, bacteria associated with diseased plants. Development and introduction of techniques for growing microorganisms (fungi and bacteria) in pure culture by Brefeld, Koch, Petri, and others (1875–1912) contributed greatly to plant pathology. In 1887, Koch's "postulates," which must be satisfied before a particular microorganism isolated from a diseased plant can be accepted as the cause of the disease and not be an unrelated contaminant, had a profound effect on plant pathology. Similarly, improvements in compound microscopes and in plant tissue-staining techniques allowed histopathological and cytological studies of infected plants that revealed the location of the pathogens (mostly fungi, nematodes, and bacteria) in relation to the infected plant cells and tissues. After 1940, the electron microscope made it possible to visualize and describe most viruses and, after 1970, helped detect and describe the mollicutes and viroids.

During the descriptive phase of plant pathology, many observations were also made and reported concerning the biology of the microorganisms involved. Most reports dealt with the types of spores produced by fungal pathogens, the means of spread of pathogens, the location of their survival during winter, the kinds of host plants infected, and so on. Quite often, such observations were correlated with the prevailing environmental conditions, such as rain and temperature, and with differences in disease severity among the various hosts. Different types of control practices, mostly cultural but also some chemical ones, were tried for various diseases. The discovery that sprays with Bordeaux mixture could control the downy mildew of grape encouraged experimentation with this and some other compounds for the control of many diseases on almost all crops.

The Experimental Phase

As the importance of plant diseases and of plant pathology as a new discipline and new profession began to be recognized in the late 1800s, scientists began to be hired as plant pathologists and to be added to the various USDA and state agricultural experiment stations. These scientists began to experiment in all areas of plant pathology. Although new diseases and pathogens continued to be discovered and described, plant pathologists began to ask questions and to design experiments to answer them about how pathogens enter their host plants, multiply, and spread within the plant; the mechanisms of host plant cell death and breakdown; pathogen sporulation; spore dispersal, overwintering, oversummering, and germination; vector involvement; and the effect of environment on disease development, among others. They also began noticing and studying variability among plant species and varieties in disease expression and loss. As knowledge accumulated, experimentation also grew rapidly on ways to control plant diseases and to avoid or reduce the losses from them.

The Etiological Phase

The etiological phase of plant pathology involved observations and experiments aimed at proving the causes (etiology) of specific plant diseases. Although the etiological phase began with the proof of pathogenicity of the late blight fungus on potatoes and of the rust and smut fungi of cereals, etiological studies were facilitated and accelerated greatly by the development of techniques for the pure culture of fungi and bacteria and by the necessity to satisfy Koch's postulates for every disease. Numerous reports in the late 1890s and in the first third of the 20th century dealt with descriptions of the symptoms of thousands of mostly fungal plant diseases on all types of hosts, of efforts to isolate and culture the suspected pathogens, and of subsequent experiments to prove the pathogenicity of the isolated, suspected pathogens. Many of these reports often included information on the losses estimated to be caused by the disease and on experiments about ways that could control the disease.

The etiological phase resumed, continued, and accelerated as new types of pathogens, such as viruses, phytoplasmas, fastidious bacteria, protozoa, and viroids, were discovered. Although the methodologies had to be adapted to the size and properties of each type of pathogen, the goal and the result remained the determination of the etiology of the disease. The etiological phase often depended on, and benefited from, improvements in methodology and instrumentation, such as the electron microscope, special nutrient media, density gradient centrifugation, electrophoresis, the development of serological techniques, the polymerase chain reaction (PCR), and the development of DNA probes and other nucleic acid tests and tools.

The Search for Control of Plant Diseases

As mentioned earlier, in addition to prayers and sacrifices to gods, some minor but realistic recommendations for control of plant diseases were reported in the writings of the ancient Greeks Homer (1000 B.C.), Democritus (470 B.C.), and Theophrastus (300 B.C.). It was not until the mid-1600s, however, that a species or variety was reported to be more resistant to a disease than another related species or variety, although it is assumed that, despite the absence of written reports, growers, knowingly or unknowingly, have been forever using a selection of resistant plants as a control of plant diseases. This is likely to have occurred not only because seeds from resistant and therefore healthier plants looked bigger and better than those from infected susceptible plants, but also because in severe disease outbreaks, resistant plants were the only ones surviving and, therefore, their seeds were the only ones available for planting.

The earliest use of chemicals for the control of plant diseases probably began in the late 1600s when some farmers in southern England planted wheat seed that had been salvaged from a ship wreck; they noticed that far fewer wheat plants produced from such seed were infected with smut (bunt) than wheat plants produced from other seed. This led some farmers to treat wheat seed with brine (sodium chloride solution) to control bunt. In the mid-1700s, copper sulfate was substituted for sodium chloride, and bunt control improved significantly. This treatment is still used in the poorer parts of the world, although in many countries copper sulfate has been replaced by other, more effective fungicides.

Diseases of fruit and ornamental trees were sometimes too obvious to ignore and although their cause was unknown, several cures, many of them worthless, were proposed. As mentioned earlier, it was noted around A.D. 1200 that a tree can be cured from mistletoe infections if the branch carrying the mistletoe is pruned out. In the mid-1700s, recommendations for the control of cankers included excisions of the canker and the application of grafting wax on the cut area. However, some "scientists" incorrectly recommended the use of vinegar to prevent canker on trees or the use of worthless mixtures of cow dung, lime rubbish from old buildings, wood ashes, and river sand to cure diseases, defects, and injuries of plants. In the early 1800s, lime sulfur and aqueous suspensions of sulfur were recommended for the control of mildew of fruit trees.

The Main Areas of Progress

Chemical Control of Plant Diseases

The introduction from America into Europe of the fungus causing the aggressive downy mildew disease of grape in the late 1870s stimulated a search by several investigators, especially in France, for chemicals that could control the disease. In 1885, Millardet noticed that vines sprayed with a bluish-white mixture of copper sulfate and lime retained their leaves, whereas the leaves of untreated vines were killed by the disease. After trying several combinations, Millardet concluded in that same year that a mixture of copper sulfate and hydrated lime could effectively control the downy mildew of grape. This mixture, which became known as Bordeaux mixture, was soon shown to be equally effective against the late blight of potato, other downy mildews, and many other leaf spots and blights of many different plants. For more than 100 years, Bordeaux mixture was used more than any other fungicide against a wide variety of plant diseases in all parts of the world, and even today it is one of the most widely used fungicides worldwide. The discovery of Bordeaux mixture proved that plant diseases can be controlled chemically and gave great encouragement and stimulus to the study of the nature and control of plant diseases.

In 1913, organic mercury compounds were introduced as seed treatments, and such treatments were routine until the 1960s when all mercury-containing pesticides were banned because of their toxicity. In the meantime, in 1928, Alexander Fleming (Fig. 1-34) discovered the antibiotic penicillin. This was effective against bacteria causing diseases of humans and animals but was not particularly effective against bacterial diseases of plants. Besides, the demand for use against bacterial diseases of humans and animals was so great and the antibiotic was so expensive that its use against bacterial diseases of plants was considered unlikely for at least the next 20 years. Penicillin, however, opened a new area for research in the control of plant diseases. In the meantime, in 1934, the first dithiocarbamate fungicide (thiram) was discovered, which led to the development of a series of effective and widely used fungicides, including ferbam, zineb, and maneb. Many other important protective fungicides followed. In 1965, the first systemic fungicide, carboxin, was discovered, and it was soon followed by the introduction of several other systemic fungicides, such as benomyl.

Antibiotics, primarily streptomycin, were first used to control bacterial plant diseases in 1950. Soon after, the antibiotic cycloheximide was shown to be effective against several plant pathogenic fungi. In 1967, tetra-

FIGURE 1-34 Alexander Fleming.

cycline antibiotics were shown to control plant diseases caused by mollicutes; a few years later, tetracycline was shown to control plant diseases caused by fastidious bacteria that live in the xylem of their host plants.

Appearance of Pathogen Races Resistant to Bactericides and Fungicides

In 1954, it was noticed that a few strains of bacteria causing disease in plants were resistant to certain antibiotics, and, in 1963, strains of fungal plant pathogens were found that were resistant to certain protective fungicides. It was in the 1970s, however, when the use of systemic fungicides became widespread, that new isolates/strains of numerous fungal plant pathogens appeared that were resistant to a fungicide that had previously been effective. The appearance of pathogen races resistant to chemicals prompted the development of new strategies in controlling plant diseases with fungicides and bactericides. Such strategies included the use of mixtures of fungicides, alternating compounds in successive sprays, and spraying with a systemic compound in the early stages of the disease and with a broad-spectrum compound in the later stages of the disease.

Public Concern about Chemical Pesticides

It had long been common knowledge that chemical pesticides are toxic poisons. The word pesticide itself means "pest killer." Pests, of course, include bacteria, fungi, insects, weeds, rodents, and other living things that affect humans, animals, or plants adversely. Depending on the kind of pest against which they are effective, pesticides are known as bactericides, fungicides, nematicides, insecticides, herbicides, and so on.

The public assumed at first that pesticides were toxic only against the kinds of pests at which they were aimed. Scientists and users alike felt certain that animals and humans were not affected by pesticides unless they were fed large amounts of pesticides accidentally or intentionally. For a long time, therefore, pesticides were applied liberally on fields, fruits, vegetables, stagnant waters, and even directly on animals and humans to control insects and diseases affecting them. Hundreds of pesticides were produced annually, and many of the newer pesticides were much more toxic than the earlier ones, i.e., they could kill or seriously injure microbes, pests, higher animals, and humans at a much lower concentration and faster than earlier pesticides. Some of the pesticides broke down into nontoxic or much less toxic compounds soon after they were applied and were exposed to air, sun, and moisture. Others, however, such as DDT and the chlorinated hydrocarbons, consisted of persistent molecules that resisted breakdown and remained toxic for many years or indefinitely.

A few voices of concern about using pesticides were beginning to be heard in the 1950s, but the obvious benefits from controlling insects and diseases in plants, animals, and humans were so overwhelming and the assurances of pesticide safety by scientists and pesticide industries so effective that few such concerns reached the wider public. Rachel Carson's (Fig. 1-35) book "Silent Spring," published in 1962, however, vividly described the dangers of polluting the environment with poisonous chemicals and documented several cases of bird and fish deaths to be the results of pesticides being accumulated and concentrated through the food chain. Carson's book generated a great deal of controversy but also a much greater awareness of the possible adverse effects of pesticides. Many scientists at first were quite skeptical and unconvinced of Carson's arguments. Little by little, however, many of them agreed to do research on the issue of safety of pesticides and began testing insects, earthworms, birds, fish, plants, animals, water streams, lakes, and even soil and underground water reservoirs for pesticides. To the surprise of many scientists, pesticides, particularly the persistent types, were found in many of these bodies, sometimes in fairly high concentrations. By that time (mid-1960s), air pollution by automobiles and factories, water and ground pollution with industrial wastes (chemicals, nuclear reactor byproducts), and so on were also becoming issues of concern to the public. The "Environmental Movement" was solidifying, and concerns about environmental pollution of all types began to gain momentum.

FIGURE 1-35 Rachel Carson.

By the mid-1960s, all pesticides containing mercury were banned by the U.S. government, and soon afterward DDT and chlorinated hydrocarbons were also banned. Laws were passed that prohibited the use of pesticides causing cancer in laboratory animals or mutations in microorganisms. All existing pesticides were subjected to a new, stricter review, and those found to be carcinogenic or mutagenic were banned and removed from the market. The uses of many pesticides that continued to be allowed were further reduced as to the crop, dosage, timing, and number of applications, while the interval between last application and harvest was increased. Since the mid-1980s, approximately 85–90% of the pesticides or pesticide uses previously available for plant disease control have been banned by the U.S. government or discontinued by the manufacturers, and it is likely that several of the remaining ones will be banned or withdrawn in the near future. In the meantime, the requirements for less toxic, more specific pesticides have increased, as have the costs of bringing a pesticide to the market. The costs of potential litigation for injury from pesticides have also increased greatly. Much stricter rules have been imposed on pesticide applicators, pesticide applications, and handlers of products treated with pesticides, with each restriction making it safer, but more expensive, to apply pesticides. The current or anticipated lack of a supply of effective pesticides has increased the effort to develop alternative controls. Different controls may be provided by using antagonistic microorganisms (biological control), improving old cultural practices, and developing new ones. Particularly desirable are new control methods that incorporate disease resistance into crop varieties, either by conventional breeding or through genetic engineering technologies, and using nontoxic compounds that activate the natural defenses of plants.

Alternative Controls for Plant Diseases

Concern over the potential toxicity of pesticides and over the continuing loss of appropriate, effective pesticides available for plant disease control has continued to increase since the 1970s. This has led to the reexamination and improvement of many old practices and to the development of some new cultural practices for use in controlling plant diseases. Proper cultural practices include removal of plant debris and infected plant parts, use of seed free of pathogens, crop rotation with plant species that are immune to the kinds of pathogens that affect the other rotation crops, soil fallow, reduced or no tillage, destruction of weeds, fertilization with appropriate amounts and forms of fertilizer, appropriate irrigation, adjusting the time and rate of sowing and date of harvest, and minimizing the influx of pathogen vectors into crops through border plants. The modification of cultural practices, use of resistant varieties, and monitoring of the appearance and development of plant disease epidemics that allow for a reduced use of pesticides have become the basis of "integrated management" of plant diseases.

It was reported early in the 20th century that some soils, through the microorganisms they harbor or through other means, suppress the development of certain diseases caused by soilborne pathogens. After Fleming reported in 1928 that certain fungi, such as *Penicillium*, inhibit the growth of other fungi and bacteria, plant pathologists began searching for nonpathogenic microorganisms that could be applied to plants before or after infection with a pathogen and that would antagonize the pathogen and keep it from infecting the plant. Numerous nonpathogenic microorganisms, mostly fungi and bacteria, have been found that antagonize various plant pathogenic fungi, bacteria, and nematodes, and some of them have been shown to protect the host plant from infection by the pathogen. In the early 1930s, it was shown that infection of a plant with a mild strain of a virus prevented or delayed infection of the plant by a severe strain of the same virus ("cross protection"). It has been shown more recently that even some plant pathogenic fungi and bacteria can be controlled by pretreatment of the plant with an avirulent or hypovirulent strain of the same species.

Biological control of plant diseases with antagonistic microorganisms is practiced to a rather limited extent. The first such control was obtained in 1963 and involved inoculation of the surface of stumps of freshly cut pines with spores of a nonpathogenic fungus (*Phleviopsis gigantea*) that protected them from infection by the fungus (*Heterobasidion annosum*) that causes root and butt rot of pines. In 1972, control of the crown gall bacterium was obtained by preinoculating seeds or roots of transplants of stone fruit trees with a related but nonpathogenic bacterium, and control of the tobacco mosaic virus in tomato fields was obtained by preinoculating tomato seedlings with a nonpathogenic strain of the virus produced by mutating the virus artificially. Experimentally, biological control can be obtained against many plant pathogenic fungi and bacteria infecting foliage or roots in the field or fruits in storage, and also against some nematodes, but field applications are still mostly ineffective. The control of viral diseases by cross protection is used in the tristeza disease of citrus and in some other virus diseases. A new and promising type of biological control of viral diseases, discovered in the late 1980s, uses the introduction of one or several appropriate viral genes into host plants through genetic engineering and expression of these genes by the host. These genes then prevent or delay infection of the plant by the virus.

Another recent, very exciting and promising means of plant disease control is through the use of pathogenic microorganisms or chemical compounds that cause tiny necrotic lesions in the treated plant and, by so doing, activate the defenses of the whole plant against subsequent infections by pathogens of the same or different types. This has been called systemic acquired (or induced or activated) resistance. In the early 1990s, nontoxic chemical compounds called plant defense activators were synthesized that, when applied to plants, activate the systemic defenses of plants against pathogens without causing necrotic lesions. The first such compound, named Actigard, was market tested with considerable success in 1996.

Interest in the Mechanisms by Which Pathogens Cause Disease

Once it became apparent that fungi and other microorganisms were the causes rather than the results of plant disease, efforts began to understand the mechanisms by which microorganisms cause disease. In 1886, deBary, working with the Sclerotinia rot disease of carrots (Fig. 1-36) and other vegetables, noted that host cells were killed in advance of the invading hyphae of the fungus and that juice from rotted tissue could break down healthy host tissue, whereas boiled juice from rotted tissue had no effect on healthy tissue. DeBary concluded that the pathogen produces enzymes and toxins that degrade and kill plant cells from which the fungus can then obtain its nutrients. In 1905, cytolytic enzymes were reported by L. R. Jones to be involved in several soft rot diseases of vegetables caused by bacteria. In 1915, it was reported that the pectic enzymes produced by fungi (Fig. 1-37A) play a significant role in their ability to cause disease on plants, but it was not until the 1940s that cellulases were implicated in plant disease development.

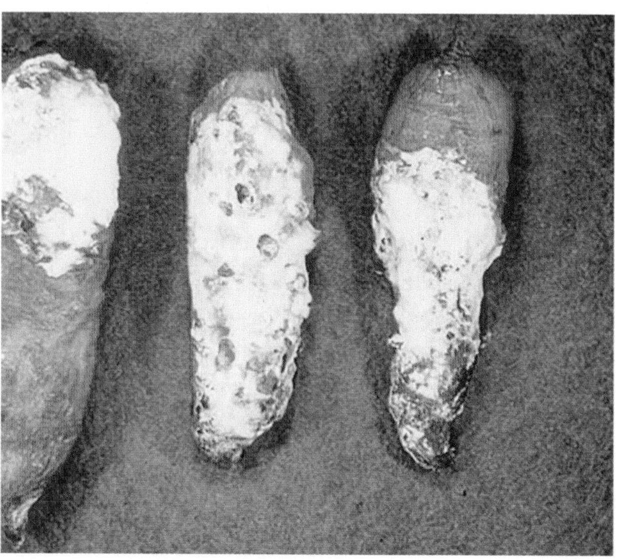

FIGURE 1-36 Sclerotinia white mold of carrots.

After deBary, many attempted to show that most plant diseases, particularly vascular wilts and leaf spots, were caused by toxins secreted by the pathogens, but those claims could not be confirmed. A 1925 suggestion that the bacterium *Pseudomonas tabaci*, the cause of the wildfire disease of tobacco, produces a toxin that is responsible for the bacteria-free chlorotic zone ("halo") (Fig. 1-37B) surrounding the bacteria-containing necrotic leaf spots was confirmed in 1934. The wildfire toxin was the first toxin to be isolated in pure form in the early 1950s. In 1947, a species of the fungus *Helminthosporium* (*Bipolaris*), which attacked and caused blight only on oats of the variety Victoria and its derivatives, was shown to produce a toxin named victorin. This toxin could induce the symptoms of the disease only on the varieties susceptible to the fungus. Many other bacterial and fungal toxins were subsequently detected and identified. The toxins exhibited several distinctive mechanisms of action, each affecting specific sites on mitochondria, chloroplasts, plasma membranes, specific enzymes, or specific cells such as guard cells. In addition, several detailed biochemical studies were carried out to elucidate the mechanisms by which toxins affect or kill plant cells or by which cells of resistant plants avoid or inactivate them.

Early observations that in many diseases the affected plants showed stunting, whereas in others they showed excessive growth, tumors, and other growth abnormalities (Fig. 1-37C), led many investigators to suspect imbalances of levels of growth regulators in diseased plants. In 1926, E. Kurosawa showed that the excessive growth of rice seedlings (Fig. 1-37D) infected with the fungus *Gibberella* could also be produced by treating healthy seedlings with sterile culture filtrates of the fungus. In 1939, the growth regulator produced by the fungus was identified and named gibberellin. By the late 1950s, numerous plant pathogenic fungi and bacteria were shown to produce the plant hormone indoleacetic acid (IAA). In the mid-1960s, a cytokinin was shown to be produced by the bacterium that causes the fasciation (leafy gall) disease of peas and other plants, and the symptoms of the disease could also be reproduced by treating the plants with kinetin, which is an animal-derived cytokinin. In the late 1970s and in the 1980s, detailed studies were made of the mechanisms of disease induction in the *Agrobacterium tumefaciens*-induced crown gall disease of many plants.

FIGURE 1-37 Chemical weapons used by pathogens in causing disease. (A) Apple infected with gray mold and showing the action of the pectinolytic enzymes ahead of the fungal pathogen. (B) Halo around lesions on tomato leaf show the presence of toxin produced by the bacterial pathogen. (C) Formation of crown gall as a result of excessive amounts of growth regulators produced by the crown gall bacterial pathogen. (D) Excessive growth of rice seedlings is the result of excessive production of gibberellin growth regulators by the fungal pathogen. [Photographs courtesy of (B) R. J. McGovern, (C) University of Florida and (D) R. K. Webster, University of California.]

These studies showed that the bacterium introduces into plant cells a specific part of transforming DNA (T-DNA) of its transformation-inducing plasmid (Ti plasmid). This DNA becomes incorporated into and is transcribed by the plant cell. The T-DNA contains several genes, one of which codes for IAA and one for a cytokinin. When these genes are expressed by the plant cell, the growth regulators they produce lead to uncontrolled enlargement and division of affected plant cells. Depending on the relative concentration of the two growth regulators, the infection may result in the production of unorganized galls (tumors), partially organized teratomas, or hairy roots.

From the mid-1950s until about 1980, a great many studies were carried out on the effect of infection on the respiration of host cells and on the possible role of altered respiration in plant defenses, and resistance, to infection. Similarly, numerous studies were carried out on the types of host cell enzymes that may be activated on infection, the types and amounts of metabolites (substances) accumulating following infection, and, particularly, the types and amounts of phenolic compounds and phenol-oxidizing enzymes produced following infection. These studies provided a great deal of information on many of the biochemical reactions that go on in plant cells following infection but did not entirely explain the mechanisms by which plants defend themselves against pathogens.

From the early 1970s onward, many studies have been devoted to the elucidation of the numerous metabolic changes associated with the hypersensitive response, i.e., the localized defense reaction of a resistant plant to a pathogen. In the hypersensitive response, numerous enzymes, known as plant pathogenesis-related (PR) proteins, are activated. Some of the PR proteins induce the synthesis of ethylene, which is a plant hormone able to induce many stress responses; some induce the production of oxidative enzymes and proteins involved in cell wall modification and strengthening against pathogen invasion; some synthesize antimicrobial compounds such as phytoalexins; and some are enzymes that attack and dissolve components of the cell wall of the pathogen or are proteinase inhibitors that neutralize specific enzymes of the pathogen. Information on such proteins is, potentially, of great practical significance for possible use to genetically engineer plants, which, upon infection, will produce sufficient amounts of appropriate pathogenesis-related proteins that will result in protecting the plants from becoming diseased.

The Concept of Genetic Inheritance of Resistance and Pathogenicity

In 1894, Eriksson showed that the cereal rust fungus *Puccinia graminis* consists of different biological races that cannot be distinguished morphologically but differ in their pathogenicity to their cereal host; for example, some of them being able to attack wheat, but not the other cereals, such as oats and rye.

In 1902, H. M. Ward recognized the necrotic defense reaction, which E. C. Stakman later (1915), studying it in the cereal rusts, called the "hypersensitive response." In 1964, Z. Klement and colleagues recognized that the hypersensitive response also operates against bacterial plant pathogens. In 1972, a similar necrotic or hypersensitive response was described in animals and was called apoptosis (= falling out); this research showed the existence of many common features in the defense reactions of plants and animals.

In 1905, Biffen reported that the resistance of two wheat varieties and their progeny to a rust fungus was inherited in a Mendelian fashion. In 1909, Orton, working with the Fusarium wilts of cotton, watermelon, and cowpea, distinguished among disease resistance, disease escape, and disease endurance (tolerance). In 1911, Barrus showed that there is genetic variability within a pathogen species; i.e., different pathogen races are restricted to certain varieties of a host species. Soon after, Stakman and colleagues (1914) established that morphologically indistinguishable races of a pathogen within a pathogen species differ in their ability to attack certain varieties. The pathogen races can be distinguished by their ability to infect different varieties within a set of host differential varieties (Fig. 1-38). Their work helped explain why a variety that was resistant in one geographic area was susceptible in another,

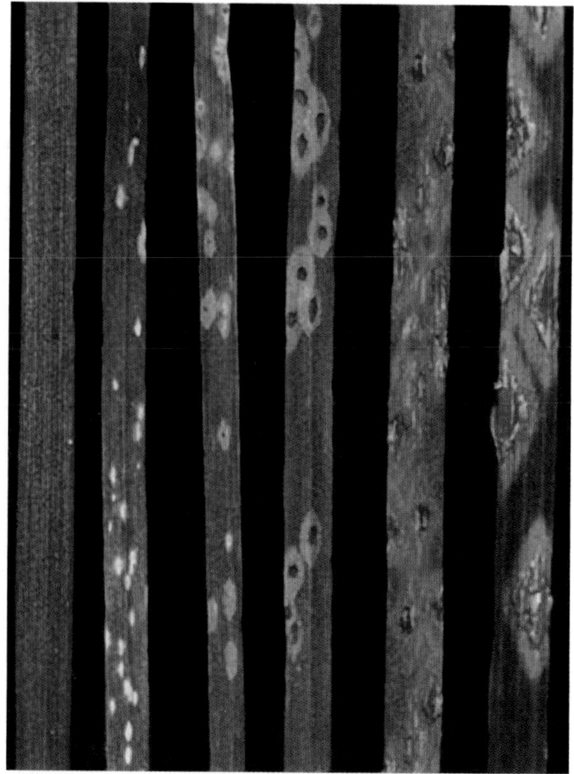

FIGURE 1-38 Differential reaction of leaves of wheat varieties to a race of wheat rust. This test is used to monitor the appearance of new rust races. (Photograph courtesy of USDA.)

why resistance changed from year to year, and why resistant varieties suddenly became susceptible. In all cases the change was due to the presence or appearance of a different physiological race of the pathogen.

The genetics of disease resistance and susceptibility remained obscure until 1946 when Flor (Fig. 1-39A), working with the rust disease of flax, showed that for each gene for resistance in the host there was a corresponding gene for avirulence in the pathogen and for each gene for virulence in the pathogen there was a gene for susceptibility in the host plant (a gene-for-gene relationship).

In 1963, Vanderplank (Fig. 1-39B) suggested that there are two kinds of resistance: one, known as vertical resistance, is controlled by a few "major" resistance genes and is strong but is effective only against one or a few specific races of the pathogen, and the other, known as horizontal resistance, is determined by many "minor" resistance genes and is weaker but is effective against all races of a pathogen species. It has been proposed that each major or minor gene for resistance represents one or several steps in a series of biochemical reactions and that it usually operates in conjunction with several other genes. Together, these genes enable the plant to produce certain types of plant cell substances and structures that interfere with, or inhibit, the growth, multiplication, or survival of the attacking pathogen, and in that way they inhibit, or stop, the development of disease. Some of the plant defense structures and substances exist before the plant comes into contact with the pathogen, but the most effective defense structures and substances are produced in response to attack by the pathogen.

In 1946, E. Gaümann proposed that in many host–pathogen combinations plants remain resistant through hypersensitivity; i.e., the attacked cells are so sensitive to the pathogen that they and some adjacent cells die immediately and in that way they isolate or cause the death of the pathogen. In the early 1960s, it was proposed that, in some cases, disease resistance is brought about by phytoalexins, i.e., antimicrobial plant substances that either are absent or are present at nondetectable levels in healthy plants, but accumulate to high levels in response to attack by a pathogen.

FIGURE 1-39 (A) H. H. Flor. (B) J. E. Vanderplank.

The genetic inheritance of pathogenicity in pathogens has been shown to parallel, and to mirror, that of resistance in plants, as mentioned previously. Some pathogen genes for virulence and even more genes for avirulence have been isolated, and the sequences as well as the products (enzymes, toxins, inhibitors, growth regulators) of several of these genes are also known.

Epidemiology of Plant Disease Comes of Age

Epidemiological observations, i.e., observations concerning the increase of disease within plant populations and how such increases relate to environmental factors, were recorded with many plant diseases as the latter began to be reported. Little effort was made, however, to correlate and utilize such information in controlling plant diseases. From studies of the apple scab disease, Mills in 1944 developed a table listing the duration of rain required at each temperature for apple buds, leaves, and fruit to become infected by the ever-present apple scab fungus. He and others then could use this information to predict whether infection would take place and whether, therefore, control measures (fungicides) should be applied.

It was in 1963, however, that Vanderplank (Fig. 1-39B), through the book "Plant Diseases: Epidemics and Control," established epidemiology as an important and interesting field of plant pathology. In his book, Vanderplank discussed the principles and variables in plant disease epidemics, stated the difference in the development and control of monocyclic and polycyclic pathogens, and described the general structure and patterns of epidemics. A few years later, modeling of plant diseases was introduced, which, through analysis of information on the host, the pathogen, and their interactions, collected at various points in time and under varying environmental conditions, could predict the course of an epidemic. In 1969, the first computer simulation program of plant disease epidemics was published for the fungal-induced early blight disease of tomato and potato. The simulation program was developed by modeling each stage of the life cycle of the pathogen as a function of various environmental conditions designed to stimulate the pathogen. Since the mid-1970s, disease modeling and computer simulation of epidemics have been developed for many diseases and, together with newly developed disease-monitoring instrumentations, have been used in plant disease-forecasting systems. Disease forecasting has become an important component of integrated pest management (IPM) and has helped reduce the amounts of pesticides applied to crops without reducing yields.

PLANT PATHOLOGY TODAY AND FUTURE DIRECTIONS

Molecular Plant Pathology

Since 1980, great emphasis has been placed on determining the specific molecule and the "genetic connection" of any substance involved in disease development. Because viruses and bacteria are small in size and because a great deal of background information is available on them, more molecular studies have been carried out with them than with the much larger fungi and nematodes. Already the number, location, size, sequence, and function of most or all genes of many viruses are known in detail. Many of these genes have been excised from the virus and have been transferred either to host plants, to which they often convey resistance, or into bacteria, in which they are expressed and the proteins they code for are isolated and studied. Similar transfers have been accomplished with a few bacterial and fungal genes coding for certain pathogenesis-related proteins.

The beginnings of molecular plant pathology can probably be traced to the isolation by W. Stanley in 1935 of the tobacco mosaic virus as a crystalline protein, which he believed to be infectious. Although 2 years later it was shown that the protein also contained a small amount of RNA, it was not until 1956, when Gierrer and Schramm showed that the ribonucleic acid and not the protein of tobacco mosaic virus was responsible for the infection of plant cells and for the reproduction of complete virus particles. In the meantime, in 1941 Beadle and Tatum showed that one gene codes for one enzyme. The following year (1942) Flor showed that a single gene is responsible for pathogenicity in the flax rust fungus and that the rust fungus gene corresponds to a single gene for resistance in the flax plant (the **gene-for-gene concept**). In 1953, Watson and Crick showed that DNA exists in a double helix and their discovery impacted greatly all of biology. In the mid-1960s, studies of tobacco mosaic virus led to the full elucidation of the genetic code according to which specific base triplets of DNA (and RNA) code for a certain amino acid. This was followed by the description in the 1970s through the 1990s of all the genes of tobacco mosaic and of many other viruses.

By the mid-1970s, the studies of *A tumefaciens* revealed that the T-DNA of its Ti plasmid contained several genes of which two, coding for growth regulators, were responsible for the production of tumors (galls) by the infected plants. It was later shown that the two genes could be removed and replaced with one or more genes from other organisms such as plants, other bacteria, viruses, and even animals, genes that could be transferred into and expressed (translated) by the plant cells. This discovery made possible the introduction of foreign genes into plants at will and, combined with tissue culture, which made possible the production of whole plants from single cells, it ushered in the era of genetic engineering of plants. Subsequently, it was discovered that foreign DNA can be introduced into plant cells in several ways, including using viruses as vectors, bombarding plant cells with foreign DNA, and growing plant cells in the presence of foreign DNA. Several viral genes coding for the coat protein or other structural or nonstructural proteins, and some noncoding regions, have been engineered into plants, and many of them have been shown to make the plant more or less resistant to the virus. Also, some bacterial and fungal genes, coding for enzymes that break down the cell wall of the pathogen, have been engineered into plants and have provided the plant with resistance to these pathogens.

In 1984, P. Albersheim and colleagues identified the molecule in the cell wall of the oomycete *Phytophthora megasperma* that acts as the elicitor of the defense response in its soybean host. It was shown later that the elicitor accomplishes this by interacting with a receptor molecule on the plant cells. In the same year, the first avirulence gene was isolated from the bacterium *Pseudomonas syringae* pv. *glycinea* by B. J. Staskawicz and colleagues. These two discoveries helped launch research that improved our understanding of pathogen virulence and plant disease resistance greatly. In 1986, bacterial hypersensitive response protein (hrp) genes were discovered. It was thought at first that the hrp genes were required for bacterial pathogenicity and production of the hypersensitive response; it is known now that they affect the transport of proteins in pathogenic bacteria and also the transport of bacteria into plant cells.

The first practical results of molecular plant pathology in improving disease resistance came in 1986 when R. Beachy and colleagues obtained tobacco plants resistant to tobacco mosaic virus (TMV) by transforming them; i.e., introducing into them the coat protein gene of the virus in a way that the plants could express the gene and produce the virus protein. Such transformed plants are called transgenic, and the resistance they acquire is called pathogen-derived resistance. In 1989, M. B. Dickman and P. E. Kolattukudi transformed a fungus, that normally could enter host plants only through wounds, with a cloned gene coding for the enzyme cutinase. That enzyme enabled the fungus to penetrate host plants directly through the cuticle, thereby proving that cutinases play a role in the direct penetration of some plants by fungi. Two years later, in 1991, R. Broglie and co-workers showed that plants transformed with the gene that codes for chitinase exhibit enhanced resistance to disease by fungi that contain chitin in their cell walls. In the meantime, in 1990, R. Cheim and colleagues obtained transgenic tobacco plants that expressed increased disease resistance by transforming them with the gene for stilbene synthetase, the enzyme that synthesizes a phytoalexin.

Discoveries in molecular plant pathology came fast and furious in the 1990s. The concept of systemic acquired resistance (SAR) burst onto the scene through the discovery of D. F. Klessig and colleagues and J. Ryals and co-workers that salicylic acid, a relative of aspirin, is associated with SAR. The first fungal avirulence gene (avr9) was isolated from *Cladosporium fulvum* by P. J. G. M. De Wit, while the first plant resistance gene (Hm-1) was isolated from corn by S. P. Briggs and J. D. Walton. The latter also showed that Hm-1 operates by producing a protein that detoxifies the host-selective toxin of the pathogen *Cochliobolus carbonum*. The only resistance gene conferring resistance in tomato to a bacterial pathogen through the hypersensitive response was isolated by G. B. Martin and colleagues in 1993. In subsequent years, dozens of plant disease resistance genes were isolated from many plants. All these genes shared a leucine-rich repeat in the protein they coded for. Tomato plants transformed by B. Baker and co-workers with the tobacco plant resistance gene N, which makes tobacco resistant to tobacco mosaic virus, were also made resistant to the virus, proving that at least some resistance genes may function in species other than the one in which they normally occur. Furthermore, it was shown by V. M. Williamson and colleagues (1998) that a single cloned disease-resistance gene from tomato can confer resistance to both a nematode pathogen and an insect. It was also shown during this period (T. Shiraishi *et al.*, 1992) that plant pathogens produce proteins that actively suppress the defense reactions of their host plants. In addition, the avirulence proteins of some pathogens contain signals that allow these proteins not only to be introduced into plant cells, most likely through the bacterial hrp protein system, but also to move into and function in the plant nucleus.

A new type of defense against pathogens was unveiled when it was discovered that many organisms, including plants, fungi, and animals, are capable of "RNA silencing," i.e., of regulating genes based on targeting and degrading sequence-specific RNAs. In plants, RNA silencing has been shown to serve as a defense against virus infections. As would be expected, however, many plant viruses carry genes that encode proteins that suppress the silencing of their RNA by the plant. RNA silencing can be induced experimentally and targeted to a single specific gene or to a family of related genes. It is believed that RNA silencing genes will soon play an important role in engineering resistance into plants.

Advances in molecular plant pathology have also provided a new set of diagnostic tools and techniques that are used to detect and identify pathogens even when they are present in very small numbers or in mixtures with other closely related pathogens. Such tools include detection with monoclonal antibodies, analysis of isozymes or of fatty acid profiles of pathogens, analysis of fragments of their nucleic acids produced by specific enzymes, calculation of percentages of hybridization of their nucleic acids, and determination of nucleotide sequences of the nucleic acids of the pathogens. Since the mid-1980s, segments of DNA (probes), complementary to specific segments of the nucleic acid of the microorganisms, have been labeled with radioactive isotopes or with color-producing compounds and are used extensively for the detection and identification of plant pathogens. Numerous techniques, often referred to by their acronyms, have been developed and are used; some of them are better suited for diagnosing one or more types of pathogens. For at least some pathogens, PCR, with selected differential random sequences of different species, can be effective for the detection and identification of each of these species. At other tests, PCR of sequence segments of rDNA internal transcribed spacer (ITS) regions are used or PCR of other genes or spacers of the fungal DNA is carried out. The product is then differentiated by digestion with restriction enzymes and gel electrophoresis and detection of differential random fragment length polymorphisms (RFLP) or use of PCR together with DNA hybridization in a reverse dot blot hybridization (RDBH) assay using PCR of selected RAPD markers. Reverse transcription PCR (RT-PCR) or immunocapture RT-PCR (IC/RT-PCR), direct binding PCR (DB-PCR), and a combination of PCR and enzyme-linked immunosorbent assay (ELISA) tests are often used successfully, especially for viruses.

An area of molecular plant pathology that is going to pay multiple dividends in the future is that of genomics, i.e., sequencing of the entire genomes of plants and their pathogens. Already, the genomes of the experimental plant *Arabidopsis thalliana*, of several plant viruses and viroids, and of the plant pathogenic bacteria *Ralstonia solanacearum* and *Xylella fastidiosa*, the white rot fungus *Phanerochaete chrysosporium*, and the model nematode *Caenorhabditis elegans* have been sequenced in their entirety. Significant progress has already been made in sequencing the entire genomes of the very destructive plant pathogenic fungi *Magnaporthe grisea*, cause of rice blast; *Ustilago maydis*, cause of corn smut; *Cochliobolus heterostrophus*, another pathogen of corn; *Botrytis cinerea*, the gray mold of many fruits and vegetables; *Fusarium graminearum*, cause of head scab of wheat; and *Phytophthora infestans*, cause of the blight of potato and of many other pathogens of crops. Once the genomes have been sequenced, it will be easier to locate, identify, compare, isolate, and manipulate the genes for pathogenicity in the pathogens and of resistance in their host plants, as well as manipulate the introduction of them into specific locations of the plant genome where they would be most effective.

The molecular phase of plant pathology is expected to develop a great deal more and to make contributions in ways that we can hardly imagine at present. One area in which molecular plant pathology is expected to contribute greatly and to provide tremendous benefits is the area of detection, identification, isolation, modification, transfer, and expression of genes for disease resistance from one plant to another. Several such resistance genes have already been identified, isolated, transferred into susceptible plants, and, when expressed, made the plants resistant. The possibility that molecular plant pathology can modify and combine resistance genes makes likely the future utilization of resistance genes from unrelated plants or from other organisms, and perhaps even the synthesis of artificial genes for resistance for incorporation into crop plants. The practical implications of such developments cannot be overestimated, as they are likely to revolutionize the control of plant diseases by providing us with cultivars that can resist disease in the presence of the pathogen, without the need to use any pesticides.

ASPECTS OF APPLIED PLANT PATHOLOGY

BOX 12 Plant biotechnology — the promise and the objections

Plant biotechnology can be defined as the use of tissue culture and genetic engineering techniques to produce genetically modified plants that exhibit new or improved desirable characteristics. The desirable characteristics include, among others, better yields, better quality, and greater resistance to adverse factors, including diseases, pests, and environmental conditions such as freezes, drought, and salinity. Plant biotechnology also makes possible the production in plants of useful proteins coded by microbial, animal, or human genes. Plant biotechnology has shown that all of these goals are attainable, at least in the kinds of plants on which they have been attempted. The number of crop, ornamental, and forest plants that have been modified genetically and released by university and industry scientists around the world is in the thousands and continues to grow.

There are numerous cases in which plant biotechnology is used successfully to produce crop plants that avoid or resist certain plant pathogens. Some plants have been rendered resistant to specific pathogens by genetically engineering (transforming) them with isolated specific genes that provide resistance against these pathogens. Transformed plants become resistant by coding for enzymes that mobilize other enzymes that carry out numerous defensive functions, such as breaking down the structural compounds of the pathogen. Several of the enzymes produce compounds in the plant that are toxic to or otherwise inhibit the growth and spread of the pathogen both through the plant and to other plants. Other plants have been transformed with animal (mouse) genes that code for antibodies (plantibodies) against a coat protein of the pathogen. Genetic engineering has been particularly effective in producing plants resistant to viruses by incorporating viral genes in the crop plants that code for virus coat protein, for altered movement protein, or by incorporating in the plant noncoding segments of virus nucleic acid or even segments of the nonsense strand of the virus nucleic acid. Many of these crop plants have been tested for resistance in the field with excellent results.

Practical examples of successful genetic engineering of disease-resistant plants include melon, squash, tomato, tobacco, and papaya crops that are protected from a variety of viral diseases. The success of genetically engineered papaya for resistance to papaya ringspot virus has saved the papaya as a crop in Hawaii and in the Far East (Fig. 1-40). Numerous other cases are still under development. For example, engineering tobacco with a chimeric transgene containing sequences from two different viruses (turnip mosaic and tomato spotted wilt) resulted in new plants resistant to both viruses. Similarly, engineering tomato plants with a truncated version of the gene coding for the DNA replicase of one of the very destructive geminiviruses resulted in plants resistant not only to the virus from which the transgene was obtained, but also to three other viruses. In other work, potato plants engineered with a chimeric gene encoding two insect proteins exhibiting antimicrobial activities showed significant resistance to the late blight oomycete and their tubers were protected in storage from infection by the soft rot-causing bacteria. In other work, raspberry plants engineered with the gene coding for the common plant polygalacturonase-inhibiting protein (PGIP) became resistant to the gray mold fungus *Botrytis cinerea*, although the transgene in raspberry, but not in other

FIGURE 1-40 Increased resistance to disease through biotechnology. Comparison of "Sunrise" papaya plants susceptible to *papaya ringspot virus* (PRSV) surrounding a block of the genetically similar "Rainbow" papaya plants that had been transformed for resistance to PRSV. Both "Sunrise" and transgenic "Rainbow" plants were inoculated by natural PRSV inoculum. (A) "Sunrise" (left) and transgenic "Rainbow" (right) plants 9 (B) 18, and (C) 23 months after transplanting. (D) Aerial photograph of the "Rainbow" block 28 months after transplanting, by which time the "Sunrise" plants surrounding the "Rainbow" block are almost totally destroyed by the virus, whereas the transgenic "Rainbow" plants remained free of virus, look healthy, and produced excellent yields. [Photographs courtesy of Ferreira (2002). *Plant Dis.* **86**, 101–105.]

plants, is expressed only in immature green fruit.

In addition to helping us engineer plants resistant to disease, molecular biology and biotechnology have made possible the development and use of nontoxic chemical substances that, when applied to plants externally, stimulate the plants and elicit the activation of their natural defense mechanisms, i.e., activation of the localized defense mechanism (hypersensitive response) and systemic-aquired resistance (SAR). Two such chemical substances that have been proven effective and are used commercially are Actigard, where one application increases the plants' resistance against some bacterial and some fungal diseases for several weeks, and Messenger, derived from the fire blight bacterium gene coding for the protein harpin, which elicits a hypersensitive response and SAR in plants. Messenger, which also promotes plant growth, is effective against a variety of diseases of several crops, including strawberry, tomato, and cotton.

In transforming plants for disease resistance or for any other characteristic, it is necessary to modify their nucleic acid by adding genetic material from another plant or, rarely, from an animal or a pathogen. In most cases, these nucleic acids are or become active, producing in the plant compounds that may be toxic to pathogens or pests and, possibly, to humans. In addition, some of this nucleic acid may find its way, through cross-pollination or through transfer by microorganisms, into weeds or other wild plants, making these plants also resistant to the pathogen or pest. Several kinds of plants have been engineered to produce toxins against certain insects; to produce vaccines against certain human pathogens; to produce animal or human growth hormones; or to produce pharmaceutical compounds

continued

that can be used to treat diseases of humans and animals. The fear by some people that some or all of these products will get into the human diet or in the animal food chain and cause allergies and other adverse health effects has resulted in significant unfavorable publicity for such products and for biotechnology. That type of publicity has, in turn, led many large buyers to refuse to buy and use products produced by genetically modified organisms (GMO). Following the adverse publicity, several governments, especially in Europe, passed laws and raised barriers to the importation of products derived from genetically modified organisms.

In addition to the argument against introducing into crops, through genetic engineering, new proteins that may cause allergic reactions in some people, there have also been arguments against biotechnology because it takes possession of, patents, and monopolizes genetic material that was previously available and free to everybody; it replaces the numerous sustainable local varieties with a few genetically engineered ones, the seed of which the farmers must buy from large companies every year; it threatens the development of pests and pathogens that can resist or overcome the transformed resistant crops; it threatens to lead to the use of larger amounts of herbicides with crops like those made herbicide resistant while the weeds are still susceptible; it threatens unknown numbers of nontarget organisms that may be affected adversely by the protein; it threatens to upset the plant balance, and through it the entire biotic balance of the environment, by having such new genes transferred naturally to nontarget plants and their proteins, harmless or not, consumed by microorganisms, animals, and humans unaccustomed to such proteins; it threatens the occurrence of accidents in which crops transformed for the production of pharmaceuticals, vaccines, and so on become mixed with edible crops.

BOX 13 Food safety

In recent years, food safety has been threatened by a number of events and developments that allow foodborne microorganisms pathogenic to humans, e.g., the bacteria *Salmonella*, *Listeria*, *Escherichia coli* strain *0157:h17*, the protozoa *Cyclospora*, *Cryptosporidium*, and *Giardia*, and the *hepatitis A virus*, to reach and contaminate our food in a variety of ways. These include (a) increased processing of fresh plant produce (e.g., fruit juices, fruit or vegetable purees, cole slaw, fruit sections and cut-up vegetables for salads in bulk or in plastic bags) that may sometimes contain produce that carries a significant amount of food-spoiling bacteria and mycotoxin-producing fungi; (b) inadequate food processing procedures that allow survival of human pathogens in the processed product; (c) long storage of foods that encourages the development of pathogenic microorganisms; (d) application to fruit and vegetable fields of improperly aged or poorly treated manure that carries human pathogens; (e) application on the plants of irrigation water that may be carrying one or many of the aforementioned human pathogens due to contamination by humans and animals through run-off of waste waters, etc.; (f) unacceptable hygiene of harvesters, handlers, and packers after using the toilet that results in the contamination of fruits and vegetables with human pathogens; and (g) the presence of pets, livestock, and wildlife animals, some of which may carry human pathogens on their bodies or in their feces to fruits and vegetables. To these should be added the ever-increasing shipment of food items among various geographical points of a country and worldwide, which may greatly multiply and expand the effects of a local contamination of food products.

One of the main effects of fears about food safety is economic. Not only is it costly to take all measures necessary to secure food safety, but there is also the fear and cost of rejection of produce shipments at the point of destination. Similarly, there is the possibility of refusal of buyers to purchase produce from farms that do not meet the buyer's food safety standards. In the United States and other developed countries, many of the large buyers of food products for their mills, processing factories, or chain stores demand third-party audits of farms by certified specially trained individuals and consulting firms regarding the employment by the farm of all necessary precautions in the type of manure they may be using, the quality of water used for irrigation, the health and hygiene of their workers and plant handlers, and so on. Also, to avoid unjustified accusations of offering contaminated produce, farmers are or will soon be expected to have a trace-back system in place. This will happen by identifying all produce leaving the farm as to origin and date of packing so that if contamination is found in the produce in the marketplace, the source will be easy to identify and appropriate measures may be taken. Also, it will become necessary to keep food safety records, such as documenting worker training sessions, recording the results of water tests, details of manure applications, if any, of dates, methods, and rates of irrigation, and so on, as well as of disease outbreaks among the farm workers. To protect themselves from purchasing contaminated produce, buyers of large quantities will test or have the produce tested with serological and molecular-based diagnostic techniques that can already detect, for example, as few as three *Salmonella* cells per 25 grams of naturally contaminated food.

In addition to the aforementioned types of contamination of food with pathogens, there are the additional threats of contamination with pathogenic microorganisms that are resistant to antibiotics, such as streptomycin and tetracyclines used in plants, as well as in humans and animals; the presence in the food of genetically engineered plants that contain genes for chemicals toxic to insects, such as the *Bacillus thuringiensis* toxin; genes for antibiotics against other human pathogens; genes for activating defensive mechanisms of plants, often through the production of proteins and phenolic compounds that make the plants resistant to insects, diseases, and to herbicides; genes for edible or otherwise delivered human vaccines and antibodies (plantibodies) against human pathogens; genes for unrelated proteins that may be allergenic in some individuals; and even genes for producing plastic. There is fear in some segments of the population, especially in developed countries, that although some of these genes are introduced into inedible plants such as tobacco, plants with such genes will intentionally or accidentally find their way into foods and feed and will affect adversely the health of animals and humans. Many large produce distributors or retailing companies and manufacturers of food products simply refuse to buy any produce that comes from genetically modified organisms (GMOs), plants, or animals. Molecular-based diagnostic tests have also been developed that detect introduced genes that may not have been declared as being present.

Since the horrendous terrorist attack in New York and Washington, DC, in September of 2001 and the subsequently declared war against terrorists wherever they exist, there is an added fear of having food contaminated intentionally by terrorists. Contamination could be carried out with human pathogenic microorganisms, such as those mentioned earlier or with others, e.g., the bacterium causing the disease anthrax, or with toxic substances. Contamination of produce can be done while the latter is still in the field, in transit, or in grocery stores. There is also fear of having the drinking water or the water used for irrigation of fruits and vegetables contaminated intentionally by terrorists with pathogenic microorganisms or with toxic substances that will then find their way to humans via the food distribution system. This subject is discussed further in the following section.

BOX 14 Bioterrorism, agroterrorism, biological warfare, etc. who, what, why?

Bioterrorism is loosely defined here as the use, or threat of use, of biological agents, mainly pathogenic microorganisms that could infect people and cause disease and, thereby, instil fear and terror in all of the populace. Bioterrorism may differ from biological warfare in that the latter is usually directed against enemy armies and its purpose is to incapacitate or kill enemy soldiers, whereas in bioterrorism the purpose is to frighten and terrorize civilian populations, although casualties in large numbers may or may not occur. The most vivid example of bioterrorism occurred in the fall of 2001 when persons in various positions in politics and the television news media in New York and Washington received letters through the mail containing spores of the bacterium *Bacillus anthracis*, the cause of the severe and often deadly anthrax disease. It became apparent at the time that the perpetrators of the anthrax bioterrorism, or others, could easily expand to other forms of bioterrorism by either contaminating agricultural products such as vegetables, milk, or meat on the farm or in the store with microorganisms pathogenic to humans, which would scare buyers away from such products (agroterrorism), or by spreading selected plant pathogenic microorganisms on certain crops, e.g., cereals, potatoes, and corn, which they could infect and destroy to various extents, thereby causing devastating losses that would further increase the fear of the people.

Biological warfare has been talked about for several decades and many of the larger countries have been producing and stockpiling pathogenic microorganisms, such as the anthrax bacterium, for potential use against the army of an enemy country with which they might go to war. At the same time, however, several countries have been experimenting with and stockpiling microorganisms that can infect and destroy important staple food crops for certain countries, e.g., rice, potatoes, wheat, or beans, which could affect the availability of food and thereby survival of the people, or at least, their will to fight and prolong the war. This type of agricultural biological warfare has revolved around important pathogens of such crops, e.g., *Magnaporthe grisea*, the fungus causing the blast disease of rice; *Phytophthora infestans*, the oomycete causing the late blight of potato; and *Puccinia graminis*, the fungus causing the rust diseases of wheat and other small grains.

As the specialization of crops in each area increases and as our knowledge of diseases of such crops increases, it becomes evident that such areas or countries become extremely vulnerable to agroterrorism or agrosabotage. This happens even if, or especially if, they grow relatively small areas of such specialty crops, e.g., bananas, citrus, coffee, and cacao, which are the main export crop and the main source of foreign currency for these countries. For each area producing such a crop there are pathogens of the crop elsewhere that, if introduced, could destroy the crop for the year to come and, possibly, forever. The pathogens that would be used on such clonal, genetically uniform, perennial crops are likely to be insect-vectored bacteria, phytoplasmas, or viruses. Such pathogens can be introduced into a field as a few bacteria- or virus-carrying insect vectors that would feed on and infect some of the plants and then, in the same or in subsequent years, multiply and spread the pathogen they carry to more plants over a continually expanding area.

WORLDWIDE DEVELOPMENT OF PLANT PATHOLOGY AS A PROFESSION

As mentioned earlier, plant pathology had its origins in plant pathological observations and studies made by botanists, naturalists, and physicians in Europe in the mid- to late 1800s. Soon after, plant pathological activity shifted primarily to the United States, where it has remained at a high level to date.

The students of the first, self-made, plant pathologists began to be hired as plant pathologists by state agricultural experiment stations, by the federal Department of Agriculture, and by universities at which they taught courses in plant pathology. In 1891, the plant pathologists in the Netherlands formed the Netherlands Society of Plant Pathology and began publishing the *Netherlands Journal of Plant Pathology* in 1895. In 1908, the plant pathologists in the United States organized into the American Phytopathological Society, and they too decided to publish a journal of plant pathology in which they could present the results of their own research and could read about the work of their colleagues. The journal, named *Phytopathology*, began to be published in 1911 as an international journal of plant pathology. The Phytopathological Society of Japan was founded in 1916, and its journal began to be published in 1918. In subsequent decades, plant pathologists formed associations and began publishing plant pathological journals in several other countries, e.g., Canada (1930) and India (1947). In the second half of the 20th century, plant pathologists in nearly 50 more countries organized into professional associations; some of them, as in Brazil, published their own national journals, whereas others formed multinational unions, e.g., the Latin American Phytopathological Association, or published a regional journal such as *Phytopathologia Mediterranea*. In 1968, an International Society of Plant Pathology was formed and it held the first International Congress of Plant Pathology in London that same year. By the end of the 20th century most or all countries have one or more plant pathologists, although in many developing countries that person is an administrator of some kind or a professor at a university. Nevertheless, in many parts of the world, plant pathology is generally unknown or rarely practiced, and losses from plant diseases in developing countries are still great.

International Centers for Agricultural Research

In the mid-1940s, the Rockefeller Foundation, in cooperation with the Mexican government, established a program in Mexico for interdisciplinary research on basic food crops such as wheat, corn, potatoes, and beans. That program was so successful in improving crops and in training personnel in the technologies that similar Rockefeller Foundation programs were established in Colombia, Chile, and India. It soon became apparent, however, that it would not be possible to have such programs in every developing country; rather, it would be preferable to have a few international centers concentrating on one or a few basic crops. So, with the cooperation of the local governments and funding from the Rockefeller and the Ford foundations, the International Rice Research Institute (IRRI) was established in the Philippines in 1960, the International Maize and Wheat Improvement Center (CIMMYT) in Mexico in 1966, the International Institute of Tropical Agriculture (IITA) in Nigeria in 1968, and the International Center of Tropical Agriculture (CIAT) in Colombia in 1969 (Fig. 1-41).

The success of these centers suggested the need for additional ones. As the finances required to operate the earlier and the new centers were beyond the means of the Ford and the Rockefeller foundations, they, in collaboration with the World Bank, set up a consortium of potential donors interested in financing international agricultural research. The consortium, known as the Consultative Group on International Agricultural Research (CGIAR), consists of wealthy countries, development banks, and other foundations and agencies. The CGIAR receives help in determining research priorities from a technical advisory committee, which consists of 13 scientists and economists. Additional centers established by the consultative group include the International Crops Research Institute for the Semi-Arid Tropics (ICRISAT) in India in 1972 and the International Potato Center (CIP) in Peru, also in 1972. A similarly operating center but not funded by the consultative group, namely the Asian Vegetable Research and Development Center (AVRDC) in Taiwan, was also established in 1972. More recent centers include the International Center for Agricultural Research in the Dry Areas (ICARDA) in Syria, the West Africa Rice Development Association (WARDA) in Gold Coast, and some others (Fig. 1-41): IFPRI, International Food Policy Research Institute; ISNAR, International Service for National Agricultural Research; IPGRI, International Plant Genetic Resources Institute; ILRI, International Livestock Research Institute; ICRAF, International Center for Research in Agroforestry; IIMI, International Irrigation Management Institute; CIFOR, Center for International Forestry Research; and ICLARM, International Center for Living Aquatic Resources Management.

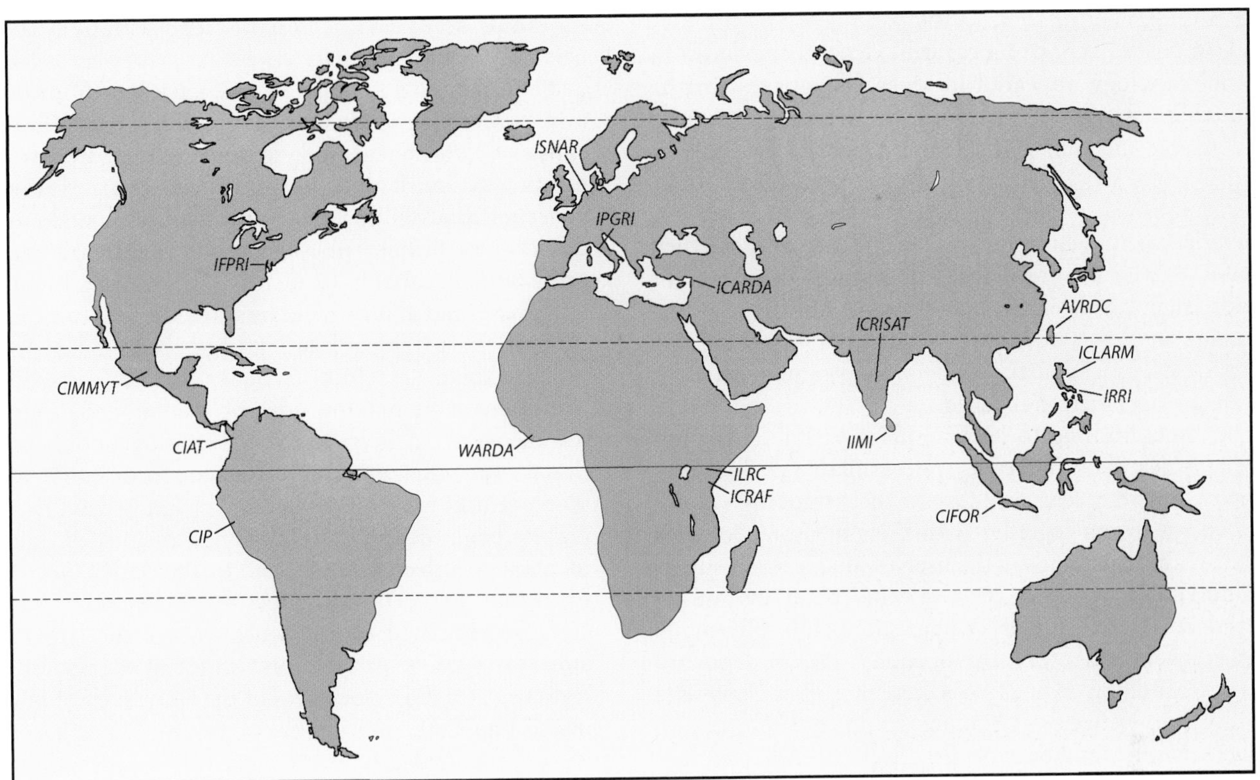

FIGURE 1-41 The global agricultural research system.

Each of the aforementioned centers studying plants includes several plant pathologists working on diseases of the specific crop(s) studied by the center. The contributions of the resident plant pathologists to the study of these diseases and to the development of disease-resistant cultivars and other controls against the diseases of these crops have been truly great. These pathologists have also helped train many other scientists not only of the host country, but from many other developing countries attempting to grow these crops, have taught plant pathology courses in universities with which their center is affiliated, and have generally helped to significantly reduce losses of crops caused by plant diseases.

The need for plant pathology has always been particularly great in tropical countries primarily because the tropical climate (hot and usually humid) favors the survival and multiplication of pathogens throughout the year, as well as the prolonged or continuous presence of primary and alternate hosts and large numbers of active vectors such as insects. Tropical climates also favor multiple and continuous infections by pathogens, which often lead to devastating epidemics. These problems in tropical countries are further compounded by low educational levels and lack of funds for carrying out effective plant disease control programs. Moreover, tremendous losses from disease occur in the tropics in all types of produce after harvest because many harvested products are already infected or contaminated while still in the field and also because harvested products often rot in storage or transit due to lack of appropriate decontamination and lack of any kind of refrigeration. It is not surprising, therefore, that so many of the international centers for agricultural research have been established in the tropics, nor that their contributions have had a big and immediate impact on reducing losses from disease. Much more, however, remains to be done.

Trends in Teaching and Training in Plant Pathology

The first course in plant pathology was offered at Harvard University by M. A. Farlow in 1875. In the early 1900s, departments of plant pathology began to be established at some of the larger universities, often as departments of botany and plant pathology. The early courses were, by necessity, primarily descriptive of the diseases of various types of crops (vegetables, fruit trees, field crops), in addition to providing information on the development of some of the pathogens and diseases and on possible control measures. General textbooks in plant pathology appeared in several languages. In the United States the main textbooks were those by Duggar (1906), Stevens and Hall (1921), Heald (1926, 1943), and Walker (1950). In the meantime, specialized books were published on plant pathogenic fungi and, later on, bacteria, viruses, and nematodes and the diseases they cause, as well as on all types of diseases of groups of crops, such as vegetables, field crops, and fruit crops. Starting in the 1960s, more specialized books on the physiology, biochemistry, epidemiology, and genetics of plant diseases were published.

Students training to become plant pathologists took as many relevant courses as were available at their university, but they learned most of their trade by watching and working together with their mentor–professor plant pathologist and by themselves, under some supervision, doing research on a specific plant disease or pathogen. Such studies, when successful, eventually earned them a doctor of philosophy (Ph.D.) degree in plant pathology, which indicates that they have the ability, knowledge, and training to do research, i.e., to solve scientific, and possibly practical, problems in plant pathology. This type of training continues to date except that, because of the tremendous increase in the amount of knowledge in plant pathology, students specialize a great deal more in what they learn and do. This has been particularly evident in the years after 1985 during which molecular plant pathology has attracted many of the students working toward their Ph.D. in plant pathology. Most of the holders of a Ph.D. in plant pathology find jobs as professors in colleges or universities, or as researchers in universities, government, or industry. Some develop their own business as private practitioners or consultants to growers. A few, usually one or two per state, work as extension plant pathologists in state land grant universities and experiment stations, where they are responsible for transferring plant pathology information from plant pathology researchers to growers and county agents, visiting crop fields and identifying diseases, identifying diseases in plant samples sent in by growers, and developing and disseminating disease control recommendations.

Similar but less extensive and intensive course work and research training can lead to a master of science (M.S.) degree in plant pathology. This enables the holder to work for the same agencies as the Ph.D. holders but with reduced responsibility and benefits. Several departments of plant pathology also offer bachelor of science (B.S.) degrees in plant pathology, which serve either as intermediate steps for advanced degrees or enable the holders to work in university, government, and industry laboratories, for various types of agribusinesses as chemical, seed, etc., company representatives, or as private practitioners.

Plant pathology, unlike its sister sciences of medicine and veterinary medicine, deals with plant diseases caused by pathogens and, to some extent, by environmental factors. It does not have teaching and training programs that will produce practitioners similar to the general practitioner physicians and veterinarians, i.e., professionals capable of identifying all types of causes of disease and injury to plants and of making recommendations to control or manage these. Such practitioners (plant doctors) would also be trained in identifying and making control recommendations for insects, weeds, damage by animal wildlife, and the nutritional and other environmental conditions that affect plant health. Development of a program leading to a professional doctor of plant medicine or doctor of plant health degree, similar to the M.D. (doctor of medicine) and D.V.M. (doctor of veterinary medicine) degrees, had been discussed since the late 1980s and was offered for the first time by the College of Agriculture and Life Sciences of the University of Florida in the year 2000.

Plant Disease Clinics

For many years, most states operated a plant disease clinic through their department of plant pathology. Growers, county extension agents, and home owners would send diseased plants, soil from areas with diseased plants, and sometimes insects to the plant disease clinic and the pathogen or insect would be identified and control measures would be recommended, all free of charge. At first, the plant disease clinics were set up rather informally and were supervised by the extension plant pathologist, with most of the diagnoses made by advanced plant pathology graduate students assisted significantly by more junior graduate students. Early plant disease clinics were equipped primarily with surface sterilants, dissecting scopes, microscopes, culture dishes and test tubes, and nutrient media for culturing fungi and bacteria. Later, much of the day-to-day operation of plant disease clinics was turned over to M.S. or Ph.D. plant pathologists hired specifically for that purpose. At the same time, nematode isolation from roots or soil and plant nematode identification became integral functions of the plant disease clinics. Virus disease identification was still made by host symptomatology alone, but some host range tests for diagnostic purposes were carried out.

Since the 1970s, every state has at least one plant disease clinic and some have several; e.g., Florida has four plant disease clinics. In addition to state-funded plant disease clinics, in some states there may be one or more privately run plant disease clinics and, in a few states, a plant disease clinic may also be operated by the state department of agriculture. Today's plant disease clinics often have one scientist with an advanced degree and one or more laboratory assistants; they are also equipped for viral disease diagnosis through host range tests, serological tests, cell inclusion identification, electron microscopy of plant sap, and dot-blot assays of radioactive or color-producing DNA probes. Plant disease clinics also have modern computers with databases and expert systems for disease and pathogen identification, computerized distance diagnostic systems that transmit plant disease images directly from the field to an expert diagnostician, CD videodisc capabilities, and e-mail for transmitting the results of diagnosis and the recommendations for control to their clientele. Also, however, due to increased costs for these tests and services, plant disease clinics in many states have now established fees that must be paid by all commercial users and home owners submitting samples of diseased plants for diagnosis.

The Practice and Practitioners of Plant Pathology

The science of plant pathology has been and continues to be developed primarily by highly specialized professors or researchers who have advanced, usually doctorate, degrees. For many discoveries, considerable contributions are made by graduate students who are themselves working toward M.S. or Ph.D. degrees at departments of plant pathology, botany, or biology and at agricultural experiment stations.

The practice of plant pathology, however, is carried out at a much lower scientific and professional level. Medicine and veterinary medicine also have Ph.D.-holding scientists who do research. These scientists advance the respective sciences at various universities and research centers. In addition, however, both medicine and veterinary medicine have numerous highly trained practicing physicians (doctors of medicine) and veterinarians (doctors of veterinary medicine) who are the practitioners of each science. They diagnose the ailments and prescribe treatments for humans and animals, both individuals and populations. In contrast, plant pathology has few well-trained practicing plant pathologists.

In general, most states have one or two extension plant pathologists. Their duty is to (a) transfer the information developed by the researchers in the state and elsewhere to county extension personnel and to growers and (b) demonstrate its effectiveness to those who need it, i.e., the growers. The same extension plant pathologists are expected to be able to diagnose all diseases on all types of plants, regardless of their cause, and to recommend measures for their control. The extension plant pathologists also train the county extension agents, who usually have little formal education or training in plant pathology, so that they can diagnose and offer recommendations for the control of plant diseases common in their county. Many states have a plant disease clinic to which samples of diseased plants or plant parts are sent by growers, home owners, and county agents for diagnosis and control recommendations. In some of the most agriculturally oriented states, a few persons, who usually have varying levels of education and training in plant pathology (B.S., M.S., or Ph.D.), offer their services as private practitioners (plant doctors) to individual growers or groups of growers, or they operate their own private plant disease clinics. Much of the time, however, growers receive information on plant diseases and recommendations for plant disease control from salesmen of pesticides, seeds, or fertilizers, and from other professionals (agronomists, horticulturists, entomologists, etc.) who may have little or no education and training in plant pathology.

Under the present conditions, therefore, most growers often receive rather limited, delayed, or inaccurate information on the kinds and development of diseases affecting their crops and, similarly, incomplete and sometimes inaccurate information about their control. As a result, plant diseases are often detected late and are sometimes misdiagnosed, and frequently the wrong pesticides or excessive dosages of pesticides are recommended and applied for their control. The amount of crop losses to plant diseases, therefore, and possibly contamination of the environment with pesticides as well, is often greater than need be.

Certification of Professional Plant Pathologists

When a professional such as a physician, veterinarian, lawyer, or engineer offers his or her services to individuals, the individuals expect the professional to have appropriate education and training that meet or exceed certain professional and ethical standards. At the same time, the professional and the public also expect that no person who does not meet such a standard will be allowed to provide such services: the professionals because they do not want such persons to compete for business with them and the public because they want to be certain that the person to whom they go for such services can actually provide them correctly. These two expectations are generally guaranteed through the licensing programs operated by each state and country.

Since the 1960s and 1970s, many states have required the licensing of pest control advisers, pesticide applicators, etc. In addition, several professional societies, such as the American Society of Agronomy, the Soil Science Society of America, the Crop Science Society of America, and the Entomological Society of America, have established professional certification programs that resulted in certified agronomists, certified soil scientists, certified crop scientists, certified entomologists, and so on.

A proposal for establishing an American registry of professional plant pathologists was submitted to the American Phytopathological Society in 1980, but it was not approved until 1991. The following year, a certified professional plant pathologist program was developed that set professional and ethical standards. A board of six plant pathologists, named by the American Phytopathological Society, was authorized to review and compare the credentials (course work, experience, references) of each applicant with the standard and to determine their eligibility to become certified professional plant pathologists. Because there were already many practicing plant pathologists (private consultants) when the certification program came into being, the standards for certification were set so that it would include most of them. The standards include a B.S. degree in plant pathology and 5 years of professional experience, a M.S. in plant pathology and 3 years of professional experience; or a Ph.D. in plant pathology and 1 year of professional experience. The board also set a curriculum that would enable new students to become certified professional plant pathologists. In addition, the board set standards for continual education and training so that certified professional plant pathologists can keep abreast of new information, techniques, conditions, regulations, and requirements in the area of plant health management.

BOX 15 Plant pathology as a part of plant medicine: the doctor of plant medicine program

In the last two decades, considerable efforts have been made to broaden the concepts of both plant health and plant protection. The American Phytopathological Society, realizing the need for such a broader concept, launched a new electronic journal called "Plant Health Progress," which publishes articles on all facets of plant health.

It has become apparent, however, that trained professionals are needed who can deal with the whole health of the plant and give recommendations for its maintenance or restoration. Such professionals would be able to diagnose all causes of plant problems, be they pathogens (fungi, bacteria, viruses, nematodes, parasitic algae and parasitic higher plants, protozoa, etc.), insects, mites, vertebrate (birds, field mice, deer) and invertebrate (snails, slugs) wildlife, weeds, soil conditions, weather extremes, pollutants, and so on, and to recommend strategies for their management or control. To develop such a broad expertise in plant protection, however, it is necessary that qualified graduates in a biological or agricultural science attend a 3- to 4-year profes-sional graduate degree program. The University of Florida's College of Agriculture and Life Sciences created such a program in 1999 and accepted its first graduate students in the fall semester of 2000. The Doctor of Plant Medicine (DPM) program, as it is called, had 14 students the first year, 15 the second year, and 10–14 students per year thereafter.

The degree is called Doctor of Plant Medicine rather than Doctor of Plant Health because it parallels the other two doctorates in the health professions, those of medicine (MD) and of veterinary medicine (DVM), in so many aspects that its goals and functions are easier to understand by this name. In addition, just like the MD and the DVM, the DPM is a professional, practitioner's degree, not a research degree as is the Ph.D. None of these degrees (MD, DVM, DPM) are replacing the Ph.D.s in their respective areas. Instead they provide a mechanism by which the information generated by the researcher Ph.D.s is used for the corresponding clientele (humans, animals, plants), the ailments of which are diagnosed and managed or controlled. Also, just like MD and DVM students, DPM students do several projects that involve mainly applied-type research and write appropriate reports, but they do not do research on a single project and do not write a thesis or dissertation.

The DPM program accepts students who have graduated with a bachelor's or a master's degree, preferably, but not necessarily, in a biological or agricultural discipline. Entering students must meet all criteria other graduate students (for Ph.D. or M.S. degrees) must meet. DPM students take 90 credits of graduate courses in the appropriate academic departments, most courses with laboratories, generally being the same courses taken by the graduate students of each department or discipline. About 65 of these credits are in required courses, a minimum of 18 in plant science, including courses in crop production, soils and crop nutrition, and weed science, 17 in entomology, 18 in plant pathology, 5 in nematology, 2 in acarology, 2 in wildlife that damage plants, 5 in plant pest management, and courses in agribusiness management, marketing, and agricultural law. The elective credits may be used by the student to specialize in a commodity area of his/her choice (e.g., agronomic crops, horticultural crops, ornamental crops and/or turf, forestry and/or urban forestry, education courses for college teaching, etc.).

In addition to the 90 credits of courses, DPM students must also do 30 credits of internships or practicums by spending appropriate lengths of time (2–3 credits each) in the soil analysis laboratory, the plant disease clinic, the nematode assay laboratory, the insect identification laboratory, and the weed identification laboratory. The students may also elect to do internships by working side by side with the extension weed scientist, horticulturist, plant pathologist, or entomologist, or they may elect to do an internship at an agricultural experiment station, at an agrichemical or seed company, or working side by side with an experienced crop consultant. The location of internships may vary from local to international. The entire curriculum is expected, although not required, to be completed within 3 or 4 years. Part-time students may take considerably longer.

Upon completion of the program, DPM graduates receive the doctorate degree and are fully educated and trained plant doctors who can identify just about anything, living and nonliving, that causes damage to plants and can provide quick and correct recommendations for their management or control. Their education, training, expertise, and the doctorate degree qualify them for a variety of well-paying jobs within the United States and internationally, including private practitioners as crop consultants; working for large farms or agribusinesses; working for the state or federal extension service (as county agents, IPM coordinators, pesticide information coordinators, etc.), for state or federal regulatory agencies [e.g., the Animal and Plant Health Inspection Service (APHIS), the Plant Protection and Quarantine (PPQ) Service, ship and airport inspectors, etc.); working for agrichemical, seed, and large food companies such as Del Monte and Campbell, teaching various biological courses at 2- and 4-year colleges and universities; and working for mid- to large size municipalities.

PLANT PATHOLOGY'S CONTRIBUTION TO CROPS AND SOCIETY

Some Historical and Present Examples of Losses Caused by Plant Diseases

Plant diseases affect the existence, adequate growth, and productivity of all kinds of plants and thereby affect one or more of the basic prerequisites for a healthy, safe life for humans. This happened since the time humans gave up their dependence on wild game and fruits and became more stationary, domesticated, and began to practice agriculture more than 6000 years ago. Destruction of food and feed crops by diseases has been an all too common occurrence in the past. It has resulted in malnutrition, starvation, migration, and death of people and animals on numerous occasions, several of which are well documented in history. Similar effects are observed annually in developing agrarian societies in which families and nations are dependent for their sustenance on their own produce. In more developed societies, losses from diseases in food and feed produce result primarily in financial losses and higher prices. It should be kept in mind, however, that loss of any amount of food or feed because of plant diseases means there is less available in the world economy. Considering the chronically inadequate amounts and distribution of food available, rich people and rich countries will be able to acquire such foodstuffs from wherever they are available, whereas many poor people somewhere in the world will be worse off because of these losses, and will go hungry.

Some examples of plant diseases that have caused severe losses in the past are shown in Tables 1-2 and 1-3.

Plant Diseases and World Crop Production

There are no dependable surveys of numbers of humans living on the earth before the year 1900. It is estimated, however, that there were about 300 million people living on the earth in the year A.D. 1, 310 million in A.D. 1000, 400 million in A.D. 1500, and 1.3 billion in A.D. 1900. During the 20th century there has been a dramatic explosion in the human population. Despite recent efforts to reduce the rate of population growth, the number of new humans added to the world population each year and the additional demands for food, energy, and other resources from our planet are frightening. Thus, the world population in 1993 was about 5.57 billion, and, at the present rate of 1.7% annual growth, it was expected to be 6.2 billion by the year 2000, be 7.1 billion by the year 2010, and be 8.5 billion by 2025. Currently, the world population increases by 1 billion every 11 years (see Fig. 1-42).

Paradoxically, the developing countries, in which from 50 to 80% of the population is engaged in agriculture, have the lowest agricultural output, their people are living on a substandard diet, and they have the highest population growth rates (2.64%). Because of the current distribution of usable land and population, of educational and technical levels for food production, and of general world economics, it is estimated that even today some 2 billion people suffer from hunger, malnutrition, or both. To feed these people and the additional millions to come in the next few years, all possible methods of increasing the world food supply are currently being pursued, including (1) expansion of crop acreages, (2) improved methods of cultivation, (3) increased fertilization, (4) use of improved varieties

TABLE 1-2
Examples of Severe Losses Caused by Plant Diseases

Disease	Location	Comments
Fungal		
1. Cereal rusts	Worldwide	Frequent severe epidemics; huge annual losses
2. Cereal smuts	Worldwide	Continuous, although lesser, losses on all grains
3. Ergot of rye and wheat	Worldwide	Infrequent, poisonous to humans and animals
4. Late blight of potato	Cool, humid climates	Annual epidemics, e.g., Irish famine (1845–1846)
5. Brown spot of rice	Asia	Epidemics, e.g., the great Bengal famine (1943)
6. Southern corn leaf blight	U.S.	Historical interest, epidemic 1970, $1 billion lost
7. Powdery mildew of grapes	Worldwide	European epidemics (1840s–1850s)
8. Downy mildew of grapes	U.S., Europe	European epidemic (1870s–1880s)
9. Downy mildew of tobacco	U.S., Europe	European epidemic (1950s–1960s); epidemic in North America (1979)
10. Chestnut blight	U.S.	Destroyed almost all American chestnut trees (1904–1940)
11. Dutch elm disease	U.S., Europe	Destroying American elm trees (1918 to present)
12. Pine stem rusts	Worldwide	Causing severe losses in many areas
13. Dwarf mistletoes	Worldwide	Serious losses in many areas
14. Coffee rust	Asia, South America	Destroyed all coffee in southeast Asia (1870s–1880s) since 1970 present in South and Central America
15. Banana leaf spot or Sigatoka disease	Worldwide	Great annual losses
16. Rubber leaf blight	South America	Destroys rubber tree plantations
17. Fusarium scab of wheat	North America	Severe losses in wet years
Viral		
18. Sugar cane mosaic	Worldwide	Great losses on sugar cane and corn
19. Sugar beet yellows	Worldwide	Great losses every year
20. Citrus tristeza (quick decline)	Africa, Americas	Millions of trees being killed
21. Swollen shoot of cacao	Africa	Continuous heavy losses
22. Plum pox or sharka	Europe, North America	Spreading severe epidemic on plums, peaches, apricots
23. Barley yellow dwarf	Worldwide	Important on small grains worldwide
24. Tomato yellow leaf curl	Mediterranean countries, Caribbean Basin, U.S.	Severe losses of tomatoes, beans, etc.
25. Tomato spotted wilt virus	Worldwide	On tomato, tobacco, peanuts, ornamentals, etc.
Bacterial		
26. Citrus canker	Asia, Africa, Brazil, U.S.	Caused eradication of millions of trees in Florida in 1910s and again in the 1980s and 1990s
27. Fire blight of pome fruits	North America, Europe	Kills numerous trees annually
28. Soft rot of vegetables	Worldwide	Huge losses of fleshy vegetables
Phytoplasmal		
29. Peach yellows	Eastern U.S., Russia	Historical, 10 million peach trees killed
30. Pear decline	Pacific coast states and Canada (1960s), Europe	Millions of pear trees killed
Nematode diseases		
31. Root knot	Worldwide	Continuous losses on vegetables and most other plants
32. Sugar beet cyst nematode	Northern Europe, Western U.S.	Continuous severe annual losses on sugar beets
33. Soybean cyst nematode	Asia, North and South America	Continuous serious losses on soybean

of crops, (5) increased irrigation, and (6) improved crop protection.

Crop Losses to Diseases, Insects, and Weeds

There is no doubt that the first five of the aforementioned measures must provide the larger amounts of food needed. Crop protection from pests and diseases can only reduce the amount lost after the potential for increased food production has been attained by proper utilization of all means possible. Crop protection, of course, has been important in the past and is important now. For example, it is estimated that in the Untied States alone, despite the control measures practiced, each year, crops worth $9.1 billion are lost to diseases,

TABLE 1-3
Additional Diseases Likely to Cause Severe Losses in the Future

Disease	Comments
Fungal	
1. Late blight of potato and tomato	New mating type of fungus spreading worldwide
2. Downy mildew of corn and sorghum	Just spreading beyond southeast Asia
3. Karnal bunt of wheat	Destructive in Pakistan, India, Nepal; since the 1980s introduced into Mexico and in the 1990s into U.S.
4. Soybean rust	Spreading from southeast Asia and from Russia; already in Hawaii, Puerto Rico, and South America
5. Monilia pod rot of cacao	Very destructive in South America; spreading elsewhere
6. Chrysanthemum white rust	Important in Europe, Asia, and recently in California
7. Sugar cane rust	Destructive in the Americas and elsewhere
8. Citrus black spot	Severe in Central and South America
9. Sweet orange scab	Severe in Australia
Viral	
10. African cassava mosaic	Destructive in Africa; threatening Asia and the Americas
11. Streak disease of maize (corn)	Spread throughout Africa on sugar cane, corn, wheat, etc.
12. Hoja blanca (white tip) of rice	Destructive in the Americas so far
13. Bunchy top of banana	Destructive in Asia, Australia, Egypt, Pacific islands
14. Rice tungro disease	Destructive in southeast Asia
15. Bean golden mosaic	Caribbean basin, Central America, Florida
16. Tomato yellow leaf curl.	East Mediterranean, Caribbean, the Americas
17. Plum pox	Destructive in Europe, spreading into U.S.
Bacterial	
18. Bacterial leaf blight of rice	Destructive in Japan and India; spreading
19. Bacterial wilt of banana	Destructive in the Americas; spreading elsewhere
20. Pierce's disease of grape	Deadly in southeast U.S.; spreading into California
21. Citrus variegation chlorosis	Destructive in Brazil; spreading
22. Citrus greening disease	Severe in Asia; spreading
Phytoplasmal	
23. Lethal yellowing of coconut palms	Destructive in Central America; spreading into U.S.
Viroid	
24. Cadang-cadang disease of coconut	Killed more than 15 million trees in the Philippines to date
Nematode	
25. Burrowing nematode	Severe on banana in many areas and citrus in Florida
26. Red ring of palms	Severe in Central America and the Caribbean
27. Pinewood nematode	Widespread and becoming severe in North America

$7.7 billion to insects, and $6.2 billion to weeds. Crop protection, however, becomes even more important in an intensive agriculture, where increased fertilization, genetically uniform high-yielding varieties, increased irrigation, and other methods are used. Crop losses to diseases and pests not only affect national and world food supplies and economies but also affect individual farmers even more, whether they grow the crop for direct consumption or for sale. Because operating expenditures for the production of the crop remain the same in years of low or high disease incidence, harvests lost to disease and pests lower the net return directly.

The amount of each crop lost to pests varies with the crop (e.g., 23.4% for fruits, 34.5% for cereals, 55.0% for sugar cane). Crop loss varies with the type of climate (warm, humid, rainy, dry, etc.), the particular year, availability of pesticides, availability of trained personnel, and educational level of growers. Also, the importance of each kind of pest (diseases, insects, weeds) varies with the crop. Generally, diseases, which are more difficult to detect, identify, and control on time, cause losses of about 14% of the crop; insects, if left unchecked, would cause tremendous losses but because they can be detected, identified, and controlled on time with effective insecticides cause losses of about 10% of the crop; and weeds, which still are poorly controlled in much of the world because of unavailability of herbicides due to cost, cause losses of about 12% of the crop. The total crop loss from diseases and pests is estimated at about 36% or one-third of the potential production of the world. To these losses should be added 6–12% postharvest losses to pests, which brings the total (preharvest

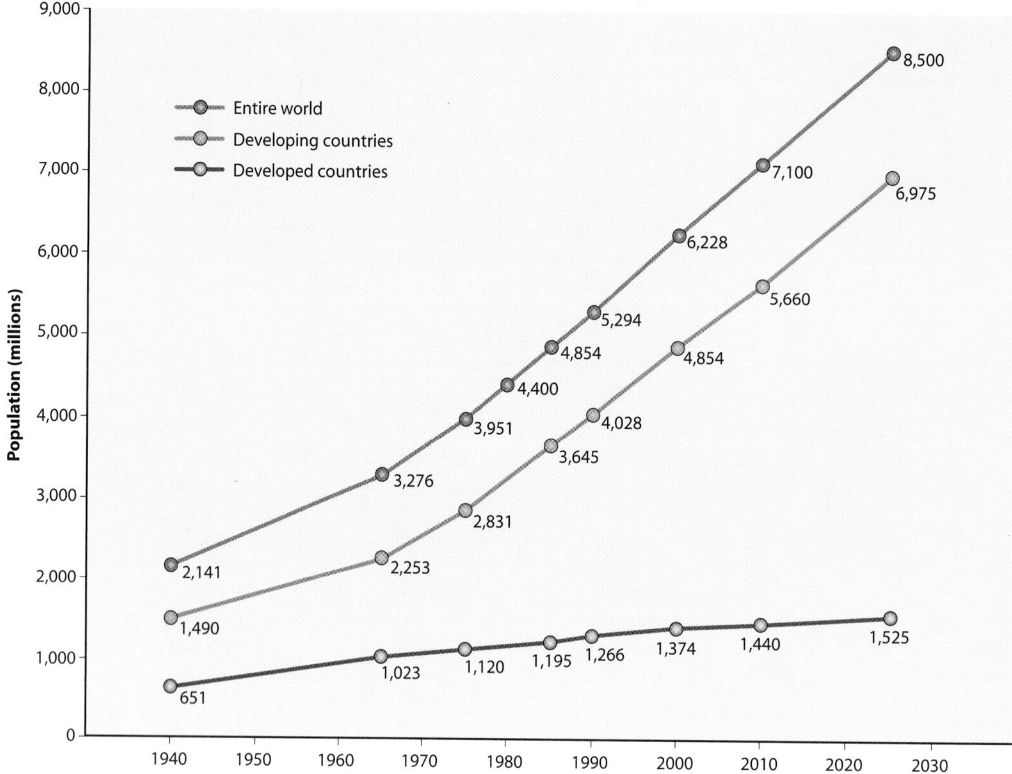

FIGURE 1-42 Real and projected population changes from 1940 to 2000 and to the year 2025. The rates of population growth were estimated for the years 1975 to 2000 and, for this graph, were assumed unchanged to the year 2025.

and postharvest) food losses to pests in the United States to about 40% and for the entire world to about 45% of all food crops. These losses occur, of course, despite all types of pest controls used. This is indeed a huge loss of needed food. It is apparent that losses are much greater in developing areas than they are in more developed ones. Another point that can be made is that insects cause much greater losses than diseases in developing countries, especially in Asia, because insects are controlled much more easily in developed countries than in developing ones, whereas losses caused by diseases seem to be as great in developed countries as they are in developing countries.

Crop losses caused by diseases, insects, and weeds become particularly striking and alarming when one considers their distribution among countries of varying degrees of development. In developed areas (Europe, North America, Australia, New Zealand, Japan, Israel, and South Africa), in which only 8.8% of the population is engaged in agriculture, the estimated losses and percentages of losses are considerably lower than those in developing countries, i.e., the rest of the world, in which 56.8% of the population is engaged in agriculture. The situation becomes particularly painful if one considers the fact that developing countries, which have much greater populations than developed countries, produce relatively less food and fiber and suffer much greater losses to plant diseases and to other pests. Taking into account the kinds of crops grown in temperate climates, where most developed countries are, and in the tropics, where developing countries are located, the total percentage losses differ considerably with the continent, as shown in Table 1-4. What is disheartening is that the more recent estimates by Oerke *et al.* (1994) indicate that the proportion of crop produce lost to diseases, insects, and weeds has actually increased in all continents (Table 1-4), despite presumably better and more widely used control materials and methods.

It is estimated that the total annual production for all agricultural crops worldwide is about $1500 billion (U.S. dollars, 2002). Of this, about $550 billion worth of produce is lost annually to diseases, insects, and weeds. An additional loss of about $455 billion would occur annually, but is averted by the use of various crop protection practices. Approximately $38 billion is spent annually for pesticides alone (fungicides, insecticides, herbicides), primarily in western Europe and in North America.

TABLE 1-4
Percentage of All Produce (1967 Estimate) and of Eight Major Crops (1994 Estimate) Lost to Diseases, Insects, and Weeds by Continent or Region[a]

	Produce lost to diseases, insects, and weeds (%)	
Continent or region	1967 estimate[b]	1994 Estimate[c]
Europe	25	28.2
Oceania	28	36.2
North and Central America	29	31.2
Russia and China	30	40.9
South America	33	41.3
Africa	42	48.9
Asia	43	47.1

[a]Reprinted from Oerke *et al.* (1994). The crops included are rice, wheat, barley, maize, potatoes, soybeans, cotton, and coffee.

[b]From H. H. Cramer (1967).

[c]The average worldwide loss to diseases, insects, and weeds was estimated at 42.1%.

Pesticides and Plant Diseases

The weed killers used increasingly in cultivated fields may cause injury to cultivated crop plants directly, but they also influence several soil pathogens and soil microorganisms antagonistic to pathogens. Other chemicals, such as fertilizers, insecticides, and fungicides, alter the types of microorganisms that survive and thrive in the soil, which sometimes leads to a reduction in the number of useful predators and antagonistic microorganisms of pathogens or their vectors. The use of fungicides and other pesticides specific against a particular pathogen often leads to increased populations and disease severity caused by other pathogens not affected by the specific pesticide. This occurs even with some rather broad-spectrum systemic fungicides that control most but not all pathogens, e.g., benomyl. Where such fungicides are used regularly and widely, some fungi, such as *Pythium*, that are not affected by them, may become more important as pathogens than when other more general fungicides are used.

The use of pesticides to control plant diseases and other pests had been, for many years since the mid–1950s, increasing steadily at an annual rate of about 14% (Fig. 1-43A). By 1999, nearly 2.6 billion kilograms (5.7 billion lbs) of active ingredients of pesticides were used per year worldwide at an annual cost of nearly $36 billion (Figs. 1-43B and 1-43C). In the United States alone, more than 550 million kg (1244 million lbs) of pesticides worth $11.2 billion (Figs. 1-43B–1-43E) were used in 1999. The relative amounts of active ingredient of herbicides, insecticides, fungicides, and other pesticides used in the United States and the world in 1998 or 1999 are shown in Figs. 1-43B–1-43E. Up to 1995, about 35% of all pesticides were applied in the United States and Canada, 45% in Europe, and the remaining 20% in the rest of the world. In the last several years, the use of pesticides has begun to decline in the United States and Europe, but as more countries become developed and can afford to buy pesticides, the use of pesticides in developing countries continues to increase sharply.

A large industry of pesticide research, production, and marketing has developed in the United States and some of the other countries. There are also hundreds of thousands of people who apply pesticides on crops as needed. The amount of pesticides applied on crops and the number of pesticide applicators varies considerably from region to region. This depends on the size of agriculture in the region, the climate of the region, and the kinds of crops grown in each. The Environmental Protection Agency has grouped the United States into 10 agricultural regions (Fig. 1-44A) and has estimated that the number of private pesticide applicators (i.e., individual farmers) and of commercial pesticide applicators varies from about 10,000 in some regions (No.1, New England states) to more than 300,000 in other regions (No.4, southeastern United States) (Fig. 1-44B).

There is little doubt that pesticide use has increased the yields of crops in most cases in which they have been applied. The cost of production, distribution, and application of pesticides is, of course, another form of economic loss caused by plant diseases and pests (Table 1-4). Furthermore, such huge amounts of poisonous substances damage the environment and food as they are spread over the crop plants several times each year. There are also the issues of worker protection from exposure to pesticides and poisonings of workers and consumers from pesticides.

Public awareness of the direct, indirect, and cumulative effects of pesticides on organisms other than the pests they are intended to control has led to increased emphasis on the protection of the environment. As a result, many pesticides have been abandoned or their use has been restricted, and their functions have been taken over by other less effective or more specific pesticides or by more costly or less efficient methods of control. The effort to control diseases and other pests by biological and cultural methods is still growing while at the same time more restrictions are being imposed on the testing, licensing, and application of pesticides. The pesticide producers must provide more detailed data on the effectiveness, toxicity, and persistence of each pesticide, and the application of each pesticide must be licensed for each crop on which it is going to be applied. Further-

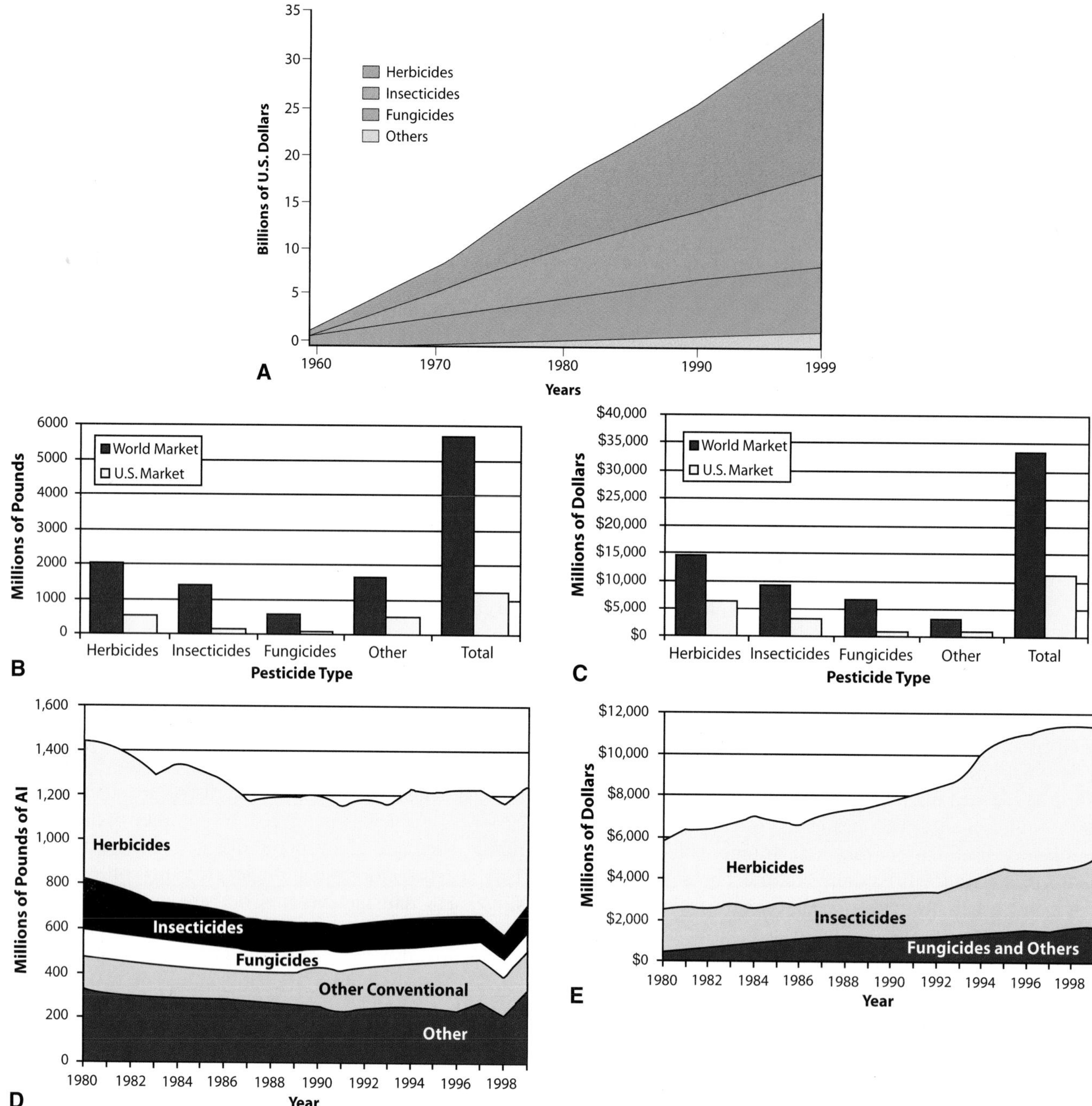

FIGURE 1-43 (A) Estimated worldwide annual sales of pesticides through 1999 in billions of dollars. Comparison of amounts of pesticides (in millions of pounds of active ingredient) used annually in the world and the United States (B) and of cost of pesticides (in millions of dollars) worldwide and the United States (C) at user level and by type of pesticide (B and C, 1999 estimates). (D) Annual usage in the United States of the various types of pesticides (in millions of pounds of active ingredient) from 1980 through 1999. (E) Cost of pesticides (in millions of dollars) spent annually in the United States from 1980 through 1999. Source: U.S. Environmental Protection Agency.

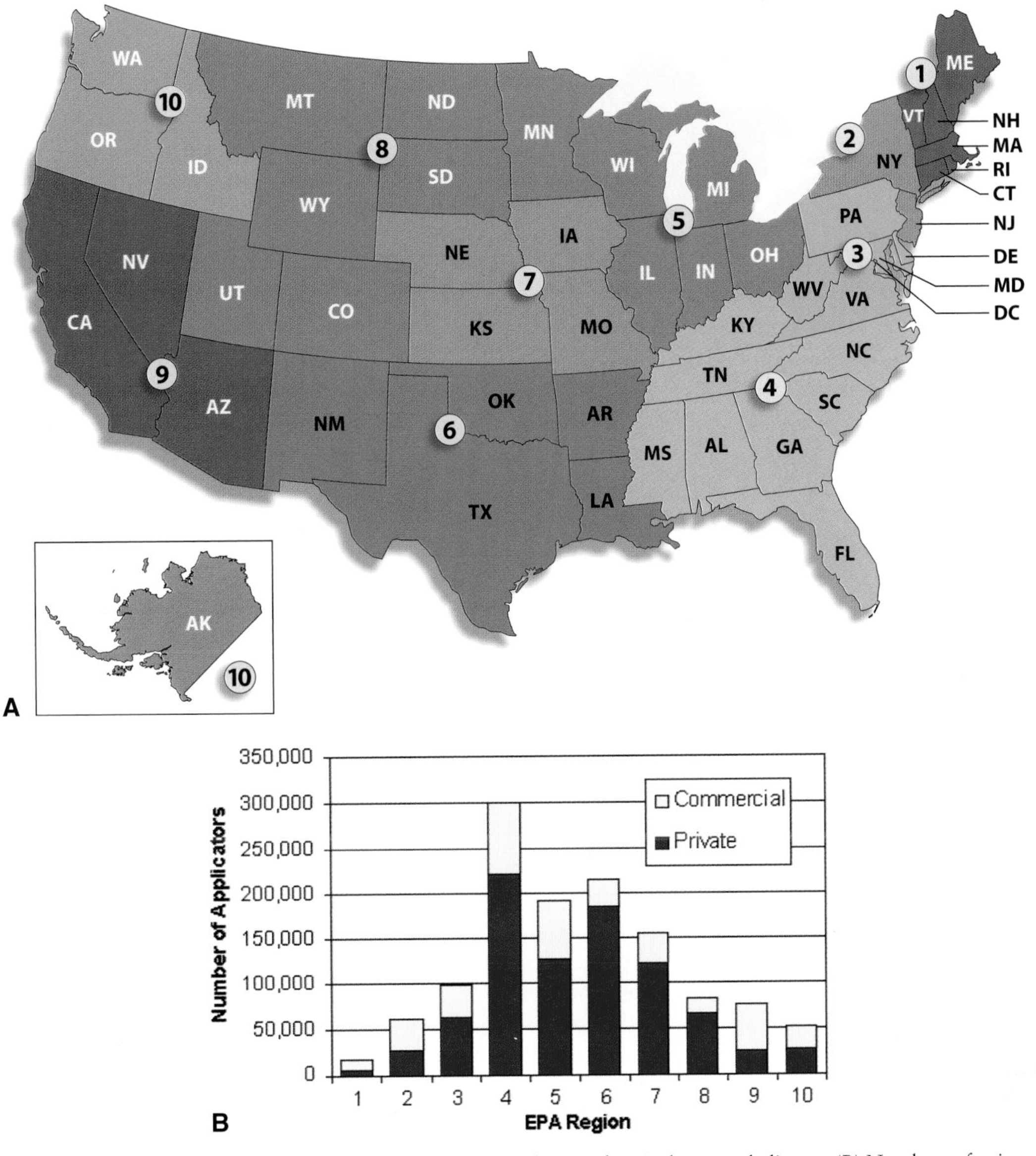

FIGURE 1-44 (A) Groups of states according to size and type of agriculture, and climate. (B) Numbers of private and commercial pesticide applicators in each region. Source: U.S. Environmental Protection Agency.

more, in some countries, each prospective commercial applicator of pesticides must pass an examination and be licensed to apply pesticides on crop plants. In some states, growers must clear with and get permission from state pest control advisors for the purchase and use of certain pesticides (prescription agriculture).

The desirability of using fewer and safer pesticides, however, is counteracted by the increasing demand of consumers over the last several decades for high-quality produce, especially fruits and vegetables free of any kind of blemishes caused by diseases or insects. A change in the attitude of consumers to demand less extravagant aesthetic quality of produce could reduce considerably the use of pesticides and the waste of perfectly wholesome foodstuffs, but such change in attitude may not occur for some time yet.

BASIC PROCEDURES IN THE DIAGNOSIS OF PLANT DISEASES

Pathogen or Environment

To diagnose a plant disease it is necessary to first determine whether the disease is caused by a pathogen or an environmental factor. In some cases, in which typical symptoms of a disease or signs of the pathogen are present, it is fairly easy for an experienced person to determine not only whether the disease is caused by a pathogen or an environmental factor, but by which one. Frequently, comparing the symptoms with those given in books that list the known diseases and their causes for specific plant hosts or in books like those of the compendia series of the American Phytopathological Society helps narrow the number of likely causes and often helps identify the cause of the disease. In most cases, however, a detailed examination of the symptoms and an inquiry into characteristics beyond the obvious symptoms are necessary for a correct diagnosis.

Infectious Diseases

In diseases caused by pathogens (fungi, bacteria, parasitic higher plants, nematodes, viruses, mollicutes, and protozoa), a few or large numbers of these pathogens may be present on the surface of the plants (some fungi, bacteria, parasitic higher plants, and nematodes) or inside the plants (most pathogens). The presence of such pathogens on or in a plant indicates that they are probably the cause of the disease. Someone with experience can detect and identify pathogens, in some cases with the naked eye or with a magnifying lens (some fungi, all parasitic higher plants, some nematodes). More frequently, identification can be accomplished only by microscopic examination (fungi, bacteria, and nematodes) (see Fig. 1-3). If no such pathogens are present on the surface of a diseased plant, then one must look for additional symptoms and, especially, for pathogens inside the diseased plant. Such pathogens are usually at the margins of the affected tissues, at the vascular tissues, at the base of the plant, and on or in its roots.

Diseases Caused by Parasitic Higher Plants

The presence of a parasitic higher plant (e.g., dodder, mistletoe, witchweed, or broomrape) growing on a plant is sufficient for diagnosis of the disease.

Diseases Caused by Nematodes

If a plant parasitic nematode is present on, in, or in the rhizosphere of a plant showing certain kinds of symptoms, the nematode may be the pathogen that caused the disease or at least was involved in the production of the disease. If the nematode can be identified as belonging to a species or genus known to cause such a disease, then the diagnosis of the disease can be made with a degree of certainty.

Diseases Caused by Fungi and Bacteria

When fungal mycelia and spores, or bacteria, are present on the affected area of a diseased plant, two possibilities must be considered: (1) the fungus or bacterium may be the actual cause of the disease or (2) the fungus or bacterium may be one of the many saprophytic fungi or bacteria that can grow on dead plant tissue once the latter has been killed by some other cause, perhaps by even other fungi or bacteria.

Fungi

To determine whether a fungus found on or in a diseased plant is a pathogen or a saprophyte, one first studies under a microscope the morphology of its mycelium, fruiting structures, and spores. The fungus can then be identified and checked in an appropriate book of mycology or plant pathology to see whether it has been reported to be pathogenic, especially on the plant on which it was found. If the symptoms of the plant correspond to those listed in the book as caused by that particular fungus, then the diagnosis of the disease is, in most cases, considered complete. If no such fungus is known to cause a disease on plants, especially one with symptoms similar to the ones under study, then the fungus found should be considered a saprophyte or, possibly, a previously unreported plant pathogen, and the search for the proof of the cause of the disease must continue. In many cases, neither fruiting structures nor spores are initially present on diseased plant tissue, and therefore no identification of the fungus is possible. For some fungi, special nutrient media are available for selective isolation, identification, or promotion of sporulation. Others need to be incubated under certain temperature, aeration, or light conditions to produce spores. With most fungi, however, fruiting structures and spores are produced in the diseased tissue if the tissue is placed in a glass or plastic "moisture chamber," i.e., a container to which wet paper towels are added to increase the humidity in the air of the container.

Bacteria and Mollicutes

Diagnosis of a bacterial disease and identification of the causal bacterium is based primarily on the symptoms of the disease, the constant presence of large numbers of bacteria in the affected area, and the absence of any other pathogens. Bacteria are small (0.8 by 1 mm), however, and although they can be seen with a compound microscope, they all resemble tiny rods and have no distinguishing morphological characteristics for identification. Care must be taken, therefore, to exclude the possibility that the observed bacteria are secondary saprophytes, i.e., bacteria that are growing in tissue killed by some other cause. Selective media are available for the selective cultivation of almost all plant pathogenic bacteria free of common saprophytes so that the genus and even some species can be identified. The easiest and surest way to prove that the observed bacterium is the pathogen is through isolation and growth of the bacterium in pure culture and, using a single colony for reinoculation of a susceptible host plant, reproducing the symptoms of the disease and comparing them with those produced by known species of bacteria. Since the late 1970s, immunodiagnostic techniques, including agglutination and precipitation, fluorescent antibody staining, and enzyme-linked immunosorbent assay, have been used to detect and identify plant pathogenic bacteria. Such techniques are quite sensitive, fairly specific, rapid, and easy to perform, and it is expected that soon standardized, reliable antisera will be available for serodiagnostic assays of plant pathogenic bacteria.

Since 1980, newer techniques have been used involving an automated analysis of fatty acid profiles of the bacteria or of the substances utilized by the bacteria for food (Biolog). Additional identification tests include comparison of the number of DNA pieces released by certain restriction enzymes, or degrees (percentages) of hybridization of the DNA of an unknown bacterium with the DNA of a known one. Some of the molecular techniques are now used for the identification of fastidious vascular bacteria.

Diseases caused by mollicutes appear as stunting of plants, yellowing or reddening of leaves, proliferation of shoots and roots, production of abnormal flowers, and eventual decline and death of the plant. Mollicutes are small, polymorphic, wall-less bacteria that live in young phloem cells of their hosts; they are generally visible only under an electron microscope and, except for the genus *Spiroplasma*, cannot be cultured on nutrient media. The diagnosis of such diseases, therefore, is based on symptomatology, graft transmissibility, transmission by certain insect vectors, electron microscopy, sensitivity to tetracycline antibiotics but not to penicillin, sensitivity to moderately high (32–35.8°C) temperatures, and, in a few cases in which specific antisera have been prepared, on serodiagnostic tests.

Diseases Caused by Viruses and Viroids

Many viruses (and viroids) cause distinctive symptoms in their hosts, and so the disease and the virus (or viroid) can be identified quickly by the symptoms. In the many other cases in which this is not possible, however, the diseases are diagnosed and the viruses are identified primarily as follows: (1) through virus transmission tests to specific host plants by sap inoculation or by grafting, and sometimes by certain insect, nematode, fungus, or mite vectors; (2) for viruses for which specific antisera are available, by using serodiagnostic tests, primarily enzyme-linked immunosorbent assays (ELISA), gel diffusion tests, microprecipitin tests, and fluorescent antibody staining; (3) by electron microscopy techniques such as negative staining of virus particles in leaf dip or purified preparations, or immune-specific electron microscopy (a combination of serodiagnosis and electron microscopy); (4) by microscopic examination of infected cells for specific crystalline or amorphous inclusions, which usually are diagnostic of the group to which the virus belongs; (5) through electrophoretic tests, useful primarily for detection and diagnosis of viroids and of nucleic acids of viruses; and (6) via hybridization of commercially available radioactive DNA complementary to a certain virus DNA or RNA, or viroid RNA, with the DNA or RNA present in plant sap and attached to a membrane filter (immunoblot).

Diseases Caused by More Than One Pathogen

Quite frequently a plant may be attacked by two or more pathogens of the same or different kinds and may develop one or more types of disease symptoms. It is important to recognize the presence of the additional pathogen(s). Once this is ascertained, the diagnosis of the disease(s) and the identification of the pathogen(s) proceed as described earlier for each kind of pathogen.

Noninfectious Diseases

If no pathogen can be found, cultured, or transmitted from a diseased plant, then it must be assumed that the disease is caused by an abiotic environmental factor. The number of environmental factors that can cause disease in plants is almost unlimited, but most of them affect plants by interfering with normal physiological processes. Such interference may be a result of an excess of a toxic substance in the soil or in the air, a lack of an essential substance (water, oxygen, or mineral nutrients), or a result of an extreme in the conditions supporting plant life (temperature, humidity, oxygen, CO_2, or light). Some of these effects may be the result of normal conditions (e.g., low temperatures) occurring at the wrong time or of abnormal conditions brought about naturally (flooding or drought) or by the activities of people and their machines (air pollutants, soil compaction, and weed killers).

The specific environmental factor that has caused a disease might be determined by observing a change in the environment, e.g., a flood or an unseasonable frost. Some environmental factors cause specific symptoms on plants that help determine the cause of the malady, but most of them cause nonspecific symptoms that, unless the history of the environmental conditions is known, make it difficult to diagnose the cause accurately.

Identification of a Previously Unknown Disease: Koch's Rules (Postulates)

When a pathogen is found on a diseased plant, the pathogen is identified by reference to special manuals; if the pathogen is known to cause such a disease and the diagnostician is confident that no other causal agents are involved, then the diagnosis of the disease may be considered completed. If, however, the pathogen found seems to be the cause of the disease but no previous reports exist to support this, then the steps described on page 27 under Koch's postulates are taken to verify the hypothesis that the isolated pathogen is the cause of the disease

Selected References

Alexopoulos, C. J., Mims, C. W., and Blackwell, M. (1996). "Introductory Mycology," 4th Ed. Wiley, New York.

Anonymous (1994). "Common Names of Plant Diseases." APS Press, St. Paul, MN.

Barger, G. (1931). "Ergot and Ergotism." Gurnev & Jackson, London.

Barrus, M. F. (1911). Variation in varieties of bean in their susceptibility to anthracnose. *Phytopathology* **1**, 190–195.

Bawden, F. C., Pirie, N. W., Bernal, J. D., and Fankuchen, I. (1936). Liquid crystalline substances from virus-infected plants. *Nature (London)* **138**, 1051–1052.

Beijerinck, M. W. (1898). Ueber ein contagium vivum fluidum als Ursache der Fleckenkrankheit der Tabaksblatter. Verh. K. Akad. Wet. Amsterdam 65(2), 3–21; Engl. transl. by J. Johnson in Phytopathol. Classics No. 7 (1942).

Biffen, R. H. (1905). Mendel's laws of inheritance and wheat breeding. *J. Agric. Sci.* **1**, 4–48.

Bridge, P. D., Couteaudier, Y., and Clarkson, J. M., eds., (1998). "Molecular Variability of Fungal Pathogens." CAB International, Wallingford, UK.

Brown, J. F., and Ogle, H. J., eds. (1997). "Plant Pathogens and Plant Diseases." Rockvale Publications, Armidale, Australia

Brown, W. (1915). Studies in the physiology of parasitism. I. The action of Botrytis cinerea. *Ann. Bot. (Lond.)* **29**, 313–348.

Bruehl, G. W. (1991). Plant pathology, a changing profession in a changing world. *Annu. Rev. Phytopathol.* **29**, 1–12.

Burrill, T. J. (1878). Remarks made in a discussion. *Trans. Ill. State Hortic. Soc.* [N. S.] **11**, 79–80.

Carefoot, G. L., and Sprott, E. R. (1967). "Famine in the Wind." Rand McNally, Chicago, IL.

Clayton, E. E. (1934). Toxin produced by *Bacterium tabacum* and its relation to host range. *J. Agric. Res. (Washington, DC)* **48**, 411–426.

Cobb, N. A. (1914). Contributions to a science of nematology. Pt. **1**, 1–33.

Cramer, H. H. (1967). Plant protection and crop production (transl. from German by J. H. Edwards), Pflanzenschutz-Nachr. 20. Farbenfabriken Bayer AG, Leverkusen.

Cruickshank, I. A. M. (1963). Phytoalexins. *Annu. Rev. Phytopathol.* **1**, 351–374.

deBary, A. (1861). Ueber die Geschlechtsorgane von *Peronospora. Bot. Z.* **19**, 89–91.

deBary, A. (1886). Ueber einige Schlerotinien und Sclerotinienkrankheiten. *Bot. Z.* **44**, 377–387.

Diener, T. O. (1971). Potato spindle tuber "virus." IV. A replicating, low molecular weight RNA. *Virology* **45**, 411–428.

Doi, Y., Teranaka, M., Yora, K., and Asuyama, H. (1967). Mycoplasma or PLT group-like microorganisms found in the phloem elements of plants infected with mulberry dwarf, potato witches broom, aster yellows, or paulownia witches broom. *Ann. Phytopathol. Soc. Jpn.* **33**, 259–266.

Dollet, M. (1984). Plant diseases caused by flagellate protozoa (*Phytomonas*). *Annu. Rev. Phytopathol.* **22**, 115–132.

Duncan, J. M., and Torrance, L. (1992). "Techniques for the Rapid Detection of Plant Pathogens." Blackwell, Oxford.

Flor, H. H. (1946). Genetics of pathogenicity in *Melampsora lini. J. Agric. Res. (Washington, DC)* **73**, 335–357.

Food and Agriculture Organization (FAO) (1993). "Production Yearbook." FAO, Rome.

Gäumann, E. (1946). Types of defensive reactions in plants. *Phytopathology* **36**, 624–633.

Gierrer, A., and Schramm, G. (1956). Infectivity of ribonucleic acid from tobacco mosaic virus. *Nature (London)* **177**, 702–703.

Goto, M. (1992). "Fundamentals of Bacterial Plant Pathology." Academic Press, San Diego.

Harper, G., *et al.* (2002). Viral sequences integrated into plant genomes. *Annu. Rev. Phytopathol.* **40**, 119–136

Heptig, G. H. (1974). Death of the American chestnut. *For. Hist.* **18**, 60–67.

Hillocks, R. J., and Waller, J. M., eds. (1998). "Soilborne Diseases of Tropical Crops." CABI, Wallingford, UK.

Holliday, P. (1989). "A Dictionary of Plant Pathology." Cambridge Univ. Press, New York.

Holmes, F. O. (1929). Local lesions in tobacco mosaic. *Bot. Gaz. (Chicago)* **87**, 39–55.

Horsfall, J. G., and Cowling, E. B., eds. (1977–1980). "Plant Disease," Vols. 1–5. Academic Press, New York.

Ivanowski, D. (1892). Ueber die Mosaikkrankheiten der Tabakspflanzen. St. Petersb. Acad. Imp. Sci. Bull. [4] 35(3), 67–70; Engl. transl. by J. Johnson in Phytopathological Classics No. 7 (1942).

Jacobsen, B. J., and Paulus, A. O. (1990). The changing role of extension plant pathologists. *Annu. Rev. Phytopathol.* **28**, 271–294.

Jones, D. R., ed. (1999). "Diseases of Banana, Abacá and Enset." CABI Publishing, Wallingford, UK.

Keen, N. T. (2000). A century of plant pathology: A retrospective view on understanding host–parasite interactions. *Annu. Rev. Phytopathol.* **38**, 31–48.

Keitt, G. W. (1959). History of plant pathology. *In* "Plant Pathology" (J. G. Horsfall and A. E. Dimond, eds.), Vol. 1, pp. 61–97. Academic Press, New York.

Klinkowski, M. (1970). Catastrophic plant diseases. *Annu. Rev. Phytopathol.* **8**, 37–60.

Lafont, A. (1909). Sur la presence d'un parasite de la classe des flagelles dans le latex de *l'Euphorbia pilulifera*. *C. R. Seances Soc. Biol. Ses Fil.* **66**, 1011–1013.

Lee, N., D'Souza, C. A., and Kronstad, J. W. (2003). Of smuts, blasts, mildews, and blights: cAMP signaling in phytopathogenic fungi. *Annu. Rev. Phytopathol.* **41**, 399–428.

Lucas, J. A. (1998). "Plant Pathology and Plant Pathogens," 3rd. Ed. Blackwell Science, Oxford, UK.

Madden, L. V., and Wheelis (2003). The threat of plant pathogens as weapons against U.S. crops. *Annu. Rev.* **41**, 155–176.

Matthews, R. E. F. (1991). "Plant Virology," 3rd Ed. Academic Press, San Diego.

Mayer, A. (1886). Ueber die Mosaikkrankheit des Tabaks. Landwirtsch. Vers.-Stn. **32**, 451–467; Engl. transl. by J. Johnson in Phytopathological Classics No. 7 (1942).

McGee, D. C., ed. (1997). "Plant Pathogens and the Worldwide Movement of Seeds." The American Phytopathological Society, St. Paul, MN.

Meehan, F., and Murphy, N. (1947). Differential phytotoxicity of metabolic by-products of *Helminthosporium victoriae*. *Science* **106**, 270–271.

Micheli, P. A. (1729). "Nova Plantarum Genera." Florence.

Millardet, P. M. A. (1885). Sur l'histoire du traitement du mildiou par le sulfate de cuivre. J. Agric. Prat. 2, 801–805; Engl. transl. by F. J. Scheiderhan in Phytopathological Classics No. 3 (1933).

Murray, T. D., Parry, D. W., and Cattlin, N. D. (1998). "A Colour Handbook of Diseases of Small Grain Cereal Crops." Manson Publishing, London, UK.

Needham, T. (1743). Concerning certain chalky tubulous concretions, callec malm; with some microscopical observations on the farina of the red lilly, and of worms discovered in smutty corn. *Philos. Trans. R. Soc. Lond.* **42**.

Oerke, E.-C., Dehne, H.-W., Schönbeck, F., and Weber, A. (1994). "Crop Production and Crop Protection: Estimated Losses in Major Food and Cash Crops."Elsevier, Amsterdam.

Orton, W. A. (1909). "The Development of Farm Crops Resistant to Disease." USDA, Washington, DC.

Parris, G. K. (1968). "A Chronology of Plant Pathology." Johnson & Sons, Strakville, Mississippi.

Prevost, B. (1807). "Memoire sur la cause immediate de la carie ou charbon des bles, et de plusieurs autres maladies des plantes, et sur les preservatifs de la carie." Paris; Engl. transl. by G. M. Keitt in Phytopathological Classics No. 6 (1939).

Prusiner, S. B. (1982). Novel proteinaceous infectious particles cause scrapie. *Science* **216**, 136–144.

Prusiner, S. B. (1995). The prion diseases. *Sci. Am.* January 1995, 48–57.

Schaad, N. W., *et al.* (2003). Advances in molecular-based diagnostics in meeting crop biosecurity and phytosanitary issues. *Annu. Rev. Phytopathol.* **41**, 305–324.

Singh, U. S., Mukhopadhyay, A. N., Kumer, J., and Chaube, H. S. (1992). "Plant Diseases of International Importance," Vols. 1–4. Prentice Hall, Englewood Cliffs, NJ.

Singh, U. S., Kohmoto, K., and Singh, R. P. (1995). "Pathogenesis and Host Specificity in Plant Diseases: Histopathological, Biochemical, Genetic, and Molecular Bases," Vols. 1–3. Pergamon/Elsevier, Amsterdam.

Sinha, K. K., and Bhatnagar, D., eds. (1998). "Mycotoxins in Agriculture and Food Safety." Dekker, New York.

Smith, E. F. (1899). Dr. Alfred Fischer in the role of pathologist. Zentralbl. Bakteriol., Abt. 2, Naturwiss.: Allg. Landwirtsch, Tech. Mikrobiol. **5**, 810–817.

Smith, E. F., and Townsend, C. O. (1907). A plant tumor of bacterial origin. *Science* **25**, 671–673.

Stahel, G. (1931). Zur kenntnis der Siebrohrenkrankheit (Phloemnekrose) des kaffeebaumes in Surinam. I, II. Phytopathol. Z. **4**, 65–82, 539–544.

Stakman, E. C. (1914). A study in cereal rusts: Physiological races. *Stn. Bull. Univ. Minn. Agric. Exp.* 138.

Stanley, W. M. (1935). Isolation of a crystalline protein possessing the properties of tobacco-mosaic virus. *Science* **81**, 644–645.

Theophrastus (370–286 b.c.). "Enquiry into Plants" (Engl. transl. by Sir Arthur Hort), Vols. 1 and 2. Harvard Univ. Press, London/New York, 1916.

Thurston, H. D. (1973). Threatening plant diseases. *Annu. Rev. Phytopathol.* **11**, 27–52.

Thurston, H. D. (1998). "Tropical Plant Diseases," 2nd Ed. APS Press, St. Paul, MN.

Tillet, M. (1755). "Dissertation sur la cause qui corrompt et noircit les grans de ble dans les epis; et sur les moyens de prevenir ces accidents. Bordeaux; Engl. transl. by H. B. Humphrey in Phytopathological Classics No. 5 (1937).

U.S. Department of Agriculture (1965). "Losses in Agriculture," Agric. Handb. No. 291. USDA, Washington, DC.

Vanderplank, J. E. (1963). "Plant Diseases: Epidemics and Control." Academic Press, New York.

von Schmeling, B., and Kulka, M. (1966). Systemic fungicidal activity of 1,4-oxanthiin derivatives. *Science* **152**, 659.

Windsor, I. M., and Black, L. M. (1972). Clover club leaf: A possible rickettsial disease of plants. *Phytopathology* **62**, 1112.

Woodham-Smith, C. (1962). "The Great Hunger, Ireland 1845–1849." Harper & Row, New York.

chapter two

PARASITISM AND DISEASE DEVELOPMENT

PARASITISM AND PATHOGENICITY
77

HOST RANGE OF PATHOGENS
78

DEVELOPMENT OF DISEASE IN PLANTS
79

STAGES IN THE DEVELOPMENT OF DISEASE: THE DISEASE CYCLE
80

INOCULATION – PREPENETRATION PHENOMENA – PENETRATION – INFECTION – DISSEMINATION OF PATHOGEN – OVERWINTERING AND/OR OVERSUMMERING OF PATHOGENS
80

RELATIONSHIPS BETWEEN DISEASE CYCLES AND EPIDEMICS
102

The pathogens that attack plants belong to the same groups of organisms that cause diseases in humans and animals. Moreover, plants are attacked by a number of other plants. With the exception of some insect-transmitted plant pathogens, however, which cause diseases in both their host plants and their insect vectors, none of the pathogen species that attack plants is known to affect humans or animals.

Infectious diseases are those that result from infection of a plant by a pathogen. In such diseases, the pathogen can grow and multiply rapidly on diseased plants, it can spread from diseased to healthy plants, and it can cause additional plants to become diseased, thereby leading to the development of a small or large epidemic.

PARASITISM AND PATHOGENICITY

An organism that lives on or in some other organism and obtains its food from the latter is called a **parasite.** The removal of food by a parasite from its host is called **parasitism.** A **plant parasite** is an organism that becomes intimately associated with a plant and multiplies or grows at the expense of the plant. The removal by the parasite of nutrients and water from the host plant usually reduces efficiency in the normal growth of the plant and becomes detrimental to the further development and reproduction of the plant. In many cases, parasitism is intimately associated with **pathogenicity**, i.e., the ability of a pathogen to cause disease, as the ability of the parasite to invade and become established in the host generally results in the development of a diseased condition in the host.

In some cases of parasitism, as with the root nodule bacteria of legume plants and the mycorrhizal infection of feeder roots of most flowering plants, both the plant and the microorganism benefit from the association. This phenomenon is known as **symbiosis**.

In most plant diseases, however, the amount of damage caused to plants is often much greater than would be expected from the mere removal of nutrients by the parasite. This additional damage results from substances secreted by the parasite or produced by the host in response to stimuli originating in the parasite. Tissues affected by such substances may show increased respiration, disintegration or collapse of cells, wilting, abscission, abnormal cell division and enlargement, and degeneration of specific components such as chlorophyll. These conditions in themselves do not seem directly to improve the welfare of the parasite. It would appear, therefore, that the damage caused by a parasite is not always proportional to the nutrients removed by the parasite from its host. **Pathogenicity**, then, is the ability of the parasite to interfere with one or more of the essential functions of the plant, thereby causing disease. Parasitism frequently plays an important, but not always the most important, role in pathogenicity.

Of the large number of groups of living organisms, only a few members of a few groups can parasitize plants: fungi, bacteria, mollicutes, parasitic higher plants, parasitic green algae, nematodes, protozoa, viruses, and viroids. These parasites are successful because they can invade a host plant, feed and proliferate in it, and withstand the conditions in which the host lives. Some parasites, including viruses, viroids, mollicutes, some fastidious bacteria, nematodes, protozoa, and fungi causing downy mildews, powdery mildews, and rusts, are **biotrophs**, i.e., they can grow and reproduce in nature only in living hosts, and they are called **obligate parasites**. Other parasites (most fungi and bacteria) can live on either living or dead hosts and on various nutrient media, and they are therefore called **nonobligate parasites**. Some nonobligate parasites live most of the time or most of their life cycles as parasites, but, under certain conditions, may grow saprophytically on dead organic matter; such parasites are **semibiotrophs** and are called **facultative saprophytes**. Others live most of the time and thrive well on dead organic matter (**necrotrophs**) but, under certain circumstances, may attack living plants and become parasitic; these parasites are called **facultative parasites**. Usually no correlation exists between the degree of parasitism of a pathogen and the severity of disease it can cause, as many diseases caused by weakly parasitic pathogens are much more damaging to a plant than others caused even by obligate parasites. Moreover, certain pathogens, e.g., slime molds and those causing sooty molds, can cause disease by just covering the surface of the plant without parasitizing the plant.

Obligate and nonobligate parasites generally differ in the ways in which they attack their host plants and procure their nutrients from the host. Many nonobligate parasites secrete enzymes that bring about the disintegration of the cell components of plants, and these alone or with the toxins secreted by the pathogen result in the death and degradation of the cells. The invading pathogen then utilizes the contents of the cells for its growth. Many fungi and most bacteria act in this fashion, growing as necrotrophs on a nonliving substrate within a living plant. This mode of nutrition is like that of saprophytes. However, all obligate (and some nonobligate) parasites do not kill cells in advance but get their nutrients either by penetrating living cells or by establishing close contact with them. The association of these pathogens with their host cells is an intimate one and results in continuous absorption or diversion of nutrients, which would normally be utilized by the host, into the body of the parasite. The depletion of nutrients, however, although it restricts the growth of the host and causes symptoms, does not always kill the host. In the case of obligate parasites, death of the host cells restricts the further development of the parasite and may result in its death.

Parasitism of cultivated crops is a common phenomenon. In North America, for example, more than 8,000 species of fungi cause nearly 100,000 diseases, and at least 200 bacteria, about 75 mollicutes, more than 1,000 different viruses and 40 viroids, and more than 500 species of nematodes attack crops. Although about 2,500 species of higher plants are parasitic on other plants, only a few of them are serious parasites of crop plants. A single crop, e.g., the tomato, may be attacked by more than 40 species of fungi, 7 bacteria, 16 viruses, several mollicutes, and several nematodes. This number of diseases is average as corn has 100, wheat 80, and apple and potato each are susceptible to about 80–100 diseases. Fortunately, however, in any given location, only a fraction of the diseases affecting a crop are present and, in any given year, only a small number of these diseases become severe.

HOST RANGE OF PATHOGENS

Pathogens differ with respect to the kinds of plants that they can attack, with respect to the organs and tissues that they can infect, and with respect to the age of the organ or tissue of the plant on which they can grow.

Some pathogens are restricted to a single species, others to one genus of plants, and still others have a wide range of hosts, belonging to many families of higher plants. Some pathogens grow especially on roots, others on stems, and some mainly on the leaves or on fleshy fruits or vegetables. Some pathogens, e.g., vascular parasites, attack specifically certain kinds of tissues, such as phloem or xylem. Others may produce different effects on different parts of the same plant. With regard to the age of plants, some pathogens attack seedlings or the young tender parts of plants, whereas others attack only mature tissues.

Many obligate parasites are quite specific as to the kind of host they attack, possibly because they have evolved in parallel with their host and require certain nutrients that are produced or become available to the pathogen only in these hosts. However, many viruses and nematodes, although obligate parasites, attack many different host plants. Nonobligate parasites, especially root, stem, and fruit-attacking fungi, usually attack many different plants and plant parts of varying age, possibly because these pathogens depend on nonspecific toxins or enzymes that affect substances or processes found commonly among plants for their attack. Some nonobligate parasites, however, produce disease on only one or a few plant species. In any case, the number of plant species currently known to be susceptible to a single pathogen is smaller than the actual number in nature, as only a few species out of thousands have been studied for their susceptibility to each pathogen. Furthermore, because of genetic changes, a pathogen may be able to attack hosts previously immune to it. It should be noted, however, that each plant species is susceptible to attack by only a relatively small number of all known plant pathogens.

DEVELOPMENT OF DISEASE IN PLANTS

A plant becomes diseased in most cases when it is attacked by a pathogen or when it is affected by an abiotic agent. Therefore, in the first case, for a plant disease to occur, at least two components (plant and pathogen) must come in contact and must interact. If at the time of contact of a pathogen with a plant, and for some time afterward, conditions are too cold, too hot, too dry, or some other extreme, the pathogen may be unable to attack or the plant may be able to resist the attack, and therefore, despite the two being in contact, no disease develops. Apparently then, a third component, namely a set of environmental conditions within a favorable range, must also occur for disease to develop. Each of the three components can display considerable variability; however, as one component changes it affects the degree of disease severity within an individual plant and within a plant population. For example, the plant may be of a species or variety that may be more or less resistant to the pathogen or it may be too young or too old for what the pathogen prefers, or plants over a large area may show genetic uniformity, all of which can either reduce or increase the rate of disease development by a particular pathogen. Similarly, the pathogen may be of a more or less virulent race, it may be present in small or extremely large numbers, it may be in a dormant state, or it may require a film of water or a specific vector. Finally, the environment may affect both the growth and the resistance of the host plant and also the rate of growth or multiplication and degree of virulence of the pathogen, as well as its dispersal by wind, water, vector, and so on.

The interactions of the three components of disease have often been visualized as a triangle (Fig. 2-1), generally referred to as the "**disease triangle**." Each side of the triangle represents one of the three components. The length of each side is proportional to the sum total of the characteristics of each component that favor disease. For example, if the plants are resistant, the wrong age, or widely spaced, the host side — and the amount of disease — would be small or zero, whereas if the plants are susceptible, at a susceptible stage of growth, or planted densely, the host side would be long and the potential amount of disease could be great. Similarly, the more virulent, abundant, and active the pathogen, the longer the pathogen side would be and the greater the potential amount of disease. Also, the more favorable the environmental conditions that help the pathogen (e.g., temperature, moisture, and wind) or that reduce host resistance, the longer the environment side would be and the greater the potential amount of disease. If the three components of the disease triangle could be quantified, the area of the triangle would represent the amount of disease in a plant or in a plant population. If any of the three components is zero, there can

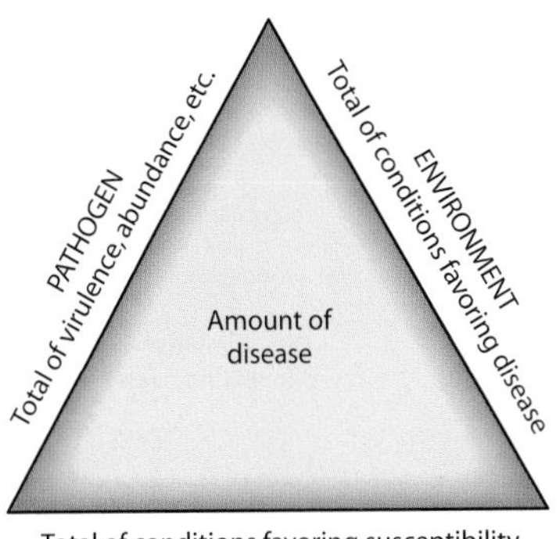

FIGURE 2-1 The disease triangle.

be no disease. The disease triangle is also represented as a triangle with the words of the three components (host plant, pathogen, environment) placed at the peaks of the triangle rather than along its sides.

STAGES IN THE DEVELOPMENT OF DISEASE: THE DISEASE CYCLE

In every infectious disease a series of more or less distinct events occurs in succession and leads to the development and perpetuation of the disease and the pathogen. This chain of events is called a **disease cycle.** A disease cycle sometimes corresponds fairly closely to the **life cycle** of the pathogen, but it refers primarily to the appearance, development, and perpetuation of the disease as a function of the pathogen rather than to the pathogen itself. The disease cycle involves changes in the plant and its symptoms as well as those in the pathogen and spans periods within a growing season and from one growing season to the next. The primary events in a disease cycle are inoculation, penetration, establishment of infection, colonization (invasion), growth and reproduction of the pathogen, dissemination of the pathogen, and survival of the pathogen in the absence of the host, i.e., overwintering or oversummering (overseasoning) of the pathogen (Fig. 2-2). In some diseases there may be several **infection cycles** within one disease cycle.

Inoculation

Inoculation is the initial contact of a pathogen with a site of plant where infection is possible. The pathogen(s) that lands on or is otherwise brought into contact with the plant is called the **inoculum.** The inoculum is any part of the pathogen that can initiate infection. Thus, in fungi the inoculum may be spores (Figs. 2-3A–2-3C), **sclerotia** (i.e., a compact mass of mycelium), or fragments of mycelium. In bacteria, mollicutes, protozoa, viruses, and viroids, the inoculum is always whole individuals of bacteria (Fig. 2-3D), mollicutes, protozoa, viruses, and viroids, respectively. In nematodes, the inoculum may be adult nematodes, nematode juveniles, or eggs. In parasitic higher plants, the inoculum may be plant fragments or seeds. The inoculum may consist of a single individual of a pathogen, e.g., one spore or one multicellular sclerotium, or of millions of individuals of a pathogen, e.g., bacteria carried in a drop of water. One unit of inoculum of any pathogen is called a **propagule.**

Types of Inoculum

An inoculum that survives dormant in the winter or summer and causes the original infections in the spring or in the autumn is called a **primary inoculum,** and the infections it causes are called **primary infections.** An inoculum produced from primary infections is called a **secondary inoculum** and it, in turn, causes **secondary infections.** Generally, the more abundant the primary inoculum and the closer it is to the crop, the more severe the disease and the losses that result.

Sources of Inoculum

In some fungal and bacterial diseases of perennial plants, such as shrubs and trees, the inoculum is produced on the branches, trunks, or roots of the plants. The inoculum sometimes is present right in the plant

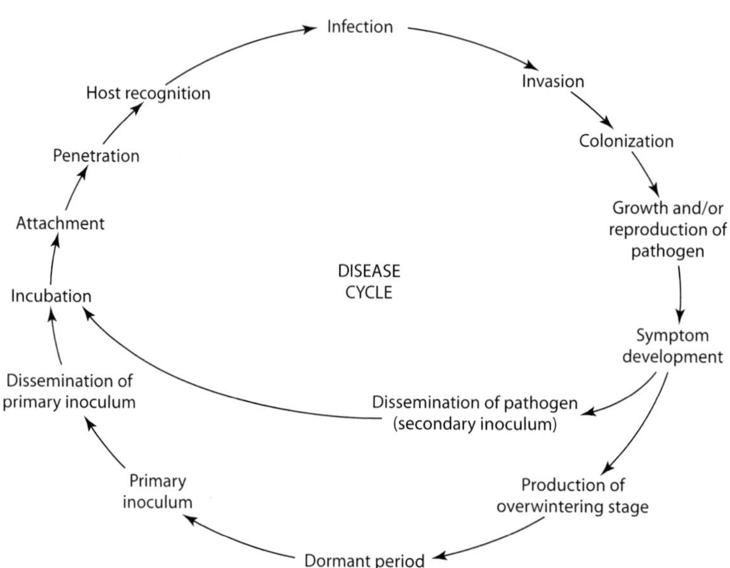

FIGURE 2-2 Stages in development of a disease cycle.

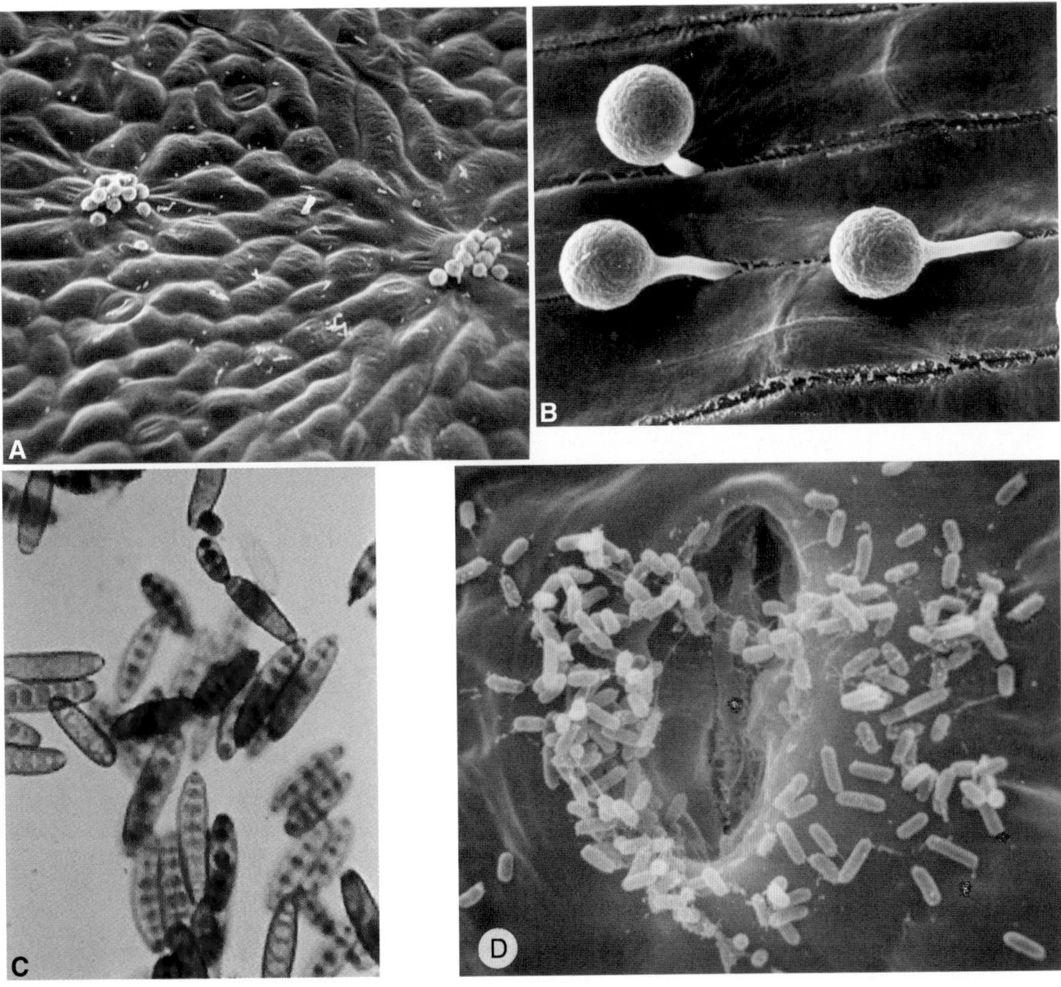

FIGURE 2-3 Types of inoculum and ways in which some pathogens enter a host plant. (A) Two groups of zoospores of the grape downy mildew oomycete have gathered over two leaf stomata. (B) Encysted zoospores of the soybean root rot pathogen *Phytophthora sojae* germinating and penetrating the root. (C) Mitospores (conidia) of a fungus that causes a corn leaf spot disease. (D) Bacteria of *Pseudomonas syringae* that causes bacterial spot and canker of stone fruits are seen in and surrounding a stoma of a cherry leaf. [Photographs courtesy of (A) D. J. Royle, (B) C. W. Mims and K. Enkerli, University of Georgia, and (D) E. L. Mansvelt, Stellenbosch, South Africa.]

debris or soil in the field where the crop is grown; other times it comes into the field with the seed, transplants, tubers, or other propagative organs or it may come from sources outside the field. Outside sources of inoculum may be nearby plants or fields or fields many miles away. In many plant diseases, especially those of annual crops, the inoculum survives in perennial weeds or alternate hosts, and every season it is carried from them to the annual and other plants. Fungi, bacteria, parasitic higher plants, and nematodes either produce their inoculum on the surface of infected plants or their inoculum reaches the plant surface when the infected tissue breaks down. Viruses, viroids, mollicutes, fastidious bacteria, and protozoa produce their inoculum within the plants; such an inoculum almost never reaches the plant surface in nature and, therefore, it can be transmitted from one plant to another almost entirely by some kind of vector, such as an insect.

Landing or Arrival of Inoculum

The inoculum of most pathogens is carried to host plants passively by wind, water, and insects. An airborne inoculum usually gets out of the air and onto the plant surface not just by gravity but by being washed out by rain. Only a tiny fraction of the potential inoculum produced actually lands on susceptible host plants; the bulk of the produced inoculum lands on things that cannot become infected. Some types of inoculum in the soil, e.g., zoospores and nematodes, may be attracted to the host plant by such substances as sugars and amino acids diffusing out of the plant roots. Vector-transmitted pathogens are usually carried to their host plants with an extremely high efficiency.

Prepenetration Phenomena

Attachment of Pathogen to Host

Pathogens such as mollicutes, fastidious bacteria, protozoa, and most viruses are placed directly into cells of plants by their vectors and, in most cases, they are probably immediately surrounded by cytoplasm, cytoplasmic membranes, and cell walls. However, almost all fungi, bacteria, and parasitic higher plants are first brought into contact with the external surface of plant organs. Before they can penetrate and colonize the host, they must first become attached to the host surface (Figs. 2-3–2-6). Attachment takes place through the adhesion of spores, bacteria, and seeds through adhesive materials that vary significantly in composition and in the environmental factors they need to become adhesive. Disruption of adhesion by nontoxic synthetic compounds results in failure of the spores to infect leaves.

The propagules of these pathogens have on their surface or at their tips mucilaginous substances consisting of mixtures of water-insoluble polysaccharides, glycoproteins, lipids, and fibrillar materials, which, when moistened, become sticky and help the pathogen adhere to the plant. In some fungi, hydration of the spore by moist air or dew causes the extrusion of preformed mucilage at the tip of the spore that serves for the immediate adherence of the spore to the hydrophobic plant surface and resistance to removal by flowing water. However, in powdery mildew fungi, which do not require free water for infection, adhesion is accomplished by release from the spore of the enzyme cutinase, which makes the plant and spore areas of attachment more hydrophilic and cements the spore to the plant surface. In other cases, propagule adhesion requires on-the-spot synthesis of new glycoproteins and it may not reach maximum levels until 30 minutes after contact. In some fungi causing vascular wilts, spores fail to adhere after hydration but become adhesive after they are allowed to respire and to synthesize new proteins.

How exactly spores adhere to plant surfaces is not known, but it seems to either involve a very specific interaction of the spore with a host plant surface via lectins, ionic interactions, or hydrophobic contact with the plant cuticle, or involve stimulation of the spore by physical rather than chemical signals. The extracellular matrix surrounding the propagules of many pathogens contains several enzymes, including cutinases, which are expected to play an important role in spore attachment. In any case, the act of attachment often seems necessary for the subsequent transmission of signals for germ tube extension and production of infection structure. It is now clear that many proteins of the fungal cell wall, in addition to their structural role, play an important role in the adhesion of fungi, as well as in the host-surface perception by the fungus.

Spore Germination and Perception of the Host Surface

It is not clear what exactly triggers spore germination, but stimulation by the contact with the host surface, hydration and absorption of low molecular weight ionic material from the host surface, and availability of nutrients play a role. Spores also have mechanisms that prevent their germination until they sense such stimulations or when there are too many spores in their vicinity. Once the stimulation for germination has been received by the spore, the latter mobilizes its stored food reserves, such as lipids, polyoles, and carbohydrates, and directs them toward the rapid synthesis of cell membrane and cell wall toward the germ tube formation and extension (Figs. 2-4 and 2-5). The germ tube is a specialized structure distinct from the fungal mycelium, often growing for a very short distance before it differentiates into an appressorium. The germ tube is also the structure and site that perceives the host surface and, if it does not receive the appropriate external stimuli, the germ tube remains undifferentiated and, when the nutrients are exhausted, it stops growing. When appropriate physical and chemical signals, such as surface hardness, hydrophobicity, surface topography, and plant signals, are present, germ tube extension and differentiation take place.

The perception of signals from plant surfaces by pathogenic fungi (Fig. 2-6) seems to be the result of signaling pathways mediated by cyclic adenosine monophosphate (cAMP) and mitogen-activated protein kinase (MAPK), which have been implicated in regulating the development of infection-related phenomena in many different fungi. In response to a signal from the host plant, e.g., the presence of a hydrophobic plant surface, which transmits a cue for appressorium formation, the fungus perceives the extracellular signal and its transmission via the plasma membrane and, as a first step, it accumulates intracellular signaling molecules and induces a phosphorylation cascade. In some fungi, the receptor of the signal is a protein in the plasma membrane of the fungal spore. Transmission of the cAMP signal proceeds via the cAMP-dependent activity of

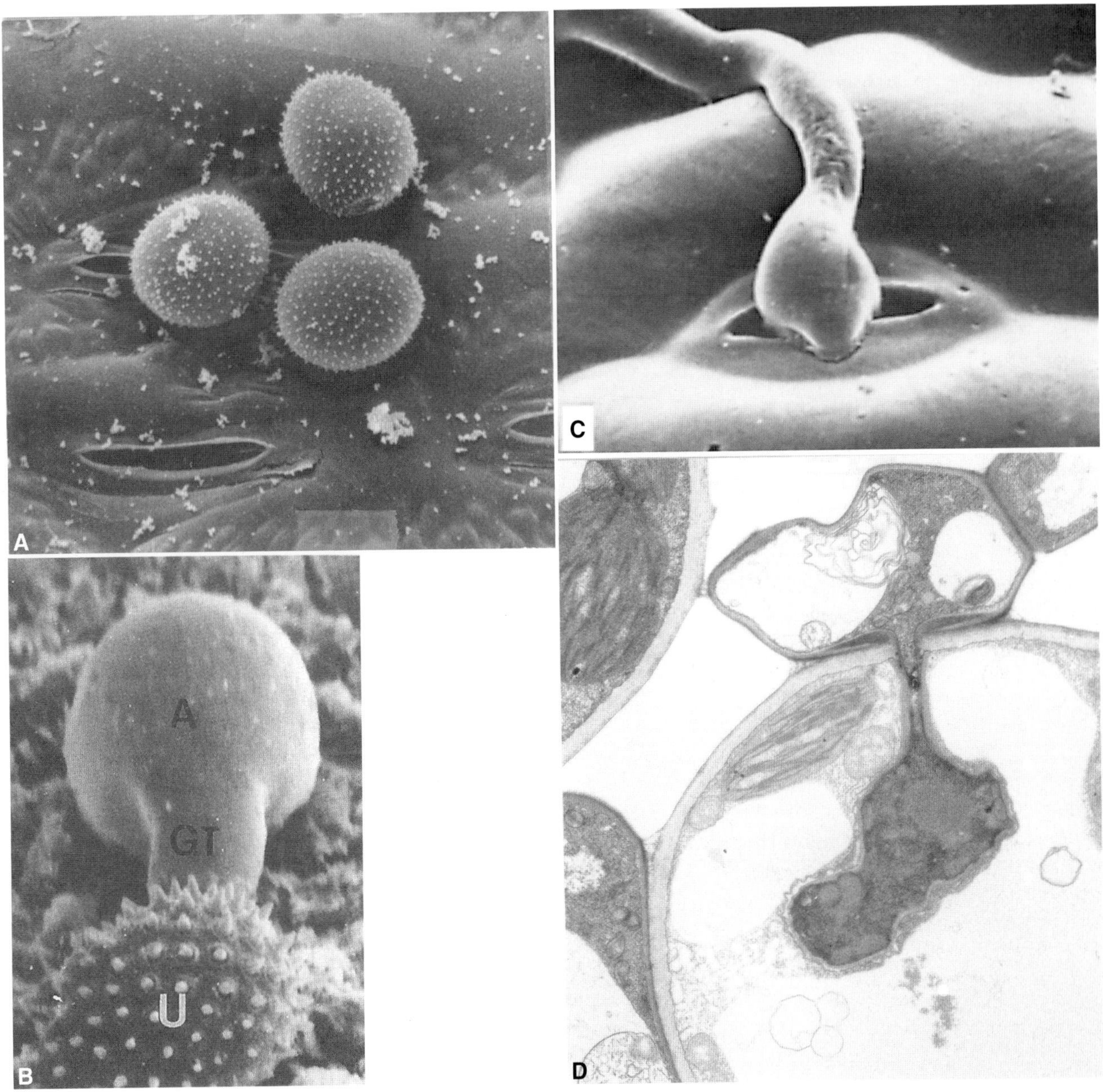

FIGURE 2-4 Methods of germination and penetration by fungi. (A) Uredospores of a rust fungus on a grass leaf next to open stomata. (B) A rust uredospore (U) that has germinated and produced a dome-like appressorium. (C) Uredospore germination, germ tube elongation, and appressorium penetration through a stoma. (D) A haustorium of a rust fungus inside a host cell. (E) A spore of the apple black rot fungus that has germinated directly into mycelium. (F) Two multicellular conidia of *Alternaria* sp. (G) A germinating conidium of *Alternaria* with a germ tube covered with extracellular material. [Photographs courtesy of (A) Plant Pathology Department, University of Florida, (B and C) W. K. Wynn and (D) C. W. Mims, University of Georgia, (E) J. Rytter and J. W. Travis, Pennslyvania State University, (F and G) Mims *et al.* (1997). *Can. J. Bot.* 75, 252–260.]

(continued on next page)

protein kinase A (= PKA) and subsequent phosphorylation of target proteins. The major activity of PKA in developing germ tubes is the mobilization of carbohydrates and lipids to the appressorium site and is, therefore, pivotal to the production of functional appressoria. In some fungi, cAMP signaling is required for the initiation of appressorium development, at which time intracellular cAMP concentrations rise during differentiation of conidia and emergence of the appressorium germ tube. Subsequently, cAMP levels fall as the germ tube extends and, if more cAMP is added at this point, further development of the germ tube is inhibited.

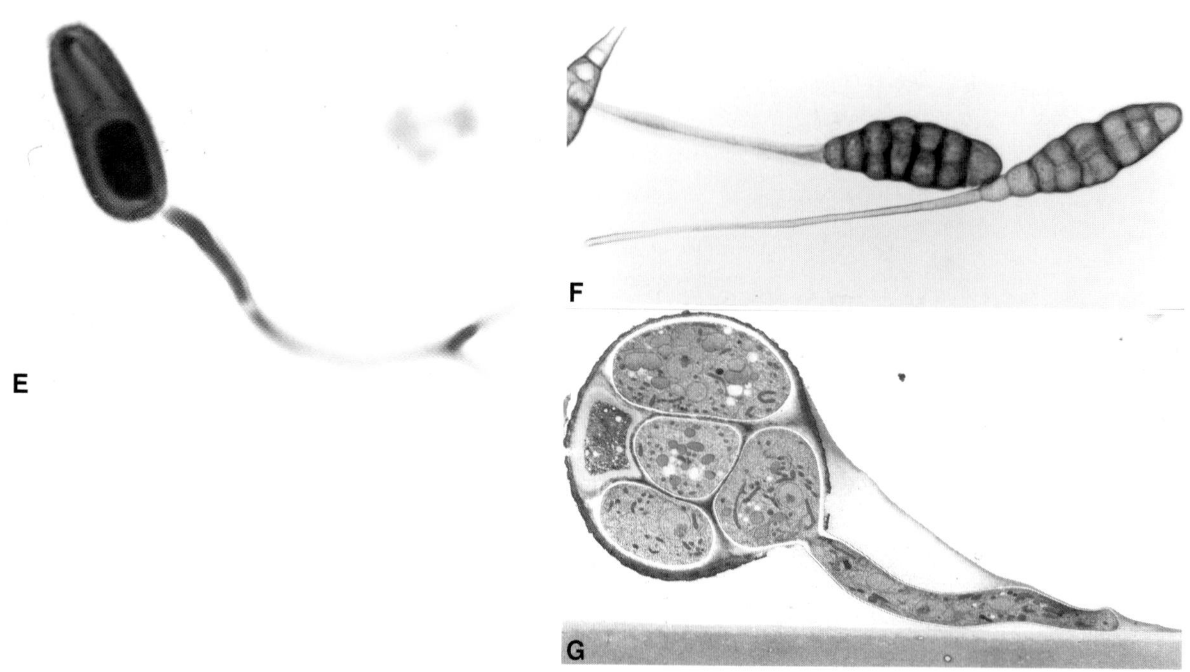

FIGURE 2-4 (*Continued*)

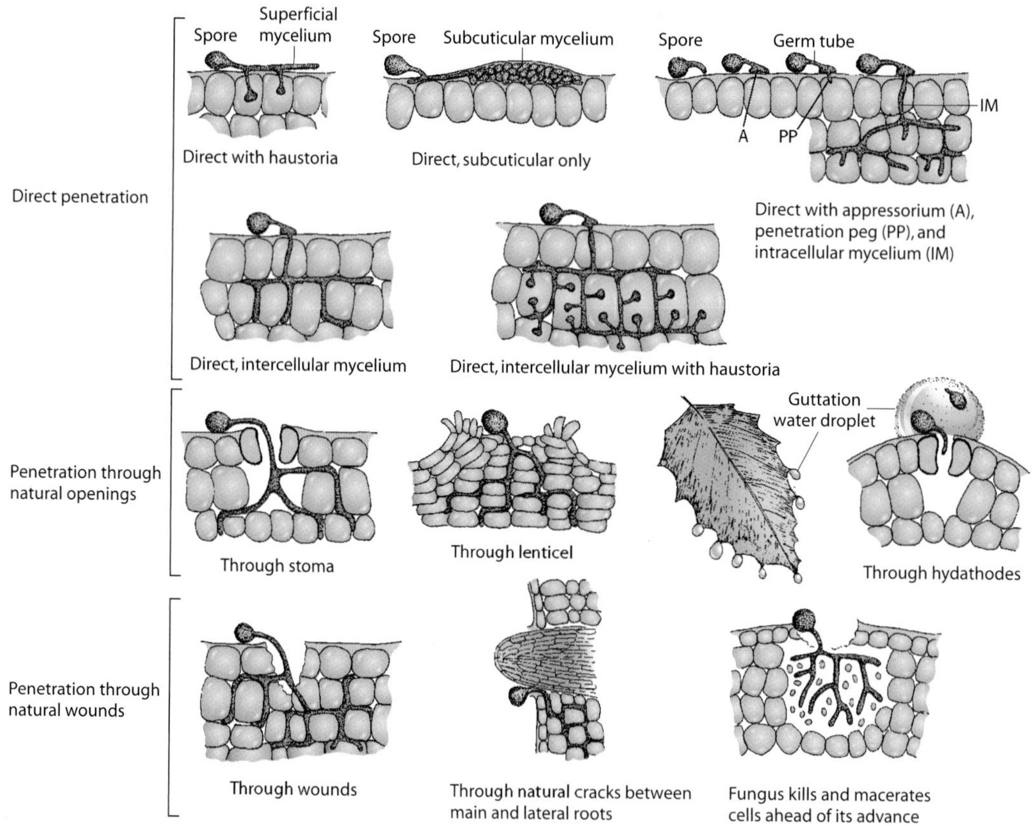

FIGURE 2-5 Methods of penetration and invasion by fungi.

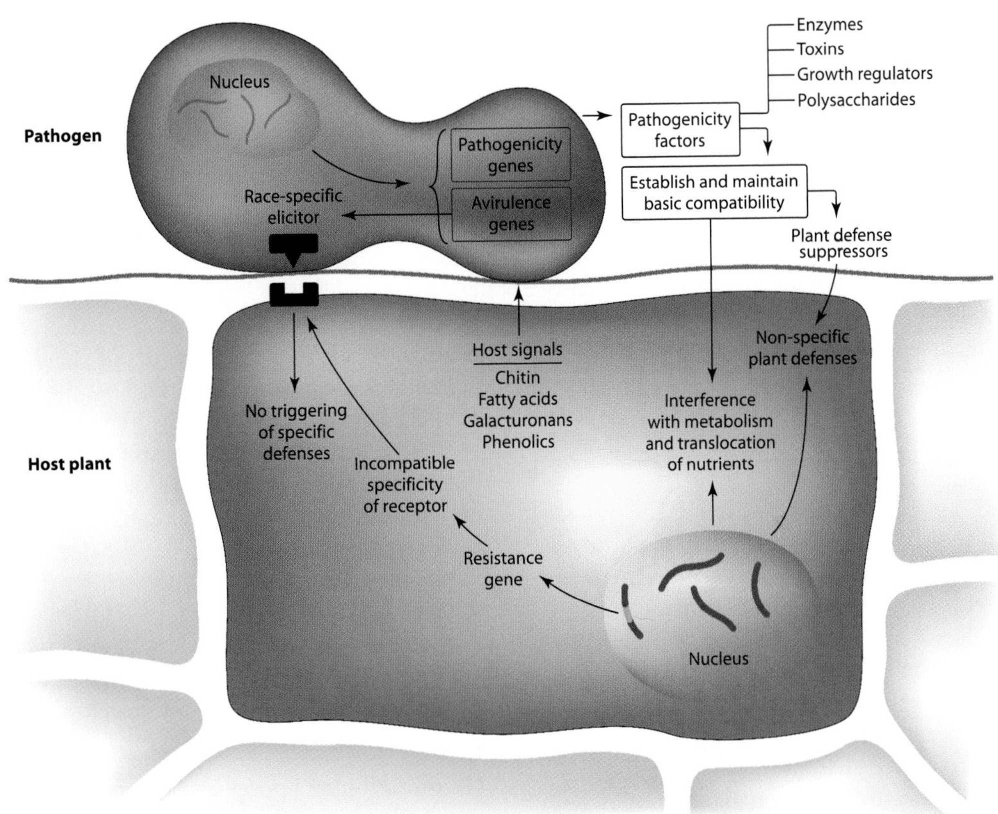

FIGURE 2-6 Establishment of infection in a compatible reaction between a pathogen and its host plant.

Signaling pathways for infection-related development are also achieved through mitogen-activated protein kinases (MAPKs) and their upstream regulatory kinases. All of these together comprise a functional unit that transmits input signals from the periphery of the cell to the cell nucleus to elicit the expression of appropriate genes. A MAP kinase, K1 or P1, regulates appressorium formation in response to a signal from the plant surface but it is also required for invasive growth or viability in its host plant.

After attachment of the propagule to the host surface, as spores and seeds germinate, germ tubes also produce mucilaginous materials that allow them to adhere to the cuticular surface of the host, either along their entire length or only at the tip of the germ tube. In regions of contact with the germ tube, the structure of the host cuticle and cell walls often appears altered, presumably as a result of degradative enzymes contained in the mucilaginous sheath.

Appressorium Formation and Maturation

Once appressoria are formed, they adhere tightly to the leaf surface (Figs. 2-4 and 2-9). Subsequently, appressoria secrete extracellular enzymes, generate physical force, or both to bring about penetration of the cuticle by the fungus. Appressoria must be attached to the host plant surface strongly enough to withstand the invasive physical force applied by the fungus and to resist the chemical action of the enzymes secreted by the fungus. Appressoria of some fungi contain lipids, polysaccharides, and proteins. Fungi that produce melanin-pigmented appressoria produce a narrow penetration hypha from the base of the appressorium and use primarily physical force to puncture the plant cuticle with that hypha.

The size of the turgor pressure inside an appressorium has been measured and found to be 40 times greater than the pressure of a typical car tire. The turgor pressure of an appressorium is due to the enormous accumulation of glycerol in the appressorium, which, due to its high osmotic pressure, draws water into the cell and generates hydrostatic pressure that pushes the thin hypha (appressorial penetration peg) outward through the host cuticle. Mobilization of spore-stored products to the developing appressorium and glycerol biosynthesis in it is regulated by the cAMP signaling pathway, whereas the initial movement of lipid and glycogen reserves to the developing appressorium was also found to be regulated by the K1 MAP. This indicates that the maturation of appressoria and their specific biochemical activity are intimately associated with genetic control of the initial development of appressoria.

The production of penetration hyphae by appressoria, or directly from germ tubes, is not well understood at the genetic level. Production of the penetration peg requires the localization of actin to the hyphal tip and rapid biosynthesis of the cell wall as the hypha grows through the cuticle and the layers of the epidermal cell walls. Production of penetration hyphae appears to be regulated by a MAP kinase pathway.

Recognition between Host and Pathogen

It is still unclear how pathogens recognize their hosts and vice versa. It is assumed that when a pathogen comes in contact with a host cell, an early event takes place that triggers a fairly rapid response in each organism that either allows or impedes further growth of the pathogen and development of disease. The nature of the "early event" is not known with certainty in any host–parasite combination, but it may be one of many biochemical substances, structures, and pathways. These may include specific host signal compounds or structures, or specific pathogen elicitor molecules, and either of them may induce specific actions or formation of specific products by the other organism (Fig. 2-6).

Host components acting as signals for recognition by and activation of pathogens are numerous. They may include fatty acids of the plant cuticle that activate production by the pathogen of the cutinase enzyme, which breaks down cutin; galacturonan molecules of host pectin, which stimulate the production of pectin lyase enzymes by the fungus or bacterium; certain phenolic compounds, such as strigol, which stimulate activation and germination of propagules of some pathogens; and isoflavones and other phenolics, amino acids, and sugars released from plant wounds that activate a series of genes in certain pathogens leading to infection. A host plant may also send cues for recognition by some of its pathogens by certain of its surface characteristics such as ridges or furrows, hardness, or release of certain ions such as calcium.

Pathogen components that act as elicitors of recognition by the host plant and subsequent mobilization of plant defenses are still poorly understood. Elicitor molecules may be released from attacking pathogens before or during entry into the host, and they may have a narrow host range, e.g., the elicitins. Some elicitors may be components of the cell surface of the pathogen (e.g., β-glucans, chitin, or chitosan) that are released by the action of host enzymes (e.g., β-glucanase and/or chitinase) and have broad host ranges; some may be synthesized and released by the pathogen after it enters the host in response to host signals. The latter elicitors include the harpin proteins of bacteria that induce development of the hypersensitive response, certain hydroxylipids, and certain peptides and carbohydrates that induce specific host defense responses such as the production of phytoalexins. Elicitors are considered as determinants of pathogen avirulence, as by their presence they elicit the hypersensitive (resistance) response and initiation of transcription of the plant genes that encode the various components of the defense response. These defense measures by the host plant, in turn, result in the pathogen appearing as avirulent.

When the initial recognition signal received by the pathogen favors growth and development, disease may be induced; if the signal suppresses pathogen growth and activity, disease may be aborted. However, if the initial recognition elicitor received by the host triggers a defense reaction, pathogen growth and activity may be slowed or stopped and disease may not develop; if the elicitor either suppresses or bypasses the defense reaction of the host, disease may develop.

Germination of Spores and Seeds

Almost all pathogens in their vegetative state are capable of initiating infection immediately. Fungal spores and seeds of parasitic higher plants, however, must first germinate (Figs. 2-4 and 2-5). Spores germinate by producing a typical mycelium (Figs. 2-4E and 2-4G) that infects and grows into host plants or they produce a short germ tube that produces a specialized infectious structure, the haustorium (Figs. 2-4B–2-4D). In order to germinate, spores require a favorable temperature and also moisture in the form of rain, dew, or a film of water on the plant surface or at least high relative humidity. The moist conditions must last long enough for the pathogen to penetrate or else it desiccates and dies. Most spores can germinate immediately after their maturation and release, but others (so-called resting spores) require a dormancy period of varying duration before they can germinate. When a spore germinates it produces a germ tube, i.e., the first part of the mycelium, that can penetrate the host plant. Some fungal spores germinate by producing other spores, e.g., sporangia produce zoospores and teliospores produce basidiospores.

Spore germination is often favored by nutrients diffusing from the plant surface; the more nutrients (sugars and amino acids) exuded from the plant, the more spores germinate and the faster they germinate. In some cases, spore germination of a certain pathogen is stimulated only by exudates of plants susceptible to that particular pathogen. In other cases, spore germination may be inhibited to a lesser or greater extent by materials released into the surrounding water by the plant, by substances contained within the spores themselves, especially when the spores are highly concentrated ("quorum sensing"), and by saprophytic microflora present on or near the plant surface.

Fungi in soil coexist with a variety of antagonistic microorganisms that cause an environment of starvation and of toxic metabolites. As a result, spores of many soilborne fungi are often unable to germinate in some soils, and this phenomenon is called **fungistasis**, or their germ tubes lyse rapidly. Soils in which such events occur are known as **suppressive soils**. Fungistasis, however, is generally counteracted by root exudates of host plants growing nearby, and the spores are then able to germinate and infect.

After spores germinate, the resulting germ tube must grow, or the motile secondary spore (zoospore) must move, toward a site on the plant surface at which successful penetration can take place (Figs. 2-3A and 2-3B). The number, length, and rate of growth of germ tubes, or the number and mobility of motile spores, may be affected by physical conditions, such as temperature and moisture, by the kind and amount of exudates the plant produces at its surface, and by the saprophytic microflora.

The growth of germ tubes in the direction of successful penetration sites seems to be regulated by several factors, including greater humidity or chemical stimuli associated with such openings as wounds, stomata, and lenticels; thigmotropic (contact) responses to the topography of the leaf surface, resulting in germ tubes growing at right angles to cuticular ridges that generally surround stomata and thus eventually reaching a stoma; and nutritional responses of germ tubes toward greater concentrations of sugars and amino acids present along roots. The direction of movement of motile spores (zoospores) is also regulated by similar factors, namely chemical stimuli emanating from stomata, wounds, or the zone of elongation of roots, physical stimuli related to the structure of open stomata, and the nutrient gradient present in wound and root exudates.

Seeds germinate by producing a radicle, which either penetrates the host plant directly or first produces a small plant that subsequently penetrates the host plant by means of specialized feeding organs called haustoria. Most conditions described earlier as affecting spore germination and the direction of growth of germ tubes also apply to seeds. Haustoria are also produced by many fungi.

Hatching of Nematode Eggs

Nematode eggs also require conditions of favorable temperature and moisture to become activated and hatch. In most nematodes, the egg contains the first juvenile stage before or soon after the egg is laid. This juvenile immediately undergoes a molt and gives rise to the second juvenile stage, which may remain dormant in the egg for various periods of time. Thus, when the egg finally hatches, it is the second-stage juvenile that emerges, and it either finds and penetrates a host plant or undergoes additional molts that produce further juvenile stages and adults.

Once nematodes are in close proximity to plant roots, they are attracted to roots by certain chemical factors associated with root growth, particularly carbon dioxide and some amino acids. These factors may diffuse through soil and may have an attractant effect on nematodes present several centimeters away from the root. Nematodes are generally attracted to roots of both host and nonhost plants, although there may be some cases in which nematodes are attracted more strongly to the roots of host plants.

Penetration

Pathogens penetrate plant surfaces by direct penetration of cell walls, through natural openings, or through wounds (Figs. 2-3–2-5). Some fungi penetrate tissues in only one of these ways, others in more than one. Bacteria enter plants mostly through wounds, less frequently through natural openings, and never directly through unbroken cell walls (Fig. 2-5). Viruses, viroids, mollicutes, fastidious bacteria, and protozoa enter through wounds made by vectors, although some viruses and viroids may also enter through wounds made by tools and other means. Parasitic higher plants enter their hosts by direct penetration. Nematodes enter plants by direct penetration and, sometimes, through natural openings (Fig. 2-10).

Penetration does not always lead to infection. Many organisms actually penetrate cells of plants that are not susceptible to these organisms and that do not become diseased; these organisms cannot proceed beyond the stage of penetration and die without producing disease.

Direct Penetration through Intact Plant Surfaces

Direct penetration through intact plant surfaces is probably the most common type of penetration by fungi, oomycetes, and nematodes and the only type of penetration by parasitic higher plants. None of the other pathogens can enter plants by direct penetration.

Of the fungi that penetrate their host plants directly, the hemibiotrophic, i.e., nonobligate parasitic ones, do so through a fine hypha produced directly by the spore or mycelium (Figs. 2-3B, 2-5, and 2-8), whereas the obligately parasitic ones do so through a penetration peg produced by an **appressorium** (Figs. 2-4B–2-4D and 2-9). The fine hypha or appressorium is formed at the point of contact of the germ tube or mycelium with a plant surface. The fine hypha grows toward the plant surface and pierces the cuticle and the cell wall through mechanical force and enzymatic softening of the cell wall substances. Most fungi, however, form an appressorium at the end of the germ tube, with the appressorium usually being bulbous or cylindrical with a flat surface in contact with the surface of the host plant (Figs. 2-4, 2-9Ab, and 2-9B). Then, a **penetration peg** grows from the flat surface of the appressorium toward the host and pierces the cuticle and the cell wall. The penetration peg grows into a fine hypha generally much smaller in diameter than a normal hypha of the fungus, but it regains its normal diameter once inside the cell. In most fungal diseases the fungus penetrates the plant cuticle and the cell wall, but in some, such as apple scab (Fig. 2-11A), the fungus penetrates only the cuticle and stays between the cuticle and the cell wall.

Parasitic higher plants also form an appressorium and penetration peg at the point of contact of the radicle with the host plant, and penetration is similar to that in fungi. Direct penetration in nematodes is accomplished by repeated back-and-forth thrusts of their stylets. Such thrusts finally create a small opening in the cell wall; the nematode then inserts its stylet into the cell or the entire nematode enters the cell (Fig. 2-12).

Penetration through Wounds

All bacteria, most fungi, some viruses, and all viroids can enter plants through various types of wounds (Fig. 2-5). Some viruses and all mollicutes, fastidious vascular bacteria, and protozoa enter plants through wounds made by their vectors. The wounds utilized by bacteria and fungi may be fresh or old and may consist of lacerated or killed tissue. These pathogens may grow briefly on such tissue before they advance into healthy tissue. Laceration or death of tissues may be the result of environmental factors such as wind breakage and hail; animal feeding, e.g., by insects and large animals; cultural practices of humans, such as pruning, transplanting, and harvesting; self-inflicted injuries, such as leaf scars; and, finally, wounds or lesions caused by other pathogens. Bacteria and fungi penetrating through wounds germinate or multiply in the wound sap or in a film of rain or dew water present on the wound. Subsequently, the pathogen invades adjacent plant cells or it secretes enzymes and toxins that kill and macerate the nearby cells.

The penetration of viruses, mollicutes, fastidious bacteria, and protozoa through wounds depends on the deposition of these pathogens by their vectors in fresh wounds created at the time of inoculation. All four types of pathogens are transmitted by certain types of insects. Some viruses are also transmitted by certain nematodes, mites, and fungi. Some viruses and viroids are transmitted through wounds made by human hands and tools. In most cases, however, these pathogens are carried by one or a few kinds of specific vectors and can be inoculated successfully only when they are brought to the plant by these particular vectors.

Penetration through Natural Openings

Many fungi and bacteria enter plants through stomata, and some enter through hydathodes, nectarthodes, and lenticels (Figs. 2-3, 2-4, 2-5, and 2-7). Stomata are most numerous on the lower side of leaves. They measure about 10–20 by 5–8 μm and are open in the daytime but are more or less closed at night. Bacteria present in a film of water over a stoma and, if water soaking occurs, can swim through the stoma easily (Fig. 2-3D) and into the substomatal cavity where they can multiply and start infection. Fungal spores generally germinate on the plant surface, and the germ tube may then grow through the stoma (Figs. 2-3A, 2-4B, and 2-5). Frequently, however, the germ tube forms an appressorium that fits tightly over the stoma, and usually one fine hypha grows from it into the stoma (Figs. 2-4 and 2-5). In the substomatal cavity the hypha enlarges, and from it grow one or several small hyphae that actually invade the cells of the host plant directly or by means of haustoria (Fig. 2-5). Although some fungi can apparently penetrate

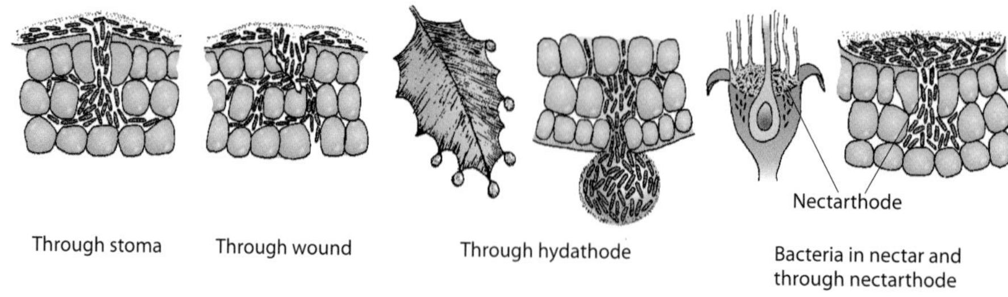

FIGURE 2-7 Methods of penetration and invasion by bacteria.

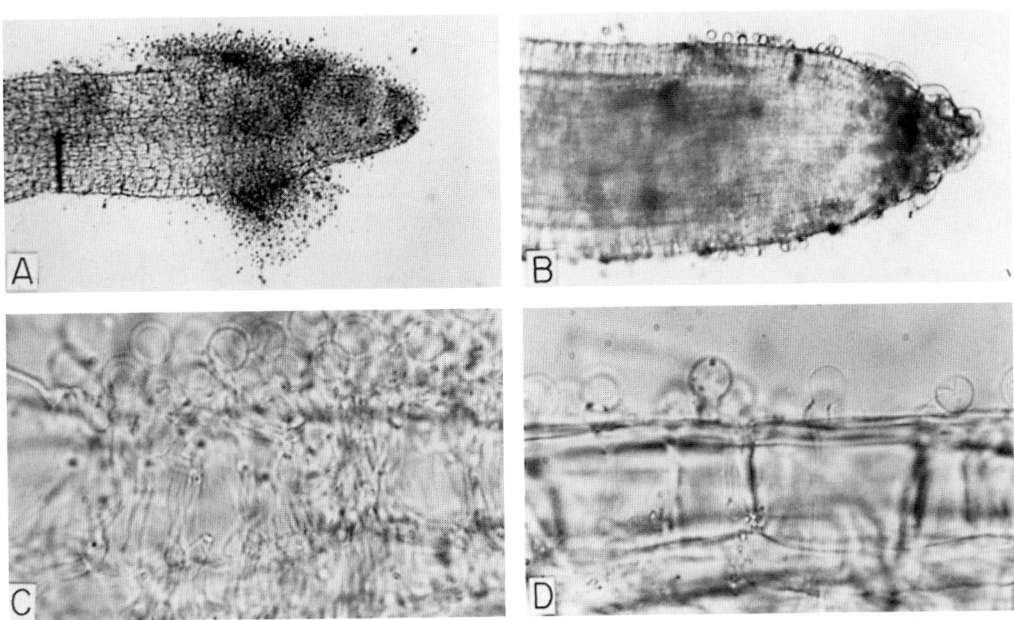

FIGURE 2-8 Attraction of zoospores of *Phytophthora cinnamomi* to roots of susceptible (A and C) and resistant (B and D) blueberry varieties, and infection of the roots by the zoospores. (A and B) Attraction of zoospores to roots 1 hour after inoculation. (C and D) Infection and colonization of the root after 24 hours are greater in the susceptible highbush blueberry (A and C) than in the more resistant rabbit-eye blueberry (B and D). (Photographs courtesy of R. D. Milholland.)

even closed stomata, others penetrate stomata only while they are open. Certain fungi, e.g., the powdery mildew fungi, may grow over open stomata without entering them.

Hydathodes are more or less permanently open pores at the margins and tips of leaves; they are connected to the veins and secrete droplets of liquid, called guttation drops, containing various nutrients (Fig. 2-5). Some bacteria use these pores as a means of entry into leaves, but few fungi seem to enter plants through hydathodes. Some bacteria also enter blossoms through the nectarthodes or nectaries, which are similar to hydathodes (Fig. 2-7).

Lenticels are openings on fruits, stems, and tubers that are filled with loosely connected cells that allow the passage of air. During the growing season, lenticels are open, but even so, relatively few fungi and bacteria penetrate tissues through them, growing and advancing mostly between the cells (Fig. 2-5). Most pathogens that penetrate through lenticels can also enter through wounds, with lenticel penetration being apparently a less efficient, secondary pathway.

Infection

Infection is the process by which pathogens establish contact with susceptible cells or tissues of the host and procure nutrients from them. Following infection, pathogens grow, multiply, or both within the plant tissues and invade and colonize the plant to a lesser or greater extent. Growth and/or reproduction of the pathogen (colonization) in or on infected tissues are actually two concurrent substages of disease development (Fig. 2-2).

Successful infections result in the appearance of symptoms, i.e., discolored, malformed, or necrotic areas on the host plant. Some infections, however, remain latent, i.e., they do not produce symptoms right away but at a later time when the environmental conditions or the stage of maturity of the plant become more favorable.

All the visible and otherwise detectable changes in the infected plants make up the **symptoms** of the disease. Symptoms may change continuously from the moment of their appearance until the entire plant dies or they may develop up to a point and then remain more or less unchanged for the rest of the growing season. Symptoms may appear as soon as 2 to 4 days after inoculation, as happens in some localized viral diseases of herbaceous plants, or as late as 2 to 3 years after inoculation, as in the case of some viral, mollicute, and other diseases of trees. In most plant diseases, however, symptoms appear from a few days to a few weeks after inoculation.

The time interval between inoculation and the appearance of disease symptoms is called the **incubation period.** The length of the incubation period of various diseases varies with the particular pathogen–host

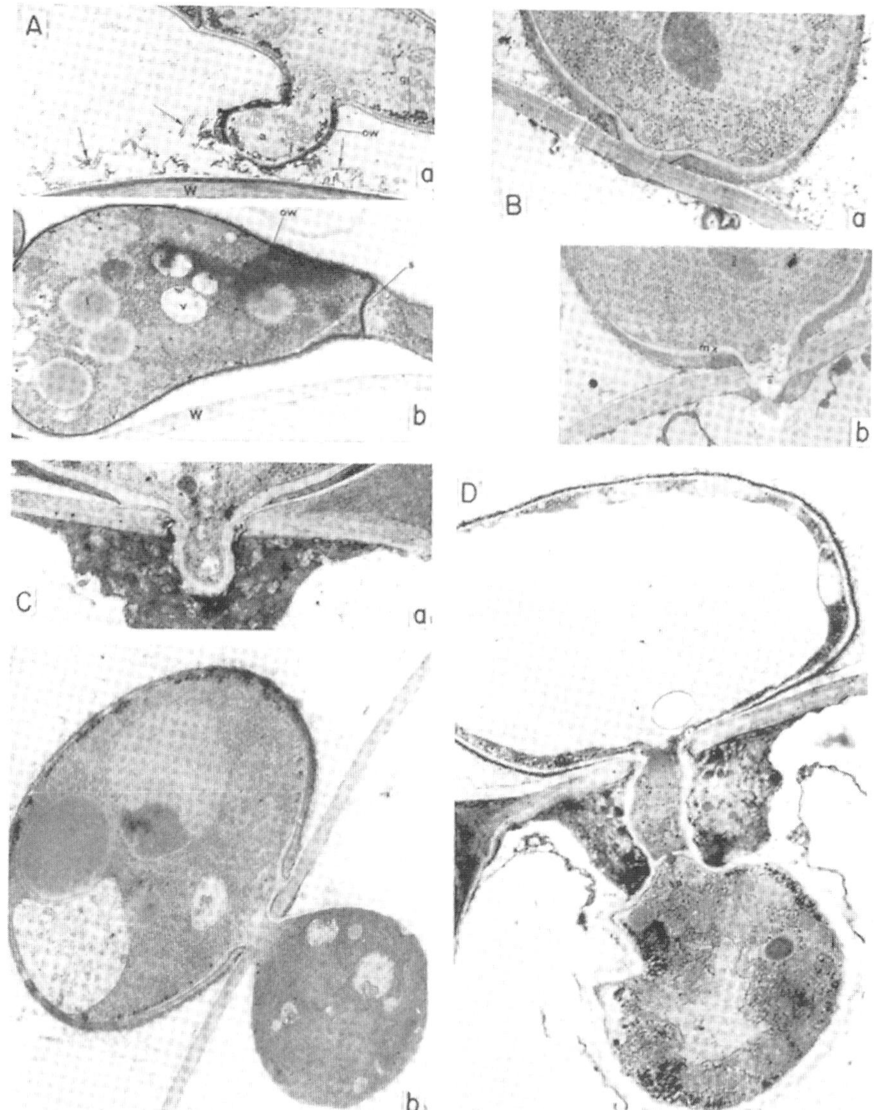

FIGURE 2-9 Electron micrographs of direct penetration of a fungus (*Colletotrichum graminicola*) into an epidermal leaf cell. (A) (a) Developing appressorium from a conidium. Note wax rods (arrows) on leaf surface. (b) Mature appressorium separated by a septum from the germination tube. (B) (a) Formation of penetration peg at the central point of contact of appressorium with the cell wall. (b) Structures in the penetration peg, which has already penetrated the cell wall, and papilla produced by the invaded cell. (C) Development of infection hypha. (a) Infection peg penetrating the papilla. (b) Appressorium and swollen infection hypha after penetration. (D) On completion of penetration and establishment of infection, the appressorium consists mostly of a large vacuole and is cut off from the infection hypha by a septum. (Photographs courtesy of D. J. Politis and H. Wheeler.)

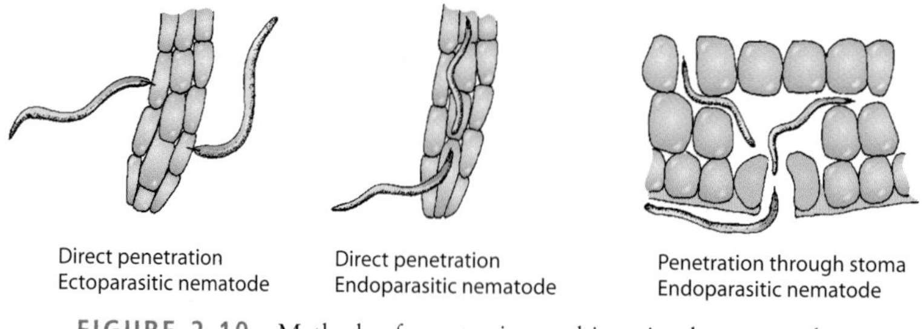

FIGURE 2-10 Methods of penetration and invasion by nematodes.

combination, with the stage of development of the host, and with the temperature in the environment of the infected plant.

During infection, some pathogens obtain nutrients from living cells, often without killing the cells or at least not for a long time; others kill cells and utilize their contents as they invade them; and still others kill cells and disorganize surrounding tissues. During infection, pathogens release a number of biologically active substances (e.g., enzymes, toxins, and growth regulators) that may affect the structural integrity of the host cells or their physiological processes. In response, the host reacts with a variety of defense mechanisms, which result in varying degrees of protection of the plant from the pathogen.

As mentioned earlier, for a successful infection to occur it is not sufficient that a pathogen comes in contact with its host; rather, several other conditions must also be satisfied. First of all, the plant variety must be susceptible to the particular pathogen and at a susceptible stage. The pathogen must be in a pathogenic stage that can infect immediately without requiring a resting (dormancy) period first, or infective juvenile stages or adults of nematodes. Finally, the temperature and moisture conditions in the environment of the plant must favor the growth and multiplication of the pathogen. When these conditions occur at an optimum, the pathogen can invade the host plant up to the maximum of its potential, even in the presence of plant defenses, and, as a consequence, disease develops.

Invasion

Various pathogens invade hosts in different ways and to different extents (Figs. 2-4, 2-5, 2-9, and 2-12). Some fungi, such as those causing apple scab and black spot of rose, produce mycelium that grows only in the area between the cuticle and the epidermis (subcuticular colonization) (Fig. 2-11A); others, such as those causing powdery mildews, produce mycelium only on the surface of the plant (Fig. 2-11B) but send haustoria into the epidermal cells. Most fungi spread into all the tissues of the plant organs (leaves, stems, and roots) they infect, either by growing directly through the cells as an **intracellular mycelium** or by growing between the cells as an **intercellular mycelium** (Figs. 2-11C and 2-11D). Fungi that cause vascular wilts invade the xylem vessels of plants (Fig. 2-11E).

Bacteria invade tissues intercellulary, although when parts of the cell walls dissolve, bacteria also grow intracellularly. Bacteria causing vascular wilts, like the vascular wilt fungi, invade the xylem vessels (Fig. 2-11E). Most nematodes invade tissues intercellularly, but some can invade intracellularly as well (Fig. 2-12). Many nematodes do not invade cells or tissues at all but feed by piercing epidermal cells with their stylets.

Viruses, viroids, mollicutes, fastidious bacteria, and protozoa invade tissues by moving from cell to cell intracellularly. Viruses and viroids invade all types of living plant cells, mollicutes and protozoa invade phloem sieve tubes and perhaps a few adjacent phloem parenchyma cells, and most fastidious bacteria invade xylem vessels and a few invade only phloem sieve tubes.

Many infections caused by fungi, bacteria, nematodes, viruses, and parasitic higher plants are local, i.e., they involve a single cell, a few cells, or a small area of the plant. These infections may remain localized throughout the growing season or they may enlarge slightly or very slowly. Other infections enlarge more or less rapidly and may involve an entire plant organ (flower, fruit, leaf), a large part of the plant (a branch), or the entire plant.

Infections caused by fastidious xylem- or phloem-inhabiting bacteria, mollicutes, and protozoa and natural infections caused by viruses and viroids are **systemic**, i.e., the pathogen, from one initial point in a plant, spreads and invades most or all susceptible cells and tissues throughout the plant. Vascular wilt fungi and bacteria invade xylem vessels internally, but they are usually confined to a few vessels in the roots, the stem, or the top of infected plants; only in the final stages of the disease do they invade most or all xylem vessels of the plant. Some downy mildew pathogens and some fungi, primarily among those causing smuts and rusts, also invade their hosts systemically, although in most cases the older mycelium degenerates and disappears and only the younger mycelium survives in actively growing plant tissues.

Growth and Reproduction of the Pathogen (Colonization)

Individual fungi and parasitic higher plants generally invade and infect tissues by growing on or into them from one initial point of inoculation. Most of these pathogens, whether inducing a small lesion, a large infected area, or a general necrosis of the plant, continue to grow and branch out within the infected host indefinitely so that the same pathogen individual spreads into more and more plant tissues until the spread of the infection is stopped or the plant is dead. In some fungal infections, however, while younger hyphae continue to grow into new healthy tissues, older ones in the already infected areas die out and disappear so that a diseased plant may have several points where separate units of the mycelium are active. Also, fungi causing vascular wilts often invade plants by producing and releasing spores within the vessels, and as the spores are carried

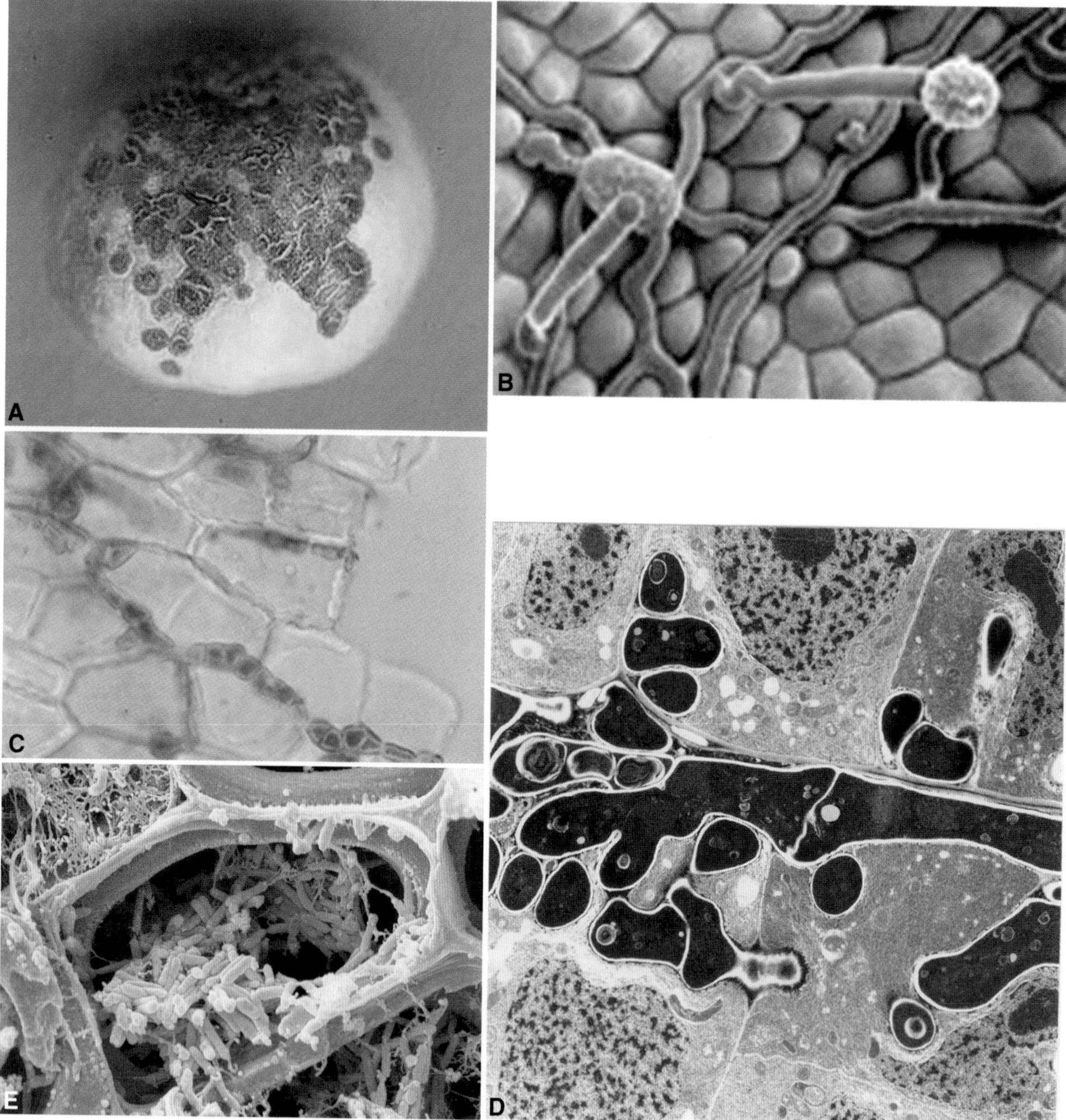

FIGURE 2-11 Types of invasion of pathogens in infected plants. (A) In apple scab disease, the pathogenic fungus grows only between the cuticle and the epidermal cells of leaves and fruit. (B) In powdery mildews the fungal mycelium grows only on the surface of host plants, but sends haustoria into the epidermal cells. (C) In many diseases the fungal mycelium (stained red here) grows only intercellularly (between the cells). (D) Hyphae of the smut fungus *Ustilago* in an infected leaf. (E) In bacterial vascular diseases, bacteria grow in and may clog the xylem vessels. [Photographs courtesy of (A) University of Oregon, (B) G. Celio, APS, (D) Mims *et al.* (1992). *Intern. J. Plant Sci.* **153**, 289–300, and (E) E. Alves, Federal University of Lavras, Brazil.]

in the sap stream they invade vessels far away from the mycelium, germinate there, and produce a mycelium, which invades more vessels.

All other pathogens, namely bacteria, mollicutes, viruses, viroids, nematodes, and protozoa, do not increase much, if at all, in size with time, as their size and shape remain relatively unchanged throughout their existence. These pathogens invade and infect new tissues within the plant by reproducing at a rapid rate and increasing their numbers tremendously in the infected tissues. The progeny may then be carried passively into new cells and tissues through plasmodesmata (viruses and viroids only), phloem (viruses, viroids, mollicutes, some fastidious bacteria, protozoa), or xylem (some bacteria); alternatively, as happens with protozoa and nematodes (Fig. 2-12) and somewhat with bacteria, they may move through cells on their own power.

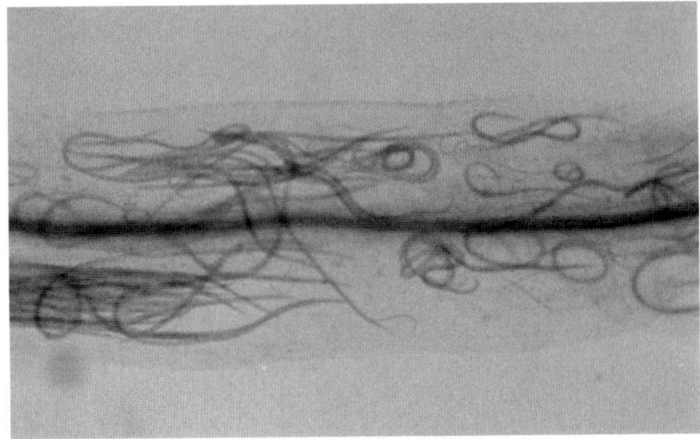

FIGURE 2-12 Alfalfa shoot invaded by plant parasitic nematodes (*Ditylenchus dipsaci*). (Photograph courtesy of J. Santo.)

Plant pathogens reproduce in a variety of ways (see Fig. 1-3 in Chapter 1). Fungi reproduce by means of spores, which may be either asexual (**mitospores**, i.e., products of mitosis, roughly equivalent to the buds on a twig or the tubers of a potato plant), or sexual (**meiospores**, i.e. products of meiosis, roughly equivalent to the seeds of plants). Parasitic higher plants reproduce just like all plants, i.e., by seeds. Bacteria and mollicutes reproduce by fission in which one mature individual splits into two equal, smaller individuals. Viruses and viroids are replicated by the cell, just as a page placed on a photocopying machine is replicated by the machine as long as the machine is operating and paper supplies last. Nematodes reproduce by means of eggs.

The great majority of plant pathogenic fungi and oomycetes produce a mycelium only within the plants they infect. Relatively few fungi and oomycetes produce a mycelium on the surface of their host plants, but most powdery mildew fungi produce a mycelium only on the surface of, and none within, their hosts (Figs. 2-13A–2-13C). The great majority of fungi and oomycetes

FIGURE 2-13 Means of reproduction of fungi and bacteria. (A–E) Mycelium [white material on leaf (A, B)], chains of conidia (C), and cleistothecium (B and D) (containing four asci, each containing ascospores) on the leaf surface. (E) Apple trees having numerous branches killed by the fire blight bacterium. (F) Large numbers of bacteria inside a xylem vessel of a bacterial wilt-infected plant. [Photographs courtesy of (A and B) D. Legard, University of Florida, (C) D. Mathre, Montana State University, (D) M. Hoffman, Oregon State University, (E) A. Jones, Michigan State University, and (F) B. Bruton, USDA, Lane, Oklahoma.]

(continued on next page)

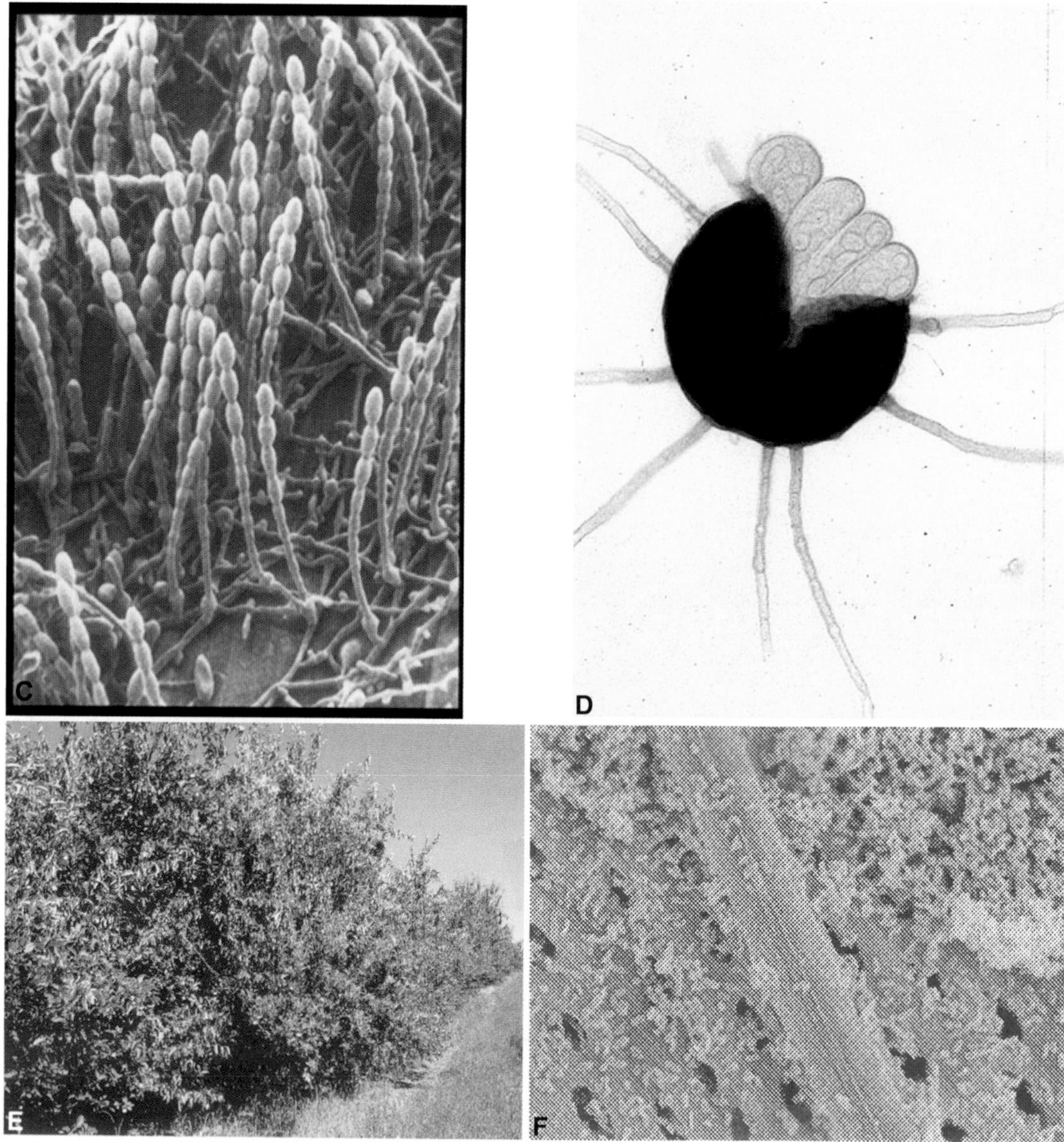

FIGURE 2-13 *(Continued)*

produce spores on, or just below, the surface of the infected area of the host, and the spores are released outward into the environment. Plant pathogenic plasmodiophoromycetes, however, such as the clubroot pathogen and fungi causing vascular wilts, produce spores within the host tissues, and these spores are not released outward until the host dies and disintegrates. Parasitic higher plants produce their seeds on aerial branches, and some nematodes lay their eggs at or near the surface of the host plant. Bacteria reproduce between or, in xylem- or phloem-inhabiting bacteria, within host cells (Fig. 2-13F), generally inside the host plant; they come to the host surface only through wounds, cracks, stomata, and so on. Viruses, viroids, mollicutes, protozoa, and fastidious bacteria reproduce only inside cells and apparently do not reach or exist on the surface of the host plant.

The rate of reproduction varies considerably among the various kinds of pathogens, but in all types, one or a few pathogens can produce tremendous numbers of individuals within one growing season. Some fungi produce spores more or less continuously (Fig. 2-14), whereas others produce them in successive crops. In either case, several thousand to several hundreds of

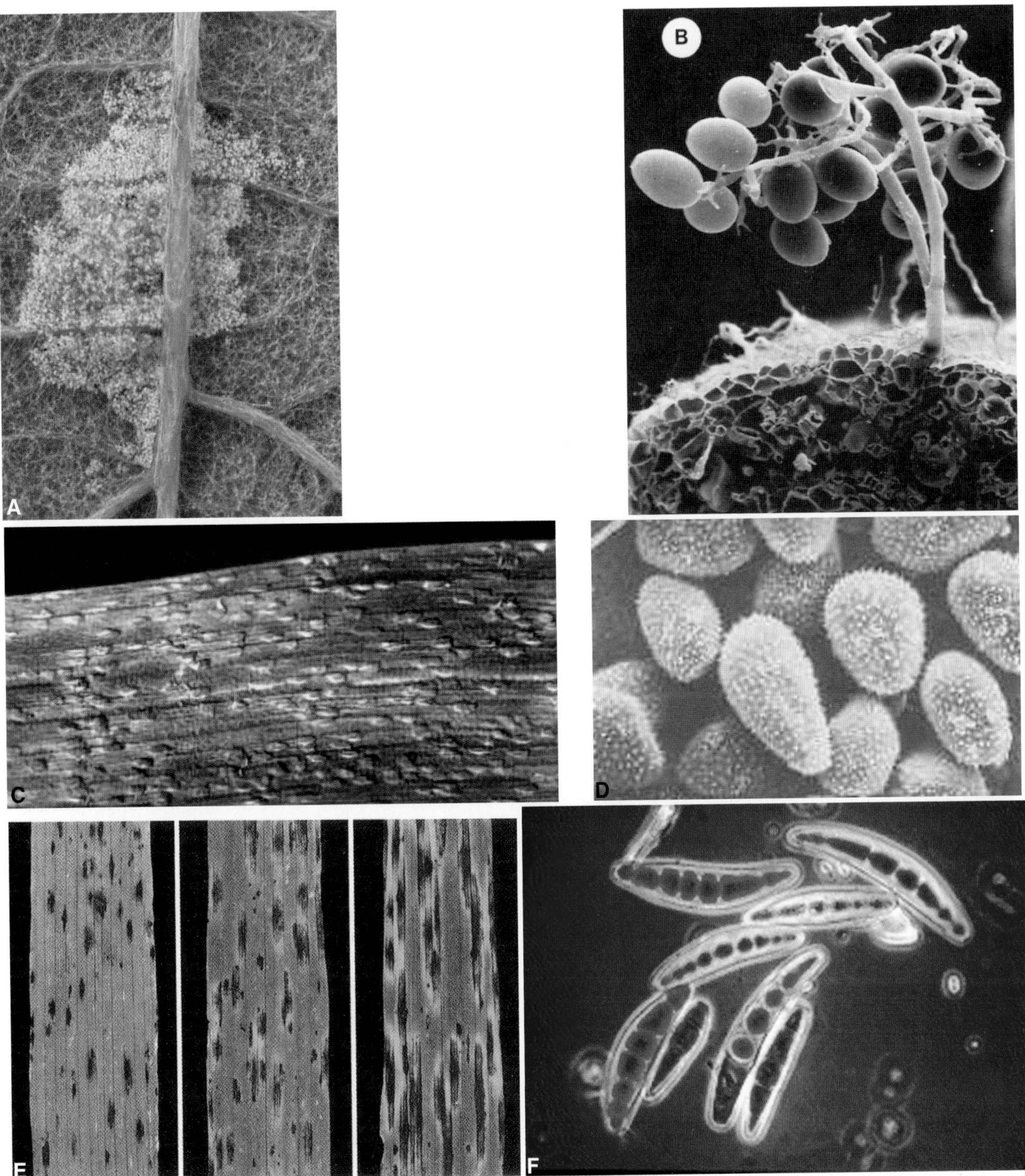

FIGURE 2-14 Invasion and reproduction of oomycete and fungal plant pathogens. Sporangiophores and sporangia (A) on the underside of a grape leaf infected with the grape downy mildew pathogen *Plasmopara viticola* and (B) on the root of a lettuce plant infected with *Plasmopara lactucae-radicis*. (C) A wheat leaf showing numerous infection lesions (uredia) of the leaf rust fungus. (D) Uredospores of the soybean rust. (E) Leaves of three barley varieties showing infection lesions, the severity (number and size) of which are inversely proportional to the degree of resistance of each variety to the fungal pathogen. (F) Spores of the fungus *Cochliobolus* that cause leaf spot on barley. [Photographs courtesy of (A) J. Rytter and J. W. Travis, Pennsylvania State University, (B) M. E. Stanghellini, University of California, Riverside, and (E) B. Steffenson, University of Minnesota.]

thousands of spores may be produced per square centimeter of infected tissue. Even small specialized sporophores can produce millions of spores, and the number of spores produced per diseased plant is often in the billions or trillions (Fig. 2-14). The number of spores produced in an acre of heavily infected plants, therefore, is generally astronomical, and enough spores are released to land on every conceivable surface in the field and the surrounding areas, enough to easily inoculate with a heavy inoculum every plant in the area.

Bacteria reproduce rapidly within infected tissues (Fig. 2-13F). Under optimum nutritional and environmental conditions (in culture), bacteria divide (double their numbers) every 20 to 30 minutes, and, presumably, bacteria multiply just as fast in a susceptible plant as long as nutrients and space are available and the temperature is favorable. Millions of bacteria may be present in a single drop of infected plant sap so the number of bacteria per plant must be astronomical. Fastidious bacteria and mollicutes appear to reproduce more slowly than typical bacteria; although they spread systemically throughout the vascular system of the plant, they are present in relatively few xylem or phloem vessels, and the total number of these pathogens in infected plants is relatively small. This also seems to be true for protozoa.

Viruses and viroids reproduce within living host cells, with the first new virus particles being detectable several hours after infection. Soon after that, however, virus particles accumulate within the infected living cell until as many as 100,000 to 10,000,000 particles may be present in a single cell. Viruses and viroids infect and multiply in most or all living cells of their hosts, and it is apparent that each plant may contain innumerable individuals of these pathogens.

Nematode females lay about 300 to 500 eggs, about half of which produce females that again lay 300 to 600 eggs each. Depending on the climate, the availability of hosts, and the duration of each life cycle of the particular nematode, a nematode species may have from two to more than a dozen generations per year. If even just half of the females survived and reproduced, each generation time would increase the number of nematodes in the soil by more than a hundred fold. Thus, the buildup of nematode populations within a growing season and in successive seasons is often quite dramatic.

Dissemination of the Pathogen

A few pathogens, such as nematodes, oomycetes, zoosporic fungi, and bacteria, can move short distances on their own power and thus can move from one host to another one very close to it. Fungal hyphae can grow between tissues in contact and sometimes through the soil toward nearby roots for a few to many centimeters. Both of these means of dissemination, however, are quite limited, especially in the case of zoospores and bacteria.

The spores of some fungi are expelled forcibly from the sporophore or sporocarp by a squirting or puffing action that results in the successive or simultaneous discharge of spores up to a centimeter or so above the sporophore. The seeds of some parasitic plants are also expelled forcibly and may arch over distances of several meters.

Almost all dissemination of pathogens responsible for plant disease outbreaks, and even for disease occurrences of minor economic importance, is carried out passively by such agents as air and insects (Figs. 2-13–2-15). To a lesser extent, water, certain other animals, and humans may be involved (Fig. 2-15).

Dissemination by Air

Spores of most oomycetes and most fungi and the seeds of most parasitic plants are disseminated by air currents that carry them as inert particles to various distances. Air currents pick up spores and seeds off the sporophores (Figs. 2-13A–2-13E, 2-14, and 2-16) or while they are being expelled forcibly or are falling at maturity. Depending on the air turbulence and velocity, air currents may carry the spores upward or horizontally in a way similar to that of particles contained in smoke. While airborne, some of the spores may touch wet surfaces and get trapped; when air movement stops or when it rains, the rest of the spores land or are "washed out" from the air and are brought down by the raindrops. Most of the spores, of course, land on anything but a susceptible host plant. Also, the spores of many fungi are actually too delicate to survive a long trip through the air and are therefore successfully disseminated through the air for only a few hundred or a few thousand meters. The spores of other fungi, however, particularly those of the cereal rusts, are very hardy and occur commonly at all levels and at high altitudes (several thousand meters) above infected fields. Spores of these fungi are often carried over distances of several kilometers, even hundreds of kilometers, and in favorable weather may cause widespread epidemics. Some fungi can spread into new areas quite rapidly and may cause severe epidemics over large areas, including entire continents, within a few years. This happened, for example, in the airborne pathogens of sugar cane smut in the Americas (Fig. 2-18) and of barley stripe rust in South America (Fig. 2-15).

Air dissemination of other pathogens occurs rather infrequently and only under special conditions, or indi-

FIGURE 2-15 Means of dissemination of fungi and bacteria.

rectly. For example, bacteria causing fire blight of apple and pear produce fine strands of dried bacterial exudate containing bacteria, and these strands may be broken off and disseminated by wind. Bacteria and nematodes present in the soil may be blown away along with plant debris or soil particles in the dust. Wind also helps in the dissemination of bacteria, fungal spores, and nematodes by blowing away rain splash droplets containing these pathogens, and wind carries away insects that may contain or are smeared with viruses, bacteria, mollicutes, protozoa, or fungal spores. Finally, wind causes adjacent plants or plant parts to rub against one another, which may help the spread by contact of bacteria, fungi, some viruses and viroids, and possibly some nematodes.

Dissemination by Water

Water is important in disseminating pathogens in three ways. (1) Bacteria, nematodes, and spores and mycelial fragments of fungi present in the soil are disseminated by rain or irrigation water that moves on the surface or through the soil. (2) All bacteria and the spores of many fungi are exuded in a sticky liquid (Figs. 2-16A, 2-16B, and 2-16D) and depend on rain or (overhead) irrigation water, which either washes them downward or splashes them in all directions, for their dissemination (3) Raindrops or drops from overhead irrigation pick up the fungal spores and any bacteria present in the air and wash them downward, where some of them may land on susceptible plants. Although water is less important than air in the long-distance transport of pathogens, the water dissemination of pathogens is more efficient for nearby infections, as the pathogens land on an already wet surface and can move or germinate immediately.

Dissemination by Insects, Mites, Nematodes, and Other Vectors

Insects, particularly aphids, leafhoppers, and whiteflies, are by far the most important vectors of viruses, whereas leafhoppers are the main vectors of mollicutes, fastidious bacteria, and protozoa. Each one of these pathogens is transmitted, internally, by only one or a few species of insects during feeding and movement of the insect vectors from plant to plant. Specific insects also transmit certain fungal, bacterial, and nematode pathogens, such as the fungus causing Dutch elm disease, the bacterial wilt of cucurbits, and the pine wilt nematode. In all diseases in which the pathogen is carried internally or externally by one or a few specific vectors, dissemination of the pathogen depends, to a large extent or entirely, on that vector. In many diseases, however, such as bacterial soft rots, fungal fruit rots, anthracnoses, and ergot, insects become smeared with various kinds of bacteria or sticky fungal spores as they move among plants. The insects carry these pathogens externally from plant to plant and deposit them on the plant surface or in the wounds they make on the plants during feeding. In such diseases, dissemination of the pathogen is facilitated by but is not dependent on the vector. Insects may disseminate pathogens over short or long

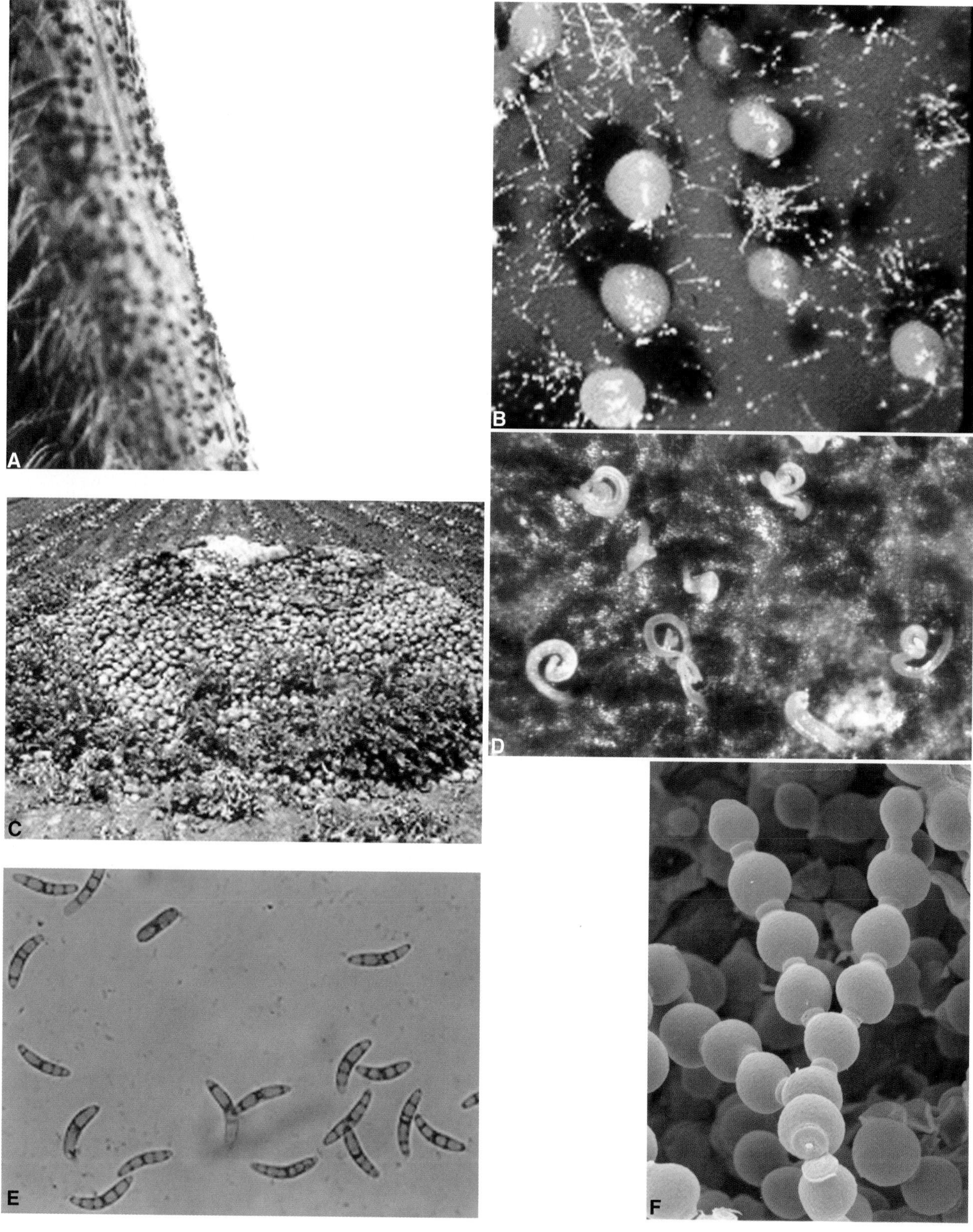

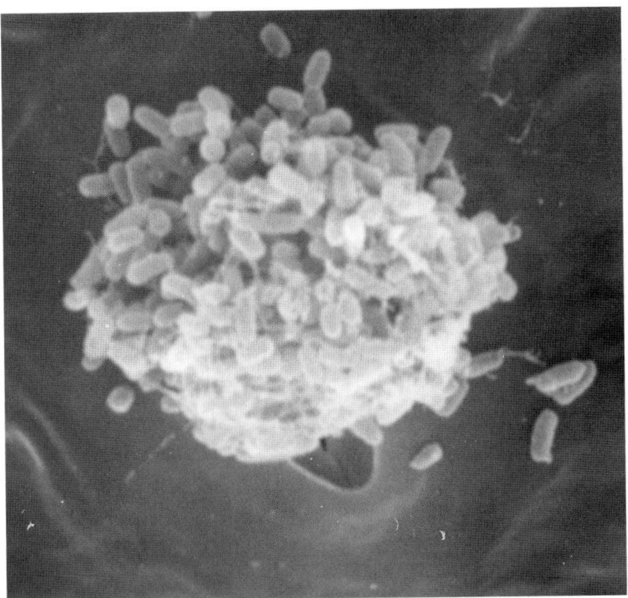

FIGURE 2-17 *Pseudomonas syringae* bacteria exuding through the stoma of an infected cherry leaf (2500X). (Photograph courtesy of E. L. Mansvelt, Stellenbosch, South Africa.)

distances, depending on the kind of insect, the insect–pathogen association, and the prevailing weather conditions, particularly wind.

A few species of mites and nematodes can transmit internally several viruses from plant to plant. In addition, mites and nematodes probably carry externally bacteria and sticky fungal spores with which they become smeared as they move on infected plant surfaces.

Almost all animals, small and large, that move among plants and touch the plants along the way can disseminate pathogens such as fungal spores, bacteria, seeds of parasitic plants, nematodes, and perhaps some viruses and viroids. Most of these pathogens adhere to the feet or the body of the animals, but some may be carried in contaminated mouthparts.

Finally, some plant pathogens, e.g., the zoospores of some fungi and certain parasitic plants, can transmit viruses as they move from one plant to another (zoospores) or as they grow and form a bridge between two plants (dodder).

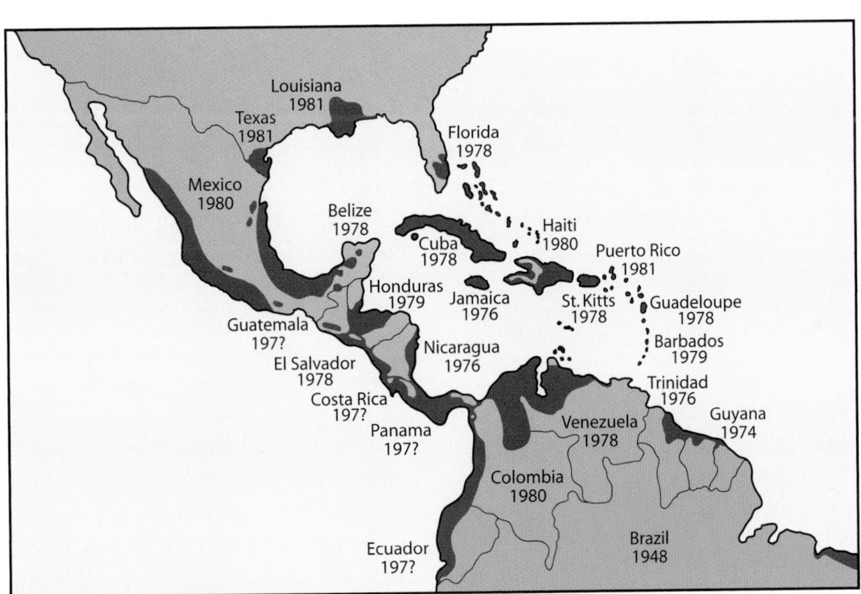

FIGURE 2-18 Map of the rapid spread of sugarcane smut, caused by the fungus *Ustilago scitaminea*, from its first sighting in Guyana in 1974 throughout the Caribbean islands, Central America, and the United States by 1981. [From Comstock *et al.* (1983). *Plant Dis.* **67**, 452–457.]

FIGURE 2-16 Fungal spore production, overwintering, and dissemination. (A) Pycnidia containing conidia produced on the stem of an infected plant. (B) Conidia oozing out of pycnidia after the latter absorbed rainwater. (C) Pile of cull potatoes in which many pathogens, such as the late blight oomycete, *Phytophthora infestans*, overwinter and are subsequently carried from the cull piles to potato fields. (D) Tendrils of conidia produced from hydrated bark-embedded pycnidia of the apple white rot fungus, *Botryosphaeria obtusa*. (E) Spores of the canker-causing fungus *Nectria*. (F) Chains of conidia of *Monilinia* sp. [Photographs courtesy of (A, B, and E) R. Cullen, University of Florida, (C) Plant Pathology Department, University of Wisconsin, (D) J. Rytter and J. W. Travis, Pennsylvania State University, and (F) and Mims *et al.* (1999). *Mycologia* **91**, 499–509.]

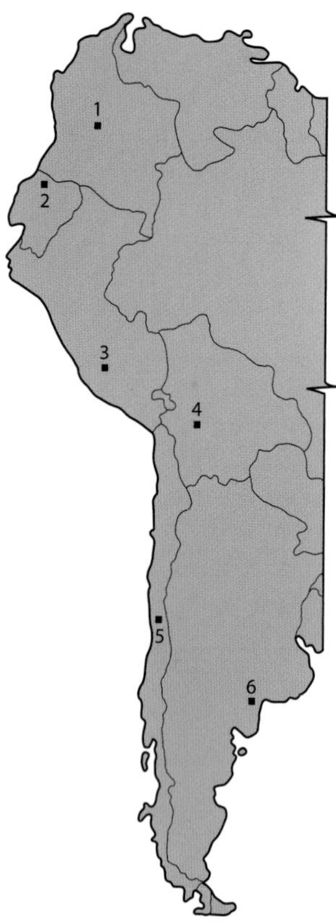

FIGURE 2-19 Map of the spatial and temporal spread of barley stripe rust, caused by the fungus *Puccinia striiformis f. sp. hordei*, in South America. The sequence of sightings are 1, Colombia 1975; 2, Ecuador 1976; 3, Peru 1977; 4, Bolivia 1978; 5, Chile 1980; and 6, Argentina 1982. [From Dubin and Stubbs (1986). *Plant Dis.* 70, 141–144.]

Dissemination by Pollen, Seed, Transplants, Budwood, and Nursery Stock

Some viruses are carried in the pollen of plants infected with these viruses and, when virus-carrying pollen pollinates a healthy plant, the virus may infect not only the seed produced from such pollination, which will then grow into a virus-infected plant, it may also infect the plant that was pollinated with the virus-carrying pollen.

Many pathogens are present on or in seeds, transplants, budwood, or nursery stock and are disseminated by them as the latter are transported to other fields or are sold and transported to other areas near and far. Dissemination of pathogens through seed, transplants, and so on is of great practical importance because it introduces the pathogen along with the plant at the beginning of the growth season and enables the pathogen to multiply and be disseminated by all the other means of spread discussed. It is also important because it brings pathogens into new areas where they may have never existed before.

Dissemination by Humans

Human beings disseminate all kinds of pathogens over short and long distances in a variety of ways. Within a field, humans disseminate some pathogens, such as tobacco mosaic virus, through the successive handling of diseased and healthy plants. Other pathogens are disseminated through tools, such as pruning shears, contaminated when used on diseased plants (e.g., pear infected with fire blight bacteria), and then carried to healthy plants. Humans also disseminate pathogens by transporting contaminated soil on their feet or equipment, using contaminated containers, and using infected transplants, seed, nursery stock, and budwood as mentioned previously. Finally, humans disseminate pathogens by importing new varieties into an area that may carry pathogens that have gone undetected, by traveling throughout the world, and by importing food or other items that may carry harmful plant pathogens. Examples of the role of humans as a vector of pathogens can be seen in the introduction into the United States of the fungi causing Dutch elm disease and white pine blister rust and of the citrus canker bacterium, in the introduction in Europe of the powdery and downy mildews of grape, and, more recently, in the rapid spread of sorghum ergot almost throughout the world (Fig. 2-20).

Overwintering and/or Oversummering of Pathogens

Pathogens that infect perennial plants can survive in them during low winter temperatures, during the hot, dry weather of the summer, or both, regardless of whether the host plants are actively growing or are dormant at the time. Annual plants, however, die at the end of the growing season, as do the leaves and fruits of deciduous perennial plants and even the stems of some perennial plants. In colder climates, annual plants and the tops of some perennial plants die with the advent of low winter temperatures, and their pathogens are left without a host for the several months of cold weather. In hot, dry climates, however, annual plants die during the summer and their pathogens must be able to survive such periods in the absence of their hosts. Thus, pathogens that attack annual plants and renewable parts of perennial plants have evolved mechanisms by which they can survive the cold winters or dry summers that may intervene between crops or growing seasons (Fig. 2-21).

Fungi have evolved a great variety of mechanisms for persisting between crops. On perennial plants, fungi overwinter as mycelium in diseased tissues, e.g., cankers,

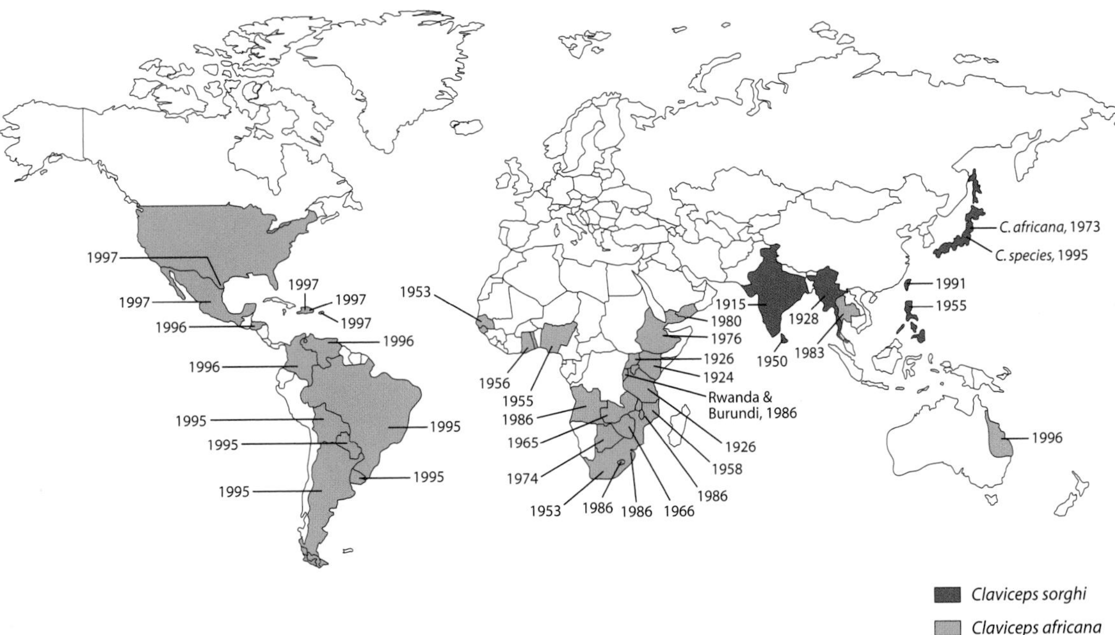

FIGURE 2-20 Map of the history of spread of ergot of sorghum, caused primarily by the fungus *Claviceps africana*, around the world. [Photograph courtesy of R. Bandyopadhyay *et al.* (1998). *Plant Dis.* **82**, 356–367.]

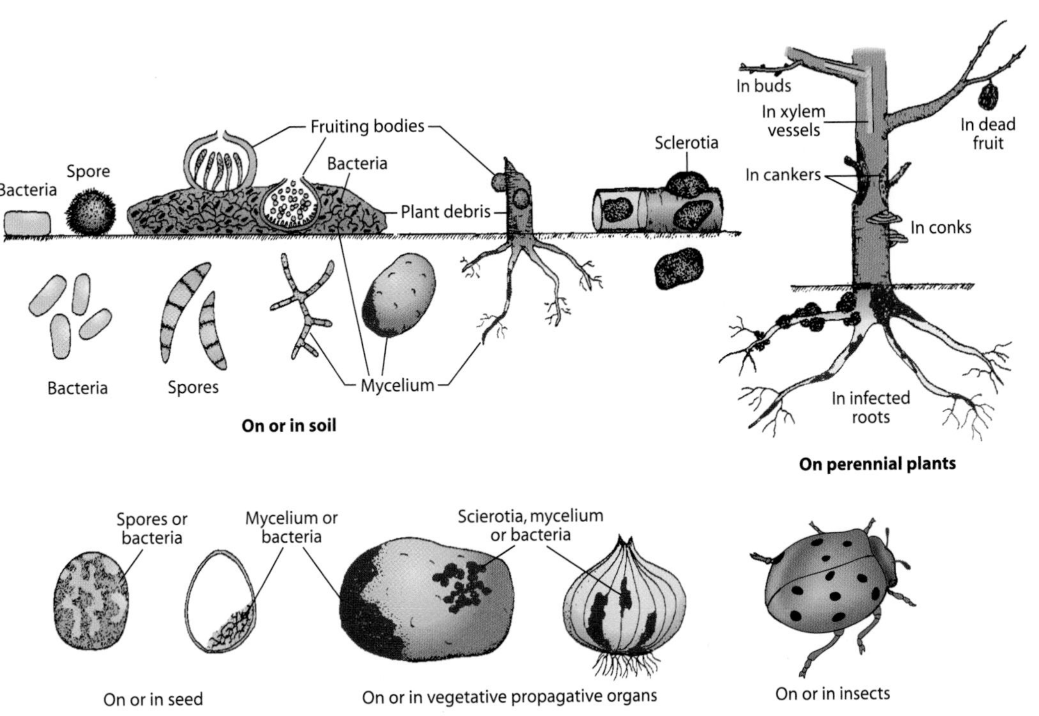

FIGURE 2-21 Forms and locations of survival of fungi and bacteria between crops.

and as spores at or near the infected surface of the plant or on the bud scales. Fungi affecting leaves or fruits of deciduous trees usually overwinter as mycelium or spores on fallen, infected leaves or fruits or on the bud scales. Fungi affecting annual plants usually survive the winter or summer as mycelium in infected plant debris, as resting or other spores and as sclerotia (hard masses of mycelium) in infected plant debris or in the soil, and as mycelium, spores, or sclerotia in or on seeds and other propagative organs, such as tubers. Some plant pathogenic oomycetes (e.g., *Pythium*) and fungi (e.g., *Fusarium, Rhizoctonia*) are **soil inhabitants**, i.e., they are able to survive indefinitely as saprophytes. Soil inhabitants are generally unspecialized parasites that have a wide host range. Other fungi are **soil transients**, i.e., they are rather specialized parasites that generally live in close association with their host but may survive in the soil for relatively short periods of time as hardy spores or as saprophytes. In some areas, fungi survive by continuous infection of host plants grown outdoors throughout the year, such as cabbage, or of plants grown in the greenhouse in the winter and outdoors in the summer. Similarly, some rust and other fungi overwinter on winter crops grown in warmer climates and move from them to the same hosts grown as spring crops in colder climates. Also, some fungi infect cultivated or wild perennial, as well as annual, plants and move from the perennial to the annual ones each growth season. Some rust fungi infect alternately an annual and a perennial host, and the fungus goes from the one to the other host and overwinters in the perennial host.

Bacteria overwinter and oversummer as bacteria in essentially the same ways as described for fungi, i.e., in infected plants, seeds, and tubers, in infected plant debris, and, for some, in the soil. Bacteria survive poorly when present in small numbers and free in the soil but survive well when masses of them are embedded in the hardened, slimy polysaccharides that usually surround them. Some bacteria also overwinter within the bodies of their insect vectors.

Viruses, viroids, mollicutes, fastidious bacteria, and protozoa survive only in living plant tissues such as the tops and roots of perennial plants, the roots of perennial plants that die to the soil line in the winter or summer, vegetative propagating organs, and the seeds of some hosts. A few viruses survive within their insect vectors, and some viruses and viroids may survive on contaminated tools and in infected plant debris.

Nematodes usually overwinter or oversummer as eggs in the soil and as eggs or nematodes in plant roots or in plant debris. Some nematodes produce juvenile stages or adults that can remain dormant in seeds or on bulbs for many months or years. Finally, parasitic higher plants survive either as seeds, usually in the soil, or as their infective vegetative form on their host.

RELATIONSHIPS BETWEEN DISEASE CYCLES AND EPIDEMICS

Some pathogens complete only one, or even part of one, disease cycle in 1 year and are called monocyclic, or single-cycle, pathogens (Fig. 2-22). Diseases caused by monocyclic pathogens include the smuts, in which the fungus produces spores at the end of the season (these spores serve as primary — and only — inoculum for the following year); many tree rusts, which require two alternate hosts and at least 1 year to complete one disease cycle; and many soilborne diseases, e.g., root rots and vascular wilts. In root rots and vascular wilts, the pathogens survive the winter or summer in decaying stems and roots or in the soil, infect plants during the growth season, and, at the end of the growth season, produce new spores in the infected stems and roots. These spores remain in the soil and serve as the primary inoculum the following growth season. In monocyclic pathogens the primary inoculum is the only inoculum available for the entire season, as there is no secondary inoculum and no secondary infection. The amount of inoculum produced at the end of the season, however, is greater than that present at the start of the season and so in monocyclic diseases the amount of inoculum may increase steadily from year to year.

In most diseases, however, the pathogen goes through more than one generation per growth season, and such pathogens are called polycyclic, or multicyclic, pathogens (Fig. 2-22). Polycyclic pathogens can complete many (from 2 to 30) disease cycles per year, and with each cycle the amount of inoculum is multiplied manyfold. Polycyclic pathogens are disseminated primarily by air or airborne vectors (insects) and are responsible for the kinds of diseases that cause most of

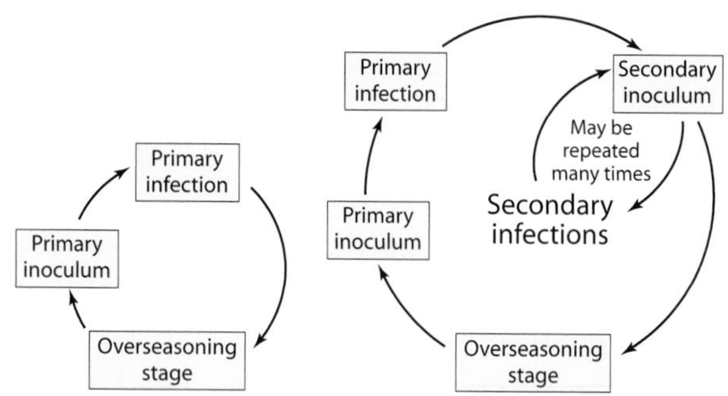

FIGURE 2-22 Diagrams of (left) monocyclic and (right) polycyclic plant diseases. Monocyclic diseases lack secondary inoculum and secondary infections during the same year.

the explosive epidemics on most crops, e.g., downy mildews, late blight of potato, powdery mildews, leaf spots and blights, grain rusts, and insect-borne viruses. In polycyclic fungal pathogens, the primary inoculum often consists of the sexual (perfect) spore or, in fungi that lack the sexual stage, some other hardy structure of the fungus such as sclerotia, pseudosclerotia, or mycelium in infected tissue. The number of sexual spores or other hardy structures that survive and cause infection is usually small, but once primary infection takes place, large numbers of asexual spores (secondary inoculum) are produced at each infection site and these spores can themselves cause new (secondary) infections that produce more asexual spores for more infections.

In some diseases of trees, e.g., fungal vascular wilts, phytoplasmal declines, and viral infections, the infecting pathogen may not complete a disease cycle, i.e., it may not produce inoculum that can be disseminated and initiate new infections, until at least the following year and some may take longer. Such diseases are basically monocyclic, but if they take more than a year to complete the cycle, they are called polyetic (multiyear). There are pathogens, however, such as those causing several rusts of trees and the mistletoes, that take several years to go through all the stages of their life cycle and to initiate new infections. These pathogens and the diseases they cause are clearly polyetic. Although polyetic pathogens may not cause many new infections over a given area within a single year and their amount of inoculum does not increase greatly within a year, because they survive in perennial hosts they have the advantage that, at the start of each year, they have almost as much inoculum as they had at the end of the previous year. Therefore, the inoculum may increase steadily (exponentially) from year to year and may cause severe epidemics when considered over several years. Examples of such diseases are Dutch elm disease, cedar apple rust, white pine blister rust, and citrus tristeza.

Whether the pathogen involved in a particular disease is monocyclic, polycyclic, or polyetic has great epidemiological consequences because it affects the amount of disease caused by the specific pathogen within a given period of time. The rate of inoculum or disease increase (r) has been calculated for many diseases and varies from 0.1 to 0.5 per day for polycyclic foliar diseases, such as southern corn leaf blight, potato late blight, grain rusts, and tobacco mosaic, to 0.02 to 2.3 per year for polyetic diseases of trees such as dwarf mistletoe of conifers, Dutch elm disease, chestnut blight, and peach mosaic. These values of r signify an increase in the amount of inoculum or disease (number of plants infected, amount of plant tissue infected, and so on) from 10 to 50% per day for polycyclic foliar diseases and from 2 to 230% per year for polyetic diseases of trees such as those listed earlier.

Selected References

Daly, J. M. (1984). The role of recognition in plant disease. *Annu. Rev. Phytopathol.* **22**, 273–307.

Dixon, R. A., Harrison, M. J., and Lamb, C. J. (1994). Early events in the activation of plant defense responses. *Annu. Rev. Phytopathol.* **32**, 479–501.

Ellingboe, A. H. (1968). Inoculum production and infection by foliage pathogens. *Annu. Rev. Phytopathol.* **6**, 317–330.

Emmett, R. W., and Parbery, D. G. (1975). Appressoria. *Annu. Rev. Phytopathol.* **13**, 147–167.

Hancock, J. G., and Huisman, O. C. (1981). Nutrient movement in host-pathogen systems. *Annu. Rev. Phytopathol.* **19**, 309–331.

Horsfall, J. G., and Cowling, E. B., eds. (1977–1980). "Plant Disease," Vols. 1–5. Academic Press, New York.

Kosuge, T., and Nester, E. W., eds. (1984). "Plant-Microbe Interactions: Molecular and Genetic Perspectives," Vol. 1. Macmillan, New York.

Kwon, Y. H., and Epstein, L. (1993). A 90 KD glycoprotein associated with adhesion of *Nectria haematococca* macroconidia to substrata. *Mol. Plant-Microbe Interact.* **6**, 481–487.

Leong, S. A., Allen, C., and Triplett, E. W., eds. (2002). "Biology of Plant-Microbe Interactions," Vol. 3. APS Press, St. Paul, MN.

Littlefield, L. J., and Heath, M. C. (1979)."Ultrastructure of Rust Fungi." Academic Press, New York.

Meredith, D. S. (1973). Significance of spore release and dispersal mechanisms in plant disease epidemiology. *Annu. Rev. Phytopathol.* **11**, 313–342.

Nielsen, K. A., *et al.* (2000). First touch: An immediate response to surface recognition in conidia of *Blumeria graminis*. *Physiol. Mol. Plant Pathol.* **56**, 63–70.

Perfect, S. E., and Green, J. R. (2001). Infection structures of biotrophic and hemibiotrophic fungal plant pathogens. *Mol. Plant Pathol.* **2**, 101–108.

Petrini, O., and Ouellette, G. B. (1994). "Host Wall Alterations by Parasitic Fungi." APS Press, St. Paul, MN.

Price-Jones, E., Carver, T., and Gurr, S. J. (1999). The roles of cellulase enzymes and mechanical force in host penetration by *Erysiphe graminis* f.sp. *hordei*. *Physiol. Mol. Plant Pathol.* **55**, 175–182.

Romantschuk, M. (1992). Attachment of plant pathogenic bacteria to plant surfaces. *Annu. Rev. Phytopathol.* **30**, 225–243.

Royle, D. J., and Thomas, G. G. (1973). Factors affecting zoospore responses towards stomata in hop downy mildew (*Pseudoperonospora humuli*) including some comparisons with grapevine downy mildew (*Plasmopara viticola*). *Physiol. Plant Pathol.* **3**, 405–417.

Schäfer, W. (1994). Molecular mechanisms of fungal pathogenicity to plants. *Annu. Rev. Phytopathol.* **32**, 461–477.

Schulze-Lefert, P., and Panstruga, R. (2003). Establishment of biotrophy by parasitic fungi and reprogramming of host cells for disease resistance. *Annu. Rev. Phytopathol.* **41**, 641–667.

Schuster, M. L., and Coyne, D. P. (1974). Survival mechanisms of phytopathogenic bacteria. *Annu. Rev. Phytopathol.* **12**, 199–221.

Singh, U. S., Singh, R. P., and Kohmoto, K. (1995). "Pathogenesis and Host Specificity in Plant Diseases: Histochemical, Biochemical, Genetic and Molecular Bases." Pergamon/Elsevier, Tarrytown, NY.

Stanley, M. S., *et al.* (2002). Inhibition of fungal spore adhesion by zosteric acid as the basis for a novel, nontoxic crop protection technology. *Phytopathology* **92**, 378–383.

Sugui, J. A., Leite, B., and Nicholson, R. L. (1998). Partial characterization of the extracellular matrix released onto hydrophobic surfaces by conidia and conidia germlings of *Colletotrichum graminicola*. *Physio. Mol. Plant Pathol.* **52**, 411–425.

Takano, Y., *et al.* (2001). Proper regulation of cyclic AMP-dependent protein kinase is required for growth, conidiation, and appressorium function in the anthracnose fungus *Colletotrichum lagenarium*. *Mol. Plant-Microbe Inter.* **14**, 1149–1157.

Tucker, S. L., and Talbot, N. J. (2001). Surface attachment and prepenetration stage development by plant pathogenic fungi. *Annu. Rev. Phytopathol.* **39**, 385–417.

Tyler, B. M. (2002). Molecular basis of recognition between *Phytophthora* pathogens and their hosts. *Annu. Rev. Phytopathol.* **40**, 137–167.

Vanderplank, J. E. (1975). "Principles of Plant Infections." Academic Press, New York.

Whitehead, N. A., and Salmond, G. P. C. (2000). Quorum sensing and the role of diffusible signaling in plant-microbe interactions. *In* "Plant-Microbe Interactions" (G. Stacey and N. T. Keen, eds.), pp. 43–92. APS Press, St. Paul, MN.

Wynn, W. K. (1981). Tropic and taxic responses of pathogens to plants. *Annu. Phytopathol.* **19**, 237–255.

chapter three

EFFECTS *OF* PATHOGENS *ON* PLANT PHYSIOLOGICAL FUNCTIONS

INTRODUCTION
105

EFFECT OF PATHOGENS ON PHOTOSYNTHESIS
106

EFFECT OF PATHOGENS ON TRANSLOCATION OF WATER AND NUTRIENTS IN THE HOST PLANT
106

EFFECT OF PATHOGENS ON HOST PLANT RESPIRATION
115

EFFECT OF PATHOGENS ON PERMEABILITY OF CELL MEMBRANES
118

EFFECT OF PATHOGENS ON TRANSCRIPTION AND TRANSLATION
118

EFFECT OF PATHOGENS ON PLANT GROWTH
119

EFFECT OF PATHOGENS ON PLANT REPRODUCTION
121

INTRODUCTION

While pathogens infect plants in the course of their obtaining food for themselves, depending on the kind of pathogen and on the plant organ and tissue they infect, pathogens interfere with the different physiological function(s) of the plant and lead to the development of different symptoms. Thus, a pathogen that infects and kills the flowers of a plant interferes with the ability of the plant to produce seed and multiply. A pathogen that infects and kills part or all of the roots of a plant reduces the ability of the plant to absorb water and nutrients and results in its wilting and death. Similarly, a pathogen that infects and kills parts of the leaves or destroys their chlorophyll leads to reduced photosynthesis, growth, and yield of the plant, and so forth. In most cases the relationship between the symptoms of the plant and the physiological functions affected is obvious and understandable. In other cases, however, the relationship of the two is more complex and the explanation is not always straightforward.

EFFECT OF PATHOGENS ON PHOTOSYNTHESIS

Photosynthesis is the basic function of green plants: it enables them to transform light energy into chemical energy, which they can utilize in all cell activities. Photosynthesis is the ultimate source of nearly all energy used in all living cells, plant or animal, as all activities of living cells, except photosynthesis, expend the energy provided by photosynthesis. In photosynthesis, carbon dioxide from the atmosphere and water from the soil are brought together in the chloroplasts of the green parts of plants and, in the presence of light, react to form glucose with a concurrent release of oxygen:

$$6CO_2 + 6H_2O \xrightarrow[\text{chlorophyll}]{\text{light}} C_6H_{12}O_6 + 6O_2$$

In view of the fundamental position of photosynthesis in the life of plants, it is apparent that any interference by pathogens with photosynthesis results in a diseased condition in the plant. That pathogens do interfere with photosynthesis is obvious from the chlorosis they cause on many infected plants, from the necrotic lesions or large necrotic areas they produce on green plant parts, and from the reduced growth and amounts of fruits produced by many infected plants.

In leaf spot, blight, and other kinds of diseases in which there is destruction of leaf tissue, e.g., in cereal rusts and fungal leaf spots (Figs. 3-1A–3-1C), bacterial leaf spots (Fig. 3-1D), viral mosaics (Fig. 3-1E) and yellowing and stunting diseases (Fig. 3-1F), or in defoliations, photosynthesis is reduced because the photosynthetic surface of the plant is lessened. Even in other diseases, however, plant pathogens reduce photosynthesis, especially in the late stages of diseases, by affecting the chloroplasts and causing their degeneration. The overall chlorophyll content of leaves in many fungal and bacterial diseases is reduced, but the photosynthetic activity of the remaining chlorophyll seems to remain unaffected. In some fungal and bacterial diseases, photosynthesis is reduced because the toxins, such as tentoxin and tabtoxin, produced by these pathogens inhibit some of the enzymes that are involved directly or indirectly in photosynthesis. In plants infected by many vascular pathogens, stomata remain partially closed, chlorophyll is reduced, and photosynthesis stops even before the plant eventually wilts. Most virus, mollicute, and nematode diseases also induce varying degrees of chlorosis and stunting. In the majority of such diseases, the photosynthesis of infected plants is reduced greatly. In advanced stages of disease, the rate of photosynthesis is no more than one-fourth the normal rate.

EFFECT OF PATHOGENS ON TRANSLOCATION OF WATER AND NUTRIENTS IN THE HOST PLANT

All living plant cells require an abundance of water and an adequate amount of organic and inorganic nutrients in order to live and to carry out their physiological functions. Plants absorb water and inorganic (mineral) nutrients from the soil through their root system. These substances are generally translocated upward through the xylem vessels of the stem and into the vascular bundles of the petioles and leaf veins, from which they enter the leaf cells. Minerals and part of the water are utilized by the leaf and other cells for the synthesis of the various plant substances, but most of the water evaporates out of the leaf cells into the intercellular spaces and from there diffuses into the atmosphere through the stomata. However, nearly all organic nutrients of plants are produced in the leaf cells, following photosynthesis, and are translocated downward and distributed to all the living plant cells by passing, for the most part, through the phloem tissues. When a pathogen interferes with the upward movement of inorganic nutrients and water or with the downward movement of organic substances, diseased conditions result in the parts of the plant denied these materials. The diseased parts, in turn, will be unable to carry out their own functions and will deny the rest of the plant their services or their products, thus causing disease of the entire plant. For example, if water movement to the leaves is inhibited, the leaves cannot function properly, photosynthesis is reduced or stopped, and few or no nutrients are available to move to the roots, which in turn become starved and diseased and may die.

Interference with Upward Translocation of Water and Inorganic Nutrients

Many plant pathogens interfere in one or more ways with the translocation of water and inorganic nutrients

FIGURE 3-1 Ways in which pathogens reduce photosynthetic area and, thereby, photosynthesis in plants. (A) Spots on barley leaves caused by the fungus *Rhynchosporium sp.* (B) Nearly complete destruction of pumpkin leaves infected heavily with the downy mildew oomycete *Pseudoperonospora cubensis.* (C) Countless tiny lesions on stems and leaves of wheat plant infected with the stem rust fungus *Puccinia graminis f.sp. tritici.* is. (D) Angular leaf spots on cucumber leaf caused by the bacterium *Pseudomonas lacrymans.* (E) Reduced chlorophyll in yellowish areas of virus-infected plants, such as cowpea infected with *cowpea chlorotic mottle virus* or (F) by stunting and yellowing of rice plants infected with the *rice tungro virus.* [Photographs courtesy of (A) Plant Pathology Department, University of Florida, (B) T. A. Zitter, Cornell University (C) I. Evans and (D) R. J. Howard, W.C.P.D., and (F) H. Hibino.]

through plants. Some pathogens affect the integrity or function of the roots, causing them to absorb less water; other pathogens, by growing in the xylem vessels or by other means, interfere with the translocation of water through the stem; and, in some diseases, pathogens interfere with the water economy of the plant by causing excessive transpiration through their effects on leaves and stomata.

Effect on Absorption of Water by Roots

Many pathogens, such as damping-off fungi (Fig. 3-2A), root-rotting fungi and bacteria (Figs. 3-2B–3-2D), most nematodes, and some viruses, cause an extensive destruction of the roots before any symptoms appear on the aboveground parts of the plant. Some bacteria and nematodes cause root galls or root knots (Figs. 3-2E and 3-2F), which interfere with the normal absorption of water and nutrients by the roots. Root injury affects the amount of functioning roots directly and decreases proportionately the amount of water absorbed by the roots. Some vascular parasites, along with their other effects, seem to inhibit root hair production, which reduces water absorption. These and other pathogens also alter the permeability of root cells, an effect that further interferes with the normal absorption of water by roots.

Effect on Translocation of Water through the Xylem

Fungal and bacterial pathogens that cause damping off, stem rots (Fig. 3-3A), and cankers (Fig. 3-3B) may reach the xylem vessels in the area of the infection and, if the affected plants are young, may cause their destruction and collapse. Cankers in older plants, particularly older trees (Fig. 3-3B), may cause some reduction in the translocation of water, but, generally, do not kill plants unless the cankers are big or numerous enough to encircle the plant. In vascular wilts, however (Figs. 3-3C–3-3F), reduction in water translocation may vary from little to complete. In many cases, affected vessels may be filled with the bodies of the pathogen (Figs. 3-4A–3-4D) and with substances secreted by the pathogen (Figs. 3-5D and 3-5E) or by the host (Fig. 3-5C) in response to the pathogen and may become clogged (Figs. 3-4A and 3-4C and 3-5C–3-5E). Whether destroyed or clogged, the affected vessels cease to function properly and allow little or no water to pass through them. Certain pathogens, such as the crown gall bacterium (*Agrobacterium tumefaciens*), the clubroot protozoon (*Plasmodiophora brassicae*), and the root-knot nematode (*Meloidogyne* sp.), induce gall formation (Figs. 3-2E and 3-2F) in the stem, roots, or both. The enlarged and proliferating cells near or around the xylem exert pressure on the xylem vessels, which may be crushed and dislocated, thereby becoming less efficient in transporting water.

The most typical and complete dysfunction of xylem in translocating water, however, is observed in the vascular wilts (Figs. 3-3 and 3-5) caused by the fungi *Ceratocystis*, *Ophiostoma*, *Fusarium*, and *Verticillium* and bacteria such as *Pseudomonas*, *Ralstonia*, and *Erwinia*. These pathogens invade the xylem of roots and stems and produce diseases primarily by interfering with the upward movement of water through the xylem. In many plants infected by these pathogens the water flow through the stem xylem is reduced to a mere 2 to 4% of that flowing through stems of healthy plants. In general, the rate of flow through infected stems seems to be inversely proportional to the number of vessels blocked by the pathogen and by the substances resulting from the infection. Evidently more than one factor is usually responsible for the vascular dysfunction in the wilt diseases. Although the pathogen is the single cause of the disease, some of the factors responsible for the disease syndrome originate directly from the pathogen, whereas others originate from the host in response to the pathogen. The pathogen can reduce the flow of water through its physical presence in the xylem as mycelium, spores, or bacterial cells (Figs. 3-4A–3-4C and 3-5B) and by the production of large molecules (polysaccharides) in the vessels (Figs. 3-5D and 3-5E). In most host–pathogen combinations, the destruction of xylem vessels by fungi (Fig. 3-3A) results in the collapse and death of the plant, as does the invasion of xylem vessels by fungi (Figs. 3-3C and 3-3D) or bacteria (Figs. 3-3E and 3-3F and 3-5A–3-5F). In host combinations with the fastidious bacterium *Xylella fastidiosa*, growth, multiplication, and spread of bacteria in xylem vessels are slower and, instead of causing wilting and rapid death of the plant, a scorching of the margins of the leaves (Fig. 3-4D) and several other symptoms occur, but rarely does the plant die quickly. In all cases, however, in infected hosts the flow of water is reduced through reduction in the size or collapse of vessels due to infection, development of tyloses (Figs. 3-5C and 3-5E) in the vessels, release of large molecule compounds in the vessels as a result of cell wall breakdown by pathogenic enzymes (Figs. 3-5D and 3-5E), and reduced water tension in the vessels due to pathogen-induced alterations in foliar transpiration.

Effect on Transpiration

In plant diseases in which the pathogen infects the leaves, transpiration is usually increased. This is the result of destruction of at least part of the protection

FIGURE 3-2 Examples of reduction of water absorption by plants. (A) Destruction of roots of young seedlings by the damping-off oomycete *Pythium sp.* (B) Roots and stems of pepper plants killed by *Phytophthora sp.* (C) Wheat roots at different stages of destruction by the take-all fungus *Gaeumannomyces tritici.* (D) Infection of crown and roots of corn plant with the fungus *Fusarium.* (E) Numerous galls caused by the bacterium *Agrobacterium tumefaciens* on roots of a cherry tree. (F) Root knot galls caused by the nematode *Meloidogyne sp.* on roots of a cantaloupe plant. [Photographs courtesy of (A) Plant Pathology Department, University of Florida, (B) K. Pernezny, University of Florida, (C) W. McFadden, W.C.P.D., (D) Plant Pathology Department, Iowa State University, (E) Oregon State University, and (F) B. D. Bruton, USDA, Lane, Oklahoma.]

FIGURE 3-3 Examples of reduction of upward translocation of water and mineral nutrients by (A) the stem of a cantaloupe plant infected with the fungus *Phomopsis* sp. (B) Canker on an almond tree caused by the fungus *Ceratocystis fagacearum.* (C) Vascular wilt of tomato caused by the fungus *Fusarium.* (D) Discolored vascular tissues of a tomato stem infected with the same fungus. (E) Wilted tomato plants infected with the vascular bacterium *Ralstonia solanacearum.* (F) Discolored vascular tissues of a tomato plant infected with the same bacterium. [Photographs courtesy of (A) B. D. Bruton, USDA, Lane, Oklahoma, (B) B. Teviotdale, Kearney Agricultural Center, Parlier, California, (C,E, and F) Department of Plant Pathology, University Florida, and (D) L. McDonald, W.C.P.D.]

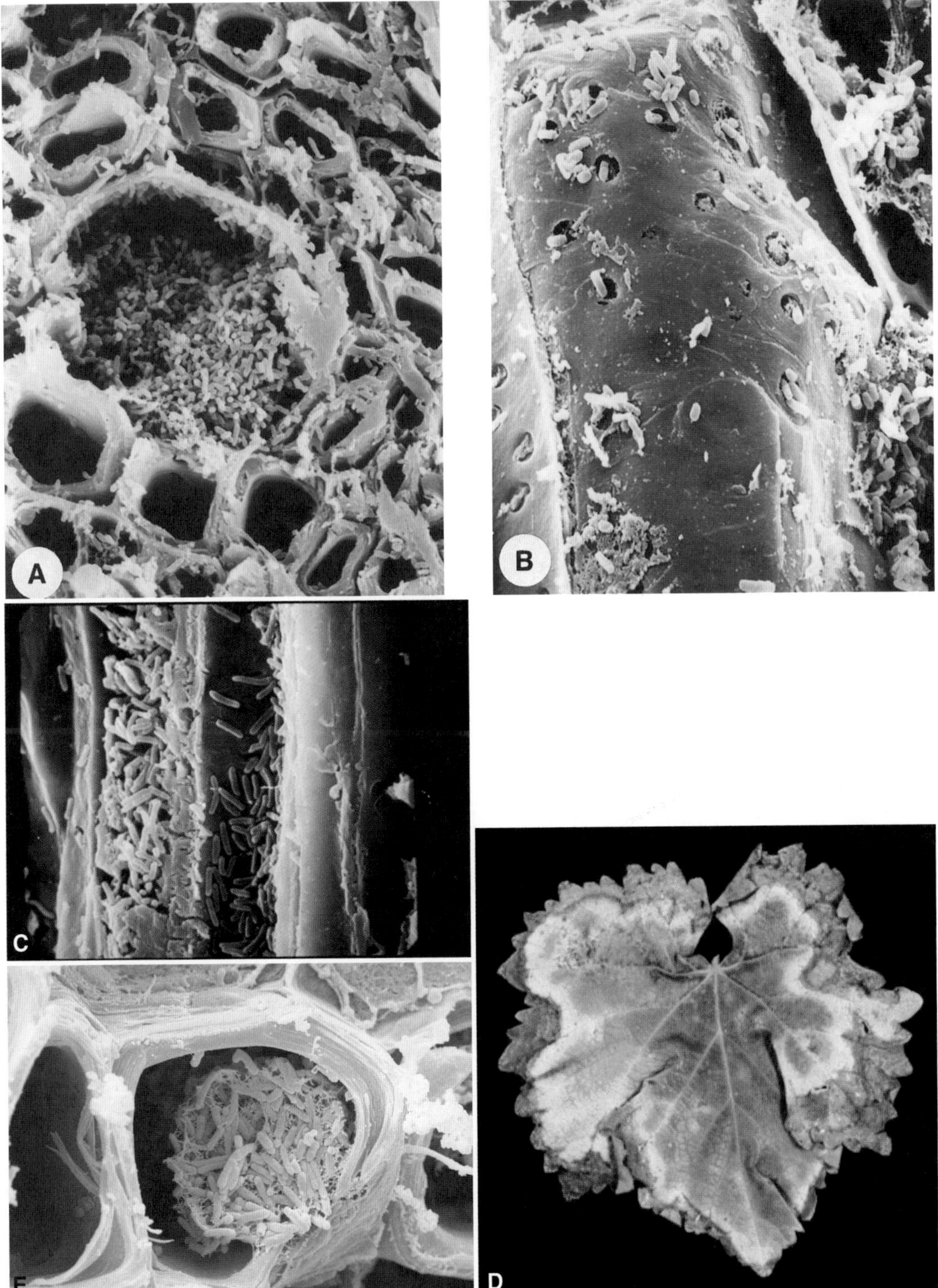

FIGURE 3-4 (A) *Pseudomonas* bacteria clogging a xylem vessel of a young plant shoot. (B) Bacteria moving from one vessel to another and to adjacent parenchyma cells through xylem pits. (C) Bacteria of the xylem-inhabiting *Xylella fastidiosa* in a vessel of a grape plant. (D) Marginal scorching of a grape leaf from a plant infected with *X. fastidiosa*, the cause of Pierce's disease of grape. (E) *Xylella* bacteria in a cross section of a xylem vessel of an infected grape leaf. [Photographs courtesy of (A and B) E. L. Mansvelt, I. M. M. Roos, and M. J. Hattingh (1500×), (C) D. Cooke, provided by E. Hellman, Texas A&M University, (D) E. Hellman, and (E) E. Alves, Federal University of Lavras, Brazil.]

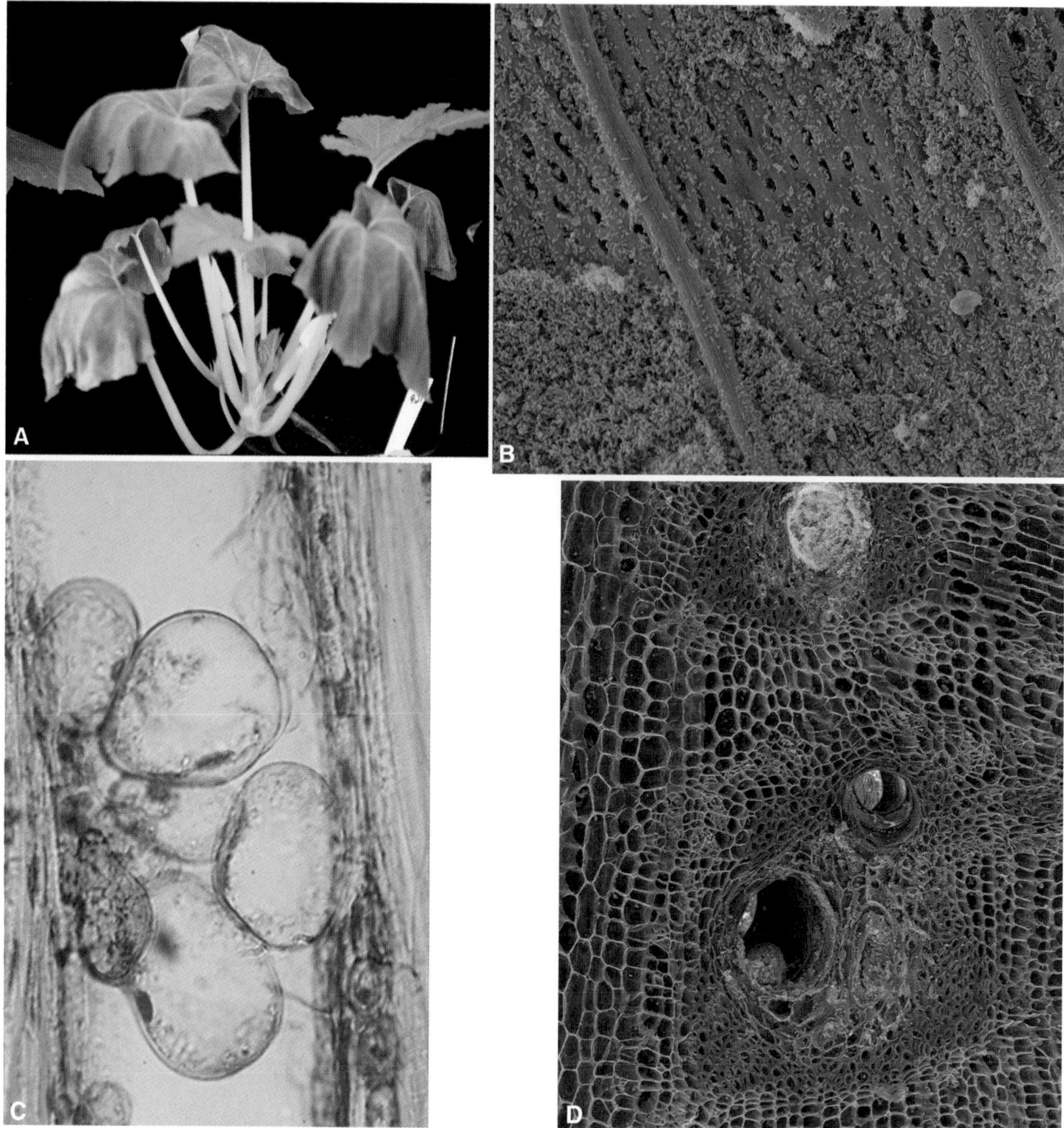

FIGURE 3-5 (A) Young squash plant showing early symptoms of vascular wilt caused by the bacterium *Erwinia tracheiphila*. (B) *E. tracheiphila* bacteria lining up the inside wall of a xylem vessel. (C) Tyloses in a xylem vessel. (D) Tyloses and gummy polysaccharides partially or totally clogging up xylem vessels of a squash plant. (E) Several xylem vessels totally clogged with gummy polysaccharides. (F) Cantaloupes in a field where the plants had been killed by the bacterium *E. tracheiphila*. [Photographs courtesy of (A,B,D,E, and F) B. D. Bruton, USDA, Lane, Oklahoma, and (C) D. M. Elgersma.]

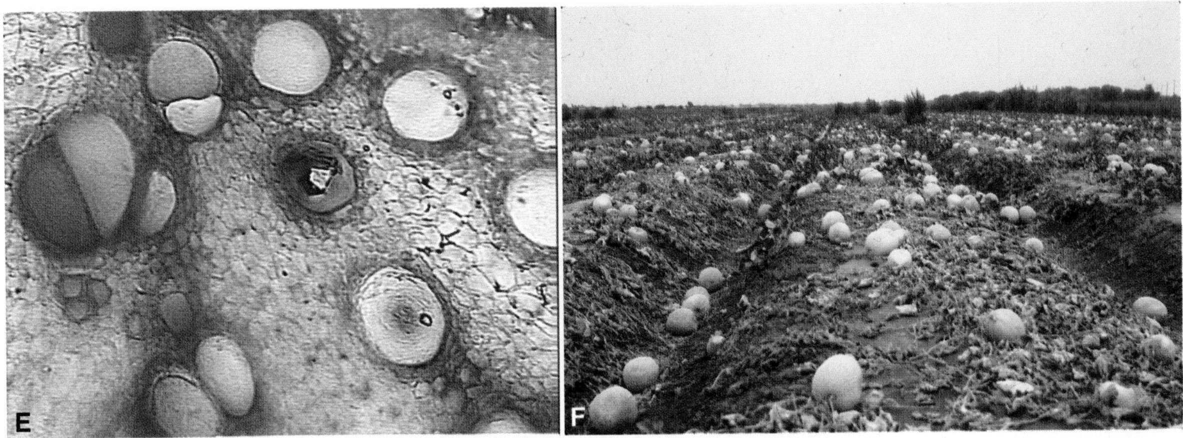

FIGURE 3-5 *(Continued)*

afforded the leaf by the cuticle, an increase in the permeability of leaf cells, and the dysfunction of stomata. In diseases such as rusts, in which numerous pustules form and break up the epidermis (Figs. 3-6A and 3-6B), in most leaf spots (Fig. 3-6E), in which the cuticle, epidermis, and all the other tissues, including xylem, may be destroyed in the infected areas, in the powdery mildews, in which a large proportion of the epidermal cells are invaded by the fungus (Fig. 3-6C), and in apple scab (Fig. 3-6D), in which the fungus grows between the cuticle and the epidermis—in all these examples, the destruction of a considerable portion of the cuticle and epidermis results in an uncontrolled loss of water from the affected areas. If water absorption and translocation cannot keep up with the excessive loss of water, loss of turgor and wilting of leaves follow. The suction forces of excessively transpiring leaves are increased abnormally and may lead to collapse or dysfunction of underlying vessels through the production of tyloses and gums.

Interference with Translocation of Organic Nutrients through the Phloem

Organic nutrients produced in leaf cells through photosynthesis move through plasmodesmata into adjoining phloem elements. From there they move down the phloem sieve tubes (Fig. 3-7) and eventually, again through plasmodesmata, into the protoplasm of living nonphotosynthetic cells, where they are utilized, or into storage organs, where they are stored. Thus, in both cases, the nutrients are removed from "circulation." Plant pathogens may interfere with the movement of organic nutrients from the leaf cells to the phloem, with their translocation through the phloem elements, or, possibly, with their movement from the phloem into the cells that will utilize them.

Obligate fungal parasites, such as rust and mildew fungi, cause an accumulation of photosynthetic products, as well as inorganic nutrients, in the areas invaded by the pathogen. In these diseases, the infected areas are characterized by reduced photosynthesis and increased respiration. However, the synthesis of starch and other compounds, as well as dry weight, is increased temporarily in the infected areas, indicating translocation of organic nutrients from uninfected areas of the leaves or from healthy leaves toward the infected areas.

In stem diseases of woody plants in which cankers develop (Figs. 3-8A–3-8C), the pathogen attacks and remains confined to the bark for a considerable time. During that time the pathogen attacks and may destroy the phloem elements in that area, thereby interfering with the downward translocation of nutrients. In diseases caused by phytoplasmas, as well as in diseases caused by phloem-limited fastidious bacteria, bacteria exist and reproduce in the phloem sieve tubes (Fig. 3-8D), thereby interfering with the downward translocation of nutrients. In several plants propagated by grafting a variety scion onto a rootstock, infection of the combination with a virus (e.g., infection of an apple or stone-fruit rootstock with *tomato ringspot virus*) leads to formation of a necrotic plate at the points of contact of the hypersensitive scion variety with the rootstock (Fig. 3-8E), which leads to the death of the scion. However, infection of a pear scion grafted on an oriental rootstock with the pear decline phytoplasma, or of a citrus variety propagated on sour rootstock with the citrus tristeza virus, results, in both cases, in the necrosis of a few layers of cells of each rootstock in contact with the tolerant variety. In these cases, the rootstock is the component of the scion/stock combination that is

FIGURE 3-6 Ways by which pathogens cause increased transpiration in infected plants. (A) The wheat leaf rust pathogen *Puccinia recondita* produces innumerable lesions (uredia) on wheat leaves and causes millions of breaks in the leaf epidermis through which transpiration goes on uncontrollably. (B) Uredospores breaking the epidermis and emerging from the surface of an infected leaf. (C) Grape berries infected with the powdery mildew fungus *Uncinula necator*, the mycelium of which penetrates and forms haustoria in almost every epidermal cell. (D) The apple scab fungus *Venturia inaequalis* grows between the cuticle and the epidermis, causing the cuticle to break in numerous places, allowing transpiration to occur. (E) Tomato leaves with numerous lesions caused by the fungus *Septoria sp.* and through which excessive transpiration occurs. [Photographs courtesy of (A and E) W.C.P.D., (B) E. A. Richardson and C. W. Mims, University of Georgia, (C) J. Travis and J. Rytter, Plant Pathology Department, Pennsylvania State University, and (D) K. Mohan, University of Idaho.]

hypersensitive to and becomes killed by the appropriate pathogen.

In some virus diseases, particularly the leaf-curling type and some yellows diseases, starch accumulation in the leaves is mainly the result of degeneration (necrosis) of the phloem of infected plants (Fig. 3-8F), which is one of the first symptoms. It is also possible, however, at least in some virus diseases, that the interference with translocation of starch stems from inhibition by the virus of the enzymes that break down starch into smaller, translocatable molecules. This is suggested by the observation that in some mosaic diseases, in which there is no phloem necrosis, infected, discolored areas of leaves contain less starch than "healthy," greener areas at the end of the day, a period favorable for photosynthesis, but the same leaf areas contain more starch than the "healthy" areas after a period in the dark, which favors starch hydrolysis and translocation. This suggests not only that virus-infected areas synthesize less starch than healthy ones, but also that starch is not degraded and translocated easily from virus-infected areas, although no damage to the phloem is present.

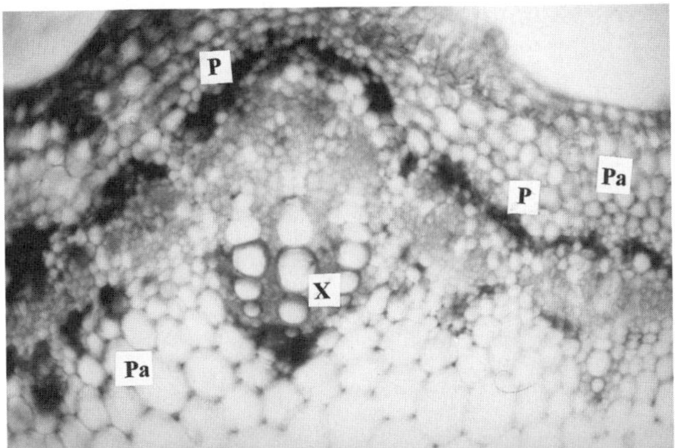

FIGURE 3-7 Necrosis of the phloem (P) in stems or petioles of plants is a common effect of viruses, such as the *tobacco ringspot virus,* on cowpea plants. As a result, roots starve and the plant declines (100×). Pa, parenchyma cells; X, xylem vessels.

EFFECT OF PATHOGENS ON HOST PLANT RESPIRATION

Respiration is the process by which cells, through the enzymatically controlled oxidation (burning) of the

FIGURE 3-8 Examples of diseases in which the pathogen interferes with the downward translocation of organic nutrients. (A) Young canker caused by the fungus *Nectria* in which the bark of the branch has been invaded and killed by the fungus. (B) Two advanced *Nectria* cankers in which both the phloem and a great deal of the xylem have been killed by the fungus. (C) Blister canker on a pine tree in which the bark and phloem have been killed by the fungus *Cronartium ribicola.* (D) Phytoplasmas filling a phloem sieve element block the downward translocation of photosynthates. (E) The graft union of a pear grafted on oriental pear rootstocks, which results in the death of pear phloem. (F) Potato tuber showing vein necrosis caused by the *potato leaf roll virus.* [Photographs courtesy of (A) USDA Forest Service, (B) A. Jones, Plant Pathology Department, Michigan State University, (C) Oregon State University, and (F) Cornell University.

(continued on next page)

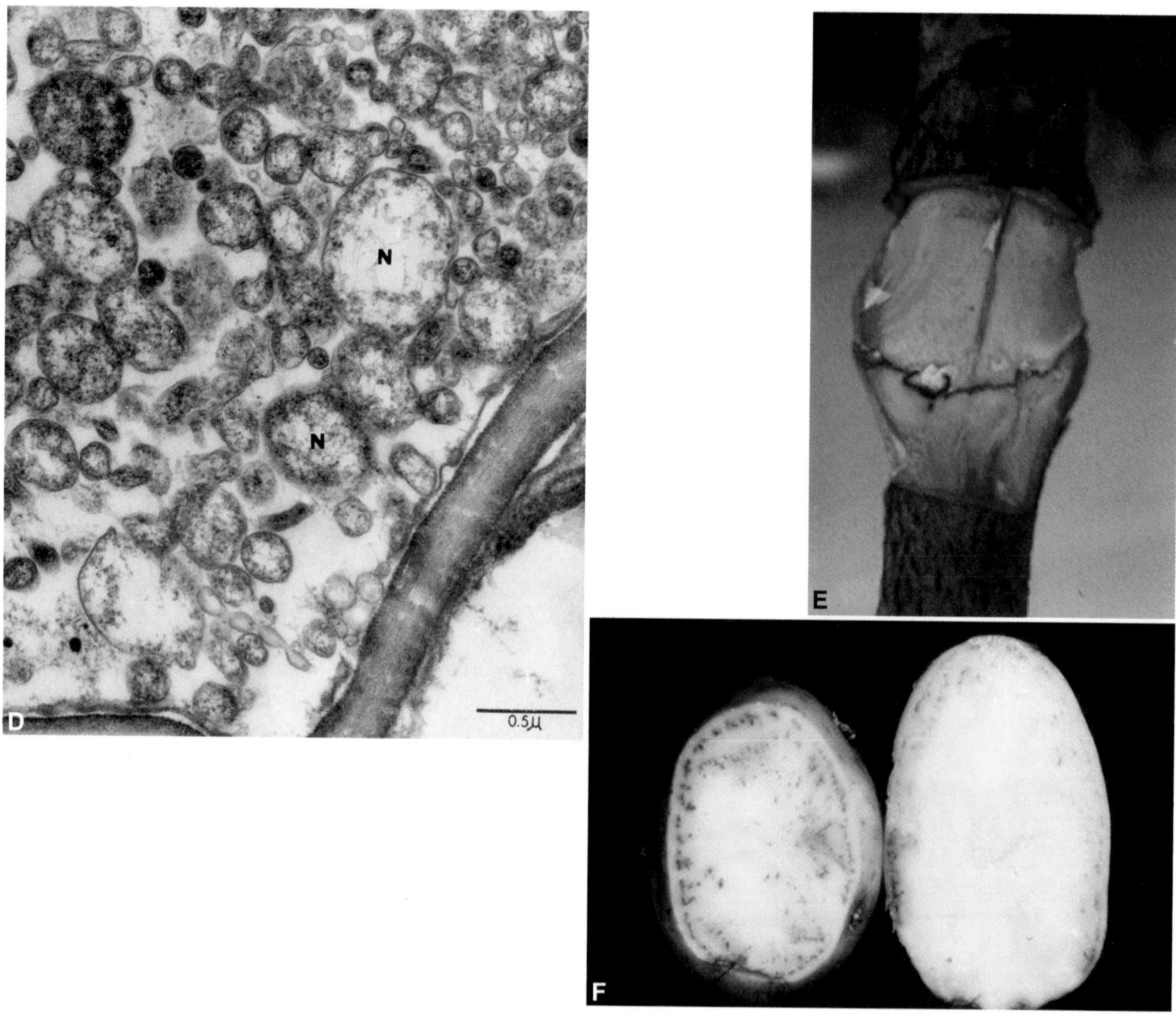

FIGURE 3-8 *(Continued)*

energy-rich carbohydrates and fatty acids, liberate energy in a form that can be utilized for the performance of various cellular processes. Plant cells carry out respiration in, basically, two steps. The first step involves the degradation of glucose to pyruvate and is carried out, either in the presence or in the absence of oxygen, by enzymes found in the ground cytoplasm of the cells. The production of pyruvate from glucose follows either the **glycolytic pathway**, otherwise known as **glycolysis**, or, to a lesser extent, the **pentose pathway**. The second step, regardless of the pathway, involves the degradation of pyruvate, however produced, to CO_2 and water. This is accomplished by a series of reactions known as the **Krebs cycle**, which is accompanied by the so-called **terminal oxidation** and is carried out in the mitochondria only in the presence of oxygen. Under normal (aerobic) conditions, i.e., in the presence of oxygen, both steps are carried out, and one molecule of glucose yields, as final products, six molecules of CO_2 and six molecules of water,

$$C_6H_{12}O_6 + 6O_2 \rightarrow 6CO_2 + 6H_2O$$

with a concomitant release of energy (678,000 calories). Some of the energy is lost, but almost half is converted to 20–30 reusable high-energy bonds of adenosine triphosphate (ATP). The first step of respiration contributes two ATP molecules per mole of glucose, and the second step contributes the rest. Under anaerobic conditions, however (i.e., in the absence of oxygen), pyruvate cannot be oxidized; instead it undergoes **fermentation** and yields lactic acid or alcohol. Because the main process of energy generation is cut off, for the cell to secure the necessary energy a much greater rate of glucose utilization by glycolysis is required in the absence of oxygen than is in its presence.

The energy-storing bonds of ATP are formed by the attachment of a phosphate (PO_4) group to adenosine diphosphate (ADP) at the expense of energy released from the oxidation of sugars. The coupling of the oxidation of glucose with the addition of phosphate to ADP to produce ATP is called **oxidative phosphorylation**. Any cell activity that requires energy utilizes the energy stored in ATP by simultaneously breaking down ATP to ADP and inorganic phosphate. The presence of ADP and phosphate in the cell, in turn, stimulates the rate of respiration. If, however, ATP is not utilized sufficiently by the cell for some reason, there is little or no regeneration of ADP and respiration is slowed down. The amount of ADP (and phosphate) in the cell is determined, therefore, by the rate of energy utilization; this rate, in turn, determines the rate of respiration in plant tissues.

The energy produced through respiration is utilized by the plant for all types of cellular work, such as accumulation and mobilization of compounds, synthesis of proteins, activation of enzymes, cell growth and division, defense reactions, and a host of other processes. The complexity of respiration, the number of enzymes involved in respiration, its occurrence in every single cell, and its far-reaching effects on the functions and existence of the cell make it easy to understand why the respiration of plant tissues is one of the first functions to be affected when plants are infected by pathogens.

Respiration of Diseased Plants

When plants are infected by pathogens, the rate of respiration generally increases. This means that affected tissues use up their reserve carbohydrates faster than healthy tissues would. The increased rate of respiration appears shortly after infection — certainly by the time of appearance of visible symptoms — and continues to rise during the multiplication and sporulation of the pathogen. After that, respiration declines to normal levels or to levels even lower than those of healthy plants. Respiration increases more rapidly in infections of resistant varieties, in which large amounts of energy are needed and used for rapid production or mobilization of the defense mechanisms of the cells. In resistant varieties, however, respiration also declines quickly after it reaches its maximum. In susceptible varieties, in which no defense mechanisms can be mobilized quickly against a particular pathogen, respiration increases slowly after inoculation, but continues to rise and remains at a high level for much longer periods.

Several changes in the metabolism of the diseased plant accompany the increase in respiration after infection. Thus, the activity or concentration of several enzymes of the respiratory pathways seems to be increased. The accumulation and oxidation of phenolic compounds, many of which are associated with defense mechanisms in plants, are also greater during increased respiration. Increased respiration in diseased plants is also accompanied by an increased activation of the pentose pathway, which is the main source of phenolic compounds. Increased respiration is sometimes accompanied by considerably more fermentation than that observed in healthy plants, probably as a result of an accelerated need for energy in the diseased plant under conditions in which normal aerobic respiration cannot provide sufficient energy.

The increased respiration in diseased plants is apparently brought about, at least in part, by the uncoupling of oxidative phosphorylation. In that case, no utilizable energy (ATP) is produced through normal respiration, despite the use of existing ATP and the accumulation of ADP, which stimulates respiration. The energy required by the cell for its vital processes is then produced through other less efficient ways, including the pentose pathway and fermentation.

The increased respiration of diseased plants can also be explained as the result of increased metabolism. In many plant diseases, growth is at first stimulated, protoplasmic streaming increases, and materials are synthesized, translocated, and accumulated in the diseased area. The energy required for these activities derives from ATP produced through respiration. The more ATP is utilized, the more ADP is produced and further stimulates respiration. It is also possible that the plant, because of the infection, utilizes ATP energy less efficiently than a healthy plant. Because of the waste of part of the energy, an increase in respiration is induced, and the resulting greater amount of energy enables the plant cells to utilize sufficient energy to carry out their accelerated processes.

Although oxidation of glucose via the glycolytic pathway is by far the most common way through which plant cells obtain their energy, part of the energy is produced via the pentose pathway. The latter seems to be an alternate pathway of energy production to which plants resort under conditions of stress. Thus, the pentose pathway tends to replace the glycolytic pathway as the plants grow older and differentiate and it tends to increase on treatment of the plants with hormones, toxins, wounding, starvation, and so on. Infection of plants with pathogens also tends, in general, to activate the pentose pathway over the level at which it operates in the healthy plant. Because the pentose pathway is not linked directly to ATP production, the increased respiration through this pathway fails to produce as much utilizable energy as the glycolytic pathway and is, therefore, a less efficient source of energy for the functions of the diseased plant. However, the pentose pathway is the main source of phenolic compounds, which play important roles in the defense mechanisms of the plant against infection.

EFFECT OF PATHOGENS ON PERMEABILITY OF CELL MEMBRANES

Cell membranes consist of a double layer of lipid molecules in which many kinds of protein molecules are embedded, parts of which usually protrude on one or both sides of the lipid bilayer (Fig. 5-2). Membranes function as permeability barriers that allow passage into a cell only of substances the cell needs and inhibit passage out of the cell of substances needed by the cell. The lipid bilayer is impermeable to most biological molecules. Small water-soluble molecules such as ions (charged atoms or electrolytes), sugars, and amino acids flow through or are pumped through special membrane channels made of proteins. In plant cells, because of the cell wall, only small molecules reach the cell membrane. In animal cells and in artificially prepared plant protoplasts, however, large molecules or particles may also reach the cell membrane and enter the cell by endocytosis, in which a patch of the membrane surrounds and forms a vesicle around the material to be taken in, brings it in, and releases it inside the cell. Disruption or disturbance of the cell membrane by chemical or physical factors alters (usually increases) the permeability of the membrane with a subsequent uncontrollable loss of useful substances, as well as the inability to inhibit the inflow of undesirable substances or excessive amounts of any substances.

Changes in cell membrane permeability are often the first detectable responses of cells to infection by pathogens, to most host-specific and several nonspecific toxins, to certain pathogen enzymes, and to certain toxic chemicals, such as air pollutants. The most commonly observed effect of changes in cell membrane permeability is the loss of **electrolytes**, i.e., of small water-soluble ions and molecules from the cell. Electrolyte leakage occurs much sooner and at a greater rate when the host–pathogen interaction is incompatible, and the host remains more resistant than when the host is susceptible and develops extensive symptoms (Fig. 3-9). It is not certain, however, whether the cell membrane is the initial target of pathogen toxins and enzymes and whether the accompanying loss of electrolytes is the initial effect of changes in cell membrane permeability

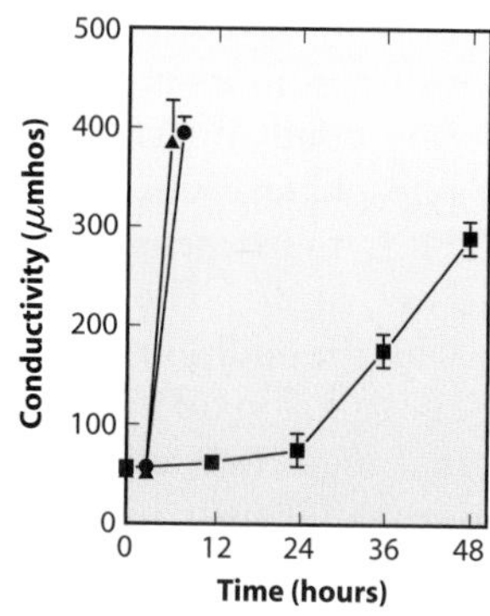

FIGURE 3-9 Levels of conductivity measuring the leakage of electrolytes released from leaves of pepper plants inoculated with three races of the bacterium *Xanthomonas campestris* pv. *vesicatoria*. (■) Release of electrolytes occurred later and at a slower rate when leaves were inoculated with a virulent race of the bacterium. (●,▲) Disruption of membranes and electrolyte leakage occurred much earlier, and at a much greater rate, when leaves were inoculated with two bacterial races carrying avirulence genes that triggered the hypersensitive response in plants carrying the corresponding resistance genes. [From Minsavage *et al.* (1990), *Mol. Plant-Microbe Interact.* 3, 41–47.]

or whether the pathogen products actually affect other organelles or reactions in the cell, in which case cell permeability changes and loss of electrolytes are secondary effects of the initial events. If pathogens do, indeed, affect cell membrane permeability directly, it is likely that they bring this about by stimulating certain membrane-bound enzymes, such as ATPase, which are involved in the pumping of H^+ in and K^+ out through the cell membrane, by interfering with processes required for the maintenance and repair of the fluid film making up the membrane, or by degrading the lipid or protein components of the membrane by pathogen-produced enzymes.

EFFECT OF PATHOGENS ON TRANSCRIPTION AND TRANSLATION

Transcription of cellular DNA into messenger RNA and translation of messenger RNA to produce proteins are two of the most basic, general, and precisely controlled processes in the biology of any normal cell (Fig. 3-10). The part(s) of the genome involved and the level and timing of transcription and translation vary with the stage of development and the requirements of each cell. Nevertheless, disturbance of any one of these processes, by pathogens or environmental factors, may cause drastic, unfavorable changes in the structure and function of the affected cells by its effect on the expression of genes.

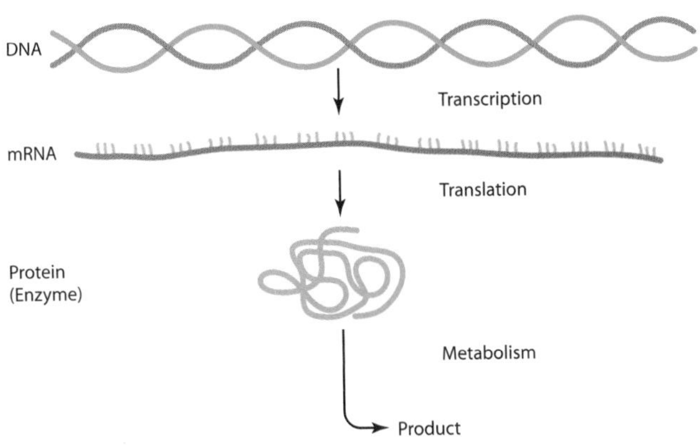

FIGURE 3-10 Transcription and translation processes.

Effect on Transcription

Several pathogens, particularly viruses and fungal obligate parasites, such as rusts and powdery mildews, affect the transcription process in infected cells. In some cases, pathogens affect transcription by changing the composition, structure, or function of the chromatin associated with the cell DNA. In some diseases, especially those caused by viruses, the pathogen, through its own enzyme or by modifying the host enzyme (RNA polymerase) that makes RNA, utilizes the host cell nucleotides and machinery to make its own (rather than host) RNA. In several diseases, the activity of ribonucleases (enzymes that break down RNA) is increased, perhaps by formation in infected plants of new kinds of ribonucleases not known to be produced in healthy plants. Finally, in several diseases, infected plants, particularly resistant ones, seem to contain higher levels of RNA than healthy plants, especially in the early stages of infection. It is generally believed that greater RNA levels and, therefore, increased transcription in cells indicate an increased synthesis of substances involved in the defense mechanisms of plant cells.

Effect on Translation

Infected plant tissues often have increased activity in several enzymes, particularly those associated with the generation of energy (respiration) or with the production or oxidation of various phenolic compounds, some of which may be involved in (defense) reactions to infection. Although a certain amount of some of these enzymes (proteins) may be present in the cell at the time of infection, several are produced *de novo*, necessitating increased levels of transcription and translation activity. Increases in protein synthesis in infected tissues have been observed primarily in hosts resistant to the pathogen and reach their highest levels in the early stages of infection, i.e., in the first few minutes and up to 2–20 hours after inoculation. If resistant tissues are treated before or during infection with inhibitors of protein synthesis, their resistance to the pathogen is reduced. These observations suggest that much of the increased protein synthesis in plants attacked by pathogens reflects the increased production of enzymes and other proteins involved in the defense reactions of plants.

EFFECT OF PATHOGENS ON PLANT GROWTH

It is easily understood and expected that pathogens that destroy part of the photosynthetic area of plants and cause significantly reduced photosynthetic output often result in smaller growth of these plants and smaller yields. Similarly, pathogens that destroy part of the roots of a plant or clog their xylem or phloem elements, thereby severely interfering with the translocation of water and of inorganic or organic nutrients in these plants, often cause a reduction in size and yields by these plants and, sometimes, their death. In many plant diseases, however, infected tissues or entire plants increase or reduce abnormally in size without a clear-cut explanation of how these changes are brought about. It is apparent that growth regulators affecting plant cell division and enlargement are involved, but very little is known about the specific compounds and mechanisms involved or the genes that control these events.

Some of the most common diseases in which pathogens cause obvious abnormal growth of their hosts' organs and tissues include clubroot of crucifers caused by the plasmodiophoromycete *Plasmodiophora brassicae*; alfalfa wart caused by the fungus *Physoderma alfalfae*, potato wart caused by the fungus *Spongospora subterranea*; peach leaf curl (Fig. 3-11A) and plum pockets (Fig. 3-11B) caused by the fungus *Taphrina sp.*, black knot canker of cherry caused by *Dibotryon morbosum* (Fig. 11-67A), Sphaeropsis gall of stone fruits caused by *Sphaeropsis sp.*; corn smut caused by *Ustilago maydis* (Figs. 5-16C and 11-144A–11-144C), dwarf bunt of wheat caused by *Tilletia contraversa* (Fig. 11-148), leaf gall of azalea caused by *Exobacidium azaleae* (Fig. 3-16A), and several rusts of pine trees caused by *Cronartium sp.* (Figs. 5-16D and 11-143). Some bacterial pathogens also cause abnormal growths such as crown gall (Fig. 3-11E) of many hosts and hairy root of apple caused by *Agrobacterium tumefaciens and A. rhizogenes*, respectively, olive knot and oleander gall caused by *Pseudomonas savastanoi*, and leafy gall of several hosts caused by *Rhodococcus sp.* (Fig. 5-17D).

A
B
C
D
E
F

Some characteristic effects on plant growth are caused by the phloem inhabiting phytoplasmas. Some phytoplasma-infected plants produce shoots that are yellowish, short, and bushy and are known as witches' brooms. Some phytoplasmas may cause stunting of their host and induce flower petals to become green as if they were leaves (known as phyllody). Nematodes are responsible for the very common root knot (Fig. 3-11F) of most cultivated plants caused by *Meloidogyne sp*.

The most frequent and unusual effects on plant growth are those caused by viruses (and viroids). Many viruses cause stunting (Fig. 3-11D) or dwarfing of infected plants, whereas others cause rolling or curling of leaves, abnormally shaped fruit, etc. Some viruses cause abnormalities even in the same leaf (Fig. 3-11C) where part of the leaf is thinner than normal and the rest is thicker than normal. Some viruses cause plants to produce galls on their root, stems, or leaves. Some induce pitting on the roots or stems of infected plants (Fig. 14-42E). How the various viruses bring about these effects on their respective hosts is not known.

EFFECT OF PATHOGENS ON PLANT REPRODUCTION

Pathogens that attack various organs and tissues of plants weaken and often kill these organs or tissues, thereby weakening the plants. As a result, such plants remain smaller in size, may produce fewer flowers, and may set fewer fruit and seeds; the latter may be of inferior vigor and vitality and, therefore, if planted, they may produce fewer and weaker new plants. In addition to these indirect effects of pathogens on plant reproduction, many pathogens have a direct adverse effect on plant reproduction because they attack and kill the flowers, fruit, or seed directly, or interfere and inhibit their production, or the pathogens interfere directly or indirectly with the propagation of their host plant.

One of the most common ways by which pathogens interfere with the reproduction of their host is by infecting and killing the flowers of the host, as happens, for example, with the brown rot of stone fruits caused by the fungus *Monilinia sp.* (Figs. 3-12A and 3-12B), the bacterial canker and gummosis of stone fruit trees caused by *Pseudomonas syringae*, and the fireblight disease of pears and apples caused by the bacterium *Erwinia amylovora.* In some diseases, e.g., in the post-bloom fruit drop of citrus, the fruit, soon after set, drops prematurely as a result of infection by the anthracnose fungus *Colletotrichum acutatum.* Similarly, plums drop prematurely from trees infected with the *plum pox virus.* In several plant diseases, especially in grain crops, the pathogen interferes directly with the reproduction of the plant host by killing the embryo, that would have produced the seed, and replacing the contents of the seed with its own fruiting structure or its own spores. Examples of such diseases are ergot of grains (Fig. 3-12C), caused by the fungus *Claviceps purpurea;* corn smut (Fig. 3-12D); and the covered (Fig. 3-12E) and loose smuts of the various cereals caused by *Tilletia and Ustilago sp.*, respectively. Finally, in some diseases caused by viruses, phytoplasmas, or phloem-limited bacteria, no flowers are produced or those produced are sterile, and therefore few or no fruit and seed are produced.

FIGURE 3-11 Effect of pathogens on plant growth. (A) Leaf curling and (B) fruit enlargement by the leaf curl fungus *Taphrina deformans* on peach and plum, respectively. (C) Leaf malformations caused by the *common bean mosaic virus* on bean and (D) a healthy and a plant showing stunting caused by the *maize streak virus* on corn (D). (E) Galls along the root and stem of a euonymus plant caused by the crown gall bacterium *Agrobacterium tumefaciens* and (F) galls along the roots of a plant caused by the root knot nematode *Meloidogyne sp.* [Photographs courtesy of (A and B) Oregon State University, (C) R. Provvidenti, Cornell University, (D) D. Coyne, Intrn. Inst. Trop. Agric., (E) R. Forster, Univ. of Idaho and (F) W. Crow, University of Florida.]

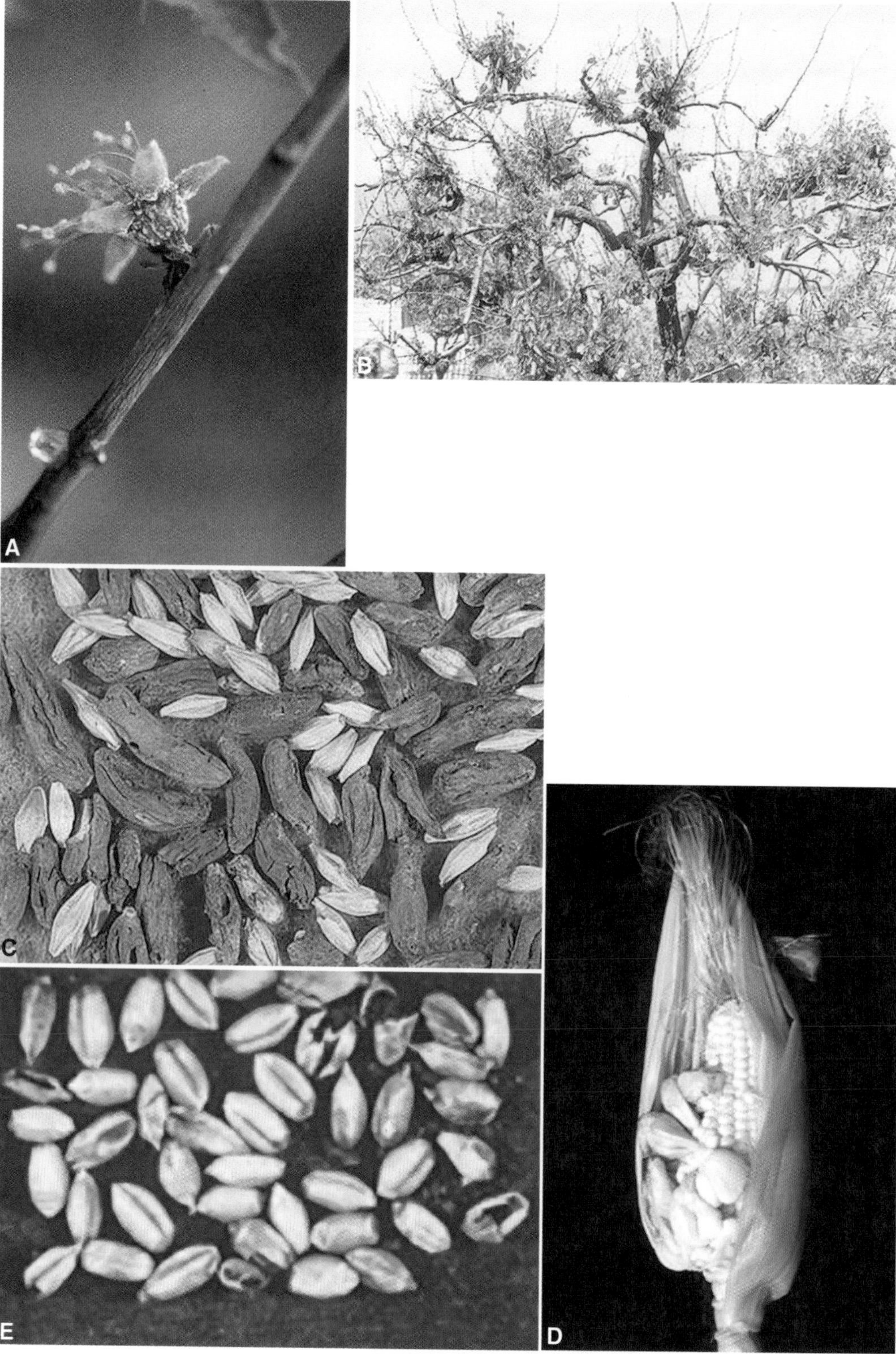

FIGURE 3-12 Ways in which pathogens affect plant reproduction. (A) Close-up of a flower and (B) macroscopic view of an apricot tree, the flowers of which have been killed by the brown rot fungus *Monilinia fructicola*. (C) A mixture of barley kernels (whitish-yellow) and ergot sclerotia (the larger black bodies) produced by the ergot fungus *Claviceps purpurea* on the heads of grain crops in place of healthy kernels. (D) Ear of corn having some of the corn kernels replaced by galls containing spores of the fungus *Ustilago maydis*. (E) A mixture of intact healthy wheat kernels and somewhat darker, broken wheat kernels filled with spores of the common bunt (covered smut) fungus *Tilletia* sp. [Photographs courtesy of (A and B) I. MacSwain, Oregon State University, (C) G. Munkvold, Iowa State University, (D) T. Zitter, Cornell University, and (E) J. Riesselman, USDA, Montana State University.]

Selected References

Allen, R. J. (1942). Changes in the metabolism of wheat leaves induced by infection with powdery mildew. *Am. J. Bot.* **29**, 425–435.

Arseniuk, E., Foremska, E., Óral, T. G., *et al.* (1999). Fusarium head blight reactions and accumulation of deoxynivalenol (DON) and some of its derivatives in kernels of wheat, triticale and rye. *J. Phytopathol.* **147**, 577–590.

Bassanezi, R. B., Amorim, L., Filho, A. B., *et al.* (2001). Accounting for photosynthetic efficiency of bean leaves with rust, angular leaf spot and anthracnose to assess crop damage. *Plant Pathol.* **50**, 443–452.

Beckman, C. H. (1987). "The Nature of Wilt in Plants." APS Press, St. Paul, MN.

Bowden, R. L., and Rouse, D. I. (1991). Effects of *Verticillium dahliae* on gas exchange of potato. *Phytopathology* **81**, 293–301.

Clover, *et al.* (1999). The effects of beet yellows virus on the growth and physiology of sugar beet (*Beta vulgaris*). *Plant Pathol.* **48**, 129–145.

Culver, J. N., Lindbeck, A. G. C., and Dawson, W. O. (1991). Virus-host interactions: Induction of chlorotic and necrotic responses in plants by tobamoviruses. *Annu. Rev. Phytopathol.* **29**, 193–217.

Eckardt, N. A. (2001). A calcium-regulated gatekeeper in phloem sieve tubes. *Plant Cell* **13**, 989–992.

Ellis, M. A., Ferree, D. C., and Spring, D. E. (1981). Photosynthesis, transpiration, and carbohydrate content of apple leaves infected by *Podosphaera leucotricha*. *Phytopathology* **71**, 392–395.

Goodman, R. N., Kiraly, Z., and Wood, K. R. (1986). "The Biochemistry and Physiology of Plant Disease." Univ. of Missouri Press, Columbia.

Hancock, J. G., and Huisman, O. C. (1981). Nutrient movement in host-pathogen systems. *Annu. Rev. Phytopathol.* **19**, 309–331.

Horsfall, J. G., and Cowling, E. B. (1978). "Plant Disease," Vol. 3. Academic Press, New York.

Livne, A., and Daly, J. M. (1966). Translocation in healthy and rust-infested beans. *Phytopathology* **56**, 170–175.

Manners, J. M., and Scott, K. H. (1983). Translational activity of polysomes of barley leaves during infection by *Erysiphe graminis* f. sp. *hordei*. *Phytopathology* **73**, 1386–1392.

Matteoni, J. A., and Sinclair, W. A. (1983). Stomatal closure in plants infected with mycoplasmalike organisms. *Phytopathology* **73**, 398–402.

McGrath, M. T., and Pennypacker, S. P. (1990). Alteration of physiological processes in wheat flag leaves caused by stem rust and leaf rust. *Phytopathology* **80**, 677–686.

Nelson, P. E., and Dickey, R. S. (1970). Histopathology of plants infected with vascular bacterial pathogens. *Annu. Rev. Phytopathol.* 8, 259–280.

Pennypacker, B. W., *et al.* (1990). Analysis of photosynthesis in resistant and susceptible alfalfa clones infected with *Verticillium alboatrum*. *Phytopathology* **80**, 1300–1306.

Samborski, D. J., Rohringer, R., and Kim, W. K. (1978). Transcription and translation in diseased plants. In "Plant Disease" (J. G. Horsfall and E. B. Cowling, eds.), Vol. 3, pp. 375–90. Academic Press, New York.

Shtienberg, D. (1992). Effects of foliar diseases on gas exchange processes: A comparative study. *Phytopathology* **82**, 760–765.

Chapter four

GENETICS OF PLANT DISEASE

INTRODUCTION
125

GENES AND DISEASE
126

VARIABILITY IN ORGANISMS-MECHANISMS OF VARIABILITY: GENERAL: MUTATION – RECOMBINATION – GENE AND GENOTYPE FLOW AMONG PLANT PATHOGENS – POPULATION GENETICS, GENETIC DRIFT, AND SELECTION – LIFE CYCLES – REPRODUCTION – MATING SYSTEMS – OUT-CROSSING-PATHOGEN FITNESS
128

SPECIALIZED MECHANISMS OF VARIABILITY: IN FUNGI: HETEROKARYOSIS – PARASEXUALISM-VEGETATIVE INCOMPATIBILITY-HETEROPLOIDY
131

IN BACTERIA. HORIZONTAL GENE TRANSFER
132

GENETIC RECOMBINATION IN VIRUSES
133

LOSS OF PATHOGEN VIRULENCE IN CULTURE
133

STAGES OF VARIATION IN PATHOGENS
134

TYPES OF PLANT RESISTANCE TO PATHOGENS
134

TRUE RESISTANCE: PARTIAL, QUANTITATIVE, POLYGENIC, OR HORIZONTAL RESISTANCE
136

R-GENE RESISTANCE, MONOGENIC, OR VERTICAL RESISTANCE
136

APPARENT RESISTANCE: DISEASE ESCAPE – TOLERANCE TO DISEASE
137

GENETICS OF VIRULENCE IN PATHOGENS AND OF RESISTANCE IN HOST PLANTS
139

THE GENE-FOR-GENE CONCEPT
140

THE NATURE OF RESISTANCE TO DISEASE
142

PATHOGENICITY GENES IN PLANT PATHOGENS
142

GENES INVOLVED IN PATHOGENESIS AND VIRULENCE BY PATHOGENS
142

PATHOGENICITY GENES IN PLANT PATHOGENIC FUNGI CONTROLLING: *PRODUCTION OF INFECTION STRUCTURES – DEGRADATION OF CUTICLE AND CELL WALL – SECONDARY METABOLITES: PHYTOANTICIPINS, PHYTOALEXINS – FUNGAL TOXINS – PATHOGENICITY SIGNALING SYSTEMS*
144

PATHOGENICITY GENES IN PLANT PATHOGENIC BACTERIA CONTROLLING: *ADHESION TO PLANT SURFACES – SECRETION SYSTEMS – ENZYMES THAT DEGRADE CELL WALLS – BACTERIAL TOXINS AS PATHOGENICITY FACTORS – EXTRACELLULAR POLYSACCHARIDES AS PATHOGENICITY FACTORS – BACTERIAL REGULATORY SYSTEMS AND NETWORKS – SENSING PLANT SIGNALING COMPONENTS – OTHER BACTERIAL PATHOGENICITY FACTORS*
146

PATHOGENICITY GENES IN PLANT VIRUSES: – *FUNCTIONS ASSOCIATED WITH THE COAT PROTEIN-VIRAL PATHOGENICITY GENES*
149

NEMATODE PATHOGENICITY GENES
150

GENETICS OF RESISTANCE THROUGH THE HYPERSENSITIVE RESPONSE
151

PATHOGEN-DERIVED ELICITORS OF DEFENSE RESPONSES IN PLANTS AVIRULENCE (AVR) GENES: ONE OF THE ELICITORS OF PLANT DEFENSE RESPONSES CHARACTERISTICS OF AVR GENE-CODED PROTEINS: – STRUCTURE OF AVR GENE PROTEINS – FUNCTION OF AVR GENE PROTEINS ROLE OF *AVR* GENES IN PATHOGENICITY AND VIRULENCE
151

HRP GENES AND THE TYPE III SECRETION SYSTEM: ALSO PATHOGENICITY GENES
155

RESISTANCE (R) GENES OF PLANTS
155

EXAMPLES OF R GENES – HOW DO R GENES CONFER RESISTANCE? – EVOLUTION OF R GENES – OTHER PLANT GENES FOR RESISTANCE TO DISEASE
156

SIGNAL TRANSDUCTION BETWEEN PATHOGENICITY GENES AND RESISTANCE GENES
159

SIGNALING AND REGULATION OF PROGRAMMED CELL DEATH
160

GENES AND SIGNALING IN SYSTEMIC ACQUIRED RESISTANCE
161

EXAMPLES OF MOLECULAR GENETICS OF SELECTED PLANT DISEASES: – THE POWDERY MILDEW DISEASE – *MAGNAPORTHE GRISEA*, THE CAUSE OF RICE BLAST – *FUSARIUM*, THE SOILBORNE PLANT PATHOGEN – *USTILAGO MAYDIS* AND CORN SMUT
161

BREEDING OF RESISTANT VARIETIES NATURAL VARIABILITY IN PLANTS – BREEDING AND VARIABILITY IN PLANTS – BREEDING FOR DISEASE RESISTANCE: SOURCES OF GENES FOR RESISTANCE – TECHNIQUES USED IN CLASSICAL BREEDING FOR RESISTANCE – SEED, PEDIGREE, AND RECURRENT SELECTION
165

TISSUE CULTURE AND GENETIC ENGINEERING TECHNIQUES
168

GENETIC TRANSFORMATION OF PLANT CELLS FOR DISEASE RESISTANCE
169

ADVANTAGES AND PROBLEMS IN BREEDING FOR VERTICAL OR HORIZONTAL RESISTANCE
169

VULNERABILITY OF GENETICALLY UNIFORM CROPS TO PLANT DISEASE EPIDEMICS
170

INTRODUCTION

The genetic information of all organisms, i.e., the information that determines what an organism can be and can do, is encoded in its deoxyribose nucleic acid (DNA). In RNA viruses, of course, it is encoded in their ribose nucleic acid (RNA). In all organisms, most DNA is present in the chromosome(s). In prokaryotes, such as bacteria and mollicutes, which lack an organized, membrane-bound nucleus, there is only one chromosome and it is present in the cytoplasm, whereas in eukaryotes, i.e., all other organisms except viruses, there are several chromosomes and they are present in the nucleus. Many prokaryotes, however, and some of the lower eukaryotes also carry smaller circular molecules of DNA called **plasmids** in the cytoplasm. Plasmid DNA also carries genetic information but multiplies and moves independently of the chromosomal DNA. Furthermore, all cells of eukaryotic organisms carry DNA in their mitochondria. Plant cells, in addition to nuclear and mitochondrial DNA, also carry DNA in their chloroplasts (Fig. 4-1).

Genetic information in DNA is encoded in a linear fashion in the order of the four bases (A, adenine; C, cytosine; G, guanine; and T, thymine). Each triplet of adjacent bases codes for a particular amino acid. A **gene** is a stretch of a DNA molecule, usually of about 100 to 500 or more adjacent triplets, that codes for one protein molecule or, in a few cases, one RNA molecule (Fig. 4-2). In eukaryotes, the coding region of a gene is often interrupted by noncoding stretches of DNA called **introns** (Fig. 4-3). When a gene is active, i.e., is expressed, one of its DNA strands is used as a template and is transcribed into an RNA strand. Some genes code only for an RNA and that RNA is either a transfer RNA (tRNA) or a ribosomal RNA (rRNA). Most genes encode proteins, however, and the transcription product is a messenger RNA (mRNA). The mRNA then becomes attached to ribosomes, which, with the help of tRNAs, translate the base sequence of the mRNA strand into a specific sequence of amino acids that folds into a specific shape and forms a particular protein. Different genes code for different proteins. Some of the proteins are part of the structure of cell membranes, but most act as enzymes. Proteins give cells and organisms their characteristic properties, such as shape, size, and color; determine what kinds of chemical substances are produced by the cell; and regulate all activities of cells and organisms.

Of course, not all genes in a cell are expressed at all times, as different kinds of cells at different times have different functions and needs. Which genes are turned on, when they are turned on or off, and for how long they stay on are all regulated by additional stretches of DNA called **promoters, enhancers, silencers**, or **terminators**. These act as signals for genes to be expressed or to stop being expressed or they act as signals for the production of RNAs and proteins that themselves act as inducers, promoters, and enhancers of gene expression or as repressors and terminators of gene expression. In many cases of host–pathogen interaction, genes in the one organism are triggered to be expressed by a substance produced by the other organism. For example, genes for cell wall-degrading enzymes in the pathogen are apparently induced by the presence of monomers or oligomers of host cell wall macromolecules that are substrates for these enzymes. Also, genes for defense reactions in the host, e.g., the production of phytoalexins, apparently are triggered to expression by certain signal compounds activated by inducer molecules (elicitors) produced by the pathogen.

GENES AND DISEASE

When different plants, such as tomato, apple, or wheat, become diseased as a result of infection by a pathogen, the pathogen is generally different for each kind of host plant. Moreover, the pathogen is often specific for that particular host plant. Thus, the fungus *Fusarium oxysporum* f. sp. *lycopersici*, which causes tomato wilt, attacks only tomato and has absolutely no effect on

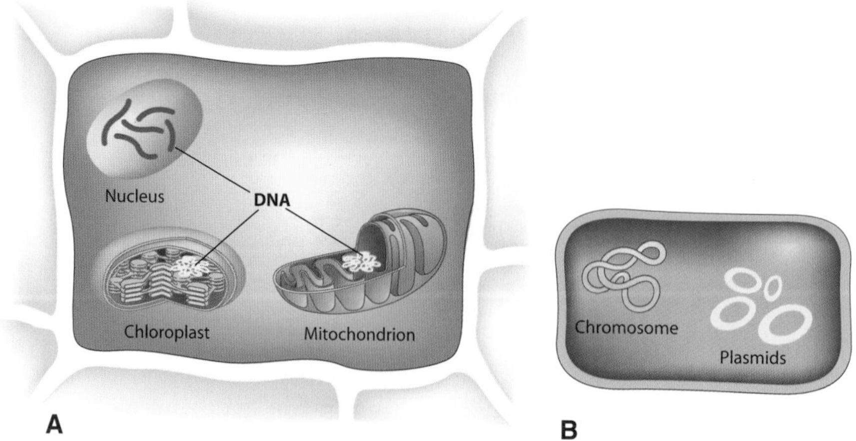

FIGURE 4-1 Location and arrangement of the genetic material in (A) eukaryotic (plant) cells and (B) prokaryotic (bacterial) cells.

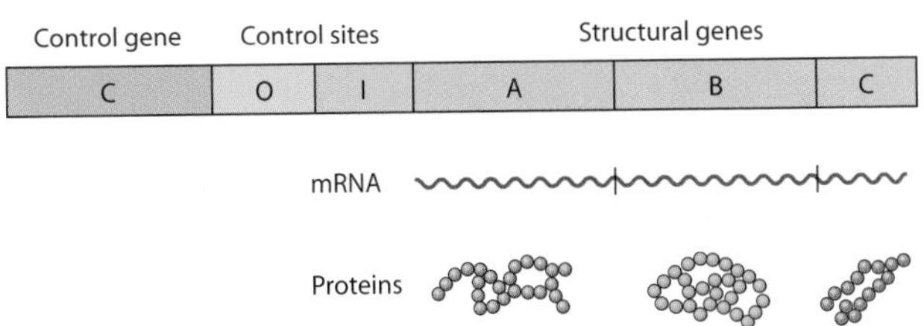

FIGURE 4-2 Gene structure, control, and expression in prokaryotes.

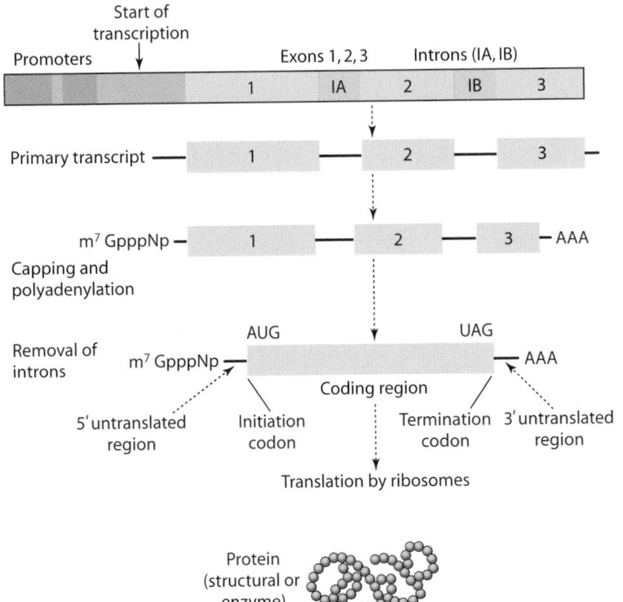

FIGURE 4-3 Gene structure, control, and expression in eukaryotes.

apple, wheat, or any other plant. Similarly, the fungus *Venturia inaequalis*, which causes apple scab, affects only apple, whereas the fungus *Puccinia graminis* f. sp. *tritici*, which causes stem rust of wheat, attacks only wheat. What makes possible the development of disease in a host is the presence in the pathogen of one or more genes for pathogenicity, for specificity, and for virulence against the particular host.

The gene(s) for virulence in a pathogen is usually specific for one or a few related kinds of plants that are hosts to the pathogen. Also, the genes and gene combinations that make a plant susceptible, i.e., a host to a particular pathogen, are present only in that one kind of plant and possibly a few related kinds of plants. All plants also have preformed and induced defenses that provide resistance against most pathogens. The specificity of microbial virulence genes that condition growth and disease on particular plants explains why a pathogen that is virulent on one kind of plant is not able to attack other kinds of plants and why a plant that is susceptible to one pathogen is not susceptible to all other pathogens of other host plants. This is known as nonhost resistance (Figs. 4-4 and 4-5).

Of course, a few pathogens are able to attack many kinds, sometimes hundreds, of host plants. Such pathogens tend to be necrotrophs and can attack so many hosts apparently because they either have many diverse genes for virulence or, more likely, because their genes of virulence somehow have much less plant specificity than those of the commonly more specialized pathogens. Each species of plant, however, seems to be susceptible to a fairly small number of different pathogens, usually less than a hundred for most plants.

Despite the many pathogens that can infect them, sometimes a few and many times countless numbers of individuals of a single plant species, such as corn, wheat, or soybean, survive in huge land expanses year after year. These plants survive either free of disease or with only minor symptoms, even though most of the other plants in the field have been killed (Fig. 4-5) and their pathogens are often widespread among the surviving plants. Why are all the plants not attacked by their pathogens, and why are those that are attacked not usually killed by the pathogens? The answer is complex, but basically it happens because plants, through evolution or through systematic breeding, have acquired, in addition to the genes that make them susceptible to a pathogen, one or usually numerous genes for resistance that protect the plants from infection or from severe disease. When a new gene for resistance to a pathogen

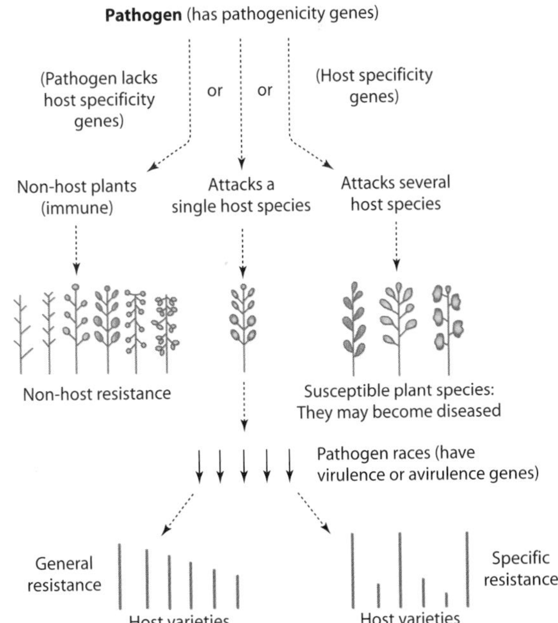

FIGURE 4-4 General interactions of a pathogen with its host and nonhost plants.

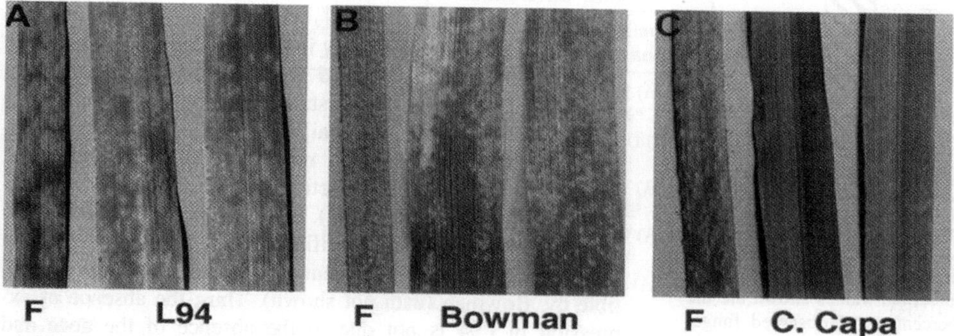

FIGURE 4-5 Infection types of two seedling leaves from each of three barley cultivars 10 days after artificial inoculation with the inappropriate wheat leaf rust fungus *Puccinia triticina*. Wheat cultivar F was used as the susceptible control. Only the cultivar C. Capa behaved as a nonhost. Infection of the others was bridged by a pathogen presumably but apparently not limited to wheat. [Photograph courtesy of Feuillet *et al.* (2003). *Mol. Plant-Microbe Interact.* **16**, 626–633.]

appears or is introduced into a plant, the plant becomes resistant to all or most of the previously existing individuals of the pathogen. Such pathogens contain one and usually more than one gene for virulence, but if they do not contain the additional new gene for virulence that is required to overcome the effect of the new resistance gene in the plant, they cannot infect the plant and the plant remains resistant. Thus, even one new gene for resistance to a pathogen can protect plants that have the gene from becoming infected by all or most preexisting races of the pathogen — at least for several months and possibly for several years.

It has been the experience of researchers with numerous host–pathogen combinations, however, that, after a new gene for resistance to a pathogen is introduced into a crop variety and that variety is planted in the fields, a new population (race) of the pathogen appears that contains a new gene for virulence that enables the pathogen to attack the crop plants containing the new gene for resistance. How did this new population of pathogens acquire the new gene for virulence? In most cases the new gene had already been present earlier at low levels, or by mutation, but only in a few pathogen individuals. New genes can arise randomly and suddenly *de novo* through mutations, or by rearrangement of the genetic material of the pathogens through the ever-ongoing events of genetic variability in organisms. Such pathogen individuals may have been but a tiny proportion of the total pathogen population and were undetected before plants with the new resistance gene were planted widely. After such plants were introduced, however, the new resistance gene excluded all other pathogen individuals except the few containing the new gene for virulence, which could attack these plants. Exclusion of the pathogens that lacked the new gene allowed the few that carried the gene to multiply and take over.

VARIABILITY IN ORGANISMS

One of the most dynamic and significant aspects of biology is that characteristics of individuals within a species are not "fixed," i.e., they are not identical but vary from one individual to another. As a matter of fact, all individuals produced as a result of a sexual process, such as the children of one family, are expected to be different from one another and from their parents in a number of characteristics, although they retain most similarities with them and belong to the same species. This is true oomycetes and of fungi produced from sexual spores such as oospores, ascospores, and basidiospores; of parasitic higher plants produced from seeds; and of nematodes produced from fertilized eggs, as well as of cultivated plants produced from seeds. Even bacteria have mechanisms for the transfer of genetic information. When individuals are produced asexually, the frequency and degree of variability among the progeny are reduced greatly, but even then certain individuals among the progeny will show different characteristics. Because of the astronomical number of individuals produced by microorganisms asexually, the total amount of variability produced by at least some microorganisms is probably as great and possibly greater than the total variability found in microorganisms reproducing sexually. This is the case in the overwhelmingly asexual reproduction of fungi by means of conidia, zoospores, sclerotia, and uredospores, and in bacteria, mollicutes, and viruses.

MECHANISMS OF VARIABILITY

In host plants and in pathogens, such as most fungi, parasitic higher plants, and nematodes, which can, and usually do, reproduce by means of a sexual process, variation in the progeny is introduced primarily through segregation and recombination of genes during the meiotic division of the zygote. Bacteria too, and even viruses, exhibit variation that seems to be the result of a sexual process. In many fungi, heteroploidy and certain parasexual processes lead to variation. However, all plants and all pathogens, especially bacteria, viruses, and fungi, and probably mollicutes, can and do produce variants by means of mutations in the absence of any sexual process.

General Mechanisms of Variability

Two mechanisms of variability, namely mutation and recombination, occur in both plants and pathogens.

Mutation

A mutation is a more or less abrupt change in the genetic material of an organism, which is then transmitted in a hereditary fashion to the progeny. Mutations represent changes in the sequence of bases in the DNA either through substitution of one base for another or through addition or deletion of one or many base pairs. Additional changes may be brought about by amplification of particular segments of DNA to multiple copies; by insertion or excision of a transposable element, i.e., a movable DNA segment, into a coding or regulatory sequence of a gene; and by inversion of a DNA segment. On average, one mutation occurs for every million copies of a gene per generation. Since the average fungus genome consists of about 10,000 genes, one cell in a hundred could be a mutant or, stated differently, there are many mutants in every colony of a fungus or a bacterium, etc. Mutation at a locus that codes for an enzyme can result in an allele that produces an altered form of the enzyme, often called an allozyme. Mutations occur spontaneously in nature in all living organisms: those that produce only sexually or only asexually and those that reproduce both sexually and asexually. Mutations in single-celled organisms, such as bacteria, in fungi with a haploid mycelium, and in viruses, may be expressed immediately after their occurrence. Most mutations, however, are usually recessive; therefore, in diploid or dikaryotic organisms, mutations can remain unexpressed until they are brought together in a homozygous condition.

Mutations for virulence probably occur no more frequently than mutations for any other inherited characteristics, but given the great number of progeny produced by pathogens, it is probable that large numbers of mutants differing in virulence from their parent are produced in nature every year. In addition, considering that only a few genetically homogeneous varieties of each crop plant are planted continuously over enormous land expanses for a number of years, and considering the difficulties involved in shifting from one variety to another on short notice, the threat of new, more virulent, mutants appearing and attacking a previously resistant variety is a real one. Moreover, once a new factor for virulence appears in a mutant, this factor will take part in the sexual or parasexual processes of the pathogen and may produce recombinants possessing virulence quite different in degree or nature from that existing in the parental strains.

Because plants and pathogens contain genetic material (DNA) outside the cell nucleus in the form of organelle or plasmid DNA or even as double-stranded RNA, mutations in the extranuclear DNA are just as common as those in the nuclear DNA and affect whatever characteristics are controlled by the extranuclear DNA. Because the inheritance of characteristics controlled by extranuclear DNA (cytoplasmic inheritance) does not follow the Mendelian laws of genetics, mutations on that DNA are more difficult to detect and characterize. Through mutations in extrachromosomal DNA, many pathogens acquire (or lose) the ability to carry out a physiological process that they could not (or could) before. Cytoplasmic inheritance presumably occurs in all organisms except viruses and viroids, which lack cytoplasm. Three types of adaptations brought about by changes in the genetic material of the cytoplasm have been shown in pathogens: Pathogens may acquire the ability to tolerate previously toxic substances, to utilize new substances for growth, and to change their virulence toward host plants. Several characteristics of plants are also inherited through extranuclear DNA, including resistance or susceptibility to infection by certain pathogens.

Recombination

Recombination occurs primarily during the sexual reproduction of plants, fungi, and nematodes whenever two haploid (1N) nuclei, containing genetic material that may differ in many loci, unite to form a diploid (2N) nucleus, called a zygote. The zygote, sooner or later, divides meiotically and produces new haploid cells (gametes, spores, mycelium). Recombination of genetic factors (different genes or alleles of the same genes) occurs during the meiotic division of the zygote as a result of genetic crossovers in which parts of chromatids (and the genes they carry) of one chromosome of a pair are exchanged with parts of chromatids of the other chromosome of the pair. Recombination of the genes of two parental nuclei takes place in the zygote, and the eventual haploid nuclei or gametes resulting after meiosis are different both from the gametes that produced the zygote and from one another. Over time, an organism may accumulate several alleles of a gene that code for slightly different forms of an enzyme, called isozymes. Such enzymes are controlled by genes at different loci and function under slightly different conditions of temperature, pH, etc. Recombination can also occur during mitotic cell division in the course of growth of an individual and it is thought to account for a significant amount of genetic exchange in fungi. In the fungi, haploid nuclei or gametes often divide mitotically to produce haploid mycelium and spores, which results in genetically different groups of relatively homogeneous individuals that may produce large populations asexually until the next sexual cycle.

Gene and Genotype Flow among Plant Pathogens

Gene flow is the process by which certain alleles (genes) move from one population to another geographically separated population. In plant pathology, gene flow is very important because it deals with the movement of virulent mutant alleles among different field populations. High gene flow in a pathogen increases the size of the population and of the geographical area in which its genetic material occurs. Therefore, pathogens that show a high level of gene flow generally have greater genetic diversity than pathogens that show a low level of gene flow. In pathogens reproducing only asexually, in which no recombination occurs, entire genotypes can be transferred from one population to another. This is known as genotype flow. Pathogens that produce hardy spores or other propagules, such as rust and powdery mildew fungi, that can spread over long distances, can distribute their genomes over large areas, sometimes encompassing entire continents. However, soil-borne fungi and nematodes move slowly and are present in small areas and their level of genetic flow is limited. With all types of pathogens, however, their gene flow can be affected significantly by human agricultural practices and by intercontinental travel and commerce. In general, pathogens with a high level of gene flow or genotype flow are much more effective and pose a greater threat to agriculture than pathogens with a low level of gene flow. Also, because asexual spores and propagules contain an already well-adapted and selected set of alleles, such propagules, through their genotype flow, pose a greater threat in enlarging the area of their adaptation than sexual propagules through their gene flow.

The frequency of alleles of importance in a population is affected by gene flow from other populations. Its magnitude depends on the number of incoming outside individuals into the population compared to the size of the population, as well as the number of different alleles brought into the population by outside individuals. Usually, allele frequencies in small populations adjacent to large ones are influenced strongly by gene flow than under any different conditions. Gene flow between distant populations is generally sporadic unless it is facilitated by intervening populations that act as stepping stones for the pathogen. The effect of gene flow is to reduce genetic differences between populations, thereby preventing or delaying the evolution of the populations in different geographical areas into separate species of the pathogen.

Population Genetics, Genetic Drift, and Selection

The size of a population affects the frequency of survival of mutants and thereby the diversity of genes in the population. Populations of most organisms in a geographic area may not be large enough to ensure that each variant will have progeny in the next generation so that random effects would occur during the transmission of genetic traits to new generations. This is known as random genetic drift. Because mutation rates are generally low (about one in a million), large populations are expected to have more mutants than small populations (a population of one million would have one mutant, another one of one billion would have 1,000 mutants). It is obvious that it is more likely that the one mutant of the smaller population will be lost than the 1,000 mutants of the larger population, i.e., in small populations, genetic drift results in a loss of alleles over time. In plant pathology, pathogens that exist in large populations have a greater potential for evolution than pathogens that exist in small populations. Large populations increase the probabilities that new mutants with greater fitness will emerge within a host, will be able to multiply in it, and will spread to a new host before the mutation is lost through genetic drift. Also, cultural practices, including chemical control, which regularly severely reduce pathogen populations in the field, are less diverse and much slower to adapt than populations that are allowed to maintain high populations year round.

Selection is a directional process by which the fittest variants in a particular environment increase their frequency in the population (positive selection), whereas less fit variants decrease their frequency (negative selection). As a result of selection in a population large enough for all variants to have progeny in the next generation, the frequency of a variant at equilibrium provides an estimate of the fitness of the variant. Selection results in a decrease in the diversity within a population, but it may cause an increase in the diversity between populations. Selection is affected by almost every factor in the life cycle of a pathogen, whether related to the pathogen itself, to its host, its vector if any, and to the environment.

Life Cycles: Reproduction — Mating Systems — Outcrossing

Life cycles of various plant pathogens vary considerably, being most complex in some oomycetes and fungi. While life cycles are very simple, and basically asexual, in bacteria and in mitosporic fungi, in most fungi and in oomycetes they can involve a strictly haploid life cycle, a haploid–dikaryotic life cycle, a haploid–diploid one, a diploid one, or an asexual one. The kind of life cycle and the mating system followed affect the opportunities and limitations for genetic diversity (gene or genome diversity) and evolution of each particular pathogen. As a brief example we will mention the wheat stem rust fungus *Puccinia graminis* f. sp. *tritici*, which in the haploid state infects barberry while in the diploid state it infects wheat. Reproduction can be sexual, asexual, or both. The mating system is important only in relation to the sexual component of reproduction and can vary from inbreeding to outcrossing. In asexual pathogen populations, genotype diversity is more significant than gene diversity, whereas sexual pathogen populations show more gene diversity. Therefore, pathogens undergoing any type of recombination pose a greater threat than pathogens that undergo little or no recombination. The result of this is that the recombining pathogen population can put together new combinations of virulence genes or alleles as fast as breeders can put together genes for resistance and, therefore, pyramiding resistance genes in plants may not be as effective a strategy for as long as plant breeders hoped it would. Also, pathogens that outcross, through which more new genotypes are created, pose a greater threat to crops than inbreeding pathogens.

Pathogen Fitness

Fitness is the ability of a pathogen to survive and reproduce. The fitness of a pathogen or parasite can be quantified by measuring its reproductive rate, rate of multiplication, efficiency of infection, and amount of disease caused (aggressiveness). Fitness seems to be the driving force in the stability and evolution of a pathosystem in agriculture. In a freely mating system, excess virulence genes in a pathogen population constitute a genetic load or drag so that future selection favors genotypes free of excess genes. Even the presence of excess genes for virulence imposes a fitness penalty to the pathogen. Therefore, a mutation from avirulence to virulence occurs only if it is needed to overcome an R gene for resistance, i.e., only if it is absolutely necessary for the pathogen to survive. So, for a specific interaction between a pathogen with an avirulence gene and a host with a matching R gene for resistance, a mutation to virulence will occur because it increases the fitness of the pathogen to survive while the R gene is present. If, however, the mutation from avirulence to virulence gene carries a fitness penalty, the pathogen will suffer from reduced fitness on the host in the absence of the R gene. Many genes coding for fitness attributes or for virulence also encode the avirulence or host recognition function. Therefore, if loss of the function for avirulence is associated with a cost to fitness, represented by k, then the reduced fitness of the gene should appear on both host varieties, the one with and the one without the corresponding R gene resistance ($1-k$). It has been suggested that if this is true, then the greater the cost of fitness (the greater the value of k), the more durable the resistance of the variety is likely to be. Although some experimental data support this hypothesis, others are inconclusive.

Specialized Mechanisms of Variability in Pathogens

Certain mechanisms for generating variability appear to operate only in certain kinds of organisms or to operate in a rather different manner than those described as general mechanisms of variability. These specialized mechanisms of variability include heterokaryosis, heteroploidy, and parasexualism in fungi; conjugation, transformation, and transduction in bacteria; and genetic recombination in viruses.

Sexual-like Processes in Fungi

Heterokaryosis

Heterokaryosis is the condition in which, as a result of fertilization or anastomosis, cells of fungal hyphae or parts of hyphae contain two or more nuclei that are genetically different. For example, in Basidiomycetes, the dikaryotic state may differ drastically from the haploid mycelium and spores of the fungus. In *P. graminis tritici*, the fungus causing stem rust of wheat, the haploid basidiospores can infect barberry but not wheat, and the haploid mycelium can grow only in barberry; however, the dikaryotic aeciospores and uredospores can infect wheat but not barberry, and the dikaryotic mycelium can grow in both barberry and wheat. Heterokaryosis also occurs in other fungi, but its importance in plant disease development in nature is not known.

Parasexualism

Parasexualism is the process by which genetic recombinations can occur within fungal heterokaryons. This comes about by the occasional fusion of the two nuclei and formation of a diploid nucleus. During multiplication, crossing-over occurs in a few mitotic divisions and results in the appearance of genetic recombinants as the diploid nucleus progressively and rapidly loses individual chromosomes to revert to its haploid state. Considering that fungi exist and grow primarily as adjacent hyphae that may form heterokaryons as a result of anastomoses or fertilization, the frequency of parasexualism and therefore of genetic variability through parasexualism may equal or surpass that brought about by sexual reproduction.

Vegetative Incompatibility

In many fungi, vegetative hyphae of the same colony, or of two colonies of the same species, coming in contact with each other, often fuse, and the fusion is called hyphal anastomosis. If, however, hyphae coming in contact belong not to different strains of the fungus but of the same species, no fusion of hyphae takes place and the phenomenon is called vegetative incompatibility (or somatic or heterokaryon incompatibility). In only a few filamentous fungi, such as the species *Thanatephorus cucumeris*, the telomorph of *Rhizoctonia solani*, does fusion incompatibility occur between distantly related strains that appear to be different species, but when it does occur, it prevents both vegetative fusion and sexual fusion and, thereby, does not allow the exchange of genetic material. It has been suggested, therefore, that perhaps the different fusion incompatibility groups constitute different biological species still unrecognized within the broad species of *T. cucumeris*.

When hyphae from two colonies that belong to different postfusion incompatibility groups meet, the hyphae fuse, but subsequently the protoplasm in the two fused hyphal compartments and some adjacent ones is destroyed and a demarcation zone of sparse and sometimes dark mycelium is produced. Such postfusion incompatibility is the result of interaction between two alleles of the same vegetative compatibility (v-c) locus and is called allelic incompatibility. Vegetative incompatibility appears to be a defense mechanism that protects individuals from harmful nuclei, mitochondria, plasmids, and viruses that could reach them from other cells through anastomosis.

Heteroploidy

Heteroploidy is the existence of cells, tissues, or whole organisms with numbers of chromosomes per nucleus that are different from the normal 1N or 2N complement for the particular organism. Heteroploids may be haploids, diploids, triploids, or tetraploids or they may be aneuploids, i.e., have one, two, three, or more extra chromosomes or are missing one or more chromosomes from the normal euploid number (e.g., N + 1). Heteroploidy is often associated with cellular differentiation and represents a normal situation in the development of most eukaryotes. In several studies, spores of the same fungus were found to contain nuclei with chromosome numbers ranging from 2 to 12 per nucleus and also diploids and polyploids. Because it has been shown that the expression of different genes is proportional to, inversely proportional to, or unaffected by dosage, obviously the existence of heteroploid cells or heteroploid whole individuals of some pathogens increases the degree of variability that can be exhibited by these pathogens. Heterploidy has been observed repeatedly in fungi and has been shown to affect the growth rate, spore size and rate of spore production, hyphal color, enzyme activities, and pathogenicity. It has been shown, for example, that some heteroploids, such as diploids of the normally haploid fungus *Verticillium alboatrum*, the cause of wilt in cotton, lose the ability to infect cotton plants even when derived from highly virulent haploids. How much of the variability in pathogenicity in nature is due to heteroploidy is still unknown.

Sexual-like Processes in Bacteria and Horizontal Gene Transfer

New biotypes of bacteria seem to arise with varying frequency by means of at least three sexual-like processes (Fig. 4-6). It is probable that similar processes occur in mollicutes. (1) **Conjugation** occurs when two compatible bacteria come in contact with one another and a small portion of the chromosome or plasmid from one bacterium is transferred to the other through a conjugation bridge or pilus. (2) In **transformation**, bacterial cells are transformed genetically by absorbing and incorporating in their own cells genetic material secreted by, or released during rupture of, other bacteria. (3) In **transduction**, a bacterial virus (phage) transfers genetic material from the bacterium in which the phage was produced to the bacterium it infects next. The transfer of genetic information in this manner is not always limited to members of the same species or even genus (**vertical inheritance**). For example, gram-negative bacteria can transmit genetic material readily across species; *Agrobacterium* transmits genes across kingdom barriers to plants. Such events are called **horizontal gene transfers.**

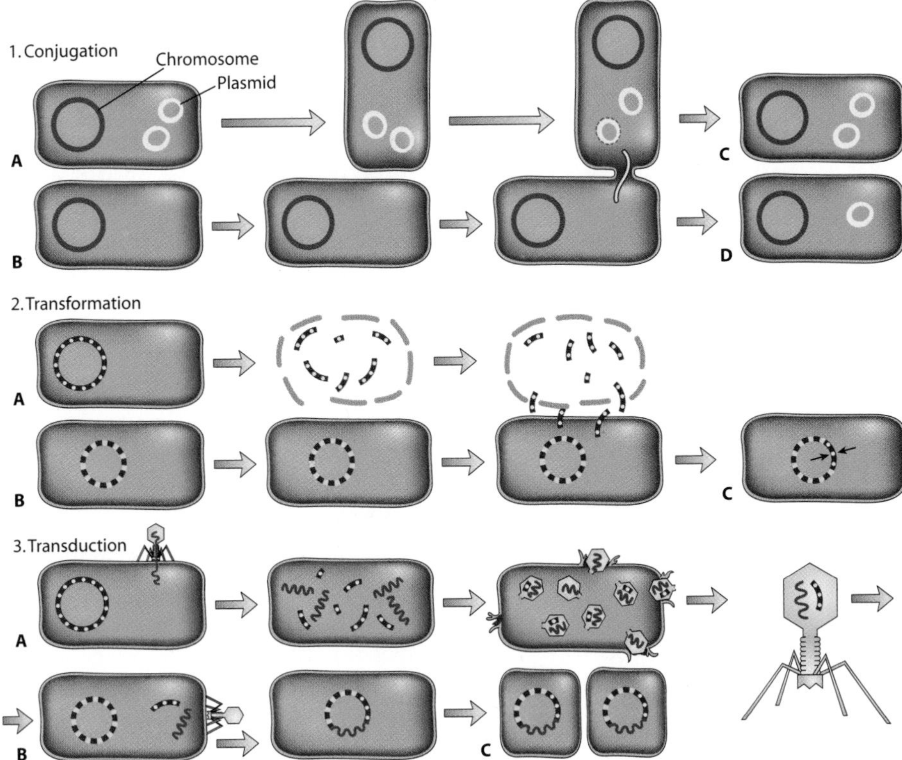

FIGURE 4-6 Mechanisms of variability in bacteria through sexual-like processes.

Genetic Recombination in Viruses

When two strains of the same virus are inoculated into the same host plant, one or more new virus strains are recovered with properties (virulence, symptomatology, and so on) different from those of either of the original strains introduced into the host. The new strains probably are recombinants, although their appearance through mutation, not hybridization, cannot always be ruled out. In multipartite viruses consisting of two, three, or more nucleic acid components, new virus strains may also arise in host plants or vectors from recombination of the appropriate components of two or more strains of such viruses.

Loss of Pathogen Virulence in Culture

The virulence of pathogenic microorganisms toward one or all of their hosts often decreases when the pathogens are kept in culture for relatively long periods of time or when they are passed one or more times through different hosts. If the culturing of the pathogen is prolonged sufficiently, the pathogen may lose virulence completely. Such partial or complete loss of virulence in pathogens is sometimes called **attenuation**, and it has been shown to occur in bacteria, fungi, and viruses. Pathogens that have experienced partial or complete loss of virulence in culture or in other hosts are often capable of regaining part or all of their virulence if they are returned to their hosts under proper conditions. Sometimes, however, the loss of virulence may be irreversible. "Loss" of virulence in culture, or in other hosts, seems to be the result of selection of individuals of less virulent or avirulent pathogen strains that happen to be capable of growing and multiplying in culture, or in the other host, much more rapidly than virulent ones. After several transfers in culture or the other hosts, such attenuated individuals largely, or totally, overtake and replace the virulent ones in the total population so that the pathogen is less virulent or totally avirulent. On reinoculation of the proper host, isolates in which the virulent individuals have been totally replaced by avirulent ones continue to be avirulent, and therefore loss of pathogenicity is irreversible. However, on reinoculation of the proper host with isolates in which at least some virulent individuals survived through the transfers in culture or the other host, the few surviving virulent individuals infect the host and multiply, often in proportion to their virulence. The virulent individuals increase in number with each subsequent inoculation while at the same time nonvirulent individuals are reduced or eliminated with each reinoculation.

STAGES OF VARIATION IN PATHOGENS

The entire population of a particular organism on the earth, e.g., a fungal pathogen, has certain morphological and other phenotypic characteristics in common and makes up the **species** of pathogen, such as *Puccinia graminis*, the cause of stem rust of cereals. Some individuals of this species, however, attack only wheat, barley, or oats, and these individuals make up groups that are called **varieties** or **special forms** (*formae specialis*) such as *P. graminis* f. sp. *tritici* or *P. graminis tritici*, *P. graminis hordei*, and *P. graminis avenae* (Table 4-1). Even within each special form, however, some individuals attack some of the varieties of the host plant but not others, some attack another set of host plant varieties, and so on, with each group of such individuals making up a race. Thus, there are more than 200 races of *P. graminis tritici* (race 1, race 15, race 59, and so on). Occasionally, one of the offspring of a race can suddenly attack a new variety or can cause severe symptoms on a variety that it could barely infect before. This individual is called a **variant**. The identical individuals produced asexually by the variant make up a **biotype**. Each race consists of one or several biotypes (race 15A, 15B, and so on).

The appearance of new pathogen biotypes may be very dramatic when the change involves the host range of the pathogen. If the variant has lost the ability to infect a plant variety that is widely cultivated, this pathogen simply loses its ability to procure a livelihood for itself and will die without even making its existence known to us. If, however, the change in the variant pathogen enables it to infect a plant variety cultivated because of its resistance to the parental race or strain, the variant individual, being the only one that can survive on this plant variety, grows and multiplies on the new variety without any competition and soon produces large populations that spread and destroy the heretofore resistant variety. This is the way the resistance of a plant variety is said to be "broken down," although it was the change in the pathogen, not the host plant, that brought it about.

TYPES OF PLANT RESISTANCE TO PATHOGENS

Plants are resistant to certain pathogens because they belong to taxonomic groups that are outside the host range of these pathogens (nonhost resistance), because they possess genes for resistance (R genes) directed against the avirulence genes of the pathogen (true, race-specific, cultivar-specific, or gene-for-gene resistance), or because, for various reasons, the plants escape or tolerate infection by these pathogens (apparent resistance).

Each kind of plant, e.g., potato, corn, or orange, is a host to a small and different set of pathogens that make up a small proportion of the total number of known plant pathogens. In other words, each kind of plant is a nonhost to the vast majority of known plant pathogens. Nonhosts are completely resistant to pathogens of other plants, usually even under the most favorable conditions for disease development (nonhost resistance). The same species of plants, however, that are nonhosts to most

TABLE 4-1
Stages of Variation in Plants and Pathogens and Characteristics by Which They Are Distinguished

Distinguishing characteristics	Fungi	Bacteria	Viruses	Nematodes	Plants
Morphology and biochemistry	Genus	Genus	Genus (formerly group)	Genus	Genus
	↓	↓	↓	↓	↓
Morphology and biochemistry	Species	Species	Virus name (species)	Species	Species
	↓	↓	↓	↓	↓
Host	Variety or special form	Variety or pathovar	Type[a]	Race or	Variety or cultivar
	↓	↓	↓		
Differential varieties or symptoms	Race	Race	Strain	biotype or pathotype	↓
	↓	↓	↓		
Localized field population	Isolate	Isolate	Isolate	or strain	
	↓	↓	↓	↓	↓
Clonal population	Single spore-derived biotype	Single colony-derived strain	Single local lesion isolate	Individual nematode	Clone

[a]Sometimes strain is used instead of type.

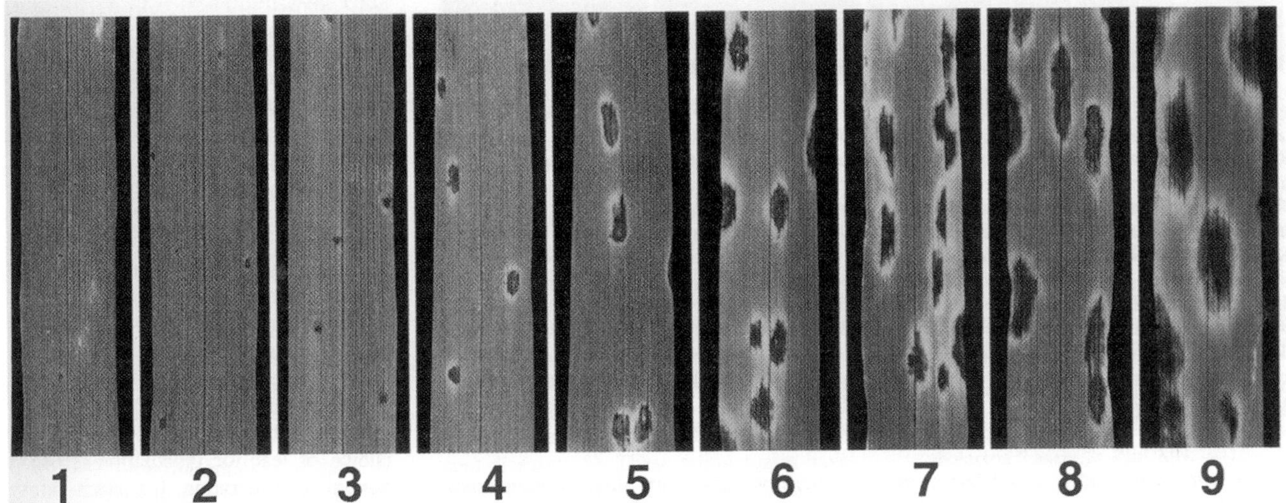

FIGURE 4-7 An infection rating scale of barley seedling leaves inoculated with the same isolate of the spot blotch fungus *Cochliobolus sativus*. Seedlings 1, 2, and 3 indicate low compatibility between hosts and the pathogen, whereas seedlings 4 and 5 show intermediate compatibility and seedlings 6, 7, 8, and 9 show high compatibility (susceptibility). [Photograph courtesy of Fetch and Steffenson (1999). *Plant Dis.* 83, 213–217.]

pathogens are susceptible, to a lesser or greater extent, to their own pathogens. Moreover, each plant species exhibits specific susceptibility toward each of its own pathogens while it exhibits complete or nonhost resistance to all other pathogens (Figs. 4-4 and 4-5).

Even within a species of plant that is susceptible to a particular species of pathogen, however, there is considerable variation in both the susceptibility of the various plant cultivars (varieties) toward the pathogen (Figs. 4-7 and 4-8) and the virulence of the various pathogen races toward the plant variety. The genetics of such host–pathogen interactions are of considerable biological interest and of the utmost importance in developing disease control strategies through breeding for resistance.

The variation in susceptibility to a pathogen among plant varieties is due to different kinds and, perhaps, different numbers of genes for resistance that may be present in each variety. The effects of individual resistance genes vary from large to minute, depending on the importance of the functions they control. A variety that is very susceptible to a pathogen isolate obviously has no effective genes for resistance against that isolate. The same variety, however, may be resistant to another pathogen isolate obtained from infected plants of another variety.

	Pathogen isolate	
	1	2
Plant variety	Susceptible	Resistant

Lack of susceptibility to the second isolate would indicate that the plant variety, which had no genes for resistance against the first pathogen isolate, has one or more genes for resistance against the second isolate. If the same plant variety is inoculated with more pathogen isolates, obtained from still different plant varieties, it is possible that the variety would be susceptible to some of them but not susceptible (and thus would be resistant) to the other isolates. The latter case would again show that the variety has one or more genes for resistance against each of the isolates to which it is resistant. Although the resistance against some of the isolates might be the result of the same genes for resistance in the variety, it is likely that the variety also contains several genes for resistance, each specific against a particular pathogen isolate.

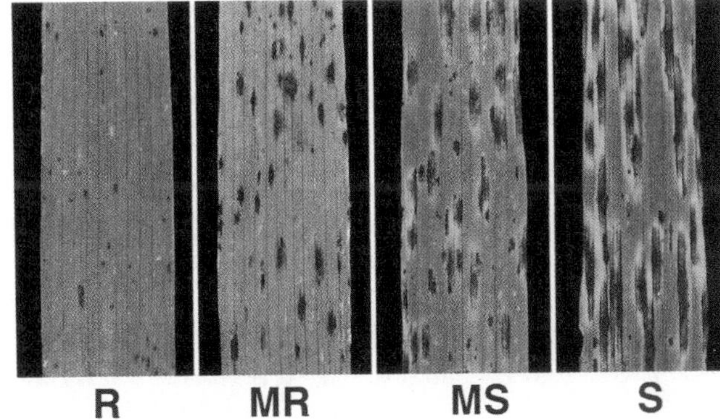

FIGURE 4-8 An infection response rating scale for leaves of adult barley plants inoculated with the same isolate of the spot blotch fungus *Cochliobolus sativus*. Rankings are R, resistant; MR, moderately resistant; MS, moderately susceptible; and S, susceptible. [Photograph courtesy of Fetch and Steffenson (1999). *Plant Dis.* 83, 213–217.]

True Resistance

Disease resistance that is controlled genetically by the presence of one, a few, or many genes for resistance in the plant is known as true resistance. In true resistance, the host and the pathogen are more or less incompatible with one another, either because of lack of chemical recognition between the host and the pathogen or because the host plant can defend itself against the pathogen. The various defense mechanisms are either already present or are activated in response to infection by the pathogen. There are two kinds of true resistance: partial, also called quantitative, polygenic, or horizontal resistance and R gene resistance, also called race specific, monogenic, or vertical resistance.

Partial, Quantitative, Polygenic, or Horizontal Resistance

All plants have a certain, but not always the same, level of possibly unspecific resistance that is effective against each of their pathogens. Such resistance is sometimes called partial, race nonspecific, general, quantitative, polygenic, adult-plant, field, or durable resistance, but in the past it was referred to most commonly as horizontal resistance.

Partial resistance is probably controlled by several genes, thereby the name **polygenic** or **multigene resistance**. There are, however, several examples of quantitative and nonrace-specific resistance that are determined by single genes, often R gene homologs. Also, in many cases where genetic analyses were performed, a limited number of genes, usually fewer than four to five, often one or two, are sufficient to explain most of the resistance. In many cases, one of these genes alone may be rather ineffective against the pathogen and may play a minor role in the total horizontal resistance (**minor gene resistance**). The several genes involved in partial resistance seem to exert their influence by controlling the numerous steps of the physiological processes in the plant that provide the materials and structures that make up the defense mechanisms of the plant. The partial resistance of a plant variety toward all races of a pathogen may be somewhat greater, or smaller, than those of other varieties toward the same pathogen (Figs. 4-7 and 4-8), but the differences are usually small and insufficient to routinely distinguish varieties (**nondifferential resistance**). In addition, partial resistance is affected by and may vary considerably more than R gene resistance under different environmental conditions. Generally, partial resistance does not protect plants from becoming infected. Instead it slows down the development of individual infection loci on a plant, thereby slowing down the spread of the disease and the development of epidemics in the field (Figs. 4-7–4-9). Some degree of partial resistance is present in plants regardless of whether monogenic resistance is present. However, although it is clear that partial resistance is inherited quantitatively, it is believed that the individual genes contributing to "partial" resistance may, in fact, be qualitatively identical to the genes of monogenic resistance.

R Gene Resistance, Race-Specific, Monogenic, or Vertical Resistance

Many plant varieties are quite resistant to some races of a pathogen while they are susceptible to other races of the same pathogen. In other words, depending on the race of the pathogen used to infect a variety, the variety may appear strongly resistant to one pathogen race and susceptible to another race (**race specific**) under a variety of environmental conditions. Such resistance differentiates clearly between races of a pathogen, as it is effective against specific races of the pathogen and ineffective against others (Figs. 4-9 and 4-10). Such resistance is

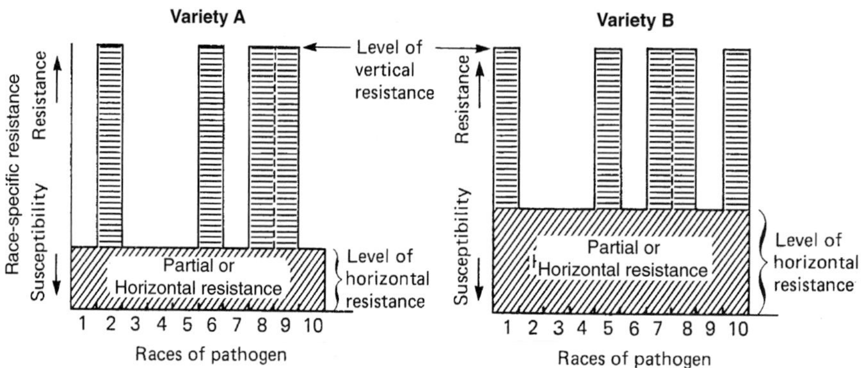

FIGURE 4-9 Levels of horizontal and vertical resistance of two plant varieties toward 10 races of a pathogen. [After Vanderplank (1984).]

FIGURE 4-10 *Brassica napus* plants following inoculation with an isolate of *turnip mosaic virus*. (Left) Susceptible. (Right) Resistant by a single dominant (R) gene. [Photograph courtesy of Walsh and Jenner (2002).]

sometimes called strong, major, race-specific, qualitative, or differential resistance, but it was more commonly referred to in the past as **vertical resistance**.

Race-specific resistance is always controlled by one or a few genes (thereby the names **monogenic** or **oligogenic resistance**). These genes, referred to as R genes, control a major step in the recognition of the pathogen by the host plant and therefore play a major role in the expression of resistance. In the presence of race-specific resistance, the host and pathogen appear incompatible (Fig. 4-9). The host may respond with a hypersensitive reaction, may appear immune, or may inhibit pathogen reproduction. Often, race-specific resistance inhibits the initial establishment of pathogens that arrive at a field from host plants that lack, or have different, major genes for resistance. Race-specific resistance inhibits the development of epidemics by limiting the initial inoculum or by limiting reproduction after infection.

Complete resistance may be provided by a single resistance gene. Often, it is desirable to combine, or **pyramid**, more than one resistance gene (R1R2, R1R3, R1R2R3) in the same plant, which then is resistant to all the pathogen races to which each of the genes provides resistance. A plant species may have as many as 20 to 40 resistance genes against a particular pathogen, although each variety may have only one or a few of these genes. For example, wheat has 20 to 40 genes for resistance against the leaf rust fungus *Puccinia recondita*, barley has a similar number of genes against the powdery mildew fungus *Erysiphe graminis hordei*, and cotton has almost as many against the bacterium *Xanthomonas campestris* pv. *malvacearum*. Each gene for resistance, such as R2, makes the plant resistant to all the races of the pathogen that contain the corresponding gene for avirulence. This pathogen race and its avirulence gene (A2), however, are detected because the pathogen attacks plants that lack the particular gene for resistance (R2).

Whether partial or race specific, true resistance is generally controlled by genes located in the plant chromosomes in the cell nucleus. There are, however, several plant diseases in which resistance is controlled by genetic material contained in the cytoplasm of the cell. Such resistance is sometimes referred to as **cytoplasmic resistance**. The two best-known cases of cytoplasmic resistance occur in the southern corn leaf blight caused by *Bipolaris (Helminthosporium) maydis* and the yellow leaf blight caused by *Phyllosticta maydis*. Resistance in these is conferred by the lack of a gene in mitochondria of normal cytoplasm of various types of corn that encodes a receptor for the host-specific toxin produced by each pathogen. The presence of such a gene in mitochondria of Texas male-sterile cytoplasm makes all corn lines with Texas male-sterile cytoplasm susceptible to these pathogens.

Varieties with race-specific (monogenic or oligogenic) resistance generally show complete resistance to a specific pathogen under most environmental conditions, but a single or a few mutations in the pathogen may produce a new race that may infect the previously resistant variety. On the contrary varieties with partial (polygenic) resistance are less stable and may vary in their reaction to the pathogen under different environmental conditions, but a pathogen will have to undergo many more mutations to completely break down the resistance of the host. As a rule, a combination of major and minor genes for resistance against a pathogen is the most desirable makeup for any plant variety.

Apparent Resistance

In any area and almost every year, limited or widespread plant disease epidemics occur on various crop plants. Under certain conditions or circumstances, however, some very susceptible plants or varieties of these crops may remain free from infection or symptoms and thus appear resistant. The apparent resistance to disease of plants known to be susceptible is generally a result of disease escape or tolerance to disease.

Disease Escape

Disease escape occurs whenever genetically susceptible plants do not become infected because the three factors necessary for disease (susceptible host, virulent pathogen, and favorable environment) do not coincide and interact at the proper time or for sufficient duration. Plants may, for example, escape disease from soilborne pathogens because their seeds germinate faster or their seedlings harden earlier than others and before the temperature becomes favorable for the pathogen to attack them. Some plants escape disease because they are susceptible to a pathogen only at a particular growth stage (young leaves, stems, or fruits; at blossoming or fruiting; at maturity and early senescence); therefore, if the pathogen is absent or inactive at that particular time, such plants avoid becoming infected. For example, young tissues and plants are infected and affected much more severely by *Pythium*, powdery mildews, and most bacteria and viruses than older ones. However, fully grown, mature, and senescent plant parts are much more susceptible to certain other pathogens, such as *Alternaria* and *Botrytis*, than their younger counterparts. Plants may also escape disease because of the distance between fields, the number and position of plants in the field, the spacing of plants in a field, and so on.

In many cases, plants escape disease because they are interspersed with other types of plants that are insusceptible to the pathogen and because the amount of inoculum that reaches them is much less than if they were in monocultural plantations; because their surface hairs and wax repel water and pathogens suspended in it; because their growth habit is too erect or otherwise unfavorable for pathogen attachment and germination; or because their natural openings, such as stomata, are at a higher level than the rest of the leaf surface or open too late in the day, by which time the leaves are dry and the germ tubes of spores, such as of *Puccinia graminis*, have desiccated. In plant diseases in which pathogens penetrate primarily through wounds caused by heavy winds and rain, dust storms, and insects, lack of such wounds allows disease escape. Also, plants that are unattractive or resistant to the vector of a pathogen escape infection by that pathogen.

Factors that affect the survival, infectivity, multiplication, and dissemination of the pathogen are also likely to allow some plants to escape disease. Such factors include the following: absence or poor growth of the pathogen at the time the susceptible plant stage is available; destruction or weakening of the pathogen by hyperparasites or by antagonistic microflora at the place of production or at the infection court; misdirection to or trapping of the pathogen by other plants; and lack of pathogen dissemination because winds, rain, or vectors are absent.

Several environmental factors play crucial roles in plant disease escape in almost every location. Temperature, for example, determines the geographical distribution of most pathogens, and plants growing outside the range of that temperature escape disease from such pathogens. Most commonly, however, plant disease escape increases in temperature ranges that favor plant growth much more than they do the growth of the pathogen. For example, many plants escape disease from *Pythium* and *Phytophthora* if the temperature is high and the soil moisture low, whereas some low temperature crops, such as wheat, escape similar diseases from *Fusarium* and *Rhizoctonia* if the temperature is low. Temperatures outside certain ranges inhibit the sporulation of fungi as well as spore germination and infection, thereby increasing the chances for disease escape. Low temperatures also reduce the mobility of many insect vectors or pathogens, allowing more plants to escape disease.

Lack of moisture caused by low rainfall or dew or low relative humidity is probably the most common cause of disease escape in plants. Plants in most dry areas or during dry years remain generally free of apple scab, late blight, most downy mildews, and anthracnoses because these diseases require a film of water on the plant or high relative humidity in almost every stage of their life cycle. Similarly, in dry soils such diseases as clubroot of crucifers and damping off induced by *Pythium* and *Phytophthora* are quite rare because such soils inhibit the production and activity of the motile spores of these pathogens. However, with some diseases, such as common scab of potato caused by *Streptomyces scabies*, plants escape disease in irrigated or wet soils because the plants can defend themselves better in the absence of water stress and because these pathogens are lysed or otherwise inhibited by microorganisms favored by high moisture. Many trees are also in a better position to defend themselves and to escape damage by the canker-causing fungus *Leucostoma* sp. (*Cytospora* sp.) in years in which sufficient rainfall or irrigation provides adequate soil moisture in late summer and early fall.

Some other environmental factors also allow plants to escape disease. For example, wind may increase disease escape by blowing from the wrong direction at the right time, thus carrying spores and vectors away from the crop plants, and by drying up plant surfaces quickly before the pathogen has time to germinate and infect. Also, soil pH increases disease escape in a few diseases, e.g., in crucifers from *Plasmodiophora brassicae* at high pH and in potatoes from *Streptomyces scabies* at low pH, in both cases because the particular pH inhibits survival and growth of the pathogen.

In general, many plant diseases are present some years in some areas and absent on the same kinds of plants in other years or in nearby areas. This suggests that in these areas or years the plants remain free of disease not because they are resistant, but because they escape disease. Earliness is often bred into many wheat and potato varieties to help them escape disease from the rusts and the late blight, respectively. Lateness, rapid growth, resistance to bruising, unattractiveness to vectors, and tolerance to low temperatures are also often bred into crop varieties to help them escape specific diseases. These and many other characteristics, of course, are those that make up horizontal resistance. It is true that there is a wide common area between horizontal resistance and disease escape in which the one leads to the other or the two appear identical. Escape from disease depends on environmental conditions, as well as on heritable characteristics in the host and the pathogen, and is often entirely controlled by the environment. Escape from disease, moreover, is a manageable quality, and farmers, through many of the most common cultural practices, actually aim at helping plants escape disease. Such practices include using disease-free, vigorous seed, choosing the proper soil, planting date, depth of sowing, and distance between plants and between fields, utilizing proper crop rotation, sanitation (rouging, pruning, and so on), interplantings, and multilines, attending to insect and vector control, and several others.

Tolerance to Disease

Tolerance to disease is the ability of plants to produce a good crop even when they are infected with a pathogen. Tolerance results from specific, heritable characteristics of the host plant that allow the pathogen to develop and multiply in the host while the host, either by lacking receptor sites for or by inactivating or compensating for the irritant excretions of the pathogen, still manages to produce a good crop. Tolerant plants are, obviously, susceptible to the pathogen, but they are not killed by it and generally show little damage. The genetics of tolerance to disease are not understood; neither is its relationship, if any, to horizontal resistance. Tolerant plants, whether because of exceptional vigor or a hardy structure, probably exist in most host–parasite combinations. Tolerance to disease is observed most commonly in many plant–virus infections in which mild viruses, or mild strains of virulent viruses, infect plants such as potato and apple systemically and yet cause few or no symptoms and have little discernible effect on yield. Generally, however, although tolerant plants produce a good crop even when they are infected, they produce an even better crop when they are not infected.

GENETICS OF VIRULENCE IN PATHOGENS AND OF RESISTANCE IN HOST PLANTS

Infectious plant diseases are the result of the interaction of at least two organisms, the host plant and the pathogen. The properties of each of these two organisms are governed by their genetic material, the DNA, which is organized in numerous segments making up the genes.

It has been known for more than a century that the host reaction, i.e., the degree of susceptibility or resistance to various pathogens, is an inherited characteristic. This knowledge has been used quite effectively in breeding and distributing varieties resistant to pathogens causing particular diseases. The ability of pathogens to inherit their infection type, however, i.e., the degree of pathogen virulence or avirulence, has been studied intensively only since the 1940s. It is now clear that pathogens consist of a multitude of races, each different from others in its ability to attack certain varieties of a plant species but not other varieties. Thus, when a variety is inoculated with two appropriately chosen races of a pathogen, the variety is susceptible to one race but resistant to the other. Conversely, when the same race of a pathogen is inoculated on two appropriately chosen varieties of a host plant, one variety is susceptible while the other is resistant to the same pathogen (Table 4-2). This clearly indicates that, in the first case, one race possesses a genetic characteristic that enables it to attack the plant, while the other race does not, and, in the second case, that the one variety possesses a genetic characteristic that enables it to defend itself against the pathogen so that it remains resistant, while the other variety does not. When several varieties are inoculated separately with one of several races of the pathogen, it is again noted that one pathogen race can infect a certain group of varieties, another race can infect another group of varieties, including some that can and some that cannot be infected by the previous race, and so on (Table 4-2).

Studies of the inheritance of resistance versus susceptibility in plants prove that single genes control resistance and their absence allows susceptibility. Studies of the inheritance of avirulence versus virulence in pathogens prove that single genes control avirulence and their absence allows virulence. Studies of their interac-

TABLE 4-2
Possible Reactions of Two (Left) and Four (Right) Varieties of a Plant to Two (Left) and Four (Right) Races of a Pathogen[a]

Plant variety	Pathogen race 1	Pathogen race 2	Plant variety	Pathogen race 1	Pathogen race 2	Pathogen race 3	Pathogen race 4
A	–	+	A	–	+	+	+
B	+	–	B	+	–	–	+
			C	–	+	–	+
			D	+	–	+	–

[a]Plus signs indicate susceptible (compatible reaction, infection); minus signs indicate resistant (incompatible reaction, noninfection).

tions prove that R genes in the plant are specific for avr genes in the pathogen. Thus, varieties possessing certain genes for resistance react differently against the various pathogen races and their genes for avirulence. The progeny of these varieties react to the same pathogens in exactly the same manner as the parent plants, indicating that the property of resistance or susceptibility against a pathogen is genetically controlled (inherited). Similarly, the progeny of each pathogen causes on each variety the same effect that was caused by the parent pathogens, indicating that the property of virulence or avirulence of the pathogen on a particular variety is also genetically controlled (inherited).

It thus appears that, under favorable environmental conditions, the outcome — infection (susceptibility) or noninfection (resistance) — in each host–pathogen combination is predetermined by the genetic material of the host and of the pathogen. The number of genes determining resistance or susceptibility varies from plant to plant, as the number of genes determining virulence or avirulence varies from pathogen to pathogen. In most host–pathogen combinations the number of genes involved and what they control are not yet known. In some diseases, however, particularly those caused by oomycetes, such as potato late blight, fungi, such as apple scab, powdery mildews, tomato leaf mold, and cereal smuts and rusts, and also in several viral and bacterial diseases of plants, considerable information regarding the genetics of host–pathogen interactions is available.

The Gene-for-Gene Concept

The coexistence of host plants and their pathogens side by side in nature indicates that the two have been evolving together. Changes in the virulence of the pathogens appear to be continually balanced by changes in the resistance of the host, and vice versa. In that way, a dynamic equilibrium of resistance and virulence is maintained, and both host and pathogen survive over considerable periods of time. The stepwise evolution of virulence and resistance can be explained by the **gene-for-gene concept**, according to which for each gene that confers virulence to the pathogen there is a corresponding gene in the host that confers resistance to the host, and vice versa.

The gene-for-gene concept was first proved in the case of flax and flax rust, but it has since been shown to operate in many other rusts, in the smuts, powdery mildews, apple scab, late blight of potato, and other diseases caused by fungi, as well as in several diseases caused by bacteria, viruses, parasitic higher plants, nematodes, and even insects. Generally, but not always, in the host the genes for resistance are dominant (R), whereas genes for susceptibility, i.e., lack of resistance, are recessive (r). In the pathogen, however, genes for avirulence, i.e., inability to infect, are usually dominant (A) whereas genes for virulence are recessive (a). Thus, when two plant varieties, one carrying the gene for resistance R to a certain pathogen and the other lacking the gene R for resistance, i.e., carrying the gene for susceptibility (r) to the same pathogen, are inoculated with two races of the pathogen, one of which carries the gene for avirulence A against the resistance gene R and the other of which carries the gene for virulence (a) against the resistance gene R, the gene combinations and reaction types shown in Table 4-3 and Fig. 4-11 are possible.

Each gene in the host can be identified only by its counterpart gene in the pathogen, and vice versa. Of the four possible gene combinations, only the AR interaction is incompatible (resistant), i.e., the host has a certain gene for resistance (R) that recognizes the corresponding specific gene for avirulence (A) of the pathogen. In the Ar combination, infection results because the host lacks genes for resistance (r) and so the pathogen can attack it with its other genes for virulence (after all, it is a pathogen on this host). In aR, infection results because although the host has a gene for resistance, the pathogen lacks the gene for avirulence that is recognized specifically by this particular gene for resistance and therefore no defense mechanisms (resistance) are activated. Finally, in the ar interaction, infection results because the plant has no resistance (r) and the pathogen, being a pathogen and therefore virulent (a), attacks it.

It is thought that genes for resistance appear and accumulate first in hosts through evolution and that they coexist with nonspecific genes for pathogenicity which evolve in pathogens. Genes for pathogenecity exist in pathogens against all host plants that lack

TABLE 4-3

Quadratic Check of Gene Combinations and Disease Reaction Types in a Host–Pathogen System in Which the Gene-for-Gene Concept for One Gene Operates[a]

	Resistance or susceptibility genes in the plant	
Virulence or avirulence genes in the pathogen	R (resistant) dominant	r (susceptible) recessive
A (avirulent) dominant	AR (−)	Ar (+)
a (virulent) recessive	aR (+)	ar (+)

[a]Minus signs indicate incompatible (resistant) reactions and therefore no infection. Plus signs indicate compatible (susceptible) reactions and therefore infection develops.

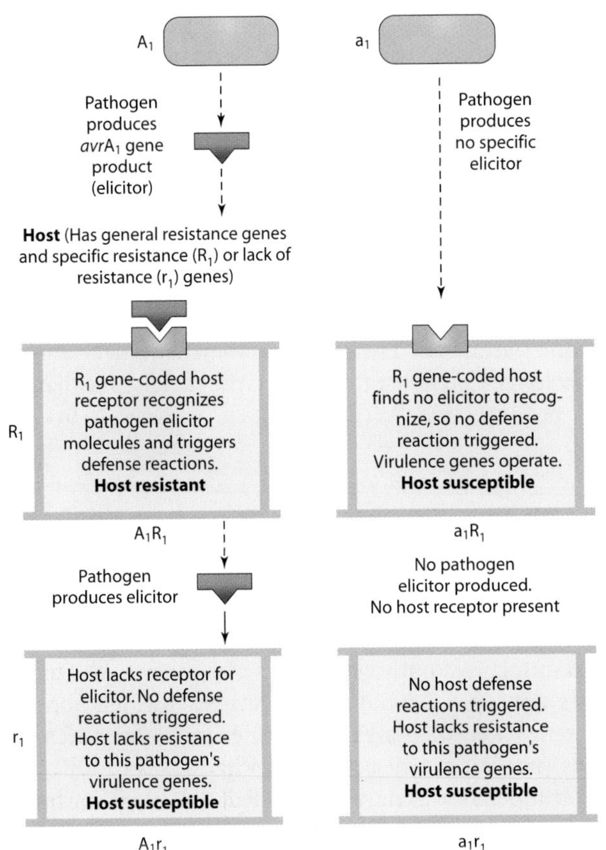

FIGURE 4-11 Basic interactions of pathogen avirulence (A)/virulence (a) genes with host resistance (R)/susceptibility (r) genes in a gene-for-gene relationship and final outcomes of the interactions.

specific resistance. When a specific gene for resistance appears in or is bred into the host, the gene enables the host to recognize the product of a particular gene for virulence in the pathogen. That pathogen gene is then thought of as the "avirulence" gene (*avr*A) of the pathogen that corresponds to the plant resistance gene R. The change in the function of the pathogen gene is because subsequent recognition of the *avr*A gene product (the elicitor molecule) by the receptor coded by the R gene triggers the hypersensitive response reaction in the plant that keeps the plant resistant. A new gene for virulence that attacks the existing gene for resistance appears by mutation of an existing avirulence gene, which then avoids gene-for-gene recognition, and the resistance of the host breaks down. Plant breeders then introduce another gene for resistance (R) in the plant, which recognizes the protein of the new gene for virulence of the pathogen and extends the resistance of the host beyond the range of the new gene for virulence in the pathogen. This produces a variety that is resistant to all races that have an avirulence gene corresponding to the specific gene for resistance until another gene for virulence appears in the pathogen. When a variety has two or more genes for resistance (R_1, R_2, . . .) against a particular pathogen, it means that each corresponds to one, two, or more genes of former virulence (and now avirulence) in the pathogen (a_1, a_2, . . .), each of which, once recognized by one of the genes for resistance in the host, subsequently functions as an avirulence gene. The gene combinations, and disease reaction types, of hosts and pathogens with two genes for resistance or virulence in corresponding loci, respectively, are shown in Table 4-4.

Table 4-4 makes clear several points. First, susceptible (r_1r_2) plants lacking genes for resistance are attacked by all races of the pathogen, regardless of the virulence (aa) or avirulence genes (A_1A_2) carried by the pathogen. Second, pathogen races or individuals designated a_1a_2, i.e., which lack genes for avirulence (A_1A_2) for each gene for resistance of the host (R_1R_2), can infect all plants that have any combination of these genes (R_1R_2, R_1r_2, r_1R_2), as the a_1a_2 pathogen produces no elicitor molecules capable of triggering the host defense response. When a pathogen has one of the two genes for virulence (a_1 or a_2), i.e., it lacks one of the two genes for avirulence (A_1 or A_2), then it can infect plants that have the corresponding gene for resistance (R_1 or R_2, respectively) but not plants that have a gene for resistance corresponding to a gene for avirulence in the pathogen (e.g., pathogen with genes A_1a_2 infects plant with r_1R_2 but not those with R_1r_2 because R_1 can recognize the avr gene A_1 and triggers defenses against it).

The gene-for-gene concept has been demonstrated repeatedly, and both pathogen avirulence genes and plant resistance genes have been isolated. Plant breeders apply the gene-for-gene concept every time they incorporate a new resistance gene into a desirable variety that becomes susceptible to a new strain of the pathogen. With the diseases of some crops, new resistance genes must be found and introduced into old varieties at relatively frequent intervals, whereas in others a single gene confers resistance to the varieties for many years.

TABLE 4-4

Complementary Interaction of Two Host Genes for Resistance and the Corresponding Two Pathogen Genes for Virulence and Their Disease Reaction Types

		Resistance (R) or susceptibility (r) genes in the plant			
		R_1R_2	R_1r_2	r_1R_2	r_1r_2
Virulence (a)	A_1A_2	–	–	–	+
or avirulence (A)	A_1a_2	–	–	+	+
genes in the	a_1A_2	–	+	–	+
pathogen	a_1a_2	+	+	+	+

The Nature of Resistance to Disease

A microorganism is pathogenic, i.e., it is a pathogen, because it has the genetic ability to infect another organism and to cause disease. Either a plant is immune to a pathogen, i.e., it is not attacked by the pathogen even under the most favorable conditions, or it may show various degrees of resistance ranging from near immunity to complete susceptibility. Resistance may be conditioned by a number of internal and external factors that operate to reduce the chance and degree of infection. The first step in any infection is recognition of the host by the pathogen and perhaps the opposite, some type of recognition of the pathogen by the host. Therefore, absence of a recognition factor(s) in the host could help it avoid infection by a particular pathogen. Generally, any heritable characteristic of the plant that contributes to localization and isolation of the pathogen at the points of entry, to reduction of the harmful effects of toxic substances produced by the pathogen, or to inhibition of the reproduction and, thereby, further spread of the pathogen, contributes to the resistance of the plant to disease. As a result, in most plant diseases, the pathogen is usually localized after varying degrees of invasion and colonization of host tissues. Indeed, there are only a few diseases in which the pathogen is allowed to advance unchecked throughout the plant and to kill the entire plant. Furthermore, any heritable characteristic that enables a particular variety to complete its development and maturation under conditions that do not favor the development of the pathogen also contributes to resistance (disease escape).

The contribution of genes conditioning resistance in the host seems to consist, primarily, of providing the genetic potential in the plant for the development of one or more of the morphological or physiological characters that contribute to disease resistance (including those described in Chapter 6 in the sections on structural and biochemical defense). The mechanisms by which genes control the physiological processes that lead to disease resistance or susceptibility are not yet clear, but they are, presumably, no different from the mechanisms controlling any other physiological process in living organisms.

It is thought that for the production of an inducible enzyme or of a fungitoxic substance needed for defense, a stimulant (elicitor), either secreted by the pathogen or caused by the activities of a pathogen, reacts with a receptor molecule of a host cell. This then transmits signals to other host cell molecules, activating plant defenses. However, if a pathogen mutant appears that does not secrete the particular elicitor that activates the defense reaction, the pathogen infects the host without opposition and so causes disease. In the latter case, the resistance of the host is said to have broken down, but it is actually bypassed by the pathogen rather than broken down. Other possible, although unproved, ways by which a pathogen could "break down" the resistance of a host are through a mutation in the pathogen that enables it to produce a substance that can react with and neutralize the defensive toxic substance of the host that is directed against the pathogen and through a mutation in the pathogen that would eliminate or block its own receptor site on which the host defensive substance becomes attached. The pathogen then can operate in the presence of that substance and of the defense mechanism that produces it.

Pathogenicity Genes in Plant Pathogens

Genes Involved in Pathogenesis and Virulence by Pathogens

Plant-infecting pathogens possess several classes of genes that are essential for causing disease (pathogenicity genes) or for increasing virulence on one or a few hosts (virulence genes).

Pathogenicity factors encoded by "pathogenicity genes" (*pat*) and "disease-specific genes" (*dsp*) are those involved in steps crucial for the establishment of disease (Fig. 4-12). Such genes include those essential for recognition of host by pathogen, attachment of the pathogen to the plant surface, germination and formation of infection structures on the plant surface, penetration of the host, and colonization of the host tissue. Genes involved in the synthesis and modification of the lipopolysaccharide cell wall of gram-negative bacteria may help condition the host range of the bacteria.

Some plant cell wall-degrading enzymes (e.g., cutinases), some toxins (e.g., victorin and HC toxin), hormones (e.g., indole acetic acid and cytokinin), polysaccharides, proteinases, siderophores, and melanin are produced by pathogens in pathogen–plant interactions in which they are essential for the pathogen to infect and cause disease on its host. In those cases, therefore, such factors function as pathogenicity factors. In other plant–pathogen systems the same compounds are helpful but not essential for disease induction and development. In these cases, these compounds are considered virulence factors. There is almost an unlimited number of virulence factors produced by pathogens. They include, in addition to many cell wall-degrading enzymes, toxins, hormones, and polysaccharides, almost all molecules or structures, e.g., amylases, lipases, signaling molecules such as homoserine lactone exopolysaccharides, and flagella. These compounds or structures may be present on the pathogen surface or

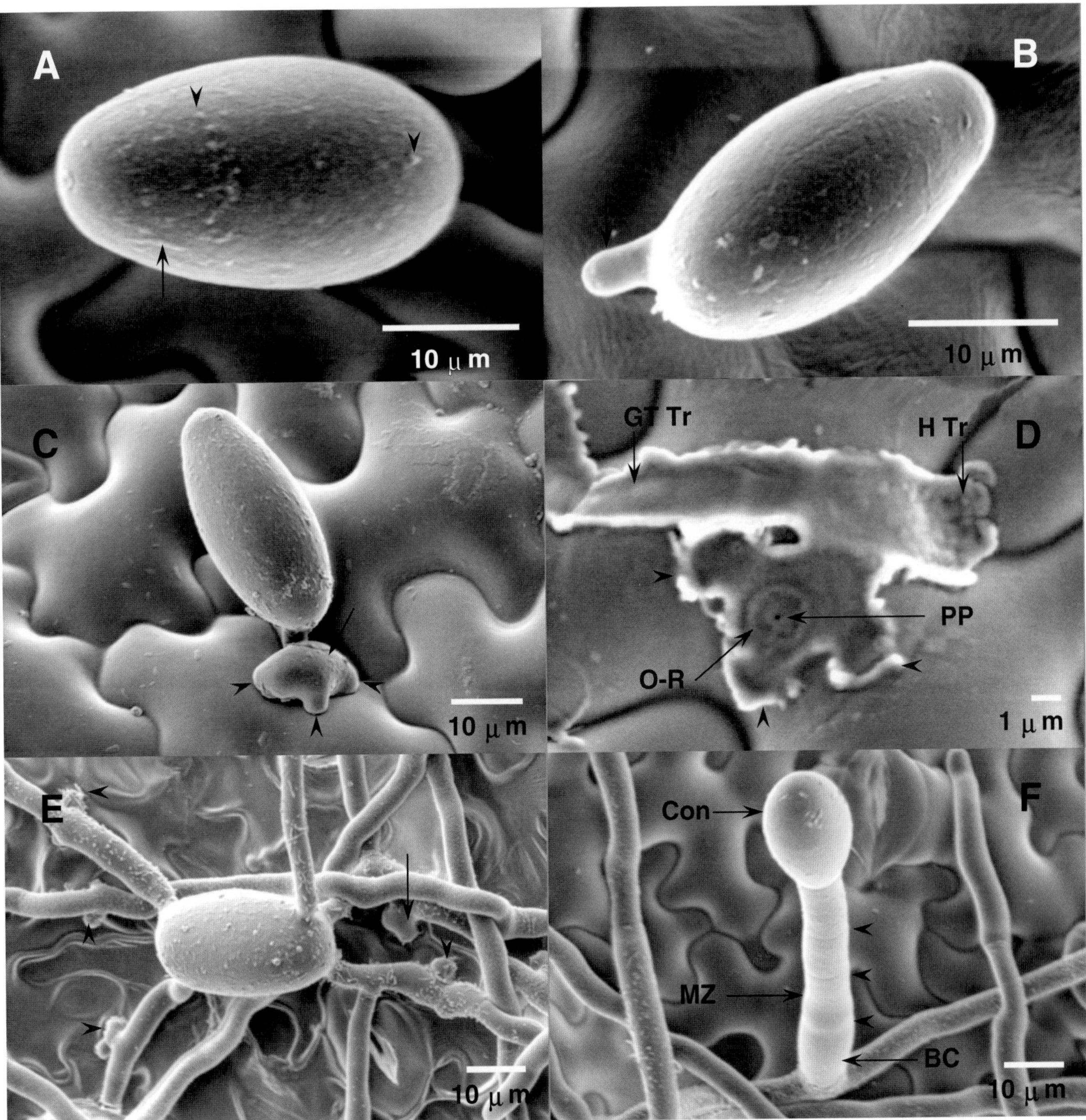

FIGURE 4-12 Electron micrographs of the infection stages of a tomato leaf by a conidium of the powdery mildew fungus *Oidium neolycopersici.* (A) Conidium. (B) Conidium with germ tube. (C) An appressorium forming at the end of the germ tube 10 hours postinoculation. (D) Imprint left on leaf after peeling germ tube and appressorium. A circular hole in the center of the appressorium shows the penetration pore made by the penetration peg. (E) Mycelium and pairs of hyphal appressoria. (F) A conidiophore bearing a conidium. [Photo courtesy S.J. Gurr, from *Can. J. Bot.* 73: (Supp 1), 5632–5639, 1995]

translocated to the extracellular environment of the pathogen and, in a variety of ways, could influence growth of the pathogen in the plant.

Plant pathogens employ diverse strategies to infect their host plants. Depending on the type of pathogen and on the infection process followed by each of them, pathogens utilize various genes that enable them to adhere to their host, form infection structures, penetrate the host, break down host wall macromolecules, produce toxins, neutralize host defenses, obtain nutrients from the host, move through the host, reproduce in the host, disseminate from host to host, respond to the environment, and so on.

Pathogenicity genes are genes that make a particular (micro)organism a pathogen, i.e., capable of causing disease. Disruption of a pathogenicity gene results in a complete loss or drastic reduction of disease symptoms. It should be noted here that virulence/avirulence genes act on top of the general pathogenicity of the pathogen and, in some cases, may have additional roles in disease. The most important types of pathogenicity genes of the main kinds of plant pathogens are discussed briefly.

Pathogenicity Genes of Fungi

Plant pathogenic fungi utilize a variety of ways and means (chemotropism, thigmotropism) to recognize and adhere to their host plant. Depending on whether the fungi enter the plant through wounds, stomata, or through direct penetration they may need to degrade the cuticle and the cell wall, for which they may need to form specialized structures, such as appressoria. Once inside the plant, the fungus may obtain nutrients without killing cells (biotroph), it may kill cells through its toxins and feed off the contents of dead cells (necrotroph), or it may act as a biotroph in early stages of infection but as a necrotroph later on.

Pathogenicity Genes Controlling Production of Infection Structures

Many fungi produce appressoria that help them penetrate epidermal cells. Appressoria contain glycerol for creating a high turgor pressure that allows the penetration peg to puncture the plant epidermal cells. Appressorial walls of *Magnaporthe grisea* and *Colletotrichum* species contain melanin that prevents glycerol from leaking out. Melanin-deficient mutants are unable to generate turgor pressure and are, therefore, nonpathogenic. Melanin biosynthesis is carried out by at least three structural genes, all of which are essential for pathogenicity of both fungi.

Several genes are involved in appressorial development, which is under both environmental and genetic control. For example, in the rice blast disease, caused by *M. grisea*, several genes have been shown to control appressorial development. One such gene, hydrophobin (mpg1), is essential for appressorial formation and, when disrupted, the fungus not only has reduced pathogenicity, it produces 100 times fewer conidia. Transcription of the mpg1 gene is controlled by three regulatory genes, two of which are also involved in the regulation of nitrogen metabolism. Another gene expressed in spores of *M. grisea* resembles transcription factors, but its disruption leads to the production of defective conidia and impaired appressorium formation, both of which cause loss of pathogenicity. A still different gene, pth11, encodes a protein that is embedded in the cell membrane and apparently enables the fungus to recognize the host surface and to form normal appressoria; disruption of that gene makes the fungus unable to do either.

Pathogenicity Genes Controlling Degradation of Cuticle and of Cell Wall

It is assumed that enzymes that degrade cell walls, cutin, pectin, and other physical structures are essential for pathogenicity. These enzymes, however, are often encoded by multigene families or by more than one gene that are not related, which results in functional redundancy of such enzymes. As a result, the disruption of one such gene does not eliminate pathogenicity of the pathogen because the other genes that encode the same enzyme fill in the need for the enzyme. Functional redundancy among virulence genes appears to be an emerging theme in explaining the degree of severity in many diseases. In addition, cell wall-degrading enzymes through their action often release oligosaccharides and cell wall proteins that can elicit or suppress the defense responses of the host plant. For example, a mutant of the elicitor enzyme xylanase II, the enzymatic activity of which was reduced 1,000-fold, still elicited a defense response in tomato and tobacco.

Cutins. These are hardy polymers that cover most external plant surfaces. They are degraded by cutinases. Cutinases are, most likely, pathogenicity factors for those fungi that need to penetrate the host surface directly. There is a whole family of cutinase multigenes and, therefore, most attempts to prove that they are essential for pathogenicity through gene disruption have been unsuccessful. The cutinase from *Fusarium solani* f. sp. *pisi*, however, when disrupted, led to mutants that had no pathogenicity.

Pectins. These consist of mixtures of primarily polygalacturonic acid with branches of many sugars. They occur in plant cell walls and in middle lamellae. Pectins exist in numerous forms and are degraded by enzymes such as pectin lyase, polygalacturonase, and pectin methylesterase, all of which appear to play a pathogenicity role in some fungi. Pectinases, however, are also encoded by multigene families, and proof of their significance as essential pathogenicity factors is difficult. Nevertheless, disruption of the gene encoding a pectate lyase in *Colletotrichum* sp. produced mutants that had reduced pathogenicity on avocado fruit.

However, disruption of a pectin lyase gene in *Alternaria* sp., *Glomerella* sp., *and Cryphonectria parasitica* had no effect on its pathogenicity, whereas disruption of a pectinase gene in *Botrytis* reduced the pathogenicity of the fungus on tomato and on apple. In a different case, disruption of either the pectin-inducible pectate lyase or the plant-inducible pectate lyase in *F. solani* pv. *pisi* had no effect on the pathogenicity of the fungus. When, however, both pectate lyase genes in *Fusarium* were disrupted at the same time, all mutants showed reduced pathogenicity. In a still different case, insertion and expression of a polygalacturonase gene in a strain of *Aspergillus flavus,* that lacked polygalacturonase, enabled the fungus to produce larger lesions on cotton bolls. Several other types of genes coding for cell wall-degrading enzymes, such as pectinases, glucanases, and xylanases, have been cloned and subsequently disrupted and their effects studied. Most disruptions failed to induce a loss of pathogenicity in the pathogen, but some gave mixed results. Little is known about the role in pathogenesis of cellulases, ligninases, or hemicellulases.

Pathogenicity Genes Controlling Secondary Metabolites

In addition to needing genes for producing infection structures and for degrading structural obstacles, fungal pathogens need genes that will help them overcome the many secondary metabolites plants produce, some of which have antimicrobial properties and help protect the plant against attack. Secondary metabolite compounds produced constitutively are called phytoanticipins, whereas those produced in response to attack by a pathogen are called phytoalexins. Pathogens respond to these chemical defenses of the host plant through genes that help pathogens avoid them, degrade them, alter their physiology, or through other mechanisms.

Phytoanticipins. They include primarily the saponins avenacin and tomatine. Saponins are glycosides with soap-like properties that can disrupt membranes. One saponin, avenacin A-1, is localized in the epidermis of oat roots but not of wheat roots. The fungus *Gaeumannomyces graminis* var. *avena* can infect oats because it has a gene that codes for the enzyme avenacinase, which degrades the saponin. When the avenacinase gene is disrupted, however, the avenacin-less mutants of the fungus fail to infect oats while they can still infect wheat, which does not produce avenacin. Another saponin, α-tomatine, is produced in tomato and has antimicrobial activity against many fungi. The fungus *Septoria lycopersici*, however, carries a gene similar in sequence to the avenacinase gene that encodes the enzyme tomatinase, which degrades the saponin tomatine. Disruption of the tomatinase gene, however, did not reduce the pathogenicity of *Septoria* on tomato, possibly because the fungus has other enzymes that can degrade the saponin. The latter happens in the oat—*Stagonospora avenae* interaction in which the fungus has three genes encoding for enzymes that can degrade the particular saponin.

Cyanogenic Glycosides and Glycosinolates. These compounds are separated in the plant from the enzymes that can degrade them. Upon wounding of a plant, these compounds and their enzymes mingle and interact, producing cyanide, isocyanates, nitriles, and thiocyanates, all toxic against all organisms and also to fungi. Their role, however, in pathogenesis of fungi and how the latter defend themselves, are not known.

Phytoalexins. Phytoalexins have been known for several decades to be produced by plants under attack but few fungal enzymes have been found that degrade them during fungal attack. One such enzyme is pisatin demethylase, which is produced by the fungus *Nectria haematococca* and degrades the pea phytoalexin pisatin. Pisatin demethylase is encoded by one of six such genes of the fungus but disruption of the gene caused only a slight reduction in pathogenicity. However, disruption of one out of four fungal genes that detoxify the phytoalexin maakiain from chickpea resulted in a reduction of pathogenicity, whereas the insertion of additional copies of the same gene in the pathogen isolates resulted in greater disease severity.

Some fungal genes protect the fungus and its pathogenicity even after it is growing inside the plant. Numerous such genes are involved in the efflux and influx of fungal molecules into the plant. Disruption of such a gene in *M. grisea* resulted in loss of pathogenicity. Because the same gene is induced by toxic drugs and by the rice phytoalexin sakuranetin, perhaps it plays a role in the efflux of plant metabolites from the fungus.

Because some fungal pathogenicity genes, when mutated, result in auxotrophic strains, it is apparent that levels of nutrients can affect the ability of fungi to colonize plants. It has been known for many years that auxotrophy is linked to a lack of pathogenicity in the corn smut fungus *Ustilago maydis*, whereas adenine auxotrophs of the apple scab fungus *Venturia inaequalis* are nonpathogenic on apple. Similarly, auxotrophs of *Fusarium sp.* in arginine and of *Stagonospora sp.* in ornithine decarboxylase also lost their ability to cause disease.

Pathogenicity Genes Controlling Fungal Toxins

Some fungi produce toxins that can disrupt host cellular functions or kill cells before or during infection.

Some toxins are nonspecific, i.e., they damage plants not attacked by the pathogen, whereas other toxins are host specific, i.e., they only damage plants that are attacked by the pathogen. The cellular targets of four host-specific fungal toxins, and possible mechanisms of action that lead to programmed cell death of their host plant cells have been studied. The HC toxin acts in the nucleus where it inhibits histone deacetylation, brings about changes in gene expression and prevents synthesis of antifungal compounds by the plant. The *Alternaria alternata* (AAL) toxin inhibits synthesis of the endoplasmic reticulum (ER) enzyme ceramide synthase; it catalyzes the formation of ceramide from phytosphingosine, both of which in animals and probably in plants can alter the signal transduction activity of the protein kinase. The T toxin reacts with the protein Urf13p of the T-cms mitochondria membrane and causes the formation of pores in it, leading to a loss of H^+ and other ions, and to cell death. Finally, victorin inhibits the enzyme glycine decarboxylase of the photorespiratory cycle and leads to the cleavage of RUBISCO, through which products of the oxygenase reaction are exchanged among the chloroplast (Cp), mitochondrion (Mit), and peroxisome (Px), leading to cell death.

Each toxin requires the participation of several genes for its biosynthesis. The genes that control the biosynthesis of toxins are often clustered together. Disruption of toxin genes in *Cochliobolus* shows that a fungus with an altered toxin profile can still be pathogenic. However, disruption of genes involved in the biosynthesis of *A. alternata* host-specific toxins resulted in reduced pathogenicity. The host-specific toxin produced by the wheat tan spot fungus *Pyrenophora tritici-repentis* is essential for pathogenicity of the fungus, as nonpathogenic toxin-minus mutants of the fungus regained their pathogenicity when they were transformed with the gene encoding the toxin.

Trichothecin are toxic metabolites (mycotoxins) produced by several species of the fungus *Gibberella (Fusarium)* and by the fungus *Myrothecium roridum*. Disruption of the gene that controls the first step in trichothecin biosynthesis resulted in reduced pathogenicity on most, but not all, hosts. Up to 11 genes have been found involved in trichothecin biosynthesis and not all the steps have been studied.

Pathogenicity Signaling Genes Used by Plant Pathogenic Fungi

Fungi, like plants and other organisms, use signaling genes that respond to changes in the environment and set off signaling cascades that alter the expression of their genes. Fungal signaling genes include the G-protein-coding genes, mitogen-activated protein (MAP) kinase genes, and cyclic AMP-dependent protein kinase genes. When such signaling genes are disrupted by mutation, the fungus loses all or most pathogenicity and exhibits a loss or reduction in several other processes, such as growth rate, mating, conidia production, and toxin production. Genes that are part of the signal transduction pathways belong to gene families such as the G protein and MAP kinase ones. In the example of *M. grisea*, three G-protein genes and three MAP kinase genes have been cloned and tested through disruption. Several but not all of the resulting mutants lost pathogenicity.

Genes in signaling pathways seem to code for the same amino acid sequences in the various fungi, but the signaling pathways and their interconnections seem to be different in various fungi. As a result, disruption of one of these genes may cause different effects. For example, disruption of the PMK1 gene of *M. grisea* reduced appressoria formation and lost the ability to infect through a wound but had no effect on mycelium and conidia formation. The CMK1 gene from *Colletotrichum lagenarium* could complement a PMK mutant of *M. grisea* and could restore its pathogenicity. Disruption of the CMK1 gene also reduced the appressorial formation and pathogenicity when inoculated through wounds but, in addition, reduced the melanization of appressoria, conidial production, and conidial germination. Disruption of the homologous gene CHK1 of *Cochliobolus heterostrophus* produced mutant strains that had reduced pathogenicity and, in addition, were infertile. Some signaling genes, in addition to controlling pathogenicity, are also involved in the mating processes in fungi. For example, the basidiomycetous fungi *Ustilago maydis* and *U. hordei* are pathogenic on plants only in a dikaryotic state obtained after two complementary strains mate. The gene loci a and b that control recognition and mating also, in an indirect way, control pathogenicity.

Pathogenicity Genes in Plant Pathogenic Bacteria

Plant pathogenic bacteria enter the intercellular spaces of plants through wounds and/or natural openings, such as stomata. Therefore, bacteria do not need to penetrate the plant surface but they must have ways to adhere.

Bacterial Adhesion to Plant Surfaces

Most bacteria do not need adhesion mechanisms except perhaps when they are moving through the xylem and phloem. The crown gall bacterium *Agrobacterium*, however, requires attachment to plant surface receptors as the first step in the transfer of T-DNA and development of disease symptoms. The attachment requires three components: a glucan molecule, the synthesis and export of which requires three genes; genes for the synthesis of cellulose; and the *att* region of the bacterial genome that contains several genes for attachment. In addition to these genes, *Agrobacterium* also contains numerous other genes with homology to genes of mammalian pathogens for adhesins and for pilus biosynthesis.

Several other plant pathogenic bacteria also have genes that encode proteins likely to be involved in attachment and aggregation. Thus, *Ralstonia solanacearum, Pseudomonas, Xanthomonas,* and *Xylella* have as many as 35 genes homologous to type IV pili genes, which in *Xanthomonas* and *Pseudomonas* is involved in cell-to-cell aggregation and protection from environmental stress, whereas in *Xylella* type IV pili are necessary for the establishment of an aggregated bacterial population in the turbulent environment of the xylem by adhering to the vessels in conjunction with components such as polysaccharides. *Xylella, Xanthomonas*, and *Ralstonia*, all colonizing plant vessels at some stage of infection, also contain additional adhesin gene homologs and homologs of hemagglutinin-related genes found in many bacteria pathogenic to mammals.

Bacterial Secretion Systems

Secretion systems are essential pathogenicity tools for bacteria because they make possible the translocation of bacterial proteins and other molecules into host plant cells. Five forms of secretion systems are recognized on the basis of the proteins that form them. Type I-SS is present in almost all plant pathogenic bacteria and carries out the secretion of toxins such as hemolysins, cyclolysin, and rhizobiocin. They consist of ATP-binding cassette (ABC) proteins and are involved in the export and import of a variety of compounds through energy provided by the hydrolysis of ATP. Type II-SS is common in gram-negative bacteria and is involved in the export of various proteins, enzymes, toxins, and virulence factors. Proteins are exported in a two-step process: First as unfolded proteins to the periplasm via the Sec pathway across the inner membrane, then as processed and folded proteins through the periplasm and across the outer membrane via an apparatus consisting of 12–14 proteins encoded by a cluster of genes. *Ralstonia* and *Xanthomonas*, which have two type II-SS per cell, use them for secretion outside the bacterium of virulence factors such as pectinolytic and cellulolytic enzymes. *Xylella* and *Agrobacterium* have one type II-SS per cell and, actually, *Agrobacterium* has the genes for only the first step of protein transport across the inner membrane, using type IV-SS for the rest.

Type III-SS is the most important in terms of pathogenicity of the bacteria in the genera *Pseudomonas, Xanthomonas,* and *Ralstonia*. The primary function of type III-SS is the transport of effector proteins across the bacterial membrane and into the plant cell. Genes that encode protein components of the type III-SS apparatus have a two-third similarity at the amino acid level and such genes are called hypersensitive response conserved (Hrc) genes. Genes that encode the transported proteins, especially the surface exposed ones, have only 35% amino acid similarity. Among the effector proteins in *R. solanacearum* are some avr homologs, most of which are similar to *Pseudomonas avr* genes. In addition to *avr* genes, *Pseudomonas, Ralstonia,* and *Xanthomonas* have effector proteins that are similar to ankyrin-related and leucin-rich proteins found in plants, humans, and insects.

Type IV-SS transports macromolecules from the bacterium to the host cell. The proteins transferred are very similar to those responsible for the mobilization of plasmids among bacteria. The *Agrobacterium tumefaciens virB* operon encodes 11 proteins that form an organized structure and are involved in the transfer of the T-DNA strand from the bacterium to the plant cell cytoplasm. The transporting structure stretches from the bacterial inner membrane through the outer membrane and terminates in a pilus-like structure that protrudes from the bacterial cell. The type V-SS autotransporter is found in *Xylella* and *Xanthomonas* and contains genes that encode surface-associated adhesins. Similar autotransporters exist in mammalian pathogens and are important for adhesion to epithelial cells.

Pathogenicity of Bacterial Enzymes That Degrade Cell Walls

Plant cell walls are composed of three major polysaccharides: cellulose, hemicellulose, and pectins and, in woody and some other plants, lignin. The number of genes encoding cell wall-degrading enzymes varies greatly in the different plant pathogenic bacteria: Soft-rotting erwinias produce a wider range of enzymes able to degrade plant cell wall components than any other plant pathogenic bacteria. The enzymes include pectinases, cellulases, proteases, and xylanases. Pectinases are believed to be the most important in pathogenesis, as they are responsible for tissue maceration by degrading the pectic substances in the middle lamella and, indirectly, for cell death. Four main types of pectin-degrading enzymes are produced, three (pectate lyase (Pel), pectin lyase (Pnl), and pectin methyl esterase (Pme)) with a high (~8.0) pH optimum, and one polygalacturonase, with a pH optimum of ~6. All are present in many forms or isoenzymes, each encoded by independent genes. For example, *E. chrysanthemi* produces five major Pel groups arranged into two families and at least three minor Pel groups induced preferentially in plant tissue and arranged into three other families. In contrast, *E. carotovora* produces three major Pels, an intercellular Pel, and several minor plant-induced Pels.

The expression of *Erwinia* genes encoding pectic enzymes and isozymes is sequential. This suggests that the genes are regulated separately. In addition, there are global regulatory systems, like the quorum-sensing system, so as to maximize the activity of the main enzymes. Because of the large number of pectinases involved, disruption of the gene encoding any one of the enzymes is not sufficient to stop cell maceration. Maceration symptoms develop when a soft rot erwinia population reaches a cell density-dependent regulatory, or quorum-sensing, system for extracellular enzymes. Enzyme production is switched on when both numbers of bacteria and the bacteria-secreted inducer homoserine lactose (HSL) have reached a critical level. Disruption of the HSL gene or addition of a gene encoding an enzyme that breaks down HSL leads to the production of mutants with reduced pathogenicity. Presumably, quorum sensing allows the bacteria to multiply within host tissue without triggering host resistance responses, such as the production of phytoalexins. In general, cell wall-degrading enzymes are considered to play a role in pathogenesis by facilitating penetration and tissue colonization, but they are also virulence determinants responsible for symptom development once growth of the bacteria has been initiated.

Some Xanthomonads, e.g., *Xanthomonas campestris* pv. *campestris*, the cause of black rot of crucifers, have genes for two pectin esterases and polygalacturonases, four pectate lyases, five xylanases, and nine cellulases. *X. citri* has no pectin esterases, one less pectate lyase, and three fewer cellulases. Because pectin esterases are important in tissue maceration, their absence in the citrus canker bacterium and presence in the crucifer rot bacterium may explain the symptoms of the two diseases. Other poor pectinolytic bacteria include *A. tumefaciens*, which has only four genes encoding pectinases of any type, and *Xylella*, which has only one gene coding for a polygalacturonase.

Bacterial Toxins as Pathogenicity Factors

Toxins have been known for a long time to play a central role in parasitism and pathogenesis of plants by several plant pathogenic bacteria. *Pseudomonas syringae, P. syringae* pv. *tomato*, and *P. syringae* pv. *maculicola* are primarily associated with production of the phytotoxin coronatine. Coronatine functions primarily by suppressing the induction of defense-related genes, but, as happens with most bacterial phytotoxins, it does not seem to be essential for pathogenicity by all strains.

The bacterium *P. syringae*, along with its pathovars, produces several pathotoxins, including syringomycin.

Albicidins, produced by *Xanthomonas albilineans*, block the replication of prokaryotic DNA and the development of plastids, thereby causing chlorosis in emerging leaves. Albicidins interfere with host defense mechanisms and thereby the bacteria gain systemic invasion of the host plant.

Extracellular Polysaccharides as Pathogenicity Factors

Extracellular polysaccharides (EPS) play an important role in pathogenesis of many bacteria by both direct intervention with host cells and by providing resistance to oxidative stress. In the bacterial wilt of solanaceous crops caused by *Ralstonia solanacearum*, EPS1 is the main virulence factor of the disease. EPS1 is a polymer composed of a trimeric repeat unit consisting of *N*-acetyl galactosamine, deoxy-L-galacturonic acid, and trideoxy-D-glucose. At least 12 genes are involved in EPS1 biosynthesis. EPS1 is produced by the bacterium in massive amounts and makes up more than 90% of the total polysaccharide. EPS likely causes wilt by occluding the xylem vessels and by causing them to rupture from the high osmotic pressure. The main component of EPS in the fire blight bacterium *Erwinia amylovora* is amylovoran, which is biosynthesized and regulated by several clusters of genes. Disturbance of production of amylovoran eliminates pathogenicity in the mutant.

Bacterial Regulatory Systems and Networks

Some plant pathogenic bacteria, such as *R. solanacearum*, the cause of wilt and soft rot diseases of solanaceous and other crops, as well as a successful soil inhabitant, have developed specialized systems of complex regulatory cascades and networks. These systems sense the different environments in which bacteria find themselves and trigger dramatic changes in their physiology by global shifts in gene expression linked to the primary network that fine-tunes virulence and pathogenicity gene expression. The majority of the network components are transcriptional regulators that consist of a transmembrane sensor kinase protein. The protein binds a specific signal molecule and, in response, its kinase transfers a phosphate group from ATP to its partner response regulator in the cytoplasm. This activates the response regulator, which turns on transcription of its targets.

Virulence and pathogenicity genes of *R. solanacearum* are regulated by a complex network of which the core is the phenotype conversion (Phc) system. The system consists of gene PhcA, a lysine-rich type transcriptional regulator, and the products of the operon phcBRSQ, which control levels of active PhcA depending on cell density or crowding. Cells that contain high levels of active PhcA produce large amounts of major virulence factors, such as EPS1 and some exoenzymes, and are very virulent. When PhcA is inactivated, the bacterial cells become quite avirulent and produce almost no EPS1 and exoproteins; instead they activate genes that produce polygalacturonase, siderophores, the Hrp secretion apparatus, and swimming motility. So the PhcA gene acts as a switch mechanism that sometimes promotes the expression of one set of genes while repressing another set, and other times does the opposite. The levels of PhcA in bacterial cells are controlled by the level of 3-OH palmitic acid methyl ester reached in the cells in response to cell density or confinement. The more dispersed the cells, the lower the concentration of 3-OH PAME in the cells, the less the activation of PhcA, and the more the activation of genes for siderophores, swimming motility, etc. When the cells are confined and dense in plant tissues, the concentration of 3-OH PAME builds up, PhcA activation increases, and genes coding pathogenicity and virulence factors (PES I, cell wall-degrading enzymes) are also activated. How 3-OH PAME activates PhcA is not yet known.

Sensing Plant Signaling Components

Agrobacterium tumefaciens has a two-component regulatory system that senses and reacts to the presence of susceptible cells. The system components are a membrane sensor protein, VirA, and a cytoplasmic response regulator protein, VirG. The two components react to exudates of wounded plant cells and initiate transcriptional activation of the *vir* genes. Expression of *vir* genes follows activation of the VirA transmembrane sensor protein by exuding phenolics such as lignin and flavonoids, and especially the phenolic acetosyringone. A number of gene groups are involved in further steps of infection. Mutants lacking these genes totally or greatly lose pathogenicity.

Other Bacterial Factors Related to Pathogenicity

Several other components of the bacterial cell or released by the bacteria appear to play roles as pathogenicity factors. Lipopolysaccharide (LPS) components of the outer cell wall of gram-negative bacteria play a role in the pathogenicity of erwinias. Proof of this is provided by the activation of pathogenesis-related proteins, such as glucanases (Fig. 4-13) in infected plants, and the fact that disruption of the LPS gene in the bacteria reduces their virulence and that protein–LPS complexes from bacteria inhibit the hypersensitive response (HR).

Catechol and hydroxamate siderophores appear to be virulence determinants for erwinias. In the fire blight bacterium *E. amylovora*, its siderophore protects the bacteria by interacting with H_2O_2 and inhibiting the generation of toxic oxygen species.

The peptide methionine sulfoxide reductase, which protects and repairs bacterial proteins against active oxygen damage, is essential for the expression of full virulence of the bacteria.

hrp genes and *avr* genes are associated with the expression of pathogenicity and host specificity and they exist in clusters. *hrp* genes encode proteins called harpins or pilins and are used to make a type III protein secretion system that is used to deliver Avr proteins across the walls and plasma membrane of living plant cells. Avr proteins and, to a lesser extent, harpins induce rapid cell death, which leads to HR; as a result, the infection by the bacteria in incompatible interactions fails. Avr proteins seem to also play a role in compatible host/bacteria interactions. *avr* genes usually determine host specificity at the pathovar and the species level. The role of *hrp* genes in the pathogenesis of soft-rotting erwinias is debatable.

Pathogenicity Genes in Plant Viruses

Viruses have a limited number of genes, but by utilizing the same genetic material in more than one way, viruses are very capable pathogens. All viruses have genes that encode one or more coat proteins that protect its nucleic acid, one or more nucleic acid replicases that produce innumerable copies of its genome, and one or more movement proteins that help the movement of the virus from cell to cell and long distance through the phloem. Several viruses have additional genes involved in virus transmission by vectors or in other ways, production of cellular inclusions, etc. Although all of these proteins are coded by the virus but are produced by the host plant, viruses also utilize host proteins for the essential functions of transcription and movement.

Viral Pathogenicity Functions Associated with the Coat Protein (CP)

Coat proteins of various viruses function in practically every aspect of viral multiplication and dissemination.

Virus Disassembly. Virus disassembly is essential for virus multiplication and the coat protein plays a central role in it. Destabilization of the weaker 5′ end CP RNA releases a few CP subunits, allowing ribosomes to bind to the exposed 5′ end of the RNA and initiate translation of the RNA replicase(s). Active translation provides the force needed to remove the CP subunits. The RNA replicase then interacts with the 3′ end of the RNA to initiate the (–) RNA strand, thereby uncoating the rest of the virus.

Virus Assembly. Virion assembly initiates at the RNA origin of assembly and proceeds in both directions of the RNA.

Virus Movement. Coat proteins apparently interact directly with movement proteins (MP). Some viruses require CP for long distance but not for cell-to-cell movement of the virus. Mutations to the CP in even a single specific amino acid inhibit the systemic infection of host plants. Other viruses absolutely require CP for even cell-to-cell movement, whereas the movement of still other viruses seems to be unaffected by the absence of CP.

Viral Genome Activation. Virus RNAs within the genera *Alfamovirus* and *Ilarvirus* require that unless a few molecules of CP are present, they cannot cause infection on their hosts. CP is probably necessary for the replication of negative-sense RNA viruses.

Symptoms. CPs can modify the symptoms caused by viruses in plants. Minor modifications of the genes of plant viruses, including the CP gene, can result in significant changes in symptomatology. In some cases, changes in a single amino acid result in dramatic changes in symptoms, ranging from stopping host development to death of the host.

Elicitor of Defense Responses. An important aspect of disease induction by a virus is the ability of the virus to neutralize or overcome the defense responses of the host. The resistance of plants to disease is via the hypersensitive response, which blocks further spread of the virus by programmed death of the infected and adjacent plant cells. Plant viral CPs generally act as elicitors of the plant defense response.

CP-Mediated Resistance in Transgenic Plants. Translatable or nontranslatable portions of CP gene sequences used to make transgenic plants confer resistance to the plant to subsequent challenge inoculation with the same or other viruses.

Viral Pathogenicity Genes

It can be concluded from the aforementioned discussion that the coat protein gene of most viruses plays one or many important pathogenicity roles for the virus. There are not enough genes in the genome of any virus to have separate genes for each of its various necessary functions that provide for its survival, multiplication, and spread. The gene encoding the nucleic acid replicase of the virus is obviously essential because without it there would be no virus. The movement protein-encoding gene is a virulence/pathogenicity gene because it enhances the multiplication and spread of the virus to other cells and plants. The same can be said for the gene(s) that encode proteins that make it possible for the virus to be acquired and then transmitted to other plants by one of the vector insects, nematodes, fungi, and so on.

Nematode Pathogenicity Genes

Nematodes attack plants by penetrating mostly root cells through their stylet. They secrete saliva that liquefies the cell contents that they absorb and move on. They enter the roots and move about in them, or they anchor themselves onto some root cells that become specialized and serve as feeder cells for the nematodes. Nematode secretions have been suspected to contain substances that nematodes use to attack their host plants and bring about a successful infection. These substances are presumably involved in hatching, in self-defense, in movement through plant tissue, and in the establishment and maintenance of a feeding site. Nematode secretions derive from several body structures, including the cuticle, amphids, and esophageal gland cells.

Cuticle Secretions

The surface of the cuticle of the infective juvenile is covered with a protein that binds to retinol and the linolenic and linoleic fatty acids, and inhibits the modification of these compounds by lipoxygenases. Peroxidation of linolenic acid by lipoxygenases is one of the steps in the synthesis of jasmonic acid, which is a signal transducer of systemic plant defenses. Also, peroxidation of lipids by lipoxygenases leads to the generation of reactive oxygen species in plants. Therefore, the protein secreted at the nematode cuticle, by inhibiting the lipoxygenase activities, downregulates and protects the nematode from the defense responses by the plant. The production of reactive oxygen species would also be a hostile environment for the nematodes, as are peroxiredoxins, which are a family of peroxidases that remove hydrogen peroxide produced at the nematode/plant interface. Superoxide dismutase, a scavenger of free oxygen radicals, is also produced in cuticle secretions.

Amphid Secretions

The role of amphids and their secetions in development of disease is not yet clear but all indications are that they play a major role in feeding site formation and maintenance. Two genes encoding two small proteins have been cloned from the amphids, but the role of the proteins in disease development is still not known.

Esophageal Gland Secretions

The esophageal glands in nematodes have for years been recognized as a major source of proteins that play a role in the parasitism of the nematode. Two sequences, one homologous to a hymenopteran venom allergen and the other homologous to a cellulose-binding cellulase-like protein, have been identified. Numerous other genes have been identified and their proteins are being studied.

Although more than 25 major resistance genes (R genes) against nematodes have been found in plants, no products encoded by nematode avirulence genes have been isolated. Of course, not all resistance to nematodes is provided by R genes.

Genetics of Resistance through the Hypersensitive Response

As mentioned previously, the hypersensitive response is a localized self-induced cell death at the site of infection of a host plant by a race or strain of a pathogen that cannot develop extensively in this particular resistant plant cultivar. Thus, the plant species as a whole may be a host to the pathogen species, but individual cultivars (varieties) of the plant may be hosts (susceptible) or nonhosts (resistant) to a particular race or strain of the pathogen. Resistance through the hypersensitive response has been shown to be the result of gene-for-gene systems in which an avirulence (*avr*) gene in the pathogen corresponds to a resistance (R) gene in the host plant. Such gene-for-gene systems that provide resistance through the hypersensitive response occur in diseases caused by obligate intracellular pathogens, such as viruses and mollicutes, as well as in diseases caused by obligate and facultative pathogens, such as bacteria, fungi, and nematodes. Whatever the type of pathogen, it is believed that resistance through the hypersensitive response is the result of recognition by the plant of specific signal molecules, the **elicitors**, produced by the avirulence genes of the pathogen and recognized by R gene-coded specific receptor molecules in the plant. Such recognition causes the activation of a cascade of host genes, which result in a burst of oxidative reactions, disruption of cell membranes, and release of phenolic and other toxic compounds, which then lead to the hypersensitive response, programmed cell death, inhibition of pathogen growth, and thereby resistance (Fig. 4-13). It also leads to the activation of numerous other defense-related genes that result in other types of resistance, including horizontal resistance and systemic acquired resistance.

Pathogen-Derived Elicitors of Defense Responses in Plants

Pathogen-produced elicitors that trigger defense responses in plants include a wide variety of molecules that seem to have little in common. Some elicitors are host specific, i.e., they induce defense responses leading to disease resistance only in specific host varieties, as is the case with elicitors produced by *avr* genes interacting with a matching R resistance gene in a host plant. Most elicitors are general or limited specificity elicitors in that they signal the presence of a potential pathogen to both host and nonhost plants, although some general elicitors are recognized by a small number of plants (Table 4-5).

In nature, the elicitor molecule either reacts directly with the receptor protein encoded by the resistance gene R, or releases compounds or reacts with another host protein (endogenous elicitors), which then interacts with the R-coded receptor.

Avirulence (*avr*) Genes: One of the Elicitors of Plant Defense Responses

Avirulence (*avr*) genes, first identified by H. H. Flor in the 1950s, were only rather recently isolated from bacteria (1984) and fungi (1991), but since then numerous bacterial and fungal *avr* genes have been identified. The *avr* genes make a pathogen avirulent, that is unable to induce disease on a specific variety of the host plant because their protein product warns the plant of the presence and impending attack by the pathogen and the host plant then mobilizes its defenses and blocks infection by the pathogen. In this way, *avr* genes, by warning the host and thereby inhibiting infection by the pathogen, determine the host range of the pathogen at the species and at the race-variety level.

As the gene-for-gene concept implies, in the majority of cases a matching dominant resistance gene (R) in the

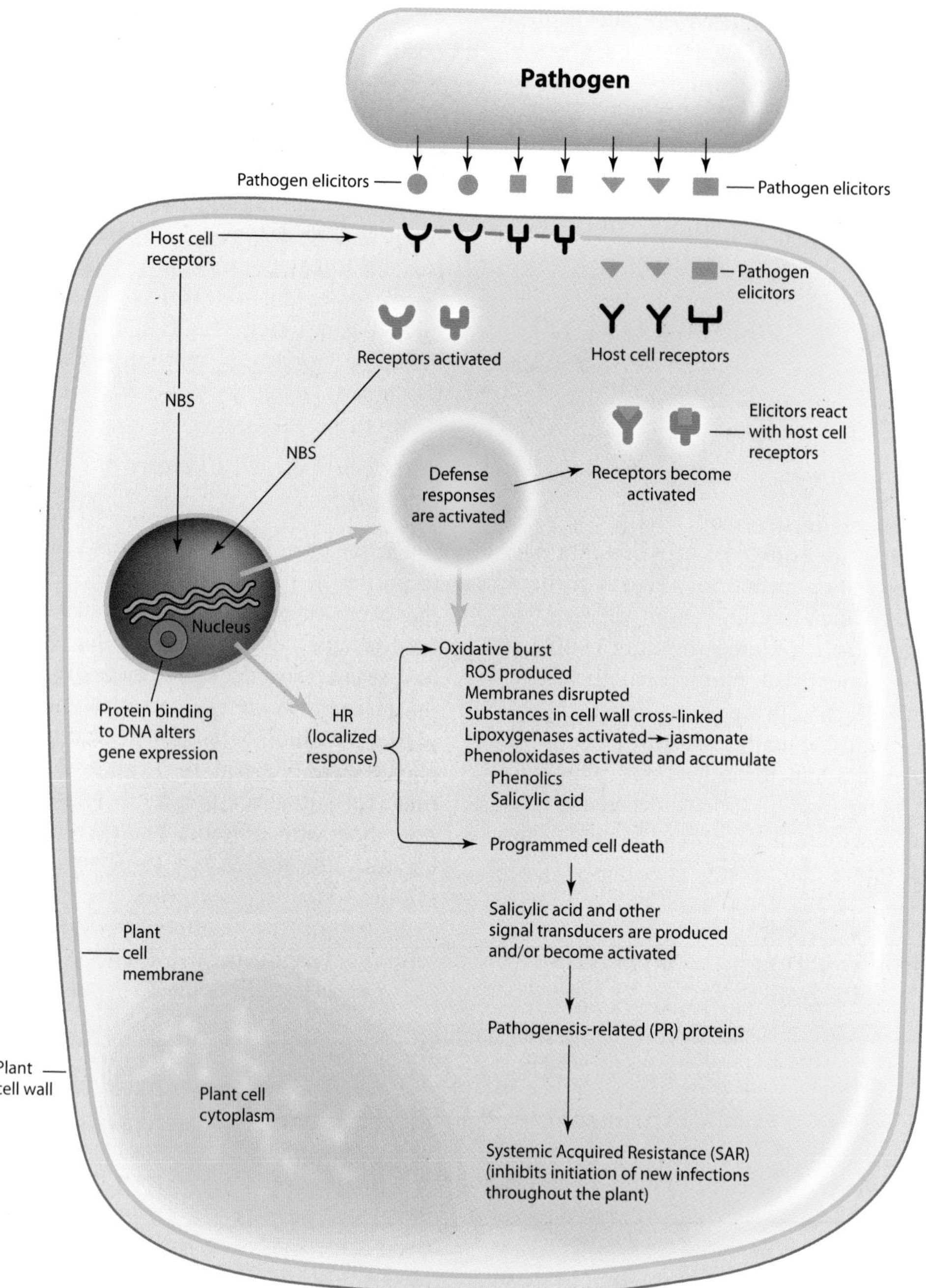

FIGURE 4-13 Basic events in an incompatible host–pathogen interaction: Elicitors from pathogen interact with plant cell receptors. Signal transductions activate hypersensitive (host defense) responses that lead to programmed cell death and systemic acquired resistance.

resistant host corresponds to each avirulence gene in the pathogen. In some cases, however, because two independent resistance (R) genes may correspond to a single *avr* gene, there apparently are genes-for-gene interactions as well. Some *avr* genes, when transferred artificially to other pathovars, are active in the new pathovars, making the recipient pathogen unable to infect their previously susceptible hosts and, instead, causing the hypersensitive response in these plants. In some host–pathogen systems, *avr* genes determine not only which cultivars of a species the pathogen can attack, but also which plant species it can attack. For example, an *avr* gene (*avrBsT*) in the tomato-infecting group of strains of the bacterium *Xanthomonas campestris* pv. *vesicatoria*, the pathogen of bacterial spot in tomato and pepper, enables the bacterium to induce

TABLE 4-5

General elicitors
Glucans, produced by *Phytophthora and Pythium*, derived from oomycete cell wall, induce phytoalexins
Chitin oligomers, by higher fungi, from chitin of fungal cell wall, induce phytoalexins and lignification
Pectin oligomers, by fungi and bacteria, from degraded cell wall, inhibit proteins and defense genes
Harpins, by several gram-negative bacteria, part of type III secretion, cause HR and defense gene response
Flagellin, by gram-negative bacteria, part of flagellum, cause callose formation and defense gene response
Glycoproteins, by *Phytophthora*, induce phytoalexin production and defense gene response
Glycopeptide fragments, by yeast, activate defense genes and ethylene production
Ergosterol, by various fungi, the main sterol of higher fungi, causes alkalinization in cell cultures
Bacterial toxins, such as coronatine of *P. syringae*, toxin, disturbs salicylic acid, mimics jasmonic acid, and induces defense genes and defense compounds
Sphinganine, the fumonisin analog, by *F. moniliforme*, toxin in necrotrophs, disturbs sphingolipid use, induces defense genes and programmed cell death (PCD)
Race-specific elicitors
avr gene products, Avr proteins, by fungi and bacteria, in some cases promoting virulence, HR, and PCD
Elicitins, by *Phytophthora and Pythium*, scavengers of sterol, induce HR in tobacco
Enzymes, e.g., endoxylanase, by *Trichoderma viride*, fungal enzymes, induce defense genes and HR
Viral proteins, e.g., viral coat proteins, by TMV, structural component, HR in tobacco, tomato
Protein or peptide toxins, e.g., victorin, by *Cochliobolus victoriae*, toxin for host, induces PCD in oat
Syringolids (acyl glycosides), by *P. syringae pv. syringae*, signal compound for bacterium, HR in soybean, carrying the Rpg4 resistance gene

the hypersensitive response on all cultivars of pepper. Loss of *avrBsT* from such tomato-infecting strains allows these strains to cause disease on normally resistant pepper cultivars.

Several avirulence genes and the proteins they code have been identified in and isolated from plant pathogenic fungi. These include especially the genes *avr2, avr4, and avr9* of strains of the fungus *Cladosporium fulvum* that are avirulent on tomato varieties carrying, respectively, the resistance loci Cf-2, Cf4, and Cf-9; and the gene *avr*Pi-ta of the rice blast fungus, *Magnaporthe grisea,* which confers avirulence to rice varieties containing the resistance gene Pi-ta. Similarly, several viral *avr* genes and their avr proteins have been obtained and studied, including those of the coat protein of *potato virus X (PVX)*, the coat protein of *turnip crinkle virus (TCV)*, and the replicase protein of *tobacco mosaic virus (TMV)*.

Characteristics of avr *Gene-Coded Proteins*

The gene-for-gene model stipulates that for every dominant gene determining resistance in the host plant, there is a matching dominant gene in the pathogen that conditions avirulence. The biochemical basis for explaining the gene-for-gene concept is the elicitor–receptor model according to which an avirulence (*Avr*) gene of a pathogen encodes an elicitor (Avr) protein that is recognized by a receptor protein encoded by the matching resistance (R) gene of the host plant.

The simplest way of recognition would be if the pathogen-produced elicitor interacted with the protein encoded by the matching resistance gene of the host. Recognition of the elicitor protein by the host plant leads to activation of a cascade of defense responses, which often include cell death around the infection site. The death of cells around the point of infection is known as the hypersensitive response and is characteristic of gene-for-gene-based resistance.

Unlike R proteins, Avr proteins encoded by pathogen *Avr* genes share few common characteristics. Because most *Avr* genes continue to exist within a pathogen population, it would seem that in addition to acting as avirulence factors, *Avr* genes probably have some additional function that is beneficial to the pathogen. From the few *Avr* genes for which a clear function for the pathogen has been demonstrated, it has now become generally accepted that their proteins carry out two functions, one of them being a contribution toward the virulence of the pathogen. Such a contribution appears to come about by the Avr proteins interacting with specific plant proteins, known as virulence targets, involved, for example, in host metabolism or in plant defense. Interaction of Avr proteins with such targets could enhance the availability of nutrients for the pathogen or could suppress defense responses by the host plant. To date, the AvrD protein, produced by the *AvrD* gene of the bacterial spot of tomato pathogen *P. syringae* pv. *tomato,* is the only Avr protein for which a biochemical function has been clearly defined. This function is the ability of the AvrD protein to direct the synthesis of low molecular weight syringolide elicitors, which elicit the hypersensitive response on soybean. A syringolide-binding protein has been identified in resistant soybean plants, possibly representing the protein of the matching R gene of the host plant.

Proteins coded by pathogen *avr* genes (Avr proteins) seem to have some features in common. Avr proteins seem to be generally hydrophilic and, therefore, water soluble, lacking stretches of hydrophobic amino acids that would enable them to be anchored in cell membranes. Avr proteins also lack stretches of amino acids known as "signal sequences" that would allow the proteins to be secreted into the external medium by the general secretory pathway. It appears, therefore, that *avr* gene proteins are produced and are either localized in the pathogen cytoplasm or they are secreted through membrane pores formed by proteins coded for by hypersensitive response and pathogenicity (*hrp*) genes, known as Hrp proteins (harpins). If they are secreted externally, the Avr proteins may act directly as elicitors. If they are localized in the pathogen cytoplasm, the *avr* gene proteins may act enzymatically to produce an elicitor molecule that is transported freely through the bacterial envelope. In either case, the elicitor reacts directly or indirectly with the product of the corresponding plant resistance R gene (Figs. 4-10 and 4-11).

Structure of avr *Gene Proteins*

Although *avr* genes are quite different, some of them have common structural characteristics that allow grouping of *avr* genes into distinct families. The best known *avr* gene group is the *Xanthomonas avr* gene family, called *pth* (for pathogenicity) genes by some. Members of this gene family are found in different species and pathovars of the bacterium *Xanthomonas*. They encode proteins that, in their central part, have from 13 to 23 copies of a nearly identical 34 amino acid repeat unit. *avr/pth* genes cause the hypersensitive response and are also required for the induction of angular leaf spot symptoms of cotton and for citrus canker disease. Elicitation of these very different symptoms (leaf spots, cankers, the HR) is determined by a single or a few amino acid differences in the repetitive regions of these genes.

Among fungal avr proteins, the *Cladosporium fulvum*-encoded avr2 is a cysteine-rich protein of 78 amino acids that has a signal peptide of 20 amino acids for external targeting; the *Cf* avr4 protein consists at first of a 135 amino acid preprotein, which upon secretion is processed at both ends, resulting in an 86 amino acid protein; and the *Cf* avr9 protein at first consisting of a precursor protein of 63 amino acids, which is further processed into a 28 amino acid peptide. All three *Cf* avr proteins are secreted in the apoplastic space of tomato leaves, are localized in the plasma membrane, and contain an extracellular leucine-rich region (LRR), a transmembrane domain, and a short cytoplasmic tail. The *Magnaporthe grisea*-encoded avr-Pi-ta protein consists of 223 amino acids but is processed into a 176 amino acid protein that has homology to zinc-dependent metalloproteases. The Pi-ta avr protein is cytoplasmic and contains a nuclear-binding site (NBS) and a leucine-rich carboxyl terminus. The viral avr proteins elicit corresponding plant resistance R genes that encode cytoplasmic proteins. These proteins consist, in the case of PVX and TCV, of either LZ-NBS-LRR domains or, as in TMV, of TIR-NBS-LRR domains (LZ, leucine zipper; TIR, toll interleukin 1 receptor).

Function of avr *Gene Proteins*

So far, the functions of only one *avr* gene, *avrD*, have been determined. The *avrD* gene is present in the bacterium *P. syringae* pv. *tomato*, but ArvD alleles are present in soybean *P. syringae* pv. *glycinea* and other hosts. *avrD* encodes syringolide elicitors, which react with the receptor protein of R gene, Rpg4 of soybean, and confers avirulence on soybean. It has no effect on the virulence of the bacterium.

The function of fungal avr proteins is not known with certainty. The timing and location of their expression suggest a role in the infection process, but so far no virulence function has been reported for most such proteins. In the case of the avrPi-ta protein, direct interaction was detected between the mature avrPi-ta protein and the leucine-rich domain of the Pi-ta R gene protein. This finding is the first experimental evidence consistent with the proposed model that avr proteins interact directly with the corresponding R proteins.

In the case of *tobacco mosaic virus*, causing the hypersensitive response in *Nicotiana sylvestris* tobacco carrying the N^1 gene for resistance, the avirulence function and thereby the elicitation of hypersensitive response seem to reside in the presence of certain amino acids on the coat protein of the virus: N^1 gene-containing plants transformed with only the gene of such TMV elicitor coat proteins, without inoculation with the virus, exhibited the hypersensitive response in the form of reduced growth, chlorotic and necrotic patches, and eventual collapse of entire leaves. Plants transformed with mutant weakly eliciting or nonelicitor coat proteins expressed respectively weaker or no hypersensitive response. In at least some viral infections then, the viral coat protein, which is produced within the cell, appears to function as a specific elicitor that activates the hypersensitive response in plant cultivars that carry the corresponding R gene for that virus.

Role of avr *Genes in Pathogenicity and Virulence*

Most *avr* genes tested so far play no role in pathogenicity or virulence of the pathogen, as even when *avr* genes are inactivated by mutation, susceptible hosts continue to be susceptible. Some *avr* genes, however, e.g., the *avrBs2* gene from the bacterium *X. campestris* pv. *vesicatoria*, encode proteins that are also necessary for pathogenicity. This is shown by the fact that this *avr* gene is present in all strains of this pathovar, whereas mutants lacking the *avr* gene lose pathogenicity on all susceptible hosts but do not gain virulence on any previously resistant hosts. However, several *avr* genes, such as the *pthA* gene from *X. citri* and *avrb6* from *X. campestris* pv. *malvacearum*, both members of the *Xanthomonas avr/pth* gene family, encode proteins that act as pathogenicity or virulence factors. For example, they enhance the virulence of a weakly pathogenic leaf-spotting strain of *X. citrumelo*, enabling it to cause canker-like lesions on its host; they may act as pathogenicity factors, e.g., *pthA* is required for the pathogenicity of *X. citri* on citrus to cause the typical citrus canker disease; and act as avirulence genes, e.g., by causing *pthA*-transformed strains of *X. phaseoli* and *X. campestris* pv. *malvacearum* that, respectively, infect bean or cotton, but not citrus, to cause the hypersensitive response on their respective hosts bean and cotton while remaining nonpathogenic on citrus.

The role of fungal *avr* genes in pathogenicity and virulence of the pathogens involved is mostly unclear. In some cases, avr proteins seem to react with other proteins that play an intermediate role in transmitting the signals for plant defense. In a few cases, as in the avr Pi-ta protein, they seem to interact directly with the R protein and to set off a cascade of defense reactions. In viruses, a certain segment of a particular coat or replicase protein seems to interact with the host R gene. Most of these statements, however, need further experimentation to support their validity.

hrp *Genes and the Type III Secretion System: Another Class of Pathogenicity Genes in Bacteria*

The *hrp* (hypersensitive response and pathogenicity) genes, found only in gram-negative bacteria so far, are additional bacterial genes that seem to be essential for some bacteria to be able to cause visible disease on a host plant, to induce a hypersensitive response on certain plants that are normally not infected by the bacteria, and to enable bacteria to multiply and reach high numbers in a susceptible host. Most bacterial species have two distinct clusters of *hrp* genes. The larger *hrp* gene cluster consists of six to nine transcription units, with each transcription unit coding for several (1 to 12) proteins. The transcription of *hrp* genes is controlled by the presence of certain nutrients, by other bacterial regulatory genes, and by so far unknown signal molecules of plant origin.

Several *hrp* gene-coded proteins, called harpins, seem to be localized in the bacterial cell membrane (Fig. 4-11). There they may be involved in forming a type III secretory apparatus involved in the outward translocation of bacterial Avr or Hrp proteins that could interact with components of host plant cells. Some *hrp* genes also code for an ATPase enzyme that may play a role in energizing the secretory apparatus.

In some bacteria, e.g., in *P. syringae*, a single promoter gene controls the expression of both *hrp* and *avr* genes, including the production of a harpin, a secretion system for harpins, and the *avr* products that elicit the hypersensitive plant response and affect the host range of the pathogens. The coregulation of both *hrp* and *avr* genes suggests that the final effectors of these genes may act together to determine the final outcome of the plant–bacterium interaction.

Resistance (R) Genes of Plants

As mentioned earlier, despite the many and different kinds of plant pathogens that come in contact with a plant, in most cases, plants remain resistant to disease because they are not hosts to the vast majority of pathogens (nonhost resistance). What makes a plant nonhost to most pathogens and host to a small number of pathogens (usually about 50–100) is still not known. Even when a plant is a host (i.e., is susceptible) to a certain pathogen, some varieties of the plant may be susceptible, or more susceptible, to the pathogen, whereas others may be resistant, or more resistant, to the pathogen. This depends on the kind and number of resistance genes present in the plant, the prevailing environmental conditions, and other factors. Even when a plant becomes attacked and diseased by a pathogen, however, a number of defense response (resistance) genes are activated. As a result, in most cases, the plant manages to limit the spread of the pathogen into a smaller or larger spot, lesion, canker, and so on through defense compounds and structures that block the further expansion of the pathogen. In a number of cases, however, plant varieties are resistant to certain pathogen races because they possess specific resistance (R) genes that enable the plant to remain resistant to pathogens carrying the corresponding avirulence (*avr*) genes.

So far, a number of plant R genes and pathogen *Avr* genes have been cloned and characterized. The proteins encoded by R genes are quite similar and are classified according to certain structural characteristics they have and according to their localization in the plant cell (Fig. 4-14). All R proteins except two contain a domain rich in the amino acid leucine (LRR, leucine rich repeats), which is thought to take part in protein–protein (e.g., elicitor–receptor) interactions. Depending on where in the plant cell the R protein LRR reside, they have either cytoplasmic LRRs or extracytoplasmic LRRs. The R proteins that have a cytoplasmic LRR domain also have a nucleotide-binding site (NBS) and some of them have a zipper-like domain of leucine molecules known as coiled coil, or have a domain of Toll/interleukin 1 recep-

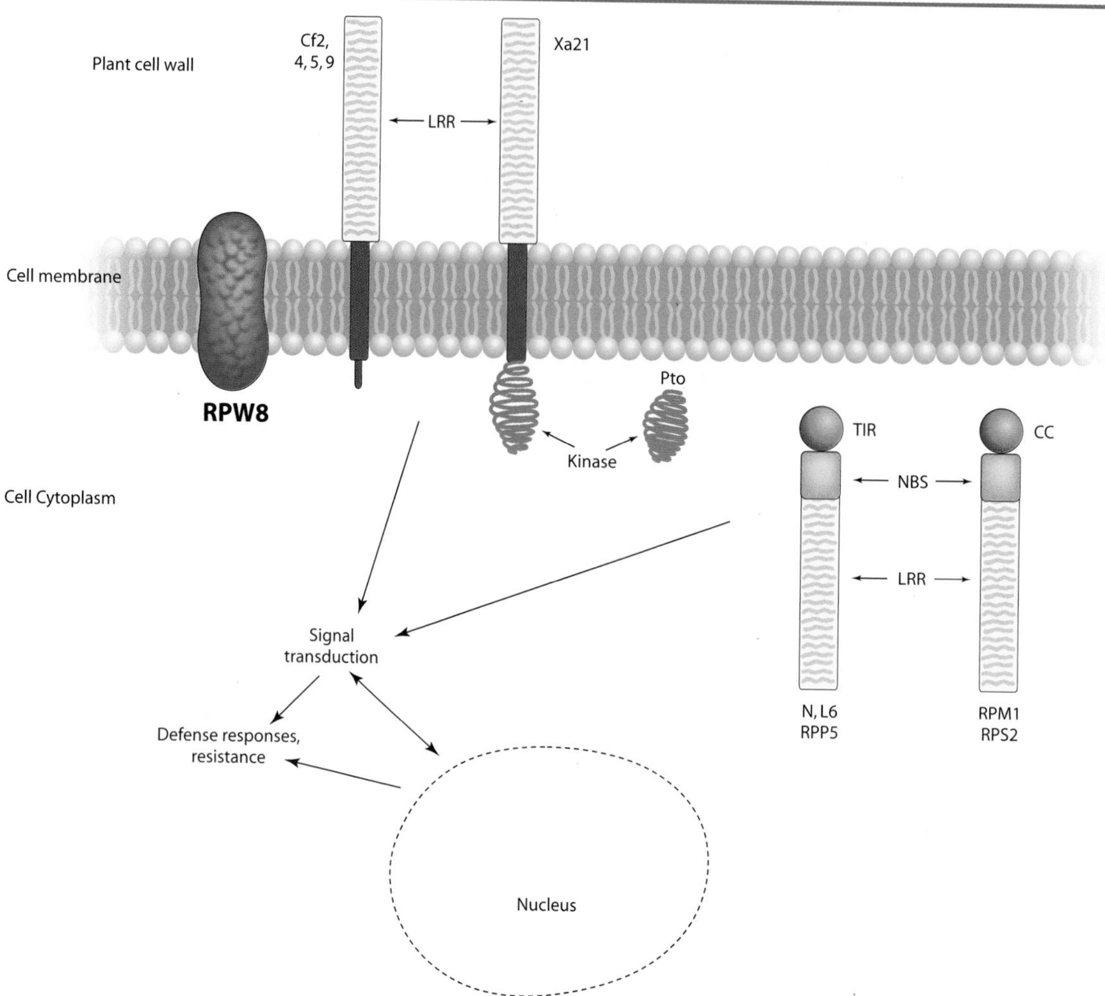

FIGURE 4-14 Schematic diagram of the structure and cell location of the six types of R-coded receptor proteins. Three types have transmembranous domains, while the other three are membrane-associated cytoplasmic proteins. LRR, leucine-rich repeats; NBS, nucleotide-binding site. TIR, Toll–interleukin-1 resistance receptor domain; CC, coiled coil with leucine zipper domain. Genes listed are tomato *Cf*-2, -4, -5, -9, rice Xa21, tomato Pto, tobacco N, flax L6, Arabidopsis RPM1, RPS2, RPP5, and the Arabidopsis broad-spectrum gene RPW8.

tor (TIR). A different kind of R gene named RPW8 has been found in *Arabidopsis*. RPW8 is different in that it confers resistance to a broad range of powdery mildew pathogens instead of a specific pathogen race. The RPW8 protein is located in the plant cell membrane but its mode of action is not known yet. The R proteins that have an extracytoplasmic LRR domain contain a transmembrane region, and some of them also contain a cytoplasmic domain that acts as a protein kinase. Although the structure of R proteins predicts a role for them in signal transduction, it is not clear how these proteins initiate defense responses.

Examples of R Genes

In 1992, the first R gene, the maize *Hm1* gene, was located, isolated, and sequenced, and its function was described at the molecular level. The *Hm1* R gene makes corn plants of certain varieties resistant to race 1 of the fungus *Cochliobolus carbonum*, which causes a leaf spot disease on susceptible corn varieties. Race 1 of *C. carbonum*, the asexual stage of which is *Bipolaris (Helminthosporium) carbonum*, produces a host-specific toxin, the HC toxin. The toxin is a pathogenicity factor for race 1 because the latter must produce HC toxin if it is to infect the corn varieties that lack the *Hm1* gene and are susceptible to the fungus. However, in corn varieties resistant to race 1, expression of the *Hm1* gene results in the production of an enzyme called HC toxin reductase. This enzyme reduces and thereby detoxifies the HC toxin and in that way keeps the plants free from infection by the fungus. If the HC toxin gene of some race 1 isolates is inactivated artificially, these isolates lose the ability to infect corn varieties that do not carry the *Hm1* gene and, therefore, the genetics of this host–pathogen system are not quite the same as in the typical gene-for-gene systems.

Within 3 years after isolation of the *Hm1* gene, more than a dozen plant R genes that conform to the classic gene-for-gene relationship were isolated from plants, sequenced, and transferred and expressed in other, susceptible, plants. The first such gene was the *Pto* gene of tomato, so called because it confers resistance in tomato to the bacterial speck-causing strains of *P. syringae* pv. *tomato* that carry the avirulence gene *avrPto*. The protein encoded by the *Pto* R gene appears to be a serine–threonine protein kinase, an enzyme suspected to play a role in signal transduction leading to the hypersensitive response. The *Pto* R gene appears to be one of five to seven homologous R genes that exist as a cluster on one of the tomato chromosomes.

Some of the other R genes isolated from plants include the tomato *Cf2, Cf4, Cf5, and Cf9* genes, which confer resistance to the leaf mold-causing fungus *Cladosporium fulvum* races 2, 4, 5, and 9 that carry the avirulence genes *avr2, avr4, avr5*, and *avr9*, respectively; the tobacco N^1 gene, which confers resistance to TMV; the flax L^6 gene, which confers resistance to the rust fungus *Melampsora lini* race 6 carrying the *avr6* gene; the rice *Xa21* gene, which confers resistance to many races of the leaf-spotting bacterium *Xanthomonas oryzae*; and several *Arabidopsis* R genes (Table 4-6).

How Do R Genes Confer Resistance?

The mechanisms by which R genes bring about disease resistance to a plant against a specific pathogen are not yet understood. It is believed that the elicitor molecule produced by an *avr* gene of the pathogen is recognized by a specific plant receptor encoded by an R gene. What happens next is mostly speculation. Following recognition of the elicitor by the receptor molecule, one or more kinase enzymes may become activated, which then amplify the signal by phosphorylating, and thereby energizing, other kinases and other enzymes. This leads to a cascade of biochemical reactions that, in ways that are still unclear, result in the hypersensitive response and, thereby, localized host resistance at the point of attack by the pathogen. Of course, in many cases, the hypersensitive response is followed by the development of various levels of systemic acquired resistance (SAR), which is expressed in the vicinity of attack as well as in distant parts of the plant.

Evolution of R Genes

It is thought that when a plant was first attacked by a new pathogen strain, the plant probably had some genes encoding nonspecific receptor molecules that enabled the activation of defense responses to wounding and to pathogens in general but that it lacked any R genes to the new pathogen (Fig. 4-15). This pathogen, therefore, was able to cause considerable damage to the plant and possibly killed many of the susceptible plants. Plants exhibiting greater or lesser general resistance survived and multiplied to proportional extents. When, during the evolutionary race for survival of the plant from the pathogen, a resistance (R_1) gene evolved, e.g., by modification of one of the general resistance genes, and that gene allowed the plant to recognize one of the initial steps of infection by the new pathogen (race 1) and to resist infection, such an individual plant and its progeny (variety 1) were selected for survival and so the plant and the R_1 gene survived and multiplied. This might have happened, for example, by modification of one of the receptors involved in activating plant defenses against pathogens in general. Thus, the modified receptor 1 product of the R_1 gene recognizes specifically a particular compound (elicitor 1) produced by a pathogen gene, which gene, as a result, behaves like an avirulence (*avr1*) gene. Pathogens carrying this *avr1* gene (race 1) cannot survive on such R_1 gene-carrying plants. If, however, in time, a mutation affects the *avr1* gene of race 1 of the pathogen, which gene until now was the cause of its avirulence, the gene and the avirulence are destroyed. As a result, the new offspring of the pathogen become virulent again, capable of attacking the so-far resistant variety 1 of the plant. This new virulent pathogen population could be called race 2. The host plant (variety 1) is now susceptible to race 2, which

TABLE 4-6

Classes of Plant R Gene Proteins

Class	Function	Example of R gene
I	Membrane–associated, transcription regulating, mediating broad-spectrum resistance	RPW8
II	Cytoplasmic signal-transducing serine–threonine protein kinase	*Pto*
III	Extracellular LRRs with transmembrane anchor	*Cf-2–Cf-9*
IV	Extracellular LRRs, with a transmembrane receptor and a cytoplasmic serine–threonine kinase	Xa21
V	Cytoplasmic, membrane associated. Contain LRRs, NBS, and TIR domains	RPP5, N^1, $L6_{6,RRPP}$
VI	Also cytoplasmic, membrane associated. Contain LRRs, NBS, and a coiled coil domain	RPM1, RPS2

infects and may kill many plants. Soon, however, through survival pressure and selection, an R_2 gene evolves that encodes a new or further modified receptor 2 that recognizes a different compound (elicitor 2) produced by the *avr* gene of individuals of the pathogen race 2. This gene, then, becomes the *avr2* gene conferring avirulence to the pathogen because it is recognized by the R_2 gene of the plant. In this way, numerous, diverse R genes have evolved in a plant host to counteract corresponding virulence genes in the various races of one of its pathogens. This gene-for-gene interaction has occurred in a stepwise fashion over time and continues to date (Fig. 4-15).

The evolutionary process just described is supported by the fact that most of the R genes studied so far seem to be present in tandem arrays of multiple (up to 10 or more) related R genes: They exhibit different specificities but behave as though they are alleles of a single gene that cannot be separated during recombination or exist as a clustered gene family. The various R genes isolated so far appear to have a portion (about 20%) of their nucleotide sequences identical, whereas a larger portion (about 50%) of the nucleotide sequences are similar. Such relationships among R genes may indicate an important mechanism by which plants, by reshuffling preexisting coding information, can respond more quickly to attack by a new pathogen by reformulating existing R genes into new R genes that then produce new specific receptors. The latter are needed, of course, to recognize one of the diverse molecules produced by pathogens, which in any case, because of their extremely large populations, change at a much greater frequency than plants can produce R genes. Besides, the change of a pathogen from avirulence to virulence is caused by the loss of an avirulence gene through a loss of function mutation on that gene, an event much more likely to happen than the positive production of a new receptor on a plant by a newly formed R gene (Fig. 4-16).

Other Plant Genes for Resistance to Disease

As mentioned earlier, how many and what types of genes make a plant nonhost, and therefore resistant, to a pathogen are unknown. It is possible that nonhost resistance is due to a lack of host recognition factors so that the pathogen is not triggered to express its pathogenicity factors on such a plant. Alternatively, it is possible that the nonhost plant carries numerous R gene-coded receptors, one or more of which quickly recognize and fend off the pathogen, or, probably, some entirely different mechanisms are responsible for nonhost resistance.

R genes, as discussed, are responsible for recognition by certain plant varieties of specific elicitors produced by certain pathogen races. That recognition results in the production of signal molecules, some of which trigger a cascade of localized reactions, leading to the hypersensitive response and, through the plant-induced death of the affected cell, to localized resistance. Such spectacular, easily identified R gene-dependent resistance is rather rare in natural genetically heterogeneous plant populations, but has been introduced into culti-

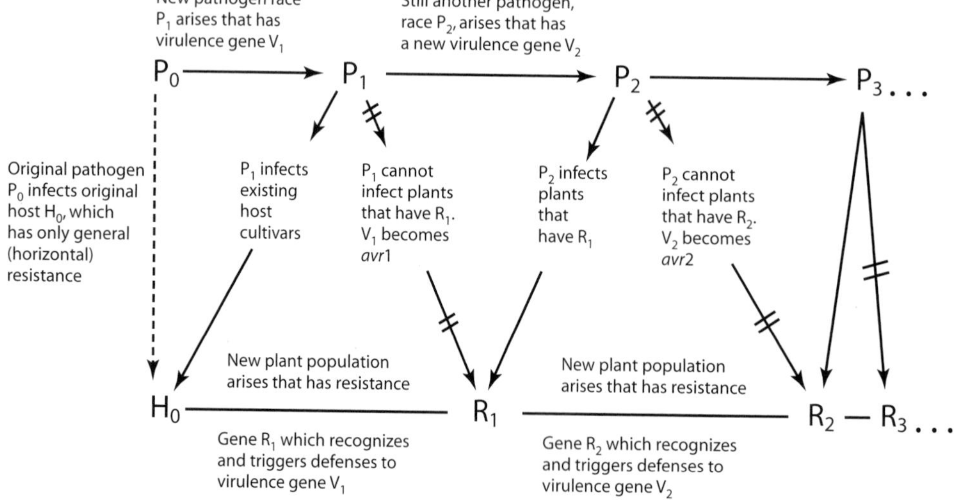

FIGURE 4-15 Steps in the evolution of genes for virulence, resistance, and avirulence. Note that race 1 pathogens (P_1) can still infect hosts carrying only the original resistance or R_2 resistance; they cannot infect plants with R_1 resistance. Also, plants with R_1 resistance are only resistant to P_1 pathogens that carry the V_1 (*avr1*) gene. R_1-carrying plants are susceptible to the original pathogen population (P_0) and to other pathogen races, e.g., to P_2.

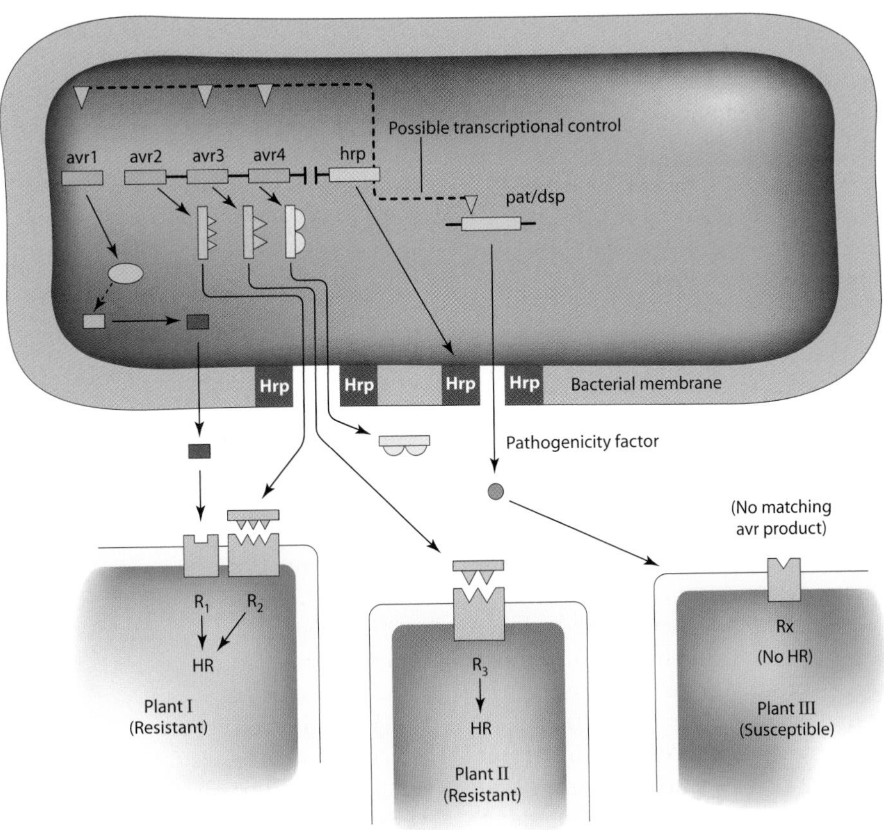

FIGURE 4-16 A simplified scheme of hypothetical molecular interactions between *avr* and *hrp* genes of a pathogenic bacterium and the R genes of two resistant and one susceptible plant. In this diagram, the avr1 product induces an intracellular enzyme to produce an elicitor that moves freely through the bacterial envelope. The products of *avr*2, *avr*3, and *avr*4, as well as effector proteins transmitted through the type III secretion system, move through membrane pores formed by the proteins (harpins) of *hrp* genes and act as elicitor molecules on receptors encoded by corresponding R genes. The pathogenicity/disease specificity (*pat/dsp*) genes are likely producers of effector proteins. From Van Gijsegen *et al.* (1995).

vated crops by breeding and is now quite common in commercial crops.

During development of the hypersensitive response, some of the signal molecules act on other signal molecules that transmit the alarm to other cells and to most distal parts of the plant. There, they trigger the activation of additional defense response genes called systemic acquired resistance genes. These genes mobilize the host defenses throughout the plant and are effective against new infections by the same pathogen and also against infections by unrelated pathogens.

The most common types of resistance genes in plants in natural populations, and quite often in cultivated crops, are numerous "minor" genes for resistance. These may affect superficial or internal, structural or biochemical defenses, preexisting or induced on or after infection. Such minor genes are probably quite numerous in all plants. They are triggered into action by signal compounds produced by the pathogen or by the infected cells and, in most cases, through their actions, produce defenses that manage to halt the advance of the pathogen and colonization of the host to a small lesion on whatever plant organ is attacked. Such minor defense genes do not always appear to effectively defend plants from pathogens, primarily because the pathogens can overcome their hosts by the sheer number of small lesions they cause on the plants. Nevertheless, in most cases, these genes manage to halt the pathogen to a small lesion in each individual infection.

Signal Transduction between Pathogenicity Genes and Resistance Genes

Induced defenses of plants against pathogens are regulated by networks of interconnecting signaling pathways in which the primary components are the plant signal molecules salicylic acid (SA), jasmonic acid (JA), ethylene (ET), and probably nitric oxide (NO). In many host/pathogen interactions, plants react to attack by pathogens with enhanced production of these substances while a distinct set of gene-to-gene resistance defense-related genes is activated and attempts to block the infection. Also, an exogenous application of SA, JA, ET, or NO to the plant often results in a higher level of resistance.

Salicylic acid reacts with several plant proteins, including the two major H_2O_2-scavenging enzymes catalase and ascorbate peroxidase, and with a chloroplast SA-binding protein, which also has antioxidant activity. The main components of the SA-mediated pathway leading to disease resistance appear to be constitutively expressed genes encoding pathogenesis-related (PR) proteins. Some of these genes also activate the JA- and ET-mediated pathways, leading to induction of the gene encoding defensin. However, NO synthase activity also increases dramatically upon inoculation of resistant but not of susceptible plants. NO induces the expression of PR-1 and the early defense gene phenylalanine lyase (PAL). Production of SA occurs within the NO-mediated pathway downstream of NO. As with SA, NO also reacts with and inhibits the activity of the enzymes aconitase, catalase, and ascorbate peroxidase. SA is not generally required for action of resistance genes R in determining resistance at the infection site, but in at least some plants, SA is required at the primary infection site and in distal secondary tissues for the establishment and maintenance of SAR. Current thinking, however, has HR based on the interplay and mutual positive feedback regulation of reactive oxygen intermediates (ROI) and SA-dependent signals. These, however, may not be the only signals required to set the HR cell death threshold. ROI and NO, generated independently during the oxidative burst, also collaborate to initiate HR. A balance between hydrogen peroxide, derived from dismutation of superoxide, and NO is required for HR. As a result, superoxide is a key regulator that can either convert NO into inert ONOO— or be dismutated to H_2O_2, keeping in mind that superoxide dismutase is induced rapidly by SA.

It is not known how plants integrate signals produced by different defense response pathways into specific defense responses. It is known, however, that the defense pathways dependent on SA, JA, ET, or NO affect each other's signaling either positively or negatively. This is called "cross talk" between pathways. Cross talk provides a regulatory potential for activating several resistance mechanisms in various combinations at once and may play a role in selecting for activation a particular defense pathway over others available. Due to negative cross talk, however, it is often assumed that SA-dependent defenses are often mutually exclusive with JA/ET–dependent defenses.

Signaling and Regulation of Programmed Cell Death

The hypersensitive response, which results in localized, very rapid cell death at the site of attempted pathogen ingress, is found in nearly all defense responses: Those mediated by one or more R genes, by nonhost resistance, and in many cases of polygenic or quantitative resistance. Interaction between elicitor and receptor molecules immediately leads to signal transduction during rapid ion flux. This results in alkalinization of the extracellular apoplast, formation of reactive oxygen intermediates, production of nitric oxide, activation of signaling cascades involving MAP kinase pathways, and transcriptional activation of a broad range of defense genes. This transcriptional reprogramming results in the production or release of antimicrobial compounds, or in the generation of signaling molecules that will act at distal points of the plant to establish systemic acquired resistance.

The extent of cell death during an HR can vary from one cell to tens of cells at the point of infection. Again, not all disease-resistant reactions lead to cell death. Depending on the "efficiency" of the R protein, resistance may be achieved without HR or, if less efficient, an R protein may require more ion flux and thus initiate HR. It should be kept in mind, however, that SA, JA, ET, and NO play a role not only in HR and programmed cell death, but also in decisions about cell growth. This strongly suggests that the relationship between SA and ROI and the genes they regulate is pivotal among signals that determine whether cells live or die. Considering that there are mutants in several species that, with their own plant genes, initiate cell death in the absence of any pathogen, HR can be considered as "programmed cell death." Such mutants could be thought of as representing a step along normal disease-resistant response pathways. Alternatively, they could be thought of as representing changes of normal metabolism that the cell senses and interprets as a commitment to rapid cell death by which it silences the renegade cell forever. Once the plant commits to the cell death pathway, it must be able to stop cell signals that might propagate cell death to neighboring cells, especially since plants have no scavengers that can engulf the corpses of cells killed by programmed cell death.

Some light on the mechanism of programmed cell death has been shed by the discovery of a recessive allele [the lesion simulating disease (lsd1)]. This allele leads to a lowered threshold for signals derived from pathogens, ROI, or SA and entering the disease resistance pathway. Disruption of this gene leads to mutant plants that are unable to stop the spread of cell death once it has started. Local applications of low concentrations of the signal molecule SA, of any of the other chemicals that activate "systemic acquired resistance," of pathogenic bacteria or fungi, or a shift to nonpermissive long-day conditions initiate foci of dead cells. These quickly become "runaway cell death" (rcd), which spreads beyond the initial site of infection and kills the entire inoculated leaf. Cell death, however, does not spread beyond the treated leaf. It has been shown that lesions form in leaves of treated lsd1-carrying plants as a result of accumulation of extracellular superoxide, to which these cells are extra sensitive, and cell death is initiated. This leads to subsequent superoxide formation in live neighboring cells, which leads to further superoxide formation and spread and to runaway cell death. At least one necrotrophic plant pathogenic fungus, *Botrytis cinerea*, attacks and induces runaway cell death in plants carrying the lsd1 gene. Upon infection, the fungus releases hydrogen peroxide or superoxide, which is converted rapidly by superoxide dismutase into hydrogen peroxide. This activates and stimulates the plant HR pathway. Thus, the fungus, by usurping the HR signaling and programmed cell death, subsequently invades the dying tissue and then continues to colonize the plant by mimicking the HR signals.

Genes and Signaling in Systemic Acquired Resistance

Systemic acquired resistance in plants is a secondary resistance response induced after a hypersensitive response to avirulent pathogens. The signal for SAR may be generated within 4–6 hours from inoculation. SA could be detected in the phloem by 8 hours after inoculation, and increases in SA occurred in the phloem of the leaf above the inoculated one within 12 hours from inoculation of the lower leaf. Expression of SAR occurred within 24 hours from inoculation. By that time the entire plant contained greatly increased levels of SA, even when the inoculated leaf had been removed before any SA increase had been detected. Plants transformed with the nahG gene, which codes for the enzyme salicylate hydroxylase, that breaks down SA to the simple phenolic catechol, cannot accumulate SA and cannot express SAR. Also, plants with suppressed phenylalanine lyase activity, a compound that is a precursor to SA, were more susceptible to infection.

External application of SA on plant tissues induces resistance to disease. At the same time, several suspected defense genes are induced systemically by the SA treatment, just as they are induced by various pathogens.

The finding that catalase binds to SA led to considering catalase as the compound that induces levels of resistance along with tissue necrosis and accumulation of PR-1. It was also shown that both H_2O_2 and SA are in the same signaling pathway, but that SA acts downstream of H_2O_2. More recently, nitric oxide has been shown to be an additional signal for the expression of defense. Application of NO on plants releases agents that induce the accumulation of phytoalexins, whereas inhibition of NO synthesis increases the susceptibility of plants. NO and SA both induce PR-1, but only NO induces PAL and accumulation of SA. These and other observations prove that NO acts through induction of SA.

The establishment of SAR follows production and accumulation of the systemic signal salicylic acid at the primary infection site, and in both local and systemic tissues. This leads to activation of numerous effector genes, the proteins of some of which are known as pathogenesis-related (PR) proteins. Concerted expression of these genes results in broad resistance to diverse pathogens. It should be kept in mind, however, that SAR is just one component of the defense responses and that possibly several defense pathways are essential for the full expression of pathogen-induced resistance.

One of the first steps toward SAR is overexpression of the NIM1/NPR1 gene, the protein of which is essential for transduction of the SA signal. This protein is translocated to the nucleus, where, in the presence of SA, nuclear localization of the genes results in regulated expression. How this gene regulates expression of other genes is not known, but it has been shown that its protein contains two domains that are involved in protein–protein interactions. The same gene has been found to interact with a subfamily of transcription factors that have been implicated in regulating SA-mediated gene expression.

Examples of Molecular Genetics of Selected Plant Diseases

The Powdery Mildew Disease

Powdery mildew fungi are obligate plant pathogens that attack approximately 10,000 species of plants belonging to more than 1600 genera. As obligate biotrophs, powdery mildew fungi obtain their nutrients from living cells of their host plants through specialized feeding organs, the haustoria. Powdery mildews evolved effective secretive ways of feeding and pathogenesis, effective counterdefense mechanisms that neutralize the host's defenses, or effective pathways for scrambling defense signaling. Numerous pathogen and host genes become involved in each of the steps in a successful infection, including recognition of host and pathogen, adhesion of fungal spores to host surfaces, spore germination, appressorial initiation and development, penetration peg development, peg penetration into host cell, haustorial initiation and development, neutralization of host defenses, removal of nutrients from host cell, hyphal growth, and sporulation (Fig. 4-12).

Halting of powdery mildew attacks by the host can be accomplished by single dominant loci of varying strength, such as R resistance genes; by single host genes that mutated to a recessive loss of function, such as barley Mlo and the Arabidopsis EDR1 and PMR1-PMR4 genes; or by the combined, additive effects from many genes.

The barley Mla locus is a race-specific R locus that confers resistance to at least 32 *Blumeria graminis* f. sp. *hordei* (Bgh) resistance specificities. Mla occupies a 240-kb chromosome section adjacent to eight nucleotide-binding, leucine-rich repeat (NB_LRR) type R gene homologs. Several other groups (MLA1-MLA12) of resistance specificities have been found with genes encoding coiled coil (CC)-NB-LRR type R proteins more than 90% identical and having four common introns. Some of the gene specificity domains have different but overlapping regions in the LRR domain that determine avirulenve (*Avr*) gene specificity. Three *avr* genes were found to be linked and to be located at an interval of about 8 kb.

Another barley gene, Rar1, was found to be needed for the resistance triggered by a subset of R specificities encoded by Mla, and also for powdery mildew R genes located on other chromosomes. Similarly, disruption of the Arabidopsis homolog, AtRAR1, produced mutant plants that had no resistance conferred by R genes against the downy mildew oomycete *Peronospora parasitica* or the bacterium *Pseudomonas syringae*. Furthermore, gene silencing of the RAR1 homolog of *Nicotiana benthamiana* destroyed the function of the tobacco N gene against *tobacco mosaic virus*. This points out the conserved function of RAR1 in resistance to diseases caused by pathogens of different taxonomic groups and in plants of different families. Actually, not only is RAR1 highly conserved in many plant species, homologs of it are also found in other eukaryotic organisms, including animals. Another highly conserved eukaryotic protein is SGT1. R proteins have varying requirements for RAR1/SGT1 that range from a total dependency on RAR1 or STG1 to dependency on both proteins to complete independence from both.

A resistance locus RPW8 is unusual in that it confers dominant resistance to several different powdery mildew species. Two RPW8 clustered sequence-related genes encode novel proteins about 17–20 kDa. These have several hydrophobic stretches found in transmembrane domains and are each able to provide resistance against several powdery mildews. Cellular events observed in PRW8-mediated resistance are similar to those in race-specific resistance, including an oxidative burst, a HR-like programmed cell death, and induction of PR-1 proteins. The similarities even include sharing regulatory components such as RAR1, and a requirement for the accumulation of salicylic acid. Depletion of SA reduced partially the race-specific resistance. Mutants of gene EDS1, which presumably encodes a lipase, and of NDR1, eliminate preferentially the race-specific resistance triggered by the intracellular R proteins composed of TIR-NB-LRR and CC-NB-LRR. It is proposed that RPW8 proteins act as compatibility factors that make possible the delivery of one or several powdery mildew pathogenicity factors into host cells. RPW8 leads to the exposure of conserved pathogen-associated molecular patterns that are recognized by acceptors of pattern recognition.

Barley plant mutants with homozygous alleles (mlo) of the Mlo gene are resistant to all tested isolates of Bgh and show increased susceptibility to the rice blast fungus *Magnaporthe grisea*. It should be noted that the mutants do not express defense responses constitutively, as proven by the absence of expression of PR genes in non-challenged plants. During leaf senescence, however, they cause the leaves to develop spontaneous spots of dead mesophyll cells and accelerated pigment removal. Mlo, therefore, appears to change defense responses to Bgh and to *M. grisea* in opposite directions and to negatively regulate certain events during leaf senescence.

A special feature of resistance of mlo to Bgh is that fungal pathogenesis stops at the time the penetration process through the cell wall is complete and does not lead to a hypersensitive response. This typically happens in most R gene-triggered responses. Instead, attempts by the fungus to enter plant cells of susceptible (Mlo) and resistant (mlo) plants cause the cells to remodel their cell wall beneath the fungal appressoria and to produce ring-shaped cell wall appositions. Although some feel that cell wall appositions serve as a scaffold and facilitate fungal pathogenesis, they most likely lead to structural reinforcement of the cell wall. In addition, cell wall appositions are resistant to cell wall-degrading enzymes and are sites of accumulation of hydrogen peroxide and other reactive oxygen species, as well as several phenolic compounds.

The MLO protein is an unusual transmembrane protein, about 60 kDa, and has seven transmembrane helices. It was the first of a family of proteins unique to plants with more than a dozen members each in rice and Arabidopsis. It shares common properties with animal and yeast G proteins, but MLO defense modulation to Bgh functions independently from G-protein silencing of single cell genes.

Magnaporthe grisea, *the Cause of Rice Blast*

Rice blast is one of the most severe diseases of rice. *M. grisea* has seven chromosomes and a genome size of

40 Mb, with approximately 9,000 genes. The pathogen is a haploid ascomysete that produces conidia on aerial conidiophores emerging from the center of lesions. The conidia consist of three cells. Each conidium contains an adhesive glycoprotein that, when wet, sticks tightly to the leaf surface. The conidium germinates rapidly from one of the terminal cells and attempts to penetrate the leaf surface. Within about 4 hours, the apex of the germ tube becomes swollen and flattened, the nucleus divides mitotically, and one daughter nucleus migrates into the appressorium being formed at the leaf surface. The appressorium differentiates by having thickened cell walls and a layer of melanin laid in the appressorium cell wall. These structural additions and the presence of glycerol increase the turgor pressure of the appressorium so that the penetration peg produced at the bottom of the flattened appressorium penetrates the cuticle and the cell wall and enters the cell. New disease lesions become apparent about 4 days after inoculation.

Numerous genes encoding proteins that act during contact and adhesion of the spore with the host and during appressorium formation have been identified. A hydrophobin protein (MPG1) produced in large amounts during appressorium formation helps appressoria to recognize hydrophobic surfaces. Disruption of the gene reduces appressorial formation on hydrophobic surfaces. The addition of cAMP to such mutants overcomes its handicap, indicating an efficient transmission of a surface signal so that appressoria can form via cAMP. A possible mechanism of transmission of the signal is through receptor PTH11, a possible membrane protein that has been identified. Mutants missing *Pth11* do not form appressoria and cannot infect plants. Another gene, mag B, encodes an inhibitory Gα protein that also affects appressorium development as mutants fail to produce appressorium and infection. However, the addition of AMP to the mutants restores the ability of the mutant. Mitogen-activated protein kinases also affect appressorium morphogenesis. There is a central signaling pathway that involves the protein PMK1. This influences appressorium development a great deal and mutants of it cannot produce appressoria or cause infection.

After it is formed, the appressorium develops enormous internal turgor pressure due to the glycerol it contains. During conidial germination, glycogen and lipids are degraded and, under the control of PMK1, they translocate rapidly to the germ tube tip. Lipolysis takes place rapidly during the generation of turgor pressure. In addition, spores contain trehalose, which seems to be required for turgor pressure and to be important for fungal infections. The mechanism by which turgor pressure is transformed into plant cuticle and cell wall penetration is not known yet. Gene *PLS1* encodes a tetraspanning protein, an unusual membrane-spanning protein, and seems to play a role in the regulation of penetration peg emergence.

Fusarium, *the Soilborne Plant Pathogen*

The genus *Fusarium* is a soilborne, necrotrophic, plant pathogenic fungus with many species that cause serious plant diseases around the world. *F. oxysporum* causes primarily vascular wilts on many crops, whereas numerous species, especially *F. solani*, cause root and stem rots and rots of seeds that are accompanied by the production of mycotoxins. A *Fusarium* species causing disease in immunocompromised human patients has been reported.

Fusarium oxysporum consists of more than 120 *formae specialis* according to the hosts they infect. Each of these can be subdivided into physiological races, each showing a characteristic pattern of virulence on differential host varieties. A gene-for-gene relationship appears to exist in many of the fungus race–host variety interactions. The fungus can survive in the soil as mycelium or as spores in the absence of its hosts. If a host is present, mycelium from germinating spores penetrates the host roots, enters the vascular system (xylem) in which it moves and multiplies, and causes the host to develop wilting symptoms. For the fungus to be successful in infecting the plant, it must mobilize different sets of genes for early plant–host signaling, attachment to root surface, enzymatic breakdown of physical barriers, defense against antifungal compounds of the host, and inactivation and death of host cells by fungal toxins.

Soil pH changes result in a transcription factor that activates alkaline-expressed genes and inhibits acid-expressed genes and thereby affect fungal cell growth, development, and possibly pathogenicity. Similarly, flavonoids and phytoalexins released by plant roots greatly affect the germination of fungal spores.

The early signals in plant–fungus recognition include transcription factor CTF1β. This mediates a constitutively expressed and starvation-activated cutinase gene (*cut2*) that release a few monomers of cutin from the plant. This triggers transcription of CTF1α. This mediates rapid activation of the fungal gene *cut1* and the latter secretes an extracellular cutinase that serves as a virulence factor.

Root attachment and penetration are under the control of a mitogen-activated protein kinase (MAPK). Once in contact with the root, the fungus needs to penetrate the cell walls. Several genes coding for cell wall-degrading pectinase and cellulase enzymes are activated sequentially. Pectin methyl esterases, pectin lyases, and polygalacturonases have been detected in vascular and other tissues and the respective genes have been identified. Several of the many genes coding for hemicellulases and xylanases have also been found and isolated.

Once inside the plant, the fungus comes in contact with preexisting antimicrobial substances (phytoanticipins), such as the saponins α-tomatine in tomato and potato, α-chalconine and α-solanine in potato, and avenacin in oats. Different *formae specialis* of the fungus show inducible extracellular enzyme activities that cleave these substances into nontoxic molecules. Two other phytoanticipins, benzoxazolinone, produced by Gramineae, and acetophenone, produced by carnations, are also broken down by appropriate enzymes encoded by genes of the respective *Fusarium* special forms.

The fungus is also equipped to detoxify phytoalexins, as has been shown with the pea phytoalexin pisatin. Depending on their pisatin-demethylating ability, naturally occurring *Fusarium* isolates are either incapable of degrading pisatin (Pda2) or degrade it slowly (PdaL) or fast (PdaH), and their degrading ability paralleled their ability to cause disease. A gene, PDA1, responsible for the production of pisatin was identified some time ago, and five more pea pathogenicity (PEP) genes have been discovered as a cluster on the same chromosome as PDA1. Each of these genes alone increased virulence of the fungus on pea.

The fungus also adopts itself to the presence of lower levels of toxic materials. This is done by sterol-deficient mutants, which being resistant to saponins, react with sterols of the fungal cell membrane, or by developing a mechanism, that reduces pisatin retention in their cells.

Species of *Fusarium* not only inactivate toxic substances produced by the host, they also produce toxins of their own that increase their virulence. Some of the toxins, such as enniatin and fusaric acid, are phytotoxins, i.e., they are toxic to plants, whereas others, the mycotoxins, such as trichothecins and fumonisins, are toxic to animals. Disruption of production of enniatin by *F. avenacearum* and of fumonicin B1 by *F. moniliforme* greatly reduced the ability of the respective mutants to cause disease on potato tubers and maize seedlings, respectively.

Some species of *Fusarium*, e.g., *F. solani*, reproduce both sexually and asexually, whereas others, e.g., *F. oxysporum,* reproduce only asexually. Sexual reproduction, which leads to formation of a heterokaryon, is controlled by a set of *het* loci. The products of these loci lead to either vegetative compatibility or vegetative incompatibility, which leads to cell lysis after fusion of the hyphae. The mating type (MAT) is conferred by alternative alleles at the MAT locus. The latter consists of two functionally distinct alleles, MAT-1 and MAT-2. They encode proteins that bind to DNA, functioning as transcriptional regulators of genes required for sexual reproduction. The mating response is activated via a MAP kinase signal transduction pathway. In heterothallic *Fusarium* species, the MAT locus has three genes in MAT1-1 and one at MAT1-2. In homothallic species, all four genes are present close together on the same chromosome. In the asexual *F. oxysporum* species, field isolates contained either the MAT1-1 or the MAT1-2 genes, and the genes were highly similar to those of heterothallic species.

Ustilago maydis *and Corn Smut*

The genetics of the *U. maydis*–maize pathosystem has been studied extensively, especially as it pertains to fungal mating, morphogenesis, and fungal–plant interactions. The fungus begins its life cycle as a saprophytic haploid basidiospore that may produce short haploid mycelium and more cells by budding. Haploid cells can fuse and form a stable dikaryon if they carry different alleles of both the genetic loci a and b. Cell fusion is controlled by the mating-type locus, which has two alternative forms, a1 and a2. These control the cell/cell recognition and fusion events. After cell fusion, the subsequent steps in pathogenic development are controlled by the alleles in locus b. Production of a stable filamentous dikaryon and pathogenicity requires that the fungus be heterozygous at the multiallelic b locus. Mating compatible haploids produces a dikaryon, which is the pathogenic cell type. This is filamentous and an obligate biotroph. While heterozygosis at the b locus is required for pathogenicity, once mating has occurred the locus is no longer needed for pathogenicity, but its presence seems to slightly help the rate of gall formation. The dikaryotic filamentous hypha enters the plant cuticle and cell wall directly and causes a localized infection on maize plants that leads to the formation of large galls on any of the aboveground plant parts. Hyphae grow in the gall tissue intra- and intercellularly. When galls mature, nuclei of the dikaryon fuse and form the diploid teliospores. The teliospores disperse and germinate the following spring, producing promycelia (basidia), which undergo meiosis and produce budding haploid basidiospores.

Mating and pathogenicity are controlled by the master control genes a and b. At the locus, there are two distinct allelic sequences, a1 and a2. The a locus possesses two tightly linked genes, *mfa* and *pra* that encode, respectively, secreted pheromone and pheromone receptors that span the membrane. The pheromone encoded by *mfa* and the pheromone receptor of the *pra* genes of the opposite a mating type interact with each other, signaling the production by the *prfl gene* of a transcription factor that links the pheromone response pathway with the expression of the b locus and thus to pathogenicity. The *prfl* gene protein can activate at least two kinase enzymes and is required for pathogenicity of the fungus due to its essential role in the regulation of the b mating type genes.

The b mating type locus encodes two proteins (bEast and bWest) that interact when produced by one of the 25 different alleles of each. The b locus controls events after cell fusion necessary for establishment of the infectious filamentous dikaryon and pathogenicity. Such interaction between bE1 and bW2 allele products establishes a novel regulatory protein that triggers formation of the infectious dikaryon. A switch controlled by a protein kinase dependent on cyclic AMP is important in the pathogenicity of *U. maydis*. Therefore a greater amount of PKA is required for initial plant infection and less for transition to gall formation and perhaps sporulation. Gall formation per se is not enough to trigger teliospore formation. One gene (hgl1) encodes a protein that is a transcription factor. Mutants of that gene produce large galls in maize kernels but the galls remain white because they do not form teliospores. Production of indole acetic acid (IAA) by the fungus has been suspected to be a factor in corn smut, and iad1, a gene encoding acetaldehyde hydrogenase, which converts indole-3-acetaldehyde to IAA, was isolated from *U. maydis*. Fungus mutants in this and another IAA gene produced a variety of IAA levels and a varying percentage of infective progeny.

BREEDING OF RESISTANT VARIETIES

The value of resistance in controlling plant disease was recognized in the early 1900s. Advances in the science of genetics and the obvious advantages of planting a resistant instead of a susceptible variety made the breeding of resistant varieties possible and desirable (Fig. 4-10). The more recent realization of the dangers of polluting the environment through chemical control of plant diseases gave additional impetus and importance to the breeding of resistant varieties. Thus, breeding resistant varieties, which is but one part of broader plant breeding programs, is more popular and more intensive today than it ever was in the past. Its usefulness and importance are paramount in the production of food and fiber. Nevertheless, some aspects of plant breeding, and of breeding resistant varieties in particular, have shown certain weaknesses and have allowed some plant disease epidemics to occur that could not have developed were it not for the uniformity created in crops through plant breeding.

Natural Variability in Plants

Today's cultivated crop plants are the result of selection, or selection and breeding, of plant lines that evolved naturally in one or many geographic areas of the world over millions of years. The evolution of plants from their ancient ancestors to present-day crop plants has occurred slowly and has produced countless genetically diverse forms of these plants. Many such plants still exist as wild types at the point of origin or in areas of natural spread of the plant. Although these plants may appear as useless remnants of evolution that are not likely to play a role in any future advances in agriculture, their diversity and survival in the face of the various pathogens that affect the crop in question indicate that they carry numerous genes for resistance against these pathogens.

Since the beginning of agriculture, some of the wild plants in each locality have been selected and cultivated and thus produced numerous cultivated lines or varieties. The most productive of these varieties were perpetuated in each locality from year to year, and those that survived the local climate and the pathogens continued to be cultivated. Nature and pathogens eliminated the weak and susceptible ones, while the farmers selected the best yielders among the survivors. Surviving varieties had different sets of major and minor genes for resistance. In this fashion, the selection of crop plants continued wherever they were grown, with people in each locality independently selecting varieties adapted to the local environment and resistant to local pathogens. Thus, numerous varieties of each crop plant were cultivated throughout the world and, by their own genetic diversity, contributed to make the crop locally adapted but, overall, genetically nonuniform and, thereby, safe from any sudden outbreak of a single pathogen over a large area.

Effects of Plant Breeding on Variability in Plants

During the 20th and 21st centuries, widespread, intensive, and systematic efforts have been made and continue to be made by plant breeders throughout the world to breed plants that combine the most useful genes for higher yields, better quality, uniform size of plants and fruit, uniform ripening, cold hardiness, and disease resistance. In searching for new useful genes, plant breeders cross existing, local, cultivated varieties with one another, with those of other localities, both here and abroad, and with wild species of crop plants from wherever they can be obtained. Furthermore, plant breeders often attempt to generate additional genetic variation by treating their plant material with mutagenic agents. More recently, plant breeders have been generating greater genetic variability and modifying or accelerating plant evolution in certain directions by various genetic engineering techniques. Using such techniques, plant breeders can introduce genetic material (DNA) into plant cells directly via ballistic devices, via vectors (such as *A. tumefaciens*), or via protoplast fusion. Breeders can also obtain plants with different characteristics through culture and regeneration of somatic plant cells, by diploidization of haploid plants, and so on.

The initial steps in plant breeding generally increase the variability of genetic characteristics of plants in a certain locality by combining in such plants genes that were more or less widely separated by distance before. As breeding programs advance, however, and as several of the most useful genes are identified, subsequent steps in breeding tend to eliminate variability by combining the best genes in a few cultivated varieties and leaving behind or discarding plant lines that seem to have no usefulness at the time. In a short time a few "improved" varieties replace most or all others over large expanses of land. The most successful improved varieties are also adopted abroad and, before too long, some of them become popular worldwide and replace the numerous but commercially inferior local varieties. Occasionally, even the wild types themselves may be replaced by such a variety. Thus, Red Delicious apples, Elberta peaches, certain dwarf wheat and rice varieties, certain genetic lines of corn and potatoes, one or two types of bananas, and sugar cane are grown in huge acreages throughout the world. In almost every crop, relatively few varieties make up the great bulk of the cultivated acreage of the crop throughout a country or throughout the world. The genetic base of these varieties is often narrow, especially as many of them have been derived from crosses of the same or related ancestors. These few varieties are used so widely because they are the best available, they are stable and uniform, and therefore everybody wants to grow them. At the same time, however, because they are so widely cultivated, they carry with them not only the blessings but also the dangers of uniformity. The most serious of these dangers is the vulnerability of large uniform plantings to sudden outbreaks of catastrophic plant disease epidemics.

Plant Breeding for Disease Resistance

Most plant breeding is done for the development of varieties that produce greater yields or better quality. While such varieties are being developed, they are tested for resistance against some of the most important pathogens present in the area where the variety is developed and where it is expected to be cultivated. If the variety is resistant to these pathogens, it may be released to growers for immediate production. If, however, it is susceptible to one or more of the pathogens, the variety is usually shelved or discarded (Fig. 4-17); sometimes it is released for production if the pathogen can be controlled by other means, such as with chemicals, but more often it is subjected to further breeding in an attempt to incorporate into the variety genes that would make it resistant to the pathogens without changing any of its desirable characteristics.

Sources of Genes for Resistance

The source of genes for resistance is the same gene pool of the crop that provides genes for every other inherited characteristic, namely, other native or foreign commercial varieties, older varieties abandoned earlier or discarded breeders' stock, wild plant relatives, and, occasionally, induced mutations. Often, genes of resistance are present in the varieties or species normally grown in the area where the disease is severe and in which the need for resistant varieties is most pressing. With most diseases, a few plants remain virtually unaffected by the pathogen, although most or all other plants in the area may be severely diseased. Such survivor plants are likely to have remained healthy because of resistant characteristics present in them (Fig. 4-18).

If no resistant plants can be found within the local population of the species, plants of the same species from other areas and plants of other species (cultivated or wild) are checked for resistance. If resistant plants are found, they are crossed with the cultivated varieties in an effort to incorporate the resistance genes of the other species into the cultivated varieties. With some diseases, such as late blight of potatoes, it has been necessary to look for resistance genes in species growing in the area where the disease originated. Presumably, plants existing in those areas managed to survive the long, continuous presence of the pathogen because of their resistance to it.

Techniques Used in Classical Breeding for Disease Resistance

The same methods used to breed for any heritable characteristic are also used for breeding for disease resistance and depend on the mating system of the plant (self- or cross-pollinated). Breeding for disease resistance, however, is considerably more complicated. The reason is that resistance can be assayed only by making the plants diseased, i.e., by employing another living and variable organism that must interact with the plants. In recent years, however, molecular markers associated with resistance-related enzymes, phenolics, and other compounds have been used in effectively selecting for resistance in place of inoculating the plants with the pathogen, at least in the early stages of breeding. Breeding for resistance is also complicated because resistance

FIGURE 4-17 Examples of resistant and susceptible corn plants. (A) Leaves of resistant (left) and susceptible (right) plants infected with the corn leaf blight fungus. (B) Lesions of gray leaf spot on a corn plant. (C) A resistant corn hybrid (left) and a resistant one to gray leaf spot caused by the fungus *Cercospora zeae-maydis*. (Photo A courtesy USDA, B & C courtesy R. Asiedu, International Instit. Tropical Agriculture).

may not be stable and may break down under certain conditions. For these reasons, several more or less sophisticated systems of screening for resistance have been developed. These screening systems include (1) precise conditions for inoculating the plants with the pathogen, (2) accurate monitoring and control of the environmental conditions in which the inoculated plants are kept, and (3) accurate assessment of disease incidence (percentage of plants, leaves, or fruits infected) and disease severity (proportion of the total area of plant tissue affected by disease). The following techniques are the main ones used for breeding disease resistance.

Seed, Pedigree, and Recurrent Selection

Mass selection of seed from the most highly resistant plants surviving in a field where natural infection occurs regularly is a simple method but improves plants only slowly. Moreover, in cross-pollinated plants there is no control of pollen source.

In pure line or pedigree selection, individual highly resistant plants and their progenies are propagated separately and are inoculated repeatedly to test for resistance. This method is easy and most effective with self-pollinated crops, but it is quite difficult with cross-pollinated ones.

FIGURE 4-18 While most of these staked yam plants were killed or nearly killed by the yam anthracnose fungus *Colletotrichum gloeosporioides*, several plants survive, despite the overwhelming amount of fungus inoculum around them due to genes for resistance they carry.

In recurrent selection or backcrossing, a desirable but susceptible variety of a crop is crossed with another cultivated or wild relative that carries resistance to a particular pathogen. The progeny is then tested for resistance, and the resistant individuals are backcrossed to the desirable variety. This is repeated several times until the resistance is stabilized in the genetic background of the desirable variety. This method is time-consuming and its effectiveness varies considerably with each particular case. It can be applied somewhat more easily in cross-pollinated than in self-pollinated crops.

Other Techniques

Other classical breeding techniques for disease resistance include the use of F_1 hybrids of two different but homozygous lines carrying different genes for resistance, which allows one to take advantage of the phenomenon of heterosis (hybrid vigor); use of natural or artificially induced (UV light, X rays) mutants that show increased resistance; and change of the number of chromosomes in a plant and production of euploids (4N, 6N) or aneuploids ($2N \pm 1$ or 2 chromosomes) using chemicals such as colchicine and by radiation.

Breeding for Resistance Using Tissue Culture and Genetic Engineering Techniques

Advances in plant tissue culture include meristem tip propagation, callus and single cell culture, haploid plant production, and protoplast isolation, culture, transformation, fusion, and regeneration into whole plants. These advances have opened up a whole new array of possibilities and methodologies for plant improvement, including improvement of plant resistance to infection by pathogens. The potential of these techniques is further augmented by combination with molecular technologies (genetic engineering). Genetic engineering techniques allow the detection, isolation, modification, transfer, and expression of single genes, or groups of related genes, from one organism to another. Several tissue culture techniques, e.g., regeneration of whole plants from calluses, single cells, protoplasts, and microspores or pollen, lead by themselves to plants showing greater variability in many characteristics, including resistance to disease. Selection of the best among such plants and subsequent application of classical breeding techniques make possible the production of improved plants with greater efficiency and at a much greater rate. The application of genetic engineering technologies in plant improvement depends on the kinds of plant tissue culture with which one is working, but it increases their potential tremendously by enabling plant scientists to pinpoint cell genes with specific functions and to transfer them into new cells and organisms.

Tissue Culture of Disease-Resistant Plants

Tissue culture of disease-resistant plants is particularly useful with clonally propagated plants such as strawberries, apples, bananas, sugar cane, cassava, and potatoes. Prolific plantlet production from meristem and other tissue cultures facilitates the rapid propagation of plants with exceptional (resistant) genotypes, especially in those crops not propagated easily by seed. An even greater use of tissue culture is for the production of pathogen-free stocks of clonally propagated susceptible plants.

Isolation of Disease-Resistant Mutants from Plant Cell Cultures

Plants regenerated from culture (calluses, single cells, or protoplasts) often show considerable variability (**somaclonal variation**), much of it useless or deleterious. However, plants with useful characteristics may also emerge. For example, when plants were regenerated from leaf protoplasts of a potato variety susceptible to both *Phytophthora infestans* and *Alternaria solani*, some of them (5 of 500) were resistant to *A. solani* and some (20 of 800) were resistant to *P. infestans*. Similarly, plants exhibiting increased resistance to disease caused by *Cochliobolus* and *Ustilago* were obtained from tissue cultures of sugar cane.

Production of Resistant Dihaploids from Haploid Plants

Immature pollen cells (microspores), and less often megaspores, of many plants can be induced to develop into haploid (1N) plants in which single copies (alleles) of each gene are present in all sorts of combinations. By vegetative propagation and proper screening for disease resistance, the most highly resistant haploids can be selected. These haploids can be subsequently treated with colchicine, which results in diploidization of the nuclei, i.e., doubling of the number of chromosomes and the production of dihaploid plants homozygous for all genes, including genes for resistance.

Increasing Disease Resistance by Protoplast Fusion

Protoplasts from closely related and even from unrelated plants, under proper conditions, can be made to fuse. The fusion produces **hybrid cells** containing the nuclei (chromosomes) and the cytoplasm of both protoplasts or it might result in **cybrid cells** containing the nucleus of one cell and the cytoplasm of the other cell. Generally, hybrids of unrelated cells sooner or later abort or may produce calluses, but they do not regenerate plants. In combinations of more or less related cells, however, although many or most of the chromosomes of one of the cells are eliminated during cell division, one or a few chromosomes of that cell survive and may be incorporated in the genome of the other cell. In this way, plants with more chromosomes and thereby new characteristics can be regenerated from the products of protoplast fusion. Protoplast fusion is particularly useful between protoplasts of different, highly resistant haploid lines of the same variety or species. Protoplast fusion of such lines results in diploid plants that combine the resistance genes of two highly resistant haploid lines.

Genetic Transformation of Plant Cells for Disease Resistance

Genetic material (DNA) can be introduced into plant cells or protoplasts by several methods. Such methods include direct DNA uptake, microinjection of DNA, liposome (lipid vesicle)-mediated delivery of DNA, delivery by means of centromere plasmids (minichromosomes), use of plant viral vectors, and, most importantly, bombardment of cells with tiny spheres carrying DNA and by use of the natural gene vector system of *A. tumefaciens*, the cause of crown gall disease of many plants. In all of these methods, small or large pieces of DNA are introduced into plant cells or protoplasts, and the DNA may be integrated in the plant chromosomal DNA. When the introduced DNA carries appropriate regulatory genes recognized by the plant cell or is integrated near appropriate regulatory genes along plant chromosomes, the DNA is "expressed," i.e., it is transcribed into mRNA, which is then translated into protein.

So far, only microprojectile bombardment and the *Agrobacterium* system have been used successfully to introduce into plants specific new genes that were then expressed by the plant. This was accomplished by isolating several genes of interest from plants or pathogens and splicing them into appropriate plasmids. These were subsequently used to coat the surface of tiny spheres, which were bombarded into plant cells or were introduced into a disarmed Ti plasmid of *Agrobacterium*; bacteria were then allowed to infect appropriate other plants. On infection, about one-tenth of the DNA of the plasmid, containing the new gene, is transferred to the plant cell and is incorporated into the plant genome. There, the new gene replicates during plant cell division and is expressed along with the other plant genes.

To date, several dozen R genes for disease resistance have been isolated, and several kinds of plants have been transformed genetically for disease resistance. This has been accomplished for fungal, bacterial, and viral host–pathogen combinations. In addition, viral, bacterial, fungal, or plant genes, when introduced into plants via genetic engineering techniques, provided various degrees of resistance (pathogen-derived resistance) in the plant to the pathogen from which the gene or DNA fragment was obtained and also to other pathogens. It is generally expected that breeding for disease resistance will quickly profit greatly from the application of techniques in genetic engineering.

Genetic engineering of plants for disease resistance is now used in practice with several crops. The best-documented cases involve plants engineered for resistance to viruses, such as cucurbits engineered for resistance to *cucumber mosaic*, *watermelon mosaic*, and *zucchini yellow mosaic viruses*; papaya engineered for resistance to *papaya ringspot virus*; potato engineered for resistance to *potato leaf roll* and *potato Y viruses*; and wheat engineered for resistance to *wheat streak mosaic virus*. More examples and details of genetic engineering of plants for disease resistance can be found in Chapter 6.

Advantages and Problems in Breeding for Vertical or Horizontal Resistance

Resistance may be obtained by incorporating one, a few, or many resistance genes into a variety. Some of these genes may control important steps in disease development and may therefore play a major role in disease resistance. Other genes may control peripheral events of lesser importance in disease development and, therefore, play a relatively minor role in disease resistance. Obviously, one or a few major role genes could be sufficient to make a plant resistant to a pathogen (R-gene, monogenic, oligogenic, or vertical resistance). However, it would take many minor effect genes to make a plant resistant (polygenic or horizontal resistance). More importantly, whereas a plant with vertical resistance may be completely resistant to a pathogen, a plant with horizontal resistance is never completely resistant or completely susceptible. Furthermore, vertical resistance is easy to manipulate in a breeding program, including the application of genetic engineering techniques, and therefore is often preferred to horizontal resistance. However, both vertical and horizontal resistances have their advantages and limitations.

Vertical resistance is aimed against specific pathogens or pathogen races. Vertical resistance is most effective when (1) it is incorporated in annual crops that are easy to breed, such as small grains; (2) it is directed against pathogens that do not reproduce and spread rapidly, such as *Fusarium*, or pathogens that do not mutate very frequently, such as *Puccinia graminis*; (3) it consists of "strong" genes (R-genes) that confer complete and long-term protection to the plant that carries it; and (4) the host population does not consist of a single genetically uniform variety grown over large acreages. If one or more of these, and several other, conditions are not met, vertical resistance becomes short lived, i.e., it breaks down as a result of the appearance of new pathogen races that can bypass or overcome it.

Horizontal resistance confers incomplete (partial) but more durable protection: it does not break down as quickly and suddenly as most vertical resistance. Horizontal resistance involves many host physiological processes that act as mechanisms of defense and that are beyond the limits of the capacity of the pathogen to change, i.e., beyond the probable limits of its variability. Horizontal resistance is present universally in wild and domesticated plants and operates against all races of a pathogen, including the most pathogenic ones. Varieties with horizontal (polygenic, general, or nonspecific) partial resistance remain resistant much longer than varieties with vertical (oligogenic or specific) resistance, but the level of resistance in plants with horizontal resistance is much lower than in plants with vertical resistance.

Because varieties with vertical resistance are often attacked suddenly and rapidly by a new virulent race and lead to severe epidemics, various strategies have been developed to avoid these disadvantages. In some crops this has been accomplished through the use of multilines or by pyramiding. **Multilines** are mixtures of individual varieties (lines or cultivars) that are agronomically similar but differ in their resistance genes. **Pyramiding** consists of using varieties that are derived from crossing several to many varieties that contain different resistance genes and then selecting from them those that contain the mixtures of genes. Multilines and pyramiding have been developed mostly in small grains against the rust fungi, but their use is likely to increase in these and other crops as the control of plant diseases with specific resistance and with chemicals becomes more risky or less acceptable.

Incorporating genes for resistance from wild or unsatisfactory plants into susceptible but agronomically desirable varieties is a difficult and painstaking process involving repeated crossings, testings, and backcrossings to the desirable varieties. The feasibility of the method in most cases, however, has been proved repeatedly. Through breeding, varieties of some crops have been developed in which genes for resistance against several different diseases have been incorporated.

Vulnerability of Genetically Uniform Crops to Plant Disease Epidemics

Varieties with even complete vertical resistance do not remain resistant forever. The continuous production of mutants and hybrids in pathogens sooner or later leads to the appearance of races that can infect previously resistant varieties. Sometimes, races may exist in an area in small populations and avoid detection until after the introduction of a new variety or virulent races of the pathogen existing elsewhere may be brought in after introduction of the resistant variety. In all cases, widespread cultivation of a single, previously resistant variety provides an excellent substrate for the rapid development and spread of the new race of the pathogen, and it usually leads to an epidemic. Thus, genetic uniformity in crops, although very desirable when it concerns horticultural characteristics, is undesirable and often catastrophic when it occurs in the genes of resistance to diseases.

The cultivation of varieties with genetically uniform disease resistance is possible and quite safe if other means of plant disease control, such as chemical, are possible. Thus, a few fruit tree varieties, such as Red Delicious apples, Bartlett pears, Elberta peaches, and navel oranges, are cultivated throughout the world in the face of numerous virulent fungal and bacterial pathogens that would destroy them in a short time were it not for the fact that the trees are protected from the pathogens by numerous chemical sprays annually. Even such varieties, however, suffer tremendous losses when affected by pathogens that cannot be controlled with chemicals, as in the case of fire blight of pears, pear decline, and tristeza disease of citrus.

Another case in which varieties with genetically uniform disease resistance are not likely to suffer from severe disease epidemics is when the resistance is aimed against slow-moving soil pathogens such as *Fusarium* and *Verticillium*. Aside from the fact that some pathogens normally produce fewer races than others, even if new races are produced at the same rate, soilborne pathogens lack the dispersal potential of airborne ones. As a result, a new race of a soilborne pathogen would be limited to a relatively small area for a long time, and although it could cause a locally severe disease, it would not spread rapidly and widely to cause an epidemic. The slow spread of such virulent new races of soilborne pathogens allows time for the control of the disease by other means or the replacement of the variety with another one resistant to the new race.

Genetic uniformity in plant varieties becomes a serious disadvantage in the production of major crops because of the potential danger of sudden and widespread disease epidemics caused by airborne or insectborne pathogens in the vast acreages in which each of these varieties is often grown. Several examples of epidemics that resulted from genetic uniformity are known and some of them have already been mentioned. Southern corn leaf blight was the result of the widespread use of corn hybrids containing the Texas male-sterile cytoplasm; the destruction of the 'Ceres' spring wheat by race 56 of *Puccinia graminis*; and of 'Hope' and its relative bread wheats by race 15B of *P. graminis*, were all the result of replacement of numerous genetically diverse varieties by a few uniform ones. *The Cochliobolus* (*Helminthosporium*) blight of Victoria oats was the result of replacing many varieties with the rust-resistant Victoria oats; coffee rust destroyed all coffee trees in Ceylon because all of them originated from uniform susceptible stock of *Coffea arabica*; and tristeza continues to destroy millions of orange trees in South, Central, and North America because they were propagated on hypersensitive resistant sour orange rootstocks.

Despite these and many other well-known examples of plant disease epidemics that occurred because of the concentrated cultivation of genetically uniform crops over large areas, crop production continues to depend on genetic uniformity. A few varieties of each crop used to, and for some crops still, make up the bulk of the cultivated crop over as vast an area as the United States. Although a relatively large number of varieties are available for each crop, only a few varieties, often two or three, are grown in more than half the acreage of each crop, and in some they make up more than three-fourths of the crop. For example, two pea varieties make up almost the entire pea crop of the country (96%), i.e., about 400,000 acres, and two varieties account for 42% of the sugar beet crop, i.e., about 600,000 acres. The figures become even more spectacular when one considers the most popular varieties of the truly large acreage crops. Thus, although six corn varieties (hybrids) account for 71%, or 47 million acres, one of them alone accounts for 26%, or 17 million acres. Similarly, six varieties of soybean account for about 24 million acres of that crop, and most of these varieties share common ancestors.

It is apparent that several hundreds of thousands or several million acres planted to one variety present a huge opportunity for the development of an epidemic. The variety, of course, is planted so widely because it is resistant to existing pathogens. However, this resistance puts extreme survival pressure on the pathogens over that area. It takes one "right" change in one of the zillions of pathogen individuals in the area to produce a new virulent race that can attack the variety. When that happens it is a matter of time — and, usually, of favorable weather — before the race breaks loose, the epidemic develops, and the yield of the variety is destroyed or reduced below acceptable economic levels. In some cases the appearance of the new race is detected early and the variety is replaced with another one, resistant to the new race, before a widespread epidemic occurs; this, of course, requires that varieties of a crop with a different genetic base are available at all times. For this reason, most varieties must usually be replaced within about 3 to 5 years from the time of their widespread distribution.

In addition to the genetic uniformity within one variety, plant breeding often introduces genetic uniformity to several or all cultivated varieties of a crop by introducing one or several genes in all of these varieties or by replacing the cytoplasm of the varieties with a single type of cytoplasm. Induced uniformity through introduced genes includes, for example, the seedless condition in grapes and watermelons, the dwarfism gene in the dwarf wheat and rice varieties, the monogerm gene in sugar beet varieties, the determinate gene in tomato varieties, and the stringless gene in bean varieties. Uniformity through replacement of the cytoplasm occurred, of course, in most corn hybrids in the later 1960s when the Texas male-sterile cytoplasm replaced the normal cytoplasm. Cytoplasmic uniformity is also employed commercially in several varieties of sorghum, sugar beet, and onions; it is studied in wheat and is also present in cotton and cantaloupe. Neither the introduced genes nor the replacement cytoplasm, of course, makes the plant less resistant to diseases, but if a pathogen appears that is favored by or can take advantage of the characters controlled by that gene or other genes linked to it or by genes in that cytoplasm, then the stage is set for a major epidemic. That this can happen was proved by the southern corn leaf blight epidemic of 1970 and by the susceptibility of dwarf wheats to new races of *Septoria* and *Puccinia*, of tomatoes with the determinate gene to *Altenaria*, and others.

In more recent years, efforts have been made to plant a smaller percentage of the total acreage of a crop with a few selected varieties, but for most crops and most areas that acreage is still too great. For example, in the mid-1990s the top six soybean varieties and the top nine wheat varieties made up only 41 and 34% of the total soybean and wheat acreage, respectively. However, the three most popular cotton varieties made up 54% of the total cotton acreage. Furthermore, the four most popular potato varieties in each of the 11 leading potato-producing states accounted for 63 to 100% of the total potato crop in any one of these states, and the most popular barley varieties in the top six barley-producing states accounted for 44 to 94% of the total crop in each state.

Selected References

Agrawal, A. A., Tuzun, S., and Bent, E., eds. (1999). "Induced Plant Defenses against Pathogens and Herbivores." APS Press, St. Paul, MN.

Agrios, G. N. (1980). Escape from disease. *In* "Plant Disease" (J. G. Horsfall and E. B. Cowling, eds.), Vol. 5, pp. 17–37. Academic Press, New York.

Bauer, D. W., *et al.* (1995). *Erwinia chrysanthemi* Harpin$_{Ech}$: An elicitor of the hypersensitive response that contributes to soft rot pathogenesis. *Mol. Plant-Microbe Interact.* **8**, 484–491.

Bent, A. F., *et al.* (1994). RPS2 of *Arabidopsis thaliana*: A leucine-rich repeat class of plant disease resistance genes. *Science* **265**, 1856–1860.

Birch, P. R., and Whisson, S. C. (2001). *Phytophthora infestans* enters the genomics era. *Mol. Plant Pathol.* **2**, 257–263.

Bisgrove, S. R. (1994). A disease resistance gene in *Arabidopsis* with specificity for two different pathogen avirulence genes. *Plant Cell* **6**, 927–933.

Boucher, C. A., Gough, C. L., and Arlat, M. (1992). Molecular genetics of pathogenicity determinants of *Pseudomonas solanacearum* with special emphasis on hrp genes. *Annu. Rev. Phytopathol.* **30**, 443–461.

Bushnell, W. R., and Rowell, J. B. (1981). Suppressors of defense reactions: A model for roles in specificity. *Phytopathology* **71**, 1012–1014.

Cao, H., Baldini, R. L., and Rahme, L. G. (2001). Common mechanisms for pathogens of plants and animals. *Annu. Rev. Phytopathol.* **39**, 259–284.

Carlyle, M. J., Watkinson, S. C., and Gooday, G. W. (2001). "The Fungi." Academic Press, San Diego, CA.

Charles, T. C., Jin, S., and Nester, E. W. (1992). Two-component sensory transduction systems in phytobacteria. *Annu. Rev. Phytopathol.* **30**, 463–484.

Crute, I. R. (1992). From breeding to cloning (and back again?): A case study with lettuce downy mildew. *Annu. Rev. Phytopathol.* **30**, 485–506.

Culver, J. N. (2002). Tobacco mosaic virus assembly and disassembly: Determinants in pathogenicity and virulence. *Annu. Rev. Phytopathol.* **40**, 287–308.

Davis, *et al.* (2000). Nematode parasitism genes. *Annu. Rev. Phytopathol.* **38**, 365–396.

deWit, P. J. G. M. (1992). Molecular characterization of gene-for-gene systems in plant-fungus interactions and the application of avirulence genes in control of plant pathogens. *Annu. Rev. Phytopathol.* **30**, 391–418.

Dinesh-Kumar, S. P., *et al.* (1995). Transposon tagging of tobacco mosaic virus resistance gene N: Its possible role in the TMV-N mediated signal transduction pathway. *Proc. Natl. Acad. Sci. USA* **92**, 4175–4180.

Dixon, R. A., Harrison, M. J., and Lamb, C. J. (1994). Early events in the activation of plant defense responses. *Annu. Rev. Phytopathol.* **32**, 479–501.

Dubin, H. J., and Rajoram, S. (1982). The CIMMYT's international approach to breeding disease-resistant wheat. *Plant Dis.* **66**, 967–971.

Eastgate, J. A. (2000). *Erwinia amylovora:* The molecular basis of fireblight disease. *Mol. Plant Pathol.* **1**, 325–329.

Flor, H. H. (1971). Current status of the gene-for-gene concept. *Annu. Rev. Phytopathol.* **9**, 275–296.

Fraser, R. S. S. (1990). The genetics of resistance to plant viruses. *Annu. Rev. Phytopathol.* **28**, 179–200.

Fry, W. E., *et al.* (1992). Population genetics and intercontinental migrations of *Phytophthora infestans. Annu. Rev. Phytopathol.* **30**, 107–130.

Gabriel, D. W., and Rolfe, B. G. (1990). Working models of specific recognition in plant-microbe interactions. *Annu. Rev. Phytopathol.* **28**, 365–391.

Garcia-Arenal, F., Fraile, A., and Malpica, J. M. (2001). Variability and genetic structure of plant virus populations. *Annu. Rev. Phytopathol.* **39**, 157–186

Gebhardt, C., and Valkonen, J. P. T. (2001). Organization of genes controlling disease resistance in the potato genome. *Annu. Rev. Phytopathol.* **39**, 79–102.

Gheysen, G., and Fenoll, C. (2002). Gene expression in nematode feeder sites. *Annu. Rev. Phytopathol.* **40**, 191–219.

Gilchrist, D. G. (1998). Programmed cell death in plant disease: The purpose and promise of cellular suicide. *Annu. Rev. Phytopathol.* **36**, 393–414.

Gold, S. E. (2003). *Ustilago* pathogenicity. *In* "Plant-Microbe Interactions (G. Stacey and N. T. Keen, eds.), Vol. 6, pp. 147–172. APS Press, St. Paul, MN.

Grandel, A., Romeis, T., and Kamper, Jorg. (2000). Regulation of pathogenic development in the corn smut fungus *Ustilago maydis. Mol. Plant Pathol.* **1**, 61–66.

Hahn, M. G. (1996). Microbial elicitors and their receptors in plants. *Annu. Rev. Phytopathol.* **34**, 387–411.

Hammerschmidt, R. (1999). Phytoalexins: What have we learned after 60 years? *Annu. Rev. Phytopathol.* **37**, 285–306.

He, S. Y. (1998). Type III protein secretion systems in plant and animal pathogenic bacteria. *Annu. Rev. Phytopathol.* **36**, 363–392.

Heath, M. C. (1991). The role of gene-for-gene interactions in the determination of host species specificity. *Phytopathology* **81**, 127–130.

Hennecke, H., and Verma, D. P. S., eds. (1991). "Advances in Molecular Genetics of Plant-Microbe Interactions," Vol. 1. Kluwer, Dordrecht, The Netherlands.

Hooker, A. L. (1974). Cytoplasmic susceptibility in plant disease. *Annu. Rev. Phytopathol.* **12**, 167–179.

Hull, R. (2002). "Matthews' Plant Virology," 4th Ed. Academic Press, San Diego.

Idnurm, A., and Howlett, B. J. (2001). Pathogenicity genes of phytopathogenic fungi. *Mol. Plant Pathol.* **2**, 241–255.

Innes, R. W. (2001). Targeting the targets of type III effector proteins secreted by phytopathogenic bacteria. *Mol. Plant Pathol.* **2**, 109–115.

Johnson, R., and Jellis, G. J., eds. (1993). "Breeding for Disease Resistance." Kluwer, Hingham, MA.

Keen, N. T. (1982). Specific recognition in gene-for-gene host parasite systems. *In* "Advances in Plant Pathology" (D. S. Ingram and P. H. Williams, eds.), Vol. 1 pp. 35–82. Academic Press, London.

Kessmann, H., *et al.* (1994). Activation of systemic acquired disease resistance in plants. *Eur. J. Plant Pathol.* **100**, 359–369.

Kistler, H. C., and Miao, V. P. W. (1992). New modes of genetic change in filamentous fungi. *Annu. Rev. Phytopathol.* **30**, 131–152.

Kiyosawa, S. (1982). Genetics and epidemiological modeling of breakdown of plant disease resistance. *Annu. Rev. Phytopathol.* **20**, 93–117.

Lamb, C. J. (1994). Plant disease resistance genes in signal perception and transduction. *Cell* **76**, 419–422.

Leach, J. E., and White, F. F. (1996). Bacterial avirulence genes. *Annu. Rev. Phytopathol.* **34**, 153–179.

Leach, J. E., *et al.* (2001). Pathogen fitness penalty as a predictor of durability of disease resistance genes. *Annu. Rev. Phytopathol.* **39**, 187–224.

Leong, S. A., Allen, C., and Triplett, E. W., eds. "Biology of Plant-Microbe Interactions," Vol. 3. Intrn. Soc. Molec. Plant-Microbe Interact., St. Paul, MN.

Lindgren, P. B. (1997). The role of *hrp* genes during plant-bacterial interactions. *Annu. Rev. Phytopathol.* **35**, 129–152.

Lindgren, P. E., Peet, R. C., and Panopoulos, N. J. (1986). Gene cluster of *Pseudomonas syringae* pv. *phaseolicola* controls pathogenicity on bean plants and hypersensitivity on nonhost plants. *J. Bacteriol.* **168**, 512–522.

Marathe, R., *et al.* (2002). The tobacco mosaic virus resistance gene N. *Mol. Plant Pathol.* **3**, 167–172.

Martin, G. B., *et al.* (1993). Map-based cloning of a protein kinase gene conferring disease resistance to tomato. *Science* **262**, 1432–1436.

McDermott, J. M., and McDonald, B. A. (1993). Gene flow in plant pathosystems. *Annu. Rev. Phytopathol.* **31**, 353–373.

McDonald, B. A., and Linde, C. (2002). Pathogen population genetics, evolutionary potential, and durable resistance. *Annu. Rev. Phytopathol.* **40**, 349–379.

Mindrinos, M., *et al.* (1994). The *Arabidopsis thaliana* disease resistance gene RPS2 encodes a protein containing a nucleotide-binding site and leucine-rich repeats. *Cell* **78**, 1089–1099.

Minsavage, G. V., *et al.* (1990). Gene-for-gene relationships specifying disease resistance in *Xanthomonas campestris* pv. *Vesicatoria*–pepper interactions. *Mol. Plant-Microbe Interact.* **3**, 41–47.

Montesano, M., Brader, G., and Palva, E. T. (2003). Pathogen-derived elicitors: Searching for receptors in plants. *Mol. Plant Pathol.* **4**, 73–83.

National Academy of Sciences (1972). "Genetic Vulnerability of Major Corps." Natl. Acad. Sci., Washington, DC.

Nene, Y. L. (1988). Multiple disease resistance in grain legumes. *Annu. Rev. Phytopathol.* **26**, 203–217.

Nester, E. W. (2000). DNA and protein transfer from bacteria to eukaryotes — the *Agrobacterium* story. *Mol. Plant Pathol.* **1**, 87–90.

Nester, E. W., and Verma, D. P. S., eds. (1993). "Advances in Molecular Genetics of Plant-Microbe Interactions," Vol. 2. Kluwer Academic, Dordrecht, The Netherlands.

Panstruga, R., and Schulze-Lefert, P. (2002). Live and let live: Insights into powdery mildew disease and resistance. *Mol. Plant Pathol.* **3**, 495–507.

Pedley, K. F., and Martin, G. B. (2003). Molecular basis of Pto-mediated resistance to bacterial speck disease in tomato. *Annu. Rev. Phytopathol.* **41**, 215–244.

Perombelon, M. C. M. (2002). Potato diseases caused by soft rot erwinias: An overview of pathogenesis. *Plant Pathol.* **51**, 1–12.

Rivas, S., and Tomas, C. M. (2002). Recent advances in the study of tomato Cf resistance genes. *Mol. Plant Pathol.* **3**, 277–282.

Roncero, M. I. G., *et al.* (2003). *Fusarium* as a model for studying virulence in soilborne plant pathogens. *Physiol. Mol. Plant Pathol.* **62**, 87–98.

Ryals, J., *et al.* (1995). Signal transduction in systemic acquired resistance. *Proc. Natl. Acad. Sci. USA* **92**, 4202–4205.

Schäfer, W. (1994). Molecular mechanisms of fungal pathogenicity in plants. *Annu. Rev. Phytopathol.* **32**, 461–477.

Singh, U. S., Singh, R. P., and Kohmoto, K., eds. (1995). "Pathogenesis and Host Specificity in Plant Diseases: Histopathological, Biochemical, Genetic and Molecular Bases," Vols. 1–3. Pergamon, Elsevier Science, Tarrytown, NY.

Stacey, G., and Keen, N. T., eds. (2000). "Plant–Microbe Interactions," Vol. 5. APS Press, St. Paul, MN.

Stacey, G., and Keen, N. T., eds. (2003). "Plant–Microbe Interactions," Vol. 6. APS Press, St. Paul, MN.

Staskawicz, B. J., *et al.* (1995). Molecular genetics of plant disease resistance. *Science* **268**, 661–667.

Staskawicz, B. J., *et al.* (2001). Common and contrasting themes of plant and animal diseases. *Science* **292**, 2285–2289.

Sticher, L., Mauch-Mani, B., and Metraux, J. P. (1997). Systemic acquired resistance. *Annu. Rev. Phytopathol.* **35**, 235–270.

Suzuki, M., *et al.* (1995). Point mutations in the coat protein of cucumber mosaic virus affect symptom expression and virion accumulation in tobacco. *J. Gen. Virol.* **76**, 1791–1799.

Thomas, P. L. (1991). Genetics of small-grain smuts. *Annu. Rev. Phytopathol.* **29**, 137–148.

Tijan, R. (1995). Molecular machines that control genes. *Sci. Am.* **February**, 54–61.

Toth, I. K., *et al.* (2003). Soft rot erwiniae: From genes to genomes. *Mol. Plant Pathol.* **4**, 17–30.

Tyler, B. M. (2002). Molecular basis of recognition between *Phytophthora* pathogens and their hosts. *Annu. Rev. Phytopathol.* **40**, 137–167.

Tzvira, T., and Citovsky, V. (2000). From host recognition to DNA integration: The function of bacterial and plant genes in the *Agrobacterium*-plant cell interaction. *Mol. Plant Pathol.* **1**, 201–212.

Valent, B., and Chumley, F. G. (1991). Molecular genetic analysis of the rice blast fungus, *Magnaporthe grisea*. *Annu. Rev. Phytopathol.* **29**, 443–467.

Vanderplank, J. E. (1982). "Host–Pathogen Interaction in Plant Disease." Academic Press, New York.

Vanderplank, J. E. (1984). "Disease Resistance in Plants," 2nd Ed. Academic Press, Orlando, FL.

Van Gisegem, F., Genin, S., and Boucher, C. (1995). *hrp* and *avr* genes, key determinants controlling the interactions between plants and gram-negative phytopathogenic bacteria. *In* "Pathogenesis and Host Specificity in Plant Diseases" (U. S. Singh, R. P. Singh, and K. Kohmoto, eds.), Vol. 1, pp. 273–292. Elsevier Science, Tarrytown, NY.

Van Sluys, M. A., *et al.* (2002). Comparative genomic analysis of plant-associated bacteria. *Annu. Rev. Phytopathol.* **40**, 169–189.

Walsh, J. A., and Jenner, C. E. (2002). Turnip mosaic virus and the quest for durable resistance. *Mol. Plant Pathol.* **3**, 289–300.

Walton, J. D., and Panacione, D. G. (1993). Host specific toxins and disease specificity: Perspectives and progress. *Annu. Rev. Phytopathol.* **31**, 275–303.

Whitham, S., *et al.* (1994). The product of the tobacco mosaic virus resistance gene N: Similarity to Toll and the interleukin-1 receptor. *Cell* **78**, 1105–1115.

Willis, D. K., Rich, J. J., and Hrabak, E. M. (1991). *hrp* genes of phytopathogenic bacteria. *Mol. Plant-Microbe Interact.* **4**, 132–138.

Wolfe, M. S., and McDermott, J. M. (1994). Population genetics of plant-patogen interactions: The example of the *Erysiphe graminis-Hordeum vulgare* pathosystem. *Annu. Rev. Phytopathol.* **32**, 89–113.

Xu, J.-R., and Xue, C. (2002). Time for a blast: genomics of *Magnaporthe grisea. Mol. Plant Pathol.* **3**, 173–176.

Zeigler, R. S. (1998). Recombination in *Magnaporthe grisea. Annu. Rev. Phytopathol.* **36**, 249–275.

chapter five

HOW PATHOGENS ATTACK PLANTS

MECHANICAL FORCES EXERTED BY PATHOGENS ON HOST TISSUES
177

CHEMICAL WEAPONS OF PATHOGENS
179

ENZYMES IN PLANT DISEASE
180

ENZYMATIC DEGRADATION OF CELL WALL SUBSTANCES
180

CUTICULAR WAX – CUTIN – PECTIC SUBSTANCES – CELLULOSE – CROSS-LINKING GLYCANS (HEMICELLULOSES) – SUBERIN – LIGNIN – CELL WALL FLAVONOIDS CELL WALL STRUCTURAL PROTEINS
180

ENZYMATIC DEGRADATION OF SUBSTANCES CONTAINED IN PLANT CELLS
189

PROTEINS – STARCH – LIPIDS
189

MICROBIAL TOXINS IN PLANT DISEASE
190

TOXINS THAT AFFECT A WIDE RANGE OF HOST PLANTS
190

TABTOXIN – PHASEOLOTOXIN – TENTOXIN – CERCOSPORIN – OTHER NON-HOST-SPECIFIC TOXINS
191

HOST-SPECIFIC OR HOST-SELECTIVE TOXINS
193

VICTORIN, OR HV TOXIN – T TOXIN [*COCHLIOBOLUS (HELMINTHOSPORIUM) HETEROSTROPHUS* RACE T TOXIN] – HC TOXIN – *ALTERNARIA ALTERNATA* TOXINS – OTHER HOST-SPECIFIC TOXINS
194

GROWTH REGULATORS IN PLANT DISEASE
196

POLYSACCHARIDES
201

DETOXIFICATION OF LOW MOLECULAR WEIGHT ANTIMICROBIAL MOLECULES
201

PROMOTION OF BACTERIAL VIRULENCE BY *AVR* GENES
202

ROLE OF TYPE III SECRETION IN BACTERIAL PATHOGENESIS
202

SUPPRESSORS OF PLANT DEFENSE RESPONSES
202

PATHOGENICITY AND VIRULENCE FACTORS IN VIRUSES AND VIROIDS
203

The intact, healthy plant is a community of cells built in a fortress-like fashion. Plant cells consist of cell wall, cell membranes, and cytoplasm, which contains the nucleus and various organelles (Fig. 5-1) and all the substances for which the pathogens attack them. The cytoplasm and the organelles it contains are separated from each other by membranes that carry various types of proteins embedded in them (Fig. 5-2). The plant surfaces that come in contact with the environment either consist of cellulose, as in the epidermal cells of roots and in the intercellular spaces of leaf parenchyma cells, or consist of a cuticle that covers the epidermal cell walls, as is the case in the aerial parts of plants. Often an additional layer, consisting of waxes, is deposited outside the cuticle, especially on younger parts of plants (Fig. 5-3).

Pathogens attack plants because during their evolutionary development they have acquired the ability to live off the substances manufactured by the host plants, and some of the pathogens depend on these substances for survival. Many substances are contained in the protoplast of the plant cells, however, and if pathogens are to gain access to them they must first penetrate the outer barriers formed by the cuticle and/or cell walls. Even after the outer cell wall has been penetrated, further invasion of the plant by the pathogen necessitates the penetration of more cell walls. Furthermore, the plant cell contents are not always found in forms immediately utilizable by the pathogen and must be broken down to units that the pathogen can absorb and assimilate. Moreover, the plant, reacting to the presence and activities of the pathogen, produces structures and chemical substances that interfere with the advance or the existence of the pathogen; if the pathogen is to survive and to continue living off the plant, it must be able to overcome such obstacles.

Therefore, for a pathogen to infect a plant it must be able to make its way into and through the plant, obtain nutrients from the plant, and neutralize the defense reactions of the plant. Pathogens accomplish these activities mostly through secretions of chemical substances that affect certain components or metabolic mechanisms of

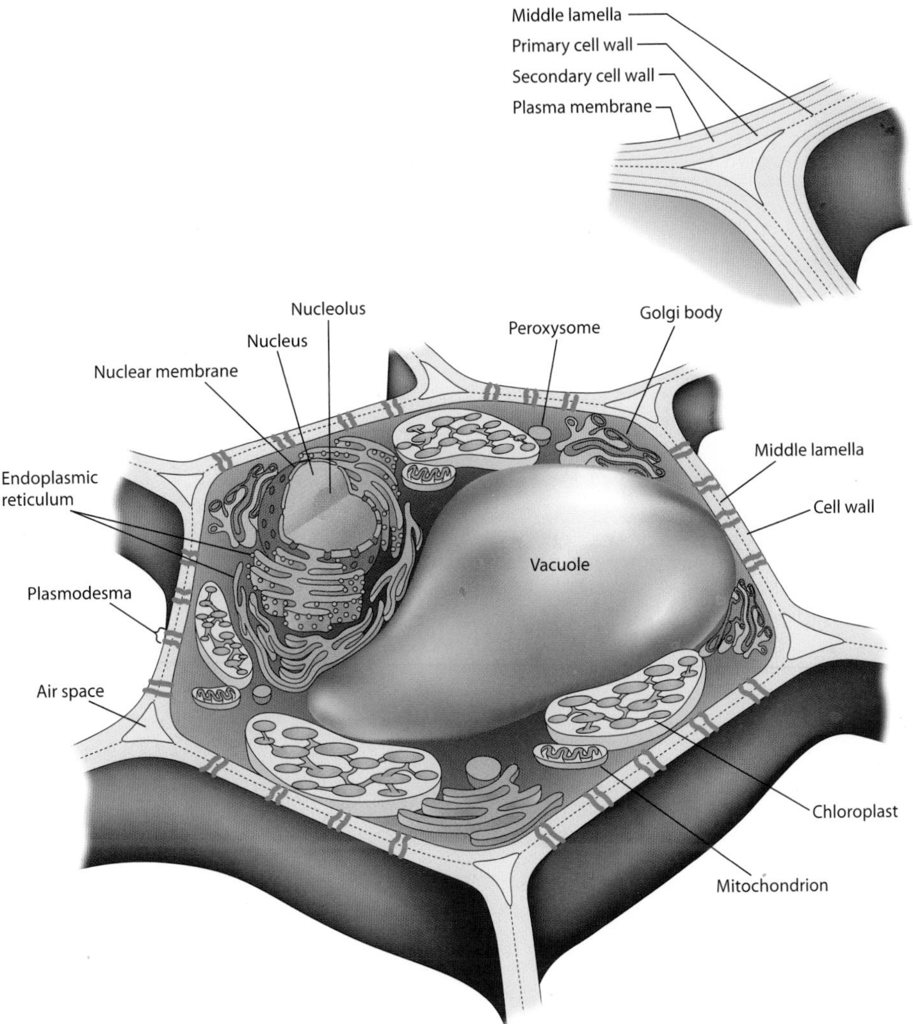

FIGURE 5-1 Schematic representation of a plant cell and its main components.

their hosts. Penetration and invasion, however, seem to be aided by, or in some cases be entirely the result of, the mechanical force exerted by certain pathogens on the cell walls of the plant.

MECHANICAL FORCES EXERTED BY PATHOGENS ON HOST TISSUES

Plant pathogens are, generally, tiny microorganisms that cannot apply a "voluntary" force to a plant surface. Only some fungi, parasitic higher plants, and nematodes appear to apply mechanical pressure to the plant surface they are about to penetrate. The amount of pressure, however, may vary greatly with the degree of "presoftening" of a plant surface by enzymatic secretions of the pathogen.

For fungi and parasitic higher plants to penetrate a plant surface, they must, generally, first adhere to it. Hyphae and radicles are usually surrounded by mucilaginous substances, and their adhesion to the plant seems to be brought about primarily by the intermolecular forces developing between the surfaces of plant and pathogen on close contact with the adhesive substances and with one another. In some cases an adhesion pad forms from the spore when it comes in contact with a moist surface, and cutinase and cellulase enzymes released from the spore surface help the spore adhere to the plant surface. Spores of some fungi carry adhesive substances at their tips that, on hydration, allow spores to become attached to various surfaces.

After contact is established, the diameter of the tip of the hypha or radicle in contact with the host increases and forms the flattened, bulb-like structure called the appressorium (Figs. 2-4 and 2-5). This increases the area of adherence between the two organisms and securely fastens the pathogen to the plant. From the appressorium, a fine growing point, called the penetration peg,

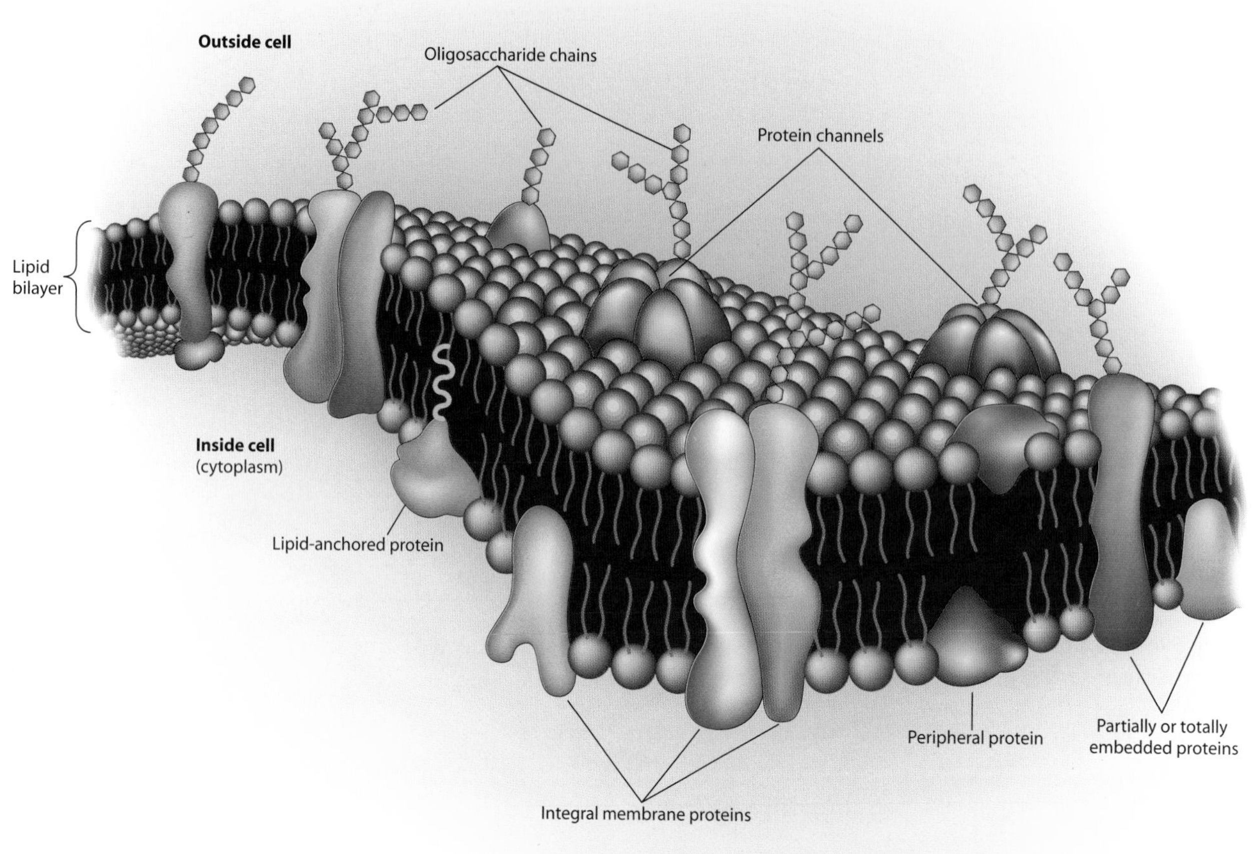

FIGURE 5-2 Schematic representation of a portion of a cell membrane and of the arrangement of protein molecules in relation to the membrane.

arises and advances into and through the cuticle and cell wall. In some fungi, such as *Alternaria*, *Cochliobolus*, *Colletotrichum*, *Gaeumannomyces*, *Magnaporthe*, and *Verticillium*, penetration of the plant takes place only if melanin (dark pigment) accumulates in the appressorial cell wall. It appears that melanin produces a rigid structural layer and, by trapping solutes inside the appressorium, causes water to be absorbed. This increases the turgor pressure in the appressorium and, thereby, the physical penetration of the plant by the penetration peg. If the underlying host wall is soft, penetration occurs easily. When the underlying wall is hard, however, the force of the growing point may be greater than the adhesion force of the two surfaces and may cause separation of the appressorial and host walls, thus averting infection. Penetration of plant barriers by fungi and parasitic higher plants is almost always assisted by the presence of enzymes secreted by the pathogen at the penetration site, resulting in the softening or dissolution of the barrier. It was found, for example, that while appressoria of some powdery mildew fungi developed a maximum turgor pressure of 2–4 MPa, approximately sufficient to bring about host cell penetration, two cellulases were also present: one primarily at the tip of the appressorial germ tube and the other at the tip of the primary germ tube.

While the penetration tube is passing through the cuticle, it usually attains its smallest diameter and

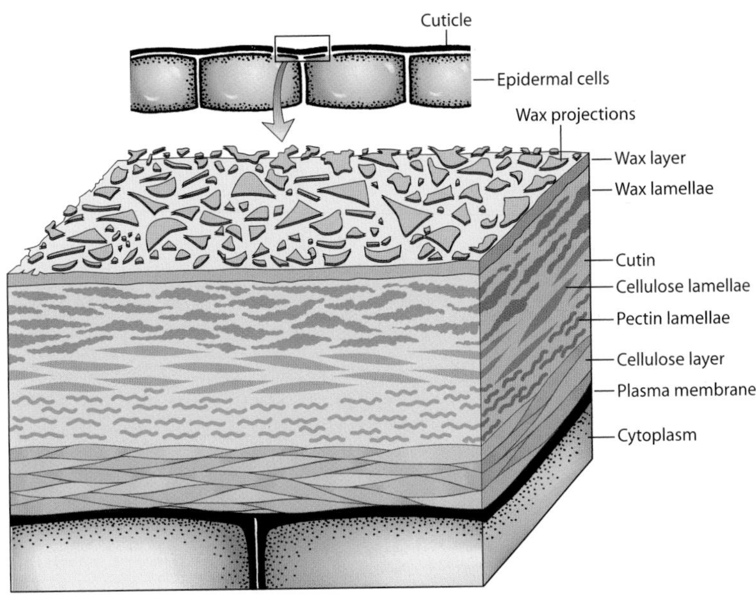

FIGURE 5-3 Schematic representation of the structure and composition of the cuticle and cell wall of foliar epidermal cells. [Adapted from Goodman *et al.* (1967).]

appears thread-like. After penetration of the cuticle, the hyphal tube diameter often increases considerably. The penetration tube attains the diameter normal for the hyphae of the particular fungus only after it has passed through the cell wall (see Figs. 2-5 and 2-9 in Chapter 2).

Nematodes penetrate plant surfaces by means of the stylet, which is thrust back and forth and exerts mechanical pressure on the cell wall (Fig. 2-10). The nematode first adheres to the plant surface by suction, which it develops by bringing its fused lips in contact with the plant. After adhesion is accomplished, the nematode brings its body, or at least the forward portion of its body, to a position vertical to the cell wall. With its head stationary and fixed to the cell wall, the nematode then thrusts its stylet forward while the rear part of its body sways or rotates slowly round and round. After several consecutive thrusts of the stylet, the cell wall is pierced, and the stylet or the entire nematode enters the cell.

Once a fungus or nematode has entered a cell, it generally secretes increased amounts of enzymes that presumably soften or dissolve the opposite cell wall and make its penetration easier. Mechanical force, however, probably is brought to bear in most such penetrations, although to a lesser extent.

Considerable mechanical force is also exerted on host tissues from the inside out by some pathogenic fungi on formation of their fructifications in the tissues beneath the plant surface. Through increased pressure, the sporophore hyphae, as well as fruiting bodies, such as pycnidia and perithecia, push outward and cause the cell walls and the cuticle to expand, become raised in the form of blister-like proturberances, and finally break.

CHEMICAL WEAPONS OF PATHOGENS

Although some pathogens may use mechanical force to penetrate plant tissues, the activities of pathogens in plants are largely chemical in nature. Therefore, the effects caused by pathogens on plants are almost entirely the result of biochemical reactions taking place between substances secreted by the pathogen and those present in, or produced by, the plant.

The main groups of substances secreted by pathogens in plants that seem to be involved in production of disease, either directly or indirectly, are enzymes, toxins, growth regulators, and polysaccharides (plugging substances). These substances vary greatly as to their importance in pathogenicity, and their relative importance may be different from one disease to another. Thus, in some diseases, such as soft rots, enzymes seem to be by far the most important, whereas in diseases such as crown gall, growth regulators are apparently the main substances involved. However, in the *Bipolaris* blight of Victoria oats, the disease is primarily the result of a toxin secreted in the plant by the pathogen. Enzymes, toxins, and growth regulators, probably in that order, are considerably more common and probably more important in plant disease development than polysaccharides. It has also been shown that some pathogens produce compounds that act as suppressors of the defense responses of the host plant.

Among the plant pathogens, all except viruses and viroids can probably produce enzymes, growth regulators, and polysaccharides. How many of them produce toxins is unknown, but the number of known toxin-producing plant pathogenic fungi and bacteria increases each year. Plant viruses and viroids are not known to produce any substances themselves, but they induce the host cell to produce either excessive amounts of certain substances already found in healthy host cells or substances completely new to the host. Some of these substances are enzymes, and others may belong to one of the other groups mentioned earlier.

Pathogens produce these substances either in the normal course of their activities (constitutively) or when they grow on certain substrates such as their host plants (inducible). Undoubtedly, natural selection has favored the survival of pathogens that are assisted in their parasitism through the production of such substances. The presence or the amount of any such substance produced, however, is not always a measure of the ability of the pathogen to cause disease. It must also be kept in mind that many substances, identical to those produced by pathogens, are also produced by the healthy host plant.

In general, plant pathogenic enzymes disintegrate the structural components of host cells, break down inert food substances in the cell, or affect components of its membranes and the protoplast directly, thereby interfering with its functioning systems. Toxins seem to act directly on protoplast components and interfere with the permeability of its membranes and with its function. Growth regulators exert a hormonal effect on the cells and either increase or decrease their ability to divide and enlarge. Polysaccharides seem to play a role only in the vascular diseases, in which they interfere passively with the translocation of water in the plants.

Enzymes in Plant Disease

Enzymes are generally large protein molecules that catalyze organic reactions in living cells and in solutions. Because most kinds of chemical reaction that occur in a cell are enzymatic, there are almost as many kinds of enzymes as there are chemical reactions. Each enzyme, being a protein, is coded for by a specific gene. Some enzymes are present in cells at all times (constitutive). Many are produced only when they are needed by the cell in response to internal or external gene activators (induced). Each type of enzyme often exists in several forms known as isozymes that carry out the same function but may vary from one another in several properties, requirements, and mechanism of action.

Enzymatic Degradation of Cell Wall Substances

Usually, the first contact of pathogens with their host plants occurs at a plant surface. Aerial plant part surfaces consist primarily of cuticle and/or cellulose, whereas root cell wall surfaces consist only of cellulose. Cuticle consists primarily of cutin, more or less impregnated with wax and frequently covered with a layer of wax. The lower part of cutin is intermingled with pectin and cellulose lamellae and lower yet there is a layer consisting predominantly of pectic substances; below that there is a layer of cellulose. Polysaccharides of various types are often found in cell walls. Proteins of many different types, both structural, e.g., elastin, which helps loosen the cell wall, and extensin, which helps add rigidity to the cell wall, some enzymes, and some signal molecules that help receive or transmit signals inward or outward, are normal constituents of cell walls. Finally, epidermal cell walls may also contain suberin and lignin. The penetration of pathogens into parenchymatous tissues is facilitated by the breakdown of the internal cell walls, which consist of cellulose, pectins, hemicelluloses, and structural proteins, and of the middle lamella, which consists primarily of pectins. In addition, complete plant tissue disintegration involves the breakdown of lignin. The degradation of each of these substances is brought about by the action of one or more sets of enzymes secreted by the pathogen.

Cuticular Wax

Plant waxes are found as granular, blade, or rod-like projections or as continuous layers outside or within the cuticle of many aerial plant parts (Fig. 5-4). The presence and condition of waxes at the leaf surface affect the degree of colonization of leaves and the effect varies with the plant species. Electron microscope studies suggest that several pathogens, e.g., *Puccinia hordei*, produce enzymes that can degrade waxes. Another fungus, *Pestalotia malicola*, which attacks fruit of Chinese quince, grows on, within, and beneath the fruit cuticle. Fungi and parasitic higher plants, however, apparently can penetrate wax layers by means of mechanical force alone.

Cutin

Cutin is the main component of the cuticle. The upper part of the cuticle is admixed with waxes, whereas its lower part, in the region where it merges into the outer walls of epidermal cells, is admixed with pectin and cellulose (see Fig. 5-3). Cutin is an insoluble polyester of C_{16} and C_{18} hydroxy fatty acids.

Many fungi and a few bacteria have been shown to produce cutinases and/or nonspecific esterases, i.e.,

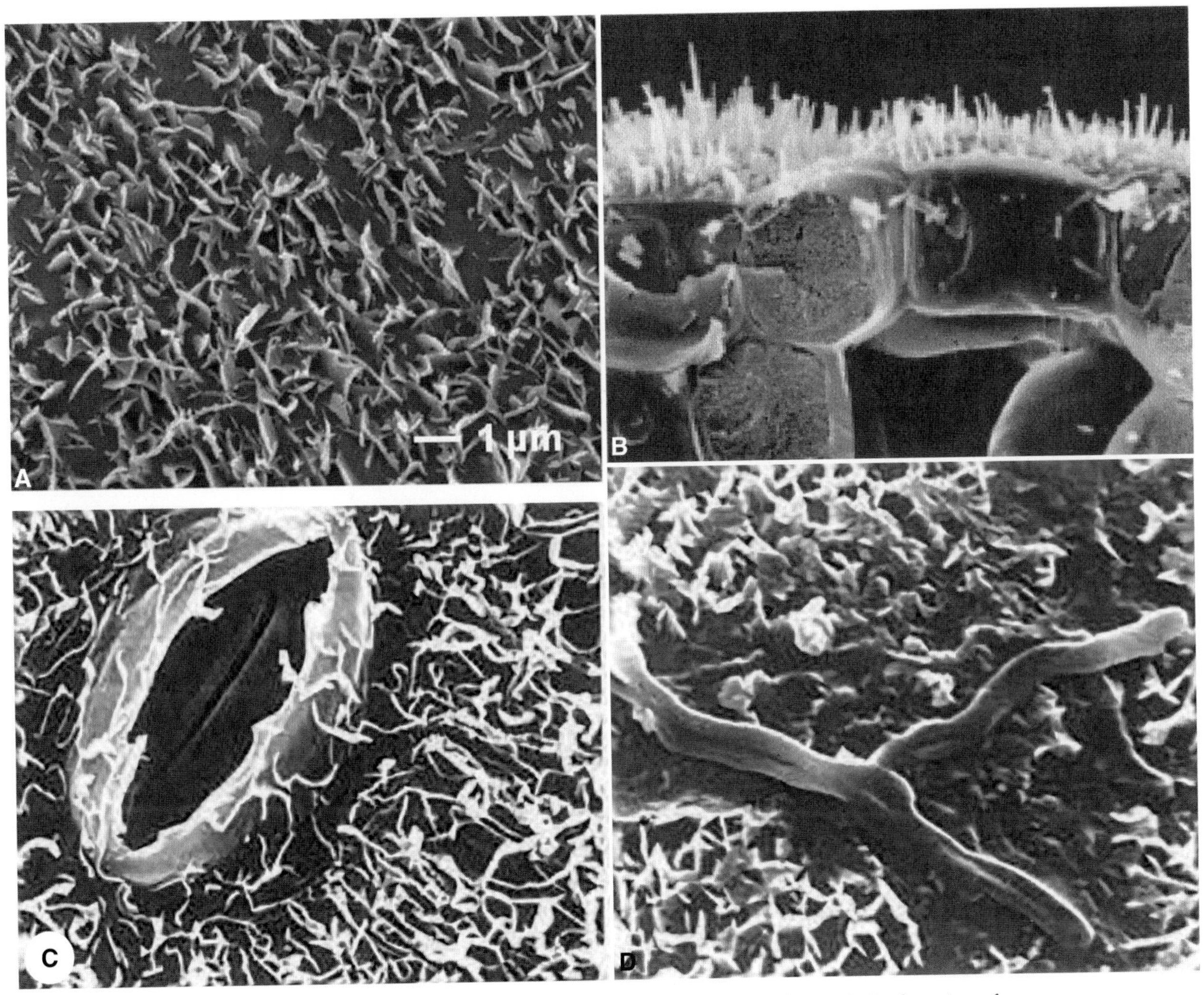

FIGURE 5-4 Morphology of cuticular wax projections on different leaf surfaces. (A) Surface view of wax on corn leaf. (B) Wax projections as seen in cross section of leaf. (C) Wax projections surrounding a stoma. (D) Wax degraded along the passage of fungal mycelium. [Photographs courtesy of (A) L. M. Marcell and G. A. Beattie, Iowa State University, (B) H. V. Davis, United Kingdom, (C and D) P. V. Sangbusen, Hamburg.]

enzymes that can degrade cutin. Cutinases break cutin molecules and release monomers (single molecules) as well as oligomers (small groups of molecules) of the component fatty acid derivatives from the insoluble cutin polymer.

Fungi that penetrate the cuticle directly seem to constantly produce low levels of cutinase, which on contact with cutin releases small amounts of monomers. These subsequently enter the pathogen cell, trigger further expression of the cutinase genes, and stimulate the fungus to produce almost a thousand times more cutinase than before (Fig. 5-5). Cutinase production by the pathogen, however, may also be stimulated by some of the fatty acids present in the wax normally associated with cutin in the plant cuticle. However, the presence of glucose suppresses expression of the cutinase gene and reduces cutinase production drastically.

The involvement of cutinase in the penetration of the host cuticle by plant pathogenic fungi is shown by several facts. For example, the enzyme reaches its highest concentration at the penetrating point of the germ tube and at the infection peg of appressorium-forming fungi. Inhibition of cutinase by specific chemical inhibitors, or by antibodies of the enzyme applied to the plant surface, protects the plant from infection by fungal pathogens. Also, cutinase-deficient mutants show reduced virulence but become fully virulent when cutinase is added on the plant surface. In the brown rot of stone fruits, caused by the fungus *Monilinia fructicola*, fungal cutinase activity seems to be inhibited greatly by

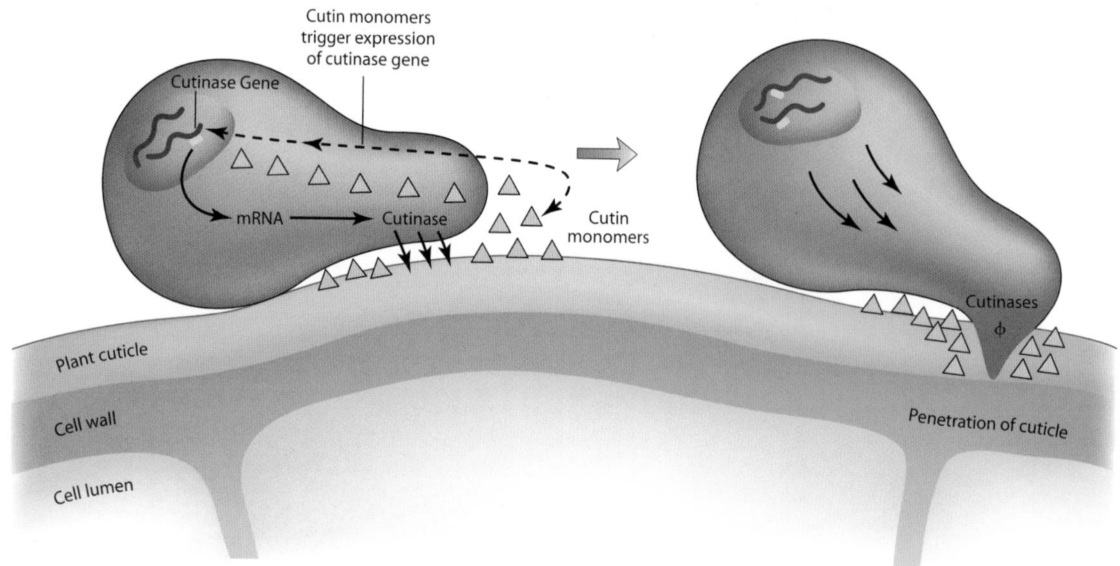

FIGURE 5-5 Diagrammatic representation of cuticle penetration by a germinating fungus spore. Constitutive cutinase releases a few cutin monomers from the plant cuticle. These trigger expression of the cutinase genes of the fungus, leading to the production of more cutinase(s), which macerates the cuticle and allows penetration by the fungus.

phenolic compounds such as chlorogenic and caffeic acids, which are abundant in epidermal cells of young fruit and the fruit is resistant to infection. As the fruit matures, the concentration of these compounds declines sharply, cutinase activity increases, and the fruit is penetrated by the fungus. Moreover, fungi that infect only through wounds and do not produce cutinase acquire the ability to infect directly if a cutinase gene from another fungus is introduced into them and enables them to produce cutinase. Pathogens that produce higher levels of cutinase seem to be more virulent than others. At least one study has shown that the germinating spores of a virulent isolate of the fungus *Fusarium* produced much more cutinase than those of an avirulent isolate of the same fungus and that the avirulent isolate could be turned into a virulent one if purified cutinase was added to its spores. The fungus *Botrytis cinerea*, the cause of numerous types of diseases on many plants, produces a cutinase and a lipase, both of which break down cutin. In the presence of antilipase antibodies, fungal spores failed to penetrate the cuticle and lesion formation was inhibited, indicating that lipase activity is required in at least the early stages of host infection.

Pectic Substances

Pectic substances constitute the main components of the middle lamella, i.e., the intercellular cement that holds in place the cells of plant tissues (Fig. 5-6). Pectic substances also make up a large portion of the primary cell wall in which they form an amorphous gel filling the spaces between the cellulose microfibrils (Fig. 5-7).

Pectic substances are polysaccharides consisting mostly of chains of galacturonan molecules interspersed with a much smaller number of rhamnose molecules and small side chains of galacturonan, xylan, and some other five carbon sugars. Several enzymes degrade pectic substances and are known as **pectinases** or **pectolytic enzymes** (Fig. 5-8). Some of them, e.g., the **pectin methyl esterases**, remove small branches off the pectin chains. Pectin methyl esterases have no effect on the overall chain length, but they alter the solubility of the pectins and affect the rate at which they can be attacked by the chain-splitting pectinases. The latter cleave the pectic chain and release shorter chain portions containing one or a few molecules of galacturonan. Some chain-splitting pectinases, called **polygalacturonases**, split the pectic chain by adding a molecule of water and breaking (hydrolyzing) the linkage between two galacturonan molecules; others, known as **pectin lyases**, split the chain by removing a molecule of water from the linkage, thereby breaking it and releasing products with an unsaturated double bond (Fig. 5-8). Polygalacturonases and pectin lyases occur in types that either can break the pectin chain at random sites (**endopectinases**) and release shorter chains, or can break only the terminal linkage (**exopectinases**) of the chain and release single units of galacturonan. The rhamnose and other sugars that may be forming part or branches of the pectin chain

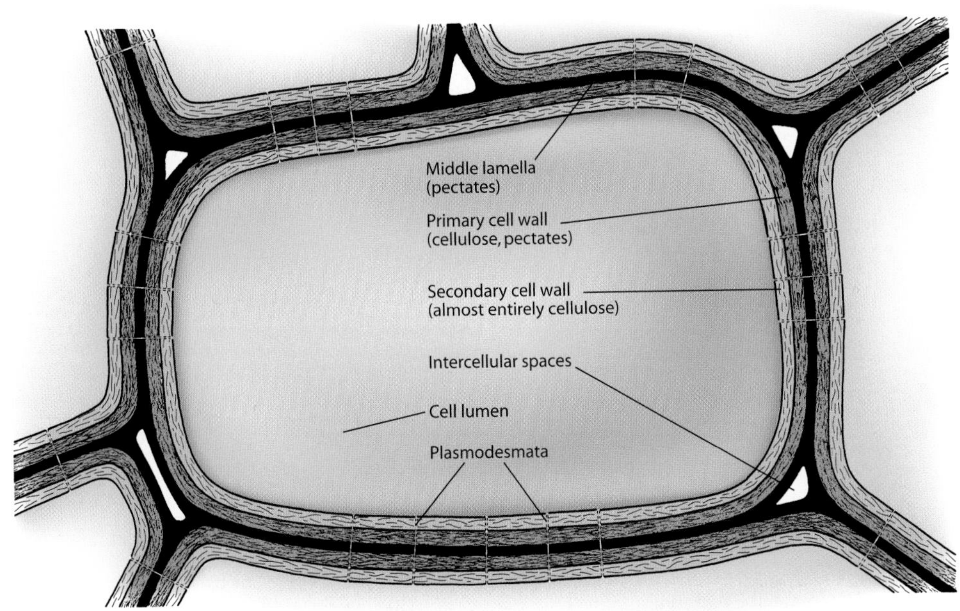

FIGURE 5-6 Schematic representation of the structure and composition of plant cell walls.

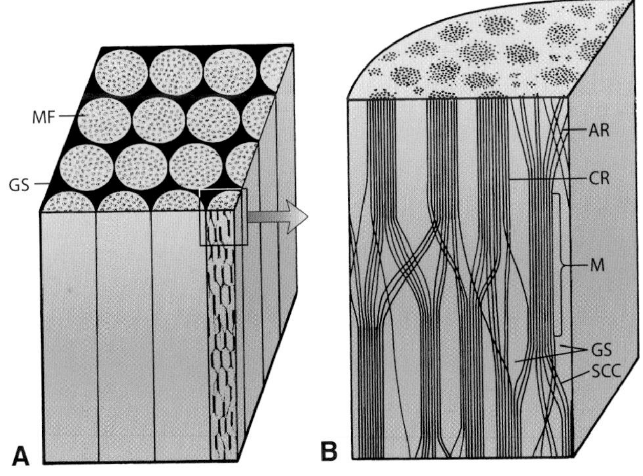

FIGURE 5-7 Schematic diagram of the gross struture of cellulose and microfibrils (A) and of the arrangement of cellulose molecules within a microfibril (B). MF, microfibril; GS, ground substance (pectin, hemicelluloses, or lignin); AR, amorphous region of cellulose; CR, crystalline region; M, micelle; SCC, single cellulose chain (molecule). [Adapted from Brown *et al.* (1949).]

are hydrolyzed by other enzymes that recognize these molecules.

As with cutinases, and with other enzymes involved in the degradation of cell wall substances, the production of extracellular pectolytic enzymes by pathogens is regulated by the availability of the pectin polymer and the released galacturonan units. The pathogen seems to produce at all times small, constitutive, base-level amounts of pectolytic enzymes that, in the presence of pectin, release from it a small number of galacturonan monomers, dimers, or oligomers. These molecules, when absorbed by the pathogen, serve as inducers for the enhanced synthesis and release of pectolytic enzymes (substrate induction), which further increase the amount of galacturonan monomers, etc. The latter are assimilated readily by the pathogen, but at higher concentrations they act to repress the synthesis of the same enzymes (catabolite repression), thus reducing production of the enzymes and the subsequent release of galacturonan monomers. The production of pectolytic enzymes is also repressed when the pathogen is grown in the presence of glucose. However, in some resistant host–pathogen combinations, pectolytic enzymes seem to elicit the plant defense response through the release from the cell wall of pectic fragments that function as endogenous elicitors of the defense mechanisms of the host.

Pectin-degrading enzymes have been shown to be involved in the production of many fungal and bacterial diseases, particularly those characterized by the soft rotting of tissues. Various pathogens produce different sets of pectinases and their isozymes. In some diseases, e.g., the bacterial wilt of solanaceous crops caused by *Ralstonia solanacearum*, pectinolytic enzymes collectively are absolutely essential for disease to develop, although some of them individually seem to not be required for disease but rather for accelerated colonization and enhanced aggressiveness by bacteria. In black rot of cantaloupe caused by the fungus *Didymella bryoniae*, there is a highly positive correlation between the

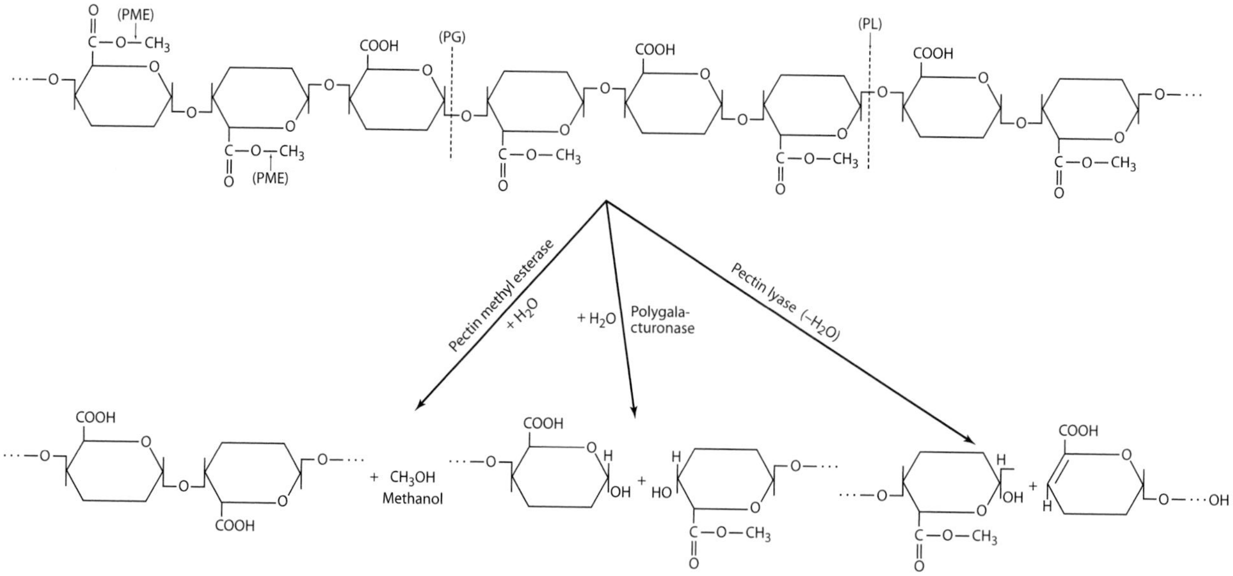

FIGURE 5-8 Degradation of a pectin chain by the three types of pectinases into modified and smaller molecules.

size of the rotting tissue lesion and the total fungal polygalacturonase activity in the rotting tissue.

In some *Colletotrichum*-caused anthracnoses, the fungus produces one pectin lyase that is a key virulence factor in disease development. The amount and activity of the enzyme and the amount of disease increase as the pH at the infection site increase to 7.5–8.0. The fungus maintains the high pH at the infection area by secreting ammonia. Inoculation of nonhost species in the presence of ammonia-releasing compounds enhances pathogenicity to levels similar to those caused by the compatible fungal and host species. Ammonia secretion by the fungus is a virulence factor for the fungus. Pectin–degrading enzymes are produced and play a role in the ability of nematodes, such as the root knot nematode, *Meloidogyne javanica*, for the penetration of root tissues, movement between plant cells along the middle lamella, and possibly in the formation of tee multinucleate giant cells on which the nematode feeds throughout the rest of its life. Some of these enzymes seem to affect the virulence of the pathogen on different hosts, i.e., they affect the degree of host specialization of the pathogen. Pectic enzymes are produced by germinating spores and, apparently, acting together with other pathogen enzymes (cutinases and cellulases), assist in the penetration of the host.

Pectin degradation results in liquefaction of the pectic substances that hold plant cells together and in the weakening of cell walls. This leads to tissue maceration, i.e., softening and loss of coherence of plant tissues and separation of individual cells, which eventually die (Fig. 5-9). The weakening of cell walls and tissue maceration undoubtedly facilitate the inter- or intracellular invasion of the tissues by the pathogen. Pectic enzymes also provide nutrients for the pathogen in infected tissues. Pectic enzymes, by the debris they create, seem to be involved in the induction of vascular plugs and occlusions in the vascular wilt diseases (Fig. 5-11). Although cells are usually killed quickly in tissues macerated by pectic enzymes, how these enzymes kill cells is not yet clear. It is thought that cell death results from the weakening by the pectolytic enzymes of the primary cell wall, which then cannot support the osmotically fragile protoplast, and the protoplast bursts.

Cellulose

Cellulose is also a polysaccharide, but it consists of chains of glucose (1–4) β-D-glucan molecules. The glucose chains are held to one another by a large number of hydrogen bonds. Cellulose occurs in all higher plants as the skeletal substance of cell walls in the form of microfibrils (see Figs. 5-7, 5-10, and 5-12). Microfibrils, which can be perceived as bundles of iron bars in a reinforced concrete building, are the basic structural units (matrix) of the wall, even though they account for less than 20% of the wall volume in most meristematic cells. The cellulose content of tissues varies from about 12% in the nonwoody tissues of grasses to about 50% in mature wood tissues to more than 90% in cotton fibers. The spaces between microfibrils and between micelles or cellulose chains within the microfibrils may be filled with pectins and hemicelluloses and probably some lignin at maturation. Although the bulk of cell wall polysaccharides is broken down by numerous enzymes produced by fungi and bacteria, a portion of them

FIGURE 5-9 Involvement of pectolytic enzymes in disease development. Peach tissues infected with the brown rot fungus *Monilinia fructicola* while still on the tree (A) and by *Rhizopus* sp. at harvest (B and C) are macerated by the pectinases of the fungus and subsequently turn brown due to the oxidation of phenolic compounds released during maceration. Subsequent loss of water results in shrinking of the fruit. (D) Potato tuber, part of which has been macerated by the enzymes of the fungus *Fusarium* and subsequently has lost some of the water. An onion bulb (E) and a potato tuber (F) macerated by the enzymes of the fungus *Botrytis* and the bacterium *Erwinia*, respectively. [Photographs courtesy of (A) D. Ritchie, North Carolina State University, (D) P. Hamm, Oregon State University, (E) K. Mohan, University of Idaho, and (F) R. Rowe, Ohio State University.]

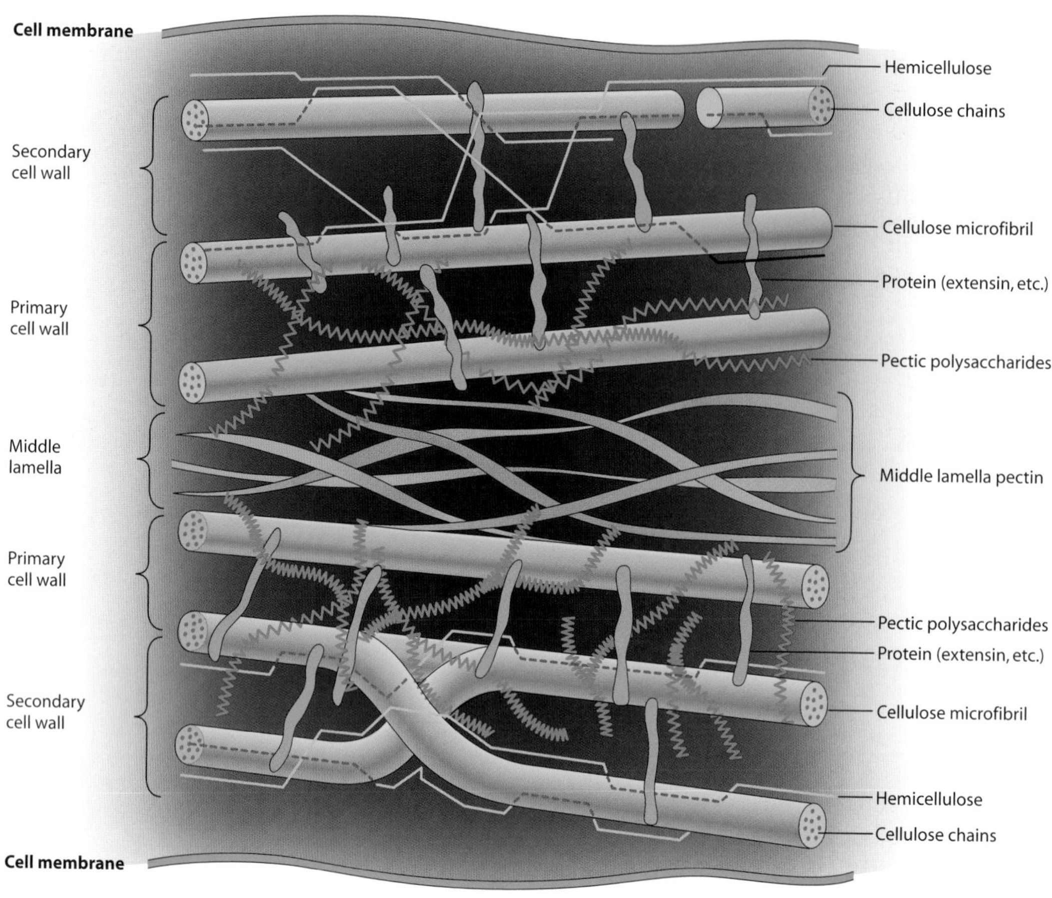

FIGURE 5-10 Schematic diagram of morphology and arrangement of some cell wall components.

appears to be broken down by nonenzymatic oxidative systems, such as activated oxygen and hydroxyl radicals (OH) produced during plant–fungus interactions. **Callose** differs from cellulose in that it consists of (1–3) β-D-glucan chains that can form duplexes and triplexes. Callose is normally made by a few cell types but is made by most cells following wounding and during attempted penetration by invading fungal hyphae.

The enzymatic breakdown of cellulose results in the final production of glucose molecules. The glucose is produced by a series of enzymatic reactions carried out by several cellulases and other enzymes. One cellulase (C1) attacks native cellulose by cleaving cross-linkages between chains. A second cellulase (C2) also attacks native cellulose and breaks it into shorter chains. These are then attacked by a third group of cellulases (Cx), which degrade them to the disaccharide cellobiose. Finally, cellobiose is degraded by the enzyme β-glucosidase into glucose.

Cellulose-degrading enzymes (cellulases) have been shown to be produced by several phytopathogenic fungi, bacteria, and nematodes and are undoubtedly produced by parasitic higher plants. Saprophytic fungi, mainly certain groups of basidiomycetes, and, to a lesser degree, saprophytic bacteria cause the breakdown of most of the cellulose decomposed in nature. In living plant tissues, however, cellulolytic enzymes secreted by pathogens play a role in the softening and disintegration of cell wall material (Figs. 5-11 and 5-12). They facilitate the penetration and spread of the pathogen in the host and cause the collapse and disintegration of the cellular structure, thereby aiding the pathogen in the production of disease. Cellulolytic enzymes may further participate indirectly in disease development by releasing, from cellulose chains, soluble sugars that serve as food for the pathogen and, in the vascular diseases, by liberating into the transpiration stream large molecules from cellulose, which interfere with the normal movement of water. In the bacterial wilt of tomato, production of an endocellulase by the bacterium was required for the latter to be pathogenic and induce the disease.

Cross-Linking Glycans (Hemicelluloses)

Cross-linking glycans, known earlier as hemicelluloses, are complex mixtures of polysaccharide

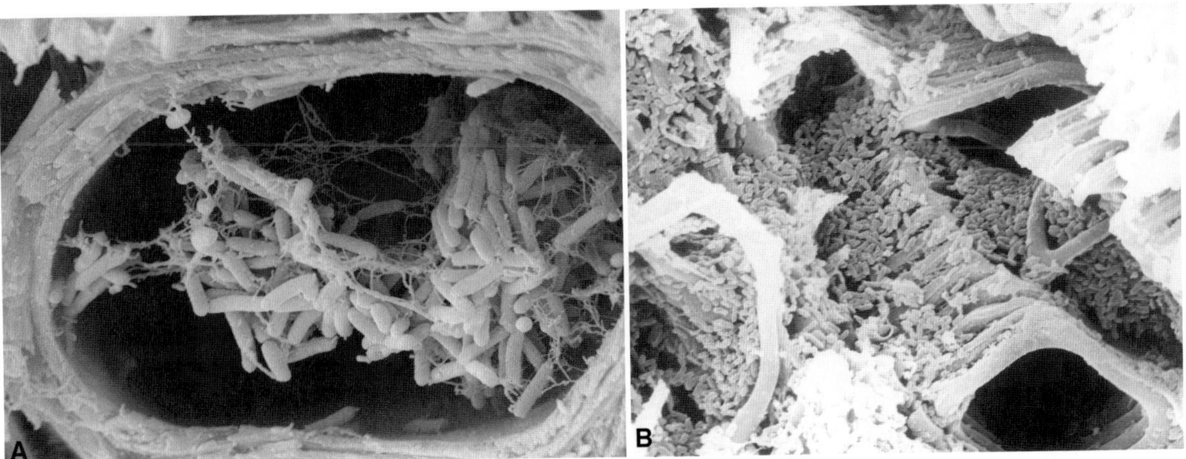

FIGURE 5-11 (A) Xylella bacteria in xylem vessel of citrus leaf. (B) Close-up of cell breakdown and maceration of pectic substances and celluloses of parenchyma cells and xylem vessels caused by enzymes secreted by bacteria of the genus *Pseudomonas*. Only the lignin-impregnated rings of xylem vessels remain intact. 1500×. [Photographs courtesy of (A) E. Alves, Federal University of Lavras, Brazil, and (B) E. L. Mansvelt, I. M. M. Roos, and M. J. Hattingh.]

polymers that can hydrogen-bond to and may cover and link cellulose microfibrils together (Figs. 5-10 and 5-12). Their composition and frequency seem to vary among plant tissues, plant species, and with the developmental stage of the plant. Cross-linking glycans are a major constituent of the primary cell wall and may also make up a varying proportion of the middle lamella and secondary cell wall. Hemicellulosic polymers include primarily xyloglucans and glucuronoarabinoxylans, but also glucomannans, galactomannans, arabinogalactans, and others. Xyloglucan, for example, is made of glucose chains with terminal branches of smaller xylose chains and lesser amounts of galactose, arabinose, and fucose. Cross-linking glycans link the ends of pectic polysaccharides and various points of the cellulose microfibrils.

The enzymatic breakdown of hemicelluloses appears to require the activity of many enzymes. Several hemicellulases seem to be produced by many plant pathogenic fungi. Depending on the monomer released from the polymer on which they act, the particular enzymes are called xylanase, galactanase, glucanase, arabinase, mannase, and so on. The nonenzymatic breakdown of hemicelluloses by activated oxygen, hydroxyl, and other radicals produced by attacking fungi also occurs. Despite the fact that fungal pathogens produce these enzymes and oxidative agents, it is still not clear how they contribute to cell wall breakdown or to the ability of the pathogen to cause disease.

Suberin

Suberin is found in certain tissues of various underground organs, such as roots, tubers, and stolons, and in periderm layers, such as cork and bark tissues. Suberins are also formed in response to wounding and to pathogen-induced defenses of certain organs and cell types. Typical suberization occurs, for example, in cut potato tubers where browning and encrustation develop in the form of multilamellar areas consisting of alternating polyaliphatic and polyaromatic layers. These layers are impermeable and help strengthen the cell wall and limit water loss through the wound. The aliphatic layer is composed of long chain (20 carbons or more) lipid substances, plus some specialized fatty acids, and is located between the primary cell wall and the plasmalemma. The polyaromatic layer consists of building blocks containing substances derived from hydroxycinnamic acid and is located in the cell wall. The polyaromatic layer also contains several phenolic compounds, such as chlorogenic acid, that act as local disinfectants. Although plants obviously produce enzymes that synthesize suberin, it is not known whether or how pathogens break it down during infection.

Lignin

Lignin is found in the middle lamella, as well as in the secondary cell wall of xylem vessels and the fibers that strengthen plants. It is also found in epidermal and occasionally hypodermal cell walls of some plants. The lignin content of mature woody plants varies from 15 to 38% and is second only to cellulose in abundance.

Lignin is an amorphous, three-dimensional polymer that is different from both carbohydrates and proteins in composition and properties. The most common basic structural unit of lignin is a phenylpropanoid:

FIGURE 5-12 (A and B) Cellulases, produced by the corn stalk rot fungus *Fusarium* sp., have broken down cellulosic walls of corn cells but did not affect the lignified vascular bundles. (C and D) Ligninases of the basidiomycete fungus *Phellinus* have caused complete disintegration and discoloration of the heartwood in the pine trunk (C) and of the roots and lower stem of the tree, causing it to topple over (D). [Photographs courtesy of (A and B) G. Munkvold, Iowa State University, (C) R. L. Anderson, USDA Forest Service, and (D) R. L. James, USDA Forest Service.]

(benzene ring)—C—C—C

where one or more of the carbons have a —OH, —OCH_3, or =O group. Lignin forms by oxidative condensation (C—C and C—O bond formation) between such substituted phenylpropanoid units. The lignin polymer is perhaps more resistant to enzymatic degradation than any other plant substance (Figs. 5-11 and 5-12).

It is obvious that enormous amounts of lignin are degraded by microorganisms in nature, as is evidenced by the yearly decomposition of all annual plants and a large portion of perennial plants. It is generally accepted, however, that only a small group of microorganisms is capable of degrading lignin. Actually, only about 500 species of fungi, almost all of them basidiomycetes, have been reported so far as being capable of decomposing wood. About one-fourth of these fungi (the brown rot fungi) seem to cause some degradation of lignin but cannot utilize it. Most of the lignin in the world is degraded and utilized by a group of basidiomycetes called white rot fungi. It appears that white rot fungi secrete one or more enzymes (ligninases), which enable them to utilize lignin (Fig. 5-12).

In addition to wood-rotting basidiomycetes, several other pathogens, primarily several ascomycetes and imperfect fungi and even some bacteria, apparently produce small amounts of lignin-degrading enzymes and cause soft rot cavities in wood they colonize. However, it is not known to what extent the diseases they cause are dependent on the presence of such enzymes.

Cell Wall Flavonoids

Flavonoids are a large class of phenolic compounds that occur in most plant tissues and, especially, in the vacuoles. They also occur as mixtures of single and polymeric components in various barks and heartwoods. Among the various functions of flavonoids, some act as signaling molecules for certain functions in specific plant/microbe combinations. Many of them, however, are inhibitory or toxic to pathogens and some of them, e.g., medicarpin, act as phytoalexins and are involved in the inducible defense in plants against fungi. It is important, therefore, that pathogens be able to survive in the presence of various flavonoids in cell walls or they must be able to neutralize them or to break them down. Little is known how pathogens accomplish this, although the joining of phenolics with sugar molecules (glycosylation) seems to neutralize the toxicity of many phenolics.

Cell Wall Structural Proteins

Cell walls consist primarily of polysaccharides, i.e., cellulose fibers embedded in a matrix of hemicellulose and pectin, but structural proteins, in the form of glycoproteins, may also form networks in the cell wall (Fig. 5-2). Four classes of structural proteins have been found in cell walls. Three of them are known by the most abundant amino acid they contain: **hydroxyproline-rich glycoproteins** (HRGPs), **proline-rich proteins** (PRPs), and **glycine-rich proteins** (GRPs). The fourth class is **arabinogalactan proteins** (AGPs). Each of these protein groups is coded by a large multigene family. Upon their production they are inserted in the endoplasmic reticulum and, through signal peptides they encode, they are targeted to the cell wall through the secretory pathway. One of the HRGP proteins is **extensin**, which makes up only 0.5% of the cell wall mass in healthy tissue but increases to 5 to 15% of the wall mass on infection with fungi and helps add rigidity to the cell wall. Another group of cell wall proteins are the **lectins**, which bind to specific sugar molecules. The role of all of these groups of proteins is not clear, but they are thought to accumulate in response to elicitor molecules released by fungi and to play a role in the plant defense response. The breakdown of structural proteins is presumably advantageous to invading pathogens and is thought to be similar to that of proteins contained within plant cells. This is discussed later.

Enzymatic Degradation of Substances Contained in Plant Cells

Most kinds of pathogens live all or part of their lives in association with or inside the living protoplast. These pathogens obviously derive nutrients from the protoplast. All the other pathogens — the great majority of fungi and bacteria — obtain nutrients from protoplasts after the latter have been killed. Some of the nutrients, e.g., sugars and amino acids, are molecules sufficiently small to be absorbed by the pathogen directly. Some of the other plant cell constituents, however, such as starch, proteins, and fats, can be utilized only after degradation by enzymes secreted by the pathogen.

Proteins

Plant cells contain innumerable different proteins, which play diverse roles as catalysts of cellular reactions (enzymes) or as structural material (in membranes and cell walls). Proteins are formed by the joining together of numerous molecules of about 20 different kinds of amino acids:

$$\underset{\substack{|\\ NH_2}}{H-CH}-CO\,\boxed{OH + H}\!-\!\overset{\substack{R-CH-COOH\\ |}}{NH} \longrightarrow \underset{\substack{|\\ NH_2}}{H-CH}-CO-\overset{\substack{R-CH-COOH\\ |}}{NH} + H_2O$$

Amino Acids and Protein

All pathogens seem to be capable of degrading many kinds of protein molecules. The plant pathogenic enzymes involved in protein degradation are similar to those present in higher plants and animals and are called **proteases** or **proteinases** or, occasionally, peptidases.

Considering the paramount importance of proteins as enzymes, constituents of cell membranes, and structural components of plant cell walls, the degradation of host proteins by proteolytic enzymes secreted by pathogens can profoundly affect the organization and function of the host cells. The nature and extent of such effects, however, have been investigated little so far and their significance in disease development is not known.

Starch

Starch is the main reserve polysaccharide found in plant cells. Starch is synthesized in the chloroplasts and, in nonphotosynthetic organs, in the amyloplasts. Starch is a glucose polymer and exists in two forms: amylose, an essentially linear molecule, and amylopectin, a highly branched molecule of various chain lengths.

Most pathogens utilize starch, and other reserve polysaccharides, in their metabolic activities. The degradation of starch is brought about by the action of enzymes called **amylases**. The end product of starch breakdown is glucose and it is used by the pathogens directly.

Lipids

Various types of lipids occur in all plant cells, with the most important being **phospholipids** and **glycolipids**, both of which, along with protein, are the main constituents of all plant cell membranes. The latter form a hydrophobic barrier that is critical to life by separating cells from their surroundings and keeping organelles such as chloroplasts and mitochondria intact and separate from the cytoplasm. **Oils** and **fats** are found in many cells, especially in seeds where they function as energy storage compounds; **wax lipids** are found on most aerial epidermal cells. The common characteristic of all lipids is that they contain fatty acids, which may be saturated or unsaturated.

Several fungi, bacteria, and nematodes are known to be capable of degrading lipids. Lipolytic enzymes, called **lipases**, **phospholipases**, and so on, hydrolyze liberation of the fatty acids from the lipid molecule. The fatty acids are presumably utilized by the pathogen directly. But Some of them, before or after hyperoxidation by plant lipoxygenases or active oxygen species, provide signal molecules for the development of plant defenses and also act as antimicrobial compounds that inhibit the pathogen directly.

Microbial Toxins in Plant Disease

Living plant cells are complex systems in which many interdependent biochemical reactions are taking place concurrently or in a well-defined succession. These reactions result in the intricate and well-organized processes essential for life. Disturbance of any of these metabolic reactions causes disruption of the physiological processes that sustain the plant and leads to the development of disease. Among the factors inducing such disturbances are substances that are produced by plant pathogenic microorganisms and are called toxins. Toxins act directly on living host protoplasts, seriously damaging or killing the cells of the plant. Some toxins act as general protoplasmic poisons and affect many species of plants representing different families. Others are toxic to only a few plant species or varieties and are completely harmless to others. Many toxins exist in multiple forms that have different potency.

Fungi and bacteria may produce toxins in infected plants as well as in culture medium. Toxins, however, are extremely poisonous substances and are effective in very low concentrations. Some are unstable or react quickly and are bound tightly to specific sites within the plant cell.

Toxins injure host cells either by affecting the permeability of the cell membrane (Fig. 5-2) or by inactivating or inhibiting enzymes and subsequently interrupting the corresponding enzymatic reactions. Certain toxins act as antimetabolites and induce a deficiency for an essential growth factor.

Toxins That Affect a Wide Range of Host Plants

Several toxic substances produced by phytopathogenic microorganisms have been shown to produce all or part of the disease syndrome not only on the host plant, but also on other species of plants that are not normally attacked by the pathogen in nature. Such toxins, called nonhost-specific or nonhost-selective toxins. These toxins increase the severity of disease caused by a pathogen, i.e., they affect the virulence of the pathogen, but are not essential for the pathogen to cause disease, i.e., they do not determine the pathogenicity of the pathogen. Several of these toxins, e.g., tabtoxin and phaseolotoxin, inhibit normal host enzymes, thereby leading to increases in toxic substrates or to depletion of needed compounds. Several toxins affect the cellular transport system, especially H^+/K^+ exchange at the cell membrane. Some, e.g., tagetitoxin, act as inhibitors of transcription in cell organelles, such as the chloroplasts. Others, e.g., cercosporin, act as photosensitizing agents, causing the peroxidation of membrane lipids.

Tabtoxin

Tabtoxin is produced by the bacterium *Pseudomonas syringae*; pv. *tabaci*, which causes the wildfire disease of tobacco; by strains of pv. tabaci occurring on other hosts such as bean and soybean; and by other pathovars (subspecies) of *P. syringae*, such as those occurring on oats, maize, and coffee. Toxin-producing strains cause necrotic spots on leaves, with each spot surrounded by a yellow halo (Figs. 5-13A and 5-13B). Sterile culture filtrates of the organism, as well as purified toxin, produce symptoms identical to those characteristic of wildfire of tobacco not only on tobacco, but in a large number of plant species belonging to many different families (nonhost-specific toxin!). Strains of *P. syringae* pv. *tabaci* sometimes produce mutants that have lost the ability to produce the toxin (they become Tox^-). Tox^- strains show reduced virulence and cause necrotic leaf spots without the yellow halo. Tox^- strains are indistinguishable from *P. angulata*, the cause of angular leaf spot of tobacco, which is now thought to be a nontoxigenic form of *P. syringae* pv. *tabaci.*

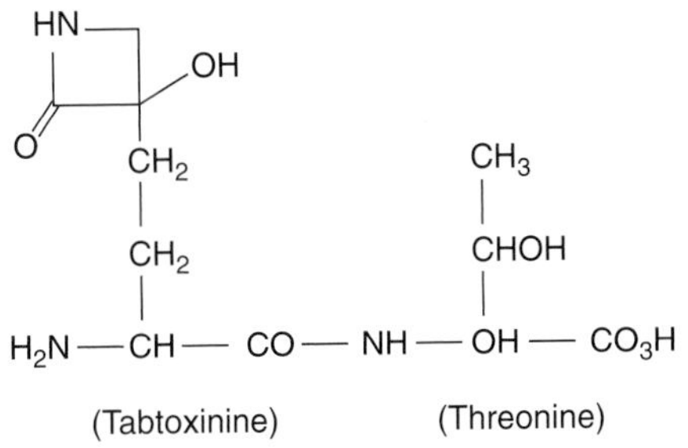

HN, OH, O, CH_2, CH_2, CH_3, CHOH

H_2N — CH — CO — NH — OH — CO_3H

(Tabtoxinine) (Threonine)

Tabtoxin

Tabtoxin is a dipeptide composed of the common amino acid threonine and the previously unknown amino acid tabtoxinine. Tabtoxin as such is not toxic, but in the cell it becomes hydrolyzed and releases tabtoxinine, which is the active toxin. Tabtoxin, through tabtoxinine, is toxic to cells because it inactivates the enzyme glutamine synthetase, which leads to depleted glutamine levels and, as a consequence, accumulation of toxic concentrations of ammonia. The latter uncouples photosynthesis and photorespiration and destroys the thylakoid membrane of the chloroplast, thereby causing chlorosis and eventually necrosis. The effects of the toxin lead to a reduced ability of the plant to respond actively to the bacterium.

Phaseolotoxin

Phaseolotoxin is produced by the bacterium *Pseudomonas syringae* pv. *phaseolicola*, the cause of halo blight of bean (Fig. 5-13C) and some other legumes. The localized and systemic chlorotic symptoms produced in infected plants are identical to those produced on plants treated with the toxin alone so they are apparently the results of the toxin produced by the bacteria. Infected plants and plants treated with purified toxin also show reduced growth of newly expanding leaves, disruption of apical dominance, and accumulation of the amino acid ornithine.

Phaseolotoxin is a modified ornithine–alanine–arginine tripeptide carrying a phosphosulfinyl group. Soon after the tripeptide is excreted by the bacterium into the plant, plant enzymes cleave the peptide bonds and release alanine, arginine, and phosphosulfinylornithine. The latter is the biologically functional moiety of phaseolotoxin. The toxin affects cells by binding to the active site of and inactivating the enzyme ornithine carbamoyltransferase, which normally converts ornithine to citrulline, a precursor of arginine. By its action on the enzyme, the toxin thus causes the accumulation of ornithine and depleted levels of citrulline and arginine. Phaseolotoxin, however, seems to also inhibit pyrimidine nucleotide biosynthesis, reduce the activity of ribosomes, interfere with lipid synthesis, change the permeability of membranes, and result in the accumulation of large starch grains in the chloroplasts. Phaseolotoxin plays a major role in the virulence of the pathogen by interfering with or breaking the disease resistance of the host toward not only the halo blight bacterium, but also several other fungal, bacterial, and viral pathogens.

Tentoxin

Tentoxin is produced by the fungus *Alternaria alternata* (previously called *A. tenuis*), which causes spots and chlorosis (Fig. 5-13D) in plants of many species. Seedlings with more than one-third of their leaf area chlorotic die, and those with less chlorosis are much less vigorous than healthy plants.

Tentoxin is a cyclic tetrapeptide that binds to and inactivates a protein (chloroplast-coupling factor) involved in energy transfer into chloroplasts. The toxin also inhibits the light-dependent phosphorylation of ADP to ATP. Both the inactivation of the protein and the inhibition of photophosphorylation are much greater in plant species susceptible to chlorosis after tentoxin treatment than in species not sensitive to the toxin. In sensitive species, tentoxin interferes with normal chloroplast development and results in chlorosis by disrupting chlorophyll synthesis, but it is not certain that these effects are solely related to tentoxin binding to the chloroplast-coupling factor protein. An additional but apparently unrelated effect of tentoxin on sensitive plants is that it inhibits the activity of polyphenol

FIGURE 5-13 Symptoms caused by nonhost-selective toxins. Early (A) and semiadvanced (B) symptoms of young tobacco leaves showing spots caused by the bacterium *Pseudomonas syringae* pv. *tabaci*. The chlorotic halos surrounding the necrotic white spots are caused by the tabtoxin produced by the bacterium. (C) Leaf spots and halos caused by the toxin phaseolotoxin produced by the bacterium *Pseudomonas phaseolicola*, the cause of halo blight of bean. (D) Leaf spots and chlorosis caused by the *Alternaria alternata* toxin. [Photographs courtesy of (A, B, and D) Reynolds Tobacco Co. and (C) Plant Pathology Department, University of Florida.]

oxidases, enzymes involved in several resistance mechanisms of plants. Both effects of the toxin, namely stressing the host plant with events that lead to chlorosis and suppressing host resistance mechanisms, tend to enhance the virulence of the pathogen. The molecular site of action of tentoxin, however, and the exact mechanism by which it brings about these effects are still unknown.

Cercosporin

Cercosporin is produced by the fungus *Cercospora* and by several other fungi. It causes damaging leaf spot and blight diseases of many crop plants, such as Cercospora leaf spot of zinnia (Fig. 5-14A) and gray leaf spot of corn (Fig. 5-14B).

Cercosporin is unique among fungal toxins in that it is activated by light and becomes toxic to plants by generating activated species of oxygen, particularly single oxygen. The generated active single oxygen destroys the membranes of host plants and provides nutrients for this intercellular pathogen. Cercosporin is a photosensitizing perylenequinone that absorbs light energy, it is converted to an energetically activated state and then reacts with molecular oxygen and forms activated oxygen. The latter reacts with lipids, proteins, and nucleic acids of plant cells and severely damages or kills the plant cells, thereby enhancing the virulence of the pathogen. The

FIGURE 5-14 Leaf spots on zinnia (A) and gray leaf spots on corn (B) caused by the photosensitizing toxin cercosporin, produced by different species of the fungus *Cercospora*. [Photographs courtesy of (A) Plant Pathology Department, University of Florida and (B) G. Munkvold, Iowa State University.]

ability of fungal spores and mycelium to survive the general toxicity of cercosporin is due to the production by the fungus of pyridoxine (vitamin B_6). Pyridoxine reacts with single oxygen atoms and is currently neutralized during that reaction.

Other Nonhost-Specific Toxins

Numerous other nonhost-specific toxic substances have been isolated from cultures of pathogenic fungi and bacteria and have been implicated as contributing factors in the development of the disease caused by the pathogen. Among such toxins produced by fungi are fumaric acid, produced by *Rhizopus* spp. in almond hull rot disease; oxalic acid, produced by *Sclerotium* and *Sclerotinia* spp. in various plants they infect and by *Cryphonectria parasitica*, the cause of chestnut blight; alternaric acid, alternariol, and zinniol produced by *Alternaria* spp. in leaf spot diseases of various plants; ceratoulmin, produced by *Ophiostoma ulmi* in Dutch elm disease; fusicoccin, produced by *Fusicoccum amygdali* in the twig blight disease of almond and peach trees; ophiobolins, produced by several *Cochliobolus* spp. in diseases of grain crops; pyricularin, produced by *Pyricularia grisea* in rice blast disease; fusaric acid and lycomarasmin, produced by *Fusarium oxysporum* in tomato wilt; and many others. Other nonhost-specific toxins produced by bacteria are coronatine, produced by *P. syringae* pv. *atropurpurea* and other forms infecting grasses and soybean; syringomycin, produced by *P. syringae* pv. *syringae* in leaf spots of many plants; syringotoxin, produced by *P. syringae* pv. *syringae* in citrus plants; and tagetitoxin, produced by *P. syringae* pv. *tagetis* in marigold leaf spot disease. One family of toxins essential for pathogenicity, is the thaxtomins, produced by species of the bacterium *Streptomyces* that cause root and tuber roots. Thaxtomins cause dramatic plant cell hypertrophy and/or seedling stunting by altering the development of primary cell walls and the ability of the cells to go through normal cell division cycles.

Host-Specific or Host-Selective Toxins

A **host-specific** or **host-selective** toxin is a substance produced by a pathogenic microorganism that, at physiological concentrations, is toxic only to the hosts of that pathogen and shows little or no toxicity against nonsusceptible plants. Most host-specific toxins must be present for the producing microorganism to be able to cause disease. So far, host-specific toxins have been shown to be produced only by certain fungi (*Cochliobolus*, *Alternaria*, *Periconia*, *Phyllosticta*, *Corynespora*, and *Hypoxylon*), although certain bacterial polysaccharides from *Pseudomonas* and *Xanthomonas* have been reported to be host specific.

Victorin, or HV Toxin

Victorin, or Hv-toxin, is produced by the fungus *Cochliobolus (Helminthosporium) victoriae.* This fungus appeared in 1945 after the introduction and widespread use of the oat variety Victoria and its derivatives, all of which contained the gene V_b for resistance to crown rust disease. *C. victoriae* infects the basal portions of susceptible oat plants and produces a toxin that is carried to the leaves, causes a leaf blight, and destroys the entire plant. All other oats and other plant species tested were either immune to the fungus and to the toxin or their sensitivity to the toxin was proportional to their susceptibility to the fungus. Toxin production in the fungus is controlled by a single gene. Resistance and susceptibility to the fungus, as well as tolerance and sensitivity to the toxin, are controlled by the same pair of alleles, although different sets of these alleles may be involved in cases of intermediate resistance. The toxin not only produces all the external symptoms of the disease induced by the pathogen, but it also produces similar histochemical and biochemical changes in the host, such as changes in cell wall structure, loss of electrolytes from cells, increased respiration, and decreased growth and protein synthesis. Moreover, only fungus isolates that produce the toxin in culture are pathogenic to oats, whereas those that do not produce toxin are nonpathogenic.

Victorin has been purified and its chemical structure has been determined to be a complex chlorinated, partially cyclic pentapeptide. The primary target of the toxin seems to be the cell plasma membrane where victorin seems to bind to several proteins. The possible site of action of victorin seems to be the glycine decarboxylate complex, which is a key component of the photorespiratory cycle. Considerable evidence, however, indicates that victorin functions as an elicitor that induces components of a resistance response that include many of the features of hypersensitive response and lead to programmed cell death.

*T Toxin [*Cochliobolus (Helminthosporium) heterostrophus *Race T Toxin]*

T toxin is produced by race T of *C. heterostrophus (Bipolaris maydis)*, the cause of southern corn leaf blight (Fig. 5-15A). Race T, indistinguishable from all other *C. heterostrophus* races except for its ability to produce the T toxin, appeared in the United States in 1968. By 1970, it had spread throughout the corn belt, attacking only corn that had the Texas male-sterile (Tms) cytoplasm. Corn with normal cytoplasm was resistant to the fungus and the toxin. Resistance and susceptibility to *C. heterostrophus* T and its toxin are inherited maternally (in cytoplasmic genes). The ability of *C. heterostrophus* T to produce T toxin and its virulence to corn with Tms cytoplasm are controlled by one and the same gene. T toxin does not seem to be necessary for the pathogenicity of *C. heterostrophus* race T, but it increases the virulence of the pathogen.

T toxin is a mixture of linear, long (35 to 45 carbon) polyketols, the most prevalent having the following formula:

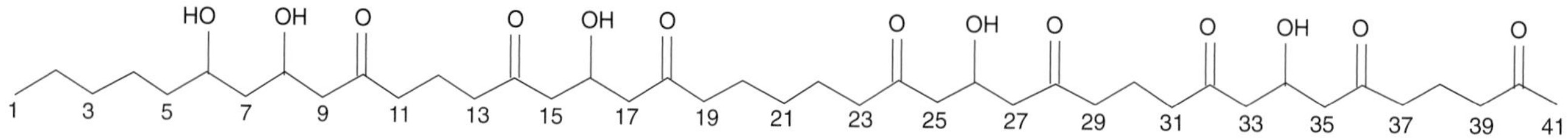

HO OH O O OH O O OH O O OH O O

1 3 5 7 9 11 13 15 17 19 21 23 25 27 29 31 33 35 37 39 41

T toxin

The T toxin apparently acts specifically on mitochondria of susceptible cells, which are rendered nonfunctional, and inhibits ATP synthesis. The T toxin reacts with a specific receptor protein molecule (URF13) that is located on the inner mitochondrial membrane of sensitive mitochondria. It is now thought that plants exhibiting cytoplasmic male sterility of the Texas type have a slight rearrangement in their mitochondrial DNA comprising gene *T-urf13* that codes for the production of protein URF13. This gene and its protein are absent from maize lines with normal cytoplasm. When the T toxin is present, protein URF13 forms pores in the inner mitochondrial membrane of maize lines with cytoplasmic male sterility. The pores cause loss of mitochondrial integrity, i.e., loss of selective permeability of the mitochondrial membrane, and disease.

HC Toxin

Race 1 of *Cochliobolus (Helminthosporium) carbonum (Bipolaris zeicola)* causes northern leaf spot and ear rot disease in maize. It also produces the host-specific HC toxin, which is toxic only on specific maize lines. Two other races of the fungus do not produce toxin but infect corn around the world, although they cause smaller lesions. The mechanism of action of HC toxin is not known, but this is the only toxin, so far, for which the biochemical and molecular genetic basis of resistance against the toxin is understood. Resistant corn lines have a gene (Hm1) coding for an enzyme called HC toxin reductase that reduces and thereby detoxifies the toxin. Susceptible corn lines lack this gene and, therefore, cannot defend themselves against the

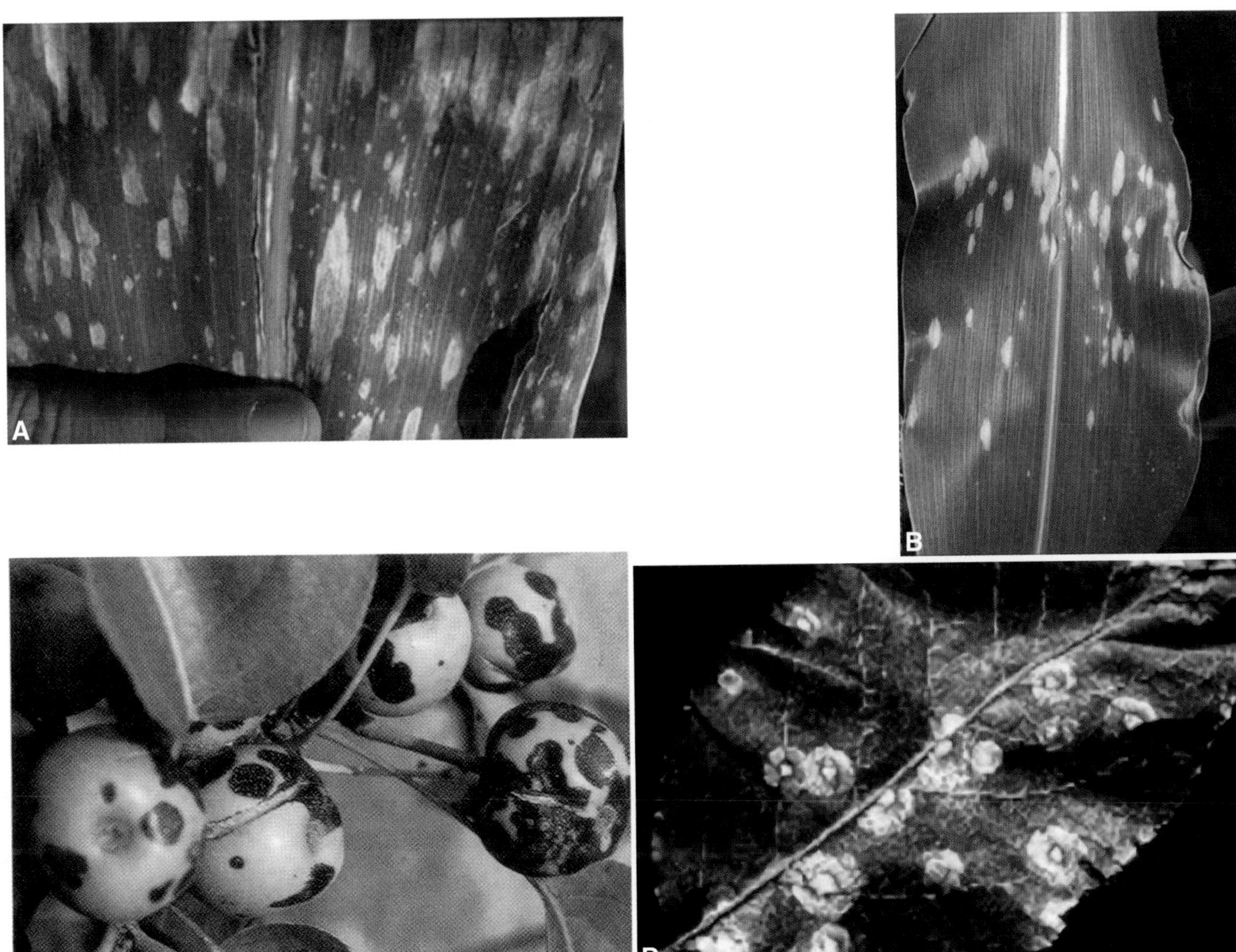

FIGURE 5-15 Symptoms caused by host-selective toxins. (A) Southern corn leaf blight symptoms caused by two race T of the fungus *Cochliobolus (Helminthosporium) heterostrophus* and its toxin, T toxin, on a corn plant containing Texas male-sterile cytoplasm. (B) Northern corn leaf spot symptoms caused by the fungus *Cochliobolus carbonum* and its toxin, HC toxin, on corn. (C) Fruit spots on Japanese pear caused by one of the strains of the fungus *Alternaria alternata* and its toxin, AK toxin. (D) Leaf spots caused by the AM toxin produced by another strain of the fungus *A. alternata* and its toxin, AM toxin, on apple leaves. [Photographs courtesy of (A) C. Martinson and (B) G. Munkvold, Iowa State University, (C) T. Sakuma, and (D) J. W. Travis, Pennsylvania State University.]

toxin. Experimental findings suggest that the HC toxin is not actually toxic in itself, but rather acts as a virulence factor by preventing initiation of the changes in gene expression that are necessary for the establishment of induced defense responses, i.e., it acts as a suppressor of defense responses.

Alternaria alternata *Toxins*

Several pathotypes of *Alternaria alternata* attack different host plants and on each they produce one of several multiple forms of related compounds that are toxic only on the particular host plant of each pathotype. Some of the toxins and the hosts on which they are produced and affect are the AK toxin causing black spot on Japanese peat fruit (Fig. 5-15C), the AAL toxin causing stem canker on tomato, the AF toxin on strawberry, the AM toxin on apple, the ACT toxin on tangerine, the ACL toxin on rough lemon, and the HS toxin on sugar cane.

As an example of *A. alternata* toxins, the AM toxin is produced by the apple pathotype of *A. alternata*, known previously as *A. mali*, the cause of alternaria leaf blotch of apple (Fig. 5-15D). The toxin molecule is a cyclic depsipeptide and usually exists as a mixture of three forms. The toxin is extremely selective for susceptible apple varieties, whereas resistant varieties can tolerate more than 10,000 times as much toxin without showing symptoms. The AM toxin causes plasma membranes of susceptible cells to develop invaginations, and cells show a significant loss of electrolytes. The initial toxic effect of the toxin seems to occur at the interface between the cell wall and the plasma membrane. However, the AM toxin also causes rapid loss of chlorophyll, suggesting that this toxin has more than one site of action.

Other Host-Specific Toxins

At least two other fungi produce well-known host-specific toxins: *Periconia circinata* produces peritoxin (PC toxin), which causes sorghum rot in sorghum root rot disease; *Mycosphaerella (Phyllosticta) zeae-maydis* produces the PM toxin (T toxin) in corn that has Texas male-sterile cytoplasm; and *Pyrenophora tritici-repentis* produces the Ptr toxin, which causes the tan spot of wheat. Another fungus, *Corynespora cassiicola*, produces the CC toxin in tomato. Toxins produced by some other fungi, e.g., *Hypoxylon mammatum* on poplar and *Perenophora teres* on barley, seem to be species selective rather than host specific. In addition, there are the SV toxins produced by *Stemphylium vesicarium* on European pear and destruxin B from *A. brassicae* on brassicas.

Growth Regulators in Plant Disease

Plant growth is regulated by a small number of groups of naturally occurring compounds that act as hormones and are generally called growth regulators. The most important growth regulators are auxins, gibberellins, and cytokinins, but other compounds, such as ethylene and growth inhibitors, play important regulatory roles in the life of the plant. Growth regulators act in very small concentrations and even slight deviations from the normal concentration may bring about strikingly different plant growth patterns. The concentration of a specific growth regulator in the plant is not constant, but it usually rises quickly to a peak and then declines quickly as a result of the action of hormone-inhibitory systems present in the plant. Growth regulators appear to act, at least in some cases, by promoting the synthesis of messenger RNA molecules. This leads to the formation of specific enzymes, which in turn control the biochemistry and the physiology of the plant.

Plant pathogens may produce more of the same growth regulators as those produced by the plant or more of the same inhibitors of the growth regulators as those produced by the plant. They may produce new and different growth regulators or inhibitors of growth regulators. Alternatively, they may produce substances that stimulate or retard the production of growth regulators or growth inhibitors by the plant.

Whatever the mechanism of action involved, pathogens often cause an imbalance in the hormonal system of the plant and bring about abnormal growth responses incompatible with the healthy development of a plant. That pathogens can cause disease through the secretion of growth regulators in the infected plant or through their effects on the growth regulatory systems of the infected plant is made evident by the variety of abnormal plant growth responses they cause, such as stunting, overgrowths, rosetting, excessive root branching, stem malformation, leaf epinasty, defoliation, and suppression of bud growth. The most important groups of plant growth regulators, their function in the plant, and their role in disease development, where known, are discussed next.

Auxins

The auxin occurring naturally in plants is indole-3-acetic acid (IAA). Produced continually in growing plant tissues, IAA moves rapidly from the young green tissues to older tissues, but is destroyed constantly by the enzyme indole-3-acetic acid oxidase, which explains the low concentration of the auxin.

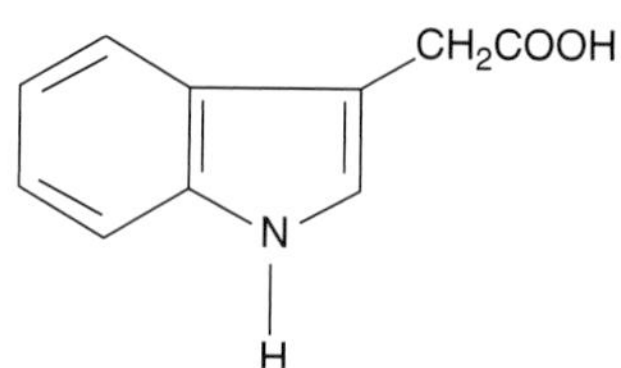

Indole-3-acetic acid

The effects of IAA on the plant are numerous. It is required for cell elongation and differentiation, and absorption of IAA to the cell membrane also affects the permeability of the membrane. The compound causes a general increase in the respiration of plant tissues and promotes the synthesis of messenger RNA and, subsequently, of proteins/enzymes as well as structural proteins.

Increased auxin (IAA) levels occur in many plants infected by fungi, bacteria, viruses, mollicutes, and nematodes, although some pathogens seem to lower the auxin level of the host. Thus, the basidiomycete *Exobasidium azaleae* causing azalea leaf and flower gall (Fig. 5-16A), the protozoon causing clubroot of cabbage (*Plasmodiophora brassicae*) (Fig. 5-16E), the bacterium

FIGURE 5-16 Plant diseases showing symptoms caused by the excessive production of growth regulators (primarily auxins) by the pathogen. (A) Enlarged and deformed leaf and flower gall of azalea caused by infection of the fungus *Exobasidium azaleae*. (B) Leafy gall produced on a sweet pea plant as a result of infection by the bacterium *Rhodococcus fascians*. (C) Corn ear and tassel showing numerous small galls as a result of infection by the corn smut fungus *Ustilago maydis*. (D) Western pine gall caused by the fungus *Cronartium* sp. (E) Cabbage roots enlarged grotesquely following infection with the clubroot pathogen *Plasmodiophora brassicae*. A few normal, thin roots are still present. (F) Root galls on bean plant infected with the root-knot nematode *Meloidogyne* sp. [Photographs courtesy of (A and B) Oregon State University, (C) K. Mohan, Idaho State University, (D) E. Hansen, Oregon State University, (E) University of Minnesota, and (F) R. T. MacMillan, University of Florida.]

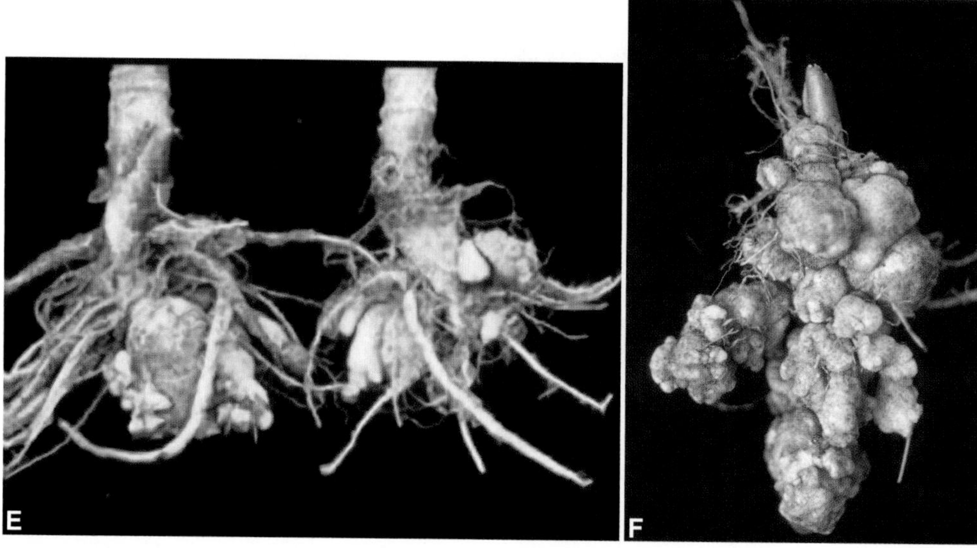

FIGURE 5-16 *(Continued)*

A. tumefaciens causing crown gall (Fig. 5-17A) and the one causing leafy gall of sweet pea and other plants (Fig. 5-16B), the fungi causing corn smut (*Ustilago maydis*) (Fig. 5-16C), cedar apple rust (*Gymnosporangium juniperi-virginianae*), banana wilt (*Fusarium oxysporum f. cubense*), pine western gall rust (Fig. 5-16D), the root-knot nematode (*Meloidogyne* sp.) (Fig. 5-16F), and others not only induce increased levels of IAA in their respective hosts, but are themselves capable of producing IAA. In some diseases, however, increased levels of IAA are wholly or partly due to the decreased degradation of IAA through the inhibition of IAA oxidase, as has been shown to be the case in several diseases, including corn smut and stem rust of wheat.

The production and role of auxin in plant disease have been studied more extensively in some bacterial diseases of plants. *Ralstonia solanacearum*, the cause of bacterial wilt of solanaceous plants, induces a 100-fold increase in the IAA level of diseased plants compared with that of healthy plants. How the increased levels of IAA contribute to the development of wilt of plants is not yet clear, but the increased plasticity of cell walls as a result of high IAA levels renders the pectin, cellulose, and protein components of the cell wall more accessible to, and may facilitate their degradation by, the respective enzymes secreted by the pathogen. An increase in IAA levels seems to inhibit the lignification of tissues and may thus prolong the period of exposure of the nonlignified tissues to the cell wall-degrading enzymes of the pathogen. Increased respiratory rates in the infected tissues may also be due to high IAA levels, and because auxin affects cell permeability, it may be responsible for the increased transpiration of the infected plants.

In **crown gall**, a disease caused by the bacterium *A. tumefaciens* on more than a hundred plant species, galls or tumors develop on the roots, stems (Figs. 3-2E, 3-11E, and 5-17A), leaves, ears, tassels, and petioles of host plants. Crown gall tumors develop when crown gall bacteria enter fresh wounds on a susceptible host. Immediately after wounding, cells around the wound produce various phenolic compounds and are activated to divide. *Agrobacterium* bacteria do not invade cells but attach to cell walls, and, in response to phenolic compounds such as acetosyringone and other signals, they become activated and begin processing the DNA in their Ti plasmid (for tumor-inducing plasmid) (Fig. 5-17). During the intense cell division of the second and third days after wounding, the plant cells are somehow conditioned and made receptive to a piece of bacterial plasmid DNA (called T-DNA, for tumor DNA). Proteins coded by genes in the T-DNA virulence (Vir) region cut out a single strand of the T-DNA from the Ti plasmid and transfer it into the plant cell nucleus as a T-DNA–protein complex. The T-DNA then becomes integrated into the nuclear plant DNA (chromosomes) and some of its genes are expressed and lead to the synthesis of auxins and cytokinins, which transform normal plant cells into tumor cells. Tumor cells subsequently grow and divide independently of the bacteria, and their

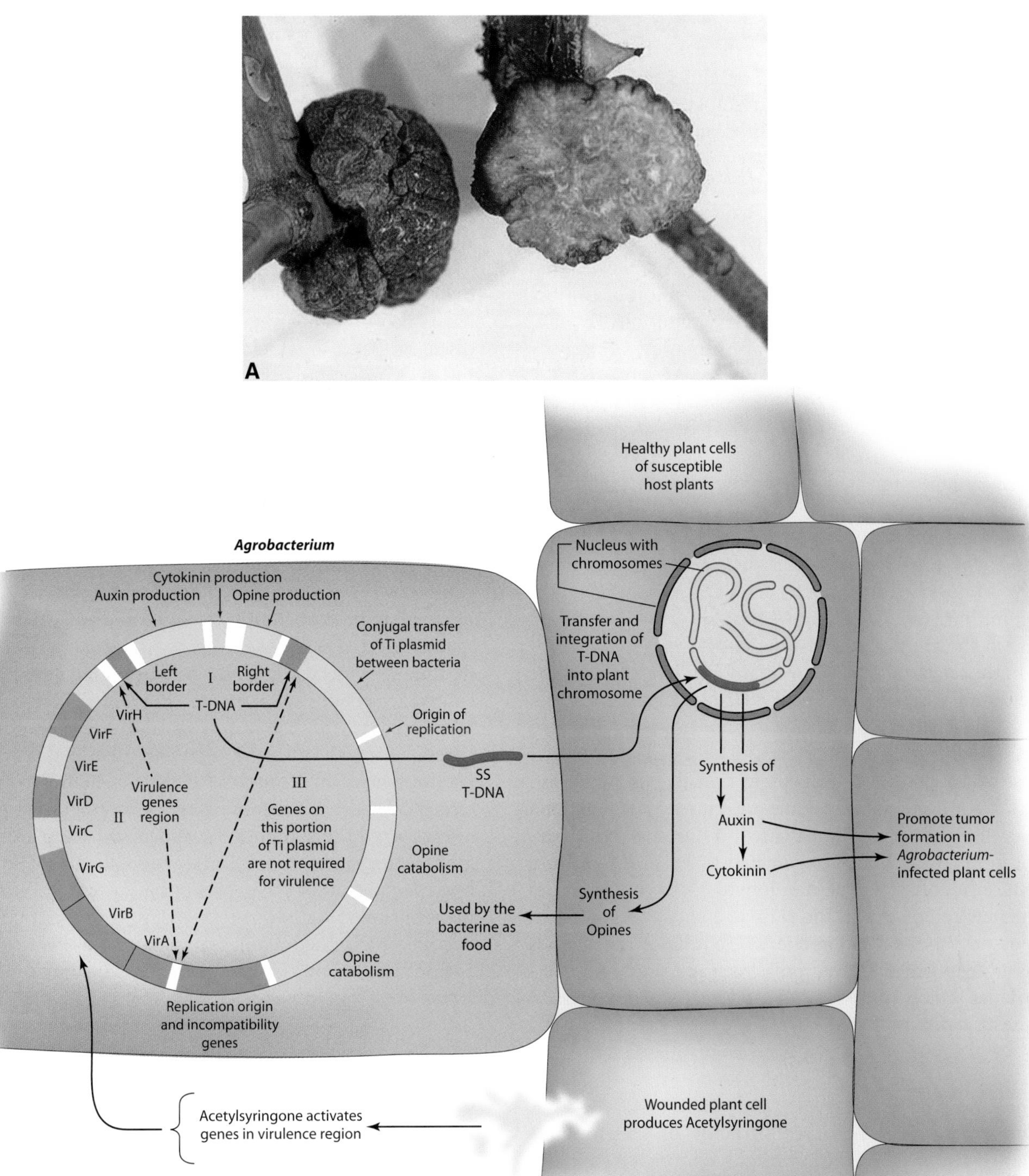

FIGURE 5-17 (A) External and cross-sectional view of crown gall on a rose stem caused by the bacterium *Agrobacterium tumefaciens*. (B) Schematic representation of the structure of Ti plasmid of the bacterium and of the transfer, integration, and expression of T-DNA in an infected plant that results in the production of crown gall tumors. Genes A, B, D, and G are needed for tumor formation on any susceptible plant species. Genes C, E, F, and H affect the host plant range and/or the size of tumors caused by the bacterium. The functions of the proteins of virulence genes are as follows: A, receptor of wound signal; B, codes for proteins that form membrane pores; C, enhances transfer of T-DNA; D, codes for proteins that nick T-DNA at its borders, help transport T-DNA across membranes, and carry signal compounds to the nucleus; E, protects T-DNA from nuclease enzymes and also carries nuclear localization signals; F, may increase host range of tumor induction; G, activates other virulence genes; H, protects the bacterium from toxic plant compounds. The entire diagram presents a simplified scheme of interaction of gene products of host cells and T-DNA that lead to the production of a gall. [Photograph (A) courtesy of Oregon State University.]

organization, rate of growth, and rate of division can no longer be controlled by the host plant.

The integrated T-DNA also contains genes that code for substances known as opines. Transformed plant cells produce opines, which can be used only by the intercellularly growing crown gall bacteria as a source of food. Although the increased levels of IAA and cytokinins of tumor cells are sufficient to cause the autonomous enlargement and division of these cells once they have been transformed to tumor cells, high IAA and cytokinin levels alone cannot cause the transformation of healthy cells into tumor cells. What other conditions or substances are involved in the transformation of healthy cells into tumor cells is not known.

In the **knot disease** of olive, oleander, and privet, another hyperplastic disease caused by the bacterium *Pseudomonas savastanoi*, the pathogen produces IAA, which induces infected plants to produce galls. The more IAA a strain produces, the more severe the symptoms it causes. Strains that do not produce IAA fail to induce the formation of galls. The bacterial genes for IAA production are in a plasmid carried in the bacterium, but some IAA synthesis is also carried out by a gene in the chromosome of the bacterium.

In the **leafy gall disease** of many plants caused by the bacterium *Rhodococcus fascians*, leafy galls are produced that consist of centers of shoot overproductions and shoot growth inhibition. The bacterium exists mostly at the surface of the plant tissues, but it can also grow internally in the plant. Auxin, cytokinins, and other hormonal substances are produced by the bacterium in cultured and by infected tissues. Signals from bacteria involved in the development of symptoms initiate new cell divisions and formation of shoot meristem in tissues already differentiated. The bacterial signals originate in genes located on a linear plasmid and exert activities much more unique and more complex than those of cytokinins alone.

Gibberellins

Gibberellins are normal constituents of green plants and are also produced by several microorganisms. Gibberellins were first isolated from the fungus *Gibberella fujikuroi*, the cause of the foolish seedling disease of rice (Figure 1-37D). The best-known gibberellin is gibberellic acid. Compounds such as vitamin E and helminthosporol also have gibberellin-like activity.

Gibberellins have striking growth-promoting effects. They speed the elongation of dwarf varieties to normal sizes and promote flowering, stem and root elongation, and growth of fruit. Such *elongation* resembles in some respects that caused by IAA, and gibberellin also induces IAA formation. Auxin and gibberellin may also act synergistically. Gibberellins seem to activate genes that have been previously "turned off." The foolish seedling disease of rice, in which rice seedlings infected with the fungus *Gibberella fujikuroi* grow rapidly and become much taller than healthy plants, is apparently the result, to a considerable extent at least, of the gibberellin secreted by the pathogen.

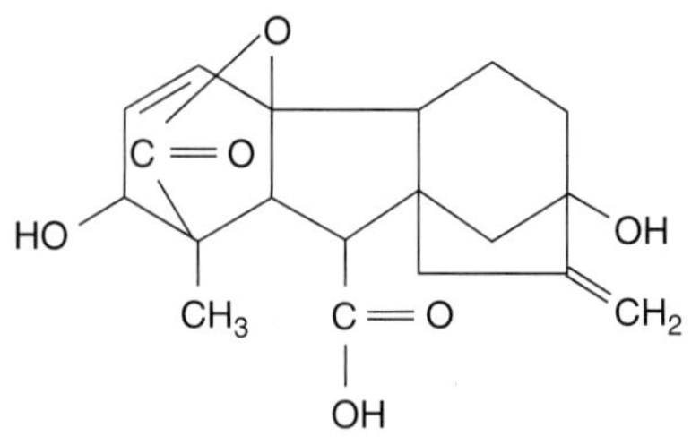

Gibberellins acid

Although no difference has been reported so far in the gibberellin content of healthy and virus- or mollicute-infected plants, spraying of diseased plants with gibberellin overcomes some of the symptoms caused by these pathogens. Thus, stunting of corn plants infected with corn stunt spiroplasma and of tobacco plants infected with severe etch virus was reversed after treatment with gibberellin. Axillary bud suppression, caused by prunus dwarf virus (PDV) on cherry and by leaf curl virus on tobacco, was also overcome by gibberellin sprays. The same treatment also increased fruit production in PDV-infected cherries. In most of these treatments the pathogen itself does not seem to be affected and the symptoms reappear on the plants after gibberellin applications are stopped. It is not known, however, whether the pathogen-caused stunting of plants is actually due to reduced gibberellin concentration in the diseased plant, especially since the growth of even healthy plants is equally increased after gibberellin treatments.

Cytokinins

Cytokinins are potent growth factors necessary for cell growth and differentiation. In addition, they inhibit the breakdown of proteins and nucleic acids, thereby causing the inhibition of senescence, and they have the capacity to direct the flow of amino acids and other nutrients through the plant toward the point of high cytokinin concentration. Cytokinins occur in very small concentrations in green plants, in seeds, and in the sap stream. The first compound with cytokinin activity to be identified was kinetin, which, however, was isolated from herring sperm DNA and does not occur naturally in plants. Several cytokinins, e.g., zeatin and isopentenyl adenosine (IPA), have since been isolated from plants.

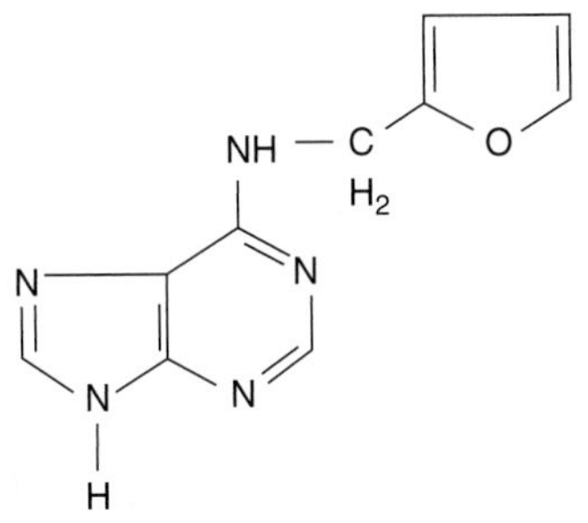

Kinetin

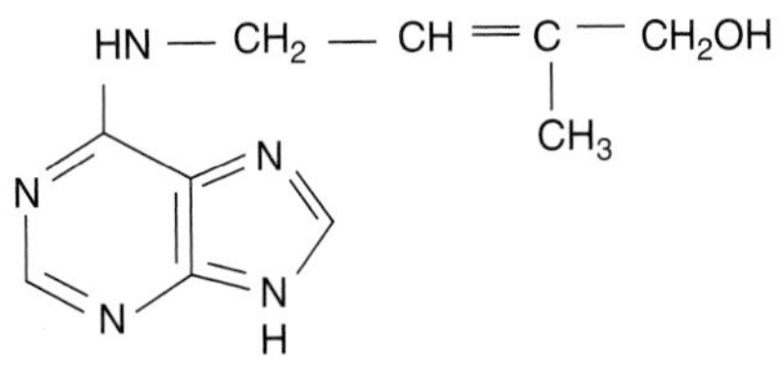

Zeatin

Cytokinins act by preventing genes from being turned off and by activating genes that have been previously turned off. The role of cytokinins in plant disease has just begun to be studied. Cytokinin activity increases in clubroot galls, in crown galls, in smut and rust galls, and in rust-infected bean leaves. In the latter, cytokinin activity seems to be related to both the juvenile feature of the green islands around the infection centers and the senescence outside the green island. However, cytokinin activity is lower in the sap and in tissue extracts of cotton plants infected with verticillium wilt and in plants suffering from drought. A cytokinin is partly responsible for several bacterial galls of plants, such as "leafy" gall disease of sweet pea caused by the bacterium *Rhodococcus (Corynebacterium) fascians*, and for the witches' broom diseases caused by fungi and mollicutes.

Treating plants with kinetin before or shortly after inoculation with a virus seems to reduce the number of infections in local lesion hosts and to reduce virus multiplication in systematically infected hosts.

Ethylene: $CH_2{=}CH_2$

Produced naturally by plants, ethylene exerts a variety of effects on plants, including chlorosis, leaf abscission, epinasty, stimulation of adventitious roots, and fruit ripening. Ethylene also causes increased permeability of cell membranes, which is a common effect of infections. However, ethylene production in infected tissues often parallels the formation of phytoalexins and the increased synthesis or activity of several enzymes or signal compounds that may play a role in increasing plant resistance to infection. Never-the-less it has not been shown that ethylene actually provides resistance. Ethylene is produced by several plant pathogenic fungi and bacteria. In the fruit of banana infected with *Ralstonia solanacearum*, the ethylene content increases proportionately with the (premature) yellowing of the fruit, whereas no ethylene can be detected in healthy fruits. Ethylene has also been implicated in the leaf epinasty symptom of the vascular wilt syndromes and in the premature defoliation observed in several types of plant diseases. In Verticillium wilt of tomato, the presence of ethylene at the time of infection inhibits disease development, whereas the presence of ethylene after infection has been established enhances Verticillium wilt development.

Polysaccharides

Fungi, bacteria, nematodes, and possibly other pathogens constantly release varying amounts of mucilaginous substances that coat their bodies and provide the interface between the outer surface of the microorganism and its environment. Exopolysaccharides appear to be necessary for several pathogens to cause normal disease symptoms either by being directly responsible for inducing symptoms or by indirectly facilitating pathogenesis by promoting colonization or by enhancing survival of the pathogen.

The role of slimy polysaccharides in plant disease appears to be particularly important in wilt diseases caused by pathogens that invade the vascular system of the plant. In vascular wilts, large polysaccharide molecules released by the pathogen in the xylem may be sufficient to cause a mechanical blockage of vascular bundles and thus initiate wilting (Figures 3-3E,F and 3-5D,E). Although such an effect by the polysaccharides alone may occur rarely in nature, when it is considered together with the effect caused by the macromolecular substances released in the vessels through the breakdown of host substances by pathogen enzymes, the possibility of polysaccharide involvement in the blockage of vessels during vascular wilts becomes obvious.

Detoxification of Low Molecular Weight Antimicrobial Molecules

Several kinds of low molecular weight antimicrobial molecules are present in plants or are produced by them in response to infection by pathogens. Some of the most common constitutive such substances are the **saponins**, which include the avenacins and the tomatines. Saponins are glycosylated triterpenoid or steroid alkaloid molecules that provide plants with some degree of protection against fungal pathogens. Saponins are thought to provide antifungal protection by forming complexes with cell membranes, leading to the formation of pores and loss of membrane integrity.

Avenacins are produced in oat roots and leaves and they protect oats from the root-infecting fungus *Gaeumannomyces graminis* while it infects the other cereals that contain no avenacins. A strain of the fungus that infects oats, *G. graminis f. sp. avenae*, produces the avenacin-detoxifying enzyme avenacinase, which is required for pathogenicity on oats. Also, the fungus *Stagonospora avenae* can infect oat leaves, despite the fact that they contain avenacins, by secreting at least three enzymes that degrade and detoxify the avenacins. Another saponin, tomatine, is present in tomatoes, which are protected from infection by some fungi that lack the tomatinase enzyme needed for tomatine detoxification. The fungus *Septoria lycopersici* produces tomatinase and infects tomato plants. Mutants of this fungus, however, that do not produce tomatinase were sensitive to tomatine but could still grow in its presence. They could cause lesions on tomato leaves that actually had more dying mesophyll cells and greater activity of a defense-related enzyme. It is not clear whether this behavior of the host is the result of differences between the mutants and the normal strains or whether the production of tomatinase helps suppress some mechanism(s) of plant defense. In *Botrytis cinerea*, all but 1 of 13 isolates could detoxify tomatine and could severely infect tomato, while one strain that was more sensitive to tomatine was also much less aggressive on tomato.

Promotion of Bacterial Virulence by *avr* Genes

avr genes in bacteria are thought to encode or to direct the production of molecules that are recognized by the host plant and elicit the rapid induction of defense responses on resistant host plants. Their prevalence among pathogens, however, suggests that they may provide some advantage to the pathogen in addition to warning host plants that they are about to be attacked. It has been proposed, therefore, and been demonstrated in many plant–bacteria combinations, that the proteins (Avr proteins) coded for by *avr* genes promote pathogen growth and disease development in susceptible hosts. How Avr proteins accomplish that is not known, but they have been shown to interfere with the resistance mediated by the *avr* gene. Because the Avr proteins are coded for by the *avr* genes, it is apparent that *avr* genes can modify the signaling of host defense pathways in resistant hosts. In some cases, in the absence of a resistance R gene, the particular *avr* gene acts as a virulence factor that not only promotes growth of the particular bacterium in several hosts, including some that exhibit varying degrees of resistance, but transgenic plants that express the *avr* gene actually exhibit enhanced susceptibility to the pathogen and/or aggressiveness of the pathogen. Different *avr* genes, however, even of the same bacterial pathogen, contribute different degrees of susceptibility/aggressiveness to bacteria that provide these genes. This shows that the particular Avr proteins function inside the host plant cell and promote bacterial virulence.

Role of Type III Secretion in Bacterial Pathogenesis

Although the primary determinants of pathogenicity and virulence in many bacteria are secreted enzymes such as pectin lyases, cellulases, and proteases that macerate plant tissues of many species, it is now known that in at least *Erwinia* bacteria, the genes for hypersensitive reaction and pathogenicity (*hrp* genes) determine the potential secondary pathogenesis. In plant pathogens, *hrp* genes code for a type III secretion machinery, which is thought to transport bacterial effector proteins directly into the host cell. *hrp* genes exist in clusters of about 20 genes, one of which codes for a constituent of an outer membrane, whereas many others code for the core secretion machinery, for regulatory genes, for harpins, for the Hrp-pilin, which in some bacteria is required for type III secretion to function, for avirulence (*avr*) genes, and so on. In nonmacerating bacteria *Pseudomonas*, *Ralstonia*, and *Xanthomonas* and in the fire blight bacterium *Erwinia amylovora*, *hrp* genes are essential for virulence and elicitation of a hypersensitive response.

Suppressors of Plant Defense Responses

It has been shown that at least some plant pathogenic fungi, e.g., *Puccinia graminis* f. sp. *tritici*, which causes stem rust of wheat, and *Mycosphaerella pinodes*, which causes a leaf spot on pea, produce substances called **suppressors** that act as pathogenicity factors by suppressing the expression of defense responses in the host plant. The defense suppressor of the wheat stem rust fungus has been found in the fungus germination fluid and in the intercellular fluid of rust-infected wheat leaves. This suppressor interacts with the wheat cell plasma membrane and reduces binding of the pathogen's 67-kDa glycoprotein elicitor of host defenses to the plasma membrane. In this way, the suppressor molecule suppresses the activity of phenylalanine lyase (PAL) and the normal development of defense responses. The pea-infecting fungus produces two suppressors in the spore germination fluid. Both suppressors are glycopeptides, counteract the elicitor of phytoalexin biosynthesis, and temporarily suppress the expression of all defense reactions of the host plant. The *Mycosphaerella* suppressors seem to reduce the proton-pumping activity of the host cell membrane ATPase and thereby temporarily lower the ability of the cell to function and to defend itself. A different mechanism of suppression of plant defense responses has been reported in the ergot disease of rye caused by the fungus *Claviceps purpurea*. In that disease the fungus produces the enzyme catalase, which reacts with and neutralizes the hydrogen peroxide that is produced as one of the first defense responses of plants against infecting pathogens. The fungal catalase concentration is greatest at hyphal walls and hyphal surfaces and is secreted by the fungus into the host apoplast at the host–pathogen interface, where the host H_2O_2 is produced. By inactivating active oxygen species produced by the host through catalase, the fungus suppresses the host defenses.

Pathogenicity and Virulence Factors in Viruses and Viroids

Until recently, little was known about the intrinsic factors of viruses and viroids that determine their pathogenicity and/or virulence. Viruses have a few, usually less than 10, genes, yet they are very capable pathogens. This requires that viral genes and gene products have multitask functions. Some of the most basic functions viral genes control are infectivity on a particular host, replication of the virus, movement of the virus from cell to cell, long-distance transport of the virus in the plant, transmissibility of the virus from plant to plant, and production of the coat protein of the virus. All of these functions are necessary for the pathogenicity and survival of the virus, although the variation in the degree most of these functions are carried out affects the virulence of the virus, i.e., the level of disease and symptoms it can cause in a host plant, rather than its pathogenicity, i.e., its ability to infect a plant.

Plant viruses have no genes that allow them to produce macerating enzymes, toxins, growth regulators, or other biologically active compounds by which to affect plant cells. However, different viruses manage to induce the plant to develop symptoms that appear to be the result of action and interaction of numerous such compounds present in the cell, despite the fact that no such compound can be found in infected cells. How viruses cause disease remains, therefore, pretty much a mystery but some facts are beginning to emerge.

One of the most important proteins coded by viruses that plays an important role in their pathogenicity and virulence is their coat protein. In addition to protecting the viral nucleic acid from external damaging factors, the coat protein plays important roles in practically everything pertaining to viral replication and dissemination. Thus, the coat protein plays a role in host recognition, uncoating and release of the nucleic acid, assistance in replication of the nucleic acid, movement of the virus between cells and organs, movement of the virus via a vector between plants, and modification of symptoms. Again, little is known on the mechanisms by which the coat protein affects these functions.

Another viral protein that has been studied extensively is the so-called movement protein, which enables viruses to move between cells and/or through the phloem system of the plant by altering the properties of plasmodesmata. However, some movement proteins not only open movement channels for the virus, they also block a defense molecule, the suppressor of virus silencing by the plant cell activated by the viral infection. Some viroids seem to form complexes with certain host proteins that help the viroids pass through plasmodesmata and with plant lectins that help viroids move through the phloem of host plants.

Selected References

Alfano, J. R., and Guo, M. (2003). The *Pseudomonas syringae* Hrp (type III) protein secretion system: Advances in the new millenium. *In* "Plant-Microbe Interactions" (G. Stacey and N. T. Keen, eds.), Vol. **6**, pp. 227–258.

Bai, J., Choi, S.-H., Ponciano, G., *et al.* (2000). *Xanthomonas oryzae* pv. *oryzae* avirulence genes contribute differently and specifically to pathogen aggressiveness. *Mol. Plant-Microbe Interact.* **13**, 1322–1329.

Barras, F., van Gijseman, F., and Chatterjee, A. K. (1994). Extracellular enzymes and pathogenesis of soft-rot erwinias. *Annu. Rev. Phytopathol.* **32**, 201–234.

Belding, R. D., *et al.* (2000). Relationship between apple fruit epicuticular wax and growth of *Peltaster fructicola and Leptodontidium elatius*, two fungi that cause sooty blotch disease. *Plant Dis.* **84**, 767–772.

Blanchette, R. A. (1991). Delignification by wood-decay fungi. *Annu. Rev. Phytopathol.* **29**, 381–398.

Bostock, R. M., Wilcox, S. M., Wang, G., *et al.* (1999). Suppression of *Monilinia fructicola* cutinase production by peach fruit surface phenolic acids. *Physiol. Mol. Plant Pathol.* **54**, 37–50.

Brown, H. P., Panshing, A. J., and Forsaith, C. C. (1949). "Textbook of Wood Technology," Vol. 1. McGraw-Hill, New York.

Burr, T. J., and Otten, L. (1999). Crown gall of grape: Biology and disease management. *Annu. Rev. Phytopathol.* **37**, 53–80.

Callaway, A., Giesman-Cookmeyer, D., Gillock, E. T., *et al.* (2001). The multifunctional capsid proteins of plant RNA viruses. *Annu. Rev. Phytopathol.* **39**, 419–460.

Cao, H., Baldini, R. L., and Rahme, L. G. (2001). Common mechanisms for pathogens of plants and animals. *Annu. Rev. Phytopathol.* **39**, 259–284.

Carzaniga, R., et al. (2002). Localization of melanin in conidia of *Alternaria alternata* using phage display antibodies. Mol. Plant-Microbe Interact. **15**, 216–224.

Chen, Z., Kloek, A. P., Boch, J., *et al.* (2000). The *Pseudomonas syringae avrRpt2* gene product promotes pathogen virulence from inside plant cells. *Mol. Plant-Microbe Interact.* **13**, 1312–1321.

Comai, L., Surico, G., and Kosuge, T. (1982). Relations of plasmid DNA to indoleacetic acid production in different strains of *Pseudomonas syringae* pv. *savastanoi*. *J. Gen. Microbiol.* **128**, 2157–2163.

Culver, J. N. (2002). Tobacco mosaic virus assembly and disassembly: Determinants in pathogenicity and resistance. *Annu. Rev. Phytopathol.* **40**, 287–308.

Cutler, D. F., Alvin, K. L., and Price, C. E., eds. (1982). "The Plant Cuticle." Linn. Soc. Symp. Ser. No. 10. Academic Press, London.

Daly, J. M., and Deverall, B. J., eds. (1983). "Toxins in Plant Pathogenesis." Academic Press, New York.

Dardick, C. D., Golem, S., and Culver, J. N. (2000). Susceptibility and symptom development in *Arabidopsis thaliana* to tobacco mosaic virus is influenced by virus cell-to-cell movement. *Mol. Plant-Microbe Interact.* **13**, 1139–1144.

Daub, M. E., and Ehrenshaft, M. (2000). The photoactivated *Cercospora* toxin *Certosporin*: Contributions to plant disease and fundamental biology. *Annu. Rev. Phytopathol.* **38**, 461–490.

Davis, E. L., Hussey, R. S., Baum, T. J., *et al.* (2000). Nematode parasitism genes. *Annu. Rev. Phytopathol.* **38**, 365–396.

Denny, T. P. (1995). Involvement of bacterial polysaccharides in plant pathogenesis. *Annu. Rev. Phytopathol.* **33**, 173–197.

Dow, M., Newman, M.-A., and von Roepenack, E. (2000). The induction and modulation of plant defense responses by bacterial lipopolysaccharides. *Annu. Rev. Phytopathol.* **38**, 241–261.

Doyle, E. A., and Lambert, K.N. (2002). Cloning and characterization of an esophageal-gland-specific pectate lyase from the root-knot nematode *Meloidogyne javanica*. *Mol. Plant-Microbe Interact.* **15**, 549–556.

Durbin, R. D. (1991). Bacterial phytotoxins: Mechanism of action. *Experientia* **47**, 776–783.

Fridborg, I., *et al.* (2003). TIP, a novel host factor linking callose degradation with the cell-to-cell movement of *potato virus* X. *Mol. Plant-Paras. Interact.* **16**, 132–140.

Fritig, B., and LeGrand, M. (1993). "Mechanisms of Plant Defense Responses." Kluwer, Dordrecht, The Netherlands.

Fry, B. A., and Loria, R. (2002). Thaxtomin A: Evidence for a plant cell wall target. *Physiol. Mol. Plant Pathol.* **60**, 1–8.

Gevens, A., and Nicholson, R. L. (2000). Cutin composition: A subtle role for fungal cutinase? *Physiol. Mol. Plant Pathol.* **57**, 43–45.

Gheysen, G., and Fenoll, C. (2002). Gene expression in nematode feeding sites. *Annu. Rev. Plant Pathol.* **40**, 191–219.

Glenn, A. E., Gold, S. E., and Bacon, C. W. (2002). *Fdb1* and *Fdb2*, *Fusarium verticillioides* loci necessary for detoxification of preformed antimicrobials from corn. *Mol. Plant-Microbe Interact.* **15**, 91–101.

Goethals, K., Vereecke, D., Jaziri, M., *et al.* (2001). Leafy gall formation. *Annu. Rev. Plant Pathol.* **39**, 27–52.

Gold, S. E. (2003). *Ustilago* pathogenicity. *In* "Plant-Microbe Interactions" (G. Stacey and N. T. Keen, eds.), Vol. **6**, pp. 147–172.

Gold, S. E., Garcia-Pedrajas, M. D., and Martinez-Espinoza, A. D. (2001). New (and used) approaches to the study of fungal pathogenicity. *Annu. Rev. Phytopathol.* **39**, 337–368.

Goodman, R. N., Kiraly, Z., and Zaitlin, M. (1967)." The Biochemistry and Physiology of Infectious Plant Disease." Van Nostrand-Reinhold, Princeton, NJ.

Graniti, A., *et al.*, eds. (1989). "Phytotoxins and Plant Pathogenesis." Springer-Verlag, Berlin.

Hammerschmidt, R. (1999). Phytoalexins: What we have learned after 60 years? *Annu. Rev. Phytopathol.* **37**, 285–306.

Harper, G., Hull, R., Lockhart, B., *et al.* (2002). Viral sequences integrated into plant genomes. *Annu. Rev. Phytopathol.* **40**, 119–136.

He, S. Y. (1998). Type III protein secretion systems in plant and animal pathogenic bacteria. *Annu. Rev. Phytopathol.* **36**, 363–392.

Henson, J. M., Butler, M. J., and Day, A. W. (1999). The dark side of the mycelium: Melanins of phytopathogenic fungi. *Annu. Rev. Phytopathol.* **37**, 447–471.

Hiriart, J.-P., Aro, E.-M., and Lehto, K. (2003). Dynamics of the VIGS-mediated chimeric silencing of the *Nicotiana benthamiana* ChIH gene and of the *tobacco mosaic virus* vector. *Mol. Plant-Microb. Interact.* **16**, 99–106.

Hooykaas, P. J. J., and Beijersbergen, A. G. M. (1994). The virulence system of *Agrobacterium tumefaciens*. *Annu. Rev. Phytopathol.* **32**, 157–179.

Horsfall, J. G., and Cowling, E. B., eds. (1979). "Plant Disease," Vol. 4. Academic Press, New York.

Hussey, R. S. (1989). Disease-inducing secretions of plant-parasitic nematodes. *Annu. Rev. Phytopathol.* **27**, 123–141.

Jahr, H., Dreier, J., Meletzus, D., *et al.* (2000). The endo-β-1,4-glucanase CelA of *Clavibacter michiganensis* subsp. *michiganensis* is a pathogenicity determinant required for induction of bacterial wilt of tomato. *Mol. Plant-Microbe Interact.* **13**, 703–714.

Keen, N. T. (2000). A century of plant pathology: A retrospective view on understanding host–parasite interactions. *Annu. Rev. Phytopathol.* **38**, 31–48.

Kolattukudy, P. E. (1985). Enzymatic penetration of the plant cuticle by fungal pathogens. *Annu. Rev. Phytopathol.* **23**, 223–250.

Kosuge, T., and Nester, E. W., eds. (1984). "Plant-Microbe Interactions: Molecular and Genetic Perspective," Vol. 1. Macmillan, New York.

Kuriger, W. E., and Agrios, G. N. (1977). Cytokinin levels and kinetin-virus interactions in tobacco ringspot virus-infected cowpea plants. *Phytopathology* **67**, 604–609.

Manulis, S., and Barash, I. (2003). Molecular basis for transformation of an epiphyte into a gall-forming pathogen, as exemplified by *Erwinia herbicola* pv. *gypsophilae*. *In* "Plant Microbe Interactions" (G. Stacey and N. T. Keen, eds.), Vol. **6**, pp. 19–52.

Marcell, L. M., and Beattie, G. A. (2002). Effect of leaf surface waxes on leaf colonization by *Pantoea agglomerans* and *Clavibacter michiganensis*. *Mol. Plant-Microbe Interact.* **15**, 1236–1244

Markham, J. E., and Hille, J. (2001). Host-selective toxins as agents of cell death in plant-fungus interactions. *Mol. Plant Pathol.* **2**, 229–239.

Martin-Hernandez, A. M., Dufresne, M., Hugouvieux,V., *et al.* (2000). Effects of targeted replacement of the tomatinase gene on the interaction of *Septoria lycopersici* with tomato plants. *Mol. Plant-Microbe Interact.* **13**, 1301–1311.

Morrissey, J. P., Wubeen, J. P., and Osbourn, A. E. (2000). *Stagonospora avenae* secretes multiple enzymes that hydrolyze oat leaf saponins. *Mol. Plant-Microbe Interact.* **13**, 1041–1052.

Navarre, D. A., and Wolpert, T. J. (1999). Effects of light and CO_2 on victorin-induced symptom development in oats. *Physiol. Mol. Plant Pathol.* **55**, 237–242.

Otani, H., Kohnobe, A., Kodama, M., *et al.* (1998). Production of a host-specific toxin by germinating spores of *Alternaria brassicicola*. *Physiol. Mol. Plant Pathol.* **52**, 285–295.

Owens, R. A., Backburn, M., and Ding, B. (2001). Possible involvement of the phloem lectin in long-distance viroid movement. *Mol. Plant-Microbe Interact.* **14**, 905–909.

Petrini, O., and Ouellette, G. B. (1994). "Host Wall Alterations by Parasitic Fungi." APS Press, St. Paul, MN.

Prusky, D., McEvoy, J. L., Leverentz, B., *et al.* (2001). Local modulation of host pH by *Colletotrichum* species as a mechanism to increase virulence. *Mol. Plant-Microbe Interact.* **14**, 1105–1113.

Pryce-Jones, E., Carver, T., and Gurr, S. (1999). The role of cellulase enzymes and mechanical force in host penetration by *Erysiphe graminis* f. sp. *hordei*. *Physiol. Mol. Plant Pathol.* **55**, 175–182.

Quidde, T., Osbourn, A. E., and Tudzynski, P. (1998). Detoxification of α-tomatine by *Botrytis cinerea*. *Physiol. Mol. Plant Pathol.* **52**, 151–165.

Rantakari, W., Virtaharju, O., Vähämiko, S., *et al.* (2001). Type III secretion contributes to the pathogenesis of the soft-rot pathogen *Erwinia carotovora*: Partial characterization of the *hrp* gene cluster. *Mol. Plant-Microbe Interact.* **14**, 962–568.

Ream, W. (1989). *Agrobacterium tumefaciens* and interkingdom genetic exchange. *Annu. Rev. Phytopathol.* **27**, 583–618.

Robison, M. M., Griffith, M., Paulls, K. P., *et al.* (2001). Dual role for ethylene in susceptibility of tomato to Verticillium wilt. *J. Phytopathol.* **149**, 385–388.

Schäfer, W. (1994). Molecular mechanisms of fungal pathogenicity to plants. *Annu. Rev. Phytopathol.* **32**, 461–477.

Schell, M. A. (2000). Control of virulence and pathogenicity genes of *Ralstonia solanacearum* by an elaborate sensory network. *Annu. Rev. Phytopathol.* **38**, 263–292.

Schouten, A., Tenberge, K. B., Vermeer, J., *et al.* (2002). Functional analysis of an extracellular catalase of *Botrytis cinerea*. *Mol. Plant Pathol.* **3**, 227–238.

Siedow, J. N., *et al.* (1995). The relationship between the mitochondrial gene T-urf13 and fungal pathotoxin sensitivity in maize. *Biochim. Biophys. Acta* **1271**, 235–240.

Sijmons, P. C., Atkinson, A. J., and Wyss, U. (1994). Parasitic strategies of root nematodes and associated host cell responses. *Annu. Rev. Phytopathol.* **32**, 235–259.

Singh, U. S., Singh, P. R., and Kohmoto, K. (1994). "Pathogenesis and Host Specificity in Plant Diseases: Histopathological, Biochemical, Genetic and Molecular Bases," Vols. 1–3. Pergamon/Elsevier, Tarrytown, NY.

Stall, R. E., and Hall, C. B. (1984). Chlorosis and ethylene production in pepper leaves infected by *Xanthomonas campestris* pv. *vesicatoria*. *Phytopathology* **74**, 373–375.

Sugui, J. A., et al. (1997). Association of *Pestalotia malicola* with the plant cuticle: Visualization of the pathogen and detection of cutinase and non-specific esterase. *Physiol. Mol. Plant Pathol.* **52**, 213–221.

Taylor, J. L. (2003). Transporters involved in commuication, attack or defense in plant-microbe interactions. *In* "Plant-Microbe Interactions" (G. Stacey and N. T. Keen, eds.), Vol. 6, pp. 97–146.

Tenllado, F., Barajas, D., Vargas, M., *et al.* (2003). Transient expression of homologous hairpin RNA causes interference with plant virus infection and is overcome by a virus encoded suppressor of gene silencing. *Mol. Plant-Microbe Interact.* **16**, 149–158.

Valette-Collet, O., Cimerman, A., Reignault, P., *et al.* (2003). Disruption of *Botrytis cinerea* pactin methylesterase gene *Bcpme1* reduces virulence on several host plants. *Mol. Plant-Microbe Interact.* **16**, 360–367.

Van Wezel, R., Liu, H., Tien, P., *et al.* (2001). Gene C2 of the monopartite geminivirus *tomato yellow leaf curl virus*-China encodes a pathogenicity determinant that is localized in the nucleus. *Mol. Plant-Microbe Interact.* **14**, 1125–1128.

Vereecke, D., Temmerman, W., Jaziri, M., *et al.* (2003). Toward an understanding of the *Rhodococcus fascians*-plant interaction. *In* "Plant-Microbe Interactions" (G. Stacey and N. T. Keen, eds.), Vol. 6, pp. 53–80.

Wolpert, T. J., Dunkle, L. D., and Ciuffetti, L. M. (2002). Host-selective toxins and avirulence determinants: What's in a name? *Annu. Rev. Phytopathol.* **40**, 251–286.

Wu, Y.-Q., and Hohn, B. (2003). Cellular transfer and chromosomal integration of T-DNA during *Agrobacterium tumefaciens*-mediated plant transformation. *In* "Plant-Microbe Interactions" (G. Stacey and N. T. Keen, eds.), Vol. **6**, pp. 1–18.

Yamada, T. (1993). The role of auxins in plant disease development. *Annu. Rev. Phytopathol.* **31**, 253–273.

Zhang, J. X., and Bruton, B. D. (1999). Relationship of developmental stage of cantaloupe fruit to black rot susceptibility and enzyme production by *Didymella bryoniae*. *Plant Dis.* **83**, 1025–1032.

Chapter six

HOW PLANTS DEFEND THEMSELVES AGAINST PATHOGENS

WHATEVER THE PLANT DEFENSE OR RESISTANCE, IT IS CONTROLLED BY ITS GENES: NON-HOST RESISTANCE – PARTIAL OR HORIZONTAL RESISTANCE – R GENE OR VERTICAL RESISTANCE
208

PREEXISTING STRUCTURAL AND CHEMICAL DEFENSES: PREEXISTING DEFENSE STRUCTURES – PREEXISTING CHEMICAL DEFENSES – INHIBITORS RELEASED BY THE PLANT IN ITS ENVIRONMENT – INHIBITORS PRESENT IN PLANT CELLS BEFORE INFECTION
210

DEFENSE THROUGH LACK OF ESSENTIAL FACTORS: LACK OF RECOGNITION BETWEEN HOST AND PATHOGEN – LACK OF HOST RECEPTORS AND SITES FOR TOXINS – LACK OF ESSENTIAL SUBSTANCES FOR THE PATHOGEN
212

INDUCED STRUCTURAL AND BIOCHEMICAL DEFENSES: RECOGNITION OF THE PATHOGEN BY THE HOST PLANT – TRANSMISSION OF THE ALARM SIGNAL TO HOST DEFENSE PROVIDER – SIGNAL TRANSDUCTION
213

INDUCED STRUCTURAL DEFENSES: CYTOPLASMIC – CELL WALL DEFENSE STRUCTURES – HISTOLOGICAL DEFENSE STRUCTURES: CORK LAYERS – ABSCISSION LAYER – TYLOSES – GUMS NECROTIC STRUCTURAL DEFENSE REACTION: THE HYPERSENSITIVE RESPONSE
214

INDUCED BIOCHEMICAL DEFENSES: INDUCED BIOCHEMICAL NON-HOST RESISTANCE – INDUCED BIOCHEMICAL DEFENSES IN PARTIAL OR HORIZONTAL RESISTANCE: *FUNCTION OF GENE PRODUCTS – MECHANISMS OF QUANTITATIVE RESISTANCE – EFFECT OF TEMPERATURE*
217

INDUCED BIOCHEMICAL DEFENSES IN THE HYPERSENSITIVE RESPONSE RESISTANCE: THE HYPERSENSITIVE RESPONSE – GENES INDUCED DURING EARLY INFECTION – FUNCTIONAL ANALYSIS OF DEFENSE GENES – CLASSES OF R GENE PROTEINS – RECOGNITION OF PATHOGEN AVR PROTEINS BY THE HOST – HOW DO R AND *AVR* GENE PRODUCTS ACTIVATE PLANT RESPONSES? – SOME EXAMPLES OF PLANT DEFENSE THROUGH R GENES AND THEIR MATCHING *AVR* GENES: *THE RICE PI-TA GENE – THE TOMATO CF GENES – THE TOMATO BS2 GENE – THE ARABIDOPSIS RPM1 GENE – THE CO-FUNCTION OF TWO OR MORE GENES* DEFENSE INVOLVING BACTERIAL TYPE III EFFECTOR PROTEINS – ACTIVE OXYGEN SPECIES, LIPOXYGENASES, AND DISRUPTION OF CELL MEMBRANES – REINFORCEMENT OF HOST CELL WALLS WITH STRENGTHENING MOLECULES – PRODUCTION OF ANTIMICROBIAL SUBSTANCES IN ATTACKED HOST CELLS – PATHOGENESIS-RELATED PROTEINS – DEFENSE THROUGH PRODUCTION OF SECONDARY METABOLITES – PHENOLICS SIMPLE PHENOLIC COMPOUNDS – TOXIC PHENOLICS FROM NONTOXIC PHENOLIC GLYCOSIDES – ROLE OF PHENOL-OXIDIZING ENZYMES IN DISEASE RESISTANCE – PHYTOALEXINS
221

DET\OXIFICATION OF PATHOGEN TOXINS BY PLANTS
236

IMMUNIZATION OF PLANTS AGAINST PATHOGENS: DEFENSE THROUGH PLANTIBODIES – RESISTANCE THROUGH PRIOR EXPOSURE TO MUTANTS OF REDUCED PATHOGENICITY
236

SYSTEMIC ACQUIRED RESISTANCE: INDUCTION OF PLANT DEFENSES BY ARTIFICIAL INOCULATION WITH MICROBES OR BY TREATMENT WITH CHEMICALS
237

DEFENSE THROUGH GENETICALLY ENGINEERING DISEASE-RESISTANT PLANTS: WITH PLANT-DERIVED GENES – WITH PATHOGEN-DERIVED GENES
242

DEFENSE THROUGH RNA SILENCING BY PATHOGEN-DERIVED GENES
244

Each plant species is affected by approximately 100 different kinds of fungi, bacteria, mollicutes, viruses, and nematodes. Frequently, a single plant is attacked by hundreds, thousands, and, in leafspot diseases of large trees, probably hundreds of thousands of individuals of a single kind of pathogen. Although such plants may suffer damage to a lesser or greater extent, many survive all these attacks and, not uncommonly, manage to grow well and to produce appreciable yields.

In general, plants defend themselves against pathogens by a combination of weapons from two arsenals: (1) structural characteristics that act as physical barriers and inhibit the pathogen from gaining entrance and spreading through the plant and (2) biochemical reactions that take place in the cells and tissues of the plant and produce substances that are either toxic to the pathogen or create conditions that inhibit growth of the pathogen in the plant. The combinations of structural characteristics and biochemical reactions employed in the defense of plants are different in different host–pathogen systems. In addition, even within the same host and pathogen, the combinations vary with the age of the plant, the kind of plant organ and tissue attacked, the nutritional condition of the plant, and the weather conditions.

WHATEVER THE PLANT DEFENSE OR RESISTANCE, IT IS CONTROLLED BY ITS GENES

One concept that must be made clear at the outset is that whatever the kind of defense or resistance a host plant employs against a pathogen or against an abiotic agent, it is ultimately controlled, directly or indirectly, by the genetic material (genes) of the host plant and of the pathogen (Fig. 6-1).

Nonhost Resistance

A plant may find it easy to defend itself, i.e., to stay resistant (immune) when it is brought in contact with a pathogenic biotic agent to which the plant is not a host. This is known as nonhost resistance and is the most common form of resistance (or defense from attack) in nature. For example, apple trees are not affected by pathogens of tomato, of wheat, or of citrus trees because the genetic makeup of apple is in some way(s) different from that of any other kinds of host plants, which, of course, are attacked by their own pathogens. However, apple can be attacked by its own pathogens, which, in turn, do not attack tomato, wheat, citrus, or anything else. Similarly, the fungus that causes powdery mildew on wheat (*Blumeria graminis* f. sp. *tritici*) does not infect barley and vice versa, the fungus that causes powdery mildew on barley (*B. graminis* f. sp. *hordei*) does not infect wheat, and so on. All such unsuccessful plant/pathogen interactions are thought to represent nonhost resistance. It has been shown recently however, that in at least some related pairings, e.g., the wheat, powdery mildew fungus inoculated on barley, the fungus produces haustoria and the host reacts by producing hydrogen peroxide (H_2O_2), cell wall appositions under the appressoria, and a hypersensitive response in which epidermal cells die rapidly in response to fungal attack.

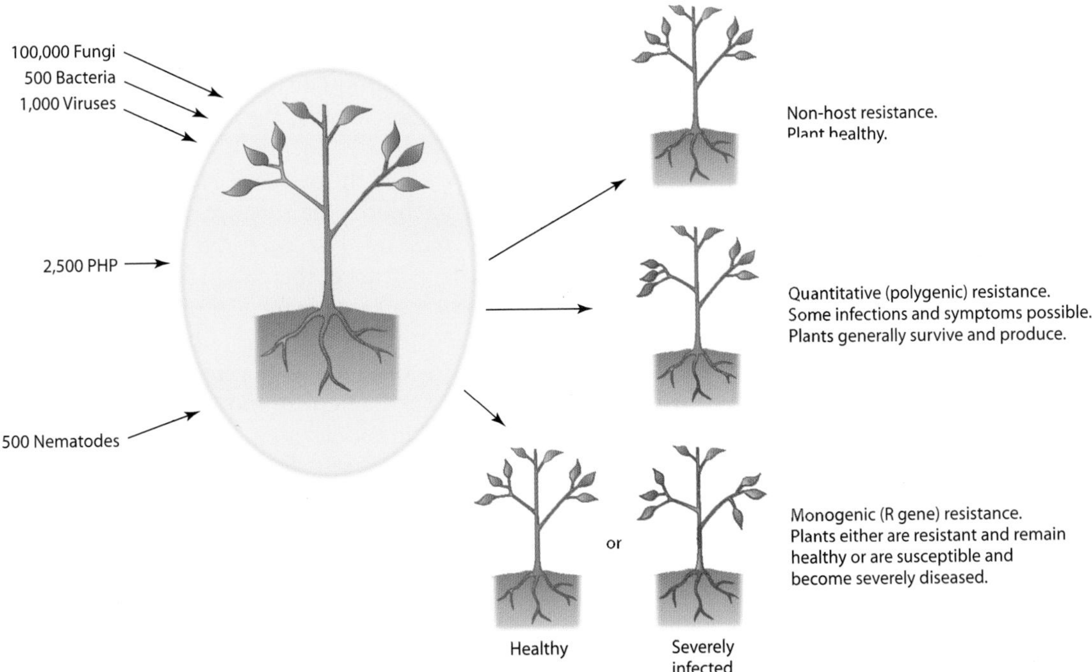

FIGURE 6-1 Types of reaction of plants to attacks by various pathogens in relation to the kind of resistance of the plant.

Partial, Polygenic, Quantitative, or Horizontal Resistance

Each plant, of course, is attacked by its own pathogens, but there is often a big difference in how effectively the plant can defend itself (how resistant the plant is) against each pathogen. Even when conditions for infection and disease development are favorable, a plant, upon infection with a particular pathogen, may develop no disease, only mild disease, or severe disease, depending on the specific genetic makeup of the plant and of the pathogen that attacks it. Many genes are involved in keeping a plant protected from attack by pathogens. Many of these genes provide for the general upkeep and well-being functions of plants, but plants also have many genes whose main functions seem to be the protection of plants from pathogens. Some of the latter plant genes code for chemical substances that are toxic to pathogens or neutralize the toxins of the pathogens, and these substances may be present in plants regardless of whether the plant is under attack or not. Plants also have genes that produce and regulate the formation of structures that can slow down or stop the advance of a pathogen into the host and cause disease. These structures can also be present in a plant throughout its life or they may be produced in response to attack by one of several pathogens or following injury by an abiotic agent. Preexisting defense structures or toxic chemical substances, and many of those formed in response to attack by a pathogen or abiotic agent, are important in the defense of most plants against most pathogens.

When a pathogen attacks a host plant, the genes of the pathogen are activated, produce, and release all their weapons of attack (enzymes, toxins, etc.) against the plants that they try to infect. With the help of different combinations of preexisting or induced toxic chemical substances or defense structures, most plants manage to defend themselves partially or nearly completely. Such plants show sufficient resistance that allows them to survive the pathogen attacks and to produce a satisfactory yield. This type of defense or resistance is known as polygenic, general, or quantitative resistance because it depends on many genes for the presence or formation of the various defense structures and for preexisting or induced production of many substances toxic to the pathogen. This type of resistance is present at different levels against different pathogens in absolutely all plants and is also known as partial, quantitative, horizontal, multigenic, field, durable, or minor gene resistance.

Most plants depend on general resistance against their pathogens, especially nonobligate parasites, e.g., the semibiotrophic or nectrotrophic oomycetes *Pythium* and *Phytophthora*, the fungi *Botrytis*, *Fusarium*, *Sclerotinia*, and *Rhizoctonia*, and most bacteria, nematodes, and so on. In at least some polygenic plant–pathogen combinations, such as the early blight of tomato caused by the necrotrophic fungus *Alternaria solani*, the more resistant the varieties are, the higher the constitutive concentration and the more rapid the accumulation in them of pathogen-induced pathogenesis related (PR) proteins, than in susceptible varieties. These PR proteins include some of the specific antifungal isozymes of chitinase and β-1,3-glucanase. Also, total enzyme preparations from resistant varieties were able to release elicitors of the hypersensitive response (HR) (see later) from purified fungal cell walls, whereas enzymes from susceptible varieties could not. Furthermore, partially purified chitinases from tomato leaves could release HR elicitors from germinating *A. solani* spores but not from mature intact cell walls. This suggests that, perhaps, constitutively produced hydrolytic enzymes may act as a mechanism of elicitor release in tomato resistance to the early blight disease. Quantitative resistance has also been shown to increase in transgenic plants carrying introduced R genes and matching avirulence genes, even though the latter do not express the hypersensitive cell death.

Race-Specific, Monogenic, R Gene, or Vertical Resistance

In many plant–pathogen combinations, especially those involving biotrophic oomycetes (downy mildews), fungi (powdery mildews, rusts), and many other fungi, e.g., *Cochliobolus*, *Magnaporthe*, *Cladosporium*, many bacteria, nematodes, and viruses, defense (resistance) of a host plant against many of its pathogens is through the presence of matching pairs of juxtaposed genes for disease in the host plant and the pathogen. The host plant carries one or few resistance genes (R) per pathogen capable of attacking it, while each pathogen carries matching genes for avirulence (A) for each of the R genes of the host plant. As explained in some detail later, the avirulence gene of the pathogen serves to trigger the host R gene into action. This then sets in motion a series of defense reactions that neutralize and eliminate the specific pathogen that carries the corresponding (matching) gene for avirulence (A), while the attacked and a few surrounding cells die. This type of defense or resistance is known as race-specific, hypersensitive response (HR), major gene, R gene, or vertical resistance. However, some R genes, e.g., Xa21 of rice, do not induce a visible HR.

PREEXISTING STRUCTURAL AND CHEMICAL DEFENSES

Preexisting Defense Structures

The first line of defense of a plant against pathogens is its surface, which the pathogen must adhere to and penetrate if it is to cause infection. Some structural defenses are present in the plant even before the pathogen comes in contact with the plant. Such structures include the amount and quality of wax and cuticle that cover the epidermal cells, the structure of the epidermal cell walls, the size, location, and shapes of stomata and lenticels, and the presence of tissues made of thick-walled cells that hinder the advance of the pathogen on the plant.

Waxes on leaf and fruit surfaces form a water-repellent surface, thereby preventing the formation of a film of water on which pathogens might be deposited and germinate (fungi) or multiply (bacteria). A thick mat of hairs on a plant surface may also exert a similar water-repelling effect and may reduce infection.

A thick cuticle may increase resistance to infection in diseases in which the pathogen enters its host only through direct penetration. Cuticle thickness, however, is not always correlated with resistance, and many plant varieties with cuticles of considerable thickness are invaded easily by directly penetrating pathogens.

The thickness and toughness of the outer wall of epidermal cells are apparently important factors in the resistance of some plants to certain pathogens. Thick, tough walls of epidermal cells make direct penetration by fungal pathogens difficult or impossible. Plants with such walls are often resistant, although if the pathogen is introduced beyond the epidermis of the same plants by means of a wound, the inner tissues of the plant are invaded easily by the pathogen.

Many pathogenic fungi and bacteria enter plants only through stomata. Although the majority of pathogens can force their way through closed stomata, some, like the stem rust of wheat, can enter only when stomata are open. Thus, some wheat varieties, in which the stomata open late in the day, are resistant because the germ tubes of spores germinating in the night dew desiccate due to evaporation of the dew before the stomata begin to open. The structure of the stomata, e.g., a very narrow entrance and broad, elevated guard cells, may also confer resistance to some varieties against certain of their bacterial pathogens.

The cell walls of the tissues being invaded vary in thickness and toughness and may sometimes inhibit the advance of the pathogen. The presence, in particular, of bundles or extended areas of sclerenchyma cells, such as are found in the stems of many cereal crops, may stop the further spread of pathogens such as stem rust fungi. Also, the xylem, bundle sheath, and sclerenchyma cells of the leaf veins effectively block the spread of some fungal, bacterial, and nematode pathogens that cause various "angular" leaf spots because of their spread only into areas between, but not across, veins. Xylem vessels seem to be involved more directly in the resistance and susceptibility to vascular diseases. For example, xylem vessel diameter and the proportion of large vessels were strongly correlated with the susceptibility of elm to Dutch elm disease caused by the fungus *Ophiostoma novo-ulni*.

Preexisting Chemical Defenses

Although structural characteristics may provide a plant with various degrees of defense against attacking pathogens, it is clear that the resistance of a plant against pathogen attacks depends not so much on its structural barriers as on the substances produced in its cells before or after infection. This is apparent from the fact that a particular pathogen will not infect certain plant varieties even though no structural barriers of any kind seem to be present or to form in these varieties. Similarly, in resistant varieties, the rate of disease development soon slows down, and finally, in the absence of structural defenses, the disease is completely checked. Moreover, many pathogens that enter nonhost plants naturally or that are introduced into nonhost plants artificially, fail to cause infection, although no apparent visible host structures inhibit them from doing so. These examples suggest that defense mechanisms of a chemical rather than a structural nature are responsible for the resistance to infection exhibited by plants against certain pathogens.

Inhibitors Released by the Plant in Its Environment

Plants exude a variety of substances through the surface of their aboveground parts as well as through the surface of their roots. Some of the compounds released by certain kinds of plants, however, seem to have an inhibitory action against certain pathogens. **Fungitoxic exudates** on the leaves of some plants, e.g., tomato and sugar beet, seem to be present in sufficient concentrations to inhibit the germination of spores of fungi *Botrytis* and *Cercospora*, respectively, that may be present in dew or rain droplets on these leaves. Similarly, in the case of onion smudge, caused by the fungus *Colletotrichum circinans*, resistant varieties generally have red scales and contain, in addition to the red pigments, the phenolic compounds protocatechuic acid and catechol. In the presence of water drops or soil moisture containing conidia of the onion smudge fungus on the surface of red onions, these two fungitoxic substances diffuse into the liquid, inhibit the germination of the conidia, and cause them to burst, thus protecting the plant from infection. Both fungitoxic exudates and inhibition of infection are missing in white-scaled, susceptible onion varieties (Fig. 6-2). It was noticed that applications of acibenzolar-*S*-methyl (ASM) on sunflower reduced infection by the rust fungus *Puccinia helianthi* through the reduction of spore germination and appressorium formation. It was subsequently shown that ASM accomplished this by increasing the production and secretion by the plant on the leaf surface of coumarins and other toxic phenolics that inhibit spore germination and appressorium formation on the leaf surfaces on which they are present.

FIGURE 6-2 Onion smudge, caused by the fungus *Colletotrichum circinans*, develops on white onions but not on colored ones, which, in addition to the red or yellow pigment, also contain the phenolics protocatechuic acid and catechol, both of which are toxic to the fungus. (Photograph courtesy of G. W. Simone.)

Inhibitors Present in Plant Cells before Infection

It is becoming increasingly apparent that some plants are resistant to diseases caused by certain pathogens because of one or more inhibitory antimicrobial compounds, known as phytoanticipins, which are present in the cell before infection. Several **phenolic compounds**, **tannins**, and some fatty acid-like compounds such as **dienes**, which are present in high concentrations in cells of young fruits, leaves, or seeds, have been proposed as responsible for the resistance of young tissues to pathogenic microorganisms such as *Botrytis*. For example, increased 9-hexadecanoic acid in cutin monomers in transgenic tomato plants led to resistance of such plants to powdery mildew because these cutin monomers inhibit the germination of powdery mildew spores. Many such compounds are potent inhibitors of many hydrolytic enzymes, including the pectolytic-macerating enzymes of plant pathogens. As the young tissues grow older, their inhibitor content and their resistance to infection decrease steadily. Strawberry leaves naturally contain **(+)-catechin**, which inhibits infection by *Alternaria alternata* by blocking the formation of infection hyphae from haustoria although it allows both spore germination and appressoria formation. Several other types of preformed compounds, such as the saponins (glycosylated steroidal or triterpenoid compounds) **tomatine** in tomato and **avenacin** in oats, not only have antifungal membranolytic activity, they actually exclude fungal pathogens that lack enzymes (saponinases) that break down the saponin from infecting the host. In this way, the presence or absence of saponin in a host and of saponinase in a fungus determines the host range of the fungus.

In addition to the simple molecule antifungal compounds listed earlier, several preformed plant proteins have been reported to act as inhibitors of pathogen proteinases or of hydrolytic enzymes involved in host cell wall degradation, to inactivate foreign ribosomes, or to increase the permeability of the plasma membranes of fungi.

For example, in a number of plants there is a family of low molecular weight proteins called phytocystatins that inhibit cysteine proteinases carried in the digestive system of nematodes and are also secreted by some plant pathogenic fungi. Constitutively present or transgenically introduced phytocystatins in plants reduce the size of nematode females and the number of eggs produced by females, thereby providing effective or significant control of several plants to root knot, cyst, reniform, and lesion nematodes.

Another type of compounds, the lectins, which are proteins that bind specifically to certain sugars and occur in large concentrations in many types of seeds, cause lysis and growth inhibition of many fungi. However, plant surface cells also contain variable amounts of hydrolytic enzymes, some of which, such as glucanases and chitinases, may cause the breakdown of pathogen cell wall components, thereby contributing to resistance to infection. The importance of either of these types of inhibitors to disease resistance is not currently known, but some of these substances are known to increase rapidly upon infection and are considered to play an important role in the defense of plants to infection.

DEFENSE THROUGH LACK OF ESSENTIAL FACTORS

Lack of Recognition between Host and Pathogen

A plant species either is a host for a particular pathogen, e.g., wheat for the wheat stem rust fungus, or it is not a host for that pathogen, e.g., tomato for wheat stem rust fungus. How does a pathogen recognize that the plant with which it comes in contact is a host or nonhost? Plants of a species or variety may not become infected by a pathogen if their surface cells lack specific **recognition factors** (specific molecules or structures) that can be recognized by the pathogen. If the pathogen does not recognize the plant as one of its host plants, it may not become attached to the plant or may not produce infection substances, such as enzymes, or structures, such as appressoria, penetration pegs, and haustoria, necessary for the establishment of infection. It is not known what types of molecules or structures are involved in the recognition of plants and pathogens, but it is thought that they probably include various types of oligosaccharides and polysaccharides, and proteins or glycoproteins. Also, it is not known to what extent these recognition phenomena are responsible for the success or failure of initiation of infection in any particular host–pathogen combination.

Lack of Host Receptors and Sensitive Sites for Toxins

In host–pathogen combinations in which the pathogen (usually a fungus) produces a host-specific toxin, the toxin, which is responsible for the symptoms, is thought to attach to and react with specific receptors or sensitive sites in the cell. Only plants that have such sensitive receptors or sites become diseased. Plants of other varieties or species that lack such receptors or sites remain resistant to the toxin and develop no symptoms.

Lack of Essential Substances for the Pathogen

Species or varieties of plants that for some reason do not produce one of the substances essential for the survival of an obligate parasite, or for development of infection by any parasite, would be resistant to the pathogen that requires it. Thus, for *Rhizoctonia* to infect a plant it needs to obtain from the plant a substance necessary for formation of a hyphal cushion from which the fungus sends into the plant its penetration hyphae. In plants in which this substance is apparently lacking, cushions do not form, infection does not occur, and the plants are resistant. The fungus does not normally form hyphal cushions in pure cultures but forms them when extracts from a susceptible but not a resistant plant are added to the culture. Also, certain mutants of *Venturia inaequalis*, the cause of apple scab, which had lost the ability to synthesize a certain growth factor, also lost the ability to cause infection. When, however, the particular growth factor is sprayed on the apple leaves during inoculation with the mutant, the mutant not only survives but it also causes infection. The advance of the infection, though, continues only as long as the growth factor is supplied externally to the mutant. In some host–pathogen combinations, disease develops but the amount of disease may be reduced by the fact that certain host substances are present in lower concentrations. For example, bacterial soft rot of potatoes, caused by *Erwinia carotovora* var. *atroseptica*, is less severe on potatoes with low-reducing sugar content than on potatoes high in reducing sugars.

INDUCED STRUCTURAL AND BIOCHEMICAL DEFENSES

Recognition of the Pathogen by the Host Plant

Early recognition of the pathogen by the plant is very important if the plant is to mobilize the available biochemical and structural defenses to protect itself from the pathogen. The plant apparently begins to receive signal molecules, i.e., molecules that indicate the presence of a pathogen, as soon as the pathogen establishes physical contact with the plant (Fig. 6-3).

Pathogen Elicitors

Various pathogens, especially fungi and bacteria, release a variety of substances in their immediate environment that act as nonspecific elicitors of pathogen recognition by the host. Such nonspecific elicitors include toxins, glycoproteins, carbohydrates, fatty acids, peptides, and extracellular microbial enzymes such as proteases and pectic enzymes. In various host–pathogen combinations, certain substances secreted by the pathogen, such as *avr* gene products, *hrp* gene products, and suppressor molecules, act as specific pathogen elicitors of recognition by the specific host plant. In many cases, in which host enzymes break down a portion of the polysaccharides making up the pathogen surface or pathogen enzymes break down a portion of the plant surface polysaccharides, the released oligomers or monomers of the polysaccharides act as recognition elicitors for the plant.

Host Plant Receptors

The location of host receptors that recognize pathogen elicitors is not generally known, but several of those studied appear to exist outside or on the cell membrane, whereas others apparently occur intracellularly. In the powdery mildew of cereals, a soluble carbohydrate that acts as an elicitor from the wheat powdery mildew fungus *Blumeria graminis* f. sp. *tritici* is recognized by a broad range of cereals (barley, oat, rye, rice, and maize) in which it induces the expression of all defense-related genes tested and also induced resistance to subsequent attacks with the fungus. The elicitor alone, in

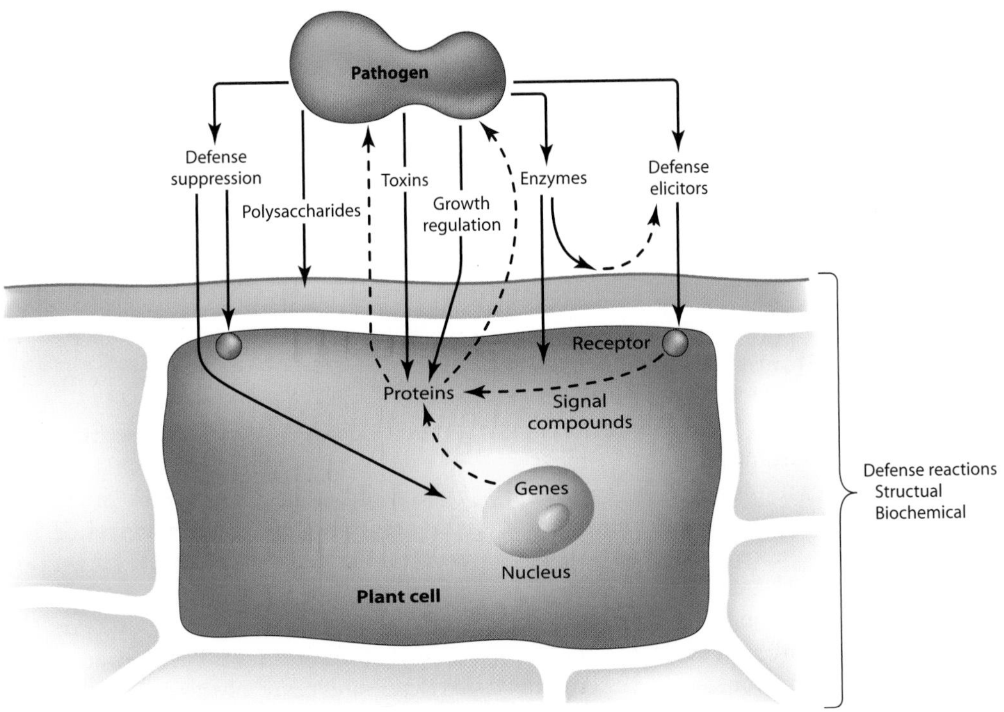

FIGURE 6-3 Schematic representation of pathogen interactions with host plant cells. Depending on its genetic makeup, the plant cell may react with numerous defenses, which may include cell wall structural defenses (waxes, cutin, suberin, lignin, phenolics, cellulose, callose, cell wall proteins) or biochemical wall, membrane, cytoplasm, and nucleus defense reactions. The latter may involve bursts of oxidative reactions, production of elicitors, hypersensitive cell death, ethylene, phytoalexins, pathogenesis-related proteins (hydrolytic enzymes, β-1,3-glucanases, chitinases), inhibitors (thionins, proteinase inhibitors, thaumatin-like proteins), and so on.

absence of the powdery mildew fungus, did not induce a hypersensitive response but it did induce an accumulation of thaumatin-like proteins in the various cereals.

Mobilization of Defenses

Once a particular plant molecule recognizes and reacts with a molecule (elicitor) derived from a pathogen, it is assumed that the plant "recognizes" the pathogen. Following such recognition, a series of biochemical reactions and structural changes are set in motion in the plant cell(s) in an effort to fend off the pathogen and its enzymes, toxins, etc. How quickly the plant recognizes the (presence of a) pathogen and how quickly it can send out its alarm message(s) and mobilize its defenses determine whether hardly any infection will take place at all (as in the hypersensitive response) or how much the pathogen will develop, i.e., how severe the symptoms (leaf spots, stem, fruit, or root lesions, etc.) will be, before the host defenses finally stop further development of the pathogen.

Transmission of the Alarm Signal to Host Defense Providers: Signal Transduction

Once the pathogen-derived elicitors are recognized by the host, a series of alarm signals are sent out to host cell proteins and to nuclear genes, causing them to become activated, to produce substances inhibitory to the pathogen, and to mobilize themselves or their products toward the point of cell attack by the pathogen. Some of the alarm substances and signal transductions are only intracellular, but in many cases the signal is also transmitted to several adjacent cells and, apparently, the alarm signal is often transmitted systemically to most or all of the plant.

The chemical nature of the transmitted signal molecules is not known with certainty in any host–pathogen combination. Several types of molecules have been implicated in intracellular signal transduction. The most common such signal transducers appear to be various protein kinases, calcium ions, phosphorylases and phospholipases, ATPases, hydrogen peroxide (H_2O_2), ethylene, and others. Systemic signal transduction, which leads to systemic acquired resistance, is thought to be carried out by salicylic acid, oligogalacturonides released from plant cell walls, jasmonic acid, systemin, fatty acids, ethylene, and others. Some natural or synthetic chemicals, such as salicylic acid and the synthetic dichloroisonicotinic acid, also activate the signaling pathway that leads to systemic acquired resistance against several diverse types of plant pathogenic viruses, bacteria, and fungi.

INDUCED STRUCTURAL DEFENSES

Despite the preformed superficial or internal defense structures of host plants, most pathogens manage to penetrate their hosts through wounds and natural openings and to produce various degrees of infection. Even after the pathogen has penetrated the preformed defense structures, however, plants usually respond by forming one or more types of structures that are more or less successful in defending the plant from further pathogen invasion. Some of the defense structures formed involve the cytoplasm of the cells under attack, and the process is called **cytoplasmic defense reaction**; others involve the walls of invaded cells and are called **cell wall defense structures**; and still others involve tissues ahead of the pathogen (deeper into the plant) and are called **histological defense structures**. Finally, the death of the invaded cell may protect the plant from further invasion. This is called the **necrotic** or **hypersensitive defense reaction** and is discussed here briefly, with more detailed treatment a little later.

Cytoplasmic Defense Reaction

In a few cases of slowly growing, weakly pathogenic fungi, such as weakly pathogenic *Armillaria* strains and the mycorrhizal fungi, that induce chronic diseases or nearly symbiotic conditions, the plant cell cytoplasm surrounds the clump of hyphae and the plant cell nucleus is stretched to the point where it breaks in two. In some cells, the cytoplasmic reaction is overcome and the protoplast disappears while fungal growth increases. In some of the invaded cells, however, the cytoplasm and nucleus enlarge. The cytoplasm becomes granular and dense, and various particles or structures appear in it. Finally, the mycelium of the pathogen disintegrates and the invasion stops.

Cell Wall Defense Structures

Cell wall defense structures involve morphological changes in the cell wall or changes derived from the cell wall of the cell being invaded by the pathogen. The effectiveness of these structures as defense mechanisms seems to be rather limited, however. Three main types of such structures have been observed in plant diseases. (1) The outer layer of the cell wall of parenchyma cells coming in contact with incompatible bacteria swells and produces an amorphous, fibrillar material that surrounds and traps the bacteria and prevents them from multiplying. (2) Cell walls thicken in response to several pathogens by producing what appears to be a cellulosic material. This material, however, is often infused with phenolic substances that are cross-linked and further increase its resistance to penetration. (3) Callose **papillae** are deposited on the inner side of cell walls in response to invasion by fungal pathogens (see Figs. 2-8C and 2-8D). Papillae seem to be produced by cells within minutes after wounding and within 2 to 3 hours after inoculation with microorganisms. Although the main function of papillae seems to be repair of cellular damage, sometimes, especially if papillae are present before inoculation, they also seem to prevent the pathogen from subsequently penetrating the cell. In some cases, hyphal tips of fungi penetrating a cell wall and growing into the cell lumen are enveloped by cellulosic (callose) materials that later become infused with phenolic substances and form a sheath or lignituber around the hypha (Fig. 6-4).

Histological Defense Structures

Formation of Cork Layers

Infection by fungi or bacteria, and even by some viruses and nematodes, frequently induces plants to form several layers of cork cells beyond the point of infection (Figs. 6-5 and 6-6), apparently as a result of stimulation of the host cells by substances secreted by the pathogen. The cork layers inhibit further invasion by the pathogen beyond the initial lesion and also block the spread of any toxic substances that the pathogen may secrete. Furthermore, cork layers stop the flow of nutrients and water from the healthy to the infected area and deprive the pathogen of nourishment. The dead tissues, including the pathogen, are thus delimited by the cork layers

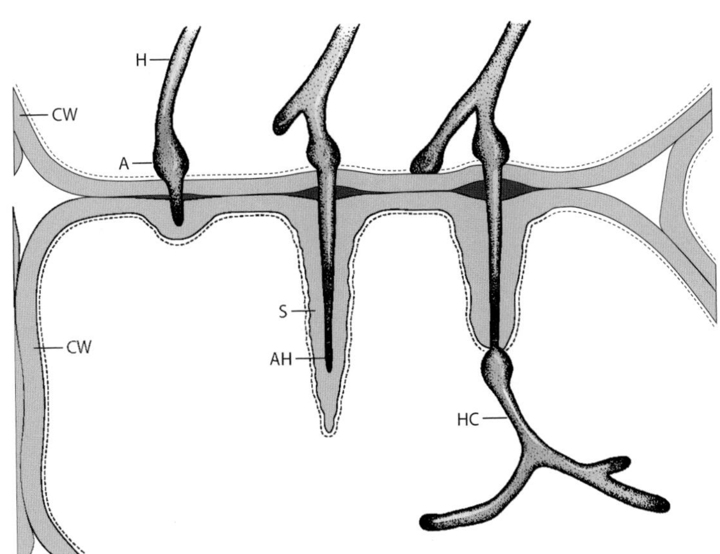

FIGURE 6-4 Formation of a sheath around a hypha (H) penetrating a cell wall (CW). A, appressorium; AH, advancing hypha still enclosed in sheath; HC, hypha in cytoplasm; S, sheath.

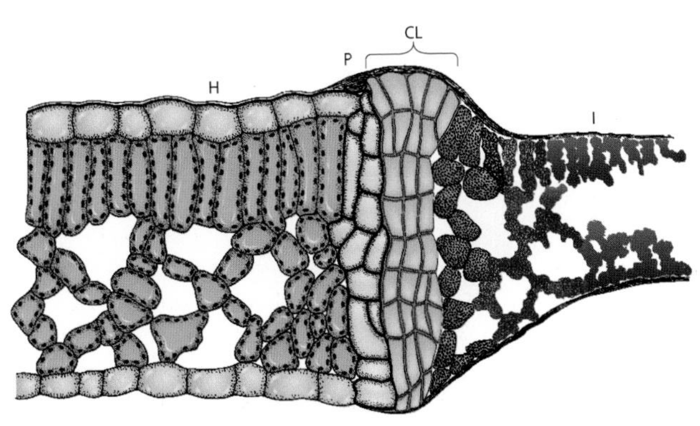

FIGURE 6-5 Formation of a cork layer (CL) between infected (I) and healthy (H) areas of leaf. P, phellogen. [After Cunningham (1928). *Phytopathology* 18, 717–751.]

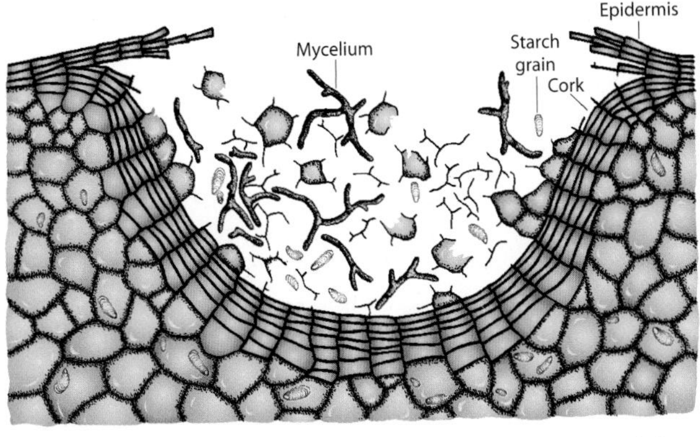

FIGURE 6-6 Formation of a cork layer on a potato tuber following infection with *Rhizoctonia*. [After Ramsey (1917). *J. Agric. Res.* 9, 421–426.]

and may remain in place, forming necrotic lesions (spots) that are remarkably uniform in size and shape for a particular host–pathogen combination. In some host–pathogen combinations the necrotic tissues are pushed outward by the underlying healthy tissues and form scabs that may be sloughed off, thus removing the pathogen from the host completely. In tree cankers, such as those caused by the fungus *Seiridium cardinale* on cypress trees, resistant plant clones restrict growth of the fungus by forming ligno-suberized boundary zones, which included four to six layers of cells with suberized cell walls. In contrast, susceptible clones have only two to four layers of suberized cells and these are discontinuous, allowing repeated penetration by the fungus past the incomplete barrier.

Formation of Abscission Layers

Abscission layers are formed on young, active leaves of stone fruit trees after infection by any of several fungi, bacteria, or viruses. An abscission layer consists of a gap formed between two circular layers of leaf cells surrounding the locus of infection. Upon infection, the middle lamella between these two layers of cells is dissolved throughout the thickness of the leaf, completely cutting off the central area of the infection from the rest of the leaf (Fig. 6-7). Gradually, this area shrivels, dies, and sloughs off, carrying with it the pathogen. Thus, the plant, by discarding the infected area along with a few yet uninfected cells, protects the rest of the leaf tissue from being invaded by the

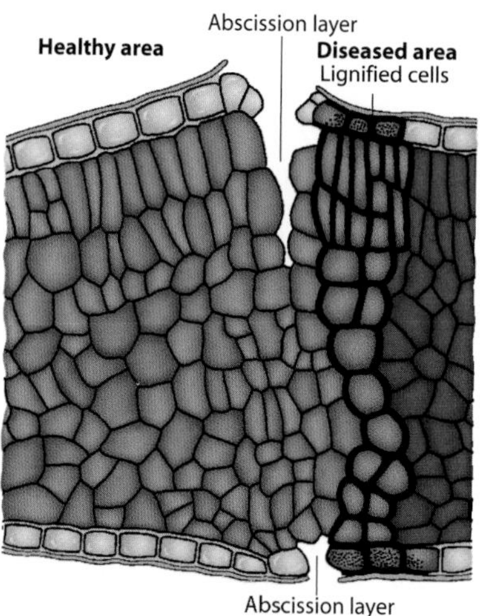

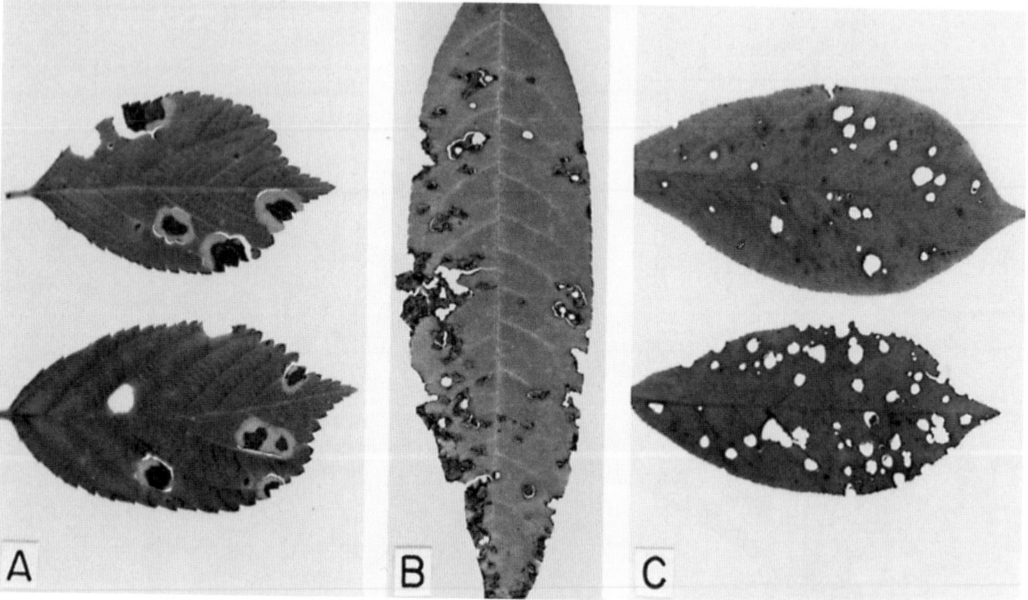

FIGURE 6-7 Schematic formation of an abscission layer around a diseased spot of a *Prunus* leaf. [After Samuel (1927).] (A–C) Leaf spots and shot holes caused by *Xanthomonas arboricola* pv. *pruni* bacteria on (A) ornamental cherry leaves; characteristic broad, light green halos form around the infected area before all affected tissue falls off, (B) on peach, and (C) on plum. The shot hole effect is particularly obvious on the plum leaves.

pathogen and from becoming affected by the toxic secretions of the pathogen.

Formation of Tyloses

Tyloses form in xylem vessels of most plants under various conditions of stress and during invasion by most of the xylem-invading pathogens. Tyloses are overgrowths of the protoplast of adjacent living parenchymatous cells, which protrude into xylem vessels through pits (Fig. 6-8). Tyloses have cellulosic walls and may, by their size and numbers, clog the vessel completely. In some varieties of plants, tyloses form abundantly and quickly ahead of the pathogen, while the pathogen is still in the young roots, and block further advance of the pathogen. The plants of these varieties remain free of and therefore resistant to this pathogen. Varieties in which few, if any, tyloses form ahead of the pathogen are susceptible to disease.

Deposition of Gums

Various types of gums are produced by many plants around lesions after infection by pathogens or injury. Gum secretion is most common in stone fruit trees but occurs in most plants. The defensive role of gums stems from the fact that they are deposited quickly in the intercellular spaces and within the cells surrounding the locus of infection, thus forming an impenetrable barrier that completely encloses the pathogen. The pathogen then becomes isolated, starves, and sooner or later dies.

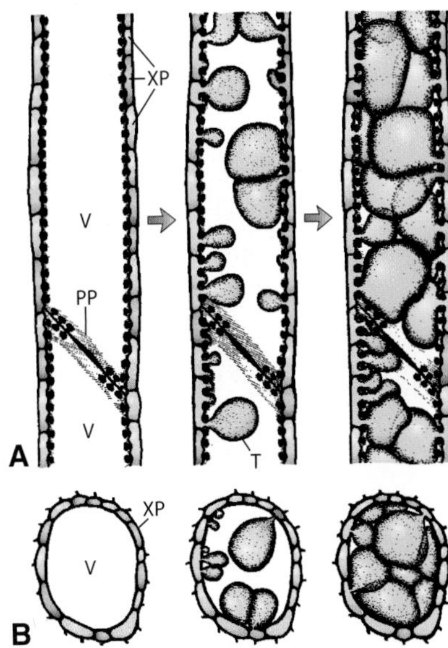

FIGURE 6-8 Development of tyloses in xylem vessels. Longitudinal (A) and cross section (B) views of healthy vessels (left) and of vessels with tyloses. Vessels at right are completely clogged with tyloses. PP, perforation plate; V, xylem vessel; XP, xylem parenchyma cell; T, tylosis.

Necrotic Structural Defense Reaction: Defense through the Hypersensitive Response

The hypersensitive response is considered a biochemical rather than a structural defense mechanism but is described here briefly because some of the cellular responses that accompany it can be seen with the naked eye or with the microscope. In many host–pathogen combinations, as soon as the pathogen establishes contact with the cell, the nucleus moves toward the invading pathogen and soon disintegrates. At the same time, brown, resin-like granules form in the cytoplasm, first around the point of penetration of the pathogen and then throughout the cytoplasm. As the browning discoloration of the plant cell cytoplasm continues and death sets in, the invading hypha begins to degenerate (Fig. 6-9). In most cases the hypha does not grow out of such cells, and further invasion is stopped. In bacterial infections of leaves, the hypersensitive response results in the destruction of all cellular membranes of cells in contact with bacteria, which is followed by desiccation and necrosis of the leaf tissues invaded by the bacteria.

Although it is not quite clear whether the HR is the cause or the consequence of resistance, this type of necrotic defense is quite common, particularly in diseases caused by obligate fungal parasites and by viruses (Fig. 6-10A), bacteria (Fig. 6-10B), and nematodes. Apparently, the necrotic tissue not only isolates the parasite from the living substance on which it depends for its nutrition and, thereby, results in its starvation and death, but, more importantly, it signifies the concentration of numerous biochemical cell responses and antimicrobial substances that neutralize the pathogen. The faster the host cell dies after invasion, the more resistant to infection the plant seems to be. Moreover, through the signaling compounds and pathways developed during the hypersensitive response, the latter serves as the springboard for localized and systemic acquired resistance.

INDUCED BIOCHEMICAL DEFENSES

Induced Biochemical Nonhost Resistance

As mentioned earlier, nonhost resistance is the resistance that keeps a plant protected from pathogens that are, through evolution, incompatible with that host. Although the nature of nonhost resistance is unknown, for a pathogen it can be as big a gap to bridge as the difference between the features of a potato plant and an oak tree, or as close as the difference between the features of potato and tomato, or barley and wheat. It appears, however, that in some plant/pathogen

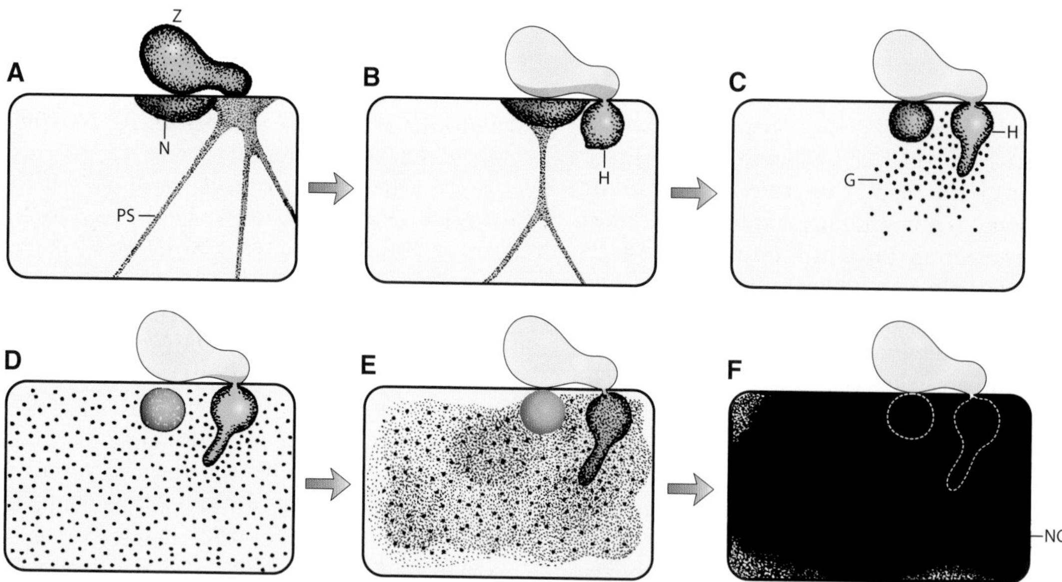

FIGURE 6-9 Stages in the development of the necrotic defense reaction in a cell of a very resistant potato variety infected by *Phytophthora infestans*. N, nucleus; PS, protoplasmic strands; Z, zoospore; H, hypha; G, granular material; NC, necrotic cell. [After Tomiyama (1956). *Ann. Phytopathol. Soc. Jpn.* **21**, 54–62.]

FIGURE 6-10 (A) Hypersensitive response (HR) expressed on leaves of a resistant cowpea variety following sap inoculation with a strain of a virus that causes local lesions (in this case, *alfalfa mosaic virus*). The virus remains localized in the lesions. (B) Tobacco leaf showing typical hypersensitive responses (white areas) 24 hours after injection with water (A) or with preparations of bacterial strains B, C, and D. Strain (B), which does not infect tobacco, and (C), which carries a *hrp* (hypersensitive response and pathogenicity) gene, both induced the hypersensitive response, whereas the third strain (D), a mutant of C that lacked the *hrp* gene, did not. [From Mukherjee *et al.* (1997). *Mol. Plant-Microbe Interact.* **10**, 462–471.]

interactions of taxonomically unrelated plants (e.g., potato and oak or oak and wheat), nonhost resistance is controlled by constitutive defenses and/or defenses induced by nonspecific stimuli in a nonspecific manner. Such defenses include physical topography and the structures present on the plant, the presence of toxic or the absence of essential compounds, and so on. In other plant/pathogen combinations, in which the plants are taxonomically related (e.g., potato and tomato, barley and wheat), nonhost resistance involves primarily inducible defenses elicited by the recognition of pathogen-specific molecules. Some cases of nonhost resistance, however, seem to be controlled by a single gene.

Some examples of questionable nonhost resistance include the resistance of the nonhost pea to the *Pseudomonas syringae* pv. *syringae* bacterium, which infects bean but not pea. The reaction occurs when that bacterium carries a gene that is responsible for elicitation of a potentially defensive response in the normally nonhost pea, that is expressed as a visible hypersensitive response. In another example, the potato late blight fungus *Phytophthora infestans*, normally does not infect the tobacco species *Nicotiana benthamiana*. The nonhost resistance of the tobacco species, however, is lost if the pathogen does not carry an "avirulence-like gene," which produces a protein that elicits cell death in the tobacco. This is unique in that in other plant/pathogen combinations, the absence of a single "nonhost avirulence gene" does not make the nonhost plant susceptible. It would appear, therefore, that if the cell death response to the elicitor controlled by the avirulence gene really contributes to resistance, then the nonhost resistance in such situations is controlled by more than one component. In still another case, nonhost resistance in some cereals [wheat to powdery mildew strains from another cereal (barley), or in barley to *Puccinia* rust races from wheat], involves similar gene-for-gene interactions and nonhost resistance occurs through defense mechanisms involving recognition of an elicitor and development of a hypersensitive response. Disease resistance does not always involve pathogen recognition events, but, especially in polygenic or quantitative resistance, it may involve directly various structural or chemical defense mechanisms. This also happens in some cases of nonhost resistance, e.g., in oat roots to the wheat fungus *Gaeumannomyces graminis* f. sp. *tritici*, while they are susceptible to the oat fungus *G. graminis* f. sp. *avenae*. The nonhost resistance of oat roots to the wheat fungus is caused by the presence of the saponin compound avenacin in the oat roots, which is toxic to the fungus. This compound is also toxic to the oat fungus, but the latter produces an enzyme that detoxifies the saponin in oat roots and can infect them. The nonhost resistance to the wheat fungus, however, is compromised in saponin-deficient mutants in which the wheat fungus causes a successful infection. This shows that nonhost resistance in some plant/microbe interactions is caused by a direct defense mechanism rather than by recognition events.

In all these examples, the pathogen or the host is already closely related and nearly fully adopted to the characteristics of nonhost resistance presented to it. In less related plants or pathogens, however, in which true nonhost resistance is found routinely, it is more likely to be the result of effective nonspecific defenses such as physical characteristics and nonspecific responses to wounding and damage done by the pathogen during attempted invasion than to defenses elicited by specific recognition events. There is also, however, the case of pathogens that have alternate hosts, such as wheat stem rust and barberry and cedar apple rust on apple and cedar. These are, perhaps, interesting from an evolutionary point of view because, presumably, before the second of the alternate hosts that became a host, it was surely a nonhost. How the rust fungus bridged the two taxonomically extremely different hosts is not known. The change in ploidy (from haploid to diploid and back to haploid) was probably involved, but how the fungus broke the nonhost resistance of the other host and how it used the nonresistant host as a completely cooperative host is still a mystery.

The present consensus is that plants that exhibit nonhost resistance against pathogens of other plants do not need to carry resistance genes that recognize these pathogens because they carry genes that provide the plants with nonspecific defenses that are fully effective in protecting the plant from these pathogens. However, it may be possible that nonhost resistance, along with polygenic and monogenic host resistance, forms a continuum of resistance that begins to overlap as the taxonomic (evolutionary) distance between host and nonhost plants becomes closer and results in a complex and continuous network of plant/pathogen interactions.

Induced Biochemical Defenses in Quantitative (Partial, Polygenic, General, or Horizontal) Resistance

In quantitative (partial, polygenic, multigenic, general, field, durable, or horizontal) resistance, plants depend on the action of numerous genes, expressed constitutively or upon attack by a pathogen (induced resistance). These genes provide the plants with defensive structures or toxic substances that slow down or stop the advance of the pathogen into the host tissues and reduce the damage caused by the pathogen. Quantitative resistance is particularly common in diseases caused by nonbiotrophic pathogens. Quantitative resistance may vary considerably, in some cases being specific against some of the strains of a pathogen, in others being effective against all strains of a pathogen, or providing resistance against more than one pathogen. Genes for quantitative resistance are present and provide a basal level of resistance to all plants against all pathogens regardless of whether the plant also carries major (or R) genes against a particular pathogen.

Function of Gene Products in Quantitative Resistance

Unlike most major (or R) genes involved in monogenic resistance, which appear to code for components that help the host recognize the pathogen and to subsequently express the hypersensitive response, genes for quantitative resistance seem to be involved directly in the expression or production of some sort of structural or biochemical defense. Quantitative resistance defenses are basically the same ones that follow the hypersensitive response in monogenic resistance; in quantitative resistance, however, defenses generally do not follow a hypersensitive response and cell death because the latter do not usually occur in quantitative resistance. Genes involved in quantitative resistance are present in the same areas of plant chromosomes that contain the genes involved in defense responses, such as the production of phenylalanine ammonia lyase, hydroxyproline-rich glycoproteins, and pathogenesis–related proteins. The defenses in quantitative resistance, however, develop slower and perhaps reach a lower level than those in the race-specific (R gene) resistance. Quantitative resistance is also affected much more by changes in the environment, mostly of changes in temperature during the various stages of development of resistance.

Mechanisms of Quantitative Resistance

Studies of defense mechanisms in diseases with quantitative resistance are few and far between. For example, in the early blight of tomato caused by the fungus *Alternaria solani*, all resistant tomato lines had higher constitutive levels of the pathogenesis-related proteins chitinase and β-1,3-glucanase than the susceptible lines. Also, preparations of constitutive enzymes from quantitatively resistant, but not from susceptible, tomato plants could release elicitors of plant cell death, and possibly of a hypersensitive response, from the cell walls of the fungus. These results show that, in this host–plant interaction, the defense responses involve the production of higher levels of pathogenesis-related proteins in resistant plants, and the same plants may also induce the pathogen to produce elicitor molecules that potentiate a more aggressive defense response through the induction of cell death and a hypersensitive-like response. The latter defenses are produced in a manner not unlike that in a specific host–pathogen interaction, but in the absence of host R genes. In the quantitatively controlled resistance of the soybean–*Phytophthora* interaction, soybean tissues actually caused the release of phytoalexin elicitors from the cell walls of the fungus, again showing that the plant can play an important role in forcing the release of defense-triggering signals from the pathogen. Finally, when five cabbage varieties of different resistance levels were inoculated with a strain of the cabbage black rot bacterium *Xanthomonas campestris* pv. *campestris*, two varieties were resistant, one was partially resistant, and two were susceptible. In all varieties there was an increase in the total oxidant activity of peroxidase and superoxide dismutase, accumulation of peroxidases, and lignin deposition. The increases, however, were greater and generally occurred earlier in resistant than in susceptible varieties. However, activity of the antioxidant catalase decreased in both resistant and susceptible varieties, but it decreased more in the resistant variety. The resistant varieties also produced new isozymes of peroxidase and superoxide dismutase that were not produced by the susceptible variety. These results suggest that in the cabbage–*X. campestris* pv. *campestris* system there is a multilevel resistance similar to a hypersensitive response, although the onset of this response was delayed when compared to the classical HR. In barley leaves infected with the fungus *Drechslera teres*, as many as eight pathogenicity-related proteins with thaumatin-like activity were detected.

Effect of Temperature on Quantitative Resistance

Quantitative resistance is often affected greatly by the temperature in the environment. This effect, however, is not unique to plants with quantitative resistance, as even in plants with monogenic (R) gene resistance, the resistance of the host may be changed drastically by changes in temperature. For example, in R resistance-carrying wheat, a change in temperature from 18 to 30°C changes the reaction of wheat plants carrying the Sr6 R gene from rust resistant to rust susceptible. Also, resistance to rust and powdery mildew was increased in pea and barley, respectively, by low-temperature hardening of these grain crops. However, a brief "heat shock" may cause a brief period of susceptibility of wheat plants to rust, while it induces resistance to powdery mildew in barley and to cucumber scab, caused by the fungus *Cladosporium cucumerinum*, in cucumber, in which it also causes an increase in peroxidase activity. There are numerous reports of different plants synthesizing a variety of pathogenesis-related (PR) proteins in response to abiotic (low temperature, drought, pollution, wounding) as well as to biotic (fungi, bacteria, etc.) stresses. Some of the PR proteins include PR-1, PR-2 (β-1,3-glucanases), PR-3 (chitinases), and PR-5 (thaumatin-like proteins), as well as peroxidases. Stressed plants also increase the production of phenylalanine ammonia lyase (PAL), which is involved in the production of phytoalexins.

In a detailed study of the effect of cold hardening of wheat on its quantitative resistance to infection by the snow mold fungi, it was found that cold hardening increases the resistance of wheat to snow mold and also induces changes in the expression (activity) of genes associated with PR proteins and other defense responses, some of them associated with induced systemic resistance. The most abundant PR proteins produced were chitinase, followed by PAL, β-1,3-glucanase, PR-1, and peroxidase. Similar PR proteins were produced by plants receiving cold treatment only, but the level of these proteins was lower and appeared later than when the plants were also infected by the snow mold fungi. It is apparent, therefore, that this biotic stress induces resistance and that the resistance is further augmented by the fungal infection. This type of resistance has characteristics similar to those of pathogen- and salicylic acid-induced resistance, including the expression of PR genes and further enhancement of defense-associated genes following the infection by a pathogen.

It should be noted in the aforementioned paragraphs that all plants produce PR and other defense-associated proteins constitutively and/or following induction by biotic and abiotic agents. In some host/pathogen combinations the level of constitutively produced PR proteins can be correlated with the level of partial resistance of the cultivars to the pathogen. There is no proof, however, that this correlation is meaningful, especially since some varieties lack the constitutive production of certain PR proteins and yet the plants exhibit partial resistance. It is possible, of course, that plants in the latter varieties have a means of upregulating PR gene expression upon infection that the other varieties lack. As was mentioned already, quantitative resistance depends (a) on the preexisting and induced structural and biochemical defenses provided by dozens and, probably, hundreds of defense-associated genes, (b) on PR proteins, which may provide another significant portion of the overall defenses, and (c) on the possible ability of PR proteins to potentiate a more aggressive response by plant cells to the pathogen invasion by inducing the pathogen to release molecules eliciting host defenses in the absence of a gene-for-gene relationship between host and pathogen.

INDUCED BIOCHEMICAL DEFENSES IN THE HYPERSENSITIVE RESPONSE (RACE-SPECIFIC, MONOGENIC, R GENE, OR VERTICAL) RESISTANCE

The Hypersensitive Response

The hypersensitive response, often referred to as HR, is a localized induced cell defense in the host plant at the site of infection by a pathogen (Fig. 6-10A). HR is the result of quick mobilization of a cascade of defense responses by the affected and surrounding cells and the subsequent release of toxic compounds that often kill both the invaded and surrounding cells and, also, the pathogen. The hypersensitive response is often thought to be responsible for limiting the growth of the pathogen and, in that way, is capable of providing resistance to the host plant against the pathogen. An effective hypersensitive response may not always be visible when a plant remains resistant to attack by a pathogen, as it is possible for the hypersensitive response to involve only single cells or very few cells and thereby remain unnoticed. Under artificial conditions, however, injection of several genera of plant pathogenic bacteria into leaf tissues of nonhost plants results in the development of a hypersensitive response. The artificially induced HR consists of large leaf sectors becoming water soaked at first and, subsequently, necrotic and collapsed within 8 to 12 hours after inoculation (Fig. 6-10B). The bacteria injected in the tissues are trapped in the necrotic lesions and generally are killed rapidly. The HR may occur whenever virulent strains of plant pathogenic bacteria are injected into nonhost plants or into resistant varieties and when avirulent strains are injected into susceptible cultivars. Although not all cases of resistance are due to the hypersensitive response, HR-induced resistance has been described in numerous diseases involving obligate parasites (fungi, viruses, mollicutes, and nematodes), as well as nonobligate parasites (fungi and bacteria).

The hypersensitive response is the culmination of the plant defense responses initiated by the recognition by the plant of specific pathogen-produced signal molecules, known as elicitors. Recognition of the elicitors by the host plant activates a cascade of biochemical reactions in the attacked and surrounding plant cells and leads to new or altered cell functions and to new or greatly activated defense-related compounds (Fig. 6-11). The most common new cell functions and compounds include a rapid burst of reactive oxygen species, leading to a dramatic increase of oxidative reactions; increased ion movement, especially of K^+ and H^+ through the cell membrane; disruption of membranes and loss of

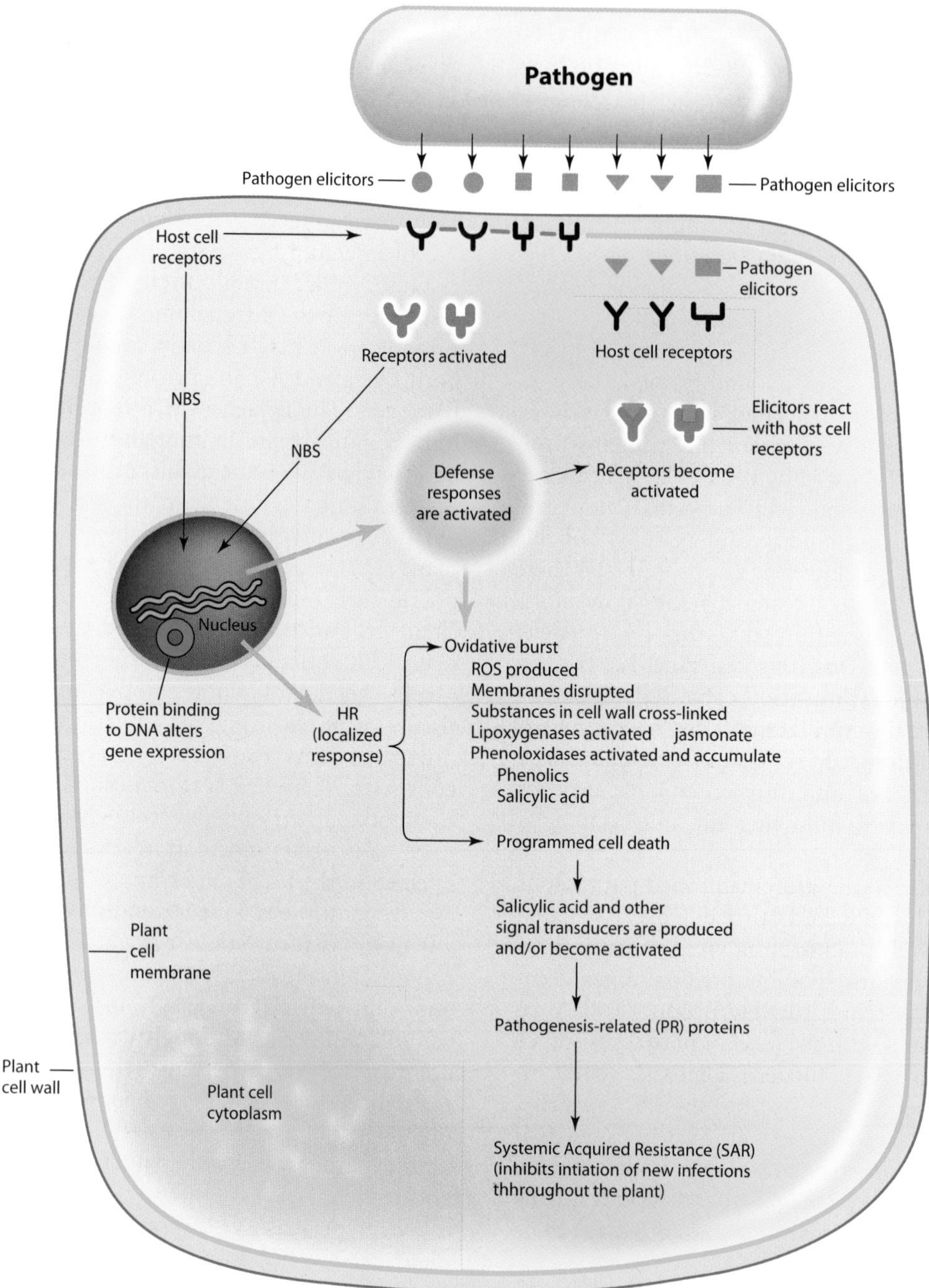

FIGURE 6-11 Diagram of the hypothetical steps in the hypersensitive response defense of plants following interaction of an elicitor molecule produced by a pathogen avirulence gene with a receptor molecule produced by the matching host R gene.

cellular compartmentalization (Fig. 6-12); cross-linking of phenolics with cell wall components and strengthening of the plant cell wall; transient activation of protein kinases (wounding-induced and salicylic acid-induced kinases); production of antimicrobial substances such as phenolics (phytoalexins); and formation of antimicrobial so-called pathogenesis-related proteins such as chitinases.

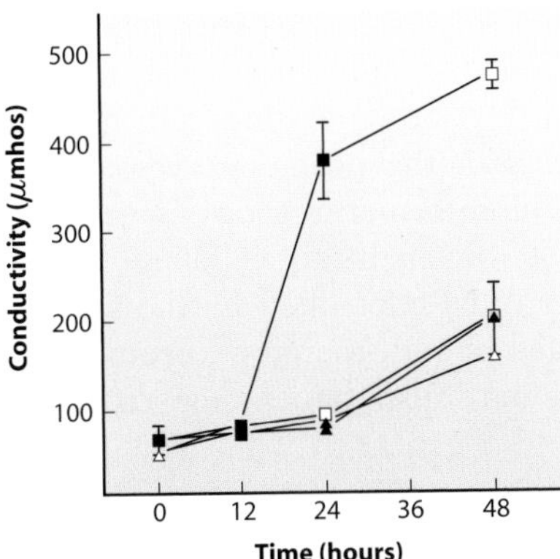

FIGURE 6-12 Disruption of cell membranes leads to a dramatic increase in cell electrolyte leakage, measured by increased current conductivity. This occurs when a resistant variety (■) containing an R gene is inoculated with pathogens containing an avirulence gene corresponding to the R gene. Same variety inoculated with a pathogen lacking the avirulence gene (□); another variety, susceptible to both pathogens (△, ▲). [From Whalen *et al.* (1993). *Mol. Plant-Microbe Interact.* 6, 616–627.]

The hypersensitive response occurs only in specific host–pathogen combinations in which the host and the pathogen are incompatible, i.e., the pathogen fails to infect the host. It is thought that this happens because of the presence in the plant of a resistance gene (R), which recognizes and is triggered into action by the elicitor molecule released by the pathogen. The pathogen-produced elicitor is, presumably, the product of a pathogen gene, which, because it triggers the development of resistance in the host that makes this pathogen avirulent, is called an avirulence gene. For several pathogens, primarily bacteria, avirulence genes have been isolated and the proteins coded by them have been identified. The first avirulence gene product to be identified was the protein of the avirulence gene D (*arv*D) of the bacterium *Pseudomonas syringae* pv. *glycinea*. This was shown to be an enzyme involved in the synthesis of substances known as syringolides. The latter have the ability to elicit the hypersensitive response in soybean varieties that carry the resistance gene D complementary to *avr*D of the bacterium.

More than 20 resistance (R) genes have been isolated from a variety of plants such as corn, tomato, tobacco, rice, flax, and *Arabidopsis*, a model plant used for experimental purposes. The corn R gene Hm1 for northern leaf spot codes for an enzyme that inactivates the HC toxin of the fungus *Cochliobolus carbonum*, the cause of northern leaf spot of corn, whereas the tomato gene Pto, that confers resistance to the tomato speck-causing bacterium *Pseudomonas syringae* pv. *tomato*, codes for a protein kinase enzyme that most likely plays a role in signal transduction by triggering other enzymes into action. The functions of the proteins encoded by most other R genes are not known with certainty, but most of them contain domains, such as leucine-rich repeats, found in proteins involved in protein–protein interactions. Proteins coded by the tobacco R gene, which protects against tobacco mosaic virus, and the *Arabidopsis* R gene, which protects against a leaf-spotting bacterium, appear to be present in the plant cell cytoplasm and, therefore, probably recognize pathogen elicitors that reach the cytoplasm. However, the protein encoded by the tomato R gene *Cf*-9, which provides resistance against race 9 of the leaf mold fungus *Cladosporium fulvum*, and the rice R gene XA21, which provides resistance against many races of the leaf-spotting bacterium *Xanthomonas oryzae*, are transmembrane receptor-like proteins with a short anchor and a protein kinase, respectively. The last two R gene products, therefore, apparently recognize pathogen-produced molecules as they approach or come in contact with the plant cell membrane.

Genes Induced during Early Infection

Through recent methodology [suppression subtractive hybridization (SSH), cDNA library construction, expressed sequence tag (EST) determination, large-scale DNA sequencing, and DNA microarrays], it is now possible to detect and identify numerous plant genes (or ESTs) and their organization, including those induced during compatible or incompatible interactions between plant pathogens and their hosts. DNA microarrays, especially, can provide extremely useful information on the expression patterns of thousands of genes in parallel. Earlier studies, for example, of a compatible interaction of *Phytophthora infestans* and potato, 43 genes appeared to be induced, 10 of which showed increased activity as a result of the infection. Some of them were homologous to genes already known to be activated during infection, e.g., for β-1,3-glucanase, some have homology to enzymes involved in detoxification, and some code for proteins that had not been reported earlier to be induced by infection. When genes expressed by rice seedlings 48 hours after inoculation with the fungus *Magnaporthe grisea* were examined, of 619 randomly selected clones, 359 expressed sequence tags that had not been described before. When 124 of 260 ESTs that showed moderate and high similarity were organized according to their suspected function, the largest group (21%) contained (24) stress or defense response genes. When looked at from a different angle, many of the genes were new and not described previously, but several had been described before and were known to be involved in the infection process; one, for example, being the rice peroxidase gene, which is expressed during the infection of rice with the bacterial blight pathogen *Xanthomonas oryzae* pv. *oryzae*.

In more recent studies, almost 2,400 genes of Arabidopsis were examined for transcriptional changes that may occur after inoculation with the incompatible fungal pathogen *Alternaria brassicicola* or after treatment with defense signaling compounds such as salicylic acid (SA), methyl jasmonate (MJ), or ethylene. More than 700 of the genes exhibited transcriptional changes in response to one or more of the treatments. Based on similarity of the sequences of these genes to known gene sequences, the majority of the activated genes were already known, but an additional 106 genes were also activated. Treatments with salicylic acid and methyl jasmonate activated 192 and 221 genes, respectively, but they also repressed the transcription of 131 and 96 genes, respectively. Of the identified genes that were activated, a number of them are involved in the oxidative burst, in antimicrobial defense, cell wall modification, phytoalexin production, and defense signal transduction. There appears to be a high level of interaction among signaling pathways regulated by pathogen infection or by treatment with SA, MJ, or ethylene. For example, of 2,375 ESTs analyzed simultaneously, 169 were regulated by more than one pathway. Of these, 55 genes were coinduced and 28 genes were corepressed by SA and MJ in local tissue, but only 6 genes were coinduced in both local and systemic tissue.

Functional Analysis of Plant Defense Genes

Expression of dozens or hundreds of genes at a particular physiological state, such as at a certain time interval after inoculation with a pathogen or a related treatment, implies the involvement of these genes in that physiological state. Determination, however, of which specific gene is responsible for a certain function requires that the study of the function of each gene be carried out individually. This is a very difficult task, partly because of the large number of genes contributing to the same function and because many of the same functions are carried out by several different genes. Also, several plant gene families consist of 100 or more members, and in some gene families related to transcription factors, most of the genes are particularly associated with defense responses. Nevertheless, candidate genes identified in microarray experiments can be subjected to detailed functional analysis *in planta* through several strategies, including posttranscriptional silencing, overexpression of genes, gene knockout experiments using insertional mutagenesis via transposon or T-DNA, through promoter trap strategies, and others.

The generation of transgenic plants for the functional analysis of genes is both time-consuming and may show high variation of transgene expression. The identification of transcription factors and their binding sites in the promoter regions of defense-related genes is also critical for understanding how defense gene expression is regulated. It is now possible to identify novel regulatory elements in the promoter regions of coregulated genes with bioinformatics tools. Genes that participate in the same biochemical, cellular, or developmental processes may be controlled by the same sets of transcription factors and, therefore, promoter sequences of such genes may also have some common regulatory sequences.

Classes of R Gene Proteins

The various plant R genes, regardless of the type of pathogen (bacterial, fungal, or viral) to which they confer resistance, have many structural similarities. It appears that most, if not all, R genes exist as clustered gene families. So far, depending on structure and function, R genes can be subdivided into five classes (Fig. 4-14, Table 4-5) (The R-like gene *Hm1*, which encodes a detoxifying enzyme, does not fit and does not follow the gene-for-gene concept.) (1) R genes, like *Pto*, encode a serine–threonine protein kinase that plays a role in signal transduction. (2) R genes, like *Xa21* of rice, which encode a transmembrane protein rich in extracellular leucine repeats and a cytoplasmic serine–threonine kinase, function as receptors of kinase-like proteins and transmit the signal to phosphokinases for further amplification. (3) R genes, like the tobacco N^1 gene, the flax L^6 gene, and the RPP5 *Arabidopsis* gene, encode proteins that are cytoplasmic. These cytoplasmic proteins, in addition to leucine-rich repeats, also have a site that binds to nucleotides (NBS) and a domain (TIR) with significant homology to the Toll/interleukin 1 receptor; such proteins may serve as receptors that activate the translocation of a transcription factor from the cytoplasm to the nucleus where it activates transcription of the genes related to hypersensitive response. (4) Another group of cytoplasmic R proteins also have LRR and NBS, but have a coiled coil domain that contains a putative leucine zipper domain, such as in RPS2 and RPM1. (5) R genes, like the tomato *Cf2*–*Cf9* genes, encode proteins that consist primarily of leucine-rich repeats and are located outside the cell membrane but are attached to the membrane with a transmembrane anchor. Such R gene-coded proteins may serve as receptors for the extracellular or intracellular elicitor molecules produced as the result of expression of the corresponding *avr* gene. For example, in the case of *avr9*, the elicitor molecule is a peptide consisting of 22 amino acids and binds to the receptor product of the *Cf9* R gene. A potential sixth class of R proteins may be coded by Arabidopsis genes RPW8.1 and RPW8.2, which individually provide resistance against a broad range of powdery mildew pathogens. RPW8 proteins have limited homology to NBS-LRR proteins, but induce localized, salicylic acid-dependent defenses similar to those induced by R genes that control specific resistance, with the important difference that RPW8 genes induce broad resistance.

Depending on their structural characteristics, plant receptors can be classified under different categories, such as receptor-like protein kinases (RLKs), histidine kinase receptors, and receptors with different numbers of transmembrane domains. The most important receptors in relation to their recognition of a pathogen are RLKs, of which, apparently, there are hundreds in each plant species. RLKs have an extracellular domain that seems to be involved in signal recognition, a transmembrane domain, and a cytoplasmic kinase domain, which may be the one that initiates a cascade of signal transduction in the cell. All the RLKs studied so far are of the serine–threonine type and, depending on the structural characteristics of the extracellular domain, the receptor-like protein kinases have been subdivided into different categories (Fig. 6-13). The variety of RLKs and the large number of them present in plants suggest that RLKs may be involved in the recognition of many and variable stimuli, in addition to those in plant–pathogen interactions. For example, some RLKs are the products of R genes, e.g., Xa21 from rice that confers resistance to the bacterium *Xanthomonas oryzae* pv. *oryzae*; several R genes actually encode cytoplasmic proteins that are related to RLKs, such as the kinase encoded by the Pto gene, which is involved in resistance against *P. syringae*. Several RLKs are involved in the plant defense responses to pathogen attacks. Some RLKs are induced by oxidative stress, salicylic acid, and pathogen attack, wounding, and bacterial infection. Furthermore, there are RLKs that structurally resemble pathogenesis-related (PR) proteins, chitinase, or have lectin-like motifs. By far the best-studied receptor system for a general pathogen elicitor is the flagellin receptor, which seems to be very similar in both plant and animal systems.

Recognition of Avr Proteins of Pathogens by the Host Plant

Although the number of R genes for which the matching *Avr* gene has been cloned is increasing steadily, in very few of the studied host–pathogen interactions has it been shown that there is a direct interaction of R and *Avr* gene products. In many host–pathogen relationships there is no physical interaction between R and Avr proteins and it appears that the recognition of Avr proteins

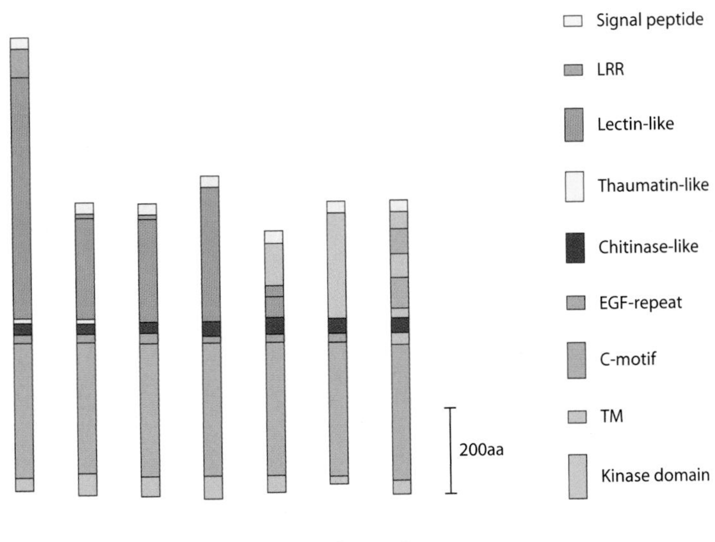

FIGURE 6-13 Schematic diagram of plant receptor-like protein kinases (RLKs) that may be involved in the recognition of elicitors and signaling of plant responses. All contain a serine–threonine kinase domain while their extracellular domains resemble different sequence motifs. (A) Leucine-rich repeats containing Xa21 from rice. (B) Leucine-like AthLecRK1 from Arabidopsis. (C) PR protein thaumatin-like PR5K from Arabidopsis. (D) PR protein chitinase-like from tobacco. (E) Epidermal growth factor-like WAK1 from Arabidopsis. (F) Dissimilar from known sequences RLK10 from wheat. (G) Bimodal cysteine motif-exhibiting StPRKs from potato. [From Montesano *et al.* (2003) *Plant Pathol.* 4, 73–79.]

by R proteins is indirect, i.e., through at least one-third component to which the Avr protein binds and is recognized. This implies that a correlation exists between the binding affinity of the Avr protein for the third component and the level of its HR-inducing activity. It is speculated that the third component may be a coreceptor of the Avr protein or possibly the virulence target of the Avr protein. Binding of the Avr protein to its virulence target serves as a signal to the R gene, which acts as a "guardian" of this virulence target and which then initiates the defense responses and defeat of the pathogen. However, absence of binding by the R protein will result in a lack of defense responses, leading to susceptibility of the host and victory of the pathogen. Of course, if the third component is indeed a virulence target, one would expect a correlation between the Avr proteins' contribution to virulence and its HR-inducing activity.

How Do R and *Avr* Gene Products Activate Plant Defense Responses?

It is assumed that once the R proteins recognize, directly or indirectly, the Avr proteins, they activate signaling networks that lead to resistance responses. Although several components of the signaling network have been identified, the mechanisms by which the R gene products and the so-far identified signaling components activate the host plant defense responses are still poorly understood.

The fact that R proteins share structural similarities suggests that, following recognition of the pathogen protein, the host plants use common signal transduction pathways. This is supported by the fact that resistance responses activated by various R proteins are similar. Such responses commonly include rapid ion fluxes, generation of superoxide and nitric oxide, and a hypersensitive response that includes localized cell death. It is also known that there are several signaling components that are utilized by more than one R proteins.

Some Examples of Plant Defense through R genes and their Matching *Avr* Genes

The Tomato Pto Gene

In many cases, the predicted structures of known R proteins provide some clues as to how the different protein classes may operate as receptors of *Avr* gene products and as generators and transducers of defense signals. For example, the Pto R gene of tomato, which confers resistance to the bacterium *P. syringae* pv. *tomato* (*Fig. 6-14*), codes for a cytoplasmic protein kinase that appears to interact directly with the bacterial avrPto protein that is delivered by the bacterium directly into the plant cell cytoplasm. The Pto kinase protein can interact with several other proteins, including another kinase and some that have homology to transcription factors. Some of these transcription factors possess a DNA-binding domain that recognizes a sequence present in the pro-

FIGURE 6-14 *Xanthomonas* bacteria (A) and tomato bacterial speck symptoms on tomato leaf (B) and fruit (C). (Photographs courtesy of R. J. McGovern.)

moters of genes that encode ethylene-induced defense-related proteins such as PR proteins. For example, when one of the transcription factor genes is overexpressed in a Pto R gene plant, the avrPto-mediated hypersensitive response is enhanced, which shows that the Pto protein can activate several distinct signaling pathways simultaneously. It has been shown, however, that the expression of Pto requires the presence and expression of another gene, *Prf*, which is located within the Pto gene cluster. *Prf* also encodes an LZ-NBS-LRR protein whose role in plant defense is still unknown. More recent work indicates that, perhaps, Pto is not the true R gene, but encodes the virulence target of AvrPto. The AvrPto–Pto complex is then recognized by the true R protein of the *Prf* gene, which is, presumably, "guarding' the virulence target. It appears that currently available data support an indirect recognition of AvrPto by *Prf* rather than a direct recognition of AvrPto by Pto; therefore, the interaction between AvrPto and Pto should not be considered an example of direct interaction of an Avr with its R gene but rather as interaction between an Avr protein and its virulence target.

The Tobacco N Gene

The class of cytoplasmic TIR-NBS-LRR R proteins appears in diseases caused by biotrophic fungi, bacteria, viruses, nematodes, and insects. All three domains of the N gene protein are required for proper N function. In the tobacco mosaic virus (TMV) disease (Fig. 6-15), replicase proteins of the virus confer avirulence to the virus in cultivars carrying the N gene. The N gene encodes a cytoplasmic TIR-NBS-LRR protein. The TMV genome encodes two replicase proteins, and a region of each of these proteins, which serves as the helicase of the virus, can induce a HR in tobacco carrying the N gene. The helicase function of the protein is not required for the avirulence function of the replicase. Whether recognition of the replicase protein by the N protein is direct or indirect is still unknown as is the signaling pathway for development of the defense responses. In other virus–plant combinations studied, avirulence is conferred by a portion of the viral coat protein to a host that carries matching R genes for resistance. No further information of how defense responses are triggered is available.

The Rice Pi-ta Gene

Of the fungal avr proteins, some of *Magnaporthe grisea*, the cause of rice blast on rice (Figs. 6-16A and 6-16B), and of *Cladosporium fulvum*, the cause of leaf mold on tomato, have been elucidated best. The rice blast fungus carries the avirulence gene avr-Pi-ta effective on rice cultivars carrying the resistance gene Pi-ta. Pi-ta encodes a cytoplasmic protein that contains an NBS domain and a leucine-rich carboxyl terminus. Direct interaction has been detected between the Avr-Pi-ta protein and the leucine-rich domain of Pi-ta. This is the first experimental evidence that an AVR protein interacts directly with its R protein. The predicted protease activity of AVR-Pi-ta is required for its avirulence function. How the AVR-Pita/Pi-ta interaction leads to defense responses is still unknown.

The Tomato Cf Genes

In the tomato leaf mold disease, strains of the fungus *C. fulvum* carrying any of the genes Avr2, Avr4, or Avr9 confer avirulence to tomato plants carrying the

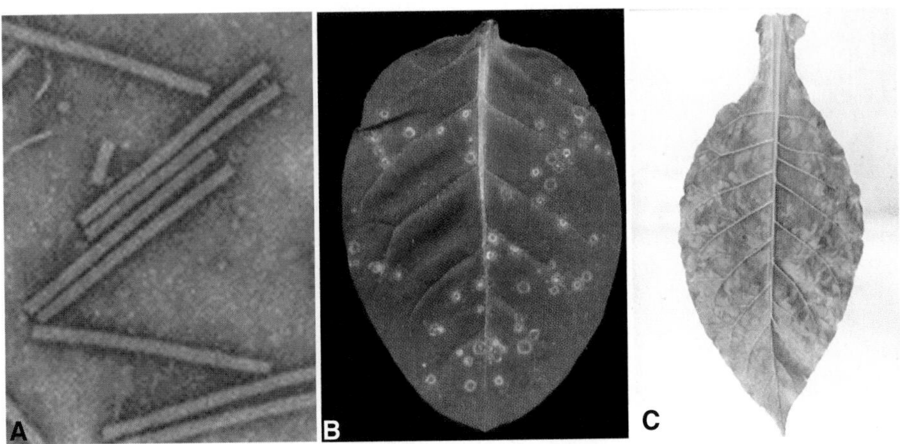

FIGURE 6-15 (A) Particles of *tobacco mosaic virus*. (B) Local lesions (hypersensitive response) on a resistant tobacco leaf. (C) Systemic mosaic symptoms on a leaf of a compatible (susceptible) tobacco plant.

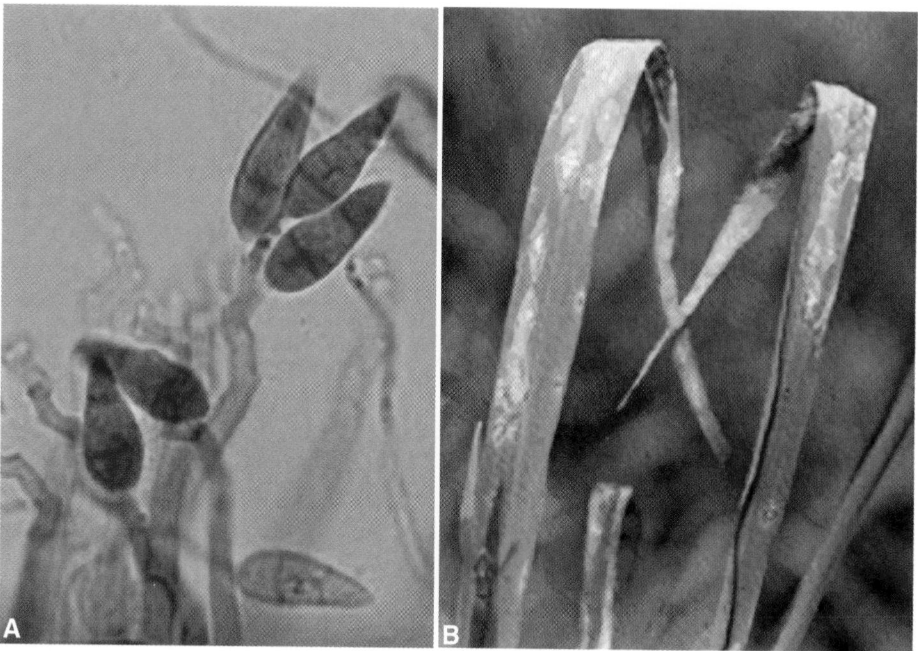

FIGURE 6-16 (A) Conidia of the rice blast fungus *Magnaporthe grisea.* (B) Individual lesions and further development of rice blast on a susceptible plant. [Photographs courtesy of (A) T. E. Freeman, University of Florida, and (B) J. Kranz, University of Giessen, Germany.]

matching resistance R genes *Cf2*, *Cf4*, *or* *Cf9*. Avr2 encodes an extracellular cysteine-rich protein that is secreted by the fungus during growth in the apoplastic space of tomato leaves. No virulence function has been detected in the Avr2. The Cf2 protein consists of a signal peptide, an extracellular LRR region, a transmembrane region, and a short cytoplasmic tail that has no homology to known signaling motifs. The Avr2 protein is recognized by Cf2 extracellularly. Cf2 specifically requires another gene, Rcr3, in order to mediate its resistance, but Rcr3 is not required for Cf5- or Cf9-mediated resistance. As these genes are more than 90% genetically identical, they seem to activate the same defense signaling pathway after the elicitor is recognized. Thus, Rcr3 might represent the third component required for the recognition of AVR2 by Cf2. If Rcr3 indeed binds to AVR2, then Rcr3 must be at least partially extracellular. Another *C. fulvum* avirulence gene confers resistance to tomato cultivars carrying the R gene Cf9. The Cf9 R protein is localized in the plasma membrane but resembles the Cf2 R protein in most respects. The AVR9 protein, also produced in the apoplastic space of tomato leaves, encodes a protein that is processed to a 28 amino acid peptide. The AVR9 protein does not have a virulence function, but because the expression of Avr9 is induced under reduced nitrogen conditions, perhaps the gene plays a role in nitrogen metabolism of the fungus. No specific binding of the proteins of Avr9 and Cf9 genes was detected, although there is a high-affinity binding site for AVR9 in plasma membranes of tomato and other solanaceous plants. It has been suggested that perhaps these binding sites are the third component required for recognition of AVR9 by Cf9.

The Tomato Bs2 Gene

Of the other bacterial avr proteins, the AvrBs2 of *Xanthomonas campestris* pv. *vesicatoria* on pepper and several avr proteins produced by various pathovars of *Pseudomonas syringae* on their specific hosts, are the best studied so far. In the *X. campestris* pv. *vesicatoria*/pepper combination, the Bs2 codes for an NBS-LRR protein that has a hydrophobic N terminus. In addition to conferring resistance to peppers with the Bs2 R gene, the avrBs2 gene, which was shown to be highly conserved among different strains of X. *campestris* pv. *vesicatoria* and among other pathovars of *X. campestris*, is needed for full virulence of the bacterium on susceptible hosts. The avrBs2 encodes a mainly hydrophilic protein, of which the C-terminal half has homology with enzymes that synthesize or hydrolyze phosphodiester linkages, but whether this relates to its role in virulence is not known. There is a correlation between reduced virulence in susceptible hosts and in HR-inducing activity exhibited by various bacterial strains, and this may indicate indirect recognition of AvrBs2 by Bs2 after the AvrBs2 protein binds to its virulence target. Recently, however, a mutant strain was found that could not trigger a resistant response in plants carrying Bs2 and yet it showed no reduction in its virulence in susceptible plants. Since this observation appears to uncouple the virulence and the avirulent functions of AvrBs2, it is not likely that recognition of AvrBs2 occurs after binding to its virulence target.

The Arabidopsis RPM1 Gene

The avrRpm1 gene of *P. syringae* pv. *maculicola* confers avirulence to the bacterium on pea, bean, soybean, and Arabidopsis but is also required for virulence of the same bacterium on Arabidopsis. Recognition of the AvrRpm1 in Arabidopsis requires the presence of the RPM1 gene. This gene encodes a peripheral membrane protein with LZ-NBS-LRR that probably resides at the cytoplasmic face of the plasma membrane. The RPM1 gene also confers resistance to *P. syringae* pv. *glycinea* expressing the avrB gene. The proteins encoded by avrRpm1 and avrB do not share homology except for an N-terminal eukaryotic consensus sequence for two fatty acids, myristic and palmitic. These sequences of AvrRpm1 and AvrB are required for the expression of full virulence and for localization of these proteins at the plasma membrane of the host cell. These observations suggest that AvrRpm1 and AvrB proteins are recognized by the RPM1 protein at the cytoplasmic face of the plasma membrane. It has been shown that recognition of both AvrRpm1 and AvrB by RPM1 requires the presence of RPM1-interacting protein 4 (RIN4), which is also probably localized at the plasma membrane. In the absence of RPM1, AvrRpm1 and AvrB form a complex with RIN4, which is predicted to be their virulence target, as it is a negative regulator of defense responses. These defense responses may be repressed after AvrRpm1 and AvrB bind to RIN4. In uninfected cells, RIN4 is present as a complex with RPM1. These observations support the suggestion that recognition of AvrRpm1 and AvrB by RPM1 is indirect and that the third component required for recognition is the virulence target RIN4.

The Cofunction of Two or More Genes

In many cases, expression of resistance mediated by several R proteins requires the presence of certain other genes. The proteins of these genes have the property to associate with a complex containing an ubiquitin ligase, which brings about ubiquitylation of certain other proteins. When substrate proteins become polyubiquitylated, they are targeted for degradation by the 26S proteasome. According to one theory, because the proteins targeted for degradation can be resistance regulators, degradation and removal of suspected negative regulators of resistance actually activate and set in motion the resistance responses. However, it is possible that monoubiquitylation regulates protein localization and the activity of several kinases and transcription factors and, therefore, the complex of ubiquitin with the other gene products mediates the translocation or activation of resistance regulators.

Defense Involving Bacterial Type III Effector Proteins

Most pathogenic bacteria have three types of secretion systems by which they secrete exoenzymes and other pathogenicity factors. The type I secretion system allows bacteria to secrete proteases from the cytoplasm to the extracellular space of the bacterium in a single step. Type I secretion plays a minor role in pathogenicity. The type II secretion system makes it possible for bacteria to secrete pathogenicity determinants like pectinases and cellulases and is essential for pathogenicity. The type II system employs a two-step mechanism for secretion. First, proteins are exported to the periplasm of bacteria. Then, a structure forms that spans the periplasmic compartment and the outer membrane and proteins marked by a special signal sequence are channeled through. The type II system is regulated in part by a quorum-sensing mechanism.

The type III secretion system (TTSS) consists of a set of 15 to 20 proteins associated with the bacterial cell membrane and making up the secretion apparatus that delivers or translocates host-specific "effector" proteins from the bacteria into their host plant cells (Fig. 6-17). The membrane-bound proteins are common to most kinds of bacteria that have type III secretion systems, whereas the proteins injected by them into their host cells are specific for that host plant. By translocating these bacterial "effector" proteins into their host cells, the TTSS interferes with host cell signal transduction and other cellular processes, thereby enhancing the virulence of bacteria in susceptible host cells. During delivery, a chaperon protein is bound to each "effector" protein that apparently protects the effector protein from premature interactions with other proteins. The type III secretion system occurs in all or most gram-negative pathogenic bacteria (*Erwinia*, *Pseudomonas*, *Xanthomonas*, *Ralstonia*, *Pantoea*), including those causing disease in humans and animals.

Proteins delivered into nonhost plant cells by type III secretion systems can elicit a hypersensitive response. For this reason, the TTSS is known as the hypersensitive response and pathogenicity (hrp) system. Most type III effectors from plant pathogenic bacteria were first identified as the products of typical avirulence (*avr*) genes. In bacteria, avirulence genes are defined as genes

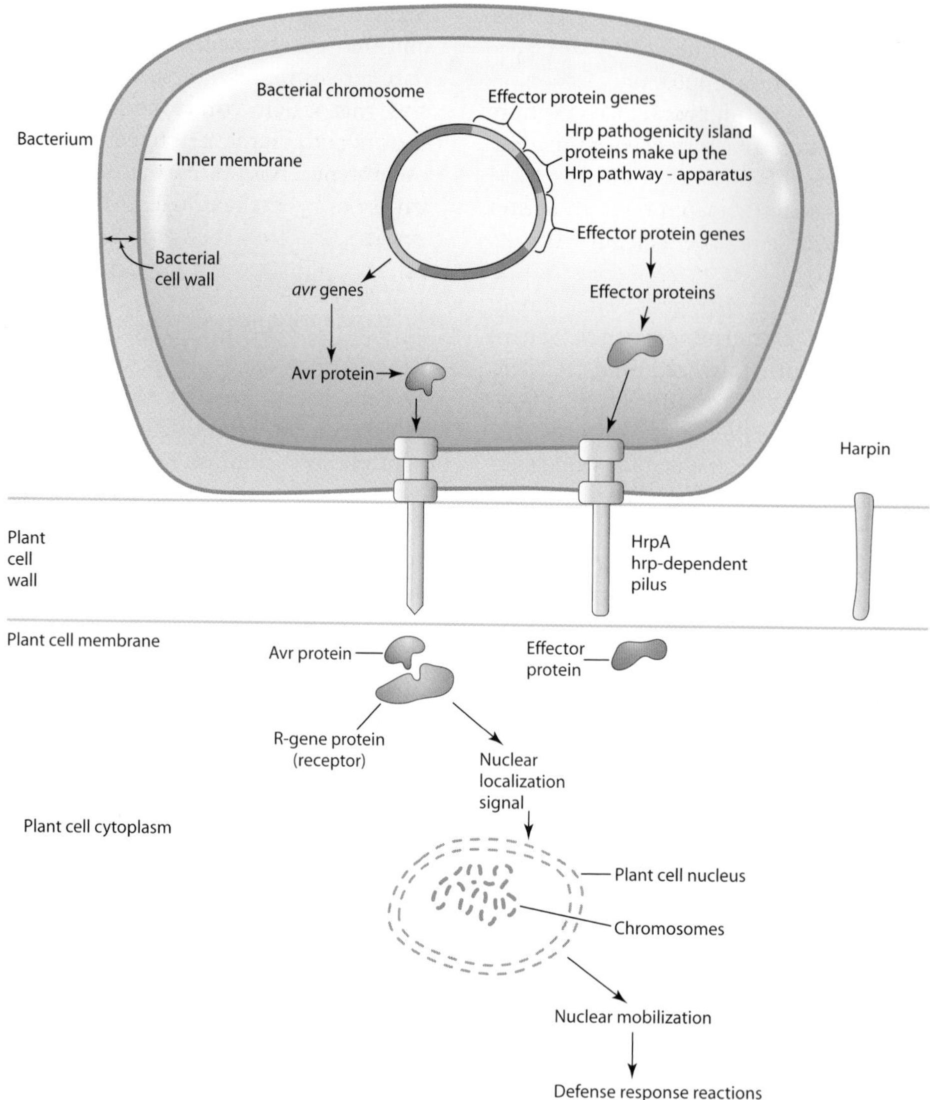

FIGURE 6-17 Diagrammatic representation of the hrp or type III secretion system in bacteria.

that can convert a normally virulent bacterial strain that infects a specific host to an avirulent one in regard to that particular host. Avirulence is usually manifested as appearance of an HR reaction on a resistant host. Because the induction of HR depends on the presence and reaction of R genes and matching hrp genes, it was thought, and later proven, that the products of hrp genes are secreted by the TTSS directly into the cytoplasm of R gene-containing host cells and lead to the induction of a hypersensitive response, i.e., cell death. It has been shown, however, that many *avr* genes that normally contribute to the defense of the host plant by being the elicitors of the HR, in the TTSS they also contribute to virulence of the bacterium by promoting more severe symptoms produced by the plant, more bacteria growing inside the leaf, and more bacteria escaping to the leaf surface. In most cases, the contribution of *avr* genes to virulence is small. Because, however, the secretion of effectors is essential for pathogenicity, it is apparent that bacteria secrete multiple effector proteins and that, therefore, they contribute to virulence in an incremental quantitative or partially redundant manner. *Xanthomonas* and *Pseudomonas* bacteria colonize the intercellular spaces (apoplast) of leaf surfaces where there are plenty of nutrients for the bacteria but where water may be a limiting factor. So, the bacteria would benefit from a susceptibility response involving leakage of water from the host cells (symplast) to the intercellular spaces (apoplast). However, the plant would benefit from a defense response, such as cell wall thickening, that would deprive the infection site from water. For the bacteria to continue to grow, they must avoid inducing host defense responses, suppress host defense responses successfully, or both.

Harpin protein, produced by the bacterium *P. syringae* pv. *syringae*, is the hrp-dependent protein that differs from most Avr proteins in that, when injected in the leaf apoplast, it can induce a hypersensitive-like response (i.e., cell death). This implies that Harpin might function on the outside of the plant cell. The Harpin HR-like response differs from HRs induced by Avr genes in that it does not depend on a matching R gene. Harpin can associate with liposomes and with bilayer membranes on which it apparently forms pores; through the pores, then, water and nutrients move out of the cell to the apoplast or, more likely, the pores serve as openings so that other types of effectors can be translocated into the cells.

Several *avr* genes have been implicated in the suppression of host defenses. Thus, the *P. syringae* pv. *phaseolicola RW60* strain can be converted by the *avrPphF* gene from avirulent to virulent on a particular bean cultivar (A) in which it suppresses the hypersensitive response. Interestingly, the same *avr* gene in the same bacterial strain (RW60) increases the HR induced by RW60 in another bean cultivar (B). This enhancement of the HR by the first *avr* gene can be suppressed by another *avr* gene. Because the suppression of HR by these genes is host specific, this points to a molecular "arms race" between bacterial effectors and host targets. It appears that the target of one effector (e.g., AvrPphC) is the R gene product, which will detect a second effector (e.g., AvrPphF). Other examples of suppression of defense responses by type III effectors are known, suggesting that such genes may interfere with the induction of defense responses at a level similar to the infection by the avirulent and the virulent pathogens.

Active Oxygen Species, Lipoxygenases, and Disruption of Cell Membranes

The plant cell membrane consists of a phospholipid bilayer in which many kinds (Figure 5-2) of protein and glycoprotein molecules are embedded. The protein molecules are often organized in groups, some of which form channels on the membrane and allow ions and metabolites to enter and exit the cell. The cell membrane in the form of endoplasmic reticulum and organelles compartmentalizes the cell into areas in which specific compounds are kept separated from others and certain biochemical reactions take place. In addition, the cell membrane is an active site for the induction of defense mechanisms; e.g., it serves as the anchor of R gene-coded proteins that recognize the elicitors released by the pathogen and subsequently trigger the hypersensitive response.

The attack of cells by pathogens, or exposure to pathogen toxins and enzymes, often results in structural and permeability changes of the cell membrane. These changes are generally thought to be an expression of susceptibility and disease development. In many host–pathogen combinations, however, particularly those involving the hypersensitive response, some membrane changes play a role in the defense against invasion by the pathogen. The most important membrane-associated defense responses include (1) the release of molecules important in signal transduction within and around the cell and, possibly, systemically through the plant; (2) the release and accumulation of reactive oxygen "species" and of lipoxygenase enzymes; and (3) as a result of the loss of compartmentalization, activation of phenol oxidases and oxidation of phenolics (Figures 4-13, 6-11).

In many host–fungus interactions, one of the first events detected in attacked host cells, or cells treated artificially with fungal elicitors, is the rapid and transient generation of activated oxygen species, including superoxide (O_2^-), hydrogen peroxide (H_2O_2), and hydroxyl radical (OH). The generation of superoxide and of other reactive oxygen species as defense response happens most dramatically in localized infections, but it also occurs in general and systemic infections, as well as in plants treated with chemicals that induce systemic acquired resistance. These highly reactive oxygen species are thought to be released by the multisubunit NADPH oxidase enzyme complex of the host cell plasma membrane. They appear to be released in affected cells within seconds or minutes from contact of the cell with the fungus or its elicitors and reach a maximum activity within minutes to a few hours.

The activated oxygen species trigger the hydroperoxidation of membrane phospholipids, producing mixtures of lipid hydroperoxides. The latter are toxic, their production disrupts the plant cell membranes, and they seem to be involved in normal or HR-induced cell collapse and death. Active oxygen species may also be involved in host defense reactions through the oxidation of phenolic compounds into more toxic quinones and into lignin-like compounds. The presence of active oxygen species, however, also affects the membranes and the cells of the advancing pathogen either directly or indirectly through the hypersensitive response of the host cell. The production of reactive oxygen species in affected but surviving nearby cells is kept under control by the radical scavenger enzymes superoxide dismutase, catalase, ascorbate peroxidase, etc. Several isoenzymes of each of these are produced, with different ones of them appearing at different stages after inoculation.

The oxygenation of membrane lipids seems to involve various lipoxygenases as well. These are enzymes that catalyze the hydroperoxidation of unsaturated fatty acids, such as linoleic acid and linolenic acid, which have been released previously from membranes by phospholipases. The lipoxygenase-generated hydroperoxides formed from such fatty acids, in addition to disrupting the cell membranes and leading to HR-induced cell collapse of host and pathogen, are also converted by the cell into several biologically active molecules, such as jasmonic acid, that play a role in the response of plants to wounding and other stresses. Jasmonic acid, for example, which is the precursor of the wound hormone traumatin, appears to induce numerous protein changes and acts as a signal transducer of the defense reaction in plant–pathogen interactions.

Reinforcement of Host Cell Walls with Strengthening Molecules

In several plant diseases caused by fungi, the walls of cells that come in contact with the fungus produce, modify, or accumulate several defense-related substances that reinforce the resistance of the wall to invasion by the pathogen. Among the defensive substances produced or deposited in plant cell walls being invaded by fungi are callose, glycoproteins such as extensin that are rich in the amino acid hydroxyproline, phenolic compounds of varying complexity including lignin and suberin, and mineral elements such as silicon and calcium. Some of these substances are also produced or deposited in defensive cell wall structures such as the papillae. Many of these substances form complex polymers and also react and cross-link with one another, thereby forming more or less insoluble cell wall structures that confine the invading fungus and prevent the further development of disease. Of course, in cases in which the host lacks resistance or exhibits incomplete resistance, apparently the host, with or without interference by fungal secretions, fails to produce reinforcing compounds or produces them too slowly to be effective and the fungus manages to invade the cell.

Production of Antimicrobial Substances in Attacked Host Cells

Pathogenesis-Related Proteins

Pathogenesis-related proteins, often called PR proteins, are a structurally diverse group of plant proteins that are toxic to invading fungal pathogens. They are widely distributed in plants in trace amounts, but are produced in much greater concentration following pathogen attack or stress. PR proteins exist in plant cells intracellularly and also in the intercellular spaces, particularly in the cell walls of different tissues. Varying types of PR proteins have been isolated from each of several crop plants. Different plant organs, e.g., leaves, seeds, and roots, may produce different sets of PR proteins. Different PR proteins appear to be expressed differentially in their hosts in the field when temperatures become stressful, low or high, for extended periods.

The several groups of PR proteins have been classified according to their function, serological relationship, amino acid sequence, molecular weight, and certain other properties. PR proteins are either extremely acidic or extremely basic and therefore are highly soluble and reactive. At least 14 families of PR proteins are recognized. The better known PR proteins are PR1 proteins (antioomycete and antifungal), PR2 (β-1,3-glucanases), PR3 (chitinases), PR4 proteins (antifungal), PR6 (proteinase inhibitors) (Fig. 6-19), thaumatine-like proteins, defensins, thionins, lysozymes, osmotinlike proteins, lipoxygenases, cysteine-rich proteins, glycine-rich proteins, proteinases, chitosanases, and peroxidases. There are often numerous isoforms of each PR protein in various host plants.

Although healthy plants may contain trace amounts of several PR proteins, attack by pathogens, treatment with elicitors, wounding, or stress induce transcription of a battery of genes that code for PR proteins (Fig. 6-18). This occurs as part of a massive switch in the overall pattern of gene expression, during which normal protein production nearly ceases. The signal compounds responsible for induction of PR proteins include salicylic acid, ethylene, xylanase, the polypeptide systemin, jasmonic acid, and probably others (Fig. 6-11).

The significance of PR proteins lies in the fact that they show strong antifungal and other antimicrobial activity (Figure 6-19). Some of them inhibit spore release and germination, whereas others are associated with strengthening of the host cell wall and its outgrowths and papillae. Some of the PR proteins, e.g., β-1,3-glucanase and chitinase, diffuse toward and affect (break down) the chitin-supported structure of the cell walls of several but not all plant pathogenic fungi, whereas lysozymes degrade the glucosamine and muramic acid components of bacterial cell walls. Lipoxygenases and lipid peroxidases generate antimicrobial metabolites as well as secondary signal molecules such as jasmonic acid. Structurally similar defensins also occur in mammals, birds, and insects. Plant defensins, which are basic cysteine-rich peptides, have antimicrobial activity and accumulate through the ethylene and jasmonic acid pathway. Plants genetically engineered to express chitinase genes show good resistance against the soilborne fungus *Rhizoctonia solani*. Tobacco plants treated with lipopolysaccharides obtained from the outer wall of gram-negative bacteria produced several PR proteins and exhibited enhanced defense responses in tobacco against *Phytophthora nicotianae*, including the production of a systemic response in the leaves of plants inoculated through the roots. Signal molecules that induce PR protein synthesis seem to be transported systemically to other parts of the

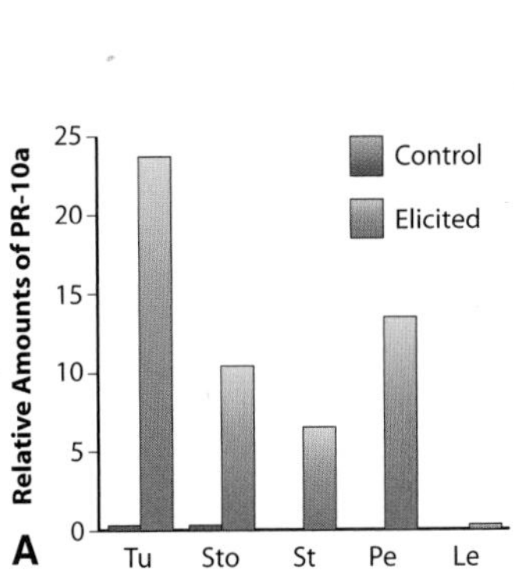

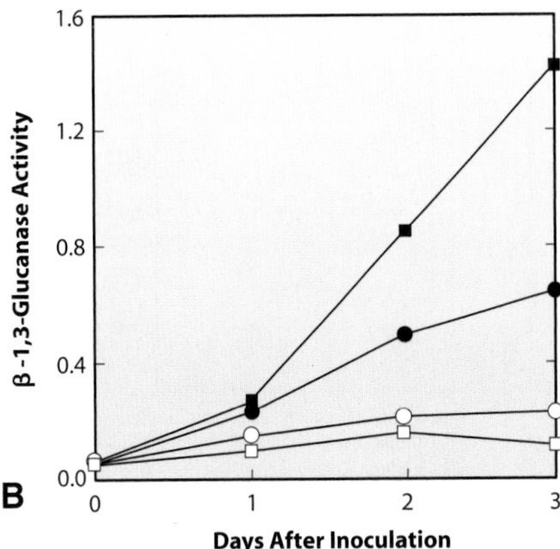

FIGURE 6-18 (A) Production and accumulation of a pathogenesis-related protein (PR10a) in potato tissues either untreated (control) or elicited by treating cut surfaces with a homogenate of the late blight fungus *Phytophthora infestans* and incubating for 4 days. Tu, tuber; Sto, stolon; St, stem; Pe, petiole; Le, leaf. [From Constabel and Brisson (1995). *Mol. Plant-Microbe Interact.* 8, 104–113.] (B) Levels of activity of the antifungal protein β-1,3-glucanase in the intercellular fluid of barley leaves, either left uninoculated (□, ○) or inoculated with the powdery mildew fungus *Erysiphe graminis* f. sp. *hordei* (■, ●). The two barley varieties are nearly isogenic, except that one (□, ■) carries an additional resistance gene that makes it resistant, whereas the other (○, ●) is susceptible. [From Jutidamrongphan *et al.* (1991). *Mol. Plant-Microbe Interact.* 4, 234–238.]

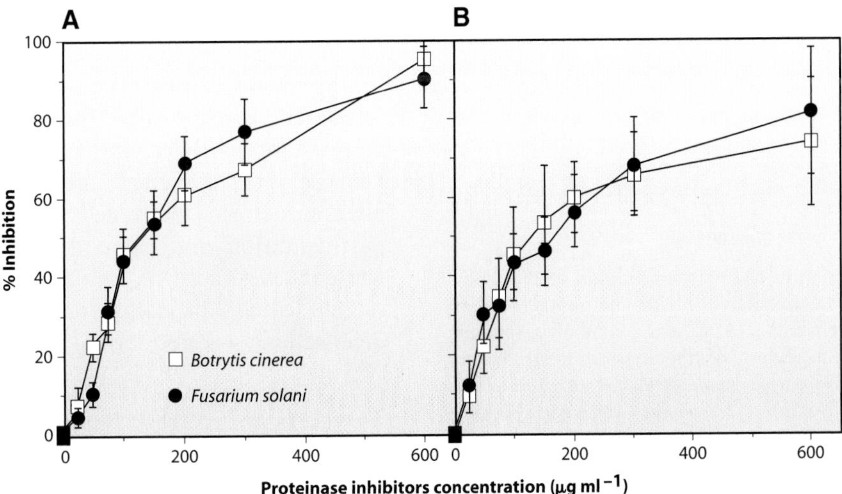

FIGURE 6-19 Inhibition of (A) spore germination and (B) germ tube elongation of fungi *Botrytis cinerea* and *Fusarium solani*, which do not infect cabbage, by proteinase inhibitors obtained from young cabbage leaves. The inhibitors caused leakage of the cellular contents of these fungi. The cabbage fungal pathogen *Alternaria brassicicola* was not affected by these proteinase inhibitors. [From Lorito *et al.* (1994). *Mol. Plant-Microbe Interact.* 7, 525–527.]

plant and to reduce disease initiation and intensity in those parts for several days or even weeks.

Defense through Production of Secondary Metabolites: Phenolics

Simple Phenolic Compounds

It has often been observed that certain common phenolic compounds that are toxic to pathogens are produced and accumulate at a faster rate after infection, especially in a resistant variety of plant relative to a susceptible variety. Chlorogenic acid, caffeic acid, and ferulic acid are examples of such phenolic compounds (Fig. 6-20). In peach, chlorogenic acid is present in quite high concentration both in immature fruit and in fruit of varieties resistant to the brown rot disease caused by the fungus *Monilinia fructicola*. The fruit is resistant in both cases, not because of the toxicity of the acid

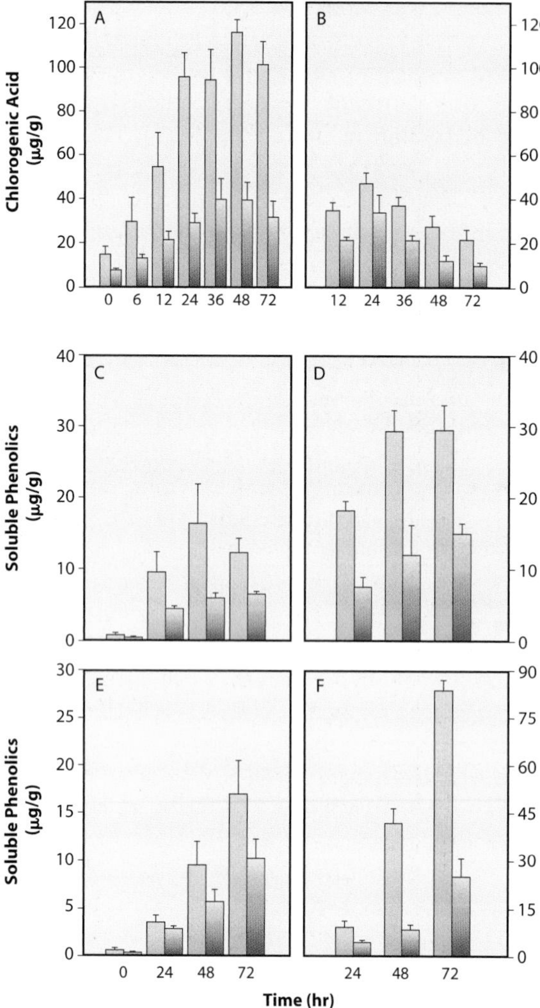

FIGURE 6-20 Production of chlorogenic acid and other soluble and wall-bound phenolics in normal (white bars) and transgenic (dark bars) potato tubers after wounding (A, C, and E) and after spraying with arachidonic acid, an elicitor of the hypersensitive defense response (B, D, and F). Transgenic plants produced an enzyme that inactivates tryptophan, a precursor of phenolics and lignin. Chlorogenic acid was increased by wounding but not by elicitation. Soluble and wall-bound phenolics increased after wounding and even more following treatment with the elicitor, but the increase was smaller in the transgenic tubers (dark bars) than in the normal tubers. Accordingly, the transgenic tubers in these treatments were more susceptible to infection when inoculated with zoospores of *Phytophthora infestans* than the treated normal plants. [From Yao *et al.* (1995). *Plant Cell* 7, 1787–1799.]

to the causal fungus, but rather because it inhibits the production of fungal enzymes that cause degradation of host tissue. In date palm tree roots, cell wall-bound hydroxybenzoic acid and sinapic acid increased 11–12 times as much in cultivars resistant to *Fusarium* than they did in susceptible cultivars. In plants such as vetch (*Vicia sativa*), resistance to the higher parasitic plant *Orobanche aegyptiaca* appears to result from higher levels of free and bound phenolics, lignin and peroxidase activity produced in the roots of resistant varieties following infection, compared to susceptible ones. In cacao infected with the witches' broom fungus *Crinipellis perniciosa*, infected young stems contain 7–8 times as much caffeine, which inhibits growth of the fungus in culture, than healthy stems. In another polygenic disease, the black sigatoka disease of banana caused by the fungus *Mycosphaerella fijiensis*, plant defenses included an activation of phenylalanine ammonia lyase and a subsequent accumulation of phenolic compounds. It also caused early activation of a banana response to the fungal compound trihydroxytetralone (THT), which, in resistant varieties, caused necrotic microlesions and elicitation of infection-induced defense reactions leading to incompatibility (resistance) between the pathogen and the host plant. In susceptible varieties, however, the fungus produced necrotizing levels of THT only at the later stages of pathogenesis after a compatible interaction had been established and typical symptoms had developed. Although some of the common phenolics may each reach concentrations that could be toxic to the pathogen, it should be noted that several of them appear concurrently in the same diseased tissue, and it is possible that the combined effect of all fungitoxic phenolics present, rather than that of each one separately, is responsible for the inhibition of infection in resistant varieties. It has even been proposed that because of the universal uniform or strategic location of phenolics-storing plant cells, these cells can, by decompartmentation and rapid oxidation of their phenolic contents, self-sacrifice, leading to the first line of defense — cell death — or leading to the production of a slower defense line — a peridermal defense layer.

Toxic Phenolics from Nontoxic Phenolic Glycosides

Many plants contain nontoxic glycosides, i.e., compounds consisting of a sugar (such as glucose) joined to another, often phenolic, molecule. Several fungi and bacteria are known to produce or to liberate from plant tissues the enzyme glycosidase that can hydrolyze such complex molecules and release the phenolic compound from the complex. Some of the released phenolics are quite toxic to the pathogen, especially after further oxidation, and appear to play a role in the defense of the plant against infection.

Role of Phenol-Oxidizing Enzymes in Disease Resistance

The activity of many phenol-oxidizing enzymes (polyphenol oxidases) is generally higher in the infected tissue of resistant varieties than in infected susceptible ones or in uninfected healthy plants. The importance of polyphenol oxidase activity in disease resistance probably stems from its property to oxidize phenolic compounds to quinones, which are often more toxic to microorganisms than the original phenols. It is reasonable to assume that an increased activity of polyphenol oxidases will result in higher concentrations of toxic products of oxidation and therefore in greater degrees of resistance to infection. A complex interaction occurs during fruit ripening in which levels of lipoxygenases increase and break down diene, a compound that is present in young, immature fruit and is toxic to fungi. These events normally result in infection (loss of resistance) of the ripening fruit. In some fruit, however, elicitors from nonpathogenic fungi stimulate production of the phenolic compound epicatechin, which inhibits the activity of lipoxygenases. As a result, epicatechin decreases degradation of the antifungal diene, thereby preventing decay of the ripening fruit by anthracnose fungi.

Another phenol oxidase enzyme, peroxidase, both oxidizes phenolics to quinones and generates hydrogen peroxide. The latter not only is antimicrobial in itself, but it also releases highly reactive free radicals and in that way further increases the rate of polymerization of phenolic compounds into lignin-like substances. These substances are then deposited in cell walls and papillae and interfere with the further growth and development of the pathogen.

Phytoalexins

Phytoalexins are toxic antimicrobial substances produced in appreciable amounts in plants only after stimulation by various types of phytopathogenic microorganisms or by chemical and mechanical injury. Phytoalexins are produced by healthy cells adjacent to localized damaged and necrotic cells in response to materials diffusing from the damaged cells. Phytoalexins are not produced during compatible biotrophic infections. Phytoalexins accumulate around both resistant and susceptible necrotic tissues. Resistance occurs when one or more phytoalexins reach a concentration sufficient to restrict pathogen development. Most known phytoalexins are toxic to and inhibit the growth of fungi pathogenic to plants, but some are also toxic to bacteria, nematodes, and other organisms. More than 300 chemicals with phytoalexinlike properties have been isolated from plants belonging to more than 30 families. The chemical structures of phytoalexins produced by plants of a family are usually quite similar; e.g., in most legumes, phytoalexins are isoflavonoids, and in the Solanaceae they are terpenoids. Most of the phytoalexins are produced in plants in response to infection by fungi, but a few bacteria, viruses, and nematodes have also been shown to induce the production of phytoalexins. Some of the better studied phytoalexins include phaseollin in bean (Fig. 6-21); pisatin in pea; glyceollin in soybean, alfalfa, and clover; rishitin in potato; gossypol in cotton; and capsidiol in pepper.

Phytoalexin production and accumulation occur in healthy plant cells surrounding wounded or infected cells and are stimulated by alarm substances produced and released by the damaged cells and diffusing into the adjacent healthy cells. Most phytoalexin elicitors are generally high molecular weight substances that are constituents of the fungal cell wall, such as glucans, chitosan, glycoproteins, and polysaccharides. The elicitor molecules are released from the fungal cell wall by host plant enzymes. Most such elicitors are nonspecific, i.e., they are present in both compatible and incompatible races of the pathogen and induce phytoalexin accumulation irrespective of the plant cultivar. A few phytoalexin elicitors, however, are specific, as the accumulation of phytoalexin they cause on certain compatible and incompatible cultivars parallels the phytoalexin accumulation caused by the pathogen races themselves. Although most phytoalexin elicitors are thought to be of pathogen origin, some elicitors, e.g., oligomers of galacturonic acid, are produced by plant cells in response to infection or are released from plant cell walls after their partial breakdown by cell wall-degrading enzymes of the pathogen.

The formation of phytoalexins in a susceptible (compatible) host following infection by a pathogen seems, in some cases, to be prevented by suppressor molecules produced by the pathogen. The suppressors seem to also be glucans or glycoproteins, or one of the toxins produced by the pathogen.

The mechanisms by which phytoalexin elicitors, phytoalexin production, phytoalexin suppressors, genes for resistance or susceptibility, and the expression of resistance or susceptibility are connected are still not well understood. Several hypotheses have been proposed to explain the interconnection of these factors, but much more work is needed before a satisfactory explanation can be obtained.

Species or races of fungi pathogenic to a particular plant species seem to stimulate the production of generally lower concentrations of phytoalexins than nonpathogens. For example, in the case of pisatin production by pea pods inoculated with the pathogen *Ascochyta pisi*, pea varieties produce concentrations of pisatin that are approximately proportional to the resistance of the variety to the pathogen. When the same pea variety is inoculated with different strains of the fungus, the concentration of pisatin produced is approximately

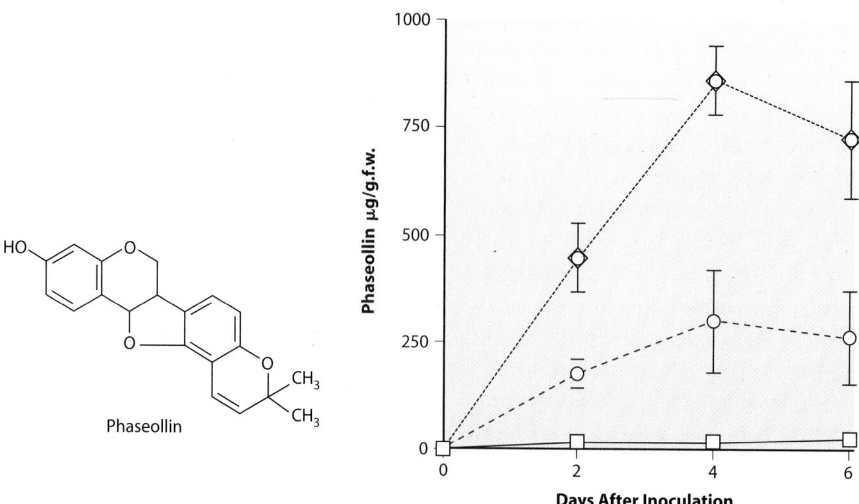

FIGURE 6-21 Levels of the phytoalexin phaseollin produced at infection sites in bean pods following inoculation with three races of the halo blight bacterium *Pseudomonas syringae* pv. *phaseolicola*. Virulent race 6 (□) infects without causing a defense response nor production of the phytoalexin. The same race 6 was transformed with an avirulence gene corresponding to resistance gene R2 (○) and with an avirulence gene to R3 (◇), and the transformants induced visibly different hypersensitive responses and also different levels of phytoalexin. [From Mansfield *et al.* (1994). *Mol. Plant-Microbe Interact.* 6, 726–739.]

inversely proportional to the virulence of each particular fungal strain inoculated on the pea variety. Also, in soybean plants infected with the fungus *Phytophthora megasperma* f. sp. *glycinea*, inoculations of fungal races on incompatible host cultivars resulted in earlier accumulations and higher concentrations of the phytoalexin glyceollin than inoculations of fungal races on compatible cultivars. It has been suggested that the higher concentrations of glyceollin in incompatible host–pathogen combinations are the result of reduced biodegradation rather than increased biosynthesis of the phytoalexin. In some host-pathogen systems, however, e.g., in the bean/*Colletotrichum lindemuthianum* and the potato/*Phytophthora infestans* systems, the respective phytoalexins, such as phaseollin and rishitin, reach equal or higher concentrations in compatible (susceptible) hosts compared to incompatible (resistant) ones.

However, pathogenic races or species of fungi seem to be less sensitive to the toxicity of the phytoalexin(s) produced by their host plant than nonpathogenic fungi. It has been suggested that pathogens may have an adoptive tolerance mechanism that enables them to withstand higher concentrations of the host phytoalexin after earlier exposures to lower concentrations of the phytoalexin. It is known, however, that many pathogenic fungi can metabolize the host phytoalexin into a nontoxic compound, thereby decreasing the toxicity of the phytoalexin to the pathogen. It is also known that numerous pathogenic fungi are successful in causing disease, although they are sensitive to or unable to metabolize the host phytoalexins. Furthermore, some fungi that can either degrade or tolerate certain phytoalexins are unable to infect the plants that produce them.

In general, it appears that phytoalexins may play a decisive or an auxiliary role in the defense of some hosts against certain pathogens, but their significance, if any, as factors of disease resistance in most host–pathogen combinations is still unknown.

DET\OXIFICATION OF PATHOGEN TOXINS BY PLANTS

In at least some of the diseases in which the pathogen produces a toxin, resistance to disease is apparently the same as resistance to the toxin. Detoxification of at least some toxins, e.g., HC toxin and pyricularin, produced by the fungi *Cochliobolus carbonum* and *Magnaporthe grisea*, respectively, is known to occur in plants and may play a role in disease resistance. Some of these toxins appear to be metabolized more rapidly by resistant varieties or are combined with other substances and form less toxic or nontoxic compounds. The amount of the nontoxic compound formed is often proportional to the disease resistance of the variety.

Resistant plants and nonhosts are not affected by the specific toxins produced by *Cochliobolus*, *Periconia*, and *Alternaria*, but it is not yet known whether the selective action of these toxins depends on the presence of receptor sites in susceptible but not in resistant varieties, on detoxification of the toxins in resistant plants, or on some other mechanism.

IMMUNIZATION OF PLANTS AGAINST PATHOGENS

Defense through Plantibodies

In humans and animals, defenses against pathogens are often activated by natural or artificial immunization, i.e., by a subminimal natural infection with the pathogen or by an artificial injection of pathogen proteins and other antigenic substances. Both events result in the production of antibodies against the pathogen and, thereby, in subsequent prolonged protection (immunity) of the human or animal from infection by any later attacks of the pathogen.

Plants, of course, do not have an immune system like that of humans and animals, i.e., they do not produce antibodies. In the early 1990s, however, transgenic plants were produced that were genetically engineered to incorporate in their genome, and to express, foreign genes, such as mouse genes that produce antibodies against certain plant pathogens. Such antibodies, encoded by animal genes but produced in and by the plant, are called plantibodies. It has already been shown that transgenic plants producing plantibodies against coat proteins of viruses, e.g., *artichoke mottle crinkle virus*, to which they are susceptible, can defend themselves and show some resistance to infection by these viruses. It is expected that, in the future, this type of plant immunization will yield dividends by expressing animal antibody genes in plants that will produce antibodies directed against specific essential proteins of the pathogen, such as viral coat proteins and replicase or movement proteins, and fungal and bacterial enzymes of attack.

Whole antibodies or fragments of antibodies can be expressed easily in plants following integration of a transgene into the plant genome, or by transient expression of the gene using viral vectors, infiltration of the gene by *Agrobacterium*, or through biolistics. Plants such as tobacco, potato, and pea have been shown to be good producers of antibody for pharmaceutical purposes. Plants have been shown to produce functional antibodies that can be used to increase the resistance of plants against specific pathogens. So far, functional plantibodies, produced by plants against specific plant pathogens, that have been shown to increase the resistance of the host plant to that pathogen include the following: Plantibodies to *tobacco mosaic virus* in tobacco decreased infectivity of the virus by 90%; to *beet necrotic yellow vein virus*, also in tobacco, provides a partial protection against the virus in the early stages of infection and against development of symptoms later on; to stolbur phytoplasma and to corn stunt spiroplasma, also in tobacco, which remained free from infection for more than two months. However, attempts to engineer plantibody-mediated resistance to plant parasitic nematodes have been unsuccessful so far. Generally, however, the expression of complete or fragment antibodies in plants has been only partially effective or mostly ineffective so far. Plantibody-derived resistance appears mostly as a delay in the development of disease and, barring a breakthrough, it does not appear that it will become an effective means of plant disease control in the near future.

Resistance through Prior Exposure to Mutants of Reduced Pathogenicity

Inoculation of avocado fruit with a genetically engineered, reduced pathogenicity strain of the anthracnose fungus *Colletotrichum gloeosporioides*, which does produce an appressorium, results in delayed decay of the fruit. Such an inoculation brings about increased levels of biochemical defense indicators, such as H^+-ATPase activity, reactive oxygen species, phenylalanine ammonia lyase, the natural antioxidant phenol epicatechin, the antifungal compound diene, and eventual fruit resistance with delay of fruit decay. However, inoculation of fruit with a similar mutant strain that does not produce an appressorium causes no activation of early signaling events and no fruit resistance. It would appear that initiation of the early signaling events that affect fruit resistance depends on the ability of the pathogen to interact with the fruit and initiate its defense mechanisms during appressorium formation.

SYSTEMIC ACQUIRED RESISTANCE

Induction of Plant Defenses by Artificial Inoculation with Microbes or by Treatment with Chemicals

As discussed earlier, plants do not naturally produce antibodies against their pathogens, and most of their biochemical defenses are inactive until they are mobilized by some signal transmitted from an attacking pathogen. It has been known for many years, however, that plants develop a generalized resistance in response to infection by a pathogen or to treatment with certain natural or synthetic chemical compounds.

Induced resistance is at first localized around the point of plant necrosis caused by infection by the pathogen or by the chemical, and it is then called local

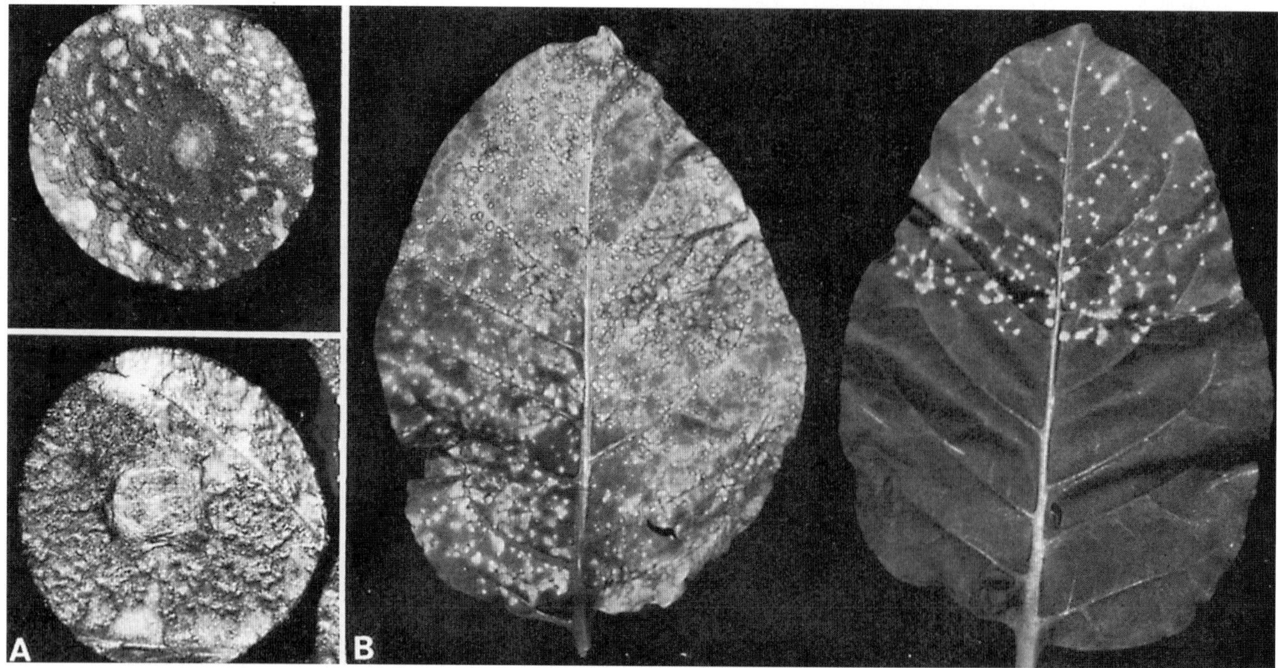

FIGURE 6-22 (A) Development of local acquired resistance to tobacco mosaic virus (TMV) around a local lesion caused by the same virus on a resistant tobacco variety. When the same leaves were reinoculated with TMV seven days later, no new lesions formed near the original one because of local acquired resistance (top), but when they were reinoculated with a different virus, no zone free of lesions remained (bottom). (B) The upper (tip) half of the leaf at the right was inoculated with TMV, and seven days later both leaves were inoculated with the same virus over their entire surface. The leaf at the left developed numerous local lesions throughout, whereas the previously half-inoculated leaf at the right developed almost no additional lesions because of acquired local and systemic resistance. [From Ross (1961). *Virology* **14**, 329–339 and 340–358.]

acquired resistance (Fig. 6-22A). Subsequently, resistance spreads systemically and develops in distal, untreated parts of the plant and is called systemic acquired resistance (Fig. 6-22B). It is known now that several chemical compounds, e.g., salicylic acid, arachidonic acid, and 2,6-dichloroisonicotinic acid, may induce localized and systemic resistance in plants at levels not causing tissue necrosis. Jasmonic acid is another type of compound, derived primarily from oxidation of fatty acids, that leads to systemic acquired resistance, often in cooperation with salicylic acid and ethylene, leading to the production of defensins. Probenazole, a synthetic chemical used in Asia for the control of rice blast disease caused by the fungus *Magnaporthe grisea*, has been shown to act upstream from the salicylic acid transcribing gene and, thereby, causing accumulation of salicylic acid. Probenazole induces systemic acquired resistance in rice against rice blast, in tomato against the bacterial pathogen *P. syringae* pv. *tabaci*, and in tobacco against the *tobacco mosaic virus*. Similarly, riboflavin was shown to induce systemic acquired resistance but it activates it in a distinct manner not involving salicylic acid. Such chemicals may be effective in inducing resistance in plants when they are applied through the roots, as a foliar spray (Fig. 6-23), or by stem injection. Local acquired resistance is induced, for example, in a 1 to 2 mm zone around local lesions caused by tobacco mosaic virus on hypersensitive tobacco varieties and probably in other host–pathogen combinations. Local acquired resistance results in near absence of lesions immediately next to the existing lesion and in smaller and fewer local lesions developing farther out from the existing local lesions when inoculations are made at least 2–3 days after the primary infection. Local acquired resistance may play a role in natural infections by limiting the number and size of lesions per leaf unit area.

Systemic acquired resistance acts nonspecifically throughout the plant and reduces the severity of disease caused by all classes of pathogens, including normally virulent ones. It has been observed in many dicot and monocot plants, but has been studied most in cucurbits, solanaceous plants, legumes, and gramineous plants following infection with appropriate fungi, bacteria, and viruses. Systemic acquired resistance is certainly produced in plants following expression of the hypersensitive response (Fig. 6-24). Localized infections of young plants, e.g., cucumber with a fungus (*Colletotrichum lagenarium*), a bacterium (*Pseudomonas lachrymans*), or a virus (*tobacco necrosis virus*), lead within a few

FIGURE 6-23 Induced resistance in *Arabidopsis* plants sprayed with water (A, C, D), salicylic acid (B), or 2,6-dichloroisonicotinic acid (INA) and inoculated with spores of *Peronospora parasitica* five (A, B) or four (C–F) days later. At six (A, B) or ten (C–F) days after inoculation, individual leaves revealed numerous oomycete structures in heavily infected H_2O-treated leaves and almost no oomycete structures in INA-treated leaves. Plants in A–C and E are normal, whereas those in D and F were transformed with a gene that blocks the accumulation of salicylic acid, indicating that INA can induce resistance in the absence of salicylic acid accumulation. C, conidiophores. [Photographs courtesy of J. A. Ryals, Ciba Agric. Biotechnology. A and B from Uknes *et al.*, *Mol. Plant-Microbe Interact.* 6, 692–698; C–F from Vernooij *et al.* (1995). *Mol. Plant-Microbe Interact.* 8, 228–234.]

days' time to broad-spectrum, systemic acquired resistance to at least 13 diseases caused by fungi, bacteria, and viruses. A single inducing infection protects cucumber from all pathogens tested for 4 to 6 weeks; when a second, booster inoculation is made 2 to 3 weeks after the primary infection, the plant acquires season-long resistance to all tested pathogens. The degree of systemic acquired resistance seems to correlate well with the number of lesions produced on the induced leaf until a saturation point is reached. Systemic acquired resistance, however, cannot be induced after the onset of flowering and fruiting in the host plant.

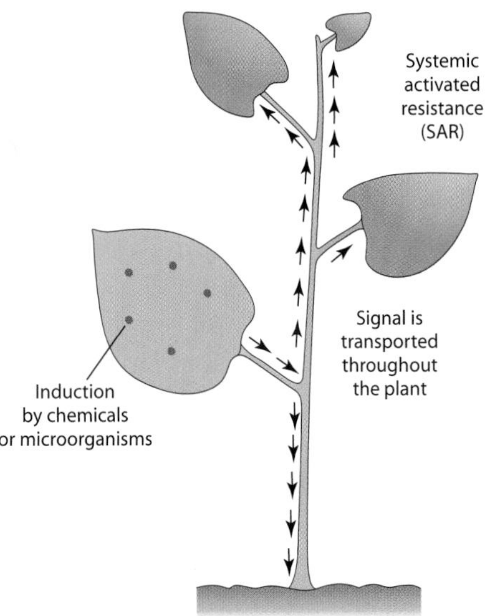

FIGURE 6-24 Principle of systemic activated (or acquired) resistance. A leaf treated with certain chemicals or with pathogens causing necrotic lesions produces a signal compound(s) that is transported systemically throughout the plant and activates its defense mechanisms, making the entire plant resistant to subsequent infections.

Systemic acquired resistance is characterized by the coordinate induction in uninfected leaves of inoculated plants of at least nine families of genes now known as systemic acquired resistance genes. Products of several SAR genes, e.g., β-1,3-glucanases, chitinases, cysteine-rich proteins related to thaumatin, and PR-1 proteins, have direct antimicrobial activity or are closely related to classes of antimicrobial proteins. The set of SAR genes that are induced in a plant may vary with the plant species. Although systemic acquired resistance does not affect spore germination and appressorium formation, penetration is reduced drastically in systemically induced resistant tissue, probably as a result of formation beneath the appressoria of papilla-like material that becomes impregnated quickly with lignin and silicon. In some host–pathogen systems, systemic acquired resistance is characterized by the induction of peroxidase and lipoxygenase activities that lead to the production of fatty acid derivatives, which exhibit strong antimicrobial activity. In plants exhibiting systemic acquired resistance in response to plant defense activators such as salicylic acid, bacterial growth and multiplication are reduced drastically (Fig. 6-25), although salicylic acid is tolerated by the bacteria at concentrations much higher than those found in the treated plant.

The mechanism of signal transduction in triggering systemic acquired resistance is still being studied. Salicylic acid seems to be involved in both the hypersensitive response and the systemic acquired resistance but may not be the signal that induces systemic acquired resistance (Fig. 6-26). Salicylic acid is present in the phloem of plants after the primary inoculation but before the onset of acquired resistance; its concentration levels correlate with the induction of PR proteins. External application of salicylic acid activates the same sets of SAR genes that are expressed after SAR induction by pathogens. Nevertheless, other evidence suggests that a signal other than salicylic acid is responsible for the systemic expression of systemic acquired resistance, but salicylic acid must be present for the real signal to be

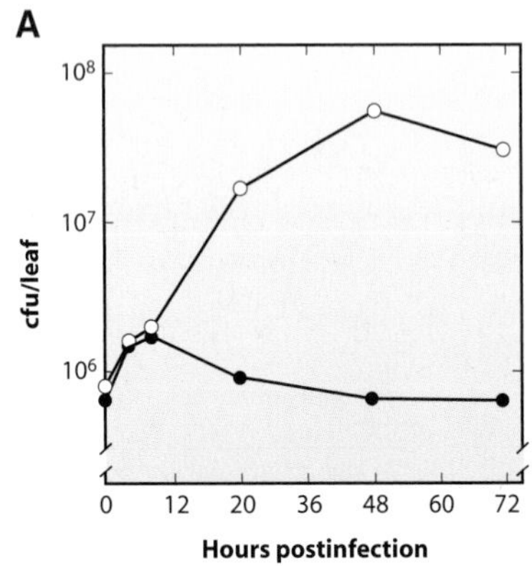

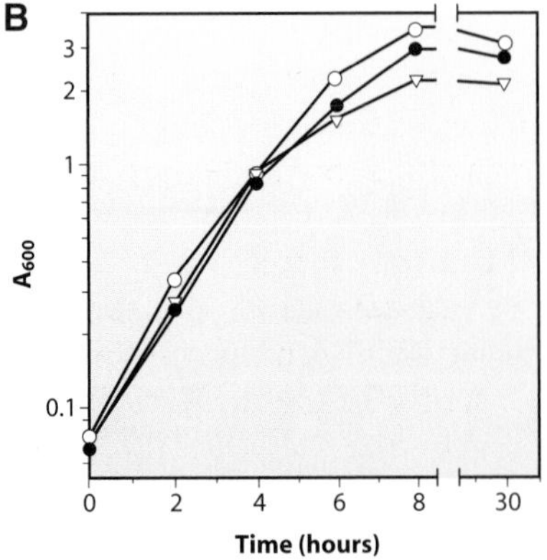

FIGURE 6-25 (A) Inhibition of growth and multiplication of *Erwinia carotovora* bacteria in inoculated leaves of tobacco seedlings growing in a medium containing 1 m*M* salicylic acid (●) or without salicyclic acid (○). cfu, colony-forming units (bacteria). Control leaves were nearly macerated 12 hours after inoculation, whereas salicylic acid-treated leaves had one small local lesion at the point of inoculation. (B) Lack of inhibition of growth and multiplication of the same bacteria in culture by various concentrations (0, 1, and 5 mM) of salicylic acid, indicating that the effect in A is caused by the plant defenses activated by salicylic acid and not by the salicylic acid itself. [From Palva *et al.* (1994) *Mol. Plant-Microbe Interact.* 7, 356–363.]

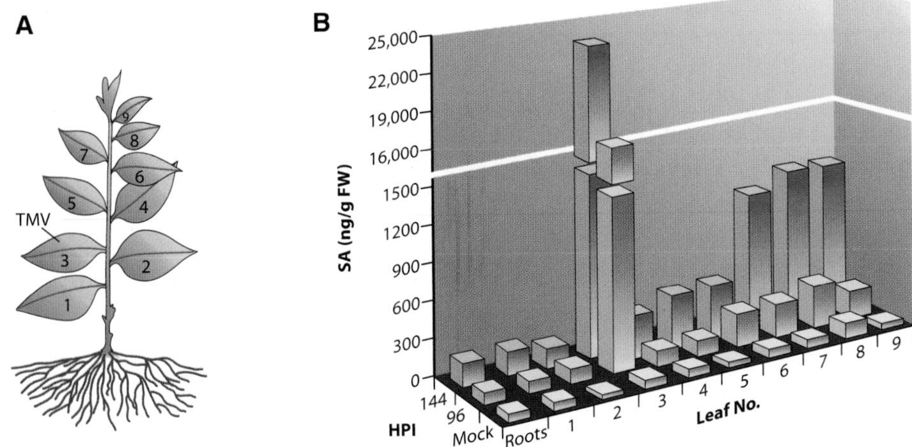

FIGURE 6-26 Salicylic acid accumulation throughout a 6-week-old tobacco plant after inoculation of a single leaf with a strain of tobacco mosaic virus that causes local lesions only and no systemic infection. (A) Inoculated leaf 3 in relation to other leaves and roots of the plant. (B) Concentrations of total salicyclic acid (SA, in nanograms per gram fresh weight) in the inoculated leaf (leaf 3) and in uninoculated roots and leaves at 96 and 144 hours postinoculation (HPI). MOCK, inoculation without virus. [From Shulaev *et al.* (1995) *Plant Cell* 7, 1691–1701.]

transduced into gene expression and acquired resistance. It had been reported earlier that salicylic acid reacts with an oxidative enzyme (catalase) and generates reactive oxygen radicals. This had been suggested as a mechanism by which the plant cell reacts to salicylic acid signaling and induces systemic acquired resistance (Fig. 6-27). This notion, however, is no longer accepted.

The onset of systemic acquired resistance in Arabidopsis is controlled by a single gene, NPR1, which also affects local acquired resistance, i.e., the ability of plants to restrict the spread of virulent pathogen infections. Disruption of the gene produces mutant plants that fail to respond to a variety of SAR-inducing treatments, they display minimum expression of pathogenesis-related genes, and they exhibit increased susceptibility to infections by allowing lesions to grow and spread much more than in nonmutant plants. The NPR1 gene encodes a novel protein that contains ankyrin repeats and these repeats are needed for NPR1 to function. Also, when the NPR1 gene was inserted into a mutant that had lost the NPR1 gene, the mutant not only reacquired the responsiveness to SAR induction in terms of expression of PR genes and resistance to infection, the mutant transgenic plants actually became more resistant to infection by the bacterium *P. syringae* even in the absence of SAR induction. It was further shown that induction of NPR1 leads to overexpression of the NPR1-coded protein and this, in turn, induces the expression of numerous downstream pathogenesis-related genes. NPR1 seems to confer resistance to some bacterial and oomycete diseases in a dosage-dependent manner. The increased resistance provided by the overexpression of NPR1 seems to occur without any detrimental effects on the plants.

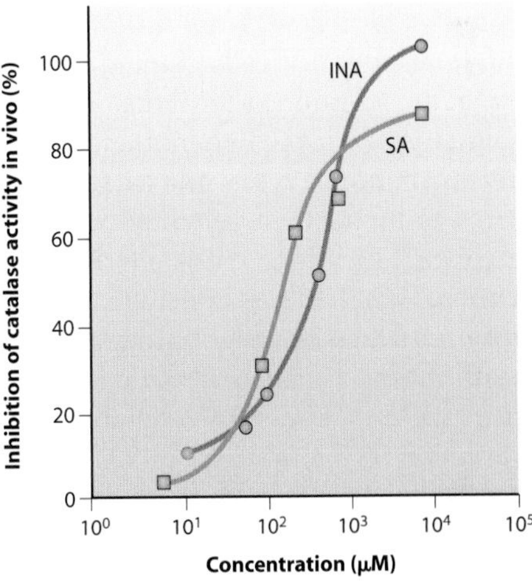

FIGURE 6-27 Inhibition of catalase activity by the plant defense-promoting compound salicyclic acid (SA) and the *in vivo*-produced active form of isonicotinic acid (INA). Such inhibition in resistant plants was earlier thought to result in the accumulation of active oxygen radicals and in the hypersensitive defense response. [From Conrath *et al.* (1995). *Proc. Natl. Acad. Sci. USA* **92**, 7143–7147.]

The induction of systemic acquired resistance through external application of salicylic acid raised the very important question of whether salicylic acid or other chemical compounds could be used to artificially induce systemic acquired resistance in plants against their numerous pathogens. Unfortunately, externally applied salicylic acid is not translocated efficiently in the plant and, in addition, salicylic acid is strongly phytotoxic when applied at even slightly higher levels above the level required for efficacy. Therefore, salicylic acid per se has not been considered for use as a practical solution for disease control.

So far, in addition to salicylic acid, derivatives of isonicotinic acid and benzothiazoles have been shown to induce systemic acquired resistance in plants against a variety of pathogens. As a matter of fact, the benzothiazole (BTH) is being used commercially. When the three compounds were used separately to protect barley against the powdery mildew fungus, they did so by inducing differential expression of a number of newly identified defense response genes, including genes encoding a lipoxygenase, a thionin, an acid phosphatase, a Ca^{2+}-binding protein, a serine proteinase inhibitor, a fatty acid desaturase, and several other proteins whose function had not been determined. Of the three chemicals, INA and BTH were more potent inducers of both gene expression and resistance. In experiments in which cowpea seeds were treated with BTH and were then inoculated with the anthracnose fungus *Colletotrichum destructivum*, the young cowpea plants were effectively protected from infection through a hypersensitive response of cells coming in contact with the pathogen. In addition, the plants showed a rapid transient increase of the phenoloxidizing enzymes phenylalanine ammonia lyase and chalcone isomerase while there was an early, accelerated accumulation of the phytoalexins kievitone and phaseollidin and of several other proteins. It was concluded that BTH protects cowpea seedlings by potentiating an early defense response rather than by altering the constitutive resistance of the tissues. The SAR-activating compounds induce expression of the same set of SAR genes that are induced either by salicylic acid or by various infectious agents and, in addition, seem to prime or sensitize plants to respond faster and with additional defense reactions than those characteristic of SAR genes. Isonicotinic acid, however, functions even in transgenic plants that are unable to accumulate salicylic acid. Apparently, therefore, isonicotinic acid triggers the signal transduction pathway that leads to SAR by acting either at the same site as salicylic acid or downstream from it.

Salicylic acid and isonicotinic acid are true SAR activators because not only do they induce resistance to the same spectrum of pathogens and induce expression of the same genes as pathogens, but these chemicals have no antimicrobial activity. Several other chemical compounds, such as the fungicides fosethyl-Al, metalaxyl, and triazoles, appear to have some resistance-inducing activity. The fungicide–bactericide probenazole is only slightly toxic *in vitro*, but induces various defense responses in rice plants, including an oxidative burst and appearance of reactive oxygen radicals, as well as significant accumulation of antimicrobial factors such as fungitoxic unsaturated fatty acids. A large number of other compounds, and also many microorganisms, have been tested for their ability to induce systemic acquired resistance in plants, but so far none has proved effective. This area of research, however, has a tremendous commercial potential, and therefore the search for SAR-inducing compounds is likely to continue and, actually, to increase.

DEFENSE THROUGH GENETICALLY ENGINEERING DISEASE-RESISTANT PLANTS

With Plant-Derived Genes

The number of plant genes for resistance (R genes) that have been isolated is increasing rapidly. The first plant gene for resistance to be isolated was the *Hml* gene of corn in 1992, which codes for an enzyme that inactivates the HC toxin produced by the leaf spot fungus *Cochliobolus carbonum*. In 1993, the Pto gene of tomato was isolated; this gene encodes a protein kinase involved in signal transduction and confers resistance to strains of the bacterium *P. syringae* pv. *tomato* that carry the avirulence gene *avrPto*. In 1994, four additional plant genes for resistance were isolated: the *Arabidopsis* RPS2 gene, which confers resistance to the strains of *P. syringae* pv. *tomato* and *P. syringae* pv. *maculicola* that carry the avirulence gene *avrRpt2*; the tobacco N gene, which confers resistance to tobacco mosaic virus; the tomato Cf9 gene, which confers resistance to the races of the fungus *Cladosporium fulvum* that carry the avirulence gene *avr9*; and the flax L^6 gene, which confers resistance to certain races of the rust fungus *Melampsora lini* carrying the avirulence gene *avr6*. The last five plant resistance genes are triggered into action by the corresponding avirulence genes of the pathogen, the products of which serve as signals that elicit the hypersensitive response in the host plant. Several more plant resistance genes have since been isolated. Some of these genes appear to provide plant resistance to pathogens expressing one or the other of two unrelated *avr* genes of the pathogen. It is expected that these and many other R genes, which are likely to be isolated in the years to come, will be used extensively in genetically engineering transgenic plants that will be resistant to many of the races of the pathogens that affect these plants.

In addition to these specific plant genes, several other plant genes encoding enzymes or other proteins (PR proteins) found widely among plants have been shown to confer resistance to transgenic plants in which they are expressed. For example, tobacco plants transformed with a chitinase gene from bean became resistant to infection by the soilborne fungus *Rhizoctonia solani* but not to infection by the oomycete *Pythium aphanidermatum*, the cell walls of which lack chitin. In other experiments, constitutive expression of a PR chitinase gene from rice in transgenic rice and cucumber plants made the rice plants more resistant to *R. solani* and the cucumber plants more resistant to *Botrytis cinerea*. Similarly, transgenic tobacco plants expressing a PR1 protein gene were resistant to the blue mold oomycete *Peronospora tabacina*, and plants expressing the systemic acquired resistance gene SAR8.2 were resistant to the black shank oomycete *Phytophthora parasitica*. Also, transgenic soybean plants expressing a wheat gene for oxalate oxidase, which oxidizes oxalic acid, a pathogenicity factor for the soybean stem rot fungus *Sclerotinia sclerotiorum*, confers resistance to soybean by exhibiting its highest activity of oxalate oxidation in cell walls proximal to the site of pathogen attack. Moreover, transgenic potato plants expressing the gene for the antibacterial enzyme T4 lysozyme exhibited resistance to the soft rot and black leg caused by the bacterium *Erwinia carotovora* pv. *atroseptica*. Also, transgenic tobacco and potato plants expressing a gene from pokeweed (*Phytolacca* sp.) that codes for an antiviral, ribosome-inactivating protein exhibited resistance against several potato and other viruses. Plants are also aided in their defense from pathogens by plant-produced, ribosome-inactivating proteins (RIPs) that inhibit foreign protein synthesis in the cell without interfering with their own ribosomes. RIP genes also show synergism with PR protein genes when the two are expressed concurrently in the same plant.

Because mixtures of pathogenicity-related proteins are more effective as antimicrobials than each of them tested separately, it was soon shown that transgenic plants (tobacco) expressing both the chitinase and the β-1,3-glucanase were significantly more resistant to the fungi *Cercospora nicotianae* and *R. solani*, as was tomato to *Fusarium oxysporum* f. sp. *lycopersici*, than plants expressing either of the genes alone. Equally effective in providing plant resistance to fungi were hydrolytic enzymes, such as chitinase and glucanase, obtained organisms other than plants, such as the soilborne bacterium *Seratia marcescens*, or the human enzyme lysozyme. Other PR proteins, such as the defensins, a group of cysteine-rich, defense-related antimicrobial peptides constitutively present in the plasma membrane of most plant species, provide enhanced resistance to different pathogens.

Modification of existing plant genes that govern the external or internal cell surface receptor to which the virus binds may result in an inability of the virus to bind and to replicate in the cell and may lead to resistance or immunity. To these must also be added the induction of resistance in potato and tobacco transgenic plants transformed, respectively, with a mouse gene coding for an enzyme involved in the synthesis of an interferon-like compound and with a mouse gene coding for an antibody (plantibody) against the coat protein of a plant virus (*artichoke mottle crinkle virus*).

Additional mechanisms of enhancing the resistance of a plant with plant-derived genes include genetic engineering of plants with R genes that provide appropriate plant resistance or an elicitor molecule that triggers it; engineering plants with genes that overexpress one or more genes that regulate the systemic acquired resistance of the plant so that it (SAR) can be kept high continually and against a variety of pathogens; and by changing a previously compatible defense reaction to an incompatible (resistant) one through insertion of a resistance gene. Engineering plants with constitutive genes that trigger or enhance the accumulation of pathogenesis-related (PR) proteins, with genes such as stilbene synthase. This enzyme triggers the production of certain phytoalexins that subsequently reduce infection, e.g., of tobacco by *Botrytis cinerea*, by 50%. Or engineering plants with defective or less active genes that reduce the level of activity of calmodulin and of catalase, thereby leading to the production of continuously high levels of active oxygen species (H_2O_2), as well as the activated expression of PR proteins. Other types of plant genes engineered into plants for disease resistance include the lectin genes, which prevent plant infection by nematodes, and defensin genes that deter plant attacks by fungi. The use of known resistance genes, e.g., of Pto, Cf-9, and N, that protect certain tomato varieties from a bacterial spot, tomato from fungal black mold, and tobacco from mosaic virus, respectively, to confer resistance to different plants has, generally, not been successful. It appears that when a gene that confers strong resistance in one host is isolated and transferred to a different plant separated from its original genetic background, it is not able to confer resistance to the new plant.

With Pathogen-Derived Genes

In 1986, it was shown for the first time that tobacco plants transformed (genetically engineered) to express the coat protein gene of TMV showed various degrees of resistance to subsequent inoculation with the same virus. Once the TMV coat protein gene was integrated in the tobacco genome, it was carried through the seed and behaved like any other tobacco gene. Since then, numerous other crop plants, especially solanaceous ones such as tobacco, tomato, and pepper; legumes such as alfalfa; grains such as barley, corn, oats, and rice; cucurbits such as cucumber, cantaloupe, and squash; and several other plants (papaya, impatiens, etc.), have been transformed with the coat protein gene of one or more of the viruses that infect them. The viruses from which the coat protein genes were obtained represent most of the virus groups.

In the vast majority of cases, transgenic plants show quite high levels of resistance to the virus from which the coat protein gene was derived and, in many cases, to other more or less related viruses. In some cases the transgenic plants were resistant to the virus if they were inoculated mechanically but not if inoculated by the specific vector of the virus, whereas in others the plants remained resistant even when inoculated by their aphid or fungus vector. In some cases, plants were transformed concurrently with as many as three viruses, the coat protein genes of which had been introduced in tandem into one location of the plant genome; such transgenic plants exhibited resistance to all three viruses.

Transgenic plants transformed with viral genes other than the coat protein gene often exhibit even higher levels of resistance to the virus providing the gene(s) and to, perhaps, additional viruses. Quite often the transferred genes either are portions of genes or are mutated artificially and thereby inactivated genes, so that they can be reproduced and expressed by the plant but do not produce a functional gene product that might aid a virus on infection. For example, highly resistant transgenic tobacco plants have been produced by transformation with modified virus replicase-coding genes of several viruses. Also, tobacco plants transformed with the TMV gene coding for the movement protein or for a dysfunctional movement protein are resistant to TMV and to several other viruses. Resistance to viruses has also been induced in plants transformed with viral genes coding for proteases needed for processing the viral nucleic acid, in plants transformed with small defective or satellite nucleic acids, and even in plants transformed with untranslatable or antisense segments of the viral nucleic acid.

Resistance to nonviral pathogens has also been increased through the engineering of plants with appropriate genes from pathogenic or nonpathogenic fungi and also from insects and other animals. For example, potato plants engineered to express the H_2O_2-generating glucose oxidase gene from the fungus *Aspergillus niger* continually produce high levels of peroxide ions in the apoplast of the plant cells, thereby increasing the resistance of the potato plants to the oomycete causing late blight (*Phytophthora infestans*), and the fungi causing early blight (*Alternaria solani*), and Verticillium wilt (*Verticillium dahliae*). The resistance of potato plants to the bacterial soft rot disease (caused by *Erwinia carotovora*), of tobacco plants to several fungal and bacterial diseases, and of apple plants to fire blight disease (caused by the bacterium *E. amylovora*) was increased when the plants were transformed with a hen, human, or T4 bacteriophage gene for lysozyme, which hydrolyzes the pteridoglycan layer of the bacterial cell wall and inhibits fungal and bacterial growth. Similarly, potato and apple plants transformed with the chitinase gene obtained from the fungus *Trichoderma harzianum,* which is used as a biocontrol agent against many plant pathogenic fungi, the walls of which it hydrolyzes with its chitinases, showed resistance to the potato early blight and to potato gray mold (caused by *Botrytis cinerea*), whereas the apple trees showed increased resistance to the apple scab disease (caused by the fungus *Venturia inaequalis*). Furthermore, tobacco, potato, apple, and pear plants showed increased resistance when transformed with certain genes; some genes were obtained from insects and code for antibacterial proteins, such as cecropins, which are lytic peptides that make pores in and cause lysis of bacterial cell membrane; or transformed with the genes coding for the antimicrobial proteins known as attacins, which inhibit the synthesis of the outer membrane protein in gram-negative bacteria. Such genes increased resistance bacterial wildfire of tobacco (caused by *P. syringae* pv. *tabaci*), of potato to bacterial black leg (caused by *E. carotovora* subsp. *atroceptica*), and of apple and pear to fire blight (caused by *E. amylovora*).

There is every expectation that the area of inducing plant resistance to pathogens through genetic transformation with pathogen-derived genes will grow and improve rapidly. Such genetic engineering strategies will provide an excellent additional tool for plant disease control.

DEFENSE THROUGH RNA SILENCING BY PATHOGEN-DERIVED GENES

RNA silencing is a type of gene regulation that, in plants, serves as an antiviral defense. RNA silencing is based on targeting specific sequences of RNA and degrading them. RNA silencing occurs in a broad range of eukaryotic organisms, including plants, fungi, and animals. While plants use RNA silencing to defend themselves against viruses, the viruses, in turn, encode proteins by which they attempt to suppress the silencing of their RNA. The consensus is that RNA silencing is one of the many interconnected pathways for RNA surveillance and cell defense.

RNA silencing was first observed in transgenic plants transformed with viral genes providing "pathogen-derived resistance." It was noticed then that sense orientation genes in the transgenic plant interfered with the expression of both the transgenes themselves and related endogenous genes of the plant. Because of the concurrent suppression of both genes, RNA silencing was at first called "cosuppression." RNA silencing is due to a process that occurs after transcription (posttranscriptional gene silencing) of the RNA and involves targeted mRNA degradation. Clues of its existence came from the discovery that plants carrying viral transgenes were resistant to related strains of the virus that replicate in the cytoplasm, which meant that silencing occurs in the cytoplasm rather than the nucleus. The nucleotide sequence specificity of the RNA depends on the sequence of 21–25 nucleotides of antisense RNA produced directly or indirectly from sense transgenes, or from dsRNA. The dsRNA is a trigger or an intermediate in the cleaving into small (21–25 nucleotides), sense or antisense RNAs called small interfering (siRNAs). siRNAs act as guides that direct the RNA degradation machinery [the RNA-induced silencing complex (RISC)] to the target RNAs.

The main events in RNA silencing, as understood at this point in time, include the following steps (Fig. 6-28): A plant or viral gene is inserted in the plant DNA where it is expressed and produces messenger RNA (mRNA). The viral gene may also be able to do that without being inserted in the plant genome. RNA viruses routinely produce double-stranded RNA (dsRNA), and RNA from some abnormal genes doubles up upon itself and forms "aberrant" dsRNA. Both dsRNAs are cleaved by an enzyme called "Dicer" into small interfering RNAs about 21–25 nucleotides long. The siRNA fragments split into individual ssRNAs and these combine with proteins and produce an RNA-induced silencing complex (RISC). This complex captures mRNAs that complement each short RNA sequence. RNAs with a nearly perfect match of their sequence with that of small RNA are sliced into useless small fragments. RNAs with less perfect sequence matches cause the RISC complex to block the movement of the ribosomes on the mRNA so that the mRNA is not translated and does not produce a protein, thereby silencing that RNA.

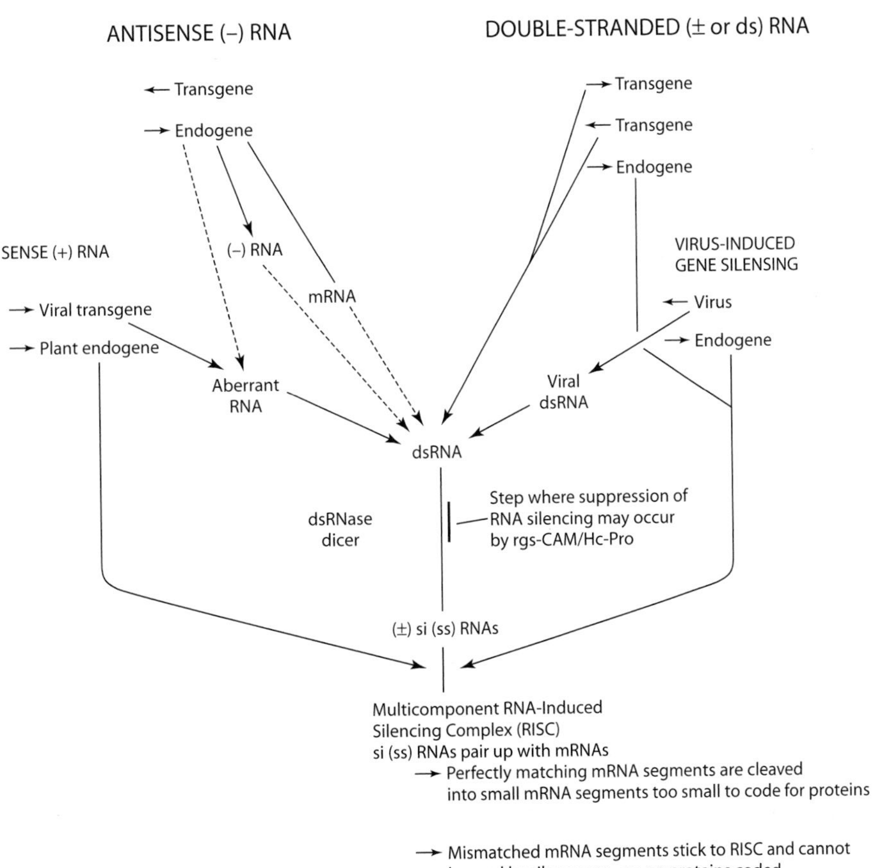

FIGURE 6-28 Diagram of the steps, some of them hypothetical, involved in the in-cell silencing of transgene, endogene, and viral RNA as a mechanism of plant defense. [Modified from Vance and Vaucheret (2001). *Science* **222**, 2277–2280.]

RNA silencing produces exceptionally strong virus resistance in transgenic plants. Such plants have neither detectable accumulation of virus in their inoculated leaves nor can this resistance be overcome with high-titer inocula. Once RNA silencing of the transgene is established, all RNAs homologous to the transgene, including those from an infecting virus, are degraded. Also, although RNA silencing is triggered locally, it can spread throughout the plant via a mobile silencing signal. The movement of the silencing signal in the plant parallels that of the virus, moving at first from cell to cell and then entering the phloem and from there spreading out to parenchyma cells again. The parallel movement of the virus and RNA-silencing signal may represent a race between the two, with the out-come of the race being a successful infection if the virus moves faster and becomes established first, or resistance, i.e., lack of infection, if RNA silencing becomes established first.

It was later shown that plant viruses could also induce RNA silencing. It was further shown that virus-induced gene silencing (VIGS) could be directed to either transgenes in the plant or endogenous genes of the plant. As a result, plant viruses could both induce RNA silencing and could be targeted for RNA silencing by transgenes. VIGS, however, is rather mild, transient, and restricted to regions around the veins. RNA silencing has not yet been reported to occur in plant DNA viruses, both the ssDNA geminiviruses and the reverse-transcribing dsDNA viruses. All DNA viruses, however, seem to have the potential to induce gene silencing in the nucleus and in the cytoplasm, as they produce multiple copies of viral DNA genomes in the nucleus, show illegitimate integration of viral DNA into host chromosomes that mimics transgene transformation for such viruses, and generate a great deal of viral RNAs in the cytoplasm.

Suppressors of RNA Silencing

Soon after the discovery of RNA silencing, it was discovered that many plant viruses encode proteins that suppress RNA silencing. The suppressors are structurally diverse and seem to have undergone repeated evolution steps in their attempt to keep up with developments in RNA silencing. One suppressor, the helper component-proteinase of potyviruses, is so effective in suppressing viral RNA silencing that it actually increases the accumulation of several unrelated plant viruses and is, possibly, responsible for the many potyvirus-associated synergistic diseases of plants. The same suppressor prevents both virus-induced and transgene-induced RNA silencing and can even reverse an already established RNA silencing of a transgene. The suppression induced by the potyvirus suppressor to a transgene-induced RNA silencing can be reversed at a step at which the accumulation of siRNAs is eliminated, but it cannot eliminate the mobile silencing signal. Another suppressor, *the potato virus X* p25 protein, is much less effective in suppressing RNA silencing and it apparently targets and interferes with systemic silencing.

In addition to the suppression of RNA silencing by virus-encoded proteins, RNA silencing can also be suppressed by certain host genes. Some of these genes are expressed in transgenic plants, in plants following infection with certain viruses, and in transgenic plants carrying the potyvirus suppressor protein. These observations suggest that the host–coded suppressor acts as a relay for the potyvirus suppressor-mediated suppression of post-transcriptional gene silencing or that the potyvirus suppressor-induced suppression of silencing perhaps takes place via activation of the host-induced suppressor protein and its unknown target protein.

Selected References

Aist, J. R. (1976). Papillae and related wound plugs of plant cells. *Annu. Rev. Phytopathol.* **14**, 145–163.

Baker, C. J., and Orlandi, E. W. (1995). Active oxygen in plant pathogenesis. *Annu. Rev. Phytopathol.* **33**, 299–321.

Balmori, E., *et al.* (2002). Sequence, tissue, and delivery-specific targeting of RNA during posttranscriptional gene silencing. *Mol. Plant-Microbe Interact* **15**, 753–763.

Beachy, R. N., Loesch-Fries, S., and Tumer, N. E. (1990). Coat protein mediated resistance against plant viruses. *Annu. Rev. Phytopathol.* **28**, 451–474.

Beyer, K., *et al.* (2001). Identification of potato genes induced during colonization by *Phytophthora infestans*. *Mol. Plant Pathol.* **2**, 125–134.

Bol, J. F., Linthorst, H. J. M., and Cornelissen, B. J. C. (1990). Plant pathogenesis-related proteins induced by virus infection. *Annu. Rev. Phytopathol.* **28**, 113–138.

Bostock, R. M., and Stermer, B. A. (1989). Perspectives on wound healing in resistance to pathogens. *Annu. Rev. Phytopathol.* **27**, 343–371.

Bowles, D. J. (1990). Defense-related proteins in higher plants. *Annu. Rev. Biochem.* **59**, 873–907.

Bowyer, P., *et al.* (1995). Host range of a plant pathogenic fungus determined by a saponin detoxifying enzyme. *Science* **267**, 371–374.

Broglie, K., *et al.* (1991). Transgenic plants with enhanced resistance to the fungal pathogen *Rhizoctonia solani*. *Science* **254**, 1194–1197.

Burrows, P. R., *et al.* (1998). Plant-derived enzyme inhibitors and lectins for resistance against plant parasitic nematodes in transgenic crops. *Pesticide Sci.* **52**, 176–183.

Cao, H., *et al.* (1997). The Arabidopsis NPR1 gene that controls systemic acquired resistance encodes a novel protein containing ankyrin repeats. *Cell* **88**, 57–63.

Cao, H., Li, X., and Dong, X. (1998). Generation of broad-spectrum disease resistance by overexpression of an essential regulatory gene in systemic acquired resistance. *Proc. Natl. Acad. Sci. USA* **95**, 6531–6536.

Cao, H., Baldini, R. L., and Rahme, L. G. (2001). Common mechanisms for pathogens of plants and animals. *Annu. Rev. Phytopathol.* **39**, 259–284.

Chamnonpol, S., *et al.* (1998). Defense activation and enhanced pathogen tolerance induced by H_2O_2 in transgenic tobacco. *Proc. Natl. Acad. Sci. USA* **95**, 5818–5823.

Chen, Z., *et al.* (2000). The *Pseudomonas syringae avrRpt2* gene product promotes pathogen virulence from inside plant cells. *Mol. Plant-Microbe Interact.* **13**, 1312–1321.

Coventry, H. S., and Dubery, I. A. (2001). Lipopolysaccharides from *Burkholderia cepacia* contribute to an enhanced defensive capacity in *Nicotiana tabacum*. *Physiol. Mol. Plant Pathol.* **58**, 149–158.

Dantan-Gonzalez, E., *et al.* (2001). Actin monoubiquitylation is induced in plants in response to pathogens and symbionts. *Mol. Plant-Paras. Interact.* **14**, 1267–1273.

De Lorenzo, G., D'Ovidio, R., and Cervone, F. (2001). The role of polygalacturonase-inhibiting proteins (PGIPs) in defense against phytopathogenic fungi. *Annu. Rev. Phytopathol.* **39**, 313–336.

Dengl, J. L., and Jones, J. D. G. (2001). Plant pathogens and integrated defense responses to infection. *Nature* **411**, 826–833.

DeWit, P. J. G. M. (1992). Molecular characterization of gene-for-gene systems in plant-fungus interactions and the application of avirulence genes in control of plant pathogens. *Annu. Rev. Phytopathol.* **30**, 391–418.

De Jaeger, G., *et al.* (2000). The plantibody approach: Expression of antibody genes in plants to modulate metabolism or to obtain pathogen resistance. *Plant Mol. Biol.* **43**, 419–428.

Dixon, R. A., Harrison, M. J., and Lamb, C. J. (1994). Early events in the activation of plant defense responses. *Annu. Rev. Phytopathol.* **32**, 479–501.

Do, H. M., *et al.* (2003). Expression of peroxidase-like genes, H_2O_2 production, and peroxidase activity during the hypersensitive response to *Xanthomonas campestris* pv. *vesicatoria* in *Capsicum annuum*. *Mol. Plant-Microbe Interact.* **16**, 196–205.

Donaldson, P. A., *et al.* (2001). Soybean plants expressing an active oligomeric oxalate oxidase from the wheat gf-2.8 gene are resistant to the oxalate-secreting pathogen *Sclerotinia sclerotiorum*. *Physiol Mol. Plant Pathol.* **59**, 297–307.

Dong, H., and Beer, S. V. (2000). Riboflavin induces disease resistance in plants by activating a novel signal transduction pathway. *Phytopathology* **90**, 801–811.

Dow, M., Newman, M.-A., and von Roepenack. (2000). The induction and modulation of plant disease responses by bacterial lipopolysaccharides. *Annu. Rev. Phytopathol.* **38**, 241–262.

Fischer, R., *et al.* (1999). Toward molecular farming in the future: Moving from diagnostic and antibody production in microbes to plants. *Biotechnol. Appl. Biochem.* **30**, 101–108.

Gaudet, D. A., *et al.* (2000). Expression of plant defense-related (PR_protein) transcripts during hardening and dehardening of winter wheat. *Physiol. Mol. Plant Pathol.* **57**, 15–24.

Gilchrist, D. G. (1998). Programmed cell death in plant disease: The purpose and promise of cellular suicide. *Annu. Rev. Phytopathol.* **36**, 393–414.

Gonzalves, D. (1988). Control of *papaya ringspot virus* in papaya. *Annu. Rev. Phytopathol.* **36**, 415–438.

Hammerschmidt, R. (1999). Phytoalexins: What have we learned after 60 years? *Annu. Rev. Phytopathol.* **37**, 285–306.

He, S. Y. (1998). Type III protein secretion systems in plant and animal pathogenic bacteria. *Annu. Rev. Phytopathol.* **36**, 363–392.

Heath, M. C. (2000). Multigenic disease resistance and the basis of host genotype specificity. *Physiol Mol. Plant Pathol.* **57**, 189–190.

Heath, M. C. (2001). Non-host resistance to plant pathogens: Non-specific defense or the result of specific recognition events? *Physiol. Mol. Plant Pathol.* **58**, 53–54.

Hennin, C., Diederichsen, E., and Hofte, M. (2002). Resistance to fungal pathogens triggered by the Cf9-Avr response in tomato and oilseed rape in the *absence* of hypersensitive cell death. *Mol. Plant Pathol.* **3**, 31–41.

Honee, G. (1999). Engineered resistance against fungal plant pathogens. *Eur. J. Plant Pathol.* **105**, 319–326.

Horsfall, J. G., and Cowling, E. B., eds. (1980). "Plant Disease," Vol. 5. Academic Press, New York.

Hückelhoven, R., Dechert, C., and Kogel, K.-H. (2001). Non-host resistance of barley is associated with a hydrogen peroxide burst at sites of attempted penetration by wheat powdery mildew fungus. *Mol. Plant Pathol.* **2**, 199–205.

Hutcheson, S. W. (1998). Current concepts of active defense in plants. *Annu. Rev. Phytopathol.* **36**, 59–90.

Innes, R. W. (2001) Targeting the targets of type III effector proteins secreted by phytopathogenic bacteria. *Mol. Plant Pathol.* **2**, 109–115.

Jones, J. B., Stall, R. E., and Bouzar, H. (1998). Diversity among xanthomonads pathogenic on pepper and tomato. *Annu. Rev. Phytopathol.* **36**, 41–58.

Joosten, M. H. A. J., and de Wit, P. J. G. M. (1999). The tomato — *Cladosporium fulvum* interaction: A versatile experimental system to study plant-pathogen interactions. *Annu. Rev. Phytopathol.* **37**, 335–368.

Kazan, K., *et al.* (2001). DNA microarrays: New tools in the analysis of plant defense responses. *Mol. Plant Pathol.* **2**, 177–185.

Keen, N. T. (2000). A century of plant pathology: A retrospective view on understanding host-parasite interactions. *Annu. Rev. Phytopathol.* **38**, 31–48.

Kessman, H., *et al.* (1994). Induction of systemic acquired disease resistance in plants by chemicals. *Annu. Rev. Phytopathol.* **32**, 439–459.

Kogel, K. H., and Hückelhoven, R. (1999). Superoxide generation in chemically activated resistance of barley in response to inoculation with the powdery mildew fungus. *J. Phytopathol.* **147**, 1–4.

Kuć, J. (1995). Phytoalexins, stress metabolism, and disease resistance in plants. *Annu. Rev. Phytopathol.* **33**, 275–297.

Latunde-Dada, A. O., and Lukas, J. A. (2001). The plant defense activator acibenzolar-S-methyl primes cowpea (*Vigna unguculata*) seedlings for rapid induction of resistance. *Physiol. Mol. Plant Pathol.* **58**, 199–208.

Lawrence, C. B., *et al.* (2000). Constitutive hydrolytic enzymes are associated with polygenic resistance of tomato to *Alternaria solani* and may function as an elicitor release mechanism. *Physiol. Mol. Plant Pathol.* **57**, 211–220.

Leong, S. A., Allen, C., and Tripplett, E. W., eds. (2002). "Biology of Plant-Microbe Interactions," Vol.3. *Intern. Soc. Plant-Microbe Interactions, St. Paul, MN.*

Lomonossoff, G. P. (1995). Pathogen-derived resistance to plant viruses. *Annu. Rev. Phytopathol.* **33**, 323–343.

Luderer, R., and Joosten, M. H. A. (2001). Avirulence proteins of plant pathogens: Determinants of victory and defeat. *Mol. Plant Pathol.* **2**, 355–364.

Marathe, R., *et al.* (2002). The tobacco mosaic virus resistance gene, N. *Mol. Plant Pathol.* **3**, 167–172.

Marcell, L. M., and Beattie, G. A. (2002). Effect of leaf surface waxes on leaf colonization by *Pantoea aglomerans* and *Clavibacter michiganensis*. *Mol. Plant-Microbe Interact.* **15**, 1236–1244.

Martin-Hernandez, A. M., *et al.* (2000). Effects of targeted replacement of the tomatinase gene on the interaction of *Septoria lycopersici* with tomato plants. *Mol. Plant-Microbe Interact.* **13**, 1301–1311.

McDonald, B. A., and Linde, C. (2000). Pathogen population genetics, evolutionary potential, and durable resistance. *Annu. Rev. Phytopathol.* **40**, 349–379.

Merighi, M., *et al.* (2003). The HrpX/HrpY two–component system activates *hrpS* expression, the first step in the regulatory cascade controlling the Hrp regulon in *Pantoes stewartii subsp. stewartii*. *Mol. Plant-Microbe Interact.* **16**, 238–248.

Moore, C. J., *et al.* (2001). Dark green islands in plant virus infection are the result of posttranscriptional gene silencing. *Mol. Plant-Microbe Interact.* **14**, 939–946.

Nakashita, H., *et al.* (2002). Probenazole induces systemic acquired resistance in tobacco through salicylic acid accumulation. *Physiol. Mol. Plant Pathol.* **61**, 197–203.

Nandi, A., *et al.* (2003). Ethylene and jasmonic acid signaling affect the NPR1-independent expression off defense genes without impacting resistance to *Pseudomonas syringae* and *Peronospora parasitica* in the Arabidopsis ssi1 mutant. *Mol. Plant-Microbe Interact.* **16**, 588–599.

Neu, C., Keller, B., and Feuillet, C. (2003). Cytological and molecular analysis of the *Hordeum vulgare-Puccinia triticina* nonhost interaction. *Mol. Plant-Parasite Interact.* **16**, 626–633.

Nicholson, R. L., and Hammerschmidt, R. (1992). Phenolic compounds and their role in disease resistance. *Annu. Rev. Phytopathol.* **30**, 369–389.

Nicholson, R. L., and Wood, K. V. (2001). Phytoalexins and secondary products, where are they and how can we measure them? *Physiol. Mol. Plant Pathol.* **59**, 63–69.

Nizan–Koren, R., *et al.* (2003). The regulatory cascade that activates the Hrp regulon in *Erwinia herbicola* pv. *gypsophilae*. *Mol. Plant-Microbe Interact.* **16**, 249–260.

Okinaka, Y., *et al.* (2002) Microarray profiling of *Erwinia chrysanthemi* 3937 genes that are regulated during plant infection. *Mol. Plant-Microbe Interact.* **15**, 619–629.

Prats, E., Rubiales, D., and Jorrin, J. (2002). Acibenzolar-S-methyl–induced resistance to sunflower rust (*Puccinia helianthi*) is associated with an enhancement of coumarins on foliar surface. *Physiol Mol. Plant Pathol.* **60**, 155–162.

Preston, G. M. (2000). *Pseudomonas syringae* pv. *tomato*: The right pathogen, of the right plant, at the right time. *Mol. Plant Pathol.* **1**, 263–275.

Qiu, W., Park, J.-W., and Scholthof (2002). Tombusvirus P19-mediated suppression of virus-induced gene silencing is controlled by genetic and dosage features that influence pathogenicity. *Mol. Plant-Microbe Interact.* **15**, 269–280.

Ramalingam, J., *et al.* (2003). Candidate defense genes from rice, barley, and maize and their association with qualitative and quantitative resistance in rice. *Mol. Plant-Microbe Interact.* **16**, 14–24.

Rantakari, A., *et al.* (2001). Type III secretion contributes to the pathogenesis of the soft-rot pathogen *Erwinia carotovora*: Partial characterization of the *hrp* gene cluster. *Mol. Plant-Microbe Interact.* **14**, 962–968.

Rauyaree, P., *et al.* (2001). Genes expressed during early stages of rice infection with the rice blast fungus *Magnaporthe grisea*. *Mol. Plant Pathol.* **2**, 347–354.

Riedle-Bauer, M. (2000). Role of reactive oxygen species and antioxidant enzymes in systemic virus infections of plants. *J. Phytopathol.* **148**, 297–302.

Romeis, T., *et al.* (2000). Early signaling events in the Avr9/Cf-9-dependent plant defense response. *Mol. Plant Pathol.* **1**, 3–8.

Ryals, J., Uknes, S., and Ward, E. (1994). Systemic acquired resistance. *Plant Physiol.* **104**, 1109–1112.

Saitoh, H., *et al.* (2001). Production of antimicrobial defensin in *Nicotiana benthamiana* with a *potato virus X* vector. *Mol. Plant-Microbe Interact.* **14**, 111–115.

Schell, M. A. Control of virulence and pathogenicity genes of *Ralstonia solanacearum* by an elaborate sensory network. *Annu. Rev. Phytopathol.* **38**, 263–292.

Schweizer, P., *et al.* (2000). A soluble carbohydrate elicitor from *Blumeria graminis* f. sp. *tritici* is recognized by a broad range of cereals. *Physiol. Mol. Plant Pathol.* **56**, 157–167.

Shirasu, K., and Schulze-Lefert, P. (2000). Regulators of cell death in disease resistance. *Plant Mol. Biol.* **44**, 371–385.

Song, W.-Y., *et al.* (1995). A kinase-like protein encoded by the rice disease resistance gene Xa21. *Science* **270**, 1804–1806.

Spanos, K. A., *et al.* (1999). Responses in the bark of *Cupressus sempervirens* clones artificially inoculated with *Seiridium cardinale* under field conditions. *Eur. J. Plant Pathol.* **29**, 135–142.

Staskawicz, B. J., *et al.* (2001). Common and contrasting themes of plant and animal diseases. *Science* **292**, 2285–2289.

Ton, J., *et al.* (2002). Differential effectiveness of salicylate-dependent and jasmonate/ethylene-dependent induced resistance in Arabidopsis. *Mol. Plant-Microbe Interact.* **15**, 27–34.

Tavladoraki, P., Benvenuto, E., *et al.* (1993). Transgenic plants expressing a functional single-chain Fv antibody are specifically protected from virus attack. *Nature* (*London*) **366**, 469–472.

Trognitz, F., *et al.* (2002). Plant defense genes associated with quantitative resistance to potato late blight in *Solanum phureja* x dihaploid *S. tuberosum* hybrids. *Mol. Plant-Microbe Interact.* **15**, 587–598.

Tyler, B. M. (2002). Molecular basis of recognition between *Phytophthora* pathogens and their hosts. *Annu. Rev. Phytopathol.* **40**, 137–168.

Vance, C. P., Kirk, T. K., and Sherwood, R. T. (1980). Lignification as a mechanism of disease resistance. *Annu. Rev. Phytopathol.* **18**, 259–288.

Vance, V., and Vaucheret, H. (2001). RNA silencing in plants: Defense and counterdefense. *Science* **292**, 2277–2280.

Van Etten, H. D., Matthews, D. E., and Matthews, P. S. (1989). Phytoalexin detoxification: Importance for pathogenicity and practical implications. *Annu. Rev. Phytopathol.* **27**, 143–164.

Van Loon, L. C., Bakker, P. A. H. M., and Pieterse (1998). Systemic resistance induced by rhizosphere bacteria.

Van Loon, L. C., and Van Strien, E. A. (1999). The families of pathogenesis-related proteins, their activities, and comparative analysis of PR-1 type proteins. *Physiol. Mol. Plant Pathol.* **55**, 85–97.

Wilson, T. M. A. (1993). Strategies to protect crop plants against viruses: Pathogen-derived resistance blossoms. *Proc. Natl. Acad. Sci. USA* **90**, 3134–3141.

Xiao, S., *et al.* (2001) Broad spectrum mildew resistance in *Arabidopsis thaliana* mediated by RPW8. *Science* **291**, 118–120.

Xin, L., *et al.* (1999). Identification and cloning of a negative regulator of systemic acquired resistance, SNI1, through a screen for suppressors of npr1-1. *Cell* **98**, 329–339.

Xu, Y., *et al.* (1994). Plant defense genes are synergistically induced by ethylene and methyl jasmonate. *Plant Cell* **6**, 1077–1085.

Yakoby, N., *et al.* (2002). The analysis of fruit protection mechanisms provided by reduced-pathogenicity mutants of *Colletotrichum gloeosporioides* obtained by restriction enzyme mediated integration. *Phytopathology* **92**, 1196–1201.

Yamamoto, M., *et al.* (2000). (+)-Catechin acts as an infection-inhibiting factor in strawberry leaf. *Phytopathology* **90**, 595–600.

Yao, N., *et al.* (2002). Apoptotic cell death is a common response to pathogen attack in oats. *Mol. Plant-Microbe Interact.* **15**, 1000–1007.

Yu, I.-C., Parker, J., and Bent, A. F. (1998). Gene-for-gene disease resistance without the hypersensitive response in *Arabidopsis dnd1* mutant. *Proc. Natl. Acad. Sci. USA* **95**, 7819–7824.

Ziegler, A., and Torrance, L. (2002). Applications of recombinant antibodies in plant pathology. *Mol. Plant Pathol.* **3**, 401–407.

chapter seven

ENVIRONMENTAL EFFECTS ON THE DEVELOPMENT OF INFECTIOUS PLANT DISEASE

EFFECT OF TEMPERATURE
251

EFFECT OF MOISTURE
253

EFFECT OF WIND
257

EFFECT OF LIGHT
257

EFFECT OF SOIL PH AND SOIL STRUCTURE
257

EFFECT OF HOST–PLANT NUTRITION
257

EFFECT OF HERBICIDES
262

EFFECT OF AIR POLLUTANTS
262

Plant diseases occur in all parts of the world where plants grow. They are more common and more severe, however, in humid to wet areas with cool, warm, or tropical temperatures (Fig. 7-1). Plants in dry areas may not be subjected to as many severe fungal, bacterial, or nematode diseases, but they are often attacked severely by powdery mildew fungi, by xylem-inhabiting fastidious bacteria, by phloem-inhabiting phytoplasmas and fastidious bacteria, and by viruses transmitted by certain insect vectors.

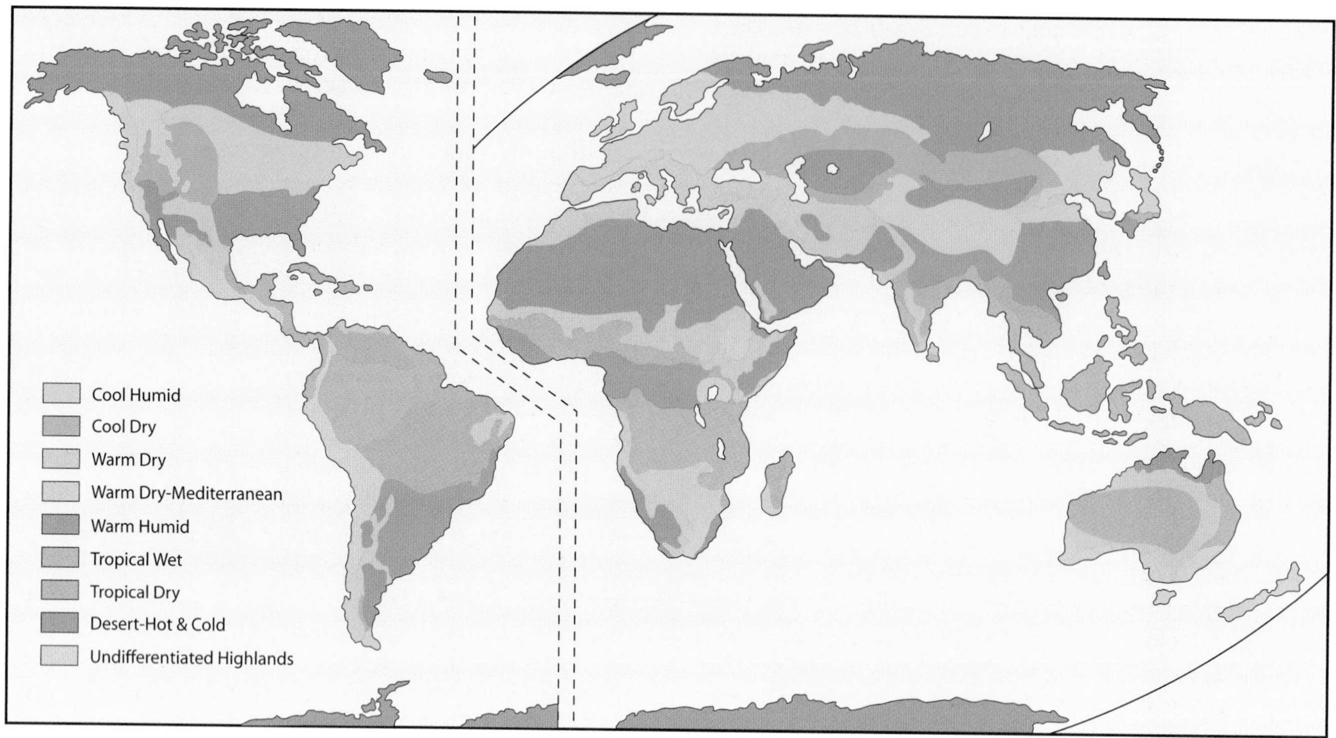

FIGURE 7-1 World map of agricultural climates. [From Duckham and Masefield (1970), "Farming Systems of the World." Chatto and Windus, London.]

Although all pathogens, all perennial plants, and, in warmer climates, many annual plants are present in the field throughout the year, almost all diseases, in all but a few very hot, dry areas, occur only, or develop best, during the warmer part of the year. Also, it is common knowledge that most diseases appear and develop best during wet, warm days and nights and that plants fertilized heavily with nitrogen are attacked much more severely by some pathogens than less fertilized plants. These general examples clearly indicate that the environmental conditions prevailing in both air and soil, after contact of a pathogen with its host, may affect the development of the disease greatly. Actually, environmental conditions frequently determine whether a disease will occur. The environmental factors that affect the initiation and development of infectious plant diseases most seriously are temperature and moisture on the plant surface (Fig. 7-2). Soil nutrients also play an important role in some diseases and, to a lesser extent, light and soil pH. These factors affect disease development through their influence on the growth and susceptibility of the host, on the multiplication and activity of the pathogen, or on the interaction of host and pathogen as it relates to the severity of symptom development.

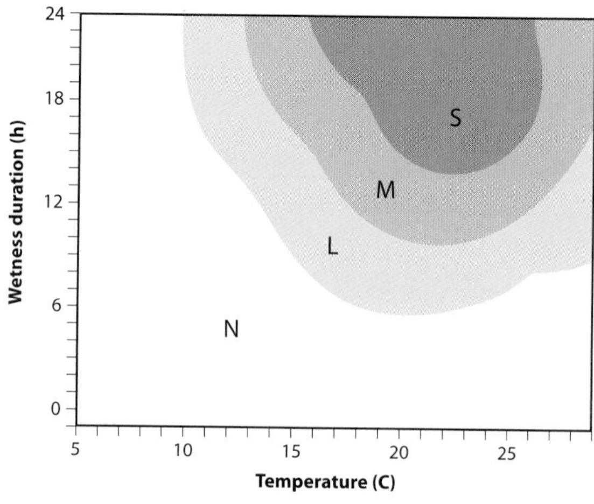

FIGURE 7-2 Various combinations of temperature and wetness duration at the fruit surface result in different severity levels of brown spot disease of pears, caused by the fungus *Stemphylium vesicarium*. N, none; L, light (20–40%); M, moderate (40–70%); S, severe (>70%). [From Montesinos *et al.* (1995). *Phytopathology* 85, 586–592.]

As mentioned previously, for a disease to occur and to develop optimally, a combination of three factors must be present: susceptible plant, infective pathogen, and favorable environment. However, although plant susceptibility and pathogen infectivity remain essentially unchanged in the same plant for at least several days, and sometimes for weeks or months, the environmental conditions may change more or less suddenly and to various degrees. Such changes may drastically influence the development of diseases in progress or the initiation of new diseases. Of course, a change in any environmental factor may favor the host, the pathogen, or both or it may be more favorable to one than it is to the other. As a result, the expression of disease will be affected accordingly. Plant diseases generally occur over a fairly wide range of the various environmental conditions. Nevertheless, the extent and frequency of disease occurrence, as well as the severity of the disease on individual plants, are influenced by the degree of deviation of each environmental condition from the point at which disease development is optimal.

EFFECT OF TEMPERATURE

Plants, as well as pathogens, require certain minimum temperatures to grow and carry out their activities (Fig. 7-1). In temperate regions, the low temperatures of late fall, winter, and early spring are below the minimum required by most pathogens. Therefore, diseases are not, as a rule, initiated during that time, and those in progress generally come to a halt. With the advent of higher temperatures, however, pathogens become active and, when other conditions are favorable, they can infect plants and cause disease. For example, in many canker diseases of perennial plants caused by fungi such as *Nectria*, *Leucostoma* (*Cytospora*), the oomycete *Phytophthora* or by bacteria such as *Pseudomonas*, infections begin and develop primarily in early spring or in the fall. The reason is that during these periods the temperatures are high enough for these fungi to grow well (Fig. 7-3) but are too low to allow optimum host development. Development of the same diseases stops during

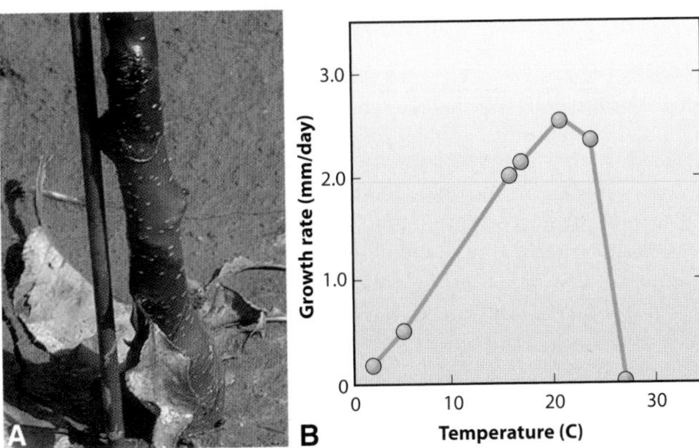

FIGURE 7-3 (A) Two cankers on a stem of a young pear tree caused by the oomycete *Phytophthora*. (B) Effect of temperature on the rate of growth of the canker-causing oomycete *Phytophthora syringae*. [Courtesy of (A) R. Regan, Oregon State University and (B) Bostock and Doster (1985). *Plant Dis.* 69, 568–571.]

the winter when temperatures are too low for both host and pathogen, and it is quite reduced during the summer months when host growth and host defenses are at their optimum.

Pathogens differ in their preference for higher or lower temperatures. Some fungi grow much faster at lower temperatures (Fig. 7-3) than others (Fig. 7-4), and there may be significant differences among races of the same fungus (Fig. 7-4). Temperature affects the number of spores formed in a unit plant area (Figs. 7-5A and 7-5B) and the number of spores released in a given time period (Figs. 7-5A, 7-5C, and 7-9). As a result, many diseases develop best in areas, seasons, or years with cooler temperatures, whereas others develop best where and when relatively high temperatures prevail. Thus, some species of the fungi *Typhula* and *Fusarium*, which cause snow mold of cereals and turf grasses, thrive only in cool seasons or cold regions. Also, the late blight pathogen *Phytophthora infestans* is most serious in the northern latitudes; in the subtropics it is serious only during the winter. Many diseases, such as the brown rot of stone fruits caused by *Monilinia fructicola*, are favored by relatively high temperatures and are limited in range to areas and seasons in which such temperatures are prevalent. Several diseases, such as the fusarial wilts, many anthracnoses caused by *Colletotrichum*, and the bacterial wilts of solanaceous plants caused by *Ralstonia solanacearum*, are favored by high temperatures and are limited to hot areas, being particularly severe in the subtropics and tropics.

The effect of temperature on the development of a particular disease after infection depends on the specific host–pathogen combination. The most rapid disease development, i.e., the shortest time required for the completion of an infection cycle, usually occurs when the temperature is optimum for the development of the pathogen but is above or below the optimum for the development of the host. At temperatures much below or above the optimum for the pathogen, or near the optimum for the host, disease development is slower. Thus, for stem rust of wheat, caused by *Puccinia graminis tritici*, the time required for an infection cycle (from inoculation with uredospores to the formation of new uredospores) is 22 days at 5°C, 15 days at 10°C, and 5 to 6 days at 23°C. Similar time periods for the completion of an infection cycle are required in many other diseases caused by fungi, bacteria, and nematodes. Because the duration of an infection cycle determines the number of infection cycles and, therefore, the number of new infections in one season, it is clear that the effect of temperature on the prevalence of a disease in a given season may be very great.

If the minimum, optimum, and maximum temperatures for the pathogen, the host, and the disease are

FIGURE 7-4 (A) Root and crown rot of tomato plant caused by *Fusarium oxysporum*. (B) Aboveground symptoms of wilting and death of tomato plants affected by such root and crown rot. (C) Effect of temperature on growth of *F. oxysporum* and difference in the growth of some of its races at the same temperatures. [Courtesy of (A and B) R. J. McGovern, University of Florida and (C) Swanson and van Gundy (1985). *Plant Dis.* 69, 779–781.]

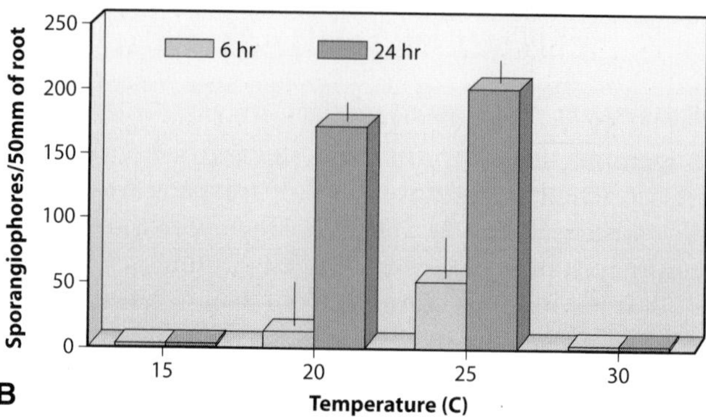

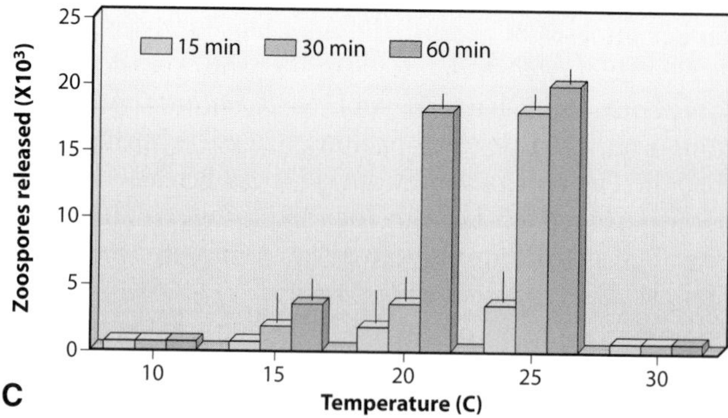

FIGURE 7-5 (A) Lettuce leaves showing symptoms of downy mildew caused by *Bremia lactucae*. (B) Effect of 6- and 24-hour temperature exposures on the production of sporangiophores (top) on infected roots and of 15-, 30-, and 60-minute temperature exposures on zoospores released from sporangia (bottom) of the same oomycete. [Courtesy of University Florida and (B) Stanghellini *et al.* (1990). *Plant Dis.* 74, 173–178.]

about the same, the effect of temperature in disease development is apparently through its influence on the pathogen. The latter becomes so activated at the optimum temperature that the host, even at its optimum growth rate, cannot contain it.

In many diseases, the optimum temperature for disease development seems to be different from those of both pathogen and host. Thus, in the black root rot of tobacco, caused by the fungus *Thielaviopsis basicola*, the optimum temperature range for disease is 17 to 23°C, that for tobacco growth is 28 to 29°C, and that for the pathogen is 22 to 28°C. Evidently, neither the pathogen nor the host grow well at 17 to 23°C, but the host grows so much more poorly and is so much weaker than the pathogen that even the weakened pathogen can cause maximum disease development. In the root rots of wheat and corn caused by the fungus *Gibberella zeae*, the maximum disease development on wheat occurs at temperatures above the optima for the development of both the pathogen and wheat, but on corn it occurs at temperatures below the optima for the pathogen and corn. Considering that wheat grows best at low temperatures whereas corn grows best at high temperatures, it would appear that the more severe damage to wheat at high temperatures and to corn at low temperatures is due to a disproportionately greater weakening of the plants than of the pathogen at the unfavorable temperatures.

The effect of temperature on virus diseases of plants is much more unpredictable. In virus inoculation experiments in the greenhouse, temperature determines not only the ease with which plants can become infected with a virus, but also whether a virus multiplies in the plant and, if it does, the type of symptoms produced. The severity of the disease may vary greatly in various virus–host combinations depending on the temperature during certain stages of the disease. In the field, temperature, probably in combination with sunlight, seems to determine the seasonal appearance of symptoms in the various virus diseases of plants. Viruses producing yellows or leaf-roll symptoms are most severe in the summer, whereas those causing mosaic or ring spot symptoms are most pronounced in the spring. New growth produced during the summer on mosaic- or ring spot-infected plants usually shows only mild symptoms or is completely free from symptoms.

It is now becoming clear that temperatures, high or low, operate by affecting the genetic machinery of the cell by favoring or inhibiting the expression of certain genes involved in disease resistance or susceptibility. For example, cold hardening increases the resistance of cereals and grasses to the snow mold disease caused by the fungus *Microdochium nivale*, partly by causing an increase in sucrose synthetase and, upon infection, in a more rapid production by the plant of pathogenesis-related proteins. However, exposure of barley leaves to 50°C for one minute resulted in induced resistance against the powdery mildew fungus *Blumeria graminis* f. sp. *hordei* by causing an oxidative burst in the plant, production of cell wall-bound proteins, and stoppage of fungal growth after appressorium formation.

EFFECT OF MOISTURE

Moisture, like temperature, influences the initiation and development of infectious plant diseases in many interrelated ways. It may exist as rain or irrigation water on the plant surface or around the roots, as relative humidity in the air, and as dew. Moisture is indispensable for the germination of fungal spores and penetration of the host by the germ tube. It is also indispensable for the activation of bacterial, fungal, and nematode pathogens before they can infect the plant. Moisture, in such forms as splashing rain and running water, also plays an important role in the distribution and spread of many of these pathogens on the same plant and on their spread from one plant to another. Finally, moisture increases the succulence of host plants and thus their susceptibility to certain pathogens, which affects the extent and severity of disease.

The occurrence of many diseases in a particular region is closely correlated with the amount and distribution of rainfall within the year (Fig. 7-1). Thus, late blight of potato, apple scab, downy mildew of grapes, and fire blight are found or are severe only in areas with high rainfall or high relative humidity during the growing season. Indeed, in all of these and other diseases, the rainfall determines not only the severity of the disease, but also whether the disease will even occur in a given season (Fig. 7-6). In fungal diseases, moisture affects fungal spore formation, longevity, and particularly the germination of spores, which requires a film of water covering the tissues. In many fungi, moisture also affects the liberation of spores from the sporophores, which, as in apple scab, can occur only in the presence of moisture. The number of infection cycles per season of many fungal diseases is closely correlated with the number of rainfalls per season, particularly of rainfalls that are of sufficient duration to allow establishment of new infections. Thus in apple scab, for example, continuous wetting of the leaves, fruit, and so on for at least 9 hours is required for any infection to take place even at the optimum range (18 to 23°C) of temperature for the pathogen. At lower or higher temperatures the minimum wetting period required is higher, e.g., 14 hours at 10°C and 28 hours at 6°C. Similar conditions are required for the initiation and development of

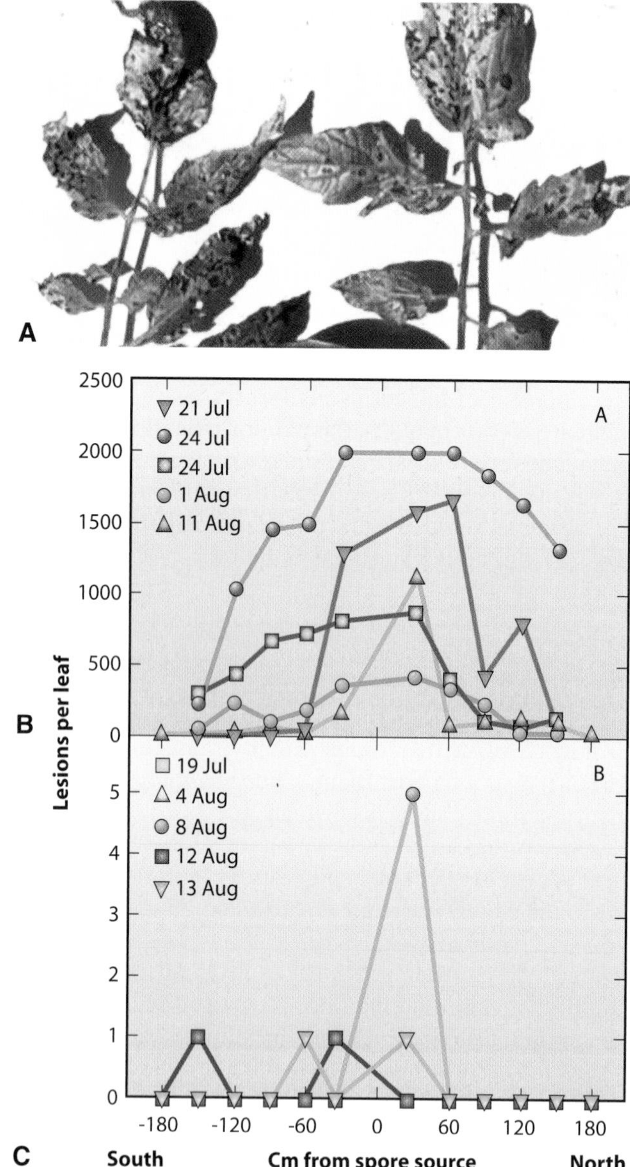

FIGURE 7-6 (A) Foliar symptoms of tomato infected by *Septoria lycopersici*. (B and C) Effect of moisture (rain) on the number of *Septoria* leaf spots per leaf on tomato plants planted at 30-centimeter intervals from an inoculated row (0) on five dates in the presence (A) and absence (B) of rainfall. The amounts of the five successive rainfalls in (A) on the dates indicated were 19.1, 30.0, 11.9, 4.1, and 8.6 millimeters [Courtesy of University of Florida and (B and C) Parker *et al.* (1995). *Plant Dis.* 79, 148–152.]

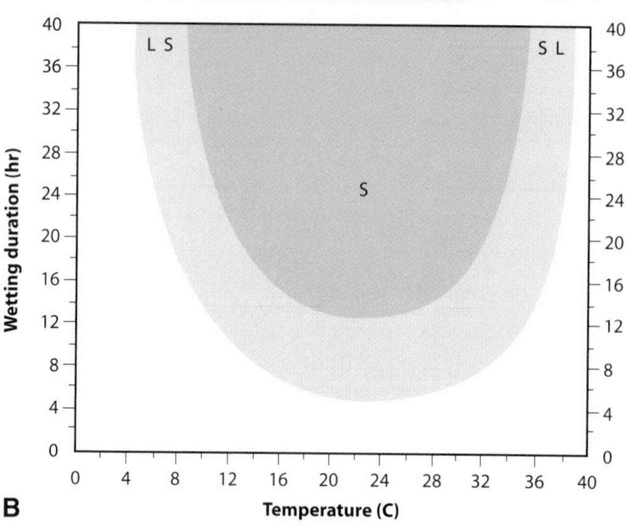

FIGURE 7-7 (A) Apple leaf spot caused by *Alternaria mali*. (B) Leaf wetness and temperature requirements for the leaf-spotting fungus *A. mali* to cause light (L) or severe (S) infection. (Photographs courtesy of T. B. Sutton, North Carolina State University.)

infections in many other diseases (Figs. 7-7 and 7-8). If the length of the wetting period is less than the minimum required for the particular temperature, the pathogen fails to establish itself in the host and fails to produce disease.

Most fungal pathogens require free moisture on the host or high relative humidity in the atmosphere for spore release (Fig. 7-9) or for germination of their spores. Most pathogens become independent of outside moisture once they can obtain nutrients and water from the host. Some pathogens, however, such as those causing late blight of potato and the downy mildews (Fig. 7-9A), must have high relative humidity or free moisture in the environment throughout their development. In these diseases, although spores may be released following a short leaf-wetness period (Figs. 7-9B and 7-9C), the growth and sporulation of the pathogen, and the production of symptoms, come to a halt as soon as dry, hot weather sets in. All these activities resume only when it rains again or after the return of humid weather.

Although most fungal and bacterial pathogens of aboveground parts of plants require a film of water to infect hosts successfully, spores of the powdery mildew

FIGURE 7-8 (A) Cedar-apple rust, caused by the fungus *Gymnosporangium juniperi-virginianae,* produces "cedar apples" on cedar. (B) Cedar-apple rust leaf spots on apple leaf resulting from basidiospores produced on cedar-apple telial horns. (C) Formation of basidiospores occurs when the temperature–leaf wetness point is at the transition line between the clear and shaded area of the diagram. If the temperature–leaf wetness point is within the shaded area, spore germination has occurred and infection is likely. [Photographs courtesy of University of Florida, (B) J. A. Christensen, Texas A&M University, and (C) Seem and Russo (1984). *Plant Dis.* **68**, 656–660.]

fungi can germinate, penetrate, and cause infection even when there is only high relative humidity in the atmosphere surrounding the plant. In powdery mildews, spore germination and infection are actually lower in the presence of free moisture on the plant surface than they are in its absence. In some of them, the most severe infections take place when the relative humidity is rather low (50 to 70%). In these diseases, the amount of disease is limited rather than increased by wet weather, as indicated by the fact that powdery mildews are more common and more severe in the drier areas of the world. The relative importance of powdery mildews decreases as rainfall increases. In high rainfall areas and periods, other diseases become more prevalent.

In many diseases affecting underground parts of plants, such as roots, tubers, and young seedlings, e.g., in the *Pythium* damping off of seedlings and seed decays, the severity of the disease is proportional to the amount of soil moisture and is greatest near the saturation point. The increased moisture seems to affect primarily the pathogen, which multiplies and moves (zoospores in the case of *Pythium*) best in wet soils. Increased moisture may also decrease the ability of the host to defend itself through a reduced availability of oxygen in water-logged soil and by lowering the temperature of such soils. Many other soil pathogens [e.g., *Phytophthora* (Fig. 7-10A), *Rhizoctonia*, *Sclerotinia*, and *Sclerotium*], some bacteria (e.g., *Erwinia* and

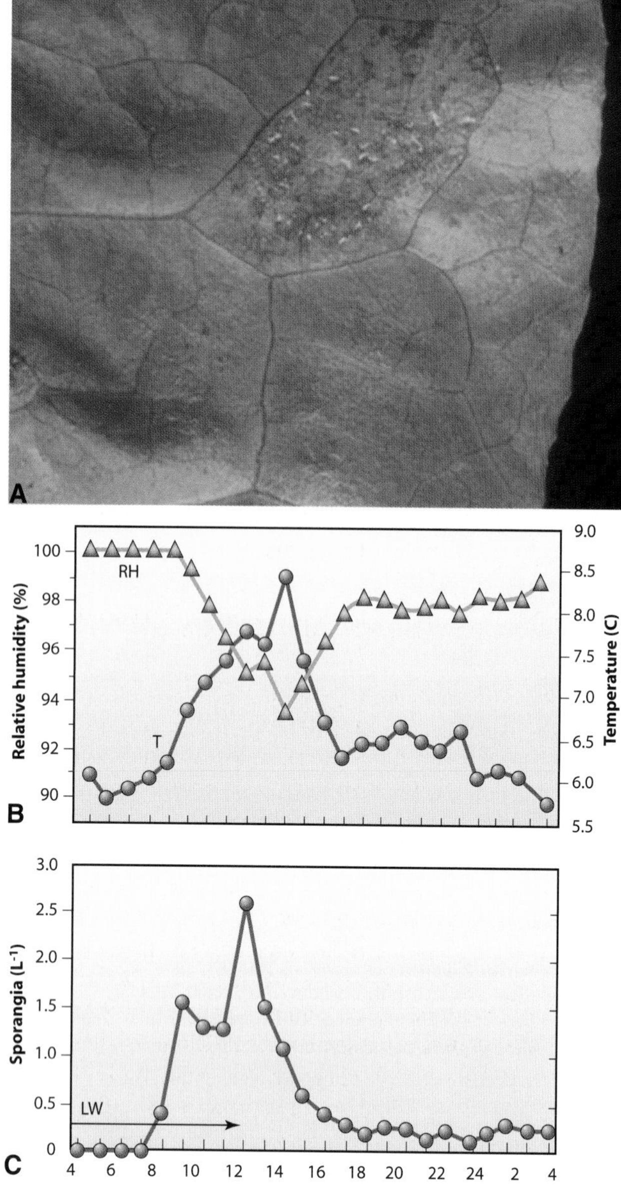

FIGURE 7-9 (A) A large and some smaller areas on a downy mildew–infected lettuce leaf producing sporangiophores and sporangia. (B) Relationship of temperature (T), relative humidity (RH), and spore release by *Bremia lactucae*, the cause of a downy mildew of lettuce, in a 24-hour period following a leaf wetness (LW) period (arrow). [Photographs courtesy of University of Florida and (B) Scherm and van Bruggen (1995). *Phytopathology* 85, 552–555.]

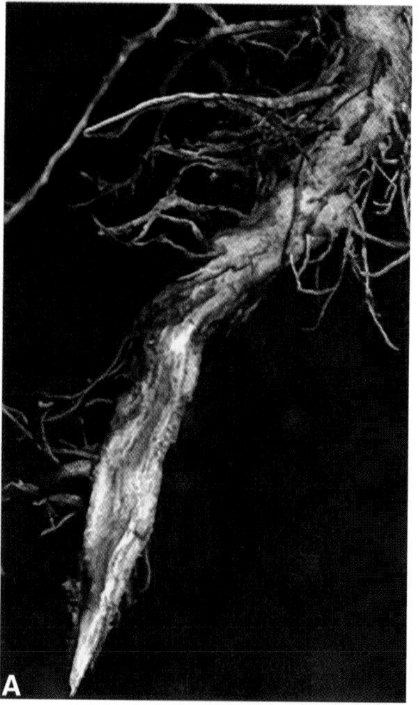

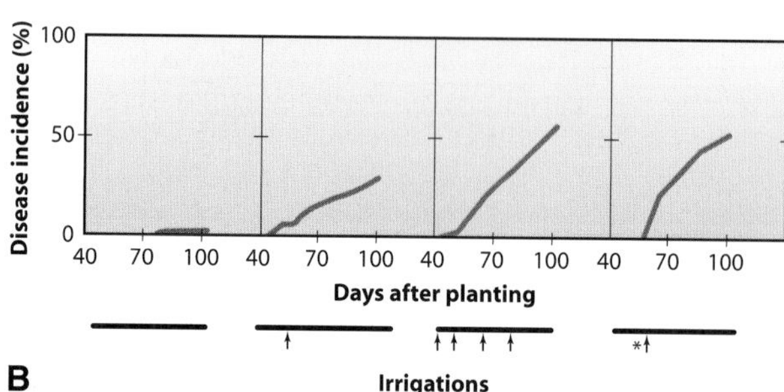

FIGURE 7-10 (A) Root rot symptoms caused by *Phytophthora sp.* (B) Development of *Phytophthora* root rot in a susceptible safflower variety under various irrigation schedules in the field. There was no rainfall. Arrows show times of surface irrigations. The asterisk before the arrow at the right field denotes plant stress before irrigation. (B) From Duniway (1982). *In* "Biometeorology in IPM" (Hatfield and Thomason, eds.). Academic Press, New York.

Pseudomonas), and most nematodes usually cause their most severe symptoms on plants when the soil is wet but not flooded (Fig. 7-10B). Several other fungi, e.g., *Fusarium solani*, which is the cause of dry root rot of beans, *Fusarium roseum*, the cause of seedling blights, and *Macrophomina phaseoli*, the cause of charcoal rot of sorghum and of root rot of cotton, grow fairly well in rather dry environments. Apparently that characteristic enables them to cause more severe diseases in drier soils on plants that are stressed by insufficient water. Vascular wilts caused by the fungus *Verticillium* and canker diseases of forest trees and seedlings caused by fungi are significantly more severe when the plants suffer from water stress. Similarly, *Streptomyces scabies*, which causes the common scab of potatoes, becomes most severe in soils drying out after wetting.

Most bacterial diseases, and also many fungal diseases of young tender tissues, are particularly favored by high moisture or high relative humidity. Bacterial pathogens and fungal spores are usually disseminated in water drops splashed by rain, in rainwater moving from the surfaces of infected tissues to those of healthy ones, or in free water in the soil. Bacteria penetrate plants through wounds or natural openings and cause severe disease when present in large numbers. Once inside the plant tissues, the bacteria multiply faster and are more active during wet weather, probably because the plants, through increased water absorption and resulting succulence, can provide the high concentrations of water that favor bacteria. The increased bacterial activity in wet weather produces greater damage to tissues. This damage, in turn, helps release greater numbers of bacteria onto the plant surface, where they are available to start more infections if the wet weather continues.

EFFECT OF WIND

Wind influences infectious plant diseases primarily by increasing the spread of plant pathogens and the number of wounds on host plants and, to a smaller extent, by accelerating the drying of wet surfaces of plants. Most plant diseases that spread rapidly and are likely to assume large epidemic proportions are caused by pathogens such as fungi, bacteria, and viruses that are spread either directly by the wind or indirectly by insect vectors that can themselves be carried over long distances by the wind. Some spores, e.g., basidiospores, and some conidia, and also zoosporangia, are quite delicate and do not survive long-distance transport in the wind. Others, e.g., uredospores and many kinds of conidia, can be transported by the wind for many kilometers. Wind is even more important in disease development when it is accompanied by rain. Wind-blown rain helps release spores and bacteria from infected tissue and then carries them through the air and deposits them on wet surfaces of plants, which, if susceptible, can be infected immediately. Wind also injures plant surfaces while they are blown about and rub against one another or through wind-blown sand; this facilitates infection by many fungi and bacteria and also by a few mechanically transmitted viruses. Wind, however, sometimes helps prevent infection by accelerating the drying of the wet plant surfaces on which fungal spores or bacteria may have landed. If the plant surfaces dry before penetration has taken place, any germinating spores or bacteria present on the plant are likely to desiccate and die, and no infection will occur.

EFFECT OF LIGHT

The effect of light on disease development, especially under natural conditions, is far less than that of temperature or moisture. Several diseases are known in which the intensity and the duration of light may either increase or decrease the susceptibility of plants to infection and also the severity of the disease. In nature, however, the effect of light is limited to the production of more or less etiolated plants as a result of reduced light intensity. This usually increases the susceptibility of plants to nonobligate parasites, for example, of lettuce and tomato plants to *Botrytis* or of tomato to *Fusarium*, but decreases their susceptibility to obligate parasites, for example, of wheat to the stem rust fungus *Puccinia*.

Reduced light intensity generally increases the susceptibility of plants to virus infections. Plants kept in the dark for 1 or 2 days before sap inoculation with a virus produce more local lesions (i.e., infections) than plants kept in the normal light–dark regime. This has become a routine procedure in many laboratories. Generally, keeping plants in the dark affects the sensitivity of plants to virus infection if it precedes inoculation with the virus, but it seems to have little or no effect on symptom development if it occurs after inoculation. However, low light intensities following inoculation tend to mask the symptoms of some diseases. In these diseases, symptoms are much more severe when the plants are grown in normal light than when they are shaded.

EFFECT OF SOIL PH AND SOIL STRUCTURE

The pH of the soil is important in the occurrence and severity of plant diseases caused by certain soilborne pathogens. For example, the clubroot of crucifers caused by *Plasmodiophora brassicae* is most prevalent and severe at about pH 5.7, whereas its development drops sharply between pH 5.7 and 6.2 and is completely checked at pH 7.8. On the contrary, the common scab of potato caused by *S. scabies* can be severe from pH 5.2 to 8.0 or above, but its development drops sharply below pH 5.2. It is obvious that such diseases are most serious in areas in which soil pH favors the particular pathogen. In these and many other diseases, the effect of soil acidity (pH) seems to be principally on the pathogen. In some diseases, however, a weakening of the host through altered nutrition that is induced by the soil acidity may affect the incidence and severity of the disease.

Soil factors other than pH may also influence the development of plant diseases. For example, the cotton root rot fungus (*Phymatotrichopsis omnivora*) affects many hosts, e.g., peach trees (Fig. 7-11A), and grows best at high pH (pH 7.2–8.0). The fungus, however, exists only in the southwestern United States and northern Mexico (Fig. 7-11B), where the soils contain relatively high concentrations of calcium carbonate.

EFFECT OF HOST–PLANT NUTRITION

Nutrition affects the rate of growth and the state of readiness of plants to defend themselves against pathogenic attack.

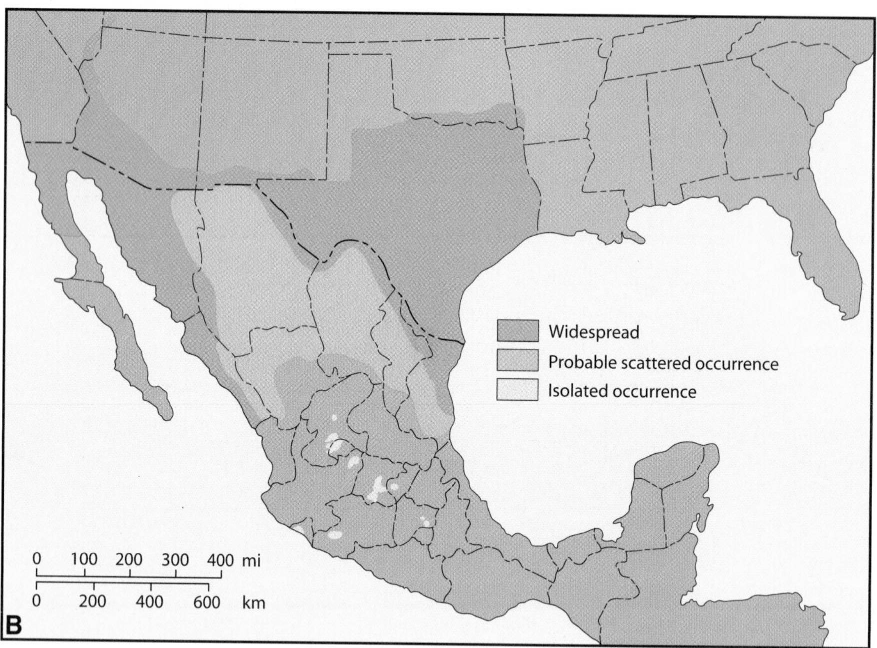

FIGURE 7-11 (A) Root rot and death of peach trees caused by the cotton root rot fungus *Phymatotrichopsis omnivora*. (B) Distribution of the cotton root rot fungus in North America is limited to areas with soils high in calcium carbonate (and high pH) and high temperatures. [Photographs courtesy of (A) R. B. Hines, New Mexico State University and (B) R. G. Percy (1983). *Plant Dis*. **67**, 981–983.]

Nitrogen abundance results in the production of young, succulent growth, a prolonged vegetative period, and delayed maturity of the plant. These effects make the plant more susceptible to pathogens that normally attack such tissues and for longer periods. Conversely, plants suffering from a lack of nitrogen are weaker, slower growing, and faster aging. Such plants, therefore, are susceptible to pathogens that are best able to attack weak, slow-growing plants. It is known, for example, that fertilization with large amounts of nitrogen increases the susceptibility of pear to fire blight (*Erwinia amylovora*) and of wheat to rust (*Puccinia*) and powdery mildew (*Erysiphe*). It has also been shown that *Cercospora* diseases of cereals, such as corn gray leaf spot, rice brown leaf spot, and the Sigatoka disease of banana, increase in severity with increasing nitrogen fertilization. The reduced availability of nitrogen may increase the susceptibility of tomato to *Fusarium* wilt, of many solanaceous plants to *Alternaria solani* early blight and *Ralstonia solanacearum* wilt, of sugar beets to *Sclerotium rolfsii*, and of most seedlings to *Pythium* damping off.

It is possible, however, that it is not the amount of nitrogen but the form of nitrogen (ammonium or nitrate) that is available to the host or pathogen that affects disease severity or resistance. Of numerous root rots, wilts, foliar diseases, and so on treated with either form of nitrogen, almost as many decreased or increased in severity when treated with a source of ammonium nitrogen as did when treated with a source of nitrate nitrogen. Each form of nitrogen, however, had exactly the opposite effect on a disease (i.e., decrease or increase in severity) than the other form of nitrogen. For example, *Fusarium* spp., *P. brassicae*, *S. rolfsii*, *Pyrenochaeta lycopersici*, and the diseases they cause (root rots and wilts, clubroot of crucifers, damping off and stem rots, and corky root rot, respectively) increase in severity when an ammonium fertilizer is applied (Fig. 7-12). Alternatively, *P. omnivora*, *Gaeumannomyces graminis*, and *S. scabies*, and the diseases they cause (cotton root rot, take-all of wheat, and scab of potato, respectively) are favored by nitrate nitrogen. The effect of each nitrogen form appears to be associated with soil pH influences. Diseases increased by ammonium nitrogen are generally more severe at acid pH, whereas those increased by nitrate nitrogen are generally more severe at neutral to alkaline pH. Ammonium ions (NH_4^+) are absorbed by the roots through exchange with H^+ released by the roots to the surrounding medium, thus reducing soil pH.

Because of the profound effects of nitrogen on growth, nitrogen nutrition has been studied the most extensively in relation to disease development. Studies with other elements, however, such as phosphorus, potassium, and calcium, and also with micronutrients have indicated similar relationships between levels of the particular nutrients and susceptibility or resistance to certain diseases.

Phosphorus has been shown to reduce the severity of take-all disease of barley (caused by *G. graminis*) and potato scab (caused by *S. scabies*) but to increase the severity of cucumber mosaic virus on spinach and of leaf and glume blotch of wheat caused by *Septoria*. Phosphorus seems to increase resistance either by improving the balance of nutrients in the plant or by accelerating the maturity of the crop and allowing it to escape infection by pathogens that prefer younger tissues.

Potassium has also been shown to reduce the severity of numerous diseases, including stem rust of wheat, early blight of tomato, and gray leaf spot and stalk rot of corn, although high amounts of potassium seem to increase the severity of rice blast (caused by *Magnaporthe grisea*), corn gray leaf spot (caused by *Cercospora zeae-maydis*), and root knot (caused by the nematode *Meloidogyne incognita*). Potassium seems to have a direct effect on the various stages of pathogen establishment and development in the host and an indirect effect on infection by promoting wound healing. Potassium also increases resistance to frost injury and thereby reduces infection that commonly begins in frost-killed tissues. In addition, potassium delays maturity and senescence in some crops and during these periods

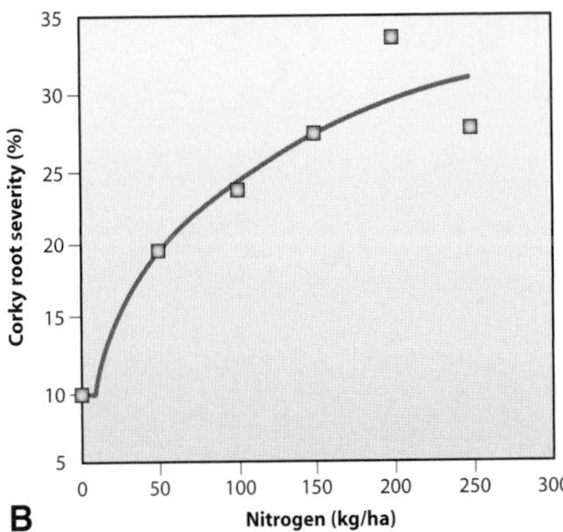

FIGURE 7-12 (A) Corky root of tomato caused by *Pyrenochaeta lycopersici*. (B) Effect of amount of nitrogen (ammonium nitrate) applied to the soil on the severity (percentage of root length infected) of corky root of tomato. [Photographs courtesy of (A) R. J. McGovern, University of Florida and (B) Worneh and van Bruggen (1994). *Phytopathology* 84, 688–694.]

infection by certain facultative parasites can be severely damaging.

Calcium reduces the severity of several diseases caused by root and stem pathogens, such as the fungi *Rhizoctonia*, *Sclerotium*, and *Botrytis*, the wilt fungus *Fusarium oxysporum*, and the nematode *Ditylenclus dipsaci*, but it increases the black shank disease of tobacco (caused by *Phytophthora parasitica* var. *nicotianae*) and the common scab of potato (caused by *S. scabies*). The effect of calcium on disease resistance seems to result from its effect on the composition of cell walls and their resistance to penetration by pathogens.

A reduction in disease levels was also observed when levels of certain micronutrients were increased. For example, application of **iron** to the soil reduced Verticillium wilts of mango and of peanuts. Foliar applications of iron compounds reduced the severity of silver leaf of deciduous fruit trees (caused by *Chondrostereum purpureum*). **Copper** applications to the soil significantly reduced take-all and ergot diseases (caused by the fungi *G. graminis* and *Claviceps purpurea*, respectively), as well as stem melanosis (caused by the bacterium *Pseudomonas chicorii*) in wheat and barley. Similarly, applications of **manganese** reduced potato scab and late blight of potato and stem rot (caused by *Sclerotinia sclerotiorum*) of pumpkin seedlings, but the addition of **magnesium** increased the severity of corn leaf blight caused by *Cochliobolus heterostrofus*, whereas applications of molybdenum reduced late blight of potato and *Ascochyta* blight of beans and peas. The severity of other diseases, however, was raised by the presence of higher levels of these micronutrients, e.g., Fusarium wilt of tomato by increased iron or manganese and tobacco mosaic of tomatoes by increased manganese.

In recent years, the addition of **silicon** to the soil or to the nutrient solution supplied to greenhouse plants has been shown to reduce diseases. Field applications of various grades of silicon increased the amount of silicon taken up by the plants (Fig. 7-13A) and reduced the amount of disease in rice such as brown spot of rice

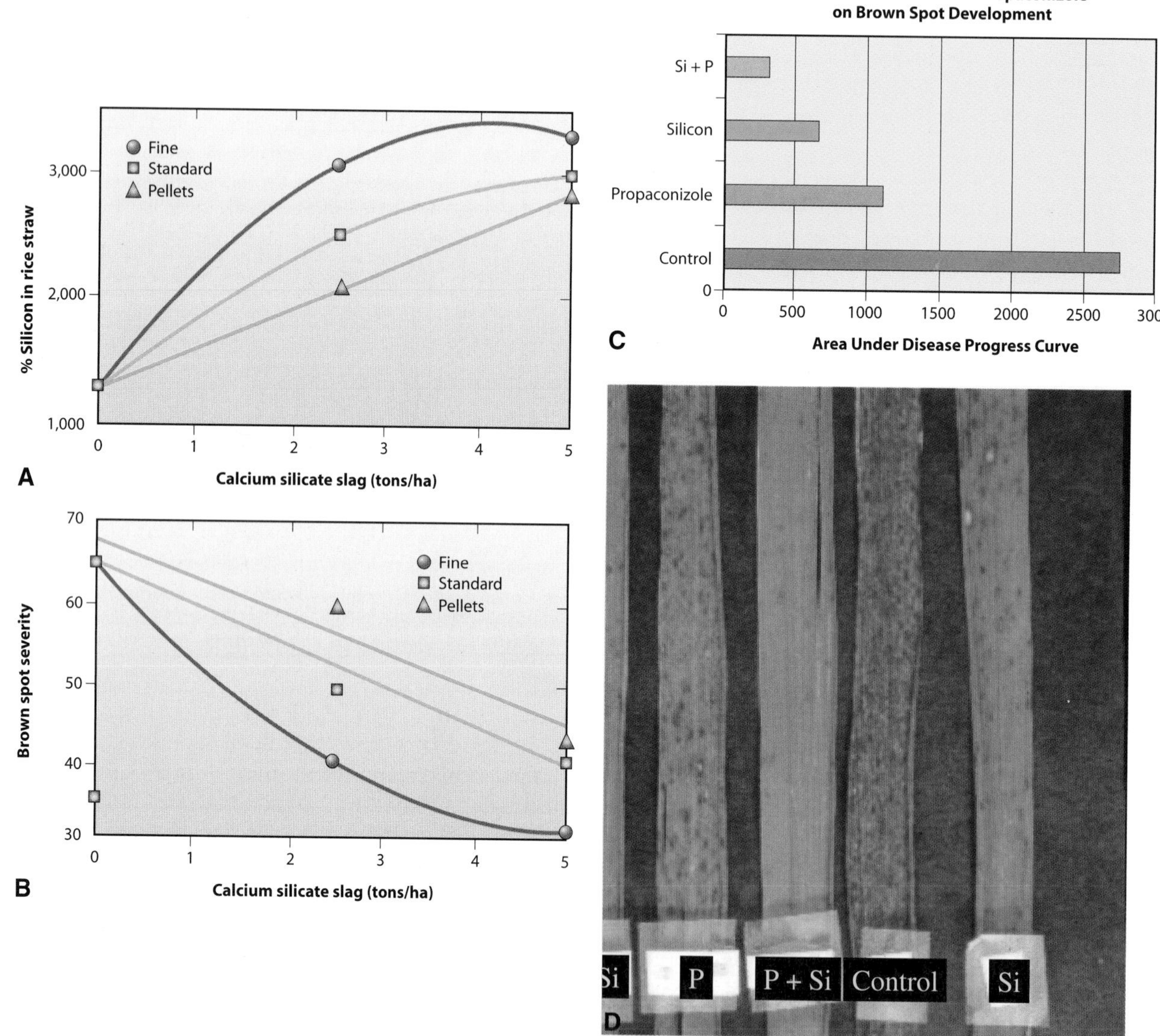

FIGURE 7-13 (A) Relationship between calcium silicate slag grades and quantity to the concentration of silicon in rice straw. (B) Reduction of severity of the brown spot disease of rice caused by the fungus *Cochliobolus miyabeanus*. (C and D) Comparison of brown spot reduction by silicon and fungicide application. [Courtesy of (A and B) Datnoff (1992). *Plant Dis.* **76**, 1011–1013 and (C and D) L. E. Datnoff, University of Florida.]

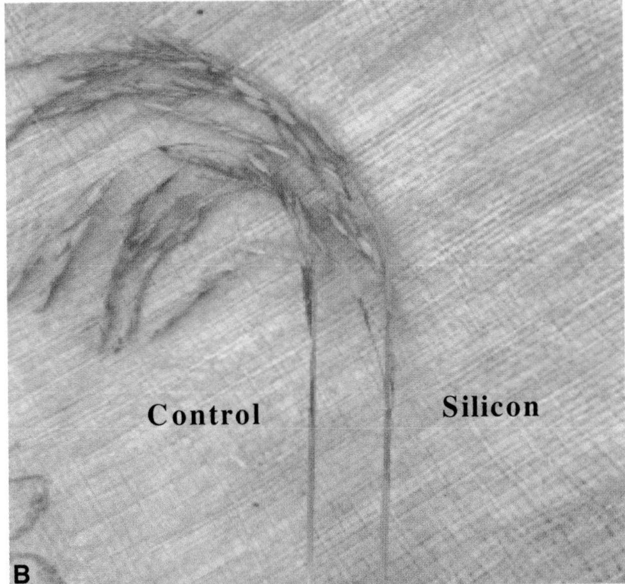

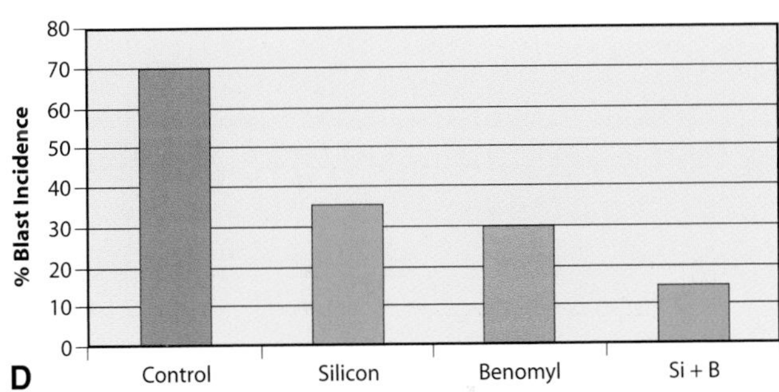

FIGURE 7-14 (A) Characteristic lesions on rice leaf (A) and panicle neck rot and blast (B) caused by the rice blast fungus *Magnaporthe grisea*. (C) Aerial photograph of a rice field, the distal half of which (+silicon) received 3 tons/acre calcium silicate slag, while the proximal half (–silicon) did not. Note the brown discoloration of the proximal half due to infection by the rice blast disease and some rice brown spot, while the distal half receiving silicon shows little or no browning of plants. (D) Graph showing the effect of silicon alone or in combination with the fungicide benomyl on the incidence of rice blast in comparison with an untreated rice field (control). (Courtesy of L. E. Datnoff, University of Florida.)

(Figs. 7-13B and 7-13C) caused by *Cochliobolus miyabeanus*, of rice blast (Fig. 7-14) caused by the fungus *M. grisea*, and of rice sheath blight caused by *Rhizoctonia solani*. The addition of silicon to the soil reduced brown spot more than application of a fungicide (Figs. 7-13C and 7-13D), reduced rice blast comparable to that of a fungicide application (Figs. 7-14C and 7-14D), and reduced rice sheath blight by at least 50% not only in susceptible but also in the resistant varieties (Figs. 7-15A and 7-15B). In greenhouse applications, silicon reduced disease levels, for example, of cucumber powdery mildew and cucumber root rot caused by the fungus *Sphaerotheca fuligena* and the oomycete *Pythium ultimum*, respectively, and of wheat powdery mildew caused by *Blumeria graminis* f. sp. *tritici*. In the latter, epidermal cells of silicon-treated plants produced specific defense reactions upon inoculation with the powdery mildew fungus, including the formation of papilla, production of callose, and release of phenolic compounds that accumulated along the cell wall and affected the integrity of the pathogen.

In diseases caused by phytoplasmas and spiroplasmas, such as maize bushy stunt and corn stunt, respectively, diseased plants took up less nutrients than healthy plants regardless of the level of availability of soil water, and spiroplasma-infected plants suppressed particularly the uptake of Mg from the soil.

In general, plants receiving a balanced nutrition, in which all required elements are supplied in appropriate amounts, are more capable of protecting themselves

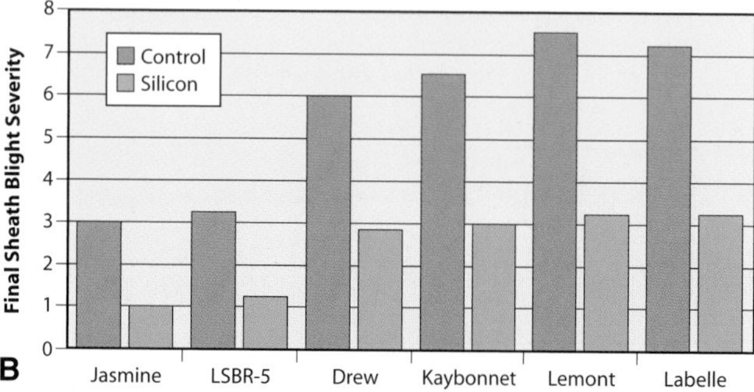

FIGURE 7-15 (A) Symptoms of rice sheath blight on rice leaves and stems caused by *Rhizoctonia solani*. (B) Graph showing the reduction of sheath blight severity by silicon added to fields planted to rice varieties of various resistances to sheath blight. Even the most resistant varieties benefited from the addition of silicon to the soil. (Courtesy of L. E. Datnoff, University of Florida.)

from new infections and of limiting existing infections than plants to which one or more nutrients are supplied in excessive or deficient amounts. However, even balanced nutrition may affect the development of a disease when the concentration of all the nutrients is increased or decreased beyond a certain range.

EFFECT OF HERBICIDES

Herbicide use is common and widespread in agriculture. In many cases, herbicides have been shown to increase the severity of certain diseases on crop plants, for example, of *R. solani* on sugar beets and cotton, *Fusarium* wilt of tomatoes and cotton, and *Sclerotium* stem rots of various crops. In other plant–pathogen combinations, herbicides appear to decrease disease, for example, *Aphanomyces euteiches* root rot of peas, *Pseudocercosporella herpotrichoides* foot rot of wheat, and *Phytophthora* collar rot of various crops. Herbicides apparently act on plant diseases either directly or indirectly. The direct effects may include stimulation or retardation of the growth of the pathogen or an increase or decrease in the susceptibility of the host. Indirect effects include an increase or decrease in the activity of soil microflora, elimination or selection of the pathogen by certain additional or alternate hosts, or alteration of the microclimate of the crop plant canopy (e.g., change in humidity).

EFFECT OF AIR POLLUTANTS

Air pollutants cause various types of direct symptoms on plants exposed to high levels of pollutants. In infectious plant diseases, both the plant and the pathogen are exposed to the same levels of pollutants, but it is not yet clear whether the presence of a particular pollutant causes an increase or a decrease in the severity of the disease caused by the pathogen alone. It appears, however, that some air pollutants, such as ozone, may affect a pathogen and sometimes the disease it causes. For example, with the rusts on oats and wheat, ozone reduces the growth of uredia and of hyphal growth and also the number of uredospores produced on ozone-injured leaves, whereas with the powdery mildew of barley, the rate of infection is reduced if the exposure to ozone is early but is increased if exposure occurs late. With nonobligate parasites, ozone may increase the percentage of diseased leaf area of wheat by *Drecslera* fungus, infection of potato leaves by *Botrytis* occurs only on ozone-injured leaves, and in the *Lophodermium* needle blight of pine, ozone exposure increases the severity of the needle blight. Similarly, the bacteria *Pseudomonas glycinea,* infecting soybean, and *Xanthomonas alfalfae,* infecting alfalfa, caused a smaller number of lesions on plants exposed to ozone than on unexposed ones.

Selected References

Altman, J., and Campbell, C. L. (1977). Effect of herbicides on plant diseases. *Annu. Rev. Phytopathol.* **15**, 361–385.

Ayres, P. G. (1984). The interaction between environmental stress injury and biotic disease physiology. *Annu. Rev. Phytopathol.* **22**, 53–75.

Belanger, R. R., *et al.* (1995). Soluble silicon: Its role in crop and disease management of greenhouse crops. *Plant Dis.* **79**, 329–336.

Belanger, R. R., Benhamou, N., and Menzies, J. G. (2003). Cytological evidence of an active role of silicon in wheat resistance to powdery mildew (*Blumeria graminis* f. sp. *tritici*). *Phytopathology* **93**, 402–412.

Caldwell, P. M., et al. (2002). Assessment of the effects of fertilizer applications on gray leaf spot and yield of maize. *Plant Dis.* **86**, 859–866.

Carver, T. L. W., et al. (1998). Silicon deprivation enhances localized autofluorescent responses and phenylalanine ammonia-lyase activity in oat attacked by *Blumeria graminis*. *Physiol Mol. Plant Pathol.* **52**, 245–257.

Chupp, C. (1928). Club root in relation to soil alkalinity. *Phytopathology* **18**, 301–306.

Colhoun, J. (1973). Effects of environmental factors on plant disease. *Annu. Rev. Phytopathol.* **11**, 343–364.

Colhoun, J. (1979). Predisposition by the environment. *In* "Plant Disease" (J. G. Horsfall and E. B. Cowling, eds.), Vol. 4, pp. 75–92. Academic Press, New York.

Datnoff, L. E., Deren, C. W., and Snyder, C. H. (1997). Silicon fertilization for disease management of rice in Florida. *Crop Prot.* **16**, 525–531.

Datnoff, L. E., Snyder, G. H., and Deren, C. W. (1992). Influence of silicon fertilizer grades on blast and brown spot development and on rice yields. *Plant Dis.* **76**, 1011–1013.

Datnoff, L. E., Snyder, G. H., and Korndorfer, G. H. (2001). "Silicon in Agriculture." Elsevier Science, B. V., Amsterdam.

de Oliveiro, E., *et al.* (2002). Growth and nutrition of mollicute-infected maize. *Plant Dis.* **85**, 945–949.

Dickson, J. G. (1923). Influence of soil temperature and moisture on the development of seedling blight of wheat and corn caused by *Gibberella saubinetii*. *J. Agric. Res.* **23**, 837–870.

Engelhard, A. W., ed. (1989). "Soilborne Plant Pathogens: Management of Diseases with Macro- and Microelements." APS Press, St. Paul, MN.

Fitt, B. D. L., McCartney, H. A., and Walklate, P. J. (1989). The role of rain in dispersal of pathogen inoculum. *Annu. Rev. Phytopathol.* **27**, 241–270.

Gallegly, M. E., Jr., and Walker, J. C. (1949). Plant nutrition in relation to disease development. V. *Am J. Bot.* **36**, 613–623.

Gunifer, B. M., Touchton, J. T., and Johnson, J. W. (1980). Effects of phosphorus and potassium fertilization on *Septoria* glume blotch of wheat. *Phytopathology* **70**, 1196–1199.

Hepting, G. H. (1963). Climate and forest diseases. *Annu. Rev. Phytopathol.* **1**, 31–50.

Huber, D. M., and Watson, R. D. (1974). Nitrogen form and plant disease. *Annu. Rev. Phytopathol.* **12**, 139–165.

Huber, L., and Gillespie, T. J. (1992). Modeling leaf wetness in relation to plant disease epidemiology. *Annu. Rev. Phytopathol.* **30**, 553–577.

Jones, L. R., Johnson, J., and Dickson, J. G. (1926). Wisconsin studies upon the relation of soil temperature to plant diseases. *Res. Bull. Wis. Agric. Exp. Stn.* **71**.

Kassanis, B. (1957). Effect of changing temperature on plant virus diseases. *Adv. Virus Res.* **4**, 169–186.

Lévesque, C. A., and Rahe, J. E. (1992). Herbicide interactions with fungal root pathogens, with special reference to glyphosate. *Annu. Rev. Phytopathol.* **30**, 579–602.

MacKenzie, D. R. (1981). Association of potato early blight, nitrogen fertilizer rate and potato yield. *Plant Dis.* **65**, 575–577.

McElhaney, R., Alvarez, A. M., and Kado, C. I. (1998) Nitrogen limits *Xanthomonas campestris* pv. *campestris* invasion of the host xylem. *Physiol. Mol. Plant Pathol.* **52**, 15–24.

Manning, W. J., and Von Tiedemann, A. (1995). Climate change: Potential effects of increased atmospheric carbon dioxide, ozone, and ultraviolet-B radiation on plant disease. *Environ. Pollut.* **88**, 219–246.

Miller, P. R. (1953). The effect of weather on diseases. *In* "Plant Diseases," pp. 83–93. U.S. Dept. of Agriculture, Washington, DC.

Palti, J. (1981). "Cultural Practices and Infectious Crop Diseases." Springer-Verlag, Berlin.

Populer, C. (1978). Changes in host susceptiblity with time. *In* "Plant Disease" (J. G. Horsfall and E. B. Cowling, eds.), Vol. 2, pp. 239–262. Academic Press, New York.

Schnathorst, W. C. (1965). Environment relationships in the powdery mildews. *Annu. Rev. Phytopathol.* **3**, 343–366.

Schoeneweiss, D. F. (1981). The role of environmental stress in diseases of woody plants. *Plant Dis.* **65**, 308–314.

Shaner, G. (1981). Effects of environment on fungal leaf blights of small grains. *Annu. Rev. Phytopathol.* **19**, 273–296.

Thomas, F. M., Blank, R., and Hartmann, G. (2002). Abiotic and biotic factors and their interactions as causes of oak decline in central Europe. *Forest Pathol.* **32**, 277–307.

Wallace, H. R. (1989). Environment and plant health: A nematological perception. *Annu. Rev. Phytopathol.* **27**, 59–75.

Zhonghua M., Morgan, D. P., and Michailides, T. J. (2001). Effects of water stress on *Botryosphaeria* blight of pistachio caused by *Botryosphaeria dothidea*. *Plant Dis.* **85**, 745–749.

chapter eight

PLANT DISEASE EPIDEMIOLOGY

THE ELEMENTS OF AN EPIDEMIC
266

HOST FACTORS THAT AFFECT THE DEVELOPMENT OF EPIDEMICS: LEVELS PF RESISTANCE OR SUSCEPTIBILITY – DEGREE OF GENETIC UNIFORMITY – TYPE OF CROP – AGE OF PLANTS
267

PATHOGEN FACTORS THAT AFFECT DEVELOPMENT OF EPIDEMICS: LEVELS OF VIRULENCE – QUANTITY OF INOCULUM NEAR HOSTS – TYPE OF REPRODUCTION OF THE PATHOGEN – ECOLOGY OF THE PATHOGEN – MODE OF SPREAD OF THE PATHOGEN
269

ENVIRONMENTAL FACTORS THAT AFFECT DEVELOPMENT OF EPIDEMICS – MOISTURE – TEMPERATURE
271

EFFECT OF HUMAN CULTURAL PRACTICES AND CONTROL MEASURES: SITE SELECTION AND PREPARATION – SELECTION OF PROPAGATIVE MATERIAL – CULTURAL PRACTICES – DISEASE CONTROL MEASURE – INTRODUCTION OF NEW PATHOGENS
272

MEASUREMENT OF PLANT DISEASE AND OF YIELD LOSS
273

PATTERNS OF EPIDEMICS
274

COMPARISON OF EPIDEMICS
276

DEVELOPMENT OF EPIDEMICS
277

MODELING OF PLANT DISEASE EPIDEMICS
278

COMPUTER SIMULATION OF EPIDEMICS
280

FORECASTING PLANT DISEASE EPIDEMICS: EVALUATION OF EPIDEMIC THRESHOLDS – EVALUATION OF ECONOMIC DAMAGE THRESHOLD – ASSESSMENT OF INITIAL INOCULUM AND OF DISEASE – MONITORING WEATHER FACTORS THAT AFFECT DISEASE DEVELOPMENT
281

NEW TOOLS IN EPIDEMIOLOGY: MOLECULAR TOOLS – GIS REMOTE SENSING – IMAGE ANALYSIS – INFORMATION TECHNOLOGY
283

EXAMPLES OF PLANT DISEASE FORECAST SYSTEMS: FORECASTS BASED ON: AMOUNT OF INITIAL INOCULUM – ON WEATHER CONDITIONS FAVORING – DEVELOPMENT OF SECONDARY INOCULUM – ON AMOUNTS OF INITIAL AND SECONDARY INOCULUM
285

RISK ASSESSMENT OF PLANT DISEASE EPIDEMICS
287

DISEASE-WARNING SYSTEMS
287

DEVELOPMENT AND USE OF EXPERT SYSTEMS IN PLANT PATHOLOGY
288

DECISION SUPPORT SYSTEMS
289

When a pathogen spreads to and affects many individuals within a population over a relatively large area and within a relatively short time, the phenomenon is called an epidemic. An epidemic has been defined as any increase of disease in a population. A similar definition of an epidemic is the dynamics of change in plant disease in time and space. The study of epidemics and of the factors that influence them is called epidemiology. Epidemiology is concerned simultaneously with populations of pathogens and host plants as they occur in an evolving environment, i.e., the classic disease triangle. As a result, epidemiology is also concerned with population genetics of host resistance and with the evolutionary potential of pathogen populations to produce pathogen races that may be more virulent to host varieties or more resistant to pesticides. Epidemiology, however, must also take into account other biotic and abiotic factors, such as an environment strongly influenced by human activity, particularly as it relates to disease management.

Plant disease epidemics, sometimes called epiphytotics, occur annually on most crops in many parts of the world. Most epidemics are more or less localized and cause minor to moderate losses. Some epidemics are kept in check naturally, e.g., by changes in weather conditions. Others are kept in check by chemical sprays and other control measures. Occasionally, however, some epidemics appear suddenly, go out of control, and become extremely widespread or severe on a particular plant species. Some plant disease epidemics, e.g., wheat rusts, southern corn leaf blight (Fig. 8-1), and grape downy mildew, have caused tremendous losses of produce over rather large areas. Others, e.g., chestnut blight (Fig. 1-8), Dutch elm disease, and coffee rust, have threatened to eliminate certain plant species from entire continents. Still others have caused untold suffering to humans. The Irish potato famine of 1845–1846 was caused by the Phytophthora late blight epidemic of potato, and the Bengal famine of 1943 was caused by the *Cochliobolus* (*Helminthosporium*) brown spot epidemic of rice.

THE ELEMENTS OF AN EPIDEMIC

Plant disease epidemics develop as a result of the timely combination of the same elements that result in plant disease: susceptible host plants, a virulent pathogen, and favorable environmental conditions over a relatively long period of time. Humans may unwittingly help initiate and develop epidemics through some of their activities, e.g., by topping or pruning plants in wet weather. More frequently, humans may stop the initiation and development of epidemics by using appropriate control measures under situations in which epidemics would almost certainly occur without human intervention. Thus, the chance of an epidemic increases when the susceptibility of the host and virulence of the pathogen are greater, as the environmental conditions approach the optimum level for pathogen growth, reproduction, and spread, and as the duration of all favorable combinations is prolonged or repeated.

To describe the interaction of the components of plant disease epidemics, the disease triangle, which is

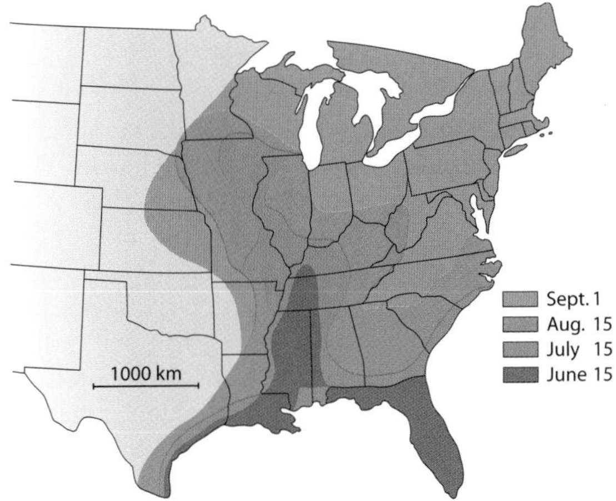

FIGURE 8-1 Development and northward spread of the southern corn leaf blight epidemic, caused by *Cochliobolus heterostrophus* (*Bipolaris maydis*), in the United States from June 15 to September 1, 1970. [From Zadoks and Schein (1979).]

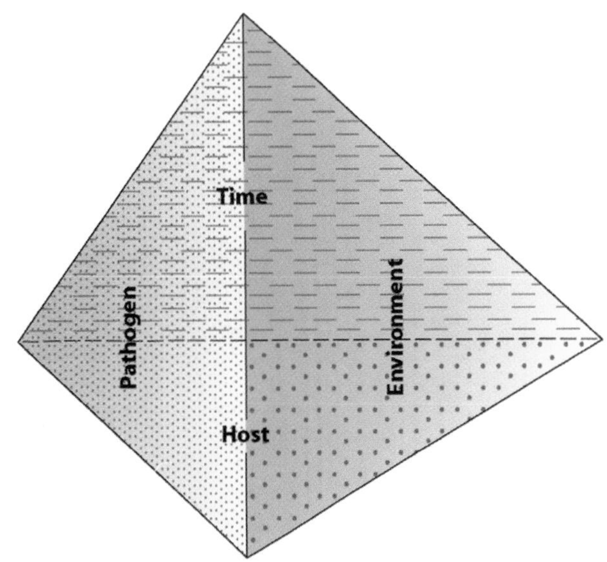

FIGURE 8-2 The disease tetrahedron.

discussed in Chapter 2 and describes the interaction of the components of plant disease, can be expanded to include time and humans. Indeed, the amount of each of the three components of plant disease and their interactions in the development of disease are affected by a fourth component: time. Both the specific point in time at which a particular event in disease development occurs and the length of time during which the event takes place affect the amount of disease. The interaction of the four components can be visualized as a tetrahedron, or pyramid, in which each plane represents one of the components. This figure is referred to as the disease tetrahedron or disease pyramid (Fig. 8-2). The effect of time on disease development becomes apparent when one considers the importance of the time of year (i.e., the climatic conditions and stage of growth when host and pathogen may coexist), the duration and frequency of favorable temperature and rains, the time of appearance of the vector, the duration of the infection cycle of a particular disease, and so on. If the four components of the disease tetrahedron could be quantified, the volume of the tetrahedron would be proportional to the amount of disease on a plant or in a plant population.

Disease development in cultivated plants is also influenced greatly by a fifth component: humans. Humans affect the kind of plants grown in a given area, the degree of plant resistance, the numbers planted, time of planting, and density of the plants. By the resistance of the particular plants they cultivate, humans also determine which pathogens and pathogen races will predominate. By their cultural practices, and by the chemical and biological controls they may use, humans affect the amount of primary and secondary inoculum available to attack plants. They also modify the effect of environment on disease development by delaying or speeding up planting or harvesting, by planting in raised beds or in more widely spaced beds, by protecting plant surfaces with chemicals before rains, by regulating the humidity in produce storage areas, and so on. The timing of human activities in growing and protecting plants may affect various combinations of these components to a considerable degree, thereby affecting the amount of disease in individual plants and in plant populations greatly. The human component has sometimes been used in place of the component "time" in the disease tetrahedron, but it should be considered a distinct fifth component that influences the development of plant disease directly and indirectly.

In Fig. 8-3, host, pathogen, and environment are each represented by one of the sides of the triangle, time is represented as the perpendicular line arising from the center of the triangle and humans as the peak of the tetrahedron whose base is the triangle and height is the length of time. In this way, humans interact with and influence each of the other four components of an epidemic, thereby increasing or decreasing the magnitude of the epidemic. Sometimes, of course, humans themselves can be affected to a greater or lesser extent by plant disease epidemics.

HOST FACTORS THAT AFFECT THE DEVELOPMENT OF EPIDEMICS

Several internal and external factors of particular host plants play an important role in the development of epidemics involving those hosts.

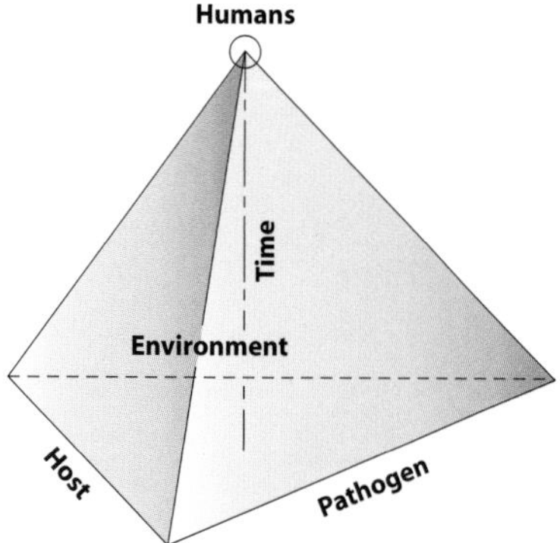

FIGURE 8-3 Schematic diagram of the interrelationships of the factors involved in plant disease epidemics.

Levels of Genetic Resistance or Susceptibility of the Host

Obviously, host plants carrying race-specific (vertical) resistance do not allow a pathogen to become established in them, and thus no epidemic can develop (Fig. 8-4). Host plants carrying partial (horizontal) resistance will probably become infected, but the rate at which the disease and the epidemic will develop depends on the level of resistance and the environmental conditions. Susceptible host plants lacking genes for resistance against the pathogen provide the ideal substrate for establishment and development of new infections. Therefore, in the presence of a virulent pathogen and a favorable environment, susceptible hosts favor the development of disease epidemics.

Degree of Genetic Uniformity of Host Plants

When genetically uniform host plants, particularly with regard to the genes associated with disease resistance, are grown over large areas, a greater likelihood exists that a new pathogen race will appear that can attack their genome and result in an epidemic. This phenomenon has been observed repeatedly, for example, in the *Cochliobolus* (*Helminthosporium*) blight on Victoria oats and in southern corn leaf blight (Fig. 8-1) on corn carrying Texas male-sterile cytoplasm. For similar reasons of genetic uniformity, the highest rates of epidemic development generally occur in vegetatively propagated crops, intermediate rates in self-pollinated crops, and the lowest rates in cross-pollinated crops. This explains why most epidemics develop rather slowly in natural populations, where plants of varying genetic makeup are intermingled.

Type of Crop

In diseases of annual crops, such as corn, vegetables, rice, and cotton, and in foliar, blossom, or fruit diseases of trees and vines, epidemics generally develop much more rapidly (usually in a few weeks) than they do in diseases of branches and stems of perennial woody crops such as fruit and forest trees. Some epidemics of fruit and forest trees, e.g., tristeza in citrus, pear decline, Dutch elm disease, and chestnut blight, take years to develop.

Age of Host Plants

Plants change in their reaction (susceptibility or resistance) to disease with age. The change of resistance with age is known as ontogenic resistance. In some plant–pathogen combinations, e.g., *Pythium* damping off and root rots, downy mildews, peach leaf curl, systemic smuts, rusts, bacterial blights, and viral infections, the hosts (or their parts) are susceptible only during the

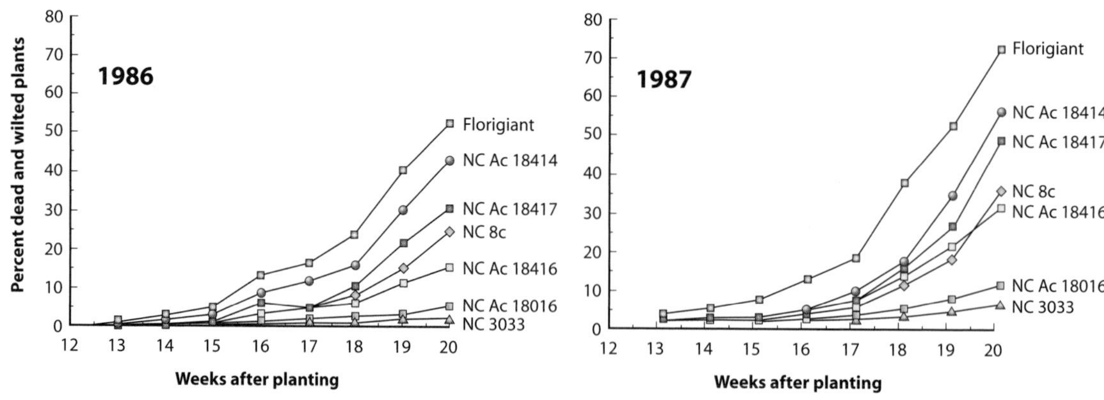

FIGURE 8-4 Development of *Cylindrocladium* black rot, caused by the fungus *C. crotalariae*, on susceptible (Florigiant), resistant (NC3033), and intermediate peanut varieties. The various genotypes maintain their resistance rankings in both years (1986, 1987) and at all inoculum density levels tested. [From Culbreath *et al.* (1991).]

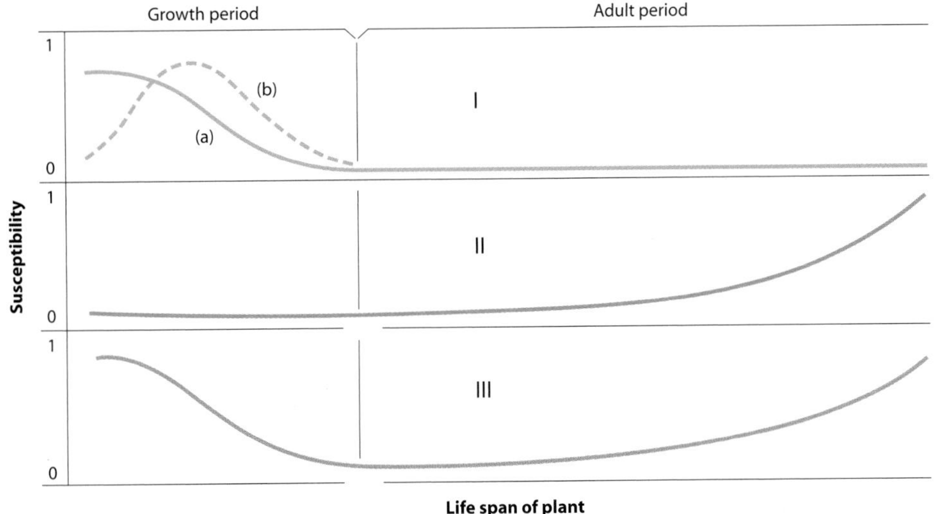

FIGURE 8-5 Change of susceptibility of plant parts with age. In pattern I, plants are susceptible only in the stages of maximum growth (Ia) or in the earliest stages of growth (Ib). In pattern II, plants are susceptible only after they reach maturity, and susceptibility increases with senescence. In pattern III, plants are susceptible while very young and again after they reach maturity. [After Populer (1978).]

growth period and become resistant during the adult period (adult resistance) (Figs. 8-5Ia and 8-5Ib). With several diseases, such as rusts and viral infections, plant parts are actually quite resistant to infection while still very young, become more susceptible later in their growth, and then become resistant again before they are fully expanded (Figs. 8-5, pattern Ib, and 8-6). In other diseases, such as infections of blossoms or fruit by *Botrytis*, *Penicillium*, *Monilinia*, and *Glomerella*, and in all postharvest infections, plant parts are resistant during growth and the early adult period but become susceptible near ripening (Fig. 8-5II). In still other diseases, such as potato late blight (caused by *Phytopthora infestans*) and tomato early blight (caused by *Alternaria solani*), a stage of juvenile susceptibility during the growth period of the plant is followed by a period of relative resistance in the early adult stage and then susceptibility after maturity (Fig. 8-5III).

Apparently then, depending on the particular plant–pathogen combination, the age of the host plant at the time of arrival of the pathogen may affect considerably the development of infection and of an epidemic.

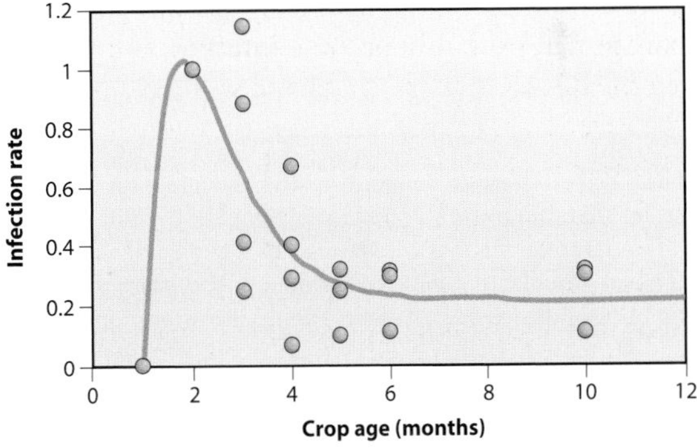

FIGURE 8-6 Effect of crop age on rate of infection. Cassava plantings of different ages exposed to the whitefly-transmitted African cassava mosaic geminivirus show increased resistance to infection as they age. [From Fargette and Vie (1994). *Phytopathology* 84, 378–382.]

PATHOGEN FACTORS THAT AFFECT DEVELOPMENT OF EPIDEMICS

Levels of Virulence

Virulent pathogens capable of infecting the host rapidly ensure a faster production of larger amounts of inoculum, and, thereby, disease, than pathogens of lesser virulence.

Quantity of Inoculum near Hosts

The greater the number of pathogen propagules (bacteria, fungal spores and sclerotia, nematode eggs, virus-infected plants, etc.) within or near fields of host plants, the more inoculum reaches the hosts and at an earlier time, thereby increasing the chances of an epidemic greatly (Fig. 8-7).

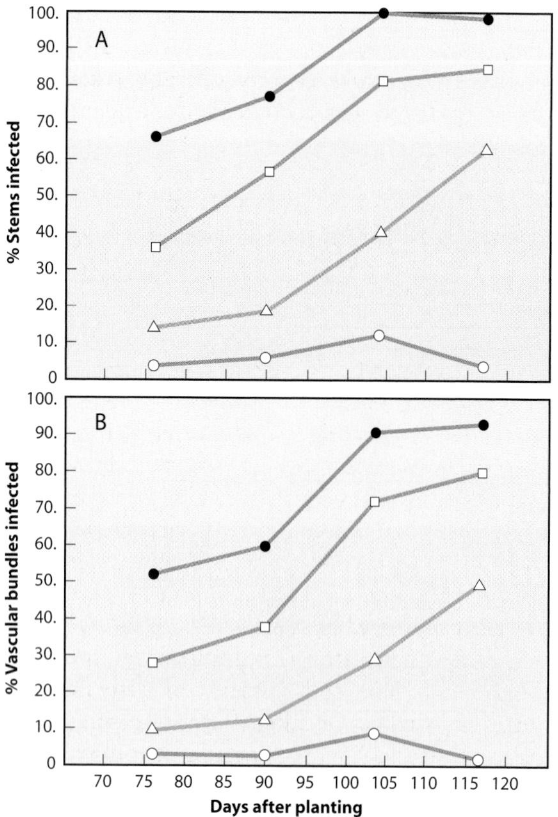

FIGURE 8-7 Effect of amount of soil inoculum of *Verticillium dahliae* on the amount of vascular wilt on potato plants at various dates after planting. Disease is expressed as a percentage of stems (A) and of main vascular bundles (B) infected at the base of the plants. ○, no pathogen detected; △, 1–5 propagules per gram (ppg); □, 6–10 ppg; and ●, more than 10 ppg. [From Nicot and Rouse (1987). *Phytopathology* 77, 1346–1355.]

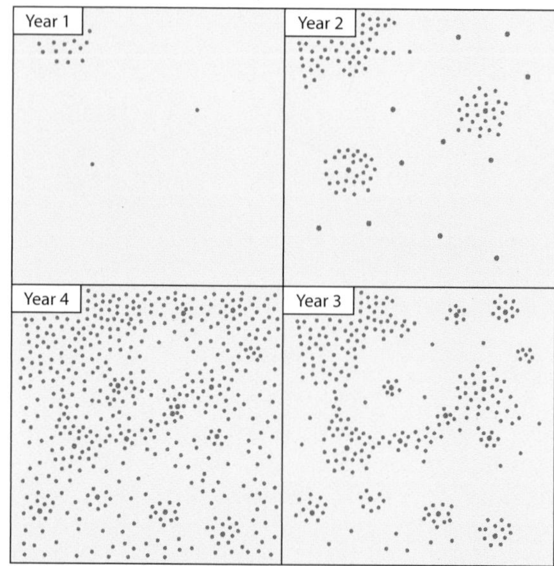

FIGURE 8-8 Schematic representation of a polyetic epidemic caused in a crop in a field by a soil pathogen over a 4-year period.

Type of Reproduction of the Pathogen

All pathogens produce many offspring. Some of them, such as most fungi, bacteria, and viruses, produce a great many offspring, while a few fungi, all nematodes, and all parasitic plants produce relatively small numbers of offspring. Some plant pathogenic fungi, bacteria, and viruses have short reproduction cycles and therefore are polycyclic, i.e., they can produce many generations in a single growing season. Polycyclic pathogens include fungi that cause rusts, mildews, and leaf spots and are responsible for most of the sudden, catastrophic plant disease epidemics in the world. Some soil fungi, such as *Fusarium* and *Verticillium*, and most nematodes usually have one to a few (up to four) reproductive cycles per growing season. For these latter pathogens, the smaller number of offspring and, especially, the conditions of their dispersal limit their potential to cause sudden and widespread epidemics in a single season. Nevertheless, they often cause localized, slower developing epidemics (Fig. 8-8). Several pathogens, such as the smuts and several short-cycle rusts, require an entire year to complete a life cycle (monocyclic pathogens) and can therefore cause only one series of infections per year. In such monocyclic diseases, the inoculum builds up from one year to the next, and the epidemic is usually polyetic, i.e., it develops over several years. Similarly, epidemics caused by pathogens that require more than a year to complete a reproductive cycle are slow to develop. Examples are cedar-apple rust (2 years), white pine blister rust (3–6 years), and dwarf mistletoe (5–6 years). As a result of overlapping of the polyetic generations, however, even such pathogens each year produce more inoculum and cause a series of infections that lead to long-term epidemics.

Ecology of the Pathogen

Some pathogens, such as most fungi and all parasitic higher plants, produce their inoculum (spores and seeds, respectively) on the surface of the aerial parts of the host. From there, spores and seeds can be dispersed with ease over a range of distances and can cause widespread epidemics. Other pathogens, such as vascular fungi and bacteria, mollicutes, viruses, and protozoa, reproduce inside the plant. In this case, spread of the pathogen is rare or impossible without the help of vectors. Therefore, such pathogens can cause epidemics only when vectors are plentiful and active. Still other pathogens, such as soilborne fungi, bacteria, and nematodes, produce their inoculum on infected plant parts in the soil, within which the inoculum disperses slowly and presents little danger for sudden or widespread epidemics.

Mode of Spread of the Pathogen

The spores of many plant pathogenic fungi, such as those causing rusts, mildews, and leaf spots, are released into the air and can be dispersed by air breezes or strong winds over distances varying from a few centimeters up to several kilometers. These kinds of fungi are responsible for the most frequent and most widespread epidemics. In terms of their ability to cause sudden and widespread epidemics, the next most important group of pathogens includes those whose inoculum is carried by airborne vectors. Many of the viruses are transmitted by aphids, whiteflies, and some other insects. Mollicutes and fastidious bacteria are transmitted by leafhoppers, plant hoppers, or psyllids. Some fungi (such as the cause of Dutch elm disease), bacteria (such as the cause of bacterial wilt of cucurbits), and even nematodes (such as the cause of pine wilt disease) are disseminated primarily by beetles. Pathogens that are transmitted by wind-blown rain (primarily fungi causing diseases such as anthracnoses and apple scab, and most bacteria) are almost annually responsible for severe but somewhat localized epidemics within a field, a country, or a valley. Pathogens carried with the seed or other vegetative propagative organs (such as tubers or bulbs) are often placed in the midst of susceptible plants, but their ability to cause epidemics depends on the effectiveness of their subsequent transmission to new plants. Finally, pathogens present in and spreading through the soil, because of the physical restrictions imposed by the soil, are generally unable to cause sudden or widespread epidemics but often cause local, slow-spreading diseases of considerable severity (Fig. 8-9A). When such primarily soil fungi, however, also produce wind-disseminated spores, the latter can spread considerable distances and can cause epidemics destructive over considerable areas (Fig. 8-9B).

ENVIRONMENTAL FACTORS THAT AFFECT DEVELOPMENT OF EPIDEMICS

The majority of plant diseases occur wherever the host is grown but, usually, do not develop into severe and widespread epidemics. The concurrent presence in the same areas of susceptible plants and virulent pathogens does not always guarantee numerous infections, much less the development of an epidemic. This fact dramatizes the controlling influence of the environment on the development of epidemics. The environment may affect the availability, growth stage, succulence, and genetic susceptibility of the host plants. It may also affect the survival, vigor, rate of multiplication, sporulation, and ease, direction, and distance of dispersal of the pathogen, as well as the rate of spore germination and penetration. In addition, the environment may affect the number and activity of the vectors of the pathogen. The most important environmental factors that affect the development of plant disease epidemics are moisture, temperature, and the activities of humans in terms of cultural practices and control measures.

Moisture

As discussed in Chapter 7, abundant, prolonged, or repeated high moisture, whether in the form of rain, dew, or high humidity, is the dominant factor in the development of most epidemics of diseases caused by oomycetes

FIGURE 8-9 (A) Lettuce heads infected by soilborne sclerotia of *Sclerotinia sclerotiorum*. (B) Large field of lettuce heads killed by infections with airborne ascospores of the same fungus. [Photographs courtesy K. V. Subbarao, *Plant Dis.* 82: 1068–1078 (1998)].

and fungi (blights, downy mildews, leaf spots, rusts, and anthracnoses), bacteria (leaf spots, blights, soft rots), and nematodes. Moisture not only promotes new succulent and susceptible growth in the host, but, more importantly, it increases sporulation of fungi (Figs. 7-6A and 7-8) and multiplication of bacteria. Moisture facilitates spore release by many fungi (Figs. 7-7 and 7-9) and the oozing of bacteria to the host surface, and it enables spores to germinate and zoospores, bacteria, and nematodes to move. The presence of high levels of moisture allows all these events to take place constantly and repeatedly and leads to epidemics. In contrast, the absence of moisture for even a few days prevents all of these events from taking place so that epidemics are interrupted or stopped completely. Some diseases caused by soilborne pathogens, such as *Fusarium* and *Streptomyces*, are more severe in dry than in wet weather, but such diseases seldom develop into important epidemics. Epidemics caused by viruses and mollicutes are affected only indirectly by moisture, primarily by the effect that higher moisture has on the activity of the vector. Moisture may increase the activity of some vectors, as happens with the fungal and nematode vectors of some viruses, or it may reduce the activity of the vectors, as happens with the aphid, leafhopper, and other insect vectors of some viruses and mollicutes. The activity of these vectors is reduced drastically in rainy weather.

Temperature

Epidemics are sometimes favored by temperatures higher or lower than the optimum for the plant because they reduce the plant's level of partial resistance. At certain levels, temperatures may even reduce or eliminate the race-specific resistance of host plants. Plants growing at such temperatures become "stressed" and predisposed to disease, provided the pathogen remains vigorous.

Low temperature reduces the amount of inoculum of oomycete fungi, bacteria, and nematodes that survives cold winters. High temperature reduces the inoculum of viruses and mollicutes that survives hot summer temperatures. In addition, low temperatures reduce the number of vectors that survive the winter. Low temperatures occurring during the growing season can reduce the activity of vectors.

The most common effect of temperature on epidemics, however, is its effect on the pathogen during the different stages of pathogenesis, i.e., spore germination (Figs. 7-8 and 7-9) or egg hatching, host penetration, pathogen growth (Figs. 7-3 and 7-4) or reproduction, invasion of the host, and sporulation (Fig. 7-5). When temperature stays within a favorable range for each of these stages, a polycyclic pathogen can complete its infection cycle within a very short time (usually in a few days). As a result, polycyclic pathogens can produce many infection cycles within a growing season. Because the amount of inoculum is multiplied manyfold (perhaps 100 times or more) with each infection cycle and because some of the new inoculum is likely to spread to new plants, more infection cycles result in more plants becoming infected by more and more pathogens, thus leading to the development of a severe epidemic.

In reality, moisture and temperature must be favorable and act together in the initiation and development of the vast majority of plant diseases and plant disease epidemics.

EFFECT OF HUMAN CULTURAL PRACTICES AND CONTROL MEASURES

Many activities of humans have a direct or indirect effect on plant disease epidemics, some of them favoring and some reducing the frequency and the rate of epidemics.

Site Selection and Preparation

Low-lying and poorly drained and aerated fields, especially if near other infected fields, tend to favor the appearance and development of epidemics.

Selection of Propagative Material

The use of seed, nursery stock, and other propagative material that carries various pathogens increases the amount of initial inoculum within the crop and favors the development of epidemics greatly. The use of pathogen-free or treated propagative material can reduce the chance of epidemics greatly.

Cultural Practices

Continuous monoculture, large acreages planted to the same variety of crop, high levels of nitrogen fertilization, no-till culture, dense plantings (Fig. 8-10), overhead irrigation, injury by herbicide application, and poor sanitation all increase the possibility and severity of epidemics.

Disease Control Measures

Chemical sprays, cultural practices (such as sanitation and crop rotation), biological controls (such as using

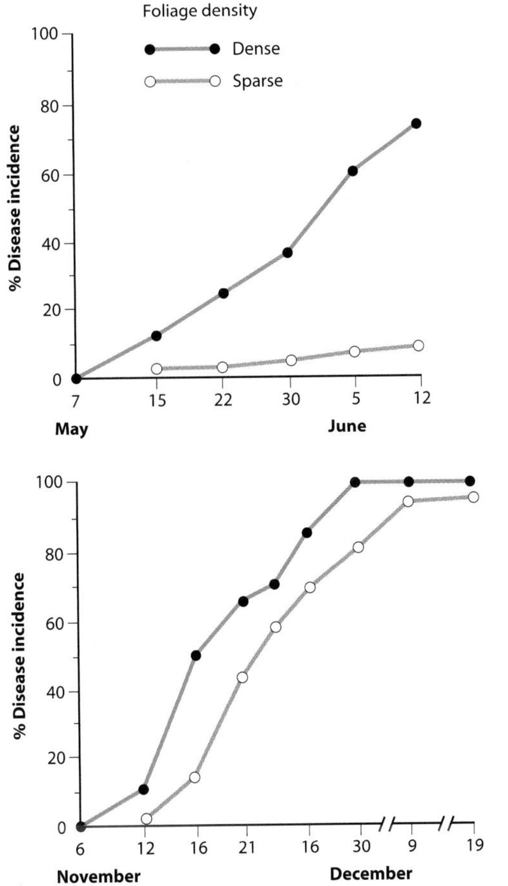

FIGURE 8-10 Effect of foliage density on development of *Phytophthora infestans* during a period of partly favorable weather (May–June) and of very favorable weather (November–December). [From Rotem and Ben-Joseph (1970). *Plant Dis. Rep.* 54, 768–771.]

resistant varieties), and other control measures reduce or eliminate the possibility of an epidemic. Sometimes, however, certain controls, e.g., the use of a certain chemical or planting of a certain variety, may lead to selection of virulent strains of the pathogen that either are resistant to the chemical or can overcome the resistance of the variety and thus lead to epidemics.

Introduction of New Pathogens

The ease and frequency of worldwide travel have also increased the movement of seeds, tubers, nursery stock, and other agricultural goods. These events increase the possibility of introducing pathogens into areas where the hosts have not had a chance to evolve resistance to these pathogens. Such pathogens frequently lead to severe epidemics. Examples are chestnut blight, Dutch elm disease, and citrus canker caused by the bacterium *Xanthomonas campestris* pv. *citri*.

MEASUREMENT OF PLANT DISEASE AND OF YIELD LOSS

When measuring disease, one is interested in measuring (1) the **incidence** of the disease, i.e., the number or proportion of plant units that are diseased (i.e., the number or proportion of plants, leaves, stems, and fruit that show any symptoms) in relation to the total number of units examined; (2) the **severity** of the disease, i.e., the proportion of area or amount of plant tissue that is diseased; and (3) the **yield loss** caused by the disease, i.e., the proportion of the yield that the grower will not be able to harvest because the disease destroyed it directly or prevented the plants from producing it (the yield loss is the difference between **attainable yield** and **actual yield**).

Measuring disease incidence is relatively quick and easy, and this measurement is the one that is used commonly in epidemiological studies to measure the spread of a disease through a field, region, or country. In a few cases, such as cereal smuts, neck blast of rice, brown rot of stone fruits, and the vascular wilts of annuals, disease incidence has a direct relationship to the severity of the disease and yield loss because each diseased plant or fruit is a total loss. However, in many other diseases (such as most leaf spots, root lesions, and rusts) in which plants are counted as diseased whether they are exhibiting a single lesion or hundreds of lesions, disease incidence may have little relationship to the severity of the disease or to yield loss. Although severity and yield loss are of much greater importance to the grower than disease incidence, their measurement is more difficult and, in some cases, not possible until too late in the development of an epidemic.

Disease severity is usually expressed as the percentage or proportion of plant area or fruit volume destroyed by a pathogen (Figs. 8-11 and 8-12). More often, disease assessment scales from 0 to 10 or 1 to 4 are used to express the relative proportions of affected tissue at a particular point in time. Yield loss due to disease is measured at a specific growth stage, from sequential disease assessments at several stages of a crop's growth, or by determining the **area under a disease progress curve** (AUDPC), i.e., the area between the disease progress curve and the *X* axis of the graph. The area under the disease progress curve is used to summarize the progress of disease severity and is calculated by a formula that takes into account the number of times the disease severity was evaluated, the disease severity at each evaluation time, and the time duration of the epidemic.

Yield loss almost always results in **economic loss** from disease. Economic loss occurs whenever economic returns from the crop decrease because of reduced yields

(Fig. 8-13), because of the cost of agricultural activities undertaken to reduce damage to the crop, or both. In managing plant diseases, however, the grower can justify applying disease control measures only when the incremental costs of control are generally smaller than the increase in crop returns. The level of disease, i.e., the amount of plant damage, at which control costs just equal incremental crop returns is called the **economic threshold** of the disease. The economic threshold of a crop–pathogen system varies with the tolerance level (**damage threshold**) of the crop, which depends on the growth stage of the crop when attacked, crop management practices, environment, shifts in pathogen virulence, and new control practices. The economic threshold also varies with changing commodity prices and control costs.

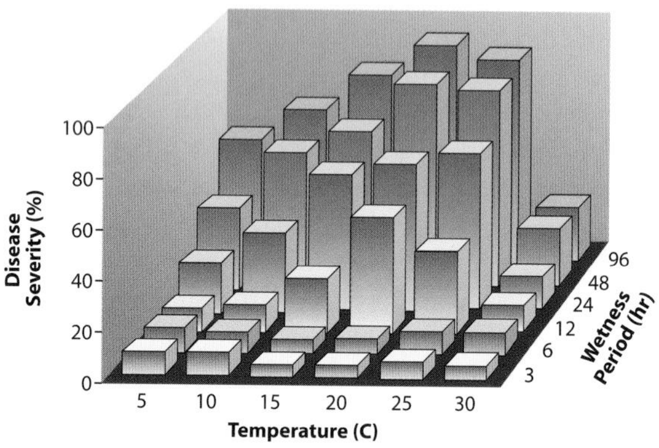

FIGURE 8-11 Development of *Ascochyta* blight of chickpea, caused by the fungus *Asochyta rabiei*, at different temperatures and leaf wetness durations. [From Trapero-Casas and Kaiser (1992). *Phytopathology* 82, 589–596.]

PATTERNS OF EPIDEMICS

Interactions of the structural elements of epidemics, as influenced over time by factors of the environment and by human interference, are expressed in patterns and rates. The pattern of an epidemic in terms of the numbers of lesions, the amount of diseased tissue, or the numbers of diseased plants is given by a curve, called the **disease–progress curve**, that shows the progress of the epidemic over time. The point of origin and the shape of a disease–progress curve reveal information about the time of appearance and amount of inoculum, changes in host susceptibility during the growing period, recurrent weather events, and the effectiveness of cultural and control measures. Disease–progress curves, because they are affected by weather, variety, and so on, vary somewhat with location and time, but they are generally characteristic for some groups of diseases. For example, a saturation-type curve is characteristic for monocyclic diseases, a sigmoid curve is characteristic for polycyclic diseases, and a bimodal curve is characteristic for diseases affecting different organs (blossoms, fruit) of the plant (Fig. 8-14). Knowledge of disease–progress curves also allows disease forecasting and selection of the best control strategy for the particular disease and time.

The progress of an epidemic in space, in terms of changes in the number of lesions, the amount of diseased tissue, and the number of diseased plants as it spreads over distance, is called its spatial pattern, i.e., the arrangement of disease entities relative to each other and to the area of cultivation of the crop. Spatial patterns of epidemics are influenced by the dispersal of the pathogen, i.e., the process of movement of individuals of the pathogen in and out of the host population or

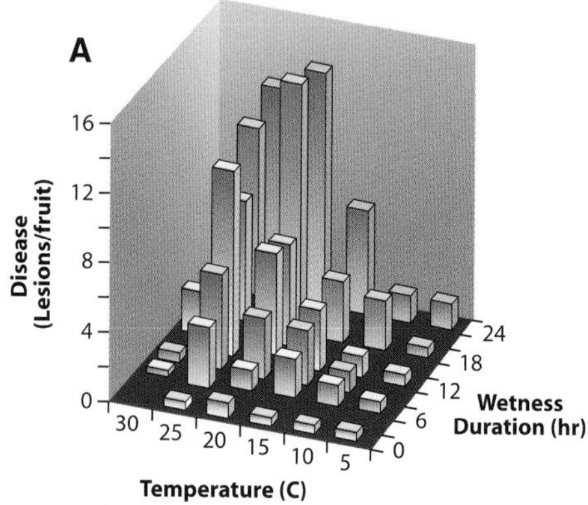

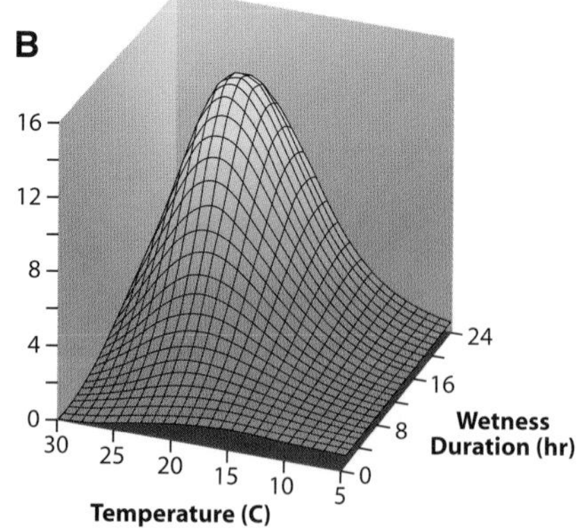

FIGURE 8-12 Severities of brown spot disease of pear, caused by the fungus *Stemphylium vesicarium*, at various combinations of temperature and wetness duration. (A) Experimental data. (B) Response surface diagram based on a model predicting the number of lesions per fruit at corresponding combinations. [From Montesinos *et al.* (1995). *Phytopathology* 85, 586–592.]

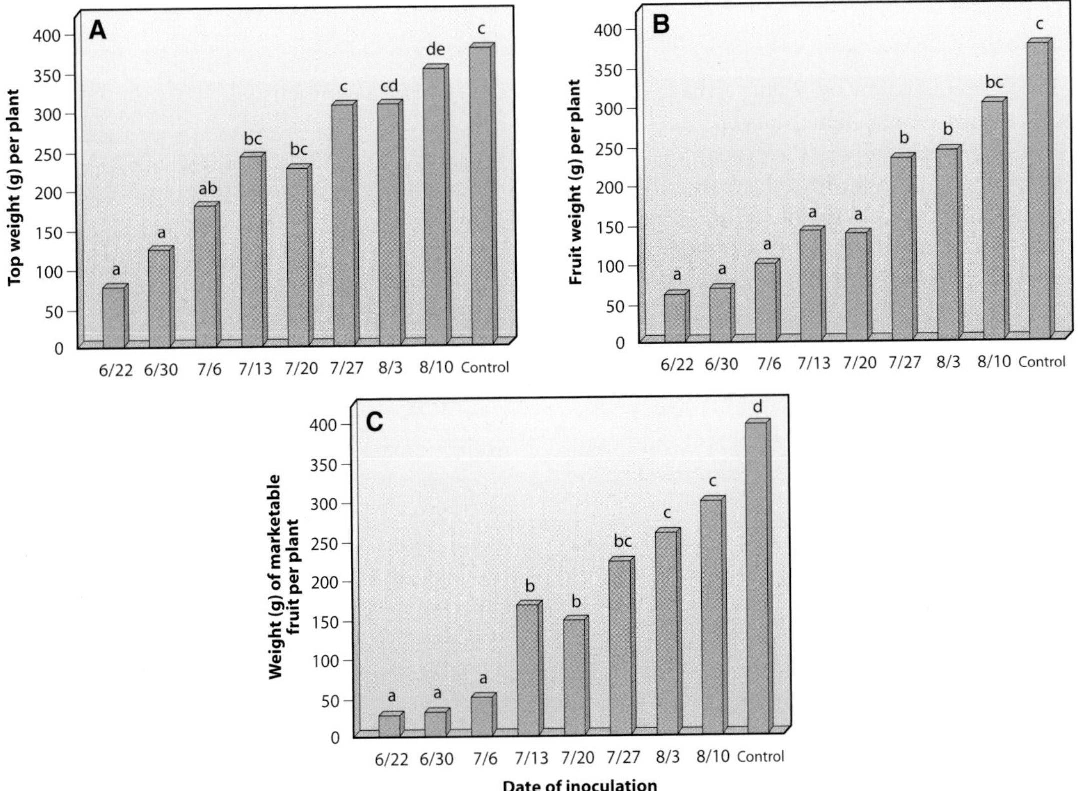

FIGURE 8-13 Average weight of tops minus fruit (A), of fruit (B), and of marketable fruit (C) of pepper plants inoculated with cucumber mosaic virus at different dates and of uninoculated control plants. [From Agrios *et al.* (1985). *Plant Dis.* **69**, 52–55.]

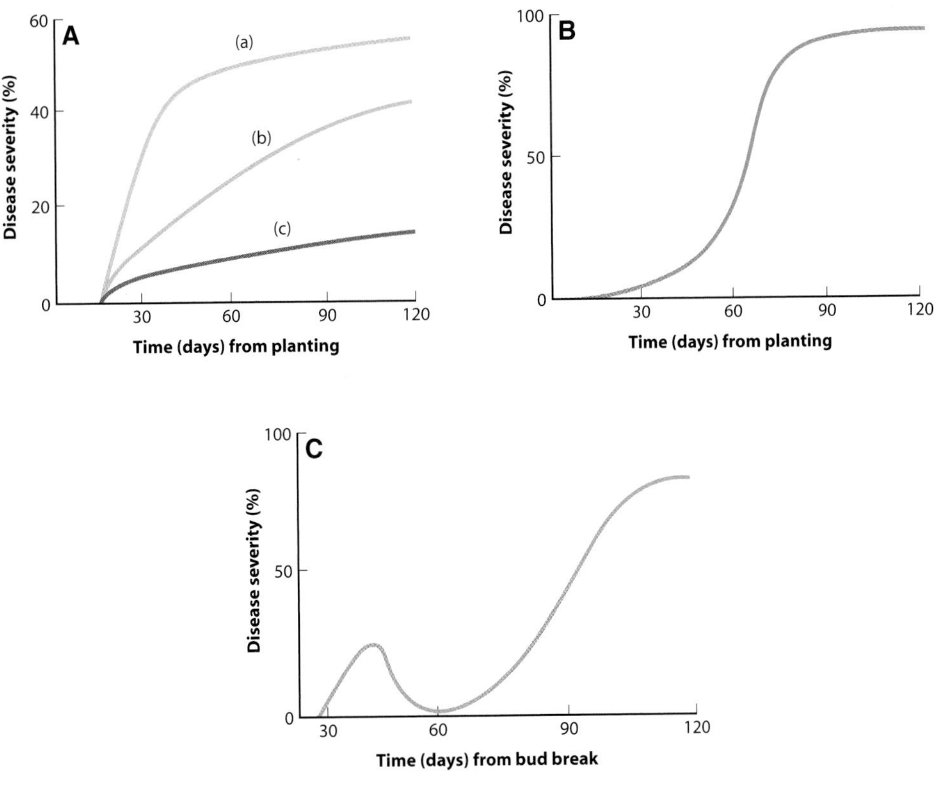

FIGURE 8-14 Schematic diagrams of disease–progress curves of some basic epidemic patterns. (A) Three monocyclic diseases of different epidemic rates. (B) Polycyclic disease, such as late blight of potato. (C) Bimodal polycyclic disease, such as brown rot of stone fruits, in which the blossoms and the fruit are infected at different, separate times.

population area, and is given by a curve that is called the **dispersal** or **disease–gradient curve**. Because the amount of disease is generally greater near the source of inoculum and decreases with increasing distance from the source, most disease–gradient curves are quite similar, at least in the early stages of the epidemic. The number of diseased plants and the severity of disease decrease steeply within short distances of the source and less steeply at greater distances until they reach zero or a low background level of occasional diseased plants (Fig. 8-15).

From data collected at various time intervals and used to plot the disease–progress curve of a disease (Fig. 8-16), one can obtain the epidemic rate of the disease, i.e., the rate of growth of the epidemic. The epidemic rate, generally designated *r*, is the amount of increase of disease per unit of time (per day, week, or year) in the plant population under consideration. The patterns of epidemic rates are given by curves called rate curves, and these curves are different for various groups of diseases (see Fig. 8-16). In some diseases, e.g., the late blight of potato, the rate curves are symmetrical (bell shaped) (Fig. 8-16A). In some diseases, e.g., in apple scab and most downy mildews and powdery mildews, the rate curves are asymmetrical, with the epidemic rate being greater early in the season (Fig. 8-16B) because of the greater susceptibility of young leaves. In still other diseases, the rate curves are asymmetrical, with the epidemic rate being greater late in the season (Fig. 8-16C). This is observed in the many diseases, e.g., *Alternaria* leaf blights and *Verticillium* wilts, that start slowly but accelerate markedly as host susceptibility increases late in the season.

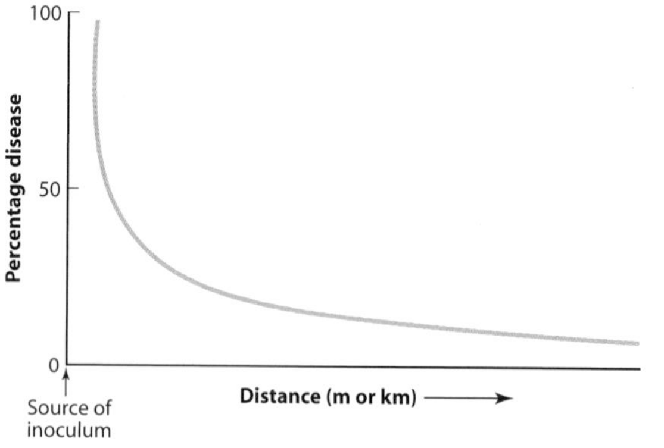

FIGURE 8-15 Schematic diagram of a disease–gradient curve. The percentage of disease and the scale for distance vary with the type of pathogen or its method of dispersal, being small for soilborne pathogens or vectors and larger for airborne pathogens.

COMPARISON OF EPIDEMICS

For better comparison of epidemics of the same disease at different times, different locations, or under different management practices or to compare different diseases, the patterns obtained for disease–progress curves and disease–gradient curves are frequently transformed mathematically into straight lines. The slopes of these lines can then be used to calculate epidemic rates.

In monocyclic diseases, the amount of inoculum does not increase significantly during the season. In such diseases, therefore, the rate of disease increase is affected only by the inherent ability of the pathogen to induce disease and by the ability of the environmental factors and cultural practices to influence host resistance and the virulence of the pathogen.

In contrast, the initial inoculum for diseases caused by polycyclic pathogens, although extremely important, has relatively less importance than the number of infection cycles in the final disease outcome (Fig. 8-17). Pathogens that have many infection cycles also have numerous opportunities to interact with the host. Therefore, the same factors mentioned earlier, namely the inherent ability of the pathogen to induce disease, environmental factors, host resistance, and cultural practices, have an opportunity to influence the dispersal, penetration, multiplication, size of lesion, rate of lesion formation, and rate and amount of sporulation, but they can do that not once but several times during the same growth season. The continuous or, sometimes, intermit-

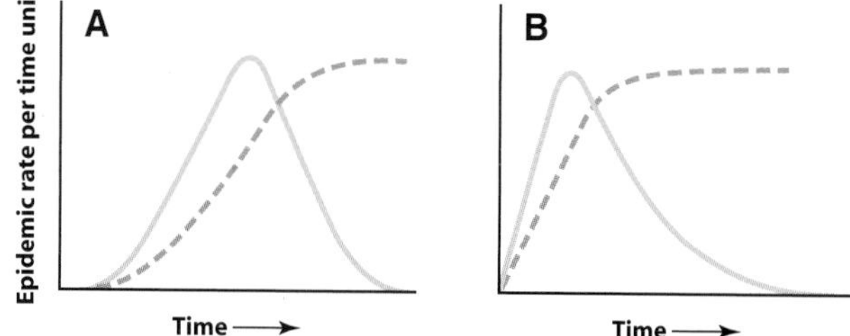

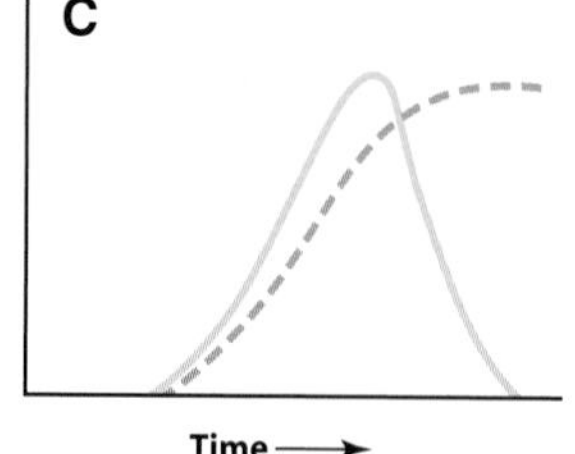

FIGURE 8-16 Schematic diagrams of epidemic rate curves of diseases with a symmetrical epidemic rate (A), with a high epidemic rate early in the season (B), and with a high epidemic rate late in the season (C). Dashed curves indicate possible disease–progress curves that may be produced in each case from the accumulated epidemic rate curves.

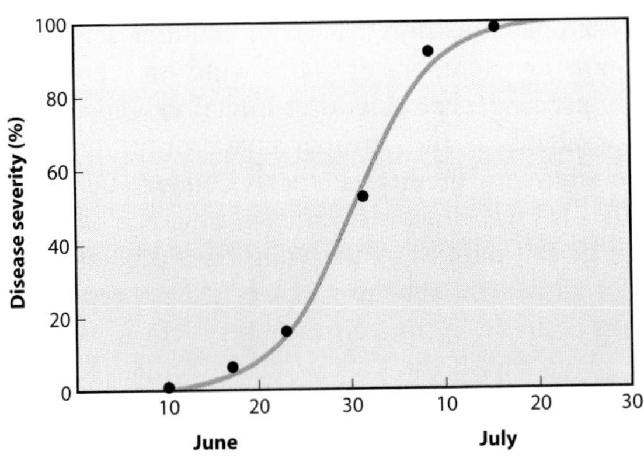

FIGURE 8-17 Predicted (—) and observed (●) disease progress curve of sunflower rust caused by the fungus *Puccinia helianthii*. [From Shtienberg and Vintal (1995). *Phytopathology* **85**, 1388–1393.]

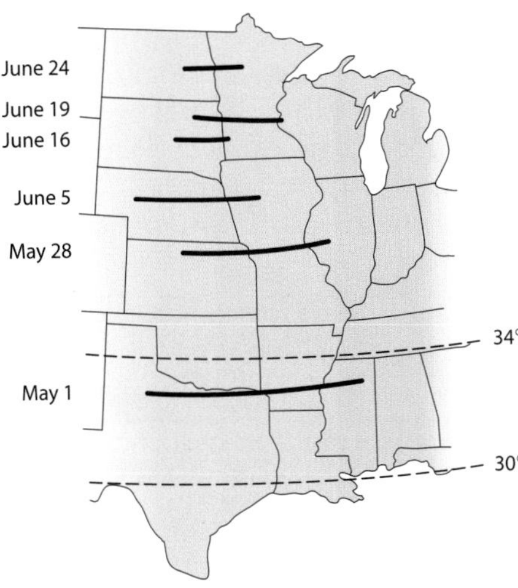

FIGURE 8-18 Normal annual advance of wheat stem rust across the United States. The fungus *Puccinia graminis tritici* generally overwinters south of the 30°N parallel and in trace amounts as far north as the 34°N parallel. In the spring it moves northward at the rates shown by the dates at left. [From Roelfs (1986).]

tent increase of the amount of inoculum and disease may result in highly variable infection rates for individual short-term intervals during the growth season, and quite variable epidemic rates for the entire season.

The epidemic rate for polycyclic diseases is usually calculated per day or per week rather than per year, which is the way it is calculated for monocyclic diseases. In general, the epidemic rate (r) for polycyclic diseases is much greater than the rate of epidemic increase (r_m) for monocyclic diseases. For example, the r_m for *Verticillium* wilt of cotton is 0.02 units per day and is 1.60 units per year for *Phymatotrichum* root rot of cotton. In contrast, the epidemic rate r for potato late blight is 0.3–0.5 units per day, is 0.3–0.6 units per day for wheat stem rust and 0.15 units per day for cucumber mosaic virus.

In addition to the epidemics caused by monocyclic and polycyclic pathogens, there are also polyetic epidemics. Pathogens causing polyetic epidemics are present for one year or more in the infected plant before they produce effective inoculum, e.g., some fungal wilts and viral and mollicute diseases of trees. Because of the perennial nature of their hosts, polyetic diseases behave basically as polycyclic diseases with a lower r. This happens because there are as many diseased trees and almost as much inoculum at the beginning of a year as at the end of the previous one, and both increase over the years, causing slower but just as severe epidemics. Some well-known polyetic epidemics are chestnut blight ($r = 0.3$–1.2 units per year) and elm yellows (phytoplasma) (0.6 units per year).

DEVELOPMENT OF EPIDEMICS

For a disease to become significant in a field, particularly if it is to spread over a large area and develop into a severe epidemic, specific combinations of environmental factors must occur either constantly or repeatedly, and at frequent intervals, over a large area. Even in a single, small field that contains the pathogen, plants almost never become severely diseased from just one set of favorable environmental conditions. It takes repeated infection cycles and considerable time before a pathogen produces enough individuals to cause an economically severe epidemic in the field (Fig. 8-1). Once large populations of the pathogen are available, however, they can attack, spread to nearby fields, and cause a severe epidemic in a very short time, often in just a few days.

A plant disease epidemic can occur in a garden, a greenhouse, or a small field, but "epidemic" generally implies the development and rapid spread of a pathogen on a particular kind of crop plant cultivated over a large area, such as a large field, a valley, a section of a country, the entire country, or even part of a continent (Figs. 8-18 and 8-19). Therefore, the first component of a plant disease epidemic is a large area planted to a genetically uniform crop plant, with the plants and the fields being close together. The second component of an epidemic is the presence or appearance of a virulent pathogen. Such cohabitations of host plants and pathogens occur, of course, daily in countless locations. Most of these, however, cause local diseases of varying severity, destroy crop plants to a limited extent, and do not develop into epidemics. Epidemics develop only when the combinations and progression of the right sets of conditions occur. These include appropriate temperature, moisture, and wind or insect vector coinciding with the susceptible stage or stages of the plant and with the production,

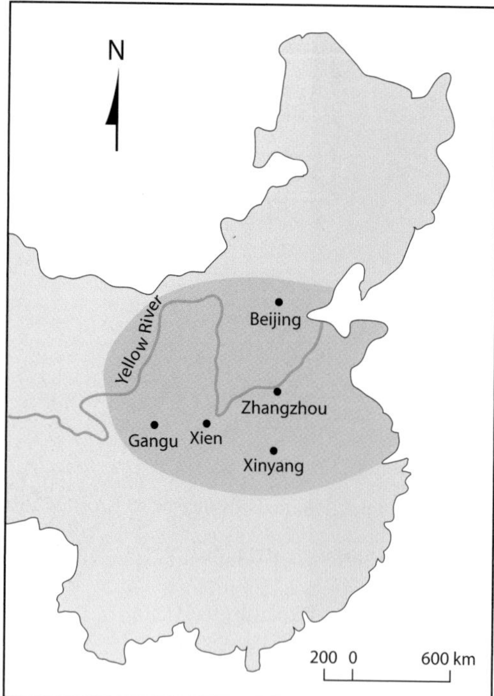

FIGURE 8-19 Annual occurrence of wheat stripe rust, caused by *Puccinia striiformis*, in northern China (shaded area). Gangu is the source region of the epidemic, and the other regions are in the dispersion area. The prevalent air currents are eastward in the fall and northward in the spring. [From Yang and Zeng (1992). *Phytopathology* **82**, 571–576.]

spread, inoculation, penetration, infection, and reproduction of the pathogen.

Thus, for an epidemic to develop, the small amount of original or primary inoculum of the pathogen must be carried by wind or vector to some of the crop plants as soon as they begin to become susceptible to that pathogen. The moisture and temperature must then be appropriate for germination or infection to take place. After infection, the temperature must be favorable for rapid growth and reproduction of the pathogen (short incubation period, short infection cycle) so that numerous new spores will appear as quickly as possible. The moisture (rain, fog, dew) then must be sufficient and should last long enough for the abundant release of spores. Winds of the proper humidity and velocity, blowing toward the susceptible crop plants, must then pick up the spores and carry them to the plants while the latter are still susceptible. Most plant disease epidemics spread from south to north in the northern hemisphere and from north to south in the southern hemisphere. Because the warmer weather and growth seasons also move in the same direction, the pathogens constantly find plants in their susceptible stage as the season progresses.

In each new location, however, the same set of favorable moisture, temperature, and wind or vector conditions must be repeated so that infection, reproduction, and dispersal of the pathogen can occur as quickly as possible. Furthermore, these conditions must be repeated several times within each location so that the pathogen can multiply, increasing the number of infections it causes on the host plants. These repeated infections usually result in the destruction of almost every plant within the area of an epidemic (Fig. 8-20), although the uniformity of the plants and the size of the area of cultivation, along with the prevailing weather, determine the final spread of the epidemic.

Fortunately, the most favorable combinations of conditions for disease development do not occur very often over very large areas; therefore, spectacular plant disease epidemics that destroy crops over large areas are relatively rare. However, small epidemics involving the plants in a field or a valley occur quite frequently. With many diseases, e.g., potato late blight, apple scab, and cereal rusts, the environmental conditions seem usually to be favorable, and disease epidemics would occur every year were it not for the control measures (chemical sprays, resistant varieties, and so on) employed annually to avoid such epidemics.

MODELING OF PLANT DISEASE EPIDEMICS

An epidemic is a dynamic process. It begins on one or a few plants and then, depending on the kind, magnitude, and duration of environmental factors that influence the host and pathogen, increases in severity and spreads over a larger geographic area until it finally dies down. Epidemics come to a stop when all host plants are killed by the pathogen, become resistant to the pathogen as they age, or are harvested. In many cases, epidemics slow down or come to a stop when the weather turns dry or unseasonably cold. In many ways, the appearance, development, and spread of epidemics resemble those of hurricanes. In both cases, humans have been extremely interested in determining the elements and conditions that initiate each, the conditions that influence the rate of increase and the direction of their path, and the conditions that bring about their demise. For both phenomena, observations, measurements, mathematical formulas, and computers are used extensively to study the development and to predict the size, path, and time of attack in any given location.

Each plant disease epidemic, e.g., of stem of wheat, late blight of potato, apple scab, or downy mildew of grape, follows a predictable course in each location each year. The course of the epidemic varies with the host

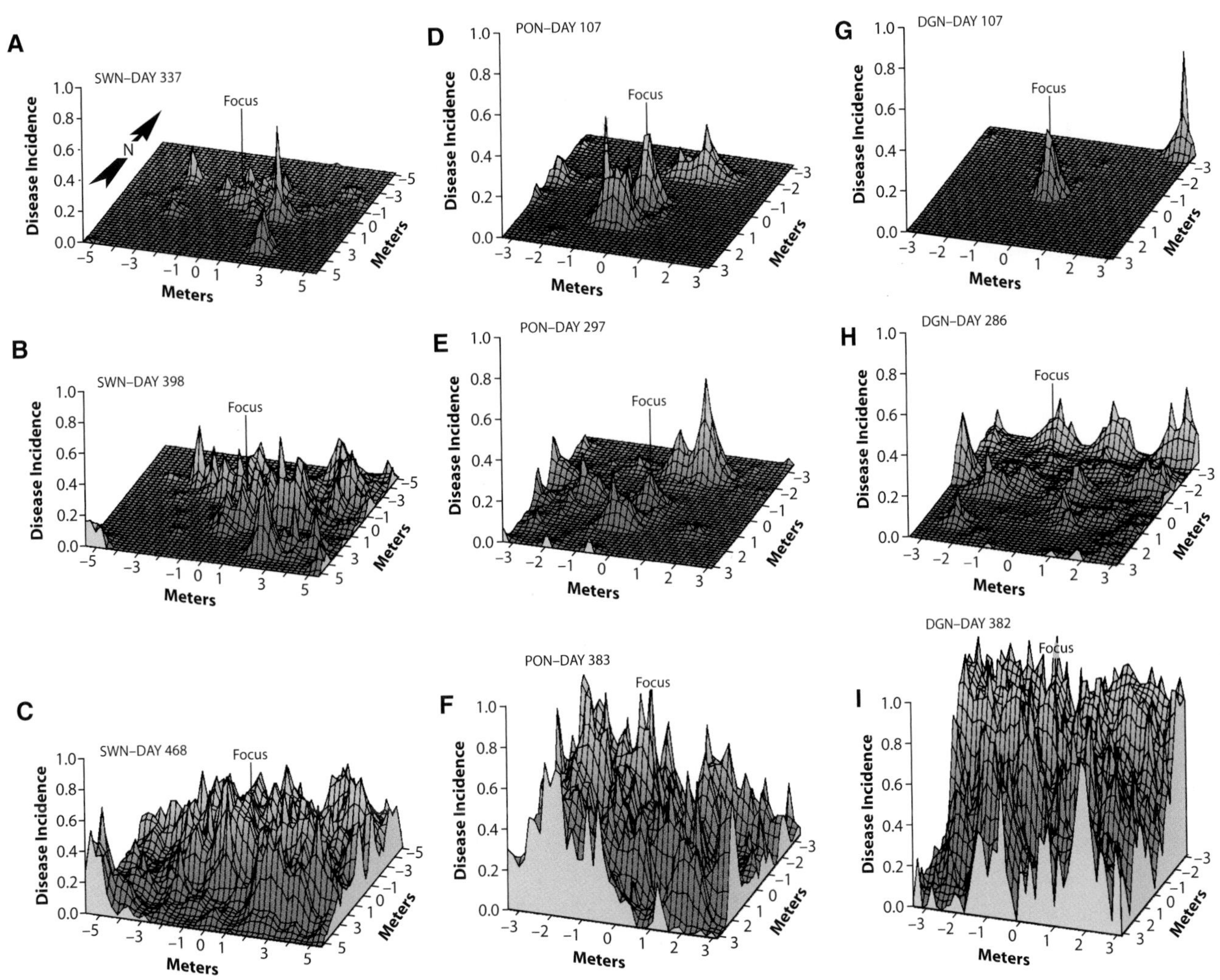

FIGURE 8-20 Development and spread of citrus canker disease, caused by the bacterium *Xanthomonas campestris* pv. *citri*, from a single inoculated plant (focus) in three citrus nurseries on the indicated days after inoculation. SWN, Swingle rootstock nursery; PON, Pinable orange nursery; DGN, Duncan grapefruit nursery. Citrus canker developed fastest in the Duncan nursery and slowest in the Swingle nursery. [From Gottwald *et al.* (1989). *Phytopathology* 79, 1276–1283.]

varieties and pathogen races present, with the amount of pathogen inoculum present at the beginning of the epidemic, and with the moisture levels and temperature ranges during the epidemic (Fig. 8-21). The more information we have about each of the components of an epidemic and about each of its subcomponents at any given moment, the better we can understand and describe the epidemic, and the better we can predict its direction and severity at some later point in time or some other place. The ability to predict the direction and severity of an epidemic, of course, has important practical consequences: it allows us to determine whether, and when, to intervene with control measures. Moreover, it often allows us to determine what types of disease management strategies can be employed to slow down, or entirely prevent, the disease in a particular location.

In an effort to improve our ability to understand and predict the development of an epidemic, plant pathologists since the late 1960s have been developing models of potential epidemics of the most common and serious diseases. The construction of a model takes into account all of the components and as many of the subcomponents of a specific plant disease for which there is information for quantitative treatment, i.e., for treatment by mathematical formulas. The models constructed are generally crude simplifications of real epidemics, roughly analogous, for example, to model toy cars or airplanes as they compare to real cars and airplanes. As with model toys, however, one can get a better picture and understanding of the real thing as the model depicts more and more parts, as the accuracy of the proportions of these parts increases, and as the number of the parts

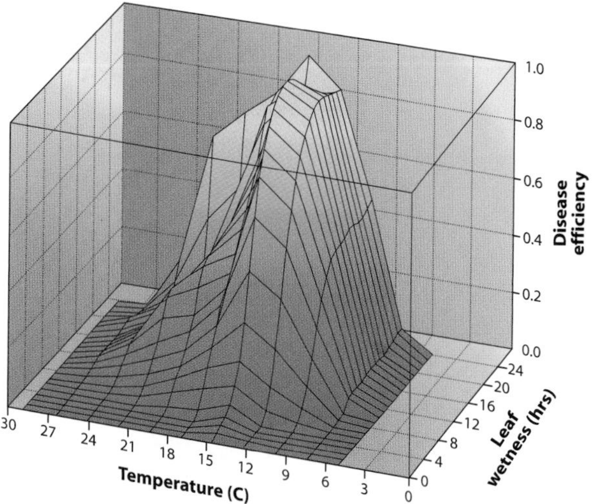

FIGURE 8-21 Model describing the effect of temperature and leaf wetness duration on the ability of the bean rust fungus *Uromyces appendiculatus* to cause disease. The maximum disease reached at 15°C and 24 hours of leaf wetness is given the maximum value of 1.0. [From Berger *et al.* (1995). *Phytopathology* 85, 715–721.]

that are interlocked and move increases. The closer the resemblance of the model to the real thing, the better we can visualize and understand the functions of the real thing by observing the model. In modeling plant disease epidemics, each component and subcomponent of the epidemic may be considered equivalent to one of the parts of the toy model; moreover, just as more accurately measured and fitted parts make for a more exact toy model, the more accurately the real subcomponents of an epidemic are measured and fitted together, the more accurately they describe the epidemic. When we have enough information about the values of the subcomponents of an epidemic at different stages and under different conditions, we can then develop a mathematical equation or equations — a mathematical model — that describes the epidemic.

Analysis of mathematical models of epidemics of specific plant diseases provides a great deal of information regarding the amount and efficacy of the initial inoculum, the effects of the environment, the disease resistance of the host, the length of time that host and pathogen may interact, and the effectiveness of various disease management strategies. Attempts to verify models of epidemics with actual observations and experimentation point out areas in which more knowledge is needed, and such analyses therefore indicate the directions in which further studies of the particular disease should be pursued.

In developing a plant disease model, a database of information is developed about as many of the components of a plant disease as possible. The database contains information on the crop, the disease, the pathogen, the location of the weather station, and sensor(s) relative to the crop and the crop canopy. The database also contains information on the input variables such as measured environmental variables, including temperature, precipitation, relative humidity and leaf wetness; calculated environmental variables such as degree hours or dew points; host variables such as crop growth stage, variety, and other host factors; and pathogen variables such as inoculum potential, spore maturity, and other pathogen factors. The mathematic relationship that describes the interaction between the environment, host and pathogen variable, and the disease is described as the model and is presented as an equation, as a graph, as a table, or as a simple statement. The information, if available, is obtained from the literature, or else it is developed through experimentation. Because plant disease models are developed for specific climates and regions, a model not developed in a specific area must be tested and validated for the specific location for one or more seasons to verify that it will work in this location.

COMPUTER SIMULATION OF EPIDEMICS

The availability of computers has allowed plant pathologists to write programs that allow the simulation of epidemics of the most important plant diseases. One of the first computer simulation programs, called EPIDEM, was written in 1969 and resulted from modeling each stage of the life cycle of a pathogen as a function of the environment. EPIDEM was designed to simulate epidemics of early blight of tomato and potato caused by the fungus *Alternaria solani*. Subsequently, computer simulators were written for Cercospora blight of celery (CERCOS), for *Mycosphaerella* blight of chrysanthemums (MYCOS), for southern corn leaf blight caused by *Cochliobolus (Helminthosporium) maydis* (EPICORN), and for apple scab caused by *Venturia inaequalis* (EPIVEN). A more general and more flexible plant disease simulator, called EPIDEMIC, was written primarily for the stripe rust of wheat but could be modified easily for other host–pathogen systems. Computer simulation programs are now available for numerous plant diseases.

In a computer simulation of an epidemic, the computer is given data describing the various subcomponents of the epidemic and control practices at specific points in time (such as at weekly intervals). The computer then provides continuous information regarding not only the spread and severity of the disease over time, but also the final crop and economic losses likely to be caused by the disease under the conditions of the epidemic as given to the computer.

Computer simulation of epidemics is extremely useful as an educational exercise for students of plant pathology and also for farmers so that they can better understand and appreciate the effect of each epidemic subcomponent on the final size of their crop loss. Computer simulations of epidemics are, however, even more useful in actual disease situations. There, they serve as tools that can evaluate the importance of the size of each epidemic subcomponent at a particular point in time of the epidemic by projecting its effect on the final crop loss. By highlighting the subcomponents of an epidemic that are most important at a particular time, the simulation serves to direct attention to management measures that are effective against these particular epidemic subcomponents. In subsequent evaluations of the epidemic, the computer evaluates not only the current status of the disease, but also the effectiveness of the applied management measures in controlling the epidemic.

FORECASTING PLANT DISEASE EPIDEMICS

Being able to forecast plant disease epidemics is intellectually stimulating and also an indication of the success of modeling or computer simulation of particular diseases. Foremost, however, it is extremely useful to farmers in the practical management of crop disease. Disease forecasting allows the prediction of probable outbreaks or increases in intensity of disease and, therefore, allows us to determine whether, when, and where a particular management practice should be applied. In managing the diseases of their crops, growers must always weigh the risks, costs, and benefits of each of numerous decisions. For example, they must decide whether or not to plant a certain crop in a particular field. Growers must also decide whether to buy more expensive propagating stock free of virus and other pathogens or whether they can "get by" with untested stock. Quite often, growers must decide whether to plant seed of a more expensive or less-yielding but resistant variety rather than seed of a high-yielding but susceptible variety that needs to be protected by chemical sprays. Most frequently, farmers need forecasts that will help them determine whether a plant infection is likely to occur so they can decide whether to spray a crop right away or to wait for several more days before they spray. If disease forecasting allows them to wait, they can reduce the amounts of chemicals and labor used without increasing the risk of losing their crop.

To develop a plant disease forecast, one must take into account several characteristics of the particular pathogen, host, and, of course, environment. In general, for most monocyclic diseases (such as root rot of peas and Stewart's wilt of corn) and for a few polycyclic diseases that may have a large amount of initial inoculum (such as apple scab), disease development may be predicted by assessing the amount of the initial inoculum. For polycyclic diseases (such as late blight of potato) that have a small amount of initial inoculum but many infection cycles, disease development can best be predicted by assessing the rate of occurrence of the infection cycles. For diseases in which both the amount of initial inoculum and the number of disease cycles are large, e.g., beet yellows, both factors must be assessed for the accurate prediction of disease epidemics. Such assessments, however, are often difficult or impossible, and, despite considerable improvements in equipment and methods, assessments of initial inoculum or rapidity of infection cycles are seldom accurate.

Disease Diagnosis: The Key to Forecasting of any Plant Disease Epidemic

Plants in a field are rarely attacked by a single kind of pathogen. More often than not, leaf spots and blotches caused by abiotic factors or bacteria may be present along with spots and blotches caused by fungi. Such symptoms may be confused with those caused by the pathogen in question and may be difficult to diagnose accurately. Such difficulty is especially likely early in the development of a disease when accurate diagnosis is needed most for determining if a threshold for development of an epidemic has been reached and appropriate instructions for its management must be issued. Inaccurate diagnosis of the pathogen in question as being present in the crop early, while in reality it is not, will lead to premature recommendation to spray and therefore to additional and unnecessary fungicide applications. However, misdiagnosis of the real pathogen as something else of lesser importance is likely to miss the opportunity to take appropriate management measures early in the development of the epidemic and to make it much more difficult and expensive to prevent the epidemic from developing and causing serious losses.

Evaluation of Epidemic Thresholds

It is always desirable for the grower to have flexibility in timing fungicide applications according to the progress of an epidemic. In diseases characterized by numerous localized infections (foliar diseases), epidemics are generally characterized by three parameters: disease incidence in individual plants, disease incidence in individual organs (usually leaves), and disease severity (percentage infected leaf area) in leaves. These parameters mark different phases of disease development. In the early stages of disease, disease incidence in plants may increase rapidly but disease severity on individual plants is low. In the second phase of the epidemic, i.e., disease incidence in leaves, there is a small increase in disease severity along with an increase in disease incidence in leaves. Depending on the specific disease, when a percentage (e.g., 1–50%) of plants and a percentage (e.g., 1–25%) of leaves show disease incidence, these are taken as the epidemic threshold in the first two phases of the epidemic for the application of fungicides to stop or slow the development of the epidemic. During the third phase of a disease, disease severity is likely to increase rapidly (up to 2–50% per week). During this phase, fungicides are applied according to disease severity assessment, the dictates of weather conditions (rainfall, relative humidity, temperature measured daily, and providing a daily infection value), and continue as long as there is healthy tissue on the plants that needs to be protected while the crop is not yet ready for harvest.

Evaluation of Economic Damage Threshold

Although it is fairly easy to determine the epidemic threshold, in many plant diseases the threshold for a fungicide application is reached late in the season, which results in disease severity remaining low and yield not being affected. Therefore, in order to apply fungicides only when needed, one must evaluate the tolerance level of disease severity at harvest. This tolerance level, known as economic damage threshold, is the highest disease severity level that does not decrease economic profits. The economic damage threshold is obtained by studying a disease–loss relationship of disease severity at harvest and the final value of the produce and then determining the point beyond which disease severity at harvest decreases economic profits.

Assessment of Initial Inoculum and of Disease

It is often difficult or impossible, in the absence of the host, to detect small populations of most pathogens. Inoculum propagules of soilborne pathogens, such as fungi and nematodes, are estimated after extraction or trapping from soil. Airborne fungal spores and insect vectors are estimated by trapping them in various devices.

Usually it is easier to assess the amount of inoculum present by measuring the number of infections produced on a host within a certain period of time. Even in the presence of a host, however, it is often difficult to find and measure a small amount of disease. Furthermore, in many diseases there is an incubation period during which the host is infected but shows no symptoms. Aerial photography, using films sensitive to near-infrared radiation, has made possible both earlier detection and sharper delineation of diseased areas in crop fields (due to the reduced reflectance of diseased foliage tissues that are occupied by water or pathogen cells). However, for many diseases, by the time aerial photography detects diseased areas in fields, yield loss has already occurred.

Monitoring Weather Factors That Affect Disease Development

Monitoring weather factors during a plant disease epidemic presents enormous difficulties. Difficulties arise from the need for the continuous monitoring of several different factors (temperature, relative humidity, leaf wetness, rain, wind, and cloudiness) at various locations in the crop canopy or on plant surfaces in one or more fields. In the past, measurements were made with mechanical instruments that measured these environmental variables roughly or infrequently and recorded data inconveniently as ink traces on chart paper. Since the 1970s, however, several types of electronic sensors have been developed that produce electrical outputs recorded easily by computerized data loggers. Such computerized sensors are now prevalent in parts of the United States and of other countries and have improved studies of weather in relation to disease greatly and have facilitated the acceptance and use of predictive systems for disease control on the farm.

In most parts of the world, however, several types of traditional and battery-operated electrical instruments are used to measure various weather factors. Temperature measurements are made with various types of thermometers, hygrothermographs, thermocouples, and especially with thermistors (the latter are semiconductors whose electrical resistance changes considerably with temperature). Relative humidity measurements are made with a hygrothermograph (which depends on the contraction and expansion of human hair in relation to relative humidity changes), with a ventilated psychrometer (consisting of a wet and dry bulb thermometer or a wet and dry thermistor), or with an electrode-bonding sulfonated polystyrene plate (whose resistance changes logarithmically with relative humidity). Leaf wetness is monitored with string-type sensors that constrict when moistened or slacken when dry and either leave an ink trace in the process or close or break an electrical circuit. Several types of electrical wetness sensors are available that can be either clipped onto leaves or placed among the leaves; they detect and measure the duration of rain or dew because either of the latter helps close the circuit between two pairs of electrodes. Rain, wind, and cloudiness (irradiance) are still measured by traditional instruments (rain funnels and tipping-bucket gauges for rain, cups and thermal anemometers for wind speed, vanes for wind direction, and pyranometers for irradiance). Several of these instruments, however, have become adapted for electronic monitoring.

In modern weather-monitoring systems, the weather sensors are connected to data-logging devices. Data may be read on a digital display or be transmitted to a cassette tape recorder or a printer. From the cassette, data may be transferred to a microcomputer. There they may be viewed, processed in several computer languages, organized into separate matrices for each weather variable, plotted, and analyzed. Depending on the particular disease model used, accurate weather information provides the most useful basis to predict sporulation and infection and therefore provides the best warning to time disease management practices, such as the application of fungicides.

The cost of purchase of automated weather systems (AWS) and the required time for operation and maintenance discourage their use by individual farmers, leading to the development of low-cost, automated weather instruments or stand-alone packages or to the creation and sharing of regional automated weather systems.

NEW TOOLS IN EPIDEMIOLOGY

The study of plant disease epidemiology has been facilitated greatly by new methods and new equipment that make possible studies of aspects of plant disease that were impossible or very difficult to study earlier. Some of the equipment and instruments that have contributed to modern epidemiology have been listed already. Some of the methods and other equipment that have been used to great advantage in plant disease epidemiology include the following.

Molecular Tools

The most important of these are the development and use of genetic (DNA) probes that allow the definitive detection and identification of a plant pathogen within or on the surface of a plant tissue, in a mixture with other microorganisms, and even in the vicinity of the host plant. The detection and identification of a pathogen by its genetic probe, however, are made immensely more effective through the use of the polymerase chain reaction (PCR) technique, which amplifies greatly a specific fragment of DNA present on a probe and produces millions of copies of it. These copies are then abundant enough to be detected, identified, and studied by conventional or other molecular tools. Random amplified polymorphic DNA (RAPD) markers are often used to detect genetic similarities among pathogenic strains known to show genetic heterogeneity and can also be used easily for designing sequence characterized amplified region (SCAR) markers for detecting the pathogen in infected plant tissue. The significance of the contributions of these, and some other, molecular techniques in epidemiology lies in the fact that they can detect pathogen arrival much earlier than could be detected before, thereby allowing the grower time to get ready and to apply whatever management treatment is most effective against the pathogen. Moreover, these techniques can detect any new mutant pathogens early that could either attack plant varieties they could not attack before or they may tolerate the fungicide to which they were sensitive before and thus produce a new resistant race. Detection of such changes in pathogens is of paramount importance in epidemiology because such changes in pathogens make useless and necessitate immediate revision of any previous predictions about the development of the epidemic and recommendations for management of the disease.

Geographic Information System

The geographic information system (GIS) is a computer system that can be installed on any recent model desktop computer and is capable of assembling, storing, manipulating, and displaying data that are referenced by geographic coordinates. GIS is adaptable to operations of any size, and data can be used at any scale from a single field to an agricultural region. It is used to better understand and manage the environment, including the understanding and management of plant disease epidemics. GIS techniques allow one to make connections between events based on geographic proximity, connections that are essential to the understanding and management of epidemics but which often go unrecognized without GIS. GIS techniques can even incorporate disease forecasting systems, although the time and cost for it may be prohibitive. However, as high-resolution weather forecast data are often available, the development of plant disease epidemics can be predicted by knowing their dependency on some critical weather variable and from estimated geographic distribution of the pathogen inoculum within a GIS framework. GIS is often used for the spatial and temporal analysis of disease development over relatively large geographic areas and helps determine the role and relative importance of various parts of these areas in the initiation and development of the epidemic.

Global Positioning System

The global positioning system (GPS) consists of a hand-held device that is coordinated with a global system of man-made satellites and, depending on the accuracy and coordination, provides quite accurate readings of the coordinates of the position of the device. GPS enables one to pinpoint an individual tree or a specific area or areas of the field that are affected by a pathogen, which then can be visited and examined again periodically for incremental advance of the symptoms. Similarly, the selected trees or areas could be treated with the appropriate pesticide or other treatment wherever the pathogen is present without the need to treat the entire field. GPS can also be used to apply pesticides, plant nutrients, and so on in only the areas of the field that are infested with the pathogen or in areas deficient in a particular micro- or macronutrient. Elimination of the pathogen from the field by early detection and treatment is often effective in not allowing the pathogen to cause an epidemic in the field and beyond.

Geostatistics

Geostatistics consist of various "geostatistical" techniques that are applied in plant disease epidemiology to characterize quantitatively spatial patterns of disease development or the development of pathogen populations in space and over time. These techniques have the capability to take into account the characteristics of spatially distributed variables whether they are random or systematic. In addition to being able to detect spatial connections, geostatistical techniques can also be used for studying continuous and discrete variables. Geostatistical techniques do not require as exacting asumptions of stationarity as do other spatial autocorrelation techniques. The spatial dependence or connection can be analyzed with semivariograms. The latter quantify spatial dependence by determining the variation between samples.

Remote Sensing

Remote sensing usually refers to the use of instruments for measuring electromagnetic radiation reflected or emitted from an object. The instruments record reflected or emitted radiation in the ultraviolet, visible, or infrared part of the spectrum. The instruments used for remote sensing may be hand-held, ground-based cameras with films and filters, digital cameras, video systems, and radiometers or they may be carried on balloons, aircraft, and satellites. The various remote-sensing instruments store data obtained from field situations, and data are then printed out and are analyzed directly or by transferring them to a computer and creating visual images of data (Fig. 8-22).

Image Analysis

Image analysis refers to photography and electronic image analysis, usually of large areas of fields or of mountains. The images or photographs are taken through aerial photography, ground-based sensor data, and satellite-borne and airborne sensors. Airborne mul-

FIGURE 8-22 An epidemic of sudden death of oak caused by *Phytophthora ramorum* in California as seen by aerial photography. (Photograph courtesy of P. Svihra, University of California.)

tispectral scanning is studied and used widely for the surveillance of plant diseases, pests, and environmental stresses in agriculture. Often, infrared light or light of other wavelengths is used for the detection of the onset and progress of a plant disease among the crop plants in the field or among the fruit or forest trees in a mountain. Plants and trees, when infected with various pathogens or subjected to other stresses, turn light green, then chlorotic (yellowish), and then brown and have different reflectances from the healthy plant. These colors or shades of colors become more distinct when photographed with the wavelengths mentioned previously than when photographed with the normal visible light spectrum. More importantly, such photographs can be examined and analyzed with specific equipment that not only better distinguishes such disease-discolored plants, but can also provide a count of the newly infected plants as well as measure the changes in intensity of the images of previously diseased plants. In that way, image analysis can provide a measure of the severity of the disease in each plant or area of infected plants and, by repetition of the photography at regular intervals, provide a measure of the rate of progress of the disease.

Information Technology

This technology involves primarily the use of computers alone or in combination with other electronic devises. They help collect data on plant diseases at various levels and various locations in a continuous manner. Data are either stored or are organized, integrated, and analyzed in tremendous quantities and at hitherto unimaginable speeds and eventually are used to produce visual images or written reports and recommendations. Electronic information technology can, above all, describe and display spatial patterns of characteristics of different pathogens, such as their genotypes, at the scale of an agricultural region.

EXAMPLES OF PLANT DISEASE FORECAST SYSTEMS

Generally, it is useful to have the maximum amount of information that is available about a disease before venturing to predict its development. In many cases, however, one or two of the factors that affect disease development predominate so much that knowledge of them is often sufficient for the formulation of a reasonably accurate forecast. Thus, forecasting systems of several plant diseases use the amount of the initial inoculum as the criterion. Such diseases include Stewart's wilt of corn, blue mold of tobacco, fire blight of apple and pear, pea root rot, and other diseases caused by soilborne pathogens such as *Sclerotium* and cyst nematodes. Forecasting systems of diseases such as the late blight of potato, *Cercospora* and other leaf spots and the downy mildew of grape use the number of infection cycles or the amount of secondary inoculum as the criterion. Forecasting systems of still other diseases, e.g., apple scab, black rot of grape, cereal rusts, *Botrytis* leaf blight and gray mold, and sugar beet yellows, use the amount of the initial inoculum and the number of infection cycles or the amount of secondary inoculum as criteria.

Forecasts Based on Amount of Initial Inoculum

In Stewart's wilt of corn [caused by the bacterium *Erwinia (Pantoea) stewartii*], the pathogen survives the winter in the bodies of its vector, the corn flea beetle. Therefore, the amount of disease that will develop in a growing season can be predicted if the number of vectors that survived the previous winter is known, as that allows an estimation of the amount of inoculum that also survived the previous winter. Corn flea beetles are killed by prolonged low winter temperatures. Therefore, when the sum of the mean temperatures for the three winter months December, January, and February at a given location is less than −1°C, most of the beetle vectors are killed and so there is little or no bacterial wilt during the following growth season. Warmer winters allow greater survival of beetle vectors and proportionately more severe wilt outbreaks the following season.

In the downy mildew (blue mold) of tobacco (caused by the oomycete *Peronospora tabacina*), the disease in most years is primarily a threat to seedbeds in the tobacco-producing states. When January temperatures are above normal, blue mold can be expected to appear early in seedbeds in the following season and to cause severe losses. However, when January temperatures are below normal, blue mold can be expected to appear late in seedbeds and to cause little damage. If the disease is expected in seedbeds, control measures can be taken to prevent it from becoming established, and subsequent control in the field is made much easier. Since 1980, a supplementary blue mold warning system has been operated in North America by the Tobacco Disease Council and the Cooperative Extension Service. The warning system keeps the industry aware of locations and times of appearance and spread of blue mold and helps growers with the timing and intensity of controls.

In pea root rot (caused by the oomycete *Aphanomyces euteiches*) and in other diseases caused by soilborne fungi and some nematodes, the severity of the disease in a field during a growing season can be predicted by winter tests in the greenhouse. In these tests, susceptible plants are planted in the greenhouse in soil taken from the field in question. If the greenhouse tests show that severe root rot develops in a particular soil, the field from which the soil was obtained is not planted with the susceptible crop. However, fields whose soil samples allow the development of little or no root rot can be planted and can be expected to produce a crop reasonably free of root rot. With some soilborne pathogens, such as fungi *Sclerotium* and *Verticillium* and the cyst nematodes *Heterodera* and *Globodera*, the initial inoculum can be assessed directly by isolating the fungal sclerotia and nematode cysts and then counting them per gram of soil. The greater the number of propagules, the more severe the disease produced.

In fire blight of apple and pear (caused by the bacterium *Erwinia amylovora*), the pathogen multiplies much more slowly at temperatures below 15°C than at temperatures above 17°C. In California, a disease outbreak can be expected to occur in the orchard if the daily average temperatures exceed a "disease prediction line" obtained by drawing a line from 16.7°C on March 1 to 14.4°C on May 1. Therefore, when such conditions occur, application of a bactericide during bloom is indicated to prevent an epidemic.

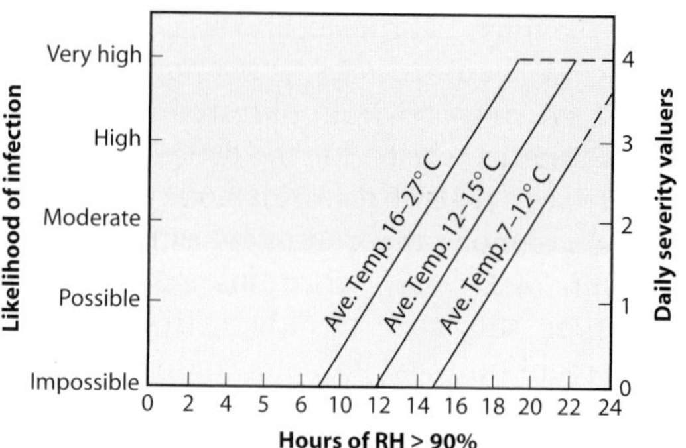

FIGURE 8-23 Relationship of the duration of high relative humidity periods and average temperature during such periods to the likelihood of potato infection by the late blight fungus *Phytophthora infestans*. The daily severity values are arbitrary values given by the relative humidity–temperature relationship; they correspond to the likelihood of infection shown at left and are used to recommend spray schedules with BLITECAST. [From MacKenzie (1981). *Plant Dis.* 65, 394–399.]

Forecasts Based on Weather Conditions Favoring Development of Secondary Inoculum

In late blight of potato and tomato (caused by the oomycete *Phytophthora infestans*), the initial inoculum is usually low and generally too small to detect and measure directly. Even with low initial inoculum, the initiation and development of a late blight epidemic can be predicted with reasonable accuracy if the moisture and temperature conditions in the field remain within certain ranges favorable to the fungus. When constant cool temperatures between 10 and 24°C prevail and the relative humidity remains over 75% for at least 48 hours or is at least 90% for 10 hours each day for 8 days, infection will take place and a late blight outbreak can be expected from 2 to 3 weeks later. If, within that period and afterward, several hours of rainfall, dew, or relative humidity close to the saturation point occur, they will serve to increase the disease and will foretell the likelihood of a major late blight epidemic (Fig. 8-23).

Computerized predictive systems have been developed for epidemics of late blight and several other diseases; in some such systems, e.g., BLITECAST for late blight (Fig. 8-20); FAST (for forecasting *Al. solani* on tomatoes); TOMCAST (for tomato forecaster) for tomato early blight, *Septoria* leaf spot, and anthracnose; and PLAM for peanut leaf spot, moisture and temperature are monitored continuously. From this information weather severity values are calculated, infection and disease severity values are predicted, and recommendations are issued to growers as to when to begin spraying. More recent refinements in late blight forecasting include, in addition to data on moisture and temperature, information on the level of resistance of the potato variety to late blight and the effectiveness of the fungicide used. Information on all these parameters is, of course, very useful in the formulation of recommendations for fungicide applications.

Several leaf spots, such as those caused by the fungi *Cercospora* on peanuts and celery and *Exserohilum* (*Helminthosporium*) *turcicum* on corn, can be predicted by taking into account the number of spores trapped daily, the temperature, and the duration of periods with relative humidity near 100%. An infection period is predicted if high (95–100%) relative humidity lasts for more than 10 hours, and growers are then urged to apply chemical sprays immediately.

Forecasts Based on Amounts of Initial and Secondary Inoculum

In apple scab (caused by the fungus *Venturia inaequalis*), the amount of initial inoculum (ascospores) is usually large and is released over a period of 1 to 2 months following bud break. Infections from the primary inoculum must be prevented with well-timed fungicide applications during blossoming, early leafing, and fruit development; otherwise, the entire crop is likely to be lost. After primary infections, however, secondary inoculum (conidia) is produced, which multiplies itself manyfold with each succeeding generation. The pathogen can infect wet leaf or fruit surfaces at a range of temperatures from 6 to 28°C. The length of time that leaves and fruit need to be wet, however, is much shorter at optimum temperatures than at either extreme (9 hours at 18–24°C versus 28 hours at 6 to 28°C). By combining temperature and leaf wetness duration data, the apple scab forecast system can predict not only whether an infection period will occur, but also whether the infection periods will result in light, moderate, or severe disease (Fig. 8-24). Such information, collected and analyzed by individuals or by weather-sensing microcomputers, is used to make recommendations to growers. The latter are advised of the need and timing of fungicide application and about the kind of fungicide (protective or eradicant) that should be used to control the disease.

In wheat leaf and stem rusts (caused by fungi *Puccinia recondita* and *Puccinia graminis*), short (1–2 week) forecasts of subsequent disease intensity can be obtained by taking into account disease incidence, stage of plant growth, and spore concentration in the air.

In many insect-transmitted virus diseases of plants (e.g., barley yellow dwarf, cucumber mosaic virus, and sugar beet yellows), the likelihood, and sometimes the severity, of epidemics can be predicted. This is accomplished by determining the number of aphids, especially viruliferous ones, coming into the field at certain stages of the host growth. A number of the aphids caught in traps placed in the field are tested for virus by allowing them to feed on healthy plants or by analyzing them for virus serologically with the ELISA technique or with nucleic acid probes. The more numerous the viruliferous aphids and the earlier they are detected, the more rapid and more severe will be the virus infection. Such predictions can be improved by taking into account late winter and early spring temperatures, which influence the population size of the overwintering aphid vectors.

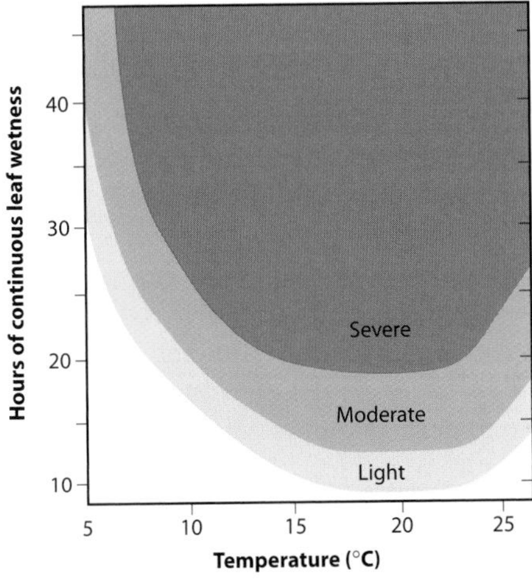

FIGURE 8-24 Relationship of temperature and duration of leaf wetness to the occurrence and severity of apple scab caused by the fungus *Venturia inaequalis*. [From Mills (1944). *Cornell Ext. Bull.* 630.]

RISK ASSESSMENT OF PLANT DISEASE EPIDEMICS

The risk of development of a plant disease into an epidemic is the probability that a certain intensity of incidence or severity of the disease will be reached. For example, a possible risk of tomato early blight can be estimated as 10% incidence with 85% probability. However, the risk of plant disease can also be determined as the probability, e.g., 90%, that the maximum possible incidence of a disease being about 60%, will not be reached. Numerous host, pathogen, and environmental factors must be taken into account in assessing the risk of development of a particular plant disease: history of the disease in the field from previous years, resistance of planted varieties, presence and amount of primary inoculum, period of susceptibility of the host, prevailing weather conditions (temperature, rainfall, relative humidity) during periods of susceptibility, availability and cost of effective control measures, and so on. Since in most cases information on all of these parameters remains fairly constant from year to year, one needs to concentrate primarily on estimating as well as possible the starting inoculum of the pathogen and, subsequently, in following closely changes in temperature and moisture, appearance of first signs of the disease in the field, and predictions of weather changes in the near future. When all the parameters, constant and variable (temperature, rainfall, relative humidity), are known, or estimated from the best data available, a knowledgeable person can project with some certainty the likely risk of the disease developing up to a certain level of severity. Risk assessment is sometimes expressed as percentages of obtaining certain values of disease severity; more often, however, it is expressed as low, moderate, or high risk of reaching those disease severity values. Nevertheless, risk assessment provides a timely warning to the grower who subsequently responds with appropriate urgency in applying effective and sufficient management measures.

DISEASE-WARNING SYSTEMS

In many states and countries, different types of warning systems are in place for one or more important plant diseases. The purpose of these systems is to warn farmers of the impending onset of an infection period or to inform them that an infection period has already occurred so that they can take immediate appropriate control measures to stop recent infections from developing or prevent further infections from occurring.

In most cases, the warning system begins with a grower, an extension agent, or a private consultant surveying certain fields on a regular basis or when the weather conditions are likely to favor maturation of the primary inoculum or appearance of the particular disease. When mature inoculum (such as ascospores in apple scab) or traces of disease (e.g., in potato late blight) are found, the county extension office is notified. The county extension office in turn notifies the state extension plant pathologist, who collates all reports about the disease from around the state and by electronic mail (e-mail), telephone, fax, or in writing notifies all concerned county agents (pest alert). They, in turn, by e-mail, radio, telephone, or letter, notify all farmers in their county. For diseases of potential regional or national epidemic consequences, the state extension plant pathologist notifies the federal plant disease survey office of the U.S. Department of Agriculture, which in turn notifies all extension plant pathologists in adjacent and other states that may be affected by that plant disease.

Since the mid-1970s, computerized warning systems have been in use for certain diseases in some states. Some of them (such as BLITECAST) use centrally located computers that process weather data either collected on the farm by individual growers and transmitted electronically or phoned in when certain weather conditions prevail, or at certain intervals. The computer then processes the data, determines whether an infection period is imminent, likely to occur, or cannot occur, and makes a recommendation to the grower as to whether to spray and what materials to apply.

After 1980, small special-purpose computers have been used that have field sensors and can be mounted on a post in a farmer's field. Such units (such as the apple scab predictor) monitor and collect data in the field on temperature, relative humidity, duration of leaf wetness, and rainfall amounts, analyze the data automatically, make predictions of disease occurrence and intensity, and, on the spot, make recommendations for disease control measures. The same unit can be used for any disease for which a prediction program is available, in which case either the unit can be reprogrammed or the program circuit boards can be interchanged. Predictions from such units are obtained by using a simplified keyboard and display right in the field or the unit can be linked to a personal computer if additional processing of data is desired.

DEVELOPMENT AND USE OF EXPERT SYSTEMS IN PLANT PATHOLOGY

Expert systems are computer programs that try to equal and, better yet, surpass the logic and ability of an expert professional in solving problems, the solutions of which require experience, knowledge, judgment, and complex interactions. The dependability of an expert system is proportional to the knowledge of the expert(s) who produced it. Expert systems can use data in almost any format and can suggest a solution to the problem; they can even use incomplete or incorrect data, as long as the degree of certainty of data is quantified by the expert and is included in the knowledge base. Expert systems in plant pathology are used frequently for diagnostic purposes, i.e., identifying the cause of a disease by the symptoms and related observations. Several expert systems, however, incorporate the decision-making process of the expert and advise producers in making disease management decisions. By incorporating infection models of the important diseases of a crop into the knowledge base of the computer, the expert system can advise growers of disease potentials on the basis of the actual occurrence of infection periods and provide pesticide recommendations and suggestions for pesticide amounts and timing of application.

The development of even simple expert systems is quite complex, but advances in computers and increasing familiarity with their use are making the development and use of expert systems increasingly attractive. In their simplest form, expert systems utilize a bank of data pertinent to the problem stored in the computer and also a knowledge base inputed by the expert(s) and consisting of one or more "IF conditions" followed by a conclusion or action (THEN action) and, finally, a recommendation. In addition to the requirement of being familiar with computer programming, the most important part of creating an expert system is the quality (expertise) of the expert(s) providing the knowledge that is inputed in the system. This knowledge of the expert(s) is then represented in a form that can be converted into computer code. Once a prototype expert system is generated, it is first tested for logic and accuracy. Usually, the expert system is also reviewed and, if necessary, revised by other experts; subsequently, it is tested with the intended users, and additional revisions are made before the expert system is released for use. Even after an expert system is released to its final users, it must be revised and updated regularly.

BLITECAST (1975), which is a computerized forecasting system for potato late blight, and the computer-based apple scab predictive system (1980) are considered to be the precursors to expert systems. The first expert system in plant pathology was developed in

1983 to diagnose nearly 20 soybean diseases in Illinois. Since then, expert systems have been developed for the diagnosis or management of diseases of tomato (TOM), grape (GrapES), wheat (CONSELLOR), peach and nectarine (CALEX), apple (POMME, the Penn State apple orchard consultant PSAOC), wheat (MoreCrop), and others.

An example of an "expert" advisory system is MoreCrop, which stands for "Managerial Options for Reasonable Economical Control of Rusts and Other Pathogens." MoreCrop is designed to provide disease management options in different geographic regions and agronomic zones of the Pacific Northwest using the vast information available on wheat diseases as well as advances in computers. The components of MoreCrop and their functional relationships are understood. Some of the frames ("windows") of the program show the wheat diseases about which one should be concerned. Brief information about each disease, suggestions for disease control through seed treatment and foliar spray, timing of sprays, spray label restrictions, and which diseases can or cannot be controlled through a particular treatment are shown in relevant frames.

Expert systems are used primarily, but not exclusively, with high value horticultural crops that require frequent application of pesticides as part of their disease and pest management, usually in response to site-specific weather conditions. Although expert systems are aimed for use by growers of such crops, they are also used by individuals, such as county agents and pesticide distributors, who influence grower decisions.

DECISION SUPPORT SYSTEMS

A fully developed decision support system (DSS) is supposed to collect, organize, and integrate all types of information related to the production of a crop, to subsequently analyze and interpret the information, and to eventually recommend the most appropriate action or action choices. Decision support systems for plant disease management may be very simple, e.g., a data processing device, fairly complex, e.g., a computerized expert system, or extremely complex, including automated weather and combinations of decision aids and expert systems, as well as multidisciplinary teams of knowledge specialists. Numerous DSS systems available are aimed to assist practitioners in the field, including county agents, crop consultants, growers, and others. Many of them have plant disease management modules, such as WISDOM for potatoes by the University of Wisconsin, RADAR for apples by the University of Maine, PAWS for several crops by the Washington State University, and another one, Fieldwise.com, used on several crops on the west coast. Of the many available DSS systems, relatively few are used because they address only specific disease problems, they are too complex to operate, or for other reasons. Cooperation among universities, growers, and industry has resulted in the development of the Penn State apple orchard consultant in the United States, while in Australia, development of the AusVit DSS for grapes came about through the cooperation of several state departments of agriculture, universities, grower organizations, and private industry. It is apparent that the development and usage of DSS will become more regional rather than local. The continuing demise of the family farm and the increase in large farms, however, are expected to increase the use of DSS systems significantly.

Selected References

Anonymous (1983). Symposium on estimating yield reduction of major food crops of the world. *Phytopathology* **73**, 1575–1600.

Anonymous (1995). Epidemiology, crop loss assessment, phytopathometry: A collection of papers *Can. J. Plant Pathol.* **17**, 95–189.

Aylor, D. E. (1998). The aerobiology of apple scab. *Plant Dis.* **82**, 838–849.

Bailey, J. E., Johnson, G. L., and Toth, S. J. (1994). Evolution of a weather-based peanut leaf spot spray advisory in North Carolina. *Plant Dis.* **78**, 530–535.

Bandyopadhyay, R., *et al.* (1998). Ergot: A new disease threat to sorghum in the Americas and Australia. *Plant Dis.* **82**, 356–367.

Berger, R. D. (1977). Application of epidemiological principles to achieve plant disease control. *Annu. Rev. Phytopathol.* **15**, 165–183.

Berger, R. D., *et al.* (1995). A simulation model to describe epidemics of rust of *Phaseolus* beans. I. Development of the model and sensitivity analysis. *Phytopathology* **85**, 715–721 and 722–727.

Bertschinger, L., Keller, E. R., and Gessler, C. (1995). Development of EPIVIT, a simulation model for contact- and aphid-transmitted potato viruses. *Phytopathology* **85**, 801–814 and 815–819.

Biggs, A. R., and Grove, G. G. (1998). Role of the worldwide web in extension plant pathology: Case studies in tree fruits and grapes. *Plant Dis.* **82**, 452–464.

Bolkan, H. A., and Reinert, W. R. (1994). Developing and implementing IPM strategies to assist farmers: An industry approach. *Plant Dis.* **78**, 545–550.

Campbell, C. L., and Madden, L. V. (1990). "Introduction to Plant Disease Epidemiology." Wiley, New York.

Coakley, S. M. (1988). Variation in climate and prediction of disease in plants. *Annu. Rev. Phytopathol.* **26**, 163–181.

Cu, R. M., and Line, R. F. (1994). An expert advisory system for wheat disease management. *Plant Dis.* **78**, 209–215.

Davis, J. M. (1987). Modeling the long-range transport of plant pathogens in the atmosphere. *Annu. Rev. Phytopathol.* **25**, 169–188.

Day, P. R., ed. (1977). The genetic basis of epidemics in agriculture. *Ann. N.Y. Acad. Sci.* **287**.

De Wolf, E. D., Madden, L. V., and Lipps, P. E. (2003). Risk assessment models for wheat Fusarium head blight epidemics based on within-season weather data. *Phytopathology* **93**, 428–442.

Duveiller, E. (1994). A pictorial series of disease assessment keys for bacterial leaf streak of cereals. *Plant Dis.* **78**, 137–141.

Ferreira, S. A., *et al.* (2002). Virus coat protein transgenic papaya provides practical control of papaya ringspot virus in Hawaii. *Plant Dis.* **86**, 101–105.

Fetch, T. G., and Steffenson, B. J. (1999). Rating scales for assessing infection responses of barley infected with *Cochliobolus sativus*. *Plant Dis.* **83**, 213–217.

Francl, L. F., and Neher, D. A., eds. (1997). "Exercises in Plant Disease Epidemiology." APS Press, St. Paul, MN.

Freeman, S., Katan, T., and Shabi, E. (1998). Characterization of *Colletotrichum* species responsible for anthracnose diseases of various fruits. *Plant Dis.* **82**, 596–605.

Fry, W. E., *et al.* (1993). Historical and recent migrations of *Phytophthora infestans*: Chronology, pathways, and implications. *Plant Dis.* **77**, 653–661.

Fry, W. E., and Goodwin, S. B. (1997). Re-emergence of potato and tomato late blight in the United States and Canada. *Plant Dis.* **81**, 1349–1357.

Gaunt, R. E. (1995). The relationship between plant disease severity and yield. *Annu. Rev. Phytopathol.* **33**, 119–144.

Gilligan, C. A. (1983). Modeling of soilborne pathogens. *Annu. Rev. Phytopathol.* **21**, 45–64.

Gleason, K. M. L., *et al.* (1995). Disease-warning systems for processing tomatoes in eastern North America: Are we there yet? *Plant Dis.* **79**, 113–121.

Horsfall, J. G., and Cowling, E. B., eds. (1978). "Plant Disease," Vol. 2. Academic Press, New York.

Huber, L., and Gillespie, T.-J. (1992). Modeling leaf wetness in relation to plant disease epidemiology. *Annu. Rev. Phytopathol.* **30**, 553–577.

Huismann, O. C. (1982). Interactions of root growth dynamics to epidemiology of root-invading fungi. *Annu. Rev. Phytopathol.* **20**, 303–327.

Jones, A. L., Fisher, P. D., Seem, R. C., *et al.* (1984). Development and commercialization of an in-field microcomputer delivery system for weather-driven predictive models. *Plant Dis.* **68**, 458–473.

Jones, D. G., ed. (1998). "The Epidemiology of Plant Diseases." Kluwer Academic, Dordrecht, Boston.

Kobayashi, T., *et al.* (2001). Detection of rice panicle blast with multispectral radiometer and the potential of using airborne multispectral scanners. *Phytopathology* **91**, 316–323.

Krause, R. A., Massie, L. B., and Hyre, R. A. (1975). BLITECAST, a computerized forecast of potato late blight. *Plant Dis. Rep.* **59**, 95–98.

Leonard, K. M., and Fry, W. E., eds. (1986). "Plant Disease Epidemiology," Vol. 1. Macmillan, New York.

Leonard, K. J., and Fry, W. E. (1989). "Plant Disease Epidemiology," Vol. 2. McGraw-Hill, New York.

Luo, Y., Morgan, D. P., and Michilidis, T. J. (2001). Risk analysis of brown rot blossom blight of prune caused by *Monilinia fructicola*. *Phytopathology* **91**, 759–768.

MacHardy, W., and Sondei, J. (1981). Weather-monitoring instrumentation for plant disease management programs and epidemiological studies. *N. H. Agric. Exp. Stn. Bull.* **519**, 1–40.

MacKenzie, D. R. (1981). Scheduling fungicide applications for potato late blight with BLITECAST. *Plant Dis.* **65**, 394–399.

Madden, L. V., and Hughes, G. (1995). Plant disease incidence: Distributions, heterogeneity, and temporal analysis. *Annu. Rev. Phytopathol.* **33**, 529–564.

Magarey, R. D., *et al.* (2001). Site-specific weather information without on-site sensors. *Plant Dis.* **85**, 1216–1226.

Magarey, R. D., *et al.* (2002). Decision support systems: Quenching the thirst. *Plant Dis.* **86**, 4–14.

McMullen, M., Jones, R., and Gallenberg, D. (1997). Scab of wheat and barley: A re-emerging disease of devastating impact. *Plant Dis.* **81**, 1340–1348.

Milgroom, M. G., and Peever, T. L. (2003). Population biology of plant pathogens: The synthesis of plant disease epidemiology and population genetics. *Plant Dis.* **87**, 608–617.

Mundt, C. C. (1995). Models from plant pathology on the movement and fate of new genotypes of microorganisms in the environment. *Annu. Rev. Phytopathol.* **33**, 467–488.

Naidu, R. A., *et al.* (1999). Groundnut rosette: A virus disease affecting groundnut production in sub-Saharan Africa. *Plant Dis.* **83**, 700–709.

Nelson, M. R., *et al.* (1999). Applications of geographic information systems and geostatistics in plant disease epidemiology and management. *Plant Dis.* **83**, 308–319.

Nesmith, W. C. (1984). The North American blue mold warning system. *Plant Dis.* **68**, 933–936.

Nilsson, H.-E. (1995). Remote sensing and image analysis in plant pathology. *Annu. Rev. Phytopathol.* **33**, 489–527.

Parker, S. K., Gleason, M. L., and Nutter, F. W., Jr. (1995). Influence of rain events on spatial distribution of Septoria leaf spot of tomato. *Plant Dis.* **79**, 148–152.

Phipps, P. M., Deck, S. H., and Walker, D. R. (1997). Weather-based crop and disease advisories for peanuts in Virginia. *Plant Dis.* **81**, 236–244.

Plumb, R. T., and Thresh, J. M., eds. (1983). "Plant Virus Epidemiology: The Spread and Control of Insect-Borne Viruses." Blackwell, Oxford.

Populer (1978). *In* "Plant Disease" (J. G. Horsfall and E. B. Cowling, eds.), Academic Press, New York.

Rapilly, F. (1979). Yellow rust epidemiology. *Annu. Rev. Phytopathol.* **17**, 59–73.

Sall, M. A. (1980). Epidemiology of grape powdery mildew: A model. *Phytopathology* **70**, 338–342.

Scott, P. R., and Bainbridge, A., eds. (1978). "Plant Disease Epidemiology." Blackwell, Oxford.

Roelfs, A. P. (1986). *In* "Plant Disease Epidemiology" (K. J. Leonard and W. E. Fry, eds.), Vol. 1, pp. 129–150, Macmillan, New York.

Seem, R. C. (1984). Disease incidence and severity relationships. *Annu. Rev. Phytopathol.* **22**, 133–150.

Subarao, K. V. (1998). Progress toward integrated management of lettuce drop. *Plant Dis.* **82**, 1068–1078.

Travis, J. W., and Latin, R. X. (1991). Development, implementation and adoption of expert systems in plant pathology. *Annu. Rev. Phytopathol.* **29**, 343–360.

Travis, J. W., *et al.* (1992). A working description of the Penn State Apple Orchard Consultant, an expert system. *Plant Dis.* **76**, 545–554.

Udin, L. S., *et al.* (1990). Disease-prediction and economic models for managing tomato spotted wilt virus disease in lettuce. *Plant Dis.* **74**, 211–216.

Vanderplank, J. E. (1963). "Plant Diseases: Epidemics and Control." Academic Press, New York.

Vanderplank, J. E. (1975). "Principles of Plant Infection." Academic Press, New York.

Van der Zwet, T., *et al.* (1994). Evaluation of the MARYBLYT computer model for predicting blossom blight on apple in West Virginia and Maryland. *Plant Dis.* **78**, 225–230.

Verreet, J. A., and Hoffmann, G. M. (1990). A biologically oriented threshold decision model for control of epidemics of *Septoria nodorum* in wheat. *Plant Dis.* **74**, 731–738.

Verreet, J. A., Klink, H., and Hoffmann, G. M. (2000). Regional monitoring for disease prediction and optimization of plant protection measures: The IPM wheat model. *Plant Dis.* **84**, 816–826.

Ward, J. M. J., *et al.* (1999). Gray leaf spot: A disease of global importance in maize production. *Plant Dis.* **83**, 884–895.

Windels, C. E., *et al.* (1998). A Cercospora leaf spot model for sugar beet: In practice by an industry. *Plant Dis.* **82**, 716–726.

Wolf, P. F. J., and Verreet, J. A. (2002). An integrated pest management system in Germany for the control of fungal diseases in sugar beet: The IPM sugar beet model. *Plant Dis.* **86**, 336–344.

Wrather, J. A., *et al.* (1995). Soybean disease loss estimates for the southern United States, 1974–1994. *Plant Dis.* **79**, 1076–1079.

Zadoks, J. C. (1984). A quarter century of disease warning. *Plant Dis.* **68**, 352–355.

Zadoks, J. C. (2001). Plant disease epidemiology in the twentieth century: A picture by means of selected controversies. *Plant Dis.* **85**, 808–816.

Zadoks, J. C., and Schein, R. D. (1979). "Epidemiology and Plant Disease Management." Oxford Univ. Press, London.

chapter nine

CONTROL OF PLANT DISEASES

CONTROL METHODS THAT EXCLUDE THE PATHOGEN FROM THE HOST: QUARANTINES AND INSPECTIONS: CROP CERTIFICATION – EVASION OR AVOIDANCE OF PATHOGEN – USE OF PATHOGEN-FREE PROPAGATING MATERIAL: SEED – VEGETATIVE PROPAGATING MATERIALS EPIDERMAL COATINGS
295

CONTROL METHODS THAT ERADICATE OR REDUCE PATHOGEN INOCULUM – CULTURAL METHODS: HOST ERADICATION – CROP ROTATION – SANITATION CREATING UNFAVORABLE CONDITIONS – PLASTIC TRAPS AND MULCHES –
298

BIOLOGICAL METHODS: SUPPRESSIVE SOILS
303

ANTAGONISTIC MICROORGANISMS: FOR SOILBORNE PATHOGENS – FOR AERIAL PATHOGENS – MECHANISMS OF ACTION CONTROL THROUGH TRAP PLANTS – THROUGH ANTAGONISTIC PLANTS
305

PHYSICAL METHODS: CONTROL BY HEAT TREATMENT – SOIL STERILIZATION BY HEAT – SOIL SOLARIZATION – HOT-WATER – HOT-AIR – LIGHT WAVELENGTHS – DRYING – RADIATION – TRENCH BARRIERS
310

CHEMICAL METHODS: SOIL TREATMENT – FUMIGATION-DISINFESTATION OF WAREHOUSES – CONTROL OF INSECT VECTORS
312

DISEASE CONTROL BY IMMUNIZING, OR IMPROVING THE RESISTANCE OF, THE HOST – CROSS PROTECTION – INDUCED RESISTANCE: SYSTEMIC ACQUIRED RESISTANCE – PLANT DEFENSE ACTIVATORS – IMPROVING THE GROWING CONDITIONS – USE OF RESISTANT VARIETIES
314

CONTROL THROUGH USE OF TRANSGENIC PLANTS THAT: TOLERATE ABIOTIC STRESSES – CARRY SPECIFIC PLANT GENES FOR RESISTANCE – CARRY GENES CODING FOR ANTI-PATHOGEN COMPOUNDS – CARRY NUCLEIC ACIDS THAT LEAD TO PATHOGEN GENE SILENCING – CARRY COMBINATIONS OF RESISTANCE GENES – PRODUCE ANTIBODIES AGAINST THE PATHOGEN – USE TRANSGENIC BIOCONTROLS
319

DIRECT PROTECTION OF PLANTS FROM PATHOGENS – BIOLOGICAL CONTROLS: FUNGAL ANTAGONISTS OF: HETEROBASIDION (FOMES) ANNOSUM – CHESTNUT BLIGHT – SOILBORNE DISEASES – DISEASES OF AERIAL PLANT PARTS WITH FUNGI – POS THARVEST DISEASES BACTERIAL ANTAGONISTS OF: SOILBORNE DISEASES – DISEASES OF AERIAL PLANT PARTS WITH BACTERIA – POSTHARVEST DISEASES – BACTERIA-MEDIATED FROST INJURY – VIRAL PARASITES OF PLANT PATHOGENS
322

BIOLOGICAL CONTROL OF WEEDS
328

DIRECT PROTECTION BY CHEMICAL CONTROLS – METHODS OF APPLICATIONS: FOLIAGE SPRAYS AND DUSTS – SEED TREATMENT – SOIL TREATMENT – TREATMENT OF TREE WOUNDS – CONTROL OF POSTHARVEST DISEASES
329

TYPES OF CHEMICALS USED FOR PLANT DISEASE CONTROL: INORGANIC: SULFUR COMPOUNDS – CARBONATES – PHOSPHATES AND PHOSPHONATES – FILM-FORMING ORGANIC CHEMICALS: CONTRACT PROTECTIVE FUNGICIDES: DIHIOCARBAMATES-MISCELLANEOUS – SYSTEMIC FUNGICIDES: HETEROCYCLIC COMPOUNDS – ACYLALANINES – BENZIMIDAZOLES – OXANTHIINS – ORGANOPHOSPHATES – PYRIMIDINES – TRIZOLES – STROBILURINS OR QOI FUNGICIDES MISCELLANEOUS SYSTEMICS – MISCELLANEOUS ORGANICS – ANTIGIOTICS-OILS – ELECTROLYZED OXIDIZING WATER – GROWTH REGULATORS – NEMATICIDES: HOLOGENATED HYDROCARBONS-ORGANOPHOSPHATES – ISOTHIOCOYANATES – CARBAMATES – MISCELLANEOUS NEMATICIDES
338

MECHANISMS OF ACTION OF CHEMICALS USED TO CONTROL PLANT DISEASES – RESISTANCE OF PATHOGENS TO CHEMICALS – RESTRICTIONS ON CHEMICAL CONTROL OF PLANT DISEASES
345

INTEGRATED CONTROL OF PLANT DISEASES: IN A PERENNIAL CROP – IN AN ANNUAL CROP
348

In addition to being intellectually interesting and scientifically justified, the study of the symptoms, causes, and mechanisms of development of plant diseases has an extremely practical purpose: it allows for the development of methods to combat plant diseases. So, control increases the quantity and improves the quality of plant products available for use. Methods of control vary considerably from one disease to another, depending on the kind of pathogen, the host, the interaction of the two, and many other variables. In controlling diseases, plants are generally treated as populations rather than as individuals, although certain hosts (especially trees, ornamentals, and, sometimes, other virus-infected plants) may be treated individually. With the exception of trees, however, the damage or loss of one or a few plants is usually considered insignificant. Control measures are generally aimed at saving the populations rather than a few individual plants.

Most serious diseases of crop plants appear on a few plants in an area year after year, spread rapidly, and are difficult to cure after they have begun to develop. Therefore, almost all control methods are aimed at protecting plants from becoming diseased rather than at curing them after they have become diseased. Few infectious plant diseases can be controlled satisfactorily in the field by therapeutic means.

The various control methods can be classified as regulatory, cultural, biological, physical, and chemical, depending on the nature of the agents employed. **Regulatory control measures** aim at excluding a pathogen from a host or from a certain geographic area. Most **cultural control methods** aim at helping plants avoid contact with a pathogen, creating environmental conditions unfavorable to the pathogen or avoiding favorable ones, and eradicating or reducing the amount of a pathogen in a plant, a field, or an area. Most **biological** and some cultural **control methods** aim at improving the resistance of the host or favoring microorganisms antagonistic to the pathogen. A new type of biological control involves the transfer of genetic material (DNA) into plants and the generation of **transgenic plants** that exhibit resistance to a certain disease(s). Finally, **physical** and **chemical methods** aim at protecting the plants from pathogen inoculum that has arrived, or is likely to arrive, or curing an infection that is already in progress. Some recent (1995) and still mostly experimental chemicals operate by activating the defenses of the plant (systemic acquired resistance) against pathogens.

Epidemiological studies, in addition to elucidating the development of diseases in an area over time, can also help determine how effective various controls might be for a particular disease. In general, excluding or reducing the initial inoculum is most effective for the management of monocyclic pathogens. Controls such as crop rotation, removal of alternate hosts, and soil fumigation reduce the initial inoculum. With polycyclic pathogens, the initial inoculum can be multiplied many times during the growing season. Therefore, a reduction in the initial inoculum must usually be accompanied by another type of control measure (such as chemical protection or horizontal resistance) that also reduces the infection rate. Many controls, e.g., excluding a pathogen from an area, are useful for both monocyclic and polycyclic pathogens.

CONTROL METHODS THAT EXCLUDE THE PATHOGEN FROM THE HOST

As long as plants and pathogens can be kept away from one another, no disease will develop. Many plants are grown in areas of the world where certain pathogens are still absent. They are, therefore, free of the diseases caused by such pathogens.

To prevent the import and spread of plant pathogens into areas from which they are absent, national and state laws regulate the conditions under which certain crops susceptible to such pathogens may be grown and distributed between states and countries. Such regulatory control is applied by means of quarantines, inspections of plants in the field or warehouse, and occasionally by voluntary or compulsory eradication of certain host plants. Furthermore, plants are sometimes grown exclusively, especially for seed production, in areas from which a pathogen is largely or entirely excluded by unfavorable climatic conditions such as low rainfall and low relative humidity or by lack of vectors. This type of exclusion is called **avoidance** or **evasion**.

Quarantines and Inspections

When plant pathogens are introduced into an area in which host plants have been growing in the absence of the pathogen, such introduced pathogens may cause much more catastrophic epidemics than the existing endemic pathogens. This happens because plants that develop in the absence of a pathogen have no opportunity to select resistance factors specific against the pathogen and are, therefore, unprotected and extremely vulnerable to attack. Also, no microorganisms antagonistic or competing with the pathogen are likely to be present, while, on the other hand, the pathogen finds a large amount of available susceptible tissue on which it can feast and multiply unchecked. Some of the worst plant disease epidemics, e.g., the downy mildew of grapes in Europe and the bacterial canker of citrus, chestnut blight, Dutch elm disease, and soybean cyst nematode in the United States, are all diseases caused by pathogens that were introduced from abroad. It has been estimated, for example, that if soybean rust were introduced into the United States it would result in losses to consumers and other sectors of the U.S. economy of several billion dollars per year. Numerous other pathogens exist in many parts of the world but not yet in the United States, and they would most likely cause severe diseases to crops and great economic losses if they were to enter the country.

To keep out foreign plant pathogens and to protect U.S. farms, gardens, and forests, the Plant Quarantine Act of 1912 was passed by Congress. This act prohibits or restricts entry into or passage through the United States from foreign countries of plants, plant products, soil, and other materials carrying or likely to carry plant pathogens not known to be established in this country. Similar quarantine regulations exist in most other countries. Because plant scientists, plant breeders, and agricultural industries need to bring into the country plant germplasm on a more or less continuing basis, a National Plant Germplasm Quarantine Center has been established in Glendale, Maryland, near Washington, DC, where all introductions are kept and tested for certain pathogens for 1 to 4 years before they are released.

Experienced inspectors stationed at all points of entry into the country enforce quarantines of produce likely to introduce new pathogens. Plant quarantines are already credited for the interception of numerous foreign plant pathogens and, thereby, with saving the country's plant world from potentially catastrophic diseases. Plant quarantines are considerably less than foolproof, however, because pathogens may be introduced in the form of spores or eggs on unsuspected carriers, and latent infections of seeds and other plant propagative organs may exist even after treatment. Various steps taken by plant quarantine stations, such as growing plants under observation for certain times before they are released to the importer, repeated serological tests of seed lots (mostly through ELISA), nucleic acid tests involving DNA probes and polymerase chain reaction (PCR) amplification of specific pathogen DNA sequences, and inspection of imported nursery stock in the grower's premises, tend to reduce the chances of introduction of harmful pathogens. With the annual imports of flower bulbs from Holland, U.S. quarantine inspectors may visit the flower fields in Holland and inspect them for certain diseases. If they find the field to be free of these diseases, they issue inspection certificates allowing the import of such bulbs into the United States without further tests.

Similar quarantine regulations govern the interstate, and even intrastate, sale of nursery stock, tubers, bulbs, seeds, and other propagative organs, especially of certain crops such as potatoes and fruit trees. The movement and sale of such materials within and between states are controlled by the regulatory agencies of each state.

Crop Certification

Several voluntary or compulsory inspection systems are in effect in various states in which appreciable amounts of nursery stock and potato seed tubers are produced. Growers interested in producing and selling disease-free plants submit to a voluntary inspection or indexing of their crop in the field and in storage by the state regulatory agency, experiment station personnel, or others. If, after certain procedures recommended by the inspecting agency are carried out, the plant material is found to be free of certain, usually virus, diseases, the inspecting agency issues a certificate indicating that the plants are free from these specific diseases, and the grower may then advertise and sell the plant material as disease free — at least from the diseases for which it was tested.

Evasion or Avoidance of Pathogen

For several plant diseases, control depends largely on attempts to evade pathogens. For example, bean anthracnose, caused by the fungus *Colletotrichum lindemuthianum*, and the bacterial blights of bean, caused by the bacteria *Xanthomonas phaseoli* and *Pseudomonas phaseolicola*, are transmitted through the seed. In most areas where beans are grown, at least a portion of the plants and the seeds become infected with these pathogens. However, in the dry, irrigated regions of the western United States, the conditions of low humidity are unsuitable for these pathogens and therefore the plants and their seeds are more likely to be free of them. Using western-grown seeds free of these pathogens is the main recommendation for control of these diseases. Similarly, to produce potato seed tubers free of viruses, potatoes are grown in remote locations in the cooler, northern states (Maine, Wisconsin, Idaho, and others) and at higher elevations, where aphids, the vectors of these viruses, are absent or their populations are small and can be controlled.

In many cases, a susceptible crop is planted at a great enough distance from other fields containing possibly diseased plants so that the pathogen would not likely infect the crop. This type of **crop isolation** is practiced mostly with perennial plants, such as peach orchards isolated from chokecherry shrubs or trees infected with the X-disease phytoplasma. Also, during much of the 20th century, banana production in Central America depended on evading the fungus *Fusarium oxysporum* f. *cubense*, the cause of fusarium wilt (Panama disease) of banana, by moving on to new, previously uncultivated fields as soon as older banana fields became infested with *Fusarium* and yields became unprofitable.

Growers carry out numerous activities aimed at helping the host evade the pathogen. Such activities include using vigorous seed, selecting proper (early or late) planting dates and proper sites, maintaining proper distances between fields and between rows and plants, planting wind break or trap crops, planting in well-drained soil, and using proper insect and weed control. All these practices increase the chances that the host will remain free of the pathogen or at least that it will go through its most susceptible stage before the pathogen reaches the host.

Use of Pathogen-Free Propagating Material

When a pathogen is excluded from the propagating material (seed, tubers, bulbs, nursery stock) of a host, it is often possible to grow the host free of that pathogen for the rest of its life. Examples are woody plants affected by nonvectored viruses. In most crops, if the host can be grown free of the pathogen for a considerable period of its early life, during which the plant can attain normal growth, it can then produce a fairly good yield despite a potential later infection. Examples are crops affected by vectored viruses and phytoplasmas and by fungal, bacterial, and nematode pathogens.

There is absolutely no question that every host plant and every crop grow better and produce a greater yield if the starting propagating material is free of pathogens, or at least free of the most important pathogens. For this reason, every effort should be made to obtain and use pathogen-free seed or nursery stock, even if the cost is considerably greater than for propagating material of unknown pathogen content.

All types of pathogens can be carried in or on propagating material. True seed, however, is invaded by relatively few pathogens, although several may contaminate its surface. Seed may carry internally one of a few fungi (such as those causing anthracnoses and smuts), certain bacteria causing bacterial wilts, spots, and blights, and one of several viruses (tobacco ring spot in soybean, bean common mosaic, lettuce mosaic, barley stripe mosaic, squash mosaic, and prunus necrotic ring spot). However, vegetatively propagated material such as buds, grafts, rootstocks, tubers, bulbs, corms, cuttings, and rhizomes are expected to carry internally almost every virus, viroid, phytoplasma, protozoon, and vascular fungus or bacterium present systemically in the mother plant, in addition to any fungi, bacteria, and nematodes that may be carried on these organs externally. Some nematodes may also be carried internally in some belowground propagating organs (tubers, bulbs, corms, and rhizomes) and in or on the roots of nursery stock.

Pathogen-Free Seed

Seed that is free of fungal, bacterial, and some viral pathogens is usually obtained by growing the crop and producing the seed in (1) an area free of or isolated from the pathogen, (2) an area not suitable for the pathogen (e.g., the arid western regions of the United States where bean seed is produced usually free of anthracnose and bacterial blights), or (3) an area not suitable for the vector of the pathogen (e.g., the northern or high-altitude fields where aphids, the vectors of many viruses, are absent or rare).

It is very important, and with seed-transmitted and aphid-borne viruses it is indispensable, that seed be essentially free of the pathogen, especially virus. Because, if carried in the seed, the pathogen will be present in the field at the beginning of the growth season, and even a small proportion of infected seeds is sufficient to provide enough inoculum to spread and infect many plants early, thus causing severe losses. It has been shown, for example, that to control lettuce mosaic virus, only seed lots that contain less than one infected seed per 30,000 lettuce seeds must be used. For this purpose, seed companies have their lettuce seed tested for lettuce mosaic virus every year. In past years, seeds were tested (indexed) by growing out hundreds of thousands of lettuce seedlings in insect-proof greenhouses, observing them over several weeks for lettuce mosaic symptoms, and attempting to transmit the virus from suspect plants to healthy plants. Later, indexing was done by inoculating a local lesion indicator plant (in this case *Chenopodium quinoa*) with sap from ground samples of groups of seeds and observing it for virus symptoms. Since the 1980s, testing for lettuce mosaic virus in seed is done with serological techniques, particularly with ELISA, which is faster, more sensitive, and less expensive than the other methods.

Testing seed for fungal and bacterial pathogens is done by symptomatology, microscopically, and by culturing the pathogen on general or selective nutrient media. For detection and identification of bacteria, serological tests are also being used with increasing frequency and accuracy.

In the 1990s, the sensitivity and accuracy of detection and identification of all types of pathogens was increased greatly by the use of techniques employing the polymerase chain reaction. PCR allows amplification of minute amounts of pathogen DNA in the sample by using DNA primers specific to the particular pathogen. The amplified DNA is then easier to detect by the various nucleic acid tests. If seed free of fungal and bacterial pathogens cannot be obtained by other means, fermentation or hot water (50°C) treatment of the seed can free it from the pathogen. An example of fermentation treatment is tomato seed freed from *Xanthomonas campestris* pv. *vesicatoria*, the cause of bacterial spot of tomato. Hot water treatments are used to free cabbage seed from *Xanthomonas campestris* pv. *campestris*, the cause of black rot of cabbage, and from *Leptosphaeria maculans* (*Phoma lingam*), the cause of black leg of cabbage. Also, hot water treatment frees seed of wheat and other cereals from *Ustilago* sp., the cause of loose smuts of cereals.

Pathogen-Free Vegetative Propagating Materials

Vegetative propagating material free of pathogens that are distributed systemically throughout the plant (viruses, viroids, mollicutes, fastidious bacteria, and some wilt-inducing fungi and bacteria) is obtained from mother plants that had been tested and shown to be free of the particular pathogen or pathogens. To ensure continuous production of pathogen-free buds, grafts, cuttings, rootstocks, and runners of trees, vines, and other perennials, the mother plant is indexed for the particular pathogen at regular (1- to 2-year) intervals. Indexing is usually done by taking grafts or sap from the plant and inoculating susceptible indicator plants to observe possible symptom development. Furthermore, the new plants must be grown in pathogen- and vector-free soil and then be protected from airborne vectors of the pathogen if they are to remain free of the pathogen for a considerable time. Indexing of mother plants for viruses (and some mollicutes) is now done in several states for most pome, stone, and citrus fruits, as well as for grapes, strawberries, raspberries, and several ornamentals, such as roses and chrysanthemums. Some viruses are now indexed by serological (ELISA) or nucleic acid tests rather than via bioassay. It is anticipated that before too long most perennial plants will be produced from pathogen-free propagating material, many of them through tissue culture.

For certain crops, such as potato, complex certification programs have evolved to produce pathogen-free seed potatoes. In every U.S. state where seed potatoes are produced, they must meet a slightly varying maximum allowable tolerance for various diseases (Table 9-1). The initial mother plants that test free of these diseases are propagated for a few years by state agencies in isolated farms, usually at high altitudes, where aphids are absent or rare. The plants and the tubers are inspected and tested repeatedly each season to ensure continued freedom from each pathogen. When enough pathogen-free seed potatoes are produced, they are turned over to commercial seed potato producers, who further multiply them and finally sell them to farmers. While in the fields of the commercial seed producers, the potato plants are inspected repeatedly, infected plants are rogued, and insect vectors are controlled. For the seed to be "certified," the plants in the field must show disease levels no higher than those allowed by the particular state (Table 9-1). In several certification programs, samples of the harvested tubers

TABLE 9-1
Maximum Tolerances for Diseases in Certified Seed Potatoes Allowed in Various States

Disease	Tolerance levels allowed (%)
Leafroll virus	0.5–1
Mosaic viruses	1–2
Spindle tuber viroid	0.1–2
Total virus content	0.5–3
Fusarium and/or *Verticillium* wilt	1–5
Ring rot (*Corynebacterium sepedonicum*)	0
Root knot (*Meloidogyne* sp.)	0–0.1
Late blight (*Phytophthora infestans*)	0

are sent to a southern state, where they are grown during the winter and checked further for symptoms. In some states, serological tests (ELISA) or nucleic acid tests are now replacing some of the bioassays.

With some crops, such as carnation and chrysanthemum, greenhouse growers need cuttings free of the vascular wilt-causing fungi *Fusarium* and *Verticillium* each time, but it is almost impossible to keep these two fungi from the production beds. It was noted early, however, that short cuttings taken from the tips of rapidly growing shoots were usually free of either of these fungi, and this became a common practice to control these diseases.

Sometimes it is impossible to find even a single plant of a variety that is free of a particular pathogen, especially viruses. In that case, one or a few healthy plants are initially obtained by tissue culture of the upper millimeter or so of the growing meristematic tip of the plant, which most viruses do not invade.

In some cases, healthy plants can be obtained from virus-infected plants by eliminating the virus through heat treatment. Dormant plant material, such as budwood, dormant nursery trees, and tubers, is usually treated with hot water at temperatures ranging from 35 to 54°C, with treatment times lasting from a few minutes to several hours. Actively growing plants are sometimes placed in growth chambers and treated with hot air, which allows better survival of the plant and more likely elimination of the pathogen than hot water. Temperatures of 35 to 40°C seem to be optimal for air treatment of growing plants. For hot-air treatment, the infected plants are usually grown in growth chambers for varying periods, generally lasting 2 to 4 weeks. Some viruses, however, require treatment for 2 to 8 months, whereas others may be eliminated in just one week. All mollicutes, all fastidious bacteria, and many viruses can be eliminated from their hosts by heat treatment, but for some viruses, such treatment has not always been dependable.

With crops such as strawberries and orchids, once one or a few pathogen-free plants have been obtained by any of the aforementioned methods, they are subsequently used as foundation material from which thousands, hundreds of thousands, and even millions of pathogen-free plants are produced through tissue culture techniques in the laboratory. These plants are later set out in the greenhouse or the field before they are sold to growers or retailers as pathogen-free plants at a premium price. This method of production of pathogen-free plants is now used even with nonsystemic bacterial and fungal pathogens, e.g., for managing strawberry anthracnose crown rot caused by the fungi *Colletotrichum fragariae* and *C. acutatum*.

Exclusion of Pathogens from Plant Surfaces by Epidermal Coatings

Successful results at controlling diseases of aboveground parts of plants have been obtained in experiments in which the plants were sprayed with compounds that form a continuous film or membrane on the plant surface and inhibit contact of the pathogen with the host and penetration of the host. Such a high-quality lipid membrane forms, for example, when plants are sprayed with a water emulsion of dodecyl alcohol. The membrane permits diffusion of oxygen and carbon dioxide but not of water. The membrane is not washed off easily by rain and remains intact for about 15 days. The film, therefore, being antitranspirant, conserves water and increases yields. It also protects plants such as cucumber, tomato, beets, wheat, and rice from several diseases such as powdery mildews and leaf and stem blights. So far, however, epidermal coatings have not been used for the commercial control of plant diseases. Similarly, kaolin-based films have proven effective in protecting apple shoots from becoming infected with the bacterial disease fire blight, caused by *Erwinia amylovora*, and apple fruit from powdery mildew, caused by the fungus *Podosphaera leucotricha* (Fig. 9-1). It also significantly protects grapevines from becoming infected with Pierce's disease, caused by the bacterium *Xylella fastidiosa*, by interfering with its transmission by the vector glassy winged sharpshooter, *Homalodisca coagulata*.

CONTROL METHODS THAT ERADICATE OR REDUCE PATHOGEN INOCULUM

Many different types of control methods aim at eradicating or reducing the amount of pathogen present in an area, a plant, or plant parts (such as seeds). Many such methods are cultural, i.e., they depend primarily on

FIGURE 9-1 Protection of apple shoots and fruit from, respectively, fire blight, caused by the bacterium *Erwinia amylovora*, and powdery mildew, caused by the fungus *Podosphaera leucotricha*, through exclusion of the pathogens from the apple tissues by kaolin-based particle films. (A, left) Untreated apple shoot inoculated with fire blight bacteria became infected and developed fire blight, while the shoot at right, which was treated with the kaolin-based film before inoculation, remained healthy. (B) Apple fruit treated with the film preparation remained free of powdery mildew, as shown by the fruit area from which the film was removed for observation. (C) Typical severe powdery mildew infection on untreated apple. (D) Apple protected from powdery mildew with fungicide sprays. (Photographs courtesy of D. M. Glenn, from Glenn *et al.*, Plant Health Progress, 2001.)

certain actions of the grower, such as host eradication, crop rotation, sanitation, improving plant growing conditions, creating conditions unfavorable to pathogens, polyethylene mulching, trickle irrigation, ecofallow, and, sometimes, reduced tillage farming. Some methods are physical, i.e., they depend on a physical factor such as heat or cold. Examples are soil sterilization, heat treatment of plant organs, refrigeration, and radiations. Several methods are chemical, i.e., they depend on the use and action of a chemical substance to reduce the pathogen. Examples are soil treatment, soil fumigation, and seed treatment with chemicals. Some methods are biological, i.e., they use living organisms to reduce the pathogen inoculum. Examples are the use of trap crops and antagonistic plants against nematodes, use of amendments that favor microflora antagonistic to the pathogen, and use of antagonistic microorganisms. The latter apparently inhibit the growth of the pathogen by producing antibiotics, by attacking and parasitizing the pathogen directly, or by competing for sites on the plant.

Cultural Methods That Eradicate or Reduce the Inoculum

Host Eradication

When a pathogen has been introduced into a new area despite a quarantine, a plant disease epidemic frequently follows. To prevent such an epidemic, all the host plants infected by or suspected of harboring the pathogen may have to be removed and burned. This eliminates the pathogen and prevents greater losses from the spread of the pathogen to additional plants. Beginning in 1915, this type of host eradication controlled the bacterial canker of citrus in Florida and other southern states, where more than three million trees had to be destroyed. Another outbreak of citrus canker occurred in Florida in 1984, and, by 1992, the disease was apparently brought under control through the painful destruction of millions of nursery and orchard trees in that state. In 1995, citrus canker was again found in Florida, but only on trees in a residential area of Miami. Immediately, an area of approximately 100 square miles was placed under quarantine, and eradication of all infected and all nearby trees, mostly in home gardens or yards, was undertaken; the disease, however, has continued to spread among trees in nearby cities and towns and its eradication has become extremely difficult, if not impossible. In a different disease, since the 1970s, a campaign to contain and eradicate witchweed (*Striga asiatica*) in the eastern Carolinas in the United States has been successful. However, attempts by several European countries to eradicate fire blight of apple and pear (caused by the bacterium *Erwinia amylovora*) and plum pox virus of stone fruits, of the United States to eradicate plum pox virus, and attempts by several South American countries to eradicate coffee rust (caused by the fungus *Hemileia vastatrix*) have not been successful, and the pathogens continue to spread. Host eradication (roguing) is also carried out routinely in many nurseries, greenhouses, and fields to prevent the spread of numerous diseases by eliminating infected plants that provide a ready source of inoculum within the crop.

Certain pathogens of annual crops, e.g., cucumber mosaic virus, overwinter only or mainly in perennials, usually wild plants. Eradication of the host in which the pathogen overwinters is sometimes enough to eliminate completely or to reduce drastically the amount of inoculum that can cause infections the following season. In some crops, such as potatoes, pathogens of all types may overwinter in infected tubers that are left in the field. Many such tubers produce infected plants in the spring that allow the pathogen to come above ground, from where it can be spread further by insects, rain, and wind. Eradication of such volunteer plants helps greatly to reduce the inoculum of these pathogens. Also, in warmer areas, volunteer plants of a crop, e.g., tomato, grow during periods between plantings of the crop. Such volunteers become infected by various pathogens, e.g., tomato mottle and tomato yellow leaf curl viruses, during the crop-free season and serve as reservoirs for the pathogens that are again spread into and cause disease once the cultivated crop is planted.

Some pathogens require two alternate hosts to complete their full life cycles. For example, *Puccinia graminis tritici* requires wheat and barberry, *Cronartium ribicola* requires pine and currant (*Ribes*), and *Gymnosporangium juniperi-virginianae* requires cedar and apple. In these cases, eradication of the wild or economically less important alternate host interrupts the life cycle of the pathogen and leads to control of the disease. This has been carried out somewhat successfully with stem rust of wheat and white pine blister rust through eradication of barberry and currant, respectively. However, due to other factors, both diseases are still widespread and often cause severe losses. In cases like cedar-apple rust, in which both hosts may be important, control through eradication of the alternate host is impractical.

Crop Rotation

Soilborne pathogens that infect plants of one or a few species or even families of plants can sometimes be reduced in the soil by planting, for 3 or 4 years, crops belonging to species or families not attacked by the particular pathogen. Satisfactory control through crop rotation is possible with pathogens that are **soil invaders**, i.e., survive only on living plants or only as long as the host residue persists as a substrate for their saprophytic existence. When the pathogen is a **soil inhabitant**, however, i.e., produces long-lived spores or can live as a saprophyte for more than 5 or 6 years, crop rotation becomes less effective or impractical. In the latter cases, crop rotation can still reduce populations of the pathogen in the soil (e.g., *Verticillium*) (Fig. 9-2), and appreciable yields from the susceptible crop can be obtained every third or fourth year of the rotation.

In some cropping systems the field is tilled and left fallow for a year or part of the year. During fallow, debris and inoculum are destroyed by microorganisms with little or no replacement. In areas with hot summers, fallow allows greater heating and drying of the soil, which leads to a marked reduction of nematodes and some other pathogens. Other cropping systems utilize herbicides and reduced tillage and fallow (ecofallow). In some such systems, certain diseases, e.g., stalk rot of grain sorghum and corn, caused by *Fusarium moniliforme*, have been reduced dramatically. In contrast, other diseases, such as Septoria leaf blotch of wheat and wheat and barley scab, have been increased.

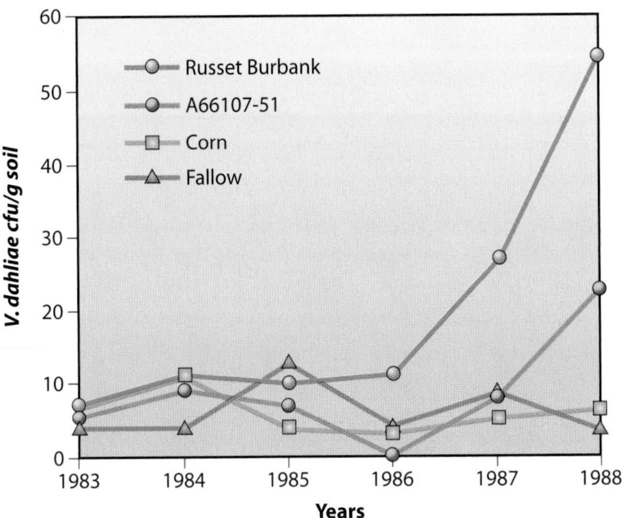

FIGURE 9-2 Effect of continuous cropping, fallow, and 5-year corn crop rotations on the development of *Verticillium dahliae* populations in a susceptible (Russet Burbank) and a resistant potato variety (A66107-51). cfu, colony-forming units. [From Davis *et al.* (1994). *Phytopathology* 84, 207–214.]

Sanitation

Sanitation consists of all activities aimed at eliminating or reducing the amount of inoculum present in a plant, a field, or a warehouse and at preventing the spread of the pathogen to other healthy plants and plant products. Thus, plowing under infected plants after harvest, such as leftover infected fruit, stems, tubers, or leaves, helps cover the inoculum with soil and speeds up its disintegration (rotting) and concurrent destruction of most pathogens carried in or on them. Similarly, removing infected leaves of house or garden plants helps remove or reduce the inoculum. Pruning infected plants (Fig. 9-3A) or infected or dead branches and then removing infected fruit and any other plant debris that may harbor the pathogen help reduce the inoculum and do not allow the pathogen to grow into still healthy parts of the tree. Such actions reduce the amount of disease that will develop later. In some parts of the world, infected crop debris of grass seed and rice crops is destroyed by burning, which reduces or eliminates the surface inoculum of several pathogens.

By washing their hands before handling certain kinds of plants, such as tomatoes, workers who smoke may reduce the spread of *tobacco mosaic virus*. Also, frequently disinfesting knives used to cut propagative stock, such as potato tubers, and disinfesting pruning shears between trees reduce the spread of pathogens through such tools. Washing the soil off farm equipment before moving it from one field to another may also help prevent the spread of any pathogens present in the soil. Similarly, by washing, often with chlorinated water, the produce (Fig. 9-3B), its containers, and the walls of storage houses, the amount of inoculum and subsequent infections may be reduced considerably.

FIGURE 9-3 Control of plant diseases through sanitation. (A) Pruning, bagging, and removal of apple and pear nursery trees infected with fireblight. (B) Washing harvested tomatoes with chlorinated water. [Photograph courtesy of (A) P. S. McManus, from P. S. McManus and V. O. Stockwell, Plant Health Progress, 2001.]

Creating Conditions Unfavorable to the Pathogen

Stored products should be aerated properly to hasten the drying of their surfaces and inhibit germination and infection by any fungal or bacterial pathogens present on them. Similarly, spacing plants properly in the field or greenhouse prevents the creation of high-humidity conditions on plant surfaces and inhibits infection by certain pathogens, such as *Botrytis* and *Peronospora tabacina*. Good soil drainage also reduces the number and activity of certain oomycete pathogens (e.g., *Pythium*) and nematodes and may result in significant disease control. The appropriate choice of fertilizers or soil amendments may also lead to changes in the soil pH, which may unfavorably influence the development of the pathogen. Flooding fields for long periods or dry fallowing may also reduce the number of certain pathogens in the soil (e.g., *Fusarium*, *Sclerotinia sclerotiorum*, and nematodes) by inducing starvation, lack of oxygen, or desiccation.

In the production of many crops, particularly containerized nursery stock, using composted tree bark in the planting medium has resulted in the successful control of diseases caused by several soilborne pathogens, e.g., *Phytophthora*, *Pythium*, and *Thielaviopsis* root rots, *Rhizoctonia* damping-off and crown rot, *Fusarium* wilt, and some nematode diseases of several crops, especially of Easter lily, poinsettia, and rhododendron. Part of the suppressive effect is apparently a result of the release from the bark of certain substances that exhibit direct fugicidal activity; additional suppression is exerted by other substances that promote the growth and activity of other microorganisms that compete with or are antagonistic to the plant pathogens (Figs. 9-4 and 9-5).

Polyethylene Traps and Mulches

Many plant viruses, such as cucumber mosaic virus, are brought into crops, such as peppers, by airborne aphid vectors. When vertical, sticky, yellow polyethylene sheets are erected along the edges of susceptible crops, a considerable number of aphids are attracted to and stick to the plastic. This is done primarily to trap and monitor incoming insects, but to some extent it also reduces the amount of virus inoculum reaching the crop. However, if reflectant aluminum or black, whitish-gray, or colored polyethylene sheets are used as mulches between the plants or rows in the field, incoming aphids, thrips, and possibly other insect vectors are repelled and misled away from the field. As a result, fewer virus-carrying vectors land on the plants and fewer plants become infected with the virus (Fig. 9-6). Reflectant mulches, however, cease to function as soon as the crop canopy covers them.

FIGURE 9-4 Biological control of *Pythium* root rot in poinsettia plants planted in potting mixes and inoculated with *Pythium ultimum*. (A) Aboveground appearance of plants grown in nonsuppressive, slightly decomposed dark-colored sphagnum peat mix (left) and in *Pythium*-suppressive, almost undecomposed light-colored sphagnum peat mix (right). (B) Root rot severity of poinsettia plants planted as in (A) left (top), as in (A) right (middle), and in a blend of composted pine bark and dark sphagnum peat (bottom). (C) Aerial view of a yard waste composting plant. [Photographs courtesy of H. A. J. Hoitink; photographs B and C are from Hoitink *et al.* (1991).]

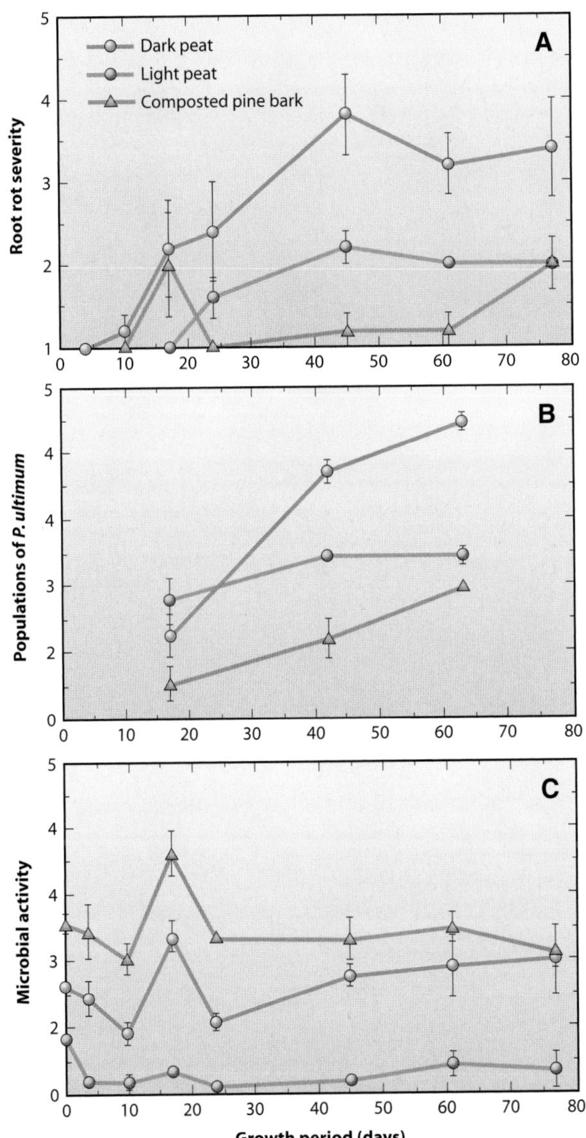

FIGURE 9-5 Effect of kind of potting mix on (A) *Pythium* root rot severity in potted poinsettias, (B) size of *Pythium ultimum* populations, and (C) level of microbial activity in the potting mix. Composted pine bark resulted in the least root rot, lowest *Pythium* populations, and greatest microbial activity. [From Boehm, and Hoitink (1992). *Phytopathology* 82, 259–264.]

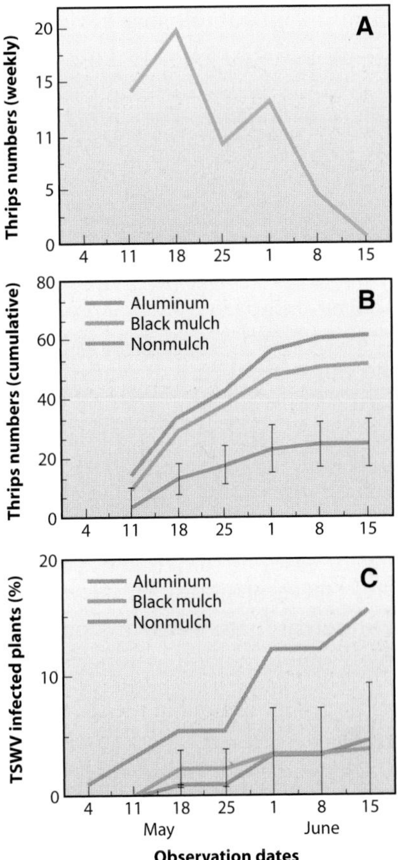

FIGURE 9-6 Relationship of (A) thrips influx (per square inch), (B) cumulative thrips numbers, and (C) percentage of tomato spotted wilt virus-infected plants as affected by aluminum and by black mulch. [From Greenough, Black, and Bond (1990). *Plant Dis.* 74, 805–808.]

Biological Methods That Eradicate or Reduce the Inoculum

Biological control of pathogens, i.e., the total or partial destruction of pathogen populations by other organisms, occurs routinely in nature. There are, for example, several diseases in which the pathogen cannot develop in certain areas either because the soil, called **suppressive soil**, contains microorganisms antagonistic to the pathogen or because the plant that is attacked by a pathogen has also been inoculated naturally with antagonistic microorganisms before or after the pathogen attack. Sometimes, the antagonistic microorganisms may consist of avirulent strains of the same pathogen that destroy or inhibit the development of the pathogen, as happens in **hypovirulence** and **cross protection.** In some cases, even higher plants reduce the amount of inoculum either by trapping available pathogens (trap plants) or by releasing into the soil substances toxic to the pathogen. Agriculturalists have increased their efforts to take advantage of such natural biological antagonisms and to develop strategies by which biological control can be used effectively against several plant diseases. Biological antagonisms, although subject to numerous ecological limitations, are expected to become an important part of the control measures employed against many more diseases.

Although the aforementioned measures attempt to control plant pathogens through the use of other microorganisms, plant pathologists have also been using specialized plant pathogens for the biological control of weeds, both terrestrial and aquatic. Biological control of weeds through pathogens that infect, damage, and sometimes kill weeds is a very promising area of plant pathology.

Suppressive Soils

Several soilborne pathogens, such as *Fusarium oxysporum* (the cause of vascular wilts), *Gaeumannomyces graminis* (the cause of take-all of wheat), *Phytophthora cinnamomi* (the cause of root rots of many fruit and forest trees), *Pythium* spp. (a cause of damping-off), and *Heterodera avenae* (the oat cyst nematode), develop well and cause severe diseases in some soils, known as **conducive soils**, whereas they develop much less and cause much milder diseases in other soils, known as **suppressive soils**. The mechanisms by which soils are suppressive to different pathogens are not always clear but may involve biotic and/or abiotic factors and may vary with the pathogen. In most cases, however, it appears that they operate primarily by the presence in such soils of one or several microorganisms antagonistic to the pathogen. Such antagonists, through the antibiotics they produce, through lytic enzymes, through competition for food, or through direct parasitizing of the pathogen, do not allow the pathogen to reach high enough populations to cause severe disease.

Numerous kinds of antagonistic microorganisms have been found to increase in suppressive soils; most commonly, however, pathogen and disease suppression has been shown to be caused by fungi, such as *Trichoderma*, *Penicillium*, and *Sporidesmium*, or by bacteria of the genera *Pseudomonas*, *Bacillus*, and *Streptomyces*. Suppressive soil added to conducive soil can reduce the amount of disease by introducing microorganisms antagonistic to the pathogen. For example, soil amended with soil containing a strain of a *Streptomyces* species antagonistic to *Streptomyces scabies*, the cause of potato scab, resulted in potato tubers significantly free from potato scab (Fig. 9-7A). Suppressive, virgin soil has been used, for example, to control *Phytophthora* root rot of papaya by planting papaya seedlings in suppressive soil placed in holes in the orchard soil, which was infested with the root rot oomycete *Phytophthora palmivora*. However, in several diseases, continuous cultivation (monoculture) of the same crop in a conducive soil, after some years of severe disease, eventually leads to reduction in disease through increased populations of microorganisms antagonistic to the pathogen. For example, continuous cultivation of wheat or cucumber

FIGURE 9-7 (A) Biological control of potato scab caused by the bacterium *Streptomyces scabies* with a suppressive strain of another *Streptomyces* species. Tubers at left were harvested from soil treated with the biocontrol agent; tubers at right were harvested from soil not amended with the biocontrol agent (B,C). Minimal incidence of lettuce drop, caused by the fungus *Sclerotinia sclerotiorum*, in a field (B) in which broccoli residue had been plowed under the previous year compared to extensive lettuce drop in a field (C) in which no broccoli residue had been incorporated in the soil. [Photographs courtesy of (A) L. Kinkel, University of Minnesota and (B and C) K. V. Subbarao (1998). *Plant Dis.* **82**, 1068–1078.]

leads to reduction of take-all of wheat and of *Rhizoctonia* damping-off of cucumber, respectively. Similarly, continuous cropping of the watermelon variety 'Crimson Sweet' allows the buildup of antagonistic species of *Fusarium* related to that causing *Fusarium* wilt of watermelon with the result that *Fusarium* wilt is reduced rather than increased. Such soils are suppressive to future disease development. That suppressiveness is due to antagonistic microflora can be shown by pasteurization of the soil at 60°C for 30 minutes, which completely eliminates the suppressiveness.

A sort of "soil suppressiveness" develops after appropriate crops are plowed under as soil amendments. Such crops, usually in the crucifer family, provide material and the time required for biological destruction of pathogen inoculum by resident antagonists in the soil. For example, significant control of lettuce drop, caused by the fungus *Sclerotinia sclerotiorum*, occurs when broccoli plants have been incorporated in the soil compared to the amount of disease in fields not receiving such treatment (Figs. 9-7B and 9-7C).

Reducing Amount of Pathogen Inoculum through Antagonistic Microorganisms

Soilborne Pathogens

The mycelium and resting spores (oospores) or sclerotia of several phytopathogenic soil oomycetes and fungi such as *Pythium*, *Phytophthora*, *Rhizoctonia*, *Sclerotinia*, and *Sclerotium* are invaded and parasitized (**mycoparasitism**) or are lysed (**mycolysis**) by several fungi, which as a rule are not pathogenic to plants. Several nonplant pathogenic oomycetes and fungi, including some chytridiomycetes and hyphomycetes, and some pseudomonad and actinomycetous bacteria infect the resting spores of several plant pathogenic fungi. Among the most common mycoparasitic fungi are *Trichoderma* sp., mainly *T. harzianum*. The latter fungus has been shown to parasitize mycelia of *Rhizoctonia* (Fig. 9-8) and *Sclerotium*, to inhibit the growth of many oomycetes such as *Pythium*, *Phytophthora*, and other fungi, e.g., *Fusarium* and *Heterobasidion* (*Fomes*), and

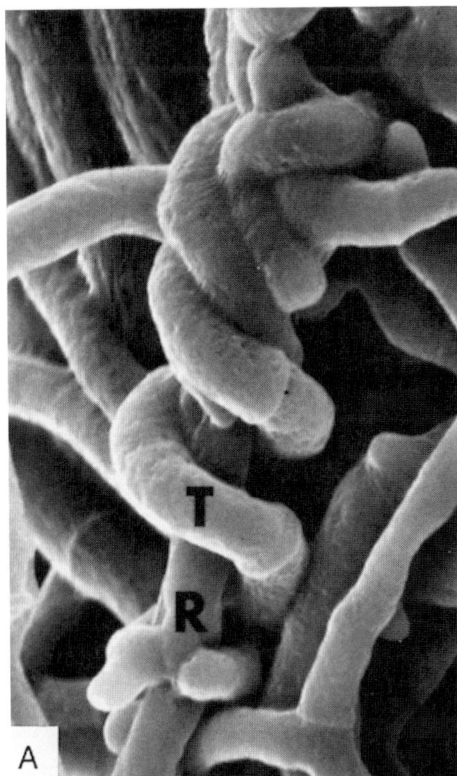

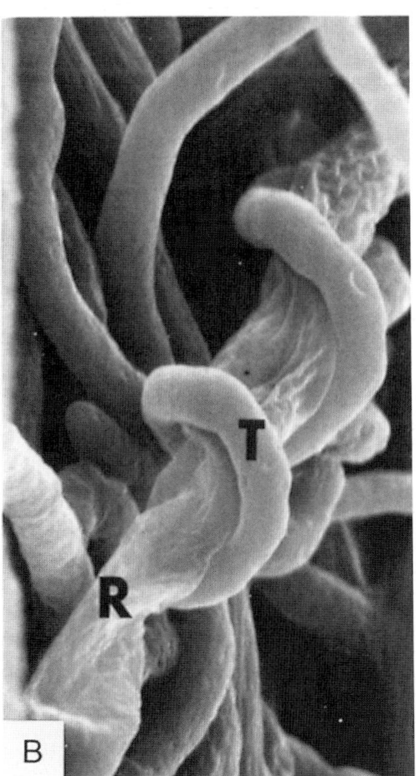

FIGURE 9-8 Effect of the biological control fungus *Trichoderma harzianum* on the plant pathogenic fungus *Rhizoctonia solani*. (A) Hyphae of *Trichoderma* (T) form dense coils and tightly encircle hyphae of *Rhizoctonia* (R) within 2 days after inoculation. (Magnification: 6000×.) (B) By 6 days after inoculation, *Rhizoctonia* hyphae show loss of turgor and marked cell collapse, whereas *Trichoderma* hyphae continue to look normal. (Magnification: 5000×.) [From Benhamou and Chet (1993). *Phytopathology* **83**, 1062–1071.]

to reduce the diseases caused by most of these pathogens. Other common mycoparasitic fungi are *Laetisaria arvalis* (*Corticium* sp.), a mycoparasite and antagonist of *Rhizoctonia* and *Pythium*; also, *Sporidesmium sclerotivorum*, *Gliocladium virens*, and *Coniothyrium minitants*, all destructive parasites and antagonists of *Sclerotinia sclerotiorum* and all effectively controlling several of the *Sclerotinia* diseases; and *Talaromyces flavus*, which parasitizes *Verticillium* and controls *Verticillium* wilt of eggplant. Also, some *Pythium* species parasitize species of *Phytophthora* (Fig. 9-9) and other species of *Pythium*. Several yeasts, e.g., *Pichia gulliermondii*, also parasitize and inhibit the growth of plant pathogenic fungi such as *Botrytis* and *Penicillium* (Fig. 9-10).

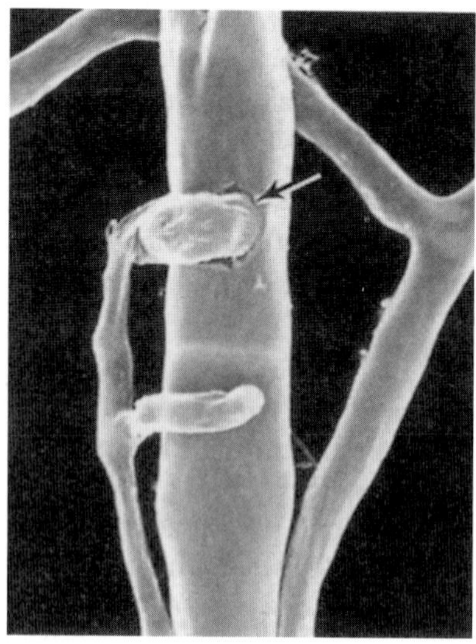

FIGURE 9-9 Hypha of a nonpathogenic species of *Pythium* (*P. nunn*) penetrating (arrow) a hypha of the pathogenic fungus *Phytophthora*. (Photograph courtesy of R. Baker.)

In addition to fungi, bacteria of the genera *Bacillus*, *Enterobacter*, *Pseudomonas*, and *Pantoea* have been shown to parasitize and/or inhibit the pathogenic oomycetes *Phytophthora* sp., *Pythium* sp, and the fungi *Fusarium* (Fig. 9-11), *Sclerotium ceptivorum*, and *Gaeumannomyces tritici*; the mycophagous nematode *Aphelenchus avenae* parasitizes *Rhizoctonia* and *Fusarium*; and the amoeba *Vampyrella* parasitizes the pathogenic fungi *Cochliobolus sativus* and *Gaeumannomyces graminis*.

Plant pathogenic nematodes are also parasitized by other microorganisms. For example, *Meloidogyne javanica* and *Pratylenchus* sp. nematodes are parasitized by the bacterium *Pasteuria (Bacillus) penetrans* (Figs. 9-12A–9-12D). Cysts of the soybean cyst nematode *Heterodera glycines* are parasitized by the fungus *Verticillium lecanii* (Fig. 9-12E); the root-knot nematode

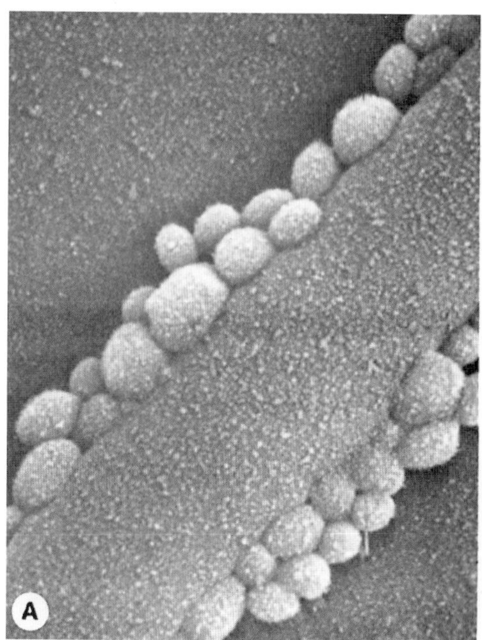

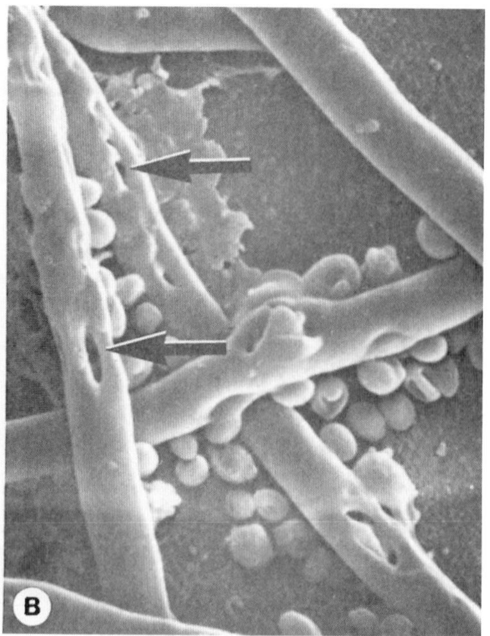

FIGURE 9-10 Attachment of the yeast biocontrol agent *Pichia guilliermondii* on hyphae of the plant pathogenic fungi *Botrytis cinerea* (A) and *Penicillium expansum* (B). Pitting is evident on the hyphae of both fungi, and, after longer interaction, numerous holes develop in the hyphal cell walls (arrows in B). (Magnification: 2350×.) [From Wisniewski, Biles, and Droby (1991), *in* "Biological Control of Postharvest Diseases of Fruits and Vegetables" (Wilson and Chalntz, eds.), pp. 167–183. USDA, ARS-92.]

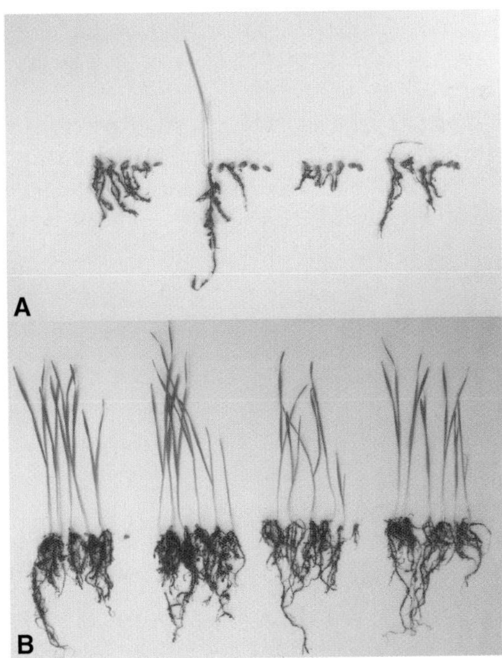

FIGURE 9-11 Biological control of wheat seedling blight caused by *Fusarium culmorum* through seed treatment with bacteria of the *Pantoea* sp. isolate MF626. (A) Extremely poor stands of wheat seedlings grown from untreated seeds and (B) normal healthy seedlings produced by treated seeds. [Photographs courtesy of P. M. Johansson, from Johansson, Johnsson, and Gerhardson (2003). *Plant Pathol.* 52, 219–227.]

Meloidogyne sp. is parasitized by fungi *Dactylella*, *Arthrobotrys* (Figs. 9-13A–9-13D), *Paecilomyces* (Fig. 9-13E), and *Hirsutella* sp.; whereas the dagger nematode *Xiphenema* and the cyst nematodes *Heterodera* and *Globodera* are parasitized by nematophagous fungi *Catenaria auxiliaris* (Fig. 9-13F), *Nematophthora gynophila*, *Verticillium chlamydosporium*, and *Hirsutella* sp.

Aerial Pathogens

Many other fungi have been shown to antagonize and inhibit numerous fungal pathogens of aerial plant parts. For example, *Chaetomium* sp. and *Athelia bombacina* suppress *Venturia inaequalis* ascospore and conidia production in fallen and growing leaves, respectively. *Tuberculina maxima* parasitizes the white pine blister rust fungus *Cronartium ribicola*; *Darluca filum* and *Verticillium lecanii* parasitize several rusts; *Ampelomyces quisqualis* parasitizes several powdery mildews; *Tilletiopsis* sp. parasitizes the cucumber powdery mildew fungus *Spaerotheca* fuligena; and *Nectria inventa* and *Gonatobotrys simplex* parasitize two pathogenic species of *Alternaria*.

Mechanisms of Action

The mechanisms by which antagonistic microorganisms affect pathogen populations are not always clear, but they are generally attributed to one of four effects: (1) direct parasitism or lysis and death of the pathogen (Fig. 9-14), (2) competition with the pathogen for food, (3) direct toxic effects on the pathogen by antibiotic substances released by the antagonist, and (4) indirect toxic effects on the pathogen by volatile substances, such as ethylene, released by the metabolic activities of the antagonist.

Many of the antagonistic microorganisms mentioned earlier are naturally present in crop soils and exert a certain degree of biological control over one or many plant pathogens regardless of human activities. Humans, however, have been attempting to increase the effectiveness of antagonists either by introducing new and larger populations of antagonists, e.g., *Trichoderma harzianum* and *Pasteuria penetrans*, in fields where they are lacking and/or by adding soil amendments that serve as nutrients for, or otherwise stimulate growth of, the antagonistic microorganisms and increase their inhibitory activity against the pathogen. Unfortunately, although both approaches are effective in the laboratory and in the greenhouse, neither has been particularly successful in the field. New microorganisms added to the soil of a field cannot compete with the existing microflora and cannot maintain themselves for very long. Also, soil amendments, so far, have not been selective enough to support and build up only the populations of the introduced or existing antagonists. Thus, their potential for eventual disease control is quite limited. There are several cases of successful biological control of plant pathogens when the antagonistic microorganism is used for direct protection of the plant from infection by the pathogen.

Control through Trap Plants

If a few rows of rye, corn, or other tall plants are planted around a field of beans, peppers, or squash, many of the incoming aphids carrying viruses that attack the beans, peppers, and squash will first stop and feed on the peripheral taller rows of rye or corn. Because most of the aphid-borne viruses are nonpersistent in the aphid, many of the aphids lose the bean-, pepper-, or squash-infecting viruses by the time they move onto these crops. In this way, trap crops reduce the amount of inoculum that reaches a crop.

Trap plants are also used against nematodes, although in a different way. Some plants that are not actually susceptible to certain sedentary plant-parasitic nematodes produce exudates that stimulate eggs of these

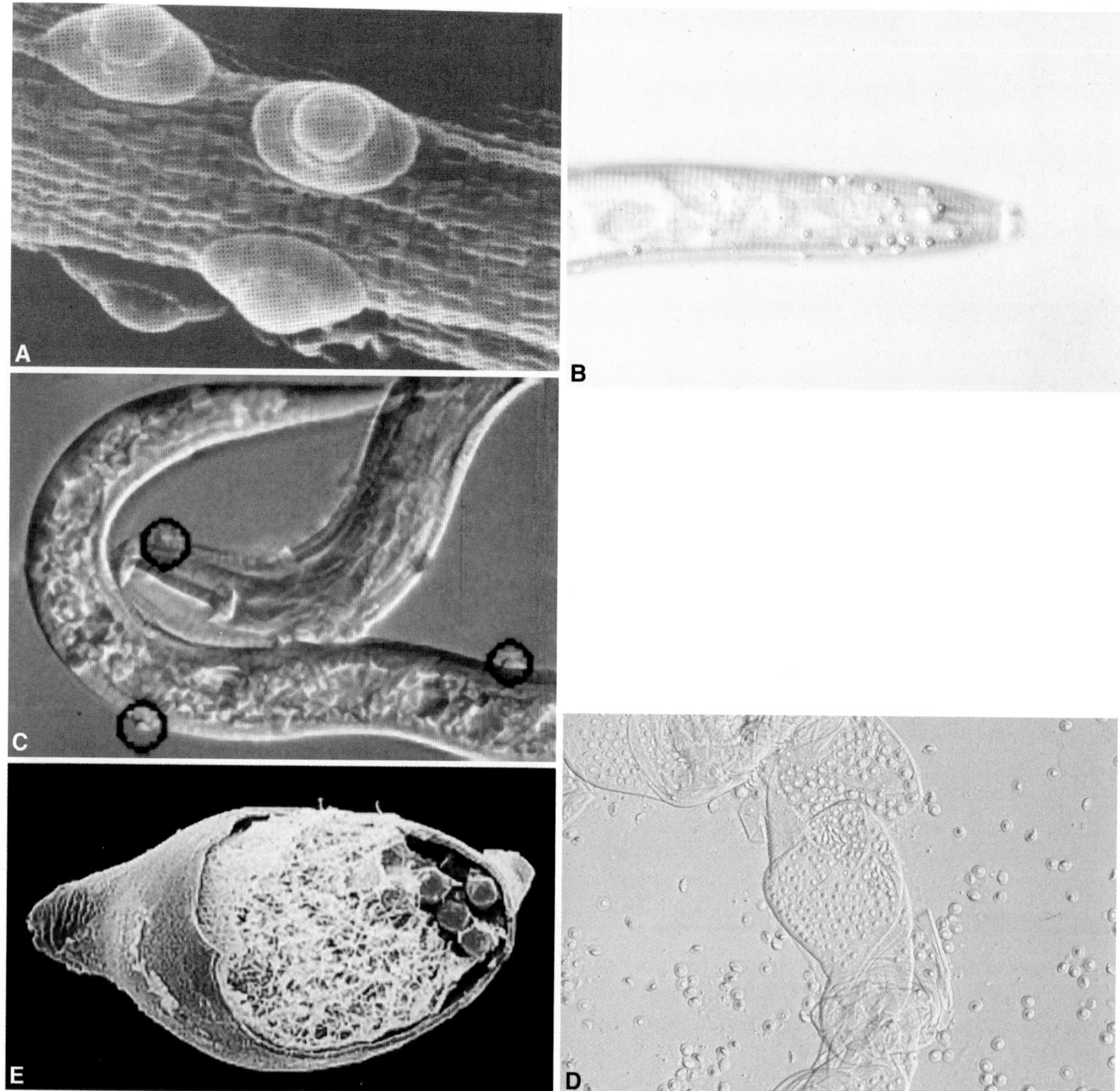

FIGURE 9-12 Biological control of nematodes. In (A, B, and C) *Meloidogyne* juveniles and (D) *Pratylenchus sp.* are attacked by the bacterium *Pasteuria penetrans* and in (E) a *Heterodera* cyst by the fungus *Verticillium lecanii.* [Photographs courtesy of (A) K. B. Nguyen, (B and D) R. M. Sayre, and (C and E) D. J. Chitwood.]

nematodes to hatch. The juveniles enter these plants but are unable to develop into adults and eventually they die. Such plants are also called **trap crops.** By using trap crops in a crop rotation program, growers can reduce the nematode population in the soil. For example, *Crotalaria* plants trap the juveniles of the root-knot nematode *Meloidogyne* sp. and black nightshade plants (*Solanum nigrum*) reduce the populations of the golden nematode *Heterodera rostochiensis.* Similar results can be obtained by planting highly susceptible plants, which after infection by the nematodes are destroyed (plowed under) before the nematodes reach maturity and begin to reproduce.

Unfortunately, trap plants have not given a sufficient degree of disease control to offset the expense and risk involved with their use. Therefore, they have been little used in the practical control of nematode diseases of plants.

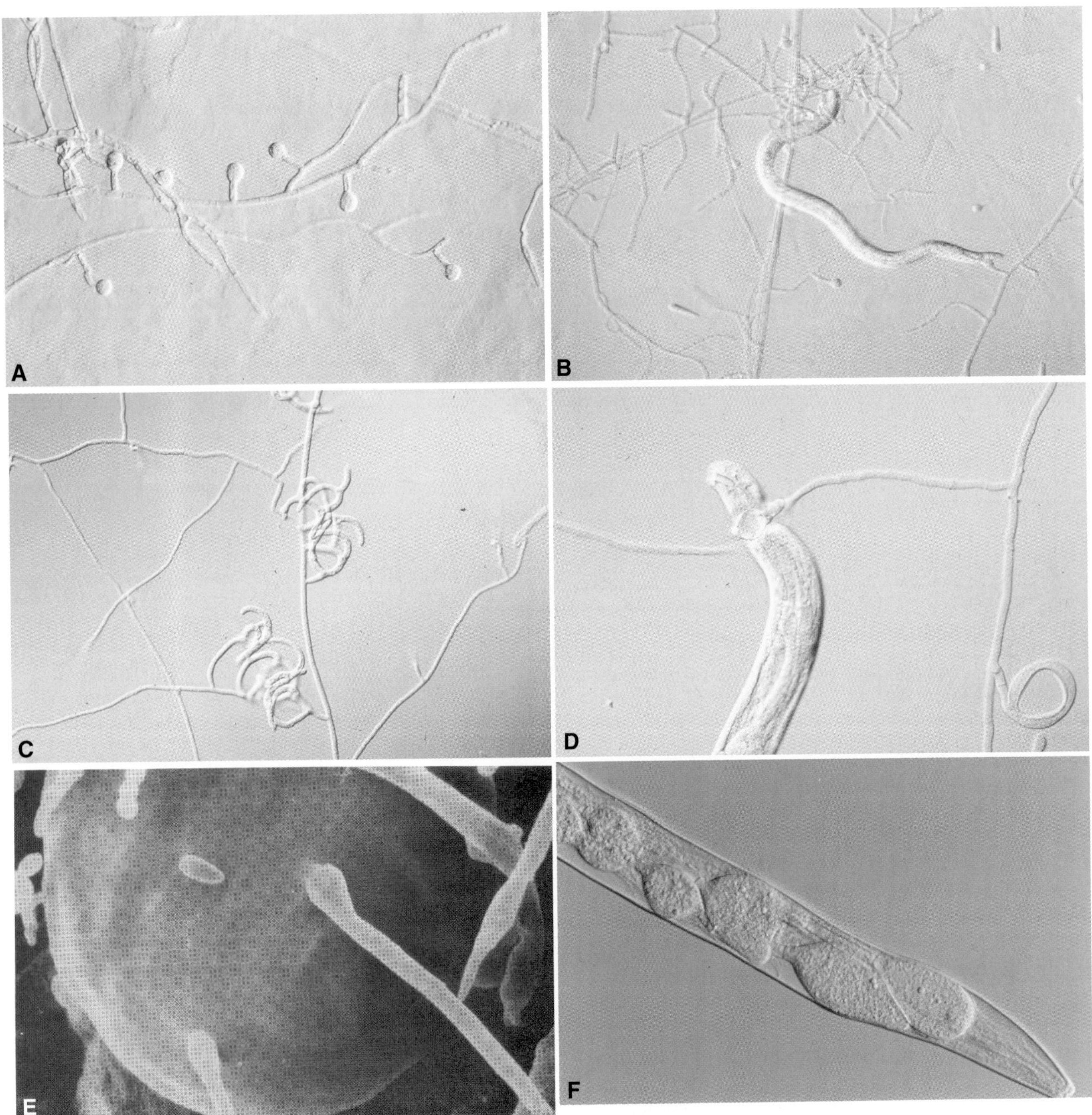

FIGURE 9-13 Biological control of nematodes with fungi. (A) Fungus with adhesive knobs and (B) a nematode trapped by such knobs. (C) *Arthrobotrys* fungus with adhesive branches and (D) a nematode trapped by the constricting ring of adhesive hyphae. (E) Egg of *Meloidogyne* invaded by the fungus *Paecilomyces*. (F) *Xiphinema* nematode invaded by zoosporangia of the fungus *Catenaria*. [Photographs courtesy of (A–D, and F) B. A. Jaffee and (E) R. M. Sayre.]

Control through Antagonistic Plants

A few kinds of plants, e.g., asparagus and marigolds, are antagonistic to nematodes because they release substances in the soil that are toxic to several plant-parasitic nematodes. When interplanted with nematode-susceptible crops, antagonistic plants decrease the number of nematodes in the soil and in the roots of the susceptible crops. Antagonistic plants, however, are not used on a large scale for the practical control of nematode diseases of plants for the same reasons that trap plants are not used.

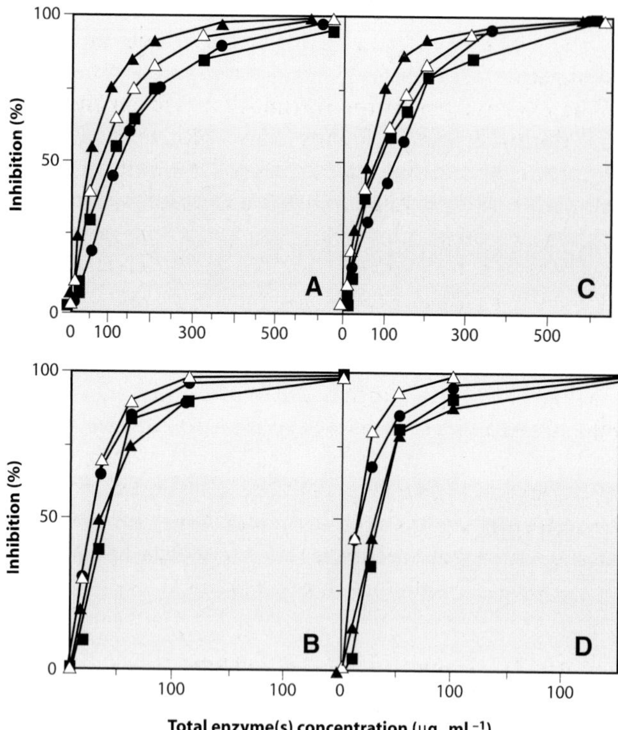

FIGURE 9-14 Inhibitory effect of fungal cell wall-degrading chitinolytic and glucanolytic enzymes obtained from the antagonistic fungus *Trichoderma harzianum* on spore germination (A and B) and germ tube elongation (C and D) of the plant pathogenic fungus *Botrytis cinerea.* (A and C) Single enzyme treatment and (B and D) treatment with a mixture of equal parts of all four enzymes. Note the much lower enzyme concentration needed in B and D. ▲, endochitinase; ●, chitobiosidase; Δ, *N*-acetylglucosaminidase; and ■, glucan 1,3-β-glucosidase. [From Lorito *et al.* (1994). *Phytopathology* 84, 398–405.]

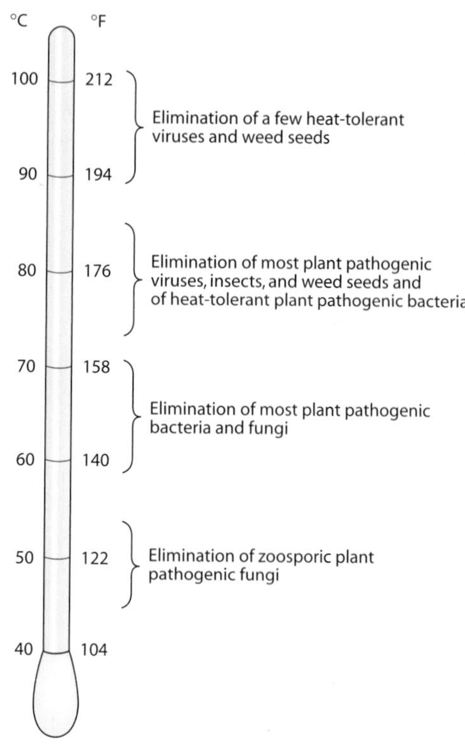

FIGURE 9-15 Temperatures (in °C and °F) at which various types of pathogens, insects, and weed seeds are eliminated from soil, seeds, and other propagative organs following exposure for 30 minutes.

Physical Methods That Eradicate or Reduce the Inoculum

The physical agents used most commonly in controlling plant diseases are temperature (high or low), dry air, unfavorable light wavelengths, and various types of radiation. With some crops, cultivation in glass or plastic greenhouses provides physical barriers to pathogens and their vectors and in that way protects the crop from some diseases. Similarly, plastic or net covering of row crops may protect the crop from infection by preventing pathogens or vectors from reaching the plants.

Control by Heat Treatment

Soil Sterilization by Heat

Soil can be sterilized in greenhouses, and sometimes in seed beds and cold frames, by the heat carried in live or aerated steam or hot water. The soil is steam sterilized either in special containers (soil sterilizers), into which steam is supplied under pressure, or on the greenhouse benches, in which case steam is piped into and is allowed to diffuse through the soil. At about 50°C, nematodes, some oomycetes, and other water molds are killed, whereas most plant pathogenic fungi and bacteria, along with some worms, slugs, and centipedes, are usually killed at temperatures between 60 and 72°C. At about 82°C, most weeds, the rest of the plant pathogenic bacteria, most plant viruses in plant debris, and most insects are killed (Fig. 9-15). Heat-tolerant weed seeds and some plant viruses, such as tobacco mosaic virus (TMV), are killed at or near the boiling point, i.e., between 95 and 100°C. Generally, soil sterilization is completed when the temperature in the coldest part of the soil has remained for at least 30 minutes at 82°C or above, at which temperature almost all plant pathogens in the soil are killed. Heat sterilization of soil can also be achieved by heat produced electrically rather than supplied by steam or hot water.

It is important to note, however, that excessively high or prolonged high temperatures should be avoided during soil sterilization. Not only do such conditions destroy all normal saprophytic microflora in the soil, but they also result in the release of toxic levels of some (e.g., manganese) salts and in the accumulation of toxic levels of ammonia (by killing the nitrifying bacteria before they kill the more heat-resistant ammonifying bacteria), which may damage or kill plants planted afterward.

Soil Solarization

When clear polyethylene is placed over moist soil during sunny summer days, the temperature at the top 5 centimeters of soil may reach as high as 52°C compared to a maximum of 37°C in unmulched soil. If sunny weather continues for several days or weeks, the increased soil temperature from solar heat, known as solarization, inactivates (kills) many soilborne pathogen fungi, nematodes, and bacteria near the soil surface, thereby reducing the inoculum and the potential for disease (Fig. 9-16).

Hot-Water Treatment of Propagative Organs

Hot-water treatment of certain seeds, bulbs, and nursery stock is used to kill any pathogens with which they are infected or which may be present inside seed coats, bulb scales, and so on, or which may be present in external surfaces or wounds. In some diseases, seed treatment with hot water was for many years the only means of control, as in the loose smut of cereals, in which the fungus overwinters as mycelium inside the seed where it could not be reached by chemicals. Similarly, treatment of bulbs and nursery stock with hot water frees them from nematodes that may be present within them, such as *Ditylenchus dipsaci* in bulbs of various ornamentals and *Radolpholus similis* in citrus rootstocks.

The effectiveness of the method is based on the fact that dormant plant organs can withstand higher temperatures than those their respective pathogens can survive for a given time. The temperature of the hot water used and the duration of the treatment vary with the different host–pathogen combinations. Thus, in the loose smut of wheat the seed is kept in hot water at 52°C for 11 minutes, whereas bulbs treated for *D. dipsaci* are kept at 43°C for 3 hours.

It has been reported that a short (15 seconds) treatment of melon fruit with hot (59 ± 1°C) water rinse and brushes resulted in a significant reduction of fruit decay while maintaining fruit quality after prolonged storage. Treated fruit had less soil, dust, and fungal spores at its surface while many of its natural openings in the epidermis were partially or entirely sealed.

Hot-Air Treatment of Storage Organs

Treatment of certain storage organs with warm air (curing) removes excess moisture from their surfaces and hastens the healing of wounds, thus preventing their infection by certain weak pathogens. For example, keeping sweet potatoes at 28 to 32°C for 2 weeks helps the wounds to heal and prevents infection by *Rhizopus* and by soft-rotting bacteria. Also, hot-air curing of harvested ears of corn, tobacco leaves, and so on removes most moisture from them and protects them from attack by fungal and bacterial saprophytes. Similarly, dry heat treatment of barley seed at 72°C for 7 to 10 days

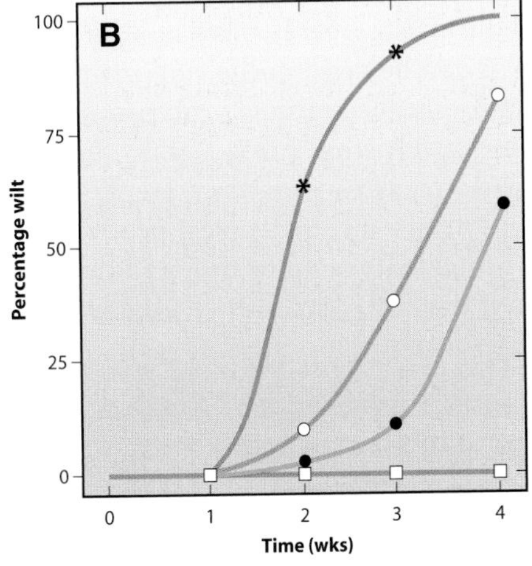

FIGURE 9-16 (A) Soil solarization in Cote d' Ivoire in Africa. Soil removed from holes to solarize before being replaced. (B) Effect of soil solarization on *Fusarium* wilt of watermelon. ∗, infested, nonsolarized soils; ○, infested soil solarized for 30 days; ●, infested soil solarized for 60 days; □, noninfested, nonsolarized soil. [From Martyn and Hartz (1986). *Plant Dis.* 70, 762–766.]

eliminates the leaf streak- and black chaff-causing bacterium *Xanthomonas campestris* pv. *translucens* from the seed with negligible reduction of seed germination.

Control by Eliminating Certain Light Wavelengths

Alternaria, *Botrytis*, and *Stemphylium* are examples of plant pathogenic fungi that sporulate only when they receive light in the ultraviolet range (below 360 nm). It has been possible to control diseases on greenhouse vegetables caused by several species of these fungi by covering or constructing the greenhouse with a special ultraviolet (UV)-absorbing vinyl film that blocks the transmission of light wavelengths below 390 nanometers.

Drying Stored Grains and Fruit

All grains, legumes, and nuts carry with them a variety and number of fungi and bacteria that can cause decay of these organs in the presence of sufficient moisture. Such decay, however, can be avoided if seeds and nuts are harvested when properly mature and then are allowed to dry in the air or are treated with heated air until the moisture content is reduced sufficiently (to about 12% moisture) before storage. Subsequently, they are stored under conditions of ventilation that do not allow buildup of moisture to levels (about 12%) that would allow storage fungi to become activated. Fleshy fruits, such as peaches and strawberries, should be harvested later in the day, after the dew is gone, to ensure that the fruit does not carry surface moisture with it during storage and transit, which could result in decay of the fruit by fungi and bacteria.

Many fruits can also be stored dry for a long time and can be kept free of disease if they are dried sufficiently before storage and if moisture is kept below a certain level during storage. For example, grapes, plums, dates, and figs can be dried in the sun or through warm air treatment to produce raisins, prunes, and dried dates and figs, respectively, that are generally unaffected by fungi and bacteria as long as they are kept dry. Even slices of fleshy fruits such as apples, peaches, and apricots can be protected from infection and decay by fungi and bacteria if they are dried sufficiently by exposure to the sun or to warm air currents.

Disease Control by Refrigeration

Refrigeration is probably the most widely used and the most effective method of controlling postharvest diseases of fleshy plant products. Although low temperatures at or slightly above the freezing point do not kill any of the pathogens that may be on or in the plant tissues, they do inhibit or greatly retard the growth and activities of all such pathogens, thereby reducing the spread of existing infections and the initiation of new ones. Most perishable fruits and vegetables should be refrigerated as soon as possible after harvest, transported in refrigerated vehicles, and kept refrigerated until they are used by the consumer. Regular refrigeration of especially succulent fruits and vegetables is sometimes preceded by a quick hydrocooling or air cooling of these products, aimed at removing the excess heat carried in them from the field as quickly as possible to prevent the development of any new or latent infections. The magnitude of disease control through refrigeration and its value to growers and consumers is immense.

Disease Control by Radiation

Various types of electromagnetic radiation, such as UV light, X rays, and γ rays, as well as particulate radiation, such as α particles and β particles, have been studied for their ability to control postharvest diseases of fruits and vegetables by killing the pathogens present on them. Some satisfactory results were obtained in experimental studies using γ rays to control postharvest infections of peaches, strawberries, and tomatoes by some of their fungal pathogens. Unfortunately, with many of these diseases the dosage of radiation required to kill the pathogen may also injure the plant tissues on which the pathogens exist. Also, this method of treatment of foodstuffs, although found safe and properly licensed by the USDA, is vigorously opposed by certain segments of the population. So far, no plant diseases are controlled commercially by radiation.

Trench Barriers against Root-transmitted Tree Diseases

Several vascular wilt and other diseases of trees are transmitted from tree to tree through contact or natural grafts between roots of adjacent trees. Several types of water permeable and nonpermeable materials are available and seem to be effective in blocking root contact. Water-permeable materials seem to be superior. However, the cost is high for either type of material and only high-value landscape trees may justify such treatment.

Chemical Methods that Eradicate or Reduce the Inoculum

Chemical pesticides are generally used to protect plant surfaces from infection or to eradicate a pathogen that has already infected a plant. A few chemical treatments, however, are aimed at eradicating or greatly reducing the inoculum before it comes in contact with the plant. They include soil treatments (such as fumigation), disinfestation of warehouses, sanitation of handling equipment, and control of insect vectors of pathogens.

Soil Treatment with Chemicals

Soil to be planted with vegetables, strawberries, ornamentals, trees, or other high-value crops, such as tobacco, is frequently treated with chemicals for control primarily of nematodes but occasionally also of soilborne fungi, such as *Fusarium* and *Verticillium*, weeds, and bacteria. Certain fungicides are applied to the soil as dusts, liquid drenches, or granules to control damping-off, seedling blights, crown and root rots, and other diseases. In fields where irrigation is possible, the fungicide is sometimes applied with the irrigation water, particularly in sprinkler irrigation. Fungicides used for soil treatments include metalaxyl, diazoben, pentachloronitrobenzene (PCNB), captan, and chloroneb, although the last two are used primarily as seed treatments. Most soil treatments, however, are aimed at controlling nematodes, and the materials used are volatile gases or produce volatile gases (fumigants) that penetrate the soil throughout (fumigate). Some nematicides, however, are not volatile but, instead, dissolve in soil water and are then distributed through the soil.

Fumigation

The most promising method of controlling nematodes and certain other soilborne pathogens and pests in the field has been through the use of chemicals usually called fumigants. Some of them, including chloropicrin, methyl bromide, dazomet, and metam sodium, either volatilize as they are applied to the soil or decompose into gases in the soil. These materials are general-purpose preplant fumigants; they are effective against a wide range of soil microorganisms, including nematodes, many fungi, insects, certain bacteria, and weeds. Contact nematicides, such as fensulfothion, carbofuran, ethoprop, and aldicarb, are of low volatility, are effective against nematodes and insects, and can be applied before and after planting of many crops that are tolerant to these chemicals.

Nematicides used as soil fumigants are available as liquids under pressure, liquids, emulsifiable concentrates, and granules. These materials are applied to the soil either by spreading the chemical evenly over the entire field (broadcast) or by applying it only to the rows to be planted with the crop (row treatment). In both cases the fumigant is applied through delivery tubes attached at the back of tractor-mounted chisel-tooth injection shanks or disks spaced at variable widths and usually reaching six inches below the soil surface. The nematicide is sealed in the soil instantly by a smoothing and firming drag or can be mixed into the soil with disk harrows or rototillers. Highly volatile nematicides are covered immediately with polyethylene sheeting (Fig. 9-16), which should be left in place for at least 48 hours. When small areas are to be fumigated, the most convenient method is through injection of the chemical with a hand applicator under a tarp that has been placed over the area. The edges of the tarp are covered with soil prior to injection of the chemical. Applications may also be made by placement of small amounts of granules in holes or furrows six inches deep, 6 to 12 inches apart, which should be covered immediately with soil. In all cases of preplant soil fumigation with phytotoxic nematicides, several days to two weeks must elapse from the time of treatment to seeding or planting in the field to avoid plant injury.

In the abovementioned types of nematicide application, only a small portion of the soil and its microorganisms immediately come in contact with the chemical. The effectiveness of the fumigants, however, is based on their diffusion in a gaseous state through the pores of the soil throughout the area in which nematode and other pest control is desired. The distance the vapors move is influenced by the size and continuity of soil pores, by the soil temperature (the best range is between 10 and 20°C), by soil moisture (best at about 80% of field capacity), by the type of soil (more material is required for soils rich in colloidal or organic matter), and by the properties of the chemical itself. Nematicides with low volatility, such as carbofuran, do not diffuse through the soil to any great extent and must be mixed with the soil mechanically or by irrigation water or rainfall. Except for the highly volatile methyl bromide and chloropicrin, most nematicides can be applied in irrigation water when it is provided as trickle soaks or drenches, but only low-volatility nematicides can be applied through overhead sprinkler systems.

In practice, chemical nematode control in the field is generally obtained by preplant soil fumigation with one of the nematicides applied only before planting. These chemicals are nonspecific, i.e., they control all types of nematodes, although some nematodes are harder to control than others no matter what the nematicide. Chloropicrin, methyl bromide, dazomet, and metam sodium are expensive, broad-spectrum nematicides that must be covered on application either with tarps (the first two) or with water or through soil (the others). All nematicides are extremely toxic to humans and animals and should be handled with great caution.

Disinfestation of Warehouses

Stored products can be protected from becoming infected by pathogens left over in the warehouse from previous years by first cleaning thoroughly the storage rooms and by removing and burning the debris. The walls and floors are washed with bleach, a copper sulfate solution (1 pound in 5 gallons of water), or some other sanitizing agent. Warehouses that can be closed airtight and in which the relative humidity can be kept at nearly 100% while the temperature is between 25 and 30°C can be fumigated effectively with chloropicrin (tear gas) used at 1 pound per each 1,000 cubic feet. In all cases the fumigants should be allowed to act for at least 24 hours before the warehouse doors are opened for aeration.

Control of Insect Vectors

When the pathogen is introduced or disseminated by an insect vector, control of the vector is as important as, and sometimes easier than, the control of the pathogen itself. Application of insecticides for the control of insect carriers of fungus spores and bacteria has been fairly successful and is a recommended procedure in the control of several such insect-carried pathogens.

In the case of viruses, mollicutes, and fastidious bacteria, however, of which insects are the most important disseminating agents, insect control has been helpful in controlling the spread of their diseases only when it has been carried out in the area and on the plants on which the insects overwinter or feed before they enter the crop. Controlling such diseases by killing the insect vectors with insecticides after they have arrived at the crop has seldom proved adequate. Even with good insect control, enough insects survive for sufficiently long periods to spread the pathogen. Nevertheless, appreciable reduction in losses from certain such diseases has been obtained by controlling their insect vectors, and the practice of good insect control is always desirable.

Success in reducing virus transmission by insects has been achieved by interfering with the ability of the aphid vector to acquire and to transmit the virus rather than by killing the insects. The interference is provided by spraying the plants several times each season with a fine-grade mineral oil. Such oil seems to have little effect on the probing and feeding behavior of the aphids and is not particularly toxic to the aphids, but it interferes with the transmission of nonpersistent, semipersistent, and even some persistent aphid-borne viruses. Control of aphid-borne viruses by oil sprays has been successful with some viruses (e.g., *cucumber mosaic virus* on cucumber and pepper and *potato virus Y* on pepper).

DISEASE CONTROL BY IMMUNIZING, OR IMPROVING THE RESISTANCE OF, THE HOST

Unlike humans and animals, plants lack an antibody-producing system and cannot be immunized by vaccination the way humans can. Through genetic engineering, however, scientists have introduced and expressed in plants genes from mice coding for the production of antibodies against certain plant pathogens, mostly viruses, with which the mice had been injected artificially. Although plants so engineered do produce antibodies, called plantibodies, against specific plant pathogens, it is not yet known whether such plantibodies will effectively protect the plant from becoming diseased by that pathogen.

Inoculation of plants with certain pathogens often leads to temporary or nearly permanent "immunization" of the plants, i.e., induced plant resistance to a pathogen to which the plants are normally susceptible. Some of these treatments involve only viruses and are known as cross protection; others may involve different kinds of pathogens and are known as induced or systemic-acquired resistance (SAR). Systemic acquired resistance can also be induced in plants against a variety of diverse pathogens by treating the plants with certain chemical compounds, such as salicylic acid and dichloroisonicotinic acid (INA), and certain benzothiazoles.

The resistance of many plants to many, mostly viral, pathogens has been improved significantly by introducing and expressing in the plant, by genetic engineering techniques, genes or other DNA segments obtained from the pathogen (pathogen-derived resistance). Some of these genes code for the production of certain structural or nonstructural proteins essential to the pathogen, but it is not clear how production of these pathogen proteins by the plant improves the resistance of the plant to the pathogens. The resistance of some plants was also improved significantly by incorporating in the plant, through genetic engineering, genes obtained from plants, other pathogens, or other organisms that code for the production of enzymes, peptides, or toxins interfering with infection by the pathogen. In some cases, the resistance of a host plant to a particular pathogen can be improved by simply improving the growing conditions (fertilizer, irrigation, drainage, and so on) of the host.

By far the most common improvement of host resistance to almost any pathogen, however, is brought about by improving the genetic resistance of the host, i.e., by breeding and using resistant varieties. In the mid-1990s, genetic engineering technology made possible the isolation of individual resistance (R) genes from resistant plants and the transfer of such genes into susceptible plants in which they induce the hypersensitive (resistant) response. It is expected that this approach for improving resistance in susceptible plants, combined with conventional plant breeding, will provide one of the most effective tools for controlling plant diseases.

Cross Protection

The term cross protection specifically applies to the protection provided to a plant by infection with a mild strain of a virus from subsequent infection by a more severe strain of the same virus that normally causes more severe symptoms. This appears to be a general phenomenon among virus strains, although some mild strains are much more effective than others in protecting plants from infection by severe strains. Its application in controlling virus diseases has met with some success in cross protecting tomatoes with mild strains of *tobacco mosaic virus*, citrus with mild strains of *citrus tristeza virus*, and papaya with mild strains of *papaya ring spot virus*. Cross protection has not gained widespread use because appropriate mild strains of viruses are often not available, and mild strains are not effective against all severe strains present in different localities; also, the method for field crops is laborious. Finally, in addition, there is a danger of mutations toward new, more virulent strains, double infections, and the spread to other crops in which virulence might be greater. Besides, in perennial crops, such as citrus trees, the cross protection ability of mild strains seems to be lost after a few years, apparently because distribution of the mild strain within trees is only partial and new severe strains can become established in mild strain-free areas, become distributed within the tree, and cause the disease.

Induced Resistance: Systemic Acquired Resistance

There are many examples in which plants infected with one pathogen become more resistant to subsequent infection by another pathogen and, also, of plants becoming resistant to a pathogen if they have been inoculated with the same pathogen at an earlier growth stage at which they are resistant to the pathogen. (There are, however, also examples of plants being more susceptible to a pathogen if they are already infected with another pathogen.) For example, bean and sugar beet inoculated with virus exhibit a greater resistance to infection by certain obligate fungal pathogens causing rusts and powdery mildews than virus-free plants. Also, in tobacco, *tobacco mosaic virus* induces a systemic resistance not only to itself, but also to unrelated viruses, to oomycetes like *Phytophthora nicotianae*, to bacteria like *Pseudomonas tabaci*, and even to certain aphids! Inversely, inoculation of tobacco with a root lesion-inducing fungus such as *Chalara elegans* (*Thielaviopsis basicola*) or a leaf lesion-inducing bacterium such as *Pseudomonas syringae* induces systemic resistance to TMV. Additional examples of induced resistance include fire blight of pear, in which induction was by inoculation with a nonpathogenic bacterium, and cucurbit anthracnose, in which resistance to the fungus *Colletotrichum lagenarium Ralstonia* was induced by inoculating the plants while young with the same fungus.

It later became apparent that resistance to pathogens could be induced in their hosts by treating (rubbing, infiltrating, or injecting) the latter with naturally occurring compounds obtained from the pathogen, such as the coat protein of TMV, a proteinaceous component or a glycoprotein fraction from a bacterium (*Ralstoni solanacearum*), a lipid component from an oomycete (*Phytophthora infestans*), or a polysaccharide such as chitosan from a fungus. Resistance to pathogens could also be induced by treating plants with unrelated natural compounds such as a water-soluble fraction of a nonpathogenic bacterium, a polysaccharide from a nonpathogenic fungus, or a proteinaceous compound isolated from an unrelated plant, all inducing complete resistance to TMV infection and to several other diseases.

Plant Defense Activators

Even more significantly, resistance in plants to several viruses, such as TMV; fungi, oomycetes such as *Peronospora tabacina*; and bacteria, such as *Pseudomonas syringae*, can be induced with several types of synthetic compounds applied by injection into the plant, spraying onto leaves, or absorption through the petiole or through the roots (see Figs. 5-16 and 5-17). Such synthetic chemicals reported as effective inducers of SAR against several pathogens of all kinds include salicylic acid (a derivative of which, acetylsalicyclic acid, is the common aspirin) and INA:

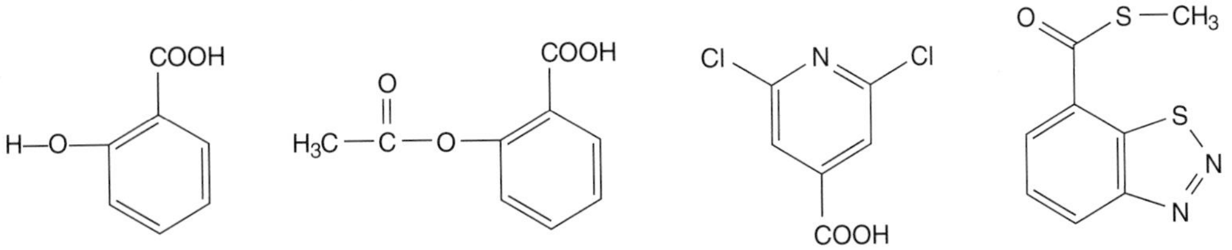

Salicylic acid

Acetylsalicylic acid (aspirin)

Dichloroisonicotinic acid (INA)

Benzothiadiazole (Actigard, BTH)

The scientific developments in the area of induced systemic acquired resistance are quite exciting and certainly promising. Few diseases are currently controlled commercially by the mechanisms of induced resistance listed here. Still, a great deal of work is going on in these areas, including work by private industry interested in developing and marketing compounds that would induce SAR to plant disease. The first such compound, benzothiadiazole, known as Actigard, is quite effective against numerous types of diseases of many diverse crops (Table 9-2). A derivative of benzothiadiazole called Acibenzolar-S-methyl (ASM) and sold as Blockade for the control of downy mildews of leafy vegetables activates the same defense responses as the biological inducers of systemic acquired resistance.It is expected that several truly effective compounds will be available commercially before too long. Such compounds represent a new class of chemical plant activators that have no antimicrobial activity but mimic the biological induction of systemic acquired resistance in both dicot and monocot crops against many diverse types of, although not all, pathogens. β-Aminobutyric acid has been reported as such a compound. It has also been shown that treatment of harvested fruit with low dosages of UV-C light (254 nm) causes an almost immediate activation and expression of certain defense-response genes and a rapid induction of chitinase, β-1,2-glucanase, and phenylalanine lyase enzymes controlled by such genes, indicating that the UV light treatment acts as an elicitor of induced systemic resistance in treated fruit. Similarly, silicon supplied as a nutrient solution seems to protect plants from powdery mildew infection by inducing localized defense reactions including papilla formation, callose production, and accumulation of phenolics along the cell wall, while plants not receiving silicon exhibit little or no such defense reactions (Fig. 9-17).

Early research also shows that some bacteria, especially several species of *Bacillus*, when added to various plants as seed treatments, soil drenches, or as transplant dips, promote growth of the plants and induce systemic resistance in them to several fungal pathogens.

Improving the Growing Conditions of Plants

Cultural practices aiming at improving the vigor of the plant often help increase its resistance to pathogen attack. Thus, proper fertilization, field drainage, irrigation, proper spacing of plants, and weed control improve the growth of plants and may have a direct or indirect effect on the control of a particular disease. The most important measures for controlling *Leucostoma* (*Valsa*) canker of fruit and other trees, for example, are adequate irrigation and proper fertilization of the trees. Seed priming with inorganic salts (osmopriming) or with a fine silicate clay (solid matrix priming) has been used effectively for controlled hydration of seeds about to be planted and allows vigorous growth of seedlings and increased resistance of seedlings to *Pythium*-caused damping-off.

TABLE 9-2
Plant Activators Available Commercially in the United States as of 2003

Name	Source	Target pathogen(s)	Crop(s)	Application
Actigard	Benzothiadiazole	Many, various	Tobacco, tomato, lettuce, spinach	Drench, spray
Bion WG50	Benzodiathiazole derivative			
Acibenzolar-S-methyl (ASM)				
Blockade	Synthetic compound	Downy mildews	Leafy vegetables	Spray
Actinovate	*Streptomyces lydicus*	Soilborne	Greenhouse, nursery, turf	Drench
AQ10 Biofungicide	*Ampelomyces quisqualis* isolate M-10	Powdery mildew	Grapes, apples, cucurbits, strawberries, ornamentals, tomatoes	Spray
Aspire	*Candida oleophilal-182*	*Botrytis* sp., *Penicillium* sp.	Citrus, pome fruits	Postharvest drench, spray
Awaiting registration				
Serenade	*Bacillus subtilis* strain QST716	Many, various	Fruits, vegetables, others	Spray
YieldShield	*B. pumilus GB34*	Root disease fungi	Soybean	Seed treatment
Messenger	*Erwinia amylovora* harpin protein	Many	Field, ornamentals, vegetables	Drench, spray
Oxycom	Synthetic salicylic acid plus oxygen generator	Many	Many	Spray, drench

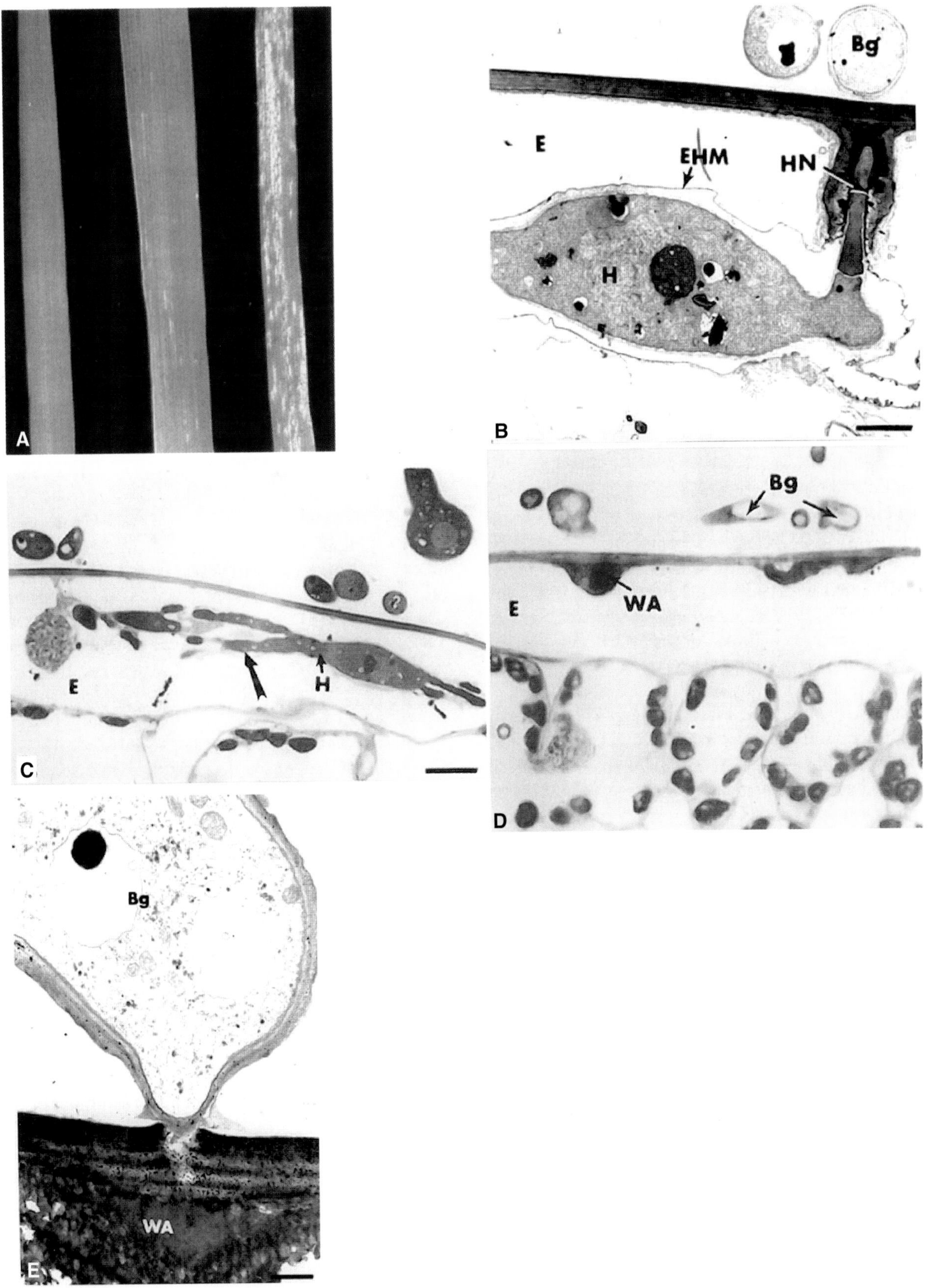

FIGURE 9-17 Enhancement of localized plant cell defenses by silicon (Si) nutrition. (A) Effect of Si on powdery mildew of wheat caused by *Blumeria graminis* f. sp. *tritici*: left, control leaf, no Si; no inoculation; middle, inoculated leaf from Si-fertilized plant; and right, inoculated leaf on plant not receiving Si. (B) The fungus (Bg) has penetrated the epidermal cell wall of Si- leaf as well as the elongated collar around the haustorial neck (HN), and the haustorium (H) develops into finger-like lobes (C). In Si+ leaves (D and E), wall appositions (WA) forming in cells below the appressorium prevent the fungus from penetrating the cells. [Photographs courtesy of R. R. Belanger, from Belanger, Benhamou, and Menzies (2003). *Phytopathology* **93**, 402–412.]

Use of Resistant Varieties

The use of resistant varieties is the least expensive, easiest, safest, and one of the most effective means of controlling plant diseases in crops. Cultivation of resistant varieties not only eliminates losses from disease, but also eliminates expenses for sprays and other methods of disease control and avoids the addition of toxic chemicals to the environment that would otherwise be used to control plant diseases. Moreover, for many diseases, such as those caused by vascular pathogens and viruses, that often cannot be controlled adequately by other means, and for others, such as cereal rusts, powdery mildews, and root rots, that in most countries are economically impractical to control in other ways, the use of resistant varieties provides a means of producing acceptable yields without any pesticides.

Varieties of crops resistant to some of the most important or most difficult to control diseases are made available to growers by federal and state experiment stations and by commercial seed companies. More than 85% of the total agricultural acreage in the United States is planted with varieties that are resistant to one or more diseases. With some crops, such as small grains and alfalfa, varieties planted because they are resistant to a certain disease(s) make up 95 to 98% of the crop. Growers and consumers alike have gained the most from the use of varieties resistant to fungi causing rusts, smuts, powdery mildews, and vascular wilts, but several other kinds of fungal diseases and many diseases caused by viruses, bacteria, and nematodes are also controlled through resistant varieties.

Resistant varieties have been used in only a few cases for disease control in fruit and forest trees. For example, some apple varieties resistant to apple scab are now available. Examples of forest tree diseases managed with resistance are white pine blister rust and fusiform rust of pine. The limited use of resistance in controlling tree diseases stems from the difficulty of replacing quickly susceptible varieties with resistant ones and keeping the resistant ones from being attacked by new races of the pathogen that are likely to develop over the long life span of trees.

It is always preferable to use varieties that have both vertical (initial inoculum-limiting) and horizontal (rate-limiting) resistance, and most resistant varieties have both types of resistance. Many of them carry only one or a few (two or three) genes of vertical resistance and an unspecified number of genes of horizontal resistance. Such varieties are, of course, resistant only to some of the races of the pathogen, and, if the pathogen is airborne and new races can be brought in easily, as happens with cereal rusts, powdery mildews, downy mildews, and *Phytophthora infestans*, new races virulent to the resistant variety appear and become widespread. As the new race takes over, the resistance of the old variety is said to break down, and the old variety must be replaced with another variety that has different genes for resistance. As a result, varieties with vertical resistance need to be replaced periodically, e.g., every 3, 5, or 10 years. How quickly varieties must be replaced depends on the genetic plasticity of the pathogen, the particular gene or combination of genes involved, the degree and manner of deployment of the gene(s), and the favorableness of the weather conditions toward disease development. It is expected that genetic engineering technology will allow for a quick transfer of individual genes or combinations of genes for resistance into susceptible crop varieties, thereby reducing the time required to develop a new resistant variety as happens today through conventional breeding alone.

Several techniques are employed to increase the useful "life span" of a resistant variety. To begin with, varieties are tested for resistance against as many pathogens and as many races of the pathogens as are available. Second, before they are released, varieties are grown and tested for resistance in as many locations as possible. For some important crops, such as cereals, new varieties are often tested in many countries and continents. Local breeding programs may carry out additional adaptation breeding to incorporate resistance to local pathogens. In this way, the released varieties are resistant to many races of the important pathogens that exist in most places.

Even after a resistant variety is released, however, measures can be taken to prolong its resistance. Any management strategy such as sanitation, seed treatment, or fungicide application that reduces the exposure of the variety to large pathogen populations (inoculum pressure) is likely to increase its useful life span. For slowly dispersing pathogens (such as soilborne pathogens), rotation of varieties with different sources of resistance reduces pathogen populations compatible with each variety so that each variety can last longer. For large area crops, such as wheat and rice, varieties could last longer against airborne pathogens (e.g., stem rust) if they were each deployed in one of three or four regional zones of the epidemiological region (Fig. 9-18). In this way, even if a new race that could attack a variety in one region did appear, it could not spread to the other varieties in the other regions because they have a different set of genes for resistance than the one whose resistance broke down. A still different approach involves the use of varietal mixtures or multilines. Multilines are composed of numerous near isogenic lines, each possessing a different gene for vertical resistance and, therefore, are resistant to a large proportion of the pathogen population. This results in overall reduction of the pathogen reproduction rate, which reduces the rate of disease increase

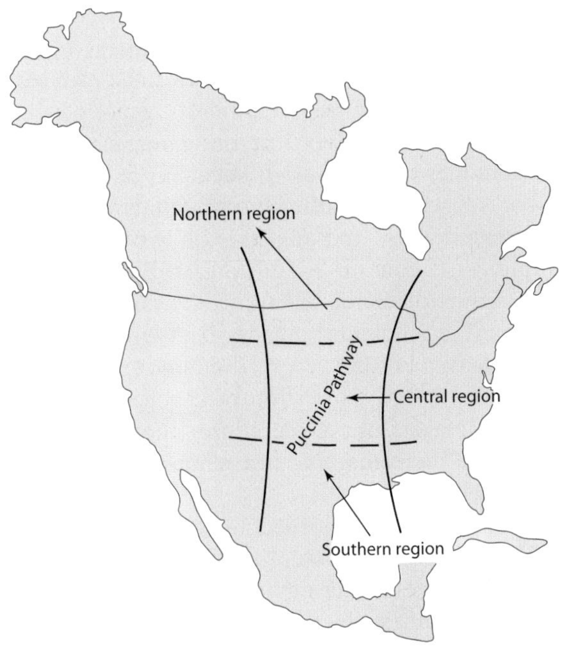

FIGURE 9-18 Pathway of stem rust of wheat in North America and deployment of wheat varieties carrying different genes for resistance in the southern, central, and northern regions to stop the spread of the same pathogen races from south to north. [From Frey *et al.* (1973).]

and the inoculum pressure on each of the other varieties. If one of the isogenic lines is attacked severely by a race of a pathogen, the following year the isoline is substituted in the multiline seed with a different isogenic line that can resist the new race.

Control through Use of Transgenic Plants Transformed for Disease Resistance

Modern DNA technology has made it possible to engineer transgenic plants that are transformed with genes for tolerance of adverse environmental factors, for resistance against specific diseases, or with genes coding for enzymes such as chitinases and glucanases directed against certain groups of pathogens, such as oomycetes, fungi, viruses, and bacteria, or with nucleic acid sequences that lead to gene silencing of pathogens.

Transgenic Plants That Tolerate Abiotic Stresses

Many types of plants have been transformed with genes that enable these plants to tolerate one or more abiotic stresses well beyond the normal range of these plants. For example, eggplant transformed with the bacterial gene coding for mannitol phosphodehydrogenase are tolerant against osmotic stress induced by salt, against drought, and against low (chilling) temperatures. Overexpression of the rice gene coding for glutamine *S*-transferase by transforming the plants with a maize ubiquitin promoter enabled the plants to tolerate lower temperature and to germinate better under submergence in water. However, upon transformation of rice plants with two genes from wheat, both genes increased the tolerance of the transgenic rice plants to stresses caused by dehydration or salt. Drought tolerance in tobacco plants was increased by introducing the yeast gene coding for trehalose phosphate synthase into the genome of tobacco chloroplasts, while introduction into the nuclear genome resulted in stunted and sterile plants. Finally, the woody plant Japanese persimmon (*Diospyros kaki*) was made more tolerant to salt stress by transforming it with a bacterial gene that codes for choline oxidase.

Transgenic Plants Transformed with Specific Plant Genes for Resistance

There are numerous crops in which plant genes for specific pathogens have been isolated from resistant plants, transferred into susceptible plants, and expressed in these plants. Provided that all the necessary supporting genes are also transferred and expressed in the new host, some of the formerly susceptible plants now behave as resistant ones. Such resistant plants are subsequently cloned and multiplied, each producing a distinctive line or variety of plant that is resistant to the specific pathogen. Examples of crops transformed with plant resistance genes include hybrid rice transformed with the rice gene Xa21 coding for resistance of rice to the bacterial blight caused by *Xanthomonas oryzae* pv. *oryzae*. Transgenic plants expressing that gene displayed high, broad-spectrum resistance to *X.oryzae* pv. *oryzae* races while they maintained the high-quality agronomic characteristics. The same gene Xa21 was also transferred into elite Indica rice varieties and the transgenic plants exhibited high levels of resistance to the bacterial blight. When the resistance gene DRR206 from pea was transferred into canola, the transgenic canola plants exhibited resistance to blackleg disease, caused by the fungus *Leptosphaeria maculans*, decreased seedling mortality caused by the root pathogen *Rhizoctonia solani*, and resulted in smaller leaf lesions caused by *Sclerotinia sclerotiorum*. Similarly, in creeping bentgrass plants transformed with the *Arabidopsis thaliana* gene PR5K, which codes for a protein kinase receptor whose extracellular domain is similar to the pathogenesis-related (PR) proteins of the PR5 family, transgenic plants showed resistance to the fungus *Sclerotinia homeocarpa* causing the dollar spot disease, with the resistance appearing as a delay in disease symptoms from 29 to 45 days. When tobacco and many other plants were transformed with animal antiapoptotic genes, the plants became resistant to necrotrophic pathogens and to abiotic stresses such as heat, cold, salt, and drought, whereas transgenic plants in which this gene was destroyed were not protected against the pathogens.

Transgenic Plants Transformed with Genes Coding for Antipathogen Compounds

Genes coding for several pathogenesis-related (PR) proteins, such as chitinase and some glucanases, have been isolated, cloned, and expressed in plants, thereby interfering with the development of certain groups of pathogens and providing resistance to affected plants. Examples of plants transformed with genes coding for antipathogen compounds include peanut plants transformed with antifungal genes that reduced the incidence of Sclerotinia blight, caused by *Sclerotinia minor*, by 36% compared to susceptible nontransgenic plants. Transgenic rose plants expressing a chitinase transgene from rice showed a 13–43% decrease in symptoms caused by the fungus *Diplocarpon rosae*, the cause of rose blackspot disease. Transgenic broccoli plants expressing an endochitinase gene obtained from the biocontrol fungus *Trichoderma harzianum* had 14–200 times the endochitinase activity of that of controls and showed significantly less severe symptoms than nontransgenic plants. Similarly, transgenic cotton and tobacco plants expressing the glucose oxidase gene obtained from the biocontrol fungus *Talaromyces flavus* showed significant resistance of seedlings to the *Rhizoctonia* fungus and partial resistance to *Verticillium*, but not to *Fusarium*. The glucose oxidase generates hydrogen peroxide, which, unfortunately, is toxic to both pathogens and plants. Transgenic *A. thaliana* plants expressing one or both of the genes coding for a cysteine protein inhibitor or, to a smaller extent, a cowpea trypsin (a serine) protein inhibitor, protect plants significantly from infection by several types of nematodes, especially the reniform nematode *Rotyenchus reniformis*. Similar increases in resistance have been shown for transgenic tobacco plants overexpressing a glutamate decarboxylase gene, which makes the plants resistant to the root knot nematode; for transgenic tobacco and potato expressing the bacterial gene ubiC, which leads to accumulation of toxic 4-hydroxybenzoic acid glucosides; for canola (*Brassica napus*) plants expressing an antimicrobial peptide, which makes the plants resistant to the blackleg disease caused by the fungus *L. maculans*, and for peanut plants transformed with genes coding for antifungal enzymes, which make plants resistant to *S. sclerotiorum*.

Transgenic Plants Transformed with Nucleic Acids That Lead to Resistance and to Silencing of Pathogen Genes

Inserting segments of viral or other nucleic acids into plant genomes often leads to silencing of genes of the virus or subsequent pathogens that have homologous sequences, thereby making the plants resistant. For example, insertion of a nontranslatable coat protein coding sequence of the *tobacco etch virus* (TEV) produces transgenic plants that develop symptoms on inoculated leaves but the rest of the plant remains free of symptoms. Similarly, insertion of a gene for a double-stranded RNase from a yeast into the genome of pea plants made the transgenic pea plants resistant to multiple viruses.

There are several well-documented cases of successful transformation of a susceptible into a resistant crop through genetic engineering with parts of the genome of the virus. The first such case was of course tobacco transformed with the coat protein gene of *tobacco mosaic virus*, which then became resistant to that virus. Other cases include squash transformed with the coat protein genome of *cucumber mosaic*, *squash mosaic*, and *watermelon mosaic viruses* and papaya transformed with the coat protein gene of *papaya ring spot virus* (Fig. 9-19). Some other susceptible plants made resistant by transforming them with the nucleic acid coding for the coat protein include soybean transformed with the CP nucleic acid of *soybean mosaic virus*, tobacco with the *tobacco vein mottling virus*, cantaloupe with the CP nucleic acids of *cucumber mosaic virus*, *zucchini yellow mosaic virus*, and *watermelon mosaic virus-2*; of potato with *potato leafroll virus*; tomato to *cucumber mosaic virus*; squash to *squash mosaic virus*; citrus to *citrus tristeza virus*; chili peppers to *cucumber mosaic virus* and *tobacco mosaic virus*; and oilseed rape to *turnip mosaic virus*.

In some cases, other types of the viral nucleic acid were used to transform the plant. For example, peanut plants were transformed and made resistant to the *tomato spotted wilt virus* by transferring into them an antisense nucleocapsid gene sequence, whereas transgenic potato plants were made highly resistant to *potato virus Y* by incorporating an antisense orientation of its P1 gene in the potato genome. In still other cases, plants are transformed by introduction into their genome of the viral replicase gene, as, for example, in transgenic wheat infected with *wheat streak mosaic virus*, in potato with *potato leafroll virus*, and so on.

Successful transformation of plants for resistance to virus diseases has been obtained through the use of viral replicase genes, as in the cases of *tomato yellow leaf curl virus* and *cucumber mosaic virus* in tomato, *potato leafroll virus* and *potato virus Y* in potato; through the

FIGURE 9-19 Plant disease control through genetic engineering. Progression of papaya ring spot disease, caused by *papaya ringspot virus*, on normal susceptible (left) and transgenic plants (right) at 9 (A), 18 (B), and 23 (C) months after transplanting. (D) Severe debilitation of normal plants but excellent growth of the transgenic plants after 28 months in a block of transgenics surrounded by normal plants. [Photographs courtesy of S. A. Ferreira, from Ferreira *et al.* (2003). *Plant Dis.* **86**, 101–105.]

use of the movement protein genes of *raspberry bushy dwarf virus* in raspberry, *tobacco mosaic virus* in tomato, *cymbidium mosaic virus* in dendrobium; of genes coding for viral inclusions in plant cells; and of genes coding for a variety of other nonstructural proteins of viruses.

Transformation of plants such as wheat, potato, pea, tobacco, and walnut with nonviral genes has also been successful in making the transgenic plants resistant to several viruses. Some of the nonviral genes that have resulted in virus resistance include the gene for double-stranded RNase (dsRNase) from *Schizosaccharomyces pombe*, tobacco resistance gene N, mouse protein kinase, tobacco systemic-acquired resistance gene 8.2, and others.

In some cases, resistance to viruses is mediated by small, defective interfering DNAs or RNAs that occur naturally in plant cells or are produced after inoculation with a DNA or RNA virus or transgene. These small DNAs and RNAs are responsible for virus silencing as well as silencing of other genes of infected cells. Generally, a transgene is supposed to be stable. However, but inactivation of transgene activity can occur either a lack of transcription, probably as a result of DNA methylation of promoter regions and condensation of chromatin, or, most likely, by instability of the transcripts (posttranscriptional silencing) as a result of initiation of a RNA degradation process affecting transgene RNAs and RNAs homologous to it.

Transgenic Plants Transformed with Combinations of Resistance Genes

Combining a host gene for resistance with pathogen-derived defense genes or with genes coding for antimicrobial compounds provides for a broad and effective resistance in many host/pathogen combinations. This has been shown with the combination of a tobacco host gene and a *tobacco vein mottling virus* coat protein gene, which showed broad and effective resistance to potyviruses in tobacco; with the combination of the Sw-5 tomato gene for resistance to *tomato spotted wilt virus* (TSWV) and the nucleocapsid (N) protein gene of TSWV in transgenic plants, the latter showed high levels of resistance to several but not all TSWV strains. Transgenic rice plants carrying a host gene plus a promoter and first intron of the maize ubiquitin gene not only were resistant to the rice blast (*Magnaporthe grisea*) disease, they also became tolerant to several abiotic stresses, such as salt, submergence, and hydrogen peroxide.

Transgenic Plants Producing Antibodies against the Pathogens

Plants lack an antibody-making machinery, but, as mentioned elsewhere, DNA technology has made it possible to transform plants with additional genes that make possible the production of functional recombinant antibodies. Such plant-produced antibodies (plantibodies) can be full antibody molecules, Fab fragments, or single chain Fv (scFv) fragments produced against specific plant viruses and they are targeted against the same viruses. They have been fully expressed in leaves and seeds of some plants and accumulate in intercellular spaces, chloroplasts, and the lumen of the endoplasmic reticulum. scFv fragments are also expressed in intracellular spaces, which allows them to be more effective in improving the resistance of the plant against the specific virus. Several plant viruses have been shown to be suppressed in transgenic plants transformed with genes that enable them to produce full antibodies or fragments of single-chain variable regions of antibodies. Such viruses include *tobacco mosaic virus*, *potato virus X*, *potato virus Y*, and *clover yellow vein virus*. More work is needed before this method of plant disease control becomes truly effective and widely practiced.

Control through Use of Transgenic Biocontrol Microorganisms

Although the mechanisms by which biocontrol organisms affect the pathogens against which they are used are not well understood, it has become apparent that at least some of them produce antibiotics toxic to the pathogens, some produce enzymes that attack structural features, e.g., the cell wall of pathogens, some compete with the pathogen for space, nutrients, and water, and so on. Genetic engineering techniques have been used to add new genes or to enhance the genetic makeup of the biocontrol organism so that it may better attack the pathogen. Such genes include plant or microbe genes that code for toxins, enzymes, and other compounds that affect the pathogen adversely, or regulator genes that overexpress appropriate biocontrol genes already present in that organism.

DIRECT PROTECTION OF PLANTS FROM PATHOGENS

Most pathogens are endemic in an area, e.g., the apple scab fungus *Venturia inaequalis*, the crown gall bacterium *Agrobacterium tumefaciens*, and the *cucumber mosaic virus*; others are likely to arrive annually from warmer areas, e.g., the wheat stem rust fungus *Puccinia graminis*. If experience has shown that none of the other methods of control is likely to prevent a major epidemic, then the plants must be protected directly from infection by such pathogens that are likely to arrive on the plant surfaces in rather large numbers. Direct protection of plants from pathogens can be achieved in a few cases by biological controls (fungal and bacterial antagonists). Primarily, however, direct control of plant pathogens is achieved with chemical control measures, i.e., the use of chemicals for foliar sprays and dusts, seed treatments, treatment of tree wounds, and control of postharvest diseases of produce. Obviously, the value of the crop must be large enough to justify application of these control measures.

Direct Protection by Biological Controls

Biological control practices for direct protection of plants from pathogens involve the deployment of antagonistic microorganisms at the infection court before or after infection takes place. The mechanisms employed by biocontrol organisms in weakening or destroying the plant pathogens they attack are primarily their ability to parasitize the pathogens directly, production of antibiotics (toxins) against the pathogens, their ability to compete for space and nutrients and to survive in the presence of other microorganisms, production of enzymes that attack the cell components of the pathogens, induction of defense responses in the plants they surround, metabolism of plant produced stimulants of pathogen spore germination, and possibly others. Although thousands of microorganisms have been shown to interfere with the growth of plant pathogens in the laboratory, greenhouse, or field and to provide some protection from the diseases caused by them, strains of relatively few microorganisms have been registered and are available commercially for use so far. The most commonly used microorganisms include three fungi: *Gliocladium virens*, sold as GlioGard for the control of seedling diseases of ornamental and bedding plants; *Trichoderma harzianum*, sold as F-Stop, for the control of several soilborne plant pathogenic fungi; and *Trichoderma harzianum/T. polysporum*, sold as BINAB T, for the control of wood decays. The other three commercially available microorganisms are bacteria: *Agrobacterium radiobacter* K-84, sold as Gallex or Galltrol for use against crown gall; *Pseudomonas fluorescens*, sold as Dagger G for use against *Rhizoctonia* and *Pythium* damping-off of cotton; and *Baccillus subtilis*, sold as Kodiak and used as a seed treatment. Although the actual use of these biological control products is rather limited, it is expected that these and other such products will find wide acceptance and will fill a real need in the not too distant future. Table 9-3 lists the various biocontrol products that have been available commercially to date.

Fungal Antagonists

Biocontrol of Heterobasidion (Fomes) annosum *by* Phleviopsis (Peniophora) gigantea

Heterobasidion annosum, the cause of root and butt rot of conifer trees (Fig. 9-20A), infects freshly cut pine stumps and then spreads into the roots. Through the

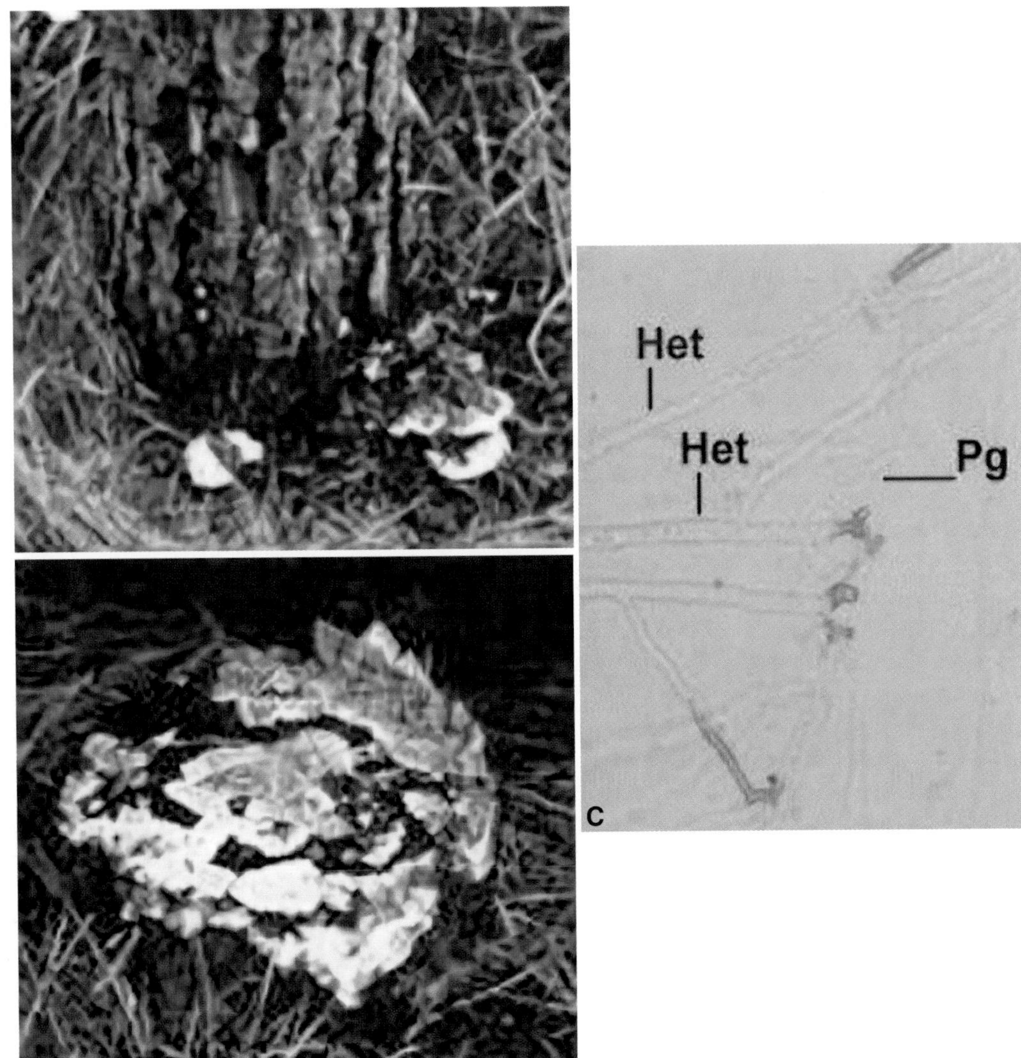

FIGURE 9-20 Biological control of a tree root rot disease. (A) Mushroom of the fungus *Heterobasidion annosum* growing at the base of pine tree killed by the fungus. (B) Stub of infected tree trunk has been inoculated with the biocontrol fungus *Phlebiopsis* (*Peniophora*) *gigantea*, which grows into the dead roots and kills the pathogen, thereby preventing it from spreading into adjacent healthy trees. (C) Interaction and killing of *Heterobasidion* mycelium by the mycelium of *Phlebiopsis*. [Photographs courtesy of (A–B) J. Rishbeth and (C) F. E. O. Ikediugwu.]

TABLE 9-3
Biocontrol Products Produced by Bacteria or Fungi and Available Commercially in the USA as of 2003

Name	Source	Target pathogen(s)	Crop(s)	Application
Bacterial				
Galltrol	*Agrobacterium radiobacter* strain 84	*A. tumefaciens* crown gall	Fruit and ornamental nursery stock grapes, brambles	Slurry to seeds, seedlings, drench
Nogall	*A. radiobacter* strain K1026	*A. tumefaciens* crown gall	Fruit, nut, and ornamental nursery stock	Suspension, drench
Companion	*Bacillus subtilis* str. GB03, other	*Pytium, Phytophthora, Fusarium, Rhizoctonia*	Many in greenhouse and nursery	Drench at planting time
HiStick N/T	*B. subtilis* str. MBI600	*Fusarium, Rhizoctonia, Aspergillus*	Legumes	Slurry to seeds
Kodiak	*B. subtilis GB03*	*Rhizoctonia solani, Fusarium, Alternaria, Aspergillus*	Cotton, legumes	Slurry to seeds
Deny	*Burkholderia cepacia,* Wisc.	*Pythium, Rhizoctonia, Fusarium,* several nematode.	Legumes, cotton, grain crops	Seed treatment
Intercept	*B. cepacia*	*R. solani, Fusarium, Pythium*	Maize, vegetables, cotton	Seed treatment, drench
BioJect Spot-Less	*Pseudomonas aureofaciens*	Dollar spot, anthracnose, *Pythium*, pink snow mold	Turf, other	Overhead irrigation
Bio-save 10LP, 110	*P. syringae*	Postharvest *Botrytis, Mucor, Penicillium, Geotrichum*	Pome fruit, citrus, cherries, potatoes	Drench, dip, spray
BlightBan A506	*P. fluorescence* A506	Frost damage, *Erwinia amylovora*, russeting bacteria	Pome and stone fruits, potatoes, tomatoes, strawberries	Spray
Dagger G	*P. fluorescens*	*Rhizoctonia, Pythium*	Field crops, vegetables	Seed treatment
Cedomon	*P. chlororaphis*	Barley, oat leaf spots, *Fusarium*	Grain cereals	Seed treatment
Fungal				
AQ10 Biofungicide	*Ampelomyces* quisqualis M-10	Powdery mildews	Apples, grapes, ornamentals, cucurbits strawberries, tomatoes	Spray
Aspire	*Candida oleophila* I-182	*Botrytis, Penicillium*	Citrus, pome fruit	Drench, drip, spray
Biotox C	Nonpathogenic *F. oxysporum*	*F. oxysporum*	Basil, carnation, tomatoes, cyclamen	Drench
Fusaclean	Nonpathogenic F. oxysporum	*F. oxysporum*	Basil, carnation, tomatoes, cyclamen	Drench
Contans WG, Intercept WG	*Coniothyrium* minitans	*Sclerotinia sclerotiorum, S. minor*	Many crops. All soils	Spray
DiTera Biocontrol	*Myrothecium verrucaria*	Parasitic nematodes	Cole crops, grape, ornamentals, turf, trees	Soil application
Polygandron	*Pythium oligsndrum*	*Pythium ultimum*	Sugar beet	
Primastop	*Gliocladium catenulatum*	Soilborne pathogens causing rots and wilts	Ornamentals, vegetables, tree crops	Drench, spray, irrigated water
RootShield, Plant Shield, T-22 Planter box	*Trichoderma harzianum,* Rifai strain — KRL_AG2(T-22)	*Pythium, Rhizoctonia, Fusarium*	Tree, shrub, ornamental, transplants, cabbage, tomato, cucumber	Mixed w/soil, soil drench
F-Stop	*T. harzianum*	*Rhizoctonia, Pythium*	Ornamental and food crops	Seed treatment
SoilGard (GlioGard)	*Gliocladium (Trichoderma) virens GL-21*	*Rhizoctonia solani, Pythium*	Ornamental and food crops, greenhouses, nurseries	Slurry, seed treatment
BINAB T	*T. harzianum/ T. polysporum*	Wood decay fungi	Trees	Spray, wound
Promote	*T. harzianum* and *T. viride*	*Pythium, Rhizoctonia, Fusarium*	Transplants, trees	
Rotstop	*Phlebia gigantea*	*Heterobasidion annosum*	Trees	
Trichodex	*T. harzianum*	*Colletotrichum, Monilin., Plasmopara Rhizop. Sclerotinia*	Various	
Trichopel,	*T. harzianum*, and *T. viride*	*Armillaria*, Botryosphaerim, *Fusarium*	Various	
Trichoject		*Nectria, Phytpphthora, Pythium, Rhizoctonia*		

root contacts, it then spreads into the roots of standing trees, which it kills. If the stump surface is inoculated with oidia of the fungus *Phleviopsis* (*Peniophora*) *gigantea* immediately after the tree is felled, *Phleviopsis* occupies the cut surface (Fig. 9-20B) and spreads through the stump into the lateral roots. There, it competes with successfully and excludes or replaces the pathogenic *Heterobasidion* (Fig. 9-20C) in the stump, thereby protecting nearby trees. Oidia are applied to the cut surface either as a water suspension or as a powder or they are added to the lubricating oil placed on the chain saw and are thus deposited on the surface as it is cut.

Biocontrol of Chestnut Blight with Hypovirulent Strains of the Pathogen

Chestnut blight, caused by the fungus *Cryphonectria* (*Endothia*) *parasitica*, is controlled naturally in Italy and artificially in France through inoculation of cankers caused by the normal pathogenic strains of the fungus, with hypovirulent strains of the same fungus. The hypovirulent strains carry virus-like double-stranded RNAs (dsRNAs) that apparently limit the pathogenicity of the virulent strains. The dsRNAs apparently pass through mycelial anastomoses (fusions) from the hypovirulent to the virulent strains, the latter are rendered hypovirulent, and the development of the canker slows down or stops. Hypovirulent strains are also being tested in the United States, but so far the control of chestnut blight has been limited to experimental trees. It appears that U.S. strains are more variable than European ones and require multiple hypovirulence factors, therefore, the interaction is more strain specific.

Biological Control of Soilborne Diseases

Principal fungi used as biological control agents against soilborne diseases include the two mentioned earlier as being used commercially, namely *Gliocladium virens* and *Trichoderma harzianum*. They are used in potting mixes, are mixed with soil, or are used as solid matrix in seed-priming treatments. They are effective against damping-off diseases of ornamentals and vegetables caused by the oomycetes *Pythium* and *Phytophthora*, by the fungi *Botrytis* (Fig. 9-21) and *Sclerotium*, and some other fungi. In addition, *Sporidesmium sclerotivorum*, *Coniothyrium minitants*, *Talaromyces flavus*, and others give experimental control of some diseases caused by *Sclerotinia*, *Rhizoctonia*, and *Verticillium*. Some species of *Pythium*, such as *Pythium nunn* and *P. oligandrum*, protect potted ornamentals and vegetables from plant pathogenic species of *Pythium*, whereas *T. flavus* and binucleate Rhizoctonias protect plants from the pathogenic *Rhizoctonia solani*. Experimental control of several other soilborne diseases has been obtained with many other fungi. Finally, Fusarium wilts of several crops, such as celery, cucumber, and sweet potato, caused by the respective formae specialis of *Fusarium oxysporum* have been reported to be controlled successfully by inoculating transplants or cuttings with nonpathogenic strains of the same fungus. Some of these strains have been isolated from the vascular tissue of host plants that remained healthy while nearby plants had been killed by the wilt-inducing strains of the fungus. It is believed that the nonpathogenic strains not only compete with the pathogenic ones in the rhizosphere and for infection sites, but they also enhance the resistance of the host toward the pathogenic strains.

FIGURE 9-21 Comparison of biocontrol and other control methods of begonia plants inoculated with *Botrytis cinerea* and kept under optimum conditions for development of the disease. From left to right: Un, untreated control; CaCl, calcium chloride; Fung, fungicide (chlorothalonil) treatment; and T382, treatment with the biocontrol agent *Trichoderma hamatum* strain T382 inoculated into the potting mix. (Photograph courtesy of H. A. J. Hoitink.)

Roots of most plants form a symbiotic relationship with certain kinds of zygomycete, ascomycete, and basidiomycete fungi that exist as mycorrhizae. Mycorrhizae colonize roots intercellularly (ectomycorrhizae) or intracellularly (endomycorrhizae). Although mycorrhizae obtain organic nutrients from the plant, they benefit the plant by promoting nutrient uptake and enhancing water transport by the plant, thus increasing growth and yield, and sometimes by providing the plant with considerable protection against several soilborne pathogens. Mycorrhizae have been shown to provide considerable protection to pine seedlings from *Phytophthora cinnamomi*, to tomato and Douglas fir seedlings from *F. oxysporum*, to cotton from *Verticillium* wilt and the root-knot nematode, and to soybean from *Phytophthora megasperma* and *Fusarium solani*. Commercial preparations of certain mycorrhizal fungi are available for promoting growth of the host plants; however, problems of production, specificity, and application of mycorrhizae to plants remain and, therefore, they are not used with the goal of protecting plants from their pathogens.

Biological Control of Diseases of Aerial Plant Parts with Fungi

Many filamentous fungi and yeasts have been shown to be effective antagonists of fungi infecting the aerial parts of plants. For example, inoculation of postbloom, dead tomato flowers with conidia of *Cladosporium herbarum* or *Penicillium* sp. almost completely suppressed the subsequent infection of developing fruits by *Botrytis cinerea*. Similarly, spraying spores of common bark saprophytes, such as *Cladosporium* sp. and *Epicoccum* sp., and of the soil fungus *Trichoderma* on pruning cuts of fruit trees has prevented infection by canker-causing pathogens such as *Nectria galligena* and *Leucostoma* (*Cytospora*) sp. Sprays with *Trichoderma* in the field also reduced Botrytis rot of strawberries and of grapes at the time of harvest and in storage. Sclerotinia head rot of sunflower was reduced significantly by releasing into the field honeybees that had been previously contaminated heavily with spores of the biocontrol fungi *Trichoderma* spp., which the honeybees delivered promptly to the flowers. Several foliar diseases have also been reduced significantly (by more than 50%) when the leaves were sprayed with spores of common phylloplane fungi, e.g., *Alternaria*, *Cochliobolus*, *Septoria*, *Colletotrichum*, and *Phoma* or with spores of hyperparasites, e.g., the cucumber powdery mildew fungus *Sphaerotheca fuliginea* with spores of *Ampelomyces quisqualis* or *Tilletiopsis*, the wheat leaf rust fungus *Puccinia triticina* with spores of *Darluca filum*, and the carnation rust fungus with *Verticillium lecanii*. None of the aforementioned nor any of the numerous other known cases of fungal antagonism by fungi are used as yet for the practical control of any disease of aerial plant parts.

Biological Control of Postharvest Diseases

Postharvest rots of several fruits could be reduced considerably by spraying the fruit with spores of antagonistic fungi and saprophytic yeasts at different stages of fruit development, or by dipping the harvested fruit in the inoculum. For example, yeast treatments reduced postharvest rotting of peach and apple. Also, significant reduction of citrus green mold (caused by *Penicillium digitatum*) was obtained by treating the fruit with antagonistic yeasts (Fig. 9-22A) or the fungal antagonist *Trichoderma viride*, whereas preharvest and postharvest Botrytis rot of strawberries was reduced by several sprays of *Trichoderma* spores on strawberry blossoms and young fruit. Penicillium rot of pineapple was reduced considerably by spraying the fruit with nonpathogenic strains of the pathogen. Similarly, several antagonistic yeasts protected grapes and tomatoes from Botrytis, Penicillium (Fig. 9-18B), and Rhizoctonia rots. One such yeast, *Candida saitoana*, controlled postharvest decay of apple fruit by inducing systemic resistance in apple fruit while at the same time increasing chitinase and β-1,3-glucanase activities in the fruit. In addition, the yeast *Candida oleophila* was approved for postharvest decay control in citrus and apples under the trade name Aspire.

Bacterial Antagonists

Biocontrol of Soilborne Diseases

Crown gall of pome, stone, and several small fruits (grapes, raspberries) and ornamentals (rose and euonymous) is caused by the bacterium *Agrobacterium tumefaciens*. Crown gall can be controlled commercially by treating the seeds, seedlings, and cuttings with Galltrol, a suspension of strain K84 of the related but nonpathogenic bacterium *Agrobacterium radiobacter*. Control is based on production by strain K84 of a bacteriocin, called agrocin 84, which is an antibiotic specific against related bacteria. The bacteriocin selectively inhibits most pathogenic agrobacteria that arrive at surfaces occupied by strain 84. Because strains of *A. tumefaciens* insensitive to agrocin 84 were produced as a result of the natural transfer of the K84 gene for resistance to agrocin 84, a new strain (K-1026) was produced from K84 through genetic engineering so that the latter (K-1026) lacks the ability to transfer its resistance gene to pathogenic *Agrobacterium* strains.

Treatment of seeds such as cereals, sweet corn, and carrots with water suspensions, slurries, or powders containing the bacteria *Bacillus subtilis* strain A13 or *Streptomyces* sp. has protected the plants against root pathogens and has resulted in better growth and yield of these crops.

Pseudomonas rhizobacteria, primarily of the *P. fluorescens*, *P. putida*, *P. cepacia*, and *P. aureofaciens* groups, applied to seeds, seed pieces, and roots of plants have resulted in less damping-off, less soft rot, and consistent increases in growth and yield in several crops. Formulations of two of the aforementioned bacteria are

FIGURE 9-22 Biological control of postharvest diseases of fruit. (A) Oranges treated with yeasts (right) remained healthy, whereas oranges not treated with yeast developed extensive decay following inoculation with *Penicillium*. (B) Apples 3 months after they had been wounded and inoculated with fungi *Penicillium* and *Botrytis* with (right) or without (left) treatment with the biocontrol bacterium *Pseudomonas syringae (*BioSave 110). (C) Peaches at left were protected from infection by the brown rot fungus (*Monilinia fructicola*) by prior treatment with the nonpathogenic bacterium *Bacillus subtilis*. Untreated peaches (right) became severely rotten within 8 days from inoculation with *M. fructicola* [Photographs courtesy of (A and C) C. L. Wilson and (B) W. Janisiewicz, USDA.]

sold commercially: *B. subtilis* is sold as Kodiak and *P. fluorescens* is sold as Dagger G. All have given good results in experimental trials but have, in general, given inconsistent results in large-scale trials. For example, in some experiments, treated potato seed tubers produced from 5 to 33% greater yield; treated sugar beet seeds produced 4 to 6 tons more of sugar beets per hectare, corresponding to an increase of from 955 to 1,227 kg sugar per acre; treated radish seeds produced from 60 to 144% more root weight than untreated ones; and treated wheat seed planted in soil infested with *Gaeumannomyces graminis* var. *tritici* (take-all of wheat) produced 27% more yield than untreated seed. The most common soilborne diseases controlled by soilborne bacteria are damping-off and root rot diseases caused by the oomycetes *Pythium* and *Phytophthora* and by fungi *Rhizoctonia*, *Fusarium*, and *Gaeumannomyces*. *Bacillus cereus*, especially *B. cereus* strain UW85, provides effective biocontrol of damping-off diseases of legumes. Three fluorescent pseudomonads and a species of *Pantoea*, used alone or together as a seed treatment for wheat, reduced seedling death by *Fusarium culmorum* and increased crop stand and yield equal to that of fungicides; an isolate of *Pantoea* increased yields by an average of 200 kg per acre. *Pasteuria penetrans*, however, parasitizes and controls the root-knot nematode, and unspecified bacteria inhibit hatching of nematode eggs. The mechanism(s) by which these plant growth-promoting rhizobacteria increase yield is not clear. It appears, however, that inhibition of harmful, toxic microorganisms and of soilborne pathogens by antibiotics, or by competition for iron, is involved in at least some of the determinants of their effectiveness.

Biological Control of Diseases of Aerial Plant Parts with Bacteria

Numerous bacteria, most of them saprophytic gram-negative bacteria of the genera *Erwinia*, *Pseudomonas*, and *Xanthomonas* and a few of the gram-positive genera *Bacillus*, *Lactobacillus*, and *Corynebacterium*, are found on aerial plant surfaces, particularly early in the growing season. Some pathogenic bacteria, such as *Pseudomonas syringae* pv. *syringae*, *P. syringae* pv. *morsprunorum*, *P. syringae* pv. *glycinea*, *Erwinia amylovora*, and *E. carotovora*, also live epiphytically (on the surface) on leaves, buds, and so on before they infect and cause disease. In several cases, spraying leaf surfaces with preparations of saprophytic bacteria or with avirulent strains of pathogenic bacteria has reduced considerably the number of infections caused by bacterial and fungal pathogens. For example, fire blight of apple blossoms, caused by *E. amylovora*, was partially controlled with sprays of *Erwinia herbicola*; and bacterial leaf streak of rice, caused by *Xanthomonas translucens* ssp. *oryzicola*, was reduced with sprays of isolates of *Erwinia* and of *Pseudomonas*.

Several cases of added epiphytic bacteria inhibiting plant infections by fungi are known. For example, spraying grass plants with *Pseudomonas fluorescens* reduced infection by *Drechslera (Helminthosporium) dictyoides*, and spraying with *Bacillus subtilis* reduced infection of apple leaf scars by *Nectria galligena* and of grape by *Eutypa lata*. Similarly, spraying of peanut or tobacco plants with *Pseudomonas cepacia* or *Bacillus* sp. reduced *Cercospora* and *Alternaria* leaf spot on these hosts, respectively. None of these biological controls is used in practice to control any disease of aerial plant parts so far.

Biocontrol of Postharvest Diseases

Pseudomonas bacteria protected lemons from Penicillium green mold and pear from various storage rots. Two *Pseudomonas syringae* strains have been approved for postharvest decay control in citrus, apples (Fig. 9-22B), and pears under the trade name Bio-Save. When several kinds of stone fruits, namely peaches, nectarines, apricots, and plums, were treated after harvest with suspensions of the antagonistic bacterium *Bacillus subtilis*, they remained free of brown rot, caused by the fungus *Monilinia fructicola*, for at least nine days (Fig. 9-22C). *Bacillus subtilis* also protected avocado fruit from storage rots.

Biocontrol with Bacteria of Bacteria-Mediated Frost Injury

Frost-sensitive plants are injured when temperatures drop below 0°C because ice forms within their tissues. Small volumes of pure water can be supercooled to −10°C or below without ice formation, provided no catalyst centers or nuclei are present to influence ice formation. It has been shown, however, that certain strains of at least three species of epiphytic bacteria (*P. syringae*, *P. fluorescens*, and *E. herbicola*), which are present on many plants, serve as ice nucleation-active catalysts for ice formation at temperatures as high as −1°C. Such bacteria usually make up a small proportion (0.1–10%) of the bacteria found on leaf surfaces. By isolating, culturing, mass producing, and applying non-ice nucleation-active bacteria antagonistic to ice-nucleation-active bacteria on the plant surfaces, it has been possible to reduce and replace large numbers of ice nucleation-active bacteria on treated plant surfaces with non-ice nucleation-active bacteria. This treatment protects frost-sensitive plants from injury at temperatures at which untreated plants may be severely injured.

Viral Parasites of Plant Pathogens

All pathogens — fungi, bacteria, mollicutes, and nematodes — are attacked by viruses. So far, however, for bacterial pathogens, viruses have been tested as possible biological controls only. **Bacteriophages** or **phages** (bacteria-destroying viruses) are known to exist in nature for most plant pathogenic bacteria. Successful experimental control of several bacterial diseases was obtained when the bacteriophages were mixed with the inoculated bacteria, when the plants were first treated with bacteriophages and then inoculated with bacteria, and when the seed was treated with the phage. In practice, however, not one bacterial disease is controlled effectively by bacteriophages. Also, no plant disease caused by a bacterium has been cured yet by treatment with phage after the disease has developed.

Inoculation of plants with mild strains of viruses to protect them from more severe strains is, of course, an effective biocontrol strategy, as discussed earlier in the section on cross protection. It is possible, however, to also significantly protect plants from some viruses, and even from some viroids, by preinoculating the plants with mild strains of the virus containing a satellite RNA (satRNA). Many satellite RNAs act as parasites of the viruses that reproduce the satRNA by competing with them, reducing the concentration the viruses can reach in plants, and thereby often causing milder symptoms and less damage to the plant. Although cases of fungal hypovirulence, such as that in the chestnut blight fungus *Cryphonectria parasitica*, are caused by what appear to be viral double-stranded RNAs, it has not yet been possible to transmit such RNAs other than by mycelial anastomosis and so their usefulness as biocontrol agents is limited.

BIOLOGICAL CONTROL OF WEEDS

Weeds, i.e., wild plants that thrive where they are not wanted, not only are a serious nuisance in lawns and gardens, they also clog waterways, displace useful plants from pastures, and, most importantly, grow among cultivated crops and compete with them for nutrients, water, and light. By so doing, weeds cause world crop losses worth approximately $150 billion annually, which is equal to about one-third of all crop losses in the world, with the other losses being caused by insects (about $135 billion) and diseases (about $190 billion).

Weeds used to be managed entirely by plowing, hoeing, or pulling them up by hand. After World War II, however, several chemical herbicides were discovered, and their use increased rapidly and dramatically. By 1990, more than $12 billion was spent annually on herbicides, a sum equal to that spent on pesticides used for the control of all other pests and diseases. The widespread use of herbicides and other pesticides, however, raised many concerns about food, water, and farm worker safety, and intensive efforts began to find alternative control methods. An alternative weed control strategy is biological control through the use of natural microorganisms (and insects) that infect and damage or kill weeds.

Microorganisms used as biocontrol agents of weeds are generally fungi pathogenic to specific weeds, to which they cause significant damage or death. Weed pathogens are isolated from infected weeds locally or elsewhere in the world. They are then grown and multiplied in culture and are tested for their efficiency in infecting and damaging or killing the specific weed in the laboratory, the greenhouse, and in the field. They are also tested for their specificity, i.e., inability to infect cultivated hosts of related plant genera and families. In addition to fungi pathogenic to weeds, numerous bacteria and several viruses infecting weeds have been tested as weed biocontrol agents.

Of the several hundred, perhaps thousands, of pathogen–weed host combinations tested, several dozen of them were shown to be quite effective. Some of the weed pathogens are now sold commercially. For example, the fungus *Colletotrichum gloeosporioides*, sold as Collego, is effective against the northern jointvetch weed (*Aeschynomene virginica*) growing in rice and soybean fields in several states; and *Phytophthora palmivora*, sold as DeVine, is effective against the milkweed vine weed (*Morrenia adorata*) growing in citrus orchards of Florida. Two other weed pathogens are expected to be marketed in the near future: *Alternaria cassiae*, to be sold as Casst, is effective against the weed sicklepod (*Cassia obtusifolia*) growing in peanut and soybean crops in the southeastern United States (Fig. 9-23); and another strain of *C. gloeosporioides*, to be sold as BioMal, is effective against the weed roundleaf mallow (*Malva pusilla*) growing in small-grain fields in North America. A few other, among the many, promising pathogens of weeds that are likely to be used as biocontrol agents against these weeds in the near future include *Phomopsis amaranticola* against pigweed (*Amaranthus* spp.) (Fig. 9-23), *Colletotrichum dematium* f. sp. *crotalariae* against showy *crotalaria* (*Crotalaria spectabilis*) (Fig. 9-23), *Alternaria helianthi* against cocklebur (*Xanthium pennsylvanicum*), *Alternaria macrospora* against spurred anoda (*Anoda cristata*), *Colletotrichum coccodes* against velvetleaf (*Abutilon theophrasti*), and *Cercospora rodmanii* against the waterweed water hyacinth (*Eichhornia crassipes*) (Fig. 9-24). Many other fungi (*Ascochyta*, *Bipolaris*, *Fusarium*, *Phoma*, *Puccinia*, *Sclerotinia*, etc.) have been studied extensively as mycoherbicides of various weeds.

Considerable effort has been made since the mid-1990s toward discovering fungi and bacteria that produce natural phytotoxic substances that can damage or kill weeds when sprayed on them. Also, efforts have been made either to increase the pathogenicity or to limit the host range of a given mycoherbicide on its weed host through genetic engineering techniques. Efforts are also being made to promote the use of mixtures of mycoherbicides for controlling several important weeds in a crop or waterway at once; to determine the best time of application of a mycoherbicide; and, not unlike the approach with fungicides, to determine how many applications of a mycoherbicide will be required for satisfactory control of weeds in a crop.

Biological control of weeds, so far, has not received sufficient attention from the industry or the grower clientele because herbicides are available that are usually broad spectrum, relatively inexpensive, and dependably effective. However, as environmental concerns increase the constraints placed on herbicides, there may be few available herbicides or their uses may be limited drastically. At the same time, as more and more effective biocontrol agents are discovered and better formulations and methods of application of the biocontrol agents on weeds are developed, it may not be too long before the biological control of weeds becomes not only accepted, but also an effective and, indeed, the preferred method for weed control.

DIRECT PROTECTION BY CHEMICAL CONTROLS

One of the most common means of controlling plant diseases in the field, in the greenhouse, and, sometimes, in storage is through the use of chemical compounds

FIGURE 9-23 (Left column) Biological control of the weed sicklepod [*Senna* (*Cassia*) *obtusifolia*] by the fungus *Alternaria cassiae*. (A) Typical field symptoms on sicklepod seedlings 10 days after treatment with fungal spores. (B) Treated seedlings become defoliated and nearly always are killed (two seedlings at right, compared to two controls). (C) Soybean field infested heavily with sicklepod, showing square area where the weed was almost completely eliminated within 7 days following treatment with *A. cassiae* spores. (Right column) Biological control of pigweed (*Amaranthus* spp.) and showy crotalaria (*Crotalaria spectabilis*) with fungi. (D) Necrotic leaf spots and (E) stem lesions on pigweed caused by the fungus *Phomopsis amaranticola*. (F) Healthy showy crotalaria seedlings (left) and seedlings killed by the biocontrol agent fungus *Colletotrichum dematium* f. sp. *crotalariae* (right). (Photographs courtesy of R. Charudattan, University of Florida.)

FIGURE 9-24 Biological control of the aquatic weed water hyacinth (*Eichhornia crassipes*) with the fungus *Cercospora rodmanii*. (A) Necrotic areas on leaf from plants treated with fungal spores. (B) Water hyacinth plants in a plot treated with fungicide that excluded the fungus but allowed insect predators of the weed to feed. (C) Water hyacinth plants treated with the fungus and the insect predators but no insecticide or fungicide. Plots shown in both B and C were photographed 8 weeks after treatment with the fungus and the insect biocontrol agents. (Photographs courtesy of R. Charudattan, University of Florida.)

that are toxic to the pathogens. Such chemicals either inhibit germination, growth, and multiplication of the pathogen or are outright lethal to the pathogen. Depending on the kind of pathogens they affect, the chemicals are called fungicides, bactericides, nematicides, viricides, or, for the parasitic higher plants, herbicides. Some chemicals are broad-spectrum pesticides, i.e., they are toxic to all or most kinds of pathogens, whereas others affect only a few or a single specific pathogen. About 60% of all the chemicals (mostly fungicides) used to control plant diseases is applied to fruit and about 25% to vegetables.

Most of the chemicals are used to control diseases of the foliage and of other aboveground parts of plants. Others are used to disinfest and/or protect from infection seeds, tubers, and bulbs. Some are used to disinfest the soil, others to disinfest warehouses, to treat wounds, or to protect stored fruit and vegetables from infection. Still others (insecticides) are used to control the insect vectors of some pathogens.

In earlier years, chemicals applied on plants or plant organs worked by modifying reactive groups of numerous enzymes and, therefore, were confined to the plant surfaces and could not act as postinfection fungicides without causing phytotoxicity. Such chemicals could not stop or cure a disease after it had started. The great majority of these older chemicals are effective only in the plant area to which they have been applied (local action) and are not absorbed or translocated by the plants. Many new chemicals, however, do have a therapeutic (eradicant) action, and several are absorbed and translocated systemically by the plant (systemic fungicides and antibiotics). Of the three main groups of systemic fungicides, benzimidazoles and sterol demethylation inhibitors require specific action between the chemical inhibitor and a fungel component. This is most often achieved by an optimized fit of inhibitors into their fungal target sites. In contrast, the third group, strobilurin-related fungicides, are strong inhibitors of respiration and their mechanisms of specificity are secondary responses, such as alternative respiration or detoxification.

Methods of Application of Chemicals for Plant Disease Control

Chemicals used to control plant diseases are applied directly to plants or to the soil with the help of various types of equipment (Figs. 9-25–9-27 and 9-31).

Foliage Sprays and Dusts

Chemicals applied as sprays or dusts on the foliage of plants are usually aimed at control of fungus (and oomycete) diseases and, to a lesser extent, control of bacterial diseases. Most fungicides and bactericides are **protectants** and must be present on the surface of the plant in advance of the pathogen in order to prevent infection. Their presence usually does not allow fungus spores to germinate or the chemicals may kill spores on

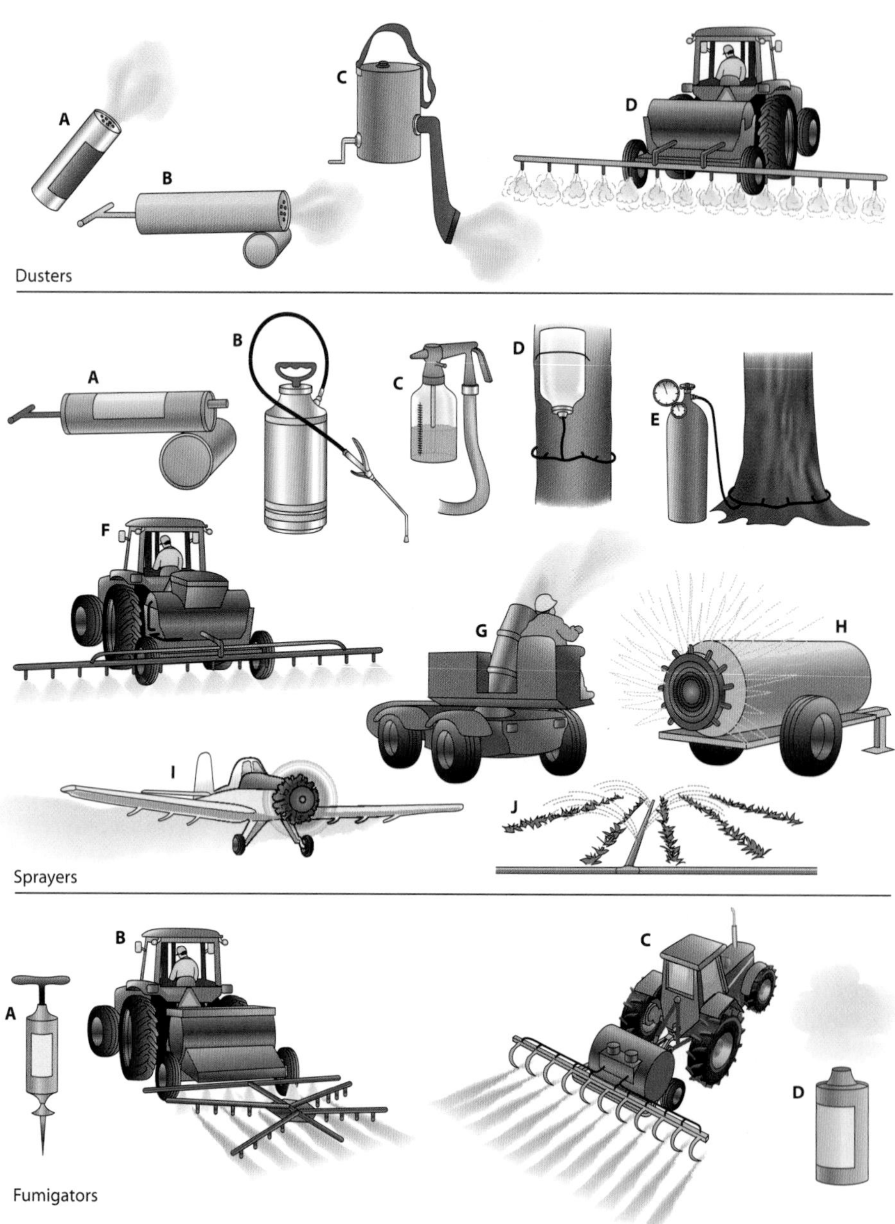

FIGURE 9-25 Various types of equipment used for the control of plant diseases by dusting, spraying, injection, or fumigation. Dusters: portable dusters (A–C) and tractor-mounted duster (D). Sprayers: portable sprayers (A–C), tree injection gravity-flow apparatus (D), apparatus for tree injection under pressure (E), tractor-mounted sprayers for annuals (F) and for trees (G and H), airplane spraying (or dusting) (I), and spraying through the irrigation system (J). Fumigators: handgun fumigator (A), tractor-mounted gravity-flow (B) or pump-driven injectors (C), and fumigation can for greenhouse or warehouse (D).

FIGURE 9-26 Application of fungicides (and insecticides) in a fruit tree orchard using a tractor-mounted, high-pressure, high-volume air blast sprayer. (Photograph courtesy of R. J. McGovern, University of Florida.)

FIGURE 9-27 Methods of fungicide spray application and some results. (A) Multinozzle vegetable and other row crop sprayer. (B) Spraying or dusting by airplane. (C) Development of severe leaf spot in a sugarbeet field sprayed with a fungicide to which the previously susceptible pathogen had developed resistance. (D) A large, tractor-operated boom duster. [Photographs courtesy of (A) and (B) USDA, (C) C. E. Windels from Windels *et al.* (1998). *Plant Dis.* **82**, 716–726 and (D) Agric. Japan, 67 (1995).]

germination. Contact of bacteria with bactericides may inhibit their multiplication or cause their death.

Some newer fungicides also have a direct effect on pathogens that have already invaded the leaves, fruit, and stem, and in this case they act as **eradicants**, i.e., they kill the fungus inside the host or may suppress the sporulation of the fungus without killing it. Some fungicides have a partial systemic action because they can be absorbed by parts of the leaf tissues and translocated internally into the whole leaf area. Several fungicides (e.g., benomyl, thiabendazole, carboxin, and matalaxyl) are clearly systemics and can be translocated internally throughout the host plant. Some bactericides, such as streptomycin, tetracyclines, and some other antibiotics, are also systemics, especially when applied by injection.

Some newer systemic fungicides, such as metalaxyl and the **sterol inhibitors** triadimefon and fenarimol, are so effective in postinfection applications that they can be used as **rescue treatments** of crops; in other words, they can be applied effectively after infection has already taken place. This use pattern is not generally recommended, however, because it is contrary to best practices for the management of pathogen resistance.

Fungicides and bactericides applied as sprays (Figs. 9-26 and 9-27) are generally more efficient in creating a protective residue layer on the plant surfaces than when applied as dusts. Neither dusts nor sprays stick well when applied during a rain. Sometimes other compounds, e.g., lime, may be added to the active chemical in order to reduce its phytotoxicity and make it safer for the plant. Compounds with a low surface tension, called **surfactants**, are often added to fungicides so that they spread better, thereby increasing the contact area between fungicide and the sprayed surface. Some compounds with good sticking ability (**stickers**) are added to increase the adherence of the fungicide to the plant surface. Finally, there are certain spreader–sticker compounds that have both properties. A newer product derived from grain by-products is added to adjuvants, allowing the plant organs to transpire without sealing the stomata.

In fields with sprinkler irrigation available, some control of foliar diseases can be obtained by applying protectant or systemic fungicides to the foliage, and somewhat to the roots, through the irrigation system (**fungigation**).

Because many fungicides and bactericides are protectant in their action, it is important that they be at the plant surface before the pathogen arrives or at least before it has had time to germinate, enter, and establish itself in the host. Because most spores require a film of water on the leaf surface or at least atmospheric humidity near saturation before they can germinate, different devices (Fig. 9-28) are used to monitor changes in weather, especially temperature and moisture. In general, sprays are more effective when they are applied before or immediately after every rain. Considering that many fungicides and bactericides are effective only on contact with the pathogen, it is important that the whole surface of the plant be covered with the chemical in order to ensure protection. Some limited redistribution of fungicides between the areas covered by spray droplets generally occurs, however. For this reason, young, expanding leaves, twigs, and fruits may have to be sprayed more often than mature tissues, as small, growing leaves may outgrow protection 3 to 5 days after spraying. The interval between sprays of mature tissue may vary from 7 to 14 days or longer, depending on the particular disease, the frequency and duration of rains, the persistence or residual life of the fungicide, and the season of the year. The same factors also determine the optimal number of sprays per season, which may vary from 0 to 15 or more (Fig. 9-29).

Since the mid-1970s, several systemic fungicides have become available, and their number, ease of application, duration of effectiveness, and even the number of diseases they control are increasing steadily. Systemic chemicals are gradually replacing many of the contact, preventive fungicides because of their effectiveness and long-lasting activity, which result in the need for only a limited number of applications to protect a crop from one or many diseases.

The number and variety of chemicals used for foliar sprays are quite large, and new chemicals, even new classes of chemicals, are being added from time to time. Some fungicides available in the past have been banned by the U.S. Environmental Protection Agency (EPA) or by the U.S. Food and Drug Administration, or they have been discontinued by the manufacturer. Many of the remaining chemicals are being reviewed for safety and efficacy, and, possibly, some of them, too, may be banned as dangerous to humans or the environment. Some fungicides are specific against certain diseases, whereas others are effective against a wide spectrum of pathogens. Sprays with these materials usually contain 0.5 to 2 pounds of the compound per hundred gallons of water, although some are applied at a few ounces and others, e.g., sulfur, at 4 to 6 pounds per 100 gallons of water. Some of the fungicides used for foliar sprays are also used for seed treatments following appropriate reformulation.

Seed Treatment

Seeds, tubers, bulbs, and roots are often treated with chemicals to prevent their decay after planting or the damping-off of young seedlings. These chemicals may

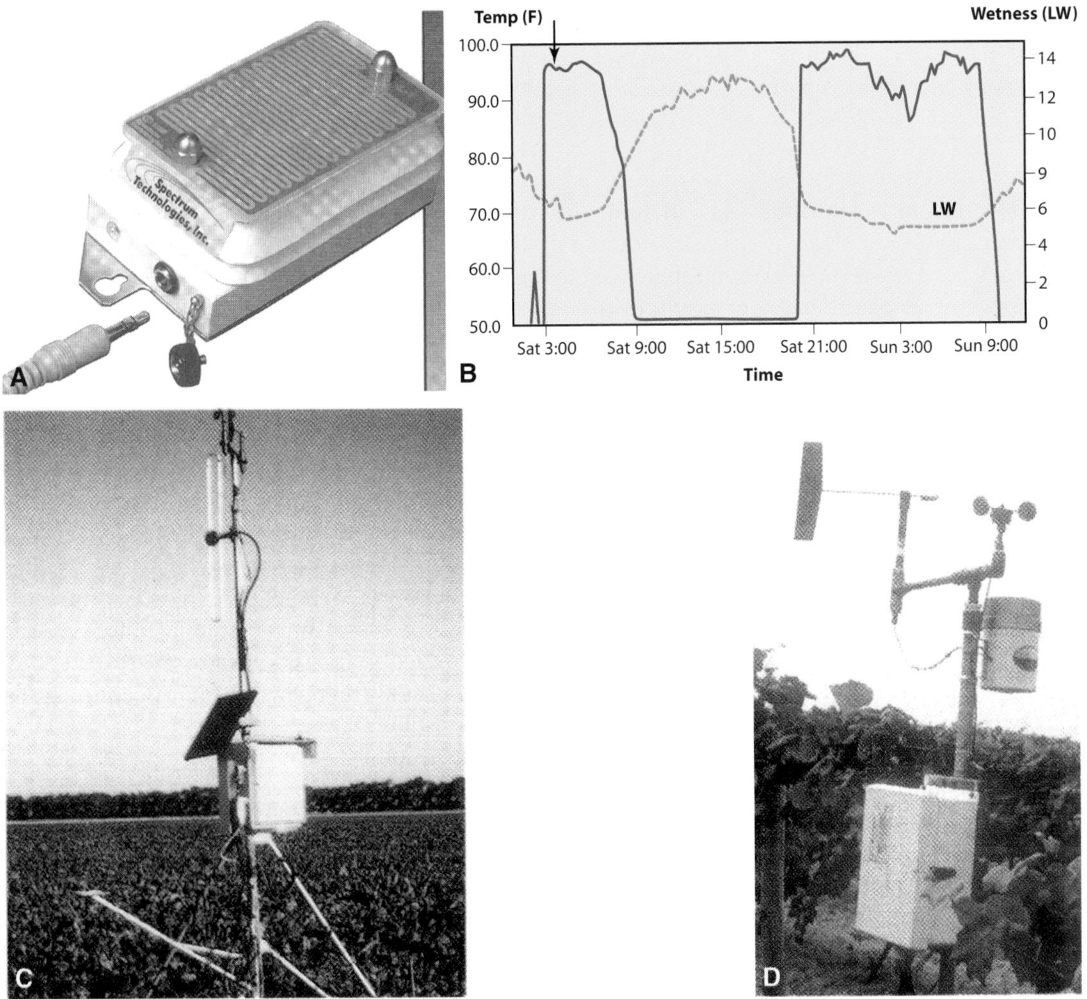

FIGURE 9-28 Weather monitoring equipment for plant disease control. (A) Temperature and surface wetness monitor. (B) A typical graph of the relationship between temperature and surface wetness during a 30-hour period. (C and D) Two types of complete weather monitoring systems (C: From Windels et al. 1998, Plant Dis., 82: 716–726.).

control pathogens carried on seeds, tubers, and so on, or existing in the soil where they will be planted. Since the mid-1970s, seeds have been treated with systemic fungicides in order to inactivate pathogens in infected seeds (e.g., carboxin for control of loose smut) or in order to provide the foliage of the developing plant with systemic protection against the pathogen (e.g., metalaxyl for the control of downy mildews of oats and sorghum and triadimenol for the control of leaf rust and *Septoria* leaf blotch of wheat, and of *Pyrenophora* net blotch of barley). Chemicals can be applied on the seed as dusts or as thick water suspensions (slurries) mixed with the seed. The seed can also be soaked in a water or solvent solution of the chemical and then allowed to dry. Tubers, bulbs, corms, and roots can be treated in similar ways, but treatments are effective mostly when the chemical is applied to protect such organs, when they are healthy, from infection in the field (Fig. 9-30) rather than eliminate pathogens that have already infected these organs.

In treating seeds or any other propagative organs with chemicals, precautions must be taken so that their viability is not lowered or destroyed. At the same time, enough chemical must stick to the seed to protect it from attacks of pathogens. When the seed is planted the chemical diffuses into the soil and disinfests a sphere of soil around the seed. In this way, the new plant will grow without being attacked at this particularly vulnerable period of growth.

Most treatments of propagative stock are with organic protectant compounds such as captan, chloroneb, maneb, mancozeb, thiram, pentachloronitrobenzene (PCNB), and the systemic compounds carboxin, benomyl, thiabendazole, metalaxyl, and triadimenol. Some chemicals may control specific diseases of some plants, whereas others are more general in their action and may control many diseases of a number of plants.

Soil Treatment

Volatile chemicals (fumigants) are often used to fumigate the soil before planting for reducing the inoculum of nematodes, fungi, and bacteria. Certain fungicides are applied to the soil as dusts, drenches, or granules to control damping-off, seedling blights, crown and root rots, and other diseases. Such fungicides include PCNB, metalaxyl, triadimefon, ethazol, and propamocarb. Some of the systemic fungicides may provide season-long control from a single preplant application. In some cases, foliar diseases (e.g., downy mildews and rusts) can be controlled by incorporating the fungicide (e.g., metalaxyl, triadimenol) into the fertilizer and applying them together before planting.

Highly volatile chemicals are applied to the soil with tractors dragging devices equipped with chisels (Figs. 9-31A and 9-31D) that release the chemical 6–12 inches deep into the soil and the treated area is covered immediately with plastic (Figs. 9-31A and 9-31B) to keep the chemical from escaping prematurely. Granular materials and low-volatility liquid pesticides either are broadcast on the soil and then disked into the soil (Fig. 9-31C) or are injected into the soil through chisels but, usually, without being covered with plastic afterward (Fig. 9-31D). Protective and systemic fungicides have also been applied to the soil (and to the foliage) through irrigation water (fungigation) for the control of soilborne diseases.

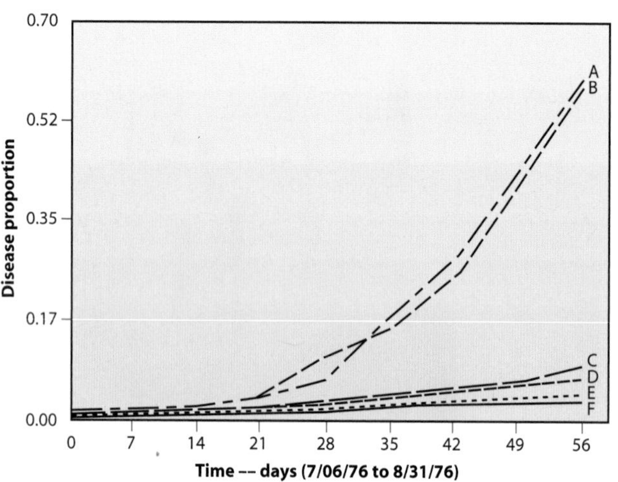

FIGURE 9-29 Proportion of tomato plant tissue infected by the early blight fungus *Alternaria solani* after application of different numbers of fungicidal sprays. A, 0 sprays; B, 1 spray; C, 3 sprays; D, 7 sprays; E, 10 sprays; and F, 13 sprays. Treatments C and D were applied on the basis of weather data; treatments E and F were applied on the basis of the earlier recommended rigid schedule. [From Madden *et al.* (1978). *Phytopathology* 68, 1354–1358.]

Treatment of Tree Wounds

Large pruning cuts and wounds made on the bark of branches and trunks accidentally or in the process of removing infections of fungi and bacteria need to be protected from drying and from becoming ports of entry for new pathogens. Drying of the margins of large tree wounds is usually prevented by painting them with shellac or any commercial wound dressing. The exposed

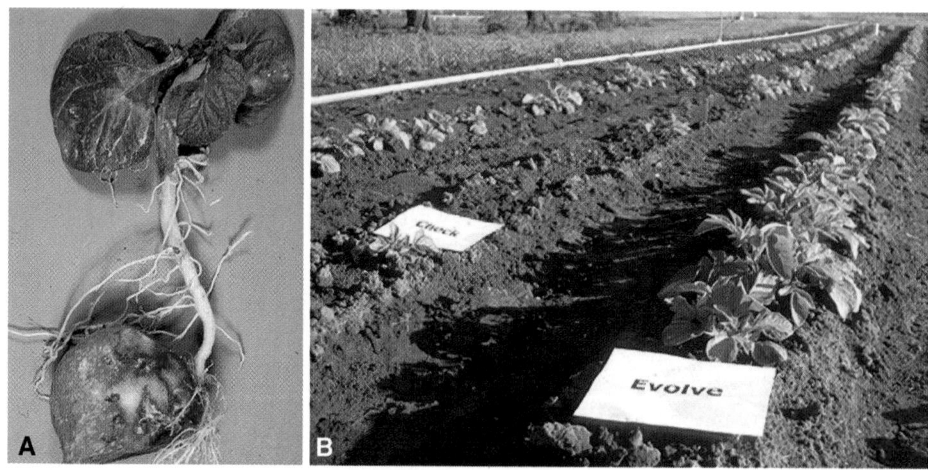

FIGURE 9-30 (A) Young potato plant growing from a late blight-infected potato tuber becomes infected and is soon killed by the late blight oomycete *Phytophthora infestans*. (B) Potato field showing the result of planting untreated tubers (center row) and healthy tubers treated with an effective fungicide. [Photographs courtesy of M. L. Powelson, from Powelson *et al.* (2002). healthy Online, Plant Health Progress.]

FIGURE 9-31 Equipment for application of soil pesticides and fumigants. (A) Tractor applying a fumigant and laying plastic over it to keep the chemical from early escape. (B) Field beds treated with a volatile chemical and covered with plastic. (C) Multidisk tractor used to incorporate nonvolatile granular chemicals in soil. (D) Broadcast chisel application of low-volatility liquid fumigants into soil. [Photographs courtesy of (A) R. T. McMillan and (B–D) D. W. Dickson, University of Florida.]

wood is then sterilized by swabbing it with a solution of either 0.5–1.0% sodium hypochlorite (10–20% Clorox bleach) or 70% ethyl alcohol. Finally, the entire wound is painted with a permanent tree wound dressing, such as a 10:2:2 mixture of lanolin, rosin, and gum; or Cerano, or Bordeaux paint, or an asphalt-varnish tree paint. Some wound dressings, such as Cerano and Bordeaux paint, are themselves disinfectants, whereas most others require the addition of a disinfectant, such as 0.25% phenyl mercuric nitrate or 6% phenol. It must be kept in mind, however, that many commercial wound dressings, especially those that are asphalt based, are phytotoxic enough to prevent, rather than promote, wound healing.

In commercial orchards, vineyards, and so on, where the number of wounds created during the annual pruning operations is too large to treat individually, the wounds are protected from infection by spraying as soon after pruning as practical with one of several fungicides, including benomyl, dichlone, and captafol.

Control of Postharvest Diseases

The use of chemicals for the control of postharvest diseases of fruits and vegetables is complicated by the fact that compounds effective against storage diseases may leave visible residues on the produce that detract from marketability. Excessive residues of some compounds may also be toxic to consumers. Many chemicals also cause injury to the products under storage conditions and give off undesirable odors.

A number of fungicides, however, have been developed that are used primarily or specifically for the control of postharvest diseases. Most of them are used as dilute solutions into which the fruits or vegetables are dipped before storage or as solutions used for the washing of fruits and vegetables immediately after harvest (Fig. 9-3B). Some chemicals, e.g., elemental sulfur, are used as dusts or crystals that undergo sublimation in storage, and others, such as SO_2, as gases.

Finally, some chemicals are impregnated in the boxes or wrappers containing the fruit. Among the compounds used for commercial control of postharvest diseases of, primarily, citrus fruits but also of other fruits, are borax, biphenyl, sodium *o*-phenylphenate, and the widely used fungicides benomyl, thiabendazole, and imazalil. Chlorinated water is used routinely in dump tanks to wash and treat tomatoes and certain vegetables in commercial packinghouses. The chlorinated water prevents the accumulation of pathogens in the water, as well as helps reduce populations on the surface of produce. In the last few years, ozonated water was shown to have similar protective effects on pear fruit. Certain other chemicals, such as elemental sulfur, sulfur dioxide, dichloran, captan, and benzoic acid, have been used mostly for the control of storage rots of stone and pome fruits, bananas, grapes, strawberries, melons, and potatoes.

Types of Chemicals Used for Plant Disease Control

Many hundreds of chemicals have been advanced to date for crop protection as fumigants, soil treatments, sprays, dusts, paints, pastes, and systemics. Some of the most important of these chemicals and some of their properties and uses are described. It should be noted that some of these chemicals have not yet been registered for any use in the United States; in addition, some are being reviewed, and their use may be canceled in the near future. The use of chemical compounds that act as plant defense activators, such as salicyclic acid (SA), isonicotinic acid (INA), phenolic acids, and the commercially available benzothiadiazole known as BTH or Actigard, which activate the natural defense of the host (systemic acquired resistance), has been described elsewhere.

Inorganic Chemicals

Copper Compounds

The Bordeaux mixture, named after the Bordeaux region of France where it was developed and used against the downy mildew of grape, is the product of reaction of copper sulfate and calcium hydroxide (hydrated lime). It was the first fungicide to be developed and still is the most widely used copper fungicide throughout the world. It controls many fungal (including oomycete) and bacterial leaf spots, blights, anthracnoses, downy mildews, and cankers. The Bordeaux mixture, however, can cause burning of leaves or russeting of fruit such as apples when applied in cool, wet weather. The phytotoxicity of Bordeaux is reduced by increasing the ratio of hydrated lime to copper sulfate. Copper is the only ingredient in the Bordeaux mixture that is toxic to pathogens and, sometimes, to plants, whereas the role of lime is primarily that of a "safener." For dormant sprays, concentrated Bordeaux is made by mixing 10 pounds of copper sulfate, 10 pounds of hydrated lime, and 100 gallons of water; it has the formula 10:10:100. The most commonly used formula for Bordeaux is 8:8:100. For spraying young, actively growing plants, the amounts of copper sulfate and hydrated lime are reduced, and the formulas used may be 2:2:100, 2:6:100, and so on. For plants known to be sensitive to Bordeaux, a much greater concentration of hydrated lime may be used, as in the formula 8:24:100.

In "fixed" or "insoluble" copper compounds, the copper ion is less soluble than that in the Bordeaux mixture. These compounds are, therefore, less phytotoxic than Bordeaux but are effective as fungicides. Fixed coppers are used for control of the same diseases as Bordeaux. Fixed coppers contain basic copper sulfate, sold as Microcop and many other names; copper oxychlorides, sold as Oxycor or C-O-C-S; copper hydroxide, sold as Kocide, Champ and Nu-Cop; copper oxides, sold as Nordox; copper ammonium carbonate, sold as Copper Count-N, Kop-R-Spray; or miscellaneous other copper sources.

Inorganic Sulfur Compounds

The element sulfur is probably the oldest fungicide known. As a dust, wettable powder, paste, or liquid, sulfur is used primarily to control powdery mildews, certain rusts, leaf blights, and fruit rots. Sulfur, in its different forms, is available under a variety of trade names, e.g., Microthiol, Disperss, and Thiolux. Sulfur may cause injury in hot (temperatures above 30°C), dry weather, especially to sulfur-sensitive plants such as tomato, melons, and grape, and when used in combination with spray oils and certain other insecticides.

By boiling lime and sulfur together, lime–sulfur, self-boiled lime–sulfur, and dry lime–sulfur are produced. These, sold as Lime Sulfur, Orthorix, Sulforix, or Polysul, are used as sprays for dormant fruit trees to control blight or anthracnose, powdery mildew, apple scab, brown rot of stone fruits, and peach leaf curl, and are sometimes used for summer control of the same diseases.

Carbonate Compounds

Sodium bicarbonate as well as bicarbonate salts of ammonium, potassium, and lithium plus 1% superfine oil were shown to be inhibitory and fungicidal to the powdery mildew fungi of roses, to several fungi infecting cucumber, to the black spot fungus of roses, to the southern blight fungus *Sclerotium rolfsii*, and to the gray mold fungus *Botrytis cinerea*.

Phosphate and Phosphonate Compounds

Spraying cucumber or grape plants with solutions of either monopotassium phosphate (KH_2PO_4) or dipotassium phosphate (K_2HPO_4) gave satisfactory control of the powdery mildew diseases of these two hosts.

Film-Forming Compounds

Film-forming compounds, such as antitranspirant polymers, mineral oils, surfactants, and kaolin-based particle films Fig. 9-1, applied on plant surfaces before inoculation with the pathogen reduce the number of infections significantly. Most of these film-forming polymers are permeable to gases, are nonphytotoxic, resist weathering for at least one week, and are biodegradable. They seem to reduce infections by altering the characteristics of the leaf surface, thereby interfering with the adhesion of the pathogen to the host and with the recognition of infection sites on the host.

Organic Chemicals

Contact Protective Fungicides

Organic Sulfur Compounds: Dithiocarbamates. Organic sulfur compounds are one of the most important, versatile, and widely used group of modern fungicides. They include thiram, ferbam, nabam, maneb, zineb, and mancozeb. They are all derivatives of dithiocarbamic acid. It is believed that dithiocarbamates are toxic to fungi mainly because they are metabolized to the isothiocyanate radical (—N=C=S). This radical inactivates the sulfhydryl groups (—SH) in amino acids and in enzymes within pathogen cells, thereby inhibiting the production and function of these compounds.

Thiram consists of two molecules of dithiocarbamic acid joined together. It is used mostly for seed and bulb treatment for vegetables, flowers, and grasses but also for the control of certain foliage diseases, such as rusts of lawns. Thiram, in various formulations, is sold under many trade names. Thiram and Tersan are two examples.

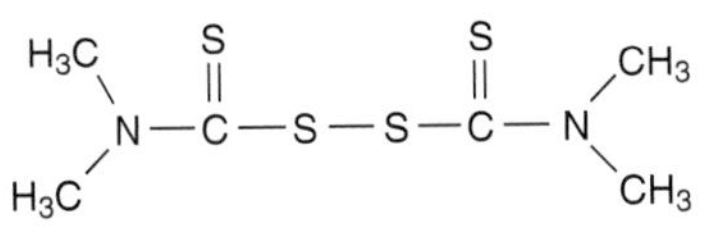

Thiram

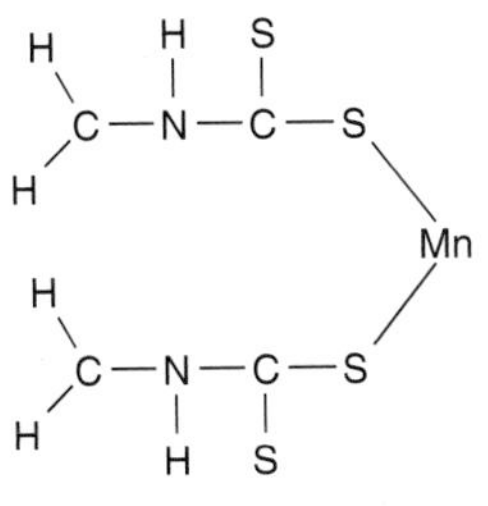

Maneb

Ferbam consists of three molecules of dithiocarbamic acid reacted to one atom of iron. Ferbam is sold as Ferbam or Carbamate and is used to control foliage diseases, especially fruit on trees and ornamentals.

Ethylenebisdithiocarbamates. Another group of dithiocarbamic acid derivatives with different molecular configurations, the ethylenebisdithiocarbamates, contains the fungicides maneb and zineb. Maneb contains manganese; it is sold as maneb and Tersan LSR. It is an excellent, broad-spectrum fungicide for the control of foliage and fruit diseases of many vegetables, especially tomato, potato, and vine crops, and of flowers, trees, turf, and some fruit. Maneb is sometimes mixed with zinc or with zinc ion and results in the formulations known as maneb zinc (sold as Manzate D) and zinc ion maneb, called mancozeb (sold as Manzate 200, Dithane M-45, and Pencozeb). The addition of zinc reduces the phytotoxicity of maneb and improves its fungicidal properties. A secondary effect of the use of mancozeb is to supply Mn and Zn to deficient plants.

Zineb is sold as Dithane Z-78. It is an excellent, safe, multipurpose foliar and soil fungicide for the control of leaf spots, blights, and fruit rots of vegetables, flowers, fruit trees, and shrubs.

Quinones. Quinones, which occur naturally in many plants and are also produced upon oxidation of plant phenolic compounds, often show antimicrobial activity and are often considered to be associated with the innate resistance of plants to disease. Only two quinone compounds, chloranil and dichlone, however, have been developed, but they are no longer used commercially as fungicides in the United States.

Aromatic Compounds. Many rather unrelated compounds that have an aromatic (benzene) ring are toxic to microorganisms, and several have been developed into fungicides and are used commercially. Most seem to inhibit production of compounds that have $—NH_2$ and —SH groups, namely amino acids and enzymes.

Pentachloronitrobenzene, sold as PCNB Terraclor, Engage, and Defend, is a long-lasting soil fungicide. It controls various soilborne diseases of vegetables, turf, and ornamentals and is applied as a dip or in the furrow at planting time. It is used primarily against *Rhizoctonia* and *Plasmodiophora*.

Dichloran (DCNA), sold as Botran and Allisan, is used as foliar and fruit fungicide or postharvest spray for diseases of vegetables and flowers caused mostly by *Botrytis*, *Sclerotinia*, or *Rhizopus*.

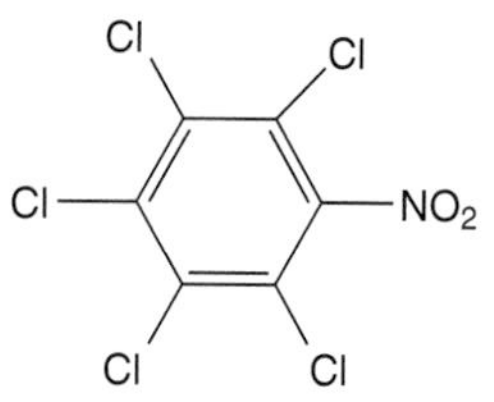

Pentachloro-nitrobenzene

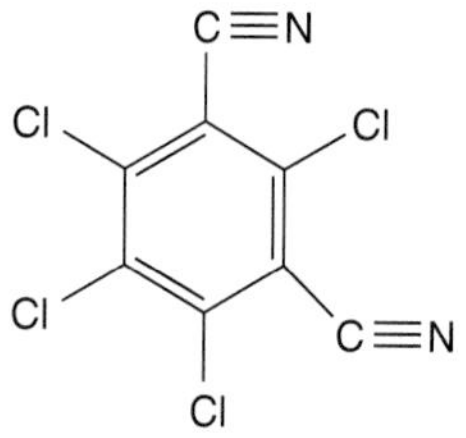

Chlorothalonil

Chlorothalonil, available as Bravo, Daconil, Exotherm Termil, and several other trade names, is an excellent broad-spectrum fungicide against many leaf spots, blights, downy mildews, rusts, anthracnoses, scabs, and fruit rots of many vegetables, field crops, ornamentals, turf, and even trees. A tablet formulation of chlorothalonil called Termil is dispersed thermally in greenhouses for the control of *Botrytis* on many ornamentals and for several leaf molds and blights of tomato.

Biphenyl, sold as biphenyl, has been used widely for control of postharvest diseases of citrus caused by *Penicillium*, *Diplodia*, *Botrytis*, and *Phomopsis*. Biphenyl is volatile and is applied by impregnating shipping materials with it; the compound then volatilizes in storage and protects the stored fruit.

Heterocyclic Compounds. Heterocyclic compounds are a rather heterogeneous group but include some of the best fungicides, e.g., captan, iprodione, and vinclozolin. Most of them also inhibit the production of essential compounds containing $—NH_2$ and —SH groups (amino compounds and enzymes).

Captan, is an excellent fungicide for the control of leaf spots, blights, and fruit rots on fruit crops, vegetables, ornamentals, and turf. It is also used as a seed protectant for agronomic crops, vegetables, flowers, and grasses. Captan has also been reported to repel "seed-pulling" birds.

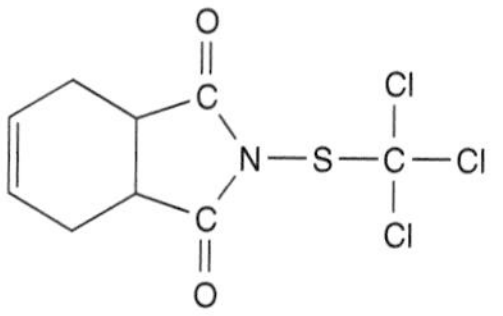

Captan

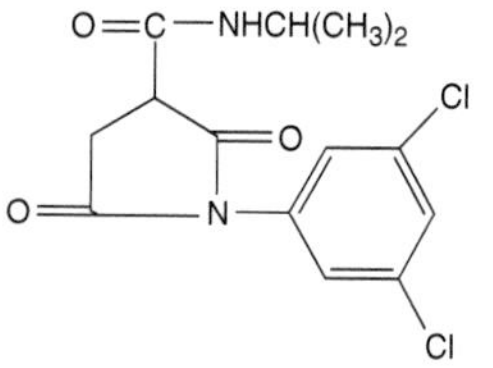

Iprodione

Iprodione, sold as Rovral, Chipco-26019, and Epic 30, is a broad-spectrum, foliage-contact fungicide. It inhibits spore germination and mycelial growth but shows mostly preventative and only early curative activity. It is effective against *Botrytis*, *Monilinia*, and *Sclerotinia* and also against *Alternaria* and *Rhizoctonia*. It is applied most often as a foliar spray and also as a postharvest dip and as a seed treatment. Iprodione is used on turf, stone fruits, grapes, peanuts, onions, lettuce, and other crops.

Flutolanil, a benzanilide, sold as Moncut, Contrast, or Prostar, is used as a protective systemic and curative fungicide against the basidiomycetes *Rhizoctonia*, *Sclerotium rolfsii*, *Corticium*, and *Typhula*. It is applied as a spray.

Vinclozolin, sold as Ornalin, Ronilan, Touché or Vorlan, is a contact, protective fungicide, effective against sclerotia-producing ascomycetes (*Botrytis*, *Monilinia*, *Sclerotinia*) and other fungi. It is used mostly as a spray on strawberries, lettuce, turf, aornamentals, and on fruit.

Organic Compounds: Systemic Fungicides

Systemic fungicides are absorbed through the foliage or roots and are translocated within the plant through the xylem. Systemic fungicides generally move upward in the transpiration stream and may accumulate at the leaf margins. A few of them, e.g., fosetyl-Al, also move downward. These fungicides are not re-exported to new growth. Some of them become translocated systemically when sprayed on herbaceous plants, but most are only locally systemic within the sprayed leaves. Many systemics are effective when applied as seed treatments, root dips, in-furrow treatments or soil drenches, and in trees when injected into the trunks.

Several systemic fungicides are currently available, and many more, belonging to many different groups of compounds, are being developed. Almost all systemic fungicides are site specific, inhibiting only one or perhaps a few specific steps in the metabolism of the fungi they control. As a result, many target fungi through simple mutation and become resistant to each frequently used systemic fungicide within a few years of introduction of the compound. For this reason, various strategies have been developed for preserving the usefulness of such chemicals. To avoid abandonment of a systemic fungicide after appearance of a pathogen strain resistant to it, the fungicide must be used in combination with another broad-spectrum contact fungicide under various schemes of application.

Acylalanines. The most important acylalanine is the fungicide metalaxyl. It is effective against the oomycetes *Pythium*, *Phytophthora*, and several of the downy mildews. It is sold as Ridomil for use in the soil and, in conjunction with a companion broad-spectrum fungicide, on foliage. It is also sold as Apron for use as a seed dressing and as Subdue for use on ornamentals and turf. Metalaxyl is one of the best systemic fungicides against oomycetes. It is widely used as a soil or seed treatment for the control of *Pythium* and *Phytophthora* seed rot and damping-off and as soil treatment for the control of *Phytophthora* stem rots and cankers in annuals and perennials and of certain downy mildews (e.g., of tobacco). It is also effective as a curative treatment if it has to be applied after infection has begun. Metalaxyl is quite water soluble and is translocated readily from roots to the aerial parts of most plants, but its lateral translocation is slight. Because the use of metalaxyl has already resulted in the appearance of strains resistant to it in some pathogens, it is recommended that it be used in combination with other, broad-spectrum fungicides.

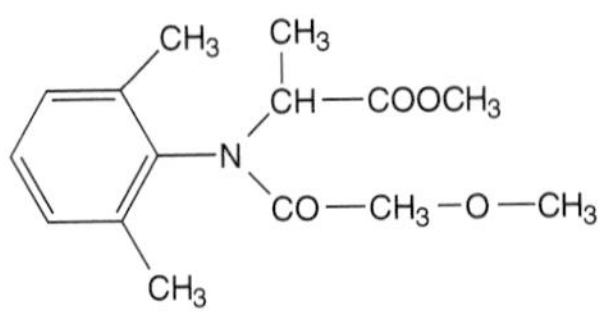

Metalaxyl

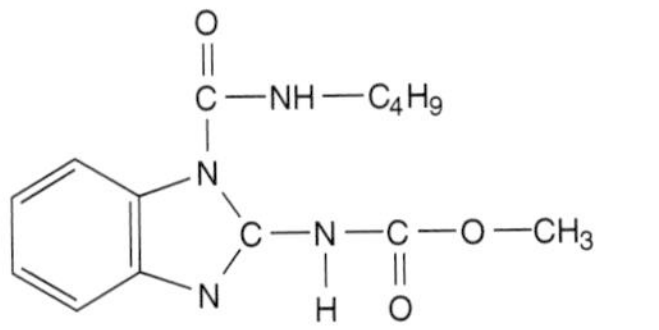

Benomyl

Benzimidazoles. They include some important systemic fungicides, such as benomyl, carbendazim, thiabendazole, and thiophanate. They are effective against numerous types of diseases caused by a wide variety of fungi. Most benzimidazoles are converted at the plant surface to methyl benzimidazole carbamate (MBC, carbendazim), and this compound interferes with nuclear division of sensitive fungi.

Benomyl is sold as Benlate, Tersan 1991, and others. It is a safe, broad-spectrum fungicide, effective against a large number of important fungal pathogens. It controls a wide range of leaf spots and blotches, blights, rots, scabs, and seed-borne and soilborne diseases. Benomyl is particularly effective for powdery mildew of all crops; scab of apples, peaches, and pecans; brown rot of stone fruits; fruit rots in general; *Cercospora* leaf spots; cherry leaf spot; black spot of roses; blast of rice; and various *Sclerotinia* and *Botrytis* diseases. It is highly active against and suppresses infection by *Rhizoctonia*, *Thielaviopsis*, *Ceratocystis*, *Fusarium*, and *Verticillium*. It has no effect on oomycetes, on some dark-spored imperfect fungi such as *Bipolaris*, *Drechslera*, and *Alternaria*, on some Basidiomycetes, and on bacteria. Benomyl may be applied as a seed treatment, foliar spray, trunk injection, root dip, or row treatment, and as a fruit dip.

Thiabendazole is sold as Mertect 340-F, Arbotect 20-S, and Decco Salt No.19. It is also a broad-spectrum fungicide and is effective against many imperfect fungi causing leaf spot diseases of turf and ornamentals and diseases of bulbs and corms. It is commonly used as a postharvest treatment for the control of storage rots of citrus, apples, pears, bananas, potatoes, and squash.

Thiophanate, under the trade name Topsin, is effective against several root and foliage fungi affecting turf grasses and vegetable crops.

Thiophanate methyl, under the trade names Fungo, Topsin M, Domain, Cavalier, Halt, etc., is a broad-spectrum preventive and curative fungicide for use on turf and as a foliar spray to control powdery and downy mildews, *Botrytis* diseases, numerous leaf and fruit spots, scabs, and rots. It is also used as a soil drench or dry soil mix to control soilborne fungi attacking bedding plants, foliage plants, and container-grown plants.

Oxanthiins. Oxanthiins were the first fungicides to be discovered as having systemic activity (1966). They include primarily carboxin and oxycarboxin and are effective against some smut and rust fungi and against *Rhizoctonia*. Oxanthiins are selectively concentrated in cells of these fungi and inhibit succinic dehydrogenase, an enzyme important in mitochondrial respiration.

Carboxin is sold as Vitavax. It is used as a seed treatment and is effective against damping-off diseases caused by *Rhizoctonia* and against the various smuts of grain crops.

Oxycarboxin is sold as Plantvax or Carbojec. It is sometimes used as a seed or foliar treatment and is effective in controlling a wide variety of rust diseases.

Flutolanil, is sold as Contrast, ProStar, and Moncut.

Nicobifen, belonging to a new anilide family of fungicides, interferes with mitochondrial respiration and energy production, moves in a translaminar and acropetal systemic manner, and controls a range of Ascomycetes on many crops.

Organophosphate Fungicides

Organophosphates include primarily fosetyl-Al, sold as Aliette, and phosphorous acid, sold as Fosphite. Aliette is very effective against foliar, root, and stem diseases caused by oomycetes such as *Phytophthora*, *Pythium*, and downy mildews in a wide variety of crops. It is applied as a foliar spray, soil drench, root dip, or postharvest dip, and in soil incorporation. Treatments may be effective for 2 to 6 months, depending on the crop. Fosetyl-Al has been reported to stimulate defense reactions and the synthesis of phytoalexins against oomycetes. Three other compounds are also included: kitazin (IBP) and edifenphos (Hinosan), both effective against rice blast and several other diseases, and pyrazophos (Afugan), which is effective against powdery mildews and *Bipolaris* and *Drechslera* diseases on various crops.

Pyrimidines. Pyrimidines include diamethirimol (Milcurb), ethirimol (Milstem), and bupirimate (Nimrod), all effective against powdery mildews of various crop plants. Fenarimol (Rubigan) and nuarimol (Trimidal) are effective against powdery mildews and also several other leaf spot, rust, and smut fungi.

Triazoles. Triazoles (conazoles or imidazoles) include several excellent systemic fungicides, such as triadimefon (Bayleton), triadimenol (Baytan), bitertanol (Baycor), difenoconazole (Divident, Score), fenbuconazole or butrizol (Indar or Enable), propiconazole (Tilt, Orbic, Banner, Alamo), etaconazole or cyprodinil (Vangard), myclobutanil (Rally, Immunox, Spectracide-Pro, Nova, Eagle, Systhane), cyproconazole (Sentinel), and tebuconazole (Elite, Folicure, Raxil, Lynx). They show long protective and curative activity against a broad spectrum of foliar, root, and seedling diseases such as leaf spots, blights, powdery mildews, rusts, smuts, and others caused by many ascomycetes, imperfect fungi, and basidiomycetes. They are applied as foliar sprays and as seed and soil treatments.

Strobilurins or QoI Fungicides

This group contains the newest and most important fungicides. The first such fungicide was isolated from the wood-rotting mushroom fungus *Strobilurus tenacellus* and was thought to help the fungus defend itself from other microbes present in rotting wood. Subsequently, chemists produced more effective and more stable strobillurin compounds. It was also determined that strobilurins have a common mode of action, i.e., all of them interfere with respiration, i.e., energy production, in the fungal cell. They do that by blocking electron transfer at the site of quinol oxidation (the Qo site) in the cytochrome bc1 complex, thereby preventing ATP formation. Strobilurins are, therefore, site-specific fungicides and as such are subject to selecting for fungicide-resistant strains of fungi and development of fungicide (strobilurin)-resistant pathogen populations.

All strobilurins are absorbed by treated leaves and other plant parts and at first are held on or within the waxy cuticle of plant surfaces. Some of the active ingredient subsequently moves into the underlying plant cells and may reach and build up again at the cuticle of the other side of leaves. Strobilurins, therefore, move trans-laminarly within leaves. In addition, some strobilurins, such as azoxystrobins, move trans-laminarly and systemically through the vascular system of the plant. Some strobilurin fungicides show growth-promoting effects on treated plants, apparently by delaying leaf senescence and having water-conserving effects. Strobilurins have been shown to be phytotoxic to plants of certain genotypes, i.e., plants of certain varieties, such as MacIntosh apples, Concord grapes, and certain sweet cherries, are sensitive to strobilurins whereas other varieties of these crops are not affected.

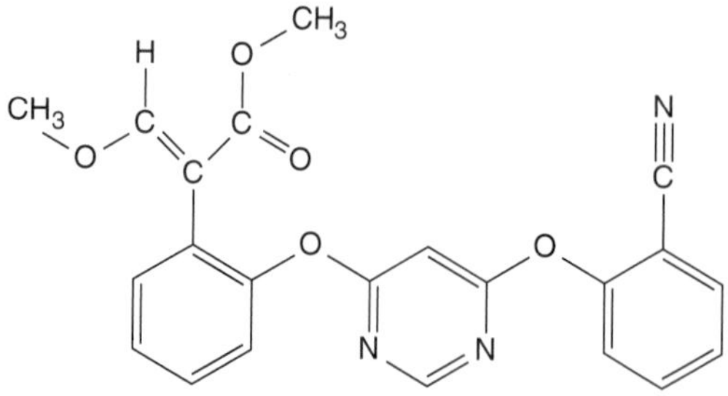

Azoxystrobin

The most important strobilurins are Azoxystrobin, sold as Abound, Heritage, or Quadris; Trifloxystrobin, sold as Flint for use on grapes, pome and stone fruits, cucurbits, fruiting vegetables, as Gem for use on citrus, rice, and sugar beets, and as Stratego for ruse on peanuts, wheat, corn and rice. Kresoxim methyl, sold as Sovran and Cygnus. Pyraclostrobin, sold as Insignia.

Strobilurins are effective against most fungal diseases of most crops. However, they seem to increase the severity of a few diseases, probably by eliminating or suppressing some naturally occurring microorganism(s) that is antagonistic to the pathogen.

Miscellaneous Systemics. Several excellent systemic fungicides of different chemical composition and affiliation are included in the miscellaneous category.

Chloroneb, sold as chloroneb, is used as a seed treatment for beans, soybeans, and cotton and as a soil fungicide for turf and ornamentals. It is sometimes applied as a seed overcoat to seed treated with standard fungicides. It does not leach from the soil.

Ethazol, sold as Truban, Terrazole, or Koban, is a seed, soil, and turf fungicide effective against damping-off and root and stem rots caused by *Pythium* and *Phytophthora*. It is often sold combined with PCNB or with thiophanate methyl (Banrot) for broader spectrum application, particularly against *Fusarium* and *Rhizoctonia*.

Imazalil, sold as Fungaflor, Flo-Pro, or Nu-Zone, is effective against many ascomycetes and imperfect fungi causing powdery mildews, leaf spots, fruit rots, and vascular wilts. It is applied as a foliar spray, a seed treatment, and as a postharvest treatment. It has excellent curative and preventive properties.

Triflumizole, another imidazole, is sold as Procure or Terraguan.

Prochloraz, also an imidazole sold as Prochloraz, is effective against ascomycetes and imperfects causing powdery mildews, leaf spots, and blights and fruit rots. It is used as a spray or a seed treatment.

Propamocarb, sold as Banol and Previcur, is effective against *Pythium*, *Phytophthora*, downy mildews, some rusts, and others. It is applied as a seedling dip, soil drench, seed treatment, soil surface spray, and foliar spray.

Triforine, sold as Funginex or Triforine, is effective against many ascomycetes and imperfect fungi causing powdery mildews, foliar and fruit spots, fruit rots, anthracnose, and some basidiomycetes causing rusts. It is used as a foliar spray.

Miscellaneous Organic Fungicides

A number of other, chemically diverse compounds are excellent protectant fungicides for certain diseases or groups of diseases.

Dodine is sold as Syllit. (Its former name, Cyprex, has been withdrawn.) It is an excellent fungicide against apple scab and also controls certain foliage diseases of cherry, strawberry, pecan, and roses. It gives long-lasting protection and is also a good eradicant. It appears to have limited local systemic action in leaves. Strains of the apple scab fungus resistant to dodine have appeared and predominate in some areas.

Fentin hydroxide, sold as Super Tin, is a broad-spectrum fungicide with activity against many leaf spots, blights, and scabs. It also has suppressant or antifeeding properties on many insects.

Fludioxonil, is used for different crops and postharvest uses.

Famoxadone, sold as Famoxate, is effective against Ascomycetes and Oomycetes on cucurbits, etc.

Oxyquinoline sulfate (as well as the benzoate and citrate salts) has been used as a soil drench to control damping-off and other soilborne diseases. An oxyquinoline–copper complex has also been used as a seed treatment, as a foliar spray against certain diseases of fruits and vegetables, and as a wood preservative for packing boxes, baskets, and crates.

Piperalin is sold as Pipron.

Zinc is sometimes used as zinc naphthenate for the disinfection and preservation of wood.

Zoxamide is sold as Busan or Gavel against the late blight of potato.

Antibiotics

Antibiotics are substances produced by one microorganism and toxic to another microorganism. Most antibiotics known to date are products of branching bacteria, such as *Streptomyces*, and some fungi, e.g., *Penicillium*, and are toxic mostly to bacteria, including fastidious bacteria, mollicutes, and also certain fungi. Chemical formulas of most antibiotics are complex and are not, as a rule, related to one another. Antibiotics used for plant disease control are generally absorbed and translocated systemically by the plant to a limited extent. Antibiotics may control plant diseases by acting on the pathogen or on the host. In many cases, the application of antibiotics to control bacterial plant diseases has led to the development of bacterial strains resistant to the antibiotic. Generally, only a few antibiotics are available for plant disease control.

Among the most important antibiotics in plant disease control are streptomycin, tetracyclines, and cycloheximide. Streptomycin is produced by the actinomycete *Streptomyces griseus*. It binds to bacterial ribosomes and prevents protein synthesis. Streptomycin or streptomycin sulfate is sold as Agrimycin and Phytomycin and as a spray shows activity against a broad range of bacterial plant pathogens causing spots, blights, and rots. Streptomycin has also been used as a soil drench, e.g., in the control of geranium foot rot caused by *Xanthomonas* sp., as a dip for potato tuber pieces used for seed against various bacterial rots of tubers, and as a seed disinfectant against bacterial pathogens of beans, cotton, crucifers, and cereals. Moreover, streptomycin is effective against several oomycetous fungi, especially *Pseudoperonospora humuli*, the cause of downy mildew of hops.

Tetracyclines are antibiotics produced by various species of *Streptomyces* and are active against many bacteria and against all mollicutes. Tetracyclines also bind to bacterial ribosomes and inhibit protein synthesis. Of the tetracyclines, Terramycin (oxytetracyline), Aureomycin (chlortetracycline), and Achromycin (tetracycline) have been used to some extent for plant disease control. Oxytetracycline is often used with streptomycin in the control of fire blight of pome fruits during blossoming (Fig. 9-32). When injected into trees infected with mollicutes or fastidious bacteria, tetracyclines stop the development of the disease and induce the remission of symptoms, i.e., the symptoms disappear and the trees resume growth as long as some tetracycline is present in the trees. Usually one injection at the end of the growing season is sufficient for normal growth of the tree during the following season.

Several more antibacterial and antifungal antibiotics are used in Japan and some other countries in Asia. Of these the most common are blasticidin, used against the rice blast fungus *Magnaporthe grisea*, and kasugamycin and polyoxin, used against rice blast and many other leaf, stem, and fruit spots.

Strobilurins were first isolated from a fungus (see earlier discussion) and as such could be classified as

FIGURE 9-32 Application of honeybee-safe antibiotic spray in a pear orchard in bloom to protect trees from fire blight caused by the bacterium *Erwinia amylovora*. [Photograph courtesy of V. O. Stockwell, from McManus and Stockwell (2001). Online, Plant Health Progress.]

antibiotics. Following their discovery, however, chemists have synthesized new active compounds not produced by fungi and so they are now discussed among the systemic fungicides.

Petroleum Oils and Plant Oils

Mineral oils of petroleum origin have been used commercially and extensively for the control of the banana black Sigatoka leaf spot disease caused by the fungus *Mycosphaerella fijiensis*, the greasy spot of citrus caused by *Mycosphaerella citri*, and the powdery mildew diseases of a variety of crops. Highly refined petroleum spray oils kill insects and mites through suffocation, are used as adjuvants with conventional pesticides, and are excellent stand-alone fungicides against some tree powdery mildews. They are also useful in programs trying to reduce the development of resistance by pathogens to strobilurin and demethylation-inhibiting fungicides. There is also some evidence that petroleum spray oils enhance the plant's resistance to infection by pathogens.

Oil obtained from seeds of several plants such as sunflower, olive, corn, and soybean gave excellent control of powdery mildew of apple when applied from 1 day before to 1 day after inoculation of the plants with the fungus. Similarly, several essential oils have been shown to reduce infection of plants by pathogens. So far, none of them is used commercially.

Electrolyzed Oxidizing Water

Some greenhouse diseases, such as powdery mildew, can be managed fairly well by spraying host plants with acidic electrolyzed oxidizing (EO) water. Such water is obtained by passing an electric current through a dilute salt solution, separating the charged products, and collecting the anode water, which is bactericidal and fungicidal due to the combined effect of low pH, high oxidation–reduction potential, and the presence in it of hypochlorous acid. EO water can be mixed with several fungicides and insecticides without losing its potency against pathogens. Research on this product is continuing.

Growth Regulators

Certain plant hormones have been shown to reduce the infection of plants by certain pathogens under experimental conditions, but none of them is used commercially. When tobacco plants were treated with maleic hydrazide, a growth retardant, the root-knot nematode *Meloidogyne* was unable to induce giant cell formation and was thereby prevented from completing its life cycle and from causing disease. Kinetin treatment of leaves, before or shortly after inoculation with virus, also reduces virus multiplication and the number and size of lesions on local-lesion hosts and postpones the onset of systemic symptoms and death of the plant. Stunting and axillary bud suppression associated with certain virus and mollicute diseases of plants can be overcome with sprays of gibberellic acid. Gibberellic acid sprays have been used somewhat for the field control of sour cherry yellows virus on cherries.

Nematicides

Many of the nematicides are broad-spectrum volatile soil fumigants that are active against not only nematodes, but also insects, fungi, bacteria, weed seeds, and almost anything else living in the soil. Several newer chemicals are nonfumigant granular or liquid substances active mostly against nematodes and insects. The four main groups of nematicides are halogenated hydrocarbons, organophosphates, isothiocyanates, and carbamates.

Halogenated Hydrocarbons

The main halogenated hydrocarbon still available for soil fumigation in some parts of the world is methyl bromide (CH_3Br). Even this, however, is scheduled for withdrawal from use in the United States after the year 2004 because it is thought to contribute to the depletion of the ozone layer in the earth's atmosphere. In some formulations, a small amount (1–2%) of chloropicrin is added to this chemical to serve as a warning agent. Mixtures of 70:30 or 50:50 of methyl bromide and chloropicrin are also used as fumigants. Chloropicrin provides better control of fungi and bacteria than methyl bromide, whereas methyl bromide is better against nematodes and weed seeds. Methyl bromide is applied to the soil by injection, after which a waiting period must be adhered to in order to allow the chemical to dissipate before planting. It kills nematodes and insects, and at higher dosages it kills soilborne pathogens and weed seeds. Methyl bromide is a broad-spectrum fumigant against soilborne pathogens and is also used for the aboveground control of dry-wood termites and for the fumigation of agricultural produce for insect control. Methyl bromide affects organisms because it is soluble in lipids and disrupts the function of membranes and nervous systems.

Organophosphate Nematicides

Organophosphates include the insecticides phorate (Thimet), disulfoton (Disyston), ethoprop (Mocap), fensulfothion (Dasanit), fenamiphos (Nemacur), isazofos (Triumph 4E), Terbufos (Counter), and a few others. Many of the organophosphates were developed initially as insecticides; however, they are taken up and are distributed systemically through the plant and are effective nematicides. They are available as water-soluble liquids or granules, have low volatility, can be applied before or after planting, and are effective only against nematodes. Most have minimal or no activity against soil fungi. Like organophosphate insecticides, these nematicides inhibit the nerve-transmitter enzyme cholinesterase and result in paralysis and ultimately death of affected nematodes.

Isothiocyanates

Isothiocyanates include metam sodium (Vapam, Busan), vorlex (Vorlex), and dazomet (Basamid, Mylone). They are active against nematodes, soil insects, weeds, and most soil fungi. Metam sodium and vorlex are applied by injection, incorporation, or irrigation into the soil at least two weeks before planting. Basamid is a granular product. They all act by releasing methylisothiocyanate, which inactivates the —SH group in enzymes.

Carbamates

Carbamates include aldicarb (Temik), carbofuran (Furadan), oxamyl (Vydate), and carbosulfan (Advantage). They are active against nematodes and soil insects, as well as some foliage insects. Available as granules or liquids of low volatility, they are easily soluble in water and can be taken up and translocated systemically by the plant. They are incorporated into the soil by disking before or after planting. They act by inhibiting the enzyme cholinesterase, causing paralysis and death of affected nematodes and insects.

Miscellaneous Nematicides

Chloropicrin (Cl_3CNO_2), the common tear gas, sold as Chlor-o-Pic, is highly volatile and is effective primarily against insects, fungi, and weed seeds. It is generally used mixed with other nematicides. 1,3-Dichloropropene is sold as Telone II, and 1,3-Dichloropropene/Chloropicrin is sold as Telone-C17.

Avermectins are a new class of natural compounds that are obtained as fermentation products of *Streptomyces avermitilis* and exhibit nematicidal properties. Avermectins are still used only for experiments.

Mechanisms of Action of Chemicals Used to Control Plant Diseases

The complete mechanisms by which the various chemicals applied to plants control plant diseases are as yet unknown for most of the chemicals. Some of the chemicals, e.g., fosetyl-Al, seem to reduce infection by increasing the resistance of the host to the pathogen, but how they do that is not clear.

The majority of chemicals are used for their toxicity directly to the pathogen and are effective as protectants at the points of entry of the pathogens or they act systemically through the plant. Such chemicals act by inhibiting the ability of the pathogen to synthesize certain of its cell wall substances; by acting as solvents of, or otherwise damaging, the cell membranes of the pathogen; by forming complexes with, and thus inactivating, certain essential coenzymes of the pathogen; or by inactivating enzymes and causing general precipitation of proteins of the pathogen. For example, sulfur interferes with electron transport along the cytochrome system of fungi, thereby depriving the cell of energy. Sulfur is reduced to hydrogen sulfide (H_2S), which is toxic to most cellular proteins and may contribute to killing the cell. Copper ion (Cu^{2+}) is toxic to all cells because it reacts with sulfhydryl (—SH) groups of certain amino acids and causes denaturation of proteins and enzymes. Many organic fungicides also are toxic because they inactivate proteins and enzymes through reaction with their —SH groups. For example, the dithiocarbamates and ethazol, when taken up by fungal cells, release thiocarbonyl (—N=C—S), which binds irreversibly with and inactivates —SH groups. Similarly, the chlorinated aromatic and heterocyclic compounds, such as PCNB, chlorothalonil, chloroneb, captan, and vinclozolin, react with —NH_2 and —SH groups and inactivate enzymes that have such groups. Furthermore, some nematicides, such as halogenated hydrocarbons, disrupt the function of membranes and nervous systems, whereas others, such as organophosphates, inhibit the nerve-transmitter enzyme cholinesterase and cause paralysis and death of nematodes.

Systemic fungicides and antibiotics are absorbed by the host, are translocated internally through the plant, and are effective against the pathogen at the infection locus both before and after infection has become established. Chemicals that can cure plants from infections that have already become established are called **chemotherapeutants**, and control of plant diseases with such chemicals is called **chemotherapy**. Once in contact with the pathogen, chemotherapeutants seem to affect pathogens in ways similar to those mentioned earlier for nonsystemic chemicals, but systemic fungicides are much more specific in that they apparently affect only one function in the pathogen rather than a variety of them. For example, oxanthiins inhibit the enzyme succinic dehydrogenase, which is essential in mitochondrial respiration, whereas benzimidazoles interfere with nuclear division by binding to protein subunits of the spindle microtubules. Moreover, the polyoxin antifungal antibiotics and the organophosphate fungicides kitazin and edifenphos act primarily by inhibiting chitin synthesis in the pathogen. As a result of such specificity, new pathogen races resistant to one or another of the systemic fungicides may appear soon after their widespread use in a location.

Several systemic fungicides have been shown to inhibit ergosterol biosynthesis and are commonly referred to as sterol inhibitors or sterol-inhibiting fungicides. Some of the sterol inhibitors include bitertanol, fenapanil, imazalil, prochloraz, triadimefon, triarimol, triforine, and etaconazole. Although these compounds have several structural similarities chemically, they do not form a homogeneous group. Ergosterol is a cellular compound that plays a crucial role in the structure and function of the membranes of many fungi, and chemicals that inhibit ergosterol biosynthesis have effective fungicidal action. Sterol-inhibiting fungicides penetrate the leaf cuticle and therefore are highly effective in curative applications after infection has already taken place.

The newest group of systemic fungicides, known as strobilurins or QoI fungicides, act by interfering with respiration, i.e., energy production, in the fungal cell. They do that by blocking electron transfer at the site of quinol oxidation (the Qo site) in the cytochrome bc1 complex, thereby preventing ATP formation. Strobilurins are, therefore, site-specific fungicides and as such are subject to selecting for fungicide-resistant strains of fungi and development of fungicide (strobilurin)-resistant pathogen populations.

Resistance of Pathogens to Chemicals

Just as human pathogens can become resistant to antibiotics and just as insects and mites resistant to certain insecticides and miticides appeared after continuous and widespread use of these chemicals, several plant pathogens have also developed strains that are resistant to certain fungicides. For many years, when only protectant fungicides such as thiram, maneb, or captan were used, no such resistant strains were observed, presumably because these fungicides affect several vital processes of the pathogen and too many gene changes would be necessary to produce a resistant strain. Resistance to some fungicides, all of which contained a benzene ring, began to appear in the 1960s when *Penicillium* strains resistant to diphenyl, *Tilletia* strains resistant to hexachlorobenzene, and *Rhizoctonia* strains resistant to PCNB were found to occur naturally. In some areas these strains became major practical problems. Later, a strain of *Venturia inaequalis* (the cause of apple scab) appeared that was resistant to dodine, and that excellent chemical became ineffective against the fungus over a large area.

Strains of *Erwinia amylovora*, the fire blight bacterium, that were resistant to the systemic antibiotic streptomycin had been known since the late 1950s (Fig. 9-33). It was the introduction and widespread use of the systemic fungicides, especially benomyl, and later metalaxyl and the strobilurins or QoI fungicides, however, that really led to the appearance of strains of numerous fungi resistant to one or more of these fungicides. In some cases, strains resistant to the fungicide appeared and became widespread after only two years of use of the chemical. To date, several of the important fungal pathogens, e.g., *Alternaria*, *Botrytis*, *Cercospora* (Fig. 9-27D), *Colletotrichum*, *Fusarium*, *Verticillium*, *Sphaerotheca*, *Mycosphaerella*, *Aspergillus*, *Penicillium*, *Phytophthora*, *Pythium*, and *Ustilago*, are known to have produced strains resistant to one or more of the systemic fungicides. It appears that resistant strains of all fungi can be expected to develop wherever single-site chemicals are used extensively for their control. This is apparently because systemic fungicides are specific in their action, i.e., they affect only one or perhaps two steps in a genetically controlled event in the metabolism of the fungus; as a result, a resistant population can arise quickly either by a single mutation or by selection of resistant individuals in a population.

The most common mechanisms by which pathogens develop resistance to various fungicides, bactericides,

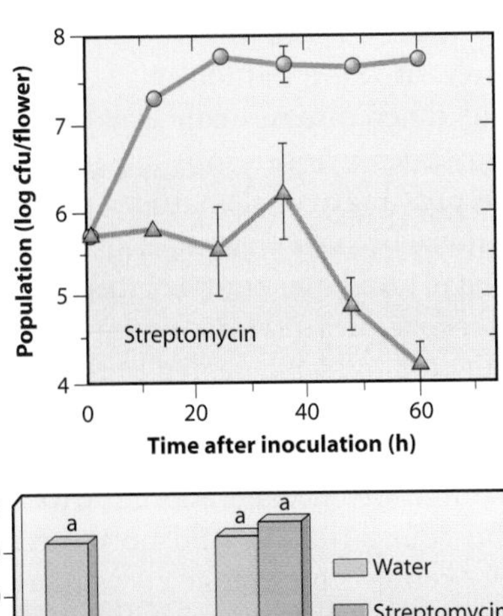

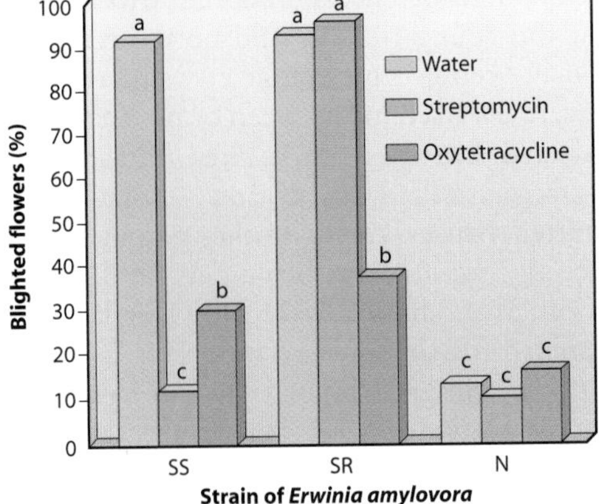

FIGURE 9-33 (Top) Multiplication of streptomycin-resistant strains (○) of the fire blight bacterium *Erwinia amylovora* at streptomycin concentrations that reduce and eventually eliminate streptomycin-susceptible strains (△) of the same bacterium. (Bottom) Incidence of blossom blight in apple flowers 7 days after inoculating with streptomycin-sensitive (SS), streptomycin-resistant (SR) strains, or water only (N) and then spraying with water, streptomycin, or oxytetracycline. Identical letters above bars show that these treatments have similar effects. [From McManus and Jones (1994). *Phytopathology* 84, 627–633.]

and so on is by (1) decreased permeability of pathogen cell membranes to the chemical, (2) detoxification of the chemical through modification of its structure or through binding it to a cell constituent, (3) decreased conversion to the real toxic compound, (4) decreased affinity at the reactive site in the cell (e.g., of benomyl to spindle protein subunits), (5) bypassing a blocked reaction through a shift in metabolism, and (6) compensation for the effect of inhibition by producing more of the inhibited product (e.g., an enzyme).

Good systemic or nonsystemic fungicides that become ineffective because of the appearance of new resistant strains often can continue to be used, and the resistant strains can still be controlled to a practical level through changes in the methods of deployment of the fungicide. This can be achieved by using mixtures of specific systemic and wide-spectrum protectant fungicides, such as benomyl or a strobilurin and either captan or dichloran or iprodione for the control of *Botrytis* or *Sclerotinia*, or matalaxyl and mancozeb for the control of downy mildews; by alternating sprays with systemic and protectant fungicides; or by spraying during half the season with systemic and the other half with protectant fungicides. In each of these schedules, the systemic or specific-action chemical carries most of the weight in controlling the disease, whereas the protectant or nonspecific chemical reduces the possibility of survival of any strains of the pathogen that may develop resistance to the systemic or specific-action chemical.

Restrictions on Chemical Control of Plant Diseases

Although most chemicals used to control plant diseases are much less toxic than most insecticides, they are, nevertheless, toxic substances, and some of them, especially the nematicides, are extremely toxic. Also, some have adverse genetic effects, causing morphological and physiological abnormalities in test animals. For this reason, a number of restrictions are imposed in the licensing, registration, and use of each chemical.

In the United States, both the Food and Drug Administration (FDA) and the Environmental Protection Agency (EPA) keep a close watch on the registration, production, and use of pesticides. It is estimated that only 1 out of 10,000 new compounds synthesized by the pesticide industry turns out to be a successful pesticide, and it takes 7 to 9 years and more than $100 million from initial laboratory synthesis to government registration and first commercial use. In the meantime, exhaustive biological tests, field tests, crop residue analyses, toxicological tests, and environmental impact studies are carried out. If the compound meets all requirements, it is then approved for use on specific food or nonfood crops for which data have been obtained. Clearance must be obtained separately for each crop and each use (seed treatment, spray, soil drench) for which the chemical is recommended.

Once a chemical is approved for a certain crop, two important restrictions on the use of the chemical must then be observed: (1) the number of days that must elapse before harvest of a crop after use of a particular chemical on the crop and (2) the amount of the chemical that can be used per application must not exceed a certain amount. If either of these restrictions is not observed, it is likely that, at harvest, the crop, especially vegetables and fruits, carries on it a greater amount than is allowed for the particular chemical and the crop then must be destroyed. Recommendations contained in bulletins published by the federal or state Cooperative Extension Service are within the tolerances established

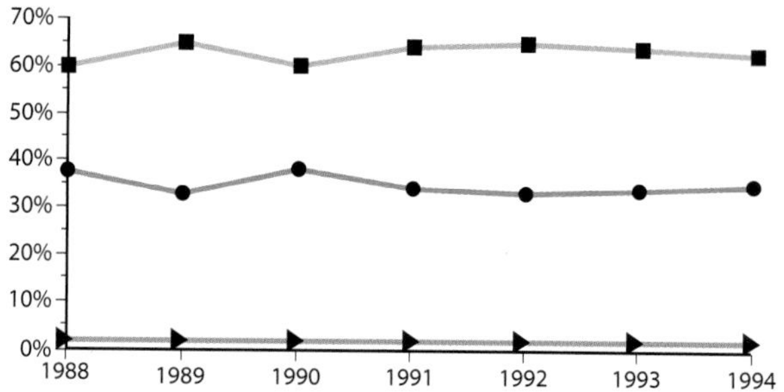

FIGURE 9-34 Results of residue monitoring on domestic and imported foods by the Food and Drug Administration from 1988 to 1994. During that period, 63 to 67% of the 11,348 samples tested had no pesticide residues (■), 33 to 37% had residues within legal limits (●), and less than 1% (►) had residues that were over the tolerances allowed by the Environmental Protection Agency.

by the FDA and EPA and should be followed carefully. As a result of all mandated and voluntary precautions regarding the use of pesticides, the great majority of foodstuffs reaching U.S. markets is either free of all pesticides or carries minimal residues within the legal limits (Fig. 9-34). Less than 1% of samples tested had residues over the allowed tolerances.

INTEGRATED CONTROL OF PLANT DISEASES

The control of plant diseases is most successful and economical when all available pertinent information regarding the crop, its pathogens, the history of disease in previous years, varietal resistance to diseases, the environmental conditions expected to prevail, locality, availability of materials, land, labor, and costs is taken into account in developing the control program. Usually, an integrated control program is aimed against all diseases affecting a crop. However, if a specific disease constitutes the major or only threat to a crop, then an integrated control program is directed against that threat, e.g., apple scab or potato late blight.

The main goals of an integrated plant disease control program are to (1) eliminate or reduce the initial inoculum, (2) reduce the effectiveness of initial inoculum, (3) increase the resistance of the host, (4) delay the onset of disease, and (5) slow the secondary cycles.

Integrated Control in a Perennial Crop

In an integrated control program of an orchard crop, such as apple, peach, or citrus, one must first consider the nursery stock to be used and the location where it will be planted. If the fruit tree is susceptible to certain viruses, mollicutes, crown gall bacteria, or nematodes, the nursery stock (both the rootstock and the scion) must be free of these pathogens. Stock free of certain viruses and other diseases can usually be bought from selected nurseries whose crops are inspected and certified. If the possibility of nematodes on the roots exists, the stock must be heat treated. The location where the trees will be planted must not be infested with fungi such as *Phytophthora*, *Armillaria*, or numerous nematodes; if it is, the field should be treated with fumigants before planting and varieties grafted on rootstocks resistant to these pathogens should be preferred. The drainage of the location should be checked and improved, if necessary. Finally, the young trees should not be planted on sites previously occupied by similar crops, particularly if the latter were diseased, or between or next to old trees that are infected heavily with canker fungi and bacteria, insect-transmitted viruses and mollicutes, pollen-transmitted viruses, or with other pathogens.

Once the trees are in place and until they begin to bear fruit, they should be fertilized, irrigated, pruned, and sprayed for the most common insects and diseases so that they will grow vigorously and free of infections. Later on, when the trees bear fruit, the care should increase, as should the vigilance to detect and control diseases that affect any part of the tree. Any trees that develop symptoms of a disease caused by a systemic pathogen, such as a virus or mollicute, should be removed as soon as possible.

Disease control measures in an orchard may begin in the winter, when weak, diseased, or dead twigs, branches, or fruit are removed during pruning operations and are buried or burned. This reduces the amount of potential fungal or bacterial primary inoculum that will start infections in the spring. In some cases, when infected leaves, fruits, or twigs on the orchard floor were raked and removed from the orchard or were sprayed with fungicides or biological control agents, the pathogen inoculum was reduced or eliminated. Pruning shears and saws should be disinfested before moving to new trees to avoid spreading any pathogens from tree to tree. Pruned trees should be sprayed as soon as possible with a fungicide, such as benomyl, to protect pruning cuts from becoming infected by canker-causing fungi.

Because many fungi and bacteria (as well as insects and mites) are activated in the spring by the same weather conditions that make buds open, a "dormant" spray, containing a fungicide–bactericide (such as the Bordeaux mixture) or a plain fungicide plus a miticide–insecticide (such as Superior oil), is applied before bud break. After that, as the buds open, the blossoms and leaves that are revealed are usually very susceptible to fungal or bacterial pathogens or both, depending on what is present in the particular area. Therefore, these organs (blossoms and leaves) must be protected with sprays containing a fungicide and/or a bactericide and, possibly, an insecticide and/or miticide that does not harm bees (Fig. 9-35). It is usually possible to find effective materials compatible with one another so that all of them can be mixed in the same tank and sprayed at once. If one compound, however, must be used to control an existing disease but is incompatible with the other compounds, then a separate spray will be needed. Because flowers appear over a period of several days and the leaves enlarge rapidly at that stage and because many fungi release their spores and bacteria ooze out most abundantly during and soon after bloom, the blossoms and leaves may have to be sprayed with a systemic fungicide. If only protectant fungicides are available, the trees must be sprayed frequently (every 3–5 days) so that they will be protected by the fungicide or bactericide (or both), especially if it rains often and the plants stay wet for many hours. Insecticides and miticides may still have to be used with the fungicide, but these insecticides must not be toxic to bees, which must be allowed to pollinate the flowers.

The frequent sprays usually continue as long as there are spores being released by fungi, or bacteria oozing out, as long as the weather stays wet, and as long as there are growing plant tissues. Combining the use of weather forecasts with disease control is most helpful. When possible, computer-aided programs predicting infection periods are employed to direct growers when spray applications should begin and when they should be applied subsequently. This allows one to do the most good in protecting the crop from disease while reducing the amount of pesticides used and the total cost of plant protection.

Once blossoming is over, young fruit appear, which may be affected by the same pathogens and insects as the flowers and leaves. If they are, the same spray schedule with the same materials continues as long as there is inoculum around. If a systemic fungicide had been used early in the season, later sprays should be made with a broad-spectrum protectant fungicide to forestall the appearance of fungicide-resistant strains of the pathogen. Often, however, new pathogens and insects may attack the fruit, and the schedule must be adjusted and materials must be included that control the new pathogens.

Usually, fruit becomes susceptible to several fruit-spotting or fruit-rotting fungi that attack fruit from the stage of early maturity through harvest and storage. Therefore, fruit must be sprayed every 10 to 14 days with materials that will control these fungi until harvest. Most fruit rots start at wounds made by insects, and therefore insect control must continue. Also, wounding of fruit during harvesting and handling must be avoided or minimized to prevent fungus infections. Fruit-picking baskets and crates must be clean and free of rotten

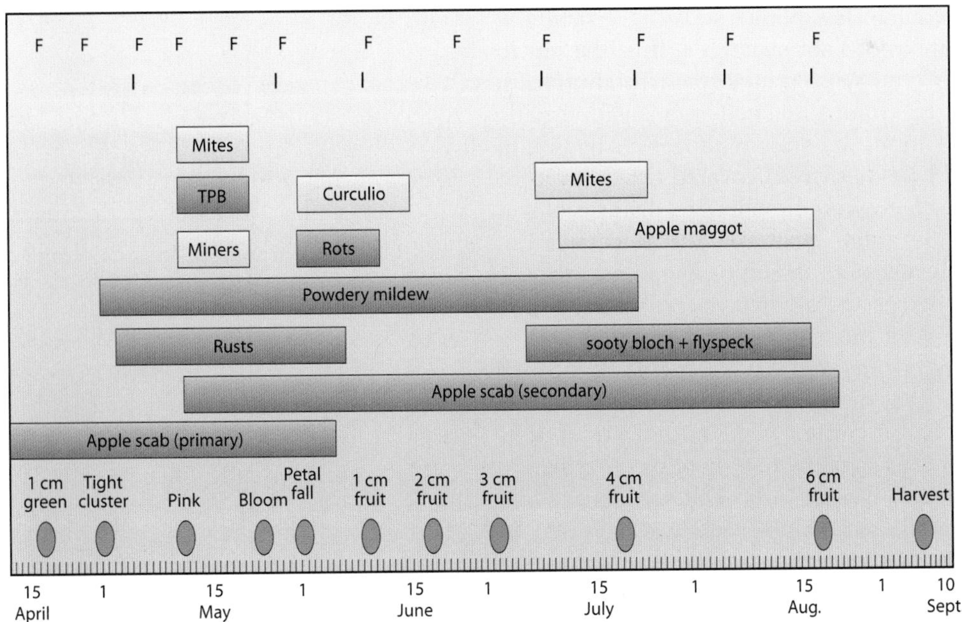

FIGURE 9-35 Number and timing of fungicide (F) and insecticide (I) applications for the control of the most important diseases (shaded bars) and insects (open bars) in apple orchards in the northeastern United States under normal weather conditions. Bars indicate periods during which a disease or pest is controlled, although it may be present during longer periods. The phases of fruit development are approximations for one variety. TPB, tarnished plant bug. [From Gadoury *et al.* (1989). *Plant Dis.* **93**, 98–105.]

debris (which may harbor fruit-rotting fungi), and the packinghouse and warehouse must also be clean, free of debris, and preferably fumigated with formaldehyde, sulfur dioxide, or some other fumigant. Harvested fruit is often washed in a water solution containing a fungicide or biological control agent to further protect the fruit during storage and transportation. Before packing, infected and injured fruit are removed and discarded. The fruit should be refrigerated during storage, transportation, and marketing, and even after it is purchased by the consumer, so that any existing infections will develop slowly and no new infections will get started.

Integrated Control in an Annual Crop

In an integrated control program of an annual crop, such as potatoes, one must again start with healthy stock and must plant it in a suitable field. Potato tuber seed may carry several viruses, the late blight fungus, ring rot bacteria, as well as several other fungi, bacteria, and nematodes. Therefore, starting with clean, disease-free seed is of paramount importance. Certified potato seed is usually free of most such important pathogens and is produced under strict quarantine and inspection rules that guarantee seed free of these pathogens. Healthy seed treated with a fungicide or biocontrol agent must then be planted in a field free of old potato tubers that may harbor some of the aforementioned pathogens. The field, as much as possible, must also be free or contain low populations of *Verticillium*, *Fusarium*, and the root-knot nematode. It is best not to follow a potato crop with another, and rotation with legumes, corn, or other unrelated crops will usually reduce the populations of potato pathogens. Any potato cull piles should be destroyed, covered, or sprayed to ensure that no *Phytophthora* sporangia will be blown from there to the potato plants in the field later on. Tubers should be cut with disinfested knives to reduce the spread of ring rot and other pathogens among seed pieces, and the seed pieces are usually treated with a fungicide, a bactericide, and an insecticide to protect them from pathogens on their surface or in the soil. The soil may have to be treated with a fumigant if it is known to be infested with the root-knot or other nematodes, *Fusarium*, or *Verticillium*. The seed pieces are planted at a date when their sprouts are expected to grow quickly, as slow-growing sprouts in cool weather are particularly susceptible to infection by *Rhizoctonia*. The field must, of course, have good drainage to reduce damping-off, seed-piece rot, and root rots.

A few weeks after young plants have emerged, under conditions of stress or high moisture, they become susceptible to attack by early blight (*Alternaria*) or late blight (*Phytophthora infestans*). If the diseases occur regularly year after year (Fig. 9-36), the grower, in addition to using resistant varieties, should start spraying with the appropriate fungicides as soon as the disease appears, or even before, and should continue the sprays, especially for late blight, throughout the season whenever the weather is cool and damp. Insecticide sprays control insects and may reduce the spread of some persistently transmitted viruses, but they usually have no effect or actually increase the spread of nonpersistently transmitted viruses. Using weather data to forecast disease appearance and development can help in spraying at the right time and in not wasting any sprays. Spraying must continue throughout the growing season as needed and as the weather dictates. Before harvest, the infected vines must be killed with chemicals to destroy late blight inoculum that could come in contact with the tubers when they are dug up. Tubers must be harvested carefully to avoid wounding that would allow storage-rot fungi such as *Fusarium* and *Pythium* and bacteria such as *Erwinia carotovora* to gain entrance into the tuber. The tubers must then be sorted and the

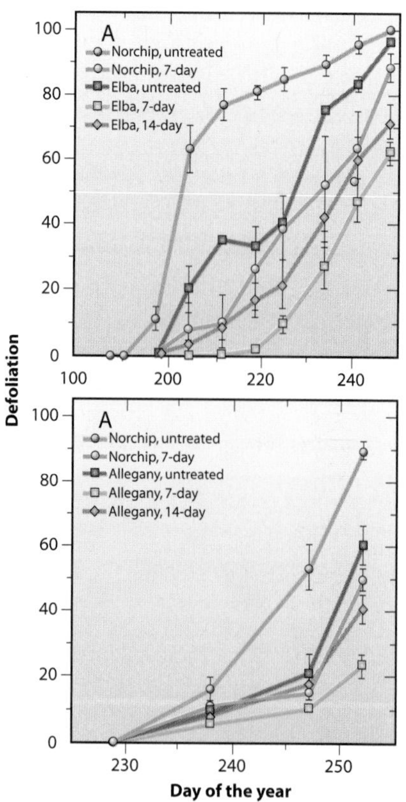

FIGURE 9-36 Effect of plant disease resistance and fungicide treatment on defoliation induced by potato early and late blights during two different years. Variety Norchip is susceptible and Elba and Allegany are moderately resistant to both diseases. [From Schtienberg *et al.* (1994). *Plant Dis.* 78, 23–26.]

damaged ones discarded. Healthy tubers are stored at about 15°C for the wounds to heal and then at about 10°C to prevent the development of fungus rots in storage. Storage rooms must of course be cleaned and disinfested before the tubers are brought in. Potato cull piles should not be kept near the field but should be either burned or buried as soon as possible.

Thus, in an integrated control program, several control methods are employed, including regulatory inspections for healthy seed or nursery crop production, cultural practices (crop rotation, sanitation, pruning), biological control (resistant varieties, biocontrol agents), physical control (storage temperature), and chemical controls (soil fumigation, seed or nursery stock treatment, sprays, disinfestation of cutting tools, crates, warehouses, and washing solutions). Each one of these measures must be used for best results, and the routine use of each of them makes all of them that much more effective.

Selected References

Adams, P. (1990). The potential of mycoparasites for biological control of plant diseases. *Annu. Rev. Phytopathol.* **28**, 59–72.

Anagnostakis, S. L. (1982). Biological control of chestnut blight. *Science* **215**, 466–471.

Andrews, J. H. (1992). Biological control in the phyllosphere. *Annu. Rev. Phytopathol.* **30**, 603–635.

Anonymous (Annually). "Biological and Cultural Control Tests." APS Press, St. Paul, MN.

Anonymous (Annually). "Fungicide and Nematicide Tests." APS Press, St. Paul, MN.

Anonymous (1992). Symposium papers on root health. *Can. J. Plant Pathol.* **14**, 76–114.

Baker, K. F., and Cook, R. J. (1974). "Biological Control of Plant Pathogens." Freeman, San Francisco.

Bau, H.-J., *et al.* (2003). Broad spectrum resistance to different geographic strains of *papaya ringspot virus* in coat protein gene transgenic papaya. *Phytopathology* **93**, 112–120.

Belanger, R. R., Benhamou, N., and Menzies, J. G. (2003). Cytological evidence of an active role of silicon in wheat resistance to powdery mildew (*Blumeria graminis f.sp. tritici*). *Phytopathology* **93**, 402–412.

Birch, R. G. (2001). *Xanthomonas albilineans* and the antipathogenesis approach to disease control. *Mole. Plant Pathol.* **2**, 1–11.

Bockus, W. W., *et al.* (2001). Success stories: Breeding for wheat disease resistance in Kansas. *Plant Dis.* **85**, 453–461.

Bokshi, A. I., Morris, S. C., and Deverall, B. J. (2003). Effects of benzothiadiazole and acetylsalicylic acid on β-1,3-glucanase activity and disease resistance in potato. *Plant Pathol.* **52**: 22–27.

Bolkan, H. A., and Reinert, W. R. (1994). Developing and implementing IPM strategies to assist farmers: An industry approach. *Plant Dis.* **78**, 545–550.

Cai, W.-Q., *et al.* (2003). Development of CMV- and TMV-resistant transgenic chili pepper: Field performance and biosafety assessment. *Mol. Breeding* **11**, 25–35.

Charrier, B., *et al.* (2000). Co-silencing of homologous transgenes in tobacco. *Mol. Breeding* **6**, 407–419.

Charudattan, R. (1990). Pathogens with potential for weed control. *In* "Microbes and Microbial Products as Herbicides" (R. E. Hoagland, ed.), pp. 132–154. Am. Chem. Soc., Washington, DC.

Charudattan, R., and Walker, H. L., eds. (1982). "Biological Control of Weeds with Plant Pathogens." Wiley & Sons, New York.

Cohen, Y. R. (2002). β-Aminobutyric acid-induced resistance against plant pathogens. *Plant Dis.* **86**, 448–457.

Cook, R. J. (1993). Making greater use of introduced microorganisms for biological control of plant pathogens. *Annu. Rev. Phytopathol.* **31**, 53–80.

Cook, R. J., and Baker, K. F. (1983). "The Nature and Practice of Biological Control of Plant Pathogens." APS Press, St. Paul, MN.

Cooksey, D. A. (1990). Genetics of bactericide resistance in plant pathogenic bacteria. *Annu. Rev. Phytopathol.* **28**, 201–219.

Costa, A. S., and Muller, G. W. (1980). Tristeza control by cross protection: A U.S.-Brazil cooperative success. *Plant Dis.* **64**, 538–541.

DeWard, M. A., *et al.* (1993). Chemical control of plant diseases: Problems and prospects. *Annu. Rev. Phytopathol.* **31**, 403–421.

Escande, A. R., Laich, F. S., and Pedraza, M. V. (2002). Field testing of honeybee-dispersed *Trichoderma* spp. to manage sunflower head rot (*Sclerotinia sclerotiorum*). *Plant Pathol.* **51**, 346–351.

Frischmuth, T., Engel, M., and Jeske, H. (1997). *Beet curly top virus* DI DNA-mediated resistance is linked to its size. *Mol. Breeding* **3**, 213–217.

Gadoury, D. M., McHardy, W. E., and Rosenberger, D. A. (1989). Integration of pesticide application schedules for disease and insect control in apple orchards of the northern United States. *Plant Dis.* **73**, 98–105.

Ghaouth, A. E., Wilson, C. L., and Callahan, A. M. (2003). Induction of chitinase, β-1,3-glucanase, and phenylalanine ammonia lyase in peach fruits by UV-C treatment. *Phytopathology* **93**, 349–355.

Ghaouth, A. E., Wilson, C. L., and Wisniewski, M. (2003). Control of postharvest decay of apple fruit with *Candida saitoana* and induction of defense responses. *Phytopathology* **93**, 344–348.

Glenn, D. M., *et al.* (2001). Efficacy of kaolin-based particle films to control apple diseases. *Plant Health Progress doi*: 10.1094/PHP-2001-0823-01-RS

Gnanamanickam, S. S., ed. (2002). "Biological Control of Crop Diseases." Dekker, New York.

Gonsalves, D. (1998). Control of *papaya ringspot virus* in papaya: A case study. *Annu. Rev. Phytopathol.* **36**, 415–437.

Graham, J. H., and Goltwald, T. R. (1991). Research perspectives on eradication of citrus bacterial diseases in *Florida. Plant Dis.* **75**, 1193–1200.

Gubba, A., *et al.* (2002). Combining transgenic and natural resistance to obtain broad resistance to tospovirus infection in tomato (*Lycopersicum esculentum* Mill). *Mol. Breeding* **9**, 13–23.

Han, J.-S. (1990). Use of antitranspirant epidermal coatings for plant protection in China. *Plant Dis.* **74**, 263–266.

Hardison, J. R. (1976). Fire and flame for plant disease control. *Annu. Rev. Phytopathol.* **14**, 355–379.

Harman, G. H. (2000). Myths and dogmas of biocontrol: Changes in perceptions derived from research on *Trichoderma harzianum* T-22. *Plant Dis.* **84**, 377–393.

Harveson, R. M., *et al.* (2002). An integrated approach to cultivar evaluation and selection for improving sugar beet profitability: A successful case study for the central high plains. *Plant Dis.* **86**, 192–204.

Harvey, J. M. (1978). Reduction of losses in fresh market fruits and vegetables. *Annu. Rev. Phytopathol.* **16**, 321–341.

Heiniger, U., and Rigling, D. (1994). Biological control of chestnut blight in Europe. *Annu. Rev. Phytopathol.* **32**, 581–599.

Hoitink, H. A. J., and Boehm, M. J. (1999). Biocontrol within the context of soil microbial communities: A substrate-dependent phenomenon. *Annu. Rev. Phytopathol.* **37**, 426–446.

Hoitink, H. A. J., Inbar, Y., and Boehm, M. J. (1991). Status of compost-amended potting mixes naturally suppressive to soilborne diseases of floricultural crops. *Plant Dis.* **75**, 869–873.

Hornby, D. (1983). Suppressive soils. *Annu. Rev. Phytopathol.* **21**, 65–85.

Hornby, E., ed. (1990). "Biological Control of Plant Pathogens." CAB Int., Oxon, UK.

Horsfall, J. G., and Cowling, E. B., eds. (1977). "Plant Disease," Vol. 1. Academic Press, New York.

Howell, C. R. (2003). Mechanisms employed by *Trichoderma* species in the biological control of plant diseases: The history and evolution of current concepts. *Plant Dis.* **87**, 4–10.

Janisiewicz, W. J., and Korsten, L. (2002). Biological control of postharvest diseases of trees. *Annu. Rev. Phytopathol* **40**, 411–441.

Janse, J. D., and Weneker, M. (2002). Possibilities of avoidance and control of bacterial plant diseases when using pathogen-tested (certified) or -treated planting material. *Plant Pathol.* **51**, 523–536.

Jarvis, W. R. (1989). Managing diseases in greenhouse crops. *Plant Dis.* **73**, 190–194.

Jarvis, W. R. (1992). "Managing Diseases of Greenhouse Crops." APS Press, St. Paul, MN.

Johansson, P. M., Johnsson, L., and Gerhardson, B. (2003). Suppression of wheat seedling diseases caused by *Fusarium culmorum* and *Microdochium nivale* using bacterial seed treatment. *Plant Pathol.* **52**, 219–227.

Jones, S. S., Murray, T. D., and Allan, R. E. (1995). Use of alien genes for the development of disease resistance in wheat. *Annu. Rev. Phytopathol.* **33**, 429–443.

Kahn, R. P. (1991). Exclusion as a plant disease control strategy. *Annu. Rev. Phytopathol.* **29**, 219–246.

Katan, J. (1981). Solar heating (solarization) of soil for control of soilborne pests. *Annu. Rev. Phytopathol.* **19**, 211–236.

Kerr, A. (1980). Biological control of crown gall through production of Agrocin 84. *Plant Dis.* **64**, 24–30.

Kerry, B. (1981). Fungal parasites: A weapon against cyst nematodes. *Plant Dis.* **65**, 390–394.

Knight, S. C., *et al.* (1997). Rationale and perspectives on the development of fungicides. *Annu. Rev. Phytopathol.* **35**, 349–372.

Ko, W.-H. (1982). Biological control of Phytophthora root rot of papaya with virgin soil. *Plant Dis.* **66**, 446–448.

Lewis, F. H., and Hickey, K. D. (1972). Fungicide usage on deciduous fruit trees. *Annu. Rev. Phytopathol.* **10**, 399–428.

Lim, P. O., *et al.* (2002). Multiple virus resistance in transgenic plants conferred by the human dsRNA-dependent protein kinase. *Mol. Breeding* **10**, 11–18.

Lindow, S. E. (1983). Methods of preventing frost injury caused by epiphytic ice-nucleation-active bacteria. *Plant Dis.* **67**, 327–333.

Lomonossoff, G. P. (1995). Pathogen-derived resistance to plant viruses. *Annu. Rev. Phytopathol.* **33**, 323–343.

MacDonald, W. L., and Fulbright, D. W. (1991). Biological control of chestnut blight: Use and limitations of transmissible hypovirulence. *Plant Dis.* **75**, 656–661.

Mathre, D. E., Cook, R. J., and Callan, N. W. (1999). From discovery to use: Traversing the world of commercializing biocontrol agents for plant disease control. *Plant Dis.* **83**, 972–983.

Mathre, D. E., Johnson, R. H., and Grey, W. E. (2001). Small grain cereal seed treatment. *Plant Health Instructor.* DOI: 10.1094/PHI-I-2001-1008-01.

McGee, D. C. (1995). Epidemiological approach to disease management through seed technology. *Annu. Rev. Phytopathol.* **33**, 445–466.

McGrath, M. T. (2001). Fungicide resistance in cucurbit powdery mildew: Experiences and challenges. *Plant Dis.* **85**, 236–245.

McInnes, T. B., Black, L. L., and Gatti, J. M. (1992). Disease-free plants for management of strawberry anthracnose crown rot. *Plant Dis.* **76**, 260–264.

McManus, P. S., *et al.* (2002). Antibiotic use in plant agriculture. *Annu. Rev. Phytopathol.* **40**, 443–465.

McSpadden Gardener, B. B., and Fravel, R. D. (2002). Biological control of plant pathogens: Research, commercialization, and application in the USA. Online. *Plant Health Progress* doi: 10.1094/PHP-2002-0510-01-RV.

Melouk, H. A., and Shokes, F. M., eds. (1995). "Peanut Health Management." APS Press, St. Paul, MN.

Norelli, J. L., Jones, A. L., and Aldwinckle, H. S. (2003). Fire blight management in the twenty-first century: Using new technologies that enhance host resistance in apple. *Plant Dis.* **87**, 756–765.

Palti, J. (1981). "Cultural Practices and Infectious Crop Diseases." Springer-Verlag, Berlin.

Pang, S.-Z., *et al.* (2000). Resistance to squash mosaic comovirus in transgenic squash plants expressing its coat protein genes. *Mol. Biol.* **6**, 87–93.

Rast, A. T. B. (1972). M11-16, an artificial symptomless mutant of tobacco mosaic virus for seedling inoculation of tomato crops. *Neth. J. Plant. Pathol.* **78**, 110–112.

Ristaino, J. B., and Thomas, W. (1997). Agriculture, methyl bromide, and the ozone hole: Can we fill the gaps? *Plant Dis.* **81**, 964–977.

Rowe, R. C., ed. (1993). "Potato Health Management." APS Press, St. Paul, MN.

Schneider, R. W., ed. (1982). "Suppressive Soils and Plant Disease." APS Press, St. Paul, MN.

Schroth, M. N., and Hancock, J. G. (1982). Disease-suppressive soil and root-colonizing bacteria. *Science* **216**, 1376–1381.

Schubert, T. S., *et al.* (2001). Meeting the challenge of eradicating citrus canker in Florida — again. *Plant Dis.* **85**, 340–356.

Shepard, J. F., and Claflin, L. E. (1975). Critical analyses of the principles of seed potato certification. *Annu. Rev. Phytopathol.* **13**, 271–293.

Shtienberg, D., *et al.* (1994). Incorporation of cultivar resistance in a reduced-sprays strategy to suppress early and late blights on potato. *Plant Dis.* **78**, 23–26.

Sikora, R. A. (1992). Management of the antagonistic potential in agricultural ecosystems for the biological control of plant parasitic nematodes. *Annu. Rev. Phytopathol.* **30**, 245–270.

Slabaugh, W. R., and Grove, M. D. (1982). Postharvest diseases of bananas and their control. *Plant Dis.* **66**, 746–750.

Sommer, N. F. (1982). Postharvest handling practices and postharvest diseases of fruit. *Plant Dis.* **66**, 351–364.

Staub, T. (1991). Fungicide resistance: Practical experience with antiresistance strategies and the role of integrated use. *Annu. Rev. Phytopathol.* **29**, 421–442.

Stirling, G. R. (1991). "Biological Control of Plant Parasitic Nematodes." CAB Int., Chichester, England.

Sumner, D. R., Doupnik, B., Jr., and Boosalis, M. G. (1981). Effect of reduced tillage and multiple cropping on plant diseases. *Annu. Rev. Phytopathol.* **19**, 167–187.

Subbarao, K. V. (1998). Progress toward integrated management of lettuce drop. *Plant Dis.* **82**, 1068–1078.

Sutton, J. C., and Peng, G. (1993). Manipulation and vectoring of biocontrol organisms to manage foliage and fruit diseases in cropping systems. *Annu. Rev. Phytopathol.* **31**, 473–493.

Te Beest, D. O., Yang, X. B., and Cisar, C. R. (1992). The status of biological control of weeds with fungal pathogens. *Annu. Rev. Phytopathol.* **30**, 637–657.

Thomson, W. T. (1993). "Agricultural Chemicals, Book IV: Fungicides," 1993–1994 Revision. Thomson, Fresno, CA.

Tepfer, M. (2002). Risk assessment of virus-resistant transgenic plants. *Annu. Rev. Phytopathol.* **40**, 467–491.

Thurston, H. D. (1991). "Sustainable Practices for Plant Disease Management in Traditional Farming Systems." Westview, Boulder, CO.

Tuite, J., and Foster, G. H. (1979). Control of storage diseases of grains. *Annu. Rev. Phytopathol.* **17**, 343–366.

Uchimiya, H., *et al.* (2002). Transgenic rice plants conferring increased tolerance to rice blast and multiple environmental stresses. *Mol. Breeding* **9**, 25–31.

Verchot-Lubicz, J., (2003). Soilborne viruses: Advance in virus movement, virus induced gene silencing, and engineered resistance. *Physiol Mol. Plant Pathol.* **62**, 55–63.

Verreet, J. A., and Hoffmann, G. M. (1990). A biologically oriented threshold decision model for control of epidemics of Septoria nodorum in wheat. *Plant Dis.* **74**, 731–738.

Vincelli, P. (2002). QoI (Strobilurin) fungicides: Benefits and risks. *The Plant Health Instructor.* DOI: 10.1094/PHI-I-2002-0809-02.

Wang, Y., and Fristensky, B. (2001). Transgenic canola lines expressing pea defense gene DRR206 have resistance to aggressive blackleg isolates and to *Rhizoctonia solani. Mol. Breeding* **8**, 263–271.

Waterworth, H. E. (1993). Processing foreign plant germ plasm at the National Plant Germplasm Quarantine Center. *Plant Dis.* **77**, 854–860.

Weller, D. M. (1988). Biological control of soilborne pathogens in the rhizosphere. *Annu. Rev. Phytopathol.* **26**, 379–407.

Wilson, C. L., and Wisniewski, M. E. (1989). Biological control of postharvest diseases of fruits and vegetables: An emerging technology. *Annu. Rev. Phytopathol.* **27**, 425–441.

Wilson, D. L., *et al.* (1994). Potential of induced resistance to control postharvest diseases of fruits and vegetables. *Plant Dis.* **78**, 837–844.

Windels, C. E., and Lindow, S. E. (1985). "Biological Control in the Phylloplane." APS Press, St. Paul, MN.

Windels, C. E., *et al.* (1998). A Cercospora leaf spot model for sugar beet: In practice by an industry. *Plant Dis.* **82**, 716–726.

Wolf, P. F. J., and Verreet, J. A. (2002). An integrated pest management system in Germany for the control of fungal leaf diseases in sugar beet: The IPM Sugar Beet Model. *Plant Dis.* **86**, 336–344.

Xiao, X. W., *et al.* (2000). Antibody-mediated improved resistance to CIYVV and PVY infections in transgenic tobacco plants expressing a single-chain variable region antibody. *Mol. Breeding* **6**, 421–431.

Young, X. B., Dowler, W. M., and Royer, M. H. (1991). Assessing the risk and potential impact of an exotic plant disease. *Plant Dis.* **75**, 976–982.

Ypema, H. L., and Gold, R. E. (1999). Kresoxim-methyl: Modification of a naturally occurring compound to produce a new fungicide. *Plant Dis.* **83**, 4–19.

Zhai, W., *et al.* (2001). Breeding bacterial blight-resistant rice with the cloned bacterial blight resistance gene Xa21. *Mol. Breeding* **8**, 263–271.

Ziegler, A., and Torrance, L. (2002). Application of recombinant antibodies in plant pathology. *Mol. Plant Pathol.* **3**, 401–407.

Zitter, T. A., and Simons, J. N. (1980). Management of viruses by alteration of vector efficiency and by cultural practices. *Annu. Rev. Phytopathol.* **18**, 289–310.

Ziv, O., and Zitter, T. A. (1992). Effects of bicarbonates and film-forming polymers on cucurbit foliar diseases. *Plant Dis.* **76**, 513–517.

part two

SPECIFIC PLANT DISEASES

chapter ten

ENVIRONMENTAL FACTORS *THAT* CAUSE PLANT DISEASES

INTRODUCTION
358

TEMPERATURE EFFECTS
358

MOISTURE EFFECTS
365

INADEQUATE OXYGEN
367

LIGHT
367

AIR POLLUTION
368

NUTRITIONAL DEFICIENCIES IN PLANTS
372

SOIL MINERALS TOXIC TO PLANTS
372

HERBICIDE INJURY
378

HAIL INJURY
380

LIGHTNING
381

OTHER IMPROPER AGRICULTURAL PRACTICES
381

THE OFTEN CONFUSED ETIOLOGY OF STRESS DISEASES
383

INTRODUCTION

Plants grow best within certain ranges of the various abiotic factors that make up their environment. Such factors include temperature, soil moisture, soil nutrients, light, air and soil pollutants, air humidity, soil structure, and pH. Although these factors affect all plants growing in nature, their importance is considerably greater for cultivated plants, which are often grown in areas that are at the margins and beyond their normal habitat and, therefore, that barely meet the requirements for normal growth. Moreover, cultivated plants are frequently grown or kept in completely artificial environments (greenhouses, homes, warehouses) or are subjected to a number of cultural practices (fertilization, irrigation, spraying with pesticides) that may affect their growth considerably.

General Characteristics

The common characteristic of abiotic, i.e., noninfectious diseases of plants, is that they are caused by the lack or excess of something that supports life. Noninfectious diseases occur in the absence of pathogens and cannot, therefore, be transmitted from diseased to healthy plants. Noninfectious diseases may affect plants in all stages of their lives (e.g., seed, seedling, mature plant, or fruit), and they may cause damage in the field, in storage, or at the market. The symptoms caused by noninfectious diseases vary in kind and severity with the particular environmental factor involved and with the degree of deviation of this factor from its normal. Symptoms may range from slight to severe, and affected plants may even die.

Diagnosis

The diagnosis of noninfectious diseases is sometimes made easy by the presence of characteristic symptoms known to be caused by the lack or excess of a particular factor on the plant (Fig. 10-1). At other times, diagnosis can be arrived at by carefully examining and analyzing several factors: the weather conditions prevailing before and during the appearance of the disease; recent changes in the atmospheric and soil contaminants at or near the area where the plants are growing; and the cultural practices, or possible accidents in the course of these practices, preceding the appearance of the disease. Often, however, the symptoms of several noninfectious diseases are too indistinct and closely resemble those caused by several viruses, mollicutes, and many root pathogens. The diagnosis of such noninfectious diseases then becomes a great deal more complicated. One must obtain proof of absence from the plant of any of the pathogens that could cause the disease, and one must reproduce the disease on healthy plants after subjecting them to conditions similar to those thought of as the cause of the disease. To distinguish further among environmental factors causing similar symptoms, the investigator must cure the diseased plants, if possible, by growing them under conditions in which the degree or the amount of the suspected environmental factor involved has been adjusted to normal.

Control

Noninfectious plant diseases can be controlled by ensuring that plants are not exposed to the extreme environmental conditions responsible for such diseases or by supplying the plants with protection or substances that would bring these conditions to levels favorable for plant growth.

TEMPERATURE EFFECTS

Plants normally grow at a temperature range from 1 to 40°C, with most kinds of plants growing best between 15 and 30°C. Perennial plants and dormant organs (e.g., seeds and corms) of annual plants may survive temperatures considerably below or above the normal temperature range of 1 to 40°C. The young, growing tissues of most plants, however, and the entire growth of many annual plants are usually quite sensitive to temperatures near or beyond the extremes of this range.

The minimum and maximum temperatures at which plants can still produce normal growth vary greatly with the plant species and with the stage of growth the plant is in during the low or high temperatures. Thus, tomato, citrus, and other tropical plants grow best at high temperatures and are injured severely when the temperature drops to near or below freezing. However, plants such as cabbage, winter wheat, alfalfa, and most perennials of the temperate zone can withstand temperatures considerably below freezing without any apparent ill effects. Even the latter plants, however, are injured and finally killed if the temperature drops too low.

A plant may also differ in its ability to withstand extremes in temperature at different stages of its growth. Thus, older, hardened plants are more resistant to low temperatures than young seedlings. Also, different tissues or organs on the same plant may vary greatly in their sensitivity to the same low temperature. Buds are more sensitive than twigs; flowers and newly formed fruit are more sensitive than leaves; and so on.

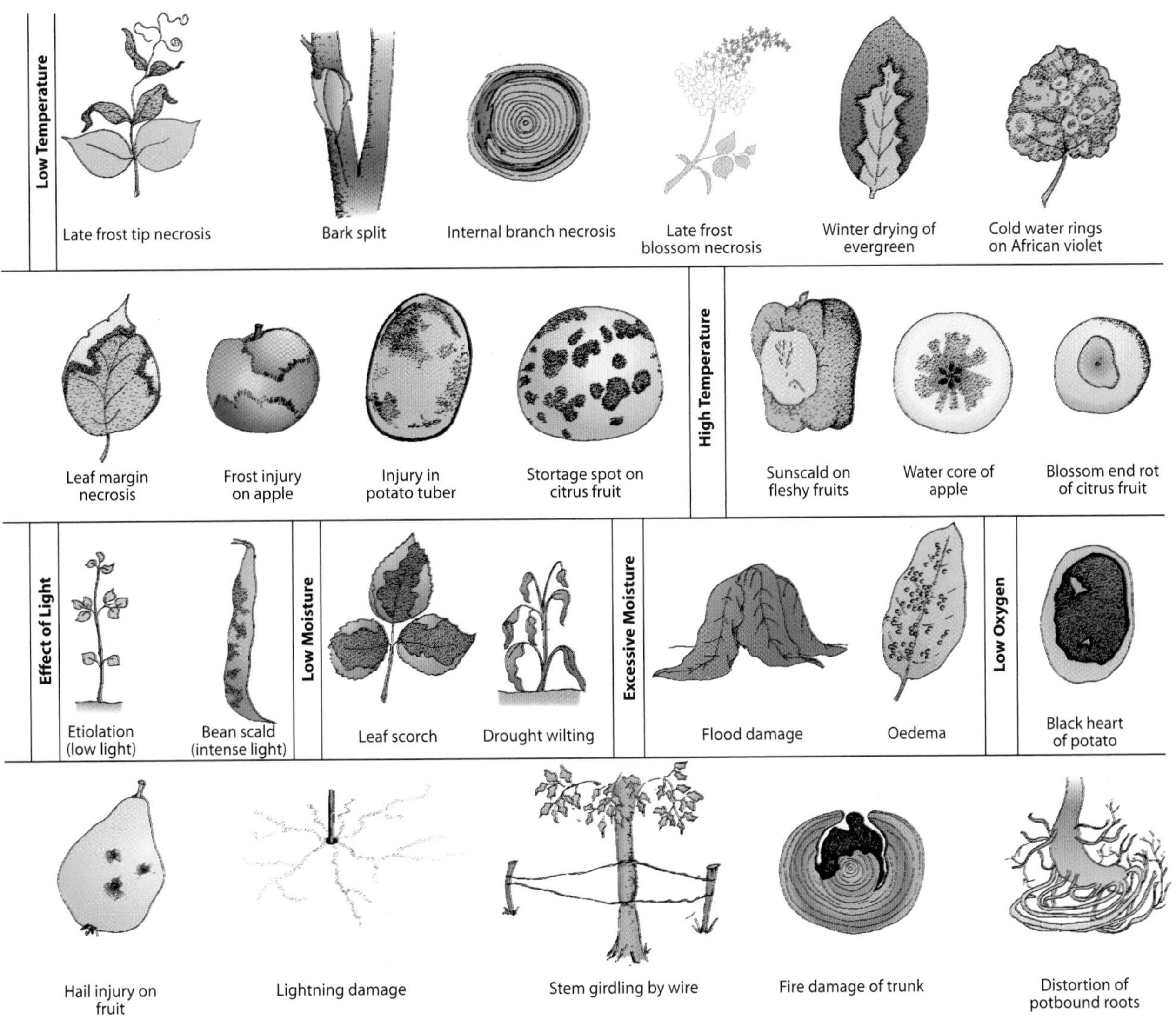

FIGURE 10-1 Various types of symptoms caused by different environmental factors.

High-Temperature Effects

Plants are generally injured faster and to a greater extent when temperatures become higher than the maximum for growth than when they are lower than the minimum. However, too high a temperature rarely occurs in nature. High temperature seems to cause its effects on the plant in conjunction with the effects of other environmental factors, particularly excessive light, drought, lack of oxygen, or high winds accompanied by low relative humidity. High temperatures are usually responsible for sunscald injuries (Figs. 10-2A and 10-2B) appearing on the sun-exposed sides of fleshy fruits and vegetables, such as peppers, apples, tomatoes, onion bulbs, and potato tubers. On hot, sunny days the temperature of the fruit tissues beneath the surface facing the sun may be much higher than that of those on the shaded side and of the surrounding air. This results in discoloration, a water-soaked appearance, blistering, and desiccation of the tissues beneath the skin, which leads to sunken areas on the fruit surface. Succulent leaves of plants may also develop sunscald symptoms, especially when hot, sunny days follow periods of cloudy, rainy weather. Irregular areas on the leaves become pale green at first but soon collapse and form brown, dry spots. This is a rather common symptom of fleshy leaved houseplants kept next to windows with a southern exposure in early spring and summer when solar rays heat the fleshy leaves excessively. Too high a soil temperature at the soil line sometimes kills young seedlings (Fig. 10-2C) or causes cankers at the crown on the stems of older plants (Fig. 10-2D). High tempera-

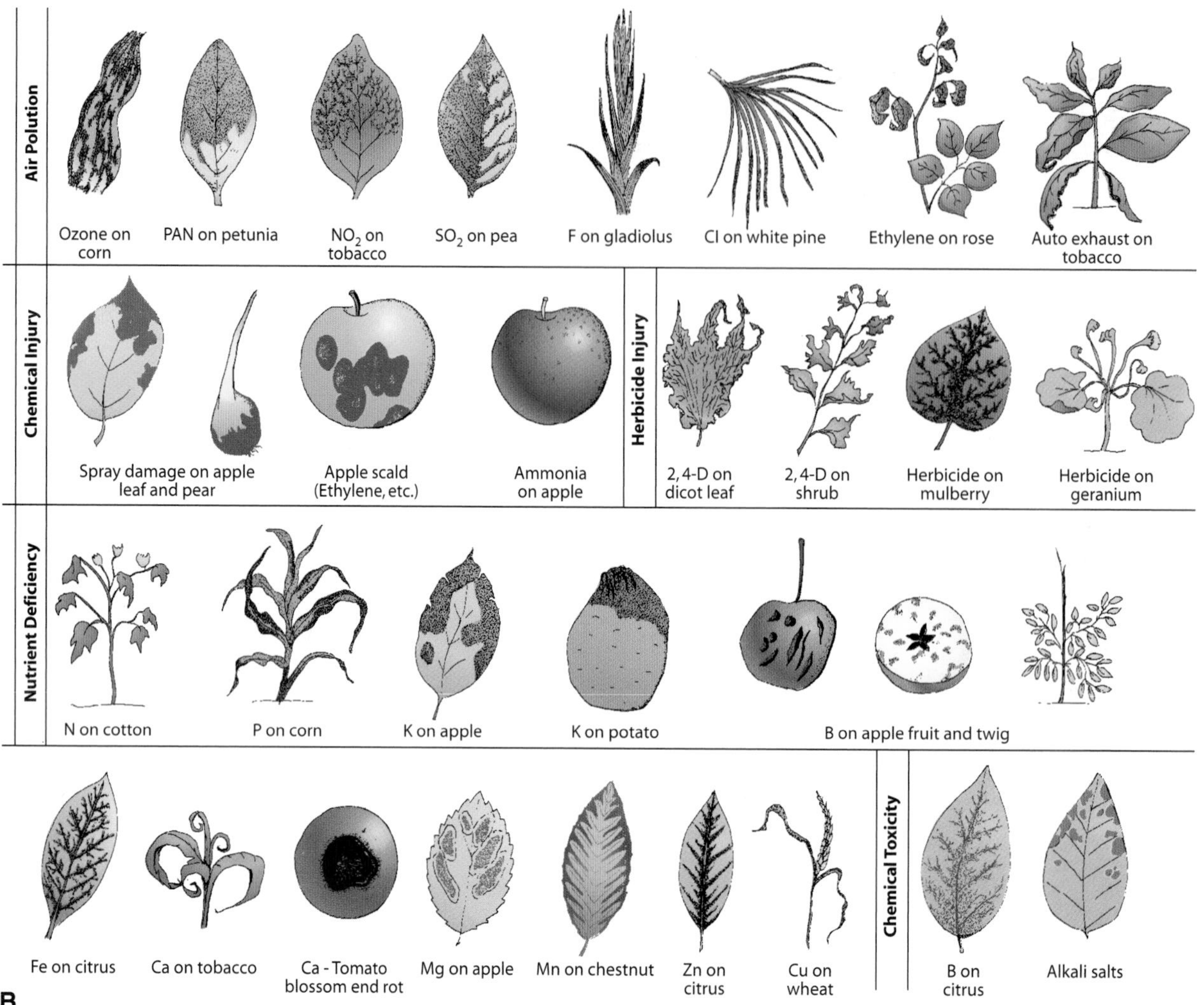

FIGURE 10-1 *(Continued)*

tures also seem to be involved in the water core disorder of apples and, in combination with reduced oxygen, in the blackheart of potatoes.

Low-Temperature Effects

Far greater damage to crops is caused by low than by high temperatures. Low temperatures, even if above freezing, may damage warm-weather plants such as corn and beans. They may also cause excessive sweetening and, on frying, undesirable caramelization of potatoes due to the hydrolysis of starch to sugars at the low temperatures.

Temperatures below freezing cause a variety of injuries to plants. Such injuries include the damage caused by late frosts to young leaves and meristematic tips (Figs. 10-3A–10-3C) or entire herbaceous plants, the frost killing of buds of peach, cherry, and other trees, and the killing of flowers, young fruit, and, sometimes, succulent twigs of most trees. Frost bands, consisting of discolored, corky tissue in a band or large area of the fruit surface, are often produced on apples, pears, and so on after a late frost (Fig. 10-3D). Low winter temperatures may kill the young roots of trees and may also cause bark splitting and canker development (Figs. 10-3E and 10-3F) on trunks and large branches, especially on the sun-exposed side, of several kinds of fruit trees. Cross sections of limbs may show a black ring or a blackheart condition in the wood. Fleshy tissues, such as tomato fruit, canola pods, and potato tubers, may be injured at subfreezing temperatures (Figs. 10-4A–10-4C). In potatoes, the injury varies depending on the degree of temperature drop and the duration of the low temperature. Early injury affects only the main vascular tissues and appears as a ring-like necrosis; injury of the finer vascular elements that are interspersed in the tuber gives the appearance of net-like necrosis. With more

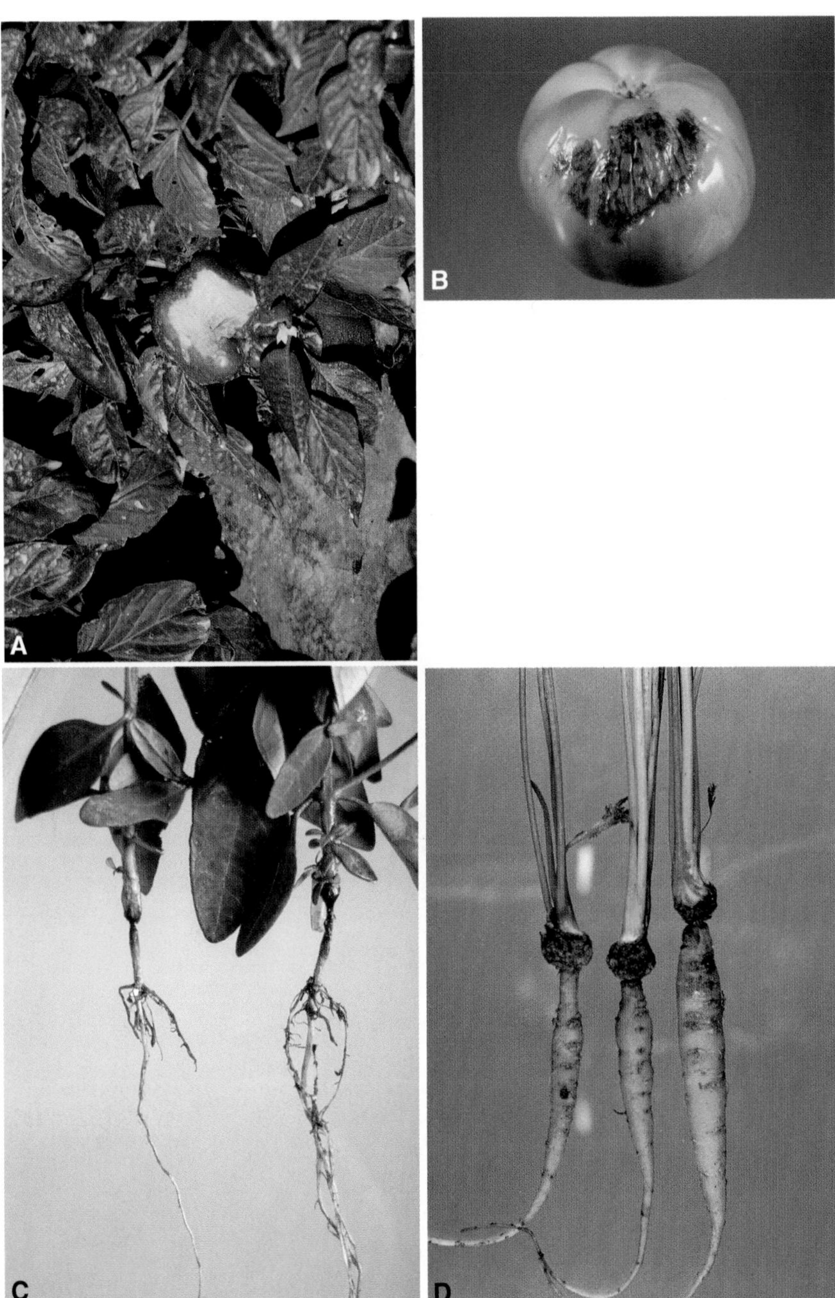

FIGURE 10-2 Some types of symptoms caused by excessively high temperatures. Sunscald damage on pepper (A) and tomato fruit (B). Damage to seedling stems (C) and to carrot (D) by the soil surface heated excessively by the sun. [Photographs courtesy of (A) R. J. McGovern, Univ. of Florida, (B, and D) I. R. Evans, W.C.P.D., and (C) E. L. Barnard, Florida Dept. Agriculture.]

FIGURE 10-3 Types of damage to plants caused by low temperatures. (A) Slight injury of tomato leaves caused by cold wind. Freeze damage on potato leaves (B), young shoots of azalea (C), russetting of apple fruit (D), cracks in forest tree branch (E), and cracking and peeling of bark in apple tree trunk (F). [Photographs courtesy of (A) R. J. McGovern, Univ. of Florida and (E) E. L. Barnard, Florida Dept. Agric.]

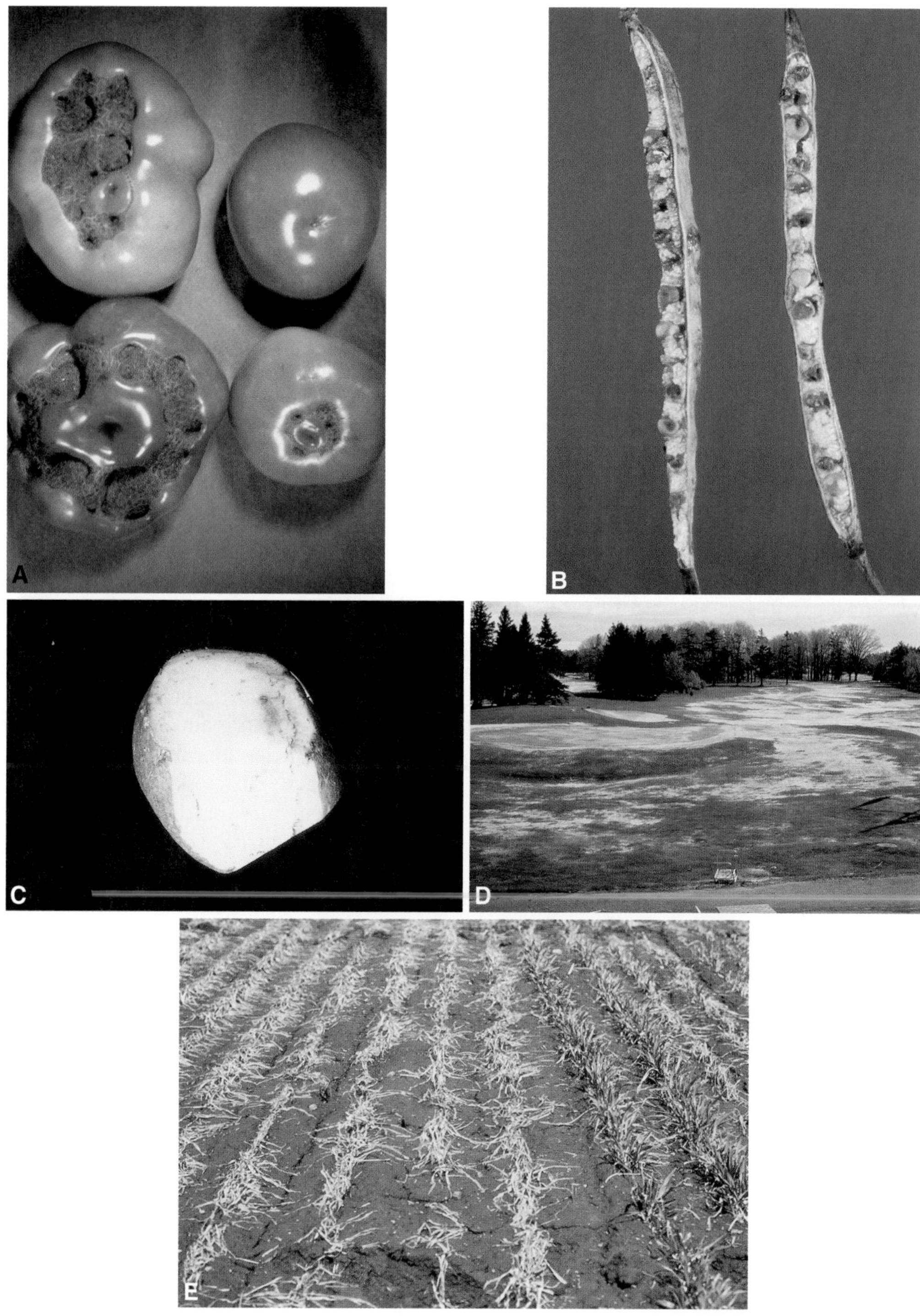

FIGURE 10-4 Additional symptoms of low-temperature damage in plants. (A) Tomato "catface," (B) killing of young seeds in canola, (C) potato freezing necrosis, (D) winter ice damage in poorly drained areas of turf, and (E) subfreezing kill of winter wheat plants. [Photographs courtesy of (A,B, and E) I. R. Evans, W.C.P.D., (C) University of Florida, and (D) S. Fustey, W.C.P.D.]

general injury, large chunks of the tuber are damaged, creating the so-called blotch-type necrosis. Subfreezing temperatures, especially in poorly drained areas where ice formation and thawing are common, may severely injure and may kill turf grass and young wheat plants (Figs. 10-4D and 10-4E).

Low-Temperature Effects on Indoor Plants

Indoor plants, whether grown in a home or a greenhouse, are particularly sensitive to low temperatures, both where they are growing and during transportation from a greenhouse or florist's shop to a home or from one home to another. Often, indoor plants are tropical plants grown far away from their normal climate. Exposure of such plants to low, not necessarily freezing, temperatures may cause stunting, yellowing, leaf or bud drop, and so on. Similarly, when grown indoors, even local plants remain in a succulent vegetative state and are completely unprepared for the stresses of low, particularly subfreezing, temperatures. Plants near windows or doors during cold winter days and, especially, nights are subject to temperatures that are much lower than those away from the window. Also, cracks or breaks in windows or the holes of electrical outlets on outside walls let in cold air that may injure the plants. A drop of night temperatures below 12°C may cause leaves and particularly flower buds of many plants to turn yellow and drop. Exposure of indoor plants to subfreezing temperatures for a few minutes or a few hours, e.g., while they are carried in the trunk of a car from the greenhouse to the house, may result in the death of many shoots and flowers or in a sudden shock to the plants from which they may take weeks or months to recover completely. Such a shock is often observed on plants that had been kept indoors and are then transplanted in the field in the spring when temperatures outdoors, although not freezing, are nevertheless much lower than those in the greenhouse. Even without the shock effect, plants growing at temperatures that are generally near the lower — or near the upper — limit of their normal range grow poorly and produce fewer and smaller blossoms and fruits.

Mechanisms of Low- and High-Temperature Injury to Plants

The mechanisms by which high and low temperatures injure plants are quite different. High temperatures apparently inactivate certain enzyme systems and accelerate others, thus leading to abnormal biochemical reactions and cell death. High temperature may also cause coagulation and denaturation of proteins, disruption of cytoplasmic membranes, suffocation, and possibly release of toxic products into the cell.

Low temperatures, however, injure plants primarily by inducing ice formation between or within the cells. The rather pure water of the intercellular spaces freezes first and normally at about 0°C, whereas the water within the cell contains dissolved substances that, depending on their nature and concentration, depress the freezing point of water several degrees. Furthermore, when the intercellular water becomes ice, more vapor (water) moves out of the cells and into the intercellular spaces, where it also becomes ice. The reduced water content of the cells depresses further the freezing point of the intracellular water. This could continue, up to a point, without damaging the cell, but below a certain temperature ice crystals form within the cell, disrupt the plasma membrane, and cause injury and death to the cell.

Ice formation in supercooled water within leaves is influenced greatly by the kinds and numbers of epiphytic bacteria that may be present on the leaves. Certain strains of some pathogenic (e.g., *Pseudomonas syringae*) bacteria and of some saprophytic bacteria, when present on or in the substomatal cavities of leaves, act as catalysts for ice nucleation. By their presence alone, such ice nucleation-active bacteria induce the supercooled water around them and in the leaf cells to form crystals, thereby causing frost injury to the leaves, blossoms, and so on at temperatures considerably higher (−1°C) than would have happened in the absence of such bacteria (approximately −5 to −10°C).

The freezing point of water in cells varies with the tissue and species of the plant; in some tissues of winter-hardy species of the north, ice probably never forms within the cells regardless of how low the temperatures become. Even when ice forms only in the intercellular spaces, cells and tissues may be damaged either by the inward pressure exerted by the ice crystals or by loss of water from their protoplasm to the intercellular spaces. This loss causes plasmolysis and dehydration of the protoplasm, which may cause coagulation. The rapidity of the temperature drop in a tissue is also important, as this affects the amount of water remaining in a cell and, therefore, the freezing point of the cell contents. Thus, a rapid drop in temperature may result in intracellular ice formation where a slow drop to the same low temperature would not. The rate of thawing may have similarly variable effects, as rapid thawing may flood the area between the cell wall and the protoplast and may cause tearing and disruption of the protoplast if the latter is incapable of absorbing the water as fast as it becomes available from the melting of ice in the intercellular spaces.

MOISTURE EFFECTS

Low Soil Moisture Effects

Moisture disturbances in the soil are probably responsible for more plants growing poorly and being unproductive annually, over large areas, than any other single environmental factor. Small or large territories may suffer from drought over time. The subnormal amounts of water available to plants in these areas may result in reduced growth, a diseased appearance, or even death of the plants. Lack of moisture may also be localized in certain types of soil, slopes, or thin soil layers underlaid by rock, clay, or sand and may result in patches of diseased-looking plants, while the immediately surrounding areas appear to contain sufficient amounts of moisture and the plants in them grow normally. Plants suffering from lack of sufficient soil moisture usually remain stunted, are pale green to light yellow, have few, small and drooping leaves, flower and fruit sparingly, and, if the drought continues, wilt and die (Fig. 10-5A). Although annual plants are considerably more susceptible to short periods of insufficient moisture, even perennial plants and trees are damaged by prolonged periods of drought and produce less growth, small, scorched leaves (Fig. 10-5B) and short twigs, dieback, defoliation, and finally wilting and death. Plants weakened by drought are also more susceptible to certain pathogens and insects.

Low Relative Humidity Effects

Lack of moisture in the atmosphere, i.e., low relative humidity, is usually temporary and seldom causes damage. When combined with high wind velocity and high temperature, however, it may lead to excessive loss of water from the foliage and may result in leaf scorching or burning, shriveled fruit, and temporary or permanent wilting of plants.

Conditions of low relative humidity are particularly common and injurious to houseplants during the winter. In modern homes and apartments, heating provides comfortable temperatures for plant growth, but it often dries the air to relative humidities of 15 to 25% which are equivalent to that of desert environments. The air is particularly dry over or near the sources of dry heat, such as radiators. Potted plants kept under these conditions not only use up the water much faster, grow poorly, and may begin to wilt sooner, but the leaves, especially the lower ones, of many kinds of plants become spotted or scorched and fall prematurely, while their flowers suddenly wither and drop off. These effects are particularly noticeable when plants are brought into such a hot, dry house directly from a cool, moist greenhouse or florist's shop. Generally, all houseplants prefer high humidity, and certain ones require high humidity if they are to grow properly and produce flowers. Therefore, houseplants should never be placed over radiators, and humidity should be increased with a commercial humidifier, by occasionally dampening the leaves with water, or by placing the pot on a brick or pebbles in a large pan of water, in a plastic case, or some other container.

High Soil Moisture Effects

Excessive soil moisture occurs much less often than drought where plants are grown. However, poor drainage or flooding of planted fields, gardens, or potted plants may result in more serious and quicker damage, or death, to plants than that from lack of moisture. Poor

FIGURE 10-5 Damage caused to plants by low water availability. (A) Wilting of pepper plants due to water stress. (B) Reduced growth, leaf scorching, and dieback of twigs of ornamental shrub due to prolonged water stress. [Photographs courtesy of (A) R. J. McGovern and (B) University of Florida.]

drainage results in plants that lack vigor, wilt frequently, and have leaves that are pale green or yellowish green. Flooding during the growth season may cause permanent wilting and death of succulent annuals within 2 to 3 days (Figs. 10-6A–C). Trees, too, are killed by waterlogging, but the damage usually appears more slowly and after their roots have been flooded continually for several weeks.

As a result of excessive soil moisture caused by flooding or by poor drainage, the fibrous roots of plants decay, probably because of the reduced supply of oxygen to the roots. Oxygen deprivation causes stress, asphyxiation, and collapse of many root cells. Wet, anaerobic conditions favor the growth of anaerobic microorganisms that, during their life processes, form substances, such as nitrites, that are toxic to plants. In addition, root cells damaged directly by the lack of oxygen lose their selective permeability and may allow toxic metals or other poisons to be taken up by the plant. Also, once parts of roots are killed, more damage is done by facultative parasites that may be favored greatly by the new environment. Thus, the wilting of the plants, which soon follows flooding, is probably the result of lack of water in the aboveground parts of plants caused by the death of the roots, although it appears that translocated toxic substances may also be involved.

In addition, many plants, particularly potted houseplants, show several symptoms that are the result of incorrect watering: either the soil is allowed to dry out too much before it is then flooded repeatedly with water or the plant is almost constantly overwatered. In either case, overwatered plants may suddenly drop their lower leaves or their leaves may turn yellow. Sometimes they develop brown or black wet patches on the leaves or stems, or the roots and lower stem may turn black and rot as a result of infection by pathogenic microorganisms encouraged by the excessive watering. Such symptoms can be avoided or corrected by watering only when the topsoil feels dry and then applying enough water to saturate thoroughly the whole mass of soil. Plants should never be watered when the soil is still wet, especially during the winter. When watering, any excess

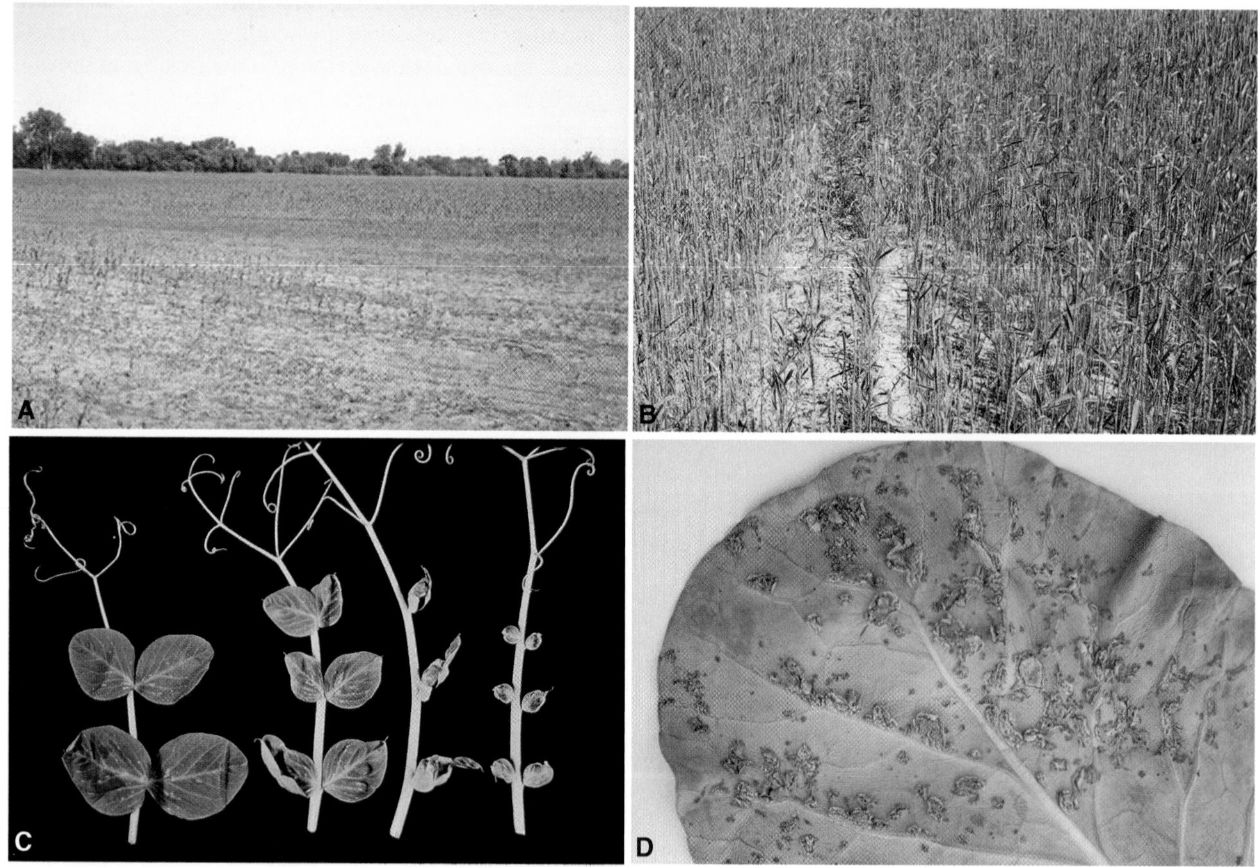

FIGURE 10-6 Damage caused to plants by excessive water. (A) Reduced growth and death of corn seedlings caused by flooding of low areas of the field. (B) Poor growth of plants following temporary flooding of the field. (C) Damage in field pea caused by water congestion. (D) Edema excrescences on the lower side of cabbage leaves caused by excessive water in the soil. [Photographs courtesy of (A) G. P. Munkvold, Iowa State Univ. (B) University of Florida, (C), W.C.P.D., and (D) D. P. Weingartner, Univ. of Florida.]

water should be drained through the drainage hole, which should always be present in the bottom of the pot. A period of dryness should not be followed with repeated heavy watering but by a gradual return to normal watering. Generally, the supply of water should be maintained as uniform as possible.

Another common symptom of houseplants, and sometimes of outdoor plants, that is caused by excessive moisture is the so-called edema [or oedema (swelling)]. Edema (Fig. 10-6D) appears as numerous small bumps on the lower side of leaves or on stems. The "bumps" are small masses of cells that divide, expand, and break out of the normal leaf surface and at first form greenish-white swellings or galls. Later, the exposed surface of the swelling becomes rusty colored and has a corky texture. Edema is caused by overwatering, especially during cloudy, humid weather, and can be avoided by reduced watering and providing better lighting and air circulation to the plant. Many other disorders are caused by excessive or irregular watering. It is known, for example, that tomatoes and some other fruits, such as cherries and grapes, grown under rather low moisture conditions at the time they are ripening often crack if they are suddenly supplied with abundant moisture by overwatering or by a heavy rainfall. Also, bitter pit of apples, consisting of small, sunken, black spots on the fruit, is the result of an irregular supply of moisture, although excessive nitrogen and low calcium fertilization also seem to be involved in bitter pit development.

INADEQUATE OXYGEN

Low oxygen conditions in nature are generally associated with high soil moisture or high temperatures. Lack of oxygen may cause the desiccation of roots of different kinds of plants in waterlogged soils, as was mentioned in the section on moisture effects. A combination of high soil moisture and high soil or air temperature causes root collapse in plants. The first condition, apparently, reduces the amount of oxygen available to the roots, whereas the other increases the amount of oxygen required by the plants. The two effects together result in an extreme lack of oxygen in the roots and cause their collapse and death.

Low oxygen levels may also occur in the centers of fleshy fruits or vegetables in the field, especially during periods of rapid respiration at high temperatures, or in storage of these products in fairly bulky piles. The best known such case is the development of the so-called blackheart of potato, in which fairly high temperatures stimulate respiration and abnormal enzymatic reactions in the potato tuber. The oxygen supply to the cells in the interior of the tuber is insufficient to sustain the increased respiration and the cells die of suboxidation. Enzymatic reactions activated by the high temperature and suboxidation go on before, during, and after the death of the cells. These reactions abnormally oxidize normal plant constituents into dark melanin pigments. The pigments spread into the surrounding tuber tissues and finally make them appear black (Figs. 10-7A and 10-7B).

LIGHT

Lack of sufficient light retards chlorophyll formation and promotes slender growth with long internodes, thus leading to pale green leaves, spindly growth, and premature drop of leaves and flowers. This condition is known as etiolation. Etiolated plants are found out-

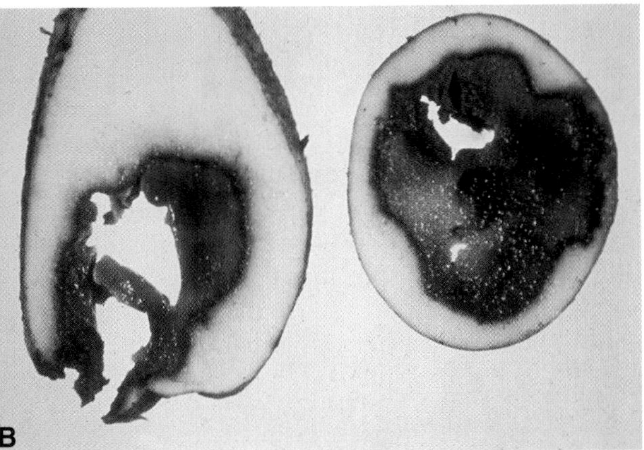

FIGURE 10-7 Damage caused by low oxygen. (A) Internal browning of potato stem end and (B) black heart of potato caused by low oxygen and high temperature in the soil. [Photographs courtesy of Plant Pathology Department, University of Florida.]

doors only when plants are spaced too close together or when they are growing under trees or other objects. Etiolation of various degrees, however, is rather common in houseplants and also in plants grown in greenhouses, seedbeds, and cold frames, where plants often receive inadequate light. Etiolated plants are usually thin and tall and are susceptible to lodging.

Excess light is rather rare in nature and seldom injures plants. Many injuries attributed to light are probably the result of high temperatures accompanying high-light intensities. Excessive light, however, seems to cause sunscald of pods of beans grown at high altitudes where, due to the absence of dust, more light of short wavelengths reaches the earth. The pods develop small, water-soaked spots that quickly become brown or reddish brown and shrink.

The amount of light is considerably more important in relation to houseplants. Some of them prefer shade or semishade during the growth season but full sunlight during the winter. Others prefer shade throughout the year, whereas still others must have sunlight all year long. As a rule, houseplants with deep green leaves prefer or tolerate shade much better than plants with colored leaves, with the latter generally doing better when they receive considerable sunlight. Most flowering houseplants grow and flower best with full exposure to sunlight at all seasons. Lack of sufficient light for any of these kinds of plants has the same effects as on outdoor plants, namely pale green leaves, spindly growth, leaf drop, few or no flowers, and flower drop. However, excessive sunlight on plants that prefer less light often results in the appearance of yellowish-brown or silvery spots on the leaves. Plants moved suddenly to an area with strikingly different light intensity than in the previous area often respond with general defoliation.

AIR POLLUTION

The air at the earth's surface consists primarily of nitrogen and oxygen (78 and 21%, respectively). Much of the remaining 1% is water vapor and carbon dioxide. The activities of humans in generating energy, manufacturing goods, and disposing of wastes result in the release of a number of pollutants into the atmosphere that may alter plant metabolism and induce disease. Air pollution damage to plants, especially around certain types of factories, has been recognized for about a century. Its extent and importance, however, have increased with continued industrialization and will, apparently, increase further with the world's increasing population and urbanization.

Air Pollutants and Kinds of Injury to Plants

Almost all air pollutants causing plant injury are gases, but some particulate matter or dusts may also affect vegetation. Some gas contaminants, such as ethylene, ammonia, and chlorine, exert their injurious effects over limited areas. Most frequently they affect plants or plant products stored in poorly ventilated warehouses in which the pollutants are produced by the plants themselves (ethylene) or result from leaks in the cooling system (ammonia).

More serious and widespread damage is caused to plants in the field by chemicals such as ozone (Fig. 10-8), sulfur dioxide, hydrogen fluoride, nitrogen dioxide, peroxyacyl nitrates, and particulates. In many localities, e.g., the Los Angeles basin, air pollutants spread into the area surrounding the source(s) of pollution, become trapped, and cause serious plant damage. More frequently, most air pollutants are transported downwind from the urban or industrial centers in which they are produced and may be carried by wind to areas that are several miles, often hundreds of miles and sometimes thousands of miles, from the source. High concentrations of or long exposure to these chemicals cause visible and sometimes characteristic symptoms (such as necrosis) on the affected plants. More important economically, however, is the fact that even when plants are exposed to dosages less than those that cause acute damage, their growth and productivity may still be suppressed by 5 to 10% because of interference by the pollutants with the metabolism of the plant. Moreover, prolonged exposure to air pollutants seems to weaken plants and to predispose them to attack by insects, by some pathogens, and by other environmental factors such as low winter temperatures. The main pollutants, their sources, and their effects on plants are given in Table 10-1.

Main Sources of Air Pollutants

Some air pollutants, such as sulfur dioxide and hydrogen fluoride, are produced as such directly from a source, such as refineries, combustion of fuel, and ore and fertilizer processing. Others, such as ozone and peroxyacyl nitrates, are produced in the atmosphere as secondary products of photochemical reactions involving NO_2, O_2, hydrocarbons, and sunlight.

Automobile exhaust in the streets and highways and exhausts of other internal combustion engines in factories and in homes are probably the most important sources of ozone and other phytotoxic pollutants. Thousands of tons of incompletely burned hydrocarbons and

TABLE 10-1
Air Pollution Injury to Plants

Pollutant	Source	Susceptible plants	Symptoms	Remarks
Ozone (O_3)	Automobile exhausts and other internal combustion engines (released NO_2 combines with O_2 in sunlight $\rightarrow O_3$). From stratosphere From lightning, from forests	Expanding leaves of all plants, especially tobacco, bean, cereals, alfalfa, petunia, pine, citrus, and corn	Stippling, mottling, and chlorosis of leaves, primarily on upper leaf surface. Spots are small to large, bleached white to tan, brown, or black (Figs. 10-8A–D). Premature defoliation and stunting occur in plants such as citrus, grapes, and vines	Enters through stomata. It is the most destructive air pollutant to plants. A major component of smog
Peroxyacyl nitrates (PAN)	Automobile exhausts and other internal combustion engines (gasoline vapors and incompletely burned gasoline $+O_3 + NO \rightarrow$ PAN)	Many kinds of plants, including spinach, petunia, tomato, lettuce, and dahlia	Causes "silver leaf" on plants, i.e., bleached white to bronze spots on lower surface of leaves that may later spread throughout leaf thickness and resemble ozone injury	Particularly severe near metropolitan areas with smog and inversion layers
Sulfur dioxide (SO_2)	Stacks of factories Automobile exhausts and other internal combustion engines	Many kinds of plants, including alfalfa, violet, conifers, pea, cotton, and bean Toxic at 0.3–0.5 ppm	Low concentrations cause general chlorosis. Higher concentrations cause bleaching of interveinal tissues of leaves (Fig. 10-8E)	Also combines with moisture and forms toxic acid droplets (acid rain)
Nitrogen dioxide (NO_2)	From oxygen and nitrogen in the air by hot combustion sources, e.g., furnaces, internal combustion engines	Many kinds of plants, including beans and tomatoes Toxic at 2–3 ppm	Causes bleaching and bronzing of plants similar to that caused by SO_2. At low concentration it also suppresses growth of plants	
Hydrogen fluoride (HF)	Stacks of factories processing ore or oil	Many kinds of plants, including corn, peach, and tulip; actively growing, especially wet leaves, are most sensitive Toxic at 0.1–0.2 ppb	Leaf margins of dicots and leaf tips of monocots turn tan to dark brown, die, and may fall from the leaf. Some plants tolerate HF up to 200 ppm	HF may evaporate or be washed out of plant and plant recovers slowly
Chlorine (Cl_2) and hydrogen chloride (HCl)	Refineries, glass factories, incineration of plastics (Fig. 10-8F)	Many kinds of plants, usually near the source Toxic at 0.1 ppm	Leaves show bleached, necrotic areas between veins. Leaf margins often appear scorched. Leaves may drop prematurely. Damage resembles that caused by SO_2	
Ethylene (CH_2CH_2)	Automobile exhausts Burning of gas, fuel oil, and coal From ripening fruit in storage	Many kinds of plants Toxic at 0.05 ppm	Plants remain stunted, their leaves develop abnormally and senesce prematurely. Plants produce fewer blossoms and fruit Fruit, e.g., apples, develop depressed, necrotic, dark areas (scald)	Ethylene is a plant hormone with numerous functions
Particulate matter (dusts)	Dust from roads, cement factories Burning of coal, etc.	All plants	Forms dust or crusty layers on plant surfaces. Plants become chlorotic, grow poorly, and may die. Some dusts are toxic and burn leaf tissues directly or after dissolving in dew or rainwater	

FIGURE 10-8 Types of damage to plants caused by air pollutants. Ozone damage to leaves of tobacco (A), potato (B), conifer (C), and sycamore (D). (E) Sulfur dioxide damage to poplar leaves. (F) Chlorine damage to horsechestnut leaves. [Photographs courtesy of Plant Pathology Department, University of Florida.]

NO_2 are released into the atmosphere daily by automobile exhausts. In the presence of ultraviolet light from the sun, this nitrogen dioxide reacts with air oxygen and forms ozone and nitric oxide. The ozone may react with nitric oxide to form the original compounds:

$$NO_2 + O_2 \xleftrightarrow{\text{sunlight}} O_3 + NO$$

In the presence of unburned hydrocarbon radicals, however, the nitric oxide reacts with these instead of ozone, and therefore the ozone concentration builds up:

$$O_3 + [NO + \text{unburned hydrocarbons from automobiles, etc.}] \rightarrow O_3 + \text{peroxyacyl nitrates}$$

Ozone can also react with vapors of certain unsaturated hydrocarbons, but the products of such reactions (various organic peroxides) are also toxic to plants. Normally, the noxious fumes produced by automobiles and other engines are swept up by the warm air currents from the earth's surface rising into the cooler air above, where the fumes are dissipated. During periods of calm, stagnant weather, however, an inversion layer of warm air is formed above the cooler air, which prevents the upward dispersion of atmospheric pollutants. The pollutants are then trapped near the ground, where, after sufficient buildup, they may seriously damage living organisms.

Peroxyacyl nitrate (PAN) injury has been observed primarily around metropolitan areas where large amounts of hydrocarbons are released into the air from automobiles. The problem is especially serious in areas such as Los Angeles and New Jersey, where the atmospheric conditions are conducive to the formation of inversion layers. Many different kinds of plants are affected by PAN compounds over large geographical areas surrounding the locus of PAN formation due to diffusion or to dispersal of the pollutant by light air currents.

How Air Pollutants Affect Plants

The concentration at which each pollutant causes injury to a plant varies with the plant and even with the age of the plant or the plant part. As the duration that the plant is exposed to the pollutant is increased, damage can be caused by increasingly smaller concentrations of the pollutant until a minimum dose-injury threshold is reached. Plant injury by air pollutants generally increases with increased light intensity, increased soil moisture and air relative humidity, and increased temperature and with the presence of other air pollutants.

In a given location, ozone fluctuates from 0.01–0.03 parts per million (ppm) in the morning to 0.05–0.10 ppm at peak sunlight intensity in early afternoon and decreases gradually afterward. There are, however, frequently days of higher O_3 concentration of up to 0.15 ppm in most rural areas, whereas in heavily populated and industrial areas such as the Los Angeles basin, O_3 peaks of 0.25 ppm are common.

Ozone injures the leaves of plants exposed for even a few hours at concentrations of 0.1 to 0.3 ppm. Ozone is taken into leaves through stomata and injures primarily palisade but also other cells by disrupting the cell membrane. Affected cells near stomata collapse and die, and white (bleached) necrotic flecks appear, first on the upper side and later on either leaf surface. Many crop plants, such as alfalfa, bean, citrus, grape, potato, soybean, tobacco, and wheat, and many ornamentals and trees, such as ash, lilac, several pines, and poplar, are quite sensitive to ozone, whereas some other crops, such as cabbage, peas, peanuts, and pepper, are of intermediate sensitivity, and some, such as beets, cotton, lettuce, strawberry, and apricot, are tolerant.

Sulfur dioxide may injure plants in concentrations as low as 0.3 to 0.5 ppm. Because sulfur dioxide is absorbed through the leaf stomata, conditions that favor or inhibit the opening of stomata similarly affect the amount of sulfur dioxide absorbed. After absorption by the leaf, sulfur dioxide reacts with water and forms phytotoxic sulfite ions. The latter, however, are oxidized slowly in the cell to produce harmless sulfate ions. Thus, if the rate of sulfur dioxide absorption is slow enough, the plant may be able to protect itself from the buildup of phytotoxic sulfites.

Peroxyacyl nitrates are also taken into leaves through stomata and cause injury at concentrations as low as 0.01 to 0.02 ppm. In large urban areas, concentrations of 0.02 to 0.03 ppm are not uncommon, and in the downtown areas of some cities, PAN concentrations of 0.05 to 0.21 have been measured. Once inside leaves, PAN attacks preferentially the spongy parenchyma cells, which collapse and are replaced by air pockets that give the leaf a glazed or silvery appearance. The symptoms on broad-leaved plants appear on the lower leaf surface, whereas monocot leaves show symptoms on both sides. Young leaves and tissues are more sensitive to PAN, and periodic exposures of leaves to PAN often cause "banding" and in some plants even margin "pinching" of leaves because of discoloration and death of the most sensitive affected cells, respectively.

Acid Rain

Normal, unpolluted rain would contain almost pure water (H_2O) in which there would be dissolved some carbon dioxide (CO_2), some ammonia (NH_3) originating from organic matter and existing in water as NH^+_4, and varying but small amounts of cations (Ca^{2+}, Mg^{2+}, K^+, and Na^+) and anions (Cl^2, SO_4^{22}). Although the pH of pure water is a neutral pH 7.0, the pH of normal, "unpolluted" rain is usually pH 5.6; in other words, rain is already acidic. Such rain, however, is considered normal, and only when the pH of rain or snow is below pH 5.6 is it considered acidic (acid rain).

Acid rain is the result of human activities, primarily the combustion of fossil fuels (oil, coal, and natural gas) and the smelting of sulfide ores. These activities release large quantities of sulfur and nitrogen oxides in the atmosphere, which when in contact with atmospheric moisture are converted to two of the strongest acids known (sulfuric and nitric) and fall to the ground in rain, snow, and fog. The pH of rain and snow over large regions of the world ranges from pH 4.0 to 4.5, which is from 5 to 30 times more acid than the lowest pH (pH 5.6) expected for unpolluted areas. The lowest rain pH values reported so far (pH 2.4 in Scotland, pH 1.5 in West Virginia, and pH 1.7 in Los Angeles) are more acidic than vinegar (pH 3.0) and lemon juice (pH 2.2). It is estimated that about 70% of the acid in acid rain is sulfuric acid, with nitric acid contributing about 30%. In addition to sulfur contained in the acids carried in the rain, it is believed that an approximately equal amount of sulfur reaches all surfaces through dry deposition of particulate sulfur. In humid or wet weather, this sulfur is also oxidized to sulfuric acid.

Acid rain exerts a variety of influences by greatly increasing the solubility of all kinds of molecules and by directly (through the low pH and the toxicity of the SO_4^{2-} and NO^{-}_3 ions) or indirectly (through the dissolved molecules) affecting many forms of life. The adverse effects of acid rain on the microorganisms, plants, and fishes of rivers and lakes have been well documented. The effects of acid rain on crop plants have been more difficult to document. Experiments in which acidic rain (pH 3.0) was applied to plants showed that, under some conditions, treated leaves developed pits, spots, and curling and that treated plants, with or without symptoms, sometimes showed reductions in dry weight. Also, more seeds of some plant species germinated when the soil in which they were planted received acid rain than when it did not, whereas the opposite was observed for other species. Experiments conducted to determine the effect of acid rain on the initiation and development of plant diseases have shown that in some diseases, such as *Cronartium fusiforme* rust of oak, only 14% as many telia formed under acid (pH 3.0) rain treatment than under a pH 6.0 rain treatment and that beans treated with acidic rain (pH 3.2) had only 34% as many nematode egg masses than they did under a pH 6.0 rain treatment. However, a bacterial disease (halo blight) and the rust disease of bean were sometimes more severe and others milder with the acidic rain than with the pH 6.0 rain. In general, although some evidence exists that acid rain causes variable amounts of damage to at least some plants, consistent quantitative data are still insufficient to determine the extent of such damage on various crops in the areas where they occur.

NUTRITIONAL DEFICIENCIES IN PLANTS

Plants require several mineral elements for normal growth. Some elements, such as nitrogen, phosphorus, potassium, calcium, magnesium, and sulfur, needed in relatively large amounts, are called major elements, whereas others, such as iron, boron, manganese, zinc, copper, molybdenum, and chlorine, needed in very small amounts, are called trace or minor elements or micronutrients. Both major and trace elements are essential to the plant. When they are present in the plant in amounts smaller than the minimum levels required for normal plant growth, the plant becomes diseased and exhibits various external and internal symptoms. The symptoms may appear on any or all organs of the plant, including leaves, stems, roots, flowers, fruits, and seeds.

The kinds of symptoms produced by deficiency of a certain nutrient depend primarily on the functions of that particular element in the plant. These functions presumably are inhibited or interfered with when the element is limiting. Certain symptoms are the same when any of several elements are deficient, but other diagnostic features usually accompany a deficiency of a particular element. Numerous plant diseases occur annually in most agricultural crops in many locations as a result of reduced amounts or reduced availability of one or more of the essential elements in the soils where the plants are grown. The presence of lower than normal amounts of most essential elements usually results in merely a reduction in growth and yield. When the deficiency is greater than a certain critical level, however, the plants develop acute or chronic symptoms and may even die. Some of the general deficiency symptoms caused by each essential element, the possible functions affected, and some examples of common deficiency disorders are listed in Table 10-2 and are shown in Figs. 10-7 and 10-8.

SOIL MINERALS TOXIC TO PLANTS

Soils often contain excessive amounts of certain essential or nonessential elements, either of which at high concentration may be injurious to the plants. Of the essential elements, those required by plants in large amounts, such as nitrogen and potassium, are usually much less toxic when present in excess than are elements required only in trace amounts, such as manganese, zinc, and boron. Even among the latter, however, some trace elements such as manganese and magnesium have a much wider range of safety than others, such as boron or zinc. Besides, not only do the elements differ in their ranges of toxicity, but various kinds of plants also differ in their susceptibility to the toxicity to a certain level of a particular element. Concentrations at which nonessential elements are toxic also vary among elements, and plants in turn vary in their sensitivity to them. For example, some plants are injured by very small amounts

TABLE 10-2
Nutrient Deficiencies in Plants

Deficient nutrient	Functions of element	Symptoms
Nitrogen (N)	Present in most substances of cells	Plants grow poorly and are light green in color. Lower leaves turn yellow or light brown and stems are short and slender (Fig. 10-9A)
Phosphorus (P)	Present in DNA, RNA, phospholipids (membranes), ADP, ATP, etc.	Plants grow poorly and leaves are bluish green with purple tints. Lower leaves sometimes turn light bronze with purple or brown spots. Shoots are short and thin, upright, and spindly (Figs. 10-9B and 10-9C)
Potassium (K)	Acts as a catalyst of many reactions	Plants have thin shoots, which in severe cases show dieback. Older leaves show chlorosis with browning of tips, scorching of margins, and many brown spots usually near the margins. Fleshy tissues show end necrosis (Figs. 10-9D, 10-9E and 10-9F)
Iron (Fe)	Is a catalyst of chlorophyll synthesis. Part of many enzymes	Young leaves become severely chlorotic, but main veins remain characteristically green. Sometimes brown spots develop. Part of or entire leaves may dry. Leaves may be shed (Figs. 10-10A and 10-10B)
Magnesium (Mg)	Present in chlorophyll and is part of many enzymes	First older, then younger leaves become mottled and chlorotic, then reddish. Sometimes necrotic spots appear. Tips and margins of leaves may turn upward and leaves appear cupped. Leaves may drop off (Figs. 10-10C and 10-10D)
Boron (B)	Not really known. Affects translocation of sugars and utilization of calcium in cell wall formation	Bases of young leaves of terminal buds become light green and finally break down. Stems and leaves become distorted. Plants are stunted. Fruit, fleshy roots or stems, etc., may crack on the surface and/or rot in the center. Causes many plant diseases, e.g., heart rot of sugar beets, brown heart of turnips, browning or hollow stem of cauliflower (Fig. 10-11A), cracked stem of celery (Fig. 10-11B), corky spot, cracked fruit of peae (Fig. 10-11C), dieback, and rosette of apples, hard fruit of citrus, top sickness of tobacco
Calcium (Ca)	Regulates the permeability of membranes. Forms salts with pectins. Affects activity of many enzymes	Young leaves become distorted, with tips hooked back and margins curled. Leaves may be irregular in shape and ragged with brown scorching or spotting. Terminal buds finally die. Plants have poor, bare root systems. Causes blossom end rot of many fruits (Figs. 10-12A–10-12C). Increases fruit (e.g., apple) decay in storage. May be responsible for tip burns in mature detached lettuce heads at high (24–35°C) temperatures
Sulfur (S)	Present in some amino acids and coenzymes	Young leaves are pale green or light yellow without any spots. Symptoms resemble those of nitrogen deficiency
Zinc (Zn)	Is part of enzymes involved in auxin synthesis and in oxidation of sugars	Leaves show interveinal chlorosis. Later they become necrotic and show purple pigmentation. Leaves are few and small, internodes are short and shoots form rosettes, and fruit production is low. Leaves are shed progressively from base to tip. It causes little leaf of apple, stone fruits, and grape, sickle leaf of cacao, white tip of corn, etc.
Manganese (Mn)	Is part of many enzymes of respiration, photosynthesis, and nitrogen utilization	Leaves become chlorotic but smallest veins remain green and produce a checked effect. Necrotic spots may appear scattered on leaf. Severely affected leaves turn brown and wither
Molybdenum (Mo)	Is essential component of nitrate reductase enzyme	Melons, and probably other plants, exhibit severe yellowing and stunting and fail to set fruit (Fig. 10-12D)
Copper (Cu)	Is part of many oxidative enzymes	Tips of young leaves of cereals wither and their margins become chlorotic (Fig. 10-12F). Leaves may fail to unroll and tend to appear wilted. Heading is reduced and heads are dwarfed and distorted and yield is reduced (Fig. 10-12F). Citrus, pome, and stone fruits show dieback of twigs in the summer, burning of leaf margins, chlorosis, rosetting, etc. Vegetable crops fail to grow

of nickel but can tolerate considerable concentrations of aluminum.

The injury occurring from the excess of an element may be slight or severe and is usually the result of direct injury by the element to the cell. However, the element may interfere with the absorption or function of another element and thereby lead to symptoms of a deficiency of the element being interfered with. Thus, excessive sodium induces a deficiency of calcium in the plant, whereas the toxicity of copper, manganese, or zinc both is direct on the plant and induces a deficiency of iron in the plant.

Excessive amounts of sodium salts, especially sodium chloride, sodium sulfate, and sodium carbonate, raise the pH of the soil and cause what is known as alkali injury. This injury varies in different plants and may range from chlorosis to stunting, leaf burn, wilting, and outright killing of seedlings and young plants. Some

FIGURE 10-9 Types of damage caused by some air pollutants. (A) Nitrogen deficiency-induced chlorosis and stunting of two wheat plants (left) compared to two healthy plants (right). Purplish coloration and stunting of phosphorous-deficient corn seedlings (B) and alfalfa leaf (C). Potassium deficiency-induced marginal leaf necrosis in alfalfa (D), in a corn seedling (E), and in a field of corn (F). [Photographs courtesy of (A,B,E, and F) Plant Pathology Department, University of Florida, (C) C. Richard, W.C.P.D., and (D) R. J. Howard, W.C.P.D.]

FIGURE 10-10 Damage caused by deficiencies of other nutrients. Iron deficiency chlorosis in an ornamental shrub (A) and in bean (B). Magnesium deficiency symptoms in cucumber leaves (C) and in palm plant (D). [Photographs courtesy of (A,B, and D) Plant Pathology Department, University of Florida and (C) I. R. Evans, W.C.P.D.]

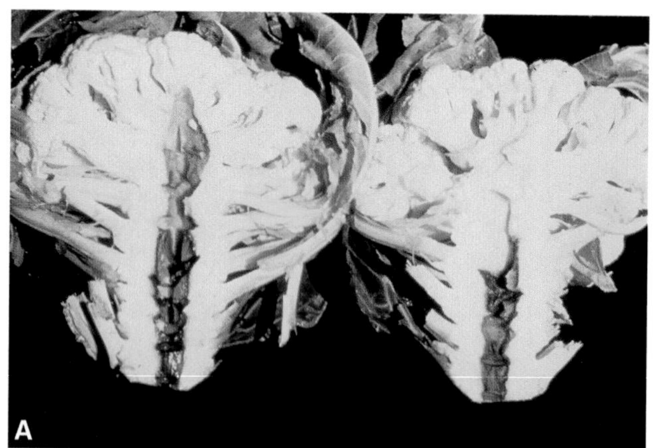

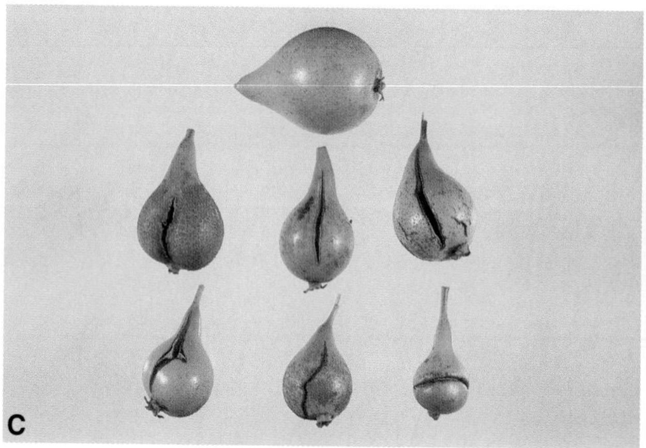

FIGURE 10-11 Boron deficiency symptoms on cauliflower (A), celery stem (B), and pears (C). Boron toxicity symptoms on pea seedlings (D). [Photographs courtesy of (A,B, and D) D. Ormrod, W.C.P.D.]

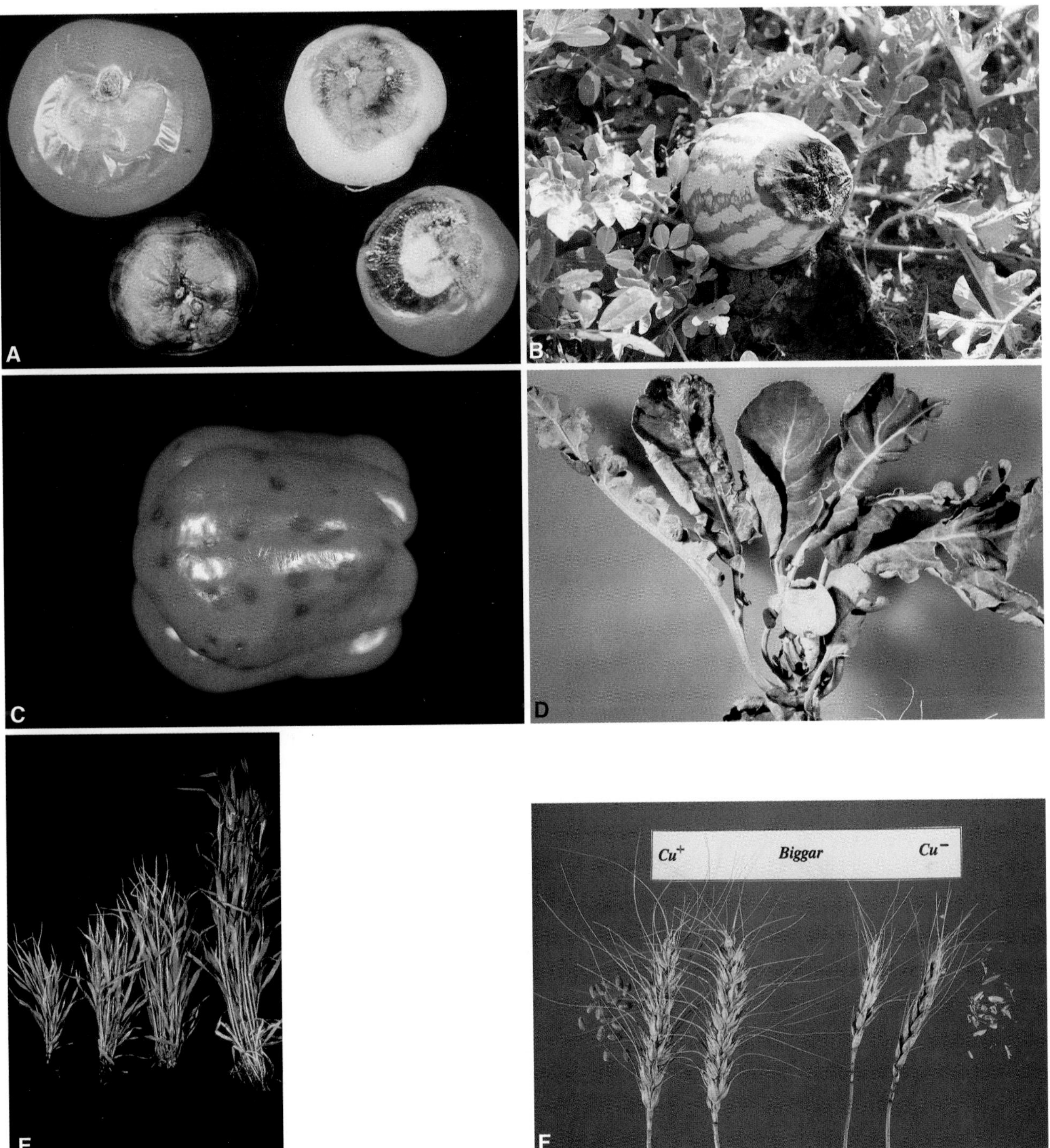

FIGURE 10-12 Additional deficiency symptoms. Calcium deficiency-induced blossom end rot in tomato fruit (A) and in watermelon (B). Pitting on pepper fruit due to calcium deficiency near harvesting (C). (D) Molybdenum deficiency symptoms in young cauliflower plant. (E) Copper deficiency-induced chlorosis and reduction of growth in wheat plants compared to healthy plant (right) and (F) reduction in head size and in number and size of kernels in copper-deficient plants. [Photographs courtesy of (A–C) University of Florida and (D–F) I. R. Evans, W.C.P.D.]

FIGURE 10-13 Soybean chlorosis caused by alkaline soils high in sodium salts and a high pH. (A) Chlorosis as it appears in a soybean field from a distance. (B) Close-up of chlorosis and death of soybean seedlings in such a field. [Photographs courtesy of E. J. Penas.]

plants, such as wheat and apple, are very sensitive to alkali injury, whereas others, such as sugar beets, alfalfa, and several grasses, are quite tolerant. In the river valleys of Nebraska, approximately 250,000 acres of alkaline land exist in which soybeans develop chlorosis or yellowing (Figs. 10-13A and 10-13B), especially in parts of such areas in which soil pH is 7.5 or higher. However, when the soil is too acidic, the growth of some kinds of plants is impaired and various symptoms may appear. Plants usually grow well in a soil pH range from pH 4.0 to 8.0, but some plants grow better at lower pH than others, and vice versa. Thus, blueberries grow well in acid soils, whereas alfalfa grows best in alkaline soils. The injury caused by low pH is, in most cases, brought about by the greater solubility of mineral salts in acid solutions. These salts then become available in concentrations that, as mentioned earlier, either are toxic to the plants or interfere with the absorption of other necessary elements and so cause symptoms of mineral deficiency.

Boron, manganese, and copper have been implicated most frequently in mineral toxicity diseases, although other minerals, such as aluminum and iron, also damage plants in acid soils. Excess boron is toxic to many vegetables and trees. Excess manganese is known to cause a crinkle-leaf disease in cotton and has been implicated in the internal bark necrosis of Red Delicious apple and in many other diseases of several crop plants. Sodium and chlorine ions also have been shown to cause symptoms of poor growth and decline such as those shown by some of the trees along roads in northern areas where heavy salting is carried out in the winter to remove ice from roads.

HERBICIDE INJURY

Some of the most frequent plant disorders seem to be the result of the extensive use of herbicides (weed killers). The constantly increasing number of herbicides in use by more and more people for general or specific weed control is creating numerous problems among those who use them, their neighbors, and those who use soil that has been treated with herbicides.

Herbicides are either specific against broad-leaved weeds [atrazine, simazine, (2,4-dichlorophenoxy) acetic acid (2,4-D), dicamba (Banvel-D)] and are applied in corn and other small grain fields and on lawns or they are specific against grasses and some broad-leaved weeds [Dacthal, trifluralin (Treflan)] and are applied in pastures, orchards, and in vegetable and truck crop fields. In addition, some herbicides are general weed or shrub killers [glyphosate (Roundup), paraquat, terbacil (Sinbar), picloram]. Most herbicides are safe as long as they are used to control weeds among the right crop plants, at the right time, at the correct dosage, and when the correct environmental conditions prevail. When any one of the aforementioned conditions is not met, abnormalities develop on the cultivated plants with which the herbicides come in contact. Affected plants show various degrees of distortion or yellowing of leaves (Figs. 10-14A–D), browning, drying and shedding of leaves, stunting (Fig. 10-14E), and even death of the plant (Fig. 10-15). Much of this damage is caused by too high doses of herbicides or by applications made too early in the season or on too cold or too hot a day or when dust or spray droplets of an herbicide are carried by the wind to nearby sensitive plants or to gardens or

FIGURE 10-14 Types of herbicide injury to plants. Foliage distortion and malformations caused by herbicides on tomato (A), cotton (B), and papaya (C). (D) Tomato fruit malformations as a result of exposure to the 2,4-D herbicide. (E) Chlorosis and stunting in lemon seedlings due to improper fumigation with methyl bromide. [Photographs courtesy of (A and C) University of Florida, (B) S. D. Eubanks, (D) I. R. Evans, W.C.P.D., and (E) J. H. Graham.]

FIGURE 10-15 (A and B) Additional types of damage caused by herbicides on single corn seedling (A) and on an entire field of corn seedlings. Spray injury on rose leaves (C) and on watermelon plants (D). [Photographs courtesy of (A and B) R. Hartzler and (C and D) University of Florida.]

fields in which plants sensitive to the herbicide are grown. Of course, direct application of the wrong pesticide in a field with a particular crop plant will kill the crop just as if it were a weed.

Use of preplant or preemergence herbicides through application to the soil before or at planting time often affects seed germination and growth of the young seedlings if too much or the wrong herbicide has been applied. Most herbicides are used up or are inactivated within a few days to a few months from the time of application; some, however, persist in the soil for more than a year. Sensitive plants planted in fields treated previously with such a persistent herbicide may grow poorly and may produce various symptoms. Also, home owners, home gardeners, and greenhouse operators often obtain what looks like good, weed-free soil from fields that, unbeknown to them, had been treated with herbicides. Such soil when used to grow potted, bench, or garden plants results in smaller, distorted, yellowish plants, which sometimes shed some or all of their leaves and either die or finally recover.

HAIL INJURY

Depending on the stage of development of the plant, the size of the hail, and duration of the hail storm, damage to crops from hail may be small, intermediate, or complete (Figs. 10-16); in the latter case, all plants are destroyed by the hail.

FIGURE 10-16 Hail injury on watermelon (A), cabbage (B), and cotton (C). [Photographs courtesy of Plant Pathology Department, University of Florida.]

LIGHTNING

Lightning is a rather rare event in most locations but it does occur and in some locations, e.g., central Florida, it occurs quite frequently. When lightning strikes a tree, the trunk or main branches may crack (Fig. 10-17A,B), tip over, or fall. Fields, however, may also be hit by lightning either directly (Figs. 10-17C, 10-17E, and 10-17F) or indirectly by hitting a taller object, such as a tree or pole (Fig. 10-17F), and then distributed to the field. In either case, plants in the field may receive an electric shock but survive it, but more frequently many plants in the path or immediate vicinity of the lightning are killed in characteristic configurations (Figs. 10-17C and 10-17D) or in a circular area (Figs. 10-17E and 10-17F).

OTHER IMPROPER AGRICULTURAL PRACTICES

As with herbicides, a variety of other agricultural practices carried out improperly may cause considerable damage to plants and significant financial losses. Almost every agricultural practice can cause damage when applied the wrong way, at the wrong time, or with the wrong materials. Most commonly, however, losses result from the application of chemicals, such as fungicides, insecticides, nematicides, and fertilizer, at too high concentrations or on plants sensitive to them. Spray injury resulting in leaf burn or spotting or russeting of fruit is common on many crop plants (Fig. 10-15C).

Excessive or too deep cultivation between rows of growing plants may be more harmful than useful

FIGURE 10-17 Lightning injury to oak (A) and palm (B) trees, to turf grass (C and D), to cabbage (E), and to corn (F) plants in the field. [Photographs courtesy of (A) R. J. McGovern and (B–F) Plant Pathology Department, University of Florida.]

because it cuts or pulls many of the plants' roots. Road or other construction often cuts a large portion of the roots of nearby trees and results in their dieback and decline. Inadequate or excessive watering may cause wilting or any of the symptoms described earlier. In the case of African violets, droplets of cold water on the leaves cause the appearance of rings and ring-like patterns reminiscent of virus ringspot diseases. Potatoes stored next to hot water pipes under the kitchen sink often develop black heart. Trees frequently grow poorly and their leaves are chlorotic, curled, or reddened because their trunk is girdled by fence wire. The roots of plants potted in pots that are too small for their size are often badly distorted and twisted and the whole plant grows poorly (Fig. 10-1).

THE OFTEN CONFUSED ETIOLOGY OF STRESS DISEASES

Diagnosis of an abiotic disease is often every bit as difficult as the diagnosis of a biotic disease. When combinations of single or multiple abiotic and biotic diseases occur on the same plant or in an entire area, however, the diagnosis of the diseases and the determination of the relative importance of each become extremely difficult and often impossible.

When plants are adversely affected by an environmental factor, such as low moisture, nutrient deficiency, air pollution, or freezing, they are generally and concurrently weakened and predisposed to infection by one or more weakly parasitic pathogens. For example, all the conditions mentioned earlier predispose annual plants to infection by the fungus *Alternaria* and many perennial plants to infection by canker-causing fungi such as *Leucostoma* (*Cytospora*) and *Botryosphaeria*. A late blossom frost is often followed by infection with *Botrytis*, *Alternaria*, or *Pseudomonas*. Herbicide injury is likely to be followed by root rots caused by *Fusarium* and *Rhizoctonia*. Flooding injury is often followed by *Pythium* root infections.

Obviously, many of the stresses discussed in this chapter are often complicated by biotic diseases that follow. As a matter of fact, many epidemic disease problems, such as stalk rot of corn, tree declines, and stand depletions in forage legumes, although thought of as being caused by one or more biotic agents, they are in reality set off by one or another of the environmental factors discussed in this chapter. Thus, stalk rot of corn, although caused by one of several common fungi (*Fusarium, Diplodia, Gibberella*), actually occurs or becomes important only under conditions of low potassium and low moisture stress in early season. Similarly, the additional stress caused by some herbicides on soybean, sugar beet, and cotton seedlings increases the susceptibility of these crops to the *Thielaviopsis basicola* and *Rhizoctonia* root rots and damping off.

A striking example of the often confused etiology of stress diseases was developed in the last 30 years in Europe, where many different forest tree species, shrubs, and herbs have been exhibiting various degrees of yellowing, reduced growth, defoliation, abnormal growth, decline, and eventually death. This widespread general decline of forests (called *waldsterben*) occurred and spread over large areas of central Europe after about 1980. Such declines seem to be triggered by the stress caused by atmospheric depositions of toxic or growth-altering air pollutants that are subsequently aggravated by additional abiotic and biotic predisposing or stress-inducing factors. The air pollutants themselves, such as ozone, cause some direct injury and reduction in photosynthesis, but the mixture of deposited acidic pollutants may also cause the acidification of soils. This may result in leaching out and therefore deficiency in certain elements, such as magnesium, or in increases in the solubility of certain toxic elements, such as aluminum, thereby causing aluminum toxicity in plants. The latter then causes necrosis of fine roots, which leads to increased moisture or nutrient stress and eventual drying out and death of trees, particularly during dry periods. In addition to the effects caused by these abiotic factors, affected trees show increased susceptibility to insects and to foliage and root pathogens such as *Lophodermium*, *Phytophthora*, and *Armillaria*, which further increase the moisture and water stress and reduce photosynthesis in the plant.

Selected References

Aiken, R. M., and Smucker, A. J. M. (1996). Root system regulation of whole plant growth. *Annu. Rev. Phytopathol.* **34**, 325–346.

Anonymous (1980–). Noninfectious or abiotic diseases and disorders. Chapters in each "Compendium of Diseases of . . ." specific crops. APS Press, St. Paul, MN.

Bennett, W. F., ed. (1993). "Nutrient Deficiencies and Toxicities in Crop Plants." APS Press, St. Paul, MN.

Carne, W. M. (1948). The non-parasitic disorders of apple fruits in Australia. *Bull. C.S.I.R.O. (Aust.)* **238**, 1–83.

Dodd, J. L. (1980). The role of plant stresses in development of corn stalk rots. *Plant Dis.* **64**, 533–537.

Eagle, D. J., Caverly, D. J., and Holly, K. (1981). "Diagnosis of Herbicide Damage to Crops." Chem. Publ., New York.

Evans, L. S. (1984). Acidic precipitation effects on terrestrial vegetation. *Annu. Rev. Phytopathol.* **22**, 397–420.

Kandler, O. (1990). Epidemiological evaluation of the development of Waldsterben in Germany. *Plant Dis.* **74**, 4–12.

Katterman, F., ed. (1990). "Environmental Injury to Plants." Academic Press, San Diego.

Krupa, S., *et al.* (2001). Ambient ozone and plant health. *Plant Dis.* **85**, 4–12.

Krupa, S. V., Pratt, G. C., and Teng, P. S. (1982). Air pollution: An important issue in plant health. *Plant Dis.* **66**, 429–434.

Lacasse, N. L., and Treshow, M., eds. (1976). "Diagnosing Vegetation Injury Caused by Air Pollution." Applied Science Associates, Washington, DC.

Laurence, J. A., and Weinstein, L. H. (1981). Effects of air pollutants on plant productivity. *Annu. Rev. Phytopathol.* **19**, 257–271.

Levitt, J. (1972). "Responses of Plants to Environmental Stresses." Academic Press, New York.

Pasternak, D. (1987). Salt tolerance and crop production: A comprehensive approach. *Annu. Rev. Phytopathol.* **25**, 271–291.

Penas, E. J., and Wiese, R. A. (1996). Soybean chlorosis management. Field Crops NebGuide G89-953-A, 9p.

Sandermann, H., Jr. (1996). Ozone and plant health. *Annu. Rev. Phytopathol.* **34**, 347–366.

Schoenweiss, D. F. (1981). The role of environmental stress in diseases of woody plants. *Plant Dis.* **65**, 308–314.

Schutt, P., and Cowling, E. B. (1985). Waldsterben, a general decline of forests in central Europe: Symptoms, development and possible causes. *Plant Dis.* **69**, 548–558.

Skelly, J. M., and Innes, J. L. (1994). Waldsterben in the forests of central Europe and eastern North America: Fantasy or reality? *Plant Dis.* **78**, 1021–1031.

Tucker, D. P. H., *et al.* (1994). Tree and Fruit Disorders. Fact Sheet HS-140, Florida Coop. Extension Service, 18p.

Wallace, T. (1961). "The Diagnosis of Mineral Deficiencies in Plants by Visual Symptoms." Stationery Office, London.

chapter eleven

PLANT DISEASES CAUSED BY FUNGI

INTRODUCTION – CHARACTERISTICS: MORPHOLOGY – REPRODUCTION – ECOLOGY – DISSEMINATION CLASSIFICATION: FUNGALLIKE ORGANISMS – THE TRUE FUNGI – IDENTIFICATION: SYMPTOMS – ISOLATION – LIFE CYCLES OF FUNGI – CONTROL OF FUNGAL DISEASES OF PLANTS
386

DISEASES CAUSED BY FUNGALLIKE ORGANISMS – BY MYXOMYCOTA (MYXOMYCETES) – BY PLASMODIOPHOROMYCETES: CLUBROOT OF CRUCIFERS
404

DISEASES CAUSED BY OOMYCETES – PYTHIUM DAMPING-OFF – PHYTOPHTHORA DISEASES: ROOT AND STEM ROTS- WAR ON PLANTS – LATE BLIGHT OF POTATOES – THE DOWNY MILDEWS: INTRODUCTION-DOWNY MILDEW OF GRAPE
409

DISEASES CAUSED BY TRUE FUNGI – BY CHYTRIDIOMYCETES – BY ZYGOMYCETES – BY ASCOMYCETES – BY BASIDIOMYCETES
433

DISEASES CAUSED BY ASCOMYCETES AND MITOSPORIC FUNGI – INTRODUCTION
439

SOOTY MOLDS – TAPHRINA LEAF CURL DISEASES – POWDERY MILDEWS
440

FOLIAR DISEASES CAUSED BY ASCOMYCETES AND DEUTEROMYCETES (MITOSPORIC FUNGI) – INTRODUCTION – ALTERNARIA DISEASES – CLADOSPORIUM DISEASES – NEEDLE CASTS AND BLIGHTS OF CONIFERS – MYCOSPHAERELLA DISEASES – BANANA LEAF SPOT OR SIGATOKA DISEASE – SEPTORIA DISEASES – CERCOSPORA DISEASES – RICE BLAST DISEASE – COCHLIOBOLUS, PHRENOPHORA AND SETOSPHAERIA DISEASES OF CEREALS
452

STEM AND TWIG CANKERS CAUSED BY ASCOMYCETES AND DEUTEROMYCETES – INTRODUCTION – BLACK KNOT OF PLUM AND CHERRY – CHESTNUT BLIGHT – NECTRIA CANKER – LEUCOSTOMA CANKER – CANKERS OF FOREST TREES
473

ANTHRACNOSE DISEASES CAUSED BY ASCOMYCETES AND DEUREROMYCETES – INTRODUCTION – BLACK SPOT OF ROSE – ELSINOE ANTHRACNOSE AND SCAB DISEASES: – GRAPE ANTHRACNOSE OR BIRD'S-EYE ROT – RASPBERRY ANTHRACNOSE – CITRUS SCAB DISEASES – AVOCADO SCAB – COLLETOTRICHUM DISEASES: OF ANNUAL PLANTS – ANTHRACNOSE OF BEANS – ANTHRACNOSE OF CUCURBITS – ANTHRACNOSE OF RIPE ROT OF TOMATO –ONION ANTHRACNOSE OR SMUDGE – STRAWBERRY ANTHRACNOSE – ANTHRACNOSE OF CEREALS AND GRASSES – A MENACE TO TROPICAL CROPS – COLLETOTRICHUM FRUIT ROTS: MANGO ANTHRACNOSE – CITRUS POSTBLOOM FRUIT DROP – BITTER ROT OF APPLE – RIPE ROT OF GRAPE – GNOMONIA ANTHRACNOSE AND LEAF SPOT DISEASES – DOGWOOD ANTHRACNOSE
483

FRUIT AND GENERAL DISEASES CAUSED BY ASCOMYCETES AND DEUTEROMYCETES – INTRODUCTION – ERGOT OF CEREALS – APPLE SCAB – BROWN ROT OF STONE FRUITS – MONOLIOPHTHORA POD ROT OF CACAO – BOTRYTIS DISEASES – BLACK ROT OF GRAPE – CUCURBIT GUMMY STEM BLIGHT AND BLACK ROT – DIAPORTHE, PHOMOPSIS, AND PHOMA DISEASES – STEM CANKER OF SOYBEANS – MELANOSE DISEASE OF CITRUS – PHOMOPSIS DISEASES – BLACK ROT OF APPLE
501

VASCULAR WILTS CAUSED BY ASCOMYCETES AND DEUTEROMYCETES – INTRODUCTION – FUSARIUM WILTS: OF TOMATO – FUSARIUM OR PANAMA WILT OF BANANA – VERTICILLIUM WILTS – OPHIOSTOMA WILT OF ELM TREES: DUTCH ELM DISEASE – CERATOCYSTIS WILTS – OAK WILT – CERATOCYSTIS WILT OF EUCALYPTUS
522

ROOT AND STEM ROTS CAUSED BY ASCOMYCETES AND DEUTEROMYCETES – INTRODUCTION – *GIBBERELLA DISEASES* : – GIBBERELLA STALK, EAR, AND SEEDLING ROT OF CORN-FUSARIUM (GIBBERELLA) HEAD BLIGHT (FHB) OR SCAB OF SMALL GRAINS – FUSARIUM ROOT AND STEM ROTS OF NON-GRAIN CROPS – TAKE-ALL OF WHEAT – THIELAVIOPSIS BLACK ROOT ROT – MONOSPORASCUS ROOT ROT AND VINE DECLINE OF MELONS – *SCLEROTINIA DISEASES OF* VEGETABLES AND FLOWERS – PHYMATOTRICHUM ROOT ROT
534

POSTHARVEST DISEASES OF PLANT PRODUCTS CAUSED BY ASCOMYCETES AND DEUTEROMYCETES – INTRODUCTION – POSTHARVEST DECAYS OF FRUITS AND VEGETABLES CAUSED BY: *ASPERGILLUS* , *PENICILLIUM* , *RHIZOPUS* , *AND MUCOR* : – *ALTERNARIA* – *BOTRYTIS* – *FUSARIUM* – *GEOTRICHUM* – *PENICILLIUM* – *SCLEROTINIA* – CONTROL OF POSTHARVEST DECAYS OF FRESH FRUITS AND VEGETABLES – POSTHARVEST DECAYS OF GRAIN AND LEGUME SEEDS – MYCOTOXINS AND MYCOTOXICOSES: ASPERGILLUS TOXINS – AFLATOXINS – FUSARIUM TOXINS – OTHER CONTROL OF POSTHARVEST GRAIN DECAYS
553

DISEASES CAUSED BY BASIDIOMYCETES
562

THE RUSTS – INTRODUCTION – CEREAL RUSTS – STEM RUST OF WHEAT AND OTHER CEREALS – RUSTS OF LEGUMES – BEAN RUST – *SOYBEAN RUST – A MAJOR THREAT TO A MAJOR CROP (BOX)* – CEDAR-APPLE RUST – COFFEE RUST – RUSTS OF FOREST TREES – WHITE PINE BLISTER RUST – FUSIFORM RUST
562

THE SMUTS – INTRODUCTION – CORN SMUT – LOOSE SMUT OF CEREALS – COVERED SMUT, OR BUNT, OF WHEAT – KARNAL BUNT OF SMALL GRAINS – LEGITIMATE CONCERNS AND POLITICAL PREDICAMENTS *(BOX)*
582

ROOT AND STEM ROTS CAUSED BY BASIDIOMYCETES – INTRODUCTION – ROOT AND STEM ROT DISEASES CAUSED BY THE "STERILE FUNGI": RHIZOCTONIA DISEASES – SCLEROTIUM DISEASES – ROOT ROTS OF TREES: ARMILLARIA ROOT ROT OF FRUIT AND FOREST TREES – *WOOD ROTS AND DECAYS* – WITCHES' BROOM OF CACAO
593

INTRODUCTION

Fungi are small, generally microscopic, eukaryotic, usually filamentous, branched, spore-bearing organisms that lack chlorophyll. Fungi have cell walls that contain chitin and glucans (but no cellulose) as the skeletal components. These are embedded in a matrix of polysaccharides and glycoproteins. A group of fungal-like organisms, the Oomycota, usually referred to as oomycetes, until about 1990 were considered to be true fungi. With a few chitin-containing exceptions, the vast majority of oomycetes have cell walls composed of glucans and small amounts of cellulose, but no chitin. The Oomycota are now members of the kingdom Chromista rather than Fungi but continue to be treated as fungi because of their many other similarities to them, at least in the way they cause disease in plants.

Most of the more than 100,000 known fungus species are strictly **saprophytic**, i.e., they live on dead organic matter, which they help decompose. Some, about 50 species, cause diseases in humans, and about as many cause diseases in animals, most of them superficial diseases of the skin or its appendages. More than 10,000 species of fungi, however, can cause disease in plants. All plants are attacked by some kinds of fungi, and each of the parasitic fungi can attack one or many kinds of plants. Some fungi, known as **obligate parasites** or **biotrophs**, can grow and multiply only by remaining, during their entire life, in association with their host plants. Others, known as **nonobligate parasites**, require a host plant for part of their life cycles but can complete their cycles on dead organic matter, or they can grow and multiply on dead organic matter as well as on living plants. Fungi that are nonobligate parasites can be **facultative saprophytes** or **facultative parasites** depending on whether they are primarily parasites or primarily saprophytes.

BOX 16 Some Interesting Facts about Fungi

Most people think that fungi are just the fluffy mildew on molded bread and the mushrooms that grow on the ground or on trees and, for some, the mushrooms one eats with pizza or with steaks. It is estimated that there are between 70,000 and 1.5 million species of fungi, most of them yet to be discovered and described. In addition to these facts, however, there are many, more or less interesting, facts about fungi.

To begin with, fungi used to be thought of as plants that did not have chlorophyll. No longer! Fungi are recognized as a separate kingdom of organisms, alongside the kingdoms of other eukaryotic (= having a nucleus) organisms, i.e., Planta (photosynthetic plants), Animalia (= ingestive animals), Chromista (roughly the multicellular algae), and Protozoa (various mostly phagotrophic (= engulfing their food) unicellular organisms. These five kingdoms of eukaryotes are in addition to the two kingdoms of prokaryotes (= lacking an organized nucleus), i.e., Archaea (preexisting ancient prokaryotes) and Bacteria (unicellular prokaryotes). Actually, in a recent taxonomic reshuffling of organisms, some of the plant pathogenic fungi known the longest, such as those causing the late blight of potato and the downy mildews, and also the clubroot of cabbage, were taken out of the kingdom Fungi and placed in the kingdoms Chromista and Protozoa, respectively.

True fungi are primarily terrestrial organisms that have mycelium and produce airborne spores, but some do produce flagella-bearing zoospores that can move in water. Although most fungi are microscopic and a few of them, the yeasts, are mostly unicellular, a group of them produce fairly large mushrooms. As a matter of fact, some of the mushroom-producing fungi are the largest living organisms of any kind. For instance, a mushroom in England reached a diameter of 170 centimeters and an estimated weight of 284 kilograms by 1996 and was still growing by about 20 centimeters in diameter per year. In Canada, a puffball reached a circumference of 300 centimeters and a weight of 22 kilograms. Until August 2000, the largest known organism to have been produced from a single spore was a fungus, *Armillaria ostoyae*, known as the honey mushroom; this fungus grew below ground through the soil and produced mushrooms through an area of about 1,500 acres near Mount Adams in the state of Washington. Then, in August 2000, in the forest east of Prairie City, Oregon, another *A. ostoyae* fungus was found that had grown and produced its mushrooms into an area 3.5 miles in diameter, an area as big as 1,600 football fields. The fungus was found to a depth of about 100 centimeters into the ground. Nobody has estimated the weight of this fungus yet, but it has been estimated that it took the fungus filaments about 2,400 years to grow from the single spore to the giant fungus size it has reached to date. That all of these mushrooms near the periphery of the fungus growth were derived from one organism was proven by the identity of the DNA in dozens of samples of mycelium and growing mushrooms within the area in question.

Fungi such as the honey mushroom, while sometimes causing considerable damage and loss by killing trees, offer an invaluable service to nature and to humans by causing tree, shrub, and other plant wood to rot and decay, thereby freeing the earth of dead wood and making room for new plants to grow. Several fungi help degrade dead nematodes and other animals.

Many fungi, however, are biotrophs and either attack plants, animals, and other organisms to which they cause disease or, in some cases, develop symbiotic associations with them. Thus, symbiotic associations between fungi and photosynthetic algae or cyanobacteria produce **lichens**, whereas symbiotic associations of fungi and the roots of higher plants result in **mycorrhizae.** There are also stem and leaf endophytic fungi that feed off the plant but, in return, provide some protection to the plant from outside factors.

One of the best known contributions of certain fungi is the production, for example, by *Penicillium*, of antibiotics, such as penicillin, that are used against bacteria pathogenic to humans and animals. There are many other even more basic contributions of fungi to the well-being of humans. A group of fungi, the yeasts, produce enzymes that, in the right substrate, make possible the production of bread from wheat and other grains, of cheese and yogurt from milk, of wine from grapes, of beer from barley, of various liquors from rye, potato, etc. Several fungi are also used for the production, accumulation, and release of certain organic acids, antibiotics, hormones, enzymes, and so on, or for the breakdown of certain carbohydrates in ways that release desirable compounds.

There are several fungi, however, that have adverse effects on humans, in addition to causing plant diseases. Some fungi cause infections of human extremities, a type of pneumonia, valley fever, etc. The spores of many fungi induce allergies and hay fever in humans, and several of them, such as *Aspergillus*, *Claviceps*, *Fusarium* and *Penicillium*, produce toxic compounds, the mycotoxins. Also, some fungi, e.g., *Amanita*, contain poisonous substances and are extremely toxic to persons eating them, often causing their death in a matter of minutes.

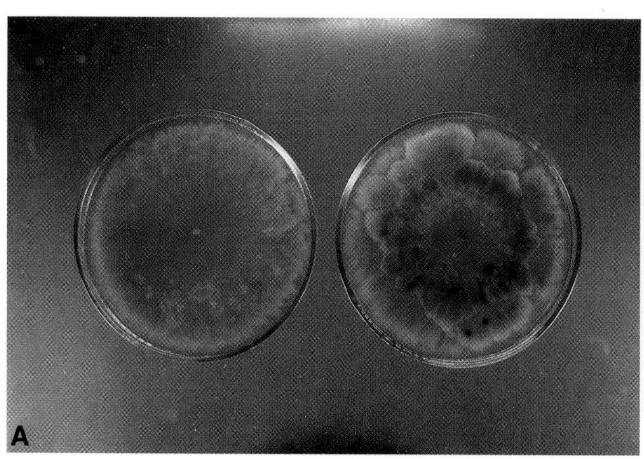

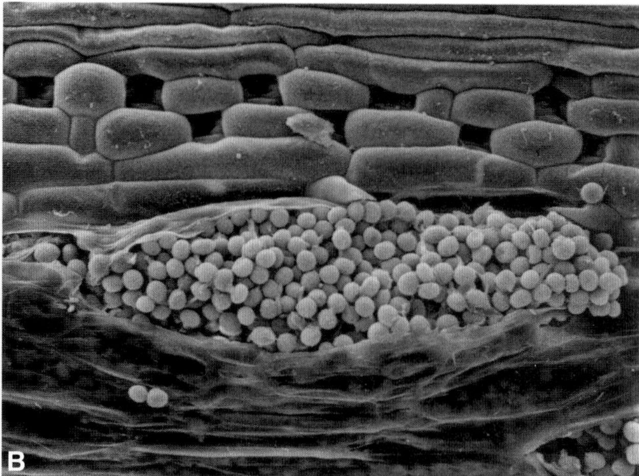

FIGURE 11-1 (A) Appearance of the vegetative body (mycelium) of two cultures of the apple-infecting fungus *Botryosphaeria* growing on a nutrient medium. (B) Fungal spores (teliospores) produced by a cereal smut-causing fungus in a leaf. [Photo (B) from Mims et al. (1998). Intern. J. Pl. Sci. 153: 289–300].

CHARACTERISTICS OF PLANT PATHOGENIC FUNGI

Morphology

Most fungi have a filamentous vegetative body called a **mycelium**. The mycelium branches out in all directions (Fig. 11-1A). The individual branches of the mycelium are called **hyphae** and are generally uniform in thickness, usually about 2 to 10 micrometers in diameter, but in some fungi may be more than 100 micrometers thick. The length of the mycelium may be only a few micrometers in some fungi, but in others it may be several meters long.

In some fungi the mycelium consists of many cells containing one or two nuclei per cell. In others the mycelium contains many nuclei, which may or may not be partitioned by cross walls (septa). Growth of the mycelium occurs at the tips of the hyphae.

Some lower fungi lack true mycelium and produce instead a system of strands of grossly dissimilar and continuously varying diameter called a **rhizomycelium**. Some microorganisms (myxomycota, plasmodiophoromycetes), formerly thought to be primitive fungi but now considered to belong to the kingdom Protozoa, instead of mycelium produce a naked, amoeboid, multinucleate body called **plasmodium**.

Reproduction

Fungi reproduce chiefly by means of spores (Figs. 11-1B and 11-2). **Spores** are reproductive bodies consisting of one or a few cells. Spores may be formed asexually, like buds produced on a twig, or as the result of sexual fertilization.

In some fungi, asexual spores are produced inside a sac called a **sporangium**. Some of these spores can swim by means of flagella and are called **zoospores**. Other fungi produce asexual spores called **conidia** by the cutting off of terminal or lateral cells from special hyphae called **conidiophores**. In some fungi, conidiophores produce short hyphae, called phialides, that produce and carry conidia endogenously, sometimes in chains. In most fungi, however, asexual spores (conidia) are produced at the tips of conidiophores, either directly on the mycelium or inside walled structures called **conidiomata**. A distinctive form of flask-shaped conidiomata is called a **pycnidium**. In some fungi, terminal or intercalary cells of a hypha enlarge, round up, form a thick wall, and separate to form **chlamydospores**.

Sexual reproduction occurs in most groups of fungi. In Zygomycetes, two cells (gametes) of similar size and appearance unite and produce a zygote called a **zygospore**. In Chytridiomycetes, motile gametes of equal or unequal size fuse to form **meiosporangia**. In some fungi, no definite gametes are produced, but instead one mycelium may unite with other compatible mycelium. In one group of fungi (Ascomycetes), sexual spores, usually eight in number, are produced within a sac-like zygote cell, the **ascus**, and the spores are called **ascospores**. In another group of fungi (Basidiomycetes), sexual spores are produced on the outside of a club-like zygote cell called the **basidium**, and the spores are called **basidiospores**. In the fungal-like Chromista called Oomycetes, gametangia of unequal size fuse to form zygotes, which are referred to as oospores.

For a large group of fungi, called mitosporic fungi (formerly known as fungi imperfecti or deuteromycetes),

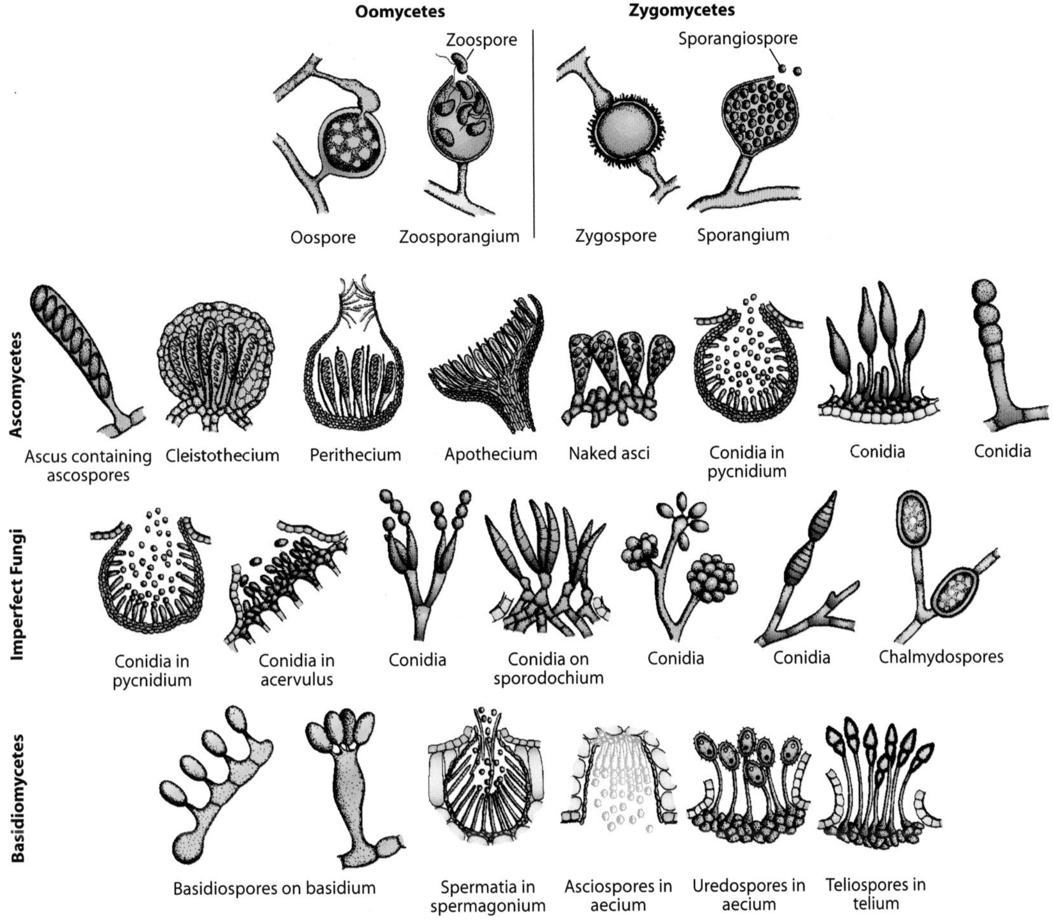

FIGURE 11-2 Representative spores and fruiting bodies of the fungal-like Oomycetes and the main groups of fungi.

no sexual reproduction is known either because they do not have one or because it has not yet been discovered. Apparently these fungi reproduce only asexually.

Ecology

Almost all plant pathogenic fungi spend part of their lives on their host plants and part in the soil or in plant debris on the soil. Some fungi are strictly **biotrophs**, i.e., they spend all of their lives on the host, and only the spores may land on the soil, where they die or remain inactive until they are again carried to a host on which they grow and multiply. Others, such as the apple scab fungus *Venturia*, are **hemibiotrophs**, i.e., they must pass part of their lives on the host as parasites and part on dead tissues of the same host on the ground as saprophytes in order to complete their life cycle in nature. The latter group of fungi, however, remains continually associated with host tissues, whether living or dead, and in nature does not grow on any other kind of organic matter. A third group of fungi are **facultative saprophytes** because they grow parasitically on the hosts, but they continue to live, grow, and multiply on the dead tissues of the host after its death and may further move out of the host debris into the soil or other decaying plant material on which they grow and multiply strictly as saprophytes. The dead plant material that they colonize need not be related at all to the host they can parasitize. These fungi are usually soil pathogens, have a wide host range, and can survive in the soil for many years in the absence of their hosts. They, too, however, may need to infect a host from time to time in order to increase their populations, as protracted and continuous growth of these fungi as saprophytes in the soil results in more or less rapid reduction in their numbers. Finally, some fungi are **facultative parasites** because they can live perfectly well in the soil or elsewhere as saprophytes, but if they happen to come in contact with a plant organ under the right conditions, they have the faculty to parasitize and cause disease on the plant.

During the parasitic phase, fungi assume various positions in relation to the plant cells and tissues. Some fungi (such as powdery mildews) grow on the plant surface but send their feeding organs (haustoria) into epidermal cells of the plant. Some, such as the apple scab causing fungus *Venturia*, grow only between the cuticle and the epidermal cells. Others grow in the intercellular spaces between the cells and may send haustoria into the cells. Still others grow between and through the cells indiscriminately. Fungi that cause vascular wilts, such as *Fusarium*, grow inside the xylem vessels of infected plants, whereas so-called endophytic fungi, growing mostly within symptomless plants, exist intercellularly in the various plant organs. Obligate parasites (biotrophs) can grow only in association with living cells, being unable to feed on dead cells. However, the mycelium of some nonobligate parasites never comes in contact with living plant cells because their macerating enzymes or toxins kill the plant cells ahead of the mycelium. In most cases, however, regardless of the position of the mycelium in the host, the reproductive bodies (spores) of the fungus are produced at or very near the surface of the host tissues to ensure their prompt and efficient dissemination.

The survival and performance of most plant pathogenic fungi depend greatly on the prevailing conditions of temperature and moisture or on the presence of water in their environment. Free mycelium survives only within a certain range of temperatures (–5 to 45°C) and in contact with moist surfaces, inside or outside the host. Most kinds of spores, however, can withstand broader ranges of both temperature and moisture and carry the fungus through the low winter temperatures and hot, dry summer periods. Spores, however, also require favorable temperatures and moisture in order to germinate. Furthermore, oomycetes and fungi producing zoospores require free water for the production, movement, and germination of the zoospores.

Dissemination

Zoospores are the only fungus structures that can move by themselves. Zoospores, however, can move for only very short distances (a few millimeters or centimeters, perhaps). Only myxomycetes, oomycetes, and chytridiomycetes produce zoospores. The great majority of plant pathogenic fungi depend for their spread from plant to plant and to different parts of the same plant on hyphal growth or chance distribution by such agents as wind, water, birds, insects, other animals, and humans. Fungi are disseminated primarily in the form of spores. Fragments of hyphae and hard masses of mycelium known as sclerotia may also be disseminated by the same agents, although to a much lesser extent.

Spore dissemination in almost all fungi is passive, although the initial discharge of spores in some fungi is forcible. The distance to which spores may be disseminated varies with the agent of dissemination. Wind is probably the most important disseminating agent of spores of most fungi and may carry spores over great distances. For specific fungi, other agents such as water or insects may play a much more important role than wind in the dissemination of their spores.

CLASSIFICATION OF PLANT PATHOGENIC FUNGI

The fungi and fungal-like organisms that cause diseases on plants are a diverse group. Because of their large numbers and diversity, only a sketchy classification of some of the most important phytopathogenic genera is presented here. Some fungal-like organisms, often referred to as lower fungi, are now considered to belong to the kingdom **Protozoa** (e.g., myxomycetes and plasmodiophoromycetes) or to the kingdom **Chromista** (also known as **Stramenopiles**) (e.g., oomycetes). True fungi, however (i.e., chytridiomycetes, zygomycetes, ascomycetes, basidiomycetes, and deuteromycetes) belong to the kingdom **Fungi**.

It should be pointed out here that, in a more recent classification scheme, organisms are no longer divided into kingdoms. Instead, five top-level taxa are recognized: two taxa, Archaea and Eubacteria, are prokaryotes, i.e., they do not have an organized nucleus; one, the Eukaryota, have an organized nucleus and include all organisms such as plants, animals, fungi, and others that we are familiar with; the Viruses; and the Viroids. All fungi and fungal-like organisms are members of the Eukaryota and belong to the following groups:

- Eukaryota
  - Mycetozoa — produce a plasmodium or plasmodium-like structures
    - Dictyosteliida — cellular slime molds
    - Myxogastria — plasmodial slime molds
  - Plasmodiophoridae — endoparasitic slime molds
  - Stramenopiles, heteroconts, with two different flagella, one having hollow tripartite hairs
    - Oomycetes — have elongated nonseptate mycelium, biflagellate zoospores in zoosporangia, oospores
  - Fungi
    - Chytridiomycota — have zoospores with a single posterior flagellum, round or elongated mycelium
    - Zygomycota — produce nonmotile asexual spores in sporangia. Resting spore is a zygospore

Ascomycota — produce sexual spores, ascospores, in asci. Produce nonmotile asexual spores (conidia)

Basidiomycota — produce sexual spores, basidiospores, externally on a basidium

This system is likely to prevail in the future. However, for purposes of continuity and easier reference to the existing literature, the earlier, up-to-now more widely used system of classification is presented in some greater detail here and is followed in the book.

Fungal-Like Organisms

I. Kingdom: Protozoa — Microorganisms that may be unicellular, plasmodial, colonial, very simple multicells, or phagotrophic, i.e., feeding by engulfing their food. The kingdom contains many microorganisms in addition to the fungal-like organisms myxomycetes and plasmodiophoromycetes

Phylum: Myxomycota — Produce a plasmodium or plasmodium-like structure

Class: Myxomycetes (slime molds) — Their body is a naked, amorphous plasmodium. They produce zoospores (swarm cells). May grow on and may cover parts of low-lying plants but do not infect plants

Order: Physarales — Saprophytic plasmodium that gives rise to crusty fructifications containing spores. They produce zoospores that have two flagella

Genus: *Fuligo*, *Mucilago*, and *Physarum* cause slime molds on low-lying plants

Phylum: Plasmodiophoromycota (the Plasmodiophoromycetes — Endoparasitic slime molds)

Order: Plasmodiophorales — Plasmodia produced within cells of roots and stems of plants. They produce zoospores that have two flagella. Obligate parasites

Genus: *Plasmodiophora*, *P. brassicae* causing clubroot of crucifers

Polymyxa, *P. graminis* parasitic on wheat and other cereals. Can transmit plant viruses

Spongospora, *S. subterranea* causing powdery scab of potato tubers

II. Kingdom: Chromista (Stramenopiles) — Unicellular or multicellular, filamentous or colonial, primarily phototrophic (micro)organisms, some with tubular flagellar appendages or with chloroplasts inside the rough endoplasmic reticulum or both. Contains brown algae, diatoms, oomycetes, and some other similar organisms

Phylum: Oomycota — Have biflagellate zoospores, with longer tinsel flagellum directed forward and a shorter whiplash flagellum directed backward. Diploid thallus, with meiosis occurring in the developing gametangia. Gametangial contact produces thick-walled sexual oospore. Cell walls composed of glucans and small amounts of hydroxyproline and cellulose

Class: Oomycetes (water molds, white rusts, and downy mildews) — Have nonseptate elongated mycelium. Produce zoospores in zoosporangia. Zoospores have two flagella. Sexual resting spores (oospores) produced by the union of morphologically different gametangia called antheridia (male) and oogonia (female)

Order: Saprolegniales — Have well-developed mycelium. Zoospores produced in long, cylindrical zoosporangia attached to mycelium. Usually several oospores in an oogonium

Genus: *Aphanomyces*, *A. euteiches* causing root rot of peas

Order: Peronosporales — Mycelium well-developed, nonseptate, branching, inter- or intracellular, often with haustoria. Zoosporangia oval or lemon shaped, borne on ordinary mycelium or on sporangiophores. Sporangia in most species germinate by producing zoospores, but in some they germinate directly and produce a germ tube. Sexual reproduction is by characteristic oogonia and antheridia that fuse and produce an oospore. Oospores germinate by giving rise to a sporangium containing zoospores or to a germ tube, which soon produces a sporagium, depending on the species

Family: *Pythiaceae* — Sporangia, usually zoosporangia, produced along somatic hyphae or at tips of hyphae of indeterminate growth and set free. Oogonia thin-walled. Facultative parasites

Genus: *Pythium*, causing damping-off of seedlings, seed decay, root rots, stem lesions, rotting of vegetable fruit and tubers on/or in the ground, and cottony blight of turf grasses

Phytophthora, *P. infestans* causing late blight of potato; several others causing mostly root and stem rots, rots of fleshy fruits and vegetables, cankers and diebacks

Family: *Peronosporaceae* (the downy mildews) — Sporangia borne on sporangiophores of determinate growth. Sporangia wind-borne. Obligate parasites

Genus: *Plasmopara*, *P. viticola* causing downy mildew of grape

Peronospora, *P. tabacina* causing downy mildew (blue mold) of tobacco

Bremia, *B. lactucae* causing downy mildew of lettuce

Pseudoperonospora, *P. cubensis* causing downy mildew of cucurbits

Peronosclerospora causing downy mildew of corn (*P. philippinensis*), of sugarcane and corn (*P. sacchari*), of sorghum (*P. sorghi*), and others

Sclerophthora causing crazy top (downy mildew) of corn

Sclerospora causing downy mildew of pearl millet and many other grasses

Family: *Albuginaceae* (the white rusts) — Sporangia borne in chains

Genus: *Albugo*, *A. candida* causing white rust of crucifers

True Fungi

Kingdom: Fungi — Produce mycelium, the walls of which contain glucans and chitin. Lack chloroplasts

Phylum: Chytridiomycota — Produce zoospores that have a single posterior flagellum

Class: Chytridiomycetes — Have round or elongated mycelium that lacks cross walls

Genus: *Olpidium*, *O. brassicae* being parasitic in roots of cabbage and other plants. Can transmit plant viruses

Physoderma, *P. maydis* causing brown spot of corn, and *P.* (= *Urophlyctis*) *alfalfae* causing crown wart of alfalfa

Synchytrium, *S. endobioticum* causing potato wart

Phylum: Zygomycota — Produce nonmotile asexual spores in sporangia. No zoospores. The resting spore is a zygospore, produced by the fusion of two morphologically similar gametes

Class: Zygomycetes (bread molds) — Saprophytic or parasites of plants, humans, and animals

Order: Mucorales — Nonmotile asexual spores formed in terminal sporangia

Genus: *Rhizopus*, causing bread molds and soft rot of fruits and vegetables

Choanephora, *C. cucurbitarum* causing soft rot of squash

Mucor, causing bread mold and storage rots of fruits and vegetables

Order: Glomales — Fungi forming vesicular–arbuscular mycorrhizae with roots, also known as endomycorrhizae. Arbuscules produced in host root. Chlamydospore-like spores produced singly in soil, in roots, or in sporocarps. Sexual reproduction rare

Genus: *Glomus*, *Acaulospora*, *Gigaspora*, *Scutellospora*

Phylum: Ascomycota (ascomycetes, sac fungi) — Most have a sexual stage (teleomorph) and an asexual stage (anamorph). Produce sexual spores, called ascospores, generally in groups of eight within an ascus. Produce asexual spores (conidia) on free hyphae or in asexual fruiting structures (pycnidia, acervuli, etc.)

I. Class: Archiascomycetes — A group of diverse fungi, difficult to characterize

Order: Taphrinales — Asci arising from binucleate ascogenous cells

Genus: *Taphrina*, causing peach leaf curl, plum pocket, oak leaf blister, etc.

II. Class: Saccharomycetes (yeasts) — Asci naked, no ascocarps produced. Mostly unicellular fungi that reproduce by budding

Genus: *Galactomyces*, causing citrus sour rot
Saccharomyces, *S. cerevisiae*, the bread yeast

III. Filamentous ascomycetes

Order: Erysiphales (the powdery mildew fungi) — Asci in fruiting bodies completely closed (cleistothecia). Mycelium, conidia, and cleistothecia on surface of host plant. Obligate parasites

Genus: *Blumeria*, causing powdery mildew of cereals and grasses
Erysiphe, causing powdery mildews of many herbaceous plants
Leveillula, causing powdery mildew of tomato
Microsphaera, one species causing powdery mildew of lilac
Oidium (anamorph only), causing powdery mildew of tomato
Podosphaera, *P. leucotricha* causing powdery mildew of apple
Sphaerotheca, *S. pannosa* causing powdery mildew of roses and peach
Uncinula, *U. necator* causing powdery mildew of grape

A. Pyrenomycetes: Ascomycetes with perithecia — Perithecia or, in some groups, cleistothecia in a stroma, immersed in a loose hyphal mat, or free. Asci have one wall

Order: Hypocreales — Stromata pale to blue, purple or brightly colored. Asci ovoid to cylindrical with an apical pore. Ascospores are spherical to needle like, one to several celled, usually discharged forcibly. Conidia produced from phialidic conidiophores. Some produce substances toxic to humans and animals. Some produce growth regulators. Some are antagonistic or parasitic on other fungi, and some are systemic parasites (endophytes) of many grain crops and grasses, making them poisonous to grazing animals

Genus: *Hypocrea*, some species of which produce the anamorphs *Trichoderma* and *Gliocladium*, which are used as biocontrol agents against several plant pathogenic fungi
Melanospora, whose anamorphs *Phialophora* and *Gonatobotrys* parasitize the mycelium of many fungi, including the important plant pathogens *Ophiostoma*, *Ceratocystis*, *Fusarium*, and *Verticillium*
Nectria, causing twig and stem cankers of trees
Gibberella, causing foot or stalk rot of corn and small grains
Claviceps, *C. purpurea* causing ergot of grain crops, which is poisonous to humans and animals, *C. sorghi*, of sorghum.
Epichloe, endophytic in grasses (its anamorph is *Acremonium*)
Balansia, endophytic in grasses and sedges
Atkinsonella, endophytic in grasses and sedges
Myriogoenospora, endophytic in grasses and sedges

Order: Microascales — Lack stromata. Most have perithecia but some have cleistothecia. Asci are globoid or ovoid, disintegrating. Acospores one-celled

Genus: *Ceratocystis*, causing oak wilt (*C. fagacearum*); cankers in stone fruit and other trees and root rot of sweet potato (*C. fimbriata*); butt rot of pineapple (*C. paradoxa*); sapstain or blue stain of cut wood surfaces (*C. coerulescens* and others)
Monosporascus, *M. cannonballus* causing root rot and collapse of cucurbits

Order: Phyllachorales — Perithecia in stroma, asci oblong to cylindrical, with pores at their tips. Ascospores of varying shapes, hyaline or dark

Genus: *Glomerella*, *G. cingulata* causing many anthracnose diseases and bitter rot of apples; its anamorphic stage is *Colletotrichum gloeosporioides*
Phyllachora, *P. graminis* causing leaf spots on grasses

Order: Ophiostomatales — Perithecia without paraphyses. Asci globose to ovoid, disintegrating. Several species are dispersed by beetles. Some species cause sapstain (blue stain) in wood

Genus: *Ophiostoma*, *O. novo-ulmi*, causing the Dutch elm disease (anamorphs are *Sporothrix* and *Graphium*)

Order: Diaporthales — Perithecia in a stroma of either fungal or plant tissue, or of hyphae on the substrate. Asci cylindrical with pores. Ascospores have one to several septa and may be hyaline to brown

Genus: *Diaporthe*, causing citrus melanose (*D. citri*), eggplant fruit rot (*D. vexans*), soybean pod and stem rot (*D. phaseolorum*); their anamorphs are species of *Phomopsis*

Gnomonia, causing anthracnose and leaf spot diseases

Gaeumannomyces, *G. graminis* causing the take-all disease of grain crops (wheat, rice, oats) and grasses

Magnaporthe, *M. grisea* causing rice blast disease; its anamorph is *Pyricularia oryzae*

Cryphonectria, *C. parasitica* causing the chestnut blight disease

Leucostoma (formerly *Valsa*), causing canker diseases of peach and other trees

Order: Xylariales — Perithecia dark, leathery, hard, sometimes embedded in a stroma Asci cylindrical to subglobose. Ascospores one to a few celled, hyaline or dark

Genus: *Hypoxylon*, *H. mammatum* causing a severe canker on poplars

Rosellinia, *R. necatrix* causing root diseases of fruit trees and vines

Xylaria, causing tree cankers and wood decay

Eutypa, *E. armeniacae* causing serious canker diseases of fruit trees and vines

B. Loculoascomycetes: Ascomycetes with ascostromata — Produce asci within locules (cavities) preformed in a stroma. Ascostroma may be monolocular (pseudothecium) or multilocular. Asci have a double wall

Order: Dothideales — Locules lack sterile hyphae and open by an apical pore. Asci ovoid to cylindrical, in fascicles. Ascospores one to several celled, hyaline to brown

Genus: *Mycosphaerella*, causing leaf spots on many plants, such as the Sigatoka diseases of banana (*M. musicola* and *M. fijiensis*), leaf spots of cereals and grasses (*M. graminicola*), and leaf spot of strawberry (*M. fragariae*); its anamorphs may be *Cercospora*, *Septoria*, and many others

Elsinoë, causing citrus scab (*E. fawcetti*), grape anthracnose (*E. ampelina*), and raspberry anthracnose (*E. veneta*)

Order: Capnodiales — Ascocarps superficial, produced in a loose mat of dark hyphae

Genus: *Capnodium*, being one of many fungi causing sooty molds on plants

Order: Pleosporales — Asci surrounded by pseudoparaphyses. Ascostroma variable

Genus: *Cochliobolus*, whose anamorphs are *Bipolaris* or *Curvularia*, causes leaf spots and root rots on grain crops and grasses

Pyrenophora, whose anamorph is *Drechslera*, causing leaf spots on cereals and grasses

Setosphaera (anamorph is *Exserohilum*), causing leaf spots on cereals and grasses

Pleospora (anamorph is *Stemphylium*), causing black mold rot of tomato

Leptosphaeria (anamorph is *Phoma*), causing black leg and foot rot of cabbage

Venturia (anamorphs are *Pollaccia* and *Spilocaea*), causing apple scab (*V. inaequalis*) and pear scab (*V. pyrina*)

Guignardia (anamorph is *Phyllosticta*), causing black rot of grapes

Apiosporina, *A. morbosa* (anamorph *Fusicladium*) causing black knot of cherries and plums

C. Discomycetes: Ascomycetes with apothecia — Ascocarps shaped like cups, saucers, or cushions and called apothecia. Asci cylindrical to ovoid, often interspersed with paraphyses. Ascospores discharged forcibly

Order: Rhytismales — Ascocarps are black, spherical, discoid, or elongate and are produced in stromata. Asci variable. Ascospores hyaline or brown, ovoid to filiform

Genus: *Hypoderma*, causing pine leaf spot (needle cast) diseases

Lophodermium, causing pine needle cast
Rhabdocline, causing Douglas fir needle cast
Rhytisma, *R. acerinum* causing tar spot of maple leaves

Order: Helotiales — Apothecia cup or disk shaped. Asci with only slightly thickened apices. Ascospores are spherical, elongate, to filiform, and have none to several septa

Genus: *Monilinia*, causing the brown rot disease of stone fruits
Sclerotinia, *S. sclerotiorum* causing the white mold or watery soft rot of vegetables
Stromatinia, *S. gladioli* causing corm rot of gladiolus
Pseudopeziza, *P. trifolii* causing alfalfa leaf spot
Diplocarpon, *D. maculatum* causing black spot of quince and pear and *D. rosae* causing black spot of roses

D. Deuteromycetes or mitosporic fungi (imperfect or asexual fungi) — Mycelium well-developed, septate, branched. Sexual reproduction and structures rare, lacking, or unknown. Asexual spores (conidia) formed on conidiophores existing singly, grouped in specialized structures such as sporodochia and synnemata, or produced in structures known as pycnidia and acervuli. The most important mitosporic fungi are listed.

Anamorphic stage	Certain or likely teleomorphic group
Genus: *Geotrichum*, *G. candidum* causing sour rot of fruits and vegetables	Saccharomycetales
	Cleistothecial ascomycetes
Penicillium, causing blue mold rot of fruits	*Talaromyces*
Aspergillus, causing bread mold and seed decays	*Eurotium*
Paecilomyces, used as biological control agent against whiteflies	*Byssochlamys*
Oidium, causing the powdery mildews	*Erysiphe*, etc.
	Perithecial ascomycetes
Chalara, causing oak wilt, tree cankers	*Ceratocystis*
Acremonium, endophytic in grasses	*Epichloe*
Sporothrix and *Graphium*, causing Dutch elm disease	*Ophiostoma*
Trichoderma, used as biocontrol agent against other fungi	*Hypocrea*
Verticillium, causing vascular wilts in many plants	*Hypocrea*
Fusarium, causing vascular wilts, root rots, stem rots, seed infections	*Gibberella*
Colletotrichum, causing anthracnoses in many plants	*Glomerella*
	Loculoascomycetes
Cercospora, causing Sigatoka disease of bananas	*Mycosphaerella*
Septoria, causing leaf spots on many crops	*Mycosphaerella*
Phyllosticta, causing black rot of grape	*Guignardia*
Alternaria, causing many leaf spots, blights	*Lewia*
Stemphylium, causing fruit rots on tomato	*Pleospora*
Bipolaris, causing leaf spots and root rots in grasses	*Cochliobolus*
Drechslera, causing leaf spots on grasses	*Pyrenophora*
Exserohilium, causing leaf spots on grasses	*Setosphaera*
Curvularia, causing leaf spots on grasses	*Cochliobolus*
Cladosporium, causing leaf mold on tomato (*C. fulvum*), and scab of peach and almond (*C. carpophilum*)	*Fulvia*, *Venturia*
Sphaeropsis, causing black rot on apple	*Botryosphaeria*
	Apothecial ascomycetes
Botrytis, *B. cinerea* causing gray mold rots on many plants	*Botryotinia*
Monilia, causing the brown rot of stone fruits	*Monilinia*
Marssonina, causing the black spot of rose	*Diplocarpon*
Entomosporium, causing a leaf and fruit spot on pear	*Diplocarpon*
Cylindrosporium, causing leaf spots on many kinds of plants	*Mycosphaerella*
Melanconium, causing the bitter rot of grape	*Greeneria*
	Basidiomycetes
Rhizoctonia, *R. solani* causing root and stem rots	*Thanatephorus*
Rhizoctonia binucleate forms	Ceratobasidiales
Sclerotium, *S. rolfsii* causing southern blight of many crops	*Aethalium*

Phylum: Basidiomycota (basidiomycetes, the club and mushroom fungi) — Sexual spores, called basidiospores, are produced externally on a club-like, one- or four-celled spore-producing structure called a basidium

Order: Ustilaginales (the smut fungi) — Basidium has cross walls or is nonseptate. It is the promycelium of the teliospore. Teliospores single or united into crusts or columns, remaining in host tissue or bursting through the epidermis. Fertilization by union of compatible spores, hyphae, etc. Only teliospores and basidiospores are produced

Genus: *Ustilago*, causing smut of corn (*U. maydis*), loose smuts of oats (*U. avenae*), of barley (*U. nuda*) and of wheat (*U. tritici*)

Tilletia, causing covered smut or bunt of wheat (*T. caries*) and Karnal bunt (partial bunt) of wheat (*T. indica*)

Urocystis, *U. cepulae* causing smut of onion

Sporisorium, causing covered kernel smut of sorghum (*S. sorghi*) and loose sorghum smut (*S. cruentum*)

Sphacelotheca, causing head smut of sorghum

Order: Uredinales (the rust fungi) — Basidium with cross walls. Sperm cells called spermatia fertilize special receptive hyphae in spermagonia. Produce two to several types of spores: teliospores, basidiospores, aeciospores, and uredospores (sometimes called "urediniospores"). Uredospores can be repeating spores. Obligate parasites

Genus: *Cronartium*, several species causing stem rusts of pines

Gymnosporangium, *G. juniperi-virginianae* causing cedar-apple rust

Hemileia, *H. vastatrix* causing coffee rust

Melampsora, *M. lini* causing rust of flax, *M. medousae* causing rust of poplars and conifers

Phakopsora, *P. pachyrrhizi* causing rust of soybeans

Phragmidium, one species causing rust of roses

Puccinia, several species causing severe rust diseases of cereals and of other plants

Uromyces, *U. appendiculatus* causing rust of beans

Order: Exobasidiales — Basidiocarp lacking: basidia produced on surface of parasitized tissue

Genus: *Exobasidium*, causing leaf, flower, and stem galls on several ornamentals

Order: Ceratobasidiales — Basidiocarp is web like, inconspicuous. Basidia without cross walls, with four prominent sterigmata

Genus: *Athelia,* the teleomorph of *Sclerotium* causing Southern blight of many plants, *S. cepivorum* causing the white rot of onions

Thanatephorus, *T. cucumeris* is the teleomorph of *Rhizoctonia solani*, causing root and stem rots, damping-off, and fruit rots in many plants

Typhula, causing typhula blight (snow mold) of turf grasses

Order: Agaricales (the mushrooms) — Basidium without cross walls, produced on radiating gills or lamellae. Many are mycorrhizal fungi

Genus: *Armillaria*, *A. mellea* and other species causing root rots of trees

Crinipellis, *C. perniciosus* causing witches'-broom of cacao

Marasmius, causing the fairy ring disease of turf grasses

Pleurotus, causing white rot on logs, tree stumps, and living trees

Pholiota, causing brown wood rot in deciduous forest trees

Order: Aphyllophorales — Basidia without cross walls produced on hymenium-forming hyphae and lining the surfaces of small pores or tubes

Genus: *Athelia* (anamorph is *Sclerotium*), causing root and stem rots of many plants

Chondrostereum, *C. purpureum* causing the silver leaf disease of trees

Corticium, one species causing the red thread disease of turf grasses

Heterobasidion, *H. annosum* causing root and butt rot of many trees

Ganoderma, causing root and basal stem rots in many trees

Inonotus, causing a heart rot of living trees and rot of dead trees and logs
Postia, causing wood and root rots of forest trees
Phellinus, causing tree root rots and cubical rots in buildings
Peniophora, causing decay in coniferous logs and pulpwood
Polyporus, causing heart rot of living trees and rot of dead trees or logs

IDENTIFICATION

The most significant fungus characteristics used for identification are spores and spore-bearing structures (sporophores) and, to some extent, the characteristics of the fungus body (mycelium). These items are examined under a compound microscope directly after removal from the specimen. The specimen is often kept moist for a few days to promote spore development. Alternatively, the fungus may be isolated and grown on artificial media and identified on the basis of spores produced on the media. For some fungi, special nutrient media have been developed that allow selective growth only of the particular fungus, allowing quick identification of the fungus.

The shape, size, color, and manner of arrangement of spores on the sporophores or in the fruiting bodies, as well as the shape and color of the sporophores or fruiting bodies, are sufficient characteristics to suggest, to one somewhat experienced in the taxonomy of fungi, the class, order, family, and genus to which the particular fungus belongs. In any case, these characteristics can be utilized to trace the fungus through published analytical, often dichotomous keys of the fungi to the genus and, finally, to the species to which it belongs. Once the genus of the fungus has been determined, descriptions of the known species are found in monographs of genera or in specific publications in research journals.

Because there are usually lists of the pathogens affecting a particular host plant, one may use such host indexes as short cuts in quickly finding names of fungus species that might apply to the fungus at hand. Host indexes, however, merely offer suggestions in determining identities, which must ultimately be determined by reference to monographs and other more specific publications.

In many fungi, hyphae in a colony or in adjacent colonies fuse (hyphal anastomosis). If the hyphae that fuse carry genetically different nuclei, the colony that is produced is a heterocaryon. Many fungi, however, have genetic systems that prevent mating between genetically identical cells. If the hyphae that come in contact belong to different strains of the same species but are of the same mating type, their encounter may result in vegetative incompatibility. Thus, the resulting vegetative incompatibility between colonies of various strains belonging to the same species is used to type the strains as belonging to different incompatibility groups constituting different biological species.

In recent years, immunoassay techniques, often involving monoclonal antibodies against specific proteins of a fungus conjugated with a fluorescent compound, have been used for the detection and identification of certain fungi.

The advent of molecular techniques, particularly of the polymerase chain reaction (PCR), of quick and inexpensive sequencing of DNA, and the accumulation of a relatively large databank of ribosomal DNA sequences have revolutionized both the lower limits of detection of pathogens and the accuracy and rapidity of their identification. These developments have made possible the detection of pathogens within plant tissues in the early stages of infection while there is still a minimal presence of the pathogen and early intervention may prevent an epidemic. They have also made possible a definitive identification of the pathogen by using DNA probes of known pathogens and, furthermore, they have made possible the quantification of the pathogen within, or in a mixture with, plant tissue, such as seed. Most DNA primers are for internal transcribed sequences of ribosomal DNA. The methodology, however, improves constantly and quickly. Much more sensitive and specific sets of primers have been developed based on families of highly repeated DNA that were 10 times more sensitive than primers directed at internal transcribed spacer sequences for ribosomal DNA.

SYMPTOMS CAUSED BY FUNGI ON PLANTS

Fungi cause local or general symptoms on their hosts and such symptoms may occur separately or concurrently or may follow one another. In general, fungi cause local or general necrosis of plant tissues, and they often cause reduced growth (stunting) of plant organs or entire plants. A few fungi cause excessive growth of infected plants or plant parts. The most common necrotic symptoms are as follows.

Leaf spots: Localized lesions on host leaves consisting of dead and collapsed cells

Blight: General and extremely rapid browning and death of leaves, branches, twigs, and floral organs
Canker: Localized necrotic lesion on a stem or fleshy organ, often sunken, of a plant
Dieback: Extensive necrosis of twigs beginning at their tips and advancing toward their bases
Root rot: Disintegration or decay of part or all of the root system of a plant
Damping-off: Rapid death and collapse of very young seedlings
Basal stem rot: Disintegration of the lower part of the stem
Soft rots and dry rots: Maceration and disintegration of fruits, roots, bulbs, tubers, and fleshy leaves
Anthracnose: Necrotic and sunken ulcer-like lesion on the stem, leaf, fruit, or flower of the host plant caused mainly by a certain group of fungi
Scab: Localized lesions on host fruit, leaves, tubers, etc., usually slightly raised or sunken and cracked, giving a scabby appearance
Decline: Progressive loss of vigor; plants growing poorly; leaves small, brittle, yellowish, or red; some defoliation and dieback present

Almost all of the aforementioned symptoms may also be associated with pronounced stunting of the infected plants. In addition, certain other diseases, such as rusts, mildews, wilts, and even those causing excessive growth of some plant organs, may cause stunting of the plant as a whole.

Symptoms associated with excessive enlargement or growth and distortion of plant parts include the following.

Clubroot: Enlarged roots appearing like spindles or clubs
Galls: Enlarged portions of plant organs (stems, leaves, blossoms, roots)
Warts: Wart-like protuberances on tubers and stems
Witches'-brooms: Profuse, upward branching of twigs
Leaf curls: Distortion, thickening, and curling of leaves

In addition to those just given, four groups of symptoms may be added.

Wilt: Generalized loss of turgidity and drooping of leaves or shoots
Rust: Many small lesions on leaves or stems, usually of a rusty color
Smut: Seed or a gall filled with the mycelium or black spores of the smut fungi
Mildew: Areas on leaves, stems, blossoms, and fruits, covered with whitish mycelium and the fructifications of the fungus

In many diseases, the fungal pathogen grows, or produces various structures, on the surface of the host. These structures may include mycelium, sclerotia, sporophores, fruiting bodies, and spores, and are called **signs**. **Signs** are distinct from **symptoms**, which refer only to the appearance of the infected plants or plant tissues. Thus, in the mildews, for example, one sees mostly the signs consisting of a whitish, downy or powdery growth of fungus mycelium and spores on the plant leaves, fruit, or stem, whereas the symptoms consist of chlorotic or necrotic lesions on leaves, fruit, and stem, reduced growth of the plant, and so on.

ISOLATION OF FUNGI (AND BACTERIA)

Many plant diseases can be diagnosed by observation with the naked eye or with the microscope, and for these the isolation of the pathogen is not necessary. There are many fungal and bacterial diseases, though, that have similar symptoms and cannot be distinguished visually from one another. In many, the pathogen cannot be identified because it is mixed with one or more contaminants, because it has not yet produced its characteristic fruiting structures and spores, or because the same disease could be caused by more than one similar-looking pathogen and perhaps by some environmental factor. In many cases in which diagnosis can be made by visual observation, isolation and identification of the pathogen are still desirable in order to verify the diagnosis. Occasionally, a disease is caused by a new, previously unknown pathogen that must be isolated and studied. Just as often, even pathogens of known diseases must be isolated from diseased plant tissues whenever a study of the characteristics and habits of these pathogens is to be undertaken. Of course, if the identity of the pathogen is suspected or determined and a specific nutrient medium that allows only the growth of that pathogen is available, then the isolation of the particular pathogen is achieved by growing a small section of infected tissue on such medium.

Preparing for Isolation

Even before attempting to isolate the causal fungus or bacterium from a diseased plant tissue, one must perform several preliminary operations, including the following.

1. Use already sterilized plastic items or sterilize glassware, such as petri dishes, test tubes, and pipettes, by dry heat (150–160°C for one hour or more), or autoclaving, or by dipping for one minute or more in 70–80% ethyl alcohol.

2. Prepare solutions for treating the surface of the infected or infested tissue to eliminate or markedly reduce surface contaminants that could interfere with the isolation of the pathogen. These solutions can be used either as a surface wipe or as a dip. The most commonly used surface sterilants are 0.5% sodium hypochlorite solution (one part household bleach to nine parts water), used both for wiping infected tissues or dipping sections of such tissues in it and for wiping down table or bench surfaces before making isolations; and 70% ethyl alcohol, which is used for leaf dips for three seconds or more. The tissues must be blotted dry with a sterile paper towel.
3. Prepare culture media on which the isolated fungal or bacterial pathogens will grow. An almost infinite number of culture media can be used to grow plant pathogenic fungi and bacteria. Some of them are entirely synthetic (i.e., made up of known amounts of certain chemical compounds) and may be quite specific (selective) for certain pathogens. Some are liquid or semiliquid and are used primarily for the growth of bacteria but also of fungi in certain cases. Most media contain an extract of a natural source of carbohydrates and other nutrients, such as potato, corn meal, lima bean, or malt extract, to which variable amounts of agar are added to solidify the medium and form a gel on or in which the pathogen can grow and be observed. The most commonly used media are potato dextrose agar (PDA), which is good for most, but not all fungi; water agar or glucose agar (1–3% glucose in water agar) for separating some oomycetes (*Pythium*) and fungi (*Fusarium*) from bacteria; V-8 and other less rich media, which encourage fungal sporulation; and nutrient agar, which contains beef extract and peptone and is good for isolating bacterial plant pathogens. Fungi can also be separated in culture from bacteria by adding 1 or 2 drops of a 25% solution of lactic acid, which inhibits the growth of bacteria, to 10 milliliters of the medium before pouring it in the plate. Solutions of culture media are prepared in flasks, which are plugged and placed in an autoclave at 120°C and 15 pounds (6.8 kg) pressure for 20 minutes (Fig. 11-3). Sterilized media are then allowed to cool somewhat and are subsequently poured from the flask into sterilized petri dishes, test tubes, or other appropriate containers. If agar was added, the medium will soon solidify and is then ready to be used for growth of the fungus or bacterium. Pouring of the culture medium into petri dishes, tubes, and so on is carried out as aseptically as possible either in a separate culture room or in a clean room free from drafts and dust. In either case, the work table should be wiped with a 10% Clorox solution, hands should be clean, and tools such as scalpels, forceps, and needles should be dipped in alcohol and flamed to prevent introduction of contaminating microorganisms. Working in a laminar flow hood greatly helps to grow the desired fungus free of airborne contaminants.

Although most fungi and most bacteria can be cultured on nutrient media with ease, some of them have specific and exacting requirements and will not grow on most commonly used nutrient media. Some groups of fungi, namely Erysiphales, causes of the powdery mildew diseases, and the oomycetes Peronosporaceae, causing downy mildews, are considered strictly obligate parasites and cannot be grown on culture media but can be grown on leaf-containing dishes. Another group of fungi, Uredinales, which cause the rust diseases of plants, were, until the late 1960s, also thought to be strictly obligate parasites. Since then, however, it has become possible to grow some stages of some rust fungi in culture by adding certain components to the media so rust fungi are no longer impossible to grow in culture, although they are, of course, obligate parasites in nature. Fastidious phloem- and xylem-limited bacteria also either are impossible to grow in culture so far or must be grown on special complex nutrient media. Of the other pathogens, only spiroplasmas have been grown in culture. None of the phytoplasmas and none of the viruses, nematodes, or protozoa have been grown on nutrient culture media so far.

Isolating the Pathogen

From Leaves

If the infection of the leaf is still in progress in the form of a fungal leaf spot or blight and if there are spores present on the surface, a few spores may be shaken loose over a petri plate containing culture medium or picked up at the point of a sterile needle or scalpel and placed on the surface of the culture medium. Also, infected tissues may be placed in a moist chamber to allow the pathogen to grow out on the tissue and then pick off spores and fruiting bodies and plate them out. If the fungus does grow in culture, isolated colonies of mycelium will appear in a few days as a result of germination of the added spores. These colonies can be transferred to separate plates, thus assuring that some plates will contain the pathogen free of contaminants.

Sometimes, isolation of the pathogen from fungal or bacterial leaf spots and blights is made by surface sterilizing the area to be cut with Clorox solution, remov-

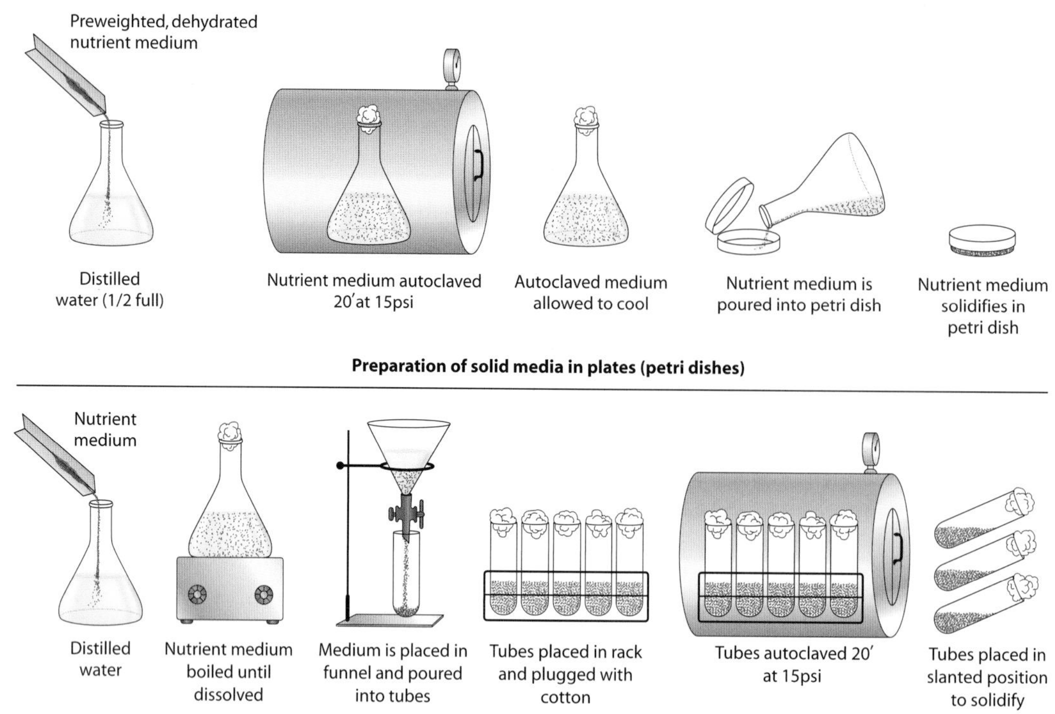

FIGURE 11-3 Preparation of solid nutrient media in plates (petri dishes) and in test tube slants.

ing a small part of the infected tissue with a sterile scalpel, and placing it in a plate containing a nutrient medium. The most common method, however, for isolating pathogens from infected leaves, as well as other plant parts, involves cutting several small sections 5 to 10 millimeters square from the margin of the infected lesion so that they contain both diseased and healthy-looking tissue (Fig. 11-4A). These are placed in one of the surface sterilant solutions, making sure that the surfaces get wet. After about 15 to 30 seconds, the sections are taken out aseptically one by one and at regular intervals (e.g., every 10–15 seconds) so that each of them has been surface sterilized for different times. The sections are then blotted dry on clean sterile paper towels or are washed in three changes of sterile water and are finally placed on the nutrient medium, usually three to five per dish. Those sections surface sterilized the shortest time usually contain contaminants along with the pathogen, whereas those surface sterilized the longest produce no growth at all because all organisms have been killed by the surface sterilant. Some of the sections left in the surface sterilant for intermediate periods of time, however, will allow only the pathogen to grow in culture in pure colonies (Fig. 11-4B). This happens because the sterilant was allowed to act long enough to kill surface contaminants but not long enough to kill the pathogen that was advancing alone from the diseased to the healthy tissue. These colonies of the pathogen are then subcultured asceptically for further study.

If fruiting structures (pycnidia, perithecia) are present on the leaf, it is sometimes possible to pick them out, drop them in the surface sterilant for a few seconds, and then plate them on the nutrient medium. This procedure, however, requires that most of the work be done under a stereoscopic microscope (binoculars) because the fruiting structures are generally too small to see with the naked eye and to handle. Fruiting structures, after surface sterilization, may also be crushed in a small drop of sterile water and then the spores in the water may be diluted serially in small tubes or dishes containing sterile water. Finally, a few drops from each tube of the serial dilution are placed on a nutrient medium, and single colonies free of contaminants develop from germinating spores obtained from some of the serial dilution tubes.

The serial dilution method is often used to isolate pathogenic bacteria from diseased tissues contaminated with other bacteria. After surface sterilization of sections of diseased tissues from the margin of the infection, the sections are ground aseptically but thoroughly in a small volume of sterile water and then part of the homogenate is diluted serially in equal volumes or 10 times the volume of the initial water. Finally, plates containing nutrient agar are streaked with a needle or loop dipped in each of the different serial dilutions, and single

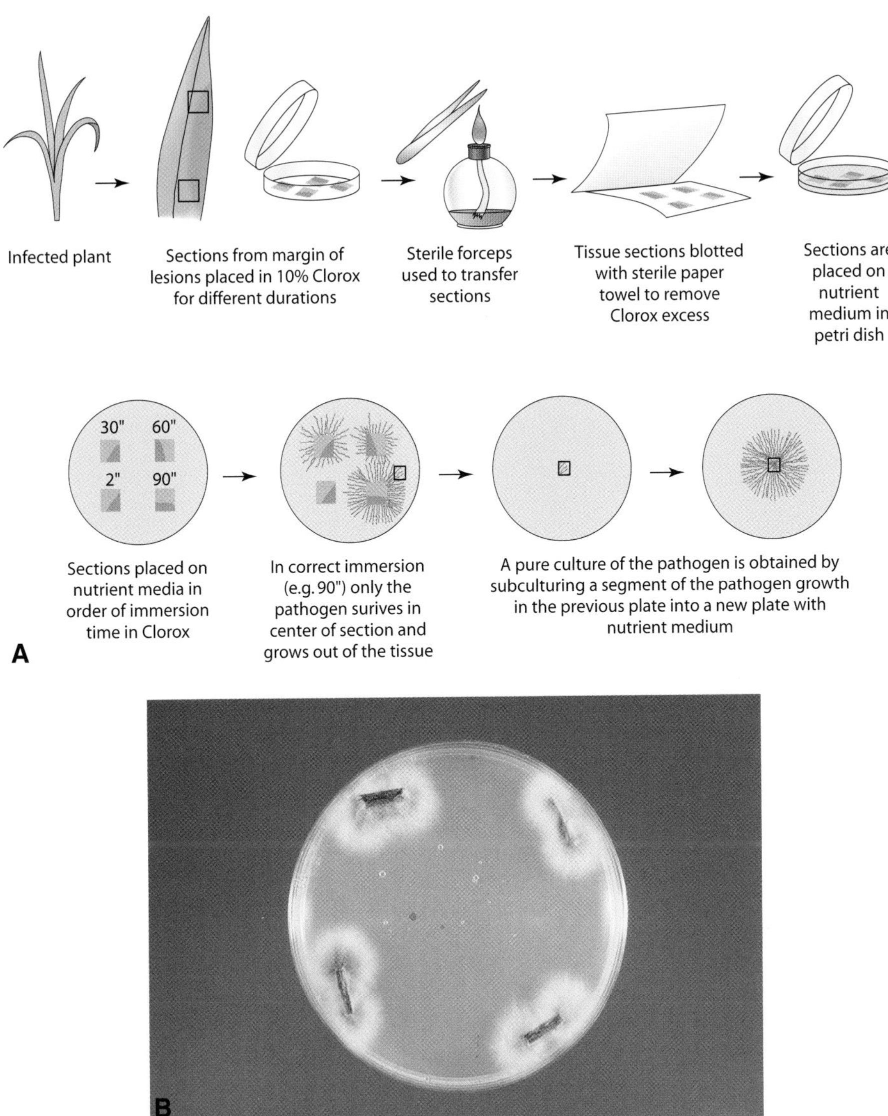

FIGURE 11-4 Isolation of fungal pathogens from infected plant tissue.

colonies of the pathogenic bacterium are obtained from the higher dilutions that still contain bacteria.

From Stems, Fruits, Seeds, and Other Aerial Plant Parts

Almost all the methods described for isolating fungal and bacterial pathogens from leaves can also be used to isolate these pathogens from superficial infections of stems, fruits, seeds, and other aerial plant parts. Entire seeds can be plated. In addition to these methods, however, pathogens can often be isolated easily from infected stems and fruits in which the pathogen has penetrated fairly deeply. This is accomplished by splitting the stem or breaking the fruit from the healthy side first and then tearing it apart toward and past the infected margin, thus exposing tissues not previously exposed to contaminants and not touched by hand or knife and therefore not contaminated. Small sections of tissue can be cut from the freshly exposed area of the advancing margin of the infection with a flamed scalpel and can be plated directly on the culture medium (Fig. 11-4B).

From Roots, Tubers, Fleshy Roots, and Vegetable Fruits in Contact with Soil

Isolating pathogens from any diseased plant tissue in contact with soil presents the additional problems of numerous saprophytic organisms invading the plant tissue after it has been killed by the pathogen. For this reason, the first step in isolating the pathogen is repeated thorough washing of such diseased tissues to remove all

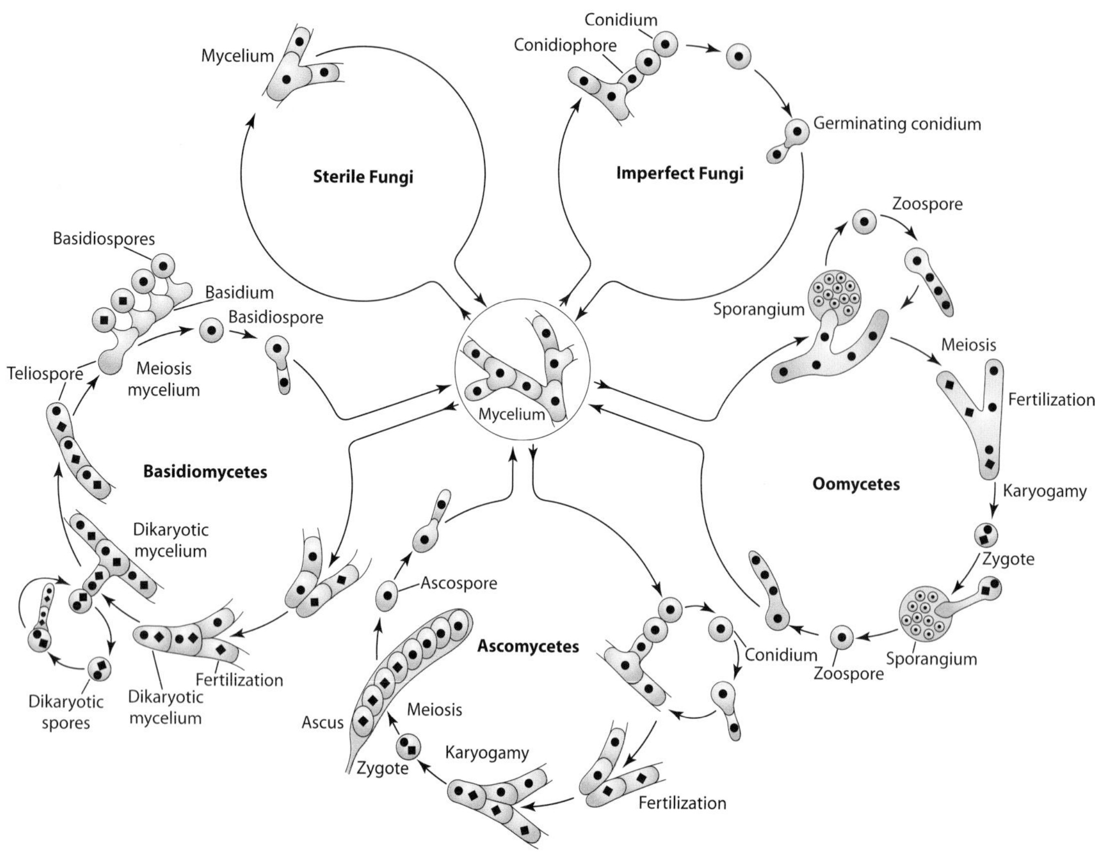

FIGURE 11-5 Schematic presentation of the generalized life cycles of oomycetes and the main groups of phytopathogenic fungi.

soil and most of the loose, decayed plant tissue in which most of the saprophytes are present. If the diseased root is small, once it is washed thoroughly, pathogens can be isolated from it by following one of the methods described for isolating pathogens from leaves. If isolation is attempted from fleshy roots or other fleshy tissues penetrated only slightly by the pathogen and showing only surface lesions, the tissue is washed free from adhering soil, and several bits of tissue from the margin of the lesions are placed in Clorox solution. The tissue sections are picked from the solution one by one, blotted or washed in sterile water, and placed on agar in petri plates. If the pathogen has penetrated deeply into the fleshy tissue, the method described earlier for stems and fruit can be used most effectively, namely breaking the specimens from the healthy side first and then tearing toward the infected area and plating bits taken from the previously unexposed margin of the rot.

LIFE CYCLES OF FUNGI

Although the life cycles of fungi and oomycetes of the different groups vary greatly, the majority of them go through a series of steps that are quite similar (Fig. 11-5). Almost all fungi have a spore stage with a simple, haploid nucleus (possessing one set of chromosomes, or 1N). The spore germinates into a hypha, which also contains haploid nuclei. The hypha may either produce simple, haploid spores again (as is always the case in the imperfect fungi) or it may fuse with another hypha to produce a fertilized hypha in which the nuclei unite to form one diploid nucleus, called a zygote (containing two sets of chromosomes, or 2N). In the oomycetes, the zygote divides to produce diploid mycelium and spores. The mycelium produces gametangia in which meiosis occurs, then fertilization, and production of the zygote. In a brief phase of most Ascomycetes, and generally in Basidiomycetes, the two nuclei of the fertilized hypha do not unite but remain separate within the cell in pairs (dikaryotic or N + N) and divide simultaneously to produce more hyphal cells with pairs of nuclei. In Ascomycetes, dikaryotic hyphae are found only inside the fruiting body, in which they become the ascogenous hyphae. In these, the two nuclei of one cell of each hypha unite into a zygote (2N), which divides meiotically to produce ascospores that contain haploid nuclei.

In Basidiomycetes, haploid spores produce haploid hyphae. On fertilization, dikaryotic mycelium (N + N) is produced, which develops into the main body of the fungus. Such dikaryotic hyphae may produce, asexually, dikaryotic spores that will grow again into a dikaryotic mycelium. Finally, however, the paired nuclei of the cells unite and form diploid nuclei. These may replicate mitotically or act as zygotes. The zygotes divide meiotically and produce basidiospores that contain haploid nuclei.

In mitosporic fungi (deuteromycetes or imperfect fungi), only the asexual cycle (haploid spore → haploid mycelium → haploid spore) is found. Even in oomycetes, however, a similar asexual cycle that can be repeated many times during each growth season is the most common by far. The sexual cycle usually occurs only once a year.

CONTROL OF FUNGAL DISEASES OF PLANTS

The endless variety and the complexity of the many diseases of plants caused by fungi and pseudofungi have led to the development of a correspondingly large number of approaches for their control. The particular characteristics of the life cycle of each fungus, its habitat preferences, and its performance under certain environmental conditions are some of the most important points to be considered in attempting to control a plant disease caused by a fungus. However, although some diseases can be controlled completely by just one type of control measure, a combination of measures is usually necessary for the satisfactory control of most diseases. An integrated approach to disease management and control is a must for most fungal diseases of plants.

The most common cultural or biologically based control measures include use of resistant plant varieties; use of pathogen-free seed or propagating stock; destruction of plant parts or refuse harboring the pathogen; destruction of volunteer plants or alternative hosts of the pathogen; use of clean tools and containers; proper drainage of fields and aeration of plants; crop rotation; and support or use of microorganisms antagonistic or pathogenic to the fungus causing the plant disease. For many fungal diseases, however, the most effective method, and sometimes the only one available for their control, is the application of chemical sprays or dusts (fungicides) on the plants, on seeds, or into the soil where the plants are to be grown. Soil-inhabiting fungi in potting mixes may be controlled by steam or electric heat and, in fields, by volatile liquids (fumigants) or solarization.

In some diseases the fungus is carried in the seed, and control can be obtained only through treatment of the seed with fungicides that are absorbed and are distributed through the plant (systemic fungicides) or through hot water. In others, control of the insect vectors may be helpful or the only available possibility. Great advances have been made toward controlling fungal diseases of plants, especially through the use of resistant varieties developed by conventional plant breeding or through genetic engineering and chemicals. As a result, these diseases are probably much easier to control than any other group of plant diseases, although the control costs and the losses caused by fungal diseases of plants are still very great.

Selected References

Ainsworth, G. C., Sparrow, F. K., and Sussman, A. S., eds. (1965–1973). "The Fungi: An Advanced Treatise," Vols. 1–4. Academic Press, New York.

Alexopoulos, C. J., Mims, C. W., and Blackwell, M. (1996). "Introductory Mycology," 4th Ed. Wiley, New York.

Arx, J. A. von. (1987). "Plant Pathogenic Fungi." J. Cramer, Berlin.

Barnett, H. L., and Hunter, B. B. (1998). "Illustrated Genera of Imperfect Fungi." 4th Ed. APS Press, St. Paul, MN.

Barr, D. J. S. (1992). Evolution and kingdoms of organisms from the perspective of a mycologist. *Mycologia* **84**, 1–11.

Baudoin, A. B. A. M., ed. (1988). "Laboratory Exercises in Plant Pathology: An Instructional Kit." APS Press, St. Paul, MN.

Buczacki, S. T., ed. (1983). "Zoosporic Plant Pathogens: A Modern Perspective." Academic Press, London.

Carlile, M. J., Watkinson, S. C., and Gooday, G. W. (2001). "The Fungi," 2nd Ed., Academic Press, San Diego.

Cole, G. T., and Kendrick, B. (1981). "Biology of Conidial Fungi," Vols. 1 and 2. Academic Press, New York.

Cummins, G. B., and Hiratsuka, Y. (1983). " Illustrated Genera of Rust Fungi." APS, St. Paul, MN.

Fergus, C. L. (1960). "Illustrated Genera of Wood Destroying Fungi." Burgess, Minneapolis, MN.

Frisvad, J. C., Bridge, P. D., and Arora, D. K. (1998). "Chemical Fungal Taxonomy." Dekker, New York.

Hanlin, R. T. (1990). "Illustrated Genera of Ascomycetes." APS Press, St. Paul, MN.

Hawksworth, D. L., Kirk, P. M., Sutton, B. C., and Pegler, D. N. (1995). "Ainsworoth and Bisb's Dictionary of the Fungi," 8th Ed. CAB International, Wallingford, UK.

Jennings, D. H., and Lysek, G. (1999). "Fungal Biology: Understanding the Fungal Lifestyle," 2nd Ed. Bios, Oxford.

Judelson, H. S., and Tooley, P. W. (2000). Enhanced polymerase chain reaction methods for detecting and quantifying *Phytophthora infestans* in plants. *Phytopathology* **90**, 1112–1119.

Kendrick, B., ed. (1971). "Taxonomy of Fungi Imperfecti." Toronto Univ. Press, Toronto.

Kronstad, J. W., ed. (2000). "Fungal Pathology." Kluwer Academic, Dordrecht, The Netherlands.

Margulis, L., Corliss, J. O., Melkonian, M. K., and Chapman, D. J., eds. (1990). "Handbook of Protoctista." Jones and Bartlett, Boston, MA.

Margulis, L., and Schwartz, K. V. (1982). "Five Kingdoms: An Illustrated Guide to the Phyla of Life on Earth." 3rd Ed. Freeman, New York.

Martin, R. R., James, D., and Levesque, C. A. (2000). Impacts of molecular diagnostic technologies on plant disease management. *Annu. Rev. Phytopathol.* 38, 207–239.

Rossman, A. Y., Palm, M. E., and Spielman, L. J. (1987). "A Literature Guide to the Identification of Plant Pathogenic Fungi." APS Press, St. Paul, MN.

Samuels, G. J., and Seifert, K. A. (1995). The impact of molecular characters on systematics of filamentous ascomycetes. *Annu. Rev. Phytopathol.* **33**, 37–67.

Scott, K. J., and Chakravorty, A. K., eds. (1982). "The Rust Fungi." Academic Press, London.

Shoemaker, R. A. (1981). Changes in taxonomy and nomenclature of important genera of plant pathogens. *Annu. Rev. Phytopathol.* **19**, 297–307.

Singleton, L. L., Mihail, J. D., and Rush, C. M., eds. (1992). "Methods for Research on Soilborne Phytopathogenic Fungi." APS Press, St. Paul, MN.

Subramanian, C. V. (1983). "Hyphomycetes; Taxonomy and Biology." Academic Press, New York.

Taylor, J. W., Jacobson, D. J., and Fisher, M. C. (1999). The evolution of asexual fungi: Reproduction, speciation, and classification. *Annu. Rev. Phytopathol.* **34**, 197–246.

Tsao, P. H. (1970). Selective media for isolation of pathogenic fungi. *Annu. Rev. Phytopathol.* **8**, 157–186.

DISEASES CAUSED BY FUNGAL-LIKE ORGANISMS

Diseases Caused by Myxomycota (Myxomycetes)

Myxomycetes, also called slime molds, are fungal-like members of the kingdom Protozoa. Their body is a plasmodium, i.e., an amoeboid mass of protoplasm that has many nuclei and no definite cell wall. At a certain point of its life cycle, the plasmodium is transformed into superficial fructifications that contain resting spores (Fig. 11-6). The slime molds produce zoospores that have two flagella (Fig. 11-7).

Myxomycetes cause disease in plants by simply growing externally on the surface of plants growing low on the ground, such as turf grasses (Fig. 11-6A), strawberries, and vegetables. They are most common in warm weather after heavy rains or watering. In some areas, all aboveground parts of plants, and even the soil between plants, may be covered by a creamy white or colored slimy growth. Later, the slimy growth changes to crusty, ash-gray, or colored fruiting structures that make the affected plants appear dull gray.

Slime molds are saprophytic. Their plasmodium creeps like an amoeba and feeds on decaying organic matter and microorganisms such as bacteria, which it simply engulfs and digests. There are many species of slime mold fungi, the most common of which are *Physarum*, *Fuligo*, *Mucilago*, and *Didymium*.

The plasmodium grows mostly in the upper layer of the soil and in the thatch. During warm, wet weather the plasmodium comes to the soil surface and creeps over low-lying vegetation, producing its crusty fruiting structures on the plant surface. The fruiting structures are sporangia filled with dark masses of powdery spores and vary in size, shape, and color depending on the species of slime mold (Figs. 11-6, 11-8, and 11-9). The spores are rubbed off the plant easily and are spread by wind, water, mowers, or other equipment and can survive unfavorable weather. In cool, humid weather,

FIGURE 11-6 (A) Turfgrass leaves covered with fructifications (sporangia) of the slime mold *Physarum*. (B) Close-up of slime mold fructifications. [Photographs courtesy of (A) D. Smith, WCPD, and (B) by R. E. Cullen, Plant Pathology Department, University of Florida.]

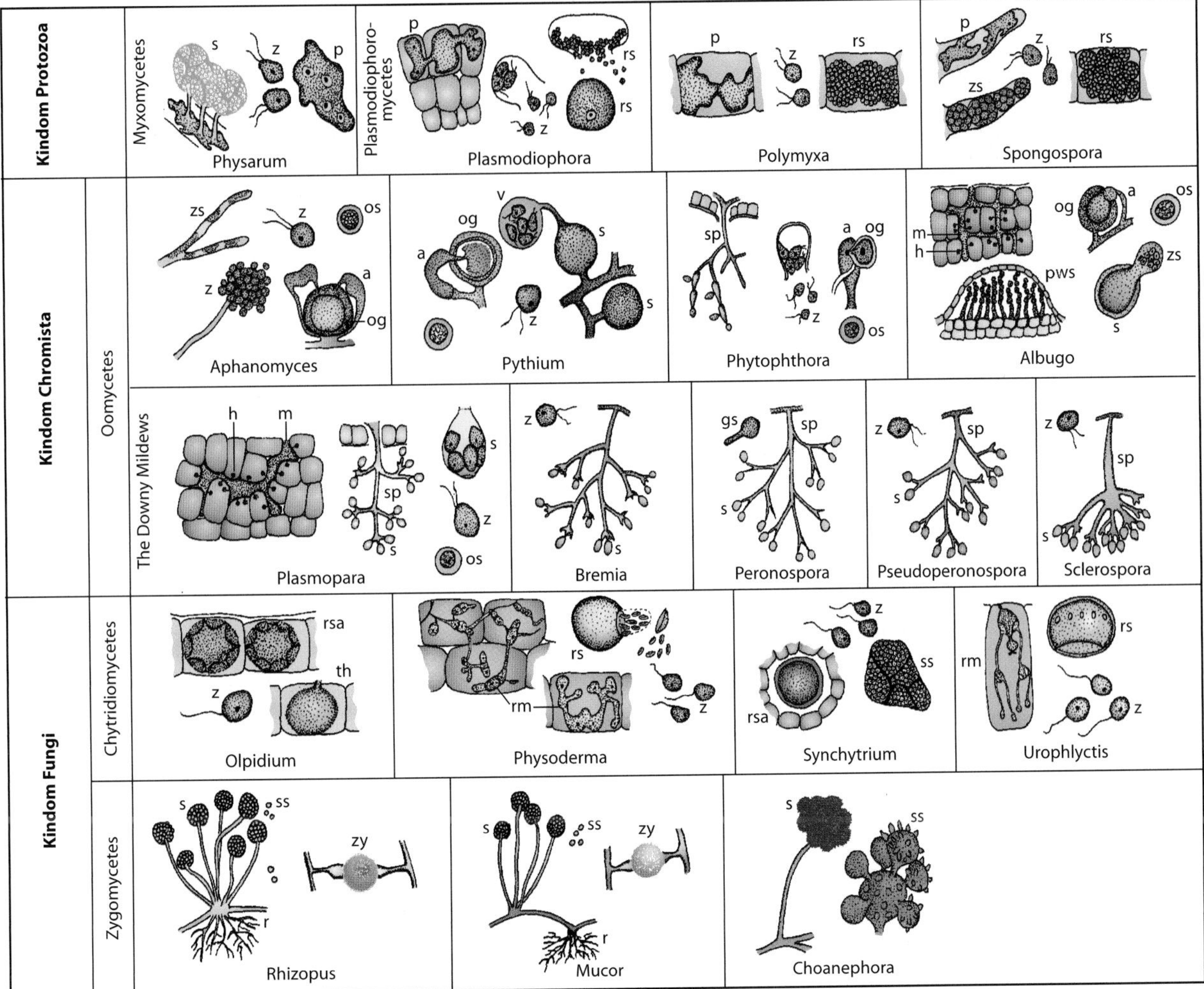

FIGURE 11-7 The most common protozoa and chromista (stramenopiles) and some of the fungi that cause disease in plants. a, Antheridium; gs, germinating sporangium; h, haustorium; m, mycelium; og, oogonium; os, oospore; p, plasmodium; pws, pustule with sporangia; rm, rhizomycelium; rs, resting spore; rsa, resting sporangium; s, sporangium; sp, sporangiophore; ss, sporangiospore; th, thallus; z, zoospore; zs, zoosporangium; zy, zygospore.

the spores absorb water, their cell wall cracks open, and a single zoospore emerges from each. The zoospores undergo various changes and unite in pairs to form amoeboid zygotes. The latter enlarge, become multinucleate, and become the plasmodium.

No control is usually necessary against slime molds. When they become too numerous and unsightly, spraying with any fungicide, such as captan or thiram, will control the slime molds.

Selected References

Alexopoulos, C. J., Mims, C. W., and Blackwell, M. (1996). "Introductory Mycology," 4th Ed. Wiley, New York.

Couch, H. B. (1995). "Diseases of Turfgrasses," 3rd Ed. Krieger, New York.

Diseases Caused by Plasmodiophoromycetes

Three Plasmodiophoromycetes cause the following common diseases of plants:

- *Plasmodiophora*, causing clubroot of crucifers
- *Polymyxa*, causing a root disease of cereals and grasses
- *Spongospora*, causing the powdery scab of potato (Fig. 11-10)

The pathogens are obligate parasites, and although they can survive in the soil as resting spores for many years, they can grow and multiply in only a few hosts. The plasmodium lives off the host cells it invades but does not kill these cells for a long time. However, in some dis-

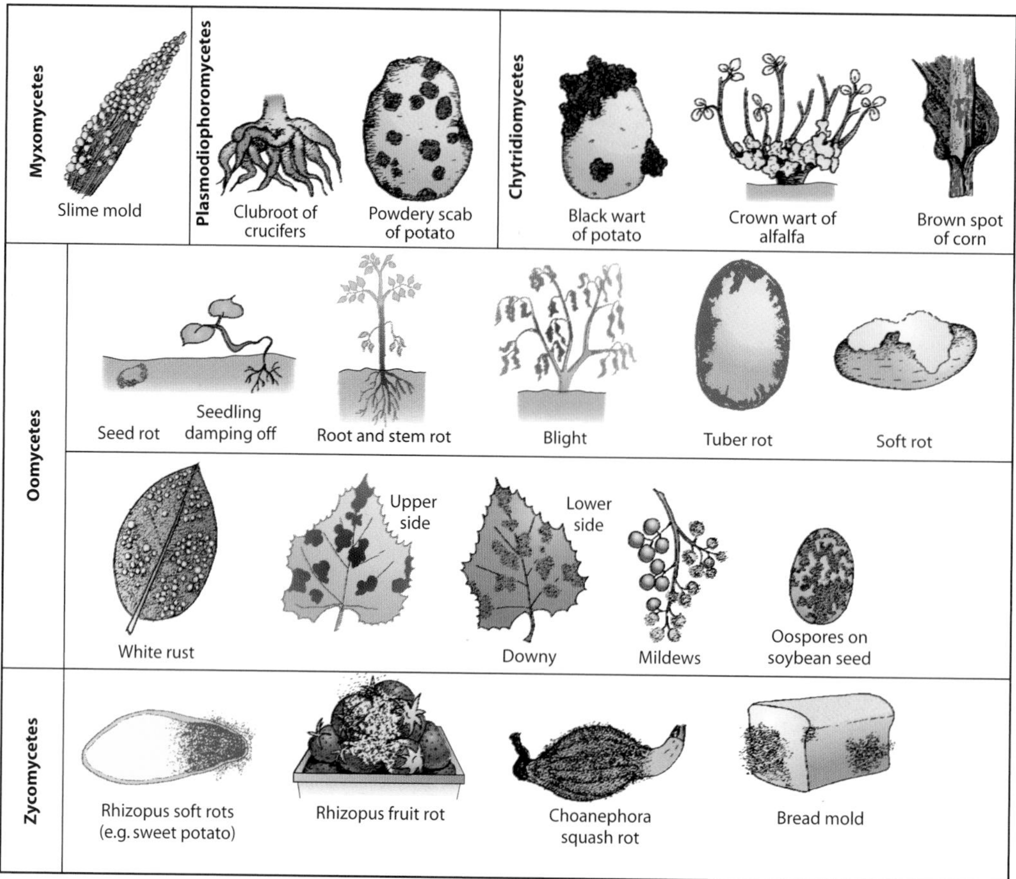

FIGURE 11-8 The most common symptoms caused by some fungal-like organisms and some fungi.

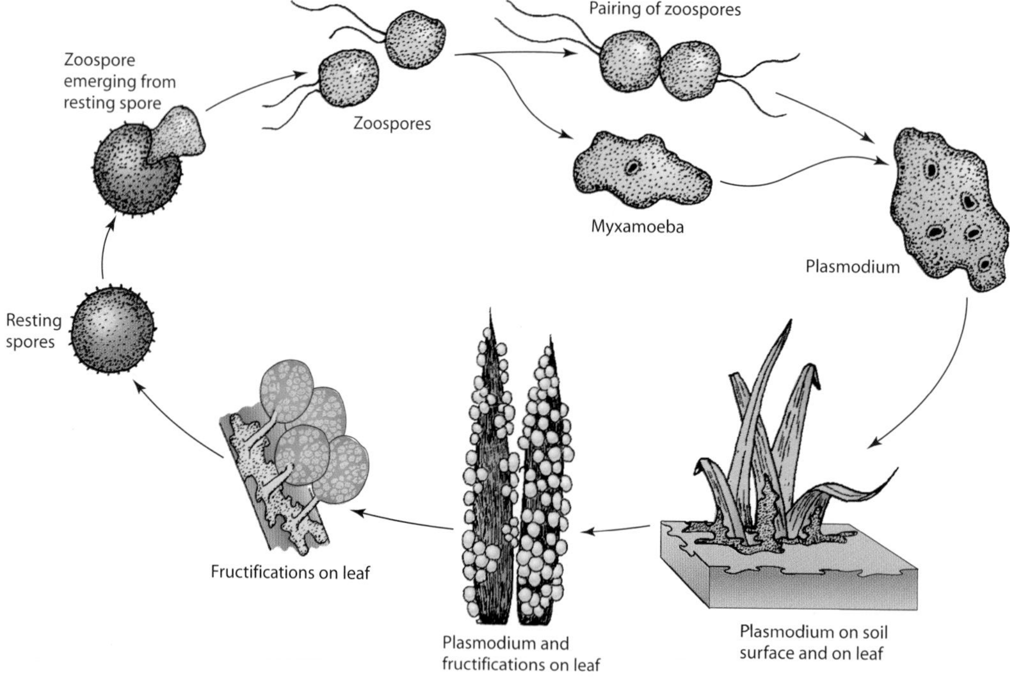

FIGURE 11-9 Life cycle of slime molds.

eases, many invaded and adjacent cells are stimulated by the pathogen to enlarge and divide, thus making more nutrients available for the pathogen. The pathogens spread from plant to plant by means of zoospores, by anything that moves soil or water containing spores, by infected transplants, and so on.

Polymyxa and *Spongospora*, in addition to the diseases they cause, can also transmit destructive plant viruses. *Polymyxa graminis* is a vector of several viruses of grain crops and of peanuts, whereas *P. betae* is a vector of beet necrotic yellow vein virus. *Spongospora* is a vector of the potato mop-top virus.

FIGURE 11-10 Potato tubers infected with powdery scab caused by *Spongospora subterranea*. (Photograph courtesy of K. Mohan, University of Idaho.)

CLUBROOT OF CRUCIFERS

The clubroot disease of cruciferous plants, such as cabbage and cauliflower, is widely distributed all over the world. Clubroot can cause serious losses to susceptible varieties. Fields once infested with the clubroot pathogen remain so indefinitely and become unfit for the cultivation of crucifers.

Symptoms

Infected plants at first have pale green to yellowish leaves. Later, infected plants show wilting in the middle of hot, sunny days, recovering during the night (Fig. 11-11A). Young plants may be killed by the disease soon after infection, whereas older plants may remain alive but become stunted and fail to produce marketable heads.

The most characteristic symptoms of the disease appear on the roots (Fig. 11-11B) as spindle-like, spherical, knobby, or club-shaped swellings. The swellings may be few and isolated or they may coalesce and cover the entire root system. The older and usually the larger clubbed roots disintegrate before the end of the season because of invasion by bacteria and other fungi.

FIGURE 11-11 (A) Field and (B) plants of cabbage infected with the clubroot disease caused by *Plasmodiophora brassicae*. [Photographs courtesy of (A) M. A. Hansen, Virginia Tech., and (B) I. R. Evans, WCPD.]

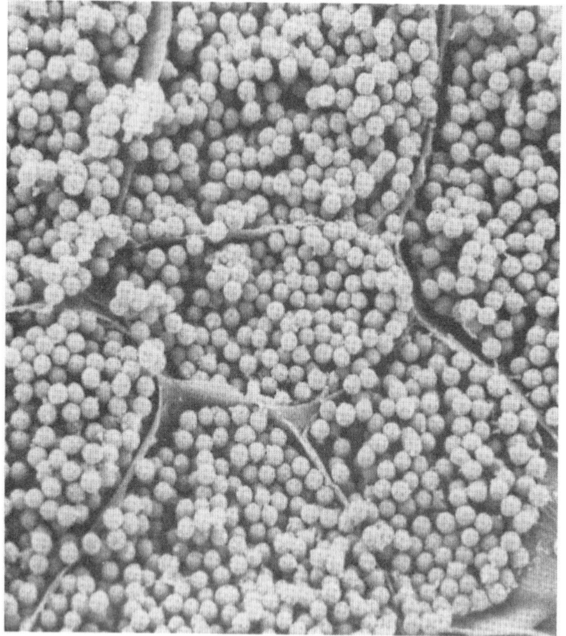

FIGURE 11-12 Scanning electron micrograph of resting spores of *Plasmodiophora brassicae* within cells of club roots. (Photograph courtesy of M. F. Brown and H. G. Brotzman.) Magnification: ×1000.

The Pathogen: Plasmodiophora Brassicae

Its body is a plasmodium. The plasmodium gives rise to zoosporangia or to resting spores (Fig. 11-12), which, on germination, produce zoospores.

Development of Disease

The single zoospore produced from resting spores penetrates root hairs and there develops into a plasmodium. After a few days, the plasmodium cleaves into multinucleate portions and each develops into a zoosporangium containing four to eight secondary zoospores. The zoospores are discharged outside the host through pores dissolved in the host cell wall. Some zoospores fuse in pairs to produce zygotes, which can cause new infections and produce new plasmodium. These zoospores penetrate young root tissues directly, whereas older, thickened roots and underground stems are penetrated through wounds. From these points of primary infection the plasmodium spreads to cortical cells and the cambium by direct penetration (Fig. 11-13). When it reaches the cambium, the plasmodium spreads in all directions in the cambium, outward into the cortex and inward toward the xylem.

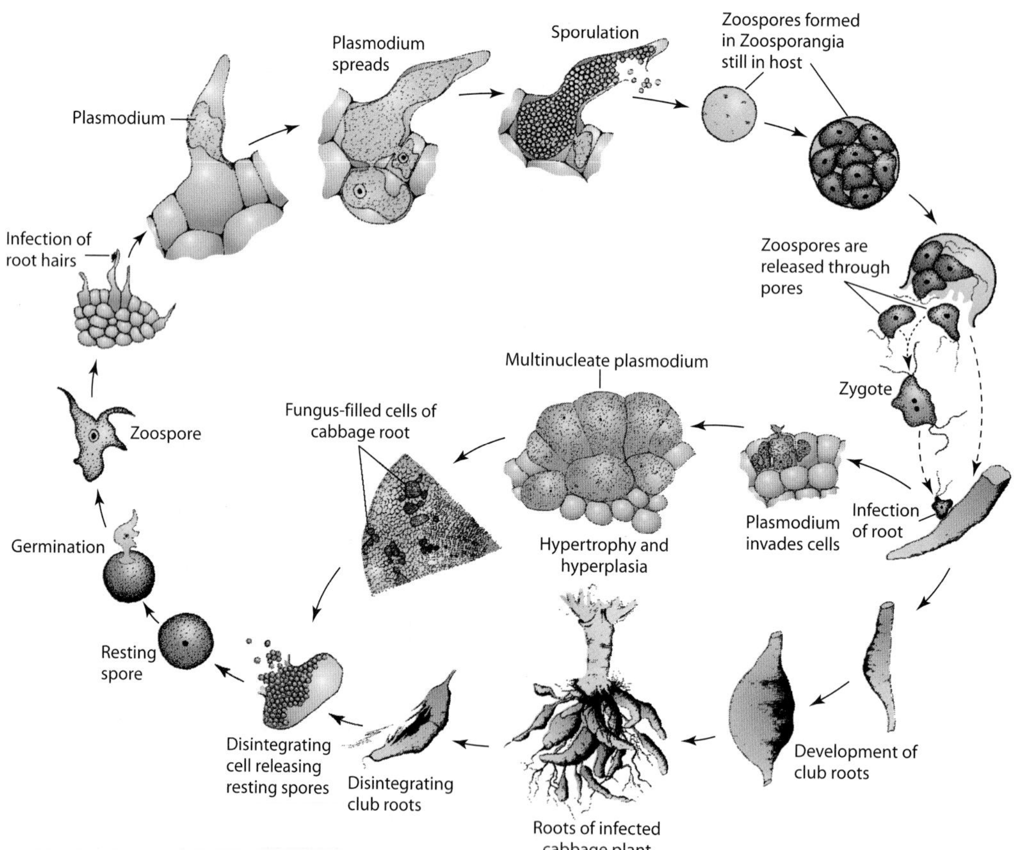

FIGURE 11-13 Disease cycle of clubroot of crucifers caused by *Plasmodiophora brassicae*.

As the plasmodia pass through cells they become established in some of the cells, which are stimulated to enlarge, divide abnormally, and become up to five or more times the normal size. The infected cells of a club occur in small groups throughout the diseased tissue, and the groups are usually separated by uninfected cells. Only rarely are all the cells of a club infected; usually, only about 30% of the tissue is occupied by plasmodium. However, even noninvaded cells of diseased tissues are stimulated to grow abnormally.

The plasmodium-infected clubs not only utilize much of the food required for the normal growth of the plant, they also interfere with the absorption and translocation of mineral nutrients and water through the root system. This results in gradual stunting and wilting of the aboveground parts of the plant. Furthermore, the rapidly growing and greatly enlarged cells of the club tissues are unable to form a cork layer at their surface and are easily ruptured and invaded by secondary, weakly parasitic microorganisms.

Control

Avoid growing cruciferous crops in fields known to be infested with the clubroot pathogen. A PCR-based assay can detect resting spores of the pathogen in field soil. If avoidance is not possible, plant cabbage and other susceptible cruciferous crops in well-drained fields that have a pH slightly above neutral (usually about pH 7.2) or in fields in which hydrated lime has been added to raise the soil to pH 7.2. At that pH, spores of the clubroot organism germinate poorly or not at all. Seedbed areas can be kept free of clubroot by treating the soil with appropriate soil fumigants approximately two weeks before planting. The clean, clubroot-free seedlings should, on transplanting, be watered with a solution of an effective fungicide.

Some varieties of cruciferous hosts are resistant to certain races of the clubroot organism and can be grown in areas infested with these races. However, significant genetic variability occurs among field isolates of the pathogen and, so far, none of the varieties are resistant to all the races of *P. brassicae*.

Selected References

Buczacki, S. T. (1983). *Plasmodiophora*: An inter-relationship between biological and practical problems. *In* "Zoosporic Plant Pathogens" (S. T. Buczacki, ed.), pp. 161–191. Academic Press, London.

Colhoun, J. (1958). Clubroot disease of crucifers caused by *Plasmodiophora brassicae*. *Commonw. Mycol. Inst. Phytopathol. Pap.* 3, 1–108.

Dobson, R. L., and Gabrielson, R. L. (1983). Role of primary and secondary zoospores of *Plasmodiophora brassicae* in the development of clubroot in Chinese cabbage. *Phytopathology* 73, 559–561.

Faggian, R., *et al.* (1999). Specific polymerase chain reaction primers for the detection of *Plasmodiophora brassicae* in soil and water. *Phytopathology* **89**, 392–397.

Harrison, J. G., Searle, R. G., and Williams, N. A. (1997). Powdery scab disease of potato: A review. *Plant Pathol.* **46**, 1–25.

Ito, S., Maehara, T. Maruno, E., *et al.* (1999). Development of a PCR-based assay for the detection of *Plasmodiophora brassicae* in soil. *J. Phytopathol.* **147**, 83–88.

Manzanares-Dauleux, M. J., Divaret, I., Baron, F., *et al.* (2001). Assessment of biological and molecular variability between and within field isolates of *Plasmodiophora brassicae. Plant Pathol.* **50**, 165–173.

Woronin, M. (1878). *Plasmodiophora brassicae.* Urheber der Kohlpflanzen-Hernie. *Jahrb. Wiss. Bot.* **11**, 548–574; Eng. Transl. by C. Chupp in *Phytopathological Classics* **4**, (1934).

Diseases Caused by Oomycetes

Oomycetes are members of the kingdom Chromista (= Stramenopila) that have mycelium containing cellulose and glucans but have no cross walls except to separate living (cytoplasmic) parts of hyphae from old parts from which the cytoplasm has been withdrawn. They produce oospores as their resting spores and zoospores or zoosporangia as their asexual spores. The most important plant pathogenic Oomycetes belong to two orders, namely Saprolegniales and Peronosporales. Of the Saprolegniales, only one genus, *Aphanomyces*, has important plant pathogens, one causing root rot diseases of many annual plants, particularly of pea and sugar beet.

The order Peronosporales includes several of the most important genera of plant pathogens known (Fig. 11-7): these are *Pythium* and *Phytophthora*, each consisting of many very important plant pathogenic species, and several genera causing downy mildews. Another genus, *Albugo*, causes the less important white rust on crucifers.

Pythium sp., one of the most common and most important causes of seed rot, seedling damping-off, and root rot of all types of plants, and also of soft rots of fleshy fruits in contact with the soil

Phytophthora sp., one causing late blight of potato and several others causing root rots, fruit rots, and blights of many other annual and perennial plants, and root and stem rots, cankers and diebacks of trees

Bremia, *Peronospora*, *Plasmopara*, and *Pseudoperonospora*, causing the very destructive diseases known as downy mildews of dicotyledonous plants, such as lettuce, tobacco, grape, and cucurbits

Peronoslerospora, *Sclerophthora*, and *Sclerospora*, causing the downy mildew diseases of monocots such as corn, sorghum, and sugarcane

Albugo, causing the common but usually not serious white rust diseases of cruciferous plants (Figs. 11-33C–11-33C-E)

Plant diseases caused by Oomycetes are basically of two types (Fig. 11-8): (1) Diseases that affect plant parts present in the soil or in contact with the soil, e.g., roots, lower stems, tubers, seeds, and fleshy fruits lying on the soil; they are caused by all the species of *Aphanomyces* and *Pythium* and by some species of *Phytophthora*. (2) Diseases that affect only or primarily aboveground plant parts, particularly the leaves, young stems, and fruits. These are caused by some species of *Phytophthora*, by all of the downy mildew, and by *Albugo*.

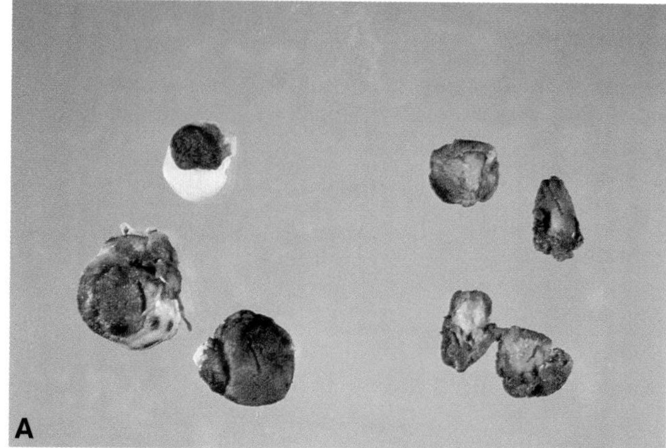

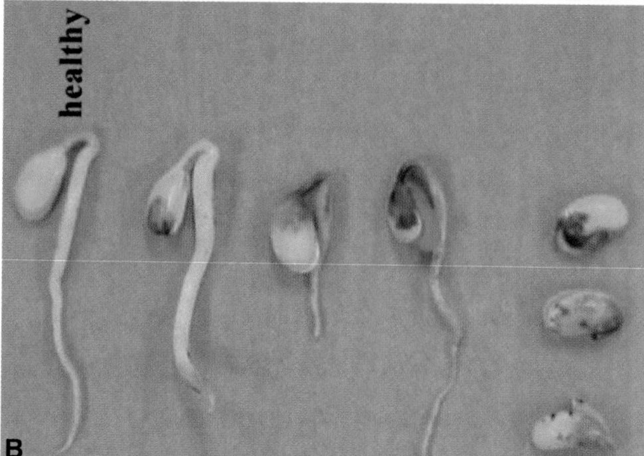

FIGURE 11-14 (A) Pythium seed rot. (B) One healthy bean seedling and several seeds and seedlings infected with *Pythium*. (C) Damping off of cucumber seedlings caused by *Pythium sp*. [Photographs courtesy of (A) R. J. Howard, WCPD, and (B) P. E. Lipps, Ohio State University.]

PYTHIUM SEED ROT, DAMPING-OFF, ROOT ROT, AND SOFT ROT

Damping-off diseases of seedlings occur worldwide in valleys and forest soils, in tropical and temperate climates, and in every greenhouse. The disease affects seeds, seedlings, and roots of all plants. In all cases, however, the greatest damage is done to the seed and seedling roots during germination either before or after emergence. Losses vary considerably with soil moisture, temperature, and other factors. Quite frequently, seedlings in seedbeds are completely destroyed by damping-off or they die soon after they are transplanted. In many instances, poor germination of seeds or poor emergence of seedlings is the result of damping-off infections in the preemergence stage. Older plants are seldom killed when infected with the damping-off pathogen, but they develop root and stem lesions and root rots, their growth may be retarded considerably, and their yields may be reduced drastically. Some species of the damping-off oomycete also attack the fleshy organs of plants, which rot in the field or in storage.

Symptoms

When seeds of susceptible plants are planted in infested soils and are attacked by the damping-off fungi, they fail to germinate, become soft and mushy, and then turn brown, shrivel, and finally disintegrate (Fig. 11-14A). Young seedlings can be attacked before emergence at any point on the plant, from which the infection spreads rapidly, the invaded cells collapse, and the seedling is overrun by the oomycete and dies (preemergence damping-off).

Seedlings that have already emerged are usually attacked at the roots and sometimes in the stems at or below the soil line. The invaded areas become water soaked and discolored and they soon collapse (Figs. 11-14B and 11-14C). The basal part of the seedling stem becomes softer and much thinner than the uninvaded parts above it; as a result, the seedling falls over on the soil. The fungus continues to invade the fallen seedling, which quickly withers and dies (postemergence damping-off). In cereals and turf grasses, the pathogen causes "Pythium blight," i.e., it invades and kills the roots and whole seedlings and even young plants, causing the appearance of numerous empty patches on the lawn or field (Figs. 11-15A–11-15C).

FIGURE 11-15 Pythium root rots and blights. Root rot of *Caladium* (A right), barley seedlings (B left), blight of turfgrass (C), and root rot and wilt of tomato (D) caused by *Pythium*. [Photographs courtesy of (A) R. J. McGovern, (C) T. E. Freeman, and (D) Plant Pathology Department, University of Florida, and (B) L. J. Piening, WCPD.]

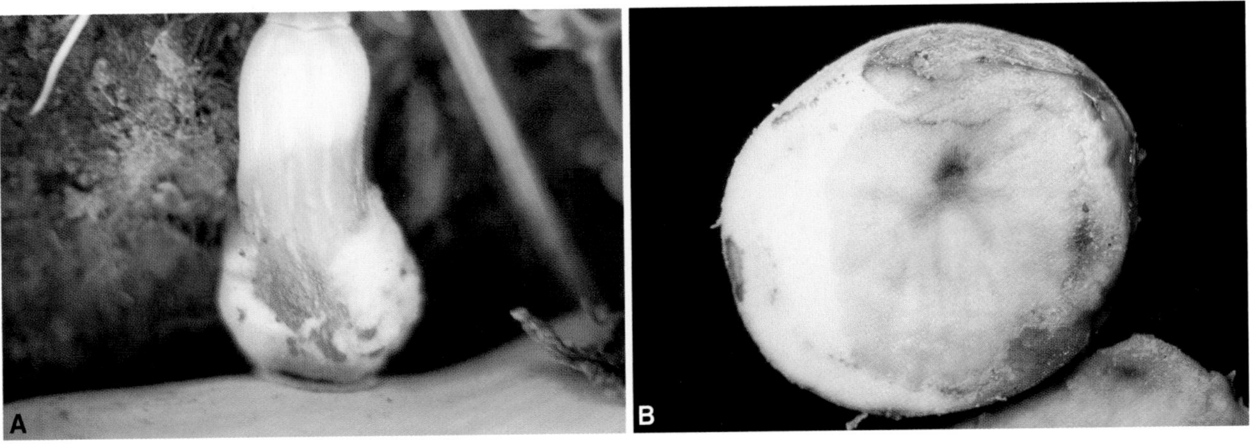

FIGURE 11-16 Soft rots of squash (A) and potato (B) caused by *Pythium*. [Photograph courtesy of (B) D. P. Weingartner, University of Florida.]

In older plants the damping-off oomycete may kill rootlets or induce lesions on the roots and stem. The lesions cause plants to become stunted and sometimes to wither or die (Fig. 11-15D).

Soft, fleshy organs of vegetables in contact with the soil, such as cucurbit fruits, green beans, and potatoes, are sometimes infected by damping-off oomycetes during extended wet periods. Such infections result in a cottony growth on the surface of the fleshy organ, while the interior turns into a soft, watery, rotten mass, called "leak" (Figs. 11-16A and 11-16B).

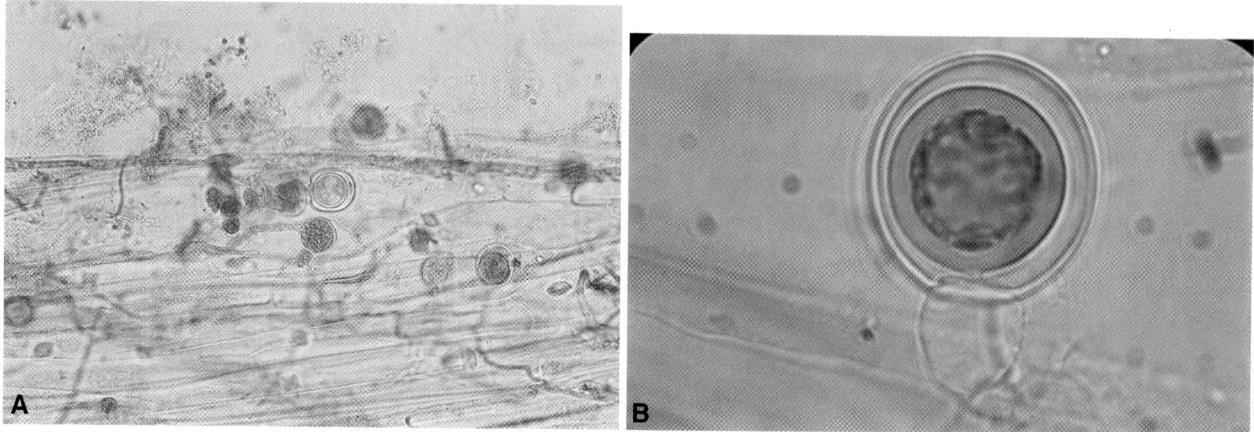

FIGURE 11-17 *Pythium* mycelium and sporangia in infected root tissue (A) and oospore (B) of *Pythium*. (Photographs courtesy of R. E. Cullen, University of Florida.)

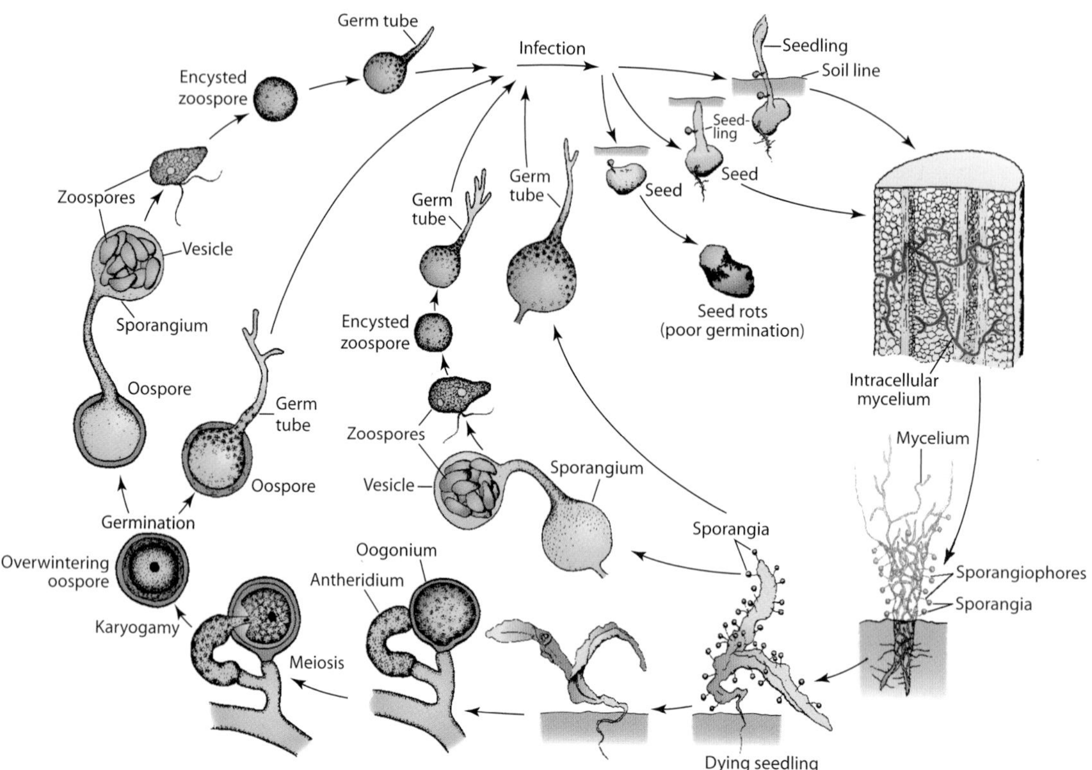

FIGURE 11-18 Disease cycle of damping-off and seed decay caused by *Pythium* sp.

The Pathogen: Pythium *spp*

Several species of *Pythium* cause pre- and postemergence damping-off. Certain other oomycetes and fungi, however, such as *Phytophthora*, *Rhizoctonia*, and *Fusarium*, often cause symptoms quite similar to those described earlier. Several more fungi, and even some bacteria, when carried in or on the seed, also cause damping-off and kill seedlings.

Pythium produces a white, rapidly growing mycelium. The mycelium gives rise to sporangia, which germinate directly by producing one to several germ tubes or by producing a short hypha at the end of which forms a balloon-like secondary sporangium called a vesicle (Figs. 11-17 and 11-18). In the vesicle, 100 or more zoospores are produced, which, when released, swarm about for a few minutes, round off to form a cyst, and then germinate by producing a germ tube. The germ tube usually penetrates the host tissue and starts a new infection, but sometimes it produces another vesicle in which several secondary zoospores are formed, and this may be repeated.

The mycelium also gives rise to spherical oogonia and club-shaped antheridia (Figs. 11-17A and 11-17B). The antheridium produces a fertilization tube, which enters the oogonium; nuclei of the antheridium move through the tube toward the nuclei of the oogonium, unite with them, and form the zygote. The fertilized oogonium produces a thick wall and is then called an oospore (Fig. 11-17B). Oospores are resistant to adverse temperatures and moisture and serve as the survival and resting stage of the fungus. Oospores germinate in a way similar to that described for sporangia. The type of germination of both sporangia and oospores is determined primarily by the temperature; temperatures above 18°C favor germination by germ tubes, whereas temperatures between 10 and 18°C induce germination by means of zoospores.

Pythium species occur in surface waters and soils throughout the world. They live on dead plant and animal materials as saprophytes or as parasites of fibrous roots of plants. The pathogen needs free water for its zoospores to swim and infect. When a wet soil is infested heavily with *Pythium*, any seeds or young seedlings in such a soil may be attacked by the pathogen.

Development of Disease

Spore germ tubes or saprophytic mycelium of *Pythium* coming in contact with seeds or seedling tissues of host plants enter by direct penetration. Pectinolytic enzymes secreted by the oomycete dissolve the pectins that hold the cells together, resulting in maceration of the tissues. The oomycete grows between and through the cells. Proteolytic enzymes break down the protoplasts of invaded cells, and, in some cases, cellulolytic enzymes cause complete collapse and disintegration of the cell walls. As a result, infected seeds and young seedlings are killed and turn into a rotten mass consisting primarily of oomycete and substances such as suberin and lignin, which this pathogen cannot break down. When the invasion of the oomycete is limited to the cortex of the belowground stem of the seedling, the latter may continue to live and grow for a short time until the lesion extends above the soil line. Then the invaded, collapsed tissues cannot support the seedling, which falls over and dies (Figs. 11-14–11-16, and 11-18).

If the infection occurs when the seedling is already well developed and has well-thickened and lignified cells, the advance of the oomycete is stopped at the point of infection, and only small lesions develop. Rootlets can be attacked at any stage of plant growth. The oomycete enters root tips and proliferates, causing a rapid collapse and death of the rootlet. Invasion of older roots is usually limited to the cortex. Relatively young or fleshy roots may be invaded and form lesions several centimeters long.

Pythium can infect fleshy vegetable fruits and other organs in the field, in storage, in transit, and in the market. Infections begin at the point of contact of the fruit with wet soil infested with the oomycete or with other infected fruit. Enzymes secreted by the oomycete macerate the tissue, which becomes soft and watery. An entire cucumber fruit may be invaded within 3 days of inoculation. As the infection progresses, sporangia begin to appear, followed by the production of oospores, inside or outside the host tissues, or both.

The disease and losses caused by *Pythium* infections are more severe when the soil is kept wet for prolonged periods, when the temperature is unfavorable (usually too low) for the host plant, when there is an excess of nitrogen in the soil, and when the same crop is planted in the same field for several consecutive years.

Control

Pythium diseases in the greenhouse can be controlled through the use of soil sterilized or pasteurized by steam or dry heat and through the use of chemically treated seed. Greenhouse benches and containers must also be sterilized or treated with an appropriate chemical solution.

So far, no commercial varieties of plants resistant to *Pythium* are available. Since the mid-1990s, experimental control of *Pythium* seed rot and damping-off has been obtained by treating the seeds with conidia of antagonistic fungi and with certain bacteria, or by incorporating conidia of antagonistic fungi into commercial soilless mixes used in greenhouses and by nursery owners.

Certain cultural practices are sometimes helpful in reducing the amount of infection. Such practices include providing good soil drainage and good air circulation among plants, planting when temperatures are favorable for fast plant growth, avoiding application of excessive amounts of nitrate forms of nitrogen fertilizers, and practicing crop rotation. Some new methods, such as osmopriming, i.e., controlled hydration, of seeds before planting, have appeared promising. For container-grown nursery crops and ornamentals, including composted tree bark as a replacement for most of the peat markedly reduced the root rots caused by *Pythium* and several other root pathogens.

In the field, seed or bulb treatment with one or more effective chemicals is the most important disease preventive measure. Some systemic fungicides, usually in combination with broad-spectrum fungicides, give excellent control of damping-off, seedling blights, and root rots caused by *Pythium* and *Phytophthora*; they can be applied as soil or seed treatment.

Seed treatment is sometimes followed by spraying of seedlings with the same or different effective fungicides than those used for seed treatment. This is especially important when the soil is infested heavily with *Pythium* or when the soil stays wet for prolonged periods during the early stages of plant growth. Cucumber seed treatment with *Pseudomonas putida* bacteria or with the mycoparasite *Verticillium lecanii* results in the systemic production of phytoalexins and other host defense reactions that protect seedlings from attack by *Pythium*. Similarly, experimental treatment of Norway spruce seedlings with methyl jasmonate induced accumulation of free salicylic acid, chitinase, and other defense responses that protected up to 75% of the seedlings from infection by *Pythium*.

Selected References

Anonymous (1974). Symposium on the genus *Pythium*. *Proc. Am. Phytopathol. Soc.* **1**, 200–223.

Buczacki, S. T., ed. (1983). "Zoosporic Plant Pathogens: A Modern Perspective." Academic Press, London.

Dick, M. W. (1990). "Keys to *Pythium*." Department of Botany, School of Plant Science, University of Reading, Reading, UK.

Hendrix, F. F., Jr., and Campbell, W. A. (1973). Pythiums as plant pathogens. *Annu. Rev. Phytopathol.* **11**, 77–98.

Martin, F. N. (1992). *Pythium*. *In* "Methods for Research on Soilborne Phytopathogenic Fungi" (L. L. Singleton, J. D. Mihail, and C. M. Rush, eds.), pp. 39–49. APS Press, St. Paul, MN.

Middleton, J. T. (1943). The taxonomy, host range and geographic distribution of the genus *Pythium*. *Torrey Bot. Club. Mem.* **20**, 1–171.

Rey, P., Benhamou, N., and Tirilly, Y. (1998). Ultrastructural and cytochemical investigation of asymptomatic infection by *Pythium* spp. *Phytopathology* **88**, 234–244.

Waterhouse, G. M. (1968). The genus *Pythium*. *Mycol. Pap.* **110**, 1–71.

Phytophthora Diseases

The name *Phytophthora* means plant destroyer, and with good reason. Species of *Phytophthora* cause a variety of devastating diseases on many different types of plants ranging from seedlings of annual vegetables or ornamentals to fully developed fruit and forest trees. Most species cause root rots, damping-off of seedlings, and rots of lower stems, tubers, and corms similar to those caused by *Pythium* spp. Others cause rots of buds or fruits, and some cause blights of the foliage, young twigs, and fruit. Some species attack only one or two species of host plants, but others may cause similar or different symptoms on many different kinds of host plants. The best known species is *Phytophthora infestans*, the cause of late blight of potatoes and tomatoes, but several other species also cause extremely destructive diseases on their hosts. *Phytophthora cactorum*, *P. cambivora*, *P. cinnamoni*, *P. citrophthora*, *P. fragariae*, *P. palmivora*, and *P. syringae* cause primarily root and lower stem rots, but also some cankers, twig blights, and fruit rots of woody ornamentals and of fruit and forest trees as well as of vegetables and other herbaceous plants. Several other species, such as *P. capsici*, *P. cryptogea*, *P. megasperma*, and *P. parasitica*, cause root, stem, and fruit rots of many vegetables, ornamentals, and field crops, but also of some woody plants.

Phytophthora Root and Stem Rots

Most species of *Phytophthora* cause root and lower stem rots on numerous species of plants (Figs. 11-19 to 11-24). The losses caused by such root and stem rots are great, especially on trees and shrubs. In many such diseases, however, the pathogen often goes undetected or unidentified. *Phytophthora*-infected plants at first show symptoms of drought and starvation, but then quickly become weakened and susceptible to attack by other pathogens or conditions that are mistakenly taken as the causes of the death of the plants.

Phytophthora root and stem rots cause damage to their hosts in nearly every part of the world where the soil becomes too wet for the good growth of susceptible plants and the temperature remains fairly low, i.e., between 15 and 23°C.

Annual plants and young seedlings of trees may be killed by the disease within a few days, weeks, or months (Fig. 11-19). In some cases the oomycete also attacks and causes partial or complete rot of the fruit, as e.g., in pepper, cucurbits (Figs. 11-20A and 11-20B), tomato (Fig. 20C), strawberry, citrus, and cacao (Fig. 11-22).

In older trees the killing of roots may be slow or rapid, depending on the amount of fungus present in the soil and the prevailing environmental conditions. As a result, older trees show sparse foliage, shorter, cupped, and yellow leaves, and dieback of twigs and branches. In some diseases, such as the collar rot of apple trees (Fig. 11-21A), foot rot of citrus trees, and root and crown rot of peach (Fig. 11-21B) and cherry trees, the oomycete invades and kills the bark of the lower stem. The oomycete also kills palm trees (Figs. 11-21C and 11-21D) by killing and causing bud rot of the only growing point at the top of palm trees. The oomycete invades and kills the bark of the lower stem of innumerable annual plants, shrubs, and trees (Figs. 11-21, 11-23, and 11-24). Infected trees increase very little in height and diameter and usually die within 3 to 10 years after infection. Fewer and smaller fruit and seeds are produced each succeeding year.

FIGURE 11-19 *Phytophthora* symptoms: Killed stem of soybean plant (A), pine seedlings in nursery killed by *Phytophthora* (B), and close-up (C) and overview of pepper plants in the field (D) killed by *Phytophthora*. [Photographs courtesy of (A) W. L. Seaman, WCPD, (B) E. L. Barnard, Florida Department of Agriculture, Division of Forestry, and (C and D) R. J. McGovern, University of Florida.]

FIGURE 11-20 *Phytophthora* symptoms on fleshy organs: Rotting of watermelon (A and B) and tomato (C) by *P. capsici* and rotting of potato (D) by *P. erythroseptica*. [Photographs courtesy of (A) B.D. Bruton, (B and C) R.J. McGovern, and (D) D.P. Weingartner, University of Florida.]

FIGURE 11-21 *Phytophthora* symptoms on trees: (A) Foot rot of citrus, (B) partial necrosis of trunk in peach tree, and bud rot of single (C) and a group (D) of palm trees. [Photographs courtesy of (A, C, and D) R.J. McGovern, University of Florida.]

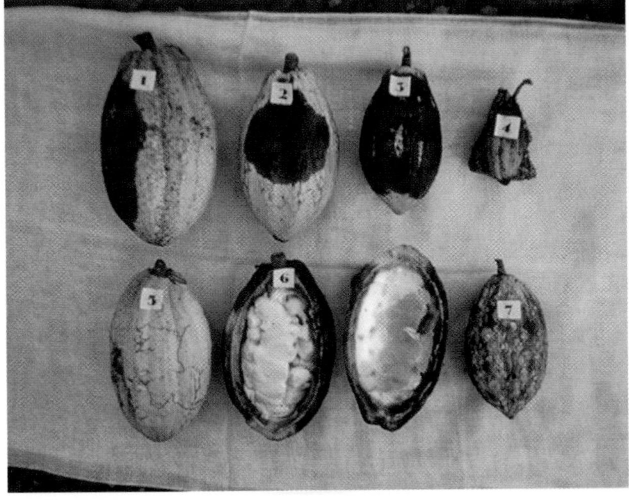

FIGURE 11-22 Phytophthora black pod of cacao. (Photograph courtesy of H.D. Thurston, Cornell University.)

BOX 17 Phytophthoras Declare War on Cultivated Plants and on Native Tree Species

The importance of *P. infestans* as destroyer of potatoes and tomatoes is recognized widely and has been reemphasized in the last decade with the appearance and rapid distribution worldwide of new more virulent strains and compatible sexual types. The latter not only allow the production and overseasoning of the oomycete in the form of hardy oospores, they also facilitate and accelerate the appearance of new more virulent strains. Furthermore, additional crops, such as peppers and cucurbits, have also been added to the list of annual plants that have become preferred hosts of Phytophthora blights. These crops are now extremely susceptible to the blight caused by *Phytophthora capsici*, a species that also appears to have produced much more virulent strains in the last decade.

Although various *Phytophthora* species have long been known to cause significant losses in many cultivated annual and perennial crops, several Phytophthoras have proven devastating against native tree species growing in different parts of the world. In the 1970s, a severe widespread epidemic of *Phytophthora cinnamomi* in the jarrah eucalyptus forests of western Australia destroyed more than 20% of the eucalyptus trees, while at the same time 60 to 70% of some important shrubs in the same forests were also infected by the oomycete. In the last several years, *P. cinnamomi* has also been reported to kill oak trees in southern and central Europe, in Argentina and Chile, in the state of Pennsylvania, in Canada, and in Alaska. More recently it has been found to be the principal cause of mortality of at least three species of native forest oaks in the Colima state of Mexico. Although present in the Pacific Northwest for many decades, in recent years, the species *Phytophthora lateralis* has been killing large numbers of Port-Orford-Cedar, also known as Lawson's cypress trees, and, to a smaller extent, Pacific yew trees. Finally, at the turn of the millennium, a new *Phytophthora* species, *P. ramorum*, was shown to be the cause of the "sudden oak death" disease that has been killing oak trees at a rapid rate (Figs. 11-23 and 11-24) in California and Oregon.

Phytophthora ramorum was known to cause diebacks of rhododendrons in Europe. In the late 1990s, however, it was found to cause bark cankers resulting in "sudden oak death" in four species of oaks in California and southern Oregon. *P. ramorum* also causes leaf and branch infections on other plants belonging to several other families, but their effects are not nearly as severe as those on oaks. Some species of infected oaks, such as the tanoaks (*Lithocarpus* spp.), exhibit drooping or wilting (Fig. 11-23D) as the first symptom, followed by the formation of bleeding cankers on infected trunks (Fig. 11-23E) that produce a reddish brown to black viscous sap. On true oaks (*Quercus* spp.), the first symptoms are bleeding cankers that form usually from above the soil line to about three meters high of the trunk and main branches. In advanced infections, especially in tanoaks, bleeding may occur up to 20 meters high of the trunk and branches. In tanoaks, the pathogen often also causes leaf spots and cankers on small twigs. After bleeding begins, infected oak trees begin to show subtle changes in the color of their foliage (Figs. 11-23A–11-23C). As the disease advances, the foliage color changes rapidly from healthy green to chlorotic yellow and finally reddish brown (Fig. 11-24). Infected trees may be isolated or in groups. They generally die within the first or second year following infection, although some trees survive for several years after they have developed bleeding cankers. In infected trees, leaves may cling to the tree for up to a year after the tree has died. The pathogen produces spores on the leaf spots it causes on several of its hosts but not on the cankers on oaks. The pathogen has also been found in the soil, in rainwater, and in downed trunks and branches. It is not known how the pathogen spreads to trees and how it enters the trunk and branches, although the involvement of insects is suspected. So far, no cure is available for the disease and the main steps recommended are to reduce water and nutrient deficiency stress on the trees.

There is no particularly good explanation of why so many Phytophthoras have become apparently reenergized against their former hosts or why they have been recognized to attack severely new hosts that apparently had been free from attack until recently. The cases of new *Phytophthora* strains with increased virulence in *P. infestans* and *P. capsici* are almost as baffling as the increased aggressiveness of *P. cinnamomi* on oaks in Mexico and the appearance of an invigorated species, *P. lateralis*, in Oregon and California. Most serious of all is the detection of a new species, *P. ramorum*, in California and Oregon where it causes an apparently new and different canker disease and severe mortality of oak trees. When the losses to *Phytophthora* of native forest trees are added to those of cultivated fruit trees such as apple, peach, citrus, and cacao and to those of annual losses to *Phytophthora* of all types of seedling plants and of vegetables and field crops, *Phytophthora* qualifies as one of the worst plant-destroying pathogens of all time.

FIGURE 11-23 Sudden oak death disease caused by *P. ramorum*. (A) Drooping and wilting of new growth. (B–D) Successive color changes in foliage of infected tree between December and June. (D) Bleeding of canker on trunk of infected tree. (F) Cankers and necrosis of trunk of infected oak tree. (Photographs courtesy of P. Svihra, University of California Cooperative Extension, Marin County.)

Continued

FIGURE 11-24 Panoramic view of oak trees killed by *P. ramorum*. (Photograph courtesy of P. Svihra, University of California Cooperative Extension, Marin County.)

On all hosts affected by Phytophthora root rot, many of the small roots are dead, while larger roots show necrotic brown lesions. On young plants or on older succulent plants the whole root system may decay, followed by a more or less rapid death of the plant. In strawberries, as in the other plants, most of the small rootlets rot away, while the larger ones turn brown beginning at the tips. In addition, in late spring, affected larger strawberry roots show a red-colored core or stele, a symptom diagnostic of the strawberry red stele root rot caused by *Phytophthora fragariae*.

In many plants, the oomycete attacks the plant at or near the soil line where it causes a water soaking and darkening of the bark on the trunk. The infected area enlarges, and if the plant is small and succulent, it may encircle the entire stem, after which the lower leaves drop and eventually the whole plant wilts (Figs. 11-20 to 11-24). On larger plants and on trees, the infected, darkened area may be on one side of the stem and becomes a depressed canker below the level of healthy bark. In early stages, the diseased bark is firm and intact, but later it becomes shrunken and cracked. The collar rot canker may spread up into the trunk (Fig. 11-21B), and sometimes the branches, or down into the root system. Invasion of the root usually begins at the crown area or at ground level. As the cankers spread and enlarge, they may girdle the trunk, limbs, or roots, causing the plant or tree to grow poorly, produce fewer and smaller fruit, show sparse foliage, and suffer dieback of twigs, finally killing the whole plant.

The various *Phytophthora* species that cause root and stem rots survive cold winters or hot, dry summers as oospores, chlamydospores, or mycelium in infected roots or stems (Figs. 2-3B and 11-25). These structures may also survive in the soil. In the spring, the oospores and chlamydospores germinate by means of zoospores, whereas the mycelium grows further and produces zoosporangia that release zoospores. The zoospores swim around in the soil water and infect roots of susceptible hosts with which they come in contact. More mycelium and zoospores are produced during wet, cool weather and spread the disease to more plants.

The control of *Phytophthora* root rots depends on planting susceptible crops in soils free of the pathogen or in soils that are light and drain well and quickly. All planting stock should be free of infection and, when available, only resistant varieties should be planted. For plants in pots, greenhouses, or seedbeds, the soil and containers should be sterilized with steam before planting.

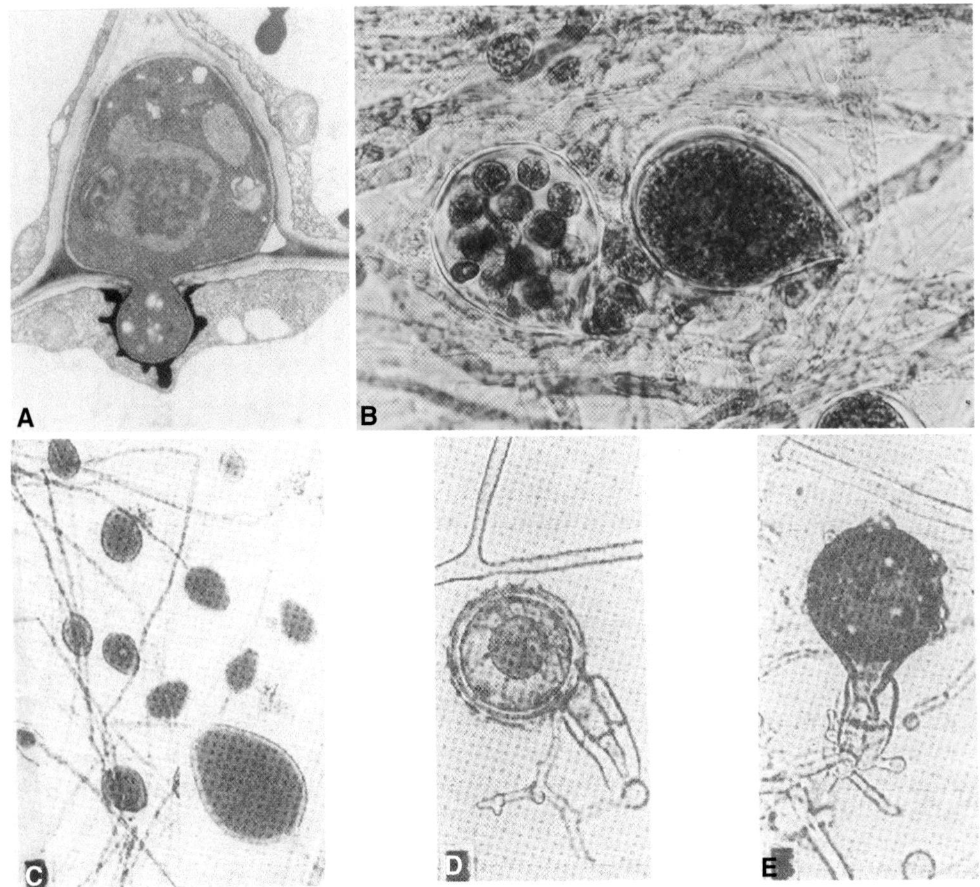

FIGURE 11-25 (A) An intercellular hypha of a *Phytophthora* sp. that has formed a haustorium in a cortical cell of a root. (B) A sporangium of *P. capsici* containing zoospores. Sporangia (C), oospore with antheridium (D), and oospore (E) of *P. cambivora*. (C–E magnified 400x.) [Photographs courtesy of (A) Mims and Enkerli, from *Can. J. Bot.* (1997). 75, 1493–1508, (B) R.J. McGovern, University of Florida, and (C–E) S. M. Mircetich, from *Phytopathology* **66**, 549–558.]

Excellent control of Phytophthora root and lower stem rots has been obtained since the mid-1990s through the use of several systemic fungicides that are applied as seed treatments, soil treatments, transplant dips, and sprays, or with drip and overhead irrigation water. Some protection of trees can be obtained by injections of selected fungicides into their trunks. Also, application of a solution of some fungicides in the soil around trees seems to inhibit the growth and activity of the oomycete greatly. Resistant varieties should always be preferred, especially for heavy, poorly drained soils. With stone fruit and other trees, resistant rootstocks and sometimes interstocks offer the most effective means of controlling foot rot or collar rot. For some crops, such as strawberries, Phytophthora root rot has also been controlled effectively through soil fumigation.

In some cases, Phytophthora root rots have been controlled by planting seedlings in suppressive soil that contains either microorganisms antagonistic to *Phytophthora* or inorganic substances toxic to the oomycete. Several fungi and bacteria have been shown to parasitize *Phytophthora* oospores or to be antagonistic to *Phytophthora*, but none of them has been effective in controlling *Phytophthora* so far. Since the mid-1990s, however, composted tree bark mixed with soil or soilless mixes used in the production of container-grown plants, in greenhouse beds, and in field experiments has reduced plant infections by *Phytophthora* significantly.

LATE BLIGHT OF POTATOES

The late blight disease of potatoes is the most devastating disease of potatoes in the world. It is most destructive, however, in areas with frequent cool, moist weather. Zones of high late blight severity include the northern United States and the east coast of Canada, western Europe, central and southern China, southeastern Brazil, and the tropical highlands. Late blight is also very destructive to tomatoes and some other members of the family Solanaceae.

Late blight may kill the foliage and stems of potato and tomato plants at any time during the growing season. It also attacks potato tubers and tomato fruits in the field, which rot either in the field or while in storage. Late blight may cause total destruction of all plants in a field within a week or two when weather is cool and wet. Even when losses in the field are small, potatoes may become infected during harvest and may rot in storage.

The historical aspects of late blight of potatoes in relation to the Irish famine and the establishment of *Phytophthora infestans* as the cause of late blight are presented on page 19–21.

Symptoms

Symptoms appear at first as water-soaked spots, usually at the edges of the lower leaves. In moist weather the spots enlarge rapidly and form brown, blighted areas with indefinite borders. A zone of white, downy mildewy growth 3 to 5 millimeters wide appears at the border of the lesions on the undersides of the leaves (Figs. 11-26A and 11-26B). Soon entire leaves are infected, die, and become limp. Under continuously wet conditions, all tender, aboveground parts of the plants blight and rot away (Figs. 11-26C and 11-26D), giving off a characteristic odor. Entire potato plants and plants in entire fields may become blighted and die in a few days or a few weeks (Fig. 11-26D). In dry weather the activities of the pathogen are slowed or stopped. Existing lesions stop enlarging, turn black, curl, and wither, and no oomycete appears on the underside of the leaves. When the weather becomes moist again the oomycete resumes its activities and the disease once again develops rapidly.

Affected tubers at first show purplish or brownish blotches consisting of water-soaked, dark, somewhat reddish brown tissue that extends 5 to 15 millimeters into the flesh of the tuber (Figs. 11-27A and 11-27B). Later the affected areas become firm and dry and somewhat sunken. Such lesions may be small or may involve almost the entire surface of the tuber without spreading deeper into the tuber interior. The rot, however, continues to develop after the tubers are harvested (Figs. 11-27A and 11-27B). Infected tubers may be subsequently covered with sporangiophores and spores of the pathogen (Figs. 11-27B and 11-27C) or become invaded by secondary fungi and bacteria, causing soft rots and giving the rotting potatoes a putrid, offensive odor.

Tomato leaves, stems, and fruit are also attacked. Entire tomato fields may be destroyed. Fruit may rot rapidly in the field or in storage (Figs. 11-28A–11-28C).

The Pathogen: Phytophthora Infestans

The mycelium produces branched sporangiophores that produce lemon-shaped sporangia at their tips (Figs. 11-27C and 11-29). At the places where sporangia are produced, sporangiophores form swellings that are characteristic for this oomycete. Sporangia germinate almost entirely by releasing three to eight zoospores at temperatures up to 12 or 15°C, whereas above 15°C sporangia may germinate directly by producing a germ tube.

The oomycete requires two mating types for sexual reproduction. Until the late 1980s, only one mating type was present in countries outside Mexico. Since then, however, both mating types have become widely distributed in most countries and, as a result, new strains of the pathogen have appeared. Some of the new strains are much more aggressive than the old ones and quickly replace them. When the two mating types grow adjacently, the female hypha grows through the young antheridium (= male reproductive cell) and develops into a globose oogonium (= female reproductive cell) above the antheridium. The antheridium then fertilizes the oogonium, which develops into a thick-walled and hardy oospore. Oospores germinate by means of a germ tube that produces a sporangium, although at times the germ tube grows directly into the mycelium.

Development of Disease

The pathogen strains that prevailed until the 1980s belonged to mating type A1 and reproduced in the absence of its compatible mating type A2, i.e., asexually. Therefore, they did not produce oospores and overwintered only as mycelium in infected potato tubers. Spread of the compatible mating type A2 from Mexico to the rest of the world has made possible the sexual reproduction of the pathogen, which results in the production of oospores in infected aboveground and belowground potato and tomato tissues. Usually, the more susceptible the potato variety the more oospores the pathogen produces per unit leaf area. Oospores may survive in the soil for 3–4 years. Such oospores not only can overwinter in the soil, they also make possible the production of new more virulent strains through genetic recombination of pathogenic characteristics of the mating strains.

During infection, a number of potato defense-related genes are induced (activated) by the pathogen, including genes coding for β-1,3-glucanase, known to be induced in many host–pathogen systems, genes coding

FIGURE 11-26 Stages in potato late blight caused by *Phytophthora infestans*. (A) Single leaf lesion with sporangiophores and sporangia. (B) Blight lesions on many leaflets. (C) Necrosis of stem. (D) Death and collapse of shoots and stem of a potato plant. (E) Death and collapse of blighted plants in the field. [Photographs courtesy of (A–C and E) D.P. Weingarten, University of Florida, and (D) K. Mohan, University of Idaho.]

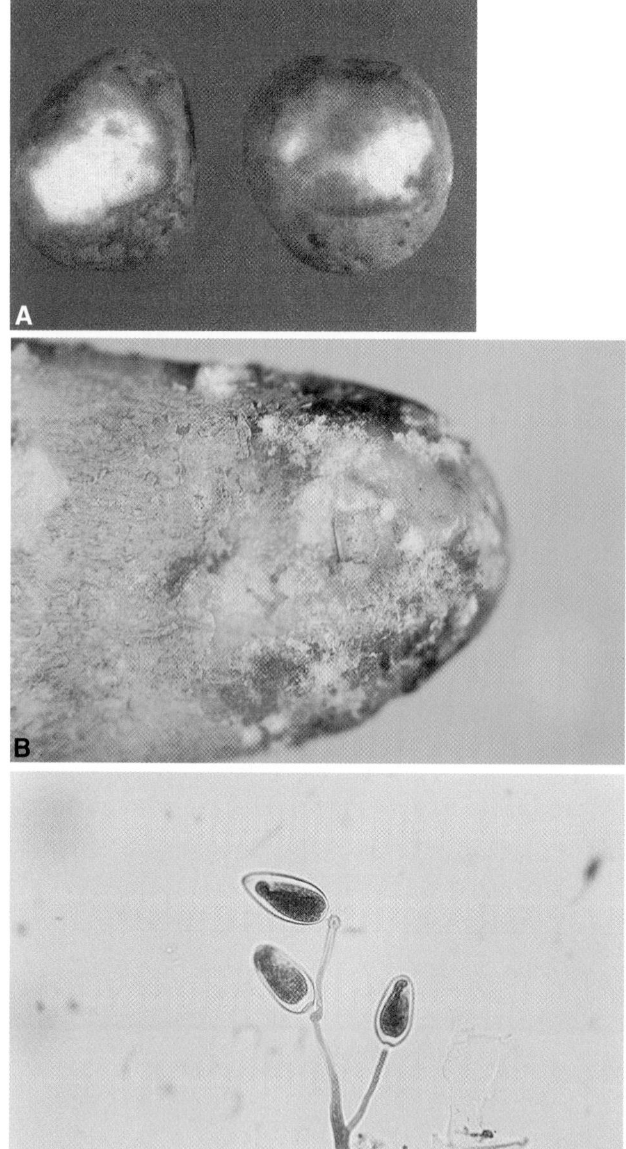

FIGURE 11-27 (A) Potato tubers rotten by the late blight disease as they appear in cross section. (B) Potato tuber infected with and producing sporangia of *Phytophthora infestans*. (C) Close-up of sporangiophore and three sporangia of the pathogen. [Photographs courtesy of (A) R. Rowe, Ohio State University, and (B and C) D.P. Weingartner, University of Florida.]

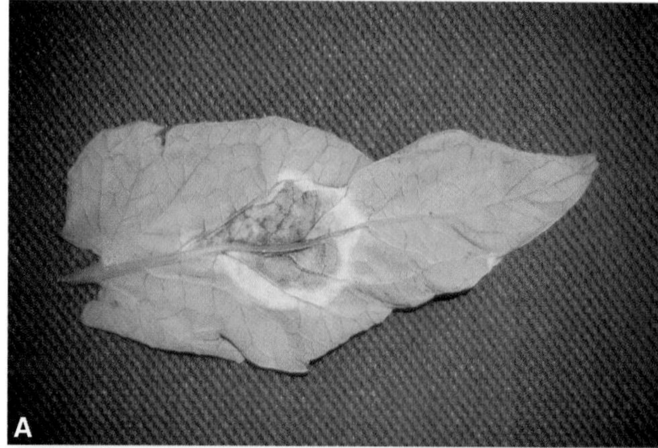

FIGURE 11-28 Symptoms of late blight on tomato leaf (A), fruit (B), and entire field of tomatoes (C) caused by *Phytophthora infestans*. [Photographs courtesy of (A) R.J. McGovern, University of Florida, (B) K. Mohan, University of Idaho, and (C), R. Jaime-Garcia, Cornell University.]

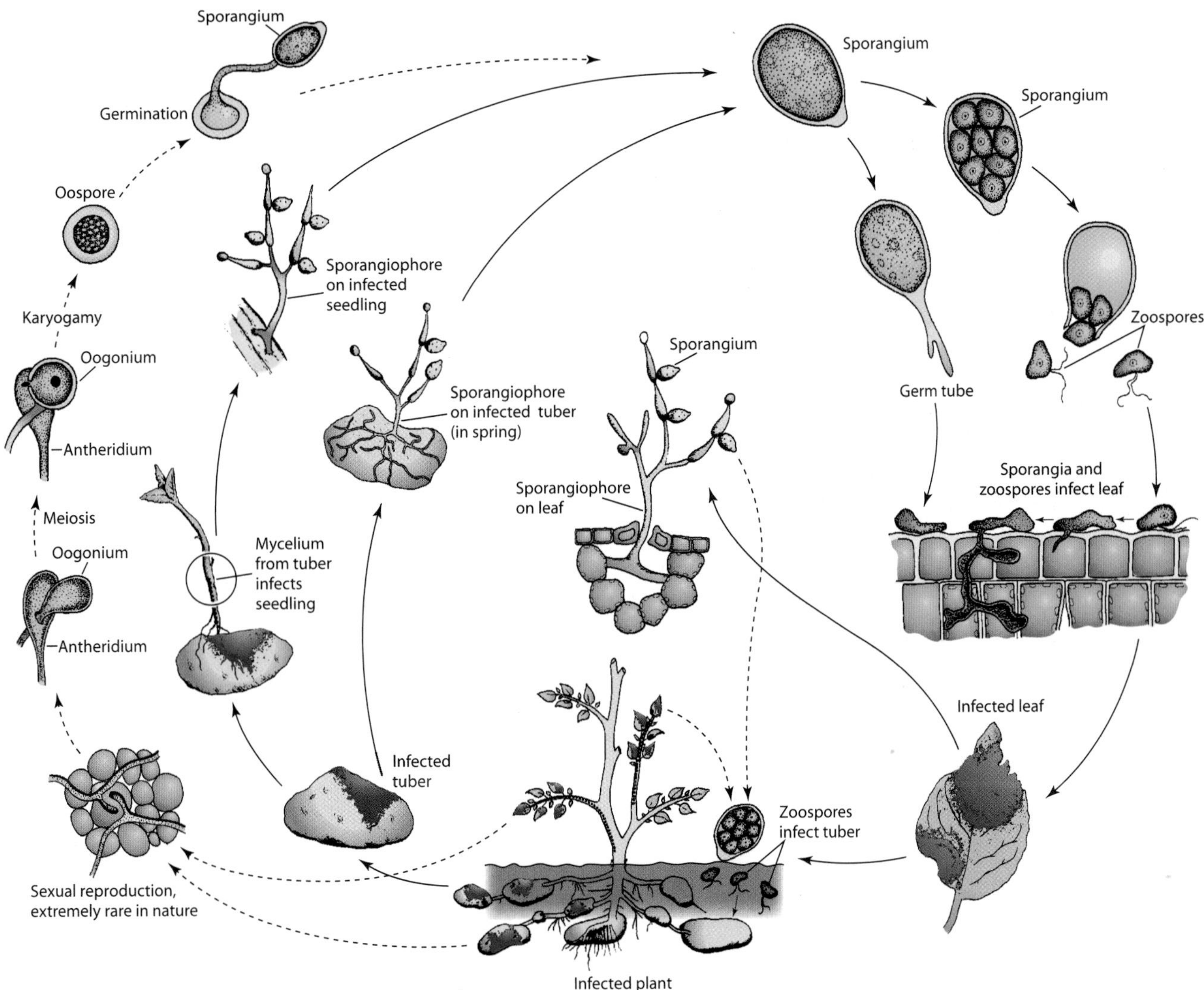

FIGURE 11-29 Disease cycle of late blight of potato and tomato caused by *Phytophthora infestans*.

for enzymes involved in detoxification, and several other types of genes involved in plant defense against pathogens.

The mycelium from infected tubers (Fig. 11-27) or from germinating oospores and zoospores spreads into shoots produced from infected or healthy tubers, causing discoloration and collapse of the cells (Fig. 11-26C). When the mycelium reaches the aerial parts of plants, it produces sporangiophores, which emerge through the stomata of the stems and leaves and produce sporangia (Figs. 11-27 and 11-29). The sporangia, when ripe, become detached and are carried off by the wind or are dispersed by rain; if they land on wet potato leaves or stems, they germinate and cause new infections. The germ tube penetrates directly or enters through a stoma, and the mycelium grows profusely between the cells, sending long, curled haustoria into the cells. Older infected cells die while the mycelium continues to spread into fresh tissue. A few days after infection, new sporangiophores emerge from the stomata of the leaves and produce numerous sporangia, which are spread by the wind and infect new plants. In cool, moist weather, new sporangia may form within four days from infection; thus, a large number of asexual generations and new infections may be produced in one growing season. Wherever the two mating types A1 and A2 are present together in the same plant tissue, fertilization may take place and oospores may be produced. The frequency of oospore formation and their role in the development of the disease within a growing season are not yet known. In any case, as the disease develops, established lesions enlarge and new ones develop, often killing the foliage and reducing potato tuber yields.

The second phase of the disease, the infection of tubers, varies between potato varieties and pathogen isolates. It begins in the field when, during wet weather, sporangia are washed down from the leaves and are carried into the soil. Emerging zoospores germinate and penetrate the tubers through lenticels or through wounds. In the tuber the mycelium grows mostly between the cells and sends haustoria into the cells. Tubers contaminated at harvest with living sporangia present on the soil or on diseased foliage may also become infected. Most of the blighted tubers rot in the ground or during storage.

The development of late blight epidemics depends greatly on the prevailing humidity and temperature during the different stages of the life cycle of the oomycete. The oomycete grows and sporulates most abundantly at a relative humidity near 100% and at temperatures between 15 and 25°C. Temperatures above 30°C slow or stop the growth of the oomycete in the field but do not kill it, and the oomycete can start to sporulate again when the temperature becomes favorable, provided, of course, that the relative humidity is sufficiently high.

Control

Late blight of potatoes can be controlled successfully by a combination of sanitary measures, resistant varieties, and well-timed chemical sprays. Only disease-free potatoes should be used for seed. Potato dumps or cull piles should be burned before planting time in the spring or sprayed with strong herbicides to kill all sprouts or green growth. All volunteer potato plants in the area, whether in the potato field or in other fields, should be destroyed, as any volunteer potato plant can be a source of late blight infection. The recent introduction of the A2 mating type and the potential for mating and production of hardy oospores that can survive the winter in the soil may drastically change our ability to control late blight by the means just described.

Only the most resistant potato varieties available should be planted. Unfortunately, most popular commercial potato varieties are more or less susceptible to late blight. The blight oomycete comprises a number of races, which differ from one another in the potato varieties that they can attack. Several potato varieties resist one or more races of the late blight oomycete. Some of them are resistant to vine infection but not to tuber infection. New varieties, derived from crosses with wild potato species, have one or more genes for resistance (R) to late blight and have withstood attack by all known races of the oomycete for a short while; however, they were attacked by other races not previously distinguished or perhaps not previously existent. Many varieties possess so-called field resistance, which is a partial resistance of varying degrees but is effective against all races of the blight oomycete. However, it is not sufficient to rely on varietal resistance to control late blight, as, in favorable weather, late blight can severely affect these varieties unless they are sprayed with a good protective fungicide. Even resistant varieties should be sprayed regularly with fungicides to eliminate, as much as possible, the possibility of becoming suddenly attacked by races of the oomycete to which they are not resistant. However, it is always advisable to use resistant varieties, even when sprays with fungicides are considered the main control strategy, because resistant varieties delay the onset of the disease or reduce its rate of development so that fewer sprays on a resistant variety may be needed to obtain a satisfactory level of control of the disease (see Figs. 8-20 and 9-28). Various computerized light forecasting systems (e.g. Blightcast) have been developed and are used.

Several broad-spectrum and systemic fungicides are used for late blight control. The new strains of the oomycete produced as recombinants of fertilization of the two mating types are resistant to some of the systemics (metalaxyl) and, therefore, sprays with such materials are ineffective against such strains. Protective spraying of foliage usually affects a considerable reduction in tuber infection. However, when partially blighted leaves and stems are surviving at harvest time, it is necessary to remove the aboveground parts of potato plants or destroy them by chemical sprays (herbicides) or mechanical means to prevent the tubers from becoming infected. Experimental but not yet practical control of the disease has been obtained by the pretreatment of tomato plants with the chemical DL-3-amino-butyric acid or preinoculation with *tobacco necrosis virus*, both of which induce systemic-acquired resistance (SAR) in the tomatoes, protecting them from late blight infection. Haustoria formation and growth of hyphae in SAR-induced leaves against *P. infestans* appear inhibited, different, and damaged. Certain pathogenesis-related proteins accumulate in the leaves of treated plants and only in plant wall papillae and in the cell walls of the oomycete pathogen. Whether these changes play a significant role in resistance to the disease is not clear.

Selected References

Andrivon, D. (1996). The origin of *Phytophthora infestans* populations present in Europe in the 1849s: A critical review of historical and scientific evidence. *Plant Pathol.* 45, 1027–1035.

Bain, H. F., and Demaree, J. B. (1945). Red stele root disease of the strawberry caused by *Phytophthora fragariae*. *J. Agric. Res.* 70, 11–30.

Baines, R. C. (1939). Phytophthora trunk canker or collar rot of apple trees. *J. Agric. Res.* 59, 159–184.

Berg, A. (1926). Tomato late blight and its relation to late blight of potato. *Bull. W. Va. Agric. Exp. Stn.* **205**, 1–31.
Brasier, C. M. (1992). Evolutionary biology of *Phytophthora*. *Annu. Rev. Phytopathol.* **30**, 153–200.
Cox, A. E., and Large, E. C. (1960). Potato blight epidemics throughout the world. *U.S. Dep. Agric. Agric. Handb.* **174**, 1–230.
Erwin, D. C., Bartnicki-Garcia, S., and Tsao, P. H., eds. (1983). "*Phytophthora*: Its Biology, Taxonomy, Ecology, and Pathology." APS Press, St. Paul, MN.
Fry, W. E., and Goodwin, S. B. (1997). Re-emergence of potato and tomato late blight in the United States. *Plant Dis.* **81**, 1349–1357.
Fry, W. E., *et al.* (1992). Population genetics and intercontinental migrations of *Phytophthora infestans*. *Annu. Rev. Phytopathol.* **30**, 107–129.
Fry, W. E., *et al.* (1993). Historical and recent migrations of *Phytophthora infestans*: Chronology, pathways, and implications. *Plant Dis.* **77**, 653–661.
Gisi, U., and Cohen, Y. (1996). Resistance to phenylamide fungicides: A case study with *Phytophthora infestans* involving mating type and race structure. *Annu. Rev. Phytopathol.* **34**, 549–572.
Hansen, E. M., *et al.* (2000). Managing Port-orford-cedar and the introduced pathogen *Phytophthora lateralis*. *Plant Dis.* **84**, 4–14.
Horner, I. J., and Wilcox, W. F. (1996). Spatial distribution of *Phytophthora cactorum* in New York apple orchard soils. *Phytopathology* **86**, 1122–1132.
Hwang, B. K., and Kim, C. H. (1995). Phytophthora blight of pepper and its control in Korea. *Plant Dis.* **79**, 221–227.
Ingram, D. S., and Williams, P. H., eds. (1991). "*Phytophthora infestans*: The Cause of Late Blight of Potato." *Adv. Plant Pathol.* 7. Academic Press, San Diego.
Ko, W. H. (1982). Biological control of Phytophthora root rot of papaya with virgin soil. *Plant Dis.* **66**, 446–448.
Krause, R. A., Massie, L. B., and Hyre, R. A. (1975). Blitecast: A computerized forecast of potato blight. *Plant Dis. Rep.* **59**, 95–98.
Lebreton, L., Lucas, J.-M., and Andrivon, D. (1999). Aggressiveness and competitive fitness of *Phytophthora infestans* isolates collected from potato and tomato in France. *Phytopathology* **89**, 679–686.
Levin, A., *et al.* (2001). Oospore formation by *Phytophthora infestans* in potato tubers. *Phytopathology* **91**, 579–585.
Madden, L. V., *et al.* (1991). Epidemiology and control of leather rot of strawberries. *Plant Dis.* **75**, 439–446.
Man in 't Veld, W. A., *et al.* (1998). Natural hybrids of *Phytophthora nicotianae* and *Phytophthora cactorum* demonstrated by isozyme analysis and random amplified polymorphic DNA. *Phytopathology* **88**, 922–929.
Newhook, F. J., and Podger, F. D. (1972). The role of *Phytophthora cinnamoni* in Australian and New Zealand forests. *Annu. Rev. Phytopathol.* **10**, 229–326.
Ristaino, J. B., and Gumpertz, M. L. (2000). New frontiers in the study of dispersal and spatial analysis of epidemics caused by species in the genus *Phytophthora*. *Annu. Rev. Phytopathol.* **38**, 541–576.
Ristaino, J. B., and Johnston, S. A. (1999). Ecologically based approaches to management of *Phytophthora* blight on bell pepper. *Plant Dis.* **83**, 1080–1089.
Rizzo, D. M., *et al.* (2001). A new Phytophthora canker disease as the probable cause of sudden oak death in California. *Phytopathology* **91**, No. 6 (Supplement), S76.
Stover, A. J., *et al.* (2002). Diagnosis and monitoring of sudden oak death. Univ. of California, Pest Alert #6.
Tainter, F. H., *et al.* (2000). *Phytophthora cinnamomi* as a cause of oak mortality in the state of Colima, Mexico. *Plant Dis.* **84**, 394–398.
Waterhouse, G. M. (1970). "The genus *Phytophthora*," 2nd Ed. Commonw. Mycol. Inst. Misc. Publ. 122.

Downy Mildews

Downy mildews are primarily foliage blights. They attack and spread rapidly in young, tender green leaf, twig, and fruit tissues. They develop and are severe when a film of water is present on the plant tissues and the relative humidity in the air is high during cool or warm, but not hot, periods. Downy mildews can cause severe losses in short periods of time.

Although even the late blight of potato and tomato looks like and is often called a downy mildew, true downy mildews are caused by a group of oomycetes that belong to the family Peronosporaceae. All species of this family are obligate parasites of higher plants and cause downy mildew diseases on most cultivated grain crops, vegetables, field crops, ornamentals, shrubs, and vines.

Downy mildews have caused spectacular and catastrophic epidemics on several crops in the past, and some of them continue to cause severe losses. The best known downy mildew is the one affecting grapes, which in the mid to late 1800s almost completely destroyed the grape and wine industry in France and most of the rest of Europe. In recent years, the downy mildew of sorghum has appeared and spread in the United States and has raised fears of future introduction of other, even more serious, downy mildews of grain crops now present in Asia and Africa. In 1979, a devastating epidemic of downy mildew (blue mold) of tobacco spread rapidly from Florida up the eastern states into New England and Canada and destroyed much of the tobacco in its path, causing losses to growers worth hundreds of millions of dollars.

Downy mildew oomycetes produce sporangia on sporangiophores that branch in ways distinctive for each oomycete. The sporangia are located at the tips of the branches. The sporangiophores are usually long, white at first, grayish to brown later, emerging in groups from the plant tissues through the stomata. Sporangiophores form a visible mat of oomycete growth on the lower side or both sides of leaves and on other affected tissues. Each sporangiophore grows until it reaches maturity and then produces its crop of sporangia, all at about the same time.

In most downy mildews, sporangia germinate by producing zoospores or, at higher temperatures, by producing germ tubes. In the genus *Bremia*, however, sporangia germinate most commonly by means of a germ tube, and in genera *Peronospora* and *Peronosclerospora* the sporangia germinate only by means of a germ tube. Whenever sporangia germinate by producing a germ tube, they are considered spores in themselves rather than sporangia, and in that case they are often called conidia. Oospores of downy mildews usually germinate by germ tubes, but in a few cases they may produce a sporangium that releases zoospores.

In most downy mildews, in which the pathogen is carried in the seed or bulb or infection takes place at the seedling or young plant stage, the pathogen routinely causes systemic shoot infection of its host. When older plants are attacked they may develop small or large localized infected areas or they may allow the oomycete to spread into young tissues and become locally systemic.

Downy mildews often cause rapid and severe losses of young crop plants still in the seedbed or in the field. They often destroy from 40 to 90% of the young plants or young shoots in the field, causing heavy or total losses of crop yields. The severity of loss depends on the prolonged presence of wet, cool weather during which the downy mildews sporulate profusely, cause numerous new infections, and spread into and rapidly kill young succulent tissues. In cool, wet weather downy mildews are often uncontrollable, checked only when the weather turns hot and dry. Since the discovery of systemic fungicides, our ability to control these diseases has improved considerably, although downy mildews are still very difficult to control.

Some of the most common or most serious downy mildew oomycetes and the diseases they cause are listed below. The structure of their sporangiophores is given in Fig. 11-7.

- *Bremia lactucae*, causing downy mildew of lettuce
- *Hyaloperonospora parasitica*, causing downy mildew of crucifers
- *Peronospora*, causing downy mildew of snapdragon (*P. antirrhini*), of onion (*P. destructor*), of spinach (*P. effusa*), of soybeans (*P. manchurica*) (Figs. 11-30C and 11-30D), mildew (blue mold) of tobacco (*P. tabacina*) (Fig. 11-30B), and of alfalfa and clover (*P. trifoliorum*)
- *Peronosclerospora*, causing downy mildew of sorghum and corn (*P. sorghi*), of corn (Figs. 11-30E and 11-30F) (*P. maydis* and *P. philippinensis*), and of corn and sugarcane (*P. sacchari*)
- *Plasmopara*, causing downy mildew of grape (*P. viticola*) (Figs. 11-31C–11-31F) and of sunflower (*P. halstedii*)
- *Pseudoperonospora*, causing downy mildew of cucurbits (*P. cubensis*) (Figs. 11-30A and 11-31A) and of hops (*P. humuli*)
- *Sclerophthora*, causing downy mildew of cereals (corn, rice, sorghum, wheat) and grasses (*S. macrospora*) (Fig. 11-30E)
- *Sclerospora*, causing downy mildew of grasses and millets (*S. graminicola*) (Fig. 11-31B)

The most important downy mildew diseases are those affecting tobacco, onion, grape, and cucurbits, but in a given year any of the downy mildews can cause catastrophic losses in their hosts.

DOWNY MILDEW OF GRAPE

Downy mildew of grape occurs in most parts of the world where grapes are grown. Some historical aspects of this disease are given on page 30–32. Although the pathogen is native to North America, where it attacks native grape vines, it does not affect them very seriously. When the oomycete, however, was introduced inadvertently into Europe in about 1875, the European or wine grape, *Vitis vinifera*, which had evolved in the absence of the downy mildew pathogen, was extremely susceptible to it and the oomycete began to spread among vineyards throughout France and most of Europe, destroying the crop and the vineyards in its path. Downy mildew is still most destructive in Europe and in the eastern half of the United States, where it may cause severe epidemics year after year and, in some years, in other humid parts of the world. Dry areas are usually free of the disease.

Downy mildew affects the leaves, fruit, and shoots of grapevines. It causes losses through killing of leaf tissues and defoliation, through production of low-quality, unsightly, or entirely destroyed grapes, and through weakening, dwarfing, and killing of young shoots. When the weather is favorable and no protection against the disease is provided, downy mildew can easily destroy 50 to 75% of the crop in one season.

Symptoms. At first, small, pale yellow, irregular spots appear on the upper surface of the leaves, and a white downy growth of the sporangiophores of the oomycete appears on the underside of the spots (Figs. 11-31C and 11-31D). Later, the infected leaf areas are killed and turn brown, while the sporangiophores of the oomycete turn gray. The spots often enlarge, coalesce to form large dead areas on the leaf, and frequently result in premature defoliation (Fig. 11-30E).

All young grapevine tissues are particularly susceptible to infection. Infected grapes are quickly covered with the downy growth (Fig. 11-31F), may become distorted or thickened, and may die. If infection takes place after the berries are half-grown, the oomycete grows mostly internally; the berries become leathery and somewhat wrinkled and develop a reddish marbling to brown coloration. In late or localized infections of shoots, the shoots usually are not killed but show various degrees of distortion.

The Pathogen: **Plasmopara viticola.** The mycelium diameter varies from 1 to 60 micrometers because the

FIGURE 11-30 Downy mildew symptoms on leaves of cantaloupe (A), cabbage (B), and soybean (C). (D) Soybean seeds encrusted with oospores of the downy mildew pathogen *Peronospora manchurica*. Sorghum downy mildew on corn caused by *Peronosclerospora sorghi* (E) and crazy top downy mildew caused by *Sclerophthora macrospora* (F). [Photographs courtesy of (A, B, and F) Plant Pathology Department, University of Florida, (C) W.L. Seaman and (D) R.G. Platford, WCPD, and (E) by H.D. Thurston, Cornell University.]

FIGURE 11-31 (A) Sporangiophores and sporangia of *Pseudoperonospora cubense*. (B) Oospores of *Peronosclerospora sorghi* in leaf tissue. Downy mildew symptoms on upper side of grape leaf (C), on lower (left) and upper side of grape leaf (D), and on grape cluster (E). (F) Grape varieties showing different resistance to leaf loss due to infection by downy mildew. [Photographs courtesy of (A and B) R.E. Cullen, University of Florida, (C) J.W. Travis, Pennsylvania State University, (D and E) E. Hellman, Texas A&M University, and (F) G. Ash, Charles Stuart University, Australia, with permission of APS.]

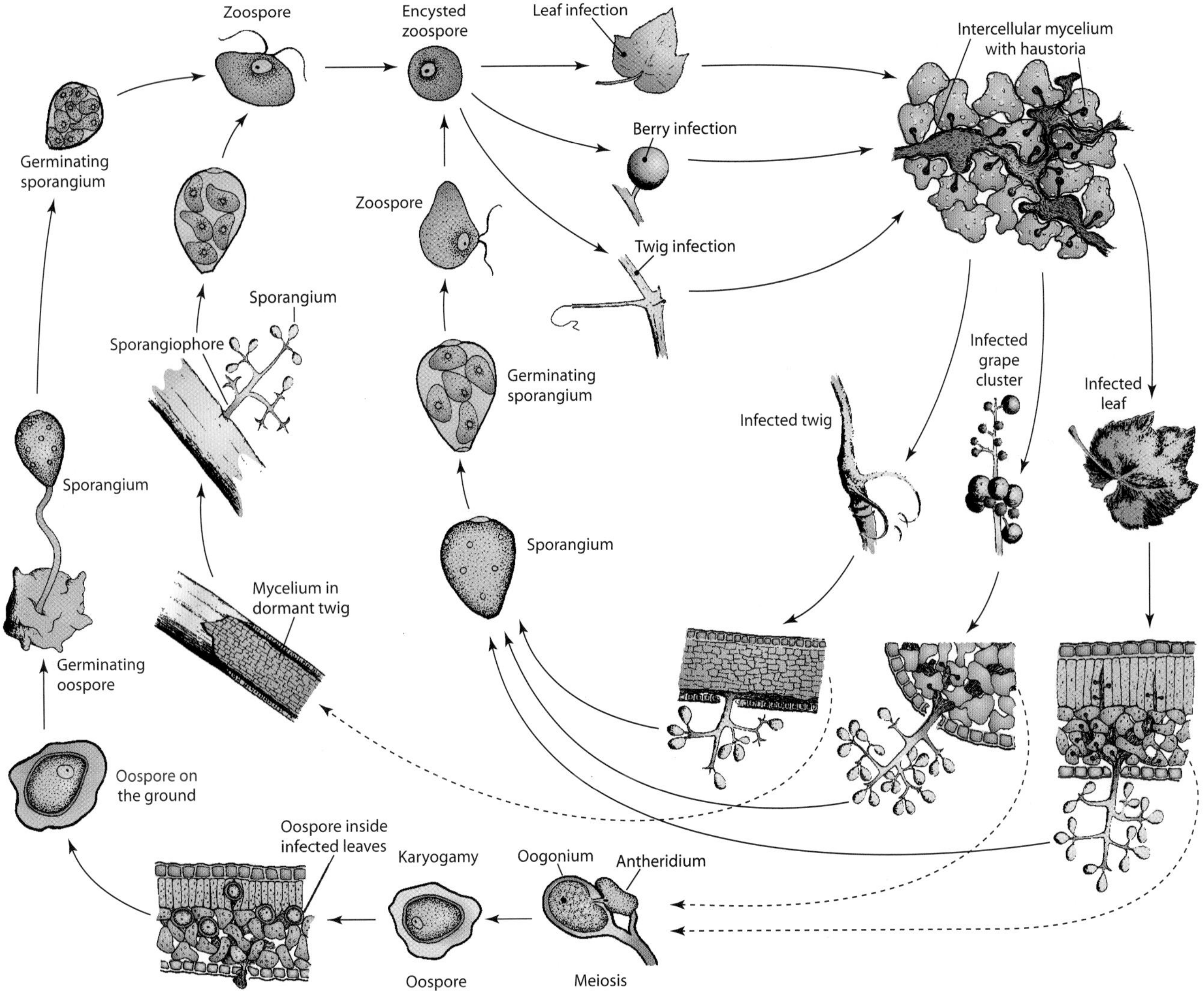

FIGURE 11-32 Disease cycle of downy mildew of grapes caused by *Plasmopara viticola*.

hyphae take the shape of the intercellular spaces of the infected tissues. Globose haustoria grow into the cells (Fig. 11-32). The mycelium produces sporangiophores on the underside of the leaves and on the stems through stomata and, in young fruit, through lenticels. Four to six or more sporangiophores arise through a single stoma. Each produces four to six branches at nearly right angles to its main stem. Each branch produces two or three secondary branches in a similar manner. At the tips of the branches, single, lemon-shaped sporangia (conidia) are produced. The oomycete also produces numerous oospores (Fig. 11-31B). It appears, however, that *P. viticola* is heterothallic, consisting of two mating types, P1 and P2, that must be present for sexual reproduction to occur.

Development of Disease. The pathogen overwinters as oospores in dead leaf lesions and shoots (Figs. 11-31B and 11-32) and, in certain areas, as mycelium in infected, but not killed, twigs. During rainy periods in the spring the oospores germinate to produce a sporangium. The sporangium or its zoospores are transported by wind or water to the wet leaves near the ground, which they infect through stomata of the lower surface (Figs. 11-31C, 11-31D, 11-32, and 11-33). Leaf hairs provide a basic protection barrier against the downy mildew pathogen, but in varieties lacking additional or different defense strategies it is overcome. The mycelium then spreads into the intercellular spaces of the leaf, and when it reaches the substomatal cavity it forms a cushion of mycelium from which sporangio-

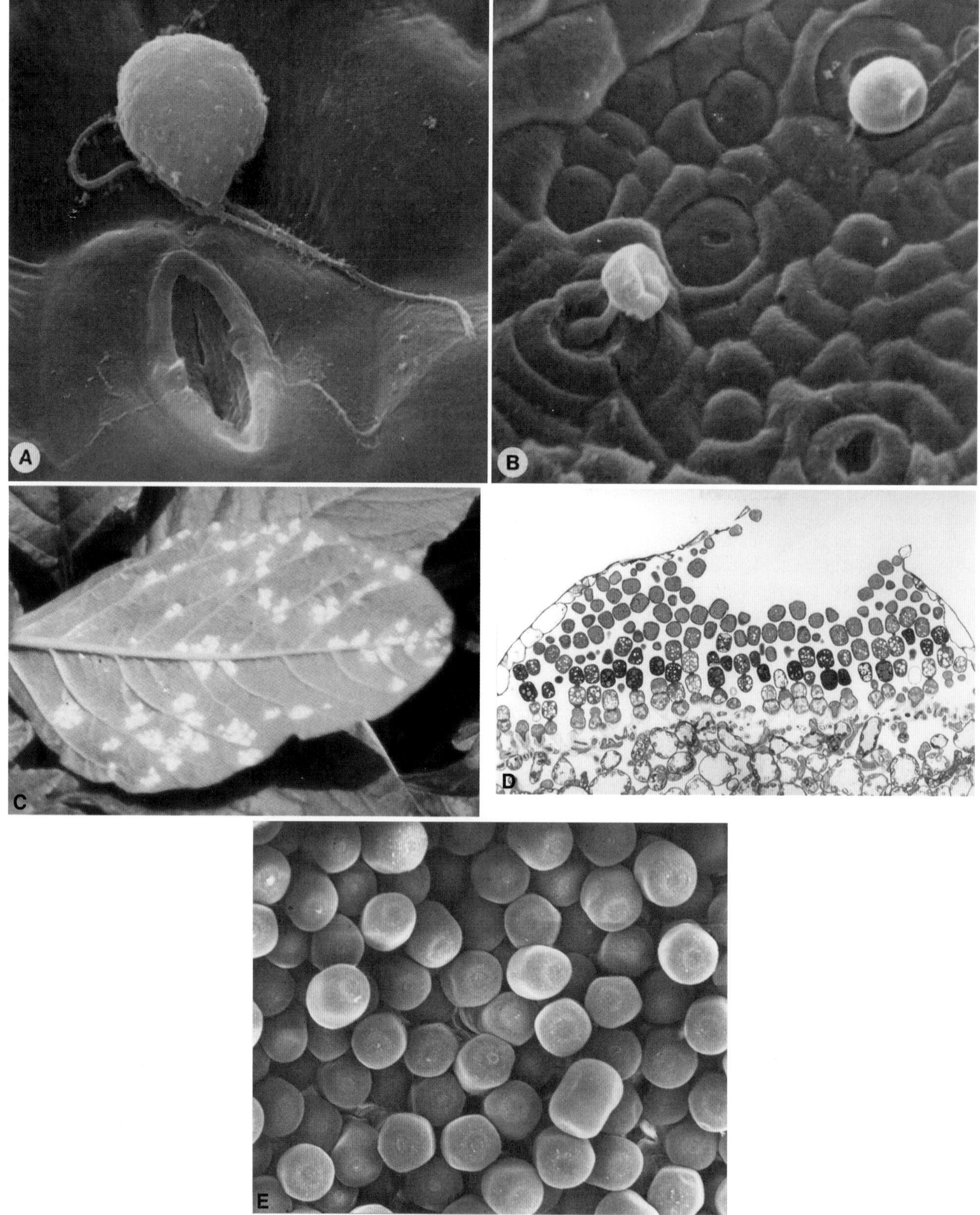

FIGURE 11-33 (A) Zoospore of *Pseudoperonospora humuli* about to settle on a stoma of a hop leaf. (B) Zoospore (upper right) settling on a leaf stoma and another (lower left) that has encysted, germinated, and produced an appressorium while penetrating through the stoma. (C) Leaf of a crucifer showing symptoms of white rust caused by *Albugo* sp. (D) Section of a white rust sorus showing zoosporangia. (E) Scanning electron micrograph of *Albugo* zoosporangia. [Photographs courtesy of (A and B) D.J. Royle and (D and E) from Mims and Richardson (2003). *Mycologia* **95**, 1–10.]

phores arise and emerge through the stoma. The sporangia may be carried by wind or rain to nearby healthy plants, germinate quickly, and produce many zoospores that cause secondary infections and thus spread the disease rapidly. A disease cycle may take from 5 to 18 days, depending on temperature, humidity, and varietal susceptibility.

In the stems, enlargement of the affected cells and the large volume of mycelium present in the intercellular spaces cause distortion and hypertrophy. Finally, the affected cells are killed and collapse, producing brown, sunken areas in the stem. In the young berries, infection is also intercellular; chlorophyll breaks down and disappears, and the cells collapse and turn brown.

At the end of the growing season the oomycete forms oospores in the infected old leaves and sometimes in the shoots and berries.

Control

Several American grape varieties show considerable resistance to downy mildew, but most European (vinifera) varieties are quite susceptible. Even the relatively resistant varieties, however, require protection through chemicals. The most effective fungicides for the control of downy mildew have been copper-based products such as the Bordeaux mixture, some broad-spectrum protective fungicides, and several systemic fungicides. The applications begin before bloom and are continued at 7- to 10-day intervals or, depending on the frequency and duration of rainfall, during the growing season. Disease prediction systems, based on the duration of leaf wetness, relative humidity, and temperature, are used to identify infection periods and to time fungicide applications. In recent years, sprays of systemic fungicides in combination with copper or broad-spectrum preventive fungicides have given excellent control of grape downy mildew.

Selected References

Crute, I. R. (1992). From breeding to cloning (and back again?): A case study with lettuce downy mildew. *Annu. Rev. Phytopathol.* 30, 485–506.

Falk, S. P., *et al.* (1996). *Fusarium proliferatum* as a biocontrol agent against grape downy mildew. *Phytopathology* **86**, 1010–1017.

Frederiksen, R. A., *et al.* (1973). Sorghum downy mildew, a disease of maize and sorghum. *Tex. Agric. Exp. Stn. Res. Monogr.* **2**, 1–32.

Gadoury, D. M., and Seem, R. C. eds. (1994). Proc. Intnl. Workshop on Grapevine Downy Mildew Modelling. N. Y. Agric. Exp. Stn. Special Report 68.

Madden, L. V., *et al.* (2000). Evaluation of a disease warning system for downy mildew of grapes. *Plant Dis.* **84**, 549–554.

McKeen, W. E., ed. (1989). "Blue Mold of Tobacco." APS Press, St. Paul, MN.

Millardet, P. M. A. (1885). (1) Traitement du mildiou et du rot. (2) Traitement du mildiou par le melange de sulphate de cuivre et de chaux. (3) Sur l'histoire du traitment du mildiou par le sulphate de cuivre. *J. Agric. Prat.* **2**, 513–516, 707–719, 801–805; Engl. trans. by F. L. Schneiderhan in *Phytopathol. Classics* **3** (1933).

Raid, R. N., and Datnoff, L. E. (1990). Loss of the EBDC fungicides: Impact on control of downy mildew of lettuce. *Plant Dis.* **74**, 829–831.

Smith, R. W., Lorbeer, J. W., and Abd-Elrazik, A. A. (1985). Reappearance and control of onion downy mildew epidemics in New York. *Plant Dis.* **69**, 703–706.

Spencer, D. M., ed. (1981). "The Downy Mildews." Academic Press, New York.

Williams, R. J. (1984). Downy mildews of tropical cereals. *Adv. Plant Pathol.* **2**, 2–103.

DISEASES CAUSED BY TRUE FUNGI

True Fungi include the Chytridiomycetes, Zygomycetes, Agcomycetes, Deuteromycetes (also known as Imperfect Fungi or as Mitosporic Fungi), and the Basidiomycetes.

Diseases Caused by Chytridiomycota (Chytridiomycetes)

The Chytridiomycetes, often referred to as chytrids, lack true mycelium. They have a round or irregularly shaped thallus, the walls of which contain chitin. They live entirely within the host cells. On maturity, the vegetative body is transformed into one or many thick-walled resting spores or sporangia.

Chytridiomycetes are water- or soil-inhabiting fungi. Because they produce zoospores, all require or are favored by free water or a film of water in the soil or on the plant surface. The class Chytridiomycetes contains three plant pathogenic genera: *Olpidium*, which infects the roots of many kinds of plants; *Synchytrium*, which causes black wart of potato (Fig. 11-34A); and *Physoderma*, which causes the crown wart of alfalfa [*P. (formerly Urophlyctis) alfalfae*] (Fig. 11-34B) and the brown spot disease of corn (*P. maydis*) (Fig. 11-34C).

These fungi survive in the soil as resting spores or in host plants as a spherical or irregularly shaped thallus. The resting spores germinate to produce one or many zoospores, which infect plant cells and either produce thalli directly and cause the typical infection, or first produce zoosporangia. The zoosporangia produce secondary zoospores, which then cause the typical infection. Abundant moisture favors the local spread of the pathogens. Over long distances the pathogens are spread in infected plant parts or on contaminated plants and in soil. Infected plant cells are not usually killed. Instead, in diseases caused by *Synchytrium* and *Physoderma alfalfae*, cells in infected tissues are stimulated to divide and enlarge excessively.

Olpidium can also transmit viruses from the hosts in which it is produced to those it infects next. *Olpidium*

FIGURE 11-34 (A) Black wart of potato caused by *Synchytrium endobioticum.* (B) Crown wart of alfalfa caused by *Physoderma alfalfae.* (C) Brown spot of corn caused by *P. zeae.* [Photographs courtesy of (A) WCPD, (B) Oregon State University, and (C) Plant Pathology Department, University of Florida.]

is a vector of at least six plant viruses, including tobacco necrosis virus and lettuce big vein virus.

Diseases Caused by Zygomycetes

Zygomycetes have well-developed mycelia without cross walls and produce nonmotile spores in sporangia; their resting spore is a thick-walled zygospore produced by the union of two morphologically similar gametes. Zygomycetes are strictly terrestrial fungi, their spores often floating around in the air, and are either saprophytes or weak parasites of plants and plant products on which they cause soft rots or molds. Some, e.g., *Rhizopus*, are opportunistic pathogens of humans.

Three genera of Zygomycetes are known to cause disease in plants or plant products (Figs. 11-7 and 11-8): (1) *Choanephora*, which attacks the withering floral

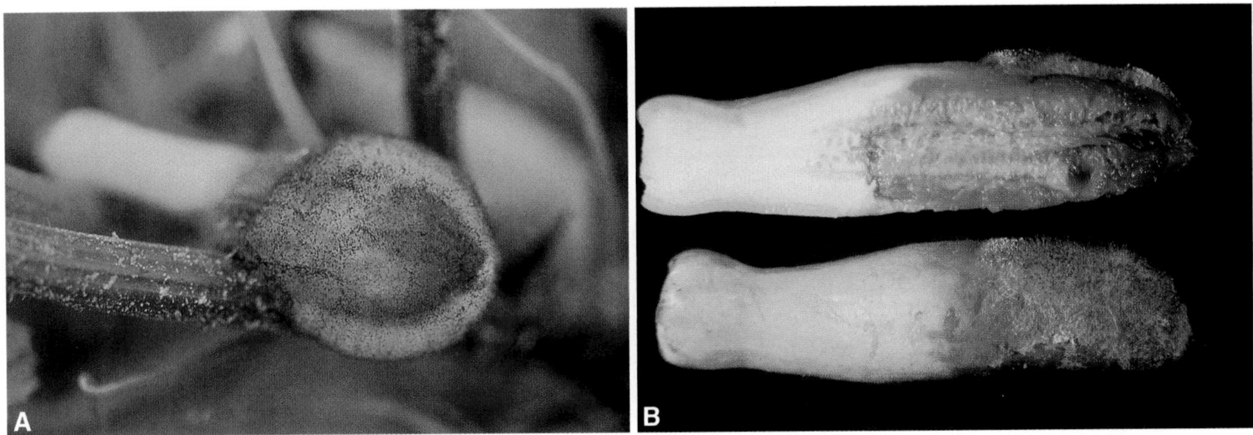

FIGURE 11-35 Choanephora wet rot of squash. (Photograph (A) courtesy of R.J. McGovern, University of Florida.)

parts of many plants after fertilization and from there invades the fruit and causes a soft rot of primarily summer squash (Fig. 11-35) but also of pumpkin, pepper, and okra; and (2) *Rhizopus* and (3) *Mucor*, both common bread mold fungi, which in addition cause soft rot of many fleshy fruits (Figs. 11-36A–11-36C), vegetables, flowers, bulbs, corms, and seeds. Other genera are fungi that become associated with roots of plants and form ectomycorrhizae, e.g., *Endogone*, or endomycorrhizae, e.g., *Glomus*, that are beneficial to plants.

Plant pathogenic Zygomycetes are weak parasites. They grow mostly as saprophytes on dead or processed plant products; even when they infect living plant tissues, they first attack injured or dead plant parts. In the latter, the fungi build up large masses of mycelium. This secretes enzymes that diffuse into the living tissue and disrupt and kill the cells. The mycelium then grows into and colonizes the tissues it killed.

RHIZOPUS SOFT ROT OF FRUITS AND VEGETABLES

Rhizopus soft rot of fruits and vegetables occurs throughout the world on harvested fleshy organs of vegetable, fruit, and flower crops during storage, transit, and marketing of these products. Among the crops affected most by this disease are sweet potatoes, strawberries (Fig. 11-36A), all cucurbits, peaches (Figs. 11-36B and 11-36C), cherries, peanuts, and several other fruits and vegetables. Corn and some other cereals are affected under fairly high moisture. Bulbs, corms, and rhizomes of flower crops, e.g., gladiolus and tulips, are also susceptible to this disease. When conditions are favorable, the disease spreads rapidly throughout the containers, and losses can be great in a short period of time (see Fig. 11-125D).

Rhizopus also causes hull rot of maturing almond fruit and necrotic areas and death of adjacent leaves and of part or all of the attached spur or shoot. Three species of *Rhizopus* also cause head rot of sunflower.

Symptoms

Infected areas of fleshy organs appear water soaked at first and are very soft. If the skin of the infected organ remains intact, the tissue loses moisture gradually until it shrivels into a mummy. More frequently, however, fungal hyphae grow outward through the wounds and cover the affected portions by producing tufts of whisker-like gray sporangiophores and sporangia (Figs. 11-35 and 11-36). The bushy growth of the fungus often spreads over the surface of the healthy portions of affected fruit and even to the surface of the containers when they become wet with the exuding liquid. Affected tissues at first give off a mildly pleasant smell, but soon yeasts and bacteria move in and a sour odor develops. When loss of moisture is rapid, infected organs finally dry up and mummify; if the loss of moisture is slow, they break down and disintegrate in a "leaky" watery rot.

The Pathogen: Rhizopus *spp.*

The mycelium of the fungus produces long, aerial sporangiophores at the tips of which black spherical sporangia develop (Figs. 11-36D and 11-37). The sporangia contain thousands of spherical sporangiospores. When the mycelium grows on a surface, it produces stolons, i.e., hyphae that arch over the surface

FIGURE 11-36 Rhizopus rot of strawberries (A), of peach externally (B), and of peach in cross section (C). Sporangiophores with sporangia (D) and zygospore (E) of *Rhizopus* sp. [Photographs courtesy of (A, D, and E) Plant Pathology Department, University of Florida.]

and at the next point of contact with the surface produce both root-like hyphae, called rhizoids, which grow toward the surface, and aerial sporangiophores bearing sporangia. From each point of contact more stolons are produced in all directions. Adjacent hyphae produce short branches called progametangia, which grow toward one another. When they come in contact, the tip of each hypha is separated from the progamentangium by a cross wall. The terminal cells are gametangia. These fuse and their nuclei pair. The cell formed by the fusion enlarges and develops a thick, black, and warty cell wall. This sexually produced spore is called a zygospore

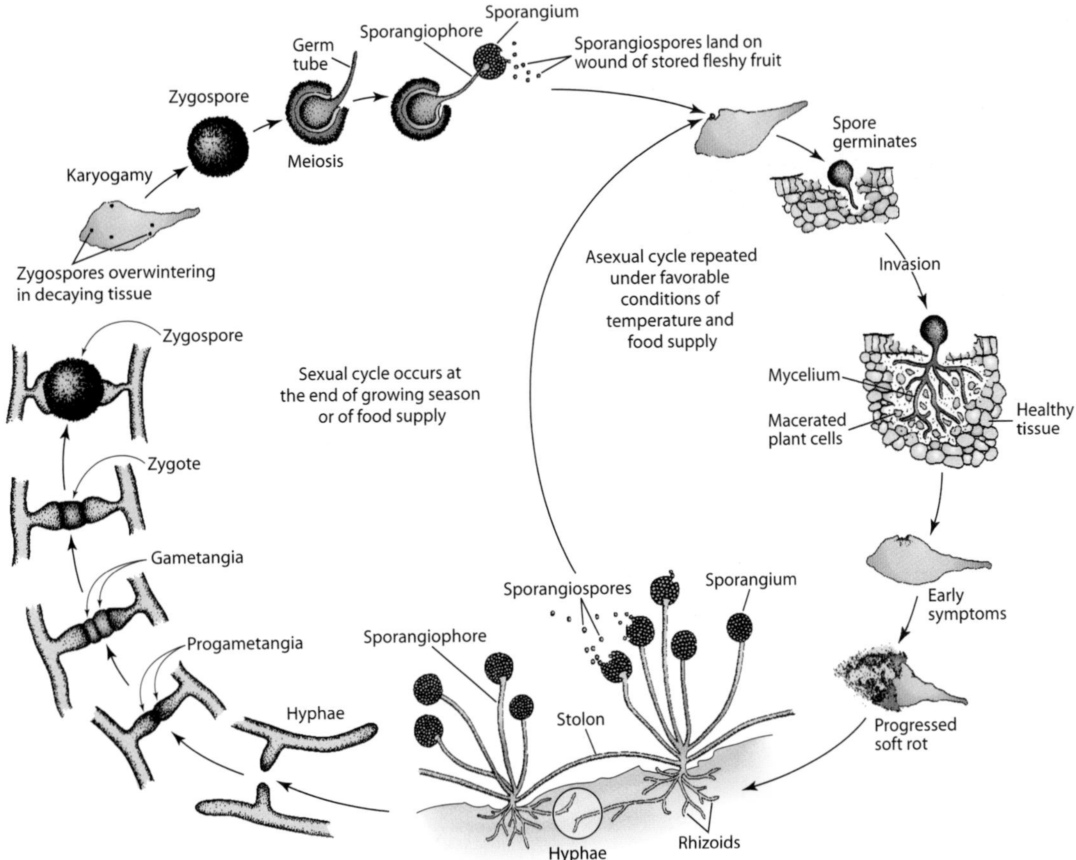

FIGURE 11-37 Disease cycle of soft rot of fruits and vegetables caused by *Rhizopus* spp.

(Figs. 11-36E and 11-37) and is the overwintering or resting stage of the fungus. When it germinates it produces a sporangiophore bearing a sporangium full of sporangiospores.

Development of Disease

Throughout the year, sporangiospores float about and if they land on wounds of fleshy fruits, roots, corms, or bulbs they germinate (Fig. 11-37). The resulting hyphae secrete pectinolytic enzymes, which break down and dissolve the pectic substances of the middle lamella that hold the plant cells in place in the tissues. This results in loss of cohesion among the cells and development of "soft rot."

The pectinolytic enzymes secreted by the fungus advance ahead of the mycelium and separate the plant cells, which are then attacked by the cellulolytic enzymes of the fungus. The cellulases break down the cellulose of the cell wall and the cells disintegrate. The mycelium does not seem to invade cells but is instead surrounded by dead cells and nonliving organic substances, the fungus living more like a saprophyte than a parasite.

The fungus continues to grow inside the tissues. When the epidermis breaks, the fungus emerges through the wounds and produces aerial sporangiophores, sporangia, stolons, and rhizoids, the latter capable of piercing the softened epidermis. In extremely fleshy fruits, the mycelium can penetrate even healthy fruit. Unfavorable temperature and humidity, or insufficient maturity of the fruit, slow down the growth and activity of the fungus. This allows some hosts to form layers of cork cells and other histological barriers that retard or completely inhibit further infection by the fungus.

When the food supply in the infected tissues begins to diminish and compatible strains are present together, zygospores are produced. Zygospores help the fungus survive periods of starvation and of adverse temperature and moisture.

Control

Avoid wounding fleshy fruits, roots, tubers, and bulbs during harvest, handling, and transportation. Discard or pack and store wounded organs separately from healthy ones.

Clean and disinfest storage containers and warehouses with a copper sulfate solution, formaldehyde, sulfur fumes, or chloropicrin.

Control temperatures of storage rooms and shipping cars. Pick succulent fruits, such as strawberries, in the morning when it is cool and keep them at temperatures below 10°C. Keep sweet potatoes and some other not so succulent organs at 25 to 30°C and 90% humidity for 10 to 14 days, during which the cut surfaces cork over and do not allow subsequent penetration by the fungus. Subsequently lower the temperature to about 12°C. Biological control of *Rhizopus* on stored peaches and nectarines has been achieved experimentally by treating them with yeasts of the genera *Candida* and *Pichia*.

Selected References

Lunn, J. A. (1977). *Rhizopus stolonifer.* CMI Descript. Pathogens. Fungi, Bacteria **524**, 1–2.

Michailides, T. J., and Spotts, R. A. (1990). Postharvest diseases of pome and stone fruits caused by *Mucor pyriformis* in the Pacific Northwest and California. *Plant Dis.* **74**, 537–543.

Srivastava, D. N., and Walker, J. C. (1959). Mechanisms of infection of sweet potato roots by *Rhizopus stolonifer*. *Phytopathology* **49**, 400–406.

Shtienberg, D. (1997). Rhizopus head rot of confectionary sunflower: Effects on yield quantity and quality and implications for disease management. *Phytopathology* **87**, 1226–1232.

Teviotdale, B. L., et al. (1996). Effects of hull abscission and inoculum concentration on severity of leaf death associated with hull rot of almond. *Plant Dis.* **80**, 809–812.

DISEASES CAUSED BY ASCOMYCETES AND DEUTEROMYCETES (MITOSPORIC FUNGI)

Ascomycetes and Mitosporic Fungi, i.e., the asexual fungi previously called Fungi Imperfecti or Deuteromycetes, are two groups of fungi that closely resemble one another: both produce a haploid mycelium that has cross walls, both produce conidia in identical types of conidiophores or fruiting bodies, and both cause the same kinds of plant diseases (leaf spots, blights, cankers, fruit spots, fruit rots, anthracnoses, stem rots, root rots, vascular wilts, or soft rots). The only difference is that Ascomycetes also produce sexual spores, known as ascospores, whereas mitosporic fungi produce all their spores through mitosis and none through meiosis and, therefore, lack sexual spores. In many Ascomycetes, however, ascospores are seldom found in nature. Therefore, such Ascomycetes reproduce, spread, cause disease, and overwinter as mycelium, conidia, or both so that they actually behave as mitosporic fungi. However, many fungi that were earlier classified as mitosporic fungi were found later to produce ascospores and were then reclassified as Ascomycetes. Many mitosporic fungi, therefore, are really Ascomycetes that have lost the need for or the ability to produce their sexual stage. Actually, analysis of DNA sequences has made possible the classification of these asexual fungi with their closest sexual relatives. Such analysis has revealed that mitosporic fungi do not constitute a natural group but have arisen from many different groups of Ascomycetes and some Basidiomycetes by the loss of sexuality and, therefore, ultimately they could all be assigned to one or the other of these two groups.

Some Ascomycetes, e.g., the anthracnose fungus *Glomerella*, although named according to the types of their sexual spores (ascospores) and fruiting bodies (perithecia), seldom produce ascospores and perithecia. Such fungi, however, routinely produce asexual spores (conidia) of a certain type, which have been classified as belonging to a particular mitosporic genus and species. Certain species of the same anthracnose fungus in the above example produce copious amounts of asexual spores (conidia) of the mitosporic or imperfect genus *Colletotrichum*. Such Ascomycetes, therefore, are often known by the name of their asexual stage, in this example, *Colletotrichum*, which name is obviously completely different from the name of the sexual stage (*Glomerella*). Usually, all species within a genus of an ascomycete produce the same type of conidia that belong to one genus of a mitosporic fungus; conversely, various species within a genus of a mitosporic fungus usually belong to one genus of an ascomycete. In many instances, however, different species within a genus of an ascomycete have asexual spores that belong to species in different genera of mitosporic fungi, and vice versa.

Ascomycetes (the sac fungi) produce sexual spores that are called **ascospores** because they form within a sac known as an **ascus**. They also produce asexual spores known as **conidia**. The ascus or sexual stage of Ascomycetes is often called the **teleomorph** or perfect stage, whereas the conidial or asexual stage is the **anamorph** or mitosporic or imperfect stage. In almost all plant pathogenic Ascomycetes, during the growing season the fungus exists as mycelium and reproduces and causes most infections with its asexual stage, i.e., conidia. The sexual or perfect stage is produced on or in infected leaves, fruits, or stems only at the end of the growing season or when the food supply is diminishing. The perfect stage is usually the overwintering stage. In many cases, however, the fungus can overwinter as mycelium and, occasionally, as conidia. Generally, ascospores act as the primary inoculum and cause the first (primary) infections in the spring of each year. The primary infections then produce conidia, which act as the secondary inoculum and cause all subsequent infections during the growing season.

The ascus in most Ascomycetes is formed as a result of fertilization of the female sex cell, called an **ascogonium**, by either an **antheridium** or a minute male sex spore called a **spermatium**. The fertilized ascogonium produces one to many **ascogenous hyphae**, the cells of which contain two nuclei, one male and one female. The cell at the tip of each ascogenous hypha develops into an ascus (Fig. 11-38), in which the two nuclei fuse to produce a zygote, which then undergoes meiosis to produce four haploid nuclei. The cell containing these nuclei elongates, and all four nuclei in most Ascomycetes undergo mitosis and produce eight haploid nuclei. Each nucleus is then surrounded by a portion of the cytoplasm and is enveloped by a wall, thus becoming a spore inside an ascus, i.e., an ascospore. There are usually eight ascospores per ascus (Fig. 11-38).

The asci in some Ascomycetes, e.g., in yeasts and leaf curl fungi, are naked (Fig. 11-38). In all other Ascomycetes the asci are produced, singly or in groups, in fruiting bodies called **ascocarps**. In some, such as the powdery mildews, the ascocarp is a completely closed spherical container called a **cleistothecium**. In others, such as most of the Pyrenomycetes, the ascocarp is more or less closed, but at maturity has an opening through which the ascospores escape; such an ascocarp is called a **perithecium**. In Loculoascomycetes (ascostromatic ascomycetes), asci are formed directly in cavities within a stroma (matrix) of mycelium and is called a **pseudothecium** or an **ascostroma**. Finally, in Discomycetes (cup

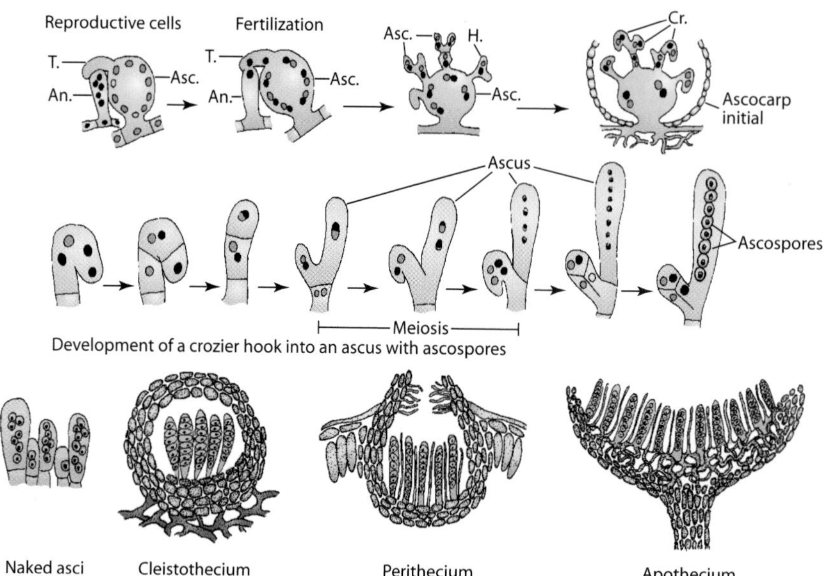

FIGURE 11-38 General scheme of sexual reproduction, ascus development, and types of ascocarps in Ascomycetes. An, antheridium; Asc, ascogonium; T, trichogyne; Asc H, Ascogenous hyphae; Cr, crozier.

fungi), asci are produced in an open, cup- or saucer-shaped ascocarp called an **apothecium** (Fig. 11-38).

Ascomycetes are identified by the characteristics of their ascocarps, asci, and ascospores (Fig. 11-39). The mitosporic fungi and those Ascomycetes that exist primarily as their mitosporic stage are identified by the conidial characteristics plus the shape of the **conidiophore** (i.e., the hypha that produces the conidium), the arrangement of the conidiophores, and the way the conidia are borne on the conidiophore (Fig. 11-40). In many cases, the conidia are borne singly or in chains at the tips of conidiophores arising from the mycelium, free from one another (Figs. 11-40 and 11-41). Some fungi produce conidiophores on a cushion-shaped stroma of mycelium, and the whole structure is called a **sporodochium;** alternatively, conidiophores are cemented together into an elongated spore-bearing structure called a **synnema.** Many fungi produce conidiophores inside a flask-shaped or globular fruiting body called a **pycnidium** (Figs. 11-40 and 11-42), whereas others produce conidia in a saucer- or cushion-shaped fruiting body called an **acervulus** (Figs. 11-40 and 11-43), which bursts through the plant surface.

Many Ascomycetes and mitosporic fungi cause a variety of diseases in all types of plants (Fig. 11-44). The most important plant pathogenic Ascomycetes and mitosporic fungi are discussed briefly later, grouped according to the general symptoms they cause on their hosts.

SOOTY MOLDS

Sooty molds appear on the leaves or stems of plants as a superficial, black growth of mycelium forming a film or crust on these plant parts (Fig. 11-45). Sooty molds may be found on all types of plants. They are most common in warm, humid weather.

Sooty molds are caused by several species of fungi of various types, but primarily dark-colored Ascomycetes of the order Capnodiales. These fungi, e.g., *Capnodium*, are not parasitic but live off honeydew, the sugary deposit forming on plant parts from the droppings of certain insects, particularly aphids and scale insects. The fungal growth is so abundant that it gives the leaf a black, sooty appearance and interferes with the amount of light that reaches the plant. This mycelium sometimes forms a black papery layer that can be peeled off from the underlying leaf. The presence of sooty mold fungi is usually of rather minor importance to the health of the plant, but it does indicate the presence of insects and may be a warning of a severe aphid or scale problem.

Sooty molds can be diagnosed easily by the fact that the black sooty mycelial growth can be completely wiped off a leaf or stem with a moistened cloth, paper, or hand, leaving a clean, healthy-looking plant surface underneath.

No control measures are applied against the sooty mold fungi. Because they grow on the excretions of insects, control of the particular insect with the appropriate insecticide or other means also results in the elimination of the sooty mold fungi.

Selected References

Alexopoulos, C. J., Mims, C. W., and Blackwell, M. (1996). "Introductory Mycology," 4th Ed. Wiley, New York.

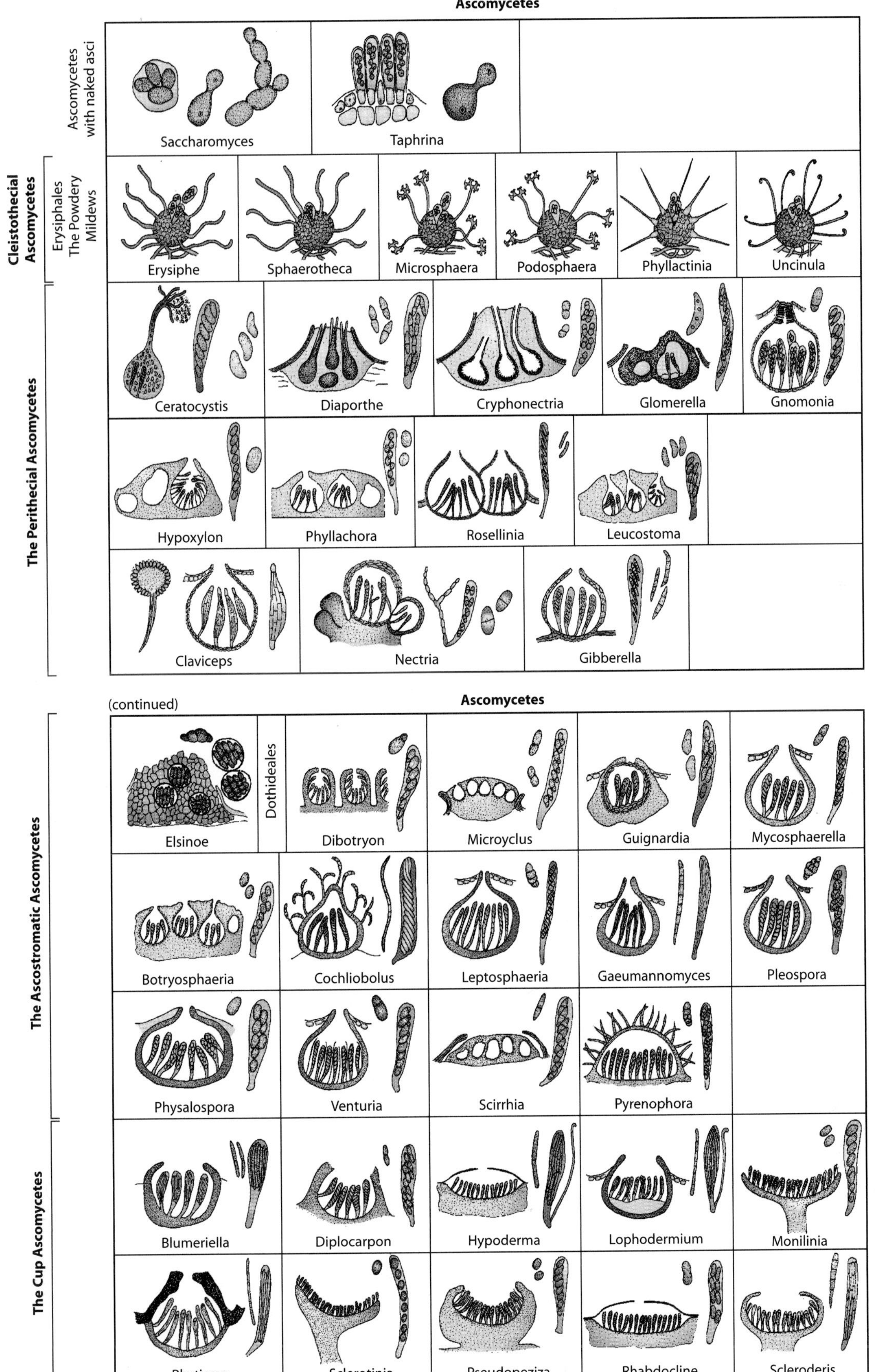

FIGURE 11-39 Morphology of fruiting bodies, asci, and ascospores of the main groups and genera of phytopathogenic Ascomycetes.

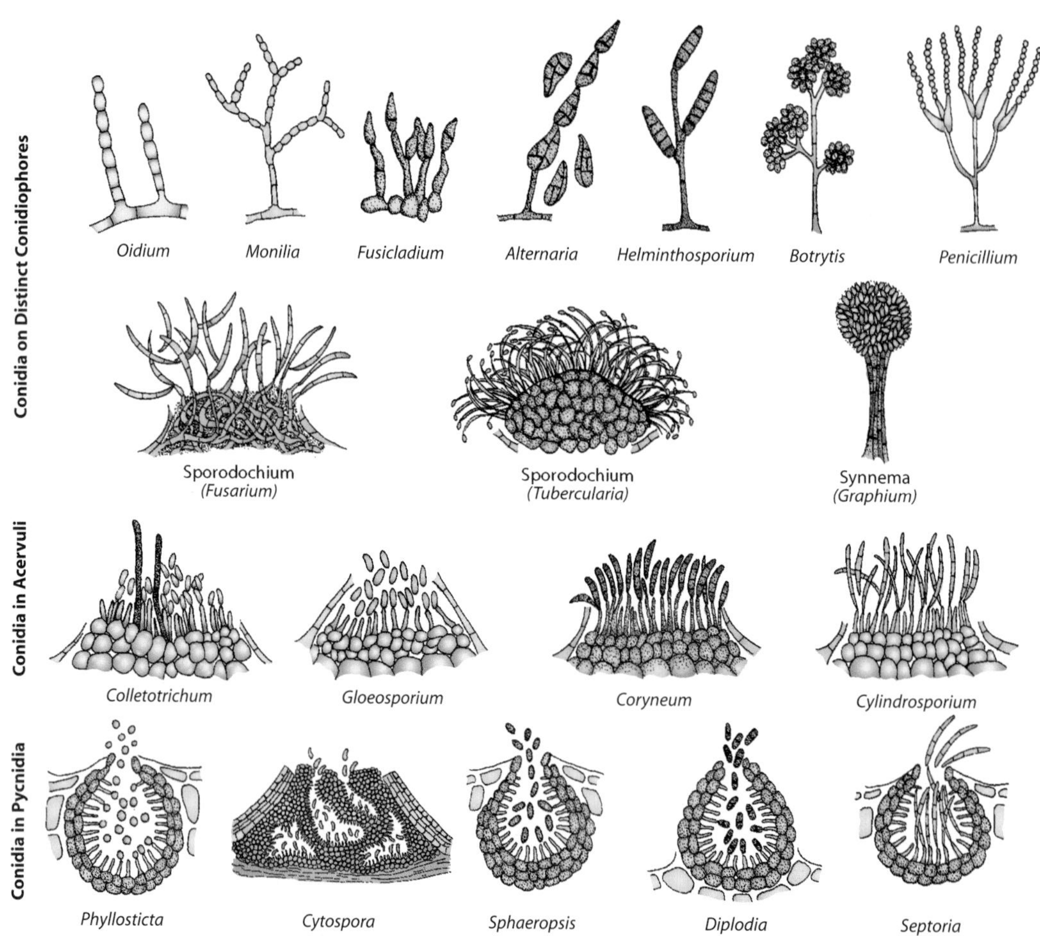

FIGURE 11-40 Types of conidia, conidiophores, and asexual fruiting bodies produced by Ascomycetes and Deuteromycetes (mitosporic fungi).

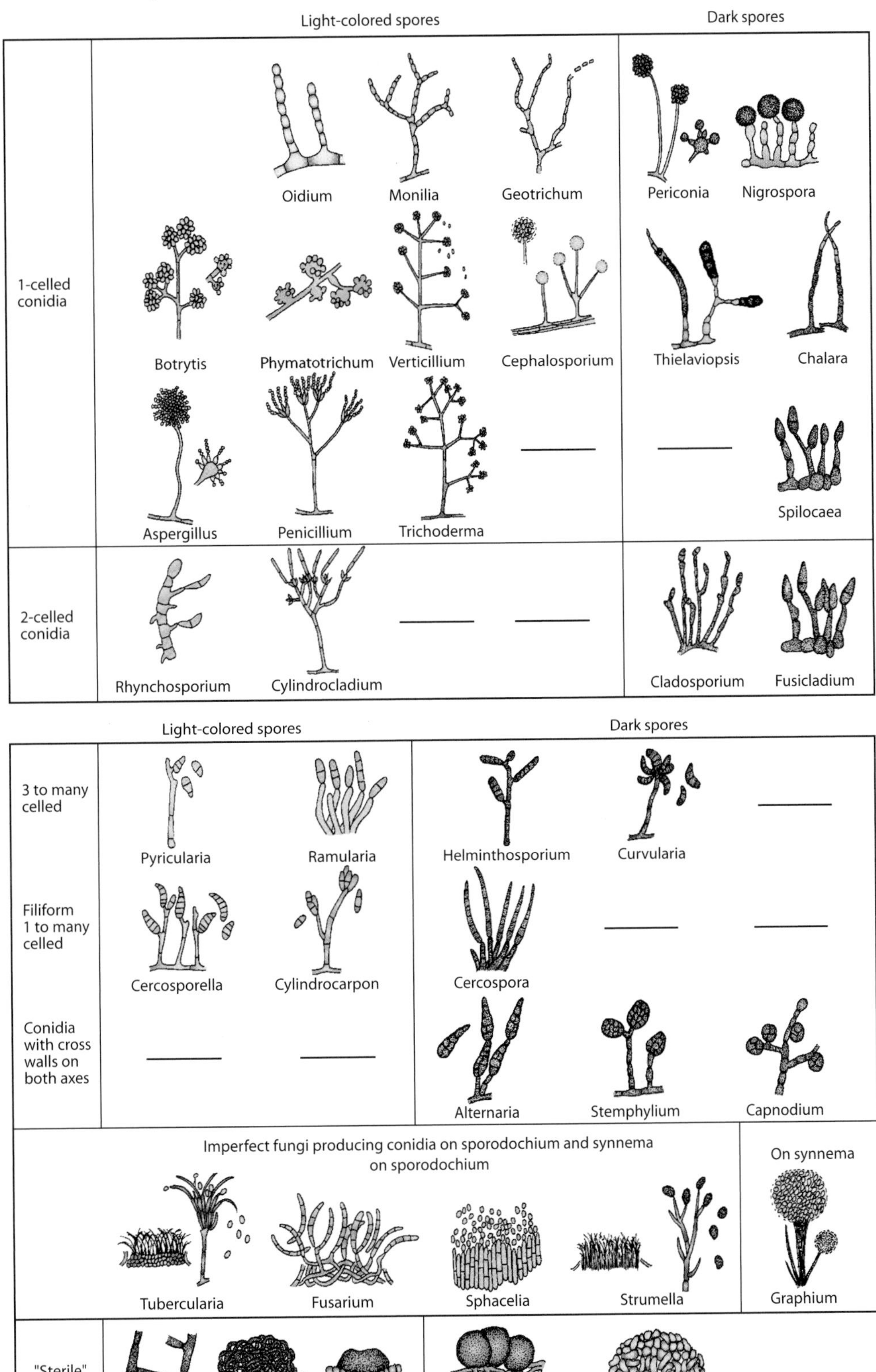

FIGURE 11-41 Grouping and morphology of conidiophores and conidia of the main genera of phytopathogenic Ascomycetes and mitosporic fungi that produce conidia on free hyphae or groups of hyphae. Also shown are mycelium and sclerotia of the two most important "sterile" fungi.

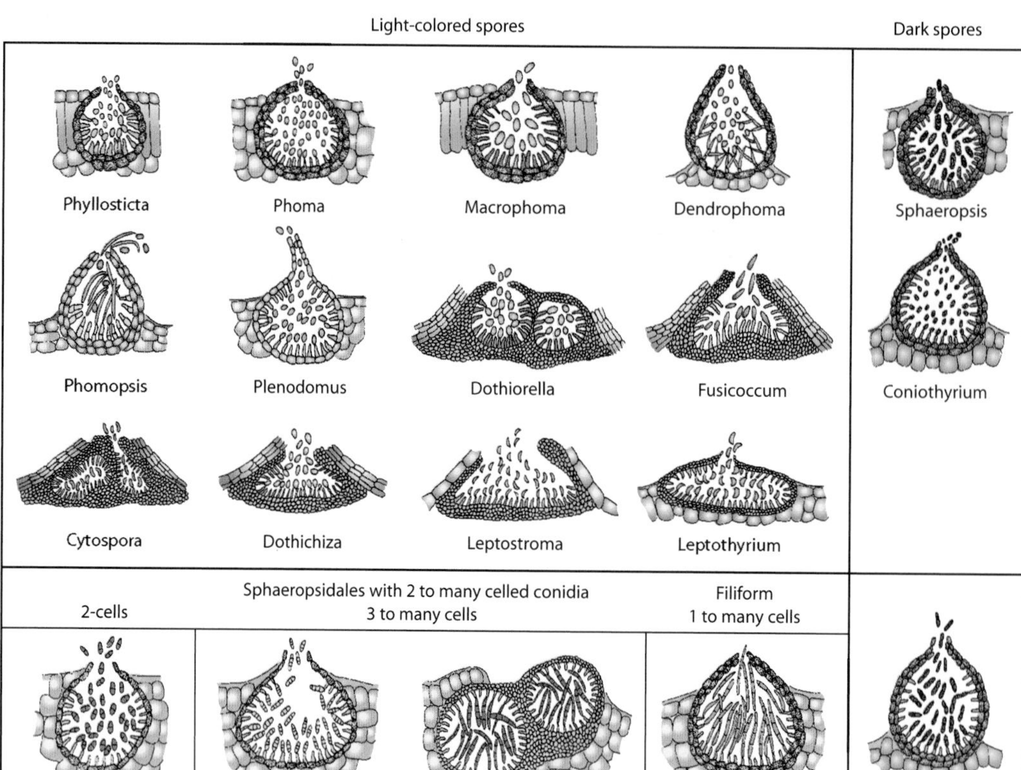

FIGURE 11-42 Morphology of pycnidia and conidia of the main genera of mitosporic fungi.

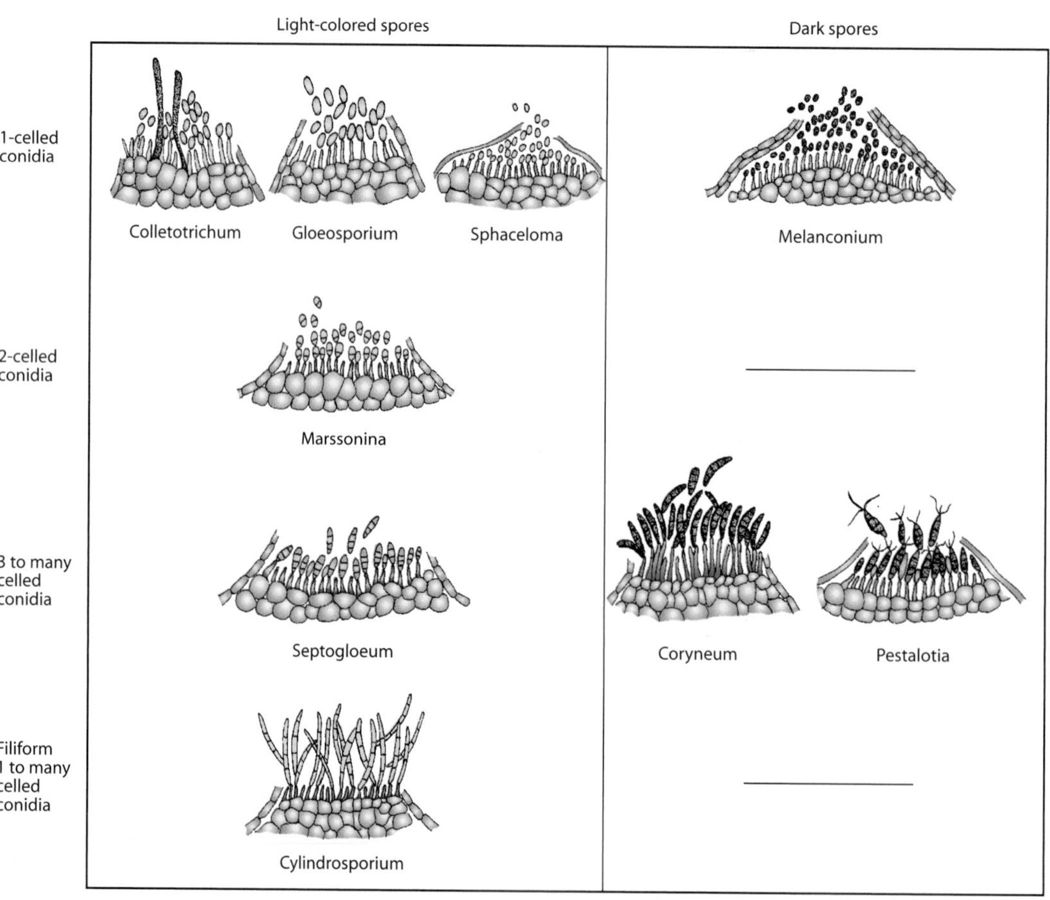

FIGURE 11-43 Morphology of acervuli and conidia produced by some genera of mitosporic fungi.

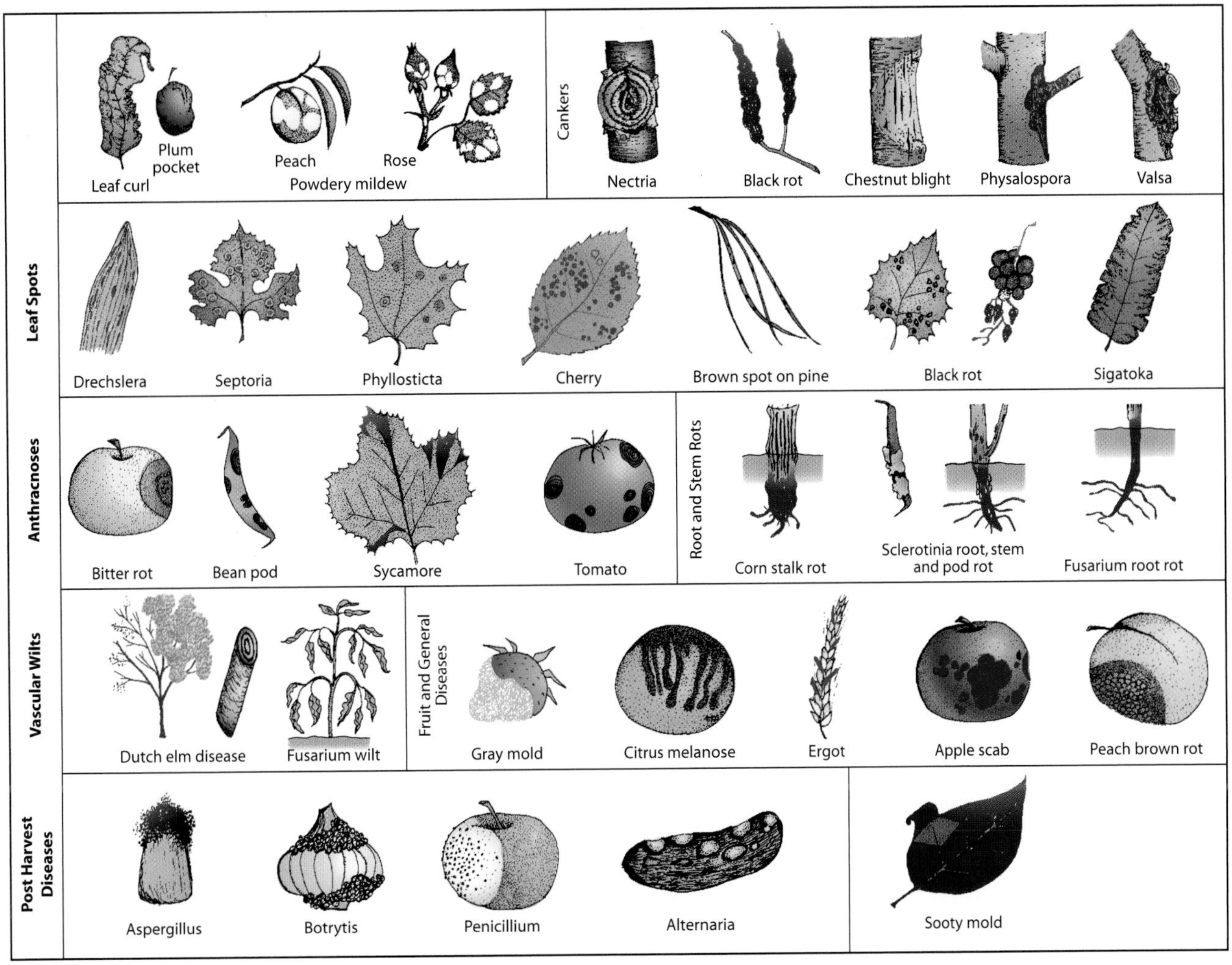

FIGURE 11-44 Common symptoms caused by some important Ascomycetes and mitosporic fungi.

FIGURE 11-45 Sooty mold on leaves of an ornamental shrub.

Barr, M. D. (1955). Species of sooty molds from western North America. *Can. J. Bot.* **33**, 497–514.

Fraser, L. (1937). The distribution of sooty-mold fungi and its relation to certain aspects of their physiology. *Proc. Linn. Soc. N. S. W.* **62**, 25–56.

Webber, H. J. (1897). Sooty mold of the orange and its treatment. *U.S. Dep. Agric. Div. Veg. Physiol. Pathol. Bull.* **13**, 1–34.

LEAF CURL DISEASES CAUSED BY *TAPHRINA*

Several species of *Taphrina* cause leaf, flower, and fruit deformation on stone fruit and forest trees, such as leaf curl on peach (Fig. 11-46A) and nectarine, plum pocket on plums (Fig. 11-46B), leaf curl and witches'-broom on cherries, and leaf blister of oak (Fig. 11-46C). The most important losses are those caused primarily on peach, nectarine, and sometimes plum.

Taphrina diseases probably occur all over the world. *Taphrina* causes defoliation of peach trees, which may

FIGURE 11-46 Peach leaf curl (A) and plum pockets (fruit in middle) at (B) caused by *Taphrina* sp. (C) Oak leaf blister caused by *T. coerulescens*. Cross section of an infected peach leaf showing naked asci of *Taphrina*. [Photographs courtesy of (A) M. Ellis, Ohio State University, (B) D.S. Wysong, University of Nebraska, (C) U.S. Forest Service, and (D) Plant Pathol. Department, University of Florida.]

lead to small fruit or fruit drop. In plum, 50% or more of the fruit may be affected and lost in years when the disease is severe. In both peach and plum, buds and twigs may also be affected, thus reducing the vitality of the tree significantly.

Symptoms

In peach and nectarine, parts of or entire infected leaves are thickened, distorted, and curled downward and inward (Fig. 11-46A). Affected leaves at first appear reddish or purplish, but later, when the fungus produces its spores on these areas, they appear reddish yellow or powdery gray, turn yellow to brown, and drop. Blossoms, young fruit, and the current year's twigs may also be attacked. Infected blossoms and fruit generally fall early in the season. The infected twigs are swollen and stunted and die during the summer.

In plum, the disease first appears on the fruit as small white blisters that enlarge rapidly as the fruit develops and soon involve the entire fruit. The fruit increases abnormally in size and is distorted (Fig. 11-46B), with the flesh becoming spongy. The seed ceases to develop, turns brown, and withers, leaving a hollow cavity. The fruit appears reddish at first, but later becomes gray and covered with a grayish powder. Leaves and twigs may also be affected, as in peach.

The Pathogen: Taphrina *spp.*

Mycelial cells of *Taphrina* in the plant contain two nuclei. These cells may develop into an ascus, usually

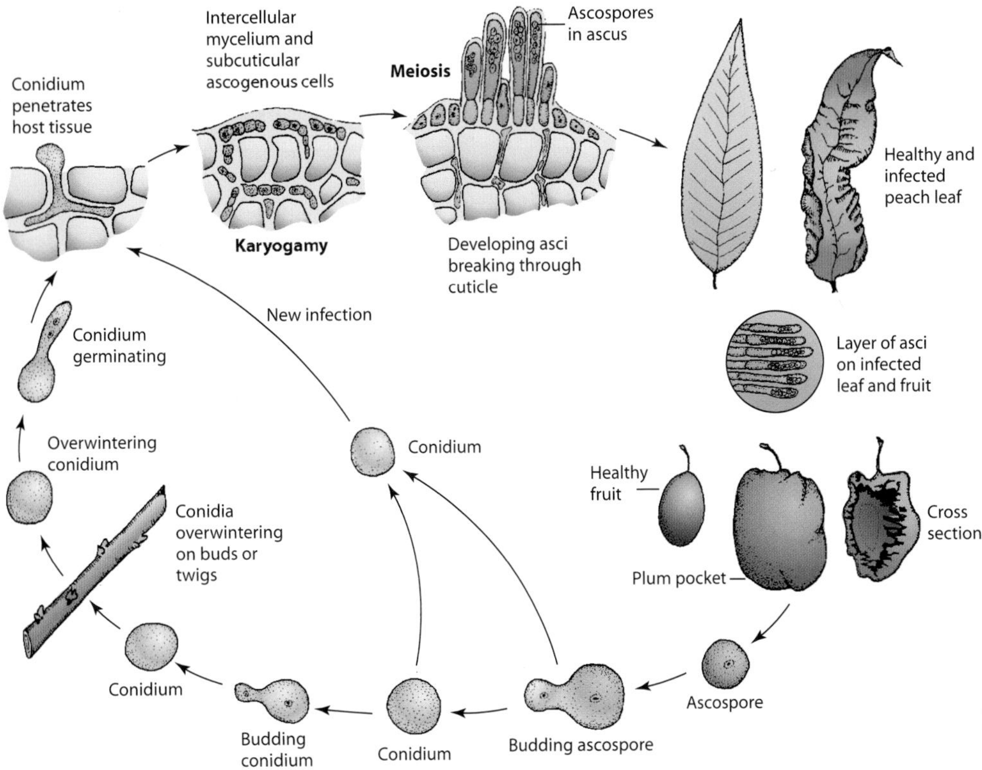

FIGURE 11-47 Disease cycle of peach leaf curl and plum pocket caused by *Taphrina* sp.

containing eight uninucleate ascospores. The ascospores multiply by budding inside or outside the ascus, producing conidia. The latter may bud again to produce more conidia or may germinate to produce mycelium. On germination, the conidial nucleus divides, and the two nuclei move into the germ tube. As the mycelium grows, both nuclei divide concurrently, producing the binucleate cells of the mycelium. Mycelial cells near the plant surface separate from one another and produce the asci (Fig. 11-46D).

Development of Disease

The fungus apparently overwinters as ascospores or thick-walled conidia on the tree, perhaps among the bud scales. In the spring, these spores are splashed or blown onto young tissues, germinate, and penetrate the developing leaves and other organs directly through the cuticle or through stomata (Fig. 11-47). The mycelium then grows between cells and invades the tissues extensively, causing excessive cell enlargement and cell division, which result in the enlargement and distortion of the plant organs. Later, numerous hyphae grow between the cuticle and the epidermis. There, their component cells separate and each produces an ascus. The asci enlarge, exert pressure on the host cuticle from below, and eventually break through to form a compact, felt-like layer of naked asci. The ascospores are released into the air, carried by wind to new tissues, and bud to form conidia. Infection occurs mainly during a short period after the buds open. All organs become resistant to infection as they grow older. Infection is favored by low temperature and a high humidity from the time of bud swell until young shoots and leaves develop, i.e., the period during which the new tissues are susceptible.

Control

Taphrina diseases are controlled easily by a single fungicide spray, preferably in late fall after the leaves have fallen or in early spring before leaf buds swell. The fungicides used most commonly are the Bordeaux mixture and chlorothalonil; the latter controls the disease if applied twice, in late fall and in early spring.

Selected References

Mix, A. J. (1949). A monograph of the genus *Taphrina*. *Univ. Kans. Sci. Bull.* 33, 1–167.

Ritchie, D. F., and Werner, D. J. (1981). Susceptibility and inheritance of susceptibility to peach leaf curl in peach and nectarine cultivars. *Plant Dis.* 65, 731–734.

Powdery Mildews

Powdery mildews are probably the most common, conspicuous, widespread, and easily recognizable plant diseases. They affect all kinds of plants except gymnosperms.

Powdery mildews appear as spots or patches of a white to grayish, powdery, mildewy growth on young plant tissues or as entire leaves and other organs being completely covered by the white powdery mildew (Figs. 11-48 and 11-49). Tiny, pinhead-sized, spherical, at first white, later yellow-brown, and finally black cleistothecia (Figs. 11-49E and 11-49F) may be present singly or in groups on the white to grayish mildew in the older areas of infection. Powdery mildew is most common on the upper side of leaves, but it also affects the underside of leaves, young shoots and stems, buds, flowers, and young fruit.

Fungi causing powdery mildews are obligate parasites: they cannot be cultured on artificial nutrient media, but recently the powdery mildew fungus of barley, *Blumeria graminis* f. sp. *hordei*, was grown in culture. They produce mycelium that grows only on the surface of plant tissues but does not invade the tissues themselves. They obtain nutrients from the plant by sending haustoria (feeding organs, Fig. 11-49B) into the epidermal cells of the plant organs. The mycelium produces short conidiophores on the plant surface (Figs. 11-49C and 11-49D). Each conidiophore produces chains of rectangular, ovoid, or round conidia that are carried by air currents. When environmental or nutritional conditions become unfavorable, the fungus may produce cleistothecia containing one or a few asci (Figs. 11-49E and 11-49F). Powdery mildew fungi, although they are common and cause serious diseases in cool or warm, humid areas, are even more common and severe in warm, dry climates. This happens because their spores can be released, germinate, and cause infection even when there is no film of water on the plant surface as long as the relative humidity in the air is fairly high. Once infection has begun, the mycelium continues to spread on the plant surface regardless of the moisture conditions in the atmosphere.

Powdery mildews are so common, widespread, and ever present among crop plants and ornamentals that the total losses, in plant growth and crop yield, they cause each year on all crops probably surpass the losses caused by any other single type of plant disease. Powdery mildews seldom kill their hosts but utilize their nutrients, reduce photosynthesis, increase respiration and transpiration, impair growth, and reduce yields, sometimes by as much as 20 to 40%.

Among the plants affected most severely by powdery mildew are the various cereals, such as wheat and barley, primarily because the chemical control of plant diseases in these crops is difficult, impractical, or not cost effective. Other crops that suffer common and severe losses from powdery mildew are the cucurbits, especially cantaloupe, squash, and cucumber; sugar beets; strawberries; clovers; many ornamentals, such as rose, begonia, dephinium, azalea, and lilac; grape; and many trees, particularly apple, catalpa, and oak.

The control of powdery mildews in grapes and some other crops depends on dusting the plants with sulfur. In cereals and several other annual crops, powdery mildew control is primarily through the use of resistant varieties. More recently, powdery mildew control has been obtained with systemic fungicides used as seed treatments or as foliar sprays. The same chemicals are used as sprays for the control of powdery mildews in other crops and in ornamentals. Several powdery mildew fungi, however, have developed resistance and are no longer controlled by some systemic fungicides. Powdery mildew on trees, such as apple, is controlled effectively with sprays of any of several sterol-inhibiting systemic fungicides. Powdery mildews have also been controlled experimentally with sprays of phosphate salt solutions and detergents or ultrafine oils and, in the greenhouse, by using blue photosensitive polyethylene sheeting. Experimentally, powdery mildew control has also been obtained through sprays with the biocontrol fungus *Ampelomyces quisqualis* and with plant activator compounds.

The powdery mildew diseases of the various crop or other plants are caused by many species of fungi of the family Erysiphaceae grouped onto several main genera. These genera are distinguished from one another by the number (one versus several) of asci per cleistothecium and by the morphology of hyphal appendages growing out of the wall of the cleistothecium. The main genera are illustrated in Fig. 11-39, and the most important diseases they cause are listed here.

Blumeria, B. graminis causing powdery mildew on cereals and grasses (Figs. 1-6 and 11-49A)

Erysiphe, E. cichoracearum causing powdery mildew of begonia, chrysanthemum, cucurbits (Fig. 11-48D), dahlia, and zinnia; *E. polygoni* of legumes, beets, crucifers, and cucurbits; *E. betae* of beets; and *E. orontii* of tomato

Leveillula, L. taurica causing powdery mildew of tomato

Microsphaera, M. alni causing powdery mildew of many shade trees and woody ornamentals

Oidium, O. neolycopersicum causing powdery mildew of tomato

Phyllactinia spp., causing powdery mildew of shade and forest trees

FIGURE 11-48 Powdery mildew symptoms on rose leaves (A) and petals (B), peach fruit (C), and squash leaf (D). Powdery mildew symptoms on dark (E) and white (F) grape bunches. [Photographs courtesy of (D) Plant Pathol. Department, University of Florida, (E) J. Travis, Pennsylvania State University, and (F) E. Hellman, Texas A&M University.]

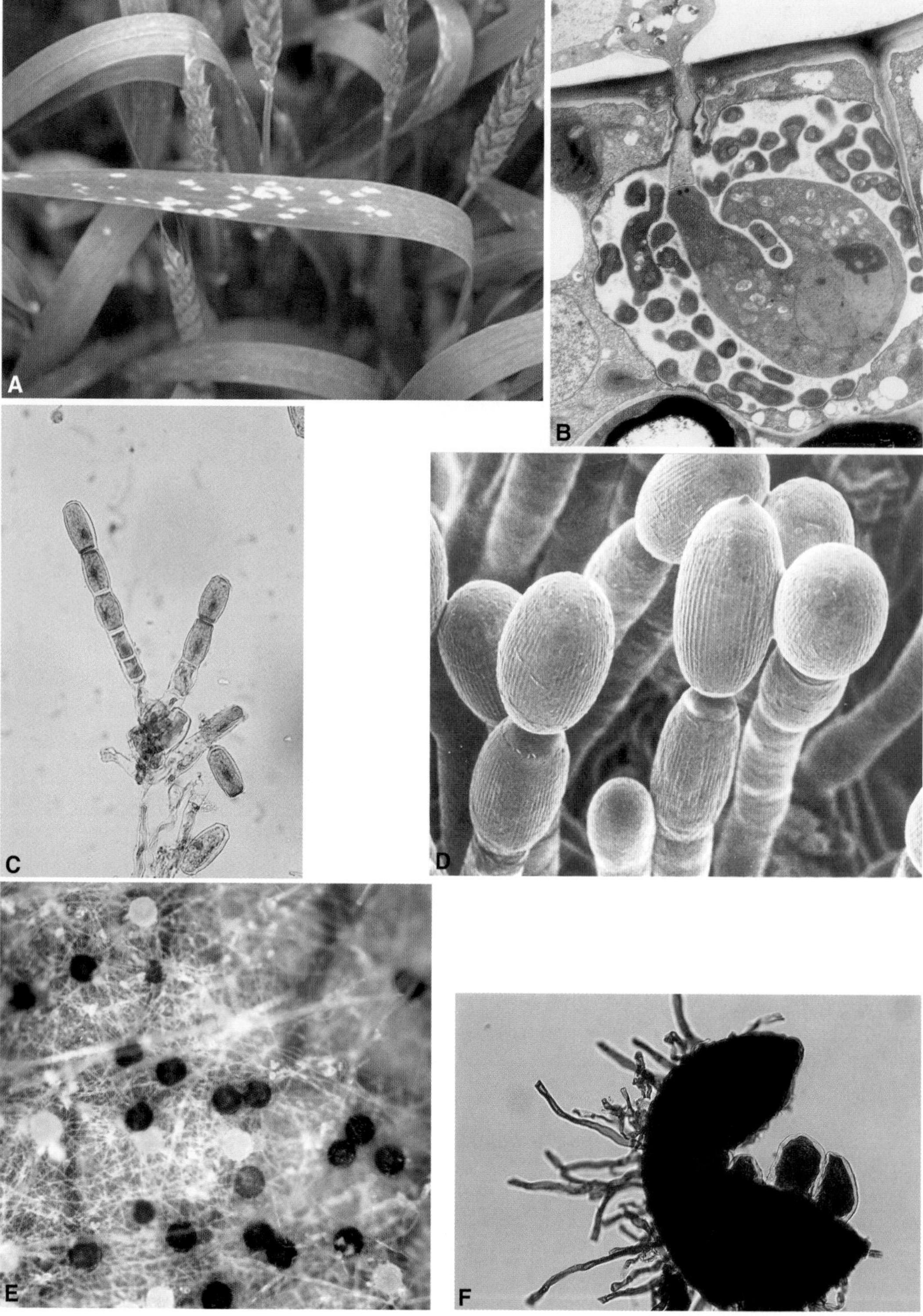

FIGURE 11-49 (A) Powdery mildew on wheat leaf. (B) A haustorium of a powdery mildew fungus inside an epidermal cell of a host leaf. (C) Conidia of a powdery mildew fungus in typical shape and arrangement in chains. (D) Scanning electron micrograph of conidia of a powdery mildew fungus. (E) Mycelium and cleistothecia of varying maturity (color) on a strawberry leaf. (F) Cleistothecium of a powdery mildew (*Erysiphe* sp.) showing two asci and mycilioid appendages. [Photographs courtesy of (A) P.E. Lipps, Ohio State University, (B and D), C.W. Mims, University of Georgia, (D) from Mims *et al.* (1995). *Phytopathology* 85, 352–358, (C and F) Plant Pathology Department, University of Florida, and (E) D.E. Legard, University of Florida.]

Podosphaera, *P. leucotricha* causing powdery mildew of apple, pear, and quince; *P. oxyacanthae*, of apricot, cherry, peach, and plum; and *P. xanthii*, of cucurbits

Sphaerotheca, *S. macularis* causing powdery mildew of strawberry (Fig. 11-49E), *S. mors-uvae* of gooseberry and currant, *S. pannosa* of peach (Fig. 11-48C) and rose (Figs. 11-48A and 11-48B), and *S. fuligena* of sugar beets

Uncinula necator, causing powdery mildew of grape (Figs. 1-23A, 1-23B, and 11-48E)

Uncinuliella flexuosa, causing powdery mildew of horsechestnut

POWDERY MILDEW OF ROSE

Powdery mildew is one of the most important diseases of roses, both in the garden and in the greenhouse. The disease appears on roses year after year and causes reduced flower production and weakening of the plants by attacking their buds, young leaves, and growing tips.

Symptoms. On young leaves the disease appears at first as slightly raised blister-like areas that soon become covered with a grayish white, powdery fungus growth (Figs. 11-48 and 11-49). As the leaves expand, they become curled and distorted. On older leaves, large white patches of fungus growth appear that cause little distortion but may eventually become necrotic.

White patches of fungus growth also appear on young, green shoots, and they may coalesce and cover the entire terminal portions of the growing shoots. Infected shoots may become arched or curved at their tip. Sometimes buds are attacked, become covered with white mildew before they open, and either fail to open or open improperly. The infection may also spread to the flower parts, which become discolored, dwarfed, and eventually die.

The Pathogen: **Sphaerotheca pannosa *f. sp.* rosae.** Powdery mildew on roses is caused by a special form of *S. pannosa*. The fungus produces white mycelium that grows on the surface of the plant tissues, sending globose haustoria into the epidermal cells (Fig. 11-50). The mycelium forms a weft of hyphae on the surface, some of which develop into short, erect, conidiophores. At the tip of each conidiophore, 5 to 10 egg-shaped conidia are produced that cling together in chains.

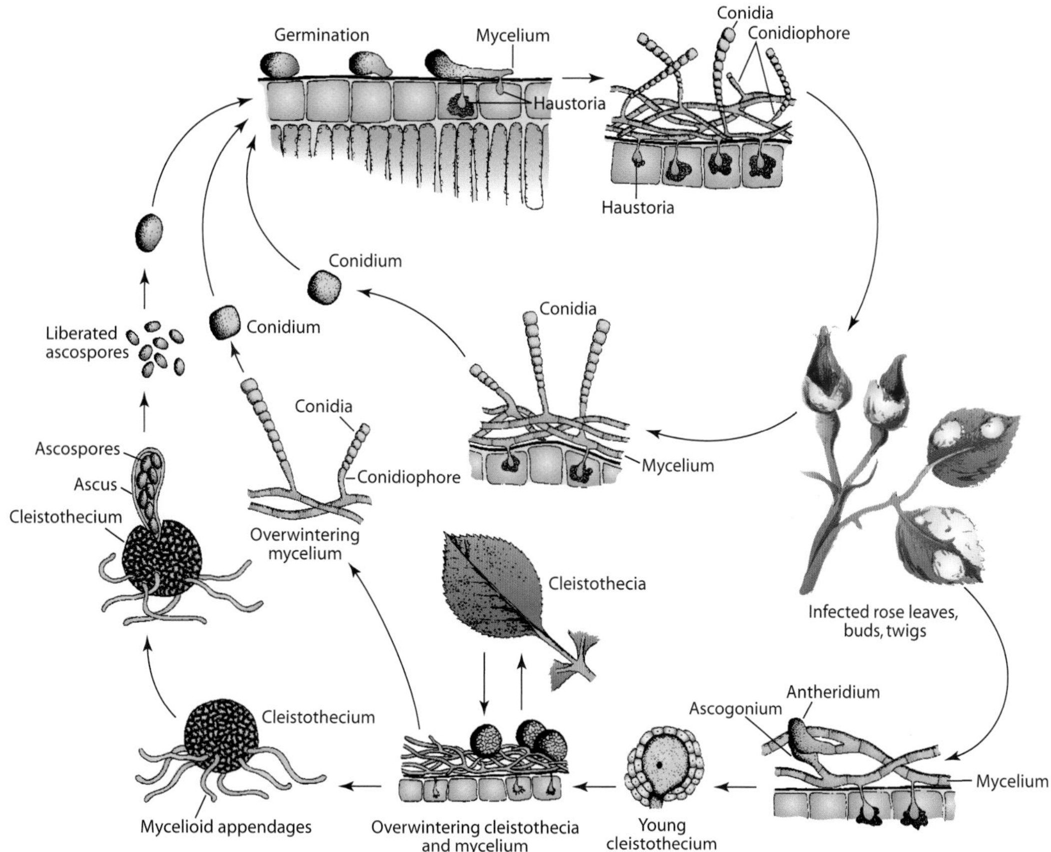

FIGURE 11-50 Disease cycle of powdery mildew of roses caused by *Sphaerotheca pannosa* f. sp. *rosae*.

With the coming of cool weather late in the season, the production of conidia ceases and cleistothecia may be formed, mainly on canes. The cleistothecia have several mycelioid appendages, i.e., hyphae arising from cells of the cleistothecium. The cleistothecia are more or less buried in the mycelial wefts on the plant tissues. The ascospores continue to develop during the fall, and in the spring they are mature and ready for dissemination. In the spring the cleistothecia absorb water and crack open. The tip of the single ascus in each cleistothecium then protrudes, bursts open, and discharges eight mature ascospores.

Development of Disease. On outdoor roses the fungus overwinters mostly as mycelium in the buds. Cleistothecia form occasionally toward the end of the season. On greenhouse roses the pathogen survives exclusively as mycelium and conidia.

Shoots arising from buds containing mycelium become infected and provide inoculum (mycelium and spores) for subsequent secondary infection and disease development on foliage and flowers. Cleistothecia, if present, discharge ascospores that also serve as primary inoculum (Fig. 11-50). Ascospores or conidia are carried by wind to young green tissues, and if temperature and relative humidity are sufficiently high the spores germinate and infect these tissues. The germ tube produces a short, fine hypha that grows directly into the epidermal cells and forms a globose haustorium by which the fungus obtains its nutrients. The germ tube, however, continues to grow and branch on the surface of the plant tissue, producing a network of superficial mycelium that sends haustoria into the epidermal cells. The absorption of nutrients from the cells depletes their food supply, weakens them, and may sometimes lead to their death. Photosynthesis in the affected areas is reduced greatly. Infection of young tissues also causes irritation and uneven growth of the affected and the surrounding cells, resulting in slightly raised areas on the leaf and distortion of the leaf. The aerial mycelium produces numerous conidia, which cause new infections on the expanding leaves and shoots. Greenhouse roses are susceptible throughout the year. In the field, however, expanding tissues seem to be the most susceptible to infection. The growth of severely infected shoots is inhibited. Infected buds often do not open. If they do open, the flowers become infected and do not develop properly.

Control. Many rose varieties show a moderately high level of resistance, but most popular varieties are highly susceptible to powdery mildew. The disease has been controlled in the past by application of sulfur or by spraying with one of several other fungicides. Sulfur may be used as a spray, as a dust, and, in the greenhouse, as a vapor. Under most conditions, weekly applications give adequate protection, but during rapid development of new growth or frequent rains, more frequent applications may be necessary. Since the early 1990s, more effective systemic fungicides have replaced many of the older fungicides in the control of powdery mildew. More recently, sprays of defense activating compounds like Actigard and of sodium bicarbonate solution and ultrafine oils have been shown to control powdery mildew of rose.

Several fungi have been reported to parasitize or antagonize the powdery mildew fungi of several crops. Although this control approach appears promising, so far it has not been developed sufficiently to be used for practical control of powdery mildews.

Selected References

Aust, H.-J., and Hoyningen-Huene, J. V. (1986). Microclimate in relation to epidemics of powdery mildew. *Annu. Rev. Phytopathol.* **24**, 491–510.

Braun, U. (1987). "A Monograph of the Erysiphales (Powdery Mildews)." Beiheft zur Nova, Hedwigia.

Everts, K. L., Leath, S., and Finney, P. L. (2001). Impact of powdery mildew and leaf rust on milling and baking quality of soft red winter wheat. *Plant Dis.* **85**, 483–429.

Horst, R. K. (1983). "Compendium of Rose Diseases." APS Press, St. Paul, MN.

LaMondia, J. A., Smith, V. L., and Douglas, S. M. (1999). Host range of *Oidium lycopersicum* on selected solanaceous species in Connecticut. *Plant Dis.* **83**, 341–344.

McGrath, M. T. (2001). Fungicide resistance in cucurbit powdery mildew: Experiences and challenges. *Plant Dis.* **85**, 236–245.

Niewoechner, A. S., and Leath, S. (1998). Virulence of *Blumeria graminis* f.sp. *tritici* on winter wheat in the eastern United States. *Plant Dis.* **82**, 64–68.

Spencer, D. M., ed. (1978). "The Powdery Mildews." Academic Press, New York.

FOLIAR DISEASES CAUSED BY ASCOMYCETES AND DEUTEROMYCETES (MITOSPORIC FUNGI)

Many Ascomycetes and mitosporic fungi cause primarily foliage diseases, but some may also affect blossoms, young stems, fruit, and even roots. Most foliar Ascomycetes reproduce by means of conidia that may overwinter; others reproduce by means of conidia during the growing season and by their perfect stage at the end of the season in which they overwinter. Some produce ascocarps and ascospores, along with conidia, throughout the growing season. The primary inoculum of these fungi, therefore, may be either ascospores or conidia, and usually originates from infected fallen or hanging leaves of the previous year.

Some of the most common Ascomycetes causing primarily foliar diseases include the following.

Cochliobolus, several species of which cause leaf spots, blights, and root rots on most cereals and grasses
Blumeriella (*Higginsia*), causing leaf spot or shot hole of cherries and plums (Fig. 11-60)
Magnaporthe, *M. grisea* causing the rice blast disease and gray leaf spot of other cereals and of turf grasses
Microcyclus, *M. ulei* causing South American leaf blight of rubber
Dothisrtroma, *D. pini* causing needle blight of pines
Elytroderma deformans, causing a leaf spot and witches'-broom of pines
Lirula, causing needle blight of spruce
Lophodermium seditiosum, causing needle blight of pines
Mycosphaerella, *M. musicola*, *M. fijiensis* causing the extremely destructive Sigatoka disease of banana, *M. graminicola* the Septoria leaf blotch of cereals, *M. fragariae* leaf spot of strawberry, *M. citri* citrus greasy spot, *M. pini* needle blights of pine, and other diseases
Pseudopeziza, causing the common leaf spot of alfalfa and clovers
Pyrenophora, several species causing leaf spots and blights on many cereals and grasses
Rhabdocline, causing needle cast of Douglas fir
Rhizosphaera, causing needle cast of spruce
Rhytisma, causing tar spot of maple and willow
Scirrhia, causing needle blights of pine

Several other Ascomycetes causing primarily foliar diseases, such as *Diplocarpon*, *Gnomonia*, and *Venturia*, could be listed here, but they either cause important additional symptoms (e.g., *Venturia* causes scab on apple fruit) or are discussed with another more cohesive group (e.g., *Gnomonia* and the related *Diplocarpon* are included in the section on anthracnose diseases).

Some of the most common mitosporic fungi causing primarily foliar but also other symptoms on a large variety of host plants are *Alternaria*, *Ascochyta*, *Cercospora*, *Cladosporium*, *Phyllosticta*, *Pyricularia*, *Septoria*, and *Stemphylium*. Many other less common imperfect (mitosporic) fungi causing leaf spots could be included here.

The foliar spots and blights caused by imperfect fungi affect numerous hosts and are quite diverse. The disease cycles and controls of these diseases are quite similar, however. Nevertheless, considerable variability may exist among diseases on different hosts, or when the diseases develop under different environmental conditions. For example, most of these fungi attack primarily the foliage of plants by means of conidia. On the infected areas, numerous conidia are produced that spread to other plants by wind, wind-blown rain, water, and insects and cause more infections. In most cases, these fungi overwinter primarily as conidia or mycelium in fallen leaves or other plant debris. Some, however, can overwinter as conidia or mycelium in or on seed of infected plants or as conidia in the soil. When perennial plants are infected, the pathogens may overwinter as mycelium in infected tissues of the plant. When these fungi are carried with the seed of annual plants, damping-off of seedlings may develop. Control of such diseases is accomplished primarily by using resistant varieties and employing fungicidal sprays or seed treatments. In some diseases, however, use of disease-free seed, removal and destruction of contaminated debris, or both may be most important.

ALTERNARIA DISEASES

Diseases caused by *Alternaria* are among the most common diseases of many kinds of plants throughout the world. They affect the leaves, stems, flowers, and fruits of primarily annual plants, especially vegetables and ornamentals, but also of trees such as citrus and apple. Total aggregate losses caused by the various Alternarias on all of their hosts rank among the highest caused by any pathogen.

Symptoms

Alternaria diseases appear usually as leaf spots and blights, but they may also cause damping-off of seedlings, stem rots, and tuber and fruit rots. Some of the diseases caused by *Alternaria* include early blight of potato and tomato (Figs. 11- 51A–D), leaf spot and fruit spot on cucurbits and onions (Figs. 11-52A and 11-52B) and on apple and citrus, fruit rot on cherry and sour cherry, core rot of apple, and rot of lemons and oranges.

The leaf spots are generally dark brown to black, often numerous and enlarging, and usually developing in concentric rings, which give the spots a target-like appearance (Figs. 11-51A–C). Lower, senescent leaves are usually attacked first, but the disease progresses upward and makes affected leaves turn yellowish, become senescent, and either dry up and droop or fall off. Dark sunken spots develop on branches and stems of plants such as tomato (Figs. 11- 51C and 11-52). Stem lesions developing on seedlings may form cankers, which may enlarge, girdle the stem, and kill the plant. In belowground parts, such as potato tubers, dark, slightly sunken lesions develop that may be up to 2 centimeters in diameter and 5 to 6 millimeters in depth.

FIGURE 11-51 Symptoms caused by *Alternaria* solani on potato (A) and tomato (B) leaves, on tomato stem (C), and on tomato fruit (D). [Photographs courtesy of (A) D.P. Weingartner, (B and C) Plant Pathology Department, University of Florida, and (D) Oregon State University.]

Alternaria may attack fruits when they approach maturity in some hosts at the blossom end but in others at the stem end or at other points through wounds (Fig. 11-51D). The spots may be small and sunken or may enlarge to cover most of the fruit, and they may be leathery and have a black, velvety surface layer of fungus growth and spores. In some fruits, such as citrus and tomato, a small lesion at the surface may indicate an extensive spread of the infection inside the fruit.

The Pathogen

Alternaria spp. have dark-colored mycelium, and in older diseased tissue they produce short, simple, erect conidiophores that bear single or branched chains of conidia (Figs. 11-52C and 11-53). Conidia are large, dark, long, or pear shaped and multicellular, with both transverse and longitudinal cross walls. Conidia are detached easily and are carried by air currents.

FIGURE 11-52 *Alternaria* symptoms on leaves of onion (A) and watermelon (B). (C) Conidia of *Alternaria* sp. (Photographs courtesy of Plant Pathology Department, University of Florida.)

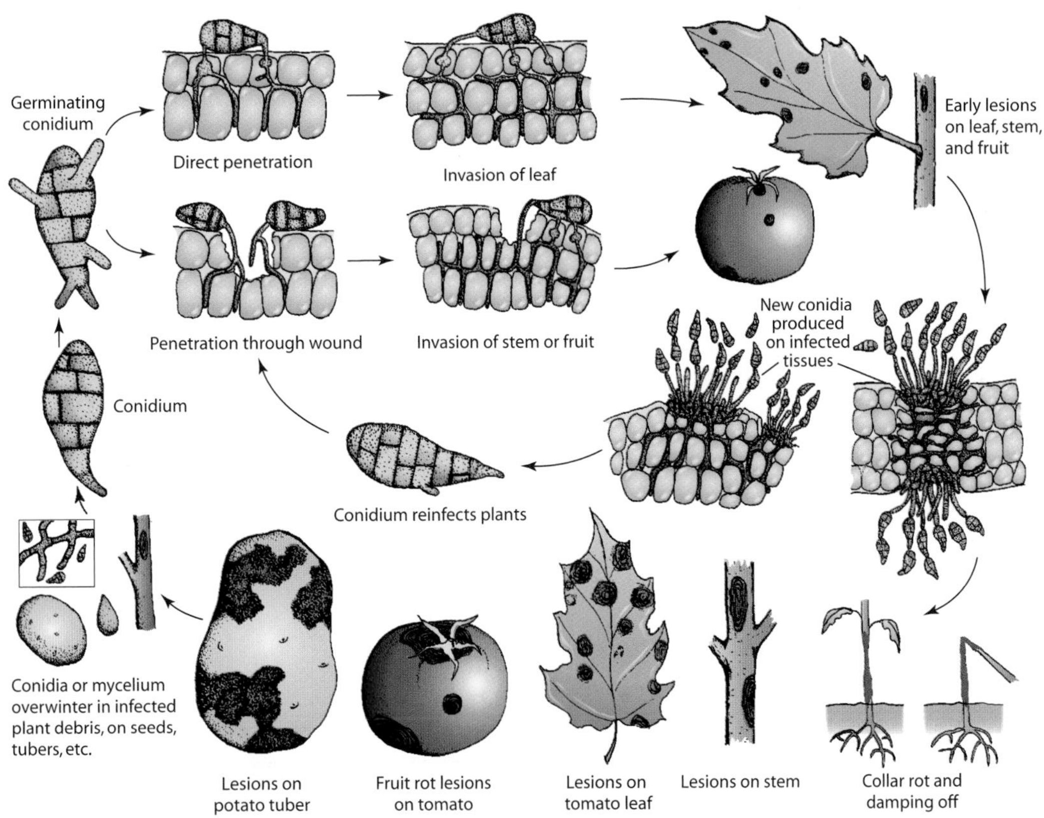

FIGURE 11-53 Development and symptoms of diseases caused by *Alternaria*.

Alternaria occurs on many plant/crop species throughout the world. Their spores are present in the air and dust everywhere and are one of the most common fungal causes of hay fever allergies. *Alternaria* spores also land and grow as contaminants in laboratory cultures of other microorganisms and on dead plant tissue killed by other pathogens or other causes. Actually, many species of *Alternaria* are mostly saprophytic, i.e., they cannot infect living plant tissues but grow only on dead or decaying plant tissues and, at most, on senescent or old tissues such as old petals, old leaves, and ripe fruit. Therefore, it is often difficult to decide whether an *Alternaria* fungus found on diseased tissue is the cause of the disease or a secondary contaminant.

Many species of *Alternaria* produce toxins. Some *Alternaria* toxins affect many different plants, whereas others are host specific.

Development of Disease

Plant pathogenic species of *Alternaria* overwinter as mycelium or spores in infected plant debris and in or on seeds (Fig. 11-53). If the fungus is carried with the seed, it may attack the seedling, usually after emergence, and cause damping-off or stem lesions and collar rot. More frequently, however, spores are produced abundantly, especially during heavy dews and frequent rains, and are blown in from infected debris or infected cultivated plants and weeds. The germinating spores penetrate susceptible tissue directly or through wounds and soon produce new conidia that are further spread by wind, splashing rain, etc. With few exceptions, *Alternaria* diseases are more prevalent in older, senescing tissues, particularly on plants growing poorly because of some kind of stress.

Control

Alternaria diseases are controlled primarily through the use of resistant varieties, disease-free or treated seed, and chemical sprays with appropriate fungicides. Adequate nitrogen fertilizer generally reduces the rate of infection by *Alternaria*. Crop rotation, removal and burning of plant debris, if infected, and eradication of weed hosts help reduce the inoculum for subsequent plantings of susceptible crops. Several mycoparasitic fungi are known to parasitize various species of *Alternaria*, but so far none has been developed into an effective biological control of the pathogen. In the greenhouse, infections by at least some *Alternaria* species can be reduced drastically by covering the greenhouse with special UV light-absorbing film, as filtering out UV light inhibits spore formation by these fungi.

CLADOSPORIUM DISEASES

Several quite diverse but important diseases are caused by different species of the mitosporic fungus *Cladosprium*. They include the leaf mold disease of tomato (Fig. 11-54A), caused by *Cladosprium fulvum* (teleomorph *Fulvia fulvum*), cucumber scab and gummosis (Fig. 11-54B) caused by *C. cucumerinum*, peach scab and twig blight (Figs. 11-54C and 11-54D) caused by *Cladosporium carpophilum*, pecan scab and leaf spot caused by *C. caryigenum*, and pod rot and blight of pea and southern pea caused by *C. cladosporioides*.

Tomato leaf mold is primarily a disease of the foliage of tomatoes grown in the greenhouse and sometimes in field-grown tomatoes in areas of high humidity. Symptoms appear first as yellowish green spots on older leaves. Later, the spots enlarge and coalesce, turn brown to black, spread to the remaining younger leaves, and may cause defoliation. Occasionally, all the other parts of the plant are attacked by the fungus.

Peach scab and twig blight, which also affects apricots and nectarines, is of major economic importance in the southeastern United States because it reduces the quality and market value of the fruit, as well as the future productivity of the trees. Symptoms appear on fruit, twigs, and leaves as rather small but numerous circular, olive-colored to black velvety spots. Spots on the fruit are usually more numerous at the stem-end half of the fruit, they often coalesce, and, when too numerous, the fruit surface below them develops cracks. The spots on shoots and young twigs are somewhat elongated and have purplish raised margins.

Cucumber scab and gummosis appear as small and sunken spots on the fruit. Such spots sometimes ooze out a rather clear fluid. The pathogen, *Cladosporium* spp., produces tall, dark, and upright conidiophores that may branch near the top. Conidia are oval, irregular to cylindrical, pale to dark brown or black and may consist of one to three cells. They give the fungus a dark, velvety appearance.

The fungus overwinters as mycelium or conidia in debris and in twig lesions. Conidia produced in these areas following periods of high humidity are airborne and waterborne and cause all infections on leaves, twigs, or fruit. Control of *Cladosporium* diseases is through sanitation and through application of appropriate fungicides.

NEEDLE CASTS AND BLIGHTS OF CONIFERS

Several ascomycetous fungi, such as *Elytroderma*, *Hypoderma*, *Lophodermium*, *Mycosphaerella* (formerly *Scirrhia*) anamorphs *Lecanosticta* and *Dothistroma*,

FIGURE 11-54 Symptoms of diseases caused by *Cladosporium fulvum* on tomato leaves (leaf mold) (A), *C. cucumerinum* on cucumber fruit (scab) (B), and *C. carpophilum* on peach twigs (C) and fruit (scab) (D). [Photographs courtesy of (A) Plant Pathology Department, University of Florida, (B) I.R. Evans, WCPD, and (C and D) P.W. Steiner, West Virginia University.]

Ploioderma, and others cause leaf diseases on pine, whereas *Rhabdocline*, *Rhizosphaera*, *Phaeocryptopus*, *Lirula*, and others infect leaves of Douglas fir, spruce, and balsam fir, respectively. All needle cast and blight diseases have certain common characteristics, although each differs from all others in some respects. The needle-like leaves of the conifers are infected by the conidia and occasionally by the ascospores of these fungi at some time during the growing season. The type and time of infection may vary with the location in which the particular species grows. The fungus enters the needle and usually causes a light green to yellow spot that sooner or later turns brown or red, encircles the needle, and kills the part of the needle beyond the spot (Fig. 11-55). The entire needle is often killed and either clings to the tree for a while, giving the tree a reddish-brown, burned appearance, or is shed, resulting in partial or total defoliation of the tree. On the infected needles, whether on the tree or on the ground, the fungus produces its conidia and, occasionally, its ascospores in perithecia, which are either released into the air or are exuded during wet weather and are washed down or splashed by the rain into other needles and trees. In some needle blights, the fungus may overwinter as mycelium in infected but still living needles, but in most cases the fungi overwinter as

FIGURE 11-55 Pine needle cast symptoms caused by *Lophodermium* sp. (A and B) and brown spot needle blight caused by *Mycosphaerela dearnessii* (C and D). [Photographs courtesy of (A and B) Plant Pathology Department, University of Florida and (C and D) E.L. Barnard, Florida Department of Agriculture, Forestry Division.]

ascospores or conidia in dead needles on the tree or on the ground.

Needle casts and blights can be destructive on mature trees, especially in plantations of a single species, which may be killed following repeated defoliations. Every year, thousands of trees are cut when dead or dying from foliage diseases. These same diseases, however, can be devastating in young or nursery trees, which they can kill by the millions in a relatively short time if the weather is favorable and no adequate control measures are taken.

Most, but not all, needle casts and blights can be controlled with fungicidal sprays, especially in the nursery and in young plantation trees. Larger trees are either cut before they die (salvage cutting) or they, too, may be protected, when possible, with fungicides applied from airplanes. In some needle diseases, two sprays either early or late in the season, when most of the infections with the particular fungus take place, are sufficient to keep the disease in check, especially in large trees. In most cases, however, nurseries must be sprayed at least every two weeks from May through October if the seedlings are to survive the needle attacks by fungi and to grow.

MYCOSPHAERELLA DISEASES

As mentioned earlier, there are many and diverse diseases, e.g., of banana, cereals, strawberries, citrus, and pines, that are caused by various species of *Mycosphaerella*. In addition, these fungi produce conidia that belong to different anamorphic genera, such as *Cercospora* and *Septoria*, each of which causes a variety of diseases of annual and perennial plants. The distinction, therefore, between these fungi and between the diseases they cause is nonexistent and is made only for discussion and teaching purposes.

BANANA LEAF SPOT OR SIGATOKA DISEASE

Banana leaf spot, or Sigatoka disease, occurs throughout the world and is one of the most destructive diseases of banana. The name Sigatoka comes from the name of the valley in the South Pacific island Fiji where the disease was first observed. It causes losses by reducing the functional leaf surface of the plant, which results in small bananas that fail to ripen and may fall.

Symptoms. The disease first appears as small, light yellow spots or streaks parallel to the side veins of leaves that unfurled about a month earlier. A few days later, the spots become 1 to 2 centimeters long and turn brown with light gray centers. Such spots soon enlarge further, the tissue around them turns yellow and dies, and adjacent spots coalesce to form large, dead areas on the leaf (Fig. 11-56). In severe infections, entire leaves die within a few weeks. Destruction of most mature leaves by the leaf spot disease may leave only a few functioning leaves; as a result, immature fruit bunches on such plants fail to fill out and ripen and may fall. If the fruit is nearing maturity at the time of heavy infection, the flesh ripens unevenly, individual bananas appear undersized and angular in shape, their flesh develops a buff pinkish color, and they store poorly.

The Pathogen. The causal fungus was *Mycosphaerella musicola*, anamorph *Pseudocercospora musae*. Since the mid-1970s, however, what used to be the common or yellow Sigatoka disease, caused by *M. musicola*, was replaced by the black Sigatoka disease caused by *M. fijiensis*, anamorph *Pseudocercospora fijiensis*. The black Sigatoka pathogen was discovered in Honduras in 1972. It causes spotting 8 to 10 days faster than *M. musicola*, and after severe outbreaks of black Sigatoka in 1973 and 1974, *M. fijiensis* replaced *M. musicola* within two years. By 1980, the black Sigatoka pathogen spread to southern Mexico and throughout Central America, where the cost of sprays to control it accounts for nearly 30% of production costs. By the turn of the century, the more severe black Sigatoka pathogen appeared to have spread to and to have replaced the common Sigatoka fungus in all important banana-producing areas of the world.

The two fungi have similar life cycles and morphology, except that *M. fijiensis* produces sporodochia in young spots and its hyphae spread from one stoma to another and cause lesions over entire leaves much more commonly than *M. musicola*. Both fungi produce spermatia in spermagonia, ascospores in perithecia, and conidia of the *Pseudocercospora* type in sporodochia. Successive abundant crops of conidia are produced on both sides of the leaf during the brown spot stage of the disease (Fig. 11-57).

Development of Disease. Conidia are spread by wind and dripping or splashing water. Release and germination of conidia depend on leaf wetness or high humidity. Perithecia are produced during warm, humid weather, and their ascospores are shot out violently in response to wetting of the perithecia. Ascospores are

FIGURE 11-56 Leaf symptoms of black Sigatoka (A) and whole plant symptoms of yellow (B) Sigatoka disease of banana caused by *Mycosphaerella fijiensis* and *M. musicola*, respectively. (Photographs courtesy of H.D. Thurston, Cornell University.)

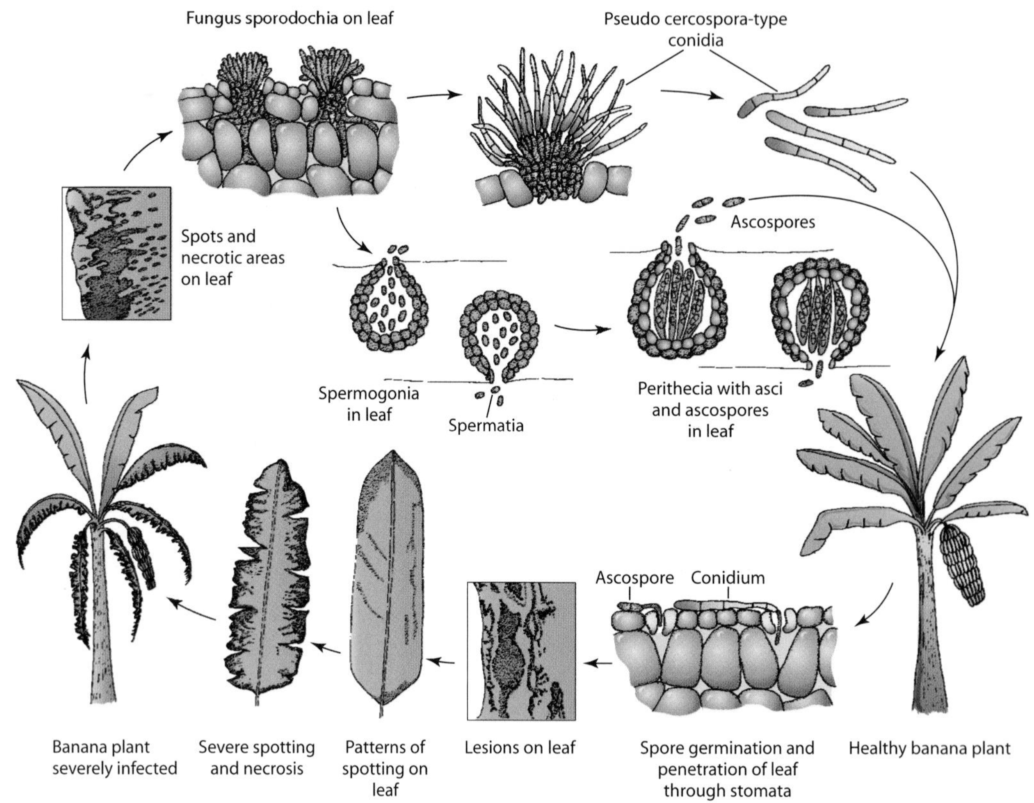

FIGURE 11-57 Development of Sigatoka disease of banana caused by *Mycosphaerella musicola* or *M. fijiensis*.

spread by air currents and are responsible for the long-distance dissemination of the disease, whereas conidia are generally the most important means of local spread of the disease. Infection by either ascospores or conidia produces the same type of spot and subsequent development of the disease (Fig. 11-57).

Control. Sigatoka diseases are controlled by a combination of measures including quarantine, sanitation, and frequent, year-round application of fungicidal sprays. For many years, the Bordeaux mixture or copper oxychloride with or without zineb was the fungicide used. Later it was shown that zineb or copper oxychloride suspended in mineral oil gave better and less expensive control than Bordeaux. To date, several other fungicides are routinely included in oil–water–fungicide emulsions for best all-around results. In some areas it is necessary to apply ground or airplane sprays every 10 to 12 days throughout the year, especially for the control of black leaf streak and black Sigatoka diseases, whereas in other areas one application every 3, 4, or even 6 weeks suffices to maintain control. The repeated exposure of the fungus to fungicides often leads to development of resistance to some of them.

SEPTORIA DISEASES

Septoria diseases occur throughout the world and affect numerous crops on which they cause mostly leaf spots and blights. The most common and serious diseases they cause are leaf blotch and glume blotch of wheat and other cereals and grasses, and leaf spots of celery, tomato, and many other vegetables, field crops, and ornamentals.

Symptoms

On vegetables and flowers, leaf spots begin as small yellowish specks that later enlarge, turn pale brown or yellowish gray, and finally dark brown, usually surrounded by a narrow yellow zone. The spots, depending on the host and fungus species, vary in size from barely visible to 1 to 2 centimeters in diameter to occasional individual spots that affect up to one-third of the leaf area. The spots may have distinct margins with a circular outline or may be very irregular with indistinct edges (Figs. 11-58A and 11-58B). In some hosts, leaves with two or three spots may turn yellow and die, whereas in others the leaves may develop numerous

FIGURE 11-58 Septoria symptoms on leaves of celery (celery late blight) (A) and soybean (brown spot) (B). *Mycosphaerella graminicola* symptoms on oat leaves (leaf blotch) (C) and glumes of wheat (glume blotch) (D) caused by different species of *Septoria*. [Photographs courtesy of (A) L. Mc Donald (WCPD), (B) Plant Pathology Department, University of Florida, (C) W. McFadden, and (D) D.E. Harder, WCPD.]

spots before they turn yellow and eventually droop and die. As the spots form, small black pycnidia appear as dots in them. The disease usually starts on the lower foliage and progresses upward.

On cereals and grasses, the leaf spots appear as light green to yellow or brown spots, first between the veins but soon becoming darker and spreading rapidly to form irregular blotches (Figs. 11-58C and 11-58D). The spots may be restricted or may coalesce and cover the entire blade and sheath, depending on the variety and humidity. The blotches often are speckled with more or less abundant, small, submerged dark pycnidia of the pathogen. In favorable weather, plants become defoliated and the fungus invades the culm, causing black necrotic lesions that result in weakened, dead, and often lodged plants. Smaller lesions with fewer pycnidia may develop on the floral bracts and on the pericarp of the kernels.

The Pathogen

The fungus, *Septoria* spp., the teleomorph of which is *Mycosphaerella*, had some of its species reclassified as *Stagonospora*, the teleomorph of which is *Phaeosphaeria*. These fungi exist as many species that affect different hosts. They produce long, filiform, colorless, one- to several-celled conidia in dark, globose pycnidia. *Septoria tritici*, the teleomorph of which is *Mycosphaerella graminicola*, and *Stagonospora nodorum*, the teleomorph of which is *Phaeosphaeria nodorum*, cause frequent and severe disease on cereals and grasses, while numerous other species cause diseases on vegetables and other crop plants.

Development of Disease

When the pycnidia become wet, they swell and the conidia are exuded in long tendrils. Conidia are spread by splashing rain, irrigation water, tools, animals, and so on. *Septoria* overwinters as mycelium and as conidia within pycnidia on and in infected seed and on diseased plant refuse left in the field (Fig. 11-59). When the fungus is carried in the seed, it produces seedling infection that may result in damping-off or provide inoculum for subsequent infections. Although all *Septoria* species require high moisture for infection and severe disease development, they can cause disease at a wide range of temperatures, e.g., between 10 and 27°C.

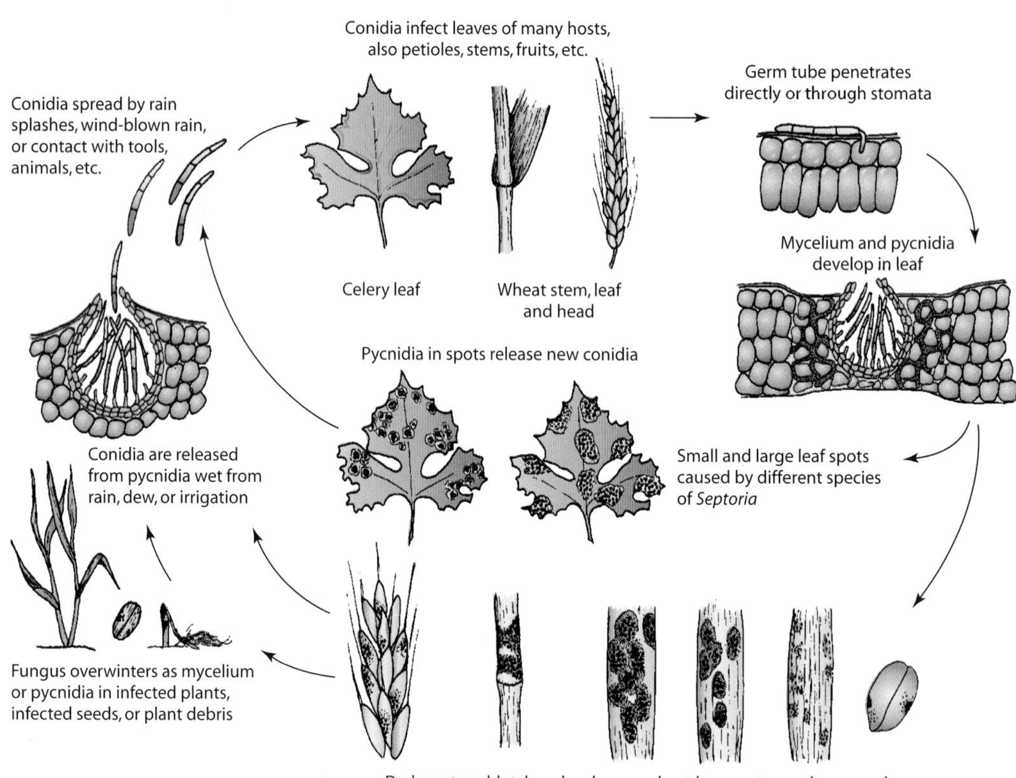

FIGURE 11-59 Development of diseases caused by *Mycosphaerella graminicola* and *Septoria* spp.

Control

Control of *Septoria* diseases depends on the use of disease-free seed in a field free of the pathogen, 2- to 3-year crop rotations, sanitation by deep plowing of plant refuse, use of resistant varieties, and chemical sprays of the plants in the seedbed and in the field. Several fungicides are available for the control of *Septoria* diseases.

CERCOSPORA DISEASES

Cercospora diseases are almost always leaf spots. The spots either stay relatively small and separate or may enlarge and coalesce, resulting in leaf blights. The diseases are generally widespread among most cereals and grasses, many field crops, vegetables, ornamentals, and trees. The Cercospora early blight of celery (Fig. 11-60), leaf spot of beets and of peanuts, leaf spot and blight of soybean (Figs. 11-60A and 11-60B), and gray leaf spot of corn are common and severe. Losses from Cercospora diseases are usually small, but in some hosts, and occasionally in others, they can be significant.

Symptoms

The leaf spots on some plants are brown, small, about 3 to 5 millimeters in diameter, and roughly circular with reddish-purple borders (Figs. 11-60A and 11-60C). Later, their centers become ashen gray, thin, papery, and brittle and may drop out, leaving a ragged hole, or the spots, if sufficiently numerous, may coalesce, causing large necrotic areas. On most hosts, the spots are irregularly circular to angular, with or without a distinct border, and often coalesce to form large blighted areas. The seed of some soybean varieties infected with certain species of *Cercospora* exhibit patches of a purplish coloration (Fig. 11-60B). In monocotyledonous plants the spots are narrow and long, usually 0.5 by 5.0 centimeters, and may coalesce and kill leaves. In humid weather the affected leaf surface on all hosts is covered with an ashen gray mold barely visible to the naked eye. In severe attacks, all the foliage is destroyed and may fall off. On fleshy plants, similar lesions are produced on stems and leaf petioles.

The Pathogen

Several species of *Cercospora* are responsible for the diseases on the various hosts. Some *Cercospora* species have *Mycosphaerella* as their teleomorph. The fungus produces long, slender, colorless to dark, straight to slightly curved, multicellular conidia on short dark conidiophores (Fig. 11-57). Conidiophores arise from the plant surface in clusters through the stomata and form conidia successively on new growing tips. Conidia are detached easily and are often blown long distances by the wind. The fungus is favored by high temperatures and therefore is most destructive in the summer months and in warmer climates. Most *Cercospora* species produce the nonspecific toxin cercosporin, which acts as a photosensitizing agent in the plant cells, i.e., it kills cells only in the light. The toxin incites the production of reactive atomic oxygen in the cells, which causes disruption of cell membranes and loss of electrolytes from cells. Although *Cercospora* spores need water to germinate and penetrate, heavy dews seem to be sufficient for infection. The pathogen overseasons in or on the seed and as minute black stromata in old infected leaves.

Control

Cercospora diseases are controlled by using disease-free seed or seed at least three years old, by which time the fungus in the seed has died; by using crop rotations with hosts not affected by the same *Cercospora* species; and by spraying the plants, both in the seedbed and in the field, with appropriate fungicides.

RICE BLAST DISEASE

The blast disease of rice occurs worldwide and is one of the most important diseases of rice, particularly where rice is irrigated or receives high amounts of rainfall and high levels of nitrogen fertilizer. Several rice blast epidemics have occurred in different parts of the world, resulting in yield losses in these areas ranging from 50 to 90% of the expected crop.

Symptoms

Rice blast affects the leaves, on which it causes diamond-shaped white to gray or reddish-brown lesions with reddish to brown borders (Fig. 11-61A); the lesions may enlarge, coalesce, and kill entire leaves. Blast also affects the leaf collar, which it may kill (thereby killing the entire leaf), and the stem nodes and occasionally the internodes, which, at heading, may result in the production of white panicles or breakage of the stem at the infected node (Figs. 11-61B and 11-61C). At heading, the fungus also attacks the panicle neck node, which is girdled. This is usually the most destructive symptom of the disease and is called the neck rot, neck blast, or panicle blast stage of the disease (Figs. 11-61B–11-61D). If infection of the panicle neck occurs early, the grains do not fill and the panicle remains erect. If the panicle neck is infected late, the grains become partially filled, and because of the weight of the grains, the base of the panicle breaks and the panicle droops. Sometimes only

FIGURE 11-60 Cercospora symptoms on soybean leaf (A), soybean seed (B) (purple stain), and celery leaf (C) showing celery early blight. (D) Cherry leaf spot caused by *Blumeriella jaapii*. [Photographs courtesy of (A and B) Plant Pathology Department, University of Florida, (C) R.T. McMillan, University of Florida, and (D) M. Ellis, Ohio State University.]

FIGURE 11-61 Rice blast symptoms on (A) rice leaves, (B) rice stalks, and (C) neck rot or blast symptoms leading to white heads. (D) Severe blasting of rice panicles in the field. (E) Conidia of the rice blast fungus *Magnaporthe (Pyricularia)* sp. [Photographs courtesy of (A and C) J. Breithaupt, FAO, (B) J. Kranz, University of Giessen, and (D) L.E. Datnoff and (E) R.E. Cullen, University of Florida.]

parts of the panicle and some glumes become infected and develop brown to black spots.

The Pathogen

The rice blast fungus pathogen has been known as *Pyricularia oryzae* but is indistinguishable from *P. grisea*, which causes gray leaf spot on other grasses. The teleomorph stage, *Magnaporthe grisea*, has not been found in nature, but it has been produced after crossing appropriate compatible isolates in the laboratory. The fungus produces simple, gray conidiophores that bear terminal, pear-shaped, mostly two-septate conidia (Fig. 11-61E). The fungus produces several toxins, e.g., pyricularin and α-picolinic acid, that seem to contribute to the development of rice blast.

Development of Disease

The pathogen overseasons as mycelium and conidia on diseased rice straw and seed, and possibly on weed hosts. In the tropics, conidia are present in the air throughout the year. The fungus produces and releases conidia during periods of high relative humidity (i.e., 90% or higher). The conidia become airborne, and on landing on rice plants, they adhere strongly through sticky mucilage they produce at their tip. When rice leaf or stem surfaces are wet, the conidia germinate and the germ tube produces an appressorium through which the fungus penetrates plant surfaces or enters through stomata. Production and accumulation of melanins in the appressorium cell wall seem to be necessary for successful penetration. Rice seedlings and young leaf and stem tissues are more susceptible than older plants and tissues. At optimum temperatures, new blast lesions appear within 4 to 5 days. In wet weather or high relative humidity, new conidia are produced and released within hours from appearance of the lesions, and this continues for several days, with most conidia being released between midnight and sunrise.

Rice blast is favored greatly by high nitrogen fertilization, prolonged leaf wetness, and night temperatures around 20°C. The pathogen exists as numerous pathogenic races, each carrying different genes for virulence. Several major genes for resistance to blast have been identified in different rice cultivars, but each resistance gene is quickly (within 2 to 3 years) overcome by appearance of new pathogen races.

Control

Control of rice blast in areas of low blast pressure is based primarily on planting resistant cultivars. Where blast epidemics are common and severe, in addition to planting resistant cultivars, which must be changed frequently, control is aided by early planting, keeping nitrogen fertilizers to the minimum necessary, and using fungicides. Many fungicides have been used with varying effectiveness over the years. More recently, systemic fungicides, which inhibit penetration by the fungus through interference with melanin production in its appressorium, have been shown to give good control of rice blast when applied as sprays and even as seed treatments. An experimental preplant soil application of silicon-containing basic ground granulated blast-furnace slag seems to reduce rice blast in some areas.

COCHLIOBOLUS, PYRENOPHORA, AND SETOSPHAERIA DISEASES OF CEREALS AND GRASSES

Cochliobolus, Pyrenophora, and Setosphaeria diseases occur throughout the world and are very common and severe on many important crop plants of the grass family. Thus, different species cause corn leaf blights; brown spot or blight of rice; and leaf spots, blights, crown rots, and root rots of wheat, barley, oats, rye, sorghum, sugarcane, and turf grasses. The total losses in grain and forage caused annually by these pathogens are staggering.

Leaf spots and blights, and also the crown and root rot diseases, caused by *Cochliobolus*, *Pyrenophora*, and *Setosphaeria* on the various hosts have many similarities, but also some significant differences. Three of them, brown spot or blight of rice, southern corn leaf blight, and Victoria blight of oats, caused sudden and catastrophic epidemics that resulted in huge crop losses, human suffering, and new approaches to disease control. All Cochliobolus, Pyrenophora, and Setosphaeria diseases destroy various percentages of the leaf area, may attack and destroy part of the stem or roots, or attack the kernels directly, in every case causing considerable yield loss.

DISEASES ON CORN

Southern corn leaf blight, caused by *Cochliobolus heterostrophus*, anamorph *Bipolaris maydis*, causes small (0.6 by 2.5 cm), tan lesions that may be so numerous that they almost cover the entire leaf (Figs. 11-62A and 11-62B). Some races of the fungus also attack the stalks, leaf sheaths, ear husks, shanks, ears, and cobs (Figs. 11-62C and 11-63). Affected kernels are covered with a black, felty mold, and cobs may rot or, if the shank is infected early, the ear may be killed prematurely and drop. Seedlings from infected kernels may wilt and

FIGURE 11-62 Symptoms of southern corn leaf blight on corn leaves (A and B) and corn leaf sheaths (C) caused by *Cochliobolus heterostrophus (Bipolaris maydis)*. (D) Northern corn leaf blight caused by *Cochliobolus carbonum (Bipolaris zeae)* showing its much larger spots. [Photographs courtesy of (A and C) Plant Pathology Department, University of Florida and (B and D) P.E. Lipps, Ohio State University.]

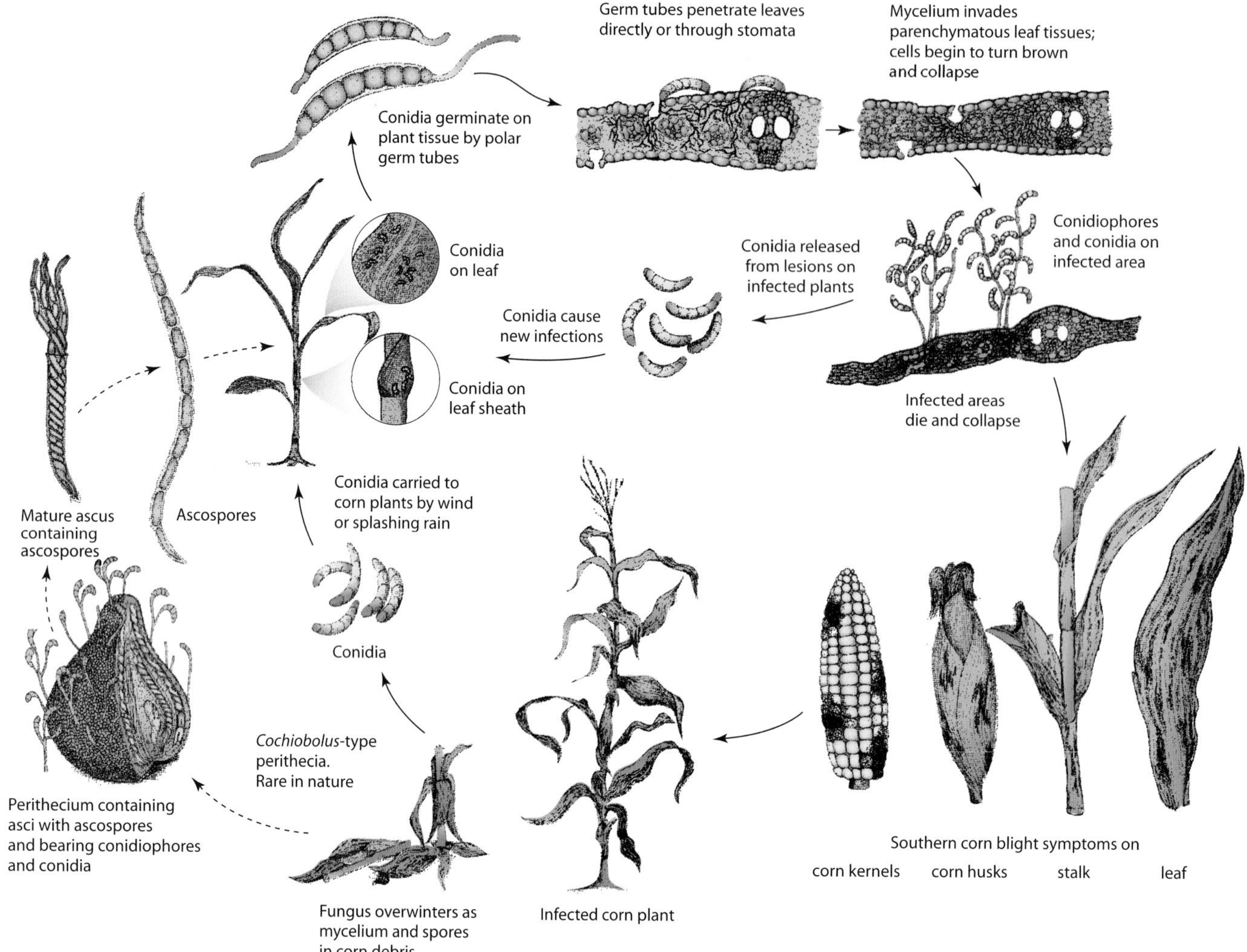

FIGURE 11-63 Disease cycle of southern corn leaf blight caused by *Cochliobolus heterostrophus* race T.

die within a few weeks of planting. A widespread epidemic caused by a new race (race T) of the southern corn leaf blight fungus occurred suddenly in 1970 on all corn hybrids containing the Texas cytoplasmic male sterility gene (used for efficient crossing and production of corn hybrids) and destroyed about 15% of all corn produced in the United States that year. The monetary value of the lost crop was estimated at $1 billion.

Northern corn leaf blight, caused by *Setosphaeria turcica*, anamorph *Exserohilum turcicum*, affects only the leaves. Lesions range in length from 2 to 15 centimeters and are 1 to 3 centimeters wide (Fig. 11-62D). The same fungus causes similar but smaller and darker spots on sorghum.

Northern corn leaf spot, caused by *Cochliobolus carbonum*, anamorph *Bipolaris zeicola*, is widespread but is important primarily on susceptible inbreds used for the production of hybrid seed. Leaf spot size varies with the race of the fungus. Some of its races also attack ears of corn, producing a black, felty mold on the kernels.

DISEASES ON RICE

Brown spot disease of rice, caused by the fungus *Cochliobolus miyabeanus*, anamorph *Bipolaris oryzae*, appears on the leaves, panicles, glumes, and grain, at first as spots (Fig. 11-64) that have a gray center and brown border. Entire glumes may be covered by several small spots or one large spot on which a dark brown, velvety layer of conidiophores and conidia is present. The fungus causes damage primarily by attacking the leaves during the seedling stage, as a result of which the plants are weakened and the yield is reduced drastically. It was such seedling infections that resulted in the Bengal famine in 1942, when approximately two million people died from starvation.

FIGURE 11-64 Brown spot of rice caused by *Cochliobolus miyabeanus*. (Photograph courtesy of Plant Pathology Department, University of Florida.)

COCHLIOBOLUS DISEASES ON WHEAT, BARLEY, AND OTHER GRASSES

Crown rot and common root rot, caused by *Cochliobolus sativus*, anamorph *Bipolaris sorokiniana*, appear as spots on seedlings, plant crowns, stems, leaves, floral parts, and kernels. Brown to black lesions develop on seedlings near the soil line and may spread into the leaves (Figs. 11-65A and 11-65B). The seedlings may be blighted and killed before or after emergence or they may survive but their growth is retarded. Crowns are infected at or just below the soil line and show a reddish-brown decay that destroys the tiller buds and advances into the root system, which it kills (Fig. 11-65E). Winter survival of root rot-infected wheat and barley plants is reduced considerably, in some cases by 10 to 30% in wheat and 20 to 60% in barley. The leaf spots frequently enlarge and coalesce to form brown, irregular stripes or blotches (Fig. 11-65D) that cover large areas of the leaf blade. Older lesions are covered with a layer of olive-colored conidiophores and conidia of the fungus. Floral parts and kernels also develop lesions or their entire surface may appear dark brown. The embryo end of the kernel is often black (Fig. 11-65C), and this symptom is characteristic of this disease and is referred to as black point. Because of crown and root rot and leaf blotch, surviving plants are shorter; the spikes may be only partially emerged and may be sterile or have poorly filled kernels. Blight of floral parts and kernels causes sterility or death of individual kernels.

In wheat and barley crown rot and root rot, *C. sativus* is often found in plants together with *Fusarium*, *Rhizoctonia*, and *Pythium*.

Spot blotch of barley and wheat, caused by *C. sativus*, anamorph *B. sorokiniana*, appears as spots or blotches on leaves (Fig. 11-65B) and also as seedling blight, root rot, and kernel blight, black point, or smudge.

PYRENOPHORA DISEASES ON WHEAT, BARLEY, AND OATS

Pyrenophora diseases on cereals are common and widespread and cause considerable losses year after year. The losses vary with the crop, the variety, and the prevailing weather conditions. The diseases seem to be more common and more severe on barley than on other crops.

Net blotch of barley (Fig. 11-66A), caused by *Pyrenophora teres*, anamorph *Drechslera teres*, produces almost square-shaped, net-like blotches near the tip of the seedling leaf. The spots later enlarge and spread along the entire leaf blade, and the leaf develops a netted appearance.

Barley stripe (Fig. 11-66B), caused by *Pyrenophora graminea*, anamorph *Drechslera graminea*, causes yellow stripes along leaf blades and sheaths of the older leaves. Later, these stripes become brown; near the end of the season, the leaves often split along the stripe lesions and become shredded. Infected plants are stunted and usually do not produce normal heads.

Tan spot of wheat (Fig. 11-66C), caused by *Pyrenophora tritici-repentis*, anamorph *Drechslera tritici-repentis*, is an important disease in the northern wheat-producing areas of the United States. Symptoms include tan lesions surrounded by a yellow halo.

Lawn and golf course grasses are attacked frequently by various *Cochliobolus*, *Pyrenophora*, and *Setosphaeria* species, which cause common and serious diseases of grasses known as leaf spots, blights, crown (foot) rots, and root rots (melting out) (Figs. 11-65A, 11-65D, and 11-65E). These diseases resemble in most respects those described earlier for small grain crops. In severe infections, leaves become completely blighted,

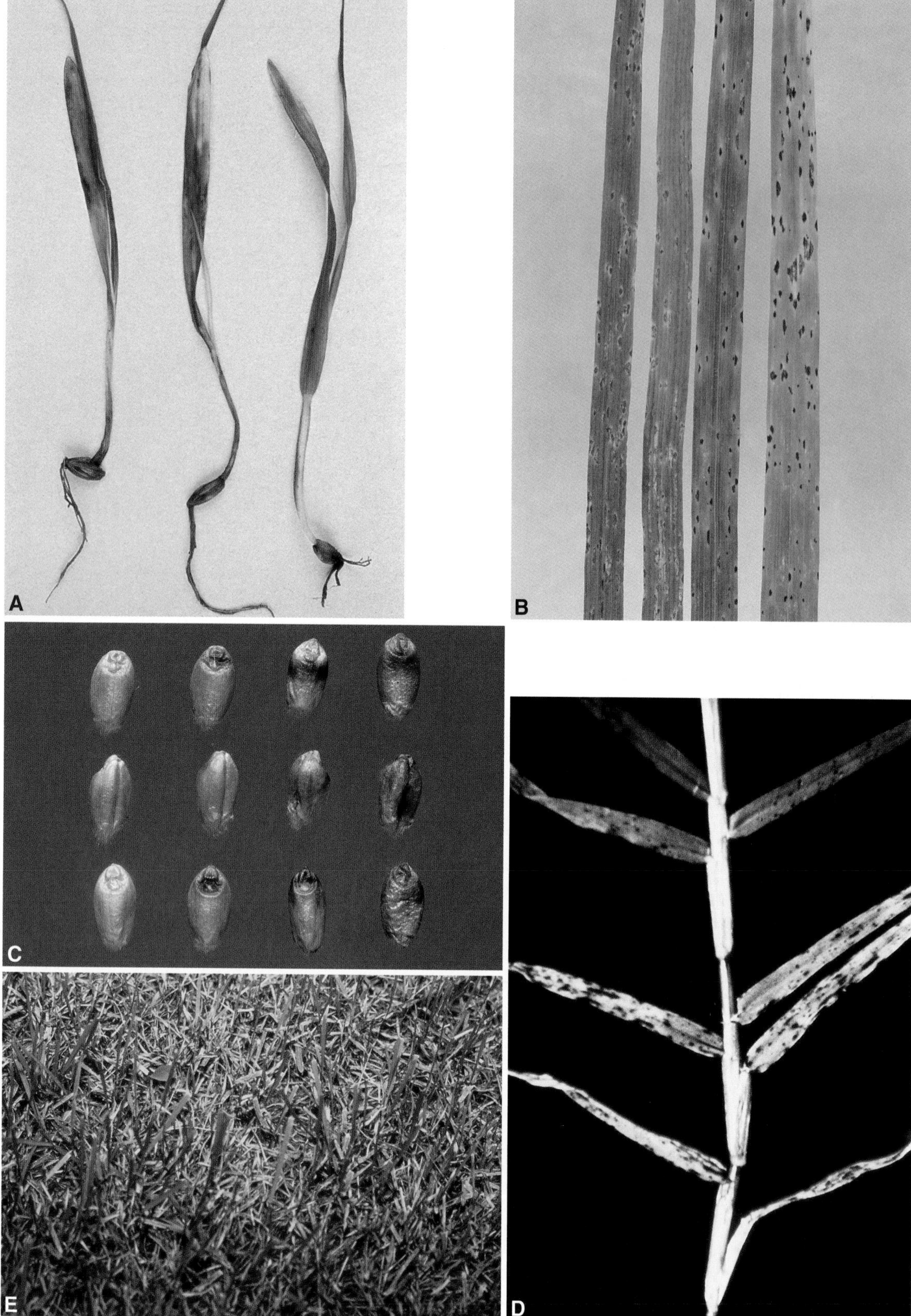

FIGURE 11-65 Common root rot of barley (A), spot blotch of wheat (B), and black point or kernel smudge of wheat (C) caused by *Cochliobolus sativus*. (D and E) Bluegrass leafspot and melting out caused by *Drechslera* sp. [Photographs courtesy of (A) I.R. Evans, (B, D, and E) S. Fushtey, and (C) L.J. Duczek, WCPD.]

FIGURE 11-66 Barley net blotch (A) caused by *Pyrenophora terres*, barley leaf stripe (B) caused by *P. graminea*, wheat tan spot (C) caused by *P. tritici-repentis*, oat stripe blight (D) caused by *Drechslera avenacea*, and Victoria oat blight (E) caused by *Bipolaris victoriae*. (F) Scanning electron micrograph of *C. heterostrophus* conidiophores and conidia on corn leaf surface. [Photographs courtesy of (A, B, and D) I.R. Evans, (C) L.J. Duczek, WCPD, (E) Plant Pathology Department, University of Florida, and (F) M.F. Brown and H.G. Brotzman.]

wither, die, and drop off. Furthermore, the perennial nature of the turf grasses and the fact that they are mowed and irrigated several times each year create additional opportunities for the diseases to spread and to become established. The diseases are usually most destructive during wet or humid weather, or where the turf is sprinkle irrigated frequently, especially late in the day. In advanced cases of infection, all plants in areas of various sizes and shapes turn yellow, then brown to straw colored, and finally are killed (melting out). The disease, if unchecked and if the weather is favorable, may spread and kill the entire turf in the area.

The pathogens in the aforementioned diseases are species of *Cochliobolus*, *Pyrenophora*, and *Setosphaeria*. These fungi produce cylindrical, dark, three to many celled (usually 5–10) conidia (Figs. 11-63 and 11-66F). The conidia are produced successively on new growing tips of dark, septate, irregular conidiophores. The fungi also produce, with more or less regularity, black perithecia containing cylindrical asci, within which are formed colorless, threadlike to pyreno-form, four- to nine-celled ascospores. Species that produce a *Cochliobolus* perfect stage have *Bipolaris* species as their anamorph (conidial stage), whereas those that produce a *Pyrenophora* perfect stage have *Drechslera* species as their anamorph. Some closely related fungi causing similar diseases produce a *Setosphaeria* perfect stage and their anamorph is *Exserohilum*.

Bipolaris conidia are slightly curved, whereas those of *Drechslera* are mostly straight and cylindrical. Conidia of *Exserohilum* are straight with pointed ends and protruding basal tip (hilum).

Development of Disease

The various *Cochliobolus*, *Pyrenophora*, and *Setosphaeria* species survive the winter as mycelium or spores in or on infected or contaminated seed, in plant debris, and in infected crowns or roots of susceptible plants. Some species of these fungi are weak parasites, but several are potent pathogens. When in the soil, however, all of them are weak as saprophytes, probably because of antagonism by soil microorganisms, especially at high nitrogen content. Many *Cochliobolus* species, e.g., *C. victoriae* and *C. heterostrophus* race T, produce potent host-specific toxins such as victorin and T-toxin that play important roles in the development of the respective diseases. Most species are favored by moderate to warm (19–32°C) temperatures and particularly by humid, damp weather. Most diseases, especially leaf spots, are retarded by dry weather, whereas crown- and root-affecting fungi may continue their invasion of diseased plants, killing the plants in irregular areas. Spread of the fungus is through the seed and infected debris. During the growing season, the fungus spreads over short distances through its numerous conidia, which may be carried by air currents, splashing rain, or by clinging to cultivating equipment, feet, animals, and so on.

Control

Control of Cochliobolus, Pyrenophora, and Setosphaeria diseases depends on the use of resistant varieties, pathogen-free seed, seed treatment with fungicides, proper crop rotation and fertilization, plowing under of infected plant debris, and fungicides. In turf grasses, control of these diseases is facilitated by mowing at the recommended maximum height, reducing or removing the accumulated dense thatch, supplying sufficient fertilizer, and irrigating quickly and sufficiently but in widely spaced (7- to 10-day) intervals. If fungicides are necessary, several of them can be applied beginning in early spring and continuing at 1- to 2-week intervals for as long as necessary to get the disease under control. Several systemic fungicides, when applied as seed treatments or, in turf grasses, with irrigation water, give good control of the root rot and several of the other symptoms and are being tested for grower application.

Selected References

Alcorn, J. L. (1988). The taxonomy of "*Helminthosporium*" species. *Annu. Rev. Phytopathol.* **26**, 37–56.

Anonymous (1970). Southern corn leaf blight. *Plant Dis. Rep.* **54** (Special Issue), 1099–1136.

Berger, R. D. (1973). Early blight of celery: Analysis of disease spread in Florida. *Phytopathology* **63**, 1161–1165.

Chupp, C. (1953). "A Monograph of the Fungus Genus *Cercospora*." Published by the author, Ithaca, New York.

Cunfer, B. M., and Ueng, P. P. (1999). Taxonomy and identification of *Septoria* and *Stagonospora* species on small grain cereals. *Annu. Rev. Phytopathol.* **37**, 267–284.

Frank, J. A. (1985). Influence of root rot on winter survival and yield of winter barley and winter wheat. *Phytopathology* **75**, 1039–1041.

Gibson, I. A. S. (1972). *Dothistroma* blight of *Pinus radiata*. *Annu. Rev. Phytopathol.* **10**, 51–72.

Gottwald, T. R. (1983). Factors affecting spore liberation by *Cladosporium carpophilum*. *Phytopathology* **73**, 1500–1505.

Jewell, F. F., Sr. (1983). Histopathology of the brown spot fungus on longleaf pine needles. *Phytopathology* **73**, 854–858.

Joosten, M. H. A. J. (1999). The tomato–*Cladosporium fulvum* interaction: A versatile experimental system to study plant-pathogen interactions. *Annu. Rev. Phytopathol.* **37**, 335–367.

Keinath, A. P. (2001). Effect of fungicide applications scheduled to control gummy stem blight on yield and quality of watermelon fruit. *Plant Dis.* **85**, 53–58.

Mackenzie, D. R. (1981). Association of potato early blight, nitrogen fertilizer rate, and potato yield. *Plant Dis.* **65**, 575–577.

Meredith, D. S. (1970). Banana leaf spot disease (Sigatoka) caused by *Mycosphaerella musicola*. *Commonw. Mycol. Inst. Phytopathol. Pap.* **11**, 1–147.

Ou, S. H. (1980). A look at worldwide rice blast disease control. *Plant Dis.* **64**, 439–445.

Padnamadhan, S. Y. (1973). The great Bengal famine. *Annu. Rev. Phytopathol.* **11**, 11–26.

Pearson, R. C., and Goheen, A. C., eds. (1988). "Compendium of Grape Diseases." APS Press, St. Paul, MN.

Rotem, J. (1994). "The Genus *Alternaria*." APS Press, St. Paul, MN.

Sherff, A. F., and MacNab, A. A. (1986). "Vegetable Diseases and Their Control," 2nd Ed. Wiley, New York.

Shipton, W. A., Boyd, W. R. J. Rosielle, A. A., and Shearer, B. L. (1971). The common *Septoria* diseases of wheat. *Bot. Rev.* **37**, 231–262.

Sing. U. S., Mukhopadhyay, A. N., Kumar, J., and Choube, H. S. (1992). "Plant Diseases of International Importance," Vol. 1. Prentice-Hall, Englewood Cliffs, NJ.

Sivanesan, A. (1987). *Graminicolous* species of *Bipolaris*, *Curvularia*, *Drechslera*, *Exserohilum* and their teleomorphs. *Mycol. Papers* **158**, 1–261.

Smiley, R. W., Dernoeden, P. H., and Clarke, B. B. (1992). "Compendium of Turfgrass Diseases," 2nd Ed. APS Press, St. Paul, MN.

Smith, J. D., Jackson, N., and Woolhouse, A. R. (1989). "Fungal Diseases of Amenity Turf Grasses." E. & F. N. Spon, London.

Sprague, R. (1944). *Septoria* diseases of *gramineae* in western United States. *Oreg. State Monogr. Stud. Bot.* **6**, 1–151.

Stover, R. H. (1980) Sigatoka leaf spots of bananas and plantains. *Plant Dis.* **64**, 750–756.

Stover, R. H. (1986). Disease management strategies and the survival of the banana industry. *Annu. Rev. Phytopathol.* **24**, 83–91.

Ullstrup, A. J. (1972). The impacts of the southern corn leaf blight epidemics of 1970–1971. *Annu. Rev. Phytopathol.* **10**, 37–50.

Taylor, J. W., Jacobson, D. J., and Fisher, M. C. (1999). The evolution of asexual fungi: Reproduction, speciation, and classification. *Annu. Rev. Phytopathol.* **37**, 197–246.

Ward, J. M. J., *et al.* (1999). Gray leaf spot: A disease of global importance in maize production. *Plant Dis.* **83**, 884–895.

Webster, R. K., and Gunnell, P. S., eds. (1992). "Compendium of Rice Diseases." APS Press, St. Paul, MN.

White, D. G. (1999)." Compendium of Corn Diseases", 3rd Ed. APS Press, St. Paul, MN.

Wiese, M. V. (1987). "Compendium of Wheat Diseases," 2nd Ed. APS Press, St. Paul, MN.

Zeigler, R. S. (1998). Recombination in *Magnaporthe grisea*. *Annu. Rev. Phytopathol.* **36**, 249–275.

Zhong, S., and Steffenson, B. J. (2001). Virulence and molecular diversity in *Cochliobolus sativus*. *Phytopathology* **91**, 469–476.

STEM AND TWIG CANKERS CAUSED BY ASCOMYCETES AND DEUTEROMYCETES (MITOSPORIC FUNGI)

Cankers are localized wounds or dead areas in the bark of the stem or twigs of woody and other plants that are often sunken beneath the surface of the bark. In some cankers, the healthy tissues immediately next to the canker may increase in thickness and appear higher than the normal surface of the stem.

Innumerable kinds of pathogens cause cankers on trees. The most common causes of tree cankers are ascomycetous fungi, although some other fungi, some bacteria, and some viruses also cause cankers. Particularly common are cankers caused by several species of the oomycete *Phytophthora*, already discussed.

The basic characteristic of cankers is that they are visible dead areas that develop in the bark and, sometimes, in the wood of the tree. Cankers generally begin at a wound or at a dead stub. From that point, they expand in all directions but much faster along the main axis of the stem, branch, or twig. Under some environmental conditions, the host may survive the disease by producing callus tissue around the dead areas and thus limiting the canker. In infections of large limbs, concentric layers of raised callus tissue may form. If, however, the fungus grows faster than the host can produce its defensive tissues, either no callus layers form and the canker appears diffuse and spreads rapidly, or the fungus invades each new callus layer and the canker grows larger each year. Young twigs are often girdled by the canker and killed soon after infection, but on larger limbs and stems cankers may become up to several meters long, although their width extends to only part of the perimeter of the limb. Eventually, however, the limb or entire tree may be killed through girdling either by the original canker or by additional cankers that develop from new infections caused by the spores from the original canker.

Cankers are generally much more serious on fruit trees such as apple and peach, which they debilitate and kill. On forest trees, with the exception of chestnut blight and a few others, cankers deform but do not kill their hosts. They do, however, reduce tree growth and the quality of lumber, result in greater wind breakage, and weaken the trees so that other more destructive wood- or root-rotting fungi can attack the trees.

Although most canker-causing fungi are Ascomycetes, only some of them produce their sexual stage regularly. The other canker fungi produce primarily conidia, usually in pycnidia embedded in the bark, and only occasionally do they produce perithecia. Some of the canker-causing fungi and their most important host plants are as follows.

Apiosporina morbosa, causing black knot of plum and cherry (Fig. 11-67A)

Botryosphaeria dothidea, causing canker on apple, peach, almond, sycamore (Fig. 11-67B) pecan, etc.

Ceratocystis fimbriata, causing canker diseases on cacao, coffee, stone fruits (Fig. 11-67C), etc.

Cryptodiaporthe populea, causing the Dothichiza canker of poplar

Cryphonectria (Endothia) parasitica, causing chestnut blight (Figs. 11-68A and 11-68B)

FIGURE 11-67 (A) Black knot canker on cherry twig. (B) Botryosphaeria cankers on sycamore tree. (C) Ceratocystis canker on trunk of almond tree. (D) Hypoxylon canker on oak stem. (E) Strumella canker on trunk of oak tree. [Photographs courtesy of (A) J.W. Pscheidt, Oregon State University, (B) E.L. Barnard, Florida Department of Agriculture, Forestry Division, (C) B. Teviotdale, University of California, and (D and E) U.S. Forest Service archives.]

FIGURE 11-68 (A and B) Trunk of young chestnut tree infected with, and covered with pycnidia of, the causal fungus *Cryphonectria parasitica.* (C) Canker on trunk of eucalyptus tree and (D) pycnidia of the causal fungus *Diaporthe cubensis.* (E) Canker on stem vine of grape caused by *Eutypa* sp. [Photographs courtesy of (A) J.W. Pscheidt, Oregon State University (B) U.S. Forest Service, (C and D) E.L. Barnard, Florida Department of Agriculture, Forestry Division, and (E) E. Hellman, Texas A&M University.]

Eutypa lata, causing canker and dieback of grape (Fig. 11-68E)
Eutypella parasitica, causing Eutypella trunk canker of maple, etc.
Fusarium circinatum, causing pitch canker of pines (Fig. 11-73)
Fusicoccum amygdali, causing twig cankers on almond and other stone fruit trees
Gremmeniella abietina, causing the Scleroderris canker of conifers
Hypoxylon mammatum, causing Hypoxylon canker of aspen (Fig. 11-67D)
Leucostoma sp. (*Valsa* sp.), causing canker of peach (Fig. 11-71), many other fruit trees, and more than 70 species of hardwood trees
Nectria galligena, causing canker of apple, pear, and many forest trees (Fig. 11-69)
Phomopsis juniperivora, causing Phomopsis blight of cedars, arborvitae, cypress, etc.
Seiridium cardinale, causing canker of cypress and Leyland cypress trees
Sirococcus clavignenti-juglandacearum, causing the devastating butternut canker of butternut (white walnut) trees (Fig. 11-74)
Urnula craterium, causing Strumella canker of forest trees (Fig. 11-67E)

The main characteristics of canker diseases caused by some of the aforementioned fungi are given here.

BLACK KNOT OF PLUM AND CHERRY

Black knot disease occurs on cultivated and wild plums and cherries, primarily in the eastern half of the United States and in New Zealand.

Symptoms

The disease appears as conspicuous, 2 to 25 centimeters long, black knotty swellings on one side of, or encircling, twigs and branches (Fig. 11-67). The knots may be several times the diameter of the limbs and make heavily infected trees appear quite grotesque. Infected plants become worthless after a few years as a result of limb death and stunting of the trees.

The Pathogen

The fungus, *Apiosporina* (= *Dibotryon*) *morbosa*, produces conidia on free hyphae and ascospores in perithecia formed in the black knots.

Development of Disease

Both conidia and ascospores are spread by wind and rain, and in early spring they can penetrate healthy and injured woody tissue of the current season's growth. Large limbs are also attacked, especially at points of developing small twigs. The fungus grows into the cambium and xylem parenchyma and along the axis of the twig. After 5 or 6 months, excessive parenchyma cells are produced and pushed outward, forming the swelling. The following spring, conidia are produced on the knot surface, giving it a temporary olive-green velvety appearance. The knots enlarge rapidly during the second summer, and perithecia in their surface layer are formed that develop during the winter and release ascospores the following spring. The knots continue to expand in following years.

Control

The disease can be controlled by pruning and burning all black knots and the destruction of black knots of all affected wild plums and cherries near the orchard. Spraying the orchard trees before and during bloom with one of several fungicides protects trees from infection.

CHESTNUT BLIGHT

After it was introduced in New York City in 1904, chestnut blight, caused by cankers on the lower trunks and larger branches of chestnut trees (Figs. 11-68A and 11-68B), spread rapidly. By 1940, it had destroyed practically all American chestnut trees throughout their natural range in the eastern third of the United States from the Canadian border south nearly to the Gulf of Mexico (see Fig. 1-8). American chestnuts killed by the blight composed 50% of the overall value of eastern hardwood timber stands. The fungus, *Cryphonectria* (*Endothia*) *parasitica*, also attacks oak and, sporadically, other trees, but not nearly as severely as it attacks the American chestnut. It is now present throughout North America, Europe, and Asia. Other species of *Cryphonectria* cause severe cankers on other forest trees (Figs. 11-68C and 11-68D). The fungus penetrates the bark of stems through wounds and then grows into the inner bark and cambium.

Symptoms

Swollen or sunken cankers develop on stems or branches of infected trees. The bark of the cankers is reddish-orange to yellow-green and is covered by pimple-like pycnidia and perithecia (Fig. 11-68). Cankers often have long cracks on their surface, may be several inches to many feet long, and eventually girdle the stem or branch, causing wilting and death of the parts beyond the canker.

FIGURE 11-69 (A) Nectria canker on apple trunk and (B) Nectria-infected branch covered with pycnidia of the fungus. (C) Nectria canker on main branches of red maple tree and (D) perithecia of *Nectria* sp. on infected part of the trunk. [Photographs courtesy of (A and B) A.L. Jones, Michigan State University and (C and D) E.L. Barnard, Florida Department of Agriculture, Forestry Division.]

The Pathogen

Chestnut blight is caused by the fungus *Cryphonectria parasitica.*

Development of Disease

The pathogen produces conidia that ooze out of pycnidia as long orange curls during moist weather and are spread by birds, crawling or flying insects, or splashing rain. The ascospores are shot forcibly into the air and may be carried by wind over long distances. The fungus survives and continues to invade and produce spores in trees or parts of trees already killed by the blight. Blighted trees almost always produce sprouts below the basal cankers, but the resulting saplings become blighted in turn by new infections.

Control

No control is available against chestnut blight, although some new systemic fungicides appear promising for isolated trees. Since the mid-1980s, several strains of the fungus that show reduced virulence (hypovirulence) have been found in Europe and the United States. All of these strains contain double-stranded RNA (dsRNA), which is the kind of RNA present in many viruses that infect fungi. When a chestnut tree canker caused by a typical virulent, dsRNA-free *C. parasitica* strain is inoculated with a hypovirulent virus-containing strain of the fungus, the virus passes through mycelial anastomoses into the mycelium and the conidia, but not into the ascospores of the virulent strain. The acquisition of the virus changes this strain into a hypovirulent one, and further development of the canker slows down or stops completely. Although this type of virus-mediated biological control of chestnut blight works well on isolated trees and in chestnut orchards in Europe, it has not been possible to use it on a large scale in the United States, especially under forest conditions. So far, no completely resistant American chestnuts have been found.

NECTRIA CANKER

Nectria canker is one of the most important diseases of apples and pears and of many species of hardwood forest trees in most parts of the world. Losses are greater in young trees because the fungus girdles the trunk or main branches, whereas in older trees only small branches are usually killed directly (Figs. 11-69A–11-69C). Cankers on the main stem of older trees, however, reduce the vigor and value or productivity of the tree, and such trees are subject to wind breakage. Nectria cankers usually develop around bud scars, wounds, and twig stubs or in the crotches of limbs. Young cankers are small, circular, brown areas. Later, the central area becomes sunken and black, while the edges are raised above the surrounding healthy bark. In many hosts and under favorable conditions for the host, the fungus grows slowly, the host produces callus tissue around the canker, and the margin of the canker cracks. Tissues under the black bark in the canker are dead, dry, and spongy, flake off, and fall out, revealing the dead wood and the callus ridge around the cavity. In subsequent years the fungus invades more healthy tissue and new, closely packed, roughly concentric ridges of callus tissue are produced every year, resulting in the typical open, target-shaped *Nectria* canker. In some hosts, however, and under conditions that favor the fungus, invasion of the host is more rapid. The bark in the cankered area is roughened and cracked but does not fall off, and the successive callus ridges are some distance apart. Some species of the fungus are associated with certain scale insects and grow profusely in insect-infested tissues. Through this association, *Nectria* species have been causing much more serious diseases, such as the "beech bark disease," than they do in the absence of the insects.

In hosts such as apple and pear, fruits are also infected and develop a circular, sunken, brown rot. White or yellowish pustules producing numerous conidia form on rotted areas. Internally, the rotted tissue is soft and has a striated appearance.

The Pathogen

The fungi, *Nectria galligena* and some related species, attack many different tree hosts. All *Nectria* species produce two-celled ascospores in brightly colored perithecia (Fig. 11-69D) on the surface of a cushion-shaped stroma, but different *Nectria* species produce different anamorphs. *Nectria galligena* produces single-celled microconidia and, more commonly, two- to four-celled, cylindrical macroconidia of the *Cylindrocarpon* type (Fig. 11-70) on small, white or yellowish or orange-pink sporodochia on the surface of the infected bark (Fig. 11-69B) or on fruit. Another *Nectria* species, *N. cinnabarina*, anamorph *Tubercularia vulgaris*, causes the Nectria twig blight of trees, especially apple, by infecting and causing cankers on small twigs, which it girdles and kills.

Development of Disease

Conidia are produced more commonly early in the season but also in the summer and early fall. They are spread by wind and by rain and perhaps by insects. Perithecia appear in the cankers in late summer and fall

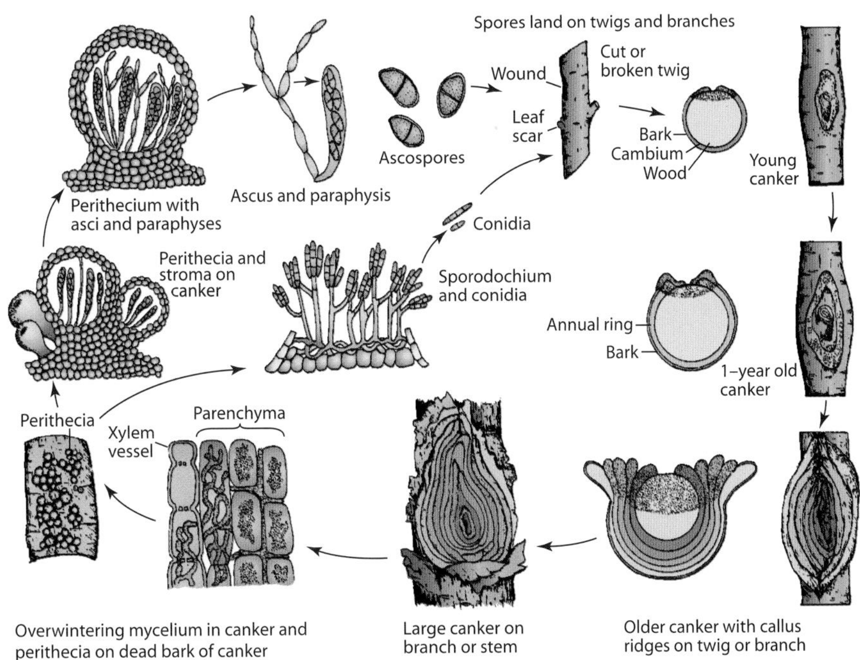

FIGURE 11-70 Disease cycle of Nectria canker caused by *Nectria galligena.*

and in the same stroma that earlier produced the conidia, which they eventually replace. The ascospores are either forcibly discharged and carried by wind or, in moist weather, ooze from the perithecium and are splashed by rain or carried by insects. Ascospores are dispersed more abundantly in late summer and fall but are also released at other times of the year (Fig. 11-70).

Control

Sanitation, i.e., removal and burning of cankered limbs or trees, is often the only control measure possible. Spraying with a fungicide such as captafol or a 8:8:100 Bordeaux mixture immediately after leaf fall helps reduce *Nectria* infections in fruit trees.

Leucostoma Canker

Leucostoma, *Valsa*, or *Cytospora* canker occurs worldwide and probably affects more species of trees than any other canker disease. More than 70 species of fruit trees, hardwood forest and shade trees, shrubs, and conifers are attacked by one of several species of the pathogen. The fungus *Leucostoma*, known previously as *Valsa*, is found most commonly in its anamorph stage, *Cytospora*, and therefore the disease is also known by these names.

Leucostoma canker is most serious on peach and other stone fruits (Fig. 11-71), but can also be serious on many other fruit, shade, or forest trees. Few orchards are free from it. Many trees are injured seriously by cankers on the trunk, in the main crotch, on the limbs, and on the branches. Infected branches of fruit trees often break from the weight of the crop or during storms. *Leucostoma* canker is most severe on trees growing under stress, such as those growing on an unfavorable site or those injured by drought or frost.

Symptoms

Infected small twigs and branches die back without showing cankers. On trunks and large branches, cankers appear at first as a gradual circular killing of the bark, which soon becomes brownish and sunken, and is often surrounded by raised callus tissue. In stone fruit trees, the diseased bark becomes dark, smelly, and oozes gum. Later, the bark shrivels and separates from the underlying wood and from the surrounding healthy bark. Small, pimple-like pycnidia appear on the dead bark. Later, the shriveled bark may slough off, exposing dead wood beneath. The cankers increase in size each year and become unsightly, rough swellings. Many twigs and branches die back as a result of cankers that girdle them completely (Figs. 11-71 and 11-72).

The Pathogen

Leucostoma cankers result mostly from infections by conidia (*Cytospora*). Perithecia and ascospores (*Leucostoma*) are not common. The pycnidia consist of

FIGURE 11-71 (A) Recent Leucostoma canker on peach twig, which exudes gum. (B) Older Leucostoma canker on peach branch. (C) Apricot tree showing dieback and decline as a result of multiple Leucostoma cankers on its twigs and branches. [Photographs courtesy of (A) A.R. Biggs, W. Virginia University, (B), J.W. Travis, Pennsylvania State University, and (C) K. Mohan, University of Idaho.]

many connecting cavities and one opening (Fig. 11-72). The spores are small, hyaline, one-celled, and slightly curved and are produced in a gelatinous matrix.

Development of Disease

During wet weather, conidia ooze out of the pycnidium (Fig. 11-72) and may be splashed by rain or may be spread by insects and humans. In moist but not rainy weather, the exuded conidia may form coiled threads of spores that dry out and harden and remain on the canker for several days or weeks. Most infections take place in late fall or early winter and in late winter or early spring. Weakened, injured trees, however, may be infected throughout the growing season. Both the mycelium and the conidia of the fungus overseason on the infected parts.

Small twigs are infected through injuries or leaf scars. In larger branches, the fungus enters through wounds of any kind. The fungus becomes established in dead bark and wood and invades the surrounding living tissues to form a canker. The fungus grows through the cells in the bark and the outer few rings of the wood.

Control

Control measures for *Leucostoma* canker include good cultural practices: watering and fertilization to

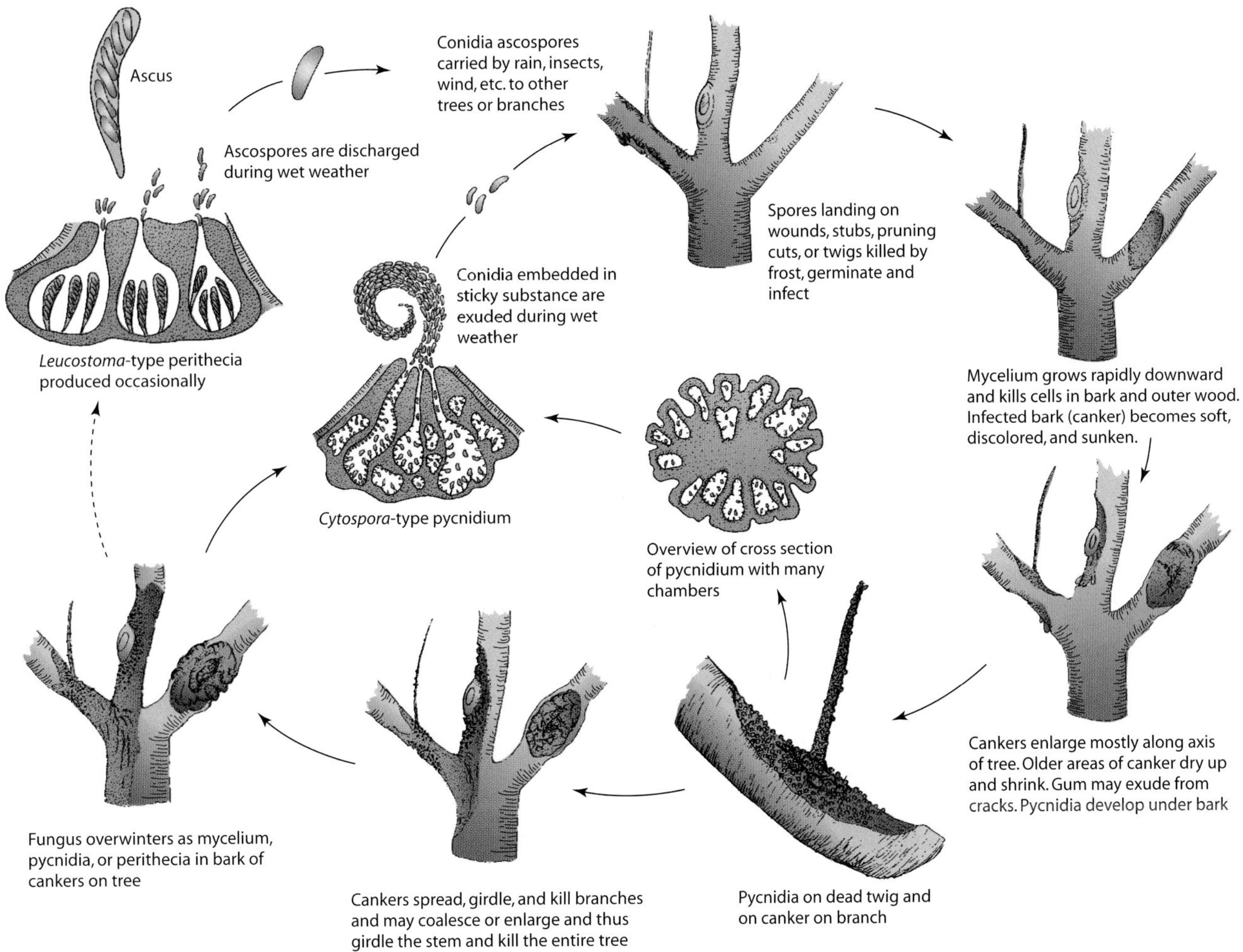

FIGURE 11-72 Disease cycle of the *Leucostoma (Valsa)* canker of peach and most other trees.

keep the trees in good vigor; avoiding wounding and severe pruning of trees; removing cankers from trunks and large branches during dry weather and treating the wound and all pruning cuts with a disinfectant and a wound dressing; removing and burning cankered and dead branches and twigs; pruning as late in the spring as possible; and spraying with one of several fungicides immediately after pruning and before it rains. These practices help but do not completely prevent canker.

CANKERS OF FOREST TREES

Although some of the cankers discussed earlier, e.g., *Nectria* and *Leucostoma*, cause cankers on forest as well as other trees, many more fungi infect and cause cankers primarily on forest trees. Some of the better known forest tree cankers are mentioned briefly here.

Hypoxylon canker. Several species of the fungus *Hypoxylon* infect the trunks and branches (Fig. 11-67D) of numerous deciduous forest tree species, especially when the trees are stressed from drought. The cankers become quite large and affected wood in trees becomes soft and weak and breaks easily in a windstorm.

Pitch canker of many pine species is caused by *Fusarium moniliforme* var. *subglutinans*. Pitch canker affects all the commercially important southern pines, but is more severe on slash pines. Heavy infections and the production of numerous cankers can kill trees (Figs. 11-73A and 11-73B). Most losses, however, come from suppression of growth. The wood beneath cankers is soaked with resin and large amounts of it flow from the affected area.

Butternut canker is caused by the fungus *Sirococcus clavigignenti-juglandacearum*. This canker has killed most butternut or white walnut (*Juglans cinerea*) trees throughout their natural range from Canada to Mississippi (Figs. 11-73C–11-73E).

Phomopsis blight, caused by *Phomopsis juniperivora*, affects various cedars, arborvitae, and cypress. Although it affects older trees by forming cankers on their

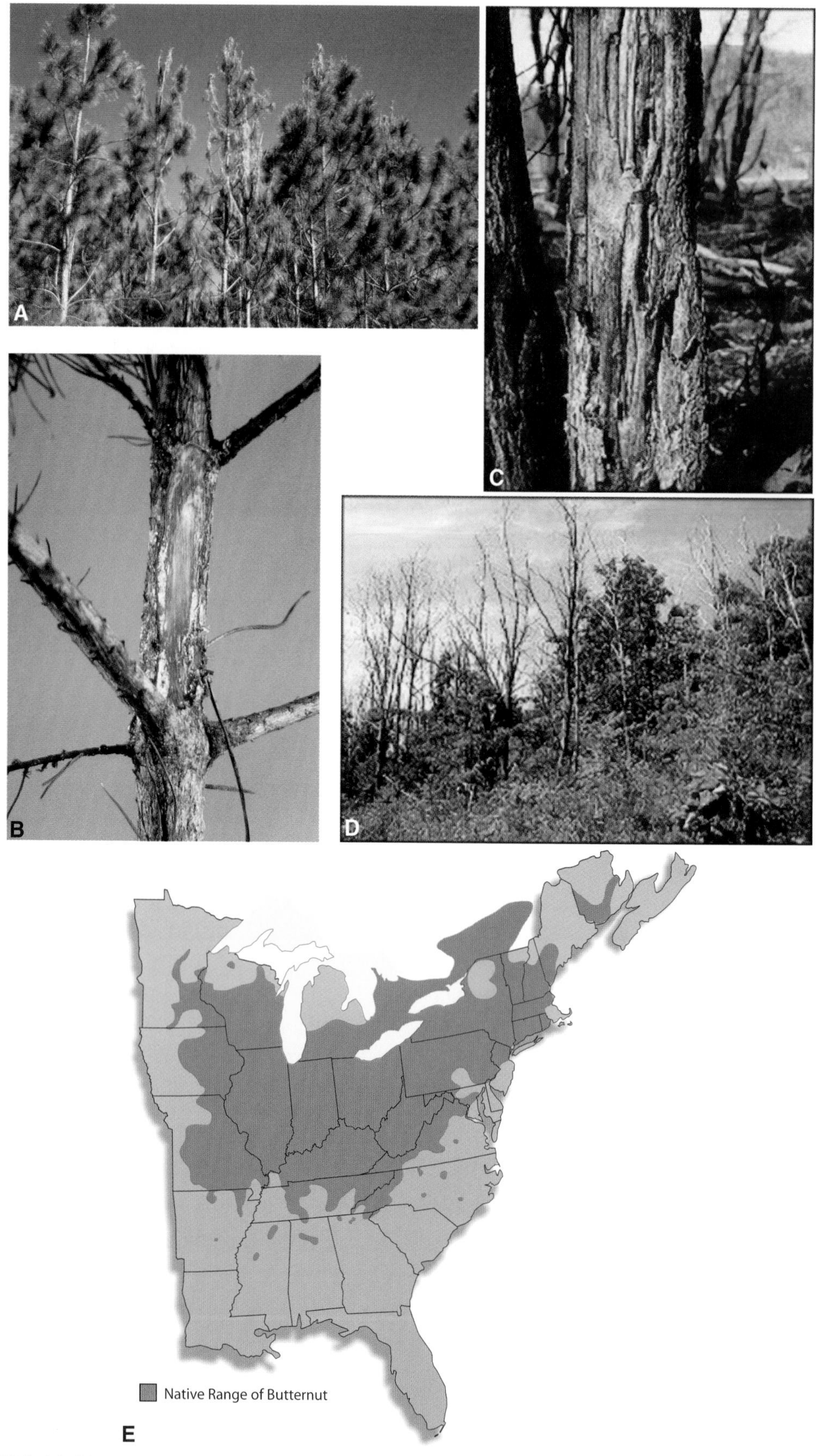

FIGURE 11-73 (A) Pine branches killed by pitch canker caused by *Fusarium subglutinans.* (B) Resin-soaked pine stem due to pitch canker infection. (C) Large canker on butternut tree caused by the fungus *Sirococcus clavigignenti juglandacearum.* (D) Butternut trees killed by the butternut canker fungus. (E) The natural range of butternut trees. [Photographs courtesy of (A and B) U.S. Forest Service and (C–E) E.L. Barnard, Florida Department of Agriculture, Forest Division.]

twigs and killing the tips of their branches, Phomopsis blight is particularly severe on nursery plants of susceptible hosts where it can kill all the plants in a nursery.

Seiridium canker, caused by the fungus *Seiridium cardinale*, has been killing Leyland cypress in the southeastern United States and cypresses in forests, plantations, and in urban settings around the world. The fungus infects and causes the formation of numerous elongated cankers on branches and stems of trees. The cankers are sunken, reddish, and ooze sap profusely. Infected branches and trees eventually wilt and die.

In most of these canker diseases, conidia or ascospores are splashed or carried to wounds of needles or bark, germinate and initiate infection. Subsequently spreads the fungus into the bark and wood of the tree and causes a canker. The cankers enlarge and encircle the twig, branch, or trunk of the host tree and either cause a blight by killing numerous of its shoots or cause a general decline and, possibly, death of the whole tree if the cankers encircle the trunk or main branches of the tree. The control of canker fungi in the nursery is though planting of genetically resistant varieties and through application of appropriate fungicides. The control of canker diseases in the forest is primarily through the use of resistant varieties.

Selected References

Biggs, A. R. (1989). Integrated approach to controlling *Leucostoma* canker of peach in Ontario. *Plant Dis.* 73, 869–874.

Davidson, A. G., and Prentice, R. M., eds. (1967). "Important Forest Insects and Diseases of Mutual Concern to Canada, the United States and Mexico." Dept. For. Urban Dev., Canada.

Gordon, T. R., Stover, A. J., and Wood, D. L. (2001). The pitch canker epidemic in California. *Plant Dis.* 85, 1128–1139.

Graniti, A. (1998). Cypress canker: A pandemic in progress. *Annu. Rev. Phytopathol.* 36, 91–114.

Houston, D. R. (1994). Major new tree disease epidemics: Beech bark disease. *Annu. Rev. Phytopathol.* 32, 75–87.

Jones, A. L., and Aldwinckle, H. S., eds. (1990). "Compendium of Apple and Pear Diseases." APS Press, St. Paul, MN.

Jones, A. L., and Sutton, T. B. (1996). "Diseases of Tree Fruits in the East". Mich. St. Univ. Extension NCR 45.

Lalancette, N., and Polk, D. F. (2000). Estimating yield and economic loss from constriction canker of peach. *Plant Dis.* 84, 941–946.

Manion, P. D. (1991). "Tree Disease Concepts," 2nd Ed. Prentice-Hall, Englewood Cliffs, NJ.

Moller, W. J., and Kassimatis, A. N. (1978). Dieback of grapevines caused by *Eutype armeniacae*. *Plant Dis. Rep.* 62, 254–258.

Ogawa, J. M., *et al.*, eds. (1995). "Compendium of Stone Fruit Diseases." APS Press, St. Paul, MN.

Roane, M. K., Griffin, G. J., and Elkins, J. R. (1986). "Chestnut Blight, Other Endothia Diseases, and the Genus *Endothia*." APS Monograph Series. APS Press, St. Paul, MN.

Schoenenweiss, D. F. (1981). The role of environmental stress in diseases of woody plants. *Plant Dis.* 65, 308–314.

Skilling, D. D. (1981). Scleroderis canker: Development of strains and potential damage in North America. *Can. J. Plant Pathol.* 3, 263–265.

Swinburne, T. R. (1975). European canker of apple (*Nectria galligena*). *Rev. Plant Pathol.* 54, 787–799.

Wainwright, S. H., and Lewis, F. H. (1970). Developmental morphology of the black knot pathogen on plum. *Phytopathology* 60, 1238–1244.

Wang, D., Iezzoni, A., and Adams, G. (1998). Genetic heterogeneity of *Leucostoma* species in Michigan peach orchards. *Phytopathology* 88, 376–381.

ANTHRACNOSE DISEASES CAUSED BY ASCOMYCETES AND DEUTEROMYCETES (MITOSPORIC FUNGI)

Anthracnoses, meaning blackenings, are diseases of the foliage, stems, or fruits that typically appear as dark-colored spots or sunken lesions with a slightly raised rim. Some cause twig or branch dieback. In fruit infections, anthracnoses often have a prolonged latent stage. In some fruit crops, the spots are raised and have corky surfaces. Anthracnose diseases of fruit often result in fruit drop and fruit rot.

Anthracnoses (from anthrax = carbon = black) are caused by fungi that produce conidia within black acervuli. Four ascomycetous fungi, *Diplocarpon*, *Elsinoe*, *Glomerella*, and *Gnomonia*, are responsible for most anthracnose diseases. They are found in nature mostly in their conidial stage and can overwinter as mycelium or conidia.

Diplocarpon, causing black spot of rose (*D. rosae*) (Figs. 11-74A and 11-74B) and leaf scorch of strawberry (*D. earliana*)

Discula, *D. destructiva* is the cause of dogwood anthracnose

Elsinoe (conidial stage: *Sphaceloma*), causing anthracnose of grape (*E. ampelina*) (Figs. 11-74C–11-74F), raspberry (*E. veneta*), (Fig. 11-75A), and scab of citrus (*E. australis* and *E. fawcettii*) (Figs. 11-75B and 11-75C), and avocado (*E. perseae*) (Fig. 11-75D)

Glomerella (conidial stage: *Colletotrichum* or *Gloeosporium*), causing anthracnose of many annual and perennial plants, bitter rot of apple, and ripe rot of grape and other fruits

Gnomonia, causing anthracnose of walnut and many forest and shade trees

Acervulus-producing mitosporic fungi used to make up a separate order (Melanconiales) but are now included in the group Coelomycetes. The most important plant pathogenic fungi that produce acervuli are *Colletotrichum* (*Gloeosporium*), *Coryneum*, *Cylindrosporium*, *Marssonina*, *Melanconium*, and *Sphaceloma*.

FIGURE 11-74 (A) Leaf symptoms of black spot of rose caused by the fungus *Diplocarpon rosae*. (B) Defoliation of rose bush as a result of black spot infection. Symptoms of grape anthracnose caused by *Elsinoe ampelina* on grape leaf (C) on young grape shoot (D) and on young grape berries on which it causes the characteristic bird's-eye spots (E and F). [Photographs courtesy of (A) J.W. Pscheidt and (B) M. Hoffer, Oregon State University, (C, D, and F) M. Ellis, Ohio State University, and (E) E. Hellman, Texas A&M University.]

Some of these are the conidial stages of Ascomycetes that cause anthracnose diseases. For some species of these fungi, however, and for *Coryneum* and *Melanconium*, no perfect stage is known. Some plant diseases caused strictly by the mitosporic stages of the fungi are as follows.

Colletotrichum (*Gloeosporium*), causing anthracnose of cereals and grasses (*C. graminicola*), anthracnose of cucurbits (*C. lagenarium*), anthracnose or fruit rot of eggplant and of tomato (*C. phomoides*), anthracnose of strawberry (*C. acutatum*), crown rot and wilt of strawberry (*C. gloeosporioides*)

FIGURE 11-75 (A) Anthracnose symptoms on cane of blackberry plant caused by *Elsinoe veneta.* Symptoms of citrus scab on orange leaves (B) and on lemon fruit (C) caused by *Elsinoe fawcettii.* (D) Scab symptoms on avocado fruit caused by *Sphaceloma perseae.* [Photographs courtesy of (A) M. Ellis, Ohio State University, (B) G. Simone, (C) R.J. MacGovern, University of Florida, and (D) T. Isakeit, Texas A&M University.]

(Fig. 11-77), red rot of sugarcane (*C. falcatum*), and onion smudge (*C. circinans*). *Colletotrichum gloeosporioides* causes many devastating fruit diseases in the tropics, such as anthracnose of citrus, fig, mango, olive, avocado, and many other plants

Coryneum (*Stigmina*), *C. beijerynki* (now *Wilsonmyces carpophilus*), causing *Coryneum* blight, shot hole, or fruit spot of stone fruits, especially peach and apricot

Greeneria uvicola, formerly *Melanconium fuligenum*, causing bitter rot of grapes

Anthracnose diseases, particularly those caused by species of *Colletotrichum* or their teleomorph *Glomerella* fungi, are very common and destructive on numerous crop and ornamental plants. Although severe everywhere, anthracnose diseases cause their most significant losses in the tropics and subtropics.

BLACK SPOT OF ROSE

The black spot of rose appears as black lesions on the leaves (Figs. 11-74A and 11-74B) and as raised, purple-red blotches on immature wood of first-year canes. The leaf spots have fringed margins and may coalesce to produce large, irregular, black lesions. The leaf tissue around the lesions turns yellow, and often entire leaves become yellow and fall off prematurely, leaving the canes almost completely defoliated.

The fungus, *Diplocarpon rosae*, produces *Marssonina*-type conidia in acervuli forming between the outer wall and cuticle of the epidermis (Fig. 11-83E) and ascospores in tiny apothecia formed in old lesions. Short conidiophores arise from a thin black stroma and give rise to successive crops of conidia. Conidia push up and rupture the cuticle. The fungus overwinters as mycelium, ascospores, and conidia in infected leaves and canes. Both kinds of spores can cause primary infections of leaves in the spring by direct penetration. The mycelium grows in the mesophyll, but within two weeks forms acervuli and conidia at the upper surface. Conidia are produced throughout the growing season and cause repeated infections during warm, wet weather.

The control of *Diplocarpon* diseases is through sanitation (e.g., removing and burning infected leaves, cutting back the canes of diseased rose plants), spraying with one of several available fungicides, or applying sulfur–copper dust. Applications should begin as soon as new leaves appear in the spring or at the first appearance of black spot on the foliage and then should be repeated at 7- to 10-day intervals or within 24 hours after each rain.

ELSINOE ANTHRACNOSE AND SCAB DISEASES

Several important anthracnose diseases are caused by the fungus *Elsinoe* on crops such as grape (*E. ampelina*) and raspberry (*E. veneta*) and also on cowpea and poinsettia. In addition, other species of *Elsinoe* cause scab diseases on citrus (*Elsinoe fawcettii* and *E. australis*) and on avocado (*E. perseae*).

Grape anthracnose or **bird's-eye rot** apparently originated in Europe from where it spread around the world. Losses from the disease can be severe, especially in humid rainy regions. Symptoms consist of 1- to 5-millimeter brown lesions on leaves, green shoots, and berries (Figs. 11-74C–11-74F). The lesions are surrounded by a dark margin, which in berries resembles a bird's eye. The disease is caused by *Elsinoe ampelina* whose anamorph is *Sphaceloma ampelinum*. The fungus produces small hyaline conidia in acervuli on the exterior of the lesions and asci in pyriform locules of ascostromata, with each ascus containing eight dark four-celled ascospores. The fungus also produces sclerotia at the edge of lesions on shoots, which serve as the main overwintering structures of the fungus. In the spring, sclerotia produce large numbers of conidia, which are disseminated by splashing rain and infect the young leaves, shoots, and berries of grapevines. The control of the disease is through planting of resistant varieties and by dormant and growing-season application of fungicides.

Raspberry anthracnose affects several *Rubus* species and probably occurs worldwide. It may cause severe losses by causing defoliation, wilting of shoots, death of fruiting canes, and making fruit unmarketable. Typical symptoms consist of reddish purple spots on canes (Fig. 11-75A) but also on other parts of the plant. The disease is caused by the fungus *Elsinoe veneta* whose anamorph is *Sphaceloma necator*. The fungus grows slowly and produces few conidia in culture. It overwinters primarily as mycelium in infected canes. In the spring, it forms a stroma beneath the epidermis, which produces acervuli with single-celled hyaline conidia. In late summer it also produces, on canes only, subepidermal ascocarps containing globose asci with oblong four-celled ascospores that mature the following spring. New infections are caused primarily by conidia in the spring. Control of the disease is primarily through cultural practices, such as avoiding excessive fertilization and overhead irrigation, improving air circulation among plants, removing infected canes and wild bramble plants from the vicinity of the planting, and applying of appropriate fungicides.

Citrus scab diseases occur in various parts of the world and can cause severe losses when they infect the fruit grown for the fresh market, but in susceptible varieties they can cause stunting of plants and can reduce the quantity and quality of fruit grown for processing. Citrus scab, or sour orange scab, caused by the fungus *Elsinoe fawcettii* (anamorph *Sphaceloma fawcettii*), is the most widespread and occurs wherever rainfall conditions are conducive to infection. Sweet orange scab, caused by *E. australis* (anamorph *S. australis*), occurs in South America, and Tyson's scab, caused by *S. fawcwttii* var. *scabiosa*, occurs on lemons in Australia. No scab diseases have been reported from California, Arizona, and from most Mediterranean countries. Citrus scab diseases cause a distortion of young shoots by producing pustules consisting of a stroma of mycelium and dead host cells, plus hyperplastic host cells that have few or no chloroplasts (Figs. 11-75B and 11-75C). Scab stromata at first are pink to light brown but later become corky and turn yellowish, grayish brown, or dark. Scab fungi produce small hyaline conidia in acervuli and, in some parts of the world (Brazil), they produce ascocarps with asci and ascospores. Scab fungi overwinter on the tree canopy. Their conidia can germinate and cause infection quickly, requiring only about 2.5 hours of wetness for initiating infection. The control of citrus scab diseases is obtained through application of appropriate fungicides.

Avocado scab occurs in the humid tropics and subtropics and causes severe yield losses through premature abscission of infected fruit and by reducing the quality of mature fruit greatly. The most striking symptoms appear on the fruit as brown to purplish brown, slightly raised fruit spots, at first oval to irregular in shape that later enlarge and coalesce and form large rough areas over the surface of the fruit (Fig. 11-75D). Lesions also form on the leaves, which eventually become crinkled and distorted. Avocado scab is caused by *Sphaceloma perseae*. The fungus produces hyaline, nonseptate conidia in acervuli. Infections occur primarily on young tissues and are favored by cool, moist weather. The control of the disease is through planting resistant varieties and through application of appropriate fungicides.

COLLETOTRICHUM DISEASES

Several species of *Colletotrichum* cause serious anthracnose diseases of numerous important annual crop and ornamental plants. Some of them produce their teleomorph, *Glomerella cingulata*, with some frequency and are sometimes referred to as Glomerella diseases. Such species also causes cankers and dieback of woody plants such as camellia and privet, bitter rot of apples, and ripe rot of grape, pears, peaches, and other fruits.

COLLETOTRICHUM ANTHRACNOSE DISEASES OF ANNUAL PLANTS

Numerous important *Colletotrichum* anthracnose diseases affect annual plants. Only a few of the most common and serious such diseases are described briefly here. They include the anthracnose of bean, cotton, cucurbits, onion, pepper, tomato, and strawberry. Severe anthracnose diseases often occur on corn and on cereals and grasses. The diseases are present wherever their hosts are grown, although they are more severe in warm to cool, humid areas. Generally, they are not a problem under dry conditions.

In **anthracnose of beans**, plants in all stages of growth are subject to anthracnose. The fungus, *Colletotrichum lindemuthianum*, is often present in or on the seed produced in infected pods. Infected seed may show yellowish to brown sunken lesions. When infected seeds are planted, many of the germinating seedlings are killed before emergence. Dark-brown, sunken lesions with pink masses of spores in the center are often present on the cotyledons of young seedlings. The fungus may destroy one or both of the cotyledons. The spores spread and infect the stem, producing more lesions. The lesions are covered with myriads of pink- to rust-colored spores. If conditions are humid, numerous lesions may girdle and weaken the stem to the point where it cannot support the top of the plant. The fungus also attacks the petioles and the veins of the underside of the leaves on which it causes long, dark-colored lesions (Fig. 11-76A). Few lesions are produced between the veins in bean, but they are rather common on plants such as cotton. In some hosts, such as sweet pea, the lesions may involve the entire leaf.

The anthracnose fungus also attacks and causes characteristic symptoms on bean pods (Fig. 11-76B) and, in cotton, on cotton bolls. On young cotton bolls, the fungus produces numerous lesions that spread, coalesce, and cover most or all their surface, with almost continuous masses of spores covering the infected area. On bean pods, small flesh- to rust-colored elongated lesions appear, which later become sunken and circular. Lesions developing on young pods may extend through the pod and even to the seed, whereas in older pods the lesions do not extend beyond the pod. As the pods mature, the pink spore masses of the lesions dry down to grayish black granulations or to small pimple-like protrusions.

Anthracnose of cucurbits, caused by *Colletotrichum orbiculare*, is probably the most destructive disease of these crops everywhere, being most severe on watermelon, cantaloupe, and cucumber. All aboveground parts of the plants are affected. On the leaves, small, water-soaked, yellowish areas appear that enlarge to 1 to 2 centimeters and become black in watermelon and brown in all other cucurbits. Infected tissues dry up and break. Lesions also develop on the petioles, which may result in defoliation of the vine; on the fruit pedicel, which cause the fruit to turn dark, shrivel, and die; and on the stem (Fig. 11-76C), which weaken or kill whole vines. The fruit becomes susceptible to infection at about the time of ripening. Circular, watery, dark, sunken lesions appear on the surface of the fruit that may be from 5 millimeters to 10 centimeters in diameter and up to 8 millimeters in depth (Fig. 11-76D). The lesions expand rapidly in the field, in transit, or in storage and may coalesce to form larger ones. The sunken lesions have dark centers, which in moist weather are filled with pink spore masses exuding from acervuli that break through the cuticle. Severely affected fruits are often tasteless or even bitter and are often invaded by soft-rotting bacteria and fungi that enter through the broken rind. The fungus overwinters in infected debris in the soil and on or in the seed.

Anthracnose or **ripe rot of tomato** and of several other vegetables and fruits causes serious losses of fruit. Occasionally, it also damages stems and foliage. Canning tomatoes are particularly susceptible to anthracnose before and after harvest, but other tomatoes, as well as eggplant and pepper (Fig. 11-76E), may be attacked in a similar manner from the time ripening begins through harvest and in storage. In early stages of tomato infection, the symptoms appear as small, circular, sunken, water-soaked spots resembling indentations caused by burnt circular objects. As the fruit softens, the spots enlarge up to 2 to 3 centimeters in diameter, and their central portion becomes dark and slightly roughened as a result of black acervuli developing just beneath the skin (Fig. 11-76F). The spots are often numerous and coalesce, leading to watery softening of the fruit and, finally, rotting of the fruit, sometimes accelerated by other invading microorganisms. Enormous numbers of conidia are present in acervuli below the skin even in the smallest spots. In later stages, pink or salmon-colored masses of spores are produced on the surface of the spots. The fungus overwinters in infected plant debris and in or on the seed. Early light infections of foliage and young stems may go unnoticed, but they enable the fungus to survive and multiply somewhat until the fruit begins to ripen and becomes susceptible to infection. High temperatures and high relative humidity or wet weather at the time of ripening favor spread of the fungus infection and often lead to destructive epidemics.

Onion anthracnose or **smudge** is caused by *Colletotrichum circinans*. Dark smudges appear on the outer scales or neck of the bulbs, primarily of white onions (Fig. 11-76G). Most colored varieties are mostly resistant except in the colorless region of the bulb neck. Smudgy spots first appear beneath the cuticle of the scale and may be scattered over the surface of the bulb; more commonly, they congregate in uniformly black,

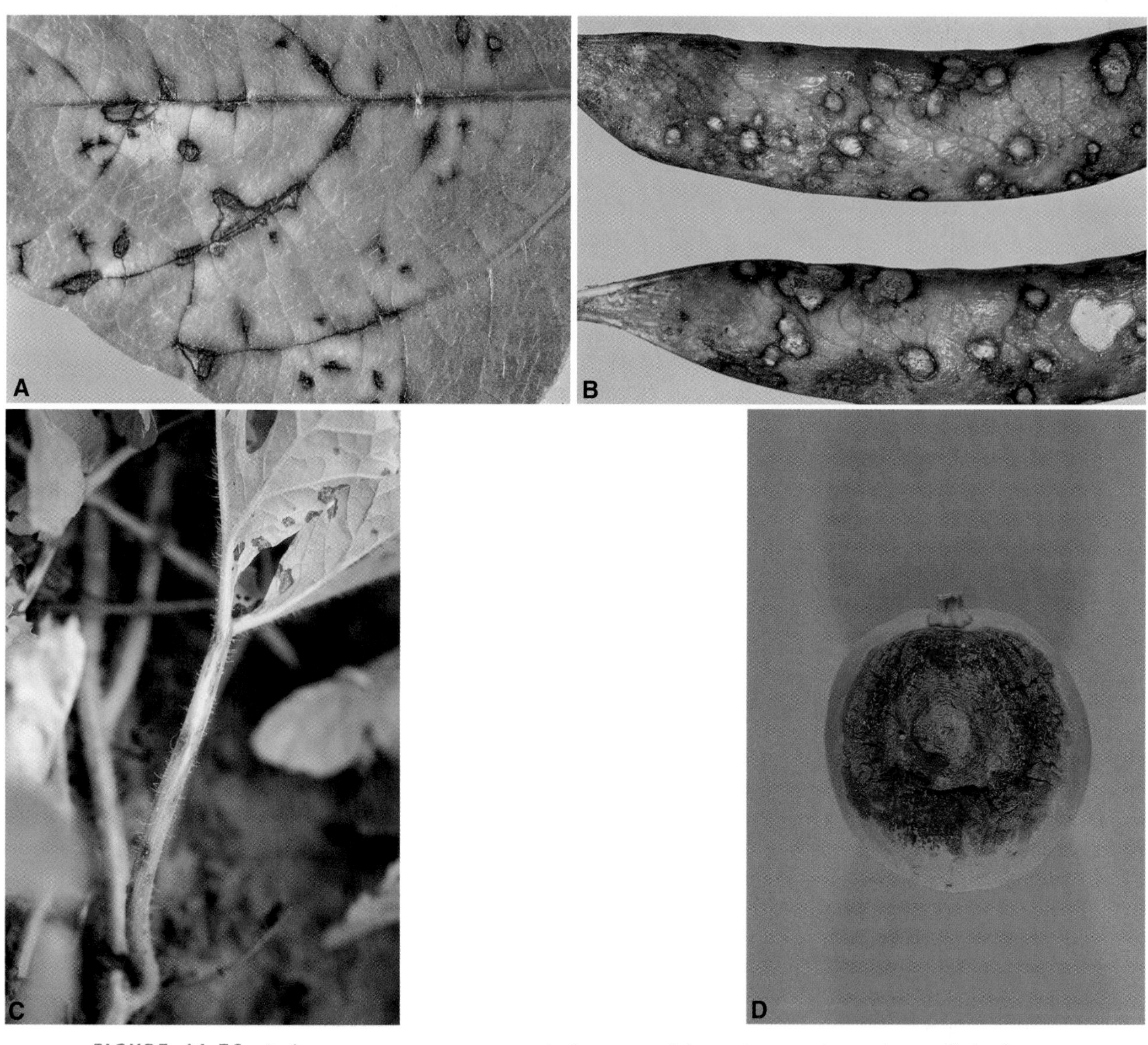

FIGURE 11-76 Anthracnose symptoms on annual plants caused by various species or forms of the fungus *Colletotrichum*. Spots and vein necrosis in bean leaf (A) and bean pod (B) caused by *C. lindemuthianum*. Lesions on stem of young watermelon plant (C) and large rotten area on acorn squash (D) caused by *C. lagenarium*. (E and F) Anthracnose symptoms on pepper and tomato fruits caused by *Colletotrichum* sp. Healthy red onions (left) and white onions infected with the onion smudge anthracnose caused by *C. circinans*. [Photographs courtesy of (A and B) W.L. Seaman, WCPD, (C) B.D. Bruton, USDA, (D, F, and G), Plant Pathology Department, University of Florida, and (E) R.J. McGovern, University of Florida.

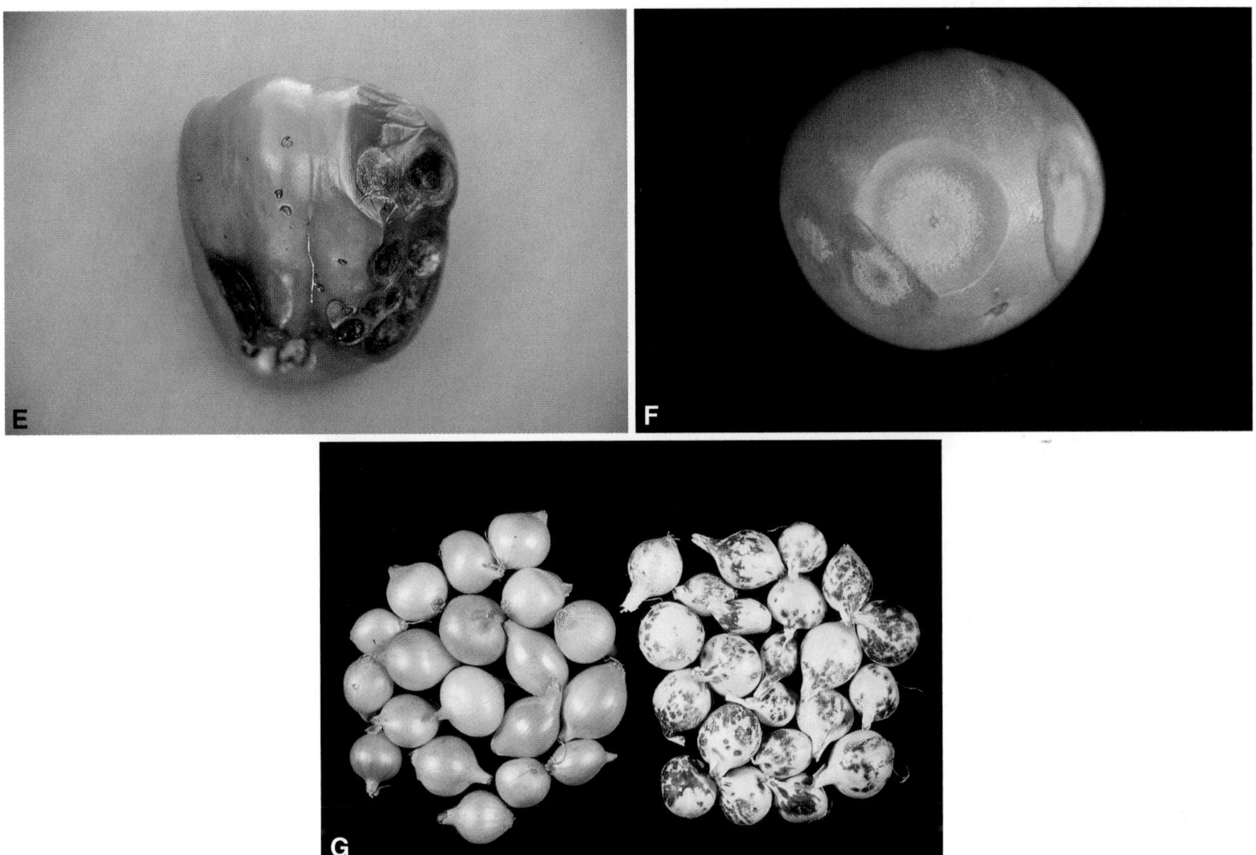

FIGURE 11-76 (*Continued*)

smudgy, circular areas or are arranged in concentric circles, with the outer one being up to 2 centimeters or more in diameter. In moist weather the fungus produces acervuli filled with cream-colored masses of conidia containing numerous black, stiff, bristle-like hairs (setae) visible with a hand lens. The fungus attacks inner, living scales only under conditions of favorable high moisture and temperature. The fungus overwinters on infected onions, on sets, and in the soil as a saprophyte.

Strawberry anthracnose, a complex of three *Colletotrichum* species, *C. acutatum*, *C. fragariae*, and *C. gloeosporioides*, causes a variety of symptoms that make anthracnose the most important disease of strawberries. *C. acutatum* causes an anthracnose fruit rot and black leaf spots, whereas *C. gloeosporioides* and *C. fragariae* infect primarily the crown of the plants and cause crown rot and wilt (Fig. 11-77). Symptoms may appear as sunken, dark lesions on petioles and stolons (runners), which may then be girdled, resulting in wilting and death of the leaf or of the daughter plant beyond the lesions on the stolon. The fungus often spreads into the crowns of young plants, which it rots, and the plants then die in the nursery or after they are transplanted in the field. Other symptoms include bud rot, flower blight, black leaf spots throughout the leaf, and irregular leaf spots on the leaf margins and tips. The fungi overseason mostly on infected or contaminated transplants. In some areas they may overseason on surviving stolons in the soil, on weed hosts, and possibly in plant debris in the soil. Fungi produce conidia in acervuli formed in the lesions. Spores are splashed by irrigation or rain to nearby plants or are carried by insects, animals, and humans moving among the plants. In warm, humid weather the disease spreads very quickly throughout a field, and effective control is impossible the rest of the season. Control measures include growing nursery plants with as little fertilizer as possible; removing all infected fruit at each harvest; practicing sanitary procedures when harvesting; using resistant cultivars when available; and applying fungicides every other day or twice per week. The most commonly used fungicides include benomyl, captan, and iprodione.

In **anthracnose of cereals and grasses**, all cereals, including corn, wheat, barley, rice, turf grasses, and pasture grasses, are attacked by *Colletotrichum graminicola* and develop symptoms of varying severity. Symptoms and losses are affected significantly by the

FIGURE 11-77 Anthracnose symptoms on strawberry: killed blossoms (A), lesions on stems (B), sunken rotten areas on fruit (C), dead stems and necrotic crown (D), and numerous strawberry plants killed by the fungus in the field (E). Fruit rot is caused by *Colletotrichum acutatum*, whereas crown rot and wilt are caused by *C. gloeosporioides* and *C. fragariae*. (Photographs courtesy of D.L. Legard, University of Florida.)

environment but can be very severe. The fungus lives saprophytically on crop residue and, although it may attack young seedlings, it usually attacks the tissues at the crown and the bases of stems of more developed plants. Infected areas first appear bleached but later become brown. Toward maturity of the plant, numerous black acervuli appear on the stems, lower leaf sheaths, and, sometimes, on the leaves and on the chaff and spikes of diseased heads (Fig. 11-83). Depending on how early the plant is attacked, the plant may show a general reduction in vigor, premature ripening or dying of the head, and shriveled grain. The fungus occasionally infects seeds and it can also overwinter as mycelium on the seed. When seed-borne, the fungus may cause root rot and crown rot of the developing plant. Anthracnose of corn and other cereals has become a major problem in areas where reduced tillage, practiced to minimize loss of soil or water, allows greater survival of inoculum of the pathogen on the crop residue.

BOX 18 Colletotrichum Anthracnoses: A Menace to Tropical Crops

Many types of plant diseases, whether caused by fungi, bacteria, nematodes, or viruses, are more severe and cause more serious losses to crops in the tropics (Fig. 11-78). There are many reasons for this, the most important being the continuous warm and humid weather that favors the growth and multiplication of the pathogens and, for pathogens whose spread depends on or is favored by vectors, on the parallel multiplication and movement of the vectors, which are

FIGURE 11-78 Anthracnose symptoms on tropical crops. (A) Large rotten lesions on papaya fruit. (B) Fructifications (black acervuli) on stems of cassava killed by the anthracnose fungus. (C) Mango fruit showing large anthracnose lesions. (D) Bananas whose point of contact with the stem has been killed by anthracnose. Three stages in the development of coffee anthracnose caused by *Colletotrichum coffeanum*: close-up of a twig in which half of the berries are rotten (E), early infection of leaves and berries still on tree (F), and fruit drop and defoliation due to anthracnose (G). (Photographs courtesy of H.D. Thurston, Cornell University.)

continued

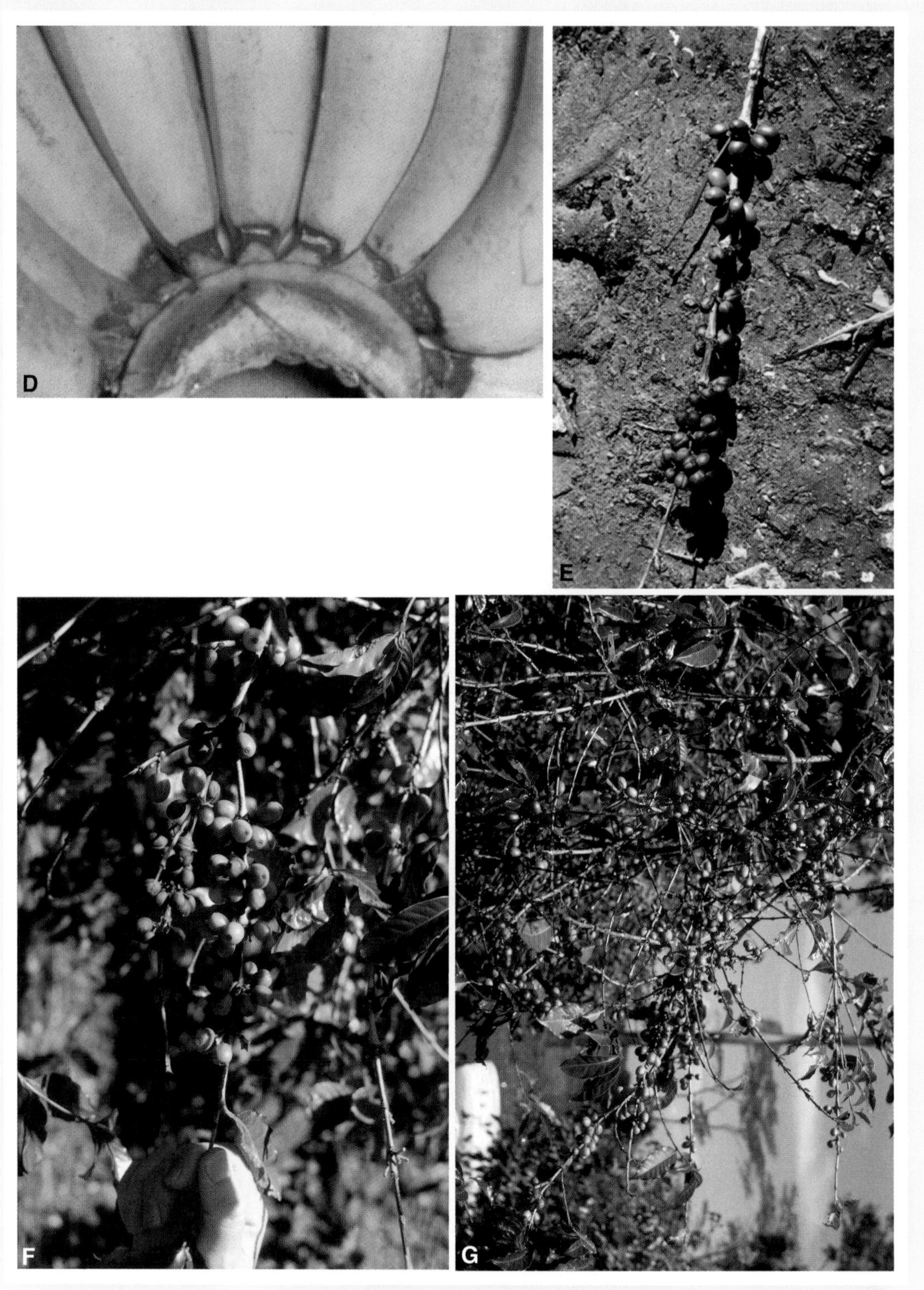

FIGURE 11-78 *(Continued)*

also favored by the same warm weather. The importance of *Colletotrichum* and of the anthracnose diseases it causes to crops in the tropics is beyond any doubt one of the most significant problems farmers and consumers have to deal with. *Colletotrichum* anthracnoses are often devastating to producers and also destroy food quickly before the consumers can use it, especially under conditions of nonexistent or poor refrigeration available to the native people living in the tropics.

A few species of *Colletotrichum* attack almost all tropical and subtropical crops and cause tremendous losses by damaging the fruit of most of them; by reducing yields through destruction of blossoms; or by affecting leaves, stems, and fruit, thereby reducing yields and quality of produced tropical fruit, root, etc. In addition, as was mentioned, even anthracnoses of not strictly tropical crops, e.g., beans, tomatoes, and peppers, are much more severe in the tropics.

Some of the most important tropical crops in which *Colletotrichum* anthracnoses cause severe losses are avocado, coffee, banana, mango, papaya, and yam. In avocado, black circular spots up to a half inch in diameter appear on ripening fruit, and the infection spreads rapidly into the flesh and causes a greenish-black decay. The infection remains latent almost until the fruit reaches the consumer. In banana, infections by *Colletotrichum gloeosporioides* are also latent but develop rapidly after harvest, causing rotting of bananas advancing internally from the point of attachment to the stalk (Fig. 11-78D) In citrus, the fungus *Colletotrichum acutatum* attacks and destroys a large percentage of the flowers, causing a postbloom drop (Fig. 11-80) of young citrus fruit, thereby directly reducing the number of fruit retained on the tree and available for growth and harvesting.

In coffee, ripening coffee berries become infected with several species of the fungus *Colletotrichum* and one of them (*C. coffeanum*) causes the very destructive green coffee berry disease (Figs. 11-78E–11-78G). Losses vary from a small percentage of berries being infected to, usually, 20 to 80% of the coffee berries becoming rotten by the fungus. In cassava, anthracnose appears as cankers on the stems (Fig. 11-78B) and the bases of leaf petioles. Affected leaves droop downward and wilt, subsequently dying and falling off. Infected shoot tips die back. In severe infections, soft parts of cassava plants become twisted. Defoliated plants and plants whose shoots have been killed fail to grow and to produce a crop of roots. In mango, anthracnose is its most important disease, affecting plants by killing inflorescences, causing spots on leaves, and, especially, by causing dark brown to black decay spots on fruit when it nears the ripening stage (Fig. 11-78C). In papaya, anthracnose appears primarily as water-soaked spots that become sunken, turn brown to black, and enlarge to 5 centimeters or more in diameter (Fig. 11-78A). Infected fruit is of much reduced quality and much of it becomes worthless and is discarded.

In yam, anthracnose appears as a leaf spot (Figs. 11-79A and 11-79B) that spreads and develops rapidly and kills leaves, shoots, and, following infection of the terminal bud, entire yam plants (Figs. 11-79C and 11-79D). The dieback or anthracnose disease is the most important disease of yam, often causing yield depressions approaching 80% of the expected crop.

FIGURE 11-79 Anthracnose effects on yam plants. Early (A) and advanced (B) infection of yam leaves by the anthracnose fungus *Colletotrichum gloeosporioides*. (C) Death and collapse of unstaked yam plants in large area of field due to infection by the anthracnose fungus. (D) Killing and destruction of most staked yam plants in a field by anthracnose infection while several resistant plants appear unaffected. (Photographs courtesy of R. Asiedu, Intl. Instit. Trop. Agric. Ibadan, Nigeria.)

continued

FIGURE 11-79 *(Continued)*

COLLETOTRICHUM FRUIT ROTS

The most important of the *Colletotrichum* fruit rots are those that occur on tropical fruits, such as avocado, bananas, citrus, coffee, mango, papaya, and others. Of the anthracnoses on temperate fruit, bitter rot of apple and ripe rot of grape are the most important.

Mango anthracnose occurs throughout the tropics where mangos are grown. It is caused by at least three species of *Colletotrichum*: *C. gloeosporioides*, *C. gloeosporioides* var. *minor*, and *C. acutatum*. The disease appears as blossom blight, as leaf blight, and, when moisture conditions are favorable, as tree dieback. Mango anthracnose is particularly severe and may destroy the total crop as a postharvest disease. Blossom blight kills individual flowers or it affects parts of or the complete inflorescence. Infected leaves develop irregular-shaped black necrotic spots that often coalesce and form large necrotic areas (Fig. 11-78C). Young twigs may also be invaded and killed, resulting in dieback of twigs. Under wet or very humid conditions, fruit become infected in the field but remain symptomless until the onset of ripening, which takes place after harvest. Fruit symptoms consist of rounded brownish-black lesions on the fruit surface. The lesions coalesce and form larger dark lesions that cover large areas of the fruit spreading downward from the stem end toward the distal end of the fruit (Fig. 11-78C). Fruit lesions are usually shallow, affecting only the peel but under favorable conditions the lesions extend into the pulp.

Mango anthracnose fungi produce abundant conidia on infected leaves, inflorescences, and on mummified aborted fruit. Conidia are spread by splashing rain and cause new infections on leaves, blossoms, and fruit. In the infected fruit in the field, the fungus remains quiescent until the fruit is harvested and ripening begins. The fungus then becomes activated and the lesions begin to develop and to enlarge. In storage, however, the fungus does not move from one fruit to the next. Conidia of *Colletotrichum* spp. produced on hosts, such as avocado, papaya, banana, and citrus, can also infect and cause the disease on mango fruit.

Citrus anthracnose refers to several nearly symptomless combinations of citrus hosts and species of *Colletotrichum*, but is best applied to the serious disease known as **citrus postbloom fruit drop**. Citrus postbloom fruit drop is caused by a slow-growing strain of *Collectotrichum acutatum*. This fungus infects citrus flowers. It produces orange to peach-colored spots on the petals (Fig. 11-80) or affects entire flower clusters. Such infections induce newly formed fruitlets to drop, leaving behind a persistent calyx (button) surrounded by distorted leaves. Postbloom fruit drop affects most citrus species in Florida, the Caribbean, and Central America. In moist weather, abundant conidia are produced in acervuli on diseased petals, which are splashed to healthy flowers by rain. In prolonged damp or rainy weather, over 90% of the blossoms may be destroyed by *Colletotrichum* within a few days. Control has been difficult in wet weather. Sprays with benomyl or captafol help reduce fruit drop.

Bitter rot of apple, caused by *Colletotrichum gloeosporioides* and by *C. acutatum*, occurs worldwide. In warm, humid weather it may cause enormous losses by destroying an entire crop of apples just a few weeks before harvest. Bitter rot symptoms usually appear when the fruit approaches its full size. The rot starts as small, dark areas that enlarge rapidly and become circular and sunken in the center. The surface of the spots is smooth and dark brown at first. When the spots are 1 to 2 centimeters in diameter, numerous acervuli-forming cushions appear concentrically near the center and fewer

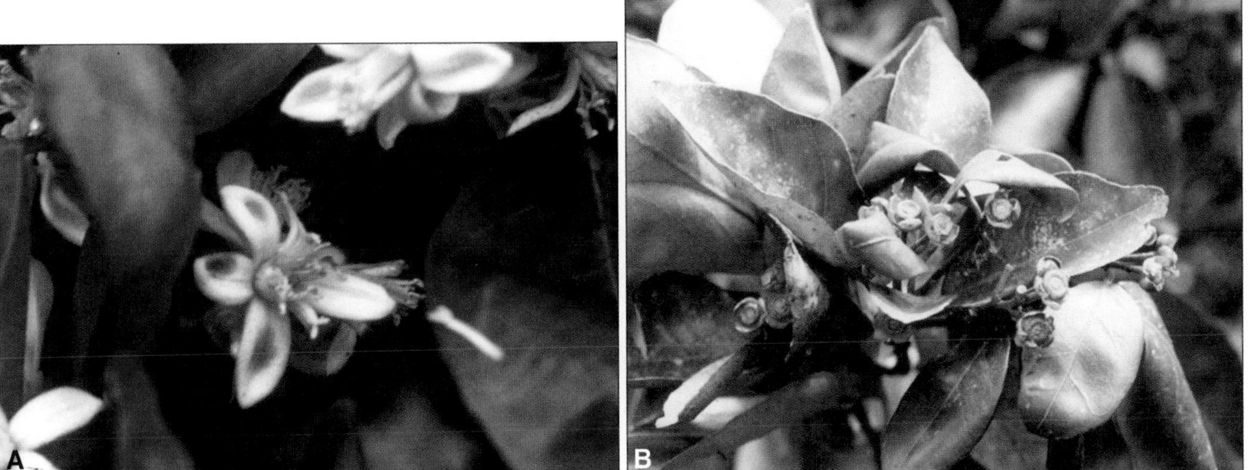

FIGURE 11-80 Postbloom fruit drop of citrus caused by *Colletotrichum acutatum.* (A) Symptoms on blossoms. (B) Fruit buttons remaining on tree after infected fruit drops off the tree. (Photographs courtesy of L.W. Timmer, University of Florida.)

FIGURE 11-81 Rotten areas on apple fruit infected with bitter rot (A) and cankers on trunk and branches of apple tree (B) caused by the fungus *Glomerella cingulata.* [Photographs courtesy of (A) Plant Pathology Department, University of Florida, and (B) Oregon State University.]

toward the edge of the spots. In humid weather, the acervuli produce creamy masses of pink-colored spores, the rotted area expands rapidly, and more rings of spore masses appear (Fig. 11-81). In older rotted areas the pink masses disappear and the tissue becomes dark brown to black, wrinkled, and sunken. The rot also spreads inward toward the apple core, and the rotted tissue may be bitter. Several spots on a fruit usually enlarge, fuse, and rot the entire fruit, which may mummify and drop or cling to the twig. Bitter rot infections fail to develop appreciably during cold storage. When, however, the fruit is marketed and kept at room temperature, bitter rot may develop very rapidly. Occasionally, bitter rot cankers may develop on the limbs.

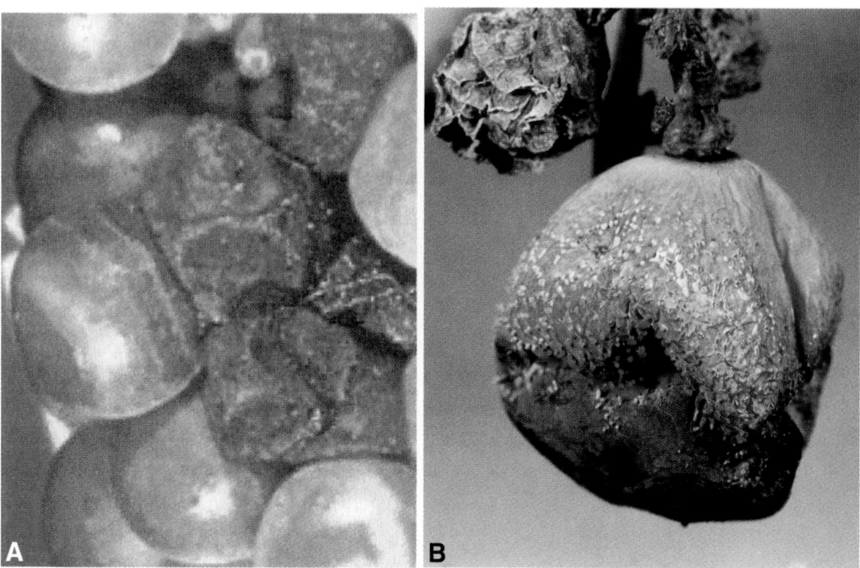

FIGURE 11-82 (A) Part of a grape cluster with berries infected by the bitter rot fungus *Greeneria uvicola*. (B) A grape berry entirely rotten and covered by fructifications (acervuli) of the causal fungus *Greeneria uvicola*. [Photographs courtesy of (A) M. Ellis, Ohio State University and (B) Plant Pathology Department, University of Florida.]

Ripe rot of grape and other fruits, caused by *Colletotrichum acutatum* and *C. gloeosporioides*, also occurs worldwide but is most serious in areas with warm, humid weather during the ripening of the fruit. Ripe rot appears when the fruit is nearly mature and may continue its destruction of fruit after it has been picked and during shipment and marketing. Symptoms begin as small spots that soon spread to over half the berry. Eventually the whole berry rots, usually in a continuous manner but sometimes marked by concentric zones, and the symptoms resemble those of bitter rot of grape (Fig. 11-82A), which is caused by another anthracnose fungus, *Greeneria uvicola* (formerly *Melanconium fuligenum*). The ripe rot-affected berry becomes more or less densely covered with numerous acervuli pustules (Fig. 11-82B) from which, in humid weather, pinkish masses of spores ooze out. Later, the spore masses become darker, almost reddish-brown. The rotted berries become sunken at the point of infection and gradually become more or less shriveled and mummified, while the pustules continue to produce spores. Infected berries often "shell" or drop off before the rot causes them to dry up.

The fungus *Glomerella* produces ascospores in asci in perithecia. Much more frequently, however, the fungus produces conidia-bearing acervuli of its anamorphs *Colletotrichum* or *Gloeosporium* spp. Anamorphs produce colorless, one-celled, ovoid, cylindrical, and sometimes curved or dumbbell-shaped conidia in acervuli (Fig. 11-83). Masses of conidia appear pink or salmon colored. Acervuli are subepidermal and break out through the surface of the plant tissue. *Colletotrichum* has been distinguished from *Gloeosporium* by the fact that *Colletotrichum* acervuli have dark, long, sterile hair-like hyphae, whereas *Gloeosporium* acervuli do not. However, this is not always so and, therefore, *Colletotrichum* and *Gloeosporium* are often considered as the same fungus. As mentioned earlier, many *Colletotrichum* species produce a *Glomerella*-perfect stage, whereas many *Gloeosporium* species have *Glomerella* or *Gnomonia* as the perfect stage.

The fungus overseasons in diseased stems, leaves, and fruit as mycelium or spores, in the seed of most affected annual hosts, and in cankers of perennial hosts (Fig. 11-84). Ascospores or conidia produced by the surviving mycelium in the spring cause primary infections. Conidia cause all secondary infections during the entire season as long as temperature and humidity are favorable. Germ tubes penetrate uninjured tissue directly. The mycelium grows intercellularly and may remain latent for some time before the cells begin to collapse and rot. The mycelium then produces acervuli and conidia just below the cuticle, which rupture the cuticle and release conidia that cause more infections. Infections of young fruit generally remain latent until the fruit is past a certain stage of development and maturity, at which point the infections develop fully.

The fungus is favored by high temperatures and humid or moist weather. Conidia are released and

FIGURE 11-83 Acervuli, cetae, and conidia of two anthracnose-causing fungi. (A) Acervuli of *Colletotrichum lagenarium* on cantaloupe fruit. (B–D) *C. graminicola* acervuli in a cross section of wheat leaf (B) showing close-up of cetae and conidia (C) and stereoscopic overview of acervulus on leaf surface (D). (E) Scanning electron micrograph of acervulus and conidia of the fungus *Marssonina* sp. [Photographs courtesy of (A) B.D. Bruton, USDA, (B and C) Plant Pathology Department, University of Florida, and (D and E) M.F. Brown and H.G. Broton.]

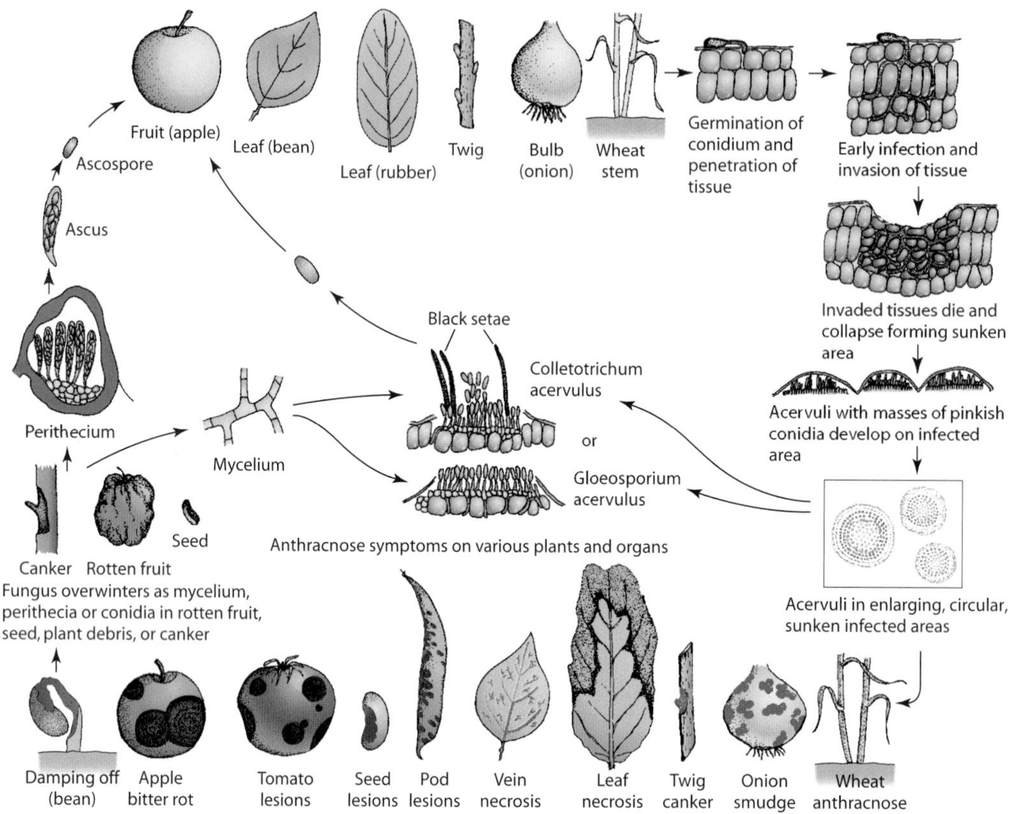

FIGURE 11-84 Disease cycle of anthracnose diseases caused by *Glomerella cingulata* and *Colletotrichum* or *Gloeosporium* sp.

spread only when the acervuli are wet and are generally spread by splashing and blowing rain or by coming in contact with insects, other animals, tools, and so on. Conidia germinate only in the presence of water and penetrate the host tissues directly (Fig. 11-84). In the beginning the hyphae grow rapidly, intercellularly and intracellularly, but cause little or no visible discoloration or other symptoms. Then, more or less suddenly, especially when fruit begins to ripen, the fungus becomes more aggressive and symptoms appear. In many hosts the fungus reaches the seed and is either carried on the seeds or, in some, may even invade a small number of seeds without causing any apparent injury to them. There is considerable variability in the kinds of host plants each species of *Colletotrichum* or *Gloeosporium* can attack, and there may be several races with varying pathogenicity within each species of the fungus.

The control of *Glomerella/Colletotrichum* diseases depends on the use of disease-free seed grown in arid areas or use of treated seed; crop rotation of hosts; use of resistant varieties when available; removal and burning of dead twigs, branches, and fruit infected with the fungus in woody plants; and, finally, spraying with appropriate fungicides.

GNOMONIA ANTHRACNOSE AND LEAF SPOT DISEASES

Various species of *Gnomonia* attack mostly forest and shade trees on which they cause symptoms primarily on leaves, e.g., elm and hickory, or on the leaves and young shoots or twigs, e.g., oak, sycamore (Figs. 11-85A and 11-85B), and walnut. *Gnomonia* diseases are also favored by wet, humid weather. The fungus overwinters primarily in fallen leaves or on infected twigs as immature perithecia. The perithecia mature and produce ascospores in the spring, which cause most primary infections on young leaves and twigs; then conidia, generally of the *Gloeosporium* type, are produced in acervuli and cause all subsequent infections. Both ascospores and condia are disseminated only during rainy weather.

Sycamore anthracnose is the most important disease of sycamore (*Platanus* spp.). The disease may kill the tips of small, 1-year-old twigs before leaf emergence and may kill the buds before they open (Fig. 11-85C). The most frequently observed symptom is the sudden death of expanding shoots and young leaves, which has often been confused with frost damage. Later, the fungus may

FIGURE 11-85 Symptoms of sycamore anthracnose caused by the fungus *Gnomonia veneta*. (A) Vein necrosis on leaves, (B) cankers on twigs, and (C) defoliation, dieback, and decline of entire trees. Spots on leaves (D) and flower petals (E) of dogwood trees caused by the dogwood anthracnose fungus *Disculla destructiva*. [Photographs courtesy of (A, D, and E) Plant Pathology Department, University of Florida and (B and C) U.S. Forest Service.]

cause irregular brown areas adjacent to the midrib, veins, and leaf tips. In moist weather, small, cream-colored acervuli form on the underside of dead leaf tissue along the veins. In some years, sycamores are completely defoliated by this disease in the spring and produce new leaves in the summer. From the buds or leaves the fungus spreads into the twigs, on which it produces cankers, or spreads through and kills small twigs and forms cankers on the branches around the bases of the dead twigs (Fig. 11-85B). In trees affected severely with anthracnose for several successive years, many branches may die (Fig. 11-85C). The fungus overwinters as mycelium and as immature perithecia in twig cankers and in fallen leaves. As in the other anthracnose diseases, this too is favored by rainy weather and rather cool temperatures.

Control of *Gnomonia* diseases in trees under forest conditions is not economically feasible. In shade and ornamental trees, however, burning of leaves, pruning of infected twigs, fertilization, and watering help reduce the disease. Valuable trees should be sprayed with appropriate fungicides two to four times at 10- to 14-day intervals starting as soon as the buds begin to swell (sycamore) or soon after the buds open.

DOGWOOD ANTHRACNOSE

Dogwood anthracnose is a relatively new disease, having been reported from the northwestern United States and Canada in the mid-1970s, from the eastern seaboard states from New York to Georgia in the mid-1980s, and by the mid-90s it had spread westward to Michigan, Indiana, Kentucky, etc. The disease may have been disseminated by people bringing in infected trees as ornamentals, which subsequently served as a source of inoculum for forest trees. In some areas of Maryland, for example, dogwood tree mortality increased from 0 to 17% between 1988 and 1992. The survival of dogwood trees has been seriously threatened by this disease.

Dogwood anthracnose appears as reddish or brown necrotic lesions on the leaves and flower petals (Figs. 11-85D and 11-85E). The leaves appear blighted and may cling to the twigs until the following spring, at which time twig dieback also becomes apparent. Any new shoots on the twigs are also killed and from them the fungus spreads and forms cankers on the branches at the base of the shoots. Dogwood anthracnose is caused by the mitosporic fungus *Discula destructiva*. It produces conidia in acervuli that develop on infected leaves and twigs and exude sticky masses of pale to lightly colored conidia. The fungus overwinters as mycelium and conidia on infected twigs and clinging leaves. Conidia are spread by splashing rain and possibly by insects and birds. Rains seem to favor both the spread of the fungus and the infection of new trees. Control/management of the disease is difficult, if not impossible, especially under forest conditions. Proper irrigation and fertilization of ornamental dogwoods seem to help, as does an early season application of appropriate fungicides.

Selected References

Arauz, L. F. (2000). Mango anthracnose: Economic impact and current options for integrated management. *Plant Dis.* **84**, 600–611.

Bailey, J. A., and Jeger, M. J., eds. (1992). "*Colletotrichum*: Biology, Pathology, and Control." Commonw. Agric. Bur., Farnham Royal, Bucks, UK.

Baker, K. F. (1948). The history, distribution, and nomenclature of the rose blackspot fungus. *Plant Dis. Rep.* **32**, 260–274, 397.

Bergstrom, G. C., and Nicholson, R. L. (1999). The biology of corn anthracnose: Knowledge to exploit for improved management. *Plant Dis.* **83**, 596–608.

Bowen, K. L., and Poark, R. S. (2001). Management of black spot of rose with winter fungicide treatment. *Plant Dis.* **85**, 393–398.

Brook, P. J. (1977). *Glomerella cingulata* and bitter rot of apple. *N. Z. J. Agric. Res.* **20**, 547–555.

Daughtrey, M. L., and Hibben, C. R. (1994). Dogwood anthracnose: A new disease threatens two native *Cornus* species. *Annu. Rev. Phytopathol.* **32**, 61–73.

Daykin, M. E., and Milholland, R. D. (1984). Ripe rot of muscadine grape caused by *Colletotrichum gloeosporioides* and its control. *Phytopathology* **74**, 710–714.

Ellis, M. A., *et al.* (1991). "Compendium of Raspberry and Blackberry Diseases and Insects." APS Press, St. Paul, MN.

Freeman, S., Katan, T., and Shabi, E. (1998). Characterization of *Colletotrichum* species responsible for anthracnose diseases of various fruits. *Plant Dis.* **82**, 596–605.

Horst, R. K. (1983). Black spot. *In* "Compendium of Rose Diseases" (R. K. Horst, ed.), pp. 7–10, The American Phytopathological Society, St. Paul, MN.

Howard, C. M., Mass, J. L., Chandler, C. K., and Albreghts, E. E. (1992). Anthracnose of strawberry caused by the *Colletotrichum* complex in Florida. *Plant Dis.* **76**, 976–981.

Kendrick, J. B., Jr., and Walker, J. C. (1948). Anthracnose of tomato. *Phytopathology* **32**, 247–260.

Maas, J. L., ed. (1998). "Compendium of Strawberry Diseases," 2nd Ed. APS Press, St. Paul, MN.

Pearson, R. C., and Goheen, A. C. (1988). "Compendium of Grape Diseases." APS Press, St. Paul, MN.

Ploetz, R. C., Nishijima, W., Robrach, K., Zentmyer, G. A., and Ohr, H., eds. (1994). "Compendium of Tropical Fruit Diseases." APS Press, St. Paul, MN.

Politis, D. J., and Wheeler, H. (1973). Ultrastructural study of penetration of maize leaves by *Colletotrichum graminicola*. *Physiol. Plant Pathol.* **3**, 465–471.

Prusky, D., Freeman, S., and Dickman, M. B., eds. (2000). *Colletotrichum*: Host Specificity, Pathology, and Host-Parasite Interaction. APS Press, St. Paul, MN.

Redlin, S. C. (1991). *Discula destructiva* sp. nov., cause of dogwood anthracnose. *Mycologia* **83**, 633–642.

Ridings, W. H., and Clayton, C. N. (1970). *Melanconium fuligineum* and the bitter rot disease of grapes. *Phytopathology* **60**, 1203–1211.

Sumner, D. R. (1995). Smudge. In "Compendium of Onion and Garlic Diseases" (H. F. Schwartz and S. K. Mohan, eds.), pp. 29–30. APS Press, St. Paul, MN.

Sutton, T. B., and Shane, W. W. (1983). Epidemiology of the perfect stage of *Glomerella cingulata*. *Phytopathology* **73**, 1179–1183.

Thompson, D. C., and Jenkins, S. F. (1985). Pictorial assessment key to determine fungicide concentrations that control anthracnose development on cucumber cultivars with varying resistance levels. *Plant Dis.* **69**, 833–836.

Timmer, L. W., Garnsey, S. M., and Graham, J. H., eds. (2000). "Compendium of Citrus Diseases," 2nd Ed. APS Press, St. Paul, MN.

Weber, G. F. (1973). "Bacterial and Fungal Diseases of Plants in the Tropics." Univ. of Florida Press, Gainesville, FL.

Wellman, F. L. (1972). "Tropical American Plant Disease," pp. 236–273. Scarecrow Press, Metuchen, NJ.

Whiteside, J. O., Garnsey, S. M., and Timmer, L. W., eds. (1988). "Compendium of Citrus Diseases." APS Press, St. Paul, MN.

FRUIT AND GENERAL DISEASES CAUSED BY ASCOMYCETES AND DEUTEROMYCETES (MITOSPORIC FUNGI)

The diseases discussed in this section are found most commonly on the fruit or cause most of their damage by their effect on the fruit, but they may affect other parts of the plant as well. Most of these fungi produce ascospores in perithecia and conidia on free hyphae, but they differ from one another in life cycles and in the diseases they cause. The most common ascomycetous fungi and the most important diseases they cause are the following.

- *Claviceps*, *C. purpurea*, causing ergot of cereals and grasses
- *Diaporthe*, *D. citri*, causing melanose of citrus fruits, *D. phaseolorum* causing stem canker of soybeans
- *Botryosphaeria* (*Physalospora*), causing black rot, frogeye leaf spot, and canker of apple (*P. obtusa*), other species causing canker and dieback of many temperate and tropical trees, such as oak, willow, citrus, cacao, coconut, rubber, and tropical forest trees
- *Didymella*, *D. bryoniae* causing the gummy stem blight of cucurbits
- *Guignardia*, *G. bidwellii* causing the black rot of grape
- *Venturia inaequalis*, causing apple scab
- *Monilinia*, three species causing brown rot of stone fruits
- *Moniliophthora*, causing the Monilia pod rot of cacao

The most common mitosporic fungus causing fruit and general diseases on plants is *Botrytis*, causing blossom blights and fruit rots but also damping-off, stem cankers or rots, leaf spots, and tuber, corm, bulb, and root rots of many vegetables, flowers, small fruits, and other fruit trees. Another mitosporic fungus causing a variety of diseases is *Phomopsis*.

ERGOT OF CEREALS AND GRASSES

Ergot occurs worldwide, most commonly on rye and pearl millet, less often on wheat and certain wild and cultivated grasses, and rarely on barley and oats. An ergot disease affecting corn occurs in Mexico. The disease, caused by the fungus *Claviceps purpurea*, destroys some 5 to 10% of the grains in infected heads, but its main importance is due to the fact that it replaces the grains with fungal sclerotia, which are poisonous to humans and animals that eat bread or feed containing sclerotia (Figs. 11-86 and 11-87; see also Fig. 1-29).

Another species of the ergot fungus, *C. africana*, had been known to infect sorghum but it seemed to be present only in Africa, India, and Japan. Then, in 1995, this ergot fungus was found to attack sorghum in Brazil and in 1996 in Australia. By 1997, the new sorghum fungus had spread throughout South and Central America, the Caribbean islands, and most of the southern United States up to Kansas and Nebraska. Losses to sorghum from the new ergot can be significant, especially in cool weather. Since the 1960s, sorghum production around the world has depended on hybrid sorghum seed, which has improved yields by 300 to 500%. Male-sterile lines used to produce hybrid seed, however, are extremely susceptible to the new ergot and losses of 10–80%, and sometimes total losses, have been observed. Because sorghum is the fifth most important crop in the world, being cultivated in 45 million acres, and because almost all sorghum is produced with hybrid seed, destruction by the new ergot of the hybrid seed parents causes immeasurable losses from lack of hybrid seed to plant.

Symptoms

The first symptoms appear as creamy droplets of a sticky liquid exuding from young florets of infected heads. The droplets are soon replaced by a hard, horn-shaped, purplish-black fungal mass a few millimeters in diameter and 0.2 to 5.0 centimeters long. These are the sclerotia or ergots of the fungus that grow in place of the kernel and consist of a hard compact mass of fungal tissue (Fig. 11-86).

The Pathogen

Ergot is caused by *Claviceps purpurea*, *C. sorghi*, *C. africana*, and other *Claviceps* species. There is considerable genetic variation among isolates of *C. purpurea* and of the other species. The fungus overwinters as sclerotia on or in the ground or mixed with the seed. In the spring, about the time cereals are in bloom, sclerotia on or near the surface of the soil germinate by forming from 1 to 60 flesh-colored stalks 0.5 to 2.5 centimeters tall (Figs. 11-86, 11-87, and 11-88). The tip of each stalk produces a spherical head at the periphery of which develop numerous perithecia, each containing many asci. Each ascus contains eight long, multicellular ascospores.

FIGURE 11-86 Ergot of grain crops. (A) Ergot sclerotia on wheat head. (B) Wheat kernel transformed into ergot sclerotium (left) and healthy kernel of wheat (right). (C) Healthy barley kernels and kernels transformed into ergot sclerotia. (D) Two ergot sclerotia, one producing several stalks and perithecia-bearing stroma. [Photographs courtesy of (A) D.T. Atkinson, (B) R. Tekauz and (C) I.R. Evans, WCPD, and (D) WCPD.]

Development of Disease

Ascospores are carried by wind or by insects to young open flowers, where they germinate and infect the ovaries directly or by way of the stigma. During infection, the fungus secretes the enzyme catalase, which apparently plays a role in development of the ergot disease by suppressing the defenses of the host. Within about a week, the fungus in the ovary forms sporodochia that produce conidia of the *Sphacelia* type. The conidia exude from the young florets as creamy droplets known as the "honeydew" stage (Fig. 11-87A). The honeydew attracts insects, which become smeared with the conidia of the fungus and carry them to healthy flowers, which the conidia infect. Conidia are also spread to flowers by splashing rain. Gradually, infected ovaries, instead of producing normal seed, become replaced by a hard mass of fungal mycelium, which eventually forms the characteristic ergot sclerotium (Fig. 11-86). The sclerotia mature about the same time as the

FIGURE 11-87 Ergot on sorghum caused by *Claviceps sorghi.* (A) Yellowish-brown droplets of conidia—containing honeydew produced by ergot-infected sorghum flowers. Ergot sclerotia produced on sorghum by *C. africana* (B) and *C. sorghi* (C). (Photographs courtesy of R. Bandyopadhyay, with permission from APS.)

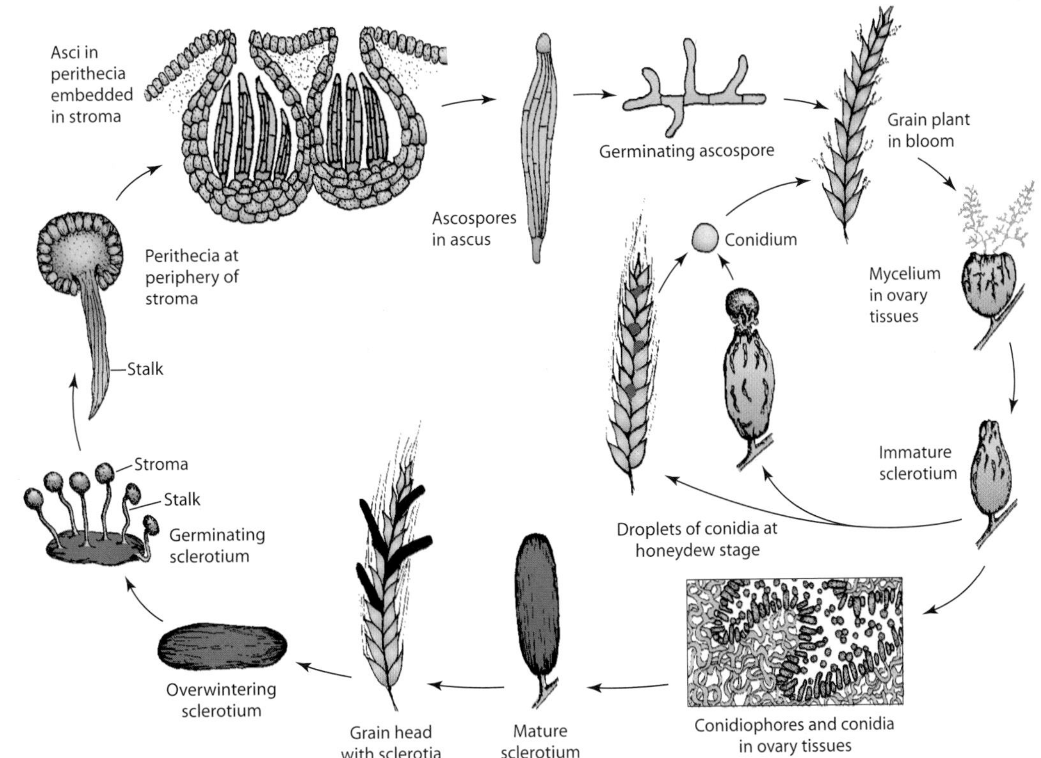

FIGURE 11-88 Disease cycle of ergot of grains caused by *Claviceps purpurea.*

healthy seeds and either fall to the ground, where they overwinter, or are harvested with the grain and may be returned to the land with the seed (Fig. 11-86C).

Although ergotism is no longer common in humans, ergot may be involved in many otherwise unexplainable poisonings of humans, and ergotism certainly continues to be of economic importance as an animal disease. Grain containing more than 0.3% by weight of the ergot sclerotia can cause ergotism, and therefore such grain may not legally be sold and milled for flour and human consumption. Also, it is costly and quite often difficult to remove enough sclerotia to meet the legal standards, particularly in poorer countries, and the remaining traces are often toxic to livestock. Moreover, feeding livestock with cleanings from contaminated grain or grazing in pastures that have infected grass heads can lead to reproductive failure or gangrene of the peripheral parts of the animals.

Control

The control of ergot depends entirely on using clean seed or seed that has been freed from ergot. Sclerotia may be removed from seed by machinery or by soaking contaminated seed for three hours in water and then floating off the sclerotia in a solution of about 18 kilograms of salt in 100 liters of water. Ergot sclerotia do not survive for more than a year and do not germinate if buried deep in the ground. Therefore, deep plowing or crop rotation with a noncereal for at least a year helps eliminate the pathogen from a particular field. Wild grasses should be mowed or grazed before flowering to prevent production of ergot sclerotia on them and avoid poisoning of livestock through them, and also to prevent the spread of the fungus to cultivated cereals and grasses. Ergot on turf grasses grown for seed can be controlled with fungicides such as flusilazole. Control of ergot in hybrid seed-producing nurseries is possible with four to five preventive sprays of fungicides at 5- to 7-day intervals, but only if the inoculum pressure is relatively low.

APPLE SCAB

Apple scab exists worldwide but is more severe in areas with cool, moist springs and summers. In the United States it is most serious in the north central and northeastern states. Similar scab diseases affect pears (*V. pyrina*) and hawthorns (*V. inaequalis* sp.f. *pyracanthae*).

Scab is the most important disease of apples. Its primary effect is reduction of the quality of infected fruit. Scab also reduces fruit size or results in premature fruit drop, defoliation, and poor fruit bud development for the next year, and it reduces the length of time infected fruit can be kept in storage. Losses from apple scab may be 70% or more of the total fruit value. In most apple-producing areas, no marketable fruit can be harvested if scab control measures are not taken.

Symptoms

At first, light, olive-colored, irregular spots appear on the lower surface of sepals or young leaves of the flower buds. Soon after, the lesions become olive green to gray with a velvety surface. Later, the lesions appear metallic black in color and may be slightly raised. Lesions on older leaves generally form on the upper surface of the leaves (Fig. 11-89A). Lesions may remain distinct or they may coalesce. Leaves infected young remain small and curled and may later fall off. Occasionally, small scab spots are produced on twigs and blossoms.

Infected fruit develop circular scab lesions, velvety and olive green at first but later becoming darker, scabby, and sometimes cracked (Figs. 11-89B–11-89D). The cuticle of the fruit is ruptured at the margin of the lesions. Fruit infected early become misshapen and cracked, and frequently drop prematurely. Fruit infected when approaching maturity form only small lesions, which, however, may develop into dark scab spots during storage.

The Pathogen: Venturia Inaequalis

The mycelium in living tissues is located only between the cuticle and the epidermal cells. There, it produces short, erect, brownish conidiophores that give rise to several, one- or two-celled, *Spilocaea*-type conidia of rather characteristic shape (Fig. 11-90). In dead leaves the mycelium grows through the leaf tissues and produces ascogonia and antheridia; following fertilization, pseudothecia form. The latter are dark brown to black with a slight beak and an opening. Each pseudothecium contains 50 to 100 asci, each with eight ascospores consisting of two cells of unequal size (Fig. 11-89E).

Development of Disease

The pathogen overwinters in dead leaves on the ground (Fig. 11-90) as immature pseudothecia. Pseudothecia complete their growth in late winter and spring, and ascospores mature as the weather becomes favorable for growth and development of the host (Fig. 11-89F). Pseudothecia and asci mature sequentially. Some ascospores mature before the apple buds start to open in the spring, but most mature in the period during which the fruit buds open.

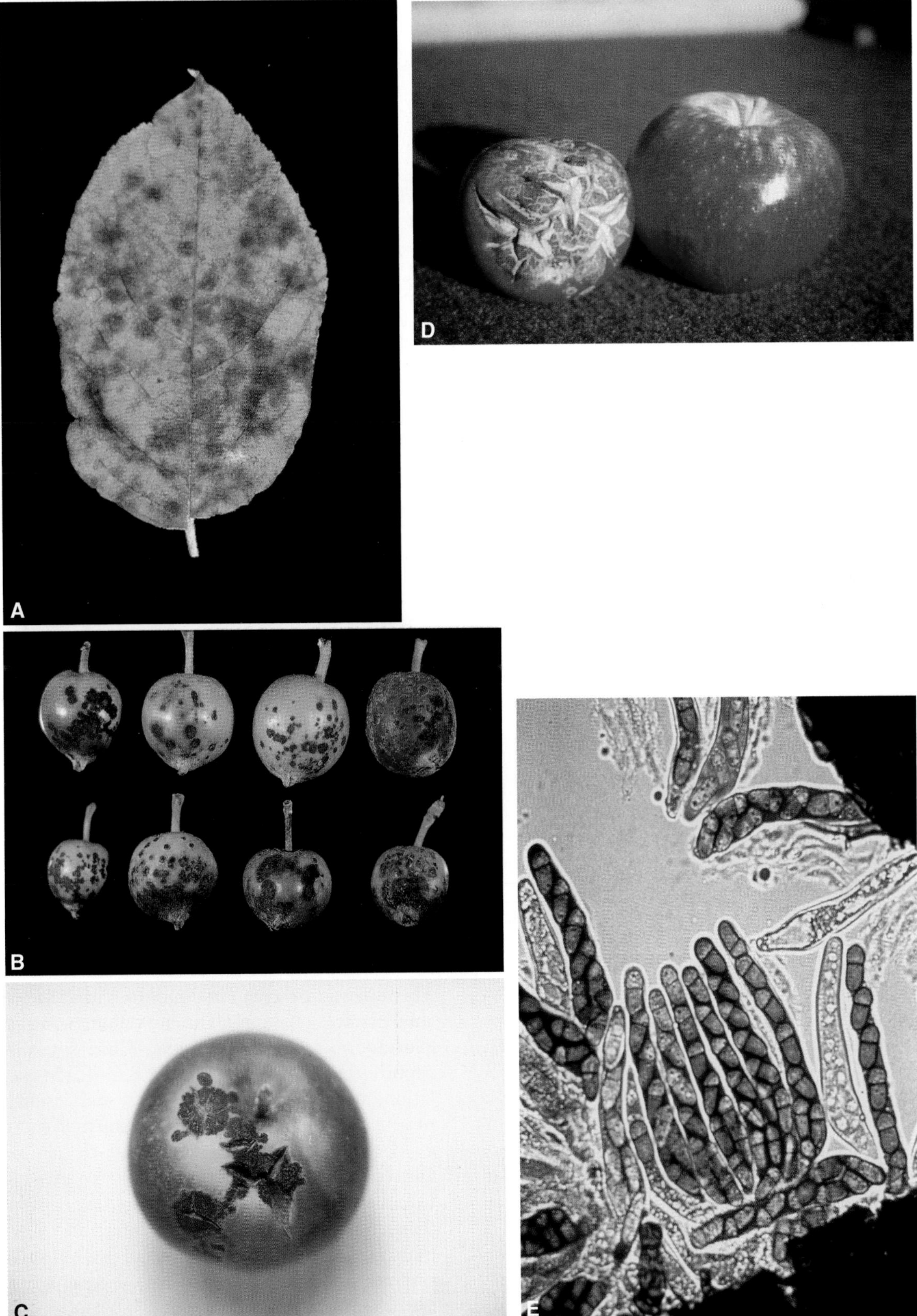

FIGURE 11-89 Apple scab symptoms caused by *Venturia inaequalis* on leaves (A), young fruit (B), and mature fruit (C and D). (E) Open perithecium of the fungus showing asci and ascospores. [Photographs courtesy of (C and D) D.R. Cooley, University Massachusetts and (E) D. Aylor, Connecticut Agricultural Experiment Station.]

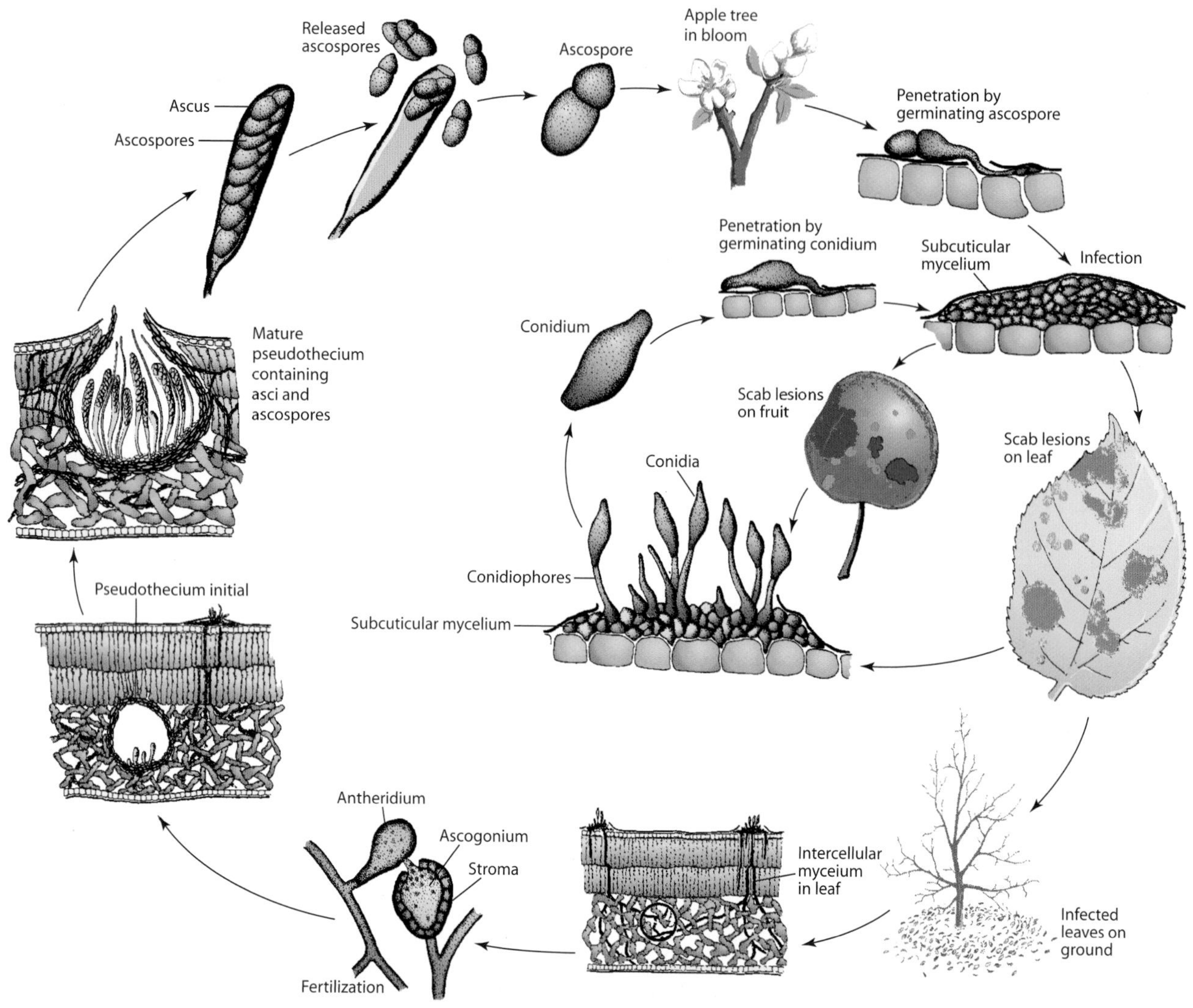

FIGURE 11-90 Disease cycle of apple scab caused by *Venturia inaequalis*.

When pseudothecia become thoroughly wet in the spring, the asci forcibly discharge the ascospores into the air; air currents may carry them to susceptible green apple tissues. Ascospore discharge may continue for 3 to 5 weeks after petal fall.

Ascospores germinate and cause infection when kept wet at temperatures ranging from 6 to 26°C. For infection to occur, the spores must be continuously wet for 28 hours at 6°C, for 14 hours at 10°C, for 9 hours at 18–24°C, or for 12 hours at 26°C (see Fig. 8-24).

On germination on an apple leaf or fruit, the ascospore germ tube pierces the cuticle and grows between the cuticle and the outer cell wall of the epidermal cells. At first the epidermal cells and later the palisade and the mesophyll cells show a gradual depletion of their contents, eventually collapsing and dying. The fungus, however, continues to remain largely in the subcuticular position. The mycelium soon produces enormous numbers of conidia, which push outward, rupture the cuticle, and, within 8 to 15 days of inoculation, form the olive-green, velvety scab lesions. During or after a rain, conidia may be washed down or blown away to other leaves or fruit on which they germinate and cause infection in the same way ascospores do. Conidia continue to cause infections during wet weather throughout the growing season. Infections, however, are more abundant during cool, wet periods of spring, early summer, and fall, while they are infrequent or absent in the dry, hot summer weather.

After infected leaves fall to the ground, the mycelium invades the interior of the leaf and forms pseudothecia, which carry the fungus through the winter.

Control

Several apple varieties resistant to scab are available, but many popular ones are moderately to highly susceptible. It appears that all apple cultivars are susceptible to some apple scab fungus isolates and resistant to others. A number of fungi are antagonistic to the apple scab fungus, and some of them decrease ascospore production when applied to scab-infected apple leaves on the orchard floor. So far, no effective practical biological control of apple scab has been developed. Introducing endochitinase genes from fungi into apple increased the resistance of apple to scab, but it also reduced vigor of the plant. Shredding of apple leaf litter or treating them with urea in the fall reduced the risk of scab by about 65%. Apple scab, however, can be controlled thoroughly by timely sprays with the proper fungicides.

For an effective apple scab control program, apple trees must be sprayed or dusted diligently before, during, or immediately after a rain from the time of budbreak until all the ascospores are discharged from the pseudothecia. If these primary infections from ascospores are prevented, there will be less need to spray for scab during the remainder of the season. If primary infections do develop, spraying will have to be continued throughout the season. In most areas, the application of fungicides for scab control begins when buds show a slight green tip and a rainy period is sufficiently long at the existing temperature to produce an infection. Sprays are repeated every 5 to 7 days, or according to rainfall, until petal fall. After petal fall, and depending on the success of the control program to that point, sprays are usually repeated every 10 to 14 days for several more times (see Fig. 9-35).

Since the early 1990s, considerable progress has been made in developing simple or computerized apple scab prediction systems of spore release and infection for scheduling fungicide applications for scab control. All the systems are based on the interactions among temperature, amount and duration of rainfall, and duration of leaf wetness, on the one hand, and the period required for the pathogen to initiate infection on the other. The accuracy and dependability of these models vary considerably under different local conditions.

Several fungicides give excellent control of apple scab. Some of them protect a plant from becoming infected, but they cannot cure an infection, whereas some can stop infections that may have started. In some areas, new strains of *Venturia inaequalis* have now appeared that are resistant to several of the systemic fungicides. These chemicals, therefore, can no longer be relied on to control the disease by themselves; rather, they must be applied in combination with one of the broad-spectrum fungicides.

BROWN ROT OF STONE FRUITS

Brown rot occurs wherever stone fruits are grown and there is sufficient rainfall during the blossoming and fruit ripening periods. It affects peaches, cherries, plums, apricots, and almonds with about equal severity.

Losses from brown rot result primarily from fruit rotting in the orchard, but serious losses may also appear during transit and marketing of the fruit. Yields may also be reduced by destruction of the flowers during the blossom blight stage of the disease. In severe infections, 50 to 75% of the fruit may rot in the orchard, and the remainder may become infected before it reaches the market.

Symptoms

The first symptoms of the disease appear on the blossoms (Fig. 11-91A) and may involve the entire flower and its stem. In humid weather the infected organs are covered with the grayish-brown conidia of the fungus and later shrivel and dry up, with the rotting mass clinging to the twigs for some time. At the base of infected flowers, small, sunken cankers develop on twigs around the flower stem, which sometimes they encircle and cause twig blight. In humid weather, gum and also gray tufts of conidia appear on the bark surface.

Fruit symptoms appear when the fruit approaches maturity as small, circular, brown spots that spread rapidly in all directions. Ash-colored tufts of conidia break through the skin of the infected areas and appear on the fruit surface (Fig. 11-91B). One large or several small rotten areas may be present on the fruit, which finally becomes completely rotted and either dries up into a mummy or remains hanging from the tree (Figs. 11-91C and 11-91D). Sometimes small cankers also develop on twigs or branches bearing infected fruit.

The Pathogen: Monilinia (Sclerotinia) Fructicola, M. Laxa, and M. Fructigena

The mycelium produces chains of elliptical *Monilia*-type conidia (Figs. 11-91E and 11-92) on hyphal branches arranged in tufts (sporodochia). The fungus also produces microconidia (spermatia) in chains on bottle-shaped condiophores. The spermatia do not germinate, but seem to be involved in fertilization of the fungus. The sexual stage, the apothecium, originates from pseudosclerotia formed in mummified fruit buried partly or wholly in the soil or debris. More than 20 apothecia may form on one mummy. The inside or upper surface of the apothecium is lined with thousands of asci interspersed with sterile hyphae (paraphyses). Each ascus contains eight single-celled ascospores.

FIGURE 11-91 (A–D) Symptoms of brown rot of stone fruits caused by *Monilinia fructicola*. (A) Blossom blight and twig canker. (B) Brown rot of a peach still on the tree. (C) Brown rot of peaches developed in storage (right) while peaches on the left were protected by appropriate fungicides. (D) (E) Conidia of *M. fructicola*. (Photographs courtesy of D.F. Richie, North Carolina State University.)

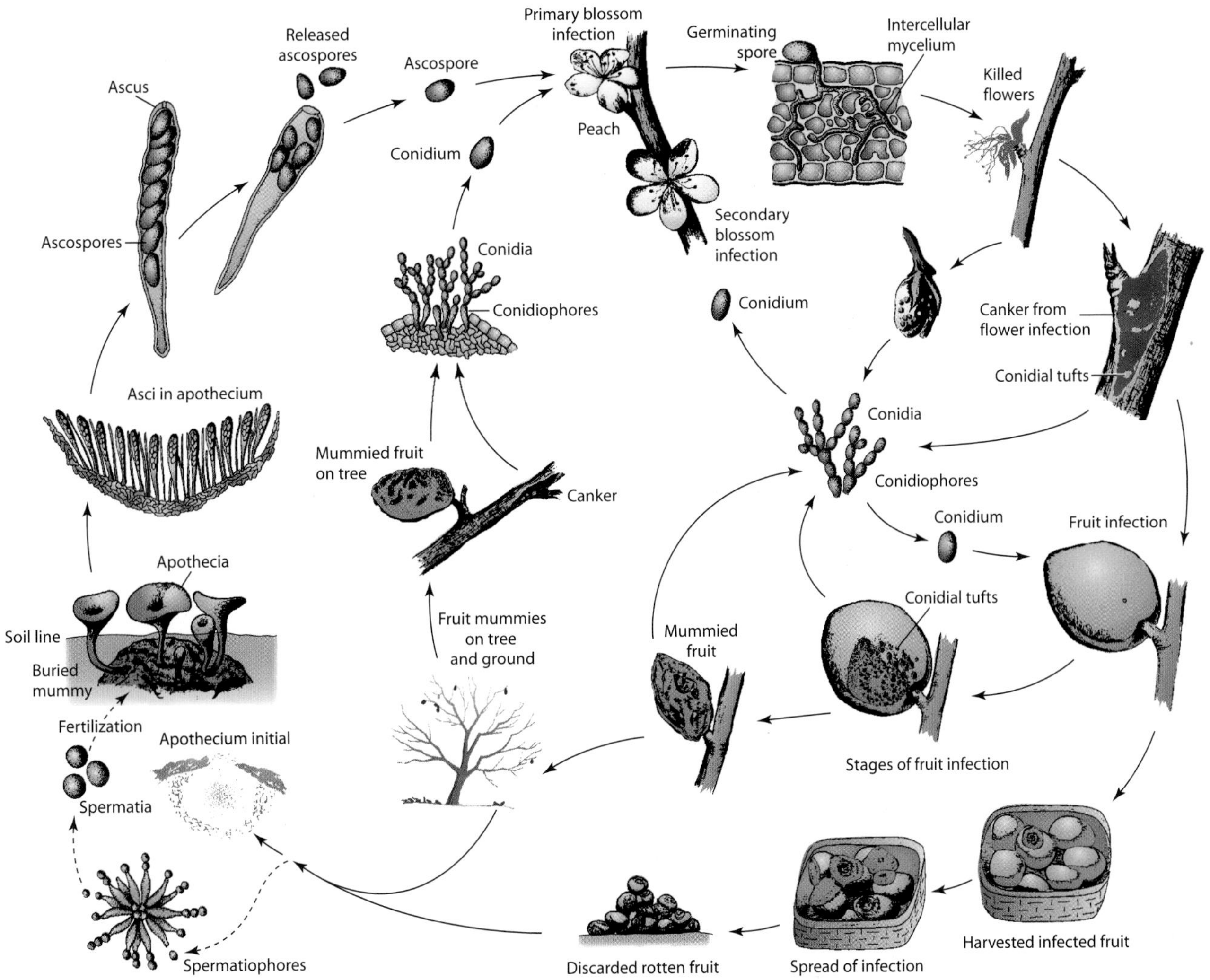

FIGURE 11-92 Disease cycle of brown rot of stone fruits caused by *Monilinia fructicola*.

Development of Disease

The pathogen overwinters as mycelium in mummified fruit on the tree and in cankers of affected twigs or as pseudosclerotia in mummies in the ground (Fig. 11-92). In the spring the mycelium in mummified fruit on the tree and in twig cankers produces new conidia, whereas the pseudosclerotia in mummified fruit buried in the ground produce apothecia, which form asci and ascospores.

Both conidia and ascospores can cause blossom infections. The conidia are windblown or may be carried to floral parts by rainwater splashes or insects. The ascospores are forcibly discharged by the ascus, forming a whitish cloud over the apothecium. Air currents then carry the ascospores to the flowers. Temperature and wetness duration play critical roles in the number of flowers that will become infected (Fig. 11-93). Conidia and ascospores germinate and can cause infection within

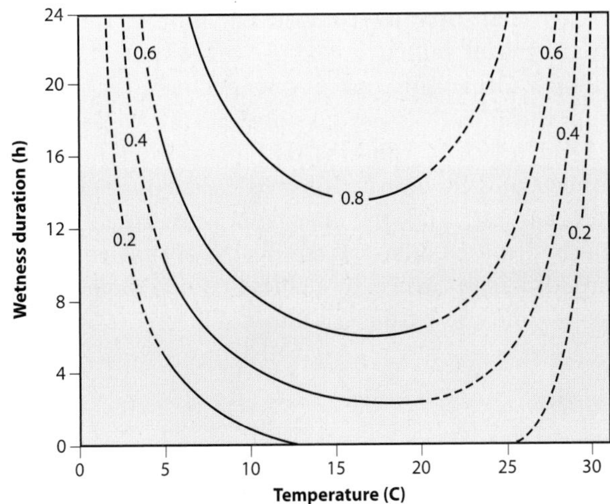

FIGURE 11-93 Predicted effect of temperature and wetness duration on disease incidence (0.8 = 80%) of cherry blossoms infected by *Monilinia laxa*. —, measured; —, suggested by the model. [From Tamm, Minder, and Flückiger (1995). *Phytopathology* 85, 401–408.]

a few hours. The mycelium, especially in humid weather, produces numerous conidial tufts on the rotten, shriveled floral parts from which new masses of conidia are released. In the meantime, the mycelium advances rapidly into the blossom petioles and into the fruit spurs and the twigs, where a depressed, reddish-brown, shield-shaped canker forms. The canker may encircle the twig, which then becomes girdled and dies. The surface of the canker is soon covered with conidial tufts, and conidia from these serve as inoculum for fruit infection later in the season when the fruit begins to ripen.

Because ascospores and new conidia are short lived, the gap between the time of blossom infection and when ripening fruit can become infected is bridged by conidia formed on twig cankers during humid weather in the summer. In addition, conidia produced on infected flowers of late-blooming stone fruits may be carried to and infect the fruit of early-ripening stone fruit species or varieties.

Conidia usually penetrate fruit through wounds made by insects, twig punctures, or hail, but in some cases they also gain access through stomata or directly through the cuticle. The fungus grows intercellularly at first and secretes enzymes that cause maceration and browning of the infected tissues. The fungus invades the fruit quite rapidly while it also produces conidial tufts on the already rotted area. The new conidia may be carried away and infect more fruit. The entire fruit may become completely rotten within a few days, and it either clings to and hangs from the tree or falls to the ground. Fruit falling to the ground soon after infection usually disintegrates through the action of saprophytic fungi and bacteria. Fruit left hanging on the tree loses moisture, shrivels, and becomes a dry, distorted mummy consisting of the remains of the fruit cells held in place by mycelial threads interwoven into a hard rind. Mummy fruit falling to the ground is not affected by soil microorganisms and may persist there for two years or more.

Fruit infection can also take place after harvest, in storage, and in transit. Infected fruit continues to rot after harvest, and the mycelium can attack directly healthy fruit in contact with infected ones. Healthy fruit may also be attacked by conidia at any time between harvest and use by the consumer.

Control

Brown rot of stone fruits can be controlled best by completely controlling the blossom blight phase of the disease. This can be done by spraying two to four times with an effective fungicide from the time the blossom buds show pink until the petals fall. Several fungicides are excellent for brown rot control. Resistant strains of the brown rot fungus have developed to systemic fungicides; therefore, these chemicals are generally used in combination with one of the broad-spectrum fungicides, such as captan or sulfur.

Twigs bearing infected blossoms or cankers should be removed as early as possible to reduce the inoculum available for fruit infections later in the season, and for overwintering.

To control brown rot in ripening fruit, fungicides are applied to the trees a few weeks before harvest, and applications continue weekly or biweekly until just before harvest. Because most infections of immature fruit and many of mature fruit originate in wounds made by insect punctures, the control of insects will also help control the disease.

To prevent infections at harvest and during storage and transit, fruit should be picked and handled with the greatest care to avoid punctures and skin abrasions on the fruit, which enable the brown rot fungus to gain entrance more easily. All fruit with brown rot spots should be discarded. Postharvest brown rot (Fig. 11-91D) can be reduced by dipping or drenching fruit in an appropriate fungicidal solution before storing and by hydrocooling or cooling fruit in air before refrigeration at 0 to 3°C. Biological control of postharvest brown rot has been obtained with several fungi, but it still needs additional work and is not yet used commercially.

MONILIOPHTHORA POD ROT OF CACAO

Monilia pod rot of cacao, caused by the fungus *Moniliophthora roreri*, anamorph *Monilia roreri*, is one of the most serious diseases of cacao in the Americas. It destroys from less than 25% in some regions to 100% of the crop in other regions. The pathogen completes its entire life cycle on the pods on the tree. Infected pods develop conspicuous bumpy swellings on their surface, which are subsequently covered by *Monilia*-type conidia (Fig. 11-94). Detection and removal of infected pods before sporulation of the fungus are essential for management of the disease.

BOTRYTIS DISEASES

Botrytis diseases are probably the most common and most widely distributed diseases of vegetables, ornamentals, fruits, and even some field crops throughout the world. They are the most common diseases of greenhouse-grown crops. Botrytis diseases appear primarily as blossom blights (Fig. 11-95A) and fruit rots (Figs. 11-95B, 11-96A, and 11-96C), but also as

FIGURE 11-94 Cacao pod rot caused by *Moniliophthora roreri*. (A) Early infection of pod. (B) Clump of cacao pods infected severely with the pod rot disease. (Photographs courtesy of Intl. Instit. Trop. Agric. Ibadan, Nigeria.)

damping-off, stem cankers (Fig. 11-95C) or rots, leaf spots, and tuber, corm, and bulb rots (Figs. 11-96B and 11-96D). Under humid conditions, the fungus produces a noticeable gray-mold fruiting layer on the affected tissues that is characteristic of *Botrytis* diseases. Some of the most serious diseases caused by *Botrytis* include gray mold of strawberry, grapes and of many vegetables (Fig. 11-95), calyx end rot of apples (Figs. 11-96A and 11-96B), onion blast and neck rot (Figs. 11-96C and 11-96D), blight or gray mold of many ornamentals, bulb rot of amaryllis, corm rot of gladiolus, and others. *Botrytis* also causes secondary soft rots of fruits and vegetables in storage, transit, and market.

Symptoms

In the field, blossom blights often precede and lead to fruit rots and stem rots. The fungus becomes established in flower petals, particularly when they begin to age, and there it produces abundant mycelium (Figs. 11-95A–11-95D). In cool, humid weather the mycelium produces large numbers of conidia, which may cause further infections. The mycelium grows and invades the inflorescence, which becomes covered with a whitish-gray or light brown cobweb-like mold. The fungus then spreads to the pedicel, which rots and lets the buds and flowers lop over. The fungus later moves from the petals into the fruit and causes a blossom end rot of the fruit, which advances and may destroy part or all of the fruit. Infected fruit and succulent stems become soft, watery, and light brown (Figs. 11-95B, 11-95C, and 11-96). As the tissue rots, the epidermis cracks open and the fungus fruits abundantly. Flat black sclerotia may appear on the surface or are sunken within the wrinkled, dry tissue.

Damping-off of seedlings due to *Botrytis* occurs primarily in cold frames, where the humidity is high. It also occurs in the field if the seed is contaminated with sclerotia of the fungus or if fungus mycelium or sclerotia are present in the soil.

Some species of *Botrytis* cause leaf spots on their hosts, e.g., on gladiolus, onion, and tulip. The spots are small and yellowish at first but later become larger, whitish gray or tan, and sunken, coalesce, and frequently involve the entire leaf.

Stem lesions usually appear on succulent stems or stalks. They may spread through the stalk and cause it to weaken and break over at the point of infection (Fig. 11-95C). In wet weather the diseased parts become covered with a grayish-brown coat of fungus spores. Sclerotia may also be produced on infected stems.

Infection of belowground parts, such as bulbs, corms, tubers, and roots, may begin while these organs are still

FIGURE 11-95 Symptoms of gray mold disease caused by *Botrytis* sp. on different plants and organs. (A) Infected petunia flowers. (B) Gray mold on strawberries. (C) Gray mold and stem canker on tomato. (D) Gray mold of lettuce. (E) Gray mold or bunch rot of grapes. (F) Rotting and shrinking of berries in part of grape bunch due to *Botryris* infection. [Photographs courtesy of (A) R.J. McGovern and (B) D.L. Legard, University of Florida, (C) Plant Pathology Department, University of Florida, (D) J.W. Travis, Pennsylvania State University, and (E and F) E. Hellman, Texas A&M University.]

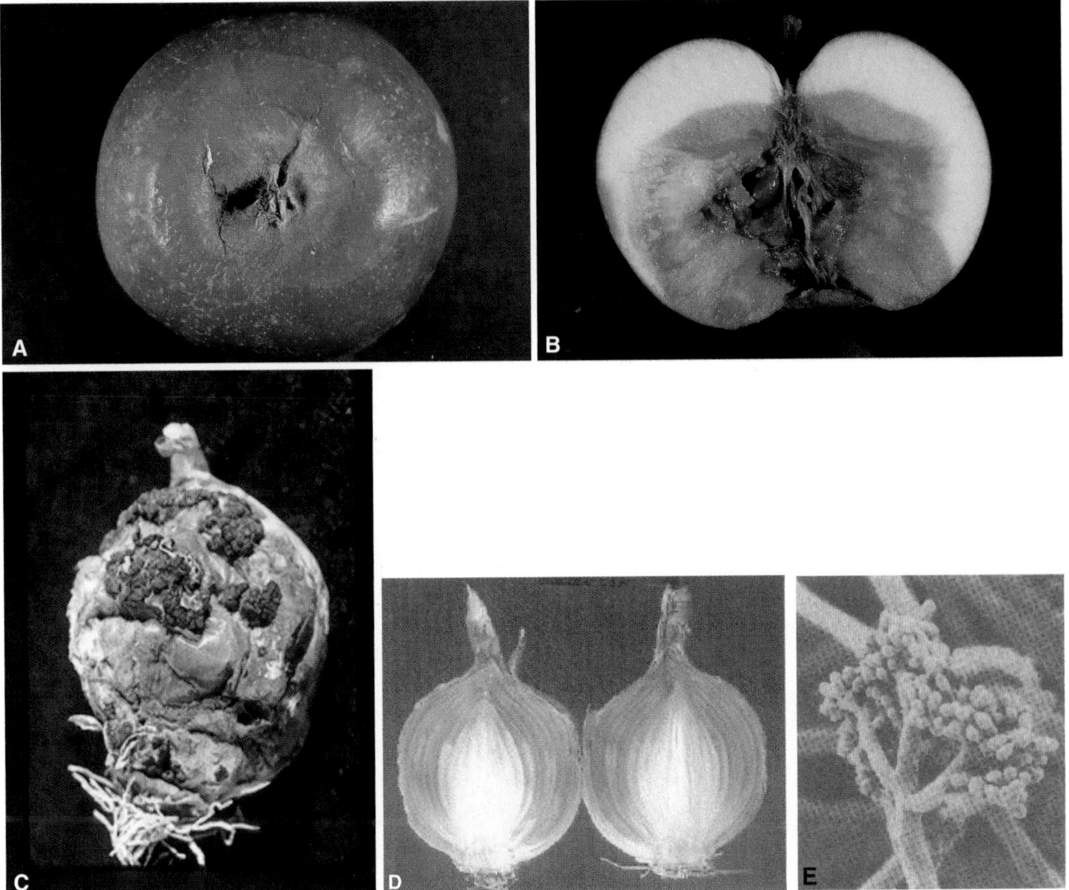

FIGURE 11-96 Symptoms caused by *Botrytis*. Blossom end rot infection of apple as it appears externally (A) and in cross section (B). Onion infected with *Botrytis* showing external sclerotia (C) and internal rotting (D). (E) Scanning electron micrograph of conidiophore and conidia of *Botrytis*. [Photographs courtesy of (C and D) K. Mohan, University of Idaho and (E) M.F. Brown and H.G. Brotzman.]

in the ground or at harvest. Infected tissues usually appear soft and watery at first, but later they turn brown and become spongy or corky and light in weight. Black sclerotia are often found on the surface or intermingled with the rotted tissues and mycelium (Figs. 11-96C and 11-97).

The Pathogen

The pathogen, *Botrytis cinerea* and a few other species, produces abundant gray mycelium and long, branched conidiophores that have rounded apical cells bearing clusters of colorless or gray, one-celled, ovoid conidia (Fig. 11-96E). The conidiophores and clusters of conidia resemble a grape-like cluster. Conidia are released readily in humid weather and are carried by air currents. The fungus frequently produces black, hard, flat, irregular sclerotia (Fig. 11-97). Some species of *Botrytis* occasionally produce a *Botryotinia* perfect stage in which ascospores are produced in an apothecium.

Development of Disease

Botrytis overwinters in the soil as mycelium in decaying plant debris and as sclerotia (Fig. 11-97). The fungus does not seem to infect seeds, but it can be spread with seed contaminated with sclerotia the size of the seed or with bits of plant debris infected with the fungus. The fungus requires cool (18–23°C), damp weather for best growth, sporulation, spore release and germination, and establishment of infection. The pathogen is active at low temperatures and causes considerable losses on crops kept for long periods in storage, even if the temperatures

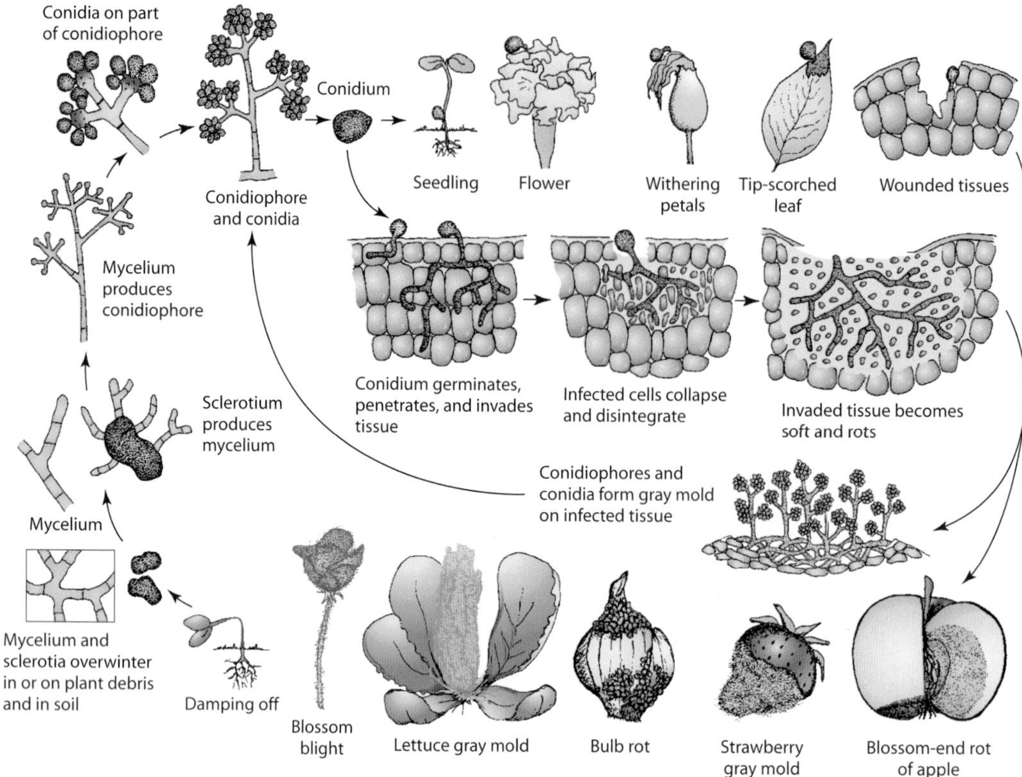

FIGURE 11-97 Disease cycle of *Botrytis* gray mold diseases.

are between 0 and 10°C. Germinating spores penetrate tissues through wounds and produce mycelium on old flower petals, dying foliage, dead bulb scales, and so on. *Botrytis* sclerotia usually germinate by producing mycelial threads that can infect directly, but in a few cases sclerotia germinate by producing apothecia and ascospores.

Control

The control of *Botrytis* diseases is aided by the removal of infected and infested debris from the field and storage rooms and by providing conditions for proper aeration and quick drying of plants and plant products. In greenhouses, humidity should be reduced by ventilation and heating. Storage organs such as onion bulbs can be protected by keeping them at 32 to 50°C for 2 to 4 days to remove excess moisture and then keeping them at 3°C in as dry an environment as possible. Biological control of *Botrytis* gray mold was obtained by spraying the flowers or fruits with spore suspensions of certain antagonistic fungi and with mixtures of several biocontrol fungi and bacteria, but this is not used in practice yet. Control of *Botrytis* in the field through chemical sprays has been only partially successful, especially in cool, damp weather. Sprays with a number of broad-spectrum or systemic fungicides give excellent control of *Botrytis* on a wide variety of crops. *Botrytis* strains resistant to several systemics and even to some broad-spectrum fungicides have been found in various crops sprayed with these chemicals. Therefore, the use of different fungicides and fungicide combinations is recommended to reduce the appearance and establishment of resistant strains.

BLACK ROT OF GRAPE

Black rot of grape is probably the most serious disease of grapes where it occurs. In favorable weather, the crop may be destroyed completely, either through direct rotting of the berries or through blasting of the blossom clusters.

Symptoms

The disease causes numerous red necrotic spots on leaves in late spring (Figs. 11-98A and 11-98B). Later, as the spots enlarge, they appear brown to grayish-tan,

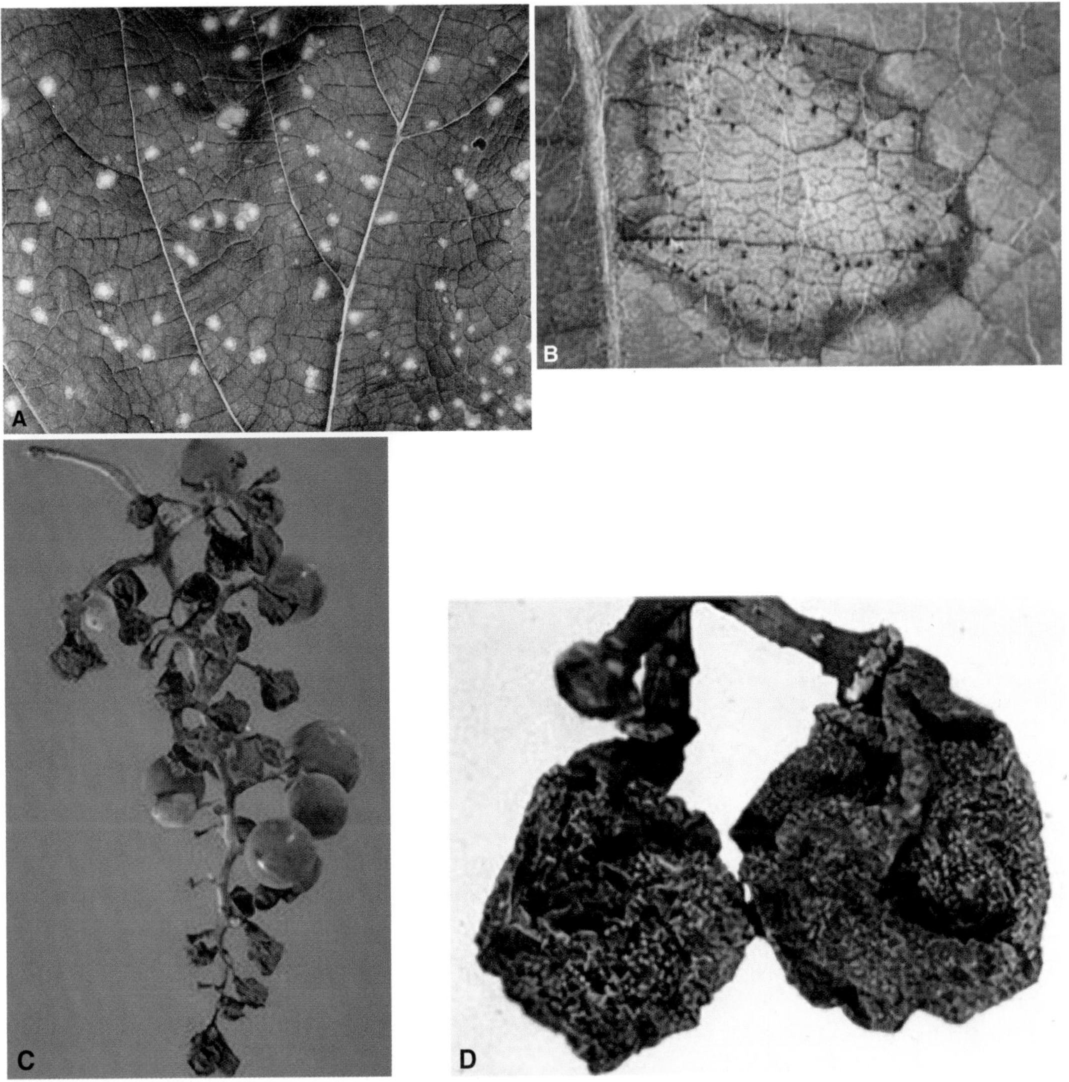

FIGURE 11-98 Symptoms of black rot of grape caused by *Guignardia bidwellii* on grape leaf (A), single spot showing pycnidia (B), rotting and drying of infected grape berries (C), and production of fungus perithecia on rotten, shriveled berries. [Photographs courtesy of (A) Plant Pathology Department, University of Florida and (B–D) M. Ellis, Ohio State University.]

with black margins. Black, dot-like, *Phyllosticta*-type pycnidia are formed on the upper side of the spots in leaves, tendrils, leaf and flower stalks, and leaf veins. Spots begin to appear on berries when the berries are about half grown (Fig. 11-98C). These spots are at first whitish but are soon surrounded by a rapidly widening brown ring. The central area of the spot remains flat or becomes depressed, and dark pycnidia appear near the center. The whole berry soon becomes rotten and shrivels, and it becomes black as the surface becomes studded with numerous black pycnidia (Fig. 11-98D).

The Pathogen

The fungus *Guignardia bidwellii*, anamorph *Phyllosticta ampelicida*, in addition to conidia-bearing pycnidia, also produces ascospore-containing perithecia in rotten, mummified fruit. The perithecia supposedly develop from transformed pycnidia.

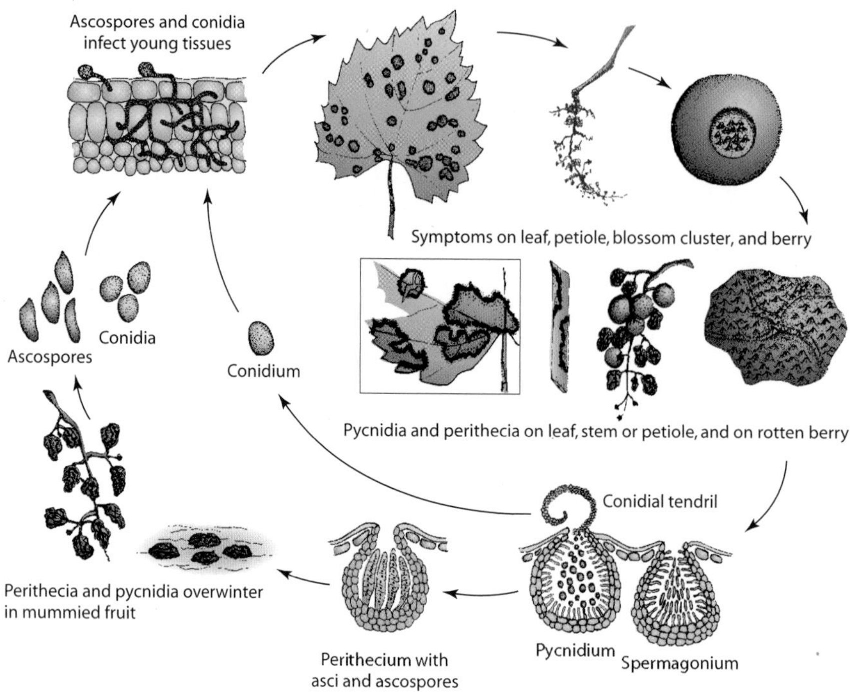

FIGURE 11-99 Disease cycle of black rot of grapes caused by *Guignardia bidwellii.*

Development of Disease

The fungus overwinters mostly as ascospores in perithecia, but because conidia can also survive the winter in most locations where grapes grow, both ascospores and conidia can cause primary infections in the spring (Figs. 11-98 and 11-99). The release of ascospores and conidia takes place only when the perithecia and pycnidia become thoroughly wet; whereas ascospores are shot out forcibly and may then be carried by air currents, conidia are exuded in a viscid mass from which they can be washed down or splashed away by rain. Ascospores may be discharged continually through the spring and summer, although most of them are discharged in the spring. Primary infections, whether from ascospores or conidia, take place on young, rapidly growing leaves and on fruit pedicels. In the ensuing spots, pycnidia are produced rapidly, and these pycnidia provide the conidia for the secondary infections of berries, stems, and so on.

Control

The control of grape leaf spot and black rot depends primarily on timely sprays of grapevines with fungicides. Sprays just before bloom, immediately after bloom, and a few to 14 days later give good control of the disease. Berries become naturally resistant to infection 3–5 weeks after bloom. Some of the fungicides are also effective against other grape diseases and are used when downy mildew or powdery mildew is also to be controlled. Where black rot has been severe, another fungicide application should be made during the first part of June, i.e., when shoot growth has reached about 25–30 centimeters in length, which in the northern states happens about two weeks before bloom.

CUCURBIT GUMMY STEM BLIGHT AND BLACK ROT

Cucurbit gummy stem blight, often accompanied by leaf spot and black fruit rot, is probably worldwide in distribution. In favorable weather the pathogen infects all parts of all cucurbits. Usually, however, it attacks the leaves and stems of watermelons, cucumber, cantaloupe, and the fruit of squash and pumpkin. When the fungus is carried in the seed it also causes damping-off, killing the seedlings.

Symptoms

On the leaves, petioles, and stems, pale brown or gray spots develop (Figs. 11-100A and 11-100B). On the stems, the spots usually start at the joints, become elongated and cracked, and exude amber-colored gummy sap (Fig. 11-100C). Spots on the leaves, stems, and petioles enlarge and make the leaves turn yellow and die. Even

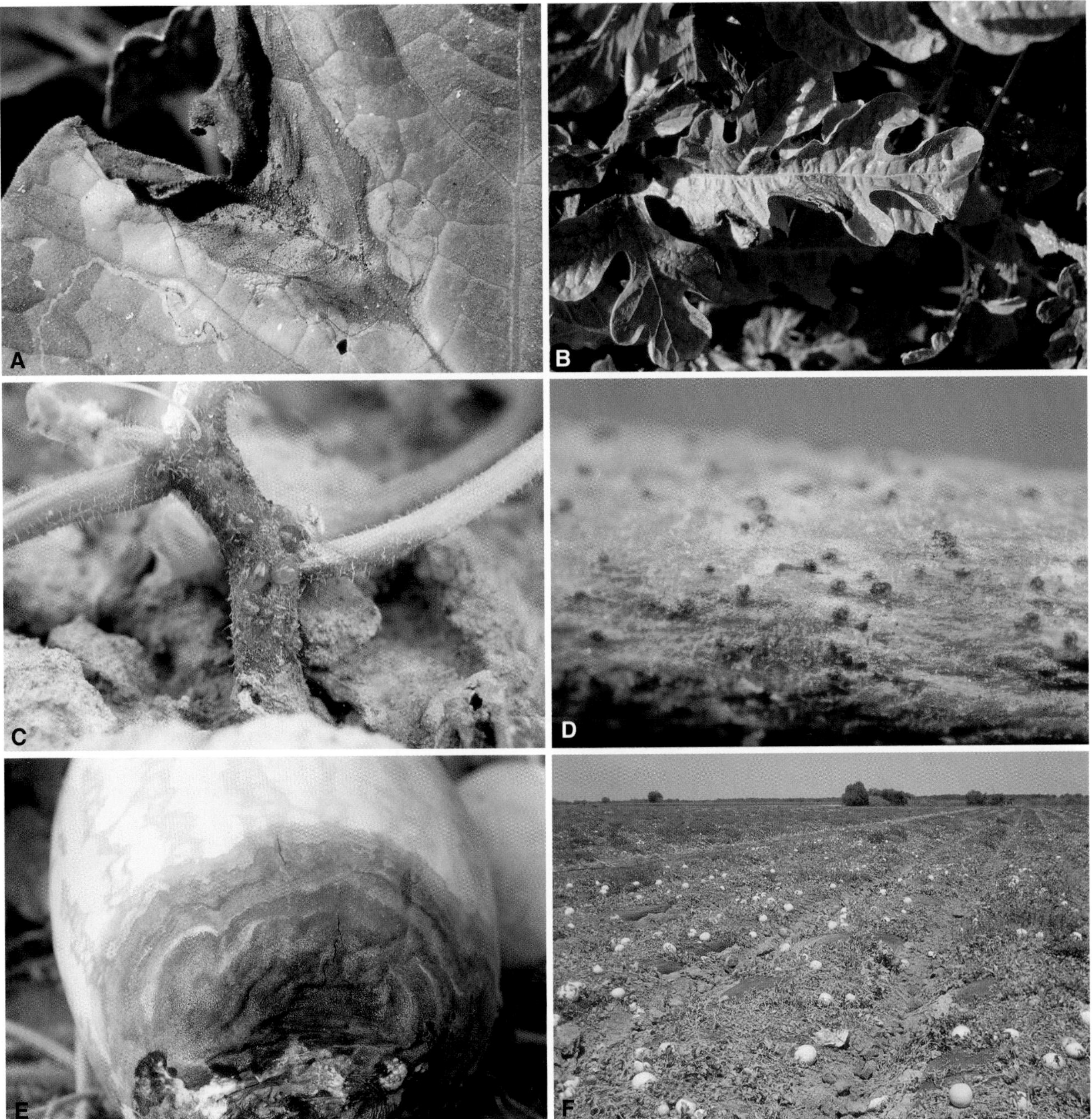

FIGURE 11-100 Symptoms of gummy stem blight of cucurbits caused by *Didymella bryoniae*. Leaf spots and blotches on cantaloupe (A) and watermelon (B). (C) Necrotic gummy stem of infected cantaloupe. (D) Pycnidia of the pathogen on infected stem. (E) Large rotten area on infected watermelon. (F) Field of honeydew melon killed by gummy stem blight. [Photographs courtesy of B.D. Bruton, USDA, and (B, D, and E) Plant Pathology Department, University of Florida.]

the whole plant may wilt and die, and plants in large areas of fields collapse and die (Fig. 11-100F). On the fruit, spots appear at first as yellowish, irregularly circular areas that later turn gray to brown and may have a droplet of gummy exudate in the center. The spot finally turns black. In some kinds of squash the lesions are superficial and spread over much of the fruit surface. Very often, however, especially in storage, the fungus penetrates the rind and spreads throughout the squash and into the seed cavity (Fig. 11-100E). On all the spots, whether on the leaf, stem, or fruit, the fungus produces closely spaced groups of pale-colored pycnidia (Fig. 11-100D) and dark, globular perithecia that are sometimes arranged in rings and are visible with the naked eye.

The Pathogen

The fungus *Didymella bryoniae*, anamorph *Phoma cucurbitacearum*, produces conidia and ascospores, both of which are short-lived after they are released.

Development of Disease

The fungus generally overwinters in diseased plant refuse as chlamydospores and in or on the seed. Thus, either spores or infected seed can result in primary infections. The subsequently profusely-produced conidia cause the secondary infections. Cantaloupe fruit had the greatest amount of decay, and the fungus produced the most polygalacturonase when cantaloupes were infected at 10 days of age than at any subsequent stage. Cucurbit plants seem to be predisposed to infection by this fungus by previous infestation with beetles or infection with powdery mildew. The striped cucumber beetle, in addition, appears to serve as a vector of *D. bryoniae* among cucurbit plants in the field.

Control

The control of black rot of cucurbits is difficult, requiring the use of clean or treated seed, long crop rotations, and frequent applications of fungicides. Good control of leaf and stem infections reduces fruit infections both in the field and in storage. However, further care is needed to avoid infection in storage. Wounding of stored fruit must be avoided. Curing of squash at 23 to 29°C for two weeks to heal the wounds and subsequent storage at 10 to 12°C is very helpful. If the inoculum present in the field is heavy, dipping the squash fruit in formaldehyde or Clorox before curing and storage is also helpful.

DIAPORTHE, PHOMOPSIS, AND PHOMA DISEASES

Several species of the fungus *Diaporthe* cause serious diseases on several crop plants. The most important *Diaporthe* diseases are stem canker of soybeans and melanose of citrus fruit.

Stem canker of soybeans is caused by *Diaporthe phaseolorum* var. *caulivora*. The disease is present in Europe and in North America, and on susceptible varieties it can cause losses of up to 50% of the crop. Symptoms consist of cankers of various sizes that result in the girdling of the stem at one or more shoots and the death of the plant beyond the cankers with a proportionate reduction in yield. The fungus produces perithecia with asci and ascospores readily on infected senescent and dead plant tissues (Fig. 11-101A), but it rarely produces pycnidia and conidia. Seeds of infected plants are often moldy (Fig. 11-101B). The fungus overwinters as perithecia in plant debris, and ascospores cause the first infections of young stems in the spring and all subsequent infections. The disease is managed by planting resistant varieties, by appropriate crop rotations, and by application of appropriate fungicides, such as benzimidazoles.

The **melanose disease of citrus** is caused by *Diaporthe citri*, anamorph *Phomopsis citri*. The fungus causes spots on the leaves, which may sometimes be severe but are generally of minor economic importance, and a superficial but objectionable blemish on the fruit (Figs. 11-101C and 11-101D). The fruit blemish, depending on its severity, may consist of scattered specks, solid dark patches, or roughened and cracked blemished areas known as mudcake melanose. Fruit infected young may remain small and fall off prematurely. In years of severe infection, particularly when twigs have been damaged by frost, the fungus infects and can kill young twigs. The fungus produces ascospores in perithecia and conidia in pycnidia on dead twigs. Conidia, spread by splashing rain, are the main inoculum, while ascospores may play a role in long-distance spread of the disease. The control of melanose is difficult and can be achieved only partially through pruning of dead twigs, if the trees are small, or through sprays of appropriate fungicides.

Phomopsis diseases appear primarily as cankers that kill twigs and small branches of ornamental shrubs and trees, as well as of some fruit trees, and as sunken lesions of various sizes on eggplant fruit. The most important Phomopsis diseases are **Phomopsis cane and leaf spot of grapes, tip blight of junipers** (Figs. 11-102A and 11-102B), **Phomopsis canker of peach, Phomopsis leaf blight of strawberries, Phomopsis stem rot of cantaloupe** (Fig. 11-102C), and **Phomopsis blight of eggplant** (Fig. 11-102D). The causes of these diseases are different species of the mitosporic fungus *Phomopsis*. They produce only conidia in rather globose pycnidia, but each pycnidium contains two types of single-celled conidia: α conidia, which are clear and oval to fusoid, and β conidia, which are clear, long, thin, and have a characteristic bend or curve. Both types of conidia must be present for the fungus to be *Phomopsis*. Phomopsis cane and leaf spot of grape is present wherever grapes are grown and weakens vines, reduces yields and quality of grapes, and kills nursery stock. In all diseases the pathogen overwinters as mycelium and/or conidia-containing pycnidia. Conidia are spread primarily by rain splashes and cause all primary and secondary infections. Management of these diseases is by sanitation and by application of appropriate fungicides.

Phoma spp. cause numerous diseases of vegetables and other annual plants. They are often present with

FIGURE 11-101 Soybean stem and seed infected with *Diaporthe phaseolorum* showing pycnidia on the stem (A) and on infected seed (B). Citrus melanose, caused by *Diaporthe citri*, and the symptoms it causes on leaf (C) and fruit (D). [Photographs courtesy of (A and B) M.C. Shurtleff, (C) R.J. McGovern, University of Florida, and (D) Plant Pathology Department, University of Florida.]

other weak pathogens. An important Phoma disease, black leg of cabbage (Fig. 11-102E), is now better known as caused by *Leptosphaeria maculans*, one of the teleomorphs of *Phoma*. Another common Phoma disease is the Phoma disease of tomato (Fig. 11-102F).

BLACK ROT OF APPLE

Black rot of apple manifests itself as three distinct symptoms: a leaf spot called frog eye leaf spot (Fig. 11-103A); a black rot of apple fruit (Figs. 11-103B and 11-103C) on the tree and in storage; and a canker of branches and limbs (Fig. 11-103D) that may destroy the tree. The leaf spot and fruit rot are more important in the southeastern part of the United States, while the canker phase of the disease is the most important in the northeastern United States.

Symptoms

Leaf spots first appear a few weeks after petal fall and have a tan to brown center and purple margins that gave them the name "frog eye." Young fruit also become infected and develop reddish flecks that turn into purple pimples and later develop into dark-brown necrotic areas of the fruit. Later infections are usually black, firm, have an irregular shape, and a red halo; larger ones may have concentric rings. Black, pimple-like pycnidia appear on

FIGURE 11-102 Symptoms of diseases caused by *Phomopsis* sp. Phomopsis tip blight on cedar (A) and in a row of juniper shrubs (B). (C) Phomopsis stem rot of cantaloupe and (D) Phomopsis rot of eggplant. (E) Phoma diseases of cabbage (black leg and root rot) caused by *Phoma lingam (Leptosphaeria maculans)*, and (F) Phoma rot of tomato caused by *Phoma destructiva*. [Photographs courtesy of (A, B, and D) Plant Pathology Department, University of Florida, (C) B.D. Bruton, USDA, (E) R.J. Howard, WCPD, and (F) R.J. McGovern, University of Florida.]

FIGURE 11-103 Apple frog eye leaf spot (A), black rot of fruit (B and C), and canker (D) caused by *Botryosphaeria obtusa*. Perithecia of the fungus can be seen on the infected area of the fruit (C). [Photographs courtesy of (A–C) D.R. Cooley, University of Massachusetts and (D) J.W. Travis, Pennsylania State University.]

the rotten areas. Infected fruit may rot throughout, mummify, and may remain hanging on the tree. Infected areas on limbs and branches become reddish brown and slightly sunken cankers. Some cankers become several feet long and the branches are weakened and break with heavy fruit loads or strong winds.

The Pathogen

All these symptoms are caused by the fungus *Botryosphaeria obtusa*, formerly *Physalospora obtusa*. It produces conidia in pycnidia (anamorph *Sphaeropsis malorum*) and ascospores in perithecia in cankers, mummified fruit, and the bark of dead wood.

Development of Disease

The fungus overwinters as mycelium, ascospores, and conidia on the tree. In the spring and throughout the growing season, conidia and ascospores are released during rains and are splashed onto leaves, fruit, and wood where they initiate infections and cause symptoms.

Control

It is achieved by removing infected branches, twigs, and mummified fruit by pruning, burning the prunings, and spraying the trees with appropriate fungicides.

Selected References

Bandyopadhyay, R., *et al.* (1998). Ergot: A new disease threat to sorghum in the Americas and Australia. *Plant Dis.* **82**, 356–367.

Batra, L. R. (1991). World species of *Monilinia* (Fungi): Their ecology, biosystematics and control. *Mycol. Mem.* **16**, 1–246.

Becker, C. M., and Burr, T. J. (1994). Discontinuous wetting and survival of conidia of *Venturia inaequalis* on apple leaves. *Phytopathology* **84**, 372–378.

Bove, F. J. (1970). "The Story of Ergot." Karger, Basel.

Byrde, R. J. W., and Willetts, H. J. (1977). "The Brown Rot Fungi of Fruit: Their Biology and Control." Pergamon, Oxford.

Coley-Smith, J. R., Verhoeff, K., and Jarvis, W. R., eds. (1980). "The Biology of *Botrytis*." Academic Press, New York.

Ellis, M. A., Madden, L. V., and Wilson, L. L. (1984). Evaluation of an electronic apple scab predictor for scheduling fungicides with curative activity. *Plant Dis.* **68**, 1055–1058.

Forsberg, J. L. (1975). "Diseases of Ornamental Plants," Special Publ. No. 3 Rev. University of Illinois, College of Agriculture, Urbana-Champaign.

Fulton, R. H. (1989). The cacao disease trilogy: Black pod, Monilia pod rot, and witches' broom. *Plant Dis.* **73**, 601–603.

Gould, C. J. (1954). *Botrytis* diseases of gladiolus. *Plant Dis. Rep. Suppl.* **224**, 1–33.

James, J. R., and Sutton, T. B. (1982). Environmental factors influencing pseudothecial development and ascospore maturation of *Venturia inaequalis*. *Phytopathology* **72**, 1073–1080.

Jones, A. L., and Aldwinckle, H. S., eds. (1990). "Compendium of Apple and Pear Diseases." APS Press, St. Paul, MN.

Jones, A. L., Fisher, P. D., Seem, R. C., Kroon, J. C., and Van De-Motter, P. J. (1984). Development and commercialization of an in-field microcomputer delivery system for weather-driven predictive models. *Plant Dis.* **68**, 458–463.

Keinath, A. P. (2001). Effect of fungicide applications scheduled to control gummy stem blight on yield and quality of watermelon fruit. *Plant Dis.* **85**, 53–58.

Keitt, G. W., and Jones, L. K. (1926). Studies of the epidemiology and control of apple scab. *Res. Bull.-Wis. Agric. Exp. Stn.* **73**, 1–104.

McClellan, W. D., and Hewitt, W. B. (1973). Early *Botrytis* rot of grapes: Time of infection and latency of *Botrytis cinerea* in *Vitis vinifera*. *Phytopathology* **63**, 1151–1156.

MacHardy, W. E., and Gadoury, D. M. (1985). Forecasting the seasonal maturation of ascospores of *Venturia inaequalis*. *Phytopathology* **75**, 185–190.

Merwin, I. A., *et al.* (1994). Scab-resistant apples for the northeastern United States: New prospects and old problems. *Plant Dis.* **78**, 4–10.

Ogawa, J. M., and English, H. (1991). "Diseases of Temperate Zone Tree Fruit and Nut Crops." Univ. Calif. Div. Agric. Nat. Resources Publ. 3345.

Pearson, R. C., and Goheen, A. C., eds. (1988). "Compendium of Grape Diseases." APS Press, St. Paul, MN.

Pepin, H. S., and MacPherson, E. A. (1982). Strains of *Botrytis cinerea* resistant to benomyl and captan in the field. *Plant Dis.* **66**, 404–405.

Pusey, L. P., and Wilson, C. L. (1984). Postharvest biological control of stone fruit brown rot by *Bacillus subtilis*. *Plant Dis.* **68**, 753–757.

Rose, D. H., Fisher, D. F., and Brooks, C. (1937). Market diseases of fruits and vegetables: Peaches, plums, cherries and other stone fruits. *Misc. Publ.- U.S. Dep. Agric.* **228**, 1–26.

Schultz, T. R., Johnston, W. J., Golob, C. T., and Maguire, J. D. (1993). Control of ergot in Kentucky bluegrass seed production using fungicides. *Plant Dis.* **77**, 685–687.

Segall, R. H., and Newhall, A. G. (1960). Onion blast or leaf spotting caused by species of *Botrytis*. *Phytopathology* **50**, 76–82.

Vincelli, P. C., and Lorbeer, J. W. (1989). BLIGHT-ALERT: A weather-based predictive system for timing fungicide applications on onion before infection periods of *Botrytis squamosa*. *Phytopathology* **79**, 493–498.

VASCULAR WILTS CAUSED BY ASCOMYCETES AND DEUTEROMYCETES (MITOSPORIC FUNGI)

Vascular wilts are widespread, very destructive, spectacular, and frightening plant diseases. They appear as more or less rapid wilting, browning, and dying of leaves and succulent shoots of plants followed by death of the whole plant. Wilts occur as a result of the presence and activities of the pathogen in the xylem vessels of the plant. Entire plants may die within a matter of weeks, although in perennials, death may not occur until several months or years after infection. As long as the infected plant is alive, wilt-causing fungi remain in the vascular (xylem) tissues and a few surrounding cells. Only when the infected plant is killed by the disease do these fungi move into other tissues and sporulate at or near the surface of the dead plant.

There are four genera of fungi that cause vascular wilts: *Ceratocystis, Ophiostoma, Fusarium*, and *Verticillium*. Each of them causes disease on several important crop, forest, and ornamental plants. *Ceratocystis* causes the vascular wilt of oak trees (*C. fagacearum*), of cacao, and of eucalyptus. *Ophiostoma* causes the vascular wilt of elm trees, known as Dutch elm disease (*O. novo-ulmi*).

Fusarium causes vascular wilts of vegetables and flowers, herbaceous perennial ornamentals, plantation crops, and the mimosa tree (silk tree). Most of the wilt-causing *Fusarium* fungi belong to the species *Fusarium oxysporum*. Different host plants are attacked by special forms or races of the fungus. The fungus that attacks tomato is designated *F. oxysporum* f. sp. *lycopersici*; cucurbits, *F. oxysporum* f. sp. *conglutinans*; banana, *F. oxysporum* f.sp. *cubense*; cotton, *F. oxysporum* f. sp. *vasinfectum*; carnation, *F. oxysporum* f. sp. *dianthii*; and so on.

Verticillium causes vascular wilts of vegetables, flowers, field crops, perennial ornamentals, and fruit and forest trees. Two species, *Verticillium albo-atrum* and *V. dahliae*, attack hundreds of kinds of plants, causing wilts and losses of varying severity.

All vascular wilts have certain characteristics in common. The leaves of infected plants or of parts of infected plants lose turgidity, become flaccid and lighter green to greenish yellow, droop, and finally wilt, turn yellow, then brown, and die. Wilted leaves may be flat or curled. Young, tender shoots also wilt and die. In cross sections of infected stems and twigs, discolored brown areas appear as a complete or interrupted ring consisting of discolored vascular tissues. In the xylem vessels of infected stems and roots, mycelium and spores of the causal fungus may be present. Some of the vessels may be clogged with mycelium, spores, or polysaccharides produced by the fungus. Clogging is increased further by gels and gums formed by the accumulation and oxidation of breakdown products of plant cells attacked by fungal enzymes. The oxidation and translocation of some such breakdown products seem to also be responsible for the brown discoloration of affected vascular tissues. In newly infected young stems, the number of xylem vessels formed is reduced and their cell walls are thinner than normal. Often the parenchyma cells surrounding xylem vessels are stimulated by secretions of the pathogen to divide excessively, and this, combined with the thinner and weaker vessel walls, results in a reduction of the diameter or complete collapse of the vessels. In some hosts, tyloses are produced by parenchyma cells adjoining some xylem vessels. The balloon-like tyloses protrude into the vessels and contribute to their clogging. Toxins secreted in the vessels by wilt-causing fungi are carried to the leaves, in which they cause reduced chlorophyll synthesis along the veins (vein clearing) and reduced photosynthesis, disrupt the permeability of the leaf cell membranes and their ability to control water loss through transpiration, and thereby result in leaf epinasty, wilting, interveinal necrosis, browning, and death.

FUSARIUM WILTS

As mentioned earlier, *Fusarium* wilts affect and cause severe losses on most vegetables (Fig. 11-104) and flowers; several field crops, such as cotton and tobacco; plantation crops, such as banana (Figs. 11-104E–11-104G), plantain, coffee, and sugarcane; and a few shade trees. Fusarial wilts are most severe under warm soil conditions and in greenhouses. Most fusarial wilts have disease cycles and develop similar to those of the *Fusarium* wilt of tomato.

FUSARIUM WILT OF TOMATO

Fusarium wilt is one of the most prevalent and damaging diseases of tomato wherever tomatoes are grown intensively. The disease is most destructive in warm climates and warm, sandy soils of temperate regions. In the United States the disease is most severe in the southern regions and in the central states; in the northern states, it can become important only on greenhouse tomatoes.

The disease causes great losses, especially on susceptible varieties and when soil and air temperatures are rather high during much of the season. Infected plants become stunted and soon wilt and finally die. Occasionally, entire fields of tomatoes are killed or damaged severely before a crop can be harvested.

Symptoms. The first symptoms appear as slight vein clearing on the outer, younger leaflets. Subsequently, the older leaves show epinasty caused by drooping of the petioles. Plants infected at the seedling stage usually wilt and die soon after appearance of the first symptoms. Older plants in the field may wilt and die suddenly if the infection is severe and if the weather is favorable for the pathogen (Figs. 11-104A and 11-104D). More commonly, however, in older plants, vein clearing and leaf epinasty are followed by stunting of the plants, yellowing of the lower leaves, occasional formation of adventitious roots, wilting of leaves and young stems, defoliation, marginal necrosis of the remaining leaves, and finally death of the plant. Often these symptoms appear on only one side of the stem and progress upward until the foliage is killed and the stem dies. Fruit may occasionally become infected and then it rots and drops off without becoming spotted. Roots also become infected; after an initial period of stunting, the smaller side roots rot.

In cross sections near the base of the infected plant stem, a brown ring is evident in the area of the vascular bundles. The upward extent of the discoloration depends on the severity of the disease (Figs. 11-104B and 11-104C).

The Pathogen. *Fusarium oxysporum* f. sp. *lycopersici*. The mycelium is colorless at first, but with age it becomes cream-colored, pale yellow, pale pink, or somewhat purplish. The fungus produces three kinds of asexual spores (Fig. 11-105). Microconidia, which have one or two cells, are the most frequently and abundantly produced spores under all conditions, even inside the vessels of infected host plants. Macroconidia are the typical "*Fusarium*" spores; they are three to five celled, have gradually pointed and curved ends, and appear in sporodochia-like groups on the surface of plants killed by the pathogen. Chlamydospores are one- or two-celled, thick-walled, round spores produced within or terminally on older mycelium or in macroconidia. All three types of spores are produced in cultures of the fungus and probably in the soil, although only chlamydospores can survive in the soil for long.

Development of Disease. The pathogen is a soil inhabitant. Between crops it survives in infected plant

FIGURE 11-104 Fusarium wilt of tomato caused by *Fusarium oxysporum* f. sp. *Lycopersici*: wilted tomato plants in the field (A) and severe brown discoloration of vascular tissues along stem of infected plant (B). Clogged and discolored vascular tissues in cross section of watermelon stem infected with *F. oxysporum* f. sp. *niveum* (C) and infected watermelon plants wilted and dead in the field (D). (E–G) Fusarium wilt (Panama disease) of banana caused by *F. oxysporum* f. sp. *cubense*. (E) Lower leaves of infected banana plants wilt, turn brown, and die. (F) Entire infected banana plant killed by the infection. (G) Discolored vascular tissues of infected banana rhizomes. [Photographs courtesy of (A and B) R.J. McGovern, (C and D) B.D. Bruton, (E) by B. Niere, provided by D. Coyne, IITA, Ibadan, Nigeria, and (F and G) A. Silagyi, University of Florida.]

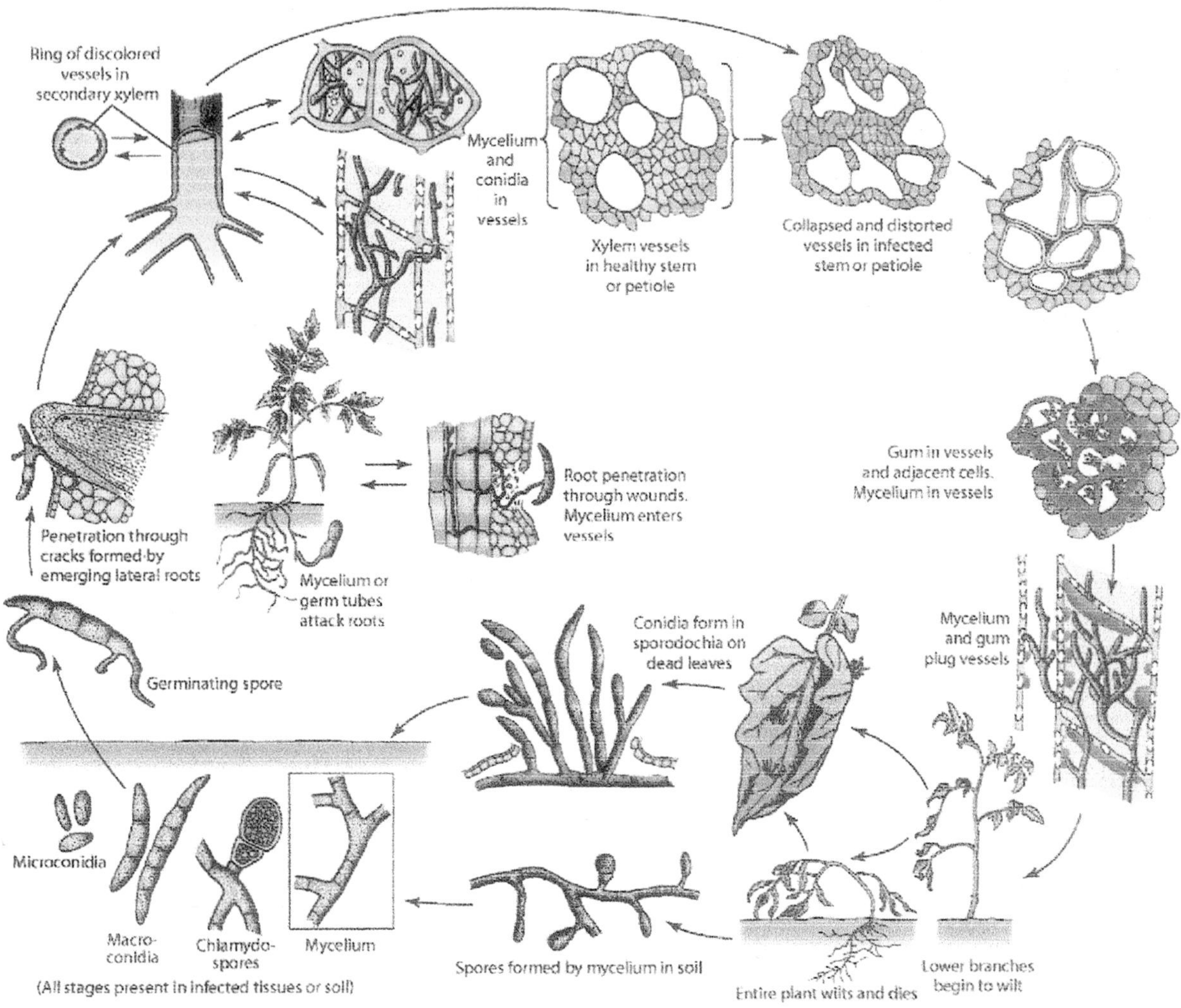

FIGURE 11-105 Disease cycle of *Fusarium* wilt of tomato caused by *Fusarium oxysporum* f. sp. *lycopersici*.

debris in the soil as mycelium and in all its spore forms but, most commonly, especially in the cooler temperate regions, as chlamydospores (Fig. 11-105). It spreads over short distances by means of water and contaminated farm equipment and over long distances primarily in infected transplants or in the soil carried with them. Usually, once an area becomes infested with *Fusarium*, it remains so indefinitely.

When healthy plants grow in contaminated soil, the germ tube of spores or the mycelium penetrates root tips directly or enters the roots through wounds or at the point of formation of lateral roots. The mycelium advances through the root cortex intercellularly, and when it reaches the xylem vessels it enters them through the pits. The mycelium then remains exclusively in the vessels and travels through them, mostly upward, toward the stem and crown of the plant. While in the vessels, the mycelium branches and produces microconidia, which are detached and carried upward in the sap stream. Microconidia germinate at the point where their upward movement is stopped, the mycelium penetrates the upper wall of the vessel, and more microconidia are produced in the next vessel. The mycelium also advances laterally into the adjacent vessels, penetrating them through the pits.

A combination of the processes discussed earlier, namely vessel clogging (Figs. 11-104B, 11-104C, and 11-104G) by mycelium, spores, gels, gums, and tyloses and crushing of the vessels by proliferating adjacent parenchyma cells, is responsible for the breakdown of the water economy of the infected plant. When the leaves transpire more water than the roots and stem can transport to them, the stomata close and the leaves wilt and finally die, followed by death of the rest of the plant. The fungus then invades all tissues of the plant extensively, reaches the surface of the dead plant, and there sporulates profusely. The spores may be disseminated to new plants or areas by wind, water, and so on.

Sometimes, when the soil moisture is high and the temperature relatively low, infected plants may produce good yields; however, in such cases the fungus may reach the fruit of the plants and penetrate or contaminate the seed. Usually, infected fruits decay and drop. If harvested, infected seeds are so light that they are eliminated in the procedures of extraction and cleaning of the seed and therefore play little role in the spread of the fungus.

Control. Use of tomato varieties resistant to the fungus is the only practical measure for controlling the disease in the field. Several such varieties are available today. The fungus is so widespread and so persistent in soils that seedbed sterilization and crop rotation, although always sound practices, are of limited value. Soil sterilization is too expensive for field application, but it should be always practiced for greenhouse-grown tomato plants. Use of healthy seed and transplants is of course mandatory, and hot-water treatment of seed suspected of being infected should precede planting.

In the last several years, biological control of *Fusarium* wilt has given encouraging results. Control may involve prior inoculation of plants with nonpathogenic strains of *F. oxysporum* or the use of antagonistic fungi, such as *Trichoderma* and *Gliocladium*, *Pseudomonas fluorescens* and *Burkholderia cepacia* bacteria, and others. However, none of the biocontrols are used in practice yet. Solar heating (solarization) of field soil by covering with transparent plastic film during the summer also reduces disease incidence. More recently, it was shown that spraying tomato plants with a suspension of zoospores of the oomycete *Phytophthora cryptogea* induces the development of systemic acquired resistance in the tomato plants, which remained free of wilt following inoculation with *F. oxysporum* f. sp. *lycopersici*. Although promising, none of these methods have been used for control of *Fusarium* wilt in practice so far.

FUSARIUM (PANAMA) WILT OF BANANA

Panama wilt of banana, caused by *Fusarium oxysporum* f.sp. *cubense*, was discovered in Australia in the late 1880s but it reached epidemic proportions in the 1950s in Panama, where it destroyed 40,000 hectares of Gros Michel bananas and was then recognized as a devastating disease and a major threat to the banana industry in Central America. Panama disease now occurs in most areas where bananas are grown.

The symptoms of Panama disease consist of yellowing of the oldest leaves or lengthwise splitting of the lower leaf sheath. Leaves may wilt and buckle at their petiole base and, later, younger leaves collapse and die (Figs. 11-104E and 11-104F). Internally, brown streaks develop on and within older leaf sheaths and these are followed by large portions of the xylem turning brick red to brown (Fig. 11-104G). In the meantime, the fungus, which enters the banana plant from the soil through the feeder roots, advances into the xylem vessels of the rhizome and from there into the pseudostem, which it colonizes, resulting in discoloration and blockage (Fig. 11-104G).

The pathogen is *F. oxysporum* f. sp. *cubense*. It produces micro- and macroconidia and chlamydospores. Several races of the pathogen are known that differ in the banana varieties they attack.

The fungus overseasons in infected plants as mycelium and in the soil mostly as chlamydospores. The latter survive in the soil for at least 20 years. The pathogen is spread primarily in infected rhizomes (suckers), which are used traditionally for the vegetative propagation of banana. Less frequently, the pathogen is spread as spores in soil, running water, and on farm equipment and machinery.

There are no easy or good controls of Panama disease. The most effective control is achieved by planting banana varieties resistant to the existing races of the pathogen. Planting pathogen-free rhizomes in pathogen-free soil is also effective. The use of tissue culture-produced propagative material free of the pathogen is helpful. Also, certain cultural practices and measures toward biological control of the disease are helpful but far from adequate.

VERTICILLIUM WILTS

Verticillium wilts occur worldwide but are most important in temperate regions. *Verticillium* attacks more than 200 species of plants, including most vegetables, flowers, fruit trees (Fig. 11-106A), strawberries (Fig. 11-106C), field crops, and shade and forest trees. *Verticillium* is also the main cause of the potato early dying disease.

The symptoms of Verticillium wilts are almost identical to those of Fusarium wilts. In many hosts and most areas, however, *Verticillium* induces wilt at lower temperatures than *Fusarium*. Moreover, the symptoms develop more slowly and often appear only on the lower or outer part of the plant or on only a few of its branches. In some hosts, Verticillium wilt develops primarily in seedlings, which usually die shortly after infection. More common are late infections, which cause upper leaves to droop and other leaves to develop irregular chlorotic patches that become necrotic. Older

FIGURE 11-106 Verticillium wilt of peach (A and B) and strawberries (C) caused by the fungus *Verticillium* sp. (D). (A and C) Wilting and death of part of the plants. (B) Discolored vascular tissue of infected peach stem. (D) The fungus *Verticillium* showing the verticillate arrangement of its conidiophores and conidia. [Photographs courtesy of (A and B) A.L. Jones, Michigan State University and (C and D) D. Legard, University of Florida.]

plants infected with *Verticillium* are usually stunted and their vascular tissues show characteristic brownish discoloration (Fig. 11-106B). *Verticillium* infection may result in defoliation, gradual wilting and death of successive branches, or abrupt collapse and death of the entire plant (Fig. 11-106). It appears that the presence of ethylene in tomato plants while they are being infected with *Verticillium* inhibits disease development, whereas the presence of ethylene at later stages of the disease enhances wilt development. Tomato plants engineered with the gene of the enzyme that cleaves the immediate precursor of ethylene developed significantly less Verticillium wilt, although the fungus was present in the plants.

When Verticillium wilt first appears in a field, it is mild and local. In subsequent years, as the inoculum builds up and as new, more virulent strains of the fungus appear, the attacks become successively more severe and widespread until the crop has to be discontinued or is replaced with resistant varieties.

Two species of *Verticillium, V. albo-atrum* and *V. dahliae*, cause Verticillium wilts in most plants. Both produce conidia that are short lived. *Verticillium dahliae* also produces microslerotia, whereas *V. albo-atrum* produces dark, thick-walled mycelium but not microsclerotia. *Verticillium albo-atrum* grows best at 20 to 25°C, whereas *V. dahliae* prefers slightly higher temperatures (25–28°C) and is somewhat more common in warmer regions. Some *Verticillium* strains show host specialization, but most of them attack a wide range of host plants. *Verticillium dahliae* overwinters in the soil as microsclerotia, which can survive up to 15 years.

Both species, however, can overwinter as mycelium within perennial hosts, in propagative organs, or in plant debris.

Verticillium penetrates young roots of host plants directly or through wounds. The fungus is spread by contaminated seed, by vegetative cuttings and tubers, b y wind, by surface water, and by soil, which may contain up to 100 or more microslerotia per gram; 6 to 50 microslerotia per gram is sufficient to give 100% infection in most susceptible crops (see Fig. 8-7). Many fields have become contaminated with *Verticillium* for the first time by planting infected potato tubers or other crops. Solanaceous crops such as potato, eggplant, and tomato increase the fungus inoculum level in the soil. However, *Verticillium* is often found in uncultivated areas, indicating that the fungus is native to the soils and can attack susceptible crops as soon as they are planted.

The control of Verticillium wilts depends on planting disease-free plants in disease-free soil, using resistant varieties, and avoiding the planting of susceptible crops where solanaceous crops have been grown repeatedly. Soil fumigation can be profitable when used to protect high-value crops, but it is too expensive on large areas.

Thermal inactivation via soil solarization is proving useful for the control of Verticillium in regions with high summer temperatures and low rainfall. Also, the use of black mulch with ammonium nitrogen fertilization seems to reduce damage from *Verticillium* on some plants, such as eggplant.

OPHIOSTOMA WILT OF ELM TREES: DUTCH ELM DISEASE

Dutch elm disease owes its name to the fact that the first widely publicized report of its occurrence on elm came in 1921 from Holland, although it had been reported in France in 1917. Since then the disease has spread throughout Europe, parts of Asia, and most of the temperate zones in North America. In the United States the disease was first found in Ohio and some states on the east coast in the early 1930s; by 1973, it had spread westward to the Pacific coast states.

Dutch elm disease is very destructive. It affects all elm species but most severely the American elm. The disease may kill branches and entire trees within a few weeks or a few years from the time of infection. Hundreds of thousands of elm trees in towns across the country die from Dutch elm disease every year. The cost of cutting down diseased and dead elm trees amounts to many millions of dollars per year. Of course, no one can estimate the value of the natural beauty destroyed by the disease in countless communities.

Symptoms

The first symptoms of Dutch elm disease appear as sudden or prolonged wilting of the leaves of individual branches or of the entire tree (Figs. 11-107A and 11-107C). Wilted leaves frequently curl, turn yellow, then brown, and finally fall off the tree earlier than normal (Fig. 11-107C). Most affected branches die immediately after defoliation. The disease usually appears first on one or several branches and then spreads to other portions of the tree. Thus, many dead branches may appear on a tree or a portion of a tree. Such trees may die gradually, branch by branch, over a period of several years or they may recover. Sometimes, however, entire trees suddenly develop disease symptoms and may die within a few weeks (Figs. 11-107A and 11-107B). Usually trees that become infected in the spring or early summer die quickly, whereas those infected in late summer are affected much less seriously and may even recover, unless they become reinfected.

When the bark of infected twigs or branches is peeled back, brown streaking or mottling appears on the outer layer of wood. In cross sections of the branch, the browning appears as a broken or continuous circle in the outer rings of the wood (Fig. 11-107D). At higher magnification, tyloses can be seen inside vessels (Fig. 11-107E) of newly infected shoots that block the upward movement of nutrients and water.

The Pathogen: Ophiostoma ulmi *and* Ophiostoma novo-ulmi

Aggressive strains of the fungus, causing recent Dutch elm disease pandemics, have been placed in a new species, *O. novo-ulmi*. The latter is separated into *Ophiostoma novo-ulmi* subspecies *novo-ulmi* and *Ophiostoma novo-ulmi* ssp. *americana*. These strains hybridize in nature and have been rapidly supplanting the previously predominant *O. ulmi* strains. The mycelium is creamy white. While in the vessels, the mycelium produces short branches on which clusters of *Sporothrix*-type conidia are formed (Fig. 11-109). In dying or dead trees, the mycelium produces mostly *Graphium*-type spores on coremia developing on bark, which is somewhat loose from the wood and in tunnels made in the bark by insects. Coremia consist of hyphae grouped into an erect, dark, solid stalk (synnema) and a colorless, flaring head to which the spores adhere, forming a sticky, glistening, whitish to yellowish droplet.

The fungus requires the contact of two sexually compatible strains for sexual reproduction. Because, frequently, only one of the mating types is found in large areas in nature, sexual reproduction is extremely rare. In the United States, for example, the fungus rarely

FIGURE 11-107 Symptoms of Dutch elm disease caused by the fungus *Ophiostoma novo-ulmi*. (A) Infected elm tree showing wilted, yellowish brown, and brown foliage as well as partially defoliated and dead twigs and branches. (B) Row of elm trees killed by the disease. (C) Early symptoms (twig wilting) of infected trees. (D) Infected elm twig showing discoloration of vascular tissues. (E) Tyloses in xylem vessels of infected trees contribute to the development of wilt. [Photographs courtesy of (A, B, and D) E.L. Barnard, Florida Department of Agriculture, Forestry Division, (C) U.S. Forest Service, and (E) D.M. Elgersma.]

reproduces sexually, but it does so rather frequently in Europe. When the two mating types do come in contact, perithecia develop. The perithecia are spherical and black, about 120 micrometers in diameter, and have a long neck (about 300–400 μm) (Fig. 11-109). Perithecia form singly or in groups and in the same areas in the bark as the coremia.

Inside the perithecium many asci develop, but as the asci mature, they disintegrate, leaving the ascospores free in the perithecial cavity. The ascospores are discharged through the neck canal and accumulate in a sticky droplet.

Development of Disease

Dutch elm disease is the result of an unusual partnership between a fungus and an insect (Fig. 11-108). Although the fungus alone is responsible for the disease, the insect is the indispensable vector of the fungus, carrying the fungus spores from infected elm wood to healthy elm trees. The insects responsible for the spread of the disease in North America are the European elm bark beetle (*Scolytus multistriatus*) and the native elm bark beetle (*Hylurgopinus rufipes*) (Fig. 11-108A). In addition to being spread by beetles, however, the fungus

FIGURE 11-108 (A) Native and European elm bark beetles that vector the Dutch elm disease fungus from diseased to healthy elm trees. (B) Galleries made on diseased or dead elm trees by egg-laying female adults and by the larvae and in which the fungus produces its spores. (C) Adult insects emerging from the galleries carry spores to healthy elm trees on which they feed and inoculate with the fungus. [Photographs courtesy of (A and B) U.S. Forest Service and (C) P. Svihra.]

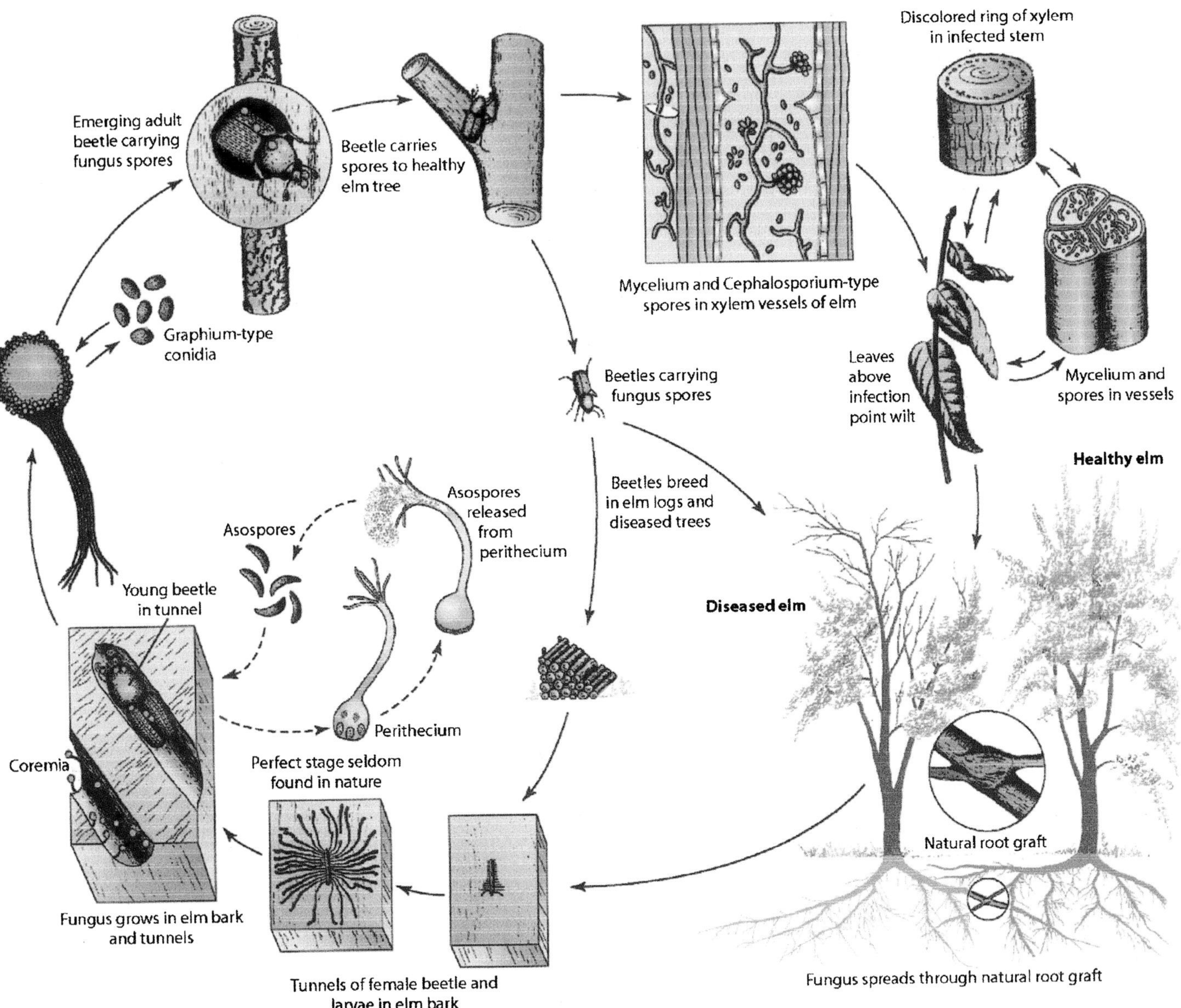

FIGURE 11-109 Disease cycle of Dutch elm disease caused by *Opiostoma novo-ulmi*.

is also spread by natural root grafts forming between adjacent trees.

The fungus overwinters in the bark of dying or dead elm trees and logs as mycelium and as spore-bearing coremia. Elm bark beetles prefer to lay their eggs in the intersurface between bark and wood of trees weakened or dying by drought or disease. The adult female beetle tunnels through the bark and opens a gallery either parallel with the grain of the wood (*Scolytus*) or at an angle or perpendicular (*Hylurgopinus*). The female lays eggs along the sides of the gallery, the eggs soon hatch, and the larvae open tunnels at right angles to the maternal gallery (Fig. 11-108B). If the tree was already infected with the fungus, the fungus produces mycelium and sticky, *Graphium*-type spores in the beetle tunnels. When the adult beetles emerge, they carry thousands of fungus spores on and in their bodies. *Scolytus* beetles feed on twigs, whereas *Hylurgopinus* beetles feed on stems 5 to 30 centimeters in diameter. As the beetles burrow into the bark and wood (Fig. 11-108B), spores are deposited in the wounded tissues of the tree, germinate, and grow rapidly into the injured bark and the wood. When the fungus reaches the large xylem vessels of the spring wood, it produces *Sporothrix*-type spores, which are carried up by the sap stream. These spores reproduce by yeast-like budding, germinate, and start new infections. The extent of symptoms in the crown is correlated with the extent of vascular invasion. In early stages of infection, the mycelium invades primarily the vessels and only occasionally tracheids, fibers, and the surrounding parenchyma cells. General invasion of tissue begins at the terminal or extensive dieback phase of the disease. Gums and tyloses are produced in the larger vessels, and sometimes isolated areas of the sapwood are blocked by a combination of gums, tyloses (Fig. 11-107E), and fungal growth. Infection also induces browning of the water-conducting vessels. Infected twigs and branches soon wilt and die.

Infections that take place in the spring or early summer allow spores to invade the long vessels of the elm springwood, through which they can be carried rapidly to all parts of the tree. If extensive vascular invasion occurs, the tree may die within a few weeks. During later infections, vascular invasion is limited to the outer, shorter vessels of the summerwood in which they move only for short distances. As a result, late infections may produce only localized infections and seldom cause serious immediate damage to the tree.

The elm bark beetles feed on living trees for only a few days; they then fly back to dying or weakened elm wood in which they construct new galleries and lay eggs. There are usually two to three generations of beetles per season. In each generation the young adult goes from dead or weakened elm trees to living, vigorous ones on which it feeds and then returns to the dead or weakened trees to lay its eggs. Therefore, once an insect becomes contaminated with fungus spores, it may carry them to either healthy or diseased wood, in both of which the fungus grows and multiplies and may contaminate all the offspring of the insect as well as any other insects that visit the infected wood.

Control

There are few resistant clones within the susceptible American elm species. Certain Asiatic species, such as the Siberian and the Chinese elm, are resistant to Dutch elm disease, but produce poor shade trees. Hybrids between various species have shown resistance in varying degrees and some look promising, but so far none of them has been planted widely or proved completely resistant.

The control of Dutch elm disease depends primarily on sanitation measures and somewhat on chemical control of the insect vectors of the fungus. Sanitation involves the removal and destruction of weakened or dead elm trees and elm logs, thus destroying the larvae contained in them or denying the insect and the fungus their overwintering habitat. Pruning out infected twigs and branches sometimes eliminates the disease. Naturally grafted roots of elm trees can be cut or killed with chemicals to prevent spread of the fungus to adjacent trees. Control of the insect vector by chemicals involves spraying the healthy elm trees while dormant and in the spring with insecticides, but spraying has been only partially effective. In some areas, trap logs and pheromone traps were tested as a means of reducing the number of insect vectors of Dutch elm disease but had little success.

Promising results for Dutch elm disease control in individual trees have been obtained with trunk or root injections of healthy or diseased elm trees with certain systemic fungicides. These fungicides have, in some cases, arrested the advance of the disease in infected trees and reduced the appearance of new infections on treated healthy trees, but they are not particularly dependable. Some protection from Dutch elm disease has also been reported when the trees were inoculated with certain *Pseudomonas* bacteria or with nonaggressive strains of the fungi *Ophiostoma* or *Verticillium*.

CERATOCYSTIS WILTS

There are three important vascular wilt diseases of trees caused by separate species of *Ceratocystis*, each affecting a different host and occurring in different parts of the world. Only a brief mention of each will be made here.

OAK WILT

At first it causes individual branches and, eventually, whole trees to wilt, become defoliated, and finally die (Figs. 11-110A–11-110D). The disease occurs in the northeastern United States but extends as far south as Texas. It affects all oaks, but red oaks are particularly susceptible. Oak wilt is caused by the fungus *Ceratocystis fagacearum*, but, as with Dutch elm disease, oak wilt is dependent for its spread on certain insects, the sap (Nitidulid) beetles. The beetles are attracted to fungal spore mats breaking out through the bark of infected trees and to tree sap coming out of wounds of any type, thereby transmitting spores from diseased to healthy trees. The fungus also spreads from one tree to adjacent ones through natural root grafts between the trees. Control of the disease is difficult and under forest conditions nearly impossible.

CERATOCYSTIS WILT OF CACAO OR MAL DE MACHETE

Infected trees show limp brown foliage first on single branches and then on the whole tree. The disease also kills the cambium and bark tissue, thereby creating a canker on the trunk or branch (Figs. 11-110E and

FIGURE 11-110 (A) Early foliage symptoms of oak wilt caused by the fungus *Ceratocystis fagacearum*. (B) Fully grown oak tree killed by oak wilt. (C) Fungal mats growing through cracks in the bark of the trunk. (D) Map showing where oak wilt occurs in North America. Symptoms of Ceratocystis wilt of cacao caused by *C. fimbriata*: wilted cacao tree (E) and discolored areas within the wood of the tree (F). [Photographs courtesy of (A–D) U.S. Forest Service and (E and F) T.C. Harrington, Iowa State University.]

11-110F). The disease, so far, has been found only in Latin America. Ceratocystis wilt of cacao is caused by the fungus *Ceratocystis fimbriata*. The fungus is likely spread from tree to tree by ambrosia beetles and by natural root grafts. No truly effective control is effective against the disease yet.

CERATOCYSTIS WILT OF EUCALYPTUS

Infected eucalyptus trees wilt and die back. Infected trees also produce typical lesions in the bark and xylem. The disease has been found recently in Congo (central Africa). Eucalyptus wilt is also caused by the fungus *C. fimbriata*.

Selected References

Anonymous (1991). Recent advances in *Fusarium* systematics: A symposium. *Phytopathology* **81**, 1043–1067.

Appel, D. N. (1995). The oak wilt enigma: Perspectives from the Texas epidemic. *Annu. Rev. Phytopathol.* **33**, 103–118.

Banfield, W. M. (1941). Distribution by the sap stream of spores of three fungi that induce vascular wilt disease of elm. *J. Agric. Res. (Washington, DC)* **62**, 637–681.

Banfield, W. M. (1968). Dutch elm disease recurrence and recovery in American elm. *Phytopathol. Z.* **62**, 21–60.

Beckman, C. H. (1987). "The Nature of Wilt Diseases of Plants." APS Press, St. Paul, MN.

Booth, C. (1971). "The Genus *Fusarium*." Commonw. Mycol. Inst., Kew, England.

Brasier, C. M. (1991). *Ophiostoma novo-ulmi* sp. nov., causative agent of current Dutch elm disease pandemics. *Mycopathologia* **115**, 151–161.

Brown, M. F., and Wyllie, T. D. (1970). Ultrastructure of microsclerotia of *Verticillium albo-atrum*. *Phytopathology* **60**, 538–542.

Chambers, L., and Corden, M. E. (1963). Semeiography of *Fusarium* wilt of tomato. *Phytopathology* **53**, 1006–1010.

Gibbs, J. N. (1978). Intercontinental epidemiology of Dutch elm disease. *Annu. Rev. Phytopathol.* **16**, 287–307.

Gordon, T R., and Martyn, R. D. (1997). The evolutionary biology of *Fusarium oxysporum*. *Annu. Rev. Phytopathol.* **35**, 111–128.

Jones, J. P., and Crill, P. (1973). The effect of Verticillium wilt on resistant, tolerant and susceptible tomato varieties. *Plant Dis. Rep.* **57**, 122–124.

Mace, M. E., Bell, A. A., and Backman, C. H., eds. (1981). "Fungal Wilt Diseases of Plants." Academic Press, New York.

Nelson, P. E., Toussoun, T. A., and Cook, R. J., eds. (1981). "*Fusarium*: Diseases, Biology, and Taxonomy." Pennsylvania State Press, University Park, PA.

Parker, K. G. (1959). Verticillium hardromycosis of deciduous tree fruits. *Plant Dis. Rep. Suppl.* **225**, 39–61.

Pegg, G. F. (1974). Verticillium diseases. *Rev. Plant Pathol.* **53**, 157–182.

Ploetz, R. C., ed. (1990). "Fusarium Wilt of Banana." APS Press, St. Paul, MN.

Pomerleau, R. (1970). Pathological anatomy of the Dutch elm disease: Distribution and development of *Ceratocystis ulmi* in elm tissues. *Can. J. Bot.* **48**, 2043–2057.

Powelson, M. L., and Rowe, R. C. (1993). Biology and management of early dying of potatoes. *Annu. Rev. Phytopathol.* **31**, 111–126.

Pullman, G. S., and DeVay, J. E. (1982). Epidemiology of *Verticillium* wilt of cotton: A relationship between inoculum density and disease progression. *Phytopathology* **72**, 549–554.

Roux, B. y. J., Wingfield, M. J., Bouillet, J.-P., *et al.* (2000). A serious new wilt disease of *Eucalyptus* caused by *Ceratocystic fimbriata* in central Africa. *Eur. J. Forest Pathol.* **20**, 175–184.

Smalley, E. B., and Guries, R. P. (1993). Breeding elms for resistance to Dutch elm disease. *Annu. Rev. Phytopathol.* **31**, 325–352.

Stipes, R. J., and Campana, R. J., eds. (1981). "Compendium of Elm Diseases." APS Press, St. Paul, MN.

Strobel, G. A., and Lanier, G. N. (1981). Dutch elm disease. *Sci. Am.* **245**, 56–66.

Wingfield, M. J., Siefert, K. A., and Webber, J. F. (1993). "*Ceratocystis* and *Ophiostoma*: Taxonomy, Ecology, and Pathogenicity." APS Press, St. Paul, MN.

ROOT AND STEM ROTS CAUSED BY ASCOMYCETES AND DEUTEROMYCETES (MITOSPORIC FUNGI)

Several ascomycetous fungi attack primarily the roots and lower stems of plants. Some, such as *Cochliobolus*, *Gibberella*, and *Gaeumannomyces*, attack only cereals and grasses. A few, such as *Sclerotinia* and *Stenocarpella* (*Diplodia*), contain some species that attack cereals and grasses while other species cause severe diseases on several vegetables and field crops. A third group of soilborne fungi, such as *Fusarium solani*, *Leptosphaeria*, *Phymatotrichopsis*, *Monosporascus*, *Acremonium*, and *Thielaviopsis*, cause root and lower stem rots of many vegetables, ornamentals, field crops, and even trees.

As a general rule, the root and stem rot diseases caused by these and by other soilborne Ascomycetes and mitosporic fungi appear on the affected plant organs at first as water-soaked areas that later turn brown to black. In some diseases, lesions are frequently covered by white fungal mycelium. The roots and stems are killed more or less rapidly, and the entire plant grows poorly or is killed. Fungi that cause these diseases are nonobligate parasites that live, grow, and multiply in the soil as soil inhabitants, usually in association with dead organic matter. These fungi are favored by high soil moisture and high relative humidity in the air. Most of them produce conidia, and some produce ascospores occasionally or regularly. Several produce sclerotia. All of the aforementioned fungi can overwinter as mycelium in infected plant tissues or debris, as sclerotia, or as spores. These stages also serve as inoculum that can be spread and start new infections. Since the mid-1980s, considerable progress has been made in the biological control of several root and stem rot fungi by treating the seed with antagonistic fungi and bacteria. Such treatments, however, are still at the experimental stage.

GIBBERELLA STALK AND EAR ROT, AND SEEDLING BLIGHT OF CORN

Stalk rots of corn are often caused by different combinations of several species of fungi and bacteria and affect plants when they are nearly mature. The fungi most commonly responsible for stalk rots in corn include several species of *Gibberella*, *Fusarium (F. verticillioides, F. proliferatum,* and *F. subglutinans)*, *Stenocarpella (Diplodia)*, *Colletotrichum graminicola*, and *Macrophomina*. The stalk rot complex often causes losses between 10 and 30%.

Gibberella diseases of corn are worldwide in distribution and cause serious losses. The most important phases of the diseases are stalk rot and ear rot (*G. zeae*). In stalk rot, lower internodes become soft and appear tan or brown on the outside while internally they may appear pink or reddish (Figs. 11-111A–11-111D). The pith disintegrates, leaving only the vascular bundles intact. The rot may also affect the roots. Stalk rot leads to a dull gray appearance of the leaves, premature death, and stalk breakage (Figs. 11-111E and 11-111F). In ear rots, ears develop a pinkish or reddish mold that often begins at the tip of the ear (Fig. 11-112). If infection occurs early, the ears may rot completely and the pinkish mold grows between the ears and the tightly adhering husks. Corn ears infected with *G. zeae* contain mycotoxins and are toxic to humans and certain animals such as hogs.

Gibberella is one of many fungi causing blight of corn seedlings. It may be carried on or in infected seed or it may attack the seed and seedling from the soil. In either case, the germinating seed may be attacked and killed before the seedling emerges from the soil or after emergence, in which case the seedling may be killed or become dwarfed and chlorotic and later die. Light brown to dark-colored lesions are usually evident on the tap and lateral roots and in the lower internode.

Two species of *Gibberella*, *G. zeae* and *G. moniliforme* (*fujikuroi*), are primarily responsible for the symptoms observed on corn and on small grains. Both fungi produce ascospores in perithecia and *Fusarium*-type conidia. Perithecia are rather rare in *G. moniliforme*. Fungi overwinter as perithecia, mycelium, or chlamydospores in infected plant debris, particularly corn stalks. In the spring, during wet, warm conditions, ascospores are released and are carried by wind to corn stalks or ears, which they penetrate directly or through wounds and cause infections. Conidia are commonly produced on infected plant parts and serve as the secondary inoculum. The diseases are favored by dry weather, which stresses young plants early in the season, but are favored by wet weather near or after silking. Also, high plant density, high nitrogen and low potassium in the plant, and early maturity of hybrids make them more susceptible to the diseases.

The control of Gibberella diseases of corn depends on the use of resistant varieties, balanced nitrogen and potassium fertilization, and lower plant density in the field. Some crop rotations help.

FUSARIUM (GIBBERELLA) HEAD BLIGHT OR SCAB OF SMALL GRAINS

Gibberella or Fusarium head blight or scab of small grains, sometimes called pink mold or white head, also occurs worldwide. Head blight or scab is usually preceded or accompanied by a seedling blight and foot rot. They are caused by the same or related fungi to those causing diseases in corn, namely *Gibberella zeae* (anamorph *Fusarium graminearum*) and perhaps some additional species such as *Fusarium culmorum*, and on barley *F. avenacearum* and *F. poae*. Symptoms of Fusarium head blight seem to be more severe in taller than in shorter wheat varieties. Losses may be as high as 50% of the yield. In some areas where corn is grown extensively, this disease makes wheat and barley production unfeasible.

Seedling blight appears as a brown cortical rot or blight either before or after emergence of the seedling above the soil line. In older plants, a foot rot develops, appearing as a browning or pronounced rotting of the basal part of the plant around the soil level and for some distance above the soil line.

Scab or head blight causes severe damage to wheat and other cereals, especially in areas such as the upper Midwest (Minnesota, North and South Dakota, Illinois, Indiana, and Ohio) and in Canada (Alberta, Manitoba, Saskachewan) that have high temperature and relative humidity during the heading and blossoming period. Infected spikelets first appear water soaked and then lose their chlorophyll and become straw colored (Fig. 11-113A). In warm, humid weather, pinkish-red mycelium and conidia develop abundantly in the infected spikelets, and the infection spreads to adjacent spikelets or through the entire head (Fig. 11-113B). Purplish perithecia may also develop on the infected floral bracts. Infected kernels become shriveled and discolored with a white, pink, or light-brown scaly appearance as a result of the mycelial outgrowths from the pericarp (Figs. 11-113C and 11-113D). As with corn, infected kernels of cereals, especially those infected with *Fusarium graminearum*, also contain mycotoxins such as deoxynivalenol that are toxic to humans, hogs, and other animals.

Control measures against small grain diseases caused by *Gibberella* are identical to those described for the same diseases of corn. A great effort is being made to

FIGURE 11-111 Corn stalk rots caused by various fungi: (A) *Gibberella* sp., (B) *Fusarium* sp., (C) *Diplodia* sp., and (D) *Microphomina* sp. (E) *Gibbberella* stalk rot of young plant. (F) Typical breakage of stalk weakened by infection. [Photographs courtesy of Plant Pathology Department, University of Florida, and (E) W.L. Seaman, WCPD.]

FIGURE 11-112 Ear rots of corn caused by various fungi: *Gibberella* sp. (A and B), *Fusarium* sp. (C), *Diplodia (Stenocarpella)* sp. (D), *Nigrospora* sp. (E), and *Trichoderma* sp. (F). [Photographs courtesy of (A) J.C. Sutton, WCPD, and (B, C, and E) M.C. Shurtleff, University of Illinois.]

FIGURE 11-113 Wheat scab or head blight caused by *Fusarium* sp. (A) Scabbed heads of wheat. (B) Pinkish spores of the fungus produced on infected glumes of wheat. (C) Shriveled and chalky kernels of wheat due to infection by *Fusarium*. (D) Healthy wheat kernels (right) and kernels from plants with varying levels of scab. [Photographs courtesy of (A and C) R.A. Martin, (B) A. Tekauz, and (D) L. Cooke, WCPD.]

find microorganisms that can be used for the biological control of wheat head blight and to develop resistance through genetic engineering of wheat.

FUSARIUM ROOT AND STEM ROTS OF NONGRAIN CROPS

Several *Fusarium* species, but primarily *F. solani* and some formae specialies of *F. oxysporum*, cause, instead of vascular wilts, rotting of seeds and seedlings (damping-off), rotting of roots, lower stems, and crowns (Figs. 11-114A–11-114E), and rots of corms, bulbs, and tubers (Fig. 11-114F). They affect many different kinds of vegetables, flowers, and field crops. These diseases occur worldwide and cause severe losses by reducing stands and the growth and yield of infected plants.

In **root rots**, such as those of bean (Fig. 11-114A), peanut, soybean, and asparagus, tap roots of young plants show a reddish discoloration that later becomes darker and larger. The discoloration may cover the tap root and the stem below the soil line without a definite margin or it may appear as streaks extending up to the soil line. Longitudinal cracks appear along the main root, whereas small lateral roots are killed. Plant growth is retarded, and in dry weather the leaves may turn yellow and even fall off. Sometimes, infected plants develop secondary roots and rootlets just below the soil line that may be sufficient to carry the plant to maturity

FIGURE 11-114 Fusarium root rots on bean plants caused by *Fusarium solani* (A) and on wheat caused by *F. culmorum*. Root and crown rot of tomato caused by *Fusarium oxysporum* f. sp. *radicis-lycopersici* (C and D), on lettuce caused by *Fusarium* sp., and on potato tubers caused by *F. solani* (F). [Photographs courtesy of (A–D and F) Plant Pathology Department, University of Florida and (E) R.T. McMillan, University of Florida.]

and to production of a fairly good crop. In many cases, however, infected plants decline and die with or without wilt symptoms.

In **stem or root and crown rots**, as in the root and crown rot of tomato (Figs. 11-114C and 11-114D) caused by *F. oxysporum* f.sp. *radicis-lycopersici*, infected plants wilt and die from rot of the roots and stem at the base of the plant. Lesions develop on the stem at or below the soil line, and their edges often are pink or red. The lesions develop inward from the outside. In some plants a brown discoloration extends into the stem for a considerable distance above the ground. In older plants, roots have often rotted and sloughed off.

Rots of bulbs, corms, and tubers by *Fusarium* can occur in the field and in storage. They are common on plants such as onion, lily, gladiolus, and potato (Fig. 11-114F). The rot often starts at wounds or through cuts formed on such tissues during harvest. Invaded bulbs and corms may show outward symptoms, although usually the basal plate, fleshy scales, and roots are brown to black, sunken, and decaying and contain mats of mycelium. The rot is generally dry and firm. The foliage turns yellow or brown and dies prematurely. Tubers usually develop small brown patches that soon enlarge, become sunken, and show concentric wrinkles that contain cavities lined with white mycelium. Eventually, parts of the tuber or entire tubers are destroyed and become hard and mummified; if it is humid, however, they are then invaded by soft rotting bacteria.

The **sudden death syndrome of soybean** is caused by blue-pigmented strains of *Fusarium solani*, now called *F. solani* f. sp. *glycines*. It occurs in almost all the soybean-producing states and in several countries in South America. Sudden death syndrome causes yield losses that depend greatly on the age of the plant at infection, variety resistance to the disease, and weather conditions, and may vary from 5 to 80%. Following infection and gradual rotting of the roots (Fig. 11-115A), the disease appears as foliar symptoms consisting of small leaf spots (Fig. 11-115B) that may later enlarge and coalesce mostly between the veins (Fig. 11-115C), defoliation, and abortion of flowers and pods. Internal root discoloration spreads outward, followed by necrosis of the taproot and lateral roots (Fig. 11-115A) and death of individual plants. Large areas in fields may be affected (Fig. 11-115D).

Fusarium root and stem rots become more severe when plants exposed to the pathogen are stressed by low temperature, by intermittent drought or excessive soil water, by herbicides, by soil compaction, and by subsurface tillage pans, which restrict root growth.

The fungus *F. solani* generally produces only asexual spores, although under certain conditions it produces its perithecial stage, *Netria haematococca*. The asexual spores are microconidia, macroconidia (Fig. 11-116), and thick-walled chlamydospores. The fungus can live on dead plant tissue and can overwinter as mycelium or spores in infected or dead tissues or seed. The fungus is already present in many soils as spores, which are spread easily by air, equipment, water, and contact.

Control of *Fusarium* rots in the greenhouse is obtained through soil sterilization and use of healthy propagative stock. There are currently no adequate control measures for these diseases in the field. Loosening compacted soil with subsoiler chisels before planting has been the most dependable method of reducing Fusarium root rot of bean. Rotation with nonsusceptible crops, ensuring good soil drainage, and using disease-free or fungicide-treated seed or other propagative stock may help reduce losses. Fertilization with the nitrate form of nitrogen also helps reduce disease, as does the use of resistant varieties when available. Treatment of propagative stock with appropriate fungicides or application of fungicide sprays on the plants has helped reduce Fusarium rots on some kinds of plants. The biological control of Fusarium root and stem rots has been attempted with some success by incorporating organic materials such as barley straw and chitin in the soil, thus favoring the increase of several fungi and bacteria antagonistic to *Fusarium*, or by treating seeds or transplants with spores of fungal antagonists, mycorrhizal fungi, or antagonistic bacteria. None of the biological controls has been used in practice so far.

TAKE-ALL OF WHEAT

Take-all is a widespread and destructive disease of wheat and of other cereals and grasses in temperate climates around the world. It is primarily a disease of the root and basal stem of winter wheat, particularly in areas of intensive, continuous cultivation of cereals. Losses may vary from negligible to 50%.

Early in the season, take-all appears as patches of poorly developed, yellowish seedlings or stunted and unthrifty plants producing few tillers (Figs. 11-117A and 11-117B). Later, affected plants ripen prematurely and produce heads that have sterile, bleached spikelets and are known as whiteheads (Figs. 11-117C and 11-117D). Infected plants are pulled easily from the soil because much of their root system has been destroyed by the fungus and the remaining few roots are short, brown-black, and brittle. The brown-black dry rot usually extends to the base of the stem up to the lower leaf bases. A dark mat of mycelium develops between the stem and the lowest leaf sheath, and the leaf sheath sometimes shows small black raised spots consisting of the necks of fungal perithecia. A diagnostic feature of the disease is the presence of thick brown strands of runner hyphae on the surface of roots (Fig. 11-117E).

FIGURE 11-115 Sudden death of soybeans syndrome caused by *Fusarium solani.* (A) Rotting of rootlets and crown; some areas of root are covered with bluish fungal growth. (B) Early symptoms of sudden death consist of yellow or white spots on leaves. (C) Intermediate sudden death symptoms consisting of brown, necrotic, mostly interveinal areas on soybean leaves. (D) Advanced field symptoms consist of almost complete defoliation of soybeans affected by the sudden death syndrome. (Photographs courtesy of X.B. Yang, Iowa State University.)

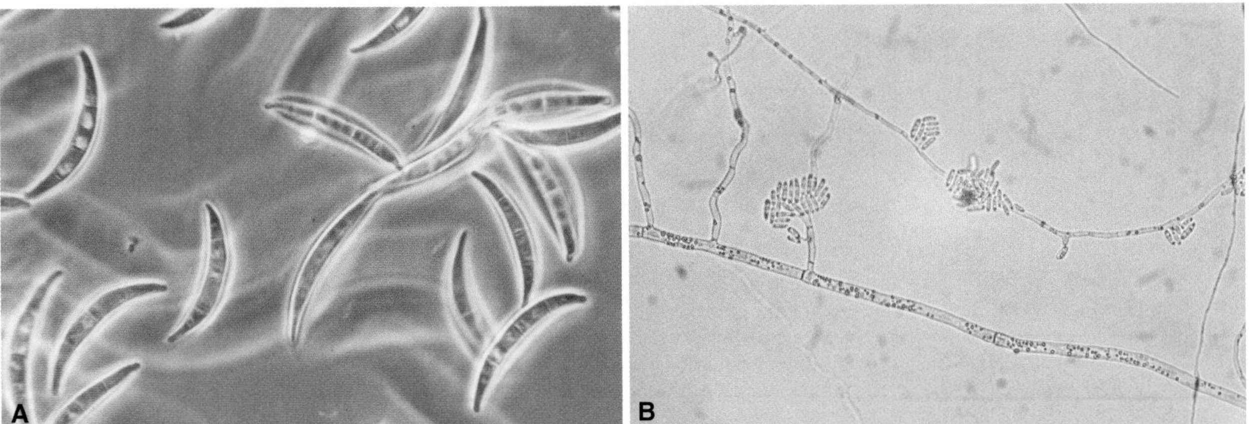

FIGURE 11-116 Macroconidia (A) and microconidia (B) of *Fusarium* sp. [Photographs courtesy of (A) R.J. McGovern and (B) R. Cullen, Plant Pathology Department, University of Florida.]

FIGURE 11-117 Take-all disease of wheat caused by *Gaeumannomyces graminis*. (A) Two healthy wheat seedlings (left) and two infected with the take-all disease (right) (B) Wheat plant whose root system and lower stem have been killed by the take-all disease. (C) Close-up of "white heads" produced by take-all-infected plants. (D) Patches of white-heads produced in a field where take-all disease was severe. (E) Superficial runner hyphae and haustorium-like feeder hyphae produced by the pathogen. [Photographs courtesy of (A and B) I.R. Evans, WCPD, (C and D) D. Mathre, Montana State University (E) R. Cullen, University of Florida.]

The pathogen of take-all is the fungus *Gaeumannomyces (Ophiobolus) graminis*. The fungus produces runner hyphae, which grow superficially on roots, and these produce lobed hyphopodia, which are short, darker, haustorium-like feeder hyphae (Fig. 11-117E) that grow toward the host. The fungus produces only one kind of spores in nature, namely ascospores in asci in perithecia. In culture, however, it also produces conidia from bottle-shaped terminal hyphal cells. The perithecia are black and embedded in basal leaf sheaths, with protruding black necks.

The fungus overwinters in infected wheat and grass plant roots and stems and in host debris. Ascospores are discharged forcibly from the asci in wet weather but rarely seem to cause infection. By far the most infections are caused by mycelium coming in contact with roots of growing plants. The superficial mycelium produces feeder hyphae that penetrate root tissues directly through pegs. The fungus invades the cortex and the vascular system but does not grow systemically through the latter. Invaded roots are killed. In young plants, the fungus extends into the crown and the base of the stem, whereas in more mature plants, its spread is slower and the fungus may remain confined to the roots. The fungus can infect plants throughout the growing season but is more active at temperatures between 12 and 18°C. Take-all is most severe in infertile, compacted, alkaline, and poorly drained soils. Its severity increases for several seasons (3–6 years) in fields cultivated continuously with wheat, but then it declines (take-all decline) and stabilizes at a lower level.

The control of take-all depends primarily on cultural practices, particularly crop rotation with nonhost plants. Other control measures include the destruction of grassy weeds and volunteer wheat plants that can harbor the fungus; application of adequate micronutrients, potassium, phosphorus, and ammonium but not nitrate-type nitrogen fertilizer; and the use of tolerant varieties, as no highly resistant ones are available. Yield losses due to take-all have been reduced by 60 to 75% through seed treatment with systemic fungicides.

A great deal of research has been carried out to discover and to develop biological controls of take-all. It was observed that some soils were suppressive to take-all, whereas others were conducive to the disease. Also, the suppressiveness could be transferred from field to field but could be eliminated by high (60°C) temperatures and by fumigation. It was later shown that take-all decline (soil suppressiveness) was brought about by root-colonizing bacteria that are antagonistic to *Gaeumannomyces* and inhibit its growth on the root surface or within the infected root. The antagonistic bacteria respond to wheat root exudates and multiply 5 to 10 times faster than other bacteria. Through their increased populations, antibiotics, siderophores, and so on, the antagonistic bacteria continue to inhibit the pathogen and eventually bring about the decline of take-all. Many bacterial strains have been found that inhibit the fungus effectively in laboratory tests. When the same bacteria are applied on the seed, however, and the seed is planted in *Gaeumannomyces*-infested soil in the greenhouse and in field plots, control is only partial and frequently fails completely. Therefore, the biological control of take-all in the field is not yet possible, but it is likely that, through research, the obstacles may be overcome.

THIELAVIOPSIS BLACK ROOT ROT

The fungus *Thielaviopsis*, mostly *T. basicola*, is a very common and important soil pathogen that causes damping of seedlings and black root rot (Fig. 11-118A) of many crops. Affected crops include many vegetables, ornamentals, and field crops, but the disease is particularly severe on beans, cucurbits, and solanaceous crops. Infected plants develop black root rot and become stunted, chlorotic, and produce reduced yields of low quality. Black root rot is due to dark-colored chlamydospores produced by the fungus on the infected roots and is diagnostic of the disease. The chlamydospores (Figs. 11-118B and 11-118C) are thick walled and are produced in chains in infected root tissue. The fungus also produces clear and cylindrical conidia in chains (Figs. 11-118B and 11-118C) on conidiophores that are expanded at the base.

MONOSPORASCUS ROOT ROT AND VINE DECLINE OF MELONS

The disease occurs in most parts of the world that have semiarid climates, relatively high summer temperatures, and soils that are saline and alkaline. Such areas include the southwestern United States, north Africa, Spain, Israel, Iran, India, Japan, and others. The disease affects primarily muskmelon and watermelon. It appears as a root rot and a sudden collapse and death of the plants in the field shortly before harvest. Losses fluctuate from year to year from 10 to 25% of the crop, but the crop of individual fields may be destroyed completely.

Symptoms

The aboveground symptoms of the disease appear as stunting, yellowing, and necrosis of the leaves in the inner crown (Fig. 11-119A). This is followed by progressive necrosis of the leaves until about 10 to 14 days before harvest when the entire canopy of the crop in a portion of or in the entire field collapses (Fig. 11-119B). The collapse leaves the still unripe fruit exposed to

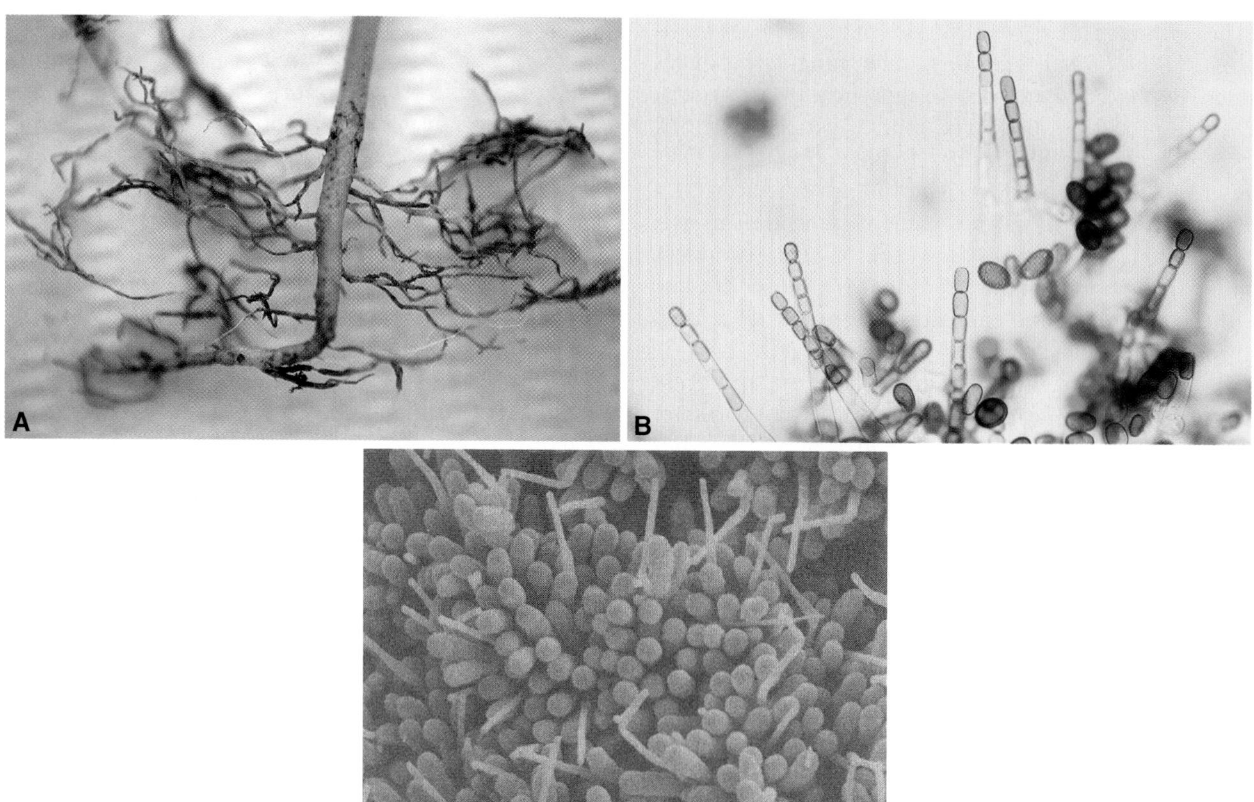

FIGURE 11-118 Black root rot disease caused by *Thielaviopsis basicola.* (A) Typical root rot symptoms in periwinkle, one of its many hosts. (B) *Thielaviopsis* conidia produced in chains and some thick-walled chlamydospores. (C) Scanning electron micrograph of chlamydospores and phialides of *T. basicola.* [Photographs courtesy of (A) R.J. McGovern, (B) R. Cullen, Plant Pathology Department, University of Florida and (C) C.W. Mims, University of Georgia, from Riggs and Mims (2000). *Mycologia* **92**, 123–129.]

intense solar radiation. The fruit fails to progress toward ripening properly and becomes unmarketable. The roots of affected plants show lesions (Figs. 11-119C and 11-119D), especially at root junctions, the feeder and secondary roots decay and slough off, and, under wet conditions, more roots rot and the tap root is killed (Figs. 11-119C and 11-119D). Some roots may have numerous perithecia embedded in the root cortex and they may occasionally be so numerous that they give the name of black pepper spot to the disease (Fig. 11-119D).

The Pathogen

The disease is caused by the fungus *Monosporascus cannonballus.* Another related species, *M. eutypoides*, seems to be responsible for the disease in southeast Asia. The fungus produces dark spherical perithecia that contain 200 or more asci each, but each ascus contains only one, spherical, cannonball-like ascospore. Ascospores (Fig. 11-119E) of *M. cannonballus* do not germinate in the laboratory. The fungus does not have an imperfect stage, i.e., it produces no conidia. The fungus grows best at high temperatures (30–35°C) and at a pH of 6 to 7, even up to pH 9.

Development of Disease

The fungus apparently survives in the soil as ascospores within or without perithecia. In the vicinity of melon roots, ascospores germinate and penetrate the roots. Feeder and secondary roots are invaded by the fungus and soon die and slough off. At this point, the plant has begun to show signs of water stress, yellowing and wilting of leaves, and may collapse and die suddenly. Infections of larger roots result in the formation of perithecia in the root cortex and the appearance of swellings on the root surface. The swellings soon turn black and finally burst open and the perithecia release the ascospores in the soil.

Control

There are no effective controls against the melon root rot and vine decline. A combination of resistant vari-

FIGURE 11-119 Monosporascus root rot and decline of cucurbit crops. Early field symptoms (A) and death and collapse of whole fields of cantaloupe plants (B) caused by infection of *Monosporascus cannonballus*. (C) Death of infected young roots and appearance of black perithecia (D) on such roots. Cannonball-like ascospores released from perithecia. (F) *M. cannonballus* mycelium parasitized by mycelium of the biocontrol fungus *Trichoderma viridae*. (Photographs courtesy of B.D. Bruton, USDA, Lane, OK.)

eties, soil fumigation, application of fungicides, and so on may reduce but do not eliminate the disease. Biological control is possible in the laboratory (Fig. 11-119F) but not, so far, in the field.

SCLEROTINIA DISEASES

Fungi of the genus *Sclerotinia*, especially *S. sclerotiorum* and *S. minor*, cause destructive diseases of numerous succulent plants, particularly vegetables and flowers. Another species, *S. homeocarpa*, causes the destructive dollar spot disease of turf grasses. *Sclerotinia* diseases occur worldwide and affect plants in all stages of growth, including seedlings, mature plants, and harvested products.

Sclerotinia Diseases of Vegetables and Flowers

Sclerotinia diseases probably affect most, if not all, annual vegetables, ornamentals, and field crops and cause huge amounts of losses both in the field and postharvest. The symptoms caused by *Sclerotinia* vary somewhat with the host or host part affected and with the environmental conditions. *Sclerotinia* diseases are known under a variety of names, such as cottony rot, white mold, watery soft rot, stem rot, drop, crown rot, and blossom blight, among others.

Symptoms and Disease Development. The most obvious and typical early symptom of *Sclerotinia* diseases is the appearance on the infected plant of a white fluffy mycelial growth in which soon afterward develop large, compact resting bodies or sclerotia (Figs. 11-120A–11-120F). The sclerotia are white at first but later become black and hard on the outside. They may vary in size from 0.5 to 1 millimeter in *S. minor* to 2 to 10 millimeters in diameter in *S. sclerotiorum*, although they are usually more flattened and elongated than spherical.

Stems of infected succulent plants (Figs. 11-120A–11-120D) at first develop pale or dark-brown lesions at their base. The lesions are often quickly covered by white cottony patches of fungal mycelium. In early stages of infection the foliage often appears normal and infected plants are easily overlooked. When the fungus grows completely through the stem and the stem rots, the foliage above the lesion wilts and dies more or less quickly. In some cases the infection may begin on a leaf and then move into the stem through the leaf. Sclerotia of the fungus may be formed internally (Fig. 11-120E) in the pith of the stem or they may be formed on the outside of the stem.

Leaves and petioles of plants such as lettuce, celery, and beets suddenly collapse and die as the fungus infects the base of the stem and the lower leaves. The fungus invades and spreads rapidly through the stem, and the entire plant dies and collapses, each leaf dropping downward until it rests on the one below. Mycelium and sclerotia usually appear on the lower surface of the outer leaves, but under moist conditions the fungus invades the plant completely and causes it to rot, producing a white, fluffy, mycelial growth over the entire plant.

Fleshy storage organs, such as lettuce, cabbage, squash, and carrots (Figs. 11-121A–11-121D), infected by *Sclerotinia* develop a white, cottony growth on their surface whether they are still in the field or in storage. Black sclerotia are formed externally (Fig. 11-121D). Invaded tissues appear darker than healthy ones and become soft and watery. If the disease develops after harvest in the storage house, the rot spreads to adjacent roots, bulbs, corms, and so on and produces pockets of rotted organs or all the organs in the crate may become infected and collapse, producing a watery soft rot covered by fungus growth.

Fleshy fruits, such as cucumber, squash, and eggplant, and seed pods of bean, are attacked by *Sclerotinia* through their closest point to the ground, at the point of their contact with the ground, or through their senescent flower parts. The fungus causes a wet rot that spreads from the tip of the fruit or pod to the rest of the organ, which eventually becomes completely rotted and disintegrates (Fig. 11-121C). The white fungal mycelium and the black sclerotia can usually be seen both externally and within the affected pods and fruits.

Flower infection is important primarily in camellias, daffodils, and narcissus. Small, watery, light brown spots appear on the petals, and these later enlarge, coalesce, and involve the entire petal. Eventually, the whole flower becomes dark brown and drops. Disintegration of the flowers occurs only in wet weather or after they have fallen, when the fungus produces abundant mycelium and sclerotia.

In turfgrasses, a related fungus, *Sclerotinia homeocarpa*, causes the dollar spot disease, called so because in low-mowed grasses, the symptoms appear as numerous small, circular, bleached-out spots the size of a quarter to a dollar. It is a persistent disease in many golf courses and in home lawns, in the latter the disease appearing as a pattern of 4- to 6-inch patches of blighted turf (Figs. 11-122A–122C).

The Pathogen. The fungus *Sclerotinia sclerotiorum* overwinters as sclerotia on or within infected tissues, as sclerotia that have fallen on the ground, and as mycelium in dead or living plants (Figs. 11-120E, 11-120F, and 11-123). In the spring or early summer, sclerotia germinate and produce slender stalks terminating at a small, disk- or cup-shaped apothecium 5 to 15 millimeters in diameter, in which asci and ascospores are

FIGURE 11-120 Symptoms caused by *Sclerotinia sclerotiorum*. Stem rot and white mold on beans (A and B), potato (C), and pepper (D). Sclerotia of the fungus inside a tomato stem (E) and germinating sclerotia producing apothecia (F). [Photographs courtesy of (A and E) K. Pernezny, (C and F) D.P. Weingartner, University of Florida, and (B and D) Plant Pathology Department, University of Florida.]

FIGURE 11-121 Sclerotinia disease symptoms. (A) Lettuce drop. (B) Cabbage rot. (C) Squash white mold and soft rot. (D) Carrot white mold, with some sclerotia on the surface. (E) Lettuce killed by *Sclerotinia sclerotiorum* in field growing continuous crops of lettuce. (F) Lettuce field destroyed by the fungus as a result of aerial spread of ascospores. [Photographs courtesy of D. Ormrod, WCPD, (B–D) I.R. Evans, WCPD, and (E and F) K. V. Subbarao, University of California, Salinas.]

FIGURE 11-122 Dollar spot disease of turfgrasses caused by *Sclerotinia homeocarpa*. (A) A few spots of diseased plants killed by the disease. (B) Close-up of one of the spots showing growth of the fungus on infected grass plants. (C) Overview of a lawn with numerous "dollar spots" in it. (Photographs courtesy of T.E. Freeman, University of Florida.)

produced. Large numbers of ascospores are discharged from the apothecia into the air over a period of 2 to 3 weeks. The ascospores are blown away, and if they land on senescent plant parts, such as old blossoms, which provide a readily available source of food, the ascospores germinate and cause infection (Fig. 11-121F). In some *Sclerotinia* species, sclerotia cause infection by producing mycelial strands that attack and infect young plant stems directly (Fig. 11-121E). Under moist conditions the latter method of infection is probably more common than the one by ascospores, although in *S. sclerotiorum* almost all infections are initiated by ascospores.

The control of *Sclerotinia* diseases depends on a number of cultural practices and on chemical sprays. Few varieties show appreciable degrees of resistance to the pathogen. In the greenhouse, soil sterilization with steam eliminates the pathogen. Susceptible crops should be planted only in well-drained soils, the plants should not be planted too close together for air drainage, and the soil should be kept free of weeds between crops. Because sclerotia remain viable in the soil for a least three years and because they do not all germinate or die out at the same time, infected fields should be planted to nonsusceptible crops, such as small grains, for at least three years before susceptible crops are planted again. In several crops, good control of the *Sclerotinia* disease has been obtained by spraying the soil or the plants with appropriate fungicides before and during their stage of susceptibility to the pathogen. Some of the newer contact and systemic fungicides give excellent control of *Sclerotinia*.

Numerous species of fungi, bacteria, insects, and other organisms have been reported to parasitize or to interfere with the growth of *Sclerotinia* spp. Encouraging results with biological control of *Sclerotinia* diseases in some crops have been obtained by incorporating the mycoparasitic fungi *Coniothyrium*

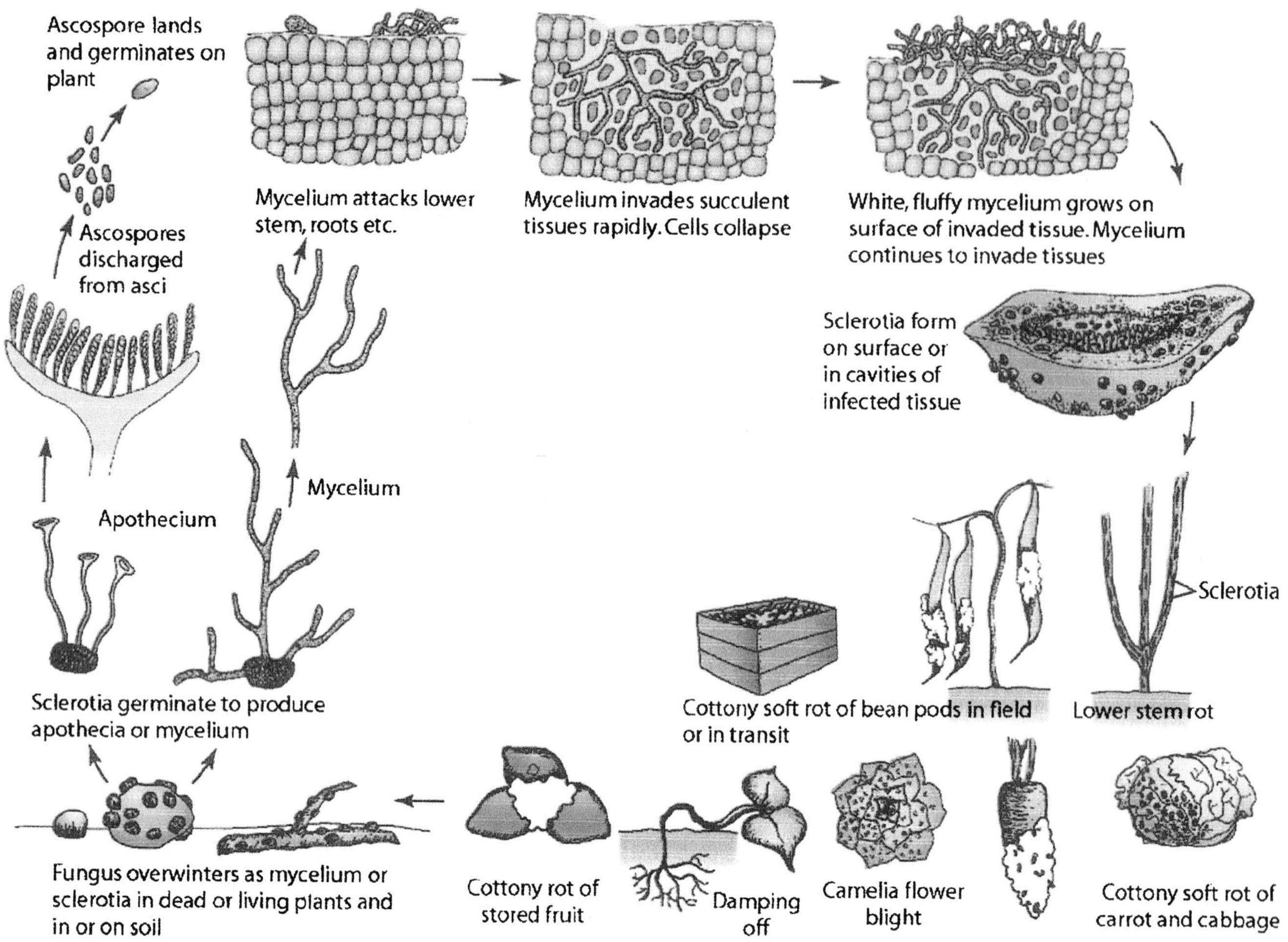

FIGURE 11-123 Development and symptoms of diseases of vegetables and flowers caused by *Sclerotinia sclerotiorum*.

minitans, *Gliocladium roseum*, *G. virens*, *Sporodesmium sclerotivorum*, and *Trichoderma viride* in *Sclerotinia*-infested soil. The mycoparasites destroy existing sclerotia or inhibit the formation of new sclerotia by the fungus and, thereby, markedly reduce the fungus population in the soil. So far, some of these biocontrols are being used by some growers on their crops but no widely accepted practical biological control recommendations have been developed.

PHYMATOTRICHUM ROOT ROT

Phymatotrichum root rot, usually called Texas root rot or cotton root rot, occurs only in the southwestern United States and Mexico (see Fig. 7-11B). It probably affects more kinds of cultivated and wild dicotyledonous plants than any other. Its hosts include many fruit, forest, and shade trees, most vegetables and flowers, ornamental shrubs, and many weeds. It causes its greatest losses in cotton in the area from Texas to Arizona and Mexico.

Infected plants appear in patches in the field (Fig. 11-124A). Leaves show yellowing, bronzing, and a slight wilting. Later they turn brown and dry but remain attached to the plant. Below the soil line, and in some plants up to 30 centimeters or more above the soil line, the bark and cambium turn brown, resulting in a firm brown rot of the root and the lower stem. The rotted roots are usually partly covered by coarse, brown, parallel strands of mycelium, and this characteristic helps diagnose the disease.

The fungus *Phymatotrichopsis omnivora* is unclassified, thought to be an ascomycete (Pezizomycete) by some and a basidiomycete by others. It produces hyphae that grow closely compressed together into thick mycelial strands (Fig. 11-124C) that have characteristic slender, cross-like side branches (Fig. 11-124D). Older strands are dark brown and have few side branches.

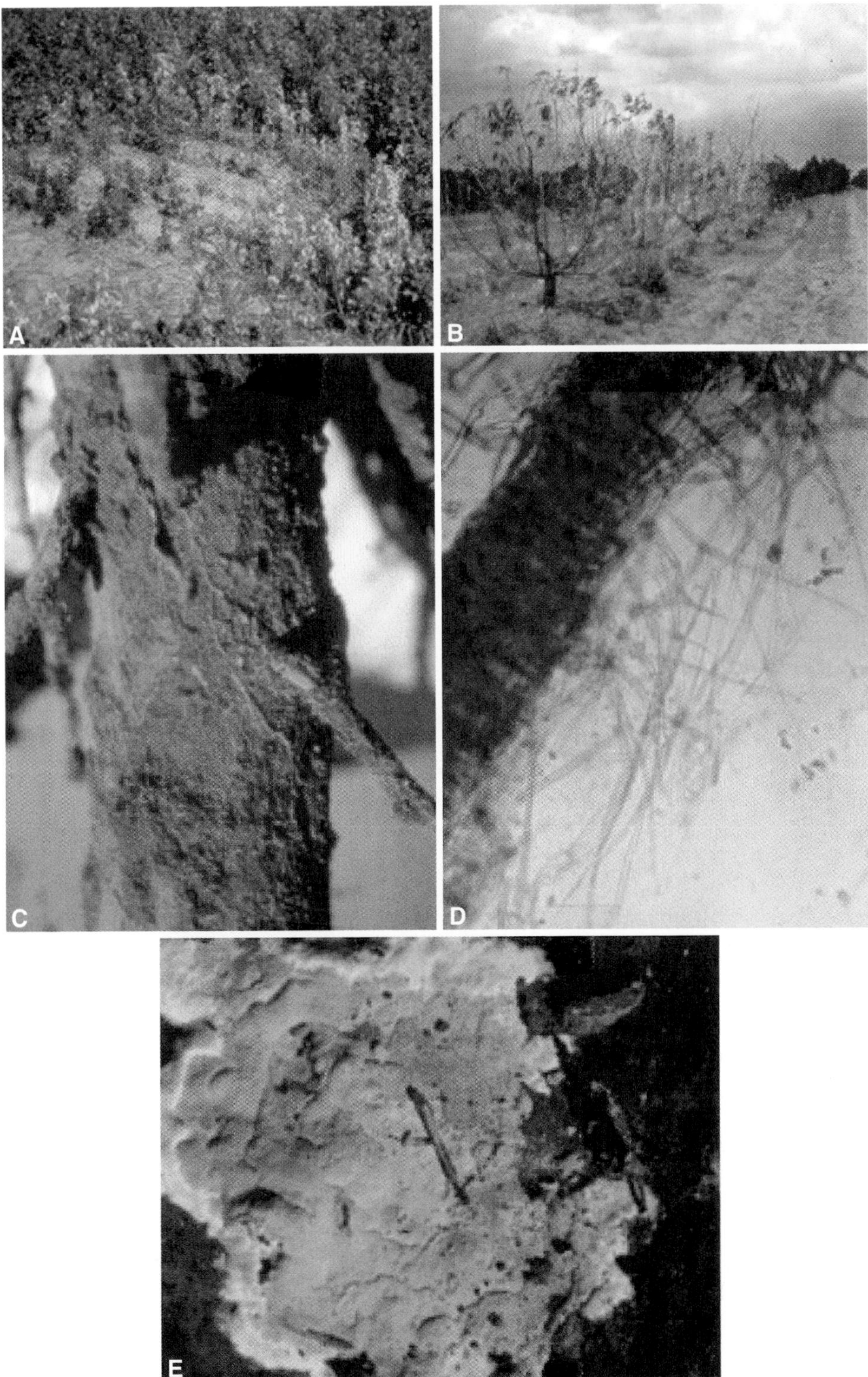

FIGURE 11-124 Symptoms of root rot diseases caused by *Phymatotrichopsis omnivorua*. (A) Alfalfa plants killed by *Phymatotrichum* in an ever-widening circle. (B) Peach trees being killed by the fungus. (C) Fungal strands on the surface of infected root. (D) Cross-shaped hyphae growing out of fungal strands. (E) A spore mat present around infected plants. (Photographs courtesy of R.B. Hine and N.P. Goldberg, New Mexico State University.)

Phymatotrichopsis produces white to tan-colored spore mats (Fig. 11-124E) on the soil around infected plants. The spore mats bear short conidiophores that have one-celled conidia that apparently do not germinate or cause infection. The fungus also produces small, brown to black sclerotia that germinate to produce mycelium. Most of the fungus mycelium and sclerotia are found in soil at depths between 30 and 75 centimeters. They can survive in the soil for five years or more.

The fungus survives best and causes considerably more damage to plants growing in alkaline, black, heavy clay soils that are poorly aerated. The fungus requires high temperature and adequate soil moisture for greatest activity, provided the soil pH is near or above neutral.

The fungus enters the plant below the soil line and then grows downward throughout the root system; on some plants, it also invades the lower stem. The fungus spreads from plant to plant through the growth of the mycelial strands and through the spread of such strands or sclerotia by farm equipment, transplants, and so on. Once introduced into an area, the fungus can survive on cultivated plants and weeds indefinitely, provided the soil and temperature conditions are favorable. The pathogen cannot stand temperatures below freezing for any appreciable amount of time, and its narrow geographic distribution seems to be the result of its high temperature and alkalinity requirements (Fig. 7-11B).

The control of Phymatotrichum root rot depends on long rotations with grain crops, weed eradication, deep and frequent plowing to keep the soil well aerated, and the use of green manure crops, such as thickly planted corn, sorghum, or legumes. The latter, on decay, favor the buildup of large populations of microorganisms that are antagonistic to *Phymatotrichopsis*. Soil fumigation is effective if applied annually and if the value of the crop justifies the expense, but it has not generally proved practical because of the rapid spread of the pathogen from deeper in the soil to the root zone once the fumigant has evaporated.

Selected References

Adams, P. B., and Fravel, D. R. (1990). Economical biological control of *Sclerotinia* lettuce drop by *Sporidesmium sclerotivorum*. *Phytopathology* **80**, 1120–1124.

Aegerter, B. J., Gordon, T. R., and Davis, R. M. (2000). Occurrence and pathogenicity of fungi associated with melon root rot and vine decline in California. *Plant Dis.* **84**, 224–230.

Anderson, A. L. (1948). The development of *Gibberella zeae* head blight of wheat. *Phytopathology* **38**, 595–611.

Anonymous (1979). Symposium on *Sclerotinia* (=*Whetzelinia*): Taxonomy, biology and pathology. *Phytopathology* **69**, 873–910.

Anonymous (1991). Symposium on recent advances in *Fusarium* systematics. *Phytopathology* **81**, 1043–1067.

Asher, M. J. C., and Shipton, P. J., eds. (1981). "Biology and Control of Take-all." Academic Press, London.

Augustin, C., Ulrich, K., Ward, E., *et al.* (1999). RAPD-based inter- and intravarietal classification of fungi of the *Gaeumannomyces–Phialophora* complex. *J. Phytopathol.* **147**, 109–117.

Bai, G., and Shaner, G. (1994). Scab of wheat: Prospects for control. *Plant Dis.* **78**, 760–766.

Bockus, W. W., and Shroyer, J. P. (1998). The impact of reduced tillage on soilborne plant pathogens. *Annu. Rev. Phytopathol.* **36**, 485–500.

Bruehl, G. W., ed. (1975). "Biology and Control of Soil-Borne Plant Pathogens." APS Press, St. Paul, MN.

Burke, D. W., and Miller, D. E. (1983). Control of Fusarium root rot with resistant beans and cultural management. *Plant Dis.* **67**, 1312–1317.

Christensen, J. J., and Wilcoxson, R. D. (1966). "Stalk Rot of Corn." Monogr. No. 3. APS Press, St. Paul, MN.

Cohen, R., *et al.* (2000). Toward integrated management of *Monosporascus* wilt of melons in Israel. *Plant Dis.* **84**, 496–505.

Cook, R. J., and Baker, K. F. (1983). "The Nature and Practice of Biological Control of Plant Pathogens." APS Press, St. Paul, MN.

Davanlou, Madsen, Madsen, and Hockenhull (1999). Parasitism of macroconidia, chlamydospores and hyphae of *Fusarium culmorum* by mycoparasitic *Pythium* species. *Plant Pathol.* **48**, 352–359.

Dodd, J. L. (1980). The role of plant stresses in development of corn stalk rots. *Plant Dis.* **64**, 533–537.

El-Tarabily, K. A., Soliman, M. H., Nassar, A. H., *et al.* (2000). Biological control of *Sclerotinia minor* using a chitinolytic bacterium and actinomycetes. *Plant Pathol.* **49**, 573–583.

Engelhard, A. W., ed. (1989). "Soilborne Plant Pathogens: Management of Diseases with Macro- and Microelements." APS Press, St. Paul, MN.

Escande, A. R., Laich, F. S., and Pedraza, M. V. (2002). Field testing of honeybee-dispersed *Trichoderma* spp. to manage sunflower head rot (*Sclerotinia sclerotiorum*). *Plant Pathol.* **51**, 346–351.

Flett, B. C., and McLaren, N. W. (2000). Incidence of *Stenocarpella maydis* ear rot of corn under crop rotation systems. *Plant Dis.* **85**, 92–94.

Gatch, E. W., and Munkvold, G. P. (2002). Fungal species composition in maize stalks in relation to European corn borer injury and transgenic insect protection. *Plant Dis.* **86**, 1156–1162.

Gatch, E. W., Hellmich, R. L., and Munkvold, G. P. (2002). A comparison of maize stalk rot occurrence in Bt and non-Bt hybrids. *Plant Dis.* **86**, 1149–1155.

Hestbjerg, H., Felding, G., and Elmholt, S. (2002). *Fusarium culmorum* infection of barley seedlings: Correlation between aggreassiveness and deoxynivalenol content. *J. Phytopathol.* **150**, 308–312.

Hilton, Jenkinson, Hollins, and Parry (1999). Relationship between cultivar height and severity of *Fusarium* ear blight in wheat. *Plant Pathol.* **48**, 202–208.

Hornby, D. (1998). "Take-All Disease of Cereals: A Regional Perspective." CAB International.

Hornby, D., *et al.* (1999). "Take-All Disease of Cereals." CABI, Wallingford, UK.

Huber, D. M., and McCay-Buis, T. S. (1993). A multiple component analysis of the take-all disease of cereals. *Plant Dis.* **77**, 437–447.

Khan, N. I., Schisler, D. A., Boehm, M. J., *et al.* (2001). Selection and evaluation of microorganisms for biocontrol of *Fusarium* head blight of wheat incited by *Gibberella zeae*. *Plant Dis.* **85**, 1253–1258.

Lumsden, R. D., and Dow, R. L. (1973). Histopathology of *Sclerotinia sclerotiorum* infections of bean. *Phytopathology* **63**, 708–715.

Lyda, S. D. (1978). Ecology of *Phymatotrichum omnivorum*. *Annu. Rev. Phytopathol.* **16**, 193–209.

Martyn, R. D., and Miller, M. E. (1996). Monosporascus root rot and vine decline: An emerging disease of melons worldwide. *Plant Dis.* **80**, 716–725.

McMullen, M., Jones, R., and Gallenberg, D. (1997). Scab of wheat and barley: A re-emerging disease of devastating impact. *Plant Dis.* **81**, 1340–1348.

Nelson, P. E., Toussoun, T. A., and Cook, R. J., eds. (1981). "*Fusarium*: Diseases, Biology and Taxonomy." Pennsylvania State Univ. Press, University Park, PA.

Pivonia, S., Cohen, R., Katan, J., *et al.* (2002). Effect of fruit load on the water balance of melon plants infected with *Monosporascus cannonballus*. *Physiol. Mol. Plant Pathol.* **60**, 39–49.

Powelson, M. L., and Rowe, R. C. (1993). Biology and management of early dying of potatoes. *Annu. Rev. Phytopathol.* **31**, 111–126.

Pritsch, C., Vance, C. P., Bushnell, W. R., *et al.* (2001). Systemic expression of defense response genes in wheat spikes as a response to *Fusarium graminearum* infection. *Physiol. Mol. Plant Pathol.* **58**, 1–12.

Reglinski, T., Whitaker, G., Cooney, J. M., *et al.* (2001). Systemic acquired resistance to *Sclerotinia sclerotiorum* in kiwifruit vines. *Physiol. Mol. Plant Pathol.* **58**, 111–118.

Ribichich, K. F., Lopez, S. E., and Vegetti, A. C. (2000). Histopathological spikelet changes produced by *Fusarium graminearum* in susceptible and resistant wheat cultivars. *Plant Dis.* **84**, 794–802.

Roy, K. W., *et al.* (1997). Sudden death syndrome of soybean. *Plant Dis.* **81**, 1100–1111.

Rush, C. M., Gerik, T. J., and Lyda, S. D. (1984). Factors affecting symptom appearance and development of *Phymatotrichum* root rot of cotton. *Phytopathology* **74**, 1466–1469.

Sherf, A. F., and MacNab, A. A. (1986). "Vegetable Diseases and Their Control." Wiley (Interscience), New York.

Singleton, L. L., Mihail, J. D., and Rush, C. M. (1992). "Methods for Research on Soilborne Phytopathogenic Fungi." APS Press, St. Paul, MN.

Subbarao, K. V. (1998). Progress toward integrated management of lettuce drop. *Plant Dis.* **82**, 1068–1078.

Willets, H. J., and Wong, J. A. L. (1980). The biology of *Sclerotinia sclerotiorum*, *S. trifoliorum* and *S. minor* with emphasis on specific nomenclature. *Bot. Rev.* **46**, 101–165.

Windels, C. E. (2000). Economic and social impacts of *Fusarium* head blight: Changing farms and rural communities in the northern grain plains. *Phytopathology* **90**, 17–21.

POSTHARVEST DISEASES OF PLANT PRODUCTS CAUSED BY ASCOMYCETES AND DEUTEROMYCETES (MITOSPORIC FUNGI)

Postharvest diseases develop on fruit and other plant products during harvesting, grading, and packing, during transportation to market and to the consumer, and while the produce is in the possession of the consumer until the moment of actual consumption or use (Figs. 11-125 and 11-126). During this period, the plant product may show symptoms of diseases that had begun in the field but remained latent; it may be subjected to environmental conditions or treatments that are harmful and therefore impair its appearance and food value; or it may be subjected to conditions that favor its attack by microorganisms, which cause a portion of it to rot. In many cases, such microorganisms also secrete toxic substances that make the remainder of the product unfit for consumption or lower its nutritional and sale value.

All types of plant products are susceptible to postharvest diseases. Generally, the more tender or succulent the exterior of the product and the greater the water content of the entire product, the more susceptible it is to injury and to infection by fungi and bacteria. Thus, succulent, fleshy fruits and vegetables, cut flowers, bulbs, and corms are often affected by postharvest diseases. The extent of damage depends on the particular product, on the disease organism or organisms involved, and on the storage conditions.

Postharvest rotting of cereal grains and of legumes is also quite common and the losses caused by it are quite large. Such losses occur primarily at the large bins or warehouses of the growers, wholesalers, or manufacturers and are seldom observed by the general public. In addition, postharvest molds and decays of bread, hay, silage, and other feedstuffs are quite common and extensive, and we all frequently have to throw away bread because it has become moldy.

Postharvest diseases destroy 10 to 30% of the total yield of crops, and in some perishable crops, especially in developing countries, they destroy more than 30% of the crop yields. Postharvest diseases usually cause great losses of fresh fruits and vegetables by reducing their quality, quantity, or both. Postharvest diseases of grains and legumes also result in the production by some infecting microorganisms of toxic substances known as **mycotoxins**. Mycotoxins are poisonous to humans and animals that consume products made from seeds infected with such microorganisms. Mycotoxins are also produced by some fungi in infected fresh fruits and vegetables, but in these cases they are generally removed when the rotten fruits or vegetables or their rotten parts are discarded before consumption. As manufacturers use bulk quantities of fresh fruits and vegetables to make fruit or vegetable juices, purees, cole slaw, baby foods, and so on, quality control of individual fruits and vegetables becomes all but economically impractical; therefore, postharvest infections and mycotoxins in bulk-prepared foods are likely to increase in the future.

Postharvest diseases are caused primarily by a relatively small number of Ascomycetes and mitorporic fungi and by a few species of Oomycetes, Zygomycetes, Basidiomycetes, and bacteria. The bacteria are primarily of the genera *Erwinia* and *Pseudomonas*. Of the Oomycetes, *Pythium* and *Phytophthora* cause only soft rots of fleshy fruits and vegetables that are usually in

FIGURE 11-125 Postharvest pathogens and diseases. (A) Conidiophore and conidia of *Aspergillus* sp. (B) Bread molded with the fungus *Penicillium* sp. (C) Conidiophores and conidia of *Penicillium* sp. (D) Peach rotting as a result of postharvest infection with *Rhizopus* sp. (E) Pear fruit rotting as a result of natural infection with *Penicillium* through a natural tiny puncture.

contact with or very near the soil and they may spread to new, healthy fruit during storage. Two Zygomycetes, *Rhizopus* and *Mucor*, affect fleshy fruits and vegetables after harvest and also stored grains and legumes, as well as prepared foods such as bread, when moisture conditions are favorable (Figs. 11-125 and 11-126). Of the Basidiomycetes, *Rhizoctonia* and *Sclerotium* cause rotting of fleshy fruits and vegetables, whereas several fungi cause deterioration of wood and wood products. The Ascomycetes and imperfect fungi that cause postharvest diseases are by far the most common and most important causes of postharvest decay, and they are discussed in some detail here.

Fungi and bacteria causing postharvest diseases usually can attack healthy, living tissue, which they disintegrate and cause to rot. Often, however, other fungi and bacteria follow them and live saprophytically on the tissues already killed and macerated by the former.

Many of the postharvest diseases of fruits, vegetables, grains, and legumes are the results of infections by pathogens in the field. Symptoms from such "field infections" may be too inconspicuous to be noticed at harvest. In fleshy fruits and vegetables, field infections continue to develop after harvest, whereas in grains and legumes they cease to develop soon after harvest. In

FIGURE 11-126 Additional postharvest pathogens and diseases. (A) Botrytis rot of strawberries. (B) Sclerotinia white mold of beans. (C) Sclerotium rot of tomato. (D) Penicillium rot of tangerines. (E) Macrophomina rot of cantaloupe. (F) Pile of tomatoes rotting from infection with various fungi. [Photographs courtesy of (A–D and F) Plant Pathology Department, University of Florida.]

fleshy fruits and vegetables, new infections may be caused in storage by the same or other pathogens, whereas in grains and legumes storage infections are usually caused by pathogens other than those causing field infections.

As with all fungal and bacterial plant diseases, postharvest diseases are favored greatly by high moisture and high temperatures. Fleshy fruits and vegetables are generally kept at high relative humidities to avoid shrinkage and therefore they are attacked easily by pathogenic microorganisms, especially when wounds, cuts, and bruises are available for penetration. However, penetration through natural openings and directly through the cuticle and epidermis, especially of fruits and vegetables in contact with infected ones, is quite common. Once a fruit or vegetable becomes infected, development and spread of the infection increase as the storage temperature increases. At lower (3 to 6°C) temperatures, pathogens and the diseases they cause develop more slowly or cease to develop at all.

Grains and legumes, however, can be kept for long periods of time because their moisture content is or can be reduced to about 12 to 14%. At such moisture contents, the fungi that cause field infections cease to grow and do not cause new infections even if the grains become remoistened. Other fungi, however, can infect grains and legumes even when their moisture content is about 13 to 15%, and the severity and spread of infection increase drastically with the slightest increase in moisture above that range. Infection of grains with a high moisture content is also favored by high temperatures. Frequently, however, the temperature of moistened infected grain rises drastically due to the heat produced from respiration of the actively growing fungi and bacteria that caused the infection.

POSTHARVEST DECAYS OF FRUITS AND VEGETABLES

Some of the most common Ascomycetes or mitosporic fungi that cause postharvest diseases are listed here.

Aspergillus, Penicillium, Rhizopus, and *Mucor*

All four of them are found commonly to cause molding of bread, whereas *Penicillium* and *Rhizopus* also cause postharvest rots of numerous kinds of wounded or senescent fruits and vegetables. *Aspergillus* (Fig. 11-125A) is found more commonly causing molding of grains and legumes; *Rhizopus* causes many fruit rots, as in peach (Fig. 11-125D) and strawberry (Fig. 11-126A), whereas *Penicillium* causes the rotting of many wounded fruit, e.g., pears (Fig. 11-125E).

Alternaria

The various species of *Alternaria* cause decay on most, if not all, fresh fruits and vegetables either before or after harvest. Symptoms appear as brown or black, flat or sunken spots with definite margins, or as diffuse, large, decayed areas that are shallow or extend deep into the flesh of the fruit or vegetable. The fungus develops well at a wide range of temperatures, even in the refrigerator, although at a slower rate. The fungus may spread into and rot tissues internally with little or no mycelium appearing on the surface, but usually a mat of mycelium that is white at first but later turns brown to black forms on the surface of the rotted area.

Botrytis

Botrytis causes the gray molds or gray mold rots of fruits and vegetables, both in the field and in storage (Figs. 11-126D and 11-126F). Almost all fresh fruits, vegetables, and bulbs are attacked by *Botrytis* in storage. Some products, such as strawberry, lettuce, onion, grape, and apple, are also attacked in the field near maturity or while green. The decay may start at the blossom or stem end of the fruit or at any wound. The decay appears as a well-defined water-soaked, then brownish, area that penetrates deeply and advances rapidly into the tissue. In most hosts and under humid conditions a grayish or brownish-gray, granular, mold layer develops on the surface of decaying areas. Gray molds are most severe in cool, humid environments and continue to develop, although slowly, even at 0°C.

Fusarium

Fusarium causes postharvest pink or yellow molds on vegetables (Fig. 126E) and ornamentals and especially on root crops, tubers, and bulbs. Low-lying crops such as cucurbits and tomatoes are also affected frequently. Contamination with *Fusarium* usually takes place in the field before or during harvest, but infection may develop in the field or in storage. Losses are particularly heavy with crops such as potatoes that are stored for long periods of time. Affected tissues appear fairly moist and light brown at first, but later they become darker brown and somewhat dry. As the decaying areas enlarge, they often become sunken, the skin is wrinkled, and small tufts of whitish, pink, or yellow mold appear. The infection of softer tissues such as tomatoes and cucurbits develops faster and is characterized by pink mycelium and pink, rotten tissues.

Geotrichum

Geotrichum causes the sour rots of citrus fruits, tomatoes, carrots, and other fruits and vegetables. Sour rot is one of the messiest and most unpleasant rots of susceptible fruits and vegetables (Fig. 11-126C). Although it may affect them at the mature green stage, it is the ripe or overripe fruits and vegetables and those kept in moisture-holding plastic bags or packages that are particularly susceptible to sour rot. The fungus occurs in soils and decaying fruits and vegetables and contaminates new ones before or during harvest. The fungus penetrates fruits, usually after harvest, at wounds of various sorts. Infected areas appear water soaked and soft and are punctured easily. The decay spreads rapidly. Later, the skin frequently cracks over the affected area and is usually filled with a white, cheesy, or scum-like development of the fungus. Also, a thin, water-soaked layer of compact, cream-colored fungal growth develops on the surface, while the whole inside becomes a sour-smelling, decayed, watery mass. Fruit flies, which are attracted to tissues affected with sour rot, spread the pathogen further. The fungus prefers high temperatures (24–30°C) and humidity but is active at temperatures as low as 2°C.

Penicillium

The various species of *Penicillium* cause the blue mold rots and the green mold rots, also known as *Penicillium* rots. They are the most common and usually the most destructive of all postharvest diseases, affecting most kinds of fruits and vegetables (Figs. 11-125B and 11-125E). On some fruits, such as citrus, some infections may take place in the field, but blue molds or green molds are essentially postharvest diseases and often account for up to 90% of decay in transit, in storage, and in the market. *Penicillium* enters tissues through wounds. However, it can spread from infected fruit in contact with healthy ones through the uninjured skin. *Penicillium* rots at first appear as soft, watery, slightly discolored spots of varying size and on any part of the fruit. The spots are rather shallow at first but quickly become deeper. At room temperature most or all of the fruit decays in just a few days. Soon a white mold begins to grow on the surface of the fruit, near the center of the spot, and starts producing spores. The sporulating area has a blue, bluish-green, or olive-green color and is usually surrounded by white mycelium and a band of water-soaked tissue. The fungus develops on spots of any size as long as the air is moist and warm. In cool, dry air, surface mold is rare, even when the fruits are totally decayed. Decaying fruit has a musty odor. Under dry conditions it may shrink and become mummified. Under moist conditions, secondary fungi and yeasts also enter the fruit, which is then reduced to a wet, soft mass.

In addition to the losses caused by the rotting of fruits and vegetables by *Penicillium*, the fungus also produces several mycotoxins, such as patulin, in the affected products, which contaminate juices and sauces made from healthy and partly rotten fruits. These and other mycotoxins and their effects are discussed later.

Sclerotinia

Sclerotinia causes the cottony rot of citrus fruits, especially lemons, and the watery soft rot of many fruits and practically all vegetables (Fig. 11-126B) except onions and potatoes. In a moist atmosphere, a soft, watery decay is produced, and the affected tissues are covered rapidly with a white, cottony growth of mycelium that is characteristic of this decay. In moist air, succulent decaying products may be completely liquefied, leak, and leave a pool of juice. In dry air the water may evaporate as fast as it is liberated by the decay, and the tissues dry down into a mummy or parchment-like remains. Cottony rot is a rapidly spreading, contact decay that attacks both green and mature fruits and vegetables. Black sclerotia, 2 to 15 millimeters long, later develop in the fungus mat. The activity of the fungus and the severity of the rot increase with temperature up to 25°C, but, once started, rotting of tissues continues at temperatures as low as 0°C.

CONTROL OF POSTHARVEST DECAYS OF FRESH FRUITS AND VEGETABLES

For some postharvest diseases, control depends on effective control of the pathogens that cause the same diseases in the field so that the crop will not be contaminated with the pathogens at harvest and subsequently in storage. The crop should be harvested and handled carefully to avoid wounds, bruises, and other injuries that could serve as ports of entry for the pathogen. Harvesting and handling of the crop should be done when the weather is dry and cool to avoid further contamination and infection. The crop should be cooled as quickly as possible to prevent the establishment of new infections and the development of existing ones. All fruits or vegetables showing signs of infection should be removed from the crop that is to be stored or shipped to avoid further spread of the disease. Storage containers, warehouse, and shipping cars should be clean and disinfected with formaldehyde, copper sulfate, or other disinfectant before use. The crop should be stored and shipped at a temperature low enough to slow down the development of infections and the physiological breakdown of the tissues but not so low as to cause chilling injuries, which then serve as ports of entry for fungi. The crop should be free of surface moisture when placed in storage and there should be adequate ventilation in storage to prevent excessively high relative humidity from building up and condensing on the fruit surface. Packaging in plastic bags should be avoided. The crop should be free of insects and other pests when placed in storage and should be kept free of them while in storage to avoid the creation of new wounds and the development of new infections. Some crops, such as sweet potatoes and onions, can be protected from some decay fungi by curing at 28 to 32°C for 10 to 14 days, which helps reduce surface moisture and heal any exposed wounds by suberization or wound periderm formation. Hot-air or hot-water treatment is sometimes used to eradicate incipient infections at the surface of some fruits.

Controlled atmosphere storage and transport employing low oxygen (5%) or increased carbon dioxide levels (5–20%) have been used to suppress respiration of both the host and the pathogen, thereby suppressing development of postharvest rots. These results are further improved by the addition of 10% carbon monoxide. Biological controls employing antagonistic microorganisms or antimicrobial metabolites isolated from such microorganisms, have been developed that are effective against some fungal and bacterial pathogens of postharvest diseases, and some of them are about to be used commercially while many more are still in the experimental stage. Gamma rays were shown to reduce storage rots of some crops. Also, spraying some crops or infiltrating some fruit with calcium chloride seems to reduce development of postharvest infections in the fruit.

Finally, postharvest decays can be controlled by the use of chemical treatments to prevent infection and suppress the development of pathogens on the surface of the diseased host. The chemicals used most commonly for such treatments include biphenyl, sodium *o*-phenylphenate, dichloran, 2-aminobutane, thiabendazole, benomyl, thiophanate-methyl, imazalil, chlorothalonil, cytovirin, triforine, captan, iprodione, vinclozolin, soda ash, and borax. They are usually applied as fungicidal wash treatments and are more effective when used hot, i.e., at temperatures between 28 and 50°C, depending on the susceptibility of the crop to injury from heat. In some crops, postharvest diseases are controlled by periodic fumigations with sulfur dioxide. Some fungicides, such as dichloran, biphenyl, acetaldehyde vapors, and some ammonia-emitting or nitrogen trichloride-forming chemicals, are used as supplementary, volatile in-package fungistats impregnated in paper sheets during storage and transport. Fungal strains resistant to one or more of the systemic fungicides are common, and precautions must be taken to include additional agents, preferably broad-spectrum fungicides, in control programs.

POSTHARVEST DECAYS OF GRAIN AND LEGUME SEEDS

Although several Ascomycetes and mitosporic fungi such as *Alternaria*, *Cladosporium*, *Colletotrichum*, *Diplodia*, *Fusarium*, and *Cochliobolus* attack grains and legumes in the field, they require too high a moisture content in the seed (24–25%) in order to grow and are, therefore, unable to grow much in grains after harvest, as grains are usually stored at a moisture content of 12 to 14%. Such fungi apparently die out after a few months in storage or are so weakened that they cannot infect new seeds; however, by then they may have had time to discolor seeds, kill ovules, weaken or kill the embryos, or cause shriveling of seeds, and they may have produced mycotoxins, i.e., fungal compounds toxic to humans and animals.

Most of the decay or deterioration of grains and legumes after harvest, i.e., during storage or transit, is caused by several species of the fungus *Aspergillus* (Fig. 11-125A). Sometimes *Penicillium* (Fig. 11-125C) infection occurs in grains or legumes stored at low temperatures at slightly above normal moisture content. *Aspergillus*, however, particularly *A. flavus*, often infects corn kernels and groundnuts while still in the field, and its incidence in the field is increased by damage to kernels by insects or other agents, by stalk rots, drought, severe leaf damage, or lodging, and by other stresses on the plant.

Each of the various species or groups of species of *Aspergillus* responsible for seed deterioration has rather definite lower limits of seed moisture content below which it will not grow. Each also has less well-defined optimum and upper limits of seed moisture content. These, however, especially the upper limit, are determined mostly by competition with associated species whose requirement for optimum moisture content coincides with the upper limit at which the former species can survive. Because of competition with field fungi or for other unknown reasons, storage fungi do not invade grains to any appreciable extent before harvest.

Aspergillus and several of the fungi that attack grains in the field often invade the embryos of seeds and cause a marked decrease in germination percentage of infected seeds used for planting or in malting barley for beer. Field and storage fungi also discolor the embryos and the seeds they kill or damage, which reduces the grade and price at which the grain can be sold. Flour containing more than 20% discolored kernels yields bread of smaller loaf volume and with an off flavor. In many cases, nearly 100% of the embryos of wheat may be infected with *Aspergillus* without yet showing discoloration. Such wheat is used routinely and unknowingly to make bread, but whether such grain ever poses a health hazard is not known. Infection of grains, hay, feeds, and cotton stored in bulk or during long shipping results in increased growth and respiration of the fungi, which causes varying degrees of heating of the material. Respiration also releases moisture, which raises the moisture in adjacent grain. Although not all spoilage of stored grains results in drastic or even detectable heating, in some materials heating from spoilage may raise the temperature up to 70°C or higher. Fungi operate at the lower moisture contents where no free water is available, and bacteria are active at the higher moisture contents.

MYCOTOXINS AND MYCOTOXICOSES

One of the more important effects of postharvest decays of fruits and vegetables, especially of seeds (Fig. 11-112), is the induction of diseases of animals and humans caused by the consumption of feeds and foods invaded by certain common fungi. These fungi produce toxic substances called mycotoxins. The diseases they cause are called **mycotoxicoses.** Ergotism (St. Anthony's fire) of humans and animals, caused by eating ergot-containing wheat and rye bread and feeds, and poisoning of humans from eating poisonous mushrooms, are classic examples of mycotoxicoses and have been known for a long time. The magnitude of the mycotoxin problem began to be appreciated during World War II, when it was noted that the consumption of moldy grain led to necroses of the skin, hemorrhage, liver and kidney failure, and death in numerous humans and animals. Similar symptoms also appeared in horses fed moldy hay. In 1960, a large number of young turkeys died in England after they were fed contaminated peanut feed. That led to intensive research on mycotoxins, which established that they are a global problem. Mycotoxins pose an ever-present threat to the health of humans and animals. When they are present in relatively high concentrations they cause acute disease symptoms. Perhaps even more serious are the chronic effects on health and productivity caused by the constant presence of subacute dosages of mycotoxins in the food and feed consumed throughout the world, particularly in developing countries.

Most mycotoxicoses are caused by the common and widespread fungi *Aspergillus*, *Penicillium*, and *Fusarium*. Some may result in severe illness and death. *Aspergillus* and *Penicillium* produce their toxins mostly in stored seeds and hay, but also on commercially processed foods and feeds, including meats, cheeses, and spices. Infection of seeds usually takes place in the field. *Fusarium* produces its toxins primarily on corn and other grains infected in the field or after they are stored. Many other common fungi that infect agricultural commodities or contaminate food produce several mycotoxins.

Mycotoxins differ in their chemical formula, in the products in, and conditions under which they are produced, in their effects on various animals and humans, and in their degree of toxicity. Several different fungi, however, produce some of the same or closely related toxins. The main, but not all, mycotoxins and some of their properties are listed here.

ASPERGILLUS TOXINS — AFLATOXINS

Aflatoxins are produced by *Aspergillus flavus* and several other species of *Aspergillus*. Aflatoxins are produced in infected cereal seeds and most legumes, but they often reach a rather low and probably nontoxic concentration (about 50 ppb). During some years, a rather high percentage (30% or more) of the corn harvest over large areas contains more than 100 ppb aflatoxin, which is five times that allowed in food for humans and in feed for sensitive animals such as chickens. However, in peanuts, cottonseed, fishmeal, Brazil nuts, and probably other seeds or nuts grown in warm and humid regions, aflatoxin is produced at high concentrations (up to 1000 ppb or more) and causes mostly chronic or occasionally acute mycotoxicoses in humans and domestic animals. Aflatoxins exist in a variety of derivatives with varying effects. Some of these toxins, when ingested with the feed by dairy cattle, are excreted in the milk in still toxic form.

The symptoms of mycotoxicoses caused by aflatoxin in animals, and presumably humans, vary widely with the particular toxin and animal species, dosage, age of the animal, and so on. Young ducklings and turkeys fed high dosages of aflatoxin become severely ill and die. Pregnant cows, calves, fattening pigs, mature cattle, and sheep fed low dosages of aflatoxin over long periods develop weakening, intestinal bleeding, debilitation, reduced growth, nausea, refusal of feed, predisposition to other infectious diseases, and may abort. Moreover, most of the ingested aflatoxin is taken up by the liver, and, in some experiments, animals given feed containing even less than the permissible amount of aflatoxin (20 ppb) almost invariably developed liver cancer.

FUSARIUM TOXINS

Three groups of toxins, zearalenones, trichothecenes, and fumonisins, are produced by several species of *Fusarium*, primarily in moldy corn.

Deoxynivalenol, also known as **vomitoxin** or **DON** is produced by the fungus *Gibberella zeae* (anamorph *Fusarium graminearum*), the cause of Gibberella ear rot of corn and of head blight (scab) of wheat. The mycotoxin at first causes reduced feeding by the animals and, thereby, slower gain or loss of weight. At higher concentrations of the mycotoxin, the animals are induced to vomit and totally refuse to eat.

Zearalenones seem to be most toxic to swine, in which they cause abnormalities and degeneration of the reproductive system, the so-called estrogenic syndrome. Female swine fed zearalenone-containing feed develop swollen vulvas bearing bleeding lesions and atrophying, nonfunctioning ovaries. They are susceptible to abortion, and piglets that are born are small and weak. Male swine show signs of feminization, namely atrophy of the testes and enlargement of the mammary glands.

Fumonisins are produced by ***Fusarium moniliforme***, which causes Fusarium ear rot of corn that affects as much as 90% of the corn fields. Fumonisins are the cause of blind staggers (equine leukoencephalomalacia) in horses, donkeys and mules, pulmonary edema in swine, and, possibly, cancer in humans.

Trichothecins (or **trichothecenes**), of which there are more than 100, are produced by species of *Fusarium* and by several other fungi. They are most toxic when fed to swine, in which they cause, among other symptoms, listlessness or inactivity, degeneration of the cells of the bone marrow, lymph nodes, and intestines, diarrhea, bleeding, and death. Other animals, however, such as cows, chicks, and lambs, are also affected.

Other Aspergillus Toxins and Penicillium Toxins

In addition to aflatoxins, species of *Aspergillus* also produce other toxins in infected grains. The same or similar toxins are also produced in grains infected by species of *Penicillium*. The most important such toxins are **ochratoxins**, which cause degeneration and necrosis of the liver and kidney, along with several other symptoms, in domestic animals. Some ochratoxins can persist in the meat of animals fed contaminated feed and can be transmitted to humans through the food chain. **Yellowed-rice toxins**, primarily citreoviridin, citrinin, and luteoskyrin, are all produced by species of *Penicillium* growing in stored rice, barley, corn, and dried fish. They cause toxicoses associated with various diseases, nervous and circulatory disorders, and degeneration of the kidneys and liver.

Tremorgenic toxins cause marked body tremors and excessive discharge of urine, followed by convulsive seizures that often end in death. They are produced by species of both *Aspergillus* and *Penicillium* infecting foodstuffs in storage and also in refrigerated foods, grains, and cereal products. **Patulin** is produced by *Penicillium* and *Aspergillus*. It causes edema and bleeding in lungs and brain, damage to kidneys, and paralysis of motor nerves and it also induces cancer in higher organisms. It is commonly found to occur naturally in foodstuffs such as fruit or juices made with fruit partly infected with *Penicillium*, in naturally molded bread and bakery products, and in most commercial apple products. Thus, patulin may constitute a serious health hazard for humans as well as for animals.

Ergotism is the oldest known mycotoxicosis. It is caused by several toxic substances contained in the sclerotia (ergots) of the ergot fungus (*Claviceps*) when they contaminate grain crops, such as rye, barley, sorghum, millet, wheat, and wild grasses, and are ingested by humans and animals. Ergotism is expressed as convulsions and limb swellings, followed by gangrene of body extremities and of burning sensations (St. Anthony's fire). Ground-up ergots have been used in the past to stop heavy bleeding, as happens, e.g., during labor or accidents.

Fescue toxicosis affects cattle and horses feeding on plants of the perennial grass tall fescue infected systemically with the fungus *Acremonium*. The fungus is an endophyte growing internally through the plant without invading its cells. The fungus actually seems to make the infected plants more resistant to stress, particularly drought. Horses eating tall fescue plants infected with the fungus show only reproductive disorders. Cattle feeding on such plants, in addition to reduced calving and lower milk production, show reduced weight gains, elevated body temperature, and rough hair coat; moreover, as in ergotism, feet or other body extremities may develop gangrene and drop off ("fescue foot").

CONTROL OF POSTHARVEST GRAIN DECAYS

The control of postharvest deterioration and spoilage by fungi of grains, legumes, fodder, and commercial feeds depends on certain precautions and conditions that must be met before and during harvest and then during storage. Provided that the crop was healthy and of high quality when harvested, its subsequent infection and spoilage in storage will be avoided if several steps are taken. (1) The moisture content is kept at levels below the minimum required for the growth of the common storage fungi. Some hardy *Aspergillus* species will grow and cause spoilage of starchy cereal seeds with a moisture content as low as 13.0 to 13.2% and of soybeans with a moisture content of about 11.5 to 11.8%. Others require a minimum moisture of 14% or more to cause spoilage. (2) The temperature of stored grain is kept as low as possible, as most storage fungi grow most rapidly at temperatures between 30 and 55°C, they grow very slowly at 12 to 15°C, and their growth almost ceases at 5 to 8°C. Low temperature also slows down the respiration of grain and prevents an increase of moisture in grain. (3) Infestation of stored products by insects and mites is kept to a minimum through the use of fumigants. This helps keep the storage fungi from getting started and growing rapidly. (4) The stored grain should not be unripe or too old; it should be clean, have good germinability, and be free of mechanical damage and broken seeds. Such grain resists infection by storage fungi that otherwise could invade weakened or cracked grain.

In addition to starting with good sound crops free of insects, or fumigating to eliminate the insects, the simplest and most common solution to maintaining grain free of storage fungi is through quick air drying and through the use of aeration systems in storage bins in which air is moved through the grain at relatively low rates of flow. The airflow removes excess moisture and heat. The flow can be regulated so that it brings the moisture content of the grain mass to the desired level and reduces the temperature to 8 to 10°C, at which insects and mites are inactive and storage fungi are almost dormant. It has been shown in recent years that certain agents, such as hydrated sodium calcium aluminosilicate, bind to mycotoxins, and if such agents are added to moldy corn before it is fed to swine the effects of any mycotoxins present are reduced considerably.

Selected References

Anonymous (1983). Symposium on deterioration mechanisms in seeds. *Phytopathology* **73**, 313–339.

Anonymous (1989). Colloquium on management of disease resistance in harvested fruits and vegetables. *Phytopathology* **79**, 1393–1390.

Barkai-Golan, R., and Phillips, D. J. (1991). Postharvest heat treatment of fresh fruits and vegetables for decay control. *Plant Dis.* **75**, 1085–1089.

Boyd, A. E. W. (1972). Potato storage diseases. *Rev. Plant Pathol.* **51**, 297–321.

CAST (1989). "Mycotoxins: Economics and Health Risks." Task Force Rep. 116. Council Agric. Sci. and Technology, Ames, IA.

Ceponis, M. J., and Butterfield, J. E. (1974). Market losses in Florida cucumbers and bell peppers in metropolitan New York. *Plant Dis. Rep.* **58**, 558–560.

Christensen, C. M. (1975). "Molds, Mushrooms, and Mycotoxins." Univ. of Minnesota Press, Minneapolis.

Coursey, D. G., and Booth, R. H. (1972). The post-harvest phytopathology of perishable tropical produce. *Rev. Plant Pathol.* **51**, 751–765.

Dennis, C., ed. (1983). "Post-Harvest Pathology of Fruits and Vegetables." Academic Press, New York.

Diener, U. L., *et al.* (1987). Epidemiology of aflatoxin production by *Aspergillus flavus*. *Annu. Rev. Phytopathol.* **25**, 249–270.

Eckert, J. W., and Ogawa, J. M. (1985). The chemical control of postharvest diseases: Subtropical and tropical fruits. *Annu. Rev. Phytopathol.* **23**, 421–454.

Food and Agriculture Organization (1981). Food loss prevention in perishable crops. *Agric. Serv. Bull. (F. A. O.)* **43**, 1–72.

Harmon, G. E., and Pfleger, F. L. (1974). Pathogenicity and infection sites of *Aspergillus* species in stored seeds. *Phytopathology* **64**, 1339–1344.

Harvey, J. M. (1978). Reduction of losses in fresh market fruits and vegetables. *Annu. Rev. Phytopathol.* **16**, 321–341.

Jones, R. K. (1979). The epidemiology and management of aflatoxins and other mycotoxins. *In* "Plant Disease" (J. G. Horsfall and E. B. Cowling, eds.), Vol. 4, pp. 381–392. Academic Press, New York.

Laidou, I. A., Thanassoulopoulos, C. C., and Liakopoulou-Kyriakides, M. (2001). Diffusion of patulin in the flesh of pears inoculated with four post-harvest pathogens. *J. Phytopathol.* **149**, 547–461.

Marasas, W. F. O., and van Rensburg, S. J. (1979). Mycotoxins and their medical and veterinary effects. *In* "Plant Disease" (J. G. Horsfall and E. B. Cowling, eds.), Vol. 4, pp. 357–379. Academic Press, New York.

McColloch, L. P., Cook, H. T., and Wright, W. R. (1968). Market diseases of tomatoes, peppers, and eggplants. *U.S. Dep. Agric. Agric. Handb.* **28**, 1–74.

Michailides, T. J., and Spotts, R. A. (1990). Postharvest diseases of pome and stone fruits caused by *Mucor pyriformis* in the Pacific Northwest and California. *Plant Dis.* **74**, 537–543.

Moline, H. E., ed. (1984). "Postharvest Pathology of Fruits and Vegetables: Postharvest Losses in Perishable Crops." Univ. of Calif. Publ. NE-87 (U.C. Bull. No. 1914).

Nelson, P. E., Desjardins, A. E., and Plattner, R. D. (1993). Fumonisins, mycotoxins produced by *Fusarium* species: Biology, chemistry and significance. *Annu. Rev. Phytopathol.* **31**, 233–252.

Pierson, C. F. (1971). Market diseases of apples, pears and quinces. *U.S. Dep. Agric. Agric. Handb.* **376**, 1–112.

Sauer, D. B., Storey, C. L., and Walker, D. E. (1984). Fungal populations in U.S. farm-stored grain and their relationship to moisture, storage times, regions, and insect infestation. *Phytopathology* **74**, 1050–1053.

Smoot, J. J., Houck, L. G., and Johnson, H. B. (1971). Market diseases of citrus and other subtropical fruits. *U.S. Dep. Agric. Agric. Handb.* **398**, 1–115.

Tuite, J., and Foster, G. H. (1979). Control of storage diseases of grain. *Annu. Rev. Phytopathol.* **17**, 343–366.

Wells, J. M., and Butterfield, J. E. (1999). Incidence of *Salmonella* on fresh fruits and vegetables affected by fungal rots or physical injury. *Plant Dis.* **83**, 722–726.

Wells, J. M., Butterfield, J. E., and Revear, L. G. (1993). Identification of bacteria associated with postharvest diseases of fruits and vegetables by cellular fatty acid composition: An expert system for personal computers. *Phytopathology* **83**, 445–455.

DISEASES CAUSED BY BASIDIOMYCETES

Basidiomycetes are fungi that produce their sexual spores, called **basidiospores,** on a club-shaped spore-producing structure called a **basidium** (Figs. 11-127 and 11-128). Most Basidiomycetes are fleshy fungi, such as the common mushrooms, the puffballs, and the shelf fungi or conks, and are either saprophytes or cause wood decay, including root and stem rots of trees (Figs. 11-128 and 11-129). Basidiomycetes, however, also include two very common and very destructive groups of plant pathogenic fungi that cause the rust and the smut diseases of plants (Figs. 11-127 and 11-129).

RUSTS

Plant rusts, caused by Basidiomycetes of the order Uredinales, are among the most destructive plant diseases. They have caused famines and ruined the economies of large areas, including entire countries. They have been most notorious for their destructiveness on grain crops, especially wheat, oats, and barley, but they also attack vegetables such as bean and asparagus, field crops such as cotton and soybeans, and ornamentals such as carnation, chrysanthemum, and snapdragon, and have caused tremendous losses on trees such as pine, apple, and coffee.

Rust fungi attack mostly leaves and stems. Rust infections usually appear as numerous rusty, orange, yellow, or even white-colored spots that rupture the epidermis. Some form swellings and even galls. Most rust infections are strictly local spots, but some may become systemic. There are about 5,000 species of rust fungi. The most important rust fungi and the diseases they cause are listed here (Figs. 11-127 and 11-129).

Puccinia, causing severe and often catastrophic diseases on numerous hosts such as the stem rust of wheat and all other small grains (*P. graminis*); yellow or stripe rust of wheat, barley, and rye (*P. striiformis*); leaf or brown rust of wheat and rye (*P. triticina*); leaf rust of barley (*P. hordei*); crown rust of oats (*P. coronata*); corn rust (*P. sorghi*); southern corn rust (*P. polysora*); sorghum rust (*P. purpurea*); and sugarcane rusts (*P. sacchari* and *P. melanocephala*). *Puccinia* also causes severe rust diseases on field crops such as cotton (*P. stakmani*);

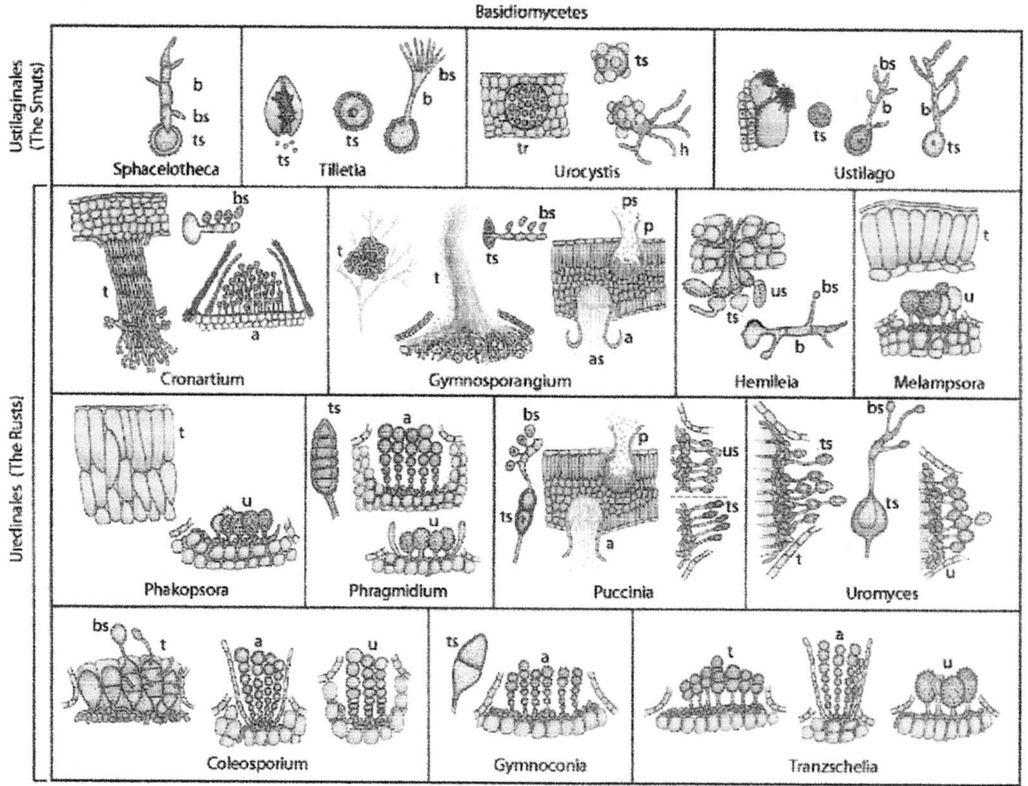

FIGURE 11-127 Basidiomycetes: some common smut and rust fungi. A, aecium; as, aeciospore; b, basidium; bs, basidiospore; h, hypha; sg, spermagonium; s, spermatium; t, telium; tr, teliosorus; ts, teliospore; u, uredium; us, uredospore.

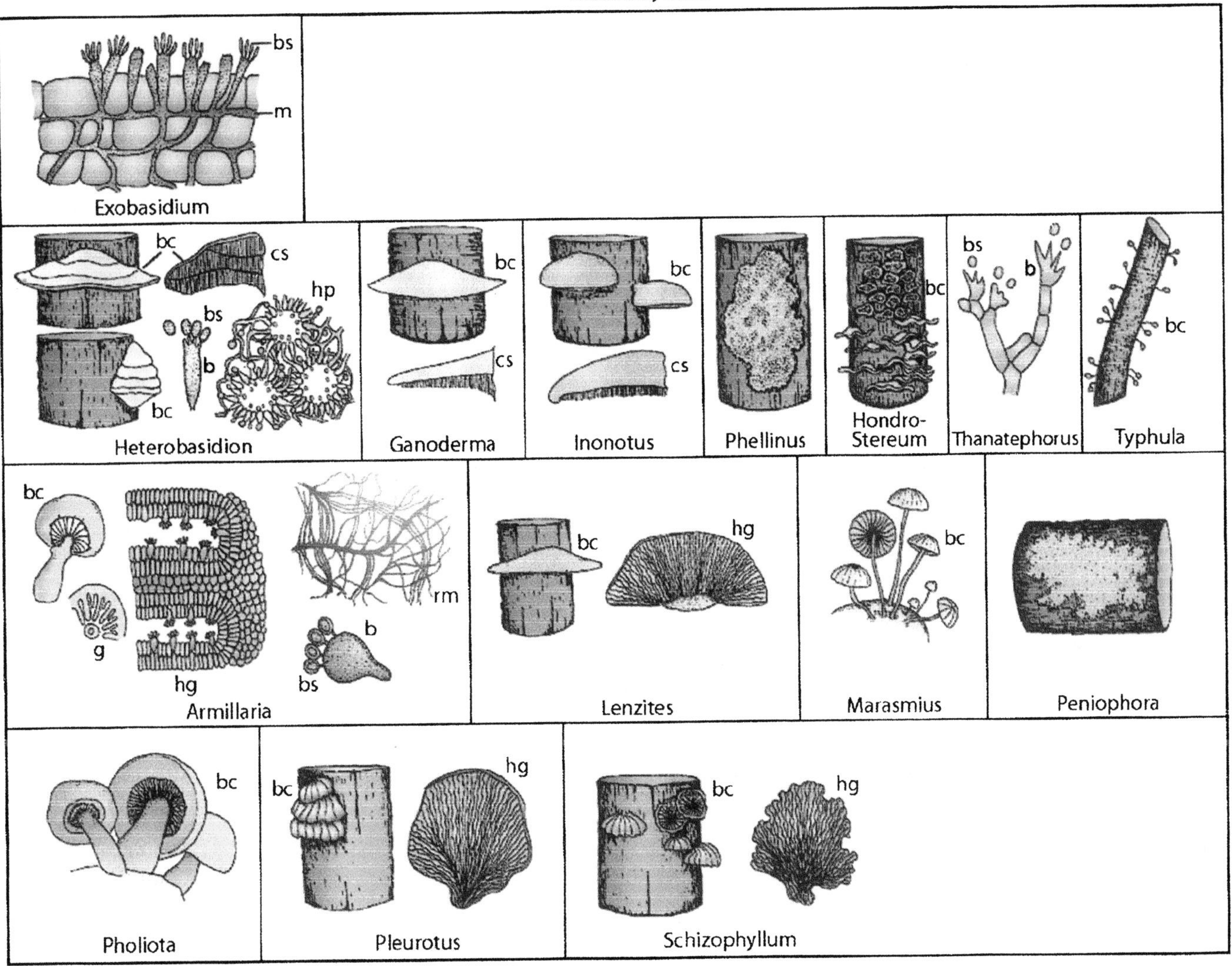

FIGURE 11-128 Basidiomycetes; some of the conk- and mushroom-forming plant pathogens. b, basidium; bc, basidiocarp; bs, basidiospore, cross section; g, gill; hg, hymenial gills; hp, hymenial pores; m, mycelium; rm, rhizomycelium.

vegetables such as asparagus (*P. asparagi*); and flowers such as chrysanthemum (*P. chrysanthemi*), hollyhock (*P. malvacearum*), and snapdragon (*P. antirhini*)

Gymnosporangium, causing cedar-apple rust (*G. juniperi-virginianae*)

Hemileia, causing the devastating coffee leaf rust (*H. vastatrix*)

Phragmidium, causing rust on roses and yellow rust on raspberry (*P. rubi-idaei*)

Uromyces, causing the rusts of legumes (*U. appendiculatus*) and of carnation (*U. caryophyllinus*)

Cronartium, causing white pine blister rust (*C. ribicola*) and fusiform rust of pines and oaks (*C. quercuum* f. sp. *fusiforme*)

Peridermium, causing western gall rust in pine (*P. harknessi*)

Melampsora, causing rust of flax (*M. lini*)

Coleosporium, causing blister rust of pine needles (*C. asterinum*)

Gymnoconia, causing orange rust of blackberry and raspberry (*G. nitens*)

Phakopsora, causing the potentially catastrophic soybean rust (*P. pachyrhizi*)

Tranzschelia, causing rust of peach (*T. discolor*)

Most rust fungi are very specialized parasites and attack only certain genera or only certain varieties of plants. Rust fungi that are morphologically identical but attack different host genera are regarded as special

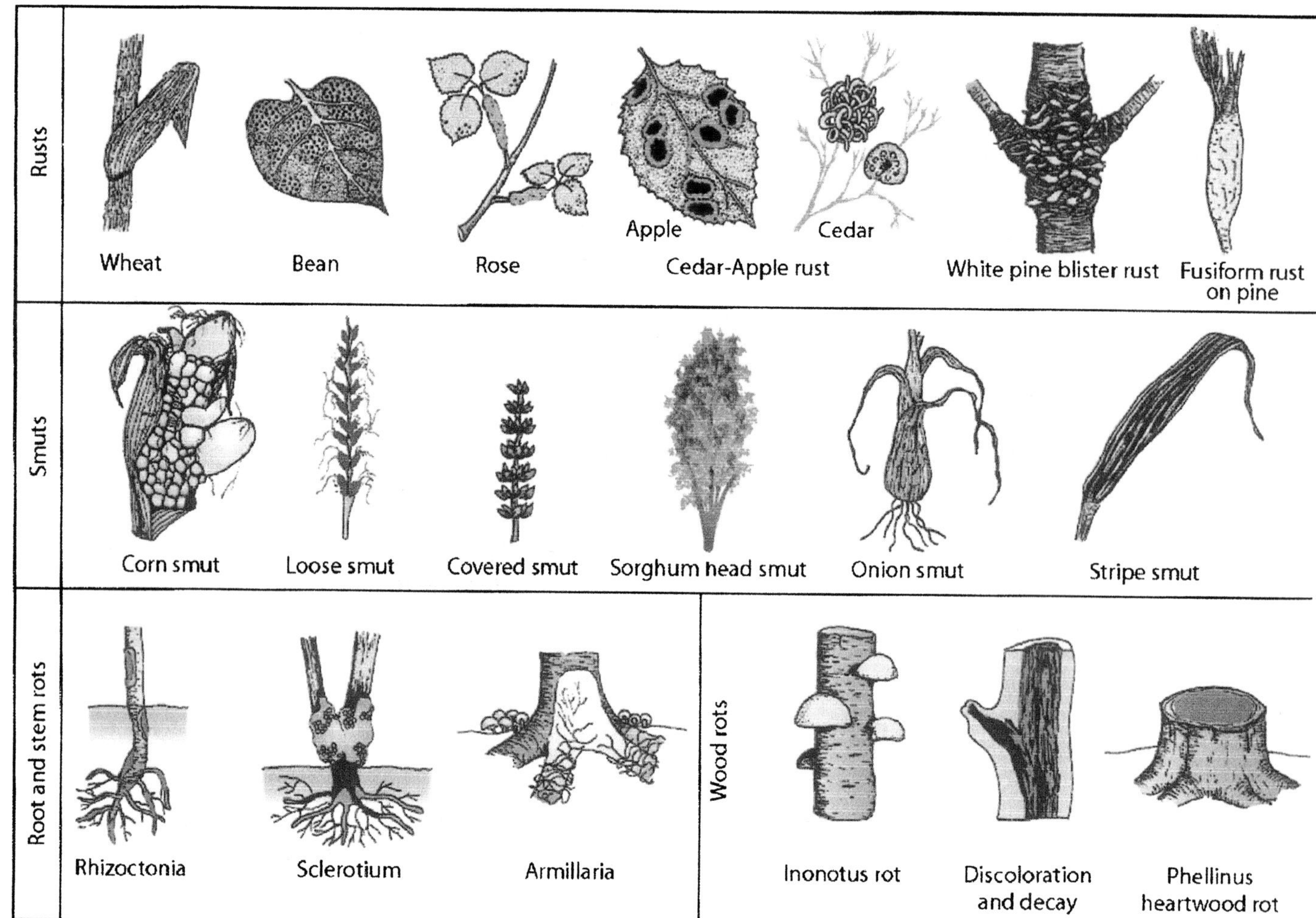

FIGURE 11-129 Common symptoms caused by Basidiomycetes.

forms (*formae specialis*), e.g., *Puccinia graminis* f. sp. *tritici* on wheat and *P. graminis* f. sp. *hordei* on barley. Within each special form of a rust there are many so-called pathogenic (physiological) races. These can attack only certain varieties within the species and can be detected and identified only by the set of differential varieties they can infect. Where sexual reproduction of the rust fungus is rare, the races are more stable over fairly long periods of time, but even so some of these fungi have as many races as those in which sexual reproduction is common.

Rust fungi are obligate parasites in nature, but some of them have now been grown on special culture media in the laboratory. Most rust fungi produce five distinct fruiting structures with five different spore forms that appear in a definite sequence (Fig. 11-130). Some of the spore stages infect one host while the others must infect and parasitize a different, alternate host. All rust fungi produce teliospores and basidiospores. Rusts caused by fungi that produce only teliospores and basidiospores are called **microcylic** or **short cycled.** Other rust fungi produce, in addition to teliospores and basidiospores, spermatia (formerly known as pycniospores), aeciopores, and uredospores (also known as urediospores or urediniospores) in that order. These are called **macrocyclic** or **long-cycled** rusts (Fig. 11-130). In some macrocyclic rusts, spermatia, uredospores, or both may be absent. Although basidiospores are produced on basidia, the other spore forms are produced in specialized fruiting structures called, respectively, spermagonia, aecia, uredia (also known as uredinia), and telia (Figs. 11-127 and 11-130).

Basidiospores, aeciospores, and uredospores can attack and infect host plants. **Teliospores** serve only as the sexual, overwintering stage, which on germination produce the basidium. The basidium, following meiosis,

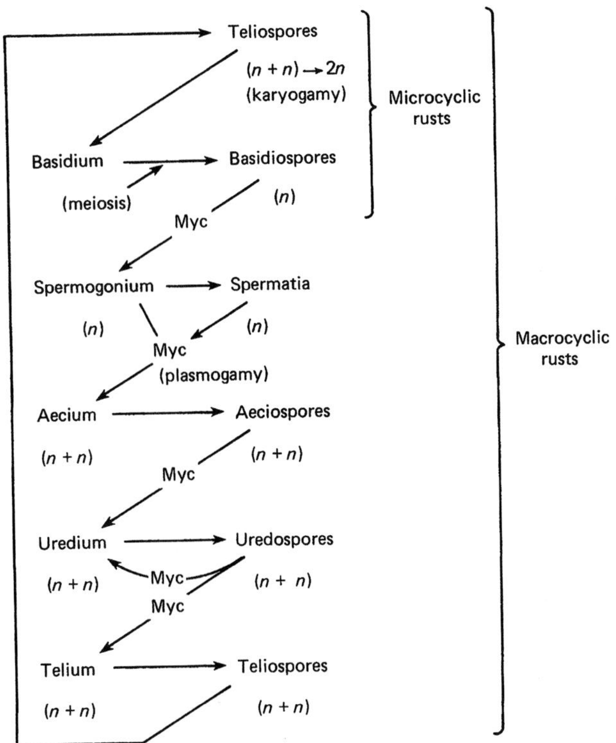

FIGURE 11-130 Kinds and sequence of spores and spore-producing structures in rust fungi along with the nuclear condition of each. Myc, mycelium.

produces four haploid basidiospores. **Basidiospores**, on infection, produce haploid mycelium that forms **spermagonia** (formerly known as pycnia), containing haploid spermatia and receptive hyphae. **Spermatia** act as male gametes and are unable to infect plants; their function is the fertilization of receptive hyphae of the compatible mating type and the subsequent production of **dikaryotic mycelium** and **dikaryotic spores**. This mycelium forms aecia that produce **aeciospores**, which on infection produce more dikaryotic mycelium that this time forms **uredia**. The latter produce **uredospores**, which also infect and produce either more uredia and uredospores or, near host maturity, telia and teliospores. The cycle is thus completed.

Some macrocyclic rusts, e.g., asparagus rust, complete their life cycles on a single host and are called **autoecious**. Others, such as stem rust of cereals, require two different or alternate hosts (e.g., wheat and barberry) for completion of their full life cycle and are called **heteroecious**.

Rust fungi spread from plant to plant mostly by windblown spores, although insects, rain, and animals may play a role. Some of their spores (uredospores) are transported over long distances (several hundred kilometers) by strong winds and, on landing (being scrubbed from the air by rain), can start new infections.

The control of rust diseases in some crops, such as grains, is achieved by means of resistant varieties. In some vegetable, ornamental, and fruit tree rusts, such as cedar-apple rust, the disease is controlled with chemical sprays. In others, e.g., white pine blister rust, control has been attempted through removal of the alternate host (*Ribes* spp.) and avoidance of high rust-hazard zones. With the discovery of several new systemic fungicides effective against rusts, such as triadimefon, triarimol, and fenapanil, the control of rust diseases of annual plants as well as trees is possible with these chemicals applied as sprays, seed dressings, soil drenches, or injections. Since the early 1990s, biological control of rust diseases has been obtained experimentally by the application of antagonistic fungi and bacteria on the surface of the plants or by prior systemic inoculation of the plants with certain viruses that make the plants more resistant to rust infection. The possibility that rust diseases will be controlled in the field with any of these biological controls, however, seems quite remote at present.

CEREAL RUSTS

Various species or special forms of *Puccinia* attack all cultivated and wild grasses, including all small grains, corn, and sugarcane. They are among the most serious diseases of cultivated plants, resulting in losses equivalent to about 10% of the world grain crop per year. Rusts may debilitate and kill young plants, but more often they reduce foliage, root growth, and yield by reducing the rate of photosynthesis, increasing the rate of respiration, and decreasing translocation of photosynthates from infected tissue, instead diverting materials into the infected tissue. The quantity of grain produced by rusted plants may be reduced greatly, and the grain produced may be of extremely poor quality, as it may be devoid of starch and may consist mostly of cellulosic materials that are of low or no nutritional value to humans. One of the most important cereal rusts is discussed here.

STEM RUST OF WHEAT AND OTHER CEREALS

Stem rust of wheat occurs worldwide and affects wheat wherever it is grown. Similar rusts affect other cultivated cereals and probably most wild grass genera and species.

The stem rust fungus attacks all the aboveground parts of the wheat plant (Figs. 11-131A and 11-131B). Infected plants usually produce fewer tillers and set fewer seeds per head, and the kernels are smaller in size, generally shriveled, and of poor milling quality and food

FIGURE 11-131 Stem rust and other rusts of wheat. (A) Uredia of wheat leaf rust almost completely covering the leaf. (B) Uredia of stem rust of wheat on leaves and stems of wheat. (C) Smaller heads (left) produced by rust-infected wheat plants resulting in smaller kernels of lower quality (D, left), compared to kernels of healthy plants (D, right). (E) Wheat stems showing numerous black telia that weaken the stems and result in lodging of plants (F). [Photographs courtesy of (A) L.D. Duczek, WCPD, and (B–F) USDA.]

value (Figs. 11-131C and 11-131D). Plants with heavily infected stems cannot support themselves and lodge (Figs. 11-131E and 11-131F). Under extreme situations, heavily infected plants may die. Heavy seedling infection of winter wheat may weaken the plants and make them susceptible to winter injury and to attack by other pathogens. The amount of losses caused by stem rust may vary from slight to complete destruction of wheat fields over large areas, sometimes encompassing several states. More than 1 million metric tons of wheat is lost to stem rust in North America annually, and during years of severe stem rust epidemics the losses are in the tens or hundreds of millions of tons. Losses from stem rust are at least as severe, and generally much more severe, in many other wheat-growing countries, particularly developing ones.

Symptoms. The pathogen causing stem rust of wheat attacks and produces symptoms on wheat and related cereals (barley, oats, rye) and grasses and on plants of common barberry (*Berberis vulgaris*) and certain other related species.

The symptoms on wheat appear as elliptical blisters or pustules, known as uredia, that develop parallel with the long axis of the stem, leaf, or leaf sheath (Figs. 11-131A, 11-131B, 11-131E, 11-132A, and 11-133E). Blisters may also appear on the neck and glumes of the wheat spike. The epidermis covering the pustules is later ruptured irregularly and pushed back, revealing a powdery mass of brick red-colored uredospores. The uredia vary in size from 1 to 3 millimeters wide by 10 millimeters long. Later in the season, as the plant approaches maturity, the pustules turn black as the fungus produces teliospores (Figs. 11-131B, 11-131E, and 132B) instead of uredospores and uredia are transformed into black telia. Sometimes telia may develop independently of uredia. Uredia and telia may exist on wheat plants in such great numbers that large parts of the plant appear to be covered with the ruptured areas, which are filled with the rust-red uredospores, the black teliospores, or both (Figs. 11-131A, 11-131B, 11-131E, and 11-132B).

On barberry (Fig. 11-132E), the symptoms appear as yellowish to orange-colored spots (Figs. 11-132F and 11-133A) primarily on the leaves. Within the spots, and in leaves generally on the upper side, appear a few minute, orange-colored bodies, the spermagonia (Figs. 11-132F and 11-132G), usually bearing a small droplet of liquid or nectar. Beneath the spermagonia, and occasionally next to them, groups of orange-yellow horn- or cup-like aecia appear (Figs. 11-133A and 11-133B). The infected host tissue is frequently swollen. The torn, whitish aecial wall usually protrudes at the margin of the aecia.

The Pathogen. Puccinia graminis. Puccinia graminis is a macrocyclic, heteroecious rust fungus producing spermagonia and aecia on barberry and uredia and telia on wheat and other cereals and grasses.

Development of Disease. In cooler regions the fungus overwinters as teliospores on infected wheat debris (Fig. 11-134). Teliospores germinate in the spring and produce a basidium on which form four basidiospores (Fig. 11-132C). The basidiospores are ejected forcefully into the air and are carried by air currents for a few hundred meters. Basidiospores landing on young barberry leaves germinate (Fig. 11-132D) and penetrate the epidermal cells. After that, the mycelium grows mostly intercellularly. Within 3 or 4 days the mycelium develops into a spermagonium (Figs. 11-132F, 11-132G, and 11-134), which ruptures the epidermis, and its opening emerges on the surface of the plant tissue. Receptive hyphae from the spermagonium extend beyond the opening, and spermatia embedded in a sticky liquid are exuded through the opening. Visiting insects become smeared with spermatia and carry them to other spermagonia. Spermatia may also be spread by rainwater or dew running off the plant surface. When a spermatium comes in contact with a receptive hypha of a compatible spermagonium, fertilization takes place. The nucleus of the spermatium passes into the receptive hypha, but it does not fuse with the nucleus already present in the latter. Instead, it migrates through the cells of the monokaryotic mycelium, dividing as it progresses to the aecial mother cells. Thus, the dikaryotic condition is reestablished, and mycelium and aeciospores formed subsequently are dikaryotic. This mycelium then grows intercellularly toward the lower side of the leaf, where it forms thick mycelial mats that develop into aecia. In the meantime, host cells surrounding the mycelium are stimulated to enlarge; this, along with the increased volume of the fungus, results in a swelling of the infected area on the lower surface of the leaf.

The aecia (Figs. 11-133A and 11-133B) form in groups and protrude considerably beyond the surface of the barberry plant. The aeciospores are produced in chains inside the aecium, and each spore contains two separate nuclei of opposite mating type. Aeciospores are released in late spring and are carried by wind to nearby wheat plants (Fig. 11-133D) on which they germinate and infect wheat stems, leaves, or sheaths through stomata. After the mycelium grows intercellularly for a while, it then grows more profusely below the surface of the wheat tissue and forms a mat of mycelium just below the epidermis. Many short sporophores and uredospores are produced that exert pressure on the epidermis, which is pushed outward and forms a uredial

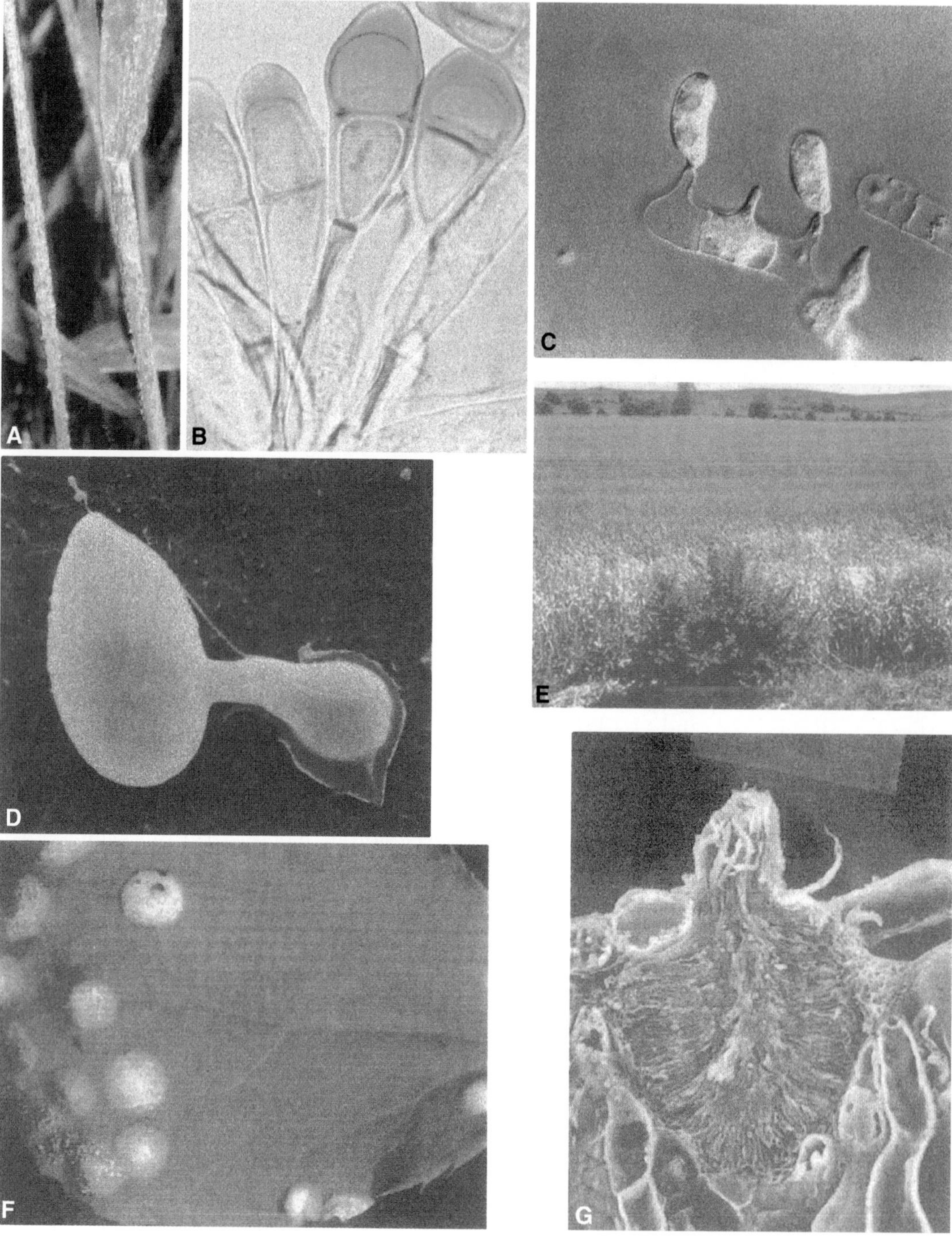

FIGURE 11-132 (A) Uredia and telia on stem rust-infected wheat plant. (B)Teliospores of stem rust of wheat. (C) Basidium with two of the four basidiospores produced by a teliospore. (D) A germinating basidiospore with a swollen appressorium covered with extracellular material. (E) Early infection of wheat plants growing next to a bush of barberry, the alternate host of the wheat stem rust fungus. (F) Spermagonia of the stem rust fungus produced on the upper surface of a barberry leaf. (G) Cross section of spermagonium of a rust fungus. [Photographs courtesy of (A) WCPD, (B) J.F. Hennen, (C, D, and G) C.W. Mims, University of Georgia, (E) Cereal Dis. Lab. Archives, and (F) D. L. Long, USDA.]

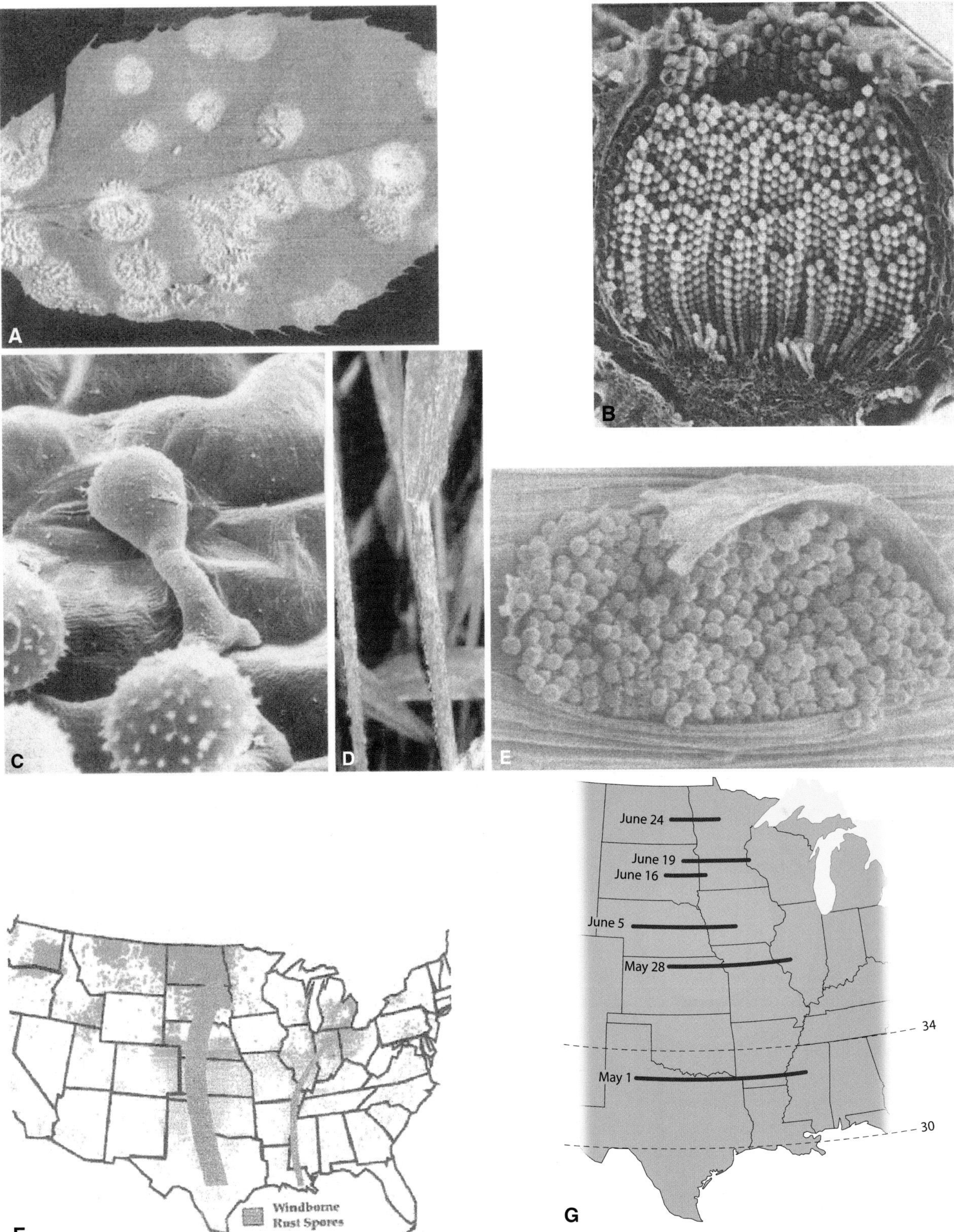

FIGURE 11-133 (A) Groups of aecia produced on the lower side of barberry leaves following fertilization of spermagonia. (B) Scanning electron micrograph of a sectioned aecium showing chains of aeciospores; aeciospores like uredospores (C), infect wheat and cause rust pustules (uredia) (C). Uredia contain uredospores (D and E); these can reinfect wheat (C and D). (E) Scanning electron micrograph of uredium and uredospores of wheat stem rust. (F) Major pathways and distribution of wheat stem rust in the United States. (G) Approximate dates on which uredospores arrive at various northern latitudes from the south. [Photographs courtesy of (A) D.L. Long, USDA, (B) C.M. Mims, University of Georgia, (C) WCPD, (E) M.F. Brown and H.G. Brotzman, and (F) M. E. Hughes and (G) A. P. Roelfs, USDA.]

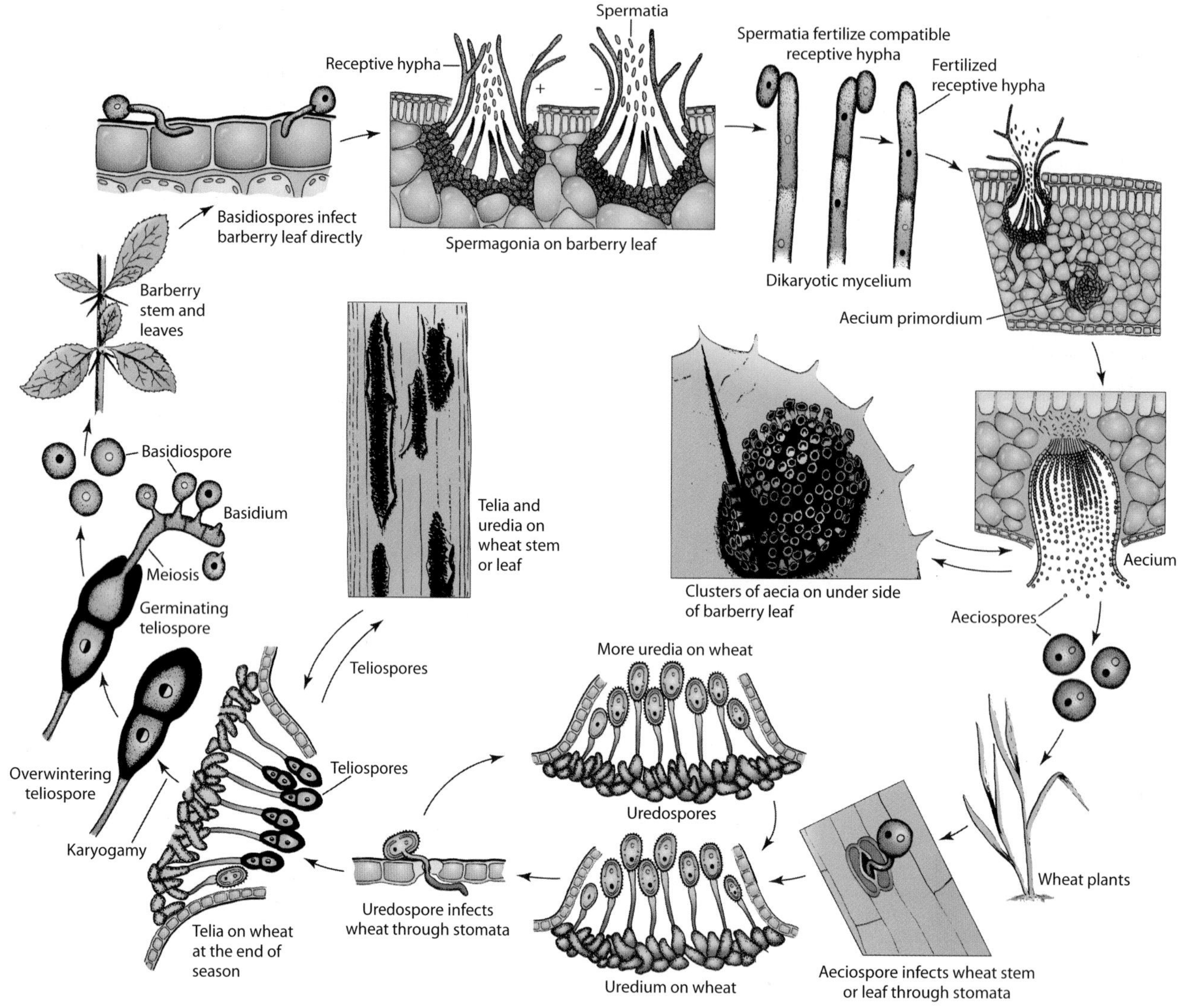

FIGURE 11-134 Disease cycle of stem rust of wheat caused by *Puccinia graminis tritici*.

pustule. Finally, the epidermis is broken irregularly, revealing several hundred thousand rust-colored uredospores, which give a powdery appearance to the uredium (Figs. 11-131A, 11-131B, 11-133C, 11-133E, and 11-134).

The uredospores are easily blown away by air currents, sometimes for many kilometers, even hundreds of kilometers, from the point of their origin (Figs. 11-133F and 11-133G). The uredospores can reinfect wheat plants in the presence of dew, a film of water, or relative humidities near the saturation point. Their germ tubes enter the plant through stomata (Fig. 11-133C). Within 8 to 10 days from inoculation the mycelium produces a new uredium and more uredospores. Uredospores cause new infections on wheat plants up to the time the plant reaches maturity. Most of the damage caused to wheat growth and yield results from such uredospore infections, which may literally cover the stem, leaf, leaf sheaths, and glumes with uredia.

Rust-infected wheat plants show increased water loss because they transpire more and because more water evaporates through the ruptured epidermis. In addition, the fungus itself removes much of the nutrients and water that would normally be used by the plant. The respiration of infected plants increases rapidly during the development of the uredia, but subsequently respiration drops to slightly below normal. Photosynthesis of diseased plants is reduced considerably due to the destruction of much of the photosynthetic area (Figs. 11-131A–11-131C) and to the interference of fungal secretions with the photosynthetic activity of the remaining green areas on the plant. The fungus also seems to interfere with normal root development and uptake of nutrients by the roots. All these effects reduce the normal number and size of seeds on the plant (Figs. 11-131C and 11-131D). The fungus also induces earlier maturity of the plant, resulting in decreased time available for the seed to fill. Heavy rust infections before or at the flowering stage of the plant are extremely damaging and may cause total yield loss (Figs. 11-131C and 11-131D), whereas if heavy infections occur later, the damage to yield is much smaller.

When the wheat plant approaches maturity or when the plant fails because of overwhelming infection, the uredia produce teliospores instead of uredospores or new telia may develop from recent uredospore infections. Teliospores do not germinate immediately and do not infect wheat; rather, they are the overwintering stage of the fungus. Teliospores also serve as the stage in which fusion of the two nuclei takes place and, after meiosis in the basidium, results in the production of new combinations of genetic characters of the fungus through genetic recombination. Several hundred races of the stem rust fungus are known to date, and new ones appear every year.

In southern regions the fungus usually overwinters in fall-sown wheat infected by uredospores produced on the previous year's crop. Heavy rust infections in warm regions in early spring not only cause heavy losses locally, but also produce uredospores that are carried northward by the warm southern winds of spring and summer and initiate infections of wheat in successively northern regions (Figs. 8-18, 11-133F, and 11-133G).

Control. The most effective, and the only practical, means of control of wheat stem rust is through the use of wheat varieties resistant to infection by the pathogen. A tremendous amount of work has been and is being done on the development of wheat varieties resistant to existing races of the fungus. The best varieties of wheat that combine rust resistance and desirable agronomic characteristics are recommended annually by the agricultural experiment stations and change periodically in order to evade the existing rust races. Much effort is now directed toward the development of varieties with general or partial resistance and toward the development of multiline cultivars.

Eradication of barberry has reduced losses from stem rust by eliminating the early season infections on wheat in areas where uredospores cannot overwinter, and by reducing the opportunity for the development of new races of the stem rust fungus through genetic recombination on barberry. This provides for greater stability in the race population of the pathogen and contributes to the success of breeding of resistant varieties.

Several fungicides can effectively control the stem rust of wheat. In most cases, however, 4 to 10 applications per season are required for complete control of the rust; because of the low income return per acre of wheat, such a control program is not economically practical. Two applications of some fungicides, coordinated with forecasts of weather conditions favoring rust epidemics, may reduce damage from stem rust by as much as 75%. These chemicals, which have both protective and eradicative properties, and therefore even two sprays, one at trace to 5% rust prevalence and the second 10 to 14 days later, can give an economically rewarding control of rust.

Certain systemic fungicides also control stem rust when applied as one or two sprays 1 to 3 weeks apart during the early stages of disease development. Seed treatments with some systemic chemicals inhibit early but not late season infections.

Damage by the stem rust fungus is usually lower in fields in which heavy fertilization with nitrate forms of nitrogen and dense seeding have been avoided.

RUSTS OF LEGUMES

Legumes, i.e., beans, broad beans, peanuts, southern peas, chick peas, lentils, and some others, provide a major portion of the food for people in many parts of the world, particularly in Central and South America and in Africa, only second in importance to the food provided by cereals. In addition, several other legumes, such as soybean, are cultivated in huge acreages, especially in the Far East and in the United States, for food, feed, oil, and numerous industrial uses, while alfalfa and various clovers are cultivated or grow naturally and are used as animal feed. All of these crops are attacked by various rust fungi that annually cause varying amounts of losses in yield, in many cases the losses being very severe. The main rust fungi involved and the crops they attack include the following.

Uromyces appendiculatus causes rust in beans
Uromyces striatus causes rust in alfalfa
Uromyces vicia-fabae causes rust in fava beans
Uromyces vignae causes rust of southern peas
Phakopsora pachyrhyzi causes rust in soybeans

BEAN RUST

It occurs worldwide but is more common and severe in humid tropical and subtropical areas. Depending on

FIGURE 11-135 Bean rust caused by *Uromyces appendiculatus*. (A) Uredia of the fungus in the upper and lower sides of bean leaves. (B) Bean plants infected heavily with rust. (C) Bean plants defoliated from severe rust infection. [Photographs courtesy of (A) R.G. Platford, WCPD, (B) H.D. Thurston, Cornell University, and (C) J. R. Steadman, University of Nebraska.]

earliness and severity of infection, it can cause almost total crop loss. Symptoms consist of numerous reddish brown circular pustules consisting of uredia about 1–2 millimeters in diameter that develop on leaves and pods. The pustules burst and release reddish brown uredospores and, later, black teliospores. Often, the tissue surrounding single large or small groups of uredia turns yellow (Figs. 11-135A, 11-135B, and 1-10). Heavily infected leaves may become shredded and parts of them may fall off (Fig. 11-135C), while the plants remain small and produce low yields.

The bean rust pathogen is the basidiomycete *Uromyces appendiculatus*. It has a macrocyclic life cycle, producing all its spore stages, the occasionally found spermagonia and aecia, and the ever-present uredia, telia, and their spores, on the same host, bean. This fungus is extremely variable, consisting of more than 300 races, often several of them found in the same field.

The control of bean rust depends on the use of bean varieties resistant to the existing races of the pathogen in the area where beans will be grown. Several fungicides give satisfactory control of bean rust. Cultural practices such as appropriate crop rotation with non-host crops, elimination of plant debris, and so on help reduce the inoculum and future infections.

BOX 19 Soybean Rust — a Major Threat to a Major Crop

Soybean rust has been known to occur in the Far East from Japan to Australia, in India, parts of Central Africa, in Central and South America, and the islands of the Caribbean Basin. Wherever it occurs it causes severe losses in yield ranging from 10 to 50%, with even higher losses, up to 80%, occurring in the more humid tropical and subtropical regions. A complicating factor for managing this disease is that it has many hosts in addition to soybean. The soybean rust fungus has been found in naturally infected plants of at least 30 species of legumes, including lima beans, cowpeas, clovers, and even perennial plants such as the kudzu vine *Pueraria lobata*. The soybean rust fungus has not yet been found in the continental United States, but in 1995 it was found in Hawaii. Considering that the fungus produces hardy, windborne uredospores, it is considered quite likely that the pathogen will be introduced into the mainland of the United States in the next few years. Soybean rust was indeed discovered in the states of Louisiana, Mississippi, and Florida in the middle of November of 2004. Since the United States annually produces approximately 70 million metric tones of soybeans, or slightly more than half the world production of soybeans, introduction of this pathogen into the United States is a major financial catastrophe to producers and worldwide consumers of soybeans alike. It has been estimated that much of the soybean-producing area in the United States would suffer losses of 10% or more, whereas the southeastern United States, which have more favorable climatic conditions for the disease, would suffer losses of about 50%, bringing the cost of the disease to producers and consumers to approximately $7.2 billion per year.

Soybean rust is caused by the basidiomycete fungus *Phakopsora pachyrhiza* and a similar but less aggressive fungus, *P. meibomiae*. It affects soybeans but also many other legumes. It causes numerous uredial lesions on both sides of leaves (Figs. 11-136A–11-136C), thereby reducing greatly the ability of leaves to carry on photosynthesis and produce high yields of soybeans. Entire fields of soybeans may be destroyed by the rust (Fig. 11-136E). Its life cycle appears to be microcyclic, producing only uredia and telia, and is completed on the same host, soybean or other legumes. Uredia produce uredospores (Fig. 11-136F) that are spread by wind and can cause infection, while the telia produce teliospores, which, however, have never been shown to germinate.

The control of soybean rust is attempted through the use of resistant varieties (Fig. 11-136D), when available, and the use of appropriate fungicides, which must be applied repeatedly for adequate control. In the United States, of course, where the pathogen has been quarantined vigorously in the past, its quarantine will continue.

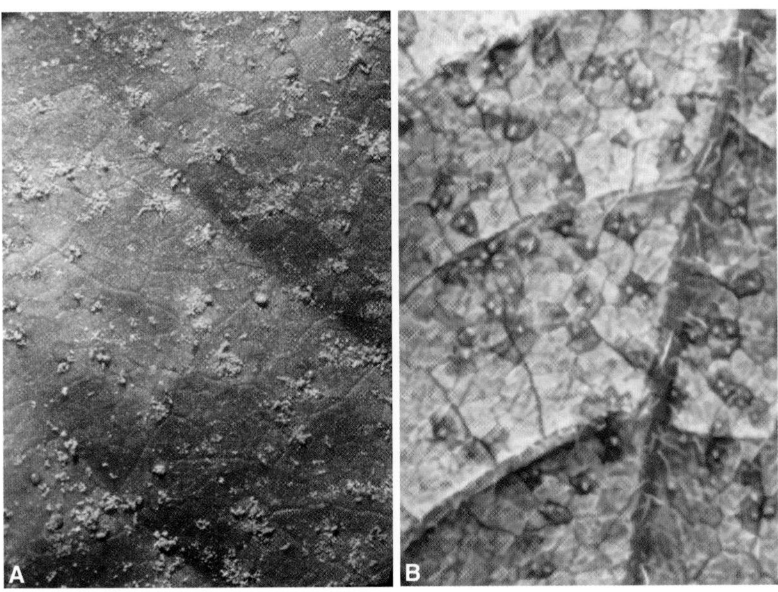

FIGURE 11-136 Soybean rust caused by *Phakopsora pachyrhizi*. Leaves with tan (A) and reddish brown (B) lesions as they appear macroscopically in the field (C). (D) Relative resistance in four soybean varieties to soybean rust. (E) Field with soybean plants infected heavily with rust. (F) Scanning electron micrograph of soybean rust uredium and uredospores. (Photographs courtesy of USDA.)

continued

FIGURE 11-136 *Continued*

Cedar-Apple Rust

Cedar-apple rust is present in North America and in Europe. It causes galls, often called cedar apples (Fig. 11-137A) that produce jelly-like horns on cedar (Fig. 11-137B), and yellow to orange-colored spots on apple leaves and fruit (Figs. 11-137C and 11-137D). It can cause considerable damage to both hosts when they are located near each other. Similar diseases affect hawthorn and quince.

The fungus, *Gymnosporangium juniperi-virginianae*, overwinters as dikaryotic mycelium in the galls on cedar trees. Cedar needles or buds are infected in the summer by windborne aeciospores from apple leaves (Fig. 11-138). The fungus grows little in the cedar needles during fall and winter; the following spring or early summer, however, galls begin to appear as small swellings on the upper surface of the needle. The fungus is present in the galls as mycelium growing between the cells. The galls enlarge rapidly and, by fall, they may be 3 to 5 centimeters in diameter, have turned brown, and become covered on the surface with small circular depressions. The cedar-apple rust fungus does not produce uredia or uredospores. The following spring, however, the small depressions on the galls absorb water during warm, wet weather, swell, and produce very conspicuous orange-

FIGURE 11-137 Cedar apple rust caused by *Gymnosporangium juniperi-virginianae.* (A) "Cedar apples" produced on cedar twigs. (B) Cedar apples develop telial horns that produce and release basidiospores. Basidiospores infect apple leaves (C) and fruit (D) and cause spots to develop. [Photographs courtesy of (A and B) E.L. Barnard, Florida Department of Agriculture, Forestry Division, and (C and D) D.R. Cooley, University of Massachusetts.]

brown, jelly-like "horns" that are 10 to 20 millimeters long (Figs. 11-137A and 11-137B). The jelly-like horns are columns of teliospores that germinate in place for several weeks and produce basidiospores that can infect apple leaves. The galls eventually die but may remain attached to the tree for a year or more.

Basidiospores are windborne and may be carried for up to 3 to 5 kilometers. Their germ tubes penetrate young apple leaves or fruit directly and produce haploid mycelium that spreads through or between the apple cells. The mycelium forms orange-colored spermagonia on the upper leaf surface and, after fertilization of the receptive hyphae by compatible spermatia, produces aecial cups on concentric rings on the lower side of leaves and on fruit. The area of the leaf where aecia are produced is swollen, and the clusters of orange-yellow aecial cups and their white cup walls stand out conspicuously. In the fruit, spermagonia appear first in the center of the spot and the aecia subsequently in the surrounding area. Infected fruit areas are usually large and flat or depressed rather than swollen (Fig. 11-137D). The aeciospores are produced in chains. They are released in the air during dry weather in late summer and are carried by wind to cedar leaves, where they start new infections.

The control of cedar-apple rust can be achieved by keeping apple and cedar trees sufficiently removed from one another so that the fungus cannot complete its life cycle. This, however, is often impossible or impractical,

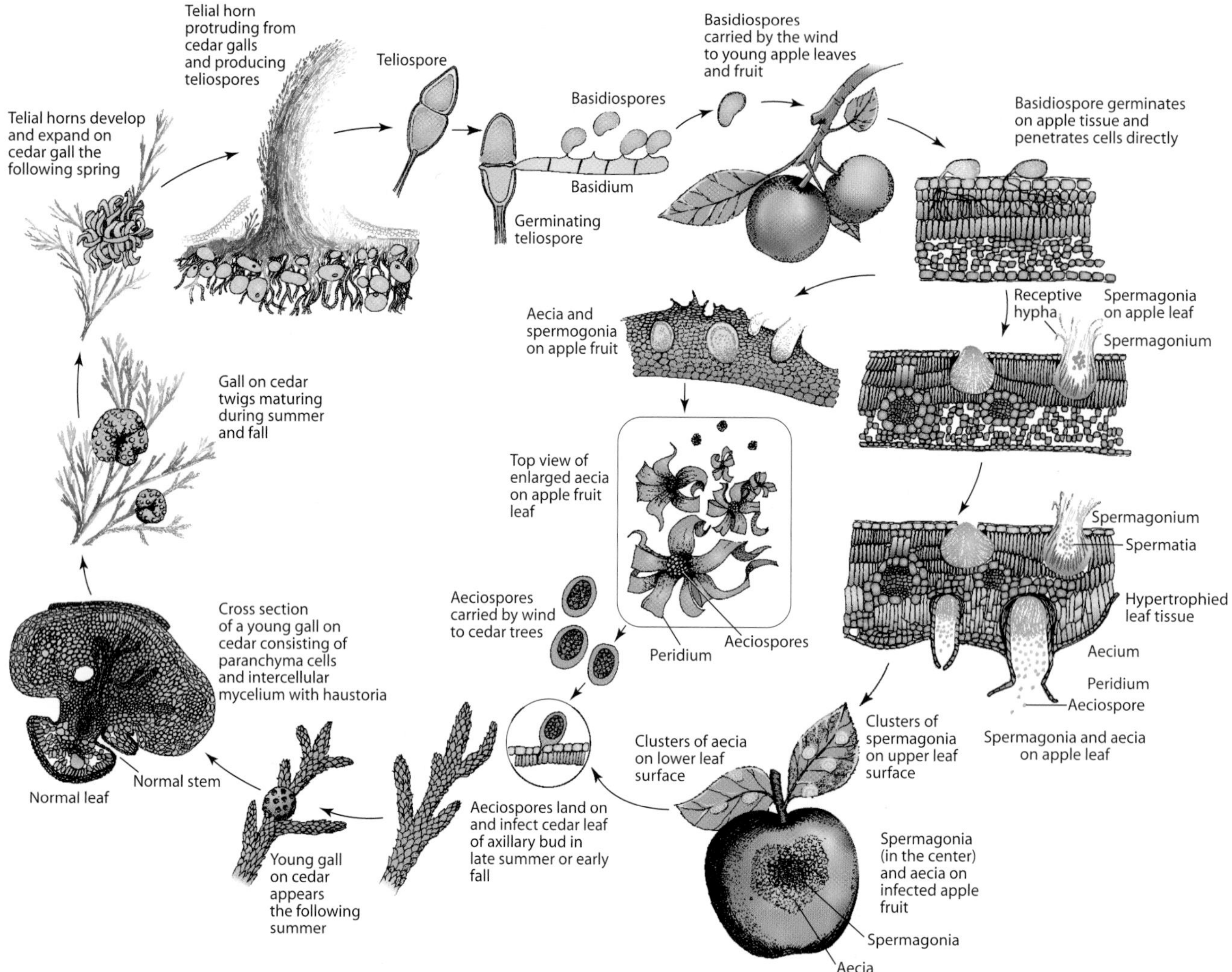

FIGURE 11-138 Disease cycle of cedar-apple rust caused by *Gymnosporangium juniperi-virginianae*.

and therefore the disease is generally controlled with chemical sprays. Many apple varieties are also quite resistant to rust.

COFFEE RUST

Coffee rust is the most destructive disease of coffee. It damages trees and reduces yields by causing premature drop of infected leaves. Coffee rust has caused devastating losses in all coffee-producing countries of Asia and Africa. It attacks all species of coffee but is most severe on *Coffea arabica*. In 1970 the disease appeared for the first time in the western hemisphere, in Brazil, and has since been steadily spreading into the world's most important coffee-producing countries of South and Central America, where all commercial coffee cultivars are susceptible to the rust.

Symptoms appear as orange-yellow powdery spots on the lower side of the leaves. The spots are circular and small, about 5 millimeters in diameter, at first, but they often coalesce and form large patches that may be 10 times as large. The centers of the spots eventually become dry and turn brownish, and the leaves fall off prematurely. Infected trees produce small yields of poor quality, and repeated infections and defoliations result in the death of trees (Figs. 11-139A–11-139C).

The fungus, *Hemileia vastatrix*, exists primarily as mycelium, uredia, and uredospores in infected leaves that they infect continuously and successively. The fungus occasionally produces teliospores, which on germination form basidiospores; the latter do not infect coffee, however, and no alternate host has so far been found. Uredospores are spread easily by wind, rain, and perhaps by insects. Spores germinate only in the pres-

FIGURE 11-139 Coffee rust caused by *Hemileia vastatrix*. Rust uredia in recent (A) and in older infections (B) of the lower side of coffee leaves. (C) Coffee tree nearly defoliated (left) due to rust infection compared to a healthy one. (Photographs courtesy of H.D. Thurston.)

ence of free water and enter leaves through the stomata of the lower surface. The mycelium grows between the leaf cells and sends haustoria into the cells. Young leaves are generally more susceptible to infection than older ones. New uredia may appear on the lower side of the leaf within 10 to 25 days from infection, depending on the climatic conditions. Once uredia develop, premature falling of infected leaves may occur at any time; sometimes even one uredium is sufficient to cause the leaf to fall. New leaves are affected after the older ones have fallen. The premature shedding of leaves weakens the trees and results in reduced yields, severe dieback of twigs, and death of trees (Fig. 11-140).

The control of coffee rust is difficult, but satisfactory results can be obtained with copper fungicides. Fungicides must be applied before and during the rainy season at 2- to 3-week intervals or less, depending on weather conditions and the severity of the attack. Systemic fungicides, which have a curative effect on developing uredial pustules, have been used in alternate applications with the copper fungicides. Sufficient tree pruning, good site selection, and use of resistant varieties help minimize losses from the rust. New races of the pathogen virulent to the new resistant varieties of the host have already appeared in some regions, however.

RUSTS OF FOREST TREES

Several species of *Cronartium* are responsible for a number of rust diseases that cause major losses in forest

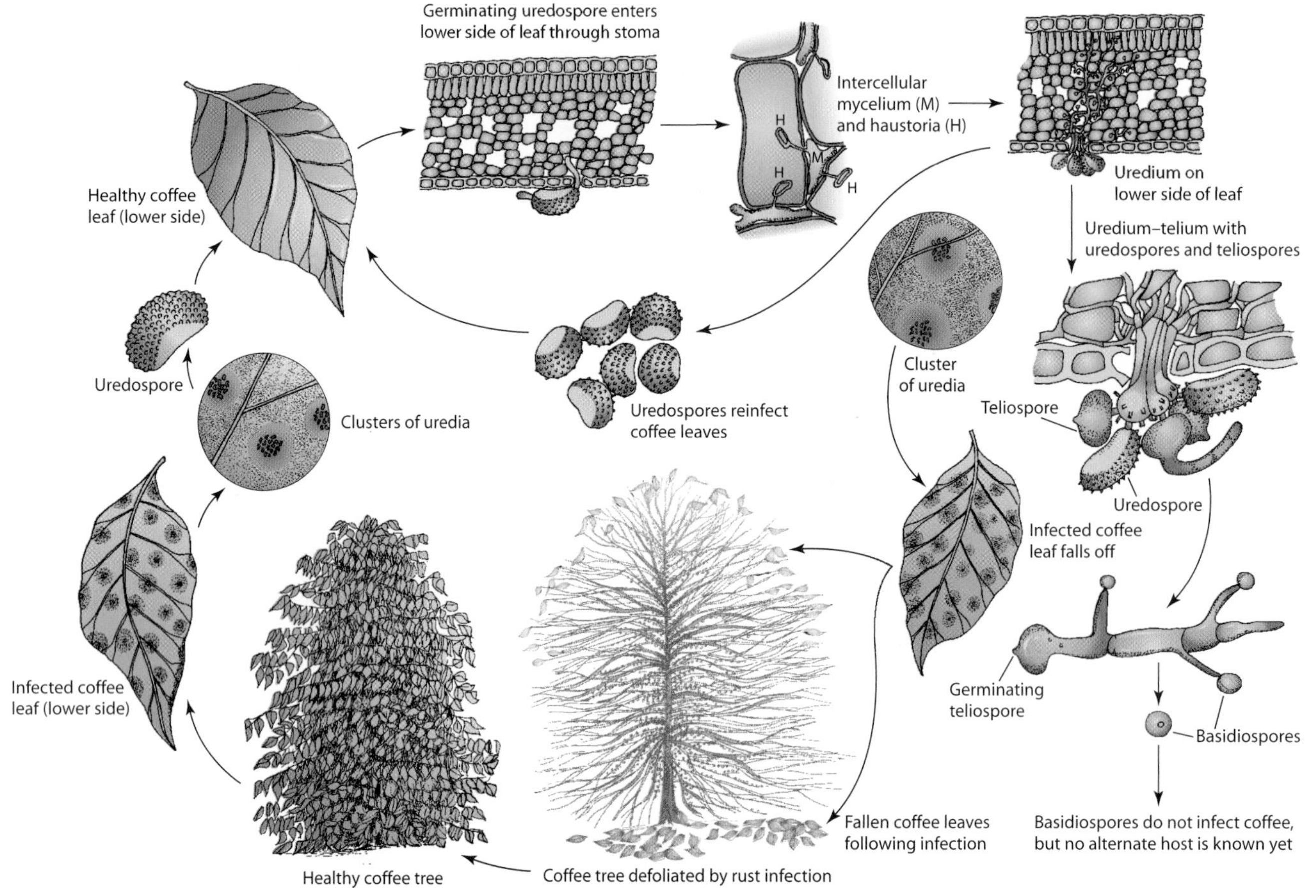

FIGURE 11-140 Disease cycle of coffee rust caused by *Hemileia vastatrix*.

trees. Some *Cronartium* species attack the main stem or branches of trees, and these are the most destructive; other species attack the needles or leaves and are less serious. All rusts, however, are especially destructive when they attack young trees in the nursery or in recently established plantations. The main economic host of the majority of forest tree rusts and the one to which they cause the most damage is pine. Some of these rusts have oak as their alternate host, but the damage to oaks is minor. Other pine rusts have as their alternate hosts various wild or cultivated shrubs or weeds.

WHITE PINE BLISTER RUST

White pine blister rust is native to Asia from where it spread to Europe and, about 1900, to North America. It is one of the most important forest diseases in North America, where it causes extensive annual pine growth loss and mortality, and, if not controlled, it makes white pine growing impossible or unprofitable. White pine blister rust is caused by the fungus *Cronartium ribicola*, which produces its spermagonia and aecia on white pine (the five-needle pines) and its uredia and telia on wild and cultivated currant and gooseberry bushes (*Ribes* spp.). Blister rust kills pines of all ages and sizes. Small pines are killed quickly, whereas larger pines may develop cankers that girdle and either kill the trees or retard their growth and weaken the stems, which then break at the canker. Infection of *Ribes* bushes causes relatively little loss through premature, partial defoliation and reduced fruit production.

Symptoms of blister rust on white pine stems or twigs appear first as small, discolored, spindle-shaped swellings surrounded by a narrow band of yellow-orange bark (Fig. 11-141A). In the swelling, small, irregular, dark brown, blister-like spermagonia appear (Fig. 11-141B), which rupture, ooze droplets full of the year's crop of spermatia, and then dry. As the swelling expands, the margin and the zone of the spermagonia expand, and the portion formerly occupied by spermagonia is the area where, a year later, aecia are produced. Aecia appear as white blisters containing orange-yellow aeciospores that push through the diseased bark. Aecial blisters soon rupture (Fig. 11-141C),

FIGURE 11-141 White pine blister rust caused by *Cronartium ribicola*. Stages in the development of a rust canker in white pine: (A) soft, yellowish bark, (B) canker feels bumpy and turns gray, (C) bumpy areas erupt, revealing spermagonia and, subsequently, aecia, (D) resin flows down the stem and hardens. (E) Ribes leaf showing uredia of the fungus. [Photographs courtesy of (A, D, and E) U.S. Forest Service and (B and C) O.C. Maloy, Washington State University.]

and the orange-yellow aeciospores are carried by the wind, sometimes for several hundred kilometers, some of them landing on and infecting *Ribes* leaves. After the aeciospores have been released, the blisters persist on the bark for several weeks, although the bark of that area dies. Resin often flows down the stem and hardens in masses (Fig. 11-141D). The fungus, however, continues to spread into the surrounding healthy bark, and the sequence of spore production and bark killing continues in subsequent years until the stem or branch is girdled and killed. The dead branches, called flags, have dead, brown needles and are visible from a distance.

On currants and gooseberries, symptoms appear on the undersides of the leaves as slightly raised, yellow-orange uredia grouped in circular or irregular spots. Uredia produce orange masses of uredospores that reinfect *Ribes*. Later, telia develop in the same or new lesions. The telia are slightly darker than the uredia and consist of brownish, hair-like structures up to 2 millimeters in height that bear the teliospores (Fig. 11-141E).

The pathogen, *Cronartium ribicola*, overwinters mostly as mycelium in infected white pines and to some extent on *Ribes*. Pines are infected only by basidiospores produced by teliospores still in the telia on the undersides of *Ribes* leaves (Fig. 11-142). Basidiospores are produced only during wet, cool periods, especially during the night, and can be carried by wind and infect pines within a few kilometers from the *Ribes* host. Basidiospores infect pine needles through stomata in late summer or early fall. Small, discolored spots may appear on the needles 4 to 10 weeks after infection. The mycelium grows down the conducting tissues of the needle and into the bark of the stem, which it reaches about 12 to 18 months after infection. Spermagonia develop on infected stems or branches in the spring and early summer 2 to 4 years after the needle infection, and aecia are produced in the spring 3 to 6 years from inoculation. Spermatia are short lived and spread over short distances by rain or insects, whereas aeciospores may live for many months and may be carried by wind over many kilometers to *Ribes* leaves. On the latter, the aeciospores germinate and infect the leaves, producing uredia and uredospores within 1 to 3 weeks after inoculation. Uredospores can reinfect *Ribes* plants again and again, producing many generations of uredospores in a single growing season. Uredospores can survive for many months and can be spread by wind for a kilometer or more, but they can infect only *Ribes*. Finally, the same mycelium that produced the uredospores begins to produce telial columns and teliospores. The latter germinate from July to October and produce short-lived basidiospores, which if blown to nearby white pines infect the needles and complete the life cycle of the fungus.

The control of white pine blister rust can be obtained by the eradication of wild and cultivated *Ribes* bushes mechanically or, better still, with herbicides. Pruning infected branches on young trees reduces stem infections and tree mortality. The most promising control for blister rust seems to be the selection and breeding of resistant trees.

FUSIFORM RUST

Fusiform rust is one of the most important diseases on southern pines, especially loblolly and slash pines. The disease is present from Maryland to Florida and west to Texas and Arkansas, where it causes tremendous losses in nurseries, young plantations, and seed orchards, leading to 20 to 60% or greater mortality of young trees. Fusiform rust is caused by *Cronartium quercuum* f. sp. *fusiforme*, which produces spermagonia and aecia on pine stems and branches and uredia and telia on oak leaves. Damage on oak is slight, mostly through occasional partial defoliation.

Symptoms on pine first appear as small, purple spots on needles and succulent shoots. Soon small galls and later spindle-shaped swellings or galls develop on branches and stems of mostly young pines. These galls may elongate from 5 to 15 centimeters per year and often encircle the stem or branch and cause it to die (Fig. 11-143A). Infection of young seedlings results in their death within a very few years, whereas infected young trees may branch excessively for a period and show a bushy growth. On older trees, stem or branch infections lead to weak, distorted swellings or, as host tissue is killed, to sunken cankers that break easily during strong winds. Yellowish masses of spermatia and later orange-yellow aeciospores appear on the galls. On oak, symptoms appear as orange pustules (uredia) and brown, hair-like columns (telia) on the underside of the leaves.

The fusiform rust pathogen, *C. quercuum* f. sp. *fusiforme*, overwinters as mycelium in the fusiform galls. From February to April, spermagonia and spermatia form, and soon aeciospores are produced on the galls. Wind carries the aeciospores to young, expanding oak leaves, which they infect. On the oak leaves, orange uredial pustules develop in a few days and produce uredospores from February to May. Uredospores can reinfect more oak leaves and produce more uredospores. The same mycelium also produces brown telia from February to June in place of uredia or in new lesions. The teliospores germinate on the telia (Fig. 11-143B), and the basidiospores produced are carried by wind to pine needles and shoots, which they infect directly. The mycelium grows first in the needles and later spreads into branches or the stem, where it induces formation of a gall.

The control of fusiform rust infections in the nursery can be obtained by frequent, twice-a-week sprays with appropriate fungicides, especially before and during cool wet weather. Some of the newer systemic fungicides give good control of fusiform rust of seedlings when applied as sprays or as seed dressings. Several fungi antagonistic to or parasitic on the pathogen are known, but no practical biological control of the disease has been developed yet. All infected seedlings should be discarded. In plantations and natural stands, limited control can be obtained against fusiform rust by avoiding planting highly susceptible slash and loblolly pines in areas of known high rust incidence and by pruning

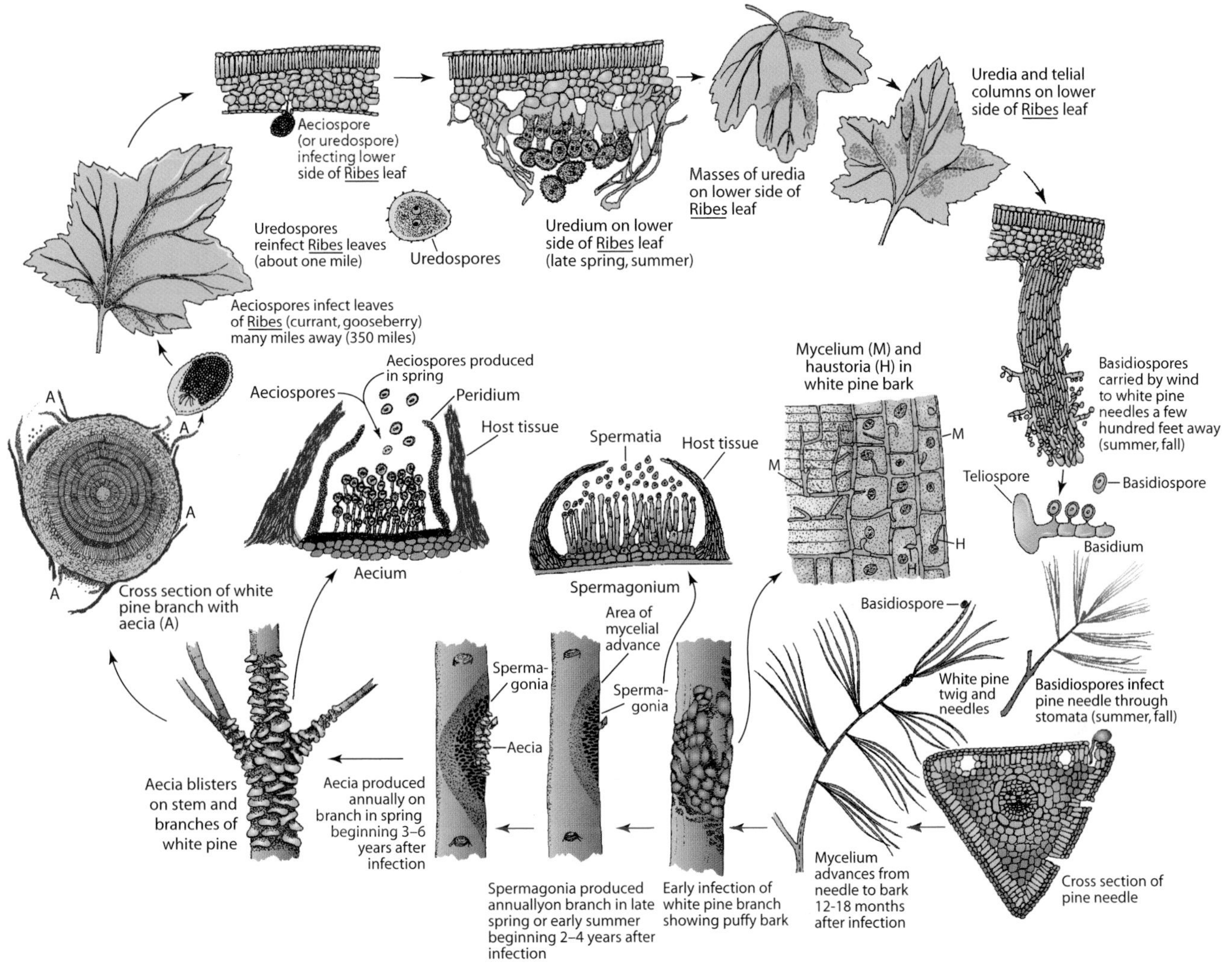

FIGURE 11-142 Disease cycle of white pine blister rust caused by *Cronartium ribicola*.

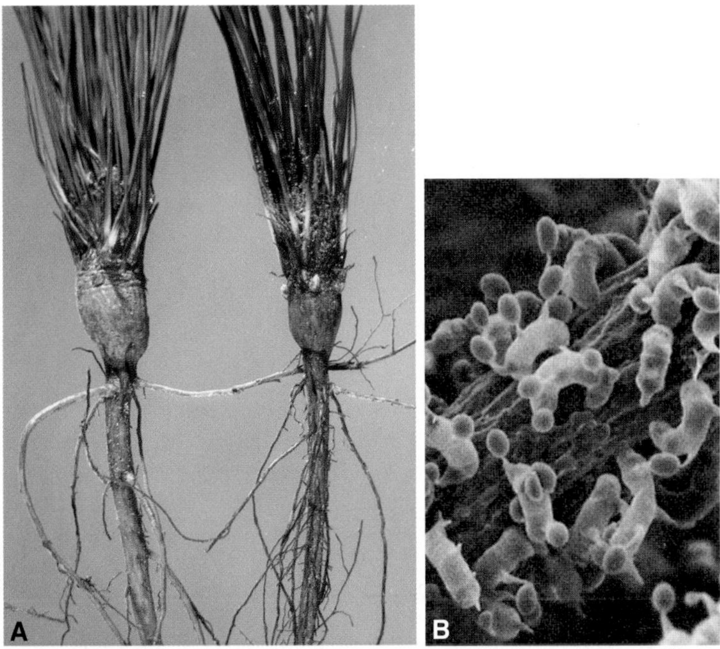

FIGURE 11-143 (A) Pine seedlings showing fusiform galls on their stem caused by *Cronartium quercuum* f. sp. *fusiforme*. (B) Scanning electron micrograph of basidia and basidiospores of the fusiform rust fungus. [Photographs courtesy of (A) E.L. Barnard, Florida Department of Agriculture, Forestry Division and (B) M.F. Brown and H.G. Brotzman.]

infected branches before the fungus reaches the trunk. As with white pine blister rust, and perhaps even more so, the control of fusiform rust is obtained through selection and breeding of resistant trees, with emphasis on trees possessing general rather than specific resistance.

Selected References

Alexopoulos, C. J., Mims, C. W., and Blackwell, M. (1996). "Introductory Mycology," 4th Ed. Wiley, New York.

Allen, R. F. (1930). A cytological study of heterothallism in *Puccinia graminis*. *J. Agric. Res. (Washington, DC)* **40**, 585–614.

Anguelova-Methar, V. S., Van Der Westhuizen, A. J., and Pretorius, Z. A. (2001). β-1,3-Glucanase and chitinase activities and the resistance response of wheat to leaf rust. *J. Phytopathol.* **149**, 381–384.

Anonymous (1981). Stakman-Craigie symposium on rust diseases. *Phytopathology* **71**, 967–1000.

Bromfield, K. R. (1984). "Soybean Rust." Monograph 11. The American Phytopathological Society, St. Paul, MN.

Browning, J. A., and Frey, K. J. (1969). Multiline cultivars as a means of disease control. *Annu. Rev. Phytopathol.* **7**, 355–382.

Crowell, I. H. (1934). The hosts, life history, and control of the cedar-apple rust fungus *Gymnosporangium juniperi-virginianae*. *J. Arnold Arbor. Harv. Univ.* **15**, 163–232.

Cummins, G. B. (1959). "Illustrated Genera of Rust Fungi." Burgess, Minneapolis, MN.

De Jesus, W. C., Jr., do Vale, F. X. R., Coelho, R. R., *et al.* (2001). Effects of angular leaf spot and rust on yield loss of *Phaseolus vulgaris*. *Phytopathology* **91**, 1045–1053.

Ellis, J., Lawrence, G., Ayliffe, M., *et al.* (1997). Advances in the molecular genetic analysis of the flax-flax rust interaction. *Annu. Rev. Phytopathol.* **35**, 271–291.

Eversmeyer, M. G., and Kramer, C. L. (2000). Epidemiology of wheat leaf and stem rust in the central Great Plains of the USA. *Annu. Rev. Phytopathol.* **38**, 491–513.

Eversmeyer, M. G., Kramer, C. L., and Browder, L. E. (1984). Presence, viability, and movement of *Puccinia recondita* and *P. graminis* inoculum in the Great Plains. *Plant Dis.* **68**, 392–395.

Everts, K. L., Leath, S., and Finney, P. L. (2001). Impact of powdery mildew and leaf rust on milling and baking quality of soft red winter wheat. *Plant Dis.* **85**, 423–429.

Frederick, R. D., Snyder, C. L., Peterson, G. L., *et al.* (2002). Polymerase chain reaction assays for the detection and discrimination of the soybean rust pathogens *Phakopsora pachyrhizi* and *P. meibomiae*. *Phytopathology* **92**, 217–227.

Gross, P. L., and Venette, J. R. (2000). Overwinter survival of bean rust urediniospores in North Dakota. *Plant Dis.* **85**, 226–227.

Hamelin, R. C., Dusabenyagasani, M. and Et-touil, K. (1998). Fine-level genetic structure of white pine blister rust populations. *Phytopathology* **88**, 1187–1191.

Hart, H. (1931). Morphologic and physiologic studies on stem-rust resistance in cereals. *Minn. Agric. Exp. Stn. Tech. Bull.* **266**, 1–75.

Hartman, G., Sinclair, J., and Rupe, J. (1999). "Compendium of Soybean Diseases." APS Press, St. Paul, MN.

Hooker, A. L. (1967). The genetics and expression of resistance in plants to rusts of the genus *Puccinia*. *Annu. Rev. Phytopathol.* **5**, 163–182.

Hovmøller, M. S., Justesen, A. F., and Brown, J. K. M. (2002). Clonality and long-distance migration of *Puccinia striiformis* f. sp. *tritici* in northwest Europe. *Plant Pathol.* **51**, 24–32.

Hunt, R. (1997). White pine blister rust. *In* "Compendium of Conifer Diseases" (E. M. Hansen and K. J. Lewis, eds.), pp. 26–27. American Phytopathological Society, St. Paul, MN.

Johnson, T., Green, G. J., and Samborski, D. J. (1967). The world situation of the cereal rusts. *Annu. Rev. Phytopathol.* **5**, 183–200.

Khan, M. A., Trevathan, L. E., and Robbins, J. T. (1997). Quantitative relationship between leaf rust and wheat yield in Mississippi. *Plant Dis.* **81**, 769–772.

Kinloch, B. B., Jr., and Dupper, G. E. (2002). Genetic specificity in the white pine-blister rust pathosystem. *Phytopathology* **92**, 278–280.

Kuchler, F., Duffy, M., Shrum, R. D., and Dowler, W. M. (1984). Potential economic consequences of the entry of an exotic fungal pest: The case of soybean rust. *Phytopathology* **74**, 916–920.

Kushalappa, A., and Eskes, A. B. (1989). Advances in coffee rust research. *Annu. Rev. Phytopathol.* **27**, 503–531.

Littlefield, L. J., and Heath, M. C. (1979). "Ultrastructure of Rust Fungi." Academic Press, New York.

Lopes, D. B., and Berger, R. D. (2001). The effects of rust and anthracnose on the photosynthetic competence of diseased bean leaves. *Phytopathology* **91**, 212–220.

Malloy, O. C. (1997). White pine blister rust control in North America: A case history. *Annu. Rev. Phytopathol.* **35**, 87–109.

McIntosh, R. A., and Brown, G. N. (1997). Anticipatory breeding for resistance to rust diseases in wheat. *Annu. Rev. Phytopathol.* **35**, 311–326.

Nagarajan, S., and Singh, D. V. (1990). Long-distance dispersion of rust pathogens. *Annu. Rev. Phytopathol.* **28**, 139–153.

Peterson, R. S., and Jewell, F. F. (1968). Status of American stem rusts of pine. *Annu. Rev. Phytopathol.* **6**, 23–40.

Powers, H. R., Schmidt, R. A., and Snow, G. A. (1981). Current status and management of fusiform rust on southern pines. *Annu. Rev. Phytopathol.* **19**, 353–371.

Rapilly, F. (1979). Yellow rust epidemiology. *Annu. Rev. Phytopathol.* **17**, 59–73.

Roelfs, A. P. (1989). Epidemiology of cereal rusts in North America. *Can. J. Plant Pathol.* **11**, 86–90.

Roelfs, A. P., and Bushnell, W. R., eds. (1985). "The Cereal Rusts," Vols. 1 and 2. Academic Press, Orlando, FL.

Scott, K. J., and Chakravorty, A. K., eds. (1982). "The Rust Fungi." Academic Press, New York.

Sillero, J. C., and Rubiales, D. (2002). Histological characterization of resistance to *Uromyces viciae-fabae* in fava bean. *Phytopathology* **92**, 294–299.

Silva, M. C., Nicole, M., Guerra-GuimarÃes, L., *et al.* (2002). Hypersensitive cell death and posthaustorial defense responses arrest the orange rust (*Hemileia vastatrix*) growth in resistant coffee leaves. *Physiol. Mol. Plant Pathol.* **60**, 169–183.

Subrahamanyam, P., Reddy, L. I., Gibbons, R. W., and McDonald, D. (1985). Peanut rust: A major threat to peanut production in the semiarid tropics. *Plant Dis.* **69**, 813–819.

Ward, H. M. (1882). Researches on the life history of *Hemileia vastatrix*, the fungus of the "coffee leaf disease." *Linn. Soc. J. (Bot.)* **19**, 229–335.

Zadoks, J. C. (1965). Epidemiology of wheat rusts in Europe. *FAO Plant Prot. Bull.* **13**, 97–108.

Zhang, L., and Dickinson, M. (2001). Fluorescence from rust fungi: A simple and effective method to monitor the dynamics of fungal growth *in planta*. *Physiol. Mol. Plant Pathol.* **59**, 137–141.

SMUTS

Plant smuts, caused by Basidiomycetes of the order Ustilaginales, occur throughout the world. There are approximately 1,200 species of smut fungi. Until the 20th century, smuts were the causes of serious grain losses that were equal to, or second only to, losses caused by the rusts. In some respects, the smuts of cereals were dreaded by farmers even more than rusts because many smuts attack the grain kernels themselves and replace the kernel contents with the black, dusty spore masses that resemble soot or smut. Thus, the reduction in yield is conspicuous and direct and the quality of the remaining yield is reduced drastically by the presence of the black smut spores on the surface of healthy kernels. In addition to the various cereals, smuts also affect sugarcane, onions, and some ornamentals such as carnation.

Most smut fungi attack the ovaries of grains and grasses and develop in them and in the fruit, i.e., the kernels of grain crops, which they destroy completely (Fig. 11-129). Several smuts, however, attack the leaves, stems, or floral parts. Some smuts infect seeds or seedlings before they emerge from the ground, and they grow internally in the seedling until they reach the inflorescence; others cause only local infections on leaves, stems, and so on. Cells in affected tissues are either destroyed and replaced by black smut spores or they are first stimulated to divide and enlarge to produce a swelling or gall of varying size and are then destroyed and replaced by the black smut spores. The spores are present in masses that may be held together only temporarily by a thin, flimsy membrane or by a more or less durable one. Smut fungi seldom kill their hosts, but in some cases infected plants may be severely stunted.

Most smut fungi produce only two kinds of spores: teliospores and basidiospores (Fig. 11-127). Teliospores are usually formed from mycelial cells along the length of the mycelium within the smut galls, and basidiospores either bud off laterally from the basidium cells or are produced as a cluster at the tip of a nonseptate basidium. Basidiospores of the smuts are not borne on sterigmata. When basidiospores germinate, the germ tubes either unite with compatible ones while still on the basidium and then infect or they penetrate tissues directly. Their haploid mycelium, however, cannot invade tissues extensively and does not cause typical infections until two compatible mycelia unite to produce dikaryotic mycelium. The latter then invades tissues inter- or intracellularly and produces the typical symptoms and the teliospores. Smut fungi also exist in many races; however, races of smut fungi are not as stable as rusts, as each generation of smut fungi on the host plant involves meiosis, i.e., genetic recombination, and this results in new races appearing constantly. The most common smut fungi and the diseases they cause are the following.

Ustilago, causing corn smut [*U. zeae* (*maydis*)], loose smut of cereals (*U. avenae*, *U. nuda*, and *U. tritici*), and sugarcane smut (*U. scitaminea*)

Tilletia, causing covered smut or bunt of wheat [*T. caries* (= *T. tritici*) and *T. laevis* (= *T. foetida*)], dwarf bunt of wheat (*T. controversa*), and Karnal bunt of wheat (*T. indica*)

Sphacelotheca, causing the sorghum smuts (*S. sorghi*, *S. cruenta*, and *S. reiliana*)

Urocystis, causing onion smut (*U. cepulae*)

Neovossia, causing kernel smut of rice (*N. barclayana*)

Entyloma, causing leaf smut of rice (*E. oryzae*)

Smuts generally overwinter as teliospores on contaminated seed, in plant debris, or in the soil. However, some smuts overwinter as mycelium inside infected kernels or in infected plants. The teliospores are not infectious but produce basidiospores, which on germination either fuse with compatible ones and then infect or penetrate the tissue and then fuse to produce dikaryotic mycelium and the typical infection. Smut fungi have only one generation per year, each infection resulting in one crop of teliospores per growing season.

The control of smuts is primarily by use of resistant varieties and seed treatment. The latter may involve either chemical dusting or dipping, if the fungus is present as teliospores on the seed surface or in the soil, or hot water if the fungus is present as mycelium inside the seed. The discovery of carboxin, thiabendazole, etaconazole, and other fungicides that are absorbed and translocated systemically by seeds and seedlings allows chemical control by seed treatment of even those smuts present as mycelium inside the seeds. Soil treatments with these and other chemicals are also useful in the control of smut diseases.

CORN SMUT

Corn smut occurs wherever corn is grown. It is more prevalent, however, in warm and moderately dry areas. Corn smut damages plants and reduces yields by forming galls on the aboveground parts of plants, including ears, tassels, stalks, and leaves. The number, size, and location of smut galls on the plant affect the amount of yield loss. Galls on the ear usually destroy it to a large extent, whereas large galls above the ear cause much greater reduction in yield than galls below the ear. Losses from corn smut range from a trace up to 10% or more in localized areas. Some individual fields of sweet corn may show losses approaching 100% from corn smut. Generally, however, over large areas and with the use of resistant varieties, losses in grain yields average about 2%.

Symptoms

When young corn seedlings are infected, minute galls form on the leaves and stems, and the seedling may remain stunted or may be killed. On older plants, infections occur on the young, actively growing tissues of axillary buds, individual flowers of the ear and tassel, leaves, and stalks (Fig. 11-144).

Infected areas are permeated by the fungus mycelium, which stimulates the host cells to divide and enlarge, thus forming galls. Galls are first covered with a greenish white membrane. Later, as the galls mature, they reach a size from 1 to 15 centimeters in diameter, and their interior darkens and turns into a mass of powdery, dark olive-brown spores. The silvery gray membrane then ruptures and exposes the millions of sooty teliospores, which are released into the air. Galls on leaves frequently remain very small (about 1–2 cm in diameter), hard, dry, and do not rupture.

The Pathogen: Ustilago zeae

The fungus produces dikaryotic mycelium, the cells of which are transformed into black, spherical, or ellipsoidal teliospores. Teliospores germinate by producing a four-celled basidium (promycelium) from each cell of which a basidiospore (sporidium) develops (Fig. 11-145).

Development of Disease

The fungus overwinters as teliospores in crop debris and in the soil, where it can remain viable for several years. In the spring and summer, teliospores germinate and produce basidiospores, which are carried by air currents or are splashed by water to young, developing tissues of corn plants. Basidiospores germinate and produce a fine hypha, which can enter epidermal cells directly. After an initial development, however, its growth stops and the hypha usually withers and sometimes dies, unless it contacts and fuses with a haploid hypha derived from a basidiospore of the compatible mating type. If fusion takes place, the resulting hypha becomes dikaryotic, enlarges in diameter, and grows into the plant tissues mostly intercellularly (Fig. 11-145). Cells surrounding the hypha are stimulated to enlarge and divide, and galls begin to form even before the fungus actually gets there.

Galls in older plants seem always to be the result of local infections. Systemic infections occur occasionally in very young seedlings. Frequently, however, only a small number of the actual local infections develop into typical, large galls, with the others remaining too small to be visible.

The mycelium in the gall remains intercellular during most of gall formation, but before sporulation, the enlarged corn cells are invaded by the mycelium, collapse, and die. The mycelium utilizes the cell contents for its further growth, and the gall then consists primarily of dikaryotic mycelium and plant cell remains. Most of the dikaryotic cells subsequently develop into teliospores and, in the process, seem to absorb and utilize the protoplasm of the other mycelial cells, which remain empty. Only the membrane covering the gall is not affected by the fungus, but finally the membrane breaks and the teliospores are released. Some of the released teliospores, if they land on young, meristematic corn tissues, may cause new infections and new galls during the same season, but most of them fall to the ground or remain in the corn debris, where they can survive for several years.

Control

No corn varieties or hybrids completely resistant to smut are known, but several corn hybrids show moderate resistance to the fungus. New pathogen races appear constantly, however; therefore partial resistance is the major type of resistance selected for in breeding programs. Sanitation measures, such as removal of smut galls before they break open, and crop rotation help where corn is grown in small, rather isolated plots but is impractical and impossible in large corn-growing areas. In some countries, e.g., Mexico, corn smuts are collected and used as food delicacies, not unlike the edible mushrooms consumed elsewhere.

LOOSE SMUT OF CEREALS

Loose smut of cereals occurs worldwide but is more abundant and serious in humid and subhumid regions. Loose smut causes damage by destroying the kernels (Fig. 11-146) of the infected plants and by smearing and thus reducing the quality of the grain of the noninfected plants on harvest. Losses from loose smut may be up to 10 or 40% in certain localities in a given year, but the overall losses in the United States are approximately 1% per year.

Symptoms

Loose smut generally does not produce discernible symptoms until the plant has produced a head. Smutted plants sometimes head earlier than healthy ones, and smutted heads are often elevated above those of healthy plants (Fig. 11-146). In an infected plant, usually all the heads and all the spikelets and kernels of each head are

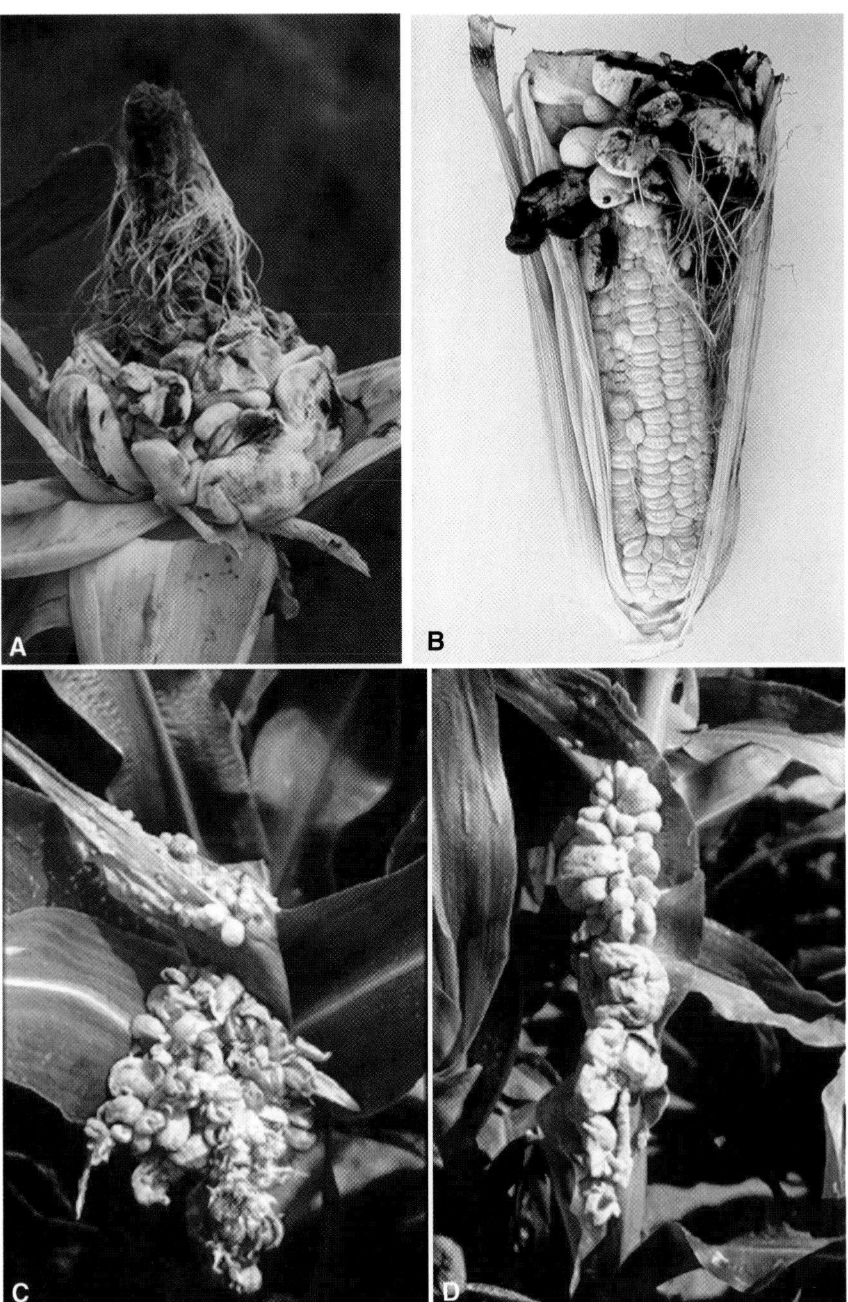

FIGURE 11-144 Corn smut caused by *Ustilago maydis*. Smut in younger (A) and older (B) ears of corn in which individual corn kernels have enlarged greatly and filled with smut spores. (C) Smut on a corn tassel and (D) smut galls on corn stem. [Photographs courtesy of (A) P. E. Lipps, Ohio State University, (B) D. Ormrod, WCPP, and (C and D) K. Mohan, University of Idaho.]

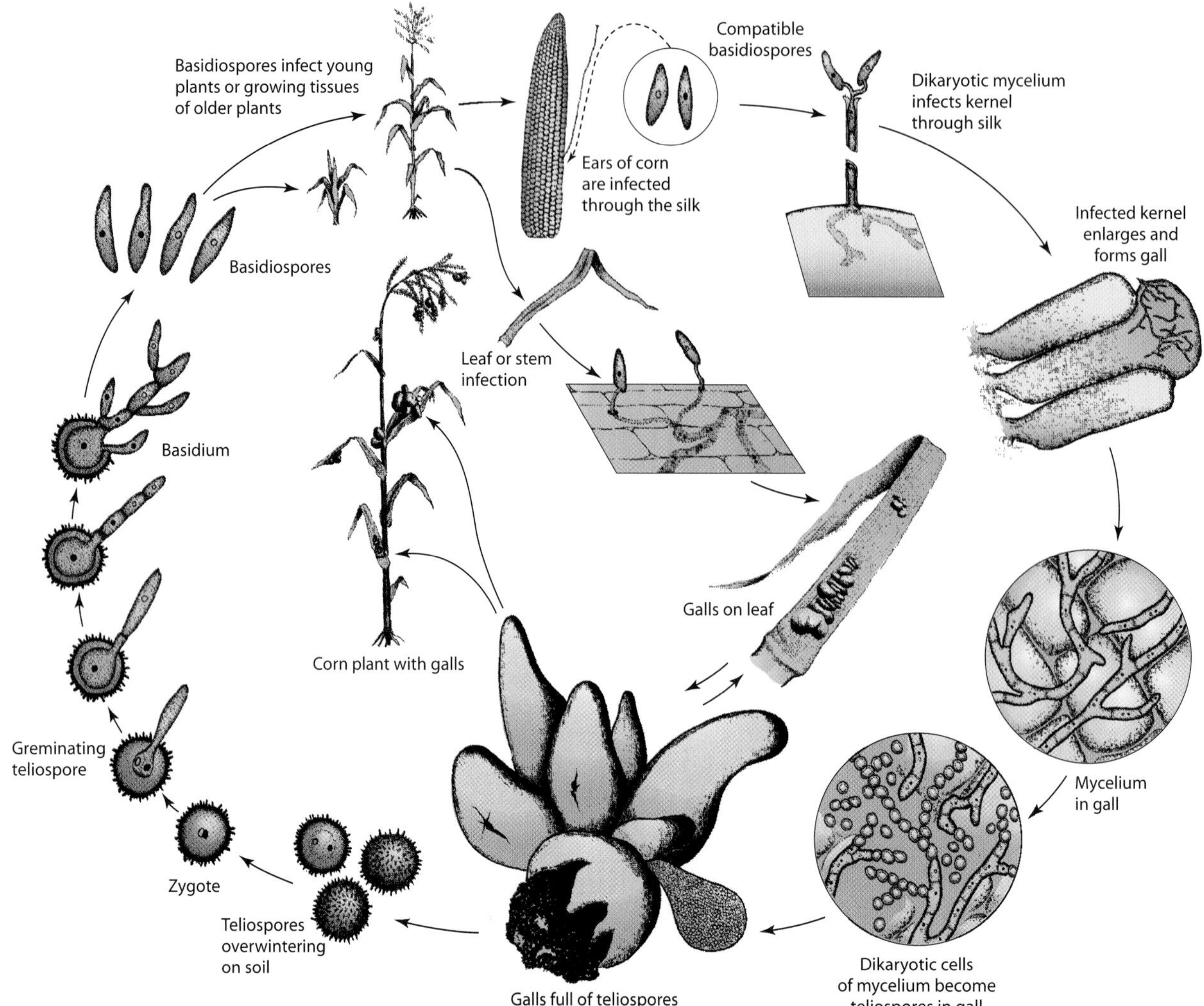

FIGURE 11-145 Disease cycle of corn smut caused by *Ustilago maydis*.

smutted, i.e., they are each transformed into a smut mass consisting of olive-green spores (Fig. 11-146). Smutted kernels are at first covered by a delicate grayish membrane, which soon bursts and sets the powdery spores free. The spores are then blown off by the wind and leave the rachis a naked stalk.

The Pathogens: Ustilago nuda and Ustilago tritici

The mycelium is hyaline during its growth through the plant, and it is hyaline changing to brown near maturity. The mycelial cells are transformed into brown, spherical teliospores, which germinate readily and produce a basidium consisting of one to four cells. The basidium produces no basidiospores, but its cells germinate and produce short, uninucleate hyphae that fuse in pairs and produce dikaryotic mycelium, which is capable of infection (Figs. 11-146 and 11-147).

Development of Disease

The pathogens overwinter as dormant mycelium in the scutellum of the cotyledon of infected kernels. When planted, infected kernels begin to germinate, and the mycelium resumes its activity and grows intercellularly through the tissues of the young seedling until it reaches the growing point of the plant (Fig. 11-147). The mycelium then follows closely the growing point of the plant, while the hyphae in the tissues of the lower stem frequently disappear. When the plant forms the head, the mycelium invades all the young spikelets, where it grows intracellularly and destroys most of the tissues of the spike, except the rachis. By this time, most infected plants are slightly taller than most healthy plants due to the stimulatory action of the pathogen. The mycelium in the infected kernels is soon transformed into teliospores, which are contained only by a delicate outer membrane

FIGURE 11-146 Loose smut of cereals. (A) Field with heads of barley infected with loose smut caused by *Ustilago nuda*. (B) Close-up of a healthy (right) and several heads of barley infected with loose smut. (C) Microscopic view of smut fungus mycelium and spores in an infected barley embryo. [Photographs courtesy of (A) P. Thomas, (B) I.R. Evans, WCPD, and (C) V. Pederson, North Dakota State University.]

of host tissue. The membranes burst open soon after maturation of the teliospores, and the spores are released and blown off by air currents to nearby healthy plants. Spore release coincides with the opening of the flowers of healthy plants. Teliospores landing on flowers germinate through formation of a basidium on which the haploid hyphae are produced. After fusion of sexually compatible haploid hyphae, the resulting dikaryotic mycelium penetrates the flower through the stigma or through the young ovary walls and becomes established in the pericarp and in the tissues of the embryo before the kernels become mature. The mycelium then becomes inactive and remains dormant, primarily in the scutellum, until the infected kernel germinates.

Control

Loose smut is now controlled by treating infected seeds with carboxin and its carboxanilide derivatives before planting. These chemicals are absorbed and act systemically in the seed or in the growing plant. Although some barley and wheat varieties are quite resistant to loose smut, most of the commercial varieties are very susceptible to it.

The best means of controlling loose smut is through the use of certified smut-free seed. Until the discovery of systemic fungicides, when seed was known to be infected with loose smut mycelium, the best way of disinfecting it was by treating it with hot water. Usually small lots of seed are treated with hot water and planted in isolated fields to produce smut-free seed to be used during the next season. The hot-water treatment consists of soaking the seed, contained in half-filled burlap bags, in 20°C water for five hours, draining it for one minute, dipping it in 49°C water for about one minute and then in 52°C water for exactly 11 minutes, and immediately afterward placing it in cold water for the seed to cool off. The seed is then allowed to dry so that it can be sown. Because some of the seed may be killed by the hot-water treatment, a higher seeding rate may be employed to offset the reduced germinability of the treated seed.

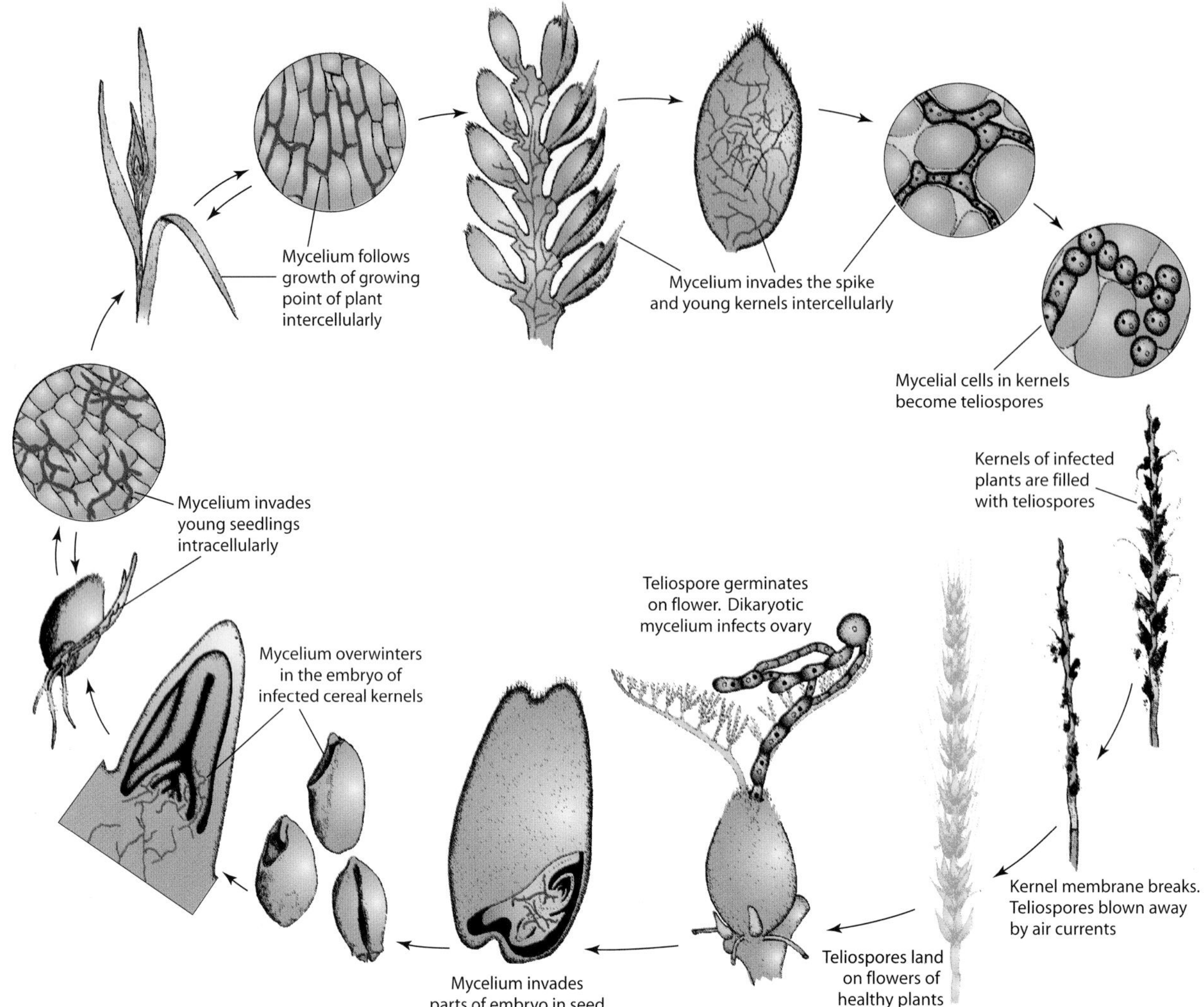

FIGURE 11-147 Disease cycle of loose smuts of barley and wheat caused by *Ustilago nuda* and *U. tritici*.

COVERED SMUT, OR BUNT, OF WHEAT

Covered smut, or bunt, or stinking smut of wheat occurs in all wheat-growing areas of the world. There are actually three kinds of bunt caused by related but different fungi: common bunt, which now is controlled easily by treating the seed with fungicides and therefore causes few losses in most developed countries; dwarf bunt, which still cannot be controlled and therefore continues to cause severe losses in many parts of the world, including the Pacific Northwest of the United States; and Karnal bunt, which, so far, occurs only in India and some other Asian countries, Mexico, and a few locations in the southwestern United States.

Bunt destroys the contents of infected kernels and replaces them with the spores of the fungus (Figs. 11-148A–11-148C). Bunt also causes slight to severe stunting of infected plants, depending on the species of bunt fungus involved. Infected plants are usually more susceptible than healthy plants to certain other diseases and to winter injury. When bunt is not controlled, it may cause devastating losses, but even with the effective control measures practiced in the United States today, the disease continues to cause severe losses. In addition, bunt causes market losses by reducing the quality, and the price, of wheat contaminated with smutted kernels or smut spores. Such wheat is discolored, has a foul odor, and is suitable for feed uses only. Bunt, moreover, results in explosions in combines and elevators during threshing or handling of smutted wheat because of the extreme combustibility of the oily smut spores in the presence of sparks from machinery (Fig. 11-148D).

FIGURE 11-148 Bunt, stinking smut, or covered smut of wheat caused by *Tilletia tritici and T. laevis.* (A) Field with stunted, covered smut-infected wheat plants and close-up of a single infected plant (A1). (B) Healthy (left) and infected (right) wheat heads showing size and direction of growth of infected kernels. (C) Healthy wheat kernels (golden yellow) mixed with black kernels filled with smut spores (teliospores). (D) Cloud of smut spore dust produced by a combine harvesting a heavily infected field. (E and F) Teliospores (smut spores) of *Tilletia tritici* and *T. laevis.* (G) A covered smut teliospore germinates by producing a basidium and eight primary sporidia that fuse and produce H-shaped structures. [Photographs courtesy of (A) R. Johnston, USDA, (B and D) L.J. Duczek, WCPD, (C) P.E. Lipps, Ohio State University, (E and F) M. Babadoost, University of Illinois, and (G) M.F. Brown and H.G. Brotzman.]

Symptoms

Plants infected with the common bunt fungi are usually a few to several centimeters shorter than healthy plants and may sometimes be only half as tall. Plants infected with the dwarf bunt fungus may be only one-fourth as tall as healthy plants and may show an increase in the number of tillers. Infected plants may appear slightly bluish green to grayish green in color, but this is not easily distinguishable.

Distinct bunt symptoms, however, are shown by the heads of infected plants. Infected heads are slimmer and are usually bluish green rather than the normal yellowish green, and their glumes seem to spread apart and form a greater angle with the main axis (Fig. 11-148B). Infected kernels are shorter and thicker than healthy ones and are grayish brown rather than the normal golden yellow or red (Figs. 11-148B and 11-148C). When mature kernels are broken, they are found to be full of a sooty, black, powdery mass of fungus spores (Figs. 11-148E and 11-148F) that give off a distinctive odor resembling that of decaying fish. During the harvest of infected fields, large clouds of spores may be released in the air (Fig. 11-148D).

The Pathogens

Tilletia caries (= *T. tritici*) and *T. laevis* (= *T. foetida*) cause the common bunt, whereas *T. controversa* causes dwarf bunt and *T. indica* causes Karnal bunt (see later). The first two species are similar in their life histories and disease development. The biology of *T. controversa* is different and somewhat similar to that of *T. indica*. The pathogens produce teliospores with different sets of wall markings.

The mycelium is hyaline. During sporulation, most cells are transformed into spherical, brownish teliospores, while the rest of the mycelial cells remain hyaline and sterile. On germination of a teliospore a basidium is produced, at the end of which 8 to 16 basidiospores develop in *T. caries* and *T. laevis*, whereas 14 to 30 basidiospores develop in the dwarf bunt fungus *T. controversa* and 32 to 128 in *T. indica*. Basidiospores, usually called primary sporidia, fuse in pairs through the production of lateral branches between compatible mating types and appear as H-shaped structures (Figs. 11-148G and 11-149). The nucleus of each primary sporidium divides, and through exchange of one of the nuclei the two fused primary sporidia become dikaryotic. When the primary sporidia germinate, they produce dikaryotic secondary sporidia. These produce dikaryotic mycelium, which can penetrate the plants and cause infection. After systemic development through the plant, the mycelium again forms teliospores in the kernels.

Development of Disease

The pathogens of common bunt overwinter as teliospores on contaminated wheat kernels and less frequently in the soil. Teliospores of the common bunt fungi are short lived in wet areas, losing viability within two years, whereas those of the dwarf bunt and Karnal bunt fungi may remain viable in any soil for at least three years and often for as long as 10 years.

When contaminated seed or healthy seed is sown in bunt-infested fields, approximately the same conditions that favor the germination of seeds favor the germination of common bunt teliospores. Teliospores of the common bunt fungi germinate readily, and as the young seedling emerges from the kernel, the teliospore on the kernel or near the seedling also germinates through the production of the basidium, primary sporidia, and secondary sporidia (Figs. 11-148 and 11-149). The secondary sporidia then germinate, and the dikaryotic mycelium they produce infects the young seedling.

Teliospores of the dwarf bunt fungus, however, germinate slowly even under optimum conditions of temperature (3–8°C) and moisture, requiring from 3 to 10 weeks for maximum germination. Persistent snow cover, providing soil surface temperatures of –2 to 2°C, is consistently correlated with high dwarf bunt incidence. Dwarf bunt infections apparently originate from teliospores germinating at or near the soil surface from December through early April. Germinating secondary sporidia penetrate the tiller initials of wheat seedlings after seedling emergence. The more tiller initials formed during the infection period, the greater the incidence of bunted plants and of bunted heads per plant. Germinating seedlings and older tillers apparently are not susceptible to infection by the dwarf bunt fungus.

After penetration, the mycelium grows intercellularly and invades the developing leaves and the meristematic tissue at the growing point of the plant. The mycelium remains dormant in the seedling during the winter; however, when the seedling begins to grow again in the spring, the mycelium resumes its growth and grows with the growing point. When the plant forms the head of the grain, the mycelium invades all parts of it even before the head emerges. As the head fills and becomes mature, the mycelial threads increase in number and soon take over and consume the contents of the kernel cells. The mycelium, however, does not affect the tissues of the pericarp of the kernel, which form a rather sturdy covering for the smutted mass they contain. At the same time, most hyphal cells are transformed into teliospores.

Smutted kernels are usually kept intact while on the plant, but break and release their spores on harvest or

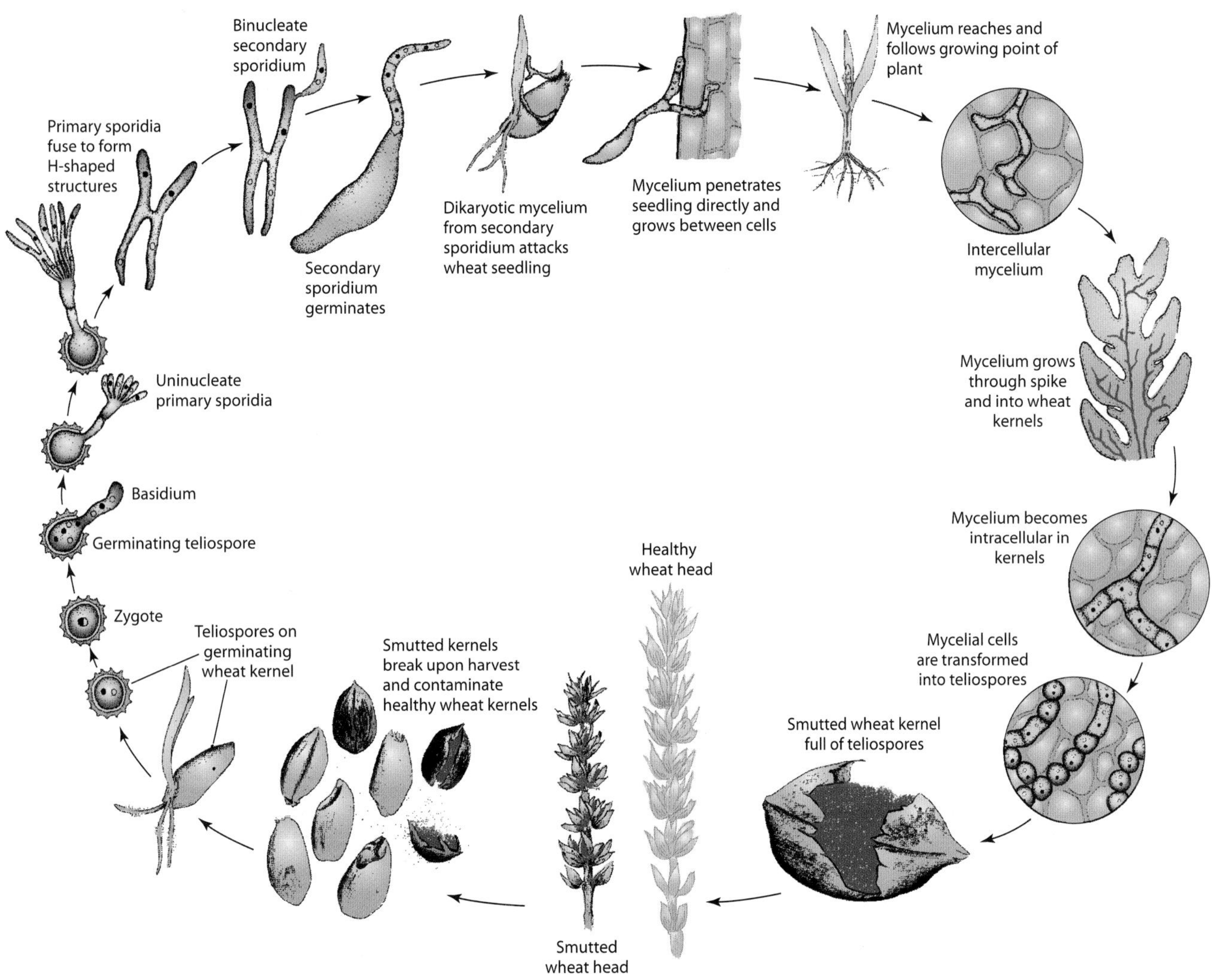

FIGURE 11-149 Disease cycle of covered smut or bunt of wheat caused by *Telletia* sp.

threshing. The liberated spores contaminate the healthy kernels and are also blown away by air currents, thus contaminating the soil.

Control

Common bunt can be controlled by using smut-free seed of a resistant variety treated with an appropriate fungicide. Contaminated seed should be cleaned to remove any unbroken, infected kernels and as many of the smut spores on the seed as possible. The seed is then treated with appropriate fungicides. In dwarf bunt, Karnal bunt, and in common bunt in drier areas, the spores survive in the soil for long periods and can cause infection of seedlings. In such cases, the most effective control is through the use of resistant cultivars, whereas seed treatments with certain systemic fungicides are moderately effective.

BOX 20 Karnal Bunt of Small Grains — Legitimate Concerns and Political Predicaments

Karnal bunt, named after the town Karnal of India where it was first observed in 1931, affects wheat and triticale (a hybrid of wheat and rye) on which it causes a covered smut (Fig. 11-150). Karnal bunt is caused by the fungus *Tilletia indica*, known previously as *Neovossia indica*. Karnal bunt is also known as partial bunt because only a portion of the wheat kernels is affected. The disease has now been reported from several other countries in Asia, in the early 1980s from Mexico, and, since 1996, from a few places in the southwestern United States. A great effort was launched in the United States to determine the extent of the area to which the Karnal bunt fungus had spread. These efforts were hampered somewhat by the fact that the teliospores of this fungus were very similar in appearance to those of the fungus *Tilletia walkeri*, the cause of ryegrass smut. This similarity led at first to several misdiagnoses and to unnecessary augmentation of the area where Karnal bunt presumably existed. By the end of 2001, Karnal bunt was found only in a few areas in Arizona, California, and Texas.

Karnal bunt has a minimal effect on yield and quality of wheat. Reported yield losses from Karnal bunt in India and in Mexico varied from 0.12 to 0.5%. In addition, the disease is managed easily through the use of clean seed treated with appropriate fungicides and application of appropriate agricultural practices. Also, the disease cannot become established in new locations that do not provide favorable climatic conditions, which, for teliospore and sporidia germination, include high relative humidity or free water and temperatures around 20°C. However, because countries presently free from the disease do not allow importation of Karnal bunt-infected wheat, the inability to export wheat contaminated with the disease has created an emergency situation in the United States, one of the largest exporters of wheat. Wheat infected with the Karnal bunt fungus is not toxic to humans or animals.

The Karnal bunt fungus overwinters as teliospores (Fig. 11-150D) on the soil. In the presence of moisture, teliospores germinate by producing a basidium

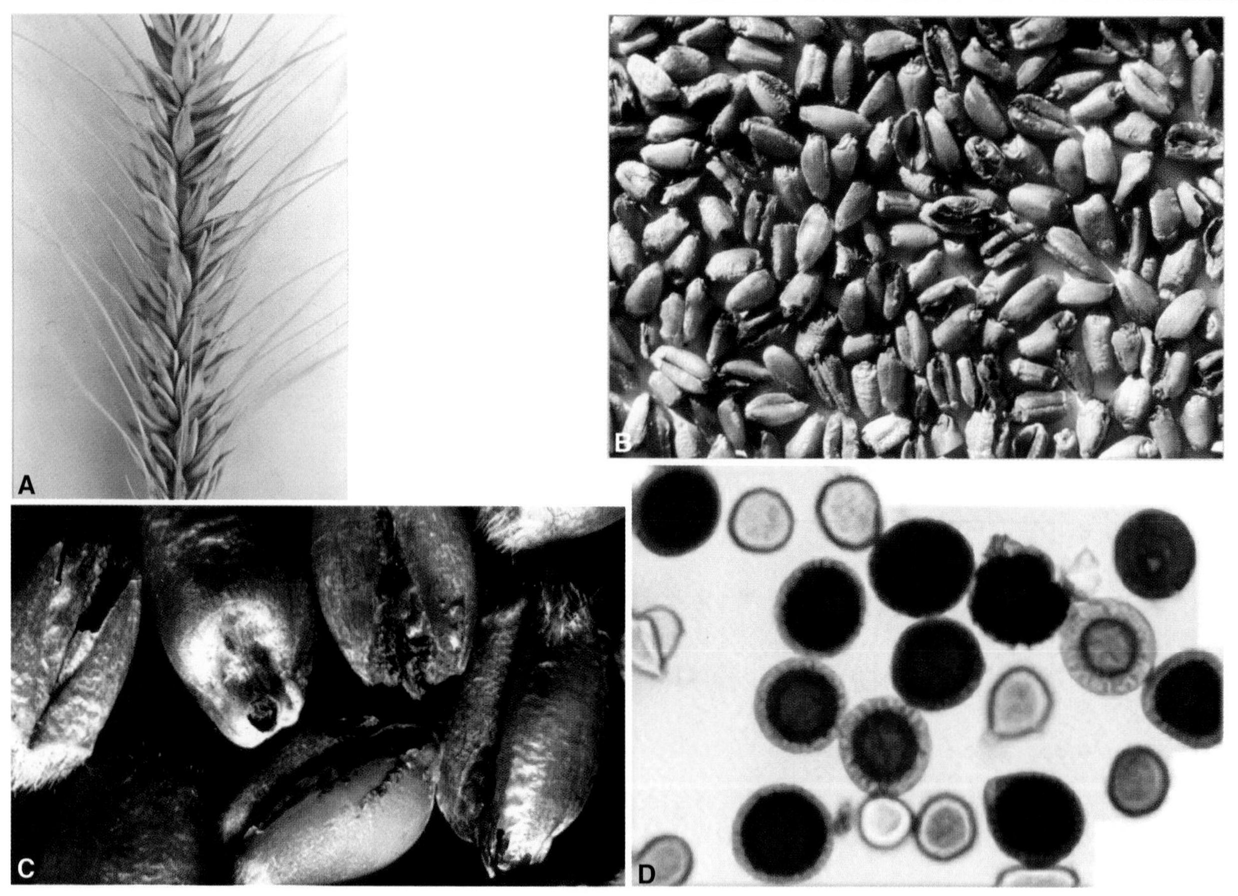

FIGURE 11-150 Karnal bunt caused by *Tilletia indica*. (A) Head of wheat containing kernels infected with Karnal bunt. Numerous wheat kernels infected with Karnal bunt (B) and close-up of a few infected kernels indicating the possible severity of infection (C). (D) Teliospores of Karnal bunt at various stages of maturity. (Photographs courtesy of USDA.)

(promycelium) that, in turn, produces as many as 180 primary sporidia. The primary sporidia germinate and produce mycelia, which then produce large numbers of secondary sporidia. At the time of flowering, both primary and secondary sporidia are blown or splashed upward on wheat plants and those that reach the plant head infect the developing kernels. The fungus is restricted to the pericarp of the kernel and there, as the kernels mature, the fungus produces large numbers of teliospores and, if the kernels rupture at harvest, the teliospores are released on the soil. Teliospores can survive in soil for five years and can survive in dry wheat seed for several years.

Karnal bunt is a disease that could, at some point, become a serious disease of wheat and other grains. So far, however, the disease causes little yield loss and is of concern only because strict international quarantines prohibit the international sale of Karnal bunt-contaminated wheat. The measures and their costs required for monitoring wheat fields in the United States for Karnal bunt are considerable, as are delays in processing and shipping of wheat internationally. This, of course, is in addition to measures taken for reducing and certifying, for export to certain countries, wheat as free of, or containing no more than a low maximum number of teliospores of the much more severe dwarf bunt disease of wheat, caused by *Tilletia controversa*. The cases of both diseases point out the need for early detection, unambiguous diagnosis, and effective monitoring of a pathogen so that it is kept from entering a country that, so far, is free from it, avoiding imposing quarantines in areas where only look-alike pathogens are present, and limiting the quarantine to areas in which the real pathogen is present.

Selected References

Christensen, J. J. (1963). Corn smut caused by *Ustilago maydis*. *Am. Phytopathol. Soc. Monogr.* **2**, 1–41.

Bonde, M. R., Peterson, G. L., Schaad, N. W., *et al.* (1997). Karnal bunt of wheat. *Plant Dis.* **81**, 1370–1377.

Fischer, G. W., and Holton, C. S. (1957). "Biology and Control of the Smut Fungi." Ronald, New York.

Gogoi, R., Singh, D. V., and Srivastava, K. D. (2001). Phenois as a biochemical basis of resistance in wheat against Karnal bunt. *Plant Pathol.* **50**, 470–476.

Hoffmann, J. A. (1982). Bunt of wheat. *Plant Dis.* **66**, 979–986.

Joshi, L. M., *et al.* (1983). Karnal bunt: A minor disease that is now a threat to wheat. *Bot. Rev.* **49**, 309–330.

Mathre, D. E., ed. (1982). "Compendium of Barley Diseases." APS Press, St. Paul, MN.

Mathre, D. E. (1996). Dwarf bunt: Politics, identification, and biology. *Annu. Rev. Phytopathol.* **34**, 67–85.

Snetselaar, K. M., and Mims, C. W. (1993). Infection of maize stigmas by *Ustilago maydis*: Light and electron microscopy. *Phytopathology* **83**, 843–850.

Thakur, R. P., Leonard, K. J., and Pataky, J. K. (1989). Smut gall development in adult corn plants inoculated with *Ustilago maydis*. *Plant Dis.* **73**, 921–925.

Thomas, P. L. (1991). Genetics of small-grain smuts. *Annu. Rev. Phytopathol.* **29**, 137–148.

Trione, E. J. (1982). Dwarf bunt of wheat and its importance in international wheat trade. *Plant Dis.* **66**, 1083–1088.

Trione, E. J., Stockwell, W. O., and Latham, C. J. (1989). Floret development and teliospore production in bunt-infected wheat, in plant and in cultured spikelets. *Phytopathology* **79**, 999–1002.

Urech, P. A. (1972). Investigations on the corn smut caused by *Ustilago maydis*. *Phytopathol. Z.* **73**, 1–26.

Warham, E. J. (1992). Karnal bunt of wheat. *In* "Plant Diseases of International Importance" (U.S. Singh, A. N. Mukhopadhyay, J. Kumar, and H. S. Chaube, eds.), Vol. 1, pp. 1–24. Prentice-Hall, Englewood Cliffs, NJ.

Whitaker, T. B., Wu, J., Peterson, G. L., *et al.* (2001). Variability associated with the official USDA sampling plan used to inspect export wheat shipments for *Tilletia controversa* spores. *Plant Pathol.* **50**, 755–760.

Wiese, M. V. (1987). "Compendium of Wheat Diseases," 2nd Ed. APS Press, St. Paul, MN.

ROOT AND STEM ROTS CAUSED BY BASIDIOMYCETES

Several Basidiomycetes cause serious plant losses by attacking primarily the roots and lower stems of plants (Fig. 11-129). Some of these fungi, e.g., *Rhizoctonia* (teleomorph: *Thanatephorus*) and *Sclerotium* (teleomorph: *Aethalium*), attack primarily herbaceous plants. Some, like *Typhula*, attack only grasses and some, like *Marasmius*, affect primarily turfgrasses. However, some other fungi, e.g., *Armillaria*, some species of *Heterobasidion*, particularly *H. annosum*, and of *Phellinus* and *Polyporus*, attack only roots and lower stems of woody plants, primarily forest trees, and certain fruit trees. Other fungi cause root or crown rot of tropical plants such as banana and sugarcane, witches'-broom of cacao (*Crinipellis* spp.), or wiry cord blights on the tops of tropical trees.

ROOT AND STEM ROT DISEASES CAUSED BY THE "STERILE FUNGI" RHIZOCTONIA AND SCLEROTIUM

Fungi *Rhizoctonia* and *Sclerotium* are soil inhabitant basidiomycetes and cause serious diseases on many hosts by affecting the roots, stems, tubers, corms, and other plant parts that develop in or on the ground. These two fungi were known as sterile fungi because for many years they were thought to produce only sclerotia and to be incapable of producing spores of any kind, either sexual or asexual. The two were distinguished from one another by the characteristics of their mycelium and by the fact that *Rhizoctonia* sclerotia have a uniform texture throughout, whereas *Sclerotium* sclerotia are internally differentiated into three areas. It is known now that at least some species within these two genera produce basidiospores as their sexual spores and, therefore, they are Basidiomycetes. However, some fungi previously thought to belong to *Sclerotium*, e.g., *S. bataticola*, causing stem blight of beans, are now known to produce conidia (*Macrophomina*), and some (e.g., *S. oryzae*) produce ascospores (*Magnaporthe*). Others, however, such as *S. cepivorum*, causing white rot of onion, still have no known spore stage. In any case, the spores of *Rhizoctonia* and *Sclerotium* either are produced only under special conditions in the laboratory or are extremely rare in nature and therefore of little value in identifying the fungus. For these reasons, these fungi continue to be considered as sterile mycelia and because, for all practical purposes, they behave as such they continue to be referred to by the names *Rhizoctonia* and *Sclerotium*.

RHIZOCTONIA DISEASES

Rhizoctonia diseases occur throughout the world. They cause losses on almost all vegetables and flowers, several field crops, turfgrasses, and even perennial ornamentals, shrubs, and trees. Symptoms may vary somewhat on the different crops, with the stage of growth at which the plant becomes infected, and with the prevailing environmental conditions. The most common symptoms on most plants are damping-off of seedlings and root rot, stem rot, or stem canker of growing and grown plants (Fig. 11-151). On some hosts, however, *Rhizoctonia* also causes rotting of storage organs (Fig. 11-152) and foliage blights or spots (Fig. 11-153), especially of foliage near the ground.

Damping-off is probably the most common symptom caused by *Rhizoctonia* on most plants it affects (Fig. 11-151E). It occurs primarily in cold, wet soils. Very young seedlings may be killed before or soon after they emerge from the soil. Thick, fleshy seedlings such as those of legumes and the sprouts from potato tubers may show noticeable brown lesions and dead tips before they are killed. After the seedlings have emerged, the fungus attacks their stem and makes it water soaked, soft, and incapable of supporting the seedling, which then falls over and dies. In older seedlings, invasion of the fungus is limited to the outer cortical tissues, which develop elongate, tan to reddish-brown lesions. The lesions may increase in length and width until they finally girdle the stem, and the plant may die; alternatively, as often happens in crucifers, before the plant dies the stem turns brownish black and may be bent or twisted without breaking, giving the disease the name wire stem (Fig. 11-151A).

A **seedling stem canker**, known as soreshin, is common and destructive in cotton, tobacco, and other seedlings that have escaped the damping-off or seedling blight phase of the disease. It develops under conditions that are not especially favorable to the disease. Soreshin lesions appear as reddish-brown, sunken cankers that range from narrow to completely girdling the stem near the soil line (Fig. 11-151). As soil temperature rises later in the season, affected plants may show partial recovery due to new root growth.

Root lesions form in seedlings and on partly grown or mature plants. Reddish-brown lesions usually appear first just below the soil line, but in cool, wet weather the lesions enlarge in all directions and may increase in size and number to include the whole base of the plant and most of the roots. This results in weakening, yellowing, and sometimes death of the plant.

On **low-lying plants** such as lettuce and cabbage, lower leaves touching the ground or close to it are attacked at the petioles and midribs. Reddish-brown, slightly sunken lesions develop and the entire leaf becomes dark brown and slimy. From the lower leaves the infection spreads upward to the next leaves until most or all leaves, and the head, may be invaded and rot, with mycelium and sclerotia permeating the tissues or nestled between the leaves.

On lawn and turfgrasses, *Rhizoctonia* causes **brown patch** (Fig. 11-153C), a disease particularly severe during periods of hot and humid or wet weather, especially with heavy dew periods. Roughly circular areas appear, ranging from a few centimeters to one or more meters in diameter, in which the grass blades become water soaked and dark at first but soon become dry, wither, and turn light brown. Diseased areas appear slightly sunken; at the border of the diseased areas, where the fungus is still active and attacking new grass blades, however, infected leaves look water soaked and dark. On damp days or in the early morning hours the areas appear as a characteristic grayish black "smoke" ring 2 to 5 centimeters wide. As the grass dries, the activity of the fungus slows down or stops and the ring disappears. Brown to black, hard, round sclerotia about 2 millimeters in diameter form in the thatch, diseased plants, and soil. In brown patch, *Rhizoctonia* usually kills only the leaf blades, and plants in the affected area begin to recover and grow again from the center outward, resulting in a doughnut-shaped diseased area.

On **fleshy, succulent stems and roots** and on tubers, bulbs, and corms, *Rhizoctonia* causes brown rotten areas that may be superficial or may extend inward to the middle of the root or stem. The rotting tissues usually decompose and dry, forming a sunken area filled with the dried plant parts mixed with fungus mycelium and sclerotia (Fig. 11-152). On **potato tubers**, *Rhizoctonia* causes "black scurf," in which small, hard, black

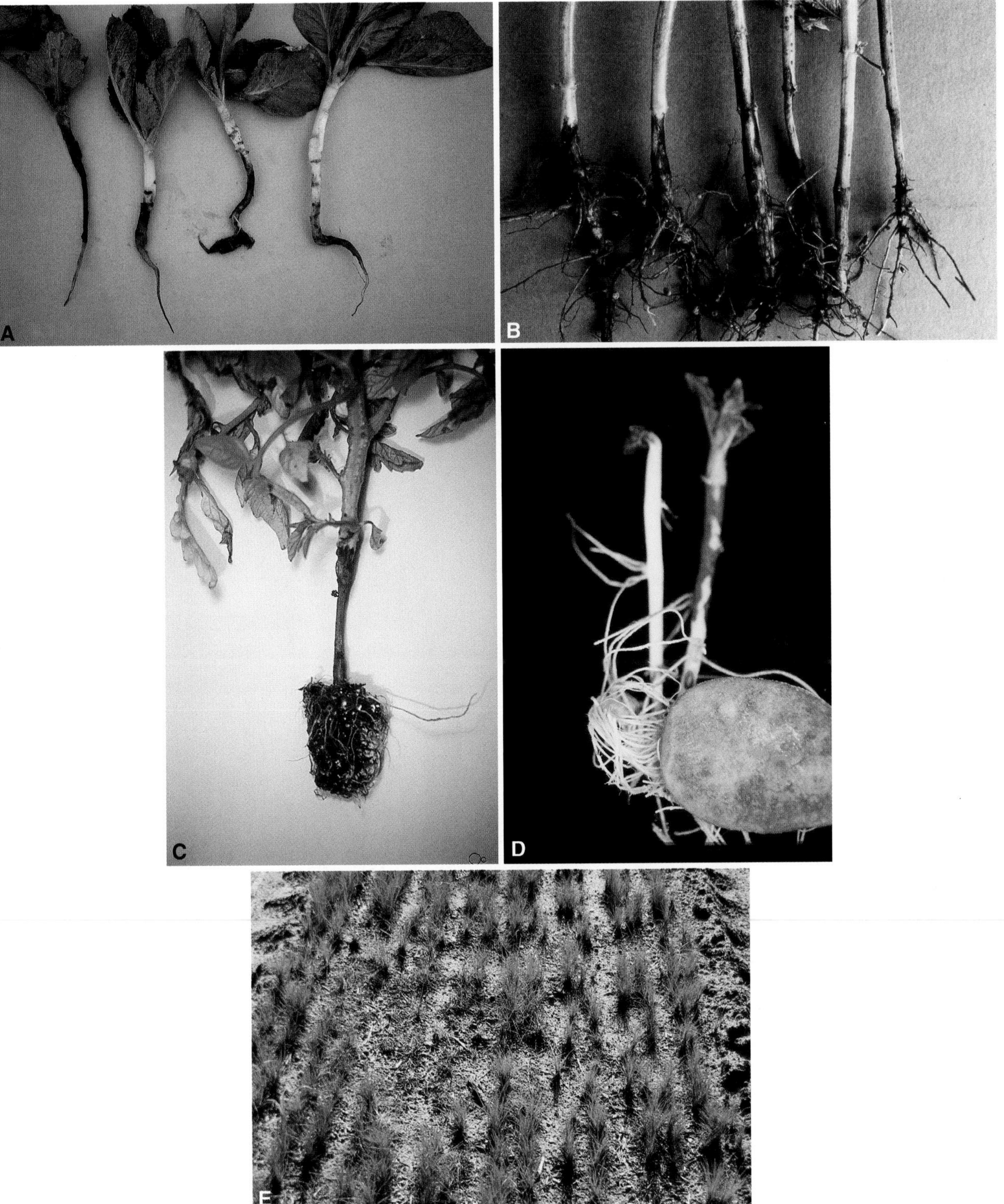

FIGURE 11-151 Symptoms of various diseases caused by *Rhizoctonia* sp. (A) Root and stem rot (wire stem) of cabbage. (B) Root and stem rot of soybeans. (C) Root and stem rot of potted tomato plant. (D) Stem rot of germinating potato tuber. (E) Patch of seedlings in pine tree nursery killed by *Rhizoctonia*. [Photographs courtesy of (A) Plant Pathology Department, University of Florida, (B) T.R. Anderson, WCPD, (C) R.J. McGovern, (D) D.P. Weingartner, and (E) E.L. Barnard, Florida Department of Agriculture, Forestry Division.]

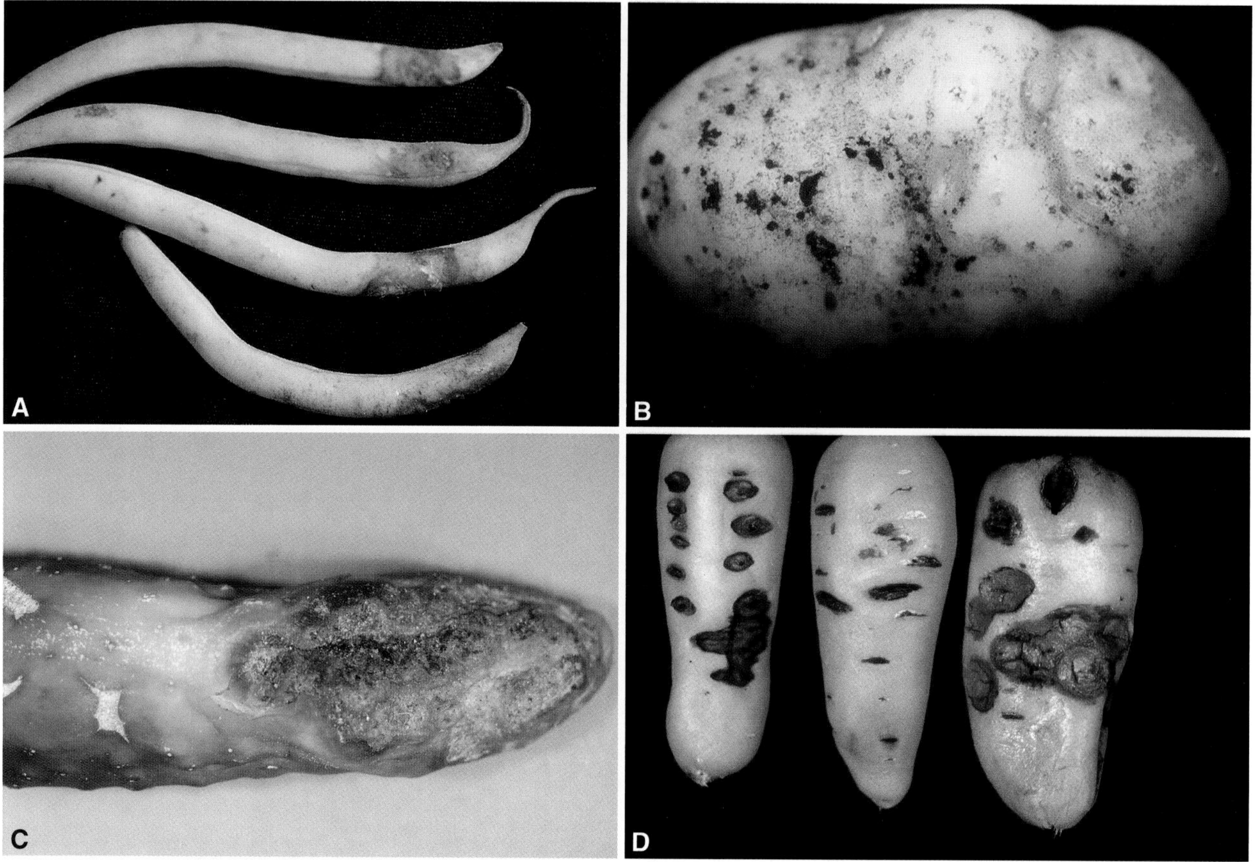

FIGURE 11-152 Rhizoctonia symptoms on soft fruits and vegetables. (A) Rot of bean pods. (B) Scarf of potato. (C) Belly rot of cucumber. (D) Crater rot of carrot. [Photographs courtesy of (A, C, and D) Plant Pathology Department, University of Florida and (B) D.P. Weingartner, University of Florida.]

sclerotia occur on the tuber surface and are not removed by washing (Fig. 11-152B), or "russeting" or "russet scab," in which the skin becomes roughened in a crisscross pattern resembling the shallow form of common potato scab.

Finally, *Rhizoctonia* causes **rots on fruits and pods** lying on or near the soil, such as cucumbers (Fig. 11-152), tomatoes, eggplants, and beans. These rots develop most frequently in wet, cool weather and appear first in the field but may continue to spread to other fruits after harvest and during transportation and storage.

In the **sheath and culm blight of rice** (Figs. 11-153A and 11-153B), one of the most serious diseases of rice and sometimes important on other cereals as well, different *Rhizoctonia* species cause large, irregular lesions that have a straw-colored center and a wide, reddish-brown margin. Seedlings and mature plants may become blighted under favorable conditions for the pathogen.

The Pathogen. *Rhizoctonia* spp. represent a large, diverse, and complex group of fungi. All *Rhizoctonia* fungi exist primarily as sterile mycelium and, sometimes, as small sclerotia that show no internal tissue differentiation. Mycelial cells of the most important species, *R. solani*, contain several nuclei (multinucleate *Rhizoctonia*), whereas mycelial cells of several other species contain two nuclei (binucleate *Rhizoctonia*). The mycelium, which is colorless when young but turns yellowish or light brown with age, consists of long cells and produces branches that grow at approximately right angles to the main hypha, are slightly constricted at the junction, and have a cross wall near the junction (Fig. 11-153D). The branching characteristics are usually the only ones available for identification of the fungus as *Rhizoctonia*. Under certain conditions the fungus produces sclerotia-like tufts of short, broad cells that function as chlamydospores, or eventually the tufts develop into rather small, loosely formed brown to black sclerotia, which are common on some hosts such as potato. As mentioned earlier, *Rhizoctonia* species infrequently produce a basidiomycetous perfect stage. The perfect stage of the multinucleate *R. solani* is *Thanatephorus cucumeris*, whereas that of binucleate *Rhizoctonia* is

FIGURE 11-153 Rice sheath blight caused by *Rhizoctonia*: early (A) and advanced (B) stages of the disease. (C) Brown patch disease of ryegrass caused by *Rhizoctonia*. (D) Typical *Rhizoctonia* mycelium showing its branching at a right angle and septa close to the branching point. [Photographs courtesy of (A and B) L.E. Datnoff, (C) T.E. Freeman, and (D) Plant Pathology Department, University of Florida.]

Ceratobasidium. A few multinucleate *Rhizoctonia* spp. (*R. zeae* and *R. oryzae*) have *Waitea* as their perfect basidiomycetous stage. The perfect stage forms under high humidity and appears as a thin, mildew-like growth on soil, leaves, and infected stems just above the ground line. Basidia are produced on a membranous layer of mycelium and have four sterigmata, each bearing one basidiospore.

It has now become evident that *Rhizoctonia solani* and other species are "collective" species, consisting of several more or less unrelated strains. The *Rhizoctonia* strains are distinguished from one another because **anastomosis** (fusion of touching hyphae) occurs only between isolates of the same **anastomosis group**. After anastomosis, which can be detected microscopically, an occasional heterokaryon hypha may be produced, under certain conditions, from one of the anastomosing cells. In the vast majority of anastomoses, however, five to six cells on either side of the fusion cells become vacuolated and die, appearing as a clear zone at the junction of two colonies. This "killing reaction" between isolates of the same anastomosis group is the expression of somatic or vegetative incompatibility. Such somatic incompatibility limits outbreeding to a few compatible pairings. The existence of anastomosis groups in *Rhizoctonia solani* represents genetic isolation of the populations in each group.

Although the various anastomosis groups are not entirely host specific, they show certain fairly well-defined tendencies: isolates of anastomosis group 1 (AG1) cause seed and hypocotyl rot and aerial (sheath) and web blights of many plant species; isolates of AG2 cause canker of root crops, wire stem on crucifers, and brown patch on turfgrasses; isolates of AG3 affect mostly potato, causing stem cankers and stolon lesions and producing black sclerotia on tubers; and isolates of AG4 infect a wide variety of plant species, causing seed and hypocotyl rot on almost all angiosperms and stem lesions near the soil line on most legumes, cotton, and sugar beets. Six more anastomosis groups are known within *R. solani* and there are many more in other *Rhizoctonia*. Recognition of the existence of anastomosis groups and of their lesser or greater host specificity has been important in determining the anastomosis group of the isolate that must be used for inoculations in breeding different crops for resistance to *Rhizoctonia* and of the propagules counted for making disease predictions for the various crops affected by that fungus.

Development of Disease. The pathogen overwinters usually as mycelium or sclerotia in the soil and in or on infected perennial plants or propagative material such as potato tubers. In some hosts the fungus may even be carried in the seed (Fig. 11-154). The fungus is present in most soils and, once established in a field, remains there indefinitely. The fungus spreads with rain, irrigation, or flood water; with tools and anything else that carries contaminated soil; and with infected or contaminated propagative materials. For most races of the fungus the optimum temperature for infection is about 15 to 18°C, but some races are most active at much higher temperatures, up to 35°C. Disease is more severe in soils that are moderately wet than in soils that are waterlogged or dry. Infection of young plants is most severe when plant growth is slow because of adverse environmental conditions for the plant.

Control. Control of *Rhizoctonia* diseases is difficult. Wet, poorly drained areas should be avoided or drained better. Disease-free seeds should be planted on raised beds under conditions that encourage fast growth of the seedling. There should be wide spaces among plants for good aeration of the soil surface and of plants. When possible, as in greenhouses and seed beds, the soil should be sterilized with steam or treated with chemicals. Drenching of soil with pentachloronitrobenzene (PCNB) helps reduce damping-off in seed beds and greenhouses. When specific races of the pathogen have built up, a 3-year crop rotation with another crop may be valuable. With most vegetables, no effective fungicides are available against *Rhizoctonia* diseases, although some other chemicals are sometimes recommended as soil drenches before planting and spraying them once or twice on the seedlings soon after emergence. On turfgrasses, fungicide applications with some contact and systemic fungicides seem to provide effective control.

Since the mid-1980s, tremendous efforts have gone into developing alternative, more effective means of control of *Rhizoctonia* diseases. Such methods include mulching of fields with certain plant materials or with photodegradable plastic, avoiding application of some herbicides that seem to increase *Rhizoctonia* diseases in certain crops, and, especially, using biological controls. Rhizoctonia is parasitized by several microorganisms, such as fungi, soil myxobacteria, and mycophagous nematodes. *Rhizoctonia* also often suffers from the so-called *Rhizoctonia* decline, which is caused by two or three infectious double-stranded RNAs. These RNAs, through anastomoses, spread from infected hypovirulent *Rhizoctonia* individuals to healthy virulent ones and reduce both their ability to cause disease and their ability to survive. Addition of these agents to *Rhizoctonia*-infested soil or to seeds, tubers, and transplants before planting in *Rhizoctonia*-infested soil reduces disease incidence and severity greatly in almost all crops. So far, however, biological controls are still at

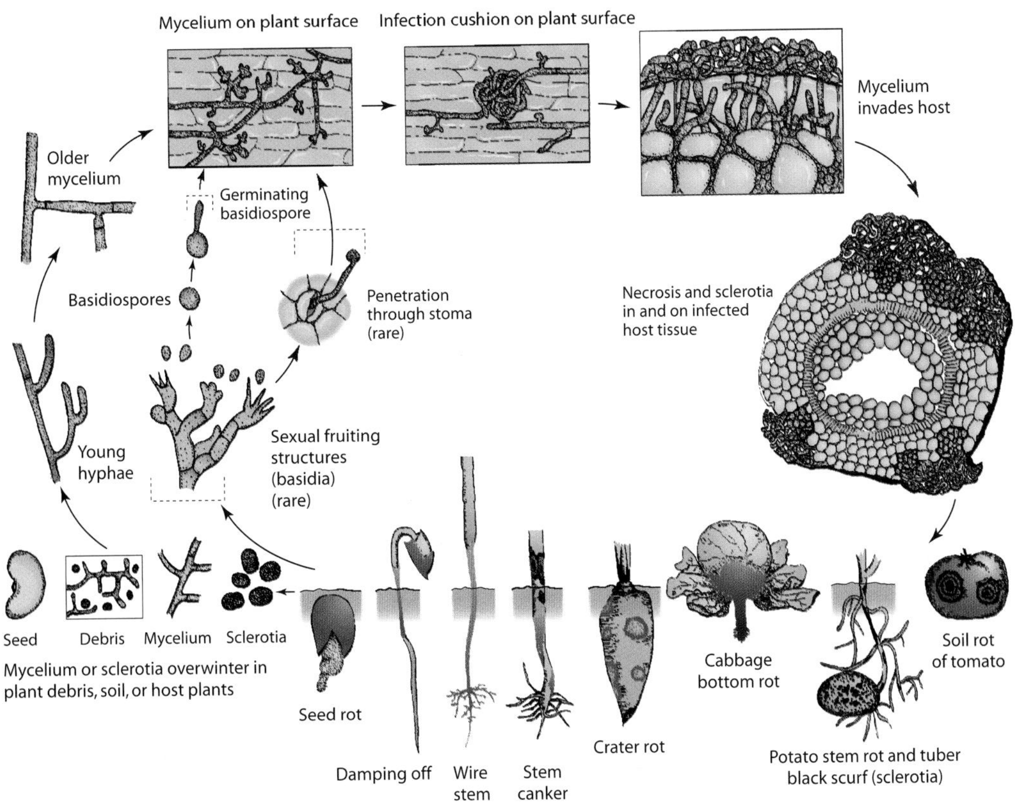

FIGURE 11-154 Disease cycle of *Rhizoctonia solani (Thanatephorus cucumeris)*.

the experimental stage and are not available for use by farmers.

SCLEROTIUM DISEASES

Sclerotium diseases occur primarily in warm climates. They cause damping-off of seedlings, stem canker, crown blight, root, crown, bulb, and tuber rot, and fruit rots (Figs. 11-155 and 11-156). *Sclerotium* frequently causes severe losses of fleshy fruits and vegetables during shipment and storage. In the United States, they are often called southern wilts or southern blights and affect a wide variety of plants, including most vegetables, flowers, legumes, cereals, forage plants, and weeds.

Symptoms. Seedlings are invaded by the fungus quickly and then die. Plants that have already developed some woody tissue are not invaded throughout, but the fungus grows into the cortex and girdles the plants slowly or quickly, which eventually die. Usually the infection begins on the succulent stem as a dark-brown lesion just below the soil line. Soon, at first the lower leaves and then the upper leaves turn yellow or wilt or die back from the tips downward. In plants with very succulent stems, such as celery, the stem may fall over, whereas in plants with harder stems, such as tomato, the invaded stem stands upright and begins to lose its leaves or to wilt (Fig. 11-155B). In the meantime, the fungus grows upward in the plant, covering the stem lesion with a cottony, white mass of mycelium (Fig. 11-155C), with the upward advance of the fungus depending on the amount of moisture present. The fungus moves even more rapidly downward into the roots and finally destroys the root system. The white mycelium is always present in and on infected tissues, and from these it grows over the soil to adjacent plants, starting new infections. Invaded stem, tuber, and fruit tissues are usually pale brown and soft but not watery. When fleshy roots or bulbs are infected (Fig. 11-156), a watery rot of the outer scales or root tissues may develop or the entire root or bulb may rot and disintegrate and be replaced by debris interwoven with mycelium. If bulbs, roots, and fruits are infected late in their development, symptoms may go unnoticed at harvest, but the disease continues as a storage rot.

On all infected tissues, and even on the nearby soil, the fungus produces numerous small roundish sclerotia of uniform size that are white when immature, becoming dark brown to black on maturity (Figs. 11-155A, 11-155C, and 11-155D). Each sclerotium is differentiated into an outer melanized rind, a middle cortex, and an innermost area of loosely arranged hyphae.

The Pathogen. The fungus, *Sclerotium* spp., produces abundant white, fluffy, branched mycelium that forms numerous sclerotia (Fig. 11-155A) but is usually sterile, i.e., does not produce spores. *Sclerotium rolfsii*, which causes the symptoms described earlier on most of the hosts, occasionally produces basidiospores at the margins of lesions under humid conditions. Its perfect stage is *Aethalium rolfsii*. As mentioned earlier, the species *S. bataticola*, which causes diseases in several different hosts, occasionally produces conidia in pycnidia and is now known as the imperfect fungus *Macrophomina phaseolina*. A third *Sclerotium* species, *S. cepivorum*, which causes the white rot disease of onion and garlic, in addition to sclerotia also produces occasional conidia on sporodochia; these conidia, however, seem to be sterile.

Development of Disease. The fungus overwinters mainly as sclerotia. It is spread by moving water, infested soil, contaminated tools, infected transplant seedlings, infected vegetables and fruits, and, in some hosts, as sclerotia mixed with the seed.

The fungus attacks tissues directly. However, the mass of mycelium it produces secretes oxalic acid and also pectinolytic, cellulolytic, and other enzymes and it kills and disintegrates tissues before it actually penetrates the host. Once established in the plants, the fungus advances and produces mycelium and sclerotia quite rapidly, especially at high moisture and high temperature (between 30 and 35°C).

Control. The control of *Sclerotium* diseases is difficult. Crop rotation, cultural practices such as deep plowing to bury fungal sclerotia in surface debris, fertilizing with ammonium-type fertilizers, applying calcium compounds, and, in some cases, application of fungicides such as PCNB to the soil before planting or in the furrow during planting provide only partial control.

In recent years, some control of *S. rolfsii* diseases has been obtained by soil solarization and by use of parasitizing and antagonistic species of fungi and bacteria. The latter are used for the treatment of seeds or other propagative organs of crops planted in *Sclerotium*-infested fields. So far, however, all such controls are at the experimental stage.

Selected References

Anderson, N. A. (1982). The genetics and pathology of *Rhizoctonia solani*. *Annu. Rev. Phytopathol.* **20**, 329–347.

Bruehl, G. W., ed. (1975). "Biology and Control of Soil-Borne Plant Pathogens." APS Press, St. Paul, MN.

Burpee, L., and Martin, B. (1992). Biology of *Rhizoctonia* species associated with turf grasses. *Plant Dis.* **76**, 112–117.

Christou, T. (1962). Penetration and host-parasite relationships of *Rhizoctonia solani* in the bean plant. *Phytopathology* **52**, 381–389.

Costanho, B., and Butler, E. E. (1978). *Rhizoctonia* decline: A degenerative disease of *Rhizoctonia solani*. II. Studies on hypovirulence and potential use in biological control. III. The association of double stranded RNA with *Rhizoctonia* decline. *Phytopathology* **68**, 1505–1519.

Dasgupta, M. K. (1992). Rice sheath blight: The challenge continues. *In* "Plant Diseases of International Importance" (U. S. Singh, A. N. Mukhopadhyay, J. Kumar, and H. S. Chaube, eds.), Vol. 1, pp. 130–157. Prentice-Hall, Englewood Cliffs, NJ.

Ellil, A. H. A. Abo (1999). Oxidative stress in relation to lipid peroxidation, sclerotial development and melanin production by *Sclerotium rolfsii*. *J. Phytopathol.* **147**, 561–566.

Hwang, J., and Benson, D. M. (2002). Biocontrol of Rhizoctonia stem and root rot of poinsettia with *Burkholderia cepacia* and binucleate *Rhizoctonia*. *Plant Dis.* **86**, 47–53.

Lees, A. K., Cullen, D. W., Sullivan, L., *et al.* (2002). Development of conventional and quantitative real-time PCR assays for the detection and identification of *Rhizoctonia solani* AG-3 in potato and soil. *Plant Pathol.* **51**, 293–302.

Lilja, A., and Rikala, R. (2000). Effect of uninucleate *Rhizoctonia* on Scots pine and Norway spruce seedlings. *Eur. J. Forest Pathol.* **30**, 109–115.

Lucas, P., Smiley, R. W., and Collins, H. P. (1993). Decline of *Rhizoctonia* root rot on wheat in soils infested with *Rhizoctonia solani* AG-8. *Phytopathology* **93**, 260–265.

Madi, L., and Katan, J. (1998). *Penicillium janczewskii* and its metabolites, applied to leaves, elicit systemic acquired resistance to stem rot caused by *Rhizoctonia solani*. *Physiol. Mol. Plant Pathol.* **53**, 163–175.

Metcalf, D. A., and Wilson, C. R. (2001). The process of antagonism of *Sclerotium cepivorum* in white rot affected onion roots by *Trichoderma koningii*. *Plant Pathol.* **50**, 249–257.

Nelson, E. G., and Hoitink, H. A. J. (1983). The role of microorganisms in the suppression of *Rhizoctonia solani* on container media amended with composted hardwood bark. *Phytopathology* **73**, 274–278.

Ogoshi, A. (1975). Grouping of *Rhizoctonia solani* and their perfect stages. *Rev. Plant Prot. Res.* **8**, 93–103.

Ogoshi, A. (1987). Ecology and pathogenicity of anastomosis and intraspecific groups of *Rhizoctonia solani* Kuhn. *Annu. Rev. Phytopathol.* **25**, 125–143.

Parmeter, J. R., Jr., ed. (1970). "*Rhizoctonia solani*, Biology and Pathology." Univ. of California Press, Berkeley.

Punja, Z. K. (1985). The biology, ecology, and control of *Sclerotium rolfsii*. *Annu. Rev. Phytopathol.* **23**, 97–127.

Savary, S., Castilla, N. P., and Willocquet, L. (2001). Analysis of the spatiotemporal structure of rice sheath blight epidemics in a farmer's field. *Plant Pathol.* **50**, 53–68.

Shew, H. D., and Melton, T. A. (1995). Target spot of tobacco. *Plant Dis.* **79**, 6–11.

Singleton, L. L., Mihail, J. D., and Rush, C. M. (1991). "Methods for Research on Soilborne Phytopathogenic Fungi." APS Press, St. Paul, MN.

Sneh, B., Burpee, L., and Ogoshi, A. (1991). "Identification of Rhizoctonia species." APS Press, St. Paul, MN.

Tu, C. C., and Kimbrough, J. W. (1978). Systematics and phylogeny of fungi in the *Rhizoctonia* complex. *Bot. Gaz. (Chicago)* **139**, 454–466.

Willocquet, L., Fernandez, L., and Savary, S. (2000). Effect of various crop establishment methods practiced by Asian farmers on epidemics of rice sheath blight caused by *Rhizoctonia solani*. *Plant Pathol.* **49**, 346–354.

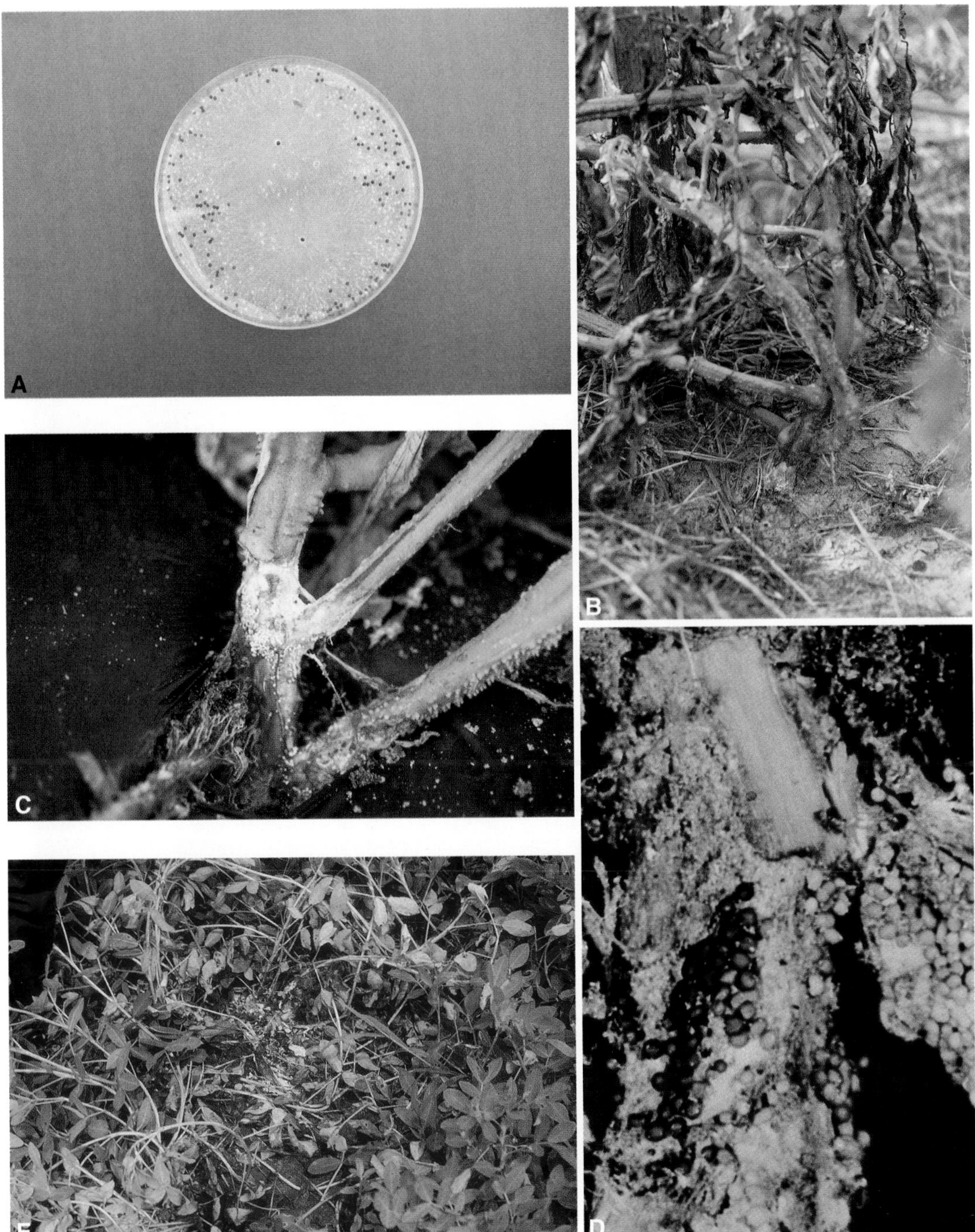

FIGURE 11-155 Symptoms of southern blight diseases caused by *Sclerotium* sp. (A) Culture of the fungus on nutrient media. Numerous spherical sclerotia can be seen. Southern blight of tomato showing lower stem rot (B and C) and sclerotia. (D) Sclerotium blight on tomato plant showing numerous sclerotia. (E) Sclerotium blight of peanuts. [Photographs courtesy of (A, B, D, and E) Plant Pathology Department, University of Florida and (C) R.J. McGovern, University of Florida.]

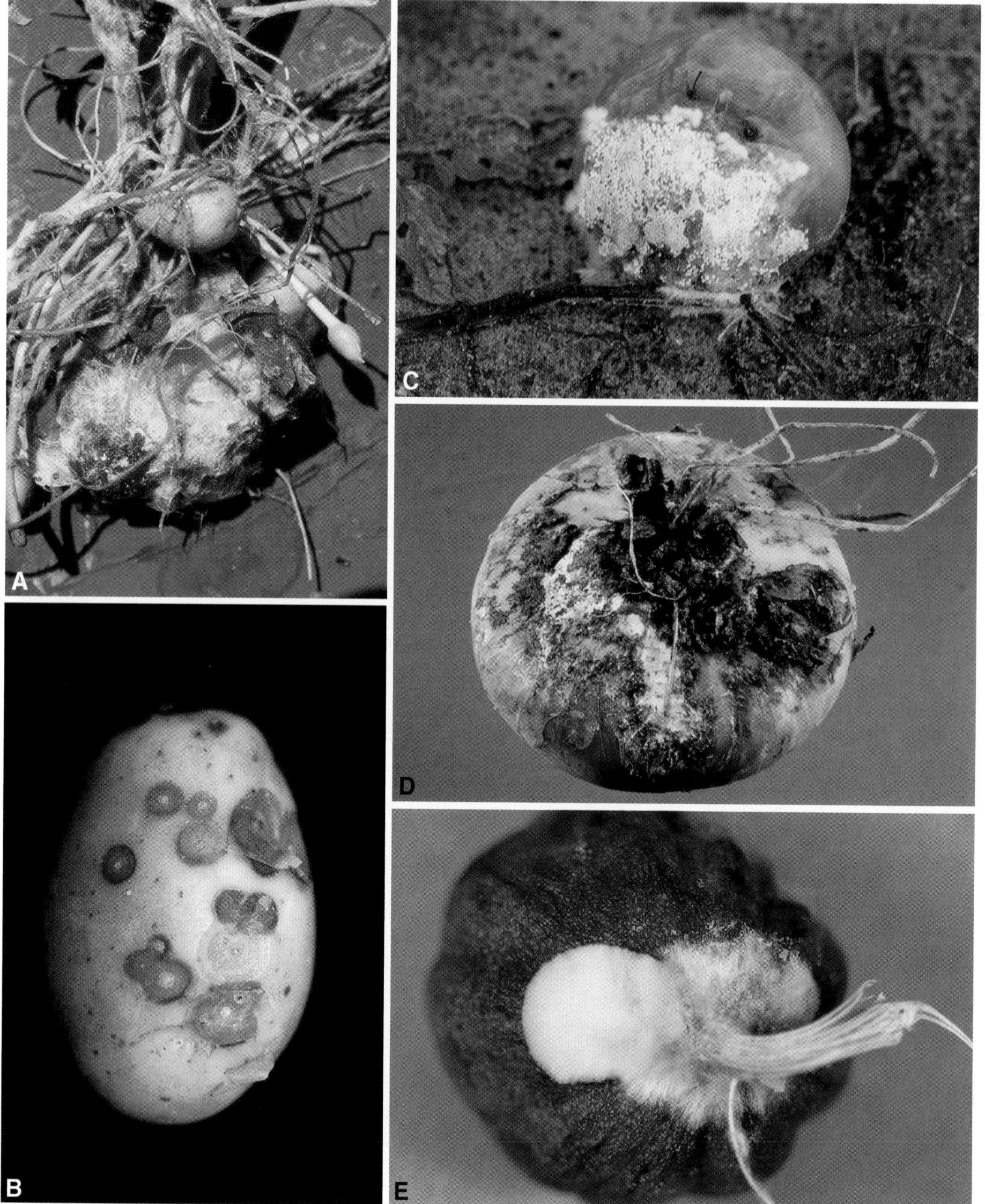

FIGURE 11-156 Sclerotium diseases of fleshy fruits and other organs. Rotting of potato seed and roots (A) and potato tubers (B). Rotting of tomato fruit (C), onion bulb (D), and squash (E). [Photographs courtesy of (A and B) D.P. Weigartner, (C and E) Plant Pathology Department, University of Florida, and (D) R.J. Howard, WCPD.]

ROOT ROTS OF TREES

ARMILLARIA ROOT ROT OF FRUIT AND FOREST TREES

Armillaria root rot occurs worldwide. It affects hundreds of species of fruit trees, vines, shrubs, and shade and forest trees, as well as other plants such as potatoes and strawberries. The disease is often known as shoestring root rot, mushroom root rot, or oak root fungus disease (Fig. 11-157). The pathogen, *Armillaria mellea* and related species, is one of the most common fungi in forest soils. The most spectacular losses occur in orchards or vineyards planted in recently cleared forest lands or in forest tree plantations, particularly in stands recently thinned. Most commonly, however, losses from

Armillaria root rot, caused primarily by *Armillaria ostoyae*, are greatest in forests, where they occur as steady but inconspicuous slow decline and death of occasional trees, with greater numbers of trees dying from this disease during periods of moisture stress or after defoliation.

Symptoms. Affected trees show symptoms similar to those caused by other root rot diseases, namely reduced growth, smaller, yellowish leaves (Fig. 11-157A), dieback of twigs and branches, and gradual or sudden death of the tree. Symptomatic trees may be scattered at first, but soon circular areas of diseased trees appear because of the spread of the fungus from its initial infection point. Diagnostic characteristics of Armillaria root rot appear at decayed areas in the bark, at the root–stem junction, and on the roots. White mycelial mats, their margins often veined and shaped like fans, form between the bark and wood (Fig. 11-157C). The mycelium may extend for a few feet upward in the phloem and cambium of the trunk and may cause white rot decay. In addition to mycelial fans, another even more characteristic sign of the disease is the formation of reddish-brown to black rhizomorphs or "shoestrings." These are cord-like threads of mycelium 1 to 3 millimeter in diameter, consisting of a compact outer layer of black mycelium and a core of white or colorless mycelium. The rhizomorphs often form a branched network of sorts on the roots or under the bark or, in severely decayed wood, some strands spread into the soil surrounding the roots (Fig. 11-157D). In areas in which the mycelium has invaded the cambium, cankers form and gum (in hardwoods) or resin (in conifers) is exuded from the infected area and flows into the soil. As the fungus gradually girdles and kills the tree at the base, infected wood changes from firm and slightly moist to somewhat soft and dry (Fig. 11-157E). At the base of dead or dying trees, a few to many honey-colored, speckled mushrooms, about 7 centimeters or more tall and with a cap 5 to 15 centimeters in diameter grow from trunks, stumps, or on the ground near infected roots (Fig. 11-157B). These are the fruiting bodies of *Armillaria*, which appear in early fall and within their radial gills produce numerous basidia and basidiospores.

Development of Disease. The fungus overwinters as mycelium or rhizomorphs in diseased trees or in decaying roots. The principal method of tree-to-tree spread of the fungus is through rhizomorphs or direct root contact. Rhizomorphs grow from roots of infected trees or from decaying roots or stumps through the soil to roots of adjacent healthy trees (Fig. 11-158). Also, pieces of rhizomorphs in infected plant debris may be carried by cultivating equipment into new areas. The fungus can spread by basidiospores. These generally colonize dead stumps or woody material first and then rhizomorphs radiating from them attack living roots directly or through wounds. When roots of trees are in contact with infected or decaying roots, mycelium may invade healthy roots appressed to diseased roots directly without forming rhizomorphs. In all cases, trees weakened from other causes are attacked much more easily by *Armillaria* than vigorous trees.

Control. The control of Armillaria root rot is usually not attempted under forest conditions. Generally, however, losses can be reduced by removing tree stumps and roots and by delaying planting, for several years, of susceptible fruit or forest trees in recently cleared forest land that had oaks or other plants favoring buildup of large amounts of *Armillaria* inoculum. Control of the disease in orchards and occasionally in forest plantations is attempted by digging a trench around infected trees and their neighbors to prevent the growth of rhizomorphs to adjacent trees and by local soil fumigation of the infested area to destroy the fungus in the soil before *Armillaria*-killed trees can be replaced.

Selected References

Bruhn, J. N., Wetteroff, J. R., Mihail, J. D., *et al.* (2000). Distribution of *Armillaria* species in upland Ozark Mountain forests with respect to site, overstory species composition and oak decline. *Eur. J. Forest Pathol.* **30**, 43–60.

Fox, R. T. V., ed. (2000). "*Armillaria* Root Rot: Biology and Control of Honey Fungus." Intercept Ltd., Andover, UK.

Morrison, D. J., and Pellow, K. W. (2002). Variation in virulence among isolates of *Armillaria ostoyae. Forest Pathol.* **32**, 99–107.

Munnecke, D. E., Kolbezen, M. J., Wilbur, W. D., and Ohr, H. D. (1981). Interactions involved in controlling *Armillaria mellea. Plant Dis.* **65**, 384–389.

O'Reilly, H. J. (1963). *Armillaria* root rot of deciduous fruits, nuts and grapevines. *Calif. Agric. Exp. Stn. Ext. Serv. Circ.* **525**, 1–15.

Rizzo, D. M., and Whiting, E. C. (1998). Spatial distribution of *Armillaria mellea* in pear orchards. *Plant Dis.* **82**,1226–1231.

Robinson, R. M., and Morrison, D. J. (2001). Lesion formation and host response to infection by *Armillaria ostoyae* in the roots of western larch and Douglas fir. *Forest Pathol.* **31**, 371–385.

Solla, A., Tomlinson, F., and Woodward, S. (2002). Penetration of *Picea sitchensis* root bark by *Armillaria mellea*, *Armillaria ostooyae* and *Heterobasidion annosum. Forest Pathol.* **32**, 55–70.

Wargo, P. M., and Kile, G. A. (1992). *Armillaria* root disease. *In* "Plant Diseases of International Importance" (A. N. Mukhopadhyay, J. Kumar, H. S. Chaube, and U. S. Singh, eds.), Vol. 4. Prentice-Hall, Englewood Cliffs, NJ.

Wargo, P. M., and Shaw, C. G., III (1985). *Armillaria* root rot: The puzzle is being solved. *Plant Dis.* **69**, 826–832.

Woodward, S. (2000). "*Armillaria* Root Rot: Biology and Control of Honey Fungus." (R. T. V. Fox, ed.). Intercept Press, Andover, UK.

FIGURE 11-157 Symptoms of root and stem rot of trees caused by *Armillaria mellea*. (A) Pine tree showing yellowing and thinning of its foliage due to nutrient and water stress caused by *Armillaria* infection. (B) *Armillaria* fruiting bodies (the mushrooms) growing out from infected roots of the tree. (C) Mycelial growth of *Armillaria* under the bark of an infected tree. (D) Thick, branching mycelial rhizomorphs of *Armillaria* growing under and over the bark of infected root and stems and toward adjacent trees. (E) Wood of infected roots and trunks becomes spongy and weak, cannot support the tree, and the latter falls over. (Photographs courtesy of U.S. Forest Service.)

WOOD ROTS AND DECAYS CAUSED BY BASIDIOMYCETES

Huge losses of timber trees in the forest and in harvested wood are caused every year by the wood-rotting Basidiomycetes (Figs. 11-160, 11-161, and 11-162). In living trees, most of the rotting is confined to the older, central wood of roots, stems, or branches sometimes referred to as heartwood (Figs. 11-159 and 11-161). Once the tree is cut, however, the outer wood, which is sometimes referred to as sapwood, is also attacked by the wood-rotting fungi. All types of wood products are also attacked by these fungi under favorable moisture conditions.

Depending on the tree part attacked, wood rots may be called root rots, root and butt rots, or stem rots. Fungi that cause tree or wood product decays grow inside the wood cells and utilize the cell wall compo-

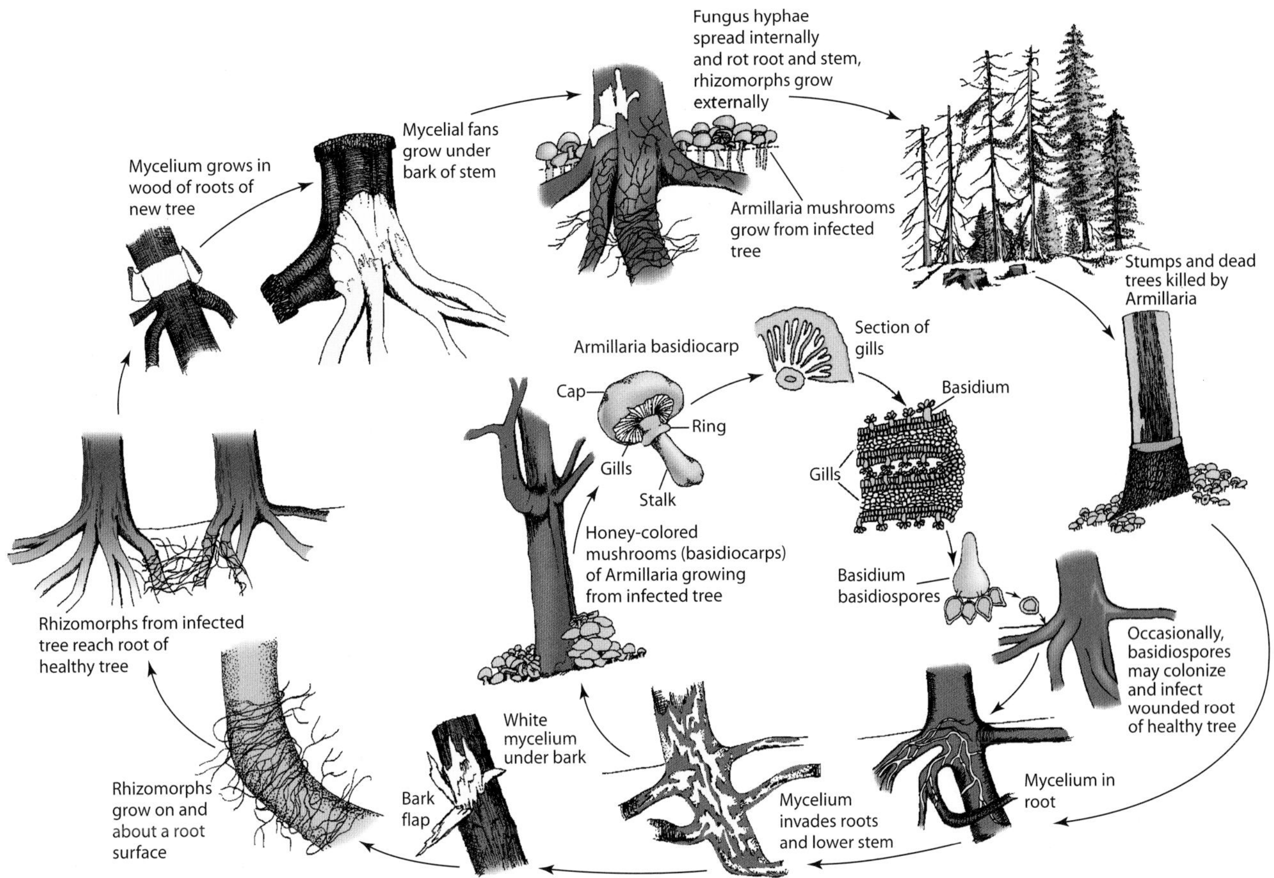

FIGURE 11-158 Disease cycle of root rots of trees caused by *Armillaria mellea.*

nents for food and energy. Some of them, the **brown-rot fungi**, attack preferably softwoods and break down and utilize primarily the cell wall polysaccharides (cellulose and hemicellulose), leaving the lignin more or less unaffected. This usually results in rotten wood that is some shade of brown and, in advanced stages, has a cubical pattern of cracking and a crumbly texture. Other wood rotters, the **white-rot fungi**, either decompose lignin and hemicellulose first and cellulose last or decompose all wood components simultaneously, in either case reducing the wood to a light-colored spongy mass (white rot) with white pockets or streaks separated by thin areas of firm wood (Figs. 11-161D, 11-161E, and 11-162D–11-162E). White-rot fungi are able to or preferably attack hardwoods normally resistant to brown-rot fungi.

It should be noted here that, in addition to the brown rots and white rots caused by Basidiomycetes, wood is also attacked by certain Ascomycetes and imperfect fungi. Some Ascomycetes, such as *Daldinia*, *Ustilina*, and *Xylaria*, cause a relatively slow white rot with variable black zone lines in and around the rotting wood, both in standing hardwood trees and in slash. In standing trees the decay is usually associated with wounds or cankers, whereas in wood pieces the decay is usually at or near a surface of wood that has high moisture content. Others, such as species of *Alternaria*, *Bisporomyces*, *Diplodia*, and *Paecilomyces*, cause the so-called **soft rots** of wood that affect the surface layers of wood pieces maintained more or less continuously at a high moisture content. Soft-rot fungi utilize both polysaccharides and lignin. They invade wood preferably through rays or vessels, from where they grow into the adjacent tracheids and invade their cell walls. Within the cell wall they produce conical or cylindrical cavities parallel to the orientation of the microfibrils; with progressing decay, the entire secondary wall is interlaced by confluent cavities. Several types of bacteria also attack wood, primarily in wood parenchyma rays, where they break down and utilize the contents and walls of the parenchyma cells, thus increasing the porosity and permeability of the wood to liquids, including solutions of fungal enzymes. Furthermore, several Ascomycetes and imperfect fungi result in the appearance of unsightly discolorations in the wood and thus reduce the quality but not the strength of the wood. Some of the **wood-**

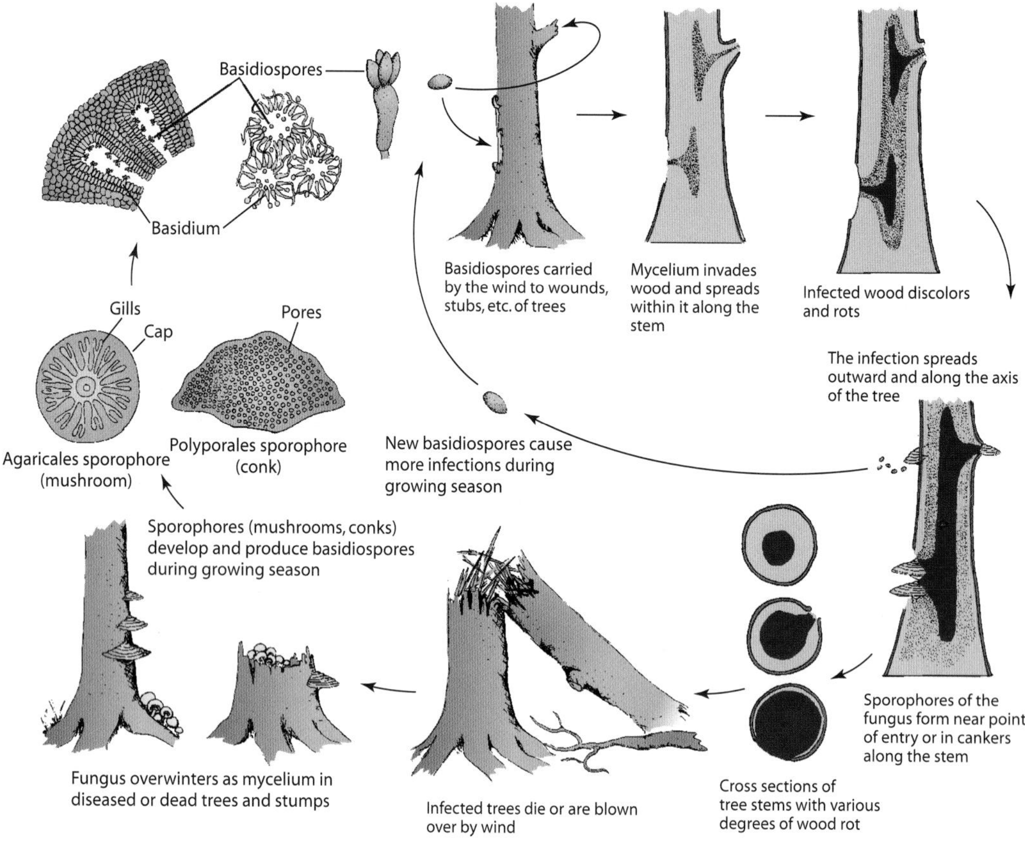

FIGURE 11-159 Disease cycle of wood-rotting fungi.

staining fungi are simply surface molds that usually grow on freshly cut surfaces of wood and impart to the wood the color of their spores; e.g., *Penicillium* stains wood green or yellow, *Aspergillus* stains wood black or green, *Fusarium* produces a red color, and *Rhizopus* causes a gray color (Figs. 11-162D and 11-162E). Other wood-staining fungi, however, usually called **sapstain** or **blue-stain fungi,** cause discoloration of the sapwood by producing pigmented hyphae that grow mainly in the ray parenchyma but can spread throughout the sapwood and cause lines of discoloration (Fig. 11-162D). Among blue-stain fungi are species of *Ceratocystis*, *Hypoxylon*, *Xylaria*, *Graphium*, *Leptographium*, *Diplodia*, and *Cladosporium*.

The bulk of wood rotting, however, is carried out by Basidiomycetes, and the most important such fungi that rot wood in standing trees or in wood products are the following.

- *Heterobasidion*, causing root and butt rot of living trees (Figs. 11-160D and 11-160E)
- *Polyporus*, a few species causing heart rot of living trees and rot of dead trees or logs
- *Inonotus*, causing a white heart rot of living trees and rot of dead trees or logs (Fig. 11-161)
- *Laetiporus*, causing a brown heart rot of many trees
- *Phellinus*, causing root rots on most conifers and many hardwood trees, and brown cubical rot in buildings and in stored lumber
- *Ganoderma*, causing root and basal rots in conifers and hardwoods, in palms, and in other tropical plantation trees (Figs. 11-160A–11-160C)
- *Chondrostereum*, causing "silver leaf" of fruit trees as a result of decay of the interior of the tree trunk and branches (*C. purpureum*), and heart rots of other trees
- *Peniophora*, causing decay in coniferous logs and pulpwood
- *Lenzites*, causing a brown cubical rot on coniferous logs, posts, and poles and a decay of hardwood slash
- *Pleurotus*, *Schizophyllum*, and *Trametes*, causing white rot in hardwoods

The development of wood rots varies with the particular fungus involved and the host tree attacked, but

FIGURE 11-160 (A) Ganoderma butt rot of palm tree. Two fruiting bodies (conks) grow out of the stem of an infected tree. (B) Top of a Ganoderma-infected palm tree showing yellowish brown foliage and general decline. (C) Ganoderma rot of trunk of citrus tree. (D) Pine trees show thinning foliage, dieback, and decline due to infection of the trees by the root- and stem-rotting fungus *Heterobasidion annosum*. Infected root (E) of one such tree (D) showing weak spongy wood. (F) Tree thoroughly rotten by *Hondrostereum purpureum* and supporting large numbers of its fructifications. (G) Fructifications (mushrooms) of the fungus *Marasmius*, the cause of fairy ring disease of turfgrasses. [Photographs courtesy of (A, B, F, and G) Plant Pathology Department, University of Florida. (C) R.J. McGovern, University of Florida, and (D and E) E.L. Barnard, Florida Department of Agriculture, Forestry Division.]

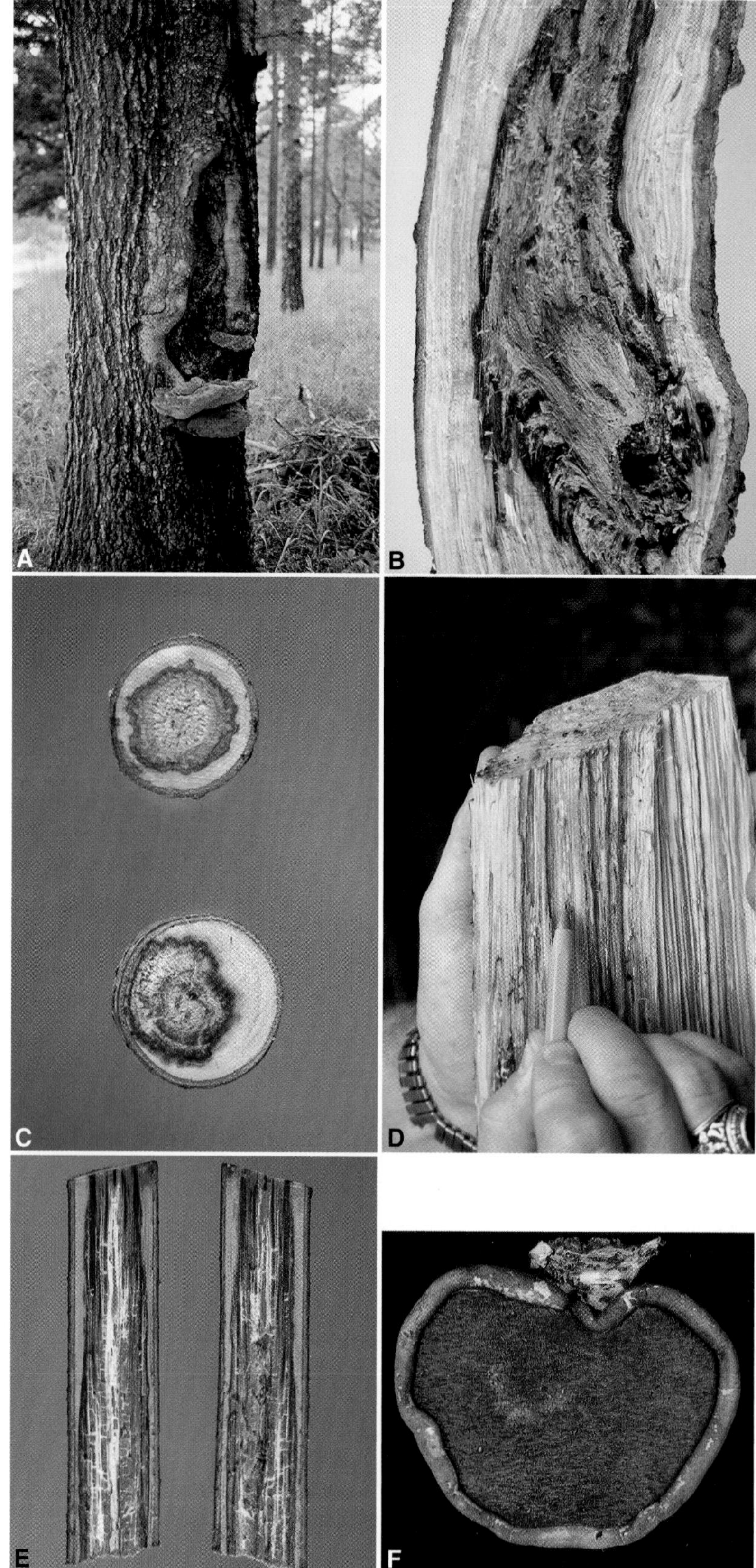

FIGURE 11-161 (A) Canker and fructifications of *Inonotus hispidus* on oak tree. (B and C) Heart rot caused by *Inonotus* on stem and branches of infected trees. Early (D) and more advanced (E) white rot of wood infected with *Inonotus* sp. (F) Sporocarp of *Inonotus betulinus*. [Photographs courtesy of (A–E) E.L. Barnard, Florida Department of Agriculture, Forestry Division.]

FIGURE 11-162 (A) Large canker originating at decaying smaller branch. (B) Remnants of tree trunk attacked by wood-rotting fungi. (C) White rot caused by *Phellinus* sp. (D) Zone lines of white rot fungus (*P. igniarius*). (E) Advanced white rot on wood caused by *Trametes* and other wood-rotting fungi.

there are many similarities (Fig. 11-159). All wood rot fungi enter trees as germinating basidiospores or as mycelium through wounds, dead branches, branch stubs, tree stumps, or damaged roots and they spread from there to the heartwood or sapwood of the tree. Wounds caused by fire and by cutting operations are the most common points of entry. Fungi develop in the wood and spread upward, downward, or both in a cylinder much faster than they do radially. In some wood rots, especially those of hardwoods originating from wounds or branch stubs, the rotten cylinder is only a few inches in diameter, forms a column no larger than the diameter of the tree at the time of injury, and may extend to one or a few meters above and below the area where the fungus entered the tree. In other wood rots, particularly those of conifers, the rotten cylinder enlarges steadily until the tree is killed or blown over by heavy winds and it may extend upward over much of the height of the tree.

The process of discoloration and decay in the wood of living trees appears to be quite complex, involving a number of successive or overlapping events. First, there must be an injury to the tree that exposes the wood as a result of a dead or broken branch, animal damage, fire burn, or mechanical scraping. The injured cells and those around them undergo chemical changes such as oxidation and become discolored. As long as the wound is open, discoloration advances toward the pith and around the tree; however, if the wound is small and occurs early in the season, a new growth ring forms and its cells act as a barrier to the discoloration process. The discoloration moves up and down within the cylinder of barrier cells but not outward into the new and subsequent growth rings.

Of course, many microorganisms are likely to land or be brought to the surface of a tree wound, and many of them begin to grow on the moist surface. Among these, however, only some bacteria and some Ascomycetes or imperfect fungi manage to survive on the discolored wood of the wound. These microorganisms do not cause wood decay, but they increase the discoloration and wetness of the wood and erode parts of the cell walls. Such wood is called **wetwood, redheart,** or **blackheart**.

Finally, however, the wood-rotting Basidiomycetes become active and begin to disintegrate and digest the cell wall components. These wood rotters attack only the tissues that have already been altered first by the chemical processes and then by the bacteria and the Ascomycetes and imperfects. Thus, wood-rotting Basidiomycetes also remain confined to the discolored column within the new growth, being unable to attack the latter. The decay in the discolored column continues until the wood is completely decomposed, but the influx of new microorganisms through the wound continues, even after the first decay fungus has caused the tissue to rot, and stops only when all tissues are completely digested.

It should be noted that the process of discoloration and decay may take from 3 to 5 to 50 or 100 years. It is most common and rapid in older, larger trees, and the older the trees, the more likely they are to contain decay columns. The discoloration and decay process starting at a particular wound need not, of course, go through to completion. Quick healing of the wound, antagonisms among the microorganisms involved, natural wood resistance, and other factors may stop the process at any stage. However, a large tree is likely to be injured many times over its long life. The events described earlier may be repeated many times after each new wound is formed, and thus more and more of the wood may be involved in the more or less continuous process of discoloration and decay. The end result is formation of a single large column or multiple columns of discolored and decayed wood.

The sporophores or conks of wood-rotting Basidiomycetes appear near the point of entry of the fungus, near the base of the tree, in cankers or swollen knots along the stem of living trees, or along the length of the tree stem after its death. The sporophores of most wood-rotting fungi, such as *Inonotus* and *Phellinus*, are formed annually and do not last for more than a year, those of *Heterobasidion* are annual or perennial, and those of *Phellinus* are perennial, adding a layer of tissue with vertical tubes and pores each year. The sporophores produce basidiophores during part or most of the growing season, and the spores are carried by wind, rain, or animals to nearby trees.

The control of wood rots and decays is impossible in the forest, but losses can be reduced by (1) management practices that reduce or eliminate the chance of introducing the fungi into healthy stands, (2) conducting logging and thinning operations in a way that minimizes breakage of branches or other wounds of trees, and (3) harvesting trees before the age of extreme susceptibility to wood rot fungi. Damage caused by wood-rotting fungi in shade and fruit trees can be prevented or minimized by avoiding or preventing wounds; by pruning dead and dying branches with a flush cut as close to the main stem as possible but without cutting the collar-like part of the stem that surrounds the base of the branch; by cleaning wounds by cutting the torn bark and shaping the wound like a vertical ellipse; and by keeping the trees in good vigor through adequate irrigation and proper fertilization. Treating large cuts or wounds with a wound dressing or tree paint has been practiced routinely in the past, but its usefulness in preventing wood discoloration and decay is questionable.

Since the early 1960s, considerable success in controlling wood rot and decays has been obtained through the treatment of tree wounds and stumps with antagonistic fungi. In the case of *Heterobasidion (Fumes) annosum* root rot and butt rot of forest trees, commercial control of the disease is obtained by applying conidia of the antagonistic fungus to freshly cut stumps by mixing the spores with the oil that constantly lubricates the chain of the chain saw used to cut the trees.

The control of discoloration or decays in lumber and wood products is usually accomplished by drying the wood or by treating it with an organic mercuric or a chlorophenate fungicide or with a mixture of the two. Wood that is likely to be in contact with soil or other moist surfaces should be treated with one of several wood preservatives, such as creosote, pentachlorophenol, copper naphthanate, and zinc chromate.

Selected References

Adaskaveg, J. E., and Ogawa, J. M. (1990). Wood decay pathology of fruit and nut trees in California. *Plant Dis.* **74**, 341–352.

Bahnweg, G., Möller, E. M., Anegg, S., *et al.* (2002). Detection of *Heterobasidion annosum* s.l. [(Fr.) Bref.] in Norway spruce by polymerase chain reaction. *J. Phytopathol.* **150**, 382–389.

Blanchette, R. A. (1991). Delignification by wood-decay fungi. *Annu. Rev. Phytopathol.* **29**, 381–398.

Boyce, J. S. (1961). "Forest Pathology." McGraw-Hill, New York.

Chase, T. E., and Ullrich, R. C. (1988). *Heterobasidion annosum*, root and butt-rot of trees. *Adv. Plant Pathol.* **6**, 501–510.

Eslyn, W. E., Kirk, T. K., and Effland, M. J. (1975). Changes in the chemical composition of wood caused by six soft-rot fungi. *Phytopathology* **65**, 473–476.

Fischer, M., and Wagner, T. (1999). RFLP analysis as a tool for identification of lignicolous basidiomycetes: European polypores. *Eur. J. Forest Pathol.* **29**, 295–304.

Gilbertson, R. L. (1980). Wood-rotting fungi of North America. *Mycologia* **72**, 1–49.

Hansen, E. M., and Goheen, E. M. (2000). *Phellinus weirii* and other native root pathogens as determinants of forest structure and process in western North America. *Annu. Rev. Phytopathol.* **38**, 515–539.

Hiratsuka, Y., and Chakravarty, P. (1999). Role of *Phialemonium curvatum* as a potential biological control agent against a blue stain fungus on aspen. *Eur. J. Forest Pathol.* **29**, 305–310.

Ihrmark, K., Zheng, J., Stenström, E., *et al.* (2001). Presence of double-stranded RNA in *Heterobasidion annosum. Forest Pathol.* **31**, 387–394.

Jacobi, W. R., *et al.* (1980). Disease losses in North Carolina forests. I. Losses in softwoods. 1973–74. II. Losses in hardwoods. 1973–74. III. Rationale and recommendations for future cooperative survey efforts. *Plant Dis.* **64**, 573–576, 576–578, and 579–581.

Levy, J. F. (1965). The soft rot fungi: Their mode of action and significance in the degradation of wood. *Adv. Bot. Res.* **31**, 323–357.

Manion, P. (1991). "Tree Disease Concepts," 2nd Ed. Prentice-Hall, New York.

Miller, Holderness, Bridge, *et al.* (1999). Genetic diversity of *Ganoderma* in oil palm plantings. *Plant Pathol.* **48**, 595–603.

Moykkynen, T., Miina, J., and Pukkala, T. (2000). Optimizing the management of a *Picea abies* stand under risk of butt rot. *Eur. J. Forest Pathol.* **30**, 65–76.

Raymer, A. D. M., and Boddy, L. (1988). "Fungal Decomposition of Wood: Its Biology and Ecology." Wiley, New York.

Roy, G., Bussières, Laflamme, G., *et al.* (2001). *In vitro* inhibition of *Heterobasidion annosum* by *Phaeotheca dimorphospora. Forest Pathol.* **31**, 395–404.

Schulze (1999). Rapid detection of European *Heterobasidion annosum* intersterility groups and intergroup gene flow using taxon-specific competitive-priming PCR. (TSCP-PCR). *J. Phytopathol.* **147**, 125–127.

Shigo, A. L. (1967). Successions of organisms in discoloration and decay of wood. *Int. Rev. For. Res.* **2**, 237–299.

Shigo, A. L. (1984). Compartmentalization: A conceptual framework for understanding how trees grow and defend themselves. *Annu. Rev. Phytopathol.* **22**, 189–214.

Shigo, A. L. (1985). Compartmentalization of decay in trees. *Sci. Am.* **252**, 96–103.

Swedjemark, G., Johannesson, H., and Stenlid, J. (1999). Intraspecific variation in *Heterobasidion annosum* for growth in sapwood of *Picea abies* and *Pinus sylvestris. Eur. J. Forest Pathol.* **29**, 249–258.

Swedjemark, Stenlid, and Karlsson (2001). Variation in growth of *Heterobasidion annosum* among clones of *Picea abies* incubated for different periods of time. *Forest Pathol.* **31**, 163–175.

Utomo, C., and Niepold, F. (2000). Development of diagnostic methods for detecting *Ganoderma*-infected oil palms. *J. Phytopathol.* **148**, 507–514.

Woodward, S., Stenlid, J., Karjalainen, R., and Hutterman, A., eds. (1998). "*Heterobasidion annosum*: Biology, Ecology, Impact and Control." CAB International, Wallingford, UK.

Zabel, R. A., and Morrell, J. J. (1992). "Wood Microbiology: Decay and Its Prevention." Academic Press, San Diego.

Witches'-Broom of Cacao

Witches'–broom disease of cacao is caused by the basidiomycete *Crinipellis perniciosa.* The disease occurs in South and part of Central America and in some Caribbean islands, i.e., most of the cocoa-growing areas of the western hemisphere. Witches'-broom causes severe yield reductions approaching 90%. In the Bahia state of Brazil, where the disease was observed for the first time in 1989, cocoa yields decreased by 60% by 1994.

During rainy periods, the pathogen produces small mushroom-like basidiocarps (Fig. 163A) that grow on witches'-brooms and on diseased cocoa pods. Basidiospores produced by the basidiocarps are spread by wind and rain and, if they land on free water on susceptible tissues of cacao, i.e., terminal and axillary buds and flowers, germinate and cause infection. Infected terminal and axillary buds produce vegetative brooms (Fig. 11-163B), whereas infected flower cushions produce smaller cushion brooms (Fig. 11-163C) that are similar to vegetative brooms, as well as diseased flowers (star blooms) and diseased chirimoya-like pods (Fig. 11-163D). Newly flashing leaves of cacao that range in size from 0.3 to 5.0 centimeters in length are also infected by the fungus. Seeds in pods infected during the first 12 weeks of their development are generally destroyed and, thereby, there are no cocoa beans produced. Pods infected after their 12th week of growth have little or no adverse effects on the development of cocoa beans.

Following penetration of the host by the mycelium, the latter ramifies and moves into the tissues intercellularly. Green brooms are produced by infected buds and flower cushions, and the pathogen grows in them, producing intercellular swollen hyphae. After several weeks, leaves on brooms begin to show necrotic areas and die. Soon after that, the stems of the brooms begin to die from the tip back. Dead brooms often remain attached to the tree or they may fall off. After 4–8 weeks of rainy weather, small mushroom-like basidiocarps appear growing out from dead brooms and from dried pods on the soil surface.

The control of witches'–broom of cacao is difficult, but several measures taken together can have a positive effect in protecting the crop. Some relatively resistant cultivars are available and should be preferred for planting. Frequent pruning of brooms and other infected material every 10–14 days during the dry period helps reduce the inoculum, especially if it is combined with petroleum oil sprays of the cuttings and the soil. Selecting trees that set pods during the dry period helps many of them escape infection by withes-broom. Protective sprays with fungicides must be repeated every seven days, making them very costly, and they are not even particularly effective.

FIGURE 11-163 Witches'-broom disease of cacao caused by *Crinipellis perniciosa*. (A) Basidiocarp of *Crinipellis perniciosa* growing out of a witches'-broom twig. (B) Witches'-broom produced by growth of axillary buds. (C) Witches'-brooms produced on flower cushions. (D) Cacao pod infected with the witches'-broom disease. (Photographs courtesy of L.H. Purdy, University of Florida.)

Selected Reference

Purdy, L. H., and Schmidt, R. A. (1996). Status of cacao witches' broom: Biology, epidemiology, and management. *Annu. Rev. Phytopathol.* 34, 573–594.

Mycorrhizae

The feeder roots of most flowering plants growing in nature are generally infected by symbiotic fungi that do not cause root disease but, instead, are beneficial to their plant hosts. The infected feeder roots are transformed into unique morphological structures called mycorrhizae, i.e., "fungus roots." Mycorrhizae, known for many years to be common in forest trees, are now considered to be normal feeder roots for most plants, including cereals, vegetables, ornamentals, and, of course, trees.

There are two types of mycorrhizae, distinguished by the way the hyphae of the fungi are arranged within the cortical tissues of the root.

Ectomycorrhizae

Ectomycorrhizal roots are usually swollen and, in some host–fungus combinations, appear considerably more

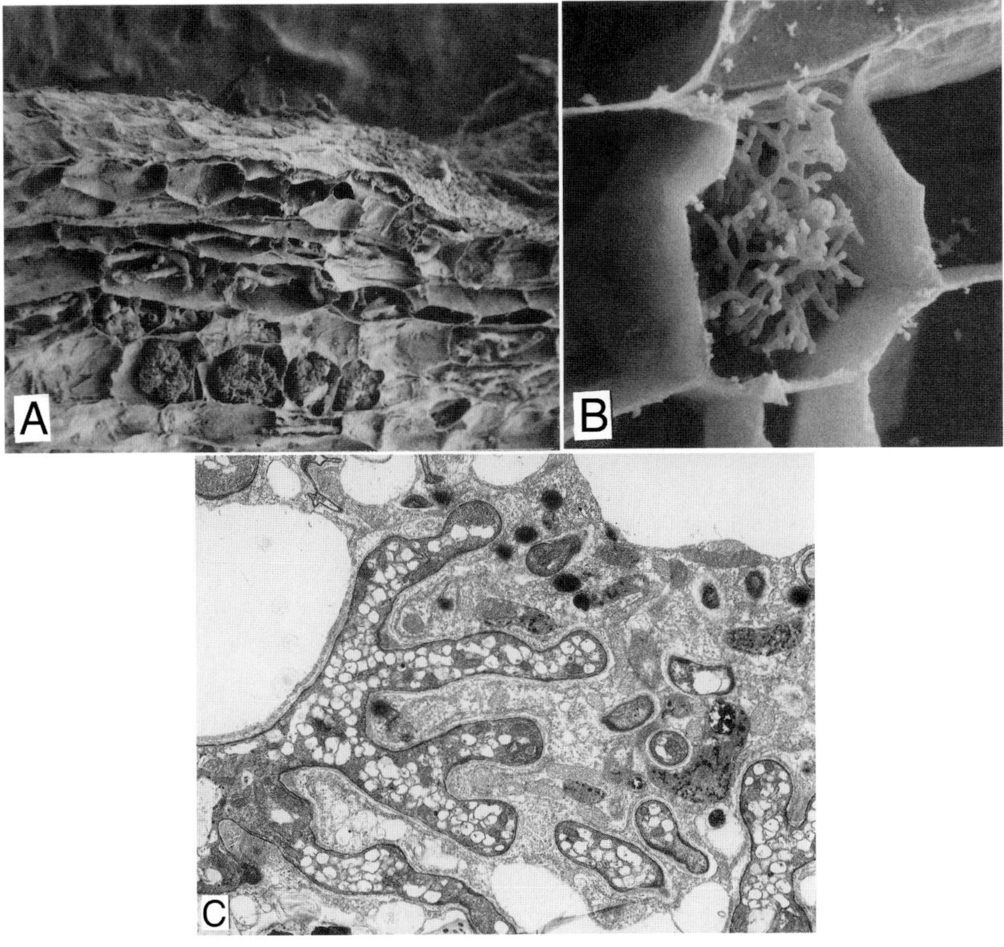

FIGURE 11-164 Vesicular–arbuscular mycorrhizae (endomycorrhizae) on yellow poplar *(Liriodendron tulipifera)* produced by *Glomus mosseae.* (A) Scanning electron micrograph of interior of mycorrhizal root showing coiled intracellular hyphae in outer cortical cells and three inner cortical cells that contain arbuscules. Some external mycelium of the fungus can be seen on the outside of the epidermis (top center). (B) Scanning electron micrograph of arbuscular morphology in a sample treated to remove host cytoplasm, which ordinarily surrounds the structure. This is a mature, viable arbuscule prior to the initiation of degenerative processes that lead to breakdown of this part of the endophyte. (C) Transmission electron micrograph of a similar arbuscule in a cortical cell. (Photographs courtesy of M.F. Brown and D.A. Kinden.)

forked than nonmycorrhizal roots. Ectomycorrhizae are formed primarily on forest trees mostly by mushroom- and puffball-producing basidiomycetes and by some ascomycetes. Spores of most ectomycorrhizal fungi are produced aboveground and are wind disseminated. Hyphae of ectomycorrhizal fungi usually produce a tightly interwoven "fungus mantle" around the outside of the feeder roots, with the mantle varying in thickness from 1 or 2 hyphal diameters to as many as 30 to 40. These fungi also enter the roots, but they only grow around the cortical cells, replacing part of the middle lamella between the cells and forming the so-called **Hartig net**. Ectomycorrhizae appear white, brown, yellow, or black, depending on the color of the fungus growing on the root.

Endomycorrhizae

By far the most common and most important mycorrhizae, endomycorrhizae externally appear similar to nonmycorrhizal roots in shape and color, but internally the fungus hyphae grow into the cortical cells of the feeder root either by forming specialized feeding hyphae (haustoria), called **arbuscules**, or by forming large, swollen, food-storing hyphal swellings, called **vesicles**. Most endomycorrhizae contain both vesicles and arbuscules and are, therefore, called **vesicular–arbuscular** mycorrhizae (Fig. 11-164). Endomycorrhizae are not surrounded by a dense fungal mantle but by a loose mycelial growth on the root surface from which hyphae and large pearl-covered zygospores or chlamydospores are produced underground. Endomycorrhizae are produced on most cultivated plants and on some forest trees mostly by zygomycetes, primarily of the genus *Glomus*, but also by other fungi, such as *Acaulospora*. Endomycorrhizae are also produced by some basidiomycetes.

Mycorrhizae apparently improve plant growth by increasing the absorbing surface of the root system and alleviating water stress on plants; by selectively absorbing and accumulating certain nutrients, especially phosphorus; by solubilizing and making available to the plant some normally nonsoluble minerals; by somehow keeping feeder roots functional longer; and by making feeder roots more resistant to infection by certain soil fungi, such as *Phytophthora*, *Pythium*, and *Fusarium*, and by nematodes. It should be kept in mind, however, that there may be many different host–fungus mycorrhizal associations, and each combination may have different effects on the growth of the plant. Some mycorrhizal fungi have a broad host range, whereas others are more specific. Also, some mycorrhizal fungi are more beneficial to a certain host than other fungi, and some hosts need and profit from association with a certain mycorrhizal fungus much more than other hosts. Mycorrhizal fungi also need the host in order to grow and reproduce; in the absence of hosts, fungi remain in a dormant condition as spores or resistant hyphae.

The symbiosis between the host plant and the mycorrhizal fungus is generally viewed as providing equal benefits to both partners. Nevertheless, it is quite probable that under certain nutritional conditions, one of the two partners may dominate and benefit more than the other. It has been suggested that the fungus is most aggressive in its invasion of root tissues when the host is growing at suboptimal nutritional levels (host defenses weak?) and the symbiotic relationship is terminated when nitrogen supply in the host reaches its optimum (host defenses at their best?). If the nitrogen supply is again reduced to deficiency levels, the fungus partner begins to dominate and forms in abundance while plant growth is suppressed.

As far as is known, mycorrhizae do not cause disease; however, an absence of mycorrhizae in certain fields results in plant stunting and poor growth, which can be avoided if the appropriate fungi are added to the plants. Also, soil fumigation often results in the eradication of mycorrhizal fungi, which in turn causes plants to remain smaller than plants growing in nonfumigated soil. The systemic fungicides metalaxyl and fosetyl-Al increase mycorrhizal infection and yield in some crop plants.

Selected References

Allen, M. F. (1992). "Mycorrhizal Functioning." Chapman & Hall, New York.

Anonymous (1988). Symposium on interactions of mycorrhizal fungi with soilborne plant pathogens and other microorganisms. *Phytopathology* **78**, 363–378.

Caron, M. (1989). Potential use of mycorrhizae in control of soilborne diseases. *Can. J. Plant Pathol.* **11**, 177–179.

Hackskaylo, E. (1971). Mycorrhizae. *Misc. Publ.-U.S. Dep. Agric.* **1189**, 1–255.

Powell, C. L., and Bagyaraj, eds. (1984). "VA Mycorrhiza." CRC Press, Boca Raton, FL.

Safir, G. R., ed. (1987). "Ecophysiology of VA Mycorrhizal Plants." CRC Press, Boca Raton, FL.

Schenck, N. C. (1982). "Methods and Principles of Mycorrhizal Research." APS Press, St. Paul, MN.

Wilcox, H. E. (1983). Fungal parasitism of woody plant roots from mycorrhizal relationships to plant disease. *Annu. Rev. Phytopathol.* **21**, 221–242.

chapter twelve

PLANT DISEASES CAUSED *BY* PROKARYOTES: BACTERIA *AND* MOLLICUTES

INTRODUCTION – PLANT DISEASES CAUSED BY BACTERIA – CLASSIFICATION AND CHARACTERISTICS OF PLANT PATHOGENIC BACTERIA – TYPES OF DISEASES – MORPHOLOGY – REPRODUCTION – ECOLOGY AND SPREAD – IDENTIFICATION OF BACTERIA – SYMPTOMS – CONTROL
616

BACTERIAL SPOTS AND BLIGHTS – INTRODUCTION – WILDFIRE OF TOBACCO – BACTERIAL BLIGHTS OF BEAN – ANGULAR LEAF SPOT OF CUCUMBER – ANGULAR LEAF SPOT OR BACTERIAL BLIGHT OF COTTON – BACTERIAL LEAF SPOTS AND BLIGHTS OF CEREALS AND GRASSES – BACTERIAL SPOT OF TOMATO AND PEPPER – BACTERIAL SPECK OF TOMATO – BACTERIAL FRUIT BLOTCH OF WATERMELON – CASSAVA BACTERIAL BLIGHT – BACTERIAL SPOT OF STONE FRUITS
627

BACTERIAL VASCULAR WILTS – INTRODUCTION – BACTERIAL WILT OF CUCURBITS – FIRE BLIGHT OF PEAR AND APPLE – SOUTHERN BACTERIAL WILT OF SOLANACEOUS PLANTS – BACTERIAL WILT OR MOKO DISEASE OF BANANA – RING ROT OF POTATO – BACTERIAL CANKER AND WILT OF TOMATO – BACTERIAL WILT (BLACK ROT) OF CRUCIFERS – STEWART'S WILT OF CORN
638

BACTERIAL SOFT ROTS – INTRODUCTION – BACTERIAL SOFT ROTS OF VEGETABLES – THE INCALCULABLE POSTHARVEST LOSSES FROM BACTERIAL (AND FUNGAL) SOFT ROTS
656

BACTERIAL GALLS – INTRODUCTION – CROWN GALL – *THE CROWN GALL BACTERIUM – THE NATURAL GENETIC ENGINEER*
662

BACTERIAL CANKERS – INTRODUCTION – BACTERIAL CANKER AND GUMMOSIS OF STONE FRUIT TREES – CITRUS CANKER
667

BACTERIAL SCABS – INTRODUCTION – COMMON SCAB OF POTATO
674

PLANT DISEASES CAUSED BY FASTIDIOUS VASCULAR BACTERIA – XYLEM-INHABITING FASTIDIOUS BACTERIA – PIERCE'S DISEASE OF GRAPE – CITRUS VARIEGATED CHLOROSIS – RATOON STUNTING OF SUGARCANE – PHLOEM-INHABITING FASTIDIOUS BACTERIA – INTRODUCTION – YELLOW VINE DISEASE OF CUCURBITS – CITRUS GREENING DISASE – PAPAYA BUNCHY TOP DISEASE
678

PLANT *DISEASES* – CAUSED BY MOLLICUTES: PHYTOPLASMAS AND SPIROPLASMAS – INTRODUCTION – PROPERTIES OF TRUE MYCOPLASMAS – PHYTOPLASMAS – SPIROPLASMAS EXAMPLES OF PLANT DISEASES CAUSED BY MOLLICUTES – ASTER YELLOWS – LETHAL YELLOWING OF COCONUT PALMS – APPLE PROLIFERATION – EUROPEAN STONE FRUIT YELLOWS – ASH YELLOWS – ELM YELLOWS (PHLOEM NECROSIS) – PEACH X-DISEASE – PEAR DECLINE – SPIROPLASMA DISEASES – INTRODUCTION – CITRUS STUBBORN DISEASE – CORN STUNT DISEASE

687

INTRODUCTION

Bacteria and mollicutes are **prokaryotes.** These are generally single-celled microorganisms whose genetic material (DNA) is not bound by a membrane and therefore is not organized into a nucleus. Their cells consist of cytoplasm containing DNA and small (70 S) ribosomes. The cytoplasm in mollicutes is surrounded by a cell membrane only, but in bacteria it is surrounded by a cell membrane and a cell wall. The cells of all other organisms (eukaryotes) contain membrane-bound organelles (nuclei, mitochondria, and — in plants only — chloroplasts). Eukaryotes also have two types of ribosomes, larger ones (80 S) in the cytoplasm and smaller ones (70 S) in mitochondria and chloroplasts. In fact, the organelles of eukaryotic cells and the prokaryotes have much in common. For example, some of the antibiotics that affect bacteria often inhibit the functions of mitochondria or chloroplasts but do not interfere with the other functions of eukaryotic plant cells.

Certain bacteria and the phytoplasmas of mollicutes (Fig. 12-1), the latter often referred to in the past as mycoplasma-like organisms (MLO), cause disease in plants. Plant pathogenic bacteria have been known since 1882; they are by far the largest group of plant pathogenic prokaryotes, cause a variety of plant disease symptoms, and are the best understood prokaryotic pathogens of plants. Even so, some types of phytopathogenic bacteria, e.g., fastidious phloem- or xylem-inhabiting bacteria, which for several years were thought to be rickettsia-like organisms (RLO), were only discovered in 1972; more of them, e.g., *Serratia*, *Sphimgomonas*, *Candidatus liberatus*, and the papaya bunchy top bacterium, are still being discovered as plant pathogens and their properties and relationships to the other plant pathogenic bacteria are still poorly understood. A general classification of plant pathogenic prokaryotes is shown.

Kingdom: Procaryotae
- Bacteria — Have cell membrane and cell wall
  - **Division:** Gracilicutes — Gram-negative bacteria
    - Class: Proteobacteria — Mostly single-celled bacteria
      - Family: Enterobacteriaceae
        - Genus: *Erwinia*, causing fire blight of pear and apple, Stewart's wilt in corn, and soft rot of fleshy vegetables
          - *Pantoea*, causing wilt of corn
          - *Serratia*, *S. marcescens*, being a phloem-inhabiting bacterium causing yellow vine disease of cucurbits
          - *Sphingomonas*, causing brown spot of yellow Spanish melon fruit
      - Family: Pseudomonadaceae
        - Genus: *Acidovorax*, causing leaf spots in corn, orchids, and watermelon
          - *Pseudomonas*, causing numerous leaf spots, blights, vascular wilts, soft rots, cankers, and galls
          - *Ralstonia*, causing wilts of solanaceous crops.
          - *Rhizobacter*, causing the bacterial gall of carrot
          - *Rhizomonas*, causing the corky root rot of lettuce
          - *Xanthomonas*, causing numerous leaf spots, fruit spots, and blights of annual and perennial plants, vascular wilts, and citrus canker
          - *Xylophilus*, causing the bacterial necrosis and canker of grapevines
      - Family: Rhizobiaceae
        - Genus: *Agrobacterium*, the cause of crown gall disease

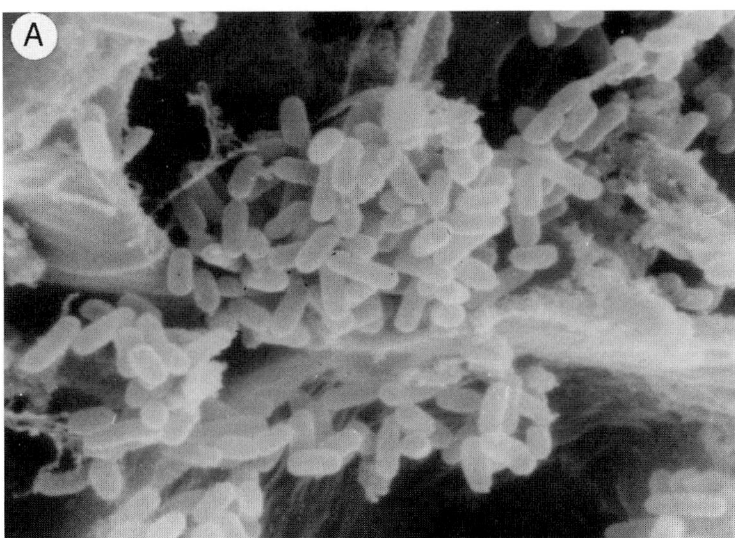

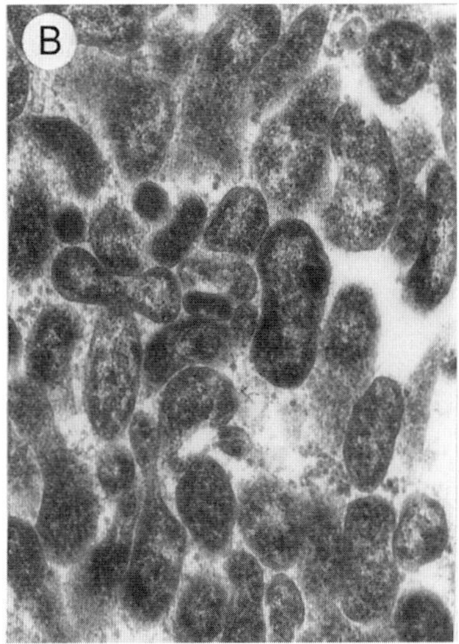

FIGURE 12-1 Plant pathogenic bacteria (A) and phytoplasmas (B) in infected plant cells. [Photographs from (A) Roos and Hattingh (1987), *Phytopathology* 77, 1246–1252; courtesy of (B) J. E. Worley, USDA.]

Rhizobium, the cause of root nodules in legumes

Family: still unnamed

Genus: *Xylella*, xylem — inhabiting, causing leaf scorch and dieback diseases on trees and vines

Candidatus liberobacter, phloem inhabiting, causing citrus greening disease

Unnamed, laticifer-inhabiting, causing bunchy top disease of papaya

Division: Firmicutes — Gram-positive bacteria

Class: Firmibacteria — Mostly single-celled bacteria

Genus: *Bacillus*, causing rot of tubers, seeds, and seedlings, and white stripe of wheat

Clostridium, causing rot of stored tubers and leaves and wetwood of elm and poplar

Class: Thallobacteria — Branching bacteria

Genus: *Arthrobacter*, causing bacterial blight of holly

Clavibacter, causing bacterial wilts in alfalfa, potato, and tomato

Curtobacterium, causing wilt in beans and other plants

Leifsonia, causing ratoon stunting of sugarcane

Rhodococcus, causing fasciation of sweet pea

Streptomyces, causing the common potato scab

Mollicutes — Have only cell membrane and lack cell wall

Division: Tenericutes

Class: Mollicutes

Family: Spiroplasmataceae

Genus: *Spiroplasma*, causing corn stunt, citrus stubborn disease

Family(ies): still unknown

Genus: *Phytoplasma*, causing numerous yellows, proliferation, and decline diseases in trees and some annuals

The taxonomy of plant pathogenic fastidious xylem-limited and phloem-limited bacteria is still unknown, and even the taxonomy of the plant pathogenic phytoplasmas, and of the spiroplasmas, is still tentative.

Because most bacteria lack distinctive morphological characteristics, their taxonomy and names are less clear and stable than in other organisms. A bacterial species is really a group of bacterial strains that share certain phenotypic and genotypic characteristics. One of these strains serves as the type strain, with the other strains of the species differing to a lesser or greater extent from the type strain. Bacterial strains may differ from one another in morphological, cultural, physiological, biochemical, or pathological characteristics. When a strain or group of strains infects a host plant not infected by the other strains of the species, that strain or group of strains comprise a pathovar (pv.) of the species.

PLANT DISEASES CAUSED BY BACTERIA

About 1,600 bacterial species are known. Most are strictly saprophytic and as such are beneficial to humans because they help decompose the enormous quantities of organic matter produced yearly by humans, animals, and factories as waste products or by the death of plants and animals. Several species cause diseases in humans, including tuberculosis, pneumonia, and typhoid fever, and a similar number cause diseases in animals, such as brucellosis and anthrax. About 100 species of bacteria cause diseases in plants. Most plant pathogenic bacteria are facultative saprophytes and can be grown artificially on nutrient media; however, fastidious vascular bacteria are difficult to grow in culture and some of them have yet to be grown in culture.

Bacteria may be rod shaped, spherical, spiral, or filamentous (threadlike). Some bacteria can move through liquid media by means of flagella, whereas others have no flagella and cannot move themselves. Some can transform themselves into spores, and the filamentous bacteria *Streptomyces* can produce spores, called conidia, at the end of the filament. Other bacteria, however, do not produce any spores. The vegetative stages of most types of bacteria reproduce by simple fission. Bacteria multiply with astonishing rapidity, and their significance as pathogens stems primarily from the fact that they can produce tremendous numbers of cells in a short period of time. Bacterial diseases of plants occur in every place that is reasonably moist or warm, and they affect all kinds of plants. Bacterial diseases are particularly common and severe in the humid tropics, but under favorable environmental conditions they may be extremely destructive anywhere.

Characteristics of Plant Pathogenic Bacteria

Morphology

Most plant pathogenic bacteria are rod shaped (Figs. 12-2 and 12-3), the only exception being *Streptomyces*, which is filamentous. In young cultures, bacteria range from 0.6 to 3.5 micrometers in diameter. In older cultures or at high temperatures, the rods may be longer, even filamentous, and they may form a club, Y, or V shape.

The cell walls of bacteria of most species are enveloped by a viscous, gummy material, which, if thin

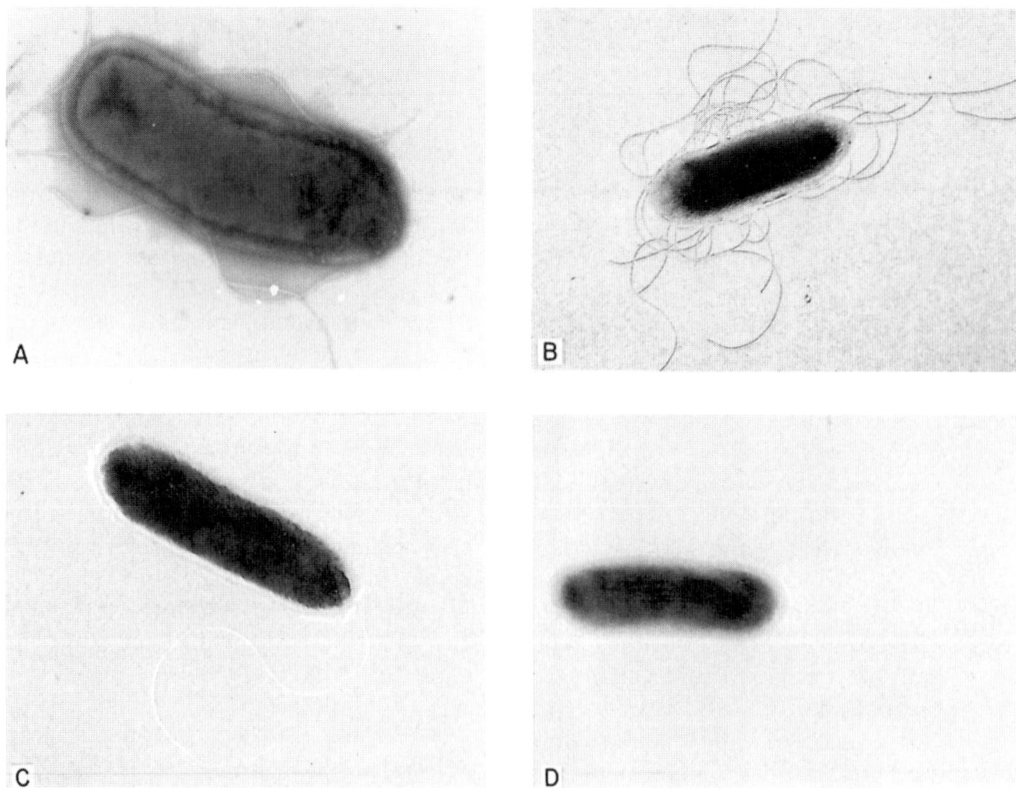

FIGURE 12-2 Electron micrographs of some of the most important genera of plant-pathogenic bacteria: (A) *Agrobacterium*, (B) *Erwinia*, (C) *Pseudomonas*, and (D) *Xanthomonas*. [Photographs courtesy of (A) R. E. Wheeler and S. M. Alcorn and (B–D) R. N. Goodman and P. Y. Huang.] Magnified 1600×.

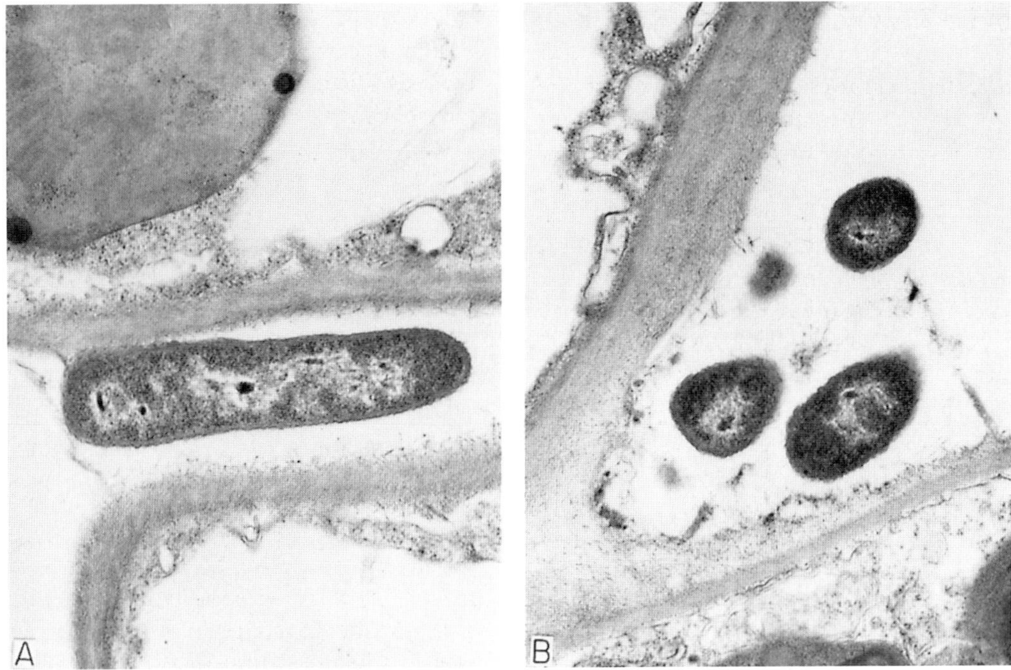

FIGURE 12-3 Electron micrographs of longitudinal (A) and cross sections (B) of bacteria (*Pseudomonas syringae* pv. *tabaci*) in the intercellular spaces of tobacco mesophyll cells. [Photographs courtesy of D. J. Politis and R. N. Goodman.] Magnified 1600×.

and diffuse, is called a **slime layer**, but if thick, forming a definitive mass around the cell, is called a **capsule**. Most plant pathogenic bacteria have delicate, threadlike *flagella*, considerably longer than the cells on which they are produced. In some bacterial species, each bacterium has only one flagellum, whereas others have a tuft of flagella at one end of the cell (**polar** flagella); still others have **peritrichous** flagella, i.e., distributed over the entire surface of the cell.

In the filamentous *Streptomyces* species, cells consist of branched threads, which usually have a spiral formation and produce conidia in chains on aerial hyphae (Fig. 12-4).

Single bacteria appear hyaline or yellowish-white under the compound microscope and are very difficult to observe in detail. When a single bacterium is allowed to grow (multiply) on the surface of or in a solid medium, its progeny soon produces a visible mass called a **colony**. Colonies of different species may vary. They may be a fraction of a millimeter to several centimeters in diameter and may be circular, oval, or irregular. Their edges may be smooth, wavy, or angular, and their elevation may be flat, raised, or wrinkled. Colonies of most species are whitish or grayish, but some are yellow. Some produce diffusible pigments into the agar that may be fluorescent with ultraviolet light.

Bacteria have thin, relatively tough, rigid cell walls and an inner cytoplasmic membrane. Gram-negative bacteria also have an outer membrane that appears to merge with the slime layer or capsule. The cell wall allows the inward passage of nutrients and the outward passage of waste matter and digestive enzymes.

All the material inside the cell wall constitutes the **protoplast**. The protoplast consists of a cytoplasmic or protoplast membrane, which determines the degree of selective permeability of the various substances into and out of the cell; the **cytoplasm**, which is the complex mixture of proteins, lipids, carbohydrates, many other organic compounds, minerals, and water; and nuclear material, which consists of a large circular **chromosome** composed of DNA. The chromosome DNA makes up the main body of the genetic material of a bacterium and appears as a spherical, ellipsoidal, or dumbbell-shaped body within the cytoplasm. Often, bacteria also have single or multiple copies of additional, smaller circular genetic material called **plasmids**. Each plasmid consists of several nonessential genes and can move or be moved between bacteria or even between bacteria and plants, as happens in the crown gall disease.

Reproduction

Rod-shaped phytopathogenic bacteria reproduce by the asexual process known as **binary fission**, or **fission**. This occurs by the inward growth of the cytoplasmic membrane toward the center of the cell, forming a trans-

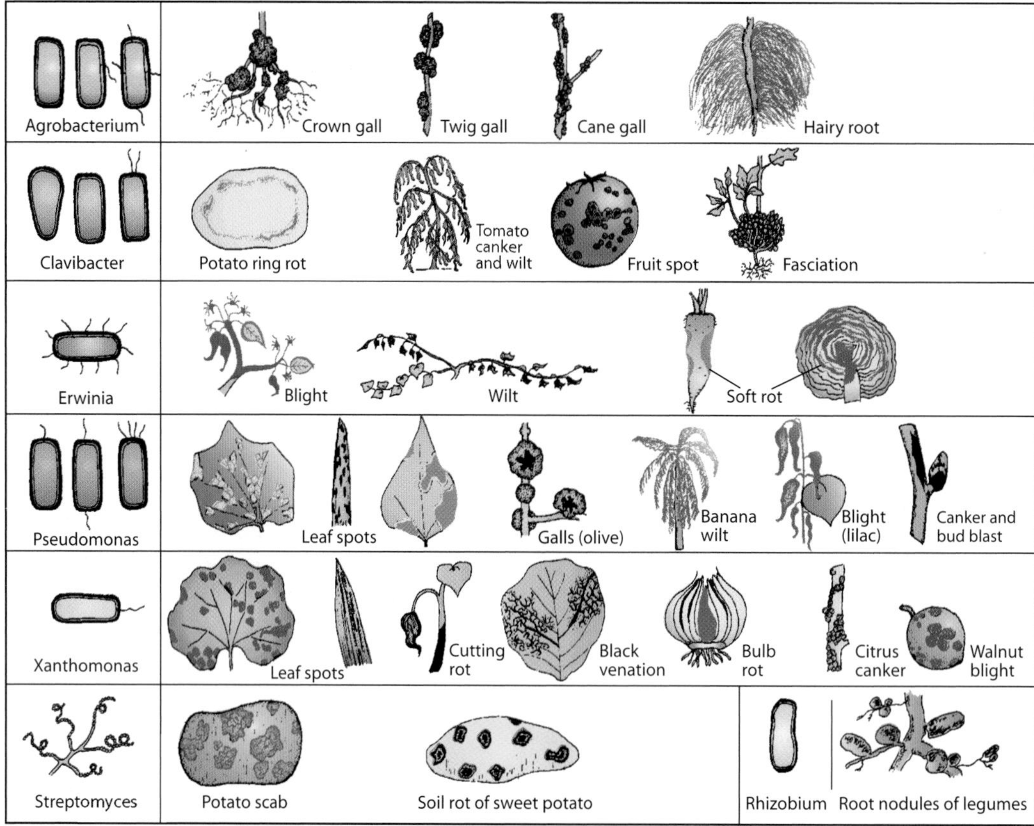

FIGURE 12-4 The most important genera of plant pathogenic bacteria and the kinds of symptoms they cause.

verse membranous partition dividing the cytoplasm into two approximately equal parts. Two layers of cell wall material, continuous with the outer cell wall, are then synthesized between the two layers of the membrane. When the formation of these cell walls is completed, the two layers separate, splitting the two cells apart.

While the cell wall and the cytoplasm are undergoing fission, the nuclear material becomes organized into a circular chromosome-like structure that duplicates itself and becomes distributed equally between the two cells formed from the dividing one. Plasmids also duplicate and distribute themselves equally in the two cells.

Bacteria reproduce at an astonishingly rapid rate. Under favorable conditions, bacteria may divide every 20 to 50 minutes, one bacterium becoming two, two becoming four, four becoming eight, and so on. At this rate, one bacterium conceivably could produce one million progeny bacteria in less than a day. However, because of the diminution of the food supply, accumulation of metabolic wastes, and other limiting factors, reproduction slows and may finally come to a stop. Nevertheless, bacteria do reach tremendous numbers in a short time, and they cause great chemical changes in their environment. It is these changes caused by large populations of bacteria that make them of such great significance in the world of life in general and in the development of bacterial diseases of plants in particular.

Ecology and Spread

Almost all plant pathogenic bacteria develop mostly in the host plant as parasites, on the plant surface, especially buds, as epiphytes, and partly in plant debris or in the soil as saprophytes. There are great differences among species, however, in the degree of their development in one or the other environment. Wherever plant pathogenic, and other, bacteria exist, they often exist as biofilms, that is, communities of idemtical or different microorganisms attached each other and/or to a solid surface.

Some bacterial pathogens, such as *Erwinia amylovora*, which causes fire blight of pear, produce their populations in the plant host, while in the soil their numbers decline rapidly and usually do not contribute to the propagation of the disease from season to season. These pathogens have developed sustained plant-to-plant infection cycles, often via insect vectors, and, either because of the perennial nature of the host or the association of the bacteria with its vegetative propagating organs or seed, they have lost the ability to survive in the soil.

Some other bacterial pathogens, such as *Agrobacterium tumefaciens*, which causes crown gall, *Ralstonia solanacearum*, which causes the bacterial wilt of solanaceous crops, and particularly *Streptomyces scabies*, which causes the common scab of potato, are rather typical **soil inhabitants**. Such bacteria build up their populations within the host plants, but these populations only gradually decline when they are released into the soil. If susceptible hosts are grown in such soil in successive years, sufficiently high numbers of bacteria could be present to cause a net increase of bacterial populations in the soil from season to season. Most plant pathogenic bacteria, however, can be considered **soil invaders**. Such bacteria enter the soil in host tissue and, because they have poor ability to compete as saprophytes, persist in the soil either as long as the host tissue resists decomposition by saprophytes or for varying durations afterward, depending on the bacterial species and on the soil temperature and moisture conditions.

When in soil, bacteria live mostly on plant material. Less often they live freely or saprophytically, or in their natural bacterial ooze, which protects them from various adverse factors. Bacteria may also survive in or on seeds, other plant parts, or insects found in the soil. On plants, bacteria often survive epiphytically, in buds, on wounds, in their exudate, or inside the various tissues or organs that they infect (Fig. 2-3D).

The dissemination of plant pathogenic bacteria from one plant to another or to other parts of the same plant is carried out primarily by water, insects, other animals, and humans (Fig. 2-15). Even bacteria possessing flagella can move only very short distances on their own power. Rain, by its washing or spattering effect, carries and distributes bacteria from one plant to another, from one plant part to another, and from the soil to the lower parts of plants. Water also separates and carries bacteria on or in the soil to other areas where host plants may be present. Insects not only carry bacteria to plants, but they inoculate the plants with the bacteria by introducing them into the particular sites in plants where they can almost surely develop. In some cases, bacterial plant pathogens also persist in insects and depend on them for their survival and spread. In other cases, insects are important but not essential in the dissemination of certain bacterial plant pathogens. Birds, rabbits, and other animals moving among plants may also carry bacteria on their bodies. Humans help spread bacteria locally by handling plants, cultural practices, and, over long distances, by transporting infected transplants or plant parts to new areas. In cases in which bacteria infect the seeds of their host plants, they can be carried in or on them for short or long distances by any of the agents of seed dispersal.

Identification of Bacteria

The main characteristics of some of the most common plant pathogenic genera of bacteria (Fig. 12-4) are as follows.

Agrobacterium

Bacteria are rod shaped, 0.8 by 1.5–3 micrometers. They are motile by means of one to four peritrichous flagella; when only one flagellum is present, it is more often lateral than polar. When growing on carbohydrate-containing media, bacteria produce abundant polysaccharide slime. The colonies are nonpigmented and usually smooth. These bacteria are rhizosphere and soil inhabitants.

Clavibacter (Corynebacterium)

Cells have the shape of straight to slightly curved rods, 0.5–0.9 by 1.5–4 micrometers. Sometimes they have irregularly stained segments or granules and club-shaped swellings. The bacteria are generally nonmotile, but some species are motile by means of one or two polar flagella. They are gram positive.

Erwinia

Bacteria are straight rods, 0.5–1.0 by 1.0–3.0 micrometers, and are motile by means of several to many peritrichous flagella. *Erwinia* are the only plant pathogenic bacteria that are facultative anaerobes. Some *Erwinia* do not produce pectic enzymes and cause necrotic or wilt diseases (the "amylovora" group), whereas other *Erwinia* have strong pectolytic activity and cause soft rots in plants (the "carotovora" group).

Pseudomonas

Pseudomonads are straight to curved rods, 0.5–1 by 1.5–4 micrometers. They are motile by means of one or many polar flagella. Many species are common inhabitants of soil or of freshwater and marine environments. Most pathogenic *Pseudomonas* species infect plants; few infect animals or humans. Plant pathogenic *Pseudomonas* species (e.g., *P. syringae*), when grown on a medium of low iron content, produce yellow-green, diffusible, fluorescent pigments.

Ralstonia

Until very recently classified as *Pseudomonas*, these resemble the latter in most respects with the important difference that its cells do not produce fluorescent pigments.

Xanthomonas

Cells are straight rods, 0.4–1.0 by 1.2–3 micrometers, and are motile by means of a polar flagellum. Growth on agar media is usually yellow, and most are slow growing. All species are plant pathogens and are found only in association with plants or plant materials.

Streptomyces

Bacteria have the shape of slender, branched hyphae without cross walls, 0.5–2 micrometers in diameter. At maturity the aerial mycelium forms chains of three to many spores. On nutrient media, colonies are small (1–10 millimeters in diameter) at first with a rather smooth surface but later with a weft of aerial mycelium that may appear granular, powdery, or velvety. The many species and strains of *Streptomyces* produce a wide variety of pigments that color the mycelium and the substrate; they also produce one or more antibiotics active against bacteria, fungi, algae, viruses, protozoa, or tumor tissues. All species are soil inhabitants. They are gram positive.

Xylella

Cells are mostly single, straight rods, 0.3 by 1–4 micrometers, producing long filamentous strands under some cultural conditions. Colonies are small, with smooth or finely undulated margins. Nutritionally fastidious, *Xylella* require specialized media; their habitat is xylem of plant tissue. They are gram negative, nonmotile, aflagellate, strictly aerobic, and nonpigmented.

The genus *Streptomyces* can be distinguished easily from other bacterial genera because of its much-branched, well-developed mycelium and curled chains of conidia. Identification of bacteria belonging to the rod-shaped genera, however, is a much more complex and difficult process. It can be made by taking into consideration not only visible characteristics such as size, shape, structure, and color, but also such obscure properties as chemical composition, serological reactions, ability to use certain nutrients, enzymatic action, pathogenicity to plants, and growth on selective media.

The shape and size of bacteria of a given species can vary in culture with the age of the culture, the composition and pH of the medium, temperature, and staining method. Under given conditions, however, the predominating form, size, and arrangement of cells in a pure culture are quite apparent, and they are important and reliable characteristics. The presence, number, and arrangement of flagella on the bacterial cell are also determined, usually after the flagella have been stained.

The chemical compositions of certain substances in bacterial cells can be detected with specific staining techniques. Information about the presence or absence of such substances is used for the identification of bacteria. **Gram's staining reaction** differentiates bacteria into gram-positive and gram-negative types. In this reaction, bacteria fixed on a glass slide are treated with a crystal violet solution for 30 seconds, rinsed gently, treated with iodine solution, and rinsed again with water and then alcohol. Gram-positive bacteria retain the violet-iodine stain combination because it forms a complex with certain components of their cell wall and cytoplasm. Gram-negative bacteria have no affinity for the stain combination, which is therefore removed by the alcohol rinse, and bacteria remain as nearly invisible as before. Of the rod-shaped phytopathogenic bacteria, only the genera *Clavibacter* and *Curtobacterium*, and the relatively unimportant plant pathogens of the genera *Arthrobacter*, *Bacillus*, and *Rhodococcus*, are gram positive. *Agrobacterium*, *Erwinia*, *Pseudomonas*, *Ralstonia*, *Xanthomonas*, and *Xylella* are gram negative.

Bacteria are also distinguished by the **substances that they can or cannot use for food** and by the **kinds of enzymes produced** when the bacteria are grown on certain media. Over a hundred characteristics of a bacterium can be determined by these tests, and the profiles for each bacterium are often used in numerical taxonomy of bacteria.

Phytopathogenic bacteria are also tested for their **pathogenicity** on various species and varieties of host plants. This test, for practical purposes, may be sufficient for tentative identification of the bacterium.

In many cases, the effort to establish the identity of an isolated bacterium begins with observation of the **external symptom**, e.g., a plant appears wilted (Fig. 12-5A) or the spots on the leaves are surrounded by a halo (Fig. 12-5B). The next step is observation of some of the easier **internal symptoms**, e.g., the wilted plant shows discoloration of the vascular system (Fig. 12-5A), so the wilt is caused by a pathogen and not by drought. Further examination of the wilted plant can be done by placing a freshly cut wilted stem in a tube or dish of water and looking for appearance or lack of a cloudy diffusate from the stem (Fig. 12-5C), which, if present, indicates that the wilt is cause by bacteria rather than a fungus or anything else. By being familiar or comparing the literature about which bacterium causes symptoms like that observed in this particular host, one can identify the

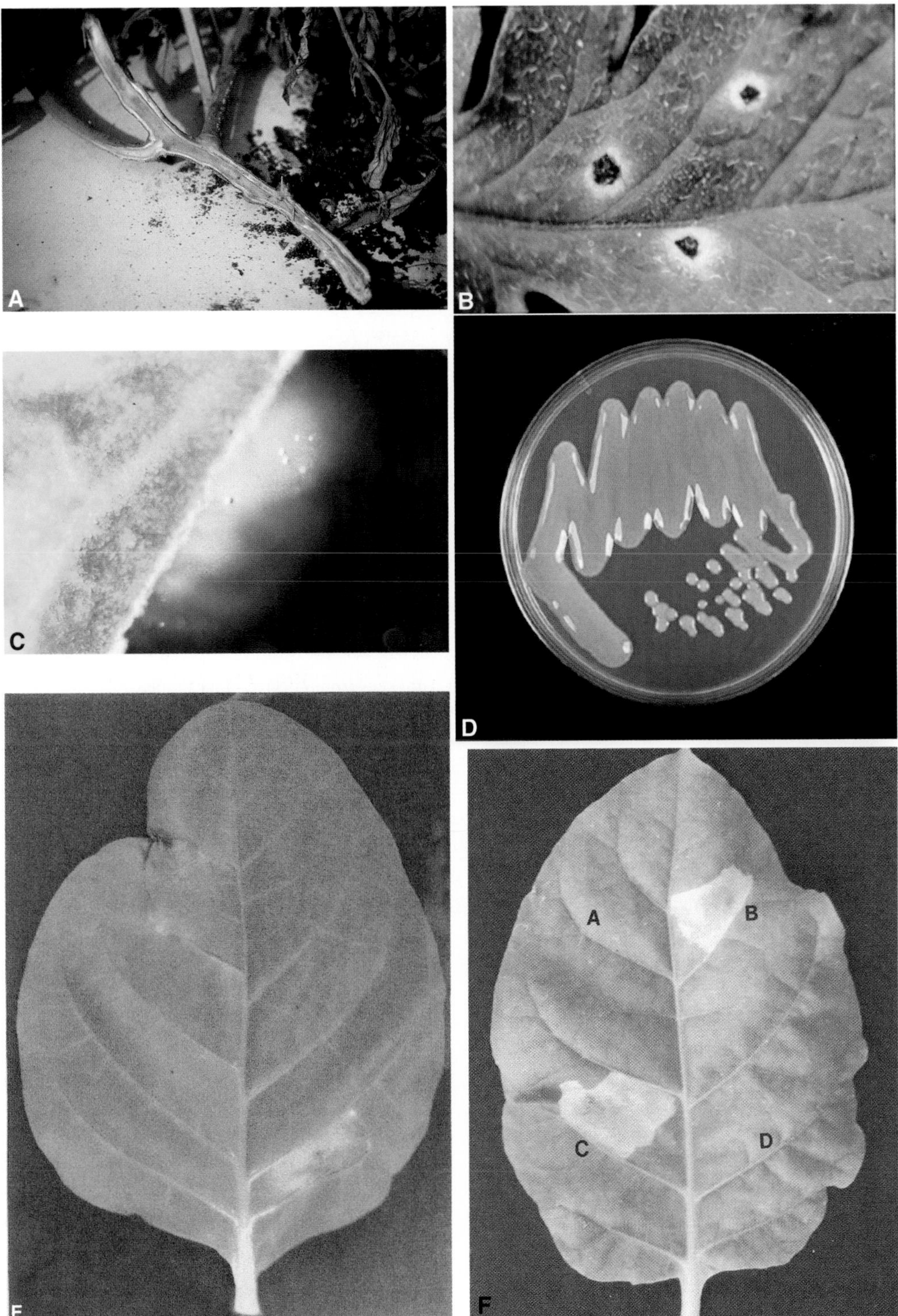

FIGURE 12-5 Some macroscopic features used to approximately determine the bacterial nature of the cause of a plant disease. (A) Brown discoloration of vascular tissues of wilting plant. (B) Halo surrounding lesions on leaf of plant. (C) Cloud-like exudate of bacteria oozing out from infected plant section placed in water. (D) Appearance characteristics of a culture and colonies of bacteria isolated from infected plant. Hypersensitive reaction tests in which injection of pathogenic bacteria into a leaf of an appropriate nonhost plant induces at first water soaking (E) and then collapse and necrosis of plant tissues (E and F), whereas injection of water at the opposite sides of the leaf at E and at A of leaf F or of a nonpathogenic bacterium (D) at leaf F induces no such reaction. [Photographs courtesy of (A and C) University of Florida, (B) R. J. McGovern, University of Florida, (D) T. R. Gottwald, USDA, Ft. Pierce, FL, and (E and F) A. Chatterjee, University of Missouri.]

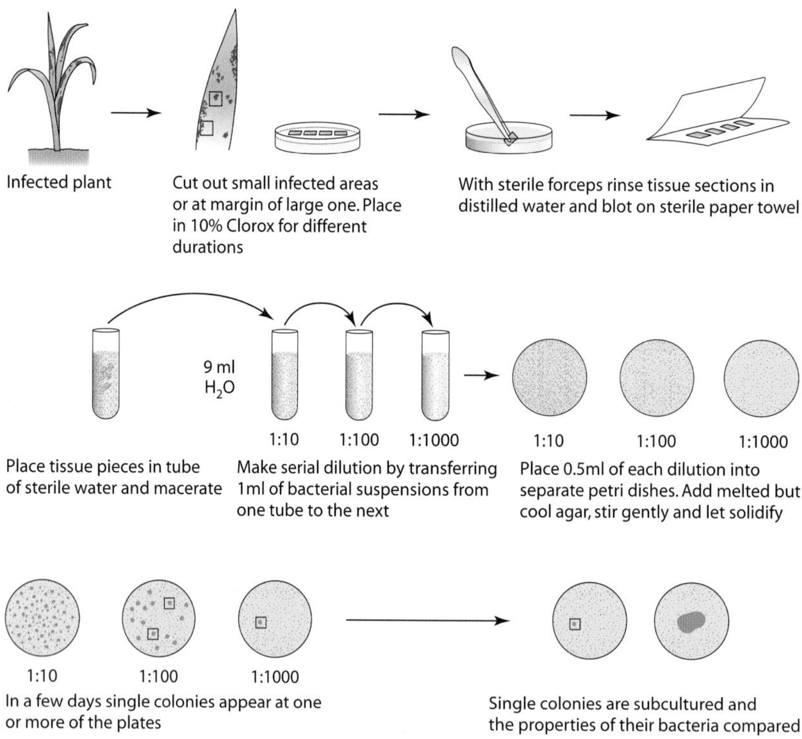

FIGURE 12-6 Isolation of bacterial pathogens from infected plant tissue.

bacterium and diagnose the disease. If further work is needed, then one cultures the bacteria and observes the shape, size, color, and so on of its culture (Figs. 12-5D and 12-5E). To make sure that the isolated bacterium is the pathogen rather than a saprophyte, a series of dilutions of the bacteria is injected into the leaves of a nonhost, such as tobacco. If nonpathogenic, the leaves show no change at the points of injection. If the bacterium is pathogenic, however, it produces a **hypersensitive response** (dead tissues around the points of injection) (Fig. 12-5F).

Serological methods, especially those employing antibodies labeled with a fluorescent compound (immunofluorescent staining), are used for the quick and fairly accurate identification of bacteria and have gained popularity in recent years. The use of serological methods is becoming widespread in plant pathology as the availability of species-specific and pathovar-specific antisera increases.

An excellent method of isolation and identification of bacteria obtained from plant tissues (Fig. 12-6) or soil is through the use of **selective nutrient media.** Selective media contain nutrients that promote the growth of a particular type of bacterium while at the same time contain substances that inhibit the growth of other types of bacteria. Positive identification usually requires more than one subculturing on selective media because seldom does only one bacterium grow on a selective medium. The available selective media for plant pathogenic bacteria are helpful for routine isolation and sometimes identification of bacterial genera and of several species and even pathovars. In the past 10 years, fairly quick distinction and identification of bacterial genera, species, and, in some cases, lower subdivisions have been made by extraction and comparison of the fatty acids present in the bacterial cell membranes (**fatty acid profile analysis**). The same bacteria grown under identical conditions also produce identical membrane proteins and identical **enzymes** and **isoenzymes**. Isolation and comparison of such structural proteins or enzymes are also used to identify bacteria (Fig. 12-7).

Similarly, bacteria are detected, identified, and their genetic relatedness measured by comparison of the **profiles of DNA bands** obtained on a separation gel following digestion (cutting up) of the bacterial chromosomal DNA with certain restriction endonucleases (Fig. 12-8). Such enzymes cut the DNA only at certain nucleotide sequences and release defined sets of DNA fragments called restriction fragment length polymorphisms (**RFLPs**). RFLP profiles may be characteristic of the bacterium and, therefore, can be used to identify the bacterium.

In other techniques, **DNA probes** are used to detect and identify bacteria. The probe consists of a comple-

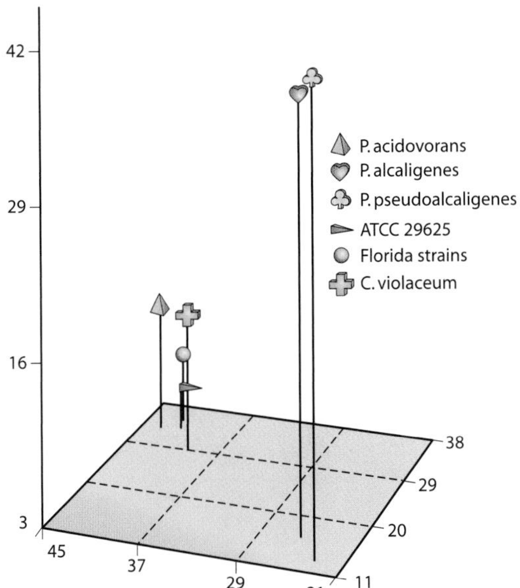

FIGURE 12-7 Proximate identification of plant pathogenic bacteria by comparison of the percentage content in certain fatty acid groups. Thus, the hitherto unknown watermelon fruit blotch bacteria (Florida strains) were shown to be related to *Acidovorax avenae* subsp. *citrulli* (formerly *P. acidovorans*). [From Cameron-Somodi *et al.* (1991). *Plant Dis.* 75, 1053–1056.]

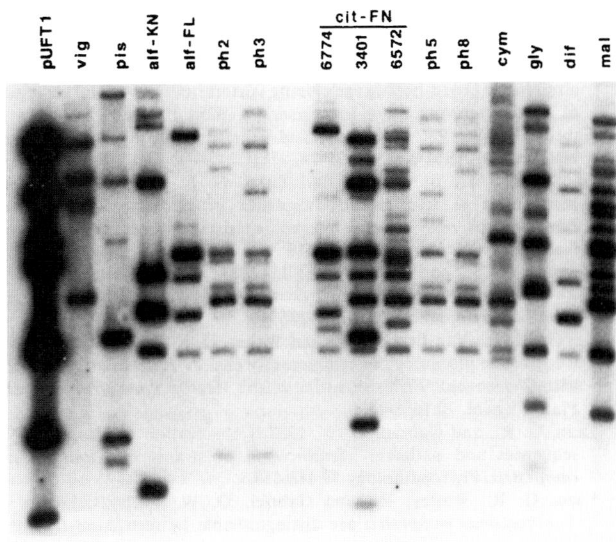

FIGURE 12-8 Identification of bacterial strains and pathovars within a species (*Xanthomonas campestris*) by isolating and digesting all their DNA with a particular nuclease enzyme and comparing the fragment profiles to those of known pathovars. [From Gabriel *et al.* (1988). *Mol. Plant/Microbe Interact.* 1, 59–65.]

mentary segment of a part of the DNA of the bacterium that exists only or primarily in that kind of bacterium, e.g., DNA of a specific toxin gene or a virulence gene of the bacterium. A radioactive element or a color-producing substance is attached to the DNA probe. Bacteria to be tested with the probe are treated so that their DNA is released onto a nylon membrane or filter and the DNA is then treated with the probe. If the probe finds its complementary DNA on the filter, it reacts (hybridizes) with it and stays on the filter even after washing. The presence of the probe is detected by its radioactive element or the color-producing (chromogenic) compound attached to it; a positive hybridization signal, of course, indicates the identity of the bacterium tested.

The availability of the **polymerase chain reaction (PCR)** technique, by which one or a few strands of DNA can be multiplied indefinitely to millions of copies, has made possible the detection, through a DNA probe, of the presence of one or a few bacteria in or on a seed or transplant or in a mixture of bacteria obtained from a plant or from the soil.

Symptoms Caused by Bacteria

Plant pathogenic bacteria induce as many kinds of symptoms on the plants they infect as do fungi. They cause leaf spots and blights, soft rots of fruits, roots, and storage organs, wilts, overgrowths, scabs, and cankers (Fig. 12-4). Any given type of symptom can be caused by bacterial pathogens belonging to several genera, and each genus may contain pathogens capable of causing different types of diseases. Species of *Agrobacterium*, however, can cause only overgrowths or proliferation of organs. However, overgrowths can also be caused by certain species of *Rhodococcus* and *Pseudomonas*. Also, the plant pathogenic species of *Streptomyces* cause only scabs or lesions of belowground crops. Species of *Rhizobium* and the related genera *Azorhizobium* and *Bradyrhizobium* are gram-negative, soil-inhabiting bacteria that induce the formation of nodules on the roots of legume plants, but these bacteria are beneficial rather than pathogenic to the plant because they fix nitrogen that is used by the plants. Parts of the DNA of the three latter genera are nearly identical to parts of the DNA of *Agrobacterium* bacteria.

Control of Bacterial Diseases of Plants

Bacterial diseases of plants are usually very difficult to control. Frequently, a combination of control measures is required to combat a given bacterial disease. Infestation of fields or infection of crops with bacterial pathogens should be avoided by using only healthy seeds or transplants. **Sanitation practices** aiming at reducing the inoculum in a field by removing and burning infected plants or branches, and at reducing the spread of bacteria from plant to plant by decontaminating tools and hands after handling diseased plants, are very important. **Adjusting fertilizing and watering** so that the plants are not extremely succulent during the period of infection may also reduce the incidence of disease. **Crop rotation** can be very effective with bacteria that have a limited host range, but is impractical and ineffective with bacteria that can attack many types of crop plants.

The use of **crop varieties resistant** to certain bacterial diseases is one of the best ways of avoiding heavy losses. Varying degrees of resistance may be available within the varieties of a plant species, and great efforts are made at crop breeding stations to increase the resistance of, or introduce new types of resistance into, currently popular varieties of plants. Resistant varieties, supplemented with proper cultural practices and chemical applications, are the most effective means of controlling bacterial diseases, especially when environmental conditions favor the development of disease.

Soil infested with phytopathogenic bacteria can be sterilized with steam or electric heat and with chemicals such as formaldehyde, but this is practical only in greenhouses and in small beds or frames. **Seed**, when infested superficially, can be disinfested with sodium hypochlorite or HCl solutions or by soaking it for several days in a weak solution of acetic acid. If seeds can remain for 2 to 3 days in fermenting juices of fruit in which they are borne, bacterial pathogens can be eliminated. When the pathogen is inside the seed coat and in the embryo, such treatments are ineffective. Treating seed with hot water does not usually control bacterial diseases because of the relatively high thermal death point of the bacteria, but treatment at 52°C for 20 minutes often considerably reduces the number of infected seeds.

The use of **chemicals** to control bacterial diseases has been, generally, much less successful than the chemical control of fungal diseases. Of the chemicals used as foliar sprays, copper compounds give the best results. However, even they seldom give satisfactory control of the disease when environmental conditions favor development and spread of the pathogen. Bordeaux mixture, fixed coppers, and cupric hydroxide are used most frequently for the control of bacterial leaf spots and blights. Bacterial strains resistant to copper fungicides, however, are quite common. Zineb, maneb, or mancozeb mixed with copper compounds is used for the same purpose, especially on young plants that may be injured by the copper compounds.

Antibiotics have been used against certain bacterial diseases with mixed results. Some antibiotics are absorbed by the plant and are distributed systemically. They can be applied as sprays or as dips for transplants. The most important antibacterial antibiotics in agriculture are formulations of streptomycin or of streptomycin and oxytetracycline. Unfortunately, bacterial races resistant to antibiotics develop soon after widespread application of antibiotics; in addition, no antibiotics are permitted on edible plant produce.

Successful practical **biological control** of the bacterial plant disease crown gall has been obtained by treating seeds or nursery stock with bacteriocin-producing antagonistic strains of *Agrobacterium*. Treatment of tubers, seeds, and so on with antagonistic bacteria and spraying of aerial plant parts with bacteria antagonistic to the pathogen have given control of various diseases under experimental conditions but have been less successful in practice.

Selected References

Beattie, W. A., and Lindow, S. E. (1995). The secret life of foliar bacterial pathogens on leaves. *Annu. Rev. Phytopathol.* **33**, 145–172.

Birch, R. G. (2001). *Xanthomonas albilineans* and the antipathogenesis approach to disease control. *Mol. Plant Pathol.* **2**, 1–11.

Botha, W. J., Serfontein, S., Greyling, M. M., *et al.* (2001). Detection of *Xylophilus ampelinus* in grapevine cuttings using a nested polymerase chain reaction. *Plant Pathol.* **50**, 515–526.

Bradbury, J. F. (1986). "Guide to Plant Pathogenic Bacteria." CAB Int. Mycol. Inst., Kew, Surrey, England.

Buonaurio, R., Stravato, V. M., and Cappelli, C. (2001). Brown spot caused by *Sphingomonas* sp. on yellow Spanish melon fruits in Spain. *Plant Pathol.* **50**, 397–401.

Buonaurio, R., *et al.* (2002). *Sphingomonas melonis* sp. nov., a novel pathogen that causes brown spots on yellow Spanish melon fruits. IJSEM.

Civerolo, E. L., Collmer, A., and Gillaspie, A. G., eds. (1987). Plant pathogenic bacteria. *In* Proceeding of the Sixth International Conference, Martinus Nijhoff, Boston.

Cooksey, D. A. (1990). Genetics of bactericide resistance in plant pathogenic bacteria. *Annu. Rev. Phytopathol.* **28**, 201–219.

Evtushenko, L. I., *et al.* (2000). *Leifsonia poae* gen. nov., sp. nov., isolated from nematode galls on *Poa annua*, and reclassification of *Corynebacterium aquaticum* Leifson *1962 as Leifsonia aquatica* gen. nov., and *Clavibacter xyli* (Davis *et al.* 1984) with two subspecies as *Leifsonia xyli* (Davis *et al.* 1984) gen. nov., comb. nov. IJSEM **50**, 371–380.

Fahy, D. C., and Persley, G. F., eds. (1983). "Plant Bacterial Diseases: A Diagnostic Guide." Academic Press, New York.

Goto, M. (1992). "Fundamentals of Bacterial Plant Pathology." Academic Press, San Diego.

He, S. Y. (1998). Type III protein secretion system in plant and animal pathogenic bacteria. *Annu. Rev. Phytopathol.* **36**, 363–392.

Hirano, S. S., and Upper, C. D. (1990). Population biology and epidemiology of *Pseudomonas syringae*. *Annu. Rev. Phytopathol.* **28**, 155–177.

Klement, Z., Bozsó, Z., Ott, P. G., *et al.* (1999). Symptomless resistant response instead of the hypersensitive reaction in tobacco leaves after infiltration of heterologous pathovars of *Pseudomonas syringae*. *J. Phytopathol.* **147**, 467–475,

Klement, Z., Rudolph, K., and Sands, D. C. (1990). "Methods in Phytobacteriology." Akademiai Kiato, Budapest.

Kushalappa, A. C., and Lui, L. H. (2001). Volatile fingerprinting (SPME-GC-FID) to detect and discriminate diseases of potato tubers. *Plant Dis.* **86**, 131–137.

Lindgren, P. B. (1997). The role of *hrp* genes during plant-bacterial interactions. *Annu. Rev. Phytopathol.* 35, 129–152.

Louws, F. J., Rademaker, J. L. W., and de Bruijn, F. J. (1999). The three Ds of PCR-based genomic analysis of phytobacteria: Diversity, detection, and disease diagnosis. *Annu. Rev. Phytopathol.* 37, 81–125.

McManus, P. S., *et al.* (2002). Antibiotic use in plant agriculture. *Annu. Rev. Phytopathol.* 40, 443–465.

Mount, M. S., and Lacey, G. H., eds. (1982). "Phytopathogenic Prokaryotes," Vols. 1 and 2. Academic Press, New York.

Newman, M.-A., von Roepenack, E., Daniels, M. and Dow, M. (2000). Lipopolysaccharides and plant responses to phytopathogenic bacteria. *Mol. Plant Pathol.* !, 25–31.

Panopoulos, N. J., and Peet, R. C. (1985). The molecular genetics of plant pathogenic bacteria and their plasmids. *Annu. Rev. Phytopathol.* 23, 381–419.

Richael, C., Lincoln, J. E., Bostock, R. M., *et al.* (2001). Caspase inhibitors reduce symptom development and limit bacterial proliferation in susceptible plant tissues. *Physiol. Mol. Plant Pathol.* 59, 213–221.

Salmond, G. P. C. (1994). Secretion of extracellular virulence factors by plant pathogenic bacteria. *Annu. Rev. Phytopathol.* 32, 181–200.

Schaad, N. W. (1979). Serological identification of plant pathogenic bacteria. *Annu. Rev. Phytopathol.* 17, 123–147.

Schaad, N. W., ed. (1980). "Laboratory Guide of Identification of Plant Pathogenic Bacteria." APS Press, St. Paul, MN.

Sigee, D. C. (1992). "Bacterial Plant Pathology: Cell and Molecular Aspects." Cambridge Univ. Press, New York.

Singh, U. S., Singh, R. P., and Kohmoto, K. (1995). "Pathogenesis and Host Specificity in Plant Diseases: Histopathological, Biochemical, Genetic and Molecular Bases," Vol. I. Elsevier, Tarrytown, NY.

Starr, M. P. (1984). Landmarks in the development of phytobacteriology. *Annu. Rev. Phytopathol.* 22, 169–188.

Swings, J. G., and Civerolo, E. L. (1993). "*Xanthomonas*." Chapman & Hall, London.

Van Sluys, M. A., *et al.* (2002). Comparative genomic analysis of plant-associated bacteria. *Annu. Rev. Phytopathol.* 40, 169–190.

Vidaver, A. K. (1982). The plant pathogenic corynebacteria. *Annu. Rev. Microbiol.* 36, 495–517.

Whitcomb, R. F., and Tully, J. G., eds. (1989). "The Mycoplasmas," Vol. 5. Academic Press, New York.

Young, J. M., Takikawa, Y., Gardan, L., and Stead, D. E. (1992). Changing concepts in the taxonomy of plant pathogenic bacteria. *Annu. Rev. Phytopathol.* 30, 67–105.

BACTERIAL SPOTS AND BLIGHTS

The most common types of bacterial diseases of plants are those that appear as spots of various sizes on leaves, stems, blossoms, and fruits. In some bacterial diseases the spots continue to advance rapidly and the diseases are then called blights. In severe infections the spots may be so numerous that they destroy most of the plant surface and the plant appears blighted or the spots may enlarge and coalesce, thus producing large areas of dead plant tissue and blighted plants. The spots are necrotic, circular or roughly circular, and in some cases are surrounded by a yellowish halo. In dicotyledonous plants the bacterial spots on some hosts are restricted by large veins, and the spots appear angular. For the same reason, bacterial spots on monocotyledonous plants appear as streaks or stripes. In humid or wet weather, infected tissue often exudes masses of bacteria that spread to new tissues or plants and start new infections. In such weather, dead leaf tissue often tears up and falls out, leaving holes that are round or irregular in shape with ragged edges.

Almost all bacterial spots and blights of leaves, stems, and fruits are caused by bacteria in the genera *Pseudomonas* and *Xanthomonas*.

Pseudomonas syringae, pathovars (pv.) causing wildfire of tobacco (*P. syringae* pv. *tabaci*), angular leaf spot of cucumber (*P. syringae* pv. *lacrymans*), halo blight of beans (*P. syringae* pv. *phaseolicola*), citrus blast, pear blast, bean leaf spot, and lilac blight (*P. syringae* pv. *syringae*), and bacterial speck of tomato (*P. syringae* pv. *tomato*)

Xanthomonas compestris, pathovars causing common blight of beans (*X. campestris* pv. *phaseoli*), angular leaf spot of cotton (*X. campestris* pv. *malvacearum*), bacterial leaf blight of rice (*X. campestris* pv. *oryzae*), bacterial blight or stripe of cereals (*X. campestris* pv. *translucens*), bacterial blight of cassava (*X. campestris* pv. *manihotis*), bacterial spots of stone fruits (*X. arboricola* pv. *pruni*) and of tomato and pepper (*X. campestris* pv. *vesicatoria*)

In bacterial spots and blights, routine diagnosis of the disease depends on the morphology of the symptoms, the absence of pathogenic fungi, and the presence of bacteria in recently infected tissue. Microscopic distinction among these pathogens is impossible, as it is among most plant pathogenic bacteria. The bacteria overwinter on infected or healthy parts, especially buds, of perennial plants, on or in seeds, on infected plant debris, on contaminated containers or tools, and on or in the soil. Their spread from the place of overwintering to their hosts and from plant to plant takes place by means of rain, runoff, rain splashes, windblown rain, direct contact with the host, insects such as flies, bees, and ants, handling of plants, and tools. Penetration takes place through stomates, hydathodes, and injuries. Water soaking of tissues during heavy rains greatly favors penetration and invasion by bacteria. Bacteria multiply on walls of host cells, which collapse after disruption of the cell membrane. The control of bacterial spots and blights can be obtained to some extent by the use of resistant varieties, crop rotation, and sanitation. Some control can be obtained by spraying several times during the period of plant susceptibility with chemicals such as copper compounds mixed with zineb, maneb, or mancozeb, antibiotics such as streptomycin and tetracyclines, and, in some cases, with plant defense activators.

WILDFIRE OF TOBACCO

Wildfire of tobacco occurs worldwide. In some regions it occurs year after year and is very destructive, whereas in others it appears sporadically. In addition to tobacco, the pathogen, *P. syringae* pv. *tabaci*, also affects soybean (Fig. 12-9A).

Wildfire causes losses in both the seedbed and field. Affected seedlings may be killed. In tobacco plants already in the field, wildfire causes large, irregular, dead areas on the leaves, which wither and fall off, making the leaves commercially worthless.

Symptoms

The first symptoms appear on the leaves of young plants in seedbeds as an advancing wet rot at the margins and tips. A water-soaked zone separates the rotting and the healthy tissues. The whole leaf area or only parts of it may rot and fall off. Some seedlings may be killed in the seedbed or after they are transplanted.

Leaves of plants in the field develop round, yellowish spots 0.5 to 1.0 centimeters in diameter. The centers of the spots quickly turn brown and are surrounded by whitish-yellow halos (Fig. 12-9). The brown spots and the halos enlarge rapidly, and in a few days they become 2 to 3 centimeters in diameter. Adjacent spots usually coalesce and form large dead areas on the leaf. In dry weather, the dead areas dry up and remain intact, but in wet weather they continue to enlarge while their centers fall off, making the leaves worthless. Spots appear less frequently on flowers, seed capsules, petioles, and stems.

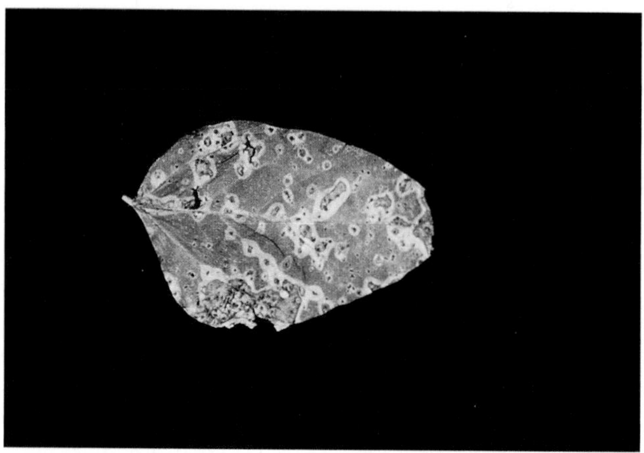

FIGURE 12-9 Wildfire symptoms on soybean leaf caused by *Pseudomonas tabaci*. Note bright halo surrounding each lesion or group of lesions. [Photograph courtesy of Plant Pathology Department, University of Florida.]

The Pathogen: **Pseudomonas syringae** ***pv.*** **tabaci**

The bacterium produces a fluorescent pigment and a potent toxin, called tabtoxin or wildfire toxin. A mere 0.05 milligrams of this toxin can produce a yellow lesion on a tobacco leaf in the absence of bacteria. The bacterium produces a hypersensitive reaction when injected into leaves of tomato and pepper.

Development of Disease

The bacterium overwinters in plant debris in the soil, in dried diseased leaves, on seed from infected seed capsules, and on contaminated seedbed covers. From these the bacteria are carried to the leaves by rain splashes or by wind (Fig. 12-10). They may also be spread during handling of the plants.

High humidity or a film of moisture must be present for infections to occur and for the development of epidemics. Water-soaked areas forming in the leaves during long rainy periods or from windblown rain are excellent infection courts for the bacterium and result in extensive lesions within 2 to 3 days. Bacteria enter the leaf through stomata, hydathodes, and wounds. Certain insects such as flea beetles, aphids, and whiteflies also act as vectors of this pathogen.

Once inside the leaf the bacteria multiply intercellularly (Fig. 12-3) at a rapid rate and secrete the wildfire toxin. The toxin spreads radially from the point of infection and results in the formation of the chlorotic halo, which consists of a zone of cells free of bacteria surrounding the bacteria-containing spot. Variants of the bacterium that do not produce tabtoxin produce a similar disease without halos, known as angular leaf spot or blackfire.

In favorable weather, bacteria continue to spread intercellularly and, through the toxin and enzymes they secrete, cause the breakdown, collapse, and death of the parenchyma cells in the leaf tissues they invade. Necrotic areas are also invaded by saprophytic bacteria and fungi, which disintegrate the tissues further. Bacteria in the disintegrated areas of the leaf fall to the ground or are carried by air currents and splashing rain to other plants.

Control

Whenever possible, only resistant varieties should be planted. Control practices should begin in the

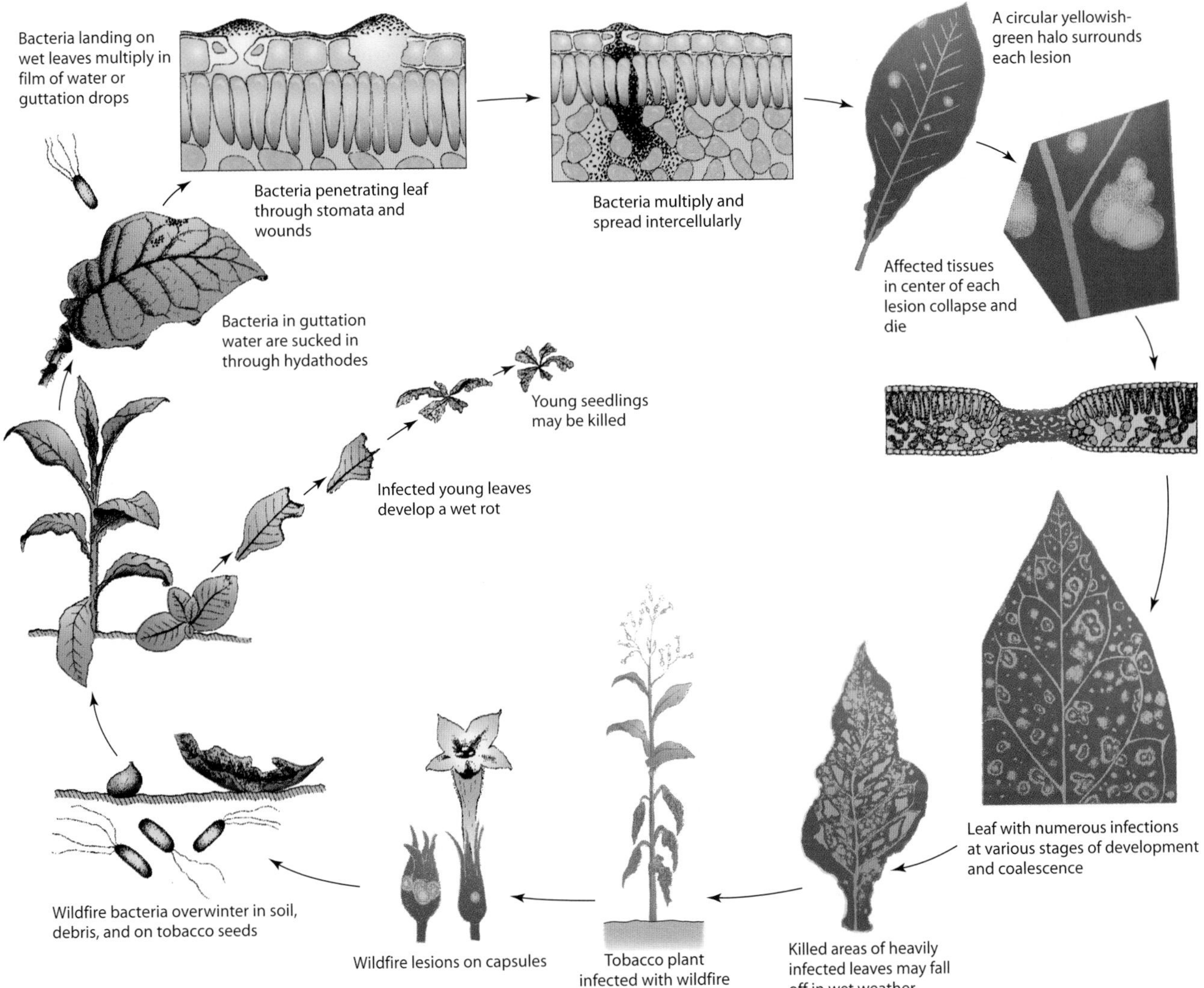

FIGURE 12-10 Disease cycle of a bacterial leaf blight, e.g., wildfire of tobacco or soybeans caused by *Pseudomonas syringae* pv. *tabaci*.

seedbed, as the disease often starts there. Only healthy seed should be used. Contaminated seed should be disinfested by soaking in a formaldehyde solution for 10 minutes. The seedbed soil should be sterilized before planting. After seedlings emerge, seedbeds should be sprayed with a copper fungicide and streptomycin. The streptomycin sprays should be continued at weekly intervals until plants are transplanted. If isolated spots of wildfire appear, the infected plants plus all healthy plants in a 25-centimeter band around them should be destroyed by drenching with formaldehyde. Only healthy seedlings should be transplanted into the field, and they should be planted only in fields that did not have a diseased crop during the previous year. Overfertilization should be avoided, as rapidly growing, succulent plants are very susceptible to the disease.

BACTERIAL BLIGHTS OF BEAN

Three blights of bean are caused by bacteria: common blight, caused by *Xanthomonas campestris* pv. *phaseoli*, halo blight caused by *Pseudomonas syringae* pv. *phaseolicola*, and bacterial brown spot caused by *P. syringae* pv. *syringae*. All three diseases occur wherever beans are grown and cause similar symptoms. In the field, the three diseases affect the leaves, pods, stems, and seeds in a similar way and are usually impossible to distinguish from one another on the basis of symptoms. Common blight seems to be more prevalent in relatively warm weather, whereas the other two blights are more prevalent in relatively cool weather.

Symptoms

The symptoms appear first on the lower sides of the leaves as small, water-soaked spots. The spots enlarge, coalesce, and form large areas that later become necrotic. Bacteria may also enter the vascular tissues of the leaf and spread into the stem. In common blight and in bacterial brown spot, the infected area, which is surrounded by a narrow zone of bright yellow tissue, turns brown and becomes rapidly necrotic. Several small spots coalesce and produce large dead areas of various shapes (Figs. 12-11A–12-11C). In halo blight, a much wider halo-like zone of yellowish tissue 10 millimeters or more in width forms outside the water-soaked area, giving the leaves a yellowish appearance (Figs. 12-11D and 12-11E). All diseases produce identical symptoms on the stems, pods, and seeds, but when a bacterial exudate is produced on them, it is yellow in common blight (*Xanthomonas*) and light cream or silver colored in halo blight and in bacterial brown spot (*Pseudomonas*).

On the stem, water-soaked, sometimes sunken lesions form that gradually enlarge longitudinally and turn brown, often splitting at the surface and emitting a bacterial exudate. Such lesions are most common in the vicinity of the first node, where they girdle the stem, usually at about the time the pods are half mature. The weighted plant often breaks at the lesion. On the pods, water-soaked spots also develop that enlarge and turn reddish with age (Fig. 12-11B). Often the vascular systems of the pod become infected, resulting in infection of the seed through its connection with the pod. Seeds may rot or may show various degrees of shriveling and discoloration depending on the timing and degree of infection. Similar symptoms are caused on pea and soybean by two different species of *Pseudomonas*.

Development of Disease

In all three bacterial blights, bacteria overwinter in infected seed and infected bean stems. From the seed, bacteria infect the cotyledons, and from these they spread to the leaves or enter the vascular system and cause systemic infection, producing stem and leaf lesions. Internally, bacteria move between cells; however, the latter collapse, are invaded and then digested, and cavities form. When in the xylem, bacteria multiply rapidly and move up or down in the vessels and out into the parenchyma. They may ooze out through splits in the tissue and may reenter stems or leaves through stomata or wounds.

Control

Control of bacteria bean blights is through the use of disease-free seed, 3-year crop rotation, and sprays with copper fungicides.

ANGULAR LEAF SPOT OF CUCUMBER

Angular leaf spot of cucumber is caused by *Pseudomonas syringae* pv. *lachrymans*. It affects the leaves, stems, and fruits of cucumber, cantaloupe, squash, and watermelon. At first, small circular spots appear on the leaves and soon become large, angular to irregular, water-soaked areas. In wet weather, droplets of bacterial ooze exude from the spots on the lower leaf surfaces. In dry weather the exudate becomes a whitish crust. Later, the infected areas die and shrink, often tearing and falling off, leaving large, irregular holes in the leaves (Fig. 12-12A). Infected fruits show small, circular, usually superficial spots (Fig. 12-12B). Affected tissues die, turn white, and crack open and then soft-rot fungi and bacteria enter and rot the whole fruit.

Bacteria overwinter on contaminated seed and in infected plant refuse. From there the bacteria are splashed to cotyledons and leaves, which they penetrate through stomata and wounds, and may move systemically to other parts of the plant. Control is obtained through the use of clean or treated seed, resistant varieties, crop rotation, and somewhat by spraying with fixed copper-containing bactericides.

ANGULAR LEAF SPOT OR BACTERIAL BLIGHT OF COTTON

Angular leaf spot of cotton is caused by *Xanthomonas campestris* pv. *malvacearum*. The disease is present wherever cotton is grown. Small, round, water-soaked spots appear on the undersides of cotyledons and young leaves and on stems of seedlings soon after emergence. Most such leaves and plants are killed. In later stages, the spots on leaves appear as angular, brown to black lesions of varying sizes (Fig. 12-13A). In some varieties, bacteria enter and kill parts of the veins and adjacent tissues (Fig. 12-13). Infected leaves of some varieties turn yellow, curl, and fall. On young stems the lesions become long and black, which has given the name "black arm" to the disease. Stem lesions sometimes girdle and kill the stems. Angular to irregular black spots also develop on young cotton bolls (Fig. 12-13B). On these, the spots become sunken,

FIGURE 12-11 Symptoms on bean leaves (A), pods (B), and whole plants (C) caused by the bean common blight bacterium *Xanthomonas phaseoli* and on bean leaves (D and E) caused by the bean halo blight bacterium *Pseudomonas phaseolicola*. Note similarity of symptoms and even of halos. [Photographs courtesy of (A and E) W. L. Seaman, W.C.P.D. and (B–D) Plant Pathology Department, University of Florida.]

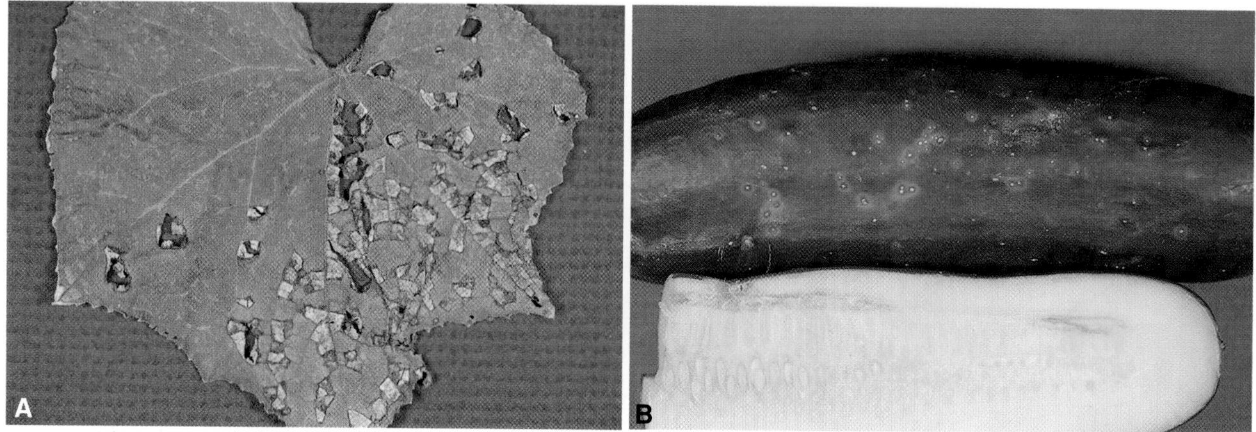

FIGURE 12-12 Angular leaf spots on cucumber leaf (A) and small circular spots with halo on cucumber fruit (B) caused by the bacterium *Pseudomonas lacrymans*. [Photographs courtesy of Plant Pathology Department, University of Florida.]

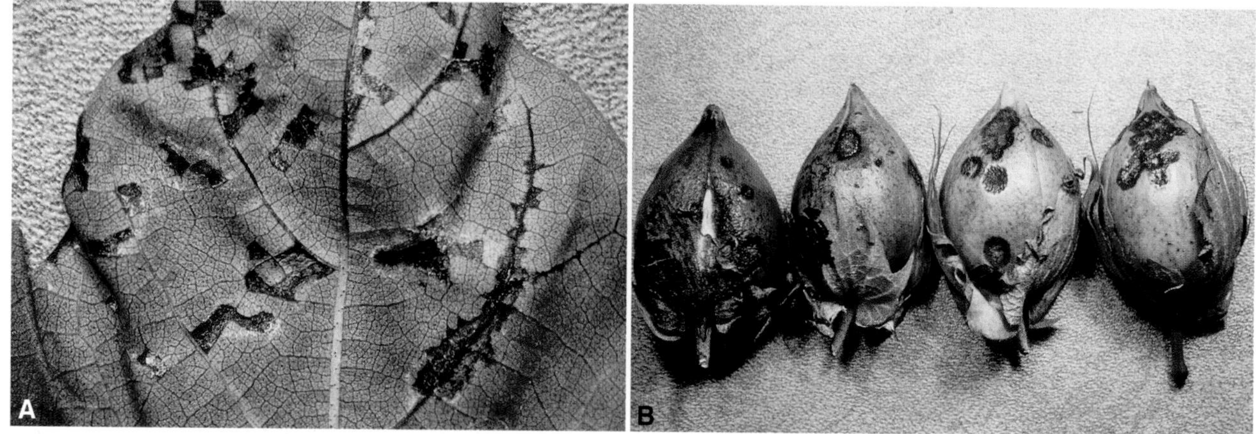

FIGURE 12-13 Angular spots and necrotic veins on cotton leaves (A) and sunken circular spots on cotton bolls (B) caused by the cotton blight bacterium *Xanthomonas campestris* pv. *malvacearum*. [Photographs courtesy of Plant Pathology Department, University of Florida.]

and in hot, humid weather the bacteria may invade and rot the bolls and cause them to drop or to become distorted.

Bacteria overwinter in or on the seed, on the lint, and on undecomposed plant debris. Control is through the use of disease-free or treated seed and resistant varieties.

BACTERIAL LEAF SPOTS AND BLIGHTS OF CEREALS AND GRASSES

Several *Pseudomonas* and *Xanthomonas* species and pathovars attack each of the cultivated cereals and wild grasses, and some of them cause severe losses to their respective hosts. The most common bacterial diseases of these crops are bacterial stripe of sorghum and corn (*P. andropogonis*), leaf blight of all cereals (*P. avenae*), red stripe and top rot of sugarcane (*P. rubrilineans*), basal glume rot of cereals (*P. syringae* pv. *atrofaciens*), halo blight of oats and other cereals (*P. syringae* pv. *coronafaciens*), bacterial blight, stripe, or streak of several cereals (*X. campestris* pv. *translucens*) (Figs. 12-14A and 12-14B), bacterial leaf blight of rice (*X. oryzae* pv. *oryzae*) (Fig. 12-14C), bacterial leaf streak of rice (Fig. 12-14D), and leaf scald of sugarcane (*X. albilineans*).

Most bacterial leaf spots and blights of cereals are probably worldwide in distribution. They cause more or less similar diseases on one or more of the cereals and grasses. Most such diseases only occasionally cause reduction in yields, but some are of major importance. The symptoms appear on leaf blades and sheaths as

FIGURE 12-14 Longitudinal lesions on leaves (A) and reddish-black lesions on glumes (B) of wheat infected with the wheat streak and black chaff bacterium *Xanthomonas campestris* pv. *transluscens*. (C) Bacterial blight of rice caused by *X. oryzae* pv. *oryzae* and (D) bacterial leaf streak caused by *X. oryzae* pv. *oryzicola*. [Photographs courtesy of (A and B) University of Idaho and (C and D) H. D. Thurston, Cornell University.]

small, linear, water-soaked areas that soon elongate and coalesce into irregular, narrow, yellowish, or brownish stripes (Fig. 12-14). Droplets of white exudate are common on the stripes. Severe infections cause leaves to turn yellow and die from the tip downward (Figs. 12-14C and 12-14D); they also retard spike elongation and cause blighting. Small lesions form on the kernels as well. The diseases develop mainly in rainy, damp weather. Bacteria overwinter on the seed and in crop residue and are spread by rain, direct contact, and insects. The main control measures are use of disease-free or treated seed and crop rotation.

BACTERIAL SPOT OF TOMATO AND PEPPER

Bacterial spot of tomato and pepper is caused by *Xanthomonas campestris* pv. *vesicatoria* and is widespread. Different strains of the bacteria cause disease on pepper and tomato, pepper only, or tomato only. They damage leaves, stems, and fruit. On the leaves, symptoms appear as small (about 3 millimeters), irregular, black, greasy lesions. Leaves with many lesions may turn yellow, may appear ragged (Figs. 12-15A and 12-15C), or may fall off. Infection of flower parts usually results in serious blossom drop. On green fruit, small, water-soaked,

FIGURE 12-15 Bacterial spot on tomato leaves (A) and fruit (B) and pepper leaves (C) and fruit (D) caused by *Xanthomonas campestris* pv. *vesicatoria*. Bacterial speck on green tomato fruit and leaves (E) and ripe tomato fruit (F) caused by *Pseudomonas syringae* pv. *tomato*. [Photographs courtesy of (A) J. A. Bartz, (B) R. T. McMillan, (C and D) Plant Pathology Department, University of Florida, (E) K. Pernezny, and (F) R. J. McGovern, all University of Florida.]

slightly raised spots appear, which sometimes have greenish-white halos, and enlarge to about 3 to 6 millimeters in diameter (Figs. 12-15B and 12-15D). Later, the halos disappear and the spots become dark brown and slightly sunken, with a scabby surface. Bacteria overwinter on seed contaminated during extraction, in infected plant debris in the soil, and on weeds and other hosts. They are spread by rain, wind, or contact and penetrate leaves and fruits through wounds and through stomata. Control of the disease depends on the use of bacteria-free seed and seedlings, resistant varieties, crop rotations, and sprays with copper fungicides tank mixed with mancozeb or maneb. The disease, however, after it appears in the field, can be controlled with copper–maneb fungicides only under reasonably dry weather.

BACTERIAL SPECK OF TOMATO

A disease called bacterial speck of tomato is similar to bacterial spot but is caused by the bacterium *Pseudomonas syringae* pv. *tomato*. Bacterial speck has become economically important throughout the world since the mid-1970s. The lesions on leaves, stems, and fruit are similar but smaller than those of bacterial spot (Figs. 12-15E and 12-15F), although they often coalesce and appear as scabby areas that, on the fruit, may cover one-fourth or more of its surface. Bacterial speck is favored by cool moist weather. Control is the same as for bacterial spot.

BACTERIAL FRUIT BLOTCH OF WATERMELON

Bacterial fruit blotch of watermelon appeared for the first time in the early 1990s. It occurs in several watermelon-producing areas of the United States, particularly in the southeast. It occurs sporadically from year to year and affects a fairly small number of fields each year. However, in affected fields, losses are usually quite high because of the disfigurement of the rind, which makes the fruit unmarketable. The fruit is safe to eat.

Symptoms

The first symptoms of watermelon fruit blotch appear first as water-soaked and then dry necrotic areas on the undersides of cotyledons and leaves (Fig. 12-16A) but these are easily overlooked or misdiagnosed. Distinctive symptoms, however, appear on mature fruit shortly before it is to be harvested. Symptoms consist of large infected areas or lesions on the rind that at first appear water soaked or oily (Figs. 12-16B and 12-16C) and are located on the top or sides of the fruit and not where the fruit touches the soil. The tissue in the oily lesions is at first as firm as in the unaffected areas and it does not extend deeper than the rind. As the disease progresses, however, the surface of the lesions becomes

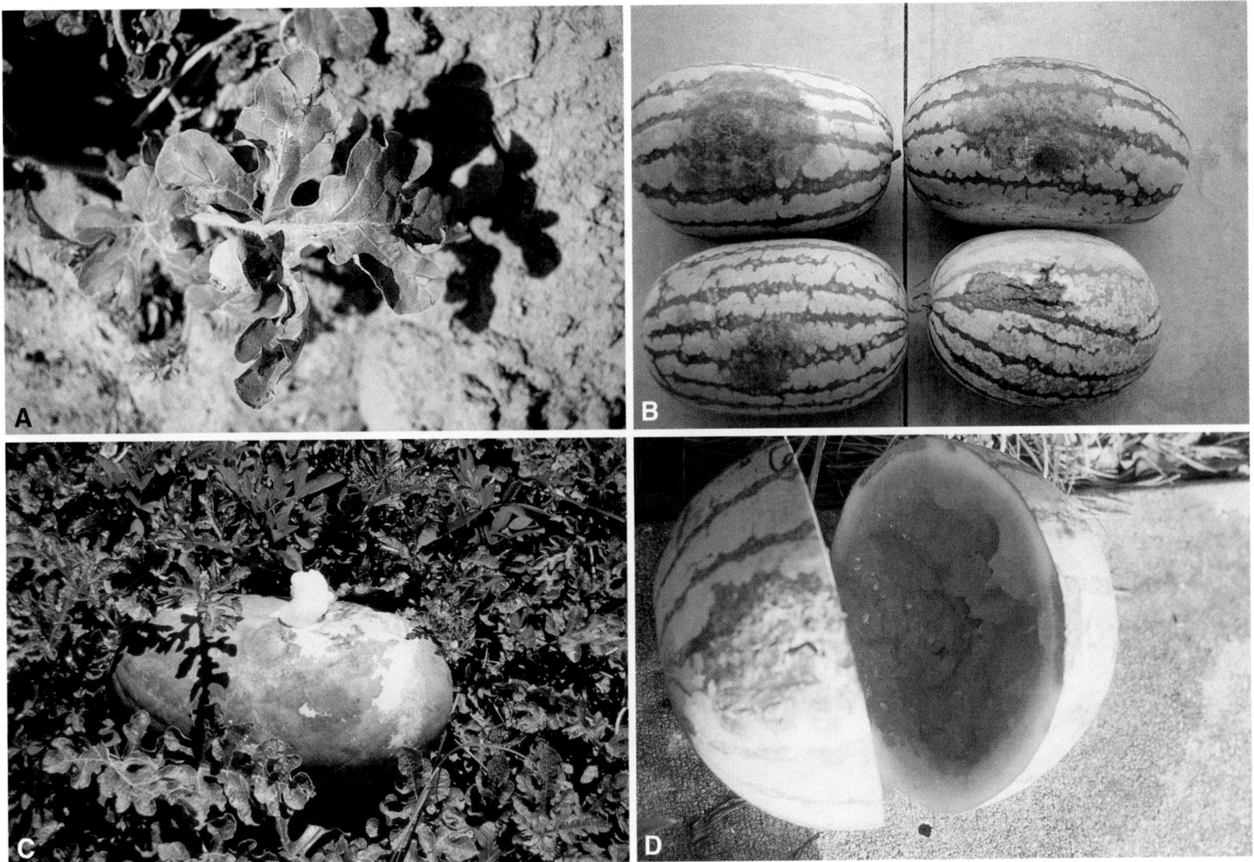

FIGURE 12-16 Watermelon fruit blotch disease caused by *Acidovorax avenae* subsp. *citrulli*. (A) Spots and vein blotches on young watermelon seedling. (B) Watermelons showing superficial oily blotches, with one also showing cracks. (C) Watermelon with large brown blotch and foam exiting the fruit, which is apparently fermenting. (D) Infected watermelon cut to show the rotting and fermentation of its contents due to invasion by secondary microorganisms. [Photographs courtesy of D. L. Hopkins, University of Florida.]

brownish and bumpy and cracks. A brown gummy ooze may develop in it. The cracks allow other microorganisms to enter the watermelon. These produce gasses may exit the fruit as small foamy eruptions (Fig. 12-16C) or noisy explosions as the inside "meat" of the watermelon liquefies (Fig. 12-16D) and rots.

Pathogen

The cause of watermelon fruit blotch is the bacterium *Acidovorax avenae* subsp. *citrulli*.

Development of Disease

The bacterium overwinters in seed from infected plants and in volunteer plants growing in fields from such fruit abandoned in the field. Even a small percentage of infected seed can produce enough bacteria to infect a large percentage of the cotyledons and leaves of young plants. Leaf infections do not seem to damage plants, but they provide bacterial inoculum for infection of the fruit. The fruit is susceptible to infection only during flowering and fruit set, but the infections remain dormant until shortly before ripening at which time bacteria become activated and cause the symptoms on the rind of the fruit. Few, if any, lesions develop on the fruit after ripening and no further infections spread after harvest during transit or storage of the fruit. The spread of the disease in the field is favored greatly by rain and overhead irrigation, both of which help spread the bacteria to more plants and fruit.

Control

The control of watermelon fruit blotch depends on the use of watermelon seed free of the fruit blotch bacteria. Rotation of watermelon fields for at least a year to non-cucurbit crops and destruction of volunteer cucurbit plants in them are essential. Do not work fields while wet. Avoid using overhead irrigation. If infected plants are found in the field, apply copper bactericides from flowering until all fruit are mature.

CASSAVA BACTERIAL BLIGHT

The disease occurs in all major cassava-producing areas of the world. Cassava losses from bacterial blight range from high to total, depending on variety, bacterial strains present, and weather conditions. In countries where the majority of the rural population relies heavily on cassava as the staple crop, bacterial blight can be devastating.

The symptoms appear at first on leaves as water-soaked lesions that enlarge, become chlorotic, and then brown necrotic areas give the leaf a wilted, blighted appearance (Fig. 12-17A). Infections soon become sys-

FIGURE 12-17 Bacterial blight of cassava caused by *Xanthomonas campestris* pv. *manihotis*. (A) Young cassava leaves and main shoot killed by the blight. (B) Leaves of older plant show numerous angular spots. Older leaves have been killed and have fallen off. Younger plant is blighted and dying. [Photographs courtesy of J. Hughes, Intl. Inst. Trop. Agric., Ibadan, Nigeria.]

temic producing vascular discoloration, stem cankers, oozing of bacteria, especially during periods of high humidity in the morning hours, and dieback. Eventually entire plants wilt and die (Fig. 12-17B).

Cassava bacterial blight is caused by the bacterium *Xanthomonas axonopodis* pv. *manihotis*.

The pathogen spreads over long distances primarily by infected cuttings. Within a field, bacteria are spread from plant to plant by rain splashes, various insects that are attracted by and then become smeared with bacteria-containing ooze, and by human hands and cultivating equipment. The disease is generally more common, more widespread, and more severe in years with above-normal rainfall.

The control of cassava bacterial blight is attempted through planting of bacteria-free cuttings, planting resistant varieties, crop rotation, and leaving the field fallow for at least six months. Also, cultural practices are used, e.g., roguing of infected plants and sterilizing tools such as knives used to cut stems, in ways that decrease rather than increase the spread of the bacteria.

BACTERIAL SPOT OF STONE FRUITS

Bacterial spot of stone fruits is caused by *Xanthomonas arboricola* (formerly *X. campestris*) pv. *pruni*. It is present in most areas where stone fruits are grown and may cause serious losses by reducing the marketability of the fruit and by weakening trees through leaf spots, defoliation, and lesions on twigs.

Symptoms appear on the leaves as circular to irregular, water-soaked spots about 1 to 5 millimeters in diameter, which later turn purple or brown (Figs 11-12A and 12C). Often halos and cracks develop around the spots, and the affected areas break away from the surrounding healthy tissue, drop out, and give a shot-ridden appearance, known as shot hole, to the leaves (Fig. 12-18A). Several spots may coalesce. Severely affected leaves turn yellow and drop. On the fruit, small, circular, brown, slightly depressed spots appear, usually on a localized area of the fruit (Fig. 12-18B). On some plums, spots are large, black and often coalesce (Fig. 12-18D). Pitting and cracking occur in the vicinity of the fruit spots and, after rainy weather, gum may exude from the injured areas. On the twigs, dark, slightly sunken lesions form usually around buds in the spring or on green shoots later in the summer.

Bacteria overwinter in twig lesions and in the buds. In the spring they ooze out and are spread by rain splashes and insects to young leaves, fruits, and twigs, which they infect through natural openings, leaf scars, and wounds. The disease is more severe on weakened trees than on vigorous ones; therefore, keeping trees in good vigor helps them resist the disease. Chemical sprays have not been effective so far.

Selected References

Abbasi, P. A., Soltani, N., Cuppels, D., *et al.* (2002). Reduction of bacterial spot disease severity on tomato and pepper plants with foliar applications of ammonium lignosulfonate and potassium phosphate. *Plant Dis.* **86**, 1232–1236.

Bonas, U., Van den Ackerveken, G., Büttner, D., *et al.* (2000). How the bacterial plant pathogen *Xanthomonas campestris* pv. *vesicatoria* conquers the host. *Mol. Plant Pathol.* **1**, 73–76.

Brinkerhoff, L. A. (1970). Variation in *Xanthomonas malvacearum* and its relation to control. *Annu. Rev. Phytopathol.* **8**, 85–110.

Clayton, E. E. (1936). Water soaking of leaves in relation to development of the wildfire disease of tobacco. *J. Agric. Res.* **52**, 239–269.

Cuppels, D. A., and Elmhirst, J. (1999). Disease development and changes in the natural *Pseudomonas syringae* pv. *tomato* populations on field tomato plants. *Plant Dis.* **83**, 759–764.

Daft, G. C., and Leben, C. (1972). Bacterial blight of soybeans: Epidemiology of blight outbreaks. *Phytopathology* **63**, 57–62.

Fahy, P. C., and Persley, G. J. (1983). "Plant Bacterial Diseases: A Diagnostic Guide." Academic Press, New York.

Feliciano, A., and Daines, R. H. (1970). Factors influencing ingress of *Xanthomonas pruni* through peach leaf scars and subsequent development of spring cankers. *Phytopathology* **60**, 1720–1726.

Fourie, D. (2002). Distribution and severity of bacterial diseases on dry beans (*Phaseolus vulgaris* L.) in South Africa. *J. Phytopathol.* **150**, 220–226.

Gitaitis, R., McCarter, S., and Jones, J. (1992). Disease control in tomato transplants in Georgia and Florida. *Plant Dis.* **76**, 651–656.

Hirano, S. S., and Upper, C. D. (1983). Ecology and epidemiology of foliar bacterial plant pathogens. *Annu. Rev. Phytopathol.* **21**, 243–269.

Jones, J. B., Stall, R. E., and Bouzar, H. (1998). Diversity among Xanthomonads pathogenic on pepper and tomato. *Annu. Rev. Phytopathol.* **36**, 41–58.

Kritzman, G., and Zutra, D. (1983). Systemic movement of *Pseudomonas syringae* pv. *lachrymans* in the stem, leaves, fruits, and seeds of cucumber. *Can. J. Plant Pathol.* **5**, 273–279.

Louws, F. J., Wilson, M., Campbell, H. L., *et al.* (2001). Field control of bacterial spot and bacterial speck of tomato using a plant activator. *Plant Dis.* **85**, 481–488.

Mew, T. W. (1987). Current status and future prospects of research on bacterial blight of rice. *Annu. Rev. Phytopathol.* **25**, 359–382.

Msikita, W., James, B., Nnodu, E., *et al.* (2000). "Disease Control in Cassava Farms." International Institute of Tropical Agriculture, Cotonou, Benin.

Pohronezny, K., *et al.* (1992). Sudden shift in the prevalent race of *Xanthomonas campestris* pv. *vesicatoria* in pepper fields in southern Florida. *Plant Dis.* **76**, 118–120.

Preston, G. M. (2000). *Pseudomonas syringae* pv. *tomato*: The right pathogen, of the right plant, at the right time. *Mol. Plant Pathol.* **1**, 263–275.

Romero, A. M., Kousik, C. S., and Ritchie, D. F. (2000). Resistance to bacterial spot in bell pepper induced by acibenzolar-*S*-methyl. *Plant Dis.* **85**, 189–194.

Shepard, D. P., Zehr, E. I., and Bridges, W. C. (1999). Increased susceptibility to bacterial spot of peach trees growing in soil infested with *Criconemella xenoplax*. *Plant Dis.* **83**, 961–963.

FIGURE 12-18 Symptoms of bacterial spot caused by *Xanthomonas arboricola* pv. *pruni*. (A) Tiny spots and holes on peach leaves. (B) Numerous small, halo-surrounded spots that later turn brown appear on infected fruit. (C) Spots on plum leaf. (D) Large coalescing spots on fruit of susceptible plum variety. [Photographs courtesy of M. Ellis, Ohio State University, (B) K. D. Hickey, Pennsylvania State University, and (C and D) K. Mohan, University of Idaho.]

Tillman, B. L., Harrison, S. A., and Russin, J. S. (1999). Yield loss caused by bacterial streak in winter wheat. *Plant Dis.* **83**, 609–614.

Verdier, Restrepo, Mosquera, *et al.* (1998). Genetic and pathogenic variation of *Xanthomonas axonopodis* pv. *manihotis* in Venezuela. *Plant Pathol.* **47**, 601–608.

Webster, D. M., Atkin, J. D., and Cross, J. E. (1983). Bacterial blights of snap beans and their control. *Plant Dis.* **67**, 935–940.

Zhao, Y., Damicone, J. P., Demezas, D. H., *et al.* (2000). Bacterial leaf spot diseases of leafy crucifers in Oklahoma caused by pathovars of *Xanthomonas campestris*. *Plant Dis.* **84**, 1008–1014.

BACTERIAL VASCULAR WILTS

Vascular wilts caused by bacteria affect mostly herbaceous plants such as several vegetables, field crops, ornamentals, and tropical plants. The bacteria and the most important vascular wilts they cause are listed.

Clavibacter (*Corynebacterium*), causing ring rot of potato (*C. michiganense* subsp. *sepedonicum*) and bacterial canker and wilt of tomato (*C. michiganense* subsp. *michiganense*)

Curtobacterium (*Corynebacterium*) *flaccumfaciens*, causing bacterial wilt of bean

Erwinia, causing bacterial wilt of cucurbits (*E. tracheiphila*), and fire blight of pome fruits (*E. amylovora*)

Pantoea, causing Stewart's wilt of corn (*P. stewartii*)

Ralstonia, causing the southern bacterial wilt of solanaceous crops and the Moko disease of banana (*R. solanacearum*)

Xanthomonas, causing black rot or black vein of crucifers (*X. campestris* pv. *campestris*)

In vascular wilts, bacteria enter, multiply in, and move through the xylem vessels of the host plants (Fig. 12-19). In the process, they interfere with the translocation of water and nutrients, resulting in the drooping, wilting, and death of the aboveground parts of the plants. In these respects, bacterial vascular wilts are similar to the fungal vascular wilts caused by *Fusarium*, *Verticillium*, *Ophiostoma*, and *Ceratocystis*. However, whereas in fungal wilts the fungi remain almost exclusively in the vascular tissues until the death of the plant, in bacterial wilts the bacteria often destroy (dissolve) parts of cell walls of xylem vessels or cause them to rupture quite early in disease development. Subsequently, the bacteria spread and multiply in adjacent parenchyma tissues at various points along the vessels, kill and dissolve the cells, and cause the formation of pockets or cavities full of bacteria, gums, and cellular debris. In some bacterial vascular wilts, e.g., those of corn and sugarcane, the bacteria, once they reach the leaves, move out of the vascular bundles, spread throughout the intercellular spaces of the leaf, and may ooze out through the stomata or cracks onto the leaf surface. Similarly, in some cases, as in the bacterial wilt of carnation, bacteria ooze to the surface of stems through cracks formed over the bacterial pockets or cavities. More commonly, however, wilt bacteria are confined primarily to the vascular elements and do not reach the plant surface until the plant is killed by the disease.

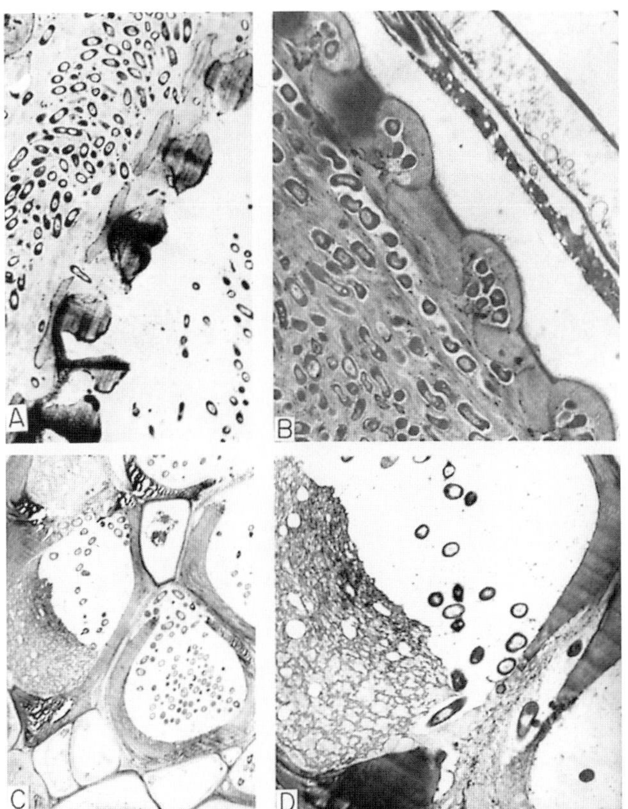

FIGURE 12-19 *Xanthomonas campestris* pv. *campestris* bacteria inside xylem vessels of leaf vein of black rot-infected cabbage. (A) Uneven distribution of bacteria in xylem vessels and passage of bacteria between adjacent xylem vessels. (B) Bacteria in xylem vessel and in bulges in interspiral regions toward the xylem parenchyma cell. (C) Bacteria-containing and bacteria-free xylem vessels. (D) A few bacteria and a mass of plugging material in invaded vessel. [Photographs courtesy of F. M. Wallis (1973). *Physiol. Plant Pathol.* 3, 371–378.]

Bacterial vascular wilts can sometimes be determined by cutting an infected stem with a sharp razor blade and then separating the two parts slowly, in which case a thin bridge of a sticky substance can be seen between the cut surfaces while they are being separated (Fig. 12-20E). Better still, small pieces of infected stem, petiole, or leaf can be placed in a drop of water and observed under the microscope, in which case masses of bacteria will be seen flowing out from the cut ends of the vascular bundles (Fig. 12-20C).

Wilt bacteria overwinter in plant debris in the soil, in the seed, in vegetative propagative material, or, in some cases, in their insect vectors. They enter the plants through wounds that expose open vascular elements and multiply and spread in the latter. They spread from plant to plant through the soil, through handling and tools, through direct contact of plants, or through insect vectors. Control of bacterial vascular wilts is difficult and depends primarily on the use of crop rotation, resistant varieties, bacteria-free seed or other propagative material, control of the insect vectors of the bacteria when such vectors exist, removal of infected plant debris, and proper sanitation.

BACTERIAL WILT OF CUCURBITS

Bacterial wilt of cucurbits occurs in the United States, Europe, South Africa, and Japan. It affects many species of the family Cucurbitaceae. Cucumber, cantaloupe, squash, and pumpkin are susceptible, whereas water-

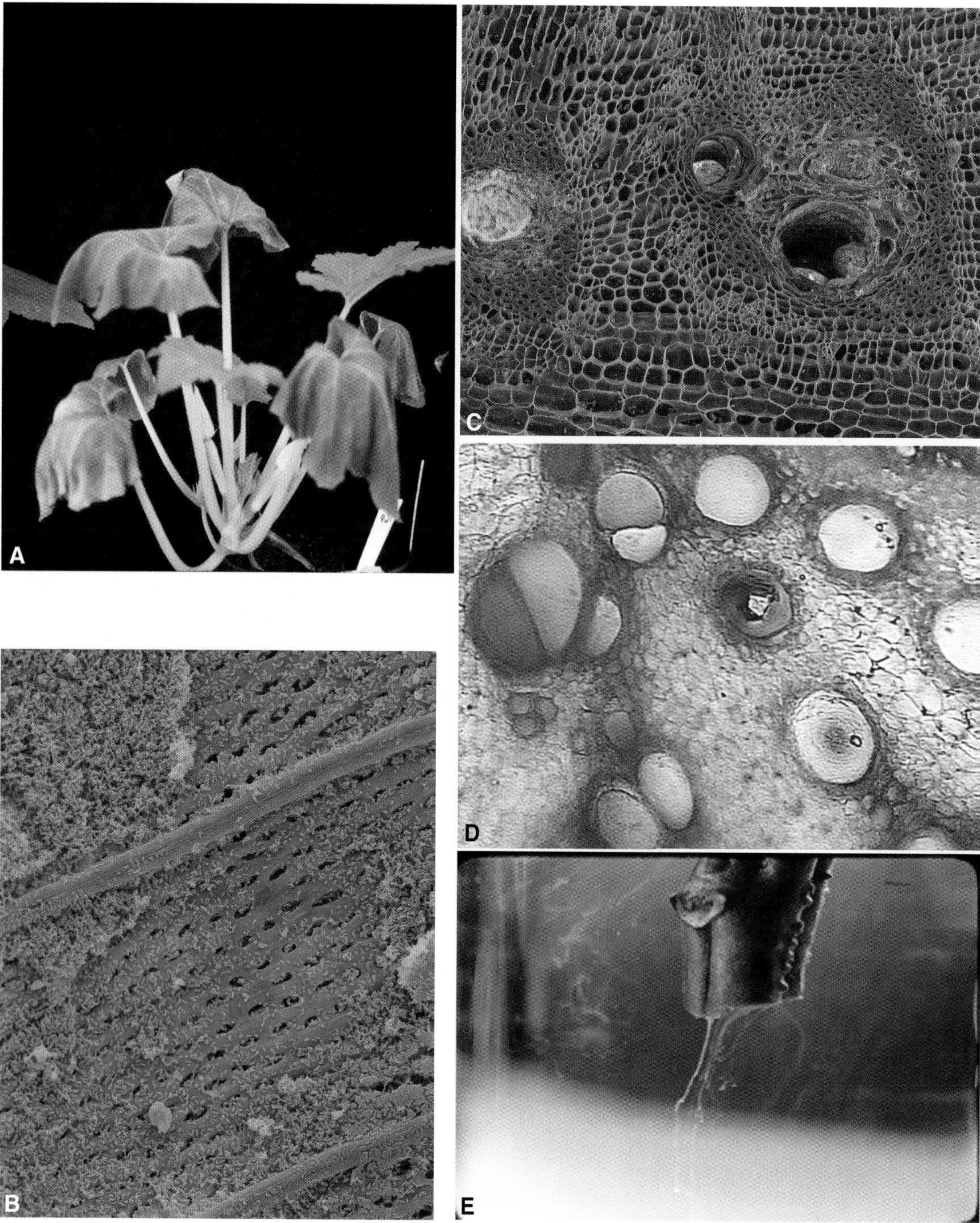

FIGURE 12-20 Bacterial wilt of cucurbits caused by *Erwinia tracheiphila*. (A) Young cantaloupe plant showing early wilt symptoms. (B) Wilt bacteria lining up much of the xylem vessel wall. (C) Most of the vessels in petiole of wilted leaf appear partially or totally occluded by a mixture of gel-like materials. (D) Almost all xylem vessels totally plugged by an almost solid mixture of gels and gums. (E) A stream of bacteria squeezed out of an infected stem. [Photographs courtesy of B. D. Bruton, USDA, Lane, OK.]

melon is resistant or immune to bacterial wilt. Affected plants develop sudden wilting of foliage and vines and finally die. Affected squash fruit develop a slime rot in storage. The severity of the disease varies from an occasional wilted plant to destruction of 75 to 95% of the crop.

Symptoms

Symptoms appear as drooping of one or more leaves of a vine followed by drooping and wilting of all the leaves of that vine (Fig. 12-20A) and, subsequently, by wilting of all leaves and collapse of all vines of the infected plant. Wilted leaves shrivel and dry up; affected stems first become soft and pale, but later they too shrivel and become hard and dry. In moderately resistant plants or under unfavorable conditions, symptoms develop slowly and may occasionally be accompanied by excessive blossoming and branching of the infected plants. Under the microscope, sections of wilted stems and petioles reveal bacteria in xylem vessels (Fig. 12-20B) and some (Fig. 12-20C) or all (Fig. 12-20D) of the xylem vessels clogged with almost solidified mixtures of polysaccharides, proteins, and so on that completely block passage of water and nutrients. When infected stems are cut and pressed between the fingers, droplets of white bacterial ooze appear on the cut surface. The viscid sap sticks to the fingers or to the cut sections, and if they are gently pulled apart the ooze forms delicate threads that may extend for several centimeters (Fig. 12-20E). The stickiness of the sap of infected plants is frequently used as a quick diagnostic characteristic of the disease.

The slime rot of stored squash progresses internally while the exterior of the fruit may appear perfectly sound. Usually, however, as the internal rot progresses there appear on the surface dark spots or blotches that coalesce and enlarge. The disease develops over several months in storage. Infected squash fruits are further invaded by soft-rot microorganisms and are completely destroyed.

The Pathogen

Erwinia tracheiphila. The bacterium survives for only a few weeks in infected plant debris. However, it survives over winter in the intestines of striped cucumber beetles (*Acalymma vittata*) and spotted cucumber beetles (*Diabrotica undecimpunctata*) (Fig. 12-21), in which it hibernates.

Development of Diseases

In the spring, the insects that carry bacteria feed and cause deep wounds on the leaves of cucurbit plants; the insects deposit bacteria in the wounds with their feces. Through the wounds, the bacteria enter the xylem vessels, multiply rapidly, and spread to all parts of the plant (Fig. 12-21). As bacteria multiply in the xylem, they and their polysaccharides obstruct the vessels, as do gum deposits and tyloses formed in the xylem elements of infected plants. Stems of wilted plants allow less than one-fifth the normal water flow, indicating that extensive plugging of the vessels is the primary cause of wilting.

Bacteria are spread by contaminated mouthparts of the striped and the spotted cucumber beetles and by some other insects. Each contaminated beetle can infect several healthy plants after one feeding on a wilted plant. Only a rather small percentage of beetles, however, become carriers of bacteria. The first wilt symptoms appear 6 or 7 days after infection, and the plant is usually completely wilted by the 15th day. Bacteria present in the vessels of infected plants die within 1 or 2 months after the dead plants dry up.

Fruit infection of squash plants usually takes place through infected vines and occasionally through beetles feeding on the blossoms and the rind of developing squash.

Control

The bacterial wilt of cucurbits can be controlled best by controlling the cucumber beetles, especially the early ones, with insecticides. To avoid squash rot in storage, only fruit from healthy plants should be picked and stored in a clean, fumigated warehouse. Resistant cucurbit varieties should be preferred to more susceptible ones.

FIRE BLIGHT OF PEAR AND APPLE

Fire blight is the most destructive disease of pear, making commercial pear growing under certain conditions impossible. Fire blight causes damage to pear and apple orchards in many parts of the world. Certain apple and quince varieties are also very susceptible to the disease. Many other plant species are affected by fire blight, including several of the stone fruits and many cultivated and wild ornamental species, but only those in the pome fruit group are affected seriously.

Fire blight may kill flowers and twigs (Figs. 12-22A–12-22D and 12-22F) and it may girdle large branches and trunks (Figs. 12-22D and 12-22E), thereby killing the trees. Young trees may be killed to the ground by a single infection in one season (Fig. 12-22G). A panoramic view of a fire blight epidemic is shown in Fig. 12-23C.

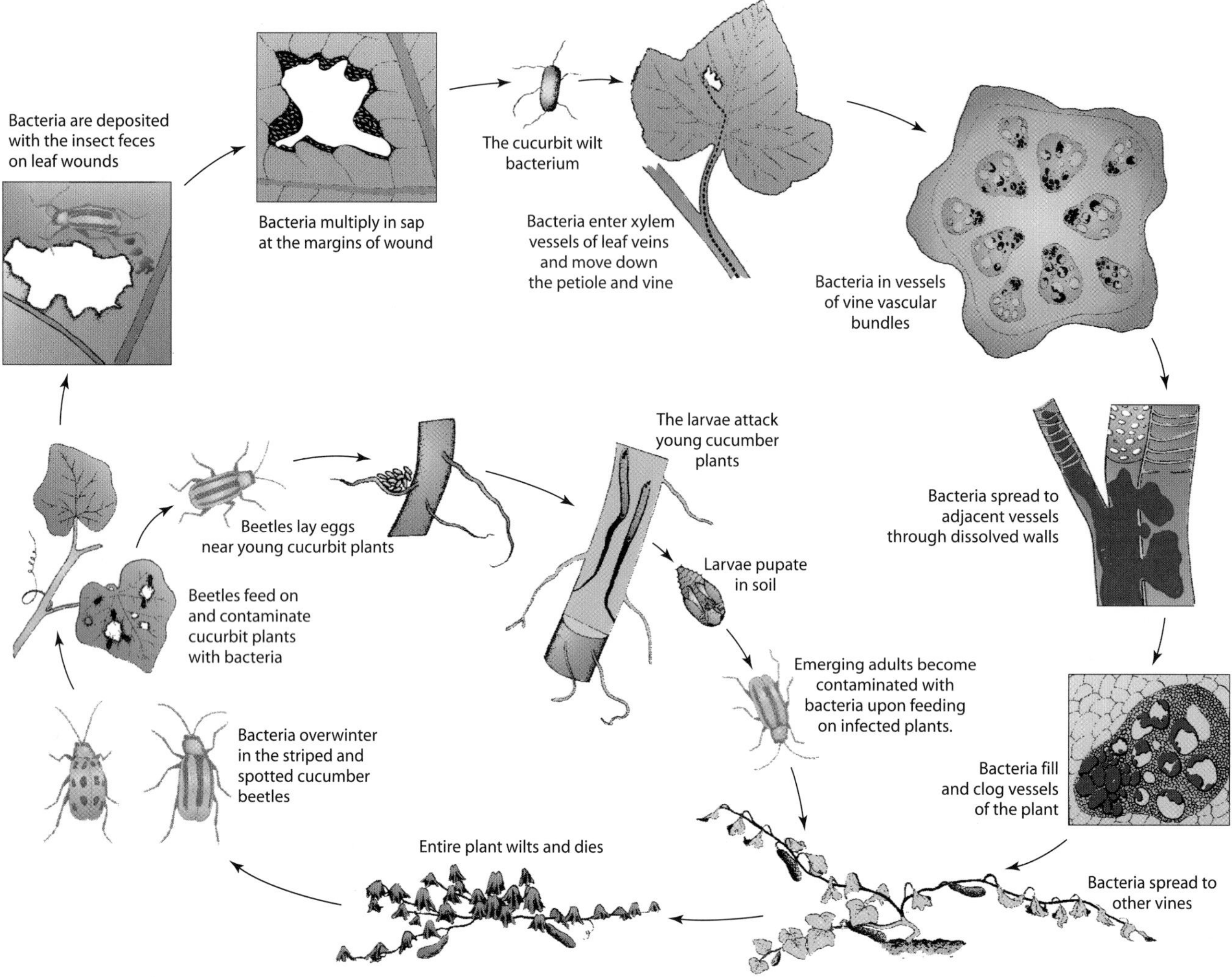

FIGURE 12-21 Disease cycle of bacterial wilt of cucurbits caused by *Erwinia tracheiphila*.

Symptoms

Infected flowers become water soaked, then shrivel, turn brownish black, and fall or remain hanging in the tree (Fig. 12-22B). Soon leaves on the same spur or on nearby twigs develop brown-black blotches along the midrib and main veins or along the margins and between the veins. As the blackening progresses, the leaves curl and shrivel, hang downward, and usually cling to the curled, blighted twigs.

Terminal twigs and suckers are usually infected directly and wilt from the tip downward. Their bark turns brownish black and is soft at first but later shrinks and hardens. The tip of the twig is hooked (Fig. 12-22A), and the leaves turn black and cling to the twig. From fruit spurs and twigs the symptoms progress down to the branches, where cankers are formed (Fig. 12-22D). The bark of cankers appears water soaked at first, later becoming darker, sunken, and dry. If the canker enlarges and encircles the branch, the part of the branch above the infection dies. If the infection stops short of girdling the branch, it becomes a dormant canker, with sunken and sometimes cracked margins (Fig. 12-22D). Bacteria can move downward internally through branches and trunks of trees, even of symptomless varieties, and may reach the rootstocks, which, if susceptible, may be killed by fire blight (Fig. 12-22E).

Infected small, immature fruit become water soaked, then turn brown, shrivel, turn black, and may cling to the tree for several months after infection (Figs. 12-22C and 12-23A).

Under humid conditions, droplets of a milky colored, sticky ooze may appear on the surface of any recently infected part (Fig. 12-23B). The ooze usually turns brown soon after exposure to the air.

FIGURE 12-22 Fire blight of apple and pear caused by *Erwinia amylovora*. (A) Infection of young shoot, which shows a shepherd's hook-like appearance. (B) Infection of blossoms and young fruit. (C) Infection of fruit and, through the pedicel, of the supporting twig. (D) Bacteria move through infected twig to the branch and cause a canker in which the bacteria overwinter. (E) Rootstock infected by bacteria moving downward through the stem or through suckers. (F) Pear tree showing typical symptoms at its crown. (G) A young apple orchard destroyed by fire blight. [Photographs courtesy of (A–F) T. Van Der Zwet, USDA, and (G) A. Jones, Michigan State University.]

(Continued next page)

FIGURE 12-22 *(Continued)*

FIGURE 12-23 (A) Pears infected with fire blight through the pedicel. (B) Infected apple fruit exuding droplets of fire blight bacteria. (C) The *Erwinia amylovora* bacterium. (D) Panoramic view of apple and pear orchard in which most of the trees were killed by fire blight. [Photographs courtesy of (A, B, and D) T. Van Der Zwet, USDA, and (C) Oregon State University.]

The Pathogen

Erwinia amylovora. It is a rod-shaped bacterium, has peritrichous flagella (Fig. 12-23C), and requires nicotinic acid as a growth factor. It is identified from the symptoms it causes and by serological tests.

Development of Disease

Bacteria overwinter at the margins of cankers and possibly in buds and apparently healthy wood tissue. In the spring, bacteria in the cankers become active again, multiply, and spread into the adjoining healthy bark. During humid or wet weather, bacterial masses exude through lenticels and cracks. The bacterial ooze appears at about the time when the pear blossoms are opening. Various insects, such as bees, flies, and ants, are attracted to the sweet, sticky, bacteria-filled exudate, become smeared with it, and spread it to the flowers they visit afterward. In some cases, bacteria are also spread from oozing cankers to flowers by splashing rain (Fig. 12-24). When the ooze dries, it often forms aerial strands that can be spread by wind and serve as inoculum.

Bacteria multiply rapidly in the nectar and, through the nectarthodes, enter the tissues of the flower. Bees visiting an infected flower carry bacteria from its nectar to all the succeeding blossoms that they visit. Once inside the flower, bacteria multiply quickly and cause death and collapse of nearby cells. Bacteria move quickly through the intercellular spaces and also through the macerated middle lamella and flower cells. In some cases, fairly large cavities form that are filled with bacteria. From the flower, bacteria move down the

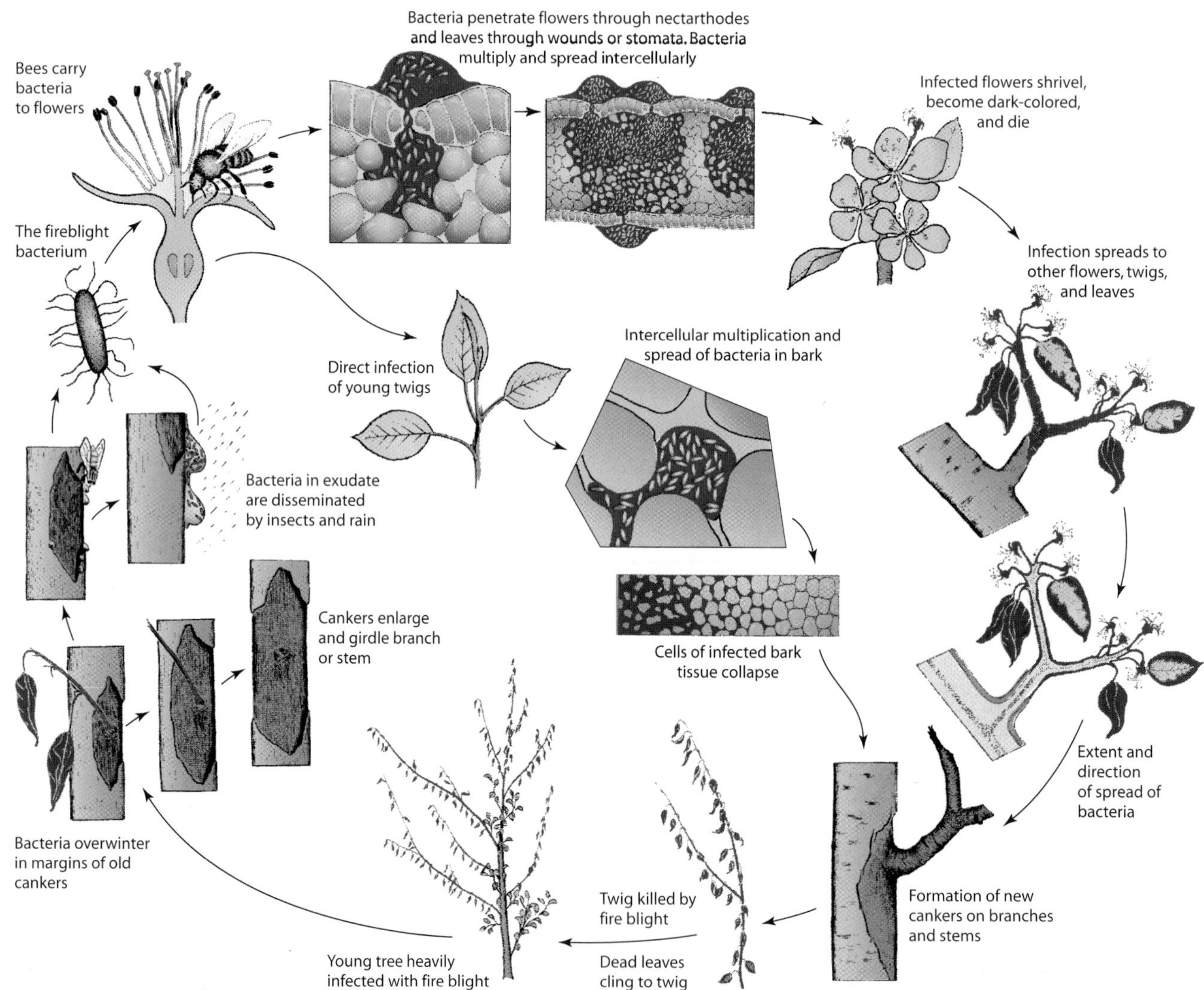

FIGURE 12-24 Disease cycle of fire blight of pear and apple caused by *Erwinia amylovora*.

pedicel into the fruit spur. Infection of the spur results in the death of all flowers, leaves, and fruit on it (Fig. 12-24).

Penetration and invasion of leaves are similar to those of flowers. Bacteria may enter through stomata and hydathodes, but usually they enter through wounds made by insects, hail storms, and so on. From the leaf, bacteria pass into the petiole and the stem. As *E. amylovora* bacteria enter tissues, they initially colonize and move through vessels, colonizing other tissues only later in the infection process. In contrast to other bacterial wilts, however, *E. amylovora* bacteria move rapidly from the vessels to other tissues, killing cells, and causing blight and canker symptoms in the process.

Young, tender twigs may be infected by bacteria through their lenticels, through wounds, and through flower and leaf infections. In the twig, bacteria travel intercellularly or through the xylem. Nearby cortical or xylem parenchyma cells collapse and break down, forming large cavities. If bacteria reach the phloem, they are carried upward to the tip of the twig and to the leaves. Invasion of large twigs and branches is restricted primarily to the cortex. Infection of succulent tissues is rapid under warm, humid conditions. Under cool, dry conditions the host forms cork layers around the infected area and limits the expansion of the canker. In susceptible varieties and during warm, humid weather, bacteria may progress from spurs or shoots into the second-year, third-year, and older growth, killing the bark all along the way.

Control

Several measures must be integrated for successful fire blight control. During the winter, all blighted twigs, branches, cankers, and even whole trees, if necessary, should be cut out about 10 centimeters below the last point of visible infection and burned. Cutting of blighted shoots in the summer can reduce the inoculum; however, bacteria are very active in the summer and should not be spread to new branches or trees. Cutting should be done about 30 centimeters below the point of visible infection. The tools should be disinfested after each cut by wiping them with a sponge soaked in 10% commercial sodium hypochlorite solution. The latter mixture can also be used to disinfect large cuts made by the removal of branches and cankers.

To reduce excessive succulence, trees should be grown in sod and should receive balanced fertilization and limited pruning. Also, a good insect control program should be followed in the postblossom period to reduce or eliminate the spread of bacteria by insects to succulent twigs.

No pear or apple varieties are immune to fire blight when conditions are favorable and the pathogen is abundant. However, moderately resistant varieties are available and should be chosen for areas where fire blight is destructive.

Satisfactory control of fire blight with chemicals can be obtained only in combination with the aforementioned measures. Dormant sprays with copper sulfate or with the Bordeaux mixture offer some, but not much, protection from fire blight. Bordeaux and streptomycin are the only effective blossom sprays. Bordeaux or streptomycin is sometimes used to control twig blight, but neither gives good control. Besides, streptomycin-resistant strains of the fire blight bacterium are encountered in many areas, making that antibiotic ineffective. In such areas, oxytetracycline has been used with some success.

In many areas, fire blight forecasting models have been developed and are used with variable success. Most models use a combination of data on temperature, rainfall or humidity, and growth stage of the tree. By forecasting when a severe outbreak of fire blight infection is likely to occur, growers are warned to begin applying bactericidal sprays as soon as such conditions are observed.

SOUTHERN BACTERIAL WILT OF SOLANACEOUS PLANTS

Southern bacterial wilt of solanaceous plants is caused by *Ralstonia solanacearum*. The disease is present in the tropics and in the warmer climates throughout the world. It causes severe losses on tobacco, tomato, potato, and eggplant in some warm areas outside the tropics. Many other hosts, however, are attacked by the disease. At least five races of the pathogen cause disease on the various hosts. One of them attacks all the solanaceous and many nonsolanaceous crops as well as some bananas, another attacks only plants in the banana family, and a third attacks potato and sometimes tobacco. Two other races cause disease in plants of little importance.

Bacterial wilt on solanaceous crops appears as a sudden wilt. Infected young plants die rapidly. Older plants first show wilting of the youngest leaves, or one-sided wilting and stunting, and finally the plants wilt permanently and die (Figs. 12-25A, 12-25B, and 12-25D). In some plants, such as tomato, excessive adventitious roots may form. The vascular tissues of stems, roots, and tubers turn brown, and in cross sections they ooze a whitish bacterial exudate (Figs. 12-25C and 12-25E). Bacterial pockets develop around the vascular bundles in the pith and in the cortex, and roots and especially tubers often rot and disintegrate (Figs. 12-25E and 12-25F) by the time the plant wilts permanently.

FIGURE 12-25 Bacterial wilt of tomato (A–C) and potato (D–F) caused by *Ralstonia solanacearum*. (A) Early and (B) later symptoms of wilt and death of infected tomato plants. (C) Brown discoloration of stem xylem of infected tomato plant. (D) Potato plant showing typical bacterial wilt symptoms. (E) Rotting of and cavity formation along the ring of vessels in an infected potato tuber. (F) Rotting and cracking of potato tubers infected with the wilt bacteria. [Photographs courtesy of (A–C) R. J. McGovern and (D–F) D. P. Weingartner, both University of Florida.]

Ralstonia solanacearum bacteria overwinter in diseased plants or plant debris, in vegetative propagative organs, such as potato tubers, on the seeds of some crops, in wild host plants, and probably in the soil. Injured or decaying infected tissues release bacteria in the soil. Bacteria are spread through the soil water, through infected or contaminated seeds, tubers, and transplants, by contaminated knives used for cutting tubers or for pruning suckers, and, in some instances, by insects. Bacteria enter plants through wounds made in roots by cultivating equipment, nematodes, insects, and at cracks where secondary roots emerge. Bacteria reach the large xylem vessels and through them spread into the plant. Along the vessels they escape into the intercellular spaces of the parenchyma cells in the cortex and pith, dissolve the cell walls, and create cavities filled with slimy masses of bacteria and cellular debris.

The control of bacterial wilt of solanaceous plants depends mostly on the use of resistant varieties, when available, and proper crop rotation or fallow. Only bacteria-free propagative material should be used, and tools, such as knives, should be disinfested when moving from one plant to another. Infested soils should be kept fallow for about a year and frequently disked during the dry season to accelerate the desiccation of plant material and the death of wilt bacteria. Experimental biological control of the disease through treatment of propagative organs with antagonistic bacteria has been obtained.

BACTERIAL WILT OR MOKO DISEASE OF BANANA

The Moko disease of banana got its name from the fact that it almost eliminated the banana relative Moko plantain in Trinidad around 1890, long before the cause of the disease was known. The Moko disease of banana now occurs throughout the tropical western hemisphere where bananas are grown.

In the Moko disease of banana, young plants wilt rapidly and die, their central leaves breaking at a sharp angle while still green (Fig. 12-26A). In older plants, first the inner leaf turns a dirty yellow near the petiole, the petiole breaks down, and the leaf wilts and dies. In the meantime, more and more of the surrounding leaves droop and die from the center outward until all the leaves bend down and dry out (Figs. 12-26B and 12-26C). Fruit growth in infected plants, if it had started, stops. Banana fingers are deformed, turn black, and shrivel (Fig. 12-26C). If the fruit was near maturity when infected, it may show no outward symptoms, but the pulp of some fingers may be discolored and decaying (Fig. 12-26E). In cross section, an infected banana pseudostem shows many discolored, yellowish brown or almost black vascular bundles, particularly in the inner leaf sheaths and in the fruit stalk (Fig. 12-26D). Pockets of bacteria and decay may be present in the pseudostem, in the rhizome, and most strikingly in individual bananas that become filled with a dark, gummy substance (Fig. 12-26E). The pulp of such bananas finally dries out into a gray, crumbly, starchy residue that pours out when the peel splits open.

The pathogen is *R. solanacearum*, race 2. When cultured on certain special media containing triphenyl tetrazolium chloride, the pathogen produces characteristic colonies (Fig. 12-26F).

The epidemiology of the disease is very similar to the other blights caused by this bacterium and described earlier. Bacteria survive in host plants and, for several months at least, in the soil. Bacteria enter roots through wounds, reach the xylem vessels and multiply, and move through them. Bacteria are transmitted through infected banana rhizomes and through contaminated tools and equipment.

The control of Moko disease is very difficult. It depends primarily on the use of resistant varieties, crop rotation, and cultural practices. Diseased and adjacent banana plants and rhizomes should be cut up and burned.

RING ROT OF POTATO

Ring rot of potato is caused by *Clavibacter michiganense* subsp. *sepedonicum*. The disease occurs and used to cause severe losses in North America and Europe. Through strict inspection and certification of potato seed tubers, the disease has almost been eliminated from seed lots, but occasional outbreaks still occur by contamination from handling or transportation equipment at the seed producers or on the farm. Infected plants usually do not show aboveground symptoms until they are fully grown or the symptoms occur so late in the season that they are overlooked or masked by senescence or other diseases. In years with cool springs and warm summers, however, one or more of the stems in a hill may appear stunted, the interveinal areas of its leaflets turn yellowish, and the leaf margins roll upward and become necrotic (Fig. 12-27A). The leaves then begin to wilt, continuing until all the leaves of the stem wilt and the stem then dies. If a wilted stem is cut at the base and is squeezed, a creamy exudate oozes out of the vascular bundles.

The disease also affects one or more tubers of each plant. Symptoms begin to develop at the stem end of the tuber and progress through the vascular tissue. When cut through, infected tubers show at first a ring of light yellow vascular discoloration (Fig. 12-27B) and some bacterial ooze that may be increased by squeezing the

FIGURE 12-26 Bacterial wilt (Moko disease) of banana caused by *Ralstonia solanacearum*. Banana plants showing different stages of bacteria wilt, including wilted foliage only (A), infection of stalk and early infection of banana fruit (B), and thorough invasion and destruction of banana fruit (C). (D) Invasion and discoloration of several vascular bundles in the banana pseudostem. (E) Early (right) and later invasion and destruction of the contents of infected bananas (left). (F) Colonies of *R. solanacearum* growing on a specialized nutrient medium. [Photographs courtesy of H. D. Thurston, Cornell University.]

tuber (Fig. 12-22). As the disease advances, a creamy yellow or light brown crumbly or cheesy rot develops in the region of the vascular ring, and if the tuber is squeezed, a soft, pulpy exudate oozes from the diseased areas while a more or less continuous ring of cavities (Fig. 12-27C) is formed by the rotting of tissues in the vascular area. Secondary, soft-rotting bacteria often invade infected tubers, and these bacteria may cause complete rot of the tuber (Fig. 12-27D) while producing a foul odor.

The characteristic morphology of *Clavibacter* cells and its gram-positive reaction, taken together with the

FIGURE 12-27 Potato ring rot disease caused by *Clavibacter michiganense* subsp. *sepedonicum*. (A) Foliage symptoms. (B) Early and (C) later symptoms of ring rot along the vessels of a potato tuber. (D) External appearance of rotting in infected tubers. [Photographs courtesy of Plant Pathology Department, University of Florida.]

host and the symptoms, are the primary diagnostic features of this disease.

Ring rot bacteria overwinter mostly in infected tubers and as dried slime on machinery, crates, and sacks. Bacteria are spread easily by knives used to cut potato seed pieces: a knife used to cut an infected tuber may infect the next 20 healthy pieces cut with it. Bacteria enter plants only through wounds and invade the xylem vessels in which they multiply profusely and may cause plugging. Bacteria also move out of the vessels into the surrounding parenchyma tissues, which they break down, and cause cavities, and then again into new vessels. Bacteria also invade the roots and cause them to deteriorate.

Potato ring rot is controlled through the use of healthy seed tubers.

BACTERIAL CANKER AND WILT OF TOMATO

Bacterial canker and wilt of tomato is caused by *Clavibacter michiganense* subsp. *michiganense*. It occurs in many parts of the world and causes considerable losses. The disease appears as spots on leaves, stems, and fruits and as wilting of the leaves and shoots (Fig. 12-28). Eventually, the whole plant wilts and collapses. Very small cankers may occur on stems and leaf veins.

The leaves on lower parts of plants often have white, blister-like spots in the margins that become brown with age and may coalesce (Fig. 12-28). Leaves wilt and curl upward and inward and later turn brown and wither but do not fall off. The wilt may develop gradually from one leaflet to the next or it may become general and destroy much of the foliage (Fig. 12-28B). On stems, shoots, and leaf stalks, light-colored streaks appear, usually at the joints of petioles and stems. Later, cracks develop in the streaks and form the cankers (Figs. 12-28C–12-28E). In humid or wet weather, slimy masses of bacteria ooze through the cracks to the surface of the stem, from which they are spread to leaves and fruits and cause secondary infections. Fruits develop small, shallow, water-soaked, white spots, the centers of which

FIGURE 12-28 Bacterial wilt and canker of tomato caused by *Clavibacter michiganense* subsp. *michiganense*. Symptoms on individual leaves (A) and on whole plants (B). (C) Browning and death of vascular tissue and stem bark (D). (E) Tomato stem cut slanted perpendicularly to show discoloration of vessels. (F) Tomato fruit showing white and brownish spots in response to infection by this bacterium. [Photographs courtesy of (A, B, D, and F) T. A. Zitter, Cornell University, (C) Plant Pathology Department, University of Florida, and (E) L. McDonald, W.C.P.D.]

later become slightly raised, tan colored, and rough. The final, bird's-eye-like appearance of the spots, which have brownish centers and white halos around them, is quite characteristic of the disease (Fig. 12-28F).

In longitudinal sections of infected stems, vascular tissues show a brown discoloration, while large cavities are present in the pith and in the cortex and extend to the outer surface of the stem, where they form the cankers (Figs. 12-28C and 12-28D). Discoloration of the vascular tissues extends all the way to the fruits, both outward toward the surface and inward toward the seeds, and small dark cavities may develop in the centers of such fruits.

Bacteria overwinter in or on seeds and, in some areas, in plant refuse in the soil. Some primary infections result from spread of the bacteria from the seed to cotyledons or leaves, but most infections result from the penetration of bacteria through wounds of roots, stems, leaves, and fruits during transplanting, from windblown rain, and from cultural practices such as tying and suckering

FIGURE 12-29 Bacterial wilt or black rot of cabbage caused by *Xanthomonas campestris* pv. *campestris*. V-shaped infected areas in close-up (A) and around several leaves of cabbage (B). (C) Dark greenish-brown discoloration in the veins of a leaf and at a cross section of the base stem of a cabbage. (D) Area of field with many cabbage plants showing symptoms of bacterial wilt. [Photographs courtesy of Plant Pathology Department, University of Florida.]

of tomatoes. Once inside the plant, bacteria enter the vascular system, move and multiply primarily in the xylem vessels, and move out of them into the phloem, pith, and cortex, where they form the large cavities that result in the cankers.

The disease is controlled through the use of bacteria-free seed, protective application of copper or streptomycin in the seed bed, and soil sterilization of the seedbeds.

BACTERIAL WILT (BLACK ROT) OF CRUCIFERS

Black rot of crucifers is caused by *Xanthomonas campestris* pv. *campestris*. The disease is present throughout the world. It affects all members of the crucifer family and often causes severe losses. The disease affects primarily the aboveground parts of plants of any age. In hosts such as turnip and radish, however, the fleshy roots may also be affected and may develop a dry rot. Infected young seedlings show dwarfing, one-sided growth, and their lower leaves droop. The first symptoms, however, usually appear in the field as large, often V-shaped, chlorotic blotches at the margins of the leaves (Figs. 12-29A and 12-29B). The chlorosis progresses toward the midrib of the leaf, while some of the veins and veinlets within the chlorotic area turn black. The affected area later turns brown and dry. In the meantime, the blackening of the veins advances to the stem and from there upward and downward to other leaves and roots. When leaves become invaded systemically from bacteria moving upward through the midvein, chlorotic areas may appear anywhere on the leaves. Infected leaves may fall off prematurely one after the other. The stem and the stalks of infected leaves in cross section show blackening of vascular tissues (Fig. 12-29C), yellow slime droplets of bacteria, and, sometimes, cavities full of bacteria in the pith and cortex. Cabbage and cauliflower heads are also invaded and discolored, as are the fleshy roots of turnip and radish. Infected areas are subsequently invaded by soft-rotting bacteria, which destroy the tissue, and a repulsive odor is given off.

Black rot bacteria overwinter in infected plant debris and on or in the seed. The bacteria infect cotyledons or young leaves through stomata, hydathodes, or wounds and spread through them intercellularly until they reach the open ends of outer vessels, which they invade. The bacteria then multiply in the vessels and spread in them throughout the plant (Fig. 12-29C), reaching even the seeds. At the same time, the xylem disintegrates in places, and the bacteria spread between the surrounding parenchyma cells. These cells are soon killed and disintegrate, and cavities are formed. Bacteria often ooze to the surface of the leaves through hydathodes or wounds and are subsequently spread by rain splashes and wind or are carried by equipment, to other leaves, which they infect. In wet, warm weather, infection develops rapidly, and visible symptoms may appear within hours (Figs. 12-29B and 12-29D).

The control of black rot depends on the use of bacteria-free seed and transplants planted in soil free of black rot for at least two years. Seed treatment with hot water (50°C for 30 minutes) and tetracycline or streptomycin helps ensure bacteria-free seed. Sprays with copper fungicides at 10-day intervals help reduce spread of the disease.

STEWART'S WILT OF CORN

It has been reported from many countries and probably exists throughout the world. In the United States the importance of the disease has declined with the availability of resistant corn hybrids, but the disease causes significant losses in developing countries.

Infection of young plants causes them to wilt rapidly or, if they survive, to produce linear pale yellow streaks with wavy margins. The streaks may extend the length of the leaf and may turn brown and desiccate. Infected plants, especially of sweet corn, become infected systemically, and may develop cavities in the pith near the soil line and bleached and dead tassels. More common and appearing usually after tasseling are streak lesions originating from the feeding sites of the corn flea beetle (*Chaetocnema pulicaria*) that die and become straw colored (Figs. 12-30A and 12-30B). Such lesions may cover entire leaves, which die and dry up.

The pathogen of Stewart's wilt of corn is *Erwinia stewartii*, which, in 1993, was renamed *Pantoea stewartii*, but the latter name has not yet been totally accepted by scientists.

The pathogen overwinters in the gut of, primarily, the corn flea beetle (Fig. 12-30C), which is also the most important vector of the bacteria from plant to plant. As the adult corn flea beetles emerge from dormancy, they feed on plants to which they carry the bacteria and deposit them in the feeding wounds where they can start new infections. The severity of the disease in a year is almost proportional to the favorable temperatures for the survival of large numbers of beetles through the previous winter. Monitoring these conditions in the winter is used to forecast the severity of the disease the following growing season. Stewart's wilt bacteria may also be carried in a small number of seed corn.

The control of Stewart's wilt of corn depends on the use of resistant corn hybrids, the use of bacteria-free seed, and, to a lesser extent, spraying plants with insecticides to control the insect vector of the bacteria.

Selected References

Abdalla, M. Y. (2001). Sudden decline of date palm trees caused by *Erwinia chrysanthemi*. *Plant Dis.* **85**, 24–26.

Bereswill, S. J., Bellemann, P., and Geider, K. (1997). Identification of *Erwinia amylovora* by growth morphology on agar containing copper sulfate and by capsule staining with lectin. *Plant Dis.* **82**, 158–164.

Coplin, D. L., Majerczak, D. R., Zhang, Y., *et al.* (2001). Identification of *Pantoea stewartii* subsp. *stewartii* by PCR and strain differentiation by PFGE. *Plant Dis.* **86**, 304–311.

Coutinho, T. A., Roux, J., Riedel, K.-H., *et al.* (2000). First report of bacterial wilt caused by *Ralstonia solanacearum* on eucalypts in South Africa. *Eur. J. Forest Pathol.* **30**, 205–210.

Deberdt, Quénéhervé, Darrasse, and Prior (2000). Increased susceptibility to bacterial wilt in tomatoes by nematode galling and the role of the *Mi* gene in resistance to nematodes and bacterial wilt. *Plant Pathol.* **48**, 408–414.

DeBoer, S. H., and Slack, S. A. (1984). Current status and prospects for detecting and controlling bacterial ring rot of potatoes in North America. *Plant Dis.* **68**, 841–844.

Eastgate, J. A. (2000). *Erwinia amylovora*: The molecular basis of fireblight disease. *Mol. Plant Pathol.* **1**, 325–329.

Genin, S., and Boucher, C. (2002). *Ralstonia solanacearum*: Secrets of a major pathogen unveiled by analysis of its genome. *Mol. Plant Pathol.* **3**, 111–118.

Gleason, M. L., Gitaitis, R. D., and Ricker, M. D. (1993). Recent progress in understanding and controlling bacterial canker of tomato in eastern North America. *Plant Dis.* **77**, 1069–1076.

Goodman, R. N., and White, J. A. (1981). Xylem parenchyma plasmolysis and vessel wall disorientation caused by *Erwinia amylovora*. *Phytopathology* **71**, 844–852.

Hartman, G. L., and Hayward, A. C., eds. (1993). Bacterial Wilt. In "Proceedings of an International Conference at Kaohsiung," Taiwan, 1992. AGIAR Proceedings.

Hayward, A. C. (1991). Biology and epidemiology of bacterial wilt caused by *Pseudomonas solanacearum*. *Annu. Rev. Phytopathol.* **29**, 65–87.

Hildebrand, M., Dickler, E., and Geider, K. (2000). Occurrence of *Erwinia amylovora* on insects in a fire blight orchard. *J. Phytopathol.* **148**, 251–256.

Hildebrand, M., Tebbe, C. C., and Geider, K. (2001). Survival studies with the fire blight pathogen *Erwinia amylovora* in soil and in a soil-inhabiting insect. *J. Phytopathol.* **149**, 635–639.

Huang, Q., and Allen, C. (2000). Polygalacturonases are required for rapid colonization and full virulence of *Ralstonia solanacearum* on tomato plants. *Physiol. Mol. Plant Pathol.* **57**, 77–83.

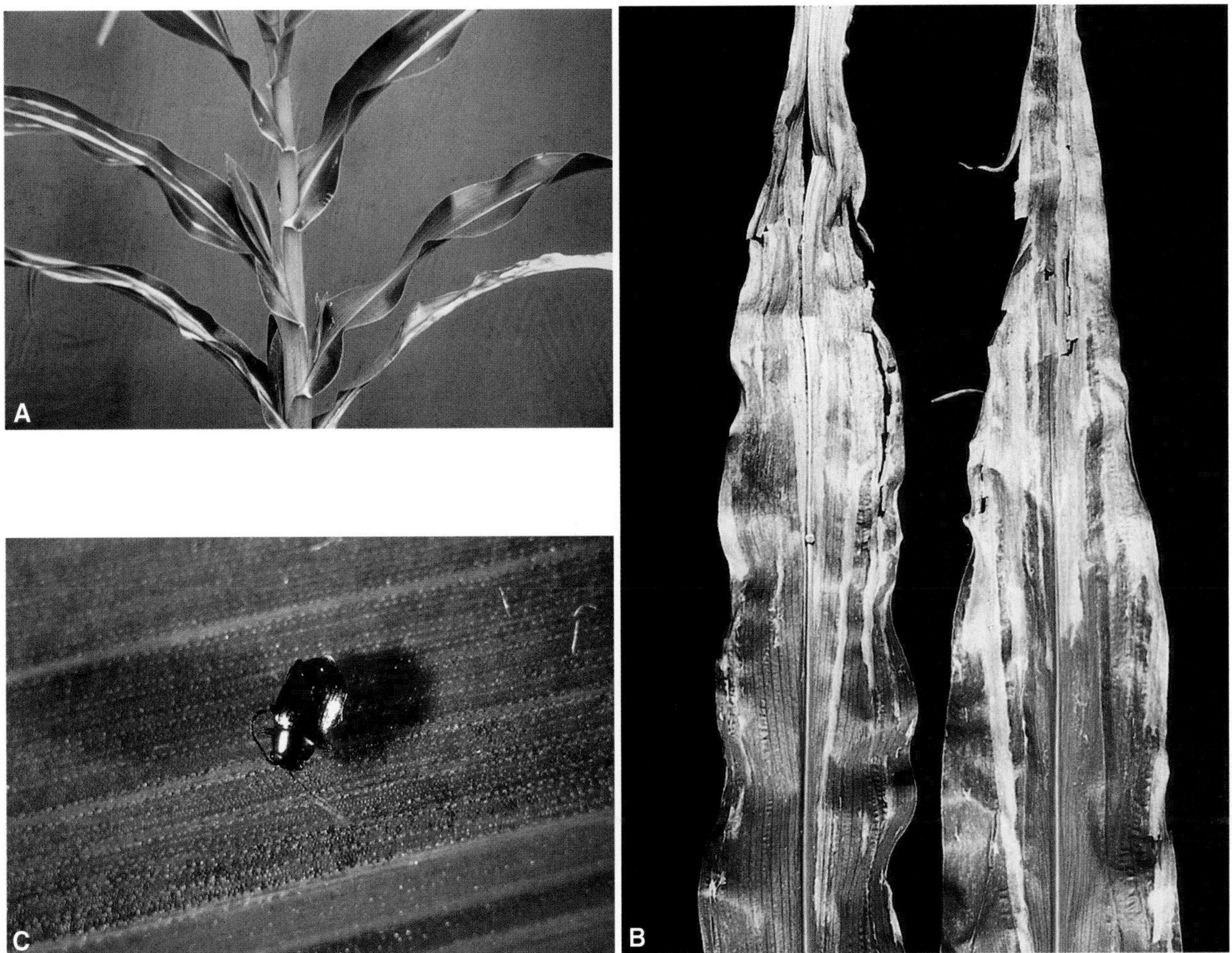

FIGURE 12-30 Bacterial wilt (Stewart's disease) of corn caused by *Erwinia (Pantoea) stewartii.* (A) Young corn plant showing leaf stripes caused by the disease. (B) Close-up of infected leaves showing stripes of dead tissue caused by bacterial wilt. (C) The corn flea beetle, which is the main vector of the bacterial wilt bacteria. [Photographs courtesy of Plant Pathology Department, University of Florida.]

Johnson, K. B., and Stockwell, V. O. (1998). Management of fire blight: A case study in microbial ecology. *Annu. Rev. Phytopathol.* **36**, 227–248.

Jones, A. L. (1992). Evaluation of the computer model MARYBLYT for predicting fire blight blossom infection on apple in Michigan. *Plant Dis.* **76**, 344–347.

Kelman, A. (1953). The bacterial wilt caused by *Pseudomonas solanacearum*. *N.C. Agric. Exp. Stn. Tech. Bull.* **99**, 1–194.

Kocks, Zadocks, and Ruissen (1999). Spatio-temporal development of black rot (*X. campestris* pv. *campestris*) in cabbage in relation to initial inoculum levels in field plots in The Netherlands. *Plant Pathol.* **48**, 176–188.

Lee, I.-M., Lukaesko, L. A., and Maroon, C. J. M. (2001). Comparison of dig-labeled PCR, nested PCR, and ELISA for the detection of *Clavibacter michiganensis* subsp. *sepedonicus* in field-grown potatoes. *Plant Dis.* **85**, 261–266.

McElhaney, R., Alvarez, A. M., and Kado, C. I. (1998). Nitrogen limits *Xanthomonas campestris* pv. *campestris* invasion of the host xylem. *Physiol. Mol. Plant Pathol.* **52**, 15–24.

McManus, P. S., and Jones, A. L. (1994). Role of wind-driven rain, aerosols, and contaminated budwood in incidence and spatial pattern of fire blight in an apple nursery. *Plant Dis.* **78**, 1059–1066.

Mitchener, P. M., Pataky, J. K., and White, D. G. (2002). Transmission of *Erwinia stewartii* from plants to kernels and reactions of corn hybrids to Stewart's wilt. *Plant Dis.* **86**, 167–172.

Mitchener, P. M., Pataky, J. K., and White, D. G. (2002). Rates of transmitting *Erwinia stewartii* from seed to seedlings of a sweet corn hybrid susceptible to Stewart's wilt. *Plant Dis.* **86**, 1031–1035.

Nakaho, K., Hibino, H., and Miyagawa, H. (2000). Possible mechanisms limiting movement of *Ralstonia solanacearum* in resistant tomato tissues. *J. Phytopathol.* **148**, 181–190.

Nelson, P. E., and Dickey, R. S. (1970). Histopathology of plants infected with vascular bacterial pathogens. *Annu. Rev. Phytopathol.* **8**, 259–280.

Pan, Y.-B., Grisham, M. P., Burner, D. M., *et al.* (1998). A polymerase chain reaction protocol for the detection of *Clavibacter xyli* subsp. *xyli*, the causal bacterium of sugarcane ratoon stunting disease. *Plant Dis.* **82**, 285–290.

Pataky, J. K., Michener, P. M., Freeman, N. D., *et al.* (2000). Control of Stewart's wilt in sweet corn with seed treatment insecticides. *Plant Dis.* **84**, 1104–1108.

Pepper, E. H. (1967). "Stewart's Bacterial Wilt of Corn," Monogr. No. 4. APS Press, St. Paul, MN.

Pradhanang, P. M., Elphinstone, J. G., and Fox, R. T. V. (2000). Sensitive detection of *Ralstonia solanacearum* in soil: A comparison of different detection techniques. *Plant Pathol.* **49**, 414–422.

Rahman, M. A., Abdullah, H., and Vanhaecke, M. (1999). Histopathology of susceptible and resistant *Capsicum annuum* cultivars infected with *Ralstonia solanacearum. J. Phytopathol.* **147**, 129–140.

Schell, M. A. (2000). Control of virulence and pathogenicity genes of *Ralstonia solanacearum* by an elaborate sensory network. *Annu. Rev. Phytopathol.* **38**, 263–292.

Singh, U. S., Mukhopadhyay, A. N., Kumar, J., and Chaube, H. S., eds. (1992). "Plant Diseases of International Importance," Vols. 1–4. Prentice-Hall, Englewood Cliffs, NJ. [Extensive chapters on black rot of crucifers and fire blight of apple and pear.]

Strider, D. L. (1969). Bacterial canker of tomato caused by *Corynebacterium michiganense*: A literature review and bibliography. *N. C. Agric. Exp. Stn. Bull.* **193**, 1–110.

Van der Zwet, T., and Keil, H. L. (1979). Fire blight: A bacterial disease of rosaceous plants. *U. S. Dept. Agric. Agric. Handb.* **510**, 1–200.

Van der Zwet, T., Zoller, B. G., and Thompson, S. V. (1988). Controlling fire blight of pear and apple by accurate prediction of the blossom blight phase. *Plant Dis.* **72**, 464–472.

Vanneste, J. L., ed. (2000). "Fire Blight: The Disease and its Causative Agent, *Erwinia amylovora*." CABI, Wallingford, UK.

Wallis, F. M. (1977). Ultrastructural histopathology of tomato plants infected with *Corynebacterium michiganense. Physiol. Plant Pathol.* **11**, 333–342.

Wallis, F. M., *et al.* (1973). Ultrastructural histopathology of cabbage leaves infected with *Xanthomonas campestris. Physiol. Plant Pathol.* **3**, 371–378.

Watterson, J. C., *et al.* (1972). Multiplication and movement of *Erwinia tracheiphila* in resistant and susceptible cucurbits. *Plant Dis. Rep.* **56**, 949–952.

Williams, P. H. (1980). Black rot: A continuing threat to world crucifers. *Plant Dis.* **64**, 736–742.

Williamson, L., Nakaho, K, Hudelson, B., *et al.* (2002). *Ralstonia solanacearum* race 2, biovar 2 strains isolated from geranium are pathogenic on potato. *Plant Dis.* **86**, 987–991.

BACTERIAL SOFT ROTS

Bacteria are invariably present whenever fleshy plant tissues are rotting in the field or in storage, and the foul smell given off by such rotting tissues is due, usually, to volatile substances released during the disintegration of plant tissues by such bacteria. Rotting tissues become soft and watery, and slimy masses of bacteria and cellular debris frequently ooze out from cracks in the tissues. In many soft rots the bacteria involved are not plant pathogenic, i.e., they do not attack living cells, rather they are saprophytic or secondary parasites, i.e., they grow in tissues already killed by pathogens and environmental causes, or in tissues so weakened or old that they are unable to resist attack by any organism. Some bacteria, however, attack living plant tissues and cause soft rots in the field or in storage.

Erwinia, the "carotovora" or "soft rot" group, causing soft rots of numerous fleshy fruits, vegetables, and ornamentals (*E. carotovora* pv. *carotovora*), and blackleg of potato (*E. carotovora* pv. *atroseptica*)

Pseudomonas, also causing soft rots of fleshy fruits and fleshy vegetables (*P. fluorescens*), such as the pink eye disease of potato, the slippery skin disease of onion, and the sour skin of onion

Bacillus, causing rotting of potatoes and tobacco leaves in storage, of tomato seedlings, and of soybeans

Clostridium, also causing rotting of potatoes and tobacco leaves in storage and the wetwood syndrome of poplar and elm

Soft-rot bacteria may survive in infected tissues, in the soil, and in contaminated equipment and containers. Some of them also overwinter in insects. They are spread by direct contact, hands, tools, soil, water, and insects. They enter plants or plant tissues primarily through wounds. Within the tissues they multiply profusely in the intercellular spaces, where they produce several kinds of enzymes that, by dissolving the middle lamella and separating the cells from one another, cause maceration and softening of affected tissues. The cells, surrounded as they are by the bacteria and their enzymes, at first lose water and their contents shrivel; finally, parts of their walls are dissolved and the cells are invaded by bacteria. The control of bacterial soft rots is difficult and depends on proper sanitation, avoiding injuries, keeping storage tissues dry and cool, assuring good insect control, and practicing crop rotation.

BACTERIAL SOFT ROTS OF VEGETABLES

Bacterial soft rots occur most commonly on fleshy storage tissues of vegetables and annual ornamentals, such as potatoes (Figs. 12-31A–12-31E), carrots (Fig. 12-32B), onions (Fig. 12-32D), iris, and fleshy fruits, such as cucumber and tomato (Fig. 12-32C), or succulent stems, stalks, or leaves, such as cabbage (Fig. 12-32A), lettuce, celery (Fig. 12-32E), and spinach. In the tropics, soft rots often develop on the fleshy stems of some plants while still in the field, e.g., in corn (Fig. 12-33A), cassava (Figs. 12-33B and 12-33C), and banana (Figs. 12-33D and 12-33E). Bacterial soft rots occur worldwide and cause serious diseases of crops in the field, in transit, and especially in storage. They cause a greater total loss of produce than any other bacterial

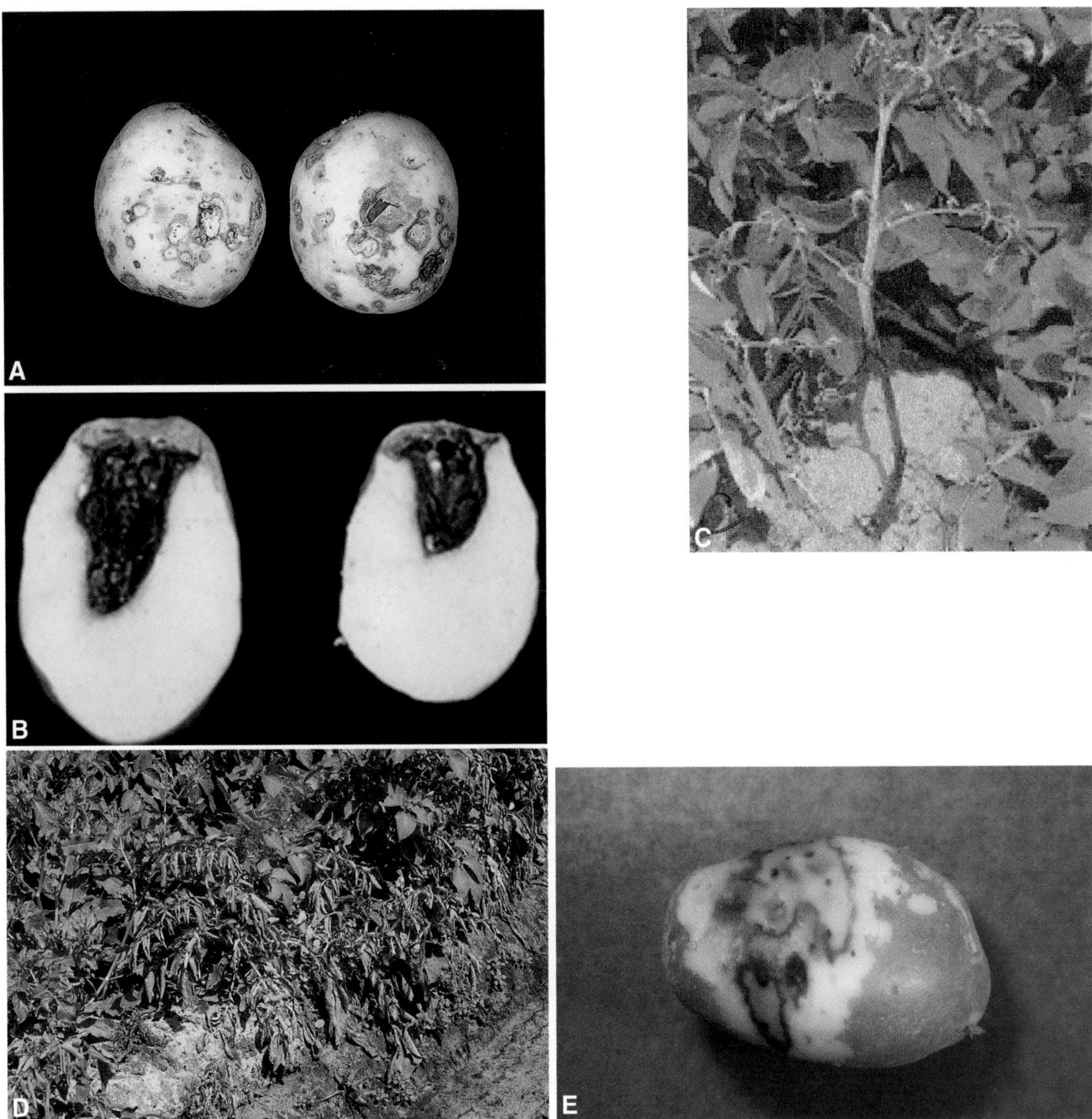

FIGURE 12-31 Bacterial soft rots of fruits and vegetables caused primarily by species of *Erwinia* and *Pseudomonas*. (A) Lenticel infection of potato tuber leading to soft rot. (B) Stem-end rot of potato tubers induced by *Erwinia carotovora* subsp. *atroseptica*, the cause of potato blackleg. (C) Potato plant infected with bacterial wilt and blackleg. (D) Potato plants in the field showing blackleg symptoms. (E) A new potato showing bacterial soft rot at harvest time. [Photographs courtesy of (A, C, and D) Plant Pathology Department, University of Florida, (B) D. P. Weingartner, and (E) R. T. McMillan, both of University of Florida.]

disease. Nearly all fresh vegetables are subject to bacterial soft rots, which may develop within a few hours in storage or during marketing. Bacterial soft rots reduce quantities of produce available for sale, reduce the quality and thus the market value of crops, and increase expenses greatly for preventive measures against soft rots.

Symptoms

Soft-rot symptoms begin as a small water-soaked lesion, which enlarges rapidly in diameter and in depth. The affected area becomes soft and mushy (Figs. 12-31, 12-32, and 12-33) while its surface becomes discolored and somewhat depressed. Tissues within the affected

FIGURE 12-32 Bacterial soft rot of vegetables on (A) cabbage head, (B) carrots, (C) tomato, (D) onion bulb, and (E) celery. [Photographs courtesy of (A, C, and E) Plant Pathology Department, University of Florida, (B) R. J. Howard, W.C.P.D. and (D) G. Q. Pelter, Washington State University.]

FIGURE 12-33 Bacterial wilt and stem rot of fleshy plants caused by *Erwinia chrysanthemi*. (A) Stem rot of corn. (B) Canker and stem rot of cassava and (C) stem rot of cassava. *Erwinia* stem rot of banana in close-up showing internal stem discoloration (E) and in the field (D). [Photographs courtesy of H. D. Thurston, Cornell University.]

region become cream colored and slimy, disintegrating into a mushy mass of disorganized plant cells and bacteria. The outer surface may remain intact while the entire contents have changed to a turbid liquid; alternatively, cracks develop and the slimy mass exudes to the surface and, in air, turns tan, gray, or dark brown. A whole fruit or tuber may be converted into a soft, watery, decayed mass within 3 to 5 days. Infected fruits and tubers of many plants are almost odorless until they collapse, and then secondary bacteria grow on the decomposing tissues and produce a foul odor. Cruciferous plants and onions, however, when infected by soft-rot bacteria, almost always give off a repulsive odor. When root crops are affected in the field, the lower parts of the stem may also become infected and watery and may turn black and shrivel, causing the plants to become stunted, wilt, and die. Infections of succulent leaves and stems are seldom important in the field. When these parts are infected in storage or in packages, however, especially in plastic containers, they rapidly become soft and disintegrate and may yield a wet, green, slimy mass within 1 or 2 days.

BOX 21 The Incalculable Postharvest Losses from Bacterial (and Fungal) Soft Rots

Everyone is familiar with the need to throw away at least a few of the strawberries purchased in a box because they are partially rotten or portions of or whole tomatoes, potatoes, peaches, cherries, bananas, spinach, celery, grapes, peppers, and onions, and almost every other flesly plant produce. This happens especially if at some point in storage they became wet or they were kept in a plastic bag (which maintains high humidity in the bag) or kept for a somewhat long time. These are the everyday experiences of everybody with postharvest soft rot, some of which are caused by fungi and some by bacteria and, in many cases, by both working together.

There are no accurate measurements of the losses, primarily of fleshy fruits, vegetables, and underground storage organs such as roots and tubers used for human consumption, to postharvest bacterial (and fungal) soft rots. They are estimated, however, to vary between 15 and 30% of the harvested crop. Of course, losses vary between crops, depending on their softness and the hardness of their peel. For example, they are much greater in strawberries, peaches, and papayas than they are in apples and watermelons. Losses also vary with the control measures taken against diseases and insects while the crops are still in the field; this affects the number of fruit, vegetables, and so on that come in from the field already infected, although some of them may not yet show visible symptoms at harvest. Losses vary with the method of harvest and the ability or effort to harvest the produce with as few injuries while cool and dry rather than while hot and wet and in ways that inflict wounds to them. Losses vary with the ability and willingness to discard from packing any and all infected or injured fruit and vegetables, which, although increases the number of those lost as culls, saves the packed ones from soft rots spreading during storage, transit, and marketing. Losses vary with the ability to provide refrigeration during storage, transit, and marketing. Finally, losses vary with the ability of the ultimate consumer to provide refrigeration to these products from the moment of purchase until the product is finally consumed.

It is obvious that avoidance of losses to soft rots depends on the crop (soft or hard), on the weather (hot or cool, wet or dry), the ability by training or having the means to control diseases and insects in the field, and the ability to provide refrigeration to the produce from the moment of harvest until the moment of consumption of the produce by the ultimate consumer. The softness or hardness of the crop is inherent to the crop and cannot, of course, be changed. There is little one can do about the weather in a location except try to take advantage of all desirable situations. The other two prerequisites for avoiding losses to soft rots, i.e., appropriate training to control things in the field and ability to provide refrigeration during all the stages after harvest, are difficult or impossible to provide in poorer, developing countries. The training of farmers in such countries is generally inadequate for appropriate controls; they often lack the materials and equipment to bring about satisfactory controls, and they most likely lack refrigeration facilities not only for storage, transit, and marketing of these products, but also in their own homes. Considering that many of the poorer, developing countries are in the tropics, where temperatures, rainfall, and relative humidity are generally high, it is easy to understand that those are the countries that suffer by far greater losses to soft rots than countries in cooler, drier — and more affluent — areas.

The Pathogens

Erwinia carotovora pv. *carotovora*, *E. chrysanthemi*, and *Pseudomonas fluorescens*. Bacteria *E. carotovora* pv. *carotovora* and *P. fluorescens* cause the most common and the most destructive soft rots. *Erwinia caratovora* pv. *atroseptica*, the cause of blackleg of potato (Figs. 12-31B–12-31D), may be thought of as a cool temperature variant of *E. caratovora* pv. *carotovora* and is restricted mostly to potatoes. *Erwinia chrysanthemi* affects many hosts and causes many of the soft rot of tropical plants while they are still growing in the field. Soft-rot bacteria can grow and are active over a range of temperatures from 5 to 35°C. They are killed with extended exposure at about 50°C.

Development of Disease

Soft-rot bacteria survive in infected fleshy organs in storage and in the field, in debris, on roots or other parts of host plants, in ponds and streams used for water irrigation, occasionally in the soil, and in the pupae of several insects (Fig. 12-34). The disease may first appear in the field on plants grown from previously infected seed pieces. Some tubers, rhizomes, and bulbs become infected through wounds or lenticels after they are set or formed in the soil. The inoculation of bacteria into fleshy organs and their further dissemination in storage and in the field are facilitated greatly by insects. Soft-rot bacteria can live in all stages of the insect. Moreover, the bodies of the insect larvae (maggots) become contaminated with bacteria when they crawl about on rotting seed pieces, carry them to healthy plants, and put them into wounds where they can cause the disease. Even when the plants or storage organs are resistant to soft rot and can stop its advance by the formation of wound-cork layers, the maggots destroy the wound cork as fast as it is formed and the soft rot continues to spread.

When soft-rot bacteria enter wounds, they feed and multiply at first on the liquids released by the broken

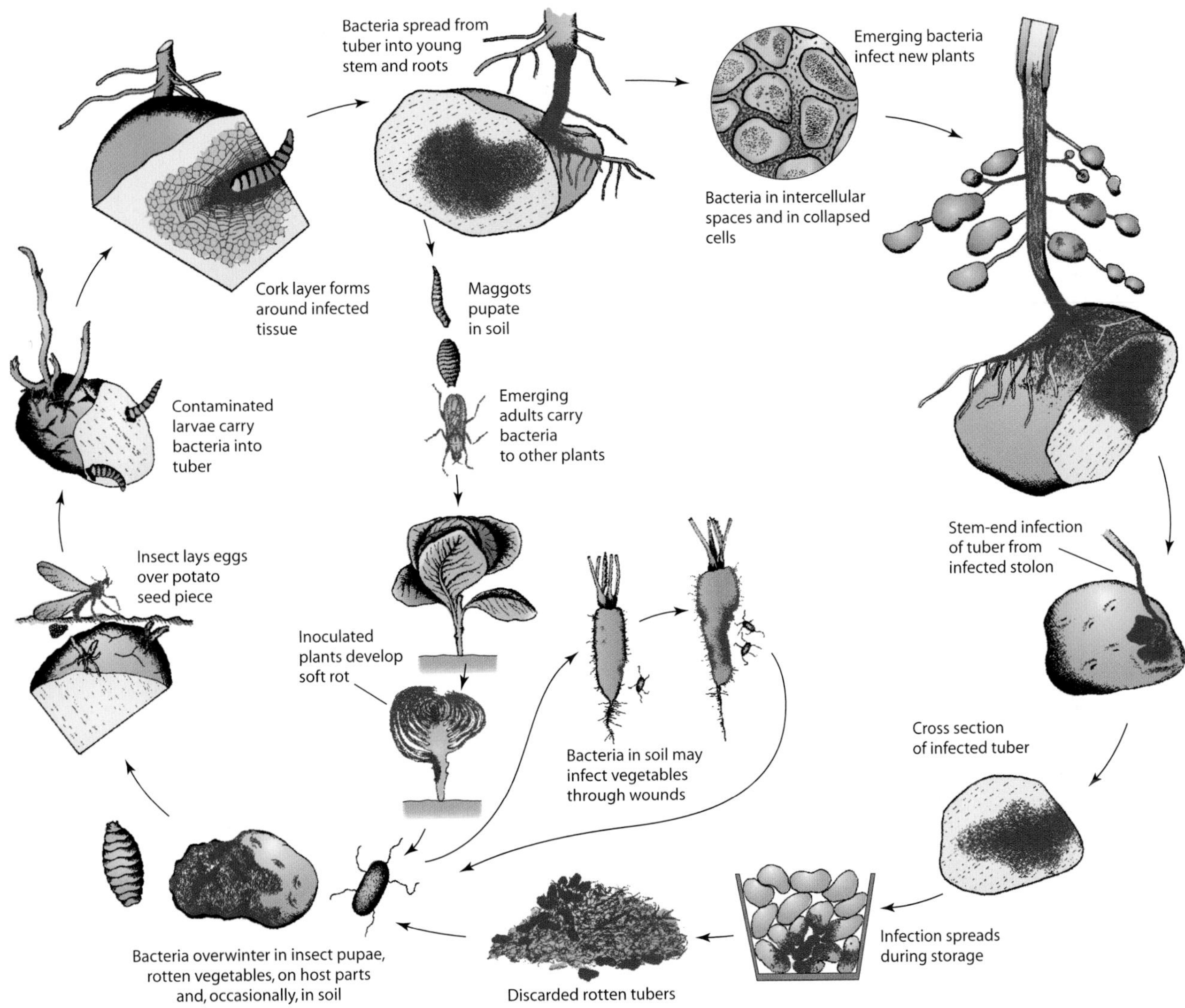

FIGURE 12-34 Disease cycle of bacterial soft rot of vegetables caused by soft-rotting *Erwinia* sp.

cells on the wound surface. There they produce increasing amounts of pectolytic enzymes that break down the pectic substances of the middle lamella and bring about maceration of the tissues. Because of the high osmotic pressure of the macerated tissue, water from the cells diffuses into the intercellular spaces; as a result, the cells plasmolyze, collapse, and die. Bacteria continue to move and to multiply in the intercellular spaces, while their enzymes advance ahead of them and prepare the tissues for invasion. The invaded tissues become soft and are transformed into a slimy mass consisting of innumerable bacteria swimming about in the liquefied substances. The epidermis of most tissues is not attacked by the bacteria; however, cracks are usually present, and the slimy mass extrudes through them into the soil or in storage, where it comes into contact with other fleshy organs, which are subsequently infected.

Control

The control of bacterial soft rots of vegetables is based almost exclusively on sanitary and cultural practices. All debris should be removed from warehouses, and the walls should be disinfested with formaldehyde or copper sulfate. Wounding of plants and their storage organs should be avoided as much as possible. Products to be stored should be dry, and the humidity and temperature of warehouses should be kept low.

In the field, plants should be planted in well-drained areas and at sufficient distances to allow adequate ventilation. Susceptible plants should be rotated with cereals or other nonsusceptible crops. Few varieties have any resistance to soft rot, and no variety is immune.

Chemical sprays are generally not recommended for the control of soft rots. Control of the insects that spread the disease reduces infections both in the field and in storage. Experimental biological control of bacterial soft rot of potatoes has been obtained by treating potato seed pieces before planting with antagonistic bacteria or with plant growth-promoting rhizobacteria.

Selected References

Barras, F., van Gijsegem, F., and Chatterjee, A. K. (1994). Extracellular enzymes and pathogenesis of soft-rot *Erwinia*. *Annu. Rev. Phytopathol.* **32**, 201–234.

DeBoer, S. H. (2002). Relative incidence of *Erwinia carotovora* subsp. *atroseptica* in stolon end and peridermal tissue of potato tubers in Canada. *Plant Dis.* **86**, 960–964.

DeBoer, S. H., and Kelman, A. (1978). Influence of oxygen concentration and storage factors on susceptibility of potato tubers to bacterial soft rot (*Erwinia carotovora*). *Potato Res.* **21**, 65–80.

Hélias, Andrivon, and Jouan (2000). Development of symptoms caused by *Erwinia carotovora* ssp. *atroseptica* under field conditions and their effects on the yield of individual potato plants. *Plant Pathol.* **49**, 23–32.

Hélias, Andrivon, and Jouan (2000). Internal colonization pathways of potato plants by *Erwinia carotovora* ssp. *atrooseptica*. *Plant Pathol.* **49**, 33–442.

Liao, C.-H., McCallus, D. E., and Wells, J. M. (1993). Calcium-dependent pectate lyase production in the soft-rotting bacterium *Pseudomonas fluorescens*. *Phytopathology* **83**, 813–818.

Palava, T. K., *et al.* (1993). Induction of plant defense response by exoenzymes of *Erwinia carotovora* subsp. *carotovora*. *Mol. Plant-Microbe Interact.* **6**, 190–196.

Pérombelon, M. C. M. (2002). Potato diseases caused by soft rot erwinias: An overview of pathogenesis. *Plant Pathol.* **51**, 1–12.

Pérombelon, M. C. M., and Kelman, A. (1980). Ecology of the soft rot *Erwinias*. *Annu. Rev. Phytopathol.* **18**, 361–387.

Smith, C., and Bartz, J. A. (1990). Variation in the pathogenicity and aggressiveness of strains of *Erwinia carotovora* subsp. *carotovora* isolated from different hosts. *Plant Dis.* **74**, 505–509.

Wells, J. M., and Butterfield, J. E. (1997). *Salmonella* contamination associated with bacterial soft rot of fresh fruits and vegetables in the marketplace. *Plant Dis.* **81**, 867–872.

BACTERIAL GALLS

Galls are produced on the stems and roots of plants infected primarily by bacteria of the genus *Agrobacterium* and by certain species of *Pseudomonas*, *Rhizobacter*, and *Rhodococcus*. The galls may be amorphous, consisting of overgrowths of more or less unorganized or disorganized plant tissues, as are most *Agrobacterium* and *Pseudomonas* galls, or they may be proliferations of tissues that develop into more or less organized, teratomorphic organs, as are some *Agrobacterium* and *Rhodococcus* galls. The bacterial species that cause galls and the main diseases they cause are the following:

Agrobacterium, causing crown gall of many woody plants, primarily stone fruits, pome fruits, willows, and grapes (*A. tumefaciens* or biovar 1), hairy root of apple (*A. rhizogenes* or biovar 2), and cane gall of raspberries and blackberries (*A. rubi*). The kind of symptoms produced is actually determined not by the species of *Agrobacterium*, but by the kind of plasmid they carry: bacteria carrying a tumor-inducing (Ti) plasmid induce crown gall, whereas bacteria carrying a root-inducing (Ri) plasmid induce hairy root symptoms. Thus, strains of all species can carry the Ti plasmid and can, therefore, cause crown gall, but so far only strains of *A. tumefaciens* and *A. rhizogenes* have been found to contain the Ri plasmid and to induce hairy root. Another species, *A. radiobacter*, carries neither plasmid and causes no disease

Pseudomonas, causing the olive knot disease and the bacterial gall or canker of oleander (*P. syringae* subsp. *savastanoi*)

Rhizobacter, causing carrot bacterial gall (*R. daucus*)

Rhodococcus, causing fasciation or leafy gall on sweet pea (*R. fascians*)

CROWN GALL

Crown gall occurs worldwide. It affects woody and herbaceous plants belonging to 140 genera of more than 60 families. In nature it is found mostly on pome and stone fruit trees, brambles, and grapes. Tumors or galls of varying size and shape form on the lower stem and main roots of the plant (Figs. 12-35A–12-35C). Infected nursery plants are unsalable. Plants with tumors at their crowns or on their main roots grow poorly and their yields are reduced. Severely infected plants or vines may die.

Crown gall tumors have certain histological similarities to human and animal cancers and, therefore, the cause and mechanism of their formation have been studied extensively. Despite the apparent similarities to cancer, however, many differences exist between crown gall of plants and malignant tumors of humans and animals.

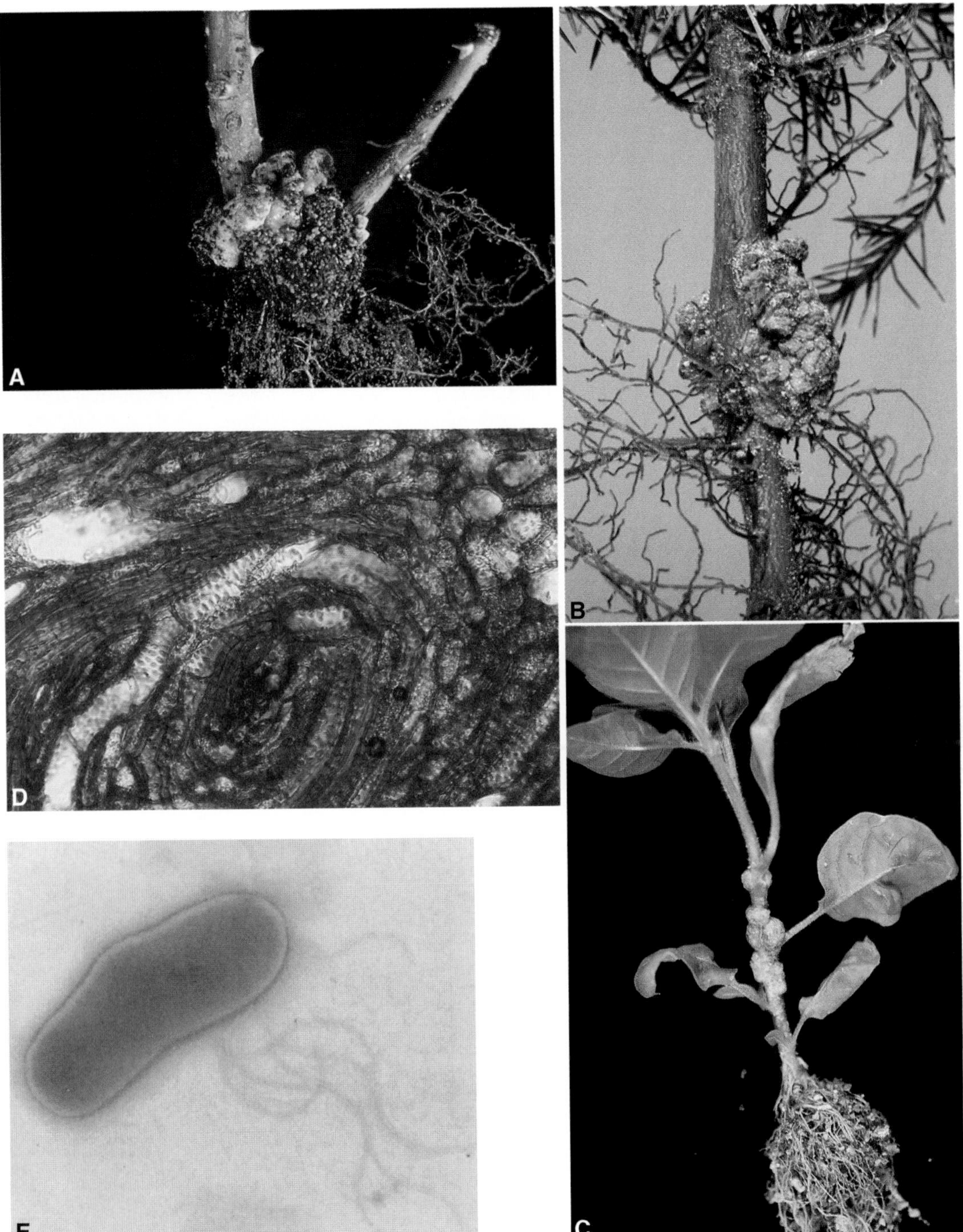

FIGURE 12-35 Crown gall disease caused by *Agrobacterium tumefaciens*. Naturally occurring crown galls on rose (A) and on spruce seedling (B). (C) Artificially induced galls in young tobacco plant by injecting crown gall bacteria into the stem. (D) Abnormal arrangement of tissues within a young crown gall. (E) The crown gall bacterium. [Photographs courtesy of (A) D. R. Cooley, University of Massachusetts, (B) E. L. Barnard, Florida Division of Forestry, and (E) R. E. Wheeler and S. M. Alcorn.]

BOX 22 The Crown Gall Bacterium — The Natural Genetic Engineer

The crown gall bacterium *Agrobacterium tumefaciens* was discovered by Erwin Smith in the 1890s and he immediately noticed that the disease it causes, crown gall, had certain similarities with human and animal cancer. Many scientists studied the physiology of the bacterium and the mechanism by which this bacterium, almost alone among all others, caused cells in infected plant tissues to divide and enlarge rather than to disintegrate and die. It was later shown that the bacterium produced an auxin, which made exposed plant cells enlarge; it also produced a cytokinin, which made exposed plant cells divide. This explained the production of the gall on infected plants except for two important observations. One was that galls were initiated only if the plants were wounded within 24 hours before inoculation with the bacteria, suggesting that cells before they could be infected had to be preconditioned by prior wounding of adjacent cells. The other very important observation was that once inoculated, plant cells began to divide and enlarge and continued to do so in the absence of the bacteria. This suggested that the affected plant cells were altered permanently by something introduced into them by the bacteria. In the meantime, studies of the bacterium itself showed that its genetic material (DNA) consisted of a circular chromosome and a much smaller, also circular plasmid.

Scientists then discovered that what the bacterium introduced into the altered plant cells was part of the DNA of its plasmid, causing the cells to grow into a tumor. For this reason, the plasmid is known as a tumor-inducing (Ti) plasmid, and the segment of the plasmid DNA that is inserted by the bacterium into the plant cell is known as T-DNA. The T-DNA carries genes for auxin and cytokinin, genes that code for small molecules called opines that the bacterium can use as food, and sequences forming the left and the right border of T-DNA. The border sequences allow the T-DNA to separate from the Ti plasmid, to be inserted into the plant cell, and to be integrated randomly in the chromosomes of the plant genome. Once it has become integrated in the plant genome, the genes of T-DNA are expressed just like other genes of the plant. The remaining DNA of the Ti plasmid codes for a number of proteins specifically required for the transfer process. The genes coding for these proteins, about 11 of them, are called virulence or *vir* genes and are activated by plant signal molecules produced following the wounding of plant cells. One of the structures coded for by the *vir* genes is a pilus, which presumably functions in the transfer of the Ti-DNA and proteins from the bacterium to the plant cell. There also appear to be some additional virulence genes located in the bacterial chromosome.

Scientists then noted that they could disarm the T-DNA so it does not cause the crown gall disease by removing the genes that code for auxin and cytokinin. They then noted that they could insert in the disarmed T-DNA genes coding for other functions, e.g., resistance to disease. The T-DNA containing these plant genes is then placed in a disarmed Ti plasmid and that is placed back into *A. tumefaciens* bacteria, which are used to inoculate the desired plant. Successfully inoculated plants express the inserted gene. Such genetically engineered plants are known as transformed plants because they now express new characteristics and, therefore, are transformed. It is now routinely possible to extract the Ti plasmid, introduce new genes (segments of DNA) from one kind of plant (such as bean) in the T-DNA, reintroduce the plasmid into *A. tumefaciens* bacteria, and, by allowing them to infect another kind of plant, such as sunflower, introduce the plasmid and the new (bean) gene into the second kind of plant (sunflower). This procedure is now used widely, and already several types of genes, including genes for disease resistance, have been transferred from one kind of plant to another by such recombinant DNA (genetic engineering) technology using *A. tumefaciens* as the vehicle. Plant biologists are using it as the main tool (vehicle) for transferring all kinds of genes between related and unrelated plants and even between other organisms, e.g., insects or viruses, plants, and animals. It subsequently became possible to inoculate plant protoplasts directly with intact or engineered Ti plasmids in the absence of the bacterium.

Symptoms

Crown gall first appears as small, round, whitish, soft overgrowths on the stem and roots, particularly near the soil line. As the tumors enlarge, their surfaces become convoluted, and the outer tissues become dark brown due to the death and decay of the peripheral cells (Fig. 12-35). The tumor may appear as an irregular swelling and may surround the stem or root or it may lie outside but close to the outer surface of the host, being connected only by a narrow neck of tissue. Some tumors are spongy and may crumble or become detached from the plant. Others become woody and hard, looking knobby or knotty, and reaching sizes up to 30 centimeters in diameter. Some tumors rot in the fall and develop again during the next growing season.

Several galls may occur on the same root or stem, continuous or in bunches. Tumors, however, can also appear on vines up to 150 centimeters from the ground, on branches of trees, on petioles, and on leaf veins. In addition to forming galls, affected plants may become stunted; they produce small, chlorotic leaves and are more susceptible to adverse environmental conditions, especially winter injury.

The Pathogen: Agrobacterium tumefaciens

This bacterium is rod shaped with a few peritrichous flagella (Fig. 12-30A). Virulent bacteria carry one to several large plasmids (small chromosome-like bodies composed of circular double-stranded DNA). One of these plasmids carries the genes for tumor induction and is called the tumor-inducing plasmid. Bacteria that lack the Ti plasmid or lose it on heat treatment are not virulent. The Ti plasmid also carries genes that determine the host range of the bacterium and the kinds of symptoms that will be produced. The most characteristic property of this bacterium is its ability to introduce part of the Ti plasmid (T-DNA) into plant cells and to transform normal plant cells to tumor cells in short periods of time. Transformed plant cells then synthesize specific chemicals called opines, which can be utilized only by bacteria that contain an appropriate Ti plasmid. This property makes the bacterium a genetic parasite, as a piece of its DNA parasitizes the genetic machinery of the host cell and redirects the metabolic activities of the host cell to produce substances used as nutrients only by the parasite.

Development of Disease

The bacterium overwinters in infested soils, where it can live as a saprophyte for several years. When host plants are growing in such infested soils, the bacterium enters the roots or stems near the ground through fairly recent wounds made by cultural practices, grafting, and insects. Once inside the tissue, bacteria occur primarily intercellularly and, through the products of the genes on the Ti plasmid, stimulate the surrounding host cells to divide at a very fast rate (Fig. 12-36). The new cells show no differentiation or orientation (Fig. 12-36D) and produce a swelling that develops into a young tumor. Bacteria are absent from the center of the tumors but can be found intercellularly in their periphery. Some tumor cells differentiate into vessels or tracheids, but they are unorganized and have little or no connection with the vascular system of the host plant. As the tumor cells increase in number and size, they exert pressure on the surrounding and underlying normal tissues, which may become distorted or crushed. Crushing of xylem vessels by tumors sometimes reduces the amount of water reaching the upper parts of a plant to as little as 20% of normal.

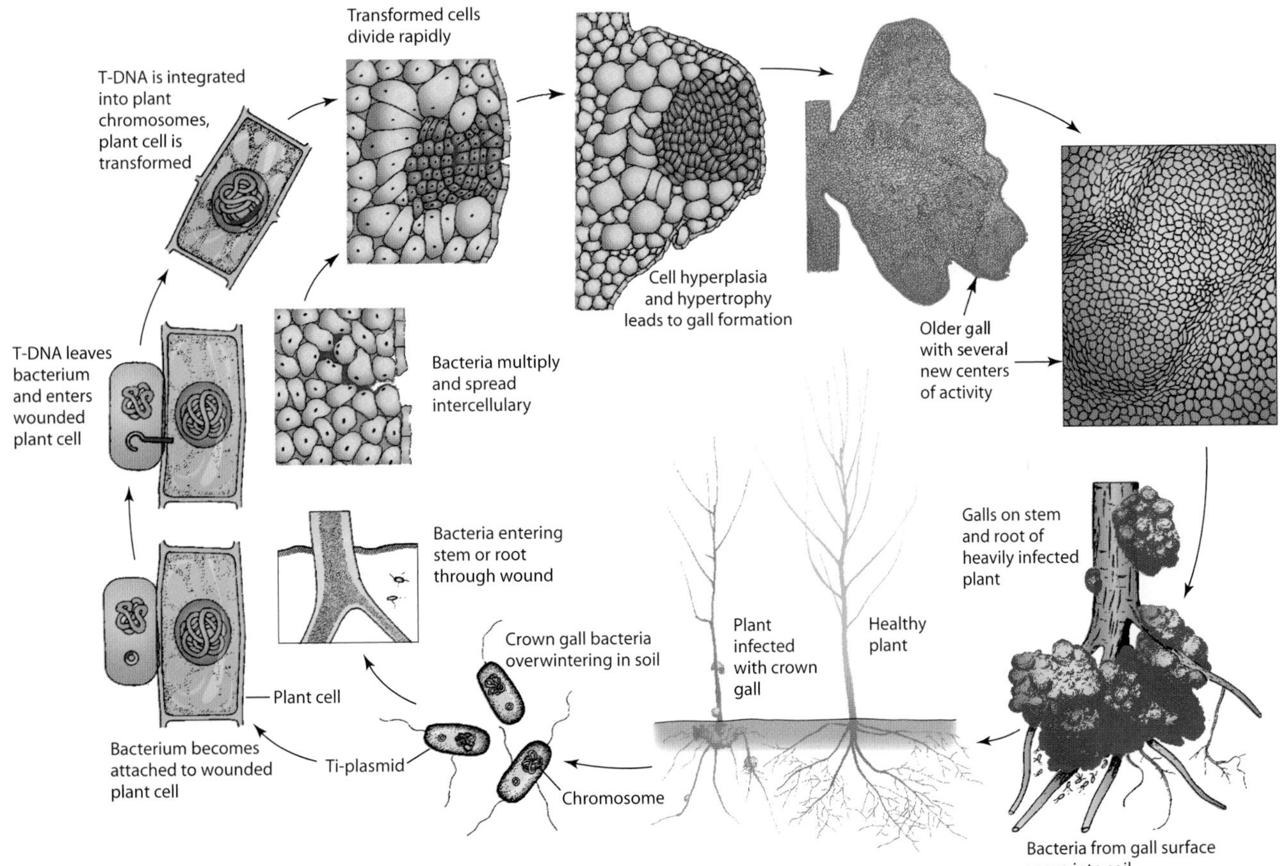

FIGURE 12-36 Disease cycle of crown gall caused by *Agrobacterium tumefaciens*.

The smooth and soft young tumors are easily injured and attacked by insects and saprophytic microorganisms. These secondary invaders cause the peripheral cell layers of tumors to decay and to discolor. Breakdown of these tissues releases crown gall bacteria into the soil, where they can be carried in the water and infect new plants.

Older tumors often become woody and hard. Because vascular bundles in the tumors are ineffective, the water and nourishment that the tumors are able to obtain can carry them only to a certain point in growth, after which enlargement stops, decay sets in, and the necrotic tissues are sloughed off. In some cases the entire tumor regresses and does not reappear. More often, however, some portion of the tumor remains alive and forms additional tumor tissue during the same or the following season.

When very young and expanding tissues are infected, in addition to the primary tumor that develops at the point of infection, secondary tumors appear at varying distances from it. The bacteria-free secondary tumors may develop at wounds made by various agents or on apparently unwounded parts of the stem, on the petiole, and even on leaf midribs or larger veins several internodes above the primary tumor. Their starting point seems to be in the xylem of the vascular bundles.

When Ti plasmid-carrying virulent bacteria are present near a recent wound, they are attracted by phenolic substances produced by the wounded cells, and in response to these substances the bacteria produce growth factors that appear to condition the plant cells for cell division and transformation. In the meantime, the induced bacteria produce an enzyme that cuts the Ti plasmid at specific sites, releasing the segment of T-DNA. The T-DNA passes into the wounded plant cells and one or several copies of it become incorporated in several places along the various chromosomes of the plant cell. Such cells express the genes present on the T-DNA. Transformed cells produce substances called opines, which can be utilized as food only by T-DNA-carrying bacteria, and also elevated amounts of indoleacetic acid (the plant hormone), cytokinins, and various enzymes. The increases in growth regulators lead to the uncontrollable growth of transformed cells.

Control

Crown gall control begins with a mandatory inspection of nursery stock and a rejection of infected trees. Susceptible nursery stock should not be planted in fields known to be infested with the pathogen. Instead, infested fields should be planted with corn or other grain crops for several years before they are planted with nursery stock. Because the bacterium enters only through relatively fresh wounds, wounding of the crowns and roots during cultivation should be avoided and root-chewing insects in the nursery should be controlled to reduce crown gall incidence. Nursery stock should be budded rather than grafted because of the much greater incidence of galls on graft than on bud unions. Growers should purchase and plant only crown gall-free trees.

Excellent biological control of crown gall is obtained by soaking germinated seeds or dipping nursery seedlings or rootstocks in a suspension of a particular strain (No. 84) of *Agrobacterium radiobacter*. This strain of bacteria is antagonistic to most strains of *A. tumefaciens*. Some control is also obtained by treating nongerminated seeds with the antagonist or by drenching the soil with a suspension of the antagonistic bacterium. It is postulated that the antagonist controls crown gall initiation by establishing itself on the surface of the plant tissues, where it produces the bacteriocin agrocin 84. This bacteriocin is inhibitory to most virulent *A. tumefaciens* strains. Unfortunately, some strains of *A. tumefaciens* inherited from strain 84 its resistance to agrocin 84. Therefore, a new strain (K-1026) is now used because it lacks the ability to transfer its resistance gene to pathogenic *Agrobacterium* strains.

Selected References

Burr, T. J., and Otten, L. (1999). Crown gall of grape: Biology and disease management. *Annu. Rev. Phytopathol.* **37**, 53–80.

Cervera, M. K., López, M. M., Navarro, L., and Peña, L. (1998). Virulence and supervirulence of *Agrobacterium tumefaciens* in woody fruit plants. *Physiol. Mol. Plant Pathol.* **52**, 67–78.

DeCleene, M., and DeLey, J. (1976). The host range of crown gall. *Bot. Rev.* **42**, 389–466.

Goethals, K., *et al.* (2001). Leafy gall formation by *Rhodococcus fascians*. *Annu. Rev. Phytopathol.* **39**, 27–51.

Hooykaas, P. J. J., and Beijersbergen, A. G. M. (1994). The virulence system of *Agrobacterium tumefaciens*. *Annu. Rev. Phytopathol.* **32**, 157–179.

Horsch, R. B., *et al.* (1985). A simple and general method for transferring genes into plants. *Science* **227**, 1229–1231.

Karimi, M., Van Montagu, M., and Gheysen, G. (2000). Nematodes as vectors to introduce *Agrobacterium* into plant roots. *Mol. Plant Pathol.* **1**, 383–387.

Kerr, A. (1980). Biological control of crown gall through production of agrocin 84. *Plant Dis.* **64**, 25–30.

Nester, E. E. W. (2000). DNA and protein transfer from bacteria to eukaryotes: The agrobacterium story. *Mol. Plant Pathol.* **1**, 87–90.

Otten, L., *et al.* (1992). Evolution of agrobacteria and their Ti plasmids: A review. *Mol. Plant-Microbe Interact.* **5**, 279–287.

Parrott, D. L., Anderson, A. J., and Carman, J. G. (2002). *Agrobacterium* induces plant cell death in wheat (*Triticum aestivum* L.). *Physiol. Mol. Plant Pathol.* **60**, 59–69.

Ream, W. (1989). *Agrobacterium tumefaciens* and interkingdom genetic exchange. *Annu. Rev. Phytopathol.* **27**, 583–618.

Smith, E. F., Brown, N. A., and Townsend, C. O. (1911). Crown gall of plants: Its cause and remedy. *U.S. Dept. Agric. Bull.* **213**, 1–215.

Süle and Burr (1998). The effect of resistance of rootstocks to crown gall (*Agrobacterium* spp.) on the susceptibility of scions in grape vine cultivars. *Plant Pathol* **47**, 84–88.

Thomashow, M. F., *et al.* (1980). Host range of *Agrobacterium tumefaciens* is determined by the Ti-plasmid. *Nature (London)* **283**, 794–796.

Tzfira, T., and Citovsky, V. (2000). From host recognition to T-DNA integration: the function of bacterial and plant genes in the *Agrobacterium*-plant cell interaction. *Mol. Plant Pathol.* **1**, 201–212.

Winans, S. C. (1992). Two-way chemical signaling in *Agrobacterium*-plant interactions. *Microbiol. Rev.* **56**, 12–31.

Zambryski, P. C. (1992). Chronicles from the *Agrobacterium*-plant cell DNA transfer story. *Annu. Rev. Plant Physiol. Plant Mol. Biol.* **43**, 465–490.

BACTERIAL CANKERS

Relatively few canker diseases of plants are caused by bacteria, but some of them are widespread and devastating. The bacteria and the most important cankers they cause are the following:

Pseudomonas, causing the bacterial canker of stone fruit and pome fruit trees (*P. syringae* pv. *syringae* and *P. syringae* pv. *morsprunorum*)

Xanthomonas, causing the bacterial canker of citrus (*X. axonopodis*, formerly *X. campestris* pv. *citri*)

In many bacterial cankers, the canker symptoms on stems, branches, or twigs are accompanied by direct symptoms on fruits, leaves, buds, or blossoms that may be at least as important in the overall effect of the disease on the plant as are the cankers. Bacterial cankers are often sunken and soft, as in the fungal cankers, but they may also appear as splits in the stem, as necrotic areas within the woody cylinder, or as scabby excrescences on the surface of the tissue.

Canker bacteria overwinter in perennial cankers, in buds, in plant refuse, and, in some plants, in or on the seed. They are spread by rain splashes or runoff water, windblown rain, handling of plants, contaminated tools, and infected plant material. Bacteria enter tissues primarily through wounds, but in young plants or early shoots they may also enter through natural openings. The control of bacterial cankers is through proper sanitation and eradication practices, through the use of bacteria-free seeds or budwood, and somewhat through several sprays with the Bordeaux mixture, other copper formulations, or antibiotics.

BACTERIAL CANKER AND GUMMOSIS OF STONE FRUIT TREES

Bacterial canker and gummosis disease affects primarily stone fruit trees and apparently occurs in all major fruit-growing areas of the world. The same pathogen or, more likely, specific pathovars of it also affect pear, apple, citrus, many annual and perennial ornamentals, some vegetables, and some small grains. The disease is also known as bud blast, blossom blast, dieback, spur blight, and twig blight.

Bacterial canker and gummosis is one of the most important diseases of stone fruit trees in many fruit-growing areas. Exact losses are difficult to assess because of serious damage to trees as well as reduction of yields. The disease causes cankers on branches and main trunks, kills young trees, and reduces the yield of or kills older ones. Tree losses from 10 to 75% have been observed in young orchards. Bacterial canker and gummosis also kills buds and flowers of trees, usually resulting in yield losses of 10 to 20% but sometimes up to 80%. Leaves and fruits are also attacked, resulting in weaker plants and in low-quality or unsalable fruit.

Symptoms

The most characteristic symptom of the disease is the formation of cankers accompanied by gum exudation (Figs. 12-37 and 12-38). Cankers usually develop at the base of infected spurs, in pruning wounds, and at the bud union. They then spread mostly upward and to a lesser extent down and to the sides. Infected areas are slightly sunken and darker brown than the surrounding healthy bark. The cortical tissues of the cankered area are bright orange to brown. Cankers are first noticed in late winter or early spring. In the spring, gum is produced in most cankers, breaks through the bark, and runs down on the surface of the limbs. Cankers that do not produce gum are softer, moister, sunken, and may have a sour smell. When the trunk or branch is girdled by a canker, the leaves show curling and drooping and turn light green and then yellow. Within a few weeks the branch or entire tree above the canker is dead (Figs. 12-37 and 12-38).

Dormant bud blast is especially serious on cherry, apricot, and pear. In some areas, great numbers of buds are killed or fail to develop. When sectioned, infected buds show brown scales and bases. Such buds eventually die (Fig. 12-38). Both flower and leaf buds are equally affected. The light bloom of infected trees is most conspicuous during full bloom.

Leaf infections appear as water-soaked spots about 1 to 3 millimeters in diameter. Later, the spots become brown, dry, and brittle and eventually fall out, giving the leaves a shot-hole or tattered appearance. Infected fruit has superficial, flat or depressed, dark brown to black spots, 2 to 10 millimeters in diameter and depth. The underlying tissue is gummy or spongy.

FIGURE 12-37 Bacterial canker of stone fruits caused by *Pseudomonas syringae* pv. *syringae* and *P. syringae* pv. *morsprunorum*. (A) Young peach shoots developing gum-producing cankers following natural infection with the bacteria. (B) Older cankers of various sizes on peach twigs. (C) Bacteria have also moved internally into the twig tissues, causing discoloration and death of tissues. (D) Bacterial canker encircling much of a peach stem and accompanied by secretion of gum. [Photograph (D) courtesy of I. MacSwann, Oregon State University.]

FIGURE 12-38 Bacterial canker on cherry. (A) Infection and killing of cherry buds by bacteria. (B) Infection moves from bud into the twig or branch where it begins to induce a canker. (C) A large bacterial canker on main branch of a cherry tree. (D) A healthy cherry tree (left) and a cherry tree with the buds or flowers killed by the bacteria (right) giving the tree temporarily a bronze color. (E) The bacterium responsible for stone fruit canker. Photos (A,B) courtesy J. W. Pscheidt, (C) I. MacSwann, and (E) H. R. Cameron, all of Oregon State University. (D) A. L. Jones, Michigan State University.

The Pathogen

The pathogens are *Pseudomonas syringae* pv. *syringae* (Fig. 12-38E) and the more specialized *P. syringae* pv. *morsprunorum*, which is restricted predominantly to cherry and plum. Most strains of *P. syringae* pv. *syringae* produce the phytotoxins syringomycins, which appear to play a role in the virulence of the pathogen. Many strains of *P. syringae* are ice nucleation active, i.e., they serve as nuclei for ice formation, and therefore cause frost injury to plants at relatively high freezing temperatures. The same bacteria also produce bacteriocins toxic against nonice nucleation-active strains, thus assuring a competitive advantage for themselves.

Development of Disease

Bacteria overwinter in active cankers, in infected buds and leaves, systemically in the xylem of some hosts, epiphytically on buds and limbs of infected or healthy trees, and possibly on weeds and on nonsusceptible hosts (Figs. 12-37, 12-38, and 12-40).

Infection of limbs usually takes place during the fall and early winter. Bacteria enter limbs through the bases of infected buds or spurs and also through pruning cuts, leaf scars, and other wounds. Bacteria move intercellularly into the bark and into the ray parenchyma of the phloem and xylem. In advanced stages of infection, bacteria break down parenchyma cells and cavities full of bacteria develop. Xylem vessels are sometimes invaded by bacteria, but the bacteria do not seem to move far through the vessels.

Cankers develop rather rapidly in the fall, slowly in the winter, and most rapidly in the period between the end of the cold weather and the beginning of rapid tree growth in the spring. Cankers on the south side of trees are usually larger due to warming by the sun during the dormant season. The advance of the canker is checked by the higher temperatures and the active growth of the host in the spring, when callus tissue forms around the canker and the canker becomes inactive. Some cankers are inactivated permanently, but others become active again the following year and continue to spread in succeeding years. Infections during the active growing season are seldom of any consequence and apparently are isolated very quickly by callus tissue.

Infections of buds originate at the base of the outside bud scales and then spread throughout the base of the bud, killing the tissues across the base and resulting in the death of the bud (Figs. 12-34 and 12-35). The bacteria sometimes spread downward and kill stem tissues around the base of the bud. Infection of buds, blossoms, and young leaves seems to be favored by frost injury to the tissues of these organs. Bacteria contribute to this by causing frost to form at somewhat higher temperatures.

Flower infection is rare and seems to occur through natural openings and through wounds. Under very humid conditions, bacteria spread through the floral parts quickly and may advance into the spur and twig, where they initiate canker formation.

Leaf infections appear on young, succulent leaves, frequently during cool, wet springs. Infection takes place through stomata. Bacteria spread intercellularly and cause collapse and death of the cells, resulting in small angular leaf spots. During wet weather, bacteria ooze out of stomata and necrotic spots (Fig. 12-39) and are

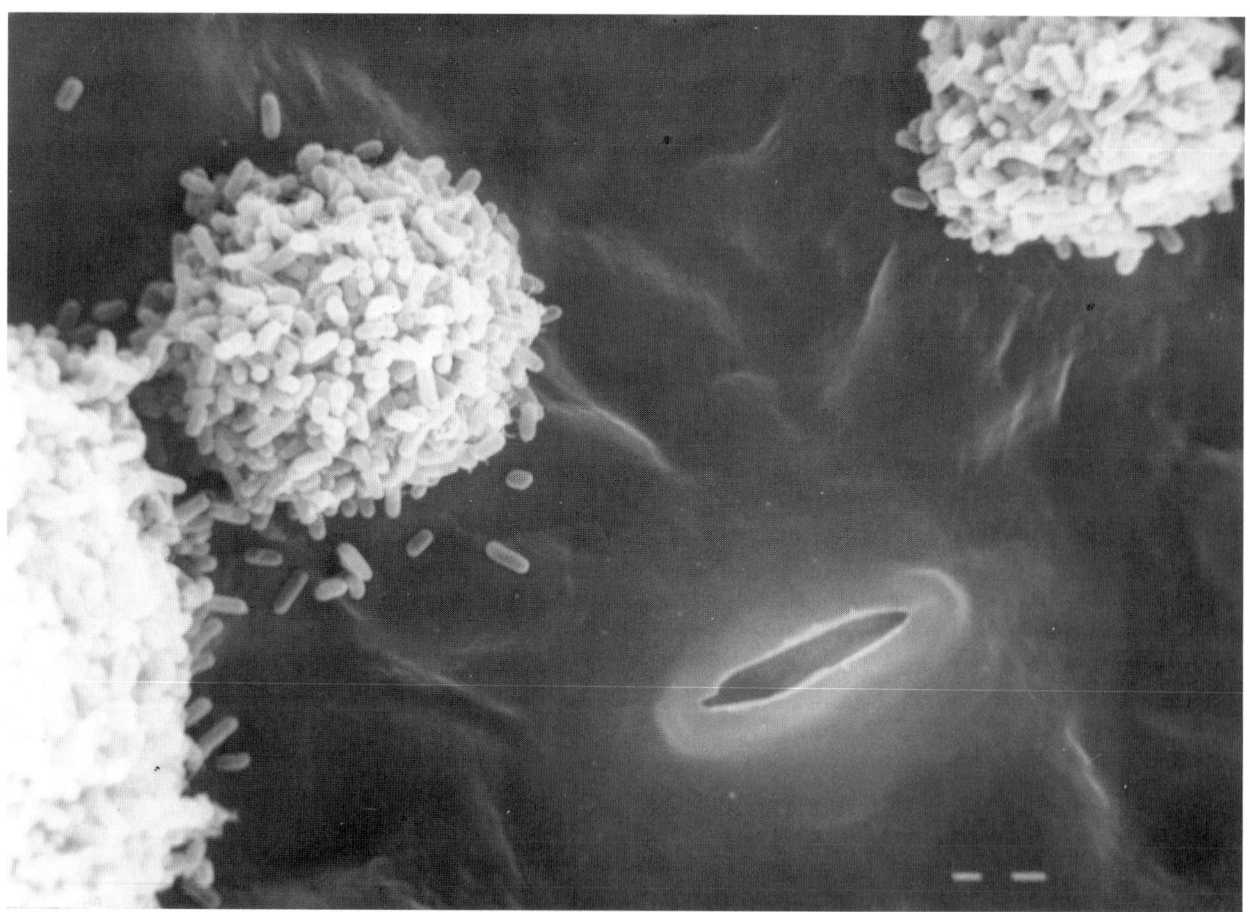

FIGURE 12-39 *Pseudomonas syringae* pv. *morsprunorum* exuding from stomata of infected cherry leaves. [Photograph courtesy of Roos and Hattingh (1983). *Phytopathol. Z.* 108, 18–25.]

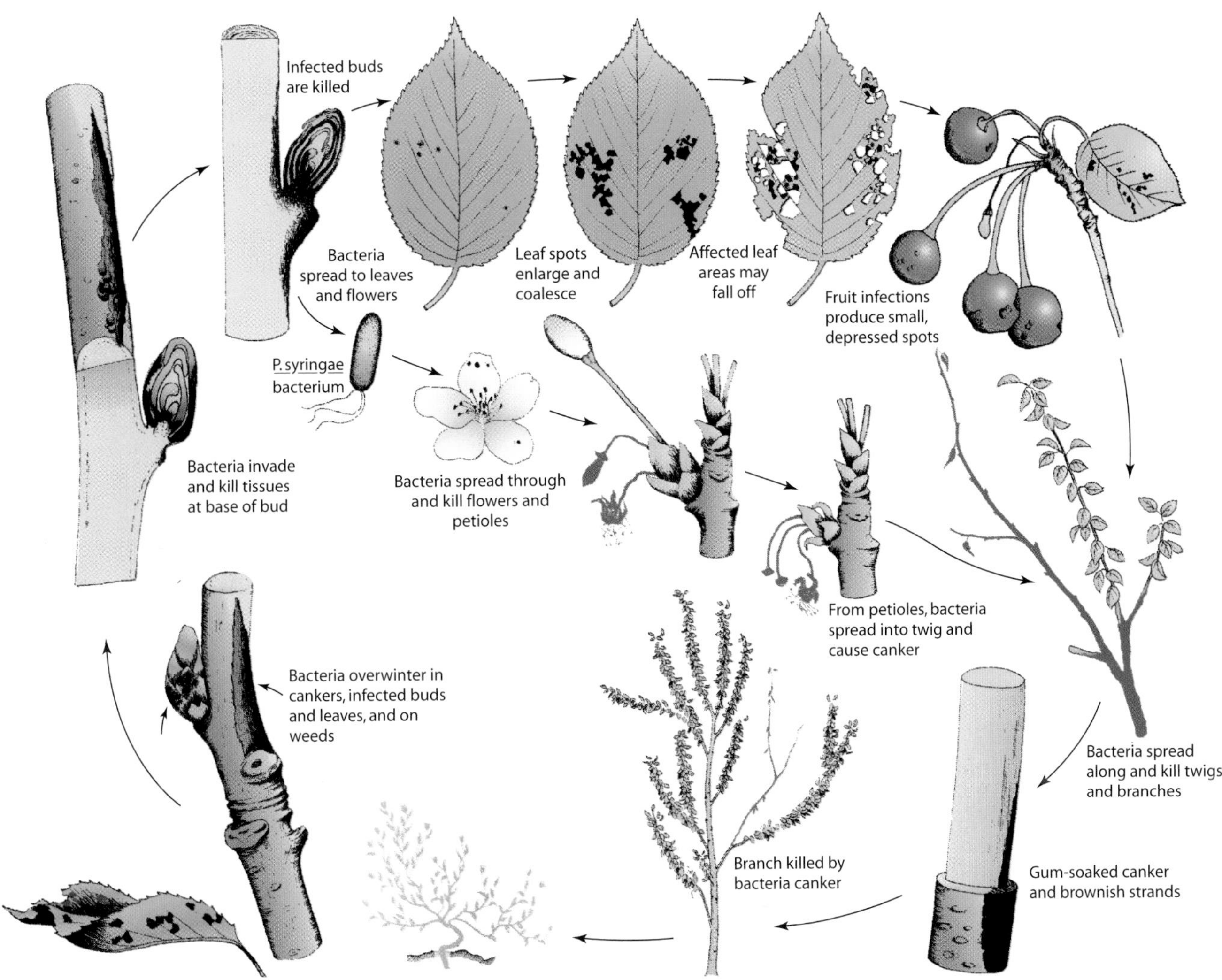

FIGURE 12-40 Disease cycle of canker and gummosis of stone fruits caused by *Pseudomonas syringae* pv. *syringae*.

spread to other leaves by direct contact, by visiting insects, and by rain and wind. As leaves mature, however, they become less susceptible, and leaf infections late in the season are rare.

Control

No complete control of bacterial canker and gummosis of fruit trees can be obtained as yet by any single method. Certain cultural practices and control measures help keep down the number and severity of infections. As a start, only healthy budwood should be used for propagation. Susceptible varieties should be propagated on rootstocks resistant to the disease and should be grafted as high as possible. Only healthy nursery trees should be planted in the orchard. Orchards should not be located in areas where trees are subjected to freeze damage, waterlogged soils, or prolonged drought.

Partial control of the canker phase of the disease both in the nursery and in the orchard is obtained with sprays of fixed copper or Bordeaux mixture in the fall and in the spring before blossoming, but pathogen strains resistant to copper exist in orchards. Cankers on trunks and large branches can be controlled by cauterization with a handheld propane burner. The flame is aimed at the canker and especially its margins for 5 to 20 seconds until the underlying tissue begins to crackle and char. The treatment is carried out in early to midspring and, if necessary, should be repeated 2 to 3 weeks later.

CITRUS CANKER

Citrus canker is one of the most feared of citrus diseases, affecting all types of important citrus crops. It causes necrotic lesions on fruit, leaves, and twigs. Losses are caused by reduced fruit quality and quantity and premature fruit drop. The disease is endemic in Japan and southeast Asia, from where it has spread to all other citrus-producing continents except Europe. In the United States, citrus canker was introduced into Florida in 1912, with infected nursery trees from Japan, and spread to all the Gulf states and beyond. It took 20 years, destruction by burning of more than a quarter million bearing trees and more than three million nursery trees, many millions of dollars in expenses, and untold inconvenience and heartaches before citrus canker was eradicated from Florida. It took 20 more years (until 1949) to eliminate it entirely from the United States. Unfortunately, a bacterial leaf spot resembling that caused by the citrus canker bacterium appeared in Florida in August 1984 and was assumed to be citrus canker. A new series of eradicating measures went into effect immediately, resulting in the destruction of at least 20 million nursery and young orchard trees through 1990. However, in 1986, the real citrus canker (Asiatic canker or canker A) was also found in Florida, and the eradications continued in areas where canker was present until 1992. After no citrus canker was found for 2 years, Florida was declared free of citrus canker in early 1994, and all regulations were suspended. Citrus canker, however, was again found in residential trees in the Miami area in October 1995, and the tree removal regulations were reinstated.

Citrus canker has been eradicated from South Africa, Australia, and New Zealand. The latest outbreak and eradication of citrus canker in Australia occurred in 1991. In South America, citrus canker was found in Brazil in 1957. It subsequently spread to Uruguay, Paraguay, and Argentina and despite attempts to eradicate it, the disease has become permanently established there. Eradication efforts in Brazil, however, have kept the large citrus-producing areas of that country free of the disease. All citrus-producing countries without canker maintain a strict prohibition on import of citrus plants and fruit from noncanker-free countries.

Symptoms

Quite similar lesions are produced on young leaves, twigs, and fruits (Fig. 12-41). The lesions at first appear as small, slightly raised, round, light green spots. Later, they become grayish white, rupture, and appear corky with brown, sunken centers. The margins of the lesions are often surrounded by a yellowish halo. The size of the lesions varies from 1 to 9 millimeters in diameter on leaves and up to 1 centimeter in diameter or length on fruits and twigs. Severe infections of leaves, twigs, and branches debilitate the tree, while severely infected fruit appear scabbed and deformed.

The Pathogen: Xanthomonas axonopodis pv. citri

Development of Disease

Bacteria overseason in leaf, twig, and fruit canker lesions. During warm, rainy weather they ooze out of lesions and, if splashed onto young tissues, bacteria enter them through stomata or wounds. Bacteria infect older tissues only through wounds. Several cycles of infection can occur on fruit; therefore, fruits often have lesions of many sizes. Free moisture and strong winds seem to greatly favor the spread of the bacteria. Citrus canker seems to be much more severe in areas in which the periods of high rainfall coincide with the period of high mean temperature (such as Florida and other Gulf Coast states), whereas it is not important in areas where high temperatures are accompanied by low rainfall (such as in the southwestern United States).

Control

In canker-free citrus-producing areas, strict quarantine measures are practiced to exclude the pathogen. When the canker bacterium is found in such an area (as it was in Florida in 1986), every effort must be made to eradicate it. This is attempted by burning all infected and adjacent trees to prevent the spread of the pathogen. In areas where the citrus canker bacterium is endemic, three to four sprays of copper are required for even partial control of the disease on susceptible trees. Generally, control on susceptible trees has not been adequate for commercial production. Because of the presence of this bacterium, 85% of citrus production in Japan is of the moderately resistant Unshiu (Satsuma) orange. On such trees, satisfactory control of citrus canker is obtained by using windbreaks, pruning diseased summer and autumn shoots, forecasting of impending epidemics, and applying copper sprays.

Selected References

Agrios, G. N. (1972). A severe new canker disease of peach in Greece. *Phytopathol. Mediterr.* **11**, 91–96.

Cameron, H. R. (1962). Diseases of deciduous fruit trees incited by *Pseudomonas syringae* van Hall. *Oreg. Agric. Exp. Stn. Bull.* **66**, 1–64.

Crosse, J. E. (1966). Epidemiological relations of the pseudomonad pathogens of deciduous fruit trees. *Annu. Rev. Phytopathol.* **4**, 291–310.

Gottwald, T. R., Sun, X., Riley, T., *et al.* (2002). Geo-referenced spatiotemporal analysis of the urban citrus canker epidemic in Florida. *Phytopathology* **92**, 361–377.

Gottwald, T. R., Timmer, L. W., and McGuire, R. G. (1989). Analysis of disease progress of citrus canker in nurseries in Argentina. *Phytopathology* **79**, 1276–1283.

Grahman, J. H., and Gottwald, T. R. (1991). Research perspectives on eradication of citrus bacterial diseases in Florida. *Plant Dis.* **75**, 1193–1200.

Gross, D. C., *et al.* (1984). Ecotypes and pathogenicity of ice-

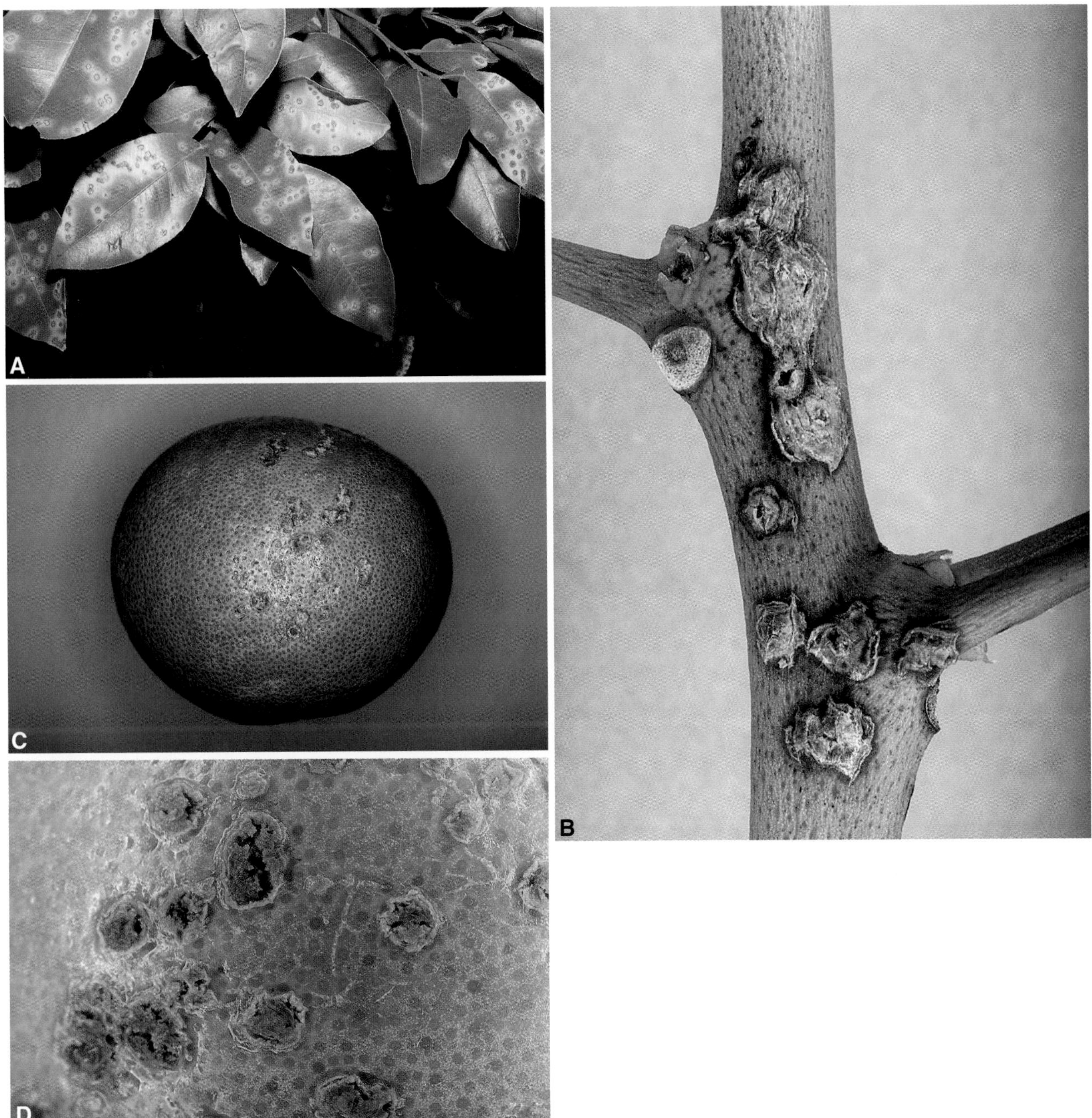

FIGURE 12-41 Citrus canker caused by *Xanthomonas axonopodis* pv. *citri*. Symptoms on leaves (A), stem (B), fruit (C), and fruit close-up (D). [Photographs courtesy of Div. Plant Industry, Florida Department of Agriculture.]

nucleation-active *Pseudomonas syringae* isolates from deciduous fruit tree orchards. *Phytopathology* **74**, 241–248.

Hattingh, M. J., Roos, I. M. M., and Mansvelt, E. L. (1989). Infection and systemic invasion of deciduous fruit trees by *Pseudomonas syringae* in South Africa. *Plant Dis.* **73**, 784–789.

Hawkins, J. E. (1976). A cauterization method for the control of cankers caused by *Pseudomonas syringae* in stone fruit trees. *Plant Dis. Rep.* **60**, 60–61.

Jones, A. L. (1971). Bacterial canker of sweet cherry in Michigan. *Plant Dis. Rep.* **55**, 961–965.

Kuhara, S. (1978). Present epidemic status and control of the citrus canker disease *Xanthomonas citri* in Japan. *Rev. Plant Prot. Res.* **11**, 132–142.

Pruvost, O., Boher, B., Brocherieux, C., *et al.* (2002). Survival of *Xanthomonas axonopodis* pv. *citri* in leaf lesions under tropical environmental conditions and simulated splash dispersal of inoculum. *Phytopathology* **92**, 336–346.

Stall, R. E., and Civerolo, E. L. (1991). Research relating to the recent outbreak of citrus canker in Florida. *Annu. Rev. Phytopathol.* **29**, 399–420.

Vigouroux, A. (1999). Bacterial canker of peach: Effect of tree winter water content on the spread of infection through frost-related water soaking in stems. *J. Phytopathol.* **147**, 553–559.

Wilson, E. E. (1933). Bacterial canker of stone fruit trees in California. *Hilgardia* **8**, 83–123.

BACTERIAL SCABS

Bacterial scabs include mainly diseases that affect belowground parts of plants and whose symptoms consist of more or less localized scabby lesions affecting primarily the outer tissues of these parts. Bacterial scabs are caused by species of *Streptomyces*. The most important ones are the common scab of potato and of other belowground crops (*S. scabies*) and the soil rot or pox of sweet potato (*S. ipomoeae*).

Scab bacteria survive in infected plant debris and in the soil and penetrate tissues through natural openings or wounds. In the tissues, bacteria grow mostly in the intercellular spaces of parenchyma cells, but these cells are sooner or later invaded by the bacteria and break down. In typical scabs, healthy cells below and around the lesion divide and form layers of corky cells. These cells push the infected tissues outward and give them the scabby appearance. Scab lesions often serve as points of entry for secondary parasitic or saprophytic organisms that may cause the tissues to rot.

COMMON SCAB OF POTATO

Common scab of potato is caused by *Streptomyces scabies* and occurs throughout the world. It is most prevalent and important in neutral or slightly alkaline soils, especially during relatively dry years. The same pathogen also affects beets, radish, and other root crops. The usually superficial blemishes on tubers and roots reduce the value rather than the yield of the crop. Severe infections may reduce yields, and deep scabs increase the waste in peeling.

Common scab of potato affects mostly the tubers. Infected tubers at first develop small, brownish, raised spots. Later, the spots usually enlarge, coalesce, and become corky. The lesions extend 3 to 4 millimeters deep in the tuber. Sometimes the lesions appear as numerous russeted areas that almost cover the tuber surface or they may appear as slight protuberances with depressed centers covered with corky tissue (Fig. 12-42).

The pathogen, *S. scabies*, is a saprophyte that can survive indefinitely in most soils except the most acidic ones. *Streptomyces scabies* consists of slender (about 1 micrometer thick), branched mycelium with few or no cross walls. The mycelium produces cylindrical spores about 0.6 by 1.5 micrometers, on specialized spiral hyphae. These hyphae develop cross walls from the tip toward their base, and, as the cross walls constrict, spores are pinched off at the tip and eventually break away. The spores produce one or two germ tubes, which develop into the mycelioid form (Fig. 12-43).

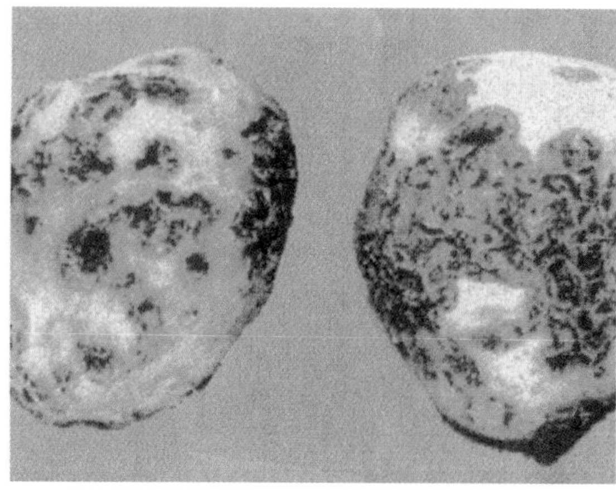

FIGURE 12-42 Potato scab caused by *Streptomyces scabies*. [Photograph courtesy of R. Loria, Cornell University.]

The pathogen is spread through soil water, by windblown soil, and on infected potato seed tubers. It penetrates tissues through lenticels, wounds, and stomata and, in young tubers, directly. Young tubers are more susceptible to infection than older ones. After penetration the pathogen apparently grows between or through a few layers of cells, the cells die, and the pathogen then derives food from them. In response to the infection, living cells surrounding the lesion divide rapidly and produce several layers of cork cells that isolate the pathogen and several plant cells. Usually, several such groups of cork cell layers are produced, and as they are pushed outward and sloughed off, the pathogen grows and multiplies in the additional dead cells, thereby allowing large scab lesions to develop. The depth of the lesion seems to depend on the host variety, on soil conditions, and on the invasion of scab lesions by other organisms, including insects. The latter apparently break down the cork layers and allow the pathogen to invade the tuber in great depth.

The severity of common scab of potato increases as the pH of the soil increases from pH 5.2 to 8.0 and decreases beyond these limits. Potato scab incidence is reduced greatly by high soil moisture during the period of tuber initiation and for several weeks afterward. Potato scab is also lower in fields after certain crop rotations and the plowing under of certain green manure crops, probably as a result of inhibition of the pathogen by antagonistic microorganisms.

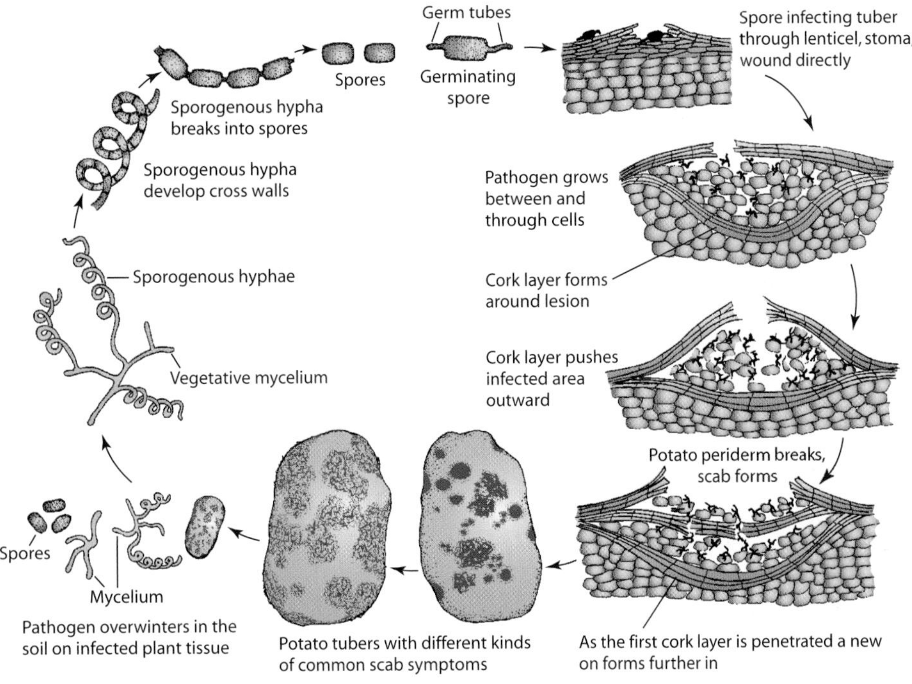

FIGURE 12-43 Disease cycle of common scab of potato caused by *Streptomyces scabies*.

Control of the common scab of potato is through the use of certified scab-free seed potatoes or through seed treatment with pentachloronitrobenzene (PCNB) or with maneb–zinc dust. If the field is already infested with the pathogen, a fair degree of disease control may be obtained by using certain crop rotations, bringing and holding the soil to about pH 5.3 with sulfur, irrigating for about six weeks during the early stages of tuber development, and using resistant or tolerant potato varieties. Biological control of scab with *Streptomyces* phage looks promising.

Selected References

Faucher, E., *et al.* (1993). Characterization of streptomycetes causing russet scab in Quebec. *Plant Dis.* 77, 1217–1220.

Jones, A. P. (1931). The histogeny of potato scab. *Ann. Appl. Biol.* **18**, 313–333.

Keinath, A. P., and Loria, R. (1989). Population dynamics of *Streptomyces scabies* and other actinomycetes as related to common scab of potato. *Phytopathology* 79, 681–687.

Levick, D. R., Evans, T. A., Stephens, C. T., and Lacy, M. L. (1985). Etiology of radish scab and its control through irrigation. *Phytopathology* 75, 568–572.

McKenna, F., *et al.* (2001). Novel *in vivo* use of polyvalent *Streptomyces* phage to disinfest *Streptomyces scabies*-infected seed potatoes. *Plant Pathol.* 50, 666–675.

Wilson, C. R. (2001). Variability within clones of potato cv. Russet Burbank to infection and severity of common scab disease of potato. *J. Phytopathol.* **149**, 625–628.

ROOT NODULES OF LEGUMES

Root nodules are well-organized structures produced on the roots of most legume plants after inoculation with certain, mostly nitrogen-fixing species of bacteria of the genus *Rhizobium* and, to a lesser extent, by bacteria of the genera *Bradyrhizobium* and *Azorhizobium*. Bacteria of the thallomycetous genus *Frankia* induce root nodule formation and fix nitrogen in a broad range of woody dicotyledonous host plants. Bacteria of another genus, *Azospirillum*, which also fix nitrogen but do not induce root nodule formation, have been isolated from the roots of many grasses, including the very important crops maize, rice, and wheat. Although they are the result of infection of legumes and other plants by bacteria, root nodules are considered a condition of symbiosis rather than of disease.

On infection, bacteria fix (trap) atmospheric nitrogen and make it available to the plant in a utilizable organic form, and the plant profits from this nitrogen more than it loses in sugars and other nutrients to the bacteria. Unfortunately, not all root nodule bacteria are beneficial to the legume host. Some strains of nodule bacteria are apparently strictly parasites, as they form nodules on the roots but fail to fix nitrogen. Therefore, the number of root nodules does not always indicate their value to the plant unless the strain of bacteria is known to be effective in fixing nitrogen. As a result, legume seeds are routinely inoculated commercially with appropriate strains of root nodule bacteria to improve plant growth and yields.

Nodules are produced on taproots as well as lateral roots of legumes and may vary in size from 1 millimeter to 2 to 3 centimeters (Fig. 12-44A). Nodules may be round or cylindrical and as large or larger than the root diameter on which they form. Their number and size vary with the plant, bacterial strain, age of infection, and so on. On herbaceous plants, nodules are fragile and short-lived, whereas on woody plants they may persist for several years. Each nodule consists of an epidermal layer, a cortical layer, and the bacteria-containing central tissue, each consisting of several layers of cells (Fig. 12-44B). Vascular bundles are present in the cortical layer just outside the central tissue. In elongated nodules, the tip of the nodule farthest away from the root consists of a zone of meristematic cells through which the nodule grows. In rounded nodules, the meristematic region is laid around the nodule except at the neck.

The Organism

Root nodule bacteria include *Rhizobium, Bradyrhizobium, Azorhizobium*, and *Frankia*. Bacteria vary in size and shape with age, with typical bacteria being rod-shaped (1.2–3.0 by 0.5–0.9 millimeter) or irregular, club-shaped forms. They have no flagella, and most are gram negative. *Frankia* bacteria are gram positive and filamentous and they produce spores. Root nodule bacteria survive in roots of susceptible legumes and, for varying periods of time, in the soil. Continued growth of the same legume in the soil tends to build up the population of nodule bacteria affecting that legume. Not all nodule bacteria affect all legumes. For example, bacteria that grow on alfalfa do not grow on clovers, beans, peas, or soybeans, and vice versa. Strains of nodule bacteria often have definite varietal preferences; e.g., some soybean bacteria work better on one or two soybean varieties than on others.

Development of Nodules

The mechanism of root nodule formation is a highly specific process and involves the interaction of many bacterial and host genes. The host roots release flavonoid compounds that serve as signal compounds and initiate the coordinated expression of bacterial genes required for nodulation (*nod* genes). Nodulation genes code for enzymes involved in the synthesis of nod factors, i.e., acylated chitin oligomers. Nod factors act as signal molecules that trigger the activation of host genes, which leads to the curling of root hairs, where penetration takes place, and to the formation of nodule meristem that produces the nodule. In addition to acting as attractants for bacteria and initiating the expression of their nodulation genes, host flavonoids also make the nodulating bacteria resistant to the host phytoalexins. Furthermore, host flavonoids are involved in nodule meristem formation, possibly by disturbing the auxin–cytokinin balance in the plant root.

Nodulation genes enable bacteria to induce nodule formation in a host-specific manner. Some of the nodulation genes, such as *nod*ABC, are essential for nodulation and are common to all nodulating bacterial species and strains. Other nodulation genes, however, are present in only certain bacterial species or strains and determine the host range of these bacteria, i.e., they determine the host species or varieties on which these bacteria can induce root nodule formation.

The observable development of root nodules begins with the direct penetration of root hairs by the nodule-inducing bacteria. Within the root hair cell, bacteria become embedded in a double-walled, tubular, mucoid sheath called an infection thread. The infection thread, which contains bacteria, penetrates into the cortical parenchyma cells and branches along the way, with terminal and lateral vesicles forming on the strands. These vesicles soon break and release the bacteria, mostly within the cells (Fig. 12-44C). The released bacteria then enlarge and become enclosed in a membrane envelope (Fig. 12-44D). These membrane-enclosed bacteria are called bacteroids. In the meantime, cortical parenchyma cells along the path of bacterial invasion begin to divide, and the invaded cells increase in size as the bacteroids appear. The increased meristematic activity and cell enlargement of cortical cells result in formation of the nodule, which grows outward from the root cortex. At the same time, differentiation of vascular tissues, both xylem and phloem, takes place in the nodule. The vascular tissues of the nodule are not connected directly with those of the root.

While the outermost tip or layer of the nodule remains meristematic and continues to grow and thus to increase the size of the nodule up to a certain point, many of the cortical cells behind the meristematic zone and in all the central tissue of the nodule are uniformly enlarged and infected with several bacteroids. In the most recently infected cells, each bacteroid is enclosed in a membrane envelope, whereas in earlier infected cells several bacteroids may be enclosed in a membrane envelope. In cells that have been infected even longer than the latter, the bacteroids lack a membrane envelope and the host cellular membrane system also deteriorates. It appears that the membraneless bacteroids, which occur in the advanced stages of infection and which increase in numbers while the nodule is still growing, lack the

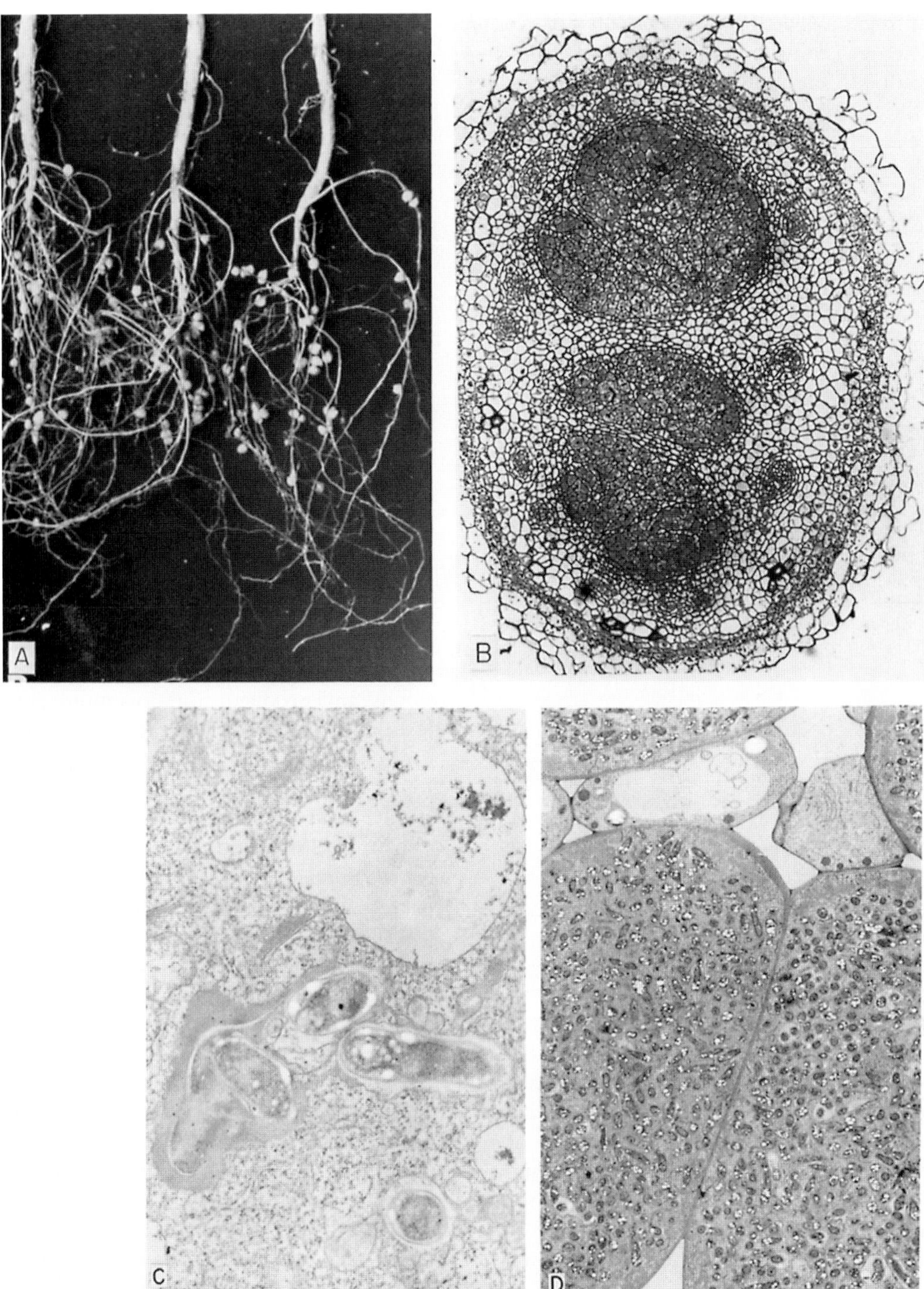

FIGURE 12-44 (A) "Healthy" soybean roots bearing numerous bacterial nodules (B) Cross section of a developing soybean nodule 12 days after inoculation. There are at least three central areas containing bacteroids, apparently resulting from several closely adjacent infections. (C and D) Electron micrographs of sections of a soybean root nodule. (C) Area of an infection thread where bacteria are apparently being released. (D) Infected and uninfected cells in a young nitrogen-fixing nodule. Membrane envelopes are visible around some bacteria. Electron-lucent granules in the bacteria consist of poly-β-hydroxybutyrate. [Photographs courtesy of (A) USDA and (B–D) B. K. Bassett and R. N. Goodman.]

ability to fix nitrogen. Therefore, the efficiency of root nodules in nitrogen fixation is proportional to the number of enveloped bacteroids they contain and not necessarily to the size of the nodules. As the nodules age, first cortical cells in the earliest infected areas and then in the entire central area of the nodule disintegrate and collapse. The bacteroids, which have by now lost their membrane envelope, either disintegrate or become intercellular bacteria and are finally released into the soil as the nodule cortex and epidermis disintegrate.

Selected References

Dénarié, J., Debellé, F., and Rosenberg, C. (1992). Signaling and host range variation in nodulation. *Annu. Rev. Microbiol.* **46**, 497–531.

Djordjevic, M. A., Gabriel, D. W., and Rolfe, B. G. (1987). *Rhizobium* — the refined parasite of legumes. *Annu. Rev. Phytopathol.* **25**, 145–168.

Palacios, R., Mora, J., and Newton, W., eds. (1993). "New Horizons in Nitrogen Fixation." Kluwer, Dordrecht, The Netherlands.

Stacey, G., Burris, R. H., and Evans, H. J., eds. (1992). "Biological Nitrogen Fixation." Chapman & Hall, New York.

Tu, J. C. (1975). Rhizobial root nodules of soybeans as revealed by scanning and transmission electron microscopy. *Phytopathology* **65**, 447–454.

Vance, C. P. (1983). *Rhizobium* infection and nodulation: A beneficial plant disease? *Annu. Rev. Microbiol.* **37**, 399–424.

Vance, C. P., and Johnson, L. E. B. (1981). Nodulation: A plant disease perspective. *Plant Dis.* **65**, 118–124.

Vande Broek, A., *et al.* (1993). Spatial-temporal colonization patterns of *Azospirillum brazilense* on the wheat root surface and expression of the bacterial *nif*H gene during association. *Mol. Plant-Microbe Interact.* **6**, 592–600.

PLANT DISEASES CAUSED BY FASTIDIOUS VASCULAR BACTERIA

The fastidious vascular bacteria that cause plant diseases cannot be grown on simple culture media in the absence of host cells, and some of them have yet to be identified, named, and classified. Fastidious phloem-limited bacteria were observed first in 1972 in the phloem of clover and periwinkle plants affected with the clover club leaf disease and later in citrus plants affected with the greening disease. More recently, in the mid-1990s, bacteria were observed in the phloem of cucurbit plants affected with the yellow vine disease and of papaya plants affected with the bunchy top disease. In 1973, fastidious xylem-limited bacteria were observed in the xylem vessels of grape plants affected with Pierce's disease and of alfalfa affected with alfalfa dwarf. Subsequently, similar organisms were observed in the xylem of plants affected with one of more than 20 other diseases, e.g., peach affected with the phony peach disease, sugarcane affected with ratoon stunting, in citrus affected with citrus variegation chlorosis, and in plum leaf scald, almond leaf scorch, and elm leaf scorch.

XYLEM-INHABITING FASTIDIOUS BACTERIA

Fastidious xylem-inhabiting bacteria are generally rod-shaped cells 0.2 to 0.5 micrometers in diameter by 1 to 4 micrometers in length. They are bounded by a cell membrane and a cell wall. They have no flagella. The cell is usually undulating or rippled (Fig. 12-46). Nearly all fastidious xylem bacteria are gram negative. Several such xylem-limited bacteria have been placed in the genus *Xylella*. *Xylella fastidiosa* has the distinction that it is the first plant pathogenic bacterium the genome of which was completely sequenced. Only xylem-inhabiting bacteria causing sugarcane ratoon stunting and Bermuda grass stunting are gram positive, and they are classified as members of the genus *Clavibacter*. All xylem-inhabiting fastidious bacteria can be grown in culture on complex nutrient media, on which they grow slowly and produce tiny (1–2 millimeter) colonies. All fastidious xylem bacteria are unable to grow on conventional bacteriological media.

All gram-negative xylem-inhabiting fastidious bacteria are transmitted by xylem-feeding insects, such as sharpshooter leafhoppers (Cicadellinae) and spittlebugs (Cercopidae). The vectors can acquire and transmit the bacteria in less than two hours. Carrier adult insects can transmit the bacteria for life but do not pass them on to progeny. So far, no insect vector is known for gram-positive xylem-inhabiting fastidious bacteria, but at least one of them, the cause of sugarcane ratoon stunting, can be transmitted mechanically by cutting implements during harvest. The symptoms of diseases caused by fastidious xylem-inhabiting bacteria often consist of marginal necrosis of leaves, stunting, and general decline and reduced yields. Such symptoms are probably caused by plugging of the xylem by bacterial cells and by a matrix material partly of bacterial and partly of plant origin. In some diseases, however, such as phony peach, no marginal leaf necrosis occurs, and in others, such as sugarcane ratoon stunting, the only diagnostic symptom is stunting and an internal discoloration of the stalk. Although fastidious vascular bacteria are sensitive to several antibiotics such as tetracyclines and penicillin, chemotherapy of infected plants in the field has proved impractical. Fastidious vascular bacteria are sensitive to high temperatures. Heat treatment of entire plants or of propagative organs, by immersing them in water kept at 45 to 50°C for two to three hours, or by keeping them in hot air at 50 to 58°C for several (4–8) hours, has

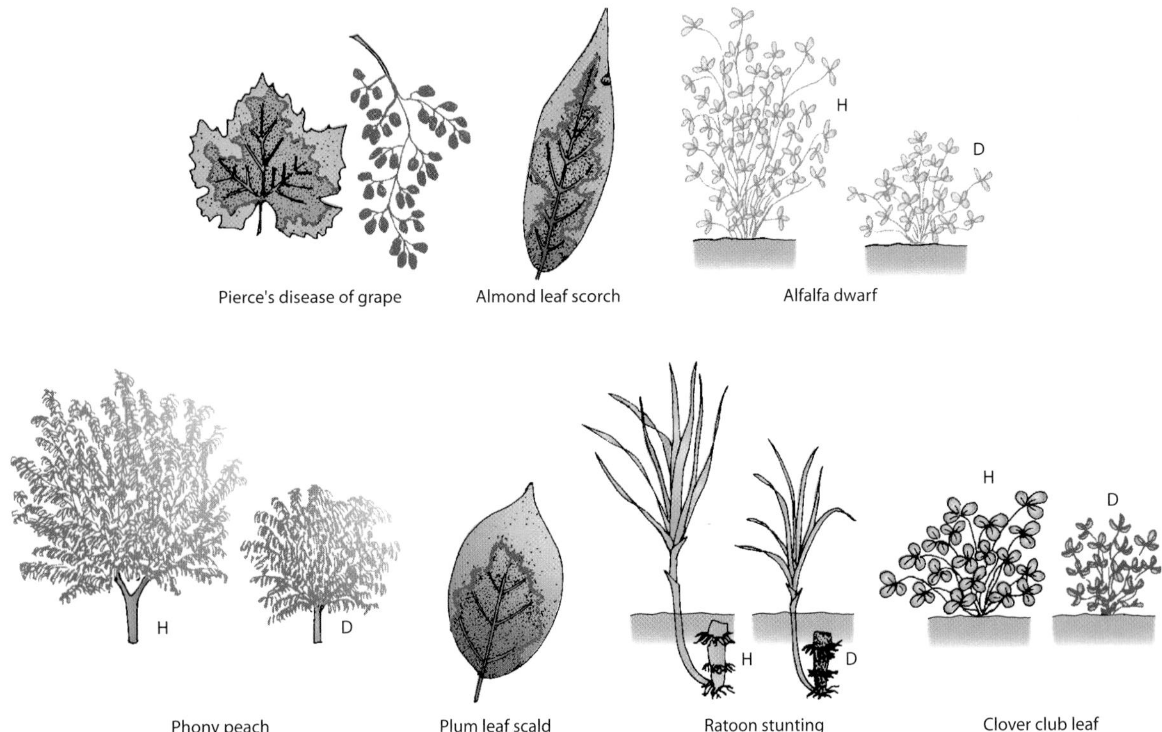

FIGURE 12-45 Symptoms caused by fastidious vascular bacteria. H, healthy plant; D, diseased plant.

cured grapevines from Pierce's disease and sugarcane from ratoon stunting disease.

Among the most important plant diseases caused by fastidious xylem-limited, gram-negative bacteria are Pierce's disease of grape, citrus variegation chlorosis, phony peach disease, almond leaf scorch, and plum leaf scald (Fig. 12-45). They are all caused by forms of the bacterium *Xylella fastidiosa*, which also causes leaf scorch diseases on elm, sycamore, oak, and mulberry. The also very important ratoon stunting disease of sugarcane is caused by the xylem-limited, gram-positive bacterium *Leifsonia xyli* (formerly *Clavibacter xyli subsp. xyli*).

PIERCE'S DISEASE OF GRAPE

Pierce's disease of grape (Fig. 12-46) is present in the southern United States from California to Florida and in Central America. In many areas the disease is endemic and no grapes can be grown because of it, whereas in other areas it breaks out as infrequent epidemics. California's vineyards had been free of Pierce's disease for many years until the introduction of the glassy winged sharpshooter (*Homalodisca coagulata*) (Fig. 12-46F), which turned out to be an efficient grapevine to grapevine vector of the disease. Many other annual and perennial kinds of plants of some 28 families, including grasses, herbs, shrubs, and trees, contain the bacteria and sometimes show symptoms of the disease. Other important diseases caused by the Pierce's disease pathogen *Xylella fastidiosa* include citrus variegated chlorosis, coffee leaf scorch (Fig. 12-47A), oak leaf scorch, oleander leaf scorch, and others. It appears that different strains of the pathogen are responsible for the diseases in the various hosts.

Infected grapevines die within a few months or may live for several years after infection. Some varieties survive infections longer than others. The disease is most severe on young, vigorous vines, especially in warm areas.

In grapes, symptoms appear as a sudden drying and scalding of much of the margin area of the leaf while the rest of the leaf is still green (Figs. 12-46A and 12-46B). Scalded areas advance toward the central area of the leaf and later turn brown. In late season, affected leaves usually drop, leaving the petioles attached to the canes. Grape clusters on vines with leaf symptoms stop growth, wilt, and dry up (Fig. 12-46C). Infected canes mature irregularly, forming patches of brown bark while the rest of the bark remains immature and green. During the following season(s), infected plants show delayed spring growth and dwarfed vines. Later in the season,

FIGURE 12-46 Pierce's disease of grape caused by *Xylella fastidiosa*. (A) Leaf scorching. (B) close-up of advanced leaf scorching. (C) Defoliation of an infected grape vine leaving petioles attached, the bark of the vine showing patchy discoloration, and the grape cluster wilted, withered, and drying up. Two of the vectors of pierce's disease, the blue green sharpshooter (D) and the glassy-winged sharpshooter (E). (F) single cell of *Xylella* showing rippled cell wall. (G) *Xylella* bacteria in xylem vessel, one undergoing binary fission. (H) *Xylella* bacteria in a tracheary element in a leaf vein. Photos (A-E) courtesy D.L. Hopkings, University of Florida. (F,G,H) courtesy H. H. Mollenhauer and D. L. Hopkins, University of Florida

leaves become scorched, and fruits wilt and dry up. Decline of the top is followed by dieback of the root system. Internally, the current-season wood of infected vines shows yellow to brown streaks both in longitudinal and in cross sections. In the same wood, gum forms in vessels and other types of cells, and tyloses develop in vessels of all sizes. Both gum and tyloses cause plugging of vessels, which results in many of the external symptoms of diseased plants.

The pathogen is *Xylella fastidiosa*, a typical fastidious xylem-inhabiting bacterium (Fig. 12-46F, 12-47A,B). It can be cultured on special nutrient media. The pathogen is transmitted by grafting and by many species of leafhoppers, such as the blue-green sharpshooter (*Graphocephala atropunctata*) and the glassy-winged sharpshooter (Figs. 12-46D and 12-46E). Leafhopper vectors acquire the pathogen after feeding on infected hosts for about two hours and may continue to transmit it to healthy hosts for the rest of their lives (Fig. 12-47C).

There is no practical control of Pierce's disease of grape in the field. All commercial grape varieties are susceptible to the disease. Drench treatments with tetracycline solutions inhibit symptom development, but such treatments are not feasible commercially. Individual plants can be freed of the pathogen by heat treatment. Such treatments, however, are of little help to the grower. The best defense is to plant in areas remote from natural reservoirs of the pathogen.

CITRUS VARIEGATED CHLOROSIS

Citrus variegated chlorosis was first reported in 1987 as affecting citrus trees in Brazil. In subsequent years, the disease spread rapidly and appeared to pose an immediate threat to the citrus industry in Brazil and possibly worldwide. A similar disease, known as "pecosita," seems to occur in Argentina. Citrus variegated chlorosis causes tree stunting, twig and branch dieback, and reduced size and quality of fruit.

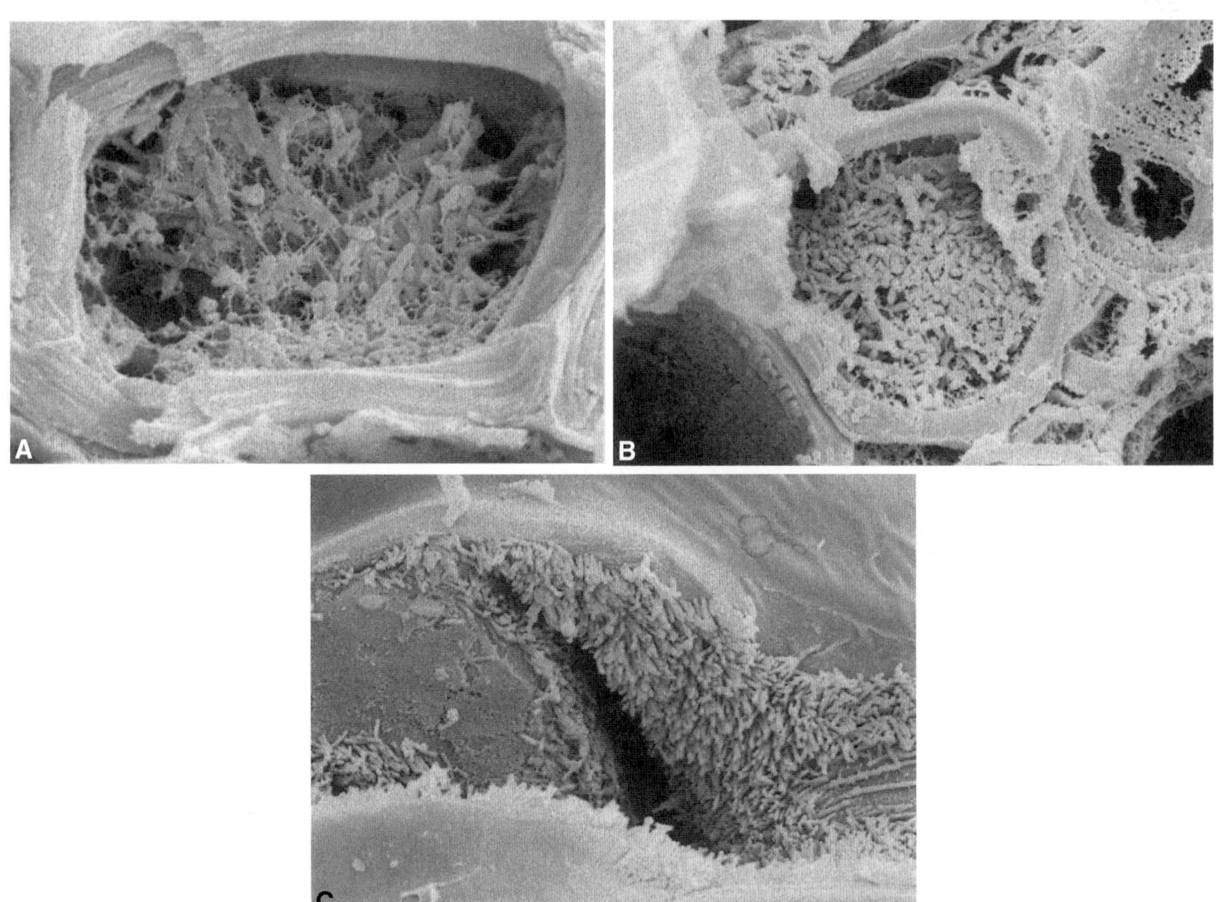

FIGURE 12-47 (A) *Xylella* bacteria exhibiting numerous thread-like connections in xylem vessel of coffee plant. (B) *Xylella* bacteria clogging a xylem vessel of a grape leaf. (C) *Xylella* bacteria in a tissue of its sharpshooter insect vector. Photos courtesy E. Alves, Federal Univ. Lavras, Brazil

FIGURE 12-48 Symptoms of citrus variegated chlorosis caused by distinct strains of *Xylella fastidiosa*. (A) Young leaves showing mottling and chlorosis. (B) Fruit of infected trees (left) is much smaller than healthy fruit and its rind is very hard. (C) Orange tree showing severe variegated chlorosis. (D) Fruit from severely affected trees is small, hard, and worthless. [Photographs courtesy of R. E. Lee, University of Florida.]

Young leaves of affected trees appear mottled and chlorotic as though they were affected by zinc deficiency (Figs. 12-48A and 12-48B). In more mature leaves, the lower sides of the chlorotic areas produce small, light brown gummy lesions that later may become dark brown, somewhat raised, and necrotic. The entire foliage of trees becomes chlorotic to yellow (Fig. 12-48C). Fruit of affected trees is smaller, often no more than one-third the diameter of healthy fruit (Figs. 12-48B and 12-48D). The fruit rind is hard to the point that it damages juicing machines, and, therefore, processing plants reject batches that contain a significant number of affected fruit. Soon after a young citrus tree becomes infected with variegation chlorosis, tree growth slows down, the tree remains stunted, and twigs and branches die back, but the trees do not die. In some cases, trees may appear to recover.

The pathogen of citrus variegation chlorosis is a strain of the xylem-limited fastidious bacterium *X. fastidiosa*. The bacterium grows in the xylem vessels of affected plants and reaches large numbers in them. The bacterium is spread by the vegetative propagation of infected budwood and, most likely, by xylem-feeding sharpshooter insects known to transmit other *X. fastidiosa* strains. The latter mode of transmission, although not yet proved, would explain the observed rapid spread of citrus variegated chlorosis within citrus orchards.

The control of citrus variegated chlorosis remains difficult. The only effective means of control to date is though the use of pathogen-free budwood in areas where the disease does not yet exist. Once introduced into an area, the disease seems to be spread rapidly to new trees by insect vectors and its control becomes impossible.

FIGURE 12-49 Sugarcane ratoon stunt disease. (A) Sugarcane planted with infected ratoons (left) and with hot-water treated cane (right). (B) Pinkish discoloration of stem at area of node due to infection by the bacterium. [Photographs courtesy of (A) H. D. Thurston, Cornell University and (B) A. G. Gillespie, USDA.]

RATOON STUNTING OF SUGARCANE

Ratoon stunting disease occurs in most and possibly all sugarcane-growing areas of the world. Losses due to stunting and unthrifty growth of infected plants usually range from 5 to 10%, but in some years losses may reach 30% or more. Losses tend to be greater in the ratoon (stubble) crops than in the crop of the first growing season.

Infected plants are slower to initiate growth, appear stunted, and may have fewer and thinner stalks, but they show no other external diagnostic symptoms (Fig. 12-49A). If plenty of water is available, infected plants may show no stunting and no loss in yield. However, infected plants often show internal symptoms. In very young shoots, symptoms consist of a pinkish discoloration just below the meristematic area of the shoot (Fig. 12-49B) above its attachment to the seed piece. In mature canes, symptoms appear as discolorations of individual vascular bundles at the nodes but not extending into the internodes. Under the microscope, affected bundles appear plugged with bacteria contained in a colored gummy substance.

The pathogen, *Leifsonia xyli* (formerly *Clavibacter xyli* subsp. *xyli*), is a gram-positive, fastidious, xylem-inhabiting coryneform bacterium 0.3 to 0.5 by 1 to 4 micrometers in size. It can be grown in culture on specialized media. The pathogen overseasons in infected sugarcane plants and propagative materials such as seed cane. New plants produced from infected seed can develop the disease. The bacterium is also spread by cutting knives and by cultivation and harvesting equipment. It has been spread and continues to be spread to different countries through infected sugarcane germplasm.

The control of ratoon stunting disease depends on the use of disease-free cane, heat treatment of suspected infected cane, sanitation of cutting equipment, and use of disease-resistant cultivars.

PHLOEM-INHABITING FASTIDIOUS BACTERIA

Fastidious vascular bacteria are generally rod-shaped cells 0.2 to 0.5 micrometer in diameter by 1 to 4 micrometers in length. They are bounded by a cell membrane and a cell wall, although in some phloem-inhabiting bacteria the cell wall appears more as a second membrane than as a cell wall. They have no flagella. The cell is usually undulating or rippled (Fig. 12-50E). Nearly all fastidious vascular bacteria are gram negative.

The symptoms of diseases caused by fastidious phloem-inhabiting bacteria often consist of leaf stunting and clubbing; in some cases they may appear as shoot proliferation and witches'-brooms and as greening of floral parts. In some of these diseases, symptoms are often mild and sometimes are followed by spontaneous recovery.

Phloem-limited bacteria are so far known to cause the very important citrus greening disease, the yellow vine disease of watermelon and other cucurbits, the bunchy top disease of papaya, and some minor diseases of clover and periwinkle.

The vectors of clover club leaf and citrus greening bacteria are leafhoppers and psyllid insects, respectively. The clover club leaf bacterium is known to multiply in its leafhopper vector and to be passed from the mother to the progeny insects through the eggs (transovarial

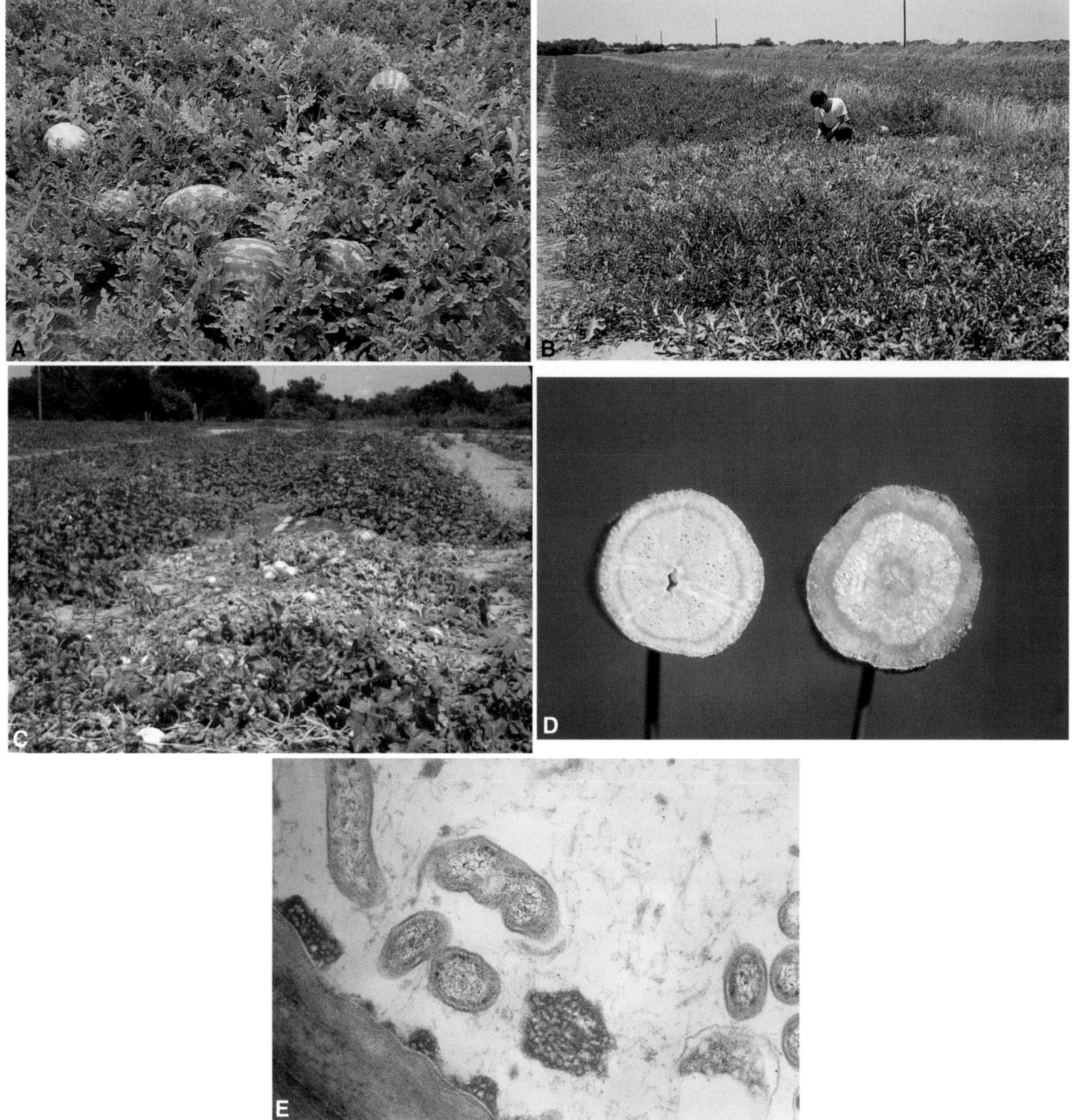

FIGURE 12-50 Yellow vine disease of cucurbits, the cause of which has been tentatively identified as the bacterium *Serratia marcescens*. Early symptoms of yellow vine in a watermelon field (A) are followed by more general yellowing (B) and death and collapse of the plants over large areas (C). Cross sections of stems of infected plants show brown discoloration of the phloem (D, right) compared to healthy plants. (E) The yellow vine bacterium, tentatively identified as *S. marcescens*, inside a phloem sieve tube. [Photographs courtesy of B. D. Bruton, USDA, Lane, OK.]

transmission). The cucurbit yellow vine bacterium is transmitted by the squash bug.

YELLOW VINE DISEASE OF CUCURBITS

Yellow vine disease was first reported in 1991 and occurs widely in several states, including Oklahoma, Texas, Tennessee, and Massachusetts. It affects several cucurbits, including watermelon, melon, squash and pumpkin. Infected plants show yellowing of leaves (Figs. 12-50A and 12-50B), discoloration of the phloem (Fig. 12-50D), and collapse of the whole plant (Fig. 12-50C). Yields may be reduced slightly or they may be destroyed completely.

FIGURE 12-51 Citrus greening disease caused by *Candidatus liberobacter asiaticum*. (A) Citrus tree affected by yellow shoot and citrus greening. (B) Leaves of greening-infected orange and lemon trees showing progressive symptoms of the disease. (C) Oranges showing delayed and abnormal coloration due to citrus greening. (C) Citrus psylla, one of the important vectors of citrus greening. [Photographs courtesy of (A, C, and D) T. R. Gottwald and S. M. Garnsey, USDA, Ft. Pierce, FL, and (B) S. P. van Vuuren, ARC-ITSC, Nelspruit, South Africa.]

The pathogen of yellow vine is a fastidious phloem-inhabiting bacterium tentatively identified as *Serratia marcescens* (Fig. 12-50E). It is spread from plant to plant by the squash bug (*Anasa tristis*).

Control measures are looking mostly toward resistance to disease in the various crops.

CITRUS GREENING DISEASE

Citrus greening or Huanglongbing is one of the most severe diseases of citrus. It has reduced yields in all types of citrus wherever it occurs in Asia, from China and the Philippines to the Arabian peninsula, and to Africa. It is one of the diseases the rest of the citrus-producing countries are guarding against and bracing for its eventual spread to them. Symptoms of citrus greening consist of smaller leaves, yellowing of the leaves of part or, usually, the entire canopy of the trees, reduced foliage, and severe dieback of twigs (Figs. 12-51A and 12-51B). The most characteristic symptoms, however, are that infected trees produce fruit that is lopsided, fails to ripen, and instead remains green (Fig. 12-51C) and imparts an unpleasant flavor to juice produced from such fruit.

The cause of citrus greening is the fastidious phloem-limited bacterium *Candidatus liberobacter asiaticus* in Asia and *C. liberobacter africanus* in Africa. The African strain does not require as high a temperature for optimum expression as the Asiatic strain. Neither of them can be cultured on artificial media.

The pathogen is spread by vegetative propagation and by two psyllid insects. The Asian strain is spread primarily by *Diaphorina citri* (Fig. 12-51D), whereas the primary vector for the African strain is *Trioza erytreae*, but both insect vectors can transmit either strain of the bacterium.

Control of citrus greening depends on exclusion of the pathogen from a citrus-producing area, use of disease-free propagating material, removal of infected trees as soon as they are detected, and attempts to control the insect vectors with insecticides or by biological control.

PAPAYA BUNCHY TOP DISEASE

Papaya bunchy top disease occurs wherever papaya is grown in the American tropics. In this area, bunchy top is one of the most economically important diseases of papaya, as infected plants produce few flowers and set few or no fruit.

Symptoms of papaya bunchy top consist of chlorosis and narrowing of leaves and elongation of petioles and internodes. Later spots and blotches appear on the petioles and stems. Petioles become rigid and extend out from the stem more horizontally than healthy ones. Leaf blades become thickened and stiff and may become chlorotic and necrotic and cup downward. Infected plants rarely produce flowers and set fruit. As the disease progresses, infected plants drop most of their leaves except for a tuft of small leaves that remain at the top of the plants (Fig. 12-52A). Unlike what happens in healthy papayas, fresh wounds in infected plants fail to exude latex. Some papaya varieties show considerable dieback when infected.

The pathogen of papaya bunchy top disease is a small, rod-shaped, gram-negative, laticifer-inhabiting

FIGURE 12-52 (A) Papaya plants showing severe bunchy top symptoms. (B) The rickettsia-like phloem-inhabiting bacterium causing the papaya bunchy top disease. [Photographs courtesy of M. J. Davis, University of Florida.]

bacterium that seems to be a member of the genus *Rickettsia*.

The papaya bunchy top pathogen is transmitted by two leafhoppers, *Empoasca papayae* and *E. stevensi*. The pathogen has been observed with the electron microscope in both infected plants and vectors transmitting the disease but not in healthy plants or in vectors that do not transmit the disease.

The control of papaya bunchy top depends on the use of resistant varieties, use of pathogen-free propagating material, and early removal of infected trees.

Selected References

Almeida, R. P. P., and Pereira, E. F. (2001). Multiplication and movement of a citrus strain of *Xylella fastidiosa* within sweet orange. *Plant Dis.* **85**, 382–386.

Avila, F. J., Bruton, B. D., Fletcher, J., *et al.* (1998). Polymerase chain reaction detection and phylogenetic characterization of an agent associated with yellow vine disease of cucurbits. *Phytopathology* **88**, 428–436.

Bextine, B., Wayadande, A., Bruton, B. D., *et al.* (2001). Effect of insect exclusion on the incidence of yellow vine disease and of the associated bacterium in squash. *Plant Dis.* **85**, 875–878.

Bruton, B., Fletcher, J., Shaw, M., *et al.* (1998). Association of a phloem-limited bacterium with yellow vine disease in cucurbits. *Plant Dis.* **82**, 512–520.

Davis, M. J., *et al.* (1980). Ratoon stunting disease of sugarcane: Isolation of the causal bacterium. *Science* **240**, 1365.

Davis, M. J., *et al.* (1996). Association of a bacterium and not a phytoplasma with papaya bunchy top disease. *Phytopathology* **86**, 102–109.

Gillaspie, A. G., Jr., and Davis, M. J. (1992). Ratoon stunting of sugarcane. *In* "Plant Diseases of International Importance: Diseases of Sugar, Forest, and Plantation Crops" (A. N. Mukhopadhyay, J. Kumar, H. S. Chaube, and U. S. Singh, eds.), Vol. 4. Prentice-Hall, Englewood Cliffs, NJ.

Goheen, A. C., Nyland, G., and Lowe, S. K. (1973). Association of a rickettsia-like organism with Pierce's disease of grapevines and alfalfa dwarf and heat therapy of the disease in grapevines. *Phytopathology* **63**, 341–345.

Hopkins, D. L. (1989). *Xylella fastidiosa*: Xylem limited bacterial pathogens of plants. *Annu. Rev. Phytopathol.* **27**, 271–290.

Hopkins, D. L., and Mollenhauer, H. H. (1973). Rickettsia-like bacterium associated with Pierce's disease of grapes. *Science* **179**, 298–300.

Hopkins, D. L., and Purcell, A. H. (2002). *Xylella fastidiosa*: Cause of Pierce's disease of grapevine and other emergent diseases. *Plant Dis.* **86**, 1056–1066.

Hoy, J. W., Grisham, M. P., and Damann, K. E. (1999). Spread and increase of ratoon stunting disease of sugarcane and comparison of disease detection methods. *Plant Dis.* **83**, 1170–1175.

Hung, T. H., Wu, M. L., and Su, H. J. (2000). Identification of alternative hosts of the fastidious bacterium causing citrus greening disease. *J. Phytopathol.* **148**, 321–326.

Hurtung, J. S., *et al.* (1994). Citrus variegated chlorosis bacterium: Axenic culture, pathogenicity, and serological relationships with other strains of *Xylella fastidiosa*. *Phytopathology* **84**, 591–597.

Lee, R. F., *et al.* (1991). Citrus variegated chlorosis: A new destructive disease of citrus in Brazil. *Citrus Ind.* **72**, 12, 13, and 15.

Leu, L. S., and Su, C. C. (1993). Isolation, cultivation and pathogenicity of *Xylella fastidiosa*, the causal bacterium of pear leaf scorch disease in Taiwan. *Plant Dis.* **77**, 642–646.

Li, W.-B., Pria, W. D., Jr., Teixeira, D. C., *et al.* (2001). Coffee leaf scorch caused by a strain of *Xylella fastidiosa* from citrus. *Plant Dis.* **85**, 501–505.

Li, W.-B., Zhou, C.-H., Pria, W. D., Jr., *et al.* (2002). Citrus and coffee strains of *Xylella fastidiosa* induce Pierce's disease in grapevine. *Plant Dis.* **86**, 1206–1210.

Nyland, G., *et al.* (1973). The ultrastructure of a rickettsialike organism from a peach tree affected with phony disease. *Phytopathology* **63**, 1275–1278.

Pierce, N. B. (1892). The California vine disease. *USDA Div. Veg. Pathol. Bull.* **2**, 1–222.

Purcell, A. H. (1982). Insect vector relationships with procaryotic plant pathogens. *Annu. Rev. Phytopathol.* **20**, 397–417.

Qin, X., Miranda, V. S., Machado, M. A., *et al.* (2001). An evaluation of the genetic diversity of *Xylella fastidiosa* isolated from diseased citrus and coffee in São Paulo, Brazil. *Phytopathology* **91**, 599–605.

Raju, B. C., and Wells, J. M. (1986). Diseases caused by fastidious xylem-limited bacteria and strategies for management. *Plant Dis.* **70**, 182–186.

Schaad, N. W., Opgenorth, D., and Gaush, P. (2002). Real-time polymerase chain reaction for one-hour on-site diagnosis of Pierce's disease of grape in early season asymptomatic vines. *Phytopathology* **92**, 721–728.

Teakle, D. S., Smith, P. M., and Steindl, D. R. L. (1973). Association of small coryneform bacterium with the ratoon stunting disease of sugarcane. *Aust. J. Agric.* **24**, 869–874.

Wells, J. M., Raju, B. C., and Nyland, G. (1983). Isolation, culture, and pathogenicity of the bacterium causing phony disease of peach. *Phytopathology* **73**, 859–862.

PLANT DISEASES CAUSED BY MOLLICUTES: PHYTOPLASMAS AND SPIROPLASMAS

In 1967, wall-less microorganisms were seen with the electron microscope in the phloem of plants infected with one of several yellows-type diseases and in insect vectors of these diseases. Such diseases, up to that moment, were thought to be caused by viruses. The new microorganisms were subsequently called mycoplasma-like organisms because of their superficial resemblance to mycoplasmas. It was later shown that these organisms are not mycoplasmas. Although all are mollicutes, i.e., prokaryotic cells without cross walls, a few of them have a helical structure and are called **spiroplasmas**. Most, however, are round to elongate but are not spiral and are now called **phytoplasmas**.

More than 200 distinct plant diseases affecting numerous types of plants have been determined to be caused by phytoplasmas. Among the diseases caused by phytoplasmas are some very destructive diseases of trees and vines, e.g., pear decline, grape yellows, coconut lethal yellowing, X disease of peach, and apple proliferation, but also diseases of herbaceous annual and

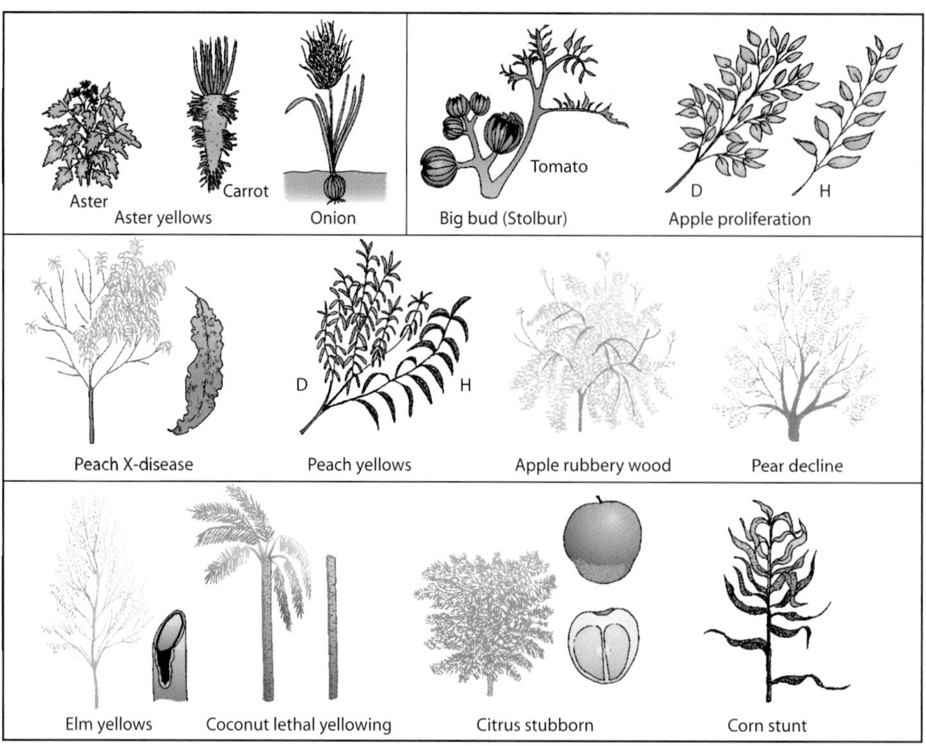

FIGURE 12-53 Symptoms caused by mollicutes. D, diseased plant; H, healthy plant.

perennial plants such as aster yellows of vegetables and ornamentals and stolbur of tomato. So far, only a few diseases, such as citrus stubborn and corn stunt, are known to be caused by spiroplasmas. The main characteristics of yellows-type diseases are a more or less gradual, uniform yellowing or reddening of the leaves, smaller leaves, shortening of the internodes and stunting of the plant, excessive proliferation of shoots and formation of witches'-brooms, greening or sterility of flowers, reduced yields, and, finally, a more or less rapid dieback, decline, and death of the plant (Fig. 12-53). Root abnormalities and necrosis often precede the aboveground symptoms.

The true nature of phytoplasmas and their taxonomic position among the lower organisms is still uncertain. Morphologically, the organisms observed in plants resemble the typical mycoplasmas found in animals and humans and those living saprophytically, but their genomes are only distantly related to true mycoplasmas. Phytoplasmas cannot be grown on artificial nutrient media and, so far, no plant disease has been reproduced on healthy plants inoculated directly with phytoplasmas obtained from diseased plants. The spiroplasmas causing citrus stubborn and corn stunt, however, have been grown on artificial nutrient media and have been shown to reproduce the disease in plants when inoculated by insects injected with the organism from culture. At present, the phytoplasmas are considered to belong to the class Mollicutes, which include the true mycoplasmas, but no family or genus has been designated for them. The citrus stubborn and the corn stunt organisms are placed in the newly created genus of mollicutes called *Spiroplasma*.

Properties of True Mycoplasmas

Mycoplasmas are prokaryotic organisms that have no cell walls. They are members of the class Mollicutes, which has one order, Mycoplasmatales. The order has three families, each with one genus: Mycoplasmataceae, genus *Mycoplasma*, Acholeplasmataceae, genus *Acholeplasma*, and Spiroplasmataceae, genus *Spiroplasma*.

As they lack a true cell wall, mycoplasmas are bounded only by a "unit" membrane. They are small, sometimes ultramicroscopic cells containing cytoplasm, randomly distributed ribosomes, and strands of nuclear material. They measure from 175 to 250 nanometers in diameter during reproduction but grow into various sizes and shapes. Shapes range from spherical or slightly ovoid to filamentous. Sometimes they produce branched mycelioid structures. The size of fully developed spherical mycoplasmas may vary from one to a few micrometers, whereas slender branched filamentous forms may range in length from a few micrometers to 150 micrometers. Mycoplasmas reproduce by budding and by binary transverse fission of cells. Mycoplasmas have no flagella, produce no spores, and are gram negative. Nearly all mycoplasmas parasitic to humans and animals and all saprophytic ones can be grown on more or less complex artificial nutrient media in which they produce minute colonies that usually have a characteristic "fried-egg" appearance. Mycoplasmas have been isolated mostly from healthy and/or diseased animals and humans suffering from diseases of the respiratory and urogenital tracts; they have been associated with some arthritic and nervous disorders of animals; and some have been found to exist as saprophytes. Most mycoplasmas are completely resistant to penicillin; however, they are sensitive to tetracycline and chloramphenicol, and some are sensitive to erythromycin and to certain other antibiotics.

Phytoplasmas

The organisms observed in plants and insect vectors, i.e., the phytoplasmas, which do not include the spiroplasmas, resemble mycoplasmas of the genera *Mycoplasma* or *Acholeplasma* in all morphological aspects. Genetically, phytoplasmas are more related to *Acholeplasma* than to *Mycoplasma*. They lack cell walls, are bounded by a "unit" membrane, and have cytoplasm, ribosomes, and strands of nuclear material. The size of their chromosomes varies from 530 kilobases of DNA to 1130 kilobases. Their shape is usually spheroidal to ovoid or irregularly tubular to filamentous, and their sizes are comparable to those of the typical mycoplasmas (Fig. 12-54).

Phytoplasmas (and spiroplasmas) are generally present in the sap of a small number of phloem sieve tubes (Fig. 12-54). The concentration of phytoplasmas in their host plants seems to vary a great deal. For example, 370 to 34,000 phytoplasma cells were found per gram plant tissue of resistant proliferation-affected apple trees and in some other trees, whereas from 220 million to 1.5 billion phytoplasma cells per gram of plant tissue were found in periwinkle plants infected with various phytoplasmas. Most plant mollicutes are transmitted from plant to plant by leafhoppers (Fig. 12-55), but some are transmitted by psyllids and plant hoppers (see Fig. 14-18). Plant mollicutes also grow in the alimentary canal, hemolymph, salivary glands, and intracellularly in various body organs of their insect vectors.

Insect vectors can acquire the pathogen after feeding on infected plants for several hours or days, or if they are injected with extracts from infected plants or vectors. More insects become vectors when feeding on young leaves and stems of infected plants than on older ones. The vector cannot transmit the mollicutes immediately after feeding on the infected plant, but it begins to transmit them after an incubation period of 10 to 45 days, depending on the temperature; the shortest incubation period occurs at about 30°C, the longest at about 10°C.

The incubation period is required for the multiplication and distribution of the mollicute within the insect (Fig. 12-55). If the mollicute is acquired from the plant, it multiplies first in the intestinal cells of the vector; it then passes into the hemolymph and infects internal organs and, eventually, the brain and the salivary glands. When the concentration of the mollicute in the salivary glands reaches a certain level, the insect begins to trans-

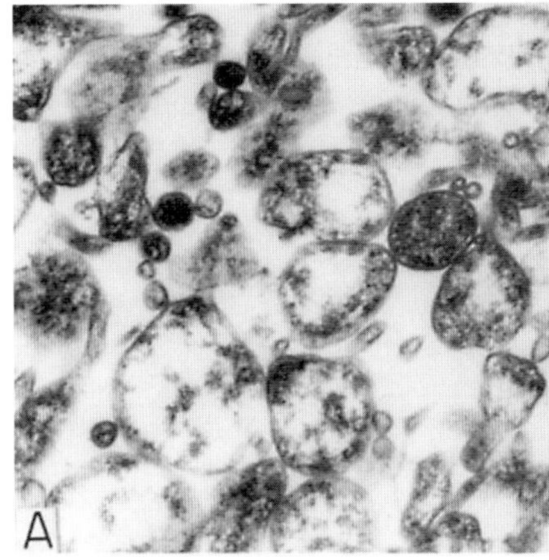

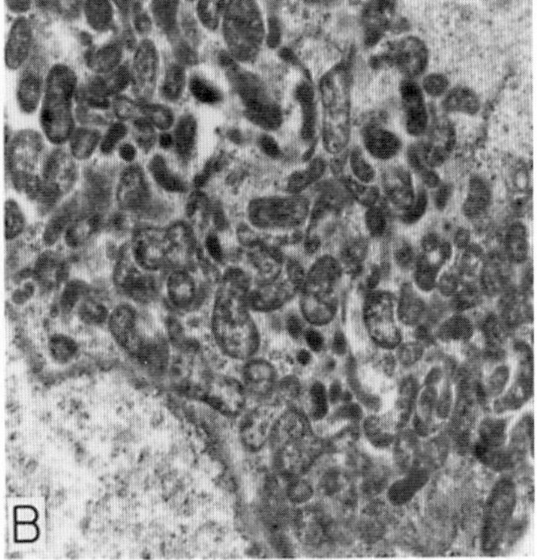

FIGURE 12-54 Aster yellows phytoplasma. (A) Older, larger phytoplasmas of aster yellows. (B) Young, active phytoplasmas. [Photographs courtesy of J. F. Worley, USDA.]

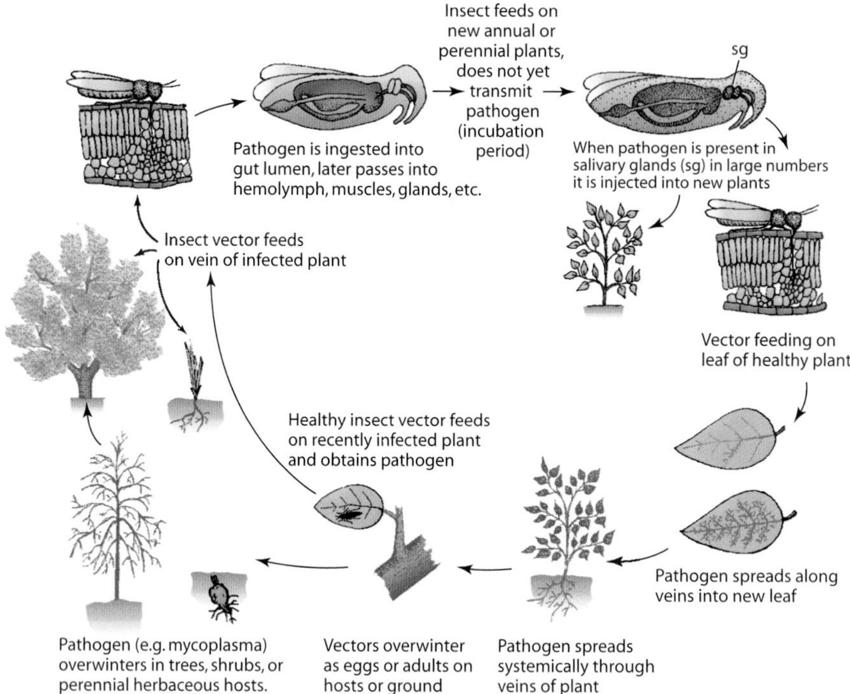

FIGURE 12-55 Sequence of events in the overwintering, acquisition, and transmission of fastidious bacteria, mollicutes, and viruses by leafhoppers and related insect vectors.

mit the pathogen to new plants and continues to do so more or less efficiently for the rest of its life. Insect vectors usually are not affected adversely by the mollicutes, but in some cases they show severe pathological effects. Mollicutes can be acquired as readily or better by nymphs than by adult leafhoppers and survive through subsequent molts, but they are not passed from the adults to the eggs and to the next generation, which, therefore, must feed on infected plants in order to become infective vectors.

Despite countless attempts by numerous investigators to culture phytoplasmas on artificial nutrient media, including the media on which all typical mycoplasmas grow, the culture of phytoplasmas has not yet been possible. Phytoplasmas, however, have been extracted from their host plants and from their vectors in more or less pure form, and for most of them, antisera, including monoclonal antibodies, have been prepared. Specific antibodies, DNA probes, RFLP profiles (Fig. 12-56), and analysis of 16 S rRNA genes with the help of the PCR amplification technique have become extremely useful in the detection and identification of the pathogen in suspected hosts and in grouping and classifying the pathogens. So far, with these techniques at least 62 phytoplasmas have been distinguished and identified and have been placed into about 14 groups. No phytoplasmas have been given accepted Latin binomials, but for a few of them tentative names (candidatus) have been proposed, e.g., Candidatus *Phytoplasma fraxinii* for ash yellows and Candidatus *Phytoplasma australasia* for Australian tomato big bud. These methods of detection and identification are also helpful in controlling these diseases through the production of pathogen-free propagating stock. Serological and nucleic acid techniques are currently replacing other methods used to detect mollicute infections. Earlier methods included indexing to sensitive hosts, fluorescent staining with either the DNA-specific stain 4,6-diamidino-2-phenylindole (DAPI) or the callose-specific stain aniline blue, or staining with the so-called Dienes' stain.

Mollicutes are sensitive to antibiotics, particularly those of the tetracycline group. When infected plants are immersed in or injected with tetracycline solutions, the symptoms, if already present, recede or disappear or, if not yet present, are delayed. Foliar and soil application are ineffective. The symptoms reappear, however, soon after treatment stops. Generally, the treatment of plants during the early phases of the disease is much more effective than treatment of plants in advanced stages of the disease. Infected plants or dormant propagative organs can be totally freed of mollicutes by heat treatment. Infected plants are kept in growth chambers at 30 to 37°C for several days, weeks, or months; dormant organs are immersed in hot water at 30 to 50°C for as

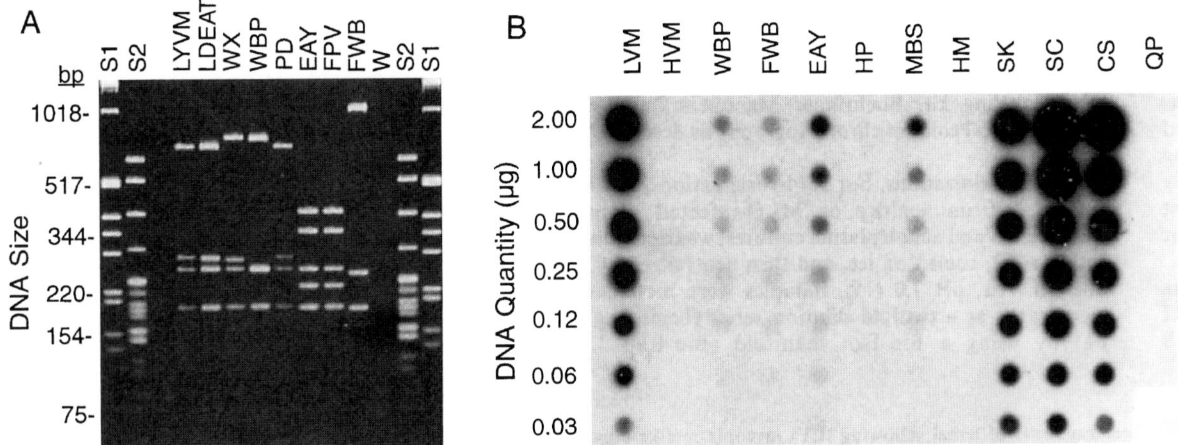

FIGURE 12-56 Detection and identification of mollicutes in plants by comparison of fragment profiles (columns) of their ribosomal DNA (A) and total DNA (B). To produce profile A, DNA was cut with a particular nuclease enzyme and the fragment profile of each isolate was compared to that of known and unknown mollicutes. To produce profile B, drops of DNA dilutions were placed on a film and reacted with an appropriate radioactive probe. The size and intensity of the dots reveal the relatedness of the tested DNA to the known probe. HVM, HP, HM, and QP are from healthy plants. [Photographs courtesy of N. Harrison, University of Florida.]

short as 10 minutes at the higher temperatures and as long as 72 hours at the lower temperatures.

Spiroplasmas

Spiroplasmas are helical mollicutes. So far, spiroplasmas are known to cause the stubborn disease in citrus plants and the brittle root disease in horseradish (*Spiroplasma citri*), stunt disease in corn plants, and a disease in periwinkle. *Spiroplasma citri* has also been found in many other dicots, such as crucifers, lettuce, and peach, and both *S. citri* and the corn stunt spiroplasma also infect their respective leafhopper vectors. Furthermore, several kinds of spiroplasmas have been shown to infect honeybees and several other insects, and several more live saprophytically on flowers and other plant surfaces and, possibly, internally in plants.

Spiroplasmas are cells that vary in shape from spherical or slightly ovoid, 100 to 240 nanometers or larger in diameter, to helical and branched nonhelical filaments. The latter are about 120 nanometers in diameter and 2 to 4 micrometers long during active growth and considerably longer (up to 15 micrometers) in later stages of growth. Unlike phytoplasmas, spiroplasms can be obtained from their host plants or their insect vectors and cultured on nutrient media (Figs. 12-57). They produce mostly helical forms in liquid media. They multiply by fission. They lack a true cell wall and are bounded by a unit membrane. The helical filaments are motile, moving by a slow undulation of the filament and probably by a rapid rotary or screw motion of the helix. There are no flagella. Colonies of spiroplasmas on agar have a diameter of about 0.2 millimeters; some have a typical fried-egg appearance, but others are granular (Fig. 12-57C). Spiroplasmas are resistant to penicillin but are inhibited by tetracycline.

Cultured plant spiroplasmas can be injected into or fed to their insect vectors, which then, on feeding on the host plants, transmit the organisms to the plants. The experimentally infected hosts develop typical symptoms of the disease.

EXAMPLES OF PLANT DISEASES CAUSED BY MOLLICUTES

Important plant diseases caused by phytoplasmas are aster yellows, apple proliferation, European stone fruit yellows, coconut lethal yellowing, elm yellows, ash yellows, grapevine yellows, peach X disease, pear decline, and many more. In addition, two diseases caused by spiroplasmas, citrus stubborn and corn stunt, are also of economic importance.

ASTER YELLOWS

Aster yellows causes general yellowing (chlorosis) and dwarfing of the plant, abnormal production of shoots, sterility of flowers, malformation of organs, and a general reduction in the quantity and quality of yield (Fig. 12-58). Losses from aster yellows vary among host crops, being greatest in carrot, in which 10 to 25% losses are rather common and occasional losses reach 80

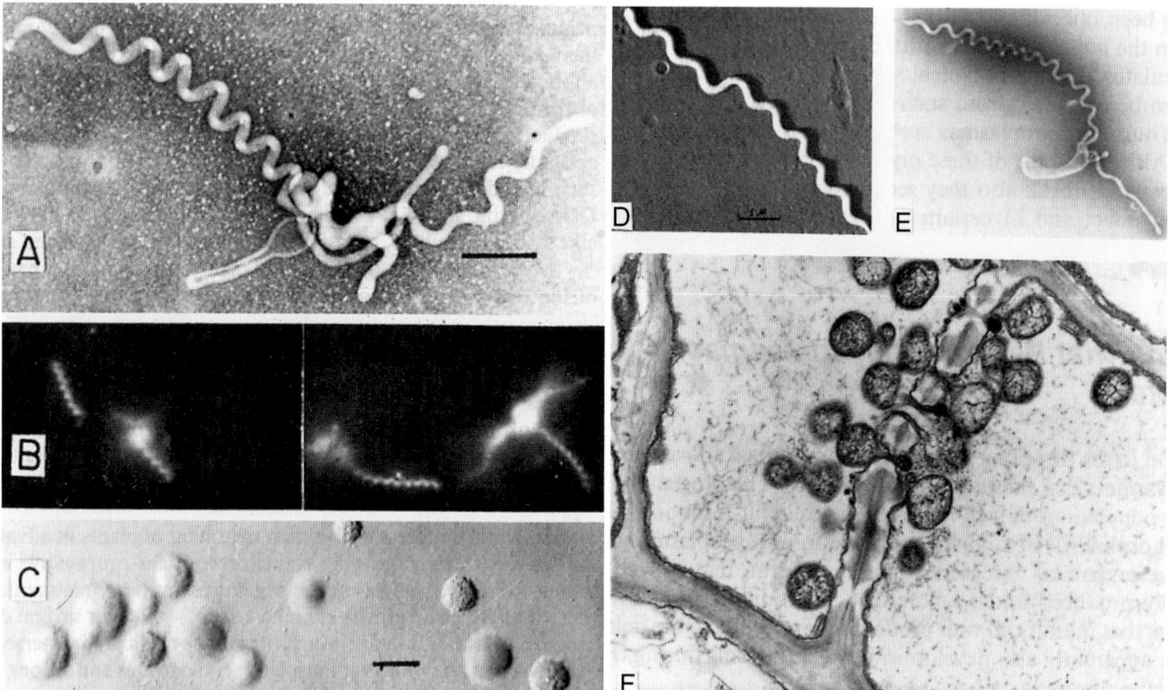

FIGURE 12-57 (Left) Corn stunt spiroplasma isolated from infected corn plants and grown on nutrient media. (Right): *Splropldsma citvi.*) (A) Typical helical morphology of spiroplasma. (B) Active spiroplasmas from liquid culture observed by dark-field microscopy. (C) Colonies of corn stunt spiroplasma on agar plates 14 days after inoculation (scale bar: 50 μm). (D) Replicative form of *Spiroplasma citri* isolated from stubborn-infected citrus. (E) *Spiroplasma citri* obtained from its leafhopper vector *Circulifer tenellus* and grown in broth culture. Note presence of bleb. (F) *Spiroplasma citri* in sieve plate in midvein of a sweet orange leaf. [[Photographs A,B,C, courtesy of T. A. Chen, Rutgers University. D,E,F, of E. C. Calavan, University of California.]

to 90% of the crop. Infected carrots, in addition to being smaller, also have an unpleasant flavor.

Although the general effects of aster yellows on host plants are similar, some hosts also produce characteristic symptoms. On carrot, symptoms appear first as a vein clearing and yellowing of the younger leaves. Infected plants then produce many adventitious shoots, and the tops look like a witches'-broom (Figs. 12-58C–12-58E). The internodes of such shoots are short, as are the leaf petioles. The young leaves are smaller and often become dry. The petioles of older leaves become twisted and break off. Later in the season, the remaining older leaves usually become bronzed and reddened. The floral parts of infected plants are deformed. Plants infected when young may die, whereas plants infected later become unsightly, have lower market value, and are difficult or impossible to harvest mechanically. The roots are predisposed to soft rots in the field and in storage.

Infected carrot roots are small, tapered, abnormally shaped, and have woolly secondary roots on which the soil clings tenaciously when the plant is pulled from the ground. In section, the xylem or core of infected carrots appears enlarged, whereas the cortex zone is much narrower than in healthy carrots. Infected carrots have an unpleasant flavor, the degree of which is proportional to the severity of the disease. In processed carrots (canned or frozen purees), the presence of even 15% of yellows-infected carrots imparts an objectionable off flavor to the entire processed product.

The aster yellows pathogen exists in several strains and strain clusters. Aster yellows is transmitted by budding or grafting and by several leafhoppers.

The phytoplasma survives in perennial ornamental, vegetable, and weed plants. A few such weeds are thistle, wild carrot, dandelion, field daisy, black-eyed Susan, and wide-leafed plantain.

The vector leafhopper acquires the phytoplasma while feeding by inserting its stylet into the phloem of infected plants and withdrawing the phytoplasma with the plant sap. After an incubation period, when the insect feeds on healthy plants it injects the phytoplasma through the stylet into the phloem of the healthy plants, where it establishes infection and multiplies. The phytoplasma moves out of the leaf and into the rest of the plant occasionally within 8 hours but

FIGURE 12-58 Aster yellows symptoms on various host plants. (A) Yellowish red foliage on potato plant. (B) Yellowing and stunting of strawberry plant. (C–E) Aster yellows-infected carrots produce smaller root and proliferation of stems (A and B), while in the field they stand out by their yellowish-red color, stunted growth, and witches'-broom appearance of their stems and leaves. [Photographs courtesy of (A) P. Koepsell, Oregon State University, (B and D) Plant Pathology Department, University of Florida, and (C and E) R. J. Howard, W.C.P.D.]

generally within 24 hours after inoculation. Infected plants usually show symptoms after 8–9 days at 25°C and 18 days at 20°C, whereas no symptoms develop at 10°C.

Aster yellows phytoplasma is limited primarily to the phloem of infected plants. Some cells adjacent to the phloem first enlarge and then die. Surviving cells begin to divide, but these too soon die. Cells surrounding the necrotic areas then begin to divide and enlarge excessively, producing abnormal sieve elements, while the phloem elements within the necrotic areas degenerate and collapse.

Several measures help reduce losses from aster yellows, although none of them will control the disease completely. Eradication of perennial weed hosts from the field and planting susceptible crops away from crops harboring the pathogen help eliminate a large source of phytoplasma inoculum. Control of the leafhopper vector in the crop and on nearby weeds with insecticides as early in the season as possible helps reduce transmission of the phytoplasma to the crop plants. Certain varieties of plants are more resistant to the disease than others, but none is immune; during severe outbreaks of the disease, they too suffer serious loses.

LETHAL YELLOWING OF COCONUT PALMS

Lethal yellowing appears as a blight that kills palm trees within 3 to 6 months after the first appearance of symptoms. The disease is present in Florida, Texas, Mexico, most Caribbean islands, in west Africa, and elsewhere. The disease was first identified in Key West in 1955 and in the next five years killed about three-fourths of the coconut palms in Key West. Lethal yellowing appeared in the Miami area of the Florida mainland in the fall of 1971, and it had killed an estimated 15,000 trees by October 1973 and 40,000 coconut palms by August 1974. By August 1975, 75% of the coconut palms in Dade County (Miami area) were reported to have been killed by, or be dying of, the lethal yellowing disease. In addition to coconut palm (*Cocos nucifera*), the disease apparently affects several other kinds of palms. All the diseased palms appear to be infected with phytoplasmas and decline and die with symptoms similar to lethal yellowing.

The first symptoms of lethal yellowing are the premature drop of coconuts of any size. Then, the next inflorescence that appears has blackened tips, almost all its male flowers are dead and black, and it sets no fruit. Soon the lower leaves turn yellow, and the yellowing progresses upward from the older to the younger leaves (Figs. 12-59A–12-59C). The older leaves then die prematurely, turn brown, and cling to the tree while the younger leaves are turning yellow (Fig. 15-59A). Before long, all the leaves die, as does the vegetative bud. Finally, the entire top of the palm falls away and leaves nothing but the tall trunk of the palm tree, which by now looks like a telephone pole (Figs. 12-59B–12-59D).

The pathogen is a phytoplasma morphologically similar to all other such organisms observed in plants. The pathogen occurs mainly in young phloem cells (Fig. 12-60). Although the disease is obviously spreading rapidly in nature, the vector is not known with certainty. The planthopper *Myndus crudus* has been implicated as one of the vectors.

Control of lethal yellowing depends on the use of genetically resistant coconut varieties and hybrids. Malayan dwarf varieties, and certain other cultivars appear to be resistant or immune to lethal yellowing, and thousands of such trees, and hybrids of Malayan dwarf with susceptible palms, are now planted to replace the other coconut palms wherever lethal yellowing exists. Sanitation measures, i.e., removal and burning of diseased palms as soon as symptoms appear, and insecticidal sprays to reduce vector populations have not reduced the spread of lethal yellowing. Control of lethal yellowing by injecting infected trees with solutions of tetracycline antibiotics is effective and economically feasible in landscape plantings, but it is too expensive for coconut-producing regions.

APPLE PROLIFERATION

Apple proliferation occurs in Europe and can cause serious losses of fruit and trees. The phytoplasma causing apple proliferation seems to be related to some of the other fruit tree phytoplasmas but it does not go to other fruit tree species.

Symptoms of apple proliferation include witches'-brooms (Figs. 12-61A and 12-61B), i.e., rosettes of leaves developing on the terminal parts of shoots as a result of growth of dormant buds in the summer; the rosettes consist of smaller leaves but longer stipules growing at a very narrow angle. Infected trees produce fewer flowers and a portion of them show phyllody, i.e., development of green leaflets in place of white petals. As a result, fruit set is often reduced at least by half and sometimes there is no fruit produced at all. Any fruit present are considerably smaller than fruit of healthy trees (Fig. 12-61C), are incompletely colored, and have poor flavor. The roots of infected trees also develop abnormally, producing abundant roots but thin ones and forming compact, felt-like masses that seem to contribute to the overall stunted growth of the tree. All these symptoms, however, vary from year to year, since due to redistribution of the phytoplasma each winter, symptoms do not occur over the entire tree and

FIGURE 12-59 Lethal yellowing of coconut palms. Symptoms begin at the lower leaves, which turn yellow (A) and later fall off while younger leaves turn yellow (B). Eventually all the leaves are killed, fall, and are followed by death of the tree bud (B and C), leaving the dead trees standing like utility poles (D). [Photographs courtesy of Plant Pathology Department, University of Florida.]

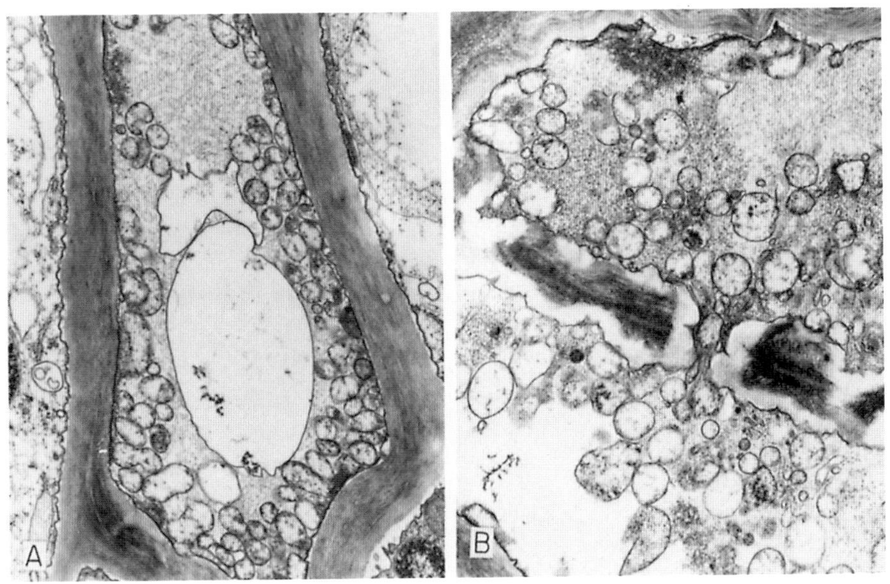

FIGURE 12-60 Lethal yellowing phytoplasmas in sieve element of infected young coconut palm inflorescence (A) and passing through a sieve-plate pore lined with callose (B). [Photographs courtesy of M. V. Parthasarathy.]

FIGURE 12-61 Apple proliferation symptoms on young twig (A), mature apple tree (B), and on reduced fruit size (C, right). (D) European stone fruit yellows symptoms on apricot, followed by death of the tree (E) within a short time. [Photographs courtesy of (A) E. Seemuller, Heidelberg, Germany, and (C–E) L. Giunchedi, University of Bologna, Italy.]

do not repeat themselves on the same branches every year.

Apple proliferation phytoplasma is spread by vegetative propagation and, presumably, by several leafhoppers. Therefore, the primary control is through the use of pathogen-free and, preferably, proliferation-resistant propagating rootstocks and scions. Insecticidal sprays during the entire growth season seem to be helpful.

EUROPEAN STONE FRUIT YELLOWS

European stone fruit yellows is the common name recently proposed for genetically related phytoplasma-caused diseases in European stone fruits. It includes the causal agents of "apricot chlorotic leaf roll" and "plum leptonecrosis." Symptoms and losses vary with the crop plant and with the strain of the pathogen. It causes serious disease on apricot, peach, and Japanese plum. European plum seems to be a symptomless carrier while cherries seem to be resistant.

Young trees are infected systemically and quickly and are killed within a year or two. In trees older than five years, the symptoms at first are localized in lower branches but then they spread to the rest of the crown and entire trees are killed rather quickly. Symptoms consist of premature opening of buds and leafing in late winter or early spring. Later on, leaf blades roll upward and turn pale green and chlorotic (Figs. 12-61D and 12-61E). Leaves remain on the tree later than usual and new buds continue to open even at freezing temperatures. Fruits produced on affected trees are smaller, may be bumpy, and drop prematurely. Fruit flesh is brown and spongy near the pit. The bark of affected trees may develop necrotic areas, which in transverse sections may appear as thin orange lines or thicker brown bands.

The pathogen of European stone fruit yellows is a phytoplasma. It is not yet certain whether the phytoplasma is spread by an insect vector(s), although the psyllid *Cacopsylla pruni* has been implicated as the vector in France. It is certainly spread by budding and grafting of scion wood onto rootstocks. Control of the disease is through the use of pathogen-free propagating materials, possibly by removal of infected trees, and through use of resistant varieties and rootstocks.

ASH YELLOWS

Ash yellows causes a significant reduction of growth of affected white ash trees, which subsequently decline and die. Green ash is also affected, but, in most areas, not as severely. Ash trees of all ages and sizes are susceptible to infection by ash yellows. Symptoms consist of reduced radial and shoot growth followed by branch dieback, sparse chlorotic foliage (Fig. 12-62A), development of sprouts chlorotic, witches broms-like on the trunk (Fig. 12-62B) and branches, cracks in the bark, early color change in the fall, and premature death of trees. The most diagnostic symptom is the appearance of witches'-brooms on the trunk and branches of infected trees, but their formation is rather inconsistent. Ash yellows is, of course, caused by a phytoplasma tentatively named *Candidatus phytoplasma fraxinii*. It is not known how the phytoplasma enters the tree and how it spreads from tree to tree, but one or more insect vectors are suspected. No control against ash yellows is attempted, especially in the forest.

ELM YELLOWS (PHLOEM NECROSIS)

Elm yellows occurs in about 20 central, eastern, and southern U.S. states. Elm yellows epidemics have killed thousands of trees in each of numerous communities.

Symptoms consist of a general decline of the tree in which the leaves droop and curl, turn bright yellow, then brown, and finally fall. Some trees are killed within a few weeks, and most trees that show symptoms in June or July die in a single growing season. Trees infected late may live through the winter, but then in the spring they produce a thin crop of small leaves and die soon after. In later stages of the disease the inner layers of peeled bark (phloem) at the base of the stem turn yellowish-brown (Fig. 12-62C) and have a faint odor of wintergreen. The latter characteristics are often used for a quick diagnosis of the disease. Discoloration of the phloem is the result of deposition of callose within the sieve tubes and then a collapse of sieve elements and companion cells. The cambium produces replacement phloem, but its cells become quickly necrotic also.

The pathogen is a phytoplasma present in the phloem of infected trees. It is transmitted from diseased to healthy trees by the leafhopper *Scaphoideus luteolus*.

Injection of tetracyclines into recently infected trees causes a remission of symptoms for several months and up to three years. Severely diseased or dead trees should be removed and burned.

PEACH X DISEASE

The X disease, including western X disease, occurs in the northwestern and northeastern parts of the United States, in Michigan and several other states, and in the adjacent parts of Canada. Where present, X disease is one of the most important diseases of peach. Affected trees become commercially worthless in 2 to 4 years (Fig. 12-62D). Young peach trees are rendered useless

FIGURE 12-62 (A) Declining ash yellows-affected tree with witches'-broom-like rosettes along its branches. (B) Close-up of rosette at the base of the trunk of ash yellows-affected ash tree. (C) Discolored phloem in trunk of elm yellows-affected elm tree. (D) Peach tree affected with X disease phytoplasma. (E) Close-up of foliar symptoms on X disease-affected peach. (F) Four healthy cherries and smaller, discolored cherries from a X disease-affected cherry tree. [Photographs courtesy of (A–C) USDA Forest Service, (D) S. Douglas, Connecticut Agricultural Experiment Station, (E) K. D. Hickey, Pennsylvania State University, and (F) Oregon State University.]

within one year of inoculation. The X disease of peach also attacks sweet and sour cherries, nectarines, and chokecherries.

The first symptoms of X disease of peach appear on the leaves of some or all branches as a slight mottle and reddish purple spots, which later die and fall out, giving a shot-hole appearance to the leaf (Fig. 12-62E). The leaves soon turn reddish and roll upward. Later, most leaves on affected branches drop, except the ones at the tips.

The fruits on affected branches usually shrivel and drop soon after the symptoms appear on the leaves. Any fruits remaining on the trees ripen prematurely, have an unpleasant taste, and are unsalable. No seeds develop in the pits of affected fruit. Infected cherries remain small, are discolored, and worthless (Fig. 12-62F). Fruits on healthy looking parts of infected trees show no signs of the disease.

The pathogen is a phytoplasma. It is transmitted by several species of leafhoppers of the genera *Colladonus* and *Scaphytopius* and, of course, by budding and grafting. Control of X disease on peach can be obtained using disease-free buds and rootstocks, by removing any X-diseased trees, and by eradicating chokecherry from the vicinity of peach orchards within about 200 meters from the orchard. Injections of tetracyclines into diseased trees result in the temporary remission of X-disease symptoms and in reduced transmission of the disease by leafhoppers that obtain the inoculum from treated trees. This control is not practiced, however, because of the costs involved, the injury caused to trees, and the possibility of antibiotic residue in the fruit.

PEAR DECLINE

Pear decline occurs in North America, in Europe, and probably in other continents. Similar decline-like disorders of pear have been observed in many countries, but their relationship to pear decline has not been established. Pear decline causes either a slow, progressive weakening and final death of trees or a quick, sudden wilting and death of trees. The disease can be extremely catastrophic. It killed more than 1 million trees in California between 1959 and 1962. Pear decline affects all pear varieties when they are grafted on rootstocks that are susceptible to the pathogen. Oriental rootstocks such as *Pyrus serotina* and *P. ussuriensis* are affected the most, but pear decline has also been observed on trees grafted on the more resistant or tolerant rootstocks *P. communis* and *P. betulaefolia*, and on quince.

Symptoms in the "slow decline" syndrome appear as a progressive weakening of the trees over many years. During this period there is little twig growth, and the leaves are few, small, pale green, and leathery and roll slightly upward. Such leaves often turn reddish in late summer and drop prematurely in the fall. Early in the disease the trees produce abundant blossoms, but as the disease progresses, the trees produce fewer blossoms, set fewer fruit, and the fruits are small. By this time, starch accumulates above the graft union but is almost absent below the union, and most of the feeder roots of the trees are dead. Eventually, despite occasional apparent improvement, the trees are killed by the disease.

In the "quick decline" syndrome the trees wilt suddenly and die within a few weeks (Fig. 12-63A). Quick decline is more common in trees grafted on the more susceptible oriental rootstocks, whereas trees grafted on other, more tolerant rootstocks usually develop the slow decline syndrome.

In slowly or quickly declining trees the current season's ring of phloem immediately below the graft union degenerates, and the degeneration becomes more pronounced as the season progresses (Fig. 12-63B). The replacement phloem produced at the graft union of diseased trees consists of narrow, small sieve tube elements rather than normal ones.

The pathogen is a phytoplasma. It can be transmitted by budding or grafting, although only about one-third of the buds seem to transmit the disease. The decline phytoplasma is also transmitted naturally by pear psylla (*Psylla pyricola*).

The most effective control of pear decline is obtained by growing disease-free pear varieties on resistant rootstocks such as *Pyrus communis* and by avoiding the highly sensitive oriental rootstocks. Control of the pear psylla vector has not been successful. Injection of a tetracycline solution in the trunk of infected trees soon after fruit harvest results in a temporary remission of symptoms. Antibiotic treatments must be repeated annually, however, or the disease will reappear.

SPIROPLASMA DISEASES

CITRUS STUBBORN DISEASE

Citrus stubborn is present in hot and dry areas such as most Mediterranean countries, the southwestern United States, Brazil, Australia, and possibly South Africa. In some Mediterranean countries and in California, stubborn is regarded as the greatest threat to the production of sweet oranges and grapefruit. Because of the slow development of symptoms and the long survival of affected trees, the spread of stubborn is insidious and its detection difficult. However, yields are reduced drastically; the trees produce fewer fruits and many of those are too small to be marketable.

FIGURE 12-63 (A) Young pear tree showing symptoms of pear decline caused by a phytoplasma. (B) Disruption of phloem at and below the graft union as a result of pear decline infection is responsible for decline symptoms.

Stubborn disease affects leaves, fruits, and stems of all commercial varieties regardless of the rootstock. Symptoms, however, vary a great deal, and frequently only a few are expressed at one time on an entire tree or parts of a tree. In general, affected trees show a bunchy, upright growth of twigs and branches, with short internodes and an excessive number of shoots (Fig. 12-64A). Some of the affected twigs die back. The trees show slight to severe stunting. The leaves are small, often mottled or chlorotic. Excessive winter defoliation is common. Affected trees bloom at all seasons, especially in the winter, but produce fewer fruits. Some of the fruit are very small and lopsided, frequently resembling acorns. Such fruit have abnormally thin rind from the fruit equator to the stylar end. The rind is often dense or cheesy. Some fruit show greening of the stylar end (Fig. 12-64B). Affected fruit tends to drop prematurely. Fruit are usually sour or bitter and have an unpleasant odor and flavor. Also, fruit from affected trees or parts of trees tend to have poorly developed and aborted seeds.

The pathogen is *Spiroplasma citri* (Fig. 12-57). It is found in the phloem. It can be cultured readily on artificial media. *Spiroplasma citri* has also been found in or transmitted to plants of many dicotyledonous families and some monocots, including most crucifers and several stone fruits, such as peach and cherry. Some infected hosts, such as pea and bean, become wilted and die, whereas most others remain symptomless.

Citrus stubborn disease is transmitted with moderate frequency by budding and grafting. It is spread naturally in citrus orchards by several leafhoppers, such as *Circulifer tenellus*, *Scaphytopius nitridus*, and *Neoaliturus haemoceps*.

The control of citrus stubborn depends on the use of spiroplasma-free budwood and rootstocks, as well as

FIGURE 12-64 (A) Healthy citrus tree (left) and tree affected with the citrus stubborn spiroplasma, *Spiroplasma citri* (right). The infected tree is stunted and has compact growth. (B) Healthy fruit (left), several infected fruit showing delayed coloration and reduced size, infected leaves showing mottling, and two healthy leaves (extreme right). [Photographs courtesy of C. N. Roistacher, California Department of Agriculture.]

early detection and removal of infected trees. Young citrus trees responded experimentally to treatment with tetracycline antibiotics, but this is not practiced commercially.

CORN STUNT DISEASE

Corn stunt occurs in the southern United States, Central America, and northern South America. The disease causes severe losses in most areas where it occurs, although disease severity varies with the variety and the stage of host development at the time of infection.

Early symptoms consist of yellowish streaks in the youngest leaves. As the plant matures, yellowing of leaves becomes more apparent and more general (Figs. 12-65A and 12-65B). Later, much of the leaf area turns reddish purple. Infected plants remain stunted due to shorter stem internodes in the part of the plant produced after infection. This gives the plants a somewhat bunchy appearance at the top. Infected plants often have more ears, but the ears are smaller and bear little or no seed. Tassels of infected plants are usually sterile. There is also a proliferation of sucker shoots and, in severe infections, of roots.

The corn stunt pathogen is the spiroplasma *Spiroplasma kunkelli* (Fig. 12-65C). It is transmitted in nature by the leafhoppers *Dalbulus elimatus*, *D. maidis*, and others. The leafhoppers must feed on diseased plants for several days before they can acquire the spiroplasma, and an incubation period of 2 to 3 weeks from the start of the feeding must elapse before the insects can infect healthy plants. A feeding period of a few minutes to a few days may be required for the insects to inoculate the healthy plants with the spiroplasma. Plants show corn stunt symptoms 4 to 6 weeks after inoculation.

Where the corn stunt spiroplasma overwinters is not known with certainty, although it was previously believed to overwinter in Johnson grass and possibly other perennial plants. In the tropics, it perpetuates itself in continuous croppings of corn.

The control of corn stunt depends on the planting of corn hybrids resistant to corn stunt.

Selected References

Abou-Jawdah, Y., Karakasian, A., and Sobh, H. (2002). An epidemic of almond witches'-broom in Lebanon: Classification and phylogenetic relationships of the associated phytoplasma. *Plant Dis.* **86**, 477–484.

Ahrens, U., Lorenz, K.-H., and Seemuller, E. (1993). Genetic diversity among mycoplasmalike organisms associated with stone fruit diseases. *Mol. Plant-Microbe Interact.* **6**, 686–691.

Bove, J. M. (1984). Wall-less prokaryotes of plants. *Annu. Rev. Phytopathol.* **22**, 361–396.

Chen, T. A., and Liao, C. H. (1975). Corn stunt spiroplasma: Isolation, cultivation, and proof of pathogenicity. *Science* **88**, 1015–1017.

Da Graca, J. V. (1991). Citrus greening disease. *Annu. Rev. Phytopathol.* **29**, 109–136.

Doi, Y., *et al.* (1967). Mycoplasma- or PLT group-like microorganisms found in the phloem elements of plants infected with mulberry dwarf, potato witches'-broom, aster yellows, or paulownia witches'-broom. *Ann. Phytopathol. Soc. Jpn.* **33**, 259–266.

Dyer, A. T., and Sinclair, W. A. (1991). Root necrosis and histological changes in surviving roots of white ash infected with mycoplasmalike organisms. *Plant Dis.* **75**, 814–819.

Fridlund, P. R., ed. (1989). "Virus and Viruslike Diseases of Pome Fruits and Simulating Noninfecting Disorders." Wash. State Univ. Coop Ext. Special Publication.

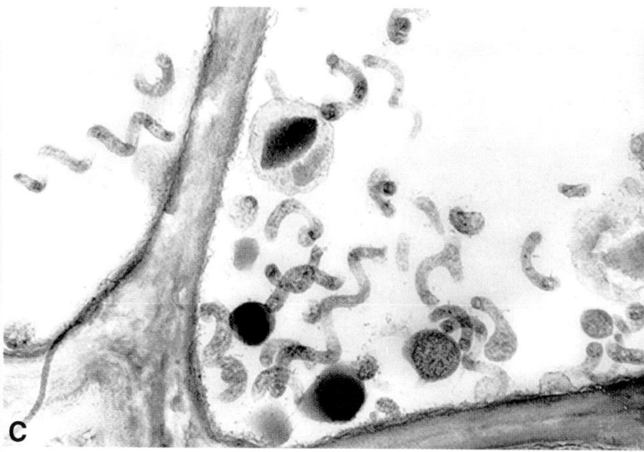

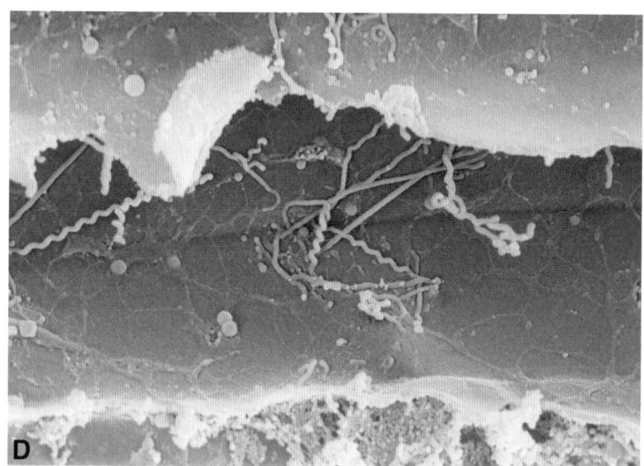

FIGURE 12-65 Corn stunt disease caused by *Spiroplasma kunkelii*. (A) All but two corn plants are infected, reddish-yellow and quite stunted. (B) The nearest plant shows extreme stunting and yellowing caused by corn stunt while plants farther away show a variety of corn stunt symptoms. (C) Portions of *S. kunkelii* in a phloem cell of a corn stunt–infected leaf. (D) Spiral cells of *S. kunkelii* in a phloem cell of an infected corn plant. [Photographs courtesy of (A–C) H. D. Thurston, Cornell University and (D) E. Alves, Federal University, Lavras, Brazil.]

Griffiths, H. M., *et al.* (1999). Phytoplasmas associated with elm yellows: Molecular variability and differentiation from related organisms. *Plant Dis.* **83**, 1101–1104.

Harrison, N. A., *et al.* (1994). Comparative investigation of MLOs associated with Caribbean and African coconut lethal decline diseases by DNA hybridization and PCR assays. *Plant Dis.* **78**, 507–511.

Hervey, G. E. R., and Schroeder, W. T. (1949). The yellows disease of carrot. *N.Y. Agric. Exp. Stn. Ithaca Bull.* **737**, 1–29.

Hibino, H., Kaloostian, G. H., and Schneider, H. (1971). Mycoplasma-like bodies in the pear psylla vector of pear decline. *Virology* **43**, 34–40.

Hiruki, C., ed. (1988). "Tree Mycoplasmas and Mycoplasma Diseases." Univ. of Alberta Press, Edmonton.

Kunkel, L. O. (1926). Studies on aster yellows. *Am. J. Bot.* **13**, 646–705.

Jaraush, W., Lansac, M., and Dosba, F. (1999). Seasonal colonization pattern of European stone fruit yellows phytoplasmas in different *Prunus* species detected by specific PCR. *J. Phytopathol.* **147**, 47–54.

Jarausch, W., *et al.* (2001). Mapping the spread of apricot chlorotic leaf roll (ACLR) in southern France and implication of *Cacopsylla pruni* as a vector of European stone fruit yellows (ESFY) phytoplasmas. *Plant Pathol.* **50**, 782–790.

Lee, I.-M., *et al.* (1993). Universal amplification and analysis of pathogen 16 S rDNA for classification and identification of mycoplasmalike organisms. *Phytopathology* **83**, 834–842.

Maramorosch, K., ed. (1973). Mycoplasma and mycoplasma-like agents of human, animal, and plant diseases. *Ann. N.Y. Acad. Sci.* **225**, 1–532.

Maramorosch, K., and Raychaudhuri, S. P. (1981). "Mycoplasma Diseases of Trees and Shrubs." Academic Press, New York.

Martin, R., *et al.* (2001). Four Spanish isolates of pear decline phytoplasma are related to other European phytoplasmas of the apple proliferation group. *J. Phytopathol.* **149**, 481–484.

Matteoni, J. A., and Sinclair, W. A. (1985). Role of the mycoplasmal disease, ash yellows, in decline of white ash in New York State. *Phytopathology* **75**, 355–360.

Nichols, C. W., *et al.* (1960). Pear decline in California. *Calif. Dept. Agric. Bull.* **49**, 186–192.

Nienhaus, F., and Sikora, R. A. (1979). Mycoplasmas, spiroplasmas, and rickettsia-like organisms as plant pathogens. *Annu. Rev. Phytopathol.* **17**, 37–58.

Prince, J. P., *et al.* (1993). Molecular detection of diverse mycoplasmalike organisms (MLOs) associated with grapevine yellows and their classification with aster yellows, X-disease, and elm yellows MLOs. *Phytopathology* **83**, 1130–1137.

Sinclair, W. A., and Griffiths, H. M. (1994). Ash yellows and its relationship to dieback and decline of ash. *Annu. Rev. Phytopathol.* **32**, 49–60.

Sinha, R. C. (1983). Relative concentration of mycoplasma-like organisms in plants at various times after infection with aster yellows. *Can. J. Plant Pathol.* **5**, 7–10.

Sinha, R. C., and Chiykowski, L. N. (1984). Purification and serological detection of mycoplasmalike organisms from plants affected by peach eastern X-disease. *Can. J. Plant Pathol.* **6**, 200–205.

Thomas, D. L. (1979). Mycoplasmalike bodies associated with lethal declines of palms in Florida. *Phytopathology* **69**, 928–934.

Whitcomb, R. F., and Tully, J. G., eds. (1989). "The Mycoplasmas," Vol. 5. Academic Press, New York.

chapter thirteen

PLANT DISEASES CAUSED BY PARASITIC HIGHER PLANTS, INVASIVE CLIMBING PLANTS, AND PARASITIC GREEN ALGAE

INTRODUCTION
705

PARASITIC HIGHER PLANTS
706

INVASIVE CLIMBING PLANTS
716

PARASITIC GREEN ALGAE
719

PLANT DISEASES CAUSED BY ALGAE
719

INTRODUCTION

More than 2500 species of higher plants are known to live parasitically on other plants. Their main common characteristic is that these parasites are vascular plants that have developed specialized organs which penetrate the tissues of other (host) vascular plants, establish connections to the host plant vascular elements, and absorb nutrients from them. These parasitic plants produce flowers and seeds and belong to several widely separated botanical families. They vary greatly in their dependence on their host plants. Some, e.g., mistletoes, have chlorophyll but no roots so they depend on their hosts only for water and minerals. Others, e.g., dodder, have little or no chlorophyll and no true roots so they depend entirely on their hosts for their existence.

Relatively few of the known parasitic higher plants cause important diseases on agricultural crops or forest trees. The most common and serious parasites belong to the following botanical families and genera:

Cuscutaceae
- Genus: *Cuscuta*, the dodders of alfalfa, onion, potato, and numerous other plants

Lauraceae
- Genus: *Cassytha*, *C. filiformis* infecting shrubs and trees in the Caribbean islands and in Florida (Figure 13-1E).

Viscaceae
- Genus: *Arceuthobium*, the dwarf mistletoes of conifers
  - *Phoradendron*, the American true mistletoes of broad-leaved trees
  - *Viscum*, the European true misteltoes

Orobanchaceae
- Genus: *Orobanche*, the broomrapes of legumes, solanaceous, and other plants

Scrophulariaceae
- Genus: *Striga*, the witchweeds of many monocotyledonous and some legume plants

The dodders and the dwarf and true mistletoes attach themselves to and parasitize aboveground parts, i.e., shoots and branches of their hosts; they cause relatively small economic losses. The witchweeds and the broomrapes attach themselves to and parasitize the roots of their host plants and cause serious economic losses. Witchweeds are one of the biggest biological hindrances in grain and corn production in Africa.

PARASITIC HIGHER PLANTS

DODDER

Dodder is widely distributed in the Americas, Europe, Africa, southern Asia, and Australia. Crops that suffer losses from dodder include alfalfa, onions, sugar beets, several ornamentals, and potatoes.

Dodder affects the growth and yield of infected plants. Losses range from slight to complete destruction of the crop in the infested areas. Names such as strangleweed, pull-down, and hellbind, by which dodder is referred to in different areas, are descriptive of the ways in which dodder affects its host plants. Dodder may also serve as a bridge for transmission of viruses from virus-infected to virus-free plants as long as both plants are infected by the same dodder plant.

Symptoms

Orange or yellow vine strands grow and entwine around the stems (Figs. 13-1A and 13-1B) and the other aboveground parts of the plants (Fig. 13-1C). The growing tips reach out and attack adjacent plants until a circle of infestation, up to 10 feet in diameter, is formed by a single dodder plant. Dodder-infested patches in the field (Figs. 13-1D and 13-1E) continue to enlarge during the growth season and, in perennial plants such as alfalfa, become larger every year. During late spring and in the summer, dodder produces massed clusters of white, pink, or yellowish flowers, which soon form seed. The infected host plants become weakened by the parasite, their vigor declines, and they produce poor yields. Many are smothered and may be killed by the parasite. As the infection spreads, several patches coalesce and form large areas covered by the yellowish vine of the parasite.

The Pathogen: Cuscuta spp

Several species of dodder exist. Some species prefer legumes, whereas others attack many other broad-leaved plants as well as legumes.

Dodder is a slender, twining plant (Fig. 13-2). The stem is tough, curling, threadlike, and leafless, bearing only minute scales in place of leaves. The stem is usually yellowish or orange in color, sometimes tinged with red or purple; sometimes it is almost white. Clusters of tiny flowers occur on the stem from early June until frost. Gray to brown seeds are produced in abundance by the flowers and mature within a few weeks after bloom.

Development of Disease

Dodder seed overwinters in infested fields or is mixed with the seed of crop plants. During the growing season the seed germinates and produces a slender yellowish shoot but no roots (Fig. 13-2). This leafless shoot rotates as though in search of a host. If no contact with a susceptible plant is made, the stem falls to the ground, where it lies dormant for a few weeks and then dies.

Dodder stems in contact with a susceptible host encircle the host plant, send haustoria into it, and begin to climb the plant. The haustoria penetrate the stem or leaf and reach into the vascular tissues, from which they absorb foodstuffs and water.

Soon after contact with the host is established, the base of the dodder shrivels and dries so that the dodder loses all connection with the ground and becomes completely dependent on the host for nutrients and water. The dodder continues to grow and expand, and its twisting tips reach out and attack adjacent plants, forming patches of infected plants. The growth of infected plants is suppressed and they may finally die.

In the meantime, the dodder plant has developed flowers and produced seeds. The seeds fall to the ground where they either germinate immediately or remain dormant until the next season. The seed may be spread to nearby areas by animals, water, and equipment, and over long distances by contaminated crop seed.

Control

Dodder is best controlled by preventing its introduction into a field by the use of dodder-free seed, by

FIGURE 13-1 Common dodder (*Cuscuta* sp.) parasitism and symptoms. Dodder stems entwined around stems of sunflower (A) and potato (B). (C) Dodder entwined around and overcoming a pepper plant. (D) Dodder covering and overcoming all watermelon plants in an area of a field. (E) A Dodder of the Lauraceae species *Cassythia filiformis* spreading over roadside shrubs and trees in Florida. [Photographs courtesy of (A) L. J. Musselman, Southern Illinois University, (B) D. P. Weingartner, University of Florida, (C) G. W. Simone, and (D) D. N. Maynard, University of Florida.]

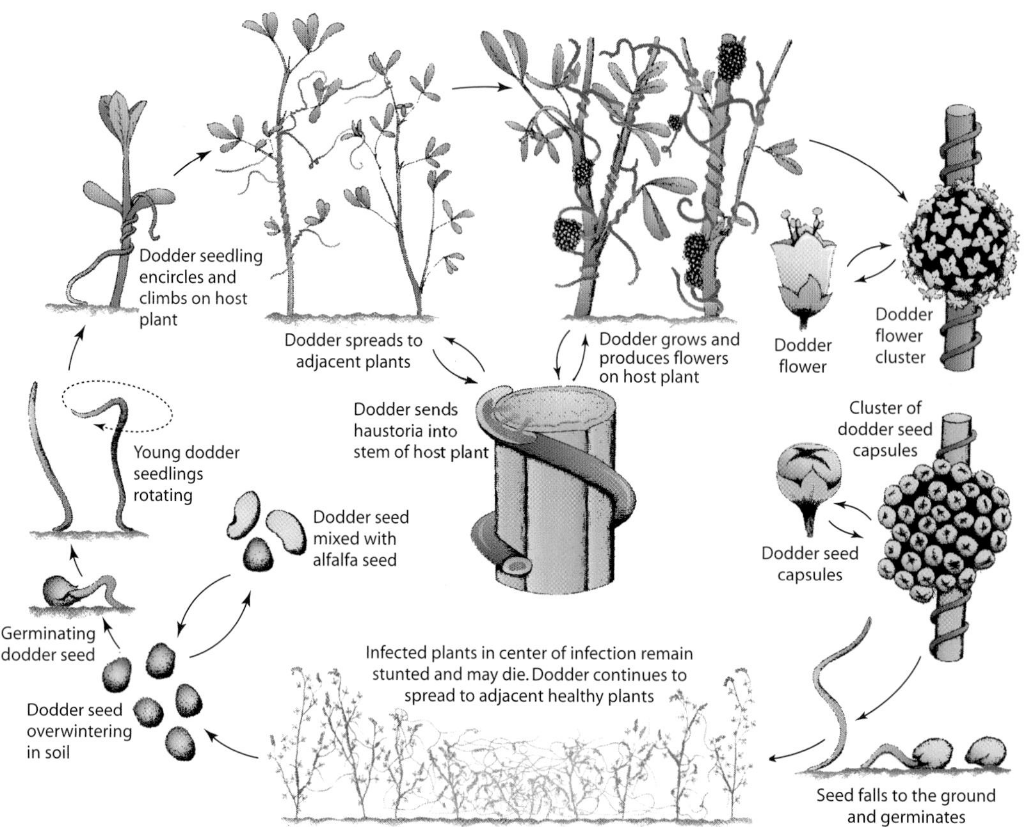

FIGURE 13-2 Disease cycle of dodder (*Cuscuta* sp.) on a plant such as alfalfa.

cleaning equipment thoroughly before moving it from dodder-infested fields to new areas, and by limiting the movement of domestic animals from infested to dodder-free fields. If dodder is already present in the field, scattered patches may be sprayed early in the season with contact herbicides. Such treatment, or cutting or burning of patches, kills both the dodder and the host plants but prevents dodder from spreading and from producing seed. When dodder infestations are already widespread in a field, dodder can be controlled by frequent tillage, flaming, and use of herbicides that kill the dodder plant on its germination from the seed but before it becomes attached to the host.

WITCHWEED

Witchweed (Fig. 13-3A) is a serious parasitic weed in Africa, Asia, and Australia. In 1956 the weed was discovered for the first time in America, in North and South Carolina. Because of effective federal and state quarantines, the spread of the parasite has been largely limited to the area of the original infestations.

Witchweed parasitizes important economic plants and is one of the most destructive pathogens in Africa. It attacks mostly monocots such as corn (Figs. 13-3A–13-3C), sorghum, millet, upland rice, and sugarcane, but also cowpeas, peanuts, other legumes (Fig. 13-3D), sweet potato, and tobacco. Infected plants become stunted and chlorotic. Heavily infected plants usually wilt and die. Losses vary and may range from slight to 100%.

Symptoms

Affected plants remain stunted, wilt, and turn yellowish (Fig. 13-3C). Death may follow these symptoms if the plants are heavily parasitized. Infected roots bear a large number of witchweed haustoria, which are attached to the root and feed on it. One to several witchweed plants may be growing above ground next to the infected plants, although roots of many more witchweed plants, which do not survive to reach the surface, may parasitize the roots of the same host (Fig. 13-3).

The Pathogen: Striga *spp*

Witchweed is a small, pretty plant. It has a bright green, slightly hairy stem and leaves and grows 15 to 30

FIGURE 13-3 Witchweed (*Striga* sp.) parasitizing plants. (A) Witchweed plant in bloom parasitizing a corn plant. (B) Groups of witchweed plants parasitizing each corn plant along a row in a field. (C) Corn plants parasitized by witchweed plants appear stressed, wilted, and stop growing. (D) A different species of *Striga* parasitizing the legume plant hairy indigo. [Photographs courtesy of L. J. Musselman, Southern Illinois University.]

centimeters high. It produces many branches both near the ground and higher on the plant. The leaves are rather long and narrow in opposite pairs (Fig. 13-4).

The flowers are small and usually red or yellowish, or white, always having yellow centers. Flowers appear just above the leaf attachment to the stem and are produced throughout the season. After pollination, seed pods or capsules develop, each containing more than a thousand tiny brown seeds. A single plant may produce from 50,000 to 500,000 seeds.

The root of witchweed is white and round in cross section. It has no root hairs, for it obtains all nutrients from the host plant through haustoria.

The life cycle of the parasite, from the time a seed germinates until the developing plant releases its first seeds, takes 90 to 120 days. Although the witchweed plant is green and can probably manufacture some of its own food, it appears that it still continues to depend on the host, not only for all its water and minerals, but for organic substances as well.

Development of Disease

The parasite overwinters as seeds, most of which require a rest period of 15 to 18 months before germination, although some can germinate without any dor-

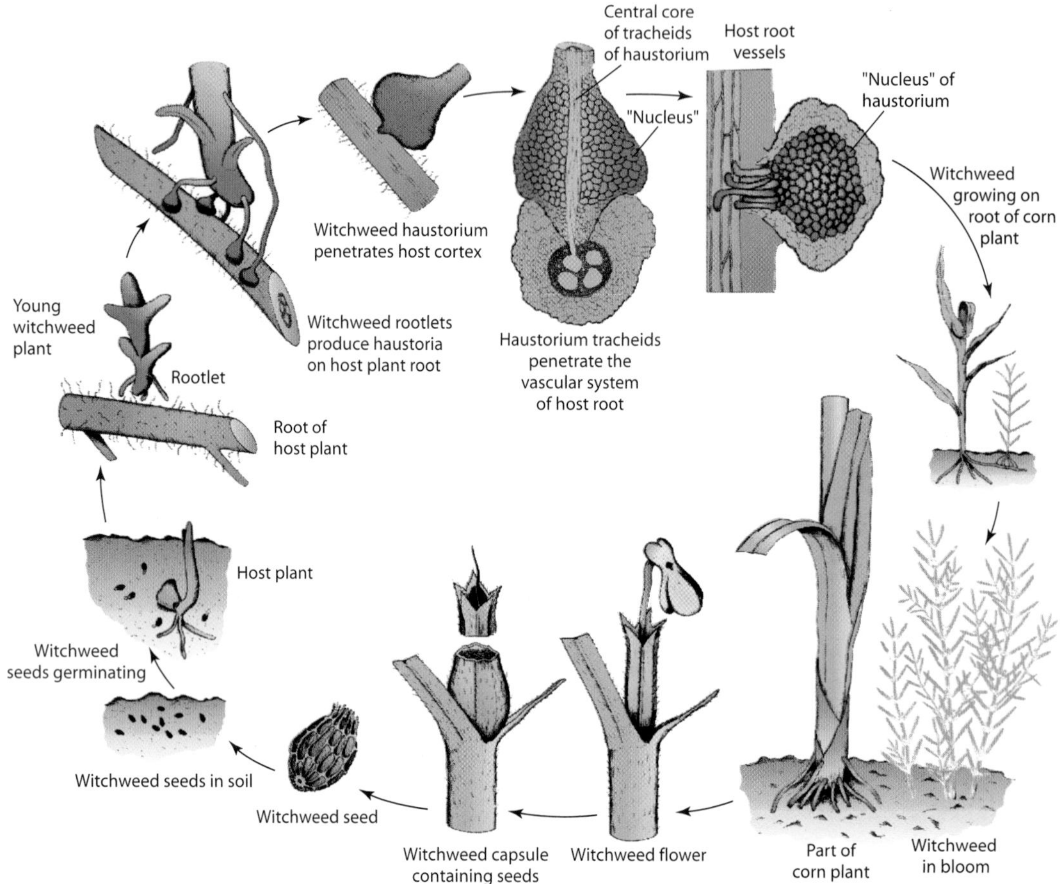

FIGURE 13-4 Disease cycle of witchweed (*Striga asiatica*) on corn.

mancy. Seeds close to host roots germinate and grow toward these roots, attracted by the exudates of the host roots. As soon as the witchweed rootlet comes in contact with the host root, its tip swells into a bulb-shaped haustorium. The haustorium dissolves and penetrates the host roots within 8 to 24 hours and advances into the roots: finally, its leading cells, usually tracheids, reach the vessels of the host roots (Fig. 13-4). The tracheids force their way into the vessel, from which they absorb water and nutrients. Although xylem vessels are present in the haustorium, no typical phloem cells develop. However, cells in the "nucleus" of the haustorium seem to connect the phloem of host and parasite. Although the chlorophyll of witchweed plants is functional, manufactured foodstuffs still move from the host plant into the parasite.

The weed produces several roots, which send more haustoria into host roots. Often, several hundred separate witchweed plants parasitize the roots of a single host plant at once, but relatively few of these survive to reach the surface because the host plant cannot support so many.

The disease spreads in the field in a circular pattern. The circle of infected plants increases year after year as the witchweed seeds spread in increasingly larger areas. The seeds are spread by wind, by water, by contaminated tools and equipment, or by contaminated soil carried on farm machinery.

Control

Witchweed is difficult to control. Introduction of witchweed should be avoided by all means. Catch crops, consisting of host plants, may be planted to force the germination of witchweed seed, and the witchweed plants then can be destroyed by plowing under or by the use of herbicides. Trap crops, consisting mostly of nonhost legumes, may be used to stimulate the germination of witchweed seeds, which, however, cannot infect the trap plants and therefore starve to death. Use of resistant cultivars, seed treatment with herbicides of differential toxicity, and use of witchweed-infecting fungi as a biological control are possible control methods under investigation. Usually, a combination of the aforementioned methods is required to prevent witchweed plants from flowering and seeding.

BROOMRAPES

Broomrapes occur in warm and dry regions worldwide. They are more common and severe in countries around the Mediterranean Sea and west Asia. They attack several hundred species of herbaceous dicotyledonous crop plants (Figs. 13-5A–13-5E). In some areas, broomrapes cause losses varying from 10 to 70% of the crop.

Symptoms

Plants affected by broomrapes usually occur in small patches and may be stunted to various degrees, depending on how early in their lives and by how many broomrapes they were infected. The broomrape pathogen,

FIGURE 13-5 Broomrapes (*Orobanche* sp.) parasitizing various plants: on fava bean (A) and on broad bean (B). *Orobanche* parasitizing the respective plants and destroying the crop in a tomato field (C), a carrot field (D), and in a broadbean field (E). [Photographs courtesy of L. J. Musselman, Southern Illinois University.]

Orobanche sp., is a whitish to yellowish-brown annual plant 15 to 50 centimeters tall. It has a fleshy stem and scale-like leaves and produces numerous pretty, white, yellow-white, or slightly purple, snapdragon-like flowers arising singly along the stem (Figs. 13-5A and 13-5B). The broomrapes produce seed pods about 5 millimeters long, each containing several hundred minute seeds.

Development of Disease

Broomrapes overwinter as seeds, which may survive in the soil for more than 10 years. Seeds germinate only when roots of certain plants grow near them, although not all these plants are susceptible to the pathogen. On germination the seed produces a radicle, which grows toward the root of the host plant, becomes attached to it, and produces a shallow cup-like appressorium that surrounds the root. From the appressorium, a mass of undifferentiated cells penetrate the host, extend to and, occasionally, into the xylem, and absorb nutrients and water from it. Some of these cells differentiate into parasite xylem vessel elements and connect the host xylem with the main vascular system of the parasite. Other undifferentiated cells become attached to phloem cells and obtain nutrients from them, which they transport back to the parasite. Soon the parasite begins to develop a stem, which appears above the soil line and looks like an asparagus shoot. Meanwhile, the original root produces secondary roots that grow outward until they come in contact with other host roots to which they become attached and subsequently infect. From these points of contact, new roots and stems of the parasite are produced and result in the appearance of the typical clusters of broomrape plants arising from the soil around infected host plants. Several such broomrapes may be growing concurrently on the roots of the same host plant. The broomrape stems continue to grow and produce flowers and seeds, which mature and are scattered over the ground in less than two months from the emergence of the stems.

Control

The control of broomrapes depends on preventing the introduction of its seeds in new areas, planting nonsusceptible crops in infested fields, frequent weeding and removal of broomrapes before they produce new seed, and, where feasible, treating the soil with an appropriate herbicide. It has been reported that flax serves as a trap crop for broomrape. Flax root exudates stimulate broomrape seeds to germinate, and these then infect flax but do not flower on it. Some plant varieties are resistant to broomrapes. Also, some fungi have been shown to parasitize *Orobanche* and may be useful for its biological control in the future.

DWARF MISTLETOES OF CONIFERS

Dwarf mistletoes occur wherever conifer trees grow. In the United States they are more prevalent and most serious in the western half of the country. The damage caused by dwarf mistletoes in coniferous forests is extensive, although not always spectacular. Trees of any age may be stunted, deformed, or killed. Their height may be reduced by 50 to 80%. Timber quality is reduced by numerous large knots and by abnormally grained, spongy wood. Seedlings and saplings, as well as trees of certain species, are frequently killed by dwarf mistletoe infections.

Symptoms

Shoots of dwarf mistletoe plants occur in tufts along the twigs, branches, and trunks of the hosts (Fig. 13-6). Infected twigs and branches develop swellings and cankers on the infected areas. Cross sections at the swellings reveal wedge-shaped haustoria of the parasite (Fig. 13-7), which grow into the bark, cambium, and xylem of the branch. Large swellings or flattened cankers may also develop on the trunks of some infected trees. Infected branches often produce witches'-brooms. Heavily infected stands contain deformed, stunted, dying, and dead trees or trees broken off at trunk cankers.

The Pathogen: Arceuthobium *spp*

In some species the shoots are up to 10 centimeters long, whereas in others they are no more than 1.5 centimeters. The dwarf mistletoe shoots may be simple or branched, and they are joined. The leaves are inconspicuous, scalelike, in opposite pairs, and of the same color as the stem. Dwarf mistletoe plants also produce a complex system of haustoria, which consists of longitudinal strands, external to and fairly parallel to the host cambium, and radial "sinkers" produced by the former and oriented radially into the phloem and xylem.

The plants are either male or female and produce flowers when they are 4 to 6 years old (Fig. 13-6A). After flowering, the male shoots die; the female shoots die after the seeds are discharged. Fruits mature 5 to 16 months after pollination of the flowers. The fruit at maturity is turgid and, on ripening, develops considerable internal pressure. When disturbed, the fruit expels the seed upward or obliquely at lateral distances up to 15 meters. The seed is covered with a sticky substance and adheres to whatever it comes in contact with. This

FIGURE 13-6 (A) Male (yellow) and female (with fruit or seed capsules) dwarf mistletoe plants growing on a pine branch. (B) Dwarf mistletoe plant parasitizing the trunk of a conifer (larch) and causing it to swell and, later, to possibly break at the point of infection. (C) A Douglas-fir tree the top branches and trunk of which have been killed by dwarf mistletoe infections. [Photographs courtesy of the USDA Forest Service.]

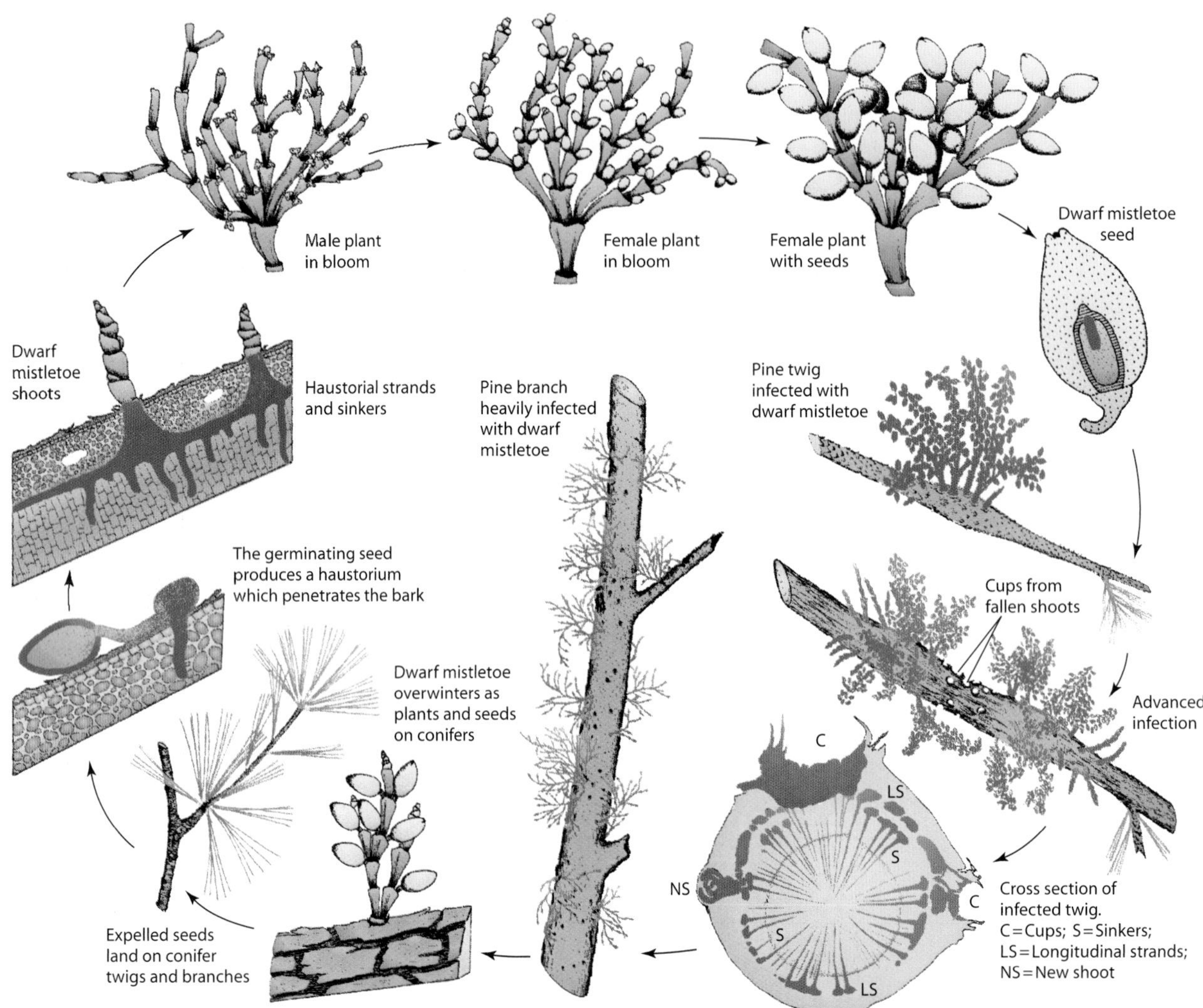

FIGURE 13-7 Disease cycle of dwarf mistletoe (*Arceuthobium* sp.) on conifers.

is the main means of spread of the parasite; occasionally, however, long-distance spread occurs when birds transport seed on their bodies.

Development of Disease

When a dwarf mistletoe seed lands on and becomes attached to the bark of a twig or a young branch of a susceptible host, it germinates and produces a germ tube or radicle. This grows along the bark surface until it meets a bud or a leafbase, at which point it produces a root-like haustorium that penetrates the bark directly and reaches the phloem and the cambium. From this haustorium develops the system of longitudinal strands and radial sinkers, all of which absorb from the host the nutrients needed for the development of the parasite (Fig. 13-7). The sinkers that reach the cambium of the host become permanently embedded in the wood as the latter is laid down each year, but they always retain their connections with the strands in the phloem. After the endophytic system is well established and developed in the host, it produces buds from which shoots develop the following year or several years later. The shoots first appear near the original point of infection, but later more shoots emerge in concentric zones of increasing diameter. The center of the infection usually deteriorates and becomes attacked easily by various decay-producing fungi. If witches'-brooms are produced on the affected area, the haustoria pervade all branches and produce mistletoe shoots along the proliferating host branches.

The parasite removes water, minerals, and photosynthates from the host and so starves and kills the portion of the branch lying beyond the point of infection. It also

FIGURE 13-8 (A) One true mistletoe plant growing on a branch of a hardwood tree. (B) A large group of true mistletoes growing on many branches of a hardwood tree. [Photograph (A) courtesy E. L. Barnard.]

saps the vitality of the branch and, when sufficiently abundant, of the whole tree. Furthermore, it upsets the balance of hormonal substances of the host in the infected area and causes excessive cell enlargement and division, with resulting swellings and deformities of various shapes on the branches. This hormonal imbalance also stimulates the normally dormant lateral buds to excessive formation of shoots, forming a dense growth of abnormal appearance. Heavy dwarf mistletoe infections weaken trees and predispose them to wood-decaying and root pathogens, to beetles, and to wind breakage.

Control

The only means of controlling dwarf mistletoes is by physical removal of the parasite. This is done either by pruning infected branches or by cutting and removing entire infected trees. Uninfected stands can be protected from dwarf mistletoe infections by maintaining a protective zone free of the parasite between the diseased stand and the stand to be protected.

TRUE OR LEAFY MISTLETOES

True or leafy mistletoes occur throughout the world, particularly in warmer climates. They attack primarily hardwood forest and shade trees but also many of the common fruit and plantation trees, such as apple and rubber, and even some gymnosperms, such as juniper and cypress. They cause serious economic losses in some areas, although not nearly as severe as those caused by the dwarf mistletoes. True mistletoes were first recognized as parasites of plants in the 12th century and have been the subject of numerous legends and traditions in Europe and North America (see p. 14–16).

The symptoms are quite similar to those caused by dwarf mistletoes. Infected areas become swollen and produce witches'-brooms (Figs. 13-8A and 13-8B). The mistletoe plants sometimes are so numerous that they make up almost half of the green foliage of the tree, and in the winter they make deciduous trees appear like evergreens (Fig. 13-8B), with the normal tree branches appearing as though they have died back. Infected trees may survive for many years; however, they show reduced growth, and portions of the tree beyond the mistletoe infection often become deformed and die.

The pathogens are *Phoradendron* sp. in most of North America and *Viscum* sp. in California, Europe, and the other continents. These mistletoes are parasitic evergreens that have well-developed leaves and stems up to 1 or 2 centimeters in diameter (Fig. 13-8). In some species of true mistletoe, however, the stems may be up to 30 centimeters or more in diameter. The height of mistletoe plants varies from a few centimeters to a meter or more. The true mistletoes produce typical green leaves that can carry on photosynthesis, usually small, dioecious flowers, and berry-like fruits containing a single seed. Instead of roots, however, true mistletoes, too, produce haustorial sinkers, which grow in branches and stems of trees and absorb water and mineral nutrients.

True mistletoes are spread by birds that eat the seed-containing berries and excrete the sticky seeds in the tops of taller trees on which they like to perch. From that point on, infection, disease development, and control of true mistletoes are almost identical to those of dwarf mistletoes. Control in isolated shade or fruit trees can be obtained by pruning of infected branches or periodic removal of mistletoe stems from the branches or trunks.

Selected References

Aigbokhan, E. I., Berner, D. K., and Musselman, L. J. (1998). Reproductive ability of hybrids of *Striga aspera* and *Striga hermonthica*. *Phytopathology* **88**, 563–567.

Ayensu, E. S., *et al.*, eds. (1984). "*Striga*: Biology and Control." International Council of Scientific Union, Paris, France.

Berner, D. K., *et al.* (1994). Relative roles of wind, crop seeds, and cattle in dispersal of *Striga* spp. *Plant Dis.* **78**, 402–406.

Berner, D. K., Kling, J. G., and Singh, B. B. (1995). *Striga* research and control: A perspective from Africa. *Plant Dis.* **79**, 652–660.

Berner, D. K., and Williams, O. A. (1998). Germination stimulation of *Striga gesnerioides* seeds by hosts and nonhosts. *Plant Dis.* **82**, 1242–1247.

Ciotola, J. M., Watson, A. K., and Hallett, S. G. (1995). Discovery of an isolate of *Fusarium oxysporum* with potential to control *Striga asiatica* in Africa. *Weed Res.* **35**, 303–309.

Dawson, J. H., Lee, W. O., and Timmons, F. L. (1969). Controlling dodder in alfalfa. *Farm. Bull.* **2211**, 1–16.

Dawson , J. H., Musselman, L. J., Wolswinkel, P., and Dopp, I. (1994). Biology and control of *Cuscuta*. *Rev. Weed Sci.* **6**, 265–317.

Dorr, I. (1987). The haustorium of *Cuscuta*: New structural results. *In* "Parasitic Higher Plants" (H. C. Webber and W. Forstreuter, eds.), pp. 163–170. Philipps Univ., Marburg-Lahm, Germany.

Eplee, R. E. (1981). *Striga's* status as a plant parasite in the United States. *Plant Dis.* **65**, 951–954.

Hawksworth, F. G., and Wiends, D. (1996). "Dwarf Mistletoes: Biology, Pathology, and Systematics." Agric. Handbook No. 709. USDA Forest Service.

Hood, M. E., *et al.* (1998). Primary haustorial development of *Striga asiatica* on host and nonhost species. *Phytopathology* **88**, 70–75.

Kuijt, J. (1969). "The Biology of Parasitic Flowering Plants." Univ. of California Press, Berkeley.

Kuijt, J. (1977). Haustoria of phanerogamic parasites. *Annu. Rev. Phytopathol.* **15**, 91–118.

Kuiper, E. A., *et al.* (1998). Tropical grasses vary in their resistance to *Striga aspera, Striga hermonthica*, and their hybrids. *Can. J. Bot.* **76**, 2131–2144.

Musselman, L. J. (1980). The biology of *Striga, Orobanche*, and other root-parasitic weeds. *Annu. Rev. Phytopathol.* **18**, 463–489.

Musselman, L. J., ed. (1987). "Parasitic Weeds in Agriculture," Vol. 1, CRC Press, Boca Raton, FL.

Nickrent, D. L. (2001). Parasitic plants of the world. *In* "Guide to the Parasitic Plants of the Iberian Peninsula and Balearic Islands" (J. A. Lopez-Saez, P. Catalan, and L. Saez, eds.). Mundi-Prensa Libros, S. A., Madrid.

Nickrent, D. L. (1998). Molecular phylogenetic and evolutionary studies of parasitic plants. *In* "Molecular Systematics of Plants" (D. Soltis, P. Soltis, and J. Doyle, eds.), Vol. II, pp. 211–241. Kluwer Academic, Boston, MA.

Pennypacker, B. W., Nelson, P. E., and Wilhelm, S. (1979). Anatomic changes resulting from the parasitism of tomato by *Orobanche ramosa*. *Phytopathology* **69**, 741–748.

Press, M. C., and Gurney, A. L. (2000). Plant eats plant: Sap-feeding witchweeds and other parasitic angiosperms. *Biologist* **47**, 189–193.

Roman, B., *et al.* (2002). Variation among and within populations of the parasitic weed *Orobanche crenata* from Spain and Israel revealed by inter simple sequence repeat markers. *Phytopathology* **92**, 1262–1266.

Sauerborn, J., *et al.* (2002). Benzothiadiazole activates resistance in sunflower (*Helianthus annuus*) to the root-parasitic weed *Orobanche cumata*. *Phytopathology* **93**, 59–64.

Shaw, C. G., and Hennon, P. E. (1991). Spread, intensification, and upward advance of dwarf mistletoe in thinned, young stands of western hemlock in southeast Alaska. *Plant Dis.* **75**, 363–367.

Sukno, S., Fernandez-Martinez, J. M., and Melero-Vara, J. M. (2001). Temperature effects on the disease reactions of sunflower to infection by *Orobanche cumana*. *Plant Dis.* **85**, 553–556.

Thoday, M. G. (1991). On the histological relations between *Cuscuta* and its host. *Ann. Bot.* **25**, 655–682.

Thomas, H., *et al.* (1999). Fungi of *Orobanche aegyptiaca* in Nepal with potential as biocontrol agents. *Biocontr. Sci. Technol.* **9**, 379–381.

Webber, H. C., and Forestreuter, W., eds. (1987). Parasitic higher plants. *In* "Proceedings of the 4th International Symposium." Philipps University, Marburg-Lahm, Germany.

INVASIVE CLIMBING PLANTS

Plant species have evolved over time a variety of structures and characteristics that we enjoy in plants wherever they grow. Their growth and properties, however, have developed in relation to the environment and other plants growing in the same or adjacent areas and have been kept in check by them. The variety, properties, and location of the various plant species are due, in part, to their separation by physical barriers such as oceans and mountains.

Over the past several centuries, however, the ability of humans to travel over oceans and mountains has enabled them to also carry with them, or to unwittingly transport, numerous plant species beyond their natural barriers to new habitats. In the new, noncultivated habitats, some introduced plant species outcompete native species for water, nutrients, and sunlight and, as a result, out-reproduce the local plant species. Consequently, before too long, certain introduced plant species become the predominant plant species over large areas of land or water, displacing the local plants. In the process, such invasive plant species disrupt the native habitat, reduce the number, size, and survival of native plants, clog waterways and lakes, and, particularly invasive plants growing as climbing vines, completely cover and block out the sunlight from plants they clime or cover, causing them to grow poorly or to die. Such plants, therefore, behave as noxious weeds. More than 300 introduced plants have invaded uncultivated areas across the United States, about 120 of them found in Florida. Some introduced plants are so invasive that, for example, in Florida, 28 of them are prohibited from possession or sale in the state. Some of the most invasive and destructive invasive plant species are as follows.

Ferns (Pteridophytes)
- Old world climbing fern — *Lygodium sp.*
- Japanese climbing fern — *Lygodium japonicum*

Monocots
- Taro — *Colocasia esculenta*
- Water lettuce — *Pistia stratiotes*
- Hydrilla — *Hydrilla verticillata*
- Cogon grass — *Imperata cylindrica*
- Torpedo grass — *Panicum repens*
- Water hyacinth — *Eichhornia crassipes*

Dicots
- Brazilian pepper — *Schinus terebinthifolius*
- Suckering Australian-pine — *Casuarina glauca*
- Chinese tallow tree — *Sapium sebiferum*
- Kudzu vine — *Pueraria montana*
- Melaleuca — *Melaleuca quinquenervia*
- Tropical soda apple — *Solanum viarum*

A couple of examples of invasive plants growing as climbing vines are described briefly.

Old World Climbing Fern

The old world climbing fern, *Lygodium microphyllum*, has climbing and twining fronds that grow up to 30 meters (90 feet) long. It produces dark brown wiry rhizomes and wiry, stem-like leaf stalk and leafy branches, and stalked, unlobed leaflets. Fertile leaflets are fringed with tiny lobes that cover the sporangia along the leaf margin.

Old world fern is native to Africa, southeast Asia, South Pacific Islands, and Australia. It was first detected in Florida in 1958. By the year 2000, old world climbing fern had spread to more than 110,000 acres of uncultivated areas in the southern half of Florida. In its path, the fern blankets and smothers native plants whether they are sawgrass standing in water, shrubby and herbaceous plants, or tree groves (Fig. 13-9).

Old world fern survives as a vine and as wiry rhizomes that can accumulate as dense mats 1 meter or more thick on soil. Spores are windborne and can germinate within 6–7 days.

The control of old world fern has been tried both mechanically and with herbicides but it is very difficult. Fire usually kills it back but does not eliminate it.

Kudzu Vine

Kudzu vine, *Pueraria montana*, is a dicot leguminous deciduous woody vine that produces tuberous roots and dark brown rope-like stems that climb up to 20 meters (65 feet) high (Fig. 13-10). Young stems are hairy, and the leaves are trifoliate and also hairy (Fig. 13-10A). It produces pretty reddish purple pea-like flowers that lead to the production of dark brown hairy pods.

FIGURE 13-9 (A) Leaves and vines of the old world climbing fern. (B) Vines and shoots of the old world climbing fern growing and almost completely covering shrubs and trees in a natural setting. [Photographs courtesy of University of Florida.]

FIGURE 13-10 (A) Leaves and flowers of kudzu vine *Pueraria montana.* (B–D) Kudzu vine plants climbing over and suffocating (blocking the light off) plants in a field (Fig. 13-9B) and on trees adjacent to it (C). (D) Kudzu plants climbing on and blanketing trees along a road. [Photographs courtesy of University of Florida.]

Kudzu vine is native to eastern Asia. It has spread to South Africa, Malaysia, and the western Pacific Islands. It was introduced into the United States as an ornamental in 1876, as a forage plant in Florida in the 1920s, and was promoted as an erosion control by the U.S. Soil Conservation Service in the 1930s. Finally, the U.S. Department of Agriculture declared Kudzu vine a weed in 1972. The vine completely engulfs nonwooded areas but it also grows over wooded areas on which it produces large impenetrable masses and completely envelops trees and other plants, killing them all by shutting out all sunlight. Kudzu vine is now widely distributed in the United States, including all the southeast, north to Massachusetts and Illinois, and west to Texas and Oklahoma. Approximately 2,000,000 acres of forest land are covered by Kudzu vine.

Kudzu vine forms new roots from stem nodes touching the ground. Thick storage roots grow as deep as 1 meter in the ground. It produces large numbers of seeds that are disseminated by animals, especially birds. The plant is drought tolerant and frosts kill only the aboveground parts of the vine. The roots are also resistant to herbicides and it can take 3–10 years of repeated treatments with herbicides before the nutrient reserves of the roots are exhausted.

Selected References

Beckner, J. (1968). *Lygodium microphyllum*, another fern escaped in Florida. *Am. Fern J.* 58, 93–94.

Bodle, M. J. (1994). Does the scourge of the South threaten the Everglades? *In* "An Assessment of Invasive Non-indigenous Species in Florida's Public Lands" (D. C. Schmitz and T. C. Brown, eds.), Technical Report No. TSS-94-100, Dept. Environm. Protection, Tallahassee, Fl.

Brown, V. M. (1984). A biosystematic study of the fern genus *Lygodium* in eastern North America. Thesis, Univ. Central Florida, Orlando.

Cronk, Q. C. B., and Fuller, J. L. (1995) "Plant Invaders." Chapman and Hall, New York.

Godfrey, R. K. (1988). Trees, Shrubs, and Woody Vines of Northern Florida and Adjacent Georgia and Alabama." Univ. of Georgia Press, Athens, GA.

Holms, L. J., *et al.* (1977). "The World's Worst Weeds: Distribution and Biology". Hawaii Univ. Press, Honolulu.

Langeland, K. A., and Craddock Burks, K., eds. (1998). "Identification and Biology of Non-Native Plants in Florida's Natural Areas." Univ. of Florida.

Moorhead, D. J., and Johnson, K. D. (1996). Controlling kudzu in CRP stands. Conserv. Res. Rept. 15, Univ. of Georgia, Athens, GA.

Nauman, C. F., and Austin, D. F. (1978). Spread of the exotic fern *Lygodium microphyllum* in Florida. *Am. Fern J.* 68, 65–66.

Roberts, R. E. (1996). The monster of Hobe Sound. *In* "Proceedings of Invasive Vines Workshop" (M. Bodle, ed.). West Palm Beach, FL.

Shores, M. (1997). The amazing story of kudzu. Univ, of Alabama, Web site: http://www.cptr.ua.edu/kudzu.htm.

PARASITIC GREEN ALGAE

Algae are the organisms, often microorganisms, other than typical land plants, that can carry on photosynthesis. Algae are sometimes considered as protists with chloroplasts. There are eight groups of algal protists. Some algae, the so-called blue-green cyanobacteria, belong to the kingdom Eubacterial Prokaryotes, but most of them, i.e., the rest, belong to the kingdom Chromista. Algae are the main producers of photosynthetic materials in aquatic ecosystems, including unstable areas such as muds, sands, and intertidal aquatic habitats. Green algae are single-celled organisms that form colonies, or multicellular, free-living organisms, all of which have chlorophyll b.

Several algae are pathogens of other organisms. For example, cyanobacteria cause the black band disease that leads to bleaching and death of coral symbionts of the algae. Many red algae are parasitic on other, mostly related, red algae. Colorless green algae of the genus *Prototheca* cause skin infections in humans. Most of the green algae are free-living organisms, but several of their genera live as endophytes of many hydrophytes to which they seem to cause little or no damage. A few genera of green algae, however, are parasitic on higher plants.

The green algal genera of *Rhodochytrium* of the family Chlorococcaceae and the genus *Phyllosiphon* of the family Phyllosiphonaceae infect numerous weeds and a few cultivated plants of relatively minor economic importance. However, green algae of the genus *Cephaleuros* of the family Tentepohliaceae are true parasites of many wild and cultivated plants and cause diseases of economic importance.

Cephaleuros green algae, especially the genus *Cephaleuros virescens*, cause leaf spots (Figs. 13-11A–13-11C) and spots on stems (Fig. 13-11D) of plants belonging to more than 200 species growing primarily in the tropics between latitudes 32°N and 32°S. These green algae also cause lesions on fruit (Figs. 13-11E and 13-11F) but less frequently. Some of the economically most important plants attacked by green algae are tea, coffee, cacao, black pepper, citrus, and mango.

Cephaleuros green algae consist of a vegetative thallus that is disc-like and is composed of cells arranged symmetrically (Figs. 13-12A–13-12C). The algal thallus produces filaments that grow mostly between the cuticle and the epidermis of host leaves but, under some conditions, the filaments also grow between the palisade and the mesophyll cells of leaves. *Cephaleuros* algae produce filaments on which zoosporangia are produced (Figs. 13-12D–13-12F). They reproduce by means of zoospores in zoosporangia, which can be disseminated by wind, rain splashes, and wind-driven rain. Zoospores can infect new leaves, shoots, and fruit of plants. Infections are much more common at the end of the rainy season. Following infection, plant cells next to the invading thallus turn yellow, while nearby cells enlarge and divide. If the plants are under stress, the infecting thallus expands, while cells in tissues invaded earlier die and produce a lesion. There may be so many lesions produced on leaves and shoots that they almost cover the entire surface.

The control of parasitic green algae, when needed, can be obtained by spraying plants that may become infected with appropriate fungicides at the time most infections occur.

PLANT DISEASES CAUSED BY ALGAE

When and where conditions allow, different types of algae that are not parasitic on plants are favored in their growth over the growth of cultivated land plants, outcompete with the latter, and prevail at the expense of the latter. What really happens is that the algae, which normally are aquatic plants, are favored by frequent and heavy rains or irrigation, by a high water table or poor drainage, and by poor air circulation or partial shade, all of which tend to keep the soil surface and the environment quite moist. In such a moist environment, and in the presence of a readily available source of nitrogen, the algae grow and multiply rapidly. At the same time, cultivated land plants grow rather poorly under such wet conditions and the algae begin to grow not only on the soil but also on the surface of leaves, shoots, and so on of such plants without withdrawing any nutrients from the land plants. When algae grow on lawns or golf course grasses, the overrun grass plants lose vigor and

FIGURE 13-11 Parasitic green algae (*Cephaleuros* sp.) symptoms usually appear as spots on leaves (A–C), but sometimes appear as spots on stems (D) and on fruits (E and F). [Photographs courtesy of University of Florida.]

appear to thin out. The thinned out areas are then colonized by algae of various shapes and colors. Most of the time, however, these algae are green or brown and appear like sheets, leaves, or cushions. Because algae contain a large percentage of water and because their vegetative body frequently contains a high amount of gelatin, the areas of lawn or golf courses that are invaded by algae become quite slippery. Many times, the algae grow and multiply so prolifically that fairly large areas of turf are covered by algae (Fig. 13-13A). The algae continue to grow, multiply, and expand outward as long as the high moisture conditions prevail. When dry weather sets in later, the algae and the plants or soil they were growing on dry up and form a caked, cracked sheet (Figs. 13-13B and 13-13C) that sometimes can almost be peeled off from the plants and the soil. Such

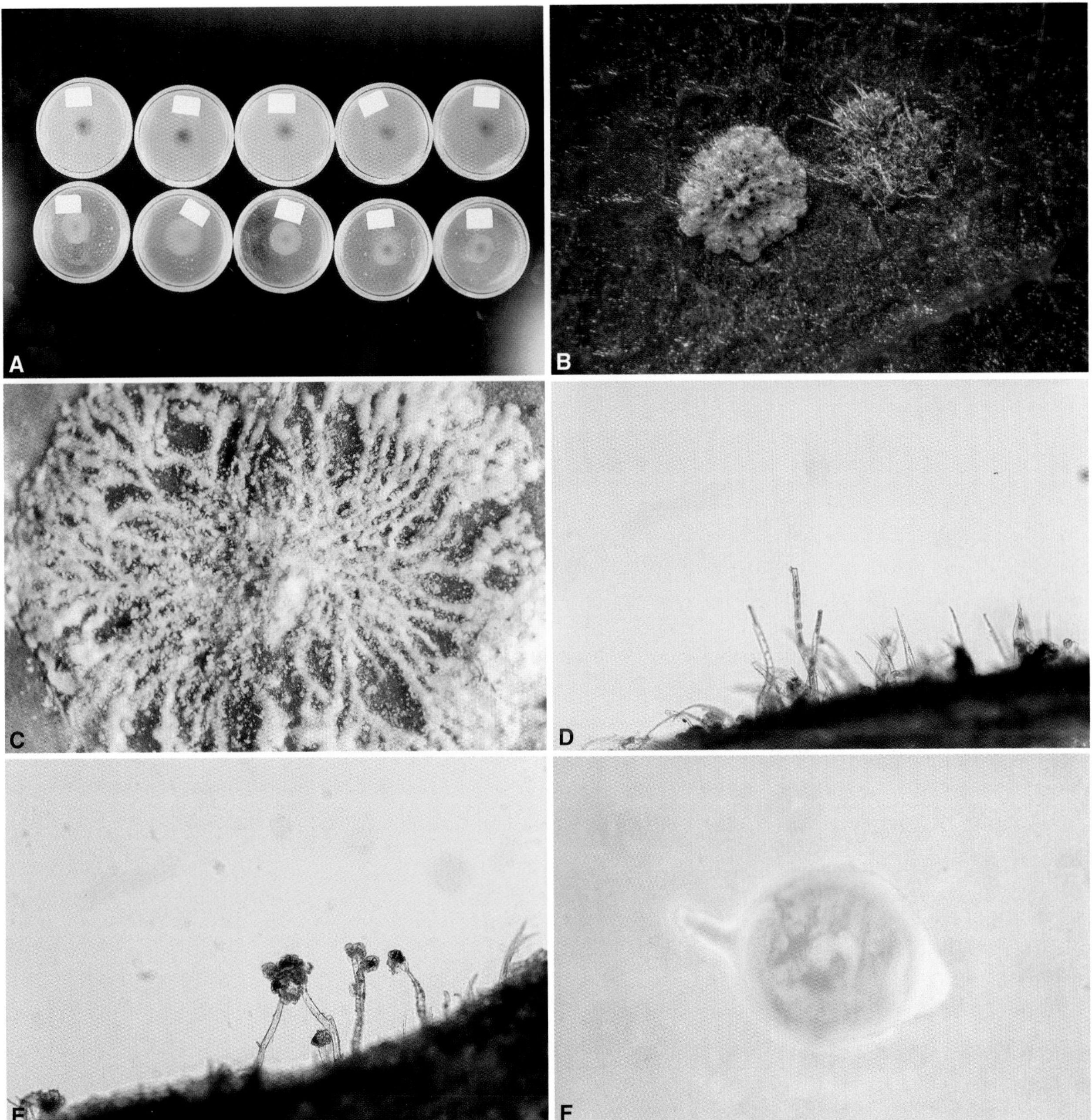

FIGURE 13-12 Colonies of the parasitic green alga *Cephaleuros virescens*. (A) Colonies grown on nutrient media kept in bright light (yellow) or in dim light (green). (B and C) Colonies under increased magnification revealing filaments. (D and E) Air filaments, some of which carry sporangia (E and F). (F) Sporangium. [Photographs courtesy of University of Florida.]

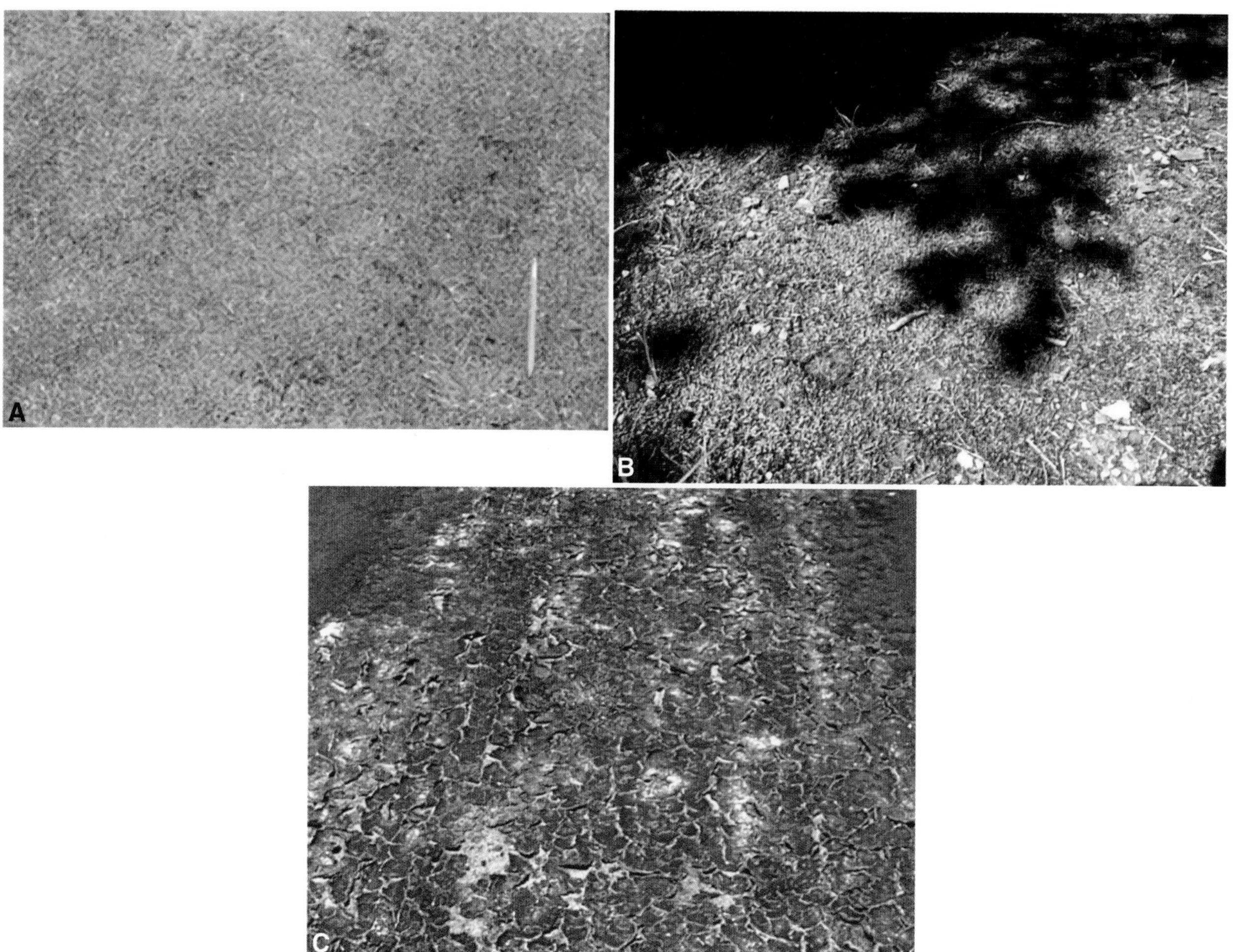

FIGURE 13-13 Turf grasses showing increasing degrees of severity of damage caused by algae. (A) Some green algae begin to grow among and to displace the turf grass. (B) Algae have eliminated grass plants from larger areas. (C) Grass has been replaced by algae, which have formed a hard layer that is cracked in innumerable areas. [Photographs courtesy of University of Florida.]

algae can be managed by reversing the high moisture conditions, if possible, reducing the availability of nitrogen, and, if needed, by spraying with approved fungicides, which, however, control the algae for a rather short period of time.

Selected References

Hood, I. A. (1985). Algal and fungal leaf spots of native plants. New Zealand Forest Service, Forest Pathology in New Zealand No. 12.

Joubert, J. J., and Rijkenberg, F. H. J. (1971). Parasitic green algae. *Annu. Rev. Phytopathol.* 9, 45–64.

chapter fourteen

PLANT DISEASES CAUSED *BY* VIRUSES

INTRODUCTION – CHARACTERISTICS OF PLANT VIRUSES: DETECTION – MORPHOLOGY – COMPOSITION AND STRUCTURE: OF VIRAL PROTEIN – OF VIRAL NUCLEIC ACID, – SATELLITE VIRUSES AND SATELLITE RNAS
724

PROPERTIES OF PLANT VIRUSES: THE BIOLOGICAL FUNCTION OF VIRAL COMPONENTS: CODING – VIRUS INFECTION AND VIRUS SYNTHESIS – TRANSLOCATION AND DISTRIBUTION OF VIRUSES IN PLANT – SYMPTOMS CAUSED BY PLANT VIRUSES – PYSIOLOGY OF VIRUS-INFECTED PLANTS – TRANSMISSION OF PLANT VIRUSES – EPIDEMIOLOGY OF PLANT VIRUSES AND VIROIDS
731

WORKING WITH AND MANAGING PLANT VIRUSES: – PURIFICATION OF PLANT VIRUSES – SEROLOGY OF PLANT VIRUSES – NOMENCLATURE AND CLASSIFICATION OF PLANT VIRUSES – DETECTION AND IDENTIFICATION – ECONOMIC IMPORTANCE – CONTROL – THE GROUPS OF PLANT VIRUSES
743

DISEASES CAUSED BY RIGID ROD SSRNA VIRUSES – DISEASES CAUSED BY TOBAMOVIRUSES: TOBACCO MOSAIC – THE CONTRIBUTION OF TOBACCO MOSAIC VIRUS TO BIOLOGY AND MEDICINE DISEASES CAUSED BY: TOBRAVIRUSES: – TOBACCO RATTLE
757

DISEASES CAUDED BY FILAMENTOUS ssRNA VIRUSES – DISEASES CAUSED BY POTEXVIRUSES – DISEASES CAUSED BY POTYVIRIDAE – DISEASES CAUSED BY POTYVIRUSES: BEAN COMMON MOSAIC AND BEAN YELLOW MOSAIC – LETTUCE MOSAIC – PLUM POX – PAPAYA RINGSPOT – POTATO VIRUS Y – SUGARCANE MOSAIC – TOBACCO ETCH – TURNIP MOSAIC – WATERMELON MOSAIC – ZUCCHINI YELLOW MOSAIC
762

DISEASES CAUSED BY CLOSTEROVIRUSES: CITRUS TRISTEZA – BEET YELLOWS – DISEASES CAUSED BY CRINIVIRUSES: LETTUCE INFECTIOUS YELLOWS
774

DISEASES CAUSED BY ISOMETRIC ssRNA VIRUSES – SEQAUIVIRIDAE: RICE TUNGRO DISEASES – LUTEOVIRIDAE: BARLEY YELLOW DWARF – POTATO LEAFROLL – BEET WESTERN YELLOWS – COMOVIRIDAE: COMOVIRUSES – NEPOVIRUSES: TOMATO RING SPOT – GRAPEVINE FANLEAF-RASPBERRY RING SPOT – BROMOVIRIDAE: CUCUMOVIRUSES: CUCUMBER MOSAIC – ILARVIRUSES: PRUNUS NECROTIC RING SPOT
779

DISEASES CAUSED BY ISOMETRIC DSRNA VIRUSES – REOVIRIDAE
792

DISEASES CAUSED BY NEGATIVE (-)ssRNA VIRUSES – BACILLIFORM – RHABDOVIRUSES – MEMBRANOUS CIRCULAR – BANYOVIRIDAE – TOSPOVIRUSES – TOMATO SPOTTED WILT – THIN, FLEXUOUS, MULTIPARTITE – TENUIVIRUSES
794

DISEASES CAUSED BY DSDNA VIRUSES – ISOMETRIC – CAULIMOVIRUSES – CAULIFLOWER MOSAIC – BACILLIFORM – BADNAVIRUSES
801

DISEASES CAUSED BY GEMINI ssDNA VIRUSES – GEMINIVIRIDAE – BEET CURLY TOP – MAIZE STREAK – AFRICAN CASSAVA MOSAIC – BEAN GOLDEN MOSAIC – SQUASH LEAF CURL – TOMATO MOTTLE – TOMATO YELLOW LEAF CURL
805

DISEASES CAUSED BY ISOMETRIC ssNDA VIRUSES CIRCOVIRIDAE – BANANA BUNCHY TOP – COCONUT FOLIAR DECAY
813

VIROIDS – DISEASES CAUSED BY VIROIDS – TAXONOMY (GROUPING) OF VIROIDS POTATO SPINDLE TUBER – CITRUS EXOCORTIS – COCONUT CADANG-CADANG
816

INTRODUCTION

A virus is a nucleoprotein that multiplies only in living cells and has the ability to cause disease. It is too small to be seen individually with a light microscope. All viruses parasitize cells and cause a multitude of diseases in all forms of living organisms. Some viruses attack humans, animals, or both and cause such diseases as influenza, polio, rabies, smallpox, acquired immunodeficiency syndrome (AIDS), and warts; others attack higher plants; and still others attack microorganisms, such as fungi and bacteria. The total number of viruses known to date exceeds 2,000, and new viruses are described almost every month. Nearly half of all known viruses attack and cause diseases in plants. One virus may infect one or dozens of different species of plants, and each species of plant is usually attacked by many different kinds of viruses. A plant may sometimes be infected by more than one kind of virus at the same time.

Although viruses behave like microorganisms in that they have genetic functions, are able to reproduce, and cause disease, they also behave as chemical molecules. At their simplest, viruses consist of nucleic acid and protein, with the protein forming a protective coat around the nucleic acid. Although viruses can take any of several forms, they are mostly rod shaped, polyhedral, or variants of these two basic structures. In each virus, there is always only RNA or only DNA and, in most plant viruses, there is only one kind of protein. Some viruses, however, may have two or more different proteins.

Viruses do not divide and do not produce any kind of specialized reproductive structures such as spores. Instead, they multiply by inducing host cells to make more virus. Viruses cause disease not by consuming cells or killing them with toxins, but by utilizing cellular substances during multiplication, taking up space in cells, and disrupting cellular processes. These in turn upset the cellular metabolism and lead to the development of abnormal substances and conditions injurious to the functions and the life of the cell or the organism.

CHARACTERISTICS OF PLANT VIRUSES

Plant viruses differ greatly from all other plant pathogens not only in size and shape, but also in the simplicity of their chemical constitution and physical structure, methods of infection, multiplication, translocation within the host, dissemination, and the symptoms they produce on the host. Because of their small size and the fact that they are transparent, viruses generally cannot be viewed and detected by the methods used for other pathogens. Cell inclusions consisting of virus particles, however, are visible by light microscopy. Viruses are not cells nor do they consist of cells.

Detection

When, from the symptoms exhibited by the plant (Figs. 14-1A–14-1D), a plant disease appears to be caused by a virus, individual virus particles are too small to be seen with the light microscope. Frequently, however, young leaf cells of virus-infected plants contain inclusion bodies of fairly distinctive shapes and sizes (Fig. 14-2). Such inclusion bodies consist of virus aggregates that can be seen with the light microscope and can be used to detect and identify the genus of the virus. Examination of cell sections or of crude sap from virus-infected plants under the electron microscope may reveal details of virus arrangement in the inclusion bodies and also independently occurring virus-like particles (Fig. 14-3). The presence of virus particles of a certain shape and size (Fig. 14-4) in a given host plant can be used for a quick identification of the virus. Particles of many viruses are not always easy to find under the electron micro-scope, however, and even when such particles are revealed, proof that the particles are of the virus that causes the particular disease requires much additional work and time.

A few plant symptoms, such as oak-leaf patterns on leaves (Fig. 14-1B) and chlorotic or necrotic ring spots, can be attributed to viruses with some degree of certainty. Some of the other symptoms shown in Fig. 14-1 can be identified by an experienced person as caused by a virus and, indeed, that some of them are caused by a certain virus. Most other symptoms caused by viruses resemble those caused by mutations, nutrient deficiencies or toxicities, insect or mite feeding damage, other pathogens, and other factors. The determination, therefore, that certain plant symptoms are caused by viruses involves the elimination of every other possible cause

FIGURE 14-1 Some of the types of symptoms caused by viruses on plants. (A) Mosaic or mottle on cowpea leaf. (B) Line pattern or mosaic on rose leaves. (C) Leaf malformation (shoe string) on squash leaves. (D) Pitting on stem of grapevine. [Photographs courtesy of (B and C) Plant Pathology Department, University of Florida.]

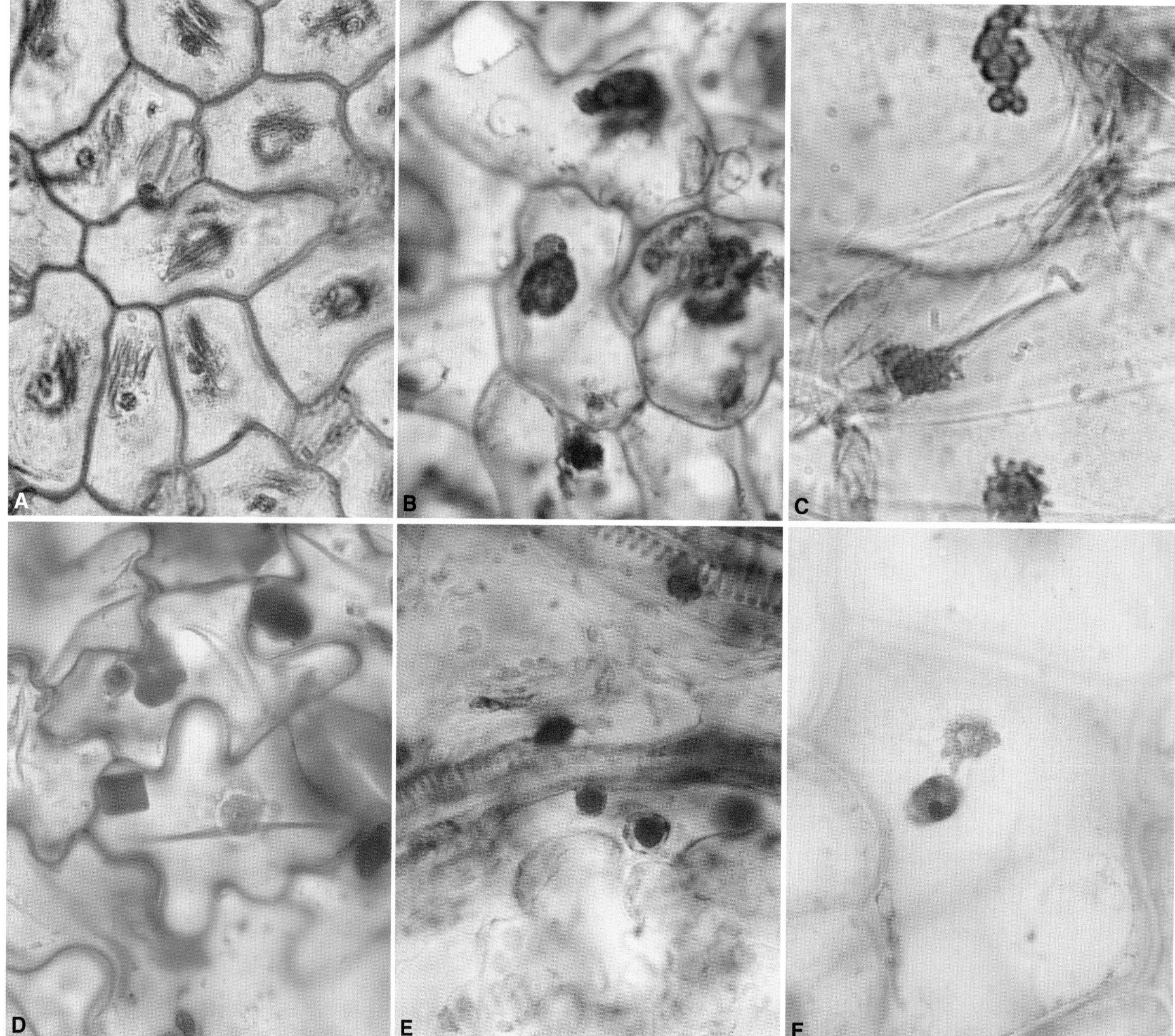

FIGURE 14-2 Cellular inclusions produced by plant cells in reaction to infection by certain viruses. The inclusions are quite specific for a virus, can be observed with a high-power compound microscope, and help identify the virus, usually to genus or family. (A) *Tobacco mosaic virus* inclusion. (B) *Bean yellow mosaic virus* inclusion. (C) *Cucumber mosaic virus*. (D) *Cowpea mosaic virus*. (E) *Tomato spotted wilt virus*. (F) *Tomato mottle virus*. (Photographs courtesy of M. Gouch, University of Florida.)

of the disease and the transmission of the virus from diseased to healthy plants in a way that would exclude transmission of any other causal agent.

The present methods of detecting plant viruses involve primarily the transmission of the virus from a diseased to a healthy plant by budding or grafting, or by rubbing leaves of healthy plants with sap from an infected plant. Certain other methods of transmission, such as by dodder or insect vectors, are also used to demonstrate the presence of a virus. Most of these methods, however, cannot distinguish whether the pathogen is a virus, a mollicute, or a fastidious vascular

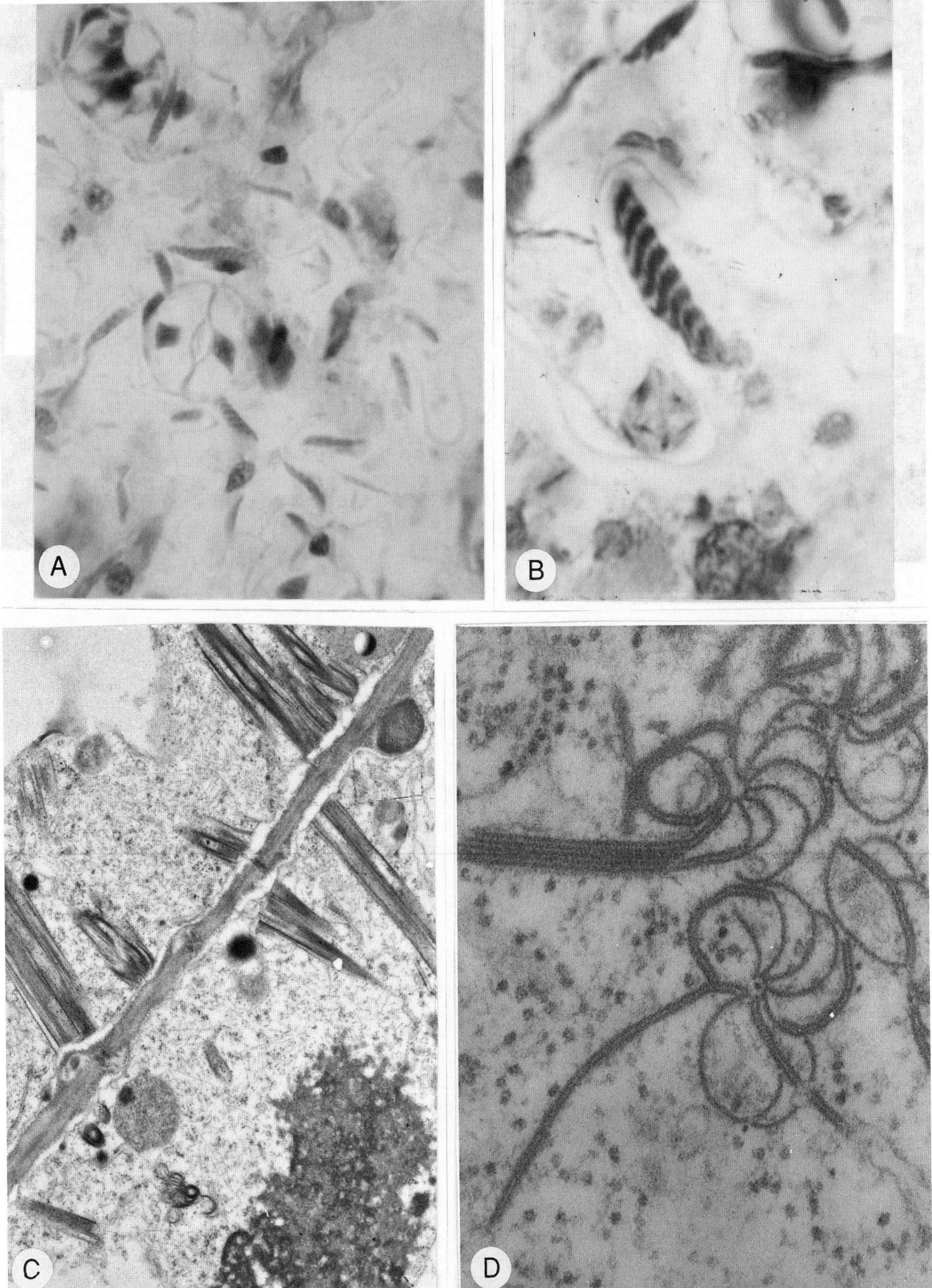

FIGURE 14-3 Cell inclusion bodies observed with an electron microscope. As mentioned in the text, they are indicative of the presence of one or more viruses and, often, are diagnostic of the genus of the virus that induces them. (A and B) Inclusion bodies in plant cells infected with potexviruses (A, 2000×; B, 8000×). (C) Cylindrical and irregular inclusion bodies induced by potyviruses (48,000×). (D) Typical pinwheel-like inclusion bodies diagnostic of potyviruses (125,000×). (Photos courtesy R. G. Christie.)

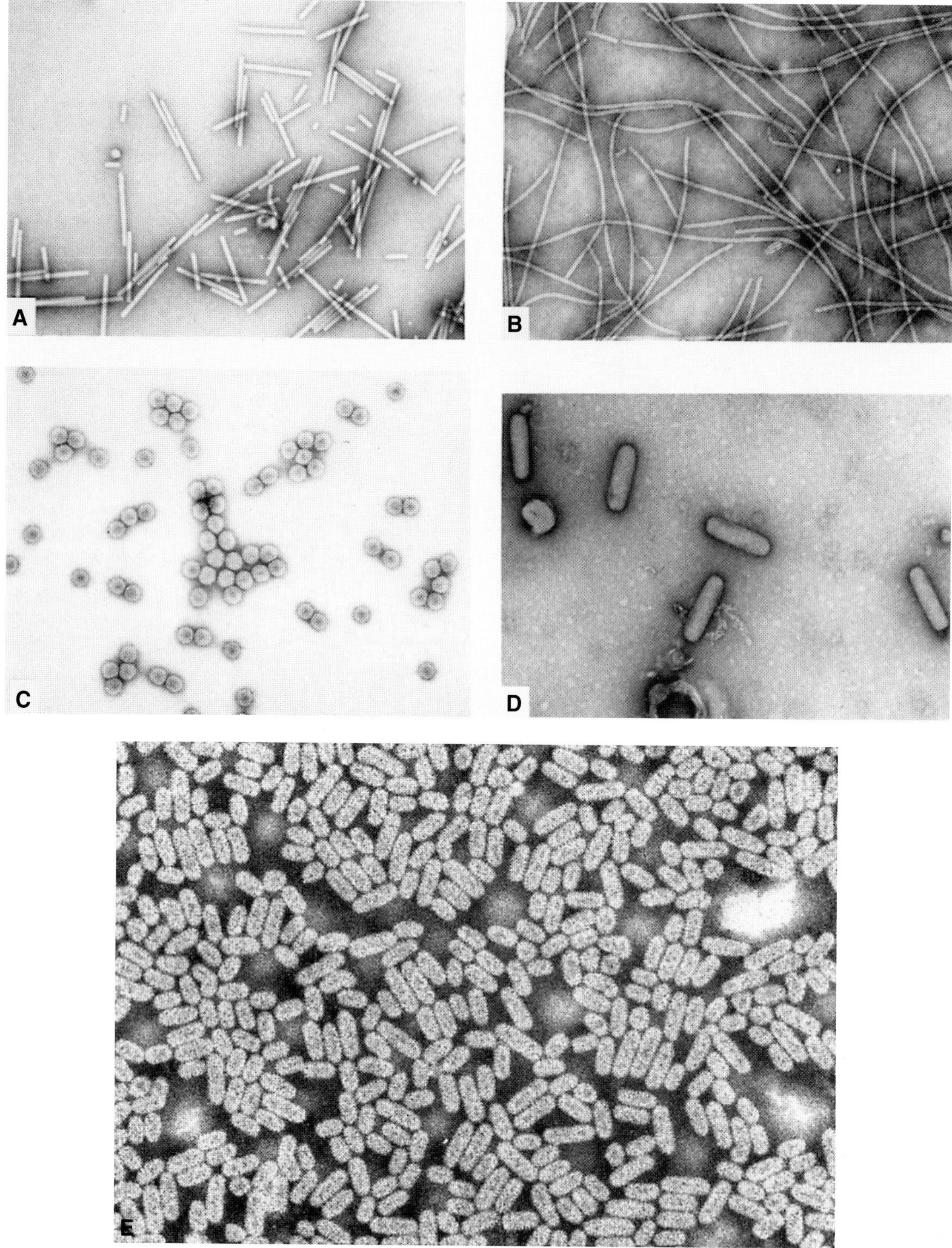

FIGURE 14-4 Electron micrographs of the various shapes of plant viruses. (A) Rod-shaped virus (*tobacco mosaic virus*) (36,000×). (B) Flexuous thread virus (*sugarcane mosaic virus*) (80,000×). (C) Isometric virus (*cowpea chlorotic mottle virus*) (100,000×). (D) Bacilliform rhabdovirus (*broccoli necrotic yellows virus*) (28,500×). (E) The various shapes and sizes of *alfalfa mosaic virus* (168,000×). [Photographs courtesy of (D) Lin and Campbell (1972). *Virology* **48**, 30–40, and (E) E. M. J. Jaspars.]

bacterium; only transmission through bacteria- and fungi-free plant sap is currently considered as proof of the viral nature of the pathogen. The most definitive proof of the presence of a virus in a plant is provided by purification, electron microscopy, and, most commonly, serology. In the past 5 to 10 years, the use of DNA or RNA probes and amplification of segments of viral nucleic acid through polymerase chain reaction (PCR) techniques have gained popularity as sensitive methods for the detection and identification of many viruses.

Morphology

Plant viruses come in different shapes and sizes. Nearly half of them are elongate (rigid rods or flexuous threads), and almost as many are spherical (isometric or polyhedral), with the remaining being cylindrical bacillus-like rods (Figs. 14-4 and 14-5). Some elongated viruses are rigid rods about 15 by 300 nanometers, but most appear as long, thin, flexible threads that are usually 10 to 13 nanometers wide and range in length from 480 to 2,000 nanometers. Rhabdoviruses are short, bacilluslike, cylindrical rods, approximately three to five times as long as they are wide (52–75 by 300–380 nm). Most spherical viruses are actually polyhedral, ranging in diameter from about 17 nanometers (tobacco necrosis satellite virus) to 60 nanometers (wound tumor virus). Tomato spotted wilt virus is surrounded by a membrane and has a flexible, spherical shape about 100 nanometers in diameter.

Many plant viruses have split genomes, i.e., they consist of two or more distinct nucleic acid strands encapsidated in different-sized particles made of the same protein subunits. Thus, some, like tobacco rattle virus, consist of two rods, a long one (195 by 25 nm) and a shorter one (43 by 25 nm), whereas others, like alfalfa mosaic virus, consist of four components of different sizes (Fig. 14-4E). Also, many isometric viruses have two or three different components of the same size but containing nucleic acid strands of different lengths. In multicomponent viruses, all of the nucleic acid strand components must be present in the plant for the virus to multiply and perform in its usual manner.

The surface of viruses consists of a definite number of protein subunits, which are arranged spirally in the elongated viruses and packed on the sides of the polyhedral particles of the spherical viruses (Fig. 14-5). In cross section, the elongated viruses appear as hollow tubes with the protein subunits forming the outer coat and the nucleic acid, also arranged spirally, embedded between the inner ends of two successive spirals of the protein subunits. In spherical viruses the visible shell consists of protein subunits, while the nucleic acid is inside the shell and is arranged in an as yet unknown manner.

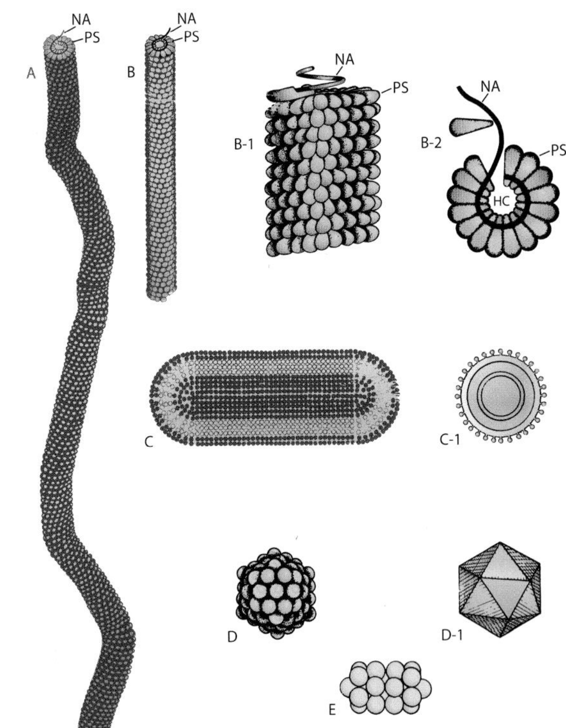

FIGURE 14-5 Relative shapes, sizes, and structures of some representative plant viruses. (A) Flexuous thread-like virus. (B) Rigid rod-shaped virus. (B-1) Side arrangement of protein subunits (PS) and nucleic acid (NA) in viruses A and B. (B-2) Cross-section view of the same viruses. HC, hollow core. (C) Short, bacillus-like virus. (C-1) Cross-section view of such a virus. (D) Isometric polyhedral virus. (D-1) Icosahedron representing the 20-sided symmetry of the protein subunits of the isometric virus. (E) Geminivirus consisting of twin particles.

Rhabdoviruses, and a few spherical viruses, are provided with an outer lipoprotein envelope or membrane. Inside the membrane is the nucleocapsid, consisting of nucleic acid and protein subunits.

Composition and Structure

Each plant virus consists of at least a nucleic acid and a protein. Some viruses consist of more than one size of nucleic acid and proteins, and some of them contain enzymes or membrane lipids.

The nucleic acid makes up 5 to 40% of the virus, protein making up the remaining 60 to 95%. The lower nucleic acid percentages are found in the elongated viruses, whereas the spherical viruses contain higher percentages of nucleic acid. The total mass of the nucleoprotein of different virus particles varies from 4.6 to 73 million daltons. The weight of the nucleic acid alone, however, ranges only between 1 and 3 million ($1–3 \times 10^6$) daltons per virus particle for most viruses, although some have up to 6×10^6 daltons and the 12 component wound tumor virus nucleic acid is approximately 16×10^6 daltons. All viral nucleic acid sizes are quite small when compared to 0.5×10^9 daltons for mollicutes and 1.5×10^9 daltons for bacteria.

Composition and Structure of Viral Protein

Viral proteins, like all proteins, consist of amino acids. The sequence of amino acids within a protein, which is dictated by the sequence of nucleotides in the genetic material, determines the nature and properties of the protein.

The protein shells of plant viruses are composed of repeating subunits. The amino acid content and sequence for identical protein subunits of a given virus are constant but vary for different viruses and even for different strains of the same virus. Of course, the amino acid content and sequence are different for different proteins of the same virus particle and even more so for different viruses. The content and sequences of amino acids are known for the proteins of many viruses. For example, the protein subunit of tobacco mosaic virus (TMV) consists of 158 amino acids in a constant sequence and has a mass of 17,600 daltons (often written as 17.6 kDa, 17.6 kd, or 17.6 K).

In TMV the protein subunits are arranged in a helix containing 16 1/3 subunits per turn (or 49 subunits per three turns). The central hole of the virus particle down the axis has a diameter of 4 nanometers, whereas the maximum diameter of the particle is 18 nanometers. Each TMV particle consists of approximately 130 helix turns of protein subunits. The nucleic acid is packed tightly in a groove between the helices of protein subunits. In rhabdoviruses the helical nucleoproteins are enveloped in a membrane.

In polyhedral plant viruses the protein subunits are packed tightly in arrangements that produce 20 (or some multiple thereof) facets and form a shell. Within this shell the nucleic acid is folded or otherwise organized.

Composition and Structure of Viral Nucleic Acid

The nucleic acid of most plant viruses consists of RNA, but a large number of viruses have been shown to contain DNA. Both RNA and DNA are long, chain-like molecules consisting of hundreds or, more often, thousands of units called nucleotides. Each nucleotide consists of a ring compound called the base attached to a five-carbon sugar [ribose (**I**) in RNA, deoxyribose (**II**) in DNA], which in turn is attached to phosphoric acid (Fig. 14-6). The sugar of one nucleotide reacts with the phosphate of another nucleotide, which is repeated many times, thus forming the RNA or DNA strand. In viral RNA, only one of four bases, adenine, guanine, cytosine, and uracil, can be attached to each ribose molecule. The first two, adenine and guanine, are purines and interact with the other two, uracil and cytosine, the pyrimidines. The chemical formulas of the bases and one of their possible relative positions in the RNA chain are shown in Fig. 14-6 (structure **III**). DNA is similar to RNA with two small, but very important differences: the oxygen of one sugar hydroxyl is missing and the base uracil is replaced by the base methyluracil, better known as thymine (**IV**). The size of both RNA and DNA is expressed either in daltons or as the number of bases [kilobases (kb) for single-stranded RNA and DNA or kilobase pairs (kbp) for double-stranded RNA and DNA], or as the number of nucleotides or nucleotide pairs.

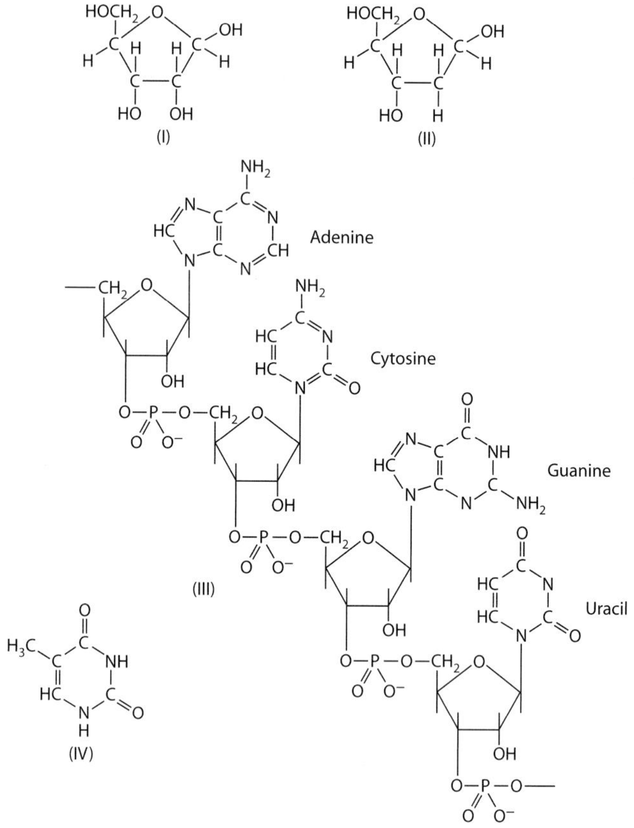

The sequence and the frequency of the bases on the RNA strand vary from one RNA to another, but they

FIGURE 14-6 Chemical formulas of ribose (I), deoxyribose (II), ribonucleic acid or RNA (III), and thymine (IV).

are fixed within a given RNA and determine its properties. Healthy cells of plants always contain double-stranded DNA and single-stranded RNA. Of the nearly 1,000 described plant viruses, most (about 800) contain single-stranded RNA, but 50 contain double-stranded RNA, 40 contain double-stranded DNA, and about 110 contain single-stranded DNA.

Satellite Viruses and Satellite RNAs

Typical viruses consist of one or more rather large strands of nucleic acid contained in a capsid composed of one or more kinds of protein molecules that can multiply and cause infection by themselves. In addition to typical viruses, however, two other types of virus-like pathogens are associated with plant diseases. **Satellite viruses** are viruses but cannot cause infection by themselves. Instead, they must always be associated with certain typical viruses (helper viruses) because they depend on the latter for multiplication and plant infection. Satellite viruses often reduce the ability of the helper viruses to multiply and cause disease; i.e., satellite viruses act like parasites of the associated helper virus. There are also **satellite RNAs**, i.e., small, linear or circular RNAs found inside virions of certain multicomponent viruses. Satellite RNAs are not related, or are only partially related, to the RNA of the virus; satellite RNAs may increase or decrease the severity of viral infections.

PROPERTIER OF PLANT VIRUSES: THE BIOLOGICAL FUNCTION OF VIRAL COMPONENTS: CODING

The protein coat of a virus not only provides a protective sheathing for the nucleic acid of the virus, but also plays a role in determining vector transmissibility of a virus and the kinds of symptoms it causes. Protein itself has no infectivity, but serves to protect the nucleic acid and its presence generally increases the infectivity of the nucleic acid.

The infectivity of viruses is strictly the property of their genomic nucleic acid, which in most plant viruses is RNA. Some viruses carry within them a transcriptase enzyme that they need in order to multiply and infect. The capability, however, of the viral RNA to reproduce both itself and its specific protein indicates that the RNA carries all the genetic determinants of the viral characteristics. The expression of each inherited characteristic depends on the sequence of nucleotides within a certain area (gene) of the viral RNA, which determines the sequence of amino acids in a particular protein, either structural or enzyme. This is called **coding** and seems to be identical in all living organisms and the viruses.

The code consists of coding units called **codons**. Each codon consists of three adjacent nucleotides and determines the position of a given amino acid in the protein being synthesized.

The amount of RNA, then, contained in each virus indicates the approximate length of, and the number of nucleotides in, the viral RNA. This in turn determines the number of codons in each RNA and, therefore, the number of amino acids that can be coded for. In some viruses, the amount of nucleic acid available for coding is increased by having some genes overlap parts of or whole other genes, or by frameshifting, i.e., reading the nucleotides in a different sequence from the first one and thereby forming entirely different codons and genes. Because the protein subunit of viruses contains relatively few amino acids (158 in TMV), the number of codons utilized for its synthesis is only a fraction of the total number of codons available (158 of 2,130 in TMV). In addition to protecting the viral nucleic acid, the coat protein in some cases affects, as mentioned already, the symptoms caused by the virus, the movement of some viruses in their hosts, and transmission of viruses by their vectors.

The remaining codons are presumably involved in the synthesis of other proteins, either structural proteins or enzymes (Fig. 14-7). One of these enzymes is called an RNA polymerase (RNA synthetase or RNA replicase) and is needed to replicate the RNA of the virus. The specific role of some proteins coded for by the viral nucleic acid is still unknown; however, some proteins have been shown to facilitate the movement of the virus through cells; others to be required for transmission of the virus by its vector; some for production of proteins needed for cleaving the nucleic acid of the virus in precise positions; and some for producing the cellular inclusion bodies observed in cells infected by viruses but whose role and function are not known.

So far, it appears that the diseased condition induced in plants by viruses is the result of the interference and disruption of normal metabolic processes in infected parenchyma or specialized cells. Such interference is caused by the mere presence and multiplication of the virus and, possibly, by the abnormal or toxic effects of additional virus-induced proteins or their products, although no such substances have been found to date.

VIRUS INFECTION AND VIRUS SYNTHESIS

Plant viruses enter cells only through wounds made mechanically or by vectors or by deposition into an ovule by an infected pollen grain.

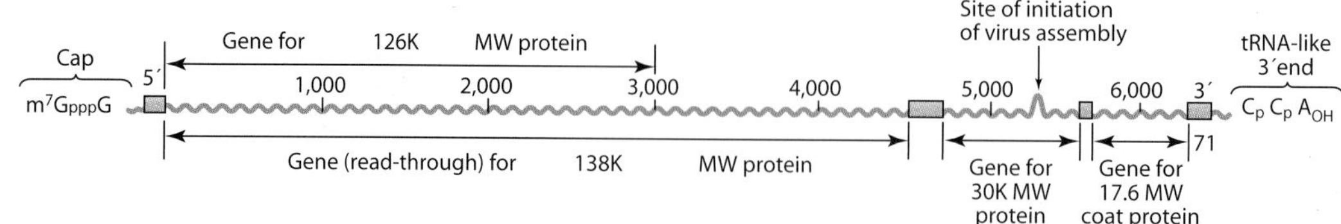

FIGURE 14-7 The 6,400 nucleotide genome of *tobacco mosaic virus* (TMV). Four genes are translated and produce proteins of 126, 183, 30, and 17.6K molecular weight, respectively. The two largest proteins function as the viral replicase(s), the 30K protein facilitates cell-to-cell movement of the virus, and the 17.6K protein makes up the coat protein of the virus. Translation of the viral genome is from left (5′ end) to right (3′ end). Four short segments of the genome (hatched boxes) are not translated. They include signals for initiation, promotion, and termination of translation. The site of the genome at which assembly with coat proteins takes place to produce complete viruses is shown, as are the 5′ end cap of the genome and the transfer RNA-like 3′ end. Numbers along the RNA indicate nucleotides.

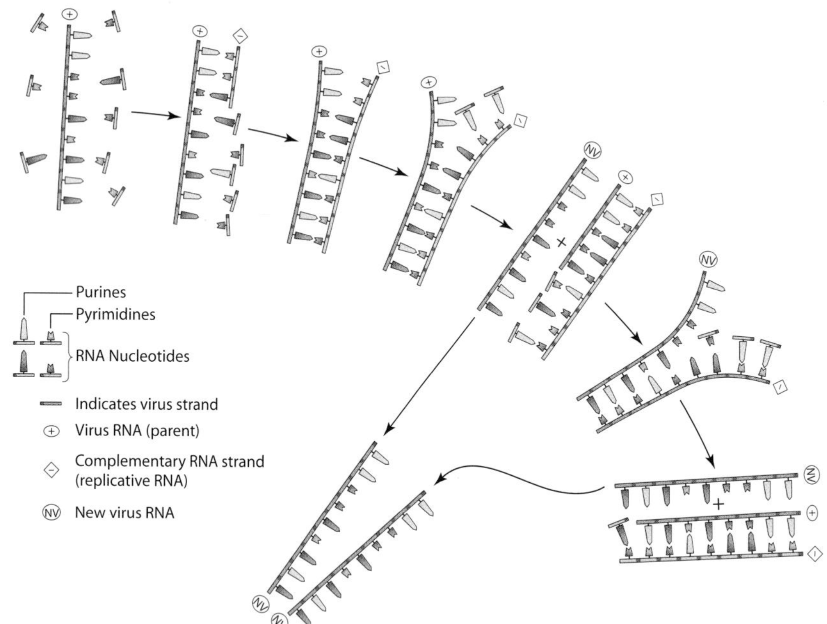

FIGURE 14-8 Schematic representation of viral RNA replication.

In a simplified replication of an RNA virus, the nucleic acid (RNA) of the virus is first freed from the protein coat. It then induces the cell to form the viral RNA polymerase. This enzyme utilizes the viral RNA as a template and forms complementary RNA. The first new RNA produced is not the viral RNA but a mirror image (complementary copy) of that RNA. As the complementary RNA is formed, it is temporarily connected to the viral strand (Fig. 14-8). Thus, the two form a double-stranded RNA that soon separates to produce the original virus RNA and the mirror image (–) strand, with the latter then serving as a template for more virus (+strand) RNA synthesis.

The replication of some viruses differs considerably from the aforementioned scheme. In viruses in which different RNA segments are present within two or more virus particles, all the particles must be present in the same cell for the virus to replicate and for infection to develop. In single-stranded RNA rhabdoviruses the RNA is not infectious because it is the (–) strand. This RNA must be transcribed by a virus-carried enzyme called transcriptase into a (+) strand RNA in the host, and the latter RNA then replicates as described earlier. In double-stranded RNA isometric viruses, the RNA is segmented within the same virus, is noninfectious, and depends for its replication in the host on a transcriptase enzyme also carried within the virus.

On infection of a plant with a double-stranded DNA (dsDNA) virus, the viral dsDNA enters the cell nucleus, where it appears to become twisted and supercoiled and forms a minichromosome. The latter is transcribed into two single-stranded RNAs: the smaller RNA is transported to the cytoplasm, where it is translated into virus-coded proteins and the larger RNA is also transported to the same location in the cytoplasm, but it becomes encapsidated by coat protein subunits and is used as a template for reverse transcription into a complete virion dsDNA. The method of replication of the single-stranded DNA (ssDNA) of plant ssDNA viruses has not yet been determined with certainty. There is some evidence, however, that the ssDNA replicates by forming a rolling circle that produces a multimeric (−) strand (see Fig. 14-64), which serves as a template for the production of multimeric (+) strands that are then cleaved to produce unit length (+) strands.

As soon as new viral nucleic acid is produced, some of it is translated, i.e., it induces the host cell to produce the protein molecules coded by its nucleic acid. Protein synthesis in healthy cells depends on the presence of amino acids and the cooperation of ribosomes, messenger RNA, and transfer RNAs. Each transfer RNA is specific for one amino acid, which it carries toward the appropriate nucleotide sequence along the messenger RNA. Messenger RNA, which is produced in the nu-cleus and reflects part of the DNA code, determines the kind of protein that will be produced by coding the sequence in which the amino acids will be arranged. The ribosomes seem to travel along the messenger RNA and to provide the energy for the bonding of the prearranged amino acids to form the protein (Fig. 14-9).

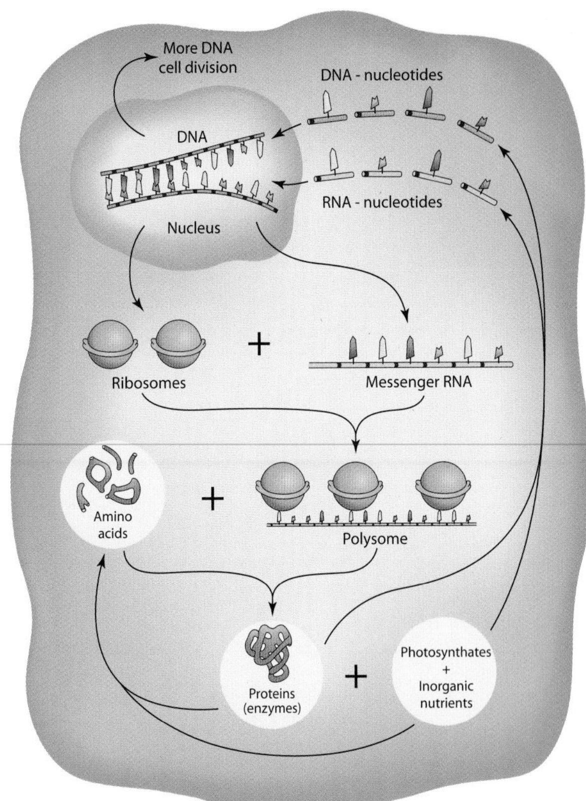

FIGURE 14-9 Schematic representation of the basic functions in a living cell.

For virus protein synthesis, the part of the viral RNA coding for the viral protein plays the role of messenger RNA. The virus utilizes the amino acids, ribosomes, and transfer RNAs of the host; however, it becomes its own blueprint (messenger RNA), and the proteins formed are for exclusive use by the virus as a coat (Fig. 14-10) or in other functions. When new virus nucleic acid and virus protein subunits have been produced, the nucleic acid organizes the protein subunits around it, and the two are assembled together to form the complete virus particle, the virion.

The site(s) of the cell in which virus nucleic acid and protein are synthesized and in which these two components are assembled to produce the virions varies with the particular genus or family of the virus. For most RNA viruses, the virus RNA, after it is freed from the protein coat, replicates itself in the cytoplasm, where it also serves as a messenger RNA and, in cooperation with the ribosomes and transfer RNAs, produces the virus protein subunits. The assembly of virions follows, also in the cytoplasm. In other viruses, e.g., those with ssDNA, the synthesis of viral nucleic acid and protein, as well as their assembly into virions, seems to take place in the nucleus, from which the virus particles are then released into the cytoplasm.

The first intact virions appear in plant cells approximately 10 hours after inoculation. The virus particles may exist singly or in groups and may form amorphous or crystalline inclusion bodies (Fig. 14-2) within the cell areas (cytoplasm, nucleus) in which they happen to be.

TRANSLOCATION AND DISTRIBUTION OF VIRUSES IN PLANTS

When a virus infects a plant, it moves from one cell to another and multiplies in most, if not all, such cells. Viruses move from cell to cell through the plasmodesmata connecting adjacent cells (Fig. 14-11). Viruses multiply in each parenchyma cell they infect. In leaf parenchyma cells the virus moves approximately 1 millimeter, or 8 to 10 cells, per day.

In all economically important viral infections, viruses reach the phloem and through it are transported rapidly

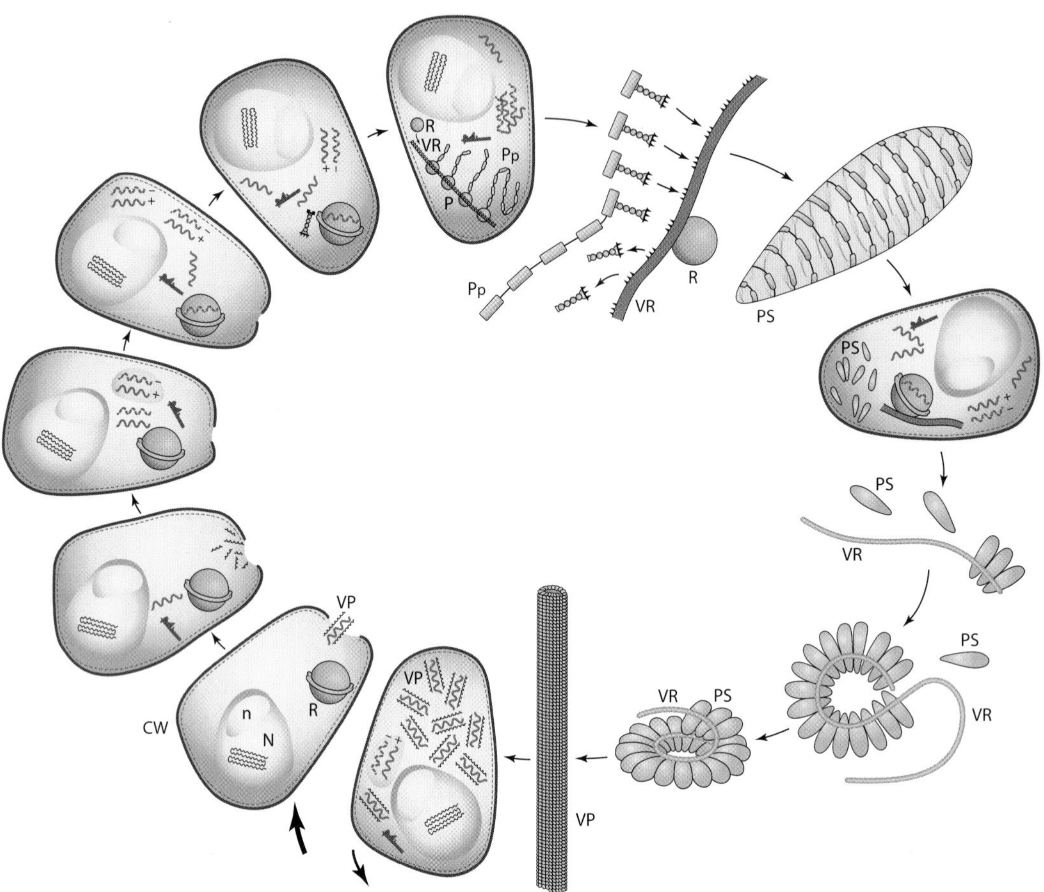

FIGURE 14-10 Sequence of events in virus infection and biosynthesis. CW, cell wall; R, ribosome; N, nucleus; n, nucleolus; P, polyribosome; Pp, polypeptide; PS, protein subunit; VP, viral particle; VR, viral RNA.

over long distances within the plant. Most viruses, however, require 2 to 5 days or more to move out of an inoculated leaf. Once the virus has entered the phloem, it moves rapidly in it toward growing regions (apical meristems) or other food-utilizing parts of the plant, such as tubers and rhizomes (Fig. 14-12). In the phloem, the virus spreads systemically throughout the plant and reenters the parenchyma cells adjacent to the phloem through plasmodesmata.

The development of local lesion symptoms is an indication of the localization of the virus within the lesion area (Fig. 14-13). In several diseases, however, the lesions continue to enlarge and, sometimes, the development of systemic symptoms follows, indicating that the virus continued to spread beyond the borders of the lesions.

In systemic virus infections, some viruses are limited to the phloem and to a few adjacent parenchyma cells. Viruses causing mosaic-type diseases are not generally tissue restricted, although there may be different patterns of localization. Mosaic virus-infected plant cells have been estimated to contain between 100,000 and 10,000,000 virus particles per cell. The systemic distribution of some viruses is quite thorough and may involve all living cells of a plant. Many viruses, however, seem to leave segments of plant tissues that are virus-free. Also, a few viruses invade all new meristem tip tissues, whereas most others leave the growing points of stems or roots of affected plants apparently free of virus.

SYMPTOMS CAUSED BY PLANT VIRUSES

Almost all viral diseases seem to cause some degree of dwarfing or stunting of the entire plant and reduction in total yield. Viruses usually shorten the length of life of virus-infected plants, although they rarely kill plants on infection. These effects may be severe and striking in appearance or they may be very slight and easily overlooked.

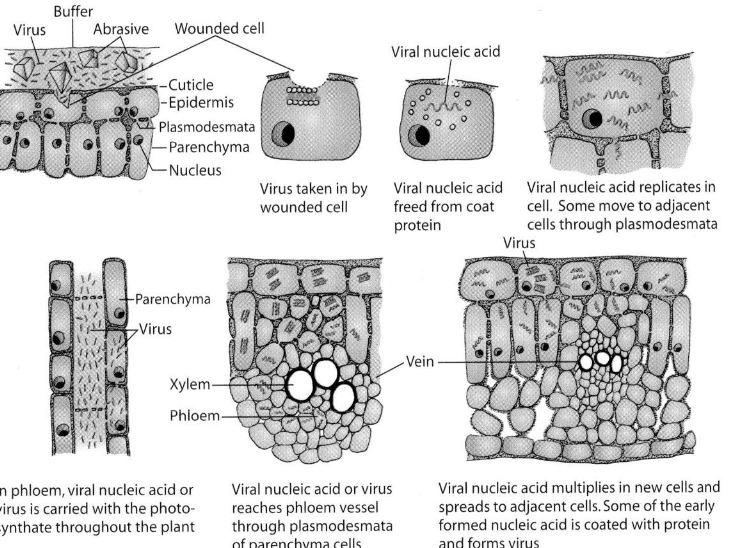

FIGURE 14-11 Mechanical inoculation and early stages in the systemic distribution of viruses in plants.

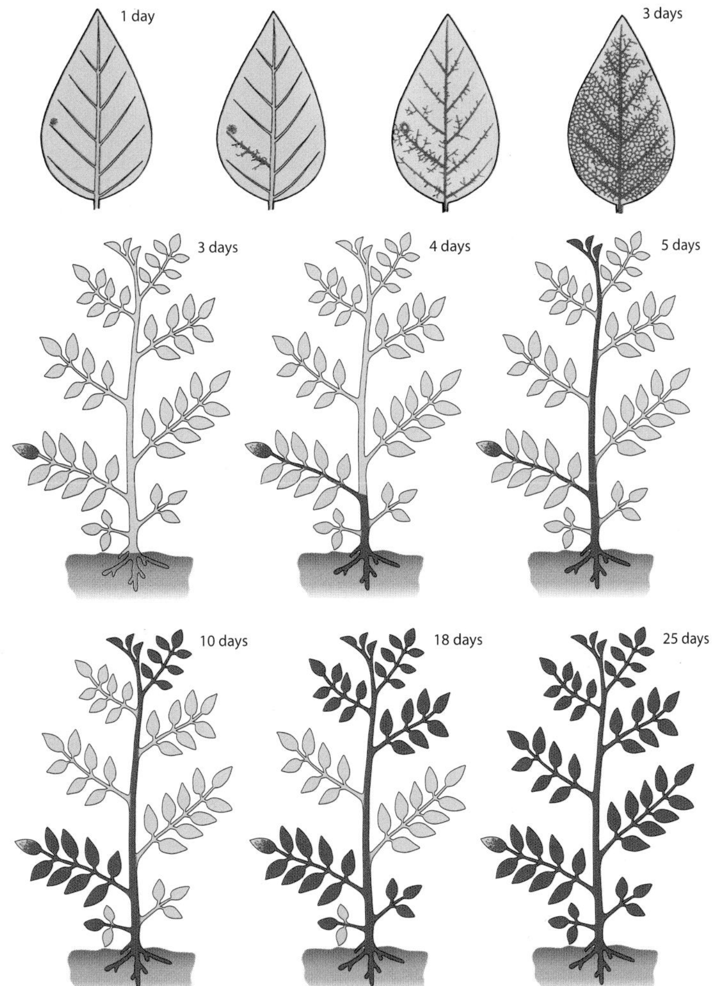

FIGURE 14-12 Schematic representation of the direction and rate of translocation of a virus in a plant. [Adapted from Samuel (1934). *Ann. Appl. Biol.* **21**, 90–11.]

FIGURE 14-13 Types of local lesion symptoms. (A) Mechanical inoculation of corn seedlings with *sugarcane mosaic virus* (SCMV) — infected sap. (B) Early lesion of SCMV infection detected by immunofluorescent microscopy. (C) Local lesions caused by SCMV on sorghum leaf. (D) Local lesions on *Chenopodium* leaf caused by *potato virus Y*. Local lesions on cowpea leaves caused by *alfalfa mosaic virus (AMV)* (E) and on tobacco leaves caused by *tomato ring spot virus* (F).

The most obvious symptoms of virus-infected plants are usually those appearing on the leaves, but some viruses may cause striking symptoms on the stem, fruit, and roots while they may or may not cause any symptom development on the leaves (Figs. 14-1 and 14-14). In almost all virus diseases of plants occurring in the field, the virus is present throughout the plant (systemic infection) and induces the formation of systemic symptoms. In many plants inoculated artificially with certain viruses, the virus causes the formation of small, chlorotic or necrotic lesions only at the points of entry (local infections), and the symptoms are called **local lesions** (Fig. 14-13). However, many viruses infect certain hosts without causing development of visible symptoms on them. Such viruses are usually called **latent viruses**, and the hosts are called **symptomless carriers**. In other cases, however, plants that usually develop symptoms on infection with a certain virus may remain temporarily symptomless under certain environmental conditions (e.g., high or low temperature), and such symptoms are called **masked**. Finally, plants may show acute severe symptoms soon after inoculation that may lead to death of young shoots or of the entire host plant; if the host survives the initial shock phase, the symptoms tend to become milder (chronic symptoms) in the subsequently developing parts of the plant, leading to partial or even total recovery. In some diseases, however, symptoms may increase progressively in severity and may result in gradual (slow) or quick decline of the plant.

The most common types of plant symptoms produced by systemic virus infections are **mosaics** and **ring spots**. Mosaics are characterized by light-green, yellow, or white areas intermingled with the normal green of the leaves or fruit or of lighter-colored areas intermingled with areas of the normal color of flowers or fruit. Depending on the intensity or pattern of discolorations, mosaic-type symptoms may be described as mottling, streak, ring pattern, line pattern, veinclearing, veinbanding, or chlorotic spotting. Ring spots are characterized by the appearance of chlorotic or necrotic rings on the leaves and sometimes also on the fruit and stem. In many ring spot diseases the symptoms, but not the virus, tend to disappear later on.

A large number of other, less common virus symptoms have been described (Fig. 14-14) and include stunt (e.g., tomato bushy stunt), dwarf (barley yellow dwarf), leaf roll (potato leafroll), yellows (beet yellows), streak (tobacco streak), pox (plum pox), enation (pea enation mosaic), tumors (wound tumor), pitting of stem (apple stem pitting), pitting of fruit (pear stony pit), and flattening and distortion of stem (apple flat limb). These symptoms may be accompanied by other symptoms on other parts of the same plant.

PHYSIOLOGY OF VIRUS-INFECTED PLANTS

Plant viruses do not contain any enzymes, toxins, or other pathogenic substances and yet cause a variety of symptoms on the host. The mere presence of viral nucleic acid or complete virus in a plant, even in large quantities, does not always cause disease symptoms. For example, some plants containing much higher concentrations of virus than others may show milder symptoms than the latter or may even be symptomless carriers. Viral diseases of plants, then, are not due primarily to the depletion of nutrients that have been diverted toward synthesis of the virus itself, but rather are due to other, more indirect effects of the virus on the metabolism of the host. These effects are brought about probably through the virus-induced synthesis of new proteins by the host, some of which are biologically active substances (enzymes, etc.) and may interfere with the normal metabolism of the host.

Viruses generally cause a decrease in photosynthesis through decreases in chlorophyll per leaf, chlorophyll efficiency, and leaf area per plant. Viruses usually cause a decrease in the amount of growth-regulating substances (hormones) in the plant, frequently by inducing an increase in growth-inhibiting substances. A decrease in soluble nitrogen during rapid virus synthesis is rather common in virus diseases of plants, and in mosaic diseases there is a chronic decrease in the levels of carbohydrates in the plant tissues.

The respiration of plants is generally increased immediately after infection with a virus. After the initial increase, however, the respiration of plants infected with some viruses remains higher, whereas with other viruses it becomes lower than that of healthy plants, and with still other viruses it may return to normal.

TRANSMISSION OF PLANT VIRUSES

Plant viruses are transmitted from plant to plant in a number of ways. Modes of transmission include vegetative propagation, mechanically through sap, through seed, pollen, dodder, and by specific insects, mites, nematodes, and fungi.

Transmission of Viruses by Vegetative Propagation

Whenever plants are propagated vegetatively by budding or grafting, by cuttings, or by the use of tubers, corms, bulbs, or rhizomes, any viruses present in the mother plant from which these organs are taken will

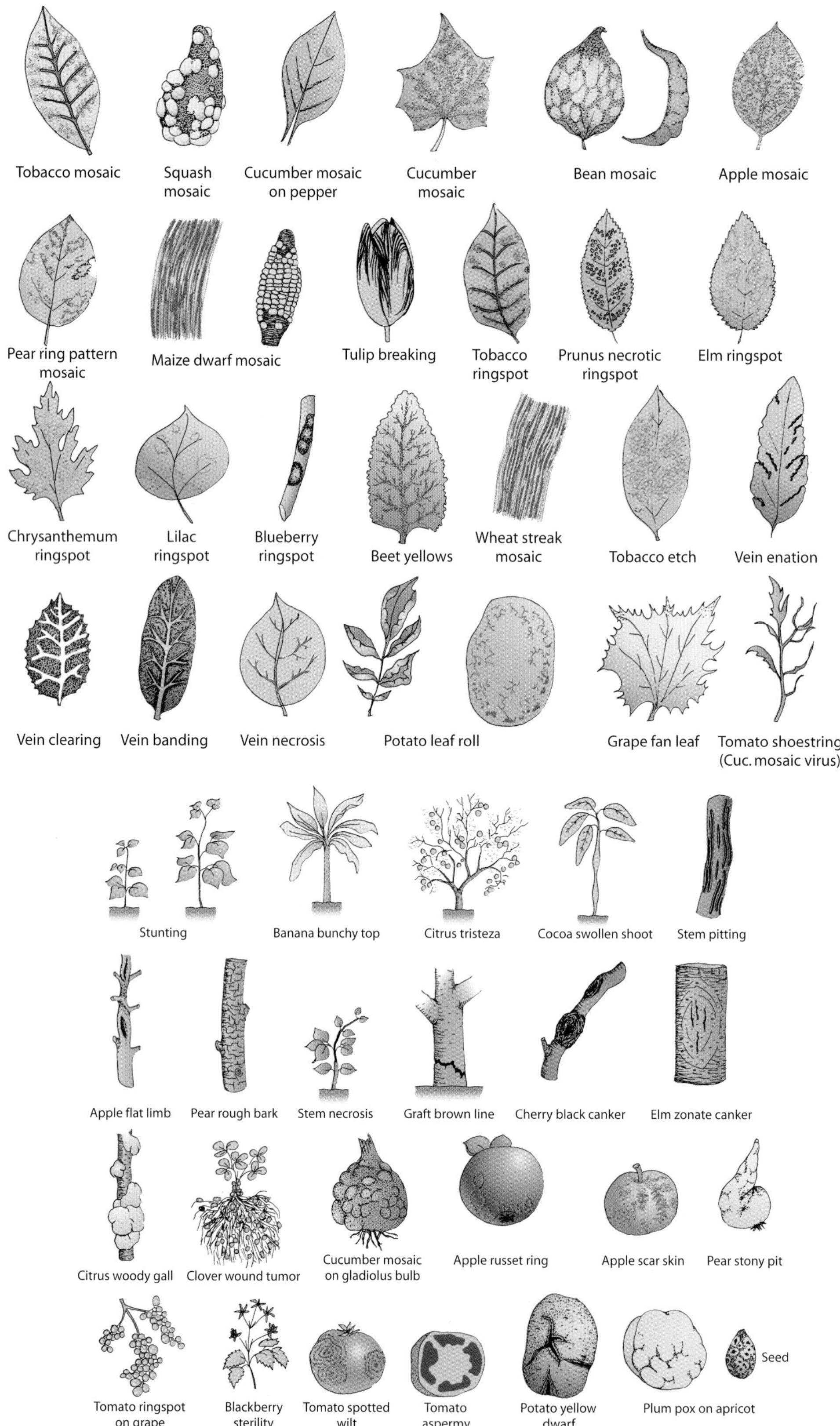

FIGURE 14-14 Types of systemic symptoms caused by viruses and viroids in plants.

almost always be transmitted to the progeny (Fig. 14-15). Considering that almost all fruit and many ornamental trees and shrubs, many field crops, such as potatoes, and most florist's crops are propagated vegetatively, this means of transmission of viruses is the most important for all these types of crop plants. In cases of propagation by budding, the presence of a virus in the bud or in the rootstock may result in an appreciable reduction of successful bud unions with the rootstock and, therefore, in poor stands.

The transmission of viruses may also occur through natural root grafts of adjacent plants, particularly trees (Fig. 14-15). For several tree viruses, natural root grafts are the only known means of tree-to-tree spread of the virus within established orchards.

Mechanical Transmission of Viruses through Sap

The mechanical transmission of plant viruses in nature by the direct transfer of sap through contact of one plant with another is uncommon and relatively unimportant. Such transmission may take place after a strong wind injures the leaves of adjacent diseased and healthy plants or when plants are wounded during cultural practices by tools, hands, or clothes, or by animals feeding on the plants, and the sap carrying virus is transferred to wounded plants (Figs. 14-16 and 14-17). Of the important plant viruses, *potato virus X*, *tobacco mosaic virus*, and *cucumber mosaic virus* are transmitted through sap in the field and may cause severe losses.

The greatest importance of mechanical transmission of plant viruses stems from its indispensability in studying the viruses that cause plant diseases. For mechanical transmission of a virus, young leaves and flower petals are ground to crush the cells and release the virus in the sap (Fig. 14-16). Often a buffer solution is added to stabilize the virus. The sap may be strained to remove tissue fragments and is then applied to the surface of leaves of young plants dusted previously with an abrasive such as Carborundum to aid in wounding of the cells. The sap is applied by rubbing the leaves gently with a cheesecloth, finger (Fig. 14-13A), glass spatula, or painter's brush dipped in the sap or by using a small sprayer. In successful inoculations, the virus enters the leaf cells through the wounds made by the abrasive or through broken leaf hairs and initiates new infections. In local-lesion hosts, symptoms usually appear within three to seven days or more, and the number of local lesions is proportional to the concentration of the virus in the sap. In systemically infected hosts, symptoms usually take 10 to 14 days or more to develop. Sometimes the same plants may first develop local lesions and then systemic symptoms.

In the mechanical transmission of viruses, the taxonomic relationship of the donor and receiving (indicator) plants is unimportant, as virus from one kind of plant, whether herbaceous or a tree, may be transmitted to dozens of unrelated herbaceous plants (vegetables, flowers, or weeds). Several viruses, however, especially of woody plants, are difficult or, so far, impossible to transmit through sap.

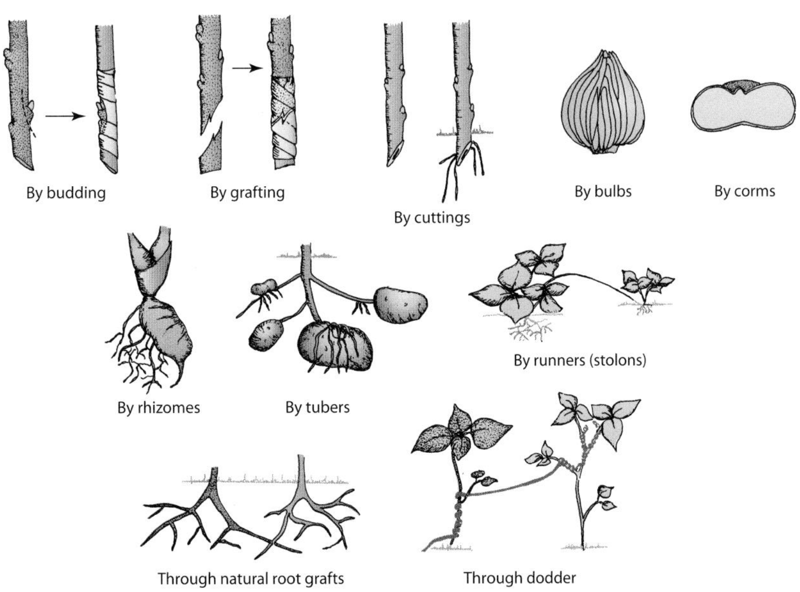

FIGURE 14-15 Transmission of viruses, mollicutes, and other pathogens through vegetative propagation, natural root grafts, and dodder.

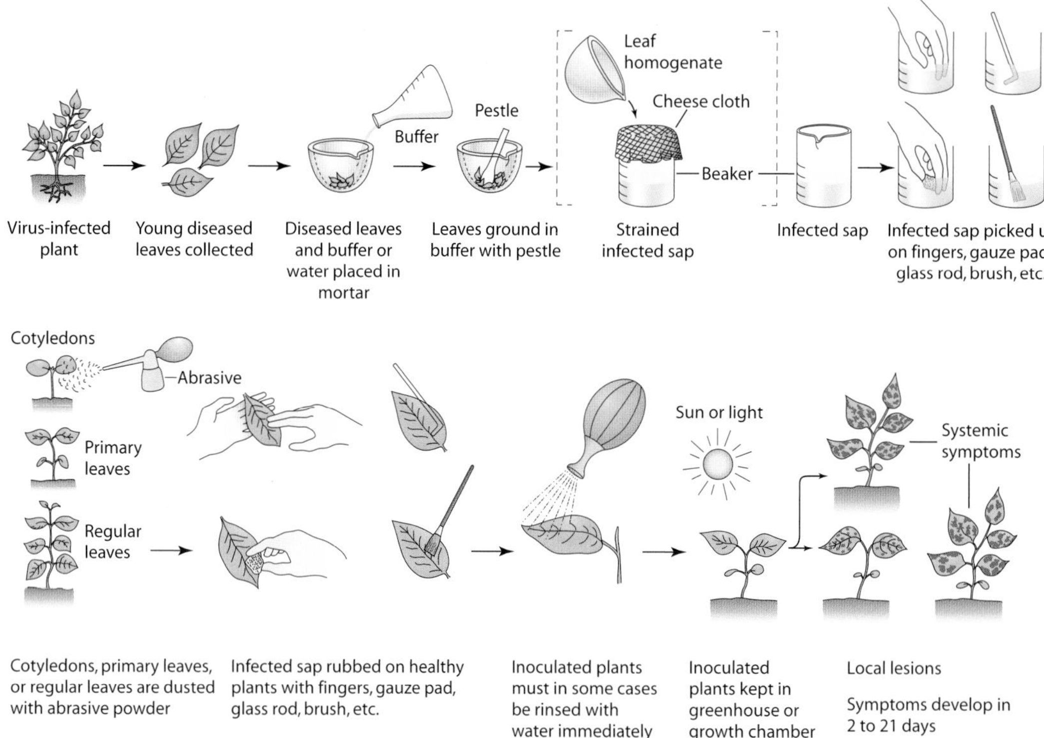

FIGURE 14-16 Typical steps in mechanical or sap transmission of plant viruses.

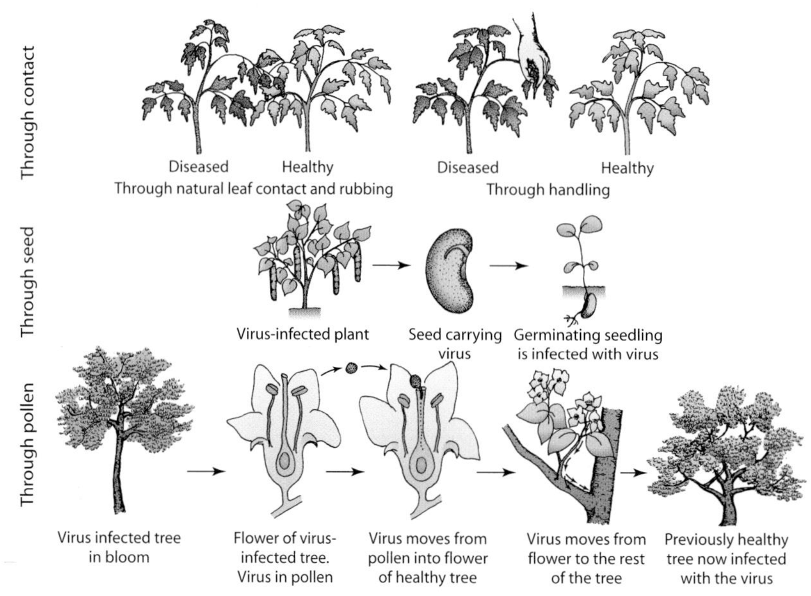

FIGURE 14-17 Plant virus transmission through direct contact of plants, handling, seed, and pollen.

Seed Transmission

More than 100 viruses are transmitted by seed to a smaller or greater extent. As a rule, only a small portion (1–30%) of seeds derived from virus-infected plants of only some hosts of the virus transmit the virus (Fig. 14-17). In some virus–host combinations, however, half or most of the seeds carry the virus, and in a few others 100% of the seeds carry the virus. The frequency of transmission varies with the host–virus combination and with the stage of growth of the mother plant when it becomes infected with the virus.

In most seed-transmitted viruses, the virus seems to come primarily from the ovule of infected plants, but several cases are known in which the virus in the seed seems to be just as often derived from the pollen that fertilized the flower. In some host–virus combinations the virus is carried in the integument of the seed and infects seedlings as they are wounded on germination.

Pollen Transmission

Virus transmitted by pollen may result in reduced fruit set, may infect the seed and the seedling that will grow from it, and, in some cases, can spread through the fertilized flower and down into the mother plant, which thus becomes infected with the virus (Fig. 14-17). Such plant-to-plant transmission of virus through pollen is known to occur, for example, in sour cherry infected with *prunus necrotic ring spot virus.*

Insect Transmission

Undoubtedly the most common and economically most important means of transmission of viruses in the field is by insect vectors. Members of relatively few insect groups, however, can transmit plant viruses (Fig. 14-18). The order Homoptera, which includes aphids (Aphididae), leafhoppers (Cicadellidae), and planthoppers (Delphacidae), contains by far the largest number and the most important insect vectors of plant viruses. Other Homoptera that transmit plant viruses are whiteflies (Aleurodidae), which transmit the usually severe geminiviruses and several other viruses, mealybugs (Coccoidae), and certain treehoppers (Membracidae). A few insect vectors of plant viruses belong to other orders, such as true bugs (Hemiptera), chewing/sucking thrips (Thysanoptera), and beetles (Coleoptera). Grasshoppers (Orthoptera) occasionally seem to carry and transmit a few viruses also. The most important virus vectors are aphids, leafhoppers, whiteflies, and thrips. These and the other groups of Homoptera, as well as true bugs, have piercing and sucking mouthparts. Beetles and

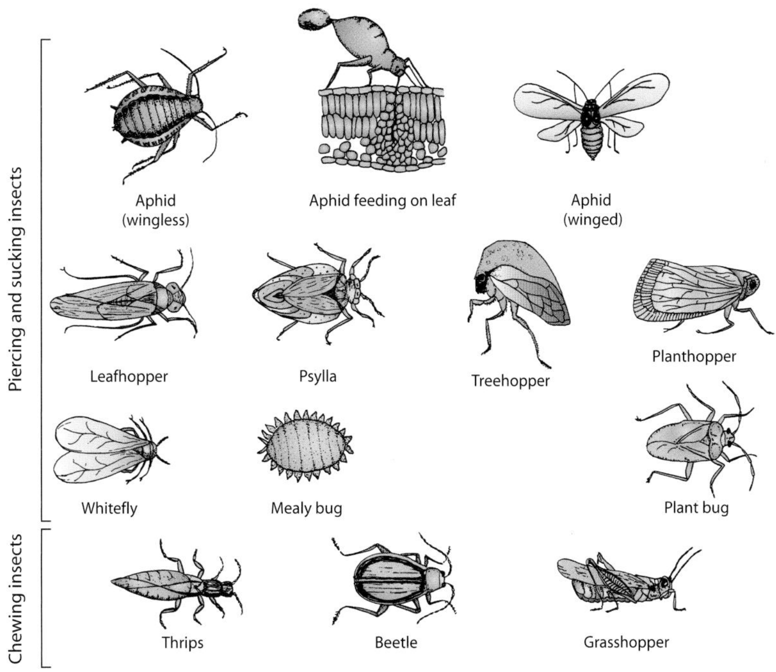

FIGURE 14-18 Insect vectors of plant viruses. Insects in second row from the top also transmit mollicutes and fastidious vascular bacteria.

grasshoppers have chewing mouthparts. Of these, the beetles are quite effective vectors of certain viruses.

Insects with sucking mouthparts carry plant viruses on their stylets — **stylet-borne viruses** — and can acquire and inoculate the virus after short feeding periods of a few seconds to a few minutes. Stylet-borne viruses persist in the vector for only a few to several hours. Therefore, they are also known as **nonpersistent viruses**. With some other viruses, the insect vectors must feed on an infected plant from several minutes or hours to a few days before they accumulate enough virus for transmission. These insects can then transmit the virus after fairly long feeding periods of several minutes to several hours. Such viruses persist in the vector for a few (1 to 4) days and are called **semipersistent viruses**. With still other viruses, the insect vectors accumulate the virus internally and, after passage of the virus through the insect tissues, introduce the virus into plants again through their mouthparts; these viruses are known as **circulative** or **persistent viruses**. Some circulative viruses may multiply in their respective vectors and are then called **propagative viruses**. Viruses transmitted by insects with chewing mouthparts (beetles) may also be circulative or may be carried on the mouthparts.

Aphids are the most important insect vectors of plant viruses and transmit the great majority (about 275) of all stylet-borne viruses. As a rule, several aphid species can transmit the same stylet-borne virus, and the same aphid species can transmit several viruses. In many cases, however, the vector–virus relationship is quite specific. Aphids generally acquire the stylet-borne virus after feeding on a diseased plant for only a few seconds (30 seconds or less) and can transmit the virus after transfer to and feeding on a healthy plant for a similarly short time of a few seconds. The length of time aphids remain viruliferous after acquisition of a stylet-borne virus varies from a few minutes to several hours, after which they can no longer transmit the virus. In aphids transmitting stylet-borne viruses, the virus seems to be borne on the tips of the stylets, it is lost easily through the scouring that occurs during the probing of host cells and it does not persist through the molt or egg. Stylet-borne viruses are said to be transmitted in a nonpersistent manner. In the few cases of aphid transmission of circulative viruses, aphids cannot transmit the virus immediately but must wait several hours after the acquisition feeding; however, once they start to transmit the virus, they continue to do so for many days after the removal of the insects from the virus source (persistent transmission).

Approximately 55 plant viruses are transmitted by leafhoppers, planthoppers, and treehoppers, including viruses with double-stranded RNA, rhabdoviruses, small isometric viruses, and some geminiviruses. Leafhopper- and planthopper-transmitted viruses cause disturbances in plants that affect primarily the region of the phloem. All such viruses are circulatory; several are known to multiply in the vector (propagative) and some persist through the molt and are transmitted through the egg stage of the vector. Most leafhopper and planthopper vectors require a feeding period of one to several days before they become viruliferous, but once they have acquired the virus, they may remain viruliferous for the rest of their lives. There is usually an incubation period of 1 to 2 weeks between the time a leafhopper or planthopper acquires a virus and the first time it can transmit it.

Mite Transmission

Primarily mites of the family Eriophyidae have been shown to transmit at least six viruses, including wheat streak mosaic and several other rymoviruses affecting cereals. These mites have piercing and sucking mouthparts (Fig. 14-19). Virus transmission by eriophyid mites seems to be quite specific, as each of these mites is the only known vector for the virus or viruses it transmits. Another virus, *peach mosaic virus*, is transmitted by mites of the family Tetranychidae.

Nematode Transmission

Approximately 20 plant viruses are transmitted by one or more species of three genera of soil-inhabiting, ectoparasitic nematodes (Fig. 14-19). Nematodes of the genera *Longidorus*, *Paralongidorus*, and *Xiphinema* transmit several polyhedral-shaped viruses known as nepoviruses, such as *grape fanleaf*, *tobacco ring spot*, and other viruses, whereas nematodes of the genera *Trichodorus* and *Paratrichodorus* transmit at least two rod-shaped tobraviruses, *tobacco rattle* and *pea early browning*. Nematode vectors transmit viruses by feeding on roots of infected plants and then moving on to roots of healthy plants. Juveniles as well as adult nematodes can acquire and transmit viruses; however, the virus is not carried through the juvenile molts or through the eggs, and, after molting, the juveniles or the resulting adults must feed on a virus source before they can transmit again.

Fungus Transmission

Root-infecting fungal-like organisms, the plasmodiophoromycetes *Polymyxa* and *Spongospora*, and the chytridiomycete *Olpidium*, transmit at least 30 plant

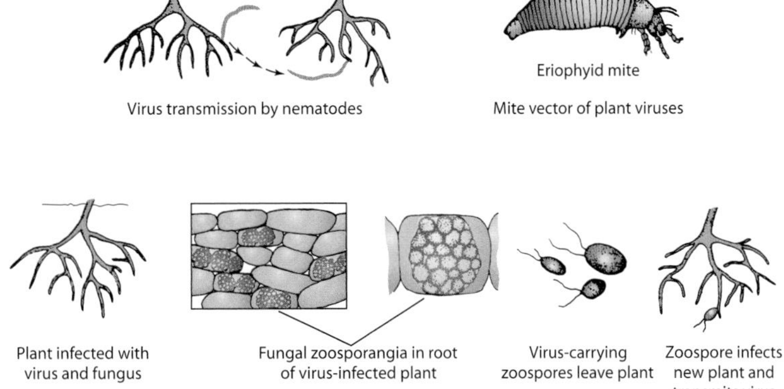

FIGURE 14-19 Transmission of plant viruses by nematodes, mites, and fungi.

viruses. Some of these viruses apparently are borne internally in, whereas others are carried externally on the resting spores and the zoospores of the fungi. On infection of new host plants, the fungi introduce the virus and cause symptoms characteristic of the virus they transmit (Fig. 14-19).

Dodder Transmission

Several plant viruses can be transmitted from one plant to another through the bridge formed between two plants by the twining stems of the parasitic plant dodder (*Cuscuta* sp.) (Fig. 14-15). A large number of viruses have been transmitted experimentally this way, frequently between plants belonging to families widely separated taxonomically. The virus is usually transmitted passively through the phloem of the dodder plant from the infected plant to the healthy one.

EPIDEMIOLOGY OF PLANT VIRUSES AND VIROIDS

Some of the methods of virus transmission (e.g., through vegetative propagation and through seed) are important primarily in the transmission of virus from one plant generation to another but play no role in the spread of virus from diseased to healthy plants of the same plant generation. By themselves, these methods of transmission result only in primary infections of plants and, therefore, only in monocyclic diseases. However, the other methods of virus transmission, particularly those involving vectors such as insects, not only bring the virus into a crop (primary infection), but also result in transmission of the virus from infected to healthy plants within the same plant generation and during the same growth season (secondary infections). The rate of secondary spread of viruses varies with the particular vector and it increases as the size of the vector population increases and as the weather, insofar as it affects the movement of the vector, remains favorable. Diseases caused by vector-transmitted viruses are polycyclic, with the number of disease cycles per season varying from a few (2–5 for nematode-transmitted viruses) to many (10–20 or more for aphid-transmitted viruses). Of course, when viruses that are transmitted by vegetative propagation or by seed are also transmitted by vectors, the availability of both modes of transmission (i.e., large primary inoculum in the crop and effective secondary virus spread by the vectors) often results in early and total infection of the crop plants with subsequent severe losses.

WORKING WITH, AND MANAGING PLANT VIRUSES: PURIFICATION OF PLANT VIRUSES

Isolation or, as it is usually called, purification of viruses is obtained most commonly by ultracentrifugation of the plant sap. This involves one to three cycles of alternate high (40,000–100,000 g or more) and low (3,000–10,000 g) speeds. Ultracentrifugation concentrates the virus and separates it from host cell components. Several modifications of the ultracentrifugation technique, particularly density-gradient centrifugation, are currently employed in virus purification with excellent results (Fig. 14-20). In all these methods, the virus is finally obtained as a colorless pellet or as a band in a test tube and may be used for infections, electron microscopy, serology, and nucleic acid studies.

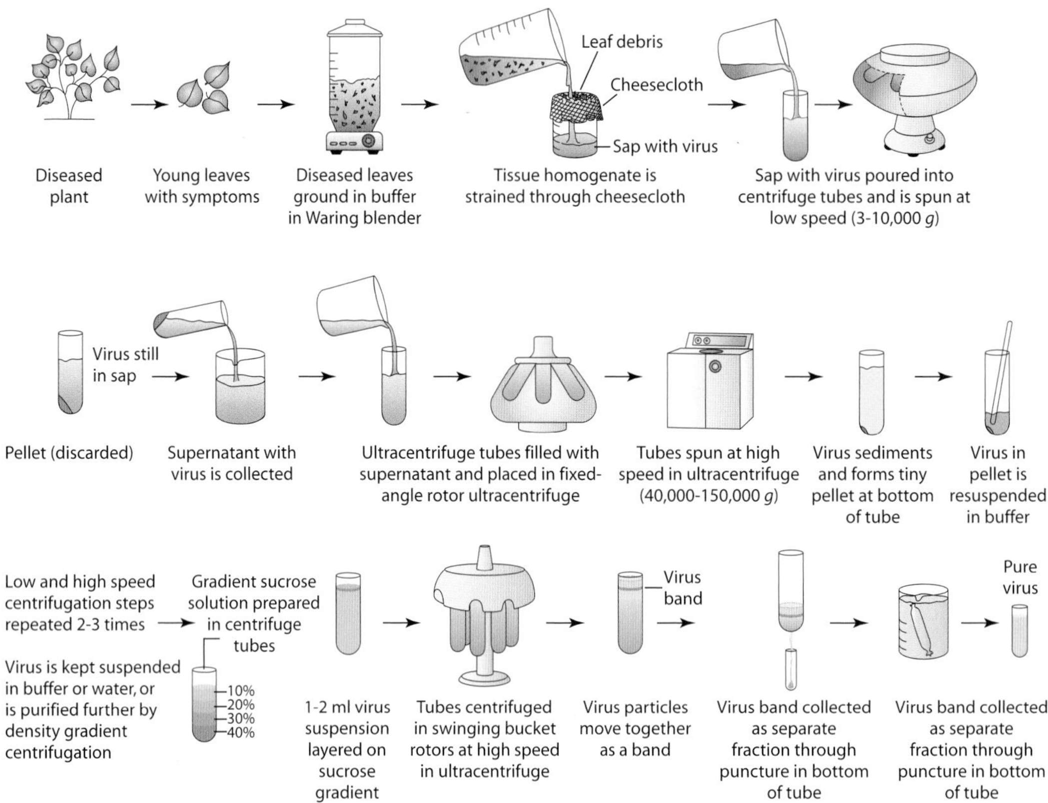

FIGURE 14-20 Steps in the purification of plant viruses.

SEROLOGY OF PLANT VIRUSES

When an **antigen**, i.e., any foreign protein, such as a virus protein, is injected into a mammal (rabbit, mouse, horse) or bird (chicken, turkey), it induces the animal to produce specific new proteins called **antibodies**. Antibodies then circulate in the blood fluid, or **serum**, of the animal. The antibodies react specifically with the **antigenic determinant** of the injected antigen, i.e., they bind to a small portion of the antigen. Each antigen, such as a virus, has many different antigenic determinants (distinct groups of 6–10 amino acids) at its surface, and because each of them prescribes the production of a different kind of antibody, the **antiserum** (serum containing antibodies) of the animal contains a mixture of many different antibodies. Such mixtures of antibodies are called **polyclonal antibodies**. Each antibody reacts with the antigen but at a different surface locality (Fig. 14-21A).

It is also possible to produce pure lines (clones) of antibodies that react only with a single antigenic determinant of a protein (or a pathogen), and such antibodies are called **monoclonal antibodies**. The production of monoclonal antibodies is possible because each cell of the immune system, e.g., of the spleen, of the animal is capable of producing many copies of only a single kind of antibody. Such cells, unfortunately, do not divide and therefore their usefulness is limited. If, however, an antibody-producing cell is fused with a mouse myeloma (cancer) cell, it produces a hybrid cell that, because of its cancerous half, can grow in culture indefinitely and, thereby, continues to produce monoclonal antibodies for a long time. Such antibody-producing hybrid cells, called **hybridomas**, can be grown in culture for months or years and produce large quantities of identical, monoclonal antibodies. Monoclonal antibodies can be obtained in high concentration and purity from the liquid of hybridoma cultures and can be used to detect, identify, and measure the antigen that induced their production. Monoclonal antibodies, however, are very specific and may not detect even strains of the same virus that happen to lack the specific antigenic determinant responsible for the monoclonal antibody. For this reason, mixtures of several monoclonal antibodies are often used in virus detection and screening tests.

The virus and its antibody are brought together in several ways, with the earliest and still quite common being the precipitin reaction. In this, the antibodies and

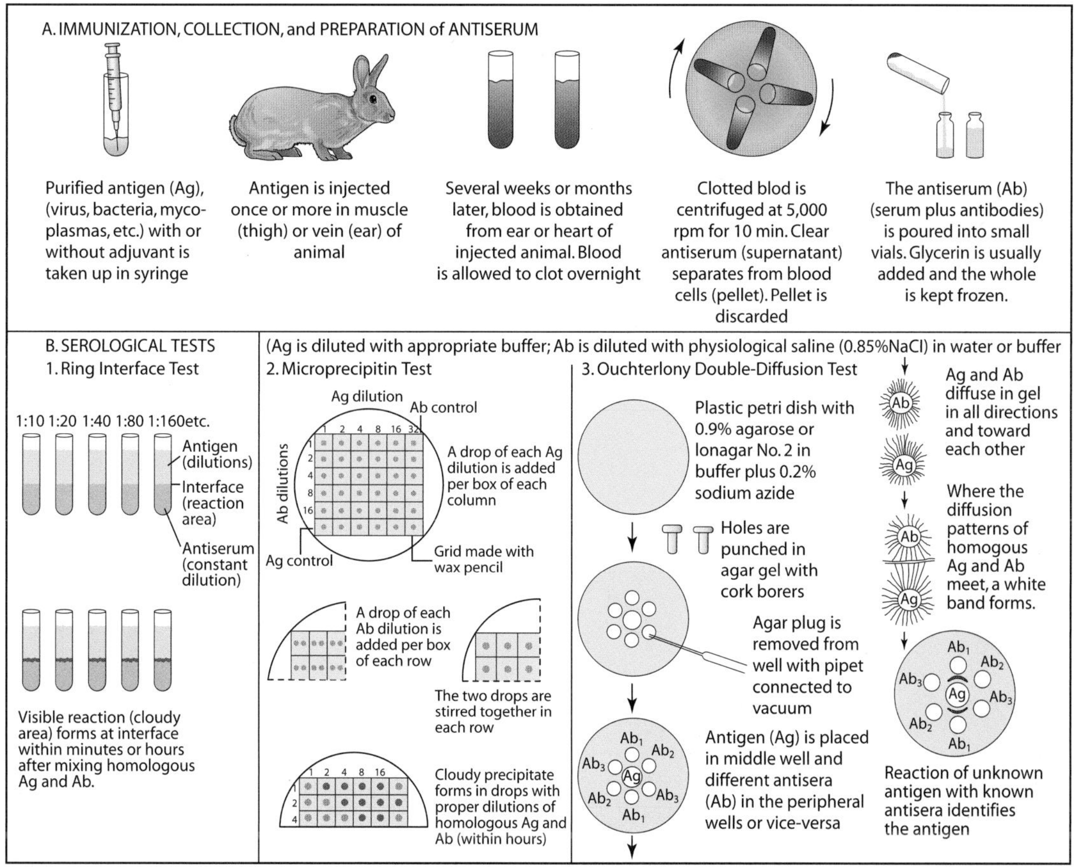

FIGURE 14-21 (A) Production of antisera. (B) Serological tests for the identification of unknown pathogens.

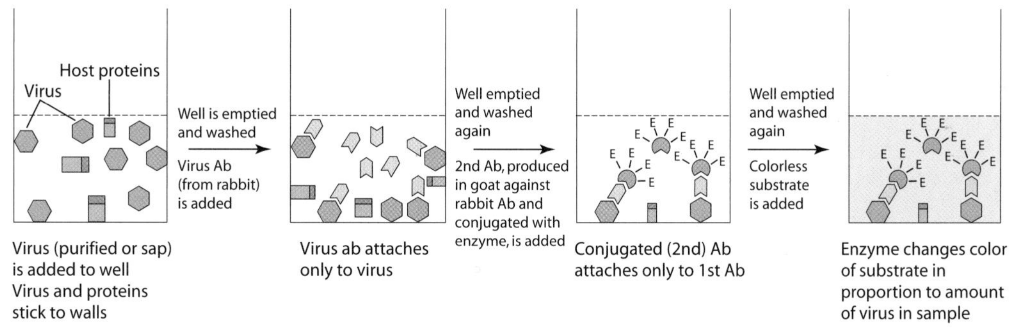

FIGURE 14-22 Schematic presentation of the steps in an indirect ELISA test.

antigens are mixed in solution — **precipitin test** in tubes or in drops on a petri dish (Fig. 14-21B, tests 1 and 2). In another test, the antigens and antibodies diffuse toward one another through an agar gel and wherever they meet in suitable concentrations they react with each other forming a whitish line or zone — **gel diffusion test** (Figs. 14-21B, test 3, and 14-23A). Sometimes the antigen is adsorbed on the surface of a large particle such as a cell, plastid, or latex sphere and these are precipitated by the addition of antibodies. This is known as the **agglutination reaction.** In all these tests the reaction of antigen and antibody becomes visible either by precipitation of the two on the bottom of the test tube or by formation of a band at the interface where the two meet.

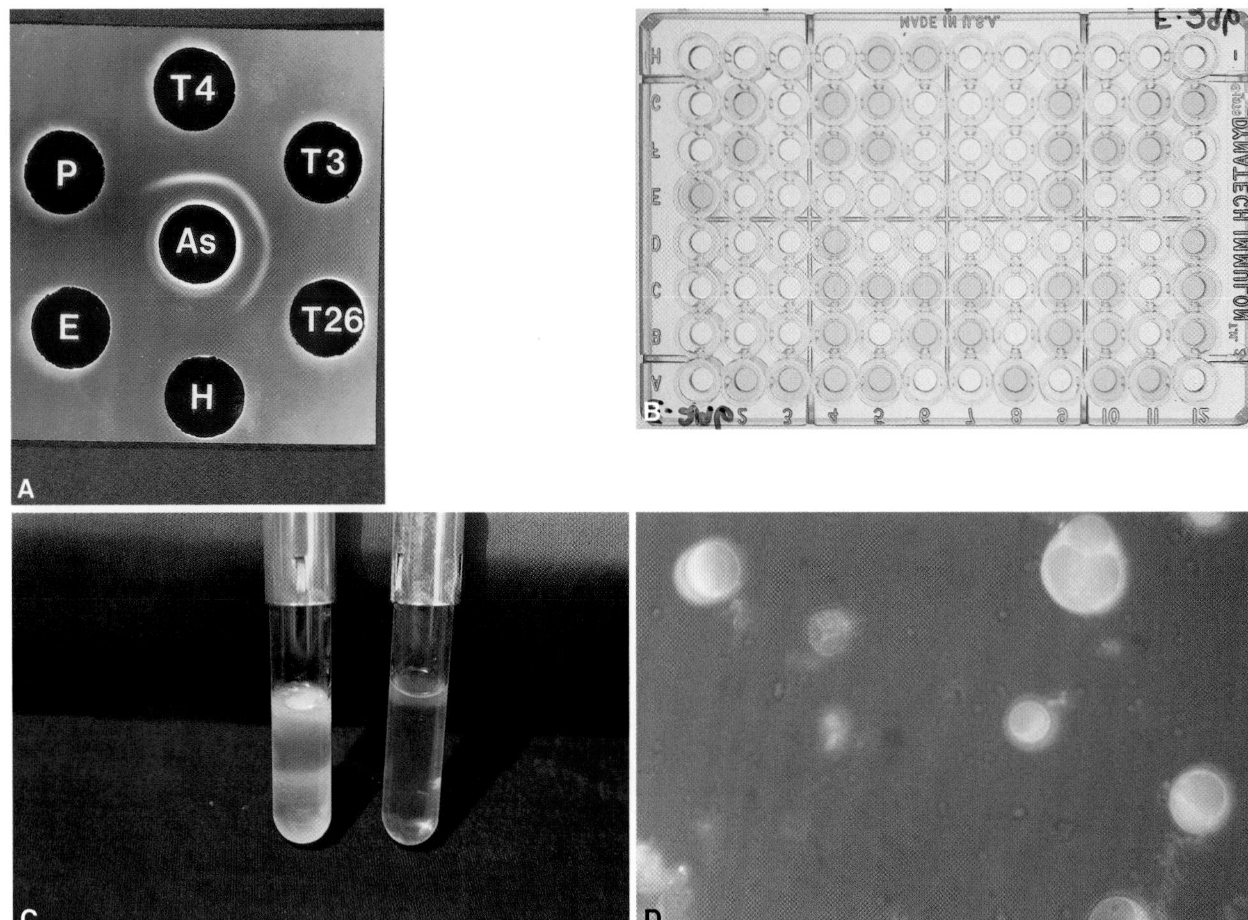

FIGURE 14-23 Detection and identification of a plant virus by different serological tests. (A) In the double diffusion test, white lines form between the antiserum (As) and the wells (T3, T4, and T26) that contain different dilutions of the virus. (B) Wells that turn yellow contain the virus against which the antiserum was produced. (C) Cloudy area formed in interface of the virus and antisera solution in a precipitin tube. (D) Bright protoplasts show infection with virus when mixed with their own antibodies, which had been conjugated with a fluorescent compound. [Photographs courtesy of (A) S. M. Garnsey and (B and C) Department of Plant Pathology, University of Florida.]

A useful serological technique called the enzyme-linked immunosorbent assay (ELISA), developed in the late 1970s, has been used widely by pathologists of all kinds and has increased tremendously the ability of plant pathologists to detect and study plant viruses and other pathogens and the diseases they cause (Figs. 14-22 and 14-23B). Several variations of the ELISA are currently in use. In the double antibody sandwich ELISA, usually referred to as **direct ELISA**, the wells (capacity 0.4 ml) of a polystyrene microtiter plate are first half-filled with and then emptied of, sequentially, (a) antibodies to the virus, (b) virus preparation or sap from an infected plant, (c) antibodies to the virus to which molecules of a particular enzyme have been attached, and (d) a substrate for the enzyme, i.e., a substance that the enzyme can break down and cause change in its color. The substrate is not emptied but is kept in the well. Within 30 to 60 minutes, the wells are "read" either visually or preferably with a colorimeter that measures the amount of color in each well. Presence of color in the well indicates that there was virus in the sample (step b, described earlier). The degree of visible coloration or the size of the reading given by the colorimeter is proportional to, and therefore a measure of, the amount of virus present in the sample.

In a variation called the **indirect ELISA** (Fig. 14-22), the sequence of steps a and b is reversed. Also, in step c the antibodies in the antibody–enzyme complex are not those against the virus; rather they are antibodies against the antibody proteins of the animal in which the virus antibody was produced (i.e., they are anti-rabbit antibodies produced in still another animal, such as a goat). All other procedures are the same. The advantage of indirect ELISA is that the same goat antirabbit antibody–enzyme complex can be used in step c of assays for any virus, as long as the first antibodies, i.e., those against the virus used in step b, were produced in rabbit.

The advantages of ELISA are as follows: the tests are extremely sensitive, large numbers of samples can be tested concurrently, only a small amount of antisera is required, results are quantitative, the procedure can be semiautomated, and the assays can be run regardless of virus morphology and virus concentration. Because of these advantages, ELISA has become one of the most popular serodiagnostic techniques, especially for multiple samples.

Two other serological techniques, each with several variations, are used by plant virologists for finding and identifying a virus present in low concentration through an electron microscope and for detecting the virus inside infected cells. In **immunosorbent electron microscopy** (ISEM), grids, prepared for electron microscopy of a virus present in low concentration or in a mixture with other viruses, are first coated with antibody to the target virus. Then, the virus sample is placed on the antibody-coated grid, and the antibodies trap the virus from the sample and concentrate it on the grid where it can be found easily with an electron microscope and identified because of its reaction with the antibodies. Identification of the virus is facilitated further by coating the virus particles already on the grid with antibodies (decoration) that make them appear quite distinctive under an electron microscope. In the **immunofluorescent staining** technique, parts of a plant leaf, whole cells, or cell sections are first "fixed," i.e., killed with acetone or other organic compounds. The fixed leaf tissues are then treated with antibodies to a virus that had been labeled previously with a compound, such as fluorescein isothiocyanate (FITC), which fluoresces under ultraviolet light. If the treated cells are infected with the virus, the virus traps the antibodies and the attached fluorescent compound. When such cells, in tissues or as protoplasts, are viewed with a microscope supplied with ultraviolet light, cells or cell parts that contain virus appear fluorescent while the rest of the cells or cell areas appear dark (see Figs. 14-13 and 14-23D).

The uses of plant virus serology are numerous. It is used to determine relationships between viruses, to identify a virus causing a plant disease, to detect virus in foundation stocks of plants, and to detect symptomless virus infections. It can also be used to measure virus quantitatively, to locate the virus within a cell or tissue, to detect plant viruses in insects, and to purify a virus.

NOMENCLATURE AND CLASSIFICATION OF PLANT VIRUSES

Many plant viruses are named after the most conspicuous symptom they cause on the first host in which they have been studied. Thus, a virus causing a mosaic on tobacco is called *tobacco mosaic virus*, whereas the disease itself is called tobacco mosaic; another virus causing spotted wilt symptoms on tomato is called *tomato spotted wilt virus* and the disease is called tomato spotted wilt, and so forth. Considering, however, the variability of symptoms caused by the same virus on the same host plant under different environmental conditions, by different strains of a virus on the same host, or by the same virus on different hosts, it becomes apparent that this system of nomenclature leaves much to be desired.

All viruses belong to the kingdom Viruses. Within the kingdom, viruses are distinguished as RNA viruses and DNA viruses, depending on whether the nucleic acid of the virus is RNA or DNA. Viruses are further subdivided depending on whether they possess one or two strands of RNA or DNA of either positive or negative sense, either filamentous or isometric. Within each of these groups there may be viruses replicating via a polymerase enzyme (+RNA or DNA viruses) or via a reverse transcriptase (–RNA or DNA viruses). Most viruses consist of nucleic acid surrounded by coat protein, but some also have a membrane attached to them. Some viruses have all their genome in one particle (monopartite viruses), but the genome of other (multipartite) viruses is divided among two, three, or, rarely, four particles. Other characteristics in the classification of viruses include the symmetry of helix in the helical viruses, or number and arrangement of protein subunits in the isometric viruses, size of the virus, and, finally, any other physical, chemical, or biological properties.

Figure 14-24 shows diagrammatically the various families and genera of plant viruses. The current nomenclature and classification scheme of plant viruses, along with the type species and the means of transmission of each virus genus, are as follows.

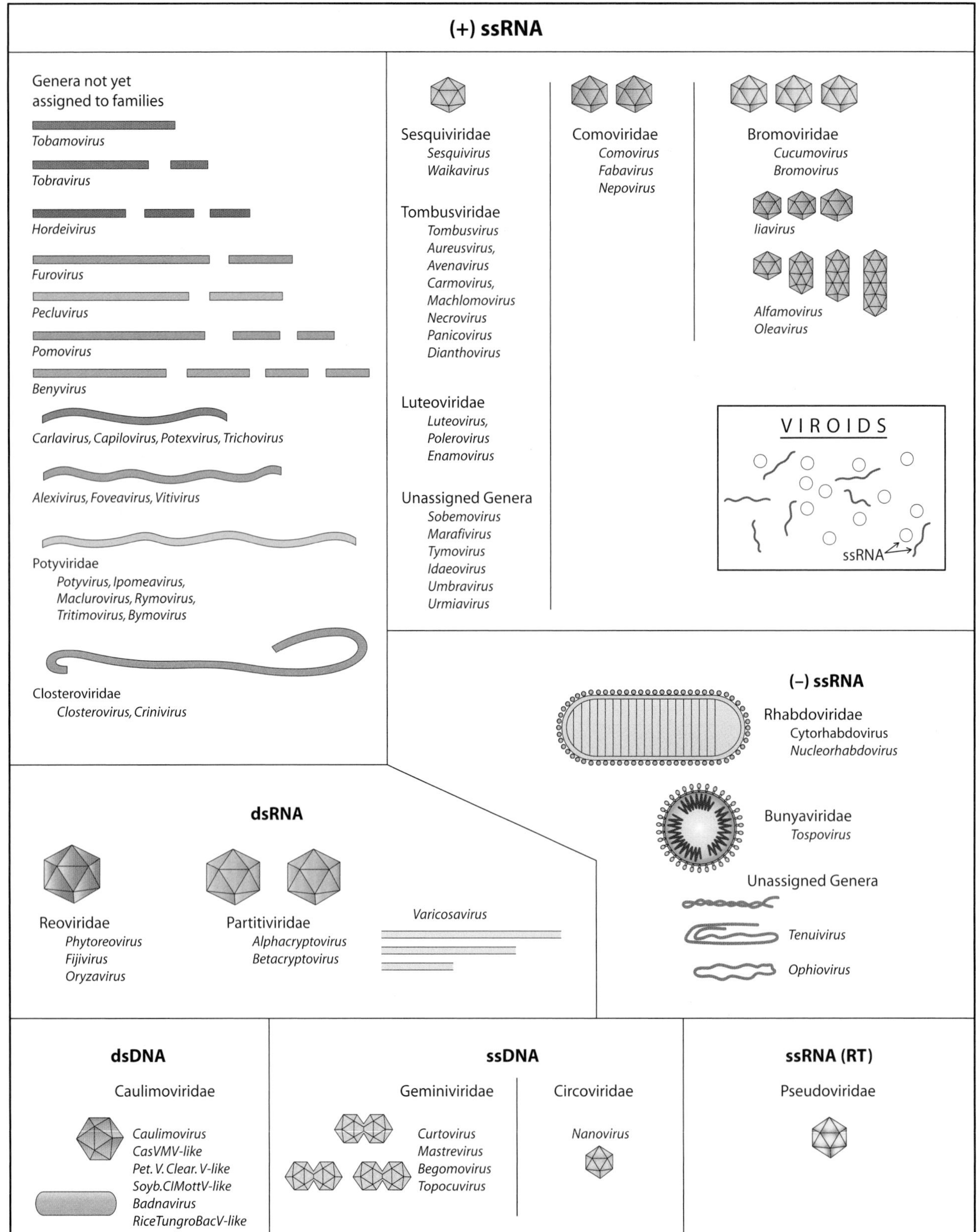

FIGURE 14-24 Schematic diagram of families and genera of viruses and of viroids that infect plants.

Kingdom: Viruses

Virus genera not yet assigned into families
RNA viruses

Single-stranded positive RNA [(+) ssRNA]

Rod-shaped particles	Family	Genus	Type species	Remarks
1 ssRNA	—	*Tobamovirus*	*Tobacco mosaic virus*	Contact transmission
2 ssRNAs	—	*Tobravirus*	*Tobacco rattle virus*	Nematode transmission
3 ssRNAs	—	*Hordeivirus*	*Barley stripe mosaic virus*	Seed transmission
2 ssRNAs	—	*Furovirus*	*Soilborne wheat mosaic virus*	Fungal transmission
	—	*Pecluvirus*	*Peanut clump virus*	Fungal and seed transmission
3 ssRNAs	—	*Pomovirus*	*Potato mop-top virus*	Fungal transmission, dicots
4 ssRNAs	—	*Benyvirus*	*Beet necrotic yellow vein virus*	Fungal transmission

Filamentous particles

1 ssRNA	—	*Allexivirus*	*Shallot virus X*	Eriophyid mite transmission
	—	*Carlavirus*	*Carnation latent virus*	
	—	*Foveavirus*	*Apple stem pitting virus*	No vector
	—	*Potexvirus*	*Potato virus X*	By contact only
	—	*Capillvirus*	*Apple stem grooving virus*	No vector. Some seed transmission
	—	*Trichovirus*	*Apple chlorotic leafspot virus*	No vector. Some seed transmission
	—	*Vitivirus*	*Grapevine virus A*	Mealybugs, scale insects, aphids

Isometric particles

1 ssRNA				
	—	*Sobemovirus*	*Southern bean mosaic virus*	Seedborne, beetles, myrids
	—	*Marafivirus*	*Maize rayado fino virus*	In Gramineae, leafhoppers
	—	*Umbravirus*	*Carrot mottle virus*	Do not code coat proteins Aphids w/ helper virus
	—	*Tymovirus*	*Turnip yellow mosaic virus*	By beetles
2 ssRNAs	—	*Idaeovirus*	*Raspberry bushy dwarf virus*	By pollen and seed

Bacilliform particles

3 ssRNAs	—	*Ourmiavirus*	*Ourmia melon virus*	No vectors known

Virus families

Filamentous viruses

1 ssRNA				
	Potyviridae	*Potyvirus*	*Potato virus Y*	Aphids, w/ helper virus
	Potyviridae	*Ipomovirus*	*Sweet pot mild mottle*	Whitefly *Bemisia tabaci*
	Potyviridae	*Macluravirus*	*Maclura mosaic virus*	Aphids
	Potyviridae	*Rymovirus*	*Ryegrass mosaic virus*	Eriophyid mites
	Potyviridae	*Tritimovirus*	*Wheat streak mosaic virus*	Eriophyid mites
	Potyviridae	*Bymovirus*	*Barley yellow mosaic virus*	Gramineae, fungal transmission

1 or 2 ssRNA				
	Closteroviridae	*Closterovirus*	*Beet yellows virus*	Aphids, mealybugs, or whiteflies
	Closteroviridae	*Crinivirus*	*Lettuce inf. yellows virus*	Whiteflies
Isometric viruses				
1 ss(+)RNA				
	Sequiviridae	*Sequivirus*	*Parsnip yellow fleck virus*	Aphids
	Sequiviridae	*Waikavirus*	*Rice tungro spherical virus*	Leafhoppers or aphids
	Tombusviridae	*Tombusvirus*	*Tomato bushy stunt virus*	Soilborne, but vector unknown
	Tombusviridae	*Aureusvirus*	*Pothos latent virus*	Soilborne
	Tombusviridae	*Avenavirus*	*Oat chlorotic stunt virus*	Soilborne
	Tombusviridae	*Carmovirus*	*Carnation mottle virus*	—
	Tombusviridae	*Dianthovirus*	*Carnation ring spot virus*	Soilborne, unknown
	Tombusviridae	*Machlomovirus*	*Maize chlorotic mottle virus*	Seed, beetles, thrips
	Tombusviridae	*Necrovirus*	*Tobacco necrosis virus*	Fungal transmission
	Tombusviridae	*Panicovirus*	*Panicum mosaic virus*	Gramineae, mechanical
	Luteoviridae	*Luteovirus*	*Barley yellow dwarf virus*	Gramineae, aphids
	Luteoviridae	*Polerovirus*	*Potato leafroll virus*	Monocot or dicot plants
	Luteoviridae	*Enamovirus*	*Pea enation mosaic virus*	Mechanically, aphids
2 ss(+)RNAs				
	Comoviridae	*Comovirus*	*Cowpea mosaic virus*	Chrysomelid beetles
	Comoviridae	*Fabavirus*	*Broad bean wilt virus*	Aphids
	Comoviridae	*Nepovirus*	*Tobacco ring spot virus*	Nematodes
3 ss(+)RNAs				
	Bromoviridae	*Bromovirus*	*Brome mosaic virus*	Beetles, mechanically
	Bromoviridae	*Cucumovirus*	*Cucumber mosaic virus*	Aphids
	Bromoviridae	*Alfamovirus*	*Alfalfa mosaic virus*	Aphids
	Bromoviridae	*Ilarvirus*	*Tobacco streak virus*	Pollen, seed
	Bromoviridae	*Oleavirus*	*Olive latent virus 2*	No vector known
dsRNA				
	Reoviridae	*Phytoreovirus*	*Wound tumor virus*	Leafhoppers
	Reoviridae	*Fijivirus*	*Fiji disease virus*	Gramineae, planthoppers
	Reoviridae	*Oryzavirus*	*Rice ragged stunt virus*	Planthoppers
	Partitiviridae	*Alphacryptovirus*	*White clover crypto. virus 1*	Nonenveloped, latent
	Partitiviridae	*Betacryptovirus*	*White clover crypto. virus 2*	Same, seed
	Partitiviridae	*Varicosavirus*	*Lettuce big-vein virus*	Fungal transmission
(−) ssRNA				
Bacilliform particles				
	Rhabdoviridae	*Cytorhabdovirus*	*Lettuce necrosis yellows virus*	Leafhoppers, planthoppers, aphids
	Phabdoviridae	*Nucleorhabdovirus,*	*Potato yellow dwarf virus*	Same
Membranous circular particles				
	Bunyaviridae	*Tospovirus*	*Tomato spotted wilt virus*	Thrips
Thin flexuous multipartite viruses				
	—	*Tenuivirus*	*Rice stripe virus*	Gramineae, planthoppers
	—	*Ophiovirus*	*Citrus psorosis virus*	No vector known
dsDNA				
Isometric	Caulimoviridae	*Caulimovirus*	*Cauliflower mosaic virus*	Aphids
	Caulimoviridae	*Soybean chlorotic mottle virus-like*		Aphids
	Caulimoviridae	*Cassava vein mosaic virus-like*		Aphids
		Petunia vein clearing virus-like		Aphids

	Caulimoviridae	*Badnavirus*	*Commelina yellow mottle virus*	Mealybugs
	Caulimoviridae	*Rice tungro bacilliform virus-like*		Leafhoppers
(+)ssDNA				
	Geminiviridae	*Mastrevirus*	*Maize streak virus*	Gramineae, leafhoppers
	Geminiviridae	*Curtovirus*	*Beet curly top virus*	Dicot, leafhoppers
	Geminiviridae	*Begomovirus*	*Bean golden mosaic virus*	2 DNAs, whiteflies
	Geminiviridae	*Topocuvirus*	*Tom. pseudocurly top vrius*	*Treehopper*
	Circoviridae	*Nanovirus*	*Subteranean clover stunt virus*	6 DNAs
ssRNA (RT)	Pseudoviridae: retrotransposons			

DETECTION AND IDENTIFICATION OF PLANT VIRUSES

Once the cause of a disease has been established as a virus, a series of tests, utilizing whatever simple or sophisticated methods and equipment are available, may be necessary to determine its identity. The host range of the virus, i.e., the hosts on which the virus induces symptoms and the kinds of symptoms produced, may help to differentiate the virus from several others. Transmission studies should indicate whether the virus is transmitted mechanically and to what hosts, or by insects and which insects (Fig. 14-25), and so on, with each new property ascertained helping to characterize the virus further. If the virus is transmitted mechanically, certain properties of the virus, such as its thermal inactivation point (i.e., the temperature required for complete inactivation of the virus in untreated crude juice during a 10-min exposure), its longevity *in vitro*, and its dilution end point (i.e., the highest dilution of the juice at which the virus can still cause infection), have been used in the past but are not reliable. If the identity of the virus is suspected, serological tests may be used, and if they are positive, a tentative identification may be made. Examination of the virus in an electron microscope and inoculation of certain plant species are also usually sufficient for a tentative identification of the virus.

In virus-like diseases of woody (and other) plants in which no pathogens have been observed so far, identification of the pathogens, which are at present presumed to be viruses, is made strictly by **indexing**. Indexing involves inoculation by grafting (Fig. 14-26) of certain plant species or varieties called **indicators**. The indicators are sensitive to specific viruses and on inoculation with these viruses develop characteristic symptoms and vice versa; i.e., development of the characteristic symptoms by an indicator identifies the virus with which the indicator was inoculated.

Because viruses are too small to be detected with the naked eye or seen through a light microscope, their presence has been detected primarily by the symptoms exhibited by the host plant (Fig. 14-1); by the symptoms induced in an indicator plant after transmission of the virus by grafting, mechanical inoculation, or by one of the vectors; by examination of young infected tissues with a microscope for cell inclusion bodies (Fig. 14-2) that may be characteristic of the virus family or genus; by electron microscopy (Figs. 14-3 and 14-4); and by one of the serological tests such as ELISA (Figs. 14-21 to 14-23) or fluorescent antibody microscopy (Fig. 14-13B). It has also been possible to detect, and even identify, RNA viruses in plants by isolating, and subsequently analyzing through electrophoresis, the dsRNA of viruses replicating in plants because healthy plants do not produce such dsRNAs. Although laborious, this technique, like the inclusion body detection technique, has the advantage that it can be used to detect known as well as unknown viruses for which no antiserum or not much information is available, and it can therefore be used for detection of even woody plant viruses.

Isolated single- or double-stranded RNA can be used further to produce complementary DNA (cDNA) to the RNA. cDNA can also be produced to single- or double-stranded DNA. The cDNA, if produced in the presence of radioactive or chromogenic (color producing) molecules, can be used for hybridization experiments with viral RNA or DNA. The viral RNA or DNA is partially

FIGURE 14-25 The most important types of insect vectors of plant viruses: (A) aphids, (B) leafhoppers, (C) whiteflies, and (D) thrips.

purified from suspected infected plants and is allowed to react with cDNA in test tubes or on nitrocellulose filter supports (dot blots). The cDNA–RNA or cDNA–DNA hybrids are detected and quantified by autoradiography or liquid scintillation counting, if radioactive, or by colorimetric techniques, if chromogenic. In addition to virus detection via the formation of cDNA–RNA hybrids, cDNA to the virus RNA can be further converted to dsDNA, which can then be cloned into suitable vectors (e.g., *Escherichia coli* bacteria) to produce almost unlimited amounts of dsDNA and, from this, cDNA probes for further hybridization experiments for virus detection (Fig. 14-27).

The PCR technique allows unlimited amplification of selected specific DNA sequences for which suitable primers (short DNA sequences) are available. Combining reverse transcription of viral RNA into DNA allows use of PCR amplification for RNA as well as DNA viruses. Once amplification of the nucleic acid is accomplished, use of labeled DNA probes or electrophoretic analysis of the PCR products allows further detection and diagnosis of the virus (Fig. 14-28).

ECONOMIC IMPORTANCE OF PLANT VIRUSES

Viruses attack all forms of life, including bacteria, fungi, and all types of plants, from herbaceous ones to trees. Plant virus diseases may damage any or all parts of a plant and may cause economic losses by reducing yields and quality of plant products. Losses may be catastrophic or may be mild and insignificant. Viruses account for a considerable portion of the losses suffered annually from diseases of the various crops.

The severity of individual virus diseases may vary with the locality and the crop variety and from one season to the next. Some virus diseases have destroyed entire plantings of certain crops in some areas, e.g., geminiviruses of tomato, plum pox, hoja blanca of rice, sugar beet yellows, and citrus tristeza. Most virus

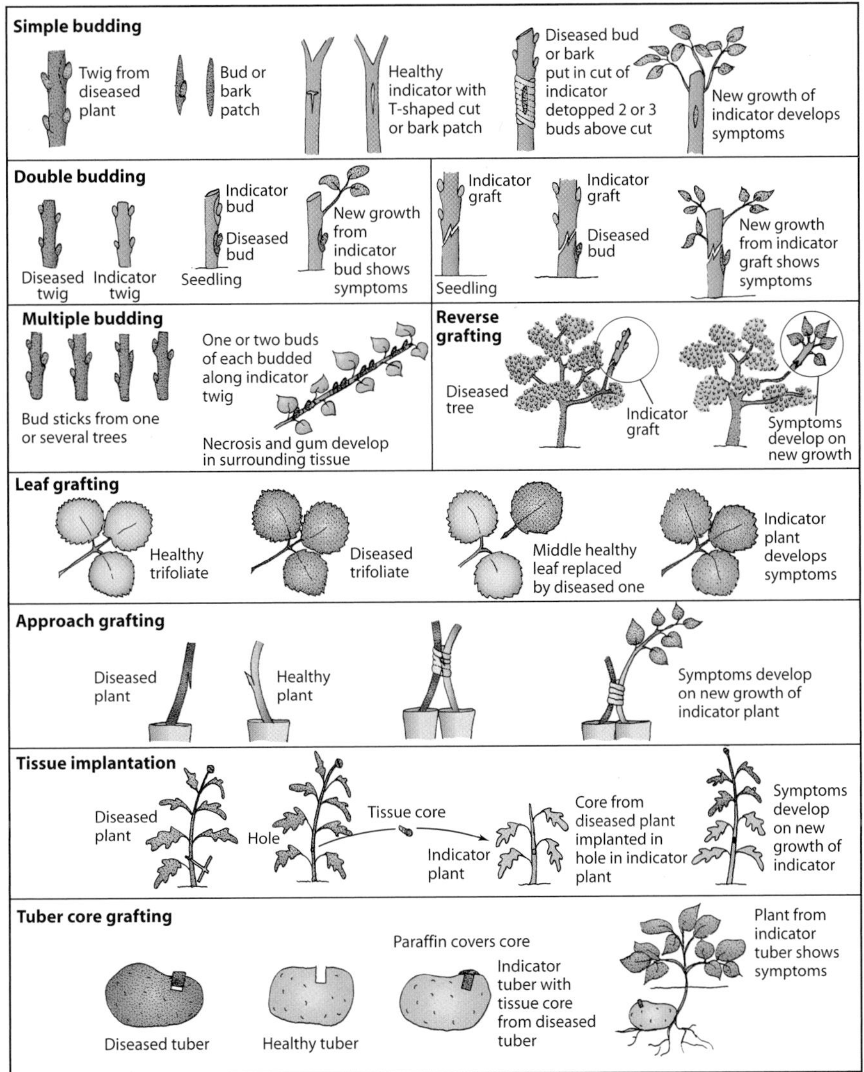

FIGURE 14-26 Indexing for viral, mollicute, and fastidious bacterial diseases.

diseases, however, occur on crops year after year and cause small to moderate but unspectacular losses. This occurs even when the viruses do not induce any visible symptoms.

CONTROL OF PLANT VIRUSES

The best way to control a virus disease is by keeping it out of an area through systems of quarantine, inspection, and certification. The existence of symptomless hosts, the incubation period after inoculation, and the absence of obvious symptoms in seeds, tubers, bulbs, and nursery stock make quarantine difficult and sometimes ineffective. Eradication of diseased plants to eliminate inoculum from the field may, in some cases, help control the disease. Plants may be protected against certain viruses by protecting them against the virus vectors. Controlling the insect vectors and removing weeds that serve as hosts may help some to control diseases. Generally, however, insect control is virtually useless for controlling insect-borne, especially aphid-transmitted, plant viruses. Losses caused by nematode-transmitted viruses can be reduced considerably by soil fumigation to control the nematodes.

The use of virus-free seed, tubers, budwood, and so on is the single most important measure for avoiding virus diseases of many crops, especially those lacking insect vectors. Periodic indexing of the mother plants producing such propagative organs is necessary to ascertain their continuous freedom from viruses. Several types of inspection and certification programs are now

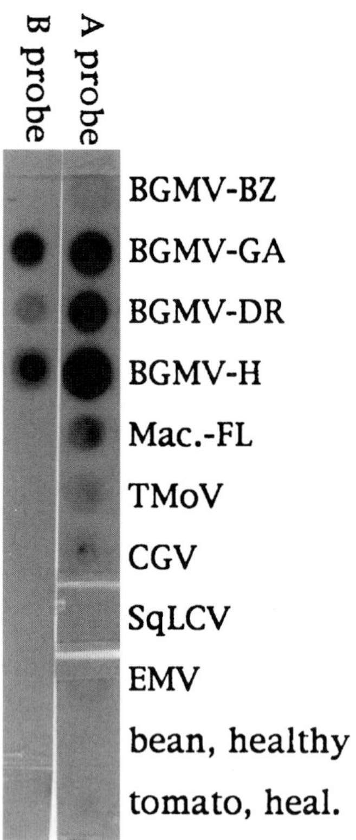

FIGURE 14-27 Use of DNA probes in a dot-blot hybridization test. Probes frrom DNAs A and B of a Florida isolate (H) of *bean golden mosaic virus* were used to detect and analyze similar viruses from other countries and also other viruses. The probes reacted and hybridized to the viruses in proportion to their relatedness to the probes. BZ, Brazil; GA, Guatemala; DR, Dominican Republic; Mac, Macroptilium weed; TMoV, *tomato mottle virus*; CGV, *cabbage geminivirus*; SqLCV, *squash leaf curl virus*; EMV, euphorbia mosaic virus. (Photograph courtesy of E. Hiebert.)

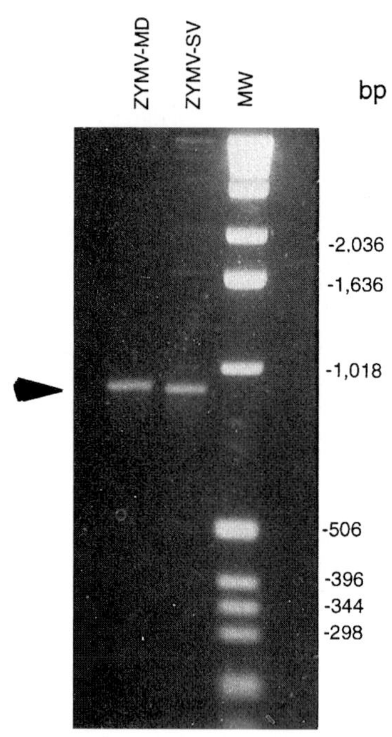

FIGURE 14-28 Detection of viral genes through the use of the polymerase chain reaction (PCR). Using previously prepared probes, the P1 genes of a mild (MD) and a severe (SV) strain of *zucchini yellow mosaic virus* were amplified by PCR. The PCR products were run on an electrophoresis gel, and the resulting bands were compared with marker DNAs of known molecular weight (MW). (Photograph courtesy of E. Hiebert.)

in effect in various states producing seeds, tubers, and nursery stock used for propagation. Serological testing of mother plants, seeds, and nursery stock for virus by the ELISA technique and, more recently, by nucleic acid techniques has helped greatly in reducing the frequency of viruses in the propagating stock of crop plants.

Although the health or vigor of host plants confers no resistance or immunity to virus disease, breeding plants for hereditary resistance to viruses is of great importance, and many plant varieties resistant to certain virus diseases have already been produced. In some host–virus combinations, the disease caused by severe strains of the virus can be avoided if the plants are inoculated first with a mild strain of the same virus, which then protects the plant from infection by the severe strain of the virus; the phenomenon is called **cross protection.** In the past 10 years, some viruses have been controlled to various extents through pathogen-derived resistance provided by introducing into plants the coat protein gene or some other segment of the genome of the virus (Fig. 14-29). Such transgenic plants transcribe and express these genes, which somehow interferes with the infection, multiplication, and disease induction by the virus. In many cases, genetic engineering of pathogen-derived resistance is the result of transcriptional or posttranscriptional virus **gene silencing,** primarily by homology of the silencing sequence, often of a small inhibitory RNA, with the silenced virus genes. Silencing, or knockout, of a plant or a viral gene can also be obtained by attaching homologous sequences or small inhibitory RNAs to appropriate satellite RNAs, which are then inoculated with their symptomless helper virus into the host plant to be protected. This is virus-induced silencing of another virus of existing plantings. Plant viruses can both initiate and be the target of gene silencing in transgenic plants, and silencing can spread systemically in the plants. It is apparent now that gene silencing is part of the normal defense system of plants against foreign nucleic acids; therefore, for viruses to become established and cause infection they must overcome this defense. Virus-induced gene silencing is a subject of a great deal of research presently being carried out.

FIGURE 14-29 Genetic engineering of plant resistance to virus infection. (A) Normal (foreground) and tomato plants (background) engineered with a truncated replicase gene of the begomovirus *tomato yellow leaf curl virus* (TYLCV). Both were inoculated as seedlings with TYLCV through viruliferous whiteflies, planted in the field, and photographed 60 days postinoculation. Normal plants became infected and remained stunted compared to beautifully growing engineered plants. (B) Normal (foreground) and transgenic tomatoes (background) transformed with a *tomato mottle begomovirus* replicase gene inoculated as seedlings with viruliferous whiteflies and then grown in the field for 60 days. Normal tomatoes show severe mottling and stunting compared to the much better growing engineered plants. (C) A commercial squash variety (background), genetically engineered for resistance to *zucchini yellow mosaic potyvirus (ZYMV)* and then inoculated mechanically with ZYMV, remained resistant and healthy looking, whereas the nontransgenic squash of the same variety and inoculated similarly with ZYMV at the front right became severely infected and stunted. Plant at front left, which is similarly transgenic with TYMV as the plant in background but inoculated with a different virus, *cucumber mosaic virus*, developed severe symptoms because, usually, the genetically engineered resistance, as happens with traditionally bred resistance, is specific to the virus for which it is obtained. [Photographs courtesy of (A and B) E. Hiebert and J. E. Polston and (C) D. E. Purcifull.]

Once inside a plant, some viruses can be inactivated by heat. Dormant, propagative organs are usually dipped in hot water (35–54°C) for a few minutes or hours, whereas actively growing plants are usually kept in greenhouses or growth chambers at 35 to 40°C for several days, weeks, or months, after which the virus in some of them is inactivated and the plants are completely healthy. Plants free of virus may also be produced from virus-infected ones by the culture of short tips (0.1 mm to 1 cm or more) of apical and root meristems, especially at elevated (28–30°C) temperatures.

No chemical substances (viricides) are yet available for controlling virus diseases of plants in the field, although some compounds, e.g., ribavirin, applied as a spray or injected into the plant reduce symptoms drastically and may eliminate the virus from the treated host plant. Foliar application of certain growth-regulating substances, such as gibberellic acid, may overcome the stunting induced by some viruses and may stimulate the growth of virus-suppressed axillary buds in virus-infected trees.

Selected References

Beachy, R. N., Loesche-Fries, S., and Turner, N. E. (1990). Coat protein mediated resistance against virus infection. *Annu. Rev. Phytopathol.* **28**, 451–474.

Bos, L. (1999). "Plant Viruses: Unique and Intriguing Pathogens — A Textbook of Plant Virology." Backhuys, Leiden, The Netherlands.

Brown, D. J. F., Robertson, W. M., and Trudgill, D. L. (1995). Transmission of viruses by plant nematodes. *Annu. Rev. Phytopathol.* **33**, 323–349.

Callaway, A., *et al.* (2001). The multifunctional capsid proteins of plant RNA viruses. *Annu. Rev. Phytopathol.* **39**, 419–460.

Campbell, R. N. (1996). Fungal transmission of plant viruses. *Annu. Rev. Phytopathol.* **34**, 87–108.

Christie, R. G., and Edwardson, J. R. (1977). Light and electron microscopy of plant virus inclusions. *Fla. Agric. Exp. Stn. Monogr.* **9**, Gainesville, FL.

Clark, M. F. (1981). Immunosorbent assays in plant pathology. *Annu. Rev. Phytopathol.* **19**, 83–106.

Colinet, D., *et al.* (1994). Identification of distinct potyviruses in mixedly-infected sweet potato by the polymerase chain reaction with degenerate primers. *Phytopathology* **84**, 65–69.

Collmer, C. W., and Howell, S. H. (1992). Role of satellite RNA in the expression of symptoms caused by plant viruses. *Annu. Rev. Phytopathol.* **30**, 419–442.

Culver, I. N. (2002). Tobacco mosaic virus assembly and disassembly: Determinants of pathogenicity and resistance. *Annu. Rev. Phytopathol.* **40**, 287–308.

Dreher, T. W. (1999). Functions of the 3′-untranslated regions of the positive strand RNA viral genomes. *Annu. Rev. Phytopathol.* **37**, 151–174.

Fedorkin, O. N., Solovyev, A. G., Yelina, N. E., *et al.* (2001). Cell-to-cell movement of potato virus X involves distinct functions of the coat protein. *J. Gen. Virol.* **82**, 449–458.

Fridlund, P. R., ed. (1989). "Virus and Viruslike Diseases of Pome Fruits and Simulating Noninfectious Disorders." Washington State Univ. Cooperative Extension, SP0003, Pullman, WA.

Garcia-Arenal, F., Fraile, A., and Malpica, J. M. (2001). Variability and genetic structure of plant virus populations. *Annu. Rev. Phytopathol.* **39**, 157–186.

Gosselè, V., *et al.* (2002). SVISS: A novel transient gene silencing system for gene function discovery and validation in tobacco plants. *Plant J.* **32**, 859–866.

Hamilton, A. J., and Baulcombe, D. C. (1999). A species of small antisense RNA in posttranscriptional gene silencing in plants. *Science* **286**, 950–952.

Hampton, R., Ball, E., and DeBoer, S. (1990). "Serological Methods for the Detection and Identification of Viral and Bacterial Plant Pathogens." APS Press, St. Paul, MN.

Harper, G., *et al.* (2002). Viral sequences integrated into plant genomes. *Annu. Rev. Phytopathol.* **40**, 119–136.

Harris, K. F., and Maramorosch, K., eds. (1977). "Aphids as Virus Vectors." Academic Press, New York.

Harris, K. F., and Maramorosch, K., eds. (1980). "Vectors of Plant Pathogens." Academic Press, New York.

Henson, J. M., and French, R. (1993). The polymerase chain reaction and disease diagnosis. *Annu. Rev. Phytopathol.* **31**, 81–110.

Hibino, H. (1996). Biology and epidemiology of rice viruses. *Annu. Rev. Phytopathol.* **34**, 249–274.

Holzberg, S., *et al.* (2002). *Barley stripe mosaic virus*-induced gene silencing in a monocot plant. *Plant J.* **30**, 315–327.

Hull, R. (1989). Movement of viruses within plants. *Annu. Rev. Phytopathol.* **27**, 213–240.

Hull, R. (1996). Molecular biology of rice tungro viruses. *Annu. Rev. Phytopathol.* **34**, 275–297.

Jan, F.-J., Pang, S.-Z., Tricoli, D. M., *et al.* (2000). Evidence that resistance in squash mosaic comovirus coat protein-transgenic plants is affected by plant developmental stage and enhanced by combination of transgenes from different lines. *J. Gen. Virol.* **81**, 2299–2306.

Johansen, E., Edwards, M. C., and Hampton, R. O. (1994). Seed transmission of viruses: Current perspectives. *Annu. Rev. Phytopathol.* **32**, 363–386.

Klahre, U., *et al.* (2002). High molecular weight RNAs and small interfering RNAs induce systemic posttranscriptional, gene silencing in plants. *Proc. Natl. Acad. Sci. USA* 1073–1078.

Lecoq, H., *et al.* (1993). Aphid transmission of a non-aphid-transmissible strain of a zucchini yellow mosaic potyvirus from transgenic plant expressing the capsid protein of plum pox potyvirus. *Mol. Plant-Microbe Interact.* **6**, 403–406.

Loebenstein, G., Lawson, R. H., and Brunt, A. A., eds. (1995). "Virus and Virus-like Diseases of Bulb and Flower Crops." J. Wiley, UK.

McLean, G. D., Garrett, R. G., and Ruesink, W. G., eds. (1986). "Plant Virus Epidemics: Monitoring, Modelling, and Predicting Outbreaks." Academic Press, Sydney.

Maramorosch, K., ed. (1993). "Plant Diseases of Viral, Viroid, Mycoplasma, and Uncertain Etiology." Westview, Boulder, CO.

Maramorosch, K., and Harris, K. F., eds. (1979). "Leafhopper Vectors and Plant Disease Agents." Academic Press, New York.

Maramorosch, K., and McKelvey, J. J., Jr., eds. (1985). "Subviral Pathogens of Plants and Animals: Viroids and Prions." Academic Press, Orlando, FL.

Maule, A. J., Escaler, M., and Aranda, M. A. (2001). Programmed responses to virus replication in plants. *Mol. Plant Pathol.* **1**, 9–15.

Mayo, M. A., and Brunt, A. A. (2001). The current state of plant virus taxonomy. *Mol. Plant Pathol.* **2**, 97–100.

Mink, G. I. (1993). Pollen and seed-transmitted viruses and viroids. *Annu. Rev. Phytopathol.* **31**, 375–402.

Pirone, T. P., and Blanc, S. (1996). Helper-dependent vector transmission of plant viruses. *Annu. Rev. Phytopathol.* **34**, 227–247.

Pogue, G. P., *et al.* (2003). Making an ally from an enemy: Plant virology and the new agriculture. *Annu. Rev. Phytopathol.* **4**, 45–74.

Roossinck, M. J. (1997). Mechanisms of plant virus evolution. *Annu. Rev. Phytopathol.* **35**, 191–208.

Savenkov, E. I., and Valkonen, J. P. T. (2002). Silencing of a viral RNA silencing suppressor in transgenic plants. *J. Gen. Virol.* **83**, 2325–2335.

Scholthof, H. B., Scholthof, K.-B. G., and Jackson, A. O. (1997). Plant virus gene vectors for transient expression of foreign proteins in plants. *Annu. Rev. Phytopathol.* **3**, 111–128.

Silhavy, D., *et al.* (2002). A viral protein suppresses RNA silencing and binds silencing-generated, 21- to 25-nucleotide double-stranded RNAs. *EMBO J.* **21**, 3070–3080.

Smith, K. M. (1972). "A Textbook of Plant Virus Diseases," 3rd Ed. Academic Press, New York.

Spiegel, S., Frison, E. A., and Converse, R. H. (1993). Recent developments in therapy and virus-detection procedures for international movement of clonal plant germplasm. *Plant Dis.* **77**, 1176–1180.

Teycheney, P.-Y., and Tepfer, M. (2001). Virus-specific spatial differences in the interference with silencing of the *chs-A* gene in non-transgenic petunia. *J. Gen. Virol.* **82**, 1239–1243.

Trepfer, M. (2002). Risk assessment of virus-resistant transgenic plants. *Annu. Rev. Phytopathol.* **40**, 467–491.

Thresh, J. M. (1982). Cropping practices and virus spread. *Annu. Rev. Phytopathol.* **20**, 193–218.

Van Regenmortel, M. H. V. (1982). "Serology and Immunochemistry of Plant Viruses." Academic Press, New York.

Zaitlin, M., and Palukaitis, P. (2000). Advances in understanding plant viruses and virus disease. *Annu. Rev. Phytopathol.* **38**, 117–143.

DISEASES CAUSED BY RIGID ROD-SHAPED ssRNA VIRUSES

Diseases Caused by Tobamoviruses: Tobacco Mosaic

Named after *tobacco mosaic virus*, the genus *Tobamovirus* contains more than a dozen rod-shaped viruses measuring 18 by 300 nanometers. Their genome consists of one positive single-stranded RNA [(+) ssRNA] of approximately 6,400 nucleotides (6.4 kb). Their protein coat consists of a single species of protein subunit arranged in a helix.

BOX 23 The Contribution of Tobacco Mosaic Virus to Biology and Medicine

Tobacco mosaic virus has contributed greatly to our understanding of not only the plant viruses and their effects in plants, but also of the viruses of humans and animals and, furthermore, of the structure and function of the genetic code in all organisms. Some of the "firsts" learned from the study of TMV are mentioned briefly here.

When Adolph Mayer began to study the tobacco mosaic disease in 1886, it was the first time that a disease was shown to be caused by a fluid free of any of the known pathogenic fungal and bacterial microbes. Then, in 1898, Beijerinck proposed that tobacco mosaic was caused by an infectious fluid, which he called a virus, free of any cellular microbe, and this changed the prevailing thinking at the time that microbes had to be cellular. TMV was the first virus to be shown (Beale, 1928) that plants infected with it contained a specific antigen. It was also the first virus that was quantified (Holmes, 1929) by the number of local lesions produced on healthy leaves by different concentrations of sap of an infected plant, although no one had any idea yet what TMV was. Then, in 1935, TMV was the first virus to be isolated in crystal form and to be reported by W. Stanley to consist of an "autocatalytic protein." The following year, 1936, Bowden and Pirie made the small but extremely important correction that TMV actually consisted mostly (95%) of protein but it also contained a small amount (5%) of ribonucleic acid (RNA). These discoveries on TMV marked the beginning of virology because, subsequently, methodologies developed to study TMV began to be applied to the study of viruses affecting humans and animals and also microbes such as bacteria. Nevertheless, TMV studies continued to lead the way. In 1939, Kausche took the first electron microscope photographs of TMV, giving the first solid evidence of what a virus looks like. Then, in the mid-1950s, Gierer and Schramm (1956) and Fraenkel-Conrat (1955), again working with TMV, demonstrated that the nucleic acid (RNA) was responsible for causing infection, whereas the protein surrounded the RNA and merely protected the RNA. In 1960, the TMV coat protein was the first virus coat protein to be fully sequenced into its 158 amino acids (Anderer, 1960; Tsugita, 1960), and the sequence of the amino acids of several natural and artificially induced mutant TMV strains was instrumental in establishing the universality of the genetic code and the chemical basis of mutation. In 1969, Takebe used TMV for the infection of suspended tobacco leaf protoplasts, thereby providing the basis for a synchronous infection system for studying virus replication. TMV was also the first plant RNA virus of which the complete genome was sequenced (Goelet, 1982) and also to which monoclonal antibodies were produced (1982). More recently, TMV was the first plant virus to be shown (Powell-Abel, 1986; Beachy, 1986) that introduction and expression of its coat protein gene in plants protected those plants against TMV. In 1987, it was shown that most or all of the TMV coat protein gene can be replaced with a foreign gene and, following inoculation into a plant, the foreign gene is expressed and, if appropriate, may increase the resistance of the plant to disease or may produce vaccines or pharmaceuticals that can be used for the control of human and animal diseases.

There are two closely related viruses of economic importance in the genus: *tobacco mosaic virus*, which infects tobacco and many other, mostly solanaceous, hosts, and *tomato mosaic virus*, which infects tomato. *Pepper green mottle virus* and *odontoglossum ring spot virus* of orchids are also commercially important. Tobamoviruses cause serious losses in their hosts by damaging the leaves, flowers, and fruits and by causing stunting of the plant. The losses are greatest when the plants are infected young. Infections at later stages of growth cause smaller losses. Tobamoviruses are easily transmitted mechanically, and in nature they are spread by incidental contact and wounding. They do not seem to be transmitted by any vectors.

Symptoms consist of various degrees of mottling, chlorosis, curling, distortion, and dwarfing of leaves (Fig. 14-30), flowers, and entire plants. In some plants, necrotic areas develop on the leaves. On tomato, leaflets may become long and pointed and, sometimes, shoestringlike. Infections of young plants reduce fruit set and may occasionally cause blemishes and internal browning on the fruit that does form. Infected cells contain virus particles (Figs. 14-4A and 14-30D) seen easily with an electron microscope and sometimes visible as crystalline aggregates or amorphous bodies with a compound microscope (Fig. 14-2A).

The pathogen is *tobacco mosaic virus* (Fig. 14-30D). The virus particle measures 18 by 300 nanometers and weighs 39 million daltons. Its protein coat consists of approximately 2,130 protein subunits, and each subunit consists of 158 amino acids. Its ssRNA consists of 6,400 nucleotides. The RNA has four open reading frames (ORF) and is translated into four proteins, one of which (17.6 kDa) is the coat protein, two (126 and 183 kDa) are components of the RNA polymerase enzyme, and the fourth (30 kDa) is associated with the cell-to-cell movement of the virus (Fig. 14-7):

```
RNA       O--------------------------ORF1--------------->-------------ORF2-------I-----ORF3-----------I-----ORF4--------I------*
Proteins      ======================ii================I==  ============  ==========
               126 kDa                        183kDa                     30 kDa       CP=17.6 kDa
```

TMV exists in numerous strains, which differ from one another in one or more characteristics.

Tobacco mosaic virus is exceptionally stable. It overseasons in infected tobacco stalks and leaves in the soil, on the surface of contaminated seeds, and for many years in cigarettes, cigars, and so on made with infected tobacco. TMV is very prevalent in many ornamentals in greenhouses and botanical gardens as a result of transmission from tobacco products. The virus is transmitted easily by handling contaminated tobacco products or implements, or infected tobacco plants, and then healthy susceptible plants. From the point of entrance (wound) the virus moves from cell to cell through plasmodesmata, multiplies in and infects each cell (Fig. 14-11), and, when it reaches the phloem, travels systemically through it and infects the entire plant.

The control of *tobacco mosaic virus* depends on sanitation and the use of resistant varieties. Sanitation includes removing infected plants and then washing hands with soap and avoiding planting susceptible hosts for two years in fields or seedbeds where a diseased crop was grown. In some countries, tomatoes in greenhouses were protected from severe strains of TMV by inoculating them while young with a mild strain of the virus. In the past 10 years promising experimental control has been obtained by genetically engineering tobacco and tomato plants with the gene coding for the TMV coat protein. Some control of TMV is also obtained by spraying the plants with or dipping them in milk, which inhibits infection by TMV.

Diseases Caused by Tobraviruses: Tobacco Rattle

The name of the genus *Tobravirus* derives from *tobacco rattle virus*, which causes significant losses in tobacco, potato, and other hosts. The virus causes necrotic areas on stems and leaf veins and a crumpled appearance on the leaves of tobacco, whereas in potato it causes necrotic areas on the leaves (Fig. 14-31A) and stem; in tubers, the latter is known as corky ring spot or spraing of potato (Figs. 14-31B and 14-31C). These diseases occur in Europe and in North and South America. Two other viruses, *pea early browning virus* and *pepper ring spot virus*, are also tobraviruses.

Tobraviruses consist of two rod-shaped particles measuring about 190 by 22 nanometers and about 80–110 by 22 nanometers. Each particle contains a positive single-stranded RNA.

FIGURE 14-30 Symptoms of tobacco mosaic infection on (A) tobacco leaf, (B) tobacco plant, and (C) tomato leaves. (D) Particles of the tobamovirus *tobacco mosaic virus*. [Photographs courtesy of (A and C) Plant Pathology Department, University of Florida and (B) E. J. Reynolds Co.]

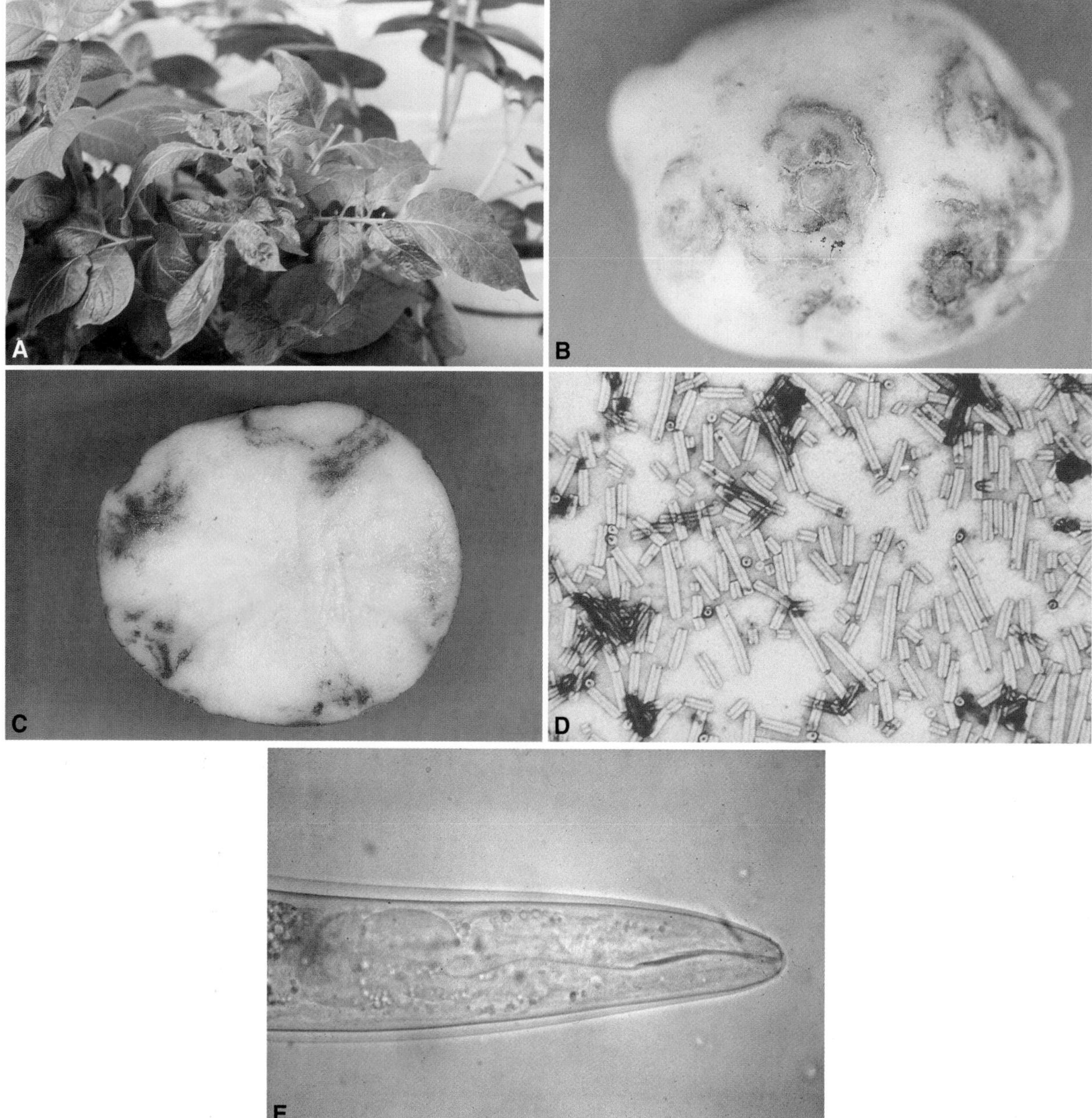

FIGURE 14-31 Foliar mosaic and necrosis (A) and tuber corky ring spot (B and C) symptoms of tobacco rattle disease on potato. (D) Particles of two lengths making up the dipartite genome of the tobravirus *tobacco rattle virus*. (E) The front part of the nematode *Trichodorus*, one of the vectors of TRV. [Photographs courtesy of (A–C) D. P. Weingartner, (D) USDA and (E) P. Lehman.]

```
RNA1 O----------------ORF1------>-----ORF2-I--3--I---4---3OH   RNA2 O-------------------------------------3OH
Proteins =================== =====   === ===                       ========= ==== ====
              194 kDa          29 kDa   29  12                         CP= 24 kDa
```

The RNA of long particles (6.8 kb) contains four genes. Two of these code for two proteins (194 and 29K) that seem to be components of the RNA polymerase enzyme, which replicates both RNAs. Another gene codes for a protein that facilitates cell-to-cell movement of the virus, and the fourth gene codes for a small protein of unknown function. The RNA of the short particles (1.8–4.5 kb) contains one gene that codes for the coat protein (CP) of both particles, and two smaller genes that code for proteins of unknown function.

Tobraviruses are transmitted in nature by nematodes of the genera *Trichodorus* and *Paratrichodorus*. The virus can persist in the vector for weeks or months but does not multiply in the vector. Tobraviruses sometimes invade only the roots of plants. Some tobraviruses, e.g., *pea early browning virus*, may be transmitted by 4 to 10% of the seed of infected plants.

Tobraviruses overseason in infected perennial cultivated or wild plants from which the nematode vector transmits it to the roots of cultivated plants. The virus multiplies in the cytoplasm of parenchyma cells and spreads through the plant from cell to cell and to some extent systemically through the phloem. Some tobraviruses produce characteristic cytoplasmic inclusions consisting of virus particles becoming arranged perpendicularly outside the mitochondria.

Diseases Caused by Furoviruses

Furovirus stands for fungus-transmitted rod-shaped viruses. They include *beet necrotic yellow vein virus*, the cause of rhizomania disease of sugar beets (Fig. 14-32A), transmitted by *Polymyxa betae*, and *soil-borne wheat mosaic virus* (Figs. 14-32B and 14-32C), transmitted by *Polymyxa graminis*. It should be noted, however, that *Polymyxa* is a plasmodiophoromycete, which are now classified as protozoa rather than fungi. The term furoviruses, therefore, is basically incorrect.

Furoviruses consist of two rod-shaped particles, each containing a positive single-stranded RNA. The particles measure from 260 to 300 nanometers and 140–160 nanometers long by 18 to 24 nanometers in diameter. The two RNAs code for nine proteins that include the RNA polymerase, the coat protein, and two proteins involved in vector transmission of the viruses.

Furoviruses cause symptoms that vary with the host. Infected plants appear in patches, they are generally stunted, and the leaves may show mottling or rings. Root systems may be reduced in size or may show excessive branching (rhizomania). Yields are reduced drastically. Furoviruses responsible for the specific diseases overseason in the resting spores of their vectors and in perennial weed and cultivated hosts. The viruses are transmitted to new hosts by viruliferous zoospores of the vectors when they infect healthy plants. The virus often seems to be limited to the roots of the infected plants. Some furoviruses (e.g., *potato mop-top virus*) seem to move systemically through the xylem rather than the phloem.

The control of furoviruses is difficult. In clean fields, only virus-free seed (such as potato tubers and peanuts) should be planted. If the virus is already present in a field, control of the vector through fumigation or by changing the pH reduces infection but is not usually economical.

Diseases Caused by Hordeiviruses

Named after *barley (Hordeum) stripe mosaic virus* (BSMV), hordeiviruses affect primarily grain crops and wild grasses. They consist of three rigid rod-shaped particles about 100 to 150 nanometers long by 20 nanometers in diameter. The longest RNA codes for the RNA polymerase for all three particle RNAs. The middle RNA codes for the coat protein of the virus and three other proteins of unknown function, whereas the short RNA codes for two proteins, one being a possible component of the viral RNA polymerase.

Hordeiviruses, with the exception of *barley stripe mosaic*, which occurs wherever barley is grown, are relatively rare in nature and cause minor losses. Infected plants show mosaics, chlorotic spots, yellow-brown stripes (Fig. 14-32D), and sometimes dwarfing, rosetting, and necrosis of plants. The virus spreads from plant to plant by contact, by pollen, and by a large percentage of seeds produced by infected plants. Virus particles occur in the cytoplasm and, sometimes, in nuclei of infected plants. Hordeiviruses overseason in infected seeds, in which they can survive for several years, and in perennial hosts. Use of virus-free seed and clean cultivation of fields generally provide good control of hordeiviruses.

Diseases Caused by Pecluviruses

Pecluvirus stands for the type species *peanut clump virus*. Each pecluvirus has two rod-shaped particles 245 and 190 nanometers long by 21 nanometers in diame-

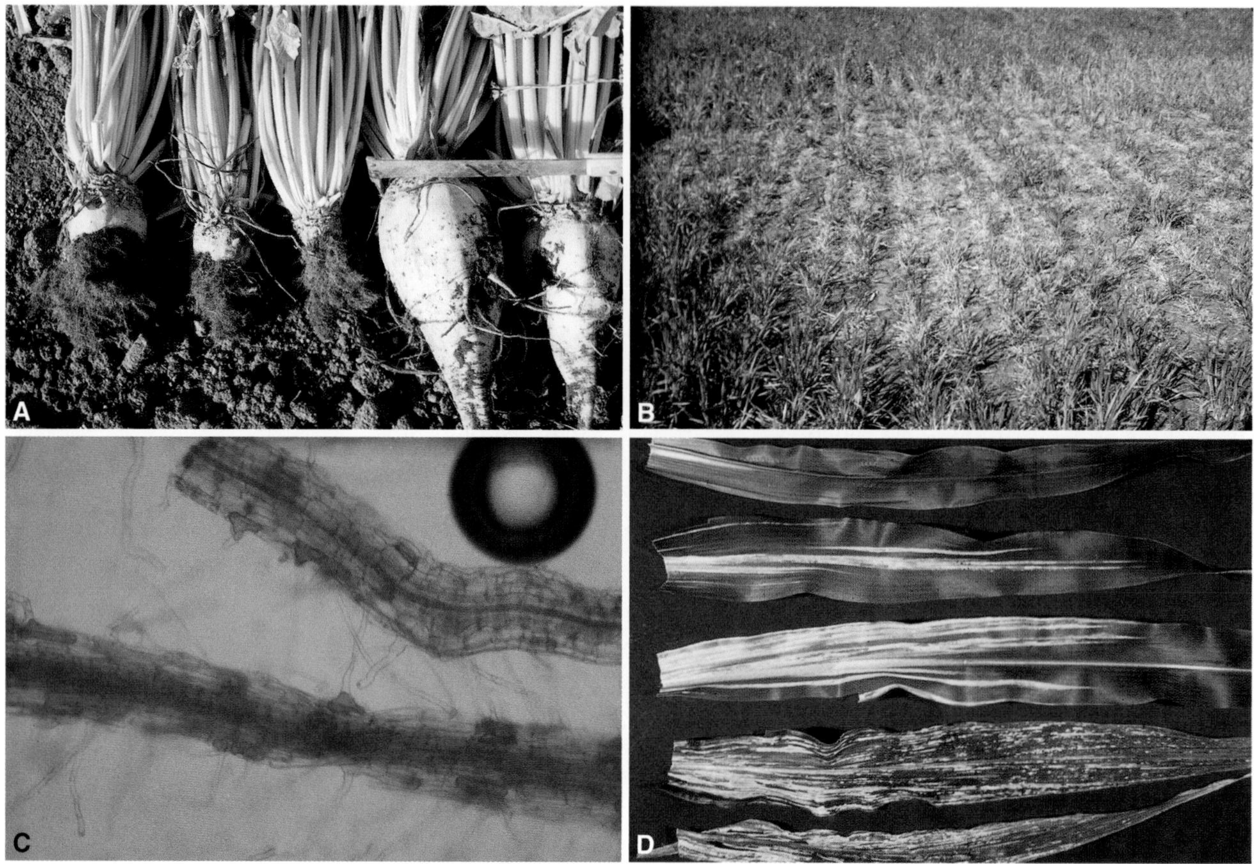

FIGURE 14-32 (A) Rhizomania of sugar beets caused by the furovirus *beet necrotic yellow vein virus*. The three beets at left grew in a naturally contaminated soil, whereas the two at right grew in fumigated soil. (B) Field of wheat infected with another furovirus, *wheat soil-borne mosaic virus*, which has severely stunted or killed the plants in a large area. (C) Roots of wheat plants infected with *Polymyxa graminis*, the fungal vector of the virus. (D) Leaf stripe symptoms on corn caused by the hordeivirus *barley stripe virus*. [Photographs courtesy of (A) G. C. Wisler and (B–D) Plant Pathology Department, University of Florida.]

ter. They are transmitted by the plasmodiophoromycete *Polymyxa graminis* and, in peanuts, by seed.

Diseases Caused by Pomoviruses

Pomoviruses are named after the type species *potato mop-top virus*. Each pomovirus consists of three rod-shaped particles 290–310, 150–160, and 65–80 nanometers long by 18–20 nanometers in diameter. Pomoviruses have narrow host ranges among the dicotyledonous plants and are transmitted by soil plasmodiophoromycetes such as *Spongospora subterranea* and *Polymyxa betae*.

Diseases Caused by Benyviruses

Benyviruses are named after their type species *beet necrotic yellow vein virus*. They consist of four rod-shaped particles 390, 265, 100, and 85 nanometers long by 20 nanometers in diameter. The two larger RNAs are responsible for infection while the other two RNAs influence transmission and symptomatology. Benyviruses are transmitted by the plasmodiophoromycete *Polymyxa betae*.

DISEASES CAUSED BY FILAMENTOUS ssRNA VIRUSES

Diseases Caused by Potexviruses

Named after *potato virus X* (PVX), potexviruses consist of a single, rather sturdy flexuous rod that is from 470 to 580 nanometers long by 11 to 13 nanometers in diameter. Their genome is a positive single-stranded RNA (5.8–7.0 kb) and they have a single species of protein subunit. The RNA codes for five proteins, including the virus RNA polymerase, the coat protein, and a cell-to-cell movement protein.

Numerous potexviruses affect many different crops worldwide. In addition to PVX, the *cymbidium mosaic virus* causes significant losses, being the most important virus of orchids. Diseases caused by potexviruses are generally some type of mosaic that results in varying degrees of stunting and reduced yields. Potexviruses produce large numbers of virus particles in the cytoplasm of infected cells. The virus particles form large aggregates visible even in the light microscope (Figs. 14-3A and 14-3B). Potexviruses lack vectors but are transmitted easily by contact of healthy plants with infected ones and while handling plants during cultivation.

Diseases Caused by Carlaviruses

Named after *carnation latent virus*, the genus *Carlavirus* contains more than 50 carlaviruses, some of which, e.g., *pea streak virus* and *poplar mosaic virus*, cause serious diseases. However, many carlaviruses cause very mild symptoms or are completely symptomless, at least in certain hosts. Some interact synergistically with other viruses and cause serious diseases, as happens when the lily symptomless virus interacts with the cucumber mosaic virus to cause the "fleck" disease. Carlaviruses consist of a single slightly flexuous rod, 610 to 700 nanometers long by 12 to 15 nanometers in diameter, and contain one positive single-stranded RNA (7.4–7.7 kb). Carlaviruses produce particles in the cytoplasm of cells, where they exist singly or in masses without forming any virus-specific inclusions. Carlaviruses are transmitted primarily by aphids or by vegetative propagative organs. Some, however, are transmitted by contact of infected and healthy plants and by handling of such plants; some are spread by whiteflies; and some are occasionally transmitted by seed.

Diseases Caused by Capilloviruses and Trichoviruses

Capilloviruses and trichoviruses have particles and histopathologies similar to those of potexviruses and carlaviruses, but they differ from both of those groups and from one another in sequence and coat protein. Capilloviruses (thin or hair-like viruses) include *apple stem grooving virus* (600–700 by 12 nm), *citrus tatter leaf virus*, and a few others. No vectors are known for capilloviruses. Trichoviruses (hair-like viruses) include *apple chlorotic leaf spot virus* (730 by 12 nm) and several other viruses. Aphids or mealybugs have been implicated as vectors of some trichoviruses, although not of apple chlorotic leaf spot virus.

Diseases Caused by Allexiviruses, Foveaviruses, and Vitiviruses

These three genera contain flexuous filamentous viruses that are about 800 nanometers long and 12 nanometers in diameter. They contain a single component of linear, positive sense ssRNA. Allexiviruses are named after their type species *shallot (Allium sp.) virus X*, have very narrow host ranges, and are transmitted in nature by eriophyid mites. Foveaviruses have as type species the *apple stem pitting virus*, infect only one or a few species of plants, and have no known vector in nature. Vitiviruses are named after the type species *grapevine (=Vitis sp.) virus A*. Each of them is restricted to a single plant species. Some vitiviruses are transmitted by mealybugs and some are also transmitted by a scale insect. One vitivirus is transmitted semipersistently by aphids.

Selected References

Abel, P. P., *et al.* (1986). Delay of disease development in transgenic plants that express the tobacco mosaic virus coat protein gene. *Science* **232**, 738–743.

Almasi, A., *et al.* (2000). BSMV infection inhibits chlorophyll biosynthesis in barley plants. *Physiol. Mol. Plant Pathol.* **56**, 227–233.

Bachard, G. D., and Costello, J. D. (2001). Immunolocalization of tobacco mosaic tobamovirus in roots of red spruce seedlings. *J. Phytophtol.* **149**, 415–419.

Brunt, A. A., and Richards, K. E. (1989). Biology and molecular biology of furoviruses. *Adv. Virus Res.* **36**, 1–32.

Choi, I.-R., Horken, K. M. K., Stenger, D. C., *et al.* (2002). Mapping of the P1 proteinase cleavage site in the polyprotein of *wheat streak mosaic virus* (genus *Tritimovirus.*). *J. Gen. Virol.* **83**, 443–450.

Clover, G. R. G., Ratti, C., and Henry, C. M. (2001). Molecular characterization and detection of European isolates of soil-borne wheat mosaic virus. *Plant Pathol.* **50**, 761–767.

"C.M.I./A.A.B. Descriptions of Plant Viruses." Carlaviruses (No. 259), carnation latent virus (No. 61), pea streak virus (No. 112), poplar mosaic virus (No. 75), apple stem grooving virus (No. 31), apple chlorotic leafspot virus (No. 30). Kew, Surrey, England.

"C.M.I./A.A.B. Description of Plant Viruses." Potexviruses (No. 200), potato virus X (No. 4). Kew, Surrey, England.

"C.M.I./A.A.B Descriptions of Plant Viruses." Tobamoviruses (No. 184), tobacco mosaic virus (No. 156), tobacco rattle virus (No. 12), pea early browning virus (No. 120), soilborne wheat mosaic virus (No. 77), peanut clump virus (No. 235), potato mop-top virus (No. 138), beet necrotic yellow vein virus (No. 144), barley stripe mosaic virus (No. 68). A series of concise publications describing individual plant viruses and virus groups by the Commonwealth Mycological Institute/Association of Applied Biologists. Kew, Surrey, England.

Dawson, W. O. (1992). Tobamovirus-plant interactions. *Virology* **186**, 359–367.

Francki, R. I. B., Milne, R. G., and Halta, T. (1985). "Atlas of Plant Viruses." Vol. 2. CRC Press, Boca Raton, FL.

Heinze, C., *et al.* (2000). Sequences of tobacco rattle viruses from potato. *J. Phytophtol.* **148**, 547–554.

Jackson, A. O., Hunter, B. G., and Gustafson, G. D. (1989). Hordeivirus relationships and genome organization. *Annu. Rev. Phytopathol.* **27**, 95–181.

Knapp, E., and Lewandowski, D. (2001). Tobacco mosaic virus, not just a single component virus anymore. *Mol. Plant Pathol.* **2**, 117–123.

Lawrence, D., and Jackson, A. O. (2001). Requirements for cell-to-cell movement of *barley stripe mosaic virus* in monocot and dicot hosts. *Mol. Plant Pathol.* **2**, 65–75.

Marathe, R., *et al.* (2002). The tobacco mosaic virus resistance gene, N. *Mol. Plant Pathol.* **3**, 167–172.

Rabentstein, F., Seifers, D. L., Schubert, J., *et al.* (2002). Phylogenetic relationships, strain diversity and biogeography of tritimoviruses. *J. Gen. Virol.* **83**, 895–906.

Rhee, Y. (2000). Cell-to-cell movement of tobacco mosaic virus: Enigmas and explanations. *Mol. Plant Pathol.* **1**, 33–39.

Richards, K. E., and Tamada, T. (1992). Mapping functions on the multipartite genome of beet necrotic yellow vein virus. *Annu. Rev. Phytopathol.* **30**, 291–313.

Rush, C. M., and Heidel, G. B. (1995). Furovirus diseases of sugar beets in the United States. *Plant Dis.* **79**, 868–875.

Scholthoff, K.-B. G., Shaw, J. G., and Zaitlin, M. (1999). "Tobacco Mosaic Virus: 100 Years of Contributions to Virology." APS Press, St. Paul, MN.

Siegrist, J., Orober, M., and Buchenauer, H. (2000). β-Aminobutyric acid-mediated enhancement of resistance in tobacco to tobacco mosaic virus depends on the accumulation of salicylic acid. *Physiol. Mol. Plant Pathol.* **56**, 95–106.

DISEASES CAUSED BY POTYVIRIDAE

The family *Potyviridae* contains six genera: *Potyvirus, Ipomovirus, Macluravirus, Tritimovirus, Rymovirus,* and *Bymovirus.* They are all flexuous filamentous viruses 11 to 15 nanometers in diameter. However, whereas most potyviridae have monopartite particles 650 to 900 nanometers long, bymoviruses have bipartite, 250 to 300 nanometers long and 500 to 600 nanometers long particles. Of the six genera, *Potyvirus* contains by far the most numerous and most important plant viruses. All potyviridae form cylindrical inclusion bodies in infected cells (Figs. 14-3C and 14-3D). Various potyviridae are transmitted in nature by a variety of vectors.

Diseases Caused by Potyviruses

Named after *potato virus Y* (PVY), potyviruses comprise the largest genus of plant viruses. It contains more than 90 confirmed potyviruses and about 90 more tentative potyvirus species. They include many of the viruses causing some of the most severe diseases of crop plants. In addition to PVY, potyviruses include the very severe *bean common mosaic virus* (BCMV), *bean yellow mosaic virus* (BYMV), *beet mosaic virus* (BtMV), *celery mosaic virus* (CeMV), *lettuce mosaic virus* (LMV), *papaya ring spot virus* (PRSV), *pepper mottle virus* (PepMV), *plum pox virus* (PPV), *soybean mosaic virus* (SoyMV), *sugarcane mosaic virus* (SCMV), *tobacco etch virus* (TEV), *turnip mosaic virus* (TuMV), *watermelon mosaic virus* (WMV1 and WMV2), *zucchini yellow mosaic virus* (ZYMV), and others. Several viruses of ornamentals, e.g., *dasheen mosaic virus* and *tulip breaking virus*, also belong in this family.

Potyviruses consist of a single flexuous rod-shaped particle 680 to 900 nanometers long by 12 nanometers in diameter (Fig. 14-4B). They have a single positive RNA species (~10 kb) and one kind of coat protein subunit. The potyvirus RNA, like RNAs of some other viruses, is joined at its 5′ end to a small protein (Vpg, for virus protein, genome linked) that seems to act as a primer for replication of the RNA, and at its 3′ end it has a polyadenylate sequence of about 190 adenine bases, the function of which is not certain but may be associated with the ability of the RNA to act as messenger RNA, i.e., to be translated into a protein(s). The main body of the potyvirus RNA is translated into one huge polyprotein of about 346,000 daltons that is subsequently cleaved at specific points to produce smaller polyproteins, which are eventually cleaved to release eight proteins (Fig. 14-33).

```
RNA O-------------------------------------------------------------------------------AAA
Proteins ======i=====i====== i ======= = ========== = ====i ==== =========i ====
          35K   63K     52K      50K  6K     71K     6K  21K  27K     58K      30K
          P1          HC-Pro     P3                  Vpg  NIa         NIb      CP
```

The 35K protein is a proteinase that cleaves the polyprotein and helps RNA binding. The 52K protein is a proteinase enzyme that cleaves the polyprotein and also has helper component activity necessary for the insect transmission of potyviruses. The activity of the 6K and 50K proteins is unknown. The 21K protein is the Vpg, i.e., the genome-linked viral protein attached to the 5′ terminus of the viral nucleic acid, which acts as a primer for replication. The 27K protein is a proteinase enzyme needed for cleaving the viral polyprotein at Gln-(Ser/Gly) bonds, whereas the 71K and the 58K proteins are components of the RNA polymerase of the virus. The 30K protein is the capsid (coat) protein. Of the proteins, the 52–71K and 6K proteins aggregate to form

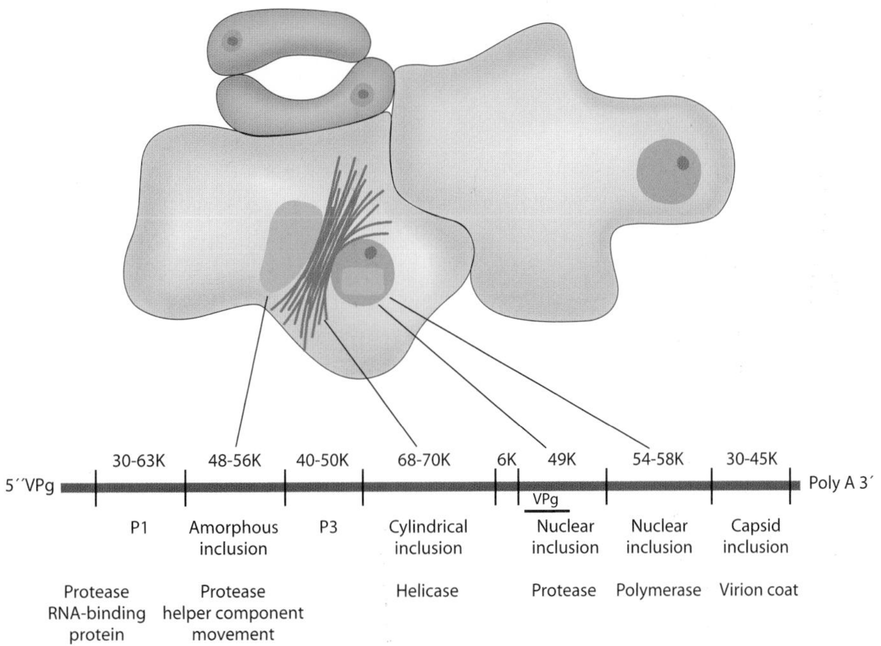

FIGURE 14-33 Generalized map of the potyviral genome. The thick, straight green line shows the arrangement of the viral genes, and above them is the size of the protein encoded by each gene. Below the line are the structural proteins and below them are the function of each protein. Arrows point to the cytoplasmic and nuclear inclusions encoded by certain genes. (Courtesy of E. Hiebert.)

FIGURE 14-34 (A) Symptoms of foliar mosaic and necrosis on potato caused by the necrotic strain of *potato virus* Y (PVY). (B) Mosaic symptoms and foliar malformations caused by PVY on pepper. [Photographs courtesy of (A) L. Brown and (B) Plant Pathology Department, University of Florida.]

characteristic cylindrical or pinwheel-like inclusions (Figs. 14-3C and 14-3D) present in plant cells infected by all potyviruses. However, the 52K protein accumulates as amorphous inclusion bodies in the cytoplasm of plants infected with some but not all potyviruses. The 21K and 27K proteins are the small and the 58K the large nuclear inclusion proteins that aggregate in the nucleus to form the nuclear inclusion body. The 30K protein is the virus coat protein and is also involved in symptom expression and insect transmission of the virus.

Potyviruses cause numerous severe diseases of plants. Most such diseases appear primarily as mosaics, mottling, chlorotic rings or color break on foliage, flowers, fruits, and stems. Many of them, however, cause severe stunting of young plants and drastically reduced yields; leaf, fruit, and stem malformations; fruit drop; and necroses of various tissues (Figs. 14-34 to 14-37).

FIGURE 14-35 Bean mosaics. (A) Common bean mosaic symptoms on bean plants. Yellow bean mosaic symptoms on leaves (B) and malformations on pods (C). (Photographs courtesy of R. Providenti.)

Potyviruses are transmitted in nature by aphids in the nonpersistent manner, and several of them are transmitted through the seed. The virus-coded helper protein of 48–52K is needed in conjunction with the capsid protein for the aphid transmission of potyviruses. Potyviruses overseason in perennial cultivated and weed hosts and, for seed-transmitted ones, in seed. Each growing season, potyviruses are transmitted by their aphid vectors from their perennial hosts, or from plants produced from virus-infected seed, to the healthy plants of the new crop. The newly infected crop plants then become a new reservoir of virus from which the aphids transmit it to additional plants. The number of infected plants in a field increases slowly at first but quite rapidly later in the season as the number of aphids and plants serving as virus reservoir increase. Often, 100% of the plants in a field become infected with the virus. The severity of the losses is proportional to the length of the time a plant has been infected, i.e., proportional to how young the plant was when it first became infected. Therefore, losses are greatest when the virus is present in perennial weeds near or within the new crop or when the virus is present in crop plants produced from virus-infected seed and interspersed among healthy crop plants because from such plants the aphids can spread the virus quickly to nearby, healthy plants.

The control of potyviruses is very difficult. Resistant varieties, when available, should be preferred. Using virus-free seed when the virus is seed transmitted is often very effective. Destroying infected volunteer plants or weeds within and around crop fields can be helpful. Planting early sometimes helps avoid the later influx of large numbers of aphid vectors and delays the age at which plants become infected. Similar delays of infection can be achieved by spraying plants with insecticides or special nontoxic oils or by applying reflective plastic covers (mulches) between plants. None of these measures, however, controls the migrating aphids; they only delay infection by a few weeks, which is sometimes sufficient to obtain a fairly good crop. For some crop viruses that have very limited host range, e.g., *lettuce mosaic* and *celery mosaic viruses*, keeping the area (e.g., valley) free of the crop (and therefore of the virus) for 2 to 3 months and subsequently using virus-free seed allow the profitable production of the crop where otherwise it would be impossible. In the 1990s, the control of several potyviruses was obtained by producing transgenic plants containing and expressing genes derived from the virus itself. Such plants transformed with the virus coat protein gene, or with mutated genes coding for the viral RNA polymerase, virus cell-to-cell movement protein, or insect transmission protein, or with other sense or antisense segments of the viral RNA, expressed various degrees of resistance to infection by potyviruses (pathogen-derived resistance). It is expected that such genetic engineering technologies will provide effective resistance in many crops to their potyviruses in the near future.

Some of the most important plant diseases caused by potyviruses are as follows.

Bean Common Mosaic and Bean Yellow Mosaic

Both bean common mosaic and bean yellow mosaic occur wherever beans are grown. Bean common mosaic affects all beans, but only beans (*Phaseolus vulgaris* and some other *Phaseolus* spp.), whereas bean yellow mosaic also affects peas, clovers, vetch, black locust, gladiolus, and yellow summer squash, among others. Both diseases are widespread in bean fields, with common mosaic being more widespread than yellow mosaic. They are often found in the same field and often on the same plants. Plants infected with either virus may show mottling, yellowing, and malformation of leaves and pods (Fig. 14-35). Infected plants may be stunted and bunchy, seeds may be aborted, smaller, or malformed, and yields may be reduced by up to 80 to 100%, depending on the plant stage of growth at the time of infection.

Both viruses, *bean common mosaic virus* and *bean yellow mosaic virus*, measure 750 by 12 nanometers and are transmitted through several aphids, most of the vectors being common to both viruses. *Bean common mosaic virus* is, moreover, readily transmitted through bean seeds; when the mother plants are infected while young, as many as 83% of their seeds may produce virus-infected plants. Seed transmission is the most important source of initial crop infection with common mosaic in bean fields. *Bean common mosaic virus* can also be transmitted by pollen. *Bean yellow mosaic virus* overseasons in one of its many cultivated and wild hosts, from which the aphids transmit it to the crop. *Bean yellow mosaic virus* is not transmitted through the seed in beans but is transmitted in about 3 to 6% of the seeds of several other legumes.

The control of bean common mosaic is obtained through the use of certified virus-free seed and through planting of bean varieties resistant to the virus. The control of bean yellow mosaic is more difficult because the virus overseasons in perennial hosts such as clovers and gladiolus and because few bean varieties show even partial resistance to some but never to all strains of the virus.

Lettuce Mosaic

The lettuce mosaic disease occurs in the United States, especially California, and Europe, and probably worldwide. *Lettuce mosaic virus*, in addition to lettuce, infects pea and sweet pea, marigold, zinnia, and weeds like groundsel and prickly sow thistle.

Lettuce mosaic symptoms consist of mottling or yellowing of the leaves, followed by distortion and marginal necrosis of leaves, dwarfing of the plant, and failure to produce a marketable lettuce head. Losses from the disease can be very severe.

Lettuce mosaic virus (750 by 12 nm) is transmitted by several species of aphids and by 1 to 8% of the seed produced by infected plants. Plants infected through the seed are the main source of virus for its subsequent transmission by aphids to other plants. The control of lettuce mosaic, therefore, depends primarily on using virus-free lettuce seed and on maintaining a period of lettuce-free cultivation in the area.

Plum Pox

Plum pox, sometimes referred to as sharka, occurs in Europe and Asia, Chile, and since 1999 in North America. It affects plum, peach, nectarine, and apricot. It causes devastating losses of fruit quantity and quality and debilitates infected trees. Plum pox, where present, is the most important disease of these trees. Leaves of infected trees show severe mottling, diffuse or bright rings, or vein yellowing and elongated line patterns (Figs. 14-36A and 14-36B). Infected plum fruits develop severe pox symptoms (Fig. 14-36C) with dark-colored rings or patches on the skin, brown or reddish discoloration in the flesh, and brown spots on the stones (pits). Most of the infected fruits fall prematurely. Peach

FIGURE 14-36 Plum pox symptoms on (A) mosaic on plum leaves, (B) line patterns on peach leaves, (C) flat, dry, pox-like areas on plum fruit, (D and E) rings and some unevenness on peach fruit, (F) rings and other discolorations and distortions and malformations on apricots, and (G) white discolorations and rings on the stones (pits) of apricots.

fruits show mottled rings and distortion (Figs. 14-36D and 14-36E), whereas apricot fruits also show rings but are more deformed, have necrotic rings and bumps (Fig. 14-36E), and contain stones that show striking whitish-yellow rings (Fig. 14-36F).

Plum pox virus (760 by 12 nm) is transmitted by budding and grafting and by several aphid species in the nonpersistent manner. The virus perpetuates itself in infected trees. The control of plum pox is extremely difficult. Planting virus-free trees in areas away from infected orchards is helpful, as is the use of resistant or tolerant varieties. Quick detection and removal of infected trees also help reduce inoculum in the orchard. Studies are underway to cross-protect trees with mild *plum pox virus* strains and by genetically engineering them to express the coat protein gene of the *plum pox virus*.

Papaya Ring Spot

The papaya ring spot disease occurs in many tropical countries and islands worldwide and in the United States in Florida, Texas, and Hawaii. Papaya ring spot is one of the most destructive diseases of papaya. In many areas, profitable papaya cultivation is impossible in the presence of papaya ring spot.

Infected trees show symptoms within 2 to 3 weeks from inoculation. Symptoms consist of intense yellow mosaic on leaves, small shoestring-like new leaves (Figs. 14-37A and 14-37B), dark green and slightly sunken rings on the fruit (Fig. 14-37C), numerous oily-looking streaks on the stem, and stunting of the plant (Fig. 14-37D). Fruits produced after infection are usually small, exhibit lichen-like lesions and ring spots, show uneven bumps, and have an unpleasant taste. Trees infected at a very young age remain stunted and never produce any fruit.

Papaya ring spot virus (800 by 12 nm), in addition to papaya, also attacks cucurbits (Figs. 14-37E and 14-37F) and used to be known as *watermelon mosaic virus 1 or* PRSV-p. Another closely related virus that infects only cucurbits used to be known as PRSV-w. The latter also causes severe losses in cucurbits but does not infect papaya and is now known as *watermelon mosaic virus*.

Papaya ring spot virus is transmitted by several species of aphids in the nonpersistent manner. Most spread by aphids is from papaya to papaya tree and is rapid, with the virus often infecting all trees in an orchard within a few months. Control of the disease is difficult. Isolation of new orchards from older ones with many infected trees and early roguing of infected trees help slow the spread of the disease. Planting papaya trees bred for tolerance to papaya ring spot is also helpful. In Hawaii, successful control of papaya ring spot has been obtained through cross protection: papaya trees are first inoculated with mild strains of *papaya ring spot virus*, and these strains protect the trees from the catastrophic effects of infection with the naturally occurring severe strains of the virus. Similar cross protection has been obtained in Hawaii and in Asia by genetically engineering the *papaya ring spot virus* coat protein gene into papaya trees. Genetically engineered resistance has been highly effective in controlling papaya ring spot in several papaya-producing parts of the world.

Potato Virus Y

Potato virus Y (PVY) occurs worldwide and is of great economic importance. It affects potato (Fig. 14-34A), pepper (Fig. 14-34B), tomato, and tobacco and causes severe losses on all these hosts. Symptoms vary from a mild to severe mottle on most hosts to a streak or "leaf-drop streak" resulting from long necrotic lesions along the veins on the underside of leaflets of some potato varieties. When present together with *potato virus X*, PVY causes "rugose mosaic," in which the plants are dwarfed and the tubers reduced in size.

Potato virus Y (730 by 11 nm) exists in nature as several distinct strains. It is transmitted through infected potato seed tubers and by at least 25 species of aphids in the nonpersistent manner. Control of PVY is difficult. Use of PVY-free potato tubers for seed certification programs is by far the most effective and most promising control measure for PVY on a worldwide basis. Potato varieties resistant to the virus and control practices that reduce PVY transmission by its aphid vectors are helpful to a limited extent. Some varieties have now been engineered to express the virus coat protein gene and are being tested for their ability to cross protect against the virus.

Sugarcane Mosaic

Sugarcane mosaic occurs worldwide, wherever sugarcane is grown. Its many strains also infect corn, sorghum, and the other Gramineae. The disease can be very severe. Symptoms appear as pale patches or blotches on the leaves, not of uniform width and not confined between the veins. Stems may show mottling or marbling, the affected areas later becoming necrotic. The stems become small and deformed; the shoots remain stunted and produce a few twisted or distorted leaves. Cane and sugar yield are reduced severely.

Two corn strains of *sugarcane mosaic virus* cause maize dwarf mosaic in the United States and Australia. They affect corn, sorghum, and several wild and cultivated grasses but apparently not sugarcane. Symptoms

FIGURE 14-37 Symptoms of *papaya ringspot virus* (PRSV) on papaya (A–D) and watermelon (E and F). (A) Papaya leaves show yellow mosaic, become narrow and flat, and fall off early. (B) Close-up of papaya leaf with mosaic. (C) Ring spots on papaya fruit. (D) Severely infected papaya trees (left) compared to unaffected trees genetically engineered for resistance. (E) PRSV-infected squash leaves showing foliar mosaic and malformations, and (F) ring spots on watermelon fruit. [Photographs courtesy of (A) M. Davis, (B and C) D. Persley, (D) D. Gonsalves, and (E and F) Plant Pathology Department, University of Florida.]

on corn and grasses develop only on plants infected early and consist of a stippled mottle, mosaic, or narrow streaks on the younger leaves (Fig. 14-38A) and shortening of upper internodes. Older leaves show no mosaic but appear yellowish-green and may have yellowish-red streaks. The corn ears remain small and incompletely filled (Fig. 14-38B). Yield in susceptible varieties may be reduced by up to 40%. Sorghum plants show mosaic followed by red striping and necrotic areas on the leaves. Maize dwarf mosaic is apparently caused by at least two, more or less host-specific strains of the *sugarcane mosaic virus* (Fig. 14-38C) that in nature infect and damage primarily corn. Strain A infects and overwinters in the perennial weed Johnsongrass, whereas strain B does not infect Johnsongrass. Both strains infect corn, sorghum, and several other annual grain crops and grasses.

Sugarcane mosaic virus (750 by 11 nm) is transmitted primarily vegetatively in sugarcane during propagation of the crop. In sugarcane and in all other grain crops, however, *sugarcane mosaic virus* is also transmitted by several aphid species in the nonpersistent manner. The virus overseasons in infected sugarcane or in appropriate perennial hosts of the specific strains. Control is possible only through the use of resistant or tolerant varieties.

FIGURE 14-38 Maize dwarf mosaic on corn caused by the maize strain of the *sugarcane mosaic virus* (SCMV). (A)Mosaic on young leaves of corn plant. (B) Mosaic, yellowing-reddening and stunting of corn plant. (C) Poorly filled ear of corn from SCMV-infected plant. (D) Electron micrograph of the virus.

Tobacco Etch

Tobacco etch occurs in North and South America. It is caused by the *tobacco etch virus*, which also infects pepper and tomato. It causes severe losses on all three hosts. Infected tobacco leaves are narrowed and show mottling and necrosis. Pepper leaves show mottling (Fig. 14-39), mosaic, and distortion; pepper fruit are distorted, and the entire plant may be stunted. Tomato plants are also stunted, and the leaves are mottled and distorted. TEV (730 by 12 nm) is transmitted by more than 10 species of aphids in the nonpersistent manner. Control is primarily through resistant varieties.

Turnip Mosaic

Turnip mosaic occurs worldwide. It affects all vegetable and ornamental crucifers. It appears as mottling, black necrotic spots, and ring spots in cabbage, cauliflower, and Brussels sprouts, whereas in the other crucifers it causes mosaic, leaf distortion, and stunting. *Turnip mosaic virus* (720 by 12 nm) is transmitted by about 50 species of aphids in the nonpersistent manner.

Watermelon Mosaic

Watermelon mosaic, caused by *watermelon mosaic virus*, occurs worldwide. It causes mosaic and mottle diseases on all cucurbits and reduces fruit production and quality (Figs. 14-40A and 14-40B). It also infects peas and other

FIGURE 14-39 Tobacco leaf showing symptoms caused by *tobacco etch virus*.

FIGURE 14-40 Mosaic on squash leaves (A) and mosaic and malformations on cucumber fruit (B) caused by *watermelon mosaic virus*. (C) Mosaic on yellow squash leaf and color reversal and swellings on yellow summer squash caused by the *zucchini yellow mosaic virus*. (Photographs courtesy of Plant Pathology Department, University of Florida.)

leguminous, malvaceous, and chenopodiaceous crop plants, ornamentals, and weeds. *Watermelon mosaic virus* (WMV-2) (760 by 12 nm) is transmitted by at least 38 species of aphids in the nonpersistent manner.

Zucchini Yellow Mosaic

Zucchini yellow mosaic probably occurs worldwide. It causes economically important diseases in zucchini squash, muskmelon, cucumber, and watermelon. Symptoms consist of severe mosaic, yellowing, shoestringing, stunting, and distortions of fruit and seed (Fig. 14-40C). The *zucchini yellow mosaic virus* (750 by 12 nm) infects many hosts experimentally, although so far in nature it has not been found in hosts other than cucurbits. It is transmitted by at least four aphid species in the nonpersistent manner.

Diseases Caused by Ipomoviruses, Macluraviruses, Rymoviruses, and Tritimoviruses

Named after the type species *sweet potato (Ipomea sp.) mild mottle virus*, ipomoviruses are 800–950 nanometers long and are transmitted by the whitefly *Bemisia tabaci* in the nonpersistent manner. Macluraviruses, named after the type species *Maclura mosaic virus*, are 650–675 nanometers long and are transmitted by aphids in the nonpersistent manner. Rymoviruses are named after their type species *rygrass mosaic virus* and are 690–720 nanometers long. They are transmitted by eriophyid mites. Tritimoviruses, the type species of which is *wheat streak mosaic virus*, are named so because they infect only grass and grain plants. Wheat streak mosaic causes severe symptoms (Figs. 14-41A and 14-41B). Tritimoviruses are also transmitted by

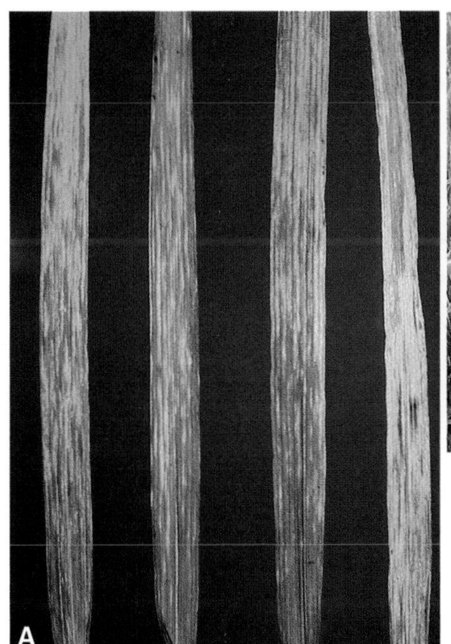

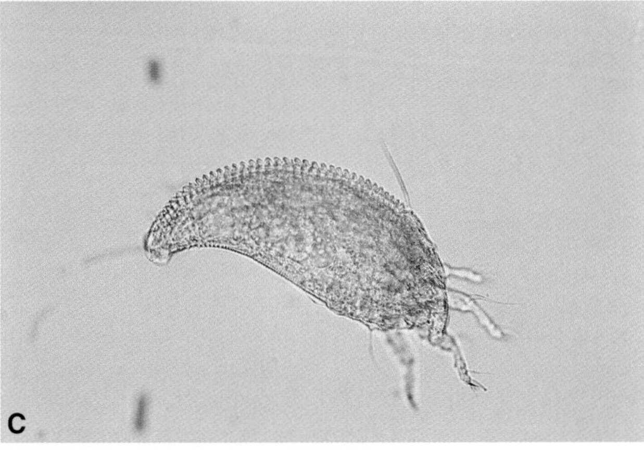

FIGURE 14-41 (A) Yellowish-orange streaks on wheat leaves caused by the tritivirus *wheat streak mosaic virus* (WSMV). (B) Wheat field with most plants infected with WSMV. (C) The eriophyte mite vector of WSMV. (Photographs courtesy of Plant Pathology Department, University of Florida.)

eriophyid mites (Fig. 14-41C), probably in a persistent manner.

Diseases Caused by Bymoviruses

Bymoviruses have particles and cytopathologies similar to those of potyviruses. However, they have their own different vectors. Bymoviruses, named after *barley yellow mosaic virus*, also affect cultivated grain crops and grasses, causing significant losses. Other bymoviruses include *oat mosaic virus*, *rice necrosis mosaic virus*, and *wheat spindle streak mosaic virus*. They are soilborne, transmitted by *Polymyxa graminis*. In addition, each bymovirus consists of two different particles, one about 500 to 600 by 12 nanometers and the other 275 to 300 by 12 nanometers.

Selected References

Clover, *et al.* The effects of beet yellows virus on the growth and physiology of sugar beet (*Beta vulgaris*). *Plant Pathol.* **48**, 129–138.

"C.M.I./A.A.B. Descriptions of Plant Viruses." Closteroviruses (No. 260), potyviruses (no. 245). A series of concise publications describing individual plant viruses and virus groups. Kew, Surrey, England.

Dougherty, W. G., and Carrington, J. C. (1988). Expression and function of potyviral gene products. *Annu. Rev. Phytopathol.* **26**, 123–143.

Edwardson, J. R., and Christie, R. G. (1991). "The Potyvirus Group." Florida Agric. Exp. Stn. Monogr. No. 16, Vols. 1–4. Gainesville, FL.

Hinrichs-Berger, *et al.* (1999). Cytological responses of susceptible and extremely resistant potato plants to inoculation with potato virus Y. *Physiol. Mol. Plant Pathol.* **55**, 143–150.

Kegler, *et al.* (2001). Hypersensitivity of plum genotypes to plum pox virus. *J. Phytopathol.* **149**, 213–218.

Kerlan, C., *et al.* (1999). Variability of potato virus Y in potato crops in France. *J. Phytopathol.* **147**, 643–651.

Khurana, S. M. P., and Garg, I. D. (1992). Potato mosaics. *In* "Plant Diseases of International Importance" (U.S. Singh *et al.*, eds.), Vol. 2, pp. 148–164. Prentice-Hall, Englewood Cliffs, NJ.

Kurstak, E., ed. (1981). "Handbook of Plant Virus Infections and Comparative Diagnosis." Elsevier, Amsterdam.

Milne, R. G., ed. (1988). "The Plant Viruses," Vol. 4. Plenum, New York.

Purcifull, D. E., and Hiebert, E. (1992). Serological relationships involving potyviral nonstructural proteins. *Arch. Virol.* (Suppl. 5), 97–122.

Schmidt, H. E. (1992). Bean mosaics. *In* "Plant Diseases of International Importance" (U.S. Singh *et al.*, eds.), Vol. 2, pp. 40–73. Prentice-Hall, Englewood Cliffs, NJ.

Shukla, D. D., Brant, A. A., and Ward, C. W. (1994). *Potyviridae*. Descriptions of Plant Viruses No. 245. Assoc. Appl. Biol., Wellesbourne, England.

Shukla, D. D., Ward, C. W., and Brunt, A. A. (1994). "Potyviruses: Biology, Molecular Structure, and Taxonomy." CAB Int., Wallingford, England.

Shukla, D. D., Ward, C. W., and Brunt, A. A. (1994). "The Potyviridae." CAB International, Wallingford, UK.

Singh, U. S., Kohmoto, K., and Singh, R. P. (1994). "Pathogenesis and Host Specificity in Plant Diseases," Vol. 3. Elsevier, Tarrytown, New York.

Yeh, S.-D., *et al.* (1988). Control of papaya ringspot virus by cross-protection. *Plant Dis.* **72**, 375–380.

DISEASES CAUSED BY CLOSTEROVIRIDAE

The term closteroviridae means "thread-like viruses." There are two genera of viruses in closteroviridae: *Closterovirus*, whose members have long, thin, very flexuous thread-like particles 1,100 to 2,000 nanometers long by 12 nanometers in diameter. They contain the largest single ssRNA genome of plant viruses; and *Crinivirus*, the genome of which is separated in two particles, 700–900 and 650–850 nanometers long and 12 nano-meters in diameter. Some closteroviridae are transmitted by aphids, some by whiteflies, and others by mealybugs. Closteroviruses include the aphid-transmitted *beet yellows virus*, *citrus tristeza virus*, and the mealybug-transmitted *grapevine leafroll-associated viruses*. All these viruses are widespread and cause very severe losses in their respective hosts. Criniviruses include the severe whitefly-transmitted *lettuce infectious yellows virus* and some other whitefly-transmitted viruses. Closteroviridae spread through their hosts systemically, but they are confined to the phloem and phloem parenchyma cells. Each virus in this family has a rather narrow host range and causes diseases of the yellows type as a result of phloem necrosis, including pitting or grooving of woody stems.

Diseases Caused by Closteroviruses

Citrus Tristeza

Tristeza occurs in almost all citrus-growing areas of the world. It affects practically all kinds of citrus plants but primarily orange, grapefruit, and lime. Severe strains of tristeza virus can cause severe losses of fruit quantity and quality and result in either a chronic or a quick decline and eventual death of infected trees. Tristeza symptoms consisting of a quick or chronic tree decline (Figs. 14-42A,B,C and 14-42E) are particularly common and severe on trees propagated on sour orange rootstocks. Millions of citrus trees have been and continue to be killed in South Africa since 1910, in Argentina and Brazil since the 1930s, and in Colombia and Spain since the 1970s. Tristeza was first reported in Florida in the 1950s, but losses became serious after severe virus strains became widespread in the 1980s. Even more severe strains and more efficient insect vectors, however, have been moving north from South

America through Central America and through the Caribbean islands, and they further threaten citrus production in the United States. In 1995, the brown citrus aphid *Toxoptera* citricida (Fig. 14-42G), considered to be the most efficient vector of severe (including stem pitting-causing) strains of citrus tristeza virus, was introduced into Florida. The following year, it spread to almost all citrus groves. This introduction poses an immediate threat to the 20 million citrus trees grafted on sour orange rootstock in Florida alone. It also threatens, however, potential catastrophic losses to the Florida and total citrus industry in the United States.

Symptoms caused by *citrus tristeza virus* on the various citrus species vary primarily with the particular

FIGURE 14-42 Citrus tristeza. (A) Orange tree on sour orange rootstock killed by quick decline. (B) Orange tree killed by the slow decline type of tristeza. (C) Orange grove in which many trees have either been killed already by tristeza or are at varying stages of decline and death by tristeza. (D) Citrus tree showing stem pitting above and necrosis at the graft union. (E) Tristeza-infected grapefruit tree showing extensive stem pitting in its trunk. (F) Small, discolored, misshapen, and poor-quality grapefruit produced by tristeza-infected trees. (G) Brown citrus aphids, the most efficient vector of the *citrus tristeza virus* (CTV). (H) An electron photograph of CTV. [Photographs courtesy of (A, C, and D) USDA, (B) R. J. McGovern, and (E–H) S. Garnsey.]

Continued

FIGURE 14-42 *(Continued)*

strain of the virus and with the rootstock on which the citrus scion is propagated. Most tristeza virus strains are mild and produce no noticeable symptoms on commercial citrus varieties; they are detected only by indexing on sensitive indicator hosts, such as Mexican lime, or by serological and nucleic acid tests. More severe strains cause a condition known as seedling yellows, consisting of severe chlorosis and dwarfing on seedlings of sour orange, lemon, and grapefruit, especially when they are kept under greenhouse conditions. In the field, young sweet orange, grapefruit, and other citrus trees growing on sour orange rootstock and inoculated with some of the severe strains of tristeza virus develop a quick decline within a few weeks. The leaves of trees developing quick decline turn yellow or brown (Fig. 14-42A) and later wilt and fall off (Fig. 14-42B) while the fruit continues to hang on the dead tree. Some severe strains, however, do not cause quick decline but instead either interfere with the growth of young trees, which remain severely stunted and fail to come into production, or cause trees to decline over several years (chronic decline), during which the trees grow poorly, become less productive, decline, and eventually die. Decline-inducing tristeza virus strains infecting citrus trees on sour orange rootstocks cause phloem necrosis at the graft union, which results in the accumulation of foodstuffs in and overgrowth of the scion above the union while few foodstuffs go through to the roots. As a result the roots grow poorly or die, causing the decline of the aboveground parts of the tree.

In addition to the mild and decline-causing strains of *citrus tristeza virus* (CTV), there are severe strains that cause stem pitting (Fig. 14-42E). Infected trees exhibit deep longitudinal pits in the wood under the bark, in trunks, in branches, and even in twigs of infected grapefruit or sweet orange trees regardless of the rootstock on which they are grafted. Actually, these strains also cause stem pitting on the rootstocks themselves. Trees with stem pitting are stunted and set less fruit, the fruit is of smaller size and of poor quality (Fig. 14-42F), the twigs are brittle and break easily, and the trees decline but do not die for many years.

The pathogen, *citrus tristeza virus*, consists of a thread-like particle 2,000 nanometers long by 12 nanometers in diameter (Fig. 14-42H). Each particle contains one positive sense single-stranded RNA consisting of 20 kilobases and a coat protein subunit with molecular weight of 25,000. The tristeza virus RNA codes for 10 to 12 proteins, but the function of several of them is still uncertain.

The largest protein (349 k) is a papain-like proteinase, methylesterase and helicase. The 25 k is the coat protein.

```
RNA O----------I----------------------------I----------I---i-------------I--------I-----------I------I--------I---I---I-----3OH
Proteins======= ================ ====== =  =======  ====  ======  === ==== == == =======
           57k      349k              33k   6k     65k    61k     27k   25k   18k  13k 20k 23k
```

Citrus tristeza virus is transmitted by budding or grafting and by several species of aphids in the semi-persistent manner, i.e., the aphids require feeding for at least 30 to 60 minutes to acquire the virus and subsequently remain viruliferous for about 24 hours. The various aphid species vary greatly in their ability to transmit CTV. The most efficient aphid vector, *Toxoptera citricida*, known as the brown citrus aphid, colonizes and affects only citrus but is 10 to 25 times more efficient as a CTV vector than any of the other aphids. Also, *T. citricida* can transmit CTV strains causing severe decline or stem pitting that the other aphid vectors do not transmit or transmit poorly. *T. citricida* occurs in most citrus-growing areas but not yet in the Mediterranean countries. In the last 20 years, this aphid had been moving northward from South America through Central America and the Caribbean islands. By 1993 it had reached Cuba. In late 1995, *T. citricida* was found in Florida and, as expected, it spread throughout most of the citrus-growing areas within the next year.

The control of citrus tristeza is difficult. Where the disease is absent, strict quarantine regulations should be enforced. Only tested budwood certified to be free of CTV should be used under all conditions, and any trees detected to carry severe strains of tristeza virus should be destroyed. If the disease already occurs in an area, considerable control can be obtained by avoiding grafting trees on sour orange and, instead, grafting on tristeza-tolerant rootstocks; using scion varieties tolerant to stem pitting also is recommended. In addition, trees can be cross protected from severe tristeza for fairly long periods by inoculating them with certain mild strains of the virus. Presently, considerable efforts are being made to genetically engineer citrus trees to express CTV genes, such as the coat protein gene, that might make the trees resistant to tristeza.

Beet Yellows

Beet yellows occurs in all major sugar beet-growing areas of the world. It causes a yellows disease in sugar beets, table beets, and spinach. The outer and middle leaves of infected plants become yellow (Fig. 14-43A), thickened, brittle, and may become necrotic. Beet production is reduced drastically, as is sugar content in the beets produced. *Beet yellows virus* is a closterovirus, 1,250 nanometers long by 12 nanometers in diameter. It is transmitted by more than 20 aphid species in the semipersistent manner.

Diseases caused by Criniviruses: Lettuce Infectious Yellows

Lettuce infectious yellows occurs in the southwestern United States and in Mexico. It affects many cultivated crops, such as lettuce (Fig. 14-43B), sugar beet, carrot, cantaloupe, melon, squash (Figs. 14-43C and 14-43D), and many weeds. Wherever it occurs it usually infects all the plants in a field and causes devastating losses usually exceeding 20 to 30% and often approaching 100%. The symptoms of lettuce infectious yellows consist of severe yellowing and/or reddening of leaves

FIGURE 14-43 (A) Symptoms of beet yellows on leaf of sugar beet caused by the *beet yellows virus*. (B) Field symptoms of lettuce infected with the crinivirus *lettuce infectious yellows virus* (LIYV). (C) Close-up of LIYV symptoms on cantaloupe and (D) symptoms of *lettuce infectious yellows* on cantaloupe in the field. (Photographs courtesy of Plant Pathology Department, University of Arizona.)

followed by stunting, rolling, and brittleness of the leaves. Infected plants remain stunted and may die.

The *lettuce infectious yellows virus* (LIYV) has a bipartite genome in two filamentous particles 700–900 and 650–850 nanometers long by 12 nanometers in diameter. LIYV has one type of protein subunit of 28 kilodaltons. The LIYV ssRNA consists of about 16 kilobases but exists in two components, one of 8.5 kilobases and the other of 7.5 kilobases. The two components together code for approximately the same number and the same kinds of proteins as those coded for by the one-component RNAs of *beet yellows virus* and *citrus tristeza virus*. Moreover, the order of the RNA genes coding for these proteins is almost identical in all these viruses.

Lettuce infectious yellows virus is transmitted by the sweet potato whitefly *Bemisia tabaci* in the semipersistent manner. Whiteflies acquire the virus after feeding for 10 minutes or more, but their efficiency increases with feeding durations up to one hour or longer. Viruliferous whiteflies can infect healthy plants for up to three days after feeding on an infected one. It was noted in California, however, that a few years after appearance of the sweet potato whitefly (*B. tabaci*) and the efficient spread of LIYV, a new whitefly biotype morphologically indistinguishable from *B. tabaci* moved in and replaced the sweet potato whitefly in nature. The new whitefly is a very poor vector of LIYV. It is now known as the silver leaf whitefly (*Bemisia argentifolii*).

Lettuce infectious yellows virus overseasons in perennial cultivated crops and weeds from which the whitefly vectors transmit it to young crop plants. LIYV epidemics follow heavy whitefly infestations of crop fields. Control of *lettuce infectious yellows virus* is very difficult and depends primarily on planting resistant crops, keeping whitefly populations down, and planting in areas and at times that will allow growth of the crop before viruliferous whiteflies arrive.

Selected References

Bodin-Ferri, M., *et al.* (2002). Systemic spread of plum pox virus (PPV) in Mariana plum GF 8-1 in relation to shoot growth. *Plant Pathol.* **51**, 142–148.

Brlansky, R. H., Howd, D. S., and Damsteegt, V. D. (2002). Histology of sweet orange stem pitting caused by an Australian Isolate of *Citrus tristeza virus*. *Plant Dis.* **86**, 1169–1174.

Cohen, S., Duffus, J. D., and Liu, H. Y. (1992). A new *Bemisia tabaci* biotype in the southwestern United States and its role in silverleaf of squash and transmission of lettuce infectious yellows virus. *Phytopathology* **82**, 86–90.

Dolja, V. V., Karasev, A. V., and Koonin, E. V. (1994). Molecular biology and evolution of closteroviruses: Sophisticated build-up of large RNA genomes. *Annu. Rev. Phytopathol.* **32**, 261–285.

Duffus, J. E., Larsen, R. C., and Liu, H. Y. (1986). Lettuce infectious yellows virus: A new type of whitefly-transmitted virus. *Phytopathology* **76**, 97–100.

Ghorbel, R., López, C, Fagoaga, C., *et al.* (2001). Transgenic citrus plants expressing the citrus tristeza virus p23 protein exhibit viral-like symptoms. *Mol Plant Pathol.* **2**, 27–36.

Hung, T. H., Wu, M., and Su, H. J. (2000). A rapid method based on the one-step reverse transcriptase-polymerase chain reaction (RT-PCR) technique for detection of different strains of citrus tristeza virus. *J. Phytopathol.* **148**, 469–475.

Lee, R. F., and Rocha-Pe±a, M. A. (1992). Citrus tristeza virus. *In* "Plant Diseases of International Importance" (U.S. Singh, *et al.*, eds.), Vol. 3, pp. 226–243. Prentice-Hall, Englewood Cliffs, NJ.

Lee, R. F., *et al.* (1992). Presence of *Toxoptera citricidus* in Central America. Citrus Industry June.

Rocha-Pe±a, M. A., *et al.* (1995). Citrus tristeza virus and its aphid vector *Toxoptera citricida*: Threats to citrus production in the Caribbean and Central and North America. *Plant Dis.* **79**, 437–445.

Rubio, L., Abou-Jawdah, Y., Lin, H.-X., and Falk, B. W. (2001). Geographically distant isolates of the crinivirus *cucurbit yellow stunting disorder virus* show very low genetic diversity in the coat protein gene. *J. Gen. Virol.* **82**, 929–933.

Rubio, L., Soong, J., Kao, J., and Falk, K. B. W. (1999). Geographic distribution and molecular variation of isolates of three whitefly-borne closteroviruses of cucurbits: Lettuce infectious yellows virus, cucurbit yellow stunting disorder virus, and beet pseudo-yellows virus. *Phytopathology* **89**, 707–711.

Sambade, A., *et al.* (2002). Comparison of viral RNA populations of pathogenically distinct isolates of citrus tristeza virus: Application to monitoring cross protection. *Plant Pathol.* **51**, 257–265.

Sedas, Haidar, Greif, *et al.* (2000). Establishment of a relationship between grapevine leafroll closteroviruses 1 and 3 by use of monoclonal antibodies. *Plant Pathol.* **49**, 80–85.

DISEASES CAUSED BY ISOMETRIC SINGLE-STRANDED RNA VIRUSES

There are numerous isometric ssRNA viruses, 26–35 nanometers in diameter, that comprise several families containing numerous genera of viruses (Fig. 14-24). One group of such viruses has its ssRNA genome contained in one isometric virion. This group includes the virus families Sequiviridae, Tombusviridae, and Luteoviridae and several genera not yet assigned to families. Another group of ssRNA viruses, the Comoviridae, has its ssRNA genome subdivided into two components, each occupying a different isometric virion. A third group of such viruses, the Bromoviridae, has three components of RNA, each contained in isometric virions of three different sizes. Finally, some genera of the Bromoviridae, such as *Ilarvirus*, *Alphamovirus*, and *Oleavirus*, have virions that vary in shape and size from quasi-isometric to bacilliform.

Diseases Caused by Sequiviridae, Genus Waikavirus

Waikaviruses, named for *rice waika* (stunting) *virus*, include the *rice tungro spherical virus* (Fig. 14-44C). Waikaviruses are isometric, about 30 nanometers in diameter, and have a single-stranded RNA genome of about 11 kilobases. The composition of their protein coat is unknown. Waikaviruses infect only certain species of grain crops and weeds. The virus particles occur in granular inclusions in the cytoplasm of phloem cells and occasionally in mesophyll cells. They are transmitted either by leafhoppers or by aphids in the semipersistent manner. Control of waikaviruses depends on the use of virus- or vector-resistant, or virus-tolerant, varieties.

Rice Tungro

Tungro is the most serious virus disease of rice in south and southeast Asia from Pakistan to the Philippines. Tungro (yellow-orange) is the result of concurrent infection by two viruses: the single-stranded RNA virus *rice tungro spherical virus* (RTSV) and the double-stranded DNA virus *rice tungro bacilliform virus* (RTBV) (Figs. 14-44A and 14-44C). Both viruses are transmitted by several leafhoppers (Fig. 14-44B), particularly *Nephotettix virescens*, in the semipersistent manner. The RTSV RNA consists of about 12.4 kilobases, which encodes a 393-kilodalton polyprotein that is cleaved into several smaller proteins. The protein coat is made of two types of protein molecules.

FIGURE 14-44 (A) Rice tungro-infected rice plants in the field showing stunting and yellow-orange coloration. (B) Female of the leafhopper vector of the tungro viruses. (C) Purified particles of the spherical (waikavirus) and bacilliform (badnavirus) viruses that together cause the rice tungro disease. (Photographs courtesy of H. Hibino.)

Tungro-infected rice plants are stunted and show mottling and yellow-orange discoloration of the leaves (Fig. 14-44A). Typical tungro symptoms can be caused by RTBV, but they are intensified by the presence of RTSV. RTSV often occurs alone but causes only very mild symptoms. The disease caused by RTSV alone was earlier known as rice waika disease and the virus as *rice waika virus*. Also, although both viruses are transmitted by leafhoppers in the semipersistent manner, only RTSV can be transmitted alone by leafhoppers, whereas RTBV transmission by leafhoppers is possible only when RTSV is also present in the donor plant.

Selected References

"C.M.I./A.A.B. Descriptions of Plant Viruses." Rice tungro spherical virus (No. 67). Kew, Surrey, England.

Gingery, R. E., and Nault, L. R. (1990). Severe maize chlorotic dwarf disease caused by double infection with mild virus strains. *Phytopathology* **80**, 687–691.

Hibino, H., *et al.* (1991). Characterization of rice tungro bacilliform and rice tungro spherical viruses. *Phytopathology* **81**, 1130–1132.

Huet, J., Mahendra, S., Wang, J., *et al.* (1999). Near immunity to rice tungro spherical virus achieved in rice by a replicase-mediated resistance strategy. *Phytopathology* **89**, 1022–1027.

Diseases Caused by Tombusviridae

Tombusviridae include eight genera of ssRNA isometric viruses 32–35 nanometers in diameter. The particles contain one species of (+)ssRNA. Individual genera can infect either monocot or dicot plants but not both. Most of these viruses are very stable and can survive in surface water or in soil from where plants can acquire them without a vector. Few viruses of genera in this family cause economically severe diseases to plants. The genera of tombusviridae and some of their most important characteristics are listed.

1. *Tombusvirus*, from its type species *tomato bushy stunt virus*. Most viruses are soilborne; some are transmitted by the chytrid fungus *Olpidium*.
2. *Aureusvirus*, a single species from pothos.
3. *Avenavirus*, also a single species in oats (*Avena*).
4. *Carmovirus*, from *carnation mottle virus*.
5. *Machlomovirus*, from *maize chlorotic mottle virus*, is restricted to Gramineae and is transmitted by seed and possibly by beetles and thrips.
6. *Necrovirus*, named after *tobacco necrosis virus A*, has a *wide* host range of mono- and dicots, generally infect roots, and are transmitted by the chytrid fungus *Olpidium*.
7. *Panicovirus*, named after *panicum mosaic virus*, affects only Gramineae and is transmitted primarily by contact.
8. *Dianthovirus*, from *carnation* (= *Dianthus*) *ring spot virus*, has genomes divided into two ssRNAs contained in virions 32–25 nanometers in diameter. Dianthoviruses seem to affect only dicots and appear to be transmitted easily through the soil but no specific vector is known.

Diseases Caused by Luteoviridae

Luteoviridae, named after the Latin word *luteus*, which means yellow, are a large group of about 30 viruses that infect plants and cause them to develop varying degrees of yellowing symptoms. All luteoviruses are confined to the phloem cells of their hosts, are present in very low concentrations, and are not transmitted by mechanical inoculation. They are transmitted by aphids in the persistent, circulative but not propagative, manner.

Luteoviruses are isometric single-stranded RNA viruses 25 to 30 nanometers in diameter. Their RNA consists of approximately 6 kilobases and seems to code for six proteins. Luteoviruses have one type of coat protein, with subunits of 22 to 23 kilodaltons. Within the luteoviridae can be distinguished four genera, some of which cause extremely severe diseases: (1) *Luteovirus*, the type species of which is *barley yellow dwarf virus* and is restricted to Gramineae: (2) *Polerovirus*, named after the type species *potato leafroll virus*, has some members that attack dicots and some that attack monocots. The *Polerovirus* also contains the very important *beet western yellows virus* (3) *Enamovirus*, named after *pea enation mosaic virus*. Actually, the disease, pea enation mosaic, is caused by the complex of one enamovirus, PEMV-1, and an umbravirus, PEMV-2. Both viruses have the same protein coat subunits coded for by the RNA of PEMV-1. Enamoviruses are transmitted mechanically and by aphids. Several additional viruses seem to belong to luteoviridae, but they have not yet been assigned to a genus.

Barley Yellow Dwarf

Barley yellow dwarf occurs throughout the world. It affects a wide variety of gramineous hosts, including barley, oats, wheat, rye, and many lawn, weed, pasture, and range grasses.

Barley yellow dwarf affects plants by causing stunting, reduced tillering, suppressed heading, sterility, and failure to fill the kernels. In some cases, entire fields are destroyed and the crops are not worth harvesting. Of the main crops, oats is the most severely affected and suffers serious losses annually. In years of barley yellow dwarf outbreaks, oat yield losses may range from 30 to 50% while barley and wheat losses range between 5 and 30%. To these losses must be added losses in quality of the grain and losses in forage crops from the resulting failure or reduced productivity of pasture, range, and meadow grasses.

Yellow dwarf-infected barley plants show yellowish, reddish, or purple areas along the margins, tips, or lamina of the older leaves. In seedling infections, leaves may emerge distorted, curled, and with serrations. Stems are shorter (Fig. 14-45). Tillering is reduced in oat and wheat plants but is excessive in severely stunted barley plants. Inflorescences of diseased plants emerge later and are smaller. Flowers are often sterile, and the number and weight of kernels are reduced. The root systems of diseased plants are reduced drastically.

Barley yellow dwarf virus (BYDV) (Fig. 14-45D) is transmitted by several aphid species. Most aphids require an acquisition feeding period of about 24 hours and an inoculation feeding period of 4 to 8 hours or more. BYDV consists of numerous strains, which differ in their relative virulence on different host varieties, in the symptoms they produce, and in their transmission by different aphid vectors.

Barley yellow dwarf virus overseasons in grass hosts, in fall-sown cereals, and in viruliferous adult aphids. The spread of the virus depends on the aphid vectors. The worst epidemics, however, develop from virus brought into cereal fields in the spring by migrating viruliferous aphids and when the spring and early summer weather is cool and moist.

FIGURE 14-45 Barley yellow dwarf symptoms on wheat plants (A) and on wheat in the field (B). (C) Barley yellow dwarf symptoms of varying severity on barley plants. (D) Purified particles of *barley yellow dwarf virus*. [Photographs courtesy of (A–C) S. M. Haber, W. C. P. D. and (D) W. F. Rochow.]

The stage of host development at the time of infection is a crucial factor in disease development. The most severe symptoms result only from infection of the annual cereals in the seedling stage. Infected seedlings often die or, if they survive, usually fail to head, and if they do, the inflorescence and entire plant are extremely small. In later stages of infection, in which the virus has progressively less time in which to affect the host, the disease severity is reduced proportionately, and only the last formed leaf may show mild symptoms. In fall-sown cereals, BYDV infections increase winter killing of plants as well as reduce yields.

The main hope for control of BYDV is the use of resistant varieties. Most of the commercial varieties of oats, barley, and wheat are susceptible to BYDV, but some are less susceptible than others. A number of varieties have been found or developed that show some tolerance or resistance to BYDV. An extensive breeding program to develop varieties of the three main cereals that can withstand heavy barley yellow dwarf epidemics is currently being carried out. Some cultural practices, such as time of sowing, can be manipulated to reduce early infection of the grain crops.

Potato Leafroll

Potato leafroll occurs worldwide. It is caused by the *potato leafroll virus* (PLRV) and affects only potato. It causes high yield losses and can be the most devastating virus of potato. It causes a prominent upward rolling of the leaves, and the plants are stunted and have a stiff

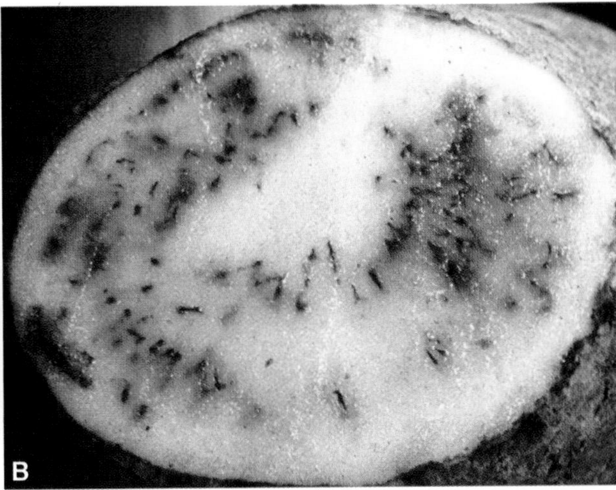

FIGURE 14-46 (A) Potato plants showing stunting and leaf rolling caused by infection with the *potato leafroll virus* (PLRV). (B) Potato tuber showing vein necrosis as a result of infection with PLRV. [Photographs courtesy of (A) Plant Pathology Department, University of Florida and (B) Plant Pathology Department, University of Idaho.]

upright growth (Fig. 14-46A). In some varieties, phloem becomes necrotic and carbohydrates accumulate in the leaves. There is phloem necrosis in tubers also (Fig. 14-46B). The virus is transmitted through infected potato seed tubers and, in the field, by more than 10 species of aphids in the persistent manner. Its control depends on the use of potato seed tubers free of the virus. Because the vector must feed for several hours to acquire the virus and for several more hours to infect the plant with the virus, some control of PLRV has been achieved through early control of the aphid vectors with insecticides.

Beet Western Yellows

Beet western yellows probably occurs worldwide. It affects sugar beets, spinach, lettuce, and many crucifers. It causes chlorosis and stunting and moderate reductions in yield. *Beet western yellows virus* is transmitted by eight species of aphids in the persistent (circulative) manner, persisting in the vector for more than 50 days.

Selected References

Burnett, P. A., ed. (1989). "Barley Yellow Dwarf Virus, the Yellow Plague of Cereals." CIMMVT, Mexico City, Mexico.

"C.M.I./A.A.B. Descriptions of Plant Viruses." Luteovirus group (No. 339), beet western yellows virus (No. 89), potato leafroll virus (No. 291), barley yellow dwarf virus (No. 32). Kew, Surrey, England.

Francki, R. I. B., Milne, R. G., and Hatta, T. (1985). "Atlas of Plant Viruses." CRC Press, Boca Raton, FL.

Gray, S. M., Smith, D., and Altman, N. (1993). Barley yellow dwarf virus isolate-specific resistance in spring oats reduced virus accumulation and aphid transmission. *Phytopathology* **83**, 716–720.

Irvin, M. E., and Thresh, J. M. (1990). Epidemiology of barley yellow dwarf: A study in ecological complexity. *Annu. Rev. Phytopathol.* **28**, 393–424.

Martin, R. R., *et al.* (1990). Evolution and molecular biology of luteoviruses. *Annu. Rev. Phytopathol.* **28**, 341–363.

Miller, W. A., and Rasochova, I. (1997). Barley yellow dwarf viruses. *Annu. Rev. Phytopathol.* **35**, 167–190.

Miller, W. A., Liu, S., and Becket, R. (2002). Barley yellow dwarf virus: Luteoviridae or Tombusviridae? *Mol. Plant Pathol.* **3**, 177–183.

Rouzé-Jouan, J., Terradot, L., Pasquer, F., *et al.* (2001). The passage of *potato leafroll virus* through *Myzus persicae* gut membrane regulates transmission efficiency. *J. Gen. Virol.* **82**, 17–23.

Smith, H. G., and Barker, H., eds. (1999). "The Luteoviridae." CABI Publ., Wallingford, CT.

Diseases Caused by Monopartite Isometric (+)ssRNA Viruses of Genera Not Yet Assigned to Families

Such genera include the following. (1) *Sobemovirus*, named after *soybean mosaic virus*, contains several viruses that cause serious losses. Several of these viruses are seed transmitted and most are transmitted by beetles. (2) *Marafivirus*, after *maize rayado fino virus*; its members are restricted to the Gramineae, are transmitted by leafhoppers, and cause severe diseases. (3) *Tymovirus*, after *turnip yellow mosaic virus*; its members affect dicot plants, are transmitted by beetles, and cause several fairly severe diseases. (4) *Idaeovirus*, the type species of which is *raspberry bushy dwarf virus*, has three genome RNAs in each particle, its members are restricted to genus *Rubus*, and it is transmitted by pollen and by seed. (5) *Ourmiavirus*, from *ourmia melon virus*, has three genome RNAs located in bacilliform particles 18 nanometers in diameter by 30, 37, 46, and 62 nanometers long. No vector of the virus is known. (6) *Umbravirus*, after *carrot mottle virus*, the members of which do not code for a coat protein but use the coat protein of some other helper virus, usually a member of Luteoviridae, which also helps them be transmitted by its aphid vector.

Diseases Caused by Comoviridae

The family *Comoviridae* contains three genera of viruses: *Comovirus*, *Fabavirus*, and *Nepovirus*. They are all isometric viruses about 30 nanometers in diameter. Their genome consists of two single-stranded RNAs each contained in a separate but identical virus particle. Some empty virus particles containing no RNA at all are also always present. The protein shell of each particle is made up of one, two, or three types of protein subunits.

```
RNA1 O---------------------------------------------------AAA
         =====i========I====i========
          32k     58k     24k     87k

RNA2 O--------------------------------------AAA
          =======I=====I===
          48/58k   37k   23k
```

The two RNAs, which have 6 to 7 and 3.4 to 4.5 kilobases, respectively, code for several proteins. The two RNAs are first translated into two large polyproteins that are then cleaved into the smaller proteins. The larger RNA codes for a 32K protein that seems to regulate another protein (the 24K protein), which is responsible for all cleavages in both polyproteins. The larger RNA also codes for the 87K polymerase, a 58K protein involved in membrane attachment of the replication complex, and a 4K protein (Vpg) attached to the 5′ end of each RNA and involved in the initiation of RNA synthesis. The shorter RNA codes for 58K and 48 K proteins, needed for viral cell-to-cell movement, and the 37K and 23K coat proteins. The 3′ end of each RNA has a short polyadenylate chain [poly(A)].

Of the *Comoviridae*, comoviruses have narrow host ranges, whereas fabaviruses and nepoviruses have wide host ranges. Viruses within each genus may cause widely different symptoms. Comoviruses are transmitted by beetles mostly of the family Chrysomelidae, whereas fabaviruses are transmitted by aphids. Most nepoviruses are transmitted by nematodes and also through a considerable portion of the seed produced by infected plants.

Diseases Caused by Comoviruses

Comoviruses, named after *cowpea mosaic virus*, affect primarily legumes (bean, cowpea, pea, soybean, clover) and a few other hosts, such as squash (*squash mosaic virus*) and radish (*radish mosaic virus*). Comoviruses cause mosaics, stunting, and malformations of varying severity. Comoviruses induce the formation of large vacuolated and crystalline inclusion bodies in the cytoplasm of infected cells.

Comoviruses are transmitted easily by mechanical inoculation and, in the field, by specific leaf-feeding beetles. Comoviruses are also transmitted through a small but significant percentage of seeds. Control of comoviruses depends primarily on the use of virus-free seed. Control of beetles with insecticides early in the season also helps reduce losses.

Selected References

"C.M.I./A.A.B. Descriptions of Plant Viruses." Comoviruses (No. 199), cowpea mosaic virus (No. 197), squash mosaic virus (No. 43), radish mosaic virus (No. 121). Kew, Surrey, England.

Diseases Caused by Nepoviruses

Nepoviruses stands for nematode-transmitted polyhedral (isometric) viruses. They are a large group of more than 30 viruses, each of which may attack many annual and perennial plants and trees. They cause many severe diseases of trees and vines. Nepovirus-infected plants often show severe shock symptoms initially or in early spring but later in the season show partial recovery during which the symptoms (chronic symptoms) are milder or disappear completely. Some of the most important nepoviruses are *tomato ring spot virus, tobacco ring spot virus* (Fig. 14-47), *cherry leaf roll virus, grapevine fanleaf virus, arabis mosaic virus*, and *raspberry ring spot virus*.

Nepovirus particles and their genomes are very similar to those of comoviruses. They are about 30 nanometers in diameter and have bipartite genomes, i.e., RNAs of 8 to 8.4 kilobases and 3.4 to 7.2 kilobases. The RNAs have a 5′ Vpg and a 3′ polyadenylate tail and their genes are similar to and arranged as in

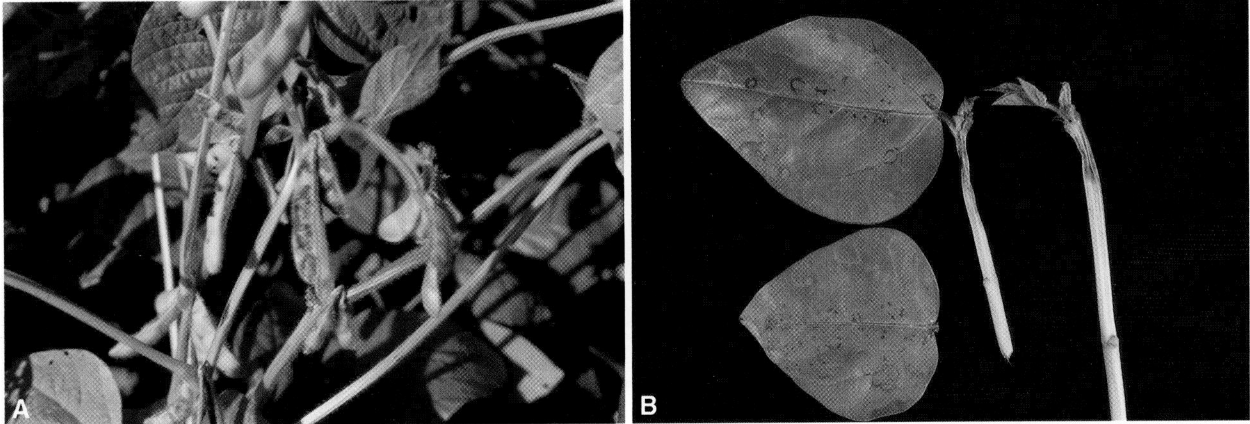

FIGURE 14-47 (A) Pod blight of soybeans caused by *tobacco ring spot virus* (TRSV). (B) Local lesions on leaves and necrosis of the top of the stem of cowpea following inoculation with TRSV. [Photograph (A) courtesy of Plant Pathology Department, University of Florida.]

comoviruses. The shell of nepoviruses, however, consists of one, two, or three types of protein subunits. Several nepoviruses contain satellite RNAs in their particles, which depend on the virus for their replication.

Nepoviruses infect parenchyma and phloem cells and can be seen as small aggregates in the cytoplasm or in vacuoles. Frequently, nepovirus particles are seen in linear arrays in tubules scattered in the cytoplasm or associated with plasmodesmata, through which they often move from cell to cell.

Nepoviruses are transmitted from plant to plant by nematodes of the genera *Longidorus*, *Paralongidorus*, and *Xiphinema*. Nematodes acquire the virus after feeding on infected hosts for several hours and they retain it and can transmit it for several months. Most nepoviruses are also transmitted through various percentages of seeds produced on infected plants. Several of them are also transmitted by pollen. Nepoviruses overwinter in perennial hosts and in seeds; during the growing season, they are transmitted to healthy annual and perennial host plants by their nematode vectors or by pollen.

The control of diseases caused by nepoviruses depends on planting only virus-free seeds and nursery plants, locating new plantings in fields free of the vector and the virus, planting crops resistant or tolerant to the virus, and fumigating the field with nematicides.

Tomato Ring Spot

Tomato ring spot is widespread in North America and has also been reported from other parts of the world. It is of minor importance to tomato production, but it infects many other hosts and causes particularly severe losses on many perennial hosts. On annual and some perennial hosts, *tomato ring spot virus* (TomRSV) causes mostly mosaic and ring spot diseases (Figs. 14-48A and 14-48B), sometimes accompanied by various degrees of systemic necrosis. On perennial hosts, however, TomRSV usually causes no distinctive symptoms on the foliage; rather, it affects the base of the plant. The virus is transmitted by the nematode *Xiphinema*. In some hosts, TomRSV is also transmitted through seed.

Many pome fruit and stone fruit varieties and rootstocks, as well as many small fruits, such as grapes, raspberries, and strawberries, are affected by TomRSV in North America; they suffer severe losses by diseases described sketchily by the names prunus stem pitting and decline, apple graft union necrosis and decline, and grapevine yellow vein disease and grapevine decline. In apple, the most common symptoms are slight stem pitting on either side of the graft union followed by gradual necrosis of the graft union (Figs. 14-48C and 14-48D). This occurs when hypersensitive apple varieties are grafted on tomato ring spot-tolerant apple rootstocks such as MM 106, which later become infected with TomRSV via nematode vectors of the virus. Eventually, affected trees show yellowing of foliage, twig dieback, and general decline and death within 3 to 5 years of the appearance of symptoms at the graft union. In *Prunus* species, there is more extensive and severe pitting of the scion or rootstock, or both, on either side of the graft union, various degrees of necrosis at the union plate, and again foliage yellowing, twig dieback, and general decline and death of the trees within 3 to 5 years. In grapevines and raspberries, the leaves may show mottling, rings, or yellow veins, the vines remain

FIGURE 14-48 *Tomato ring spot virus* (TomRSV) local (A) and systemic (B) symptoms on tobacco leaves. (C and D) Graft union necrosis symptoms caused by TomRSV on apple. (E) Stem pitting in cherry twig caused by TomRSV infection. [Photographs courtesy of (C and D) J. Halbrendt.]

stunted, fruit clusters develop poorly or not at all, and berry size may be uneven. All of these diseases, caused by the nematode-transmitted *tomato ring spot virus*, are among the most important diseases in each of the respective fruit trees or vines.

Grapevine Fanleaf

Grapevine fanleaf occurs worldwide. It affects only grapes. It occurs in many strains and causes variable symptoms but always severe losses. Depending on the virus strain, infected leaves show a green or yellow mosaic, rings, line patterns, or flecks. In most varieties, infections cause smaller, slightly asymmetric leaves, whereas in others the veins are spread abnormally, giving the leaf a fan-like appearance (Fig. 14-49). Leaves may show a chrome-yellow mottle, the mottled areas later becoming paler, then necrotic, and finally dropping, or leaves may show chrome-yellow areas along main veins of mature leaves. Canes are often deformed, having uneven internode lengths, double nodes, and pitting of the bark and wood. Fruit production is low. Many flowers shell from clusters, and small, seedless berries develop along with a few normal berries.

FIGURE 14-49 Grape fanleaf symptoms caused by *grape fanleaf virus*.

The vigor and yield of grapevines are reduced progressively, and the vines gradually degenerate and die. *Grapevine fanleaf virus* is transmitted by budding and grafting, by cuttings, and by nematodes of the genus *Xiphinema*.

Raspberry Ring Spot

Raspberry ring spot occurs primarily in northern Europe and causes major losses in yield and plant stands. It is caused by the nepovirus *raspberry ring spot virus* (RRSV) and is transmitted by nematodes of the genus *Longidorus*.

Selected References

"C. M. I./A. A. B. Descriptions of Plant Viruses." Nepoviruses (No. 185), tomato ringspot virus (No. 290), cherry leafroll virus (No. 306), arabis mosaic virus (No. 16), raspberry ringspot virus (No. 198), tomato black ring virus (No. 38). Kew, Surrey, England.

Converse, R. H., ed. (1987). "Virus Diseases of Small Fruits." USDA Agric. Handbook No. 631, Washington, DC.

Ellis, M. A., *et al.*, eds. (1987). "Compendium of Raspberry and Blackberry Diseases and Insects." APS Press, St. Paul, MN.

Frazier, N. W., ed. (1987). "Virus Diseases of Small Fruits and Grapevines." Univ. of California, Div. Agric. Sci., Berkeley.

Martelli, G. P. (1978). Nematode-borne viruses of grapevine, their epidemiology and control. *Nematol. Mediterr.* **6**, 1–27.

Rosenberger, D. A., Cummins, J. N., and Gonsalves, D. (1989). Evidence that tomato ringspot virus causes apple union necrosis and decline: Symptom development in inoculated apple trees. *Plant Dis.* 73, 262–265.

Diseases Caused by Bromoviridae

The family *Bromoviridae* contains five genera of viruses: *Bromovirus*, *Cucumovirus*, *Ilarvirus*, *Alfamovirus*, and *Oleavirus*. Virus particles of the first three genera are isometric, 26 to 35 nanometers in diameter. Two genomic RNAs (RNA1 and RNA2) are each contained in separate particles. A third RNA (RNA3) and a subgenomic one (RNA4) are contained together in a third particle. Alfamoviruses, Oleavirus, (and sometimes Ilarviruses) have four particles each. They are 18 nanometers in diameter but are mostly bacilliform, ranging in length from 30 to 57 nanometers. Three of the four particle sizes contain single copies of one of the RNAs (RNA1, RNA2, or RNA3), whereas the fourth contains two copies of RNA4.

Several members of the *Bromoviridae* are important pathogens of agronomic and horticultural crops. Many of them are distributed worldwide. Most cucumoviruses have a narrow host range within legumes and solanaceous plants, but *cucumber mosaic virus* has a very wide host range. Ilarviruses infect a wide range of mostly woody hosts. Bromoviruses infect gramineous and legume plants, and alfamoviruses infect mainly legumes.

Of the *Bromoviridae*, cucumoviruses and alfamoviruses are transmitted by many different aphids in the nonpersistent manner. Some bromoviruses have been reported to be transmitted by beetles. Ilarviruses have no vector, but some are seed transmitted and also pollen transmitted in some host species. Some cucumoviruses and alfamoviruses are also transmitted in the seed of some of their hosts.

Diseases Caused by Cucumoviruses

Cucumoviruses, named after *cucumber mosaic virus* (CMV), are a small group of viruses that include *tomato aspermy virus* (TAV) and *peanut stunt virus* (PSV). *Cucumber mosaic virus* occurs worldwide, infects more different kinds of plants than any other virus, and causes mosaics, stunting of plants, and leaf and fruit malformations (Figs. 14-50A–14-50E). *Tomato aspermy virus* affects chrysanthemum more often than tomato and is present primarily in countries where chrysanthemums are grown. TAV-infected tomato plants are stunted and bushy; fruits are small and distorted and have few seeds. *Peanut stunt virus* occurs sporadically in isolated plantings in most countries where peanuts are grown. It also affects beans, white clover, and other host plants. PSV-infected plants are severely dwarfed and produce fewer and smaller seeds that germinate poorly and produce seedlings of low vigor.

```
          RNA1                                 RNA2                            RNA3              RNA4
O-----------------------------------*  O----------------------------------*  O---------------------------*  O------------*
     ===================                    =============                    ===  =====                ==
             111k                                   97k                      30k  24.5k               24.5k
```

Cucumovirus particles are isometric, about 29 nanometers in diameter (Fig. 14-50F). The genome consists of three single-stranded RNAs of 3.4, 3.1, and 2.2 kilobases, respectively, each existing in a separate but identical particle. A fourth RNA of 1.0 kilobases, which codes for the coat protein of the virus, is generated from the smallest of the three RNAs and coexists with it in the particle. All virus particles consist of 180 identical protein subunits that have a molecular weight of about 24.5K. Many isolates of cucumoviruses also contain small single-stranded satellite RNAs of about 350 nucleotides. Some of the satellite RNAs increase and others reduce the severity of the symptoms caused by the virus. Each viral RNA codes for one protein. The two longest RNAs code for two proteins of 111K and 97K, respectively, involved in RNA replication. The shortest RNA codes for a 30K protein involved in the cell-to-cell movement of the virus. The fourth subgenomic RNA, which codes for the 24.5K coat protein, is generated from the replicative form of RNA3. The coat protein not only forms the shell of the virus, but it also determines the transmissibility of the virus from plant to plant by its aphid vectors.

Cucumoviruses overseason in perennial cultivated and wild hosts. From these, several aphid species, specific for each virus, transmit the viruses to annual and other perennial crops. All cucumoviruses are transmitted by aphids in the nonpersistent manner. Cucumoviruses are also transmitted, in at least some of their hosts, by a varying but small percentage of seeds produced on a few virus-infected plants. Cucumoviruses are easily transmitted mechanically, and some of them, e.g., *cucumber mosaic virus*, can be transmitted to a small extent by handling of the plants in the greenhouse or field. The virus infects and multiplies in phloem and parenchyma cells. Virus particles may appear scattered in the cytoplasm of infected cells or in crystalline aggregates in the cytoplasm, the vacuoles, and, possibly, the nucleus, and they may be aligned in multiple files in the cytoplasm or in single file passing through plasmodesmata.

The control of cucumoviruses is difficult. It depends primarily on breeding and use of resistant varieties, use of virus-free seed and transplants, removal of wild hosts that may carry the virus, and sprays and cultural practices that help reduce the virus inoculum or reduce or delay the aphid vectors that come and move around in the field. In the past 10 to 15 years, several alternative approaches for the control of cucumoviruses have been studied. These include use of cross protection with mild strains, transformation of plants with the coat protein gene of the virus, and use of certain of the mild satellite RNAs that, either in an inoculum applied to field-grown plants or as a transgene, can and do reduce the severity of the disease caused by the virus and the severe satellite RNAs.

Cucumber Mosaic

Cucumber mosaic is worldwide in distribution. The virus causing cucumber mosaic has, perhaps, a wider range of hosts and attacks a greater variety of vegetables, ornamentals, weeds, and other plants than any other virus. Among the most important vegetables affected by cucumber mosaic are cucumbers, gladioli, melons, squash, peppers, spinach, tomatoes, celery, beets, beans, bananas, and crucifers (Fig. 14-50).

Cucumber mosaic affects plants by causing mottling or discoloration and distortion of leaves, flowers, and fruits. Infected plants may be reduced greatly in size or they may be killed. Crop yields are reduced in quantity and are often lower in quality. Plants are seriously affected in the field as well as in the greenhouse. In some localities, one-third to one-half of the plants may be destroyed by the disease, and susceptible crops, such as summer squash, may have to be replaced by other crops.

Young seedlings are seldom attacked in the field during the first few weeks. Most general field infections occur when the plants are about six weeks old and growing vigorously. Four or five days after inoculation, the young developing leaves become mottled, distorted, and wrinkled and their edges begin to curl downward (Fig. 14-50). All subsequent growth is reduced drastically, and the plants appear dwarfed as a result of stem internodes and petioles being shorter and leaves developing to only half their normal size. Such plants produce few runners and also few flowers and fruits. Instead, they have a bunched or bushy appearance, with the leaves forming a rosette-like clump near the ground. The older leaves of infected plants develop at first chlorotic and then necrotic areas along the margins, which later spread over the entire leaf. The killed leaves hang down on the petiole or fall off, leaving part or most of the older vine bare.

Fruit produced on the plant after infection shows pale green or white areas intermingled with dark green, raised areas; the latter often form rough, wart-like projections and cause distortion of the fruit. Cucumbers

FIGURE 14-50 (A) *Cucumber mosaic virus* (CMV) symptoms of mosaic, rings, and line patterns on individual pepper leaves, (B) necrosis of pepper leaves in the field, (C) stunting of young pepper plants, (D) mottle and mosaic on squash leaf, and (E) shoestring malformation of tomato leaf. (F) Purified preparation of CMV.

produced by plants in the later stages of the disease are somewhat misshapen but have smooth gray-white color with some irregular green areas and are often called white pickles. Cucumbers infected with cucumber mosaic often have a bitter taste and on pickling become soft and soggy.

The *cucumber mosaic virus* exists in numerous strains that differ somewhat in their hosts, in the symptoms they produce, in the ways they are transmitted, and in other properties and characteristics.

The *cucumber mosaic virus* overwinters in many perennial weeds, flowers, and crop plants. Perennial weeds harbor the virus in their roots during the winter and carry it to their top growth in the spring, from which aphids transmit it to susceptible crop plants. Once a few plants have become infected with CMV, insect vectors and humans during their cultivating and handling of the plants spread the virus to many more healthy plants. Entire fields of cucurbits sometimes begin to turn yellow with mosaic immediately after the first pick has been made, indicating the ease and efficiency of transmission of CMV mechanically through sap carried on the hands and clothes of workers.

Whether the virus is transmitted by insects or through sap, it produces a systemic infection of most host plants. Older tissues and organs developed before infection are not, as a rule, affected by the virus, but young active cells and tissues developing after infection may be affected with varying severity. The virus concentration in CMV-infected plants continues to increase for several days after inoculation and then decreases until it levels off or until the plant dies.

Selected References

Agrios, G. N., Walker, M. E., and Ferro, D. N. (1985). Effect of cucumber mosaic virus inoculation at successive weekly intervals on growth and yield of pepper (*Capsicum annuum*) plants. *Plant Dis.* **69**, 52–55.

"C. M. I./A. A. B. Descriptions of Plant Viruses." Cucumber mosaic virus (No. 213), peanut stunt virus (No. 92). Kew, Surrey, England.

Crescenzy, A., *et al.* (1993). Cucumber mosaic cucumovirus populations in Italy under natural epidemic conditions and after a satellite-mediated protection test. *Plant Dis.* **77**, 28–33.

Francki, R. I. B., ed. (1985). "The Plant Viruses," Vol. 1. Plenum, New York.

Jordan, C., *et al.* (1992). Epidemic of cucumber mosaic virus plus satellite RNA in tomatoes in eastern Spain. *Plant Dis.* **76**, 363–366.

Kao, Cheng, C., and Sivakumaran, K. (2000). Brome mosaic virus, good for an RNA virologist's basic needs. *Mol. Plant Pathol.* **1**, 91–97.

Palukaitis, P., *et al.* (1992). Cucumber mosaic virus. *Adv. Virus Res.* **41**, 281–348.

Roossinck, M. J. (2001). *Cucumber mosaic virus*, a model for RNA virus evolution. *Mol. Plant Pathol.* **2**, 59–63.

Sayama, H., *et al.* (1993). Field testing of a satellite-containing attenuated strain of cucumber mosaic virus for tomato protection in Japan. *Phytopathology* **83**, 405–410.

Tolin, S. A. (1984). Peanut stunt. *In* "Compendium of Peanut Diseases" (D. M. Porter, D. H. Smith, and Rodriguez-Kabana, eds.), pp. 46–48. APS Press, St. Paul, MN.

Xue, B., *et al.* (1994). Development of transgenic tomato expressing a high level of resistance to cucumber mosaic virus strains of subgroups I and II. *Plant Dis.* **78**, 1038–1041.

Diseases Caused by Ilarviruses

Ilarviruses derive their name from their description as "isometric labile ring spot viruses," although their particles are not truly isometric and many cause symptoms other than ring spots. The type ilarvirus is *tobacco streak virus*, but more than 16 ilarviruses known have been found primarily in woody plants such as pome fruits (*apple mosaic virus*), roses (*rose mosaic virus*), stone fruits (*prunus necrotic ring spot virus, prune dwarf virus*), and citrus trees (*citrus leaf rugose virus, citrus variegation virus*), in forest trees such as elm; in shrubs such as black raspberry, lilac, and hops; and in asparagus. Ilarviruses probably occur wherever their hosts are grown, having been distributed with infected nursery stock, budwood, or seed. They cause symptoms mostly on the foliage and blossoms in the form of line patterns, ring spots, and mosaics, which are sometimes accompanied by leaf malformation and distortions. Many ilarviruses cause severe "shock" symptoms on the spring growth of their hosts, parts of which (leaves, blossoms, young twigs) may be killed by the viruses. Necrotic areas (cankers) may sometimes develop on twigs and branches. Leaves and shoots produced later may show mild or no symptoms. Trees affected with some ilarviruses may show symptoms for only one or a few years, with the virus becoming latent (symptomless) in subsequent years.

Ilarvirus particles of even the same virus are of somewhat varying shapes, with some of them being isometric or spherical with diameters ranging from 20 to 32 nanometers and some being oblong; in some viruses, some particles are isometric whereas others are bacilliform of various lengths up to 75 nanometers. Usually, the particles of ilarviruses can be separated by centrifugation into three or four classes, each of them containing one single-stranded RNA. All particles are composed of one type of coat protein subunit, the molecular weight of which is 24 to 30 K, depending on the virus. The coat protein is coded for by the smallest of the four RNAs, but this RNA is repeated in and is generated from the RNA3 negative strand by the virus replicase. All four RNAs must be present for infection to take place; however, infection does take place if the fourth smallest RNA is replaced with coat protein. The two largest RNAs code for one protein each (120 and 100 K), both of which are RNA polymerases involved in RNA replication. The third RNA codes for a 34 K protein that facilitates cell-to-cell transport of the virus. RNA4 codes for the 24 to 30K coat protein of each ilarvirus.

Ilarviruses perpetuate themselves in their perennial woody hosts and, for most of them, in a portion of the seeds produced on infected hosts. Ilarviruses have no known vectors. In addition to their transmission through vegetative propagation and by a portion of the seeds, many ilarviruses are also transmitted in the field to both the progeny (seed) and the parent plant through pollen.

Because most ilarviruses are very labile, and therefore difficult to isolate and characterize, the identity of many of them and their relationships with one another have not yet been established definitively.

Prunus Necrotic Ring Spot

Prunus necrotic ring spot occurs in all temperate regions. The disease affects most stone fruits, including sour cherry, cherry, almond, peach, apricot, and plum, their wild and flowering counterparts, and also some ornamental species such as roses.

Necrotic ring spot is the most widespread virus disease of stone fruit trees. In fruit-producing areas, almost all orchard trees in production are infected. Losses vary with the *Prunus* species or variety affected and with the time from inoculation with the virus. Bud take is also lower in combinations in which the bud or the rootstock carry the virus than when both are virus free or virus infected. The growth of virus-infected trees may be reduced by 10 to 30% or more, whereas the yield of virus-infected trees may be 20 to 60% lower than that of healthy trees. Affected trees are also susceptible to winter injury.

Infected trees or individual branches are slow to leaf out in the spring, and their leaves are small and have light green spots and dark rings 1 to 5 millimeters in diameter. Later, affected areas may become necrotic, fall out, and give a "shredded leaf" or "tatter leaf" effect (Figs. 14-51A and 14-51B). Such shock or acute symptoms are usually limited to the first leaves that unfold. Leaves formed later generally do not show marked symptoms. Affected trees, however, usually have fewer leaves and therefore have a thin appearance.

Blossoms of affected trees often are smaller and distorted, may develop chlorotic or necrotic rings or arcs, and ordinarily do not set fruit. Occasional fruits also develop small rings similar to those on the leaves.

As a rule, trees affected severely one year show few or no symptoms in subsequent years except for the thinness of foliage. If severe symptoms are present only on a few branches the first year, other branches may show striking symptoms the following year. In many areas, however, trees may continue to show symptoms for 4 to 6 years or more.

Prunus necrotic ring spot virus (PNRV) can be transmitted by budding and grafting and mechanically by rubbing sap from virus-infected tree leaves or petals onto leaves of cucumber and several other herbaceous plants. PNRV is transmitted through 5 to 70% of the seed and through pollen to seeds and to pollinated plants. No other vector of PNRV is known.

The virus overwinters in infected stone fruit trees, from which it spreads to healthy trees in the spring primarily through infected pollen. PNRV spreads slowly in orchards less than four years old but can spread rapidly in older orchards, probably because older trees have more bloom and therefore are much more subject to infection through pollen than young ones. PNRV can spread over a distance of at least 800 meters, but most infections occur within 15 meters of a known infected tree. Symptoms on trees infected by virus-

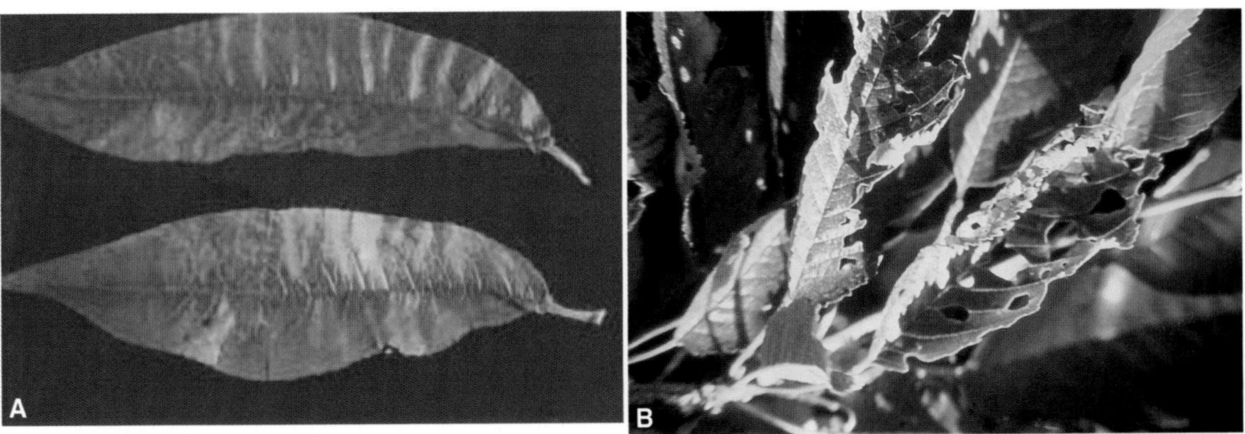

FIGURE 14-51 *Prunus necrotic ring spot virus* symptoms consisting of faint early chlorotic rings on peach leaves (A) and of advanced necrotic and fallen out rings giving a shot-hole, tattered effect on cherry leaves (B).

infected pollen usually develop in the spring one year after inoculation.

The control of necrotic ring spot of stone fruits is based almost exclusively on starting with virus-free nursery stock and on eliminating PNRV-infected *Prunus* trees from the area where the young virus-free trees are grown. Trees are tested for infection by PNRV by indexing on susceptible indicator hosts or by serological tests, especially by ELISA.

After a new orchard has been established with virus-free trees, it is necessary to remove all wild *Prunus* trees from a radius of about 200 meters around the periphery of the orchard to avoid spread of the virus into the orchard. A new orchard should not be planted next to an older one containing infected trees, and any infected trees appearing in the new orchard should be removed immediately to prevent further spread of the virus.

Selected References

Agrios, G. N., and Buchholtz, W. F. (1967). Virus effect on union and growth of peach scions on *Prunus besseyi* and *P. tomentosa* understocks. *Iowa State J. Sci.* **41**, 385–391.

"C. M. I./A. A. B. Descriptions of Plant Viruses." Ilarviruses (No. 275), apple mosaic virus (No. 83), citrus leaf rugose virus (No. 164), prune dwarf virus (No. 19), Prunus necrotic ringspot virus (No. 5). Kew, Surrey, England.

Fridlund, P. R., ed. (1989). "Virus and Viruslike Diseases of Pome Fruits and Simulating Noninfectious Disorders." SP0003, Cooperative Extension, Washington State Univ., Pullman, WA.

Garnsey, S. M. (1975). Purification and properties of citrus-leaf-rugose virus. *Phytopathology* **65**, 50–57.

Pine, T. S., ed. (1976). "Virus Diseases and Noninfectious Disorders of Stone Fruits in North America." USDA Agric. Handbook No. 437.

Sanchez-Navarro, Aparicio, Rowhani, *et al.* (1998). Comparative analysis of ELISA, nonradioactive molecular hybridization and PCR for the detection of prunus necrotic ringspot virus in herbaceous and *Prunus* hosts. *Plant Pathol.* **47**, 780–786.

Thole, V., Garcia, M.-L., Van Rossum, C. M. A., *et al.* (2001). RNAs 1 and 2 of *alfalfa mosaic virus*, expressed in transgenic plants, start to replicate only after infection of the plants with RNA 3. *J. Gen. Virol.* **82**, 25–28.

DISEASES CAUSED BY ISOMETRIC DOUBLE-STRANDED RNA VIRUSES

Viruses with isometric double-stranded RNA (dsRNA) include two families that contain genera of viruses causing disease in plants and some viruses that infect plant pathogenic fungi. Thus, the family *Reoviridae* contains the monopartite plant-infecting virus genera *Fijivirus*, *Phytoreovirus*, and *Oryzavirus* along with several genera of viruses infecting animals and humans. The other family, *Partitiviridae*, contains four bipartite virus genera. Two of these, *Alphacryptovirus* and *Betacryptovirus*, infect plants. Two others, *Partitivirus* and *Chrysovirus*, infect fungi, including several plant pathogenic ones, e.g., *Penicillium*, *Rhizoctonia*, *Gaeumannomyces*, and *Helminthosporium*.

Diseases Caused by Reoviridae

Reoviridae is a family of viruses that includes viruses that infect humans, other vertebrate animals, insects, and plants. The name *Reovirus* derives from "respiratory enteric orphan virus" because these viruses, found in the respiratory system and digestive tract of humans and animals, had not yet been associated with any disease ("orphan").

The plant *Reoviridae* contain three genera of reoviruses: *Phytoreovirus*, *Fijivirus*, and *Oryzavirus*. All but one of the plant *Reoviridae (rice dwarf virus)* cause galls or tumors on their hosts. *Phytoreovirus* includes the following species: *wound tumor virus*, which can infect several dicotyledons systemically, induces only vein enlargement in unwounded plants, but causes tumors where the plants are wounded and where new roots form (Fig. 14-52A); *rice dwarf virus*, which causes only stunting and chlorotic flecks on the plants it infects (Fig. 14-52B,); and *rice gall dwarf virus*, which causes stunting, darker leaf color, leaf distortion, galls on leaf veins, and suppression of flowering (Figs. 14-52C and 14-52D). All fijiviruses, e.g., *rice black-streaked dwarf virus* (Fig. 14-52F), *maize rough dwarf* virus, and *oat sterile dwarf virus*, and oryzaviruses, e.g., *rice ragged stunt virus* (Figs. 14-52E and 14-52G), also cause stunting, darker leaf color, leaf distortion, galls on leaf veins, and suppression of flowering (Fig. 14-52E). Most of these viruses have a rather limited host range among grass species, and although they cause severe symptoms and losses when the plants they infect are still young, the overall losses caused are moderate and generally localized. All *Reoviridae* are transmitted from plant to plant by specific leafhoppers (Fig. 14-25B) and planthoppers in the persistent, propagative manner, but only the phytoreoviruses are transmitted to new generations through the egg. Because these viruses multiply in the insect as well as in the plant, they can be considered as insect viruses as well as plant viruses.

Reoviruses are isometric with particles measuring 65 to 70 nanometers in diameter (Figs. 14-52C, 14-52F, and 14-52G). Their genome consists of 12 double-stranded RNAs in the phytoreoviruses and 10 double-stranded RNAs in the fijiviruses and the oryzaviruses. The sizes of the RNAs range from about 800 to 2,600 base pairs and each codes for a single protein of molecular weights ranging from 19 to 155K. Each reovirus also contains a transcriptase enzyme that, after removal

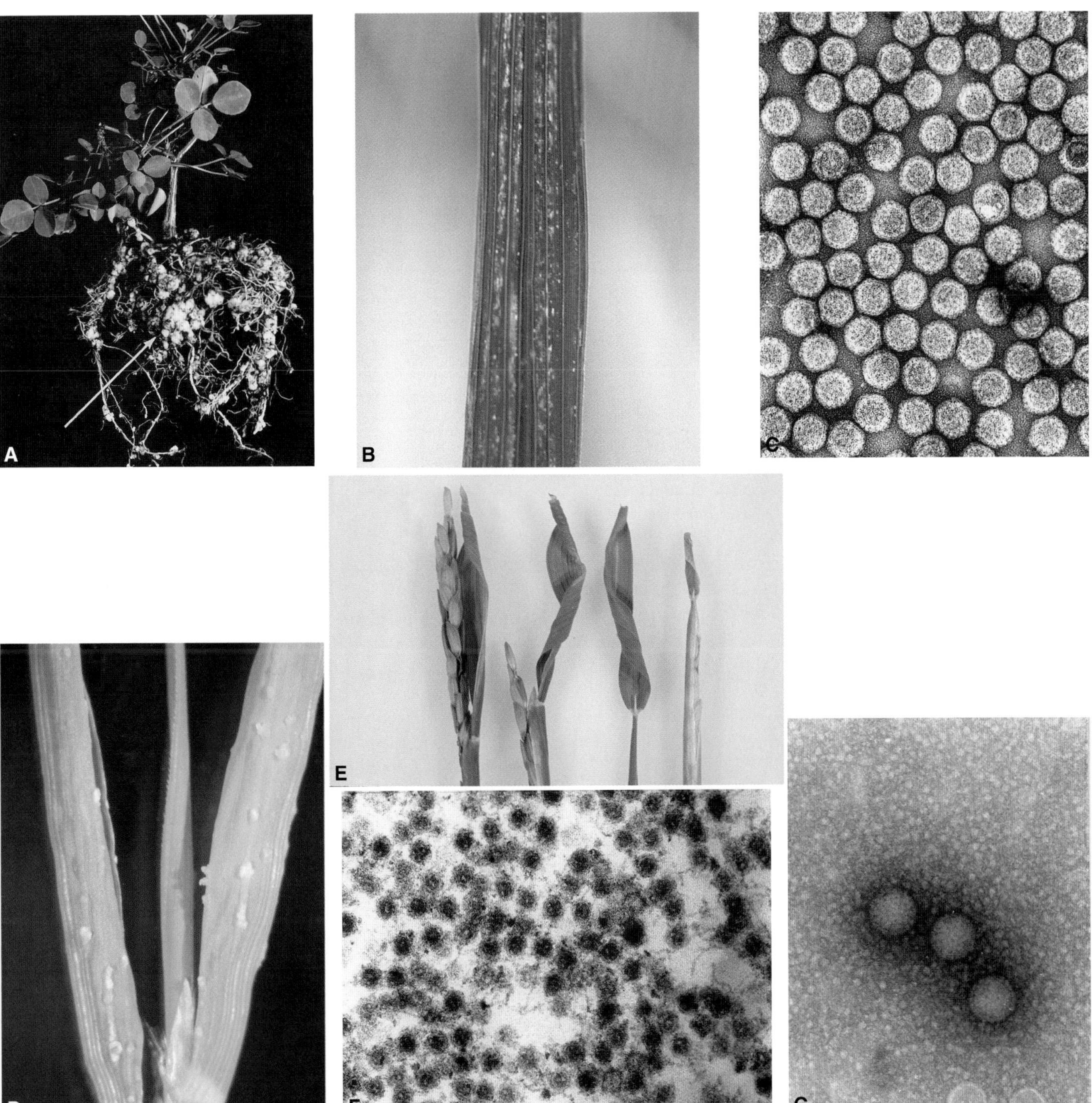

FIGURE 14-52 (A) Tumors (galls) produced on wounds of roots of sweet clover infected with *wound tumor virus*. (B) Rice leaf showing chlorotic flecks caused by rice dwarf virus. (C) *Rice gall dwarf virus* and (D) chlorotic spots, distortion, and galls on rice leaves caused by the same virus. (E) Leaf twisting and reduced flowering in rice plants infected with the *rice ragged stunt virus* (RRSV). (F) Particles of RRSV in degenerated cells of a swollen vein. (G) Three particles of the *rice black-streaked dwarf virus*. [Photographs courtesy of (A) K. Maramorosch, (B–F) H. Hibino, and (G) Y. Mikoshiba.]

of the outer protein shell of the virus on infection, uses the dsRNAs to produce the 10 or 12 equivalent single-stranded RNAs that can then be translated into proteins.

The particles of phytoreoviruses and of fijiviruses (Fig. 14-52C and 14-52G) consist of two concentric protein shells, an outer protein shell and an inner core protein shell, the latter containing double-stranded RNAs and the transcriptase enzyme. The particles of fijiviruses also have 12 spikes distributed evenly in and projecting from the surface of the virus particles. The particles of oryzaviruses have only one protein shell (the core) and also have 12 spikes, but the spikes are broader and shorter than in fijiviruses. The protein shells of phytoreoviruses contain seven different proteins, whereas those of fijiviruses contain six proteins and those of oryzaviruses contain five proteins.

In the plant, phytoreoviruses can be found in the phloem and also in other adjacent cells, including mesophyll cells. Fijiviruses and oryzaviruses are restricted to the phloem parenchyma and sieve cells. In infected cells, reoviruses occur in viroplasms, in tubular structures in the cytoplasm, or as free particles in the cytoplasm.

Reoviruses survive between crops in wild hosts, primarily grasses, in cultivated plants of overlapping croppings, in their insect vectors, and, for the phytoreoviruses only, in the eggs of their vectors. Leafhopper and plant hopper vectors acquire the virus after feeding on infected plants for a few to several hours but require an incubation period of 1 to 2 weeks before they can transmit the virus to healthy plants. The inoculation feeding period required is usually one hour to several hours. Viruliferous insects remain infective for life.

The control of reoviruses depends on: use of varieties resistant to the vector and/or to the virus; applying insecticides on seedlings before planting; cultural practices, e.g., plowing fallow paddy fields or planting vector nonhost plants, that help reduce the vector populations; planting late; and avoiding an overlap of early and late-planted rice crops.

Selected References

"C. M. I./A. A. B. Descriptions of Plant Viruses." Plant reoviruses (No. 294), maize rough dwarf virus (No. 72), oat sterile dwarf virus (No. 217), rice black-streaked dwarf virus (No. 135), rice dwarf virus (No. 102), rice gall dwarf virus (No. 296), rice ragged stunt virus (No. 248), wound tumor virus (No. 34). Kew, Surrey, England.

Francki, R. I. B., Milne, R. G., and Hatta, T. (1985). "Atlas of Plant Viruses," Vol. 1. CRC Press, Boca Raton, FL.

Nuss, D. L., and Dall, D. L. (1989). Structural and functional properties of plant reovirus genomes. *Adv. Virus Res.* 38, 249–306.

Webster, R. K., and Gunnell, P. S., eds. (1992). "Compendium of Rice Diseases." APS Press, St. Paul, MN.

DISEASES CAUSED BY NEGATIVE RNA [(−) SSRNA] VIRUSES

Negative RNA plant viruses include those belonging to the families Rhabdoviridae and Bunyaviridae, both of which also contain viruses that infect humans and animals, and to the genera *Tenuivirus* and *Ophiovirus*, which have not yet been assigned to families. All negative RNA viruses carry a transcriptase enzyme (RNA-dependent RNA polymerase) within their particle that transcribes the (−) ssRNA to (+) ssRNA so it can be translated.

Plant Diseases Caused by Rhabdoviruses

Rhabdoviruses, named after the Greek word *rhabdos*, which means a rod, are a large group of about 80 plant viruses. They have limited host ranges and have been found mostly in vegetables, weeds, and gramineous hosts. Most frequently they cause mosaics, vein clearing, yellowing, and dwarfing, and they sometimes cause malformations and necrosis of plant tissues. Some of the rhabdoviruses include *lettuce necrotic yellows virus, potato yellow dwarf virus, rice transitory yellowing virus*, and *wheat striate mosaic virus.*

Most rhabdoviruses are transmitted in the circulative, propagative manner by either leafhoppers or planthoppers or by aphids. A few rhabdoviruses have been reported to be transmitted by lace bugs or mites, but this needs to be confirmed. Acquisition feeding periods for various vectors range from one minute to several minutes. Vectors remain infective for life. Rhabdoviruses are also passed through the egg to a small percentage of the vector progeny.

The particles of rhabdoviruses are bacilliform and are the largest among plant viruses. They range in size from 50 to 95 nanometers in diameter to 200 to 500 nanometers long (Fig. 14-53).

(-) ssRNA
70k
241k G =70k M1 37k M2 54k Proteins

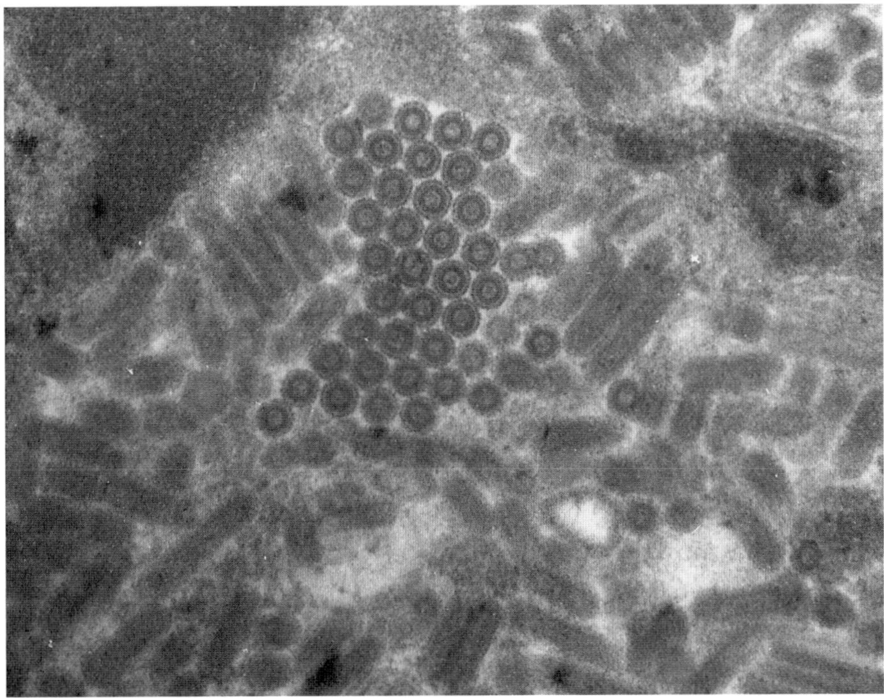

FIGURE 14-53 Particles of a rhabdovirus as seen in longitudinal and cross section within a plant cell (72,000×). (Photograph courtesy of R. G. Christie.)

Each particle is enveloped in a membrane composed of two proteins, M1 and M2 (of 19 to 45K), carrying numerous regularly-arranged projections made of a glycoprotein of 71 to 92K molecular weight. Inside the membranous envelope, rhabdoviruses have the nucleoprotein, made up of the coat protein (54 to 64K) and the (−) ssRNA containing 11 to 13 kilobases of nucleotides. The nucleoprotein is a negative single-stranded continuous molecule arranged in a helical fashion and appears as cross striations in high magnifications of virus particles. Inside or attached to the nucleoprotein are two other proteins, a large one of 241K and a nonstructural 37K protein, either or both of which may play a role as the virus transcriptase that transcribes the viral negative ssRNA into translatable positive ssRNA. A discrete mRNA for each of the six proteins of rhabdoviruses is transcribed from the one continuous (−) ssRNA of the virus. The transcriptase is also presumed to transcribe complete lengths of viral (+) ssRNA and, from these, (−) ssRNA, which is incorporated in the nucleoprotein of the virus.

In the plant, rhabdoviruses infect phloem and parenchyma cells. Particles of most rhabdoviruses are assembled in the perinuclear space of cells. Two genera of plant-infecting rhabdoviruses have been recognized so far: (1) *Cytorhabdovirus*, such as *lettuce necrotic yellows virus*, the particles of which acquire their envelope from the outer nuclear membrane (which is continuous with the endoplasmic reticulum) and accumulate in vesicles in the cytoplasm, and (2) *Nucleorhabdovirus*, such as *potato yellow dwarf virus*, the particles of which acquire their envelope from the inner nuclear membrane and accumulate in the perinuclear space. There is also a third group of rhabdoviruses in which virus particles are assembled in large viroplasms in the cytoplasm and accumulate in membrane-bound vesicles at the periphery of the viroplasm.

Selected References

Bishop, D. H. L., ed. (1980). "Rhabdoviruses," Vols. 1–3. CRC Press, Boca Raton, FL.

"C.M.I./A.A.B. Descriptions of Plant Viruses." Plant rhabdoviruses (No. 244), lettuce necrotic yellows virus (No. 26), wheat striate mosaic virus (No. 99), potato yellow dwarf virus (No. 35), rice transitory yellowing virus (No. 100). Kew, Surrey, England.

Francki, R. I. B., Milne, R. G., and Hatta, T. (1985). "Atlas of Plant Viruses." CRC Press, Boca Raton, FL.

Plant Diseases Caused by Tospoviruses

Tospoviruses, named after *tomato spotted wilt virus*, make up one of the virus genera within the family Banyaviridae, which includes several genera of viruses infecting humans and animals. Tospoviruses occur and cause plant disease epidemics primarily in tropical and subtropical regions of the world but also in many temperate regions. Their widespread distribution seems to have come about by the international movement of infected ornamentals, whereas their local abundance and severity depend on the populations of their thrips vectors. Tospoviruses have an extremely wide host range, infecting more than 500 species of ornamental, vegetable, fruit, and other annual and perennial plants in more than 50 families. Solanaceous, composite, and leguminous plants are particularly susceptible. At present, eight tospoviruses are recognized and at least five more are thought to belong to this genus. *Tomato spotted wilt virus* infects many dicots and monocots. Some of the hosts severely affected by tospoviruses are tomato, tobacco, peanut, pineapple, papaya, lettuce, dahlia, gloxinia, and impatiens (Figs. 14-54 and 14-56). Losses are proportional to the number of plants infected early by the virus, and infection rates reaching 50 to 90% are common.

The symptoms of tospoviruses vary greatly with the host affected, plant organ affected, and age of plant or organ at the time of infection. In general, however, tospovirus symptoms appear as chlorotic or necrotic rings, lines, or spots on leaves (Figs. 14-54A and 14-55A–14-55D), stems, and fruits (Figs. 14-54B and 14-54C); necrotic streaks on stems; bronzing, curling, and wilting of leaves; rings, necrotic spots, and malformations on fruits; stunting (Fig. 14-55D) and necrosis of parts or whole plants; and, generally, greatly reduced yields.

The tospovirus particles are spherical, about 80 to 110 nanometers in diameter, and are enveloped with a

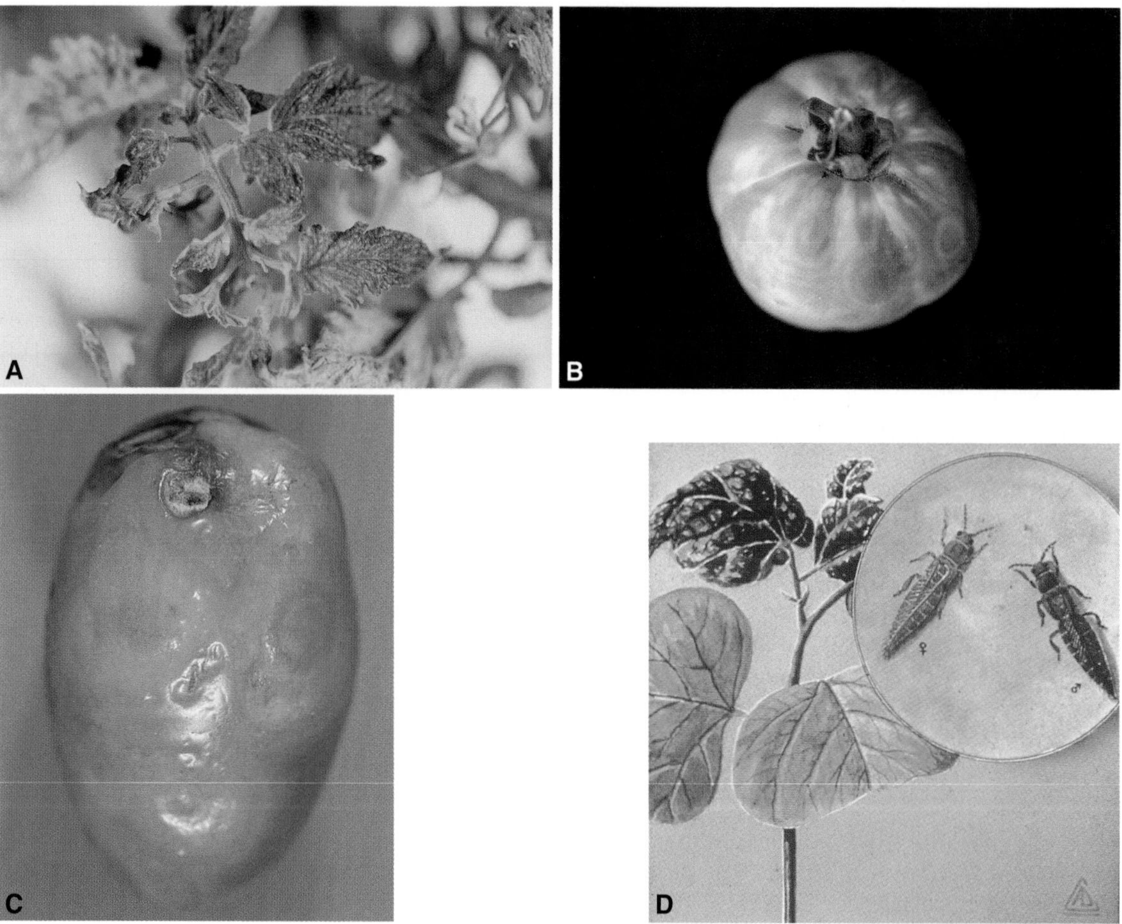

FIGURE 14-54 Tomato spotted wilt symptoms caused by the *tomato spotted wilt virus* (TSWV). (A) Tomato plant showing bronzing and necrosis. (B and C) Young and ripe tomato fruit showing rings of spotted wilt. (D) Thrips, the insect vector of TSWV. [Photographs courtesy of (A, C, and D) Plant Pathology Department, University of Florida, and (B) R. J. McGovern.]

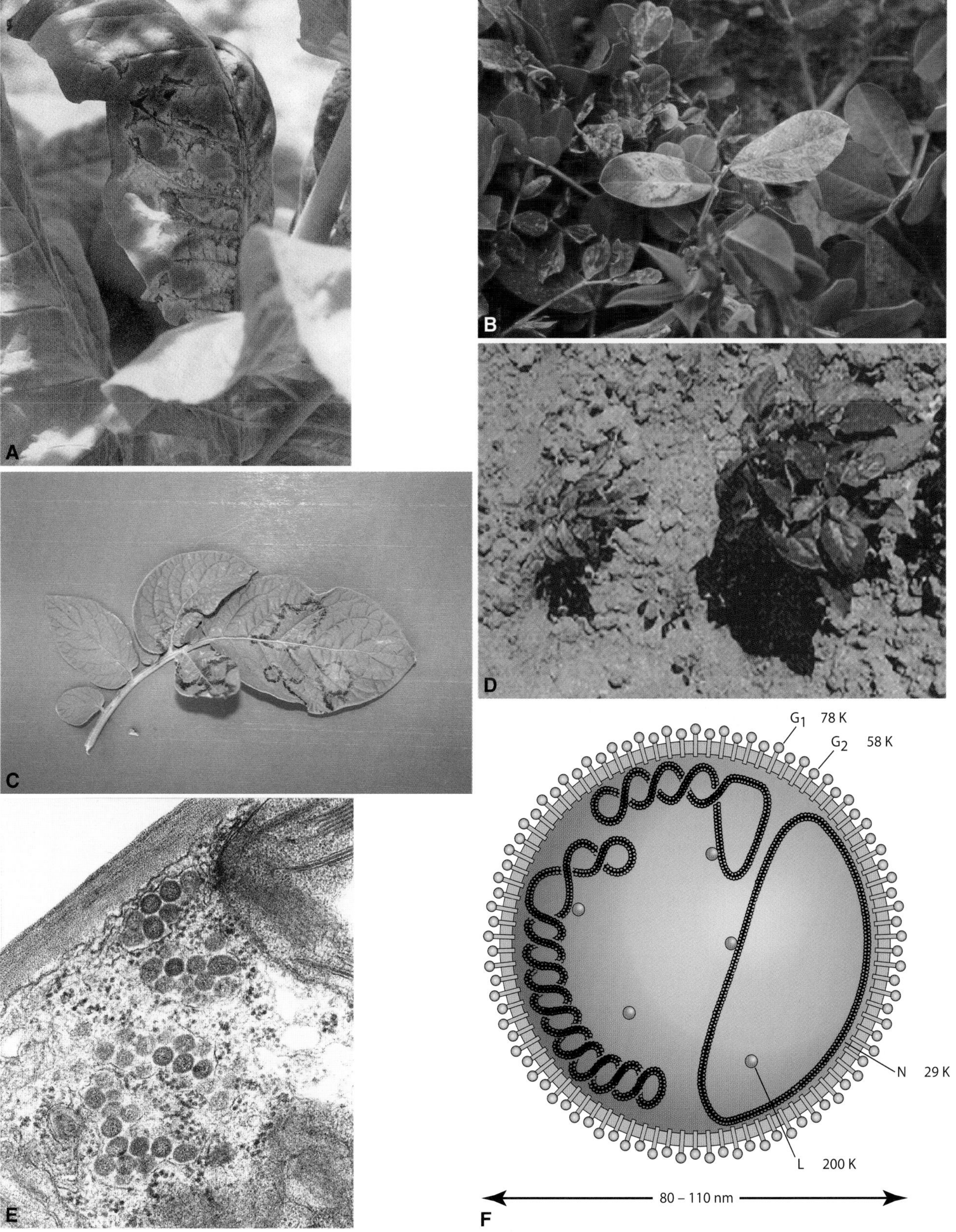

FIGURE 14-55 Tomato spotted wilt symptoms on tobacco leaf (A), peanut plant (B), potato leaves (C), and pepper plants in the field (D). (E) Membrane-bound clusters of *tomato spotted wilt virus* (TSWV) particles in a parenchyma cell (48,000×). (F) Schematic diagram of a TSWV particle. The virus genome consists of three linear (–) ssRNAs. These are tightly associated with the 29K viral coat protein subunits and form circles that may be coiled. The lipid membrane envelope contains two types of glycoproteins, G1 and G2. A few molecules of a large protein (L), possibly the viral transcriptase, are present in each viral particle. [Photographs courtesy of (A) E. Hiebert, (B and D) Plant Pathology Department, University of Florida, (C) D. P. Weingartner, (E) M. A. Petersen, and (F) R. Goldbach.]

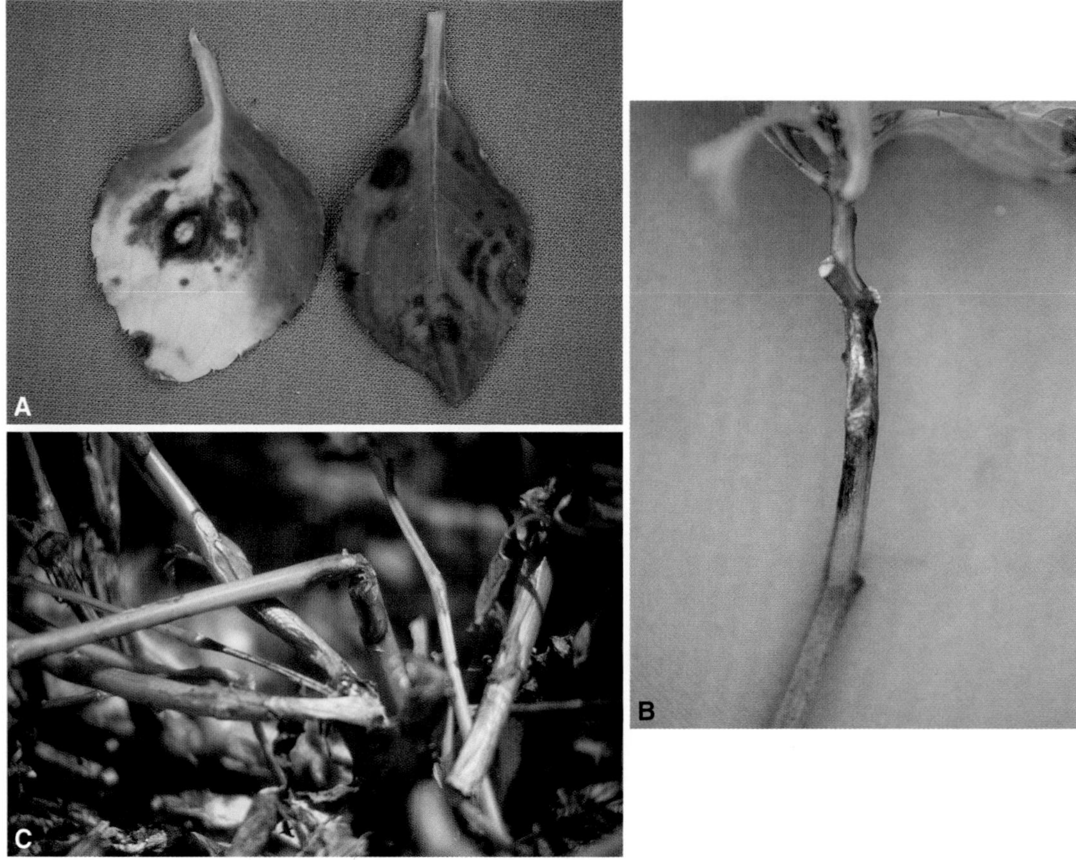

FIGURE 14-56 Symptoms caused by the other common tospovirus *impatiens necrotic spot virus* on impatiens leaves (A), stem (B), and entire plant (C). (Photographs courtesy of R. J. McGovern.)

membrane (Figs. 14-55E and 14-55F). Each virus particle contains four types of proteins: two glycoproteins (78 and 58K) partially embedded in the membrane and forming two types of external projections (Fig. 14-55F); a 29K protein associated with the three viral RNAs within the particle and forming three pseudocircular nucleoprotein structures; and a few molecules of a large (330K) protein that is thought to be the virus replicase.

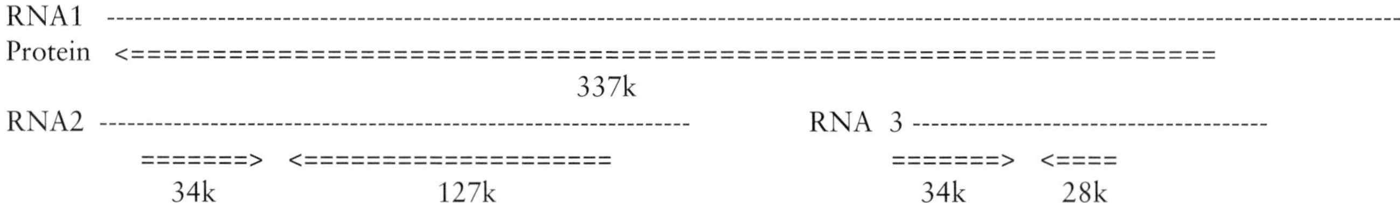

Tospoviruses have three linear ssRNAs: L is 8.7 kilobases, M is 5 kilobases, and S is 2.9 kilobases (Fig. 14-55F). The large RNA is negative sense throughout its length and, after transcription to positive sense (mRNA), codes for a 330K protein that serves as the virus replicase. The medium size RNA is ambisense (i.e., part of it is negative RNA), is transcribed as a separate segment mRNA, and codes for a large protein that is then cleaved and produces the two glycoproteins of the membrane envelope. A smaller part of the medium size RNA is already positive mRNA and on transcription and translation produces a 34K nonstructural protein of unknown function. Finally, the small RNA of *tomato spotted wilt virus* is also ambisense, part of it being positive mRNA coding for a 34K nonstructural protein of unknown function, while the remaining part is negative RNA and, on transcription to mRNA and translation, produces the 29K coat protein.

Tospoviruses are transmitted by at least seven species of thrips (Fig. 14-54D). These include the western flower thrips (*Frankliniella occidentalis*), tobacco thrips (*F. fusca*), common blossom thrips (*F. schultzei*), onion thrips (*Thrips tabaci*), and melon thrips (*T. palmi*). Tospoviruses can be acquired from infected plants by thrips larvae but not by adult thrips. Once a larva acquires the virus, however, usually after feeding on an infected leaf for 30 minutes or more, it then retains the virus through molting, pupation, and emergence so that the emerging adult thrips is viruliferous and can transmit the virus to healthy plants for the rest of its life. Inoculation feeding periods must be 30 minutes or longer. Fortunately, adult thrips, once alighting, do not move from plant to plant as much as aphids do and so transmission of tospoviruses is not as explosive as some of the aphid-borne viruses.

Tospoviruses overseason in their perennial or biennial hosts from which their thrips vectors transmit them to healthy plants. The virus spreads into phloem and parenchyma cells and multiplies in their cytoplasm. Virus particles appear to form in densely staining patches of cytoplasm (viroplasms), possibly by budding, and release from regions of two parallel membranes. Mature individual particles or groups of particles are always surrounded by irregularly shaped membranous cisternae.

The control of tospoviruses is made very difficult by the wide host range of the virus and its vectors. In a few crops, e.g., lettuce, peanut, and tomato, some resistance seems to have been identified, but much more work is needed before breeding and use of tospovirus-resistant varieties will be possible. Crops susceptible to tospoviruses should not be planted near other susceptible crops. Only tospovirus-free transplants should be planted. Roguing of infected plants may help. Plants should be monitored for thrips vectors and should be treated with insecticides to keep thrips populations to a minimum. Thrips, however, develop resistance to insecticides and move in readily from other crops so insecticidal sprays often have little or no effect on the spread of the virus. Plants with high resistance to *tomato spotted wilt virus* have been obtained by transforming them with the nucleoprotein gene of the virus itself. Such transgenic plants were resistant even when they were challenge inoculated with the virus through viruliferous thrips.

Selected References

Adkins, S. (2000). Tomato spotted wilt virus: Positive steps towards negative success. *Mol. Plant Pathol.* **1**, 151–157.

Chatzivassiliou, E. K., Peters, D., and Katis, N. I. (2002). The efficiency by which *Thrips tabaci* populations transmit *tomato spotted wilt virus* depends on their host preference and reproductive strategy. *Phytopathology* **92**, 603–609.

Cho, J. J., *et al.* (1989). A multidisciplinary approach to management of tomato spotted wilt virus in Hawaii. *Plant Dis.* **73**, 375–383.

Culbreath, A. K., Todd, J. W., and Brown, S. L. (2003). Epidemiology and management of tomato spotted wilt in peanut. *Annu. Rev. Phytopathol.* **41**, 53–76.

German, T. L., Ullman, D. E., and Mayer, J. M. (1992). Tospoviruses: Diagnosis, molecular biology, phylogeny, and vector relationships. *Annu. Rev. Phytopathol.* **30**, 315–348.

Gielen, J. J. L., *et al.* (1991). Engineered resistance to tomato spotted wilt virus, a negative strand virus. *Bio/Technology* **9**, 1363–1367.

Goldbach, R., and Peters, D. (1944). Possible causes of the emergence of tospovirus diseases. *Semin. Virol.* **5**, 113–120.

Hsu, H. T., and Lawson, R. H., eds. (1991). "Virus–Thrips–Plant Interactions of Tomato Spotted Wilt Virus," Proceedings of a USDA Workshop. USDA, ARS-87.

Ie, T. S. (1970). Tomato spotted wilt virus. *In* "C.M.I./A.A.B. Descriptions of Plant Viruses," No. 39. Kew, Surrey, England.

Jan, F.-J., Fagoaga, C., Pang, S.-Z., *et al.* (2000). A minimum length of N gene sequence in transgenic plants is required for RNA-mediated tospovirus resistance. *J. Gen. Virol.* **81**, 235–242.

Jan., F.-J., Fagoaga, C., Pang, S.-Z., *et al.* (2000). A single chimeric transgene derived from two distinct viruses confers multi-virus resistance in transgenic plants through homology-dependent gene silencing. *J. Gen. Virol.* **81**, 2103–2109.

Kato, K., Hanada, K., and Kameya-Iwaki, M. (2000). *Melon yellow spot virus*: A distinct species of the genus *Tospovirus* isolated from melon. *Phytopathology* **90**, 422–426.

Kucharek, T., *et al.* (1990). Tomato spotted wilt virus of agronomic, vegetable, and ornamental crops. Univ. of Florida, Institute of Food and Agricultural Sciences, Circular 914, Gainesville, FL.

MacKenzie, D. J., and Ellis, P. J. (1992). Resistance to tomato spotted wilt virus infection in transgenic tobacco expressing the viral nucleoprotein gene. *Mol. Plant-Microbe Interact.* **5**, 34–40.

Wilson, C. R. (2001). Resistance to infection and translocation of *tomato spotted wilt virus* in potatoes. *Plant Pathol.* **50**, 402–410.

Yeh, S. D., and Chang, T. F. (1995). Nucleotide sequence of the N-gene of watermelon silver mottle virus, a proposed new member of the genus *Tospovirus*. *Phytopathology* **85**, 58–64.

Plant Diseases Caused by Tenuiviruses

Tenuiviruses, meaning thin, filamentous viruses, cause severe diseases of gramineous hosts, especially rice and corn, in the tropic and subtropic regions. So far, four tenuiviruses have been identified: *rice hoja blanca* (white leaf) *virus, rice grassy stunt virus, rice stripe virus*, and *maize stripe virus,* and at least as many are being characterized. *Rice hoja blanca virus* (Figs. 14-57A–14-57D) occurs in the Caribbean and in Central and South America. *Rice grassy stunt virus* (Fig. 14-57E) and *rice stripe virus* (Figs. 14-57E and 14-57F) occur primarily in the rice-growing areas of south and southeast Asia, whereas *maize stripe virus* probably occurs in all tropical and subtropical maize-growing regions of the world. Each tenuivirus also infects several wild grass hosts.

Leaves of infected plants usually show chlorotic to yellowish white stripes, and young leaves may turn

FIGURE 14-57 Symptoms caused by some tenuiviruses. (A) Rice plants infected with the *rice hoja blanca virus* (RHBV) showing whitish leaves and stunting of the whole plant. (B) Sterile inflorescence of infected plant. (C) Purified preparation of RHBV particles. (D) Female (cream-colored) and male (black) planthopper vectors of RHBV. (E) Rice plants infected with the *rice grassy stunt virus* showing severe stunting, yellowing, and excessive tillering. (F) Rice plant infected with the *rice stripe virus* (RSV) showing whitish-yellow stripes on leaves, stunting, and reduced tillering. (G) Purified preparation of RSV particles (8×290–2100 nm). [Photographs courtesy of (A–D) F. Correa, (E) H. Hibino, and (F and G) K. Ishikawa.]

completely yellow or white (Figs. 14-57A, 14-57E, and 14-57G). Plants usually remain stunted, they produce few or no panicles, and flowers are absent or sterile (Fig. 14-57B). Some viruses, e.g., *rice grassy stunt virus*, induce the proliferation of tillers (Figs. 14-57A and 14-57E), whereas others, e.g., *rice stripe virus* (Fig. 14-57G), reduce tiller formation by the plants. Losses from these diseases are proportional to the number of plants becoming infected, and in some years and locations they can be very severe. Usually, however, these diseases appear sporadically, cause severe losses in localized areas, and then recede for several years before they reappear and cause severe losses again.

Tenuiviruses consist of fine, flexuous, thread-like particles. The particles are made up of one type of structural (coat) protein, which is arranged in a more or less coiled helix. Within the same helix are embedded each of the usually four (2.1, 2.4, 3.5, and 10 kb) and, in some viruses, five (1.3 kb) single-stranded ambisense RNAs of the virus. Depending on the degree of coiling (loose or supercoiled), the virus may appear to be from 3 to 12 nanometers in diameter and of varying length. Each virus, however, seems to favor a particular diameter (*rice hoja blanca virus*, 3–4 nm; *rice grassy stunt virus*, 6–8 nm; *rice stripe virus*, 8 nm; *maize stripe virus*, 12 nm). Although RNAs are linear, the virus particles usually appear circular or at least their ends seem to be coming together into a loop of some kind (Figs. 14-57C and 14-57E). The lengths of the four particles are reported to vary from 290 to 2,100 nanometers in rice stripe virus and from 950 to 1,350 nanometers in rice grassy stunt virus. In addition to the coat protein (32K), as many as three other nonstructural proteins of unknown function (19.8, 20.4, and 22.5K) have been detected in virus-infected plants.

All tenuiviruses are transmitted by one or more species of planthoppers in the circulative propagative manner. Planthopper vectors can acquire virus after feeding for a minimum of 15 minutes to 6 hours, they require a latent period of 1 to 3 weeks before they can transmit it, and they can transmit the virus for life whenever they feed on a host plant for a few minutes (5–15 min) to a few hours. All tenuiviruses except the *rice grassy stunt virus* are passed transovarially to 30 to 100% of the vector progeny. The viruses, therefore, can overseason in their insect vectors as well as in cultivated or wild hosts.

The control of tenuiviruses depends primarily on using varieties resistant to the virus or the vector. Also, planting the crop at a time when the vector population is low, avoiding overlapping of crops, and controlling the vector with insecticides help reduce disease incidence.

Selected References

Chen, C. C., *et al.* (1993). Purification, characterization, and serological analysis of maize stripe virus in Taiwan. *Plant Dis.* **77**, 367–372.

"C.M.I./A.A.B. Descriptions of Plant Viruses." Rice stripe virus (No. 269), rice hoja blanca virus (No. 299), maize stripe virus (No. 300), rice grassy stunt virus (No. 320). Kew, Surrey, England.

Falk, B. W., and Tsai, J. H. (1998). Biology and molecular biology of viruses in the genus *Tenuivirus*. *Annu. Rev. Phytopathol.* **36**, 139–163.

Ramirez, B.-C., and Haenni, A.-L. (1994). Molecular biology of tenuiviruses, a remarkable group of plant viruses. *J. Gen. Virol.* **75**, 467–475.

Sasaya, T., Ishikawa, K., and Koganezawa, H. (2001). Nucleotide sequence of the coat protein gene of *lettuce big-vein virus*. *J. Gen. Virol.* **82**, 1509–1515.

Van der Wilk, J., Dullemans, A. M., Verbeek, M., *et al.* (2002). Nucleotide sequence and genomic organization of an ophiovirus associated with lettuce big-vein disease. *J. Gen. Virol.* **83**, 2869–2877.

Webster, R. K., and Gunnell, P. S. (1992). "Compendium of Rice Diseases." APS Press, St. Paul, MN.

DISEASES CAUSED BY DOUBLE-STRANDED DNA VIRUSES

There is one family with six genera of viruses that have genomes made of double-stranded DNA. The family is called Caulimoviridae and includes all the plant viruses that replicate by reverse transcription, i.e., those that although their genome is double-stranded DNA, they produce RNA, which serves as both messenger RNA for production of proteins and as the template for reverse transcription of the viral DNA via the RNA rather than directly via the DNA. Genera of Caulimoviridae include four that are *Caulimovirus*-like, in that they have isometric viruses but differ from each other on the organization of their genome, and two genera that are badnavirus-like. The four caulimovirus-like genera are (1) *Caulimovirus*, (2) *soybean chlorotic mottle virus*-like, (3) *cassava vein mottle virus*-like, and (4) *petunia vein clearing virus*-like. The two genera that are *Badnavirus*-like in that they have bacilliform particles and also differ from each other in genome organization are (1) *Badnavirus* and (2) *rice tungro bacilliform virus*-like.

Diseases Caused by Caulimoviruses and Other Isometric Caulimoviridae

Caulimoviruses, named after *cauliflower mosaic virus*, occur collectively in many parts of the world; however, most have a limited host range and their distribution seems to also be limited to certain areas. Caulimoviruses cause mostly mottles or mosaics on certain vegetables, ornamentals, and weeds, which are accompanied by poor growth, poor quality, and reduced yields. Caulimoviruses infect plants systemically and multiply in phloem and in parenchyma cells. In the cell they are usually found in viroplasms in the cytoplasm, but sometimes also as scattered particles and as particles lined up inside plasmodesmata passing from one cell to another. Nearly 15 caulimoviruses have been reported. Some important caulimoviruses are *cauliflower mosaic virus, dahlia mosaic virus* and *carnation etched ring virus*.

Caulimovirus particles are isometric, about 50 nanometers in diameter. The protein shell consists of one type of protein molecule of 42K molecular weight. The virus genome consists of a circular double-stranded DNA (dsDNA) molecule of about eight kilobase pairs. The one circular strand (α or minus strand) of DNA, which is used for transcription, has a break of one or two nucleotides compared with the complementary β strand. The β strand has one, two, or three breaks that are not missing nucleotides but instead have a short overlap over an identical sequence at the end of the next segment (i.e., a short sequence of the β strand overlaps the end of the γ strand, and a short segment of the γ strand overlaps the end of the β strand) (Fig. 14-58).

```
              D1
DNA       -------->-I----------I—ORF1--I---ORF2-I---ORF3 I----ORF4-------I-------ORF5----------I----ORF6---------I
proteins            ====  ======== =====     ====     ========  ============   =========
                                37k     18k        15k      57k            79k                58k
```

DNA of the *cauliflower mosaic virus* (CauMV) codes for six proteins whose functions are known and two smaller ones that have no known function. The largest protein (79K) coded by gene V is the reverse transcriptase of the virus. The 58K protein, coded by gene VI, is the major protein found in CaMV viroplasms, transactivates the translation of the 35 S RNA, and seems to play a role in disease induction by the virus, symptom expression by the host, and determining the host range of the virus. The 57K protein, coded by gene IV, undergoes degradation and releases the coat protein (42K) of the virus at the site of construction of the virus shell. The 37K protein, coded by gene I, facilitates cell-to-cell movement of the virus by modifying the host cell plasmodesmata. The 18K protein, coded by gene II, seems to be responsible for aphid transmissibility of the virus and possibly for the release of virus particles by the viroplasm. Finally, the 15K protein coded by gene III binds to dsDNA and may be a structural protein carried within the virus particle.

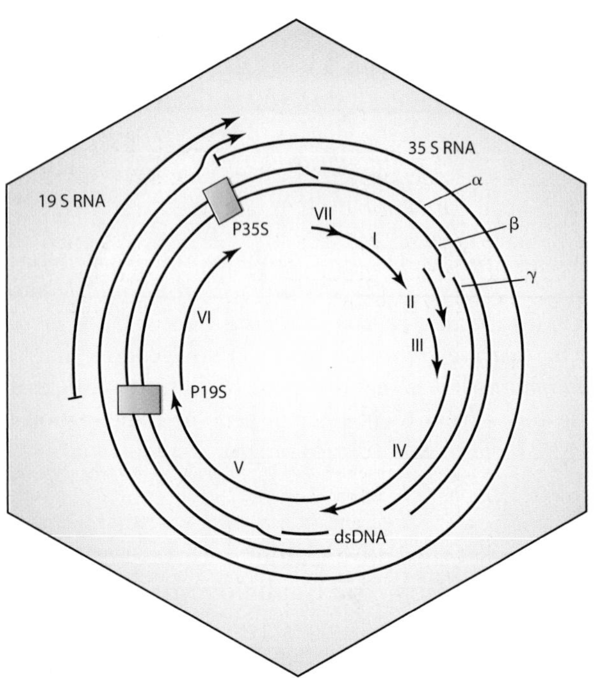

FIGURE 14-58 Shape and genomic map of *cauliflower mosaic virus*. The map shows three strands of the viral dsDNA, the 19 S and 35 S RNAs and their two promoter boxes (stippled), and several genes (open reading frames).

The replication, transcription, and translation of dsDNA viruses are quite complex. When a caulimovirus infects a plant cell, the dsDNA moves to the cell nucleus. There, the overlapping sequences are removed, the breaks are closed, and a completely circular dsDNA minichromosome forms. A host DNA-dependent RNA polymerase enzyme then transcribes two RNAs, a short one (19 S) that is subsequently translated into large amounts of viroplasm protein (58K) and a long one (35 S) that codes for and is translated sequentially into all the other proteins of the virus. It has been shown that splicing of the 35 S RNA is required to provide appropriate substrate mRNAs for protein translation and for viral infectivity. The gene for the 58K viroplasm protein is the only caulimovirus gene that is transcribed separately and early and may play a role in enhancing the translation of the other genes on the 35 S RNA.

The viral dsDNA is transcribed from the 35 S RNA by reverse transcription. The viral reverse transcriptase enzyme synthesizes a DNA minus strand (α strand) along the 35 S RNA. The enzyme stops and cleaves at stretches of purine-rich RNA, which are left at positions corresponding to nucleotides next to the β_2 and β_3

breaks of the complementary (β or plus) DNA strand. Subsequently, transcription initiated at these two primer-like RNA tracts results in the synthesis of the complementary DNA strand, thereby completing the replication of the dsDNA of the virus.

Because of its dsDNA genome, CaMV has been used as one of the best vectors of foreign DNA (genes) into plants by inserting such DNA into the genome of the virus and inoculating (transforming) plants with it. Also, the promoter for the 35 S RNA of the virus is used extensively as an effective promoter of gene replication and expression in plant transformations involving a variety of DNAs.

Caulimoviruses are transmitted in nature by many species of aphids in the nonpersistent manner. The viruses overseason in perennial hosts or in overlapping crops from which the aphids carry them into new crops. Control of caulimoviruses depends on using virus-free propagative material and on practices that help reduce or avoid high aphid populations within the crop.

Selected References

Covey, S. N., McCallum, D. G., Turner, D. S., *et al.* (2000). Pararetrovirus-crucifer interactions: Attack and defence or *modus vivendi*? *Mol. Plant Pathol.* **1**, 77–86.

"C.M.I./A.A.B. Descriptions of Plant Viruses." Caulimoviruses (No. 295), cauliflower mosaic virus (Nos. 24, 243), dahlia mosaic virus (No. 51), carnation etched ring virus (No. 182). Kew, Surrey, England.

Francki, R. I. B., Milne, R. G., and Hatta, T. (1985). "Atlas of Plant Viruses," Vol. 1. CRC Press, Boca Raton, FL.

Hull, R., Covey, S. N., and Maule, A. J. (1987). Structure and replication of caulimovirus genomes. *J. Cell Sci. Suppl.* **7**, 213–229.

Karsies, A., Merkle, T., Szurek, B., *et al.* (2002). Regulated nuclear targeting of cauliflower mosaic virus. *J. Gen. Virol.* **83**, 1783–1790.

Kiss-Laszlo, Z., Blanc, S., and Hohn, T. (1995). Splicing of cauliflower mosaic virus 35 S RNA is essential for viral infectivity. *EMBO J.* **14**, 3552–3562.

Palacios, I., Drucker, M., Blanc, S., *et al.* (2002). *Cauliflower mosaic virus* is preferentially acquired from the phloem by its aphid vectors. *J. Gen. Virol.* **83**, 3163–3171.

Pfeiffer, P., and Mesnard, J. M. (1995). The interplay of host and virus genes in the specificity and pathogenicity of cauliflower mosaic virus. *In* "Pathogenesis and Host Specificity in Plant Diseases" (R. P. Singh, U. Singh, and K. Kohmoto, eds.), Vol. 3, pp. 269–288. Elsevier, Tarrytown, NY.

Diseases Caused by Badnaviruses

Badnaviruses were given this name because they are bacilliform DNA viruses. Their particle lacks a membrane envelope (Fig. 14-59E). They differ from rhabdoviruses in that the genome of the latter is negative ssRNA and their virus particle is enveloped by a membrane. Badnaviruses are also smaller, measuring approximately 30 by 100 to 300 nanometers in size. So far, at least 12 badnaviruses have been studied in detail and have been shown to contain dsDNA (Fig. 14-59F). At least as many more nonenveloped bacilliform viruses have been reported from various host plants, but the type of their nucleic acid has not yet been determined. Badnaviruses cause diseases of varying severity in several economically important hosts, such as rice tungro (Fig. 14-59A), banana streak (Fig. 14-59B), and cacao swollen shoot (Figs. 14-59C and 14-59D). Badnaviruses have also been found in sugarcane, in other tropical crops such as taro and yucca, and in ornamentals such as canna, kalanchoe, and schefflera.

Rice tungro bacilliform virus (RTBV) (Fig. 14-44C) can infect rice plants alone, but in nature, where it is transmitted by leafhoppers, it is transmitted only in the presence of *rice tungro spherical virus*. The two viruses coexist in the vectors and in the plant and together cause one of the most destructive diseases of rice in south and southeast Asia. Both viruses are transmitted in the semipersistent manner.

Banana streak virus (BSV) occurs in many banana-growing regions and can cause significant losses. BSV DNA has been shown to integrate into the banana genome, and infections may arise from activation of such integrated BSV sequences. The virus measures 30 by 130 to 150 nanometers (Figs. 14-59E and 14-59F) and it is transmitted by mealybugs. BSV is closely related or is identical to the sugarcane bacilliform virus.

Cacao swollen shoot virus occurs in west Africa and in Ceylon. It affects cacao and cola and causes severe losses. Infected plants develop swellings in stems (Fig. 14-59C) and tap roots, necrosis of side roots, and chlorosis of leaves (Fig. 14-59D) and pods and they produce small, rounded pods that contain fewer, smaller beans. Trees decline and die or they may linger on. *Cacao swollen shoot virus* is a bacilliform virus, 142 by 27 nanometers, but has not yet been assigned to a definite taxonomic group. It is transmitted by several species of mealybugs in the semipersistent manner.

Selected References

Brunt, A. A., and Kenten, R. H. (1971). Viruses infecting cacao. *Rev. Plant Pathol.* **50**, 591–602.

"C.M.I./A.A.B. Descriptions of Plant Viruses." Cacao swollen shoot virus (No. 10). Kew, Surrey, England.

Dahal, G., Hughes, J. d'A., and Thottappilly, G. (1998). Effect of temperature on symptom expression and reliability of banana streak badnavirus detection in naturally infected plantain and banana (*Musa* spp.). *Plant Dis.* **82**, 16–21.

Delanoy, M., Salmon, M., and Kummert, J. (2002) Development of real-time PCR for the rapid detection of episomal *banana streak virus* (BSV). *Plant Dis.* **87**, 33–38.

Geering, A. D. W., Olszewski, N. E., Dahal, G., *et al.* (2001). Analysis of the distribution and structure of integrated *banana streak virus* DNA in a range of *Musa* cultivars. *Mol. Plant Pathol.* **2**, 207–213.

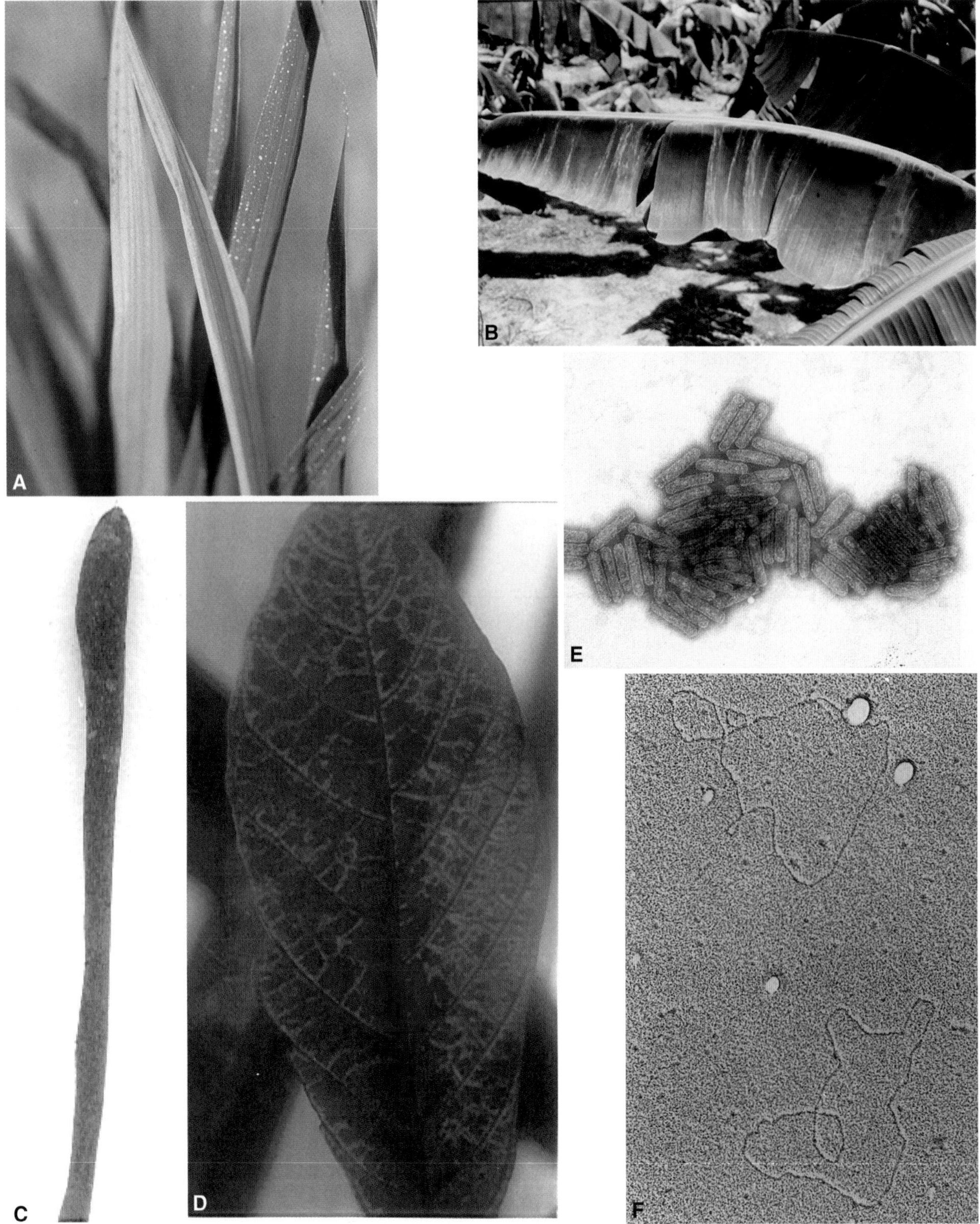

FIGURE 14-59 Symptoms and particles of some badnaviruses. (A) Rice tungro disease caused by synergism of the *rice tungro bacilliform virus* and the *rice tungro spherical virus*. (B) Banana streak disease caused by the *banana streak virus*. (C) Young stem of a cacao plant showing swelling at the tip caused by the *cacao swollen shoot virus* (CSSV). (D) Vein clearing mosaic symptoms on a cacao leaf caused by CSSV. (E) Purified particles of a badnavirus and (F) the double-stranded DNA genome of two badnavirus particles. [Photographs courtesy of (A) H. Hibino, (B) J. Hughes, IITA, (C and D) L. L. A. Ollennu, and (E and F) B. E. Lockhart.]

Geering, W., McMichael, I. A., Ditzgen, R. G., *et al.* (2000). Genetic diversity among *banana streak virus* isolates from Australia. *Phytopathology* **90**, 921–927.

Hibino, H. (1992). Tungro and waika. *In* "Compendium of Rice Diseases" (R. K. Webster and P. S. Gunnell, eds.), pp. 42–43. APS Press, St. Paul, MN.

Lockhart, B. E. L. (1990). Evidence for a double-stranded circular DNA genome in a second group of plant viruses. *Phytopathology* **80**, 127–131.

Lockhart, B. E. L. (1992). Banana streak. *In* "Compendium of Tropical Fruit Diseases" (R. C. Ploetz *et al.*, eds.), pp. 19–20. APS Press, St. Paul, MN.

DISEASES CAUSED BY SINGLE-STRANDED DNA VIRUSES

Viruses with single-stranded DNA that infect plants belong primarily to the family *Geminiviridae* and, a few of them, to the family *Circoviridae*. Viruses in both families have circular ssDNA in isometric particles, but while Circoviridae have single virions, Geminiviridae have virions that consist of two incomplete paired or twin (geminate) particles. The family Geminiviridae includes many (about 120 so far) viruses that cause numerous devastating diseases, especially in the tropics and subtropics. Circoviridae includes a few viruses of minor economic importance.

Plant Diseases Caused by Geminiviridae

All geminiviridae consist of geminate (paired or twin) particles, each appearing as the result of partial fusion together of two isometric particles (Fig. 14-60). The protein coat of geminiviridae consists of one type of protein molecule of about 28K molecular weight. The Geminiviridae family consists of four genera: (1) *Curtovirus*, named after the type species *curly top of sugar beets* (Fig. 14-60B), the genome of which consists of a single, circular ssDNA of about 2.6 to 2.8 kilobases; these viruses are transmitted by leafhoppers (Figs. 14-60C and 14-60E) in the circulative nonpropagative manner. (2) *Mastrevirus*, named after the type species *maize streak virus* (Fig. 14-60D), have genomes composed of a single component of ssDNA. With the exception of *bean* and *tobacco yellow dwarf viruses*, these viruses infect monocotyledonous plants (gramineae) on which they cause severe losses. Mastreviruses are also transmitted by leafhoppers in the circulative nonpropagative manner. (3) *Begomovirus*, named after *bean golden mosaic virus*, includes viruses transmitted by whiteflies (Figs. 14-63A and 14-63B). The genome of most of them consists of two circular ssDNAs (DNA A or 1 and DNA B or 2) of about equal size (2.4–2.8 kb each). Begomoviruses infect only dicotyledonous plants (Figs. 14-63C–14-63F). They include many geminiviruses that cause devastating losses in many crops, particularly in the tropical and subtropical regions where high populations of *Bemisia* whiteflies are common. Some of the most important begomoviruses are *bean golden mosaic virus* (Figs. 14-63C and 14-63D), *tomato mottle virus* (Fig. 14-63E), *tomato yellow leaf curl virus* (Fig. 14-63F) and *African cassava mosaic virus* (Fig. 14-65F), *squash leaf curl virus, tobacco leaf curl virus*, and *tomato golden mosaic virus*. (4) *Topocuvirus*, named after its type species *tomato pseudocurly top virus*, has a genome similar to curtoviruses but it is vectored by a treehopper rather than a leafhopper.

The construction of the genome of geminiviruses, regardless of the genus to which they belong, is quite complex. Because geminiviruses replicate by producing temporary complementary (−) ssDNA strands, both the original virus (+) ssDNA and the newly formed (−) ssDNA can have open reading frames (ORFs) that may be translated into functional proteins (>10K). Also, it should be noted that the transcription of mRNAs and subsequent translation into proteins can take place in both directions in each of the ssDNAs. For example, some single-component genome geminiviruses (*Mastrevirus*) have three ORFs in their viral (V) plus-sense ssDNA and four more ORFs in their complementary (C) minus-sense ssDNA (Fig. 14-61B), but the numbers of ORFs in each DNA can vary with the virus. Some open reading frames can overlap partially or totally as a result of frameshifts during transcription and translation of the DNA open reading frames.

Geminiviruses with two-part ssDNA genomes (*Begomovirus*) have a common 200 nucleotide sequence (Fig. 14-62). They have open reading frames on the two (+) ssDNAs (DNA A and DNA B) and, in addition, on the two (−) ssDNAs. For example, DNAs of *African cassava mosaic virus* code for 12 potentially functional proteins.

Certain open reading frames found in some geminiviruses correspond to identical or similar ORFs found in other geminiviruses, apparently because they code for common functions in all viruses, e.g., replication, movement, or vector transmissibility. However, some ORFs appear to be distinctive for the virus on which they are found.

Proteins coded for by the open reading frames include (1) the coat protein, with a size of 27 to 31K, that, in addition to providing the shell of the virus, is essential for infectivity in all single-component geminiviruses and plays a role in insect vector specificity and transmissibility of the virus; (2) the replication associated protein, which is usually the largest coded protein (40.3K); and (3) the protein(s) facilitating cell-to-cell and systemic movement of geminiviruses, which seems to be coded by DNA B of the virus.

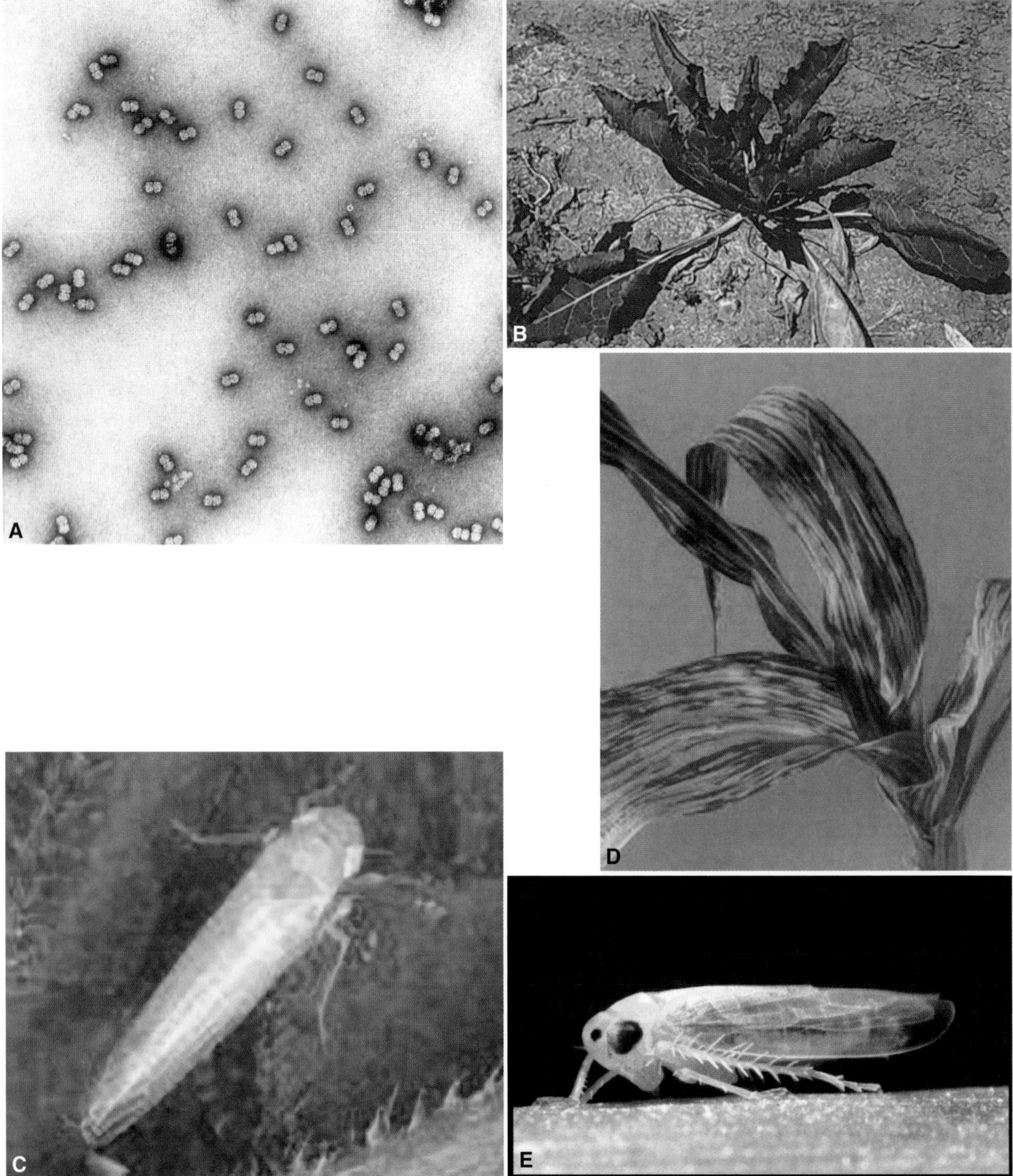

FIGURE 14-60 (A) *Bean golden mosaic virus* showing the characteristic twin particles of Geminiviridae. (B) Sugar beet plant showing curling of leaves caused by the curtovirus *beet curly top virus* (BCTV). (C) The leafhopper vector of BCTV. (D) Symptoms of maize streak caused by the mastrevirus *maize streak virus* (MSV). (E) The leafhopper vector of MSV. [Photographs courtesy of (A) E. Hiebert, University of Florida, (B) R. Harveson, University of Nebraska and (D) Institute of International Tropical Agriculture (IITA).]

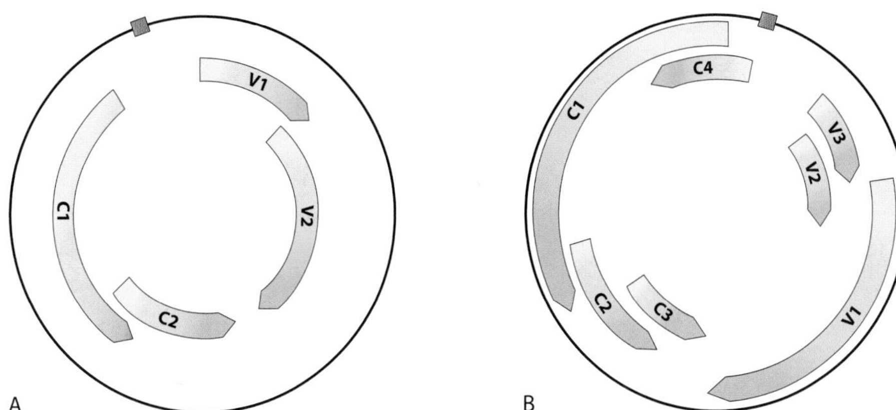

FIGURE 14-61 Organization of the genome of monopartite Geminiviridae in genera *Mastrevirus* (A) and *Curtovirus* (B). Genes (open reading frames) indicated by the letter V are encoded by the viral DNA strand, whereas genes indicated by the letter C are encoded by the complementary DNA strand. The filled small square at the top of each diagram denotes the position of a nine-nucleotide sequence that is conserved in all geminiviruses examined to date.

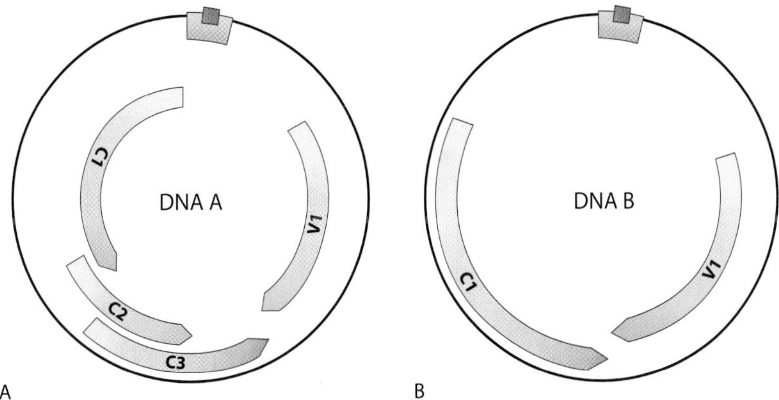

FIGURE 14-62 Organization of the genomes of the two DNAs (A and B) of bipartite geminiviruses of the genus *Begomovirus*. Open reading frames (genes) of the viral (V) strand and the complementary strand are shown, as is the "common region" (stippled box) and the highly conserved nine-nucleotide sequence (filled square) in the two DNAs.

The replication of geminivirus genomes in plant cells seems to take place through the formation of double-stranded DNA intermediates via a rolling circle mechanism. The double-stranded genomic DNA is assembled in the nucleus into nucleosomes, i.e., groups of eight histone protein molecules wrapped about by two coils of DNA, which comprise the basic unit of eukaryotic chromosome structure. Transcription of DNAs into mRNAs also takes place in the nucleus, as does the assembly of virus particles. Geminivirus particles usually accumulate in the nuclei of infected cells either in random aggregates or in crystalline arrays. In some cases, one or more dark-staining fibrillar rings appear in the nuclei of infected plants. Most geminiviruses seem to be confined to the phloem cells of their hosts, but some seem to infect and multiply in most leaf cell types.

Geminiviruses cause some of the most devastating diseases of vegetables, such as tomato, bean, and squash, and of field crops, such as sugar beets, tobacco, and corn. They are most common and catastrophic in the tropical and subtropical regions of the world because of the high populations reached by the *Bemisia* vector in these areas. Geminiviruses drastically reduce photosynthesis, plant growth, fruit set, fruit growth, and fruit quality. Losses, although they depend on the number of plants infected and on the age of plants at the time of infection, are frequently great and in years of epidemics range from 30 to 100% of the crop.

Geminivirus-infected plants often exhibit bright mottles to almost yellow mosaics, and their leaves may be curled or otherwise distorted (Figs. 14-60 and 14-63). When plants become infected young they become

FIGURE 14-63 Vectors and symptoms of begomoviruses. (A) An adult whitefly with eggs and immature ones. (B) Whiteflies almost covering the lower leaf surface of one of its many hosts. Bean plants showing early (C) and advanced (D) symptoms caused by the *golden bean mosaic virus*. The symptoms (D, right) also show greater susceptibility of the cultivar to BGMV. (E) Symptoms caused by *tomato mottle virus* on tomato. (F) Yellowish mottling and severe leaf curling in tomato plant infected with *tomato yellow leaf curl virus*. [Photographs courtesy of (A and B) Florida Department of Agriculture, (C) R. T. McMillan, (E) E. Hiebert, and (F) G. Simone, all University of Florida.]

dwarfed and bushy. Infection of older plants results in reduced new growth and fruit set, and new leaves appear mottled. Geminiviruses are spread rapidly by leafhoppers or whiteflies in the persistent manner, and usually all the plants in a field are infected by the end of the growth season. Geminiviruses overseason in their cultivated or wild hosts, in volunteer plants, and, in the tropics and subtropics, in overlapping crops and in surviving insect vectors. Control of geminiviruses is very difficult and depends primarily on measures that reduce the number of overseasoning insects and virus-infected hosts; separating susceptible crops in place and time is

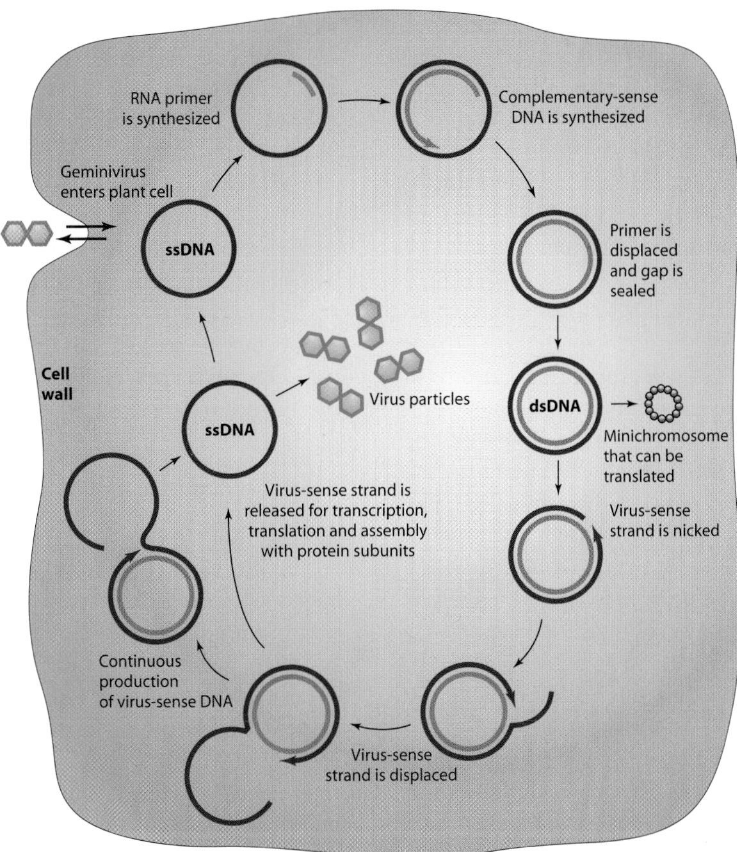

FIGURE 14-64 Schematic representation of replication of a geminivirus. Adapted from S. Stanlely (1995), *Virology* 806, 707–712.

recommended to reduce the transfer of viruliferous vectors from one crop to the other. Few plant genes for only partial resistance to geminiviruses or to their vectors are available so far. Numerous crops are now genetically engineered with genes obtained from geminiviruses in the hope of finding significant stable pathogen-derived resistance to these viruses.

Beet Curly Top

Curly top occurs primarily in the western half of North America and in several Mediterranean countries. The virus infects more than 150 species of herbaceous plants belonging to more than 50 families. It is most destructive on sugar beet, bean, tomato, melons, and spinach.

Curly top kills young plants and causes stunting, malformations, reduced yields, and lower quality in older plants. Losses have sometimes been so severe that vast areas had to be completely abandoned after years of destructive outbreaks of curly top.

Leaves of infected plants are smaller but more numerous, curl upward and inward, and their veins have swellings and spine-like protrusions (Fig. 14-60B). Such leaves later turn yellow, then brown, and die prematurely. Roots of infected plants are also severely stunted, are malformed, and are often killed. Sometimes rootlets proliferate, giving the roots a hairy appearance. In cross sections, infected roots show brownish rings, indicating degenerative changes in the vascular (phloem) tissues. In longitudinal sections, the same tissues appear as a discolored line.

Beet curly top virus is a geminivirus transmitted by the leafhopper *Circulifer tenellus* in a persistent manner. In the plant, the virus seems to be limited almost entirely to the phloem and adjacent parenchyma cells.

The virus overseasons primarily in infected perennial and biennial weeds, in perennial ornamental hosts, in annuals in the greenhouse, and occasionally in the overwintering adults of the insect vector. Insects feed on infected wild plants in the winter and spring, become viruliferous, and carry the virus to cultivated crops in late spring or summer.

In the southwestern United States, the disease has been reduced markedly in some areas by statewide and regional programs to eradicate the leafhopper by mapping and spraying the breeding ground of the leafhopper with insecticides. The most effective and most widespread means of curly top control today is through the use of resistant varieties. Several sugar beet varieties resistant to curly top are available. Resistant varieties to curly top have also been developed for tomato, bean, and other crops.

Maize Streak

Maize streak (Fig. 14-60D) is widespread and severe in the southern half of Africa, in India, and in several islands of the Indian Ocean. It appears as spots on the leaves that later develop into streaks. Plants infected young are stunted and their ears are poorly filled. The maize streak geminivirus is transmitted by several species of leafhoppers. Distinct strains of *maize streak virus* infect each of the other gramineous crops, such as wheat, rice, sugarcane, millet, and many wild grasses.

African Cassava Mosaic

African cassava mosaic occurs in all countries of sub-Saharan Africa where cassava is grown, in India, and in many islands of southern Asia. In Africa, where cassava is by far the largest source of carbohydrates for human food, African cassava mosaic is extremely widespread, affecting 80 to 100% of all cassava plants and causing yield losses of 20 to 90%. The average annual yield loss caused by *African cassava mosaic virus* (ACMV) to cassava production in Africa is estimated to be 50% of the total. This is equivalent to a loss of approximately 50 million tons of cassava roots with a market value of approximately $2 billion U.S.

Cassava plants infected with ACMV produce leaves that exhibit mild to severe mosaic symptoms (Figs. 14-65A, 14-65C, and 14-65D). If plants are infected young they tend to show more severe mosaic, their leaf blades are narrow, the entire plants remain stunted (Figs. 14-65A, 14-65B, and 14-65E), and they form small tubers that contain only a small amount of starch. In older plants, symptoms are milder or absent.

African cassava mosaic virus infects several cassava species and some wild hosts in other families. It is transmitted in nature through infected cassava stem cuttings used as propagative material and by the whitefly *Bemisia tabaci*. The whitefly acquires the virus after feeding on infected plants for about three hours; after an incubation period of at least eight hours, it transmits the virus after feeding on healthy plants for about 10 minutes.

The virus survives in infected cassava plants. Whiteflies transmit the virus from infected to healthy plants in nearby fields. Most infections in a cassava field occur by whiteflies bringing the virus from nearby cassava fields. The amount of secondary, in-field spread of ACMV is considerably smaller. The amount of disease incidence may vary from a few percent to nearly 90% of the plants in a field and reflects the size of whitefly populations moving into the field.

The control of African cassava mosaic depends on using virus-free propagative stock (cuttings), planting resistant varieties, and, where possible, planting new cassava crops away from existing cassava fields or planting only several weeks after the previous crop has been harvested and the plants destroyed.

Bean Golden Mosaic

Bean golden mosaic occurs in most tropical and subtropical areas of the New World, where it causes severe losses on beans. If plants are infected while still very young they fail to produce flowers, and yield losses may be as high as 100%. Infection of plants at later stages of development causes proportionately smaller losses. Leaves of infected plants at first show bright yellow chlorosis of the veins. This soon expands and gives the leaves a golden net-like appearance, which later becomes a striking bright golden mosaic visible from a considerable distance. New leaves produced by infected plants have stiff, leathery surfaces, curl downward, fail to expand properly, and may die (Fig. 14-63).

Bean golden mosaic virus (BGMV) exists as several rather distinct isolates. In addition to beans it also infects several other cultivated and wild legumes and also a malvaceous weed. BGMV is transmitted by the whitefly *B. tabaci* in the semipersistent manner. Some, but not all, BGMV isolates are also transmitted with difficulty by sap inoculation. The virus survives the seasons in its cultivated and wild hosts. Control of bean golden mosaic is very difficult. A few bean cultivars have some degree of resistance, but not much. Planting bean crops when seedling growth can occur at times of low whitefly populations gives the most satisfactory yields. Controlling weeds that may be potential virus reservoirs seems to be helpful, whereas attempts to control whiteflies with insecticides have generally not protected the crop from infection with the virus.

Squash Leaf Curl

Squash leaf curl occurs in southern California where it affects all cucurbits, causing severe symptoms on squash and watermelon and less severe ones on cantaloupe and cucumber. Leaves of infected plants develop thick veins and enations and become curled upward. Infected plants remain stunted, fruit is absent, small, or distorted, and

FIGURE 14-65 Symptoms of African cassava mosaic disease. (A) Healthy and infected cassava plants growing side by side. (B) Young cassava stem is killed by *cassava mosaic virus* (CsMV). (C) Upper and then lower leaves show severe mosaic and become narrower to filiform and some of them are killed. (D) Entire cassava plant remains stunted, malformed, and produces little root yield. (E) Infected susceptible cassava plants (left) grow and produce only a fraction of what healthy or resistant plants (right) do. (Photographs courtesy of D. Coynes and J. J. Hughes, IITA.)

yields are reduced drastically. Plants infected at the seedling stage are often killed by the disease. *Squash leaf curl virus* (SLCV) is transmitted by the whitefly *B. tabaci*.

Tomato Mottle

Tomato mottle causes a severe disease of tomato in Florida. Its first appearance in 1989 occurred approximately one year after a new strain of the usual vector of geminiviruses, *B. tabaci*, was observed in high numbers in Florida. *Tomato mottle virus* (TMoV) is spread readily by whiteflies, and, frequently, most (up to 95%) of the plants in a field become infected. The main symptom of TMoV-infected plants is a more or less brilliant chlorotic yellowish mottle of the foliage (Fig. 14-63E). If plants become infected young, however, they remain stunted, and the yield is reduced drastically. It is estimated that in the 1990–91 growing season of winter tomatoes, losses from TMoV for southwestern Florida exceeded $125 million.

Tomato mottle virus is a bipartite geminivirus. TMoV is transmitted by whiteflies in the persistent manner. When inoculated artificially via whiteflies, TMoV can infect some tobacco species as well as the solanaceous weed *Physalis* and the legume *Phaseolus* (common bean). In nature, however, the virus seems to survive the seasons primarily in overlapping tomato crops and, to some extent, in the perennial solanaceous weed *Solanum viarum*, commonly known as tropical soda apple. Widespread early incidence of tomato mottle within a field and the accompanying severe losses have always been associated with the presence of already infected old, or abandoned, tomato fields within a short distance from the newly planted field. This enables viruliferous whiteflies to transmit the virus from old infected plants to young plants in the new field. So far, no seed transmission of TMoV has been observed; frequently, however, tomato transplants produced in greenhouses or open fields near old or abandoned infected fields become infected with TMoV, presumably because they are within reach of viruliferous whiteflies. In such cases, TMoV is brought to the field with the infected transplants and is subsequently spread by whiteflies to other plants.

The most effective control of tomato mottle can be obtained by planting virus-free transplants at a time when there are no older tomatoes growing nearby or by planting them several miles away from fields with infected plants. This is made possible if all tomato plants are destroyed immediately after the last pick of fruit and no volunteer tomato plants are allowed to grow. Control of the whitefly vector with insecticides seems to help some if it starts early and is intensive. Whiteflies, however, quickly become resistant to the various insecticides. In addition, it takes so many frequent sprays to control them that the cost is great and the allowed limits for the number of applications and amounts of insecticide applied are reached before satisfactory control can be obtained. Some genes with a degree of resistance to the virus have been found in wild tomato species, but so far all tomato varieties are very susceptible to TMoV. In the mid-1990s, tomato varieties were engineered to carry and express certain TMoV genes that provide so-called pathogen-derived resistance resembling cross protection. To date, however, no commercial varieties are available that have such resistance.

Tomato Yellow Leaf Curl

Tomato yellow leaf curl is one of the most devastating diseases of tomato in the Middle East, southeast Asia, North and Central Africa, southern Europe, and, since 1993, the Caribbean Basin islands of Hispaniola and Jamaica. Tomato yellow leaf curl causes severe losses on all fresh market and canning tomatoes. Infected plants remain stunted while their shoots and leaves are smaller and assume an erect position. Such leaves are usually rolled upward and inward and become deformed and severely chlorotic (Fig. 14-63F). After the plant becomes infected, there is considerable drop of flowers, fruit fails to set, and no more marketable fruit is produced. Early infections almost always result in 100% yield loss.

Tomato yellow leaf curl virus (TYLCV) is a whitefly-transmitted geminivirus whose genome consists of a single circular ssDNA, whereas the genome of the other known whitefly-transmitted geminiviruses consists of two ssDNAs. TYLCV is transmitted by the whitefly *B. tabaci* in the circulative nonpropagative manner. The virus can be acquired by the vector after feeding on an infected plant for at least 15 to 30 minutes. There is a latent period of at least 21 hours after which the virus can be inoculated into a healthy plant during a feeding period of at least 15 to 30 minutes. Whiteflies remain viruliferous for about two weeks.

In addition to tomato, TYLCV infects several cultivated hosts, such as tobacco, and weeds, such as datura, belonging to five plant families. Some of these hosts serve as alternate and overseasoning hosts for the virus, which makes efforts to control the virus considerably more difficult. Control of tomato yellow leaf curl is very difficult indeed. All commercial tomato varieties used to date are susceptible, but a few genes providing a degree of resistance to TYLCV have been found in wild tomatoes and are now being tested in several tomato breeding programs. Insecticidal sprays to control the whitefly vector are only partially successful and very expensive. Separating new tomato plants from old ones (in time and space) and planting only when whitefly populations are at their minimum are the main methods that provide considerable control of this disease. In some countries, e.g., Israel, because of TYLCV, tomatoes are now produced only under fine-mesh nets and with frequent application of insecticides. Considerable efforts are presently being made to genetically engineer tomato plants to express certain genomic areas of TYLCV that seem to protect the plant from subsequent infection by the virus.

Selected References

Abouzid, A. M., Polston, J. E., and Hiebert, E. (1992). The nucleotide sequence of tomato mottle virus, a new geminivirus isolated from tomatoes in Florida. *J. Gen. Virol.* **73**, 3225–3229.

Bennet, C. W. (1971). "The Curly Top of Sugar Beet and Other Plants." Monograph No. 7, APS Press, St. Paul, MN.

Berrie, L. C., Rybicki, E. P., and Rey, M. E. C. (2001). Complete nucleotide sequence and host range of *South African cassava mosaic virus*: Further evidence for recombination amongst begomoviruses. *J. Gen. Virol.* **82**, 53–58.

Boulton, M. I. (2002). Functions and interactions of mastrevirus gene products. *Physiol. Mol. Plant Pathol.* **60**, 243–255.

Brown, J. K., and Bird, J. (1992). Whitefly-transmitted geminiviruses and associated disorders in the Americas and the Caribbean Basin. *Plant Dis.* **76**, 220–225.

"C.M.I./A.A.B. Descriptions of Plant Viruses." Maize streak virus (No. 133), beet curly top virus (No. 210), African cassava mosaic virus (No. 297), bean golden mosaic virus (No. 192). Kew, Surrey, England.

Dhar, A. K., and Singh, R. P. (1995). Geminiviruses. *In* "Pathogenesis and Host Specificity in Plant Diseases" (R. P. Singh, U. S. Singh, and K. Kohmoto, eds.), Vol. 3, pp. 289–309. Elsevier, Tarrytown, NY.

Fauquet, C., and Fargette, D. (1990). African cassava mosaic virus: Etiology, epidemiology, and control. *Plant Dis.* **74**, 404–411.

Fondong, V. N., Pita, J. S., Rey, M. E. C., *et al.* (2000). Evidence of synergism between *African cassava mosaic virus* and a new double-recombinant geminivirus infecting cassava in Cameroon. *J. Gen. Virol.* **81**, 287–297.

Fondong, V. N., Thresh, J. M., and Zok, S. (2002) Spatial and temporal spread of cassava mosaic virus disease in cassava grown alone and when intercropped with maize and/or cowpea. *J. Phytopathol.* **150**, 365–374.

Frischmuth, T., Ringel, M., and Kocher, C. (2001). The size of encapsidated single-stranded DNA determines the multiplicity of *African cassava mosaic virus* particles. *J. Gen. Virol.* **82**, 673–676.

Gafni, Y., and Epel, B. L. (2002). The role of host and viral proteins in intra- and inter-cellular trafficking of geminiviruses. *Physiol. Mol. Plant Pathol.* **60**, 231–241.

Ghanim, M., Morin, S., and Czosnek, H. (2001). Rate of *tomato yellow leaf curl virus* translocation in the circulative transmission pathway of its vector, the whitefly *Bemisia tabaci*. *Phytopathology* **91**, 188–196.

Gilbertson, R. L., *et al.* (1993). Genetic diversity in geminiviruses causing bean golden mosaic disease: The nucleotide sequence of the infectious cloned DNA components of a Brazilian isolate of bean golden mosaic geminivirus. *Phytopathology* **83**, 709–715.

Gutierrez, C. (2002). Strategies for geminivirus DNA replication and cell cycle interference. *Physiol. Mol. Plant Pathol.* **60**, 219–230.

Hall, R., ed. (1991). "Compendium of Bean Diseases." APS Press, St. Paul, MN.

Harrison, B. D., Swanson, M. M., and Fargette, D. (2002). Begomovirus coat protein: Serology, variation and functions. *Physiol. Mol. Plant Pathol.* **60**, 257–271.

Harrison, B. D. (1999) Natural genomic and antigenic variation in whitefly-transmitted Gemoniviruses (Begomoviruses). *Annu. Rev. Phytopathol.* **37**, 369–398.

Hiebert, E., Abouzid, A. M., and Polston, J. E. (1996). Whitefly-transmitted geminiviruses. *In* "Bemisia 1995: Taxonomy, Biology, Control and Management," Chap. 26, pp. 277–288. Intercept Ltd., Andover, Hants, England.

Kunik, T., *et al.* (1994). Transgenic tomato plants expressing the tomato yellow leaf curl virus capsid protein are resistant to the virus. *Bio/Technology* **12**, 500–504.

Lazarowitz, S. G. (1992). Geminiviruses: Genome structure and gene function. *Crit. Rev. Plant Sci.* **1**, 327–349.

Liu, H.-Y., Wisler, G. C., and Duffus, J. E. (2000). Particle lengths of whitefly-transmitted criniviruses. *Plant Dis.* **84**, 803–805.

Pita, J. S., Fondong, V. N., Sangare, A., *et al.* (2001). Recombination, pseudorecombination and synergism of geminiviruses are determinant keys to the epidemic of severe cassava mosaic disease in Uganda. *J. Gen. Virol.* **82**, 655–665.

Ploetz, R. C., *et al.* (1994). "Compendium of Tropical Fruit Diseases." APS Press, St. Paul, MN.

Polston, J. E., *et al.* (1990). Association of the nucleic acid of squash leaf curl geminivirus with the whitefly *Bemisia tabaci*. *Phytopathology* **80**, 850–856.

Polston, J. E., and Anderson, P. K. (1997). The emergence of whitefly-transmitted geminiviruses of tomato in the Western Hemisphere. *Plant Dis.* **81**, 1358–1369.

Saunders, K., Wege, C., Veluthambi, K., *et al.* (2001). The distinct disease phenotypes of the common and yellow vein strains of *tomato golden mosaic virus* are determined by nucleotide differences in the 3′-terminal region of the gene encoding the movement protein. *J. Gen. Virol.* **82**, 45–51.

Schnippenkoetter, W. H., Martin, D. P., and Willment, J. A. (2001). Forced recombination between distinct strains of *maize streak virus*. *J. Gen. Virol.* **82**, 3081–3090.

Swiech, R., Browning, S. Molsen, D., *et al.* (2001). Photosynthetic responses of sugar beet and *Nicotiana benthamiana* Domin. infected with beet curly top virus. *Physiol. Mol. Plant Pathol.* **58**, 43–52.

Wege, C., Saunders, K., Stanley, J., *et al.* (2001). Comparative analysis of tissue tropism of bipartite Geminiviruses. *J. Phytopathol.* **149**, 359–368.

Willment, J. A., Martin, D. P., Van der Walt, E., *et al.* (2001). Biological and genomic sequence characterization of *maize streak virus* isolates from wheat. *Phytopathology* **92**, 81–86.

Plant Diseases Caused by Circoviridae

As shown in the mid-1990s, *banana bunchy top virus* (BBTV), *subterranean clover stunt virus*, *coconut foliar decay virus*, and *fava bean necrotic yellows virus* comprise a new group distinct from all other virus groups. These viruses have been placed in the family Circoviridae (from being circular, round), genus *Nanovirus (from being small, nanos = dwarf)*. They are isometric, small (about 18–22 nm in diameter), and contain ssDNA organized in multiple (at least six, in some viruses as many as 11) circular ssDNA components. Each of them has a single open reading frame. Some nanoviruses, e.g., *banana bunchy top virus*, are transmitted by aphids in the persistent manner (Fig. 14-66). Others, e.g., *coconut foliar decay virus* (Fig. 14-67), are transmitted by plant hoppers. To date, four or five viruses are considered as possible members of this group. Multiple ssDNAs of each virus consist of 1,000 to 1,200 nucleotides each and all contain an identical stem-loop structure, a common noncoding region, and one or more open reading frames (Fig. 14-66C) coding for the various virus proteins such as coat protein, virus replicase, and virus movement protein. The coat protein of the virus has a molecular weight of about 21K.

Banana Bunchy Top

Banana bunchy top, where present, is the most important virus disease of banana and one of the few truly important diseases of that crop. It occurs in most banana-growing countries of the world except in Central and South America. It causes severe losses because infected plants produce no fruit. New leaves of infected plants develop dark green streaks on their petioles and veins while the margins become chlorotic. The leaves at the top of the plant are narrower, upright, and closer together, making the top of the plant appear bunchy (Fig. 14-66B). Depending on when the plant was

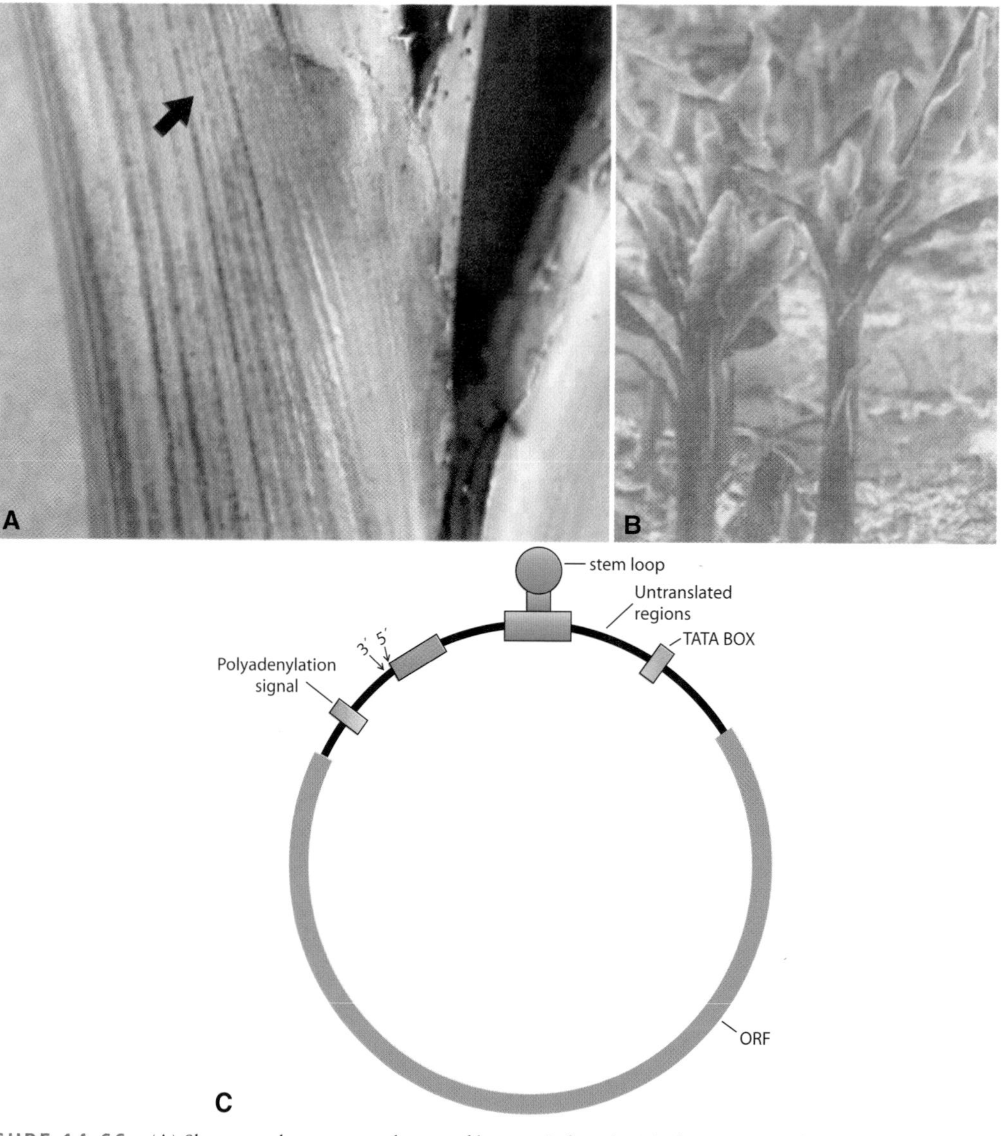

FIGURE 14-66 (A) Short streaks on young leaves of banana infected with the nanovirus *banana bunchy top virus (BBTV)*. (B) Young banana plants showing yellow leaf margins and narrow, stiff, erect leaves bunched together at the top as a result of infection by BBTV. (C) Genome organization of BBTV. The six components of the isometric ssDNA virus differ in the length of the ORF, untranslated regions, and number (1, 2, or 3) of polyadenylation signals. [Photographs courtesy of (A and B) University of Hawaii and (C) Burns *et al.* (1995). *J. Gen. Virol.* 76, 1471–1482.]

infected, the inflorescence and fruit bunch either fail to form or fail to emerge from the banana pseudostem.

Banana bunchy top virus is most severe on banana, but also affects some ornamentals such as *Canna* sp. BBTV apparently exists in several strains, which may infect some banana cultivars and its other hosts and cause only mild or no symptoms. The virus concentration in the plant is highest in the midrib, petiole, and sheath of the younger leaves and it varies with the season. Phloem cells of BBTV-infected plants produce a fluorescence specific to this disease.

Banana bunchy top virus is transmitted over long distances by propagative materials such as rhizomes, suckers, or tissue-cultured meristems and over short distances by the banana aphid *Pentalonia nigronervosa* in the persistent manner. Volunteer bananas growing in previous banana plantations are often infected with BBTV and also support large aphid populations that then transmit the virus to newly planted banana plantations. Control of banana bunchy top depends primarily on adopting cultural measures that help avoid or minimize virus infections. Such measures include quarantine to keep the virus out of a virus-free area, the use of virus-free propagative material, locating new plantations away from older infected ones, and destroying all volunteer banana plants. Roguing of infected and nearby plants seems to reduce the rate of virus spread. Attempts to control the aphid vector with insecticides have little effect on the spread of this virus.

Coconut Foliar Decay

The disease is of limited distribution at present, being found in the New Hebrides islands of the Pacific Ocean but is economically important in areas where it occurs. Trees of susceptible varieties in the field at first show yellowing in several leaflets of the intermediate fronds. The yellowing of the fronds then becomes more general, the petioles become necrotic, and the fronds die prematurely and hang downward (Figs. 14-67A and 14-67B). More fronds continue to become yellow and die while some of the oldest and some of the youngest fronds may remain green longer. Susceptible trees die within 1 or 2 years from the appearance of symptoms. Trees of resistant varieties may show remission from the disease.

The disease is caused by the nanovirus *coconut foliar decay virus* (Fig. 14-67C), which is 20 nanometers in diameter and contains circular ssDNA. The *coconut foliar decay virus* (CFDV) is transmitted by the plant hopper *Myndus taffini* in the semipersistent manner.

Selected References

Burns, T. M., *et al.* (1993). Single-stranded DNA genome organization of banana bunchy top virus. *Sixth Int. Congr. Plant Pathol.*, 312.

Burns, T. M., Harding, R. M., and Dale, J. L. (1995). The genome organization of banana bunchy top virus: Analysis of six ssDNA components. *J. Gen. Virol.* 76, 1471–1482.

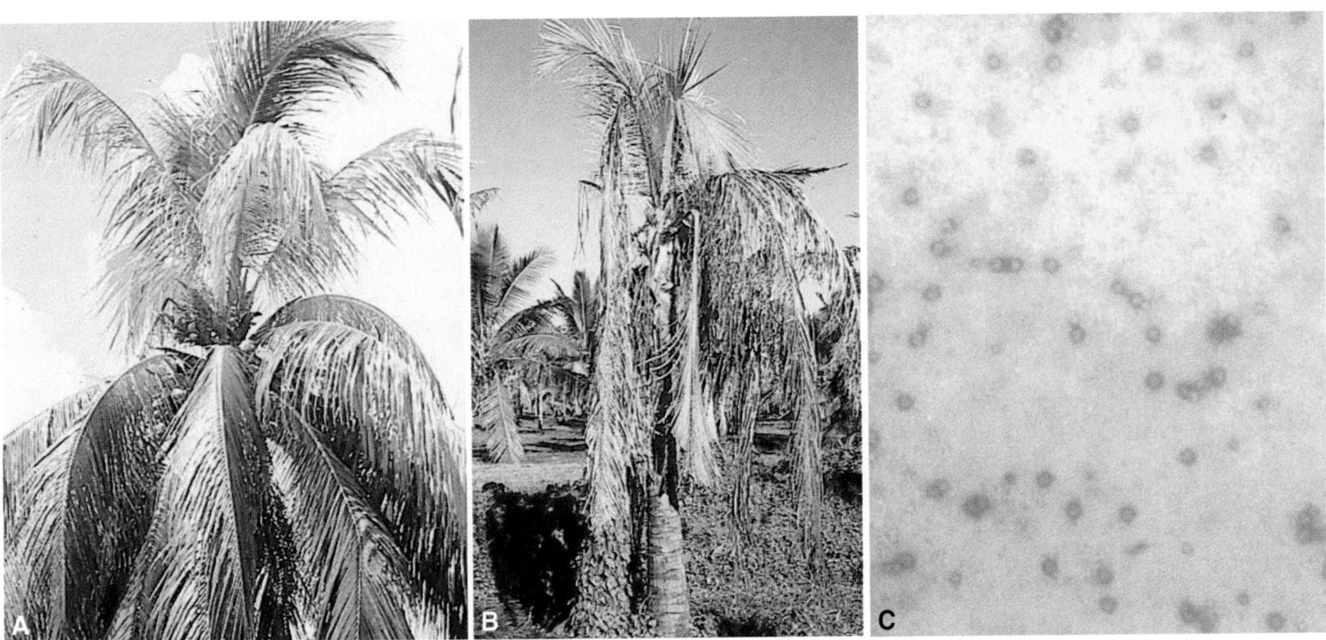

FIGURE 14-67 (A) Early stages of foliar decay disease of coconut palms caused by the *coconut foliar decay virus* (CFDV). Leaves of fronds halfway from the top turn yellow first. (B) Later in the year, leaves on more fronds are killed and finally the palm tree dies. (C) CFDV particles. (Photographs courtesy of J. W. Randles.)

Chu, P., *et al.* (1995). Non-geminated single-stranded DNA plant viruses. *In* "Pathogenesis and Host Specificity in Plant Diseases" (R. P. Singh, U. S. Singh, and K. Kohmoto, eds.), Vol. 3, pp. 311–341. Elsevier, Tarrytown, NY.

Horser, C. L., Harding, R. M., and Dale, J. L. (2001). Banana bunchy top nanovirus DNA-1 encodes the 'master' replication initiation protein. *J. Gen. Virol.* **82**, 459–464.

Kim, K.-S., and Lee, K.-W. (1992). Geminivirus-induced macrotubules and their suggested role in cell-to-cell movement. *Phytopathology* **82**, 664–669.

Merits, A., Fedorkin, O. N., Guo, D., *et al.* (2000). Activities associated with the putative replication initiation protein of coconut foliar decay virus, a tentative member of the genus *Nanovirus*. *J. Gen. Virol.* **81**, 3099–3106.

Saunders, K., Bedford, I. D., and Stanley, J. (2002). Adaptation from whitefly to leafhopper transmission of an autonomously replicating nanovirus-like DNA component associated with ageratum yellow vein disease. *J. Gen. Virol.* **83**, 907–913.

Wanitchakorn, R., Hafner, G. J., Harding, R. M., and Dale, J. L. (2000). Functional analysis of proteins encoded by banana bunchy top virus DNA-4 to -6. *J. Gen. Virol.* **81**, 299–306.

VIROIDS

Plant Diseases Caused by Viroids

To date, at least 40 plant diseases have been shown to be caused by viroids. The most important viroid plant diseases are cadang-cadang disease of coconut, potato spindle tuber, citrus exocortis, avocado sunblotch (Figs. 14-68A and 14-68B), chrysanthemum stunt (Fig. 14-68C), and apple scar skin (Figs. 14-68D and 14-68E). As the detection, separation, and identification techniques have improved greatly, many more viroids have been detected and studied. So far, no animal or human disease has been shown to be caused by a viroid. It is likely, however, that viroids will be soon implicated as the causes of several "unexplained" diseases in animals and humans and in more plants.

Viroids are small, low molecular weight ribonucleic acids that can infect plant cells, replicate themselves, and cause disease (Fig. 14-69). Viroids differ from viruses in at least two main characteristics: (1) the size of RNA in viroids, which consists of 250 to 370 bases, is much smaller compared to that in viruses, which is 4 to 20 kilobases, and (2) the fact that virus RNA is enclosed in a protein coat whereas viroids lack a protein coat and apparently exist as free (naked) RNA.

Because of their small size (250–370 nucleotides), viroids lack sufficient information to code for even one protein, even for a replicase enzyme required to replicate the viroid. The existence of viroids as free RNAs rather than as nucleoproteins necessitates the use of phenol in the sap to inactivate the plant ribonucleases and makes their visualization with an electron microscope extremely difficult even in purified preparations; indeed, in plant tissues or plant sap their detection with an electron microscope is currently impossible.

Viroids are circular, single-stranded RNA molecules with extensive base pairing in parts of the RNA strand. The base pairing results in some sort of hairpin structure with single-stranded and double-stranded regions of the same viroid and contributes to the stability of the RNA, given that it lacks a protein coat (Fig. 14-70).

It appears that, in its double-stranded form, each viroid consists of five structural regions: a left and a right terminal region, a pathogenicity region, a conserved central region, and a variable region. The terminal and pathogenicity regions determine the pathogenicity of a viroid, i.e., its ability to infect and multiply, and also the severity of the symptoms that will develop on the host plants. The severity of the symptoms, however, can be altered by changes in one or two bases in these regions. The other two regions of viroids, the conserved central region and the variable region, have not been implicated in any function of viroids.

Taxonomy (Grouping) of Viroids

The taxonomy of viroids is based on the absence in some of them (the avocado sunblotch viroid group, group A, or Avsunviroids) of a conserved central region or the presence in them (the potato spindle tuber viroid group, group B, or Pospiviroids) of a central conserved region. The avocado sunblotch viroid (ASBVd) group has only four members, whereas the potato spindle tuber viroid (PSTVd) group has all the rest of the 40 viroids. All Avsunviroids have a ribozyme activity that enables them to self-cleave their RNA multimers during viroid replication. It is also speculated that Avsunviroids replicate in chloroplasts, whereas Pospiviroids replicate in the nucleus and nucleolus. Both groups are subdivided into subgroups depending on sequence similarities in the conserved central region.

Viroids
- ASBVd group or Avsunviroids
  - Avsunviroideae
    - Avsunviroid
      - Avocado sunblotch viroid
    - Pelamoviroid
      - Chrysanthemum chlorotic mottle viroid
      - Peach latent mosaic viroid
- PSTVd group or Pospiviroids
  - Pospiviroideae

FIGURE 14-68 Diseases caused by some viroids. (A and B) Avocado sunblotch. (C) Chrysanthemum leaves showing symptoms of (from left) *chrysanthemum chlorotic mottle viroid, potato spindle tuber viroid,* and *chrysanthemum stunt viroid*; right leaf, control. (D) Apples on a tree showing severe scarring and cracking caused by the *apple scar skin viroid* (SSVd). (E) Comparison of a healthy Red Delicious apple and one infected with SSVd. [Photographs courtesy of (A and B) R. T. MacMillan and (C) R. J. McGovern.]

Pospiviroid subgroup
- Potato spindle tuber viroid
- Chrysanthemum stunt viroid
- Citrus exocortis viroid
- Columnea latent viroid
- Iresine viroid 1
- Mexican papita viroid
- Tomato apical stunt viroid
- Tomato planta macho viroid

Apscaviroid subgroup
- Apple scar skin viroid
- Apple dimple fruit viroid
- Australian grapevine viroid
- Citrus bent leaf viroid
- Citrus viroid III
- Grapevine yellow speckle viroid

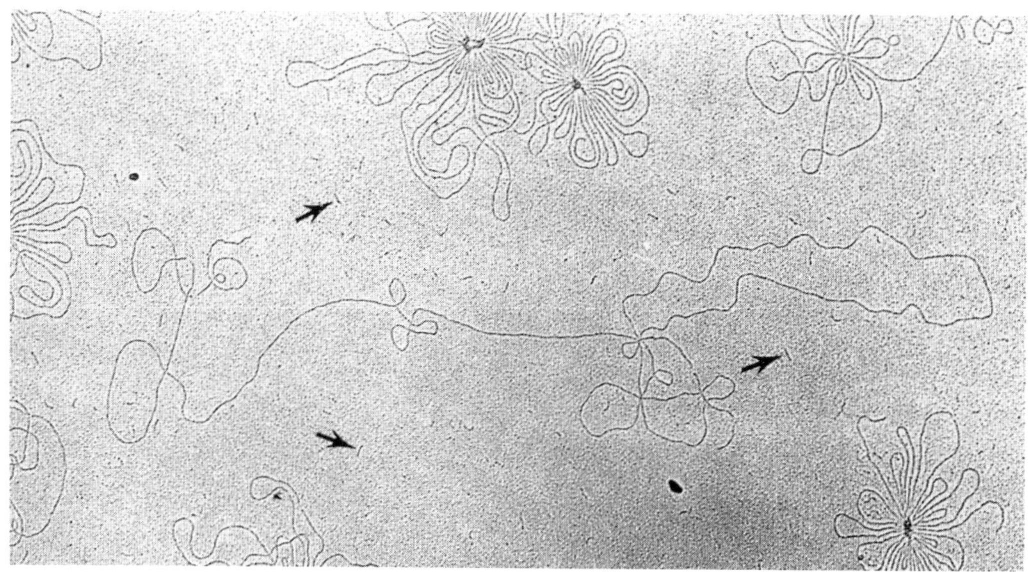

FIGURE 14-69 Electron micrograph of *potato spindle tuber viroid* particles (arrows) mixed with double–stranded DNA of a bacterial virus (bacteriophage T7) for comparison. Photograph taken by T. Koller and J. M. Sogo, supplied courtesy of T. O. Diener.

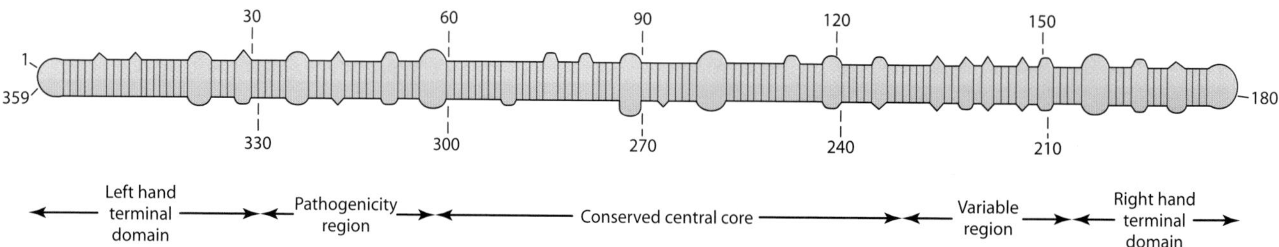

FIGURE 14-70 Likely secondary structure of the *potato spindle tuber viroid*. Its 359 bases are arranged in a way that results in extensive base pairing and greater stability of the viroid. Most known viroids appear to share common features. For instance, the left and right terminal domains are involved in viroid replication, whereas the pathogenicity region is involved in pathogenesis. The central area is the most conserved and the variable region is the least conserved among viroids.

- Grapevine yellow speckle viroid 1
- Grapevine yellow speckle viroid 2
- Pear blister canker viroid

Cocadviroid subgroup
- Coconut cadang-cadang viroid
- Citrus viroid 4
- Coconut tinangaja viroid
- Hop latent viroid

Coleviroid subgroup
- *Coleus blumei* viroid
- *C. blumei* viroid 1
- *C. blumei* viroid 2
- *C. blumei* viroid 3

Hostuviroid subgroup
- Hop stunt viroid

Unassigned viroids
- Apple fruit crinkle viroid
- Cherry small circular viroid-like RNA
- Citrus viroid Ia
- Citrus viroid II
- Citrus viroid OS
- Citrus viroid-I-LSS
- Coleus yellow viroid
- Tomato chlorotic dwarf viroid

Viroids seem to be associated with cell nuclei, particularly the chromatin, and possibly with the endomembrane system of the cell. Although viroids have many of the properties of single-stranded RNAs, when seen with the electron microscope they appear about 40 nanometers in length and have the thickness of double-stranded RNA (Fig. 14-69).

How viroids replicate themselves is still not known. Their small size is sufficient to code for a very small protein, but such a protein would be considerably smaller than known RNA polymerase (replicase) subunits and would therefore be unable to carry out replication of the viroid. In addition, viroids have been shown to be inactive as a messenger RNA in all *in vitro* protein-synthesizing systems tested. Also, no new proteins could be detected in viroid-infected plants. Evidence shows that viroids replicate by direct RNA copying in which all components required for viroid replication, including the RNA polymerase, are provided by the host. During viroid replication, the circular (+) strand of the viroid is replicated while it acts as a rolling drum producing multimeric linear strands of (–) RNA (Fig. 14-71). The linear (–) strand then serves as a template for replication of multimeric strands of (+) RNA. The (+) RNA is subsequently processed (cleaved) by enzymes that release linear, unit-length viroid (+) RNAs, which circularize and produce many copies of the original viroid RNA (Fig. 14-71).

How viroids cause disease is also not known. Viroid diseases show a variety of symptoms (Fig. 14-72) that resemble those caused by virus infections. The amount of viroids formed in cells seems to be extremely small, and it is therefore unlikely that they cause a shortage of RNA nucleotides in cells. In addition, as with viruses, many infected hosts show no obvious damage, although viroids seem to be replicated in them as much as in sensitive hosts. Moreover, as mentioned earlier, even one or two base changes at specific sites of the viroid are sufficient to change the disease from mild to severe and vice versa. Thus, viroids apparently interfere with the host metabolism in ways resembling those of viruses, but which ways are also unclear. It has been shown that both virus-specific RNAs synthesized during infection and viroid RNA *in vitro* activate a protein kinase enzyme, which in turn activates other cellular enzymes while it impedes the initiation of protein synthesis. As viroid strains that cause mild to severe plant symptoms activate the protein kinase more than 10 times as much as mild strains, it is possible that activation of the protein kinase represents the triggering event in viroid pathogenesis and in disease development by the plant.

Viroids are spread from diseased to healthy plants primarily by mechanical means, i.e., through sap carried on hands or tools during propagation or cultural practices and, of course, by vegetative propagation. Some, such as *potato spindle tuber*, *chrysanthemum stunt*, and *chrysanthemum chlorotic mottle viroids*, are transmitted through sap quite readily, whereas others, such as *citrus excortis viroid*, are transmitted through sap with some difficulty. Several viroids, e.g., those causing potato spindle tuber, cadang-cadang, tomato bunchy top, and apple scar skin, appear to be transmitted through the pollen and seed, but the rates of such transmission are usually very small. No specific insect or other vectors of viroids are known, although viroids seem to be transmitted on the mouthparts or feet of some insects.

Viroids apparently survive in nature outside the host or in dead plant matter for periods of time varying from a few minutes to a few months. Generally, they seem to overwinter and oversummer in perennial hosts, which include the main hosts of almost all known viroids. Viroids are usually quite resistant to high temperatures and cannot be inactivated in infected plants by heat treatment.

The control of diseases caused by viroids is based on the use of viroid-free propagating stock, removal and

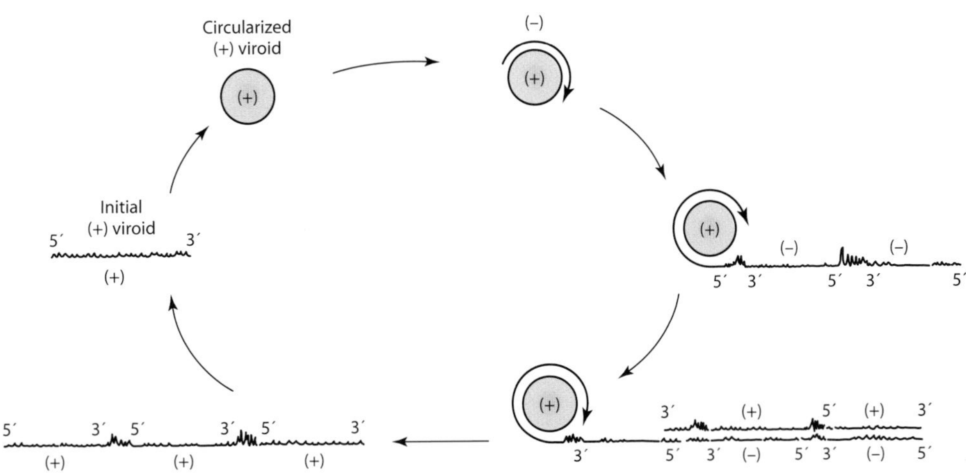

FIGURE 14-71 Schematic representation of presumed viroid replication.

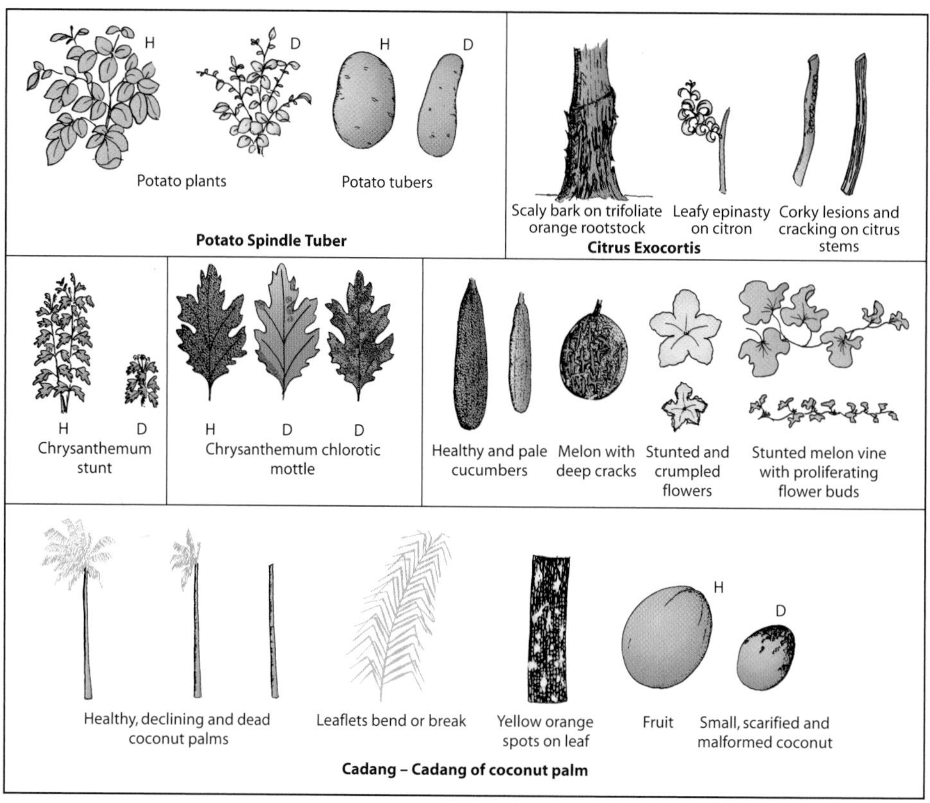

FIGURE 14-72 Types of symptoms caused by viroids. H, healthy; D, diseased.

destruction of viroid-infected plants, and washing of hands or sterilizing of tools after handling viroid-infected plants before moving on to healthy plants.

Potato Spindle Tuber

The potato spindle tuber disease occurs in North America, Russia, and South Africa. It causes severe losses, and in some regions it is one of the most destructive diseases of potatoes. It attacks all varieties and spreads rapidly. It also attacks tomato but seems to be of little economic importance in that crop.

Infected potato plants appear erect, spindly, and dwarfed (Fig. 14-73A). The leaves are small and erect, and the leaflets are darker green and sometimes show rolling and twisting. The tubers are elongated, with tapering ends (Fig. 14-73B). Tubers are smoother, but tuber eyes are more numerous and more conspicuous. Yields are reduced by 25% or more.

Potato spindle tuber viroid (PSTVd) is the first recognized viroid. PSTVd consists of 359 nucleotides. Under an electron microscope, purified PSTVd appears as short strands about 40 nanometers long and has the thickness of a double-stranded DNA (Fig. 14-69). Sap from infected plants is still infective after heating for 10 minutes at 75 to 80°C.

The viroid is mechanically transmissible and is spread primarily by knives used to cut healthy and infected potato seed tubers and during handling and planting of the crop. PSTVd seems to also be transmitted by pollen and seed and by contaminated mouthparts of several insects not normally considered as virus vectors, e.g., grasshoppers, flea beetles, and bugs. After inoculation of a tuber with PSTVd by means of a contaminated knife or of a growing plant with sap from an infected plant, the viroid replicates itself and spreads systemically throughout the plant.

Potato spindle tuber can be controlled effectively by planting only PSTVd-free potato tubers in fields free of diseased tubers that may have survived from the previous year's crop.

Citrus Exocortis

Exocortis occurs worldwide. It affects trifoliate oranges, citranges, Rangpur and other mandarin and sweet limes, some lemons, and citrons. It is important commercially when infected budwood of orange, lemon, grapefruit,

FIGURE 14-73 (A) Stunting, upright growth of shoots and rolling of leaves of potato plants caused by *potato spindle tuber viroid* infections (PSTVd). (B) Tubers of PSTVd-infected plants are often spindle shaped (at right). [Photographs courtesy of (A) T. A. Zitter and (B) H. D. Thurston, Cornell University.]

FIGURE 14-74 (A) Scaly bark symptoms on the rootstock of a citrus tree caused by the *citrus exocortis viroid*. (B) The same viroid causes leaf epinasty and bark splitting on certain clones of Etrog citron (right), which has been used for detecting and identifying this viroid. (Photographs courtesy of Plant Pathology Department, University of California, Riverside.)

and other citrus trees is grafted on exocortis-sensitive rootstocks. Such trees show slight to great reductions in growth and yields are reduced by as much as 40%.

Infected susceptible plants develop narrow, vertical, thin strips of partially loosened outer bark that give the bark a cracked and scaly appearance (Fig. 14-74) when they are about 4 to 8 years old. Infected exocortis-susceptible plants may also show yellow blotches on young infected stems, and some citrons show leaf and stem epinasty (Fig. 14-74B) along with cracking and darkening of leaf veins and petioles. In plant cells, the viroid is associated with nuclei and internal membranes of host cells and results in aberrations of the plasma membranes. All infected plants, including resistant cultivars grafted on such trees, usually appear stunted to a smaller or greater extent and have lower yields.

Citrus exocortis viroid (CEVd) consists of 371 nucleotides. CEVd is transmitted readily from diseased to healthy trees by budding knives, pruning shears, or other cutting tools, by hand, and possibly by scratching and gnawing of animals; CEVd is also transmitted by dodder and by sap to herbaceous plants. On contaminated knife blades, CEVd retains its infectivity for at least eight days. The viroid is highly resistant to heat inactivation and to almost all common chemical sterilants except sodium hypochlorite solution and ribonuclease.

Citrus exocortis viroid has been identified in the past by graft indexing on sensitive clones of Etrog citron, which develops leaf epinasty and bark splitting within a few months. In the past 10 years, CEV identification has been made by electrophoresis of infectious sap and by using radioisotope-labeled DNA probes complementary to CEV.

Coconut Cadang-Cadang

Cadang-cadang (dying) disease of coconut and other palms occurs in the Philippines, where it has killed more than 30 million coconut palms since it was first recognized in the 1930s. Even now, about 1 million palms succumb to cadang-cadang every year (Fig. 14-75B).

FIGURE 14-75 Symptoms of cadang-cadang disease of coconut palm caused by the *coconut cadang-cadang viroid* (CCCVd). (A) Leaflets from a healthy palm (right) and from a palm with late stage disease showing chlorotic spotting (left). (B) Area in one of the Philippine islands with many coconut palm trees showing early, mid, and late stages of cadang-cadang disease. (C) Purified preparation of the CCCVd. Numerous circular-form molecules of the viroid can be seen. (Photographs courtesy of J. W. Randles.)

The disease is of great economic significance to the Philippines because of the subsistence value of coconut palms to the local population as food and lumber and as a major cash crop from the export of coconuts and copra, the dried coconut "meat" from which coconut oil is extracted. A similar disease, called tinangaja, caused by a related viroid, occurs in Guam where it has killed many of the coconut palms on the island.

The symptoms of cadang-cadang in palms develop slowly over 8 to 15 years and are not particularly diagnostic of the disease unless observations are made over several years. Palm trees usually become infected with cadang-cadang after they have begun to flower. The first symptoms appear on the coconuts, which become rounded and develop scarifications on their surface, while the leaves begin to show bright yellow spots (Fig. 14-75A). Three to 4 years later, the inflorescences are killed and, as a result, no more coconuts are produced. Also, few new fronds develop, while the leaves have more and larger yellow spots, making the whole fronds appear chlorotic from a distance. Five to 7 years from the beginning of symptoms, the constantly increasing number of leaf spots gives the whole crown a yellowish or bronze color while the number and size of fronds in the crown continue to be reduced. Finally, the growing bud dies, falls off, and leaves the palm trunk standing like a telephone pole (Fig. 14-75).

In early stages of infection, the coconut cadang-cadang viroid (CCCVd) consists of 246 nucleotides, making this the smallest viroid known (Fig. 14-75C). However, it is always accompanied by a 247 nucleotide form of the identical viroid plus an additional cytosine nucleotide. In later stages of infection, two longer forms of the viroid appear and eventually replace the smaller viroids in fronds. These forms, containing 296 and 297 nucleotides, are the result of duplication of part of the right-hand end of the viroid molecule of the short forms. This pattern of changing molecular forms is unique to the *cadang-cadang viroid*, which is also the only viroid, so far, known to infect monocots and to kill its host plants. Some CCCVd-like viroids have been found to cause disease in oil palms in several islands of the southwest Pacific, but coconut palms infected with them develop only mild symptoms not typical of cadang-cadang.

The cadang-cadang viroid survives in infected coconut and possibly other palm trees. It survives in most palm tissues, including the husks and embryo of coconuts, and is transmitted through a small proportion (0.3%) of the seeds. It is also present in pollen of affected palms. It is not clear how CCCVd spreads from tree to tree. It is likely, however, that it spreads to a small extent by each of several methods: on the mouthparts of various chewing insects, mechanically on the machetes used to cut steps at the base of the palm, to dislodge the nuts, and to make cuts to the inflorescence for tapping their sugary sap, and, possibly, through infected pollen.

Cadang-cadang disease cannot yet be controlled by any available means and continues to spread outward from infected areas and into new areas of uninfected palm trees at about 500 meters per year. Eradication of infected trees and insect control have no effect on the spread of the disease, and decontamination of machetes has proved impractical. So far, no resistant coconut cultivars are available for replanting or as breeding material; breeding efforts, however, continue. Production and use of viroid-free palm seedlings whether from seed or through tissue culture are extremely important. This has become possible, easier, and faster recently through the use of electrophoresis and nucleic acid techniques that help detect the viroid in parental material, which is then excluded from further multiplication so that only viroid-free material is used for the propagation of palm trees.

Selected References

Desvignes, J. C., Grassseau, N., Boyé, R., *et al.* (1999). Biological properties of apple scar skin viroid: isolates, host range, different sensitivity of apple cultivars, elimination, and natural transmission. *Plant Dis.* **83**, 768–772.

Desvignes, J. C., Cornaggia, D., and Grasseau, N. (1999). Pear blister canker viroid: Host range and improved bioassay with two new pear indicators, Fieud 37 and Fieud 110. *Plant Dis.* **83**, 410–422.

Diener, T. O., Owens, R. A., and Hammond, R. W. (1993). Viroids: The smallest and simplest agents of infectious disease. How do they make plants sick? *Intervirology* **35**, 186–195.

Di Serio, F., and Malfitano, M. (2000). *Apple dimple fruit viroid:* Fulfillment of Koch's postulates and symptom characteristics. *Plant Dis.* **85**, 179–182.

Hadidi, A., Giunchedi, L., Shamloul, A. M., *et al.* (1997). Occurrence of peach latent mosaic viroid in stone fruits and its transmission with contaminated blades. *Plant Dis.* **81**, 154–158.

Hanold, D., and Randles, J. W. (1991). Coconut cadang-cadang disease and its viroid agent. *Plant Dis.* **75**, 330–335.

Hodgson, R. A. J., Wall, G. C., and Randles, J. W. (1998). Specific identification of coconut tinangaja viroid for differential field diagnosis of viroids in coconut palm. *Phytopathology* **88**, 774–781.

Ito, T., Ieki, H., Ozaki, K., *et al.* (2002). Multiple citrus viroids in citrus from Japan and their ability to produce exocortis-like symptoms in citron. *Phytopathology* **92**, 542–547.

Maramorosch, K. (1993). The threat of cadang-cadang disease. *Principles* **37**, 187–196.

Reanwarakorn, K., and Semancik, J. S. (1999). Correlation of hop stunt viroid variants in cachexia and xyloporosis diseases of citrus. *Phytopathology* **89**, 568–574.

Riesner, D. (1991). Viroids: From thermodynamics to cellular structure and function. *Mol. Plant-Microbe Interact.* **4**, 122–131.

Sano, T., and Singh, R. P. (1995). Avocado sunblotch viroid group. *In* "Pathogenesis and Host Specificity in Plant Diseases" (R. P. Singh, U. S. Singh, and K. Kohmoto, eds.), Vol. 3, pp. 363–371. Elsevier, Tarrytown, NY.

Semancik, J. S. (1987). "Viroids and Viroidlike Pathogens." CRC Press, Boca Raton, FL.

Singh, M., and Singh, R. P. (1995). Potato spindle viroid group. *In* "Pathogenesis and Host Specificity in Plant Diseases" (R. P. Singh, U. S. Singh, and K. Kohmoto, eds.), Vol. 3, pp. 343–362. Elsevier, Tarrytown, NY.

Symons, R. H. (1991). The intriguing viroids and virusoids: What is their information content and how did they evolve? *Mol. Plant-Microbe Interact.* **4**, 111–121.

Van Vurren, S. P., and da Graça, J. V. (2000). Evaluation of graft-transmissible isolates from dwarfed citrus trees as dwarfing agents. *Plant Dis.* **84**, 239–242.

Vivanco, J. M., Querci, M., and Salazar, L. F. (1999). Antiviral and antiviroid activity of MAP-containing extracts from *Mirabilis jalapa* roots. *Plant Dis.* **83**, 1116–1121.

Wan Chow Wah, Y. F., and Symons, R. H. (1999). Transmission of viroids via grape seeds. *J. Phytopathol.* **147**, 285–291.

Zhao, Y., Owens, R. A., and Hammond, R. W. (2001). Use of a vector based on *potato virus X* in a whole plant assay to demonstrate nuclear targeting of *potato spindle tuber viroid*. *J. Gen. Virol.* **82**, 1491–1497.

chapter fifteen

PLANT DISEASES CAUSED *BY* NEMATODES

INTRODUCTION
826

CHARACTERISTICS OF PLANT PATHOGENIC NEMATODES
827

PROPERTIES OF NEMATODES – SYMPTOMS CAUSED BY NEMATODES – HOW NEMATODES AFFECT PLANTS – INTERRELATIONSHIPS BETWEEN NEMATODES AND OTHER PLANT PATHOGENS – CONTROL OF NEMATODES
831

ROOT-KNOT NEMATODES: *MELOIDOGYNE* SPP.
838

CYST NEMATODES: *HETERODERA* AND *GLOBODERA*
842

THE CITRUS NEMATODE: *TYLENCHULUS SEMIPENETRANS*
848

LESION NEMATODES: *PRATYLENCHUS*
849

THE BURROWING NEMATODE: *RADOPHOLUS*
853

STEM AND BULB NEMATODE: *DITYLENCHUS*
858

STING NEMATODE: *BELONOLAIMUS*
860

STUBBY-ROOT NEMATODES: *PARATRICHODORUS* AND *TRICHODORUS*
863

SEED-GALL NEMATODES: *ANGUINA*
865

FOLIAR NEMATODES: *APHELENCHOIDES*
867

PINE WILT AND PALM RED RING DISEASES: *BURSAPHELENCHUS*
870

INTRODUCTION

Nematodes belong to the kingdom Animalia. Nematodes are wormlike in appearance but quite distinct taxonomically from the true worms. Most of the several thousand species of nematodes live freely in fresh or salt waters or in the soil, and feed on microorganisms and microscopic plants and animals. Numerous species of nematodes attack and parasitize humans and animals, in which they cause various diseases. Several hundred species, however, are known to feed on living plants, obtaining their food with spears or stylets (Fig. 15-1) and causing a variety of plant diseases worldwide. The annual worldwide losses caused

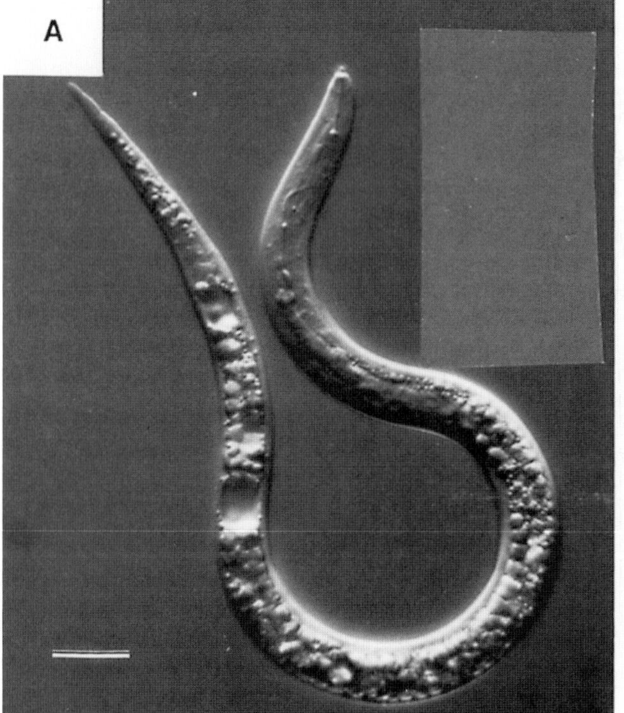

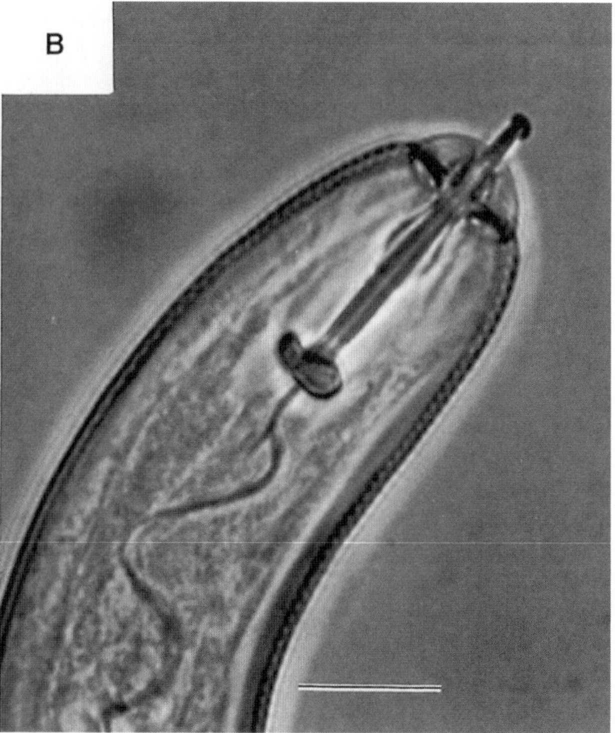

FIGURE 15-1 (A) Typical plant parasitic nematode. (B) Close-up of the head of a plant parasitic nematode showing the spear or stylet. Scale bars: 10 µm. [From McClure and von Mende (1987), *Phytopathology 77, 1463–1469.*]

by nematodes on the life-sustaining crops, which include all grains and legumes, banana, cassava, coconut, potato, sugar beet, sugarcane, sweet potato, and yam, are estimated to be about 11%; Losses for most other economically important crops (vegetables, fruits, and nonedible field crops) are about 14%, for a total of over $80 billion annually.

CHARACTERISTICS OF PLANT PATHOGENIC NEMATODES

Morphology

Plant-parasitic nematodes are small, 300 to 1,000 micrometers, with some up to 4 millimeters long, by 15–35 micrometers wide (Figs. 15-2 and 15-3). Their small diameter makes them invisible to the naked eye,

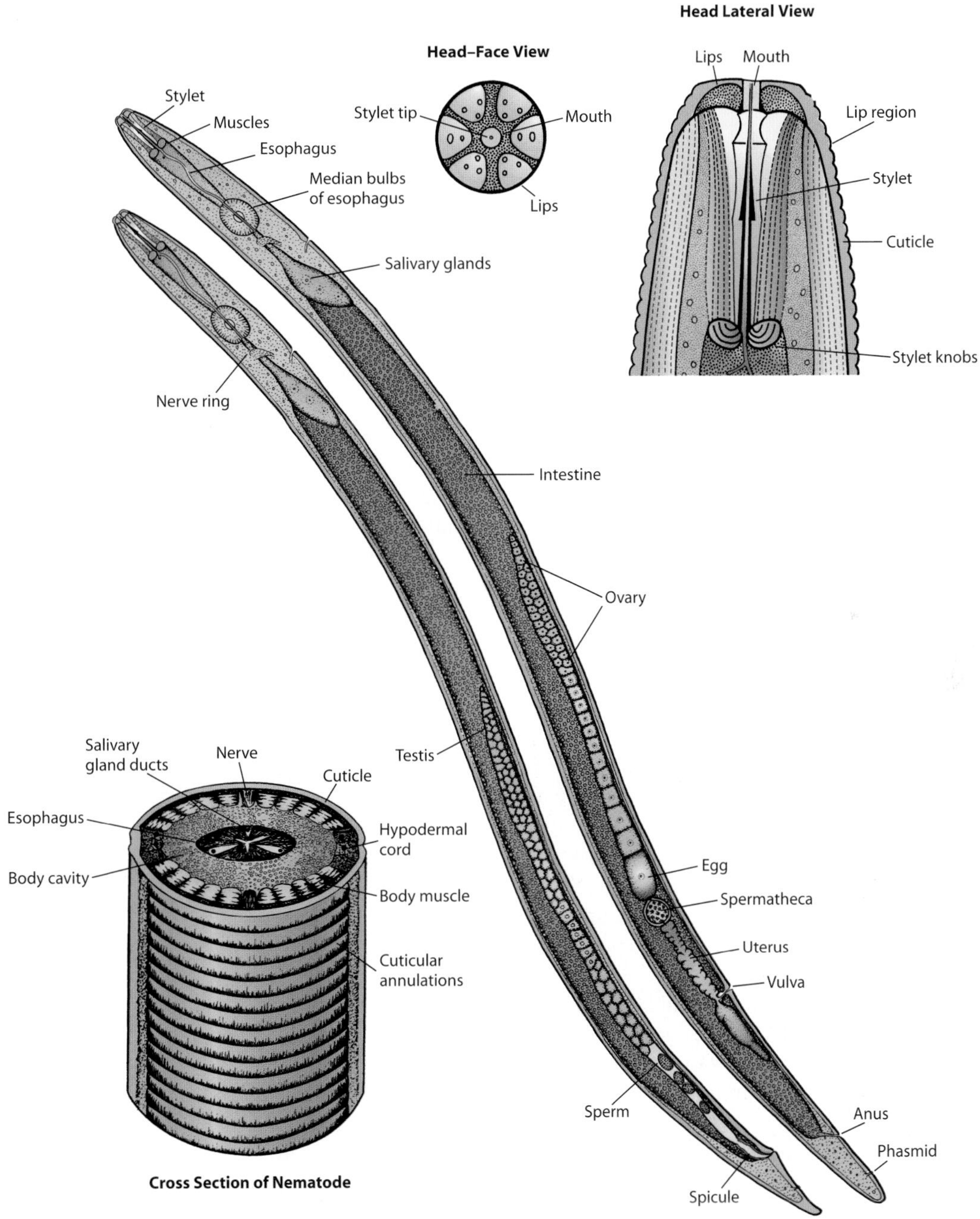

FIGURE 15-2 Morphology and main characteristics of typical male and female plant parasitic nematodes.

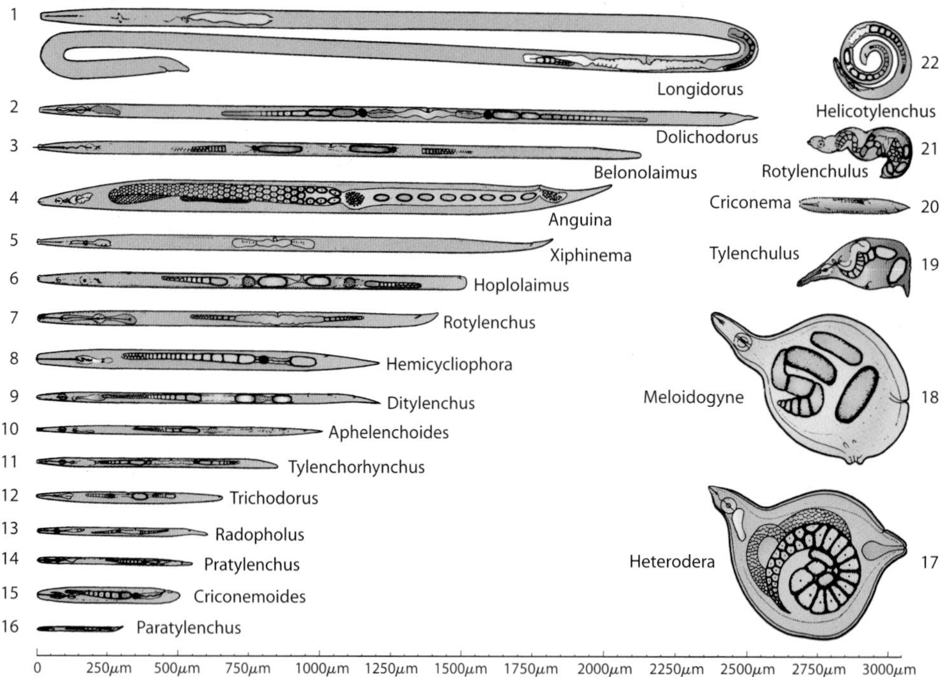

FIGURE 15-3 Morphology and related sizes of some of the most important plant parasitic nematodes.

but they can be observed easily under the microscope. Nematodes are, in general, eel shaped and round in cross section, with smooth, unsegmented bodies, without legs or other appendages. The females of some species, however, become swollen at maturity and have pear-shaped or spheroid bodies (Fig. 15-3).

Anatomy

The nematode body (Fig. 15-2) is more or less transparent. It is covered by a colorless cuticle, which is usually marked by striations or other markings. The cuticle molts when a nematode goes through the successive juvenile stages. The cuticle is produced by the hypodermis, which consists of living cells and extends into the body cavity as four chords separating four bands of longitudinal muscles. The muscles enable the nematode to move.

The body cavity contains a fluid through which circulation and respiration take place. The digestive system is a hollow tube extending from the mouth through the esophagus, intestine, rectum, and anus. Lips, usually six in number, surround the mouth. Most plant parasitic nematodes have a hollow stylet or spear (Fig. 15-1B), but a few have a solid modified spear. The spear is used to puncture holes in plant cells and through which to withdraw nutrients from the cells.

The reproductive systems of nematodes are well developed. Females have one or two ovaries, followed by an oviduct and uterus terminating in a vulva. The male reproductive structure is similar to that of the female, but there is a testis, seminal vesicle, and a terminus in a common opening with the intestine. A pair of protrusible, copulatory spicules are also present in the male. Reproduction in plant parasitic nematodes is through eggs and may be sexual or parthenogenetic. Many species lack males.

Life Cycles

The life histories of most plant parasitic nematodes are, in general, quite similar (Fig. 15-4). Eggs hatch into juveniles, whose appearance and structure are usually similar to those of the adult nematodes. Juveniles grow in size, and each juvenile stage is terminated by a molt. All nematodes have four juvenile stages, with the first molt usually occurring in the egg (Fig. 15-4B). After the final molt the nematodes differentiate into males and females. The female can then produce fertile eggs either after mating with a male or, in the absence of males, parthenogenetically.

A life cycle from egg to egg may be completed within 2 to 4 weeks under optimum environmental, especially temperature, conditions but will take longer in cooler

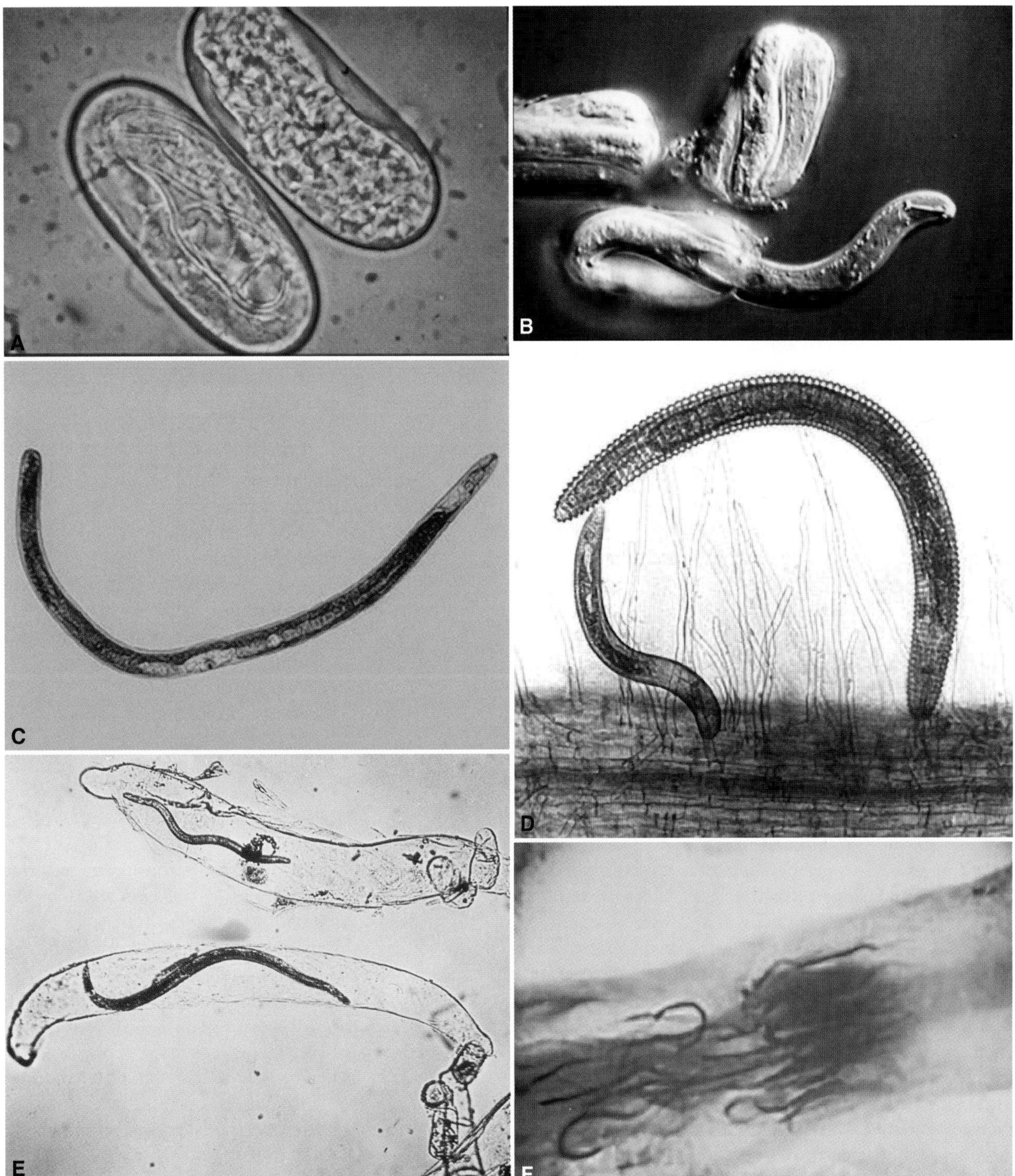

FIGURE 15-4 Stages in a life cycle and the infection process of plant parasitic nematodes. (A) Nematode eggs. (B). Nematode eggs and hatching second-stage juvenile. (C) Typical plant parasitic nematode ready to infect plant. (D) Juvenile and adult ectoparasitic ring nematodes feeding on root. (E) *Aphelenchus* nematodes present inside plant cells. (F) *Radopholus* nematodes feeding inside plant root. [Photographs courtesy of (A, C, E, and F) University of Florida, (B) U. Zunke, (D) S. W. Westcott III.]

temperatures. In some species of nematodes the first or second juvenile stages cannot infect plants and depend on the energy stored in the egg for their metabolic functions. When the infective stages are produced, however, they must feed on a susceptible host (Figs. 15-4D–15-4F) or starve to death. An absence of suitable hosts may result in the death of all individuals of certain nematode species within a few months, but in other species the juvenile stages may dry up and remain quiescent or the eggs may remain dormant in the soil for years.

Ecology and Spread

Almost all plant pathogenic nematodes live part of their lives in the soil. Many live freely in the soil, feeding superficially on roots and underground stems, and in all, even in the specialized sedentary parasites, the eggs, the preparasitic juvenile stages, and the males are found in the soil for all or part of their lives. Soil temperature, moisture, and aeration affect survival and movement of nematodes in the soil. Nematodes occur in greatest abundance in the top 15 to 30 centimeters of soil. The distribution of nematodes in cultivated soil is usually irregular and is greatest in or around the roots of susceptible plants, which they follow sometimes to considerable depths (30–150 centimeters or more). The greater concentration of nematodes in the region of host plant roots is due primarily to their more rapid reproduction on the food supply available and also to attraction of nematodes by substances released into the rhizosphere. To these must be added the so-called hatching factor effect of substances originating from the root that diffuse into the surrounding soil, markedly stimulating the hatching of eggs of certain species. Most nematode eggs, however, hatch freely in water in the absence of any special stimulus.

Nematodes spread through the soil slowly under their own power. The overall distance traveled by a nematode probably does not exceed a few meters per season. Nematodes move faster in the soil when the pores are lined with a thin film of water (a few micrometers thick) than when the soil is waterlogged. In addition to their own movement, however, nematodes can be spread easily by anything that moves and can carry particles of soil. Farm equipment, irrigation, flood or drainage water, animal feet, birds, and dust storms spread nematodes in local areas, whereas over long distances nematodes are spread primarily with farm produce and nursery plants. A few nematodes that attack the aboveground parts of plants not only spread through the soil as described earlier, but they are also splashed to the plants by falling rain or overhead watering. Some species ascend wet plant stem or leaf surfaces on their own power. Further spread takes place on contact of infected plant parts with adjacent healthy plants.

Two genera of the family Aphelenchoididae, namely *Aphelenchoides* (bud and leaf nematodes) and *Bursaphelenchus* (the pine wilt and red-ring nematodes), seldom, if ever, enter the soil. They survive instead in the tissues of the plants they infect and, for the latter, in its insect vectors.

Classification

All plant parasitic nematodes (Fig. 15-3) belong to the phylum Nematoda. Most of the important parasitic genera belong to the order Tylenchida, but a few belong to the order Dorylaimida.

Phylum: Nematoda
- Order: Tylenchida
  - Suborder: Tylenchina
    - Superfamily: Tylenchoidea
      - Family: Anguinidae
        - Genus: *Anguina*, wheat or seed-gall nematode
        - *Ditylenchus*, stem or bulb nematode of alfalfa, onion, narcissus, etc.
      - Family: Belonolaimidae
        - Genus: *Belonolaimus*, sting nematode of cereals, legumes, cucurbits, etc.
        - *Tylenchorhynchus*, stunt nematode of tobacco, corn, cotton, etc.
      - Family: Pratylenchidae
        - Genus: *Pratylenchus*, lesion nematode of almost all crop plants and trees
        - *Radopholus*, burrowing nematode of banana, citrus, coffee, sugarcane, etc.
        - *Nacobbus*, false root-knot nematode
      - Family: Hoplolaimidae
        - Genus: *Hoplolaimus*, lance nematode of corn, sugarcane, cotton, alfalfa, etc.
        - *Rotylenchus*, spiral nematode of various plants
        - *Heliocotylenchus*, spiral nematode of various plants
        - *Rotylenchulus*, reniform nematode of cotton, papaya, tea, tomato, etc.
        - *Scutellonema*, dry rot nematode of yam, cassava, etc.
      - Family: Heteroderidae
        - Genus: *Globodera*, round cyst nematode of potato
        - *Heterodera*, cyst nematode of tobacco, soybean, sugar beets, cereals

Meloidogyne, root-knot nematode of almost all crop plants
Superfamily: Criconematoidea
Family: Criconematidae
Genus: *Criconemella*, formerly *Criconema* and *Criconemoides*, ring nematode of woody plants, cause of peach tree short life
Hemicycliophora, sheath nematode of various plants
Family: Paratylenchidae
Genus: *Paratylenchus*, pin nematode of various plants
Family: Tylenchulidae
Genus: *Tylenchulus*, citrus nematode of citrus, grapes, olive, lilac, etc.
Suborder: Aphelenchina
Family: Aphelenchoididae
Genus: *Aphelenchoides*, foliar nematode of chrysanthemum, strawberry, begonia, rice, coconut, etc.
Bursaphelenchus, the pine wilt and the coconut palm or red ring nematodes
Order: Dorylaimida
Family: Longidoridae
Genus: *Longidorus*, needle nematode of some plants
Xiphinema, dagger nematode of trees, woody vines, and many annuals
Family: Trichodoridae
Genus: *Paratrichodorus*, stubby-root nematode of cereals, vegetables, cranberry, and apple
Trichodorus, stubby-root nematode of sugar beet, potato, cereals, and apple

In terms of habitat, pathogenic nematodes are either **ectoparasites**, i.e., species that do not normally enter root tissue but feed only from the outside on the cells near the root surfaces (Fig. 15-4D), or **endoparasites**, i.e., species that enter the host and feed form within (Figs. 15-4E–15-4F). Both of these can be either **migratory**, i.e., they live freely in the soil and feed on plants without becoming attached or move around inside the plant, or **sedentary**, i.e., species that, once within a root, do not move about. Ectoparasitic nematodes include the ring nematodes (sedentary) and the dagger, stubby root, and sting nematodes (all migratory). Endoparasitic nematodes include the root knot, cyst, and citrus nematodes (all sedentary), and the lesion, stem and bulb, burrowing, leaf, stunt, lance, and spiral nematodes (all somewhat migratory). Of these, the cyst, lance, and spiral nematodes may be somewhat ectoparasitic, at least during part of their lives.

ISOLATION OF NEMATODES

Plant parasitic nematodes are generally isolated from the roots of plants they infect or from the soil surrounding the roots on which they feed (Fig. 15-5). A few kinds of nematodes, however, attack aboveground plant parts, e.g., chrysanthemum foliar nematode, grass seed-gall nematode, and the stem, leaf, and bulb nematode, and these nematodes can be isolated primarily from the plant parts they infect.

Isolation of Nematodes from Soil

From a freshly collected soil sample of about 100 to 300 cm^3, the nematodes in it can be isolated either by the Baermann funnel method or by sieving.

A Baermann funnel consists of a fairly large glass funnel (12 to 15 centimeters in diameter) to which a piece of rubber tubing is attached, with a clamp placed on the tubing. The funnel is placed on a stand and filled with water. The soil sample is placed in the funnel on porous, wet-strength paper, sometimes supported by a 5- to 6-centimeter circular piece of screen, or in a beaker over which a piece of cloth is fastened with a rubber band. The beaker is then inverted in the funnel, with the cloth and all the soil being below the surface of the water, and allowed to stand overnight or for several hours. The live nematodes move actively and migrate through the cloth or porous paper into the water and sink to the bottom of the rubber tubing just above the clamp. More than 90% of the live nematodes are recovered in the first 5 to 8 milliliters of water drawn from the rubber tubing, and this sample is placed in a shallow dish for examination and, if desired, single nematode isolation.

The sieving method is based on the fact that when a small soil sample, such as $300\,cm^3$, is mixed with considerably more water, e.g., 2 liters, the nematodes float in the water and can be collected on sieves with pores of certain sizes. Thus, the soil–water mixture is stirred and then allowed to stand for 30 seconds. The liquid is poured through a 20-mesh sieve (20 holes per square inch), which holds large debris but allows the nematodes to pass into a bucket. The liquid containing the nematodes is then poured through a 60-mesh sieve, which holds the larger nematodes and some debris but lets the smaller ones pass through into another bucket. The flow through is then passed through a 200-mesh sieve, which holds the small nematodes and some debris. Both the 60- and the 200-mesh sieves are washed two or three times to remove as much of the debris as possible, and the nematodes are then washed into shallow dishes for direct examination and further isolation. For further

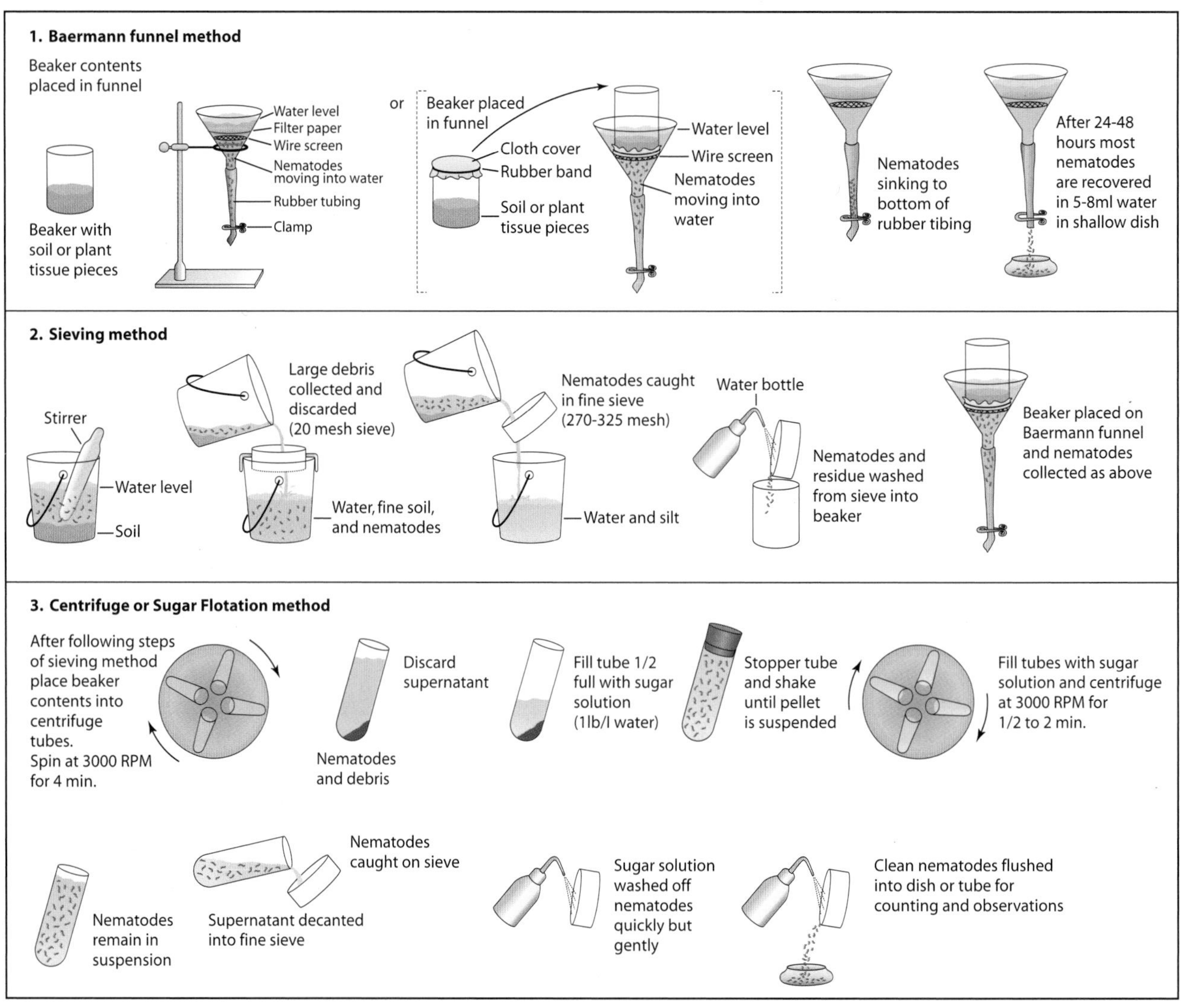

FIGURE 15-5 Methods of isolation of nematodes from soil or plant tissues.

cleaning of nematodes collected through the Baermann funnel or the sieving method, the nematodes are subjected to a combination of centrifugal flotation in a sugar solution, as shown in Fig. 15-5. A semiautomatic elutriator developed in the 1980s combines the steps just described into one continuous process, providing a much improved soil-mixing step as well as requiring less labor for operation.

Isolation of Nematodes from Plant Material

Regardless of the type of plant material containing nematodes, nematode isolation from plants begins by cutting the material into very small pieces by hand or with the use of a blender for a few seconds. The tissue is then placed in the Baermann funnel as described earlier. The nematodes leave the tissue and move into the water in the tubing, from which they are collected in a shallow dish.

SYMPTOMS CAUSED BY NEMATODES

Nematode infections of plants result in the appearance of symptoms on roots as well as on the aboveground parts of plants (Fig. 15-6). Root symptoms may appear as root lesions (Figs. 15-7A and 15-7C), root knots or root galls (Fig. 15-7E), excessive root branching, injured root tips, and, when nematode infections are accompanied by plant pathogenic or saprophytic bacteria and fungi, as root rots. The root symptoms are usually accompanied by noncharacteristic symptoms in the aboveground parts of plants (Figs. 15-7B, 15-7D, and

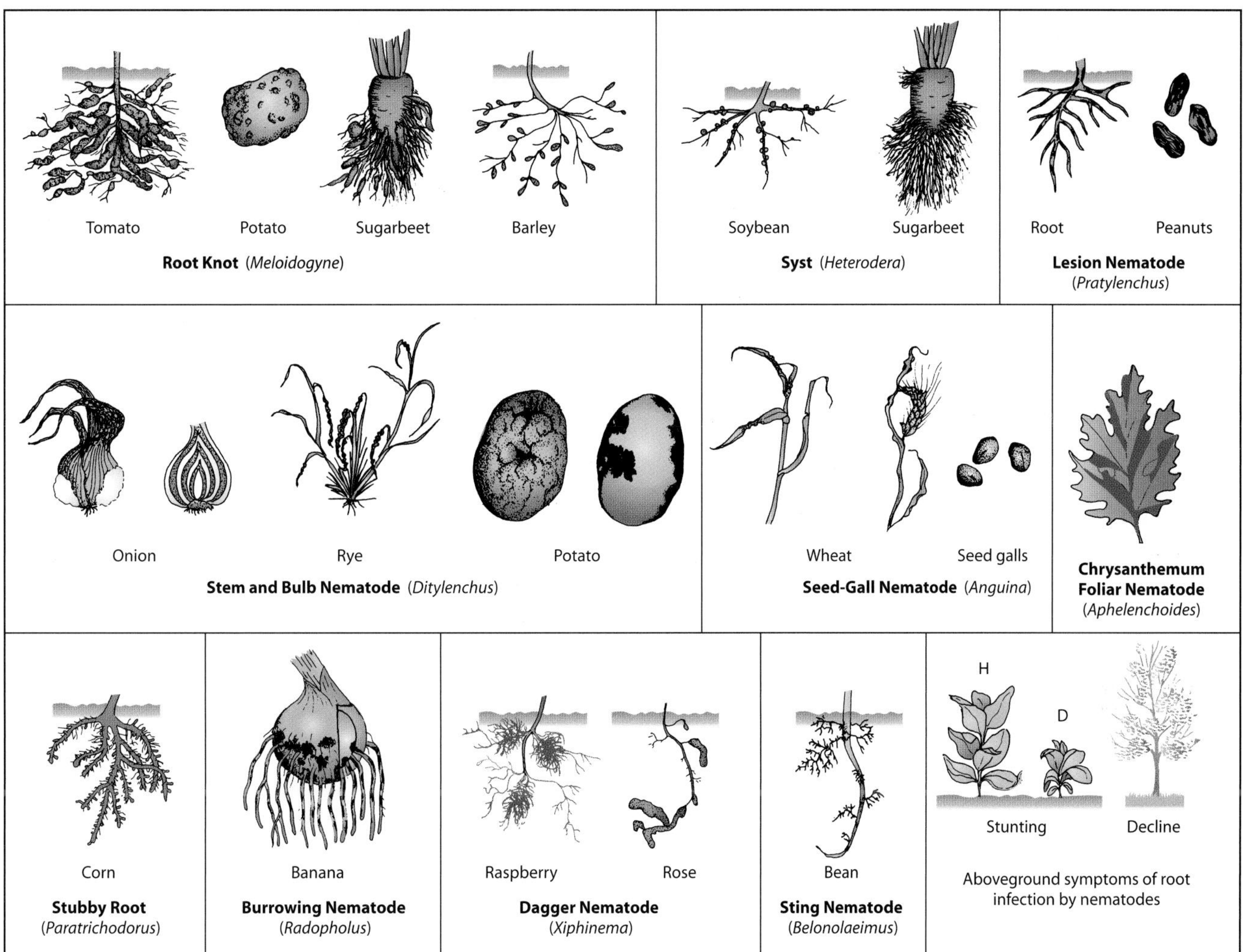

FIGURE 15-6 Types of symptoms caused by some of the most important plant-parasitic nematodes.

15-7F), appearing primarily as reduced growth, symptoms of nutrient deficiencies such as yellowing of foliage, excessive wilting in hot or dry weather, reduced yields, and poor quality of products.

Certain species of nematodes invade the aboveground portions of plants rather than the roots, and on these they cause galls, necrotic lesions and rots, twisting or distortion of leaves and stems, and abnormal development of the floral parts. Certain nematodes attack cereals or grasses and form galls full of nematodes in place of seed.

HOW NEMATODES AFFECT PLANTS

The direct mechanical injury inflicted by nematodes while feeding causes only slight damage to plants. Most of the damage seems to be caused by a secretion of saliva injected into the plants while the nematodes are feeding. Some nematode species are rapid feeders. They puncture a cell wall, inject saliva into the cell, withdraw part of the cell contents, and move on within a few seconds. Others feed much more slowly and may remain at the same puncture for several hours or days. These, as well as the females of species that become established in or on roots permanently, inject saliva intermittently as long as they are feeding.

The feeding process causes the affected plant cells to react, resulting in dead or devitalized root tips and buds, lesion formation and tissue breakdown, swellings and galls of various kinds, and crinkled and distorted stems and foliage. Some of these manifestations are caused by the dissolution of infected tissues by nematode enzymes, which, with or without the help of toxic metabolites, cause tissue disintegration and the death of cells. Others are caused by abnormal cell enlargement (hypertrophy),

FIGURE 15-7 Symptoms on plant roots and on plants in the field caused by some nematodes. (A) Lesions on and necrosis of roots. (B) Strawberry plants stressed by nematodes feeding on their roots showing stunting and death of older leaves. (C) Females and cysts of a cyst nematode on the roots of its host plant. (D) Yellowing, stunting, and death of soybean plants in a field patch infested with the soybean cyst nematode. (E) Galls on tomato root caused by the root knot nematode. (F) Stunting and death of cotton plants in patch of field infested with nematodes compared to the adjacent area treated with a nematicide. [Photographs courtesy of (A) K. R. Baker, (B) J. Noling, (C) W.C.P.D., (D) G. Tylka, (E) R. Dunn, and (F) C. Overstreet.]

by suppression of cell divisions, or by stimulation of cell division proceeding in a controlled manner and resulting in the formation of galls or of large numbers of lateral roots at or near the points of infection.

Plant diseases caused by nematodes are complex. Root-feeding species often decrease the ability of plants to take up water and nutrients from soil and thus cause symptoms of water and nutrient deficiencies in the aboveground parts of plants. In some cases, however, it is the plant–nematode biochemical interactions that impair the overall physiology of plants, as well as the role nematodes play in providing courts for entry of other pathogens, that are primarily responsible for plant injury. The mechanical damage or withdrawal of food from plants by nematodes is generally less significant but may become all important when nematode populations become very large.

INTERRELATIONSHIPS BETWEEN NEMATODES AND OTHER PLANT PATHOGENS

Although nematodes can cause diseases to plants by themselves, most of them live and operate in the soil, where they are constantly surrounded by fungi and bacteria, many of which can also cause plant diseases. In many cases an association develops between nematodes and certain of the other pathogens. Nematodes then become a part of an etiological complex, resulting in a combined pathogenic potential that sometimes appears to be far greater than the sum of the damages either of the pathogens can produce individually.

Several nematode–fungus disease complexes are known. Fusarium wilt of several plants increases in incidence and severity when the plants are also infected by root-knot, lesion, sting, reniform, burrowing, or stunt nematodes. Similar effects have also been noted in disease complexes involving nematodes and *Verticillium* wilt, *Pythium* damping off, *Rhizoctonia* and Phytophthora root rots, and in some other instances. For example, in the potato early dying syndrome, potato plants can become infected by *Verticillium dahliae* alone and may wilt and die. If, however, the plants are also infected with even small populations of the lesion nematode *Pratylenchus penetrans*, then even small amounts of fungus in the plant are activated and cause early wilting and death of the potato plant. In none of these cases is the fungus transmitted by the nematode. However, plant varieties susceptible to the respective fungi are damaged even more when the plants are infected with the nematodes, with the combined damage being considerably greater than the sum of the damages caused by each pathogen acting alone. Also, varieties ordinarily resistant to the fungi apparently become infected by them after previous infection by nematodes. The importance of nematodes in these complexes is indicated by the fact that soil fumigation aimed at eliminating the nematode but not the fungus reduces greatly the incidence and the damage caused by the fungus-induced disease.

Relatively few cases of nematode–bacteria disease complexes are known. For example, the root-knot nematode increases the frequency and severity of the bacterial wilt of tobacco caused by *Ralstonia solanacearum*, of the bacterial wilt of alfalfa caused by *Clavibacter michiganense* subsp. *insidiosum*, and of the bacterial scab of gladiolus caused by *Pseudomonas marginata*. In most of these the role of the nematode seems to be that of providing the bacteria with an infection court and to assist bacterial infection by wounding the host. However, root infection of plum trees with the ring nematode *Criconemella xenoplax* changed the physiology of the trees and resulted in the development of more extensive cankers by the bacterium *Pseudomonas syringae* pv. *syringae* on branches of nematode-infected trees than on nematode-free trees.

An interesting interaction has been established between some species of the seed-gall nematode *Anguina* and the phytopathogenic bacterium *Clavibacter toxicus*, which distorts or prevents the normal formation of grass seed heads. The bacterium also produces corynetoxins, which are among the most potent toxins produced in nature and cause lethal neurological convulsions in most domestic animals fed infected grasses and seeds. The amount of toxin in and toxicity of infected grasses seems to be proportional to the number of bacterial cells infected with a bacteriophage virus specific to this bacterium. The role of the nematode seems to be primarily that of a vector of the bacterium from plant to plant and from year to year and in facilitating entry of the bacterium into the host plant. It is not known whether corynetoxins have any effect on the nematode.

Much better known are the interrelationships between nematodes and viruses. Several plant viruses such as grapevine fanleaf virus, tomato ringspot virus, raspberry ringspot virus, and tobacco rattle virus are transmitted through the soil by means of nematode vectors. All these viruses, however, are transmitted by only one or more of five genera of nematodes: *Xiphinema*, *Longidorus*, and *Paralongidorus* transmit only polyhedral viruses, which include most of the nematode-transmitted viruses, whereas *Trichodorus* and *Paratrichodorus* transmit two rod-shaped viruses, tobacco rattle virus and pea early browning virus. These nematodes can transmit the viruses after feeding on infected plants from 1 hour to 4 days. The nematodes remain infective for periods of 2 to 4 months and sometimes even longer. All stages, juvenile and adult nematodes, can transmit viruses. Although nematodes can ingest and carry within them several plant viruses, they can only transmit certain of them to healthy plants, which suggests that there is a close biological association between the nematode vectors and the viruses they can transmit.

CONTROL OF NEMATODES

Several methods of effectively controlling nematodes are available, although certain factors, such as expense and types of crops, may influence the types of control methods employed. Control is usually attempted through cultural practices, such as use of clean planting stock, crop rotation, fallow, and cover crops; through biological control with resistant varieties (Fig. 15-8A) and certain other means, such as organic amendments and natural or genetically engineered antagonistic or parasitic bacteria (Fig. 15-8B) and fungi (Figs. 15-8C and 15-8D); through control by means of physical agents, such as tillage, heat, including solarization, and flooding; and through control with chemicals, such as various types of fumigant (Fig. 15-8E) and nonfumigant nematicides. In practice, a combination of several methods is usually employed for controlling nematode diseases of plants. Since the 1950s, nematicides have been used almost exclusively for the effective control of nematodes in high-value crops such as flowers, vegetables, strawberries, tobacco, and nursery crops. As the number of available nematicides continues to decline drastically and problems of residue toxicity (Fig. 15-8F) increase, other methods of nematode control are becoming increasingly important. Recent development of new technologies, e.g., precision agriculture, nematode identification and assessment of nematode populations, genetic engineering of host resistance, and modern advisory programs through extension or through private crop consultants, are expected to improve the accuracy of nematode diagnoses and of the risk evaluation of potential problems, thereby providing more effective management of nematodes.

Selected References

Anonymous (1972 and annually thereafter). "Commonwealth Institute of Helminthology Descriptions of Plant-Parasitic Nematodes." Commonw. Agric. Bur., Farnham Royal, Bucks, England.

Barker, K. R., and Koenning, S. R. (1998). Developing sustainable systems for nematode management. *Annu. Rev. Phytopathol.* **36**, 165–205.

Barker, K. R., Pederson, G. A., and Windham, G. L. (1998). "Plant and Nematode Interactions." ASA, CSSA, SSA Publishers, Madison, WI.

Bird, D. M., *et al.* (1999). The *Caenorhabditis elegans* genome: A guide in the post-genomic age. *Annu. Rev. Phytopathol.* **37**, 247–265.

Brown, D. J. F., Robertson, W. M., and Trudgill, D. L. (1995). Transmission of viruses by plant nematodes. *Annu. Rev. Phytopathol.* **33**, 223–249.

Brown, R. H., and Kerry, B. R., eds. (1987). "Principles and Practice of Nematode Control in Crops." Academic Press, Sydney.

Chitwood, D. J. (2002). Phytochemical based strategies for nematode control. *Annu. Rev. Phytopathol.* **40**, 221–249.

Davis, E. L., Hussey, R. L. *et al.* (2000). Nematode parasitism genes. *Annu. Rev. Phytopathol.* **38**, 365–396.

Dropkin, V. H. (1980). "Introduction to Plant Nematology." Wiley, New York.

Dropkin, V. H. (1988). The concept of race in phytonematology. *Annu. Rev. Phytopathol.* **26**, 145–161.

Duncan, L. W. (1991). Current options for nematode management. *Annu. Rev. Phytopathol.* **29**, 469–490.

Evans, K., Trudgill, D. L., and Webster, J. M., eds. (1993). "Plant Parasitic Nematodes in Temperate Agriculture." CAB Int., Wallingford, England.

Ferris, H. (1981). Dynamic action thresholds for diseases induced by nematodes. *Annu. Rev. Phytopathol.* **19**, 427–436.

Gheysen, G., and Fenol, C. (2002). Gene expression at nematode feeding sites. *Annu. Rev. Phytopathol.* **40**, 191–219.

Gowen, S. R., and Ahmad, R. (1990). *Pasteuria penetrans* for control of pathogenic nematodes. *Aspects Appl. Biol.* **24**, 25–32.

Hussey, R. S. (1989). Disease-inducing secretions of plant-parasitic nematodes. *Annu. Rev. Phytopathol.* **27**, 127–141.

Luc, M., Sikora, R. A., and Bridge, J., eds. (1990). "Plant Parasitic Nematodes in Subtropical and Tropical Agriculture." CAB Int., Wallingford, England.

McKay, A. C., and Ophel, K. M. (1993). Toxicogenic *Clavibacter/Anguina* associations infecting grass seedheads. *Annu. Rev. Phytopathol.* **31**, 151–167.

Nickle, W. R., ed. (1991). "Manual of Agricultural Nematology." Dekker, New York.

Sayre, R. M., and Walter, D. E. (1991). Factors affecting the efficacy of natural enemies of nematodes. *Annu. Rev. Phytopathol.* **29**, 149–166.

Schenk, S., and Haltzman, O. V. (1990). Evaluation of potential problems in a changing agricultural system: Nematode control in Hawaiian crops. *Plant Dis.* **74**, 837–843.

Sijmons, P. C., Atkinson, H. J., and Wyss, U. (1994). Parasitic strategies of root nematodes and associated host cell responses. *Annu. Rev. Phytopathol.* **32**, 235–259.

Stirling, G. R. (1991). "Biological Control of Plant Parasitic Nematodes." CAB Int., Wallingford, England.

Trudgill, D. L. (1991). Resistance to and tolerance of plant parasitic nematodes of plants. *Annu. Rev. Phytopathol.* **29**, 167–192.

Trudgill, D. L., and Block, V. C. (2001). Apomictic, polyphagous root-knot nematodes: Exceptionally successful and damaging biotrophic root pathogens. *Annu. Rev. Phytopathol.* **39**, 53–77.

Veech, J. A., and Dickson, D. W., eds. (1987). "Vistas on Nematology." Soc. of Nematologists, Hyattsville, Maryland.

FIGURE 15-8 Some methods of managing/controlling plant parasitic nematodes. (A) Soybean cultivars resistant (upper left) and susceptible to the cyst nematode. Biological control of the nematode by the bacterium *Pasteuria* sp. (B), by another nematode (C), and by a fungus (D). (E) Control of nematodes with chemicals covered with plastic. (F) Corn seedlings damaged (yellow, stunted) by residual nematicides applied the previous year. [Photographs courtesy of (A) J. P. Ross, (B) R. Sayre, USDA, (C) University of Florida, (D) G. Barron, and (E and F) J. Noling.]

ROOT-KNOT NEMATODES: *MELOIDOGYNE* SPP.

Root-knot nematodes occur throughout the world, especially in areas with warm or hot climates and short or mild winters, and in greenhouses everywhere. They attack more than 2,000 species of plants, including almost all cultivated plants, and reduce world crop production by about 5%. Losses in individual fields, however, may be much higher.

Root-knot nematodes damage plants by devitalizing root tips and causing the formation of swellings of the roots. These effects not only deprive plants of nutrients, but also disfigure and reduce the market value of many root crops. When susceptible plants are infected at the seedling stage, losses are heavy and may result in complete destruction of the crop. Infections of older plants may have only slight effects on yield or may reduce yields considerably.

Symptoms

Aboveground symptoms are reduced growth and fewer, small, pale green, or yellowish leaves that tend to wilt in warm weather. Blossoms and fruits are few and of poor quality. Affected plants usually linger through the growing season and are seldom killed prematurely.

Characteristic symptoms of the disease appear on the underground parts of the plants. Infected roots develop the typical root-knot galls that are two to several times as large in diameter as the healthy root (Figs. 15-9A and 15-9G). Several infections along the root give the root a rough, clubbed appearance. Roots infected by certain species of the nematode also develop a bushy root

FIGURE 15-9 Galls and other symptoms caused by the root-knot nematode on tomato (A), carrots (B), potato (C), peanuts (D), yam (F), and a dogwood tree (G). (E) Healthy yams. [Photographs courtesy of (A, C, and D) D. W. Dickson, (B) D. Ormrod, W.C.P.D., (E and F) D. Coyne, IITA, Nigeria, and (G) E. L. Barnard.]

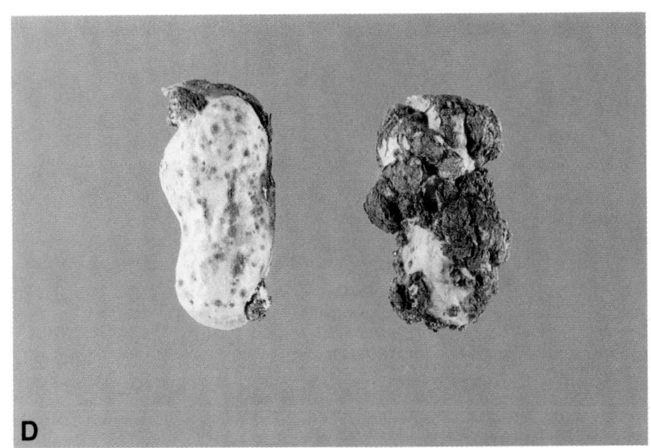

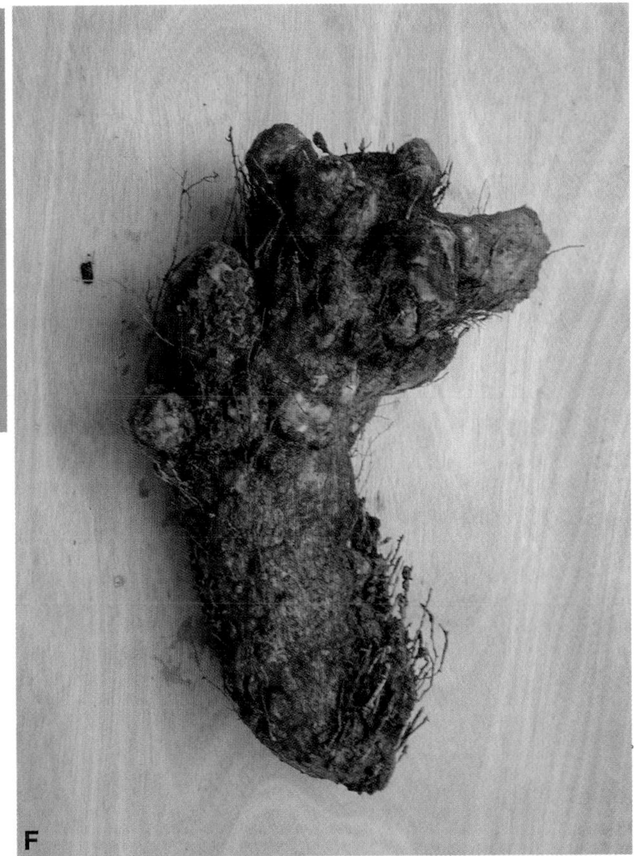

FIGURE 15-9 (*Continued*)

system (Fig. 15-9B). Usually, however, infected roots remain smaller and show necrosis and rotting, particularly late in the season. When tubers or other fleshy underground organs, such as carrots, potatoes, peanuts, and yam, are attacked, they produce small swellings over their surface, which become quite prominent and cause distortion or cracking (Figs. 15-9B–15-9D and 15-9F). Roots of trees are also attacked by the root-knot nematodes and develop galls (Fig. 15-9G) roughly proportional in size to the length of time since infection.

The Pathogen: Meloidogyne *spp.*

The male and female root-knot nematodes are easily distinguishable morphologically (Figs. 15-10 and 15-11). The males are wormlike and about 1.2 to 1.5 millimeters long by 30 to 36 micrometers in diameter.

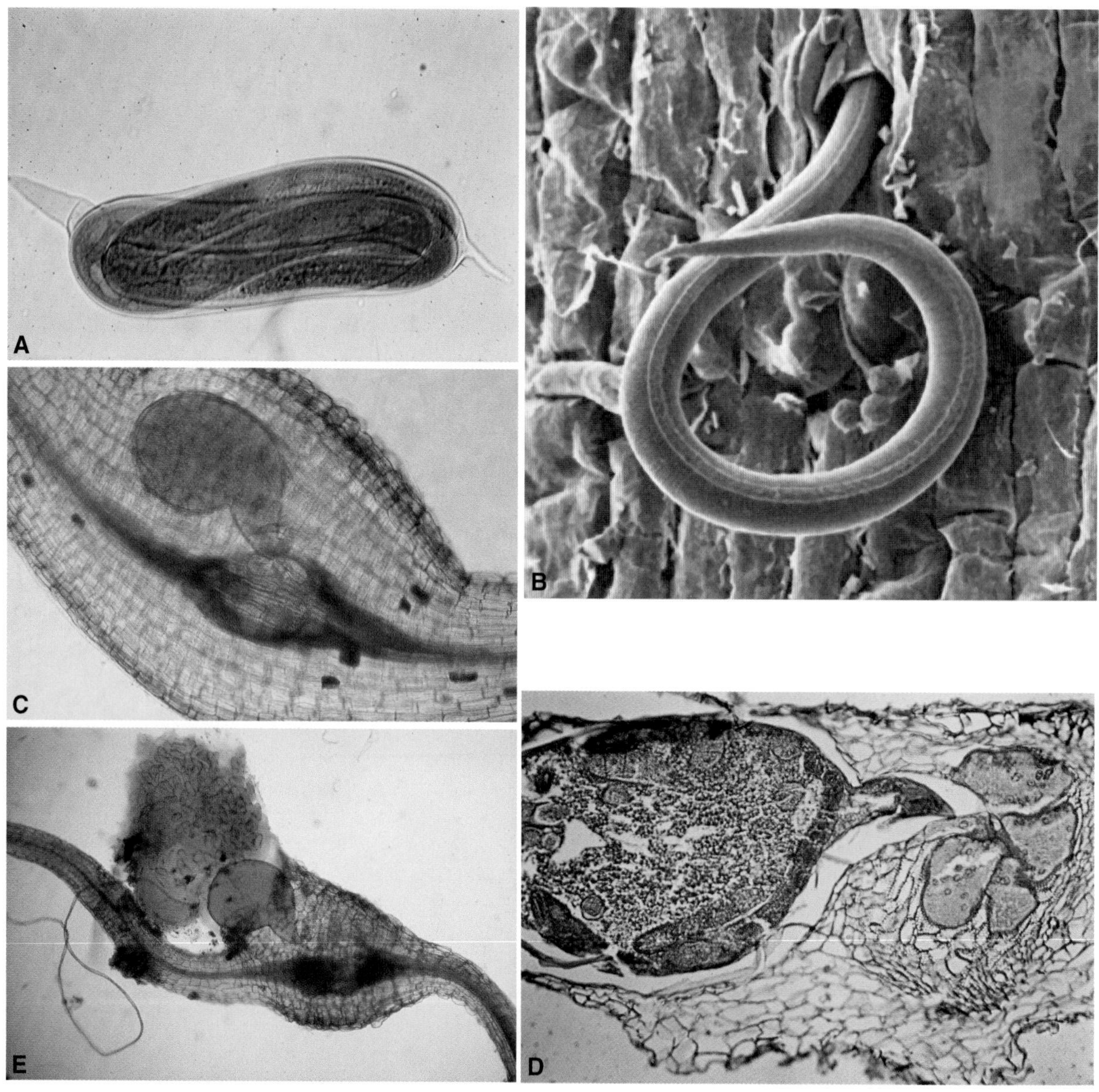

FIGURE 15-10 Stages in the life cycle of the root-knot nematode. (A) Nematode egg with second-stage juvenile ready to hatch. (B) Second-stage juvenile penetrating root tissues. (C) Female root-knot nematode in plant root causing the formation of and feeding on "giant cells." (D) Longitudinal section of *Meloidogyne* female feeding on giant cells. (E) Root-knot female laying eggs outside the root. [Photographs courtesy of (A) D. W. Dickson, (B) USDA, and (C–E) R. A. Rohde.]

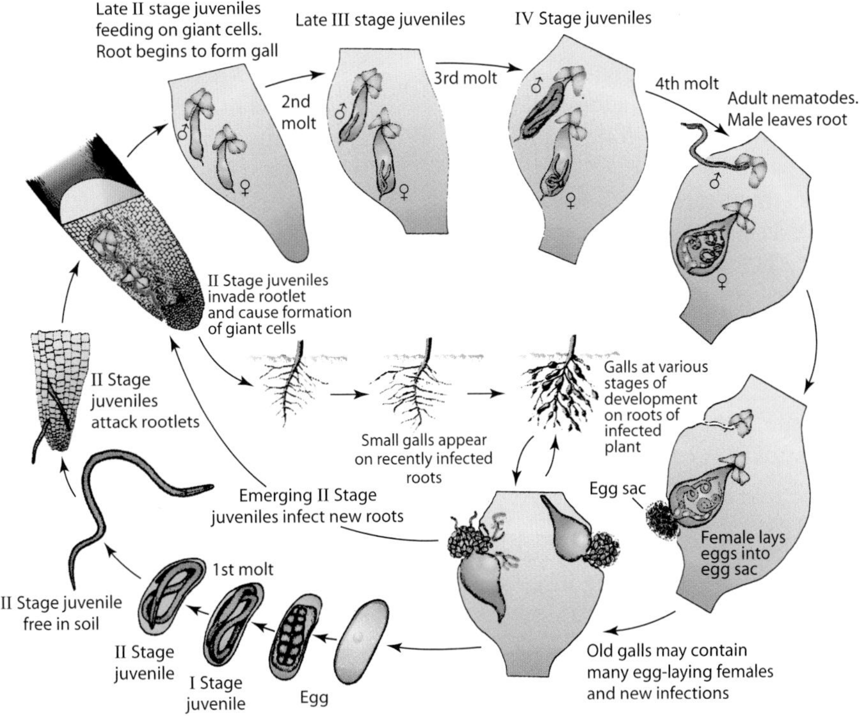

FIGURE 15-11 Disease cycle of root knot caused by nematodes of the genus *Meloidogyne.*

The females are pear shaped and about 0.40 to 1.30 millimeters long by 0.27 to 0.75 millimeters wide.

Each female lays approximately 500 eggs in a gelatinous substance. The first- and second-stage juveniles are wormlike and develop inside each egg (Fig. 15-10A). The second-stage juvenile emerges from the egg into the soil. This is the only infective stage of the nematode. If it reaches a susceptible host, the juvenile enters the root (Fig. 15-10B), becomes sedentary, and grows thick like a sausage. The nematode feeds on the cells around its head by inserting its stylet and secreting saliva into the cells. The saliva stimulates cell enlargement (Figs. 15-10C–15-10E) and also liquefies part of the contents of the cells, which are then withdrawn by the nematode through its stylet.

The nematode then undergoes a second molt and gives rise to the third-stage juvenile, which is stouter and goes through the third molt and gives rise to the fourth-stage juvenile, which can be distinguished as either male or female (Fig. 15-10C). These undergo the fourth and final molt and the male emerges from the root as the worm-like adult male, which becomes free-living in the soil, while the female continues to grow in thickness and somewhat in length and appears pear shaped. The female continues to swell and, with or without fertilization by a male, produces eggs that are laid in a gelatinous protective coat inside or outside the root tissues, depending on the position of the female (Fig. 15-10E). Eggs may hatch immediately or a few of them may overwinter and hatch in the spring.

A life cycle is completed in 25 days at 27°C, but it takes longer at lower or higher temperatures. When the eggs hatch, the infective second-stage juveniles migrate to adjacent parts of the root and cause new infections in the same root or infect other roots of the same plants or roots of other plants. Most root-knot nematodes are found in the root zone from 5 to 25 centimeters below the surface. Root-knot nematodes are spread primarily by water or by soil clinging to farm equipment or on infected propagating stock transported into uninfested areas.

Development of Disease

Second-stage juveniles enter roots behind the root tip and keep moving until they reach positions behind the growing point. There, they settle with their head in the developing vascular cylinder (Figs. 15-10 and 15-11). In older roots the head is usually in the pericycle. Cells near the path of the juveniles begin to enlarge. Two or 3 days after the juvenile has become established, some of the cells around its head begin to enlarge. Their nuclei divide, but no cell walls are laid down. The existing walls between some of the cells break down and disappear, giving rise to giant cells (Figs. 15-10C, 15-10D, and 15-11). Enlargement and coalescing of cells continues for 2 to 3 weeks, and the giant cells invade the surrounding tissues irregularly. Each gall usually contains three to six giant cells, which are due to substances contained in the saliva secreted by the nematode in the giant cells during feeding.

The giant cells attract nutrients from surrounding cells and serve as feeder cells for the nematode. The giant cells crush xylem elements already present but degenerate when nematodes cease to feed or die. In the early stages of gall development the cortical cells enlarge in size and, later, they also divide rapidly. Swelling of the root results from excessive enlargement and division of all types of cells surrounding the giant cells and from enlargement of the nematode. As the females enlarge and produce their egg sacs, they push outward, split the cortex, and may become exposed on the surface of the root (Fig. 15-10E) or remain completely covered, depending on the position of the nematode in relation to the root surface.

In addition to the disturbance caused to plants by the nematode galls themselves, damage to infected plants is frequently increased by certain parasitic fungi, which can easily attack the weakened root tissues and the hypertrophied, undifferentiated cells of the galls. Moreover, some fungi, e.g., *Fusarium*, *Rhizoctonia*, and the oomycete *Pythium*, grow and reproduce much faster in the galls than in other areas of the root, thus inducing an earlier breakdown of the root tissues.

Control

Root knot can be controlled effectively in the greenhouse with steam sterilization of the soil or soil fumigation with nematicides. In the field the best control of root knot is obtained by fumigating the soil with approved chemical nematicides. Each treatment usually gives satisfactory control of root knot for one season. In several crops, varieties resistant to root-knot nematodes are also available. Transgenic plants producing inhibitors to certain nematode proteinases have shown promising resistance to the nematode and their use may prove practical in the future. Several cultural practices, such as crop rotation, fallow soil, soil solarization, and certain soil amendments, are also helpful in reducing root-knot losses. Biological control of root knot has been obtained experimentally by treating nematode-infested soil with endospores of the bacterium *Pasteuria penetrans*, which is an obligate parasite of some plant parasitic nematodes, or with preparations of the fungus *Trichoderma harzianum*; by treating transplants or infested soils with spores of the fungus *Dactylella oviparasitica*, which parasitizes the eggs of *Meloidogyne* nematodes; and in some experiments by treating transplants or infested soils with spores of the vesicular-arbuscular mycorrhizal fungi *Gigaspora* and *Glomus*. Fairly good experimental control of root knot has also been obtained by mixing essential oils from plant spices into nematode-infested soil before planting and through an increase in plants of their local and systemic-induced resistance to root knot nematodes by mixing in the soil or spraying the plants with amino-butyric acid and other amino acids.

Selected References

Barker, K. R., Carter, C. C., and Sasser, J. N., eds. (1985). "An Advanced Treatise on *Meloidogyne*," Vol. 2. Department of Plant Pathology, North Carolina State University, and U.S. A.I.D., Raleigh, North Carolina.

Roberts, P. A. (1995). Conceptual and practical aspects of variability in root knot nematodes related to host plant resistance. *Annu. Rev. Phytopathol.* **33**, 199–221.

Sasser, J. N., and Carter, C. C., eds. (1985). "An Advanced Treatise on *Meloidogyne*," Vol. 1. North Carolina State University Graphics, Raleigh.

Sasser, J. N., *et al.* (1983). The international *Meloidogyne* project — Its goals and accomplishments. *Annu. Rev. Phytopathol.* **21**, 271–288.

Sharon, E., *et al.* (2001). Biological control of the root-knot nematode *Meloidogyne javanica* by *Trichoderma harzianum*. *Phytopathology* **91**, 687–693.

Trudgill, D. L., and Block, V. C. (2001). Apomictic, polyphagous root-knot nematodes: Exceptionally successful and damaging biotrophic root pathogens. *Annu. Rev. Phytopathol.* **39**, 53–77.

Williamson, V. M. (1998). Root-knot nematode resistance genes in tomato and their potential for future use. *Annu. Rev. Phytopathol.* **36**, 277–293.

Wishart, J., Phillips, M. S., and Block, V. C. (2002). Ribosomal intergenic spacer: A polymerase chain reaction diagnostic for *Meloidogyne chitwoodi*, *M. fallax*, and *M. hapla*. *Phytopathology* **92**, 884–892.

CYST NEMATODES: *HETERODERA* AND *GLOBODERA*

Cyst nematodes cause a variety of plant diseases, mostly in the temperate regions of the world. Some species of cyst nematodes attack only a few plant species and are present over limited geographic areas, whereas others attack a large number of plant species and are widely distributed. The round cyst nematode *Globodera rostochiensis* is known as the golden nematode and is particularly severe on potato but also on tomato and eggplant. Other common cyst nematodes and their most important hosts are *Heterodera avenae* on cereals, *H. glycines* on soybeans, *H. schachtii* on sugar beets, crucifers, and spinach, *H. tabacum* on tobacco, and *H. trifolii* on clover. The diagnostic feature of cyst nematode infections is the presence of cysts on the roots and usually the proliferation of roots and production of shallow, bushy root systems.

SOYBEAN CYST NEMATODE: *HETERODERA GLYCINES*

The soybean cyst nematode has been found in northeastern Asia, Japan, and Java, in most soybean-producing states of North America, and in Colombia and Brazil, and it continues to spread slowly to new areas. Several other legumes, such as common bean and forage legumes, and a few nonleguminous plants are also attacked by this nematode. Losses vary from slight to complete destruction of the crop. In heavily infested fields, yield is often reduced from 30 to 75%.

Symptoms

Infected soybean plants growing on sandy soils are stunted and their leaves turn yellow and fall off early. The plants bear only a few flowers and a few small seeds and they usually die (Figs. 15-7D, 15-8A, 15-12A, and 15-12F). Infected plants growing on fertile soils with plenty of moisture show only slight aboveground symptoms and produce a nearly normal yield for a year or two. In subsequent years, however, due to the tremendous buildup of nematodes in the soil, plants in these areas also become severely chlorotic and dwarfed.

The root system of infected plants appears smaller (Fig. 15-12G) and has fewer bacterial nodules than roots of healthy plants. The most characteristic symptom of the disease is the presence of female nematodes in varying stages of development and of mature cysts attached on the soybean roots (Figs. 15-12B and 15-12G). Young females are small, white, and partly buried in the root, with only part of them protruding on the surface (Fig. 15-12B). Older females are larger, almost completely on the surface of the root, and appear yellowish or brown (Figs. 15-12B and 15-12D), depending on maturity. Dead, brown cysts are also present on the roots.

The Pathogen: Heterodera glycines

The soybean cyst nematode overwinters as eggs in brown cysts (Figs. 15-12C and 15-12D), which are the leathery skins of the females, in the upper 90 to 100 centimeters of soil. The eggs contain fully developed second-stage juveniles (Fig. 15-13). When the temperature and moisture become favorable in the spring, the juveniles emerge from the cysts and infect roots of host plants. Numerous races of the pathogen are known.

After penetrating the roots, the juveniles molt and produce the next stage juveniles at 4- to 6-day intervals. The female third- and fourth-stage juvenile becomes stouter and eventually flask shaped (Fig. 15-12C), approximately 0.40 millimeters in length by 0.12 to 0.17 millimeters in width. By days 12 to 15, males and females appear.

The male is wormlike (Fig. 15-12C), about 1.3 millimeters long by 30 to 40 micrometers in diameter. The males remain in the root for a few days, during which they may or may not fertilize the females, and then they move into the soil and soon die.

The females, when fully developed, are lemon shaped, 0.6 to 0.8 millimeters in length and 0.3 to 0.5 millimeters in diameter. They are white to pale yellow at first, becoming yellowish-brown as they mature. The body cavity of the female becomes completely filled with eggs. As the female body distends during egg production, it crushes cortical cells, splits the root surface, and protrudes until it is almost entirely on the root surface. A gelatinous mass surrounds the posterior end of the females, and the nematodes deposit some of their eggs in it. Each female produces 300 to 600 eggs (Fig. 15-12D), most of which remain inside her body when the female dies. Eggs in the gelatinous matrix may hatch immediately, and the emerging second-stage juveniles may cause new infections. Finally, the old body wall, darkening to brown, becomes the cyst that persists in the soil for many years and protects the eggs in it. Approximately 21 to 24 days is required for the completion of a life cycle of this nematode.

Development of Disease

The infective second-stage juveniles penetrate young primary roots or apical meristems of secondary roots (Fig. 15-13). The juveniles pierce their stylets into and feed off cells of the cortex, the endodermis, or the pericycle, causing the enlargement of these cells. The groups of enlarged cells are called **syncytia** and serve as feeder cells for the nematode. Syncytia in contact with developing third- or fourth-stage males begin to degenerate, indicating cessation of feeding. Syncytia in contact with females degenerate after egg deposition.

Syncytia often inhibit secondary growth of both phloem and xylem. Because a short portion of a root may be attacked by many juveniles, the large number of syncytia that develop reduce the conductive elements drastically and result in poor growth and yield of soybean plants, especially under stresses of moisture.

Control

Soil fumigation of soybean cyst nematode-infested fields or soil treatment with nonfumigant nematicides temporarily increases plant growth and soybean yield,

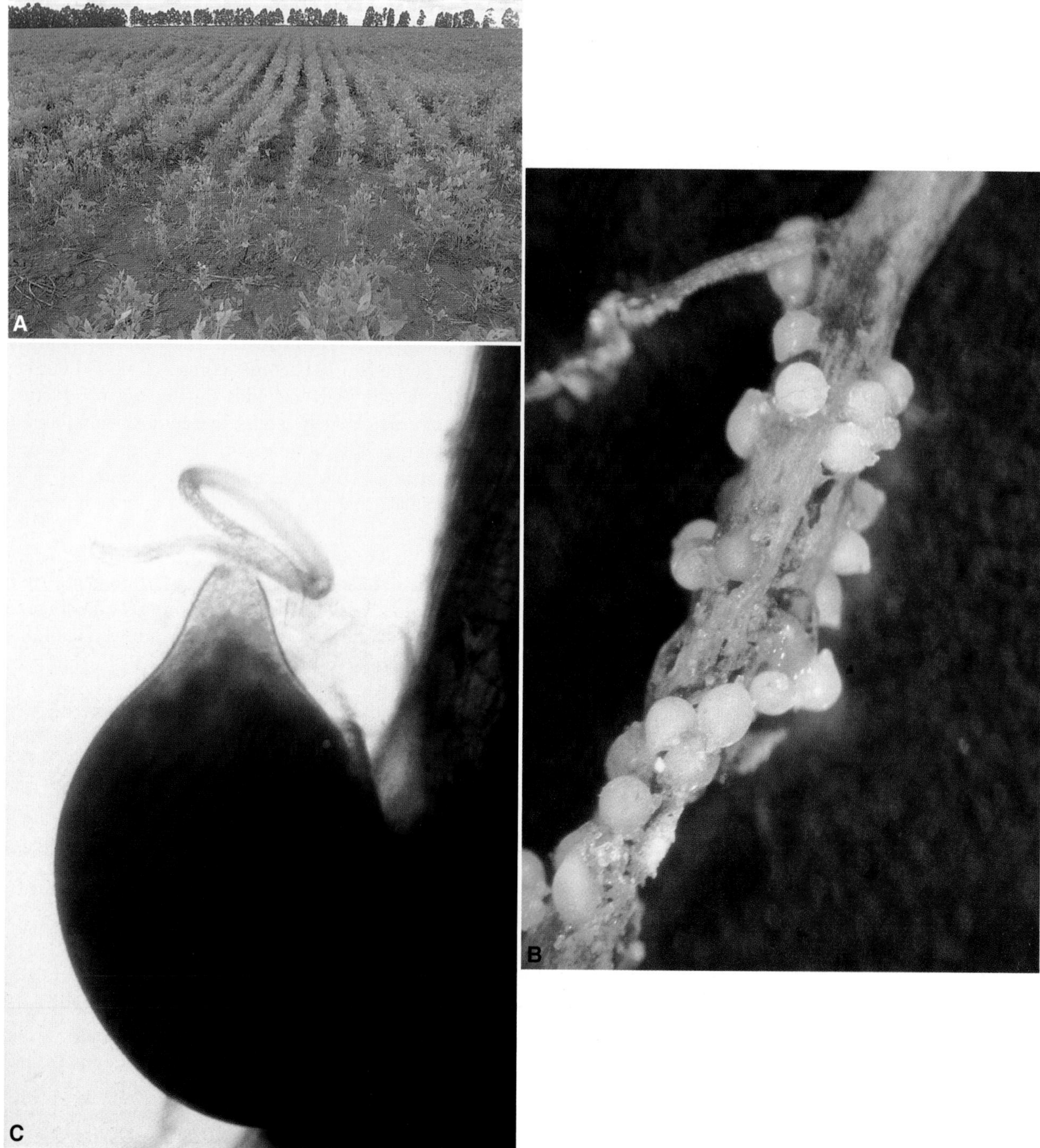

FIGURE 15-12 (A) Damage caused to a patch of soybean plants by the soybean cyst nematode (SCN). (B) Portion of soybean root with several SCN females feeding on it. (C) A flask-shaped female and a worm-like male SCN. (D) A female SCN laying eggs. (E) A female SCN parasitized by the fungus *Verticillium lecanii.* (F) Soybean plants resistant (right) and susceptible (left) to root knot. (G) Root systems from resistant (left) and susceptible (right) plants from the field at F. [Photographs courtesy of (B, F, and G) D. W. Dickson, (C) R. Huettel, (D) D. Chitwood, and (E) S. Meyer.]

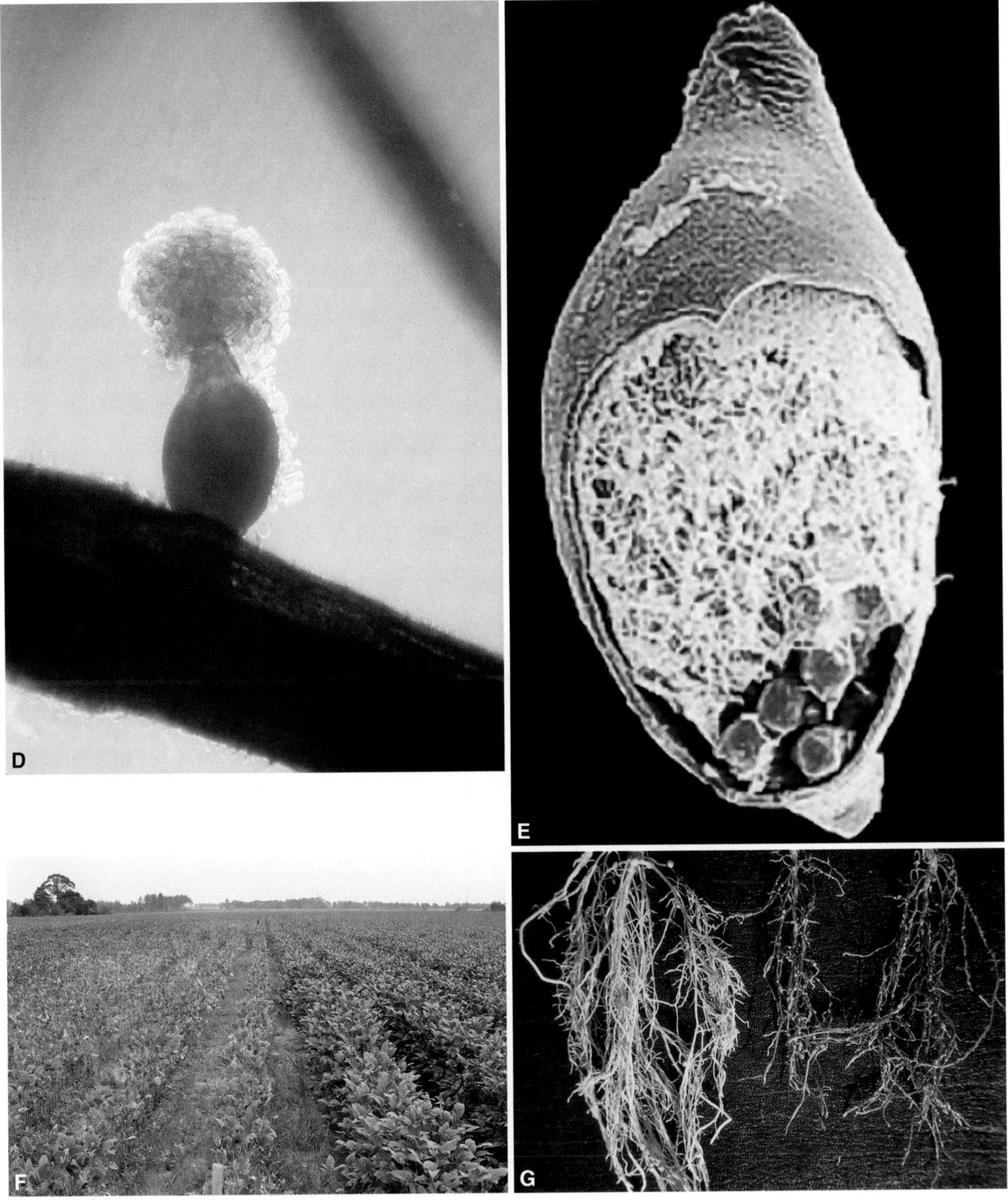

FIGURE 15-12 (*Continued*)

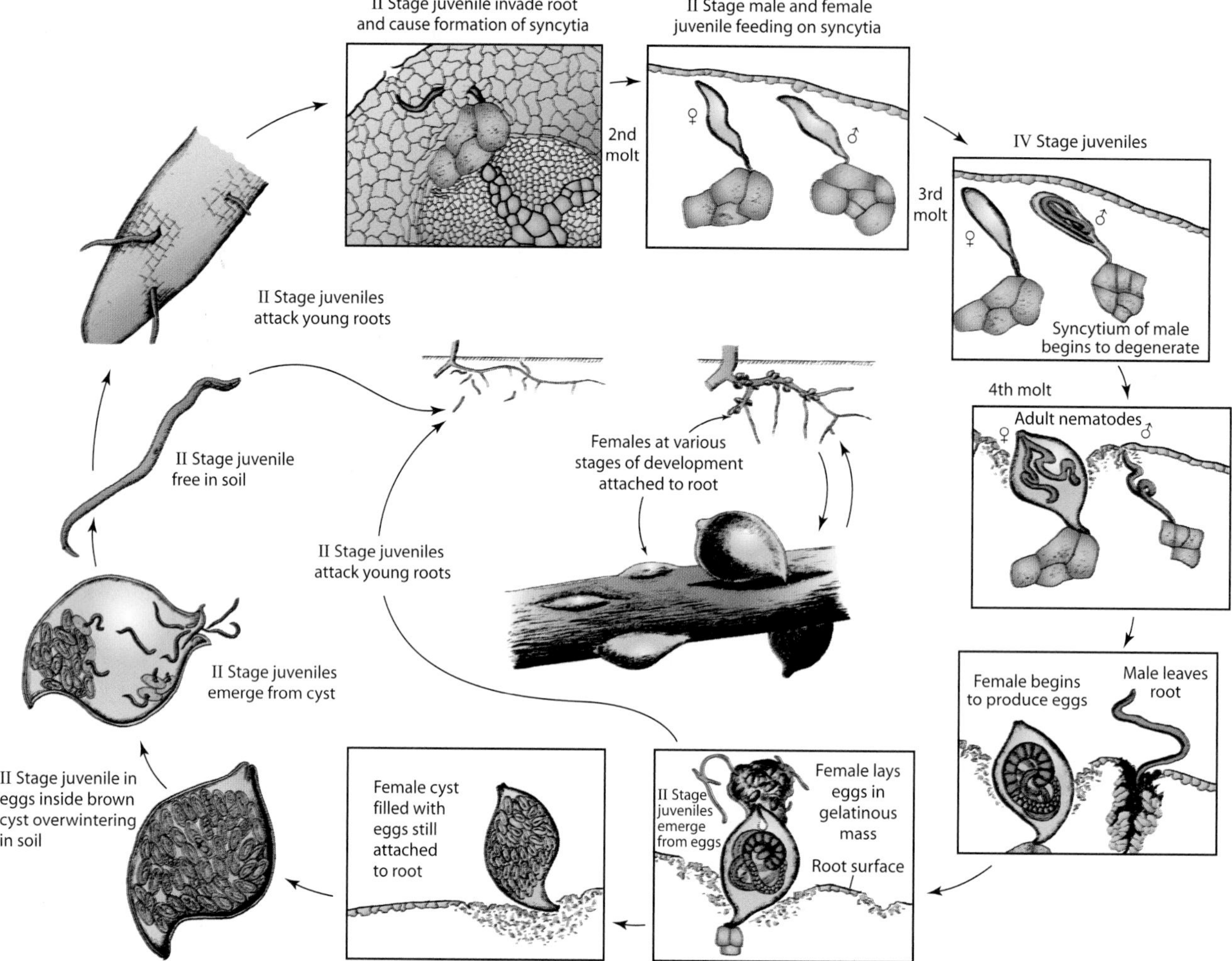

FIGURE 15-13 Disease cycle of the soybean cyst nematode *Heterodera glycines*.

but is not economically viable. Nematode cysts and juveniles, however, are almost never eradicated from a field completely by fumigation, and a small nematode population left over after fumigation can build up rapidly on soybean. In addition, the cost of fumigation per acre makes its use impractical.

The most practical method of control of the soybean cyst nematode is through the use of resistant varieties (Figs. 15-8A, 15-12F, and 15-12G) and through a 1- to 2-year crop rotation with nonhost crops, as some legumes are the only other cultivated crops that are hosts of this nematode. The effectiveness of crop rotation is increased by planting the more resistant soybean varieties, which do not allow a quick and excessive buildup of nematode populations.

Cysts and eggs of soybean and other cyst nematodes are often found infected with one of several fungi such as *Fusarium*, *Verticillium* (Fig. 15-12E), *Neocosmospora*, and *Dictyochaeta*. So far, however, none of the fungi have shown promise as biological control agents of the cyst nematodes.

SUGAR BEET NEMATODE: *HETERODERA SCHACHTII*

The sugar beet nematode occurs wherever sugar beets a re grown in North America, Europe, the Middle East, and Australia and is the most important nematode pest of sugar beet production. It also affects spinach and crucifers. The sugar beet nematode causes yield losses of 25 to 50% or more, especially in warmer climates or late-planted crops. The losses on sugar beet are mostly the result of reduced root weight; however, in warm climates the sugar content is also reduced, and, generally, the nematode aggravates losses caused by other pathogens such as *Cercospora, Rhizoctonia*, and beet viruses.

FIGURE 15-14 (A) Sugar beet field in which a large area of sugar beet plants have been severely stunted or killed by the sugar beet cyst nematode *Heterodera schachtii*. (B) A sugar beet cyst nematode laying its eggs. [Photographs courtesy of (A) R. J. Howard, W.C.P.D., and (B) D. W. Dickson.]

In fields infested with the sugar beet nematode, small to large patches of wilting or dead young plants or stunted older sugar beets appear (Fig. 15-14A). The latter have an excessive number of hair-like roots. Small white or brownish cysts of female nematodes and their eggs (Fig. 15-12B) can be seen clinging to the roots. The morphology, biology, and spread of the sugar beet nematode are similar to that of the soybean cyst nematode. Control of the sugar beet nematode in red table beets is based on several practices: early sowing so that plants can grow as much as possible at temperatures at which the nematodes are more or less inactive; crop rotations with alfalfa, cereals, or potatoes, which are not hosts of this nematode; and soil fumigation with nematicides. No sugar beet varieties of high quality resistant to this nematode are commercially available yet. Some fungi, like those listed earlier for the soybean cyst nematode, have been shown to also reduce populations of the sugar beet nematode, but none is effective as a practical biological control of the disease.

POTATO CYST NEMATODE: *GLOBODERA ROSTOCHIENSIS* AND *GLOBODERA PALLIDA*

It is also known as the "golden cyst nematode." The adult female is virtually spherical, about 450 micrometers in diameter, with a projecting neck and head (Fig. 15-15A). It affects primarily potatoes but also tomato, eggplant, and other solanaceous crops. It occurs in many parts of the world and causes severe losses. In the United States the nematode occurs only in two counties of the state of New York and in Canada only in Newfoundland and in British Columbia. Once the golden nematode infests a field, it is practically impossible to eradicate it because its eggs survive in cysts in the soil for more than 20 years. Therefore, in order to prevent the further spread of the nematode in North America, the areas now infested are under quarantine. Infected plants grow poorly, and the leaves are small and yellowish green and may wilt and die. Infected roots are smaller and a number of nematode cysts visible to the naked eye grow along their length (Fig. 15-15B). Small to large areas of infected plants appear as patches of shorter yellowish plants that have fewer and smaller tubers. Other than reduced size, tubers of infected plants show no symptoms.

Selected References

Bakker, J., *et al.* (1993). Changing concepts and molecular approaches in the management of virulence genes in potato cyst nematodes. *Annu. Rev. Phytopathol.* **31**, 171–192.

Baldwin, J. G. (1992). Evolution of cyst and noncyst-forming Heteroderinae. *Annu. Rev. Phytopathol.* **30**, 271–290.

Brodie, B. B., and Mai, W. F. (1989). Control of the golden nematode in the United States. *Annu. Rev. Phytopathol.* **27**, 443–461.

Colgrove, A. L., *et al.* (2002). Lack of predictable race shift in *Heterodera glycines*-infested field plots. *Plant Dis.* **86**, 1101–1108.

Franklin, M. T. (1972). *Heterodera schachtii*. Commonwealth Institute of Helminthology Descriptions of Plant-Parasitic Nematodes, Set 1, No. 1, pp. 1–4. St. Albans, England.

Gao, B., *et al.* (2001). Identification of putative parasitism genes expressed in the esophageal gland cells of the soybean cyst nematode *Heterodera glycines*. *Mol. Plant-Microbe Interact.* **14**, 1247–1254.

Gipson, I., Kim, K. S., and Riggs, R. D. (1971). An ultrastructural study of syncytium development in soybean roots infected with *Heterodera glycines*. *Phytopathology* **61**, 347–353.

Raski, D. J. (1950). The life history and morphology of the sugar beet nematode *Heterodera schachtii*. *Phytopathology* **40**, 135–152.

FIGURE 15-15 (A) Females (white) and cysts (brown) of the potato golden nematode. (B) White cysts feeding on potato roots. [Photographs courtesy of USDA.]

Riggs, R. D., and Wrather, J. A., eds. (1992). "Biology and Management of the Soybean Cyst Nematode." APS Press, St. Paul, MN.

Wang, J., *et al.* (2000). Soybean cyst nematode reproduction in the north central United States. *Plant Dis.* **84**, 77–82.

THE CITRUS NEMATODE: *TYLENCHULUS SEMIPENETRANS*

The citrus nematode is present and common wherever citrus trees are grown and causes the "slow decline" of citrus. In some regions, in addition to citrus, the nematode, or distinct races of it, also attacks grapevines, olive, lilac, and other plants. Infected trees show a slow decline, i.e., they grow poorly, their leaves turn yellowish and drop early (Fig. 15-16A), their twigs die back, and fruit production is gradually reduced to unprofitable levels.

The Pathogen, Tylenchulus semipenetrans

The pathogen is a semiendoparasitic sedentary nematode. The juveniles and males are wormlike, but the female body is swollen irregularly behind the neck. The nematodes measure about 0.4 millimeters long by 18 to 80 micrometers in diameter, with the larger diameters being found only in the maturing and mature females. The females bury the front end of their body in the root tissue while the rear end remains outside (Figs. 15-16B and 15-16C) and lays eggs in a gelatinous substance. The life cycle of *T. semipenetrans* is completed within 6 to 14 weeks at 24°C. The male juveniles and adults do not feed and apparently do not play a role either in the disease or in the reproduction of the nematode. The second-stage female juvenile is the only infective stage of the nematode and cannot develop without feeding, but it can survive for several years. In the soil, the citrus nematode occurs as deep as four meters.

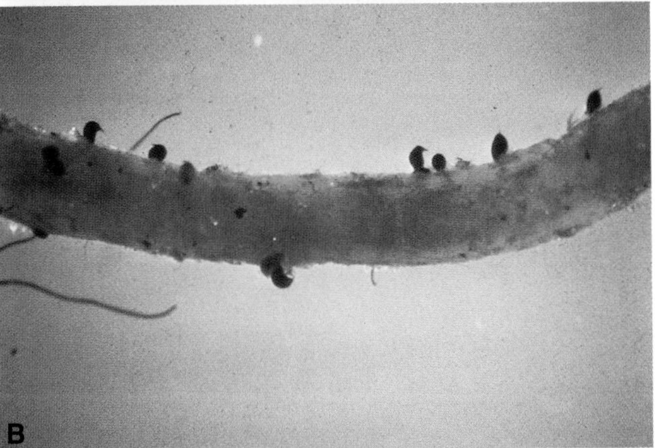

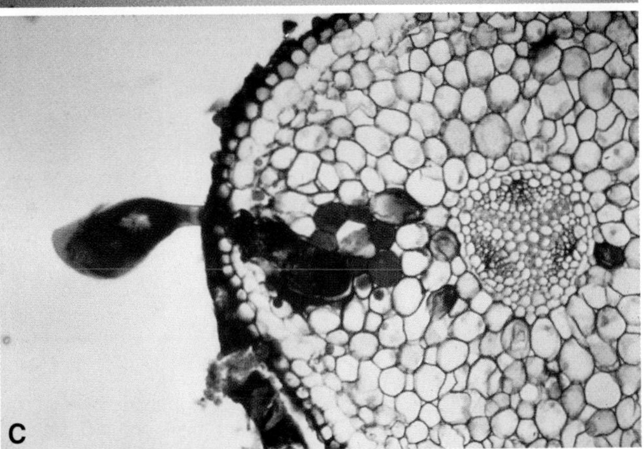

FIGURE 15-16 (A) Citrus tree showing symptoms of slow decline caused by the citrus nematode *Tylenchulus semipenetrans*. (B) Several citrus nematodes feeding on a small root of a citrus tree. (C) Cross section of a citrus root showing the head of a citrus nematode advancing into the root while the rear part of the nematode remains outside the root. [Photographs courtesy of University of Florida.]

Development of Disease

The female second-stage juveniles usually attack young feeder roots and feed on their surface cells. There, they undergo three additional molts and produce females. The young females then penetrate deeper into the cortex and may reach as deep as the pericycle (Fig. 15-16C). The head of the nematode creates a tiny cavity around it and feeds on enlarged parenchyma cells known as nurse cells. Later on, cells around the feeding site become disorganized and break down. After invasion by secondary fungi and bacteria, the affected areas turn into dark, necrotic lesions. In severe infections, 100 or more females may be feeding per centimeter of root. The females, along with soil particles that cling to the gelatinous substance of the egg mass, result in dark, bumpy, and often decayed young roots. The nematodes reach high populations in infected trees, which begin to show decline 3 to 5 years after the initial infection. When the trees show advanced stages of decline, the nematode populations also decline.

Control

The control of the citrus nematode is based on preventing its introduction into new areas by growing nursery stock in nematode-free fields and by treating nursery stock with hot water at 45°C for 25 minutes or with nematicides. Because of the great depth at which the citrus nematode can survive, soil fumigation is not always effective. Satisfactory control has been obtained by preplant fumigation or by postplant treatment with appropriate nematicides. Some citrus clones are resistant to the nematode populations of some regions but not to those of others.

Selected References

Inserra, R. N., *et al.* (1994). Citrus nematode biotypes and resistant citrus rootstocks in Florida. Nematology Circular No. 205. Florida Dept. Agric. and Consumer Services, Div. Pl. Industry.

Le Roux, H. F., Ware, A. B., and Pretorius, M. C. (1998). Comparative efficacy of preplant fumigation and postplant chemical treatment of replant citrus trees in an orchard infested with *Tulenchulus semipenetrans*. *Plant Dis*. **82**, 1323–1327.

Siddiqi, M. R. (1974). *Tylenchulus semipenetrans*. Commonwealth Institute of Helminthology Descriptions of Plant-Parasitic Nematodes, Set 3, No. 34. St. Albans, England.

Van Gundy, S. D. (1985). The life history of the citrus nematode *Tylenchulus semipenetrans*. *Nematologica* **3**, 283–294.

LESION NEMATODES: *PRATYLENCHUS*

Lesion nematodes occur in all parts of the world. They attack the roots of all kinds of plants, such as cereals and other field crops, vegetables, fruit trees, and many ornamentals. Lesion nematodes reduce or inhibit root development by forming local lesions on young roots. Roots with lesions then may rot because of secondary fungi and bacteria. As a result of the root damage, affected plants grow poorly, produce low yields, and may finally die. In potato plants, a synergism between *Pratylenchus* nematodes and the wilt fungus *Verticillium dahliae* leads to the "potato early dying syndrome" that results in reduced yields and premature death of infected plants.

Symptoms

Infected plants appear stunted and chlorotic as though they are suffering from mineral deficiencies or drought. Usually several plants are affected in one area, producing patches of yellowish-green plants that grow poorly (Figs. 15-17A and 15-17B). As the season progresses, plants appear more stunted and the foliage wilts during hot summer days and becomes yellowish brown. Such plants can be pulled easily from the soil because of the extensive destruction of the root system (Fig. 15-17C). Underground organs such as peanuts are also attacked and may be covered by dark lesions. Affected plants have drastically reduced yields and in severe infections the plants are killed.

FIGURE 15-17 Damage in fields of young corn plants (A) and cotton plants (B), on root of tobacco plant (C), and on peanut pods caused by the lesion nematode *Pratylenchus* sp. [Photographs courtesy of (A) G. Tylka, (C) University of Georgia, and (D) D. W. Dickson.]

Infected shrubs and trees show slower and less obvious damage and are rarely killed. Isolated trees or patches of trees gradually become unthrifty and produce poor crops. Their leaves are smaller, dull green or yellow, and may fall off as terminal branches die back. The patches of affected trees increase slowly in size, although this happens over a rather long period.

Infected roots at first show small, water-soaked lesions that soon turn brown to almost black. The lesions appear mainly on the young feeder roots but may appear anywhere along the roots. The lesions enlarge mostly along the root axis, but they also expand and coalesce laterally until they girdle the entire root, which they kill. Affected cortex cells in the lesions collapse, and the lesion area appears constricted. Secondary fungi and bacteria usually invade the lesions and contribute to the discoloration and rotting of the affected root areas, which may slough off. In some hosts, moderately infected plants produce adventitious roots; generally, however, the roots are discolored and stubby, and the whole root system is severely reduced by the root pruning that results from the formations of lesions (Figs. 15-17A and 15-19).

The Pathogen: Pratylenchus *sp.*

Both male and female *Pratylenchus* nematodes are wormlike, 0.4 to 0.7 millimeters long and 20 to 25 micrometers in diameter (Figs. 15-18A–15-18D). They are migratory, endoparasitic nematodes. The life cycle of the various species of *Pratylenchus* is completed within 45 to 65 days. The nematodes overwinter in

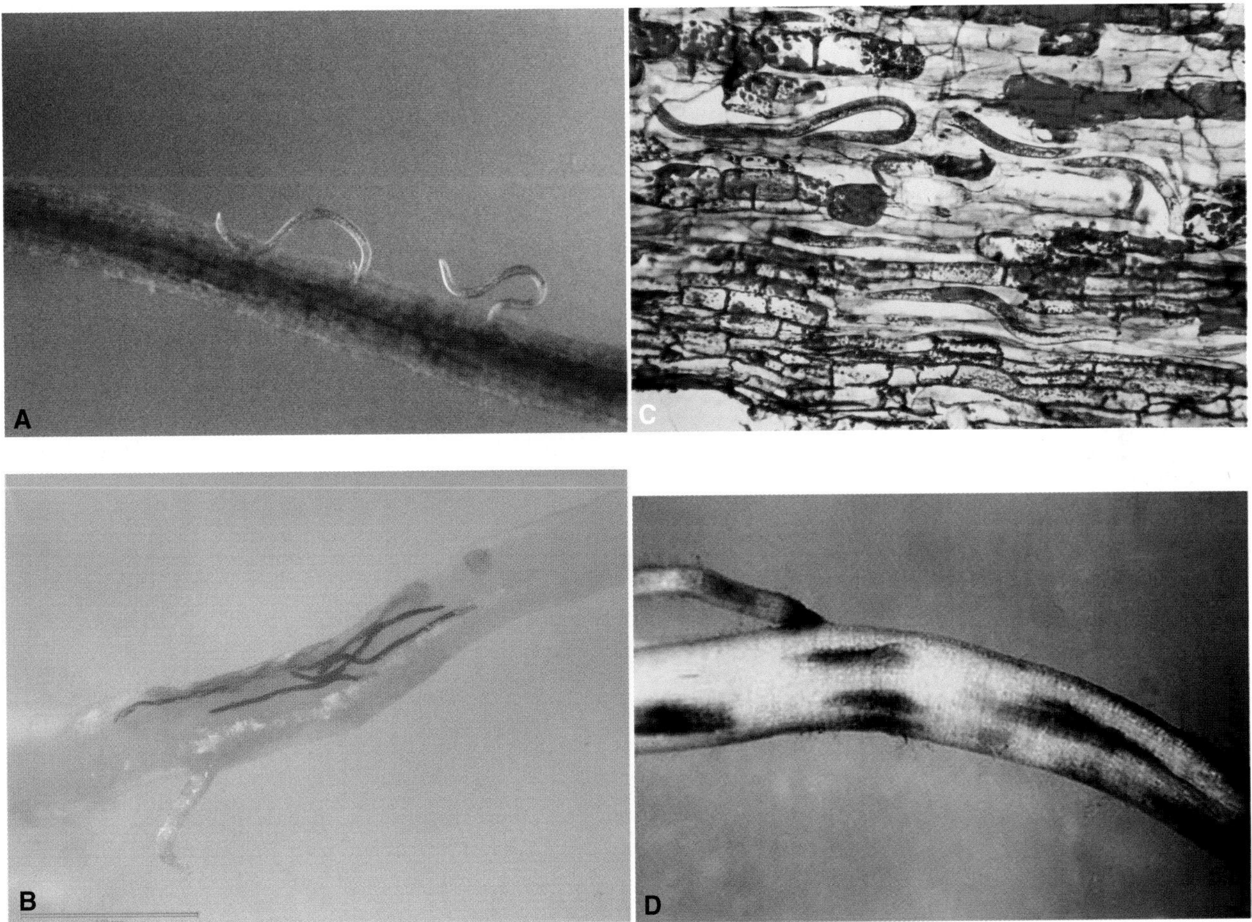

FIGURE 15-18 (A) Two *Pratylenchus* nematodes penetrating on corn root. (B) Nematodes within a tomato root iniciating a lesion. (C) Numerous *Pratylenchus* nematodes within a short segment of a root killing plant cells and leading to the formation of a lesion. (D) External appearance of lesions on young root infected with Pratylenchus nematodes. [Photographs courtesy of (A) D. Chitwood, USDA, (B) W. T. Crow and A. Hixon, (C) R. A. Rohde. (D) J. C. Townsend]

infected roots or in soil as eggs, juveniles, or adults, except for the egg-producing females, which seem to be unable to survive the winter. Adults and juveniles can infect and leave roots. The females, with or without fertilization, lay their eggs singly or in small groups inside infected roots. The eggs hatch in the roots or in the soil when released after root tissues break down. The emerging second-stage juvenile develops into the other juvenile stages and becomes an adult either in the soil or after it enters the root. When in the soil the nematodes are susceptible to drying, and during periods of drought they lie quiescent until the moisture increases and the plants resume growth.

Development of Disease

Juveniles and adult *Pratylenchus* nematodes enter roots usually in a radial direction (Figs. 15-18 and 15-19) by a persistent thrusting of the stylet and head, which seems to soften and break the cell wall. The cell walls and the cytoplasm turn light brown within a few hours after the nematodes begin feeding. The nematodes move into the cortex, where they feed and reproduce (Figs. 15-18B and 15-18C), but do not attack the endodermis. The necrosis of cortical cells follows the path of nematodes. In some hosts, only one or two cells on each side of the nematode tunnels are affected, but in others the lesion involves more than half the circumference of the root. As the feeding of the nematode on cortical cells continues, cell walls break down and cavities appear in the cortex.

Each lesion, and sometimes single host cells, is usually inhabited by more than one nematode (Figs. 15-18B,C and 15-18D). The females lay their eggs in the cortex, and frequently eggs, juveniles, and a few adults form "nests" that occur in great numbers in the cortex. As the eggs hatch, the nematodes feed on the parenchyma cells and move mostly lengthwise within the cortex, thus enlarging the lesion (Figs. 15-18 and 15-19). Some of the nematodes leave the lesion, emerge from the root, and travel to other points of the root or other roots, where they cause new infections. Necrotic cortical tissues are invaded by secondary fungi and bacteria, resulting in rotting and sloughing off of the root tissues around the point of infection and subsequent death of the distal part of the root. Thus, the reduced number of functioning roots results in reduced absorption of water and nutrients that makes the plants stunted and chlorotic.

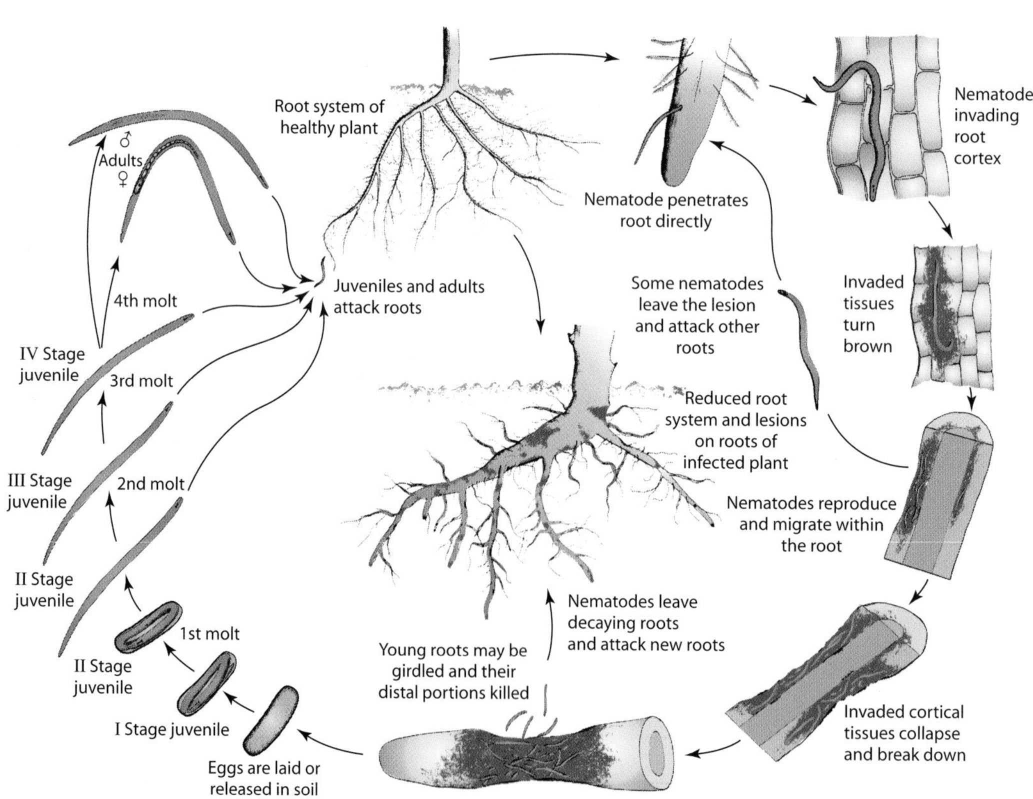

FIGURE 15-19 Disease cycle of the lesion nematode *Pratylenchus* sp.

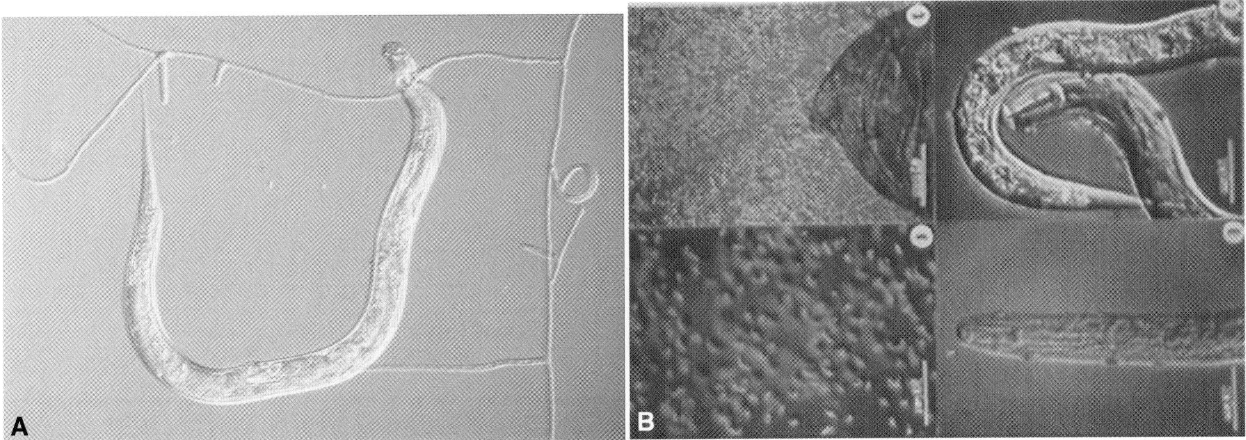

FIGURE 15-20 Biological control of *Pratylenchus* nematodes with nematode-trapping and parasitizing fungi (A) and with nematode-parasitizing *Pasteuria* bacteria (B). [Photographs courtesy of (A) B. Jaffee and (B) R. Sayre, USDA.]

Control

Lesion nematodes can best be controlled by overall or row treatment of the soil with nematicides before the crop is planted. Such treatments give good control of these nematodes, but they usually fail to eradicate them completely.

In hot and dry climates, a fairly good control of lesion nematodes can be achieved by summer fallow, which reduces the nematode populations by exposing them to heat and drying and by eliminating host plants. Control through crop rotation is rather unsuccessful because of the wide host ranges of the lesion nematodes. Several fungi and bacteria that parasitize and kill lesion nematodes are known (Figs. 15-20A and 15-20B), but none are effective as biological control agents under field conditions.

Selected References

Loof, P. A. A. (1991). The family Pratylenchidae. *In* "Manual of Agricultural Nematology" (W. R. Nickle, ed.), pp. 336–421. Dekker, New York.

MacGuidwin, A. E., and Rouse, D. I. (1990). Role of *Pratylenchus penetrans* in the potato early dying disease of Russet Burbank potato. *Phytopathology* **80**, 1077–1082.

Pinochet, J., Verdejo, S., and Marull, J. (1991). Host suitability of eight *Prunus* spp. and one *Pyrus communis* rootstocks to *Pratylenchus vulnus*, *P. neglectus*, and *P. thornei*. *J. Nematol.* **23**, 570–575.

Rebois, R. V., and Golden, R. M. (1985). Pathogenicity and reproduction of *Pratylenchus agilis* in field microplots of soybeans, corn, tomato, or corn-soybean cropping systems. Plant Dis. **69**, 927–929.

Zunke, U. (1991). Observations on the invasion and endoparasitic behavior of the root lesion nematode *Pratylenchus penetrans*. *J. Nematol.* **22**, 309–320.

THE BURROWING NEMATODE: *RADOPHOLUS*

The burrowing nematode occurs widely in tropical and subtropical regions of the world and in greenhouses in Europe. *Radopholus similis* is the most important banana root pathogen in most banana-growing areas, where it causes the so-called banana root rot, blackhead toppling disease, or decline of banana (Fig. 15-21). It also causes declines of avocado and of tea, and the yellows disease of black pepper. Furthermore, it attacks coconut, coffee and other fruit, ornamental, and forest trees, sugarcane, corn, vegetables, grasses, and weeds. Another closely related species, *Radopholus citrophilus*, causes the spreading decline of citrus in Florida (Fig. 15-22), which is now limited in distribution as a result of former quarantine and eradication programs. *Radopholus citrophilus* also infects several other cultivated crops, ornamentals, and weeds.

Symptoms

Infected banana plants grow poorly, have fewer and smaller leaves, show premature defoliation, and have smaller fruits. Often entire banana plants topple over (Figs. 15-21A and 15-21D). At first, primary banana roots show browning and cavities in the cortex, followed by deep cracks on the root surface (Fig. 15-21B). The nematodes, along with fungi and bacteria that invade the cracked roots, cause the roots to rot (Figs. 15-21B and 15-21C). As fewer short root stubs remain, they cannot anchor the plant sufficiently and the latter topples over. From the primary roots the nematodes move into the rhizome, in which they cause black, rotten areas to develop (Fig. 15-21C and 15-21E). As a

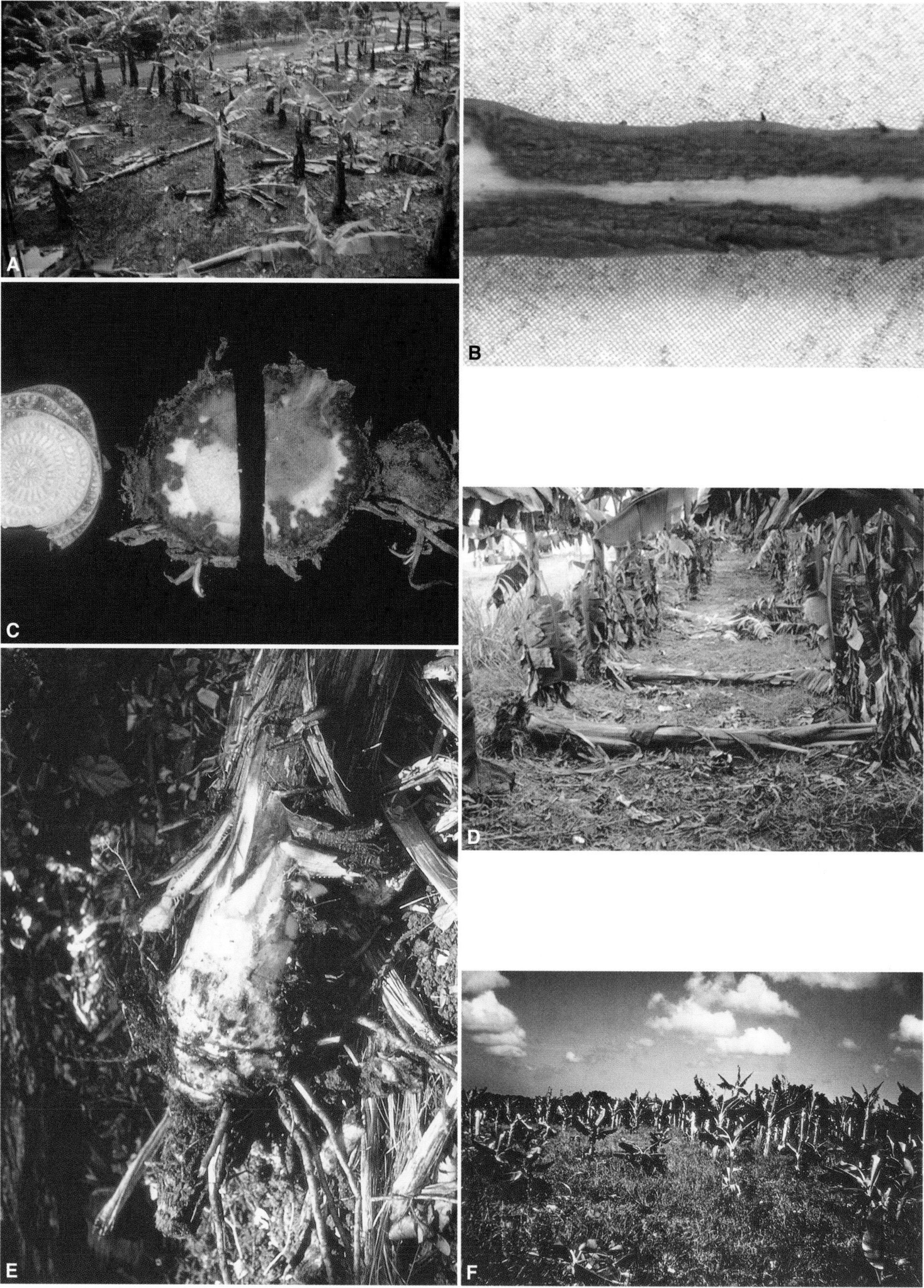

FIGURE 15-21 Symptoms of banana plants infected with the burrowing nematode *Radopholus similis*. (A and D) Most banana plants are stunted and several toppled over because roots are destroyed and provide poor anchoring. Root lesions (B) enlarge and increase destroying the root system (C) and part of the pseudostem (E). Decline has set in in this field as a result of heavy infestation with burrowing nematodes. [Photographs courtesy of (A and B) D. Coyne, IITA, Nigeria, (C and D) D. H. Thurston, and (E and F) University of Florida.]

FIGURE 15-22 (A) Declining citrus tree due to infection with the citrus-burrowing nematode *Radopholus citrophilus*. (B) A whole row of rapidly declining citrus trees due to infection by the burrowing nematode. [Photographs courtesy of University of Florida.]

result of this disease the profitable life of a banana plantation in many areas is decreased from indefinite to as short as one year (Fig. 15-21F), and the costs of annual replanting and losses in production are tremendous.

In citrus, *Radopholus* causes "the spreading decline." Blocks of affected trees have fewer and smaller leaves and fruits, and many of the twigs and branches die back (Fig. 15-22A,B). Yields of infected trees are reduced by 40 to 70 percent. Even under mild moisture stress, infected trees wilt readily, but they generally do not die and often recover temporarily after rainy periods. The symptoms of decline spread steadily to more trees each year, with the diameter of the decline area increasing approximately 10 to 20 meters per year. The symptoms on the aboveground parts follow infection of the roots by about a year. Infected feeder roots have numerous lesions and are invaded by primary and secondary fungal parasites, which result in the rotting and destruction of the feeder roots. Feeder roots seem to be attacked and destroyed most at depths of 50 cm or more, leaving less than half the feeder roots functional.

The Pathogen, Radopholus *sp.*

The pathogen, usually known as the burrowing nematode, is wormlike, about 0.65 millimeters long by 25 micrometers wide (Fig. 15-23). It spends its life and reproduces inside cavities in the root cortex, where it completes a life cycle in about 20 days. All juveniles and the females can infect roots and can emerge from the roots and spread through the soil. Most of the spread of the nematode from plant to plant, however, is through root contact or near contact. Long-distance spread of the nematode is primarily with infected plant material, such as infected banana sets. Although the nematodes infecting banana and citrus are morphologically identical, *R. similis* can attack banana but not citrus, whereas *R. citrophilus* can attack citrus as well as banana. Both can attack several other hosts. *Radopholus citrophilus*, however, is so far known to occur only in Florida; *R. similis* exists in many parts of the world.

Development of Disease

The burrowing nematode enters feeder roots and moves in the cortical parenchyma, feeding on nearby cells, destroying them, and causing the formation of cavities (Fig. 15-23). As the nematodes continue to feed, the cavities enlarge and coalesce, forming long tunnels. In banana, tunnels are limited to the cortex of the feeding roots, from which they spread into the rhizome. In citrus, the nematodes form cavities in the cortex and also in the stele, where they accumulate in the phloem and cambium, destroy them, and form nematode-filled cavities. At the same time, cells of the pericycle divide excessively and produce groups of tumor-like cells. Three to 4 weeks from infection the lesions develop one or more deep cracks. Each female lays one or a few eggs per day for many days, and as the eggs hatch, develop, and reproduce, nematode populations increase rapidly.

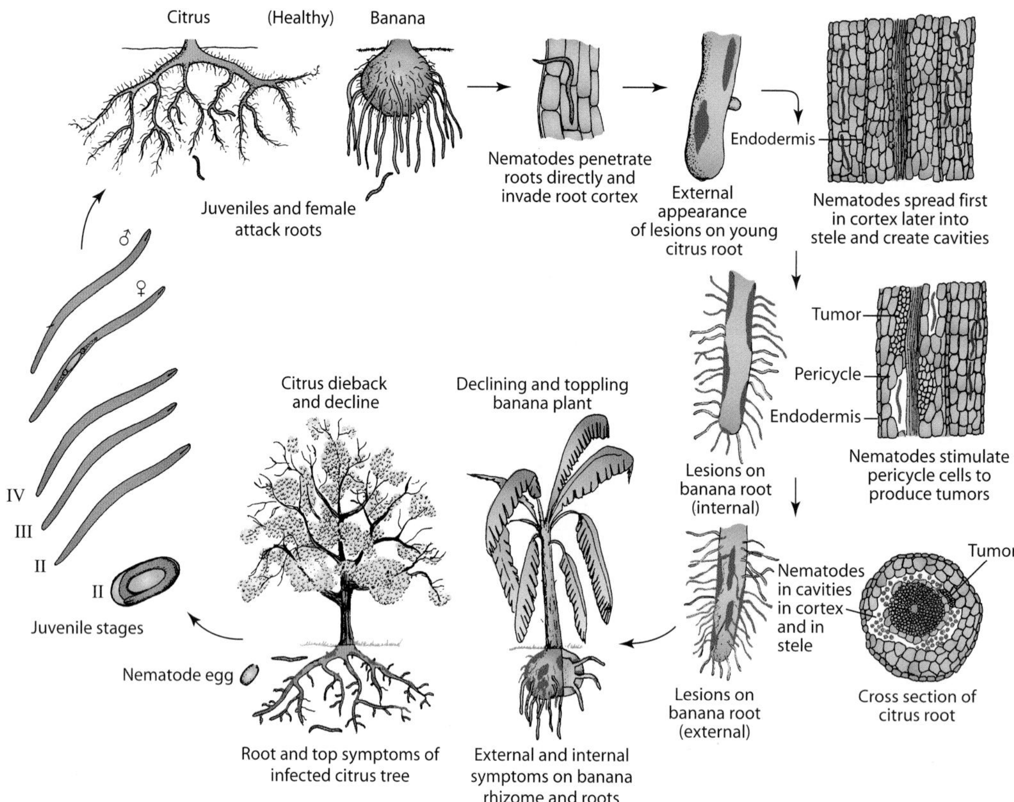

FIGURE 15-23 Disease cycle of the burrowing nematode *Radopholus* sp. in banana and citrus.

As many as 800 nematodes may be present in a single lesion. Fungi such as *Fusarium* and *Sclerotium* invade nematode-infected roots much more readily and further increase their rotting and destruction.

Control

Control of the burrowing nematode in banana can be obtained by using nematode-free plantlets produced through tissue culture; by removing discolored tissues from banana sets by paring and then dipping the sets in hot water at 55°C for 20 minutes; by flooding the field for 5 to 6 months where possible; and by soil fumigation or postplanting treatment with appropriate nematicides.

Control of the spreading decline of citrus is much more difficult and depends primarily on: (1) preventive regulatory measures that inhibit the spread and establishment of the nematode in new areas by treating nursery trees with hot water at 50°C for 10 minutes or dipping them in nematicides; (2) fumigation of decline areas with heavy doses of appropriate nematicides after removal of all declining trees and at least two rows around them; (3) use of tolerant or resistant rootstocks; and (4) control of weeds and providing trees with sufficient fertilizer and water.

Selected References

DuCharme, E. P. (1950). Morphogenesis and histopathology of lesions induced on citrus roots by *Radopholus similis*. *Phytopathology* **49**, 388–395.

Gowen, S. R., and Queneherve, P. (1990). Nematode parasites of bananas, plantains and abaca. *In* "Plant Parasitic Nematodes in Subtropical and Tropical Agriculture" (M. Luc, R. A. Sikora, and J. Bridge, eds.), pp. 431–460. CAB Int., Kew, England.

Marin, D. H., *et al.* (1998). Dissemination of bananas in Latin America and the Caribbean and its relationship to the occurrence of *Radopholus similis*. *Plant Dis.* **82**, 964–974.

Martin, D. H., *et al.* (2000). Development and evaluation of a standard method for screening for resistance to *Radopholus similis* in bananas. *Plant Dis.* **84**, 689–693.

Poucher, C., *et al.* (1967). Burrowing nematode in citrus. *Fla. Dep. Agric. Bull.* **7**, 1–63.

Sarah, J. L. (1989). Banana nematodes and their control in Africa. *Nematropica* **19**, 199–216.

Valette, C., *et al.* (1998). Histochemical and cytochemical investigations of phenols in roots of banana infected by the burrowing nematode *Radopholus similis*. *Phytopathology* **88**, 1141–1148.

Williams, K. J. O., and Siddiqui, M. R. (1973). *Radopholus similis*. Commonwealth Institute of Helminthology Descriptions of Plant-Parasitic Nematodes, Set 2, No. 27, pp. 1–4. St. Albans, England.

FIGURE 15-24 Examples of nematode diseases in some staple crops in the tropics and subtropics. (A) Poor survival and growth of bananas infested with *Radopholus similis*, *Helicotylenchus*, and *Hoplolaimus* nematodes (foreground) compared to bananas growing in a field where there was no infestation with nematodes and, instead, a mulch of plant materials was added (background). (B) Banana and (C) cassava roots with galls caused by the root-knot nematode. Yam roots (D left healthy) infected (D right and E) with the yam nematode *Scutellonema bradys*, the cause of dry root rot of yams. [Photographs courtesy of D. Coyne, IITA, Nigeria.]

BOX 24 The Added Significance of Plant Nematodes in the Tropics and Subtropics

In comparison to areas with temperate climates, crops in the tropics and subtropics suffer disproportionately from diseases and pests. This happens primarily because the growing season and, therefore, the reproduction of pathogens and pests are long or continuous, i.e., it is not interrupted by a cold winter period that normally limits reproduction and, usually, destroys and reduces the amount of inoculum available for the next growing season. In the tropics, therefore, pathogens continue to reproduce as long as there are crops growing and because many crops survive year-round in the tropics, pathogens continue to reproduce as long as there is food available for them. To this, however, should be added the lack of sufficient knowledge by the farming population regarding pathogens and diseases and their management and control and also the lack of funds for research as well as for materials, equipment, and related facilities that would allow the management or control of pathogens and pests.

Plant parasitic nematodes are no exception to the aforementioned situation. Actually, because nematodes are small, invisible, and generally exist and cause plant damage by attacking the belowground parts of plants, nematodes have been overlooked as serious pathogens even more than the aboveground-occurring pathogens and pests. Also, considering their ability to build up tremendous populations in the soil and the difficulty and expense of their control, plant parasitic nematodes are a major constraint in food production in the tropics and subtropics. Their detrimental effects are perhaps even greater in areas where soil fertility is low and moisture levels for adequate crop production are already at a critical level.

Tropical and subtropical crops are, of course, attacked by the nematodes that have worldwide distribution such as the root-knot nematode, lesion nematodes, sting nematodes, and the cyst nematodes. The root-knot nematode, for example, attacks and causes root knots on all important tropical crops that provide the main food staples such as yam (Figs. 15-9E and 15-9F), banana, and cassava (Fig. 15-24B), and the burrowing nematode attacks bananas and other crops (Figs. 15-21 and 15-24A). Often, of course, more than one kind of nematode are present in the soil and attack crop plants, together either killing or drastically reducing the size and productivity of the affected plants in relation to those of unaffected or slightly affected plants (Fig. 15-24A). Moreover, tropical crops are affected by additional types of nematodes that can cause serious losses, as is the case of the yam nematode *Scutellonema* sp., which causes dry rot of yams (Figs. 15-24D and 15-24E) and destroys huge amounts of yams every year. Great strides have been made and are being made toward identifying the nematodes affecting crops in the tropics and subtropics and toward developing methodologies that are practical and affordable under the particular circumstances. Studies on the use of nematode-free propagating material, use of resistant varieties, and crop rotation with nonhost crops have been the most successful so far.

STEM AND BULB NEMATODE: *DITYLENCHUS*

The stem and bulb nematode *Ditylenchus* occurs worldwide but is particularly prevalent and destructive in areas with temperate climate. It is one of the most destructive plant parasitic nematodes. It attacks a large number of host plants, including alfalfa (Fig. 15-25), onion (Fig. 15-26A), hyacinth (Fig. 15-26B), tulip, oat, and strawberry. Different populations or races of the stem and bulb nematode exist that have specific but often overlapping host preferences. On most crops, *Ditylenchus* causes heavy losses by killing seedlings, dwarfing plants, destroying bulbs, or making them unfit for propagation or consumption; by causing the development of distorted, swollen, and twisted stems and foliage; and, generally, by reducing yields greatly. A distinct species of *Ditylenchus*, *D. destructor*, causes the serious "potato rot" disease of potatoes (Figs. 15-26C and 15-26D).

Symptoms

In fields infested with stem and bulb nematodes, the emergence of seedlings such as onion is retarded and stands are reduced considerably. Half or more of the emerging seedlings may be diseased, appearing pale, twisted, arched, and with enlarged puffy and cracked areas along the cotyledon. Most infected seedlings die within three weeks of planting and the remainder usually die later.

Plants developing from bulbs planted in infested soil show stunting, light yellow spots, swellings ("spikkles") on the stem, shorter and curled leaves, and open lesions on the foliage. Many leaves become flaccid and are so weakened that they cannot maintain their erect growth and fall to the ground. The stem, neck, and individual scales of the bulb become softened, loose, and pale gray in color. Affected scales appear as discolored rings in cross sections of infected bulbs and as discolored,

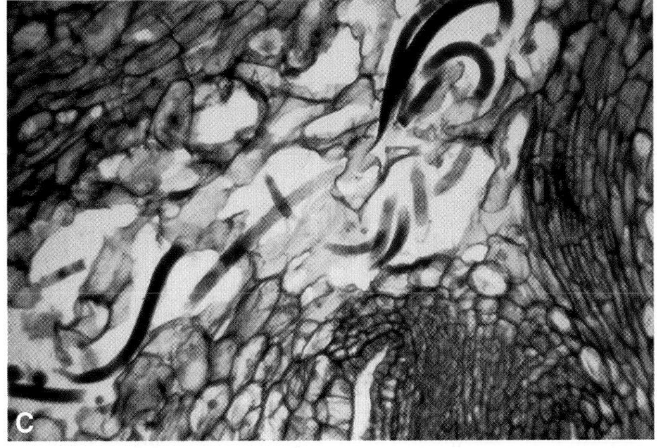

FIGURE 15-25 Damage to alfalfa plants in the field (A), individual alfalfa plants (B), and alfalfa stem tissues (C) by the stem and bulb nematode *Ditylenchus dipsaci*. [Photographs courtesy of (A and B) University of Florida and (C) E. I. Hawn, W.C.P.D.]

unequal lines in longitudinal sections. In more advanced cases, large areas or the whole bulb may be affected. Infected bulbs may also split and become malformed. In dry weather the bulbs become desiccated, odorless, and light in weight. In wet seasons a soft rot due to secondary invaders sets in, destroying the bulb and giving off a foul odor. Infected bulbs continue to decay in storage.

The Pathogen: Ditylenchus dipsaci

The nematode is 1.0 to 1.3 millimeters long and about 30 micrometers in diameter (Figs. 15-25C and 15-27). Second-stage juveniles emerge from the egg, undergo the second and third molt, and produce the preadult or infective juvenile. The latter can withstand adverse conditions of freezing and of extreme drying for long periods in fragments of plant tissue, in seeds, or in the soil. During favorable moisture and temperature the preadult juveniles become active, enter the host, pass through the fourth molt, and become males and females. The females lay 200 to 500 eggs, mostly after fertilization by the males. A complete cycle usually lasts about 19 to 25 days. Reproduction continues throughout the year. *Ditylenchus dipsaci* is an internal parasite of bulbs, stems, and leaves and passes generation after generation in these tissues, escaping to the soil only when living conditions in the plant tissues become unfavorable. When heavily infected bulbs decay, juveniles move out and sometimes accumulate about the basal plates of dried bulbs as grayish-white, cottony masses, called nematode wool, where they can remain alive for years.

Development of Disease

In germinating seeds or young seedlings, nematodes enter near the root cap or at points still within the seed and remain mostly intercellular, feeding on the parenchymatous cells of the cortex. Cells near the heads of the nematodes lose all or a portion of their contents, while cells surrounding these divide and enlarge, resulting in the development of swellings that make the seedling malformed. The epidermis of swellings often splits and allows fungi and bacteria to enter.

In young plants, nematodes enter the leaves through stomata or penetrate directly (Fig. 15-27). The nematodes usually remain and reproduce in the intercellular spaces, feeding on the nearby parenchyma cells whose contents they consume. As the bulbs enlarge, the nematodes migrate down from the leaves either intercellularly or on the surface of the leaves and enter again at the outer sheaths of the stem or neck. Heavily infected stems become soft and puffy due to the formation of large cavities through breakdown of the middle lamella and of the cells the nematodes feed on. Such stems can no longer remain rigid under the weight of the foliage and they frequently collapse. The nematodes continue to

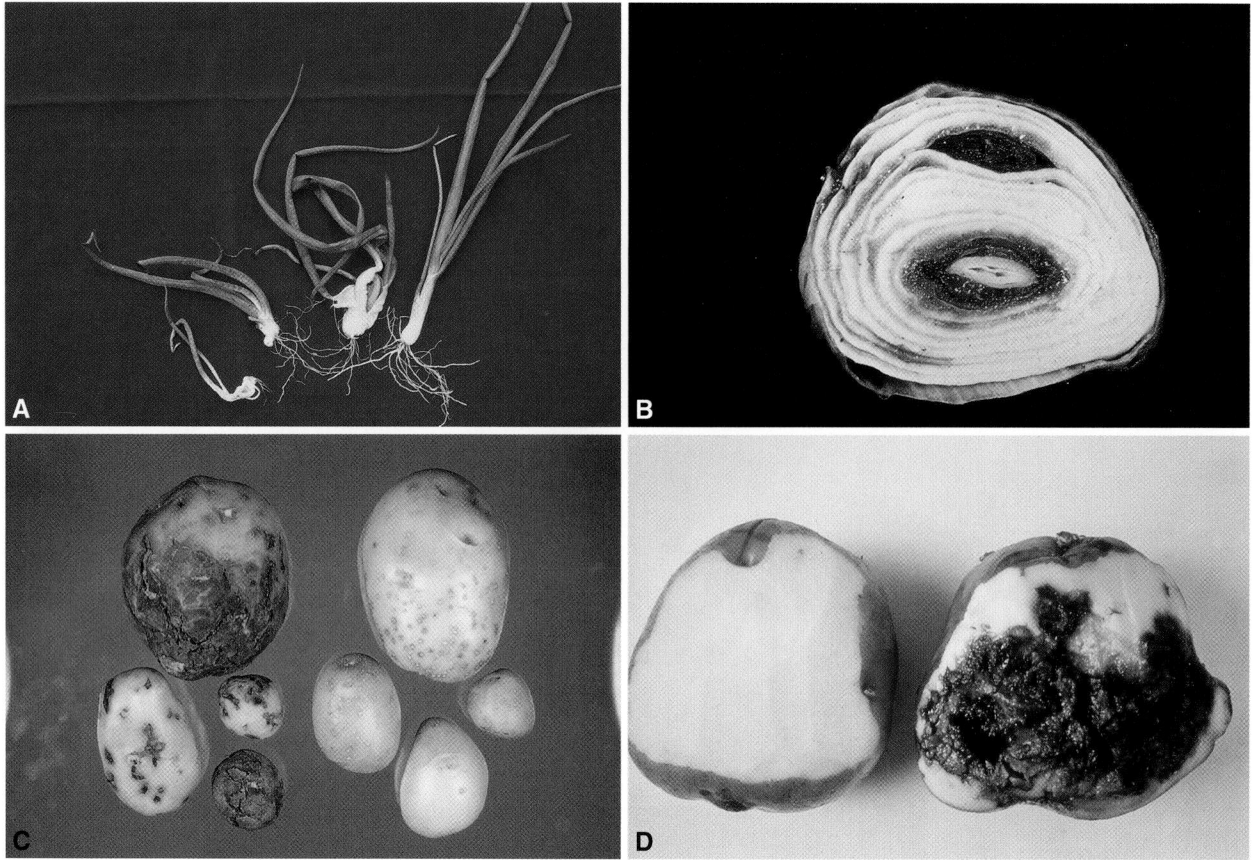

FIGURE 15-26 Damage by the stem and bulb nematode *Ditylenchus dipsaci* on young onion plants (A) and narcissus bulb (B). (C and D) Potato rot caused by *Ditylenchus destructor*. [Photographs courtesy of D. W. Dickson.]

move intercellularly through the outer scales of the bulbs. The macerated parenchyma cells have a white, mealy texture at first, but secondary invaders usually set in and cause them to turn brown. In early stages of infection the nematodes remain within individual scales; in later stages, however, they pass from one scale to the next, and thus several scales may be involved in each ring of frosty white or brownish tissue. The spread of the infection within a bulb continues in the field and in storage until, usually, the entire bulb become affected.

Control

Ditylenchus dipsaci on certain crops can be reduced by long (2–3 years at least) rotations with resistant crops. The use of nematode-free sets or seeds is extremely important. Infested seeds or bulbs can be disinfested by treating them with hot water for 1 hour at 46°C, with a nematicide in a gas-tight container, or with 0.5% formaldehyde. In the field, control can be achieved by fall fumigation of the soil, by preplant row treatment, and by treatment at or soon after planting with appropriate nematicides. In some crops, resistant cultivars provide quite satisfactory control.

Selected References

Darling, H. M., Adams, J., and Norgren, R. L. (1983). Field eradication of the potato rot nematode, *Ditylenchus destructor*: A 29-year history. *Plant Dis.* **67**, 422–423.

Hooper, D. J. (1972). *Ditylenchus dipsaci.* Commonwealth Institute of Helminthology Descriptions of Plant-Parasitic Nematodes, Set 1, No. 14, St. Albans, England.

Sturhan, D., and Brzeski, M. W. (1991). Stem and bulb nematodes, *Ditylenchus* spp. *In* "Manual of Agricultural Nematology" (W. R. Nickle, ed.), pp. 423–464. Dekker, New York.

STING NEMATODE: *BELONOLAIMUS*

Sting nematodes (Fig. 15-28A) are among the most underestimated but widespread and destructive plant

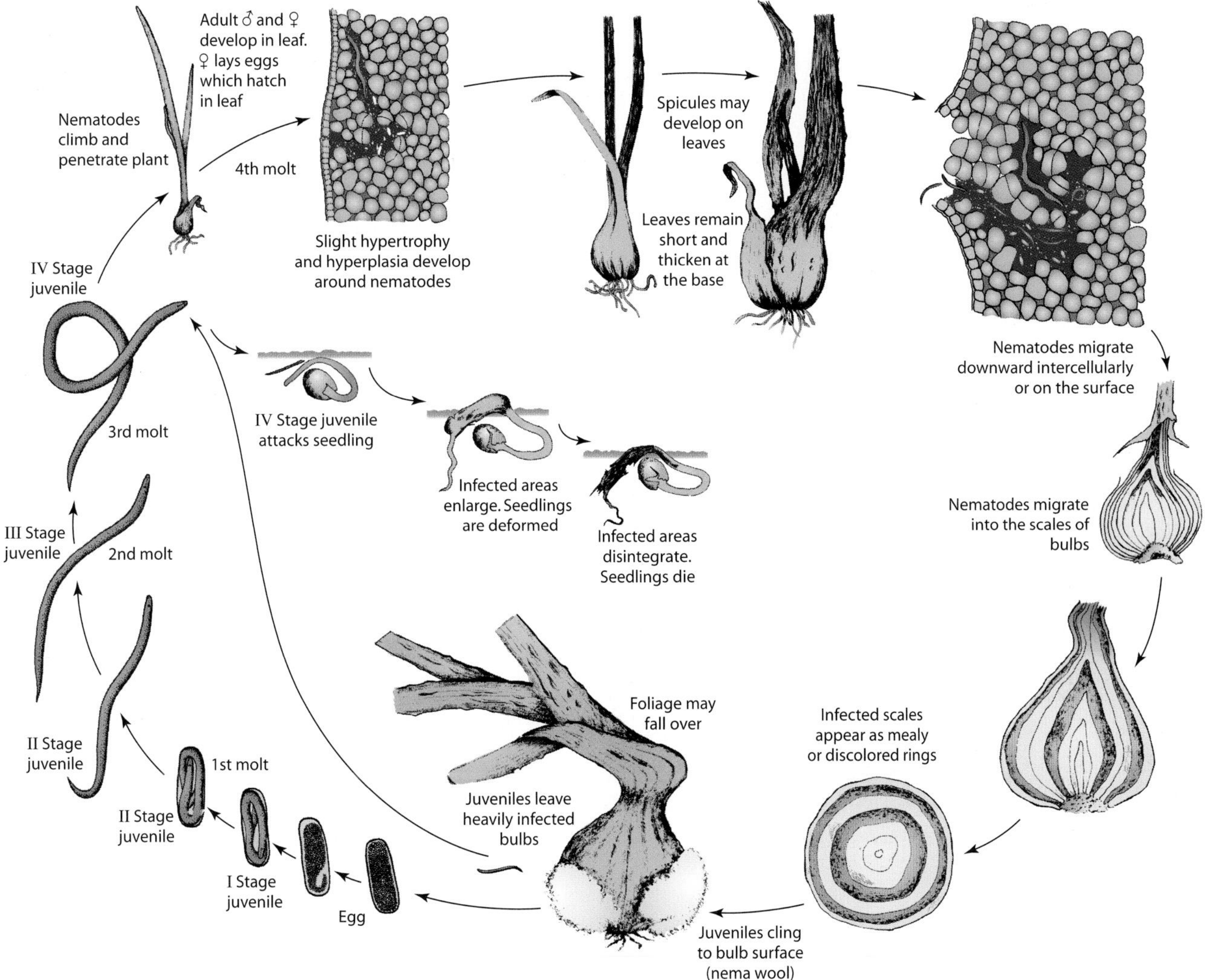

FIGURE 15-27 Disease cycle of the stem and bulb nematode *Ditylenchus dipsaci*.

parasitic nematodes. Sting nematodes are known to exist in sandy soils primarily in the coastal plains of the Atlantic and of the gulf coasts, but also in some Mid-western states and in California, in several Caribbean islands, and in Australia. They have an extremely wide host range, infecting equally effectively all grain crops, including corn (Figs. 15-28B and 15-28C) and sugarcane, turf grasses, and forage grasses; nongrass field crops such as cotton, peanut, and soybean (Fig. 15-28D); trees such as citrus, blueberries, some grapes, and pines; and many vegetables, such as beans, cucurbits, crucifers, and potato (Fig. 12-28E), strawberry (Fig. 15-28F), pepper, and many weeds. In several of these crops, sting nematodes cause serious yield losses and, when they reach high populations, they can cause complete destruction of the crop.

Symptoms

Young plants infected with sting nematodes grow poorly and then stop growing altogether. At high nematode populations such plants die. Plants becoming infected when older develop short, stubby roots that have dark, shrunken lesions, especially near the root tips. Plants with such roots may show symptoms of nutrient deficiency and subsequently remain stunted and may wilt. The symptoms usually appear aboveground as ever enlarging patches of discolored, stunted, and dead plants.

The Pathogen: Belonolaimus longicaudatus

The nematode is about 2 to 3 millimeters long and is thereby one of the longest nematodes. The nematode is

FIGURE 15-28 (A) The head region of the sting nematode *Belonolaimus longicaudatus*. (B) Typical root symptoms caused by sting nematodes on a plant (in this case, corn). (C, D, and F) Aboveground symptoms on plants caused by the sting nematode consist of patches of stunted and dead plants in corn (B), soybean (D), and strawberries (F). (E) Potato tuber showing lesions caused by the feeding of sting nematodes. [Photographs courtesy of (A) Z. Handoo, USDA, (B) D. W. Dickson, (D) University of Georgia, (E) D. P. Weingartner, and (F) J. Noling.]

an ectoparasite, moving freely in the soil and feeding by inserting its long stylet into epidermal cells of root tips, injecting enzyme — containing saliva and breaking down the cell contents, and then sucking the plant cell contents through the stylet. Affected root tips cease growing.

Sting nematodes reproduce only after mating. The fertilized female lays eggs in pairs in the soil and the eggs hatch in about five days. The emerging second stage juveniles must find a host and feed to survive and grow. They then undergo three more molts and finally produce adults. A life cycle takes 18 to 24 days to complete.

Control

The control or management of sting nematodes is difficult. Sometimes, careful crop rotation with nonhost crops is effective. Apparently, in some areas, the nematode exists as different species or races that show specificity for certain crops and, under such circumstances, crop rotation may work. Application of approved nematicides before or at planting is often effective for crops like strawberries, but may be too expensive for some crops.

Selected References

Bekal, S., and Becker, J. O. Population dynamics of the sting nematode in California turfgrass. *Plant Dis.* **84**, 1081–1084.

Huang X., and Becker, J. O. (1999). Life cycle and mating behavior of *Belonolaimus longicaudatus* in gnotobiotic culture. *J. Nematol.* **31**, 70–74.

Perry, V. G., and Rhoads, H. (1982). The genus *Belonolaimus. In* "Nematology in the Southern Region of the United States" (R. D. Riggs, ed.), pp. 144–149. Southern Cooperative Series Bulletin 276. Univ. Of Arkansas Agricultural Publications, Fayetteville, AR.

Smart, G. C., and Nguyen, K. B. (1991). Sting and awl nematodes. *In* "Manual of Agricultural Nematology" (W. R. Nickle, ed.), pp. 627–668. Dekker, New York.

STUBBY-ROOT NEMATODES: *PARATRICHODORUS* AND *TRICHODORUS*

Stubby-root nematodes occur all over the world. They attack a wide variety of plants, including cereals, vegetables, shrubs, and trees. They devitalize root tips and stop their growth. This results in reduction of the root system of plants, severe stunting and chlorosis of the whole plant, reduced yields, and poor quality of produce, but it seldom, if ever, causes death of the plant.

Symptoms

Infected plants show abnormal growth of lateral roots and proliferation of branch roots. In parasitized root tips, meristematic activity and growth stop, but cells already formed enlarge abnormally and cause swelling of the root tip (Fig. 15-29). Frequently, affected roots produce numerous lateral roots that are in turn attacked by nematodes. Repeated infections of lateral roots and their branches produce a smaller root system that lacks feeder roots and has instead short, stubby, swollen root branches (Fig. 15-30).

The Pathogen: Paratrichodorus minor

The pathogen is a small nematode about 0.65 millimeters long by 40 micrometers wide. It is an ectoparasite, feeding on the epidermal cells at or near the root-tip region, never entering the root tissue. It lays eggs in the soil, which hatch to produce juveniles and then adults. Its life cycle is completed within about 20 days (Fig. 15-30). Populations of *P. minor* build up quickly around susceptible hosts but decline in their absence or when host plants become old and do not produce new root tips. Eggs, juveniles, and adults are usually found in the soil throughout the year, although fourth-stage juveniles and eggs seem to be the stages found mostly during winter.

Development of Disease

Several species of *Paratrichodorus* and *Trichodorus* can transmit two rod-shaped plant viruses, *tobacco rattle virus* and *pea early browning virus*, from one plant to another.

When the *Trichodorus* nematode comes in contact with young roots or root tips of susceptible plants growing in infested soil, it bends its head at approximately a right angle to the root surface and punctures the cell wall with its stylet. Once inside the cell the stylet releases a viscous substance that causes the cytoplasm to aggregate around the stylet tip and the nematode ingests part of it. After that the nematode moves on to another cell within seconds or perhaps a few minutes from the beginning of feeding.

All free juvenile stages and the adults can attack plants and feed on them. Feeding is restricted to the epidermal cells at or near the root tip on older roots but encompasses the whole length of young succulent roots (Fig. 15-30).

Although one root tip may be attacked by many nematodes simultaneously or over time, the mechanical damage caused by *Paratrichodorus* and *Trichodorus* in feeding is slight and does not account for the gross changes on roots nor for the symptoms of the aboveground part of the plant. These effects seem to be the result of inhibitory or stimulatory actions of substances secreted by the nematodes into cells. These substances cause parasitized roots to lose meristematic activity at the root tip, to have no definite root cap or region of elongation, and to have a much smaller region of mitosis than that of healthy roots. Branch roots are also more abundant and closer together in infected than in healthy roots.

Control

Stubby-root nematodes can be controlled through broadcast or row application of nematicides. However, 6 to 8 weeks after treatment, stubby-root nematodes begin to reappear in the field and, if susceptible hosts

FIGURE 15-29 Damage to roots of plants caused by the stubby-root nematode *Paratrichodorus* (or *Trichodorus*) sp. [Photographs courtesy of (A–C) University of Florida.]

are present, nematode populations build up rapidly. Slow-acting nematicides retard or prevent the rapid buildup of nematodes, thus increasing the effectiveness of the treatment. Fallow or fallow and dry cultivation also help control *Paratrichodorus* and *Trichodorus*.

Selected References

Allen, M. W. (1957). A review of the nematode genus *Trichodorus* with descriptions of ten new species. *Nematologica* **2**, 32–62.

Rohde, R. A., and Jenkins, W. R. (1957). Host range of a species of *Trichodorus* and its host-parasite relationships on tomato. *Phytopathology* **4**, 295–298.

Russell, C. C., and Perry, V. G. (1966). Parasitic habit of *Trichodorus christiei* on wheat. *Phytopathology* **56**, 357–358.

Zuckerman, B. M. (1962). Parasitism and pathogenesis of the cultivated highbush blueberry by the stubby root nematode. *Phytopathology* **52**, 1017–1019.

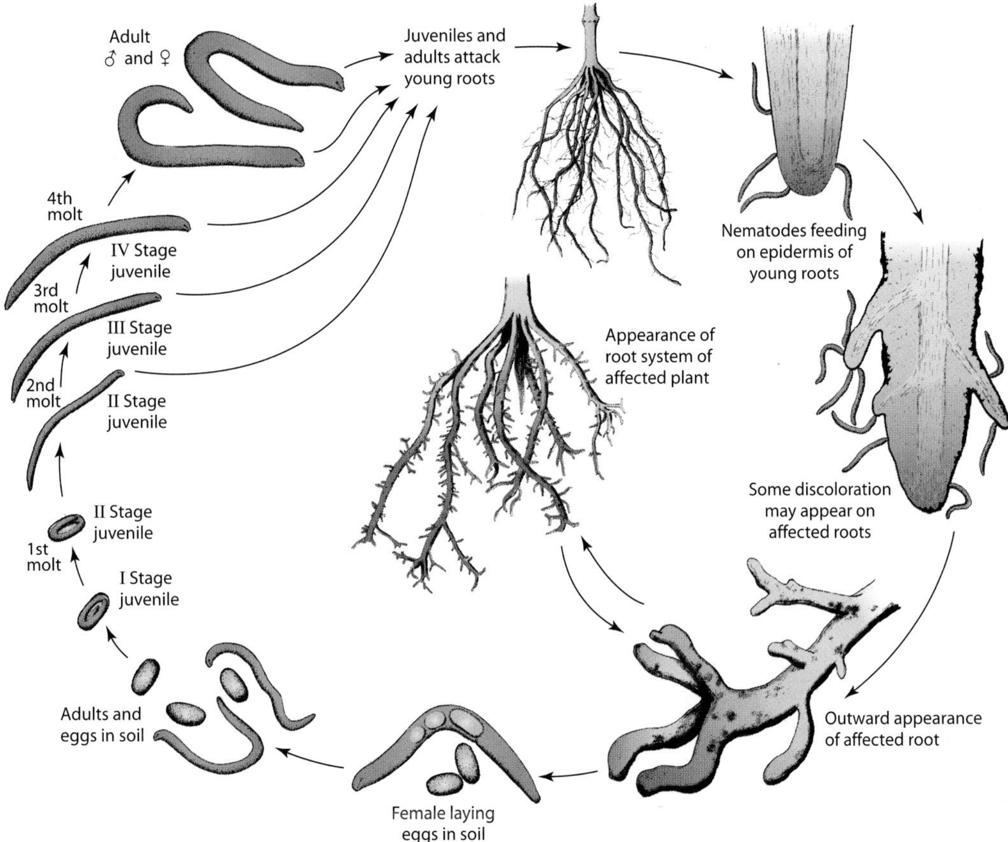

FIGURE 15-30 Disease cycle of the stubby-root nematode *Paratrichodorus minor*.

SEED-GALL NEMATODES: *ANGUINA*

Seed-gall nematodes were the first recorded plant parasitic nematodes; they were discovered in 1743, when an infected wheat seed (seed gall) was crushed in a drop of water under a microscope. Several species of *Anguina* are known and all of them cause formation of galls on seeds (Figs. 15-31A and 15-31B), leaves, and other aboveground parts of grain crops and forage grasses. The wheat seed-gall nematode is present wherever wheat is grown, but in most countries, where they use fresh and cleaned seed, this disease is quite rare. The wheat seed-gall nematode is still common, however, in eastern Europe and in parts of Asia and Africa. In Australia and in South Africa. Certain species of *Anguina* serve as vectors into seeds of certain pasture grasses of the plant pathogenic bacterium *Clavibacter toxicus*. The bacterium is often infected with a bacteriophage that induces the bacterium to produce corynetoxins. The latter are extremely toxic and cause disease and often death in sheep, cattle, horses, pigs, and so on, grazing on infected pastures or fed infected grain.

Symptoms

Symptoms caused by the seed-gall nematode appear on plants in all growth stages. Infected seedlings are more or less severely stunted and show characteristic rolling or twisting of the leaves (Fig. 15-32). A rolled leaf often traps the next emerging leaf or the inflorescence within it and causes it to become looped or bent and badly distorted. Stems are often enlarged near the base, frequently bent, and generally stunted. Diseased heads are shorter and thicker than healthy ones, and the glumes are spread farther apart by the nematode-filled seed galls (Fig. 15-31A). A diseased head may have one, a few, or all of its kernels turned into nematode galls. The galls are shiny green at first but turn brown or black as the head matures. Diseased heads remain green longer than healthy ones, and galls are shed off of the heads more readily than kernels. Mature galls are hard, dark, rounded, and shorter than normal wheat kernels (Fig. 15-31B) and often resemble cockle seeds, smutted grains, or ergot sclerotia.

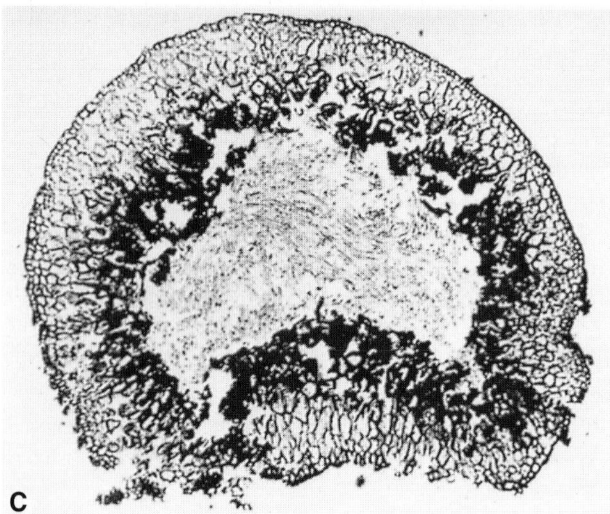

FIGURE 15-31 Damage on wheat kernels caused by the seed-gall nematode *Anguina tritici*. (A) Wheat heads, healthy (left) and infected with the seed-gall nematode showing the horizontal spreading of glumes and infected kernels. (B) Healthy and much smaller, rounded, brown-black infected kernels filled with stubby-root nematodes. (C) Cross section of infected kernel showing nematodes filling the kernel. [Photographs courtesy of (A and B) D. W. Dickson and (C) USDA.]

The Pathogen: Anguina tritici

The pathogen is a large nematode about 3.2 millimeters long by 120 micrometers in diameter. The nematode lays its eggs and produces all its juvenile stages and the adults in seed galls.

Development of Disease

The seed-gall nematode overwinters as second-stage juveniles in seed galls or in plants infected in the fall. Galls fallen to the ground or sown with the seed soften during warm, moist weather and release infective second-stage juveniles. When a film of water is present on the surface of the plants the juveniles swim upward and feed ectoparasitically on the tightly compacted leaves near the growing point, causing the leaves and stem to become malformed. When the inflorescence begins to form, the juveniles enter the floral primordia and produce the third- and fourth-stage juveniles and the adults. Each infected floral primordium becomes a seed gall and may contain 80 or more adults of both sexes. Each of the females then lays up to 2,000 eggs over several weeks within the freshly formed gall so that each gall contains 10,000 to 30,000 eggs. The adults die soon after the eggs are laid. The eggs then hatch, and the first-stage juveniles emerge; however, these soon molt and by harvest produce the second-stage juveniles, which are very resistant to desiccation and can survive in the galls for up to 30 years (Fig. 15-31C). The seed-gall nematode produces only one generation per year. The nematode is spread in infected seed.

Control

Control of the seed-gall nematode depends on the use of clean seed free of nematode-containing galls. Contaminated seed can be cleaned with modern equipment or by sieving and flotation in fresh water. Fields infested with seed-gall nematodes should not be planted to wheat or rye for at least a year. In moist weather the seed galls release second-stage juveniles; if no susceptible hosts are present, they die before they can infect and reproduce. In dry weather, however, nematodes can survive in the seed galls for many years.

Selected References

McKay, A. C., and Opfel, K. M. (1993). Toxigenic *Clavibacter/Anguina* associations infecting grass seedheads. *Annu. Rev. Phytopathol.* **31**, 151–167.

Singh, D., and Agrawal, K. (1987). Ear cockle disease (*Anguina tritici*) of wheat in Rajasthan, India. *Seed Sci. Technol.* **15**, 777–784.

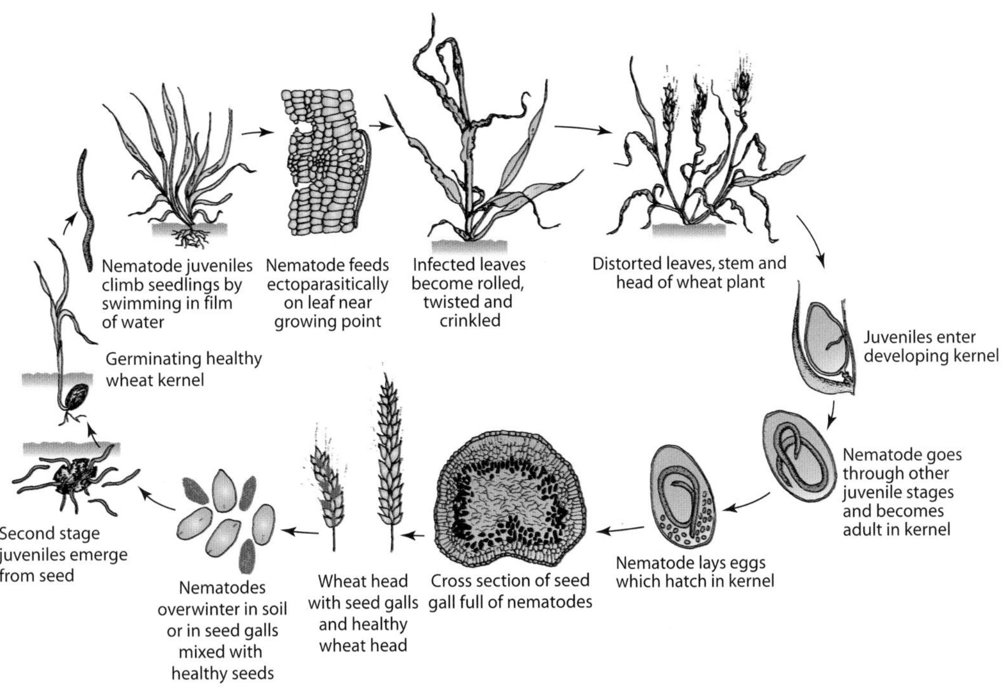

FIGURE 15-32 Disease cycle of wheat seed gall caused by *Anguina tritici*.

Southey, J. F. (1972). *Anguina tritici*. Commonwealth Institute of Helminthology Descriptions of Plant-Parasitic Nematodes, Set 1, No. 13, pp. 1–4. St. Albans, England.

Swarup, G., and Sosa Moss, C. (1990). Nematode parasites of cereals. *In* "Plant Parasitic Nematodes in Subtropical and Tropical Agriculture" (M. Luc, R. A. Sikora, and J. Bridge, eds.), pp. 109–136. CAB Int., Wallingford, England.

FOLIAR NEMATODES: *APHELENCHOIDES*

Several species of *Aphelenchoides* feed ectoparasitically and endoparasitically on aboveground plant parts. Some of the most important species are as follows: *A. ritzemabosi*, the chrysanthemum foliar nematode (Fig. 15-33A), also causes angular leaf spot of dry bean and some ornamentals (Fig. 15-33B); *A. fragariae*, the spring crimp or spring dwarf nematode of strawberry, also attacks many ornamentals such as various ferns (Figs. 15-33C and 15-33D); and *A. bessey*i, the nematode causing summer crimp or dwarf of strawberry, also causes white tip of rice.

The foliar nematode of chrysanthemums is widespread in the United States and Europe, but it occurs mostly in private gardens. It results in fairly severe losses. Foliar nematodes also attack several other plants, including aster, dahlia, zinnia, and strawberry.

Symptoms

Affected buds or growing points sometimes do not grow but turn brown or they produce short, bushy-looking plants with small and distorted leaves. As the season progresses, first the lower and then the upper leaves show small yellowish spots that later turn brownish black, coalesce, and form large blotches. At first the blotches are contained between the larger leaf veins (Figs. 15-33 and 15-34), but eventually the entire leaf is covered with spots or blotches, shrinks, becomes brittle, and falls to the ground. Defoliation, like infection, progresses from the lower to the upper leaves. Affected ray flowers fail to develop. Severely infected plants die without producing much normal foliage or many marketable flowers.

The Pathogen: Aphelenchoides ritzemabosi

The pathogen is a nematode measuring about 1 millimeter long by 20 micrometers in diameter. It may live its entire life inside leaves or at the surface of other plant organs. The female lays its eggs in the intercellular spaces of leaves. The eggs hatch and produce the four juvenile stages, and finally adults, all inside the leaf. The life cycle is completed in about 2 weeks. The foliar

FIGURE 15-33 Leaf symptoms caused by the foliar nematode *Aphelenchoides* sp. on chrysanthemum (A), Philippine violet (B), bird's nest fern (C), and maidenhair fern (D). [Photographs courtesy of (A–C) R. A. Dunn and (D) P. S. Lehman.]

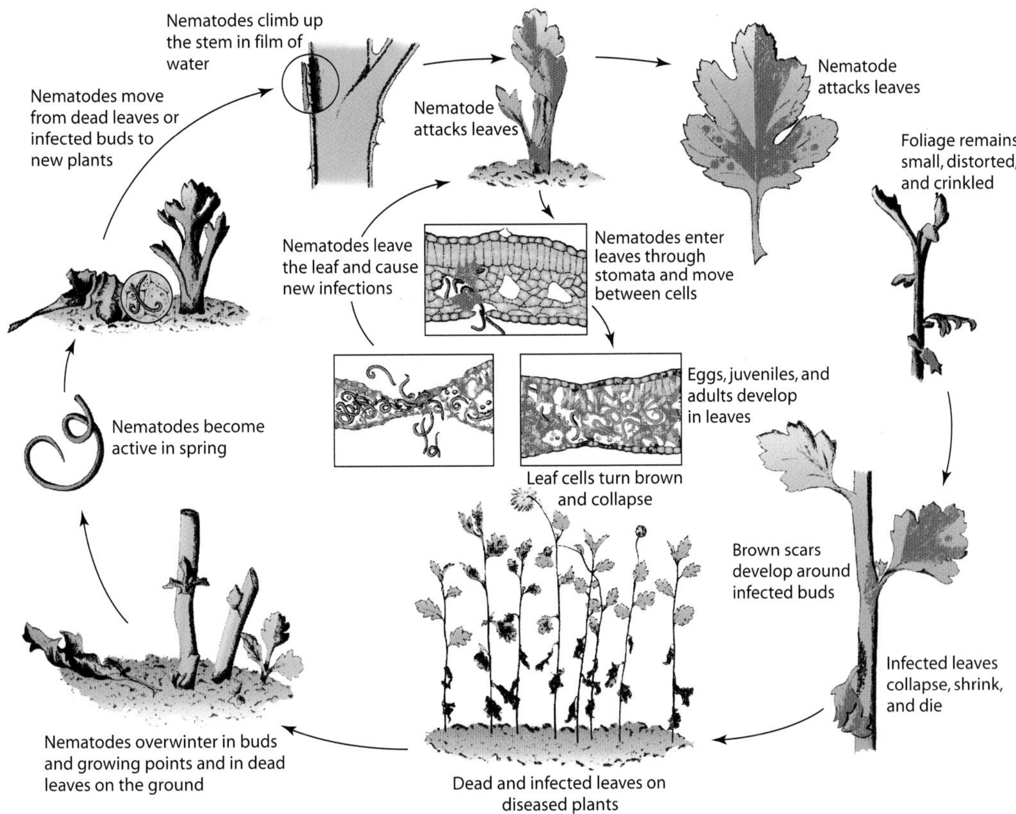

FIGURE 15-34 Disease cycle of the foliar (chrysanthemum) nematode *Aphelenchoides ritzemabosi*.

nematodes overwinter as adults in dead leaves or between the scales of buds of infected tissues.

In the spring the nematodes become activated and feed ectoparasitically on the epidermal cells of the organs in their vicinity. Thus, stem areas, petioles, and leaves near infested buds show brown scars consisting of groups of cells killed by the nematodes. In addition to direct killing of cells, the nematodes, through their secretions, cause several other symptoms: shortening of the internodes, which results in a bushy appearance of the plant; browning and failure of the shoot to grow (blindness); and development of distorted leaves.

Nematodes infest new, healthy plants by swimming up the stem when it is covered with a film of water during rainy or humid weather. They enter leaves through the stomata (Fig. 15-34). The presence of nematodes between leaf cells causes the cells to turn brown and to break down, creating large cavities in the mesophyll. In the early stage of infection the cells of the vein sheath block the extension of leaf necrosis across the veins. In advanced stages of infection, even these cells break down, and the nematodes and leaf necrosis spread over the entire leaf. Heavily infected leaves fall to the ground.

Control

Several sanitary practices are quite important in controlling the foliar nematode. The leaves and stems should be kept dry to prevent spreading of the nematodes. Cuttings should be taken only from the tops of long, vigorous branches. Suspected dormant cuttings or stools may be disinfested by dipping in hot water at 50°C for 5 minutes or at 44°C for 30 minutes. Excellent control of this nematode can be obtained by treating plants with appropriate nematicides as sprays or drenches.

Selected References

Franc, G. D., *et al.* (1996). Nematode angular leaf spot of dry beans in Wyoming. *Plant Dis.* **80**, 476–477.

French, N., and Barraglough, R. M. (1964). Observations on eelworm on chrysanthemum stools. *Plant Pathol.* **13**, 32–37.

Hesling, J. J., and Wallace, H. R. (1961). Observations on the biology of chrysanthemum eelworm *Aphelenchoides ritzemabosi* (Schwartz) Steiner in florist's chrysanthemum. I. Spread of eelworm infestation. *Annu. Appl. Biol.* **49**, 195–203, 204–209.

Siddiqi, M. R. (1974). *Aphelenchoides ritzemabosi.* Commonwealth Institute of Helminthology Descriptions of Plant-Parasitic Nematodes, Set 1, No. 32. St. Albans, England.

Siddiqi, M. R. (1975). *Aphelenchoides fragariae*. Commonwealth Institute of Helminthology Descriptions of Plant-Parasitic Nematodes, Set 1, No. 74. St. Albans, England.

PINE WILT AND PALM RED RING DISEASES: *BURSAPHELENCHUS*

Nematodes of the genus *Bursaphelenchus* cause severe wilt diseases in several tree species. These nematodes have developed specific symbiotic relationships with certain insects (beetles, weevils) which they invade and by which they are transported from diseased to healthy trees. The two most important tree wilt diseases caused by nematodes are pine wilt (Fig. 15-35), caused by *B. xylophilus*, the vectors of which are beetles of the genus *Monochamus*, and coconut red ring (Fig. 15-36), caused by *B. cocophilus*, which is vectored by the palm weevil *Rhynchophorus* sp. and the sugarcane weevil *Metamasius* sp.

PINE WILT NEMATODE: *BURSAPHELENCHUS XYLOPHILUS*

Pine wilt has been known to occur in Japan since the early 1900s. It is present in localized areas of China and, since 1979, in most of the United States, much of Canada, and Mexico. It affects, with different severity, more than 28 species of pine and several other conifers, being most severe on Scotch pine (*Pinus sylvestris*). Pine wilt is a lethal disease of pines and other forest trees. Pine wilt has caused severe losses of pines in several localities in Japan, but although it has spread widely in the Untied States, it has not yet become a significant problem. The pine wilt nematode, however, can kill whole trees or parts of them, and it is transmitted from dead to live pines by certain insects. Because of that, there is a strict embargo of forest products from North America to Europe, for example, with severe economic consequences.

Symptoms

The foliage of infected branches or whole trees suddenly becomes grayish-green, and the trees stop exuding resin from their wounds. The foliage then becomes yellowish green, and at first some, then all, of the needles turn brown (Figs. 15-35A and 15-35B). Within 4 to 6 weeks from the appearance of symptoms, the tree or branch has totally brown foliage and appears wilted, although sometimes the needles are retained without obvious droop. In many affected trees blue stain in wood is heavy (Fig. 15-35E). Infected trees invariably die (Fig. 15-35B).

The Pathogen: Bursaphelenchus xylophilus

This pathogen, also known as the pinewood nematode, is about 800 micrometers long by 22 micrometers in diameter (Fig. 15-35C). It develops and reproduces rapidly, completing a life cycle within four days during the summer. Each female lays about 80 eggs, which hatch and produce the four juvenile stages and the adults. While the tree is still living, the nematodes feed on plant cells, but after its death they feed on fungi that invade the dying or dead tree. In late stages of infections, a different form of third-stage juveniles, called the dispersal stage, appear. These have large amounts of nutritional reserves and a thick cuticle and they molt to fourth-stage dispersal juveniles. The latter are especially adapted to survive in the respiratory system of certain cerambycid beetles, by which they are transmitted to healthy trees. *Bursaphelenchus xylophilus* is mycophagous, i.e., it feeds and can complete its life cycle feeding on many kinds of fungi, e.g., *Alternaria*, *Fusarium*, and the blue stain fungi (*Ceratocystis* spp.).

Development of Disease

The pinewood nematode overwinters in the wood of infected dead trees, which also contain instars (larvae) of cerambycid beetles such as *Monochamus alternatus*. In early spring, the instars excavate small chambers in the wood in which they pupate. As the adult beetles emerge from the pupae later in the spring, large numbers of fourth-stage dispersal juveniles enter the beetles and more or less fill many of the tracheae of the respiratory system (Fig. 15-35D). The emerging adult beetles bore their way out of the wood, each carrying with it an average of 15,000 to 20,000 fourth-stage dispersal juveniles, and fly to succulent branch tips of healthy trees where they feed for several weeks. As the beetles feed by stripping the bark and reaching the cambial tissues (Fig. 15-35F), the fourth-stage dispersal juveniles emerge from the insect and enter the pine tree through the wound. Once in the plant, the dispersal juveniles undergo the final molt to produce adult nematodes, which then reproduce. The nematodes migrate to the resin canals, where they feed on the epithelial cells lining the canals and cause their death as well as the death of the surrounding parenchyma cells. The nematodes move quickly through resin canals in both the xylem and the cortex, reproduce rapidly, and, within a few weeks, build up enormous populations in the host.

The destruction of the resin canals stops all resin flow from wounds within about 10 days of inoculation. In the next three weeks, transpiration by the foliage declines rapidly and stops as the foliage loses color and the tree suddenly wilts. Nematode populations reach a

FIGURE 15-35 The pinewood or pine wilt nematode and its effects on plants and its vector. Browning, wilting, and death of individual pine tree (A) and of trees on a mountainside (B) caused by the pinewood nematode *Bursaphelenchus xylophilus* (C). (D) Fourth-stage dispersal larvae of the pinewood nematode inside a trachea of the weevil insect vector *Monochamus* sp. (E) Blue-stain fungi develop in the wood of dead pine twigs and branches and provide food for the nematode. (F) The pinewood nematode vector, *Monochamus* sp., feeding on a pine branch and depositing fourth-stage dispersal juveniles in the pine wood where they start a new infection. [Photographs courtesy of (A) R. P. Esser, (B) J. J. Witcosky, (C,F) D. R. Bergdahl, (D) E. Kondo, Saga Univ., Japan, and (E) L. D. Dwinell, USDA.]

maximum level after the death of the tree, about one month after inoculation. In later stages of the disease, as the condition of the tree deteriorates, nematode populations decline. At the same time, however, there is a gradual increase in the proportion of the dispersal third-stage juveniles in relation to the total population of the nematode in the wood. The third-stage dispersal juveniles are the resting stage of the nematode.

In the meantime, the adult *Monochamus* beetles, the vector of *B. xylophilus*, after they have fed on tender pine twigs for about one month, look for and deposit their eggs under the bark of stressed and dead pine trees, including trees showing symptoms or dying from infection by the pinewood nematode. The first two instars of the insect feed under the bark, but the third penetrates the wood, where, after a molt, it produces the fourth instar, which overwinters in the wood. In early spring, the fourth instar excavates a chamber in the wood, in which it pupates, and attracts numerous third-stage nematode juveniles all around it. The latter molt to produce fourth-stage dispersal juveniles, which infect the adult insect as soon as it emerges from the pupa, and thus the cycle is completed.

In some temperate regions, primarily pine trees stressed by various diseases and insects are attacked by the pinewood nematode but typical wilt symptoms are not usually produced.

Control

Control measures involving insecticide treatment to control the beetles, and early removal and burning of dead and dying pine trees to eliminate the breeding habitat of the nematode and of the beetle, are only moderately effective and practical only in restricted localities. Neither of these controls is possible in large forests. Affected susceptible pine species planted as shade trees should be replaced with more resistant pine species or with other types of trees.

RED RING NEMATODE: *BURSAPHELENCHUS COCOPHILUS*

Red ring of coconut palms occurs in the countries of the American tropics, from Mexico to Brazil, and in several of the southern islands of the Caribbean Basin. The disease is most common and develops rapidly in young coconut trees 3 to 10 years old; it kills them within about three months from infection. Losses from red ring can be severe. Losses of 10 to 15% of young coconut palms, and also of young and established oil palms and date palms, are common. In some areas of high infestation, losses of up to 60% of young coconut palms have been recorded.

Symptoms

The symptoms appear at first on older leaves nearest the point of infection as yellowing from the tips inward and then as browning of the leaves (Figs. 15-36A and 15-36B). Several dying or dead leaves often break close to the stem and remain hanging. Yellowing and browning then spread to progressively younger leaves and finally the whole treetop dies (Fig. 15-36B, right). In bearing trees, inflorescences wither and nuts of all stages drop prematurely. In cross sections of infected stems, an orange-red to brownish ring about 3 to 5 centimeters wide is present about 5 centimeters inside the stem periphery over the length of the stem (Fig. 15-36C). Although young, 3- to 10-year-old trees die within a few months after they become infected, older trees (>20 years old) at first produce shorter leaves and then the leaf size and number continue to decrease every year, inflorescences are aborted, and the palms become unproductive and may eventually die.

The Pathogen: **Bursaphelenchus (Formerly Rhadinaphelenchus) cocophilus**

The pathogen is about 1 millimeter long by 15 micrometers wide (Fig. 15-36D). It lays its eggs and produces all juvenile stages and the adults inside infected palm trees, completing its life cycle within 9 to 10 days. *Bursaphelenchus cocophilus* is transmitted from palm to palm by the American palm weevil *Rhynchophorus palmarum* (Fig. 15-36E), the sugarcane weevil (*Metamasius* sp.), and possibly other weevils. The weevil larvae, as they feed on red ring-infected palm tissue, swallow several hundred thousand nematode third-stage juveniles, but only a few hundred of the nematodes survive and pass through the molt, internally or externally, to the next stage weevil larva and to the adult weevil. As weevil females emerge from rotted palms, a small percentage of them carry with them third-stage juveniles of the nematode. Female weevils are attracted to red ring-diseased trees but they also lay their eggs on healthy or wounded palm trees. If the female carries red ring nematodes, it deposits them into wounds at leaf bases or internodes. The nematodes then go through repeated life cycles and spread intercellularly in the ground parenchyma cells of the stem, petioles, and roots. There they cause cell breakdown, which results in cavities, an orange-to-red discoloration, and a dry flaky texture of the diseased tissues. The discoloration extends into the leaf bases and into the petioles. Although the red ring nematodes do not invade xylem and phloem tissues, xylem vessels within the red ring develop tyloses that block the upward transport of nutrients and water. The internal symptoms generally develop before external

FIGURE 15-36 Young (A) and fully developed (B) palm trees showing yellowing, wilting, and necrosis of lower fronds caused by infection with the red ring nematode *Bursaphelenchus cocophilus*. (C) Cross section of infected palm tree trunk showing red ring where the nematodes feed and move. (D) Palm red ring nematodes. (E) The palm red ring nematode vector *Rhynchophorus palmarum*. [Photographs courtesy of (A and B) D. W. Dickson and (C–E) R. M. Giblin-Davis.]

symptoms become visible. Nematode populations in diseased trees increase rapidly at first but then they decline, slowly at first, and then sharply. Finally, about 3 to 5 months after infection, no more living red ring nematodes or any of its eggs can be found in the decomposed stem tissue of infected, dead palm trees. The nematodes, however, survive in newly infected palm trees and, briefly, in their insect vector. Generally, however, red ring is spread rather slowly to new trees.

Control

The control of red ring is difficult. Infected trees should be treated with systemic insecticides and nematicides to reduce the vector and the nematodes within the trees or such trees should be killed with herbicide or cutting to forestall nematode survival. Experiments are also underway to use insecticide-laced insect traps, which contain insect attractants derived from red ring-infested palm tissues or synthetic pheromones, to attract and to kill potential nematode-vectoring weevils, thereby controlling both the weevil and red ring.

Selected References

Dwinell, L. D. (1997). The pinewood nematode: Regulation and mitigation. *Annu. Rev. Phytopathol.* **35**, 153–166.

Giblin-Davis, R. M. (1994). Red ring disease. *In* "Compendium of Tropical Fruit Diseases" (R. C. Ploetz *et al.*, eds.), pp. 30–32. APS Press, St. Paul, MN.

Griffith, R. (1987). Red ring disease of coconut palm. *Plant Dis.* **71**, 193–196.

Griffith, R. (1992). Red ring disease of coconuts. *In* "Plant Diseases of International Importance" (U. S. Singh *et al.*, eds.), Vol. 4, pp. 258–276. Prentice-Hall, Englewood Cliffs, NJ.

Halik, S, and Bergdahl, D. R. (1994). Long-term survival of *Bursaphelenchus xylophilus* in living *Pinus sylvestris* in an established plantation. *Eur. J. Forest Pathol.* **34**, 357–363.

Ichihara, Y., Fukuda, K., and Suzuki, K. (2000). Early symptom development and histological changes associated with migration of *Bursaphelenchus xylophilus* in seedling tissues of *Pinus thunbergii*. *Plant Dis.* **84**, 675–680.

Kondo, E., *et al.* (1982). Pine wilt diseases: Nematological, entomological, and biochemical investigations. *Univ. Mo. Columbia Agric. Exp. Stn. Bull.* **SR282**, 1–56.

Mamiya, Y. (1983). Pathology of the pine wilt disease caused by *Bursaphelenchus xylophilus*. *Annu. Rev. Phytopathol.* **21**, 201–220.

Rutherford, T. A., Mamiya, Y., and Webster, J. M. (1990). Nematode-induced pine wilt disease: Factors influencing its occurrence and distribution. *For. Sci.* **36**, 145–155.

Tarès, S., *et al.* (1994). Use of species-specific satellite DNA from *Bursaphelenchus xylophilus* as a diagnostic probe. *Phytopathology* **84**, 294–298.

Wingfield, M. J. (1987). "Pathogenicity of the Pine Wood Nematode." Symposium Series. APS Press, St. Paul, MN.

chapter sixteen

PLANT DISEASES CAUSED *BY* FLAGELLATE PROTOZOA

INTRODUCTION – NOMENCLATURE, TAXONOMY, AND PATHOGENICITY OF PLANT TRYPANASOMATIDS
877

EPIDEMIOLOGY AND CONTROL OF PLANT TRYPANOSOMATIDS
878

PLANT DISEASES CAUSED BY PHLOEM-RESTRICTED TRYPANOSOMATIDS
878

"PLANT DISEASES" CAUSED BY LATICIFER-RESTRICTED TRYPANOSOMATIDS
882

"PLANT DISEASES" CAUSED BY FRUIT- AND SEED-INFECTING TRYPANOSOMATIDS
882

INTRODUCTION

Certain trypanosomatid flagellates (Fig. 16-1), belonging to the kingdom Protozoa, phylum Euglenozoa, order Kinetoplastidae, and family Trypanosomatidae, have been known to parasitize plants since the early 1900s. That flagellates may be pathogenic to their host plants was suggested several times by the investigators of these parasites, and rather good evidence was presented that some plant diseases are caused by flagellates. However, because these parasites could not be isolated in pure culture and could not be inoculated into healthy plants so that they could reproduce the disease, as Koch's rules dictate, flagellates have not yet been fully accepted as plant pathogens. Nevertheless, the pathogenicity of phytoplasmas and of some fastidi-

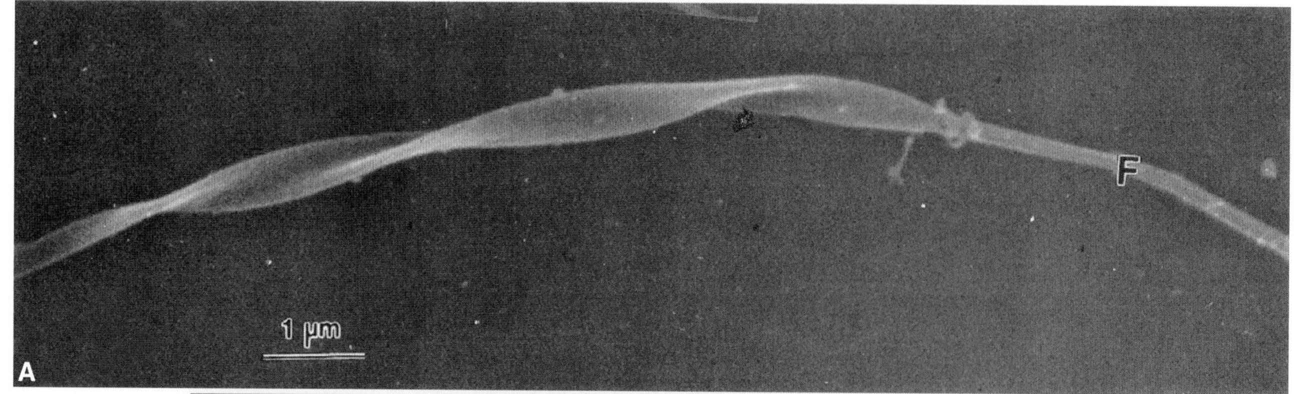

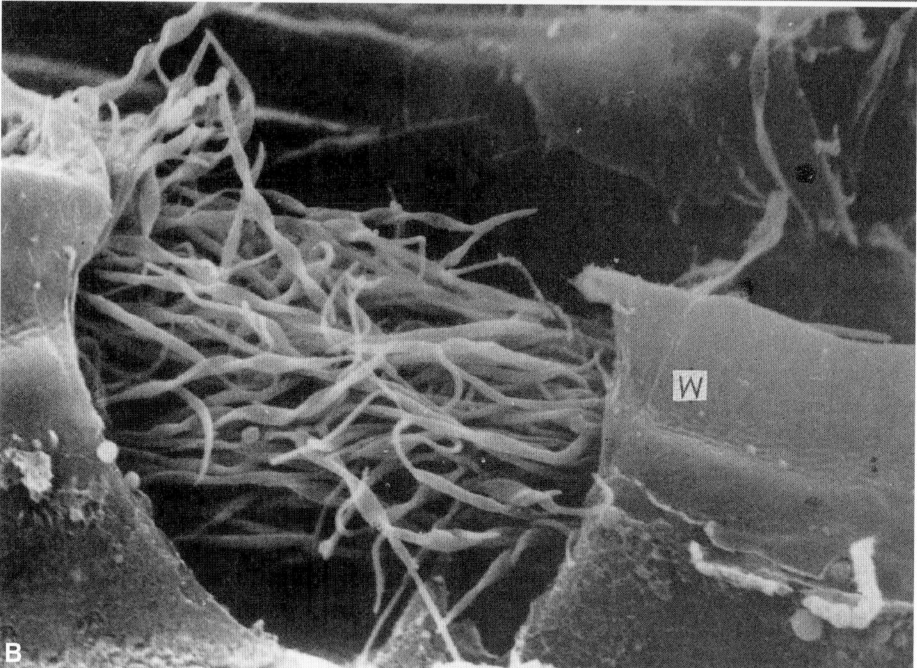

FIGURE 16-1 (A) Individual flagellate trypanosomatid protozoon (*Phytomonas frantai*) isolated from the latex of cassava plants affected with the empty root disease. F, flagellum. (B) *Phytomonas* protozoa in a phloem sieve tube of root of oil palm tree affected with sudden wilt disease. W, cell wall. [Photographs courtesy of (A) E. W. Kitajima and (B) W. de Sousa.]

ous vascular bacteria in plants is almost universally accepted, although the same Koch's rules are equally unfulfilled with these organisms as they are with flagellates. Because evidence supporting the pathogenicity of flagellates is no less compelling than that available for phytoplasmas and fastidious vascular bacteria, it is reasonable to assume that at least some flagellates are considered capable of causing disease in plants, and it is apparent that the role of flagellates, as well as the role of other protozoa, in plant pathology deserves more attention than it has received in the past.

The protozoa are mostly one-celled, microscopic organisms, generally motile, and have typical nuclei. They may live alone or in colonies and may be free living, symbiotic, or parasitic. Some protozoa subsist on other organisms, such as bacteria, yeasts, algae, and other protozoa; some saprophytically on dissolved substances in the surroundings; and some by photosynthesis as in plants. Protozoa move by flagella, by pseudopodia, or by movements of the cell itself.

Of the protozoa, apparently only the flagellates have been reported as associated with plant diseases so far, but there are no good reasons why other classes of protozoa might not be found in the future also to be parasitic on plants. Flagellates are characterized by one or more long slender flagella at some or all stages of their life cycle. Although many flagellates are saprophytic and some contain plastids with colored pigments, including functional chlorophyll, others are parasites of humans and various animals, with some causing serious diseases. The best known flagellate pathogenic to humans is the blood parasite *Trypanosoma*, the cause of sleeping sickness in Africa, which is transmitted by tsetse flies.

NOMENCLATURE OF PLANT TRYPANOSOMATIDS

Flagellate protozoa were first found to be associated with plants in 1909, in Mauritius, when Lafont reported that they parasitize the latex-bearing cells (the laticifers) of the laticiferous plant *Euphorbia* (Euphorbiaceae). To distinguish them from protozoa parasitizing humans and animals, plant protozoa were placed in a new genus, *Phytomonas*, and the one described by Lafont was named *P. davidi*. Since then several other species of *Phytomonas* have been reported from plants belonging to the families Asclepiadaceae (e.g., *P. elmassiani* on milkweed), Moraceae (e.g., *P. bancrofti* on ficus species), Rubiaceae (e.g., *P. leptovasorum* on coffee), and Euphorbiaceae (e.g., *P. fran̄tai* on cassava); unnamed species have been reported on coconut palm and on oil palm (Figs. 1-3 and 16-1) and on the ornamental plant red ginger *Alpinia purpurata* of the zingiberaceae family. All plant flagellates belong to the order Kinetoplastida, family Trypanosomatidae. In recent years, however, flagellate protozoa have been isolated from fruits such as tomato. These flagellates, although trypanosomatids, do not seem to all belong to the genus *Phytomonas*.

TAXONOMY OF PLANT TRYPANOSOMATIDS

The taxonomy of *Phytomonas* species has not yet been resolved. The genus *Phytomonas* contains promastigote trypanosomatid flagellates that are parasites and have a life cycle completed in two hosts, a plant and an insect. In plants, some *Phytomonas* species live in the phloem sieve tubes (Figs. 16-3 and 16-5) of nonlaticiferous plants, such as coconut and oil palms, red ginger and coffee, and are definitely pathogenic. Others live in the latex-containing cells of laticiferous plants and are not considered to be pathogenic, although one (*P. fran̄tai*, Fig. 16-8C) has been associated with the empty root and subsequent decline of cassava. Still other trypanosomatid flagellates, some of them *Phytomonas* species and some that may not even belong to *Phytomonas*, parasitize and cause damage only to the fruit and seed of several plants.

In addition to being the only flagellates inhabiting phloem sieve tubes of their hosts, these plant pathogenic *Phytomonas* are closely related to one another but differ from those inhabiting latex tubes or fruits and seeds in several other characteristics. The differences may be in serological relationships, the sets of isoenzymes they possess, the patterns of DNA cleavage by restriction endonuclease enzymes, the sizes of minicircles of their kinetoplastid DNA, the gene repeat sequences of 5S ribosomal RNA, the presence in them of double-stranded RNA probably of viral origin, and possibly the presence in them of virus particles. The aforementioned tests plus studies comparing the spliced leader RNA gene array employing 29 trypanosomatid isolates from coconut palms, oil palms, and red ginger, revealed that (i) the spliced leader RNA gene sequences from phloem-restricted trypanosomatids are distinct from the same sequences of latex-restricted or fruit-infecting isolates; (ii) sequences in all the phloem-restricted isolates are highly similar; (iii) the phloem-restricted isolates can nevertheless be distinguished into two main groups; and (iv) one of the two main groups of phloem-restricted isolates can be subdivided further into two subgroups, one containing only coconut isolate and the other containing a mix of palm and ginger isolates. Moreover, although all *Phytomonas* can be grown on specialized nutrient media, phloem-inhabiting *Phytomonas* must be first grown for several passages on media containing cultured insect cells before they can be grown on cell-free media.

PATHOGENICITY OF PLANT TRYPANOSOMATIDS

Many of the investigators who studied the flagellates in laticiferous plants felt that although the flagellates parasitize the plants (they live off their latex), the plants do not become diseased and, therefore, the flagellates are not pathogenic to these plants. According to some reports, however, symptoms apparently do develop in some flagellate-infected laticiferous plants, which would indicate that the flagellates are pathogenic to their hosts. This seems to be the case in the empty root disease of cassava, in which flagellate protozoa present in the laticifer ducts of roots seem to be responsible for poor development of the root system and a general chlorosis of the cassava plant.

The nonlaticiferous hosts, coffee, coconut palm, oil palm, and red ginger, are apparently infected by pathogenic *Phytomonas* species and develop characteristic internal and external symptoms and severe and economically important diseases. Flagellates apparently cause the phloem necrosis disease of coffee, the hartrot disease of coconut palm, and the marchitez sopresiva (sudden wilt) disease of oil palms. All three diseases are so far known to occur only in South America. Flagellates also cause the red ginger wilt in the island of Grenada in the Caribbean Sea.

The mechanism by which protozoa cause disease in plants is not clear. As most of the disease-inducing ones inhabit the phloem sieve tubes, it is possible that they cause disease by blocking the transport of photosynthates to the roots. Laticifer-inhabiting *Phytomonas* have been shown to produce enzymes degrading pectin and cellulose, but these enzymes have not yet been studied in other phytomonads. Fruit-inhabiting phytomonads seem to cause local damage to fruit around the point of introduction, but this may also be due to concurrent infections by fungi and bacteria.

EPIDEMIOLOGY AND CONTROL OF PLANT TRYPANOSOMATIDS

In the field, *Phytomonas* protozoa seem to be transmitted by root grafts and by insects of the families Pentatomidae, Lygaeidae, and Coreidae. Several species of the pentatomid insect genera *Lincus* (Fig. 16-6D) and *Ochlerus*, for example, have been shown to transmit trypanosomes causing the hartrot disease of coconut palms and those causing the sudden wilt (marchitez sopresiva) of oil palms. So far, no insect vector is known for ginger wilt. Because phloem-inhabiting trypanosomes can be distinguished from laticifer-inhabiting trypanosomes by the criteria mentioned earlier, it is apparent that laticiferous weeds growing on palm or coffee plantations do not serve as reservoirs for palm- or coffee-infecting *Phytomonas*. It is not known, however, if wild palms, some of which may be resistant or symptomless to *Phytomonas* trypanosomes, may serve as a reservoir and source of *Phytomonas* for cultivated palm trees.

The control of plant diseases caused by flagellate protozoa would seem to be facilitated by using pathogen-free nursery plants and planting them away from infected plants. Control of insect vectors may or may not be useful or practical.

PLANT DISEASES CAUSED BY PHLOEM-RESTRICTED TRYPANOSOMATIDS

These include the main plant diseases caused by trypanosomatids. They have several common characteristics. Coffee phloem necrosis was apparently quite widespread in northern South America but its present distribution is uncertain. The palm diseases, hartrot of coconut palm and marcitez sopresiva of oil palm, occur wherever these plants are grown north of Lima, Peru, up to Brazil, Trinidad, and Honduras. Red ginger wilt and decay have been found only in the Caribbean island of Grenada.

Phloem Necrosis of Coffee

Phloem necrosis of coffee occurs in Suriname, Guyana, and probably Brazil, San Salvador, and Colombia. It affects trees of *Coffea liberica*, *C. arabica*, and other coffee species. Infected trees show sparse, yellowing, and dropping of leaves, and as the symptoms advance gradually only the young top leaves remain on the otherwise bare branches. As the roots begin to die back, the condition of the tree worsens, and the tree eventually dies (Fig. 16-2A). Sometimes, in the beginning of the dry season, trees wilt and die within 3 to 6 weeks (Fig. 16-2B). Internally, the roots and trunk of trees show multiple divisions of cambial cells and production of a zone of smaller and shorter phloem vessels of disorderly structure right next to the wood cylinder (Figs. 16-2C and 16-2D). At this stage the bark in the roots and the trunk is firmly attached to the wood and cannot be separated from it.

The pathogen, *Phytomonas leptovasorum*, is a trypanosomatid flagellate. When symptoms first appear there are only a few, big (14–18 by 1.0–1.2 μm), spindle-shaped flagellates in the phloem (Figs. 16-3A and 16-3B). As multiple division of cambial cells and abnormal phloem production become apparent and many leaves turn yellow and fall, the flagellates are numerous, slender, and spindle shaped, 4 to 14 by 0.3 to 1.0 micrometers (Fig. 16-3C). A few shorter (2.0–3.0 μm) forms of the flagellate, called leishmania, also appear in the oldest sieve tubes. When the multiple division of cambial cells results in a multilayered sheath around the wood cylinder that extends from the roots up to 2 meters above the ground line and the tree is almost dead, there is a great abundance of small (3–4 by 0.1–0.2 μm), "spaghetti" flagellates only in the living tissues of the stem, while previously occupied cells are evacuated.

The flagellates can be traced from the roots upward into the trunk, where they seem to migrate vertically in the phloem and laterally through the sieve plates into healthy sieve tubes. They also seem to move downward into unaffected roots. Flagellates could not be found in the tree outside the areas that show multiple division.

The disease can be transmitted through root grafts but not through green branch or leaf grafts. After grafting of healthy trees with roots infected with flagellates, the flagellates can be observed in previously healthy roots within a few weeks, the tree begins to develop external symptoms 4 to 5 months later, and it then dies

FIGURE 16-2 Coffee wilt of *Coffea liberica* caused by the flagellate protozoon *Phytomonas leptovasorum*. (A) Affected tree during rainy season. Note loss of leaves and yellowing but no acute wilting. (B) Affected tree at the onset of the dry season. Note sudden wilting. (C) Cross section of abnormal phloem tissue from flagellate-affected and wilting coffee tree. (D) Cross section of healthy phloem tissue of coffee tree. C and D, 700×. [Photographs courtesy of J. H. van Emden.]

shortly afterward. The disease spreads in the field from one tree to another, and healthy trees often become infected when transplanted in areas from which a diseased tree had been removed. The vector of the disease is one or more pentatomid insects of the genus *Lincus* (Fig. 16-6D).

In recent years, coffee wilt caused by trypanosomatids has been hard to find anywhere. Coffee cultivation has almost totally been abandoned in Suriname, from where the disease had been reported in the early 1900s and up to about 1970. On addition, the new varieties grown in the big coffee-producing countries of

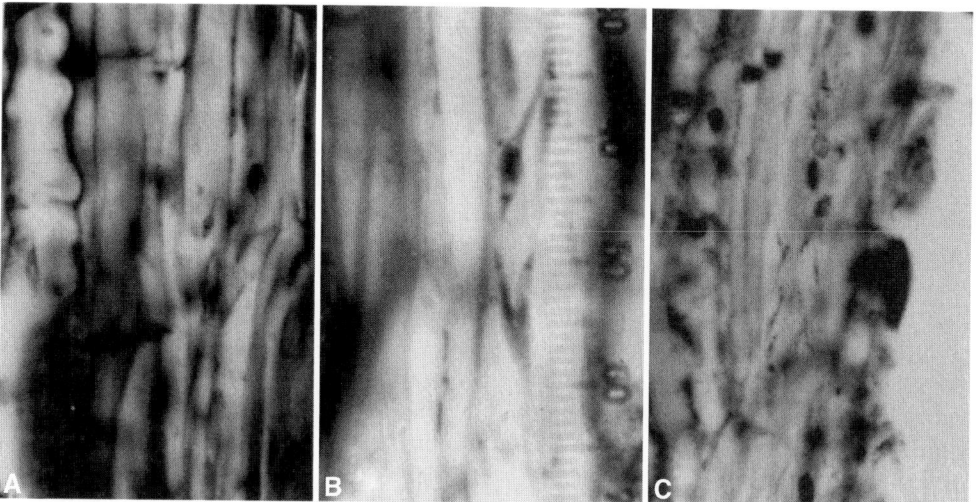

FIGURE 16-3 Flagellates associated with the coffee wilt disease. (A) Single protozoon in vascular vessel of diseased *Coffea liberica*. (B) Flagellates in vessel of *C. liberica*, one of them in the process of division. (C) Long and thin flagellates in vessels of coffee tree showing advanced symptoms of the disease. Magnified 1000×. [Photographs courtesy of J. H. van Emden.]

South and Central America are apparently resistant to the pathogen.

HARTROT OF COCONUT PALMS

Hartrot has been known in Suriname since 1906, sometimes under the more appropriate names of lethal yellowing or bronze-leaf wilt. The disease also occurs in Colombia and Ecuador and, under the local name Cedros wilt, in Trinidad. The symptoms of hartrot (Fig. 16-4) include yellowing and browning of the tips of older leaves (Fig. 16-4A) that subsequently spread to younger leaves (Figs. 16-4B–16-4D). Recently opened inflorescences are black (Figs. 16-4E and 16-4F), whereas unripe nuts of symptomatic trees fall off (Fig. 16-4G). At this stage, root tips also begin to rot. Petioles of older leaves may break, and the spear becomes necrotic. At later stages, the apical region of the crown also rots and often produces a foul odor. Trees infected with hartrot die within one to a few months of the appearance of external symptoms.

Flagellates of the genus *Phytomonas* occur in mature sieve elements of young leaves and inflorescences of hartrot-affected coconut palms. In advanced stages of the disease, 10 to 100% of the mature sieve elements contain flagellates, and many of them are plugged with flagellates, which are usually oriented longitudinally within the phloem (Fig. 16-5). The flagellates measure 12 to 18 by 1.0 to 2.5 micrometers. The number and spread of the flagellates in sieve tubes increase proportionally with the development of the disease.

Hartrot-causing protozoa are transmitted by pentatomid insects of the genera *Lincus* and *Ochlerus*. Hartrot spreads very rapidly. For example, about 15,000 coconut trees died in three years in the Cedros region of Trinidad.

SUDDEN WILT (MARCHITEZ SOPRESIVA) OF OIL PALM

Sudden wilt of oil palm is rather common and widespread in much of northern South America. It has been known since at least the 1960s in Colombia. The disease spreads rapidly through oil palm plantations and causes considerable damage by killing trees first in loci of a few to many trees (Fig. 16-6A) and then in increasingly larger areas. Symptoms begin as browning of the tips of the lower leaf leaflets. The browning subsequently spreads to the upper leaves and eventually becomes ashen gray (Figs. 16-6A and 16-6B). In the meantime, root tips also begin to die, and the whole root system deteriorates. As a result, plant growth slows down, fruit bunches discolor and rot or fall off, and within a few weeks all leaves become ashen gray and dry up and the whole tree dies (Figs. 16-6B and 16-6C). *Phytomonas* flagellates occur widely in the phloem sieve elements of roots (Fig. 16-5), leaves, and inflorescences of infected plants. These flagellates are also transmitted readily by the pentatomid insects *Lincus* (Fig. 16-6) and *Ochlerus*. Some control of sudden wilt has reportedly been obtained by spraying insecticides that control the vectors of the protozoa.

FIGURE 16-4 Hartrot of coconut palms caused by flagellate protozoa. (A) Malayan dwarf palm showing medium stage of leaf yellowing from the bottom of the tree up caused by hartrot disease. (B) A further stage of hartrot showing more browning than yellowing. (C) Medium to late stage of coconut hartrot on a plot of hybrid palm trees. (D) Late stage of hartrot on a tall palm variety. Note broken leaves and the collapsed spear. (E) Unopened inflorescence of Malayan dwarf palm showing necrotic spike tops, one of the first symptoms of the disease. (F) Inflorescence showing necrotic spike tops. (G) Ceylonese dwarf yellow palm, 4 years old, suffering from hartrot disease. Note dropping of nuts on the ground, another early symptom. [Photographs courtesy of (A–D) M. Dollet and (E–G) W. G. van Slobbe.]

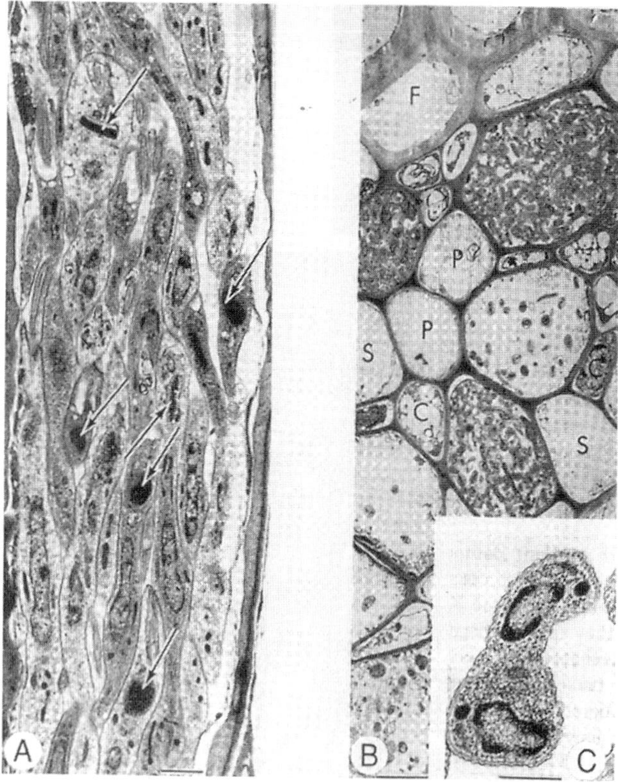

FIGURE 16-5 *Phytomonas* protozoa in phloem sieve tubes of young inflorescence of coconut palm tree affected with hartrot. Electron micrographs of longitudinal (A) and cross sections (B) of phloem cells filled with protozoa and of a cross section of a flagellate undergoing longitudinal fission (C). Arrows in A point to the DNA portion of kinetoplasts. Scale bars: 1 μm. F, fiber; P, phloem parenchyma cell; C, companion cell; S, sieve elements free of flagellates. [Photographs courtesy of M. V. Parthasarathy.]

Wilt and Decay of Red Ginger

Red ginger, *Alpinia purpurata*, an ornamental plant (Fig. 16-7A) of the family zingiberaceae has been reported to become infected with phloem-restricted trypanosomatid protozoa that cause wilt and eventually decay of the plants (Fig. 16-7). The disease on red ginger occurs in the island of Grenada of the Caribbean Sea. The red ginger trypanosomatid resembles the other phloem-limited palm-infecting trypanosomatids of Venezuela and Colombia in some respects, but no vector for it is known so far. Nevertheless, because red ginger grows near oil palm trees in Venezuela and Colombia, it is thought that the trypanosomatid infecting red ginger was probably derived from infected oil palm trees and was introduced into Grenada in infected plants by farmers who wanted to expand the cultivation of red ginger to the island.

"PLANT DISEASES" CAUSED BY LATICIFER-RESTRICTED TRYPANOSOMATIDS

Empty Root of Cassava

Trypanosomatids growing in the latex of laticiferous plants have been found in plants of at least nine families almost worldwide. Latex trypanosomatids are transmitted from plant to plant by pentatomid bugs from the genera *Lincus*, *Ochlerus*, and probably others.

Infected laticiferous plants generally show no pathological symptoms, but in the case of empty root of cassava, a laticiferous plant, symptoms are produced. The empty root disease was observed affecting certain cultivars of cassava (*Manihot esculenta*) in the Espirito Santo state of Brazil. The root system of affected plants develops poorly. Roots in general remain small and slender and contain little or no starch. The aboveground parts of infected plants show general chlorosis and decline (Fig. 16-8). The empty root disease can be transmitted by grafting. It also spreads rapidly in the field, probably by an insect vector like those mentioned earlier. Diseased plants contain numerous *Phytomonas*-like protozoa in the laticifer ducts (Fig. 16-8), but not in the phloem. Typical *Phytomonas* protozoa can be seen easily with a light microscope in latex exuded from wounds of infected plants. So far, however, it has not been proved beyond any doubt that protozoa are the cause of the empty root disease.

"PLANT DISEASES" CAUSED BY FRUIT- AND SEED-INFECTING TRYPANOSOMATIDS

Fruit Trypanosomatids

Trypanosomatids have been found to cause minor disease on tomatoes in South Africa, Spain, and Brazil. At least four genera of trypanosomatids (e.g., *Herpetomonas*, *Leptomonas*, and *Phytomonas*) have been isolated from tomato fruit but so far all are called *Phytomonas serpens*. It is possible to culture all of them in the laboratory. Also, fruit can be contaminated with trypanosomatids by all kinds of insects that feed on fruit.

In fruit and seed diseases caused by trypanosomatids, the latter are found around the point of inoculation, which is usually carried out by an insect vector. The diseases appear as localized yellow patches that may also exhibit malformations and it is within these patches that trypanosomatids can be found multiplying in the wound made by the insect vector. Only fruit damaged by the insects become infected; the infection, however, unlike systemic infections caused by phloem-restricted trypanosomes, remains localized and the mother plants,

FIGURE 16-6 (A) Sudden wilt or Marchitez sopresiva on a 5-year-old oil palm tree in north Colombia. (B) Marchitez of oil palm trees in a 2-year-old plantation east of the Andes mountains. (C) An 18-year-old palm showing sudden wilt symptoms in the Llanos region east of the Andes. (D) Adult *Lincus* insect, the most common vector of phloem-restricted trypanosomatids. [Photographs courtesy of (A–C) M. Dollet, (D) taken by R. Desmier De Chenon and provided by M. Dollet.]

FIGURE 16-7 (A) Healthy and trypanosomatid-infected red ginger plants growing in a field in the island of Grenada. (B) A lightly infected red ginger plant (left) and a severely infected plant (right) showing the bottom-to-top progress of trypanosomatid diseases of plants. [Photographs courtesy of M. Dollet.]

FIGURE 16-8 (A) A chlorotic and declining cassava plant affected with trypanosomatids and showing empty root disease. (B) Two healthy cassava roots and, above and in between them, the small, useless root of a cassava plant affected by empty root. (C) Scanning electron micrograph of *Phytomonas franзai* in a latex vessel of a cassava plant affected with the empty root disease. [Photographs courtesy of E. W. Kitajima.]

on which fruit become infected, remain free of trypanosomatids.

Selected References

Dollet, M. (1984). Plant diseases caused by flagellated protozoa (*Phytomonas*). *Annu. Rev. Phytopathol.* **22**, 115–132.

Dollet, M., *et al.* (2000). 5S ribosomal RNA gene repeat sequences define at least eight groups of plant trypanosomatids (*Phytomonas* spp.): Phloem restricted pathogens form a distinct section. *J. Eukaryot. Microbiol.* **47**, 569–574.

Dollet, M. (2001). Phloem-restricted trypanosomatids form a clearly characterized monophyletic group among trypanosomatids isolated from plants. *Int. J. Parasitol.* **31**, 459–467.

Dollet, M., Sturm, N. R., and Campbell, D. A. (2001). The spliced leader RNA gene array in phloem-restricted plant trypanosomatids (*Phytomonas*) partitions into two major groupings: Epidemiological implications. *Parasitology* **122**, 289–287.

Gargani, D., *et al.* (1992). In vitro cultivation of *Phytomonas* from latex and phloem-restricted *Phytomonas*. *Oleagineaux* **47**, 596.

Harvey, R. D., and Lee, S. B. (1943). Flagellates of laticiferous plants. *Plant Physiol.* **18**, 633–655.

Kitajima, E. W., Vainstein, M. H., and Silveira, J. S. M. (1986). Flagellate protozoon associated with poor development of the root system of cassava in the Espirito Santo state of Brazil. *Phytopathology* **76**, 638–642.

Lafont, A. (1909). Sur la presence d'un parasite de la classe des flagelles dans le latex de *l'Euphorbia pilulifera*. *C. R. Soc. Biol.* **66**, 1011–1013.

Louise, C., Dollet, M., and Mariau, D. (1986). Research into hartrot of the coconut, a disease caused by *Phytomonas* (*Trypanosomatidae*), and into its vector *Lincus* sp. (Pentatomidae) in Guiana. *Oleagineux* **41**, 437–449.

Marche, S., *et al.* (1993). RNA virus-like particles in pathogenic plant trypanosomatids. *Mol. Biochem. Parasitol.* **57**, 261–268.

McCoy, R. E., and Martinez-Lopez, G. (1982). *Phytomonas staheli* associated with coconut and oil palm diseases in Colombia. *Plant Dis.* **66**, 675–677.

Parthasarathy, M. V., van Slobbe, W. G., and Soudant, C. (1976). Trypanosomatid flagellate in the phloem of diseased coconut palms. *Science* **192**, 1346–1348.

Sanchez Moreno, M., *et al.* (1995). Isolation, *in vitro* culture, ultrastructure study, and characterization by lectin-agglutinating tests of *Phytomonas* isolates from tomatoes (*Lycopersicon escolentum*) and cherimoyas (*Anona cherimolia*) in southeastern Spain. *Parasitol Res.* **81**, 575–581.

Stahel, G. (1933). Zur Kenntnis der Siebrohren-krankheit (Phloemnekrose) des Kaffeebaumes in Surinam. III. *Phytopathol. Z.* **6**, 335–357.

Thomas, D. L., McCoy, R. E., Norris, R. C., and Espinoza, A. S. (1979). Electron microscopy of flagellated protozoa associated with marchitez sopresiva disease of African oil palm in Ecuador. *Phytopathology* **69**, 222–226.

van Emden, J. H. (1962). On flagellates associated with a wilt of *Coffea liberica*. *Meded. Landbouwhogesch. Opzoekingsstn. Staat Gent* **27**, 776–784.

Vermeulen, H. (1963). A wilt of *Coffea liberica* in Surinam and its association with a flagellate, *Phytomonas leptovasorum*. *J. Protozool.* **10**, 216–222.

Vermeulen, H. (1968). Investigations into the cause of the phloem necrosis disease of *Coffea liberica* in Surinam, South America. *Neth. J. Plant Pathol.* **74**, 202–218.

Vickerman, K., and Dollet, M. (1992). Report on the second *Phytomonas* workshop. Santa Marta, Colombia, 5–8 February 1992. *Oleagineux* **47**, 593–595.

Waters, H. (1978). A wilt disease of coconuts from Trinidad, associated with *Phytomonas* sp., a sieve tube-restricted protozoan flagellate. *Annu. Appl. Biol.* **90**, 293–302.

Glossary

Abiotic Nonliving, or caused by a nonliving agent; e.g., abiotic disease.

Acervulus A subepidermal, saucer-shaped, asexual fruiting body producing conidia on short conidiophores.

Acquired resistance Plant resistance to disease activated after inoculation of the plant with certain microorganisms or treatment with certain chemical compounds.

Active defense Defenses induced in the plant after attack by a pathogen.

Aecium A cup-shaped fruiting body of rust fungi that produces aeciospores.

Aerobic A microorganism that lives, or a process that occurs, in the presence of molecular oxygen.

Aflatoxin A mycotoxin produced by the fungus *Aspergillus flavus* and by some other fungi.

Agar A gelatin-like material obtained from seaweed and used to prepare culture media on which microorganisms are grown and studied.

Agroterrorism Terroristm caused by scaring consumers away from buying certain agricultural products such as vegetables, milk, and meat, by contaminating them on the farm or in the market with human pathogens. Also, scaring people for future shortages of food by spreading plant pathogens on crops so that terrorists reduce the amount of food produced.

Alarm signal A chemical compound, presumably produced by a host plant, in response to infection, and sent out to host cell proteins and genes that the plant activates to produce substances inhibitory to the pathogen.

Allele One of two or more alternate forms of a gene occupying the same locus on a chromosome.

Allozyme An enzyme with slightly altered properties produced by an allele of the original gene.

Alternate host One of two kinds of plants on which a parasitic fungus (e.g., rust) must develop to complete its life cycle.

Anaerobic A microorganism that lives, or a process that occurs, in the absence of molecular oxygen.

Anamorph The imperfect or asexual stage of a fungus.

Anastomosis The union of a hypha with another, resulting in intercommunication of their genetic material.

Antheridium The male sexual organ found in some fungi.

Anthracnose A disease that appears as black, sunken, leaf, stem, or fruit lesions, caused by fungi that produce their asexual spores in an acervulus.

Antibiotic A chemical compound produced by one microorganism that inhibits or kills other microorganisms.

Antibody A protein produced in a warm-blooded animal in reaction to an injected foreign antigen and capable of reacting specifically with that antigen.

Antigen A substance, usually a protein, that, when injected into a warm-blooded animal, causes the formation of an antibody.

Antiserum The blood serum containing antibodies possessed by a warm-blooded animal.

Apoplast The area outside the plasma membrane of cells, consisting of cell walls and conducting cells of the xylem, that contains the aqueous phase of intercellular solutes.

Apoptosis A common type of cell death involving a highly regulated, energy-dependent process in animals, but is quite rare in plants.

Apothecium An open cup- or saucer-shaped ascocarp of some ascomycetes.

Appressorium The swollen tip of a hypha or germ tube that facilitates attachment and penetration of the host by a fungus.

Arbuscule A branched, tuft-like haustorium, produced by certain mycorrhizal fungi inside root cells.

Area under the disease progress curve (AUDPC) The area of a graph under the line that depicts the progress of an epidemic.

Ascocarp The fruiting body of ascomycetes bearing or containing asci.

Ascogenous hyphae Hyphae arising from the fertilized ascogonium and producing the asci.

Ascogonium The female gametangium or sexual organ of ascomycetes.

Ascomycetes A group of fungi producing their sexual spores, ascospores, within asci.

Ascospore A sexually produced spore borne in an ascus.

Ascostroma The ascocarp or reproductive structure of certain ascomycetes that bears the spore sacs within cavities called locules.

Ascus A sac-like cell of a hypha in which meiosis occurs and that contains ascospores (usually eight).

Asexual reproduction Any type of reproduction not involving the union of gametes or meiosis.

Attacins Antimicrobial proteins that inhibit the synthesis of outer membrane proteins of gram-negative bacteria.

Attenuation Partial or complete loss of virulence in a pathogen.

Autoecious fungus A parasitic fungus that can complete its entire life cycle on the same host.

Auxotroph An organism partly or totally deficient on a substance, the addition of which significantly promotes the growth of the organism.

Avirulence The inability of a pathogen to infect a certain plant variety that carries genetic resistance.

Avirulent Lacking virulence.

***Avr* gene** A gene that codes for avirulence.

Avr protein The protein coded for by an *Avr* gene, acting as an elicitor of defense reactions.

Bacillus A rod-shaped bacterium.

Bactericide A chemical compound that kills bacteria.

Bacteriocins Bactericidal substances produced by certain strains of bacteria and are active against some other strains of the same or closely related species.

Bacteriophage A virus that infects bacteria and usually kills them.

Bacteriostatic A chemical or physical agent that prevents multiplication of bacteria without killing them.

Base An alkaline, usually nitrogenous organic compound, used particularly for the purine and pyrimidine moieties of the nucleic acids of cells and viruses.

Basidiomycetes A group of fungi producing their sexual spores, basidiospores, on basidia.

Basidiospore A sexually produced spore borne on a basidium.

Basidium A club-shaped structure on which basidiospores are borne.

Bioassay The use of a test organism to measure the relative infectivity of a pathogen or toxicity of a substance.

Biofilm A polysaccharide matrix in which one or more species of bacteria and fungi are embedded. May play a role in bacterial and fungal attachment, colonization, and host invasion.

Bioinformatics The accumulation of data on biological sequencing of the genome of an organism to predict gene function, protein and RNA structure, gene regulation, genome organization, etc.

Biological control Total or partial inhibition or destruction of pathogen populations by other organisms.

Biotechnology The use of genetically modified organisms and/or modern techniques and processes with biological systems for industrial production.

Bioterrorism The frightening and terrorizing of civilian populations by terrorists spreading or threatening to spread microorganisms pathogenic to humans in ways that can reach and infect the people.

Biotic Living; associated with or caused by a living organism.

Biotroph An organism that can live and multiply only on another living organism.

Biotype A subgroup within a species or race, usually characterized by the common possession of a single or a few new characters.

Blight A disease characterized by general and rapid killing of leaves, flowers, and stems.

Blotch A disease characterized by large, irregularly shaped, spots or blots on leaves, shoots, and stems.

Breeding The use of controlled reproduction to improve certain characteristics in plants and animals.

Budding A method of vegetative propagation of plants by implantation of buds from the mother plant onto a rootstock.

Bunt A disease of wheat caused by the fungus *Tilletia* in which contents of the wheat grains are replaced by odorous smut spores.

Callus A mass of thin-walled undifferentiated cells, developed as the result of wounding or culture on nutrient media.

Canker A necrotic, often sunken, lesion on a stem, branch, or twig of a plant.

Capsid The protein coat of viruses forming the closed shell or tube that contains nucleic acid.

Capsule A relatively thick layer of mucopolysaccharides that surrounds some kinds of bacteria.

Carbohydrates Foodstuffs composed of carbon, hydrogen, and oxygen (CH_2O) with the last two in a 2-to-1 ratio, as in water (H_2O).

Cecropins Antimicrobial lytic proteins that make pores in and cause the lysis of the bacterial cell membrane.

Cellulase An enzyme that breaks down cellulose.

Cellulose A polysaccharide composed of hundreds of glucose molecules linked in a chain and found in plant cell walls.

Chaperon protein A protein molecule that is attached to an "effector" protein to protect it from coming in contact with other proteins while it is being exported through a type III secretion apparatus.

Chemotherapy Control of a plant disease with chemicals (chemotherapeutants) that are absorbed and translocated internally.

Chitin A complex, N-containing carbohydrate, derived from *N*-acetyl-D-glucosamine, forming the hard outer shell of insects, crustaceans, arthropods, fungi, and some algae.

Chlamydospore A thick-walled asexual spore formed by the modification of a cell of a fungus hypha.

Chlorosis Yellowing of normally green tissue due to chlorophyll destruction or failure of chlorophyll formation.

Chronic symptoms Symptoms that appear over a long period of time.

Circulative viruses Viruses that are acquired by their vectors through their mouthparts, accumulate internally, and then are passed through tissues of the vector and introduced into plants again via the mouthparts of the vectors.

Cistron The sequence of nucleotides within a certain area of nucleic acid (DNA or RNA) that codes for a particular protein.

Cleistothecium An entirely closed ascocarp.

Clone A group of genetically identical individuals produced asexually from one individual.

Cloning A group of DNA molecules derived from one original length of DNA sequences and produced by a bacterium or virus into which it was introduced, using genetic engineering techniques, often involving plasmids.

Coding The process by which the sequence of nucleotides within a certain area of RNA determines the sequence of amino acids in the synthesis of the particular protein.

Codon A coding unit, consisting of three adjacent nucleotides, that codes for a specific amino acid.

Comparative genomics Comparisons of complete genome sequences at the whole genome level across genera, species, strains, etc.

Complementary DNA (cDNA) DNA synthesized by reverse transcriptase from an RNA template.

Conidiophore A specialized hypha on which one or more conidia are produced.

Conidium An asexual fungus spore formed from the end of a conidiophore.

Conjugation A process of sexual reproduction involving the fusion of two gametes. Also, in bacteria, the transfer of genetic material from a donor cell to a recipient cell through direct cell-to-cell contact.

Constitutive A substance, usually an enzyme, whose presence and, often, concentration in a cell remain constant, unaffected by the presence of its substrate.

Cork An external, secondary tissue impermeable to water and gases. It is often formed in response to wounding or infection.

Cross protection The phenomenon in which plant tissues infected with one strain of a virus are protected from infection by other, more severe, strains of the same virus.

Culture To artificially grow microorganisms or plant tissue on a prepared food material; a colony of microorganisms or plant cells artificially maintained on such food material.

Cuticle A thin, way layer on the outer wall of epidermal cells consisting primarily of wax and cutin.

Cutin A waxy substance comprising the inner layer of the cuticle.

Cyst An encysted zoospore (fungi); in nematodes, the carcass of dead adult females of the genus *Heterodera* or *Globodera*, which may contain eggs.

Cytokinins A group of plant growth-regulating substances that regulate cell division.

Cytoplasmic resistance Resistance controlled by genetic material present in the cell cytoplasm.

Dalton A unit of mass equaling the atomic weight of a hydrogen atom.

Damping-off Destruction of seedlings near the soil line, resulting in the seedlings falling over on the ground.

Defence activators Synthetic chemicals that, when applied to plants as sprays, injections, root treatments, etc., induce systemic acquired resistance in them to several types of pathogens.

Defensins A group of defense-related, cysteine-rich, antimicrobial peptides present in the plasma membrane of most plant species that provide resistance to different pathogens.

Denatured protein Protein whose properties have been altered by treatment with physical or chemical agents.

Density-gradient centrifugation A method of centrifugation in which particles are separated in layers according to their density.

Dieback Progressive death of shoots, branches, and roots, generally starting at the tip.

Dikaryotic Mycelium or spores containing two sexually compatible nuclei per cell. Common in the basidiomycetes.

Disease Any malfunctioning of host cells and tissues that results from continuous irritation by a pathogenic agent or environmental factor and leads to development of symptoms.

Disease cycle The chain of events involved in disease development, including the stages of development of the pathogen and the effect of the disease on the host.

Disinfectant A physical or chemical agent that frees a plant, organ, or tissue from infection.

Disinfestant An agent that kills or inactivates pathogens in the environment or on the surface of a plant or plant organ before infection takes place.

Downy mildew A plant disease in which the sporangiophores and spores of a fungus appear as a downy growth on the lower surface of leaves and stems, fruit, etc., caused by fungi in the family Peronosporaceae.

Ectoparasite A parasite feeding on a host from the exterior.

"Effector" protein A protein coded by a bacterial pathogenicity/virulence gene that is exported into the plant and interacts with an R-gene protein.

Egg A female gamete. In nematodes, the first stage of the life cycle containing a zygote or a juvenile.

Elicitors Molecules produced by a pathogen that induce a defense response by the host.

ELISA A serological test in which one antibody carries with it an enzyme that releases a colored compound.

Enation Tissue malformation or overgrowth, induced by certain virus infections.

Endoparasite A parasite that enters a host and feeds from within.

Enzyme A protein produced by living cells that can catalyze a specific organic reaction.

Epidemic A disease increase in a population; usually a widespread and severe outbreak of a disease.

Epidemic rate The amount of increase of disease per unit or time in a plant population.

Epidemiology The study of factors affecting the outbreak and spread of infectious diseases.

Epidermis The superficial layer of cells occurring on all plant parts.

Epiphytically Existing on the surface of a plant or plant organ without causing infection.

Epiphytotic A widespread and destructive outbreak of a disease of plants; epidemic.

Eradicant A chemical substance that destroys a pathogen at its source.

Eradication Control of plant disease by eliminating the pathogen after it is established or by eliminating the plants that carry the pathogen.

Expressed sequence tag (EST) Molecular landmarks that provide a profile of mRNAs and allow cloning of a large number of genes being expressed in a cell population.

Etiology of disease The determination and study of the cause of a disease.

Facultative parasite Having the ability to be a parasite.

Fermentation Oxidation of certain organic substances in the absence of molecular oxygen.

Fertilization The sexual union of two nuclei, resulting in doubling of chromosome numbers.

Filamentous Thread like; filiform.

Fission Transverse splitting in two of bacterial cells; asexual reproduction.

Fitness The ability of a pathogen to survive and reproduce.

Flagellin A receptor system for general elicitors very similar and common to plants and animals.

Flagellum A whip-like structure projecting from a bacterium or zoospore and functioning as an organ of locomotion; also called a *cilium*.

Forma specialis (f. sp.) A group of races and biotypes of a pathogen species that can infect only plants within a certain host genus or species.

Free-living Of a microorganism that lives freely, unattached, or a pathogen living in the soil, outside its host.

Fructification Production of spores by fungi; also, a fruiting body.

Fruiting body A complex fungal structure containing spores.

Fumigant A toxic gas or volatile substance that is used to disinfest soil or certain areas from various pests.

Fumigation The application of a fumigant for disinfestation of an area or soil.

Functional genomics Genetic studies focusing on the functions and interactions of genes or groups of genes that may belong to plants, pathogens, or both.

Fungicide A compound toxic to fungi.

Fungigation Application of fungicides to foliage or roots through the irrigation system.

Fungistatic A compound that prevents fungus growth without killing the fungus.

Gall A swelling or overgrowth produced on a plant as a result of infection by certain pathogens.

Gametangium A cell containing gametes or nuclei that act as gametes.

Gamete A male or female reproductive cell or nuclei within a gametangium.

Gene A linear portion of the chromosome that determines or conditions one or more hereditary characters; the smallest functioning unit of the genetic material.

Gene cloning The isolation and multiplication of an individual gene sequence by its insertion into a bacterium, which can multiply the gene as it multiplies itself.

Gene flow The process by which certain genes move from one population to another geographically separated one.

Gene for gene The concept that for each gene for virulence in a pathogen there is a corresponding gene for resistance in the host toward that pathogen.

Gene knockout The disruption of a target gene by transformation or mutation and characterization of the function of the gene by assessing the phenotype of the resulting mutant.

Gene silencing The interruption or suppression of the activity of a targeted gene that prevents it from coordinating the production of specific proteins.

Genetically modified organisms (GMOs).

Genetic drift The occurrence of random effects (mutations, etc.) in individuals of a population that affect the survival of various genetic traits in subsequent generations.

Genetic engineering Alteration of the genetic composition of a cell or organism by various procedures (transformation, protoplast fusion, etc.).

Genetic load or drag Accumulation of excess genes for any characteristic, even for virulence, that imposes a fitness penalty to the organism.

Genome sequencing The orderly reading of all the millions of nucleotides constituting the total DNA of a living organism.

Genomics Studies focusing on the analysis of whole genomes of organisms.

Genotype The genetic constitution of an organism.

Genotype flow Transfer of entire genotypes of asexually only reproducing microorganisms from one population to another.

Germ theory The proposal that infectious and contagious diseases are caused by germs (microorganisms).

Germ tube The early growth of mycelium produced by a germinating fungus spore.

Gibberellins A group of plant growth-regulating substances with a variety of functions.

G-proteins A subset of the GTPase superfamily of proteins that is concerned with the accuracy of recognition or interaction of activated receptor sites.

Grafting A method of plant propagation by transplantation of a bud or a scion of a plant on another plant; also the joining of cut surfaces of two plants so as to form a living union.

Growth regulator A natural substance that regulates the enlargement, division, or activation of plant cells.

Gum Complex polysaccharidal substances formed by cells in reaction to wounding or infection.

Gummosis Production of gum by or in a plant tissue.

Guttation Exudation of water from plants, particularly along the leaf margin.

Habitat The natural place of occurrence of an organism.

Haploid A cell or an organism whose nuclei have a single complete set of chromosomes.

Harpins, or pilins Proteins coded by *hrp* (hypersensitive response and pathogenicity) genes that are used to make type III protein secretion systems.

Haustorium A simple or branched projection of hyphae into host cells that acts as an absorbing organ.

Hectare An area of land equal to 2.5 acres.

Hemibiotrophic An organism that lives part of its life as a parasite on another organism and the other part as a sarophyte.

Herbaceous plant A higher plant that does not develop woody tissues.

Hermaphrodite An individual bearing both functional male and female reproductive organs.

Heteroecious Requiring two different kinds of hosts to complete its life cycle, pertaining particularly to rust fungi.

Heterokaryosis The condition in which a mycelium contains two genetically different nuclei per cell.

Heteroploid A cell, tissue, or organism that contains more or fewer chromosomes per nucleus than the normal 1N or 2N for that organism.

Heterothallic fungi Fungi producing compatible male and female gametes on physiologically distinct mycelia.

Homothallic fungus A fungus producing compatible male and female gametes on the same mycelium.

Hormone A growth regulator, frequently referring particularly to auxins.

Horizontal resistance Partial resistance, equally effective against all races of a pathogen.

Host A plant that is invaded by a parasite and from which the parasite obtains its nutrients.

Host range The various kinds of host plants that may be attacked by a parasite.

Hyaline Colorless; transparent.

Hybrid The offspring of two individuals differing in one or more heritable characteristics.

Hybridization The crossing of two individuals differing in one or more heritable characteristics.

Hybridoma A hybrid animal cell produced by the fusion of a spleen cell and a cancer cell and able to produce monoclonal antibodies and to multiply.

Hydathodes Structures with one or more openings that discharge water from the interior of a leaf to its surface.

Hydrolysis The enzymatic breakdown of a compound through the addition of water.

Hyperparasite A parasite parasitic on another parasite.

Hyperplasia A plant overgrowth due to increased cell division.

Hypersensitivity Excessive sensitivity of plant tissues to certain pathogens. Affected cells are killed quickly, blocking the advance of obligate parasites.

Hypertrophy A plant overgrowth due to abnormal cell enlargement.

Hypha A single branch of a mycelium.

Hypovirulence Reduced virulence of a pathogen strain as a result of the presence of transmissible double-stranded RNA.

Immune Cannot be infected by a given pathogen.

Immunity The state of being immune.

Imperfect fungus A fungus that is not known to produce sexual spores; also known as a deuteromycete or a mitosporic fungus.

Imperfect stage The part of the life cycle of a fungus in which no sexual spores are produced; the anamorph stage.

Inclusion bodies Crystalline or amorphous structures in virus-infected plant cells that are produced by and consist largely of viruses and are visible under a compound microscope.

Incubation period The period of time between penetration of a host by a pathogen and the first appearance of symptoms on the host.

Indexing A procedure to determine whether a given plant is infected by a virus or a xylem- or phloem-infecting fastidious bacterium. It involves the transfer of a bud, scion, sap, etc. from one plant to one or more kinds of (indicator) plants that are sensitive to the virus or other pathogen.

Indicator A plant that reacts to certain viruses or environmental factors with production of specific symptoms and is used for detection and identification of these factors.

Induced systemic resistance A systemic resistance in plants that is triggered by certain strains of nonpathogenic root-colonizing bacteria; its signaling requires jasmonic acid and ethylene.

Inducible or induced A substance, usually an enzyme, whose production has been or may be stimulated by another compound, often a substrate or a structurally related compound called an inducer.

Infection The establishment of a parasite within a host plant.

Infectious disease A disease that is caused by a pathogen that can spread from a diseased to a healthy plant.

Infested Containing great numbers of insects, mites, nematodes, etc. as applied to an area or field. Also applied to a plant surface, soil, container, or tool contaminated with bacteria, fungi, etc.

Injectosome In gram-positive bacteria, injectosome is a complex consisting of the ExPortal, a novel organelle that organizes the general secretory machinery (Sec), the Sec pathway for protein export, and a pore-forming cytolysin, and functions to inject signal transduction proteins into host cells.

Injury Damage of a plant by an animal, physical, or chemical agent.

Inoculate To bring a pathogen into contact with a host plant or plant organ.

Inoculation The arrival or transfer of a pathogen onto a host.

Inoculum The pathogen or its parts that can cause infection; that portion of individual pathogens that are brought into contact with the host.

Integrated control An approach that attempts to use all available methods of control of a disease or of all the diseases and pests of a crop plant for best control results but with the least cost and the least damage to the environment.

Integrated pest management The attempt to prevent pathogens, insects, and weeds from causing economic crop losses by using a variety of management methods that are cost effective and cause the least damage to the environment.

Intercalary Formed along and within the mycelium, not at the hyphal tips.

Intercellual Between cells.

Intracellular Within or through the cells.

Introns Sections of 70–140 nucleotide noncoding premessenger RNA that exist between exons and are spliced during the processing of mRNA.

Invasion The spread of a pathogen into the host.

In vivo In culture, outside the host.

In vivo In the host.

Isolate A single spore or culture and the subcultures derived from it. Also used to indicate collections of a pathogen made at different times.

Isolation The separation of a pathogen from its host and its culture on a nutrient medium.

Isozymes The different forms of an enzyme that carry out the same enzymatic reaction but require different conditions (pH, temperature, etc.) for optimum activity.

Juvenile The life stages of a nematode between the embryo and the adult; an immature nematode.

Kilobase One thousand continuous bases (nucleotides) of single-stranded RNA or DNA.

Kinase A protein enzyme that phosphorylates (adds phosphate), and thereby activating a target protein.

Latent infection The state in which a host is infected with a pathogen but does not show any symptoms.

Latent virus A virus that does not induce symptom development in its host.

Leaf spot A self-limiting lesion on a leaf.

Lectins A group of plant proteins that bind to specific carbohydrates.

Lenticel An opening in the stem of woody plants that has spongy cells at its base and allows for the exchange of gases between the plant and the atmosphere.

Leucine-rich repeats (LRR) Repetitious segments of amino acids containing multiple copies of leucine on a protein.

Life cycle The stage or successive stages in the growth and development of an organism that occur between the appearance and reappearance of the same stage (e.g., spore) of the organism.

Lipids Substances whose molecules consist of glycerin and fatty acids and sometimes certain additional types of compounds.

Local lesion A localized spot produced on a leaf upon mechanical inoculation with a virus.

LRR proteins Proteins containing leucine-rich repeats

Macroscopic Visible to the naked eye without the aid of a magnifying lens or a microscope.

Malignant Use of a cell or tissue that divides and enlarges autonomously, i.e., its growth can no longer be controlled by the organism on which it is growing.

Masked symptoms Symptoms of a irus-infected plant that are absent under certain environmental conditions but appear when the host is exposed to certain conditions of light and temperature.

Mechanical inoculation Inoculation of a plant with a virus through transfer of sap from a virus-infected plant to a healthy plant.

Meiospore A spore produced through meiosis, a sexual spore.

Melanin A dark brown to black compound found in the cell walls of some fungi and needed by them for pathogenicity.

Messenger RNA (mRNA) A chain of ribonucleotides that codes for a specific protein.

Metabolism The process by which cells or organisms utilize nutritive material to build living matter and structural components or to break down cellular material into simple substances to perform special functions.

Microarray analysis A molecular method employing large-scale hybridization of fluorescently labeled nucleic acids from biological samples to single-stranded cDNA sequences and used to study the degree of expression of thousands of genes in parallel during a certain treatment.

Micrometer (μm) A unit of length equal to 1/1000 of a millimeter.

Microscopic Very small; can be seen only with the aid of a microscope.

Middle lamella The cementing layer between adjacent cell walls; it generally consists of pectinaceous materials, except in woody tissues, where pectin is replaced by lignin.

Migratory Migrating from plant to plant.

Mildew A fungal disease of plants in which the mycelium and spores of the fungus are seen as a whitish growth on the host surface.

Millimeter (mm) A unit of length equal to 1/10 of a centimeter (cm) or 0.03937 of an inch.

Mitosporic fungi Producing spores only through mitosis (imperfect fungi or deuteromycetes).

Mold Any profuse or woolly fungus growth on damp or decaying matter or on surfaces of plant tissue.

Molecular marker A molecular characteristic (a landmark) on a piece of DNA that can be used to compare that DNA for degrees of similarity with those of other microorganisms.

Molt The shedding or casting off of the cuticle in a nematode or insect.

Monoclonal antibodies Identical antibodies produced by a single clone of lymphocytes and reacting only with one of the antigenic determinants of a pathogen or protein.

Monocyclic Having one cycle per season.

Mosaic Symptom of certain viral diseases of plants characterized by intermingled patches of normal and light green or yellowish color.

Mottle An irregular pattern of indistinct light and dark areas.

Movement protein One or more proteins of a virus that facilitate the movement of the virus through the plant and/or by the vector.

Mummy A dried, shriveled fruit.

Mutant An individual possessing a new, heritable characteristic as a result of a mutation.

Mutation An abrupt appearance of a new characteristic in an individual as the result of an accidental change in a gene or chromosome.

Mycelium The hypha or mass of hyphae that make up the body of a fungus.

Mycoplasma-like organisms Microorganisms found in the phloem and phloem parenchyma of diseased plants and assumed to be the cause of the disease; they resemble mycoplasmas in all respects except that they cannot yet be grown on artificial nutrient media. Now called phytoplasmas or spiroplasmas.

Mycoplasmas Pleomorphic prokaryotic microorganisms that lack a cell wall.

Mycorrhiza A symbiotic association of a fungus with the roots of a plant.

Mycotoxicoses Diseases of animals and humans caused by consumption of feed and foods invaded by fungi that produce mycotoxins.

Mycotoxins Toxic substances produced by several fungi in infected seeds, feeds, or foods; and capable of causing illnesses of varying severity and death to animals and humans that consume such substances.

Nanometer (nm) A unit of length equal to 1/1000 of a micrometer.

Necrotic Dead and discolored.

Nectarthode An opening at the base of a flower from which nectar exudes.

Nectrotroph A microorganism feeding only on dead organic tissues.

Nematicide A chemical compound or physical agent that kills or inhibits nematodes.

Nematode Generally microscopic, worm-like animals that live saprophytically in water or soil, or as parasites of plants and animals.

Nonhost resistance Inability of a pathogen to infect a plant because the plant is not a host of the pathogen due to lack of something in the plant that the pathogen needs or to the presence of substances incompatible with the pathogen.

Noninfectious disease A disease that is caused by an abiotic agent, i.e., by an environmental factor, not by a pathogen.

Nuclear-binding site (NBSI) A protein whose configuration of surface amino acids allows the protein to bind to and activate a protein in its nucleus.

Nucleic acid An acidic substance containing pentose, phosphorus, and pyrimidine and purine bases. Nucleic acids determine the genetic properties of organisms.

Nucleoprotein Referring to viruses: consisting of nucleic acid and protein.

Nucleoside The combination of a sugar and a base molecule in a nucleic acid.

Nucleotide The phosphoric ester of a nucleoside consisting of a base (purine or pyrimidine), a sugar, and phosphate. Nucleotides are the building blocks of DNA and RNA.

Obligate parasite A parasite that in nature can grow and multiply only on or in living organisms.

Ontogenic resistance When the degree of resistance of a plant to a pathogen varies with age and the developmental stage of the plant.

Oogonium The female gametangium of oomycetes containing one or more gametes.

Oomycete A fungus-like chromistan that produces oospores; a water mold.

Oospore A sexual spore produced by the union of two morphologically different gametangia (oogonium and antheridium).

Operon A cluster of functionally related genes regulated and transcribed as a unit.

Osmosis The diffusion of a solvent through a differentially permeable membrane from its higher concentration to its lower concentration.

Ostiole A pore-like opening in perithecia and pycnidia through which the spores escape from the fruiting body.

Ovary The female reproductive structure that produces or contains the egg.

Oxidative phosphorylation The utilization of energy released by the oxidative reactions of respiration to form high-energy ATP bonds.

Ozone (O_3) A highly reactive form of oxygen that may injure plants in relatively high concentrations.

Papilla A nipple-like protuberance of the cell wall on the inside of a cell being attacked by a fungus, apparently serving as a defense mechanism against infection.

Paraphysis A sterile hypha present in some fruiting bodies of fungi.

Parasexualism A mechanism whereby recombination of hereditary properties occurs within fungal heterokaryons.

Parasite An organism living on or in another living organism (host) and obtaining its food from the latter.

Parenchyma A tissue composed of thin-walled cells that usually leave intercellular spaces between them.

Pathogen An entity that can incite disease.

Pathogenicity The capability of a pathogen to cause disease.

Pathogenicity factors These factors are produced by pathogenicity genes, are essential, and are involved in all crucial steps in disease induction and development.

Pathogenicity genes Genes that are essential for a pathogen to be able to cause disease.

Pathovar In bacteria, a subspecies or group of strains that can infect only plants within a certain genus or species.

Pectin A methylated polymer of galacturonic acid found in the middle lamella and the primary cell wall of plants.

Pectinase An enzyme that breaks down pectin.

Penetration The initial invasion of a host by a pathogen.

Perfect stage The sexual stage in the life cycle of a fungus; The teleomorph.

Periplasm The area between the plasma membrane and the cell wall.

Perithecium The globular or flask-shaped ascocarp of the Pyrenomycetes, having an opening or pore (ostiole).

Phage A virus that attacks bacteria; also called bacteriophage.

Phenolic Applied to a compound that contains one or more phenolic rings.

Phenotype The external visible appearance of an organism.

Phloem Food-conducting tissues, consisting of sieve tubes, companion cells, phloem parenchyma, and fibers.

Phyllody Excessive production of leaves in place of shoots and blossoms.

Phytoalexin A substance that inhibits the development of a fungus on hypersensitive tissue formed when host plant cells come in contact with the parasite.

Phytoanticipins Inhibitory antimicrobial compounds present in plant cells before infection.

Phytopathogenic Term applicable to a microorganism that can incite disease in plants.

Phytoplasmas Mollicutes that infect plants and cannot yet be grown in culture, as contrasted to spiroplalsmas, which can be cultured.

Phytotoxic Toxic to plants.

Plant pathogenesis-related proteins (PR) Groups of proteins with different chemical properties produced in a cell within minutes or hours following inoculation, but all being more or less toxic to pathogens.

Plantibodies Antibodies produced in transgenic plants expressing the antibody-producing gene(s) of a mouse that had been injected previously with a pathogen (usually a virus) that infects the plant.

Plasmalemma The cytoplasmic membrane found on the outside of the protoplast adjacent to the cell wall.

Plasmid A self-replicating, extrachromosomal, hereditary circular DNA found in certain bacteria and fungi, generally not required for survival of the organism.

Plasmodesma (plural = plasmodesmata) A fine protoplasmic thread connecting two protoplasts and passing through the wall that separates the two protoplasts.

Plasmodium A naked, slimy mass of protoplasm containing numerous nuclei.

Plasmolysis The shrinking and separation of the cytoplasm from the cell wall due to exosmosis of water from the protoplast.

Plerome The plant tissues inside the cortex.

Polyclonal antibodies The usual mix of antibodies present in the serum of the blood of an animal that has been injected with a pathogen or protein that generally has many antigenic determinants.

Polycyclic Completes many (life or disease) cycles in one year.

Polyetic Requires many years to complete one life or disease cycle.

Polygenic A character controlled by many genes.

Polyhedron A spheroidal particle or crystal with many plane faces.

Polymerase An enzyme that joins single small molecules into chains of such molecules (e.g., DNA, RNA).

Polymerase chain reaction A technique that allows an almost infinite amplification (multiplication) of a segment of DNA for which a primer (short piece of that DNA) is available.

Polysaccharide A large organic molecule consisting of many units of a simple sugar.

Polysome (or polyribosome) A cluster of ribosomes associated with messenger RNA.

Population genetics The description and quantification of genetic variation in populations and its use for drawing conclusions about evolutionary processes that affect populations.

Precipitin The reaction in which an antibody causes visible precipitation of antigens.

Primary infection The first infection of a plant by the overwintering or oversummering pathogen.

Primary inoculum The overwintering or oversummering pathogen, or its spores that cause primary infection.

Probe A radioactive nucleic acid used to detect the presence of a complementary strand by hybridization.

Programmed cell death Death of specific cells of an organism, the initiation and execution of which is controlled by the organism.

Prokaryote A microorganism whose genetic material is not organized into a membrane-bound nucleus, e.g., bacteria and mollicutes.

Promoter A region on DNA or RNA recognized by the RNA polymerase in order to initiate transcription.

Promycelium The short hypha produced by the teliospore; the basidium.

Propagative virus A virus that multiplies in its insect vector.

Propagule The part of an organism, such as a spore or a bacterium, that may be disseminated and reproduce the organism.

Proteasome An extremely large protein complex that carries out most protein degradation in the nucleus and the cytoplasm.

Protectant A substance that protects an organism against infection by a pathogen.

Protein kinases Proteins that act as signal transducers and amplifiers by responding to the size of the input signal through a proportional increase in activity and corresponding cellular response.

Protein subunit A small protein molecule that is the structural and chemical unit of the protein coat of a virus.

Proteome The total of proteins produced by an organism, or produced under certain developmental or environmental conditions.

Proteomics The study of the identity and function of the proteins produced by an organism.

Protoplast A plant cell from which the cell wall has been removed. The organized living unit of a single cell; the cytoplasmic membrane and the cytoplasm, nucleus, and other organelles inside it.

Protozoa Individual organisms of the kingdom Protozoa or of the phylum Protozoa of the kingdom Protista. Among the plant pathogens, it includes Myxomycetes, Plasmodiophoromycetes, and Flagellate protozoa.

Pseudofungi A name formerly used for Myxomycetes, Plasmodiophoromycetes, and Oomycetes, all of which were thought to be fungi until about 1990, but now the first two are considered protozoa (protista) and the Oomycetes are considered chromista. All three, however, continue to be studied along with the true fungi (Chytridiomycetes, Zygomycetes, Ascomycetes, and Basidiomycetes).

Pseudothecium The ascocarp of the Loculoascomycetes (ascostromatic ascomycetes) in which asci are formed directly in cavities within a stroma (matrix) of mycelium; Pseudothecium also called an ascostroma.

Purification The isolation and concentration of virus particles in a pure form, free from cell components.

Pustule Small blister-like elevation of epidermis created as spores form underneath and push outward.

Pycnidium An asexual, spherical, or flask-shaped fruiting body lined inside with conidiophores and producing conidia.

Pycniospore Also called a spermatium. A spore produced in a pycnium (spermagonium).

Pycnium Also called a spermagonium. In some basidiomycetes, it contains spermatia and receptive hyphae.

Quarantine Control of import and export of plants to prevent spread of diseases and pests.

Quorum sensing Dependence of bacterial or spore behavior and pathogenicity on their cells reaching a certain density by sensing the concentration of certain signal molecules in their environment.

Race A genetically and often geographically distinct mating group within a species; also a group of pathogens that infect a given set of plant varieties.

Reactive oxygen radicals Oxygen species much more reactive than molecular oxygen (O_2), which, upon contact of a resistant cell with a pathogen, react with and quickly oxidize various cellular components into compounds toxic to the pathogen.

Recognition factors Specific receptor molecules or structures on the host (or pathogen) that can be recognized by the pathogen (or host).

Resistance The ability of an organism to exclude or overcome, completely or in some degree, the effect of a pathogen or other damaging factor.

Resistant Possessing qualities that hinder the development of a given pathogen; infected little or not at all.

Resting spore A sexual or other thick-walled spore of a fungus that is resistant to extremes in temperature and moisture and which often germinates only after a period of time from its formation.

Restriction enzymes A group of enzymes from bacteria that break internal bonds of DNA at highly specific points.

Reverse transcription Copying of RNA into DNA.

Rhizoid A short, thin hypha growing in a root-like fashion toward the substrate.

Rhizosphere The soil near a living root.

Ribonuclease (RNase) An enzyme that breaks down RNA.

Ribonuclic acid (RNA) A nucleic acid involved in protein synthesis; also the most common nucleic acid (genetic material) of plant viruses.

Ribosome A subcellular particle involved in protein synthesis.

Rickettsiae Microorganisms similar to bacteria in most respects but generally capable of multiplying only inside living host cells; parasitic or symbiotic.

Ring spot A circular area of chlorosis with a green center; a symptom of many virus diseases.

Rosette Short, bunchy habit of plant growth.

Rot The softening, discoloration, and often disintegration of a succulent plant tissue as a result of fungal or bacterial infection.

Russet Brownish roughened areas on skin of fruit as a result of cork formation.

Rust A disease giving a "rusty" appearance to a plant and caused by one of the Uredinales (rust fungi).

Sanitation The removal and burning of infected plant parts, decontamination of tools, equipment, hands, etc.

Saprophyte An organism that uses dead organic material for food.

Scab A roughened, crust-like diseased area on the surface of a plant organ; a disease in which such areas form.

Scion A piece of twig or shoot inserted on another by grafting.

Sclerotium A compact mass of hyphae with or without host tissue, usually with a darkened rind, and capable of surviving under unfavorable environmental conditions.

Scorch "Burning" of leaf margins as a result of infection or unfavorable environmental conditions.

Secondary infection Any infection caused by inoculum produced as a result of a primary or a subsequent infection; an infection caused by secondary inoculum.

Secondary inoculum Inoculum produced by infections that take place during the same growing season.

Secretome The total of proteins secreted by an organism or sets of proteins secreted under certain conditions.

Sedentary Staying in one place; stationary.

Selection The process by which populations of the fittest variants in a particular environment increase in frequency while those of less fit variants decrease.

Septate Having cross walls.

Septum A cross wall (in a hypha or spore).

Serology A method using the specificity of the antigen–antibody reaction for the detection and identification of antigenic substances and the organisms that carry them.

Serum The clear, watery portion of the blood remaining after coagulation.

Sexual Participating in or produced as a result of a union of nuclei in which meiosis takes place.

Shock symptoms The severe, often necrotic symptoms produced on the first new growth following infection with some viruses; also called acute symptoms.

Shot hole A symptom in which small diseased fragments of leaves fall off and leave small holes in their place.

Sieve plate Perforated wall area between two phloem sieve cells through which they are connected.

Sieve tube A series of phloem cells forming a long cellular tube through which food materials are transported.

Sign The pathogen or its parts or products seen on a host plant.

Signaling genes Genes that respond to changes in the environment and set off signaling cascades that alter the expression of the genes of the organism.

Signaling pathways The series of compounds involved in the transmission of cellular signals, often involving several protein kinases functioning in series.

Signal molecules Host molecules that react to infection by a pathogen and transmit the signal to and activate proteins and genes in other parts of the cell and of the plant so they will produce the defense reaction.

Signal transduction The means by which cells construct and deliver responses to a signal, generally involving intracellular Ca and protein kinases.

Slime molds Formerly fungi, now protozoa of the class Myxomycetes; also superficial diseases caused by these pseudofungi on low-lying plants.

Smut A disease caused by smut fungi (Ustilaginales) characterized by masses of dark, powdery and sometimes odorous spores.

Soft rot A rot of a fleshy fruit, vegetable, or ornamental in which the tissue becomes macerated by the enzymes of the pathogen.

Soil inhabitants Microorganisms able to survive in the soil indefinitely as saprophytes.

Soil solarization Attempt to reduce or eliminate pathogen populations in the soil by covering the soil with clear plastic so that sun rays will raise the soil temperature to levels that kill the pathogen.

Soil transients Parasitic microorganisms that can live in the soil for short periods.

Somaclonal variation Variability in clones generated from a single mother plant, leaf, etc., by tissue culture.

Somatic hybridization Production of hybrid cells by fusion of two protoplasts with different genetic makeup.

Sooty mold A sooty coating on foliage and fruit formed by dark hyphae of fungi that live in the honeydew secreted by insects such as aphids, mealybugs, scales, and whiteflies.

Sorus A compact mass of spores or fruiting structure found especially in rusts and smuts.

Spermagonium (formerly pycnium) A fruiting body of rust fungi in which gametes or gametangia are produced.

Spermatium (formerly pycniospore) The male gamete or gametangium of rust fungi.

Spiroplasmas Pleomorphic, wall-less microorganisms present in the phloem of diseased plants; often helical in culture and thought to be a kind of mycoplasma.

Sporagiophore A specialized hypha bearing one or more sporangia.

Sporagiospore Nonmotile, asexual spore borne in a sporangium.

Sporangium A container or case of asexual spores. In some cases it functions as a single spore.

Spore The reproductive unit of fungi consisting of one or more cells; in function, it is analogous to the seed of green plants.

Sporidium The basidiospore of smut fungi.

Sporodochium A fruiting structure consisting of a cluster of conidiophores woven together on a mass of hyphae.

Sporophore A hypha or fruiting structure bearing spores.

Sporulate To produce spores.

Stem pitting A symptom of some viral diseases characterized by depressions on the stem of the plant.

Sterigma A slender protruberance on a basidium that supports the basidiospore.

Sterile fungi A group of fungi that are not known to produce any kind of spores.

Sterilization The elimination of pathogens and other living organisms from soil, containers, etc., by means of heat or chemicals.

Strain The decendants of a single isolation in pure culture; an isolate. Also a group of similar isolates; a race. In plant viruses, a group of virus isolates having most of their antigens in common.

Stroma A compact mycelial structure on or in which fructifications are usually formed.

Stylet A long, slender, hollow feeding structure of nematodes and some insects.

Stylet borne A virus borne on the stylet of its vector; a noncirculative virus.

Substrate The material or substance on which a microorganism feeds and develops; also a substance acted upon by an enzyme.

Suppressive soils Soils in which certain diseases are suppressed because of the presence in the soil of microorganisms antagonistic to the pathogen.

Suscept Any plant that can be attacked by a given pathogen; a host plant.

Susceptibility The inability of a plant to resist the effect of a pathogen or other damaging factor.

Suseptible Lacking the inherent ability to resist disease or attack by a given pathogen; nonimmune.

Symbiosis A mutually beneficial association of two or more different kinds of organisms.

Symptom The external and internal reactions or alterations of a plant as a result of a disease.

Symptomless carrier A plant that, although infected with a pathogen (usually a virus), produces no obvious symptoms.

Syncytium A multinucleate mass of protoplasm surrounded by a common cell wall.

Synergism The concurrent parasitism of a host by two pathogens in which the symptoms or other effects produced are of greater magnitude than the sum of the effects of each pathogen acting alone.

Systemic Spreading internally throughout the plant body; said of a pathogen or a chemical.

Systemic acquired resistance Systemically activated resistance after primary infection with a necrotizing pathogen accompanied by increased levels of salicylic acid and pathogenesis-related proteins.

Teleomorph The sexual or so-called perfect growth stage or phase in fungi.

Teliospore The sexual, thick-walled resting spore of rust and smut fungi.

Telium The fruiting structure in which rust teliospores are produced.

Tissue A group of cells of similar structure that perform a special function.

Tolerance The ability of a plant to sustain the effects of a disease without dying or suffering serious injury or crop loss; also the amount of toxic residue allowable in or on edible plant parts under the law.

Toxicity The capacity of a compound to produce injury.

Toxin A compound produced by a microorganism; being toxic to a plant or animal.

Transcription Copying of a gene into RNA; also copying of a viral RNA into a complementary RNA.

Transduction The transfer of genetic material from one bacterium to another by means of a bacteriophage.

Transfer RNA The RNA that moves amino acids to the ribosome to be placed in the order prescribed by the mRNA.

Transformation The change of a cell through uptake and expression of additional genetic material.

Transgenic (or transformed) plants Plants into which genes from other plants or other organisms have been introduced through genetic engineering techniques and are expressed, i.e., produce the expected compound or function.

Translation Copying of mRNA into protein.

Translocation Transfer of nutrients or virus through the plant.

Transmission The transfer or spread of a virus or other pathogen from one plant to another.

Transpiration The loss of water vapor from the surface of leaves and other aboveground parts of plants.

Transposable element A segment of chromosomal DNA that can move around (transpose) in the genome and integrate at different sites on the chromosomes.

Tumor An uncontrolled overgrowth of tissue or tissues.

Tylosis An overgrowth of the protoplast of a parenchyma cell into an adjacent xylem vessel or tracheid.

Uredium The fruiting structure of rust fungi in which uredospores are produced.

Ubiquitin A small protein found in plants involved in the degradation of proteins.

Ubiquitination The attachment of one or more ubiquitin molecules to proteins destined for degradation and delivery to the proteasome where they are degraded.

Variability The property or ability of an organism to change its characteristics from one generation to the other.

Vascular Term applied to a plant tissue or region consisting of conductive tissue; also a pathogen that grows primarily in the conductive tissues of a plant.

Vector An animal able to transmit a pathogen. In genetic engineering, *vector (or cloning vehicle)*, a self-replicating DNA molecule, such as a plasmid or virus, used to introduce a fragment of foreign DNA into a host cell.

Vegetative Asexual; somatic.

Vegetative incompatibility Failure of the hyphae of strains of the same species of a fungus to fuse and form anastomoses.

Vertical resistance Complete resistance to some races of a pathogen but not to others.

Vesicle A bubble-like structure produced by a zoosporangium in which zoospores are released or are differentiated.

Vessel A xylem element or series of such elements whose function is to conduct water and mineral nutrients.

Virescent A normally white or colored tissue that develops chloroplasts and becomes green.

Virion A virus particle.

Viroids Small, low-molecular-weight RNA that can infect plant cells, replicate themselves, and cause disease.

Virulence The degree of pathogenicity of a given pathogen.

Virulence factors Coded for by virulence genes that are helpful but not essential for induction and development of disease.

Virulence genes Enable a pathogen to express increased virulence on only one or a few related hosts.

Virulent Capable of causing a severe disease; strongly pathogenic.

Viruliferous Said of a vector containing a virus and capable of transmitting it.

Virus A submicroscopic obligate parasite consisting of nucleic acid and protein.

Virusoid The extra-small circular RNA component of some isometric RNA viruses.

Xylem A plant tissue consisting of tracheids, vessels, parenchyma cells, and fibers; wood.

Wilt Loss of rigidity and drooping of plant parts, generally caused by insufficient water in the plant.

Witches' broom Broom-like growth or massed proliferation caused by the dense clustering of branches of woody plants.

Yellows A plant disease characterized by yellowing and stunting of the host plant.

Zoosporangium A sporangium which containing or producing zoospores.

Zoospore A spore bearing flagella and capable of moving in water.

Zygospore The sexual or resting spore of zygomycetes produced by the fusion of two morphologically similar gametangia.

Zygote A diploid cell resulting from the union of two gametes.

Index

AAL toxin, 195
Abiotic diseases. *See* Noninfectious (abiotic) diseases
Abiotic stress, 319, 383
Abscission layer, formation of, 216–217
Absorption of water by roots, 108, 109
Acervulus, 440, 444
Acetosyringone, 149
Acibenzolar-S-methyl (ASM), 211, 316
Acidovorax avenae subsp. *citrulli*, 636
Acid rain injury to plants, 371–372
ACL toxin, 195
Acquired resistance. *See* Systemic acquired resistance (SAR)
Actigard, 50, 57, 316, 338, 452
Actinovate, 316
Activated oxygen species, 231
ACT toxin, 195
Acylalanines, 340–341
Adenine, 730
Adhesion to plant surfaces, bacterial, 146–147
Aecia, 564
Aeciopores, 564
Aerial pathogens, 307
Aerial plant parts, biological control of, 326, 328
Aflatoxins, 39, 41, 559
African cassava mosaic, 67, 805, 810
AF toxin, 195
Agglutination reaction, 745
Agricultural practices, affects of improper, 381, 383
Agrimycin, 343
Agrobacterium
  adhesion, 146–147
  description of, 621
  horizontal gene transfer, 132
  radiobacter, 323
  rhizogenes, 119
  tumefaciens, 24, 50, 54, 108, 109, 119, 121, 148, 149, 198, 199, 326, 662–666
Agrocin, 326
Agrosabotage, 59
Agroterrorism, 59
Air dissemination, 96–97
Air pollution, 48, 262, 368–372
AK toxin, 195
Alarm signal, 214
Alarm substances, 214
Albersheim, P., 54
Albertus Magnus, 14, 15
Albicidins, 148
Albugo, 410, 432
Aldicarb (Temik), 313, 345
Alfalfa
  crown wart, 433, 434
  downy mildew, 428
  nematode invasion, 93, 859
  wart, 119
Alfamovirus, 150, 787
Algae
  diseases caused by, 719–722
  parasitic green, 719
Aliette (fosetyl-Al), 341
Allelic incompatibility, 132
Allexiviruses, 763
Allozyme, 129
Almonds
  cankers, 473, 474
  hull rot, 193
Alpinia purpurata, 882
Alternaria, 138, 193, 605
  alternata (AAL), 146, 191, 192, 195–196
  brassicicola, 224
  diseases caused by, 453–456, 556
  mali, 195
  penetration, 178
  solani, 168, 220, 454
Alternaric acid, 193
American Phytopathological Society, 60, 64
Amino acids, 730
Amphid secretions, role of, 151
AM toxin, 195–196
Amylases, 190
Anamorph, 439
Anastomosis, 132, 598
Anguina, 865–867
Antagonistic microorganisms, control with, 305–308
Antagonistic plants, control with, 309
Antheridium, 439
Anthracnose, fungi, 251, 398, 439, 483–500
Anthrax bacillus, 23, 26, 59
Antibiotics, 47–48, 343–344, 626
Antibodies, 744
Antigen, 744
Antigenic determinant, 744
Antimicrobial substances, pathogenesis-related proteins, 232–233
Antiserum, 744
Ants, 42
Aphanomyces euteiches, 285–286
Aphelenchoides, 830, 867–870
Aphids, 42, 45, 302, 742
Apiosporina morbosa, 473, 474

Apothecium, 440
Apparent resistance, 137–139
Apples
  bitter rot, 494–495
  black rot, 519–521
  blossom end rot, 513
  canker, 473, 667
  cedar-apple rust, 198, 255, 574–576
  chlorotic leaf spot virus, 763
  crown gall, 662
  fire blight, 42, 43, 121, 286, 299, 300, 641–647
  gray mold, 51
  hairy root, 119, 662
  mosaic virus, 790
  Nectria canker, 99, 115, 477, 478–481
  powdery mildew, 299
  proliferation, 42, 694, 696–697
  scab disease, 53, 91, 92, 113, 114, 127, 253, 286–287, 504–507
  stem grooving virus, 763
  stem pitting virus, 763
  white rot fungus, 99
Appressoria
  formation and maturation, 85–86
  penetration, 88, 144, 177–178
Apricots
  brown rot, 122
  European fruit yellows, 697
Apron (metalaxyl), 340
AQ10 Biofungicide, 316, 324
Arabidopsis, 156, 157, 224
  RPM1 gene, 229
  RPS2 gene, 242
  thalliana, 55
Arabinogalactan proteins (AGPs), 189
Arabis mosaic virus, 784
Arachidonic acid, 238
Arbuscules, 613
Arceuthobium, 712–715
Area under a disease progress curve (AUDPC), 273
Armillaria, 602–604
Aromatic compounds, 339–340
Arthrobacter, 622
Artichoke mottle crinkle virus, 237
Asclepiadaceae, 877
Ascocarps, 439
Ascochyta, 453
Ascogenous hyphae, 439
Ascogonium, 439
Ascomycetes
  anthracnoses, 483–500
  cankers, 473–476
  diseases caused by, 439–440
  foliar diseases, 452–473
  fruit and general diseases, 501–522
  morphology, 441–444
  postharvest diseases, 553–582
  reproduction in, 388
  root and stem rots, 534–553
  symptoms caused by, 445
  vascular wilts, 522–534
Ascospores, 388, 439
Ascostroma, 439
Ascus, 388, 439
Asexual fungi, 388
Ash yellows, 697, 698
Aspergillus, 40, 41, 556, 558
  flavus, 145
  toxins, 559
Aspire, 316, 324
Aster yellows, 42, 691–694
Attenuation, 133
Aureusvirus, 781
Autoecious, 565
Auxins, 196–200, 664
Avenacins, 145, 202, 211
Avenavirus, 781
Avermectins, 345
Avirulence genes (avr), 55, 141, 151
  characteristics, 153–154
  as an elicitor of plant defense responses, 151–153
  function of, 154–155
  proteins, recognition by host, 149, 225–226
  structure of, 154
  virulence promotion, 202
Avirulent, 151–153
Avocado
  anthracnose, 493
  scab, 483, 485, 486–487
avr genes. *See* Avirulence genes
Azaleas
  leaf and flower gall, 119, 120, 196–198
  powdery mildew, 11
Azorhizobiu, 676
Azoxystrobin, 342

B*acillus*, 622, 656
  anthracis, 23, 26, 59
  subtilis, 323
  thuringiensis, 58
Bacteria
  adhesion to plant surfaces, 146–147
  biocontrol products produced by, 324
  cell wall degradation, 147–148
  characteristics of, 618–626
  control of, 625–626
  description and movement of, 618
  diagnosis of diseases, 72–73
  dissemination of, 81–82, 96–100
  ecology and spread, 620–621
  effects of moisture on, 256–257
  horizontal gene transfer, 132
  identification, 621–625
  isolation, 398–402, 624
  morphology, 618–619
  pathogenicity genes, 146–149
  penetration by, 88
  phloem-inhabiting, 94
  regulatory systems and networks, 148–149
  reproduction, 96, 619–620
  resistant strains, 48
  secretion systems, 147
  sexual-like processes in, 132
  staining, 622
  strains, 617
  symptoms caused by, 625
  taxonomy, 616–617
  toxins, 148
  xylem-inhabiting, 94
Bacteria, diseases caused by, 9, 23, 24–25, 66, 618
  cankers, 667–674
  fastidious vascular bacteria, 678–687
  galls, 108, 109, 662–667
  root nodules of legumes, 675–678
  scabs, 674–675
  soft rots, 656–662
  spots and blights, 67, 627–638
  vascular wilts, 108, 638–656
Bactericides, 334
Bacteriocins, 326
Bacteriophages (phages), 328
Badnaviruses, 803
Baermann funnel, 831, 832
Baker, B., 55
Bananas
  anthracnose, 491
  bacterial wilt or Moko disease, 67, 649
  bunchy top, 67, 813, 814–815
  burrowing nematode, 853–855
  fusarium wilt (Panama disease), 296, 526
  sigatoka or leaf spot disease, 66, 234, 459–460
  streak virus, 803
  wilt, 198
Barberry, 16
  stem rust of wheat on, 567
Barley, 162
  crown and common root rots, 469, 470
  ergot, 37, 38, 502
  net blotch, 469, 471
  Pyrenophora diseases, 469, 471
  smut, 12, 583, 584, 587, 588
  spot blotch, 469
  spots, 107
  stripe, 469, 471
  stripe mosaic virus, 761
  stripe rust, 96, 100
  yellow dwarf virus, 66, 781–782
Barrus, M. F., 52
Basal stem rot, 398
Basidiomycetes, 131, 562
  reproduction in, 388
Basidiomycetes, diseases caused by
  root and stem rots, 593–603
  rusts, 562–582
  smuts, 582–593
  symptoms caused by, 564
  wood rots and decay, 604–614
Basidiospores, 86, 257, 388, 562, 565
Basidium, 388, 562, 564
Bawden, F. C., 25
Baycor (bitertanol), 342
Bayleton (triadimefon), 342
Bayram (triadimenol), 342
Beachy, R., 54
Beadle, 54

Beans
anthracnose, 296, 487, 488
bacterial blights, 296, 629–630
common mosaic virus, 121, 764, 767
fava bean necrotic yellows virus, 813
golden mosaic, 67, 805, 810
halo blight, 191, 192
root rot, 538, 539
rot, 596
rust, 13, 571–572
stem rot and white mold, 547
yellow mosaic virus, 764, 767
Bees, 42, 43
Beetles, 42–44, 530, 741–742
Beets
curly top virus, 809–809
necrotic yellow vein virus, 237, 407, 761, 762
Begomoviruses, 805
Beijerinck, M. W., 25
Belonolaimus, 860–863
Bemisia, 778, 805, 810
Benomyl, 47, 334, 341
Benyviruses, 762
Benzimidazoles, 341
Benzothiazole (BTH), 242, 338
Best western yellows, 783
Best yellows, 777
β-Aminobutyric acid, 316
β-glucanase, 86, 186
β-1,3-glucanase, 210, 220, 221, 240
Biffen, P. D., 52
BINAB T, 323, 324
Binary fission, 619–620
Biochemical defenses
induced, 217–236
used by plants, 211–212
BioJect Spot-Less, 324
Biological controls, description of, 49–50, 294, 303–305, 322–329, 626
Biological warfare, 59
BioMal, 329
Bion WG50, 316
Bio-save, 324
Bioterrorism, 59
Biotic diseases. *See* Infectious (biotic) diseases
Biotox C, 324
Biotrophs, 78, 387, 389
Biphenyl, 340
Bipolaris, 50, 468
carbonum, 156
maydis, 137, 466, 467
Bird's-eye rot, 486
Bisporomyces, 605
Bitter rot, 494–495
Blackberries
anthracnose, 485
cane gall, 662
Blackfire, 628
Black heart, 367, 610
Black knot, 473, 476
Black root rot, 543, 544
Black rot
apple, 519–521
cantaloupes, 183–184
crucifers, 653–654
cucurbits, 516–518
grapes, 514–516
Black spot
citrus, 67
roses, 91, 483, 484, 485–486
Black wart, 433, 434
Blasts, 10, 12
BlightBan, 324
Blights, 10, 13, 50
See also under type of
bacterial, 67, 627–638
fire, 24, 42, 66, 286
fungal, 66, 67, 106, 107, 398
Blight, early
celery, 463, 464
potatoes, 53, 453, 454
tomatoes, 53, 453, 454
Blight, late, 267
celery, 461
potatoes, 18, 19–21, 22, 47, 59, 66, 67, 286, 421–426
tomatoes, 67, 286, 421–426
BLITECAST, 286, 288
Blockade, 316
Blossom-end rot of fruits, 513
Blotch-type necrosis, 364
Blue mold (downy mildew), 66, 285
Blue mold rots, 557
Blue-stain fungi, 606
Blumeria graminis, 162, 208, 213, 253, 448
Blumeriella, 453, 464
Bordeaux mixture, 31, 46, 47, 338, 447, 460
Boron
deficiency diseases, 372, 373, 376
toxicity diseases, 376
Botran (dichloran), 339
Botryosphaeria, 501
dothidea, 473, 474
obtusa, 99, 521
Botrytis, 138, 145
cinerea, 22, 56, 182, 202
diseases, 510–514, 556
Bradyrhizobium, 676
Brassica napus, 137
Bravo (chlorothalonil), 340
Breeding resistant varieties, 165
advantages/disadvantages of vertical or horizontal resistance, 169–170
classical techniques, 166–167
crops and, 626
epidemics, vulnerability to, 170–172
genetic engineering techniques, 168–169
isolation of mutants, 168
protoplast fusion, 169
selection, 167–168
sources of genes for, 166
tissue culture, 168
variability affected by, 165–166
Brefeld, 45
Bremia, 252, 409, 427, 428
Briggs, S. P., 55
Brine. *See* Sodium chloride
Broglie, R., 55
Bromoviruses, 787
Broomrape, 72, 711–712
Brown patch, 594
Brown rot, 42, 43
fungi, 605
of stone fruits, 121, 122, 181–182, 185, 507–510
Brown spot
bacterial, 629
corn, 433, 434, 468–469
rice, 66
soybean, 461
Bulb nematode, 24, 858–860
Bulbs, rot, 540
Bunt. *See* Smut
Burkholderia cepacia, 526
Burrill, T. J., 24
Burrowing nematode, 67, 853–857
Bursaphelenchus, 830
cocophilus, 870, 872–874
xylophilus, 870–872
Butternut canker, 36, 481, 482
Bymoviruses, 774

Cabbage
bacterial soft rot, 658
black leg, 297, 520
black rot, 297, 653–654
clubroot, 196, 197
downy mildew, 429
rot, 548
Cacao
black pod, 414
pod rot, 67, 510, 511
swollen shoot, 66, 803
vascular wilt, 522, 532–534
witches' broom, 234, 611–612
Cadang-cadang disease, 67, 822–823
Caenorhabditis elegans, 55
Caffeic acid, 182, 233
Calcium, 259
deficiency diseases, 372, 373, 377
Callose, 186
Candidatus liberatus, 616
Candidatus liberobacter, 685–686
Cane gall, 662
Cankers
Ascomycetes and mitosporic fungi, 473–476
bacterial, 651–653, 667–674
butternut, 36
citrus, 66, 300
cypress, 36
early methods of controlling, 47
forest trees, 481–483
fungal, 108, 110, 115, 398, 473–476
Leucostoma, 479–481
Nectria, 99, 115, 477, 478–481
Cantaloupe
downy mildew, 429

Cantaloupe (*continued*)
gummy stem blight, 517
stem rot, 518, 520
Capilloviruses, 763
Capsidiol, 235
Capsule, 619
Captan, 313, 340
Carbamates, 339, 344, 345
Carbendazim, 341
Carbofuran, 313, 345
Carbonate compounds, 338–339
Carbosulfan, 345
Carboxin, 47, 334, 341
Carlaviruses, 763
Carmovirus, 781
Carnation
etched ring virus, 802
latent virus, 763
mottle virus, 781
Carrots
aster yellows, 691–692
bacterial gall, 662
bacterial soft rot, 658
crater rot, 596
mottle virus, 784
root knot, 838
Sclerotinia white mold, 50, 548
Carson, Rachel, 48
Cassava
African, mosaic, 67, 805, 810
anthracnose, 491
bacterial blight, 636–637
canker and stem rot, 659
empty root disease, 882
vein mottle virus, 801
Casst, 329
Catalase, 203
Catechin, 211
Catechol, 149
Cauliflower
mosaic virus, 801–803
Caulimoviruses, 801–803
Cedar-apple rust, 198, 255, 574–576
Cedomon, 324
Celery
bacterial soft rot, 658
blight, early, 463, 464
blight, late, 461
mosaic virus, 764
Cell(s)
components of, 176, 177
death, 160–161
enzymatic degradation of substances contained in, 189–190
membranes, disruption of, 231–232
membranes, permeability affected by pathogens, 118
Cellulases, 50, 145, 147, 148
Cellulose, 176, 184, 186
Cell walls
composition, 616
defense structures, 210, 214–215
degradation of, 144–145, 147–148
enzymatic degradation of, 180–189
flavonoids, 189
reinforcement of, 232
structural proteins, 189
Cephaleuros, 719
Ceratobasidium, 598
Ceratocystis, 606
cankers, 473, 474
fagacearum, 36, 108, 110, 522
vascular wilts, 522, 532–534
Ceratoulmin, 193
CERCOS, 280
Cercospora, 192
diseases caused by, 453, 463, 464
zeae-maydis, 167
Cercosporin, 192–193
Cereals
See also under type of
anthracnose, 484, 489, 491
bacterial leaf spots and blights, 632–633
basal glume rot, 632
downy mildew, 428
ergot, 501–504
halo blight, 632
head blight or scab, 535, 538
postharvest decays, 558–559
powdery mildew, 448, 450
rust, 52, 66, 565–571
smut, 66, 121, 584–587
snow mold, 251
sting nematode, 860–863
Certification, for plant pathologists, 63–64
Cheim, R., 55
Chemicals/chemical control of diseases
application methods, 332–338
bacteria and, 626
defense and, used by plants, 211–212
description of, 47–49, 294, 312–314, 329–348
early developments, 47–48
mechanisms of action, 345–346
public concern about, 48–49
resistance of pathogens to, 346–347
restrictions on, 347–348
types of, used to control diseases, 338–345
used by pathogens, 179–203
Chemotherapeutants, 346
Chemotherapy, 346
Cherries
black knot, 476
brown rot, 43
leaf curl and witches' broom, 445
leaf roll virus, 784
leaf spot, 464
Cherry trees
bacterial canker and gummosis, 667, 669
black knot, 119, 120
cankers, 473, 474
root knot galls, 109
Chestnut blight, 32, 33–34, 66, 193, 473, 475, 476, 478
biocontrol of, 325
Chimeric genes, 56
Chitin, 55, 86
Chitinases, 210, 212, 220, 221, 240
Chitosan, 86, 235
Chlamydospores, 388, 523
Chlorinated hydrocarbons, 48, 49
Chlorine injury to plants, 368, 369
Chlorogenic acid, 182, 233
Chloroneb, 313, 342
Chloropicrin, 313, 345
Chloroplasts, 106
Chlorothalonil (Bravo), 340, 447
Choanephora, 22, 434–435
Chondrostereum, 606
Chromista, 388, 390
Chromosome, 619
Chrysanthemum
chlorotic mottle viroid, 819
foliar nematode, 867–870
stunt viroid, 819
white rust, 67
Chytridiomycetes, 305, 388
diseases caused by, 433–434, 742
Chytridiomycota, 433
CIMMYT (International Maize and Wheat Improvement Center), 60
Circoviridae, 813–816
Circulative viruses, 742
Circulifer tenellus leafhopper, 809
Citrus
anthracnose, 483, 492–493, 494
bacterial canker, 671–673
black spot, 67
burrowing nematode, 855–856
canker, 66, 300, 671–673
exocortis viroid, 819, 820–822
foot rot, 417
greening disease, 42, 67, 685–686
leaf rugose virus, 790
melanose, 518, 519
nematode, 848–849
postbloom fruit drop, 494
quick decline, 699
scab, 483, 485, 486
sour rot, 556–557
stubborn disease, 699–701
tatter leaf virus, 763
tristeza, 49, 66, 774–777
variegated chlorosis, 67, 681–682
variegation virus, 790
Cladosporium, 606
cucumerinum, 220, 456
diseases of, 453, 456
fulvum, 55, 153, 154, 157, 456, 457
Clavibacter
description of, 621
michiganense subsp. *michiganense*, 639, 651
michiganense subsp. *sepedonicum*, 638, 649
toxicus, 865
Claviceps purpurea, 37, 39, 121, 122, 203, 501
Cleistothecium, 439
Climbing plants, invasive, 716–719
Closteroviruses, 774–777
Clostridium, 656

Clover
downy mildew, 428
subterranean stunt virus, 813
Clubroot, 108, 257, 398
in cabbage, 196, 197
of crucifers, 119, 407–409
Coat proteins (CPs), 149–150
Cobb, N. A., 24
Cochliobolus, 95, 146, 193
carbonum, 55, 156, 236, 468
diseases caused by, 453, 466–469, 470
heterostrophus, 56, 146, 267, 268, 466–468
HV toxin (victorin), 194
penetration, 178
sativus, 135, 469, 470
Coconuts/coconut palms
cadang-cadang disease, 67, 822–823
foliar decay virus, 813, 815
hartrot disease, 880
lethal yellowing of, 35–36, 42, 67, 694, 695
red ring nematode, 872–874
Coding, 731
Codons, 731
Coffee
anthracnose, 493
phloem necrosis (wilt), 878–880
rust, 66, 300, 576–577
Coiled coil, 155, 162
Cold hardening, 221, 253
Coleosporium, 563
Collego, 329
Colletotrichum, 144, 184, 251
acutatum, 121, 489, 494–498
circinans, 211, 488
destructivum, 242
diseases, 483, 484–485, 487–500
fruit rots, 494–498
gloeosporioides, 237, 329, 489–498
graminicola, 90, 489, 494
lagenarium, 146, 238
lindemuthianum, 236, 296, 487
penetration, 178
Colonization/reproduction, pathogen, 91, 93–96, 619
Comoviruses, 784
Companion, 324
Computer simulation programs, 53–54, 280–281, 286
Conducive soils, 304
Conidia, 99, 257, 388, 439, 442–444, 618
Conidiomata, 388
Conidiophores, 388, 440, 442, 443
Conifers
dwarf mistletoes, 712–715
needle casts and blights, 456–458
root and butt rot, 323, 325
Coniothyrium minitants, 306
Conjugation, 132
Consultative Group on International Agricultural Research (CGIAR), 60
Contans WG, 324
Control of plant diseases. *See* Diseases, controlling
Copper, 260
deficiency diseases, 372, 373, 377
fungicides, 338, 460, 626
Copper sulfate, 18, 31, 47
Cork layer, formation of, 215–216
Corn
bacterial stripe, 632
brown spot, 433, 434, 468–469
downy mildew, 67, 428, 429
flea beetle, 42, 44
Hml gene, 242
leaf spot, 192, 193
maize streak virus, 121
northern corn leaf blight, 468
northern corn leaf spot, 468
root rot, 109, 253
seedling blight, 535
smut, 56, 119, 120, 121, 164–165, 198, 583–584
southern leaf blight, 66, 137, 267, 268, 466–468
stalk and ear rot, 535, 536–537
stem rot, 659
Stewart's wilt, 42, 44, 285, 639, 654
streak disease, 67
stunt, 692, 701
Coronatine, 148, 193
Corynebacterium, 201, 621
Corynespora cassiicola, 196
Coryneum, 483, 485
Cotton
angular leaf spot, 630, 632
anthracnose, 487
root rot, 257
verticillium wilt, 132
Covered smut, 12, 18, 588–591
Cowpea
chlorotic mottle virus, 107
mosaic virus, 784
Crick, Francis, 54
Crinipellis perniciosa, 234, 611–612
Criniviruses, 777–779
Cronartium, 119, 577–582
ribicola, 115, 563
Crop losses. *See* Losses, from disease
Crops
breeding stations, 626
certification, 295–296
epidemics and types of, 268
isolation, 296
resistant varieties, 626
rotation, 49, 300–301, 626
Cross protection, 49, 303, 314–315, 754
Crown galls, 24, 49, 50, 51, 108, 119, 120, 121, 146, 198, 199, 326
bacterial, 662–666
Crown rots, 540
Crown wart, 433, 434
Crucifers
black rot or black vein, 653–654
clubroot, 119, 257, 407–409
Cryphonectria parasitica, 145, 193, 325, 473
Cryptodiaporthe, 473
Cucumbers
angular leaf spot, 630, 632
damping off, 410
mosaic virus, 169, 787, 788–790
rot, 596
scab and gummosis, 456, 457
Cucumoviruses, 787–790
Cucurbits
anthracnose, 484, 487
bacterial wilt, 42, 44, 639–641
downy mildew, 428, 429, 430
genetic engineered, 169
gummy stem blight and black rot, 516–518
leaf and fruit spot, 453
yellow vine disease, 684–685
Cultural control methods, 294, 300–302
Curculio beetle/weevil, 42, 43
Curly top of sugar beets, 805, 806
Curtobacterium, 622, 639
Curtoviruses, 805
Cuscuta, 10, 705, 706–708, 743
Cuticle
composition and structure, 180
defense, and role of, 210
degradation of, 144–145
nematode secretions, 150–151
Cuticular wax, 180, 181
Cutins, 86, 144, 180–182
Cutinases, 55, 181–182
Cyanogenic glycosides and glycosinolates, 145
Cybrid cells, 169
Cyclic adenosine monophosphate (cAMP), 82, 83
Cycloheximide, 47, 343
Cylindrosporium, 483
Cymbidium mosaic virus, 763
Cypress canker, 36, 216, 483
Cyprex (dodine), 343
Cysteine-rich proteins, 240
Cyst nematodes
description of, 842–848
potato, 847
soybean, 66, 843–846
sugar beet, 24, 66, 846–847
Cytokinins, 50, 200–201, 664
Cytolytic enzymes, 50
Cytoplasm, 176, 619
Cytoplasmic defense reaction, 214
Cytoplasmic inheritance, 129
Cytoplasmic resistance, 137
Cytoplasm, 616
Cytorhabdovirus, 795
Cytosine, 730
Cytospora, 138, 479

Dactylella, 842
Dagger G, 323, 324, 327
Dahlia mosaic virus, 802
Daldinia, 605
Daltons, 730
Damage threshold, 274

Damping off, 108, 109, 255, 398, 410–414
  biocontrol of, 327
  Rhizoctonia, 594
Darwin, Charles, 17
Dasheen mosaic virus, 764
Dazomet (Mylone), 313, 345
DDT, 48, 49
DeBary, Anton, 18, 20, 21, 23, 50
Decision support systems (DSS), 289
Decline, 398
  pear, 66, 699
  slow, 848
Defense responses. *See* Resistance/defense against disease
Defensins, 238
Democritus, 10, 14, 46
Deny, 324
Deoxynivalenol, 559
Deoxyribose nucleic acid. *See* DNA
Detoxification of pathogen toxins, 236–237
Deuteromycetes. *See* Mitosporic fungi
DeVine, 329
De Wit, P. J. G. M., 55
Dianthovirus, 781
Diaporthe, 501, 518–519
Diazoben, 313
Dibotryon morbosum, 119, 120
Dichloran (DCNA) (Botran/Allisan), 339
Dichloroisonicotinic acid, 238, 242
Dickman, M. B., 54–55
Didymella, 501
  bryoniae, 183–184, 518
Didymium, 404
Dieback, 398
Diener, T. O., 27, 28
Dienes, 211
Dihaploids, 169
Dikaryotic mycelium and spores, 565
Diplocarpon, 453, 483, 484
Diplodia, 605, 606
Direct penetration, 87–88, 90
Direct protection, 322–348
Discula, 483, 500
Diseases
  See also Genetics, diseases and; Infectious (biotic) diseases; Noninfectious (abiotic) diseases; Resistance/defense against disease
  agents that cause, 4
  concept of, 5–7
  cycle/development of, 79–102
  damage threshold, 274
  diagnosis of, 71–74
  dispersal, 276
  economic threshold, 274
  escape, 137–139
  forecasting, 281–283, 285–287
  gradient curve, 276
  incidence, 273
  measurement of disease and yield loss, 273–274
  methods by which pathogens cause, 50–52
  progress curves, 274, 275
  pyramid, 267
  severity, 273
  symptoms, 89
  tetrahedron, 267
  tolerance, 139
  triangle, 79
  types of, 7–8
  unknown etiology, 26
  warning systems, 287–288
Diseases, controlling
  alternative, 49–50
  biological, 49–50, 294, 303–305, 322–329
  chemical, 47–49, 294, 312–314, 329–348
  crop certification, 295–296
  crop isolation, 296
  crop rotation, 300–301
  cross protection, 49, 303, 314–315
  cultural, 294, 300–302
  direct protection, 322–348
  disinfestation of warehouses, 313–314
  epidemics and, 272–273
  fumigation, 313
  growing conditions, improving, 316
  heat treatment, 310–312
  immunization, 314
  induced resistance, 315
  inoculum, eradication or reduction of, 298–314
  insect vectors, 314
  integrated management, 49, 348–351
  methods for, 5, 46–48
  pathogen-free propagating material, 296–298
  pathogens, evasion or avoidance of, 296
  pathogens, exclusion from surfaces, 298
  physical methods, 294, 310–312
  plant defense activators, 315–316
  quarantines and inspections, 295
  regulatory measures, 294
  resistant varieties, use of, 318–319
  sanitation, 301
  soil treatment, 313
  sprays and dusts, 332–338
  systemic acquired resistance, 50
  transgenic plants, use of, 294, 319–322
  trap plants, 307–308
  traps/mulches, 302
Diseases of Cultivated Crops, Their Causes and Their Control (Kühn), 22
Disinfestation of warehouses, 313–314
Dispersal curve, 276
Dissemination, pathogen, 81–82, 96–100
  fungi, 390
DiTera Biocontrol, 324
Dithiocarbamates, 47, 339
Ditylenchus, 10, 858–860
  dipsaci, 93, 859
DNA (deoxyribose nucleic acid)
  bacteria identification and, 624–625
  cytoplasmic inheritance, 129
  discovery of double helix, 54
  double-stranded viruses, 732, 321, 801–805
  enhancers, 126
  expressed sequence tag, 223
  genetic information in, 125–126
  microarrays, 223
  molecular tools, 283
  probes, 624–625
  promoters, 126
  single-stranded viruses, 733, 805–816
  silencers, 126
  terminators, 126
  tumor (T-DNA), 198–200, 664, 665–666
  uptake methods, 169
Dodder, 10, 72, 706–708
  virus transmission by, 743
Dodine, 343
Dogwood anthracnose, 483, 500
Doi, Y., 26
DON, 559
Dothistroma, 453, 456
Double-stranded
  DNA, 321, 732, 801–805
  RNA, 245, 325
Downy mildew
  corn and, 67, 428, 429
  description of, 427–433
  grapes and, 31–32, 47, 66, 427, 428, 430–433
  pumpkins and, 107
  sorghum and, 67, 427, 428
  tobacco and, 66, 285, 427, 428
Drechslera, 469
  teres, 220
Drought, 365
Drying, control by, 312
Dry rot, 398
dsRNA, 245, 325
Duggar, 62
Dusters, 332
Dusts, 332–334
Dutch elm disease, 32, 34–35, 42, 44, 66, 193, 211, 522, 528–532
Dwarf mistletoes, 712–715

Early blight. *See* Blight, early
Economic loss, 273–274
Economic threshold, 274
Ectomycorrhizae, 612–613
Ectoparasites, 831
Edema (oedema), 366, 367
Educational and training requirements for pathologists, 61–62
Eggplant
  blight, 518
  fruit rot, 484
Electrolytes, loss of, 118
Electrolyzed oxidizing water, 344
Electron microscopy, 747
Elements, major and minor, 372

Elicitors
nonspecific, 213
pathogen-derived, 54, 86, 151–153
ELISA. *See* Enzyme-linked immunosorbent assay
Elms
bark beetles, 42, 44, 530
Dutch elm disease/wilt, 32, 34–35, 42, 44, 66, 193, 211, 522, 528–532
yellows (phloem necrosis), 697, 698
Elsinoe, anthracnoses and scabs, 483, 484, 486–487
Elytroderma, 453, 456
Endomycorrhizae, 613–614
Endoparasites, 831
Endopectinases, 182
Enniatin, 164
Entyloma, 583
Environmental factors, 7, 48–49, 137–139
air pollution, 48, 262, 368–372
epidemics and, 271–272
hail injury, 380
herbicide injury, 378–380
improper agricultural practices, 381–383
inadequate oxygen, 367
infectious diseases and, 249–262
light, 367–368
lightning, 381
moisture effects, 138, 250, 253–257, 271–272, 365–367
nutritional deficiencies as a result, 372, 373
soil minerals, toxic, 372–378
temperature effects, 138, 250, 251–253, 272, 358–364
types of symptoms caused by, 359, 360
Environmental Movement, 48
Enzyme-linked immunosorbent assay (ELISA), 55, 73, 287, 295, 297, 746–747
Enzymes
bacteria identification and, 624
degradation of cell wall substances, 180–189
degradation of substances contained in cells, 189–190
Epicatechin, 235
EPICORN, 280
EPIDEM, 280
EPIDEMIC, 280
Epidemics
comparison of, 276–277
computer simulation of, 280–281, 286
decision support systems, 289
defined, 266
development of, 277–278
elements of, 266–267
environmental factors that affect, 271–272
expert systems, 288–289
forecasting, 281–283, 285–287
host factors that affect, 267–269
human practices and, 272–273
measurement of disease and yield loss, 273–274
modeling, 53–54, 278–280
pathogen factors that affect, 269–271
patterns, 274–276
relationship between disease cycle and, 102–103
risk assessment of, 287
vulnerability of genetic crops to, 170–172
warning systems for, 287–288
Epidemiology
defined, 266
as a field, 53–54
geographic information system (GIS), 283–284
geostatistics, 284
global positioning system (GPS), 284
image analysis, 284–285
information technology, 285
molecular tools, 283
remote sensing, 284
tools, 283–285
Epiphytotics, 266
EPIVEN, 280
Eradicants, 334
Eradication of host, control by, 300
Ergot sclerotia (ergotism, Holy Fire), 16, 39, 559, 560
barley, 37, 38, 502
cereals and grasses, 501–504
rye, 37, 38, 66
sorghum, 101, 503
wheat, 37, 38, 66, 502
Eriksson, 52
Erwinia, 553
amylovara, 121, 148, 149, 286, 299, 300, 639, 641–647
carotovora, 148, 213, 656
chrysanthemi, 148
description of, 621
stewartii, 639, 654
tracheiphila, 108, 112, 639–641
Erysiphe, 448
graminis hordei, 137
Esophageal gland secretions, 151
Ethazol, 342
Ethoprop, 313
Ethylene, 52, 159–160, 201
injury to plants, 368, 369
Ethylenebisdithiocarbamates, 339
Etiolation, 367–368
Eucalyptus, vascular wilt, 522, 534
Eukaryotes
composition, 616
DNA in, 125–126
Euphorbia, 877
European fruit yellows, 697
Eutypa, 476
Eutypella, 476
Exobasidium azaleae, 119, 120, 196, 197
Exopectinases, 182
Expert systems, 288–289
Expressed sequence tags (ESTs), 223
Exserohilum, 468
Extensin, 189
Extracellular polysaccharides (EPS), 148
Fabavirus, 784
Facultative parasites, 78, 387, 389
Facultative saprophytes, 78, 387, 389
Famoxadone, 343
Farlow, M. A., 61
Fasciation (leafy gall) disease, 50
FAST, 286
Fastidious vascular bacteria
phloem-inhabiting, 683–687
symptoms caused by, 679
xylem-inhabiting, 678–683
Fats, 190
Fatty acid compounds, defense and, 211
Fatty acid profile analysis, 624
Fava bean necrotic yellows virus, 813
Fenarimol, 334
Fensulfothion, 313
Fentin hydroxide, 343
Ferbam, 47
Fermentation, 116, 297
Ferns, 717
Fertilizers, use of, 626
Ferulic acid, 233
Fescue toxicosis, 560
Fijiviruses, 792
Filamentous ssRNA viruses, 762–764
Film-forming compounds, 339
Fire blight, 24, 66, 148, 149, 639
in apples and pears, 42, 43, 121, 286, 299, 300, 641–647
Fischer, Alfred, 25
Fission, 619–620
Flagella, 618, 619
Flagellate protozoa
description of, 875–877
epidemiology and control of, 878
fruit-and-seed infecting, 882–885
laticifer-restricted, 882
nomenclature, 877
pathogenicity, 877–878
phloem-restricted, 878–882
taxonomy, 877
Flavonoids, 149, 189
Flax rust, 53, 54, 242
Fleck disease, 763
Fleming, Alexander, 47, 49
Flies, 42, 43
Flooding, damage by, 366
Flor, H. H., 53, 54, 151
Fludioxonil, 343
Fluorescent antibody microscopy, 747
Flutolanil, 340, 341
Foliar diseases, fungal, 452–473
Foliar nematodes, 867–869
Food safety, 58–59
Foot rot, 417
Ford Foundation, 60
Forecasting epidemics, 281–283, 285–287
Forest trees
cankers, 481–483
rusts, 577–582
Fosetyl-Al, 341
Foveaviruses, 763
Frankia, 676
Frog eye leaf spot, 519, 521

Frost damage, 328, 360, 362–364
Fruiting bodies, 604
Fruits
 postharvest decays, 556–558
 spot and rot, 484, 494–498
 trypanosomatids, 882, 886
Fruit trees, root rots, 602–604
F-Stop, 323, 324
Fuligo, 404
Fumaric acid, 193
Fumigants, 313
Fumigation, 313
Fumigators, 332
Fumonisins, 560
Fungal-like organisms, 391–392
Fungi, 9
 biocontrol products produced by, 324
 characteristics of, 388–390
 control of diseases, 403
 defined, 386
 diagnosis of diseases, 72
 diseases caused by, 66, 67
 dissemination, 390
 ecology and spread, 389–390
 effect of moisture on, 253–255
 effect of temperatures on, 251
 expanding role of, as causes of diseases, 21–22
 germination, invasion, and penetration by, 82, 83–84
 identification of, 397
 interesting facts about, 387
 isolation of, 398–402
 life cycles, 402–403
 morphology, 388
 pathogenicity genes, 144–146
 reproduction, 93–96, 388–389
 sexual-like processes in, 131–132
 symptoms caused by, 397–398
 taxonomy, 390–397
 toxins, pathogenicity genes controlling, 145–146
 true, 392–397
 vascular wilts caused by, 108
 virus transmission by, 742–743
Fungi, imperfecti. *See* Mitosporic fungi
Fungicides
 See also under name of
 description of, 334
 early discovery of, 47
 for soil treatment, 313
 statistics and costs of, 69–71
 sterol-inhibiting, 334
Fungigation, 334
Fungistasis, 87
Fungitoxic exudates, 211
Furoviruses, 761
Fusaclean, 324
Fusaric acid, 193
Fusarium, 102, 108, 138, 145, 535
 avenacearum, 164
 circinatum, 476
 description of, 163–164
 graminearum, 56
 moniliforme, 164, 481, 560
 oxysporum, 163, 164, 193, 198, 252, 522
 oxysporum f. sp. *cubense*, 296, 522
 oxysporum f. sp. *lycopersici*, 126–127, 522, 523
 pink or yellow molds, 556
 postharvest decays, 556
 root and stem rot, 109, 538–540
 scab, 66
 solani, 144, 145, 163, 164, 256
 toxins, 559–560
 vascular wilts, 41, 52, 110, 251, 522, 523–526
Fusicoccin, 193
Fusicoccum amygdali, 193, 476
Fusiform rust, 580–582

Gaeumannomyces
 graminis var. *avena*, 145, 202, 219
 penetration, 178
 take-all wheat disease, 540, 542–543
 tritici, 109, 219
Gallex, 323
Galls
 bacterial, 108, 109, 662–667
 fungal, 398
Galltrol, 323, 324, 326
Gametes, 388
Ganoderma, 606, 607
Gaümann, E., 53
Gel diffusion test, 745
Geminiviruses, 805–813
Gene(s)
 avirulence (avr), 55, 141, 149, 151–155
 defined, 126
 disease and, 126–128
 disease-specific, 142
 flow, 130
 horizontal gene transfer, 132
 hrp, 149, 155
 hypersensitive response (HR), 52, 53, 57, 149, 150, 151, 221–236
 induced during early infection, 223–224
 minor resistance, 136, 159
 pathogenicity, 142–151
 resistance/defense and role of, 208–210
 R gene resistance (race-specific, monogenic, or vertical), 136–137, 151, 155–158, 210, 221–236
 signaling, 146
 silencing, 320–321, 754
 structure, 126, 127
Gene-for-gene concept, 54, 140–141, 151–153
Genetically modified organisms (GMOs), 58
Genetics
 breeding resistant varieties, 165–172
 drift, 130
 engineering, 49–50, 54, 56–58, 168–169, 242–244
 recombination in, 129–130, 133
 selection, 130–131
Genetics, diseases and
 basics of, 125–128
 relationship between pathogen virulence and host plant resistance, 139–165
 resistance to disease, 52–53
 variability, mechanisms of, 128–133
 variability in organisms, 128
Genomes, split, 729
Genomics, 55–56
 viral activation, 150
Geographic information system (GIS), 283–284
Geostatistics, 284
Geotrichum, 556–557
Germination
 seed, 86–87
 spore, 82–87
Germ theory of disease, 18, 22, 26
Germ tube formation, 82, 86, 87
Gibberella, 50, 146, 200, 253
 head blight, 535
 seedling blight, 535
 stalk and ear rot, 535, 536–537
Gibberellin growth regulators, 50, 51, 200
Gierrer, A., 25, 54
Gigaspora, 842
Gliocladium, 526
 virens, 306, 322–323, 325
GlioGard, 322–323, 324
Global positioning system (GPS), 284
Globodera, 842–843, 847
Gloeosporium, 483, 484–485
Glomerella, 145, 483
Glomus, 842
Glucanases, 145, 149, 212
 β-glucanase, 86, 186
 β-1,3-glucanase, 210, 220, 221, 240
Glucans, 235
Glume blotch, 461
Glycans, 186–187
Glyceollin, 235, 236
Glycine-rich proteins (GRPs), 189
Glycolipids, 190
Glycolysis, 116
Glycolytic pathway, 116
Glycoproteins, 235
Glycosides and glycosinolates, cyanogenic, 145
Gnomonia, 453, 483
 anthracnose and leaf spot, 498–500
Gossypol, 235
G-protein-coding genes, 146
Grains. *See* Cereals
Gram staining reaction, 622
Grape berries, powdery mildew, 11, 114
Grapes
 anthracnose, 483, 484, 486
 bitter rot, 496
 black rot, 514–516
 downy mildew, 31–32, 47, 66, 428, 430–433
 fanleaf virus, 742, 786–787
 gray mold/bunch rot, 512

leaf spot, 518
phylloxera, 30–31
Pierce's disease, 36, 42, 67, 111, 679–681
powdery mildew, 30, 66, 449
ripe rot, 496, 498
Graphium, 606
Grasses
anthracnose, 484, 489, 491
bacterial leaf spots and blights, 632–633
downy mildew, 428
ergot, 501–504
leaf spots, blights and rots, 469–470, 472
Grassy stunt virus, 43, 45
Gray molds and blights, 56, 512, 556
Greece, ancient, 10–11, 14
Green algae, parasitic, 719
Greeneria uvicola, 485, 496
Greening disease, 42, 67, 685–686
Green mold rots, 557
Gremmeniella, 476
Ground pollution, 48
Growth, effect of pathogens on, 119–121
Growth regulators
effects of excessive, 50, 51
plant diseases and, 196–201, 344
Guanine, 730
Guignardia, 501, 515
Gum barrier, role in defense, 217
Gummosis, 456, 457, 667–671
Gummy stem blight, 516–518
Gymnoconia, 563
Gymnosporangium, 563
juniperi-virginianae, 198, 255

Hail injury, 380–381
Hairy root, 119, 662
Hall, 62
Halo blight, 191, 192, 629, 632
Halogenated hydrocarbons, 344–345
Haploids, 129–130, 169
Harpins, 57, 86, 149, 154, 231
Hartrot disease, 880
Haustoria, 86, 87
HC toxin, 146, 194–195, 236
reductase, 156
Head blight, 535, 538
Heald, 62
Heat injury, 359–360, 361
Heat shock, 220
Heat treatment, control by, 310–312
Helminthosporium, 50, 194
Hemibiotrophs, 389
Hemicellulases, 145
Hemiculluloses. See Glycans
Hemileia vastatrix, 300, 563, 576
Herbicides, effects of, 262, 378–380
Heterobasidion, 606
annosum, 49, 323, 325
Heterocyclic compounds, 340
Heterodera, 842
glycines, 843–846
schachtii, 846–847
Heteroecious, 565
Heterokaryosis, 131–132
Heteroploidy, 132
High-temperature effects, 359–360, 361
HiStick N/T, 324
Histological defense structures, 215–217
Hm-1 resistance gene, 55, 242
Hoja blanca, rice, 67
Holy Fire (ergot), 16, 37–39
Homer, 9, 14, 46
Homoserine lactose (HSL), 148
Hooke, Robert, 16, 21
Hops, downy mildew, 428, 432
Hordeiviruses, 761
Horizontal gene transfer, 132
Horizontal resistance, 53, 136, 169–170, 209–210, 219–221
Hosts
attachment of pathogen to, 82
epidemics and role of, 267–269
eradication, control by, 300
range, pathogen, 78–79
reaction to pathogens, 213
receptors, 212, 213–214
recognition between pathogen and, 86, 212, 213
relationship between pathogen virulence and resistance of, 139–165
Hot-air/hot-water treatment, 297, 311–312
Hrp genes. *See* Hypersensitive response and pathogenicity/protein genes
HS toxin, 195
Humans
dissemination by, 100
epidemics and role of, 272–273
HV toxin, 194
Hyaloperonospora parasitica, 428
Hybrid cells, 169
Hybridomas, 744
Hydathodes, 89
Hydrogen chloride injury, 368, 369
Hydrogen fluoride injury, 368, 369
Hydrophobin, 144, 163
Hydroxamate siderophores, 149
Hydroxyproline-rich glycoproteins (HRGPs), 189
Hypersensitive response (HR), 52, 53, 57, 149, 150, 151, 217, 218
bacteia and, 624
induced biochemical defenses in, 221–236
Hypersensitive response and pathogenicity/protein (hrp) genes
pilin, 149
protein, 54, 149, 154
secretion system, 155, 202
Hypertrophy, 833
Hyphae, 215, 388
ascogenous, 439
Hyphal anastomosis, 132, 397
Hypoderma, 456
Hypovirulence, 303
Hypoxylon, 606
mammatum, 196, 476, 481

IAA. *See* Indoleacetic acid
Ice nucleation bacteria, 364
Idaeovirus, 783–784
Ilarviruses, 150, 790–792
Image analysis, 284–285
Imazalil, 342
Immunization of plants, 237, 314
Immunofluorescent staining, 747
Immunosorbent electron microscopy (ISEM), 747
Imperfecti fungi. *See* Mitosporic fungi
Incubation period, 89, 91
Indexing, 297, 751
Indoleacetic acid (IAA), 50, 196–200
Induced resistance
biochemicals, 213–214, 217–236, 315
structural, 214–217, 315
Infection(s)
defined, 89
cycles, 80
primary and secondary, 80
structures, production of, 144
symptoms of, 89
systemic, 91
Infectious (biotic) diseases
defined, 77
diagnosis of, 72–73
environmental effects on, 249–262
types of, 8
Inhibitors, defense and, 211–212
Inoculation, 80–82
Inoculum
antagonistic microorganisms to reduce, 305–309
biological methods to eradicate or reduce, 303–305
chemical method to eradicate or reduce, 312–314
defined, 80
forecasts based on initial amounts of, 285–286
forecasts based on weather conditions and secondary, 286–287
landing or arrival of, 81–82
physical methods to eradicate or reduce, 310–312
sources of, 80–81
types of, 80
Inonotus, 606
Insecticides, statistics and costs of, 69–71
Insects
control of, 314
as vectors for disease, 42–45, 97–99
virus transmission through, 741–742
Inspections, regulatory, 295
Integrated control
annual crops and, 350–351
perennial crops and, 348–350

Integrated management, 49, 348–351
Intercellular mycelium, 91
Intercept, 324
Internal transcribed spacer (ITS) regions, 55
International centers for research, 60–61
International Institute of Tropical Agriculture (IITA), 60
International Maize and Wheat Improvement Center (CIMMYT), 60
International Rice Research Institute (IRRI), 60
International Society of Plant Pathology, 60
Intracellular mycelium, 91
Introns, 126
Invasion, 91, 92
Invasive climbing plants, 716–719
Ipomoviruses, 773
Iprodione, 340
Ireland, potato late blight, 19–21, 22
Iron, 260
  deficiency diseases, 372, 373, 375
Isoenzymes, 624
Isolation, fungal and bacteria, 398–402, 624
Isometric viruses
  double-stranded DNA, 801–805
  double-stranded RNA, 792–794
  single-stranded DNA, 805–816
  single-stranded RNA, 779–792
Isonicotinic acid (INA), 238, 242, 338
Isopentenyl adenosine (IPA), 200
Isothiocyanates, 344, 345
Isozymes, 130
Ivanowski, D., 25

Jasmonic acid, 150, 159–160, 232, 238
Jenner, William, 23
Jones, L. R., 50
Junipers, tip blight, 518, 520

Karnal bunt, wheat, 67, 592–591
Kausche, 25
Kilobase pairs, 730
Kinase, 146, 157
Kinetin, 50
Klement, Z., 52
Klessig, D. F., 55
Knot disease, 200
Koch, Robert, 19, 23
  postulates, 26–27, 45–46, 74
Kodiak, 323, 324, 327
Kolattukudi, P. E., 54–55
Krebs cycle, 116
Kudzu vine, 573, 717–718
Kühn, 22
Kurosawa, E., 50

Laetiporus, 606
Laetisaria arvalis, 306
Lafont, A., 26, 877
Late blight. *See* Blight, late
Latent viruses, 737
Laticifer-restricted trypanosomatids, 882
Leaf blight
  northern corn, 468
  southern corn, 66, 137, 267, 268, 466–468
  strawberries, 518
Leaf blotch, 461
Leaf curl, 119, 120, 121, 200, 398
  caused by *Taphrina*, 445–447
Leafhoppers, 42, 43, 45, 742
Leaf scorch, 483
Leaf spots, 50, 106, 107, 192, 193, 216, 397, 463, 464
  bacterial, 627–638
  frog eye, 519, 521
  Gnomonia, 498–500
  grapes, 518
  northern corn, 468
Leaf wetness, monitoring, 282–283
Leafy gall disease, 50, 119, 200, 662
Leathal yellowing disease, 35–36
Lecanosticta, 456
Lectins, 189, 212
Leeuwenhoek, Antonius van, 16, 23
Legumes
  postharvest decays, 558–559
  root nodules, 675–678
  rusts, 571–577
Leifsonia, 683
Lenticels, 89
Lenzites, 606
Leptographium, 606
Lesions
  local, 737
  nematodes, 849–853
  root, 594
Lethal yellowing, 35–36, 42, 67, 694, 695
Lettuce
  downy mildew, 252, 428
  drop, 304, 305, 548
  gray mold, 512
  infectious yellows virus, 777–779
  mosaic virus, 297, 764, 767
  necrotic yellows virus, 794, 795
Leucine-rich region (LRR), 154, 155–156, 162
Leucostoma, 138, 316, 476
  canker, 479–481
Leveillula, 448
Lichens, 387
Life cycles
  fungi, 402–403
  nematode, 828–830
  pathogens, 69, 131
Light
  control by eliminating, 312
  diseases caused by, 367–368
  infectious diseases and, 257
Lightning, 381, 382

Ligninases, 145
Lignin, 149, 187–189, 234
Lilacs, powdery mildew, 11
Lily symptomless virus, 763
Lime sulfur, 47
Linne', Carl von, 17
Linolenic and linoleic fatty acids, 150
Lipases, 190
Lipids, degradation of, 190
Lipopolysaccharide (LPS), 149
Lipoxygenases, 150, 231–232
Lirula, 453, 457
Local acquired resistance, 237, 238
Local lesions, 737
Loculoascomycetes, 439
Longidorus, 742
Loose smut, 12, 585–587
Lophodermium, 453, 456
Losses, from disease
  description of, 29–45, 65–69
  epidemics and yield loss, 273–274
  examples of, 65, 66, 67
  financial, 41–42, 66, 67
  postharvest, 660
  quantity and quality, 29
  statistics on, 4
Low-temperature effects, 360, 362–364
LSD, 37
Luteoviruses, 781–783
Lycomarasmin, 193
Lygodium, 717

Machlomovirus, 781
Macluraviruses, 773
Macroconidia, 523
Mad cow disease, 26
Magnaporthe
  diseases caused by, 453
  grisea, 55–56, 59, 144, 145, 146, 153, 154, 162–163, 223, 236
  penetration, 178
Magnesium, 260
  deficiency diseases, 372, 373, 375
Maize
  See also Corn
  chlorotic mottle virus, 781
  rayado fino virus, 783
  rough dwarf virus, 792
  streak virus, 121, 805, 810
  stripe virus, 799–801
Maneb, 47
Manganese, 260
  deficiency diseases, 372, 373
Mango anthracnose, 491, 494
MAPK. *See* Mitogen-activated protein kinase
Marafivirus, 783
Marchitez sopresiva, 880
Marssonina, 483
Martin, G. B., 55
Masked symptoms, 737

Mastreviruses, 805
Matalaxyl, 334
Mating systems, 131
Mayer, Adolph, 25
Mealybugs, 43
Mechanical transmission of viruses through sap, 739–740
Medicarpin, 189
Meiosis, 564
Meiosporangia, 388
Meiospores, 93
Melampsora lini, 157, 563
Melanconium, 483, 485
Melanin, 178, 466
Melanose disease of citrus, 518, 519
Meloidogyne, 108, 109, 121, 198, 838–842
  javanica, 184
Melons, *Monosporascus* root rot and vine decline, 543–546
Mercury compounds, 47, 49
Messenger, 57, 316
Messenger RNA (mRNA), 126, 245
Metabolic (biochemical) defense, 217–236
Metabolites, 52
  pathogenicity genes controlling secondary, 145
  production of secondary, 233–236
Metalaxyl, 313, 340–341
Metam sodium, 313, 345
Methyl bromide, 313
Methyluracil, 730
Micheli, Pier Antonio, 17, 21
Microconidia, 523
Microcyclus, 453
Microdochium nivale, 253
Micronutrients, diseases and, 372
Microscopes, improvements in, 46
Microsphaera, 448
Mildews, 10, 11, 398
  See also Downy mildew; Powdery mildew
  early methods of controlling, 47
Millardet, Pierre Alexis, 31, 47
Millet, downy mildew, 428
Mills, 53
Minichromosome, 733
Minor resistance genes, 136, 159
Mistletoe, 14–16, 47, 72
  dwarf, 66, 712–715
  true of leafy, 715–716
Mites, virus transmission through, 742
Mitogen-activated protein kinase (MAP or MAPK), 82, 85, 146, 163
Mitospores, 93
Mitosporic fungi
  anthracnoses, 483–500
  cankers, 473–476
  diseases caused by, 439–440
  foliar diseases, 452–473
  fruit and general diseases, 501–522
  postharvest, 553–582
  reproduction in, 388–389
  root and stem rots, 534–553
  symptoms caused by, 445
  vascular wilts, 522–534
Modeling diseases and epidemics, 53–54, 278–280
Moisture
  damage by excess of, 365–367
  damage by lack of, 365
  disease escape and, 138
  epidemics and, 271–272
  infectious diseases and, 250, 253–257
Moko disease of bananas, 649
Molecular plant pathology, 54–56
Molds
  slime, 404
  sooty, 440, 445
Mollicutes (phytoplasmas), 9, 26, 72–73
  diseases caused by, 691–701
  phytoplasmas, 689–691
  spiroplasmas, 691, 699–701
  taxonomy, 617
Molybdenum (Mo) deficiency diseases, 372, 373, 377
Monilia pod rot of cacao, 67, 510, 511
Monilinia, 99, 121, 501
  fructicola, 122, 181–182, 185, 233, 251, 507–510
Moniliophthora, 501, 510, 511
Monoclonal antibodies, 744
Monocyclic diseases, 102, 103, 270, 276
Monogenic resistance, 137, 210
Monosporascus, melon root rot and vine decline, 543–546
Moraceae, 877
MoreCrop, 289
Mosaics, 737
Movement proteins, 150
Mucilago, 404
Mucor, 554, 556
Mulches, use of, 302
Multigene resistance, 136
Multilines, 170
Mutation, 128, 129
Mycelium, 9, 86
  dikaryotic, 565
  in fungi, 388, 397
  intercellular, 91
  intracellular, 91
Mycolysis, 305
Mycoparasitism, 305
Mycoplasma-like organisms (MLO), 616
Mycoplasmas, true, 688–689
Mycorrhizae, 325–326, 387, 612–614
MYCOS, 280
Mycosphaerella, 196, 202, 203
  diseases caused by, 453, 456, 458–460
  fijiensis, 234, 453
Mycotoxins and mycotoxicosis, 39–41, 146
  postharvest diseases and, 553, 559–560
Myrothecium roridum, 146
Myxomycota, 26, 404–405
Myxomycetes, diseases caused by, 404–405
Nanoviruses, 813
Natural openings, penetration through, 88–89
Nature of Plants, The (Theophrastus), 10–11
Necrotic defense reaction, 217, 218
Necrotrophs, 78
Necrovirus, 781
Nectarthodes, 88, 89
Nectria, 99, 476, 478–481
  haematococca, 145
Nectria canker, 99, 115, 477, 478–481
Needham, T., 17–18, 23
Needle casts and blights, 456–458
Negative RNA viruses, 794–801
Nematicides, 313, 344–345
Nematodes, 23–24
  See also Cyst nematode
  anatomy, 828
  burrowing, 853–857
  characteristics of, 827–831
  chemical control, 313
  control of, 836
  cyst, 842–849
  description of, 826–827
  diagnosis of diseases, 72
  diseases caused by, 66, 67, 121
  ecology and spread, 830
  ectoparasites, 831
  endoparasites, 831
  foliar, 867–869
  hatching of eggs, 87
  how they affect plants, 833–835
  interrelationships with other pathogens, 835–836
  isolation from plant material, 832
  isolation from soil, 831–832
  lesion, 849–853
  life cycles, 828–830
  migratory, 831
  morphology, 827–828
  pathogenicity genes, 150–151
  penetration by, 90, 179
  pine wilt, 870–872
  red ring, 872–874
  reproduction, 828
  root-knot, 838–842
  seed-gall, 865–867
  sedentary, 831
  stem and bulb, 858–860
  sting, 860–863
  stubby-root, 863–865
  symptoms caused by, 832–833
  taxonomy, 830–831
  in the tropics and subtropics, 858
  virus transmission by, 742
Neovossia, 583
Nepoviruses, 742, 784–787
Net blotch, 469, 471
Netria haematococca, 540
Nicobifen, 341
Nicotiana
  benthamiana, 162
  sylvestris, 154

Nitric oxide, 159–160
Nitrogen deficiency diseases, 372, 373, 374
Nitrogen dioxide, injury by, 368, 369, 374
Nitrogen nutrition, 258–259
Nogall, 324
Nondifferential resistance, 136
Nonhost resistance, 127, 208–209
Noninfectious (abiotic) diseases
  control of, 358
  diagnosis of, 73–74, 358
  types of, 8
Nonobligate parasites, 78, 387
Nonpersistent viruses, 742
Nuclear-binding site (NBS), 154, 155
Nucleic acid, in viruses, 729–731
Nucleorhabdovirus, 795
Nutrition
  deficiencies, 372, 373–378
  infectious diseases and, 257–262

Oak
  cankers, 474
  leaf blister, 445, 446
  sudden death, 419–420
  vascular wilt, 36, 522, 532, 533
Oats
  Helminthosporium blight, 50
  leaf blotch, 461
  mosaic virus, 774
  sterile dwarf virus, 792
Obligate parasites, 78, 387
Ochratoxins, 41, 560
Oidium, 448
  neolycopersici, 143, 448
Oil palms, sudden wilt, 880
Oils, 190
Old Testament, 9, 10
Old world climbing fern, 717
Oleander gall, 119, 662
Oleaviruses, 787
Oligogenic resistance, 137
Olive knot, 119, 662
Olpidium, 433–434, 742
Onions
  anthracnose, 488–489
  bacterial soft rot, 658
  downy mildew, 428
  leaf and fruit spot, 453, 455
  rots, 540
  slippery skin, 656
  smudge, 211, 485, 488–489
  smut, 583
  sour skin, 656
  stem and bulb nematode, 860
Oomycetes
  diseases caused by, 285–286, 409–433
  symptoms caused by, 410–411
  reproduction in, 388
Oomycota, 386
Oospores, 388, 430
Ophiobolins, 193
Ophiostoma, 108
  ulmi, 35, 193, 211, 528–532
  vascular wilts, 522, 528–532
Orchids, odontoglossum ring spot virus, 758
Organelles, 616
Organophosphates, 341, 344, 345
Origin of Species by Means of natural Selection, The (Darwin), 17
Orobanche, 234
Orton, W. A., 52
Oryzaviruses, 792
Ourmia melon virus, 784
Ourmiavirus, 784
Outcrossing, 131
Overwintering/oversummering, pathogen, 100–102, 300
Oxalic acid, 193
Oxamyl, 345
Oxanthiins, 341
Oxidative phosphorylation, 117
Oxycarboxin (Plantvax)
Oxycom, 316
Oxygen, damage by lack of, 366, 367
Oxygen species, activated, 231
Oxyquinoline sulfate, 343
Ozone injury, 368, 369, 370–371

Paecilomyces, 605
Palms
  See also Coconut palms
  bud rot, 414, 417
  red ring, 67, 872–874
Panama disease, 296, 526
Panicovirus, 781
Pantoea, 327
  stewartii, 639, 654
Papaya
  anthracnose, 491
  bunchy top bacterium, 616, 686–687
  ring spot virus (PRSV), 56, 57, 169, 320, 321, 764, 769
Papillae, 215
Paralongidorus, 742
Parasexualism, 132
Parasite(s)
  defined, 77
  facultative, 78, 387, 389
  nonobligate, 78, 387
  obligate, 78, 387
Parasitic plants, 72, 88
  broomrapes, 711–712
  disease development and, 77–103
  dodder, 706–708
  green algae, 719
  mistletoes, dwarf, 712–715
  mistletoes, true of leafy, 715–716
  witchweed, 708–711
Parasitism, defined, 77
Paratrichodorus, 742, 761, 863
Parenchyma cells, 214
Partial resistance, 136, 209–210, 219–221
Particulate matter, injury by, 368, 369
Pasteur, Louis, 18, 20, 22, 23, 26
Pasteuria penetrans, 842
Pathogen-derived resistance, 54
Pathogenesis related (PR) proteins, 210, 221, 232–233
Pathogenicity
  bacteria and, 622
  defined, 77–78
  genes, 142–151
  parasitism and, 77–78
Pathogens
  aerial, 307
  attachment to host, 82
  attenuation, 133
  defined, 7
  dissemination of, 81–82, 96–100
  epidemics and role of, 269–271
  fitness, 131
  -free propagating material, 296–298
  host range of, 78–79
  inoculation, 80–82
  life cycle, 80, 131
  methods by which diseases are caused, 50–52
  methods of multiplication, 8
  monocyclic and polycyclic, 102–103, 270
  overwintering/oversummering, 100–102
  penetration by, 87–89
  recognition between host and, 86
  relationship between host plant resistance and virulence of, 139–165
  relationship of insects and, 42
  reproduction/colonization, 91, 93–96, 270–271
  resistance to, 134–139
  resistance to chemicals, 346–347
  resistant strains, 48
  shapes and sizes of, 7
  soil-borne, 163–164, 305–307
  variability in, 131–134
  virulence in culture, loss of, 133
Pathogens, attack by
  chemicals, 179–203
  enzymes, 180–190
  grow regulators, 196–201
  mechanical forces, 177–179
  reasons for, 176
  toxins, 190–196
Pathogens, effects of
  on cell membrane permeability, 118
  on photosynthesis, 106, 107
  on plant growth, 119–121
  on plant reproduction, 121–122
  on respiration in host plant, 115–118
  on transcription and translation, 118–119
  on translocation of water/nutrients in host plant, 106, 108–115
  on transpiration, 108, 113, 115
Pathovars, 152
Patulin, 557, 560
PAWS, 289

PCNB (pentachloronitrobenzene), 313, 339, 675
PCR. *See* Polymerase chain reaction
PC toxin, 196
Peaches
  brown rot, 185
  canker, 518, 668
  European fruit yellows, 697
  leaf curl, 119, 120, 445, 446, 447
  Leucostoma canker, 480
  powdery mildew, 449
  root and crown rot, 414, 417
  scab and twig blight, 456, 457
  soft rot, 435, 436
  Verticillium wilt, 526, 527
  X disease, 697–699
  yellows, 66
Peanuts
  clump virus, 761–762
  stunt virus, 787
Pears
  decline, 66, 699
  fire blight, 42, 43, 121, 258, 286, 300, 641–647
  Nectria canker, 99, 115, 477, 478–481
Peas
  early browning virus, 742, 758, 863
  fasciation (leafy gall) disease, 50
  fungus, 164
  pod rot, 456
  root rot, 285–286
  streak virus, 763
Pecluviruses, 761–762
Pectic enzymes, 50, 51
Pectic substances, 182–184
Pectinases, 144, 145, 147–148, 182
Pectins
  degradation of, 144–145
  lyases, 144–145, 147–148, 182
  methyl esterases, 163, 182
Pectolytic enzymes, 182
Penetration, pathogen, 87–89
  peg, 88, 177–179
Penicillin, 47, 387
Penicillium, 41, 49, 387
  postharvest decays, 556, 557
  toxins, 560
Peniophora, 606
Pentose pathway, 116
Peppers
  bacterial spot, 633–635
  green mottle virus, 758
  mottle virus, 764
  ring spot virus, 758
  root and stem rot, 414, 415
Peptidases, 190
Perenophora teres, 196
Periconia circinata, 196
Peridermium, 563
Perithecium, 439
Peritoxin, 196
Peritrichous flagella, 619
Permeability of cell membranes, affected by pathogens, 118
Peronosclerospora, 409, 427, 428
Peronospora, 409, 427, 428
  parasitica, 162
  tabacina, 285
Peroxidases, 150–151, 221, 234, 235
Peroxyacyl nitrate (PAN) injury, 369, 371
Persistent viruses, 742
Pestalotia malicola, 180
Pesticides
  other names for, 48
  public concern about, 48–49
  statistics and costs of, 69–71
Petri, Robert, 19, 45
Petroleum oils, 344
Petunia vein clearing virus, 801
Phaeocryptopus, 457
Phages, (bacteriophages), 328
Phakopsora, 563, 571
Phanerochaete chrysosporium, 55
Phaseollin, 235, 236
Phaseolotoxin, 191
Phellinus, 606
Phenolic compounds, 52, 211, 338
  from nontoxic glycosides, 234
  simple, 233–234
Phenolic glycocides, 234
Phenol-oxidizing enzymes, 234–235
Phenylalanine ammonia lyase (PAL), 203, 221
Phialids, 388
Phleviopsis gigantea, 49
Phloem, 93
  inhabiting bacteria, 94, 683–687
  necrosis, 115, 697, 878–880
  nutrient transport, 113, 115
  restricted trypanosomatids, 878–882
Phoma, 518–519
Phomopsis, 110, 476, 481, 483, 518–519
Phosphate compounds, 339
Phospholipases, 190
Phospholipids, 190
Phosphorus, 259
Phosphorus deficiency diseases, 372, 373, 374
Photosynthesis, effect of pathogens on, 106, 107
Phragmidium, 563
Phylactinia, 448
Phyllody, 121
Phyllosticta, 453, 515
  maydis, 137, 196
Phylloxera, 30–31
Phymatotrichopsis omnivora, 257, 550–552
Phymatotrichum root rot, 550–552
Physalospora, 521
Physarum, 404
Physical control methods, 294, 310–312
Physiology of virus-infected plants, 737
Physoderma alfalfae, 119, 433, 434
Phytoalexins, 52, 53, 55, 145, 164, 221, 235–236
Phytoanticipins, 145, 211
Phytocystatins, 212
Phytomonas, 10, 877–886
Phytomycin, 343
Phytopathology, 60
Phytophthora, 36, 138, 409, 553
  cinnamomi, 89
  cryptogea, 526
  diseases caused by, 414–427
  increase in, 418
  infestans, 56, 59, 99, 168, 219, 223, 236, 251, 286, 418, 421–426
  late blight of potato and tomato, 421–426
  megasperma, 54, 236
  ramorum, 418, 419
  root and stem rot, 109, 414–421
Phytoplasmas, 9, 26, 616, 617
  diseases from, 66, 67, 113, 115, 687, 689–691
Phytoreoviruses, 792
Pichia gulliermondii, 306
Pierce's disease, 36, 42, 67, 111, 679–681
Pigweed, 329, 330
Pilins, 149
Pines
  blister canker, 115
  blister rust, 578–580
  brown spot, 457, 458
  dwarf mistletoes, 712–715
  fusiform rust, 580–582
  needle casts and blights, 456–457
  pitch canker, 481, 482
  root and butt rot, 49
  ruts, 119
  stem rusts, 66
  western gall rust, 198
  wilt nematode, 870–872
  witches' broom, 121, 201, 398, 445
Pinewood nematode, 67
Piperalin, 343
Pisatin, 145, 164, 235
Pitch canker, 481, 482
PLAM, 286
Plant biotechnology, 56–58
Plant defense activators, 315–316
Plant Diseases: Epidemics and Control (Vanderplank), 53
Planthoppers, 742
Plantibodies, 237
Plant losses. *See* Losses, from disease
Plant pathogenesis-related (PR) proteins, 52
Plant pathologists
  certification for, 63–64
  educational and training requirements, 61–62
Plant pathology
  certification for, 63–64
  contribution to crops and society, 65–71
  defined, 4, 5
  descriptive phase, 45–46
  early history of, 8–21
  educational and training requirements, 61–62
  etiological phase, 46
  experimental phase, 46
  future for, 54–59
  international centers, 60–61
  molecular, 54–56

Plant pathology (*continued*)
  practitioners of, 63
  as a profession, 60–65
  in the 20th century, 45–54
Plant Quarantine Act (1912), 295
Plants
  basic functions in, 6
  importance of, 4
  oils, 344
Plant Shield, 324
Plasmids, 126, 619
Plasmodesmata, 94, 113, 733
Plasmodiophora, 405
  brassicae, 108, 119, 138, 196, 197, 257, 407, 408
Plasmodiophoromycetes, diseases caused by, 405, 407–409, 742
Plasmodiophoromycota, 26
Plasmodium, 388, 404
Plasmopara, 409, 428
  lactucae-radicis, 95
  viticola, 31, 95, 428–433
Pleurotus, 606
Ploioderma, 457
Plums
  black knot, 473, 476
  curculio beetle/weevil, 42, 43
  European fruit yellows, 697
  pocket, 119, 120, 445, 446, 447
  pox or sharka, 66, 67, 121, 300, 764, 767–769
Podosphaera, 299, 451
Pod rot, 67, 510, 511
Poisonous plants, 37–39
Polar flagella, 619
Pollen, dissemination by, 100
  virus transmission through, 741
Poplar mosaic virus, 763
Polyclonal antibodies, 744
Polycyclic diseases, 102, 103, 270, 276–277
Polyethylene traps and mulches, 302
Polyetic epidemics, 277
Polygalacturonase-inhibiting protein (PGIP), 56
Polygalacturonases, 182
Polygandron, 324
Polygenic resistance, 136, 209–210, 219–221
Polymerase chain reaction (PCR), 55, 283, 295, 297, 397, 625
Polymyxa, 405, 407, 742, 761
Polyphenol oxidases, 234–235
Polyporus, 606
Polysaccharides
  extracellular, 148
  role of, 201
  water translocation, 112
Pome fruits
  brown rot, 42
  fire blight, 66, 639, 641–647
Pomoviruses, 762
Postharvest diseases
  Ascomycetes and mitosporic fungi, 553–582
  bacterial, 660
  biological control of, 326, 328, 560–561
  chemical control of, 337–338, 560–561
  of fruits and vegetables, 556–558
  of grains and legumes, 558–559
  losses, 553, 660
  mycotoxins in, 553, 559–560
Potassium, 259
  deficiency diseases, 372, 373, 374
Potatoes
  bacterial soft rot, 656, 657
  black heart, 367
  blackleg, 656
  black scurf, 594, 596
  black wart, 433, 434
  blight, 56
  blight, early, 53, 453, 454
  blight, late, 18, 19–21, 22, 47, 59, 66, 67, 286, 421–426
  golden cyst nematode, 847
  leaf roll virus, 115, 169, 782–783
  mop-top virus, 761, 762
  pink eye, 656
  ring rot, 649–651
  root knot, 838
  scab (common), 256, 257, 304, 674–675
  scab (powdery), 405, 407
  soft rot, 42, 416, 656, 657
  southern bacterial wilt, 647
  spindle tuber disease, 26, 27, 28, 819, 820
  storage rotting, 540
  virus X, 153, 246, 762
  virus Y, 42, 45, 169, 764, 769
  wart, 119
  yellow dwarf virus, 794, 795
Potexviruses, 762–763
Potyviruses, 246, 764–773
Powdery mildew, 299
  apples, 299
  azaleas, 11
  description of, 161–162, 254–255, 448–452
  grape berries, 11, 114
  grapes, 30, 66
  lilacs, 11
  roses, 449, 451–452
  wheat, 11
Pratylenchus, 849–853
Precipitin test, 745
Prepenetration phenomena, 82–87
Prevost, 18, 21
Primary infections, 80
Primary inoculum, 80
Primastop, 324
Prions, 26, 27, 28, 29
Probenazole, 238, 242
Prochloraz, 343
Professional organizations/societies, 60, 64
Prokaryotes
  See also Bacteria; Mollicutes
  DNA in, 125–126, 616
  diseases caused by, 616
  taxonomy, 616–617
Proline-rich proteins (PRPs), 189
Promote, 324
Propagative viruses, 742
Propagules, 80, 86
Propamocarb, 343
Proteases, 147, 190
Protectants, 332, 334
Proteinase inhibitors, 52, 190
Proteins
  arabinogalactan, 189
  avr gene, 153–154
  cell wall structural, 189
  coat, 149–150, 731
  degradation, 189–190
  glycine-rich, 189
  hydroxyproline-rich glyco, 189
  hypersensitive response (hrp), 54, 154
  kinase, 146, 157
  kinase A (PKA), 83
  movement, 150
  pathogenesis related (PR), 210, 221, 232–233
  proline-rich, 189
  role of, 126
  subunits, 729, 730
  thaumatin-like, 221
  in viruses, 730
Protoplast, 619
Protoplast fusion, 169
Prototheca, 719
Protozoa, 10, 25–26
  See also Flagellate protozoa
  myxomycetes, 24
  taxonomy, 390
PR (pathogenesis related) proteins, 210, 221, 232–233
Prune/Prunus dwarf virus (PDV), 200, 790
Prunus necrotic ring spot virus, 790, 791–792
Prusiner, S. B., 27, 29
Pseudomonas, 108, 111, 147, 553
  angulata, 191
  description of, 621, 627, 662
  fluorescens, 323, 326, 526, 656
  lacrymans, 107, 238, 630
  phaseolicola, 296
  savastanoi, 119, 200
  syringae, 121, 122, 148, 162, 627
  syringae pv. *atropurpurea*, 193
  syringae pv. *glycinea*, 54, 154
  syringae pv. *maculicola*, 148
  syringae pv. *phaseolicola*, 191, 192, 629
  syringae pv. *syringae*, 193, 219, 629, 667
  syringae pv. *tabacci*, 50, 191, 192, 628–629
  syringae pv. *tagetis*, 193
  syringae pv. *tomato*, 148, 153, 154, 157, 635
Pseudoperonospora, 409, 428
  cubensis, 107, 428
Pseudopeziza, 453
Pseudothecium, 439, 504–506
Puccinia
  graminis f. sp. *tritici*, 13, 16, 52, 59, 107, 127, 131, 138, 251, 287, 567
  helianthi, 211

hordei, 180
recondita, 114, 137, 287
rusts, 219, 562–563
Pueraria montana, 717–718
Purification of viruses, 743–744
Purines, 730
Pycnia, 565
Pycnidia, 99, 444
Pycnidium, 388, 440
Pycniospores, 564
Pyramiding, 137, 170
Pyrenophora
diseases caused by, 453, 466, 469, 471–472
tritici-repentis, 146, 196
Pyricularia, 453, 466
grisea, 193
Pyricularin, 193, 236
Pyrimidines, 342, 730
Pythium, 69, 102, 109, 138, 306
diseases, 409, 410–414, 553

QoI fungicides, 342, 346
Quantitative resistance, 136, 209–210, 219–221
Quarantines and inspections, 295
Quinones, 339
Quorum sensing, 148

Race specific resistance, 136–137, 210, 221–236
RADAR, 289
Radiation, control by, 312
Radish mosaic virus, 784
Radopholus, 853–857
Ralstonia, 108
description of, 622
solanacearum, 55, 110, 147, 148, 149, 183, 198, 251, 639, 647
Rapid amplified polymorphic DNA (RAPD), 283
Raspberries
anthracnose, 483, 486
bushy dwarf virus, 784
cane gall, 662
gray mold, 56
ring spot virus, 787
Rate curves, 276
Ratoon stunting, 683
Reasons of Vegetable Growth (Theophrastus), 11
Receptor-like protein kinases (RLKs), 225
Recognition factors, defense and lack of, 212
Recombination, genetic, 129–130
in viruses, 133
Red ginger, wilt and decay, 882
Redheart, 610
Red ring nematode, 67, 872–874
Refrigeration, control by, 312
Regulatory control methods, 294
Remote sensing, 284
Renaissance, 16–19
Reoviridae, 792–794
Reoviruses, 792
Reproduction
bacteria, 619–620
effects of pathogens on plant, 121–122
epidemics and type of pathogen, 270
fungi, 93–96, 388–389
nematode, 828
pathogen, 91, 93–96
Rescue treatments, 334
Resistance/defense against disease
apparent, 137–139
avirulence (avr) genes, 55, 141, 149, 151–155
biochemicals, induced, 213–214, 217–236
breeding, 165–172
cell wall defense structures, 210, 214–215
chemicals, preexisting, 211–212
cytoplasmic defense reaction, 137, 214
detoxification of toxins by plants, 236–237
gene-for-gene concept, 54, 151
genes, 55, 208–210
genes and susceptibility to epidemics, 268
genetic, 52–53
genetic engineering, role of, 49–50, 54, 56–58, 242–244
histological defense structures, 215–217
horizontal, 53, 136, 169–170, 209–210, 219–221
hypersensitive response, 52, 151, 217, 221–236
immunizations of plants, 237
lack of recognition and, 212–213
nature of, 142
nonhost, 127, 208–209
partial (quantitative or polygenic), 136, 209–210, 219–221
pathogen-derived genes, 54, 243–244
plant-derived genes, 242–243
R gene resistance (race-specific or monogenic), 136–137, 151, 155–158, 210, 221–236
relationship between pathogen virulence and host plant, 139–165
RNA silencing, 244–246
strains of pathogens, 48
structural, induced, 214–217
structural, preexisting, 210–211
systemic acquired (SAR), 50, 55, 57, 157, 237–242, 315
true, 136–137
types of, 134–139
varieties of crops, 626
vertical, 53, 137, 169–170, 210
viruses and, 150
Resistant plant varieties, 318–319
Respiration
infections and affects on, 52, 737
pathogens and affects on, 115–118
Restriction fragment length polymorphisms (RFLPs), 624
R gene resistance, 136–137, 155, 210
classes of proteins, 157, 224–225
evolution of, 157–158
examples of, 156–157
induced biochemical defenses in, 221–236
mechanisms of, 157
Rhabdocline, 453, 457
Rhabdoviruses, 729, 794–795
Rhadinaphelenchus, 872
Rhizobacter, 662
Rhizobium, 676
Rhizoctonia, 102, 138, 554
diseases, 593, 594–599
solani, 132
Rhizomania, 761
Rhizomycelium, 388
Rhizopus, 22, 185, 193, 554
postharvest decays, 556
Rhizosphaera, 453, 457
Rhodococcus, 119, 622
fascians, 200, 201
soft rot of fruits and vegetables, 435–438
Rhynchosporium, 107
Rhytisma, 453
Riboflavin, 238
Ribonucleic acid. *See* RNA
Ribose, 730
Ribosomal RNA (rRNA), 126
Ribosomes, 616
Rice
bacterial leaf blight/streak, 632
black-streaked dwarf virus, 792
blast disease, 56, 59, 144, 162–163, 463, 465–466
brown spot, 66
dwarf virus, 792
gall dwarf virus, 792
gibberellin growth regulators, 50, 51
grassy stunt virus, 43, 45, 799–801
hoja blanca, 67, 799–801
leaf blight, 67
necrosis mosaic virus, 774
Pi-ta gene, 227
ragged stunt virus, 792
sheath and culm blight, 596, 597
smut, 584, 586–587
stripe virus, 799–801
transitory yellowing virus, 794
tungro bacilliform virus, 779, 801, 803
tungro disease, 67, 107, 779–780
tungro spherical virus, 779, 803
waika (stunting) virus, 779
Rickettsia-like organisms, 616
Ridomil (metalaxyl), 340
Rigid rod-shaped viruses, 757–762
Ring rot of potato, 649–651
Ring spots, 737
Rishitin, 235, 236
Risk assessment, epidemics and, 287

RNA
double stranded (dsRNA), 245, 325
filamentous ssRNA, 762–764
genetic information in, 125–126
isometric double-stranded, 792–794
isometric single-stranded, 779–792
messenger (mRNA), 126, 245
negative RNA [(-)SSRNA], 794–801
polymerase (replicase, synthetase), 731
ribosomal, 126
rod-shaped ssRNA, 757–762
satellite, 328, 731
silencing, 55, 244–246
transfer, 126
Rockefeller Foundation, 60
Rod-shaped viruses, 757–762
Romans, 14, 15
Root and stem rots
Ascomycetes and mitosporic fungi and, 534–553
Basidiomycetes, 593–603
fungal, 398, 518
Fusarium, 109, 538–540
oomycetes and, 285–286, 410–421
phymatotrichum, 550–552
Phytophthora and, 109, 414–421
sterile fungi, 593–603
of trees, 602–604
Root collapse and death, 367
Root-knot nematodes, 24, 66, 108, 109, 121, 198, 838–842
Root knots/galls, 108, 109
Root nodules of legumes, 675–678
Root lesions, 594
Root rot. *See* Root and stem rots
RootShield, 324
Roses
black spot, 91, 483, 484, 485–486
crown gall, 663
mosaic virus, 790
powdery mildew, 449, 451–452
Rotstop, 324
Rubber leaf blight, 66
Rubiaceae, 877
Rusts, 10, 14, 22, 398, 562
autoecious, 565
barley, 96, 200
basidiomycetes and, 562–582
bean, 13, 571–572
cedar-apple, 574–576
cereals, 52, 66, 565–571
chrysanthemum, 67
coffee, 66, 576–577
flax, 53, 54
forest trees, 577–582
heteroecious, 565
macrocylic/long-cycled, 564
microcylic/short cycled, 564
pine stem, 66
soybean, 67, 573–574
sugar cane, 67
wheat, 13, 16, 52, 59, 565–571
Ryals, J., 55
Rye, ergot, 37, 38, 66
Rye grass mosaic virus, 773
Rymoviruses, 773

St. Anthony's fire, 37, 559
Salicylic acid, 55, 159–160, 238, 240–242, 338
Sanitation, control by, 301, 625–626
Saponins, 145, 202, 211
Saprophyte, facultative, 78, 387, 389
Saprophytic, 386–387
Sapstain, 606
SAR. *See* Systemic acquired resistance
Satellite RNA, 328, 731
Satellite viruses, 731
Scab diseases, 398, 483, 486–487, 535
apple, 53, 91, 92, 113, 114, 127, 253, 286–287, 504–507
bacterial, 674–675
Scandinavia, 15
Schizophyllum, 606
Schramm, G., 25, 54
Scirrhia, 453, 456
Sclerenchyma cells, defense and, 210
Sclerophthora, 409, 428
Sclerospora, 409, 428
Sclerotia, 80
Sclerotinia
diseases, 546–550, 557
sclerotiorum, 193, 304, 305, 546–550
white mold, 50
Sclerotium diseases, 593, 599–601
Scutellonema, 858
Secondary infections, 80
Secondary inoculum, 80
Secretion systems, bacterial, 147
Seeds/seedlings
blights, 535
chemical treatment of, 334–336
damping off, 594
disinfected, 626
-gall nematodes, 865–867
germination, 87
pathogen-free, 296–297
rot, 410–412
stem canker, 594
virus transmission through, 741
Seiridium cardinale, 36, 216, 476, 483
Selective nutrient media, 624
Semibiotrophs, 78
Semipersistent viruses, 742
Septoria, 114, 301
diseases caused by, 453, 460–463
lycopersici, 145, 202
Sequence characterized amplified region (SCAR), 283
Sequiviruses, 779–780
Serenade, 316
Serological methods, 624
Serratia, 616, 685
Serum, 744
Setosphaeria, 466, 468
Sharka (plum pox), 66, 67, 121, 300, 764, 767–769
Sheath and culm blight, 596, 597
Sicklepod, 329, 330
Sigatoka disease, of banana, 66, 234, 459–460
Signaling
alarms, 214
components, sensing plant, 149
cyclic AMP, 146
genes, 146
G-proteins, 146
mitogen-activated protein kinase (MAP or MAPK), 82, 85, 146, 163
molecules, 158–159
regulation of programmed cell death, 160–161
in systemic acquired resistance, 161
transduction, 159–160, 214, 240–241
Signs, 398
Silencing
gene, 320–321, 754
RNA, 55, 244–246
Silent Spring (Carson), 48
Silicon, 260–261
Single-stranded DNA viruses, 805–816
Sirococcus clavigigenti-juglandacearum, 36, 476, 481, 482
Slime layer, 619
Slime molds, 404–405
fructifications, 404
life cycle, 406
Smith, Erwin, 24, 664
Smuts, 10, 21, 398, 582
barley, 12
cereals, 66, 121, 584–588
corn, 56, 119, 120, 121, 164–165, 198, 583–584
covered/bunt, 12, 18, 121, 588–591
Karnal bunt, 592–591
loose, 12, 121, 584–588
sugar cane, 96, 99
wheat, 12, 18, 21, 47, 588–591
Snapdragon downy mildew, 428
Snow mold, of cereals and turf, 251
Sobemovirus, 783
Sodium bicarbonate, 338, 452
Sodium chloride, 47
Soft rots, 50, 66, 398
bacterial, 656–662
fungal, 410–412, 435–438
in potatoes, 42
of wood, 605
Soil
-borne pathogens, 163–164, 305–307
chemical treatment of, 313, 336
conducive, 304
disease escape and, 138
fumigation, 313
infectious diseases and, 250, 257
inhabitants, 102, 300, 621
invaders, 300, 621
minerals toxic to plants, 372–378
moisture effect, low and high, 365–367

solarization, 311
sterilization, 310–311, 626
suppressive, 87, 303, 304–305
transients, 102
Soilborne diseases
bacteria biological control of, 326–327
fungi biological control of, 325–326
wheat mosaic virus, 761
SoilGard, 324
Solanaceous crops, southern bacterial wilt, 639, 647, 649
Solarization, 311
Somaclonal variation, 168
Sooty molds, 440, 445
Soreshin, 594
Sorghum
downy mildew, 67, 427, 428
ergot, 101, 503
Sour rot, 556–557
Southern bacterial wilt or blight, 601, 639, 647, 649
Soybeans
brown spot, 461
canker, 518, 519
chlorotic mottle virus, 801
cyst nematode, 66, 843–846
downy mildew, 428, 429
leaf sport, 463, 464
mosaic virus, 320, 764, 783
root and stem rot, 415
root nodules, 677
rust, 67, 573–574
stem canker, 595
sudden death syndrome, 540, 541
Spermagonia, 564, 565
Spermatia, 439, 564, 565
Sphaceloma, 483, 486
Sphaeropsis, 119, 521
Sphaerotheca, 451, 583
Sphingomonas, 616
Spiroplasma citri, 700
Spiroplasmas, 26, 73, 617, 691, 699–701
Spongospora, 405, 742
subterranea, 119 , 407
Sporangia, 86, 95, 427, 430
Sporangiophores, 95, 427, 428, 430
Sporangium, 388
Spores, 388, 397
dikaryotic, 565
germination, 82–87
Sporidesmium sclerotivorum, 306
Sporodochium, 440
Spot blotch, 469, 470
Sprayers, 332
Sprays and dusts, 332–334
Squash
leaf curl virus, 805, 810, 812
mosaic virus, 784
powdery mildew, 449
Pythium soft rot, 411
soft/wet rot, 411, 435
vascular wilt, 112
Stachybotrys chartarum, 41
Stagonospora avenae, 145, 202
Stahel, G., 26
Stakman, E. C., 52
Stalk and ear rot, 535, 536–537
Stanley, W. M., 25, 54
Starch, composition and degradation, 190
Staskawicz, B. J., 54
Stem and bulb nematode, 24, 858–860
Stem canker, 518, 519, 594
Stemphylium, 453
vesicarium, 196
Stem rot. *See* Root and stem rots
Sterile fungi, 443, 593–603
Sterilization, of soil, 311–312
Sterol-inhibiting fungicides, 334
Stevens, 62
Stewart's wilt, 44, 285, 654
Stickers, 334
Stilbene synthetase, 55
Sting nematode, 860–863
Stomata defense, role of, 210
Stone fruits
bacterial canker and gummosis, 667–671
bacterial spot, 637
brown rot, 42, 121, 122, 181–182, 185, 251, 507–510
European fruit yellows, 697
sphaeropsis gall, 119
virus, 300
Storage organs, hot-air treatment of, 311–312
Storage rots of fruits and vegetables, 656
Stramenopiles, 390
Strawberries
anthracnose, 484, 489, 490
crown rot and wilt, 484
leaf blight, 518
leaf scorch, 483
soft rot, 435, 436
Verticillium wilt, 526, 527
Streak disease, 67
Streptomyces, 193, 618
cells in, 619
description of, 622
scabies, 138, 256, 257, 304, 674–675
Streptomycin, 334, 343
Stress, abiotic, 319, 383
Striga, 708–711
asiatica, 300
Strigol, 86
Stripped (spotted) cucumber beetles, 42, 44
Strobilurins, 342, 343–344, 346
Structures, defense
induced, 214–217
preexisting, 210–211
Stubby-root nematodes, 863–864
Stylet-borne viruses, 742
Subdue (metalaxyl), 340
Suberins, 187
Sudden death
oak, 419–420
soybeans, 540, 541
Sudden wilt, 880
Sugar beets
curly top, 805, 806
cyst nematode, 24, 66, 846–847
rhizomania virus, 761
yellows, 66
Sugarcane
downy mildew, 428
leaf scald, 632
mosaic virus, 66, 764, 769, 771
ratoon stunting, 683
red rot, 485
red stripe and top rot, 632
rust, 67
smut, 96, 99
Sulfur, 47
deficiency diseases, 372, 373
dioxide injury, 368, 369, 371
Sulfur fungicides
inorganic, 338
organic, 339
Sunflowers
downy mildew, 428
rust, 211
Sunscald injury, 359, 361
Suppressive soils, 87, 303, 304–305
Suppressors
of defense response, 202–203
of RNA silencing, 246
Surfactants, 334
Sweet orange scab, 67
Sweet pea, fasciation/leafy gall, 662
Sweet potatoes
anthracnose, 493
mild mottle virus, 773
nematode, 858
soft rot, 435
Sycamores
anthracnose, 498–500
cankers, 473, 474
Symbiosis, 78
Symptomless carriers, 737
Symptoms
bacteria, 625
external, 622
fungi, 397–398
infection/disease, 89
internal, 622
masked, 737
Synchytrium, 433, 434
Synnema, 440
Syringomycin, 148, 193
Syringotoxin, 193
Systema Naturae (Linne'), 17
Systemic acquired resistance (SAR), 50, 55, 57, 157, 161
description of, 237–242, 315–316
Systemic fungicides, 340
Systemic infections, 91

Tabtoxin, 191, 628
Tagetitoxin, 193
Take-all disease, of wheat, 540, 542–543
Talaromyces flavus, 306

Tannins, 211
Tan spot, 469, 471
Taphrina, 119
  deformans, 121
  leaf curl diseases caused by, 445–447
Tatum, 54
T-DNA (tumor), 198–200, 664, 665–666
Teleomorph, 439
Telia, 564
Teliospores, 86, 564
Temperature
  disease escape and, 138
  effects of, 358–364
  epidemics and, 272
  infectious diseases and, 250, 251–253
  injury to plants, mechanisms, 364
  measurements, 282
  quantitative resistance and, 220–221
Tentoxin, 191–192
Tenuiviruses, 799–801
Terminal oxidation, 116
Tetracycline, 47–48, 334, 343
Texas root rot, 550–552
Thanatephorus cucumeris, 132
Thaumatin-like proteins, 221
Thaxtomins, 193
Theophrastus, 10–11, 14, 46
Thiabendazole (Mertect), 334, 341
Thielaviopsis basicola, 253, 543, 544
Thiophanate ethyl (Topsin), 341
Thiophanate methyl (Fungo), 341
Thiram, 47
Thoullier, 16
Thrips, 43
Thymine, 730
Tillet, M., 18, 21
Tilletia, 12, 18, 121, 122, 583
  contraversa, 119
Tip blight, 519, 520
Tissue culture, 168
Tobacco
  black root rot, 253
  downy mildew (blue mold), 66, 285, 427, 428
  etch virus (TEV), 320, 764, 772
  leaf curl virus, 805
  mosaic virus (TMV), 25, 49, 54, 153, 154, 162, 237, 320, 757–758, 759
  necrosis virus, 238, 781
  necrotic yellow vein virus, 237
  N gene, 227, 242
  rattle virus, 742, 758, 760, 761, 863
  ring spot virus, 10, 742
  southern bacterial wilt, 647
  streak virus, 790
  wildfire disease, 50, 191, 628–629
  yellow dwarf virus, 805
Tobamovirus, 757–758, 759
Tobraviruses, 742, 758, 761
Tolerance, to disease, 139
Toll interleukin receptor (TIR) 155–156
Tomatines (tomatinase), 145 , 202, 211
Tomatoes
  anthracnose or fruit/ripe rot, 484, 487–488
  aspermy virus, 787
  bacterial canker and wilt, 651–653
  bacterial soft rot, 658
  bacterial speck, 635
  bacterial spot, 633–635
  blight, early, 53, 453, 454
  blight, late, 67, 286, 421–426
  Bs2 gene, 228–229
  bushy stunt virus, 781
  Cf genes, 227–228, 242
  fusarium wilt, 110, 126–127, 252, 523–526
  golden mosaic virus, 805
  gray mold and stem canker, 512
  leaf mold, 457, 456
  mottle virus, 805, 812
  pseudocurly top virus, 805
  Pto gene, 226–227, 242
  ring spot virus, 113, 785–786
  root and crown rot, 539, 540
  root knot, 838
  rot, 416
  southern bacterial wilt, 647
  spotted wilt, 56, 66, 795–799
  yellow leaf curl, 43, 45, 66, 67, 805, 812–813
Tombusviruses, 780–781
TOMCAST, 286
Topocuviruses, 805
Topsin (thiophanate ethyl), 341
Topsin M, 341
Tospoviruses, 795–799
Toxic soil minerals, 372–378
Toxins
  See also Mycotoxins
  bacterial, as pathogenicity factors, 148
  detoxification, 236–237
  host-selective (specific), 193–196
  nonhost specific, 190–193
  in plant disease, 190–196
  sensitive sites, 212
Trace elements, 372
Trametes, 606
Transcription, effect of pathogens on, 119
Transduction, 132, 159–160, 214, 240–241
Transfer RNA (tRNA), 126
Transformation, bacterial, 132
Transformation-inducing plasmid (Ti plasmid), 51, 54
Transforming DNA (T-DNA), 51, 54
Transgenic plants, 54
antibodies produced by, 322
  coat protein mediated resistance in, 150
  control with the use of, 319–322
  genes coding and, 320
  gene silencing, 320–321
  resistance genes and, 321–322
  stress and, 319
Translation, effect of pathogens on, 119
Translocation
  of viruses, 733–734
  of water and nutrients, effect of pathogens on, 106, 108–115
Transmission
  epidemics and modes of, 271
  inoculum, 81–82
  of viruses, 737–743
Transpiration, effect of pathogens on, 108, 113, 115
Tranzschelia, 563
Trap crops, 307–308
Traps and mulches, control by, 302
Treehoppers, 742
Trees
  cankers, 473–476, 481–483
  root and stem rot, 414, 417, 419–421, 602–604
  rusts, 577–582
  wood rots and decays, 604–614
  wounds, treatment of, 336–337
Tremorgenic toxins, 560
Trench barriers, 312
Triadimefon (Bayleton), 334, 342
Triadimenol (Baytan), 342
Triazoles, 342
Trichoderma, 526
  harzianum, 305–306, 323, 325, 842
Trichodex, 324
Trichodorus, 742, 761, 863
Trichoject, 324
Trichopel, 324
Trichothecins, 41, 146, 560
Trichoviruses, 763
Triflumizole, 342
Triforine, 343
Tristeza, 49, 66, 774–777
Tritimoviruses, 773–774
Tropical plants, anthracnose, 491–493
True fungi, classification, 392–397
True resistance, 136–137
Trypanosomatids
  description of, 875–877
  epidemiology and control of, 878
  fruit-and-seed infecting, 882–885
  laticifer-restricted, 882
  nomenclature, 877
  pathogenicity, 877–878
  phloem-restricted, 878–882
  taxonomy, 877
T toxin, 146, 194
T-22 Planter Box, 324
Tulip breaking virus, 764
Tumor-inducing (Ti) plasmid, 664
Tumor (T-DNA), 198–200, 664, 665–666
Turf grasses
  dollar spot, 546, 549
  slime mold, 404
  snow mold or blight, 251
Turkey X disease, 39
Turnips
  crinkle virus (TCV), 153
  mosaic virus, 56, 137, 764, 772
  yellow mosaic virus, 783
Tylenchulus semipenetrans, 848–849
Tyloses, 112, 113
  formation of, 217
Tymovirus, 783
Type III secretion system (TTSS), 229–231

Ultracentrifugation, 743
Umbravirus, 784
Uncinula necator, 30, 114, 451
Uncinuliella flexuosa, 451
Uracil, 730
Uredinia (uredia), 564, 565
Uredospores, 95, 564, 565
Urnula, 476
Urocystis, 583
Uromyces, 571
  appendiculatus, 13, 563, 571
Ustilago sp., 12, 121, 583
  hordei, 146
  maydis, 56, 119, 120, 122, 145, 146, 164–165, 198
  nuda, 586
  scitaminea, 583
  tritici, 583, 586
Ustilina, 605

Valsa, 479
Vanderplank, J. E., 53
Variability
  affected by plant breeding, 165–166
  mechanisms of, 128–133
  in organisms, 128
  in pathogens, 131–134
  somaclonal, 168
Vascular wilts
  Ascomycetes and mitosporic fungi, 522–534
  bacterial, 108, 112, 638–656
  Ceratocystis, 532–534
  fungus, 108, 110, 256
  Fusarium, 523–526
  Ophiostoma, 528–532
  toxins in pathogens, 50
  Verticilium, 526–528
Vegetables, postharvest decays of, 556–558
Vegetative incompatibility, 132
Vegetative propagating
  materials, 297–298
  transmission of viruses, 737–739
Venturia, 453
  inaequalis, 114, 127, 145, 212, 286–287, 501, 504–507
Vermeulen, 26
Vertical inheritance, 132
Vertical resistance, 53, 137, 169–170, 210, 221–236
Verticillium
  alboatrum, 132, 522, 526–528
  dahliae, 522, 526–528
  penetration, 178
  wilts, 108
Vesicular-arbuscular, 613
Vinclozolin, 340
VirA protein, 149
VirG protein, 149
Viral diseases, 66, 67
  diagnosis of diseases, 73
Viral genes, 49–50
Viral parasites, 328
Virion, 733
Viroids, 26, 27, 28, 67, 121
  circular, 816
  diagnosis of diseases, 73
  diseases caused by, 816–824
  epidemiology, 743
  pathogenicity/virulence, 203
  properties of, 816
  replication, 819
  symptoms caused by, 820
  taxonomy, 816–818
Virulence
  in culture, loss of pathogen, 133
  epidemics and levels of, 269
  genes, 142–144
  relationship between host plant resistance and pathogen, 139–165
Viruses, 25, 27, 28, 121
  assembly, 150
  biological function of components (coding), 731
  characteristics of, 724–731
  coat proteins, 149–150, 731
  composition and structure, 729–731
  control of, 753–756
  defined, 724
  detection, 725–729, 751–752
  disassembly, 150
  economic importance, 752–753
  effects on plant physiology, 737
  effects of temperature on, 253
  epidemiology, 743
  genome activation, 150
  genetic recombination in, 133
  helper, 731
  identification, 751–752
  indexing, 751
  induced gene silencing (VIGS), 246
  infection and synthesis, 731–733
  latent, 737
  morphology, 729
  movement, 150
  nomenclature, 747–751
  nonpersistent, 742
  pathogenicity genes, 149–150
  pathogenicity/virulence, 203
  persistent (circulative), 742
  propagative, 742
  proteins, 149–150, 731
  purification, 743–744
  replication, 732–733
  satellite, 731
  semipersistent, 742
  serology, 744–747
  stylet-borne, 742
  symptoms caused by, 734–737
  taxonomy, 747–751
  translocation and distribution in plants, 733–734
  transmission and spread, 737–743
Viruses, diseases caused by
  closteroviridae, 774–779
  double-stranded DNA, 732, 801–805
  filamentous ssRNA, 762–764
  isometric double-stranded RNA, 792–794
  isometric single-stranded RNA, 779–792
  negative RNA [(-)SSRNA], 794–801
  potyviridae, 764–774
  rod-shaped ssRNA, 757–762
  single-stranded DNA, 733, 805–816
  viroids, 816–824
Vitavax (carboxin), 341
Vitiviruses, 763
Vomitoxin, 559
Vorlex, 345

Waikaviruses, 779–780
Walker, 62
Walton, J. D., 55
Ward, H. M., 52
Warehouses, disinfestation of, 313–314
Warning systems, 287–288
Warts, 398
Water dissemination, 97, 626
Water hyacinth, biological control of, 329, 331
Watermelons
  anthracnose, 487, 488
  bacterial fruit blotch, 635–636
  leaf sports, 455
  mosaic virus, 169, 764, 769, 772–773
  rot, 416
Water pollution, 48
Water stress, 365
Watery soft rot, 557
Watson, James, 54
Wax
  cuticular, 180, 181
  as a defense structure, 210
  lipids, 190
Weather, disease development and monitoring, 282–283
Weed killers, effects of, 262, 378–380
Weeds, biological control of, 329
Wetwood, 610
Wheat
  crown and root rot, 469, 470
  dwarf bunt, 119
  ergot, 37, 38, 66, 502
  fusarium scab, 66
  galls, 23
  glum blotch, 461
  head blight or scab, 56, 535, 538
  karnal bunt, 67
  leaf and glum blotch, 301
  leaf rust, 287
  powdery mildew, 11, 258, 450
  root rot, 109, 253, 469, 539
  rust, 13, 16, 52, 59, 114, 258, 565–571
  seed-gall nematode, 865–867
  smut, 12, 18, 21, 47, 122, 588–591
  soil-borne mosaic virus, 761
  spindle streak mosaic virus, 774
  spot blotch, 469, 470
  stem rust, 127, 131, 251, 287, 565–571

Wheat (*continued*)
  striate mosaic virus, 794
  streak mosaic virus, 169, 320, 773
  take-all, 540, 542–543
  tan spot, 469, 471
Whiteflies, 43, 45
White mold, 50
White pine blister rust, 578–580
White rot, 55
White rot fungi, of trees, 605
White rust, 67, 432
Wildfire disease, in tobacco, 50, 191, 628–629
Wildfire toxin, 628
Williamson, V. M., 55
Wilts, 398
  See also Fusarium wilts; Vascular wilts
  Stewart's (bacterial), 42, 44, 285, 654
  sudden, 880
Wind
  disease escape and, 138
  infectious diseases and, 257
WISDOM, 289
Witches' broom, 121, 201, 398, 445
  of cacao, 234, 611–612
Witchweed, 72, 300, 708–711
Wood rots and decays, basidiomycetes and, 604–614
Wood-staining fungi, 605–606
  sapstain or blue stain, 606
Woronin, M., 24
Wounds, penetration through, 88
Wound tumor virus, 792

Xanthomonas, 147
  albilineans, 148
  avr gene, 154
  axonopidis, 637, 667, 672
  citrumelo, 155
  description of, 622, 627
  oryzae pv. *oryzae*, 223, 225
Xanthomonas campestris, 627
  pv. *campestris*, 148, 220, 639, 653
  pv. *citri*, 155, 667
  pv. *malvacearum*, 137, 155, 630
  pv. *oryzae*, 157
  pv. *phaseoli*, 155, 296, 629
  pv. *pruni*, 637
  pv. *vesicatoria*, 152–153, 154, 633–635
X disease, peach, 697–699
Xiphinema, 742
Xylanases, 145, 147, 148
Xylaria, 605, 606
Xylella, 147, 148
  description of, 622
  fastidiosa, 36, 55, 108, 111, 679–681
Xylem, 93
  defense and, 210–211
  translocation of water through, affected by pathogens, 108, 110
Xylem-inhabiting bacteria, 94, 678–683

Yams. *See* Sweet potatoes
Yeast, 306
Yellowed-rice toxins, 560
Yellow leaf blight, 137
Yellow vine disease, 684–685
Yield loss
  attainable versus actual, 273
  measurement of epidemics, 273–274
YieldShield, 316

Zearalenones, 559–560
Zeatin, 200
Zinc, 343
  deficiency diseases, 372, 373
Zineb, 47, 460
Zoosporangia, 257, 432
Zoospores, 86, 89, 388, 432
Zoxamide, 343
Zucchini yellow mosaic virus, 169, 764, 773
Zycomycetes
  diseases caused by, 434–438
  reproduction in, 388
  symptoms caused by, 435
Zygospore, 388
Zygote, 129–130